Lemuridae
Bamboo, True and Ruffed Lemurs
5 genera • 21 spp
p. 94

Indriidae
Woolly Lemurs, Sifakas and Indri
3 genera • 19 spp
p. 98

Daubentoniidae
Aye-aye
1 genus • 1 sp
p. 102

Galagidae
Galagos
6 genera • 19 spp
p. 102

Lorisidae
Angwantibos, Pottos and Lorises
5 genera • 16 spp
p. 106

Tarsiidae
Tarsiers
3 genera • 14 spp
p. 109

Callitrichidae
Marmosets and Tamarins
6 genera • 51 spp
p. 111

Aotidae
Night Monkeys
1 genus • 11 spp
p. 118

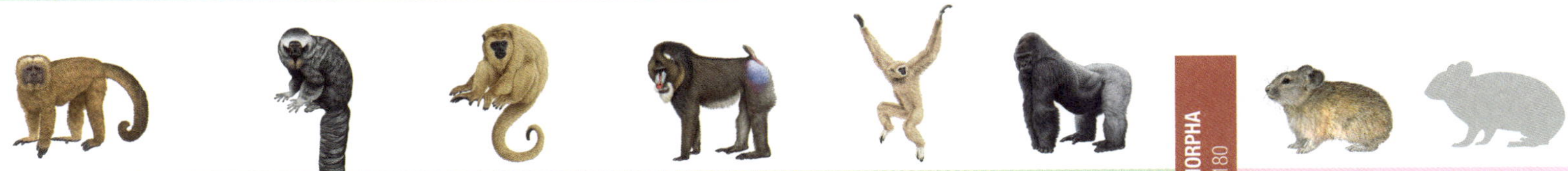

Cebidae
Squirrel Monkeys and Capuchins
3 genera • 30 spp
p. 120

Pitheciidae
Titis, Sakis and Uacaris
7 genera (†1) • 60 spp (†1)
p. 125

Atelidae
Howlers, Spider and Woolly Monkeys and Muriquis
4 genera • 23 spp
p. 137

Cercopithecidae
Old World Monkeys
23 genera • 160 spp
p. 143

Hylobatidae
Gibbons
4 genera • 20 spp
p. 173

Hominidae
Great Apes
4 genera • 8 spp
p. 178

LAGOMORPHA
p. 180

Ochotonidae
Pikas
1 genus • 34 spp
p. 180

Prolagidae
Sardinian Pika
1 genus (†1) • 1 sp (†1)
p. 700

Leporidae
Hares and Rabbits
10 genera • 62 spp
p. 184

RODENTIA
p. 191

Castoridae
Beavers
1 genus • 2 spp
p. 191

Heteromyidae
Pocket Mice, Kangaroo Mice and Kangaroo Rats
5 genera • 69 spp
p. 192

Geomyidae
Pocket Gophers
7 genera • 41 spp
p. 198

Zenkerellidae
Non-gliding Anomalure
1 genus • 1 sp
p. 202

Anomaluridae
Gliding Anomalures
2 genera • 6 spp
p. 202

Pedetidae
Springhares
1 genus • 2 spp
p. 203

Sminthidae
Birch Mice
1 genus • 19 spp
p. 204

Zapodidae
Jumping Mice
3 genera • 5 spp
p. 205

Dipodidae
Jerboas
14 genera • 38 spp
p. 206

Platacanthomyidae
Tree Mice
2 genera • 6 spp
p. 210

Spalacidae
Muroid Mole-rats
7 genera • 29 spp
p. 210

Calomyscidae
Brush-tailed Mice
1 genus • 8 spp
p. 213

Nesomyidae
Pouched Rats, Climbing Mice and Fat Mice
22 genera • 71 spp
p. 214

Cricetidae
True Hamsters, Voles, Lemmings and New World Rats and Mice
149 genera (†5) • 844 spp (†14)
p. 220

Muridae
True Mice and Rats, Gerbils and relatives
162 genera (†1) • 862 spp (†19)
p. 281

Ctenodactylidae
Gundis
4 genera • 5 spp
p. 349

Diatomyidae
Kha-nyou
1 genus • 1 sp
p. 349

Hystricidae
Old World Porcupines
3 genera • 11 spp
p. 349

Thryonomyidae
Cane Rats
1 genus • 2 spp
p. 350

Petromuridae
Noki
1 genus • 1 sp
p. 350

Heterocephalidae
Naked Mole-rat
1 genus • 1 sp
p. 351

Bathyergidae
African Mole-rats
5 genera • 25 spp
p. 351

Erethizontidae
New World Porcupines
3 genera • 18 spp
p. 353

Cuniculidae
Pacas
1 genus • 2 spp
p. 355

Caviidae
Cavies, Capybaras and Maras
7 genera • 24 spp
p. 355

Dasyproctidae
Agoutis and Acouchis
2 genera • 15 spp
p. 357

Chinchillidae
Viscachas and Chinchillas
3 genera • 7 spp (†1)
p. 358

Dinomyidae
Pacarana
1 genus • 1 sp
p. 359

Abrocomidae
Inca Rats and Chinchilla Rats
2 genera • 10 spp
p. 359

Ctenomyidae
Tuco-tucos
1 genus • 67 spp
p. 360

Octodontidae
Viscacha Rats, Degus, Rock Rats and Coruro
7 genera • 15 spp
p. 365

Echimyidae
Hutias, Coypu and South American Spiny-rats
35 genera (†7) • 116 spp (†14)
p. 366

Aplodontiidae
Mountain Beaver
1 genus • 1 sp
p. 374

Sciuridae
Tree, Flying and Ground Squirrels, Chipmunks, Prairie Dogs and Marmots
63 genera • 309 spp
p. 375

Gliridae
Dormice
9 genera • 30 spp
p. 402

All the Mammals of the World

LYNX NATURE BOOKS

All the Mammals of the World

Color illustrations by

Toni Llobet (main artist)

Ilian Velikov
Lluís Sogorb
Faansie Peacock
Francesc Jutglar
Àlex Mascarell
Blanca Martí
Jesús Rodríguez-Osorio

LYNX NATURE BOOKS

Barcelona

Recommended citation:
Lynx Nature Books. (2023). *All the Mammals of the World.* Barcelona.

Printed on Imagine Silk paper sourced from managed sustainable forests.
Printed in Barcelona by Índice Arts Gràfiques.
Legal Deposit: B 10690-2023
ISBN: 978-84-16728-66-4

PEFC Certified
This product is from sustainably managed forests, recycled and controlled sources
PEFC
PEFC/14-38-00202
www.pefc.co.uk

Project Co-ordinator
Amy Chernasky

Data Compilation
Anna Motis
Connor Burgin

Map Production
Guillem Martínez-Motis

Additional Artwork
Lluís Sogorb

French Common Names
Normand David †
Michel Gosselin

German Common Names
Rainer Hutterer
Gustav Peters

Spanish Common Names
Sociedad Española para la Conservación
y Estudio de los Mamíferos (SECEM):
Luis Javier Palomo

Asociación Española para la Conservación
y el Estudio de los Murciélagos (SECEMU):
Maria Mas
Xavier Puig

Revision
Marc Olivé
Albert Martínez-Vilalta
Michael Lockwood
Andrew Elliott

Production
Daniel Roca

Layout
Daniel Roca
Lluc Julià
Oriol Cabrero

Concept
Josep del Hoyo

Alada Books
Yolanda Aguayo
Amy Chernasky
Ana Conesa
Ana Lineros

Lynx Nature Books Editorial Council
Alex Berryman
Josep del Hoyo
Guy Kirwan
Albert Martínez-Vilalta
Daniel Roca
Jordi Sargatal

CONTENTS

Class MAMMALIA

Contributions from HMW authors and artists

First entry, author contributions; second entry, *artist contributions*

Family Tachyglossidae
S. C. Nicol
T. Llobet

Family Ornithorhynchidae
T. R. Grant
T. Llobet

Family Didelphidae
D. Astúa
T. Llobet

Family Caenolestidae
B. D. Patterson
T. Llobet

Family Microbiotheriidae
R. E. Palma & A. Valladares-Gómez
T. Llobet

Family Notoryctidae
K. P. Aplin †
T. Llobet

Family Myrmecobiidae
J. A. Friend
T. Llobet

Family Dasyuridae
A. M. Baker
T. Llobet

Family Thylacomyidae
R. I. Southgate
T. Llobet

Family Peramelidae
C. R. Dickman
T. Llobet

Family Phascolarctidae
S. M. Jackson
T. Llobet

Family Vombatidae
R. T. Wells
T. Llobet

Family Burramyidae
S. M. Jackson
T. Llobet

Family Phalangeridae
K. M. Helgen & S. M. Jackson
T. Llobet

Family Pseudocheiridae
S. M. Jackson
T. Llobet

Family Petauridae
S. M. Jackson
T. Llobet

Family Tarsipedidae
R. D. Wooller
T. Llobet

Family Acrobatidae
K. P. Aplin †
T. Llobet

Family Hypsiprymnodontidae
A. J. Dennis
T. Llobet

Family Potoroidae
M. D. B. Eldridge & G. J. Frankham
T. Llobet

Family Macropodidae
M. D. B. Eldridge & G. M. Coulson
T. Llobet

Family Tenrecidae
P. D. Jenkins
T. Llobet

Family Potamogalidae
A. Monadjem
T. Llobet

Family Chrysochloridae
W. A. Taylor, S. Mynhardt & S. Maree
T. Llobet

Family Macroscelididae
S. Heritage
T. Llobet

Family Orycteropodidae
W. A. Taylor
T. Llobet

Family Procaviidae
H. N. Hoeck
T. Llobet

Family Elephantidae
G. Wittemyer
T. Llobet

Family Dugongidae
H. D. Marsh
T. Llobet

Family Trichechidae
T. J. O'Shea
T. Llobet

Family Dasypodidae
C. M. McDonough & W. J. Loughry
T. Llobet

Family Chlamyphoridae
M. Superina & A. M. Abba
T. Llobet

Family Myrmecophagidae
A. Bertassoni
T. Llobet

Family Cyclopedidae
F. R. Miranda
T. Llobet

Family Choelopodidae
N. Moraes-Barros
T. Llobet

Family Bradypodidae
J. N. Pauli
T. Llobet

Family Ptilocercidae
M. T. R. Hawkins
T. Llobet

Family Tupaiidae
M. T. R. Hawkins
T. Llobet

Family Cynocephalidae
J. E. Janecka & M. J. Janecka
T. Llobet

Family Cheirogaleidae
C. Schwitzer, R. A. Mittermeier, E. E. Louis Jr & M. C. Richardson
F. Peacock

Family Lepilemuridae
C. Schwitzer, R. A. Mittermeier, E. E. Louis Jr & M. C. Richardson
F. Peacock

Family Lemuridae
C. Schwitzer, R. A. Mittermeier, E. E. Louis Jr & M. C. Richardson
F. Peacock

Family Indriidae
R. A. Mittermeier, C. Schwitzer, E. E. Louis Jr & M. C. Richardson
F. Peacock

Family Daubentoniidae
C. Schwitzer, R. A. Mittermeier, E. E. Louis Jr & M. C. Richardson
F. Peacock

Family Galagidae
K. A. I. Nekaris
F. Peacock

Family Lorisidae
K. A. I. Nekaris
F. Peacock

Family Tarsiidae
M. Shekelle, S. Gursky-Doyen & M. C. Richardson
F. Peacock

Family Callitrichidae
A. B. Rylands & R. A. Mittermeier
I. Velikov

Family Aotidae
E. Fernandez-Duque, M. K. Corley & A. Spence-Aizenberg
L. Sogorb & I. Velikov

Family Cebidae
A. B. Rylands, R. A. Mittermeier, B. M. Bezerra, F. P. Paim & H. L. Queiroz
L. Sogorb & I. Velikov

Family Pitheciidae
S. F. Ferrari, L. M. Veiga †, L. P. Pinto, L. K. Marsh, R. A. Mittermeier & A. B. Rylands
L. Sogorb & I. Velikov

Family Atelidae
A. B. Rylands, R. A. Mittermeier, F. M. Cornejo, T. R. Defler, K. E. Glander, W. R. Konstant, L. P. Pinto & M. Talebi
F. Jutglar & I. Velikov

Family Cercopithecidae
D. Zinner, G. H. Fickenscher, C. Roos, M. V. Anandam, E. L. Bennett, T. R. B. Davenport, N. J. Davies, K. M. Detwiler, A. Engelhardt, A. Eudey, E. L. Gadsby, C. P. Groves †, A. Healy, K. P. Karanth, S. Molur, T. Nadler, M. C. Richardson, E. P. Riley, A. B. Rylands, L. K. Sheeran, N. Ting, J. Wallis, S. S. Waters & D. J. Whittaker
I. Velikov

Family Hylobatidae
D. J. Chivers, M. V. Anandam, C. P. Groves †, S. Molur, B. M. Rawson, M. C. Richardson, C. Roos & D. J. Whittaker
I. Velikov

Family Hominidae
E. A. Williamson, F. G. Maisels, C. P. Groves †, B. I. Fruth, T. Humle, F. B. Morton, M. C. Richardson, A. E. Russon & I. Singleton
I. Velikov

Family Ochotonidae
A. A. Lissovsky
T. Llobet

Family Leporidae
S. C. Schai-Braun & K. Hackländer
T. Llobet

Family Castoridae
P. E. Busher
T. Llobet

Family Heteromyidae
D. J. Hafner
T. Llobet

Family Geomyidae
M. S. Hafner
T. Llobet

Family Zenkerellidae
S. M. Jackson
T. Llobet

Family Anomaluridae
S. M. Jackson
T. Llobet

Family Pedetidae
R. López-Antoñanzas
T. Llobet

Family Sminthidae
M. E. Holden, T. Cserkész & G. M. Musser
T. Llobet

Family Zapodidae
J. O. Whitaker Jr
T. Llobet

Family Dipodidae
J. R. Michaux & G. I. Shenbrot
T. Llobet

Family Platacanthomyidae
T. C. Giarla
T. Llobet

Family Spalacidae
R. W. Norris
T. Llobet

Family Calomyscidae
C. W. Kilpatrick
T. Llobet

Family Nesomyidae
S. M. Goodman & A. Monadjem
T. Llobet

Family Cricetidae
U. F. J. Pardiñas, P. Myers, L. León-Paniagua, N. Ordóñez Garza, J. A. Cook, B. Kryštufek, R. Haslauer, R. D. Bradley, G. I. Shenbrot, J. L. Patton, D. Ruelas, J. Brito, L. C. Bradley, E. Cuéllar Soto, J. Salazar-Bravo, E. A. Chiquito, R. Percequillo & J. R. Prado
T. Llobet

Family Muridae
C. Denys, P. J. Taylor, K. P. Aplin †, C. J. Burgin, P.-H. Fabre, R. Haslauer, J. C. Z. Woinarski, W. G. Breed & J. I. Menzies
T. Llobet

Family Ctenodactylidae
R. López-Antoñanzas
T. Llobet

Family Diatomyidae
P. D. Jenkins
T. Llobet

Family Hystricidae
E. L. Barthelmess
T. Llobet

Family Thryonomyidae
R. López-Antoñanzas
T. Llobet

Family Petromuridae
R. López-Antoñanzas
T. Llobet

Family Heterocephalidae
B. D. Patterson
T. Llobet

Family Bathyergidae
R. L. Honeycutt
T. Llobet

Family Erethizontidae
E. L. Barthelmess
T. Llobet

Family Cuniculidae
R. Samudio Jr
T. Llobet

Family Caviidae
T. E. Lacher Jr
T. Llobet

Family Dasyproctidae
J. A. Gilbert & T. E. Lacher Jr
T. Llobet

Family Chinchillidae
A. E. Spotorno & P. Valladares Faúndez
T. Llobet

Family Dinomyidae
M. A. Mares & J. K. Braun
T. Llobet

Family Abrocomidae
L. H. Emmons
T. Llobet

Family Ctenomyidae
T. R. O. Freitas
T. Llobet

Family Octodontidae
A. A. Ojeda
T. Llobet

Family Echimyidae
P.-H. Fabre, J. L. Patton & Y. L. R. Leite
T. Llobet

Family Aplodontiidae
S. S. B. Hopkins
T. Llobet

Family Sciuridae
J. L. Koprowski, E. A. Goldstein, K. R. Bennett & C. Pereira Mendes
T. Llobet

Family Gliridae
M. E. Holden, R. Juškaitis & G. M. Musser
T. Llobet

Family Erinaceidae
T. L. Best
T. Llobet

Family Soricidae
C. J. Burgin, K. He, R. Haslauer, B. I. Sheftel, P. D. Jenkins, M. Ruedi, S. Hintsche, M. Motokawa, A. Hinckley & R. Hutterer
T. Llobet

Family Talpidae
B. Kryštufek & M. Motokawa
T. Llobet

Family Solenodontidae
S. T. Turvey
T. Llobet

Family Pteropodidae
N. P. Giannini, V. Van Cakenberghe, C. J. Burgin, S. M. Tsang, T. H. Lavery, S. Hintsche, F. J. Bonaccorso, F. C. Almeida & B. P. O'Toole
I. Velikov & F. Peacock

Family Rhinopomatidae
I. Horáček
I. Velikov

Family Craseonycteridae
T. Kingston & P. Soisook
I. Velikov

Family Megadermatidae
C. M. Francis
I. Velikov

Family Rhinonycteridae
P. Benda
I. Velikov

Family Hipposideridae
A. Monadjem, P. Soisook, C. Tuneu-Corral & V. D. Thong
F. Jutglar & À. Mascarell

Family Rhinolophidae
A. M. Hutson, S. J. Rossiter, G. Csorba & C. J. Burgin
I. Velikov

Family Emballonuridae
F. J. Bonaccorso
I. Velikov

Family Nycteridae
A. Monadjem
L. Sogorb

Family Myzopodidae
S. M. Goodman
I. Velikov

Family Mystacinidae
C. A. Toth
I. Velikov

Family Noctilionidae
R. A. Medellín
I. Velikov

Family Furipteridae
J. Arroyo-Cabrales
I. Velikov

Family Thyropteridae
T. E. Lee Jr
I. Velikov

Family Mormoopidae
A. C. Pavan
I. Velikov

Family Phyllostomidae
S. Solari, R. A. Medellín, B. Rodríguez-Herrera, E. R. Dumont, S. F. Burneo, V. da C. Tavares, G. S. T. Garbino, M. A. Camacho, D. G. Tirira, B. K. Lim, J. Arroyo-Cabrales, A. Rodríguez-Durán, L. F. Aguirre, M. Tschapka & D. V. Espinosa-Martínez
I. Velikov

Family Natalidae
A. Tejedor
B. Martí

Family Molossidae
P. J. Taylor, B. K. Lim, M. Pennay, P. Soisook, L. O. Loureiro, L. M. Moras & T. Kingston
F. Jutglar, I. Velikov & L. Sogorb

Family Miniopteridae
C. Ibáñez & J. Juste
J. Rodríguez-Osorio

Family Cistugidae
M. Ruedi
I. Velikov

Family Vespertilionidae
R. Moratelli, C. J. Burgin, V. C. Cláudio, R. L. M. Novaes, A. López-Baucells & R. Haslauer
L. Sogorb, À. Mascarell, B. Martí & F. Peacock

Family Balaenidae
B. E. Curry & R. L. Brownell Jr
T. Llobet

Family Cetotheriidae
C. M. Kemper
T. Llobet

Family Balaenopteridae
T. A. Deméré
T. Llobet

Family Physeteridae
S. L. Mesnick
T. Llobet

Family Kogiidae
D. F. McAlpine
T. Llobet

Family Ziphiidae
C. D. MacLeod
T. Llobet

Family Platanistidae
G. T. Braulik
T. Llobet

Family Lipotidae
B. D. Smith
T. Llobet

Family Iniidae
V. M. F. da Silva & A. R. Martin
T. Llobet

Family Pontoporiidae
E. R. Secchi
T. Llobet

Family Monodontidae
P. R. Richard
T. Llobet

Family Delphinidae
J. Y. Wang, K. N. Riehl & S. Z. Dungan
T. Llobet

Family Phocoenidae
T. A. Jefferson
T. Llobet

Family Hippopotamidae
R. L. Lewison
T. Llobet

Family Tragulidae
E. Meijaard
T. Llobet

Family Antilocapridae
J. A. Byers
T. Llobet

Family Giraffidae
J. D. Skinner & G. Mitchell
T. Llobet

Family Cervidae
S. Mattioli
T. Llobet

Family Moschidae
C. P. Groves †
T. Llobet

Family Bovidae
C. P. Groves †, D. M. Leslie Jr, B. A. Huffman, R. Valdez, K. Habibi, P. J. Weinberg, J. A. Burton, P. J. Jarman & W. G. Robichaud
T. Llobet

Family Suidae
E. Meijaard, J. P. d'Huart & W. L. R. Oliver
T. Llobet

Family Tayassuidae
A. B. Taber, M. Altrichter, H. Beck & J. Gongora
T. Llobet

Family Camelidae
W. L. Franklin
T. Llobet

Family Equidae
D. I. Rubenstein
T. Llobet

Family Tapiridae
E. P. Medici
T. Llobet

Family Rhinocerotidae
E. Dinerstein
T. Llobet

Family Manidae
P. Gaubert
T. Llobet

Family Nandiniidae
P. Gaubert
T. Llobet

Family Prionodontidae
P. Gaubert
T. Llobet

Family Felidae
M. E. Sunquist & F. C. Sunquist
T. Llobet

Family Viverridae
A. P. Jennings & G. Veron
T. Llobet

Family Herpestidae
J. S. Gilchrist, A. P. Jennings, G. Veron, & P. Cavallini
T. Llobet

Family Eupleridae
S. M. Goodman
T. Llobet

Family Hyaenidae
K. E. Holekamp & J. M. Kolowski
T. Llobet

Family Canidae
C. Sillero-Zubiri
T. Llobet

Family Ursidae
D. L. Garshelis
T. Llobet

Family Otariidae
M. A. Webber
T. Llobet

Family Odobenidae
B. S. Stewart
T. Llobet

Family Phocidae
B. S. Stewart
T. Llobet

Family Ailuridae
F. Wei & Z. Zhang
T. Llobet

Family Mephitidae
J. W. Dragoo
T. Llobet

Family Mustelidae
S. Larivière & A. P. Jennings
T. Llobet

Family Procyonidae
R. Kays
T. Llobet

FOREWORD

The magnificent nine-volume *Handbook of the Mammals of the World* (HMW) series was completed in 2019, exactly a decade after the first volume had been published. In these nine volumes, every one of the more than 6,000 species of extant mammals is illustrated and given an individual species account with an associated distribution map. In addition, each mammal family (such as Canidae, dogs, and Felidae, cats) is discussed in great detail and lavishly illustrated with outstanding photographs. This stupendous undertaking was accomplished by drawing on the expertise and knowledge of dozens of mammalogists, each an expert in a particular group of mammals, working in harmony with a dynamic team of many other collaborators from around the world. An exceptional contribution, it most certainly was, but the size of the collection (weighing in at about 45 kg for the nine volumes!) and its resulting cost makes it difficult to handle for many mammal scientists and enthusiasts. The solution? Compress the nine volumes into a single book!

In essence, this is what this new work, *All the Mammals of the World*, represents. An innovative and unique book that contains information about each and every living species of mammal, anywhere in the world, building on the legacy of the HMW series and the subsequent two-volume taxonomic update, the *Illustrated Checklist of the Mammals of the World*. With close to 6,500 extant species of mammals currently recognized, how is this even possible? For those doubting the logistics of this enterprise, I can assure you that this experiment has in fact already been attempted successfully by Lynx Edicions (now Lynx Nature Books), with the publication in 2020 of *All the Birds of the World*. All 11,524 species of birds known at the time of publication are included in this outrageously extravagant volume.

As is to be expected from Lynx, the quality of the artwork is impeccable. Not only is each mammal species illustrated accurately, showing its distinctive features, but each image looks life-like. Additional illustrations are included for species with clearly recognized subspecies or geographic variants, and length measurements and mass are provided when available, often given as a range to show variation in size within the species. Each species has an associated map showing the known extent of its distribution, along with a concise written summary of its range, including the countries in which the species has been recorded. To assist in interpreting these maps drawn at continental (for species with wide distributional ranges) or regional scales, all major rivers are shown. A particularly useful innovation is the addition of the elevational range of the species alongside the map. Hence, a high-elevation species that occurs patchily across a mountain range can be excluded from the river valleys and other lowlands occurring within its distribution—and of course *vice versa* for lowland species occurring in and around patchy (and hence not mappable at continental scale) uplands. Finally, and very importantly, for each species the global conservation status from The IUCN Red List of Threatened Species is provided to underline the importance of conservation, as well as to encourage further study. This is exceedingly important for species that are categorized as Data Deficient, which are not on the radar for conservation efforts and may be lost before they can be protected. Typically, 10–12 species are presented per page, allowing for both the illustrations and the maps to be large enough for easy and enjoyable viewing, and the divisions between species, genera, families, and orders are clearly marked to aid comprehension and comparison.

What does all this mean? What is the relevance of this singular publication? For one thing, *All the Mammals of the World* will bring wonder and joy to untold numbers of mammal lovers and wildlife fans, who perhaps were unable to obtain copies of HMW or the *Illustrated Checklist* or have not yet heard of these publications. Even for those who own any or all of the previous tomes, this new book allows a different type of exploration of mammal diversity. For the first time ever, you can browse all the mammals of the world—from the smallest to the largest species present in every corner of the Earth—by simply paging through a single-volume book.

This new publication also includes new mammal species that were not represented in the previous works. Many new mammal (and other) species are described each year, the result of two main factors. Firstly, we are increasingly visiting remote corners of our planet that had previously been overlooked by scientists. My own recent work on Mount Nimba in the Upper Guinea rainforest zone of West Africa demonstrates this well. Over the course of less than a decade, my collaborators and I have described five new species of bat to science from this region: Rosevear's Serotine (*Pseudoromicia roseveari*; see p. 537), Isabelline Serotine (*Pseudoromicia isabella*; see p. 537), Happolds' Serotine (*Nycticeinops happoldorum*; see p. 537), Simandou Pipistrelle (*Pipistrellus simandouensis*; see p. 530), and Nimba Long-fingered Bat (*Miniopterus nimbae*; see p. 526). The importance of this becomes apparent when, by including these new species, we were able to demonstrate that this region represents an overlooked global hotspot of bat diversity and endemism. Secondly, we are now combining genetic analyses with standard morphological research to distinguish cryptic species (i.e. species that are morphologically difficult to differentiate), which were previously and incorrectly lumped together. For example, we described Mozambican Horseshoe Bat (*Rhinolophus mossambicus*; see p. 476), which is widespread in Mozambique and neighboring states, as a species genetically and morphologically distinct from Hildebrandt's Horseshoe Bat (*R. hildebrandtii*; see p. 476) occurring in East Africa. Another example of how genetic analyses have shed greater taxonomic insights is our work on the systematic relationships of African vesper bats, where we demonstrated that the species lumped in the genus *Neoromicia* (see p. 538) are in fact best represented by multiple genera, including *Laephotis* (see p. 538), *Afronycteris* (see p. 537), and *Pseudoromicia* (see p. 537) (the latter two new to science). We also showed that the tropical African vesper species previously placed in the genus *Hypsugo* (see p. 535) and later moved to the genus *Parahypsugo* are in fact closely related to *Nycticeinops schlieffeni* (see p. 536) and therefore best placed in that genus.

As a result, *All the Mammals of the World* includes more than 400 species that were not known or recognized as species at the time of publication of the relevant HMW volume. While the *Illustrated Checklist* published in 2020 aimed to update the taxonomy of the HMW series, the current book includes many taxonomic changes that have occurred during the short time since that publication, and these are summarized and the important cases explained in Appendix 3: Taxonomic changes and controversies in mammalogy (see p. 715).

However, perhaps the most important contribution made by this book is its graphic illustration of the plight of our Earth's mammals. Threatened species, easily seen by their colour-coded conservation status, appear on page after page of the book. It is a sobering thought that in the next edition of this book, some of the species may have to be moved to the Appendix on Extinct species (see p. 697). In fact, the section dedicated to extinct mammals already includes over one hundred species known to have gone extinct since 1500 AD, nearly all of them pushed over the edge by maliciousness or carelessness on the part of us humans. The irony should not be lost on us that we too are mammals, albeit naked and bipedal, in contrast to the vast majority of mammals that are quadrupedal and well furred.

At least part of the problem with conservation has been the difficulty scientists have in getting their message across to the general public and in making a broader impact beyond their own scientific community. The growing field of Science Communication, or "SciCom" as it is often abbreviated, applies to a wide set of activities that aim to connect science and society by transmitting scientific knowledge and research to non-expert audiences—including the public, industry, government, and educators—in an understandable way. The goal of this outreach is to increase public awareness and interest in science, which will have an impact on many levels. On an individual level, the hope is that such scientific engagement will produce a personal response and influence people's attitudes and behavior; on a more global scale hopefully it will affect public policy and promote actions addressing a variety of problems. For example, this book serves as an excellent SciCom tool, illustrating the tangled science of mammal taxonomy and diversity in a way that can be appreciated and understood by the general public, which, in turn, will raise awareness and interest in studying and protecting species and their habitats.

Is it too much to hope that this new book might fall into the hands of a young person who, by exploring its pages, becomes transformed by it and dedicates their life to saving our planet's amazing mammalian diversity? I do not doubt it.

Ara Monadjem
Professor, Department of Biological Sciences,
University of Eswatini, Kwaluseni, Eswatini

The Class Mammalia boasts a broad spectrum of animal types, ranging from diminutive shrews, mice, and bats to charismatic large carnivores, herbivores, marine mammals, and marsupials, and includes the largest documented animal to have ever existed, the Blue Whale (*Balaenoptera musculus*). The variety and complexity of forms and functions of mammals serve as a beacon for the importance and lure of the natural world, felt by the scientific community and the general public alike. However, unlike other more ubiquitous animals such as birds, mammals are rarely observed firsthand due to their small size, shyness, nocturnal behavior, success at hiding from humans (thanks to their remarkable hearing and olfactory senses), and lives led far from human settlements. Many of these species are also nearly indistinguishable from one another without a trained eye, making it difficult to appreciate the true diversity present in certain mammal groups. To put this into perspective, approximately 70% of all mammal species are either rodents, bats, or small insectivores, which, beyond the academic realm of study, are also the most difficult to observe and identify. The lack of awareness of these taxa leaves a major gap in knowledge and hampers conservation activities since more charismatic and easily visible species often receive far more attention from amateur naturalists, as well as from conservation organizations and policy makers. Thus, having routes for transferring scientific information from academic spheres to the general public is a key element in the conservation of these organisms and their environments and is, furthermore, the major goal of this book.

All the Mammals of the World (AMW) represents the culmination of the work contained in numerous volumes and all the research that went before it, and is also the first such book of its kind. The base of this important global project is the nine-volume *Handbook of the Mammals of the World* (HMW) series (Wilson *et al.* 2009–2019), the first work to ever cover and lavishly illustrate all mammal species, and provide information on their evolutionary relationships, natural history, ecology, and conservation status. The series was followed by the *Illustrated Checklist of the Mammals of the World* (Burgin *et al.* 2020), which updates the HMW taxonomy and provides a complete and fully illustrated checklist in a set of two volumes. The current work continues the legacy of these publications by providing a single-volume summation and an updated taxonomic arrangement of all the world's mammals in a more practical and digestible format aimed at readers with a broad-based interest in mammalian diversity. Although the HMW series and the *Illustrated Checklist* have similar goals and offer exceptional information and illustrations of all the world's mammals, both are costly, information-dense, and physically voluminous, which inevitably limits their accessibility. Thus, we hope that AMW, the first single, easy-to-use, fully illustrated volume covering every mammal species in the world, will bring the extraordinary richness and diversity of the world's mammals closer to a wider audience.

Readers familiar with other works by Lynx may be aware of "sister" projects published for birds such as the *Handbook of the Birds of the World* (del Hoyo *et al.* 1992–2013), the *HBW and BirdLife International Illustrated Checklist of the Birds of the World* (del Hoyo & Collar 2014, 2016), and *All the Birds of the World* (del Hoyo 2020). Although the current book serves a similar function as *All the Birds of the World*, there are some key distinctions between the systematics of birds and mammals that ensure that the scope of these volumes is different. *All the Birds of the World* includes all species included in each of the four major bird checklists in use at the time of preparation and so reflects the disagreements between major authorities on which species should be recognized and which species concepts should be used to define them. Unlike birds, mammals have not undergone a rigorous taxonomic vetting system and there is currently only one major authority on the world's mammals, the *Mammal Diversity Database* (MDD; mammaldiversity.org) run by the American Society of Mammalogists (Burgin *et al.* 2018). Given that there is a just a single source, the taxonomic arrangement presented here is based entirely on version 1.9 of the MDD, released on 1 April 2022 (ASM 2022), with some specific exceptions detailed in Appendix 3: Taxonomic changes and controversies in mammalogy (see p. 715). A cut-off point for applying the taxonomic criteria was necessary to allow time for processing the data and for the meticulous preparation of the layout.

There are two major caveats regarding the use of a single taxonomic checklist. Firstly, taxonomy is constantly changing, which is why the MDD releases regular updates like the taxonomic checklists published for birds (e.g. Clements *et al.* 2022), reptiles (Uetz *et al.* 2022), amphibians (AmphibiaWeb 2023), and fish (Froese & Pauly 2023). So, although this volume reflects the up-to-date taxonomy used in version 1.9 of the MDD, it does not aim to be a definitive taxonomic revision of all the world's mammals; rather, the MDD should be seen as a practical list illustrating the diversity of currently recognized mammal species. Secondly, the MDD version used to construct the taxonomic arrangement in this volume was produced using the most recent taxonomic revisions for each clade with few exceptions to avoid bias towards the opinions of the MDD curators. Thus, all taxonomic decisions made in this book reflect the opinions of the researchers working in these groups rather than the taxonomic views of the database curators (unless stated otherwise). For those interested in the specific taxonomic decisions used in this volume, it is worth consulting Appendix 3 of this book, the HMW series, the *Illustrated Checklist*, or the current published version of the MDD at mammaldiversity.org.

The constant fluctuations and discussions over the taxonomic arrangements of various organisms run the risk of making taxonomy seem unstable even to the point of irrelevancy. However, species definitions have wide-reaching implications. Species are the basal referenceable entity used in biological research including zoonotic disease research and conservation litigation (Cook *et al.* 2020, Upham *et al.* 2021, Wang *et al.* 2021). Species also serve as the foundation for developing key biodiversity infrastructure such as museum collections, databases, and publications (Cook & Light 2019). Well-supported species designations are crucial to biological research—despite the declining support for taxonomic and museum-based research, which is in turn crucial for keeping a record of the world's biodiversity (Wheeler 2020). The goal of taxonomy and the naming of species (termed "nomenclature") is twofold: 1) to produce a taxonomic arrangement that represents our current understanding of the evolutionary history of life on Earth; and 2) to maintain stability in the names and classification of life on Earth so that the scientific community, conservation and political organizations, and the general public have a common language for referring to different taxa over time and around the world. In light of this necessity, works like *All the Mammals of the World* containing an updated yet nomenclaturally consistent listing of the world's mammals can be used to make inferences about mammal diversity and conservation.

This book, although primarily focused on the dissemination of taxonomic information regarding the world's mammals among a more general audience, will also be highly valuable for professionals working in the fields of biology and mammalogy as a compendium of mammal diversity. Mammalian taxonomy is highly dynamic, with 40–60 new names described every year and other species being resurrected from synonymy due to ongoing revisionary work (see Table 1 for a summary of taxonomic changes in mammal taxonomy over time; Burgin *et al.* 2018). The increasing accessibility and power of molecular genetic analyses have stimulated this type of revisionary work throughout most major mammalian groups. For example, the silky anteater genus *Cyclopes* (see p. 82) once included a single species until a recent systematic revision using molecular and morphometric analyses split this taxon into seven distinct species (Miranda *et al.* 2018). Such integrative taxonomic work is essential for future progress and ideally incorporates data on nuclear and mitochondrial genomes, morphology, cytogenetics, distributions, and ecological and behavioral information. Future work classifying the categories of evidence supporting (or refuting) given taxonomic hypotheses will be essential in the advancement of research on mammalian taxonomy and evolutionary history.

Table 1. Total numbers of mammals recognized by major taxonomic compendia over time, comparing *Mammal Species of the World* (MSW) editions 1–3, *International Union of Conservation of Nature* (IUCN) *Red List of Threatened Species* versions 2008 and 2022-2, *ASM Mammal Diversity Database* (MDD) v1.0, and the *Illustrated Checklist of the Mammals of the World* (CMW) relative to *All the Mammals of the World* (AMW).

Taxa	MSW1 1982	MSW2 1993	MSW3 2005	IUCN 2008	MDD 2018	CMW 2020	IUCN 2022	AMW 2023
Species								
Total	4170	4631[a]	5416	5513	6495	6554	5973	6581
Extinct	NA	NA	75	79[b]	96	104	87[b]	103
Living	NA	NA	5341	5436	6399	6450	5886	6478
Domesticated	0	0	0	0	16	19	0	19
Living wild	NA	NA	5338	5436	6382	6432	5886	6459
Genera	1033	1135	1230	1226	1314	1343	1302	1340
Families	135	132	153	149	167	167	163	167
Orders	20	26	29	24	27	27	27	27

[a] Corrected total per Solari & Baker (2007)
[b] Extinct IUCN mammals include both "EX" (Extinct) and "EW" (Extinct in the Wild)

How to Use This Book

Divisions and sequence

Mammals, known by the scientific name Mammalia, are one of the four taxonomic classes of living tetrapods (along with amphibians, reptiles, and birds) and are descended from reptile-like ancestors called Synapsids. The Class Mammalia is broadly divided by reproductive method into egg-laying (Prototheria) and live-bearing (Theria) mammals, of which the latter are further divided into marsupials (mammals with pouches to house their young; Metatheria/Marsupialia) and placentals (mammals with placentas that provide nutrients for their young in the womb; Eutheria/Placentalia). The classification of mammals has changed dramatically over the past 30 years with the advent of more complex molecular genetic and statistical methods, which have enhanced our understanding of the evolutionary relationships between mammalian clades.

Like the HMW series and the *Illustrated Checklist*, the classification above family level is not highlighted in the plates of *All the Mammals of the World*, even if the sequence of the orders is influenced by their evolutionary relationships. The classification of mammals below class but above family levels are shown in Table 2, which reflects the taxonomic sequence of the orders and families in the main text. The evolutionary relationship of mammals is also presented in Figure 1, which was constructed from molecular genetic data from nearly every family of living mammal in a recent large-scale phylogenetic study of over 5900 mammal species (Upham *et al.* 2019). The sequence of this volume is based on this work as it provides the most up-to-date view of mammalian evolutionary relationships currently available. Furthermore, because of the family-level focus of this work, subfamilies and tribes, taxonomic ranks between genus and family levels, are not included; for the use of these taxonomic ranks, refer to the HMW series, *Illustrated Checklist*, or the MDD.

Readers familiar with other taxonomic or biodiversity resources may notice differences in this book regarding the placement of certain species in different genera, the inclusion of a recognized species under a different name, or even the exclusion of a species recognized elsewhere. Taxonomy is constantly in flux as more data becomes available for estimating the geographic, morphological, and genetic boundaries between species. When a species is determined to represent multiple distinct species, the resulting species are considered to have been 'split' from the parent species, whereas two or more species that are found to be synonymous (representing the same species) are 'lumped' into a single species. For example, Giraffes were previously considered to represent a single species, but were recently split into three or four species depending on the source: *Giraffa giraffa*, *G. tippelskirchi*, and *G. reticulata* were split from *G. camelopardalis* (the arrangement used here; see p. 587), although some sources lump *G. reticulata* into *G. camelopardalis,* while others continue to lump all Giraffe populations under *G. camelopardalis* (Fennessy *et al.* 2016, Bercovitch *et al.* 2017, Petzold & Hassanin 2020, Coimbra *et al.* 2021). Species are also often moved between genera and, for instance, all hairy-tailed bat species are either placed in a single genus, *Lasiurus* (the arrangement followed here; see p. 548), or in three genera, *Lasiurus*, *Aeorestes*, or *Dasypterus* (Baird *et al.* 2015, Novaes *et al.* 2018). Many cases of differential genus use are resolved once more data are produced to justify a specific arrangement, although some, like the *Lasiurus* example given above, are more subjective since the decision comes down to whether an already monophyletic group should be split into multiple genera or retained as just one for functional use. Taxonomic revisions at species level usually occur because the previously defined genus was paraphyletic or polyphyletic (see Figure 2 for graphical representation of mono-, para-, and polyphyletic groups).

The sequence of the families within each order and the species within each family are presented visually and should be "read" from left to right and from top to bottom. The species-level sequence is ordered by evolutionary relationships when molecular data are available to make such inferences, or by geographic distribution in the absence of molecular data. If the geographic distribution is used, species are ordered from north to south and from west to east. Due to the inherent complications in the layout of this work, in a small number of cases the sequence of presentation of species has been altered slightly to accommodate the design and make the most of the limited space available; we recommend readers consult the MDD for questions regarding taxonomic sequencing.

Figure 1. Family-level phylogenetic relationships of living mammals based on a mammal phylogeny of 4098 species for which DNA sequences were available (Upham *et al.* 2019). One species was chosen to represent each extant family (149 in total according to these authors' taxonomy, which differs from the more recent treatment used in this book) from the larger phylogeny, which was based on a 31-gene matrix of the 4098 species. Variants of this tree are downloadable from http://vertlife.org/phylosubsets for the different subsets of mammal species. The colored dots at each node in the tree indicate the statistical support values for each of the family-level branching relationships, quantified as probabilities of ≥ 0·95 (black, strong support), 0·94–0·75 (red, moderate support), and < 0·75 (orange, weak support). Taxa below order level are shown with branches colored according to their labels. Numbers at nodes are the mean divergence times (in millions of years); the geological timescale follows Gradstein *et al.* (2012). The 27 extant orders of mammals are labeled in capital letters on the right-hand side. Although the taxonomic sequence differs in places from the one used in this book, this does not affect the validity of the information shown here.

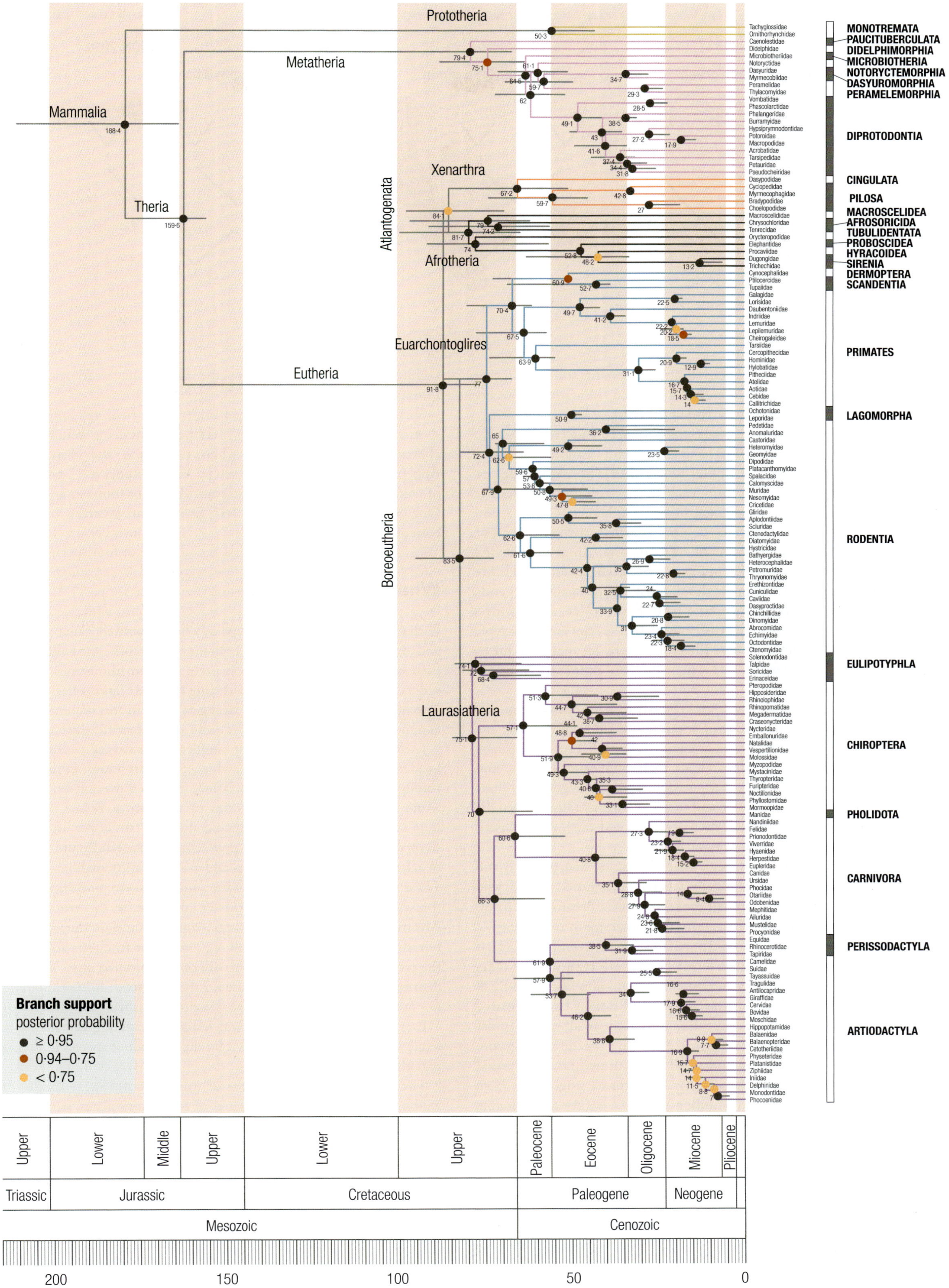

Prototheria
Metatheria
Mammalia
Theria
Xenarthra
Atlantogenata
Afrotheria
Euarchontoglires
Eutheria
Boreoeutheria
Laurasiatheria
MONOTREMATA
PAUCITUBERCULATA
DIDELPHIMORPHIA
MICROBIOTHERIA
NOTORYCTEMORPHIA
DASYUROMORPHIA
PERAMELEMORPHIA
DIPROTODONTIA
CINGULATA
PILOSA
MACROSCELIDEA
AFROSORICIDA
TUBULIDENTATA
PROBOSCIDEA
HYRACOIDEA
SIRENIA
DERMOPTERA
SCANDENTIA
PRIMATES
LAGOMORPHA
RODENTIA
EULIPOTYPHLA
CHIROPTERA
PHOLIDOTA
CARNIVORA
PERISSODACTYLA
ARTIODACTYLA
Branch support
posterior probability
≥ 0·95
0·94–0·75
< 0·75
Upper
Lower
Middle
Upper
Lower
Upper
Paleocene
Eocene
Oligocene
Miocene
Pliocene
Triassic
Jurassic
Cretaceous
Paleogene
Neogene
Mesozoic
Cenozoic
200
150
100
50
0

Figure 2. Phylogenetic trees showing examples of a monophyletic group (one that includes the most recent common ancestor and all its descendants), a paraphyletic group (one that includes a common ancestor but not all its descendants), and a polyphyletic group (one that does not include the common ancestor of all members of the taxon).

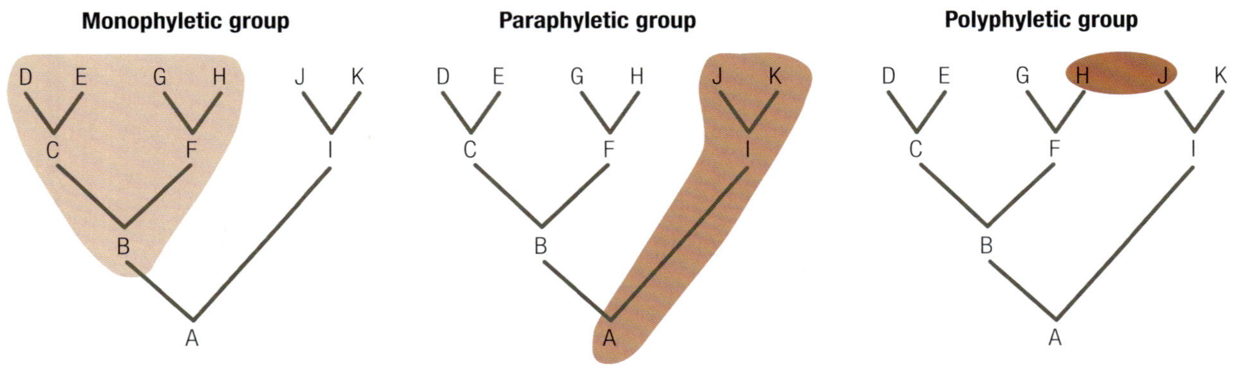

The main body of this volume includes all currently extant (still in existence) wild mammal species. However, recently extinct and domestic species are presented in separate appendices at the end of the book.

Appendix 1: Extinct species (see p. 697) covers 103 species that have become extinct roughly since 1500 AD and are currently assigned the threat category of Extinct (EX) on The IUCN Red List of Threatened Species. In addition to the family classification, each species account includes the Latin and vernacular names of the taxon in English, French, German, and Spanish, as well as a short summary describing when each species went extinct and the factors that contributed to its demise. An accompanying distribution map is included, shaded in a gray crosshatch pattern to indicate its approximate former distribution, with solid shading for small ranges and/or islands where the crosshatching would be difficult to see, along with a short phrase summarizing its former distribution. All species are monotypic unless otherwise indicated in the text. Illustrations are not included due to the lack of information and resources for many of these extinct species, such as species known exclusively from fossil material.

Appendix 2: Domestic species (see p. 711) deals with 19 domestic species defined taxonomically by the MDD. The species, divided by families, are presented with their Latin and vernacular names in English, French, German, and Spanish, as well as information about their domestication, use, and distribution. Maps are not included because domestic species generally share worldwide distributions with humans, and there are no illustrations given the variation within domestic forms (e.g. domestic dogs and cats). All the listed domestic species are considered monotypic here.

There is considerable confusion regarding the use of domesticated and wild-form names in mammals, especially when domesticated and if wild forms were once considered to represent the same species. Because wild forms were typically used in publications relating to the conservation and ecology of these species, the names applying to wild populations were recently designated as having priority over domesticated forms (which are typically the older names) if regarded as a single species (see ICZN Opinion 2027 in ICZN 2003, Gentry *et al.* 2004). Nevertheless, recent publications have suggested listing domesticated taxa as distinct species rather than as subspecies under their wild forms to avoid confusion when discussing the different forms. This arrangement is followed here and so all domesticated taxa are treated as distinct from their wild forms and any feral forms originating from domesticated stock (e.g. European Mouflon under *Ovis aries* and Dingoes under *Canis familiaris*; Gentry *et al.* 2004, Jackson *et al.* 2019). However, it must be noted that in a biological sense, domestic forms are not necessarily "true" species since they represent a selectively bred population of the wild form and are genetically identifiable as populations of the same species. They are primarily listed as distinct species here and in other publications to avoid confusion with their wild form for conservation and research purposes.

For readers interested in taxonomic matters, Appendix 3: Taxonomic changes and controversies in mammalogy (see p. 715) contains the most notable taxonomic issues dealt with in this book and includes a table highlighting the differences in the taxonomic treatment between the *Illustrated Checklist* and AMW.

Further resources are provided under three final appendices: Appendix 4: Country codes (see p. 725) has helpful lists of countries and their abbreviation codes for use when navigating the distribution texts; Appendix 5: Reference maps (see p. 729) includes a 37-page world atlas of color reference maps, with relevant details of interest; and Appendix 6: One-country endemics (see p. 767) provides an overview of the countries containing endemic species.

What is included

Each **Order** (27 in total) is marked by a header with its name in capital letters and the number of families it contains in parentheses. Unlike most other taxonomic ranks, there is no consistent suffix in the names of mammalian orders, with some, for example, ending in "-idea" and others in "-morphia". The order Primates may be the strangest of all in this naming system, as the common name for the order is the same as its scientific name.

Each **Family** of the 160 extant families is demarcated within a red-colored header giving its scientific name, which consistently ends in the suffix "-idae" throughout zoological nomenclature, followed by an English common name representative of the species in the family. Below the header, the number of genera and species included in the family are listed, as well as a short description of the habitat preferences of the family and its distribution in relation to defined territories and zoogeographic regions (see Figure 3). These zones are representative of the zoological communities associated with the seven different regions of the world: Nearctic, Neotropical, Palearctic, Afrotropical, Indo-Malayan, Australasian, and Antarctic/Subantarctic. When referring to taxa distributed in all regions except Antarctic and Subantarctic, the term "Cosmopolitan" is used; although this term often implies a global distribution, an exception is made for the Antarctic and Subantarctic as very few families have a distribution that includes this region.

Every **Genus** appears in capital letters in a gray box with its scientific name italicized, followed below it by the number of species in each genus in parentheses.

The **species** are the base units of this book and every species of extant wild mammal is presented in a series of continuous plates, with an average of 10–12 species per page, with scientific illustrations, distribution maps, and general information. The figures, map, and data attributable to a single species are delimited by a thin gray line separating each species from the others in the same genus; the members of a genus are separated from neighboring genera by a thicker gray line. If a genus has many species, its presentation may extend beyond the initial double-page where it begins, and the same genus continues until the division line marking its end. The two types of division lines are an important characteristic of this book, as they provide an easy way to navigate and interpret the relationships of species and genera to one another.

Each species is labeled with an **English common name** in bold font. Common or vernacular names of mammals vary widely and are considerably less regulated than the common names of birds. With this in mind, the names presented herein reflect the most commonly used and uncontroversial common names associated with each species (albeit with some exceptions). Additional alternative names can be consulted in the HMW series and the *Illustrated Checklist*.

One of the goals when choosing the English common names was to avoid ambiguity. For example, the common name Elk refers to *Alces alces* in Europe but to *Cervus canadensis* in North America. To avoid confusion and remove any doubt, Elk is not used as the common name for either of these two species and the names Moose (*Alces alces*) and Wapiti (*Cervus canadensis*) are employed instead. In most cases, names containing the complete name of another species are also avoided. This is often an issue when a species is split into multiple species as in the case of the Lutrine Opossum (*Lutreolina crassicaudata*). When a second species was described in the genus, *L. massoia*, the common name for that species was designated as Massoia's Lutrine Opossum. So, instead of leaving *L. crassicaudata* as Lutrine Opossum, the name was changed to Big Lutrine Opossum to make the separation between these two species clearer. An effort has also been made throughout the work to maintain consistent common names within genera or families (e.g. the use of Deermouse for all *Peromyscus* and White-toothed Shrew for all *Crocidura*).

In terms of style, for consistency all separate words within the name are capitalized, which does not include the second part of hyphenated words, including directional words (e.g. White-footed Deermouse and South-western Myotis); most names follow American English standards (Gray rather than Grey, -colored rather than -coloured) and accent marks are not included for standardization.

For newly recognized species, an English common name was generally sought by investigating preexisting sources such as subspecies names in field guides. However, if no name was found, a new name was invented based on the species' diagnostic features, distribution, or etymology of its scientific name. Preference is given to names highlighting defining features or distribution, although if no other name fits, the species is sometimes named after the person the specific epithet was named by or after.

The species' **scientific name** is given in smaller italic font. Scientific names are in Latin (or a Latinized form of another language) and consist of two parts; the first corresponds to the genus and begins with a capital letter, while the second—the specific epithet—is unique to the species and begins with a lower-case letter (e.g. *Homo sapiens*, *Homo* being the genus and *sapiens* the specific epithet). For those interested in further details about scientific names, the HMW series and *Illustrated Checklist* provide for each species the original scientific name, the authors that initially described it, and the year in which the description appeared in zoological publications.

Scientific names are meant to represent a common naming system for all languages and are governed by an organization known as the International Code of Zoological Nomenclature (ICZN), often known as "the Code". However, authorities may interpret the rules of the Code in different ways, which can lead to different spellings of specific epithets, or the use of completely different epithets between taxonomic resources. For example, the specific name for Christie's Long-eared Bat can be found as '*christii*' or '*christiei*' depending on whether the altered name spelling ('*christiei*' in this case) is considered a justified or unjustified emendation (i.e. whether the spelling change is allowed by the Code), and whether it is in prevailing usage despite being considered initially unjustified (i.e. it is used mostly in modern scientific literature); if it is in prevailing use, then an unjustified emendation

Figure 3. Zoogeographical regions

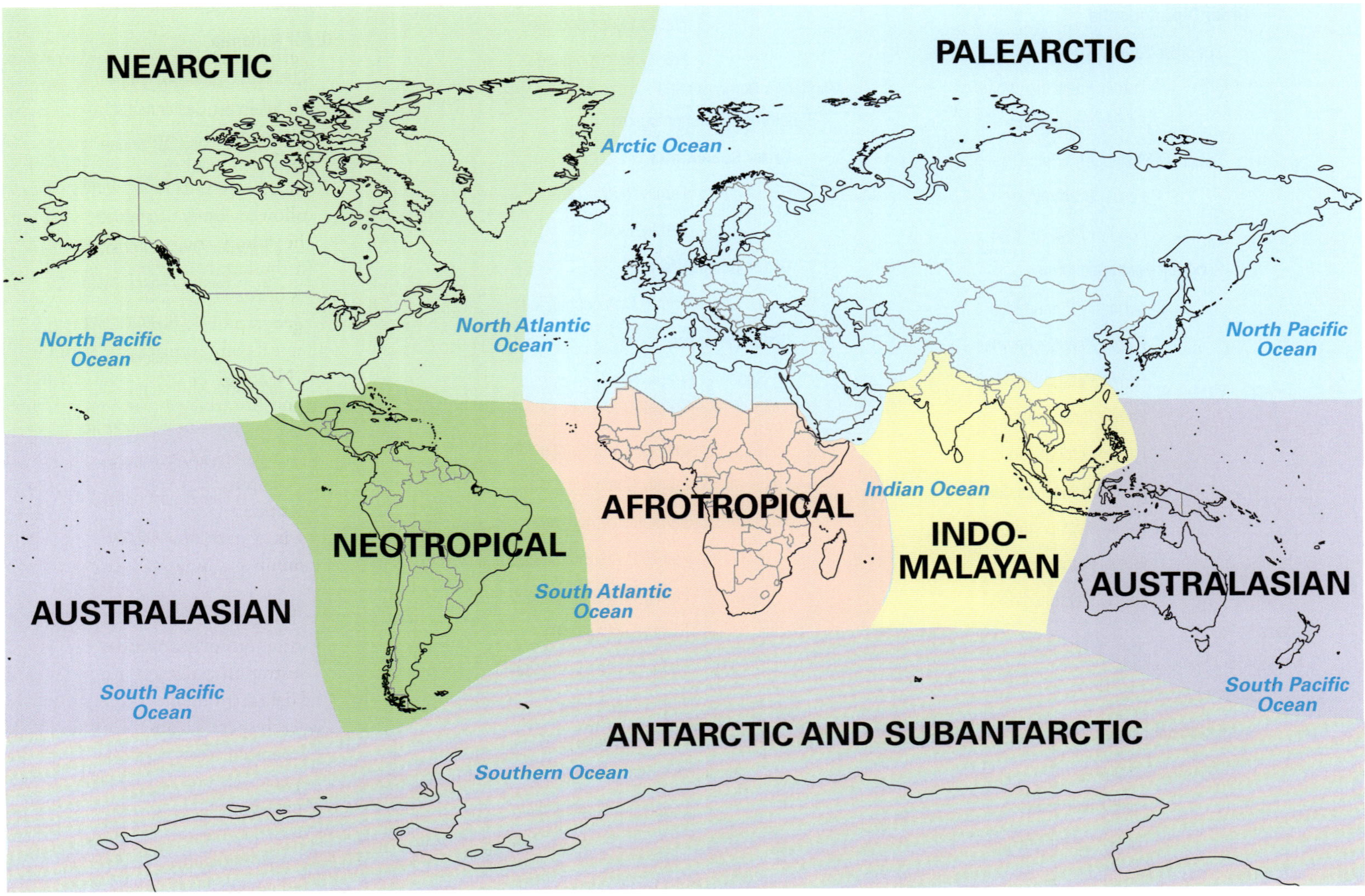

Table 2. Suprafamilial taxonomy of the world's mammals (families marked with "†" symbol are extinct)

Subclass Prototheria
- **Order Monotremata**
  - Family Tachyglossidae
  - Family Ornithorhynchidae

Subclass Theria
- Infraclass Metatheria
  - **Order Didelphimorphia**
    - Family Didelphidae
  - **Order Paucituberculata**
    - Family Caenolestidae
  - **Order Microbiotheria**
    - Family Microbiotheriidae
  - **Order Notoryctemorphia**
    - Family Notoryctidae
  - **Order Dasyuromorphia**
    - Family Thylacinidae †
    - Family Myrmecobiidae
    - Family Dasyuridae
  - **Order Peramelemorphia**
    - Family Thylacomyidae
    - Family Chaeropodidae †
    - Family Peramelidae
  - **Order Diprotodontia**
    - Suborder Vombatiformes
      - Family Phascolarctidae
      - Family Vombatidae
    - Suborder Phalangeriformes
      - Family Burramyidae
      - Family Phalangeridae
    - Suborder Macropodiformes
      - Superfamily Petauroidea
        - Family Pseudocheiridae
        - Family Petauridae
        - Family Tarsipedidae
        - Family Acrobatidae
      - Superfamily Macropodoidea
        - Family Hypsiprymnodontidae
        - Family Potoroidae
        - Family Macropodidae
- Infraclass Eutheria
  - Magnorder Atlantogenata
    - Superorder Afrotheria
      - **Order Afrosoricida**
        - Suborder Tenrecomorpha
          - Family Tenrecidae
          - Family Potamogalidae
        - Suborder Chrysochloridea
          - Family Chrysochloridae
      - **Order Macroscelidea**
        - Family Macroscelididae
      - **Order Tubulidentata**
        - Family Orycteropodidae
      - **Order Hyracoidea**
        - Family Procaviidae
      - **Order Proboscidea**
        - Family Elephantidae
      - **Order Sirenia**
        - Family Dugongidae
        - Family Trichechidae
    - Superorder Xenarthra
      - **Order Cingulata**
        - Family Dasypodidae
        - Family Chlamyphoridae
      - **Order Pilosa**
        - Suborder Vermilingua
          - Family Myrmecophagidae
          - Family Cyclopedidae
        - Suborder Folivora
          - Family Choelopodidae
          - Family Bradypodidae
  - Magnorder Boreoeutheria
    - Superorder Euarchontoglires
      - **Order Scandentia**
        - Family Ptilocercidae
        - Family Tupaiidae
      - **Order Dermoptera**
        - Family Cynocephalidae
      - **Order Primates**
        - Suborder Strepsirrhini
          - Infraorder Lemuriformes
            - Superfamily Cheirogaleoidea
              - Family Cheirogaleidae
            - Superfamily Lemuroidea
              - Family Palaeopropithecidae †
              - Family Megaladapidae †
              - Family Lepilemuridae
              - Family Archaeolemuridae †
              - Family Lemuridae
              - Family Indriidae
          - Infraorder Chiromyiformes
            - Family Daubentoniidae
          - Infraorder Lorisiformes
            - Family Galagidae
            - Family Lorisidae
        - Suborder Haplorhini
          - Infraorder Tarsiiformes
            - Family Tarsiidae
          - Infraorder Simiiformes
            - Parvorder Platyrrhini
              - Family Callitrichidae
              - Family Aotidae
              - Family Cebidae
              - Family Pitheciidae
              - Family Atelidae
            - Parvorder Catarrhini
              - Superfamily Cercopithecoidea
                - Family Cercopithecidae
              - Superfamily Hominoidea
                - Family Hylobatidae
                - Family Hominidae
      - **Order Lagomorpha**
        - Family Ochotonidae
        - Family Prolagidae †
        - Family Leporidae
      - **Order Rodentia**
        - Suborder Supramyomorpha
          - Infraorder Castorimorphi
            - Family Castoridae
            - Family Heteromyidae
            - Family Geomyidae
          - Infraorder Anomaluromorphi
            - Family Zenkerellidae
            - Family Anomaluridae
            - Family Pedetidae
          - Infraorder Myomorphi
            - Superfamily Dipodoidea
              - Family Sminthidae
              - Family Zapodidae
              - Family Dipodidae
            - Superfamily Muroidea
              - Family Platacanthomyidae
              - Family Spalacidae
              - Family Calomyscidae
              - Family Nesomyidae
              - Family Cricetidae
              - Family Muridae

Suborder Hystricomorpha

Infraorder Ctenodactylomorphi

Family Ctenodactylidae

Family Diatomyidae

Infraorder Hystricognathi

Incertae sedis

Family Hystricidae

Family Thryonomyidae

Family Petromuridae

Family Heterocephalidae

Family Bathyergidae

Superfamily Erethizontoidea

Family Erethizontidae

Superfamily Cavioidea

Family Cuniculidae

Family Caviidae

Family Dasyproctidae

Superfamily Chinchilloidea

Family Chinchillidae

Family Dinomyidae

Superfamily Octodontoidea

Family Abrocomidae

Family Ctenomyidae

Family Octodontidae

Family Echimyidae

Suborder Sciuromorpha

Family Aplodontiidae

Family Sciuridae

Family Gliridae

Superorder Laurasiatheria

Order Eulipotyphla

Family Erinaceidae

Family Soricidae

Family Talpidae

Family Nesophontidae †

Family Solenodontidae

Order Chiroptera

Suborder Pteropodiformes

Superfamily Pteropodoidea

Family Pteropodidae

Superfamily Rhinolophoidea

Family Rhinopomatidae

Family Craseonycteridae

Family Megadermatidae

Family Rhinonycteridae

Family Hipposideridae

Family Rhinolophidae

Suborder Vespertilioniformes

Superfamily Emballonuroidea

Family Emballonuridae

Family Nycteridae

Family Myzopodidae

Superfamily Noctilionoidea

Family Mystacinidae

Family Noctilionidae

Family Furipteridae

Family Thyropteridae

Family Mormoopidae

Family Phyllostomidae

Superfamily Vespertilionoidea

Family Natalidae

Family Molossidae

Family Miniopteridae

Family Cistugidae

Family Vespertilionidae

Order Artiodactyla

Suborder Whippomorpha

Infraorder Cetacea

Parvorder Mysticeti

Family Balaenidae

Family Cetotheriidae

Family Balaenopteridae

Parvorder Odontoceti

Superfamily Physeteroidea

Family Physeteridae

Family Kogiidae

Superfamily Ziphioidea

Family Ziphiidae

Superfamily Platanistoidea

Family Platanistidae

Superfamily Lipotoidea

Family Lipotidae

Superfamily Inioidea

Family Iniidae

Family Pontoporiidae

Superfamily Delphinoidea

Family Monodontidae

Family Delphinidae

Family Phocoenidae

Infraorder Ancodonta

Family Hippopotamidae

Suborder Ruminantia

Infraorder Tragulina

Family Tragulidae

Infraorder Pecora

Family Antilocapridae

Family Giraffidae

Family Cervidae

Family Moschidae

Family Bovidae

Suborder Suina

Family Suidae

Family Tayassuidae

Suborder Tylopoda

Family Camelidae

Order Perissodactyla

Suborder Hippomorpha

Family Equidae

Suborder Ceratomorpha

Family Tapiridae

Family Rhinocerotidae

Order Pholidota

Family Manidae

Order Carnivora

Suborder Feliformia

Infraorder Nandinioidea

Family Nandiniidae

Infraorder Feloidea

Family Prionodontidae

Family Felidae

Infraorder Viverroidea

Family Viverridae

Family Herpestidae

Family Eupleridae

Family Hyaenidae

Suborder Caniformia

Infraorder Canoidea

Family Canidae

Infraorder Arctoidea

Superfamily Ursoidea

Family Ursidae

Superfamily Phocoidea

Family Otariidae

Family Odobenidae

Family Phocidae

Superfamily Musteloidea

Family Ailuridae

Family Mephitidae

Family Mustelidae

Family Procyonidae

BR = Biosphere Reserve*
CA = Conservation Area*
CF = Classified Forest*
CP = Conservation Park
CZ = Conservation Zone*
NP = National Park*
NR = Nature Reserve*
PA = Protected Area*
R = Reserve (when appearing after location name)
SR = Special Reserve*
WA = Wilderness Area*
WMA = Wildlife Management Area
WS = Wildlife Sanctuary*

Additional abbreviations for Australia:

ACT = Australian Capital Territory
NSW = New South Wales
NT = Northern Territory
Qld = Queensland
SA = South Australia*
Tas = Tasmania
Vic = Victoria
WA = Western Australia*

*These cases share their abbreviations with country codes (see Appendix 4, p. 725) or each other but are made clear by context.

More specific localities and regions are not abbreviated. For example, the Southern Brown Bandicoot (*Isoodon obesulus*; see p. 50), endemic to Australia, as indicated on the map, has the following range description: "NSW, Vic, mainland SA, Kangaroo I, Nuyts Arch, Tas"; this indicates the species is found in Australia in New South Wales, Victoria, mainland South Australia, Kangaroo Island, Nuyts Archipelago, and Tasmania. Introduced and reintroduced populations are not included in the distribution maps (except for more ancient introductions, which are included in some cases and specified in the text as, for example, in the Common Fallow Deer, *Dama dama*; see p. 590), although they are listed in the range description after the main distribution if space allows (which is not always the case). For example, the range description of the Common Spotted Cuscus (*Spilocuscus maculatus*; see p. 56), reads: "ID/PG: C & S Moluccas, New Guinea (with some islands), Aru Is; introduced to Selayar I and Bismarck Arch"; this indicates that this species is found in Indonesia and Papua New Guinea in the central and south Moluccas, New Guinea including some islands, and the Aru Islands natively, but was introduced to Selayar Island and the Bismarck Archipelago. Readers interested in more detailed distributions can explore the HMW series or *Illustrated Checklist*, which provide highly detailed range descriptions for each subspecies.

To further accompany the distribution map and range description, the **elevational range** for terrestrial and aerial species is also included close to the map if available. Elevations are indicated in meters (m) and given as a range representing the approximate lower and upper limits at which the species is normally recorded (e.g. 1500–2900 m) or an approximate value, indicated with "c." when few data are available, sometimes taken directly from the species' original description. A lower limit of 0 represents sea level; a few species are present below sea level, in which case the limit is a negative number. The elevational distribution of many mammal species has not been well recorded and we hope that this will improve with further interest and study.

Final Thoughts

Given that we have produced the *Handbook of the Mammals of the World* series and the *Illustrated Checklist of the Mammals of the World*, one might wonder why we insist on publishing another tome on mammals. However, as highlighted at the beginning of this introduction, these predecessors are more geared towards the academic community and their sheer dimensions limit their accessibility. Thus, we always aimed to create a single volume that—for the first time ever—would present all the mammals of the world in a format that would make the scientific information currently available on mammals more accessible for a general audience and raise awareness of the conservation importance of many species, especially those that are less well known or rarer. It is also our hope that this book will serve as a starting point for those interested in mammalian diversity and will stimulate greater public and scientific interest in the wonderful diversity that exists within the Class Mammalia.

Connor J. Burgin
PhD Candidate, Department of Biology,
University of New Mexico, Albuquerque, New Mexico, USA

Acknowledgements

We are extremely grateful to everyone who helped in so many ways to make this book possible.

Given that this project would not have been possible without its predecessors, we would like to thank everyone who contributed to the *Handbook of the Mammals of the World* and the *Illustrated Checklist of the Mammals of the World*. Sincere thanks thus go to the Chief Editors and Associate Editors of both works: Don E. Wilson, Russell A. Mittermeier, Anthony B. Rylands, Thomas E. Lacher Jr, Wes Sechrest, and Connor J. Burgin, and Albert Martínez-Vilalta, David Leslie Jr, Andrew Elliott, Marc Olivé, Sue Ruff, David Christie, and Matthew Richardson; the expert team of authors that contributed to the HMW series; the talented artists who illustrated the full range of mammalian diversity in all its splendor, with special thanks to Toni Llobet, who single-handedly illustrated seven of the nine HMW volumes (see p. 11 for author and artist contributions by family); and the various other collaborators and staff that made this global project a success. We are also grateful for the support and collaboration of Conservation International, the International Union for the Conservation of Nature (IUCN), Global Wildlife Conservation now Re:wild, Texas A&M AgriLife Research, the East Foundation, Stephen Nash, the Margot Marsh Biodiversity Foundation, Stony Brook Medicine Anatomical Sciences, Wildlife Conservation at the Houston Zoo, the Mascort Foundation, and the Ministry of Culture of the Generalitat de Catalunya.

We would also like to take this opportunity to thank the Partners of Lynx Edicions, Ramón Mascort, Jordi Sargatal, and Josep del Hoyo, for creating this publishing house and promoting so many important projects that have become a standard for excellence, and for entrusting us with the continuation of this legacy as Lynx Nature Books. We hope they will be proud of this next steppingstone in this rich trajectory, which we will endeavor to continue far into the future. Additional thanks are due to Josep for the development of the concepts behind this trilogy of mammal publications, an important contribution to the study and conservation of mammals, and for his continued support as a consultant for this book.

We are truly grateful to the late Normand David and Michel Gosselin (Canadian Museum of Nature), who provided the French names for the series and the new names and revisions required for the book, and reviewed the relevant scientific names for spelling and gender agreement. We were extremely fortunate to be able to enjoy Normand's expert collaboration for many years; he will be greatly missed as a professional and a colleague.

The German names for the series were provided by Gustav Peters and Rainer Hutterer (LIB Museum Koenig, Bonn, Germany); Rainer prepared the new and revised names. The Spanish names for the series and the current publication were contributed by Luis Javier Palomo of the Sociedad Española para la Conservación y el Estudio de los Mamíferos (SECEM), while the bat names, including the necessary revisions and new names, were provided by Maria Mas of the Asociación Española para la Conservación y el Estudio de los Murciélagos (SECEMU).

We are also thankful for the American Society of Mammalogists' *Mammal Diversity Database* for its key input on taxonomic situations and its continued work in this field. Special thanks also go to Nathan Upham, who kindly provided the phylogeny to create Figure 1 in the Introduction based on his 2019 paper. We would also like to recognize the importance of free-access online resources (e.g. iNaturalist) and their citizen-science contributors, whose observations, photographs, and identifications create important data that improve our understanding and promote nature conservation.

Thanks are due to Ara Monadjem, author in three HMW volumes and an important bat expert and researcher, for writing the Foreword and sharing his experiences and vision; we share his hope that this book may inspire readers to make a difference in conservation!

A large, complex book like this one requires a special team of collaborators with experience, skill, and infinite patience. We have been fortunate to count on an absolute dream team: Anna Motis, Daniel Roca, Connor J. Burgin, Marc Olivé, Albert Martínez-Vilalta, Lluc Julià, Oriol Cabrero, Guillem Martínez-Motis, Michael Lockwood, and Andrew Elliott.

We would like to express our gratitude to photographer Andy Smith for his breathtaking photo of the Snow Leopard (*Panthera uncia*) on the book's cover, which enraptured all of us and reminds us of the power and—simultaneously—the fragility of our world's mammals.

Finally, we extend our sincere thanks to our families and friends, and especially to our readers, old and new, for their support and encouragement. Enjoy!

References

AmphibiaWeb. (2023). University of California, Berkeley, California. URL: https://amphibiaweb.org (download 4 May 2023).

ASM (American Society of Mammalogists). (2018). *ASM Mammal Diversity Database: v1.0.* URL: https://www.mammaldiversity.org (download 4 May 2023).

ASM (American Society of Mammalogists). (2022). *ASM Mammal Diversity Database: v1.9.* URL: https://www.mammaldiversity.org (download 4 May 2023).

Baird, A.B., Braun, J.K., Mares, M.A., Morales, J.C., Patton, J.C., Tran, C.Q. & Bickham, J.W. (2015). Molecular systematic revision of tree bats (Lasiurini): doubling the native mammals of the Hawaiian Islands. *J. Mammal.* **96(6)**: 1255–1274.

Bercovitch, F.B., Berry, P.S.M., Dagg, A., Deacon, F., Doherty, J.B., Lee, D.E., Mineur, F., Muller, Z., Ogden, R., Seymour, R., Shorrocks, R. & Tutchings, A. (2017). How many species of giraffe are there? *Curr. Biol.* **27(4)**: R136–R137.

Burgin, C.J., Colella, J.P., Kahn, P.L. & Upham, N.S. (2018). How many species of mammals are there? *J. Mammal.* **99(1)**: 1–14.

Burgin, C.J., Wilson, D.E., Mittermeier, R.A., Rylands, A.B., Lacher, T.E. Jr. & Sechrest, W. (2020). *Illustrated Checklist of the Mammals of the World.* Vols. 1–2. Lynx Edicions, Barcelona.

Clements, J.F., Schulenberg, T.S., Iliff, M.J., Fredericks, T.A., Gerbracht, J.A., Lepage, D., Billerman, S.M., Sullivan, B.L. & Wood, C.L. (2022). *The eBird/Clements checklist of Birds of the World*: v2022. URL: https://www.birds.cornell.edu/clementschecklist/ (download 4 May 2023).

Coimbra, R.T.F., Winter, S., Kumar, V., Koepfli, K.P., Gooley, R.M., Dobrynin, P., Fennessy, J. & Janke, A. (2021). Whole-genome analysis of giraffe supports four distinct species. *Curr. Biol.* **31(13)**: 2929–2938.

Cook, J.A. & Light, J.E. (2019). The emerging role of mammal collections in 21st century mammalogy. *J. Mammal.* **100(3)**: 733–750.

Cook, J.A., Arai, S., Armién, B., Bates, J., Carrion, C.A., Cortez, M.B.S., Dunnum, J.L., Ferguson, A.W., Johnson, K.M., Khan, F.A.A., Paul, D.L., Reeder, D.M., Revelez, M.A., Simmons, N.B., Thiers, B.M., Thompson, C.W., Upham, N.S., Vanhove, M.P.M., Webala, P.W., Weksler, M., Yanagihara, R. & Soltis, P.S. (2020). Integrating biodiversity infrastructure into pathogen discovery and mitigation of emerging infectious diseases. *BioScience* **70(7)**: 531–534.

Fennessy, J., Bidon, T., Reuss, F., Kumar, V., Elkan, P., Nilsson, M.A., Vamberger, M., Fritz, U. & Janke, A. (2016). Multi-locus analyses reveal four giraffe species instead of one. *Curr. Biol.* **26(18)**: 2543–2549.

Froese, R. & Pauly, D. eds. (2023). *FishBase.* URL: https://www.fishbase.org (download 4 May 2023).

Gentry, A., Clutton-Brock, J. & Groves, C.P. (2004). The naming of wild animal species and their domestic derivatives. *J. Archaeol. Sci.* **31(5)**: 645–651.

Gradstein, F.M., Ogg, J.G., Schmitz, M.D. & Ogg, G.M. (2012). *The Geologic Time Scale 2012.* Elsevier, Amsterdam, The Netherlands.

Honacki, J.H., Kinman, K.E. & Koeppl, J.W. (1982). *Mammal Species of the World: A Taxonomic and Geographic Reference.* Allen Press, Lawrence, Kansas.

del Hoyo, J. ed. (2020). *All the Birds of the World.* Lynx Edicions, Barcelona.

del Hoyo, J. & Collar, N.J. (2014). *HBW and BirdLife International Illustrated Checklist of the Birds of the World.* Vol. 1. Non-passerines. Lynx Edicions, Barcelona.

del Hoyo, J. & Collar, N.J. (2016). *HBW and BirdLife International Illustrated Checklist of the Birds of the World.* Vol. 2. Passerines. Lynx Edicions, Barcelona.

del Hoyo, J., Elliott, A., Sargatal, J. & Christie, D.A. eds. (1992–2013). *Handbook of the Birds of the World.* Vols. 1–16. Lynx Edicions, Barcelona.

ICZN (International Commission on Zoological Nomenclature). (2003). Opinion 2027 (Case 3010) – Usage of 17 specific names based on wild species which are pre-dated by or contemporary with those based on domestic animals (Lepidoptera, Osteichthyes, Mammalia): conserved. *Bull. Zool. Nomen.* **60(1)**: 81–84.

IUCN (International Union for Conservation of Nature). (2008). *The IUCN Red List of Threatened Species.* Version 2008. URL: https://www.iucnredlist.org (download 4 May 2023).

IUCN (International Union for Conservation of Nature). (2022). *The IUCN Red List of Threatened Species.* Version 2022-2. URL: https://www.iucnredlist.org (download 4 May 2023).

Jackson, S.M., Fleming, P.J.S., Eldridge, M.D.B., Ingleby, S., Flannery, T.F., Johnson, R.N., Cooper, S.J.B., Mitchell, K.J., Souilmi, Y., Cooper, A., Wilson, D.E. & Helgen, K.M. (2019). The Dogma of Dingoes—Taxonomic status of the dingo: A reply to Smith *et al. Zootaxa* **4564(1)**: 198–212.

McCarthy, T., Mallon, D., Jackson, R., Zahler, P. & McCarthy, K. (2017). *Panthera uncia.* In: *The IUCN Red List of Threatened Species.* Version 2022.2. URL: https://dx.doi.org/10.2305/IUCN.UK.2017-2.RLTS.T22732A50664030.en (download 4 May 2023).

Miranda, F.R., Casali, D.M., Perini, F.A., Machado, F.A. & Santos, F.R. (2018). Taxonomic review of the genus *Cyclopes* Gray, 1821 (Xenarthra: Pilosa), with the revalidation and description of new species. *Zool. J. Linn. Soc.* **183(3)**: 687–721.

Novaes, R.L.M., Garbino, G.S.T., Cláudio, V.C. & Moratelli, R. (2018). Separation of monophyletic groups into distinct genera should consider phenotypic discontinuities: the case of Lasiurini (Chiroptera: Vespertilionidae). *Zootaxa* **4379(3)**: 439–440.

Petzold, A. & Hassanin, A. (2020). A comparative approach for species delimitation based on multiple methods of multi-locus DNA sequence analysis: A case study of the genus *Giraffa* (Mammalia, Cetartiodactyla). *PLoS ONE* **15(2)**: e0217956.

Solari, S. & Baker, R.J. (2007). Wilson, D. E., and D. M. Reeder (eds.). 2005. MAMMAL SPECIES OF THE WORLD: A TAXONOMIC AND GEOGRAPHIC REFERENCE. 3rd edition. Johns Hopkins University Press, Baltimore, Maryland, 2142 pp. ISBN 0-8018-8221-4, 2 volumes, price (hardbound), $125.00. *J. Mammal.* **88(3)**: 824–830.

Uetz, P., Freed, P., Aguilar, R., Reyes, F. & Hošek, J. eds. (2022). *The Reptile Database.* URL: http://www.reptile-database.org (download 4 May 2023).

Upham, N.S., Esselstyn, J.A. & Jetz, W. (2019). Inferring the mammal tree: Species-level sets of phylogenies for questions in ecology, evolution, and conservation. *PLoS Biol.* **17(12)**: e3000494.

Upham, N.S., Poelen, J.H., Paul, D., Groom, Q.J., Simmons, N.B., Vanhove, M.P.M., Bertolino, S., Reeder, D.M., Bastos-Silveira, C., Sen, A., Sterner, B., Franz, N.M., Guidoti, M., Penev, L. & Agosti, D. (2021). Liberating host–virus knowledge from biological dark data. *Lancet Planet. Health* **5(10)**: e746–e750.

Wang, Y.X.G., Matson, K.D., Santini, L., Visconti, P., Hilbers, J.P., Huijbregts, M.A.J., Xu, Y., Prins, H.H.T., Allen, T., Huang, Z.Y.X. & de Boer, W.F. (2021). Mammal assemblage composition predicts global patterns in emerging infectious disease risk. *Global Change Biol.* **27(20)**: 4995–5007.

Wheeler, Q.D. (2020). A taxonomic renaissance in three acts. *Megataxa* **1(1)**: 4–8.

Wilson, D.E. & Reeder, D.M. eds. (1993). *Mammal Species of the World: A Taxonomic and Geographic Reference.* 2nd edition. Smithsonian Institution Press, Washington, D.C.

Wilson, D.E. & Reeder, D.M. eds. (2005). *Mammal Species of the World: A Taxonomic and Geographic Reference.* 3rd edition. The Johns Hopkins University Press, Baltimore, Maryland.

Wilson, D.E., Mittermeier, R.A. & Lacher, T.E. Jr. eds. (2009–2019). *Handbook of the Mammals of the World.* Vols. 1–9. Lynx Edicions, Barcelona.

Class Mammalia

Order MONOTREMATA (2 families)

Family TACHYGLOSSIDAE – Echidnas

2 genera, 4 species. Nearly all terrestrial habitats in Australia and New Guinea. Australasian Region.

Genus *ZAGLOSSUS*

(3 species)

Western Long-beaked Echidna
Zaglossus bruijnii
FR Échidné de Bruijn
DE Westlicher Langschnabeligel
ES Equidna de hocico largo occidental

CR
Monotypic

variants

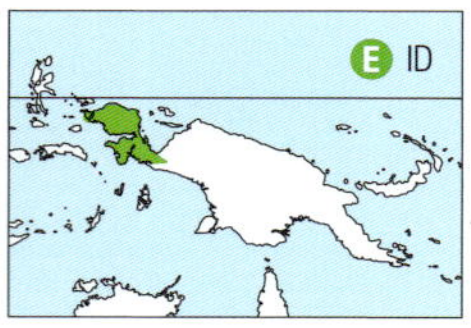

0–2500 m

W New Guinea (Vogelkop, Bomberai); possibly AU

Eastern Long-beaked Echidna
Zaglossus bartoni
FR Échidné de Barton
DE Barton-Langschnabeligel
ES Equidna de hocico largo oriental

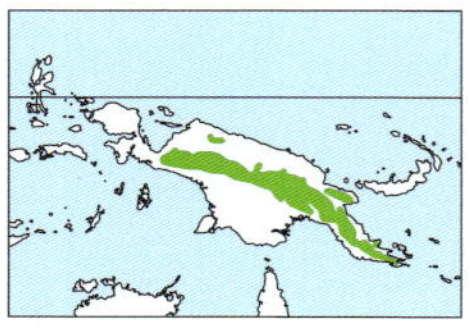

600–3200 m

ID/PG: New Guinea Highlands, Huon, Foja

ssp *bartoni*

ssp *clunius*

ssp *smeenki*

ssp *diamondi*

VU
4 sspp
HB 48–63 cm
W 4.2–9 kg

Attenborough's Long-beaked Echidna
Zaglossus attenboroughi
FR Échidné d'Attenborough
DE Attenborough-Langschnabeligel
ES Equidna de hocico largo de Attenborough

CR
Monotypic
HB c. 30 cm

New Guinea (Cyclops Mts)

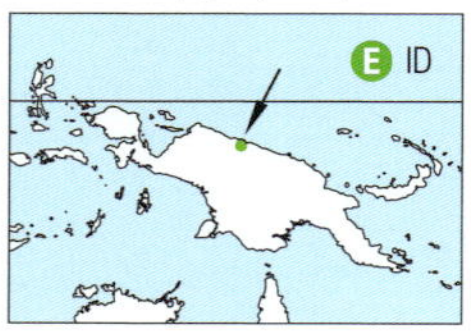

160–1700 m

Genus *TACHYGLOSSUS*

(1 species)

Short-beaked Echidna
Tachyglossus aculeatus
FR Échidné à nez court
DE Ameisenigel
ES Equidna de hocico corto

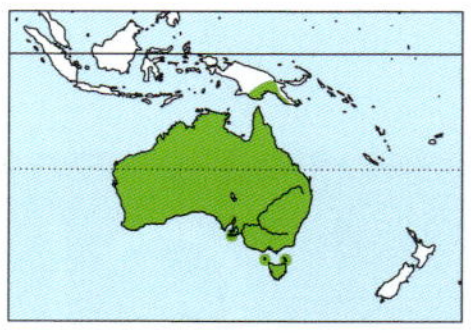

AU (with Kangaroo I, Tas and Bass Strait Is), PG (S & SE New Guinea)

ssp *acanthion*

LC
5 sspp
HB 30–45 cm
W 2–6 kg

ssp *aculeatus*

ssp *lawesii*

ssp *multiaculeatus*

ssp *setosus*

Flinders Island blond morph

Family ORNITHORHYNCHIDAE – Platypus

1 genus, 1 species. Inland freshwater bodies and associated riparian areas. Australasian Region.

Genus *ORNITHORHYNCHUS*

(1 species)

Platypus
Ornithorhynchus anatinus
FR Ornithorynque
DE Schnabeltier
ES Ornitorrinco

♀

♂

NT
Monotypic
HB 40–63 cm (M), 37–55 cm (F)
W 0.8–3 kg (M), 0.6–1.7 kg (F)

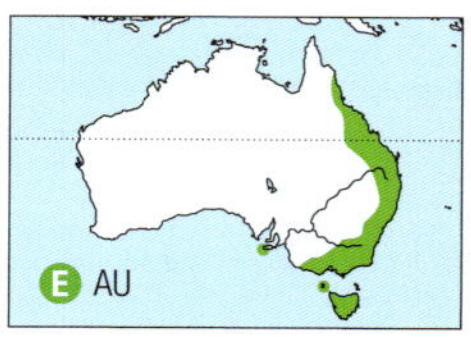

Qld (near Cooktown) S to Vic, Tas, King I; introduced to Kangaroo I

Order DIDELPHIMORPHIA (1 family)

Family DIDELPHIDAE – Opossums

18 genera, 127 species. Most habitats in the continental Americas from 50° N to 48° S. Nearctic and Neotropical Regions.

Genus *CALUROMYS*

(3 species)

Derby's Woolly Opossum
Caluromys derbianus

FR Opossum de Derby
DE Derby-Wollbeutelratte
ES Raposa lanuda de Derby

variants

LC
Monotypic
HB 22–30 cm
T 38–45 cm
W 245–370 g

S MX (Veracruz) through Central America (with Coiba I) to W CO & W EC (W Andes)

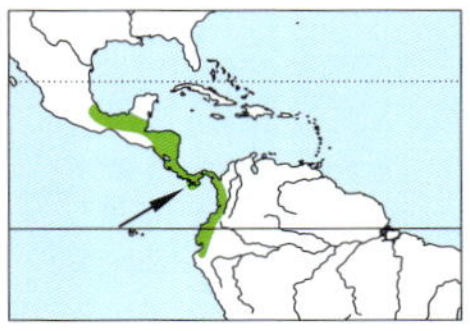

0–2500 m

Brown-eared Woolly Opossum
Caluromys lanatus

FR Opossum laineux
DE Braunohr-Wollbeutelratte
ES Raposa lanuda occidental

LC
2 sspp
HB 20–32 cm
T 33–44 cm
W 300–520 g

variants

CO, W & S VE, W & S BR, E EC, N & E PE, N & E BO, PY, NE AR (Misiones)

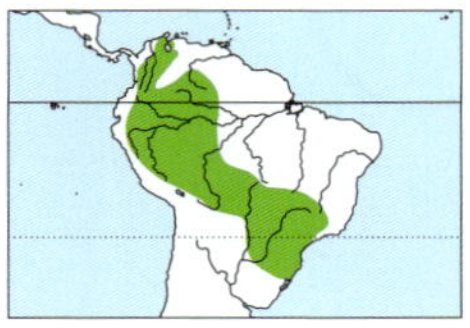

below 500 m

Bare-tailed Woolly Opossum
Caluromys philander

FR Opossum jaune
DE Gelbe Wollbeutelratte
ES Raposa lanuda oriental

LC
4 sspp
HB 16–28 cm
T 25–41 cm
W 140–390 g

at least 1600 m

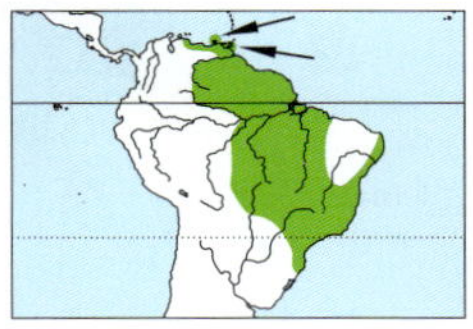

N & E VE (with Margarita I), TT (Trinidad I), Guianas, BR, E BO

Genus *CALUROMYSIOPS*

(1 species)

LC
Monotypic
HB 25–33 cm
T 31–34 cm
W 300–500 g

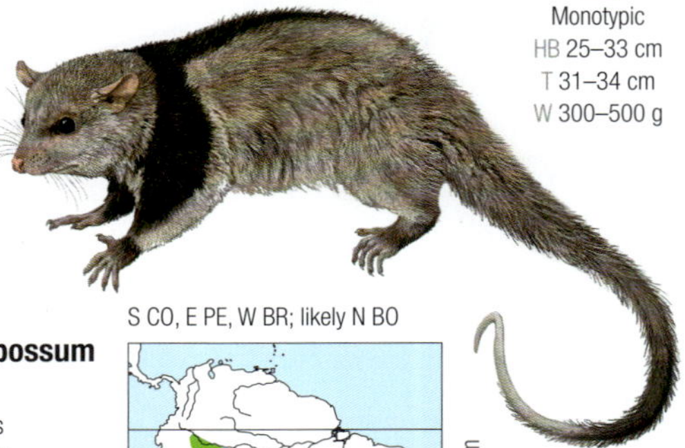

Black-shouldered Opossum
Caluromysiops irrupta

FR Opossum à épaules noires
DE Schwarzschulterbeutelratte
ES Raposa lanuda de hombros negros

S CO, E PE, W BR; likely N BO

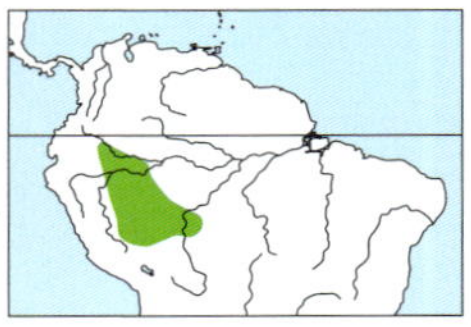

below 700 m

Genus *GLIRONIA*

(1 species)

LC
Monotypic
HB 17–21 cm
T 19.5–23 cm
W c. 130 g

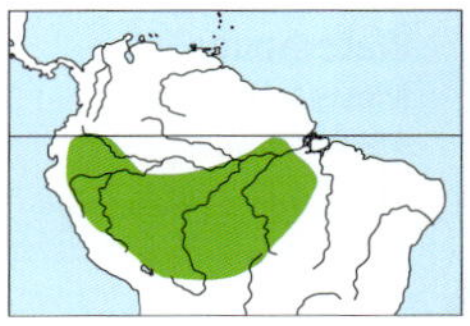

Bushy-tailed Opossum
Glironia venusta

FR Opossum à queue touffue
DE Buschschwanzbeutelratte
ES Raposa de cola peluda

S CO (Putumayo), E EC, E PE, N BO, BR (S Amazonas to Pará, S to Mato Grosso)

Genus *HYLADELPHYS*

(1 species)

Kalinowski's Opossum
Hyladelphys kalinowskii

FR Opossum de Kalinowski
DE Kalinowski-Zwergbeutelratte
ES Marmosa de Kalinowski

Guianas, Amazonian BR, E PE; possibly SE CO (Leticia) and N BO

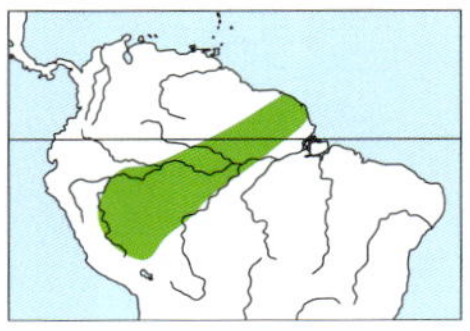

LC
Monotypic
HB 7.5–9.5 cm
T 10.2–11.7 cm
W 10–20 g

Genus *MARMOSA*

(26 species)

DD
Monotypic
HB 12.8–20 cm
T 18.3–22 cm
W 59–81 g

E EC, PE; possibly S CO

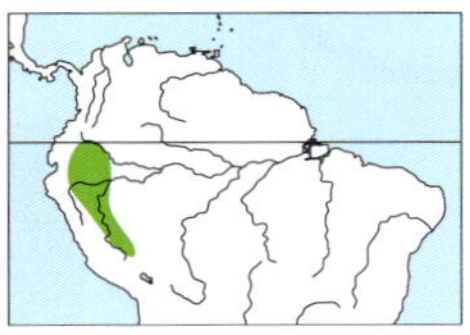

Red Mouse Opossum
Marmosa rubra

FR Opossum roux
DE Rote Zwergbeutelratte
ES Marmosa rojiza

NE
Monotypic
HB 9.8–15.6 cm
T 12.1–18.9 cm
W 24–49 g

0–1600 m

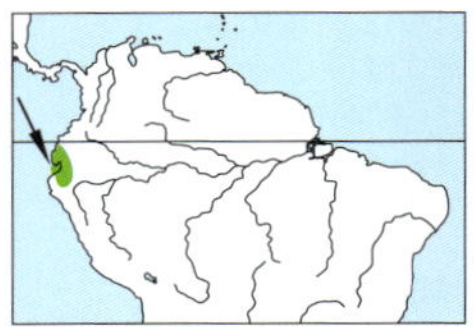

Simons's Mouse Opossum
Marmosa simonsi

FR Opossum de Simons
DE Simon-Zwergbeutelratte
ES Marmosa de Simons

W EC (with Puná I), NW PE

Panama Mouse Opossum
Marmosa isthmica

FR Opossum isthmique
DE Panama-Zwergbeutelratte
ES Marmosa de Panamá

below 1700 m

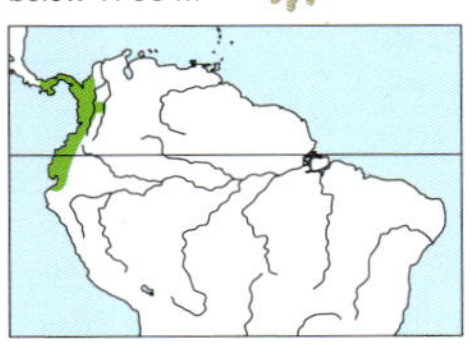

NE
Monotypic
HB 10.5–20 cm
T 15–25 cm
W 40–164 g

PA, CO (NW Caribbean lowlands, Pacific coast, W Cundinamarca), W EC

Zeledon's Mouse Opossum
Marmosa zeledoni

FR Opossum de Zeledon
DE Zeledon-Zwergbeutelratte
ES Marmosa de Zeledón

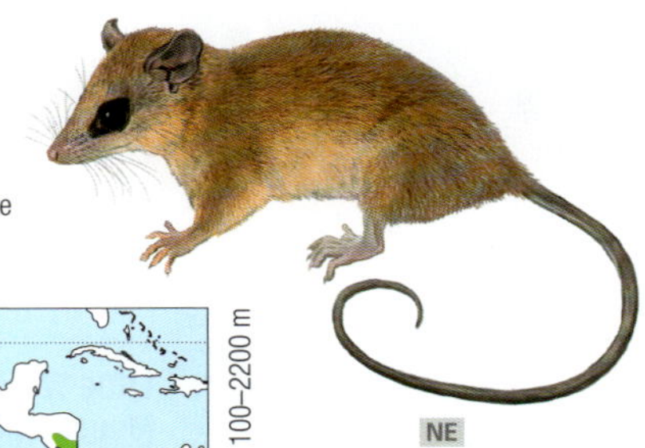

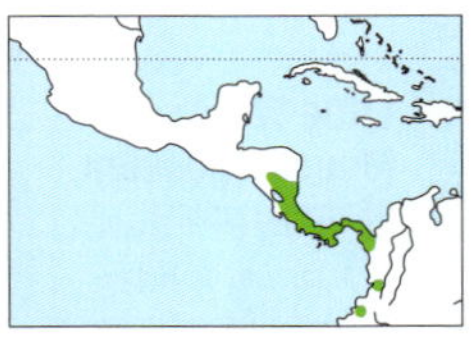

100–2200 m

NI, CR, PA, W CO (Nariño, Valle del Cauca)

NE
Monotypic
HB 11.9–16.2 cm
T 14.5–21 cm
W 28–100 g

LC
Monotypic
HB 9.4–16.8 cm
T 13–23 cm
W 24–99 g

E & SE MX (Tamaulipas) to E PA, with Roatán I (HN), Maíz Grande I (NI)

below 1600 m

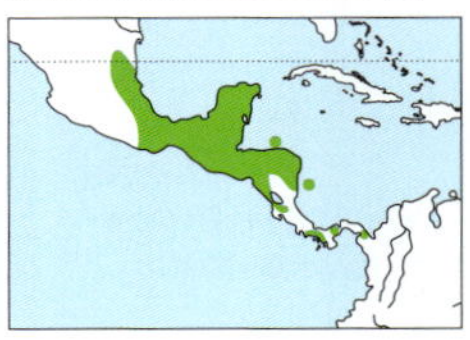

Mexican Mouse Opossum
Marmosa mexicana

FR Opossum du Mexique
DE Mexiko-Zwergbeutelratte
ES Marmosa de México

Dryland Mouse Opossum
Marmosa xerophila

FR Opossum xérophile
DE Trockenland-Zwergbeutelratte
ES Marmosa de La Guajira

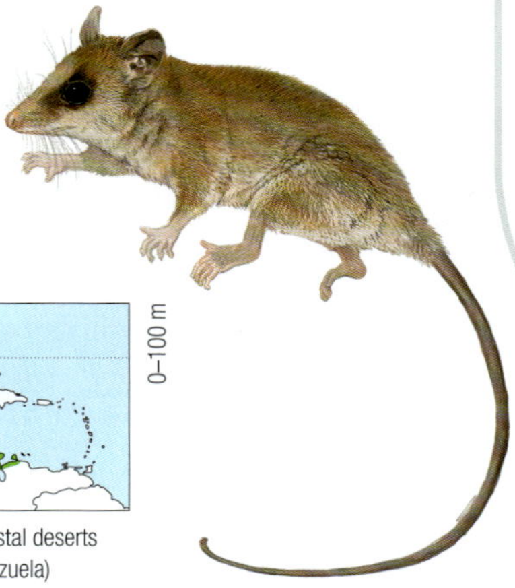

VU
Monotypic
HB 10.1–15.9 cm
T 13–18.1 cm
W 29–87 g

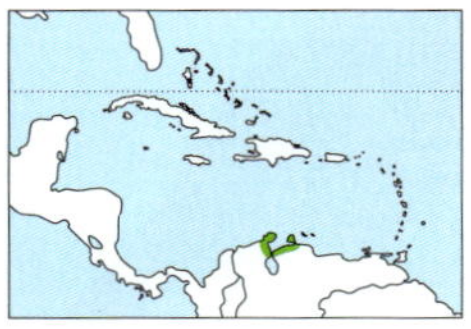

0–100 m

NE CO, NW VE (coastal deserts around Gulf of Venezuela)

Robinson's Mouse Opossum
Marmosa robinsoni

FR Opossum de Robinson
DE Robinson-Zwergbeutelratte
ES Marmosa de Robinson

0–2600 m

PA (W to Chiriquí, with del Rey and Saboga Is), N & W CO, N VE (N of R Orinoco, Margarita I), TT, GD

LC
Monotypic
HB 8.3–22 cm
T 10.1–22 cm
W 16–110 g

Anderson's Mouse Opossum
Marmosa andersoni

FR Opossum d'Anderson
DE Anderson-Zwergbeutelratte
ES Marmosa de Anderson

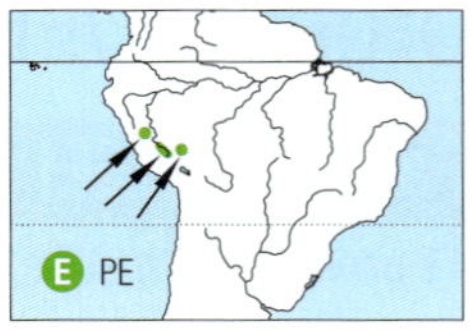

Pasco, Cusco

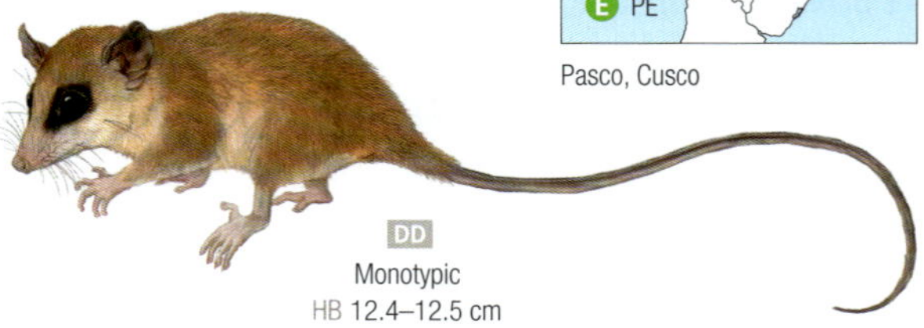

DD
Monotypic
HB 12.4–12.5 cm
T 17.5–19.3 cm
W 28–38 g

Western Amazonian Mouse Opossum
Marmosa macrotarsus

FR Opossum macrotarse
DE Großfuß-Zwergbeutelratte
ES Marmosa amazónica

NE
Monotypic
HB 10.4–13.6 cm
T 14.8–21 cm
W 23–51 g

25–1900 m

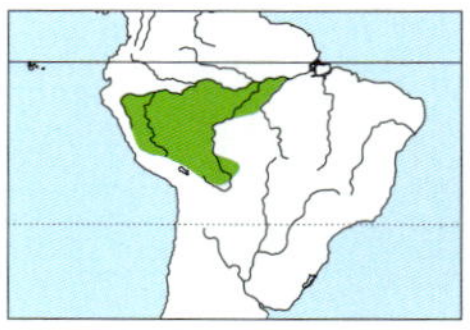

PE (S of R Solimões and R Marañón), W BR (S of R Amazon, W of R Tapajós); probably BO

Rufous Mouse Opossum
Marmosa lepida

FR Opossum de Bartlett
DE Fuchsrote Zwergbeutelratte
ES Marmosa rufa

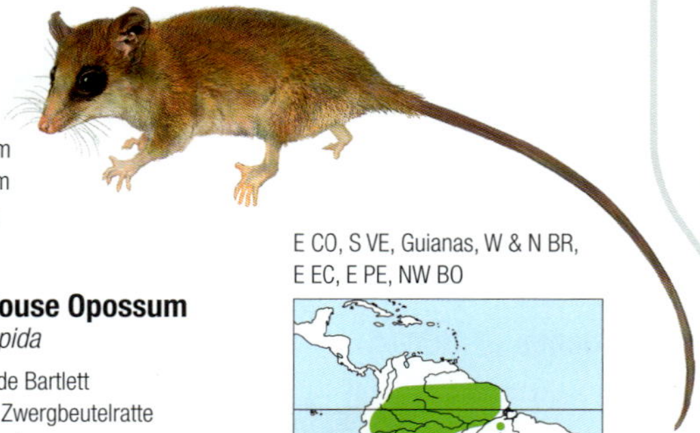

LC
Monotypic
HB 9.7–12 cm
T 14–15.9 cm
W 10–18 g

E CO, S VE, Guianas, W & N BR, E EC, E PE, NW BO

45–700 m

Tyleria Mouse Opossum
Marmosa tyleriana

FR Opossum des tépuis
DE Tepui-Zwergbeutelratte
ES Marmosa de Tyler

DD
Monotypic
HB 11.2–12.8 cm
T 14.5–17.7 cm
W 30–56 g

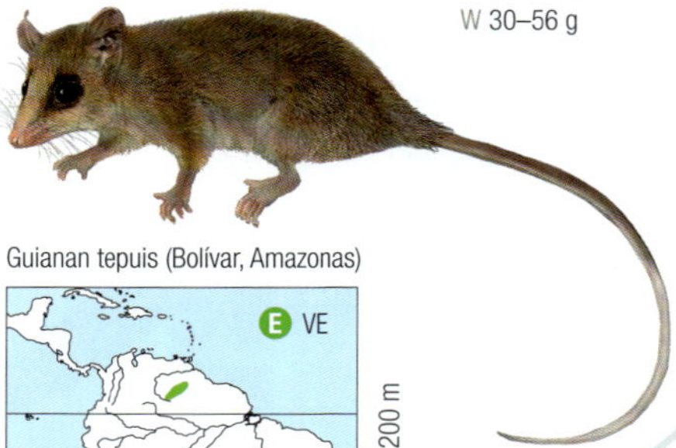

Guianan tepuis (Bolívar, Amazonas)

1300–2200 m

Waterhouse's Mouse Opossum
Marmosa waterhousii

FR Opossum de Waterhouse
DE Waterhouse-Zwergbeutelratte
ES Marmosa de Waterhouse

C, S & SE CO, W VE (Andes), E EC, N PE, W BR (W of R Negro, N of R Amazon)

50–1100 m

NE
Monotypic
HB 12.2–14.9 cm
T 17.2–22 cm
W 49–66 g

Linnaeus's Mouse Opossum
Marmosa murina

FR Opossum murin
DE Maus-Zwergbeutelratte
ES Marmosa de Linneo

LC
Monotypic
HB 9.1–24 cm
T 13.9–28 cm
W 19–100 g

VE, TT, Guianas, C & E BR

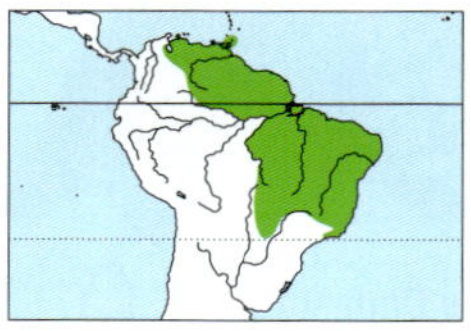

0–1350 m

North-eastern Woolly Mouse Opossum
Marmosa demerarae

FR Opossum du Demerara
DE Nordöstliche Zwergbeutelratte
ES Marmosa lanuda nororiental

LC
5 sspp
HB 15.2–21 cm
T 19.5–27 cm
W 62–130 g

below 1200 m

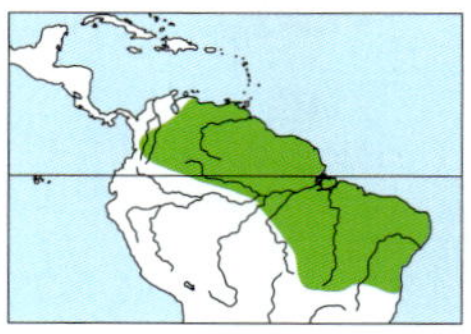

CO E to VE (Andean valleys), Guianas, BR (S to Bahia, W to Mato Grosso)

White-bellied Woolly Mouse Opossum
Marmosa constantiae

FR Opossum de Constance
DE Weißbauch-Zwergbeutelratte
ES Marmosa lanuda de vientre claro

LC
Monotypic
HB 13.4–18 cm
T 18.6–22 cm
W 68–90 g

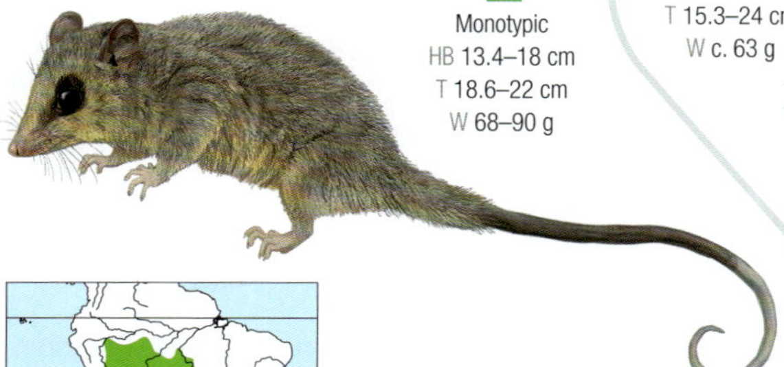

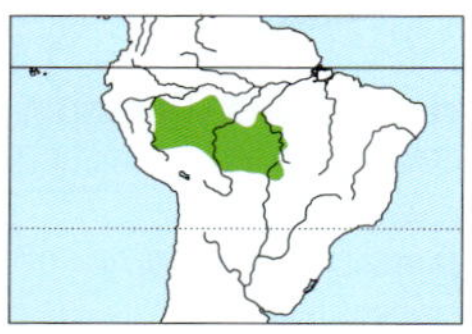

E PE, N BO, W & SW BR (Acre to Mato Grosso and Mato Grosso do Sul)

Little Woolly Mouse Opossum
Marmosa phaea

FR Opossum de Hopke
DE Kleine Zwergbeutelratte
ES Marmosa lanuda pequeña

VU
Monotypic
HB 12.1–17 cm
T 15.3–24 cm
W c. 63 g

W CO to SW EC (W slopes of Andes)

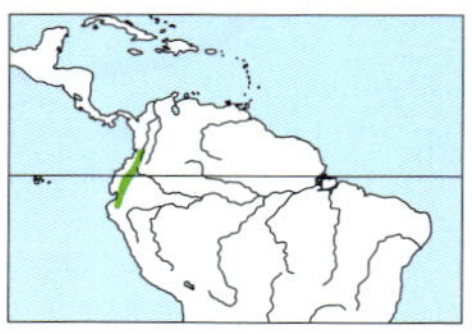

North-western Woolly Mouse Opossum
Marmosa germana

FR Opossum du Pastaza
DE Nordwestliche Zwergbeutelratte
ES Marmosa lanuda noroccidental

NE
Monotypic
HB 18.7–21 cm
T 25–28 cm
W 75–150 g

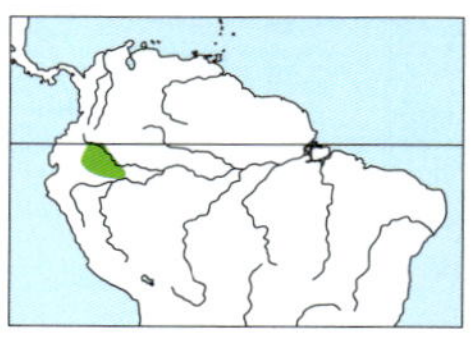

SE CO, E EC, N PE (N of R Amazon); exact range uncertain

Alston's Woolly Mouse Opossum
Marmosa alstoni

FR Opossum d'Alston
DE Alston-Zwergbeutelratte
ES Marmosa lanuda de Alston

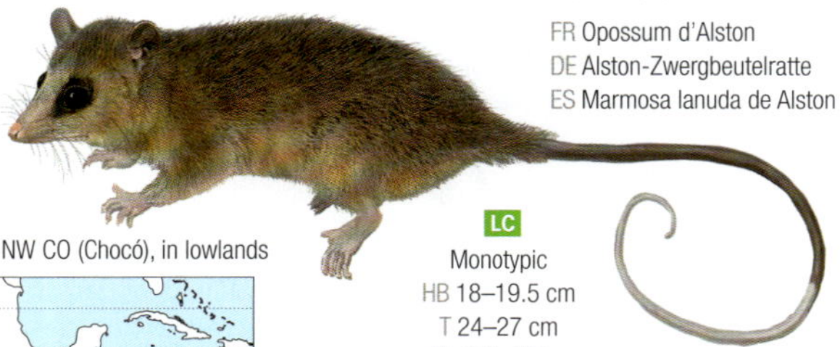

BZ S to NW CO (Chocó), in lowlands

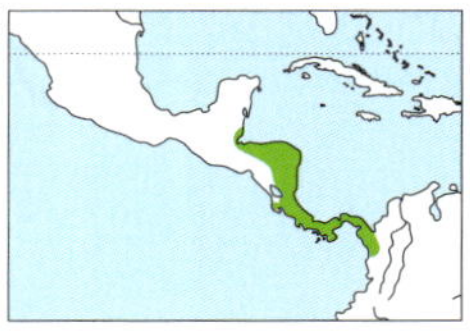

LC
Monotypic
HB 18–19.5 cm
T 24–27 cm
W 100–155 g

Nicaraguan Woolly Mouse Opossum
Marmosa nicaraguae

FR Opossum du Nicaragua
DE Nicaragua-Zwergbeutelratte
ES Marmosa lanuda de Nicaragua

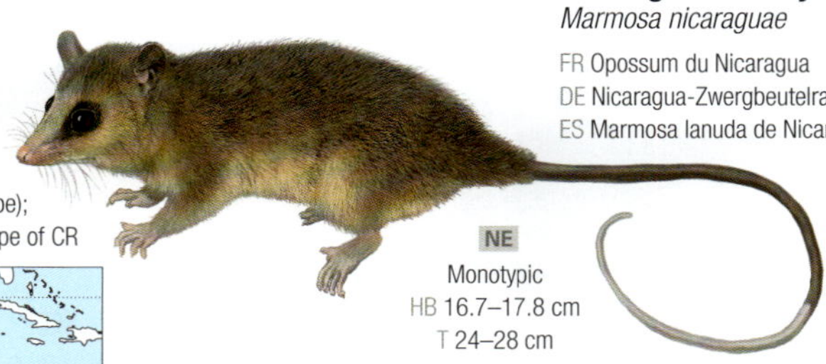

NI & CR (Caribbean slope); uncertain on Pacific slope of CR

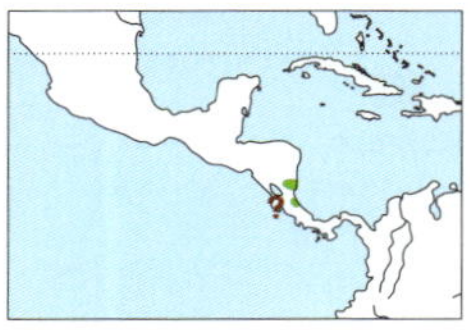

NE
Monotypic
HB 16.7–17.8 cm
T 24–28 cm

Adler's Woolly Mouse Opossum
Marmosa adleri

FR Opossum d'Adler
DE Adler-Zwergbeutelratte
ES Marmosa lanuda de Adler

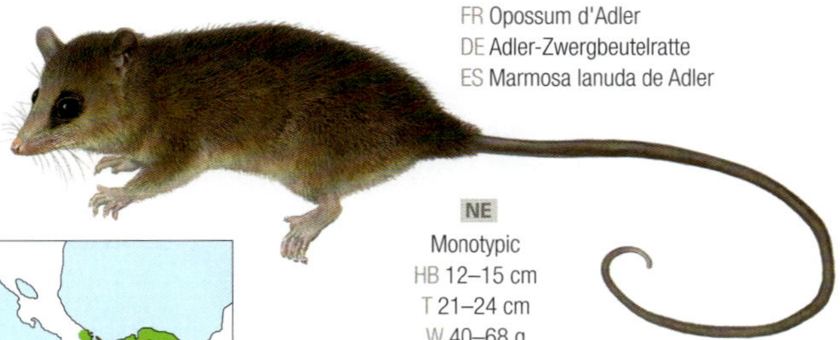

0–1500 m

NE
Monotypic
HB 12–15 cm
T 21–24 cm
W 40–68 g

South-eastern Woolly Mouse Opossum
Marmosa paraguayana

FR Opossum du Paraguay
DE Südöstliche Zwergbeutelratte
ES Marmosa lanuda meridional

E BR (S Bahia to Rio Grande do Sul) to NE AR (Misiones), E PY

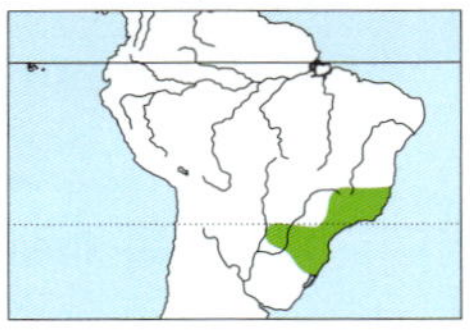

LC
Monotypic
HB 12–20 cm
T 15–26 cm
W 56–230 g

Anthony's Woolly Mouse Opossum
Marmosa perplexa

FR Opossum d'Anthony
DE Anthony-Zwergbeutelratte
ES Marmosa lanuda de Anthony

SW EC (Loja, El Oro), NW PE (Cajamarca, Tumbes)

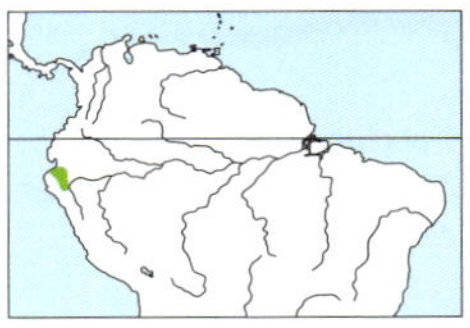

NE
Monotypic
HB 12.7–15.1 cm
T 17.8–20 cm
W 35–43 g

Jansa's Woolly Mouse Opossum
Marmosa jansae

FR Opossum de Jansa
DE Jansa-Zwergbeutelratte
ES Marmosa lanuda de Jansa

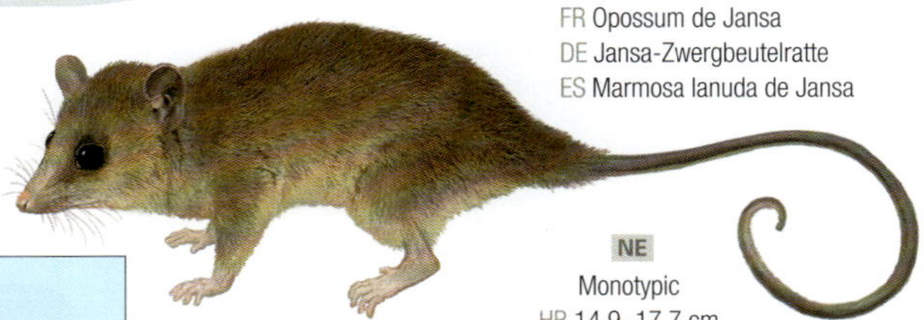

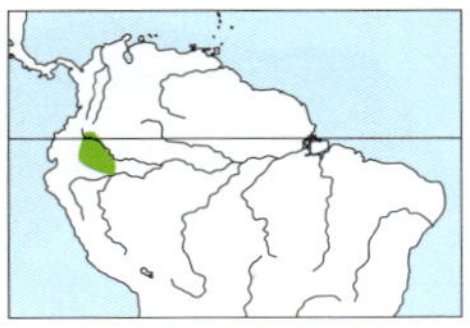

SE CO (Putumayo), E EC (Orellana, Pastaza), NE PE (Loreto)

NE
Monotypic
HB 14.9–17.7 cm
T 21–24 cm
W 56–103 g

Bolivian Woolly Mouse Opossum
Marmosa rapposa

FR Opossum rapposa
DE Bolivien-Zwergbeutelratte
ES Marmosa lanuda de Bolivia

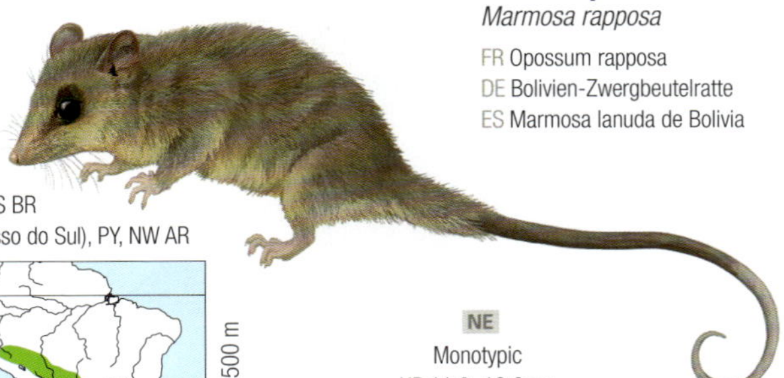

S PE, BO, S BR (Mato Grosso do Sul), PY, NW AR

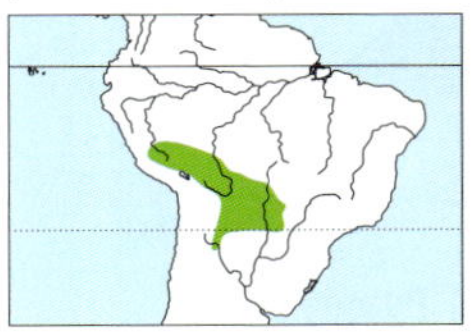

below c. 2500 m

NE
Monotypic
HB 11.8–18.6 cm
T 16.2–24 cm
W 37–132 g

Bare-tailed Woolly Mouse Opossum
Marmosa rutteri

FR Opossum de Rutter
DE Rutter-Zwergbeutelratte
ES Marmosa lanuda de Rutter

NE
Monotypic
HB 15.2–21 cm
T 21–29 cm
W 84–180 g

below c. 800 m

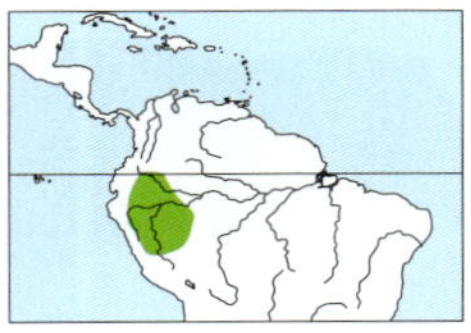

S CO, E EC, PE, W BR (W Amazonas and Acre)

Peruvian Woolly Mouse Opossum
Marmosa parda

FR Opossum du Huanuco
DE Parda-Zwergbeutelratte
ES Marmosa lanuda de Tate

R Huallaga basin (Huánuco, La Libertad)

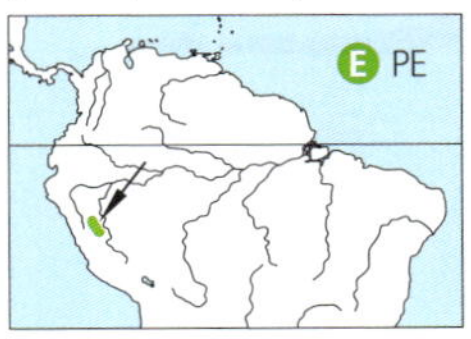

1000–1200 m

NE
Monotypic
HB 13.7–17.7 cm
T 21–27 cm
W c. 82 g

Genus *TLACUATZIN*
(5 species)

Tres Marias Gray Mouse Opossum
Tlacuatzin insularis

FR Opossum des Marias
DE Tres-Marías-Zwergbeutelratte
ES Ticuachín de Islas Marías

Marías Is, off Nayarit

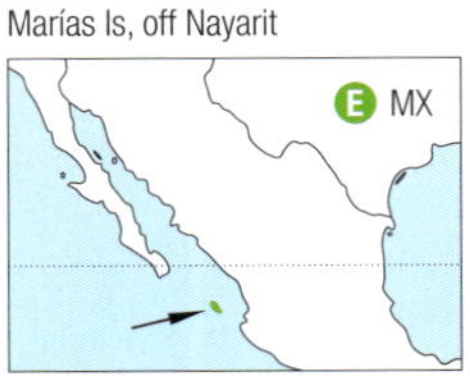

NE
Monotypic
HB 11–16 cm
T 10.5–16.7 cm
W 30–70 g

Northern Gray Mouse Opossum
Tlacuatzin sinaloae

FR Opossum du Sinaloa
DE Sinaloa-Zwergbeutelratte
ES Ticuachín septentrional

NE
Monotypic
HB 11–16 cm
T 10.5–16.7 cm
W 30–70 g

0–2100 m

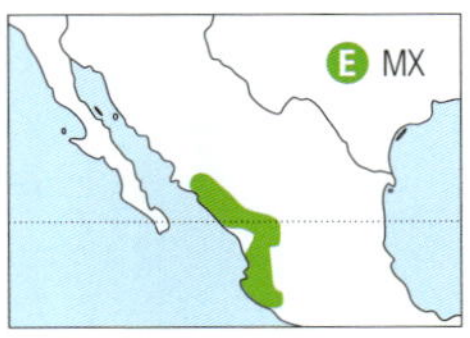

Sonora to Colima, E to Zacatecas; N & S limits uncertain

Balsas Gray Mouse Opossum
Tlacuatzin balsasensis

FR Opossum du Balsas
DE Balsas-Zwergbeutelratte
ES Ticuachín de Balsas

Jalisco E to Puebla, S to Oaxaca; E & W limits uncertain

NE
Monotypic

Tehuantepec Gray Mouse Opossum
Tlacuatzin canescens

FR Opossum cendré
DE Graue Zwergbeutelratte
ES Ticuachín de Tehuantepec

LC
Monotypic
HB 11–16 cm
T 10.5–16.7 cm
W 30–70 g

Oaxaca, W Chiapas; E & W limits uncertain

Yucatan Gray Mouse Opossum
Tlacuatzin gaumeri

FR Opossum du Yucatan
DE Yucatán-Zwergbeutelratte
ES Ticuachín de Yucatán

NE
Monotypic
HB 11–16 cm
T 10.5–16.7 cm
W 30–70 g

Yucatán, Campeche

Genus *MONODELPHIS*

(24 species)

LC
Monotypic
HB 9.7–15.8 cm
T 4.5–7 cm
W 20–60 g

0–300 m

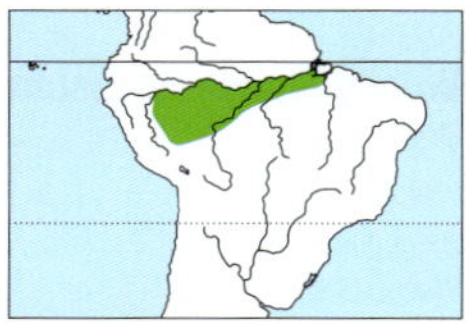

Emilie's Short-tailed Opossum
Monodelphis emiliae

FR Opossum de Boim
DE Emilie-Spitzmausbeutelratte
ES Colicorto de Emilie

E PE (S of Marañón-Ucayali junction, S to R Urubamba), N BO, BR (S of R Amazon, E to R Tocantins)

LC
Monotypic
HB 8.8–15.4 cm
T 3.7–7.6 cm
W 40–84 g

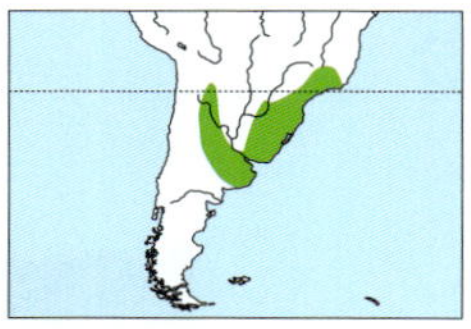

Orange-sided Opossum
Monodelphis dimidiata

FR Opossum à flancs jaunes
DE Orange-Flanken-Spitzmausbeutelratte
ES Colicorto de la Pampa

SE BR, S PY, N & E AR, UY

CR(PE)
Monotypic
HB c. 14.1 cm
T c. 6.2 cm

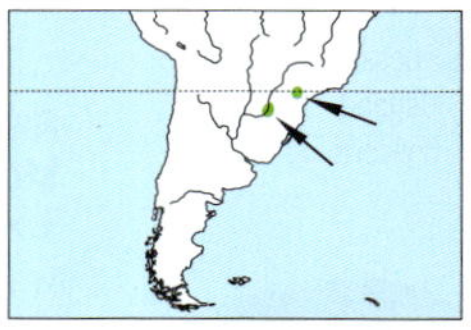

Single-striped Opossum
Monodelphis unistriata

FR Opossum à une raie
DE Einstreifen-Spitzmausbeutelratte
ES Colicorto uniestriado

SE BR (São Paulo), NE AR (Misiones)

LC
Monotypic
HB 6.3–16.3 cm
T 3.7–8.3 cm
W 23–90 g

below 1400 m

Tawny-headed Opossum
Monodelphis scalops

FR Opossum à tête rouge
DE Rotkopf-Spitzmausbeutelratte
ES Colicorto de cabeza leonada

SE BR (Espírito Santo to Santa Catarina), NE AR (Iguazú NP, Misiones)

NE
Monotypic
HB 7.6–10 cm
T 3.9–5.3 cm

1785–2800 m

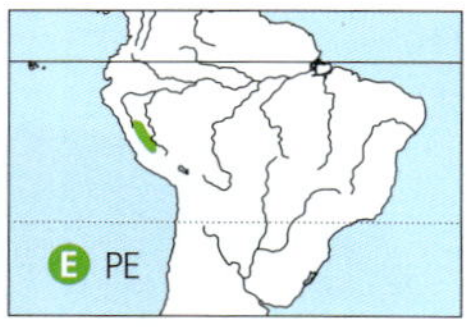

Gardner's Short-tailed Opossum
Monodelphis gardneri

FR Opossum de Gardner
DE Gardner-Spitzmausbeutelratte
ES Colicorto de Gardner

4 localities on E slope of Andes

DD
Monotypic
HB 7.5–10 cm
T 5.3–6 cm
W 15–23 g

Ihering's Three-striped Opossum
Monodelphis iheringi

FR Opossum d'Ihering
DE Ihering-Spitzmausbeutelratte
ES Colicorto de Ihering

Coastal Espírito Santo to Rio Grande do Sul; possibly NE AR (Misiones)

LC
Monotypic
HB 9–13.7 cm
T 4–6 cm
W 23–46 g

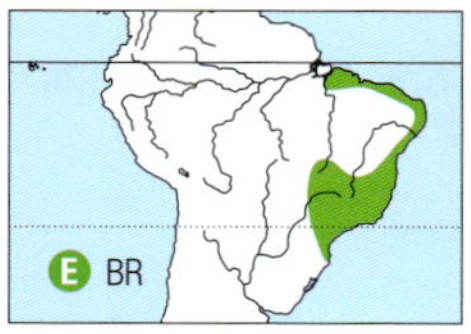

Northern Three-striped Opossum
Monodelphis americana

FR Opossum à trois raies
DE Dreistreifen-Spitzmausbeutelratte
ES Colicorto estriado

E Pará to Santa Catarina, inland W to Brasília; possibly NE AR (Misiones)

NE
Monotypic
HB 8.8–10.3 cm
T 5–5.4 cm

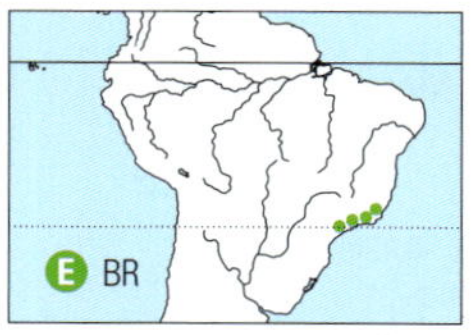

Long-nosed Short-tailed Opossum
Monodelphis pinocchio

FR Opossum pinocchio
DE Langnasen-Spitzmausbeutelratte
ES Colicorto de hocico largo

Atlantic Forest from Espírito Santo and Minas Gerais to São Paulo

LC
Monotypic
HB 6.2–10.2 cm
T 3.6–4.5 cm
W 7.5–30 g

200–1700 m

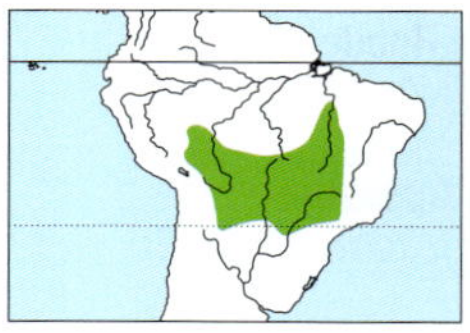

Pygmy Short-tailed Opossum
Monodelphis kunsi

FR Opossum de Kuns
DE Zwerg-Spitzmausbeutelratte
ES Colicorto pigmeo

BO (E of Andes), N PY, N AR (Salta), SC BR

Sepia Short-tailed Opossum
Monodelphis adusta

FR Opossum bistre
DE Braune Spitzmausbeutelratte
ES Colicorto andino

200–2200 m

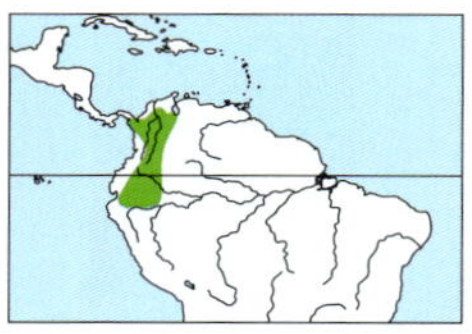

LC
Monotypic
HB 8.4–11.4 cm
T 5–6.7 cm
W 15–35 g

E PA (E Darién), CO, SW VE (Andes in Mérida and Táchira), E EC, N PE

Reig's Opossum
Monodelphis reigi

FR Opossum de Reig
DE Reig-Spitzmausbeutelratte
ES Colicorto de Reig

1100–2050 m

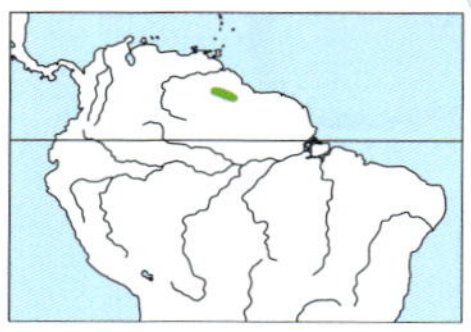

E VE (Sierra de Lema), W GY (Mt Ayanganna)

VU
Monotypic
HB c. 12.1 cm
T c. 7.4 cm

NE

Monotypic

HB 9.8–10.1 cm

T 4.2–4.9 cm

W 16–19 g

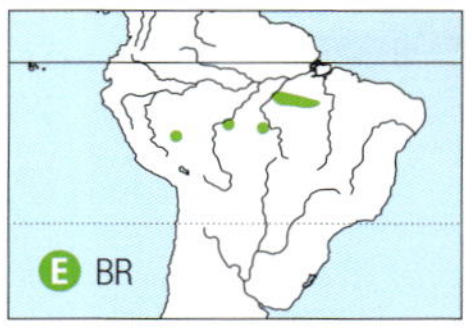

Saci Short-tailed Opossum

Monodelphis saci

FR Opossum saci

DE Saci-Spitzmausbeutelratte

ES Colicorto de Saci

Amazonia (Pará; uncertain in N Mato Grosso, N Rondônia and SE Acre)

DD

Monotypic

HB c. 12.3 cm

T c. 6.9 cm

W 45–65 g

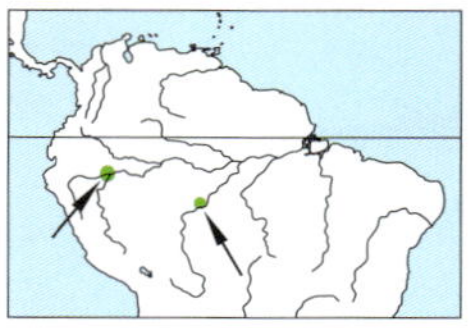

Handley's Short-tailed Opossum

Monodelphis handleyi

FR Opossum du Loreto

DE Handley-Spitzmausbeutelratte

ES Colicorto de Handley

NE PE (Loreto: R Ucayali to R Javarí), WC BR (Amazonas)

LC

Monotypic

HB c. 14.1 cm

T c. 7.2 cm

W c. 70 g

c. 355 m

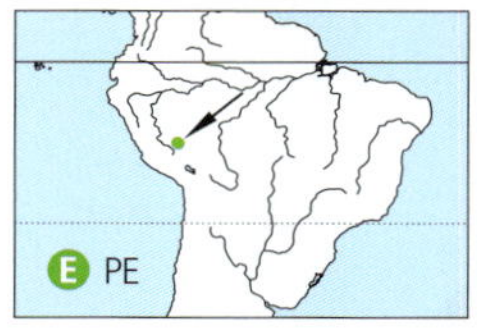

Pine's Opossum

Monodelphis ronaldi

FR Opossum de Pine

DE Pine-Spitzmausbeutelratte

ES Colicorto de Ronald

W bank of R Madre de Dios

LC

Monotypic

HB 6–6.2 cm

T 9.4–9.6 cm

W 18–29 g

1900–3200 m

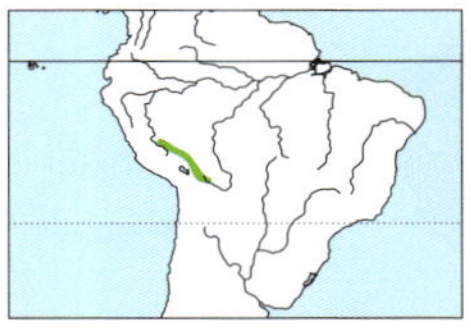

Osgood's Short-tailed Opossum

Monodelphis osgoodi

FR Opossum d'Osgood

DE Osgood-Spitzmausbeutelratte

ES Colicorto de Osgood

E Andean slopes of SE PE and W BO

LC

Monotypic

HB c. 9.6 cm

T 5.6–5.8 cm

W 24–30 g

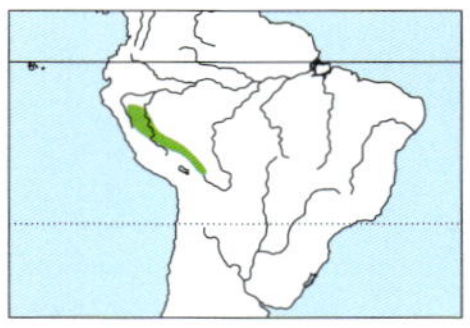

Peruvian Short-tailed Opossum

Monodelphis peruviana

FR Opossum du Pérou

DE Peru-Spitzmausbeutelratte

ES Colicorto de Perú

E Andes of C & S PE and W BO

NE

Monotypic

HB 12.7–16.2 cm

T 7.2–9.4 cm

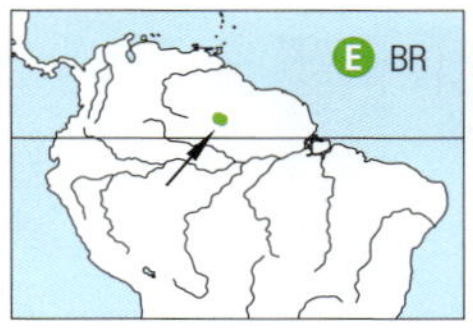

Voss's Short-tailed Opossum

Monodelphis vossi

FR Opossum de Voss

DE Voss-Spitzmausbeutelratte

ES Colicorto de Voss

2 localities in NE Roraima

LC

Monotypic

HB 11–18 cm

T 6.5–10.6 cm

W 40–150 g

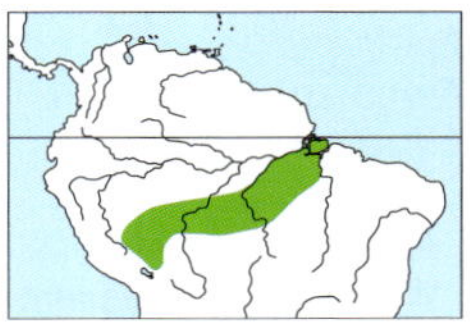

Amazonian Red-sided Opossum

Monodelphis glirina

FR Opossum à flancs roux

DE Amazonas-Spitzmausbeutelratte

ES Colicorto de flancos rojos

SE PE, N BO (La Paz, Pando), BR S of R Amazon (Acre E to R Xingu; Marajó I)

NE

Monotypic

HB c. 10.8 cm

T c. 6 cm

W c. 23 g

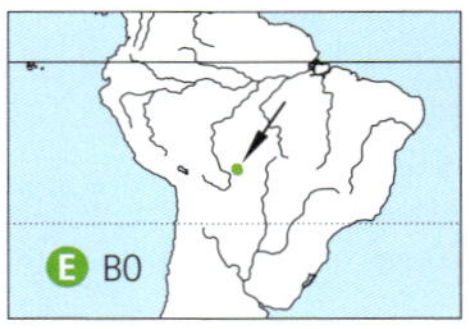

Santa Rosa Short-tailed Opossum

Monodelphis sanctaerosae

FR Opossum de Santa Rosa

DE Santa Rosa-Spitzmausbeutelratte

ES Colicorto de Santa Rosa

Santa Cruz

Gray Short-tailed Opossum

Monodelphis domestica

FR Opossum à queue courte

DE Graue Spitzmausbeutelratte

ES Colicorto gris

LC

Monotypic

HB 12.3–17.9 cm

T 4.6–9.6 cm

W 58–110 g

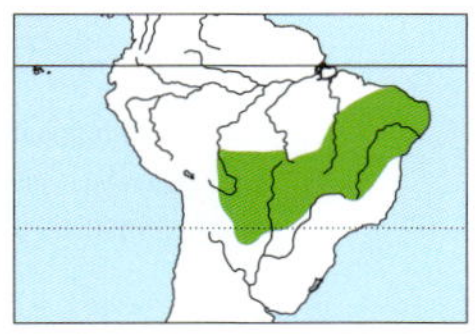

E BO, N AR (Formosa), N & C PY, C & NE BR

LC

Monotypic

HB 10.2–18.3 cm

T 5.5–10.5 cm

W 32–106 g

0–2250 m

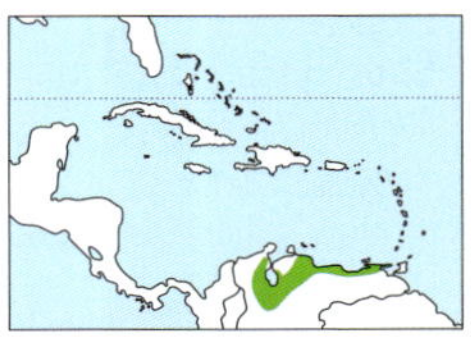

Hooded Red-sided Opossum

Monodelphis palliolata

FR Opossum à capuchon

DE Kronen-Spitzmausbeutelratte

ES Colicorto encapuchado

NE CO (Norte de Santander, Vichada), NW & N VE

North Amazon form

NE

Monotypic

HB 11.9–17 cm

T 6.5–9.5 cm

W 40–100 g

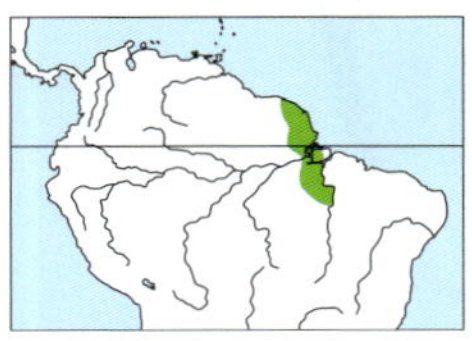

GF, N BR (Amapá, Pará S of R Amazon, Marajó I); probably SR

Touan Short-tailed Opossum

Monodelphis touan

FR Opossum touan

DE Touan-Spitzmausbeutelratte

ES Colicorto de Touan

South Amazon form

NE

Monotypic

HB 9.5–18.2 cm

T 6.5–9.8 cm

W 33–105 g

C & S GY, BR N of R Amazon (SE Roraima, E Amazonas, N Pará); probably SR

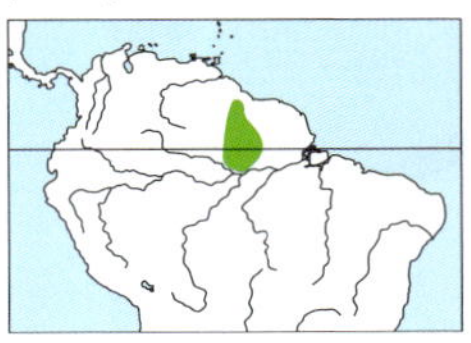

Arlindo's Short-tailed Opossum

Monodelphis arlindoi

FR Opossum d'Arlindo

DE Arlindo-Spitzmausbeutelratte

ES Colicorto de Arlindo

SE VE (S of R Orinoco), N GY, NW BR (N of R Negro, W of R Branco)

95–1080 m

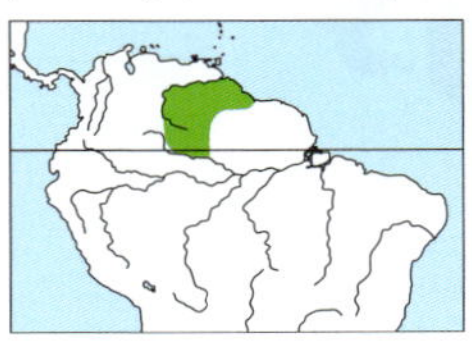

Guyanan Short-tailed Opossum

Monodelphis brevicaudata

FR Opossum à pattes rouges

DE Guiana-Spitzmausbeutelratte

ES Colicorto de Guyana

LC

Monotypic

HB 12.8–17 cm

T 6.4–9.7 cm

W 50–127 g

Guianan Brown Four-eyed Opossum
Metachirus nudicaudatus

FR Opossum quatre-yeux
DE Nacktschwanzbeutelratte
ES Filandro de Guyana

LC
Monotypic
HB 24–30 cm
T 30–38 cm
W 260–480 g

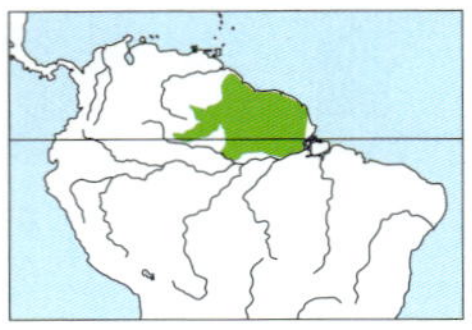

GY, SR, GF, N BR; possibly widespread

Genus *METACHIRUS*

(2 species)

Common Brown Four-eyed Opossum
Metachirus myosuros

FR Opossum myosure
DE Temminck-Nacktschwanzbeutelratte
ES Filandro pardo

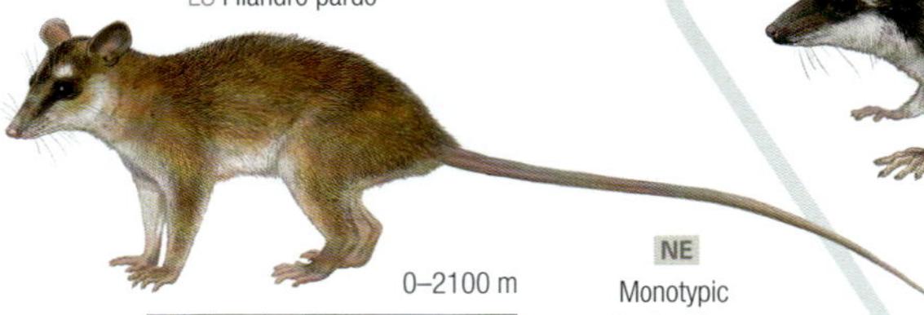

NE
Monotypic
HB 22–28 cm
T 26–34 cm
W 248–485 g

0–2100 m

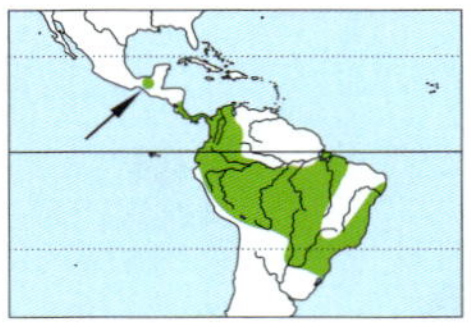

S MX (Chiapas), S NI to VE, BR, BO, PY and N AR (Formosa, Misiones)

Genus *CHIRONECTES*

(1 species)

Water Opossum
Chironectes minimus

FR Yapock
DE Schwimmbeutler
ES Zarigüeya acuática

LC
4 sspp
HB 25–40 cm
T 27–43 cm
W 510–790 g

0–1860 m

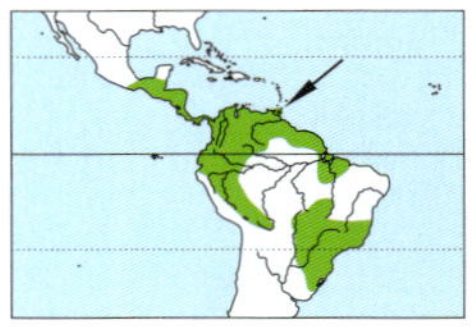

S MX (S Oaxaca, Tabasco) to C BO, N & S BR, E PY, UY (Cerro Largo), NE AR (Misiones), TT (Trinidad I)

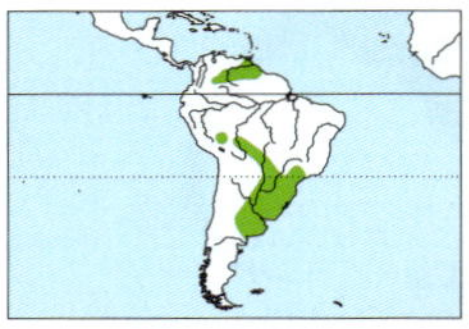

E CO, VE, NW GY, SE PE (Madre de Dios), E BO, S BR, PY, UY, E AR (S to Buenos Aires Prov)

Big Lutrine Opossum
Lutreolina crassicaudata

FR Opossum à grosse queue
DE Dickschwanzbeutelratte
ES Zarigüeya de cola gruesa

LC
2 sspp
HB 24–40 cm
T 24–36 cm
W 200–910 g

Genus *LUTREOLINA*

(2 species)

450–2000 m

Yungas forests of S BO (Chuquisaca, Tarija) and NW AR (Jujuy, Salta, Tucumán)

LC
Monotypic
HB 18.8–24 cm
T 21–26 cm
W c. 284 g

Massoia's Lutrine Opossum
Lutreolina massoia

FR Opossum de Massoia
DE Massoia-Dickschwanzbeutelratte
ES Zarigüeya de cola gruesa de Massoia

Virginia Opossum
Didelphis virginiana

FR Opossum de Virginie
DE Virginia-Opossum
ES Zarigüeya de Virginia

LC
4 sspp
HB 37–50 cm
T 29–47 cm
W 0.5–5.9 kg

0–3000 m

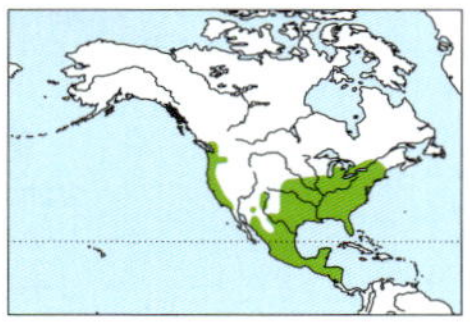

CA (British Columbia, S Ontario), US (W coast, most of E half), MX to NI; introduced populations exist

Genus *DIDELPHIS*

(6 species)

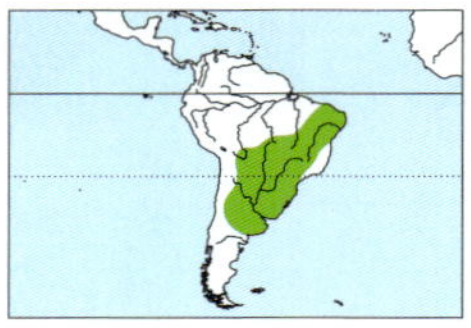

NE, C & S BR, SE BO, PY, UY, AR (S to Buenos Aires Prov in E, Monte Desert ecoregion in W)

Brazilian White-eared Opossum
Didelphis albiventris

FR Opossum à oreilles blanches
DE Südopossum
ES Zarigüeya de orejas blancas de Brasil

LC
Monotypic
HB 30–44 cm
T 29–45 cm
W 0.5–2.5 kg

Andean White-eared Opossum
Didelphis pernigra

FR Opossum des Andes
DE Anden-Opossum
ES Zarigüeya de orejas blancas andina

LC
Monotypic
HB 34–44 cm
T 32–41 cm
W 0.7–2 kg

above 1500 m

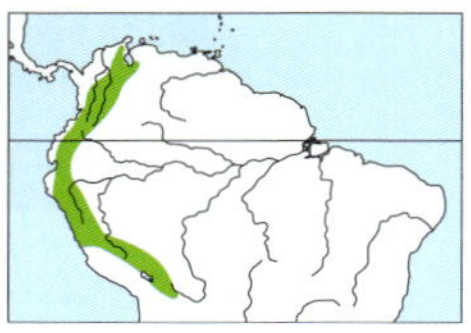

Forested Andean slopes of NW VE, CO, EC, PE, W BO

100–2200 m

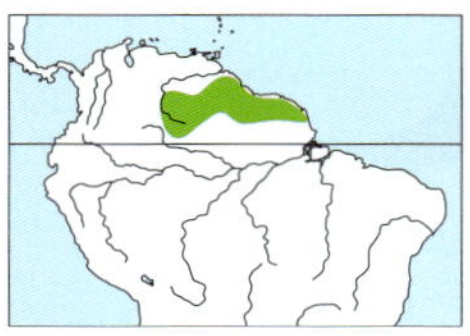

VE (S of R Orinoco), Guianas, extreme N BR (Amapá, Roraima)

Guianan White-eared Opossum
Didelphis imperfecta

FR Opossum de Guyane
DE Guiana-Opossum
ES Zarigüeya de orejas blancas de Guyana

LC
Monotypic
HB 32–39 cm
T 30–41 cm
W 0.6–1.2 kg

E BR (Pernambuco) S to SE PY and NE AR (Misiones)

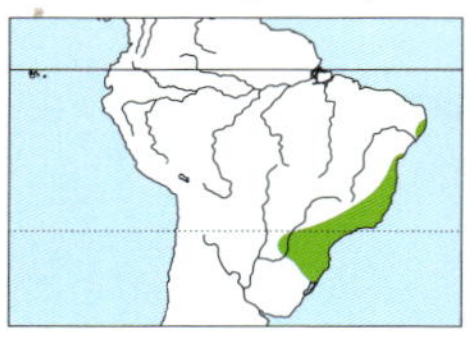

Southern Black-eared Opossum
Didelphis aurita

FR Opossum oreillard DE Großohr-Opossum
ES Zarigüeya de orejas negras meridional

LC
Monotypic
HB 31–39 cm
T 31–37 cm
W 0.7–1.5 kg

Northern Black-eared Opossum
Didelphis marsupialis

FR Opossum commun
DE Schwarzohr-Opossum
ES Zarigüeya de orejas negras septentrional

LC
Monotypic
HB 32–46 cm
T 34–46 cm
W 0.6–2.4 kg

above 1000 m

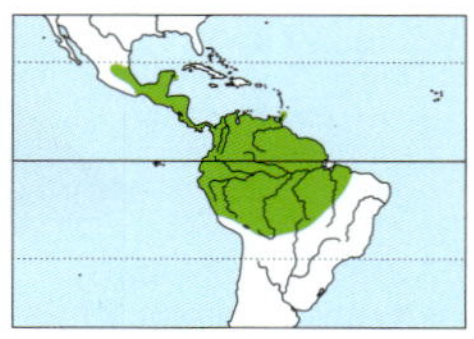

E MX (Tamaulipas) to S Amazonia, TT and (perhaps introduced) N to DM

Genus *PHILANDER*

(10 species)

LC

Monotypic

HB 26–33 cm

T 25–33 cm

W 220–910 g

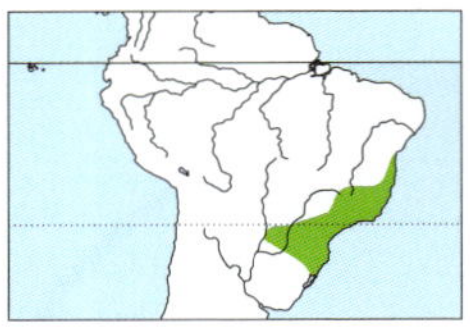

Southern Four-eyed Opossum

Philander quica

FR Opossum bridé

DE Südöstliche Vieraugenbeutelratte

ES Filandro meridional

SE BR (Bahia, Minas Gerais and Goiás inland) S to E PY and NE AR (Misiones)

Anderson's Four-eyed Opossum

Philander andersoni

FR Opossum à dos noir

DE Anderson-Vieraugenbeutelratte

ES Filandro de Anderson

LC

Monotypic

HB 22–31 cm

T 25–33 cm

W 225–600 g

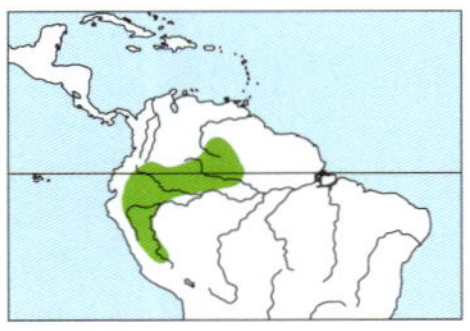

SE CO, SC VE (Bolívar, Amazonas), W BR (Amazonas, Acre), E EC, N & C PE (E of Andes)

Black Four-eyed Opossum

Philander nigratus

FR Opossum noir

DE Schwarze Vieraugenbeutelratte

ES Filandro negro

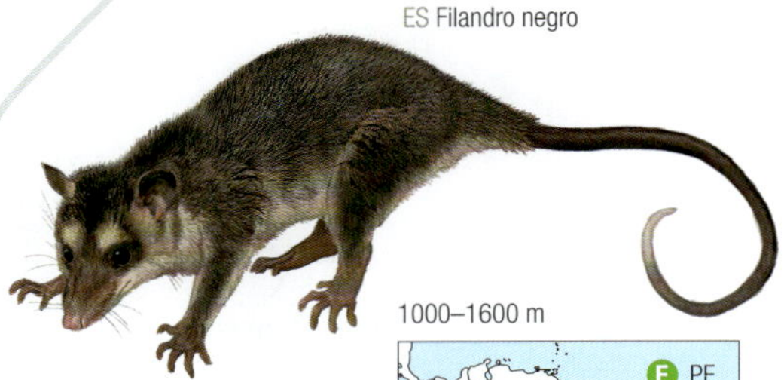

NE

Monotypic

HB 26–28 cm

T 27–30 cm

1000–1600 m

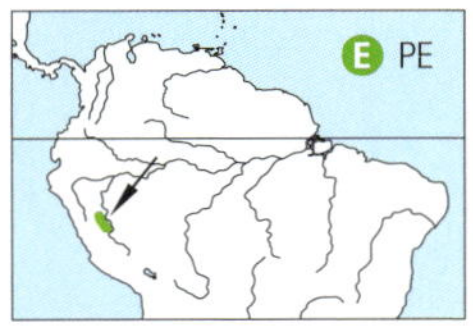

E Andes (Ayacucho, Junín)

NE

Monotypic

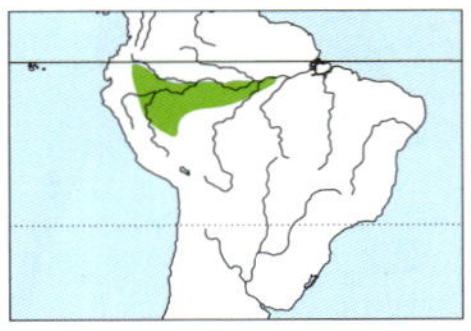

Pebas Four-eyed Opossum

Philander pebas

FR Opossum du Pebas

DE Pebas-Vieraugenbeutelratte

ES Filandro de Pebas

E EC (Orellana), E PE (S to Madre de Dios), BR (Acre, Amazonas, Pará)

Common Four-eyed Opossum

Philander canus

FR Opossum de Moyobamba

DE Gewöhnliche Vieraugenbeutelratte

ES Filandro de Moyobamba

NE

Monotypic

HB 20–33 cm

T 19–33 cm

W 200–674 g

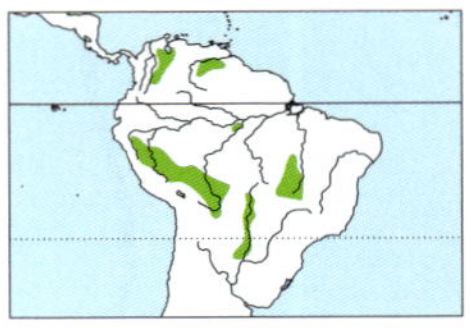

E Andes of CO, VE, PE, BO, PY, N AR (Formosa, Chaco), W & C BR (probably across Amazonia)

McIlhenny's Four-eyed Opossum

Philander mcilhennyi

FR Opossum de McIlhenny

DE McIlhenny-Vieraugenbeutelratte

ES Filandro de McIlhenny

EC PE (Loreto, Ucayali), W BR (Amazonas, Acre)

LC

Monotypic

HB 29–31 cm

T 26–38 cm

W 396–640 g

Gray Four-eyed Opossum

Philander opossum

FR Opossum gris

DE Graue Vieraugenbeutelratte

ES Filandro gris

variants

LC

Monotypic

HB 20–33 cm

T 19.5–33 cm

W 200–674 g

Guianas, E Amazonian BR; possibly widespread

LC

Monotypic

HB c. 21 cm

T c. 30 cm

W 148–350 g

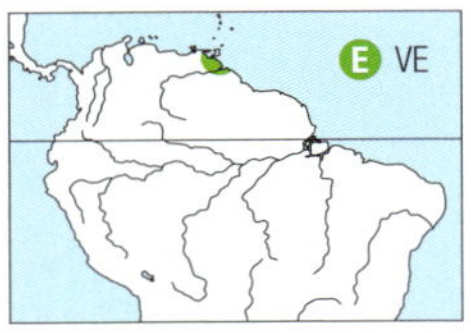

R Orinoco Delta, Gulf of Paria (E Monagas, Delta Amacuro)

Orinoco Four-eyed Opossum

Philander deltae

FR Opossum de l'Orénoque

DE Orinoko-Vieraugenbeutelratte

ES Filandro del delta del Orinoco

Dark Four-eyed Opossum

Philander melanurus

FR Opossum du Mira

DE Dunkle Vieraugenbeutelratte

ES Filandro oscuro

PA, Pacific lowlands of CO and W EC; probably CR (limits with *P. vossi* uncertain)

NE

Monotypic

HB 20–33 cm

T 19.5–33 cm

W 200–674 g

Northern Four-eyed Opossum

Philander vossi

FR Opossum d'Orizaba

DE Fahle Vieraugenbeutelratte

ES Filandro pálido

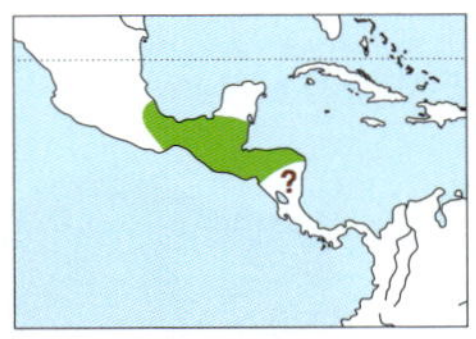

E & S MX (S of Tamaulipas), GT, BZ, SV; probably HN (limits with *P. melanurus* uncertain)

NE

Monotypic

HB 20–33 cm

T 19.5–33 cm

W 200–674 g

Genus *CHACODELPHYS*

(1 species)

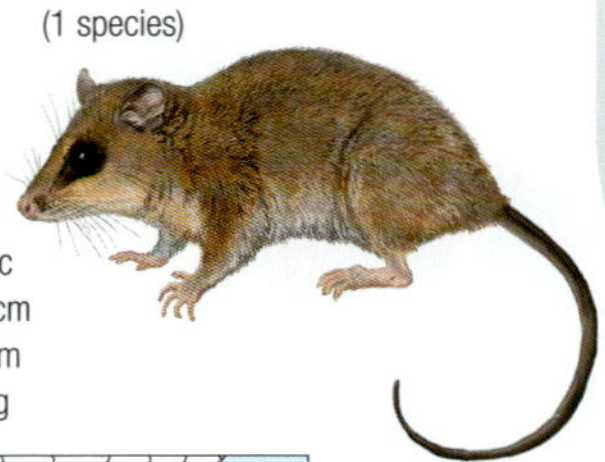

NT

Monotypic
HB c. 6.8 cm
T c. 5.5 cm
W c. 10 g

2 disjunct populations (Chaco-Formosa, Misiones)

Pygmy Opossum
Chacodelphys formosa

FR Opossum de Formosa
DE Chaco-Beutelratte
ES Zarigüeya pigmea

Agricola's Opossum
Cryptonanus agricolai

FR Opossum d'Agricola
DE Agricolas Beutelratte
ES Marmosa grácil de Agricola

DD

Monotypic
HB 7.4–9.5 cm
T 9–11.4 cm
W c. 18 g

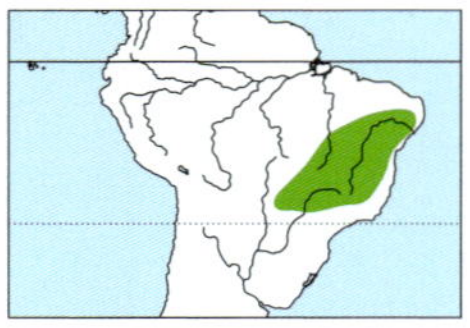

NE to SW BR (Ceará, Pernambuco to Minas Gerais and Mato Grosso do Sul)

Genus *CRYPTONANUS*

(4 species)

Chacoan Mouse Opossum
Cryptonanus chacoensis

FR Opossum du Chaco
DE Chaco-Zwergbeutelratte
ES Marmosa grácil del Chaco

LC

Monotypic
HB 7–11.4 cm
T 9.5–13.6 cm
W 10–19 g

S & SE BR (Mato Grosso do Sul, Minas Gerais, Rio Grande do Sul), S BO (Tarija), PY, N & NE AR, UY

Guaiba Mouse Opossum
Cryptonanus guahybae

FR Opossum de Guaiba
DE Guahiba-Zwergbeutelratte
ES Marmosa grácil del Guaíba

DD

Monotypic
HB 7–10.2 cm
T 9.6–13.5 cm
W 14–25 g

Rio Grande do Sul

Unduavi Mouse Opossum
Cryptonanus unduaviensis

FR Opossum de l'Unduavi
DE Unduavi-Zwergbeutelratte
ES Marmosa grácil de Unduavi

DD

Monotypic
HB 8.6–12.1 cm
T 11.2–13.5 cm
W 15–40 g

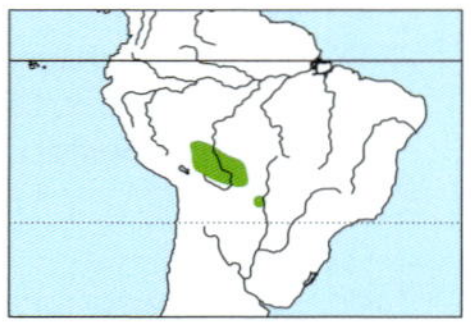

N & E BO (Pando, La Paz, Beni, Santa Cruz), N PY (Alto Paraguay)

Emilie's Opossum
Gracilinanus emiliae

FR Opossum d'Emilie
DE Emilie-Zwergbeutelratte
ES Marmosa grácil de Emilie

DD

Monotypic
HB 7.2–8.7 cm
T 12.8–15.1 cm
W 10–14 g

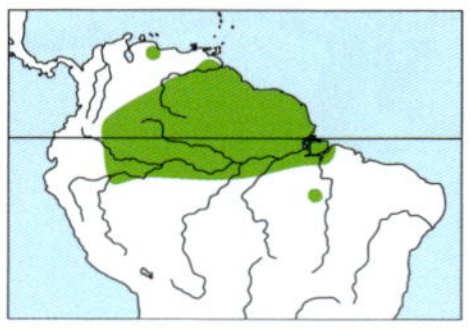

E CO, N, E & SE VE, Guianas, N BR, NE PE (Loreto)

Genus *GRACILINANUS*

(7 species)

Agile Opossum
Gracilinanus agilis

FR Opossum agile
DE Flinke Zwergbeutelratte
ES Marmosa grácil ágil

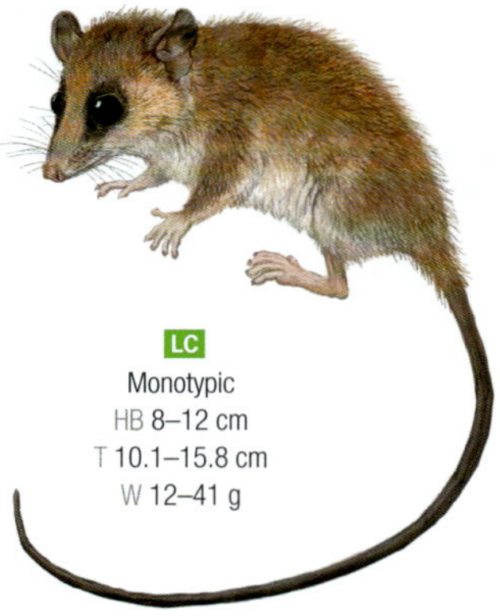

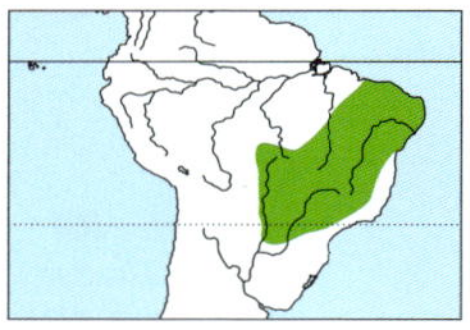

E & S BR, E PY

LC

Monotypic
HB 8–12 cm
T 10.1–15.8 cm
W 12–41 g

Peruvian Opossum
Gracilinanus peruanus

FR Opossum du Huallaga
DE Peru-Zwergbeutelratte
ES Marmosa grácil de Perú

NE

Monotypic
HB 8.5–12.1 cm
T 13–15.5 cm
W 10–49 g

below 1800 m

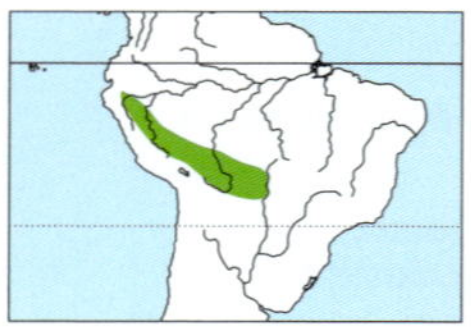

E PE, N & C BO, WC BR (Rondônia, Mato Grosso)

Aceramarca Opossum
Gracilinanus aceramarcae

FR Opossum de Bolivie
DE Aceramarca-Zwergbeutelratte
ES Marmosa grácil del Aceramarca

2530–3350 m

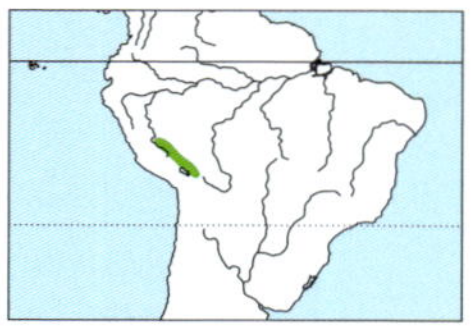

SE PE (Cusco, Puno), W BO (La Paz)

LC

Monotypic
HB 8.3–10.3 cm
T 11.2–14.5 cm
W 18.5–23 g

Brazilian Gracile Opossum
Gracilinanus microtarsus

FR Opossum du Brésil
DE Kleinfuß-Zwergbeutelratte
ES Marmosa grácil de pies pequeños

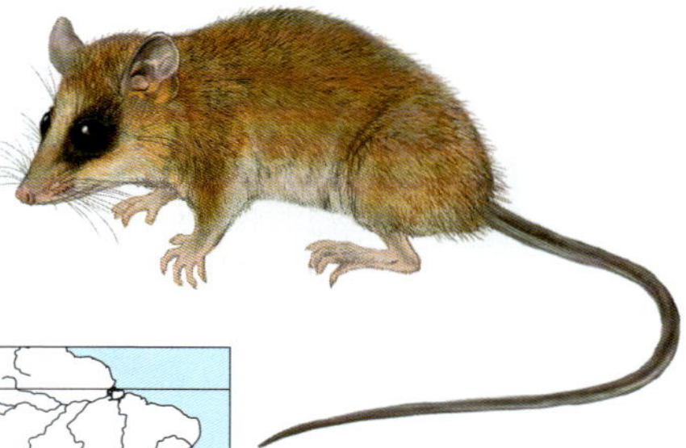

0–1800 m

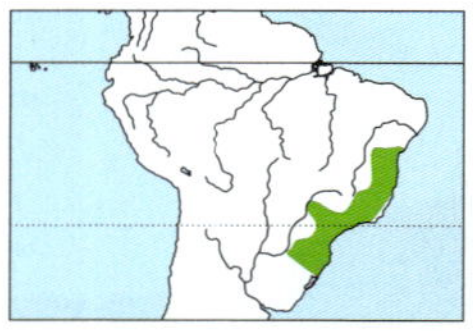

E & SE BR (C Bahia to N Rio Grande do Sul), NE AR (Misiones)

LC

Monotypic
HB 6.7–18.5 cm
T 7–17.7 cm
W 12–58 g

Wood Sprite Opossum
Gracilinanus dryas

FR Opossum des bois
DE Mérida-Zwergbeutelratte
ES Marmosa grácil duende

LC

Monotypic
HB 9–10 cm
T 13–15 cm
W c. 18 g

2300–4000 m

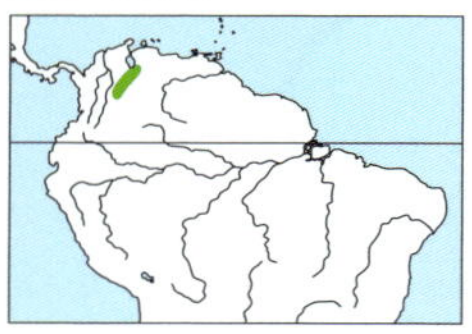

Andes of NE CO (Cundinamarca) and W VE (Trujillo, Mérida, Táchira)

Northern Gracile Opossum
Gracilinanus marica

FR Opossum de Mary
DE Kolumbien-Zwergbeutelratte
ES Marmosa grácil septentrional

LC

Monotypic
HB 9.6–11.6 cm
T 13.1–15.1 cm
W 24–30 g

0–2590 m

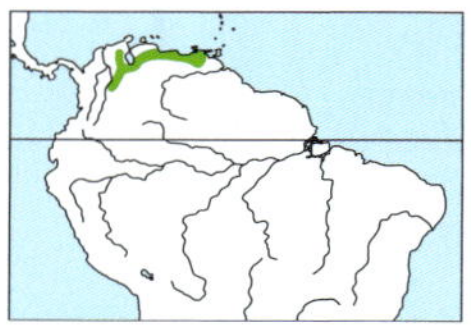

N & C CO (La Guajira to Cundinamarca), W & N VE

Genus *LESTODELPHYS*

(1 species)

Patagonian Opossum
Lestodelphys halli

FR Opossum de Patagonie
DE Patagonien-Beutelratte
ES Zarigüeya patagónica

LC

Monotypic
HB 12.3–14.4 cm
T 7.4–9.9 cm
W 60–100 g

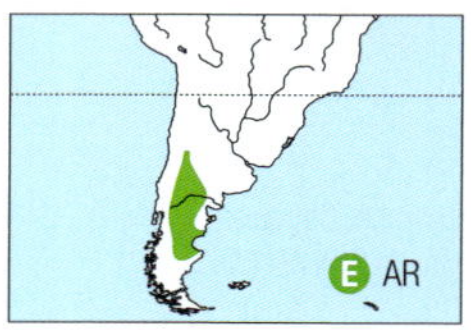

Mendoza S to Santa Cruz

Genus *THYLAMYS*

(11 species)

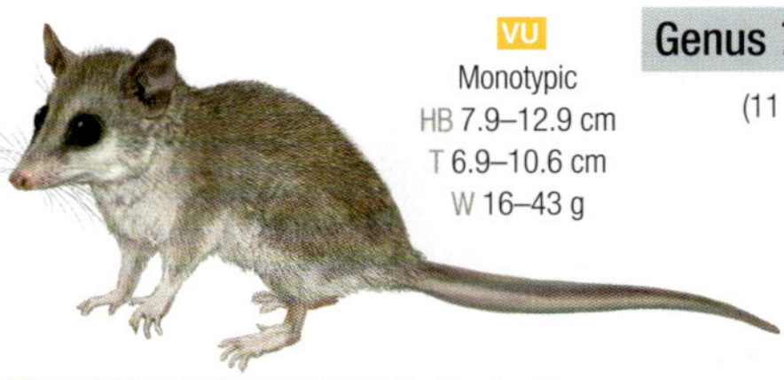

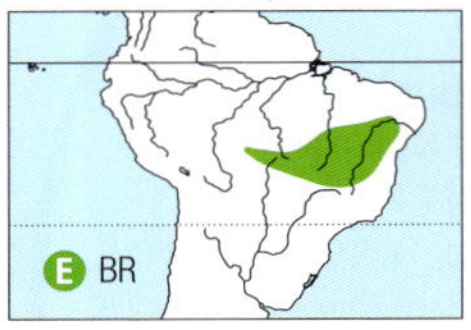

Karimi's Fat-tailed Opossum
Thylamys karimii

FR Opossum de Karimi
DE Karimi-Fettschwanzbeutelratte
ES Marmosa coligruesa de Karimi

VU
Monotypic
HB 7.9–12.9 cm
T 6.9–10.6 cm
W 16–43 g

Rondônia E to Piauí and Pernambuco, S to Minas Gerais

Dwarf Fat-tailed Opossum
Thylamys velutinus

FR Opossum velouté
DE Zwerg-Fettschwanzbeutelratte
ES Marmosa coligruesa enana

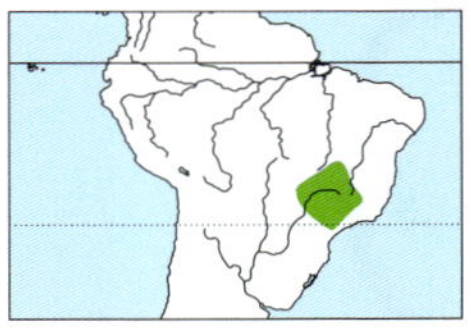

NT
Monotypic
HB 7.9–10 cm
T 6.5–9.1 cm
W 13–36 g

SC & SE BR (Goiás, Minas Gerais, São Paulo)

Elegant Fat-tailed Opossum
Thylamys elegans

FR Opossum élégant
DE Elegante Fettschwanzbeutelratte
ES Marmosa coligruesa elegante

0–1000 m

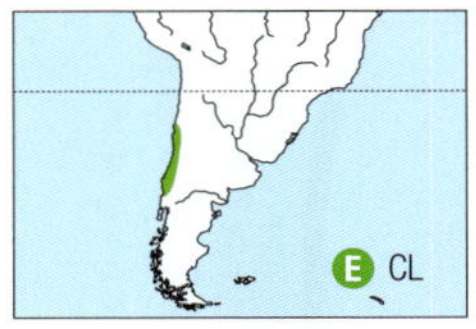

LC
Monotypic
HB 9–13.7 cm
T 10.5–14.6 cm

Coquimbo S to Los Lagos

Pallid Fat-tailed Opossum
Thylamys pallidior

FR Opossum à ventre blanc
DE Fahle Fettschwanzbeutelratte
ES Marmosa coligruesa pálida

S PE, SW BO, N CL, AR

0–4500 m

LC
Monotypic
HB 7.3–10.5 cm
T 9–11.8 cm
W 13–38 g

Tate's Fat-tailed Opossum
Thylamys tatei

FR Opossum de Tate
DE Tate-Fettschwanzbeutelratte
ES Marmosa coligruesa de Tate

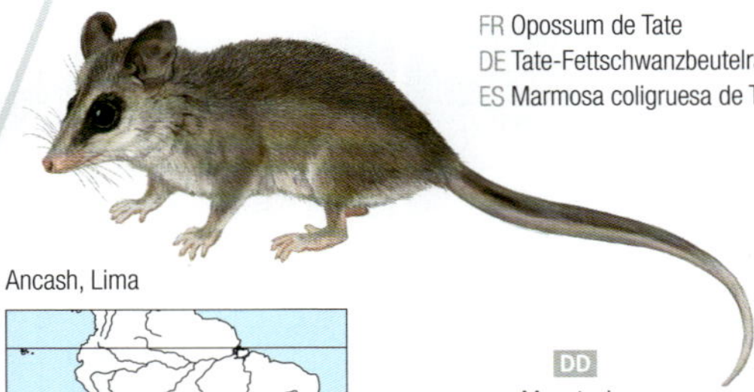

Ancash, Lima

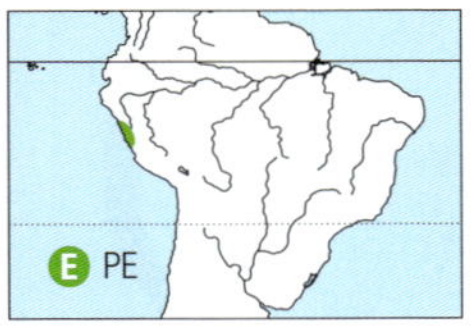

300–3000 m

DD
Monotypic
HB 10.9–12.3 cm
T 11.8–13.2 cm
W 16–54 g

Mesopotamian Fat-tailed Opossum
Thylamys citellus

FR Opossum d'Entre Rios
DE Pampas-Fettschwanzbeutelratte
ES Marmosa coligruesa de Mesopotamia

LC
Monotypic
HB 9.9–11.5 cm
T 10.2–12.5 cm
W 55–66 g

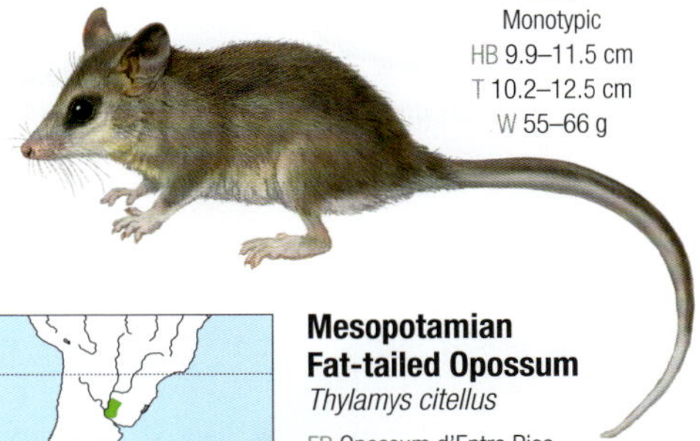

Corrientes (S of Iberá Wetlands), Entre Ríos

Dry Chaco Fat-tailed Opossum
Thylamys pulchellus

FR Opossum de Robles
DE Schöne Fettschwanzbeutelratte
ES Marmosa coligruesa del Chaco seco

LC
Monotypic
HB 6.2–9.6 cm
T 7.9–10.8 cm

Chaco, Santiago del Estero, Catamarca, San Juan

Chacoan Fat-tailed Opossum
Thylamys pusillus

FR Petit Opossum
DE Chaco-Fettschwanzbeutelratte
ES Marmosa coligruesa del Chaco

LC
Monotypic
HB 7–12 cm
T 8.5–13.4 cm
W 11–32 g

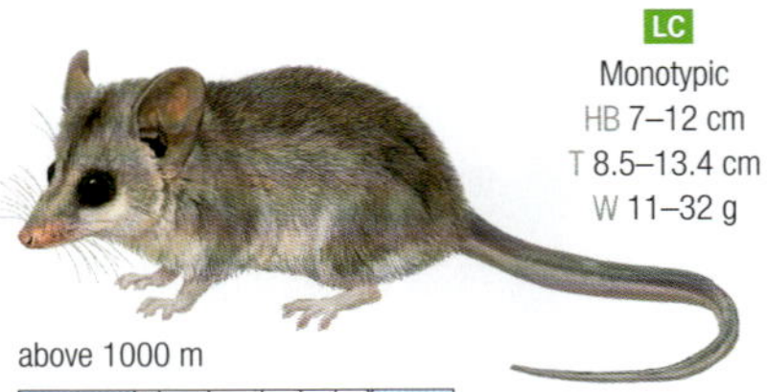

above 1000 m

SE BO (Santa Cruz), W PY, N AR (Formosa)

Paraguayan Fat-tailed Opossum
Thylamys macrurus

FR Opossum à longue queue
DE Paraguay-Fettschwanzbeutelratte
ES Marmosa coligruesa de Paraguay

NT
Monotypic
HB 10.1–13.5 cm
T 13.2–15.4 cm
W 30–57 g

S BR (Mato Grosso do Sul), E PY

Argentine Fat-tailed Opossum
Thylamys sponsorius

FR Opossum d'Argentine
DE Argentinien-Fettschwanzbeutelratte
ES Marmosa coligruesa de Argentina

LC
Monotypic
HB 8.6–11.9 cm
T 12.5–15.4 cm
W 15.5–62 g

515–3750 m

S BO (Tarija), NW AR (Jujuy, Salta, Tucumán, Catamarca)

Buff-bellied Fat-tailed Opossum
Thylamys venustus

FR Opossum charmant
DE Cinderella-Fettschwanzbeutelratte
ES Marmosa coligruesa de vientre beige

DD
Monotypic
HB 8.4–11 cm
T 11.1–13.8 cm
W 12–26 g

250–4000 m

BO (Cochabamba, Chuquisaca, Santa Cruz, Tarija), NW AR (Jujuy, Tucumán)

Genus *MARMOSOPS*

(21 species)

Junin Slender Opossum
Marmosops juninensis

FR Opossum du Junin
DE Junín-Schlankbeutelratte
ES Marmosa esbelta de Junín

VU
Monotypic
HB 10.4–11 cm
T 13.3–13.7 cm
W c. 27 g

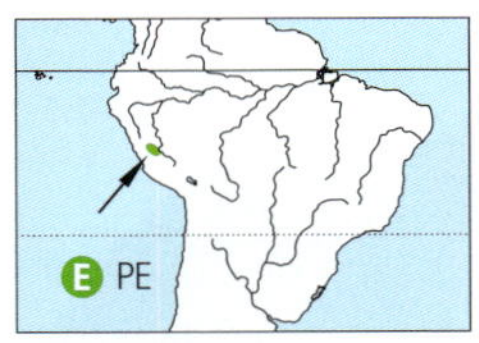

1390–2310 m

E Andean slopes (N Junín, Pasco)

Bishop's Slender Opossum
Marmosops bishopi

FR Opossum de Bishop
DE Bishop-Schlankbeutelratte
ES Marmosa esbelta de Bishop

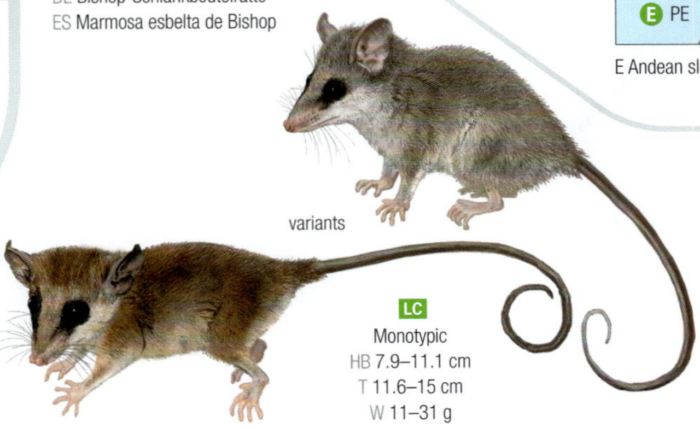

variants

LC
Monotypic
HB 7.9–11.1 cm
T 11.6–15 cm
W 11–31 g

100–1300 m

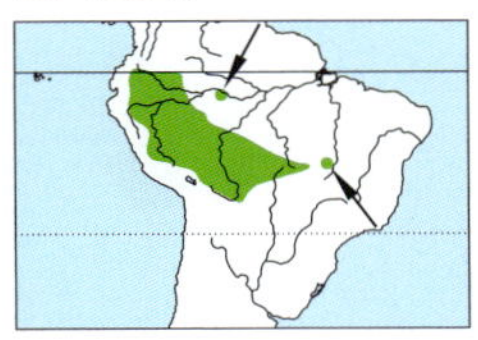

SE CO (Amazonas), E EC, E PE, W BR, N & E BO

Ojasti's Slender Opossum
Marmosops ojastii

FR Opossum d'Ojasti
DE Ojasti-Schlankbeutelratte
ES Marmosa esbelta de Ojasti

0–1850 m

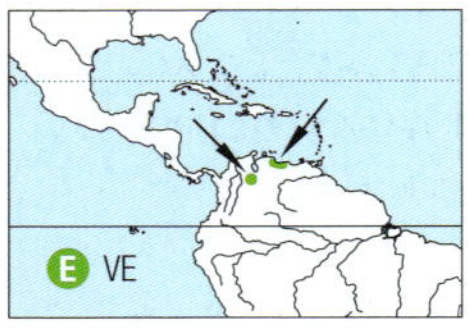

Cordillera de la Costa, S Cordillera de Mérida

NE
Monotypic
HB c. 9.7 cm
T c. 13.8 cm

Cordillera Slender Opossum
Marmosops chucha

FR Opossum d'Antioquia
DE Cordillera-Schlankbeutelratte
ES Marmosa esbelta de Antioquia

130–1400 m

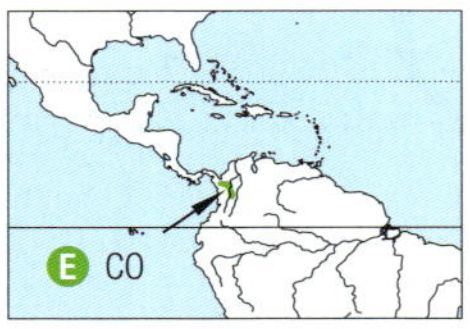

W of R Magdalena to NW terminus of W Andes Range

NE
Monotypic
HB 9.9–10.9 cm
T 13.4–14.1 cm
W 25–26 g

Rio Magdalena Slender Opossum
Marmosops magdalenae

FR Opossum du Magdalena
DE Río-Magdalena-Schlankbeutelratte
ES Marmosa esbelta de Magdalena

NE
Monotypic
HB 9.9–10.6 cm
T 12.5–13.2 cm
W c. 24 g

100–1940 m

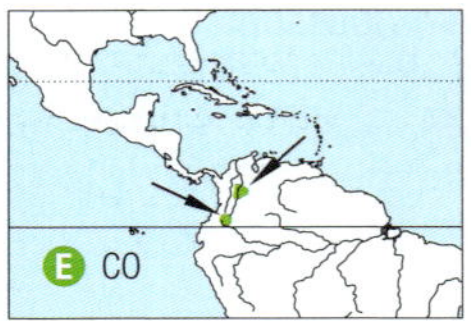

E of R Magdalena to W slope of E Andes; San Adolfo (Huila)

Carr's Slender Opossum
Marmosops carri

FR Opossum de Carr
DE Carr-Schlankbeutelratte
ES Marmosa esbelta de Carr

NE
Monotypic
HB 12.6–15.5 cm
T 15.2–18.3 cm
W 41–85 g

1000–2300 m

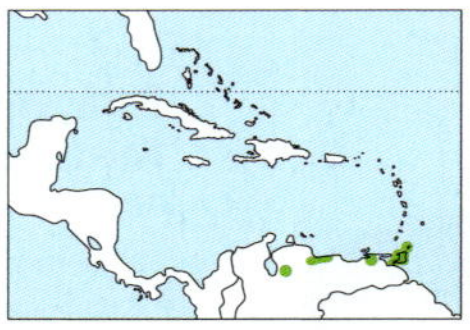

N VE (Cordillera de Trujillo, N Cordillera de la Costa), TT

Dusky Slender Opossum
Marmosops fuscatus

FR Opossum sombre
DE Dunkle Schlankbeutelratte
ES Marmosa esbelta oscura

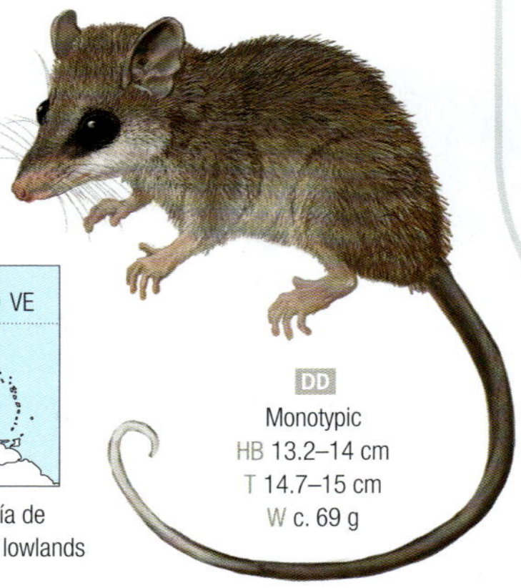

125–2350 m

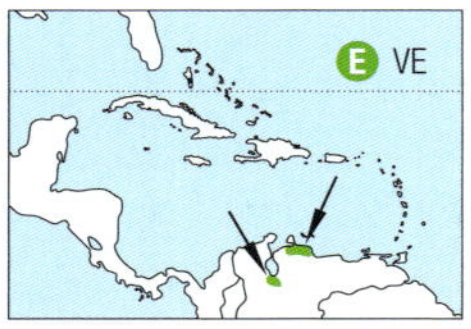

Cordillera de Mérida, Serranía de San Luis, Caribbean coastal lowlands

DD
Monotypic
HB 13.2–14 cm
T 14.7–15 cm
W c. 69 g

Panama Slender Opossum
Marmosops invictus

FR Opossum du Panama
DE Panama-Schlankbeutelratte
ES Marmosa esbelta de Panamá

LC
Monotypic
HB 10.4–12.1 cm
T 12.4–14.7 cm

450–1500 m

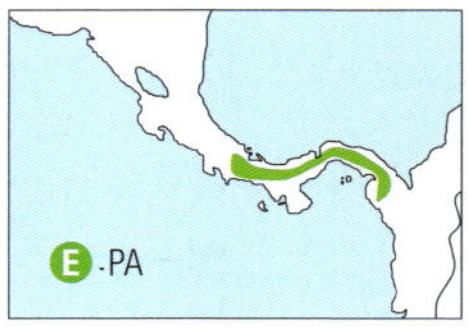

Silva's Slender Opossum
Marmosops marina

FR Opossum de Marina
DE Silva-Schlankbeutelratte
ES Marmosa esbelta de Marina

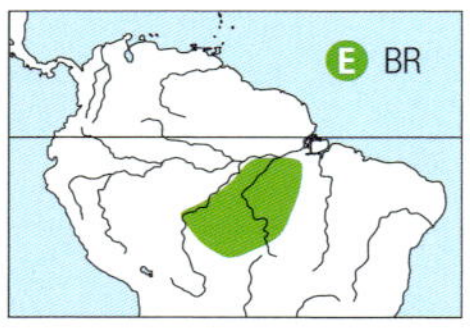

S of R Amazon, W of R Xingu (W to Rondônia, S to Mato Grosso)

NE
Monotypic
HB 6.4–11.2 cm
T 12–16 cm
W 15.1–30 g

Pinheiro's Slender Opossum
Marmosops pinheiroi

FR Opossum de Pinheiro
DE Pinheiro-Schlankbeutelratte
ES Marmosa esbelta de Pinheiro

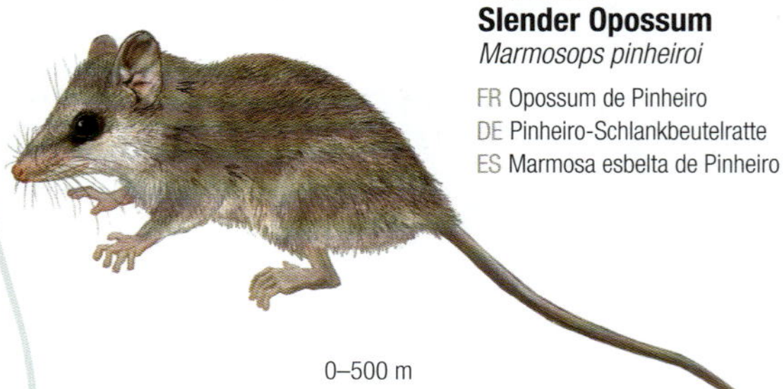

0–500 m

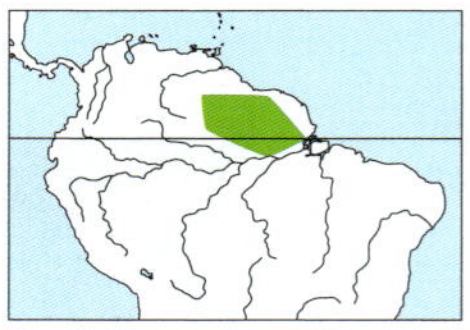

SE Venezuela (Bolívar), Guianas, N Brazil (Amapá, Roraima, Pará)

LC
Monotypic
HB 9.1–10.5 cm
T 12–14.5 cm
W 17–23 g

Handley's Slender Opossum
Marmosops handleyi

FR Opossum de Handley
DE Handley-Schlankbeutelratte
ES Marmosa esbelta de Handley

1400–1950 m

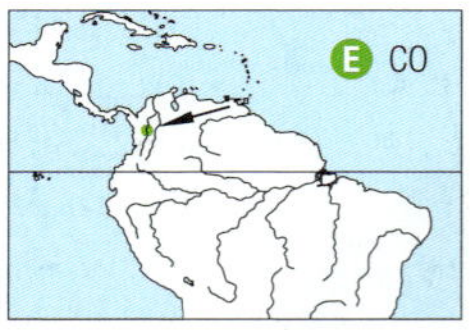

Central Andes (Antioquia)

CR
Monotypic
HB 10.4–12.2 cm
T 12.9–14.9 cm
W 21–29 g

Woodall's Slender Opossum
Marmosops woodalli

FR Opossum de Woodall
DE Woodall-Schlankbeutelratte
ES Marmosa esbelta de Woodall

NE
Monotypic
HB 9–11 cm
T 12.5–14.3 cm
W 15–28 g

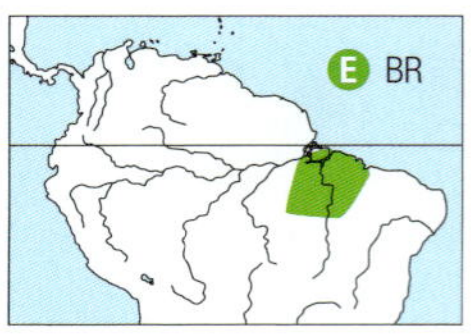

Pará (E of R Xingu, Marajó I) to Maranhão and S to Tocantins

Delicate Slender Opossum
Marmosops parvidens

FR Opossum délicat
DE Zierliche Schlankbeutelratte
ES Marmosa esbelta delicada

0–500 m

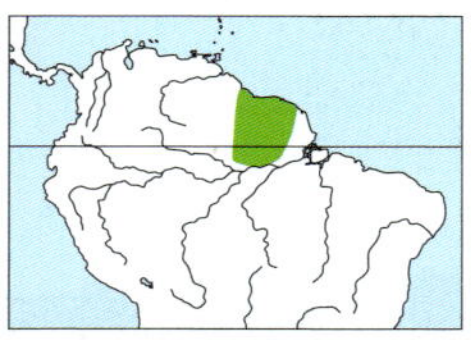

Guianas, N BR (Amazonas, NW Pará)

LC
Monotypic
HB 9.3–10.7 cm
T 14.2–16 cm
W 21–31 g

Pantepui Slender Opossum
Marmosops pakaraimae

FR Opossum des Pakaraima
DE Pantepui-Schlankbeutelratte
ES Marmosa esbelta de Pakaraima

VU
Monotypic
HB 10.4–11.6 cm
T 15.1–16.9 cm
W 25–33 g

800–1500 m

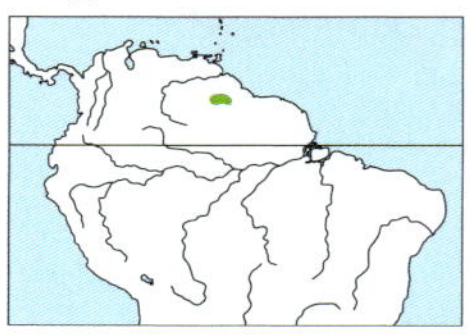

Pantepui region of SE VE and W GY

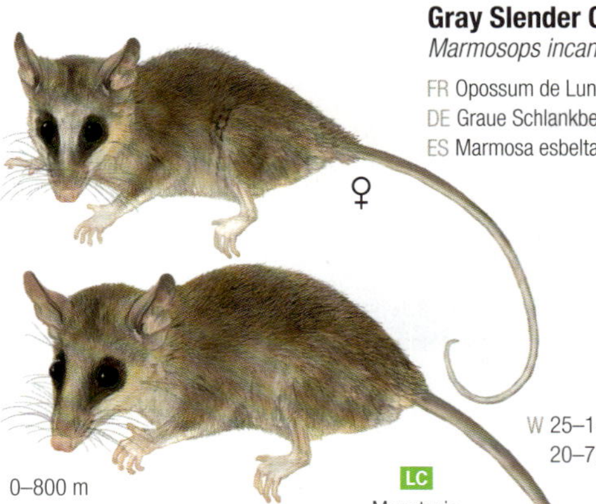

Gray Slender Opossum
Marmosops incanus
FR Opossum de Lund
DE Graue Schlankbeutelratte
ES Marmosa esbelta gris

W 25–140 g (M), 20–72 g (F)

LC
Monotypic
HB 12.7–19.4 cm (M), 10–15.8 cm (F)
T 14.6–24 cm (M), 14–20 cm (F)

0–800 m

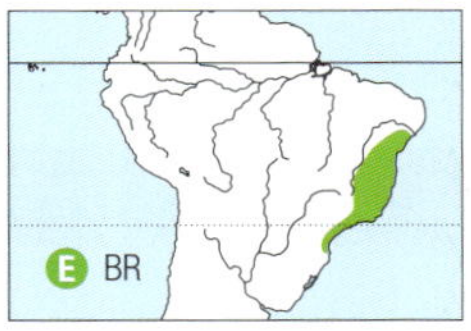

Coast from Sergipe to São Paulo, Minas Gerais

Brazilian Slender Opossum
Marmosops paulensis
FR Opossum de Sao Paulo
DE Brasilien-Schlankbeutelratte
ES Marmosa esbelta de Brasil

LC
Monotypic
HB 9.8–15.3 cm
T 14.5–21 cm
W 16–70 g

above 800 m

Minas Gerais, Rio de Janeiro, São Paulo, Paraná

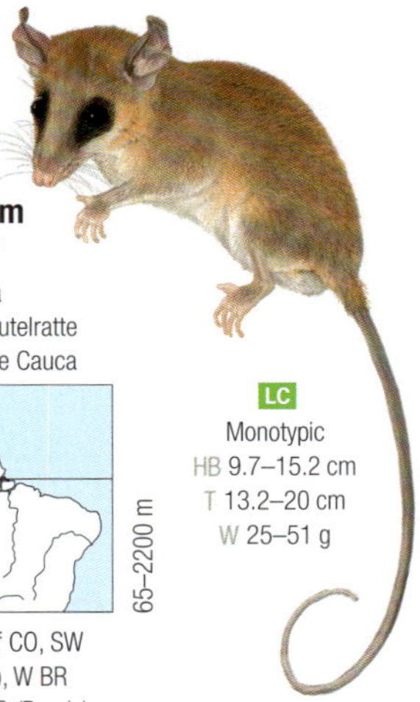

Tschudi's Slender Opossum
Marmosops caucae
FR Opossum du Cauca
DE Tschudi-Schlankbeutelratte
ES Marmosa esbelta de Cauca

LC
Monotypic
HB 9.7–15.2 cm
T 13.2–20 cm
W 25–51 g

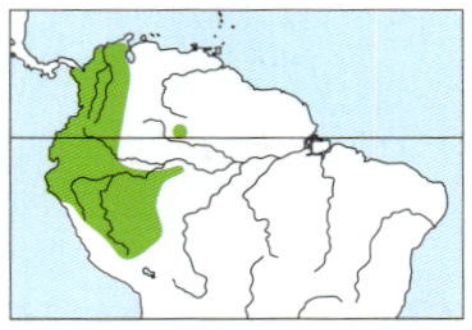

65–2200 m

E PA (Darién), Andes of CO, SW VE and PE (S to Cusco), W BR (Amazonas, Acre), N BO (Pando)

Spectacled Slender Opossum
Marmosops ocellatus
FR Opossum à lunettes DE Brillen-Schlankbeutelratte
ES Marmosa esbelta de anteojos

LC
Monotypic
HB 10.4–14 cm
T 14.6–18.6 cm
W 23–51 g

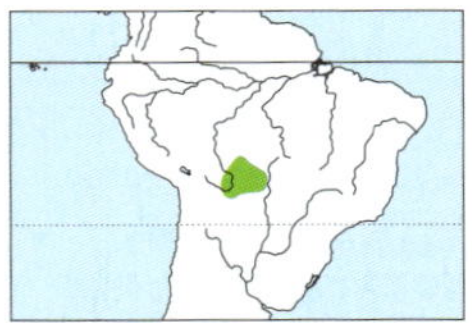

C & E BO (Santa Cruz), SW BR (Mato Grosso, Mato Grosso do Sul)

DD
Monotypic
HB 11.4–14 cm
T 15–17.7 cm
W 32–54 g

Creighton's Slender Opossum
Marmosops creightoni
FR Opossum de Creighton
DE Creighton-Schlankbeutelratte
ES Marmosa esbelta de Creighton

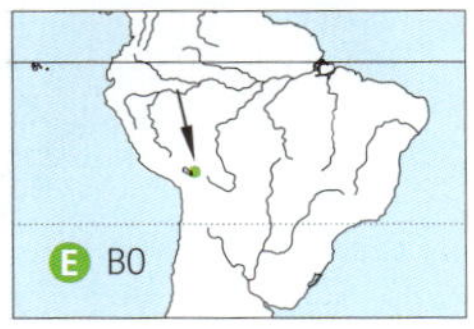

2000–3000 m

La Paz (3 close localities)

NE Monotypic
HB 10.6–13.6 cm
T 14.5–18.5 cm
W 33–54 g

Soini's Slender Opossum
Marmosops soinii
FR Opossum de Soini
DE Soini-Schlankbeutelratte
ES Marmosa esbelta de Soini

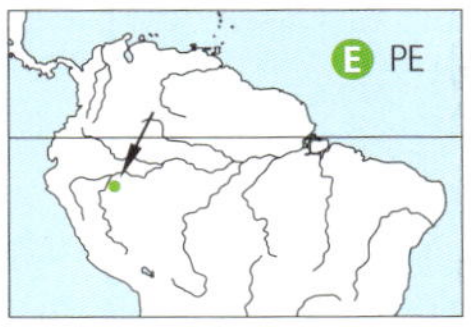

Yavarí-Ucayali interfluve

LC
Monotypic
HB 11.8–16.3 cm
T 15.4–20 cm
W 35–85 g

White-bellied Slender Opossum
Marmosops noctivagus
FR Opossum nocturne
DE Weißbauch-Schlankbeutelratte
ES Marmosa esbelta de vientre blanco

300–1500 m

W Amazon Basin of S CO (Putumayo), E EC, PE, BO, BR (S of R Amazon)

Order PAUCITUBERCULATA (1 family)

Family CAENOLESTIDAE – Shrew-opossums

3 genera, 7 species. Subtropical and montane forests and páramo grasslands. Neotropical Region.

Genus *CAENOLESTES*

(5 species)

W Andean slope in C & S EC and NW PE (N & S of Huancabamba Depression), E slope in NW PE

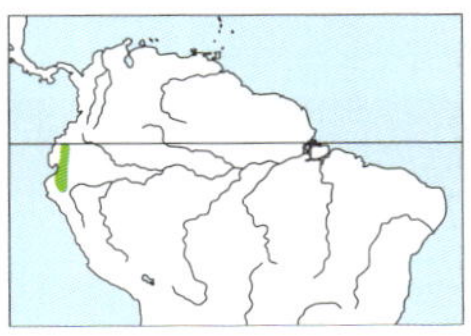

1630–3340 m

NT
Monotypic
HB 9.1–12.8 cm
T 11.8–15 cm
W 29–47 g

Gray-bellied Shrew-opossum
Caenolestes caniventer
FR Cénoleste à ventre gris
DE Graubäuchige Opossummaus
ES Musaraña marsupial de vientre gris

Sangay Shrew-opossum
Caenolestes sangay
FR Cénoleste du Sangay
DE Sangay-Opossummaus
ES Musaraña marsupial de Sangay

2300–3500 m

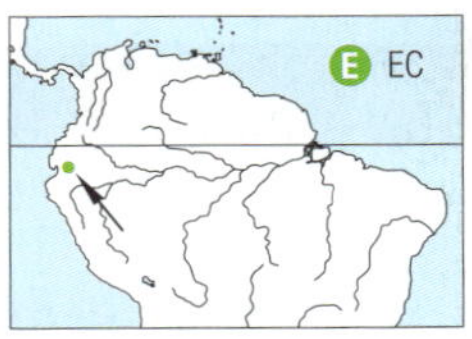

E slope of Andes along Macas–Riobamba highway

VU
Monotypic
HB 9.9–13.7 cm
T 9.5–13 cm
W 30–53 g

Condor Shrew-opossum
Caenolestes condorensis
FR Cénoleste du Condor
DE Anden-Opossummaus
ES Musaraña marsupial del Cóndor

VU
Monotypic
HB 13–14 cm
T 12.3–13 cm
W 43–48 g

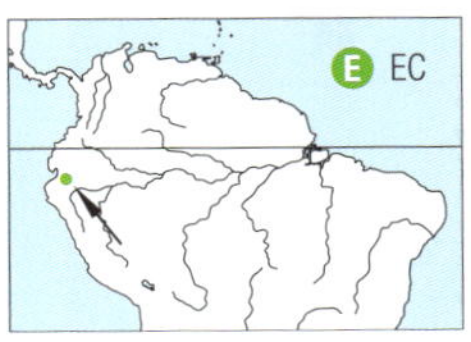

c. 2000 m

E Andean slope on SE EC (Cordillera del Cóndor, possibly in nearby Tapichala R), probably extending to PE

Dusky Shrew-opossum
Caenolestes fuliginosus
FR Cénoleste gris
DE Ecuador-Opossummaus
ES Musaraña marsupial sedosa

LC
3 sspp
HB 9.6–13.4 cm
T 10.3–13.9 cm
W 25–32 g

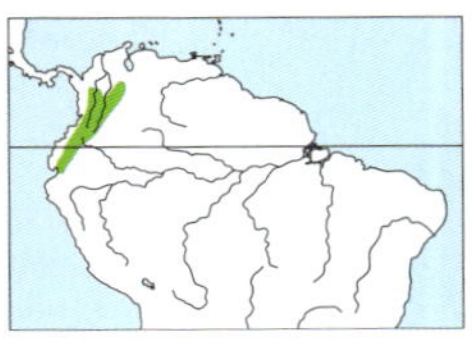

2135–4300 m

Forests and páramos of Andes of CO, extreme SW VE (Táchira) and N & C EC

Blackish Shrew-opossum
Caenolestes convelatus
FR Cénoleste noirâtre DE Schwarzbraune Opossummaus
ES Musaraña marsupial negruzca

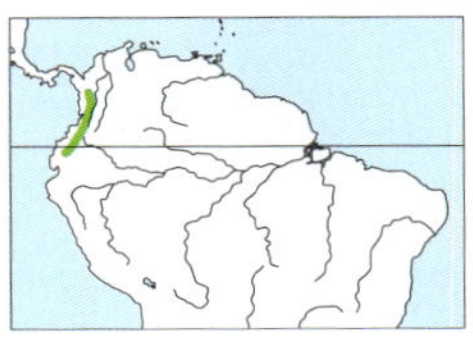

1100–3800 m

CO (W Andes), EC (W slope of Andes)

VU
2 sspp
HB 12.1–14.6 cm
T 7.2–14.1 cm
W 40–45 g

Incan Shrew-opossum
Lestoros inca

FR Cénoleste du Pérou
DE Peruanische Opossummaus
ES Musaraña marsupial andina

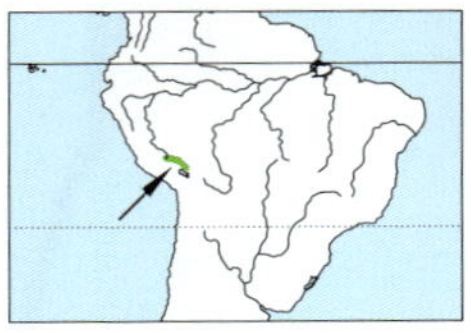

2190–3700 m

Valleys of E Andes from SE PE (Ocobamba, Cusco) to W BO (Llamachaqui, La Paz)

Genus *LESTOROS*

(1 species)

LC
Monotypic
HB 9–12 cm
T 10.5–13.5 cm
W 20–35 g

Long-nosed Shrew-opossum
Rhyncholestes raphanurus

FR Cénoleste du Chili
DE Langnasen-Opossummaus
ES Musaraña marsupial trompuda

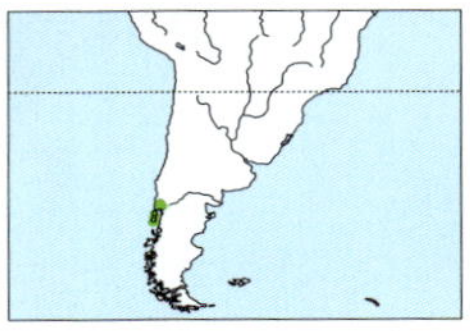

0–1135 m

SC CL (Los Lagos, with Chiloé I), AR (Nahuel Huapi NP)

Genus *RHYNCHOLESTES*

(1 species)

NT
2 sspp
HB 9.7–12.8 cm
T 6.5–8.8 cm
W 23–30 g

Order MICROBIOTHERIA (1 family)

Family MICROBIOTHERIIDAE – Monitos del Monte

1 genus, 3 species. Valdivian temperate rainforests. Neotropical Region.

Southern Monito del Monte
Dromiciops gliroides

FR Monito de montagne
DE Chiloé-Beutelratte
ES Monito del monte meridional

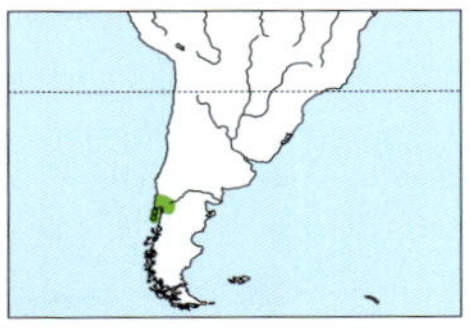

CL (Los Ríos, Los Lagos, with Chiloé I), AR (Neuquén, Río Negro, Chubut)

Genus *DROMICIOPS*

(3 species)

fattened tail

NT
Monotypic
HB 8.3–13 cm
T 9–13.2 cm
W 16–32 g

Pancho's Monito del Monte
Dromiciops bozinovici

FR Monito de Pancho
DE Pancho-Beutelratte
ES Monito del monte de Bozinovic

CL (Bío-Bío, Araucanía, probably N to S Maule), AR (Neuquén)

NE
Monotypic
HB c. 17 cm
T c. 8.6 cm
W c. 10 g

Mondaca's Monito del Monte
Dromiciops mondaca

FR Monito de Mondaca
DE Mondaca-Beutelratte
ES Monito del monte de Mondaca

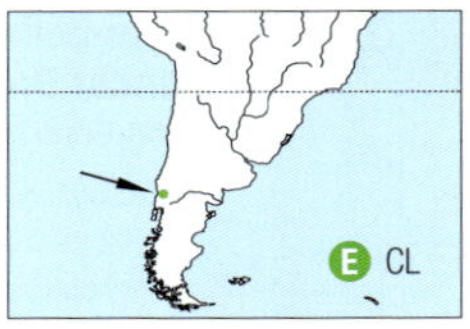

2 close localities of Coastal Cordillera (Los Ríos)

NE
Monotypic
HB c. 20 cm
T c. 11 cm
W c. 15 g

Order NOTORYCTEMORPHIA (1 family)

Family NOTORYCTIDAE – Marsupial Moles

1 genus, 2 species. Sand-dune deserts. Australasian Region.

North-western Marsupial Mole
Notoryctes caurinus

FR Kakarratul
DE Kleiner Beutelmull
ES Topo marsupial septentrional

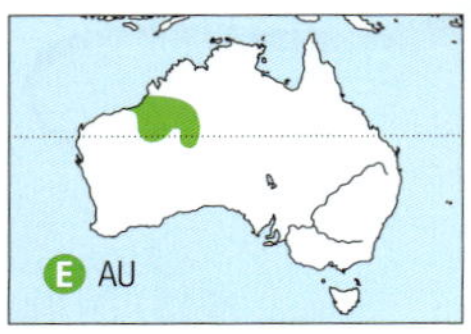

Little/Great Sandy & Gibson deserts (maybe Tanami), near coast Port Hedland to Broome (WA)

LC
Monotypic
HB 8.6–9.3 cm
T 1.6–1.8 cm
W 30–50 g

Genus *NOTORYCTES*

(2 species)

Central Desert Marsupial Mole
Notoryctes typhlops

FR Itjaritjari
DE Großer Beutelmull
ES Topo marsupial meridional

LC
Monotypic
HB 11–14 cm
T 2–2.5 cm
W 50–60 g

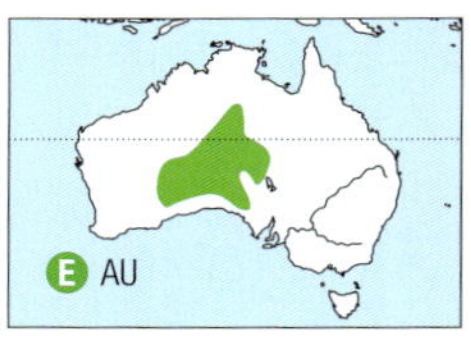

Great Victoria, Tanami & Great Sandy deserts (maybe Simpson) (S NT, SE WA, W SA)

Order DASYUROMORPHIA (2 families)

Family MYRMECOBIIDAE – Numbat

1 genus, 1 species. Eucalypt forests and woodlands and savanna woodlands. Australasian Region.

Genus *MYRMECOBIUS*

(1 species)

Numbat
Myrmecobius fasciatus
FR Numbat
DE Ameisenbeutler
ES Numbat

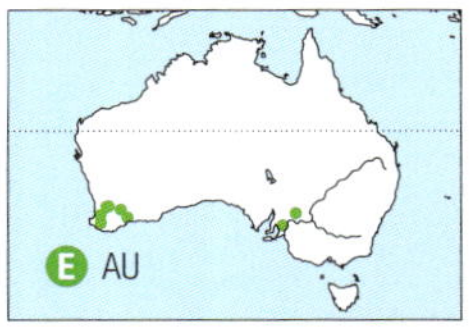

SW WA (Dryandra Woodland, Perup NR), SA, NSW; reintroduced to several sites

EN
1 extant ssp
HB 20–29 cm
T 12.5–21 cm
W 300–750 g

Family DASYURIDAE – Carnivorous Marsupials

17 genera, 76 species. Xeric deserts, shrublands and grasslands to temperate and tropical forests, woodlands, and swamps. Australasian Region.

Genus *DASYCERCUS*

(2 species)

Brush-tailed Mulgara
Dasycercus blythi
FR Mulgara à queue touffue
DE Pinselschwanz-Beutelmaus
ES Mulgara de cola de cepillo

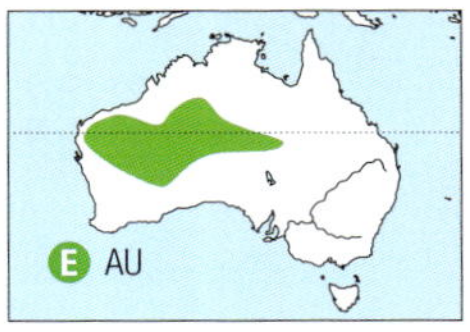

Mainly C WA and S NT, isolated locations in N SA and W Qld

LC
Monotypic
HB 12–16.5 cm
T 6–10 cm W 60–110 g

Crest-tailed Mulgara
Dasycercus cristicauda
FR Mulgara à queue en crête
DE Kammschwanz-Beutelmaus
ES Mulgara de cola crestada

NT
Monotypic
HB 12–23 cm
T 8–12.5 cm W 65–185 g

Deserts along borders of S NT, CN SA and SW Qld; formerly widespread in W & S AU

Genus *DASYKALUTA*

(1 species)

Kaluta
Dasykaluta rosamondae
FR Kaluta DE Rote Breitfuß-Beutelmaus
ES Kaluta

LC
Monotypic
HB 9–11 cm
T 5.5–7 cm
W 20–40 g

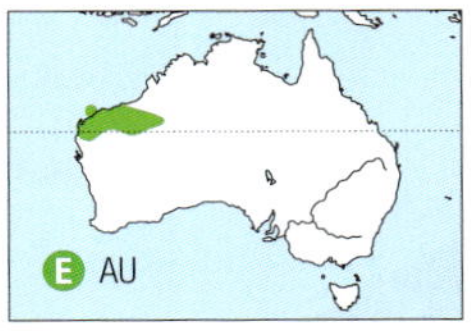

Potter I, Pilbara region, Carnarvon Basin, Little Sandy Desert (WA)

Genus *PARANTECHINUS*

(1 species)

Dibbler
Parantechinus apicalis
FR Dibbler DE Sprenkelbeutelmaus
ES Dibbler

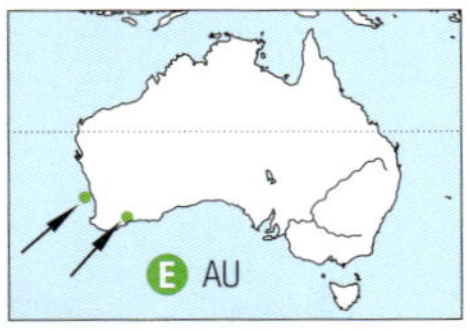

Fitzgerald River NP, Boullanger and Whitlock Is (WA); translocated to several sites of WA

EN
Monotypic
HB 14–14.5 cm
T 9.5–11.5 cm
W 40–125 g

Genus *DASYUROIDES*

(1 species)

Kowari
Dasyuroides byrnei
FR Kowari
DE Kowari
ES Kowari

SE NT, SW Qld (NW of R Diamantina), NE SA (S to Cooper Creek)

VU
2 sspp
HB 13.5–18 cm T 11–16 cm W 70–175 g

Genus *MYOICTIS*

(4 species)

Müller's Three-striped Dasyure
Myoictis melas
FR Dasyure à trois raies
DE Müller-Streifenbeutelmaus
ES Dasiuro de tres rayas de Müller

LC
Monotypic
HB 17.2–24 cm
W 172–255 g

ID/PG: locally in W & N New Guinea, Waigeo, Salawati, Batanta, Yapen

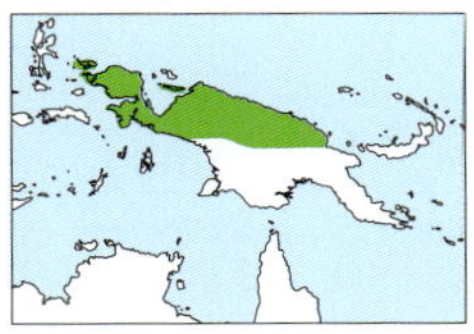

0–1800 m

Wallace's Three-striped Dasyure
Myoictis wallacii
FR Dasyure de Wallace
DE Wallace-Streifenbeutelmaus
ES Dasiuro de tres rayas de Wallace

LC
Monotypic
HB 17.6–22 cm W 192–245 g

ID/PG: Aru Is, S New Guinea (Merauke and Western Prov, R Aroa in E)

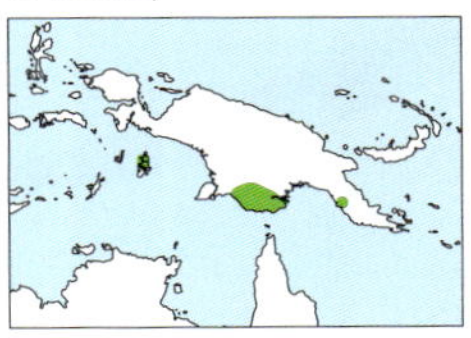

30–920 m

Woolley's Three-striped Dasyure
Myoictis leucura
FR Dasyure à queue blanche
DE Streifenbeutelmarder
ES Dasiuro de tres rayas de Woolley

C & E New Guinea (S Central Range: Mt Bosavi and Mt Sisa in W, Mt Victoria and R Vanapa in E)

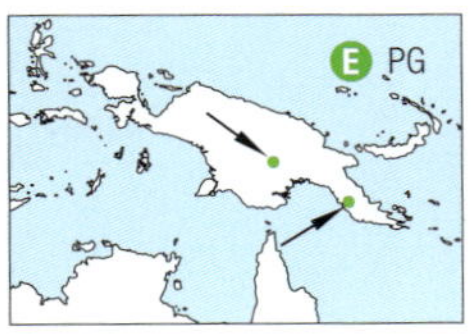

DD
Monotypic
HB 19.2–23 cm W 200–230 g

Tate's Three-striped Dasyure
Myoictis wavicus
FR Dasyure de Wau
DE Tate-Streifenbeutelmaus
ES Dasiuro de tres rayas de Tate

975–1810 m

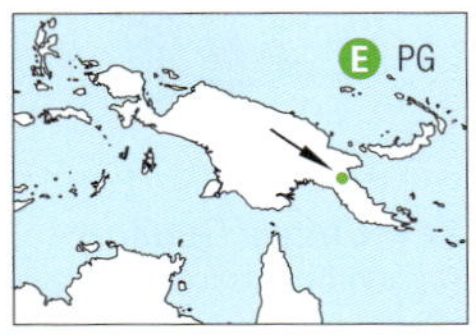

E New Guinea (Wau area, in N sice of Central Range)

DD
Monotypic
HB 14–17.5 cm
W 61–134 g

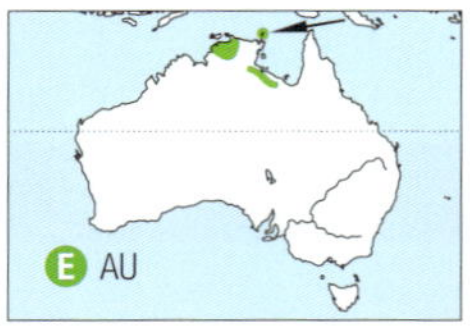

N NT (Table Top Range to Wollogorang Station, with Marchinbar I)

LC
Monotypic
HB 9–11.5 cm T 9–12.5 cm W 15–40 g

Sandstone Pseudantechinus
Pseudantechinus bilarni

FR Dasyure de Harney
DE Harney-Fettschwanz-Beutelmaus
ES Falso antequino de Bilarn

Genus *PSEUDANTECHINUS*
(5 species)

Woolley's Pseudantechinus
Pseudantechinus woolleyae

FR Dasyure de Woolley
DE Woolley-Fettschwanz-Beutelmaus
ES Falso antequino de Woolley

LC
Monotypic
HB 10.7–10.8 cm
T 7.5–8.5 cm
W 39–50 g

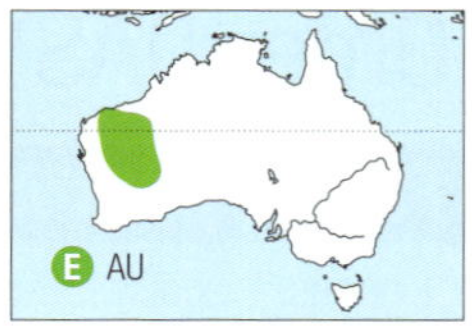

W WA (across R Fortescue Basin, W Ashburton, W Austin; recorded in Little Sandy Desert)

Fat-tailed Pseudantechinus
Pseudantechinus macdonnellensis

FR Dasyure à queue adipeuse
DE MacDonnell-Fettschwanz-Beutelmaus
ES Falso antequino de cola gruesa

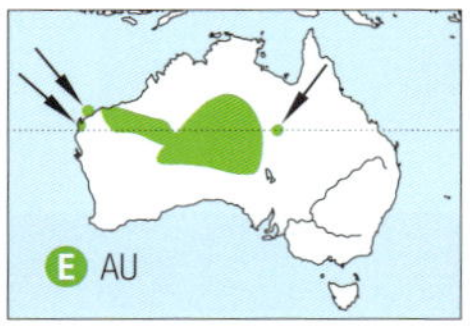

LC
Monotypic
HB 9.5–10.5 cm T 7.5–8.5 cm
W 20–45 g

WA (with Barrow I), C & S NT, N SA; isolated locality in far W Qld

Carpentarian Pseudantechinus
Pseudantechinus mimulus

FR Dasyure des termitières
DE Kleine Fettschwanz-Beutelmaus
ES Falso antequino de Carpentaria

NT (Alexandria Station, Sir Edward Pellew Group Is), Qld (near Mt Isa)

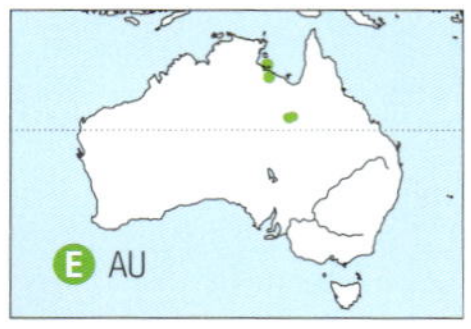

NT
Monotypic
HB 6.3–9.1 cm T 5.6–7.6 cm W 14–25 g

Ningbing Pseudantechinus
Pseudantechinus ningbing

FR Dasyure de Ningbing
DE Ningbing-Fettschwanz-Beutelmaus
ES Falso antequino de Ningbing

LC
Monotypic
HB 8–10 cm
T 7.5–9 cm
W 15–25 g

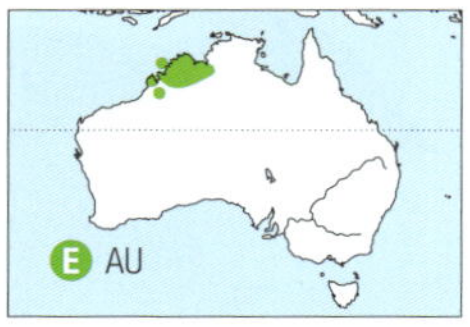

Kimberley region, Bonaparte Arch (Augustus & Heywood Is) (WA, NW NT)

Genus *NEOPHASCOGALE*
(1 species)

Speckled Dasyure
Neophascogale lorentzii

FR Dasyure de Lorentz
DE Neuguinea-Spitzhörnchenbeutler
ES Dasiuro moteado

LC
Monotypic
HB 16.6–22 cm T 17–21 cm
W c. 212 g

1200–3900 m

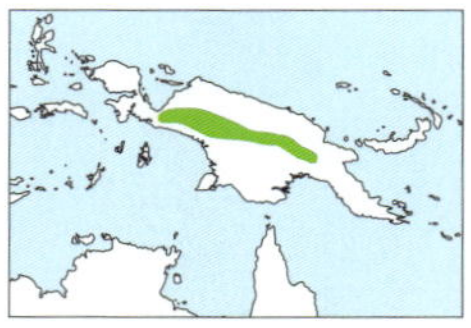

ID/PG: New Guinea Central Range (Weyland Range E to Hagen Range)

Genus *PHASCOLOSOREX*
(2 species)

Red-bellied Phascogale
Phascolosorex doriae

FR Dasyure à ventre rouge
DE Orangebauch-Streifenbeutelmaus
ES Dasiuro de vientre rojo

900–2000 m

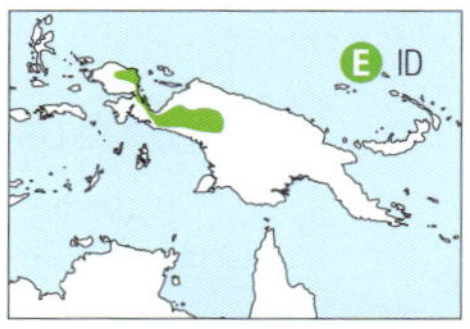

W New Guinea (Vogelkop Pen to Weyland Range, NW Central Range)

LC Monotypic
HB 15.2–23 cm T 15.2–19.1 cm

Narrow-striped Dasyure
Phascolosorex dorsalis

FR Dasyure à bande dorsale
DE Braunbauch-Streifenbeutelmaus
ES Dasiuro de banda estrecha

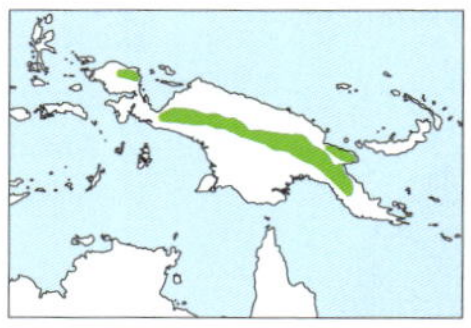

1500–3600 m

ID/PG: New Guinea (Arfak Mt, Weyland Range, Central Range, Huon)

LC
3 sspp
HB 11.7–16.8 cm
T 11.6–15.4 cm
W 132–145 g

ssp *brevicaudata*

ssp *dorsalis*

Genus *DASYURUS*
(6 species)

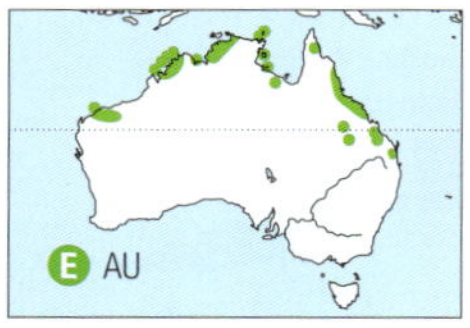

AU: patchily, including islands (W & N WA, N NT, N & E Qld)

♀

♂

EN
Monotypic
HB 27–37 cm (M), 25–31 cm (F)
T 22–34 cm (M), 20–30 cm (F)
W 0.3–1.1 kg (M), 240–690 g (F)

Northern Quoll
Dasyurus hallucatus

FR Quoll nain
DE Nördlicher Beutelmarder
ES Dasiuro septentrional

Spotted-tailed Quoll
Dasyurus maculatus

FR Quoll à queue tachetée
DE Fleckenbeutelmarder
ES Dasiuro de cola moteada

ssp *gracilis*

NT
2 sspp
HB 31–76 cm
T 28–55 cm
W 0.8–5 kg

ssp *maculatus*

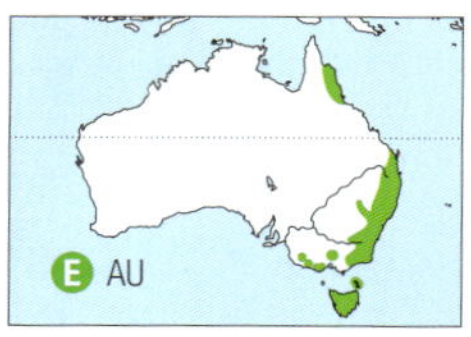

N & SE Qld, E NSW, E & S Vic, SE SA, Tas (with some Bass Strait Is)

New Guinea Quoll
Dasyurus albopunctatus

FR Quoll de Nouvelle-Guinée
DE Neuguinea-Beutelmarder
ES Dasiuro de Nueva Guinea

0–3600 m

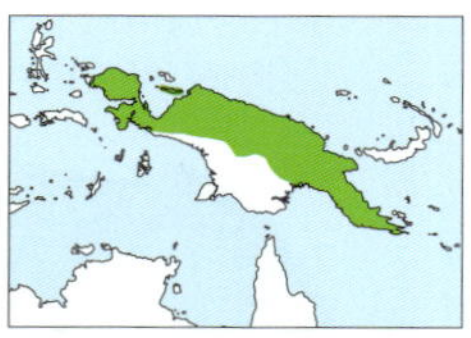

ID/PG: New Guinea (except SW lowlands), Yapen I; uncertain in Vogelkop

NT
Monotypic
HB 23–35 cm
T 21–29 cm
W 520–710 g

Eastern Quoll
Dasyurus viverrinus

FR Quoll viverrin
DE Tüpfelbeutelmarder
ES Dasiuro oriental

black morph

0–1000 m

Now only in Tas and Bruny I

EN
Monotypic
HB 32–45 cm (M), 28–40 cm (F)
T 20–28 cm (M), 17–24 cm (F)
W 0.9–1.9 kg (M), 0.7–1.1 kg (F)

Western Quoll
Dasyurus geoffroii

FR Quoll de Geoffroy
DE Geoffroy-Beutelmarder
ES Dasiuro occidental

SW WA, historically widespread; introduced to Julimar CP

NT
2 sspp
HB 26–40 cm T 21–35 cm
W 0.6–2.2 kg

Bronze Quoll
Dasyurus spartacus

FR Quoll bronzé
DE Bronzequoll
ES Dasiuro de bronce

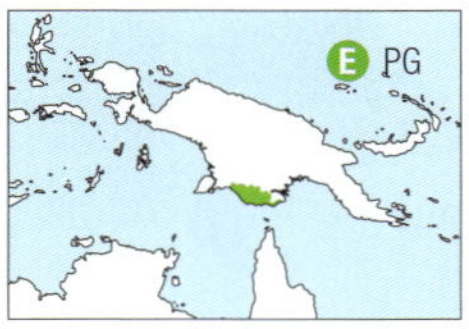

S New Guinea (Morehead area in Trans-Fly Plains); unconfirmed in Wasur NP and Tonda WMA

NT
Monotypic
HB 30–38 cm
T 25–28 cm
W c. 684 g (F)

Genus *SARCOPHILUS*

(1 species)

Tasmanian Devil
Sarcophilus harrisii

FR Diable de Tasmanie
DE Beutelteufel
ES Diablo de Tasmania

Now restricted to Tas, with Robbins I

EN
Monotypic
HB 57–65 cm T 24–26 cm W 5–14 kg

Tasmanian Dusky Antechinus
Antechinus swainsonii

FR Antéchinus de Swainson
DE Swainson-Breitfußbeutelmaus
ES Antequino oscuro de Tasmania

Genus *ANTECHINUS*

(15 species)

Forested areas in Tas

LC
Monotypic
HB 10.3–16.1 cm
T 7.7–11 cm
W 31–93 g

NE Monotypic
HB 10.5–13.3 cm (M) T 9.2–11.8 cm (M)
W 46–92 g (M)

SE Tas (Tasman Pen)

Tasman Peninsula Dusky Antechinus
Antechinus vandycki

FR Antéchinus de Van Dyck
DE Van-Dyck-Breitfußbeutelmaus
ES Antequino de Van Dyck

NE
2 sspp
HB 8.5–15 cm
T 7.6–11.4 cm
W 30–112 g

Mainland Dusky Antechinus
Antechinus mimetes

FR Antéchinus imitateur
DE Braune Breitfußbeutelmaus
ES Antequino oscuro continental

0–1800 m

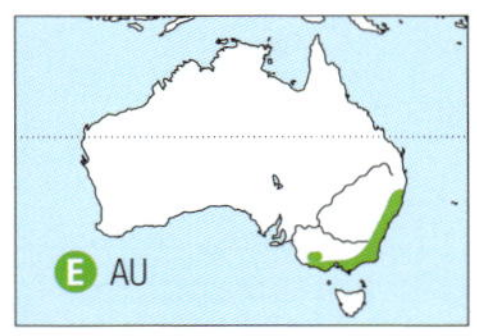

SW Vic to NE NSW (up to N of Guy Fawkes, New England NP)

LC
2 sspp
HB 9.8–14 cm
T 6.5–10 cm
W 36–129 g

E SA (Robe) to S Vic (S Gippsland, some islands), Tas, Bass Strait Is

Swamp Antechinus
Antechinus minimus

FR Antéchinus des marais
DE Sumpf-Breitfußbeutelmaus
ES Antequino palustre

NE
Monotypic
HB 10.6–14.5 cm
T 9.4–13.1 cm
W 46–120 g

780–950 m

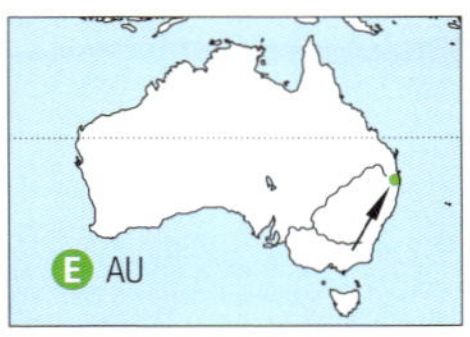

Tweed Volcano caldera (SE Qld, NE NSW)

Black-tailed Antechinus
Antechinus arktos

FR Antéchinus à queue noire
DE Schwarzschwanz-Breitfußbeutelmaus
ES Antequino de cola negra

Atherton Antechinus
Antechinus godmani

FR Antéchinus de Godman
DE Queensland-Breitfußbeutelmaus
ES Antequino de Godman

LC Monotypic
HB 10.3–15.5 cm T 9.2–14.3 cm
W 42–123 g

600–1540 m

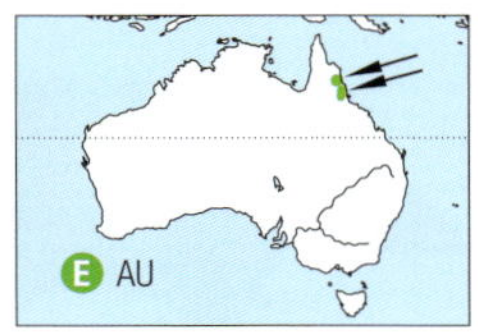

NE Qld (from Tinaroo Range, S of Cairns, to Girringun NP near Mt Thorn)

NE

Monotypic
HB 8.1–11.7 cm
T 7.9–11.7 cm
W 18–55 g

100–1000 m

Buff-footed Antechinus
Antechinus mysticus
FR Antéchinus à pattes chamois
DE Mystische Breitfußbeutelmaus
ES Antequino de patas color ante

SE Qld (Eungella NP S in scattered populations to D'Aguilar NP)

NE

Monotypic
HB 8.1–10.7 cm
T 7.8–10.8 cm
W 18–38 g

850–900 m

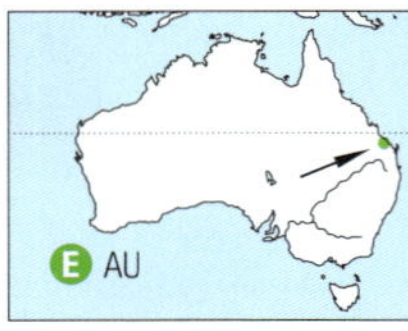

Silver-headed Antechinus
Antechinus argentus
FR Antéchinus à tête argentée
DE Silberfarbene Breitfußbeutelmaus
ES Antequino de cabeza plateada

SE Qld (escarpment at Kroombit Tops NP)

LC

Monotypic
HB 8.8–12.5 cm
T 7.2–10.4 cm
W 18–42 g

above 600 m

Rusty Antechinus
Antechinus adustus
FR Antéchinus rouilleux
DE Rostfarbene Breitfußbeutelmaus
ES Antequino herrumbroso

NE Qld (between Mt Windsor Tableland and Bluewater Range)

Cinnamon Antechinus
Antechinus leo
FR Antéchinus cannelle
DE Zimt-Breitfußbeutelmaus
ES Antequino canela

LC

Monotypic
HB 10.9–16 cm
T 12.5–14 cm
W 32–124 g

N Qld (between Iron and McIlwraith ranges)

VU

Monotypic
HB 9.6–14.8 cm
T 9.1–13.2 cm
W 25–83 g

NT (Top End, with Tiwi Is)

Fawn Antechinus
Antechinus bellus
FR Antéchinus fauve
DE Augenfleck-Breitfußbeutelmaus
ES Antequino beis

Yellow-footed Antechinus
Antechinus flavipes
FR Antéchinus à pattes jaunes
DE Gelbfüßige Breitfußbeutelmaus
ES Antequino de patas amarillas

ssp *leucogaster*

ssp *flavipes*

LC

3 sspp
HB 8.3–15.1 cm
T 6.5–11.6 cm
W 21–76 g

20–1200 m

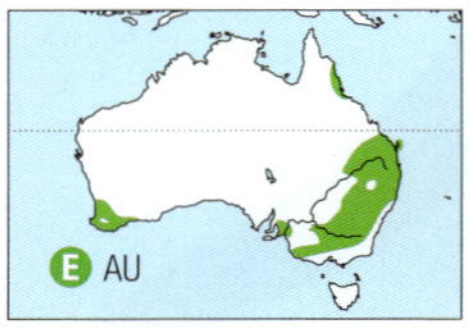

NE & SE Qld, NSW, Vic, SA, SW WA

ssp *rubeculus*

LC

Monotypic
HB 6.2–12.8 cm T 6–11.6 cm
W 16–40 g

below 2000 m

Agile Antechinus
Antechinus agilis
FR Antéchinus agile
DE Graue Breitfußbeutelmaus
ES Antequino ágil

S & SE NSW, SE Vic

Brown Antechinus
Antechinus stuartii
FR Antéchinus de Stuart
DE Stuart-Breitfußbeutelmaus
ES Antequino pardo

S Qld (from Main Range NP) to SE NSW (about Kioloa)

LC

Monotypic
HB 7.3–14.1 cm
T 6.5–12.4 cm
W 16–71 g

LC 4 sspp
HB 10.6–14.5 cm
T 10.5–17.5 cm
W 29–74 g

1400–2800 m

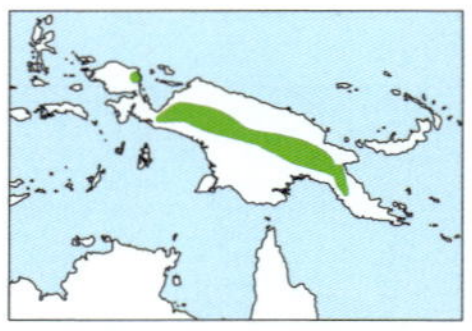

Long-nosed Dasyure
Murexia naso
FR Murexie à long museau
DE Langnasen-Neuguinea-Beutelmaus
ES Dasiuro de hocico largo

ID/PG: New Guinea (Vogelkop Pen, Central Range)

Subtropical Antechinus
Antechinus subtropicus
FR Antéchinus subtropical
DE Subtropische Breitfußbeutelmaus
ES Antequino subtropical

0–1000 m

SE Qld (from Gympie) S to NE NSW (Border Ranges NP, Dorrigo NP)

LC

Monotypic
HB 8.1–13.6 cm
T 6.4–10.6 cm
W 16–78 g

LC

2 sspp
HB 13.6–27 cm
T 14.7–28 cm
W 39–134 g

0–2200 m

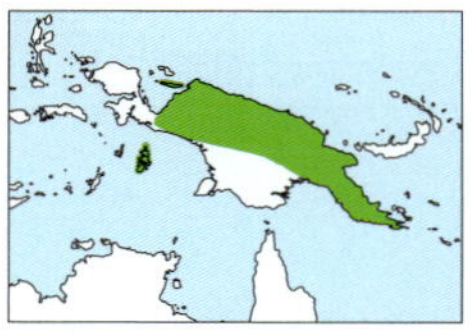

ID/PG: New Guinea, Aru Is, Yapen I

Genus *MUREXIA*

(5 species)

Short-furred Dasyure
Murexia longicaudata
FR Murexie à poil ras
DE Langschwanz-Neuguinea-Beutelmaus
ES Dasiuro de pelo corto

Black-tailed Dasyure
Murexia melanurus
FR Murexie à queue noire
DE Schwarzschwanz-Neuguinea-Beutelmaus
ES Dasiuro de cola negra

0–2800 m

LC

Monotypic
HB 9.1–16.5 cm
T 10–16.5 cm
W 17–69 g

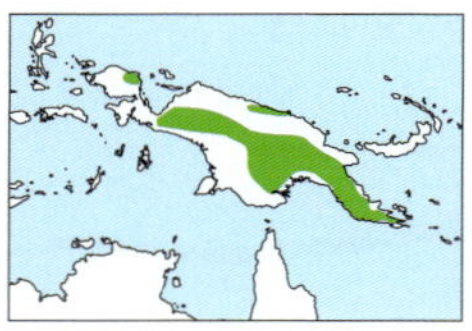

ID/PG: New Guinea (Central Range, Arfak and Torricelli Mts)

Habbema Dasyure
Murexia habbema
FR Murexie du Habbema
DE Habbema-Neuguinea-Beutelmaus
ES Dasiuro del Habbema

LC

2 sspp
HB 11–12.2 cm
T 10.9–15.7 cm
W 23–45 g

1600–3660 m

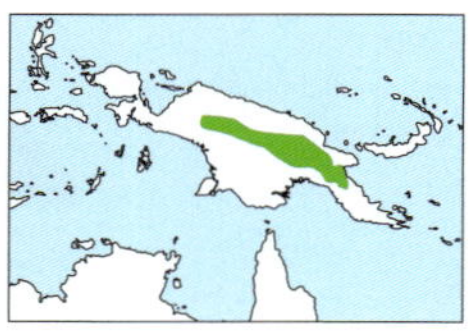

ID/PG: New Guinea (Central Range from Snow & Star Mts to N Central Prov)

NT

Monotypic
HB 12.4–17 cm
T 14–18.4 cm
W 32–102 g

SE Pen ranges of New Guinea

600–1400 m

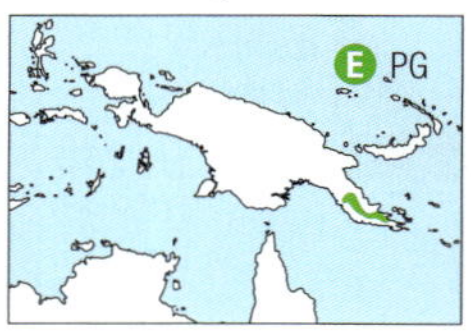

Broad-striped Dasyure
Murexia rothschildi
FR Murexie rayée
DE Rothschild-Neuguinea-Beutelmaus
ES Dasiuro de banda ancha

Genus *PHASCOGALE*

(3 species)

Red-tailed Phascogale
Phascogale calura

FR Petit Phascogale
DE Kleine Pinselschwanzbeutelmaus
ES Fascogalo de cola roja

NT
Monotypic
HB 9.3–12.2 cm
T 11.9–14.5 cm
W 38–68 g

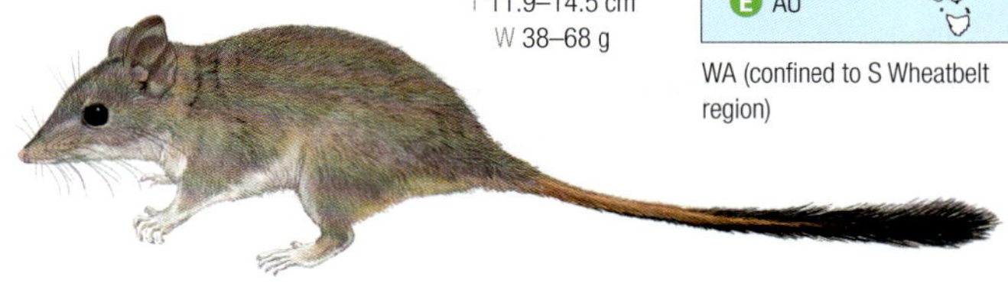

WA (confined to S Wheatbelt region)

Northern Brush-tailed Phascogale
Phascogale pirata

FR Phascogale du Nord
DE Nördliche Pinselschwanzbeutelmaus
ES Fascogalo de cola de cepillo septentrional

VU
Monotypic
HB 15–21 cm
T 18–21 cm
W 108–234 g

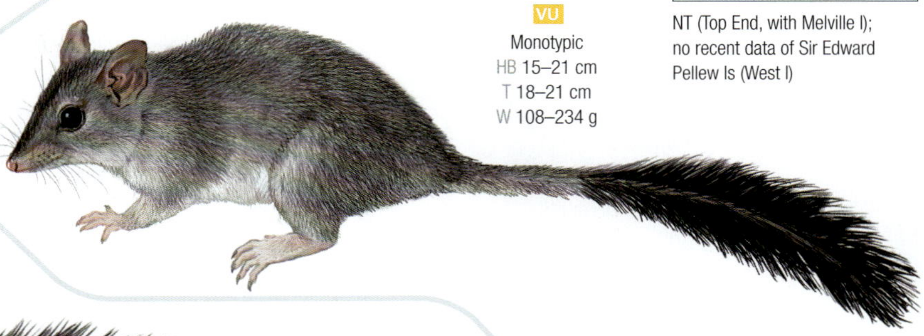

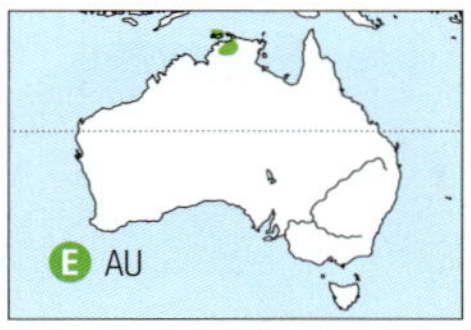

NT (Top End, with Melville I); no recent data of Sir Edward Pellew Is (West I)

Common Brush-tailed Phascogale
Phascogale tapoatafa

FR Grand Phascogale
DE Große Pinselschwanzbeutelmaus
ES Fascogalo de cola de cepillo meridional

NT
Monotypic
HB 14.8–26 cm
T 16–23 cm
W 106–311 g

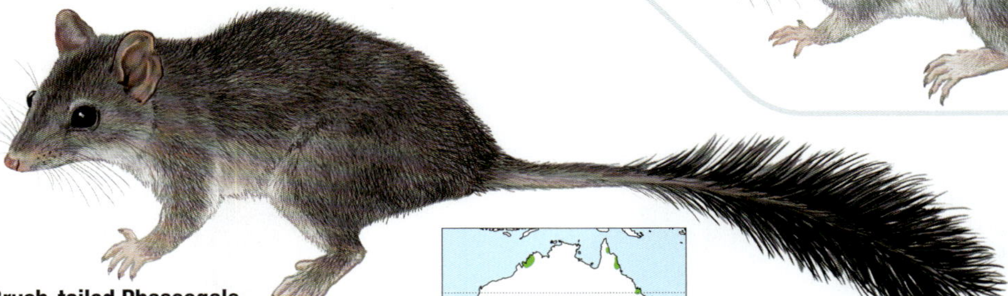

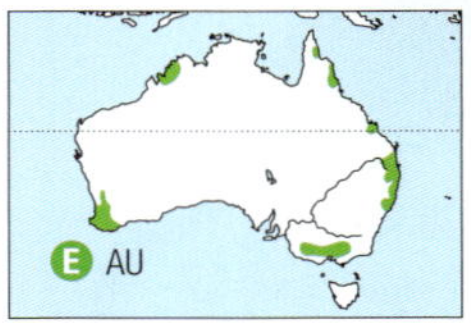

Patchily (SW & NW WA, E Qld, NE NSW, Vic); presumed extinct in SA

Genus *PLANIGALE*

(5 species)

Narrow-nosed Planigale
Planigale tenuirostris

FR Planigale à nez étroit
DE Spitznasen-Flachkopfbeutelmaus
ES Planigalo de hocico estrecho

LC
Monotypic
HB 5–7.5 cm
T 5–6.5 cm
W 4–9 g

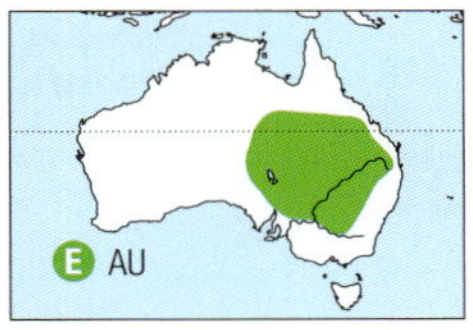

SE NT, S & W Qld, C & W NSW, C & NE SA

Papuan Planigale
Planigale novaeguineae

FR Planigale de Nouvelle-Guinée
DE Neuguinea-Flachkopfbeutelmaus
ES Planigalo de Nueva Guinea

LC
Monotypic
HB c. 7.9 cm T c. 8 cm
W c. 14.9 g

0–250 m

ID/PG: S & SE New Guinea (SE Papua, S Western & S Central prov)

Common Planigale
Planigale maculata

FR Planigale commun
DE Gefleckte Flachkopfbeutelmaus
ES Planigalo pigmeo

LC
2 sspp
HB 5.2–8.4 cm
T 4.6–6.5 cm
W 3.5–16.3 g

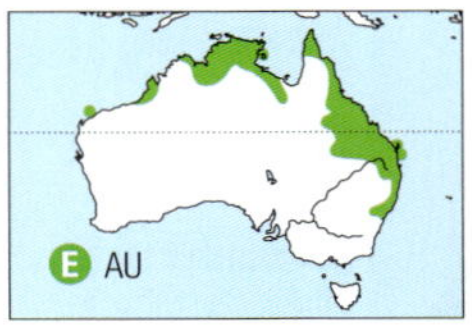

N WA, N NT (Top End, Groote Eylan), N & E Qld (with Fraser I, Bribie I), E NSW

Giles's Planigale
Planigale gilesi

FR Planigale de Giles
DE Giles-Flachkopfbeutelmaus
ES Planigalo de Giles

LC
Monotypic
HB 6–8 cm T 5.5–7 cm
W 5–16 g

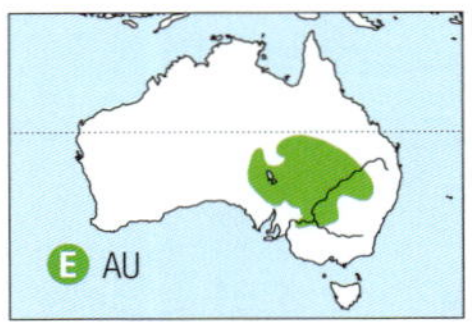

SW Qld (Boulia) to NW Vic (along R Murray), and Stuart Highway (S NT, C SA) to Moree (N NSW)

Long-tailed Planigale
Planigale ingrami

FR Planigale d'Ingram
DE Langschwanz-Flachkopfbeutelmaus
ES Planigalo de cola larga

LC
Monotypic
HB 4.9–7.1 cm
T 5.1–6.8 cm
W 2.6–6.6 g

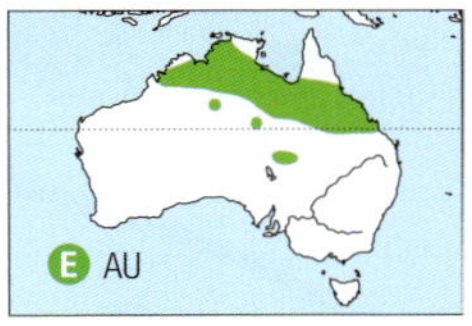

NE WA, NT, Qld, NE SA

Genus *NINGAUI*

(3 species)

Wongai Ningaui
Ningaui ridei

FR Ningaui wongai
DE Ride-Ningaui
ES Ningaui de Ride

LC
Monotypic
HB 5.7–7.5 cm
T 5.9–7.1 cm
W 6.5–10.5 g

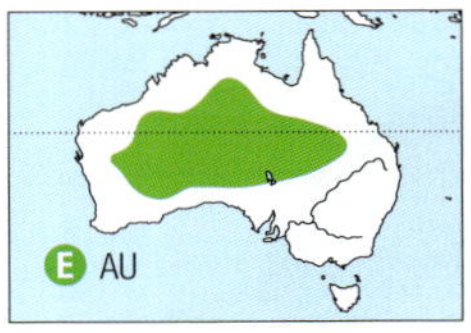

Arid regions of WA, NT, Qld, SA

Southern Ningaui
Ningaui yvonneae

FR Ningaui austral
DE Yvonne-Ningaui
ES Ningaui meridional

LC
Monotypic
HB 4.8–8.1 cm
T 5.3–7.1 cm
W 6–14 g

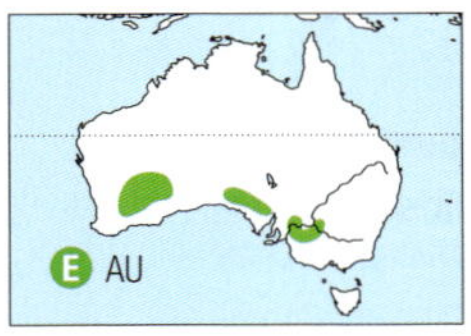

S WA, S SA, SW NSW, NW Vic

Pilbara Ningaui
Ningaui timealeyi

FR Ningaui du Pilbara
DE Pilbara-Ningaui
ES Ningaui de Pilbara

LC
Monotypic
HB 5.6–7.6 cm
T 5.7–9.5 cm
W 3.5–9.5 g

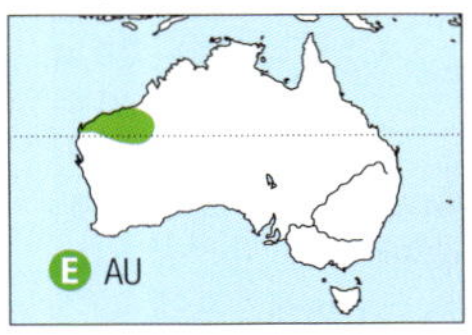

Pilbara region, Gascoyne region, Little Sandy Desert (WA)

Genus *ANTECHINOMYS*

(1 species)

Kultarr
Antechinomys laniger

FR Kultarr
DE Springbeutelmaus
ES Kultarr

LC
2 sspp
HB 7–10 cm
T 10–15 cm
W c. 30 g

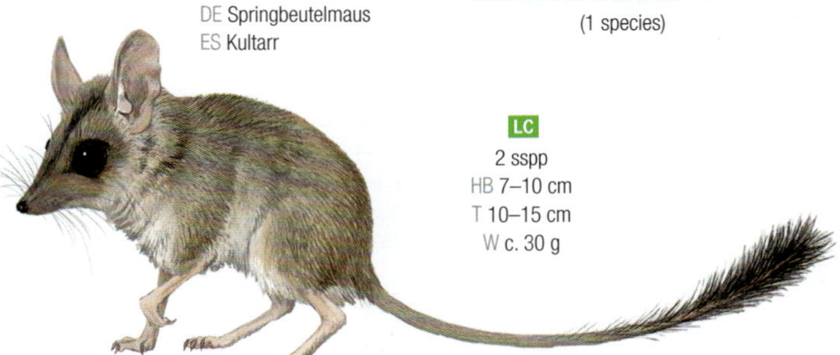

ssp *spenceri*

ssp *laniger*

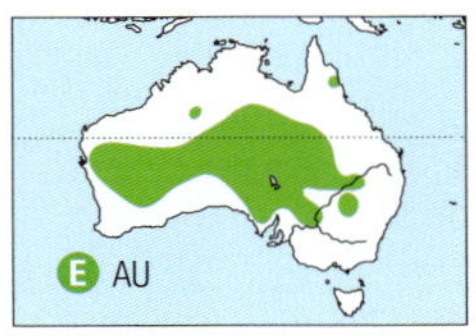

WA, NT, Qld, NSW, N SA, extreme N Vic

Genus *SMINTHOPSIS*

(20 species)

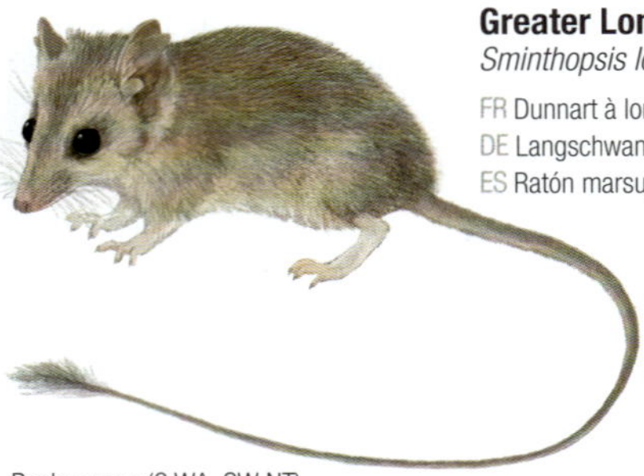

Greater Long-tailed Dunnart
Sminthopsis longicaudata

FR Dunnart à longue queue
DE Langschwanz-Schmalfußbeutelmaus
ES Ratón marsupial de cola larga

Rocky areas (C WA, SW NT)

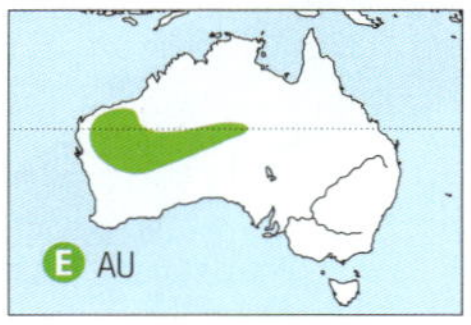

LC
Monotypic
HB 8–10 cm
T 18–22 cm
W 15–25 g

White-tailed Dunnart
Sminthopsis granulipes

FR Dunnart à queue blanche
DE Kornsohlen-Schmalfußbeutelmaus
ES Ratón marsupial de cola blanca

SW WA (Avon, Mallee, Coolgardie, Esperance Plains & Geraldton regions)

LC
Monotypic
HB 6.9–8.8 cm
T 5.6–6.6 cm
W 18–37 g

Lesser Hairy-footed Dunnart
Sminthopsis youngsoni

FR Dunnart de Youngson
DE Wüsten-Schmalfußbeutelmaus
ES Ratón marsupial pequeño de pies peludos

LC
Monotypic
HB 6.6–7.1 cm
T 6.2–6.8 cm
W 8.5–12 g

Subtropical arid regions of WA, S NT, SW Qld, extreme N SA

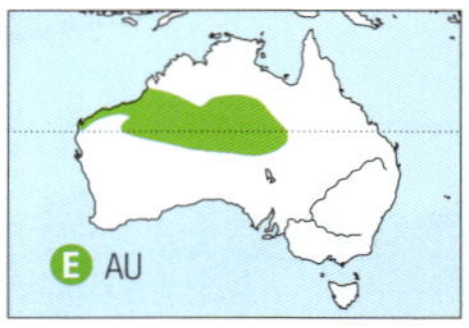

Greater Hairy-footed Dunnart
Sminthopsis hirtipes

FR Dunnart à pieds velus
DE Polster-Schmalfußbeutelmaus
ES Ratón marsupial de pies peludos

LC
Monotypic
HB 7.2–8.3 cm
T 7.2–10.1 cm
W 13–19.5 g

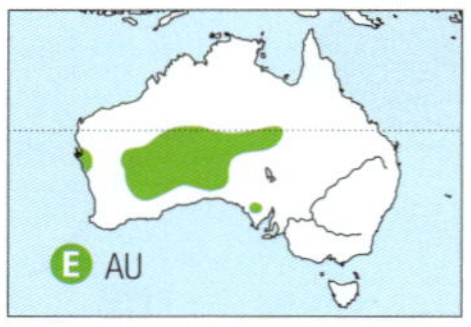

Arid and semi-arid areas of WA to SW Qld; also in Eyre Pen, S SA

Sandhill Dunnart
Sminthopsis psammophila

FR Dunnart du désert
DE Sand-Schmalfußbeutelmaus
ES Ratón marsupial del desierto

VU
Monotypic
HB 8.5–11.4 cm
T 10.6–13.3 cm
W 25–55 g

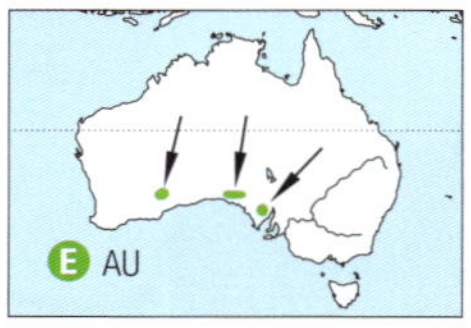

S WA (SW Great Victoria Desert), S SA (Nullarbor Res, Yellabinna Res, Eyre Pen)

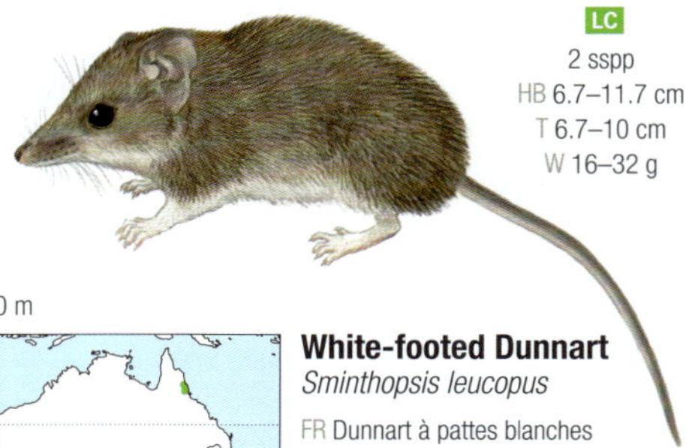

White-footed Dunnart
Sminthopsis leucopus

FR Dunnart à pattes blanches
DE Rippsohlen-Schmalfußbeutelmaus
ES Ratón marsupial de pies blancos

LC
2 sspp
HB 6.7–11.7 cm
T 6.7–10 cm
W 16–32 g

0–810 m

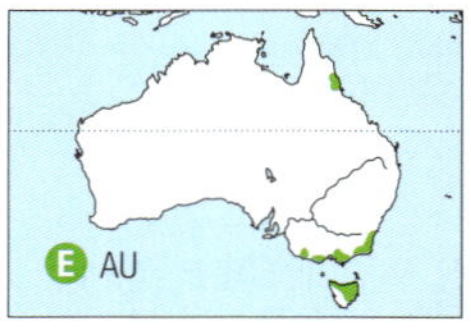

NE Qld, SW NSW, S Vic, Tas

Chestnut Dunnart
Sminthopsis archeri

FR Dunnart d'Archer
DE Archer-Schmalfußbeutelmaus
ES Ratón marsupial castaño

DD
Monotypic
HB 8.3–10.7 cm
T 8.2–10.5 cm
W 12–16 g

0–1000 m

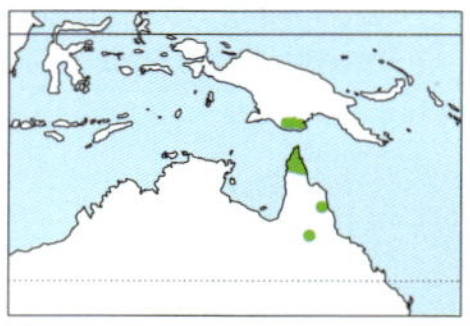

PG (S Western Prov), NE AU (N Qld)

Butler's Dunnart
Sminthopsis butleri

FR Dunnart de Butler
DE Butler-Schmalfußbeutelmaus
ES Ratón marsupial de Butler

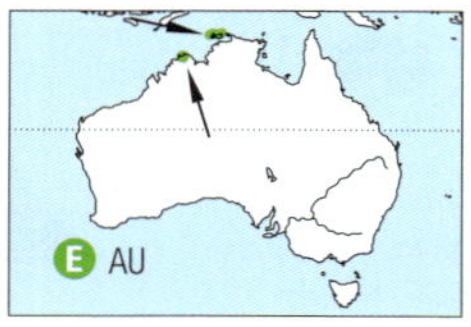

VU
Monotypic
HB c. 8.8 cm T c. 9 cm
W 15–30 g

WA (Kulamburu, Kimberley region), NT (Bathurst and Melville Is)

Common Dunnart
Sminthopsis murina

FR Petit Dunnart
DE Kleine Schmalfußbeutelmaus
ES Ratón marsupial de cola delgada

LC
2 sspp
HB 6.4–10.4 cm
T 6.8–9.9 cm
W 10–28 g

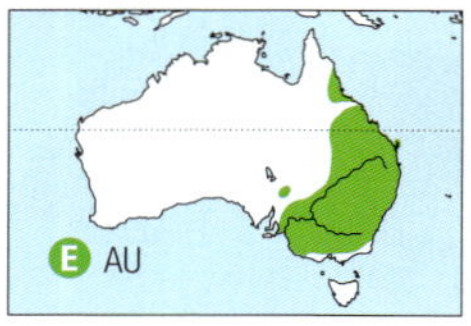

E Qld (with Fraser I), NSW, Vic, E & SE SA

Gilbert's Dunnart
Sminthopsis gilberti

FR Dunnart de Gilbert
DE Gilbert-Schmalfußbeutelmaus
ES Ratón marsupial de Gilbert

LC
Monotypic
HB 8.1–9.2 cm T 7.5–9.2 cm
W 14–25 g

below 600 m

S WA (from Darling scarp across C & S Wheatbelt region; Roe Plain, S of Nullarbor Plain)

Ooldea Dunnart
Sminthopsis ooldea

FR Dunnart d'Ooldea
DE Ooldea-Schmalfußbeutelmaus
ES Ratón marsupial de Ooldea

LC
Monotypic
HB 5.5–8.5 cm
T 6–9.5 cm
W 8–17 g

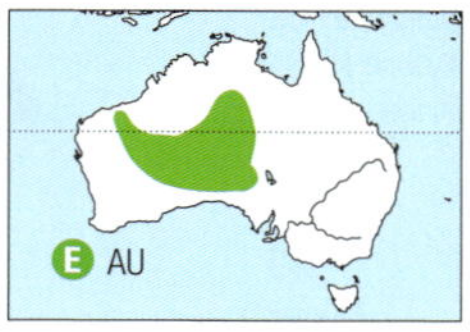

WA, NT, SA

Little Long-tailed Dunnart
Sminthopsis dolichura

FR Dunnart de Buningonia
DE Kleine Langschwänzige Schmalfußbeutelmaus
ES Ratón marsupial pequeño de cola larga

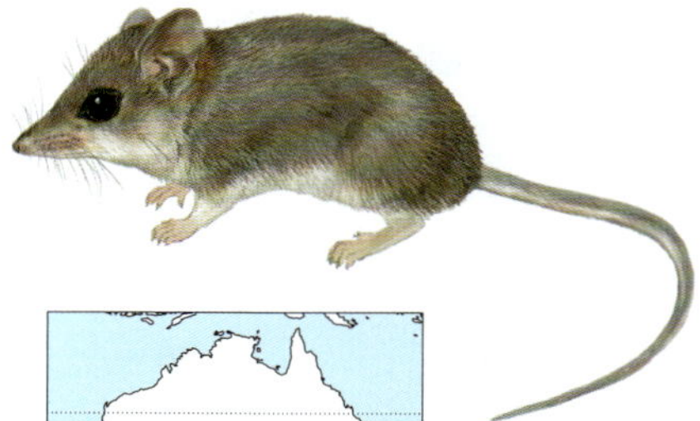

Semi-arid and arid country of SW WA and SA

LC
Monotypic
HB 6.3–9.9 cm
T 8.4–10.9 cm
W 10–21 g

Gray-bellied Dunnart
Sminthopsis fuliginosa

FR Dunnart à ventre gris
DE Graubäuchige Schmalfußbeutelmaus
ES Ratón marsupial de vientre gris

NE
3 sspp
HB 6.5–9.4 cm
T 6.5–10.6 cm
W 14–25 g

SW WA (with Boullanger I), S SA (Eyre Pen, Kangaroo I)

LC
2 sspp
HB 6–9 cm
T 4–7 cm
W 10–20 g

Less arid areas of WA, Qld, SA, NSW and Vic

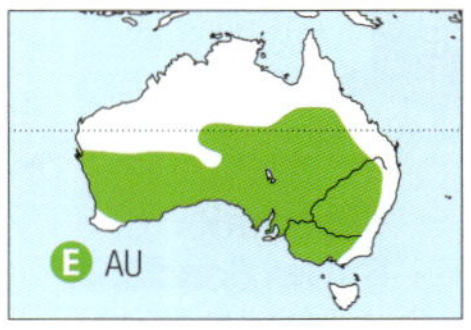

Fat-tailed Dunnart
Sminthopsis crassicaudata
FR Dunnart à grosse queue
DE Dickschwanz-Schmalfußbeutelmaus
ES Ratón marsupial de cola gruesa

Kakadu Dunnart
Sminthopsis bindi
FR Dunnart du Kakadu
DE Kakadu-Schmalfußbeutelmaus
ES Ratón marsupial de Kakadu

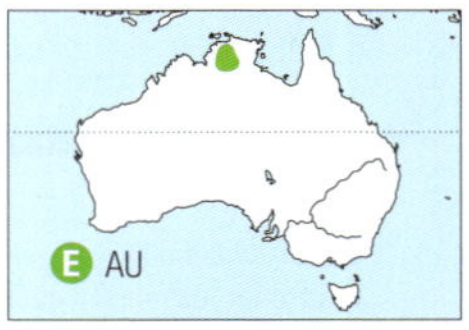

Top End (Kakadu) region of NT

NT
Monotypic
HB 5.2–8.4 cm
T 6.1–10.5 cm
W 12–20 g

Stripe-faced Dunnart
Sminthopsis macroura
FR Dunnart à face rayée
DE Streifgesichtige Schmalfußbeutelmaus
ES Ratón marsupial de cara rayada

gray morph

pale gray-brown morph

LC
Monotypic
HB 7–10 cm
T 8–11 cm
W 15–25 g

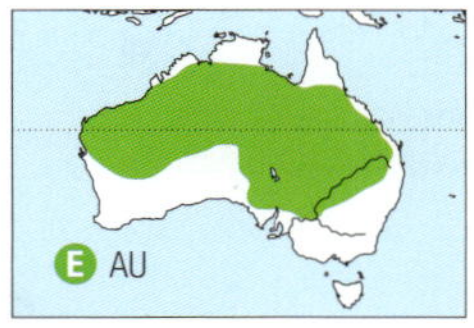

WA, NT, Qld, NSW, SA

Kimberley region of N WT

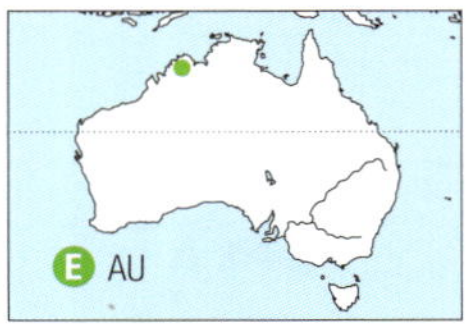

Froggatt's Dunnart
Sminthopsis froggatti
FR Dunnart de Froggatt
DE Froggatt-Schmalfußbeutelmaus
ES Ratón marsupial de Ramsay

NE
Monotypic
HB c. 7.3 cm
T c. 7.1 cm

NT, Qld

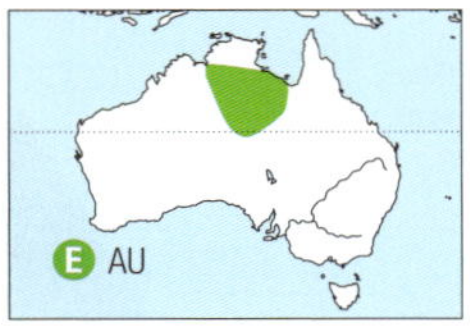

NE
Monotypic
HB 7–10 cm
T 8–11 cm
W 15–25 g

Stalker's Dunnart
Sminthopsis stalkeri
FR Dunnart de Stalker
DE Stalker-Schmalfußbeutelmaus
ES Ratón marsupial de Thomas

LC
3 sspp
HB 9–13.5 cm
T 9–13.5 cm
W 18–58 g

AU (N WA, N NT with Tiwi Is, NE Qld), ID & PG (Aru Is, S New Guinea)

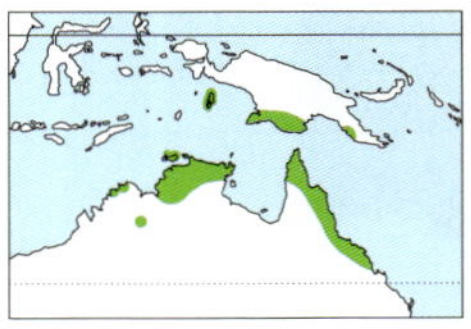

Red-cheeked Dunnart
Sminthopsis virginiae
FR Dunnart à joues rouges
DE Queensland-Schmalfußbeutelmaus
ES Ratón marsupial de mejillas rojas

NT
Monotypic
HB 11–13.5 cm
T 11–13 cm
W 40–70 g

Grassy downs country between Julia Creek and Richmond, NW Qld

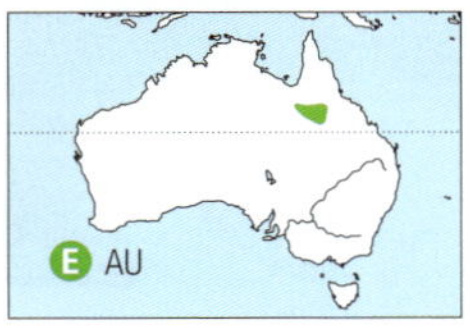

Julia Creek Dunnart
Sminthopsis douglasi
FR Dunnart de Julia Creek
DE Douglas-Schmalfußbeutelmaus
ES Ratón marsupial de Douglas

Order PERAMELEMORPHIA (2 families)

Family THYLACOMYIDAE – Bilbies

1 genus, 1 extant species. Arid and semi-arid deserts, shrublands, and grasslands. Australasian Region.

Genus *MACROTIS*

(1 species)

Greater Bilby
Macrotis lagotis
FR Grand Bilby
DE Großer Kaninchen-Nasenbeutler
ES Bilbi

VU
Monotypic
HB 30–55 cm (M), 29–39 cm (F)
T 20–29 cm (M), 20–28 cm (F)
W 1–2.5 kg (M), 0.8–1.2 kg (F)

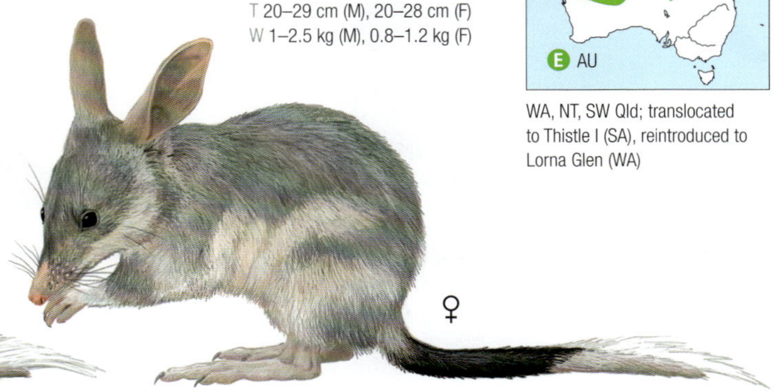

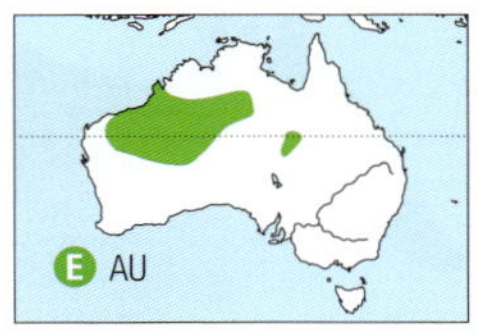

WA, NT, SW Qld; translocated to Thistle I (SA), reintroduced to Lorna Glen (WA)

Family PERAMELIDAE – Bandicoots and Echymiperas

6 genera, 21 species. Low shrublands, grasslands, and woodlands to montane tropical and deciduous forests. Australasian Region.

Genus *ISOODON*

(5 species)

LC 3 sspp
HB 28–36 cm
T 9–14.5 cm
W 0.4–1.8 kg

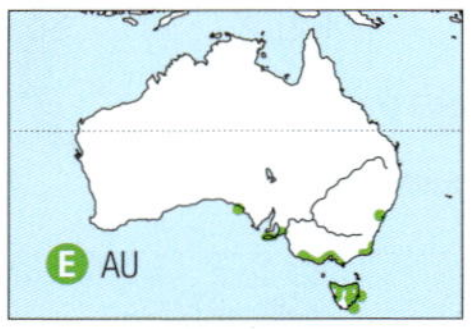

Southern Brown Bandicoot
Isoodon obesulus
FR Bandicoot obèse
DE Kleiner Kurznasenbeutler
ES Bandicut de hocico corto meridional

NSW, Vic, mainland SA, Kangaroo I, Nuyts Arch, Tas

NE
Monotypic
HB 28–36 cm
T 9–14.5 cm
W 0.4–1.8 kg

SW WA

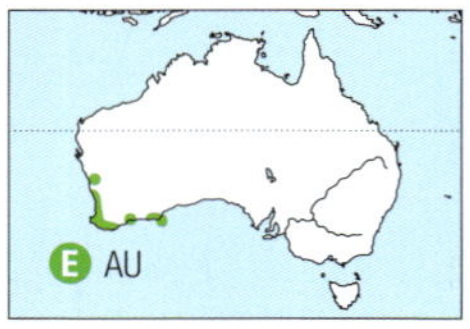

Western Brown Bandicoot
Isoodon fusciventer
FR Bandicoot quenda
DE Westlicher Kurznasenbeutler
ES Bandicut de hocico corto occcidental

NE
Monotypic

N Qld (Cape York Pen)

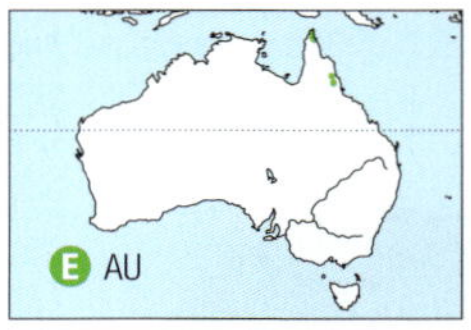

Cape York Brown Bandicoot
Isoodon peninsulae
FR Bandicoot de Cape York
DE Kap York-Kurznasenbeutler
ES Bandicut de hocico corto de Queensland

Golden Bandicoot
Isoodon auratus
FR Bandicoot doré
DE Goldbrauner Kurznasenbeutler
ES Bandicut de hocico corto dorado

ssp *barrowensis*

ssp *auratus*

VU
3 sspp
HB 19–29 cm
T 8.4–12.1 cm
W 300–670 g

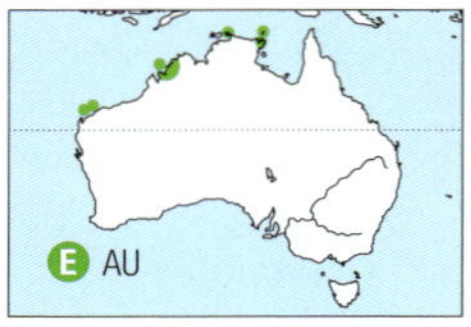

WA (NW Kimberley, some islands), NT (Cape Arnhem, Marchinbar I); introduced to some islands

Northern Brown Bandicoot
Isoodon macrourus
FR Bandicoot à longue queue
DE Großer Kurznasenbeutler
ES Bandicut de hocico corto septentrional

♀

♂

LC
3 sspp
HB 30–47 cm
T 8–21 cm
W 0.5–3.1 kg

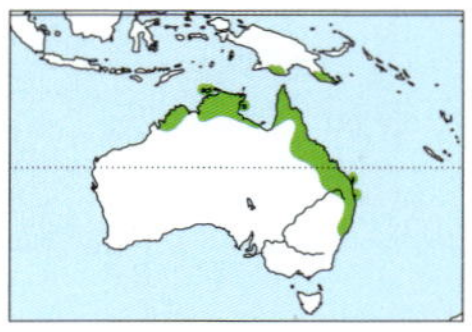

AU (N WA, N NT, Qld S to CE NSW, including many islands), S PG

Genus *PERAMELES*

(4 species)

VU
1 extant ssp
HB 17.3–23 cm
T 8.1–10.6 cm
W 165–300 g

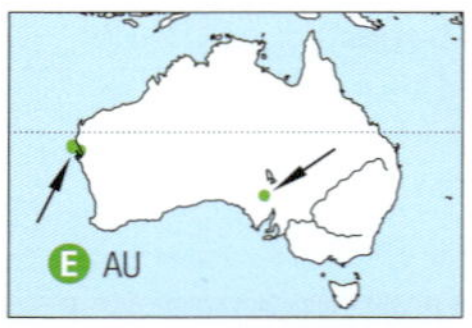

Western Barred Bandicoot
Perameles bougainville
FR Bandicoot de Bougainville
DE Streifen-Langnasenbeutler
ES Bandicut listado occidental

WA (Shark Bay), SA (reintroduced to Arid Recovery Project)

NE Monotypic
HB 31–44 cm
T 12–16 cm
W 0.52–1.3 kg

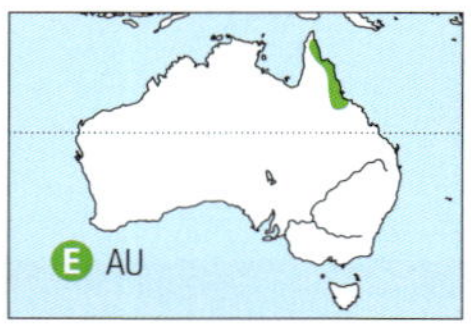

Queensland Barred Bandicoot
Perameles pallescens
FR Bandicoot du Queensland
DE Queensland-Langnasenbeutler
ES Bandicut listado de Queensland

N Qld (Wet Tropics and Cape York Pen)

LC
Monotypic
HB 31–44 cm
T 12–16 cm
W 0.52–1.3 kg

Mid-N Qld S to Vic

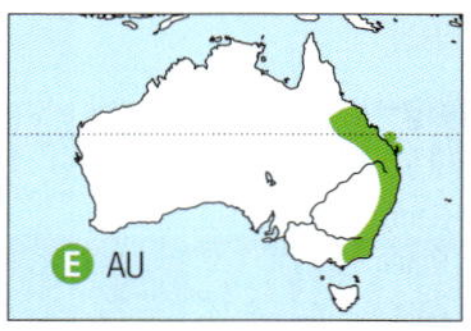

Long-nosed Bandicoot
Perameles nasuta
FR Bandicoot à long nez
DE Großer Langnasenbeutler
ES Bandicut listado de hocico largo

VU
Monotypic
HB 27–35 cm
T 7–11 cm
W 0.5–1.4 kg

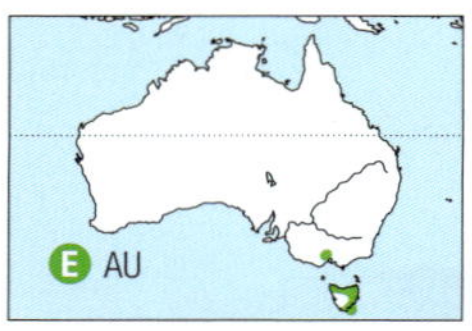

Eastern Barred Bandicoot
Perameles gunnii
FR Bandicoot de Gunn
DE Tasmanien-Langnasenbeutler
ES Bandicut listado oriental

N & E Tas (with Bruny I), S Vic

Genus *PERORYCTES*

(2 species)

50–1000 m

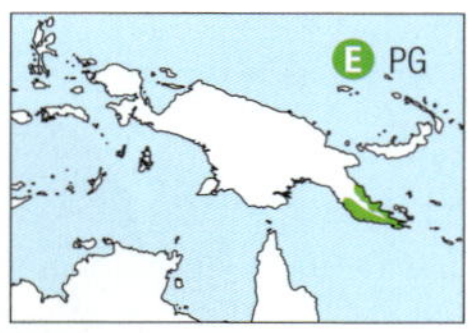

SE New Guinea Pen

♀

♂

EN
Monotypic
HB 34–56 cm
T 9.5–23 cm
W 0.9–4.9 kg

Giant Bandicoot
Peroryctes broadbenti
FR Bandicoot géant
DE Riesen-Neuguinea-Nasenbeutler
ES Bandicut gigante

Raffray's Bandicoot

Perorycytes raffrayana

FR Bandicoot de Raffray
DE Großer Neuguinea-Nasenbeutler
ES Bandicut de Raffray

ssp *raffrayana*

LC
2 sspp
HB 27–39 cm
T 13.2–17.9 cm
W 0.6–1.1 kg

ssp *rothschildi*

50–3900 m

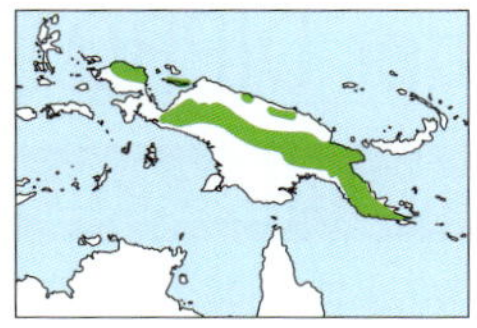

ID/PG: most of New Guinea, Yapen I

Clara's Echymipera

Echymipera clara

FR Bandicoot de Clara
DE Clara-Stachelnasenbeutler
ES Bandicut espinoso de Clara

♀

♂

LC
Monotypic
HB 27–41 cm
T 8–10.6 cm
W 0.8–1.7 kg

300–1700 m

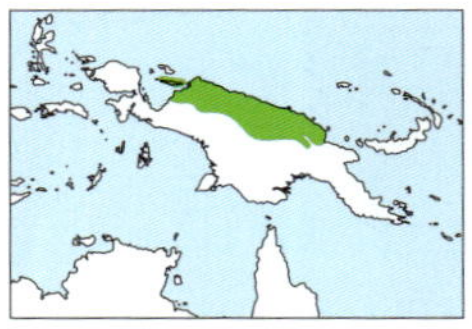

ID/PG: N lowlands of New Guinea, Yapen I

Genus *ECHYMIPERA*

(5 species)

David's Echymipera

Echymipera davidi

FR Bandicoot de David
DE Kiriwina-Stachelnasenbeutler
ES Bandicut espinoso de David

EN
Monotypic
HB 33–40 cm
T c. 10.8 cm
W c. 950 g

c. 10 m

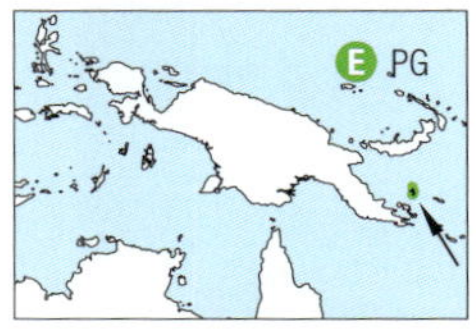

Kiriwina I (Trobriand Is), off SE New Guinea

Menzies's Echymipera

Echymipera echinista

FR Bandicoot de Menzies
DE Menzies-Stachelnasenbeutler
ES Bandicut espinoso de Menzies

40–80 m

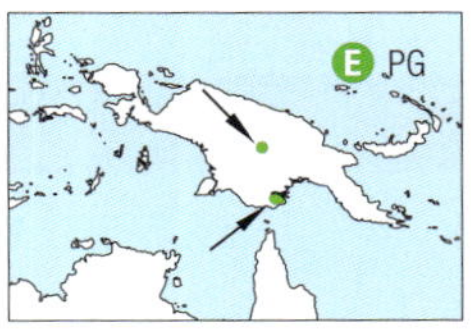

DD Monotypic
HB 33–36 cm T c. 9.7 cm
W c. 1 kg

E New Guinea (middle R Strickland, lower R Fly; unconfirmed in Mt Menawa, North Coastal Ranges)

Common Echymipera

Echymipera kalubu

FR Bandicoot épineux
DE Flachstachelnasenbeutler
ES Bandicut espinoso común

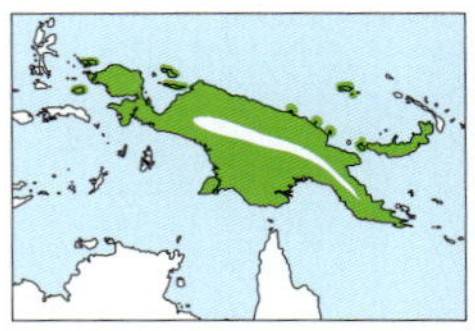

ID/PG: W Papuan Is, most of New Guinea, Geelvink Bay Is, Bismarck Arch (New Britain, Manus)

0–2000 m

ssp *philipi*
melanistic morph

ssp *kalubu*

LC 4 sspp
HB 22–38 cm
T 6.3–9.8 cm
W 0.4–1.8 kg

Long-nosed Echymipera

Echymipera rufescens

FR Bandicoot roux
DE Roter Stachelnasenbeutler
ES Bandicut espinoso de hocico largo

LC
2 sspp
HB 21–54 cm
T 7.5–11.5 cm
W 0.3–3 kg

0–2100 m

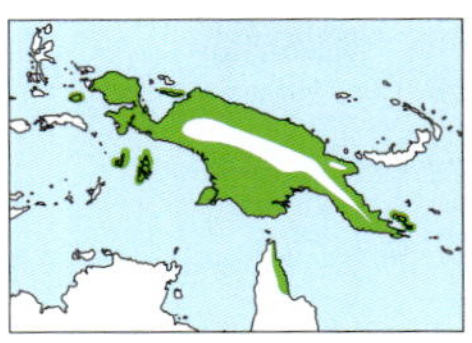

ID/PG: W Papuan Is (Misool), New Guinea (with Yapen, D'Entrecasteaux, Kai, Aru), AU (Cape York)

Genus *MICROPERORYCTES*

(4 species)

Striped Bandicoot

Microperoryctes longicauda

FR Bandicoot rayé
DE Langschwanz-Mausnasenbeutler
ES Bandicut estriado

LC 4 sspp
HB 24–30 cm
T 18–26 cm
W 350–670 g

1000–4000 m

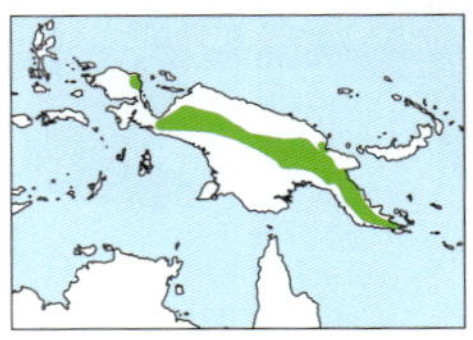

ID/PG: New Guinea (Arfak Mts, Central Range, SE Pen)

Mouse Bandicoot

Microperoryctes murina

FR Bandicoot souris
DE Weyland-Mausnasenbeutler
ES Bandicut murino

VU
Monotypic
HB 15.2–17.4 cm
T 10.5–11.1 cm
W c. 100 g

c. 2500 m

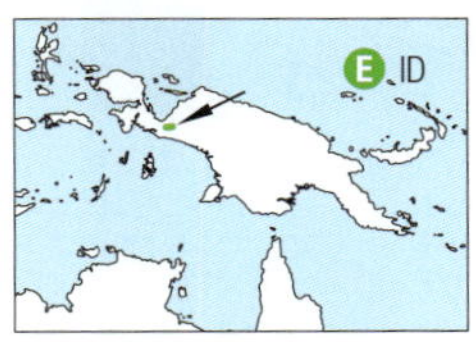

Mt Sumuri (Weyland Range), WC New Guinea

Arfak Pygmy Bandicoot

Microperoryctes aplini

FR Bandicoot d'Aplin
DE Arfak-Mausnasenbeutler
ES Bandicut de las Arfak

VU
Monotypic
HB 14.2–16 cm
T 11.5–12 cm
W c. < 100 g

1890–2200 m

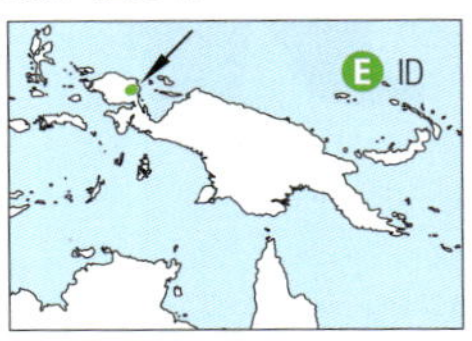

NW New Guinea (Arfak Mts)

Papuan Bandicoot

Microperoryctes papuensis

FR Bandicoot papou
DE Mura-Mausnasenbeutler
ES Bandicut de Papúa

LC
Monotypic
HB 17.5–20 cm
T 14.2–15.8 cm
W 145–184 g

1200–2650 m

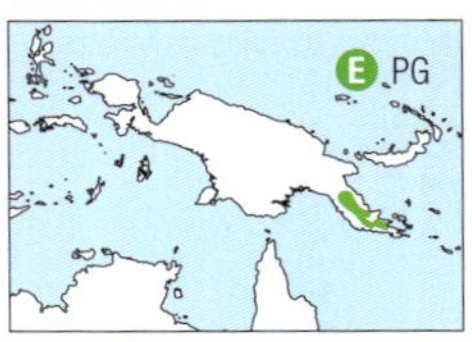

SE New Guinea (Fane village to Mt Simpson)

Genus *RHYNCHOMELES*

(1 species)

Seram Bandicoot

Rhynchomeles prattorum

FR Bandicoot de Seram
DE Seram-Nasenbeutler
ES Bandicut de Seram

EN
Monotypic
HB 24–33 cm
T 10.5–13 cm
W c. > 500 g

c. 1800 m

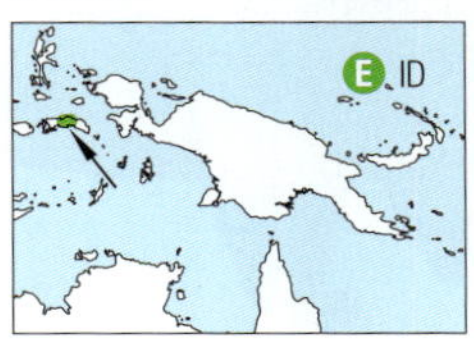

Mt Manusela, C Seram, S Moluccas

Order DIPROTODONTIA (11 families)

Family PHASCOLARCTIDAE – Koala

1 genus, 1 species. Open sclerophyll forests and woodlands. Australasian Region.

Genus *PHASCOLARCTOS*

(1 species)

Koala
Phascolarctos cinereus

FR Koala
DE Koala
ES Koala

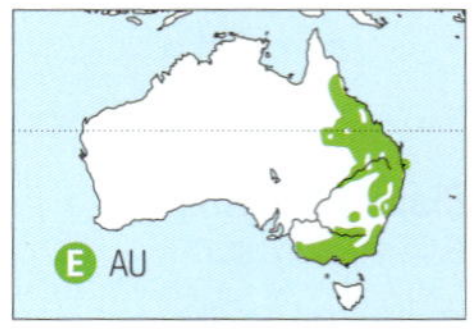

N Qld (near Cairns) patchily to SE SA; introduced to some Is (Qld, Vic, SA)

♂

♂

Victorian and South Australian form

Queensland form

♀

VU
Monotypic
HB 67–82 cm (M), 65–73 cm (F)
W 4.2–14.9 kg (M), 4.1–11 kg (F)

Family VOMBATIDAE – Wombats

2 genera, 3 species. Semi-arid shrublands and grasslands, temperate forests, and coastal heathlands. Australasian Region.

Genus *VOMBATUS*

(1 species)

ssp *ursinus*

ssp *hirsutus*

ssp *tasmaniensis*

Common Wombat
Vombatus ursinus

FR Wombat commun
DE Nacktnasenwombat
ES Uómbat

LC
3 sspp
HB 90–115 cm
W 22–40 kg

above 600 m

Mainland SE, Bass Strait Is (Flinders) and Tas

Genus *LASIORHINUS*

(2 species)

Southern Hairy-nosed Wombat
Lasiorhinus latifrons

FR Wombat du Sud
DE Südlicher Haarnasenwombat
ES Uómbat de hocico peludo meridional

NT
Monotypic
HB 84–111 cm
T 2.5–6 cm
W 17.5–36 kg

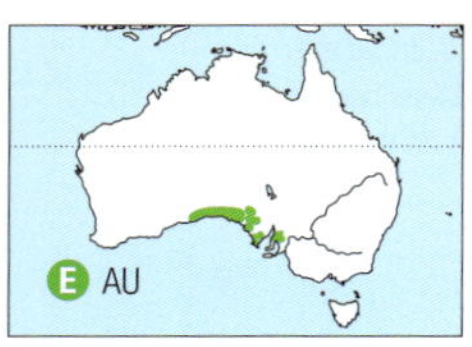

SE WA (W Nullarbor Plain), S SA (E to R Murray); introduced to Wedge I (SA)

Northern Hairy-nosed Wombat
Lasiorhinus krefftii

FR Wombat du Queensland
DE Nördlicher Haarnasenwombat
ES Uómbat de hocico peludo septentrional

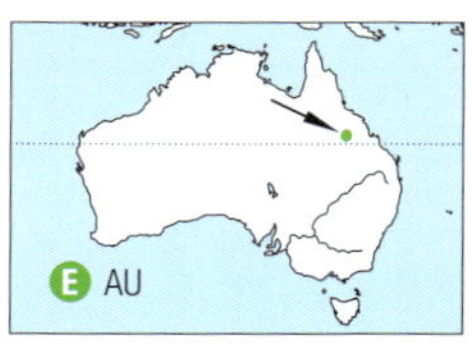

Epping Forest NP (C Qld)

CR
Monotypic
HB c. 108 cm
W c. 32 kg

Family BURRAMYIDAE – Pygmy Possums

2 genera, 5 species. Montane and lowland, rainforests to sclerophyll forests and woodlands. Australasian Region.

Genus *BURRAMYS*

(1 species)

CR
Monotypic
HB c. 11 cm
T 13.6–13.8 cm
W 30–82 g

Mountain Pygmy Possum

Burramys parvus

FR Possum-loir des montagnes
DE Bergbilchbeutler
ES Pósum pigmeo de montaña

at least c. 1200 m

NSW (Kosciuszko NP), Vic (between Mt Bogong and Mt Higginbotham, Mt Buller)

Genus *CERCARTETUS*

(4 species)

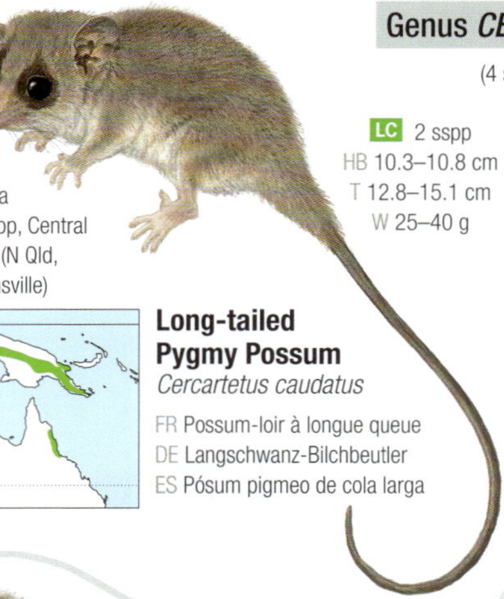

LC 2 sspp
HB 10.3–10.8 cm
T 12.8–15.1 cm
W 25–40 g

ID/PG: New Guinea highlands (Vogelkop, Central Range, Huon), AU (N Qld, Cooktown to Townsville)

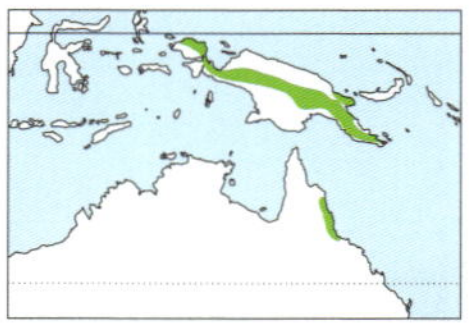

300–3450 m

Long-tailed Pygmy Possum

Cercartetus caudatus

FR Possum-loir à longue queue
DE Langschwanz-Bilchbeutler
ES Pósum pigmeo de cola larga

LC
Monotypic
HB 6.4–10.6 cm
T 5.3–10.1 cm
W 8–21 g

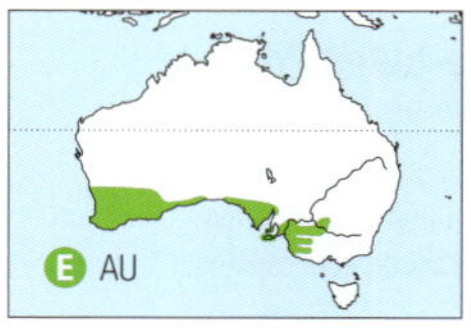

S WA, S SA (with Kangaroo I), SE SA, W Vic, SW NSW

Western Pygmy Possum

Cercartetus concinnus

FR Possum-loir de Gould
DE Dünnschwanz-Bilchbeutler
ES Pósum pigmeo occidental

NW Vic to SE SA (with Kangaroo I), Tas

Little Pygmy Possum

Cercartetus lepidus

FR Possum-loir pygmée
DE Zwergbilchbeutler
ES Pósum pigmeo pequeño

LC
Monotypic
HB 5–7.3 cm
T 6–7.5 cm
W 6–10 g

LC
2 sspp
HB 7–11 cm
T 7.5–10.5 cm
W 15–43 g

Coastal Qld–NSW border to SE SA, Tas, Bass Strait Is (King, Flinders)

0–1800 m

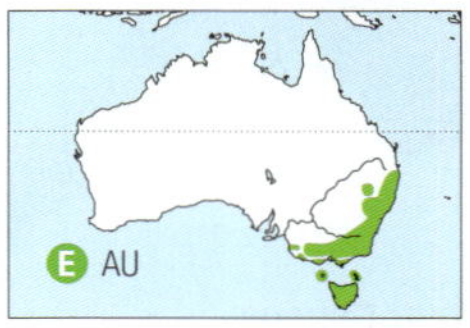

Eastern Pygmy Possum

Cercartetus nanus

FR Possum-loir de l'Est
DE Östlicher Zwergbilchbeutler
ES Pósum pigmeo oriental

Family PHALANGERIDAE – Cuscuses, Brush-tailed Possums and Scaly-tailed Possum

6 genera, 31 species. Lowland and montane tropical to sclerophyll forests and woodlands. Australasian Region.

Genus *TRICHOSURUS*

(4 species)

ssp *fuliginosus*

LC 5 sspp
HB 35–55 cm
T 25–40 cm
W 1.2–4.5 kg

ssp *vulpecula*

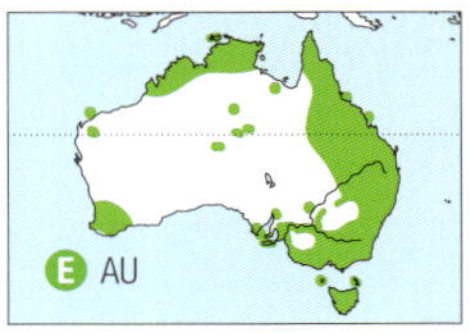

Common Brush-tailed Possum

Trichosurus vulpecula

FR Phalanger vulpin
DE Fuchskusu
ES Cuscús de cola de cepillo común

W, N, E & S mainland and several islands, Tas, Bass Strait Is; introduced to New Zealand

Coppery Brush-tailed Possum

Trichosurus johnstonii

FR Phalanger de Johnstone
DE Queensland-Kusu
ES Cuscús de cola de cepillo de Atherton

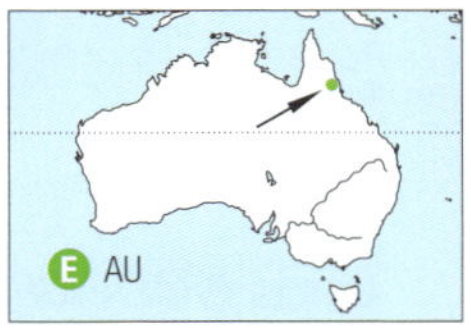

NE Qld (Atherton Tableland region, along W side from Koombooloomba to Kuranda)

NE
Monotypic
HB 35–44 cm
T 33–42 cm
W 1.2–2 kg

black morph

Short-eared Brush-tailed Possum

Trichosurus caninus

FR Phalanger canin
DE Hundskusu
ES Cuscús de cola de cepillo de orejas cortas

SE Qld to CE NSW, with Fraser I

LC
Monotypic
HB 40–55 cm
T 34–42 cm
W 2.5–4.5 kg

LC
Monotypic
HB 49–54 cm
T 34–37 cm
W 2.6–4.2 kg

S NSW (inland of Newcastle) to C & W Vic (to vicinity of Mt Cole)

Mountain Brush-tailed Possum

Trichosurus cunninghami

FR Phalanger de Cunningham
DE Cunningham-Kusu
ES Cuscús de cola de cepillo de montaña

Genus *WYULDA*

(1 species)

NT
Monotypic
HB 30–47 cm
T 28–30 cm
W 0.9–2 kg

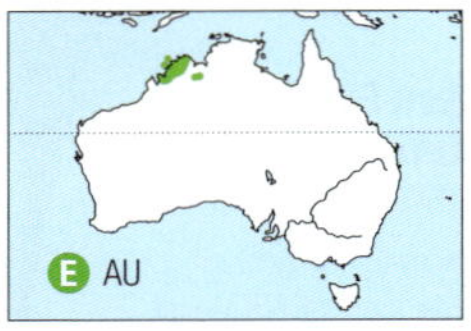

WA (Kimberley region, with Bigge and Boongaree Is)

Scaly-tailed Possum

Wyulda squamicaudata

FR Phalanger à queue écailleuse
DE Schuppenschwanzkusu
ES Cuscús de cola espinosa

Genus *AILUROPS*

(3 species)

Common Bear Cuscus

Ailurops ursinus

FR Phalanger ursin
DE Bärenkuskus
ES Cuscús ursino común

VU
3 sspp
HB 47–57 cm
T 52–58 cm
W 5–8 kg

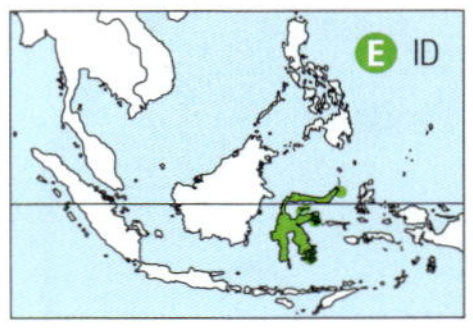

0–1000 m

Sulawesi and adjacent islands (Lembeh, Muna, Buton Is, Peleng, Togian Is, possibly Banggai)

CR
Monotypic
HB c. 45 cm
T c. 33 cm

Sangihe Is, Talaud Is

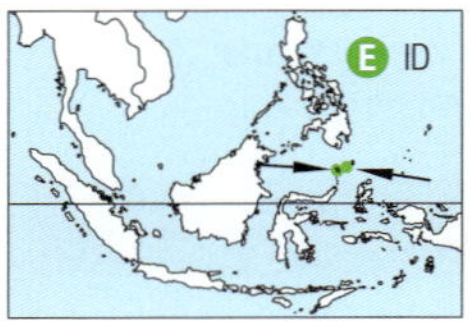

Talaud Bear Cuscus

Ailurops melanotis

FR Phalanger des Talaud
DE Talaud-Bärenkuskus
ES Cuscús ursino de las Talaud

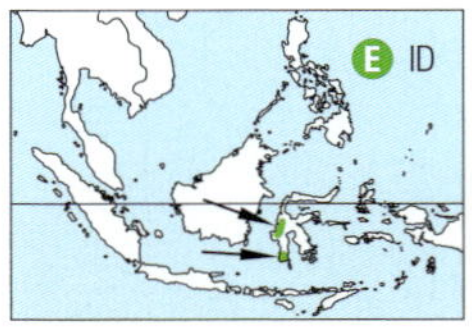

800–2000 m

C & SW Sulawesi Mts

Mountain Bear Cuscus

Ailurops furvus

FR Phalanger de Raven
DE Großer Bärenkuskus
ES Cuscús ursino de montaña

NE
Monotypic
HB 59–65 cm
T 56–62 cm
W c. 10 kg

Genus *STRIGOCUSCUS*

(2 species)

NT
2 sspp
HB 30–38 cm
T 31–37 cm
W 0.5–1 kg

Sulawesi (with Muna I)

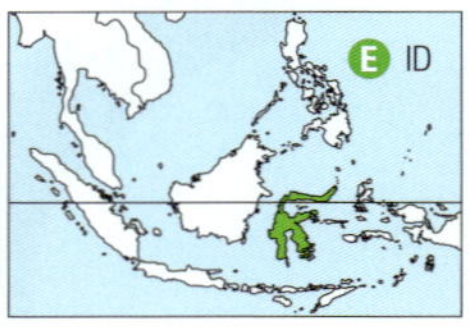

0–2000 m

Small Sulawesi Cuscus

Strigocuscus celebensis

FR Phalanger des Célèbes
DE Sulawesi-Kuskus
ES Cuscús menor de las Célebes

NE
Monotypic
HB c. 39 cm
T c. 39 cm
W c. 1 kg

Sangihe Is (Siau, Sangir)

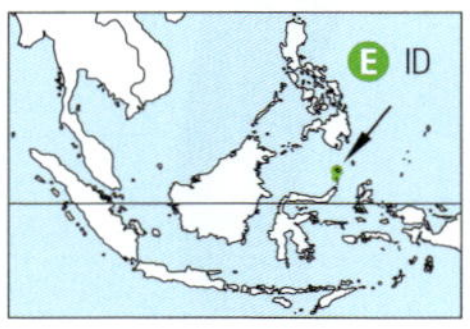

Small Sangihe Cuscus

Strigocuscus sangirensis

FR Phalanger des Sangir
DE Sangihe-Kuskus
ES Cuscús menor de las Sangihe

LC
2 sspp
HB 35–37 cm
T 25–30 cm
W 1.1–1.2 kg

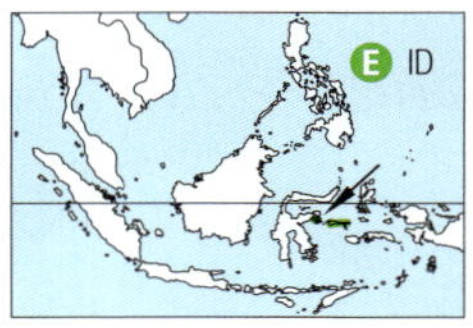

Banggai Arch (Peleng), Sula Is (Taliabu, Seho, Mangole, Sanana)

Peleng Cuscus

Phalanger pelengensis

FR Phalanger de Peleng
DE Peleng-Kuskus
ES Cuscús de Peleng

Genus *PHALANGER*

(15 species)

Moluccan Cuscus

Phalanger ornatus

FR Phalanger des Moluques
DE Molukken-Kuskus
ES Cuscús de las Molucas

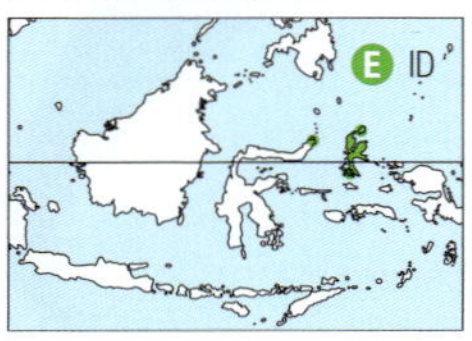

E tip of N Sulawesi, larger N Moluccan Is (Halmahera, Bacan, Morotai)

LC
Monotypic
HB 35–42 cm
T 28–36 cm
W 1.2–2.2 kg

VU
Monotypic

Blue-eyed Cuscus

Phalanger matabiru

FR Phalanger aux yeux bleus
DE Blauaugen-Kuskus
ES Cuscús de Ternate y Tidore

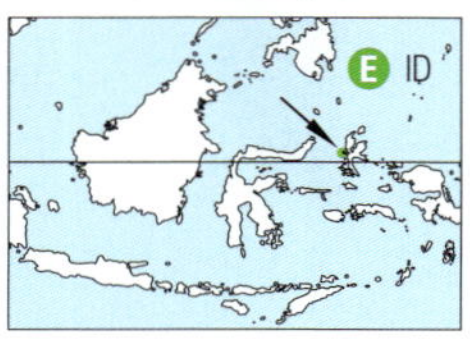

Smaller N Moluccan Is (Ternate, Tidore)

Gebe Cuscus

Phalanger alexandrae

FR Phalanger de Gebe
DE Gebe-Kuskus
ES Cuscús de Gebe

EN
Monotypic
HB c. 45 cm
T c. 36 cm
W c. 2.5 kg

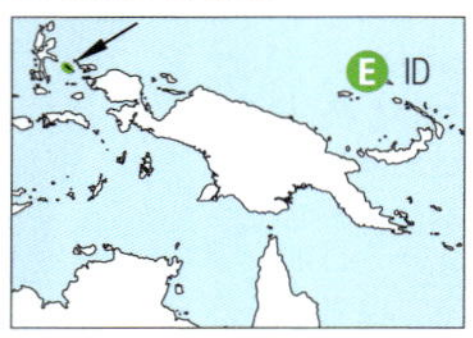

c. 300 m

Gebe I, N Moluccas (possibly on islands between Halmahera to Waigeo)

Obi Cuscus
Phalanger rothschildi

FR Phalanger d'Obi
DE Obi-Kuskus
ES Cuscús de las Obi

LC
Monotypic
HB 36–39 cm
T 30–33 cm
W 1.1–1.4 kg

orange-brown morph

gray morph

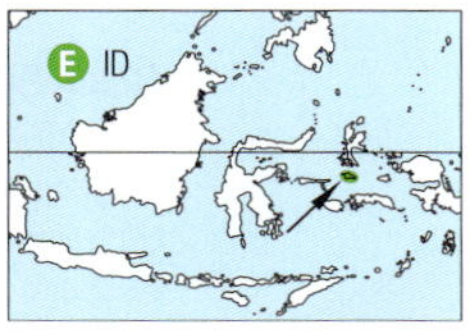

NC Moluccas (Obi, Bisa, Obilatu)

Ground Cuscus
Phalanger gymnotis

FR Phalanger gymnote
DE Gleichfarbkuskus
ES Cuscús terrestre

LC
Monotypic
HB 31–54 cm
T 29–39 cm
W 1.5–5 kg

0–2700 m

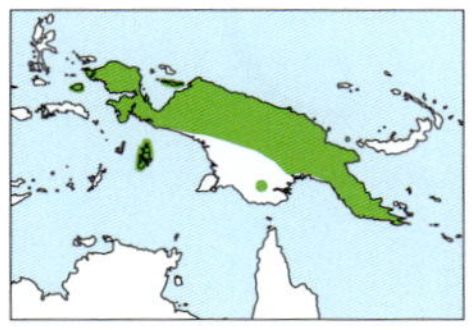

ID/PG: New Guinea mainland, off W New Guinea Is (Salawati, Misool, Yapen, Aru)

Woodlark Cuscus
Phalanger lullulae

FR Phalanger de Woodlark
DE Woodlark-Kuskus
ES Cuscús de Woodlark

variants

EN
Monotypic
HB 33–41 cm
T 27–34 cm
W 1.5–2 kg

Off SE New Guinea Is (Woodlark, Madeu, and Alcester, where perhaps introduced)

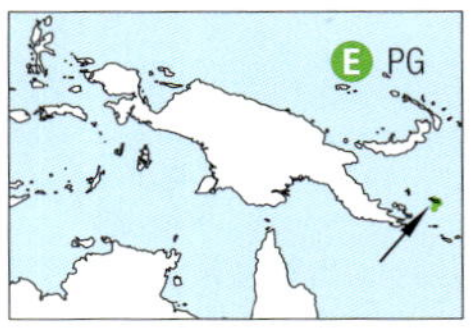

Northern Common Cuscus
Phalanger orientalis

FR Phalanger gris
DE Nördlicher Grauer Kuskus
ES Cuscús septentrional

ID/PG: C & S Moluccas, E Lesser Sundas, N New Guinea (E to Long I)

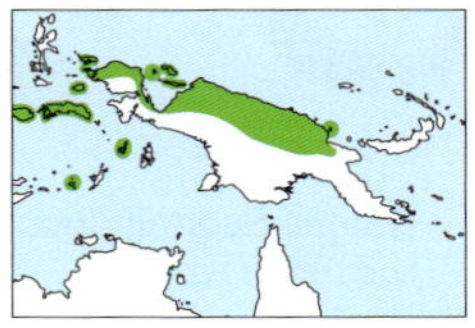

variants

LC
2 sspp
HB 30–50 cm
T 28–42 cm
W 1.6–5 kg

Eastern Common Cuscus
Phalanger intercastellanus

FR Phalanger de d'Entrecasteaux
DE Östlicher Grauer Kuskus
ES Cuscús oriental

♂

♀

LC
Monotypic
HB 35–45 cm
T 27–32 cm
W 1.4–2.2 kg

0–1250 m

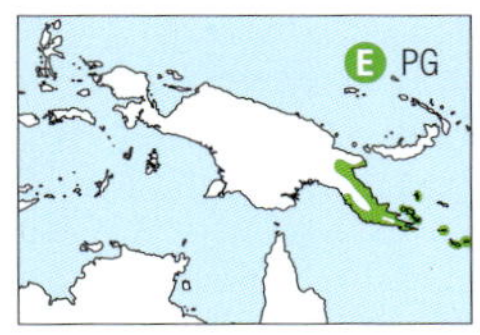

Markham Valley and Huon Gulf through SE Pen; D'Entrecasteaux, Trobriand and Louisiade Is

New Britain Cuscus
Phalanger breviceps

FR Phalanger de Nouvelle-Bretagne
DE Neubritannien-Kuskus
ES Cuscús de Nueva Bretaña

NE
2 sspp
HB 30–50 cm
T 28–42 cm
W 1.6–5 kg

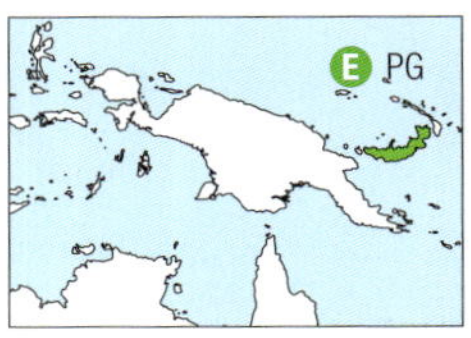

New Britain I (Bismarck Arch), likely introduced all over Bismarck Arch and Solomon Is

Southern Common Cuscus
Phalanger mimicus

FR Phalanger de Mimika
DE Südlicher Grauer Kuskus
ES Cuscús meridional

LC
Monotypic
HB 33–40 cm
T 28–35 cm
W 1.5–2.2 kg

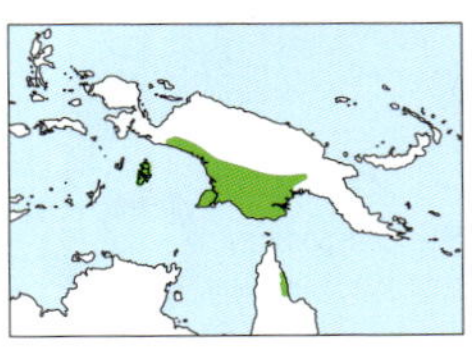

ID/PG (S New Guinea, Aru Is), AU (E Cape York Pen, Qld)

Telefomin Cuscus
Phalanger matanim

FR Phalanger de Telefomin
DE Telefomin-Kuskus
ES Cuscús de Telefomin

CR(PE)
Monotypic
HB 34–44 cm
T 27–31 cm
W 1.1–2 kg

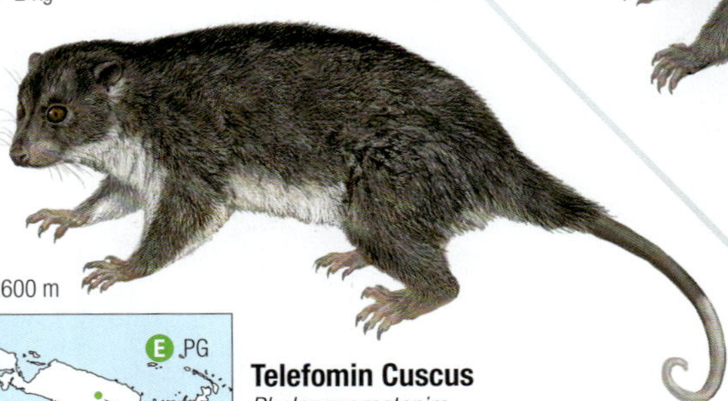

1500–2600 m

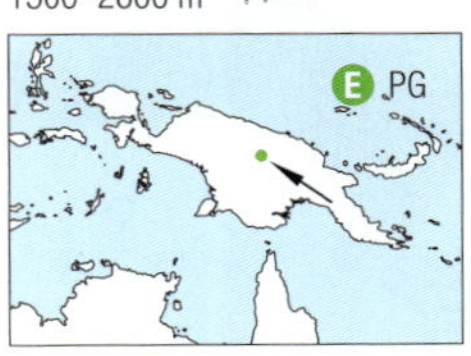

C New Guinea (Telefomin area)

Stein's Cuscus
Phalanger vestitus

FR Phalanger de Stein
DE Stein-Kuskus
ES Cuscús de Stein

LC 3 sspp
HB 35–48 cm
T 29–36 cm
W 1.4–2.4 kg

1000–2200 m

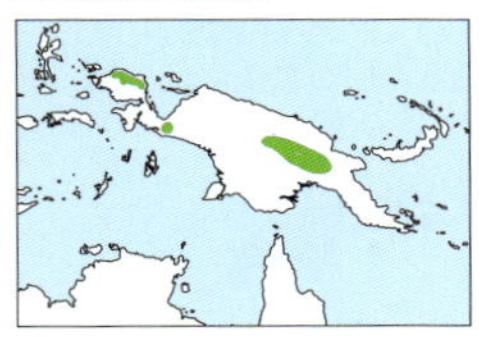

ID/PG: New Guinea (Vogelkop Pen Mts, Weyland Range, Central Range)

1500–3900 m

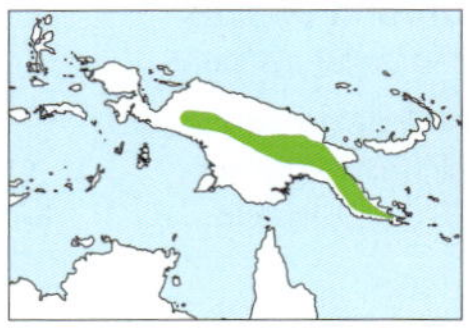

ID/PG: New Guinea Central Range; recorded in SC Papua New Guinea (Mt Bosavi area)

LC
2 sspp
HB 37–46 cm
T 27–32 cm
W 1.7–2.4 kg

Silky Cuscus
Phalanger sericeus
FR Phalanger soyeux
DE Seidenkuskus
ES Cuscús sedoso

Mountain Cuscus
Phalanger carmelitae
FR Phalanger de montagne
DE Bergkuskus
ES Cuscús de montaña

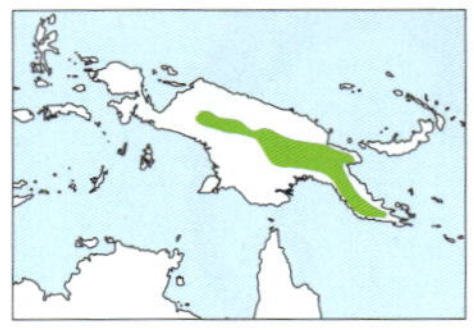

1350–3800 m

ID/PG: New Guinea Central Range (E from Wissel lakes), Huon Mts

LC
2 sspp
HB 37–43 cm
T 31–36 cm
W 1.7–2.6 kg

Genus *SPILOCUSCUS*

(6 species)

Black Spotted Cuscus
Spilocuscus rufoniger
FR Phalanger noir et roux
DE Schwarzfleck-Tüpfelkuskus
ES Cuscús rojinegro

♂

♀

0–1200 m

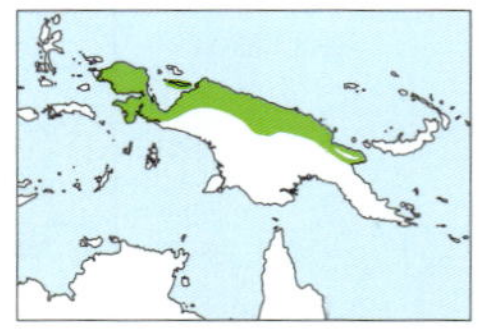

ID/PG: Vogelkop to Huon (probably Yapen I); recorded in S New Guinea

CR
Monotypic
HB 58–69 cm
T 45–65 cm
W 5.5–6.6 kg

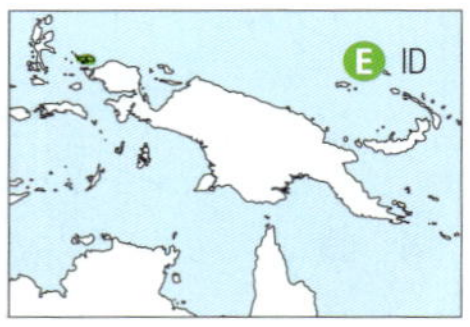

E ID

Islands off W New Guinea (Waigeo I, Gam I, perhaps Batanta I)

Waigeo Spotted Cuscus
Spilocuscus papuensis
FR Phalanger de Waigeo
DE Waigeo-Tüpfelkuskus
ES Cuscús de Waigeo

VU
Monotypic
HB 47–57 cm
T 47–56 cm
W c. 2.6 kg

♂

♀

♀

♂

NT
Monotypic
HB 31–57 cm
T 25–59 cm
W 1.9–3.2 kg

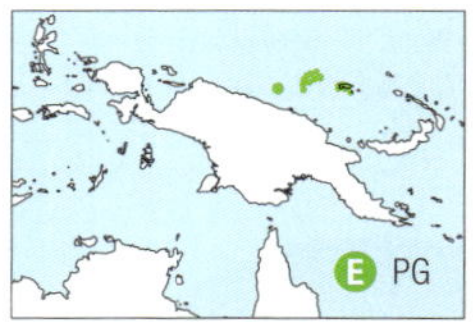

E PG

Admiralty Arch, Bismarck Arch

Admiralty Spotted Cuscus
Spilocuscus kraemeri
FR Phalanger de l'Amirauté
DE Manus-Tüpfelkuskus
ES Cuscús de las Manus

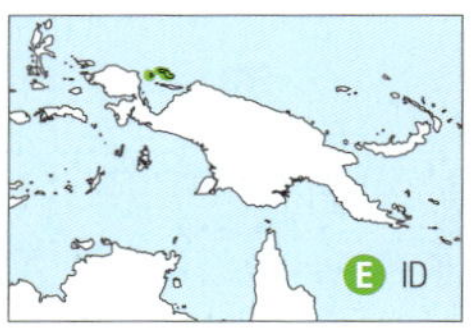

E ID

NW New Guinea Is (Numfor, Supiori, Biak)

Blue-eyed Spotted Cuscus
Spilocuscus wilsoni
FR Phalanger de Wilson
DE Blauaugen-Tüpfelkuskus
ES Cuscús moteado de ojos azules

all white morph

CR
Monotypic
HB 33–48 cm
T 28–33 cm
W 1.2–2.1 kg

spotted morph

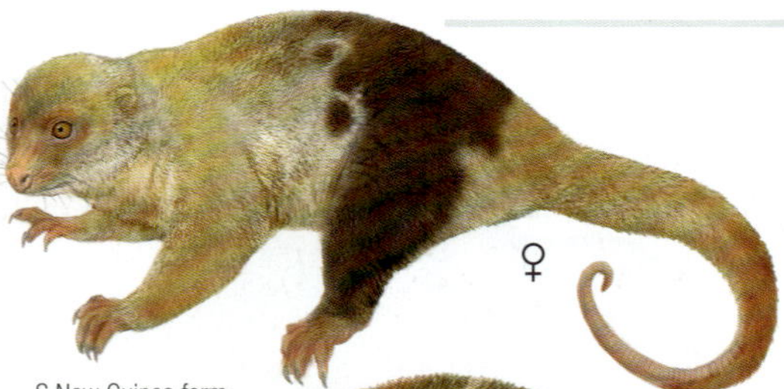

♀

S New Guinea form

W and N New Guinea form

♂

♀

LC
Monotypic
HB 42–74 cm
T 38–71 cm
W 2.4–6.1 kg

Common Spotted Cuscus
Spilocuscus maculatus
FR Phalanger tacheté
DE Tüpfelkuskus
ES Cuscús moteado de Nueva Guinea

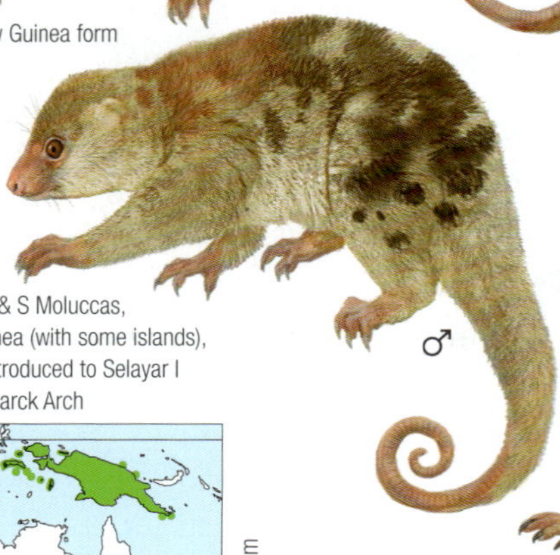

♂

ID/PG: C & S Moluccas, New Guinea (with some islands), Aru Is; introduced to Selayar I and Bismarck Arch

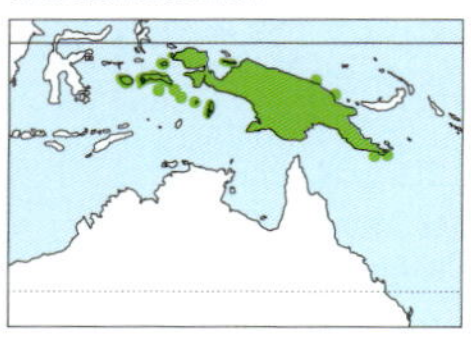

0–1400 m

♂

SW New Guinea form

♂

Moluccas gray form

Australian Spotted Cuscus
Spilocuscus nudicaudatus

FR Phalanger de Cape York
DE Kap-York-Tüpfelkuskus
ES Cuscús moteado de Australia

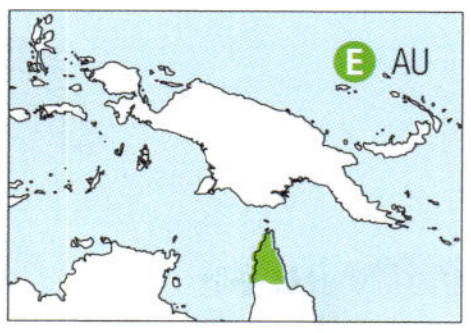

0–820 m

N Qld (Cape York Pen: N of R Stewart and R Coen-Archer system, with McIlwraith and Iron ranges)

NE Monotypic
HB 41–58 cm
T 37–48 cm W 2.7–4.9 kg

Family PSEUDOCHEIRIDAE – Ring-tailed Possums and Greater Gliders

6 genera, 20 species. Montane and lowland tropical to sclerophyll forests and woodlands. Australasian Region.

Genus *HEMIBELIDEUS*

(1 species)

Lemuroid Ring-tailed Possum
Hemibelideus lemuroides

FR Possum lémurien
DE Lemuren-Ringbeutler
ES Falangero lemuroide

N Qld

creamy white morph

NT
Monotypic
HB 31–40 cm
T 23–38 cm
W 0.8–1.2 kg

Central Greater Glider
Petauroides armillatus

FR Possum de Lumholtz
DE Mittlerer Großflugbeutler
ES Falangero planeador grande central

Mid-Qld (just N of Townsville, S to Eungella Range and near Roma)

NE Monotypic
HB 30–42 cm
T 40–53 cm
W 0.7–1.2 kg

Genus *PETAUROIDES*

(3 species)

Northern Greater Glider
Petauroides minor

FR Possum mineur
DE Nördlicher Großflugbeutler
ES Falangero planeador grande septentrional

NE
Monotypic
HB 30–40 cm
T 40–48 cm
W 700–900 g

NE Qld (just N of Cairns S to Townsville)

Southern Greater Glider
Petauroides volans

FR Possum volant
DE Südlicher Großflugbeutler
ES Falangero planeador grande meridional

SE Qld (S of from Bundaberg), NSW, Vic

variants

VU
2 sspp
HB 35–45 cm
T 45–60 cm
W 0.9–1.7 kg

Genus *PSEUDOCHIRULUS*

(8 species)

Daintree River Ring-tailed Possum
Pseudochirulus cinereus

FR Possum de Daintree
DE Grauer Ringbeutler
ES Falangero de cola anillada del río Daintree

NT
Monotypic
HB 34–37 cm
T 32–40 cm
W 0.7–1.4 kg

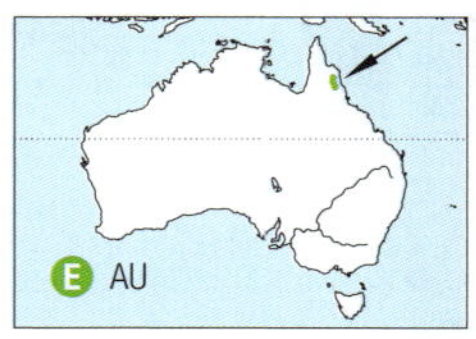

above 420 m

N Qld (Mt Thornton Peak, Mt Windsor Tableland, Mt Carbine Tableland)

Herbert River Ring-tailed Possum
Pseudochirulus herbertensis

FR Possum du Herbert
DE Herbert-River-Ringbeutler
ES Falangero de cola anillada del río Herbert

black morph

LC
Monotypic
HB 30–40 cm
T 34–47 cm
W 0.7–1.5 kg

above 350 m

N Qld (Kuranda S to Ingham)

Weyland Ring-tailed Possum
Pseudochirulus caroli

FR Possum des Weyland
DE Buntschwanz-Ringbeutler
ES Falangero de cola anillada de las Weyland

0–2200 m

ID: New Guinea (Weyland Range, R Lorentz Basin)

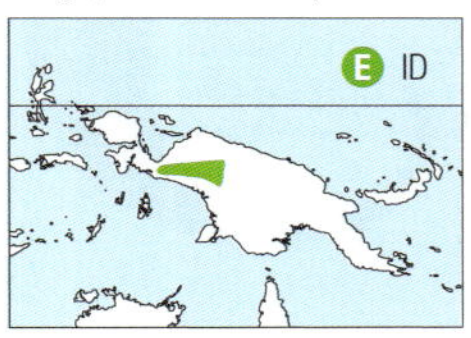

LC
2 sspp
HB 30–36 cm
T 30–37 cm
W c. 440 g (F)

LC
Monotypic
HB 19–34 cm
T 20–31 cm
W 450–835 g

500–2800 m SE New Guinea

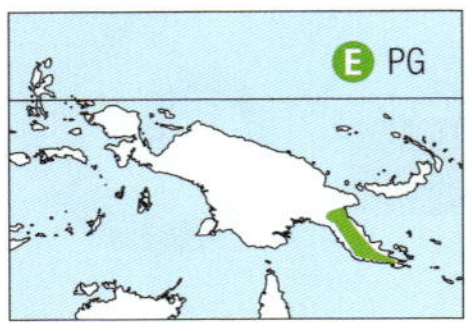

Painted Ring-tailed Possum
Pseudochirulus forbesi
FR Possum peint
DE Forbes-Ringbeutler
ES Falangero de cola anillada pintado

LC
Monotypic
HB 33–35 cm
T 23–25 cm

above 500 m

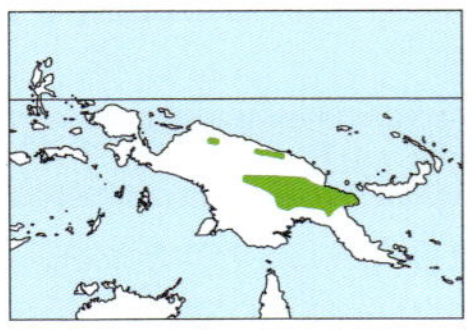

ID/PG: New Guinea (Foja, N Coastal Ranges, Central Range, Huon)

Masked Ring-tailed Possum
Pseudochirulus larvatus
FR Possum masqué
DE Masken-Ringbeutler
ES Falangero de cola anillada enmascarado

LC
Monotypic
HB 17–21 cm
T 15–19 cm
W 105–206 g

Pygmy Ring-tailed Possum
Pseudochirulus mayeri
FR Possum pygmée
DE Zwerg-Ringbeutler
ES Falangero de cola anillada pequeño

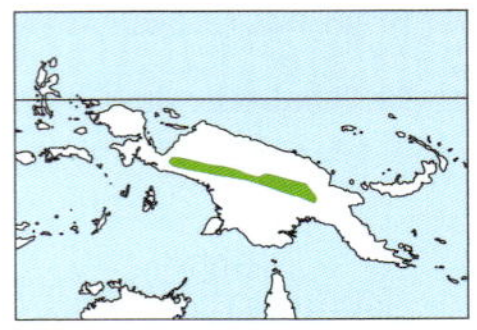

1500–3600 m

ID/PG: New Guinea (Central Range from Wissel lakes E to Mt Hagen)

VU
Monotypic
HB 21–23 cm
T 21–25 cm
W 244–305 g

Arfak Ring-tailed Possum
Pseudochirulus schlegeli
FR Possum des Arfak
DE Schlegel-Ringbeutler
ES Falangero de cola anillada de las Arfak

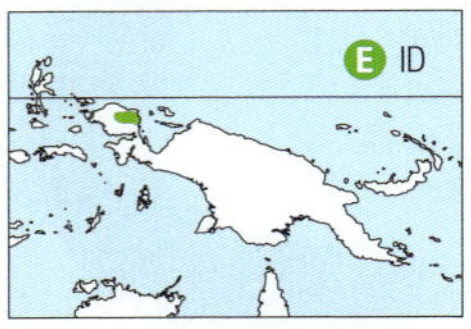

750–1900 m

New Guinea (Arfak Mts)

gray morph

LC
Monotypic
HB 20–23 cm
T 17–20 cm
W 235–380 g

rich-red morph

0–1300 m

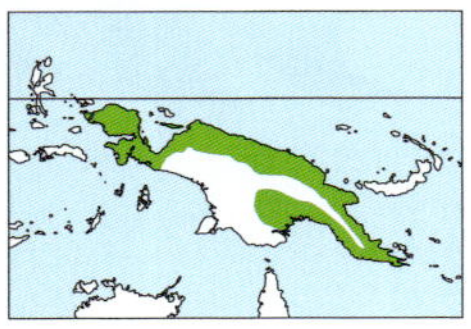

ID/PG: New Guinea (Vogelkop E to Milne Bay Prov, Salawati I, Yapen I)

Lowland Ring-tailed Possum
Pseudochirulus canescens
FR Possum cendré
DE Hundsringbeutler
ES Falangero de cola anillada de tierras bajas

SW WA (coast S of Bunbury to Waychinicup NP, in good habitat inland)

Western Ring-tailed Possum
Pseudocheirus occidentalis
FR Possum ngwayir
DE Westlicher Ringelschwanzbeutler
ES Falangero de cola anillada occidental

CR
Monotypic
HB 30–40 cm
T 30–41 cm
W 0.7–1.3 kg

Genus *PSEUDOCHEIRUS*

(2 species)

ssp *convolutor*

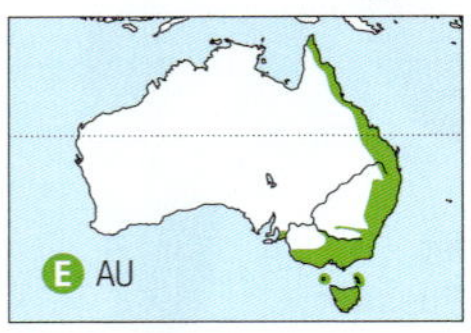

LC
4 sspp
HB 29–35 cm
T 29–36 cm
W 0.8–1.1 kg

E Qld, E NSW to extreme SE SA, Tas (with King, Furneaux Group, Maria Is); introduced to Kangaroo I

ssp *peregrinus*

ssp *cookii*

ssp *pulcher*

Eastern Ring-tailed Possum
Pseudocheirus peregrinus
FR Possum du Queensland
DE Östlicher Ringelschwanzbeutler
ES Falangero de cola anillada oriental

Genus *PETROPSEUDES*

(1 species)

LC
Monotypic
HB 32–39 cm
T 20–26 cm
W 1.3–2 kg

Rock Ring-tailed Possum
Petropseudes dahlii
FR Possum des rochers
DE Felsenringbeutler
ES Falangero rupícola de cola anillada

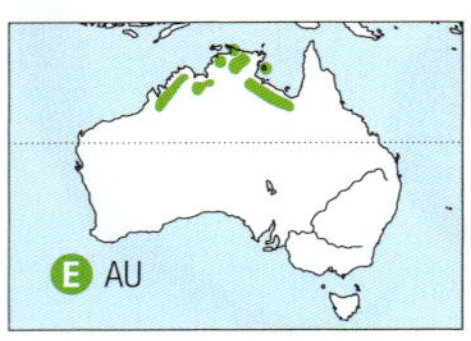

WA (Kimberley region), NT (with Groote Eylandt), NW Qld (E to Boodjamulla NP)

ID/PG: N New Guinea Mts (Vogelkop, Weyland, Foja, Cyclops, N Coastal Ranges, Yapen I)

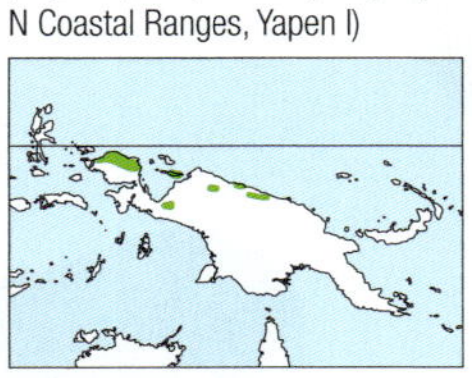

1000–1900 m

D'Albertis's Ring-tailed Possum
Pseudochirops albertisii
FR Possum de d'Albertis
DE Langhaar-Ringbeutler
ES Falangero de cola anillada de d'Albertis

NT
Monotypic
HB 29–34 cm
T 29–32 cm
W 640–875 g

Genus *PSEUDOCHIROPS*

(5 species)

Reclusive Ring-tailed Possum
Pseudochirops coronatus
FR Possum couronné
DE Großer Langhaar-Ringbeutler
ES Falangero de cola anillada solitario

NW New Guinea (Arfak Mts)

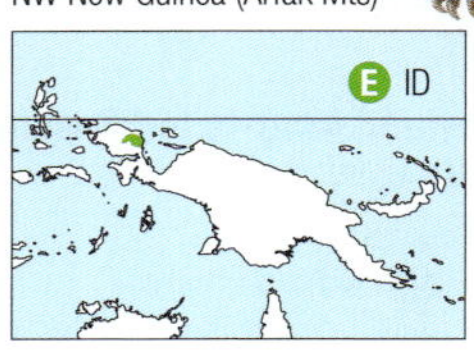

1000–2250 m

VU
Monotypic
HB 34–35 cm
T 28–33 cm
W c. 1.5 kg

LC
Monotypic
HB 36–41 cm
T 27–31 cm
W 1.3–2.2 kg

ID/PG: New Guinea (Central Range from Wissel lakes E to extreme SE)

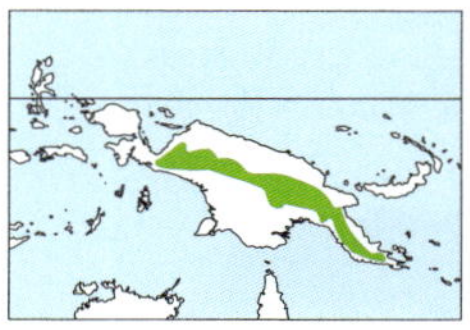

500–3995 m

Coppery Ring-tailed Possum
Pseudochirops cupreus
FR Possum cuivré
DE Kupfer-Ringbeutler
ES Falangero de cola anillada cobrizo

NT
3 sspp
HB 30–37 cm
T 26–37 cm
W 0.9–1.3 kg

ID/PG: New Guinea (Central Range to extreme SE, Huon)

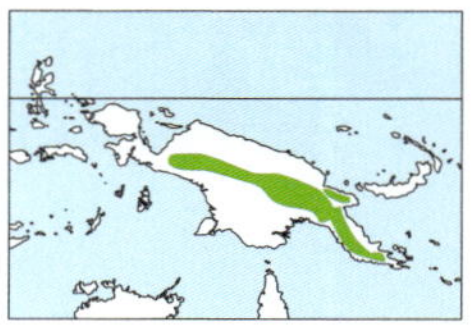

900–2900 m

Plush-coated Ring-tailed Possum
Pseudochirops corinnae
FR Possum de Corinna
DE Glanz-Ringbeutler
ES Falangero de cola anillada de felpa

Green Ring-tailed Possum
Pseudochirops archeri
FR Possum d'Archer
DE Grüner Ringbeutler
ES Falangero de cola anillada verdoso

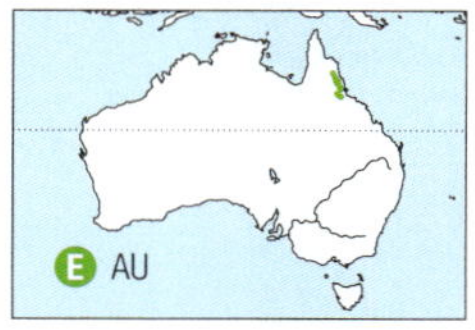

N Qld (Mt Windsor Tableland S to Paluma)

NT
Monotypic
HB 32–40 cm
T 30–36 cm
W 0.9–1.5 kg

Family PETAURIDAE – Striped Possums, Leadbeater's Possum and Lesser Gliders

3 genera, 14 species. Montane and lowland tropical to sclerophyll forests and woodlands. Australasian Region.

LC Monotypic
HB 20–26 cm
T 17–24 cm
W 320–550 g

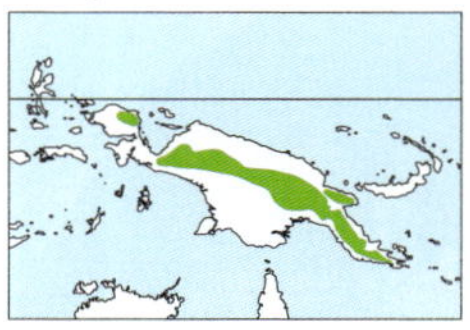

850–3050 m

Long-fingered Striped Possum
Dactylopsila palpator
FR Triok à longs doigts
DE Langfinger-Streifenbeutler
ES Falangero rayado de dedos largos

ID/PG: New Guinea Mts (Arfak Mts, Central Range to extreme SE, Huon)

Genus *DACTYLOPSILA*

(5 species)

Arfak Striped Possum
Dactylopsila kambuayai
FR Triok des Arfak
DE Arfak-Streifenbeutler
ES Falangero rayado de las Arfak

NE
Monotypic

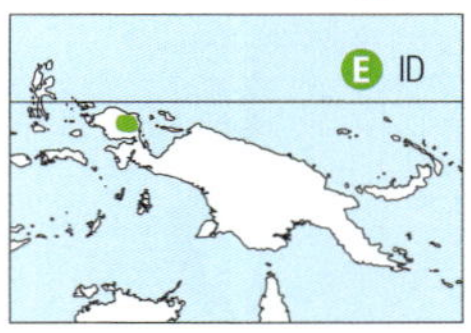

c. 860 m

NW New Guinea (Arfak Mts)

1000–2300 m

ID/PG: W & C New Guinea Central Range (Weyland, Snow and Star Mts)

LC
Monotypic
HB 20–24 cm
T 28–29 cm

Great-tailed Striped Possum
Dactylopsila megalura
FR Triok à grande queue
DE Bürstenschwanz-Streifenbeutler
ES Falangero rayado de cola grande

EN
Monotypic
HB 17–21 cm
T 26–29 cm
W c. 252 g

600–1000 m

Tate's Striped Possum
Dactylopsila tatei
FR Triok de Fergusson
DE Tate-Streifenbeutler
ES Falangero rayado de Tate

D'Entrecasteaux Is (Fergusson I), off SE New Guinea

Genus *PETAURUS*

(8 species)

Yellow-bellied Glider
Petaurus australis
FR Possum à ventre jaune
DE Großer Gleithörnchenbeutler
ES Falangero planeador de vientre amarillo

NE Qld, patchily S along coast through NSW to Vic–SA border

NT
Monotypic
HB 24–31 cm
T 38–47 cm
W 435–725 g

Genus *GYMNOBELIDEUS*

(1 species)

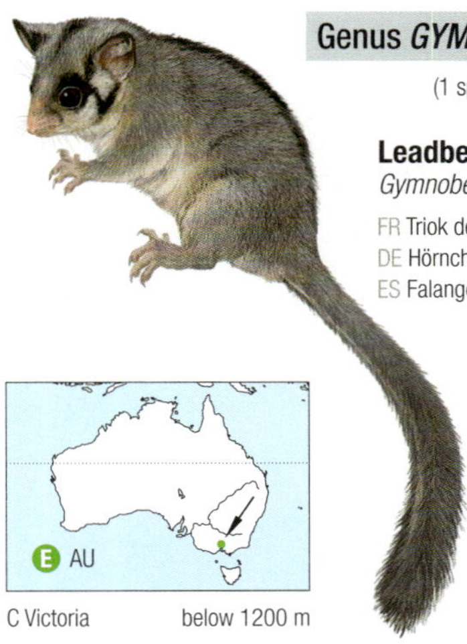

Leadbeater's Possum
Gymnobelideus leadbeateri
FR Triok de Leadbeater
DE Hörnchenbeutler
ES Falangero de Leadbeater

C Victoria below 1200 m

CR
Monotypic
HB 15–17 cm
T 15–18 cm
W 100–170 g

Torresian Striped Possum
Dactylopsila trivirgata
FR Triok rayé
DE Großer Streifenbeutler
ES Falangero rayado de Torres

ID/PG (Waigeo I, New Guinea, Yapen I, Aru Is), NE AU (Iron Range to Mt Spec, NE Qld)

LC
4 sspp
HB 24–28 cm
T 32–39 cm
W 310–545 g

CR
Monotypic
HB 25–28 cm
T 35–38 cm
W 228–332 g

300–1200 m

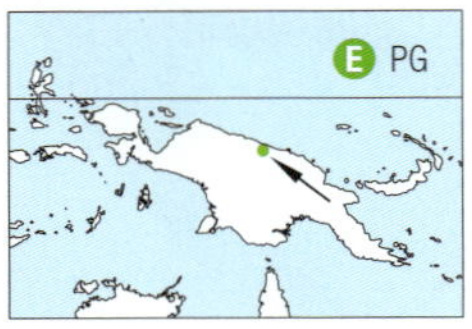

Northern Glider
Petaurus abidi

FR Possum de Ziegler
DE Nördlicher Gleithörnchenbeutler
ES Falangero planeador septentrional

New Guinea (Torricelli Mts, in Mt Somoro and adjacent areas)

LC Monotypic
HB 14–15 cm
T 14–18 cm
W 88–100 g

Biak Glider
Petaurus biacensis

FR Possum de Biak
DE Biak-Gleithörnchenbeutler
ES Falangero planeador de Biak

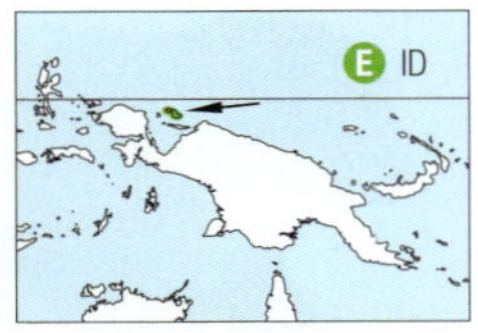

NW New Guinea (Geelvink Bay Is: Supiori, Biak, Owi)

Sugar Glider
Petaurus breviceps

FR Possum à sucre
DE Kurzkopf-Gleithörnchenbeutler
ES Falangero planeador del azúcar

E of Great Dividing Range from SE Qld (N Brisbane) to E Vic

LC
Monotypic
HB 15.5–17.5 cm
T 14.2–22 cm
W 95–155 g

ssp *notatus*

NE
2 sspp
HB 17.5–23 cm
T 17.5–23 cm
W 75–147 g

Krefft's Glider
Petaurus notatus

FR Possum de Krefft
DE Krefft-Gleithörnchenbeutler
ES Falangero planeador de Krefft

ssp *papuanus*

0–3000 m

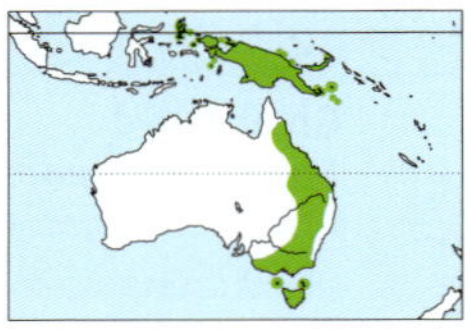

ID/PG (N & C Moluccas, New Guinea, Bismarck Arch), AU (N Qld S to SA; likely introduced to Tas)

NE
Monotypic
HB 14–21 cm
T 16.1–23 cm
W 78–159 g

Savanna Glider
Petaurus ariel

FR Possum ariel
DE Savannen-Gleithörnchenbeutler
ES Falangero planeador de sabana

WA (Kimberley coast) E to W Qld (Lawn Hill), with some islands

LC
Monotypic
HB 18–24 cm
T 22–30 cm
W 173–300 g

Squirrel Glider
Petaurus norfolcensis

FR Possum écureuil
DE Mittlerer Gleithörnchenbeutler
ES Falangero planeador mediano

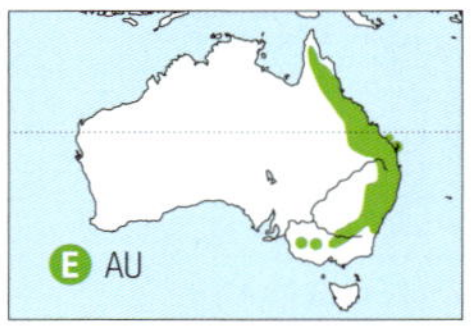

below 300 m

N Qld to C Vic, with Ricketts I and Fraser I; few records in SE SA (Western Flat and Bordertown)

Mahogany Glider
Petaurus gracilis

FR Possum à queue fine
DE Mahagoni-Gleithörnchenbeutler
ES Falangero planeador grácil

EN
Monotypic
HB 22–27 cm
T 30–39 cm
W 310–500 g

Coastal NE Qld from R Hull near Tully S to Ollera Creek

Family TARSIPEDIDAE – Honey Possum

1 genus, 1 species. Heathlands and shrublands of coastal sand-plains. Australasian Region.

Genus *TARSIPES*

(1 species)

Honey Possum
Tarsipes rostratus

FR Possum à miel
DE Honigbeutler
ES Falangero mielero

♀

♂

LC
Monotypic
HB 6.5–8.5 cm (M), 7–9 cm (F)
T 7–10.5 cm
W 7–10 g (M), 8–20 g (F)

SW WA, from Geraldton Sand-plains to Esperance Plains

Family ACROBATIDAE – Feather-tailed Gliders and Feather-tailed Possum

2 genera, 3 species. Forests, woodlands, and coastal heathlands. Australasian Region.

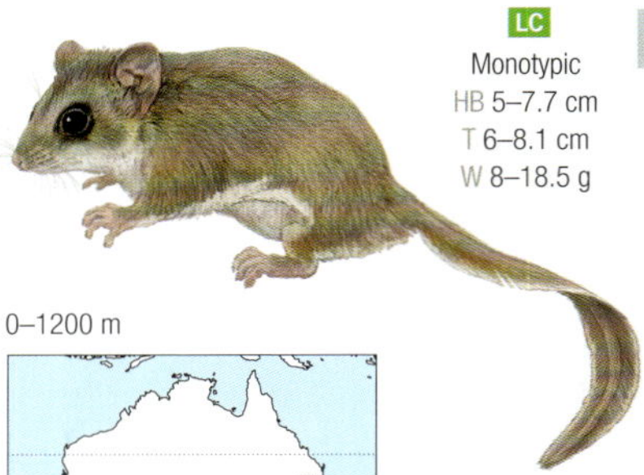

LC
Monotypic
HB 5–7.7 cm
T 6–8.1 cm
W 8–18.5 g

0–1200 m

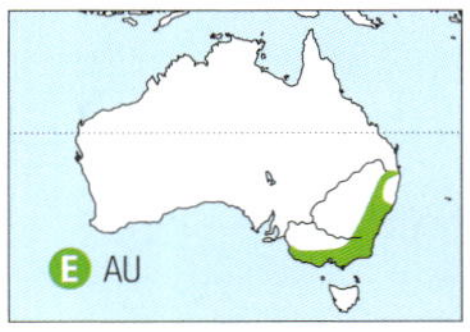

SE Qld (McPherson & Border ranges) to SE SA; absent in E Great Dividing Range (S to R Wallamba)

Narrow-toed Feather-tailed Glider
Acrobates pygmaeus

FR Acrobate pygmée
DE Schmalfuß-Federschwanz-Gleitbeutler
ES Falangero de cola plumosa de dedos estrechos

Genus *ACROBATES*
(2 species)

Broad-toed Feather-tailed Glider
Acrobates frontalis

FR Acrobate de De Vis
DE Breitfuß-Federschwanz-Gleitbeutler
ES Falangero de cola plumosa de dedos anchos

NE
Monotypic
HB 5–7.7 cm
T 6–8.1 cm
W 8–18.5 g

0–700 m

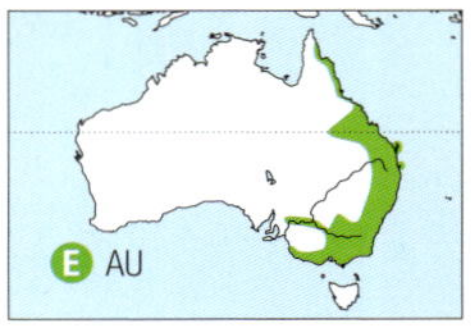

N Qld to Vic (except inland Murray Basin) and SE SA; New Guinea records probably erroneous

Genus *DISTOECHURUS*
(1 species)

Feather-tailed Possum
Distoechurus pennatus

FR Acrobate à queue pennée
DE Pinselschwanzbeutler
ES Falangero de cola plumosa

LC Monotypic
HB 9–10.8 cm
T 11.5–15.3 cm
W 30–45 g

0– 2600 m

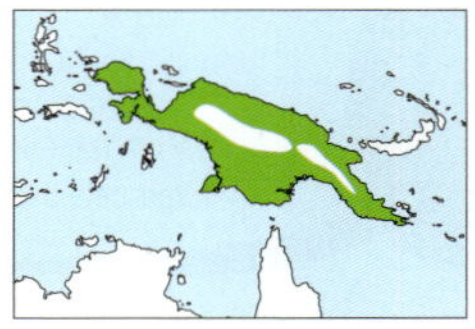

ID/PG: New Guinea, widespread throughout lowlands and montane areas

Family HYPSIPRYMNODONTIDAE – Musky Rat Kangaroo

1 genus, 1 species. Lowland and sub-alpine tropical forests. Australasian Region.

Genus *HYPSIPRYMNODON*
(1 species)

Musky Rat Kangaroo
Hypsiprymnodon moschatus

FR Bettong musquée
DE Moschusrattenkänguru
ES Rata canguro almizcleña

LC
Monotypic
HB 15–27 cm
T 12–16 cm
W 360–680 g

0–1200 m

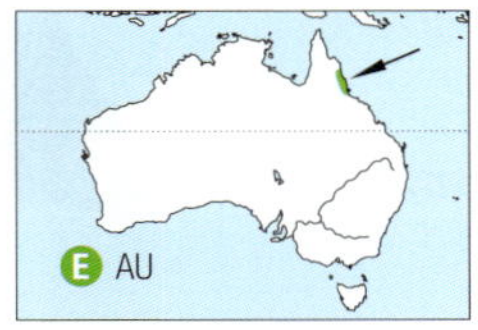

NE Qld, from Helenvale (Mt Amos) W to Ravenshoe, S to Ingham

Family POTOROIDAE – Bettongs and Potoroos

3 genera, 8 species. Xeric shrublands and heathlands to tropical and deciduous forests. Australasian Region.

Genus *AEPYPRYMNUS*
(1 species)

Rufous Bettong
Aepyprymnus rufescens

FR Bettong roussâtre
DE Rotes Rattenkänguru
ES Betong rojizo

below 700 m

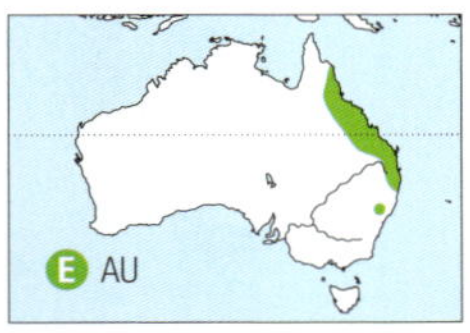

N Qld (Cooktown), S on both sides of Great Dividing Range to near Newcastle (NSW)

LC
Monotypic
HB 34–48 cm
T 31–41 cm
W 1.3–3 kg

Burrowing Bettong
Bettongia lesueur

FR Bettong de Lesueur
DE Lesueur-Bürstenrattenkänguru
ES Betong de Lesueur

NT
Monotypic
HB 28–36 cm
T 21–28 cm
W 0.7–1 kg

0–300 m

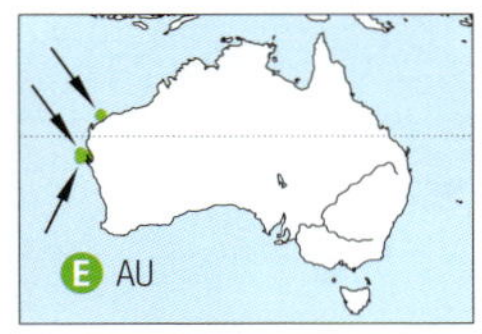

WA (Shark Bay Is, Barrow I); introduced to mainland (WA, NSW, SA)

Genus *BETTONGIA*
(4 species)

E Tas (with Maria I and Bruny I); reintroduced to Canberra (ACT)

0–1000 m

Eastern Bettong
Bettongia gaimardi

FR Bettong de Gaimard
DE Tasmanien-Bürstenrattenkänguru
ES Betong de Tasmania

NT
1 extant ssp
HB 31–33 cm
T 29–34 cm
W 1.2–2.3 kg

Brush-tailed Bettong
Bettongia penicillata

FR Bettong à queue en brosse
DE Östliches Bürstenrattenkänguru
ES Betong de cola de cepillo

CR
1 extant ssp
HB 29–36 cm
T 25–36 cm
W 0.8–1.8 kg

0–300 m

WA (Dryandra Woodland, Kingston, Perup NR, Tutanning NR); (re-)introduced elsewhere in WA, SA & NSW

Northern Bettong
Bettongia tropica

FR Bettong du Nord DE Nördliches Bürstenrattenkänguru
ES Betong septentrional

EN
Monotypic
HB 29–43 cm
T 27–35 cm
W 0.9–1.4 kg

400–1200 m

NE Qld (Mt Windsor Tableland, Mt Carbine Tableland, Lamb Range, Coane Range)

NE Vic (East Gippsland, Barry Mts), SE NSW (South East Forests NP, Yambulla State Forest)

100–1100 m

Long-footed Potoroo
Potorous longipes
FR Potorou à longues pattes
DE Langfuß-Kaninchenkänguru
ES Potoro de pies largos

VU
Monotypic
HB 38–41 cm
T 31–32 cm
W 1.6–2.2 kg

CR
Monotypic
HB 29–37 cm
T 20–24 cm
W 0.7–1.2 kg

SW WA (Mt Gardner, Two Peoples Bay NR); introduced to nearby Bald I

Gilbert's Potoroo
Potorous gilbertii
FR Potorou de Gilbert
DE Gilbert-Kaninchenkänguru
ES Potoro de Gilbert

Genus *POTOROUS*

(3 species)

ssp *apicalis*

Long-nosed Potoroo
Potorous tridactylus
FR Potorou à long nez
DE Langnasen-Kaninchenkänguru
ES Potoro de hocico largo

Patchily from SE Qld along coast to S Vic, SE SA, Tas, with many islands

ssp *tridactylus*

NT
3 sspp
HB 26–41 cm
T 18.9–26 cm
W 0.7–1.7 kg

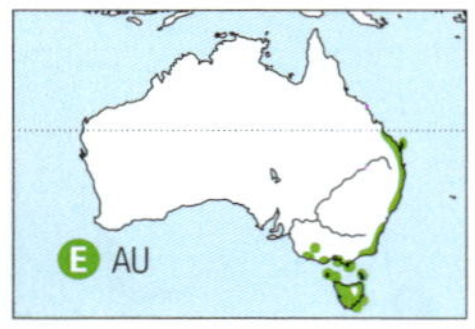

Family MACROPODIDAE – Kangaroos and Wallabies

13 genera, 63 species. Nearly all terrestrial habitats in Australia and New Guinea. Australasian Region.

Genus *LAGOSTROPHUS*

(1 species)

Banded Hare Wallaby
Lagostrophus fasciatus
FR Wallaby rayé
DE Bänderkänguru
ES Ualabí liebre de bandas

WA (Shark Bay: Dorre I, Bernier I, introduced to Faure I)

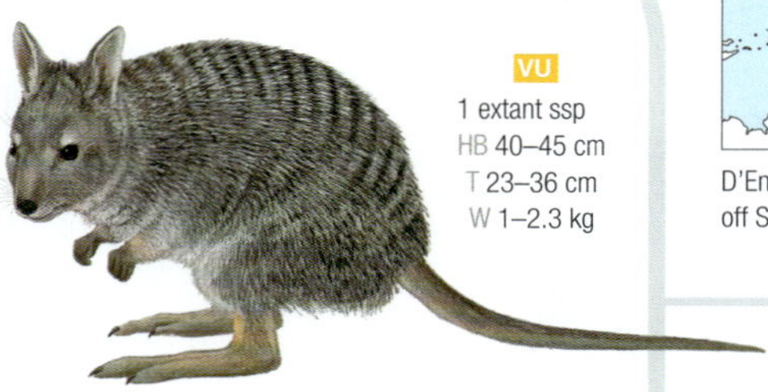

VU
1 extant ssp
HB 40–45 cm
T 23–36 cm
W 1–2.3 kg

Genus *DORCOPSIS*

(4 species)

CR
Monotypic
HB 55–69 cm
T 25–42 cm
W 3.4–7.5 kg

1000–1800 m

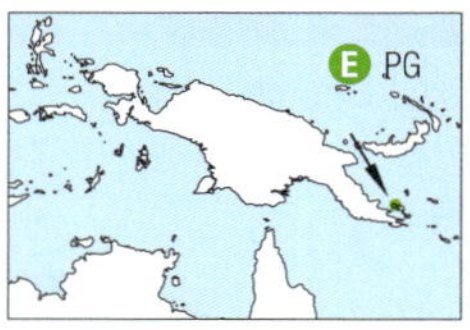

D'Entrecasteaux Is (Goodenough I), off SE New Guinea

Black Forest Wallaby
Dorcopsis atrata
FR Wallaby de Goodenough
DE Schwarzes Buschkänguru
ES Ualabí de bosque negro

White-striped Forest Wallaby
Dorcopsis hageni
FR Wallaby de Hagen
DE Hagen-Buschkänguru
ES Ualabí de bosque de rayas blancas

LC Monotypic
HB 42–60 cm
T 31–38 cm
W 5–6 kg

variants

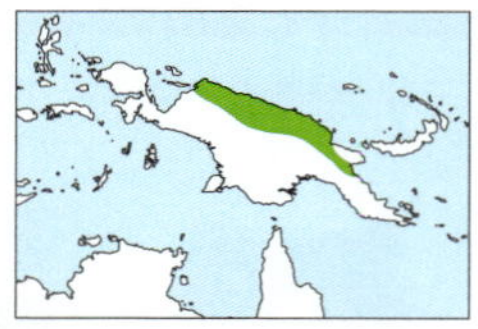

ID/PG: N coastal lowlands of New Guinea

Gray Forest Wallaby
Dorcopsis luctuosa
FR Wallaby gris DE Graues Buschkänguru
ES Ualabí de bosque gris

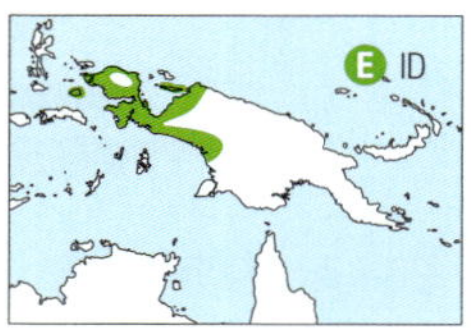

Lowlands of W New Guinea (with Misool I, Salawati I, Yapen I)

Brown Forest Wallaby
Dorcopsis muelleri
FR Wallaby de Müller
DE Braunes Buschkänguru
ES Ualabí de bosque pardo

LC
4 sspp
HB 54–77 cm
T 32–53 cm
W 5–6.8 kg (F)

ssp *luctuosa*

ssp *phyllis*

below 400 m

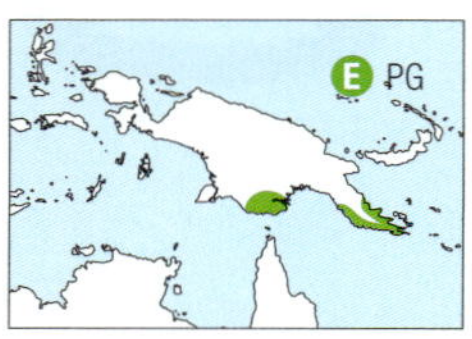

Lowlands of S & SE New Guinea

VU
2 sspp
HB 25–97 cm
T 31–39 cm
W 3.6–11.6 kg

Genus *DORCOPSULUS*

(2 species)

Macleay's Forest Wallaby
Dorcopsulus macleayi
FR Wallaby de Macleay
DE Macleay-Buschkänguru
ES Ualabí de bosque de Macleay

LC
Monotypic
HB 43–46 cm
T 31–35 cm
W 2.5–3.4 kg

1000–1800 m

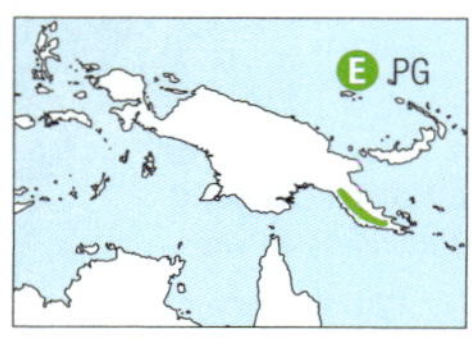

SE New Guinea Pen

800–3100 m

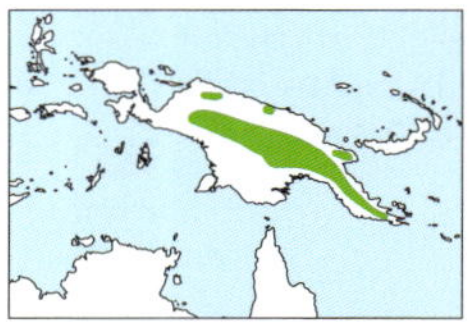

ID/PG: New Guinea (Central Range, N Coastal Ranges, Huon Pen)

NT

Monotypic

HB 31–50 cm

T 22–40 cm

W 1.5–2.3 kg

Small Forest Wallaby

Dorcopsulus vanheurni

FR Wallaby de van Heurn

DE Bergbuschkänguru

ES Ualabí de bosque pequeño

Rufous-bellied Pademelon

Thylogale billardierii

FR Thylogale à ventre roux

DE Rotbauchfilander

ES Pademelon de Tasmania

LC

Monotypic

HB 56–63 cm

T 32–42 cm

W 2.4–12 kg

0–1400 m

Tas (with Schouten I, Bruny I, De Witt I), Bass Strait Is

Genus *THYLOGALE*

(6 species)

SE Qld (from Gladstone) to SE NSW (Illawarra)

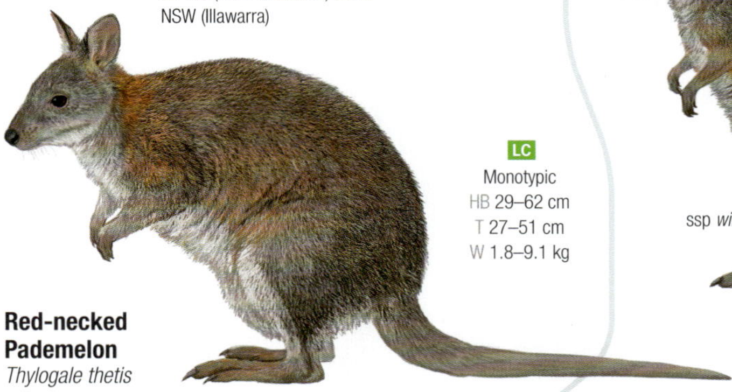

LC

Monotypic

HB 29–62 cm

T 27–51 cm

W 1.8–9.1 kg

Red-necked Pademelon

Thylogale thetis

FR Thylogale à cou rouge

DE Rothalsfilander

ES Pademelon de cuello rojo

Red-legged Pademelon

Thylogale stigmatica

FR Thylogale à pattes rousses

DE Rotbeinfilander

ES Pademelon de patas rojas

ssp *wilcoxi*

ssp *oriomo*

LC

4 sspp

HB 39–54 cm

T 30–47 cm

W 3.7–6.8 kg (M), 2.5–4.2 kg (F)

AU (coastal Qld to C NSW), PG (Trans-Fly region of S New Guinea)

ssp *coxenii*

ssp *stigmatica*

VU

2 sspp

HB 49–67 cm

T 30–52 cm

W 3–9.1 kg

New Guinea Pademelon

Thylogale browni

FR Thylogale de Brown

DE Brauner Filander

ES Pademelon de Nueva Guinea

ssp *lanata*

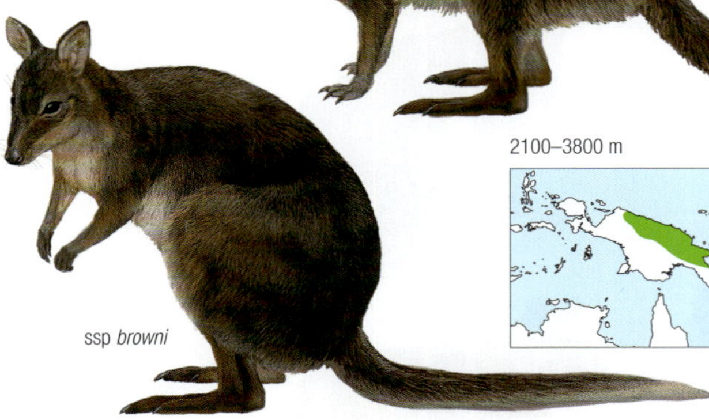

ssp *browni*

2100–3800 m

ID/PG: N & E New Guinea; introduced to Yapen I and some islands of Bismarck Arch

Genus *DENDROLAGUS*

(14 species)

Bennett's Tree Kangaroo

Dendrolagus bennettianus

FR Dendrolague de Bennett

DE Bennett-Baumkänguru

ES Canguro arborícola de Bennett

NT

Monotypic

HB 54–75 cm

T 63–84 cm

W 6.3–13.4 kg

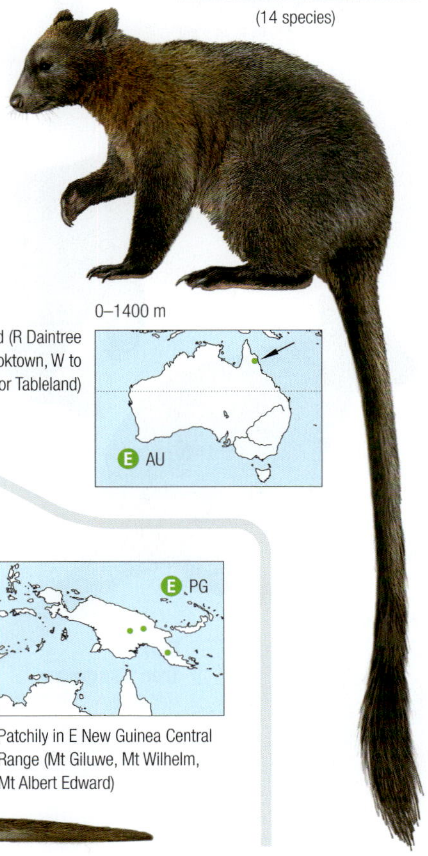

0–1400 m

NE Qld (R Daintree N to past Cooktown, W to Mt Windsor Tableland)

Dusky Pademelon

Thylogale brunii

FR Thylogale de De Bruijn

DE Neuguinea-Filander

ES Pademelon oscuro

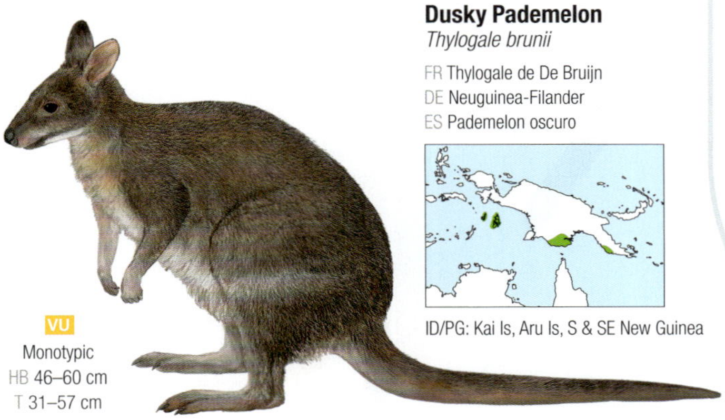

ID/PG: Kai Is, Aru Is, S & SE New Guinea

VU

Monotypic

HB 46–60 cm

T 31–57 cm

Calaby's Pademelon

Thylogale calabyi

FR Thylogale de Calaby

DE Bergfilander

ES Pademelon de Calaby

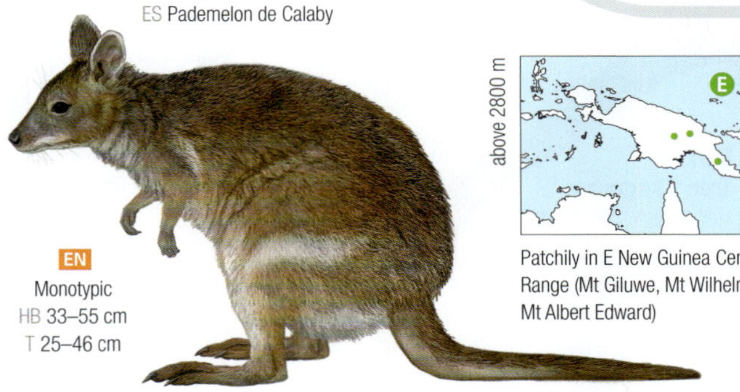

EN

Monotypic

HB 33–55 cm

T 25–46 cm

above 2800 m

Patchily in E New Guinea Central Range (Mt Giluwe, Mt Wilhelm, Mt Albert Edward)

Lumholtz's Tree Kangaroo
Dendrolagus lumholtzi

FR Dendrolague de Lumholtz
DE Lumholtz-Baumkänguru
ES Canguro arborícola de Lumholtz

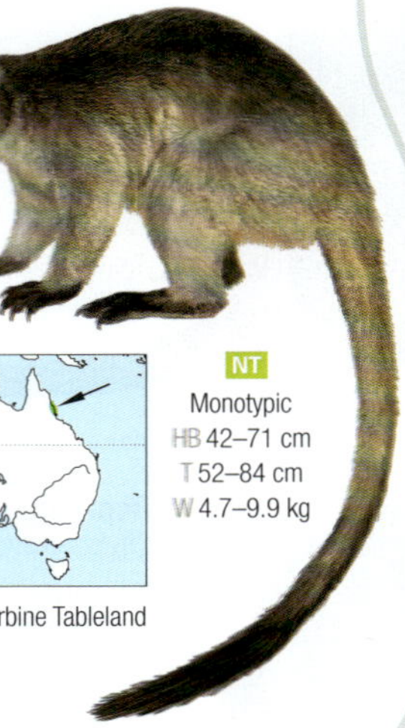

0–1600 m

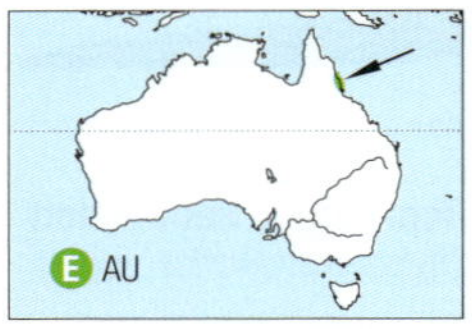

NE Qld (from Mt Carbine Tableland S to Kirrima)

NT
Monotypic
HB 42–71 cm
T 52–84 cm
W 4.7–9.9 kg

Grizzled Tree Kangaroo
Dendrolagus inustus

FR Dendrolague gris
DE Graues Baumkänguru
ES Canguro arborícola entrecano

VU
2 sspp
HB 55–89 cm
T 64–96 cm
W 7–23 kg

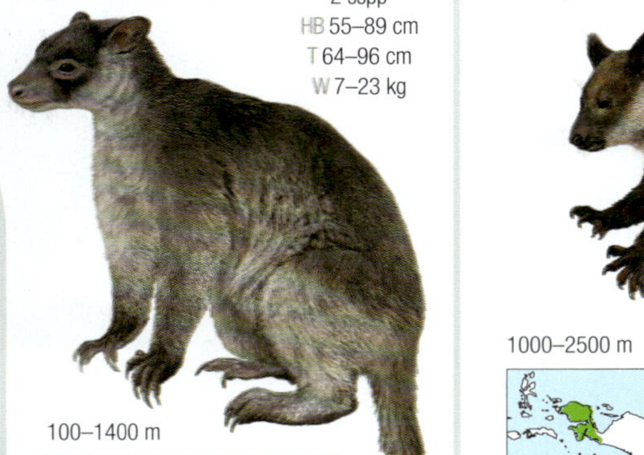

100–1400 m

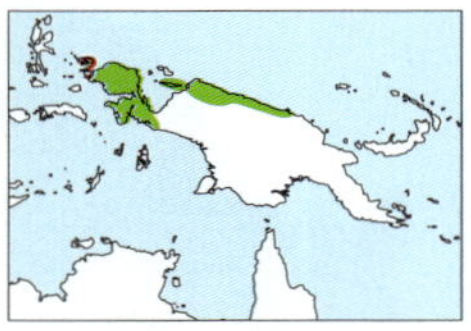

ID/PG: NW & N New Guinea (W of Wewak, with Yapen I); possibly West Papuan Is

Vogelkop Tree Kangaroo
Dendrolagus ursinus

FR Dendrolague noir
DE Bären-Baumkänguru
ES Canguro arborícola de Vogelkop

VU
Monotypic
HB 53–73 cm
T 59–72 cm

1000–2500 m

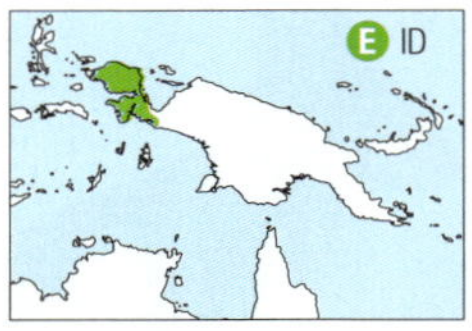

NW New Guinea (Vogelkop & Bomberai pen)

Dingiso
Dendrolagus mbaiso

FR Dendrolague dingiso
DE Schwarzweißes Baumkänguru
ES Canguro arborícola de Sudirman

EN
Monotypic
HB 66–67 cm
T 41–52 cm
W 5.4–8.6 kg

2700–3500 m

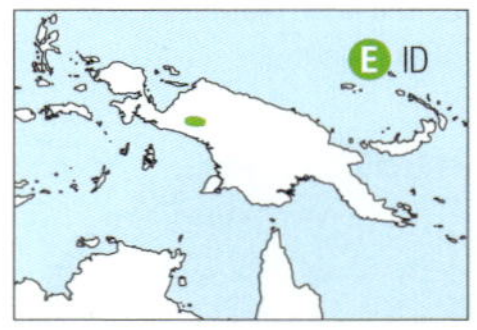

CW New Guinea (Tembagapura & Kwiyawagi areas, Snow Mts)

Doria's Tree Kangaroo
Dendrolagus dorianus

FR Dendrolague de Doria
DE Doria-Baumkänguru
ES Canguro arborícola de Doria

VU
Monotypic
HB 55–78 cm
T 45–66 cm
W 8.2–13.4 kg

600–3650 m

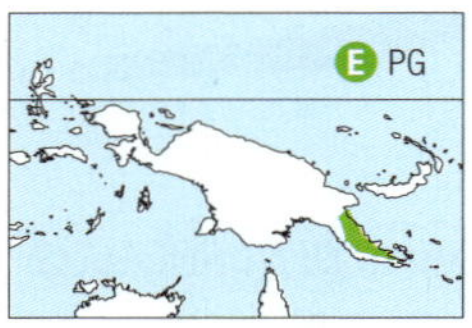

New Guinea (C mountains of SE Pen)

Wondiwoi Tree Kangaroo
Dendrolagus mayri

FR Dendrolague de Mayr
DE Wondiwoi-Baumkänguru
ES Canguro arborícola de Mayr

CR(PE)
Monotypic
HB c. 63.5 cm
T c. 57 cm
W c. 9.3 kg

c. 1600 m

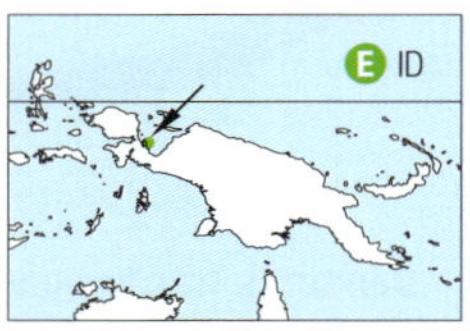

Known from Wondiwoi Pen, West Papua

Ifola Tree Kangaroo
Dendrolagus notatus

FR Dendrolague ifola
DE Ifola-Baumkänguru
ES Canguro arborícola de Ifola

900–3100 m

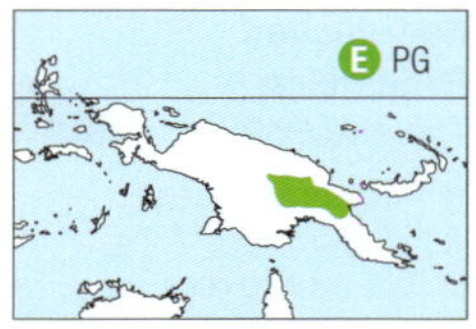

E Central Range, New Guinea

EN
Monotypic
HB 51–69 cm
T 43–56 cm
W 6–9.1 kg

Goodfellow's Tree Kangaroo
Dendrolagus goodfellowi

FR Dendrolague de Goodfellow
DE Goodfellow-Baumkänguru
ES Canguro arborícola de Goodfellow

ssp *goodfellowi*

ssp *buergersi*

EN
2 sspp
HB 50–84 cm
T 53–85 cm
W 5.6–9.5 kg

1000–2860 m

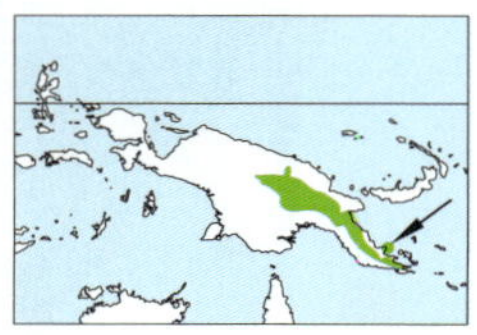

New Guinea (Central Range, SE Pen); introduced to Bismarck Arch (W New Britain)

Seri's Tree Kangaroo
Dendrolagus stellarum

FR Dendrolague de Seri
DE Seri-Baumkänguru
ES Canguro arborícola de Seri

Tembagapura form

VU
Monotypic
HB 61–70 cm
T 49–58 cm
W 6.5–9.5 kg

2600–3200 m

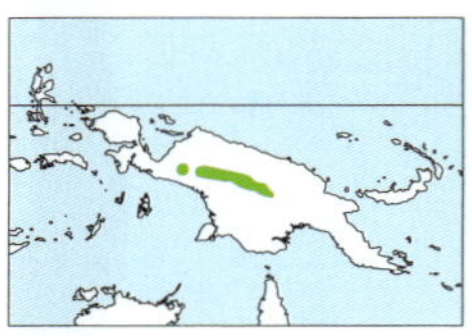

ID/PG: W Central Range (from Tembagapura to Hak-Orn region)

Scott's Tree Kangaroo
Dendrolagus scottae

FR Dendrolague de Scott
DE Scott-Baumkänguru
ES Canguro arborícola de Scott

CR
Monotypic
HB 56–75 cm
T 52–59 cm
W 6.8–11.5 kg

800–2000 m

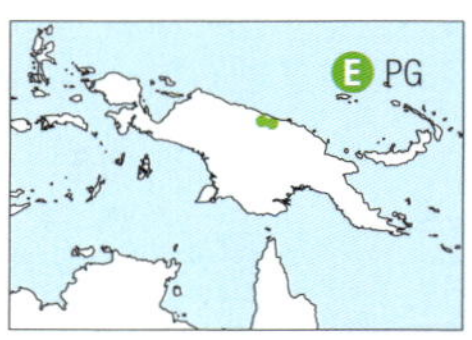

N New Guinea (Bewani & Torricelli Mts)

Golden-mantled Tree Kangaroo

Dendrolagus pulcherrimus

FR Dendrolague doré
DE Goldmantel-Baumkänguru
ES Canguro arborícola dorado

CR
HB 58–63 cm
T 71–72 cm
W 7–7.2 kg

680–1200 m

ID/PG: N New Guinea (Foja & Torricelli Mts)

Huon Tree Kangaroo

Dendrolagus matschiei

FR Dendrolague de Matschie
DE Matschie-Baumkänguru
ES Canguro arborícola de Huon

EN
Monotypic
HB 51–66 cm T 45–68 cm
W 7–10 kg

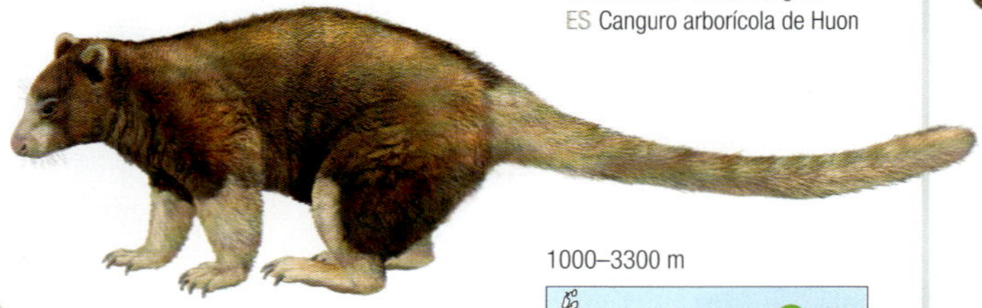

1000–3300 m

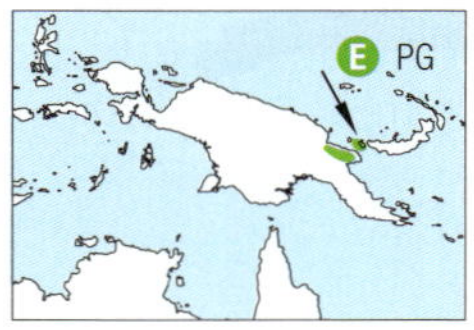

NE New Guinea (Huon Pen); likely introduced to Umboi I

Lowland Tree Kangaroo

Dendrolagus spadix

FR Dendrolague des plaines
DE Tiefland-Baumkänguru
ES Canguro arborícola de llanura

VU
Monotypic
HB 50–64 cm
T 62–72 cm
W 7–9.1 kg

below 800 m

SC New Guinea (Hinterland of Gulf of Papua)

Genus *PETROGALE*

(17 species)

Wilkins's Rock Wallaby

Petrogale wilkinsi

FR Wallaby de Wilkins
DE Wilkins-Felskänguru
ES Ualabí rupestre de Wilkins

NE
Monotypic
HB 31–57 cm
T 30–52 cm
W 1.8–5.2 kg

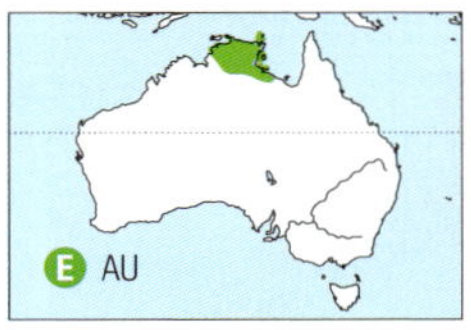

W NT (from R Daly) E to NT–Qld border (Wollogorang), also several islands; E limit uncertain

Short-eared Rock Wallaby

Petrogale brachyotis

FR Wallaby à oreilles courtes
DE Kurzohr-Felskänguru
ES Ualabí rupestre de orejas cortas

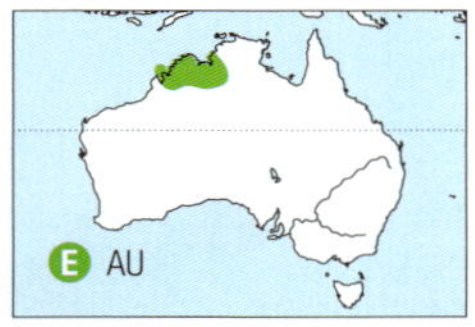

WA (Kimberley, with Koolan I), E to NW NT (R Daly)

LC
2 sspp
HB 44–52 cm
T 37–56 cm
W 2.9–6 kg

ssp *victoriae*

ssp *brachyotis*

Nabarlek

Petrogale concinna

FR Nabarlek
DE Zwergfelskänguru
ES Ualabí rupestre pequeño

EN
3 sspp
HB 29–35 cm
T 22–31 cm
W 1.1–1.7 kg

ssp *canescens*

ssp *monastria*

ssp *concinna*

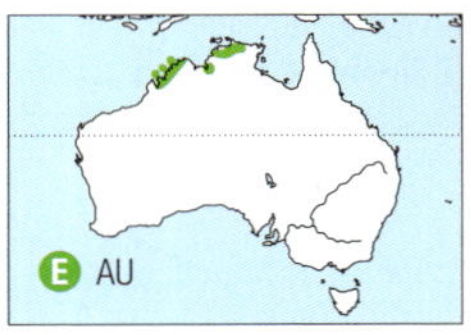

WA (Kimberley, with Borda, Augustus, Long & Hidden Is), NT (Top End, R Victoria region)

Monjon

Petrogale burbidgei

FR Wallaby de Burbidge
DE Kimberley-Felskänguru
ES Ualabí rupestre de Burbidge

NT
Monotypic
HB 31–35 cm
T 26–29 cm
W 0.9–1.4 kg

WA (NW Kimberley from Mitchell Plateau to R Charnley); also Bonaparte Arch & Boongaree I

Yellow-footed Rock Wallaby

Petrogale xanthopus

FR Wallaby à pattes jaunes
DE Gelbfuß-Felskänguru
ES Ualabí rupestre de patas amarillas

ssp *xanthopus*

ssp *celeris*

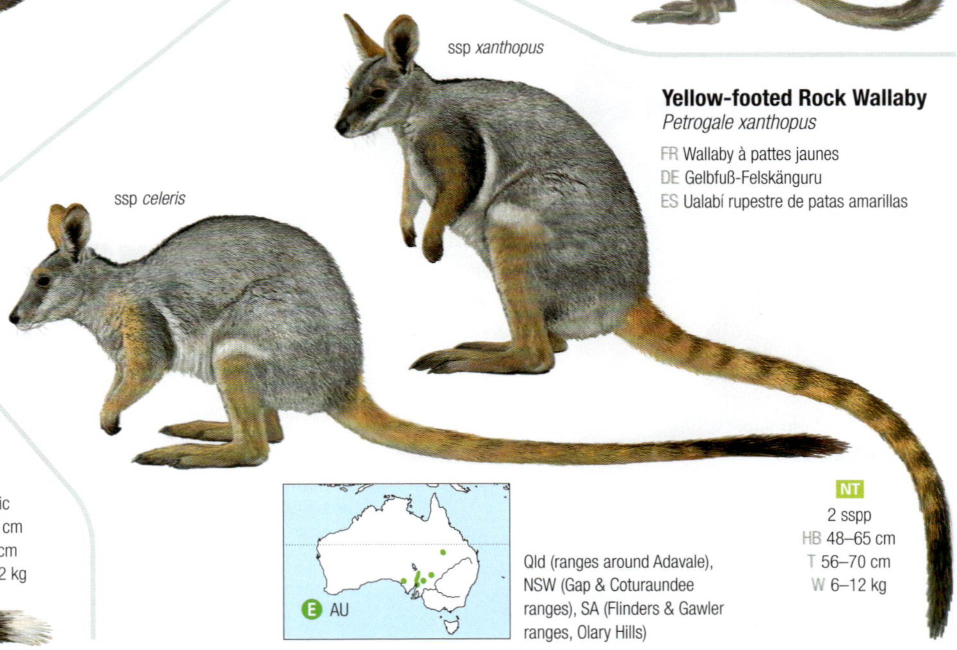

NT
2 sspp
HB 48–65 cm
T 56–70 cm
W 6–12 kg

Qld (ranges around Adavale), NSW (Gap & Coturaundee ranges), SA (Flinders & Gawler ranges, Olary Hills)

Proserpine Rock Wallaby

Petrogale persephone

FR Wallaby proserpine DE Proserpine-Felskänguru
ES Ualabí rupestre de Proserpine

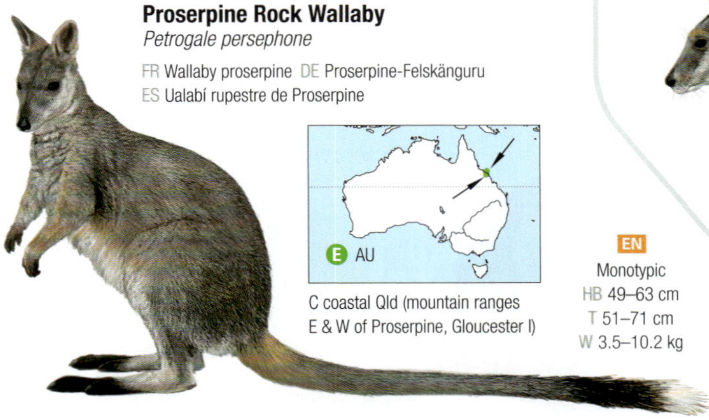

C coastal Qld (mountain ranges E & W of Proserpine, Gloucester I)

EN
Monotypic
HB 49–63 cm
T 51–71 cm
W 3.5–10.2 kg

Rothschild's Rock Wallaby
Petrogale rothschildi

FR Wallaby de Rothschild
DE Rothschild-Felskänguru
ES Ualabí rupestre de Rothschild

LC
Monotypic
HB 46–59 cm (M) T 54–70 cm
W 3.7–6.6 kg

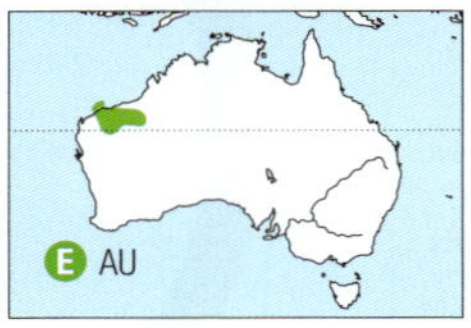

WA (Pilbara & Ashburton regions, Dampier Arch); E & S limits uncertain

Black-footed Rock Wallaby
Petrogale lateralis

FR Wallaby à pattes noires
DE Schwarzpfoten-Felskänguru
ES Ualabí rupestre de pies negros

VU
5 sspp
HB 44–57 cm
T 32–61 cm
W 2.3–6.5 kg

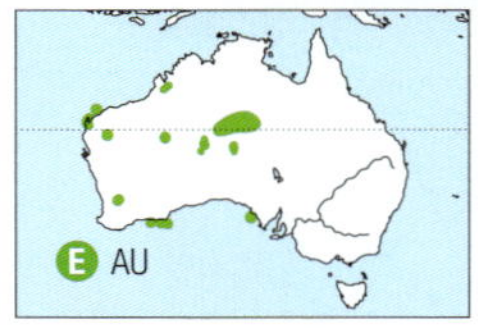

Patchily across WA, S NT and SA, with several islands; introduced to some islands of SA

NT
Monotypic
HB 39–61 cm
T 44–58 cm
W 2.7–7.1 kg

NW Qld (patchily from Mt Isa SE to Winton and NW to past Lawn Hill NP); N & W limits uncertain

Purple-necked Rock Wallaby
Petrogale purpureicollis

FR Wallaby violacé DE Purpurnacken-Felskänguru
ES Ualabí rupestre de cuello púrpura

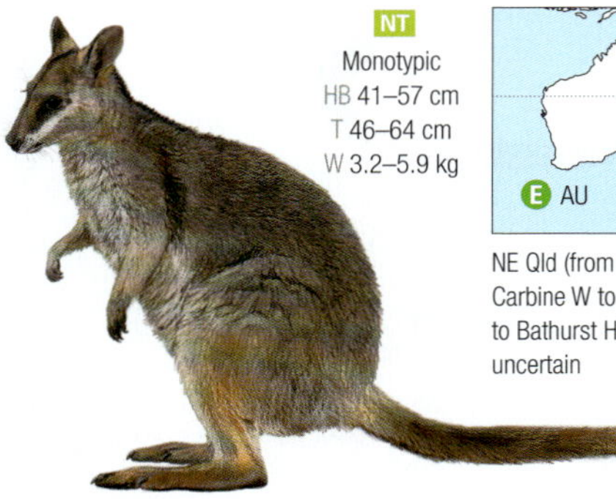

NT
Monotypic
HB 41–57 cm
T 46–64 cm
W 3.2–5.9 kg

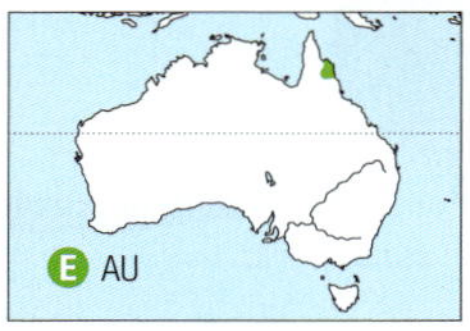

NE Qld (from R Mitchell and Mt Carbine W to "Pinnacles" and N to Bathurst Head); W & N limits uncertain

Godman's Rock Wallaby
Petrogale godmani

FR Wallaby de Godman DE Godman-Felskänguru
ES Ualabí rupestre de Godman

below c. 400 m

Qld (E Cape York Pen from R Pascoe S to Musgrave)

EN
Monotypic
HB 44–56 cm
T 46–54 cm
W 3.5–5 kg

Cape York Rock Wallaby
Petrogale coenensis

FR Wallaby de Cape York
DE Kap-York-Felskänguru
ES Ualabí rupestre del Cabo York

Unadorned Rock Wallaby
Petrogale inornata

FR Wallaby terne
DE Queensland-Felskänguru
ES Ualabí rupestre de Queensland

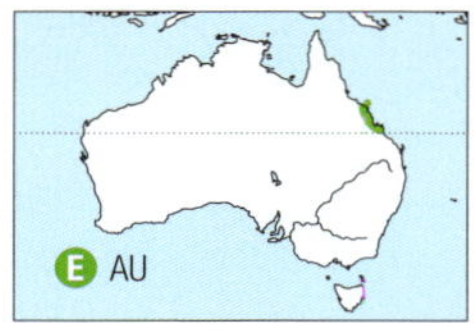

C coastal Qld (from R Burdekin at Home Hill S to R Fitzroy; also Whitsunday Is); W limit uncertain

LC Monotypic
HB 43–61 cm T 43–64 cm
W 3.1–6.6 kg

Sharman's Rock Wallaby
Petrogale sharmani

FR Wallaby de Sharman
DE Sharman-Felskänguru
ES Ualabí rupestre de Sharman

VU
Monotypic
HB 43–53 cm
T 43–54 cm
W 3–4.8 kg

NE Qld (Seaview & Coane Ranges, W of Ingham)

Allied Rock Wallaby
Petrogale assimilis

FR Wallaby de Ramsay
DE Variables Felskänguru
ES Ualabí rupestre de Ramsay

LC Monotypic
HB 44–59 cm
T 41–60 cm
W 3.1–5.9 kg

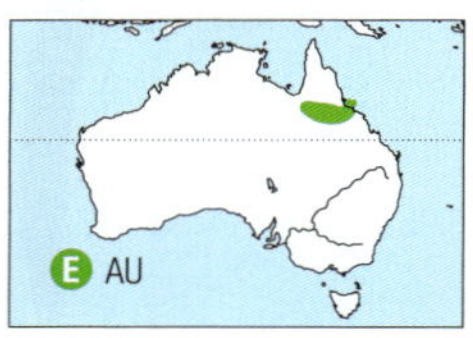

NE Qld (Croydon E to Townsville S to Hughenden & R Bowen; also Palm & Magnetic Is); W limit uncertain

Mareeba Rock Wallaby
Petrogale mareeba

FR Wallaby de Mareeba
DE Mareeba-Felskänguru
ES Ualabí rupestre de Mareeba

NT Monotypic
HB 42–60 cm
T 41–59 cm
W 2.1–6.6 kg

NE Qld (from R Mitchell and near Mt Carbine W to Mungana and Undara, S to R Burdekin); W limit uncertain

Brush-tailed Rock Wallaby
Petrogale penicillata

FR Wallaby à pinceau DE Bürsten-Felskänguru
ES Ualabí rupestre de cola de cepillo

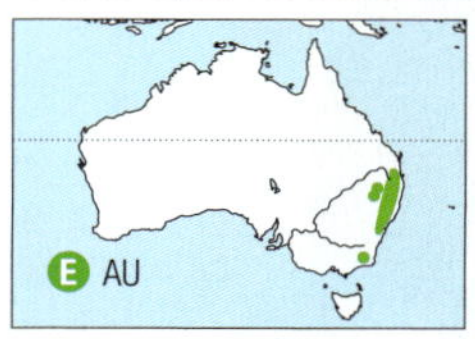

Patchily from SE Qld to E NSW (Kangaroo Valley), N NSW (Warrumbungles & Mt Kaputar), E Vic (E Gippsland)

VU
Monotypic
HB 51–61 cm
T 48–70 cm
W 4.2–10.9 kg

Herbert's Rock Wallaby

Petrogale herberti

FR Wallaby de Herbert
DE Herbert-Felsenkänguru
ES Ualabí rupestre de Herbert

LC
Monotypic
HB 47–61 cm
T 48–66 cm
W 3.1–6.7 kg

SE Qld (from R Fitzroy S to Nanango, W to beyond Rubyvale and Clermont); W limit uncertain

Genus *ONYCHOGALEA*

(2 species)

Bridled Nail-tailed Wallaby

Onychogalea frenata

FR Wallaby bridé
DE Nagelschwanzkänguru
ES Ualabí de brida

VU
Monotypic
HB 34–109 cm
T 32–57 cm
W 1.8–8.3 kg

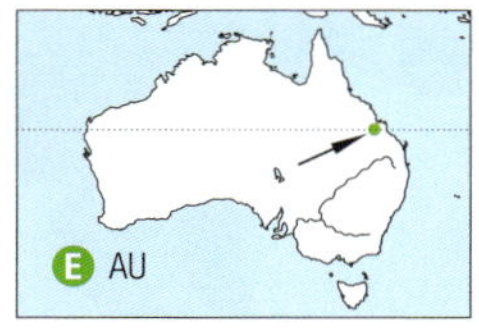

EC Qld (Taunton NP, near Dingo)

Northern Nail-tailed Wallaby

Onychogalea unguifera

FR Wallaby à queue ongulée
DE Nördliches Nagelschwanzkänguru
ES Ualabí de cola de uña

ssp *annulicauda*

LC
2 sspp
HB 49–69 cm
T 52–73 cm
W 3.9–9 kg

ssp *unguifera*

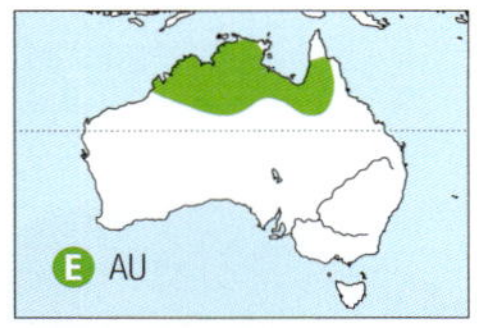

NE WA, N NT, N Qld

Genus *SETONIX*

(1 species)

Quokka

Setonix brachyurus

FR Quokka
DE Quokka
ES Ualabí de cola corta

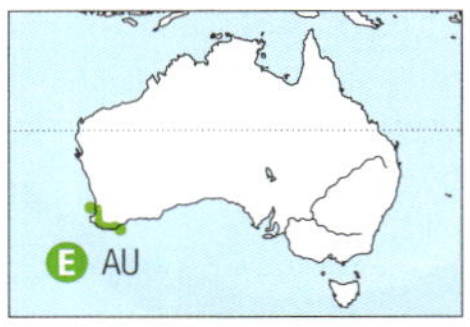

Far SW WA (E of Perth to Stirling & Green Ranges; also Rottnest & Bald Is)

VU
Monotypic
HB 39–54 cm
T 23–31 cm
W 1.6–4.2 kg

Rufous Hare Wallaby

Lagorchestes hirsutus

FR Wallaby mala
DE Zottel-Hasenkänguru
ES Ualabí liebre rojiza

VU
3 sspp
HB 31–39 cm
T 24–30 cm
W 0.8–1.9 kg

WA (Bernier I, Dorre I, introduced to Trimouille I); once in W & C mainland

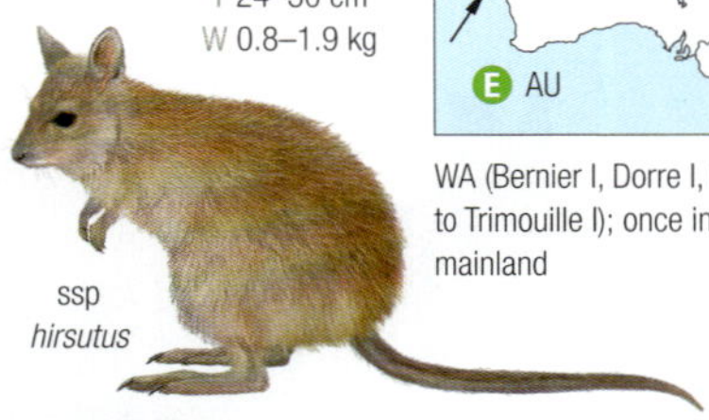

ssp *hirsutus*

ssp *bernieri*

Genus *LAGORCHESTES*

(2 species)

Spectacled Hare Wallaby

Lagorchestes conspicillatus

FR Wallaby à lunettes
DE Brillen-Hasenkänguru
ES Ualabí liebre de anteojos

LC
2 sspp
HB 39–50 cm
T 42–53 cm
W 2.1–4.8 kg

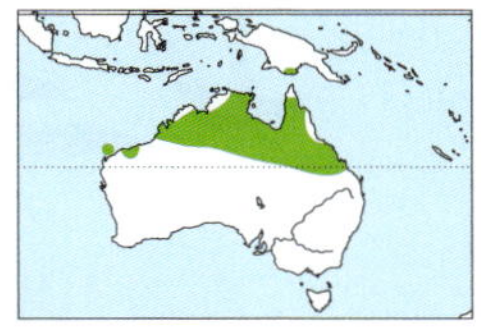

AU (N WA with Barrow I, NT, Qld), SW PG

ssp *leichardti*

ssp *conspicillatus*

Genus *MACROPUS*

(2 species)

Eastern Gray Kangaroo

Macropus giganteus

FR Grand Kangourou
DE Östliches Graues Riesenkänguru
ES Canguro gris oriental

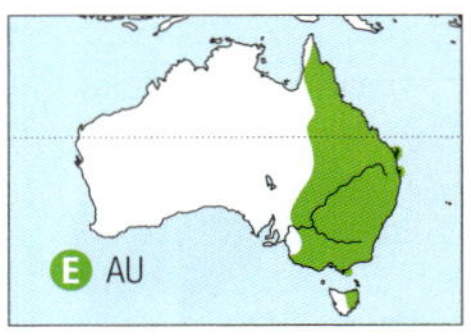

NE Qld (from Cape York Pen) to SE SA, NE Tas; some islands of Qld and Vic

LC
Monotypic
HB 97–230 cm (M), 96–186 cm (F)
T 43–109 cm (M), 44–84 cm (F)
W 19–90 kg (M), 17–42 kg (F)

Western Gray Kangaroo

Macropus fuliginosus

FR Kangourou gris
DE Westliches Graues Riesenkänguru
ES Canguro gris occidental

Kangaroo Island form

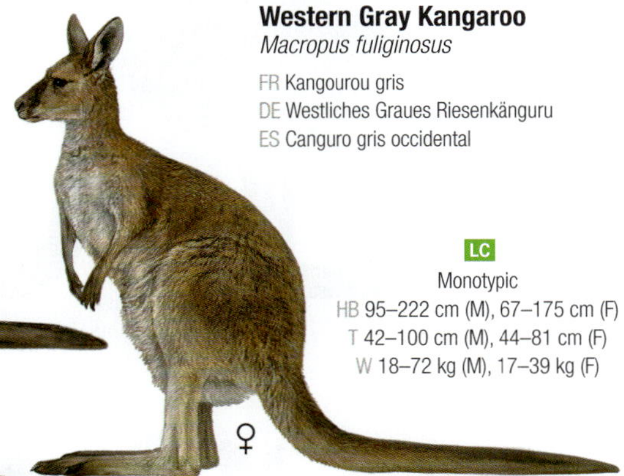

LC
Monotypic
HB 95–222 cm (M), 67–175 cm (F)
T 42–100 cm (M), 44–81 cm (F)
W 18–72 kg (M), 17–39 kg (F)

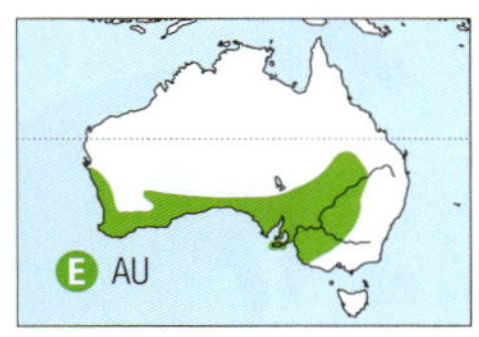

WA (from Shark Bay) to SW Qld, W NSW, W Vic; also Kangaroo I, SA

Genus *OSPHRANTER*

(4 species)

Red Kangaroo

Osphranter rufus

FR Kangourou roux
DE Rotes Riesenkänguru
ES Canguro rojo

LC
Monotypic
HB 93–140 cm (M), 74–110 cm (F)
T 71–100 cm (M), 64–90 cm (F)
W 22–92 kg (M), 17–39 kg (F)

♂

♀

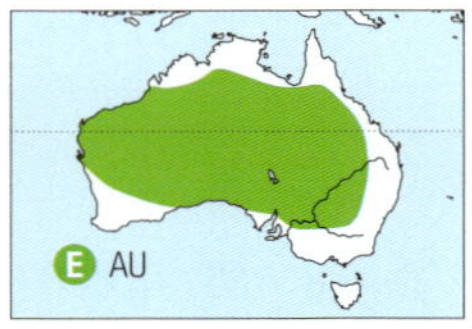

Inland and W mainland

Black Wallaroo

Osphranter bernardus

FR Wallarou noir
DE Schwarzes Bergkänguru
ES Walaró negro

NT
Monotypic
HB 59–72 cm
T 54–64 cm
W 19–22 kg (M), c. 13 kg (F)

NT (W Arnhem Land E of R South Alligator)

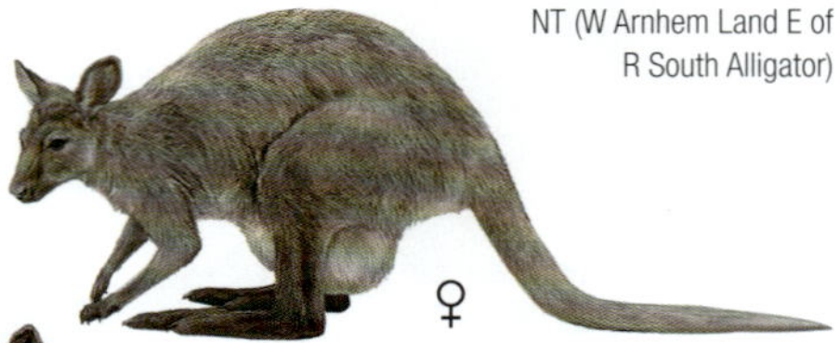

♀

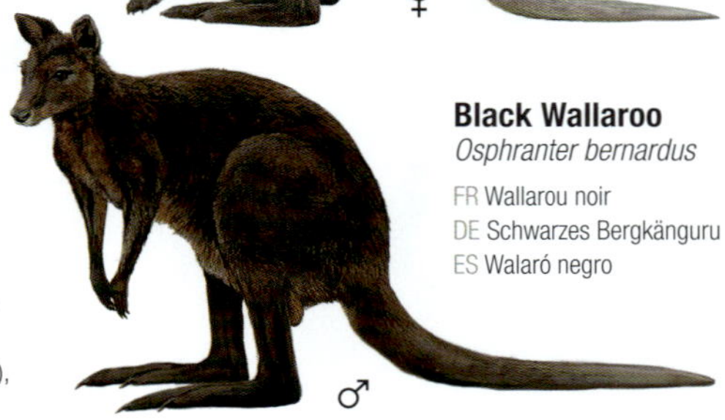

♂

Antilopine Wallaroo

Osphranter antilopinus

FR Wallarou antilope
DE Antilopenkänguru
ES Walaró antílope

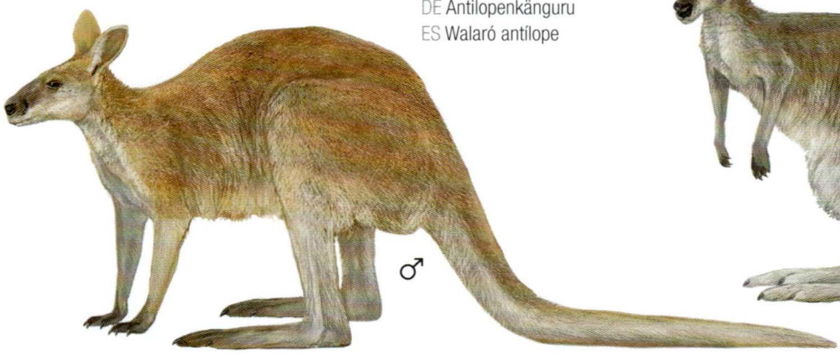

♂

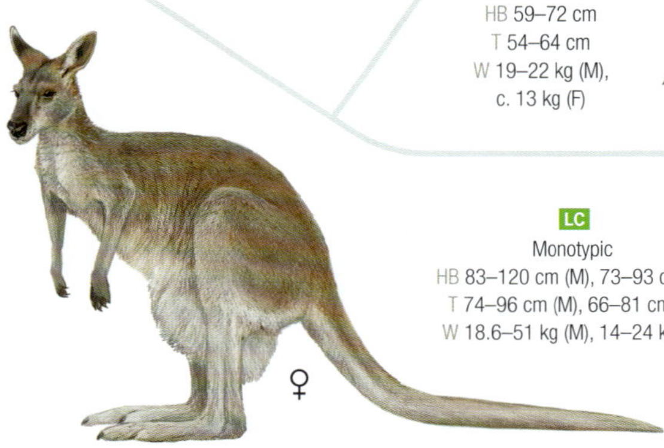

♀

LC
Monotypic
HB 83–120 cm (M), 73–93 cm (F)
T 74–96 cm (M), 66–81 cm (F)
W 18.6–51 kg (M), 14–24 kg (F)

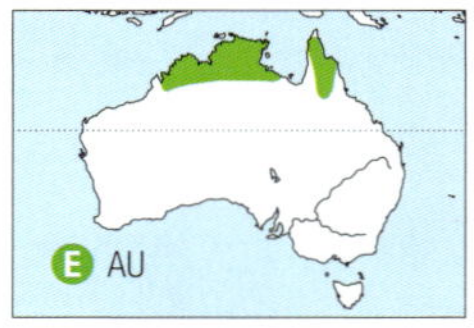

WA (SW Kimberley) to E NT (Pungalina); disjunctly in Qld (Cape York)

Common Wallaroo

Osphranter robustus

FR Wallarou commun
DE Gemeines Bergkänguru
ES Walaró común

♂
ssp *erubescens*

♂
ssp *isabellinus*

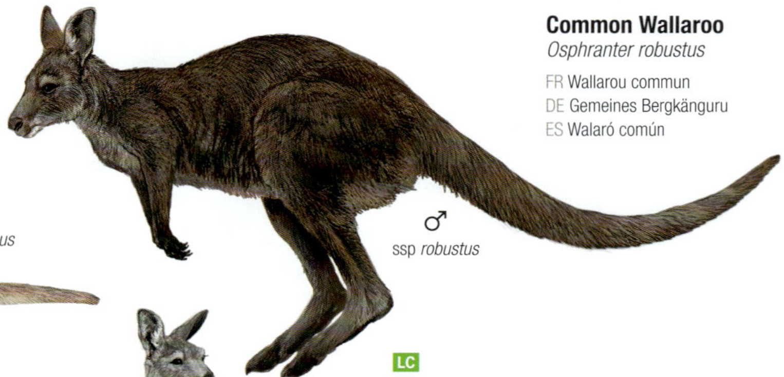

♂
ssp *robustus*

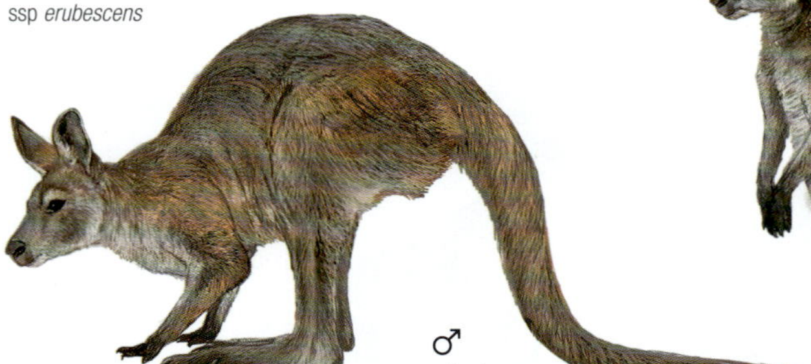

♂
ssp *robustus*
reddish-brown morph

♂
ssp *woodwardi*

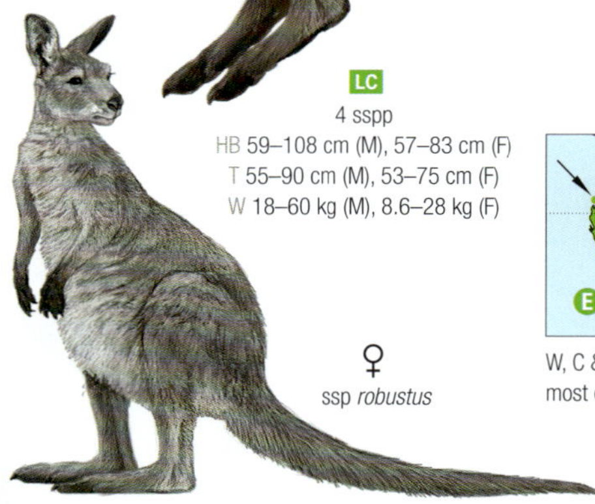

♀
ssp *robustus*

LC
4 sspp
HB 59–108 cm (M), 57–83 cm (F)
T 55–90 cm (M), 53–75 cm (F)
W 18–60 kg (M), 8.6–28 kg (F)

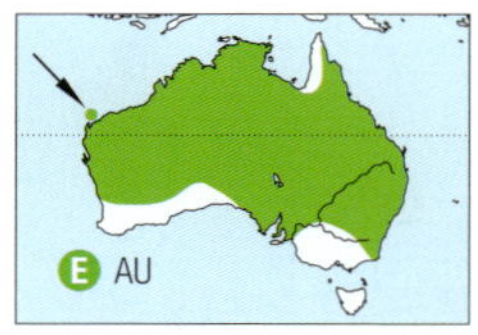

W, C & E mainland, excluding most of Vic; also Barrow I, WA

Genus *WALLABIA*

(1 species)

Swamp Wallaby

Wallabia bicolor

FR Wallaby bicolore
DE Sumpfwallaby
ES Ualabí de pantano

northern variants

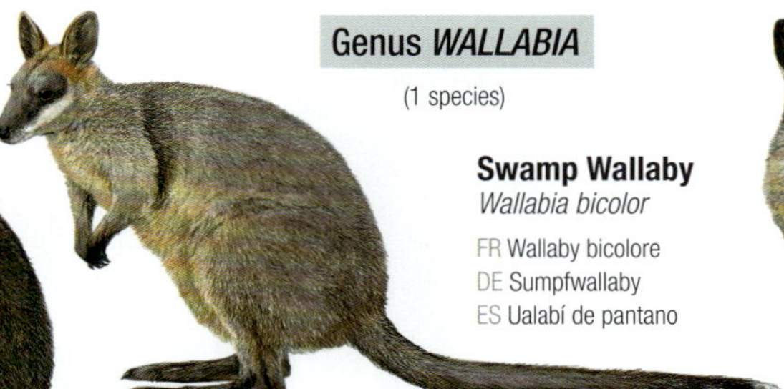

LC
Monotypic
HB 66–85 cm
T 64–86 cm
W 10–20 kg

southern form

Stradbroke I & SE Queensland golden color morph

N Qld (from Cape York, inland to Adava) to far SE SA; also some islands of Qld and Vic

Western Brush Wallaby
Notamacropus irma

FR Wallaby de Jourdan
DE Westliches Irma-Wallaby
ES Ualabí de pincel occidental

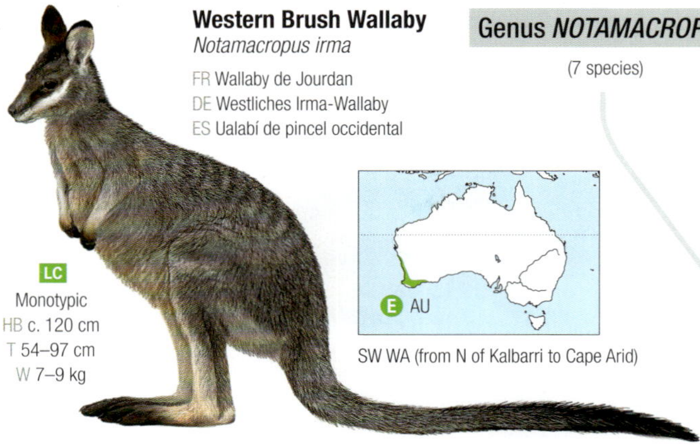

LC
Monotypic
HB c. 120 cm
T 54–97 cm
W 7–9 kg

E AU

SW WA (from N of Kalbarri to Cape Arid)

Genus *NOTAMACROPUS*

(7 species)

Red-necked Wallaby
Notamacropus rufogriseus

FR Wallaby de Bennett
DE Rotnacken-Wallaby
ES Ualabí de cuello rojo

LC
2 sspp
HB 65–92 cm
T 62–86 cm
W 11–27 kg

ssp *rufogriseus*

ssp *banksianus*

E AU

SE Qld to Far SE SA, Bass Strait Is, Tas and nearby Is (Bruny, Robbins, Walker)

Tammar Wallaby
Notamacropus eugenii

FR Wallaby tammar
DE Derby-Wallaby
ES Ualabí tammar

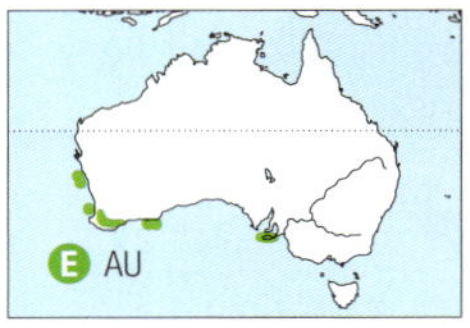

SW WA (East & West Wallabi Is, Garden I off S Perth, Middle I & North Twin Peak Is), SA (Kangaroo I)

LC
2 sspp
HB 52–68 cm
T 33–45 cm
W 3.7–10 kg

ssp *derbianus*

ssp *eugenii*

Agile Wallaby
Notamacropus agilis

FR Wallaby agile
DE Flinkwallaby
ES Ualabí ágil

LC Monotypic
HB 59–85 cm
T 59–84 cm
W 9–27 kg

♀

♂

AU (NE WA to SE Qld, with some islands) and ID/PG: S & E New Guinea, with D'Entrecasteaux Is

Whip-tailed Wallaby
Notamacropus parryi

FR Wallaby élégant
DE Hübschgesicht-Wallaby
ES Ualabí de cola de látigo

below 1400 m

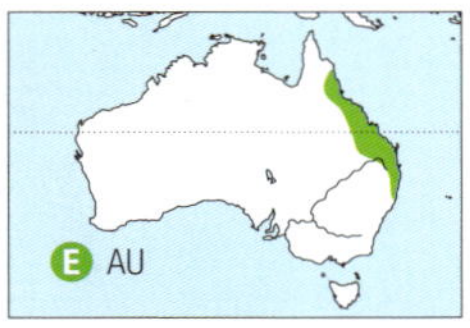

NE Qld (from Cooktown) to NE NSW (Grafton, inland to W Great Dividing Range)

LC
Monotypic
HB 61–100 cm
T 73–104 cm
W 7–26 kg

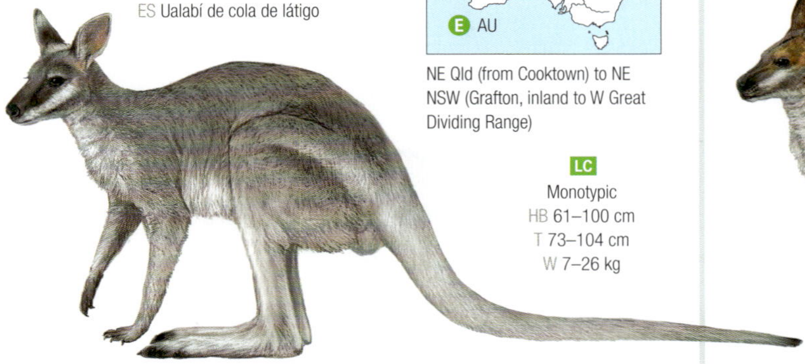

Parma Wallaby
Notamacropus parma

FR Wallaby de Parma
DE Parma-Wallaby
ES Ualabí parma

NT
Monotypic
HB 45–53 cm
T 40–54 cm
W 3.2–5.9 kg

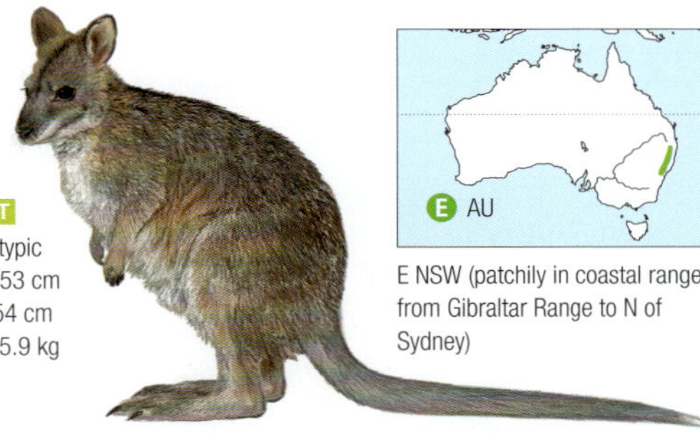

E AU

E NSW (patchily in coastal ranges from Gibraltar Range to N of Sydney)

Black-striped Wallaby
Notamacropus dorsalis

FR Wallaby à raie noire
DE Rückenstreif-Wallaby
ES Ualabí de rayas negras

LC
Monotypic
HB 48–82 cm
T 52–83 cm
W 5.2–21 kg

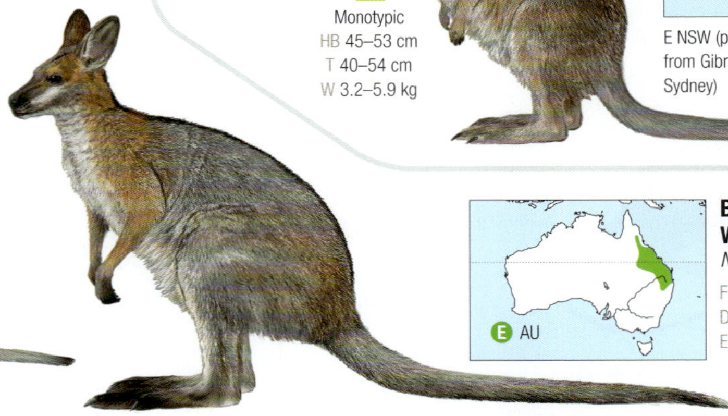

E AU

NE Qld (from Chillagoe, W to Blackall) S to NE NSW (Coonabarabran)

Order AFROSORICIDA (3 families)

Family TENRECIDAE – Tenrecs

8 genera, 31 species. Lowland to montane, humid to dry spiny forests and woodlands. Afrotropical Region.

Genus *ECHINOPS*

(1 species)

LC
Monotypic
HB 10.4–16.5 cm
T c. 1.3 cm
W 50–95 g

0–1300 m

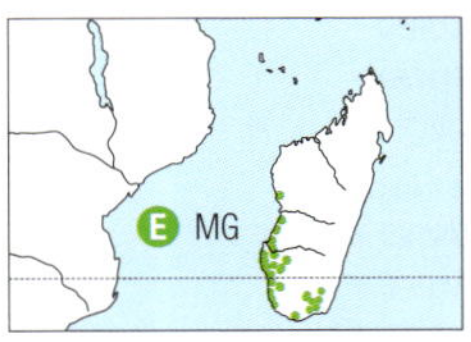

Lesser Hedgehog Tenrec
Echinops telfairi

FR Petit Tenrec-hérisson
DE Kleiner Igeltenrek
ES Tenrec erizo menor

W & SE, mainly S of R Tsiribihina

Genus *SETIFER*

(1 species)

LC
Monotypic
HB 14–23 cm
T 9–17 mm
W 108–350 g

0–2250 m

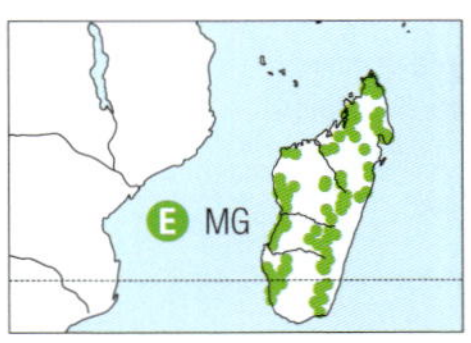

Greater Hedgehog Tenrec
Setifer setosus

FR Grand Tenrec-hérisson
DE Großer Igeltenrek
ES Tenrec erizo mayor

N, Northern & Central highlands, W, E, S

Genus *TENREC*

(1 species)

Tailless Tenrec
Tenrec ecaudatus

FR Tenrec commun
DE Großer Tenrek
ES Tenrec sin cola

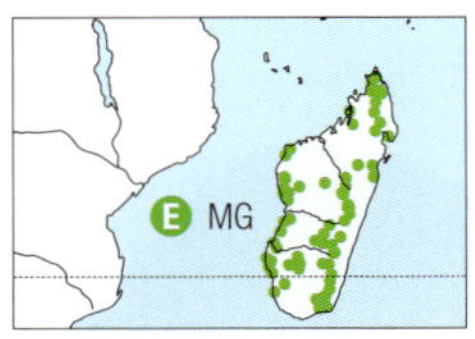

0–1680 m

N, Northern & Central highlands, W, E, S; introduced to RE, MU, SC, possibly Comoro Is

LC Monotypic
HB 18.2–35 cm
W 148–430 g

Lowland Streaked Tenrec
Hemicentetes semispinosus

FR Tenrec rayé
DE Eigentlicher Streifentenrek
ES Tenrec listado de tierras bajas

LC
Monotypic
HB 10.4–17.6 cm
W 76–108 g

0–2050 m

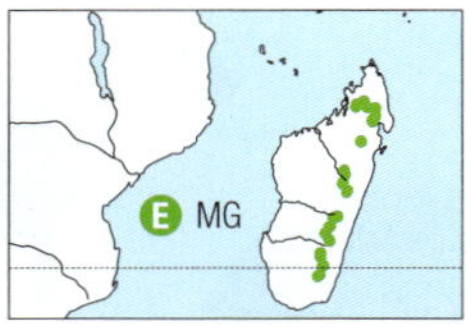

Northern & Central highlands, E

Genus *HEMICENTETES*
(2 species)

Highland Streaked Tenrec
Hemicentetes nigriceps

FR Tenrec à tête noire DE Schwarzkopf-Streifentenrek
ES Tenrec listado de tierras altas

LC
Monotypic
HB 12.4–17.4 cm
W 39–129 g

1200–2050 m

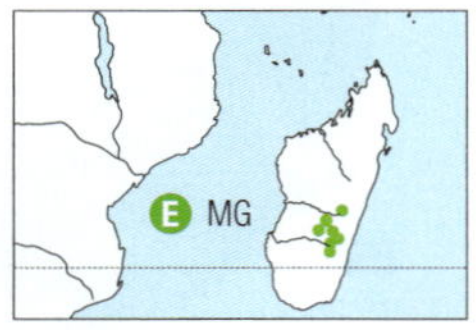

E edge of Central Highlands

Genus *GEOGALE*
(1 species)

LC
Monotypic
HB 5.1–7.6 cm
T 2.8–4.1 cm
W 6–9 g

10–870 m

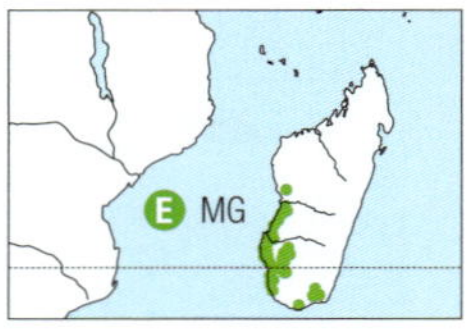

Large-eared Tenrec
Geogale aurita

FR Géogale
DE Erdtenrek
ES Tenrec de orejas grandes

W (mainly S of R Tsiribihina), S, SE; uncertain records in Central Highlands and E

Genus *ORYZORICTES*
(2 species)

LC
Monotypic
HB 9.5–12.4 cm
T 3.8–6.2 cm
W 28–59 g

20–1960 m

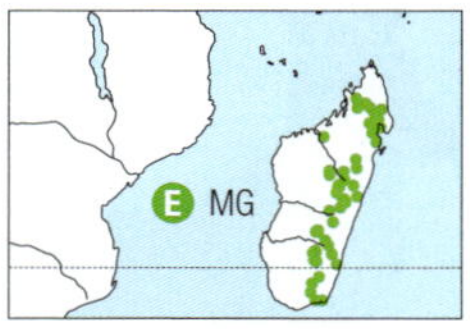

Mole-like Rice Tenrec
Oryzorictes hova

FR Oryzoricte taupe
DE Eigentlicher Reiswühler
ES Tenrec topo de los arrozales

N (with Nosy Mangabe), Northern & Central highlands, W (Marovoay region), E, SE

DD
Monotypic
HB 10.6–12.2 cm
T 4.2–4.9 cm
W 29–31 g

2050–2450 m

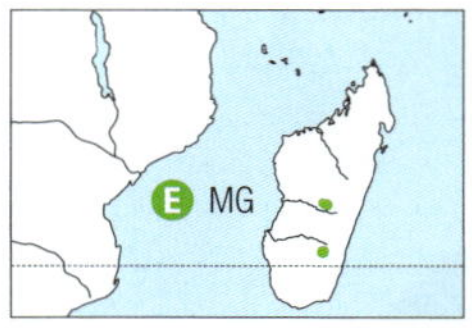

Four-toed Rice Tenrec
Oryzorictes tetradactylus

FR Oryzoricte à quatre doigts
DE Vierzehen-Reiswühler
ES Tenrec de los arrozales tetradáctilo

Central Highlands, E (Andringitra NP, other imprecise records not mapped)

Genus *NESOGALE*
(2 species)

LC
Monotypic
HB 8.7–11.6 cm
T 8.9–12.8 cm
W 21–48 g

0–2500 m

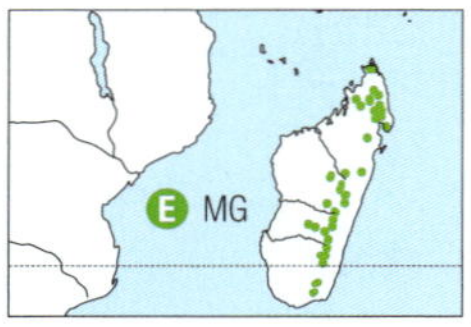

Dobson's Shrew Tenrec
Nesogale dobsoni

FR Microgale de Dobson
DE Dobson-Kleintenrek
ES Tenrec musaraña de Dobson

N, Northern & Central highlands, E, SE

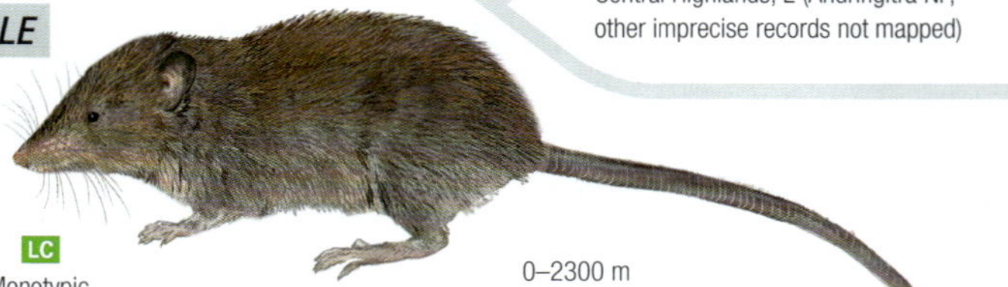

LC
Monotypic
HB 10.5–13.8 cm
T 10.3–15.8 cm
W 32–47 g

0–2300 m

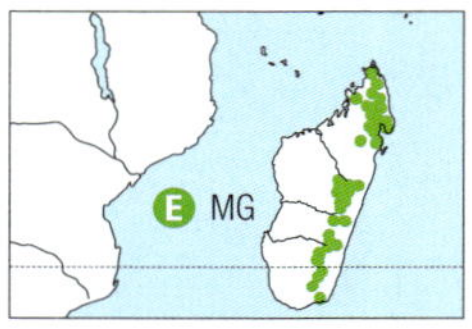

Talazac's Shrew Tenrec
Nesogale talazaci

FR Microgale de Talazac
DE Talazac-Kleintenrek
ES Tenrec musaraña de Talazac

N, Northern & Central highlands, E, SE

Genus *MICROGALE*
(21 species)

LC
Monotypic
HB 8.5–10.5 cm
T 7.5–8.8 cm
W 20–33 g

900–2000 m

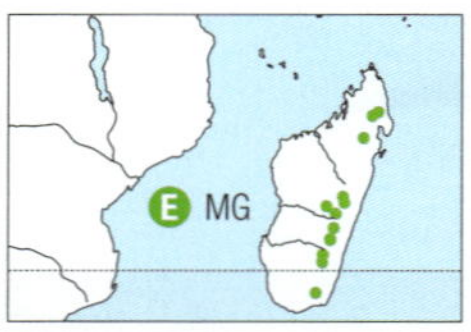

Gracile Shrew Tenrec
Microgale gracilis

FR Microgale gracile
DE Schlanker Kleintenrek
ES Tenrec musaraña grácil

N, Central Highlands, E & SE Madagascar

LC
Monotypic
HB 7.5–11.2 cm
T 5.9–8 cm
W 20–26 g

800–2000 m

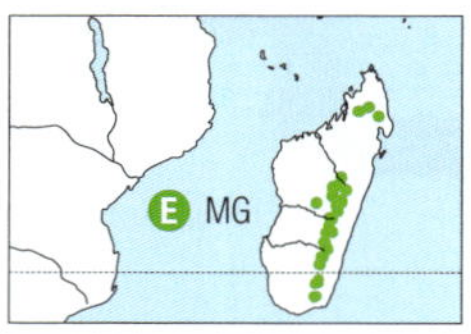

Thomas's Shrew Tenrec
Microgale thomasi

FR Microgale de Thomas
DE Thomas-Kleintenrek
ES Tenrec musaraña de Thomas

Northern (Tsaratanana and Marojejy) & Central Highlands, CE to SE

EN
Monotypic
HB 5.3–8 cm
T 4.4–5.7 cm
W 7–10 g

1000–1680 m

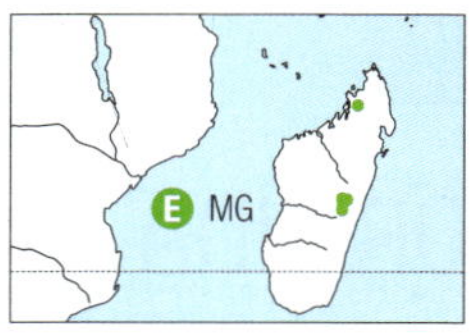

Dark Shrew Tenrec
Microgale jobihely

FR Microgale de Tsaratanana
DE Dunkler Kleintenrek
ES Tenrec musaraña oscura

NW (2 sites on SW slopes of Tsaratanana Massif), E (Analamay & Ambatovy forests, Ambatovy region)

LC
Monotypic
HB 6.8–8.7 cm
T 5.4–8.7 cm
W 12–17 g

530–2525 m

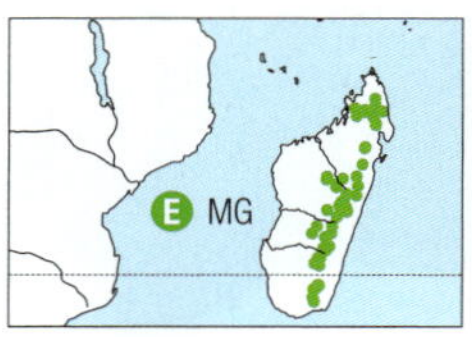

Cowan's Shrew Tenrec
Microgale cowani

FR Microgale de Cowan
DE Cowan-Kleintenrek
ES Tenrec musaraña de Cowan

N, Central Highlands, E, SE

VU
Monotypic
HB 10.6–11.4 cm
T 6.8–7.1 cm
W 30–49 g

540–1260 m

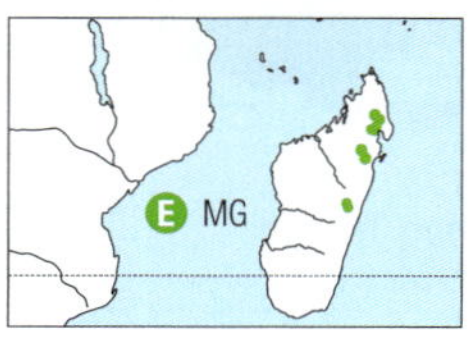

Dryad Shrew Tenrec
Microgale dryas

FR Microgale dryade
DE Dryaden-Kleintenrek
ES Tenrec musaraña de Dríade

NE & E (Anjanaharibe-Sud, Marotandrano R & Ambatovaky R, Makira Plateau and An'ala, E of R Mangoro)

LC
Monotypic
HB 7.5–10.1 cm
T 5.9–7.5 cm
W 14–26 g

595–2525 m

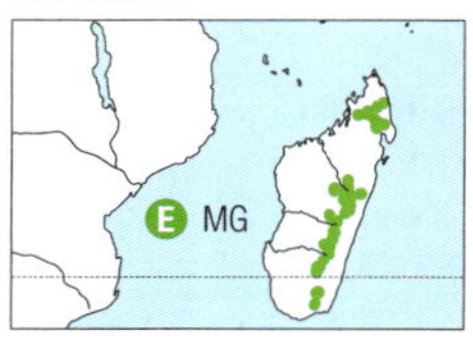

Naked-nosed Shrew Tenrec
Microgale gymnorhyncha

FR Microgale à museau nu
DE Nacktnasen-Kleintenrek
ES Tenrec musaraña de hocico desnudo

N, Central Highlands, E, SE

VU Monotypic
HB 7–8.1 cm T 5–6.2 cm
W 6–14 g

80–1050 m

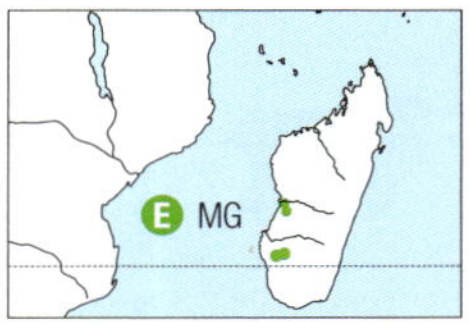

Nasolo's Shrew Tenrec
Microgale nasoloi

FR Microgale de Nasolo
DE Nasolo-Kleintenrek
ES Tenrec musaraña de Nasolo

W (Amboropotsy & Lambokely forests), SW (Vohibasia & Analavelona forests)

LC Monotypic
HB 5.9–8.1 cm T 6.9–9.4 cm
W 7–15 g

600–2500 m

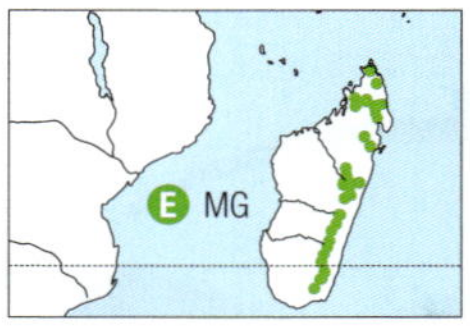

Pale Shrew Tenrec
Microgale fotsifotsy

FR Microgale pâle
DE Fahler Kleintenrek
ES Tenrec musaraña pálida

N, E, SE

LC Monotypic
HB 6.6–10.3 cm T 8.1–11.2 cm
W 14–22 g

675–2525 m

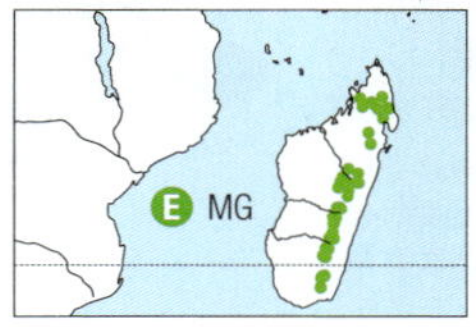

Shrew-toothed Shrew Tenrec
Microgale soricoides

FR Microgale soriçoïde
DE Spitzmaus-Kleintenrek
ES Tenrec musaraña de dientes de musaraña

N, Northern & Central highlands, E, SE

LC Monotypic
HB 6.6–8.8 cm T 3–4.5 cm
W 9–13 g

20–1150 m

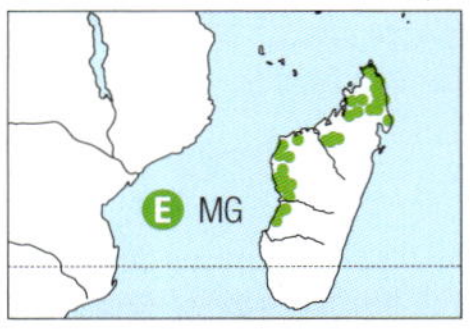

Short-tailed Shrew Tenrec
Microgale brevicaudata

FR Microgale à queue courte
DE Kurzschwanz-Kleintenrek
ES Tenrec musaraña de cola corta

NW, N, Northern Highlands, NE, CW

LC Monotypic
HB 6–8 cm T 3.3–4.3 cm
W 8–10 g

50–430 m

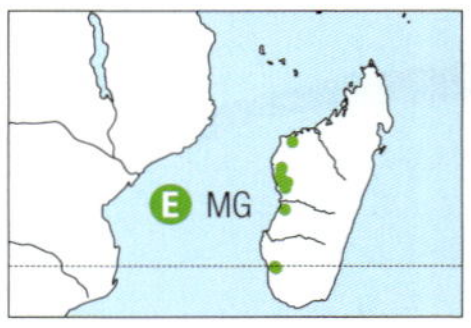

Grandidier's Shrew Tenrec
Microgale grandidieri

FR Microgale de Grandidier
DE Grandidier-Kleintenrek
ES Tenrec musaraña de Grandidier

W & SW (from Namoroka Massif S to R Onilahy)

LC Monotypic
HB 6.3–8.3 cm T 5.3–8.3 cm
W 8–14 g

530–2500 m

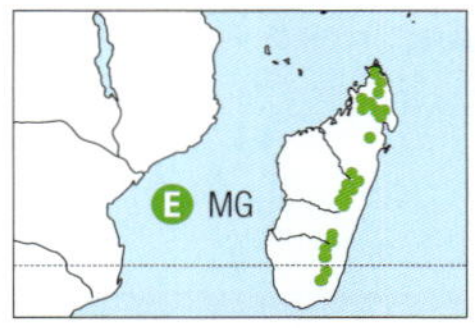

Drouhard's Shrew Tenrec
Microgale drouhardi

FR Microgale de Drouhard
DE Drouhard-Kleintenrek
ES Tenrec musaraña de Drouhard

N, E, SE

VU
Monotypic
HB 7.2–9 cm
T 9.8–11.7 cm
W 12–18 g

500–1950 m

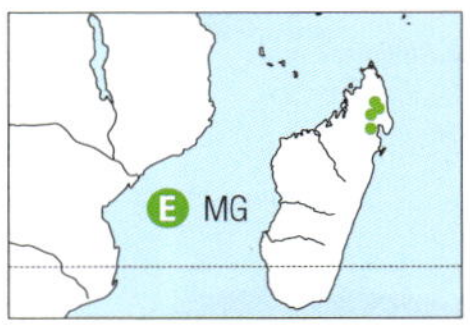

Montane Shrew Tenrec
Microgale monticola

FR Microgale des montagnes
DE Gebirgs-Kleintenrek
ES Tenrec musaraña de montaña

Northern Highlands (Marojejy & Anjanaharibe-Sud), N (Makira Sud)

Taiva Shrew Tenrec
Microgale taiva

FR Microgale des Taivas
DE Taiva-Kleintenrek
ES Tenrec musaraña de Taiva

530–2500 m

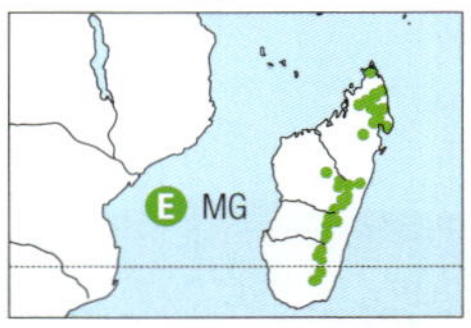

LC
Monotypic
HB 6.1–8.9 cm
T 6.6–9.5 cm
W 10–16 g

N, Northern & Central highlands, E, SE

VU
Monotypic
HB 11.6–17 cm
T 12.8–16.1 cm
W 60–110 g

450–2000 m

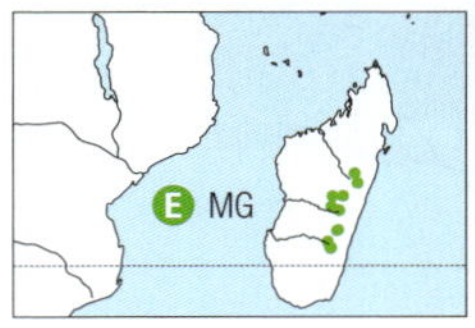

Web-footed Tenrec
Microgale mergulus

FR Microgale plongeur
DE Wassertenrek
ES Tenrec musaraña de pies palmeados

Central Highlands, E

LC Monotypic
HB 4.5–6.5 cm
T 4.7–6.6 cm
W 2–5 g

450–2050 m

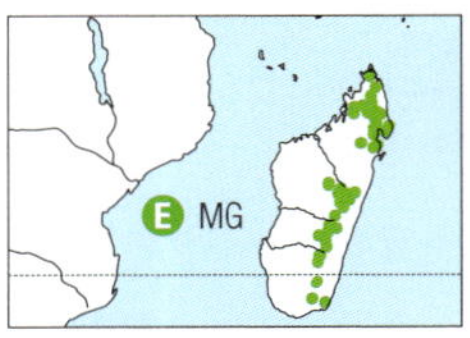

Pygmy Shrew Tenrec
Microgale parvula

FR Microgale pygmée
DE Zwergkleintenrek
ES Tenrec musaraña pigmea

N, Northern & Central highlands, E, SE

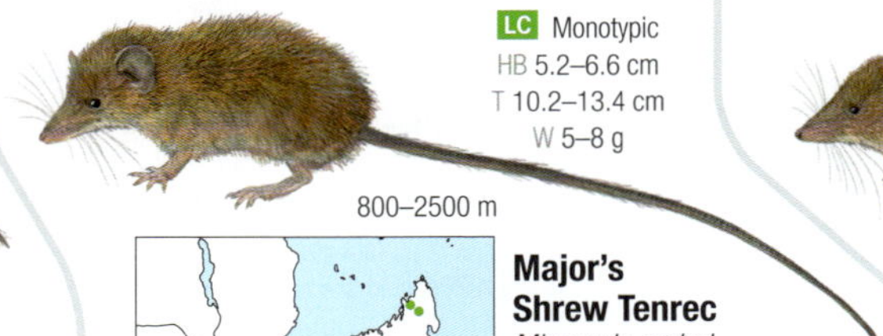

LC Monotypic
HB 5.2–6.6 cm
T 10.2–13.4 cm
W 5–8 g

800–2500 m

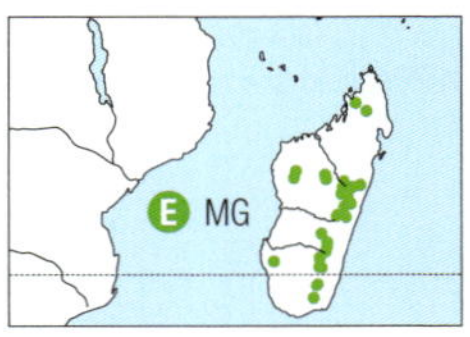

Major's Shrew Tenrec
Microgale majori

FR Microgale de Major
DE Major-Langschwanz-Kleintenrek
ES Tenrec musaraña de Major

Northern & Central highlands, W, SW, E, SE

LC Monotypic
HB 5.9–8 cm T 13.6–15.8 cm
W 6–11 g

530–2500 m

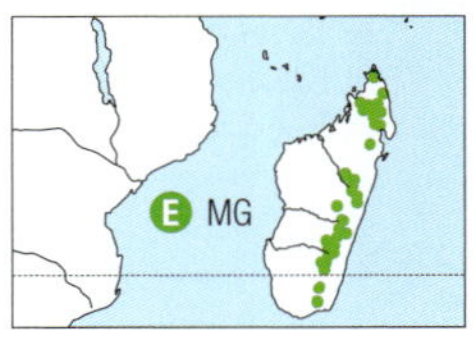

Lesser Long-tailed Shrew Tenrec
Microgale longicaudata

FR Microgale à longue queue
DE Kleiner Langschwanz-Kleintenrek
ES Tenrec musaraña de cola larga menor

N, Central Highlands, E, SE

EN Monotypic
HB 6.2–7.6 cm T 7.1–8.1 cm
W c. 5 g

c. 80 m

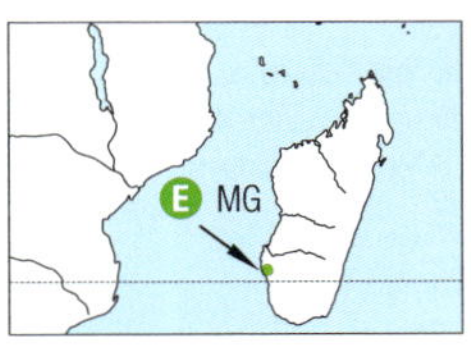

Jenkins's Shrew Tenrec
Microgale jenkinsae

FR Microgale de Jenkins
DE Jenkins-Kleintenrek
ES Tenrec musaraña de Jenkins

SW (Mikea Forest)

LC
Monotypic
HB 6.9–8.9 cm
T 14.4–17.1 cm
W 9–14 g

Greater Long-tailed Shrew Tenrec
Microgale principula

FR Grand Microgale
DE Großer Langschwanz-Kleintenrek
ES Tenrec musaraña de cola larga mayor

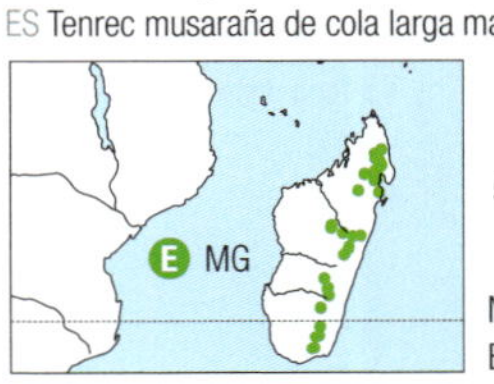

500–1875 m

NE, Central Highlands, E, SE

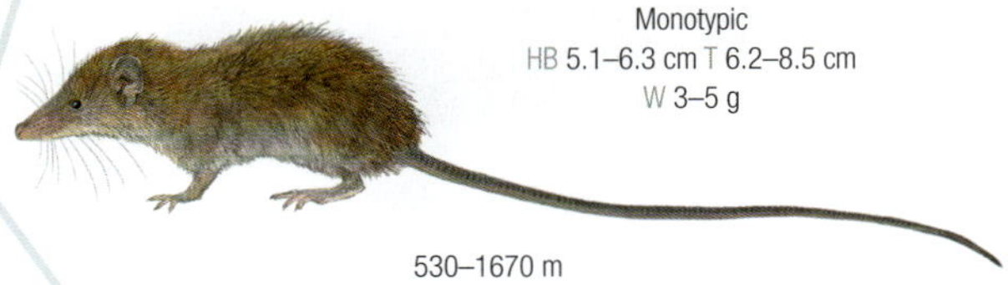

LC
Monotypic
HB 5.1–6.3 cm T 6.2–8.5 cm
W 3–5 g

530–1670 m

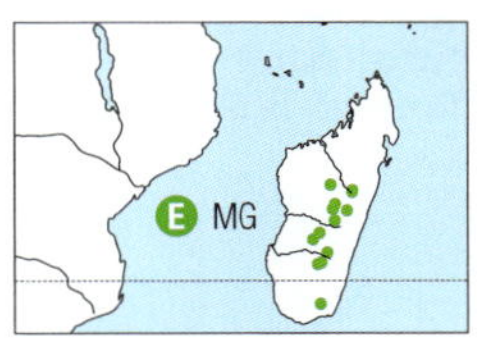

Least Shrew Tenrec
Microgale pusilla

FR Petit Microgale
DE Gnomkleintenrek
ES Tenrec musaraña mínima

Central Highlands, E, SE

Family POTAMOGALIDAE – Otter-shrews

2 genera, 3 species. Rivers and streams in lowland and sub-alpine tropical forests. Afrotropical Region.

Genus *MICROPOTAMOGALE*

(2 species)

above 200 m

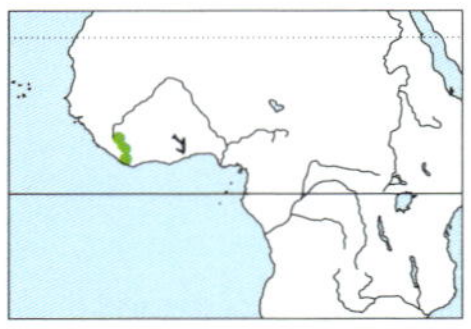

SE GN, N & E LR, E CI, in Upper Guinea rainforest zone (Mt Nimba)

Nimba Otter-shrew
Micropotamogale lamottei
FR Micropotamogale du Nimba
DE Nimba-Otterspitzmaus
ES Musaraña nutria de Nimba

VU Monotypic
HB 12–15.5 cm T 9.5–13.4 cm
W 32–95 g

800–2200 m

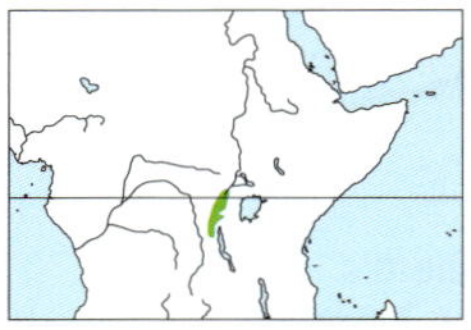

E CD, W UG, SW RW, in Albertine Rift (including Nyungwe Forest NP)

Rwenzori Otter-shrew
Micropotamogale ruwenzorii
FR Micropotamogale du Rwenzori
DE Ruwenzori-Otterspitzmaus
ES Musaraña nutria de Rwenzori

LC Monotypic
HB 13–20 cm T 12.3–15 cm
W 75–135 g

Genus *POTAMOGALE*

(1 species)

below 1800 m

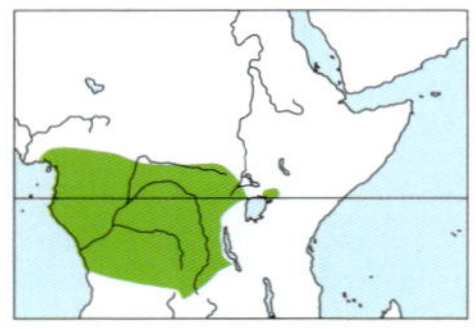

From extreme SE NG, E to S CF and W UG, S to N AO, S CD and NW ZM; also SW KE (Kakamega Forest)

Giant Otter-shrew
Potamogale velox
FR Potamogale
DE Große Otterspitzmaus
ES Musaraña nutria gigante

LC
Monotypic
HB 30–34 cm
T 23–29 cm
W 517–780 g

Family CHRYSOCHLORIDAE – Golden Moles

10 genera, 21 species. Lowland and montane subtropical and tropical deserts, grasslands, and forests. Afrotropical Region.

Genus *AMBLYSOMUS*

(5 species)

ZA (KwaZulu-Natal, Eastern Cape, Free State, Mpumalanga), LS

0–2000 m

Hottentot Golden Mole
Amblysomus hottentotus
FR Taupe-dorée hottentote
DE Hottentotten-Goldmull
ES Topo dorado de los Hotentotes

LC
5 sspp
HB 10.4–14.5 cm
W 37–85 g

ssp *hottentotus*

ssp *longiceps*

ssp *meesteri*

VU
Monotypic
HB 9–14.3 cm
W 61–98 g

2000–2100 m

Mpumalanga (Steenkampsberg around Lydenburg, Dullstroom & Belfast)

Robust Golden Mole
Amblysomus robustus
FR Taupe-dorée de Bronner
DE Robuster Goldmull
ES Topo dorado robusto

NT
Monotypic
HB 10.5–14.5 cm
W 52–86 g

1600–1800 m

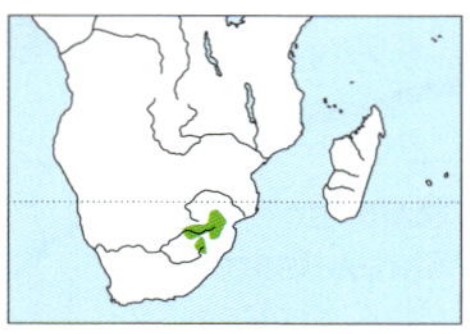

ZA (Mpumalanga, NE Free State), SZ (Piggs Peak)

Highveld Golden Mole
Amblysomus septentrionalis
FR Taupe-dorée du Highveld
DE Highveld-Goldmull
ES Topo dorado del Alto Veld

EN
Monotypic
HB 9–12 cm
W 30–34 g

490–695 m

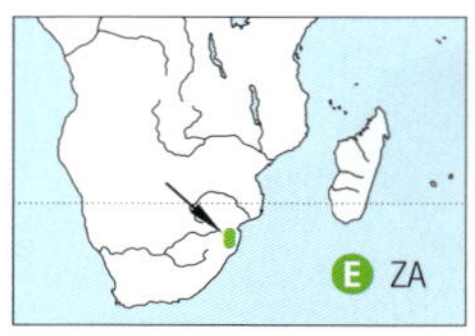

KwaZulu-Natal (Ingwavuma & Ubombo on E slopes of Lebombo Mts); possibly SE SZ (Lebombo Mts)

Marley's Golden Mole
Amblysomus marleyi
FR Taupe-dorée de Marley
DE Marley-Goldmull
ES Topo dorado de Marley

NT 2 sspp
HB 10.5–13.5 cm
W 46–64 g

10–1200 m

Western Cape (from near George) E along S Cape Mts and coastal plains to Eastern Cape (Humansdorp)

Fynbos Golden Mole
Amblysomus corriae
FR Taupe-dorée du fynbos
DE Fynbos-Goldmull
ES Topo dorado de fynbos

Genus *CALCOCHLORIS*

(1 species)

LC
3 sspp
HB 8.2–11 cm
W 15–37 g

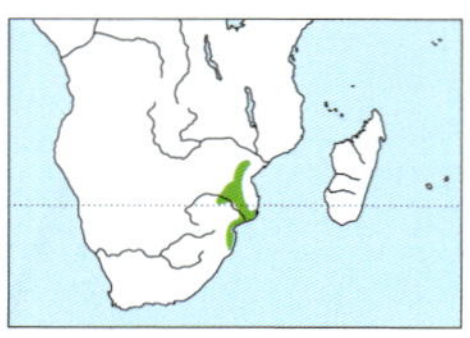

C & S MZ (Save, Changane & Limpopo rivers), marginal in SE ZW & NE ZA (Limpopo, Maputaland)

Yellow Golden Mole
Calcochloris obtusirostris
FR Taupe-dorée du Mozambique
DE Gelber Goldmull
ES Topo dorado amarillo

Genus *HUETIA*

(2 species)

DD
2 sspp
HB 6.3–12.6 cm

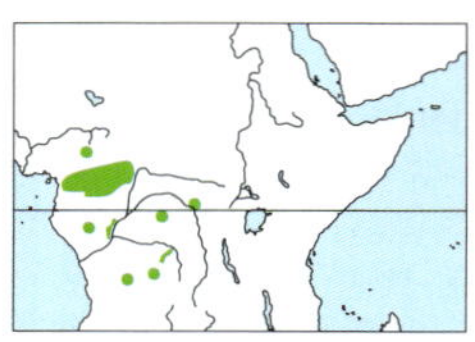

S CM, SW CF, N CG, also isolated locations in SE GA, CD and N AO

Congo Golden Mole
Huetia leucorhina
FR Taupe-dorée du Congo
DE Kongo-Goldmull
ES Topo dorado del Congo

DD
Monotypic

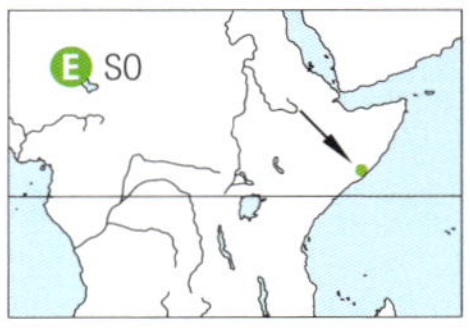

Somali Golden Mole
Huetia tytonis
FR Taupe-dorée de Somalie
DE Somalia-Goldmull
ES Topo dorado de Somalia

Known only from type locality (Guiohar)

EN
Monotypic
HB 11.1–13.2 cm
W 39–70 g

Gunning's Golden Mole
Neamblysomus gunningi
FR Taupe-dorée de Gunning
DE Gunning-Goldmull
ES Topo dorado de Gunning

Highly restricted and fragmented range in N Drakensberg (Limpopo Prov)

Genus *NEAMBLYSOMUS*

(2 species)

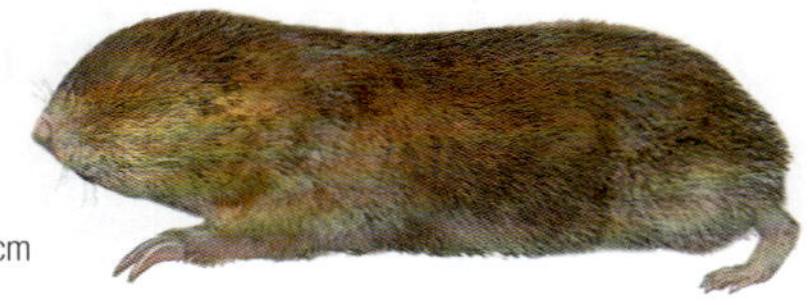

EN
Monotypic
HB 9.2–11.1 cm
W 21–46 g

Juliana's Golden Mole
Neamblysomus julianae
FR Taupe-dorée de Juliana
DE Juliana-Goldmull
ES Topo dorado de Juliana

Small and separated locations (SE Pretoria, Modimolle, SW Kruger NP)

Genus *CARPITALPA*

(1 species)

VU Monotypic
HB 11.5–14.1 cm
W 43–70 g

850–2000 m

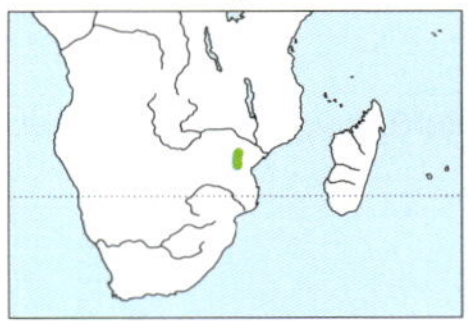

Arends's Golden Mole
Carpitalpa arendsi
FR Taupe-dorée d'Arends
DE Arends-Goldmull
ES Topo dorado de Arends

E ZW (Eastern Highlands), narrowly in W MZ

LC
4 sspp
HB 8.2–13.5 cm
W 22–54 g

ZA (Western Cape, Eastern Cape, SW KwaZulu-Natal, extreme E Free State, Mpumalanga), LS

Sclater's Golden Mole
Chlorotalpa sclateri
FR Taupe-dorée de Sclater
DE Sclater-Goldmull
ES Topo dorado de Sclater

Genus *CHLOROTALPA*

(2 species)

VU Monotypic
HB 9.5–13 cm
W 20–41 g

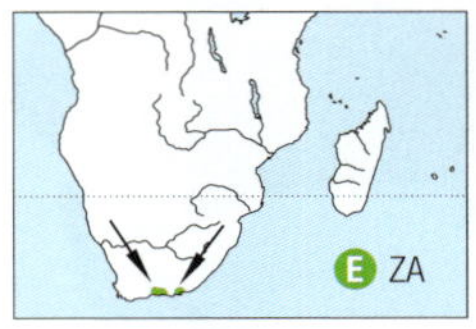

Duthie's Golden Mole
Chlorotalpa duthieae
FR Taupe-dorée de Duthie
DE Duthie-Goldmull
ES Topo dorado de Duthie

Coast of Western Cape (from George) to Eastern Cape (Port Elizabeth)

Genus *CHRYSOCHLORIS*

(3 species)

1700–3500 m

LC 5 sspp
HB 10.3–12.3 cm
W c. 42 g

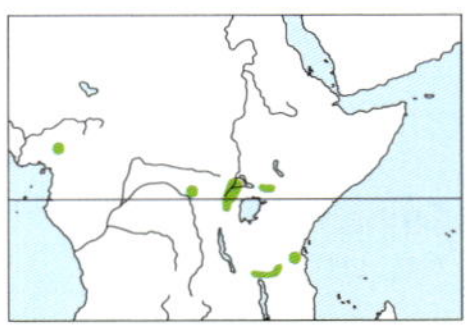

Stuhlmann's Golden Mole
Chrysochloris stuhlmanni
FR Taupe-dorée de Stuhlmann
DE Stuhlmann-Goldmull
ES Topo dorado de Stuhlmann

Patchily from CM to TZ (Mt Oku, Yambuya & Kisangani, Rwenzori Mts, Mt Elgon, Cherangani Hills, Uluguru Mts, Rungwe)

LC
Monotypic
HB 9.4–11.5 cm

Cape Golden Mole
Chrysochloris asiatica
FR Taupe-dorée du Cap
DE Kap-Goldmull
ES Topo dorado de El Cabo

Northern Cape, Western Cape

DD
Monotypic
HB c. 10.6 cm

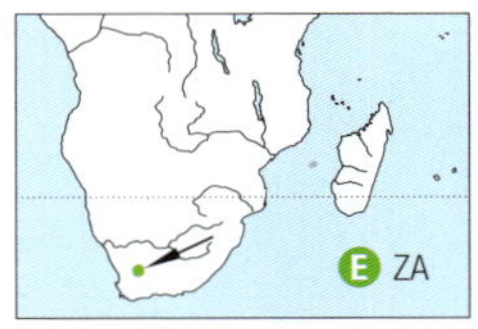

Visagie's Golden Mole
Chrysochloris visagiei
FR Taupe-dorée de Visagie
DE Visagie-Goldmull
ES Topo dorado de Visagie

Northern Cape (known only from type locality)

Genus *CHRYSOSPALAX*

(2 species)

VU 6 sspp
HB 12.7–17.5 cm
W 93–160 g

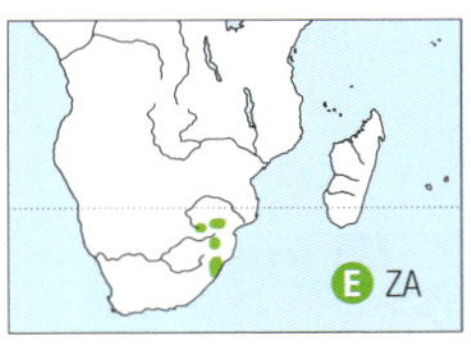

Rough-haired Golden Mole
Chrysospalax villosa
FR Taupe-dorée à poil dur
DE Rauhaar-Goldmull
ES Topo dorado de pelaje áspero

E Eastern Cape, KwaZulu-Natal, Gauteng, NC & S Mpumalanga

EN Monotypic
HB 21–23 cm
W 410–500 g

Giant Golden Mole
Chrysospalax trevelyani
FR Grande Taupe-dorée
DE Riesengoldmull
ES Topo dorado gigante

Eastern Cape (between East London and Port St Johns)

Genus *CRYPTOCHLORIS*

(2 species)

CR(PE)
Monotypic
HB 8.6–9.2 cm

De Winton's Golden Mole
Cryptochloris wintoni
FR Taupe-dorée de De Winton
DE De Winton-Goldmull
ES Topo dorado de De Winton

NW Northern Cape (known only from type locality)

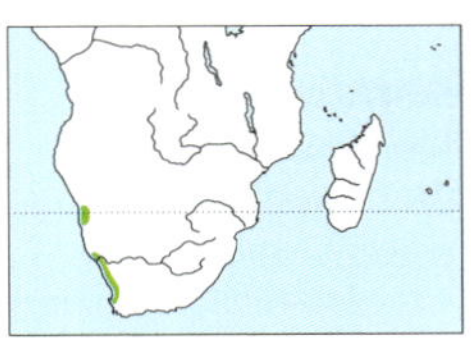

Grant's Golden Mole
Eremitalpa granti
FR Taupe-dorée de Grant
DE Wüstengoldmull
ES Topo dorado de Grant

W ZA (Strandveld Succulent Karoo), W NA (Sossusvlei, Namib Desert)

Genus *EREMITALPA*

(1 species)

LC
2 sspp
HB 6–8.6 cm
W 15–30 g

Van Zyl's Golden Mole
Cryptochloris zyli
FR Taupe-dorée de van Zyl
DE Van-Zyl-Goldmull
ES Topo dorado de van Zyl

EN
Monotypic
HB c. 8.2 cm

Northern Cape (Groenriviermond), Western Cape (Lamberts Bay); possibly N along coast to R Olifants

Order MACROSCELIDEA (1 family)

Family MACROSCELIDIDAE – Sengis

6 genera, 20 species. Xeric to mesic forests and woodlands. Palearctic and Afrotropical Regions.

Genus *RHYNCHOCYON*

(5 species)

Golden-rumped Sengi
Rhynchocyon chrysopygus
FR Sengi à croupe dorée
DE Goldsteiß-Rüsselhündchen
ES Sengi de grupas doradas

EN Monotypic
HB 22–30 cm T 21–27 cm
W 410–690 g

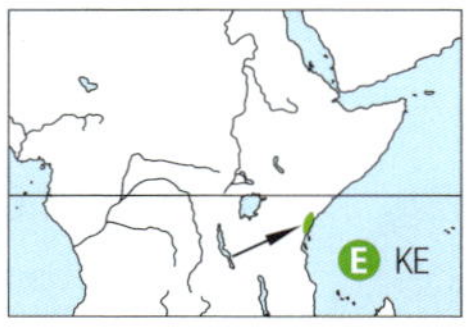

Coast from R Tana S to near Mombasa

Chequered Sengi
Rhynchocyon cirnei
FR Sengi de Cirne
DE Geflecktes Rüsselhündchen
ES Sengi estriado

LC 5 sspp
HB 23–27 cm
T 19.6–25 cm
W 320–450 g

0–2300 m

SE DC, S & SW TZ, NE ZM, MW, MZ

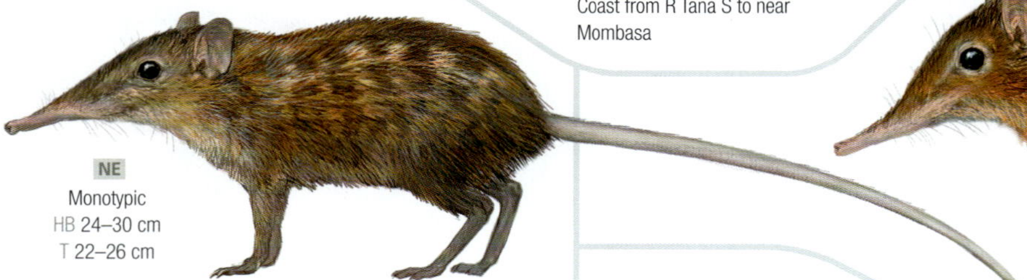

NE
Monotypic
HB 24–30 cm
T 22–26 cm

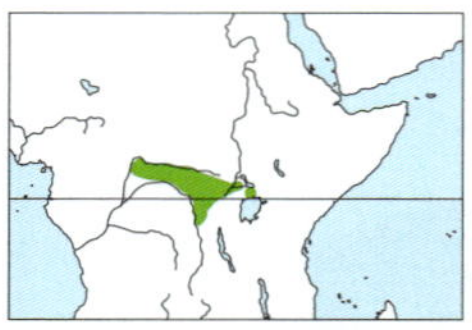

White-tailed Sengi
Rhynchocyon stuhlmanni
FR Sengi de Stuhlmann
DE Dunkles Rüsselhündchen
ES Sengi de cola blanca

N DC, UG; perhaps S CF (near Bangui)

Black-and-rufous Sengi
Rhynchocyon petersi
FR Sengi de Peters
DE Rotschulter-Rüsselhündchen
ES Sengi de Peters

LC 2 sspp
HB 25–29 cm T 22–25 cm
W c. 540 g

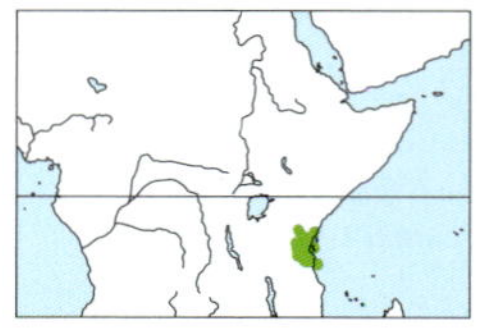

SE KE, E TZ (Eastern Arc Mts and coastal forests, Zanzibar Arch)

Gray-faced Sengi
Rhynchocyon udzungwensis
FR Sengi à face grise
DE Graugesicht-Rüsselhündchen
ES Sengi de cara gris

1000–2300 m

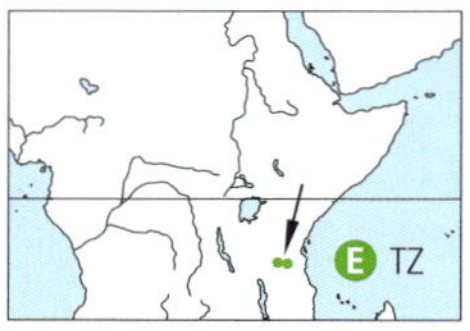

Udzungwa Mts of Eastern Arc Mts

VU
Monotypic
HB 30–32 cm
T 24–26 cm
W 658–750 g

Genus *PETROSALTATOR*

(1 species)

North African Sengi
Petrosaltator rozeti
FR Sengi de Rozet
DE Nordafrikanischer Rüsselspringer
ES Sengi norteafricano

LC 2 sspp
HB 10–12 cm
T 10.5–13.7 cm
W 40–55 g

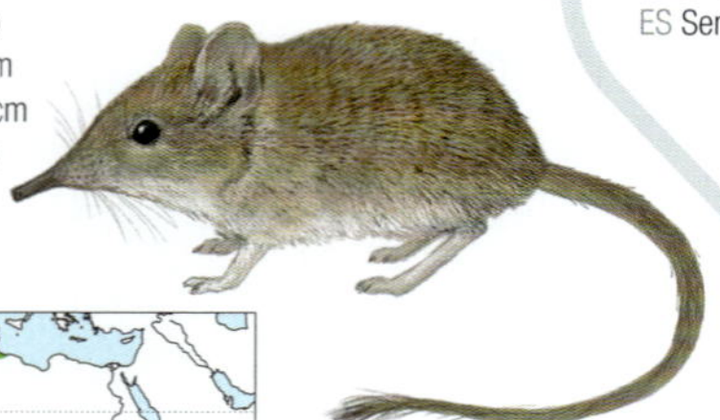

0–2700 m

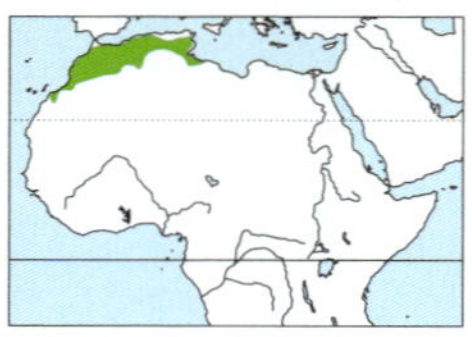

Atlas Range in EH, MA, DZ, TN, W LY

Genus *GALEGEESKA*

(2 species)

Rufous Sengi
Galegeeska rufescens
FR Sengi roux
DE Rotbraune Elefantenspitzmaus
ES Sengi bermejo

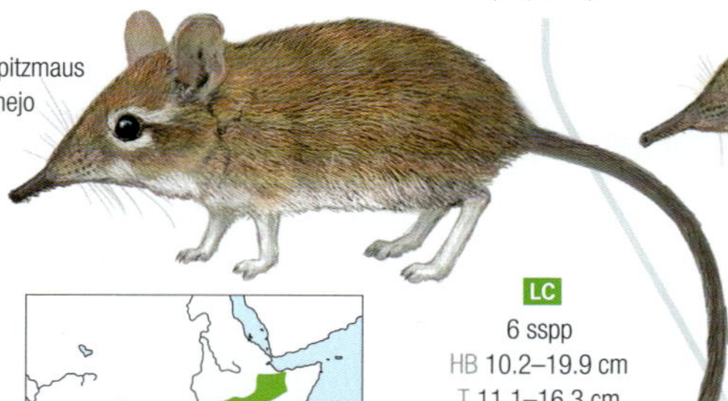

LC
6 sspp
HB 10.2–19.9 cm
T 11.1–16.3 cm
W 47–70 g

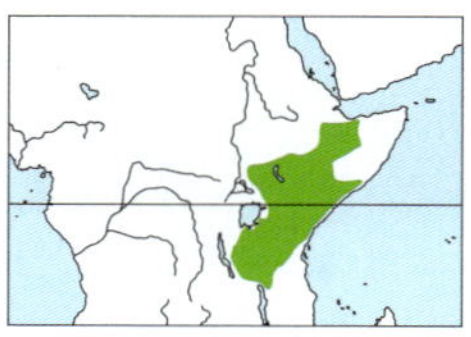

S & E ET, N & S SO, SE SS, UG, KE, TZ

Somali Sengi
Galegeeska revoilii
FR Sengi de Somalie
DE Somalia-Elefantenspitzmaus
ES Sengi de Somalia

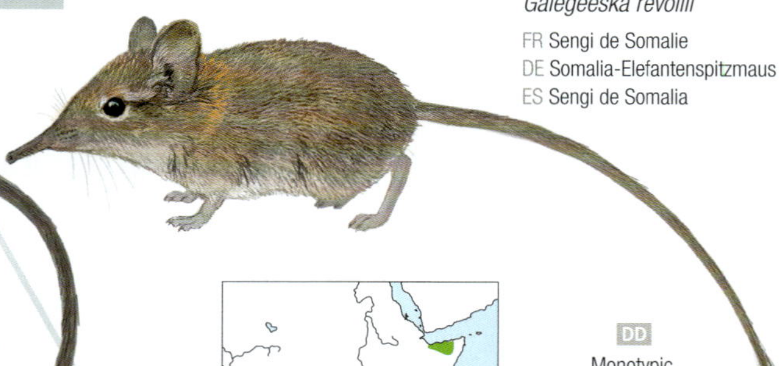

DD
Monotypic
HB 12.2–14.8 cm
T 14.4–16.7 cm

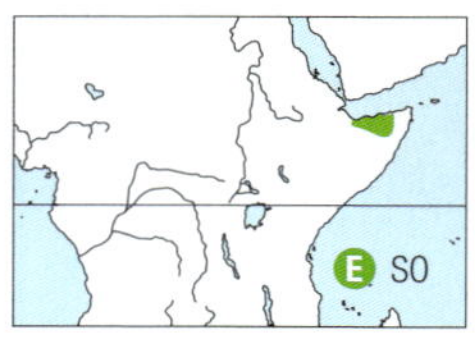

Somaliland, W Nugaal

Genus *MACROSCELIDES*

(3 species)

LC
Monotypic
HB 9.4–10.9 cm
T 9.8–13.1 cm
W 22–39 g

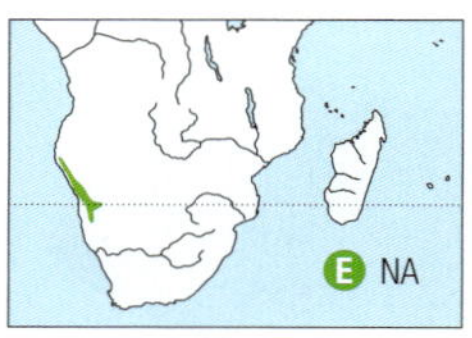

Namib Round-eared Sengi
Macroscelides flavicaudatus
FR Sengi de Namibie
DE Namib-Kurzohrrüsselspringer
ES Sengi de orejas redondas de Namib

Namib Desert, Pro-Namib

LC
Monotypic
HB 10.4–11.5 cm
T 10.7–13.4 cm
W 26–38 g

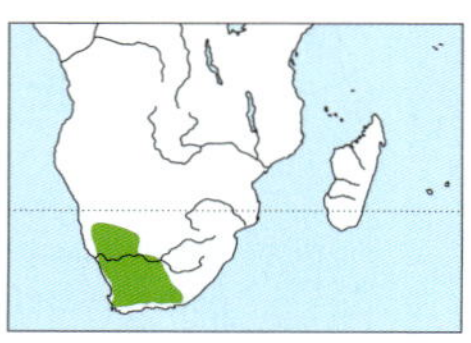

S NA, W ZA (Northern Cape, Western Cape, Eastern Cape), extreme SW BW

Karoo Round-eared Sengi
Macroscelides proboscideus
FR Sengi à oreilles courtes
DE Kurzohrrüsselspringer
ES Sengi de orejas redondas de Karoo

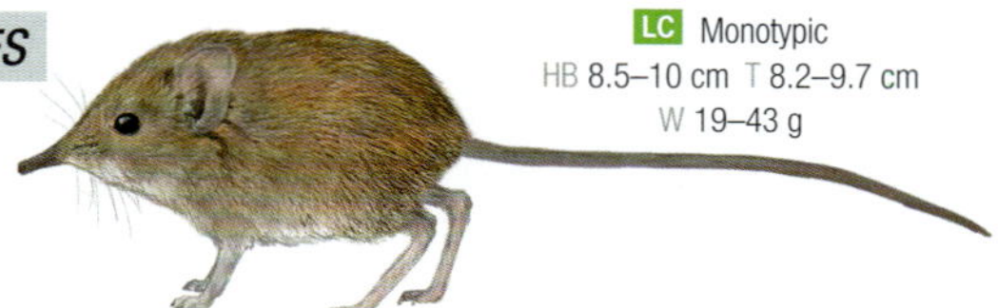

LC Monotypic
HB 8.5–10 cm T 8.2–9.7 cm
W 19–43 g

340–860 m

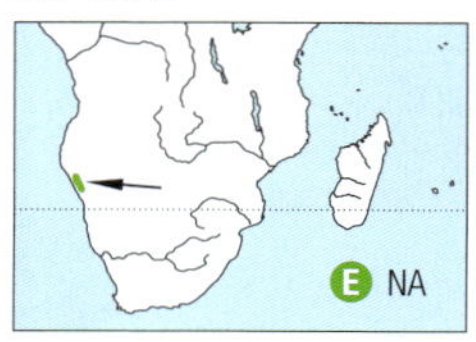

Etendeka Formation N & S of R Huab

Etendeka Round-eared Sengi
Macroscelides micus
FR Sengi de l'Etendeka
DE Etendeka-Kurzohrrüsselspringer
ES Sengi de orejas redondas de Etendeka

Genus *PETRODROMUS*

(1 species)

LC 10 sspp
HB 16.3–21 cm
T 15.6–18.7 cm
W 120–250 g

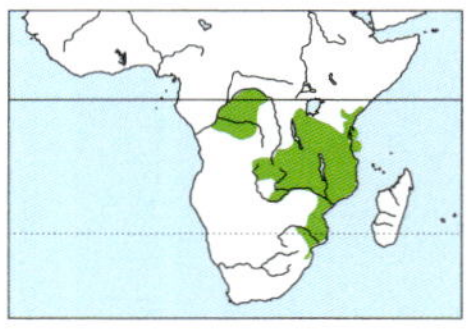

Four-toed Sengi
Petrodromus tetradactylus

FR Sengi à quatre doigts
DE Rüsselratte
ES Sengi de cuatro dedos

DC, AO, SE KE, RW, TZ (with Zanzibar Arch), E ZM, MW, MZ, E ZW, NE ZA; perhaps BI

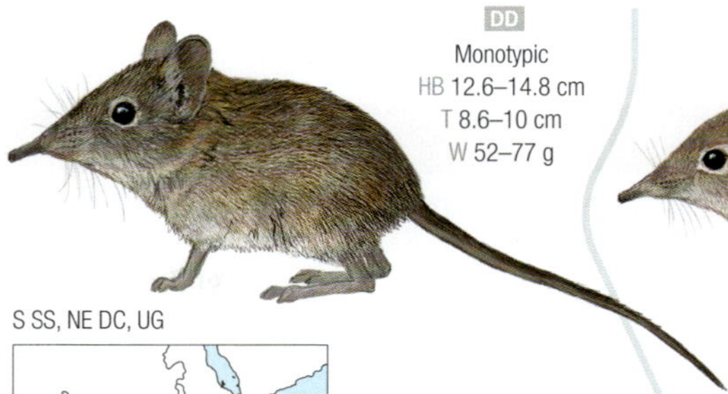

DD
Monotypic
HB 12.6–14.8 cm
T 8.6–10 cm
W 52–77 g

S SS, NE DC, UG

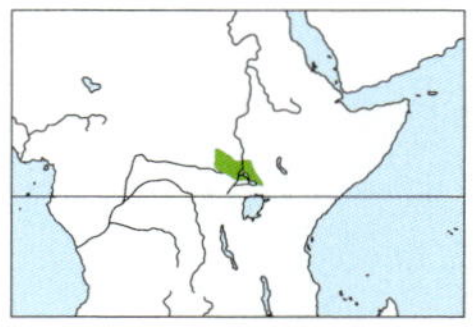

Dusky-footed Sengi
Elephantulus fuscipes

FR Sengi à pattes sombres
DE Dunkelfuß-Elefantenspitzmaus
ES Sengi de pies oscuros

Short-snouted Sengi
Elephantulus brachyrhynchus

FR Sengi à nez court
DE Kurznasen-Elefantenspitzmaus
ES Sengi de hocico corto

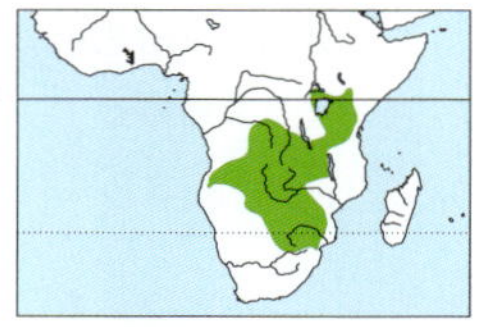

UG, KE, TZ, S DC, AO, ZM, MW, NA, N & E BW, ZW, S MZ, NE ZA, N SW

LC
Monotypic
HB 9.9–12.7 cm
T 9.5–10.8 cm
W 31–46 g

Genus *ELEPHANTULUS*

(8 species)

DD Monotypic
HB 10.5–12.5 cm
T 8.6–11.6 cm
W c. 45 g

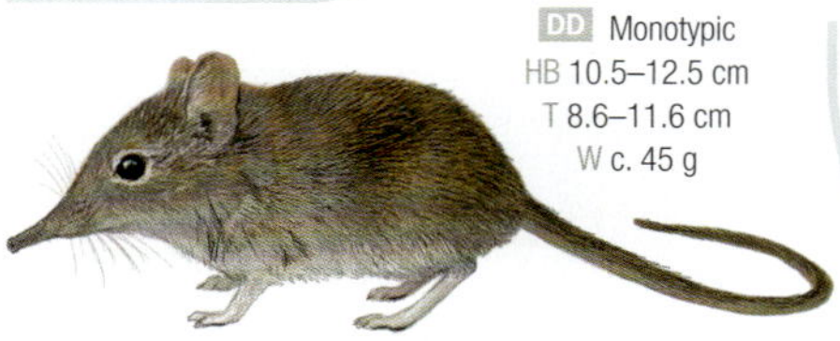

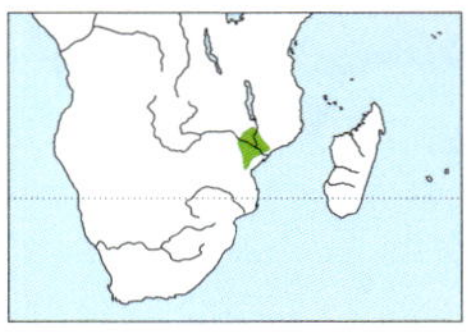

Dusky Sengi
Elephantulus fuscus

FR Sengi foncé
DE Dunkle Elefantenspitzmaus
ES Sengi oscuro

S MW, C MZ; perhaps SE ZM

LC Monotypic
HB 11.6–12.3 cm T 10.2–13.1 cm
W 34–62 g

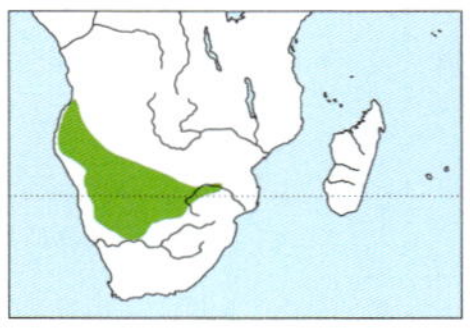

Bushveld Sengi
Elephantulus intufi

FR Sengi du bushveld
DE Trockenland-Elefantenspitzmaus
ES Sengi de Bushveld

SW AO, NA, BW, N & NE ZA, extreme S ZW

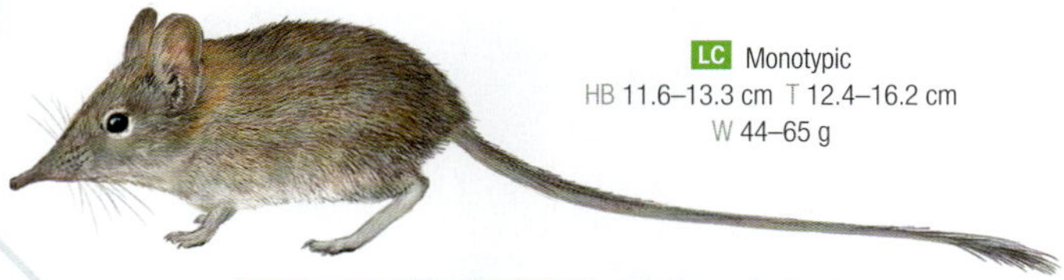

LC Monotypic
HB 11.6–13.3 cm T 12.4–16.2 cm
W 44–65 g

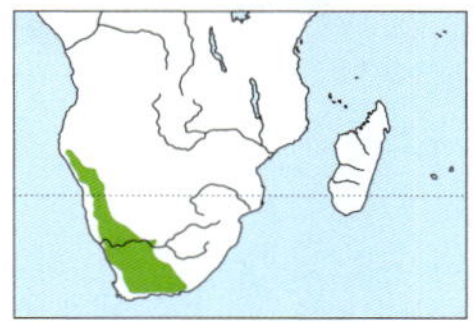

Western Rock Sengi
Elephantulus rupestris

FR Sengi des rochers
DE Westliche Elefantenspitzmaus
ES Sengi de roca occidental

W & S NA, W ZA (Northern Cape, Western Cape, Eastern Cape)

LC Monotypic
HB 11.4–13.3 cm
T 12.5–15.9 cm
W 48–67 g

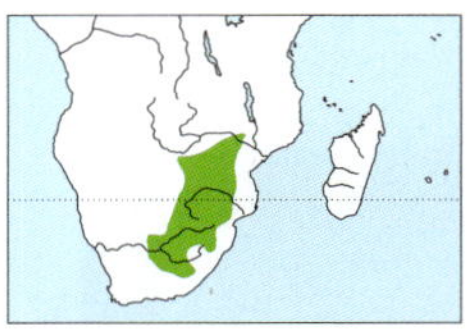

Eastern Rock Sengi
Elephantulus myurus

FR Sengi du Limpopo
DE Östliche Elefantenspitzmaus
ES Sengi de roca oriental

ZW, W MZ, E BW, N, C & E ZA, N SW; perhaps LS

LC Monotypic
HB 9.4–12.5 cm
T 11.7–16 cm
W 37–83 g

Northern Cape, Western Cape, Eastern Cape

Cape Rock Sengi
Elephantulus edwardii

FR Sengi du Cap
DE Kap-Elefantenspitzmaus
ES Sengi de roca de El Cabo

Karoo Rock Sengi
Elephantulus pilicaudus

FR Sengi du Karoo
DE Karoo-Elefantenspitzmaus
ES Sengi de roca de Karoo

DD
Monotypic
HB 11.1–13.5 cm
T 11.2–13.1 cm
W 38–59 g

below 1300 m

Upper & Lower Karoo of Nama Karoo biome

Order TUBULIDENTATA (1 family)

Family ORYCTEROPODIDAE – Aardvark

1 genus, 1 species. Semi-arid deserts to savannas and tropical and seasonal forests. Afrotropical Region.

Aardvark
Orycteropus afer

FR Oryctérope
DE Erdferkel
ES Cerdo hormiguero

Genus *ORYCTEROPUS*

(1 species)

below 3200 m

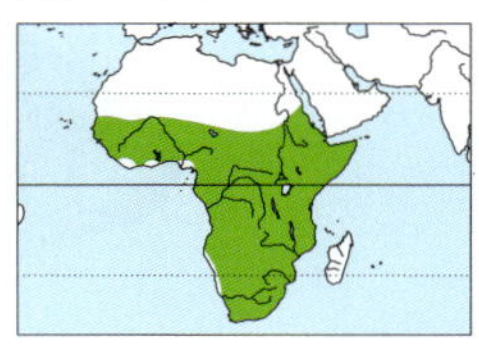

Sub-Saharan Africa, from MR E to ER, S to ZA

LC
18 sspp
HB 94–142 cm
T 44–63 cm
W 40–65 kg

Order HYRACOIDEA (1 family)

Family PROCAVIIDAE – Hyraxes

3 genera, 6 species. Lowland to montane tropical forests, woodlands, savannas, and deserts. Palearctic and Afrotropical Regions.

Genus *PROCAVIA*

(1 species)

Rock Hyrax
Procavia capensis
FR Daman des rochers
DE Klippschliefer
ES Damán de rocas

ssp *syriaca*

ssp *capensis*

LC
17 sspp
HB 39–58 cm
W 1.8–5.4 kg

0–4000 m

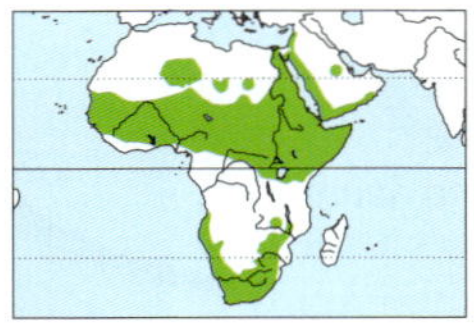

N, C & S Africa, Middle East (Syria S to Arabian Pen)

Genus *HETEROHYRAX*

(1 species)

Bush Hyrax
Heterohyrax brucei
FR Daman de Bruce
DE Buschschliefer
ES Damán de arbustos

LC
24 sspp
HB 32–56 cm
W 1.3–3.6 kg

below 3800 m

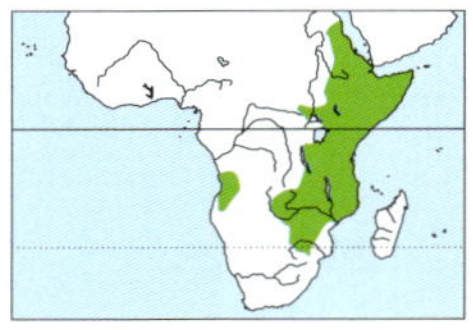

Africa from NE SD and Horn of Africa, S to ZA (Limpopo & Mpumalanga); isolated populations in AO

Genus *DENDROHYRAX*

(4 species)

Western Tree Hyrax
Dendrohyrax dorsalis
FR Daman de Beecroft
DE Westlicher Baumschliefer
ES Damán arborícola occidental

ssp *nigricans*

LC
6 sspp
HB 44–57 cm
W 1.8–4.5 kg

ssp *dorsalis*

1500–3500 m

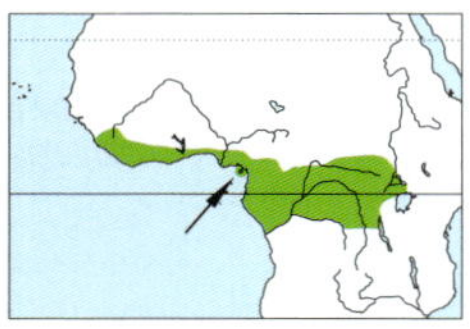

WC Africa, from Upper Guinea E to SW SS and UG (forest islands), Bioko I

Benin Tree Hyrax
Dendrohyrax interfluvialis
FR Daman du Bénin
DE Benin-Baumschliefer
ES Damán arborícola de Benín

NE
Monotypic
HB 50–55 cm

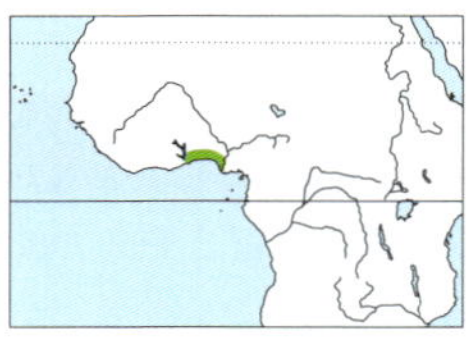

Known only between R Volta and R Niger (SE GH, S TG, S BJ, SW NG)

Southern Tree Hyrax
Dendrohyrax arboreus
FR Daman des arbres
DE Südlicher Baumschliefer
ES Damán arborícola meridional

LC
7 sspp
HB 32–60 cm
W 1.7–4.5 kg

0–4500 m

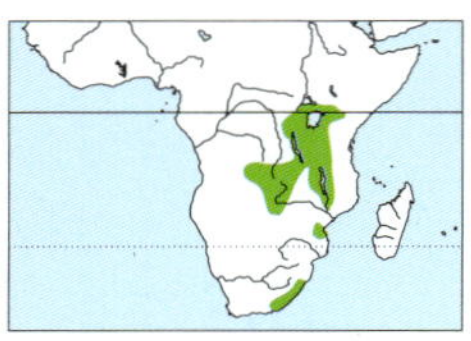

E & S DC, SW UG, RW, BI, KE, TZ, NE AO, ZM, MW, MZ, SE ZA

Eastern Tree Hyrax
Dendrohyrax validus
FR Daman du Kilimanjaro
DE Östlicher Baumschliefer
ES Damán arborícola oriental

NT
3 sspp
HB 32–60 cm
W 1.7–3 kg

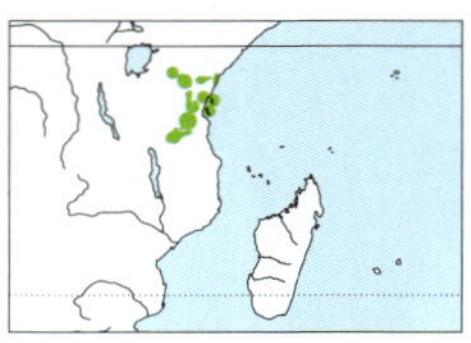

below c. 3500 m

S KE, N TZ: Eastern Arc Mts, Mt Kilimanjaro, Mt Meru, coast (with Zanzibar Arch)

Order PROBOSCIDEA (1 family)

Family ELEPHANTIDAE – Elephants

2 genera, 3 species. Most major habitats in Africa and wooded habitats in Asia. Afrotropical and Indo-Malayan Regions.

Genus *LOXODONTA*

(2 species)

African Savanna Elephant
Loxodonta africana

FR Éléphant de savane
DE Afrikanischer Savannenelefant
ES Elefante de sabana

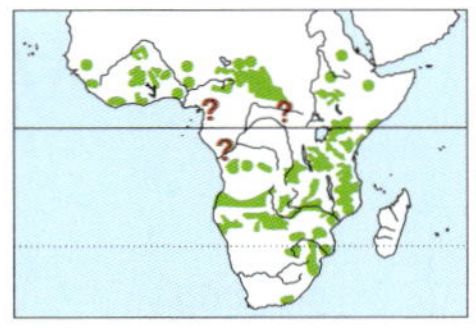

E & S Sub-Saharan Africa (from N CM, SS and ER, S to NA, BW and ZA); scattered in W (SN to NG)

EN
Monotypic
HB 600–750 cm
T 100–150 cm
W 2800–10000 kg

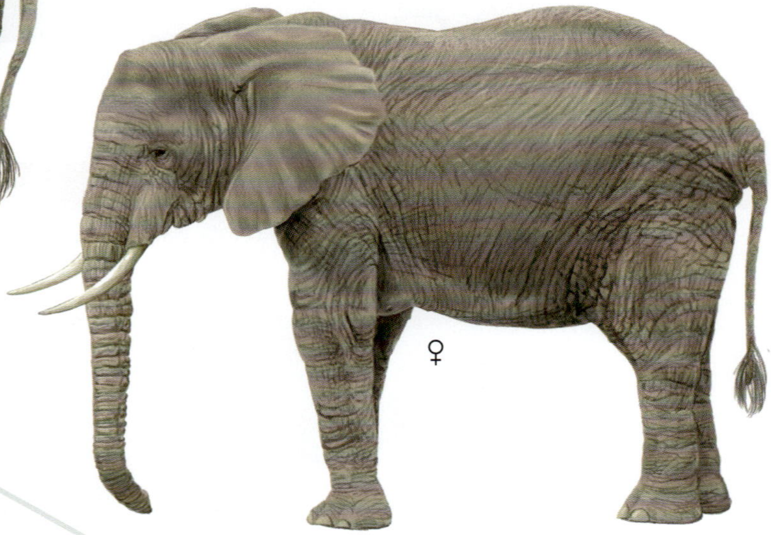

African Forest Elephant
Loxodonta cyclotis

FR Éléphant de forêt
DE Afrikanischer Waldelefant
ES Elefante de bosque

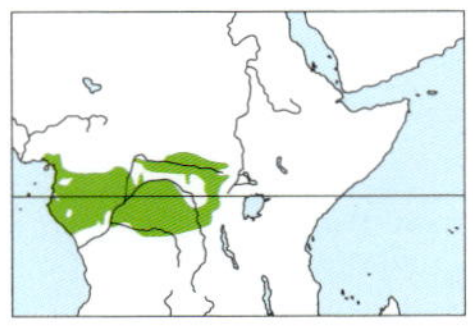

Congo Basin in C Africa (S CM, S CF, DC, GQ, GA, CG); range not well known

CR
Monotypic
HB 600–750 cm
T 100–150 cm
W 2700–6000 kg

Genus *ELEPHAS*

(1 species)

EN
3 sspp
HB 550–640 cm
T 120–150 cm
W 2700–6000 kg

Asian Elephant
Elephas maximus

FR Éléphant d'Asie
DE Asiatischer Elefant
ES Elefante asiático

0–3000 m

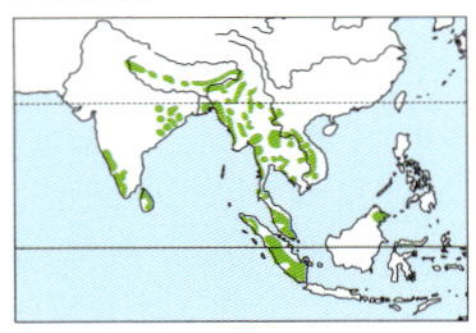

IN, LK, NP, BT, BD, S CN, MM, LA, KH, VN, TH, MY, E ID (Sumatra, NE Borneo)

Order SIRENIA (2 families)

Family DUGONGIDAE – Dugong and Steller's Sea Cow

1 extant genus, 1 extant species. Coastal and island waters with seagrass meadows. Subtropical and tropical Indo-West Pacific waters.

Genus *DUGONG*

(1 species)

Dugong
Dugong dugon
FR Dugong
DE Dugong
ES Dugón

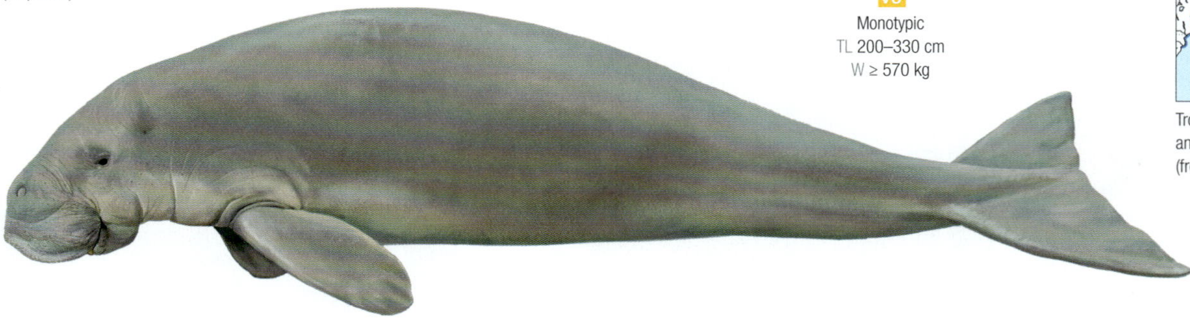

VU
Monotypic
TL 200–330 cm
W ≥ 570 kg

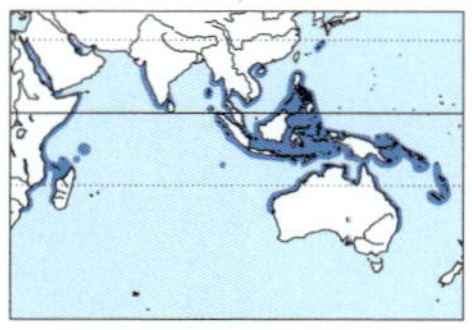

Tropical and subtropical coastal and island Indo-Pacific waters (from E Africa to VU)

Family TRICHECHIDAE – Manatees

1 genus, 3 species. Tropical and subtropical shallow coastal waters, estuaries, and river systems. Nearctic, Neotropical and Afrotropical Regions.

Genus *TRICHECHUS*

(3 species)

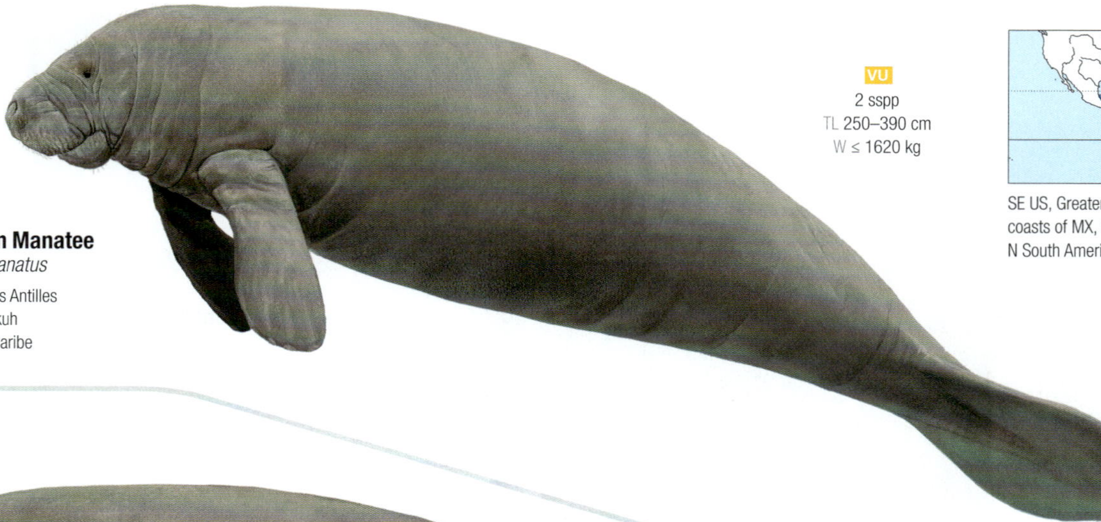

VU
2 sspp
TL 250–390 cm
W ≤ 1620 kg

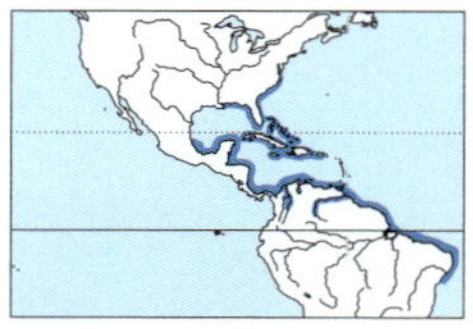

SE US, Greater Antilles, Caribbean coasts of MX, Central America and N South America (to E BR)

West Indian Manatee
Trichechus manatus
FR Lamantin des Antilles
DE Karibik-Seekuh
ES Manatí del Caribe

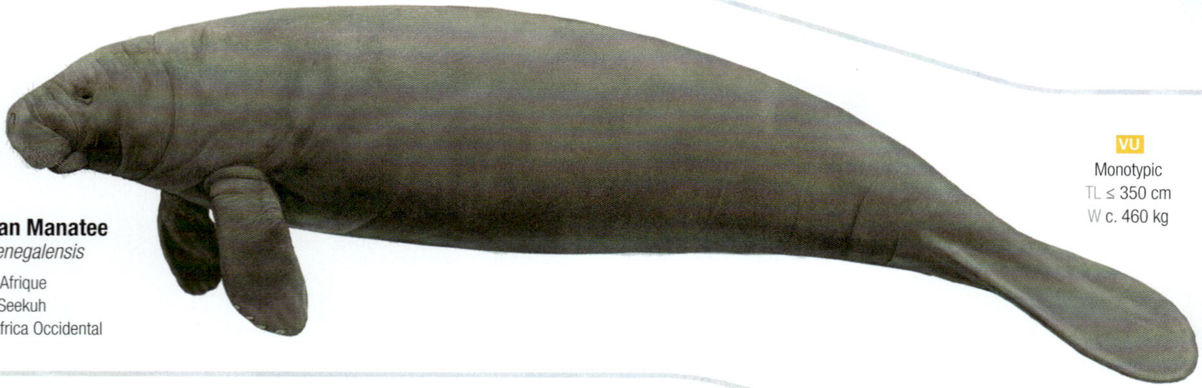

VU
Monotypic
TL ≤ 350 cm
W c. 460 kg

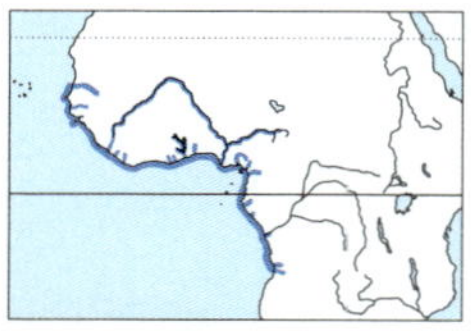

R Senegal (MR) S to R Longa (AO), and up large rivers, inland to Inner Niger Delta (ML) and L Tréné (TD)

West African Manatee
Trichechus senegalensis
FR Lamantin d'Afrique
DE Westafrika-Seekuh
ES Manatí de África Occidental

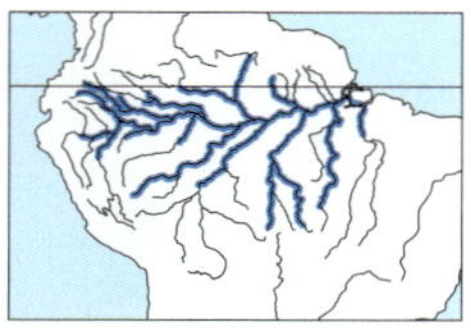

R Amazon Basin including its estuary (CO, BR, EC, PE)

VU
Monotypic
TL ≤ 300 cm
W c. 450 kg

Amazonian Manatee
Trichechus inunguis
FR Lamantin de l'Amazone
DE Amazonas-Seekuh
ES Manatí del Amazonas

Order CINGULATA (2 families)

Family DASYPODIDAE – Long-nosed Armadillos

1 genus, 8 species. Temperate and tropical forests, savannas, and riparian areas. Nearctic and Neotropical Regions.

Genus *DASYPUS*

(8 species)

East Amazonian Long-nosed Armadillo
Dasypus beniensis
FR Tatou du Rio Beni DE Ostamazonien-Gürteltier
ES Mulita del Amazonas oriental

NE
Monotypic
HB 55–56 cm
T 35–40 cm
W c. 12 kg

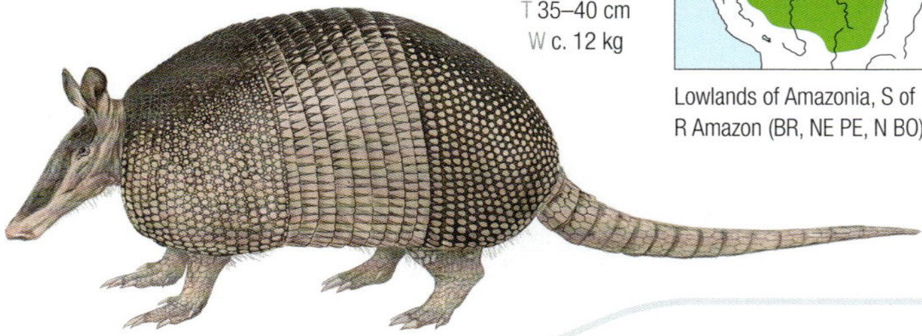

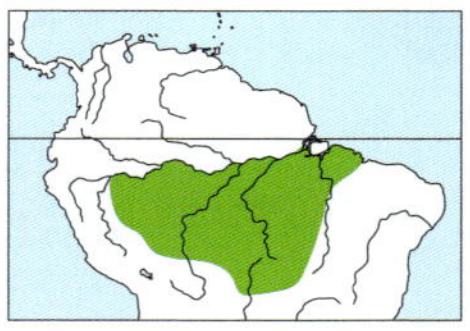

Lowlands of Amazonia, S of R Amazon (BR, NE PE, N BO)

West Amazonian Long-nosed Armadillo
Dasypus pastasae
FR Tatou du Pastaza
DE Westamazonien-Gürteltier
ES Mulita del Amazonas occidental

NE
Monotypic
HB 48–58 cm
T 37–48 cm
W 5.9–9.8 kg

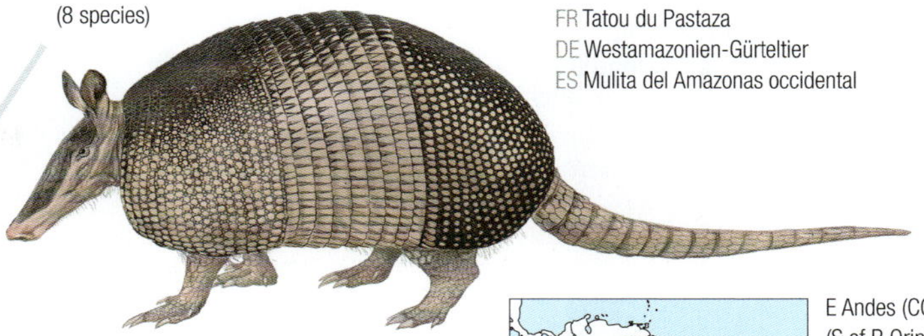

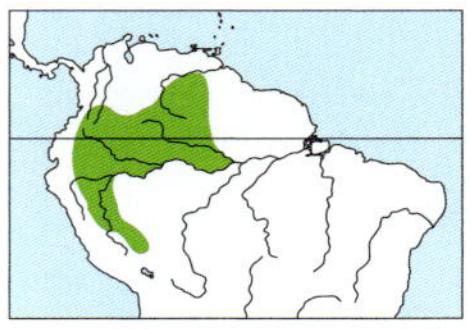

E Andes (CO, EC, PE), VE (S of R Orinoco), W Amazon Basin (BR, NW BO, N of R Madre de Dios)

Greater Long-nosed Armadillo
Dasypus kappleri
FR Tatou de Kappler DE Kappler-Gürteltier
ES Mulita de Kappler

LC
Monotypic
HB 50–90 cm
T 30–43 cm
W 4–15 kg

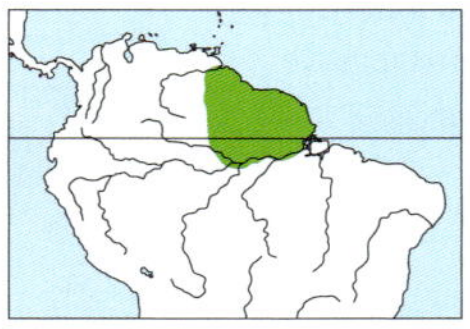

The Guiana Shield (E VE, Guianas, NE BR)

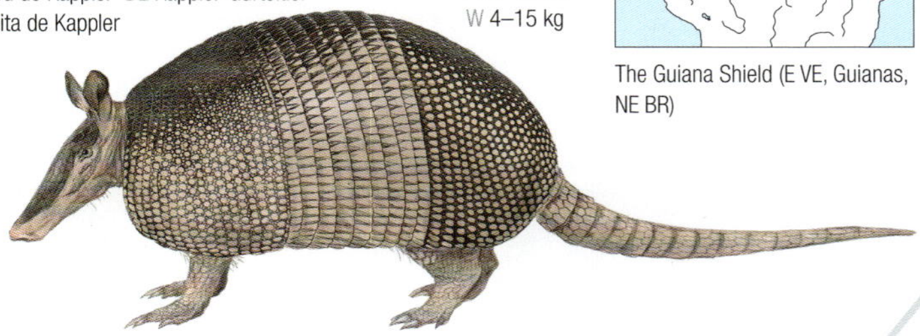

Brazilian Lesser Long-nosed Armadillo
Dasypus septemcinctus
FR Tatou à sept bandes
DE Siebenbinden-Gürteltier ES Mulita chica

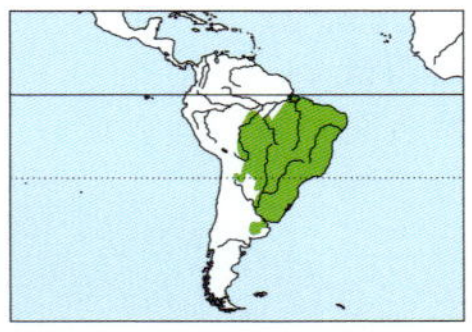

From lower Amazon Basin S to Rio Grande do Sul (BR) and UY, W to Gran Chaco (E BO, PY, N AR)

LC
3 sspp
HB 24–31 cm
T 12–17 cm
W 1–1.5 kg

ssp *hybridus*

ssp *septemcinctus*

Hairy Long-nosed Armadillo
Dasypus pilosus
FR Tatou à long museau
DE Pelzgürteltier
ES Mulita peluda

DD
Monotypic
HB 32–44 cm
T 24–31 cm
W 1–1.5 kg

2000–3500 m

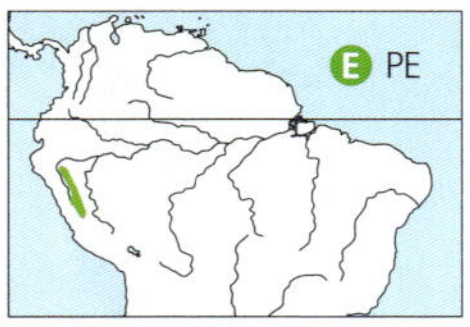

E PE

E Andean slope (Amazonas to Junín, perhaps Cajamarca)

Northern Long-nosed Armadillo
Dasypus sabanicola
FR Tatou des savanes
DE Savannen-Gürteltier
ES Mulita de Los Llanos

25–500 m

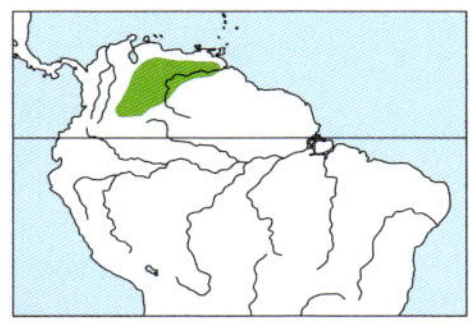

Llanos of CO and VE

NT Monotypic
HB 25–31 cm T 17–21 cm
W 1–2 kg

Yungas Lesser Long-nosed Armadillo
Dasypus mazzai
FR Tatou de Yepes
DE Yungas-Gürteltier
ES Mulita de Mazza

DD
Monotypic
HB c. 31 cm
T 18–23 cm
W 2–2.5 kg

440–1800 m

E AR

Known only from Jujuy and Salta prov

Nine-banded Armadillo
Dasypus novemcinctus
FR Tatou à neuf bandes DE Neunbinden-Gürteltier
ES Mulita de nueve bandas

LC
6 sspp
HB 36–57 cm
T 26–45 cm
W 3–6 kg

below 3000 m

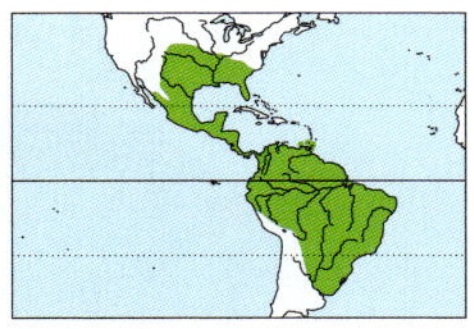

S & SE US throughout Central and South America to most of BR, BO, N & CE AR; some Caribbean islands

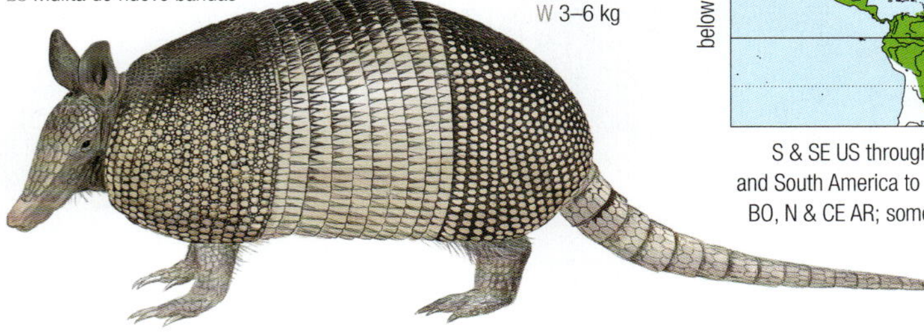

Family CHLAMYPHORIDAE – Chlamyphorid Armadillos

8 genera, 14 species. Temperate and tropical deserts, savannas, and forests. Neotropical Region.

Genus *EUPHRACTUS*

(1 species)

Six-banded Armadillo
Euphractus sexcinctus
FR Tatou à six bandes
DE Sechsbinden Gürteltier
ES Gualacate

LC
5 sspp
HB 40–50 cm
T 20–25 cm
W 3–7 kg

0–1600 m

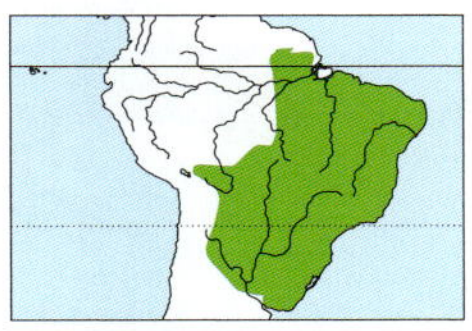

S SR, most of BR, BO, PY, UY, N AR (S to Catamarca)

Genus *CHAETOPHRACTUS*

(2 species)

LC 2 sspp
HB 20–30 cm T 8–11 cm
W 0.6–1.2 kg

0–4600 m

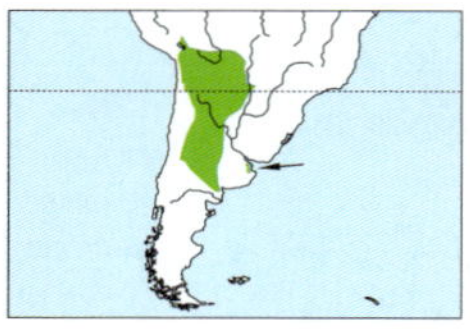

Screaming Hairy Armadillo
Chaetophractus vellerosus

FR Petit Tatou
DE Weißhaar-Gürteltier
ES Piche llorón

BO, NE CL, PY, AR (S to La Pampa, outlying population in E Buenos Aires); probably SE PE

LC
Monotypic
HB 26–40 cm
T 12–15 cm
W 2–5 kg

0–1500 m

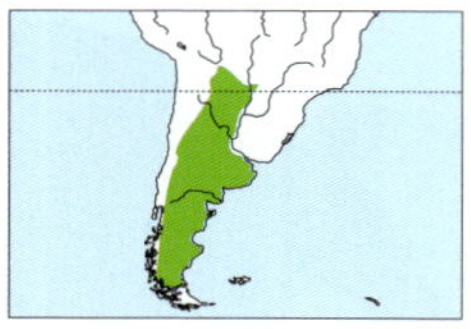

Large Hairy Armadillo
Chaetophractus villosus

FR Grand Tatou
DE Braunborsten-Gürteltier
ES Armadillo peludo

Gran Chaco (SE BO, PY, AR) and S CL (Aisén, Magallanes); introduced to Tierra del Fuego I

Genus *ZAEDYUS*

(1 species)

NT 2 sspp
HB 22–31 cm T 8–13 cm
W 0.7–1.5 kg

AR, SE CL 0–2500 m

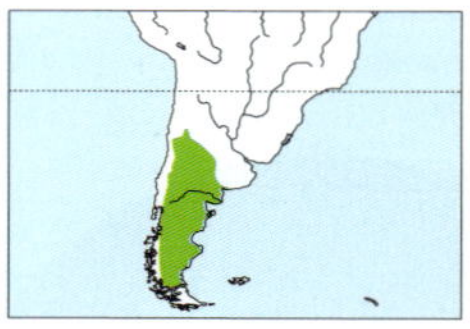

Pichi
Zaedyus pichiy

FR Pichi
DE Zwerggürteltier
ES Piche patagónico

Genus *CHLAMYPHORUS*

(1 species)

DD
Monotypic
HB 11–15 cm
T 2.5–3.5 cm
W 100–115 g

0–1500 m

From S Catamarca S to N Río Negro, SE to S Buenos Aires Prov

Pink Fairy Armadillo
Chlamyphorus truncatus

FR Tatou tronqué
DE Kleiner Gürtelmull
ES Pichiciego menor

Genus *CALYPTOPHRACTUS*

(1 species)

Greater Fairy Armadillo
Calyptophractus retusus

FR Tatou de Burmeister
DE Burmeister-Gürtelmull
ES Pichiciego chaqueño

DD
Monotypic
HB 12–17 cm
T 3–4 cm
W 90–130 g

Gran Chaco (C & SE BO, W PY, N AR, S to Santiago del Estero Prov)

Genus *PRIODONTES*

(1 species)

Giant Armadillo
Priodontes maximus

FR Tatou géant
DE Riesengürteltier
ES Armadillo gigante

VU
Monotypic
HB 75–100 cm
T 40–50 cm
W 20–60 kg

below c. 500 m

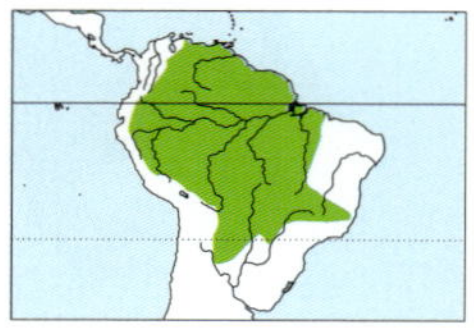

N & C South America E of Andes (CO, VE, Guianas, BR, EC, PE, BO, PY, N AR)

Genus *CABASSOUS*

(5 species)

Northern Naked-tailed Armadillo
Cabassous centralis

FR Tatou du Nord
DE Nördliches Nacktschwanzgürteltier
ES Armadillo de cola desnuda septentrional

DD
Monotypic
HB 30–38 cm
T 13–18 cm
W 2–3.5 kg

0–3000 m

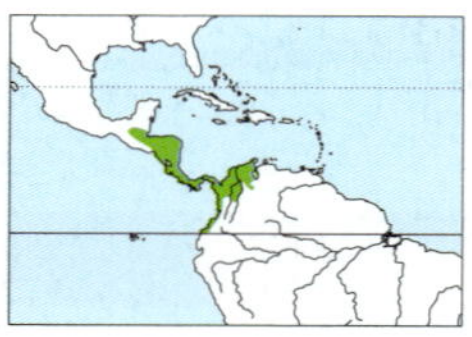

S MX (Chiapas) through Central America to NW VE, W CO and NW EC, W of Andes

Amazon Naked-tailed Armadillo
Cabassous unicinctus

FR Tatou à onze bandes
DE Südliches Nacktschwanzgürteltier
ES Armadillo de cola desnuda meridional

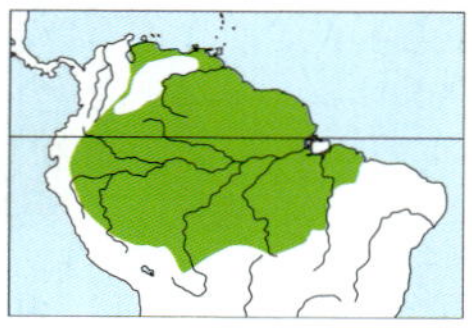

Amazon forest E of Andes in CO, VE, Guianas, EC, PE, BO and BR

LC Monotypic
HB 35–45 cm T 14–21 cm
W 2–3.8 kg

Cerrado and open areas of E BO, E PY (Amambay, San Pedro) & C BR (Mato Grosso, Mato Grosso do Sul)

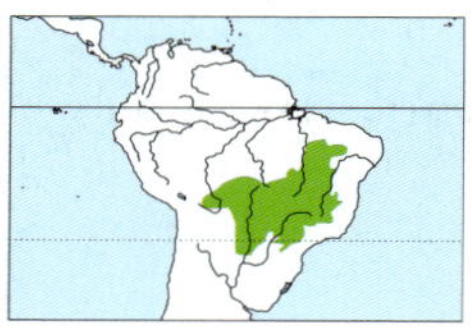

Cerrado Naked-tailed Armadillo
Cabassous squamicaudis

FR Tatou du Cerrado
DE Cerrado-Nacktschwanzgürteltier
ES Armadillo de cola escamosa

NE
Monotypic
HB 29–40 cm
T 9–14 cm
W 1.3–2.7 kg

Greater Naked-tailed Armadillo
Cabassous tatouay

FR Tatou à queue nue
DE Großes Nacktschwanzgürteltier
ES Armadillo de cola desnuda mayor

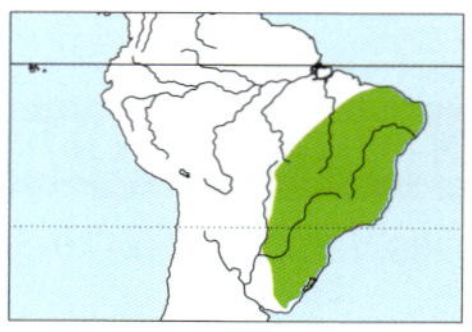

E & S BR, E PY, NE AR (Misiones, N Corrientes), C & E UY (S to N Maldonado Dep)

LC
Monotypic
HB 36–49 cm
T 15–20 cm
W 3.4–6.4 kg

NT
Monotypic
HB 30–35 cm
T 9–10 cm
W 1–1.5 kg

Chacoan Naked-tailed Armadillo
Cabassous chacoensis
FR Tatou du Chaco
DE Chaco-Nacktschwanzgürteltier
ES Armadillo de cola desnuda del Chaco

Gran Chaco of W PY and N AR (Formosa S to San Luis); unconfirmed in SE BO and BR (Mato Grosso do Sul)

VU
Monotypic
HB 23–25 cm
T 5–7 cm
W 1–1.8 kg

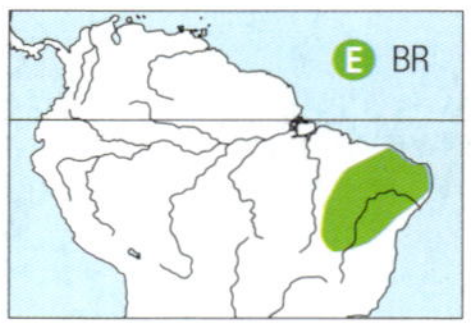

Brazilian Three-banded Armadillo
Tolypeutes tricinctus
FR Tatou du Brésil
DE Nördliches Kugelgürteltier
ES Tatú bola

NE BR: from Maranhão E to Rio Grande do Norte, S to extreme NE Goiás and extreme NW Minas Gerais

Genus *TOLYPEUTES*

(2 species)

NT
Monotypic
HB 20–25 cm
T 5–7 cm
W 1–2 kg

0–800 m

Southern Three-banded Armadillo
Tolypeutes matacus
FR Tatou du Sud
DE Südliches Kugelgürteltier
ES Mataco bola

E BO, SW BR, Gran Chaco of PY, N & C AR

Order PILOSA (4 families)

Family MYRMECOPHAGIDAE – Anteaters

2 genera, 3 species. Grasslands, savannas, and forests, avoiding the high Andes. Nearctic and Neotropical Regions.

Genus *MYRMECOPHAGA*

(1 species)

From HN throughout Central America to C South America, E of Andes; W of Andes in CO, NW EC & VE

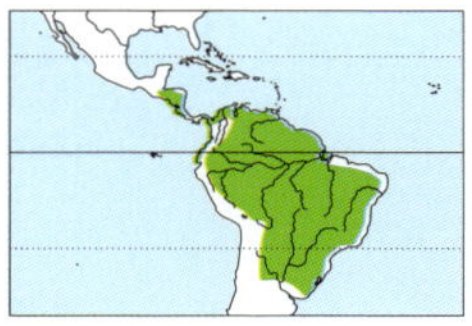

Giant Anteater
Myrmecophaga tridactyla
FR Tamanoir
DE Großer Ameisenbär
ES Oso hormiguero gigante

VU
3 sspp
HB 100–140 cm
T 60–90 cm
W 22–45 kg

Northern Tamandua
Tamandua mexicana
FR Tamandua du Mexique
DE Nördlicher Tamandua
ES Oso melero septentrional

LC
4 sspp
HB c. 56 cm
T 40–67 cm
W 3–6 kg

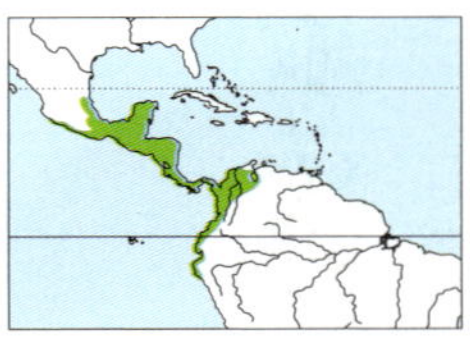

S MX (from Colima and San Luis Potosí) throughout Central America to N & W CO, NW VE, W EC and NW PE

Genus *TAMANDUA*

(2 species)

Southern Tamandua
Tamandua tetradactyla
FR Tamandua austral
DE Südlicher Tamandua
ES Oso melero meridional

E CO, VE, TT (Trinidad I), Guianas, BR, E EC, E PE, BO, PY, N AR, N UY

LC
4 sspp
HB 47–77 cm
T 40–67 cm
W 3.5–8.4 kg

Family CYCLOPEDIDAE – Silky Anteaters

1 genus, 7 species. Rainforests to mangrove forests. Nearctic (marginally) and Neotropical Regions.

Genus *CYCLOPES*

(7 species)

NE
Monotypic
HB 18.5–21 cm
T 17–22 cm
W 155–275 g

Central American Silky Anteater
Cyclopes dorsalis

FR Myrmidon du Costa Rica
DE Mittelamerika-Zwergameisenbär
ES Oso hormiguero sedoso de Centroamérica

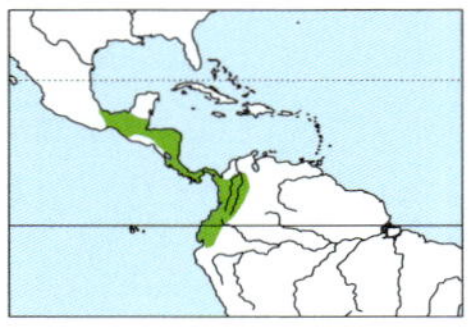

S MX to inter-Andean valleys of CO and along Pacific coast of CO and EC

LC
Monotypic
HB c. 20 cm
T 16.5–29 cm
W c. 300 g

E CO, E & S VE, TT (Trinidad I), Guianas, N & NE BR

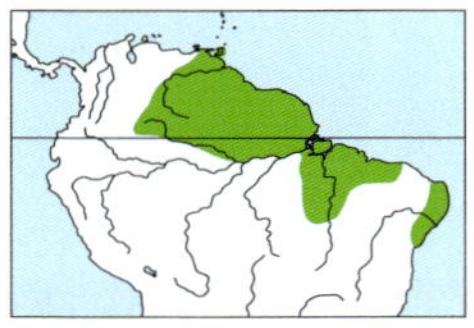

Common Silky Anteater
Cyclopes didactylus

FR Myrmidon didactyle
DE Gemeiner Zwergameisenbär
ES Oso hormiguero sedoso común

Rio Negro Silky Anteater
Cyclopes ida

FR Myrmidon du Rio Negro
DE Rio-Negro-Zwergameisenbär
ES Oso hormiguero sedoso de Río Negro

NE
Monotypic

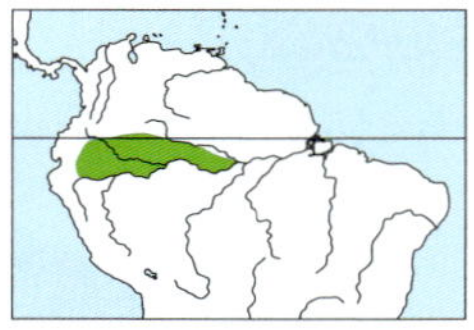

W & S of R Negro, S to R Amazon; recorded in E Andes of CO; N limit unknown

NE
Monotypic

Thomas's Silky Anteater
Cyclopes thomasi

FR Myrmidon de Thomas
DE Thomas-Zwergameisenbär
ES Oso hormiguero sedoso de Thomas

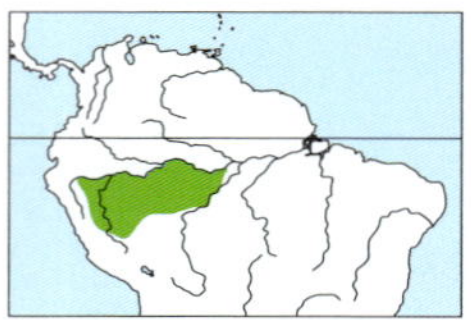

EC PE (R Juruá SW to R Ucayali) E to BR (R Madeira); E limit uncertain

Red Silky Anteater
Cyclopes rufus

FR Myrmidon roux
DE Rötlicher Zwergameisenbär
ES Oso hormiguero sedoso rojo

NE
Monotypic

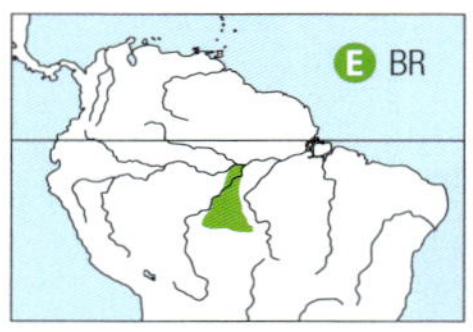

Interfluve between R Madeira and R Aripuanã; N limit possibly R Amazon, S limit unknown

NE
Monotypic

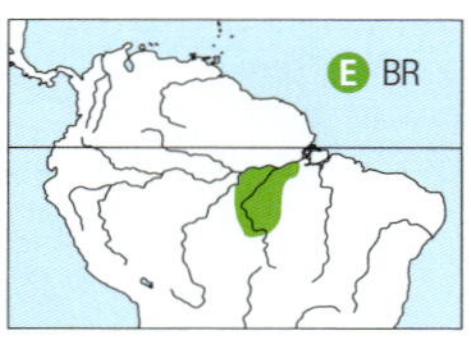

Between R Madeira and R Xingu, N to R Amazon; S limit unknown

Xingu Silky Anteater
Cyclopes xinguensis

FR Myrmidon du Xingu
DE Xingu-Zwergameisenbär
ES Oso hormiguero sedoso de Xingu

Amboro Silky Anteater
Cyclopes catellus

FR Myrmidon de Bolivie
DE Bolivien-Zwergameisenbär
ES Oso hormiguero sedoso de Amboró

Probably Andean slopes forests

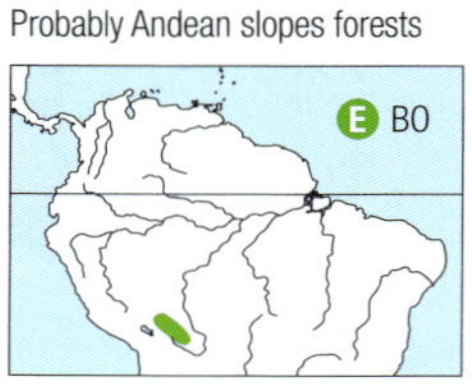

NE
Monotypic

Family CHOELOPODIDAE – Two-toed Sloths

1 genus, 2 species. Tropical moist lowland and montane forests. Neotropical Region.

Genus *CHOLOEPUS*

(2 species)

Hoffmann's Two-toed Sloth
Choloepus hoffmanni

FR Paresseux de Hoffmann
DE Hoffmann-Zweifingerfaultier
ES Perezoso didáctilo de Hoffmann

LC
5 sspp
HB 50–70 cm
T 1.4–3 cm
W 2.7–10 kg

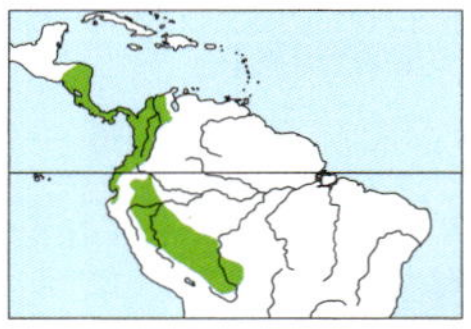

S HN to CO, NW VE, EC, SW BR Amazon (Amazonas, Acre), PE, N BO; range in BR unclear

Linnaeus's Two-toed Sloth
Choloepus didactylus

FR Paresseux à deux doigts
DE Eigentliches Zweifingerfaultier
ES Perezoso didáctilo de Linneo

LC
Monotypic
HB 54–88 cm
T 0.9–3.3 cm
W 4–11 kg

below 2400 m

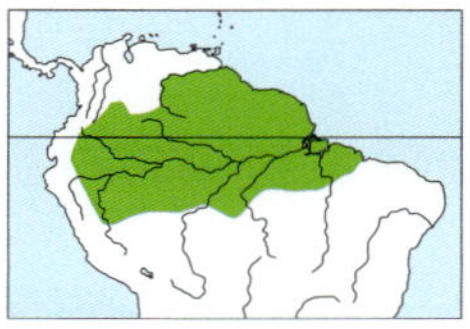

VE (S of R Orinoco), Guianas, SE CO (up to Meta & Guainía), E EC, E PE, N BR (E to Maranhão, S to Mato Grosso)

Family BRADYPODIDAE – Three-toed Sloths

1 genus, 4 species. Low-elevation tropical forests. Neotropical Region.

Genus *BRADYPUS*

(4 species)

Brown-throated Three-toed Sloth
Bradypus variegatus

FR Paresseux à gorge brune
DE Dreifinger-Faultier
ES Perezoso tridáctilo de garganta marrón

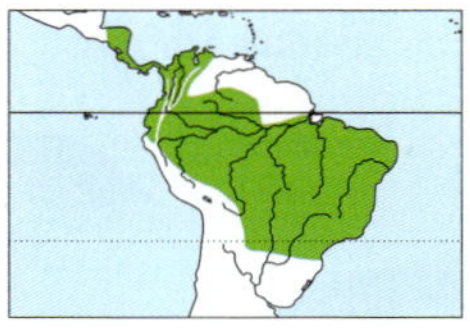

From HN throughout Central America to CO, W & S VE, EC, PE, most of BR, N BO, PY, N AR

LC
Monotypic
HB 52–54 cm
T 5.2–5.5 cm
W 3.7–6 kg

Pygmy Three-toed Sloth
Bradypus pygmaeus

FR Paresseux nain
DE Zwergfaultier
ES Perezoso tridáctilo pigmeo

CR
Monotypic
HB 48–53 cm
T 4.5–6 cm
W 2.5–3.5 kg

Escudo de Veraguas I (18 km offshore from Valiente Pen)

Maned Three-toed Sloth
Bradypus torquatus

FR Paresseux à crinière
DE Kragenfaultier
ES Perezoso tridáctilo de crín

below 1000 m

Locally in Coastal Atlantic Forests (from Sergipe & E Bahia S to N Rio de Janeiro & E Minas Gerais)

VU
Monotypic
HB 59–75 cm
T 4–5.4 cm
W 4.6–10.1 kg

Pale-throated Three-toed Sloth
Bradypus tridactylus

FR Paresseux à trois doigts
DE Weißkehl-Faultier
ES Perezoso tridáctilo de garganta pálida

LC
Monotypic
HB 45–75 cm
T 2.2–11 cm
W 3.4–6.5 kg

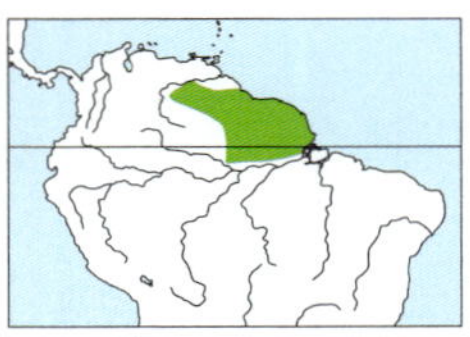

E VE (E & S of R Orinoco), Guianas, N Brazil (N of R Amazon, E to R Amazon Delta)

Order SCANDENTIA (2 families)

Family PTILOCERCIDAE – Pen-tailed Treeshrew

1 genus, 1 species. Primary rainforests, secondary forests, and occasionally gardens. Indo-Malayan Region.

Genus *PTILOCERCUS*

(1 species)

Pen-tailed Treeshrew
Ptilocercus lowii
FR Ptilocerque de Low
DE Federschwanz-Spitzhörnchen
ES Musaraña arborícola de cola plumosa

LC
2 sspp
HB 13–15 cm
T 16–20 cm
W 50–60 g

Thai-Malay Pen (from S TH), Borneo, Sumatra, some Is (S Natuna, Batu, Mentawai, Riau, Lingga, Bangka)

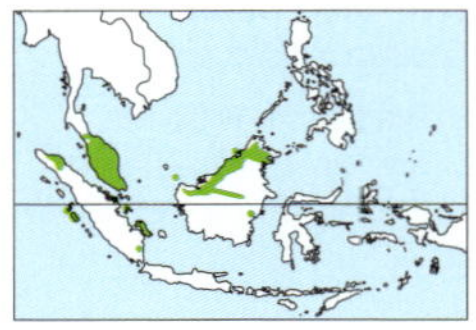

below c. 1200 m

Family TUPAIIDAE – Treeshrews

3 genera, 22 species. Mostly tropical rainforests but also scrub jungle, montane, evergreen, and deciduous forests, and plantations. Indo-Malayan Region, marginally into Palearctic Region.

Genus *DENDROGALE*

(2 species)

LC Monotypic
HB 11.5–13 cm T 11–12.5 cm

below c. 1500 m

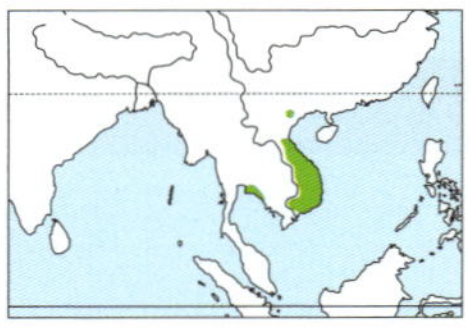

Northern Smooth-tailed Treeshrew
Dendrogale murina
FR Dendrogale d'Indochine
DE Nördliches Bergspitzhörnchen
ES Tupaya de cola lisa septentrional

C & S VN (isolated population in N), EC & SE LA, parts of SE TH, SW & E KH

DD
2 sspp
HB 12–12.5 cm
T 13.5–14 cm

900–3350 m

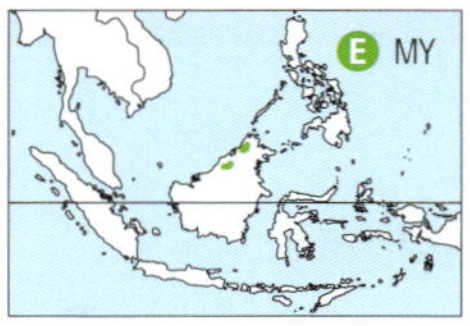

Bornean Smooth-tailed Treeshrew
Dendrogale melanura
FR Dendrogale de Bornéo
DE Borneo-Bergspitzhörnchen
ES Tupaya de cola lisa de Borneo

N & NW Borneo (Mt Dulit, Mt Mulu, Mt Murud, Mt Kinabalu, Crocker Range, Trus Madi)

Genus *ANATHANA*

(1 species)

LC
Monotypic
HB 16.5–20 cm
T 15–23 cm

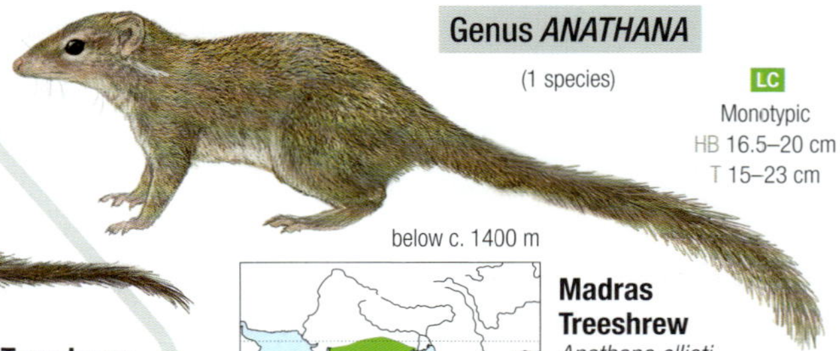

below c. 1400 m

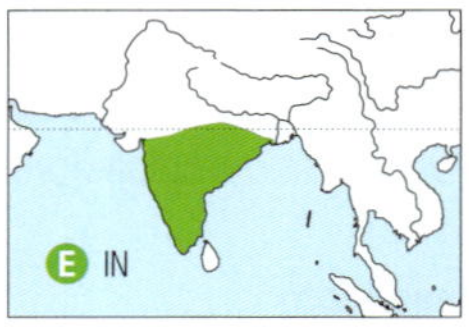

Madras Treeshrew
Anathana ellioti
FR Toupaye d'Elliot
DE Madras-Spitzhörnchen
ES Tupaya de Madrás

C & S Pen (from Satpura Hills E to SW Bihar)

Genus *TUPAIA*

(19 species)

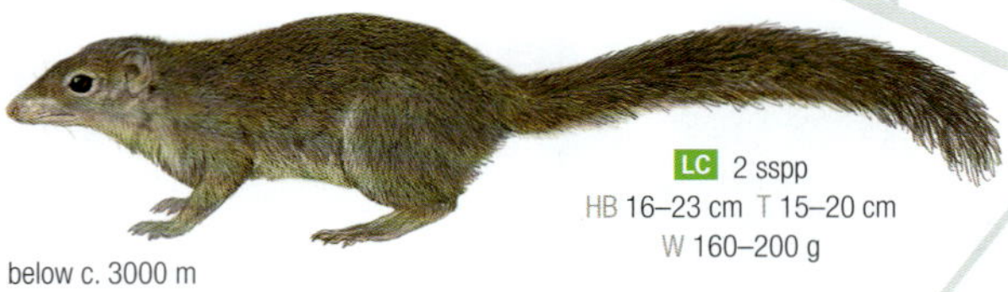

LC 2 sspp
HB 16–23 cm T 15–20 cm
W 160–200 g

below c. 3000 m

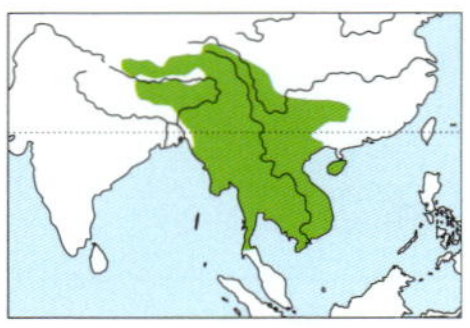

Northern Treeshrew
Tupaia belangeri
FR Toupaye de Bélanger
DE Nördliches Spitzhörnchen
ES Tupaya septentrional

From SW & SC CN (with Hainan I) and NP S through BT, NE IN, NE BD, SE Asia to N Thai-Malay Pen

below c. 1700 m

LC
4 sspp
HB 11–14 cm
T 13.5–17.5 cm
W 35–80 g

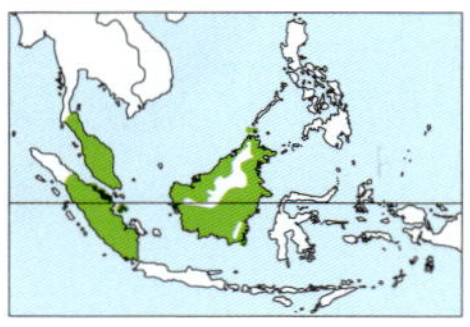

Lesser Treeshrew
Tupaia minor
FR Toupaye nain
DE Zwergspitzhörnchen
ES Tupaya menor

Thai-Malay Pen, Sumatra, Borneo, some Is (Banggi, Balambangan, Singkep, probably Laut)

Common Treeshrew
Tupaia glis
FR Toupaye commun
DE Gewöhnliches Spitzhörnchen
ES Tupaya común

LC
Monotypic
HB 13.5–21 cm
T 12.5–18 cm
W 90–190 g

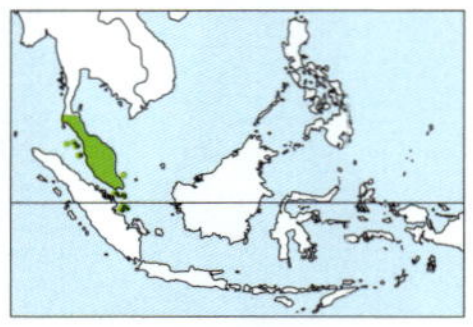

Thai-Malay Pen (with several Is off coast), Lingga Arch (Lingga, Singkep)

EN Monotypic
HB 17.5–19.5 cm T 19–23 cm

Nicobar Treeshrew
Tupaia nicobarica
FR Toupaye des Nicobar
DE Nikobaren-Spitzhörnchen
ES Tupaya de Nicobar

0–640 m

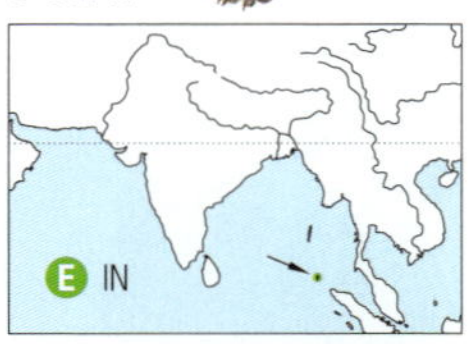

Nicobar Is (Great & Little Nicobar)

Sumatra and Batu Is (Tanahbala)

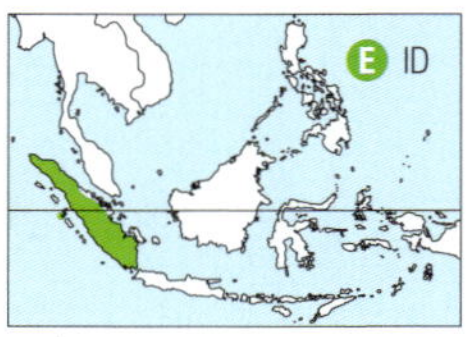

DD Monotypic
HB 17.5–20 cm
T 14–17.5 cm

Sumatran Treeshrew
Tupaia ferruginea
FR Toupaye ferrugineux
DE Rötliches Spitzhörnchen
ES Tupaya de Sumatra

Golden-bellied Treeshrew
Tupaia chrysogaster
FR Toupaye des Mentawai
DE Mentawai-Spitzhörnchen
ES Tupaya de vientre dorado

Bangka I, off SE coast of Sumatra

DD Monotypic
HB 19–21 cm T 16–19 cm

Banka Island Treeshrew
Tupaia discolor
FR Toupaye de Bangka
DE Bangka-Spitzhörnchen ES Tupaya de Banka

below 1000 m

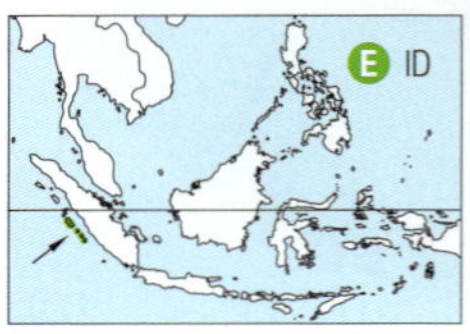

VU
Monotypic
HB 18–22 cm
T 12.5–15 cm

Mentawai Is (Sipora, North Pagai, South Pagai, probably Siberut)

below c. 1700 m

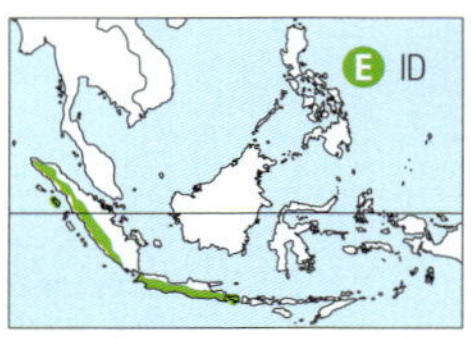

W Sumatra and Nias I, Java, Bali

LC Monotypic
HB 13–15 cm T 14.5–17.5 cm

Horsfield's Treeshrew
Tupaia javanica
FR Toupaye de Horsfield
DE Horsfield-Spitzhörnchen
ES Tupaya de Horsfield

DD Monotypic
HB c. 14.5 cm
T c. 14.5 cm

above 2000 m — W Java

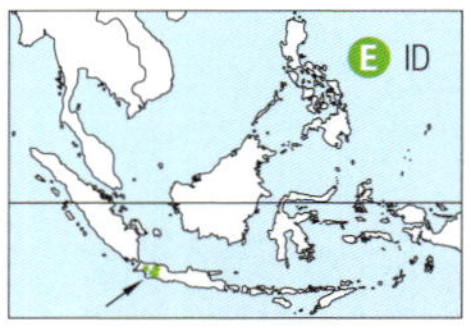

Javan Treeshrew
Tupaia hypochrysa

FR Toupaye à gorge dorée
DE Java-Spitzhörnchen
ES Tupaya de Java

LC 15 sspp
HB 17.5–21 cm
T 15–19.5 cm
W 200–300 g

below c. 1500 m

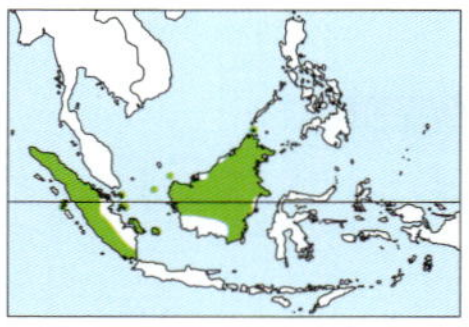

Large Treeshrew
Tupaia tana

FR Toupaye terrestre
DE Großes Spitzhörnchen
ES Tupaya grande

Sumatra, Borneo, some Is (Banyak, Batu, Bangka, Belitung, Lingga, Riau, Natuna, Labuan, Banggi, Tambelan)

Long-footed Treeshrew
Tupaia longipes

FR Toupaye à longues pattes
DE Langfuß-Spitzhörnchen
ES Tupaya de pies largos

below c. 900 m

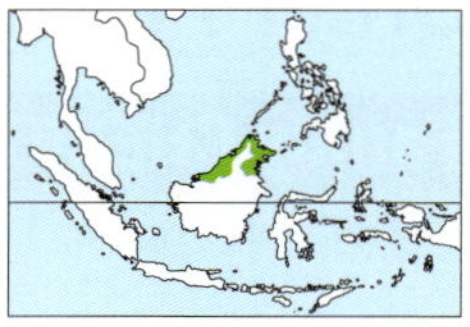

N Borneo lowlands S to R Rajang and R Kayan (MY/BN/ID)

LC Monotypic
HB 19–20 cm T 18–19 cm
W c. 165 g

below c. 1200 m

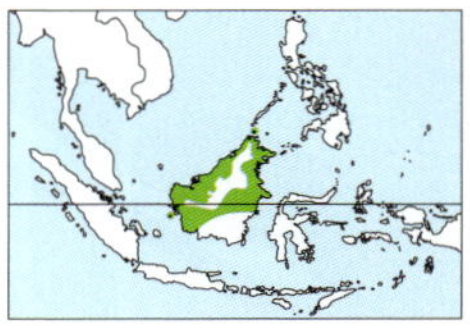

Slender Treeshrew
Tupaia gracilis

FR Toupaye grêle
DE Schlankspitzhörnchen
ES Tupaya esbelta

MY/ID: Borneo (except C highlands & SE), nearby Banggi & Karimata Is; possibly Belitung & Bangka Is

LC
Monotypic
HB 13.5–15 cm
T 15.5–18 cm
W c. 70 g

LC 3 sspp
HB 15–17.5 cm T 12–15 cm
W 110–150 g

Mountain Treeshrew
Tupaia montana

FR Toupaye de montagne
DE Hochland-Spitzhörnchen
ES Tupaya de montaña

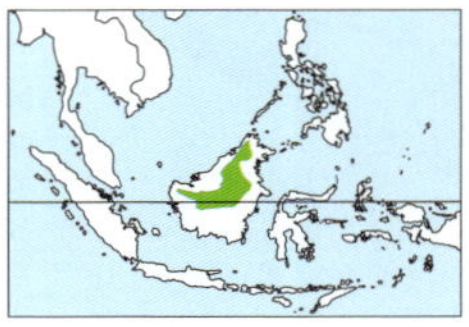

900–3200 m

Borneo: MY (Lanjak Entimau WS, Batu Song, Kalulong, Usun Apau, Dulet, Pueh & Kinabalu Mts), ID (E Kalimantan)

DD
Monotypic
HB 17.5–19.5 cm
T 14.5–15 cm

below 1200 m

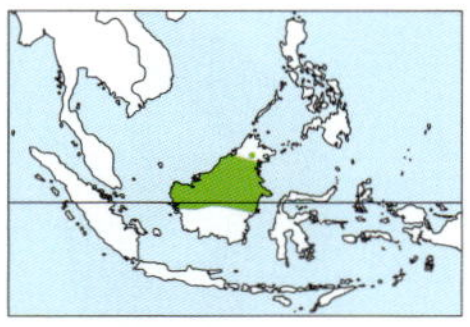

W, C & E Borneo (MY/BN/ID); in Sabah, rare along Kalimantan border

Striped Treeshrew
Tupaia dorsalis

FR Toupaye rayé
DE Streifenspitzhörnchen
ES Tupaya listada

LC 2 sspp
HB 17.4–20 cm T c. 15 cm

below c. 1000 m

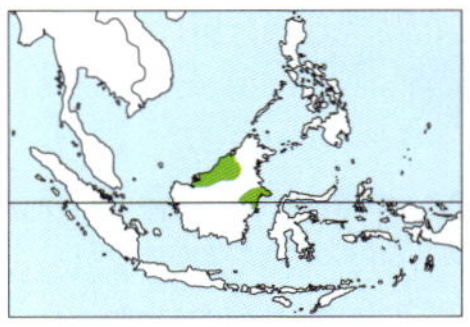

Painted Treeshrew
Tupaia picta

FR Toupaye peint
DE Tiefland-Spitzhörnchen
ES Tupaya pintada

NW & E Borneo (MY/BN/ID)

LC
Monotypic
HB 17.5–21 cm
T 17.5–20 cm

MY/ID: S Borneo, S of R Rajang and R Kayan

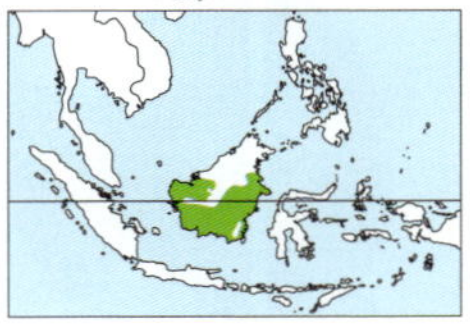

Kalimantan Treeshrew
Tupaia salatana

FR Toupaye du Kalimantan
DE Kalimantan-Spitzhörnchen
ES Tupaya de Borneo

Splendid Treeshrew
Tupaia splendidula

FR Toupaye à queue rousse
DE Rotschwanz-Spitzhörnchen
ES Tupaya espléndida

LC 5 sspp
HB 15–17.5 cm T 12–15 cm
W 110–150 g

MY/ID: Borneo, Karimata I, Natuna Is (Laut, Bunguran), Anambas Is (Kiabu)

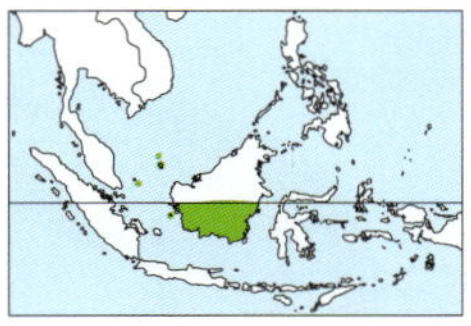

900–1200 m

Mindanao, nearby Dinagat & Siargao Is

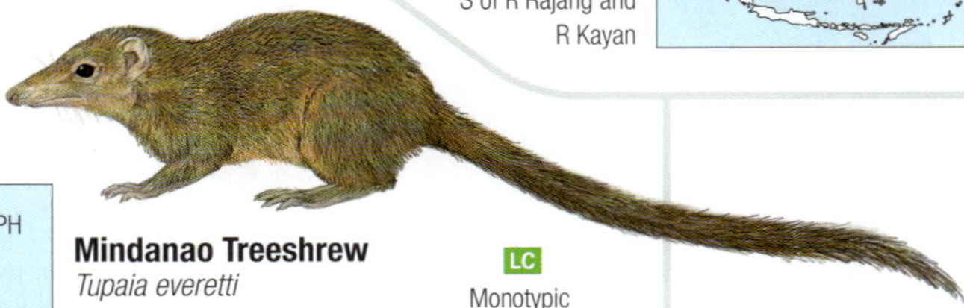

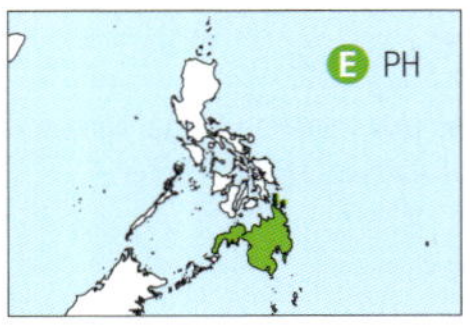

Mindanao Treeshrew
Tupaia everetti

FR Toupaye de Mindanao
DE Mindanao-Spitzhörnchen
ES Tupaya de Mindanao

LC
Monotypic
HB 17–22 cm
T 11.5–17.5 cm
W c. 350 g

0–c. 1400 m

Palawan, Balabac, Busuanga, Culion, Cuyo

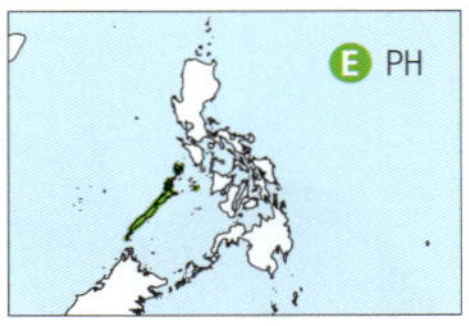

LC 2 sspp
HB 15.7–19 cm
T 15–18 cm
W 120–160 g

Palawan Treeshrew
Tupaia palawanensis

FR Toupaye de Palawan
DE Palawan-Spitzhörnchen
ES Tupaya de Palawan

Order DERMOPTERA (1 family)

Family CYNOCEPHALIDAE – Colugos

2 genera, 2 species. Primary and secondary tropical rainforests and forest plantations, with trees of sufficient height for gliding. Indo-Malayan Region.

Genus *CYNOCEPHALUS*
(1 species)

Philippine Colugo
Cynocephalus volans

FR Galéopithèque des Philippines
DE Philippinen-Gleitflieger
ES Colugo de Filipinas

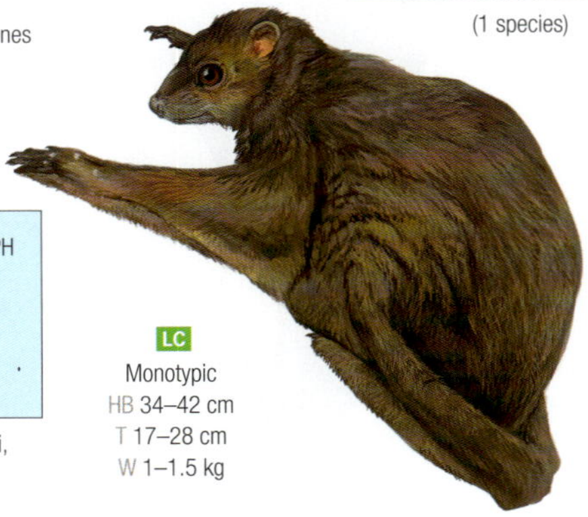

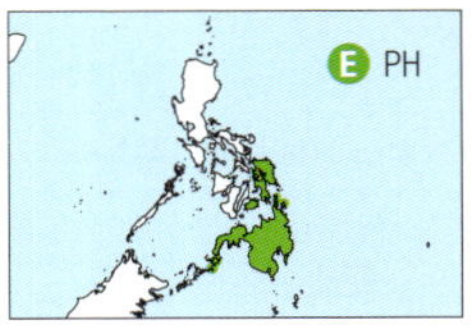

Samar, Leyte, Biliran, Maripipi, Bohol, Dinagat, Siargao, Mindanao, Basilan, Tongkil

LC
Monotypic
HB 34–42 cm
T 17–28 cm
W 1–1.5 kg

Genus *GALEOPTERUS*
(1 species)

Sunda Colugo
Galeopterus variegatus

FR Galéopithèque de Malaisie
DE Malaien-Gleitflieger
ES Colugo de Sonda

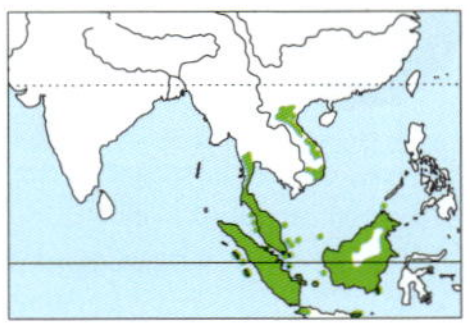

E LA, VN, SE KH, extreme SE MM, S TH, Thai-Malay Pen, Sumatra, W Java, Borneo, several is & arch

LC
4 sspp
HB 34–42 cm
T 17–28 cm
W 1.2–1.7 kg

Order PRIMATES (16 families)

Family CHEIROGALEIDAE – Mouse, Giant Mouse, Dwarf and Fork-marked Lemurs

5 genera, 42 species. Tropical lowland, humid montane, dry deciduous, and spiny forests, and scrub. Madagascar.

Genus *MICROCEBUS*

(25 species)

Gray Mouse Lemur
Microcebus murinus

FR Microcèbe murin
DE Grauer Mausmaki
ES Lémur ratón gris

LC
Monotypic
HB 12–14 cm
T 13–14.5 cm
W c. 60 g

0–800 m

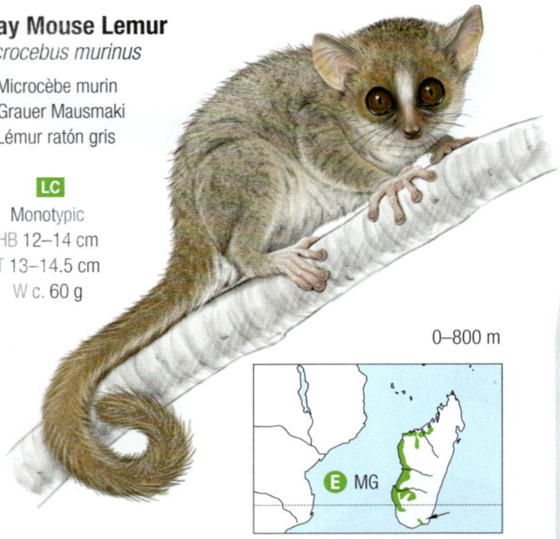

W & SE coast (S from Ankarafantsika NP; SE up to littoral forests of Mandena CZ)

Ganzhorn's Mouse Lemur
Microcebus ganzhorni

FR Microcèbe de Ganzhorn
DE Ganzhorn-Mausmaki
ES Lémur ratón de Ganzhorn

EN
Monotypic

Mandena Forest

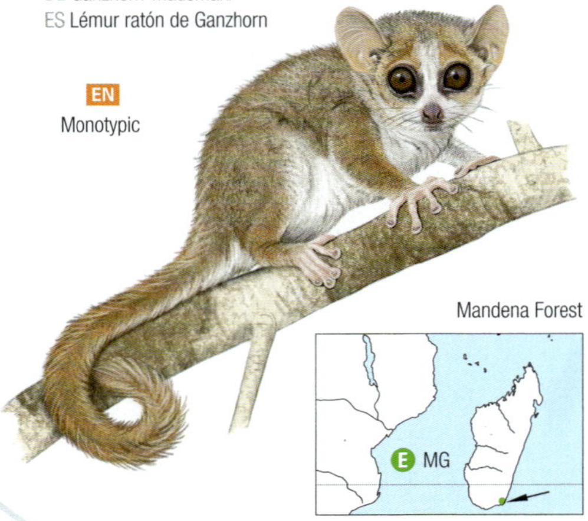

Bemnasy Mouse Lemur
Microcebus manitatra

FR Microcèbe pionnier
DE Südöstlicher Mausmaki
ES Lémur ratón meridional

CR
Monotypic

Bemanasy Forest

Gray-brown Mouse Lemur
Microcebus griseorufus

FR Microcèbe gris-roux
DE Graubrauner Mausmaki
ES Lémur ratón gris castaño

LC
Monotypic
HB 12–13 cm
T 14–15 cm
W 40–50 g

0–250 m

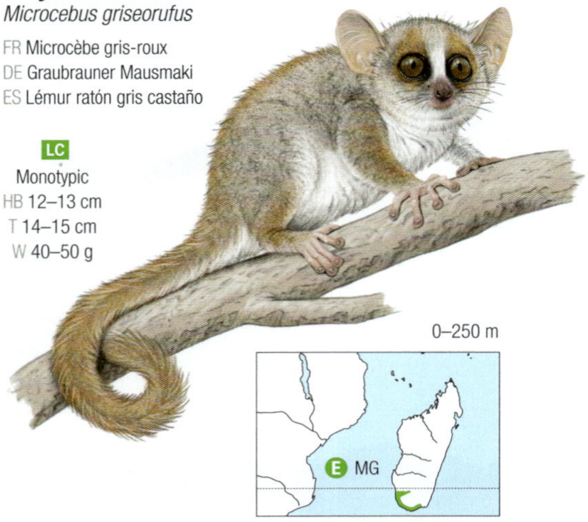

Coast from Lamboharana & Beza-Mahafaly SR, to Berenty R & Petriky in S

Ambarijeby Mouse Lemur
Microcebus danfossi

FR Microcèbe de Danfoss
DE Sofia-Mausmaki
ES Lémur ratón de Danfoss

VU
Monotypic
HB 12.7–13.6 cm
T 16.5–17.3 cm
W c. 63 g

0–780 m

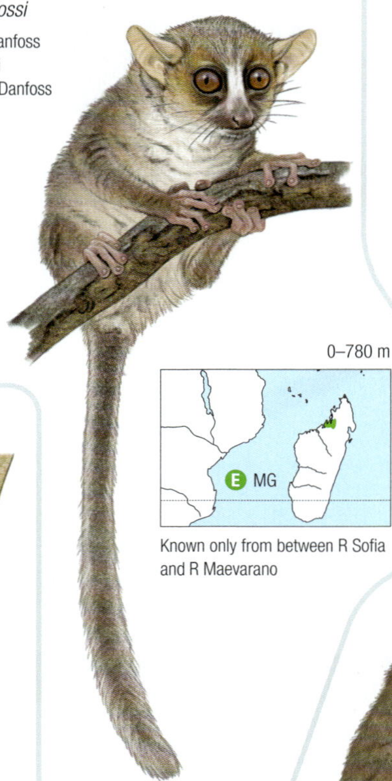

Known only from between R Sofia and R Maevarano

Bongolava Mouse Lemur
Microcebus bongolavensis

FR Microcèbe du Bongolava
DE Bongolava-Mausmaki
ES Lémur ratón de Bongolava

EN
Monotypic
HB 9–12.2 cm
T 14.7–17.4 cm
W c. 54 g

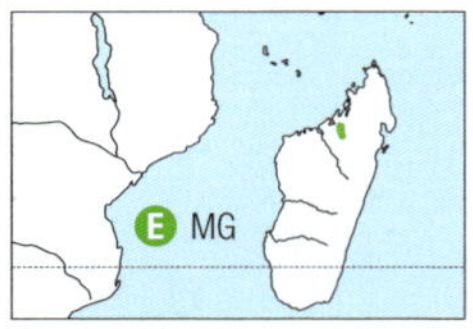

Known only from Port-Bergé, between R Mahajamba-Est and R Sofia

Golden-brown Mouse Lemur
Microcebus ravelobensis

FR Microcèbe doré
DE Goldbrauner Mausmaki
ES Lémur ratón castaño dorado

VU
Monotypic
HB 12–13 cm
T 15–17 cm
W c. 57 g

below 500 m

Ankarafantsika region; limits uncertain

Anjiahely Mouse Lemur
Microcebus macarthurii

FR Microcèbe de MacArthur
DE Anjiahely-Mausmaki
ES Lémur ratón de MacArthur

EN
Monotypic
HB 11–12 cm
T c. 14.7 cm
W c. 53 g

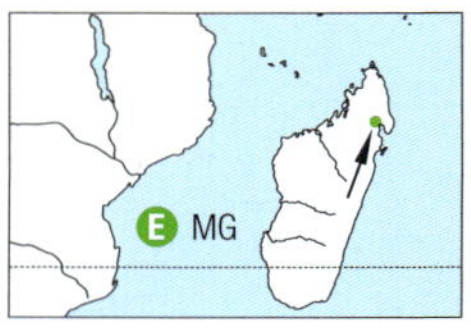

Known only from type locality (c. 26 km W of Maroantsetra)

Jonah's Mouse Lemur
Microcebus jonahi

FR Microcèbe de Jonah
DE Jonah-Mausmaki
ES Lémur ratón de Jonah

EN
Monotypic

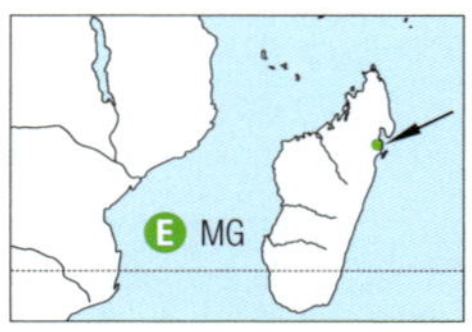

Known only from near Ambavala

Marohita Mouse Lemur
Microcebus marohita

FR Microcèbe de Marohita
DE Marohita-Mausmaki
ES Lémur ratón de Marohita

CR
Monotypic
HB 13.2–14 cm T 13.3–14.5 cm
W 64–89 g

Marohita Forest

Gerp's Mouse Lemur
Microcebus gerpi

FR Microcèbe du Gerp
DE Gerp-Mausmaki
ES Lémur ratón de Gerp

CR
Monotypic
HB c. 8.4 cm
T c. 14.8 cm
W c. 68 g

30–230 m

Known only from type locality, Sahafina Forest

Tavaratra Mouse Lemur
Microcebus tavaratra

FR Microcèbe du Nord
DE Nördlicher Mausmaki
ES Lémur ratón septentrional

20–250 m

VU
Monotypic
HB 12–14 cm
T 15–16 cm
W c. 54 g

Ankarana, Analamerana, Andavakoera & Andrafiamena regions

Anosy Mouse Lemur
Microcebus tanosi

FR Microcèbe d'Anôsy
DE Anosy-Mausmaki
ES Lémur ratón de Anosy

EN
Monotypic
HB 11.6–14 cm
T 11.5–15 cm
W 48–60 g

Manantantely & Ivorona forests

EN
Monotypic
HB c. 13 cm
T c. 12 cm
W c. 49 g

Jolly's Mouse Lemur
Microcebus jollyae

FR Microcèbe de Jolly
DE Jolly-Mausmaki
ES Lémur ratón de Jolly

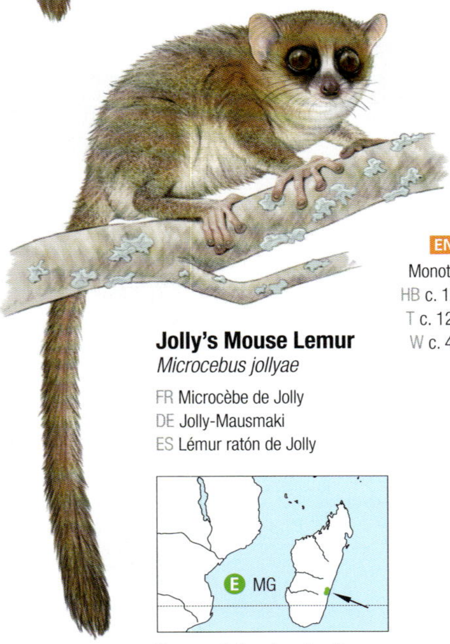

Known only from type locality (perhaps just S of R Mananara to N bank of R Mananjary)

Nosy Boraha Mouse Lemur
Microcebus boraha

FR Microcèbe de Boraha
DE Nosy-Boraha-Mausmaki
ES Lémur ratón de Santa María

DD
Monotypic

Nosy Boraha (off NE coast)

Simmons's Mouse Lemur
Microcebus simmonsi

FR Microcèbe de Simmons
DE Simmons-Mausmaki
ES Lémur ratón de Simmons

EN
Monotypic
HB c. 9.2 cm
T c. 14.2 cm
W 42–65 g

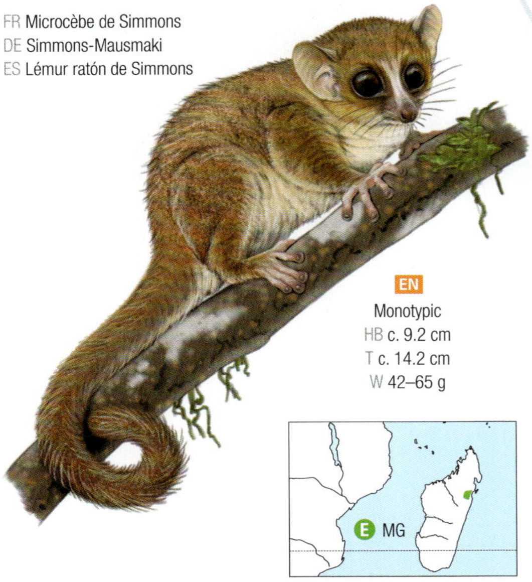

Known only from Betampona, Zahamena & Tampolo regions

Nosy Be Mouse Lemur
Microcebus mamiratra

FR Microcèbe de Claire
DE Nosy-Be-Mausmaki
ES Lémur ratón de Nosy Be

EN
Monotypic
HB c. 12.7 cm
T c. 14.3 cm
W c. 58 g

Nosy Be (principally Lokobe Strict NR) and adjacent mainland

Margot Marsh's Mouse Lemur
Microcebus margotmarshae

FR Microcèbe de Marsh
DE Antafondro-Mausmaki
ES Lémur ratón de Margot Marsh

EN
Monotypic
HB 11–12 cm
T c. 14 cm
W c. 49 g

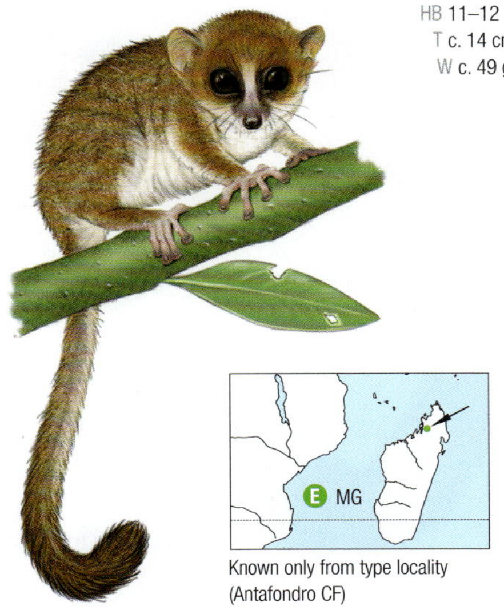

Known only from type locality (Antafondro CF)

Sambirano Mouse Lemur
Microcebus sambiranensis

FR Microcèbe du Sambirano
DE Sambirano-Mausmaki
ES Lémur ratón de Sambirano

EN
Monotypic
HB 11–12 cm
T 13.5–14.5 cm
W c. 45 g

Known only from Manongarivo SR; possibly on Ampasindava Pen & Tsaratanana Massif

Montagne d'Ambre Mouse Lemur
Microcebus arnholdi

FR Microcèbe d'Arnhold
DE Montagne-d'Ambre-Mausmaki
ES Lémur ratón de Arnhold

VU
Monotypic
HB c. 12.6 cm
T 10.6–13.6 cm
W c. 45 g

Known only from Montagne d'Ambre NP and Forêt d'Ambre SR

Mittermeier's Mouse Lemur
Microcebus mittermeieri

FR Microcèbe de Mittermeier
DE Mittermeier-Mausmaki
ES Lémur ratón de Mittermeier

NE
Monotypic
HB c. 8 cm T c. 12 cm
W c. 46 g

below 1760 m

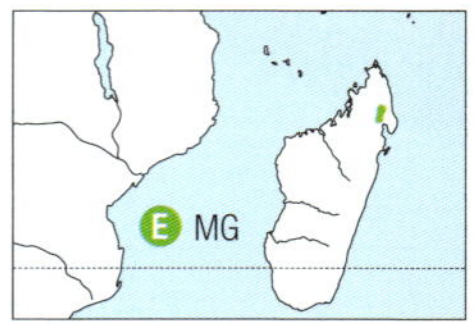

In and around Anjanaharibe-Sud SR as far NW as Bealalana

Goodman's Mouse Lemur
Microcebus lehilahytsara

NT
Monotypic
HB c. 9 cm
T c. 10 cm
W 45–48 g

FR Microcèbe de Goodman
DE Goodman-Mausmaki
ES Lémur ratón de Goodman

Andasibe area, Analamazaotra SR, Anjozorobe-Angavo PA, Mantadia NP; limits unclear

Madame Berthe's Mouse Lemur
Microcebus berthae

FR Microcèbe de Berthe
DE Madame-Berthe-Mausmaki
ES Lémur ratón de Berthe

CR
Monotypic
HB 9–9.5 cm
T 13–14 cm
W c. 31 g

0–150 m

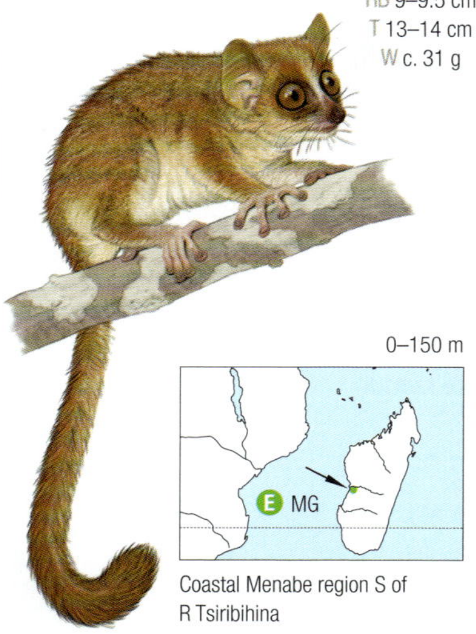

Coastal Menabe region S of R Tsiribihina

Peters's Mouse Lemur
Microcebus myoxinus

VU
Monotypic
HB 12–13 cm
T 14–15 cm
W c. 42 g

FR Microcèbe pygmée
DE Zwergmausmaki
ES Lémur ratón de Peters

0–900 m

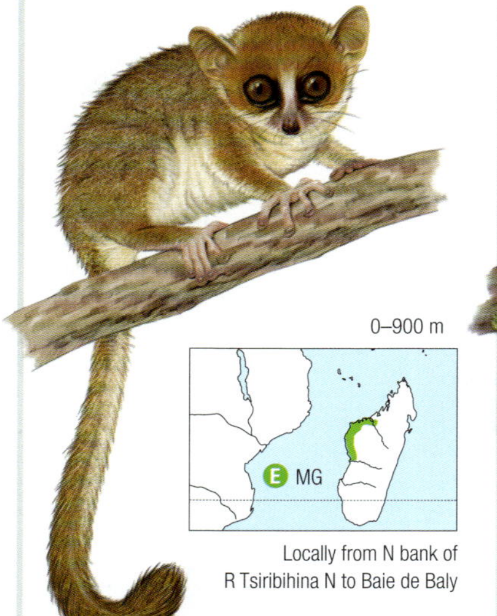

Locally from N bank of R Tsiribihina N to Baie de Baly

Rufous Mouse Lemur
Microcebus rufus

VU
Monotypic
HB c. 12 cm
T 10.9–12.5 cm
W c. 44 g

FR Microcèbe roux
DE Brauner Mausmaki
ES Lémur ratón castaño

0–2000 m

Limits unknown; perhaps from Ranomafana NP to Andringitra NP

Coquerel's Giant Mouse Lemur
Mirza coquereli

FR Microcèbe de Coquerel
DE Coquerel-Riesenmausmaki
ES Lémur ratón gigante de Coquerel

EN
Monotypic
HB c. 25 cm
T 31–32 cm
W 299–317 g

0–700 m

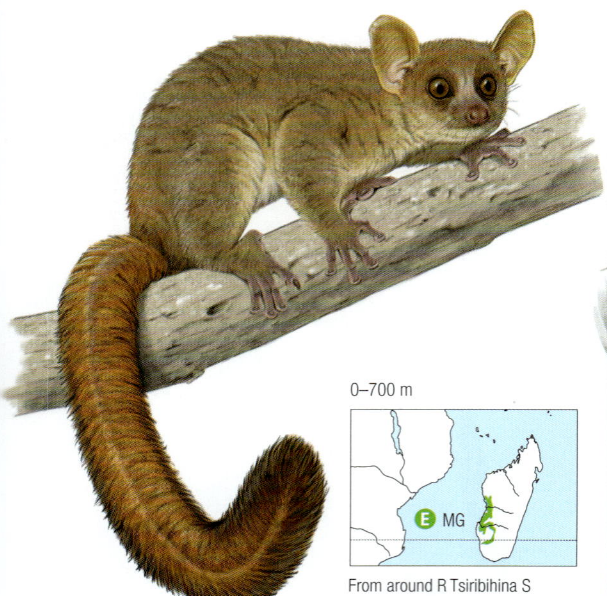

From around R Tsiribihina S R Onilay, including Zombitse-Vohibasia & Isalo NPs

Genus *MIRZA*

(2 species)

Northern Giant Mouse Lemur
Mirza zaza

FR Microcèbe zaza
DE Nördlicher Riesenmausmaki
ES Lémur ratón gigante septentrional

VU
Monotypic
HB c. 27 cm
T 27–28 cm
W c. 290 g

From Ampasindava Pen between R Mahavavy Nord and R Maevarano

Genus *ALLOCEBUS*

(1 species)

Hairy-eared Dwarf Lemur
Allocebus trichotis

FR Allocèbe
DE Büschelohrmaki
ES Lémur enano de orejas peludas

0–1600 m

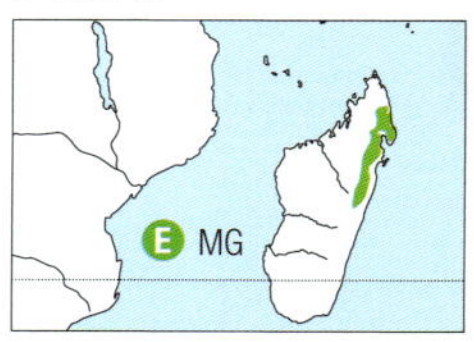

Limits unknown

EN
Monotypic
HB 13–16 cm
T 14–20 cm
W c. 79 g

Genus *CHEIROGALEUS*

(10 species)

Sibree's Dwarf Lemur

Cheirogaleus sibreei

FR Chirogale de Sibree
DE Sibree-Fettschwanzmaki
ES Lémur enano de Sibree

CR

Monotypic
HB 20–23 cm
T 22–26 cm
W c. 272 g

above 1400 m

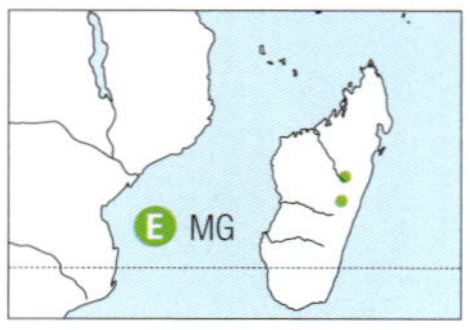

Perhaps limited by R Onive, R Mangoro, Anjozorobe and corridor to W half of Tsinjoarivo Forest

Lesser Iron-gray Dwarf Lemur

Cheirogaleus minusculus

FR Chirogale minuscule
DE Kleiner Fettschwanzmaki
ES Lémur enano minúsculo

NE

Monotypic

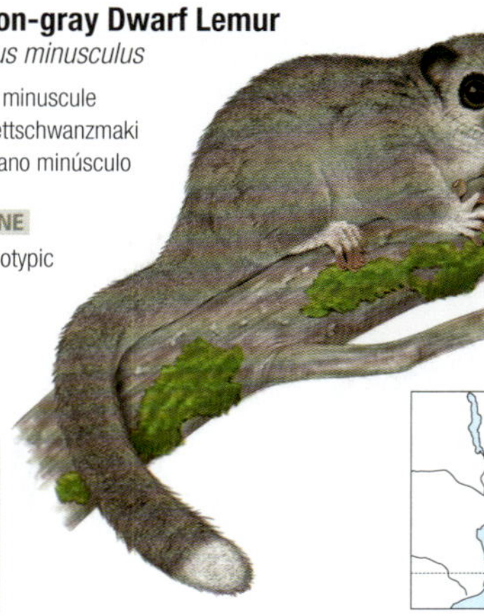

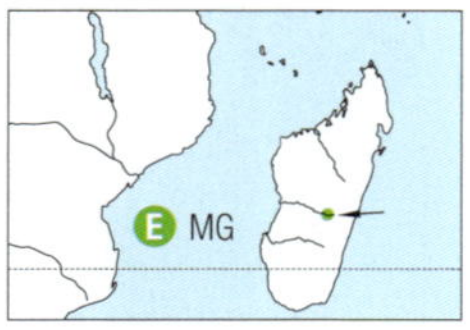

Known only from type locality (Ambositra, N of Fianarantsoa)

Ankarana Dwarf Lemur

Cheirogaleus shethi

FR Chirogale de Sheth
DE Ankarana-Fettschwanzmaki
ES Lémur enano de Ankarana

EN

Monotypic
T c. 16.3 cm
W c. 11.5 g

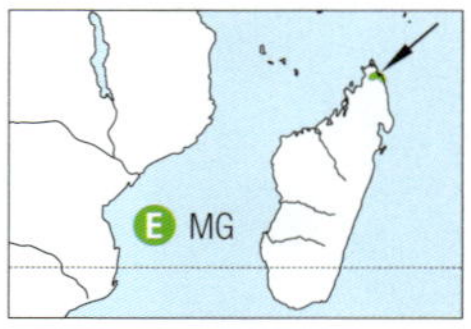

From Ankarana E to Bekaraoka

Thomas's Dwarf Lemur

Cheirogaleus thomasi

FR Chirogale de Thomas
DE Thomas-Fettschwanzmaki
ES Lémur enano de Thomas

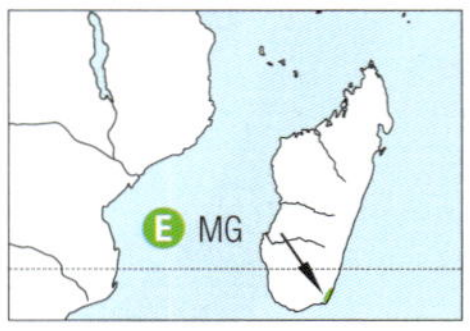

90–320 m

From St Luce to Petriky

EN

Monotypic

Fat-tailed Dwarf Lemur

Cheirogaleus medius

FR Petit Chirogale
DE Fettschwanzmaki
ES Lémur enano de cola gruesa

VU

Monotypic
HB 17–23 cm
T 19–27 cm
W c. 135 g

60–800 m

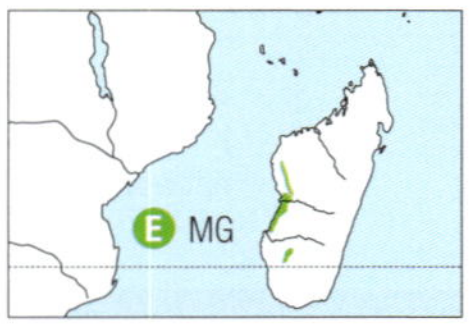

From Zombitse S to Tsingy de Bemaraha

Greater Dwarf Lemur

Cheirogaleus major

FR Grand Chirogale
DE Großer Fettschwanzmaki
ES Lémur enano mayor

VU

Monotypic
HB 28–29 cm
T 27–28 cm
W 314–414 g

0–1800 m

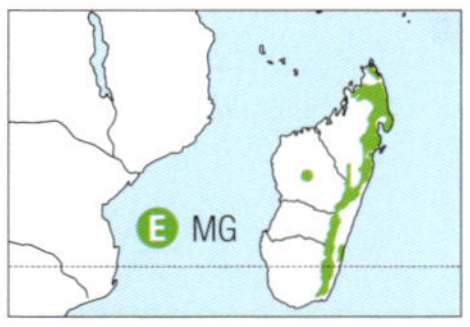

From N tip S to near Tôlanaro and littoral forest at Mandena; also CW (Vohimena & Anosy Mts)

Montagne d'Ambre Dwarf Lemur

Cheirogaleus andysabini

FR Chirogale de Sabin
DE Montagne-d'Ambre-Fettschwanzmaki
ES Lémur enano del Monte Ambre

EN

Monotypic
T c. 26 cm
W c. 28 g

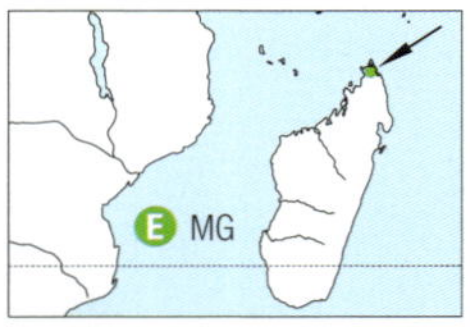

Montagne d'Ambre NP and areas around Joffreville, NW of R Irodo

Lavasoa Dwarf Lemur

Cheirogaleus lavasoensis

FR Chirogale du Lavasoa
DE Lavasoa-Fettschwanzmaki
ES Lémur enano de Lavasoa

EN

Monotypic

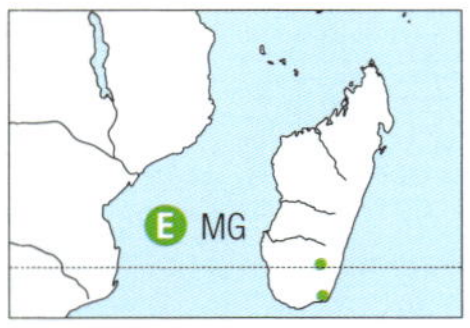

S slopes of Lavasoa Mts

Crossley's Dwarf Lemur

Cheirogaleus crossleyi

FR Chirogale de Crossley
DE Crossley-Fettschwanzmaki
ES Lémur enano de Crossley

VU

Monotypic
HB 22–26 cm
T 21–27 cm
W 370–458 g

0–1800 m

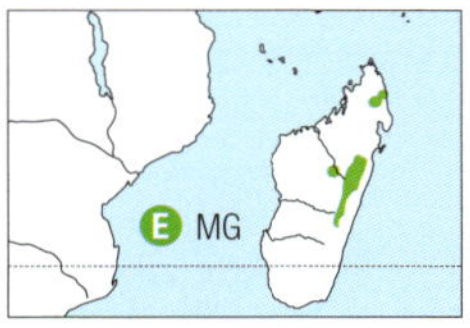

Marojejy NP, COMATSA, Makirovana-Tsihomanaomby, Ambohitromby S to Ankazomivady, Ankafobe & Ambohitantely

Groves's Dwarf Lemur

Cheirogaleus grovesi

FR Chirogale de Groves
DE Groves-Fettschwanzmaki
ES Lémur enano de Groves

DD

Monotypic
T c. 29 cm
W c. 45 g

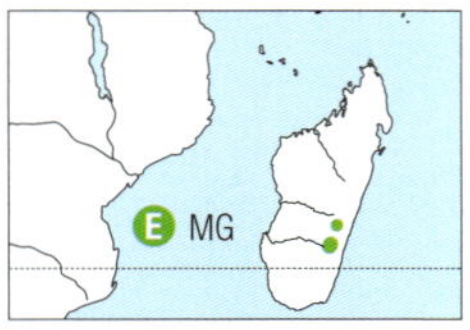

Ranomafana & Andringitra NPs, and surrounding areas; probably locally between the 2 parks

Genus *PHANER*

(4 species)

Masoala Fork-marked Lemur
Phaner furcifer

FR Phaner à fourche
DE Masoala-Gabelstreifenmaki
ES Lémur de orejas ahorquilladas de Masoala

EN
Monotypic

0–1000 m

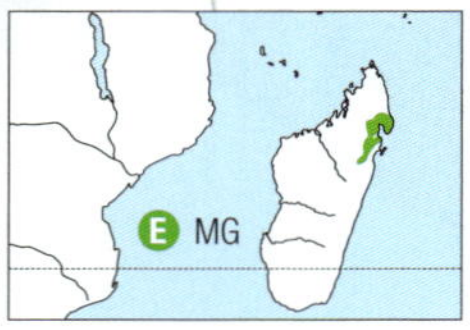

From Masoala Pen S to Toamasina

Pale Fork-marked Lemur
Phaner pallescens

FR Phaner pâle
DE Westlicher Gabelstreifenmaki
ES Lémur de orejas ahorquilladas pálido

EN
Monotypic
HB 23–29 cm
T 29–37 cm
W c. 333 g

0–800 m

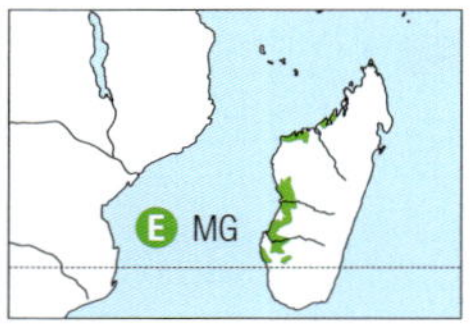

Along strip of forests from Soalala S to Toliara just S of R Fiherenana (including Mikea Forest)

Sambirano Fork-marked Lemur
Phaner parienti

FR Phaner de Pariente
DE Sambirano-Gabelstreifenmaki
ES Lémur de orejas ahorquilladas de Sambirano

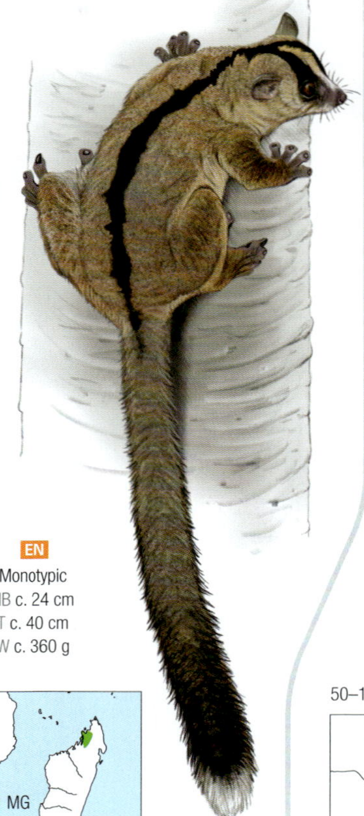

EN
Monotypic
HB c. 24 cm
T c. 40 cm
W c. 360 g

0–800 m

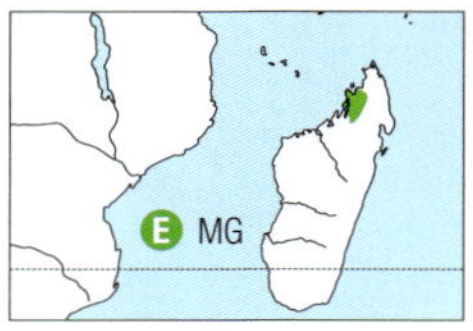

Sambirano region S of Ambanja to R Andranomalaza, with Ampasindava Pen

Montagne d'Ambre Fork-marked Lemur
Phaner electromontis

FR Phaner de la Montagne d'Ambre
DE Montagne-d'Ambre-Gabelstreifenmaki
ES Lémur de orejas ahorquilladas de Ambre

EN
Monotypic
HB c. 27 cm
T c. 34 cm
W c. 387 g

50–1000 m

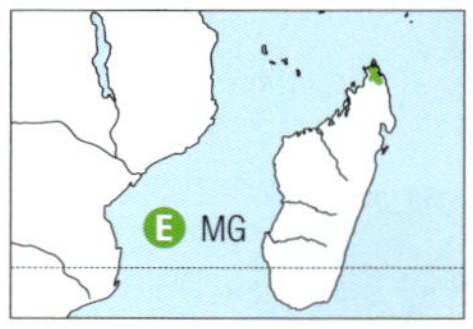

3 small populations in area of Montagne d'Ambre and Ankarana

Family LEPILEMURIDAE – Sportive Lemurs

1 genus, 25 species. Primary and secondary low- and mid-elevation evergreen and deciduous, transitional subhumid, thorn, gallery, and spiny forests, and scrub. Madagascar.

Genus *LEPILEMUR*

(25 species)

EN
Monotypic
HB 25–28 cm
T 25–29 cm
W c. 880 g

Masoala Sportive Lemur
Lepilemur scottorum

FR Lépilémur des Scott
DE Masoala-Wieselmaki
ES Lémur saltador de los Scott

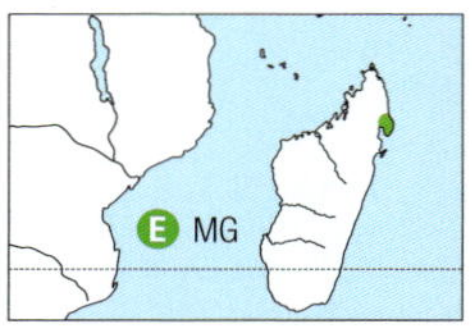

Known only from Masoala NP (Masiaposa Forest); N & E limits unclear

CR
Monotypic
HB 29–34 cm
T 27–29 cm
W c. 1 kg

Mananara-Nord Sportive Lemur
Lepilemur hollandorum

FR Lépilémur des Holland
DE Mananara-Nord-Wieselmaki
ES Lémur saltador de Mananara

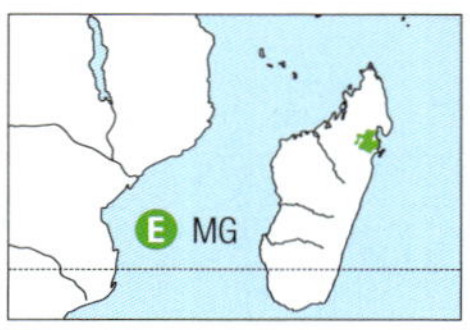

Ivontaka-Sud & Verezanantsoro parcels of Mananara-Nord BR; N & S limits unclear

VU
Monotypic
HB c. 27 cm
T c. 26 cm
W c. 950 g

Seal's Sportive Lemur
Lepilemur seali

FR Lépilémur de Seal
DE Seal-Wieselmaki
ES Lémur saltador de Seal

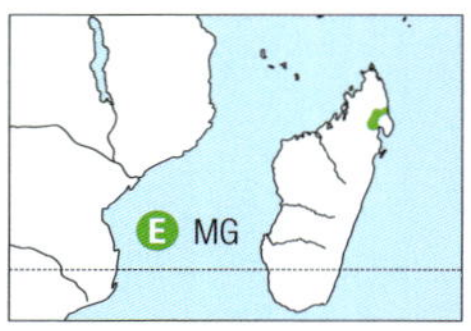

S of R Antainambalana at least as far as R Fananehana, including Makira area; N & S limits unclear

Wright's Sportive Lemur
Lepilemur wrightae

FR Lépilémur de Wright
DE Wright-Wieselmaki
ES Lémur saltador de Wright

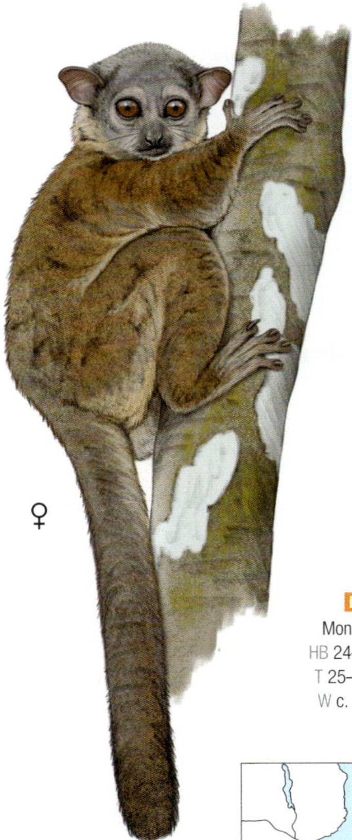

♀

♂

EN
Monotypic
HB 24–26 cm
T 25–26 cm
W c. 1.1 kg

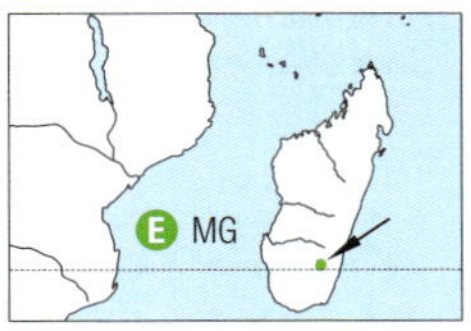

Kalambatritra SR, W of R Ionaivo, E of R Mangoky and N of R Mandrare; limits unclear

Andohahela Sportive Lemur
Lepilemur fleuretae

FR Lépilémur de Fleurette
DE Andohahela-Wieselmaki
ES Lémur saltador de Fleurette

EN
Monotypic
HB 24–26 cm
T 27–30 cm
W c. 800 g

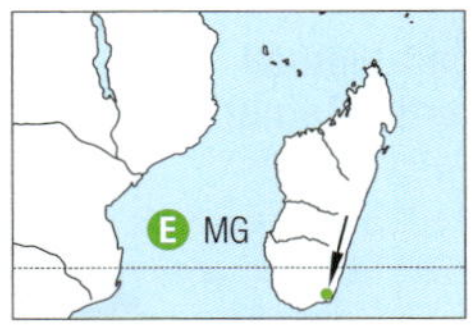

Andohahela NP (Manangotry Parcel between R Mandrare & R Manampanihy); limits unclear

Weasel Sportive Lemur
Lepilemur mustelinus

FR Lépilémur mustelin
DE Östlicher Wieselmaki
ES Lémur saltador mayor

VU
Monotypic
HB 21–25 cm
T 25–29 cm
W c. 1 kg

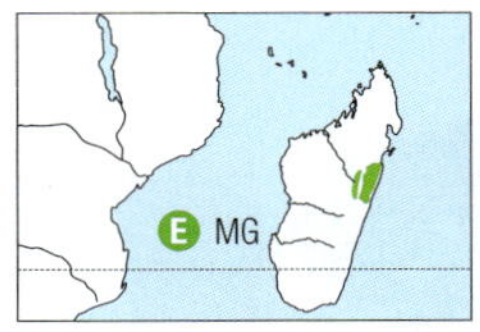

From at least R Maningory S to R Nesivolo & R Mangoro; N & S limits unclear

Betsileo Sportive Lemur
Lepilemur betsileo

FR Lépilémur des Betsileo
DE Betsileo-Wieselmaki
ES Lémur saltador de Betsileo

EN
Monotypic
HB c. 25 cm
T c. 28 cm
W 1–1.2 kg

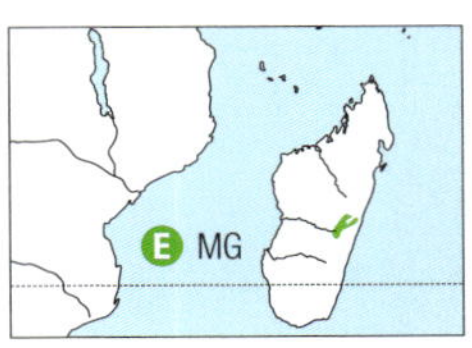

Known only from Fandriana region; N & S limits perhaps R Mangoro & R Namorona, respectively

Manombo Sportive Lemur
Lepilemur jamesorum

FR Lépilémur des James
DE Manombo-Wieselmaki
ES Lémur saltador de los James

CR
Monotypic
HB c. 26 cm
T c. 30 cm
W c. 964 g

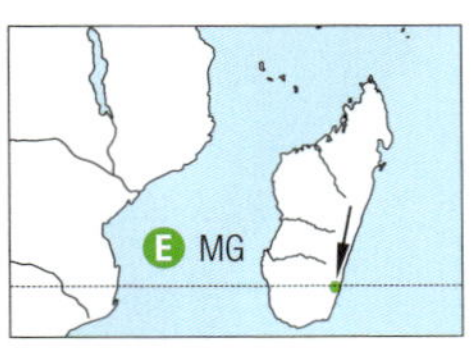

Known from Manombo SR, S of R Manampatrana and N of R Mananara; N & S limits unknown

White-footed Sportive Lemur
Lepilemur leucopus

FR Lépilémur à pieds blancs
DE Weißfuß-Wieselmaki
ES Lémur saltador de pies blancos

EN
Monotypic
HB 19–26 cm
T 22–26 cm
W c. 580 g

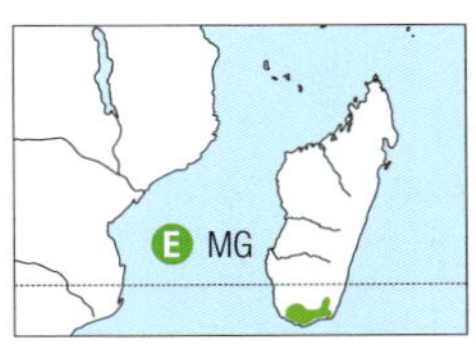

0–300 m

From S of R Onilahy in W to spiny forest part of Andohahela NP (near Tôlanaro) in E

Petter's Sportive Lemur
Lepilemur petteri

FR Lépilémur de Petter
DE Petter-Wieselmaki
ES Lémur saltador de Petter

EN
Monotypic
HB c. 23 cm
T c. 24 cm
W c. 630 g

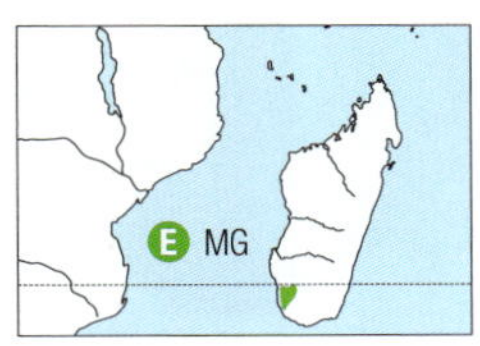

Known from Beza-Mahafaly region, S of R Onilahy and W of R Linta; N limits unclear

Bemaraha Sportive Lemur
Lepilemur randrianasoloi

FR Lépilémur de Randrianasolo
DE Randrianasolo-Wieselmaki
ES Lémur saltador de Randrianasolo

EN
Monotypic
HB c. 29 cm
T c. 28 cm
W c. 775 g

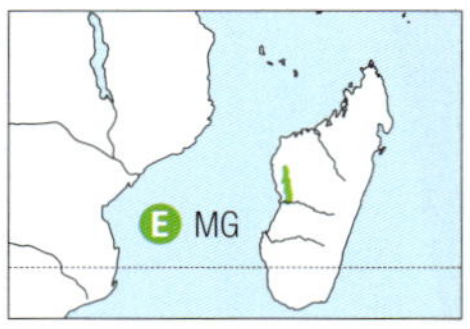

Andramasay, Tsingy de Bemaraha NP & Strict NR; probably between R Tsiribihina and R Manambolo

Antafia Sportive Lemur
Lepilemur aeeclis

FR Lépilémur d'Antafia
DE Antafia-Wieselmaki
ES Lémur saltador de Antafia

EN
Monotypic
HB 21–24 cm
T 24–25 cm
W c. 850 g

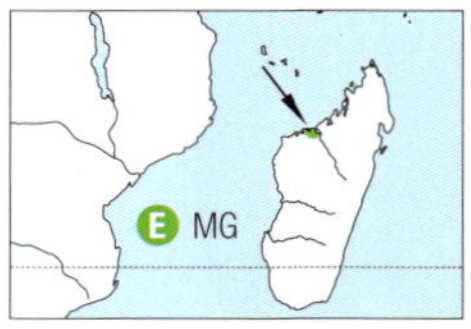

Coast NE of R Mahavavy du Sud and S of R Betsiboka; S limits unknown

CR
Monotypic
HB 24–30 cm
T 24–28 cm
W c. 771 g

Red-tailed Sportive Lemur
Lepilemur ruficaudatus

FR Lépilémur à queue rousse
DE Rotschwanz-Wieselmaki
ES Lémur saltador de cola roja

0–900 m

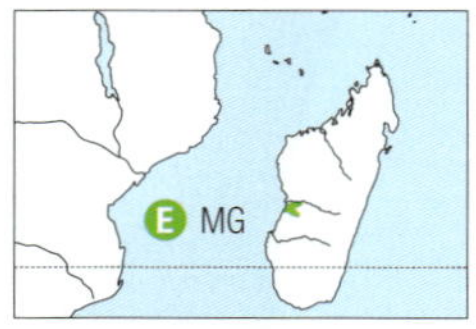

Menabe-Antimena R, Andranomena & Kasijy SR, between R Tsiribihina and R Morondava

Zombitse Sportive Lemur
Lepilemur hubbardorum

FR Lépilémur des Hubbard
DE Zombitse-Wieselmaki
ES Lémur saltador de los Hubbard

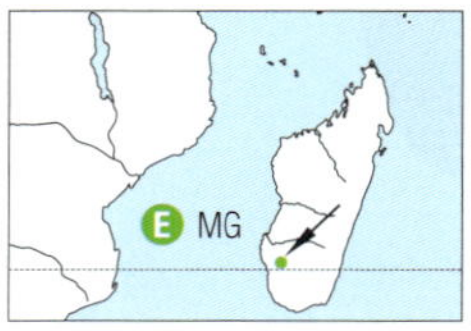

Zombitse-Vohibasia NP region, from R Fiherena S to R Onilahy; N, W & E limits uncertain

EN
Monotypic
HB 23–24 cm
T c. 24 cm
W c. 990 g

Small-toothed Sportive Lemur
Lepilemur microdon

FR Lépilémur à petites dents
DE Kleinzahn-Wieselmaki
ES Lémur saltador de dientes pequeños

EN
Monotypic
HB 27–32 cm
T 25–29 cm
W c. 1 kg

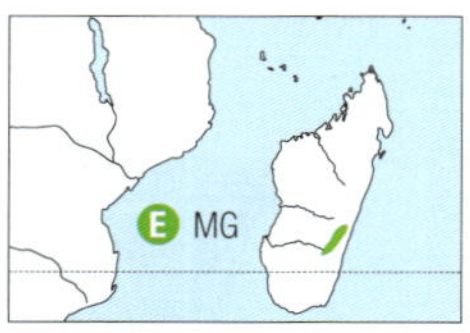

NE to SW trajectory from Ranomafana NP to Andringitra NP; limits poorly known

Ambodimahabibo Sportive Lemur
Lepilemur otto

FR Lépilémur d'Otto
DE Bongolava-Wieselmaki
ES Lémur saltador de Otto

EN
Monotypic
HB 29–30 cm
T 25–27 cm
W 853–872 g

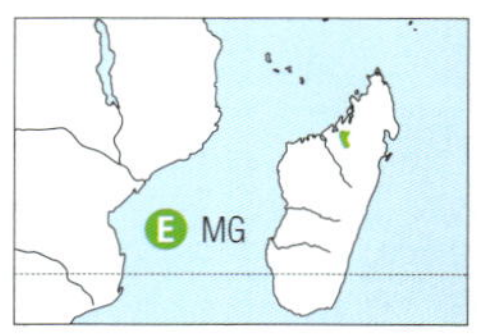

Known only from type locality, limited by R Sofia in N and R Mahajamba in S

Anjiamangirana Sportive Lemur
Lepilemur grewcockorum

FR Lépilémur des Grewcock
DE Sofia-Wieselmaki
ES Lémur saltador de los Grewcock

CR
Monotypic
HB c. 25 cm
T 28–29 cm
W c. 900 g

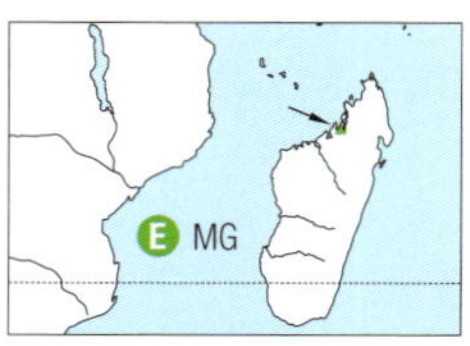

Anjiamangirana region (Anjajavy forests, between Antsohihy and Analala, S of R Maevarano, N of R Sofia)

Milne-Edwards's Sportive Lemur
Lepilemur edwardsi

FR Lépilémur de Milne-Edwards
DE Milne-Edwards-Wieselmaki
ES Lémur saltador de Milne-Edwards

EN
Monotypic
HB 26–29 cm
T 26–29 cm
W c. 965 g

up to 450 m

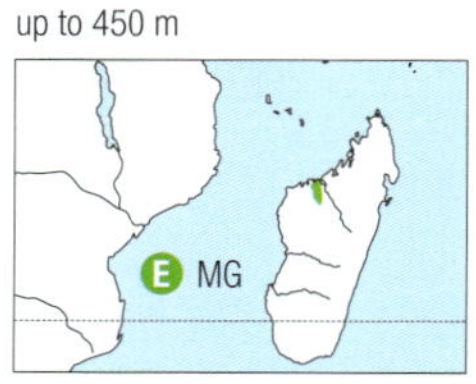

Patchily from N of R Betsiboka to R Mahajamba; N limit unclear

Sahafary Sportive Lemur
Lepilemur septentrionalis

FR Lépilémur septentrional
DE Nördlicher Wieselmaki
ES Lémur saltador de Sahafary

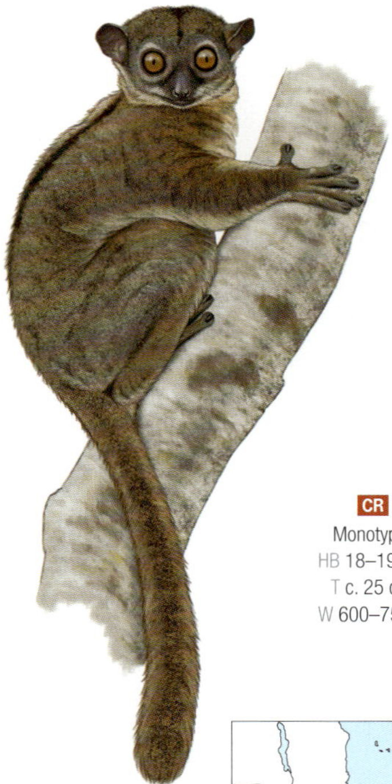

CR
Monotypic
HB 18–19 cm
T c. 25 cm
W 600–750 g

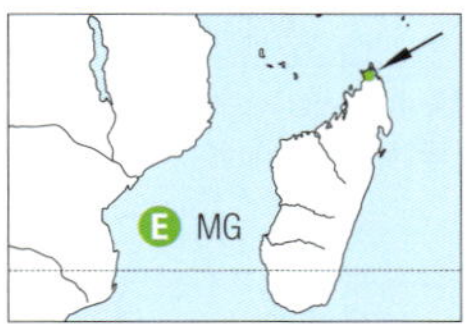

Forest patches near Madirobe & Ankarongana (Sahafary region) and Andrahona

Sahamalaza Sportive Lemur
Lepilemur sahamalaza

FR Lépilémur de Sahamalaza
DE Sahamalaza-Wieselmaki
ES Lémur saltador de Sahamalaza

CR
Monotypic
HB 19–24 cm
T c. 24 cm
W c. 720 g

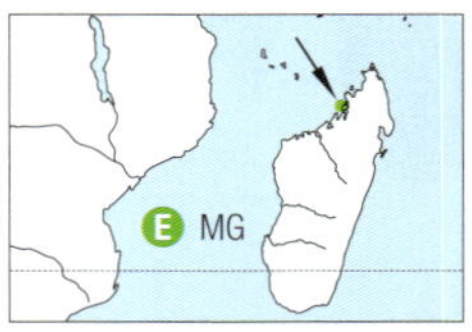

Sahamalaza Pen & adjacent mainland (likely between R Andranomalaza and R Maevarano)

Tsiombikibo Sportive Lemur
Lepilemur ahmansonorum

FR Lépilémur d'Ahmanson
DE Tsiombikibo-Wieselmaki
ES Lémur saltador de Ahmanson

CR
Monotypic
HB 24–30 cm
T 23–34 cm
W c. 550 g

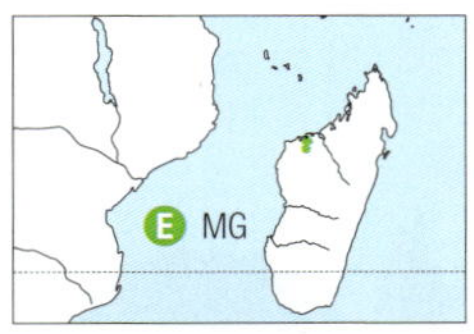

Known only from Tsiombikibo Forest, SW of R Mahavavy du Sud and near Mitsinjo; S limit unknown

Gray's Sportive Lemur
Lepilemur dorsalis

FR Lépilémur à dos gris
DE Gray-Wieselmaki
ES Lémur saltador de Gray

EN
Monotypic
HB 23–29 cm
T 25–28 cm
W c. 740 g

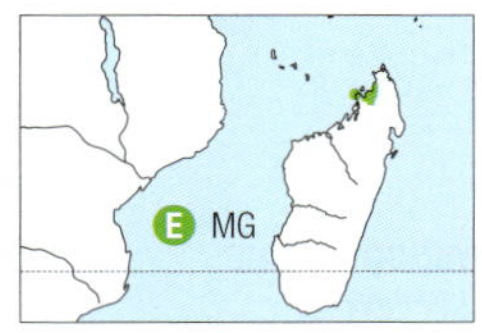

Sambirano region including Ampasindava Pen

Nosy Be Sportive Lemur
Lepilemur tymerlachsoni

FR Lépilémur des Hawk
DE Nosy-Be-Wieselmaki
ES Lémur saltador de Nosy Be

CR
Monotypic
HB c. 23 cm
T c. 25 cm
W c. 840 g

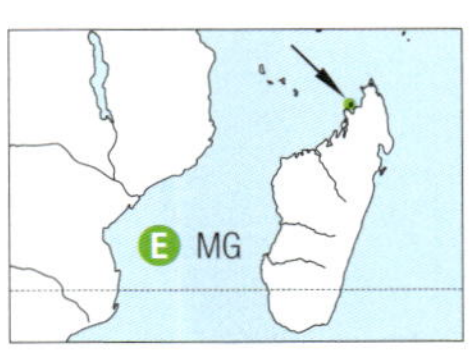

Nosy Be (Lokobe region)

Daraina Sportive Lemur
Lepilemur milanoii

FR Lépilémur de Daraina
DE Daraina-Wieselmaki
ES Lémur saltador de Daraina

EN
Monotypic
HB 22–24 cm
T 25–26 cm
W c. 711 g

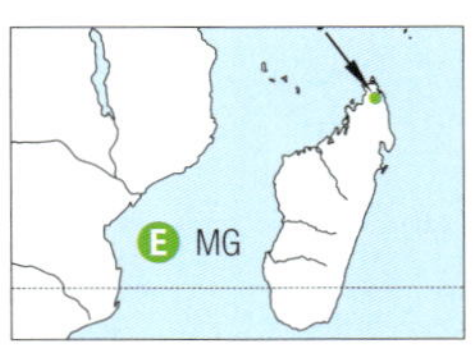

Known from Daraina region S of R Loky, also Analamerana Forest

Ankarana Sportive Lemur
Lepilemur ankaranensis

FR Lépilémur de l'Ankarana
DE Ankarana-Wieselmaki
ES Lémur saltador de Ankarana

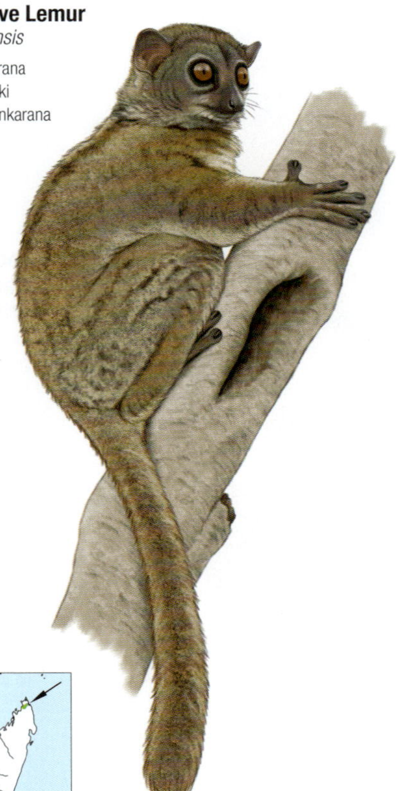

EN
Monotypic
HB c. 22 cm
T c. 27 cm
W c. 730 g

up to 1500 m

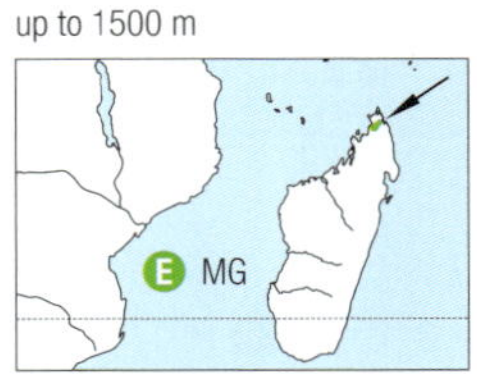

Ankarana, Andrafiamena, Analamerana and Montagne d'Ambre forests

Family LEMURIDAE – Bamboo, True and Ruffed Lemurs

5 genera, 21 species. Dry, subhumid and humid primary and degraded forests, dense bamboo and vines, reed beds, and papyrus marsh. Madagascar.

Gray Bamboo Lemur
Hapalemur griseus

FR Lémur gris
DE Grauer Bambuslemur
ES Lémur del bambú gris

VU
3 sspp
HB 28–30 cm
T 35–37 cm
W 813–967 g

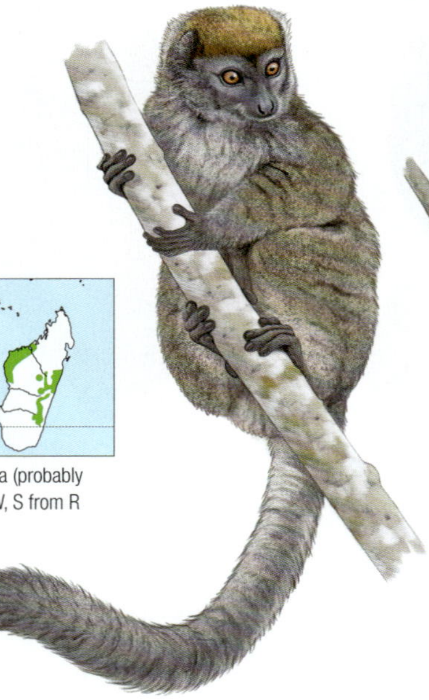

ssp *ranomafanensis*

ssp *gilberti*

ssp *griseus*

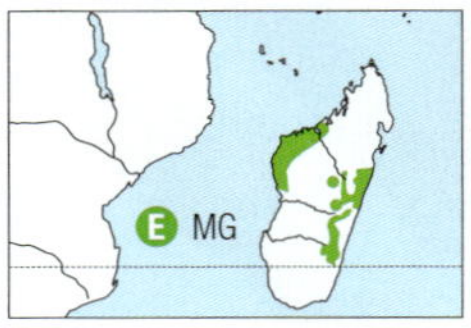

Tsingy de Bemaraha (probably to R Betsiboka) in W, S from R Onibe in E

Genus *HAPALEMUR*

(5 species)

Southern Bamboo Lemur
Hapalemur meridionalis

FR Lémur méridional
DE Südlicher Bambuslemur
ES Lémur del bambú meridional

VU
Monotypic
HB 28–30 cm
T 35–37 cm
W 840–870 g

up to 1600 m

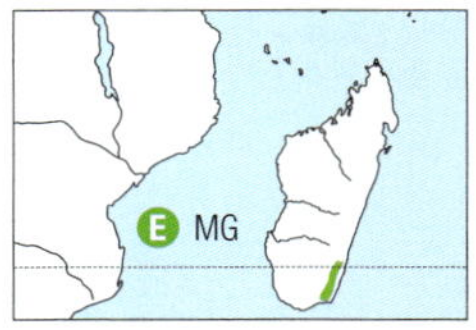

N from Mandena CZ and vicinity of Andohahela NP, limits unknown

Northern Bamboo Lemur
Hapalemur occidentalis

FR Lémur occidental
DE Nördlicher Bambuslemur
ES Lémur del bambú occidental

VU
Monotypic
HB 27–28 cm
T 36–39 cm
W 840–1200 g

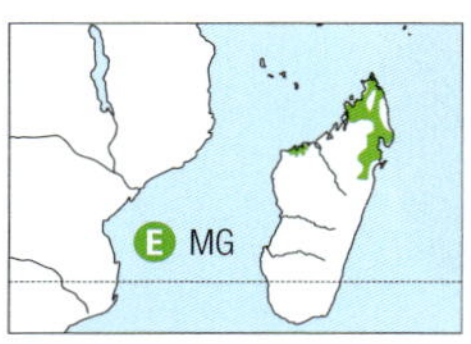

Patchily from Ankarana & Analamerana S to Tsingy de Namaroka in CW and to near L Alaotra in NE

Lac Alaotra Bamboo Lemur
Hapalemur alaotrensis

FR Lémur de l'Alaotra
DE Alaotra-Bambuslemur
ES Lémur del bambú del Lago Alaotra

CR
Monotypic
HB c. 40 cm
T c. 40 cm
W 1.2–1.3 kg

up to 750 m

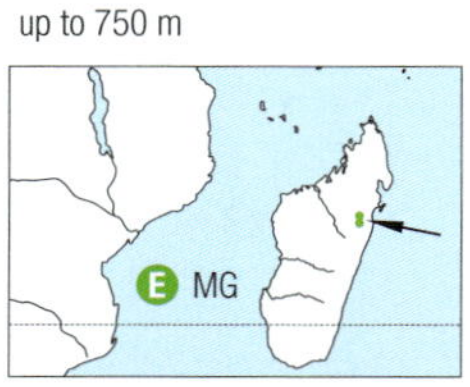

2 subpopulations in L Alaotra (around Belempona Pen in NW, along SW shores)

Golden Bamboo Lemur
Hapalemur aureus

FR Lémur doré
DE Goldener Bambuslemur
ES Lémur del bambú dorado

CR
Monotypic
HB c. 34 cm
T c. 41 cm
W 1.3–1.5 kg

625–1625 m

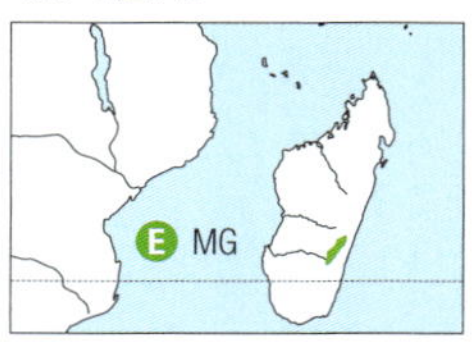

Patchily in Ranomafana NP (N to Miaranony), S to Andringitra NP; perhaps NE to Betsakafandrika

Genus *PROLEMUR*

(1 species)

Greater Bamboo Lemur
Prolemur simus

FR Lémur à nez large
DE Großer Bambuslemur
ES Lémur del bambú grande

CR
Monotypic
HB 40–42 cm
T 45–48 cm
W 2.2–2.5 kg

30–1600 m

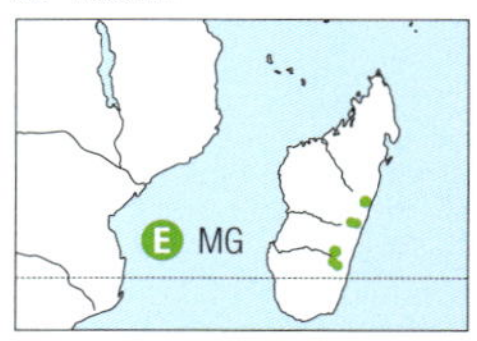

Several rainforests from Zahamena NP in N to near Midongy du Sud NP in S

Genus *LEMUR*

(1 species)

Ring-tailed Lemur
Lemur catta

FR Lémur catta
DE Katta
ES Lémur de cola anillada

E MG

N to Belo-sur-Tsiribihina, E to near Tôlanaro, inland to mountains of Andringitra (SE plateau)

EN
Monotypic
HB 39–46 cm
T 56–63 cm
W c. 2.2 kg

dark morph

Genus *EULEMUR*

(12 species)

Brown Lemur
Eulemur fulvus

FR Lémur brun
DE Brauner Maki
ES Lémur pardo

E MG

Notably disjunct populations in N, NW & CE; N limit unclear

♀

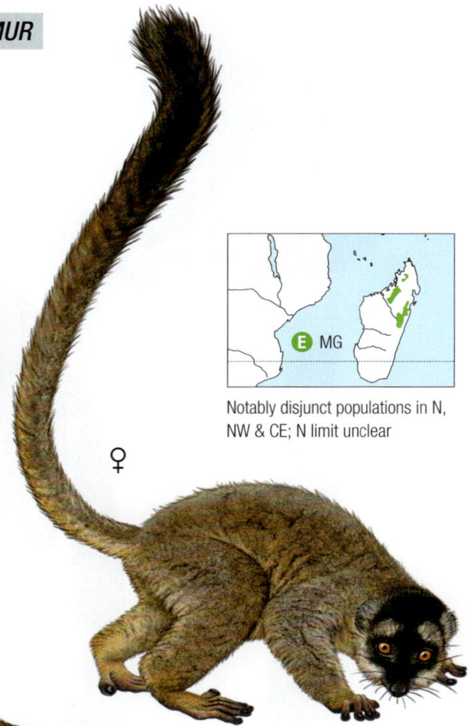

♂

VU
Monotypic
HB 43–50 cm
T 41–51 cm
W c. 1.5 kg

Rufous Brown Lemur
Eulemur rufus

FR Lémur roux
DE Nördlicher Rotstirnmaki
ES Lémur rojizo

VU
Monotypic
HB 40–48 cm
T 45–55 cm
W c. 1.6 kg

♀

♂

E MG

Patchily from R Mahavavy du Sud S at least to R Tsiribihina

Red-fronted Brown Lemur
Eulemur rufifrons

FR Lémur à front roux
DE Südlicher Rotstirnmaki
ES Lémur pardo de frente roja

VU
Monotypic
HB 40–48 cm
T 45–55 cm
W c. 2.1 kg (M), c. 2.5 kg (F)

♂

up to 1700 m

E MG

R Tsiribihina S to R Onilahy & Andringitra Massif in SW, R Mangoro & R Oniver S to R Manampatrana in SE

♀

White-fronted Brown Lemur

Eulemur albifrons

FR Lémur à front blanc
DE Weißkopfmaki
ES Lémur pardo de frente blanca

♀ ♂

VU
Monotypic
HB 39–42 cm
T 50–54 cm
W c. 1.9 kg

0–1600 m

From R Bemarivo S to Mananara-Nord region, with Masoala Pen (possibly Marojejy W to Tsaratanana)

Sanford's Brown Lemur

Eulemur sanfordi

FR Lémur de Sanford
DE Sanford-Maki
ES Lémur pardo de Sanford

♀ ♂

up to 1400 m

Ankarana, Analamerana & Montagne d'Ambre forests, disjunctly in Daraina region, S to R Manambato

EN
Monotypic
HB 38–40 cm
T 50–55 cm
W c. 1.9 kg

White-collared Brown Lemur

Eulemur cinereiceps

FR Lémur à collier blanc
DE Weißwangen-Halsbandmaki
ES Lémur pardo de cabeza gris

♀ ♂

CR
Monotypic
HB 39–40 cm
T 50–55 cm
W c. 2 kg

0–615 m

Strip from above R Manampatrana S to R Mananara, also Manombo & Agnalazaha (coast S of Farafangana)

Red-collared Brown Lemur

Eulemur collaris

FR Lémur à collier roux
DE Rotwangen-Halsbandmaki
ES Lémur pardo de collar

♀ ♂

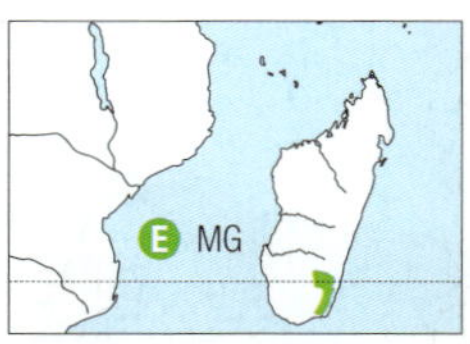

R Mananara S to near Tôlanaro, W to forests of Kalambatritra; isolated at Midongy du Sud NP & Vohipaho

EN
Monotypic
HB 39–40 cm
T 50–55 cm
W c. 2.2 kg

Black Lemur

Eulemur macaco

FR Lémur noir
DE Mohrenmaki
ES Lémur negro

♂ ♀

0–1600 m

R Mahavavy S to R Andranomalaza; Ampasindava Pen, Nosy Be & Nosy Komba, NE of Ambanja

EN
Monotypic
HB 39–45 cm
T 51–65 cm
W 1.9–2 kg

Blue-eyed Black Lemur
Eulemur flavifrons

FR Lémur aux yeux turquoise
DE Blauaugenmaki
ES Lémur negro de ojos azules

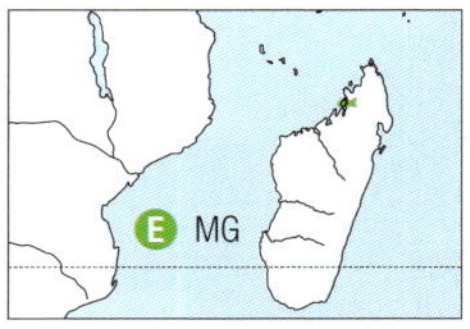

From R Andranomalaza (reported NE of river, E of Manongarivo SR) S to R Maevarano, E to R Sandrakota

CR
Monotypic
HB 39–45 cm
T 51–65 cm
W 1.8–1.9 kg

Crowned Lemur
Eulemur coronatus

FR Lémur couronné
DE Kronenmaki
ES Lémur coronado

EN
Monotypic
HB 34–36 cm
T 41–49 cm
W c. 1.3 kg

0–1400 m

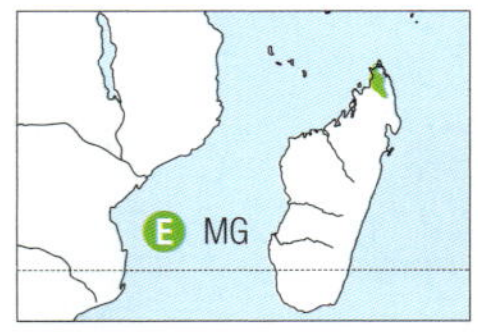

From Cap d'Ambre Pen S along E bank of R Mahavavy, E along R Manambato to just N of Sambava

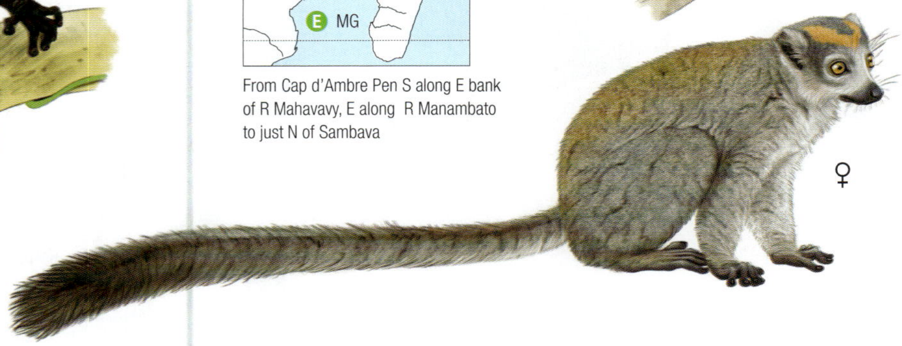

Red-bellied Lemur
Eulemur rubriventer

FR Lémur à ventre roux
DE Rotbauchmaki
ES Lémur de vientre rojo

up to 2400 m

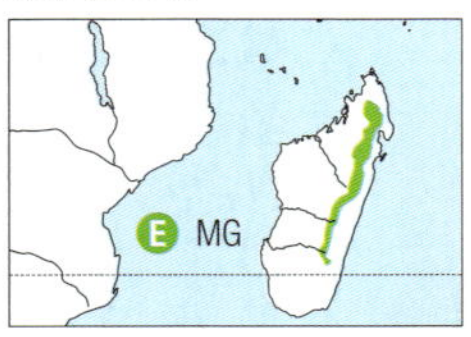

Patchily from Tsaratanana Massif S along thin strip of E rainforest to Pic d'Ivohibe SR & R Manampatran

VU
Monotypic
HB 35–40 cm
T 43–53 cm
W c. 2 kg

Mongoose Lemur
Eulemur mongoz

FR Lémur mongoz
DE Mongozmaki
ES Lémur mangosta

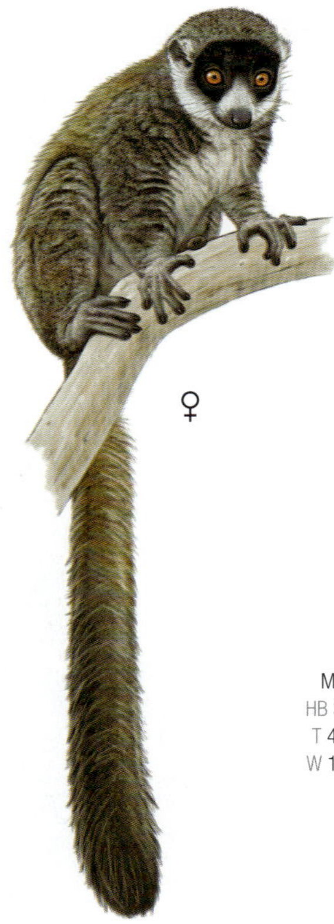

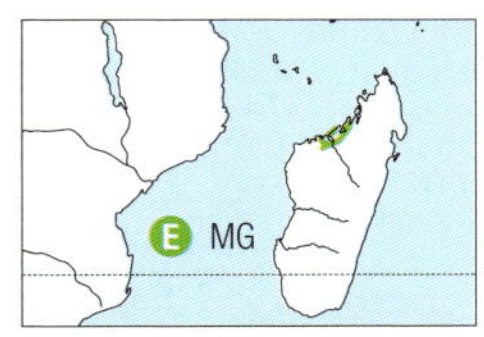

Ambato-Boéni & Ankarafantsika region, N to Bay of Narinda, S to Tsiombikibo CF & L Kinkony; introduced to Comoros

CR
Monotypic
HB 30–35 cm
T 45–48 cm
W 1.1–1.3 kg

Black-and-white Ruffed Lemur
Varecia variegata

FR Vari noir et blanc
DE Schwarzweißer Vari
ES Lémur de collar blanco y negro

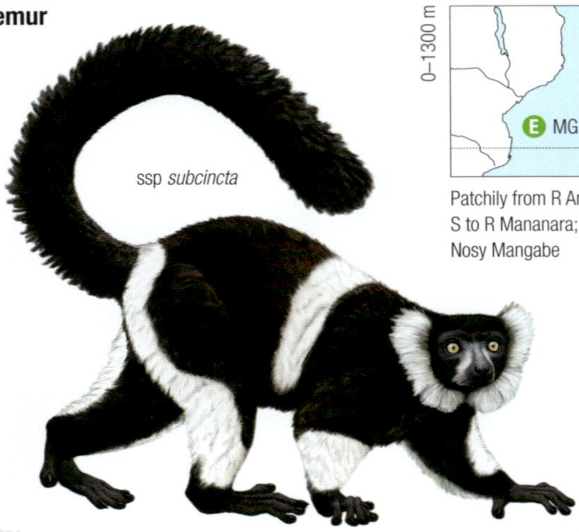

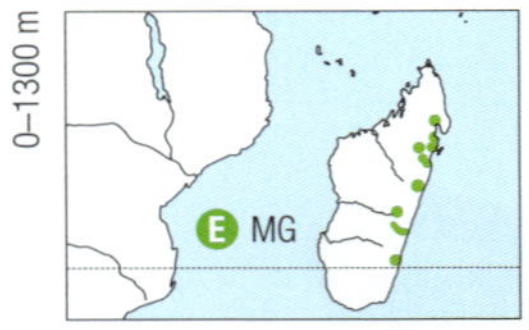

Patchily from R Antainambalana S to R Mananara; introduced to Nosy Mangabe

CR
3 sspp
HB c. 45 cm
T 60–61 cm
W c. 3.6 kg (M), c. 3.7 kg (F)

Genus *VARECIA*

(2 species)

Red Ruffed Lemur
Varecia rubra

FR Vari roux
DE Roter Vari
ES Lémur de collar rojo

CR
Monotypic
HB 50–55 cm
T 60–65 cm
W 3–3.6 kg

up to 1200 m

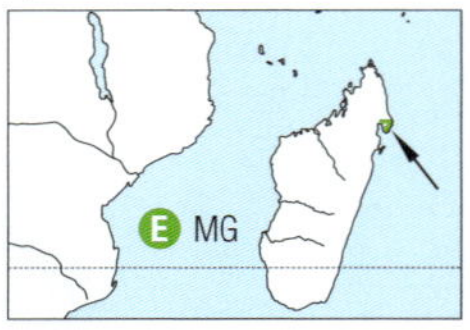

Masoala Pen and region N of Bay of Antongil; W limit unclear, perhaps N to Antalaha in past

Family INDRIIDAE – Woolly Lemurs, Sifakas and Indri

3 genera, 19 species. Tropical humid, dry, lowland, and montane forests. Madagascar.

Genus *AVAHI*

(9 species)

Eastern Woolly Lemur
Avahi laniger

FR Avahi laineux
DE Östlicher Wollmaki
ES Lémur lanudo oriental

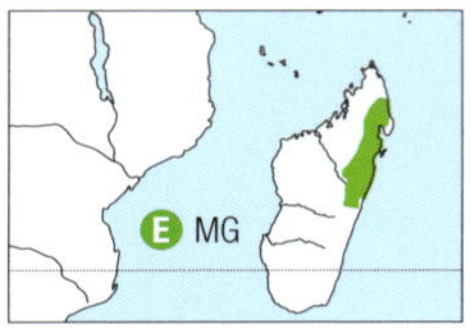

From R Bemarivo S to R Mangoro/R Nesivolo

VU
Monotypic
HB 28–32 cm
T 30–37 cm
W 1.1–1.3 kg

Masoala Woolly Lemur
Avahi mooreorum

FR Avahi des Moore
DE Masoala-Wollmaki
ES Lémur lanudo de los Moore

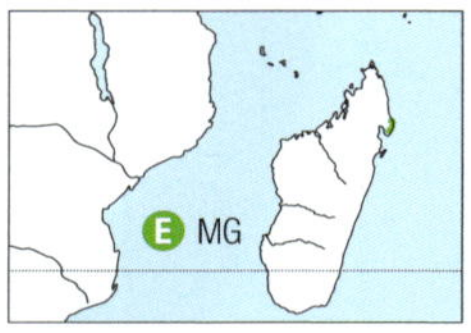

Known only from Masoala NP

EN
Monotypic
HB 28–33 cm
T 29–37 cm
W c. 920 g

Peyrieras's Woolly Lemur
Avahi peyrierasi

FR Avahi de Peyrieras
DE Ranomafana-Wollmaki
ES Lémur lanudo de Peyrieras

VU
Monotypic
HB 26–32 cm
T 28–34 cm
W 1–1.1 kg

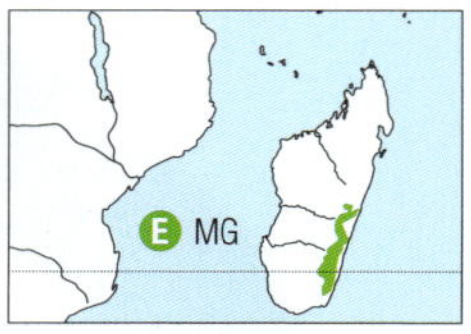

S of R Mangoro/R Nesivolo (Manara, Vatoalatsaka, Sangalampona, Mahasoarivo & Ranomafana forests)

Betsileo Woolly Lemur
Avahi betsileo

FR Avahi des Betsileo
DE Betsileo-Wollmaki
ES Lémur lanudo de Betsileo

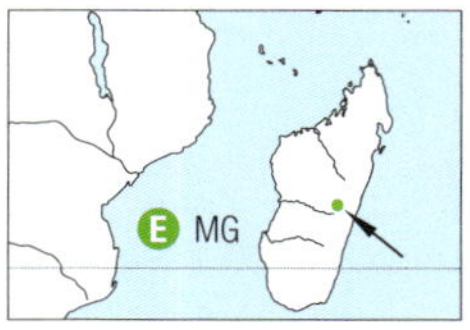

EN
Monotypic
HB 26–31 cm
T 28–34 cm
W 1–1.2 kg

Bemosary CF (Fandriana); limits perhaps from R Mangoro S to R Mananjary

Manombo Woolly Lemur
Avahi ramanantsoavanai

FR Avahi de Manombo
DE Manombo-Wollmaki
ES Lémur lanudo de Manombo

VU
Monotypic
HB 24–31 cm
T 33–40 cm
W 900–1000 g

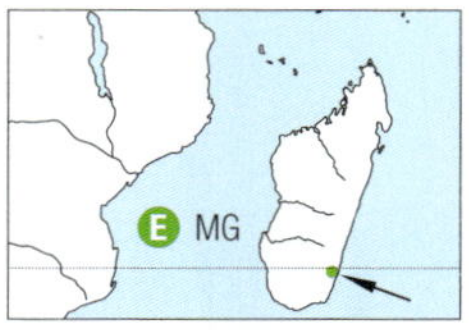

Known only from Manombo SR and Agnalazaha Forest

Southern Woolly Lemur
Avahi meridionalis

FR Avahi méridional
DE Südlicher Wollmaki
ES Lémur lanudo meridional

EN
Monotypic
HB 23–29 cm
T 30–33 cm
W 1–1.1 kg

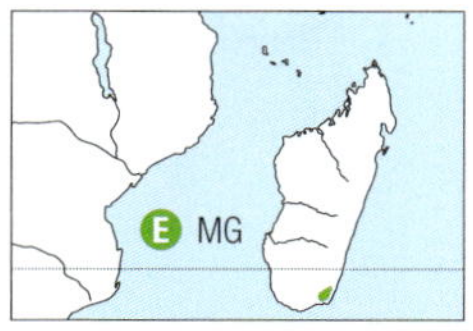

Apparently restricted to Andohahela NP and Sainte-Luce area

Western Woolly Lemur
Avahi occidentalis

FR Avahi occidental
DE Westlicher Wollmaki
ES Lémur lanudo occidental

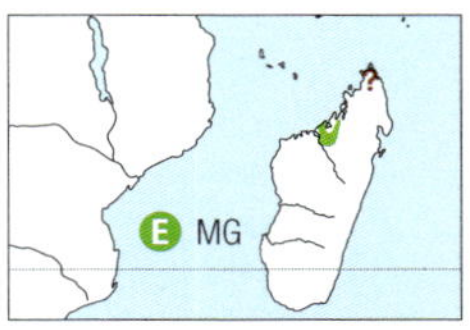

VU
Monotypic
HB 27–30 cm
T 31–38 cm
W 830–1000 g

N & E of R Betsiboka to Bay of Narinda and Ankarafantsika NP; perhaps isolated in Ankarana region

Bemaraha Woolly Lemur
Avahi cleesei

FR Avahi de Cleese
DE Bemaraha-Wollmaki
ES Lémur lanudo de Bemaraha

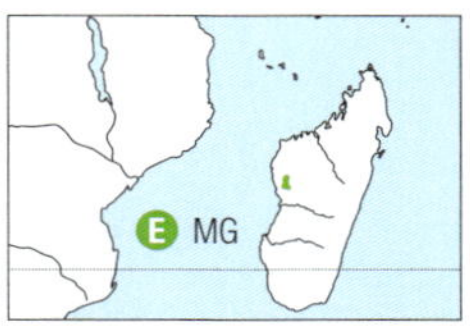

CR
Monotypic
HB 23–31 cm
T 32–36 cm
W 830–980 g

Tsingy de Bemaraha area (Ankindrodro, Ankinajao & Ambalarano forests), N of R Manambolo; N limit unclear

Sambirano Woolly Lemur
Avahi unicolor

FR Avahi unicolore
DE Sambirano-Wollmaki
ES Lémur lanudo de Sambirano

CR
Monotypic
HB 23–31 cm
T 26–30 cm
W 830–920 g

0–700 m

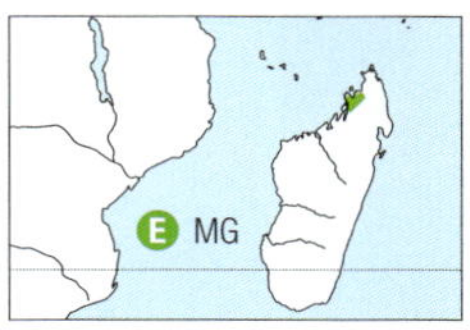

Sambirano region, including Ampasindava Pen; possibly N to R Sambirano and S to R Andranomalaza

Genus *PROPITHECUS*

(9 species)

Verreaux's Sifaka
Propithecus verreauxi

FR Sifaka de Verreaux
DE Kappensifaka
ES Sifaca de Verreaux

0–1300 m

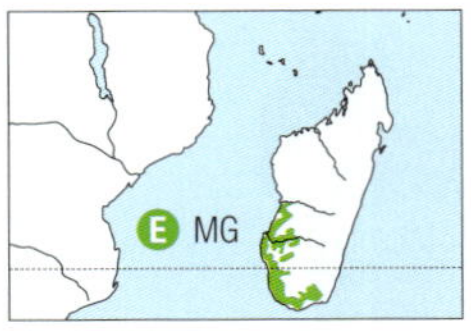

In W up to R Tsiribihina, SE to Andohahela NP (Mangatsiaka) & Nahampoana R (where probably introduced)

CR
Monotypic
HB 40–48 cm
T 50–60 cm
W c. 2.9 kg

dark morph

pale morph

Decken's Sifaka
Propithecus deckenii

FR Sifaka de der Decken
DE Von-der-Decken-Sifaka
ES Sifaca de Decken

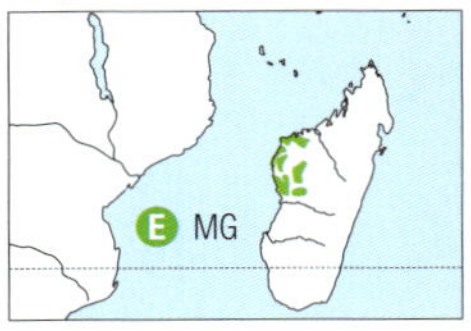

Mainly between R Manambolo and R Mahavavy; S limit does not extend to R Tsiribihina

CR
Monotypic
HB 42–48 cm
T 50–60 cm
W 2.6–2.9 kg

Crowned Sifaka
Propithecus coronatus

FR Sifaka couronné
DE Kronensifaka
ES Sifaca coronado

CR
Monotypic
HB 39–45 cm
T 48–57 cm
W 3.2–3.7 kg

0–700 m

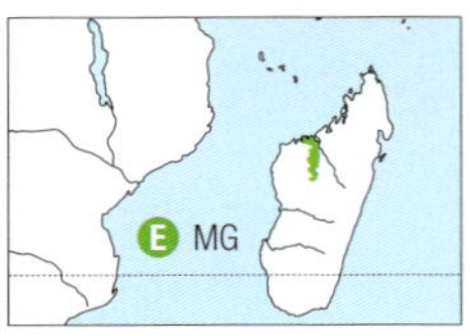

Between R Mahavavy and R Betsiboka in N, inland S to R Manambola, R Tsibirhina, R Mania & R Sakay

Coquerel's Sifaka
Propithecus coquereli

FR Sifaka de Coquerel
DE Coquerel-Sifaka
ES Sifaca de Coquerel

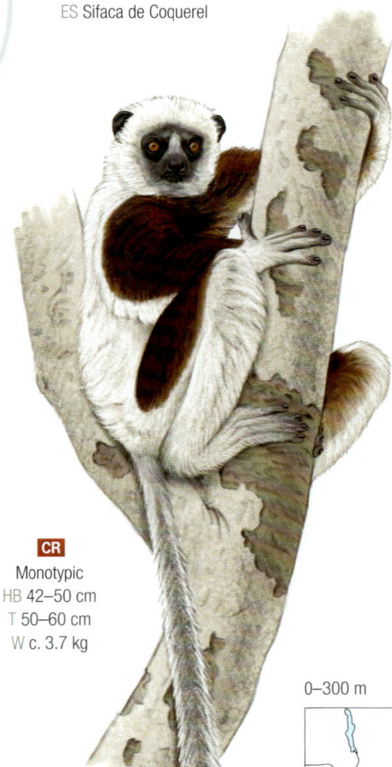

CR
Monotypic
HB 42–50 cm
T 50–60 cm
W c. 3.7 kg

0–300 m

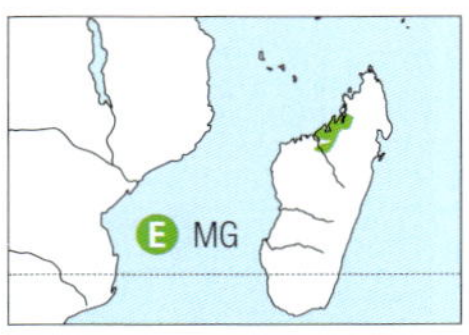

From near Bealalana in N to R Betsiboka; S to Ambato Boeni, E to near Antetemasy (W of Befandriana Nord)

Tattersall's Sifaka
Propithecus tattersalli

FR Sifaka de Tattersall
DE Goldkronensifaka
ES Sifaca de Tattersall

CR
Monotypic
HB 45–47 cm
T 42–47 cm
W 3.4–3.6 kg

50–700 m

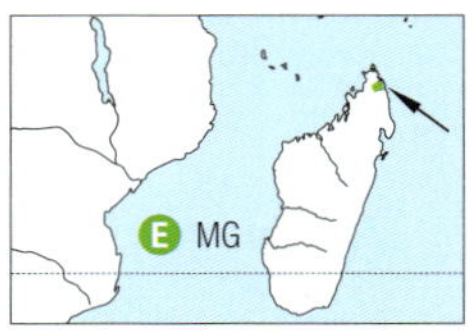

Small area in Ampandraha, Madirabe & Daraina districts bounded by R Loky in N and R Manambato in S

Diademed Sifaka
Propithecus diadema

FR Sifaka à diadème
DE Diademsifaka
ES Sifaca de diadema

CR
Monotypic
HB 50–55 cm
T 44–50 cm
W 5.7–6.8 kg

200–1600 m

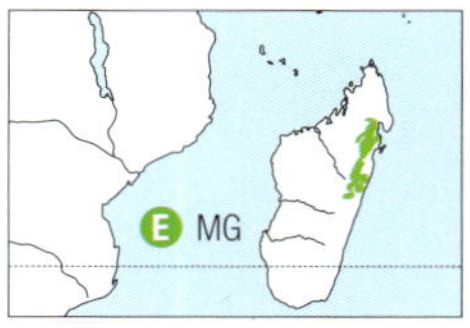

From R Mananara S to R Mangoro & R Onive; limits unknown but possibly widely distributed

Milne-Edwards's Sifaka
Propithecus edwardsi

FR Sifaka de Milne-Edwards
DE Milne-Edwards-Sifaka
ES Sifaca de Milne-Edwards

EN
Monotypic
HB 42–52 cm
T 41–48 cm
W c. 5.5 kg

600–1600 m

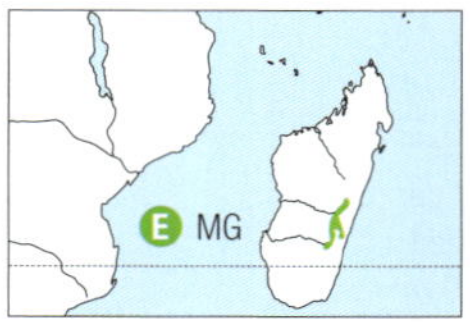

From R Mangoro & R Onive S to R Rienana in Andringitra NP

Silky Sifaka
Propithecus candidus

FR Sifaka soyeux
DE Seidensifaka
ES Sifaca sedoso

CR
Monotypic
HB 48–54 cm
T 45–51 cm
W 5–6 kg

300–1900 m

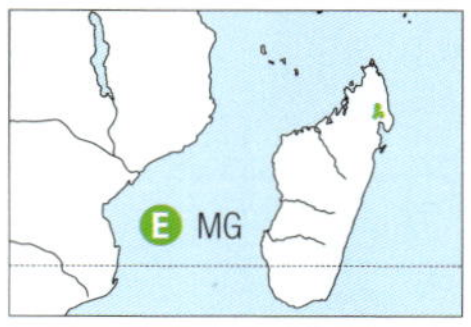

Patchily in forest belt from Marojejy NP S to Andaparaty (Makira NP), just N of R Antainambalana

Perrier's Sifaka
Propithecus perrieri

FR Sifaka de Perrier
DE Schwarzer Diademsifaka
ES Sifaca de Perrier

CR
Monotypic
HB 43–47 cm
T 42–45 cm
W 4.4–4.5 kg

0–500 m

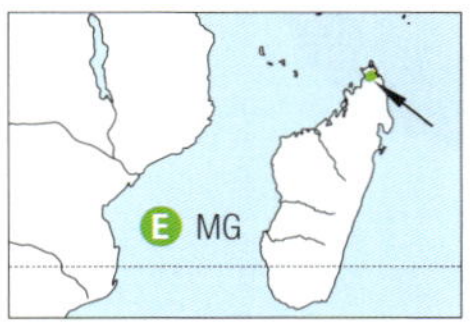

Analamerana & Andrafiamena massifs and E edge of Ankarana forests

Genus *INDRI*

(1 species)

Indri
Indri indri

FR Indri
DE Indri
ES Indri

CR
Monotypic
HB 64–72 cm
T c. 5 cm
W 5.8–9 kg

variants

0–1800 m

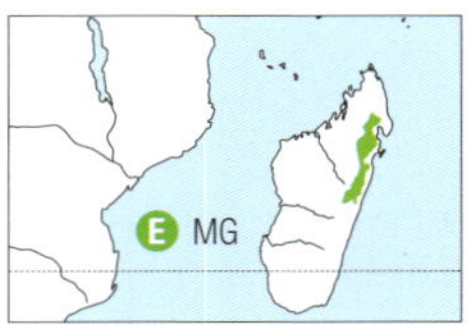

Roughly from Anjanaharibe-Sud SR S to Anosibe an'ala CF

Family DAUBENTONIIDAE – Aye-aye

1 genus, 1 species. All forest types except spiny forest. Madagascar.

Genus *DAUBENTONIA*

(1 species)

Aye-aye
Daubentonia madagascariensis
FR Aye-aye
DE Fingertier
ES Aye-aye

EN
Monotypic
HB 30–37 cm
T 44–53 cm
W 2.4–2.6 kg

0–1875 m

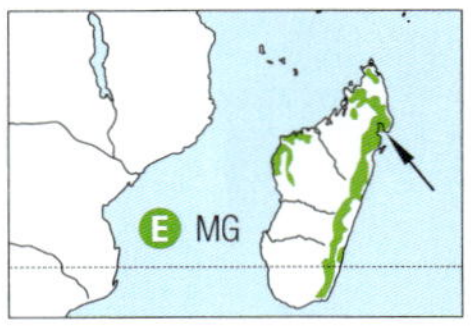

Mainly in E, N & CW but in low numbers across most coastal areas; introduced to Nosy Mangabe

Family GALAGIDAE – Galagos

6 genera, 19 species. Tropical lowland, swamp, riparian, montane, and submontane forests and forest-edge, thicket and scrub, hilly and semi-arid woodlands, savannas, and tree plantations. Afrotropical Region.

Genus *EUOTICUS*

(2 species)

Northern Needle-clawed Galago
Euoticus pallidus
FR Galago pâle
DE Nördlicher Kielnagelgalago
ES Gálago de garras septentrional

ssp *pallidus*

ssp *talboti*

NT
2 sspp
HB 16–20 cm
T 28–32 cm
W 200–260 g

up to 1600 m

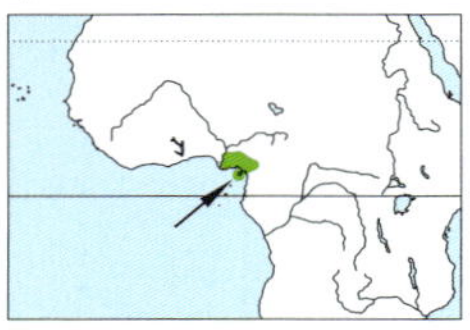

Bioko I (GQ), patchily between R Niger (SE NG) and R Sanaga (W CM)

LC
Monotypic
HB 19–23 cm
T 26–32 cm
W 270–360 g

Southern Needle-clawed Galago
Euoticus elegantulus
FR Galago élégant
DE Südlicher Kielnagelgalago
ES Gálago de garras meridional

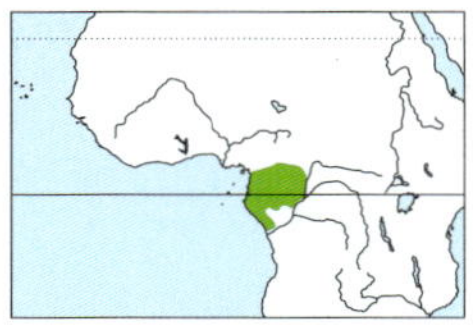

Primary forested regions of CM (S of R Sanaga), GQ, GA, CG and CD (N of R Congo & R Ubangi); probably CF

Genus *GALAGOIDES*

(3 species)

Demidoff's Dwarf Galago
Galagoides demidoff

FR Galago de Demidoff
DE Demidoff-Zwerggalago
ES Gálago pequeño de Demidoff

LC
2 sspp
HB 12–17 cm
T 14–22 cm
W 55–100 g

ssp *demidoff*

ssp *poensis*

up to 1600 m

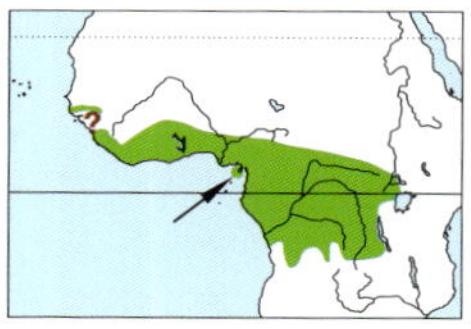

W & C Africa from SN E to SE UG and NW TZ, S to C AO, S DC and NW ZM; Bioko I (GQ)

Thomas's Dwarf Galago
Galagoides thomasi

FR Galago de Thomas
DE Thomas-Zwerggalago
ES Gálago pequeño de Thomas

LC
Monotypic
HB 14–20 cm
T 15–24 cm
W 60–150 g

up to 2690 m

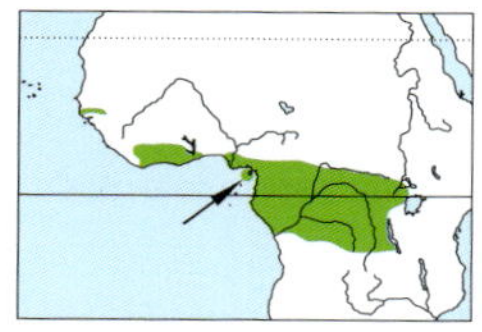

W & C Africa from S SN E to W UG and NW TZ, S to NE ZA, DC and (perhaps) AO; Bioko I (GQ); limits uncertain

Angolan Dwarf Galago
Galagoides kumbirensis

FR Galago de Kumbira
DE Angola-Zwerggalago
ES Gálago pequeño de Angola

NT
Monotypic
HB 15–17 cm
T 18–20 cm

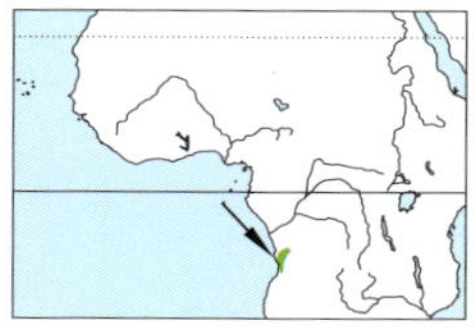

Known from 4 localities in NW AO and SW DC

Genus *PARAGALAGO*

(5 species)

Rondo Dwarf Galago
Paragalago rondoensis

FR Galago du Rondo
DE Rondo-Zwerggalago
ES Gálago pequeño de Rondo

EN
Monotypic
HB 9–12 cm
T 16–22 cm
W 35–70 g

100–900 m

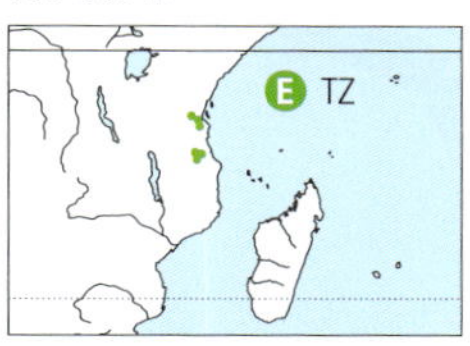

Moist-forest patches (Litipo, Rondo, Ziwani, Pugu Hills/ Kazimzumbwi, Zareninge, Pande Game R)

Mountain Dwarf Galago
Paragalago orinus

FR Galago des Usambara
DE Uluguru-Zwerggalago
ES Gálago pequeño de montaña

VU
Monotypic
HB 12.5–15 cm
T 17–19 cm
W c. 80 g

1200–2000 m

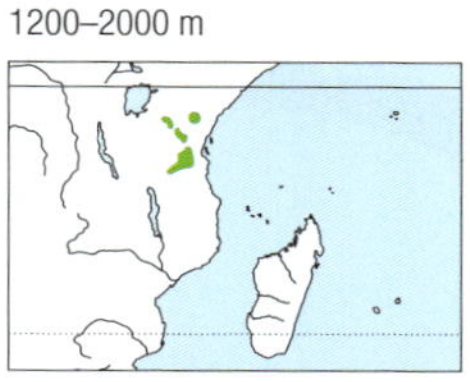

In many Eastern Arc Mts of SE KE & E TZ (with Taita Hills, West & East Usambara, Uluguru, Udzungwa, Rubeho)

Mozambique Dwarf Galago
Paragalago granti

FR Galago de Grant
DE Mosambik-Zwerggalago
ES Gálago pequeño de Mozambique

LC
Monotypic
HB 9–12 cm
T 16–22 cm
W 35–70 g

up to 1800 m

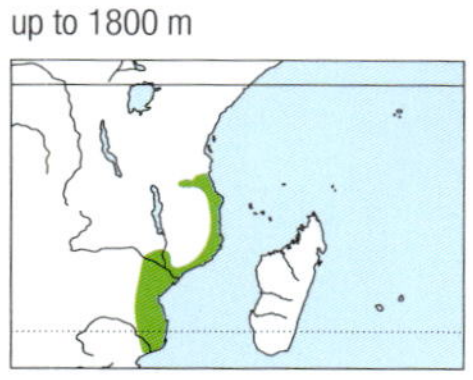

Coast from R Rufiji (SE TZ) to R Limpopo (SE MZ) and montane areas (e.g. Thyolo Mts, MW) W to Chimanimani (ZW)

Kenya Coast Dwarf Galago
Paragalago cocos

FR Galago des cocotiers
DE Kokos-Zwerggalago
ES Gálago pequeño de Kenia

LC
Monotypic
HB c. 17 cm
T c. 27 cm
W 155–180 g

0–350 m

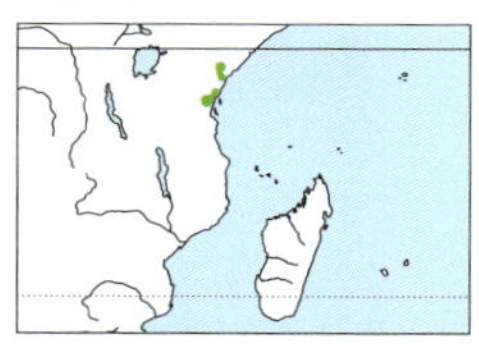

Coastal strip and gallery forests of E KE, NE TZ and possibly S SO

Tanzania Coast Dwarf Galago
Paragalago zanzibaricus

FR Galago de Zanzibar
DE Sansibar-Zwerggalago
ES Gálago pequeño de Tanzania

0–1070 m

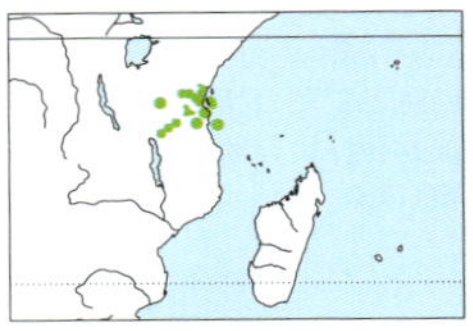

NE TZ (Udzungwa at Kihansi, Uluguru, East Usambara Mts S to R Rufiji), Zanzibar Arch (Unguja, Mafia)

ssp *udzungwensis*

ssp *zanzibaricus*

NT
2 sspp
HB 12–18 cm
T 17–24 cm
W 100–200 g

Genus *GALAGO*

(4 species)

Spectacled Lesser Galago
Galago matschiei

FR Galago de Matschie
DE Matschie-Galago
ES Gálago menor de anteojos

LC
Monotypic
HB 15–20 cm
T 19–25 cm
W 170–250 g

740–1600 m

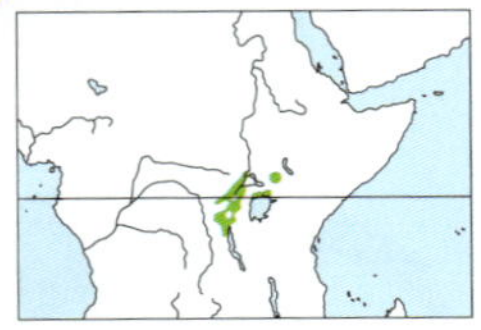

Albertine Rift of NE DC, UG, W RW & NW BI (from L Albert S to Itombwe Massif, E to R Nile)

Somali Lesser Galago
Galago gallarum

FR Galago de Somalie
DE Somali-Galago
ES Gálago menor de Somalia

LC
Monotypic
HB 13–20 cm
T 21–29 cm
W 180–250 g

150–1200 m

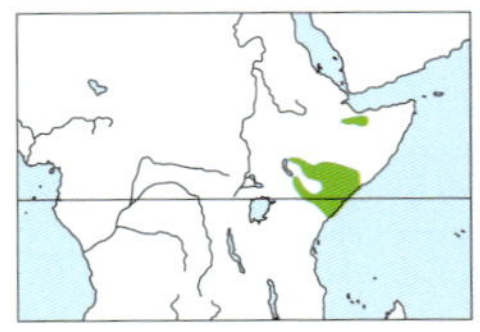

NE & S ET, E KE, NW & W SO (patchily between R Shebelle & R Tana, W to Rift Lakes); perhaps SE SS & NE UG

Northern Lesser Galago
Galago senegalensis

FR Galago du Sénégal
DE Senegal-Galago
ES Gálago menor de Senegal

LC
4 sspp
HB 13–21 cm
T 20–30 cm
W 110–300 g

ssp *braccatus*

ssp *dunni*

ssp *sotikae*

ssp *senegalensis*

up to 2400 m

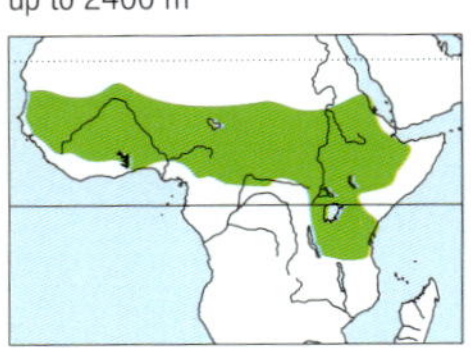

W & C Africa from SN to ER, ET (perhaps SO), S to KE and S TZ; limits uncertain

Southern Lesser Galago
Galago moholi

FR Galago moholi
DE Moholi-Galago
ES Gálago menor meridional

LC
Monotypic
HB 12–17 cm
T 16–28 cm
W 95–255 g

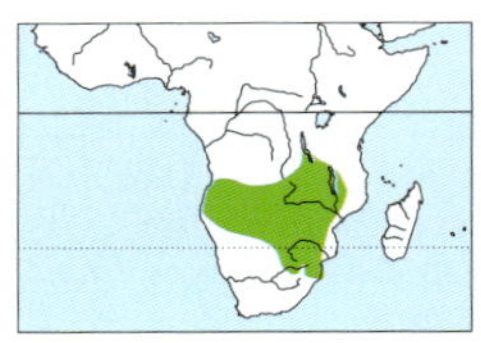

AO, SE CD, ZM, MW, W MZ, ZW, N NA, N & E BW, S to R Orange & R Limpopo (NE ZA); perhaps TZ, RW & BI

Genus *SCIUROCHEIRUS*

(3 species)

Bioko Squirrel Galago
Sciurocheirus alleni

FR Galago d'Allen
DE Bioko-Buschwaldgalago
ES Gálago ardilla de Bioko

NT
2 sspp
HB 17–28 cm
T 22–30 cm
W 200–410 g

ssp *alleni*

ssp *cameronensis*

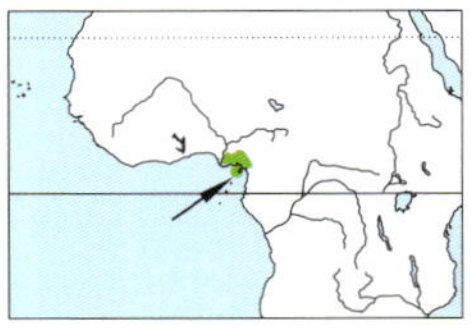

Bioko I (GQ), forested regions between R Niger (SE NG) and R Sanaga (NW CM), perhaps CI

Gabon Squirrel Galago
Sciurocheirus gabonensis

FR Galago du Gabon
DE Gabun-Buschwaldgalago
ES Gálago ardilla de Gabón

up to 800 m

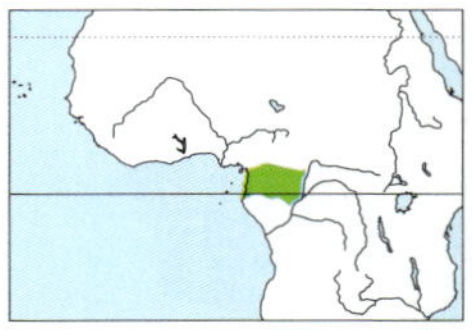

R Sanaga (W CM) to GQ, SW CF, CG & GA (S to R Ogooué), E possibly to R Congo & R Ubangi; perhaps CD & AO

LC
Monotypic
HB 16–24 cm
T 23–30 cm
W 200–500 g

Makande Squirrel Galago
Sciurocheirus makandensis

FR Galago de Makandé
DE Makande-Buschwaldgalago
ES Gálago ardilla de Makande

DD
Monotypic

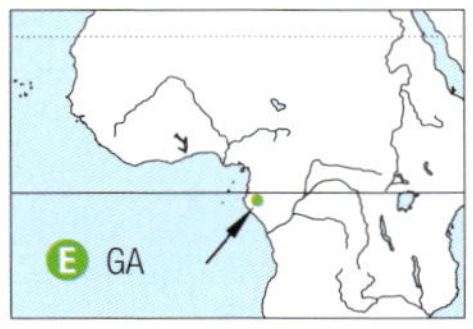

Known only from type locality in Makandé (C GA)

Genus *OTOLEMUR*

(2 species)

Thick-tailed Greater Galago
Otolemur crassicaudatus

FR Galago à queue touffue
DE Großohr-Riesengalago
ES Gálago mayor de cola gruesa

LC
4 sspp
HB 26–40 cm
T 30–50 cm
W 1.1–1.8 kg

ssp *argentatus*

ssp *monteiri*

ssp *argentatus*
melanistic morph

ssp *crassicaudatus*

ssp *kirkii*

up to 1800 m

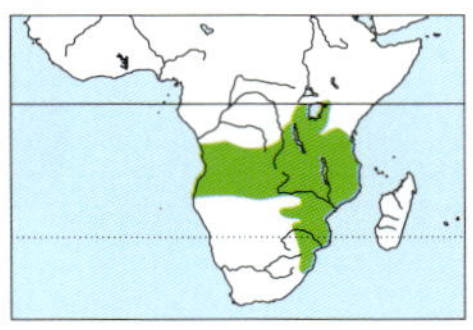

AO, S CD, UG and S KE, S to ZM, ZW, E BW, NE ZA and SZ

Garnett's Greater Galago
Otolemur garnettii

FR Galago de Garnett
DE Kleinohr-Riesengalago
ES Gálago mayor de Garnett

ssp *garnettii*

ssp *panganiensis*

ssp *kikuyuensis*

ssp *lasiotis*

0–2500 m

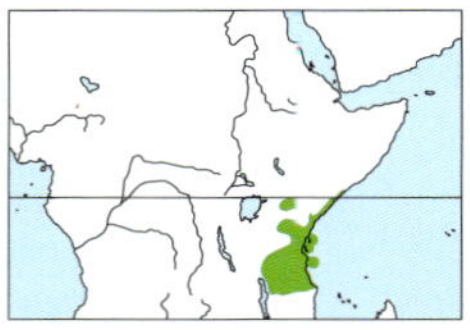

S SO (S of R Jubba), C, E & S KE, E TZ, with Zanzibar Arch (Unguja, Pemba, Mafia); perhaps N MZ & N MW

LC
4 sspp
HB 23–34 cm
T 30–44 cm
W 550–1200 g

Family LORISIDAE – Angwantibos, Pottos and Lorises

5 genera, 16 species. Swamp, lowland, flooded, riparian, semi-evergreen, dry dipterocarp, and mid-elevation to montane forests and secondary and forest-edge, coastal scrub, woodlands, and plantations. Afrotropical and Indo-Malayan Regions.

Genus *ARCTOCEBUS*

(2 species)

Calabar Angwantibo
Arctocebus calabarensis

FR Potto de Calabar
DE Bärenmaki
ES Poto dorado de Calabar

NT
Monotypic
HB 26–35 cm
T c. 1 cm
W 230–465 g

WC Africa, from E R Niger (SE NG) to R Sanaga (SW CM)

Golden Angwantibo
Arctocebus aureus

FR Potto doré
DE Gold-Bärenmaki
ES Poto dorado

LC
Monotypic
HB 22–26 cm
T 1–2 cm
W 200–270 g

CM (S R Sanaga) S & E to Congo/Ubangi river system in CF, GQ (Mbini), CG, GA and AO; perhaps W CD

Genus *PERODICTICUS*

(3 species)

West African Potto
Perodicticus potto

FR Potto de Bosman
DE Potto
ES Poto occidental

ssp *juju*

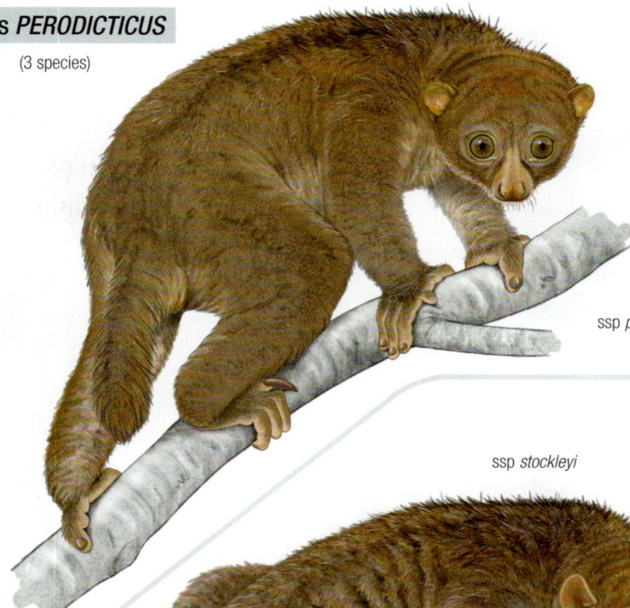

ssp *potto*

NT
2 sspp
HB c. 30 cm
T 4–6 cm
W 850–1000 g

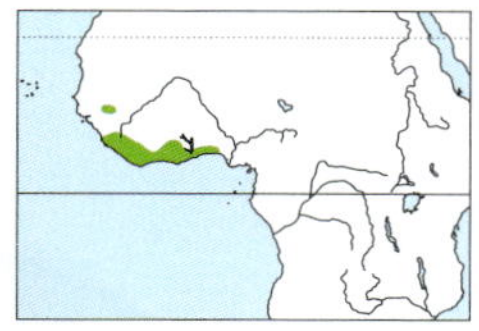

SE SN, SE GN, SL, LR, CI, GH, S TG, S BJ and SE NG; unconfirmed in GM, GW and most of GN

Central African Potto
Perodicticus edwardsi

FR Potto de Milne-Edwards
DE Kamerun-Potto
ES Poto de Milne-Edwards

LC
Monotypic
HB 32–40 cm
T c. 10 cm
W 1–1.6 kg

up to 1500 m

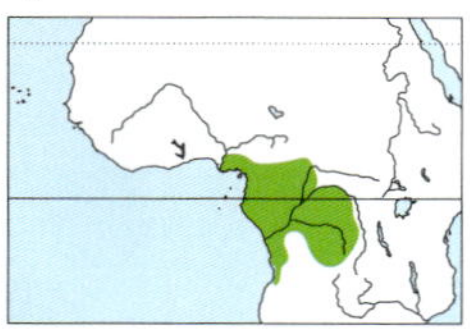

R Niger (S NG) E through CM, CF, GQ, GA, CG, CD (S of R Congo E to Irneti), S to N AO

East African Potto
Perodicticus ibeanus

FR Potto est-africain
DE Ostafrika-Potto
ES Poto oriental

ssp *stockleyi*

ssp *ibeanus*

LC
2 sspp
HB c. 35 cm
T c. 5 cm
W 1–1.2 kg

above 600 m

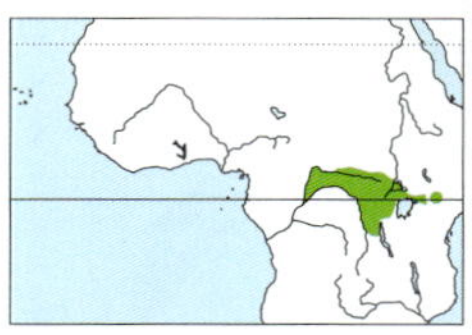

CD (E & S of R Ubangi, N of R Congo, E of R Lualaba), NW BI, RW, UG, SW KE (Kakamega & Nandi forests, Mt Kenya)

Genus *LORIS*

(2 species)

Gray Slender Loris
Loris lydekkerianus

FR Loris de Lydekker
DE Grauer Schlanklori
ES Loris esbelto gris

NT
2 sspp
HB 21–26 cm
T vestigial

ssp *lydekkerianus*

ssp *malabaricus*

up to 2000 m

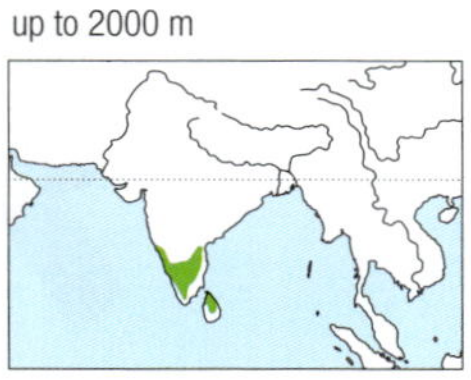

IN in Eastern & Western Ghats (Andhra Pradesh, Karnataka, Tamil Nadu, Kerala), LK

Red Slender Loris
Loris tardigradus

FR Loris grêle
DE Roter Schlanklori
ES Loris esbelto rojo

EN
Monotypic
HB 18–21 cm
T vestigial
W 85–220 g

up to 2300 m

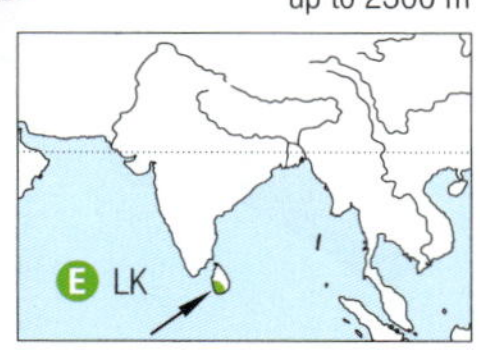

Western, Southern & Central prov

Genus *XANTHONYCTICEBUS*

(1 species)

Pygmy Slow Loris
Xanthonycticebus pygmaeus

FR Loris pygmée
DE Kleiner Plumplori
ES Loris perezoso pigmeo

EN
Monotypic
HB 19.5–23 cm
T c. 1.8 cm
W 420–424 g

summer

winter

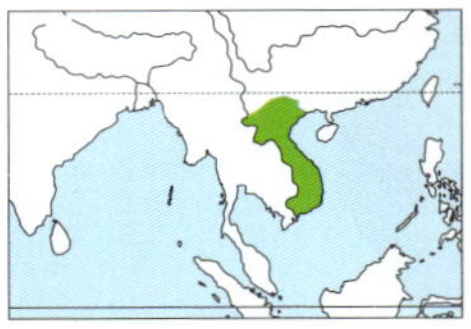

LA, VN & E KH (E of R Mekong); records from S CN (SE Yunnan) uncertain

Genus *NYCTICEBUS*

(8 species)

Bengal Slow Loris
Nycticebus bengalensis

FR Loris du Bengale
DE Bengalen-Plumplori
ES Loris perezoso de Bengala

dark morph

pale morph

EN
Monotypic
HB 34–38 cm
T vestigial
W 650–2100 g

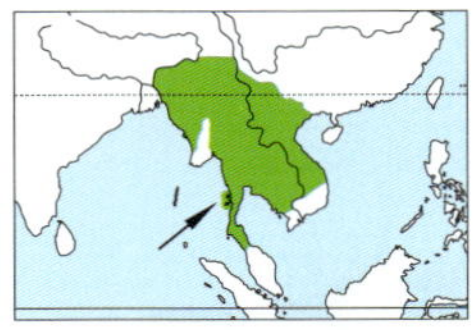

E BD, NE IN, SC CN (S Yunnan, possibly S Guangxi), MM (with Mergui Arch), TH, LA, VN, KH; perhaps N Pen MY

Malayasian Slow Loris
Nycticebus coucang

FR Loris lent
DE Sunda-Plumplori
ES Loris perezoso de Sonda

EN
2 sspp
HB 24–29 cm
T vestigial
W 483–706 g

ssp *coucang*

ssp *insularis*

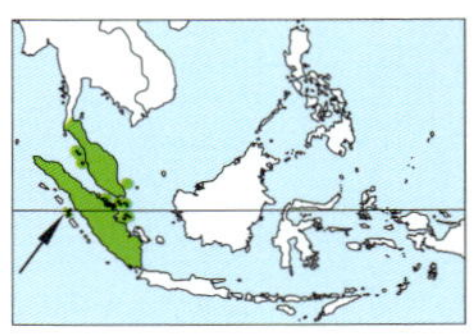

S TH, Pen MY (with Penang, Langkawi & Tioman Is), SG, ID (Sumatra, Malacca Straits Is, Riau Arch)

Hiller's Slow Loris
Nycticebus hilleri

FR Loris de Hiller
DE Hiller-Plumplori
ES Loris perezoso de Hiller

EN
Monotypic
HB 26–29 cm
T vestigial
W 532–845 g

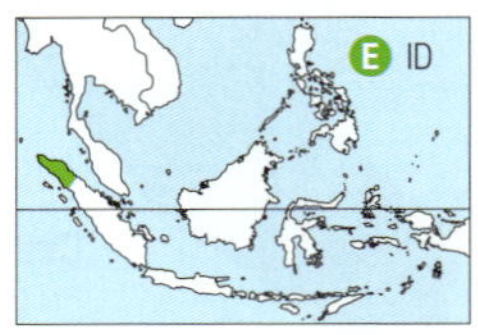

Apparently N Sumatra, mainly highlands of North Sumatra & Aceh prov

Javan Slow Loris
Nycticebus javanicus

FR Loris de Java
DE Java-Plumplori
ES Loris perezoso de Java

CR
Monotypic
HB c. 25 cm
T vestigial
W 750–1150 g

0–1800 m

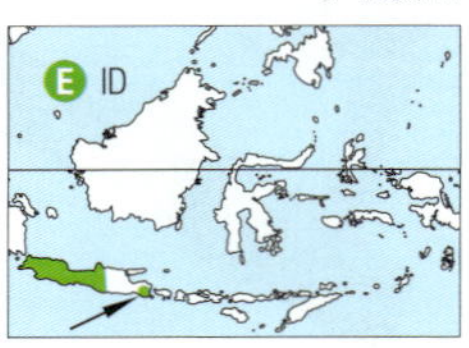

W & C Java, isolated locality in E

Philippine Slow Loris
Nycticebus menagensis

FR Loris de Menage
DE Philippinen-Plumplori
ES Loris perezoso de Filipinas

VU
2 sspp
HB 27–30 cm
T vestigial
W 265–800 g

ssp *menagensis*

ssp *natunae*

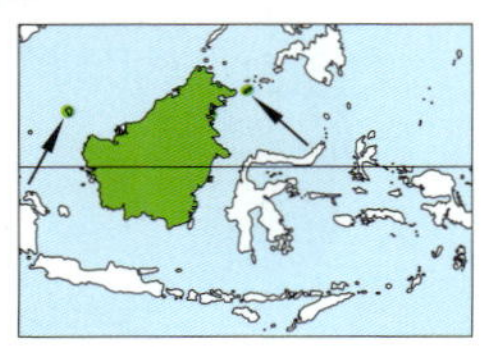

Borneo, North Natuna Is, S PH (Tawi-Tawi, Bongao, Sanga-Sanga, perhaps more Is in Sulu Arch)

Bornean Slow Loris
Nycticebus borneanus

FR Loris de Bornéo
DE Borneo-Plumplori
ES Loris perezoso de Borneo

VU
Monotypic
HB c. 26 cm
T vestigial

SC Borneo (W, S & C Kalimantan excluding extreme SW, S of R Kapuas, E to R Barito)

Bangka Slow Loris
Nycticebus bancanus

FR Loris de Bangka
DE Bangka-Plumplori
ES Loris perezoso de Bangka

CR(PE)
Monotypic
HB c. 26 cm
T vestigial

Bangka I (not reported since 1937), perhaps Belitung I; uncertain in Borneo

Kayan Slow Loris
Nycticebus kayan

FR Loris du Kayan
DE Kayan-Plumplori
ES Loris perezoso de Kayan

VU
Monotypic

N & C Borneo in MY (Sarawak, Sabah) and ID (E Kalimantan, S to R Mahakam)

Family TARSIIDAE – Tarsiers

3 genera, 14 species. Humid lowland, montane, bamboo, secondary, and degraded forests. Indo-Malayan Region.

Genus *CEPHALOPACHUS*

(1 species)

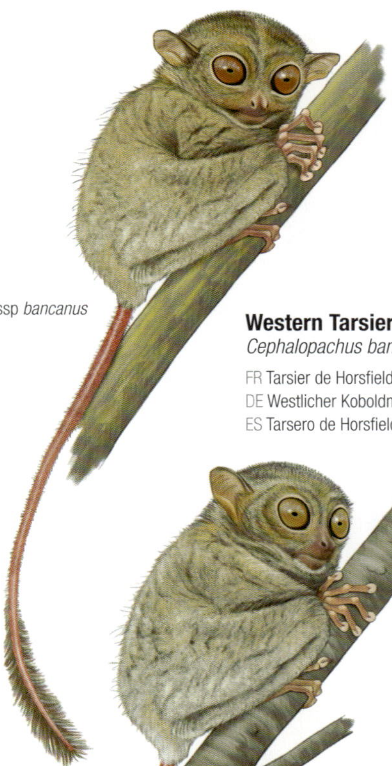

ssp *bancanus*

ssp *borneanus*

Western Tarsier
Cephalopachus bancanus

FR Tarsier de Horsfield
DE Westlicher Koboldmaki
ES Tarsero de Horsfield

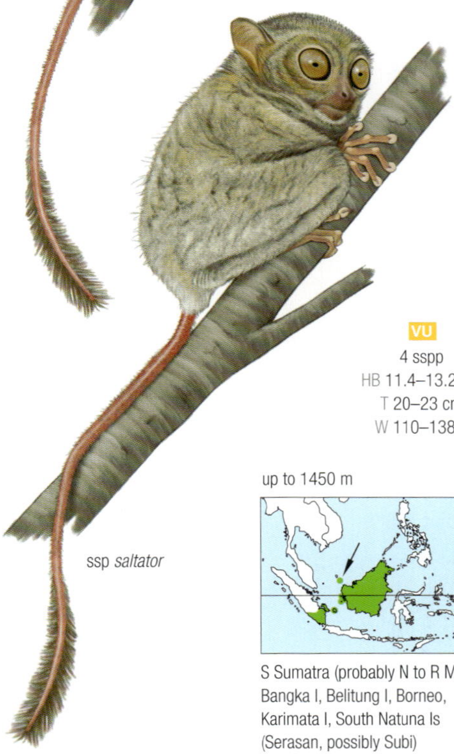

ssp *saltator*

ssp *natunensis*

VU
4 sspp
HB 11.4–13.2 cm
T 20–23 cm
W 110–138 g

up to 1450 m

S Sumatra (probably N to R Musi), Bangka I, Belitung I, Borneo, Karimata I, South Natuna Is (Serasan, possibly Subi)

Genus *CARLITO*

(1 species)

ssp *carbonarius*

ssp *syrichta*

ssp *fraterculus*

Philippine Tarsier
Carlito syrichta

FR Tarsier des Philippines
DE Philippinen-Koboldmaki
ES Tarsero de Filipinas

NT
3 sspp
HB 11.8–14 cm
T 14.7–29 cm
W 110–153 g

0–800 m

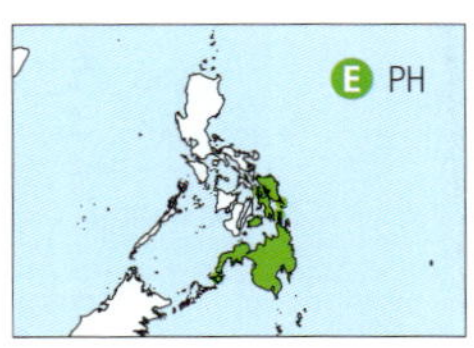

Leyte, Samar, Dinagat, Biliran, Maripipi, Bohol, Mindanao, Basilan

Genus *TARSIUS*

(12 species)

Dian's Tarsier
Tarsius dentatus

FR Tarsier de Dian
DE Diana-Koboldmaki
ES Tarsero de Dian

VU

Monotypic
HB c. 11.8 cm
T 25–27 cm
W 95–135 g

1000–1500 m

EC Sulawesi to E Pen, N to Isthmus of Palu, W at least to R Palu, S limits uncertain

Pygmy Tarsier
Tarsius pumilus

FR Tarsier pygmée
DE Zwergkoboldmaki
ES Tarsero pigmeo

EN

Monotypic
HB c. 9.7 cm
T 20–21 cm
W 48–57 g

1800–2200 m

C & S Sulawesi, fragmented on mountain tops of Rano Rano & Latimojong Mts

Lariang Tarsier
Tarsius lariang

FR Tarsier du Lariang
DE Lariang-Koboldmaki
ES Tarsero de Lariang

DD

Monotypic
HB 11.8–12.1 cm
T 24–25 cm
W 102–118 g

1100–1500 m

WC Sulawesi in R Lariang Basin near confluence with R Meweh, N to Gimpu; limits unclear

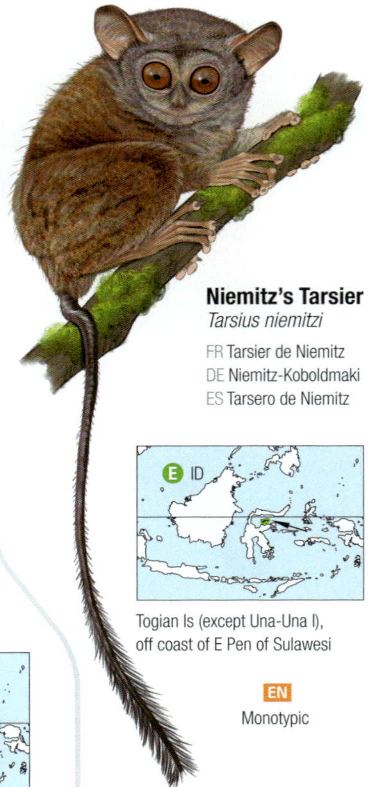

Niemitz's Tarsier
Tarsius niemitzi

FR Tarsier de Niemitz
DE Niemitz-Koboldmaki
ES Tarsero de Niemitz

Togian Is (except Una-Una I), off coast of E Pen of Sulawesi

EN

Monotypic

Peleng Tarsier
Tarsius pelengensis

FR Tarsier de Peleng
DE Peleng-Koboldmaki
ES Tarsero de Peleng

EN

Monotypic
HB 12–14 cm
T 25–27 cm

up to 520 m

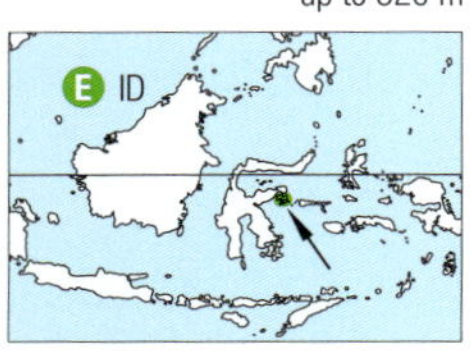

Banggai Arch (Peleng I, perhaps others), off coast of E Pen of Sulawesi

Great Sangihe Tarsier
Tarsius sangirensis

FR Tarsier des Sangir
DE Sangihe-Koboldmaki
ES Tarsero de Sangir

EN

Monotypic
HB c. 15 cm
T 30–31 cm
W 143–150 g

Sangihe Arch (Sangir I, perhaps others), off N Sulawesi

Siau Island Tarsier
Tarsius tumpara

FR Tarsier de Siau
DE Siau-Koboldmaki
ES Tarsero de Siau

CR

Monotypic
HB c. 15 cm
T c. 26 cm
W c. 104 g

Sangihe Arch (Siau I, perhaps on very small near-by islands), off N Sulawesi

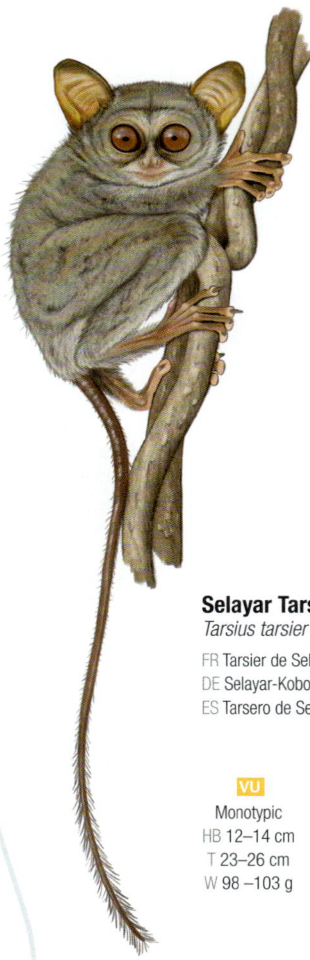

Selayar Tarsier
Tarsius tarsier

FR Tarsier de Selayar
DE Selayar-Koboldmaki
ES Tarsero de Selayar

VU

Monotypic
HB 12–14 cm
T 23–26 cm
W 98 –103 g

0–1100 m

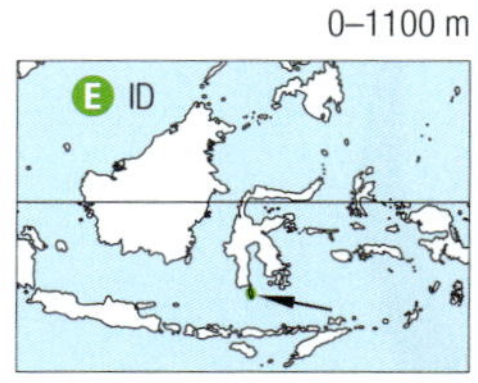

Selayar I, off SW Pen of Sulawesi

VU
Monotypic
HB 12.4–12.8 cm
T 24–26 cm
W 113–133 g

Makassar Tarsier
Tarsius fuscus

FR Tarsier de Makassar
DE Makassar-Koboldmaki
ES Tarsero de Makassar

up to c. 1500 m

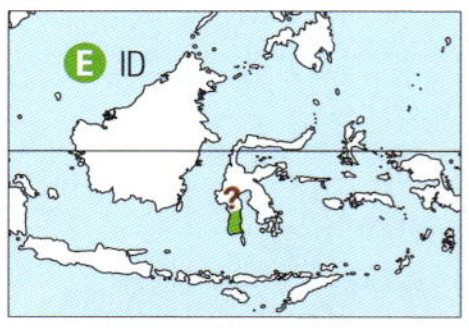

SW Pen of Sulawesi, presumably S of Lake Tempe Depression

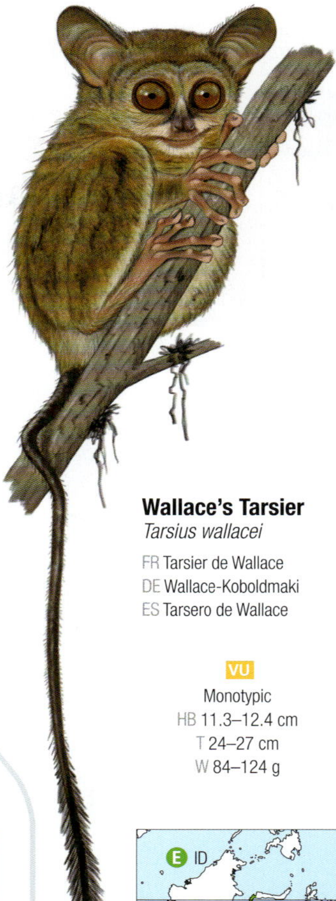

Wallace's Tarsier
Tarsius wallacei

FR Tarsier de Wallace
DE Wallace-Koboldmaki
ES Tarsero de Wallace

VU
Monotypic
HB 11.3–12.4 cm
T 24–27 cm
W 84–124 g

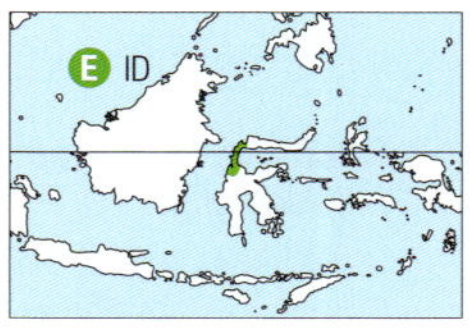

NW Sulawesi (discontinuous in Central Sulawesi Prov)

Gursky's Spectral Tarsier
Tarsius spectrumgurskyae

FR Tarsier de Gursky
DE Gursky-Koboldmaki
ES Tarsero de Gursky

VU
Monotypic

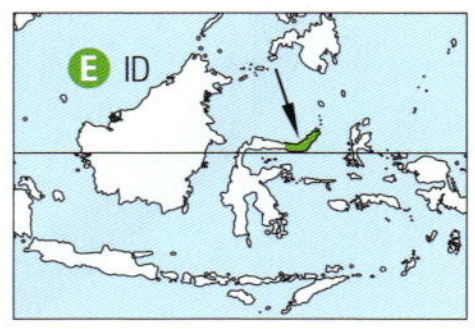

E part of NE Pen of Sulawesi from Tangkoko in N to Suwawa in W (W of Bogani Nani Wartabone NP)

Jatna's Tarsier
Tarsius supriatnai

FR Tarsier de Supriatna
DE Supriatna-Koboldmaki
ES Tarsero de Supriatna

VU
Monotypic

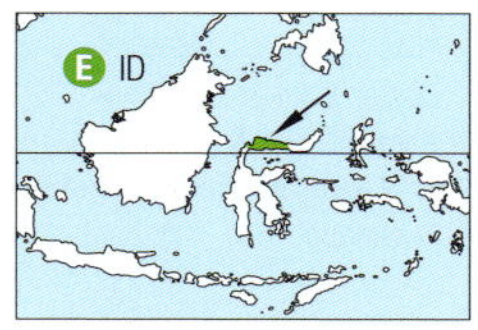

NW Pen of Sulawesi from Isthmus of Gorontalo W at least to Sejoli, probably to Ogatemuku

Family CALLITRICHIDAE – Marmosets and Tamarins

6 genera, 51 species. Humid submontane, lowland tropical, dry semi-deciduous and deciduous, gallery, and Chacoan forests, scrub, and farmland-forest mosaics. Neotropical Region.

Genus *CEBUELLA*

(2 species)

Southern Pygmy Marmoset
Cebuella niveiventris

FR Ouistiti à ventre blanc
DE Östliches Zwergseidenäffchen
ES Tití pigmeo oriental

VU
Monotypic
HB 12–16 cm
T 17–23 cm
W 85–140 g

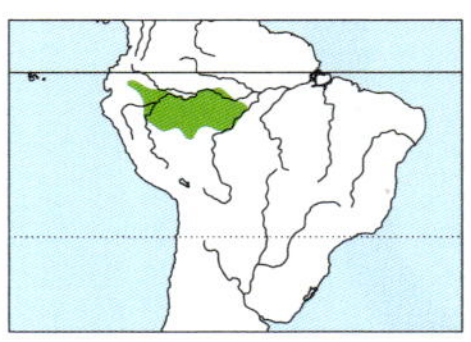

W Amazonia, N to R Napo & lower R Solimões, S to R Madeira & R Madre de Dios

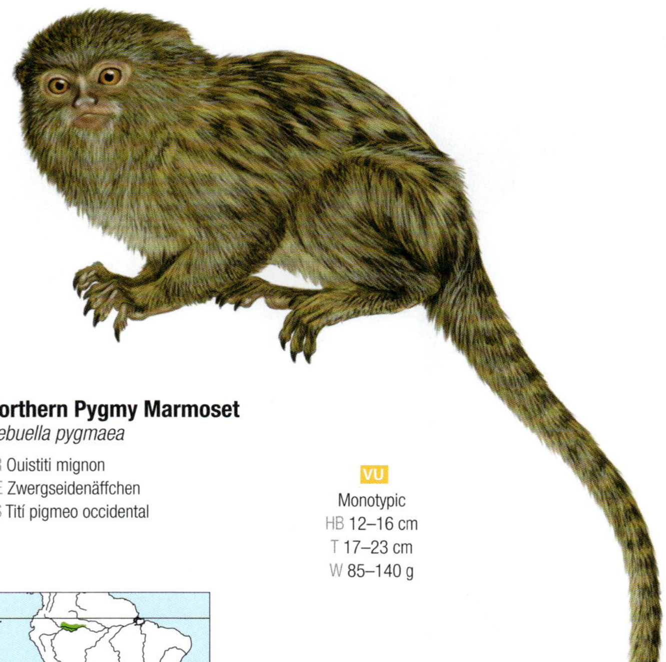

Northern Pygmy Marmoset
Cebuella pygmaea

FR Ouistiti mignon
DE Zwergseidenäffchen
ES Tití pigmeo occidental

VU
Monotypic
HB 12–16 cm
T 17–23 cm
W 85–140 g

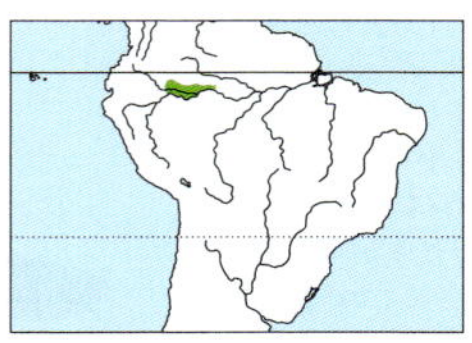

W Amazonia in S CO, W BR, NE EC & N PE (N of R Napo & Solimões, E to mouth of R Japurá)

Genus *MICO*

(16 species)

Black-crowned Dwarf Marmoset

Mico humilis

FR Ouistiti nain
DE Schwarzkronen-Seidenäffchen
ES Tití de corona negra

S from S bank of R Madeira, W of R Aripuanã, and R Roosevelt to R Manicoré; S limit unknown

LC
Monotypic
HB 16–17 cm
T 22–24 cm
W 150–185 g

Santarem Marmoset

Mico humeralifer

NT
Monotypic
HB 18–21 cm
T 31–33 cm
W 446–498 g

FR Ouistiti à camail
DE Weißschulter-Seidenäffchen
ES Tití de Santarém

S of R Amazon between R Maués and probably R Parauari in W and R Tapajós in E; S limit unknown

Satere Marmoset

Mico saterei

LC
Monotypic
HB c. 20 cm
T c. 37 cm
W c. 430 g

FR Ouistiti des Saterés
DE Sateré-Seidenäffchen
ES Tití de Sateré

Interfluvium of rivers Canumã/ Sucunduri in W and Abacaxis in E, E bank tributaries of R Madeira

Maues Marmoset

Mico mauesi

FR Ouistiti de Maués
DE Maués-Seidenäffchen
ES Tití de Maués

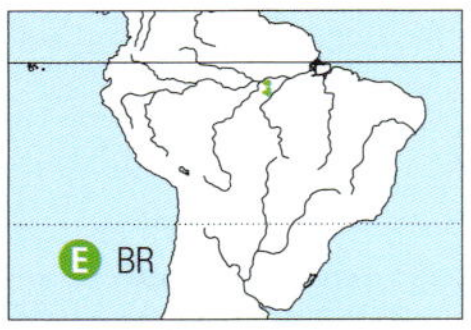

R Maués & R Parauari W to R Paraná Urariá & R Abacaxis, S to W bank of R Tapajós, W to R Sucunduri

LC
Monotypic
HB 20–23 cm
T 34–38 cm
W 315–400 g

Black-tailed Marmoset

Mico melanurus

FR Ouistiti à queue noire
DE Schwarzschwanz-Seidenäffchen
ES Tití de cola negra

450–1100 m

CS BR (Pantanal, N between R Aripuanã and R Juruena to Serra do Sucunduri), E BO (E of R Mamoré), NW PY

NT
Monotypic
HB 22–24 cm
T 30–34 cm
W c. 330 g

Rio Acari Marmoset

Mico acariensis

FR Ouistiti de l'Acari
DE Rio-Acari-Seidenäffchen
ES Tití de Acarí

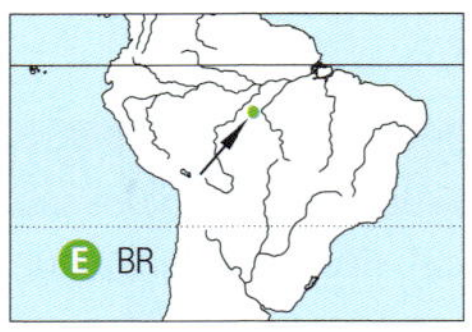

E bank of lower R Acarí and interfluvium of R Acarí in W and R Sucundurí in E; S limit uncertain

LC
Monotypic
HB c. 22 cm
T c. 38 cm
W c. 350 g

Marca's Marmoset

Mico marcai

FR Ouistiti de Marca
DE Marca-Seidenäffchen
ES Tití de Marca

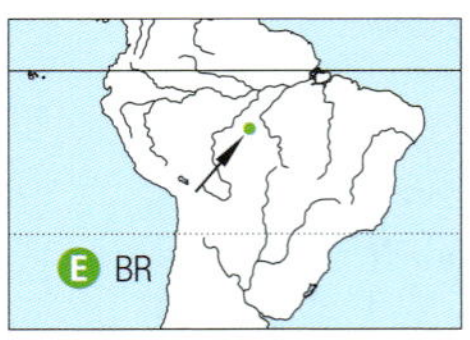

Near mouth of R Roosevelt, N along W bank of R Aripuanã, W to R Manicoré, N to E bank of R Madeira

VU
Monotypic
HB c. 22 cm
T c. 36 cm
W 280–350 g

Schneider's Marmoset

Mico schneideri

FR Ouistiti de Schneider
DE Schneider-Seidenäffchen
ES Tití de Schneider

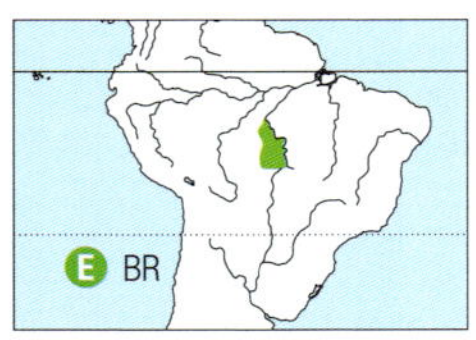

Between R Juruena and R Teles Pires, N to confluence, S to headwaters (N of Lucas do Rio Verde)

EN
Monotypic

Black-headed Marmoset
Mico nigriceps

FR Ouistiti à tête noire
DE Schwarzkopf-Seidenäffchen
ES Tití de cabeza negra

NT
Monotypic
HB 19–22 cm
T 31–33 cm
W 330–400 g

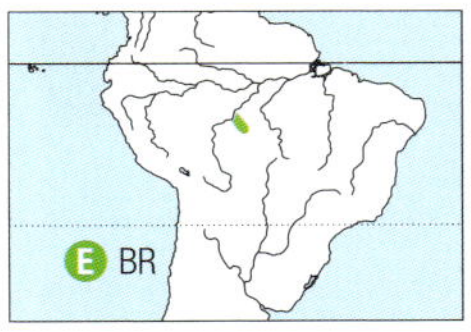

Between R dos Marmelos in N & E, R Madeira in W, and R Ji-Paraná in S

Rio Aripuana Marmoset
Mico intermedius

FR Ouistiti de l'Aripuana
DE Rio-Aripuana-Seidenäffchen
ES Tití de Aripuaná

LC
Monotypic
HB 20–24 cm
T 32–40 cm
W 280–310 g

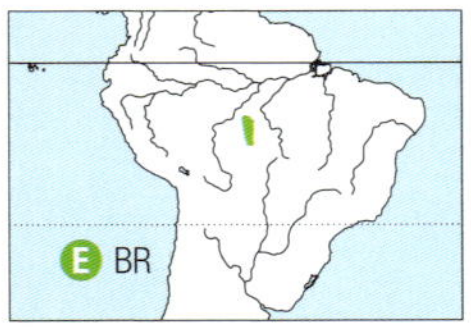

Between R Roosevelt and R Arip-uanã; S limits unknown (probably around headwaters of said rivers)

Rondon's Marmoset
Mico rondoni

FR Ouistiti de Rondon
DE Rondonia-Seidenäffchen
ES Tití de Rondon

Between R Mamoré in W, R Madeira in N, R Ji-Paraná in E, and Serra dos Pacaás Novos in S

VU
Monotypic
HB 20–24 cm
T 29–33 cm
W 250–390 g

Munduruku Marmoset
Mico munduruku

FR Ouistiti des Mundurukus
DE Munduruku-Seidenäffchen
ES Tití de los Mundurukú

W of R Jamanxim, below mouth of R Novo, possibly up to E upper R Tapajós, below mouth of R Cururú

VU
Monotypic

Silvery Marmoset
Mico argentatus

FR Ouistiti argenté
DE Silberäffchen
ES Tití plateado

LC
Monotypic
HB 20–22 cm
T 26–33 cm
W 349–406 g

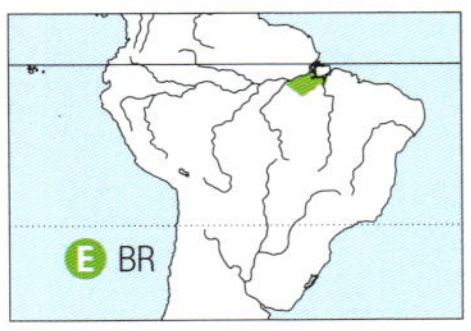

S of R Amazon, R Tapajós (S to R Cupari) E to R Tocantins

Golden-white Tassel-ear Marmoset
Mico chrysoleucos

FR Ouistiti à pieds jaunes
DE Weißgoldenes Seidenäffchen
ES Tití blanco y dorado

LC
Monotypic
HB 20–24 cm
T 32–40 cm
W 280–310 g

S of R Amazon between R Madeira and lower R Aripuanã in W and R Canumã in E

Golden-white Bare-ear Marmoset
Mico leucippe

FR Ouistiti blanc-doré
DE Goldschwanz-Seidenäffchen
ES Tití dorado

S of R Cupari, E of R Tapajós, S to E bank of R São Benedito (affluent of R Teles Pires)

LC
Monotypic
HB c. 20 cm
T c. 30 cm
W c. 330 g

Snethlage's Marmoset
Mico emiliae

FR Ouistiti de Snethlage
DE Snethlage-Seidenäffchen
ES Tití de Snethlage

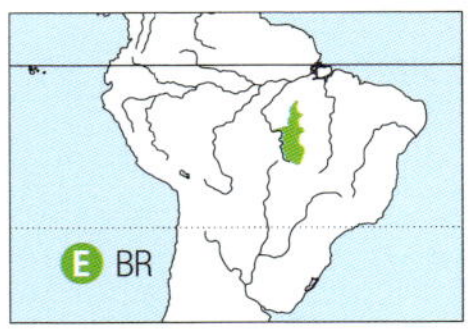

S of R Amazon between R Madeira and lower R Aripuanã in W and R Canumã in E

LC
Monotypic
HB c. 19.9 cm
T c. 30.8 cm

Genus *CALLITHRIX*

(6 species)

Buffy-tufted-ear Marmoset
Callithrix aurita

FR Ouistiti oreillard
DE Gelbohr-Büscheläffchen
ES Tití de orejas amarillas

EN

Monotypic
HB 19–25 cm
T 27–35 cm
W 400–450 g

up to 1200 m

Minas Gerais (N to Rio Doce State Park), Rio de Janeiro, São Paulo

Buffy-headed Marmoset
Callithrix flaviceps

FR Ouistiti à tête jaune
DE Gelbkopf-Büscheläffchen
ES Tití de cabeza amarilla

CR

Monotypic
HB 22–25 cm
T 30–35 cm
W c. 400 g

270–1800 m

S of R Doce S to Serra da Mantiqueira (S Espírito Santo, Minas Gerais), perhaps Rio de Janeiro

Geoffroy's Tufted-ear Marmoset
Callithrix geoffroyi

FR Ouistiti de Geoffroy
DE Weißkopf-Büscheläffchen
ES Tití de cabeza blanca

LC

Monotypic
HB 18–23 cm
T c. 29 cm
W 190–350 g

up to 500 m

S of R Jequitinhonha & R Araçuaí (Espírito Santo, E Minas Gerais, to near Rio de Janeiro border)

Wied's Marmoset
Callithrix kuhlii

FR Ouistiti de Kuhl
DE Kuhl-Büscheläffchen
ES Tití de orejas negras

VU

Monotypic
HB 20–23 cm
T 29–33 cm
W 350–400 g

Between R de Contas and R Jequitinhonha (S Bahia, NE tip of Minas Gerais)

Black-tufted-ear Marmoset
Callithrix penicillata

FR Ouistiti à pinceaux noirs
DE Schwarzbüscheläffchen
ES Tití de pincel negro

LC

Monotypic
HB 20–23 cm
T 29–33 cm
W 180–225 g

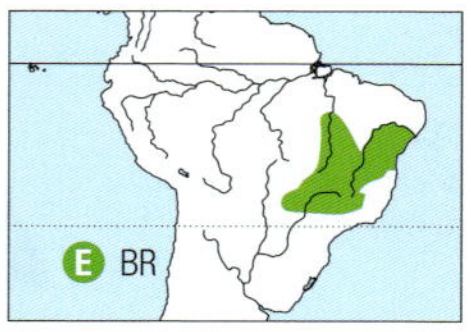

E of R Araguaia, S Maranhão & Bahia S to NE Mato Grosso do Sul & N São Paulo

Common Marmoset
Callithrix jacchus

FR Ouistiti à toupets blancs
DE Weißbüscheläffchen
ES Tití común

LC

Monotypic
HB 16–21 cm
T 24–31 cm
W 318–322 g

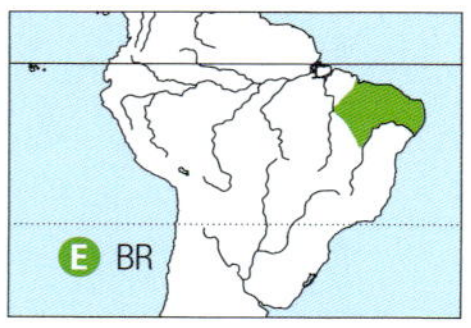

E of Maranhão (at least from R Parnaíba) and possibly NE Tocantins, S to N Bahia

Genus *CALLIMICO*

(1 species)

Goeldi's Monkey
Callimico goeldii

FR Callimico
DE Springtamarin
ES Tití de Goeldi

VU

Monotypic
HB 19–25 cm
T 26–35 cm
W 355–366 g

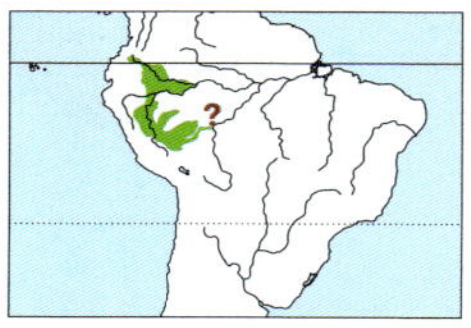

Upper Amazon in S CO (S from R Caquetá), W BR, E PE and N BO (Pando)

Genus *LEONTOPITHECUS*

(4 species)

Golden Lion Tamarin
Leontopithecus rosalia

FR Tamarin-lion doré
DE Goldgelbes Löwenäffchen
ES Tití león dorado

EN

Monotypic
HB 26–33 cm
T 32–40 cm
W 710–795 g

below 300 m

Rio de Janeiro (today largely restricted to Silva Jardim and Cabo Frio)

Black Lion Tamarin
Leontopithecus chrysopygus

FR Tamarin-lion noir
DE Schwarzes Löwenäffchen
ES Tití león negro

São Paulo (E from R Paraná, N of R Paranapanema, S of R Tieté, E to Serra de Paranapiacaba)

EN
Monotypic
HB 25–30 cm
T 36–41 cm
W 540–690 g

Golden-headed Lion Tamarin
Leontopithecus chrysomelas

FR Tamarin-lion à tête dorée
DE Goldkopf-Löwenäffchen
ES Tití león de cabeza dorada

Bahia (R de Contas S to R Jequitinhonha, on border with Minas Gerais); introduced to Rio de Janeiro

EN
Monotypic
HB 22–26 cm
T 33–39 cm
W 480–700 g

Black-faced Lion Tamarin
Leontopithecus caissara

FR Tamarin-lion à tête noire
DE Schwarzkopf-Löwenäffchen
ES Tití león de cara negra

below 40 m

Paraná, São Paulo (Superagüi I and on mainland valleys of R Sebuí & R dos Patos)

EN
Monotypic
HB c. 34 cm
T c. 40 cm
W 540–710 g

Genus *SAGUINUS*

(22 species)

Lesson's Saddle-back Tamarin
Saguinus fuscus

FR Tamarin de Lesson
DE Brauner Sattelrückentamarin
ES Tamarín de Lesson

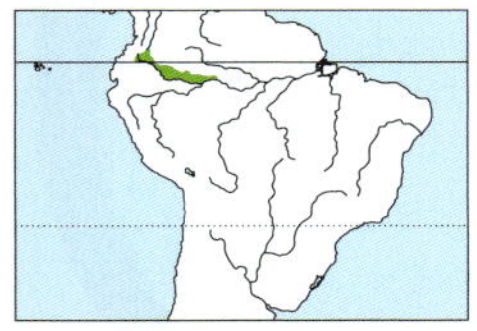

SE CO & NW BR (N of R Solimões between rivers Japurá-Caquetá and Içа-Putumayo, W to Andean foothills)

LC
Monotypic
HB 22–23 cm
T c. 35 cm
W 350–400 g

Black-mantled Tamarin
Saguinus nigricollis

FR Tamarin rouge et noir
DE Schwarzrückentamarin
ES Tamarín de dorso negro

ssp *graellsi*

ssp *nigricollis*

ssp *hernandezi*

LC
3 sspp
HB 21–25 cm
T 31–35 cm
W 420–500 g

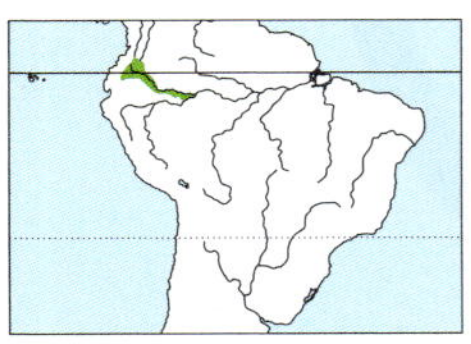

S CO (from E Andes, S of R Guayabero), NE EC, N PE & NW BR (between R Putumayo-Içá and R Napo & R Solimões)

Andean Saddle-back Tamarin
Saguinus leucogenys

FR Tamarin des Andes
DE Anden-Sattelrückentamarin
ES Tamarín andino

LC
Monotypic
HB 20–23 cm
T 30–33 cm
W 350–400 g

up to 1000 m

S of R Marañón (San Martín) to R Perené (N Junín), E to W R Huallaga and upper R Ucayali (S of R Pisqui)

Illiger's Saddle-back Tamarin
Saguinus illigeri

FR Tamarin d'Illiger
DE Schwarzkopftamarin
ES Tamarín de Illiger

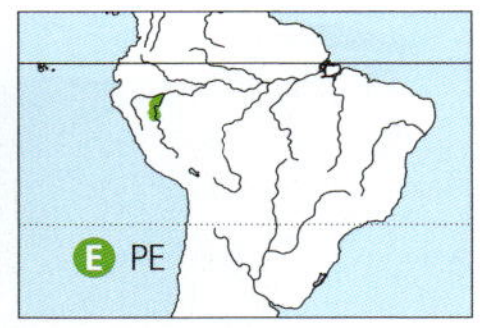

S of R Marañón, between R Huallaga and R Ucayali, also E of R Ucayali along left bank of lower R Tapiche

NT
Monotypic
HB c. 20 cm
T c. 31 cm
W c. 293 g

Red-mantled Saddle-back Tamarin
Saguinus lagonotus

FR Tamarin à manteau rouge
DE Rotschultertamarin
ES Tamarín de dorso rojo

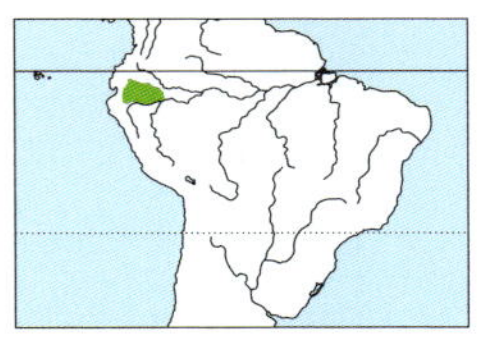

E EC & NE PE (S of R Napo & R Curaray, N of R Marañón, E to R Amazon); SW limit uncertain

LC
Monotypic
HB 22–24 cm
T 30–32 cm
W 350–400 g

Golden-mantled Saddle-back Tamarin
Saguinus tripartitus
FR Tamarin à manteau doré
DE Goldmanteltamarin
ES Tamarín de dorso dorado

NT
Monotypic
HB 22–26 cm
T 32–34 cm
W c. 400 g

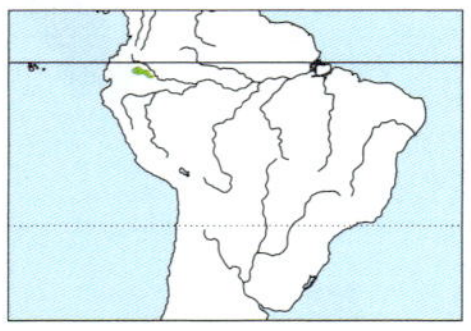

NE EC & N PE (between R Napo and R Curaray)

ssp *fuscicollis*

ssp *mura*

LC
4 sspp
HB c. 21 cm
T c. 32 cm
W c. 333 g

ssp *primitivus*

ssp *avilapiresi*

Spix's Saddle-back Tamarin
Saguinus fuscicollis
FR Tamarin de Spix
DE Braunrückentamarin
ES Tamarín de Spix

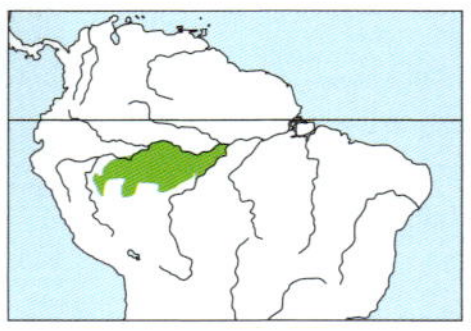

W BR & PE (S of R Javarí & R Solimões, E of R Tapiche, W of R Madeira)

Geoffroy's Saddle-back Tamarin
Saguinus nigrifrons
FR Tamarin à front noir
DE Schwarzstirntamarin
ES Tamarín de frente negra

LC
Monotypic
HB c. 21 cm
T c. 32 cm
W c. 366 g

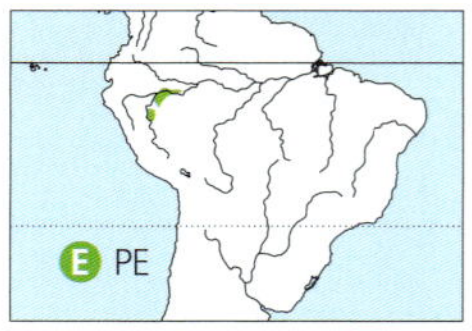

Between R Solimões and R Javarí, S to E bank of R Ucayali and R Tapiche as far as R Blanco

Cruz Lima's Saddle-back Tamarin
Saguinus cruzlimai
FR Tamarin de Cruz Lima
DE Cruz-Lima-Braunrückentamarin
ES Tamarín de Cruz Lima

LC
Monotypic
HB c. 25 cm
T c. 30 cm
W c. 390 g

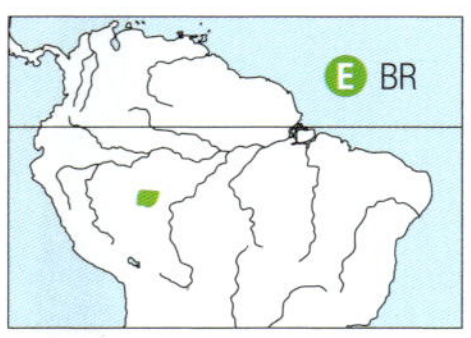

Interfluvium of R Purus and R Juruá, W bank of middle R Purus (Pauini and Boca do Acre, S Amazonas)

Weddell's Saddle-back Tamarin
Saguinus weddelli
FR Tamarin de Weddell
DE Schwarzmanteltamarin
ES Tamarín de Weddell

ssp *crandalli*

ssp *melanoleucus*

ssp *weddelli*

LC
3 sspp
HB 18–27 cm
T 25–38 cm
W 340–440 g

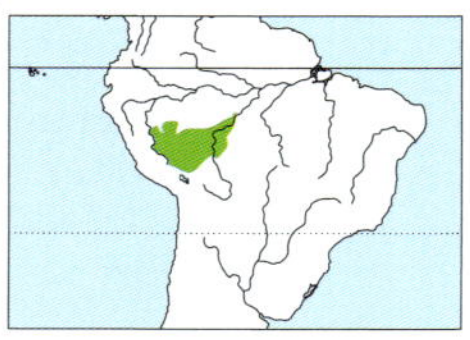

SE PE, N BO & W BR (R Ucayali E to upper R Purus, R Jamarí and R Beni)

Mustached Tamarin
Saguinus mystax
FR Tamarin à moustaches
DE Schnurrbarttamarin
ES Tamarín bigotudo

ssp *mystax*

ssp *pileatus*

ssp *pluto*

LC
3 sspp
HB 24–25 cm
T 38–39 cm
W 360–650 g

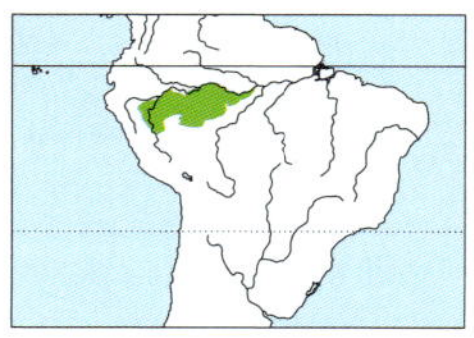

W BR (S of R Solimões, W of R Purus), NE PE (E of R Tapiche & R Ucayali, S to R Urubamba & R Inuya)

Red-bellied Tamarin
Saguinus labiatus

FR Tamarin labié
DE Rotbauchtamarin
ES Tamarín de pecho rojo

LC
3 sspp
HB 21–28 cm
T 30–38 cm
W 400–510 g

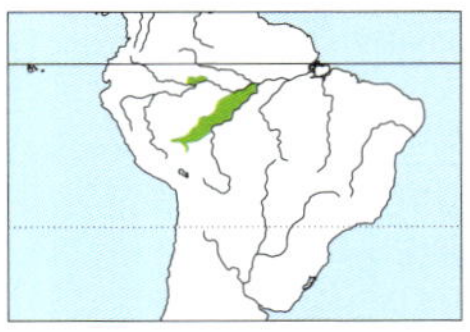

W BR, SE PE & N BO (R Tonantins to Auatí-Paraná; between R Purus and R Madeira, S to R Tahuamanú)

Emperor Tamarin
Saguinus imperator

FR Tamarin empereur
DE Kaiserschnurrbarttamarin
ES Tamarín emperador

LC
2 sspp
HB 23–26 cm
T 35–42 cm
W 400–550 g

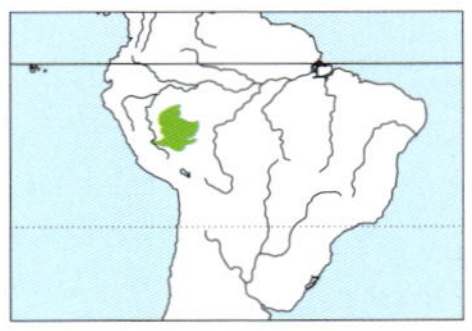

SE PE, NW BO & W BR (R Urubamba E to R Acre, N to upper R Juruá, S to R Muyumanu)

Mottled-face Tamarin
Saguinus inustus

FR Tamarin à face marbrée
DE Marmorgesichttamarin
ES Tamarín jaspeado

LC
Monotypic
HB 21–26 cm
T 33–41 cm
W c. 430 g

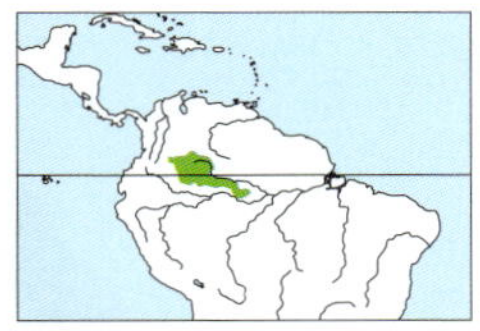

E CO (upper R Guaviare S to R Caquetá), W BR (between upper R Negro and R Japurá); N & W limits unknown

Midas Tamarin
Saguinus midas

FR Tamarin à mains dorées
DE Goldhandtamarin
ES Tamarín de Midas

LC
Monotypic
HB 24–25 cm
T 38–39 cm
W 380–500 g

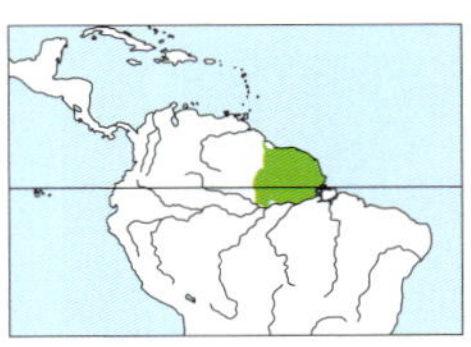

NE BR (E of R Branco & R Negro), Guianas (in GY in E bank of R Essequibo and S & E of R Rupununi)

Western Black-handed Tamarin
Saguinus niger

FR Tamarin noir
DE Mohrentamarin
ES Tamarín de manos negras occidental

VU
Monotypic
HB 21–26 cm
T 32–40 cm
W c. 431 g (M), c. 428 g (F)

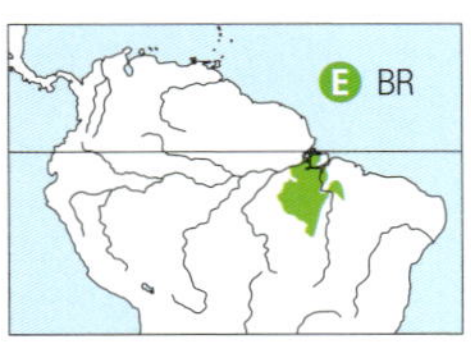

S of R Amazon, from E bank of R Xingu to W bank of R Tocantins, SW Marajó I (Pará)

Eastern Black-handed Tamarin
Saguinus ursula

FR Tamarin de Belém
DE Schwarzer Tamarin
ES Tamarín de manos negras oriental

VU
Monotypic
HB 21–26 cm
T 32–40 cm
W c. 431 g (M), c. 428 g (F)

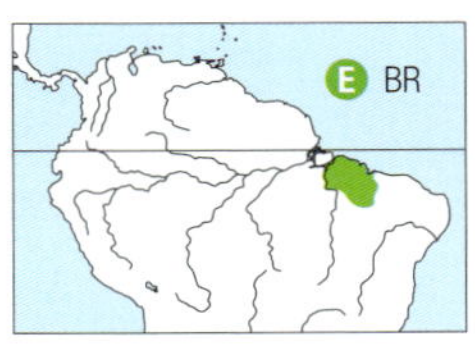

S of R Amazon, from R Tocantins to limits of Amazon forest with cerrado and caatinga (E Maranhão)

Pied Bare-faced Tamarin
Saguinus bicolor

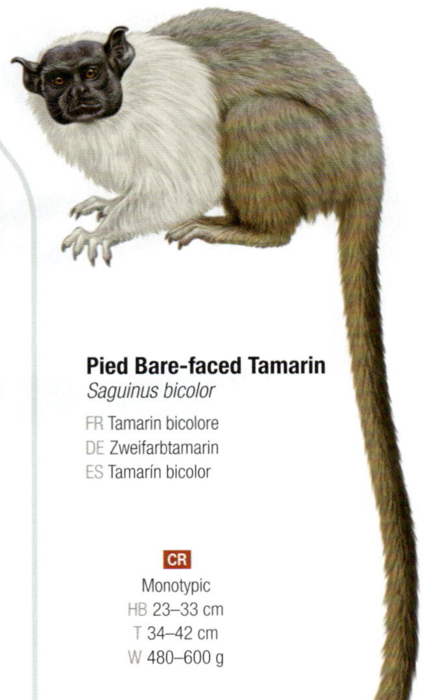

FR Tamarin bicolore
DE Zweifarbtamarin
ES Tamarín bicolor

CR
Monotypic
HB 23–33 cm
T 34–42 cm
W 480–600 g

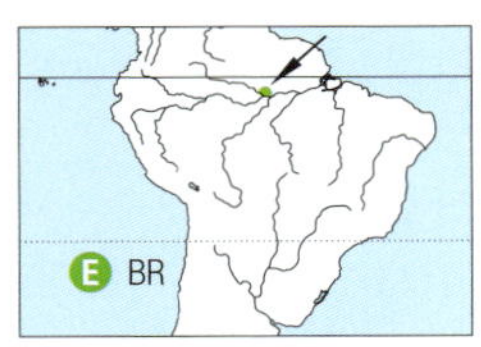

Small area N of R Amazon (from R Negro E to R Urubu, N to R Cuieiras)

ssp *martinsi*

ssp *ochraceus*

NT
2 sspp
HB 21–28 cm
T 34–42 cm
W 400–600 g

Martins's Bare-faced Tamarin
Saguinus martinsi
FR Tamarin de Martins
DE Ockermanteltamarin
ES Tamarín de Martins

N of R Amazon, between R Uatumã and R Trombetas

Cotton-top Tamarin
Saguinus oedipus
FR Tamarin pinché
DE Lisztäffchen
ES Tamarín de cabeza blanca

CR
Monotypic
HB 21–26 cm
T 33–40 cm
W 404–418 g

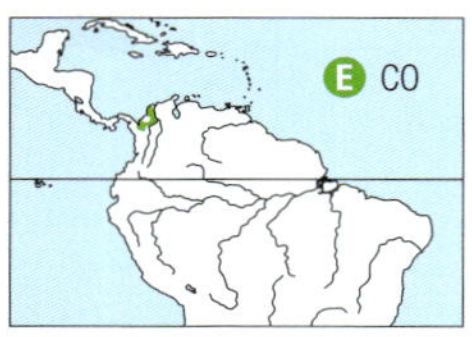

Between R Atrato and R Cauca & R Magdalena; introduced to Tayrona NP and to islands off Cartagena

Geoffroy's Tamarin
Saguinus geoffroyi
FR Tamarin de Geoffroy
DE Perückentamarin
ES Tamarín de Panamá

NT
Monotypic
HB 20–29 cm
T 32–42 cm
W 486–507 g

0–350 m

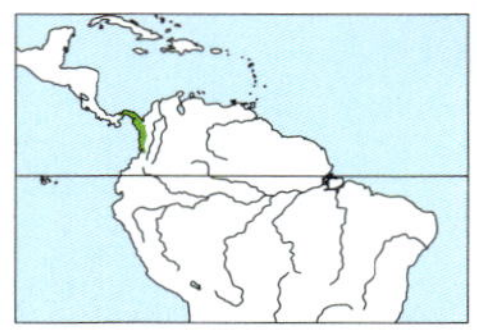

C & E PA (E of Azuero Pen, to just W of Canal Zone), NW CO (along Pacific coast, S to R San Juan)

White-footed Tamarin
Saguinus leucopus
FR Tamarin à pieds blancs
DE Weißfußtamarin
ES Tamarín de pies blancos

VU
Monotypic
HB 22–29 cm
T 35–42 cm
W c. 462 g

up to 1500 m

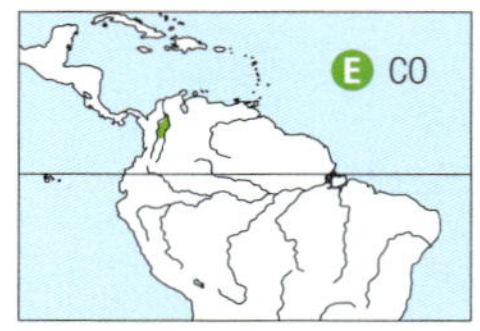

Between lower R Cauca and middle R Magdalena, S along W Magdalena at least to Mariquita, also Mompós I

Family AOTIDAE – Night Monkeys

1 genus, 11 species. Moist tropical and subtropical, submontane and montane, dry deciduous and semi-deciduous, gallery, and Chacoan forests. Neotropical Region.

Genus *AOTUS*

(11 species)

Lemurine Night Monkey
Aotus lemurinus
FR Douroucouli lémurien
DE Nördlicher Graukehl-Nachtaffe
ES Mico nocturno de vientre gris

VU
Monotypic
HB 31–32 cm
T 34–35 cm
W 800–1050 g

1000–3200 m

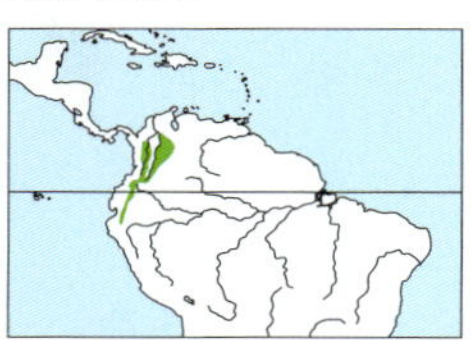

CO (upper R Cauca Valley and slopes of E Andes) to S EC; perhaps Sierra de Perijá in VE

Gray-legged Night Monkey
Aotus griseimembra
FR Douroucouli à bras gris
DE Elliot-Nachtaffe
ES Mico nocturno de patas grises

VU
Monotypic
HB c. 48 cm
T c. 42 cm
W 800–1000 g

N CO (R Magdalena Valley and N lowlands), NW VE (Sierra de Perijá and vicinity of L Maracaibo)

Panamanian Night Monkey
Aotus zonalis
FR Douroucouli du Panama
DE Panama-Nachtaffe
ES Mico nocturno de Panamá

NT
Monotypic
HB c. 30 cm
T c. 36 cm
W 889–916 g

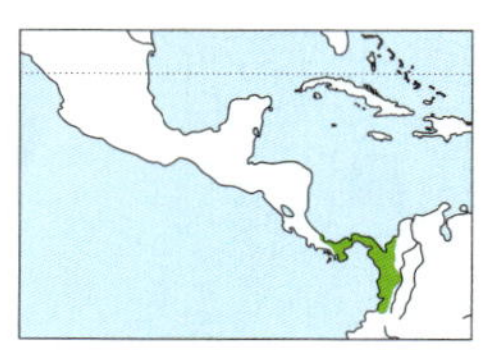

Most of PA and Pacific lowlands of NW CO; occurrence in SE CR uncertain

Brumback's Night Monkey
Aotus brumbacki

FR Douroucouli de Brumback
DE Brumback-Nachtaffe
ES Mico nocturno de Brumback

VU
Monotypic

up to 1500 m

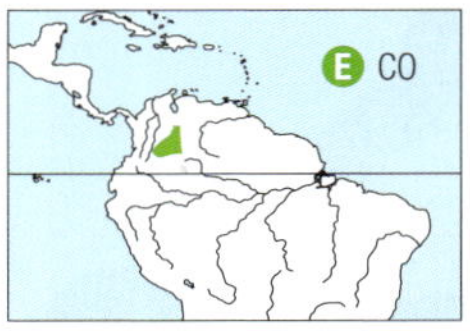

E Boyacá E to highlands of Meta; poorly known

LC
Monotypic
HB 30–38 cm
T 33–40 cm
W 736–813 g

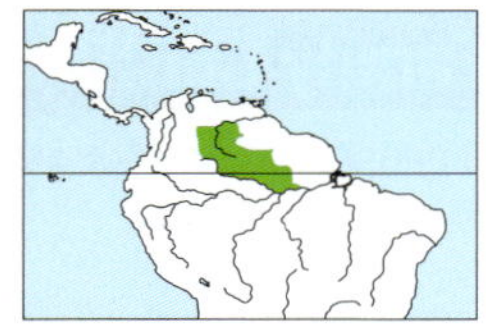

SC VE, E CO & N BR (N of R Negro, E of R Trombetas)

Humboldt's Night Monkey
Aotus trivirgatus

FR Douroucouli de Humboldt
DE Östlicher Graukehl-Nachtaffe
ES Mico nocturno de Humboldt

Spix's Night Monkey
Aotus vociferans

FR Douroucouli de Spix
DE Spix-Nachtaffe
ES Mico nocturno de Spix

LC
Monotypic
HB 35–45 cm
T 31–47 cm
W 698–708 g

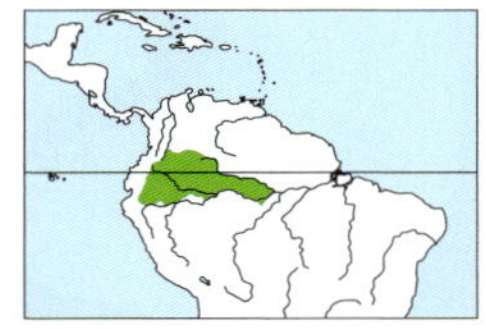

200–1550 m

SE CO, E EC, NE PE & NW BR (W of Rio Negro)

Hernandez-Camacho's Night Monkey
Aotus jorgehernandezi

FR Douroucouli de Hernandez
DE Hernández-Camacho-Nachtaffe
ES Mico nocturno de Hernández-Camacho

Believed to be on W slopes and foothills of Andes (Quindío, Risaralda)

DD
Monotypic

Andean Night Monkey
Aotus miconax

FR Douroucouli des Andes
DE Anden-Nachtaffe
ES Mico nocturno andino

800–2790 m

Small area S of R Marañón and W of R Huallaga to c. 10° S

EN
Monotypic
HB c. 39 cm
T c. 22 cm

Ma's Night Monkey
Aotus nancymai

FR Douroucouli du Loreto
DE Nancy-Ma-Nachtaffe
ES Mico nocturno de Nancy Ma

VU
Monotypic
HB 29–34 cm T 35–42 cm
W 780–794 g

60–130 m

NE PE (R Huallaga E to R Yavarí, also N of R Marañón between R Tigre and R Pastaza), W BR (S of R Solimões)

ssp *infulatus*

ssp *boliviensis*

ssp *azarae*

Azara's Night Monkey
Aotus azarae

FR Douroucouli d'Azara
DE Azara-Nachtaffe
ES Mico nocturno de Azara

LC
3 sspp
HB 33–34 cm
T c. 31 cm
W 990–1580 g

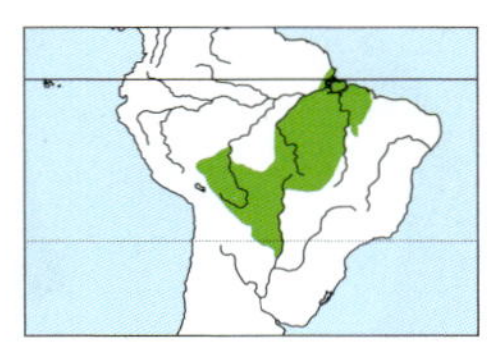

BR (SE Amapá to Maranhão, S betwen R Tapajós and R Tocantins to Pantanal), SE PE, BO, W PY, N AR

Black-headed Night Monkey
Aotus nigriceps

FR Douroucouli de Dollman
DE Schwarzkopf-Nachtaffe
ES Mico nocturno de cabeza negra

LC
Monotypic
HB 35–42 cm
T 35–44 cm
W 875–1040 g

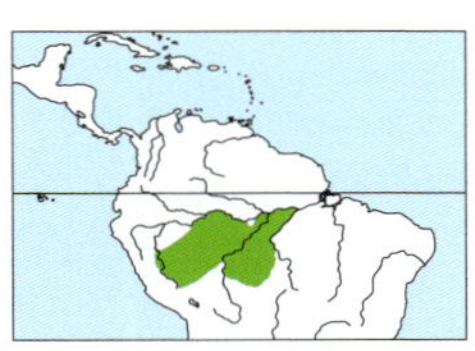

SE PE, W BR, N BO (from R Huallaga E to R Tapajós, N to R Amazon, S to R Madre de Dios & R Guaporé)

Family CEBIDAE – Squirrel Monkeys and Capuchins

3 genera, 30 species. Moist tropical and subtropical, lowland and submontane, and dry deciduous and semi-deciduous forests, scrub, and farmland-forest mosaics. Neotropical Region.

Genus *SAIMIRI*

(8 species)

Central American Squirrel Monkey
Saimiri oerstedii

FR Saimiri à dos roux
DE Mittelamerika-Totenkopfaffe
ES Mono ardilla de América Central

EN
2 sspp
HB 28–33 cm
T 33–43 cm
W 750–950 g

Pacific coast of CR (Puntarenas Prov, S from R Tulín in N Herradura Mts), SW PA (Chiriquí Prov)

Humboldt's Squirrel Monkey
Saimiri cassiquiarensis

FR Saimiri de Humboldt
DE Humboldt-Totenkopfaffe
ES Mono ardilla de Humboldt

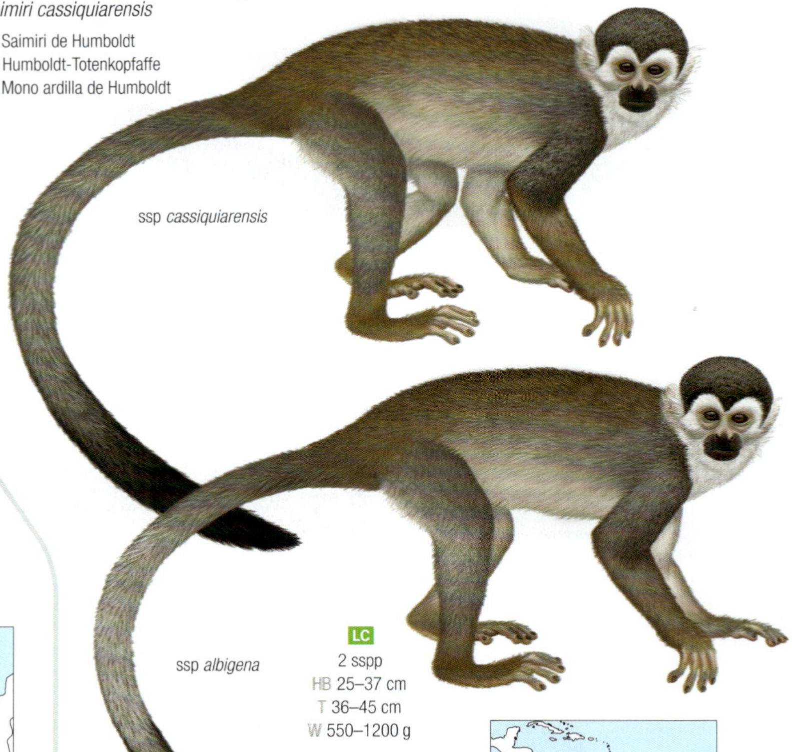

LC
2 sspp
HB 25–37 cm
T 36–45 cm
W 550–1200 g

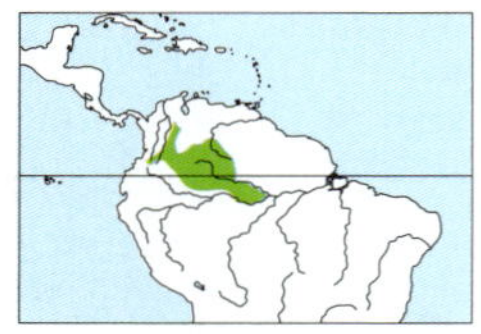

CO (E Andes S to upper R Magdalena & R Apaporis), S VE (upper R Orinoco), BR (N of R Solimões, W of R Negro)

Ecuadorian Squirrel Monkey
Saimiri macrodon

FR Saimiri d'Équateur
DE Ecuador-Totenkopfaffe
ES Mono ardilla de Ecuador

LC
Monotypic
HB 25–32 cm
T 34–44 cm
W 590–1380 g

200–1200 m

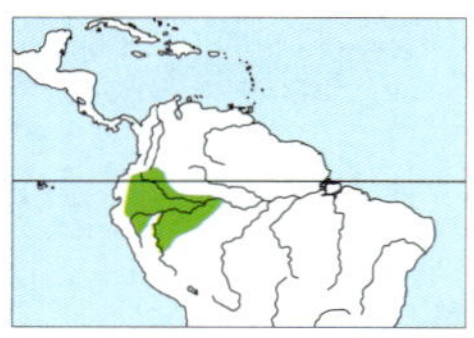

S CO, E EC, NE PE, NW BR (S of R Apaporis, E to R Japurá & R Juru, W to R Huallaga & R Ucayali, S to R Abujao)

Golden-backed Squirrel Monkey
Saimiri ustus

FR Saimiri à dos brûlé
DE Nacktohr-Totenkopfaffe
ES Mono ardilla orejudo

NT
Monotypic
HB 23–42 cm
T 31–45 cm
W 620–1200 g

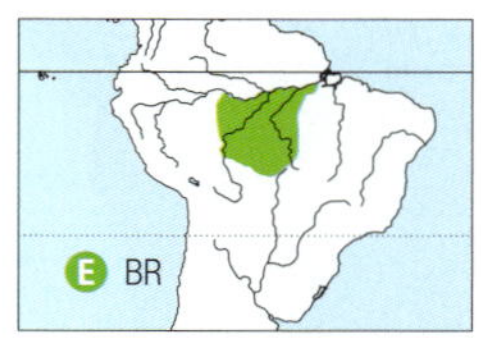

S of R Amazon, from R Tefé E to R Xingu-Iriri, S to upper R Guaporé and headwaters of R Juruena

Collins's Squirrel Monkey
Saimiri collinsi

FR Saimiri de Collins
DE Collins-Totenkopfaffe
ES Mono ardilla de Collins

LC
Monotypic
HB 25–37 cm
T 36–47 cm
W 550–1400 g

R Xingu-Iriri E to R Pindaré, S to upper reaches of R Xingu, also Marajó I

Guianan Squirrel Monkey
Saimiri sciureus

FR Saimiri écureuil
DE Guayana-Totenkopfaffe
ES Mono ardilla común

LC
Monotypic
HB 25–37 cm
T 36–47 cm
W 550–1400 g

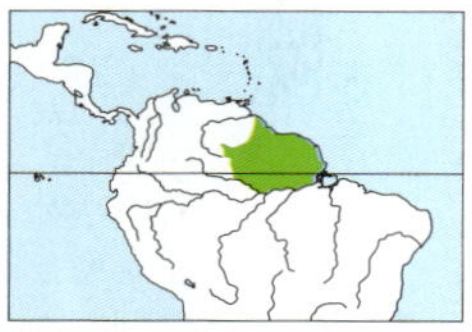

Guianas & N BR (N of R Amazon from R Negro & R Demini, E to river mouth)

EN
Monotypic
HB 22–30 cm
T 35–41 cm
W 650–950 g

Black-headed Squirrel Monkey
Saimiri vanzolinii

FR Saimiri noir
DE Dunkler Totenkopfaffe
ES Mono ardilla de Vanzolini

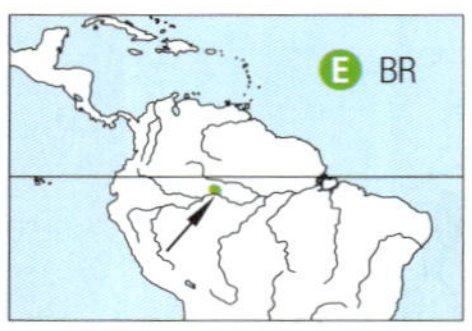

Known only in Mamirauá R, N of R Solimões

Black-capped Squirrel Monkey
Saimiri boliviensis

FR Saimiri de Bolivie
DE Schwarzkappen-Totenkopfaffe
ES Mono ardilla de Bolivia

ssp *boliviensis*

ssp *peruviensis*

LC
2 sspp
HB 26–32 cm
T 37–43 cm
W 700–1000 g

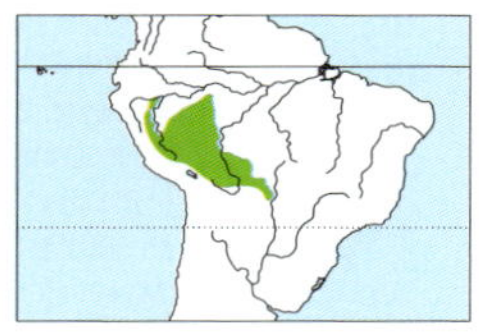

E PE (S of R Marañón), NW BR (S of R Solimões, between R Juruá and R Tefé) & N BO (W of R Guaporé)

Genus *SAPAJUS*

(7 species)

Crested Capuchin
Sapajus robustus

FR Sapajou robuste
DE Schopfkapuzineraffe
ES Capuchino crestado

From R Jequitinhonha (S Bahia) S to R Doce, E of Serra do Espinhaço (Espírito Santo, E Minas Gerais)

EN
Monotypic
HB 33–56 cm
T 43–56 cm
W 1.3–4.8 kg

Yellow-breasted Capuchin
Sapajus xanthosternos

FR Sapajou à poitrine jaune
DE Gelbbrust-Kapuzineraffe
ES Capuchino de pecho amarillo

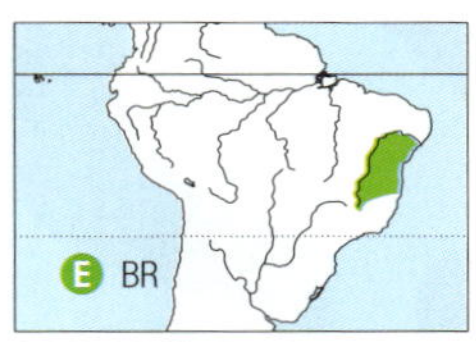

S & E of R São Francisco, S to R Jequitinhonha (Bahia, NE Minas Gerais)

CR
Monotypic
HB 36–42 cm
T 38–45 cm
W 1.3–4.8 kg

Black-horned Capuchin
Sapajus nigritus

FR Sapajou noir
DE Schwarzer Kapuzineraffe
ES Capuchino negro

ssp *nigritus*

ssp *cucullatus*

NT
2 sspp
HB 42–56 cm
T 43–56 cm (M)
W 2.6–4.8 kg

SE BR (S of R Doce & R Grande, from Minas Gerais to Rio Grande do Sul), NE AR (Misiones)

Bearded Capuchin
Sapajus libidinosus

FR Sapajou barbu
DE Rückenstreifen-Kapuzineraffe
ES Capuchino barbudo

300–600 m

W & N of R São Francisco, W to Maranhão and R Araguaia, S to N of R Grande in Minas Gerais

NT
Monotypic
HB 34–44 cm
T 38–49 cm
W 1.3–4.8 kg

Blond Capuchin
Sapajus flavius

FR Sapajou blond
DE Gelber Kapuzineraffe
ES Capuchino rubio

EN
Monotypic
HB 35–40 cm
T 37–42 cm
W 1.8–3 kg

Coast S of Rio Grande do Norte to NE Pernambuco; perhaps to N bank of R São Francisco in Alagoas

Hooded Capuchin
Sapajus cay

FR Sapajou du Paraguay
DE Azara-Kapuzineraffe
ES Capuchino de Azara

VU
Monotypic
HB 40–45 cm
T 41–47 cm
W 3–3.5 kg

up to 1500 m

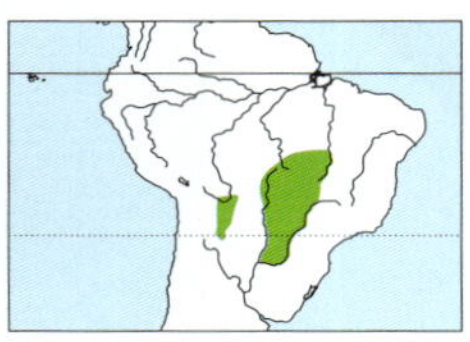

SE BO (Tarija, Santa Cruz), N AR (Jujuy, Salta, Formosa, Chaco), BR (S of Mato Grosso, W of R Paraná), E PY

Brown Capuchin
Sapajus apella

FR Sapajou brun
DE Haubenkapuzineraffe
ES Capuchino pardo

ssp *margaritae*

ssp *apella*

LC
2 sspp
HB 38–46 cm
T 38–49 cm
W 1.3–4.8 kg

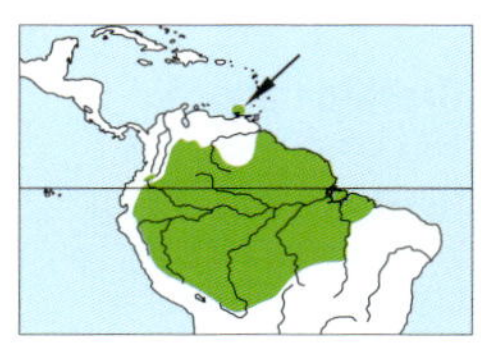

EC, CO (N to R Arauca), E PE, C & N BO (S to upper R Beni), S VE (with Margarita I), Guianas, Amazonian BR

Genus *CEBUS*

(15 species)

Marañon White-fronted Capuchin
Cebus yuracus

FR Sapajou du Maranon
DE Marañón-Kapuzineraffe
ES Capuchino del Marañón

NT
Monotypic
HB 37–43 cm
T 45–47 cm
W 2–4.7 kg

up to 2000 m

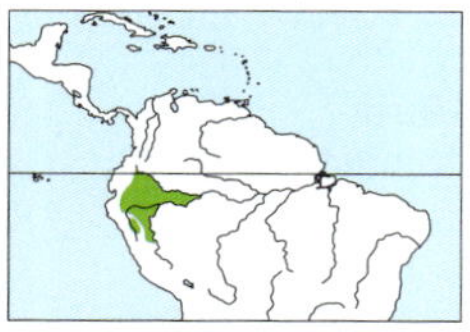

S CO, E EC, NE PE (N of R Amazon to R Putumayo, W of R Ucayali, S to R Pachitea), presumably W BR

Shock-headed White-fronted Capuchin
Cebus cuscinus

FR Sapajou ébouriffé
DE Peru-Kapuzineraffe
ES Capuchino pálido

up to 1800 m

SE PE (S bank of upper R Purus W to R Urubamba Valley) S to NW BO; poorly known

NT
Monotypic
HB 39–46 cm
T 39–47 cm
W 2.8–3 kg

Spix's White-fronted Capuchin
Cebus unicolor

FR Sapajou unicolore
DE Spix-Kapuzineraffe
ES Capuchino de Spix

NW BR (S of R Solimões, W of R Tapajós), to E PE (R Ucayali), presumably N BO (S at least to middle R Beni)

VU
Monotypic
HB 36–37 cm
T 42–46 cm

Humboldt's White-fronted Capuchin
Cebus albifrons

FR Sapajou à front blanc
DE Weißstirn-Kapuzineraffe
ES Capuchino de frente blanca

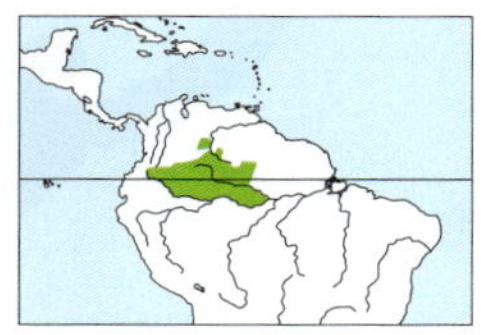

up to 2000 m

LC
Monotypic
HB 34–38 cm
T 41–46 cm
W 2.1–2.6 kg

S & E CO (E lowlands N to lower R Meta), S VE (Amazonas), NW BR (N of R Solimões, W of R Negro & R Branco)

Chestnut Weeper Capuchin
Cebus castaneus

FR Sapajou châtain
DE Kastanienbrauner Kapuzineraffe
ES Capuchino castaño

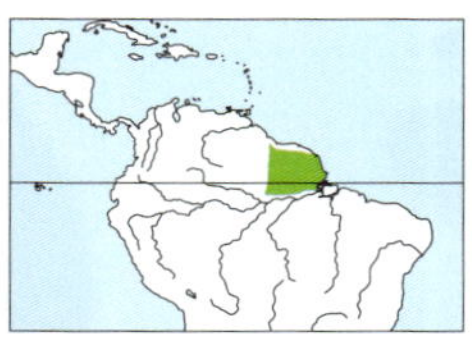

0–2000 m

Guianas (possibly E of R Essequibo), N BR (possibly limited by R Negro/R Branco in W, R Amazon in S, coast in E)

LC
Monotypic
HB 37–46 cm
T 45–55 cm
W 3–4.2 kg (M), 2.3–3 kg (F)

Guianan Weeper Capuchin
Cebus olivaceus

FR Sapajou pleureur
DE Brauner Kapuzineraffe
ES Capuchino oliva

0–2000 m

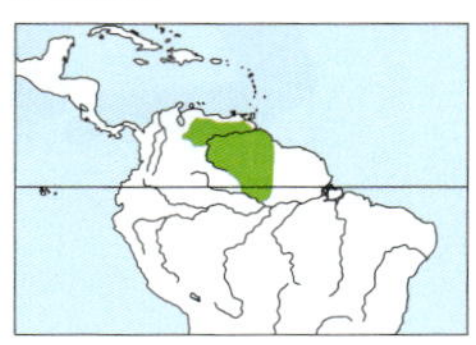

VE (E Sierra de Perijá & upper R Orinoco) E to W bank of R Essequibo (W GY), BR (R Branco/R Araca interfluve)

LC
Monotypic
HB 37–46 cm
T 45–55 cm
W 3–4.2 kg (M), 2.3–3 kg (F)

below 200 m

NE Pará, NW Maranhão; precise limits unknown

Ka'apor Capuchin
Cebus kaapori

FR Sapajou de Ka'apor
DE Ka'apor-Kapuzineraffe
ES Capuchino de Ka'apor

CR
Monotypic
HB 37–46 cm
T 40–55 cm
W 2.4–3 kg

EN
2 sspp
HB c. 42 cm
T c. 44 cm

Brown Weeper Capuchin
Cebus brunneus

FR Sapajou d'Allen
DE Venezuela-Kapuzineraffe
ES Capuchino de Venezuela

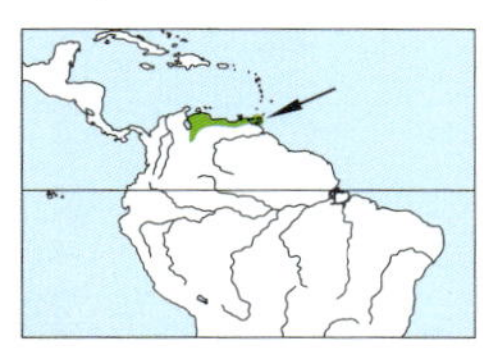

N VE (E of Sierra de Perijá, Cordillera de la Costa), TT (E & SE Trinidad I, where perhaps introduced)

VU
Monotypic
HB 37–41 cm (M)
T 39–50 cm (M)

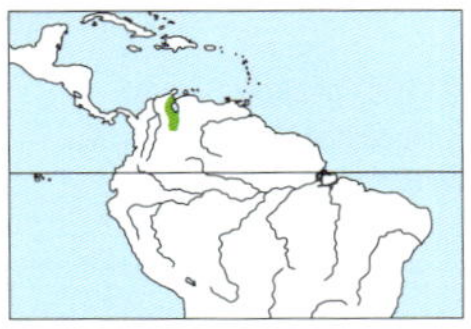

Sierra de Perija White-fronted Capuchin
Cebus leucocephalus

FR Sapajou à tête blanche
DE Weißkopf-Kapuzineraffe
ES Capuchino de Perijá

N CO (from W slope of E Andes E through low passes to R Zulia & R Catatumbo), NW VE (Zulia)

Rio Cesar White-fronted Capuchin
Cebus cesarae

FR Sapajou du Cesar
DE Rio-Cesar-Kapuzineraffe
ES Capuchino de cara blanca del Cesar

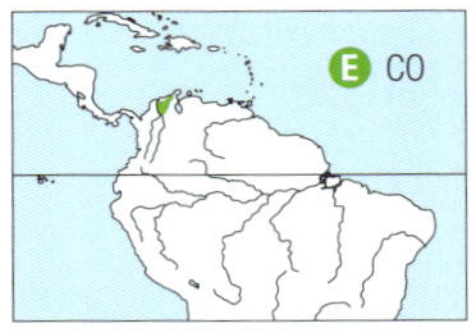

R Cesar Valley up to S & E slopes of Sierra Nevada de Santa Marta

EN
Monotypic
HB 35–41 cm
T 42–50 cm

Santa Marta White-fronted Capuchin
Cebus malitiosus

FR Sapajou de Santa Marta
DE Santa-Marta-Kapuzineraffe
ES Capuchino de Santa Marta

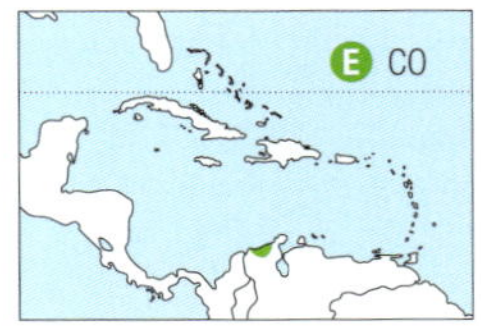

Known only from type locality; perhaps throughout lower W & N slopes of Sierra Nevada

EN
Monotypic
HB c. 46 cm
T c. 43 cm

Varied White-fronted Capuchin
Cebus versicolor

FR Sapajou varié
DE Kolumbien-Kapuzineraffe
ES Capuchino versicolor

EN
Monotypic
HB 45–50 cm
T 42–45 cm

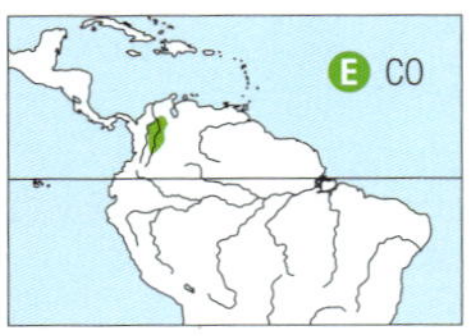

Middle R Magdalena (except W slope of E Andes), from S Magdalena S to Cundinamarca & Tolima deps

Ecuadorian White-fronted Capuchin
Cebus aequatorialis

FR Sapajou d'Équateur
DE Ecuador-Kapuzineraffe
ES Capuchino de Ecuador

CR
Monotypic
HB 35–51 cm
T 40–50 cm
W 1.2–3.6 kg

up to 2040 m

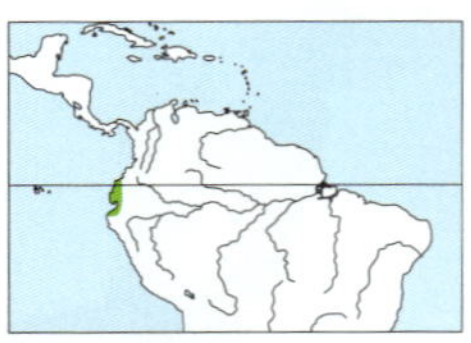

EC (lowlands W of Andes, S from R Esmeraldas-Guayllabamba), NW PE (Tumbes, S to Cerros de Amotape NP)

Colombian White-faced Capuchin
Cebus capucinus

FR Sapajou capucin
DE Weißschulter-Kapuzineraffe
ES Capuchino de cara blanca de Colombia

VU
2 sspp
HB 33–45 cm
T 35–55 cm
W 1.5–4 kg

ssp *capucinus*

ssp *curtus*

up to 2100 m

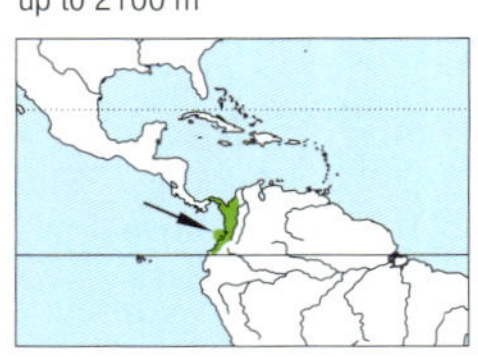

E PA, W CO (Pacific coast, with Gorgona I), NW EC (S to R Esmeraldas-Guayllabamba)

Panamanian White-faced Capuchin
Cebus imitator

FR Sapajou du Panama
DE Panama-Kapuzineraffe
ES Capuchino de cara blanca de Panamá

VU
Monotypic
HB 34–42 cm
T 43–46 cm
W 2.6–3.9 kg

0–1500 m

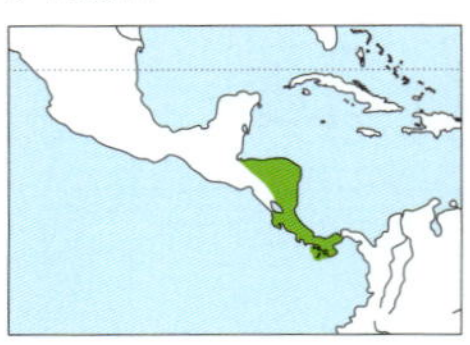

N HN, C & W NI, CR, W PA (with Coiba I and Jicarón I); unconfirmed in W BZ

Family PITHECIIDAE – Titis, Sakis and Uacaris

6 genera, 59 species. Humid submontane and lowland tropical, dry semi-deciduous and deciduous, gallery, and Chacoan forests, and scrub. Neotropical Region.

Genus *CHERACEBUS*

(5 species)

LC
Monotypic
HB 35–41 cm
T 45–48 cm
W c. 1.5 kg (M)

Yellow-handed Titi
Cheracebus lucifer

FR Titi lucifer
DE Witwenaffe
ES Tití lucifer

100–300 m

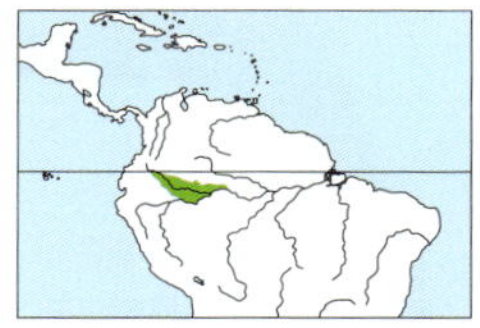

SE CO, NE EC, NE PE, NW BR (between Caquetá/Japurá rivers in N and Napo/Solimões rivers in S)

VU
Monotypic
HB 23–36 cm
T 42–49 cm
W 1.1–1.5 kg

Medem's Titi
Cheracebus medemi

FR Titi de Medem
DE Medem-Springaffe
ES Tití de Medem

100–450 m

Putumayo (E from foothills of Andes, between upper R Caquetá and R Putumayo; E limit unclear)

LC
Monotypic
HB 37–45 cm
T 44–49 cm
W 1–1.5 kg

Rio Jurua Collared Titi
Cheracebus regulus

FR Titi roitelet
DE Rotkopf-Springaffe
ES Tití de cabeza roja

100–200 m

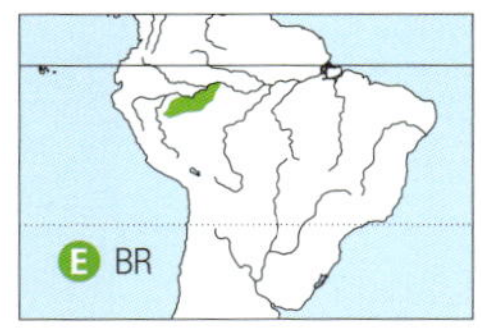

Amazonas (S of R Solimões, E of lower R Javari, W of lower R Juruá; S limit unclear)

LC
Monotypic
HB 30–40 cm
T 41–49 cm
W 1–1.5 kg

White-collared Titi
Cheracebus torquatus

FR Titi à collier
DE Halsband-Springaffe
ES Tití de collar

100–1000 m

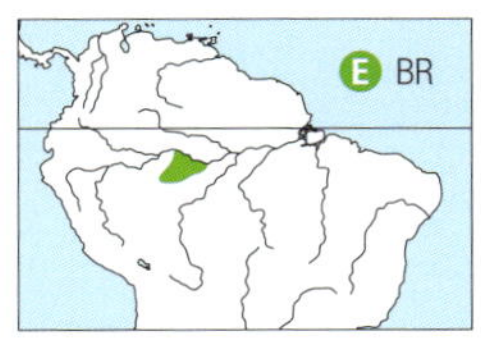

Amazonas (S of R Solimões, E of lower R Juruá, W of Purus/Tapauá river system; S limit unclear)

LC
Monotypic
HB 30–40 cm
T 41–49 cm
W 1–1.5 kg

White-chested Titi
Cheracebus lugens

FR Titi noir
DE Schwarzer Springaffe
ES Tití negro

100–1000 m

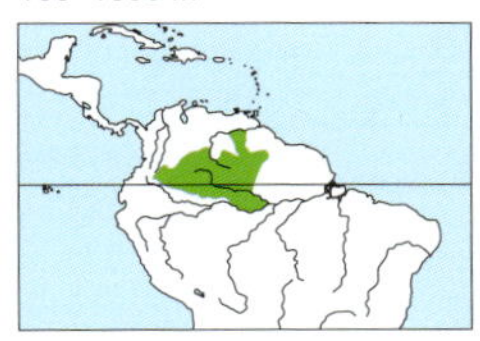

E CO, S VE, NW BR (N of R Caquetá, S of R Ventuari & middle R Orinoco to R Japurá & R Solimões, E to R Branco)

Genus *CALLICEBUS*

(5 species)

buffy morph

orange morph

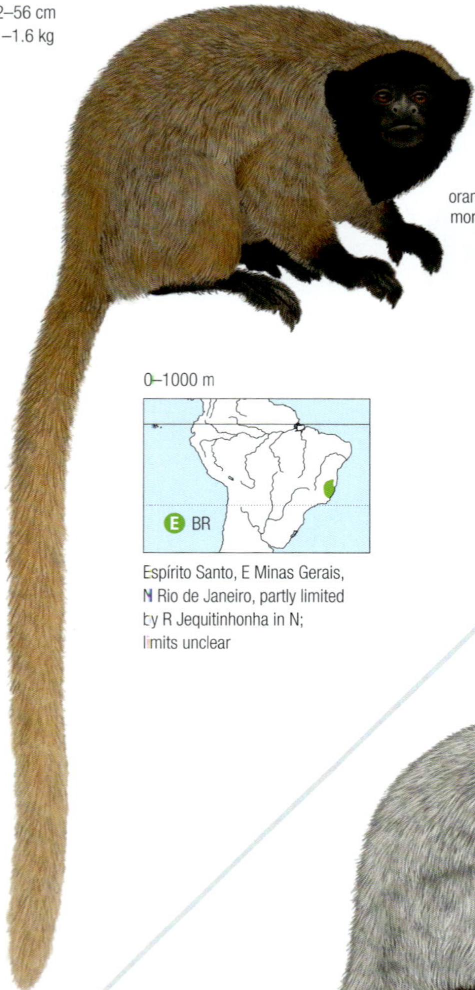

Masked Titi

Callicebus personatus

FR Titi masqué
DE Maskenspringaffe
ES Tití enmascarado

VU
Monotypic
HB 31–42 cm
T 42–56 cm
W 1–1.6 kg

0–1000 m

E BR

Espírito Santo, E Minas Gerais, N Rio de Janeiro, partly limited by R Jequitinhonha in N; limits unclear

Coimbra-Filho's Titi

Callicebus coimbrai

FR Titi de Coimbra
DE Coimbra-Filho-Springaffe
ES Tití de Coimbra

0–300 m

E BR

Sergipe (S of R São Francisco) & NE Bahia (N of R Paraguaçu), W to Atlantic Forest–caatinga transition

EN
Monotypic
HB 35–36 cm (F)
T 45–48 cm (F)
W 1–1.3 kg (F)

variants

variants

Blond Titi

Callicebus barbarabrownae

FR Titi blond
DE Barbara-Brown-Springaffe
ES Tití de Bárbara Brown

240–900 m

E BR

NC Bahia, W Sergipe (caatinga S of R São Francisco, N of R Paraguaçu)

CR
Monotypic
HB 33–36 cm (F)
T 39–43 cm (F)
W 1–1.3 kg

Black-fronted Titi
Callicebus nigrifrons

FR Titi à front noir
DE Schwarzstirn-Springaffe
ES Tití de frente negra

NT
Monotypic
HB 30–39 cm
T 45–50 cm
W 1–1.6 kg

0–1000 m

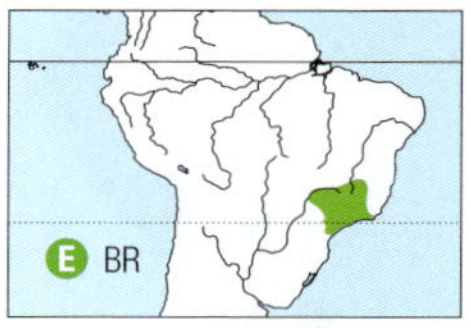

Paraná/Parnaíba rivers E to Mantiqueira & Espinhaço ranges, N to upper R São Francisco, S to R Tietê

Southern Bahian Titi
Callicebus melanochir

FR Titi à mains noires
DE Grauer Maskenspringaffe
ES Tití de manos negras

VU
Monotypic
HB 33–37 cm (F)
T 39–51 cm (F)
W c. 1.4 kg (F)

0–300 m

E Bahia, N Espírito Santo (Atlantic Forest from R Paraguaçu S to R Mucuri, W to transition with caatinga)

Genus *PLECTUROCEBUS*

(24 species)

Pale Titi
Plecturocebus pallescens

FR Titi pâle
DE Weißmantel-Springaffe
ES Tití blanco

LC
Monotypic
HB 31–36 cm
T 39–42 cm
W c. 800 g

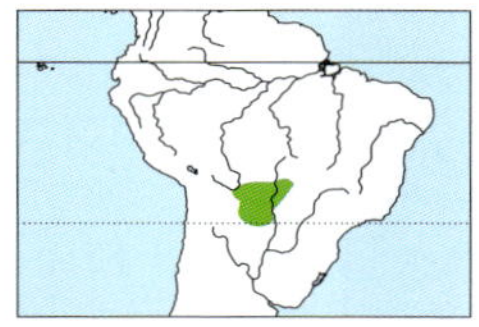

NE BO (lowland E of R Manique), W BR (S Rondônia at least to W upper R Ji-Paraná); S & E limits unclear

White-eared Titi
Plecturocebus donacophilus

FR Titi de d'Orbigny
DE Weißohr-Springaffe
ES Tití de orejas blancas

LC
Monotypic
HB 28–42 cm
T 37–46 cm
W c. 800 g

NE BO (lowland E of R Manique), W BR (S Rondônia at least to W upper R Ji-Paraná); S & E limits unclear

San Martin Titi
Plecturocebus oenanthe

FR Titi du Mayo
DE Anden-Springaffe
ES Tití andino

CR
Monotypic
HB 30–32 cm
T 38–40 cm (M), 36–38 cm (F)
W c. 800 g

750–950 m

San Martín (mainly upper R Mayo Valley but extending S possibly as far as R Huayabamba)

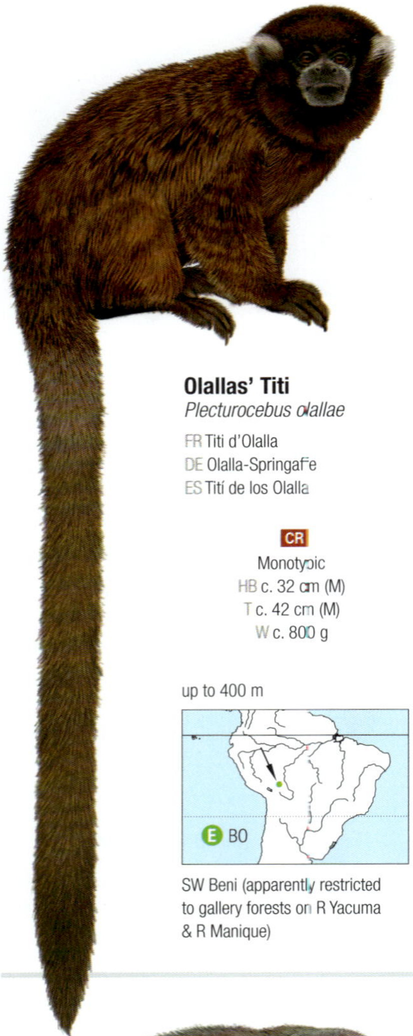

Olallas' Titi
Plecturocebus olallae

FR Titi d'Olalla
DE Olalla-Springaffe
ES Tití de los Olalla

CR
Monotypic
HB c. 32 cm (M)
T c. 42 cm (M)
W c. 800 g

up to 400 m

SW Beni (apparently restricted to gallery forests on R Yacuma & R Manique)

Rio Beni Titi
Plecturocebus modestus

FR Titi du Rio Beni
DE Rio-Beni-Springaffe
ES Tití del Beni

EN
Monotypic
HB c. 31 cm (M)
T c. 40 cm (M)
W c. 800 g

up to 400 m

SW Beni (apparently restricted to R Beni Basin, E of R Beni, W to R Manique)

Urubamba Brown Titi
Plecturocebus urubambensis

FR Titi de l'Urubamba
DE Urubamba-Springaffe
ES Tití de Urubamba

LC
Monotypic

Between R Tambo and R Urubamba and between R Manu and upper R Madre de Dios

Vieira's Titi
Plecturocebus vieirai

FR Titi de Vieira
DE Vieira-Springaffe
ES Tití de Vieira

CR
Monotypic
HB 30–35 cm
T 41–51 cm
W c. 960 g (M)

Pará, Mato Grosso (S of R Amazon, between R Xingu and R Iriri)

Red-bellied Titi
Plecturocebus moloch

FR Titi gris
DE Rotbauch-Springaffe
ES Tití de vientre rojo

LC
Monotypic
HB 27–43 cm
T 35–55 cm
W 700–1200 g

0–300 m

Pará, Mato Grosso (S of R Amazon, R Tapajós/R Juruena E to R Tocantins/R Araguaia; S limit unclear)

Groves's Titi
Plecturocebus grovesi

FR Titi de Groves
DE Groves-Springaffe
ES Tití de Groves

CR
Monotypic

Mato Grosso (R Juruena/R Arinos E to R Teles-Pires; S limits likely to Amazon forest/cerrado ecotone)

Prince Bernhard's Titi
Plecturocebus bernhardi

FR Titi de Bernhard
DE Prinz-Bernhard-Springaffe
ES Tití del Príncipe Bernardo

LC
Monotypic
HB 36–38 cm
T c. 55 cm
W 700–1200 g

S of R Madeira, from lower R Ji-Paraná E to R Aripuanã & R Roosevelt; S limits unclear

Milton's Titi
Plecturocebus miltoni

FR Titi de Milton
DE Milton-Springaffe
ES Tití de Milton

DD
Monotypic

Mato Grosso, Amazonas (interfluvial region of R Roosevelt and R Aripuanã)

Parecis Titi
Plecturocebus parecis

FR Titi des Parecis
DE Parecis-Springaffe
ES Tití de Parecis

NT
Monotypic

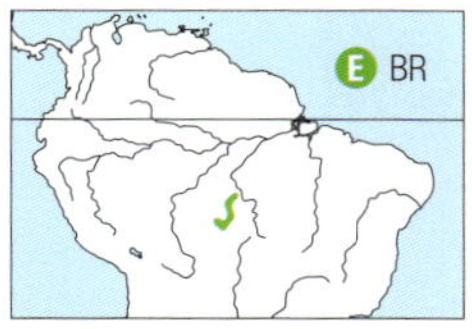

S Rondônia, W Mato Grosso (in ecotone Amazon Forest/cerrado savanna, including part of Parecis Plateau)

Ashy Titi
Plecturocebus cinerascens

FR Titi cendré
DE Dunkelgrauer Springaffe
ES Tití ceniciento

LC
Monotypic
HB 32–40 cm
T 39–48 cm
W 740–950 g

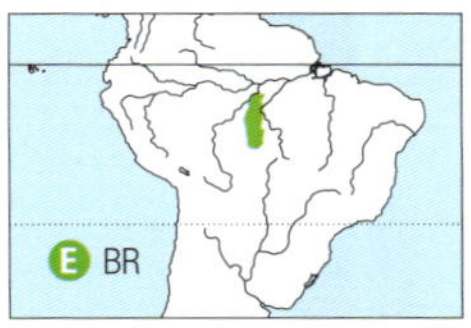

S of R Madeira, W to R Aripuanã & R Roosevelt, E to R Tapajós & R Juruena, S possibly to E R Guaporé

Hoffmanns's Titi
Plecturocebus hoffmannsi

FR Titi de Hoffmanns
DE Hoffmanns-Springaffe
ES Tití de Hoffmanns

LC
Monotypic
HB 27–36 cm
T 40–53 cm
W 920–1100 g

Amazon/Tapajós interfluvium, S of Canumã, Urariá & Ramos channels ("paranás"); W to R Abacaxis

Lake Baptista Titi
Plecturocebus baptista

FR Titi du Baptista
DE Baptistasee-Springaffe
ES Tití del Baptista

LC
Monotypic
HB 30–41 cm
T 42–50 cm
W 930–1400 g

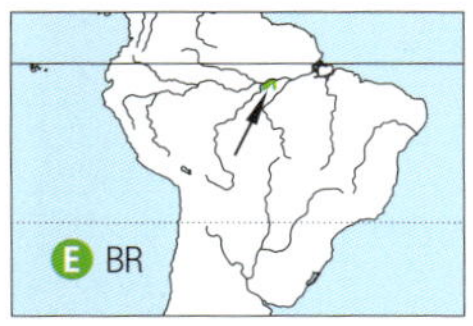

S of Amazon/Madeira river system, N of Canumã, Urariá & Ramos channels ("paranás")

Coppery Titi
Plecturocebus cupreus
FR Titi roux
DE Roter Springaffe
ES Tití cobrizo

LC
Monotypic
HB 27–41 cm
T 39–43 cm
W 1–1.2 kg (M)

100–300 m

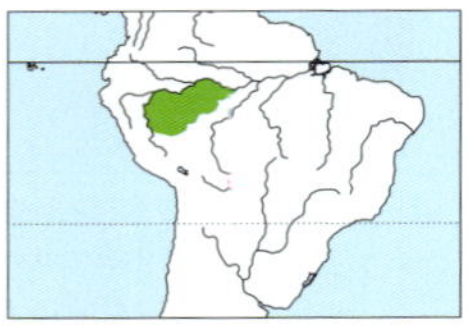

E PE (S of R Marañón/Solimões, E of R Ucayali), W BR (W of R Purus); S limit apparently foothills of Andes

Red-crowned Titi
Plecturocebus discolor
FR Titi discolore
DE Weißschwanz-Springaffe
ES Tití rojizo

LC
Monotypic
HB 28–34 cm
T 38–51 cm
W 850–1080 g

100–800 m

S CO (S of R Guamués), E EC (E from Andean foothills), PE (between R Napo and R Santiago, along R Huallaga)

Ornate Titi
Plecturocebus ornatus
FR Titi orné
DE Weißstirn-Springaffe
ES Tití ornamentado

VU
Monotypic
HB 30–36 cm
T 38–45 cm
W 1.1–1.2 kg

100–1000 m

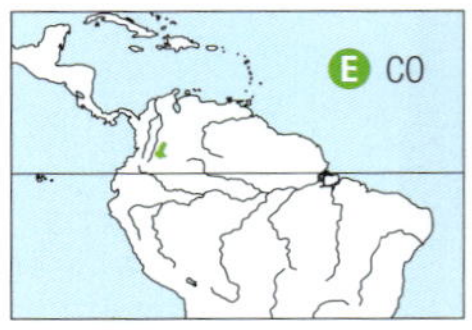

Cundinamarca (N to lower R Upía/Meta basin), Meta (Sierra de la Macarena to R Guayabero & R Guaviare)

Caqueta Titi
Plecturocebus caquetensis
FR Titi du Caqueta
DE Caquetá-Springaffe
ES Tití del Caquetá

CR
Monotypic
HB c. 35 cm (F)

190–260 m

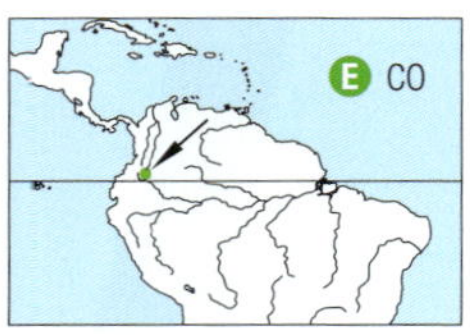

Caquetá (known between R Caquetá and R Orteguaza, W to foothills of Andes)

Brown Titi
Plecturocebus brunneus
FR Titi brun
DE Brauner Springaffe
ES Tití pardo

VU
Monotypic
HB 30–34 cm
T 37–44 cm
W c. 850 g

100–400 m

Rondônia (Mamoré/Madeira/Ji-Paraná interfluvium, limited to S in part by Serra dos Pacaás Novos)

Madidi Titi
Plecturocebus aureipalatii
FR Titi du Madidi
DE Madidi-Springaffe
ES Tití del Madidi

LC
Monotypic
HB 29–32 cm
T 48–52 cm
W 900–1000 g

100–500 m

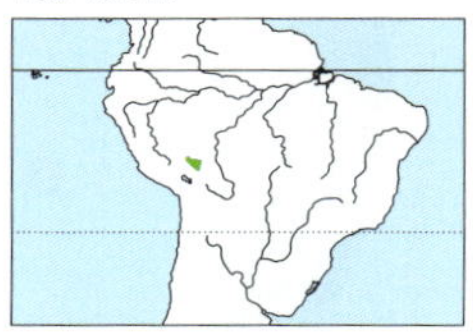

SE PE (Puno, Madre de Dios, W to R Tambopata), W BO (E to R Beni, S to foothills of Andes); limits unclear

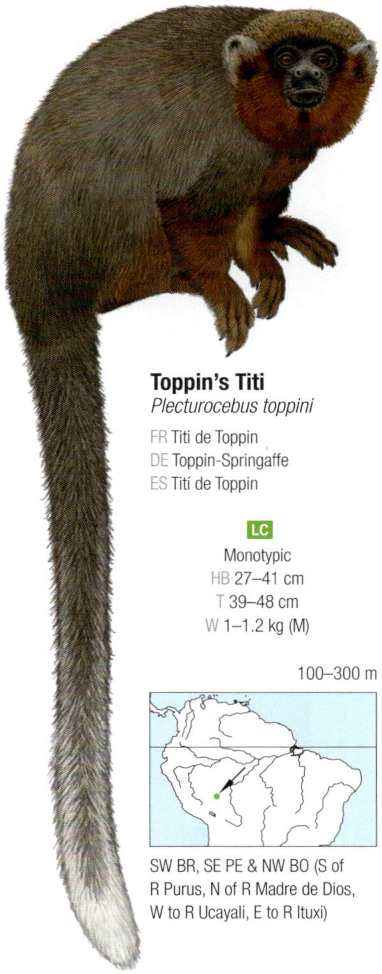

Toppin's Titi
Plecturocebus toppini

FR Titi de Toppin
DE Toppin-Springaffe
ES Tití de Toppin

LC
Monotypic
HB 27–41 cm
T 39–48 cm
W 1–1.2 kg (M)

100–300 m

SW BR, SE PE & NW BO (S of R Purus, N of R Madre de Dios, W to R Ucayali, E to R Ituxi)

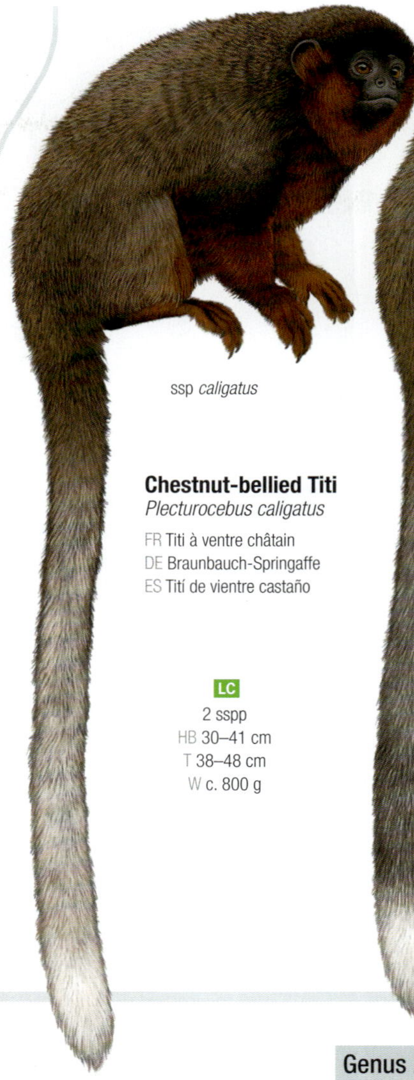

ssp *caligatus*

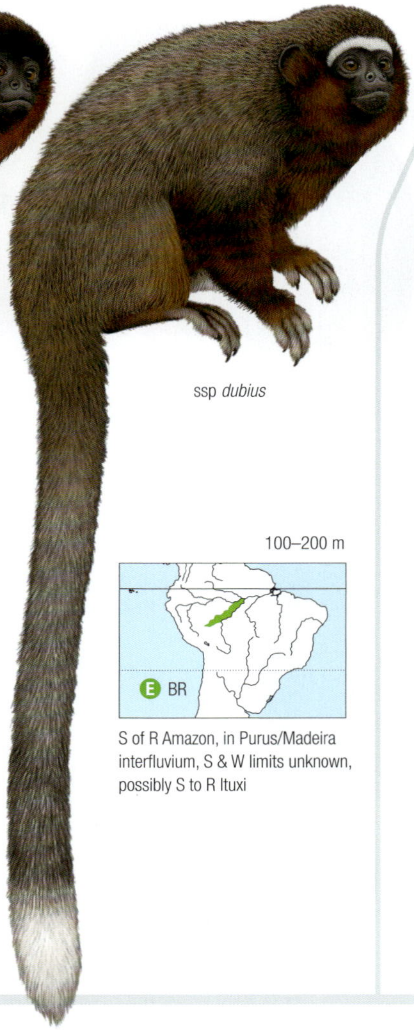

ssp *dubius*

Chestnut-bellied Titi
Plecturocebus caligatus

FR Titi à ventre châtain
DE Braunbauch-Springaffe
ES Tití de vientre castaño

LC
2 sspp
HB 30–41 cm
T 38–48 cm
W c. 800 g

100–200 m

S of R Amazon, in Purus/Madeira interfluvium, S & W limits unknown, possibly S to R Ituxi

Stephen Nash's Titi
Plecturocebus stephennashi

FR Titi de Nash
DE Stephen-Nash-Springaffe
ES Tití de Stephen Nash

DD
Monotypic
HB 27–28 cm
T c. 42 cm
W c. 780 g

100–200 m

Presumably Ipixuna/Mucuím interfluvium on S bank of R Purus, S possibly to R Madeira in Rondônia

Genus *PITHECIA*
(16 species)

♀

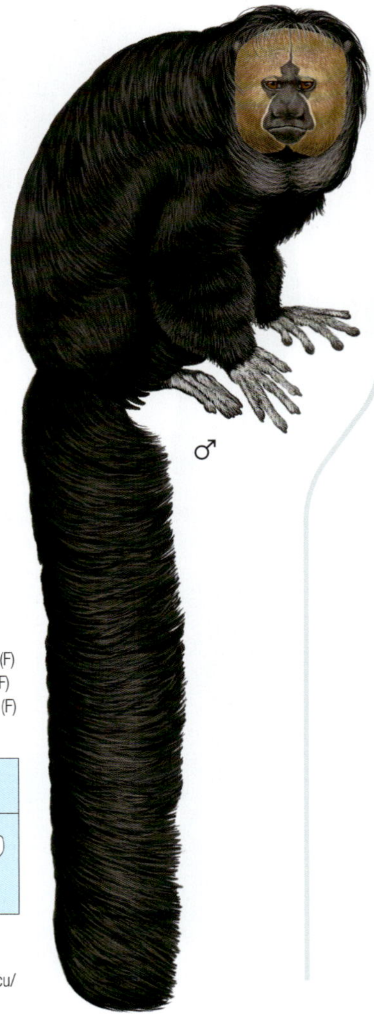

♂

Monk Saki
Pithecia monachus

FR Saki moine
DE Mönchsaffe
ES Saki cabelludo

LC
Monotypic
HB 40–48 cm (M), 37–46 cm (F)
T 30–50 cm (M), 41–48 cm (F)
W 2.5–3.1 kg (M), 1.3–2.5 kg (F)

NE PE & W BR (R Solimões/R Ucayali/lower R Yavarí/lower R Juruá interfluvium, S to Sarayacu/Serra do Divisor)

♀

♂

Hairy Saki
Pithecia hirsuta

FR Saki hirsute
DE Haariger Saki
ES Saki hirsuto

LC
Monotypic
HB 48–50 cm (M), 41–43 cm (F)
T 41–43 cm (M)

SE CO, N PE, NW BR (S of R Caquetá, N of R Napo and R Solimões, W of R Negro)

♀

♂

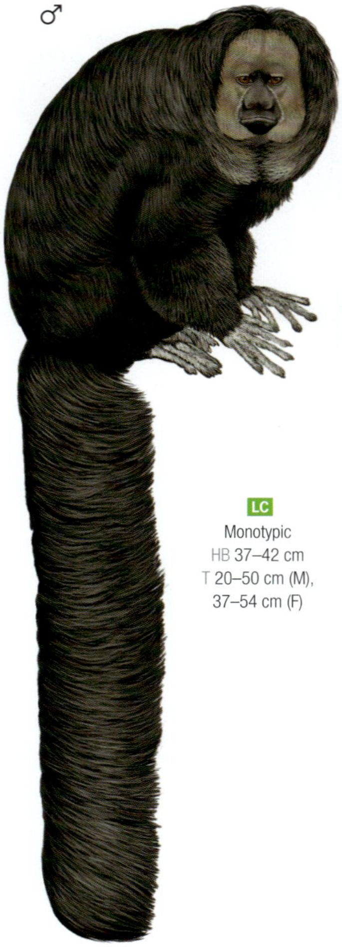

Burnished Saki
Pithecia inusta

FR Saki basané
DE Spix-Saki
ES Saki pulido

LC
Monotypic
HB 37–42 cm
T 20–50 cm (M), 37–54 cm (F)

EC PE & W BR (both sides of R Ucayali from Sarayacu S to Urubamba/Tambor split, perhaps N to R Tahuayo and R Tapiche

♀

♂

Miller's Saki
Pithecia milleri

FR Saki de Miller
DE Miller-Saki
ES Saki de Miller

VU
Monotypic
HB 33–48 cm (M), c. 39 cm (F)
T 44–49 cm (M), c. 41 cm (F)
W 2.6–2.8 kg (M), c. 2.2 kg (F)

up to 700 m

S CO (Andean foothills, NE to La Macarena in Meta, E to at least Puerto Leguízamo) & NE EC (N of R Napo)

♀

♂

Napo Saki
Pithecia napensis

FR Saki du Napo
DE Napo-Saki
ES Saki del Napo

LC
Monotypic
HB 36–56 cm
T 40–54 cm
W 2.5–3 kg

E EC (S of R Napo from Coca in W to Yasuni NP in E), N PE (W to N bank of mouth of R Curaray at R Napo)

♀

♂

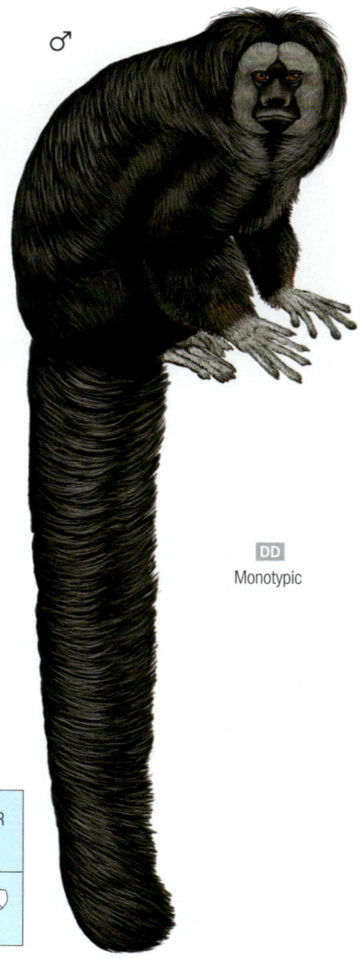

Cazuza's Saki
Pithecia cazuzai

FR Saki de Cazuza
DE Cazuza-Saki
ES Saki de Cazuza

DD
Monotypic

Known only from between lower R Japurá and R Solimões

♀

♂

Equatorial Saki
Pithecia aequatorialis

FR Saki d'Équateur
DE Äquatorialsaki
ES Saki de Ecuador

LC
Monotypic
HB 40–44 cm (M), 39–40 cm (F)
T 45–47 cm (M), c. 45 cm (F)
W c. 3 kg

S of R Napo and S of R Curaray to R Tigre in W; N limit uncertain

♀

♂

Isabel's Saki
Pithecia isabela

FR Saki d'Isabel
DE Isabel-Saki
ES Saki de Isabel

DD
Monotypic

Pacaya-Samiria R, between R Marañón in W and R Ucayali in E

♀

♂

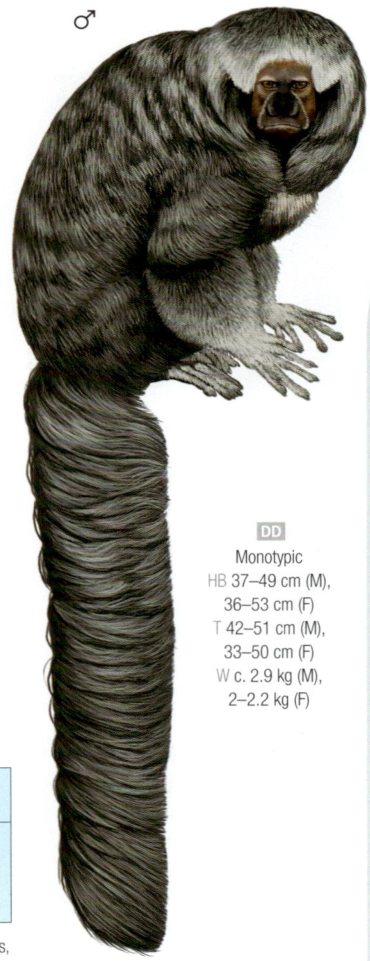

Gray's Bald-faced Saki
Pithecia irrorata

FR Saki à tête chauve
DE Nacktgesichtsaki
ES Saki gris

DD
Monotypic
HB 37–49 cm (M), 36–53 cm (F)
T 42–51 cm (M), 33–50 cm (F)
W c. 2.9 kg (M), 2–2.2 kg (F)

BR, SE PE & N BO (along R Purus, S to W of lower R Urubamba & R Madre de Dios, E limits uncertain)

♀

♂

Rylands's Bald-faced Saki
Pithecia rylandsi

FR Saki de Rylands
DE Rylands-Saki
ES Saki de Rylands

VU
Monotypic

NW BO (W Pando, perhaps NE Beni & N La Paz), SE PE (mainly N of R Madre de Dios), possibly adjacent BR

♀

♂

Mittermeier's Saki
Pithecia mittermeieri

FR Saki de Mittermeier
DE Mittermeier-Saki
ES Saki de Mittermeier

VU
Monotypic

S of R Amazon between R Madeira and R Tapajós, including R Aripuanã drainage

♀

♂

Pissinatti's Bald-faced Saki
Pithecia pissinattii

FR Saki de Pissinatti
DE Pissinatti-Saki
ES Saki de Pissinatti

DD
Monotypic

S of R Solimões, between R Purus and R Madeira; S limit unknown

♀

♂

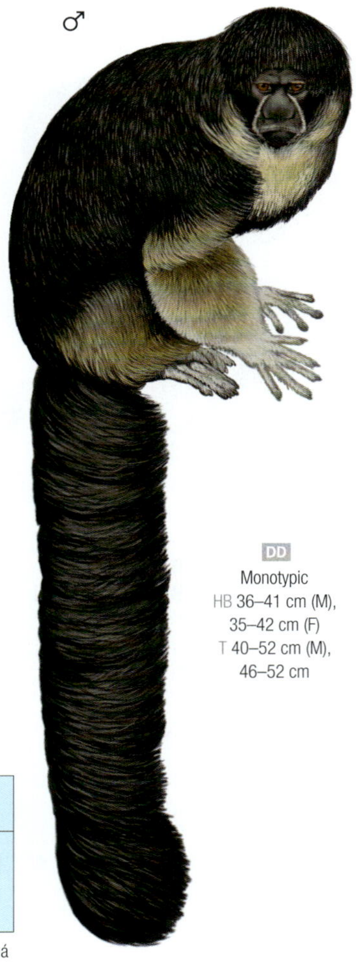

Vanzolini's Bald-faced Saki
Pithecia vanzolinii

FR Saki de Vanzolini
DE Vanzolini-Saki
ES Saki de Vanzolini

DD
Monotypic
HB 36–41 cm (M), 35–42 cm (F)
T 40–52 cm (M), 46–52 cm

Between S bank of upper R Juruá and S bank of R Tarauacá

♀

♂

Buffy Saki
Pithecia albicans

FR Saki chamois
DE Schwarzrückensaki
ES Saki beige

LC
Monotypic
HB 50–57 cm (M), 38–54 cm (F)
T 40–57 cm (M), 41–49 cm (F)
W c. 3.5 kg

S of R Solimões, between lower R Juruá and lower R Purus

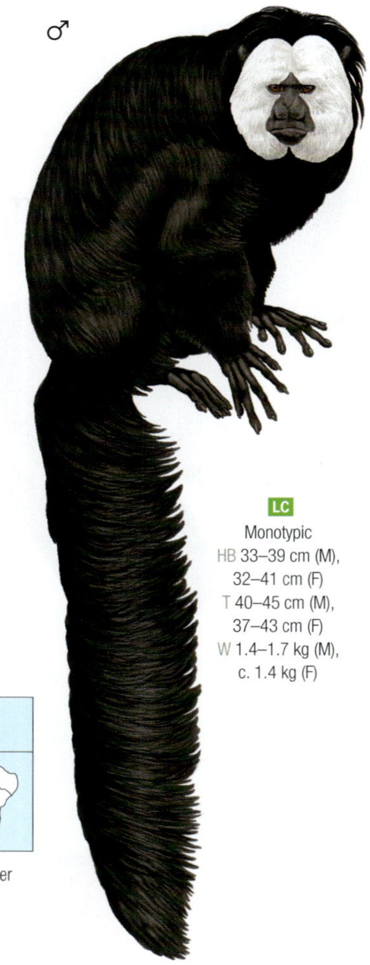

White-faced Saki
Pithecia pithecia

FR Saki à tête pâle
DE Weißkopfsaki
ES Saki de cara blanca

LC

Monotypic
HB 33–39 cm (M), 32–41 cm (F)
T 40–45 cm (M), 37–43 cm (F)
W 1.4–1.7 kg (M), c. 1.4 kg (F)

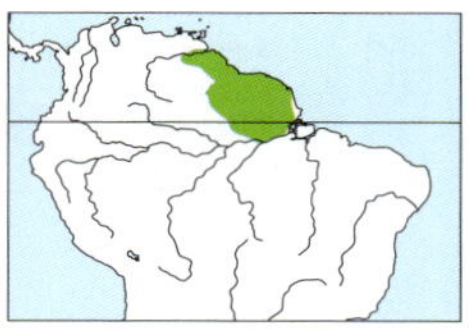

E VE (S of R Orinoco, E of lower R Caroní), Guianas, N BR (E of R Negro & R Branco, S to R Amazon)

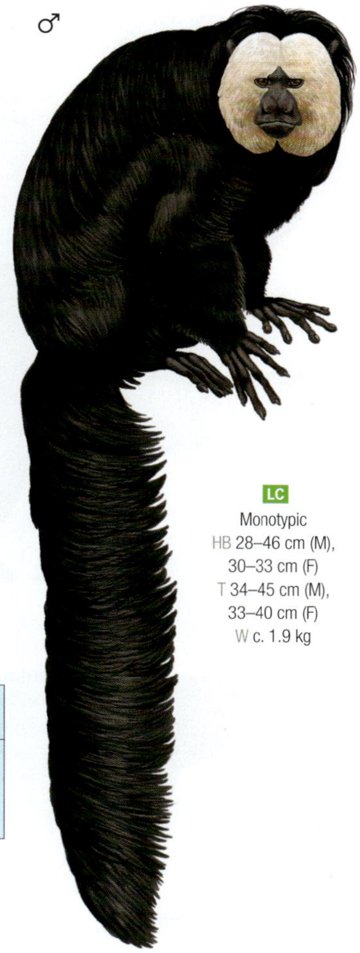

Golden-faced Saki
Pithecia chrysocephala

FR Saki à tête d'or
DE Goldgesichtsaki
ES Saki de cara dorada

LC

Monotypic
HB 28–46 cm (M), 30–33 cm (F)
T 34–45 cm (M), 33–40 cm (F)
W c. 1.9 kg

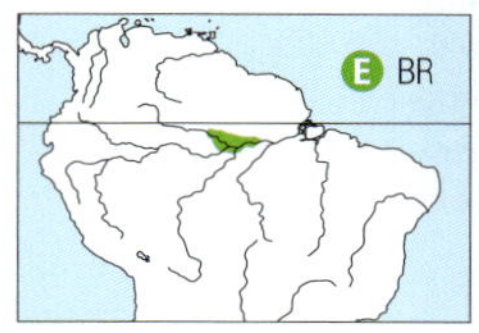

N of R Amazon, both sides of lower R Negro, E to Faro and R Nhamundá

Genus *CHIROPOTES*

(5 species)

Red-nosed Bearded Saki
Chiropotes albinasus

FR Saki à nez blanc
DE Weißnasensaki
ES Saki barbudo de nariz blanca

S of R Amazon, from R Madeira in W and R Xingu in E, to rainforest/cerrado ecotone

VU

Monotypic
HB 36–51 cm
T 36–48 cm
W 2.2–3.7 kg

Black Bearded Saki
Chiropotes satanas

FR Saki noir
DE Satansaffe
ES Saki barbudo negro

S of lower R Amazon, from R Tocantins to E limits of Amazonian rainforest

EN

Monotypic
HB 34–42 cm
T 36–42 cm
W 2–4 kg

Rio Negro Bearded Saki
Chiropotes chiropotes

FR Saki du Rio Negro
DE Rotrückensaki
ES Saki barbudo de dorso rojo

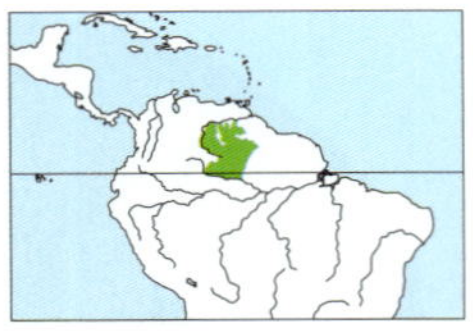

S VE & N BR: delimited by R Orinoco in N, R Branco in E, R Negro in S

LC
Monotypic
HB 35–46 cm
T 30–46 cm
W 2–4 kg

Uta Hick's Bearded Saki
Chiropotes utahicki

FR Saki de Hick
DE Uta-Hick-Saki
ES Saki barbudo de Uta Hick

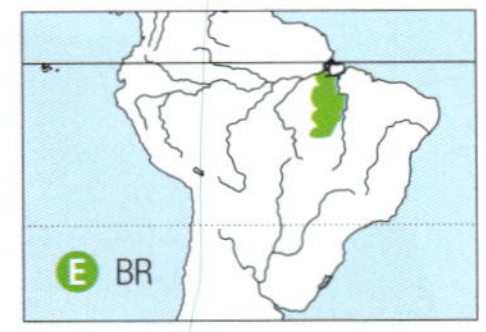

S of R Amazon, between R Xingu and Tocantins/Araguaia rivers; S limit unknown

VU
Monotypic
HB 36–42 cm
T 37–58 cm
W 2–4 kg

Guianan Bearded Saki
Chiropotes sagulatus

FR Saki à gilet
DE Guayana-Rotrückensaki
ES Saki barbudo de dorso pardo

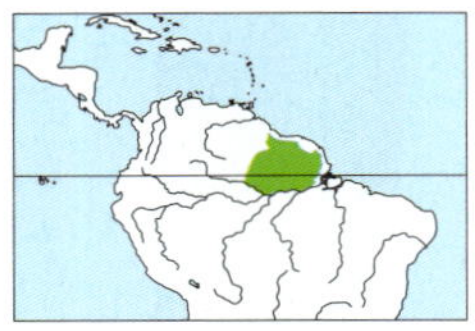

Guianas & N BR (N of R Amazon, E of R Branco)

LC
Monotypic
HB 33–46 cm
T 39–46 cm
W c. 3 kg

Genus *CACAJAO*

(4 species)

Bald Uacari
Cacajao calvus

FR Ouakari chauve
DE Rotkopfuakari
ES Uakari calvo

ssp *calvus*

VU
4 sspp
HB 36–57 cm
T 14–19 cm
W 2.3–3.5 kg

ssp *novaesi*

ssp *rubicundus*

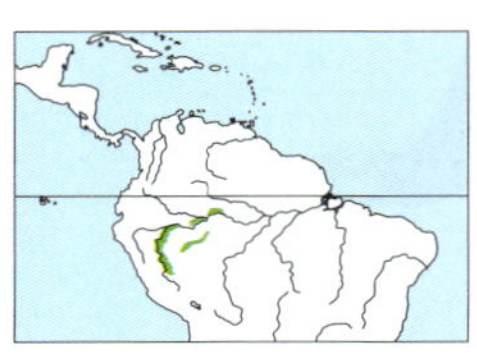

Along rivers in E PE and W BR (Japurá, Auatí-Paraná, Solimões, Juruá, Ucayali, Yavarí); poorly known

ssp *ucayalii*

VU
Monotypic
HB 42–49 cm
T 14–22 cm
W 3.1–4.5 kg

Neblina Uacari
Cacajao hosomi

FR Ouakari du Neblina
DE Schwarzkopfuakari
ES Uakari de cabeza negra

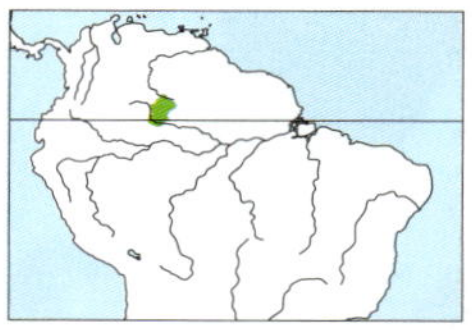

S VE and NW BR, between upper R Orinoco and upper R Negro

Araça Uacari
Cacajao ayresi

FR Ouakari d'Ayres
DE Ayres-Schwarzkopfuakari
ES Uakari del Araçá

LC
Monotypic
HB 30–40 cm
T 15–18 cm (F)
W 2–2.5 kg

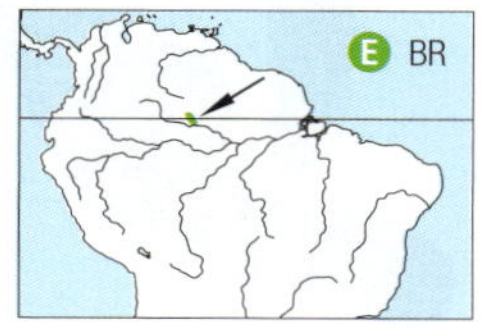

N BR, limited to small area comprising R Curuduri Basin and middle to lower R Araçá N of R Negro

Golden-brown Uacari
Cacajao melanocephalus

FR Ouakari à tête noire
DE Spix-Schwarzkopfuakari
ES Uakari de Spix

LC
Monotypic
HB 36–55 cm
T 14–15 cm
W 0.5–2.6 kg

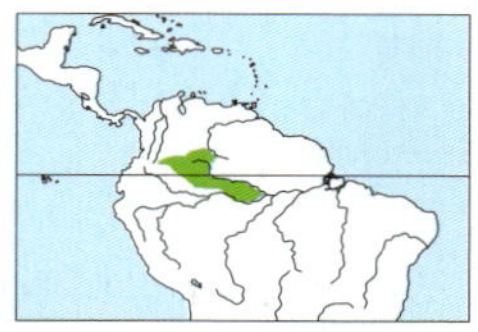

SE CO, S VE & NW BR (E of Serranía de La Macarena, S of R Casiquiare & R Negro to R Caquetá/R Japurá)

Family ATELIDAE – Howlers, Spider and Woolly Monkeys and Muriquis

4 genera, 23 species. Humid, lowland, submontane and montane, deciduous and semi-deciduous forests. Neotropical Region.

Genus *ALOUATTA*

(12 species)

Colombian Red Howler
Alouatta seniculus

FR Hurleur roux
DE Roter Brüllaffe
ES Mono aullador rojo

LC
3 sspp
HB 48–63 cm
T 52–80 cm
W 4.1–9 kg

0–3200 m

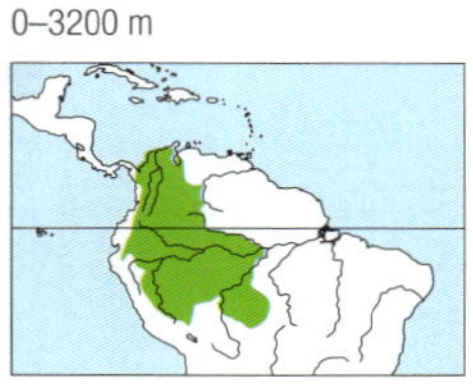

N & E CO, NW VE, E EC, E PE, W BR (S of R Negro, E to lower R Madeira and R Teles Pires); limits uncertain

Ursine Red Howler
Alouatta arctoidea

FR Hurleur ourson
DE Bärenbrüllaffe
ES Mono aullador rojo de Venezuela

LC
Monotypic
HB 45–65 cm
T 55–68 cm
W 4.5–8 kg

10–2000 m

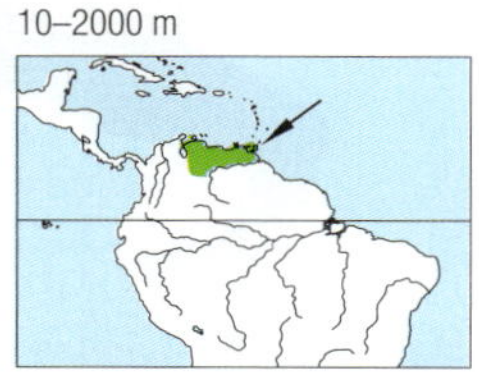

VE, TT (Trinidad I); perhaps extreme NE CO

Guianan Red Howler

Alouatta macconnelli

FR Hurleur de McConnell
DE Guayana-Brüllaffe
ES Mono aullador rojo de Guayana

LC
Monotypic
HB 48–63 cm
T 52–80 cm
W 4.1–9 kg

0–1200 m

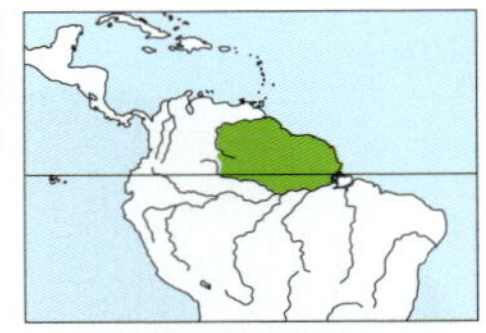

S VE (E & S of R Orinoco), Guianas, N BR (E of R Cauaburi, S to R Amazon)

up to 1000 m

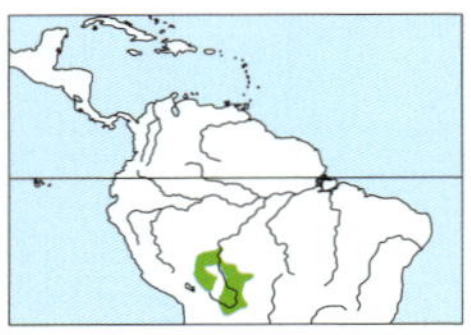

N & C BO, adjacent SE Peru

Bolivian Red Howler

Alouatta sara

FR Hurleur de Bolivie
DE Bolivien-Brüllaffe
ES Mono aullador rojo de Bolivia

NT
Monotypic
HB 54–71 cm
T 51–60 cm
W 6–9 kg

Between R Madeira and R Tapajós, S of R Amazon; outlying populations N of R Amazon and W of R Madeira

Amazon Black Howler

Alouatta nigerrima

FR Hurleur de Lönnberg
DE Amazonas-Brüllaffe
ES Mono aullador negro

LC
Monotypic
HB 48–65 cm
T 56–69 cm
W 4.9–8 kg

VU
Monotypic
HB 37–65 cm
T 45–70 cm
W 4.8–8 kg

Lower Amazonia S of R Amazon, disjunctly in Atlantic Forest (Rio Grande do Norte S to R São Francisco)

Eastern Red-handed Howler

Alouatta belzebul

FR Hurleur à mains rousses
DE Rothand-Brüllaffe
ES Mono aullador de manos rojas

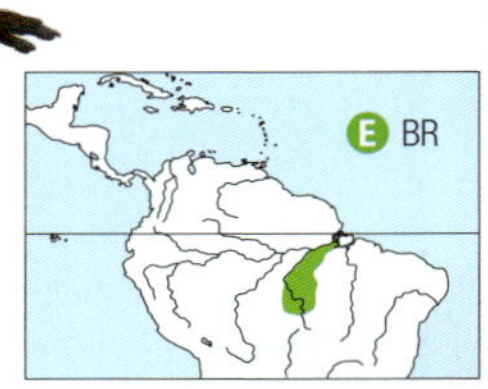

Lower Amazonia S of R Amazon (from R Tapajós & R Juruena to R Xingu & R Iriri; interfluvium Juruena/Teles Pires)

Spix's Red-handed Howler

Alouatta discolor

FR Hurleur de Spix
DE Spix-Brüllaffe
ES Mono aullador de Spix

VU
Monotypic
HB 46–91 cm
T 60–67 cm
W 4.8–8 kg

♀

♂

Maranhao Red-handed Howler

Alouatta ululata

FR Hurleur du Maranhao
DE Maranhão-Brüllaffe
ES Mono aullador de manos rojas del Maranhão

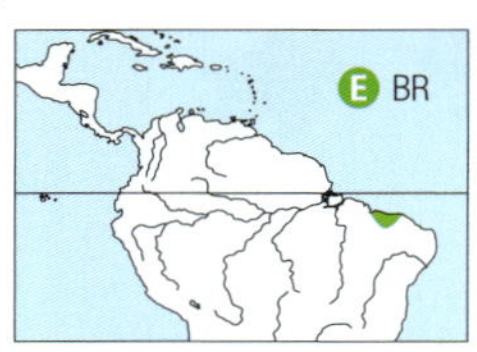

N of Maranhão, Piauí & Ceará

EN
Monotypic
HB c. 56 cm
T c. 56 cm

♂
ssp *guariba*

♂
ssp *clamitans*

♀
ssp *clamitans*

Brown Howler
Alouatta guariba

FR Hurleur brun
DE Brauner Brüllaffe
ES Mono aullador pardo

0–700 m

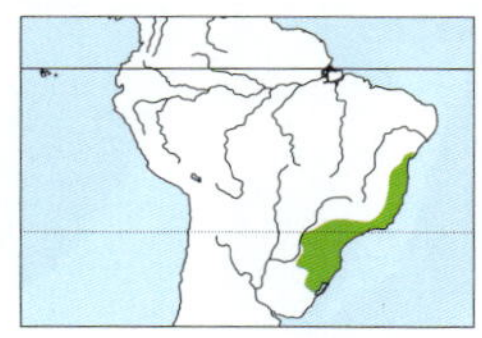

E & S BR (Bahia S to Rio Grande do Sul), NE AR (Misiones)

VU
2 sspp
HB 50–60 cm (M), 44–54 cm (F)
T 52–67 cm (M), 48–57 cm (F)
W 5.3–7.2 kg (M), 4.1–5 kg (F)

Black-and-gold Howler
Alouatta caraya

FR Hurleur noir
DE Schwarzer Brüllaffe
ES Mono aullador carayá

♀

♂

NT
Monotypic
HB 60–65 cm (M), c. 50 cm (F)
T 60–65 cm (M), 54–60 cm (F)
W 5.3–9.6 kg (M), 3.6–6.5 kg (F)

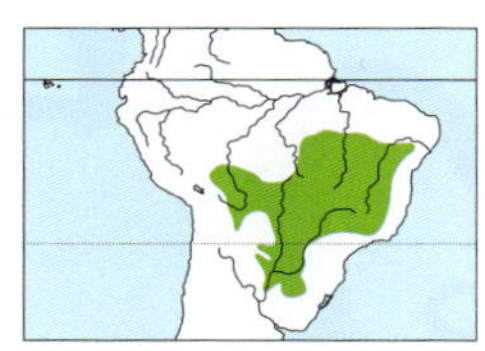

C BR, E & SE BO, E PY, N AR; perhaps NW UY

Mantled Howler
Alouatta palliata

FR Hurleur à manteau
DE Mantelbrüllaffe
ES Mono aullador de manto dorado

VU
5 sspp
HB 46–63 cm
T 55–70 cm
W 3.1–9 kg

0–2500 m

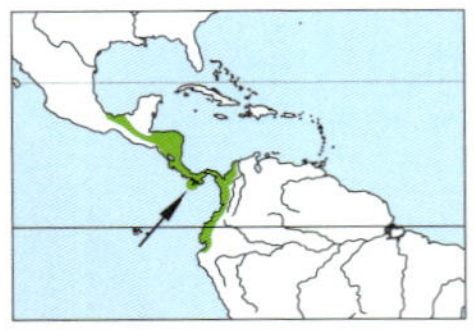

S & SE MX (S of Veracruz), NE GT to W CO, W EC & NW PE (N Tumbes); perhaps formerly in SV

ssp *trabeata*

ssp *palliata*

ssp *mexicana*

EN
Monotypic
HB 52–71 cm
T 50–69 cm
W 4.6–9.1 kg

Central American Black Howler
Alouatta pigra

FR Hurleur paresseux
DE Guatemala-Brüllaffe
ES Mono aullador del Yucatán

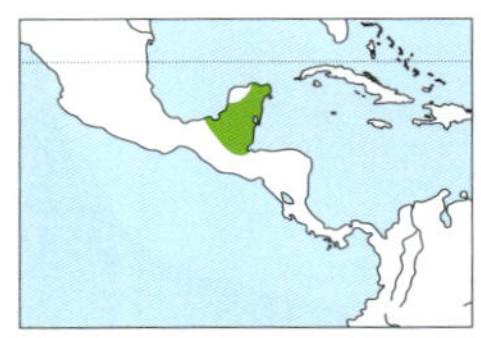

SE MX (Tabasco, Campeche, Yucatán, Quintana Roo, S Chiapas), BZ, N & C GT (S to L Izabel & El Golfete)

Central American Spider Monkey
Ateles geoffroyi

FR Atèle de Geoffroy
DE Geoffroy-Klammeraffe
ES Mono araña de América Central

Genus *ATELES*
(7 species)

E & SE MX (from E San Luis Potosí, Veracruz, E Oaxaca) S to of NW CO (Cordillera de Baudó)

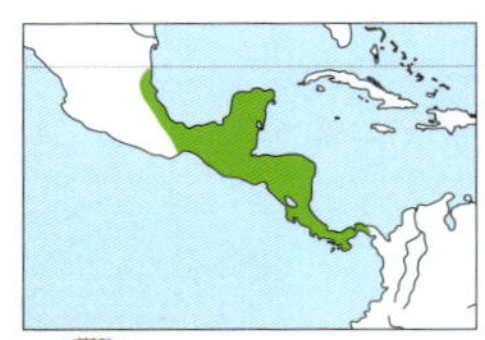

ssp *azuerensis*

ssp *frontatus*

ssp *geoffroyi*

EN
6 sspp
HB 31–63 cm
T 64–86 cm
W 6–9.4 kg

ssp *grisescens*

ssp *vellerosus*

ssp *ornatus*

Brown-headed Spider Monkey
Ateles fusciceps

FR Atèle à tête brune
DE Braunkopf-Klammeraffe
ES Mono araña de cabeza parda

EN
2 sspp
HB 30–64 cm
T 66–86 cm
W 8–9 kg

ssp *rufiventris*

100–2500 m

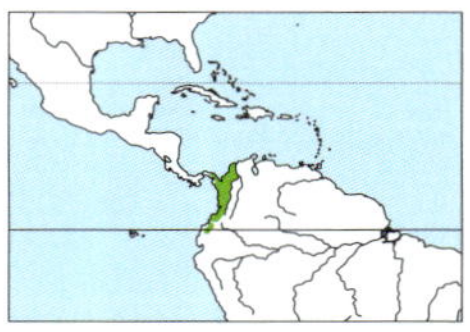

E PA (Atlantic slope), W & SW CO, NW EC (2 remnant populations W of Andes)

ssp *fusciceps*

Black Spider Monkey
Ateles chamek

FR Atèle à face noire
DE Schwarzgesicht-Klammeraffe
ES Mono araña negro

Amazonian BR S of R Amazon, E PE, N & C BO

EN
Monotypic
HB 40–60 cm
T 70–88 cm
W 5–7 kg

Red-faced Black Spider Monkey
Ateles paniscus

VU
Monotypic
HB 42–66 cm
T 64–93 cm
W 8.4–9.1 kg

FR Atèle noir
DE Rotgesicht-Klammeraffe
ES Mono araña de cara roja

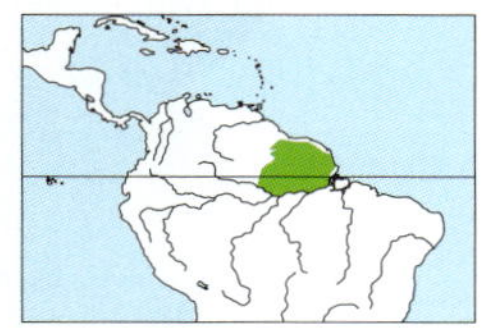

Guianas (E of R Essequibo, not in coast), NE BR (N of R Amazon, E of R Negro & R Branco); doubtful W of R Essequibo

White-whiskered Spider Monkey
Ateles marginatus

FR Atèle de Santarém
DE Weißwangen-Klammeraffe
ES Mono araña de mejillas blancas

S of R Amazon, between E bank of R Tapajós/R Teles Pires and W bank of R Xingu

EN
Monotypic
HB 35–70 cm
T 62–90 cm
W 5.8–10.4 kg

White-bellied Spider Monkey
Ateles belzebuth

FR Atèle belzébuth
DE Weißbauch-Klammeraffe
ES Mono araña de vientre amarillo

variants

EN
Monotypic
HB 46–50 cm
T 74–81 cm
W 7.9–8.3 kg

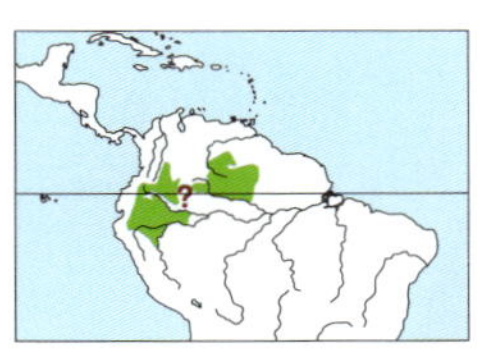

S VE, NW BR, E CO, E EC, NE PE; limits not well known

Variegated Spider Monkey
Ateles hybridus

FR Atèle varié
DE Brauner Klammeraffe
ES Mono araña jaspeado

CR
Monotypic
HB 45–50 cm
T 74–81 cm
W 7.5–10.5 kg

20–700 m

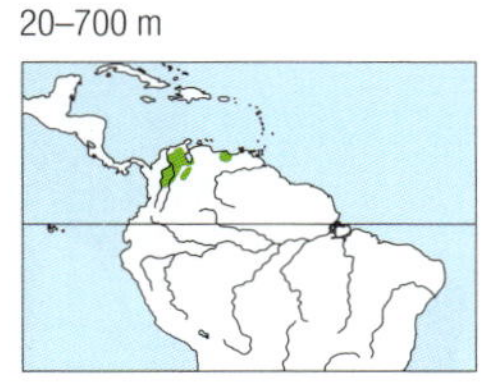

NE CO (C R Magdalena basin, Perijá Mts) to W VE (Andes; an isolated population in NC, in Guatopo NP)

Genus *LAGOTHRIX*

(2 species)

Common Woolly Monkey

Lagothrix lagothricha

FR Lagotriche de Humboldt
DE Brauner Wollaffe
ES Mono lanudo pardo

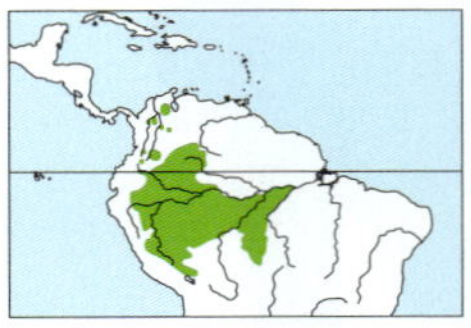

CO, NE EC, E PE, NW BO, NW BR (E to R Tapajós/R Juruena); it may occur in VE

VU
5 sspp
HB 46–65 cm
T 53–80 cm
W 5–10 kg

ssp *cana*

ssp *lugens*

ssp *lagothricha*

Yellow-tailed Woolly Monkey

Lagothrix flavicauda

FR Lagotriche à queue jaune
DE Gelbschwanz-Wollaffe
ES Mono lanudo de cola amarilla

1400–2700 m

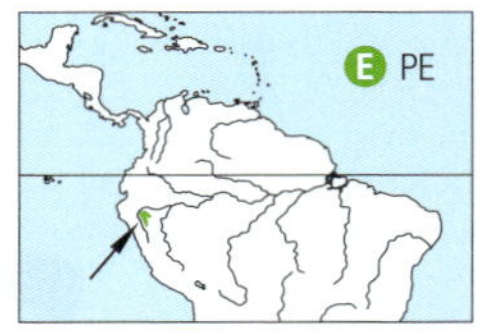

Few scattered locations on E slopes of Andes, along Marañón/Huallaga divide

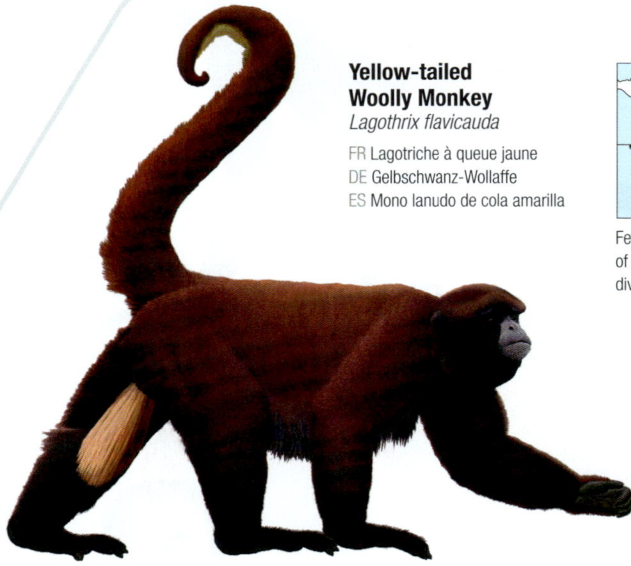

CR
Monotypic
HB 40–54 cm
T 56–63 cm
W 5.7–10 kg

Genus *BRACHYTELES*

(2 species)

Southern Muriqui

Brachyteles arachnoides

FR Muriqui du Sud
DE Südlicher Spinnenaffe
ES Muriqui meridional

40–1500 m

Through coastal Serra do Mar in Rio de Janeiro, São Paulo and NE Paraná

CR
Monotypic
HB 48–50 cm
T 73–75 cm
W 8.5–10.2 kg

Northern Muriqui

Brachyteles hypoxanthus

FR Muriqui du Nord
DE Nördlicher Spinnenaffe
ES Muriqui septentrional

250–1350 m

Atlantic Forest in Bahia, Espírito Santo, Minas Gerais and Rio de Janeiro

CR
Monotypic
HB 46–51 cm
T 73–81 cm
W 6.9–9.6 kg

Family CERCOPITHECIDAE – Old World Monkeys

23 genera, 160 species. Tropical and subtropical forests, tropical and subtropical savanna woodlands, temperate woodlands, arid shrublands, and semi-desert from sea level to 700 m. Palearctic, Afrotropical and Indo-Malayan Regions.

Barbary Macaque
Macaca sylvanus

FR Macaque de Barbarie
DE Berberaffe
ES Macaco de Berbería

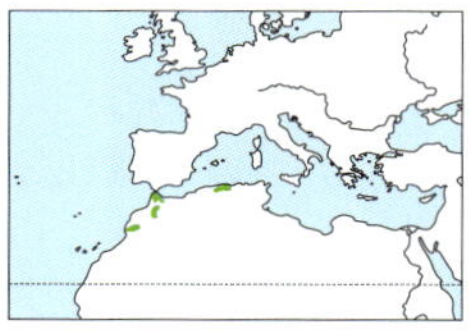

Mountains in MA (Rif, High & Middle Atlas) and DZ (Grande & Petite Kabylie); historically W TN; introduced to Gibraltar

EN
Monotypic
HB 56–63 cm
T vestigial
W 9.9–14.5 kg

Genus *MACACA*

(24 species)

Lion-tailed Macaque
Macaca silenus

FR Macaque à queue de lion
DE Bartaffe
ES Macaco de cola de león

100–1800 m

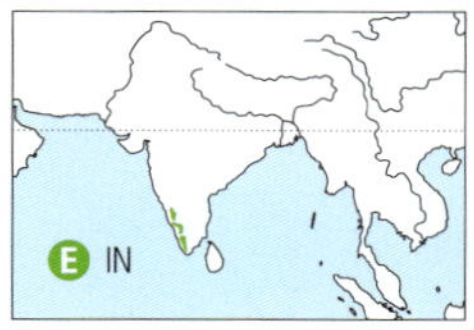

Hills of Western Ghats from Anshi Ghats in N to Kalakkad Hills in S (Karnataka, Kerala, Tamil Nadu)

EN
Monotypic
HB 42–61 cm
T 24–39 cm
W 2–10 kg

Northern Pig-tailed Macaque
Macaca leonina

FR Macaque de Birmanie
DE Nördlicher Schweinsaffe
ES Macaco de cola de cerdo septentrional

VU
Monotypic
HB 40–59 cm
T 16–25 cm
W 4.4–9.1 kg

50–2000 m

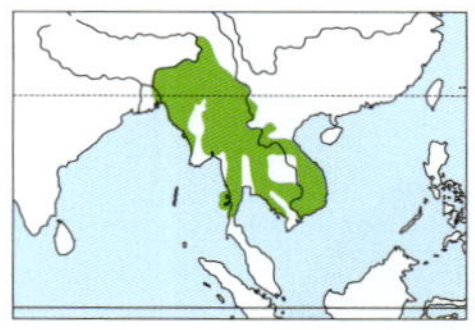

NE IN (S of R Brahmaputra), E BD, MM (with Mergui Arch), S CN (SW Yunnan), TH (S to Surat Thani/Krabi depression), LA, VN, KH

Southern Pig-tailed Macaque
Macaca nemestrina

FR Macaque à queue de cochon
DE Südlicher Schweinsaffe
ES Macaco de cola de cerdo meridional

0–1900 m

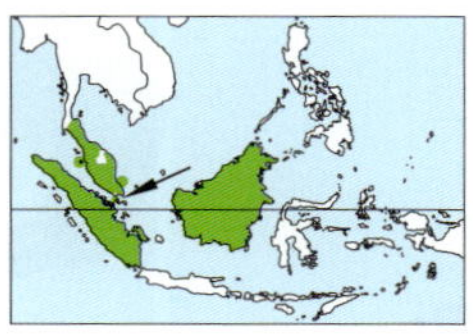

S of Surat Thani/Krabi depression in Malayan Pen (with Penang, Tioman and Batan Is), Sumatra, Bangka, Borneo

EN
Monotypic
HB 43–74 cm
T 13–25 cm
W 5.4–13.6 kg

Pagai Macaque
Macaca pagensis

FR Macaque des Pagai
DE Pagai-Makak
ES Macaco de Pagai

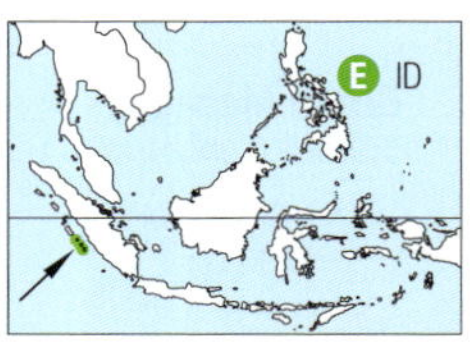

Mentawai Is (Sipora, North Pagai, South Pagai, Sinakak islet off E coast of South Pagai)

CR
Monotypic
HB 43–53 cm
T 10–16 cm
W 4.5–9 kg

Siberut Macaque
Macaca siberu

FR Macaque de Siberut
DE Siberut-Makak
ES Macaco de Siberut

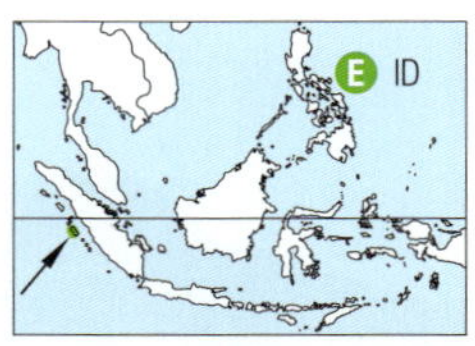

Mentawai Is (Siberut I)

EN
Monotypic
HB 40–48 cm
T 10–16 cm
W 4.5–9 kg

Crested Macaque
Macaca nigra

FR Macaque noir
DE Schopfmakak
ES Macaco crestado

CR
Monotypic
HB 44–57 cm
T 1.5–2.5 cm
W 5.5–13 kg

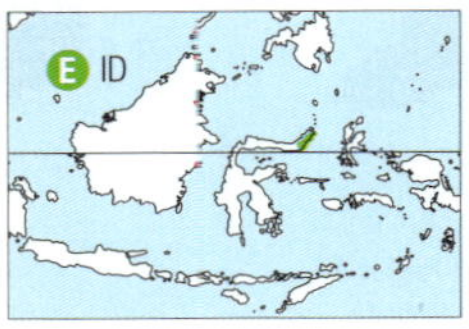

N Sulawesi (E tip of N Pen), Manado Tua and Talise Is; introduced to Moluccas Arch (Bacan I)

Gorontalo Macaque
Macaca nigrescens

FR Macaque de Gorontalo
DE Gorontalo-Makak
ES Macaco de Gorontalo

VU
Monotypic
HB 50–60 cm
T c. 2.5 cm

up to 2000 m

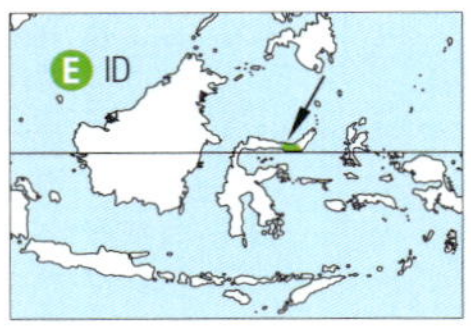

N Sulawesi (E of Gorontalo to R Onggak Dumogar in N Pen)

Heck's Macaque
Macaca hecki

FR Macaque de Heck
DE Heck-Makak
ES Macaco de Heck

VU
Monotypic
HB 50–68 cm
T 2–3 cm
W 7–10 kg

0–1800 m

NW Sulawesi (from Isthmus of Palu NE to just E of Gorontalo, in N Pen)

Tonkean Macaque
Macaca tonkeana

FR Macaque tonkéen
DE Tonkean-Makak
ES Macaco de Togian

VU
Monotypic
HB 50–68 cm
T 3–7 cm
W 8.6–12 kg

400–2500 m

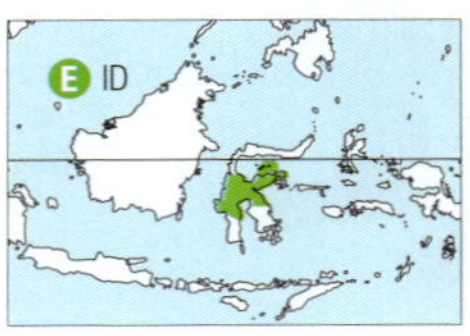

C Sulawesi (S to Latimojong, SW to Toraja highlands, SE to lakes region of SE Pen, NW to Isthmus), Togian Is

Moor Macaque
Macaca maura

FR Macaque maure
DE Mohrenmakak
ES Macaco de Célebes

EN
Monotypic
HB 46–69 cm
T 2–7 cm
W 3.8–10 kg

up to 2000 m

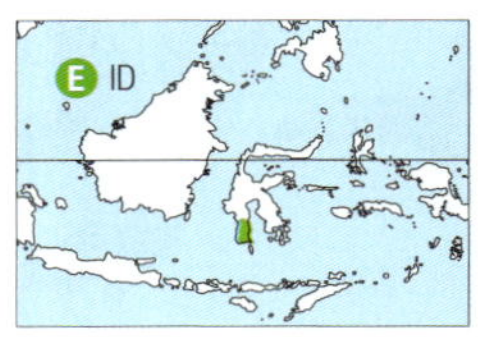

SW Sulawesi (from Bontobahari at tip of SW Pen N to Tempe Depression)

Booted Macaque
Macaca ochreata

FR Macaque à bras gris
DE Grauarm-Makak
ES Macaco calzado

VU
Monotypic
HB 48–59 cm (M), 40–48 cm (F)
T 4–6 cm (M), 3–5 cm (F)
W 6–9 kg (M), 3–5 kg (F)

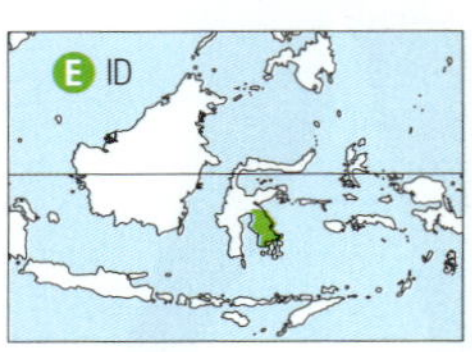

up to 800 m

SE Sulawesi, through entire SE Pen, extending to N of lakes region

Buton Macaque
Macaca brunnescens

FR Macaque de Buton
DE Buton-Makak
ES Macaco de Buton

VU
Monotypic
HB 48–59 cm (M), 40–48 cm (F)
T 4–6 cm (M), 3–5 cm (F)
W 6–9 kg (M), 3–5 kg (F)

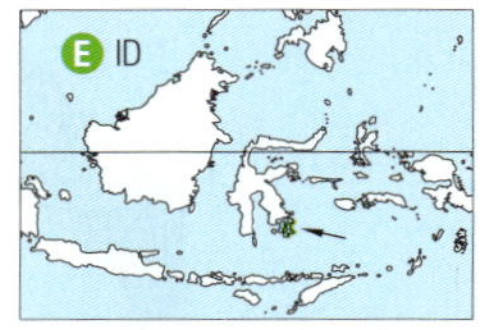

Muna I (where might be extirpated), Buton I, perhaps Pulau Labuan Blanda, off coast of S Sulawesi

Stump-tailed Macaque
Macaca arctoides

FR Macaque brun
DE Bärenmakak
ES Macaco rabón

VU
Monotypic
HB 48–65 cm
T 1.7–8 cm
W 7.5–15.5 kg

50–2800 m

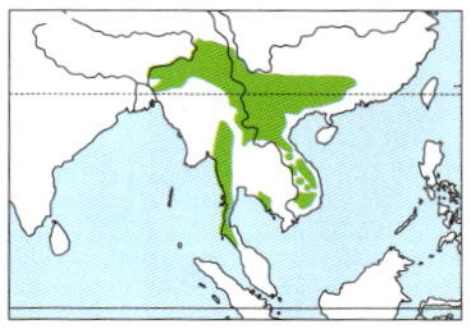

NE IN, BD (perhaps extirpated), SC CN, N MM, TH, LA, VN, KH, N Pen MY; introduced to Hong Kong

Long-tailed Macaque
Macaca fascicularis

FR Macaque crabier
DE Java-Makak
ES Macaco cangrejero

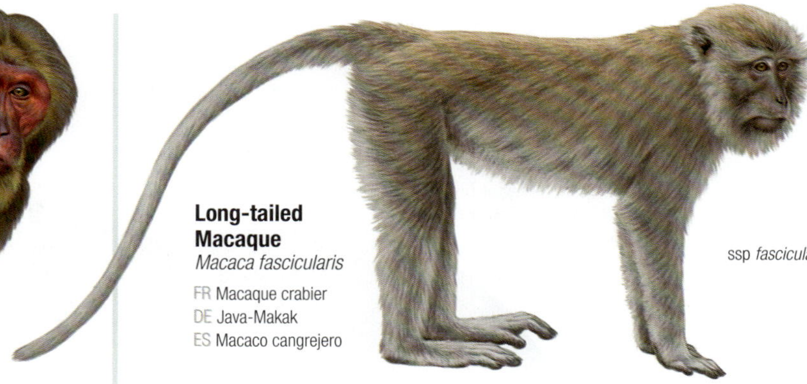

ssp *fascicularis*

ssp *condorensis*

EN
9 sspp
HB 31–63 cm
T 31–71 cm
W 2.4–12 kg

up to 2000 m

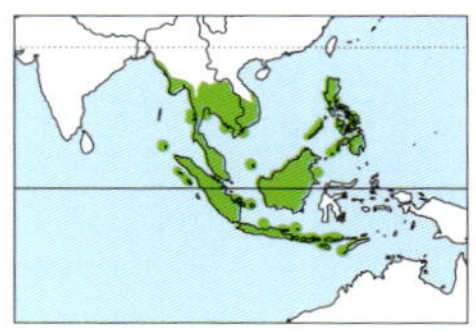

S BD, S MM, S LA, S VN, KH, TH, Thai-Malay Pen, Nicobar Is, Greater Sundas, PH; probably introduced to Nusa Penida-Timor Is

ssp *fuscus*

Rhesus Macaque
Macaca mulatta

FR Macaque rhésus
DE Rhesusaffe
ES Macaco Rhesus

LC
Monotypic
HB 37–66 cm
T 12.5–31 cm
W 3–14.1 kg

up to 4000 m

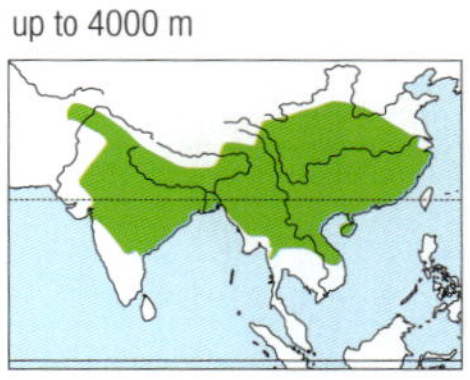

S & SE Asia from c. 36° N (AF, PK, IN & CN) S to c. 15° N (IN, TH, LA & VN), also NP, BT, BD, MM & Hainan I

ssp *fuscata*

ssp *yakui*

LC
2 sspp
HB 47–65 cm
T 8–9 cm
W 4–18.4 kg

Japanese Macaque
Macaca fuscata

FR Macaque du Japon
DE Japan-Makak
ES Macaco de Japón

S of Tsugaru Strait (Honsu, Shikoku, Kyushu and islands of Kojima, Kashima, Awaji, Shodo, Kinkazan & Yaku)

Taiwanese Macaque
Macaca cyclopis

FR Macaque de Taïwan
DE Formosa-Makak
ES Macaco de Formosa

100–3400 m

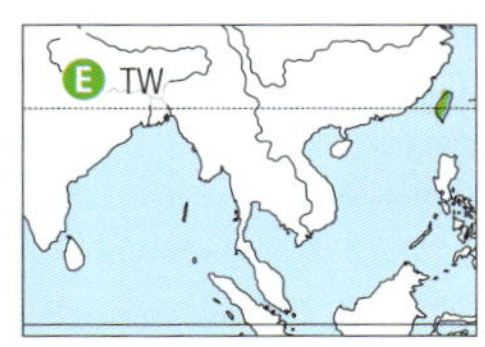

Widely in E mountains

LC
Monotypic
HB 42–65 cm
T 35–50 cm
W 5.5–18.5 kg

Toque Macaque
Macaca sinica

FR Macaque couronné
DE Ceylon-Hutaffe
ES Macaco de Sri Lanka

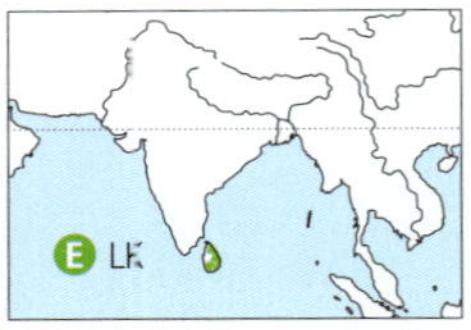

Range fragmented on Dry Zone, Wet Zone lowlands and montane areas

EN
3 sspp
HB 40–53 cm
T 46–62 cm
W 2.3–8.4 kg

ssp *aurifrons*

ssp *sinica*

Bonnet Macaque
Macaca radiata

FR Macaque à bonnet
DE Indien-Hutaffe
ES Macaco de Madras

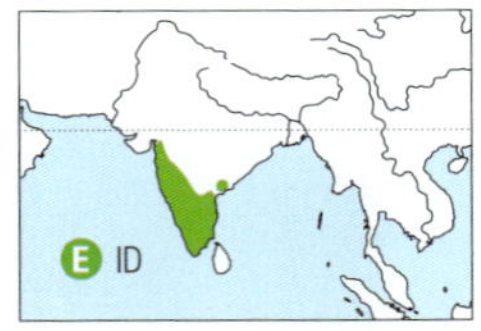

S of R Tapti in N, S of R Krishna in E; introduced to Mascarene Is

VU
2 sspp
HB 34–60 cm
T 48–69 cm
W 2.9–11.6 kg

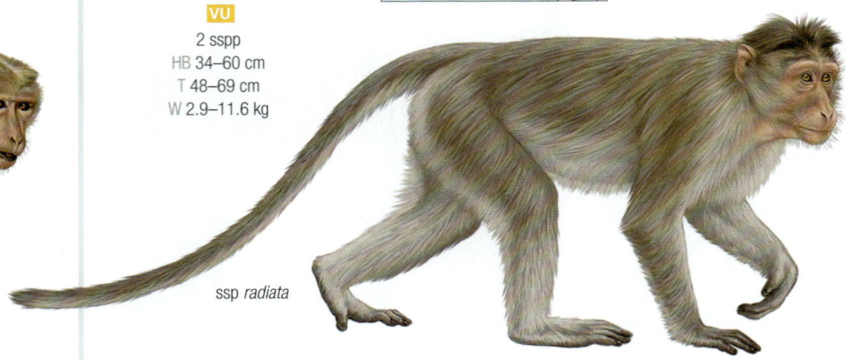

ssp *radiata*

ssp *diluta*

Assamese Macaque
Macaca assamensis

FR Macaque d'Assam
DE Assam-Makak
ES Macaco de Assam

200–3100 m

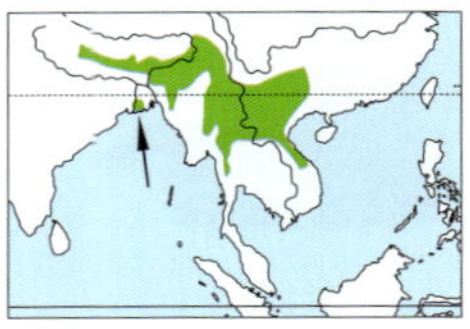

S & SE Asia in NP (E from Tipling), NE IN, BT, SC CN, N & E MM, N & W TH, LA, VN; disjunct record in coastal SW BD

NT
2 sspp
HB 43–73 cm
T 17–36 cm
W 4.9–16.5 kg

Arunachal Macaque
Macaca munzala

FR Macaque d'Arunachal
DE Arunachal-Makak
ES Macaco de Arunachal

EN
Monotypic
HB c. 57 cm
T c. 26 cm
W c. 14 kg

2000–3500 m

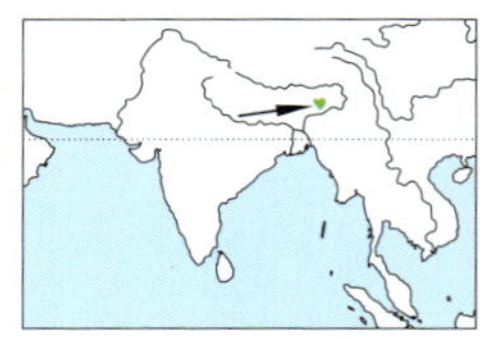

NE IN (W Arunachal Pradesh; records of Mouling NP, C Arunachal Pradesh, unconfirmed), BT; possibly Tibet

White-cheeked Macaque
Macaca leucogenys

FR Macaque à joues blanches
DE Weißwangenmakak
ES Macaco de mejillas blancas

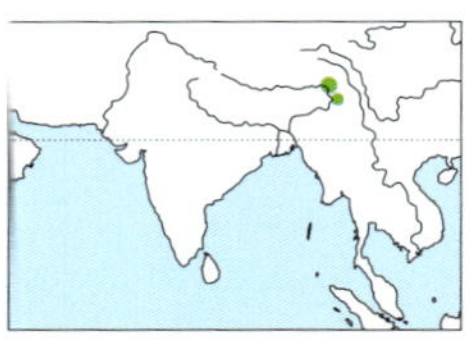

SE CN (SE Tibet in Mêdog County), NE IN (Arunachal Pradesh)

EN
Monotypic

Tibetan Macaque
Macaca thibetana

FR Macaque du Tibet
DE Tibet-Makak
ES Macaco tibetano

NT
Monotypic
HB 51–71 cm
T 4–14 cm
W 9–18.3 kg

600–2500 m

W limit in Yangtze Gorge (W & NW Sichuan), S limit in Guangxi; presence in NE IN uncertain

Genus *CERCOCEBUS*

(7 species)

Tana River Mangabey
Cercocebus galeritus

FR Mangabey de la Tana
DE Tana-Mangabe
ES Mangabey del Tana

CR
Monotypic

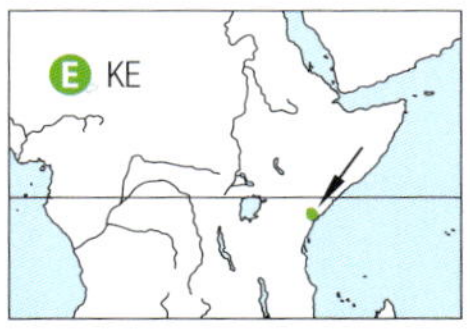

Riparian forest patches along lower R Tana and R Tana Delta

Agile Mangabey
Cercocebus agilis

FR Mangabey agile
DE Olivmangabe
ES Mangabey ágil

LC
Monotypic
HB 44–65 cm
T 45–76 cm
W 4.3–10.2 kg

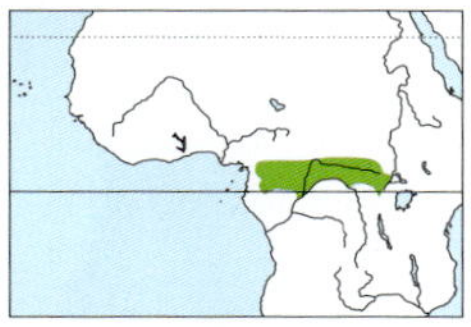

SE CM, SW CF, N CD (N of R Congo, from R Ubangi to Ituri Forest and R Semliki), NE GA (E of R Ivindo), N CG (across R Sangha)

Golden-bellied Mangabey
Cercocebus chrysogaster

FR Mangabey à ventre doré
DE Goldbauchmangabe
ES Mangabey de vientre dorado

up to 500 m

Congo Basin, S & E of R Congo (N limit could be R Lulonga, E limit probably R Lomami)

EN
Monotypic

Sanje River Mangabey
Cercocebus sanjei

FR Mangabey de la Sanje
DE Sanje-Mangabe
ES Mangabey del Sanje

EN
Monotypic

400–1600 m

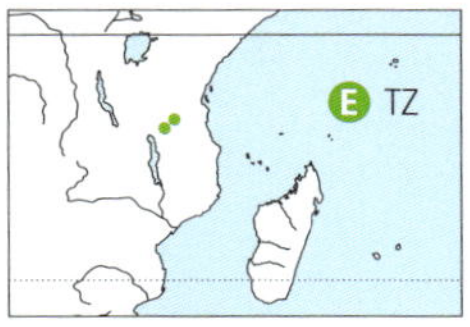

Confined to Mwanihana Forest and E slopes of Udzungwa Mts

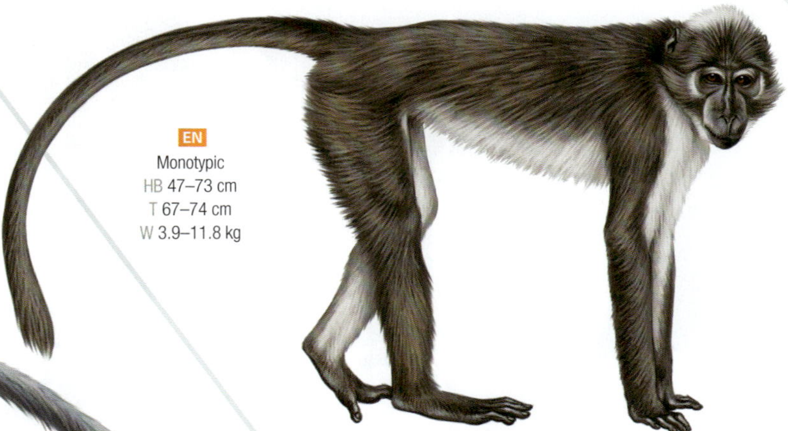

White-naped Mangabey
Cercocebus lunulatus

FR Mangabey couronné
DE Weißnackenmangabe
ES Mangabey de cuello blanco

EN
Monotypic
HB 47–73 cm
T 67–74 cm
W 3.9–11.8 kg

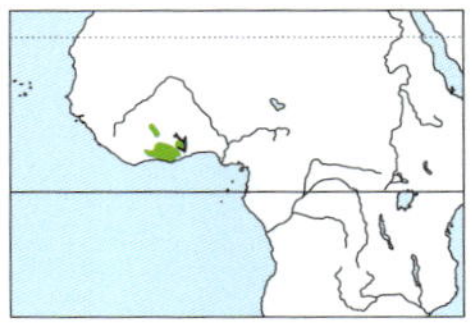

E & NE CI (from Nzo-Sassandra river system to R Volta; perhaps extirpated W of R Bandama), SW BF, SW GH

Sooty Mangabey
Cercocebus atys

FR Mangabey fuligineux
DE Rußmangabe
ES Mangabey gris

VU
Monotypic
HB 44–58 cm
T 52–64 cm
W 5.6–11.4 kg

SW SN, GW, W & SE GN, SL, LR, W CI (E boundary Nzo-Sassandra river system)

Red-capped Mangabey
Cercocebus torquatus

FR Mangabey à collier blanc
DE Rotscheitelmangabe
ES Mangabey de sombrero rojo

EN
Monotypic
HB 42–67 cm
T 46–76 cm
W 5–12.5 kg

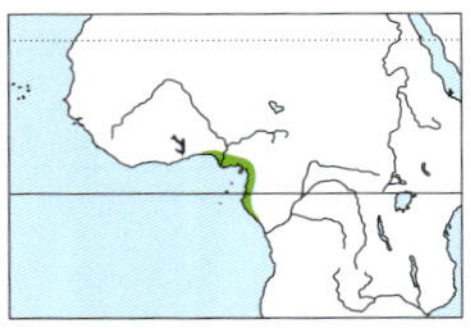

Patchy in forests in S NG, W CM, W GQ, W GA, SW CG (S to Conkouati-Douli NP); unconfirmed in BJ

Genus *MANDRILLUS*

(2 species)

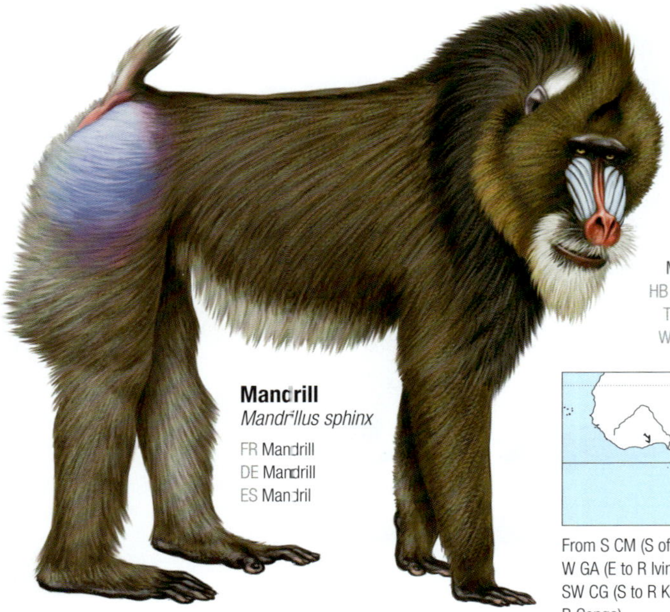

Mandrill
Mandrillus sphinx

FR Mandrill
DE Mandrill
ES Mandril

VU
Monotypic
HB 55–110 cm
T 5–10 cm
W 11–33 kg

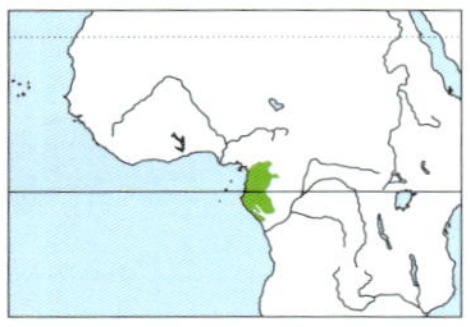

From S CM (S of R Sanaga) to GQ, W GA (E to R Ivindo & R Ogooué), SW CG (S to R Kouilou down to R Congo)

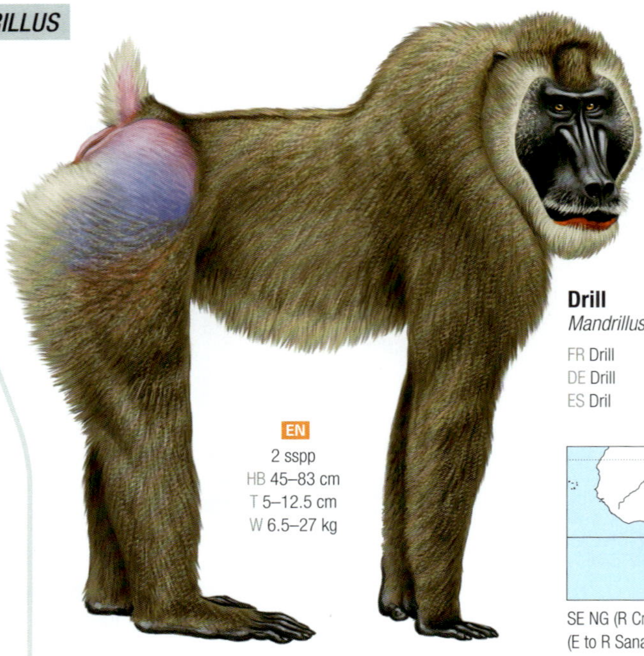

Drill
Mandrillus leucophaeus

FR Drill
DE Drill
ES Dril

EN
2 sspp
HB 45–83 cm
T 5–12.5 cm
W 6.5–27 kg

SE NG (R Cross), W CM (E to R Sanagar and its tributary R Mbam), Bioko I (GQ)

Genus *LOPHOCEBUS*

(2 species)

Gray-cheeked Mangabey
Lophocebus albigena

FR Lophocèbe à joues grises
DE Grauwangenmangabe
ES Mangabey de mejillas grises

VU
4 sspp
HB 42–73 cm
T 65–100 cm
W 4–10 kg

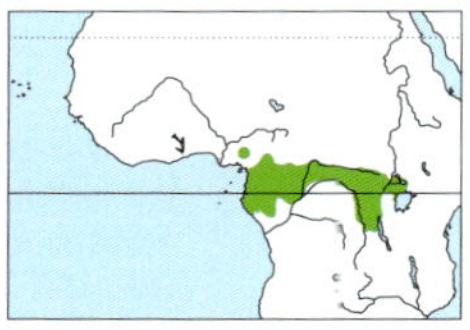

SE NG (Gashaka-Gumti NP), CM, SW CF, GQ, GA, CG, CD (S of R Ubangi, N of R Congo), W RW, W BI, UG, SS, NW TZ

Black Crested Mangabey
Lophocebus aterrimus

FR Lophocèbe noir
DE Schopfmangabe
ES Mangabey de cresta negra

VU
2 sspp
HB 40–62 cm
T 55–85 cm
W 4.5–8 kg

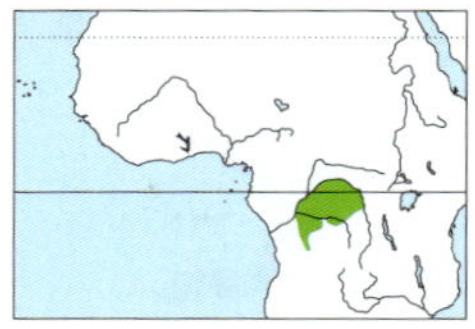

CD (SW Congo Basin, limited by left bank of R Congo), N AO

ssp *opdenboschi*

ssp *aterrimus*

Genus *RUNGWECEBUS*

(1 species)

Kipunji
Rungwecebus kipunji

FR Kipunji
DE Kipunji-Affe
ES Kipunji

EN
Monotypic
HB 85–90 cm
W 10–15 kg

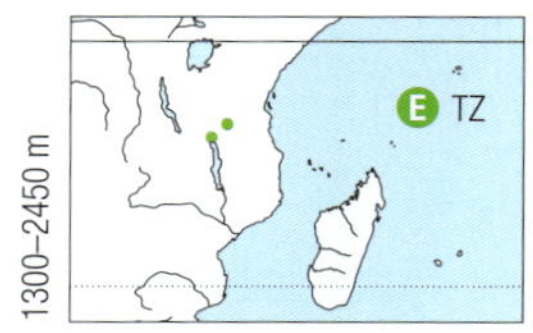

Southern Highlands and Udzungwa Mts

Genus *PAPIO*

(6 species)

Yellow Baboon
Papio cynocephalus
FR Babouin cynocéphale
DE Gelber Pavian
ES Papión dorado

LC
2 sspp
HB 51–85 cm
T 34–66 cm
W 8–28 kg

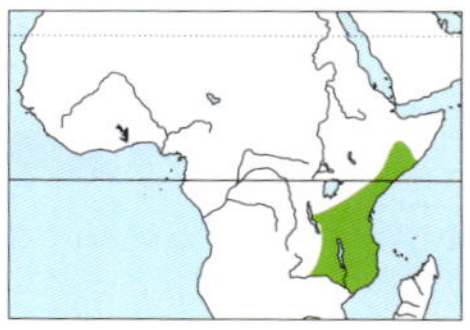

SE SO, E KE (with Manda I), TZ, MW, E ZM (E of R Luangwa), N MZ; perhaps extreme SE ET

Kinda Baboon
Papio kindae
FR Babouin de Kinda
DE Kinda-Pavian
ES Papión chico

LC
Monotypic
HB 56–58 cm
T 47–53 cm
W 10–16 kg

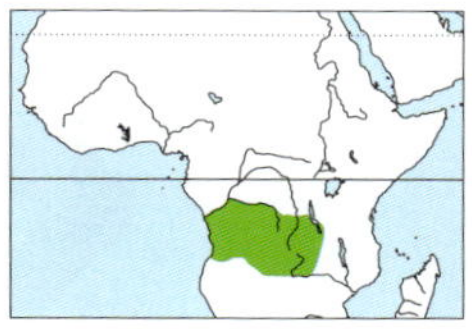

N AO, S CD, W ZM (W of R Luangwa), extreme SW TZ (possibly N to Mahale Mountains NP)

Chacma Baboon
Papio ursinus
FR Babouin chacma
DE Bärenpavian
ES Papión chacma

LC
3 sspp
HB 51–100 cm
T 37–84 cm
W 12–35 kg

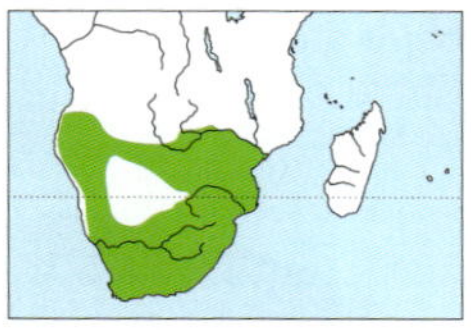

up to 3000 m

SW AO (N along coast to c. 11° 30' S), SW ZM, NA, BW, ZW, S MZ (S of R Zambezi), ZA, SZ, LS

Olive Baboon
Papio anubis
FR Babouin anubis
DE Anubis-Pavian
ES Papión oliva

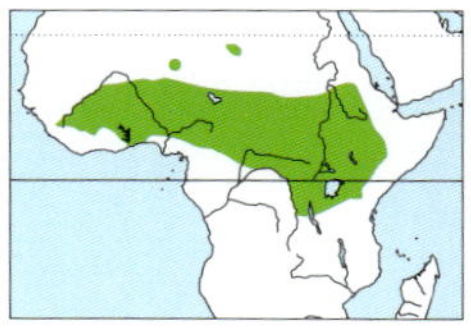

up to 2500 m

LC
Monotypic
HB 50–90 cm
T 38–60 cm
W 14–30 kg

Savanna belt from ML, GN and SL to W ER and ET, S to N & E CD and NW TZ; isolated in Sahara (Tibesti, Aïr & Ennedi massifs)

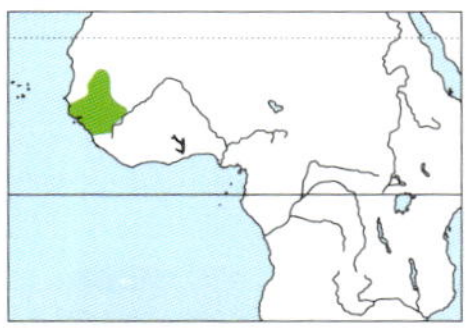

Extreme W Africa in S MR, SN, GM, GW, W ML, W GN, NW SL; occurrence in W LR highly unlikely

Guinea Baboon
Papio papio
FR Babouin de Guinée
DE Guinea-Pavian
ES Papión de Guinea

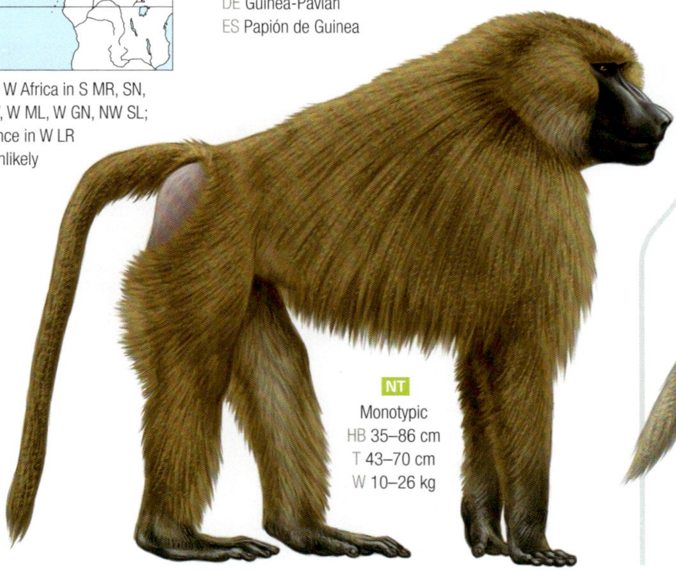

NT
Monotypic
HB 35–86 cm
T 43–70 cm
W 10–26 kg

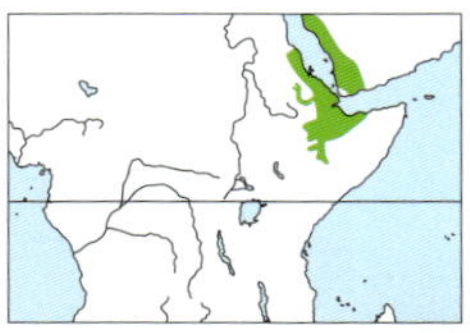

up to 3300 m

SW SA (up to Jeddah), W YE, NE SD (Red Sea Hills), E ER, NE ET, DJ, N SO

Hamadryas Baboon
Papio hamadryas
FR Babouin hamadryas
DE Mantelpavian
ES Papión sagrado

LC
Monotypic
HB 50–95 cm
T 37–60 cm
W 10–21 kg

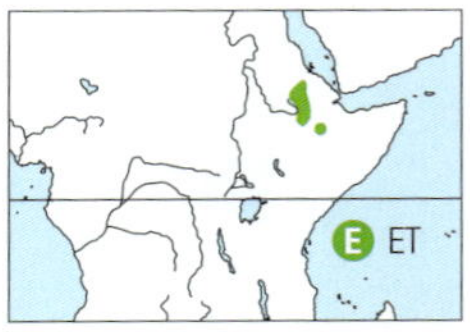

1400–4400 m

Between Blue Nile Gorge in W and upper Wabe Shebelle Valley (E of Bale Massif) in SE

Gelada
Theropithecus gelada

FR Gélada
DE Dschelada
ES Gelada

Genus *THEROPITHECUS*
(1 species)

LC
2 sspp
HB 50–75 cm
T 33–50 cm
W 12–30 kg

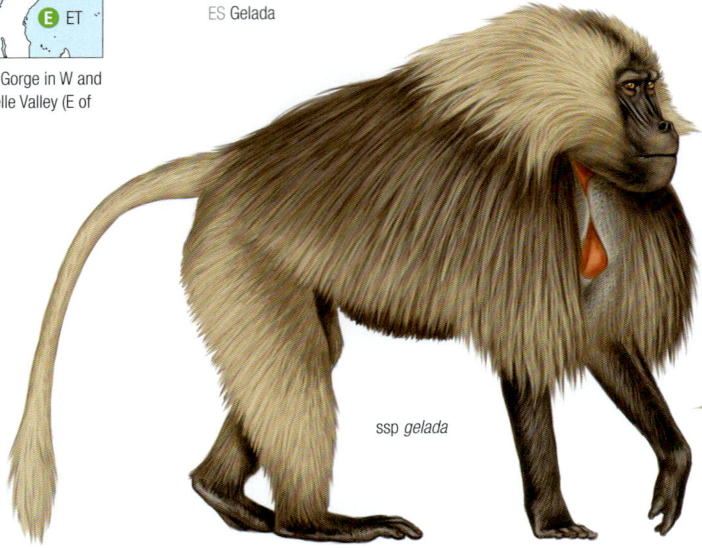

ssp *gelada*

ssp *obscurus*

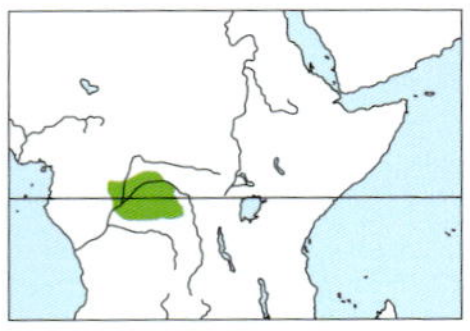

NW CD & NE CG in lowland forests on both sides of lower and middle Congo, E to R Lomami; unconfirmed in NE AO

Genus *ALLENOPITHECUS*
(1 species)

Allen's Swamp Monkey
Allenopithecus nigroviridis

FR Cercopithèque d'Allen
DE Sumpfmeerkatze
ES Mono de Allen

LC
Monotypic
HB 40–50 cm
T 36–51 cm
W 3.2–6.1 kg

Southern Talapoin Monkey
Miopithecus talapoin

FR Talapoin d'Angola
DE Südliche Zwergmeerkatze
ES Talapoin meridional

VU
Monotypic
HB 26–45 cm (M)
T c. 53 cm
W 750–1.300 g

SW CD (both sides of R Kasai and along Mebrige, Loge, Cuanza, Nhia, Cuvo & Cuango rivers), NW AO (S to c. 13° S)

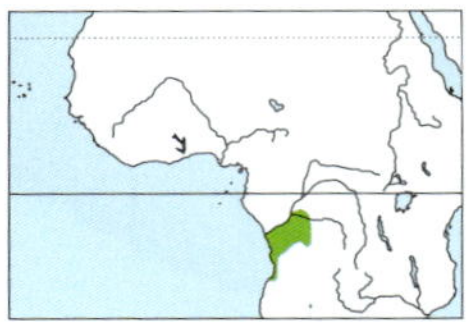

0–700 m

Northern Talapoin Monkey
Miopithecus ogouensis

FR Talapoin du Gabon
DE Nördliche Zwergmeerkatze
ES Talapoin septentrional

Genus *MIOPITHECUS*
(2 species)

NT
Monotypic
HB 28–34 cm
T 37–43 cm
W 1.1–1.4 kg

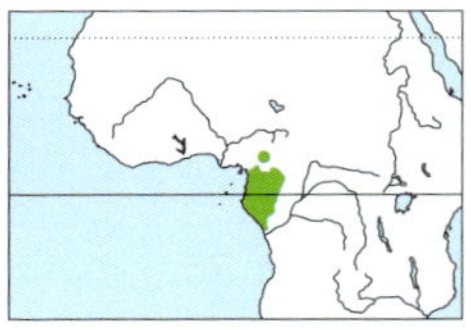

S CM (S of R Sanaga, an outlying population N of R Sanaga, at c. 6° N), GQ, GA, W CG, NW AO (Cabinda)

Common Patas Monkey
Erythrocebus patas

FR Patas roux
DE Husarenaffe
ES Mono patas

Genus *ERYTHROCEBUS*
(3 species)

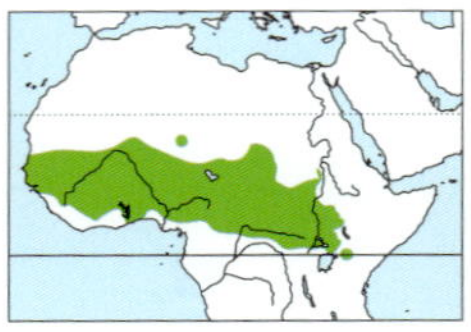

N of equatorial forests and S of Sahara from W SN and S MR E to SS and C KE; introduced to PR

ssp *patas*

NT
3 sspp
HB 48–87 cm
T 48–72 cm
W 4–13 kg

ssp *pyrrhonotus*

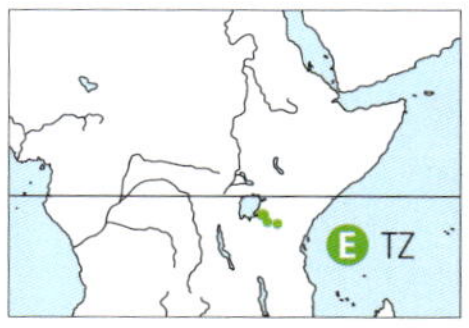

Serengeti and Athi Plains and W side of Mt Kilimanjaro; extirpated from S KE

800–1850 m

Southern Patas Monkey
Erythrocebus baumstarki

FR Patas de Baumstark
DE Südlicher Husarenaffe
ES Mono patas meridional

CR
Monotypic
HB 50–87 cm
T 50–74 cm
W 7.5–13 kg

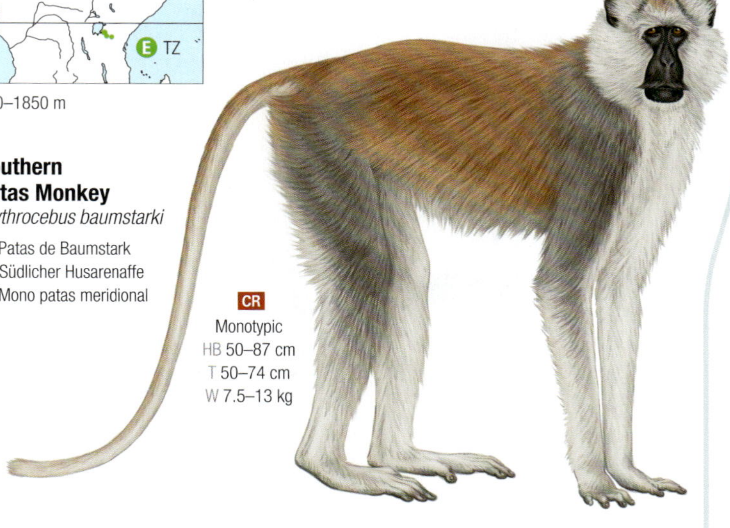

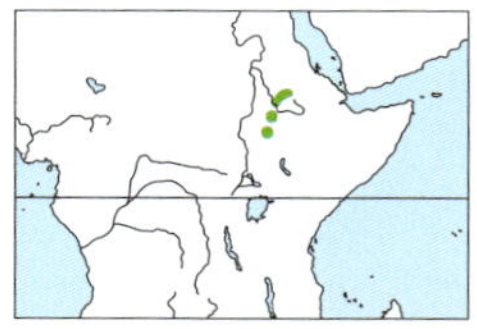

Region between R Nile and Western Ethiopian Escarpment (W ET & SE SD); probably adjacent SS

up to 1000 m

Blue Nile Patas Monkey
Erythrocebus poliophaeus

FR Patas de Heuglin
DE Heuglin-Husarenaffe
ES Mono patas de Heuglin

DD
Monotypic
HB 60–87 cm (M), 48–52 cm (F)
T 63–72 cm (M), 48–55 cm (F)
W 7–13 kg (M), 4–7 kg (F)

Genus *CHLOROCEBUS*

(7 species)

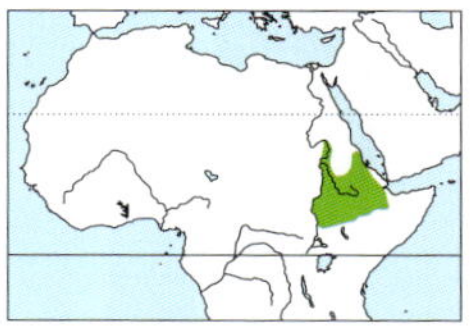

Grivet Monkey
Chlorocebus aethiops

FR Grivet d'Éthiopie
DE Äthiopien-Grünmeerkatze
ES Mono tota

E of White Nile in SE SD, E SS, ER, ET (S to R Omo, E to Rift Valley) and DJ; perhaps S EG

LC
Monotypic
HB 30–60 cm
T 41–76 cm
W 1.5–6.4 kg

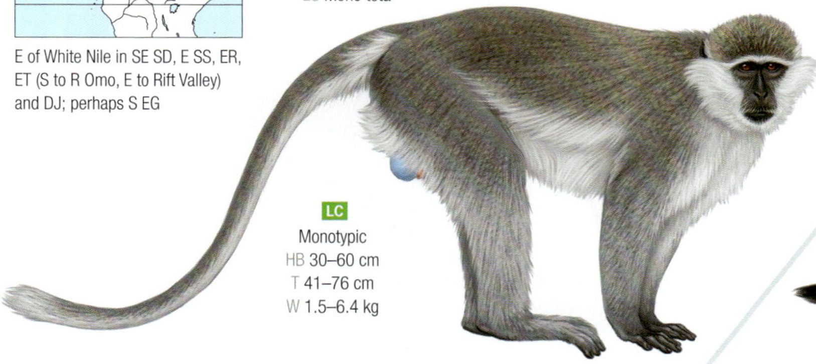

ssp *budgetti*

Tantalus Monkey
Chlorocebus tantalus

FR Vervet tantale
DE Tantalus-Grünmeerkatze
ES Vervet de Tántalo

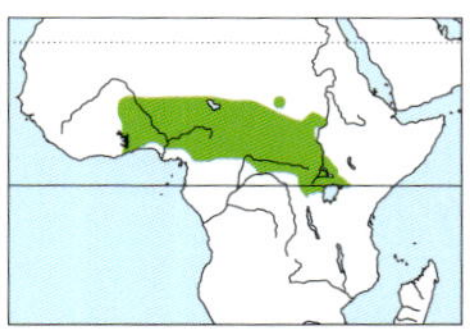

Sub-Saharan Africa from E BF and GH (E of R Volta) to W SD (W of White Nile), SS (from Jebel Marra S to Imatong Mts) and UG

ssp *marrensis*

ssp *tantalus*

LC
3 sspp
HB 30–83 cm
T 41–66 cm
W 3–6.4 kg

Green Monkey
Chlorocebus sabaeus

FR Vervet vert
DE Westliche Grünmeerkatze
ES Vervet verde

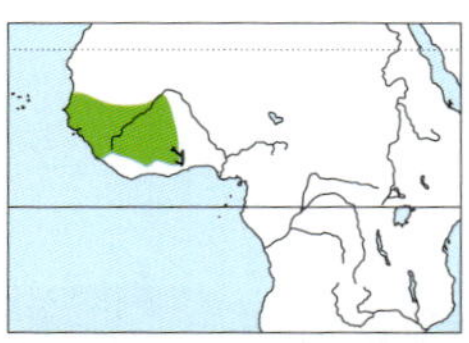

SN and GW (with Bijagos Arch) E to GH (just W of R Volta); perhaps S MR; introduced to CI and West Indies

LC
Monotypic
HB 30–60 cm
T 41–76 cm
W 1.5–6.4 kg

Malbrouck Monkey
Chlorocebus cynosuros

FR Vervet malbrouck
DE Angola-Grünmeerkatze
ES Vervet occidental

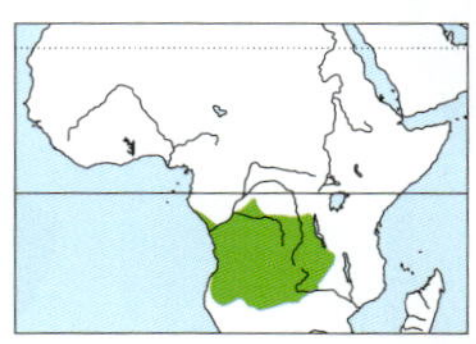

S GA, S CG, S CD, AO, ZM (W of Luangwa Valley), N NA, N BW

LC
Monotypic
HB 30–65 cm
T 35–76 cm
W 1.5–6.4 kg

Vervet Monkey
Chlorocebus pygerythrus

FR Vervet bleu
DE Südliche Grünmeerkatze
ES Vervet oriental

LC
5 sspp
HB 30–70 cm
T 41–76 cm
W 1.5–6.4 kg

ssp *hilgerti*

ssp *pygerythrus*

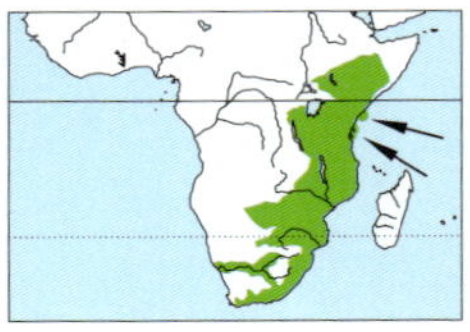
S ET (E of Rift Valley), S SO, KE (with Lamu Arch), E & SW UG, RW, BI, TZ (with Pemba and Mafia Is), ZM (E of Luangwa Valley), MW, MZ, ZW, NA, E BW, SZ, LS, much of ZA

ssp *rufoviridis*

Bale Mountains Monkey
Chlorocebus djamdjamensis

FR Grivet du Balé
DE Bale-Grünmeerkatze
ES Vervet de las Montañas Bale

2400–3000 m

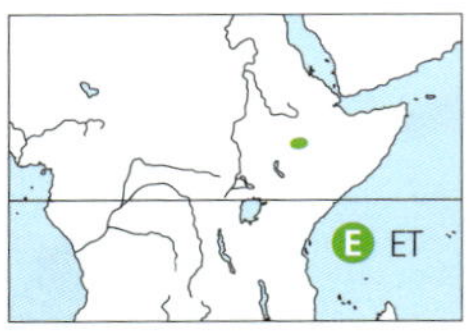

Forest fragments of Sidamo, Bale Mts, E of Rift Valley lakes of Abiata, Shalla and Zway

VU
2 sspp
HB 30–60 cm
T 41–76 cm
W 1.5–6.4 kg

up to 380 m

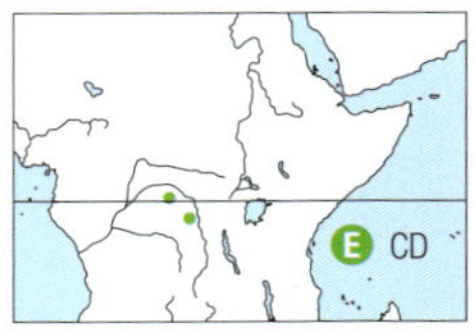

Congo Basin in Kokolopori Bonobo R and Lomami NP; exact limits uncertain

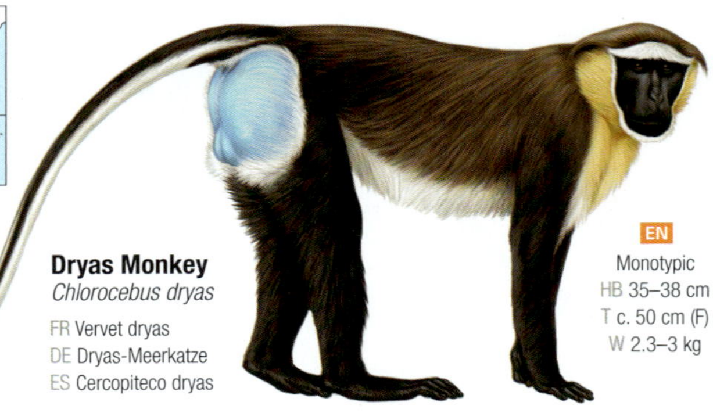

Dryas Monkey
Chlorocebus dryas

FR Vervet dryas
DE Dryas-Meerkatze
ES Cercopiteco dryas

EN
Monotypic
HB 35–38 cm
T c. 50 cm (F)
W 2.3–3 kg

Genus *ALLOCHROCEBUS*

(3 species)

VU
Monotypic
HB 45–70 cm
T 46–76 cm
W 3–10 kg

L'Hoest's Monkey
Allochrocebus lhoesti

FR Cercopithèque de L'Hoest
DE Vollbartmeerkatze
ES Cercopiteco de L'Hoest

1000–2900 m

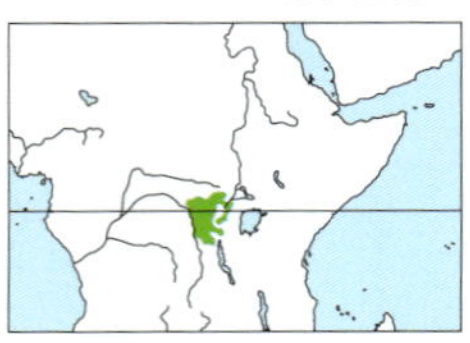
E CD (from R Lualaba E to Ituri Forest; additional population S of lower R Lindi), SW UG, RW, N BI

Preuss's Monkey
Allochrocebus preussi

FR Cercopithèque de Preuss
DE Preuss-Meerkatze
ES Cercopiteco de Preuss

SE NG & SW CM highlands (Cross River Highlands, Korup NP, Bamenda Highlands, Mt Cameroon), Bioko I (GQ)

EN
2 sspp
HB 37–70 cm
T 41–76 cm
W 2.9–10 kg

Sun-tailed Monkey
Allochrocebus solatus

FR Cercopithèque à queue de soleil
DE Gabun-Vollbartmeerkatze
ES Cercopiteco de Gabón

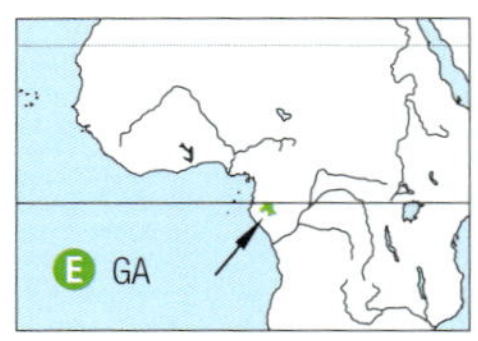

S of R Ogooué, W of R Offoué; W limit unclear but recorded E of Waka NP, S limit unknown but extends to Massif du Chaillu

NT
Monotypic
HB 50–70 cm
T 60–76 cm
W 3.5–9 kg

Genus *CERCOPITHECUS*

(19 species)

Diana Monkey

Cercopithecus diana

FR Cercopithèque diane
DE Diana-Meerkatze
ES Cercopiteco de Diana

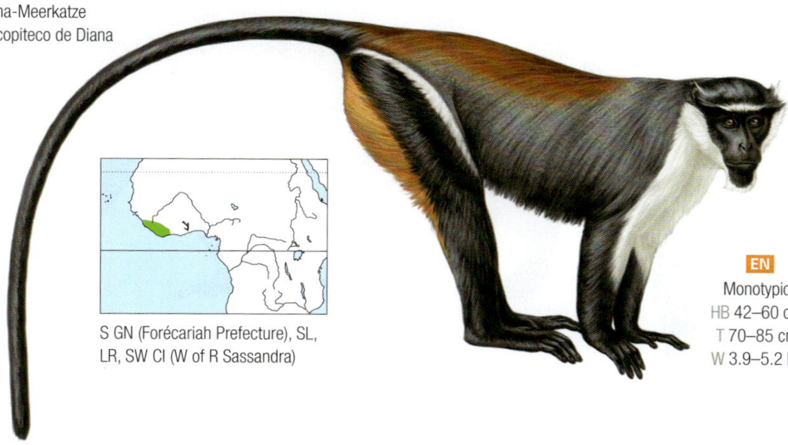

S GN (Forécariah Prefecture), SL, LR, SW CI (W of R Sassandra)

EN
Monotypic
HB 42–60 cm
T 70–85 cm
W 3.9–5.2 kg

Roloway Monkey

Cercopithecus roloway

FR Cercopithèque roloway
DE Roloway-Meerkatze
ES Cercopiteco roloway

SE CI, SW GH; possibly TG

CR
Monotypic
HB 42–60 cm
T 70–85 cm
W 3.9–5.2 kg

De Brazza's Monkey

Cercopithecus neglectus

FR Cercopithèque de De Brazza
DE Brazza-Meerkatze
ES Cercopiteco de Brazza

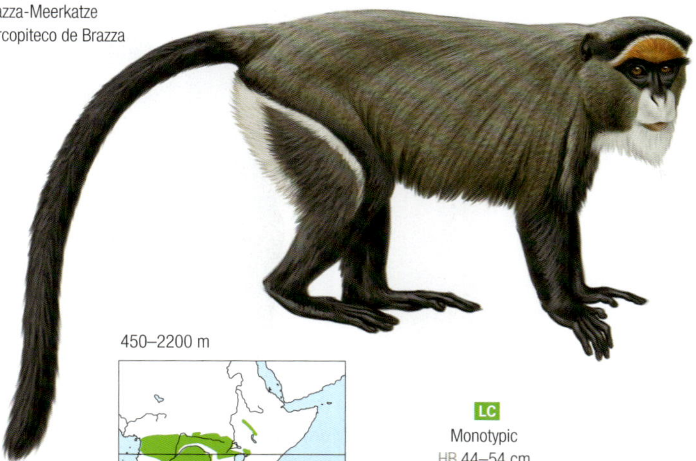

450–2200 m

S CM, S CF, GQ, GA (S to R Mpassa and its tributaries), CG (S to R Nambouli), CD, UG, KE, SW ET, NE AO

LC
Monotypic
HB 44–54 cm
T 59–70 cm
W 3.2–5.5 kg

Mona Monkey

Cercopithecus mona

FR Cercopithèque mone
DE Mona-Meerkatze
ES Cercopiteco mona

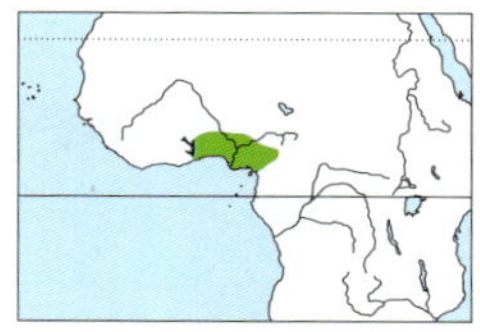

SE GH (E of R Volta), TG, BJ, NG, W CM (to a little S of R Sanaga); introduced to ST and Caribbean Is

NT
Monotypic
HB 38–63 cm
T 53–88 cm
W 2.5–7.5 kg

Campbell's Monkey

Cercopithecus campbelli

FR Cercopithèque de Campbell
DE Campbell-Meerkatze
ES Cercopiteco de Campbell

NT
Monotypic
HB 36–54 cm
T 49–85 cm
W 2.2–4.6 kg

SN (from just N of R Casamance), GM, GW (with Caravela I in Bijagos Arch), GN, SL, LR (just to R Cavally)

Lowe's Monkey

Cercopithecus lowei

FR Cercopithèque de Lowe
DE Lowe-Meerkatze
ES Cercopiteco de Lowe

VU
Monotypic
HB 40–52 cm
T 54–75 cm
W 2–5.8 kg

up to 300 m

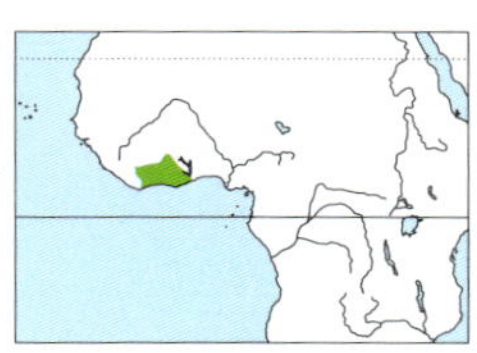

Patchily from R Cavally (W CI) E to R Volta (GH)

Crowned Monkey
Cercopithecus pogonias

FR Cercopithèque couronné
DE Kronenmeerkatze
ES Cercopiteco coronado

ssp *grayi*

ssp *nigripes*

NT

3 sspp
HB 44–57 cm
T 60–87 cm
W 3.1–4.7 kg

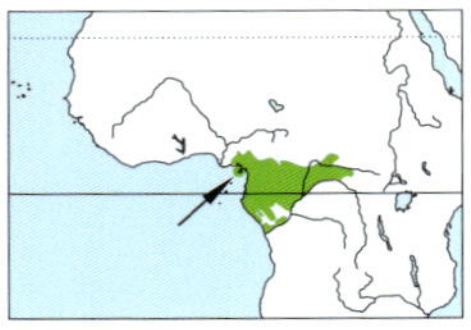

SE NG, CM, GQ (with Bioko I), GA, CG, S CF, NW CD (N of R Congo), perhaps AO (Cabinda)

ssp *pogonias*

ssp *wolfi*

ssp *elegans*

Wolf's Monkey
Cercopithecus wolfi

FR Cercopithèque de Wolf
DE Wolf-Meerkatze
ES Cercopiteco de Wolf

NT

3 sspp
HB 28–46 cm
T 60–72 cm
W 2.8–6.5 kg

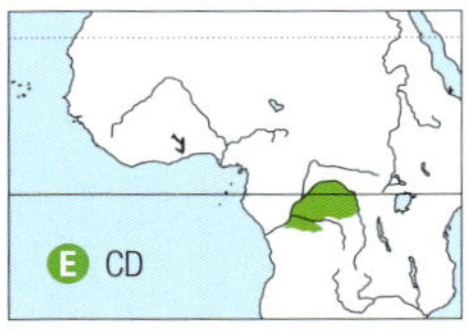

C Congo Basin from left bank of R Congo/R Lualaba S to about 7° S

Dent's Monkey
Cercopithecus denti

FR Cercopithèque de Dent
DE Dent-Meerkatze
ES Cercopiteco de Dent

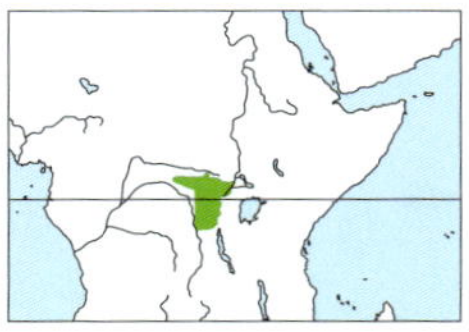

NE CD (E & N of Congo-Lualaba river system, N to R Itimbiri, S to Kasongo savanna), W UG, NW RW

LC

Monotypic
HB 40–50 cm
T 68–90 cm
W 2.8–5.7 kg

600–3300 m

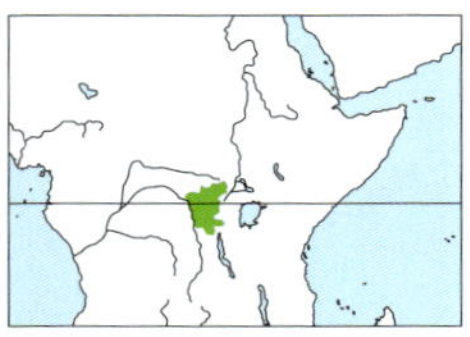

E CD (R Lualaba E to Ituri & Albertine Rift, N to Lindi-Nepoko system, S to rainforest limit), W RW (Nyungwe Forest NP); perhaps S UG and N BI

Owl-faced Monkey
Cercopithecus hamlyni

FR Cercopithèque de Hamlyn
DE Eulenkopfmeerkatze
ES Cercopiteco de Hamlyn

VU

Monotypic
HB c. 55 cm (M)
T c. 57 cm (M)
W 3.1–5.5 kg

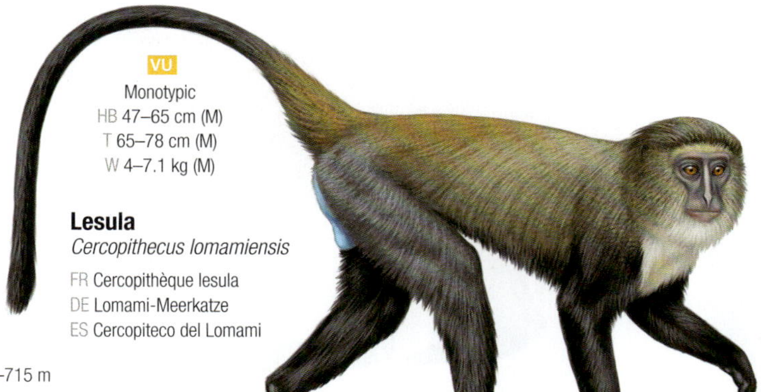

VU

Monotypic
HB 47–65 cm (M)
T 65–78 cm (M)
W 4–7.1 kg (M)

Lesula
Cercopithecus lomamiensis
FR Cercopithèque lesula
DE Lomami-Meerkatze
ES Cercopiteco del Lomami

400–715 m

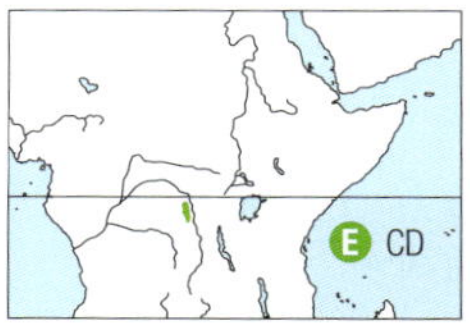

EC Congo Basin, from upper R Tshuapa (24° E) E to R Lomami (25° E), S to limits of forest (c. 3° 30′ S), N limit unclear (apparently does not occur N of 1° S)

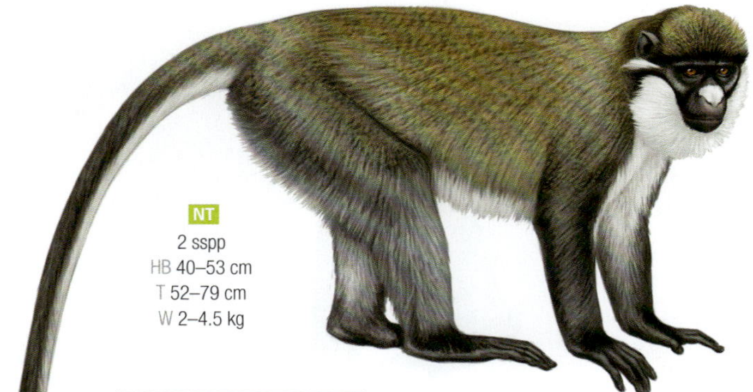

NT

2 sspp
HB 40–53 cm
T 52–79 cm
W 2–4.5 kg

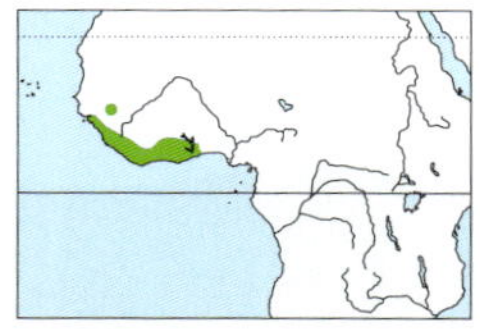

SE SN, S GW, S GN, SL, LR, CI, GH, W TG; possibly BJ

Spot-nosed Monkey
Cercopithecus petaurista
FR Cercopithèque blanc-nez
DE Kleine Weißnasenmeerkatze
ES Cercopiteco menor de nariz blanca

ssp *erythrogaster*

Red-bellied Monkey
Cercopithecus erythrogaster
FR Cercopithèque à ventre roux
DE Rotbauchmeerkatze
ES Cercopiteco de vientre rojo

ssp *pococki*

EN

2 sspp
HB 40–50 cm
T 55–70 cm
W 2–4.5 kg

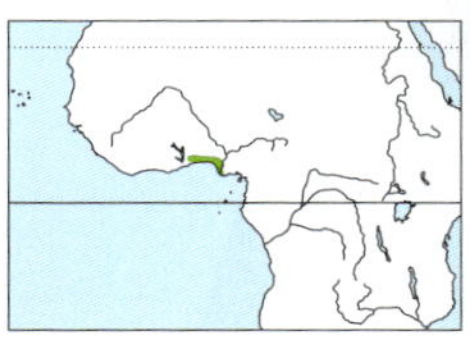

SE TG (Togodo Forest), S BJ (Lama CF and on Nigerian border, N of Porto Novo), SW & SC NG

Sclater's Monkey
Cercopithecus sclateri
FR Cercopithèque de Sclater
DE Sclater-Meerkatze
ES Cercopiteco de Sclater

below 130 m

Scattered in forest and derived savanna between lower R Niger and R Cross, also Niger Delta, N to R Eastern Aboine

EN

Monotypic
HB 40–55 cm
T 45–75 cm
W 2.5–4.5 kg

Mustached Monkey
Cercopithecus cephus
FR Cercopithèque moustac
DE Blaumaulmeerkatze
ES Cercopiteco bigotudo

ssp *cephodes*

ssp *cephus*

LC

3 sspp
HB 44–58 cm
T 60–99 cm
W 2–5 kg

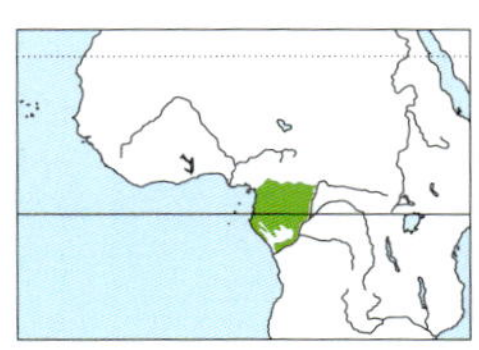

S CM (S of R Sanaga), SW CF, GQ, GA, CG, NW AO, NW DC

VU

2 sspp
HB 40–55 cm
T 46–77 cm
W 2.3–4.5 kg

Red-eared Monkey
Cercopithecus erythrotis
FR Cercopithèque à nez rouge
DE Rotohrmeerkatze
ES Cercopiteco de orejas rojas

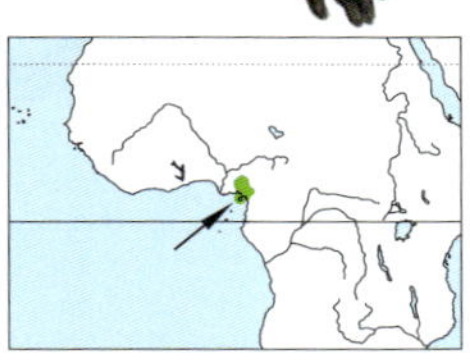

SE NG (E of R Cross), W CM (N of lower R Sanaga; small area S of R Sanaga, near its mouth), mountains on Bioko I (GQ)

Red-tailed Monkey
Cercopithecus ascanius

FR Cercopithèque ascagne
DE Rotschwanzmeerkatze
ES Cercopiteco de cola roja

LC
5 sspp
HB 34–51 cm
T 54–92 cm
W 1.8–6 kg

0–2500 m

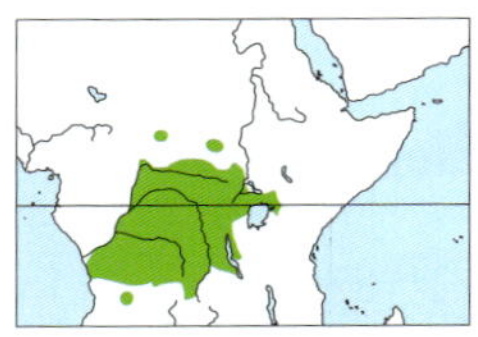

CD, UG, SW KE, RW, W BI, W TZ, NW ZM, N AO; N & W of R Ubangi in CF and S SS

Putty-nosed Monkey
Cercopithecus nictitans

FR Cercopithèque hocheur
DE Große Weißnasenmeerkatze
ES Cercopiteco de nariz blanca

NT
5 sspp
HB 43–70 cm
T 56–100 cm
W 2.7–8 kg

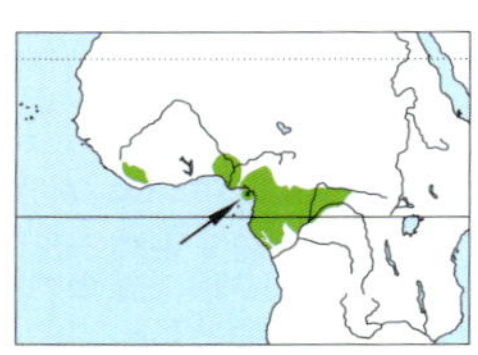

N LR, W CI, S NG, CM (extending N of R Sanaga to Cameroon Highlands), SW CF, GQ (including Bioko I), GA, CG, NW CD, perhaps NW AO

ssp *boutourlinii*

Blue Monkey
Cercopithecus mitis

FR Cercopithèque à diadème
DE Diademmeerkatze
ES Cercopiteco azul

ssp *mitis*

ssp *heymansi*

LC
16 sspp
HB 20–71 cm
T 26–95 cm
W 1.3–11.1 kg

ssp *stuhlmanni*

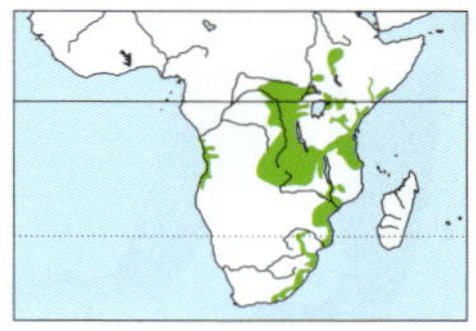

W & E AO, E Africa from SW ET, S SS and S SO through E CD, UG, KE, TZ, RW, BI, ZM, MW, MZ, ZW to E ZA

Genus *COLOBUS*
(5 species)

ssp *cordieri*

ssp *angolensis*

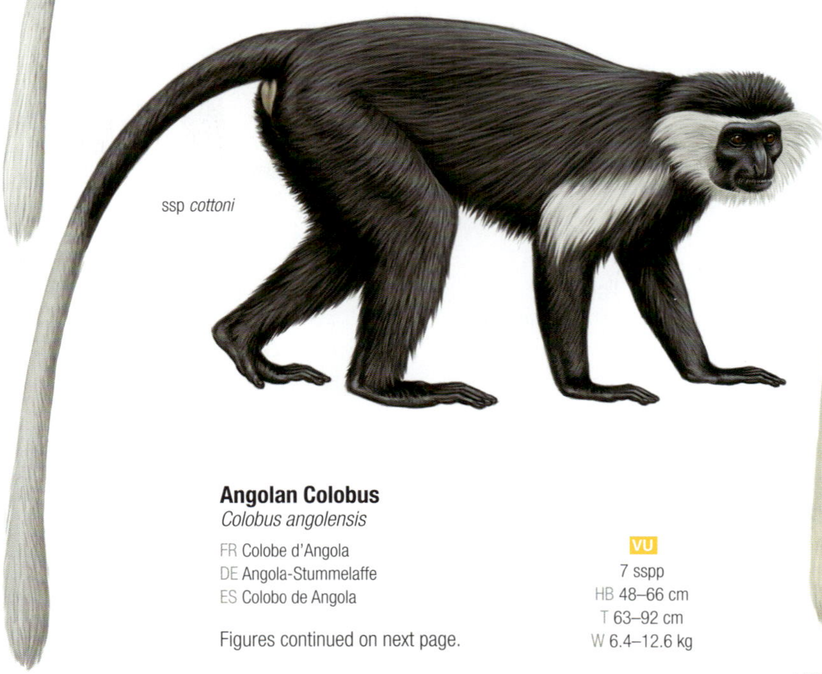

ssp *cottoni*

ssp *prigoginei*

Angolan Colobus
Colobus angolensis

FR Colobe d'Angola
DE Angola-Stummelaffe
ES Colobo de Angola

Figures continued on next page.

VU
7 sspp
HB 48–66 cm
T 63–92 cm
W 6.4–12.6 kg

up to 4000 m

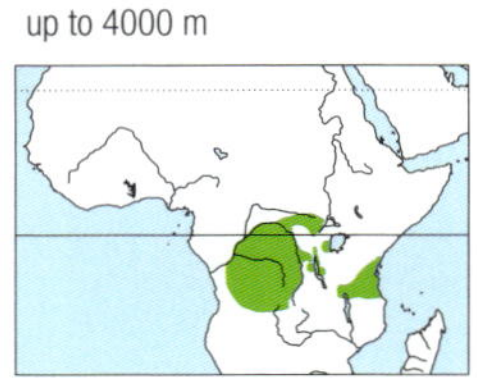

CD (S of R Congo), NE AO and NW ZM disjunctly to SW UG, W RW, W BI, SE KE, TZ, N MW, NE ZM; perhaps N MZ

Angolan Colobus

Colobus angolensis

(Continued from previous page.)

Guereza

Colobus guereza

FR Colobe guéréza
DE Guereza
ES Colobo guereza

LC
8 sspp
HB 49–75 cm
T 50–90 cm
W 5.5–13.5 kg

up to 3300 m

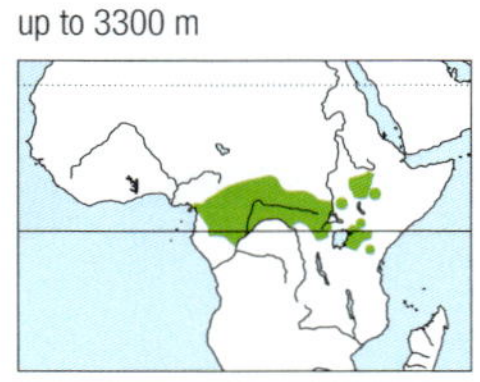

S CM, NE GA, N CG, CF, S TD, N CD (S to R Congo), SS, W & SE UG, NW RW, W & EC ET, KE, N TZ; perhaps GQ and SE NG

King Colobus
Colobus polykomos

FR Colobe noir et blanc
DE Bärenstummelaffe
ES Colobo rey

EN
Monotypic
HB 55–68 cm
T 72–100 cm
W 6.6–10 kg

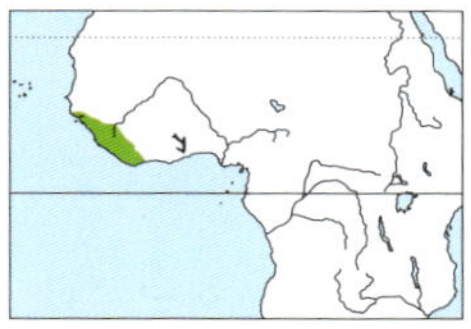

GW, GN, SL, LR, SW CI (E to R Sassandra); perhaps S SN and GM

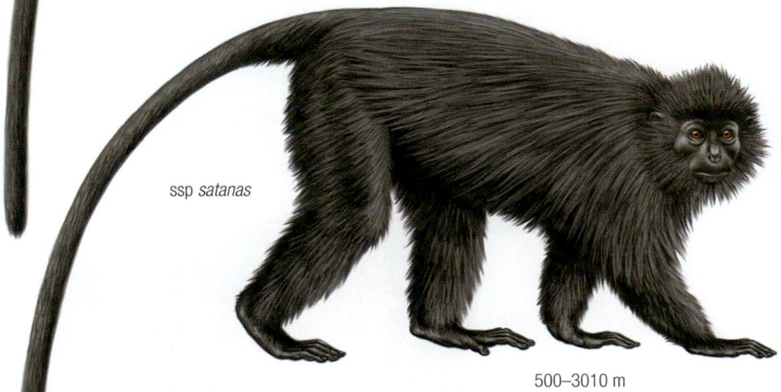

Black Colobus
Colobus satanas

FR Colobe noir
DE Schwarzer Stummelaffe
ES Colobo negro

VU
2 sspp
HB 50–77 cm
T c. 80 cm
W 10–15 kg

500–3010 m

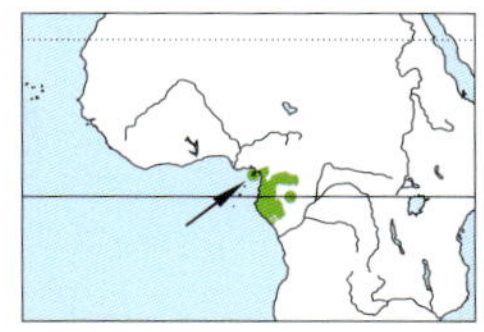

CM (S of R Sanaga), GQ (with Bioko I), W & C GA (inland to Lopé NP, uncertain E of R Ogooué/R Ivindo), W CG; perhaps AO (Cabinda)

White-thighed Colobus
Colobus vellerosus

FR Colobe de Geoffroy
DE Weißschenkel-Stummelaffe
ES Colobo de muslos blancos

CR
Monotypic
HB 56–66 cm
T 81–97 cm
W 7–9.9 kg

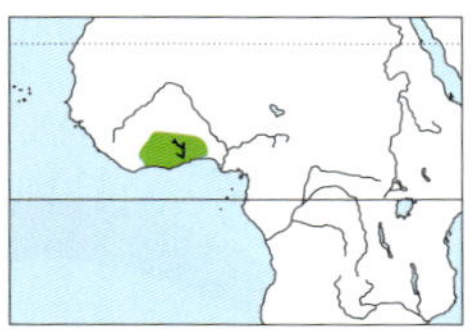

E CI (E of R Nzi-Bandama), S BF, GH, TG, BJ, SW NG

Genus *PILIOCOLOBUS*

(17 species)

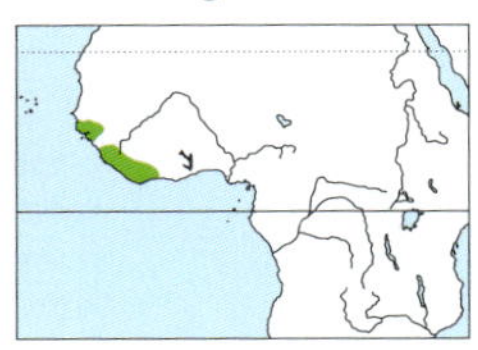

SW SN, GM, GW, NW & S GN, SL, LR, W CI (E to R Nzi-Bandama)

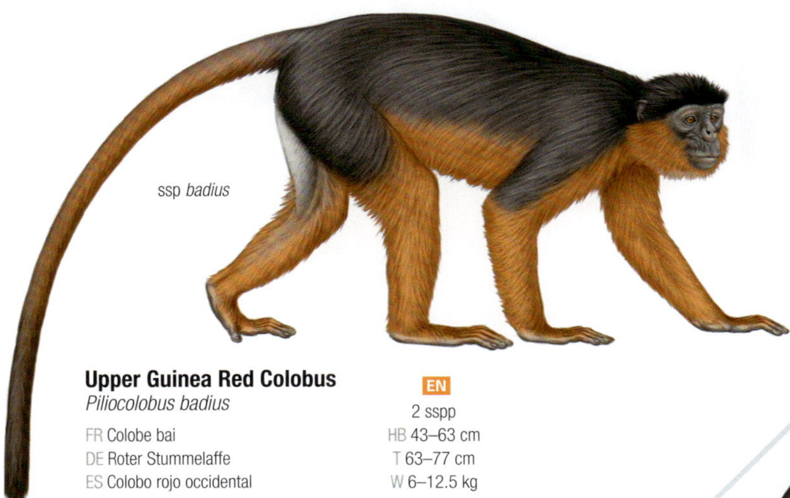

Upper Guinea Red Colobus
Piliocolobus badius

FR Colobe bai
DE Roter Stummelaffe
ES Colobo rojo occidental

EN
2 sspp
HB 43–63 cm
T 63–77 cm
W 6–12.5 kg

Miss Waldron's Red Colobus
Piliocolobus waldronae

FR Colobe de Waldron
DE Miss-Waldron-Stummelaffe
ES Colobo rojo de Ghana

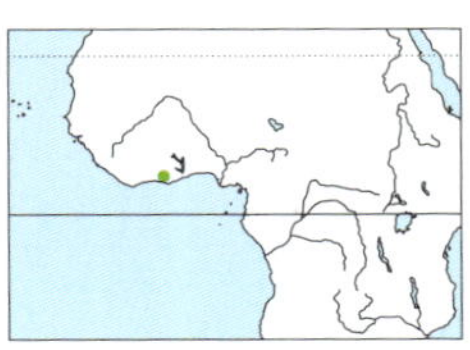

SE CI; possibly SW GH

CR
Monotypic
HB 45–67 cm
T 52–68 cm
W 5.5–6.3 kg

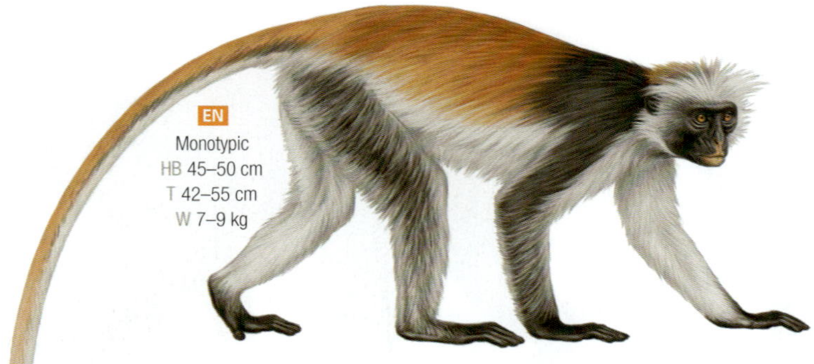

EN
Monotypic
HB 45–50 cm
T 42–55 cm
W 7–9 kg

Zanzibar Red Colobus
Piliocolobus kirkii

FR Colobe de Zanzibar
DE Sansibar-Stummelaffe
ES Colobo rojo de Zanzíbar

0–110 m

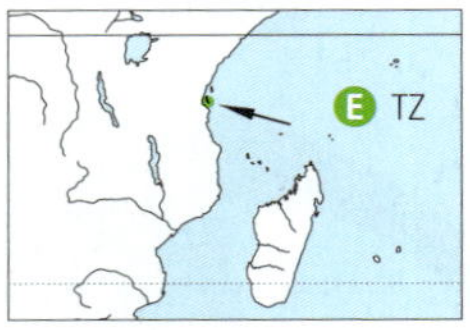

Unguja I (mainly in SE) and Muyuni and Uzi Is; introduced to nearby Pemba I

VU
Monotypic
HB 60–66 cm (F)
T 64–69 cm (F)

Udzungwa Red Colobus
Piliocolobus gordonorum

FR Colobe des Gordon
DE Uzungwe-Stummelaffe
ES Colobo rojo de Udzungwa

200–2200 m

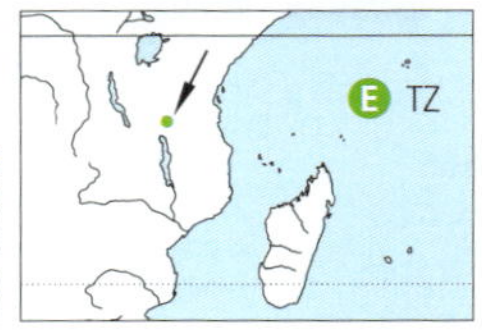

Known only from several forest patches of Udzungwa Mts

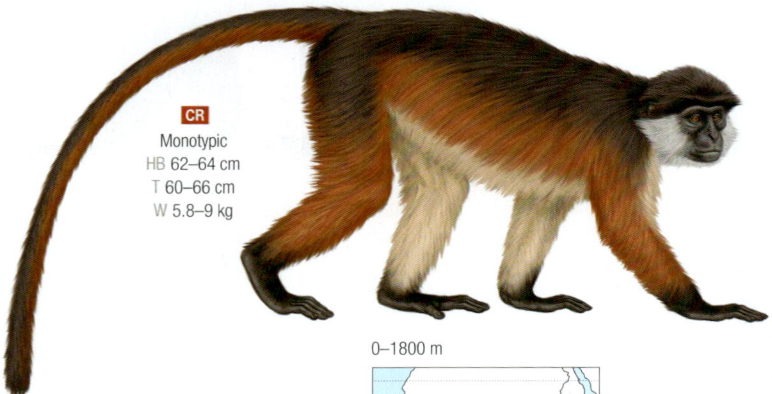

CR
Monotypic
HB 62–64 cm
T 60–66 cm
W 5.8–9 kg

Pennant's Red Colobus
Piliocolobus pennantii

FR Colobe de Pennant
DE Bioko-Stummelaffe
ES Colobo rojo de Bioko

0–1800 m

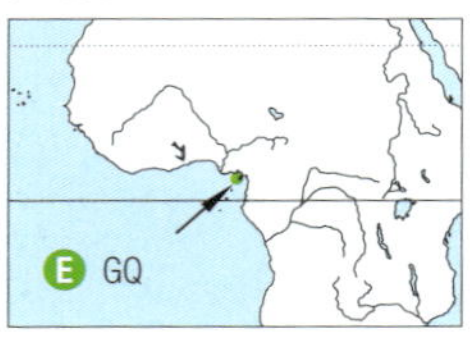

Bioko I, now restricted mainly to SW

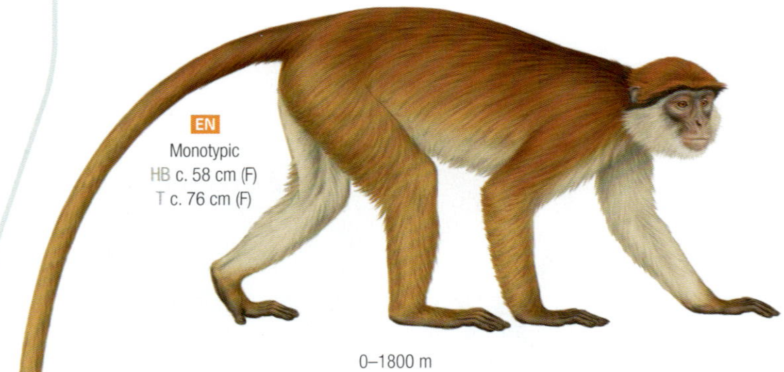

EN
Monotypic
HB c. 58 cm (F)
T c. 76 cm (F)

Bouvier's Red Colobus
Piliocolobus bouvieri

FR Colobe de Bouvier
DE Bouvier-Stummelaffe
ES Colobo rojo de Bouvier

0–1800 m

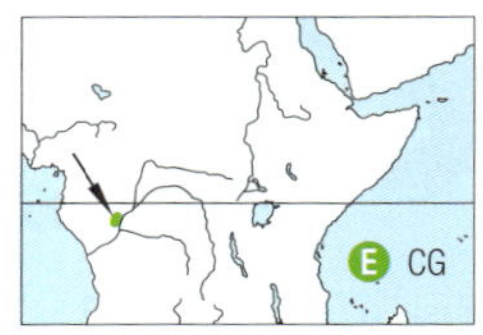

Right bank of R Congo (from mouth of R Likouala-aux-Herbe to lower R Alima); unconfirmed in Lefini R

CR
Monotypic

Niger Delta Red Colobus
Piliocolobus epieni

FR Colobe epieni
DE Nigerdelta-Stummelaffe
ES Colobo rojo del Níger

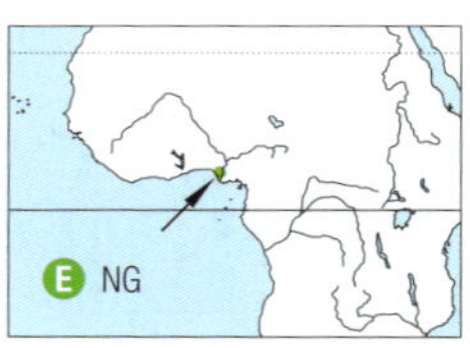

Confined to W Niger Delta (between Forcados-Nikrogha Creek and Sag-bama-Osiama-Agboi Creek)

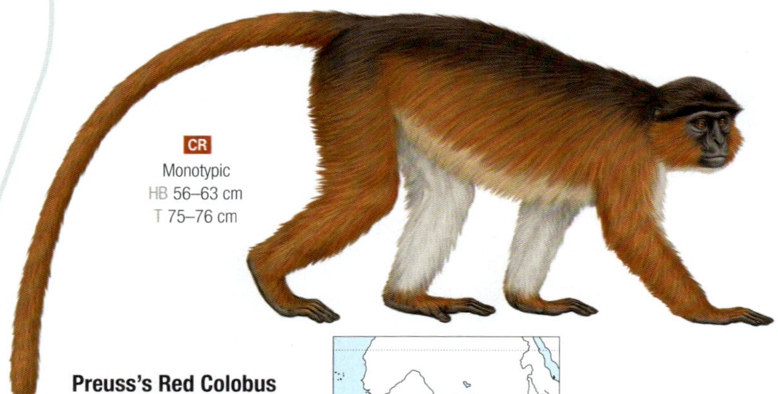

CR
Monotypic
HB 56–63 cm
T 75–76 cm

Preuss's Red Colobus
Piliocolobus preussi

FR Colobe de Preuss
DE Preuss-Stummelaffe
ES Colobo rojo de Camerún

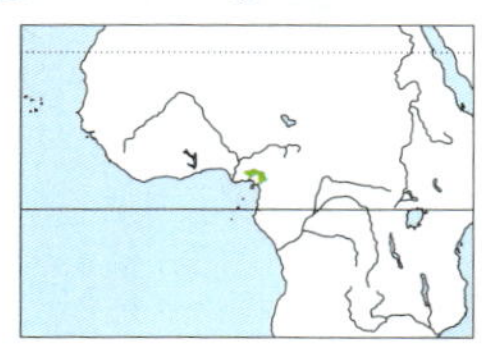

SE NG (Cross River NP), W CM (Korup NP, Yabassi, Ebo Forest, Makombe Forest)

Tana River Red Colobus
Piliocolobus rufomitratus

FR Colobe de la Tana
DE Tana-Stummelaffe
ES Colobo rojo del Tana

CR
Monotypic
HB c. 50 cm
T c. 65 cm
W c. 9 kg

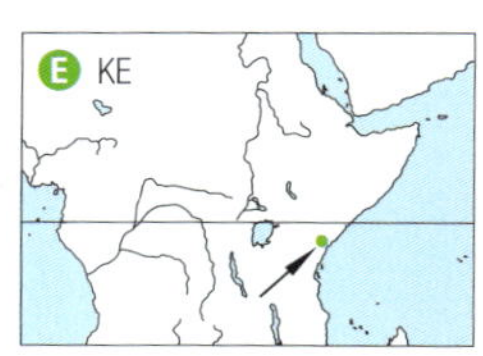

Confined to narrow stretch of forest along lower R Tana (mainly from Garsen N nearly to Wenje) and R Tana Delta

VU

Monotypic

Semliki Red Colobus
Piliocolobus semlikiensis

FR Colobe de la Semliki
DE Semliki-Stummelaffe
ES Colobo rojo del Semliki

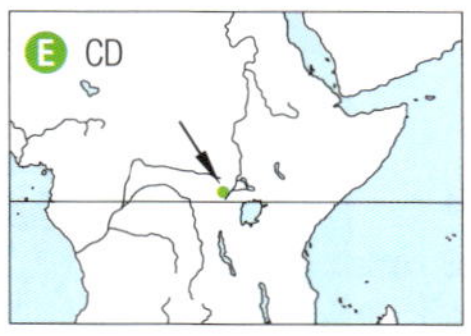

Very small area between middle R Semliki Valley and Rwenzori Mts in E

EN

Monotypic
HB 50–69 cm (M)
T 62–67 cm (M)
W 9–13 kg (M), 7–9 kg (F)

Foa's Red Colobus
Piliocolobus foai

FR Colobe de Foa
DE Luvua-Stummelaffe
ES Colobo rojo del Ouroua

900–2280 m

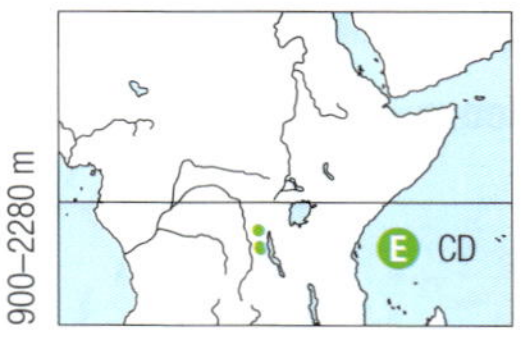

2 widely-separated locations, 1 in Itombwe Massif and 1 in L Tanganyika escarpment

EN

Monotypic
HB 50–69 cm (M)
T 62–67 cm (M)
W 9–13 kg (M), 7–9 kg (F)

Ulindi River Red Colobus
Piliocolobus lulindicus

FR Colobe de l'Ulindi
DE Ulindi-Stummelaffe
ES Colobo rojo del Ulindi

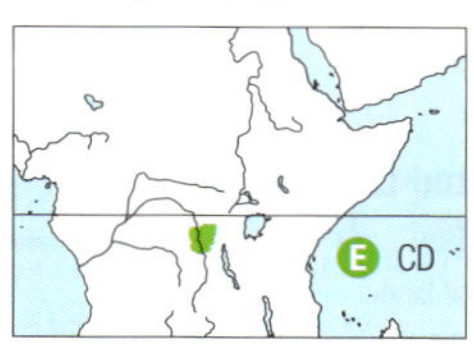

From R Congo (Lualaba) in W, E to lowland-montane transition, S to R Lukuga, N to R Lowa

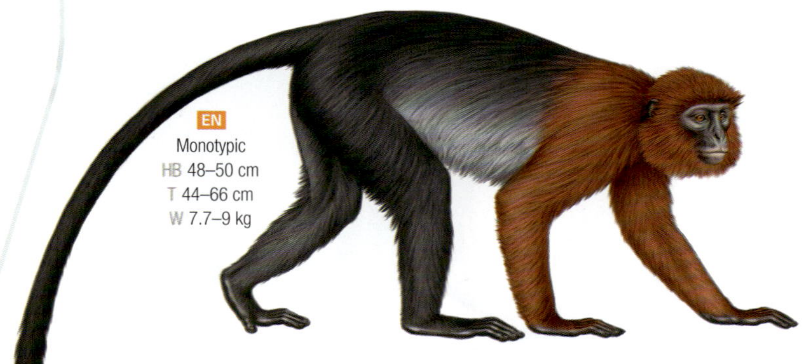

EN

Monotypic
HB 48–50 cm
T 44–66 cm
W 7.7–9 kg

Lang's Red Colobus
Piliocolobus langi

FR Colobe de Lang
DE Lualaba-Stummelaffe
ES Colobo rojo del Lualaba

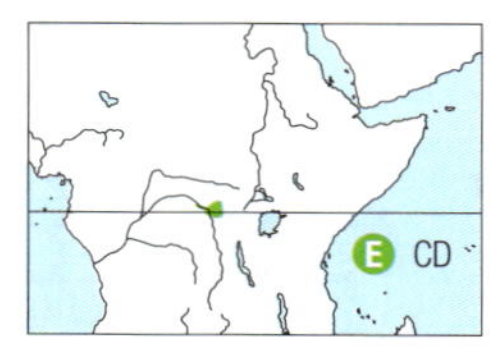

Restricted to "cul-de-sac" between Lualaba and Aruwimi-Ituri rivers

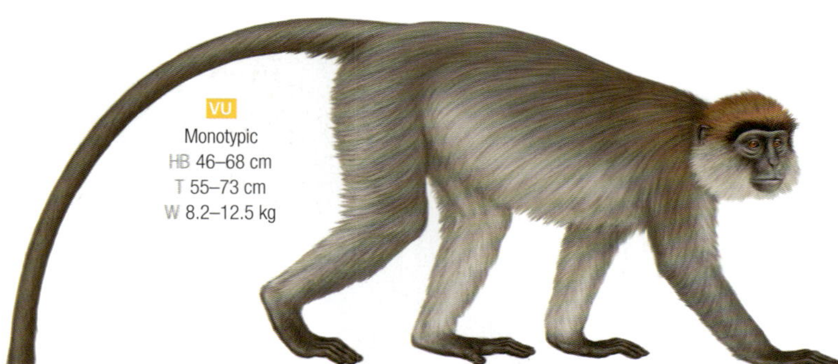

VU

Monotypic
HB 46–68 cm
T 55–73 cm
W 8.2–12.5 kg

Oustalet's Red Colobus
Piliocolobus oustaleti

FR Colobe d'Oustalet
DE Oustalet-Stummelaffe
ES Colobo rojo del Ubangui

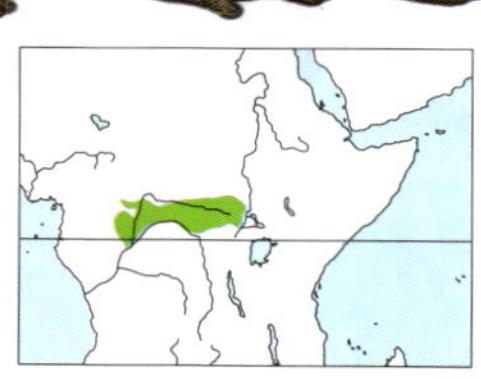

NE CG (E of R Sangha), S CF, N CD (N of R Congo, S of R Uélé, E to L Albert), extreme S SS

EN

Monotypic
W 7.5–9.2 kg

Lomami Red Colobus
Piliocolobus parmentieri

FR Colobe de la Lomami
DE Lomami-Stummelaffe
ES Colobo rojo del Lomami

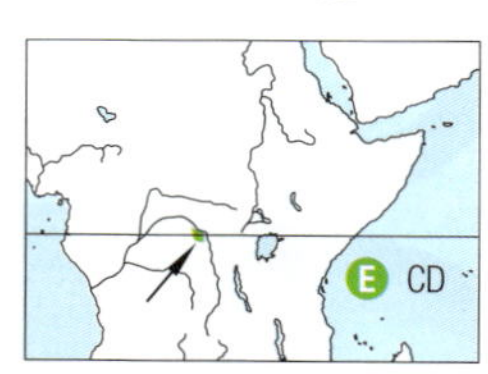

Between R Lomami and R Lualaba, S to about R Ruiki and R Lutanga

Ashy Red Colobus
Piliocolobus tephrosceles

FR Colobe d'Ouganda
DE Uganda-Stummelaffe
ES Colobo rojo de Tanzania

EN

Monotypic
HB 45–67 cm
T 65–72 cm
W 8–11.4 kg (M)

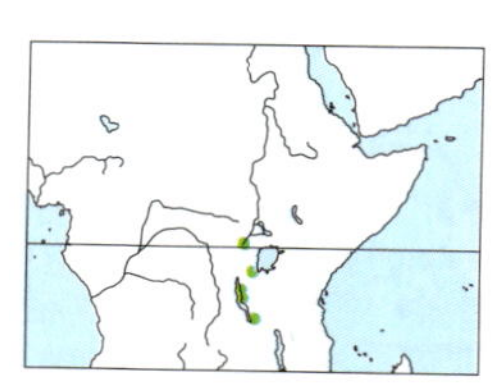

E border of W Rift Valley in SW UG (Kibale) and W TZ (Biharamulo, Gombe, Mahale Mts, L Rukwa); perhaps E CD, RW & BI

Genus *SYMPHALANGUS*

(1 species)

1500–2300 m

Extreme S TH, NW & C Pen MY, ID (W Sumatra, confined to Barisan Mts); perhaps formerly Bangka I

Siamang

Symphalangus syndactylus

FR Siamang
DE Siamang
ES Siamang

EN
Monotypic
HB 75–90 cm
W 9.1–12.7 kg

Genus *NOMASCUS*

(7 species)

Hainan Crested Gibbon

Nomascus hainanus

FR Gibbon de Hainan
DE Hainan-Schopfgibbon
ES Gibón de Hainan

♀

♂

700–1200 m

W Hainan I (once widespread, today restricted to Bawangling NR)

CR
Monotypic
HB c. 49 cm (M), c. 48 cm (F)
W 5.8–10 kg

Eastern Black Crested Gibbon

Nomascus nasutus

FR Gibbon cao-vit
DE Östlicher Schopfgibbon
ES Gibón de cresta negra oriental

500–930 m

Today restricted to a small area in NE VN (Trung Khanh District, Cao Bang Prov) and CN (Jingxi County, Guangxi)

♂

♀

CR
Monotypic

CR
2 sspp
HB 43–54 cm
W 6.9–10 kg

500–2900 m

SC CN (C & SW Yunnan), N VN (Lao Cai, Son La & Yen Bai prov), NW LA (Bokeo & Luang Namtha prov)

♀

♂

Western Black Crested Gibbon

Nomascus concolor

FR Gibbon noir
DE Westlicher Schopfgibbon
ES Gibón de cresta negra

VU
Monotypic

Semliki Red Colobus
Piliocolobus semlikiensis

FR Colobe de la Semliki
DE Semliki-Stummelaffe
ES Colobo rojo del Semliki

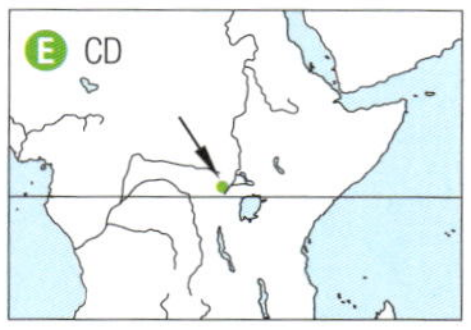

Very small area between middle R Semliki Valley and Rwenzori Mts in E

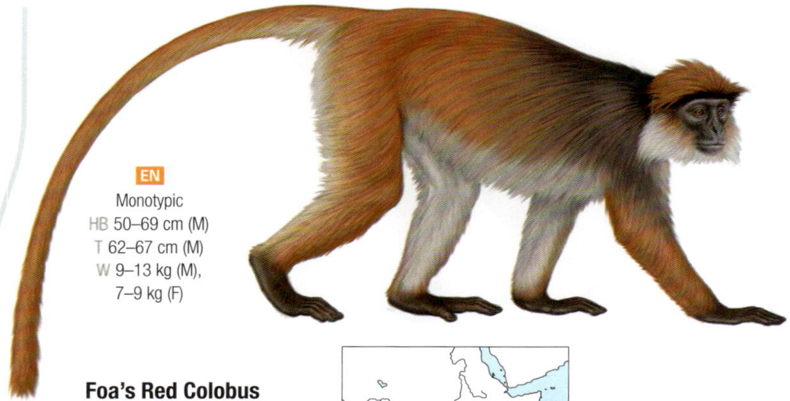

EN
Monotypic
HB 50–69 cm (M)
T 62–67 cm (M)
W 9–13 kg (M), 7–9 kg (F)

Foa's Red Colobus
Piliocolobus foai

FR Colobe de Foa
DE Luvua-Stummelaffe
ES Colobo rojo del Ouroua

900–2280 m

2 widely-separated locations, 1 in Itombwe Massif and 1 in L Tanganyika escarpment

EN
Monotypic
HB 50–69 cm (M)
T 62–67 cm (M)
W 9–13 kg (M), 7–9 kg (F)

Ulindi River Red Colobus
Piliocolobus lulindicus

FR Colobe de l'Ulindi
DE Ulindi-Stummelaffe
ES Colobo rojo del Ulindi

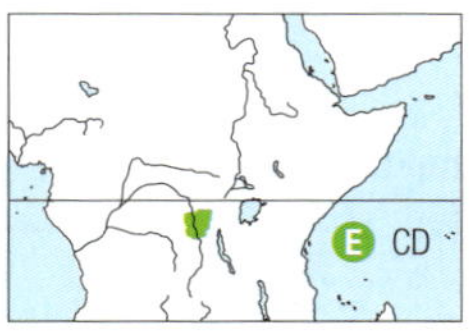

From R Congo (Lualaba) in W, E to lowland-montane transition, S to R Lukuga, N to R Lowa

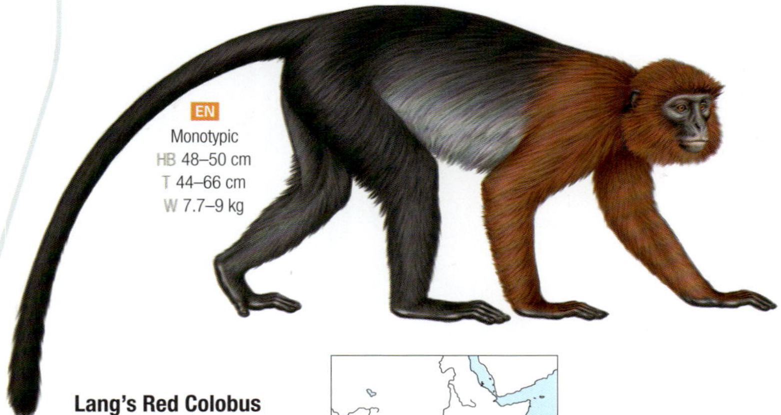

EN
Monotypic
HB 48–50 cm
T 44–66 cm
W 7.7–9 kg

Lang's Red Colobus
Piliocolobus langi

FR Colobe de Lang
DE Lualaba-Stummelaffe
ES Colobo rojo del Lualaba

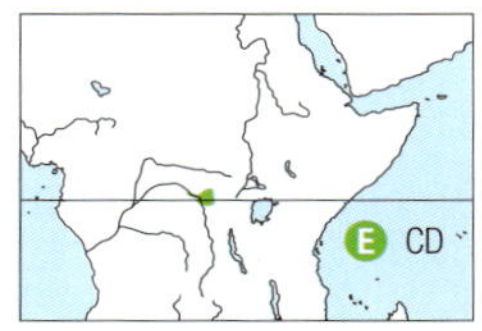

Restricted to "cul-de-sac" between Lualaba and Aruwimi-Ituri rivers

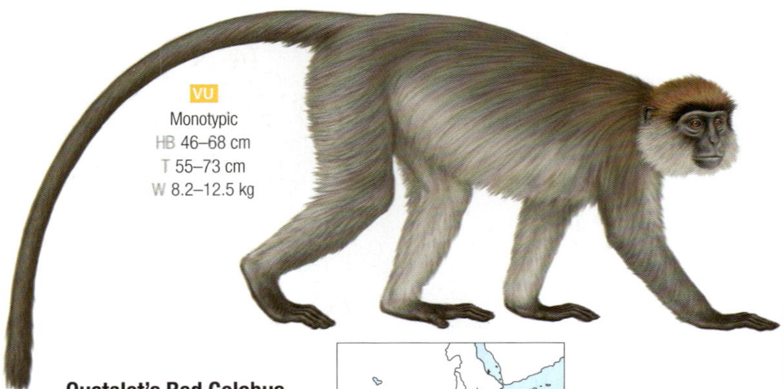

VU
Monotypic
HB 46–68 cm
T 55–73 cm
W 8.2–12.5 kg

Oustalet's Red Colobus
Piliocolobus oustaleti

FR Colobe d'Oustalet
DE Oustalet-Stummelaffe
ES Colobo rojo del Ubangui

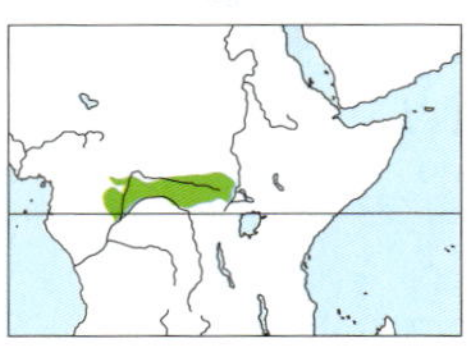

NE CG (E of R Sangha), S CF, N CD (N of R Congo, S of R Uélé, E to L Albert), extreme S SS

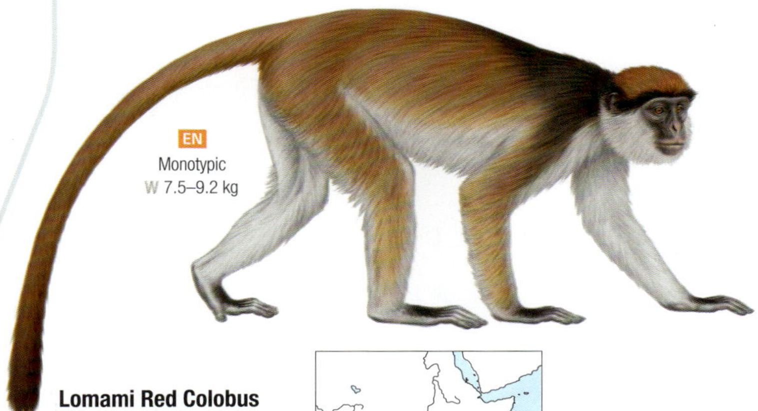

EN
Monotypic
W 7.5–9.2 kg

Lomami Red Colobus
Piliocolobus parmentieri

FR Colobe de la Lomami
DE Lomami-Stummelaffe
ES Colobo rojo del Lomami

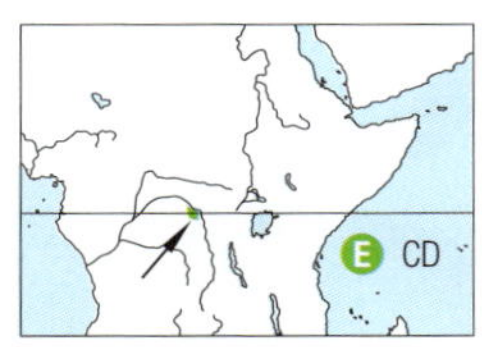

Between R Lomami and R Lualaba, S to about R Ruiki and R Lutanga

Ashy Red Colobus
Piliocolobus tephrosceles

FR Colobe d'Ouganda
DE Uganda-Stummelaffe
ES Colobo rojo de Tanzania

EN
Monotypic
HB 45–67 cm
T 65–72 cm
W 8–11.4 kg (M)

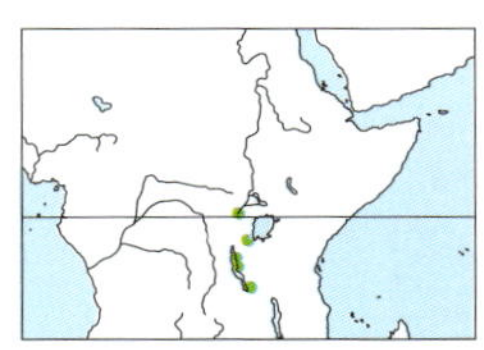

E border of W Rift Valley in SW UG (Kibale) and W TZ (Biharamulo, Gombe, Mahale Mts, L Rukwa); perhaps E CD, RW & BI

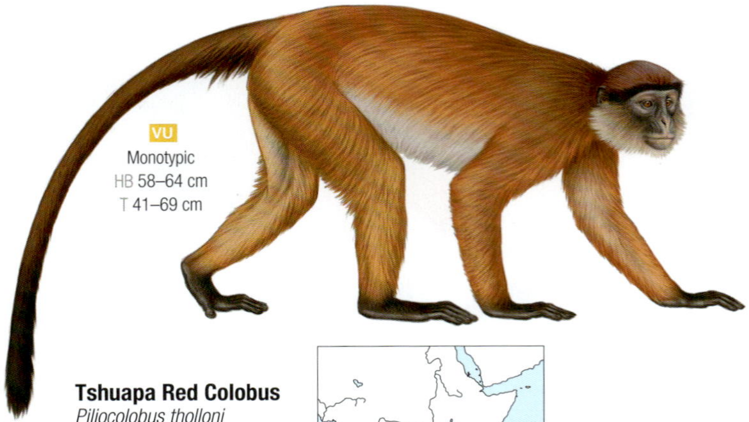

VU
Monotypic
HB 58–64 cm
T 41–69 cm

Tshuapa Red Colobus
Piliocolobus tholloni

FR Colobe de Thollon
DE Thollon-Stummelaffe
ES Colobo rojo del Congo

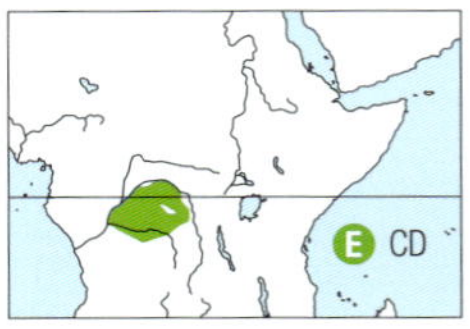

E and S of great bend of R Congo, E to R Lomami, S to R Kasai and R Sankuru

Genus *PROCOLOBUS*

(1 species)

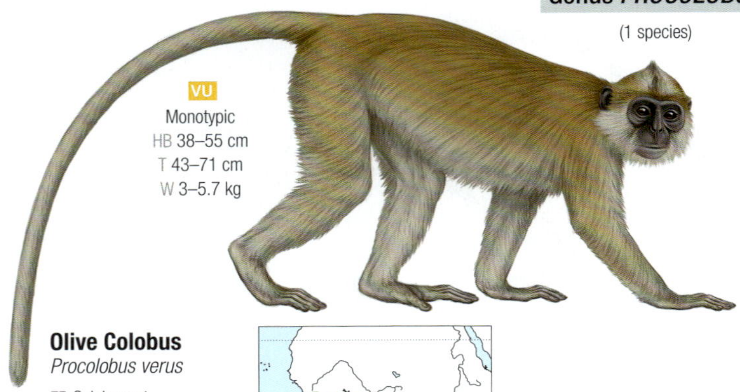

VU
Monotypic
HB 38–55 cm
T 43–71 cm
W 3–5.7 kg

Olive Colobus
Procolobus verus

FR Colobe vert
DE Grüner Stummelaffe
ES Colobo oliva

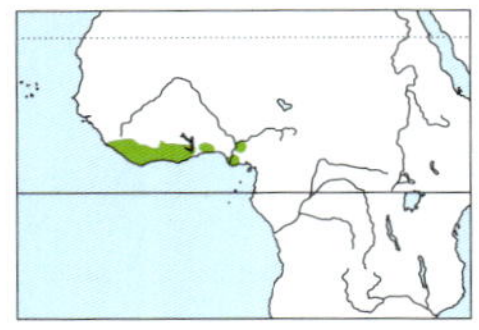

W Africa, from SE SL and S GN to GH (extending just E of R Volta), discontinuity in S TG, S BJ and S NG

Genus *PRESBYTIS*

(19 species)

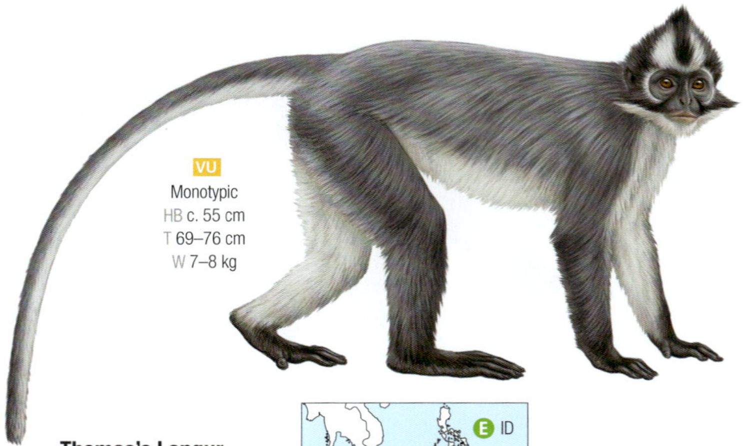

VU
Monotypic
HB c. 55 cm
T 69–76 cm
W 7–8 kg

Thomas's Langur
Presbytis thomasi

FR Langur de Thomas
DE Thomas-Langur
ES Langur de Thomas

1500–3400 m

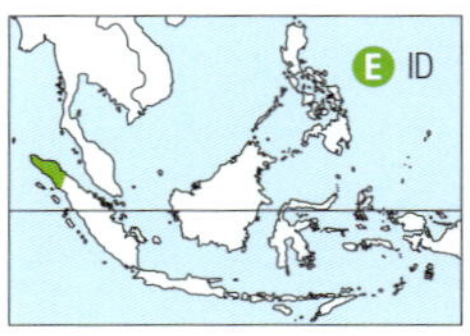

N Sumatra (Aceh Prov), mostly N of R Wampu and R Alas

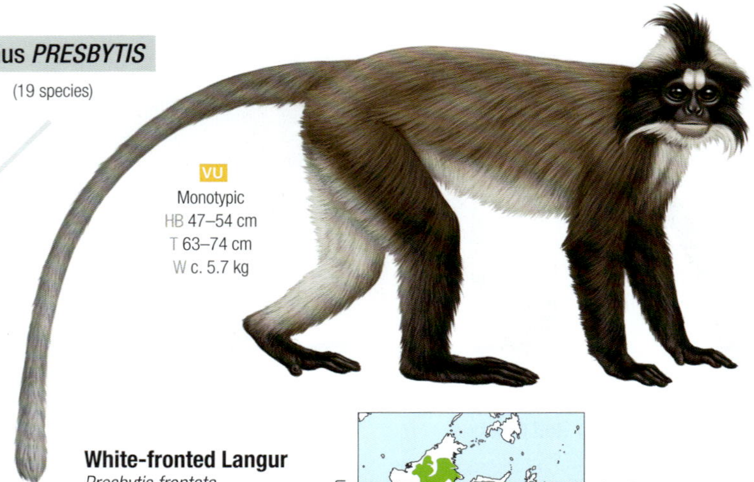

VU
Monotypic
HB 47–54 cm
T 63–74 cm
W c. 5.7 kg

White-fronted Langur
Presbytis frontata

FR Langur à front blanc
DE Weißstirnlangur
ES Langur de frente blanca

up to 1000 m

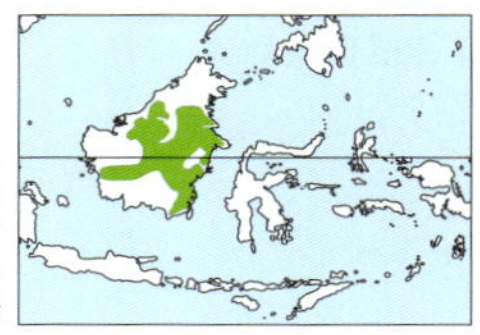

C & E Borneo: patchily from C Sarawak (MY) to most of E and few areas in W Kalimantan (ID)

Cross-marked Langur
Presbytis chrysomelas

FR Langur de Bornéo
DE Sarawak-Langur
ES Langur de Sarawak

ssp *cruciger*

CR
2 sspp
HB 43–61 cm
T 61–84 cm
W 5.9–8.2 kg

ssp *chrysomelas*

variant

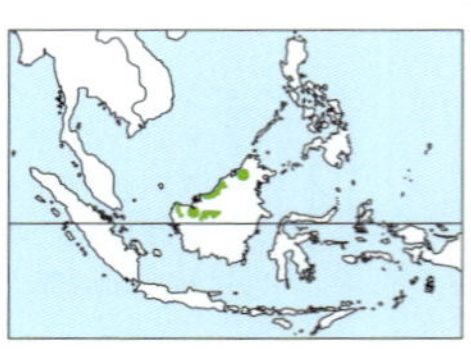

N & W Borneo: Sabah (Melalap) and Sarawak (MY) S to R Kapuas in W Kalimantan (ID)

VU
Monotypic
HB 48–56 cm
T 65–84 cm
W 5.5–7 kg

Hose's Langur
Presbytis hosei

FR Langur de Hose
DE Hose-Langur
ES Langur de Hose

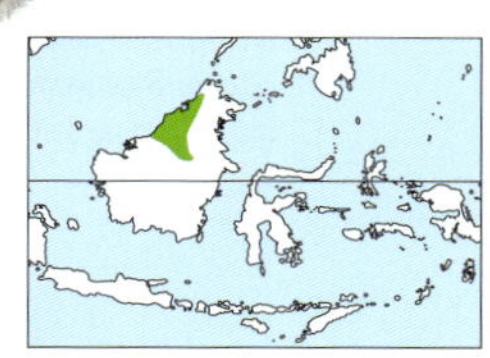

NW Borneo: NE BN, MY (W Sabah, S of Mt Kinabalu, and N Sarawak), ID (N Kalimantan)

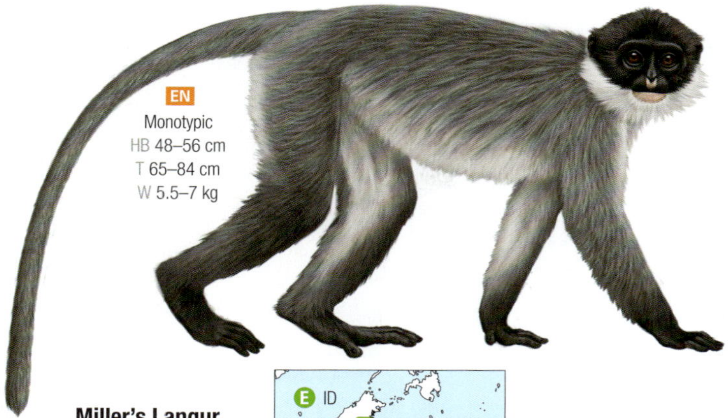

EN

Monotypic

HB 48–56 cm

T 65–84 cm

W 5.5–7 kg

Miller's Langur

Presbytis canicrus

FR Langur de Miller

DE Kalimantan-Langur

ES Langur de Miller

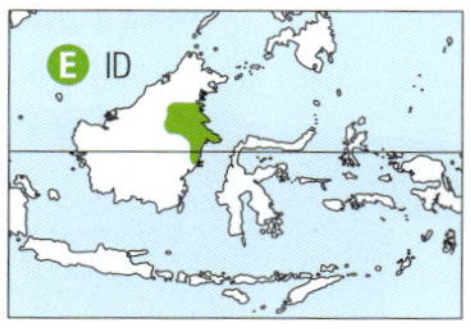

Borneo (East Kalimantan, S of R Kayan & R Sembakung, S to R Mahakam, W limit unclear)

EN

Monotypic

HB 48–56 cm

T 65–84 cm

W 5.5–7 kg

Sabah Grizzled Langur

Presbytis sabana

FR Langur de Sabah

DE Sabah-Langur

ES Langur de Sabah

N Borneo in most of Sabah, as far SW as Kalabakan; unconfirmed in North Kalimantan (ID)

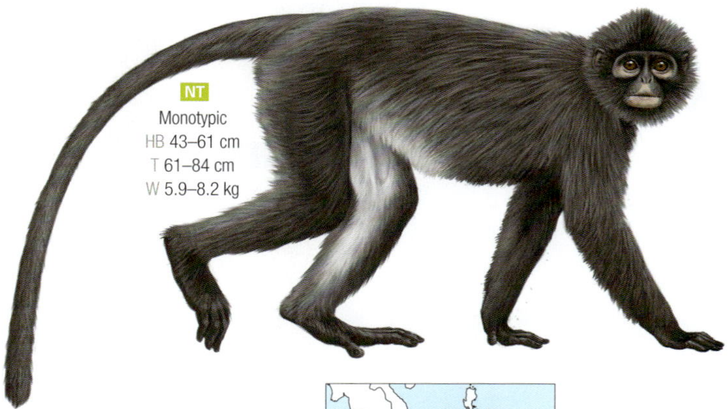

NT

Monotypic

HB 43–61 cm

T 61–84 cm

W 5.9–8.2 kg

Robinson's Banded Langur

Presbytis robinsoni

FR Langur de Robinson

DE Robinson-Bindenlangur

ES Langur de Robinson

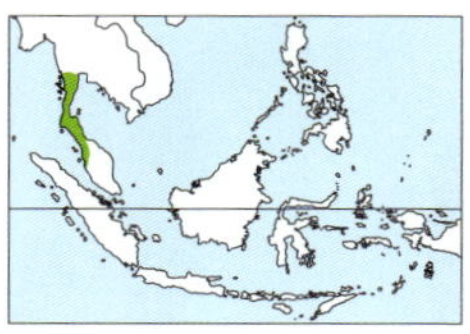

S MM, S TH (from 13° 50' N in Phetchaburi Prov), NW Pen MY (S to Larut Hills, Perak)

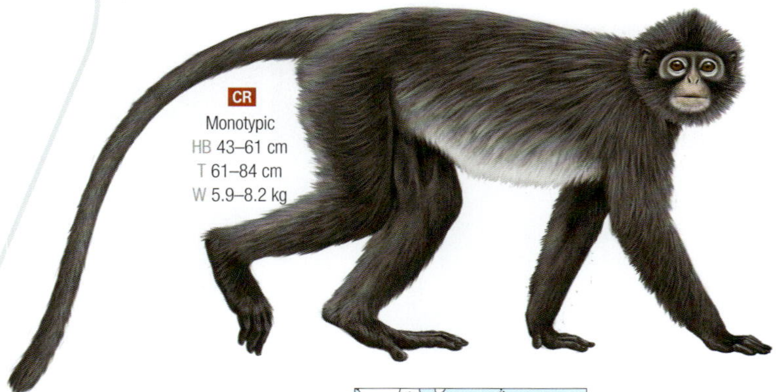

CR

Monotypic

HB 43–61 cm

T 61–84 cm

W 5.9–8.2 kg

Raffles's Banded Langur

Presbytis femoralis

FR Langur malais

DE Bindenlangur

ES Langur listado

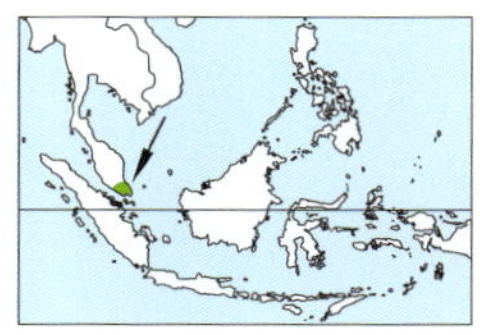

S Pen MY (extreme S Pahang and Johor), SG

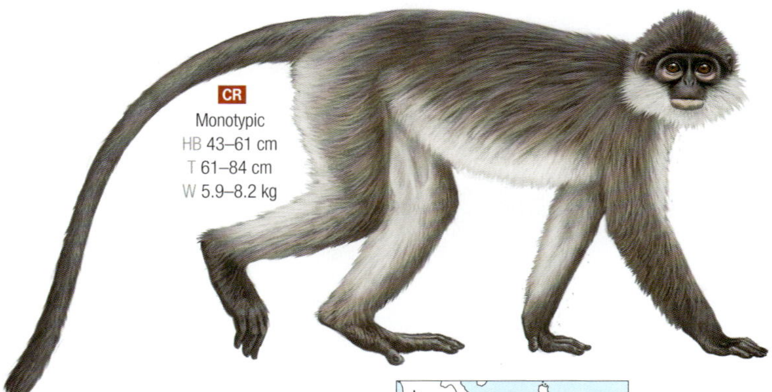

CR

Monotypic

HB 43–61 cm

T 61–84 cm

W 5.9–8.2 kg

East Sumatran Banded Langur

Presbytis percura

FR Langur du Riau

DE Ostsumatra-Bindenlangur

ES Langur de Sumatra oriental

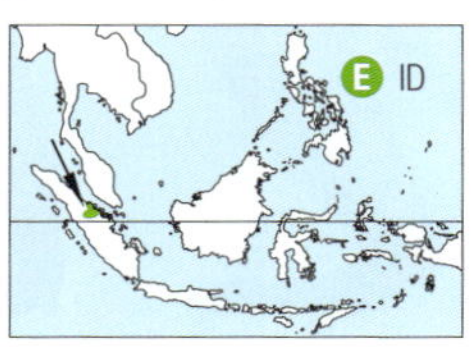

CE Sumatra, in small area between R Rokan and R Siak

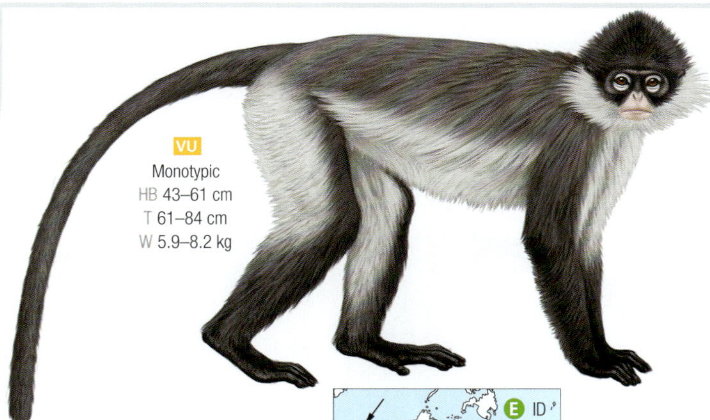

VU

Monotypic

HB 43–61 cm

T 61–84 cm

W 5.9–8.2 kg

Natuna Islands Langur

Presbytis natunae

FR Langur des Natuna

DE Natuna-Langur

ES Langur de Natuna

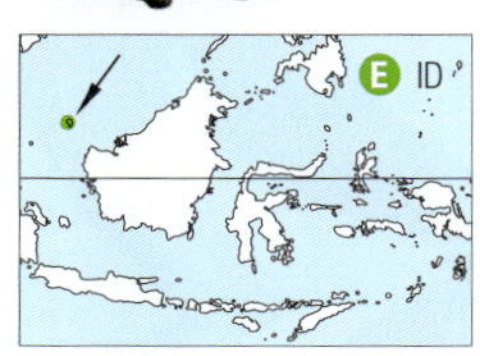

Natuna Is (Buguran I), off NW coast of Borneo

Mentawai Langur

Presbytis potenziani

FR Langur des Mentawai

DE Mentawai-Langur

ES Langur de Mentawai

CR

Monotypic

HB c. 50 cm

T c. 58 cm

W 6.4–6.5 kg

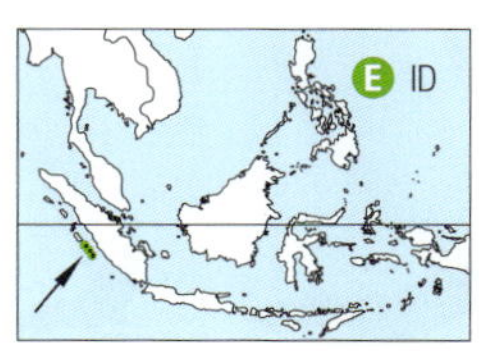

Mentawai Is (Sipora, North Pagai, South Pagai, islet of Sinakak)

EN
Monotypic
HB 48–50 cm
T 58–64 cm
W 6.4–6.5 kg

Siberut Langur
Presbytis siberu

FR Langur de Siberut
DE Siberut-Langur
ES Langur de Siberut

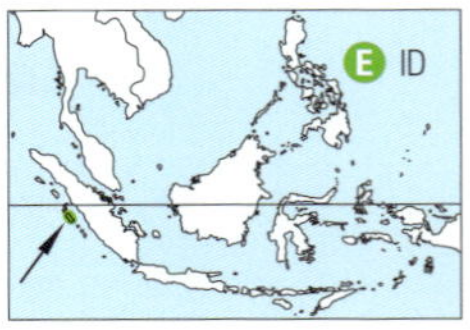

Mentawai Is (Siberut I)

Javan Langur
Presbytis comata

FR Langur de la Sonde
DE Java-Langur
ES Langur de Java

VU
2 sspp
HB 43–60 cm
T 56–72 cm
W 6.4–8 kg

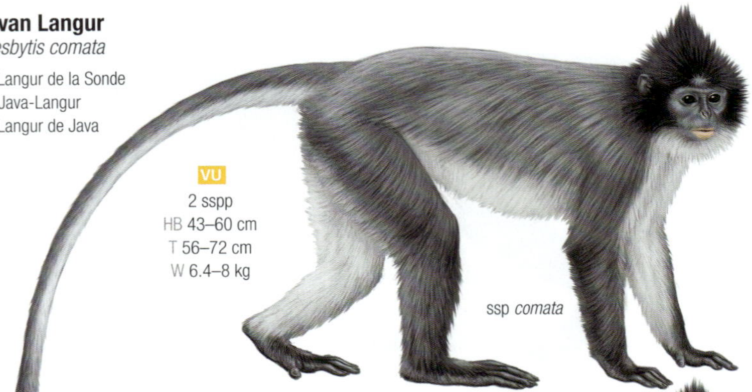

ssp *comata*

ssp *fredericae*

up to 2500 m

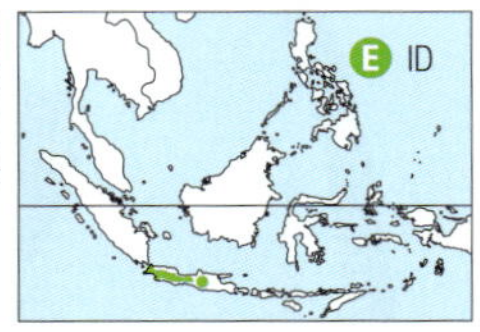

W & C Java (from Ujung Kulon to Diyeng)

VU
Monotypic
HB 42–57 cm
T 62–82 cm
W 5.8–5.9 kg

Mitered Langur
Presbytis mitrata

FR Langur mitré
DE Südlicher Kammlangur
ES Langur mitrado

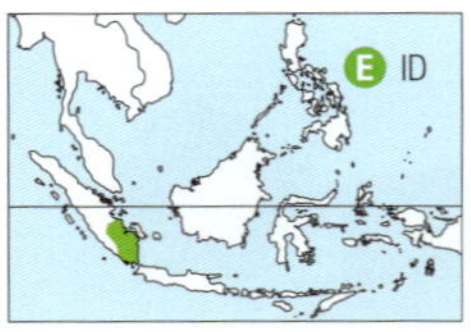

SE Sumatra E of Barisan Mts, from R Batang Hari SW to upper R Musi, S to Lampung Prov

Black-crested Sumatran Langur
Presbytis melalophos

FR Langur à crête noire
DE Rotbrauner Kammlangur
ES Langur de Sumatra

EN
Monotypic
HB 26–40 cm
T 67–86 cm
W 5.2–9 kg

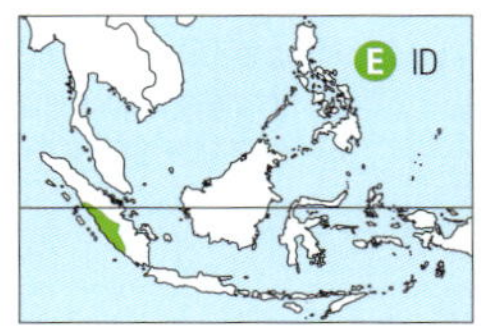

W Sumatra, from upper R Rokan (Tanangtalu) S to upper R Batang Hari and beyond along Barisan Mts

ssp *cana*

ssp *siamensis*

Pale-thighed Langur
Presbytis siamensis

FR Langur à cuisses pâles
DE Weißschenkellangur
ES Langur de muslos blancos

NT
4 sspp
HB 34–39 cm
T 67–89 cm
W 4.3–8.2 kg

ssp *paenulata*

ssp *rhionis*

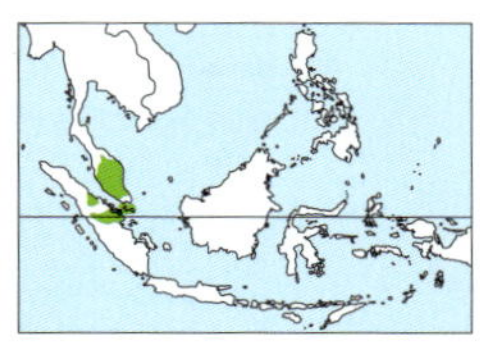

Extreme S TH, most of Pen MY, ID (CE Sumatra, Riau Arch in Kundur, Bintan, perhaps Batam and Galang)

Black Sumatran Langur
Presbytis sumatrana

FR Langur de Sumatra
DE Schwarzer Kammlangur
ES Langur negro de Sumatra

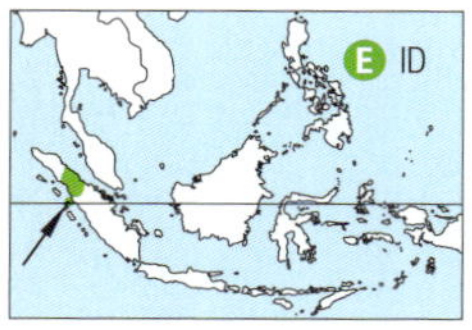

W & NC Sumatra, Batu Is (Pini I)

EN
Monotypic
HB 33–39 cm
T 64–84 cm
W 4.9–8.5 kg

Black-and-white Langur
Presbytis bicolor

FR Langur bicolore
DE Schwarzweißer Kammlangur
ES Langur bicolor

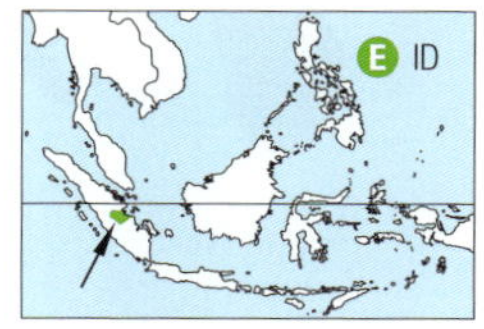

EC Sumatra (from R Indragiri S to R Batang Hari, mainly in highlands)

DD
Monotypic
HB 33–38 cm
T 64–84 cm
W 4.9–8.5 kg

ssp *carimatae*

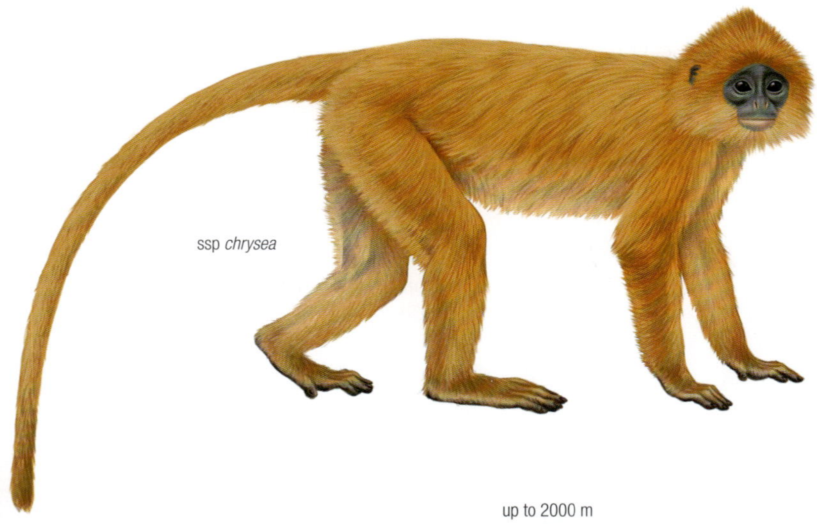

ssp *chrysea*

ssp *ignita*

Maroon Langur
Presbytis rubicunda

FR Langur rubicond
DE Maronenlangur
ES Langur granate

VU
5 sspp
HB 44–58 cm
T 63–80 cm
W 5.5–7 kg

up to 2000 m

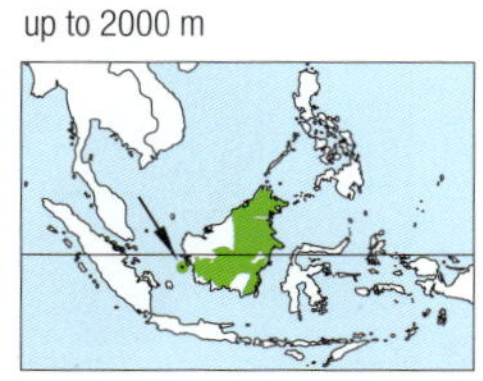

NE Borneo: Sabah and Sarawak (MY), Kalimantan and Karimata I, off W Borneo (ID); perhaps BN

ssp *rubicunda*

ssp *rubida*

Genus *PYGATHRIX*

(3 species)

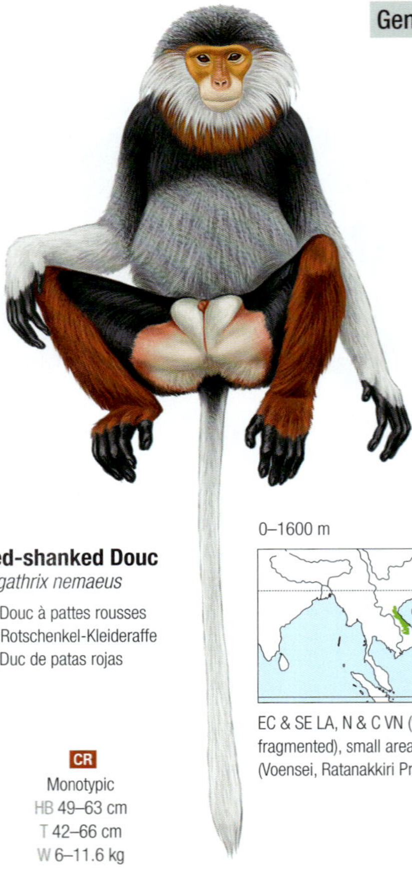

Red-shanked Douc
Pygathrix nemaeus

FR Douc à pattes rousses
DE Rotschenkel-Kleideraffe
ES Duc de patas rojas

0–1600 m

EC & SE LA, N & C VN (very fragmented), small area in NE KH (Voensei, Ratanakkiri Prov)

CR
Monotypic
HB 49–63 cm
T 42–66 cm
W 6–11.6 kg

Gray-shanked Douc
Pygathrix cinerea

FR Douc à pattes grises
DE Grauschenkel-Kleideraffe
ES Duc de patas grises

200–1500 m

C VN (Quang Nam, Kon Tum, Quang Ngai, Gia Lai and Binh Dinh prov), small area in NE KH

CR
Monotypic
HB 51–66 cm
T 54–65 cm
W 4.7–12.5 kg

CR
Monotypic
HB 54–63 cm
T 65–84 cm
W 6–11 kg

Black-shanked Douc
Pygathrix nigripes

FR Douc à pattes noires
DE Schwarzschenkel-Kleideraffe
ES Duc de patas negras

0–800 m

E KH (E of R Mekong, S of R Srepok), SW VN (from c. 14° N to S Cat Tien NP); unconfirmed in S LA

Genus *RHINOPITHECUS*

(5 species)

Tonkin Snub-nosed Monkey
Rhinopithecus avunculus

FR Rhinopithèque du Tonkin
DE Tonkin-Stumpfnase
ES Rinopiteco de Tonkin

200–1300 m

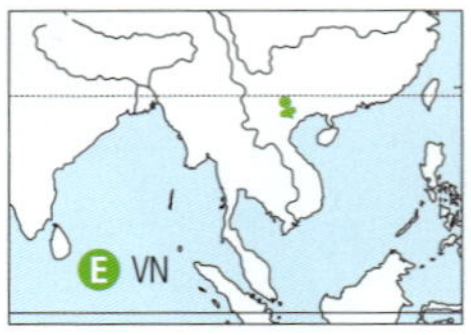

Known only from small forest patches in Ha Giang, Tuyen Quang, Bac Kan and Thai Nguyen prov

CR
Monotypic
HB 50–65 cm
T 65–92 cm
W 8.5–14 kg

Yunnan Snub-nosed Monkey
Rhinopithecus bieti

FR Rhinopithèque du Yunnan
DE Schwarze Stumpfnase
ES Rinopiteco de Yunnan

EN
Monotypic
HB 74–83 cm
T 52–75 cm
W 9.2–17 kg

3000–4700 m

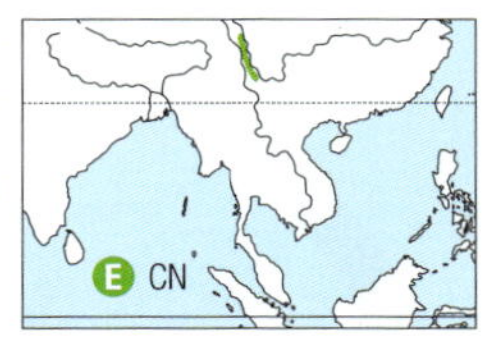

SE Tibet & NW Yunnan (fragmented populations i n Yun Ling Mts), W of R Yangtze , E of R Mekong

Guizhou Snub-nosed Monkey
Rhinopithecus brelichi

FR Rhinopithèque du Guizhou
DE Graue Stumpfnase
ES Rinopiteco de Guizhou

CR
Monotypic
HB 66–72 cm
T c. 97 cm
W 8–15 kg

570–2500 m

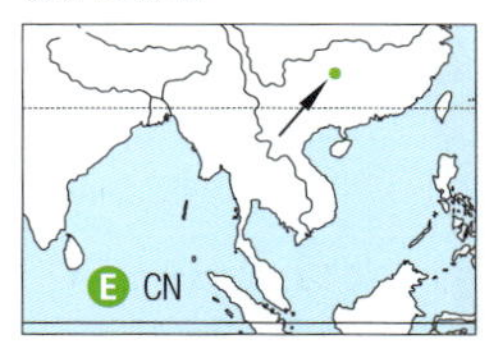

Restricted to Mt Fanjing in Wuling Mts (Guizhou)

Golden Snub-nosed Monkey

Rhinopithecus roxellana

FR Rhinopithèque de Roxellane
DE Goldstumpfnase
ES Rinopiteco dorado

EN
3 sspp
HB 47–83 cm
T 51–104 cm
W 6–19 kg

1400–3000 m

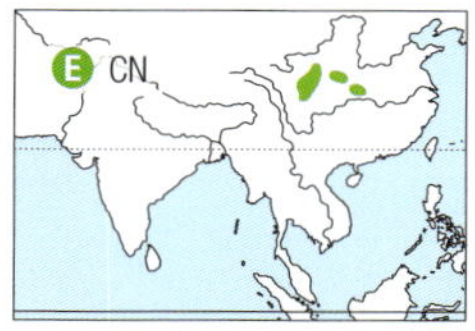

Sichuan, S Gansu, S Shaanxi, W Hubei

CR
Monotypic
HB 54–55 cm
T 68–78 cm
W 8.5–14 kg

Stryker's Snub-nosed Monkey

Rhinopithecus strykeri

FR Rhinopithèque de Stryker
DE Burma-Stumpfnase
ES Rinopiteco de Birmania

1720–3660 m

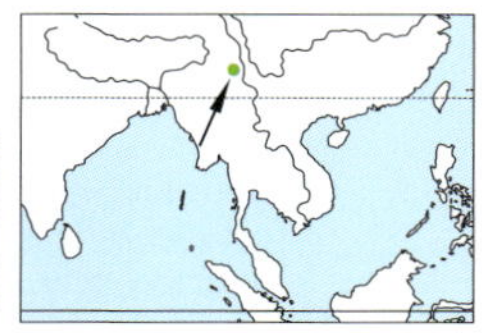

NE MM (Salween–N'mai Hka divide, around R Maw, E to Chichitago), S CN (Gaolilgongshan NR, Yunnan)

Genus *SIMIAS*

(1 species)

Pig-tailed Langur

Simias concolor

FR Langur à queue de cochon
DE Pageh-Stumpfnase
ES Langur de cola de cerdo

CR
2 sspp
HB 46–55 cm
T 14–16 cm
W 7–8.8 kg

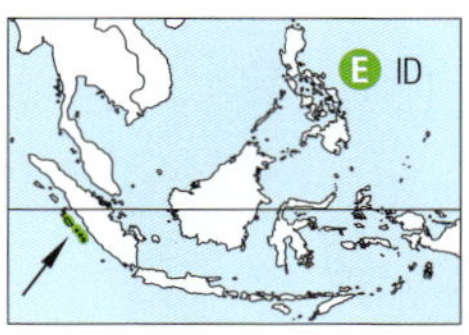

Mentawai Is (Siberut, Sipora, North Pagai, South Pagai and nearby Simalegu, Simatapi and Sinakak)

Genus *NASALIS*

(1 species)

EN
Monotypic
HB 61–76 cm
T 55–67 cm
W 10–24 kg

Proboscis Monkey

Nasalis larvatus

FR Nasique
DE Nasenaffe
ES Nasico

up to 350 m

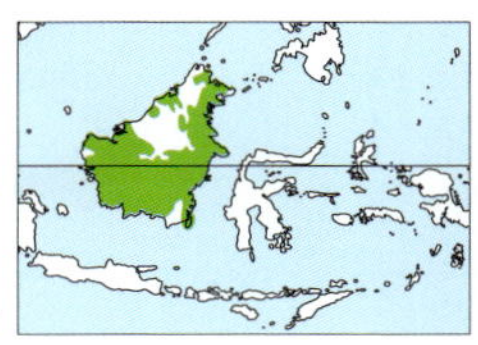

Borneo: Sabah and Sarawak (MY), BN, Kalimantan (ID), satellite islands of Berhala, Sebatik and Laut

Genus *SEMNOPITHECUS*

(8 species)

Bengal Sacred Langur

Semnopithecus entellus

FR Langur du Bengale
DE Bengalen-Hulman
ES Langur de Jánuman

LC
Monotypic
HB 45–78 cm
T 80–112 cm
W 9.5–19.5 kg

up to 400 m

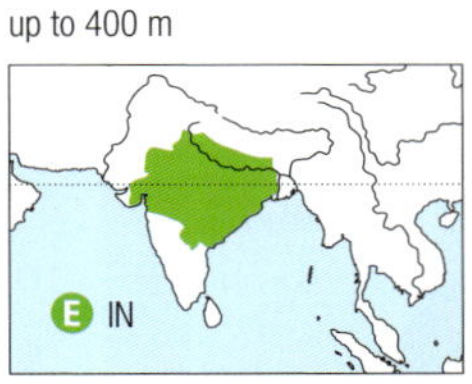

Between R Tapti (Gujarat) and R Krishna (Andhra Pradesh) to foothills of Himalaya; introduced to SW BD

Chamba Sacred Langur

Semnopithecus ajax

FR Langur du Chamba
DE Kashmir-Hulman
ES Langur de Cachemira

EN
Monotypic
HB 51–79 cm
T 72–96 cm
W 19.5–21 kg (M)

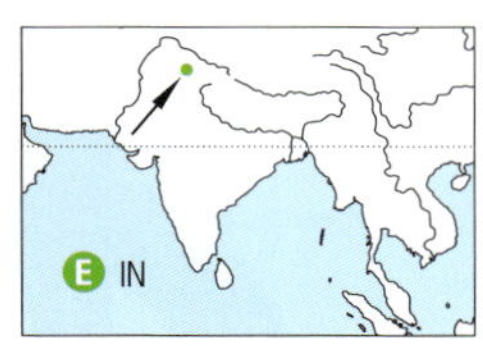

Confined to Chamba Valley (Himachal Pradesh); perhaps Kishtwar Valley (Jammu and Kashmir)

NT
Monotypic
HB 58–76 cm
T 75–99 cm
W 17–18 kg

Terai Sacred Langur
Semnopithecus hector

FR Langur du Teraï
DE Tarai-Hulman
ES Langur de Terai

150–1600 m

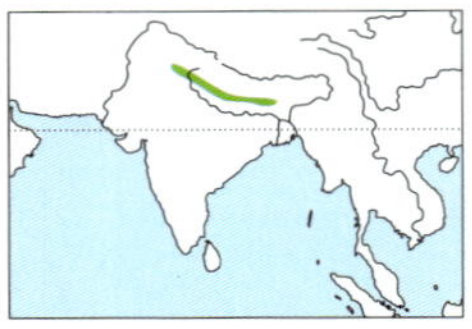

N IN (Himalayan foothills in Uttarakhand, Uttar Pradesh and West Bengal), W & C NP; perhaps SW BT (near Pankhabari)

LC
Monotypic
HB 58–76 cm
T 75–99 cm
W 17–18 kg

Nepal Sacred Langur
Semnopithecus schistaceus

FR Langur du Népal
DE Himalaya-Hulman
ES Langur de Nepal

2000–4000 m

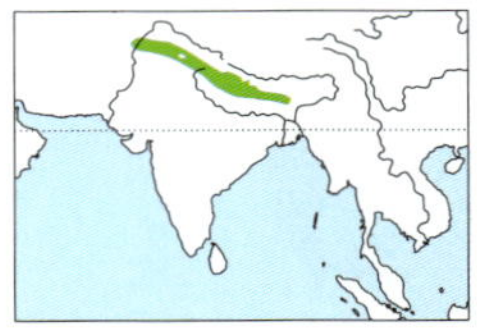

NW PK, N IN (Jammu and Kashmir E to Sikkim), S CN (Tibet), NP, W BT (E to R Sankosh); perhaps E AF

LC
3 sspp
HB 51–69 cm
T 72–109 cm
W 8.4–15.9 kg

Malabar Sacred Langur
Semnopithecus hypoleucos

FR Langur de Malabar
DE Schwarzfuß-Hulman
ES Langur de pies negros

0–1200 m

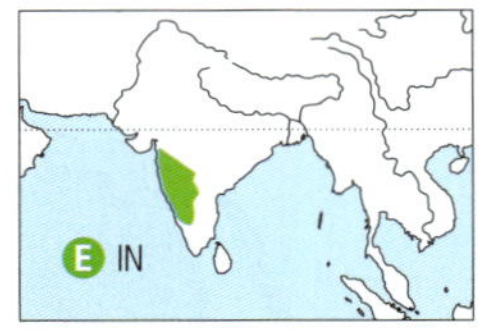

From R Tapti and R Godavari S to Deccan Plateau, all along Western Ghats

Tufted Gray Langur
Semnopithecus priam

FR Langur de Madras
DE Südlicher Hulman
ES Langur moñudo

ssp *priam*

180–1200 m

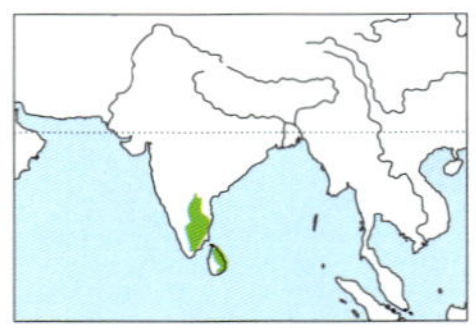

S IN (Andhra Pradesh, Karnataka, Kerala, Tamil Nadu), LK (Dry Zone)

ssp *anchises*

NT
3 sspp
HB 58–64 cm
T 66–102 cm
W 11.4–13.5 kg (M)

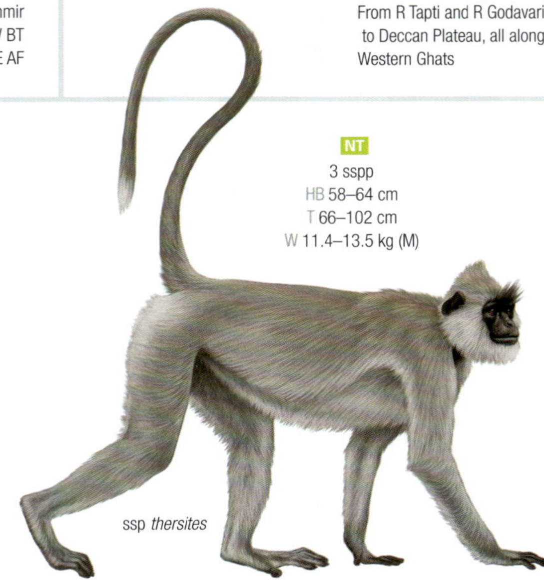

ssp *thersites*

up to 2200 m

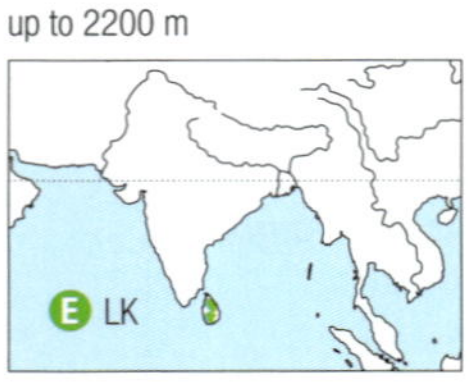

Except in dry and arid zones in E & S

Purple-faced Langur
Semnopithecus vetulus

FR Langur blanchâtre
DE Weißbartlangur
ES Langur púrpura

EN
4 sspp
HB 45–65 cm
T 62–85 cm
W 3.4–9.4 kg

ssp *monticola*

ssp *philbricki*

ssp *nestor*

ssp *vetulus*

Nilgiri Langur
Semnopithecus johnii

FR Langur des Nilgiri
DE Nilgiri-Langur
ES Langur de Nilgiri

300–2400 m

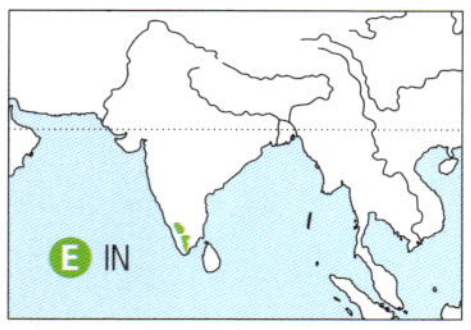

Karnataka, Kerala, Tamil Nadu

VU
Monotypic
HB 58–78 cm
T 68–96 cm
W 9.8–13.6 kg

Genus *TRACHYPITHECUS*

(22 species)

Golden Langur
Trachypithecus geei

FR Langur doré
DE Goldlangur
ES Langur dorado

EN
Monotypic
HB 49–72 cm
T 71–94 cm
W 9.5–12 kg

up to 3000 m

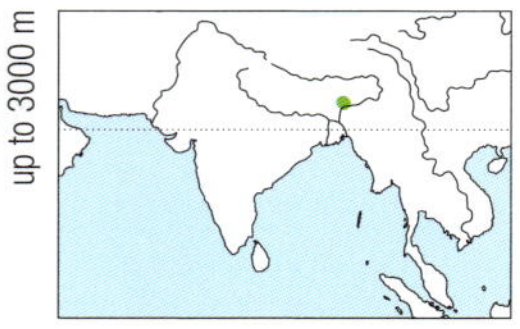

NE IN (Himalayan foothills of NW Assam), BT (N to Black Mt, E of R Sankosh; also E of R Mangde)

Capped Langur
Trachypithecus pileatus

FR Langur à capuchon
DE Kappenlangur
ES Langur encapuchado

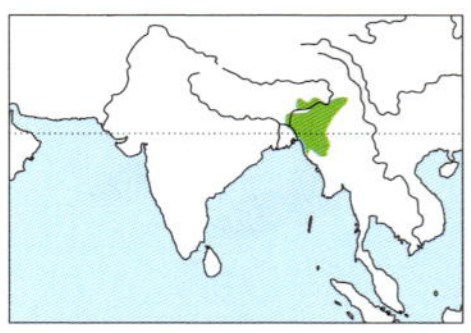

NE IN (Arunachal Pradesh, Assam, Meghalaya, Nagaland), BT, NW MM (W of R Chindwin, S to Chin Hills)

VU
3 sspp
HB 47–72 cm
T 72–104 cm
W 9.5–14 kg

ssp *tenebricus*

ssp *pileatus*

ssp *brahma*

Shortridge's Langur
Trachypithecus shortridgei

FR Langur de Shortridge
DE Shortridge-Langur
ES Langur de Birmania

EN
Monotypic
HB 67–72 cm
T 95–104 cm

200–2500 m

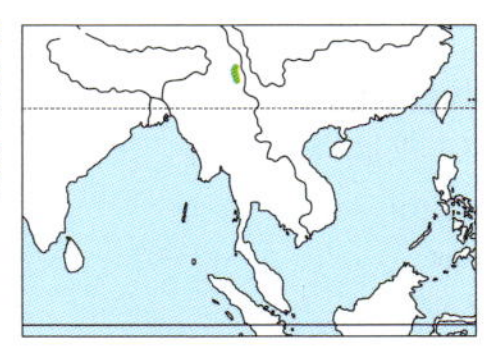

NE MM (E of R Chindwin, Kachin State), SW CN (R Dulong Valley, NW Yunnan)

East Javan Langur
Trachypithecus auratus

FR Langur noir
DE Schwarzer Haubenlangur
ES Langur de Java oriental

dark morph

VU
Monotypic
HB 44–65 cm (M)
T 61–87 cm (M)
W c. 7.1 kg (M)

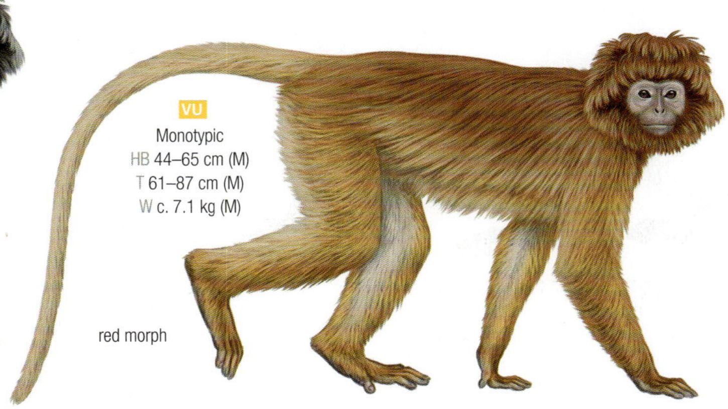

red morph

up to 3500 m

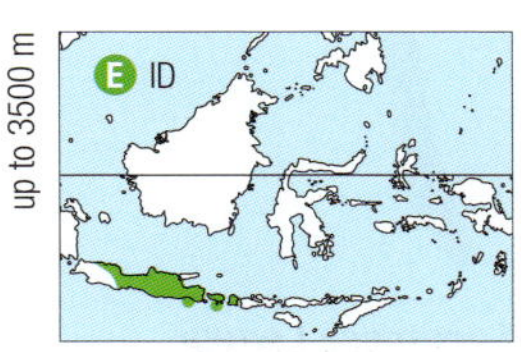

C & E Java (with Sempu I and Nusa Barung I, off S coast), Bali, Lombok

West Javan Langur

Trachypithecus mauritius

FR Langur de Griffith
DE Westlicher Haubenlangur
ES Langur de Java occidental

VU
Monotypic

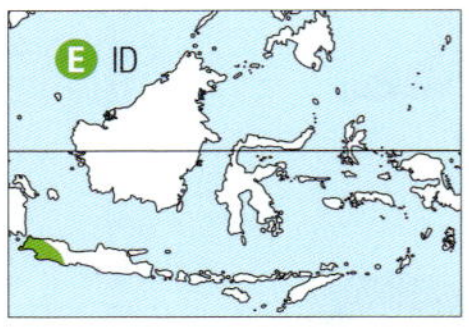

W Java (N coast W of Jakarta, inland to Bogor, Cisalak and Jasinga, S coast E to Cikaso and possibly Ciwangi)

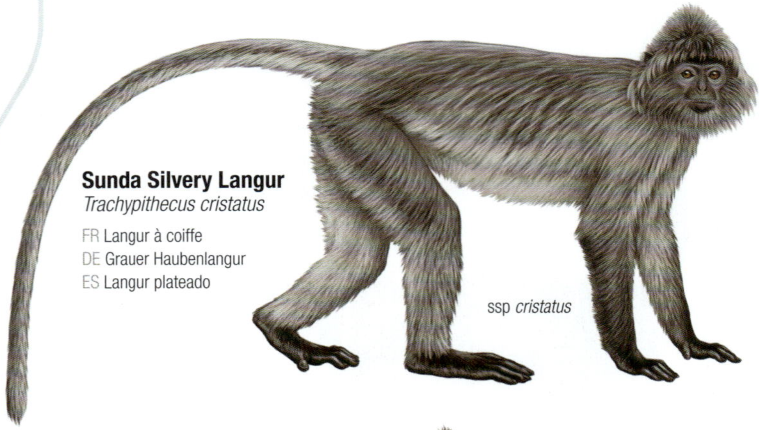

Sunda Silvery Langur

Trachypithecus cristatus

FR Langur à coiffe
DE Grauer Haubenlangur
ES Langur plateado

VU
2 sspp
HB 46–58 cm
T 66–75 cm
W 5.7–6.6 kg

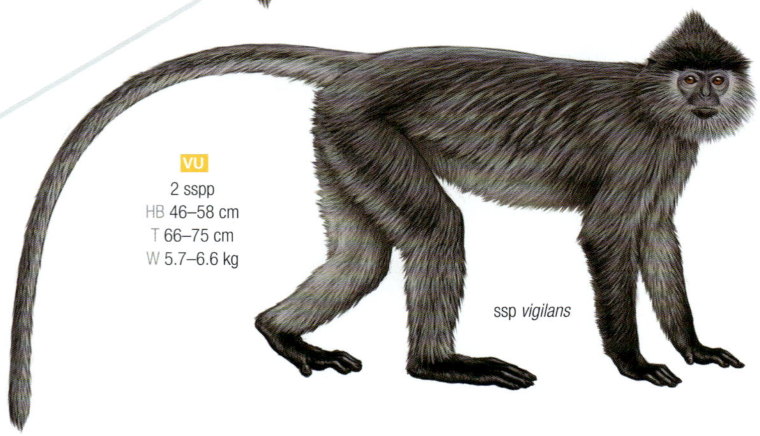

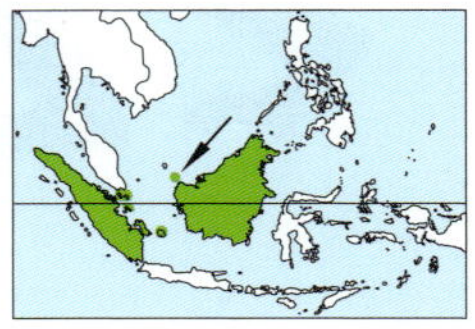

Borneo, Sumatra, Bangka, Belitung, Riau (Bintan, Combol, Sugi, perhaps Batam) and Lingga (Bakung, Lingga) arch, Natuna Is (Serasan)

NT
Monotypic
HB 46–53 cm
T 68–82 cm

Selangor Silvery Langur

Trachypithecus selangorensis

FR Langur du Selangor
DE Selangor-Langur
ES Langur de Malasia

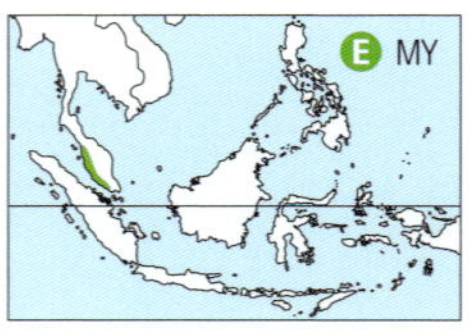

Strip along W coastal Pen from 6° N to 1° 30' N (Kedah, Perak, Selangor, Negeri Sembilan, Melaka, Johor)

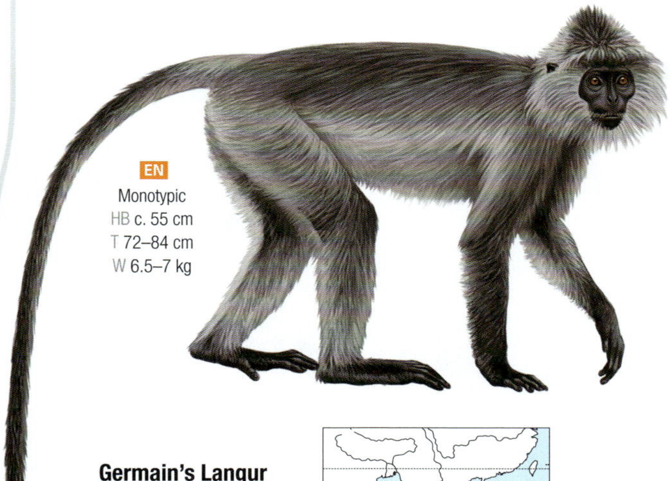

EN
Monotypic
HB c. 55 cm
T 72–84 cm
W 6.5–7 kg

Germain's Langur

Trachypithecus germaini

FR Langur de Germain
DE Germain-Langur
ES Langur de Germain

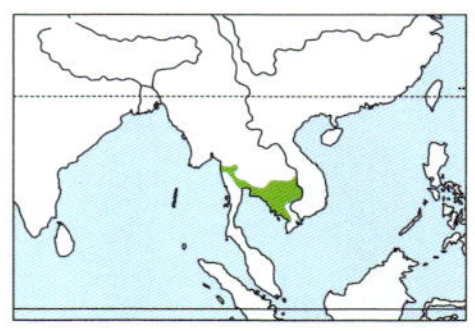

S MM, S TH, S LA, KH (W of R Mekong), S tip of VN; E limit might be R Mekong

EN
Monotypic

Annamese Langur

Trachypithecus margarita

FR Langur d'Elliot
DE Annam-Langur
ES Langur annamita

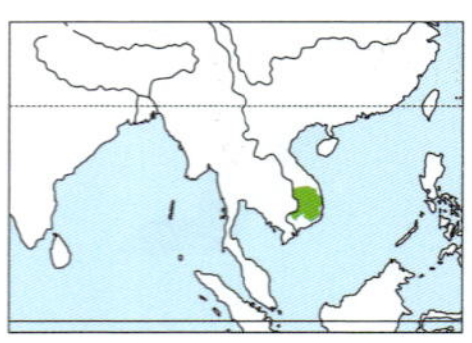

S LA (N limit 16° 23' N), SC VN (N limit 14° 30' N), E KH (Ratanakiri & Mondulkiri prov); W limit might be R Mekong

VU
Monotypic

Tenasserim Langur

Trachypithecus barbei

FR Langur de Barbe
DE Tenasserim-Brillenlangur
ES Langur de Tenasserim

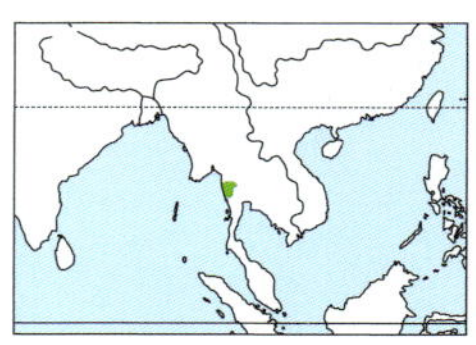

Confined to small region N of Thai-Malay Pen (E MM, W TH)

Dusky Langur
Trachypithecus obscurus

FR Langur obscur
DE Brillenlangur
ES Langur de anteojos

EN
7 sspp
HB 42–68 cm
T 57–81 cm
W 5–9.1 kg

ssp *flavicauda*

ssp *halonifer*

ssp *obscurus*

up to 1800 m

S MM (S of c. 15° 10' N, Mergui Arch), Thai-Malay Pen in SW TH and MY (including several islands)

ssp *sanctorum*

Phayre's Langur
Trachypithecus phayrei

FR Langur de Phayre
DE Phayre-Brillenlangur
ES Langur de Phayre

EN
Monotypic
HB 45–58 cm
T 65–77 cm
W 5.7–9.1 kg

E BD, NE IN (Assam, Mizoram, Tripura), W MM (SE through Arakan to Pegu)

Shan Langur
Trachypithecus melamera

FR Langur shan
DE Shan-Brillenlangur
ES Langur de Shan

NE
Monotypic
HB 55–60 cm
T 73–80 cm
W 5.7–9.1 kg (M), 6.4–7.5 kg (F)

SC CN (W Yunnan), N & E MM (Shan States and neighboring dry zone of N MM)

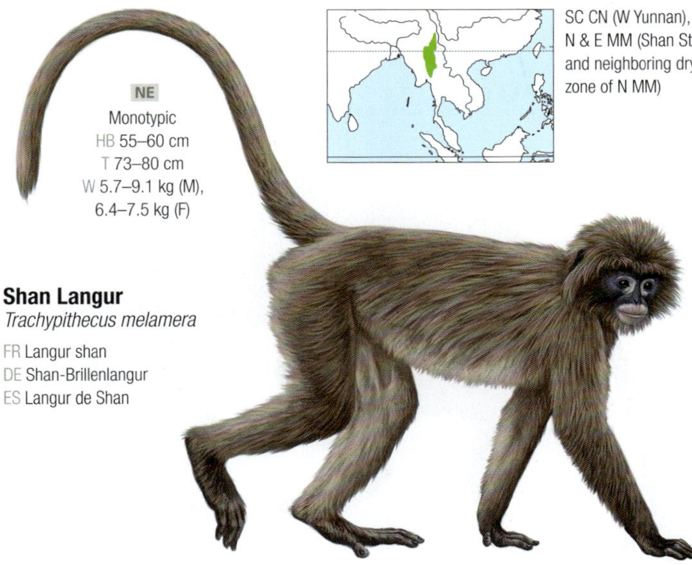

Popa Langur
Trachypithecus popa

FR Langur du Popa
DE Popa-Brillenlangur
ES Langur de Popa

CR
Monotypic
HB 50–60 cm
T 77–86 cm
W 7–8.2 kg

100–1500 m

Between R Ayeyarwaddy and R Thanlwin, also W foothills of Kayah-Karen Mts; NE limit unclear

Indochinese Gray Langur
Trachypithecus crepusculus

FR Langur gris
DE Indochina-Brillenlangur
ES Langur gris de Indochina

EN
Monotypic
HB 49–51 cm
T 82–83 cm
W 6.4–6.9 kg

SC CN (E of R Salween, S Yunnan), S MM, N TH (S to Raheng, W to Bay of Bengal), N & C LA, N VN

Cat Ba Langur
Trachypithecus poliocephalus

FR Langur à tête dorée
DE Cat-Ba-Langur
ES Langur de Cat Ba

CR
Monotypic
HB 49–59 cm
T 82–92 cm
W 6.7–9.5 kg

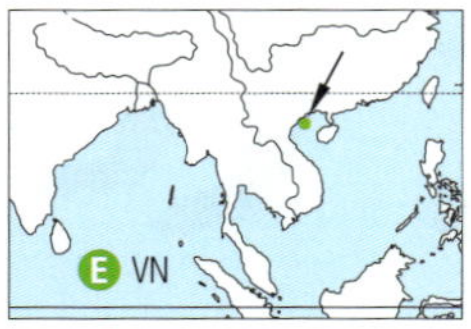

Cat Ba I in Ha Long Bay, off Quang Ninh Prov

White-headed Langur
Trachypithecus leucocephalus

FR Langur à tête blanche
DE Weißkopflangur
ES Langur de cabeza blanca

CR
Monotypic
HB 47–62 cm
T 77–89 cm
W 6.7–9.5 kg

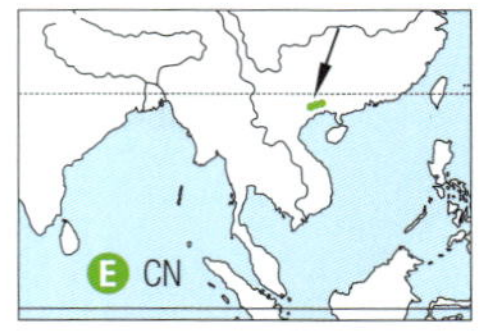

SW Guangxi; bordered in NW & N by R Zuojiang, in SW & S by R Mingjiang, in SE by Shiwan Mts

Delacour's Langur
Trachypithecus delacouri

FR Langur de Delacour
DE Delacour-Langur
ES Langur de Delacour

CR
Monotypic
HB 52–62 cm
T 81–90 cm
W 7.3–8.5 kg

up to 1000 m

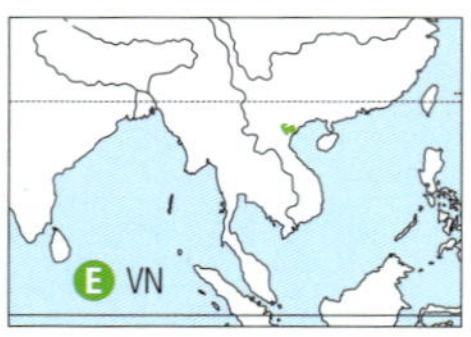

Hoa Binh, Ha Nam, Ninh Binh and Thanh Hoa prov

François's Langur
Trachypithecus francoisi

FR Langur de François
DE Tonkin-Langur
ES Langur de François

EN
Monotypic
HB 50–63 cm
T 74–90 cm
W 5.5–7.9 kg

up to 600 m

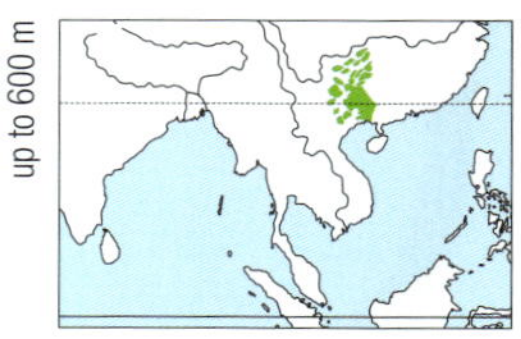

SC CN (Chongqing, Guizhou, Guangxi), N VN (Ha Giang, Cao Bang, Tuyen Quang, Bac Kan, Thai Nguyen prov)

Black Langur
Trachypithecus ebenus

FR Langur ébène
DE Khammuan-Langur
ES Langur negro

NE
Monotypic
HB c. 62 cm
T c. 94 cm
W c. 10.3 kg (M)

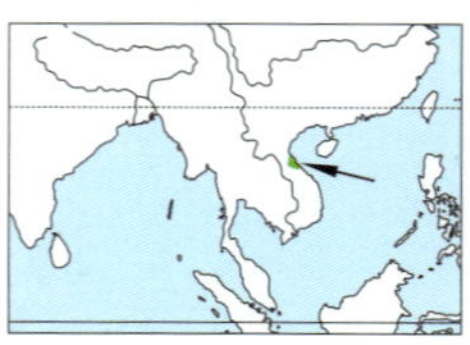

EC LA (S Khammouane Prov, scattered limestone blocks in N Savannakhet Prov), C VN (Quang Binh Prov)

Laos Langur
Trachypithecus laotum

FR Langur du Laos
DE Laos-Langur
ES Langur de Laos

EN
Monotypic
HB 46–67 cm
T 81–90 cm
W 6–8 kg

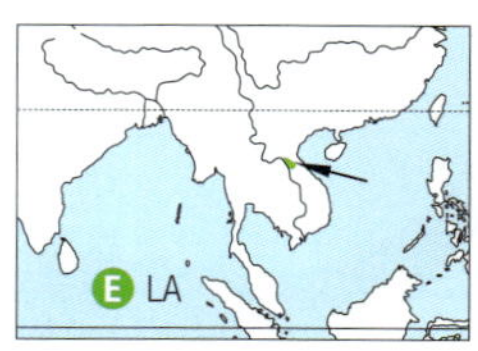

From S Bolikhamsai Prov, probably with R Nam Kading as N limit, S to C Khammouane Prov (c. 17° 40′ N)

Hatinh Langur
Trachypithecus hatinhensis

FR Langur du Hà Tinh
DE Hatinh-Langur
ES Langur de Hatinh

EN
Monotypic
HB 54–61 cm
T 78–95 cm
W 6–9 kg

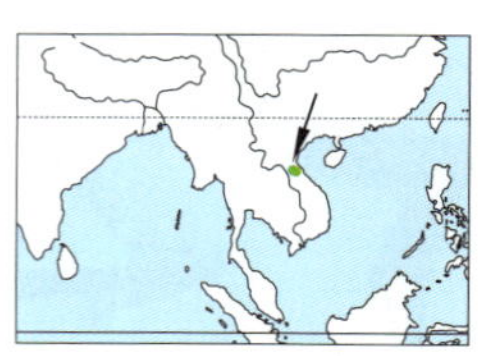

NC VN (Quang Binh and Quang Tri prov), EC LA (Khammouane Prov)

Family HYLOBATIDAE – Gibbons

4 genera, 20 species. Tropical evergreen rainforests, including lowland, hill, submontane, and tropical semi-evergreen forests. Indo-Malayan Region.

Genus *HOOLOCK*

(3 species)

Western Hoolock Gibbon
Hoolock hoolock

FR Gibbon hoolock
DE Westlicher Weißbrauengibbon
ES Gibón huloc occidental

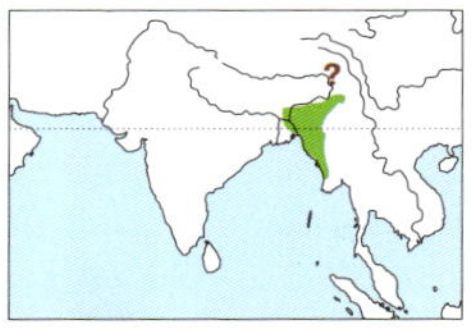

500–2700 m

BD, NE IN (S of R Brahmaputra, E of R Dibang), NW MM (W of R Chindwin); perhaps extreme SE Tibet (Medog NR)

EN
2 sspp
HB c. 81 cm
W c. 6.9 kg (M), c. 6.1 kg (F)

♀

♂

Eastern Hoolock Gibbon
Hoolock leuconedys

FR Gibbon de Groves
DE Östlicher Weißbrauengibbon
ES Gibón huloc oriental

VU
Monotypic
HB 45–65 cm
W 6–9 kg

140–2300 m

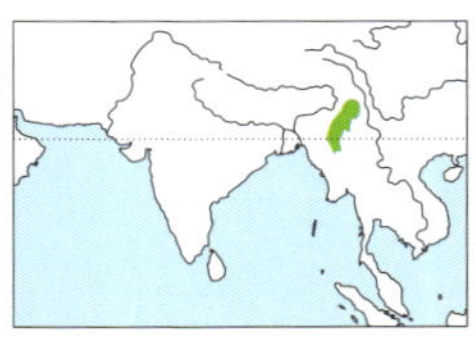

NE IN (Arunachal Pradesh and Assam), E MM (between R Chindwin and R Irrawaddy)

♀

♂

Skywalker Hoolock Gibbon
Hoolock tianxing

FR Gibbon du Gaoligong
DE Gaoligong-Weißbrauengibbon
ES Gibón de Gaoligong

EN
Monotypic

♀

♂

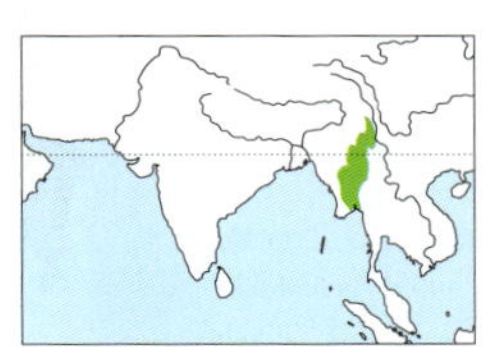

MM (between Irrawaddy-Nmai Hka and Salween rivers, historically further S at Gokteik, Shan State), CN (Mt Gaoligong, Sichuan)

Genus *HYLOBATES*

(9 species)

Pileated Gibbon

Hylobates pileatus

FR Gibbon à bonnet
DE Kappengibbon
ES Gibón crestado

up to 1500 m

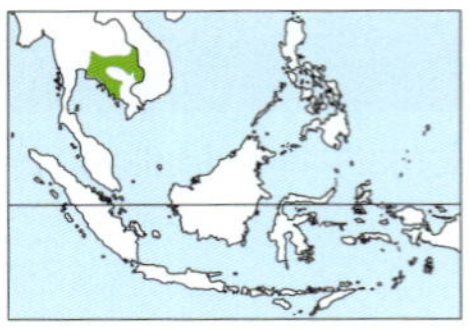

SE TH (E & S of R Mun and R Takhong), W of R Mekong in SW LA and N & W KH; perhaps formerly S VN

EN
Monotypic

Lar Gibbon

Hylobates lar

FR Gibbon lar
DE Weißhandgibbon
ES Gibón de manos blancas

up to 1200 m

SC CN (SW Yunnan), E & S MM, NW LA, W TH, Thai-Malay Pen, ID (N Sumatra)

EN
5 sspp
HB c. 41 cm (M), c. 42 cm (F)
W 4.1–7.3 kg (M), 3.9–6.1 kg (F)

Agile Gibbon

Hylobates agilis

FR Gibbon agile
DE Schwarzhandgibbon
ES Gibón ágil

EN
Monotypic
W 5–7.5 kg (M), 4.5–7.3 kg (F)

up to 1400 m

S TH (near Malaysian border, E of R Thepha), N Pen MY (from R Mudah and R Thepha S to R Perak and R Kelantan), ID (most of Sumatra)

Bornean White-bearded Gibbon

Hylobates albibarbis

FR Gibbon à barbe blanche
DE Weißbartgibbon
ES Gibón barbiblanco

EN
Monotypic
W 4.9–6.8 kg

up to 1200 m

SW Borneo (West & Central Kalimantan), S of R Kapua, W of R Barito

Southern Gray Gibbon
Hylobates muelleri

FR Gibbon de Müller
DE Müller-Gibbon
ES Gibón de Müller

EN
Monotypic
HB c. 41 cm
W 4.6–6.8 kg

variants

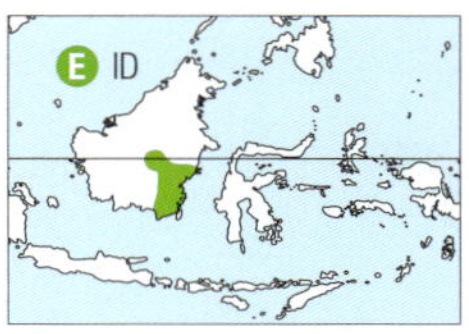

Borneo (SE Kalimantan), S of R Mahakam, E of R Barito

Western Gray Gibbon
Hylobates abbotti

FR Gibbon d'Abbott
DE Westlicher Borneo-Gibbon
ES Gibón de Abbott

up to c. 1700 m

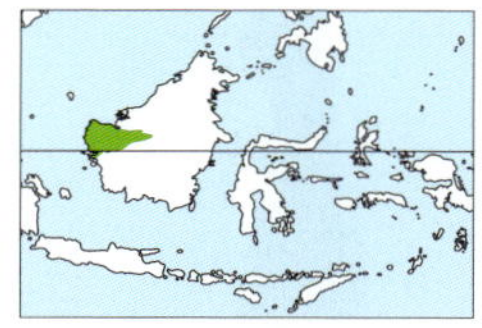

SW Borneo: S Sarawak E to Spaoh (MY), West Kalimantan N of R Kapuas (ID)

EN
Monotypic
W 5.5–6.4 kg

Eastern Gray Gibbon
Hylobates funereus

FR Gibbon de Geoffroy
DE Östlicher Borneo-Gibbon
ES Gibón de Sabah

EN
Monotypic
HB 47–49 cm
W 5–6.4 kg

up to c. 1700 m

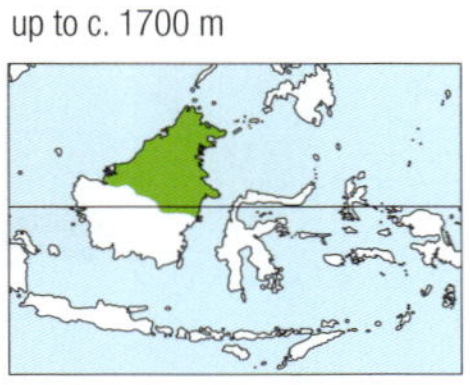

N & NE Borneo: from Sabah (MY) S to R Mahakam in East Kalimantan (ID); perhaps W to Baram District and IV Division of Sarawak (MY)

Moloch Gibbon
Hylobates moloch

FR Gibbon cendré
DE Silbergibbon
ES Gibón plateado

up to 2400 m

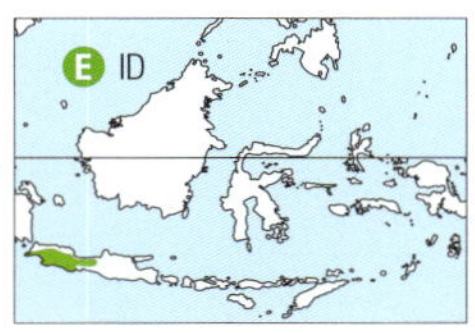

Mainly W Java (Banten & West Java prov), additional small area in C, E to Dieng Mts

EN
Monotypic
W 6.2–6.6 kg

Kloss's Gibbon
Hylobates klossii

FR Gibbon de Kloss
DE Kloss-Gibbon
ES Gibón de Kloss

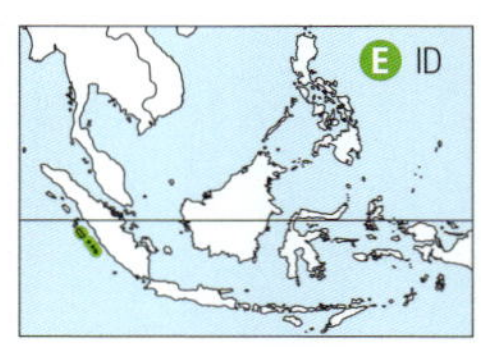

Mentawai Is (Siberut, Sipora, North Pagai, South Pagai and nearby Sinakak)

EN
Monotypic
HB 42–58 cm
W 4.4–7.8 kg

Genus *SYMPHALANGUS*

(1 species)

1500–2300 m

Extreme S TH, NW & C Pen MY, ID (W Sumatra, confined to Barisan Mts); perhaps formerly Bangka I

Siamang
Symphalangus syndactylus

FR Siamang
DE Siamang
ES Siamang

EN
Monotypic
HB 75–90 cm
W 9.1–12.7 kg

Genus *NOMASCUS*

(7 species)

Hainan Crested Gibbon
Nomascus hainanus

FR Gibbon de Hainan
DE Hainan-Schopfgibbon
ES Gibón de Hainan

700–1200 m

W Hainan I (once widespread, today restricted to Bawangling NR)

CR
Monotypic
HB c. 49 cm (M), c. 48 cm (F)
W 5.8–10 kg

♂ ♀

Eastern Black Crested Gibbon
Nomascus nasutus

FR Gibbon cao-vit
DE Östlicher Schopfgibbon
ES Gibón de cresta negra oriental

500–930 m

Today restricted to a small area in NE VN (Trung Khanh District, Cao Bang Prov) and CN (Jingxi County, Guangxi)

CR
Monotypic

♂ ♀

CR
2 sspp
HB 43–54 cm
W 6.9–10 kg

500–2900 m

SC CN (C & SW Yunnan), N VN (Lao Cai, Son La & Yen Bai prov), NW LA (Bokeo & Luang Namtha prov)

♀ ♂

Western Black Crested Gibbon
Nomascus concolor

FR Gibbon noir
DE Westlicher Schopfgibbon
ES Gibón de cresta negra

Northern White-cheeked Crested Gibbon

Nomascus leucogenys

FR Gibbon à joues blanches
DE Nördlicher Weißwangengibbon
ES Gibón de mejillas blancas septentrional

♀

♂

CR
Monotypic
HB 47–52 cm
W c. 6.4 kg (M),
c. 6.6 kg (F)

200–1650 m

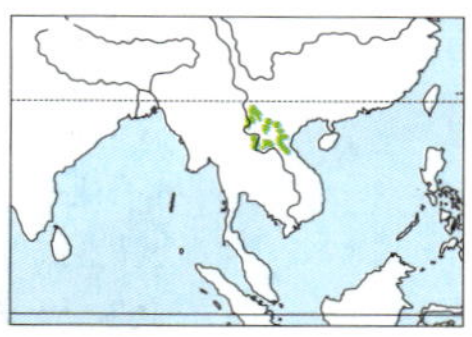

SC CN (extreme S Yunnan), N LA (S possibly to R Khading), NW VN (between R Mekong and R Black, S perhaps to R Rao Nay)

Southern White-cheeked Crested Gibbon

Nomascus siki

FR Gibbon siki
DE Südlicher Weißwangengibbon
ES Gibón de mejillas blancas meridional

CR
Monotypic
W c. 6.6 kg (M),
c. 7.9 kg (F)

♂

♀

30–1800 m

C LA (E of R Mekong, N to R Khading), C VN (N to R Rao Nay, S to R Thach Han)

EN
Monotypic
HB c. 46 cm (M),
c. 44 cm (F)
W c. 7 kg (M)

Northern Yellow-cheeked Crested Gibbon

Nomascus annamensis

FR Gibbon annamite
DE Nördlicher Gelbwangengibbon
ES Gibón de mejillas beige septentrional

50–1205 m

S LA (E of R Mekong, up to about R Banhiang), C VN (from R Thach Han S to about R Ba), NE KH (E of R Mekong, N of R Srepok)

♂

♀

Southern Yellow-cheeked Crested Gibbon

Nomascus gabriellae

FR Gibbon de Gabrielle
DE Südlicher Gelbwangengibbon
ES Gibón de mejillas beige meridional

EN
Monotypic
HB 45–49 cm (M)
46–47 cm (F)
W c. 5 kg (F)

♀

♂

100–2285 m

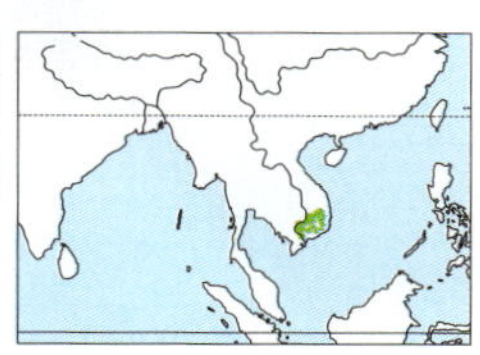

S VN (from R Ba S to Nui Ong NR), SE KH (E of R Mekong, S of R Srepok)

Family HOMINIDAE – Great Apes

4 genera, 8 species. Lowland to montane equatorial forests, swamps, and savanna-woodland mosaics. Afrotropical and Indo-Malayan Regions. *Homo sapiens* almost all terrestrial habitats and all regions.

Genus *PONGO*

(3 species)

Bornean Orangutan
Pongo pygmaeus

FR Orang-outan de Bornéo
DE Borneo-Orang-Utan
ES Orangután de Borneo

CR
3 sspp
HB 96–97 cm (M), 72–85 cm (F)
W 30–85 kg (M), 30–45 kg (F)

up to 750 m

Borneo: patchily in Sabah and Sarawak (MY) and North, East, Central & West Kalimantan (ID)

Tapanuli Orangutan
Pongo tapanuliensis

FR Orang-outan de Tapanuli
DE Tapanuli-Orang-Utan
ES Orangután de Tapanuli

CR
Monotypic

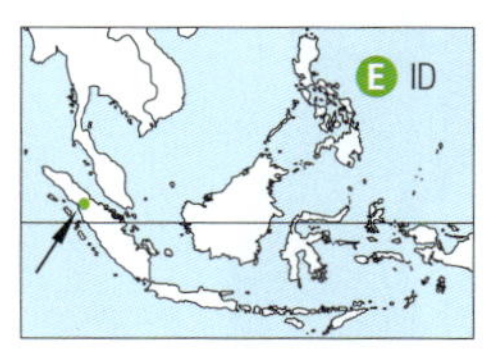

NW Sumatra (forest fragments in districts of C, N & S Tapanuli, S of L Toba)

Sumatran Orangutan
Pongo abelii

FR Orang-outan de Sumatra
DE Sumatra-Orang-Utan
ES Orangután de Sumatra

CR
Monotypic
HB 94–99 cm (M), 68–84 cm (F)
W 30–85 kg (M), 30–45 kg (F)

up to 1500 m

E ID

NW Sumatra (Aceh and North Sumatra prov), NW of L Toba

Western Gorilla
Gorilla gorilla

FR Gorille de l'Ouest
DE Westlicher Gorilla
ES Gorila occidental

Genus *GORILLA*

(2 species)

CR
2 sspp
HB 103–107 cm (M)
W 145–191 kg (M), 57–73 kg (F)

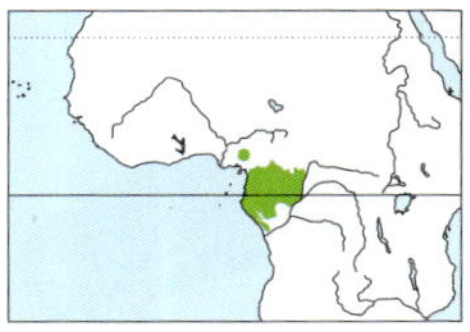

SE NG, W & S CM, SW CF, GQ, GA, CG, N AO (Cabinda)

♂

♀

ssp *beringei*

♂

♀

Eastern Gorilla
Gorilla beringei

FR Gorille de l'Est
DE Östlicher Gorilla
ES Gorila oriental

CR
2 sspp
HB 101–120 cm (M)
W 120–209 kg (M), 60–98 kg (F)

600–3800 m

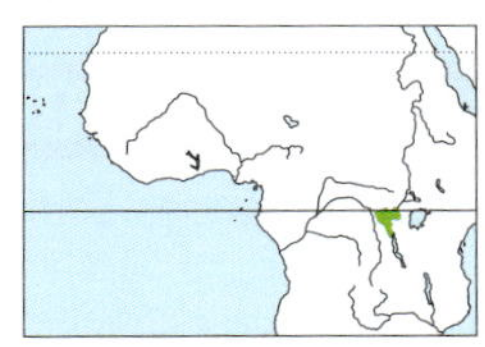

E CD (Virunga Volcanoes, Kahuzi-Biéga-Kasese, Itombwe Massif, Maiko-Tayna), SW UG (Bwindi Impenetrable NP), NW RW

ssp *graueri*

♂

♀

Human
Homo sapiens
FR Homme moderne
DE Mensch
ES Humano

Genus *HOMO*
(1 species)

NE
Monotypic
HB 150–185 cm (M), 140–170 cm (F)
W 50–99 kg (M), 45–98 kg (F)

0–6000 m

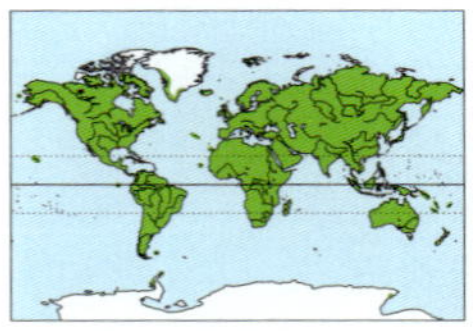

Cosmopolitan

up to 2790 m

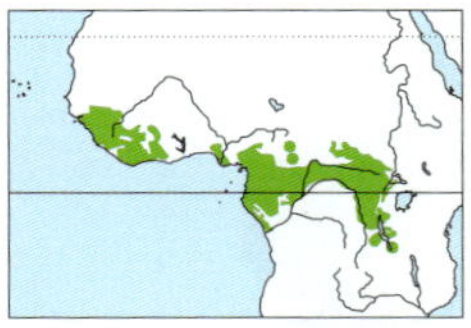

S SN and SW ML E to SW GH, also S NG and CM E to SW SS and W UG, S to N AO (Cabinda) and W TZ

EN
4 sspp
HB 70–96 cm
W 20–70 kg

Genus *PAN*
(2 species)

Chimpanzee
Pan troglodytes
FR Chimpanzé
DE Schimpanse
ES Chimpancé

Bonobo
Pan paniscus
FR Bonobo
DE Bonobo
ES Bonobo

EN
Monotypic
HB 70–83 cm
W 26–43 kg

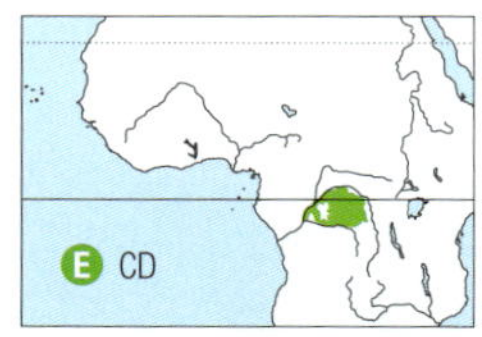

E and S of great bend of R Congo, E to R Lualaba, S to R Kasai and R Sankuru

Order LAGOMORPHA (2 families)

Family OCHOTONIDAE – Pikas

1 genus, 34 species. Semi-deserts, steppes, meadows, and forests in mountains, from elevations as low as 100 m to alpine zones at 5700 m. Holarctic Region.

Genus *OCHOTONA*
(34 species)

Steppe Pika
Ochotona pusilla
FR Pika des steppes
DE Steppenpfeifhase
ES Pica de estepa

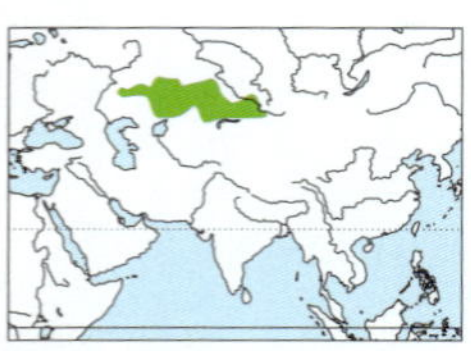

Steppe belt from Obshchy Syrt, S of Ural & Mugodzhar mts, E to W Altai (RU, KZ)

LC
Monotypic
HB 15–21 cm
W 95–270 g

American Pika
Ochotona princeps
FR Pika d'Amérique
DE Amerikanischer Pfeifhase
ES Pica americana

1200–3500 m

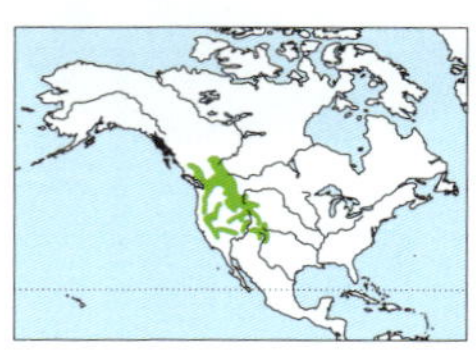

Patchily across mts of SW CA (S British Columbia, SW Alberta) and W US

LC
5 sspp
HB 16–20 cm
W 121–176 g

Collared Pika
Ochotona collaris
FR Pika à collier
DE Alaska-Pfeifhase
ES Pica de collar

500–2000 m

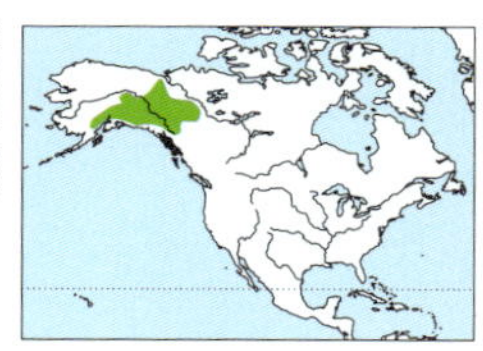

SE Alaska (US), NW CA (Yukon, Northwest Territories, NW British Columbia), W of R Mackenzie

LC
Monotypic
HB 17–19 cm
W 117–200 g

Pallas's Pika
Ochotona pallasii

FR Pika de Pallas
DE Mongolei-Pfeifhase
ES Pica de Pallas

1700–3100 m

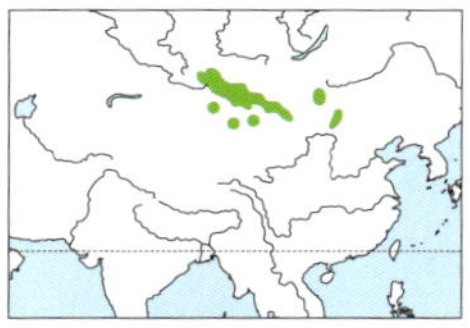

Altai Mts (S RU, N Xinjiang, W MN); disjunctly in S & E MN (Trans-Altai Gobi) & N CN (Dzungaria, Gobi)

LC
Monotypic
HB 16–23 cm
W 150–270 g

Kazakh Pika
Ochotona opaca

FR Pika du Kazakhstan
DE Kasachstan-Pfeifhase
ES Pica de Kazajistán

LC
Monotypic
HB 15–20 cm
W 100–173 g

400–1000 m

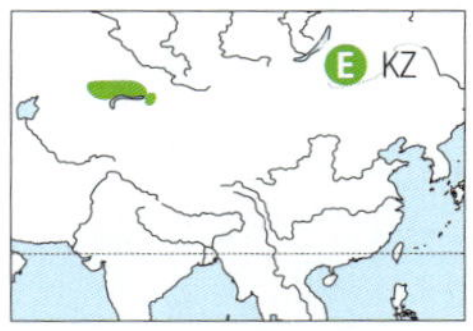

Uplands N, W & E of L Balkhash

Silver Pika
Ochotona argentata

FR Pika argenté
DE Silber-Pfeifhase
ES Pica plateada

2000–2300 m

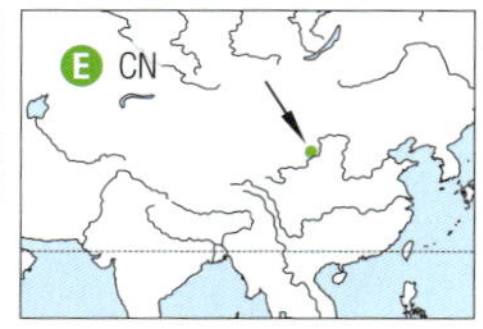

C Helan Mts (Suyukou and surroundings, Ningxia)

summer

EN
Monotypic
HB 20–24 cm
W 230–250 g

winter

Korean Pika
Ochotona coreana

FR Pika de Corée
DE Korea-Pfeifhase
ES Pica de Corea

DD
Monotypic
HB 16–20 cm
W 144–190 g

700–2500 m

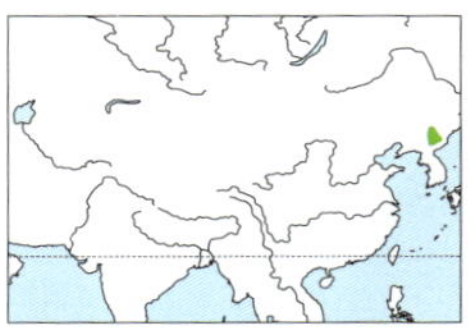

Paektu Mt and surrounding area in KP & NE CN (Jilin)

400–1300 m

CN (Greater/Lesser Khingan Ranges), RU (R Shilka & R Argun interfluves, SE Transbaikalia)

Manchurian Pika
Ochotona mantchurica

FR Pika de Mandchourie
DE Manschurischer Pfeifhase
ES Pica de Manchuria

LC
3 sspp
HB 14–22 cm
W 110–260 g

Hoffmann's Pika
Ochotona hoffmanni

FR Pika de Hoffmann
DE Hoffmann-Pfeifhase
ES Pica de Hoffmann

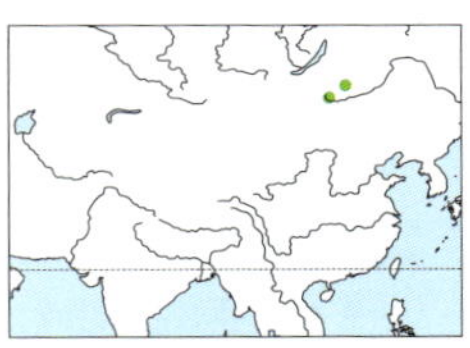

S RU (S Ermana Range in Transbaikalia), NE MN (S Ereenii Nuruu Range in Bayan Ulan Mts)

EN
Monotypic
HB 19–21 cm
W 260–310 g

Northern Pika
Ochotona hyperborea

FR Pika de Sibérie
DE Nördlicher Pfeifhase
ES Pica de Siberia

200–2000 m

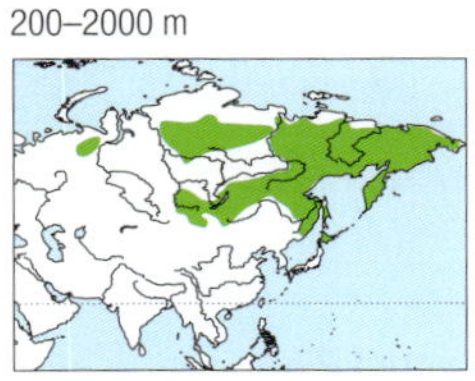

From Ural Mts across NE Eurasia (RU, MN, CN), Hokkaido (JP), Bering Sea & Sea of Okhotsk Is

LC
8 sspp
HB 13–18 cm
W 52–165 g

Turuchan Pika
Ochotona turuchanensis

FR Pika de la Touroukhan
DE Turuchan-Pfeifhase
ES Pica de Turujanks

100–1100 m

C Siberia between middle and lower R Yenisei and R Lena

LC
Monotypic
HB 16–22 cm
W 113–171 g

LC
3 sspp
HB 19–25 cm
W 150–360 g

Alpine Pika
Ochotona alpina

FR Pika de l'Altaï
DE Altai-Pfeifhase
ES Pica de Altái

400–3100 m

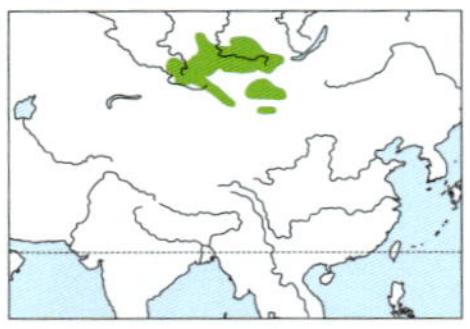

Mountain ranges in S RU, E KZ, W & C MN and NW CN (Xinjiang)

LC
3 sspp
HB 14–22 cm
W 103–190 g

Daurian Pika
Ochotona dauurica

FR Pika de Daourie
DE Daurien-Pfeifhase
ES Pica de Dauria

800–3300 m

S RU (SE Altai, S Transbaikalia), MN, N & C CN (S to Shaanxi & W Henan)

LC
Monotypic
HB 12–17 cm
W 40–115 g

Thomas's Pika
Ochotona thomasi

FR Pika de Thomas
DE Thomas-Pfeifhase
ES Pica de Thomas

2800–4100 m

NE & E Tibetan Plateau (Qilian, Burhan Budai, Amne Machin, Bayan Har Mts)

LC
3 sspp
HB 12–18 cm
W 55–105 g

Moupin Pika
Ochotona thibetana

FR Pika du Moupin
DE Tibet-Pfeifhase
ES Pica de Moupin

2100–4300 m

CN (Hengduan Mts in SE Tibet, SE Qinghai, W Sichuan & NW Yunnan), N MM

NE
Monotypic

Qionglai Pika
Ochotona qionglaiensis

FR Pika des Qionglai
DE Qionglai-Pfeifhase
ES Pica de Qionglai

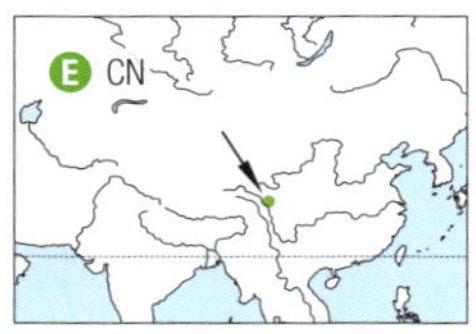

Forests in SW Sichuan

LC
3 sspp
HB 11–17 cm
W 45–95 g

Gansu Pika
Ochotona cansus

FR Pika du Gansu
DE Gansu-Pfeifhase
ES Pica de Gansu

1700–4700 m

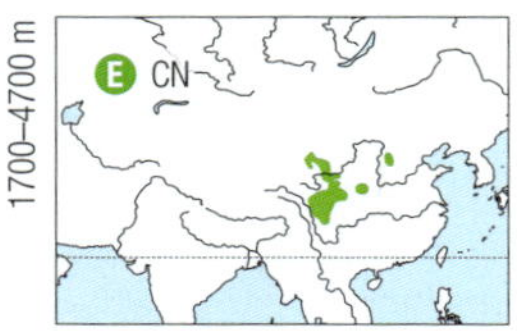

Qilian Mts, Mts NE of Tibetan Plateau, Hengduan, Qin Ling & Luyashan Mts

LC
2 sspp
HB 14–21 cm
W 90–210 g

Plateau Pika
Ochotona curzoniae

FR Pika à lèvres noires
DE Schwarzlippen-Pfeifhase
ES Pica de altiplano

3200–4800 m

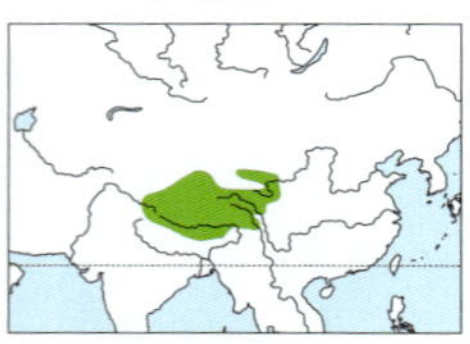

CN (Tibetan Plateau), N IN (Sikkim); probably extreme N NP

LC
2 sspp
HB 15–18 cm
W 108–135 g

Nubra Pika
Ochotona nubrica

FR Pika du Nubra
DE Nubra-Pfeifhase
ES Pica de Nubra

3100–4600 m

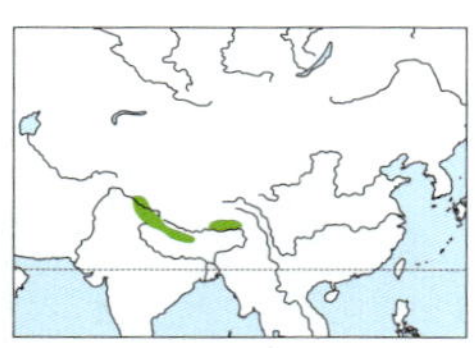

Ladakh E across S Tibetan Plateau (N IN, N NP, S CN)

NE
Monotypic
HB 14.6 cm
W 67 g

Sijin Pika
Ochotona sikimaria

FR Pika du Sikkim
DE Sikkim-Pfeifhase
ES Pica de Sikkim

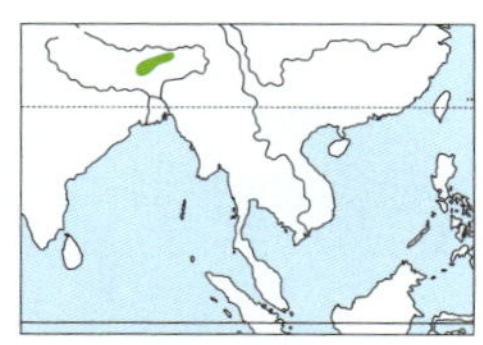

Sikkim (N IN), W BT & adjacent Tibetan Plateau (CN)

LC
2 sspp
HB 13–17 cm
W 60–110 g

Tsing-ling Pika
Ochotona syrinx

FR Pika des Qinling
DE Tsing-Ling-Pfeifhase
ES Pica de Tsing-ling

1800–3100 m

Sichuan Basin borders in Qin Ling & Daba Mts

NE
Monotypic
HB 14.8 cm
W c. 79 g

Sacred Pika
Ochotona sacraria

FR Pika de l'Emei
DE Heiliger Pfeifhase
ES Pica sagrada

1200–3100 m

Forests in SW Sichuan

NE
Monotypic
HB c. 15.1 cm
W c. 99 g

Yellow Pika
Ochotona huanglongensis

FR Pika de Huanglong
DE Gelber Pfeifhase
ES Pica amarilla

1800–3100 m

C Sichuan

Flat-headed Pika

Ochotona flatcalvariam

NE
Monotypic
HB c. 12.8 cm
W c. 71 g

FR Pika des Longmen
DE Flachkopf-Pfeifhase
ES Pica de cabeza plana

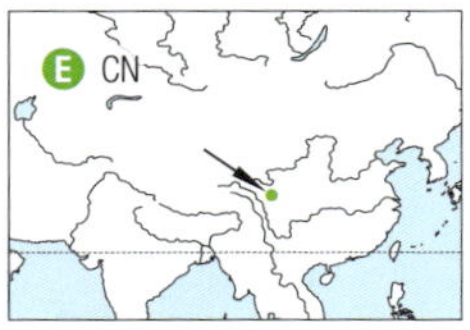

C Sichuan

Large-eared Pika

Ochotona macrotis

LC
6 sspp
HB 15–21 cm
W 160–280 g

FR Pika à oreilles longues
DE Großohr-Pfeifhase
ES Pica de orejas grandes

2500–5700 m

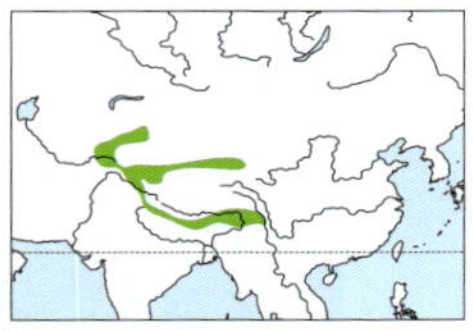

Tian Shan & Pamir-Alay Mts, Tibetan Plateau, Himalayas

Royle's Pika

Ochotona roylii

LC
2 sspp
HB 15–22 cm
W 130–200 g

FR Pika de Royle
DE Royle-Pfeifhase
ES Pica de Royle

2100–4500 m

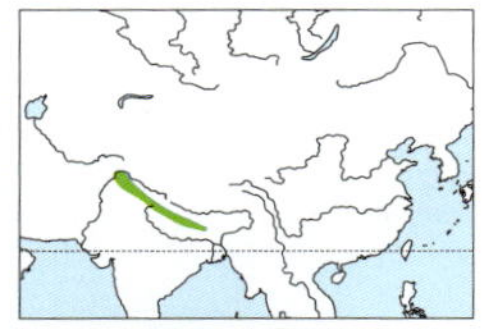

Himalayas through PK, Kashmir, N IN, NP & S Tibet

Afghan Pika

Ochotona rufescens

LC
3 sspp
HB 16–25 cm
W 155–230 g

FR Pika afghan
DE Afghanistan-Pfeifhase
ES Pica de Afganistán

200–2900 m

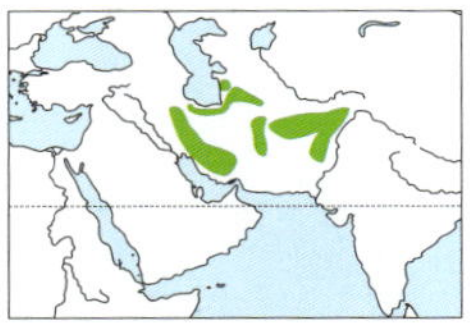

Mountains of IR, TM, AF & PK

Forrest's Pika

Ochotona forresti

LC
Monotypic
HB 15–19 cm
W 85–148 g

FR Pika de Forrest
DE Forrest-Pfeifhase
ES Pica de Forrest

2500–4400 m

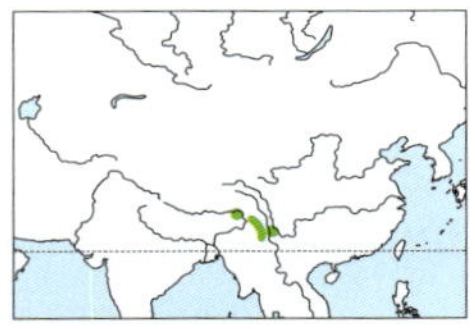

Namcha Barwa Himal (Tibet), Gaoligong Mts (CN/MM border), Mt Yulong (NW Yunnan)

Ladak Pika

Ochotona ladacensis

LC
Monotypic
HB 15–24 cm
W 100–320 g

FR Pika du Ladak
DE Ladakh-Pfeifhase
ES Pica de Ladakh

4400–5500 m

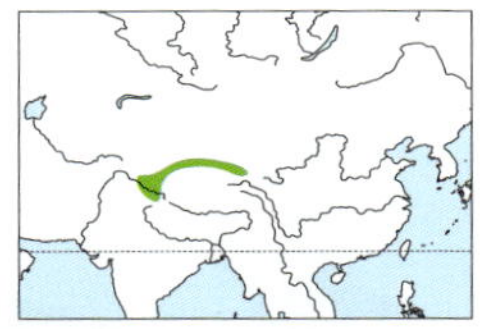

NW IN (Ladakh), Kunlun Mts in SW CN (S Xinjiang, NW Tibet, E Qinghai)

Turkestan Red Pika

Ochotona rutila

LC
Monotypic
HB 19–26 cm
W 220–320 g

FR Pika roux
DE Roter Pfeifhase
ES Pica roja de Turkestán

1800–3700 m

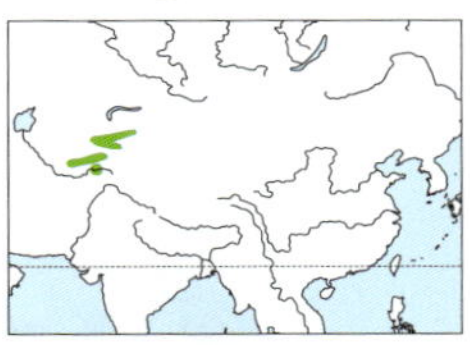

Tian Shan (from C Asia side) to Gissar ranges (SE KZ, KG, E UZ, TJ)

Kozlov's Pika

Ochotona koslowi

EN
Monotypic
HB 17–24 cm
W 100–300 g

FR Pika de Kozlov
DE Koslow-Pfeifhase
ES Pica de Kozlov

4100–4800 m

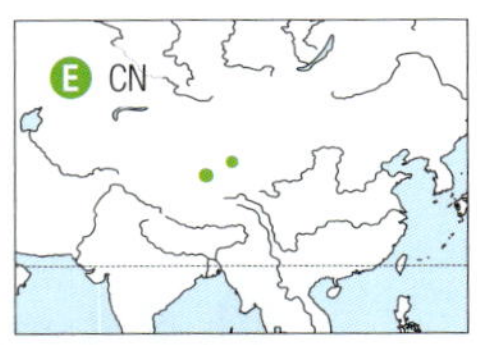

Grasslands and steppes in Kunlun Mts (S Xinjiang, Qinghai)

Chinese Red Pika

Ochotona erythrotis

LC
Monotypic
HB 17–23 cm
W 160–290 g

FR Pika à oreilles rouges
DE Rotohr-Pfeifhase
ES Pica roja de China

2000–4800 m

Upper R Yellow Basin and Qilian Mts (Qinghai, Gansu)

Glover's Pika

Ochotona gloveri

FR Pika de Glover
DE Glover-Pfeifhase
ES Pica de Glover

2200–4300 m

Upper and middle Yangtze/Mekong basins (SW Qinghai, NE Tibet, C Sichuan, NW Yunnan)

LC
2 sspp
HB 17–25 cm
W 125–350 g

Ili Pika

Ochotona iliensis

FR Pika de l'Ili
DE Ili-Pfeifhase
ES Pica de Ili

EN
Monotypic
HB 19–20 cm
W 200–250 g

2800–4100 m

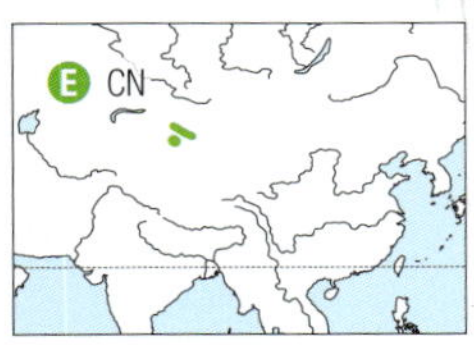

Borohoro Shan and Halke Shan in E Tian Shan (NW Xinjiang)

Family LEPORIDAE – Hares and Rabbits

10 genera, 62 species. Forests, shrubby areas, grasslands, deserts, tundra, alpine slopes, and agricultural landscapes up to 5400 m. Holarctic, Afrotropical, Indo-Malayan and Neotropical Regions.

Genus *PRONOLAGUS*

(4 species)

Jameson's Red Rock Hare
Pronolagus randensis
FR Lapin de Jameson
DE Rand-Wollschwanzhase
ES Liebre roja de Jameson

LC
3 sspp
HB 42–50 cm
T 6–13.5 cm
W 1.8–3 kg

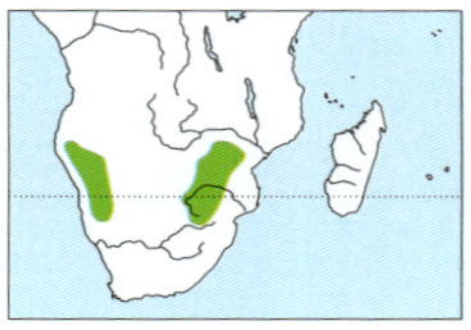

2 populations: C & NW NA (perhaps SW AO) and ZW to NE ZA, W MZ & E BW

Natal Red Rock Hare
Pronolagus crassicaudatus
FR Lapin du Natal
DE Natal-Wollschwanzhase
ES Liebre roja de Natal

LC
2 sspp
HB 46–56 cm
T 3.5–11 cm
W 2.4–3.1 kg

0–1550 m

Extreme S MZ, SZ, E & SE ZA (Mpumalanga, KwaZulu-Natal, Eastern Cape), E LS

Smith's Red Rock Hare
Pronolagus rupestris
FR Lapin de Smith
DE Rotkaninchen
ES Liebre roja de Smith

LC
4 sspp
HB 38–54 cm
T 5–11.5 cm
W 1.4–2.1 kg

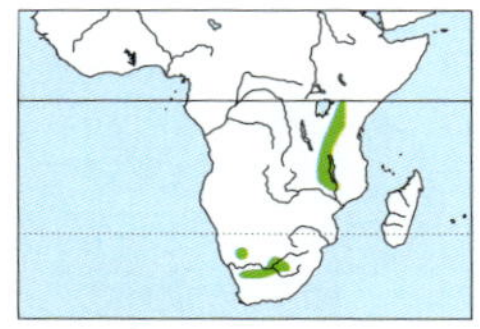

2 populations: SW KE to NE ZM and N & NW ZA; recorded in S NA

Hewitt's Red Rock Hare
Pronolagus saundersiae
FR Lapin de Hewitt
DE Hewitt-Wollschwanzhase
ES Liebre roja de Hewitt

LC
Monotypic
HB 38–54 cm
T 5–11.5 cm
W 1.4–2.1 kg

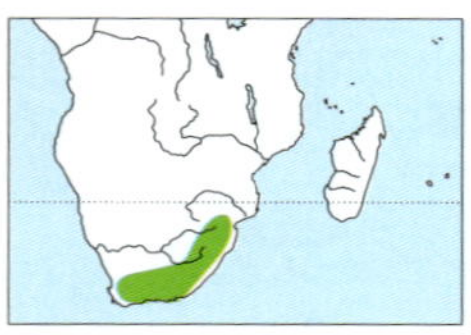

Rocky areas in ZA, LS, W SZ

Genus *POELAGUS*

(1 species)

LC
Monotypic
HB 40–61 cm
T 3.8–7 cm
W 2.3–3.2 kg

Bunyoro Rabbit
Poelagus marjorita
FR Lapin d'Afrique
DE Buschkaninchen
ES Conejo de Bunyoro

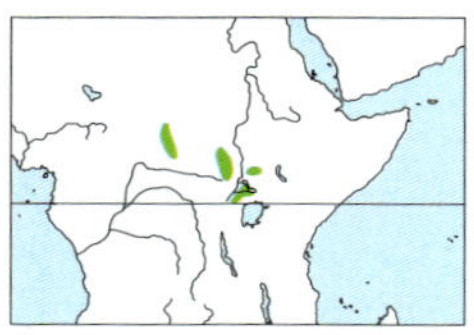

NE CF, S SS, W & C UG, NE CD

Genus *NESOLAGUS*

(2 species)

EN
Monotypic

Annamite Striped Rabbit
Nesolagus timminsi
FR Lapin d'Annam
DE Annam-Streifenkaninchen
ES Conejo rayado de Annamita

c. 200 m

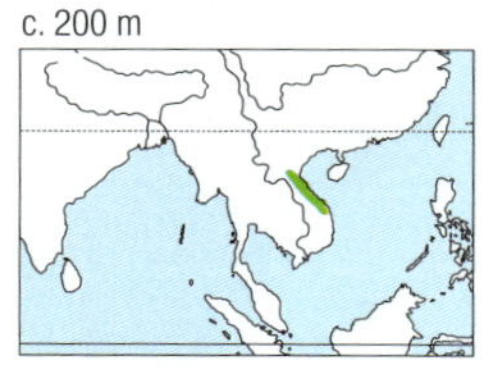

N & C Annamite Mts (LA, VN); perhaps S Annamite Mts

DD
Monotypic
HB 37–42 cm T c. 1.7 cm
W c. 1.5 kg

Sumatran Striped Rabbit
Nesolagus netscheri
FR Lapin de Sumatra
DE Sumatra-Streifenkaninchen
ES Conejo rayado de Sumatra

up to 1900 m

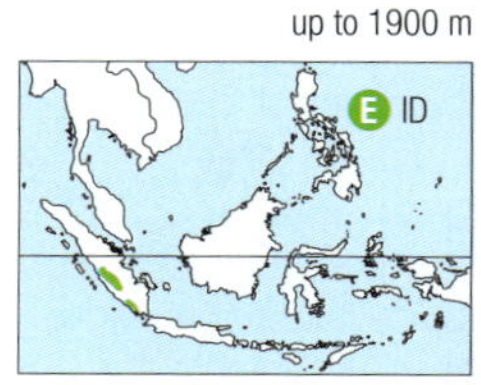

W Sumatra (Barisan Mts)

Genus *ROMEROLAGUS*

(1 species)

Volcano Rabbit
Romerolagus diazi
FR Lapin des volcans
DE Vulkankaninchen
ES Conejo de los volcanes

2800–4250 m

C Trans-Mexican Volcanic Belt (Distrito Federal, Morelos, W Puebla)

EN
Monotypic
HB 23–35 cm
T not visible
W 387–602 g

Genus *PENTALAGUS*

(1 species)

EN
Monotypic
HB 42–47 cm
T 1.1–3.5 cm
W 2–2.9 kg

Amami Rabbit
Pentalagus furnessi
FR Lapin des Amami
DE Ryukyu-Kaninchen
ES Conejo de Amami

0–695 m

Ryukyu Is (Amami Ōshima, Tokunoshima)

Genus *CAPROLAGUS*
(1 species)

Hispid Hare
Caprolagus hispidus
FR Lapin d'Assam
DE Borstenkaninchen
ES Liebre híspida

EN
Monotypic
HB c. 48 cm
T c. 5.3 cm
W 1.8–3.2 kg

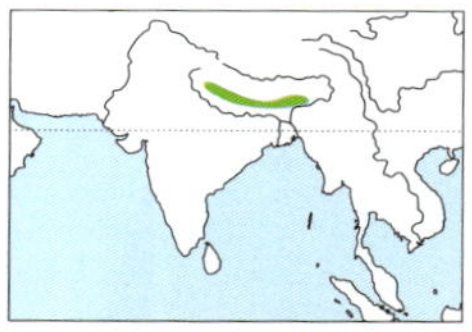
100–250 m

S Himalaya foothills in N & NE IN, S NP, NW BD; recorded in S BT

Genus *ORYCTOLAGUS*
(1 species)

European Rabbit
Oryctolagus cuniculus
FR Lapin de garenne
DE Wildkaninchen
ES Conejo europeo

EN
6 sspp
HB 36–38 cm
T 6.5–7 cm
W 1.5–3 kg

below 1500 m
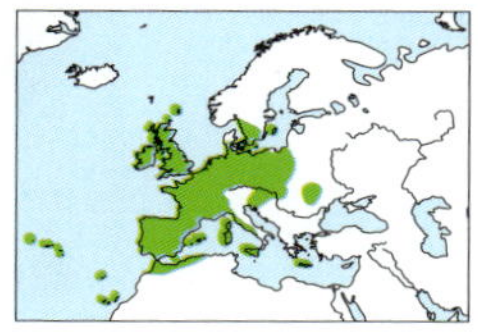

Originally Iberian Pen, W FR & N Africa; old European introductions mapped

Genus *BUNOLAGUS*
(1 species)

C Karoo region, around Victoria West, Beaufort West and Fraserburg

Riverine Rabbit
Bunolagus monticularis
FR Lapin des Bochimans
DE Buschmannhase
ES Conejo de ribera

CR
Monotypic
HB 34–47 cm
T 7–10.8 cm
W 1–1.5 kg

Genus *SYLVILAGUS*
(18 species)

Pygmy Rabbit
Sylvilagus idahoensis
FR Lapin pygmée
DE Zwergkaninchen
ES Conejo pigmeo

LC
Monotypic
HB 23–31 cm
T 1.5–2.4 cm
W 246–458 g

1370–2135 m

Most of Great Basin and close intermountain areas; reintroduced to WC Washington

1370–3540 m
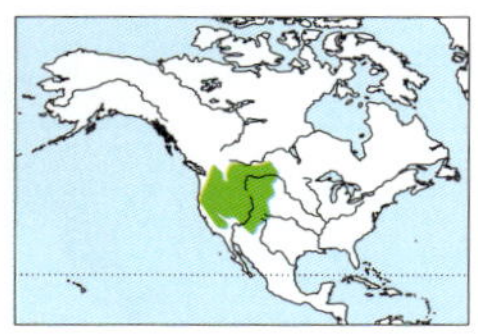

SW CA (S British Columbia to S Saskatchewan), W US (S to Arizona & New Mexico)

Mountain Cottontail
Sylvilagus nuttallii
FR Lapin de Nuttall
DE Berg-Baumwollschwanzkaninchen
ES Conejo de Nuttall

LC
3 sspp
HB 34–39 cm
T 3–5.4 cm
W 629–871 g

Coastal S Maine S to W & E Connecticut

New England Cottontail
Sylvilagus transitionalis
FR Lapin de Nouvelle-Angleterre
DE Neuengland-Baumwollschwanzkaninchen
ES Conejo de Nueva Inglaterra

VU
Monotypic
HB 39–43 cm
T 2.2–6.5 cm
W 0.8–1 kg

Western US (E just into Great Plains), N & C MX (S to Puebla & Veracruz)

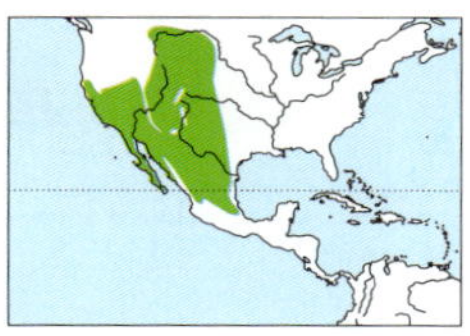
up to 1830 m

Desert Cottontail
Sylvilagus audubonii
FR Lapin d'Audubon
DE Audubon-Baumwollschwanzkaninchen
ES Conejo de desierto

LC
12 sspp
HB 37–40 cm
T 3.9–6 cm
W 511–915 g

Appalachian Cottontail
Sylvilagus obscurus
FR Lapin des Appalaches
DE Appalachen-Baumwollschwanzkaninchen
ES Conejo de los Apalaches

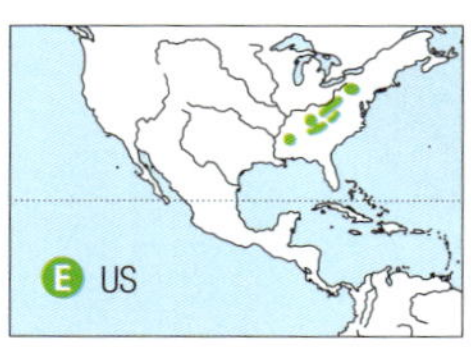

Disjunctly in N Pennsylvania, S & W along Appalachian Mts to N Alabama

NT
Monotypic
HB 39–43 cm
T 2.2–6.5 cm
W 0.8–1 kg

Brush Rabbit
Sylvilagus bachmani
FR Lapin de Bachman
DE Strauchkaninchen
ES Conejo de matorral

LC
14 sspp
HB 30–37 cm
T 1–4.4 cm
W 511–915 g

0–2070 m
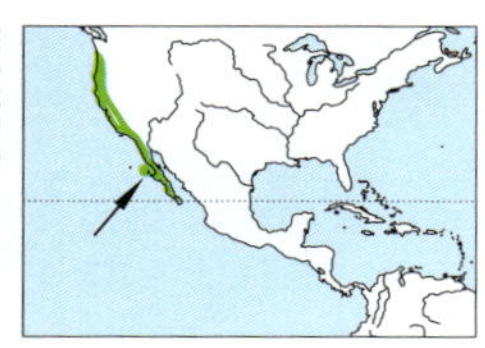

W USA (Oregon, California), NW MX (Baja California Pen, with Cedros I & San José I)

ssp *bachmani*

ssp *mansuetus*

Swamp Rabbit
Sylvilagus aquaticus

FR Lapin aquatique
DE Sumpfkaninchen
ES Conejo de ciénaga

LC
2 sspp
HB 45–55 cm
T 5–7.4 cm
W 1.6–2.7 kg

E Texas E to W North Carolina, N to SE Kansas, S Missouri & S Illinois

Marsh Rabbit
Sylvilagus palustris

FR Lapin des marais
DE Marschlandkaninchen
ES Conejo de pantano

LC
3 sspp
HB 43–44 cm
T 3.3–3.9 cm
W 1–1.6 kg

0–150 m

SE Virginia to S Alabama & Florida Pen, with Florida Keys

Mexican Cottontail
Sylvilagus cunicularius

FR Lapin du Mexique
DE Mexiko-Baumwollschwanzkaninchen
ES Conejo de México

LC
3 sspp
HB 49–52 cm
T 5.4–6.8 cm
W 1.8–2.3 kg

0–3400 m

S Sinaloa and SW Durango S & SE to SE Hidalgo, WC Veracruz and Oaxaca

Tres Marias Cottontail
Sylvilagus graysoni

FR Lapin des Marias
DE Tres-Marías-Baumwollschwanzkaninchen
ES Conejo de Tres Marías

EN
Monotypic
HB 44–47 cm
T 3.3–5 cm

0–350 m

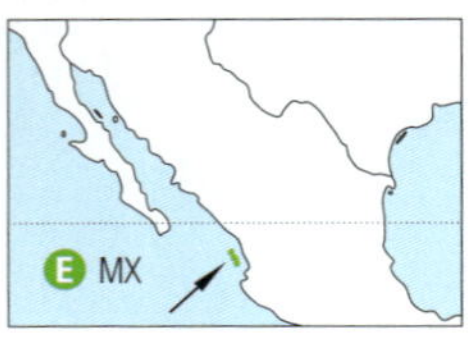

Marías Is, off Nayarit

Robust Cottontail
Sylvilagus robustus

FR Lapin des Davis
DE Texas-Baumwollschwanzkaninchen
ES Conejo de Davis

VU
Monotypic
HB 38–46 cm
T 4.7–6.7 cm
W 0.9–1.8 kg

above 1500 m

S US (SE New Mexico, SW Texas), NE MX (N Coahuila)

Manzano Mountain Cottontail
Sylvilagus cognatus

FR Lapin des Manzano
DE Manzano-Baumwollschwanzkaninchen
ES Conejo de Manzano

EN
Monotypic
HB c. 45 cm
T c. 65 cm

Coniferous forests, grasslands and scrubs in Manzano Mts (New Mexico)

Eastern Cottontail
Sylvilagus floridanus

FR Lapin à queue blanche
DE Östliches Baumwollschwanzkaninchen
ES Conejo de Florida

LC
31 sspp
HB 40–48 cm
T 2.5–6.1 cm
W 0.8–1.5 kg

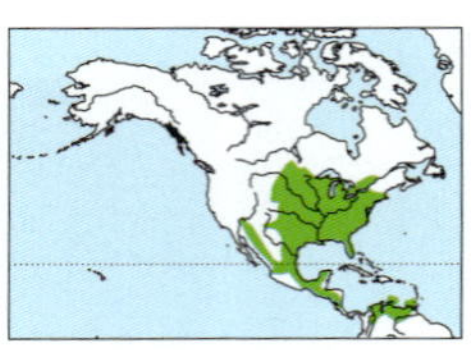

S CA, SW & E US (mainly E of Rocky Mts), Central America (MX to NW CR), CO, VE (with coastal Is)

Venezuelan Lowland Rabbit
Sylvilagus varynaensis

FR Lapin du Venezuela
DE Venezuela-Baumwollschwanzkaninchen
ES Conejo de Venezuela

DD
Monotypic
HB 42–45 cm
T 2.1–2.7 cm
W 1.5–1.8 kg

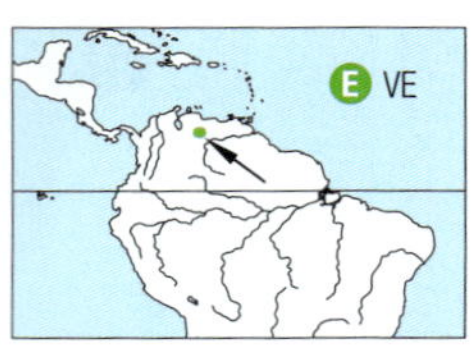

Lowlands of Barinas, Portuguesa and Guárico

Omilteme Cottontail
Sylvilagus insonus

FR Lapin d'Omiltemi
DE Omilteme-Baumwollschwanzkaninchen
ES Conejo de Omiltemi

DD
Monotypic
HB 43–50 cm
T 4–4.5 cm
W 500–950 g

2135–5280 m

Omiltemi State Ecological Park, Sierra Madre del Sur (C Guerrero)

VU
Monotypic
HB 39–45 cm
T 1.5–2.9 cm

Dice's Cottontail
Sylvilagus dicei

FR Lapin de Dice
DE Dice-Baumwollschwanzkaninchen
ES Conejo de Dice

1220–3535 m

Cordillera de Talamanca (SC CR, W PA)

ssp *incitatus*

ssp *gabbi*

Central American Tapeti
Sylvilagus gabbi

FR Lapin de Gabb
DE Mittelamerika-Tapeti
ES Conejo de Centroamérica

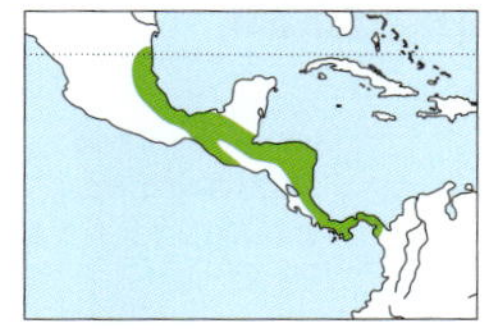

0–1525 m

E & SE MX (from S Tamaulipas) to PA (with del Rey I)

LC
3 sspp
HB 38–39 cm
T c. 2.1 cm
W 500–950 g

Common Tapeti
Sylvilagus brasiliensis

FR Lapin du Brésil
DE Tapeti
ES Conejo tropical

0–4500 m

South America in CO, EC, PE, VE, Guianas, BR (N & mainly S of R Amazon), BO, PY & N AR

EN
17 sspp
HB 38–42 cm
T 2–2.1 cm
W 500–950 g

ssp *tapetillus*

ssp *andinus*

ssp *brasiliensis*

ssp *daulensis*

ssp *apolinaris*

ssp *fulvescens*

ssp *nicefori*

ssp *parentum*

ssp *surdaster*

ssp *salentus*

ssp *sanctaemartae*

Genus *LEPUS*

(32 species)

Indian Hare
Lepus nigricollis

FR Lièvre à col noir
DE Schwarznackenhase
ES Liebre de India

LC
7 sspp
HB 33–53 cm
T 1–9 cm
W 1.8–3.6 kg

50–4500 m

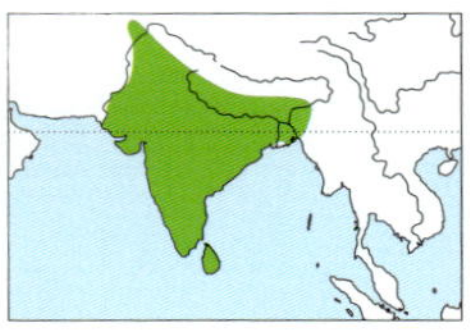

Extreme E AF, Indian subcontinent; introduced to Indian Ocean Is & New Guinea

Burmese Hare
Lepus peguensis

FR Lièvre du Pégou
DE Burmesischer-Hase
ES Liebre de Birmania

LC
2 sspp
HB 40–59 cm
T 5.5–8.4 cm
W 2–2.5 kg

below 1300 m

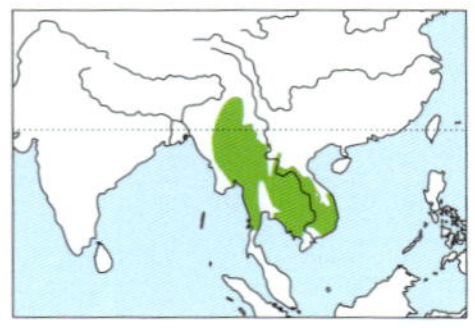

C & S MM (E from R Chindwin valley), TH, LA, KH, S VN

Hainan Hare
Lepus hainanus

FR Lièvre de Hainan
DE Hainan-Hase
ES Liebre de Hainan

EN
Monotypic
HB 35–39 cm
T 4.5–7 cm
W 1.3–1.8 kg

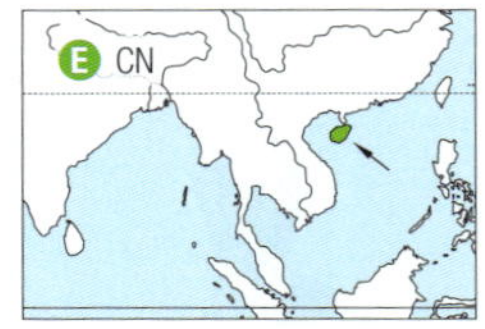

Hainan I

Ethiopian Hare
Lepus fagani

FR Lièvre d'Éthiopie
DE Äthiopischer Hase
ES Liebre de Etiopía

LC
Monotypic
HB 45–54 cm
T 5–8.2 cm

500–2500 m

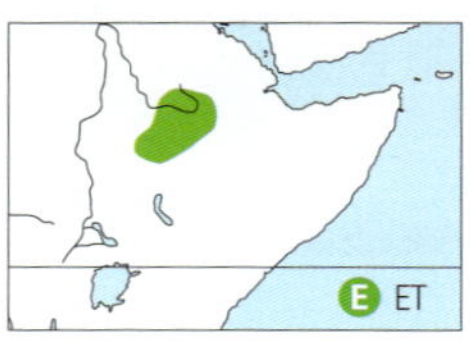

N & W highlands of ET (N of Rift Valley); perhaps adjacent E SS & NW KE

Scrub Hare
Lepus saxatilis

FR Lièvre des buissons
DE Buschhase
ES Liebre de matorral

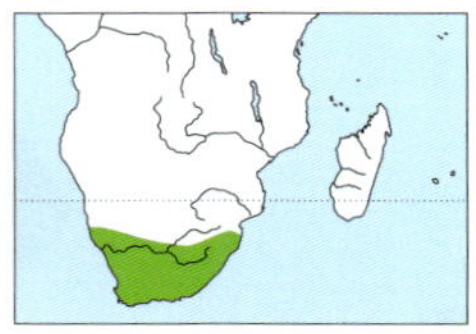

S NA, most of ZA, SZ, LS; N boundary uncertain, perhaps to S BW & S ZW

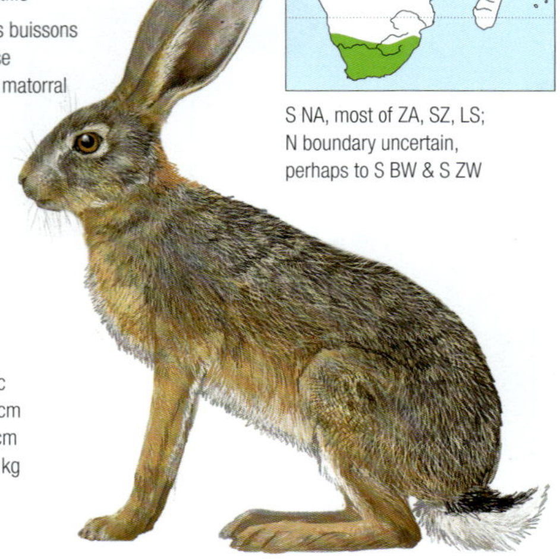

LC
Monotypic
HB 45–63 cm
T 7–12.2 cm
W 1.4–3.5 kg

African Savanna Hare
Lepus victoriae

FR Lièvre des savanes
DE Savannenhase
ES Liebre de sabana

up to 3600 m

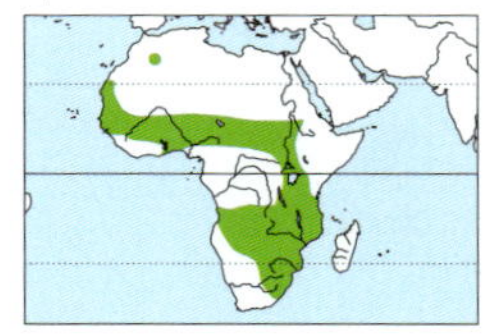

NW Africa (EH to GN), E across Sahel to SD & W ET, S through E Africa to AO & ZA; also W DZ

LC
4 sspp
HB 42–58 cm
T 6.8–12.1 cm
W 1.4–3.2 kg

Yarkand Hare
Lepus yarkandensis

FR Lièvre de Yarkand
DE Yarkand-Hase
ES Liebre de Yarkand

Steppes of Tarim Basin around edge of Taklamakan Desert (S Xinjiang)

900–1200 m

NT
Monotypic
HB 29–43 cm
T 5.5–8.6 cm
W 1.1–1.9 kg

Ethiopian Highland Hare
Lepus starcki

FR Lièvre de Starck
DE Äthiopischer Hochlandhase
ES Liebre de tierras altas de Etiopía

C highlands, especially mountains in Arussi and Bale regions

2100–4400 m

LC
Monotypic
HB 44–55 cm
T 8.2–11.5 cm
W 1.9–3 kg

Cape Hare
Lepus capensis

FR Lièvre du Cap
DE Kaphase
ES Liebre de El Cabo

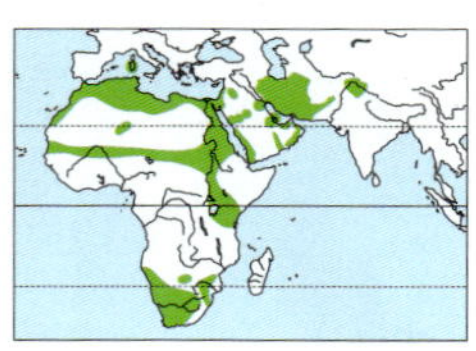

Sardinia I, patchily across Arabian Pen & Middle East, E to W Himalayas; widespread in Africa

up to 2400 m

LC
13 sspp
HB 45–55 cm
T 10–14.5 cm
W 1.7–2.5 kg

Abyssinian Hare
Lepus habessinicus

FR Lièvre d'Abyssinie
DE Abessinischer Hase
ES Liebre de Abisinia

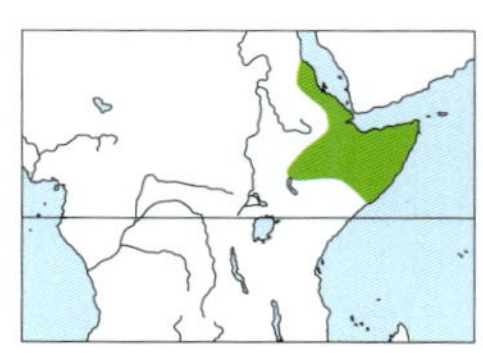

E SD, ER, DJ, E & S ET, SO (excluding S); perhaps extreme N KE

0–2500 m

LC
Monotypic
HB 40–55 cm
T 7.5–10 cm

Tolai Hare
Lepus tolai

FR Lièvre tolai
DE Tolai-Hase
ES Liebre tolai

600–4900 m

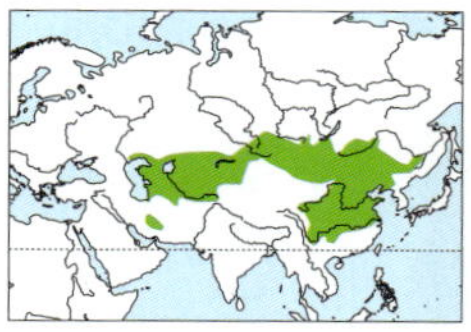

Steppes E of Caspian Sea, S to IR, E across KZ and S Siberia to MN & CN

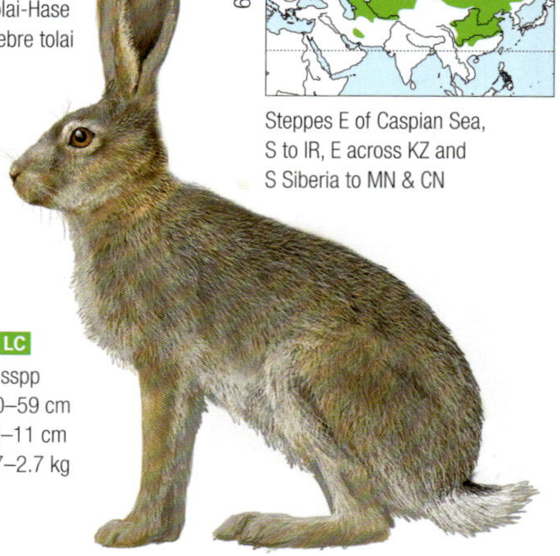

LC
7 sspp
HB 40–59 cm
T 7.2–11 cm
W 1.7–2.7 kg

Desert Hare
Lepus tibetanus

FR Lièvre du désert
DE Wüstenhase
ES Liebre de desierto

LC
5 sspp
HB 40–48 cm
T 8.7–10.9 cm
W 1.6–2.5 kg

3500–4000 m

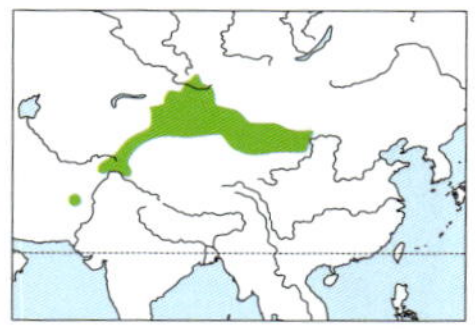

Baluchistan (PK), Upper Indus Valley N to Tian Shan, E to NC China

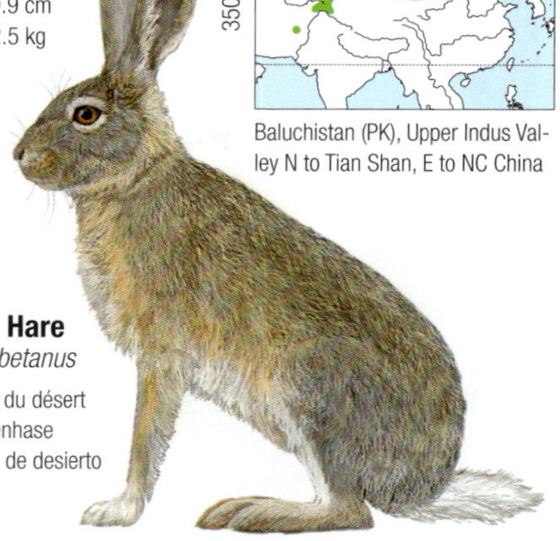

Yunnan Hare
Lepus comus

FR Lièvre du Yunnan
DE Yunnan-Hase
ES Liebre de Yunnan

1300–3200 m

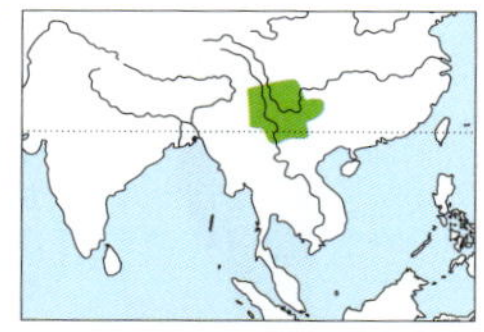

NE MM, SC CN (Yunnan, except part SW of R Mekong, SW Sichuan, W Guizhou)

LC
Monotypic
HB 32–48 cm
T 9.5–11 cm
W 1.8–2.5 kg

Woolly Hare
Lepus oiostolus

FR Lièvre laineux
DE Tibet-Wollhase
ES Liebre lanuda

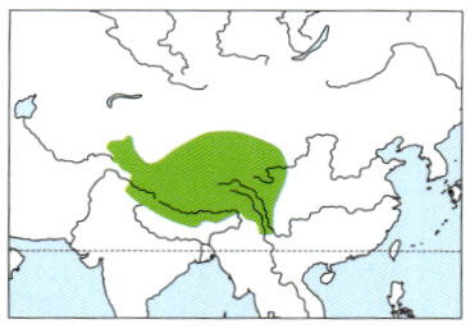

2500–5400 m

Tibetan Plateau, NW & NE IN (Jammu and Kashmir, Sikkim), N NP, perhaps N BT

LC
4 sspp
HB 40–58 cm
T 6.5–12.5 cm
W 2–4.3 kg

Snowshoe Hare
Lepus americanus

FR Lièvre d'Amérique
DE Schneeschuhhase
ES Liebre americana

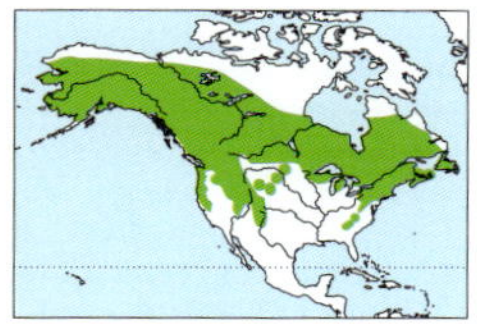

CA (except Nunavut), N US (Alaska, Sierra Nevada, Rocky Mts, Great Lakes region, Appalachian Mts)

summer

winter

LC
15 sspp
HB 36–52 cm
T 2.5–5.5 cm
W 1.1–1.6 kg

Black-tailed Jackrabbit
Lepus californicus

FR Lièvre de Californie
DE Eselhase
ES Liebre de cola negra

-85–3750 m

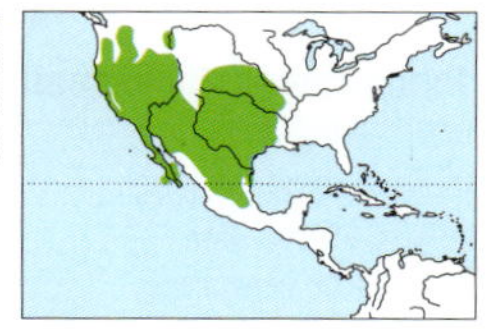

SW US (N to Washington, E to Missouri & Texas), MX (S to Mexico City, several Is)

ssp *californicus*

ssp *insularis*

LC
17 sspp
HB 52–61 cm
T 7.5–10.1 cm
W 1.5–3.6 kg

Antelope Jackrabbit
Lepus alleni

FR Lièvre d'Allen
DE Antilopenhase
ES Liebre antílope

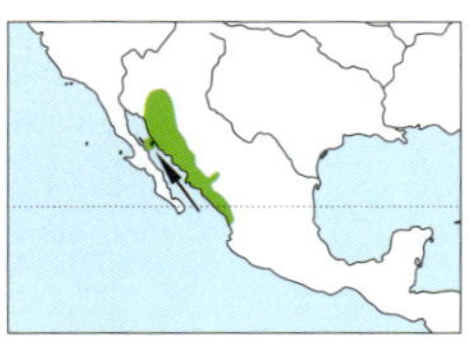

0–1500 m

SW US (SC Arizona), MX (Sonora S to NW Nayarit, Tiburón I)

LC
3 sspp
HB 55–67 cm
T 4.8–7.6 cm
W 2.7–4.5 kg

Tamaulipas Jackrabbit
Lepus altamirae

FR Lièvre d'Altamira
DE Tamaulipas-Eselhase
ES Liebre de Tamaulipas

0–150 m

Open habitats in E Tamaulipas

NE
Monotypic

White-sided Jackrabbit
Lepus callotis

FR Lièvre du Mexique
DE Weißflankenhase
ES Liebre de flancos blancos

VU
2 sspp
HB 43–60 cm
T 4.7–9.2 cm
W 1.5–3.2 kg

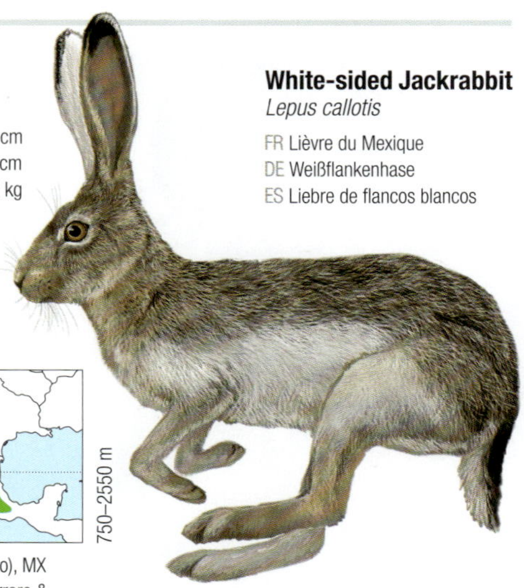

750–2550 m

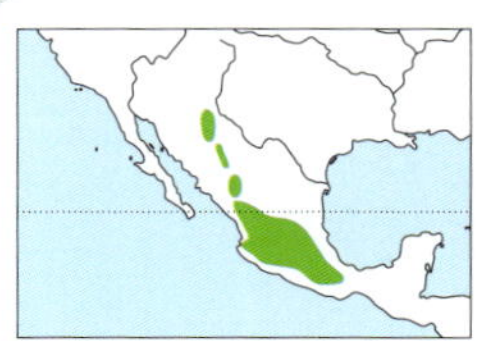

SW US (SW New Mexico), MX (Chihuahua S to N Guerrero & NW Oaxaca)

0–665 m

Small area between Salina Cruz (SE Oaxaca) and extreme W Chiapas

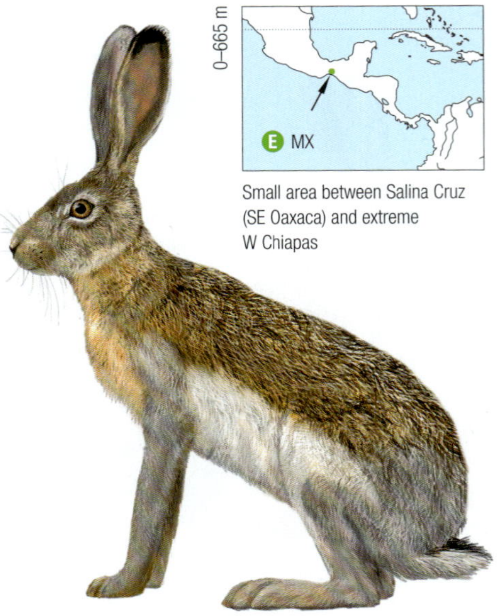

Tehuantepec Jackrabbit
Lepus flavigularis
FR Lièvre de Tehuantepec
DE Tehuantepec-Hase
ES Liebre de Tehuantepec

EN
Monotypic
HB 57–61 cm
T 6.5–9.5 cm
W 2.5–3.7 kg

0–1900 m

Iberian Pen (except N & NE), Balearic Is (Mallorca, where perhaps introduced)

Iberian Hare
Lepus granatensis
FR Lièvre ibérique
DE Andalusischer Hase
ES Liebre ibérica

LC
3 sspp
HB 44–48 cm
T 9.3–11.2 cm
W 2–2.6 kg

European Hare
Lepus europaeus
FR Lièvre d'Europe
DE Feldhase
ES Liebre europea

0–2300 m

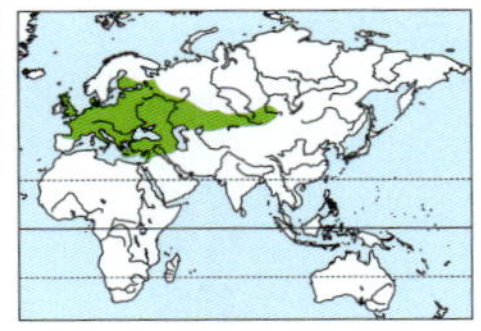

Europe, N Middle East, E to W Mongolian lowlands; old introduction to Great Britain

LC
16 sspp
HB 55–68 cm
T 7.5–14 cm
W 3.5–5 kg

Corsican Hare
Lepus corsicanus
FR Lièvre corse
DE Korsika-Hase
ES Liebre de Córcega

0–2400 m

Tuscany S to Calabria, Sicily (limits poorly known); old introduction to Corsica

VU
Monotypic
HB 55–61 cm
T 7–12 cm
W 3.5–5 kg

Broom Hare
Lepus castroviejoi
FR Lièvre de Castroviejo
DE Castroviejo-Hase
ES Liebre de piornal

VU
Monotypic
HB 49–51 cm
T 7.8–9.6 cm
W 2.7–3.2 kg

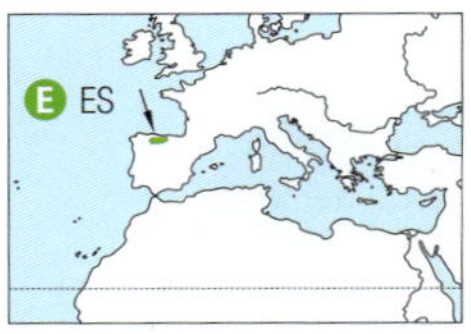

1000–1900 m

Cantabrian Range between Serra dos Ancares and Sierra de Peña Labra

30–4320 m

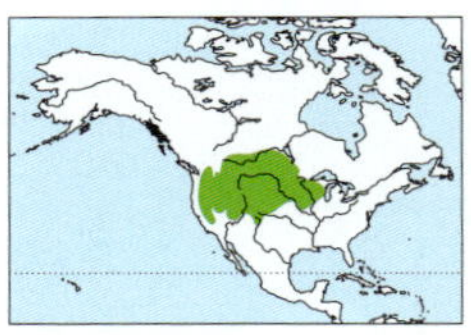

S CA (British Columbia to SW Ontario), W & C US (S to New Mexico, E to Missouri)

summer

White-tailed Jackrabbit
Lepus townsendii
FR Lièvre de Townsend
DE Präriehase
ES Liebre de cola blanca

winter

LC
2 sspp
HB 57–66 cm
T 6.6–10.2 cm
W 2.5–4.4 kg

0–900 m

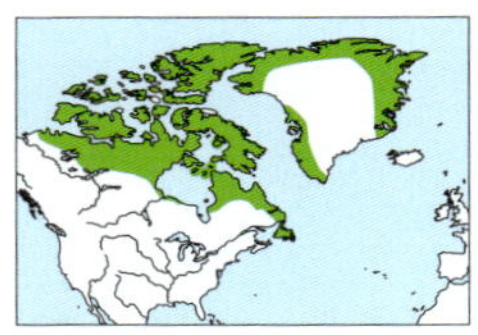

N CA (Northwest Territories E to Newfoundland, Canadian Arctic Arch), GL

Arctic Hare
Lepus arcticus
FR Lièvre arctique
DE Polarhase
ES Liebre ártica

summer

winter

LC
9 sspp
HB 56–66 cm
T 4.5–10 cm
W 2.5–6.8 kg

Alaskan Hare
Lepus othus
FR Lièvre d'Alaska
DE Alaska-Hase
ES Liebre de Alaska

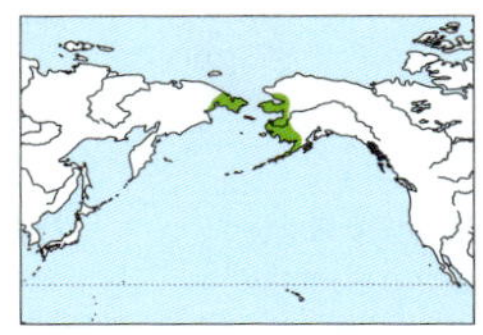

RU (Chukchi Pen, NE Siberia), US (W Alaska Pen)

summer

winter

LC
3 sspp
HB 57–62 cm
T 6.5–8.3 cm
W 3.9–4.8 kg

Mountain Hare
Lepus timidus
FR Lièvre variable
DE Schneehase
ES Liebre de montaña

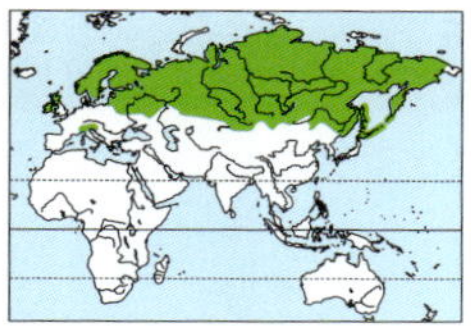

N Palearctic, disjunctly in IE, Scotland, Alps, Sakhalin, Hokkaido & Kuril Is

summer

LC
16 sspp
HB 51–55 cm
T 5.9–6.5 cm
W 2.4–3.4 kg

winter

Japanese Hare
Lepus brachyurus
FR Lièvre du Japon
DE Japan-Hase
ES Liebre de Japón

0–3000 m

Honshu (with Sado I & Oki Is), Shikoku, Kyushu (and surrounding Is)

LC
4 sspp
HB 49–54 cm
T 3.7–4.1 cm
W 2.4–2.7 kg

Manchurian Hare
Lepus mandshuricus
FR Lièvre de Mandchourie
DE Mandschurei-Hase
ES Liebre de Manchuria

300–900 m

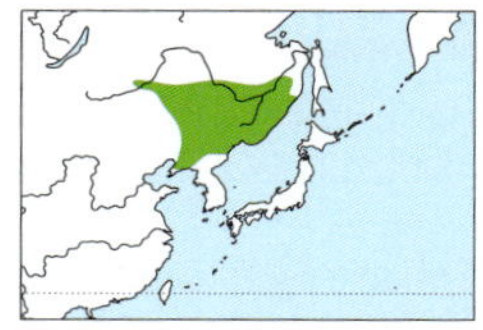

Amur Basin (E RU) S to Liaoning (NE CN); perhaps NE KP

LC
Monotypic
HB 41–54 cm
T 5–8 cm
W 1.4–2.6 kg

Korean Hare
Lepus coreanus
FR Lièvre de Corée
DE Korea-Hase
ES Liebre de Corea

LC
Monotypic
HB 43–49 cm
T 6–7.5 cm
W c. 1.7 kg

Korean Pen, NE CN (S Jilin; perhaps SE Heilongjiang and E Liaoning)

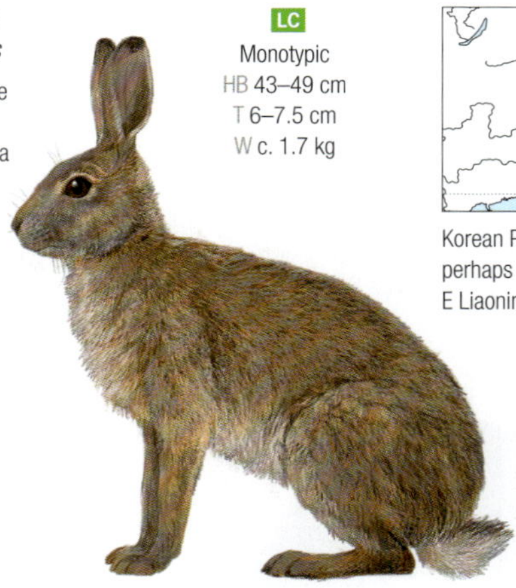

Chinese Hare
Lepus sinensis
FR Lièvre de Chine
DE China-Hase
ES Liebre de China

LC
3 sspp
HB 35–45 cm
T 4–5.7 cm
W 1–1.9 kg

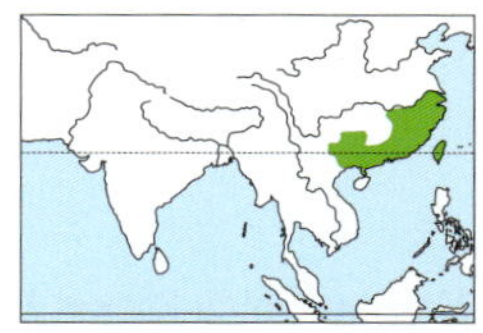

SE CN S of R Yangtze, TW; perhaps NE VN

summer

winter

Order RODENTIA (35 families)

Family CASTORIDAE – Beavers

1 genus, 2 species. All types of freshwater wetlands from streams and ponds to large rivers and lakes. Palearctic and Nearctic Regions.

Genus *CASTOR*

(2 species)

Eurasian Beaver
Castor fiber
FR Castor d'Europe
DE Eurasischer Biber
ES Castor euroasiático

LC
Monotypic
HB 80–90 cm
T 20–30 cm
W 15–20 kg

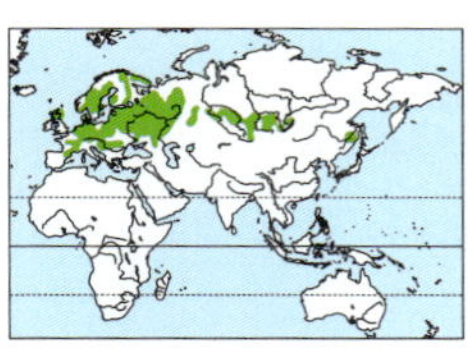

Europe from S FR (Rhone Delta) N to Scandinavia, E patchily to S Siberia, Altai, Amur Basin

North American Beaver
Castor canadensis
FR Castor du Canada
DE Kanadischer Biber
ES Castor americano

LC
Monotypic
HB 80–90 cm
T 20–30 cm
W 15–20 kg

2000–2500 m

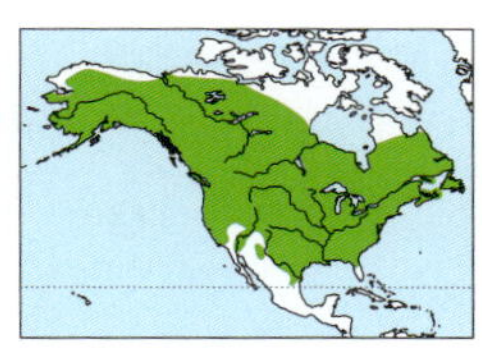

North America, from CA (R Mackenzie Delta) S across US to N MX (Tamaulipas)

Family HETEROMYIDAE – Pocket Mice, Kangaroo Mice and Kangaroo Rats

5 genera, 69 species. From arid deserts to humid tropical forests, including grasslands and chaparral, from cold temperate zones to tropics, from sea level to 3200 m. Nearctic and Neotropical Regions.

Genus *HETEROMYS*

(17 species)

Salvin's Spiny Pocket Mouse
Heteromys salvini

FR Souris-à-abajoues de Salvin
DE Salvin-Stacheltaschenmaus
ES Ratón espinoso de Salvin

up to 1500 m

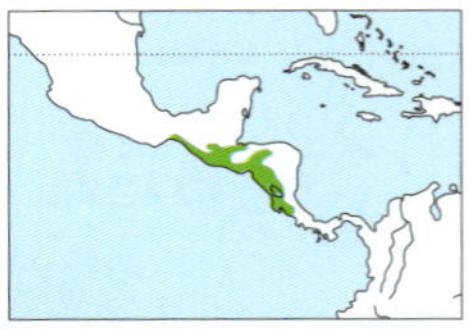

From S MX (SE Oaxaca, S Chiapas) S to C CR

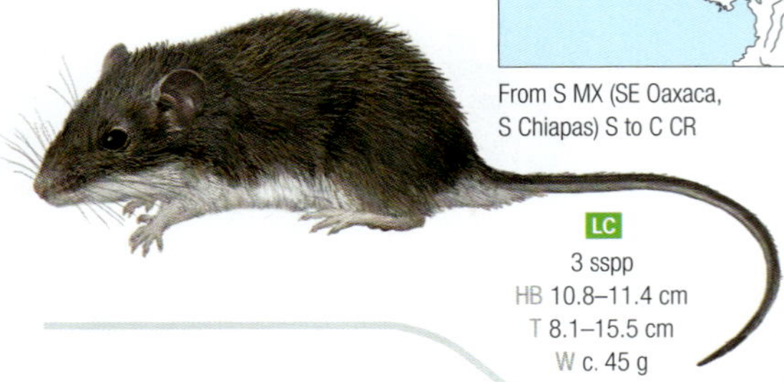

LC
3 sspp
HB 10.8–11.4 cm
T 8.1–15.5 cm
W c. 45 g

Panamanian Spiny Pocket Mouse
Heteromys adspersus

FR Souris-à-abajoues du Panama
DE Panama-Stacheltaschenmaus
ES Ratón espinoso de Panamá

below 600 m

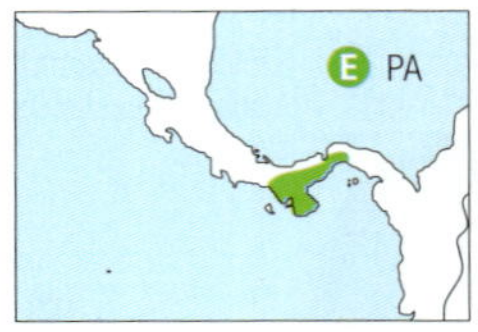

Semiarid savannas principally on Pacific side

LC
Monotypic
HB 12.3–12.7 cm
T 10.7–14.8 cm

LC
6 sspp
HB 11.9–12.5 cm
T 9.6–16.9 cm
W 34–50 g

Mexican Spiny Pocket Mouse
Heteromys irroratus

FR Souris-à-abajoues du Mexique
DE Mexiko-Stacheltaschenmaus
ES Ratón espinoso de México

0–3500 m

S US (S Texas), MX (from S Chihuahua and C Nuevo León S to Oaxaca)

NE
Monotypic
HB 23–25 cm
T 11.3–12.8 cm

Buller's Spiny Pocket Mouse
Heteromys bulleri

FR Souris-à-abajoues de Buller
DE Buller-Stacheltaschenmaus
ES Ratón espinoso de Manantlan

WC Jalisco

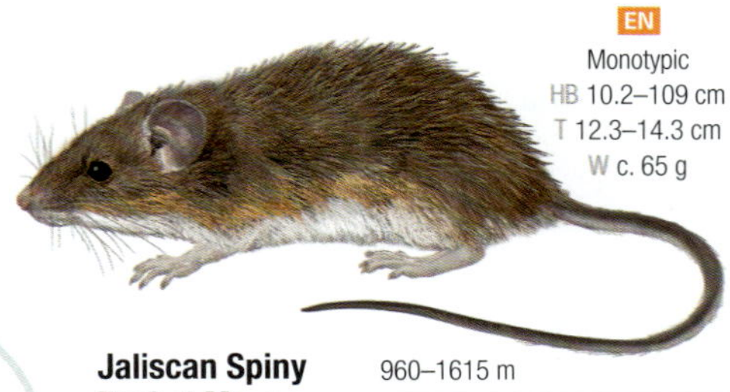

EN
Monotypic
HB 10.2–109 cm
T 12.3–14.3 cm
W c. 65 g

Jaliscan Spiny Pocket Mouse
Heteromys spectabilis

FR Souris-à-abajoues du Jalisco
DE Jalisco-Stacheltaschenmaus
ES Ratón espinoso de Jalisco

960–1615 m

SE Jalisco

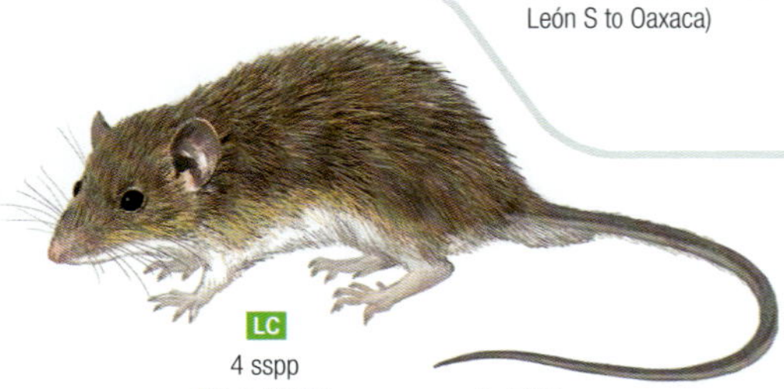

LC
4 sspp
HB c. 11 cm
T 9.1–16.8 cm
W 30–80 g

Painted Spiny Pocket Mouse
Heteromys pictus

FR Souris-à-abajoues peinte
DE Bunte Stacheltaschenmaus
ES Ratón espinoso pintado

0–2400 m

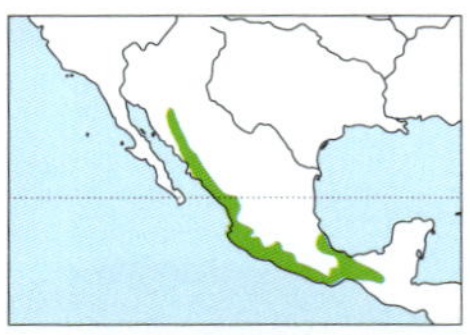

MX (coast of N Sonora to Chiapas, S Veracruz), W GT

EN
Monotypic
HB 15.1–16.2 cm
T 18.6–19.7 cm
W 60–110 g

Nelson's Spiny Pocket Mouse
Heteromys nelsoni

FR Souris-à-abajoues de Pinabete
DE Nelson-Stacheltaschenmaus
ES Ratón espinoso de Nelson

2500–2800 m

S Chiapas (S MX), adjacent W GT

Trinidad Spiny Pocket Mouse
Heteromys anomalus

FR Souris-à-abajoues de Trinidad
DE Trinidad-Stacheltaschenmaus
ES Ratón espinoso de Trinidad

LC
4 sspp
HB 12.9–13.4 cm
T 15–15.6 cm

0–2430 m

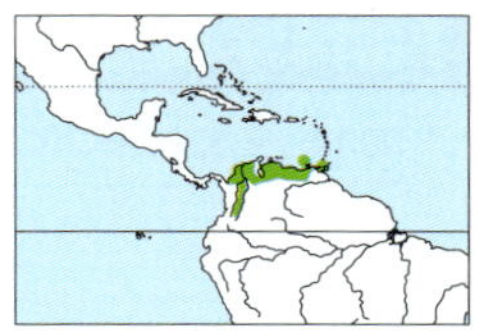

C & N CO (with Magdalena Valley), N VE (mainland, Margarita I), TT (Trinidad I)

LC
2 sspp
HB c. 13.2 cm
T c. 16.6 cm

Overlook Spiny Pocket Mouse
Heteromys catopterius

FR Souris-à-abajoues de la Costa
DE Aragua-Stacheltaschenmaus
ES Ratón espinoso de altura

350–2425 m

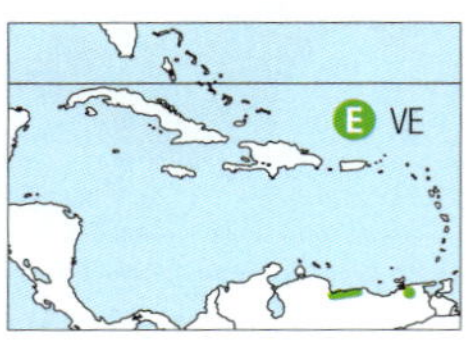

Forests in Cordillera de la Costa

Paraguana Spiny Pocket Mouse
Heteromys oasicus

FR Souris-à-abajoues de Paraguaná
DE Paraguaná-Stacheltaschenmaus
ES Ratón espinoso de Paraguaná

EN
Monotypic
HB c. 11.8 cm
T c. 11.9 cm
W c. 44 g

above 550 m

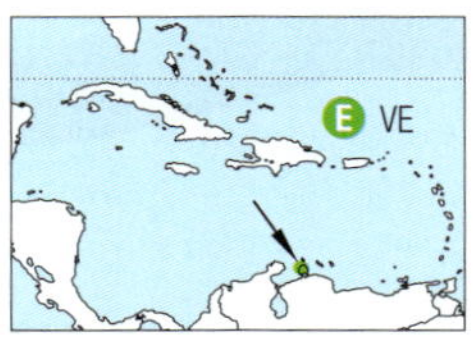

Cerro Santa Ana and Fila de Monte Cano on Paraguaná Pen (Falcón)

LC
3 sspp
HB 12–12.8 cm
T 13.6–14 cm

0–2500 m

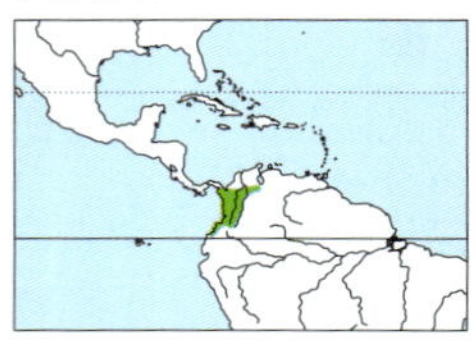

E PA, W CO (including Andes), NW EC, W VE (Cordillera de Mérida, Táchira)

Southern Spiny Pocket Mouse
Heteromys australis

FR Souris-à-abajoues australe
DE Südliche Stacheltaschenmaus
ES Ratón espinoso meridional

Ecuadorean Spiny Pocket Mouse
Heteromys teleus

FR Souris-à-abajoues d'Équateur
DE Ecuador-Stacheltaschenmaus
ES Ratón espinoso de Ecuador

VU
Monotypic
HB c. 13.5 cm
T c. 14.6 cm

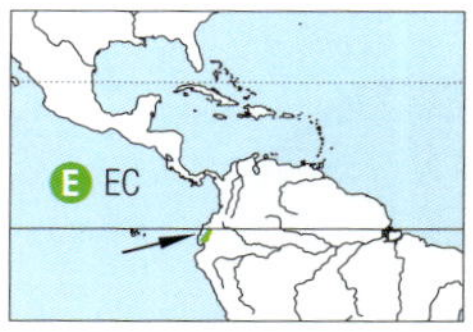

0–2000 m

W slopes of Andes (R Esmeraldas/ R Guayllabamba S to Cerros de Colonche)

Gaumer's Spiny Pocket Mouse
Heteromys gaumeri

FR Souris-à-abajoues de Gaumer
DE Gaumer-Stacheltaschenmaus
ES Ratón espinoso de Gaumer

LC
Monotypic
HB 12.3–12.5 cm
T 14.6–15.1 cm
W 43–70 g

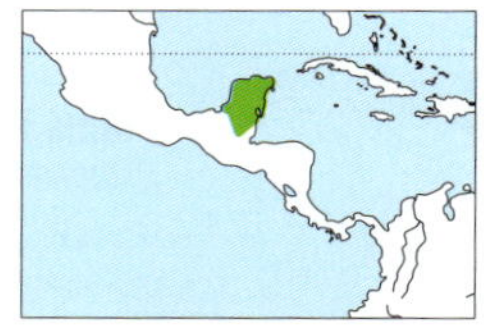

0–100 m

Forests in Yucatán Pen (SE MX, N BZ, N GT)

Desmarest's Spiny Pocket Mouse
Heteromys desmarestianus

FR Souris-à-abajoues de Desmarest
DE Desmarest-Stacheltaschenmaus
ES Ratón espinoso de Desmarest

LC
11 sspp
HB 13–13.3 cm
T 14.2–14.8 cm
W 61–83 g

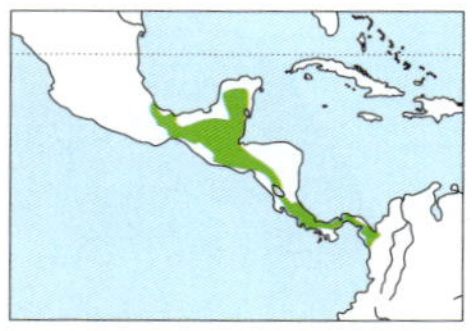

0–2400 m

Humid forests in S MX (S of Veracruz & Tabasco) to NW CO

Goldman's Spiny Pocket Mouse
Heteromys goldmani

FR Souris-à-abajoues de Chicharras
DE Goldman-Stacheltaschenmaus
ES Ratón espinoso de Goldman

NE
Monotypic
HB c. 14.7 cm
T c. 19 cm

45–1860 m

Heavily forested Pacific slope of Sierra Madre in SE MX and SW GT

Cloud-dwelling Spiny Pocket Mouse
Heteromys nubicolens

FR Souris-à-abajoues nubicole
DE Nebelwald-Stacheltaschenmaus
ES Ratón espinoso de las nubes

NE
Monotypic
HB c. 14.6 cm
T c. 17.5 cm

750–1840 m

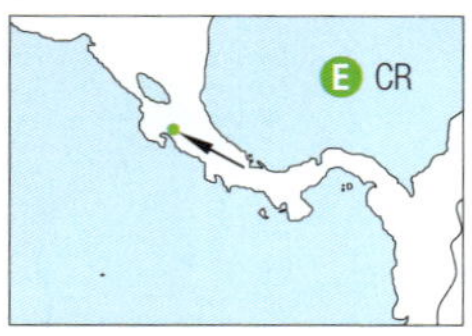

Cordillera de Tilarán and Cordillera de Guanacaste

Mountain Spiny Pocket Mouse
Heteromys oresterus

FR Souris-à-abajoues de montagne
DE Costa-Rica-Stacheltaschenmaus
ES Ratón espinoso de montaña

LC
Monotypic
HB 14.1–15.9 cm
T 16.9–17.5 cm

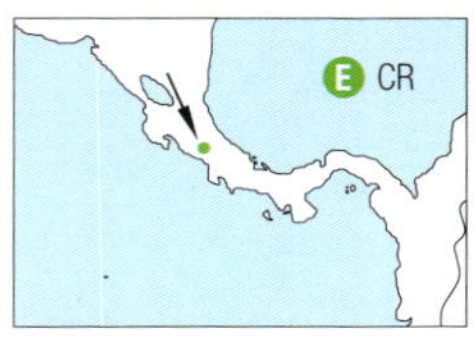

1800–2600 m

N end of Cordillera de Talamanca (Cartago & San José prov)

Genus *PEROGNATHUS*

(10 species)

Little Pocket Mouse
Perognathus longimembris

FR Petite Souris-à-abajoues
DE Kleine Seidentaschenmaus
ES Ratón de abazones pequeño

LC
16 sspp
HB c. 5.8 cm
T c. 7.4 cm W 7–12 g

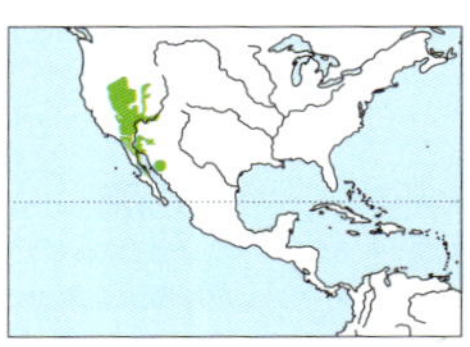

W US (from SE Oregon & W Utah) S to NW MX (N Sonora, Baja California Sur)

San Joaquin Pocket Mouse
Perognathus inornatus

FR Souris-à-abajoues du San Joaquin
DE San-Joaquin-Seidentaschenmaus
ES Ratón de abazones de San Joaquín

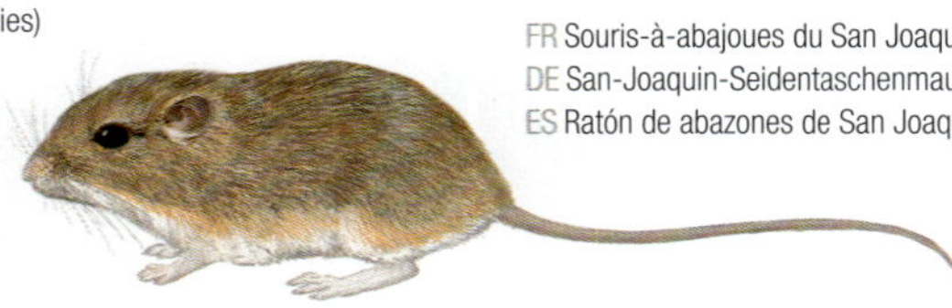

LC
3 sspp
HB c. 7.3 cm T c. 7.5 cm
W 7–12 g

CW California (Salinas, Sacramento & San Joaquin valleys, adjacent Mojave Desert)

Arizona Pocket Mouse
Perognathus amplus

FR Souris-à-abajoues d'Arizona
DE Arizona-Seidentaschenmaus
ES Ratón de abazones de Arizona

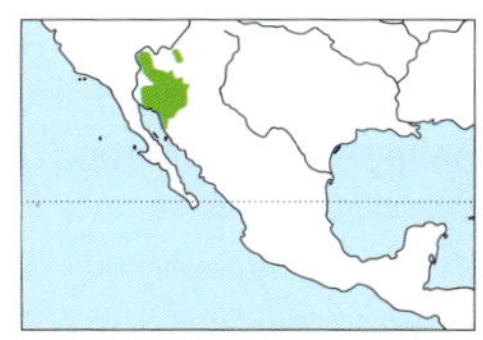

Desert scrubs in SW US (C & S Arizona) & NW MX (NW Sonora)

LC
4 sspp
HB c. 7.1 cm T c. 8.3 cm
W 9–14 g

LC
14 sspp
HB c. 6 cm T c. 5.3 cm
W 5–10 g

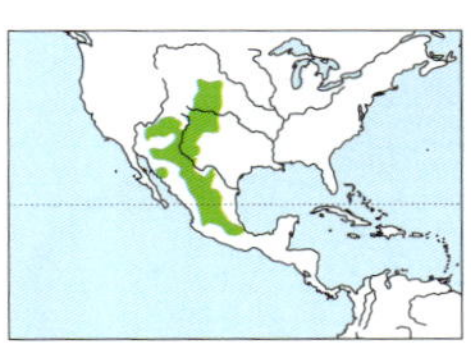

US (C & S Great Plains, intermountain West), MX (Mexican Plateau, WC Sonora)

Silky Pocket Mouse
Perognathus flavus

FR Souris-à-abajoues soyeuse
DE Seidentaschenmaus
ES Ratón de abazones sedoso

Merriam's Pocket Mouse
Perognathus merriami

FR Souris-à-abajoues de Merriam
DE Merriam-Seidentaschenmaus
ES Ratón de abazones de Merriam

S US (SW Oklahoma, E New Mexico, Texas), NE MX (S to Tamaulipas)

LC
2 sspp
HB 5.3–6 cm
T 4.2–6.1 cm
W 5–9 g

Olive-backed Pocket Mouse
Perognathus fasciatus

FR Souris-à-abajoues des plaines
DE Oliv-Seidentaschenmaus
ES Ratón de abazones de dorso oliva

up to 2500 m

SC CA (SE Alberta E to SW Manitoba), US (Great Plains S to S Colorado)

LC
2 sspp
HB c. 7.2 cm
T c. 6.5 cm
W 8–14 g

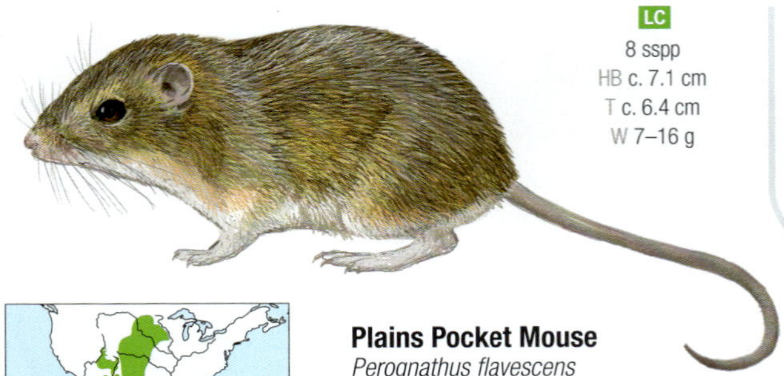

LC
8 sspp
HB c. 7.1 cm
T c. 6.4 cm
W 7–16 g

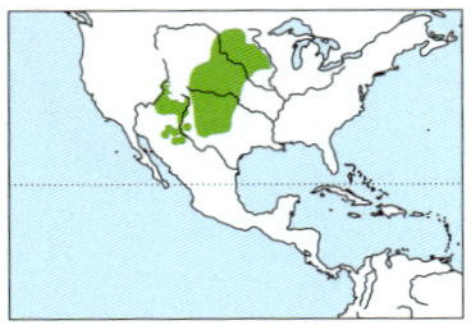

C US (from SE North Dakota & SW Minnesota) S to MX (N Chihuahua)

Plains Pocket Mouse
Perognathus flavescens

FR Souris-à-abajoues flavescente
DE Plains-Seidentaschenmaus
ES Ratón de abazones de llanura

Columbia Plateau Pocket Mouse
Perognathus parvus

FR Souris-à-abajoues des pinèdes
DE Columbia-Plateau-Seidentaschenmaus
ES Ratón de abazones de la Meseta del Columbia

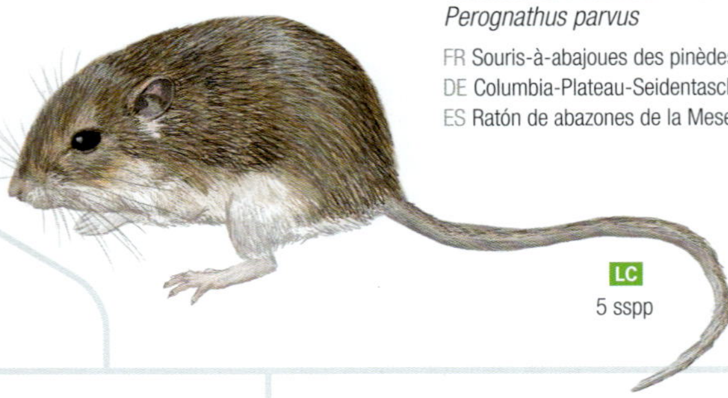

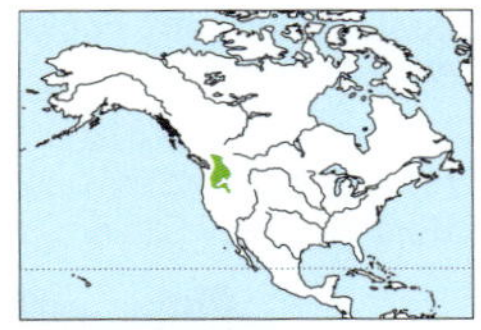

SW CA (SC British Columbia), NW US (Washington to Nevada)

LC
5 sspp

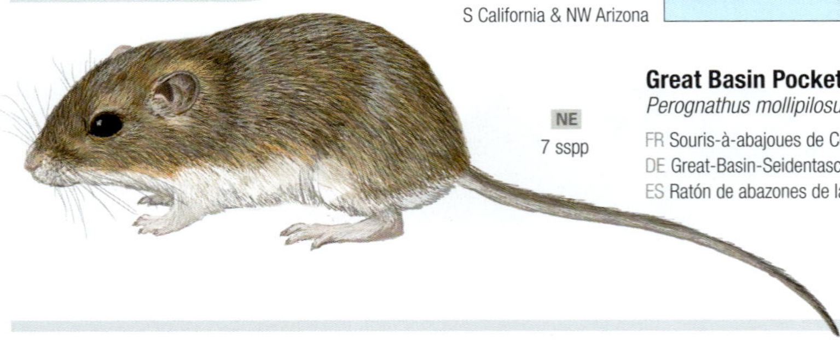

S Oregon, Idaho & SW Montana S to S California & NW Arizona

NE
7 sspp

Great Basin Pocket Mouse
Perognathus mollipilosus

FR Souris-à-abajoues de Coues
DE Great-Basin-Seidentaschenmaus
ES Ratón de abazones de la Gran Cuenca

1065–1830 m

SW California (Tehachapi Mts, also San Bernardino Mts where perhaps extinct)

White-eared Pocket Mouse
Perognathus alticola

FR Souris-à-abajoues alticole
DE Weißohr-Seidentaschenmaus
ES Ratón de abazones de orejas blancas

VU
2 sspp
HB 7.3–7.8 cm
T 7.7–8.6 cm
W 16–24 g

Genus *CHAETODIPUS*

(20 species)

Hispid Pocket Mouse
Chaetodipus hispidus

FR Souris-à-abajoues hérissée
DE Rauhaartaschenmaus
ES Ratón de abazones hirsuto

LC
3 sspp
HB 9.8–11.5 cm
T 9–11.3 cm
W 30–47 g

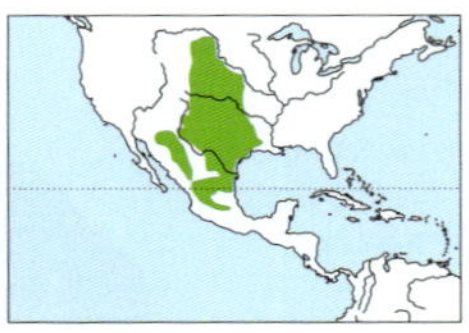

C US (Great Plains S of S North Dakota) S to C MX

LC
7 sspp
HB 7.9–8.6 cm
T 8.6–12.5 cm
W 17–25 g

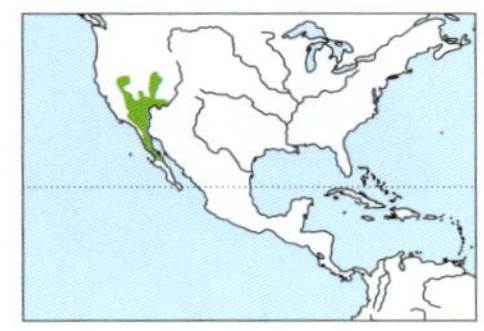

SW US (California, Nevada, Utah, Arizona), NW MX (E Baja California)

Long-tailed Pocket Mouse
Chaetodipus formosus

FR Souris-à-abajoues longicaude
DE Langschwanz-Rauhaartaschenmaus
ES Ratón de abazones de cola larga

Bailey's Pocket Mouse
Chaetodipus baileyi

FR Souris-à-abajoues de Bailey
DE Bailey-Rauhaartaschenmaus
ES Ratón de abazones de Bailey

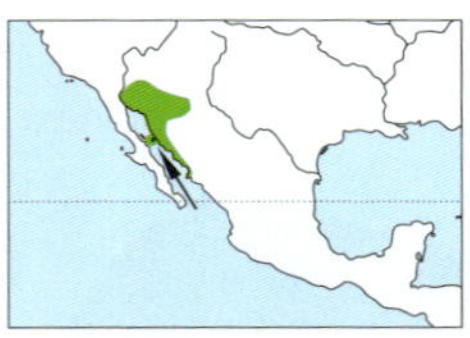

SW US (S Arizona, SW New Mexico), NW MX (Sonora, N Sinaloa, Tiburón I)

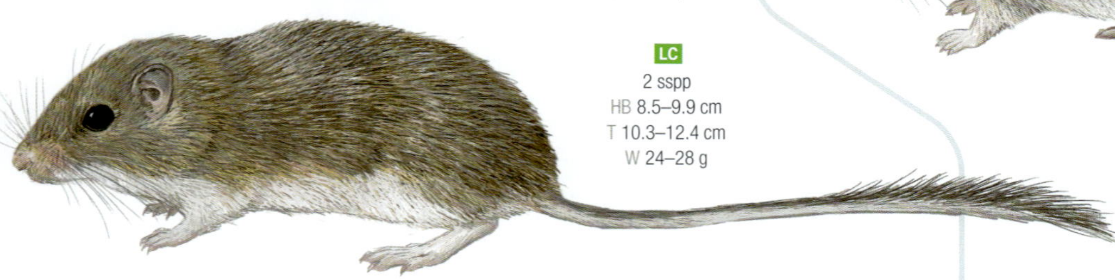

LC
2 sspp
HB 8.5–9.9 cm
T 10.3–12.4 cm
W 24–28 g

Peninsular Pocket Mouse
Chaetodipus rudinoris

FR Souris-à-abajoues de San Quintín
DE Halbinsel-Rauhaartaschenmaus
ES Ratón de abazones de Baja California

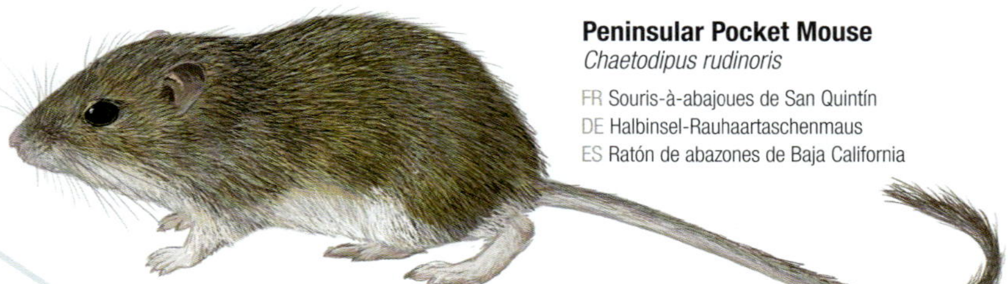

SW US (SE California), NW MX (Baja California Pen, Monserrat, Coronado, Pata & Mitlán Is)

LC
5 sspp
HB 8.5–9.9 cm
T 10.3–12.4 cm
W 24–28 g

up to 900 m

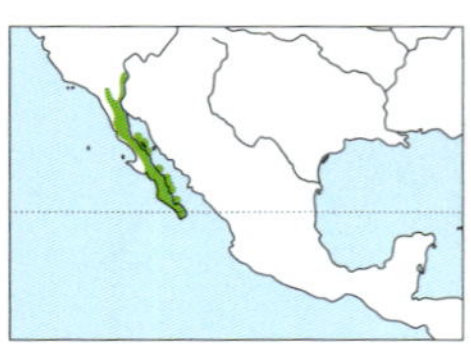

SW US (SE California, S Nevada), NW MX (Baja California Pen, nearby Is)

Spiny Pocket Mouse
Chaetodipus spinatus

FR Souris-à-abajoues épineuse
DE Stachel-Rauhaartaschenmaus
ES Ratón de abazones espinoso

LC
18 sspp
HB 7.5–9.5 cm
T 8.9–12.8 cm
W 13–18 g

California Pocket Mouse
Chaetodipus californicus

FR Souris-à-abajoues de Californie
DE Kalifornien-Rauhaartaschenmaus
ES Ratón de abazones de California

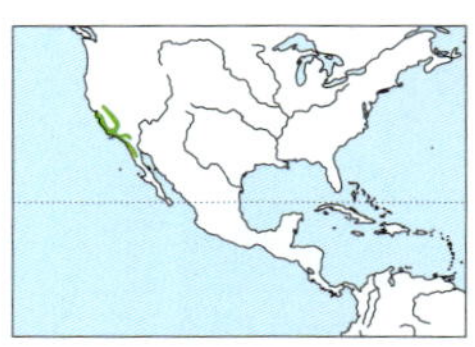

below 1500 m

W US (C & S California), NW MX (N Baja California)

LC
8 sspp
HB 8.6–9.2 cm
T 10.3–14.3 cm
W 18–29 g

San Diego Pocket Mouse
Chaetodipus fallax

FR Souris-à-abajoues de San Diego
DE Kurzohr-Rauhaartaschenmaus
ES Ratón de abazones de San Diego

LC
6 sspp
HB 8.2–8.7 cm
T 8.8–11.8 cm
W 17–22 g

0–1400 m

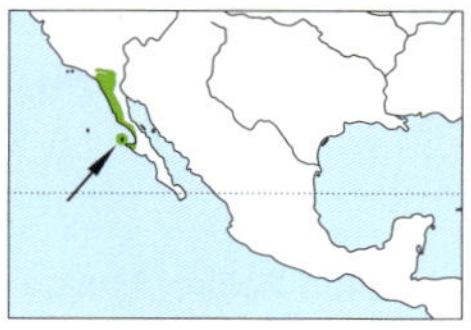

SW US (SW California), NW MX (W coast of Baja California, Cedros I)

Little Desert Pocket Mouse
Chaetodipus arenarius

FR Souris-à-abajoues des sables
DE Kleine Rauhaartaschenmaus
ES Ratón de abazones pequeño de desierto

LC
9 sspp
HB 6.7–7 cm
T 8.5–8.6 cm
W c. 11 g

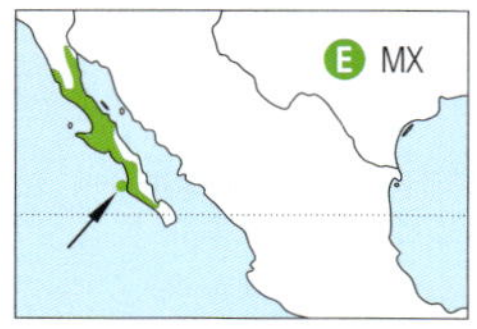

below 600 m

Baja California Pen, with Magdalena I

Dalquest's Pocket Mouse
Chaetodipus ammophilus

FR Souris-à-abajoues de Dalquest
DE Dalquest-Rauhaartaschenmaus
ES Ratón de abazones de Dalquest

NT
3 sspp
HB c. 7.2 cm
T c. 9.1 cm
W c. 14 g

below 600 m

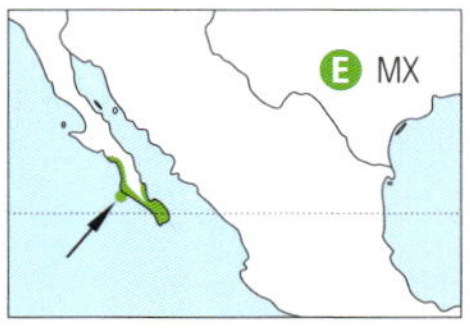

S Baja California Sur, with Santa Margarita I

LC
Monotypic
HB c. 7.2 cm
T c. 9.1 cm
W c. 14 g

Cerralvo Island Pocket Mouse
Chaetodipus siccus

FR Souris-à-abajoues de Cerralvo
DE Cerralvo-Rauhaartaschenmaus
ES Ratón de abazones de Isla Cerralvo

SE Baja California Sur (Los Planes Basin, Cerralvo I)

Desert Pocket Mouse
Chaetodipus penicillatus

FR Souris-à-abajoues du désert
DE Wüsten-Rauhaartaschenmaus
ES Ratón de abazones de desierto

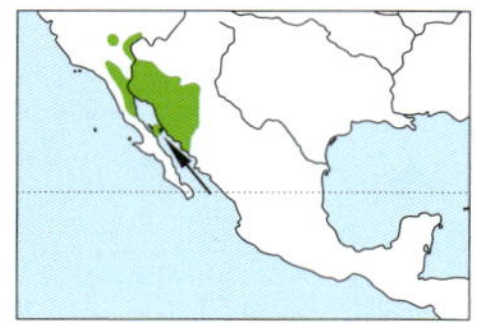

35–1585 m

US (from S Nevada & SW Utah) S to MX (NE Baja California, Sonora, Tiburón I)

LC
6 sspp
HB 7.9–8.7 cm
T 8.3–12.9 cm
W 15–19 g

LC
2 sspp
HB 7–8.1 cm T 7.7–9.8 cm
W 13–19 g

Chihuahuan Desert Pocket Mouse
Chaetodipus eremicus

FR Souris-à-abajoues du Chihuahua
DE Chihuahua-Rauhaartaschenmaus
ES Ratón de abazones de Chihuahua

below 1700 m

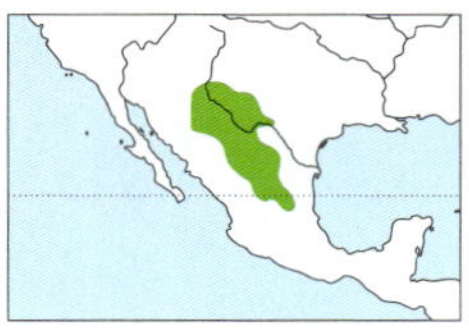

S US (from S New Mexico & SW Texas) S to N MX (to N San Luis Potosí)

Sinaloan Pocket Mouse
Chaetodipus pernix

FR Souris-à-abajoues de Sinaloa
DE Sinaloa-Rauhaartaschenmaus
ES Ratón de abazones de Sinaloa

0–90 m

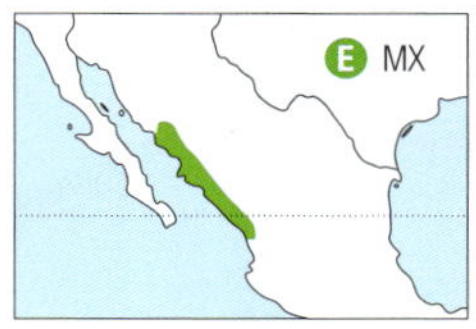

Arid coastal plains in S Sonora, Sinaloa & N Nayarit

LC
2 sspp
HB c. 7.5 cm
T c. 8.3 cm
W c. 17 g

LC
8 sspp
HB 7.3–7.6 cm T 8.4–11.2 cm
W 11–20 g

Rock Pocket Mouse
Chaetodipus intermedius

FR Souris-à-abajoues des rochers
DE Felsen-Rauhaartaschenmaus
ES Ratón de abazones de roca

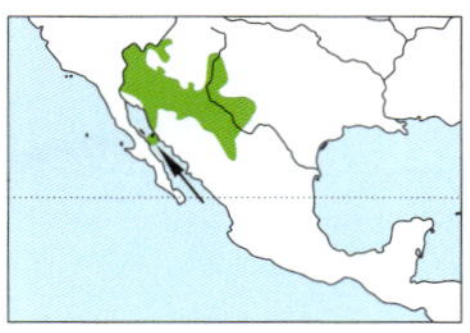

SW US (from S Utah & New Mexico) S to MX (N Sonora, N Chihuahua, Turner I)

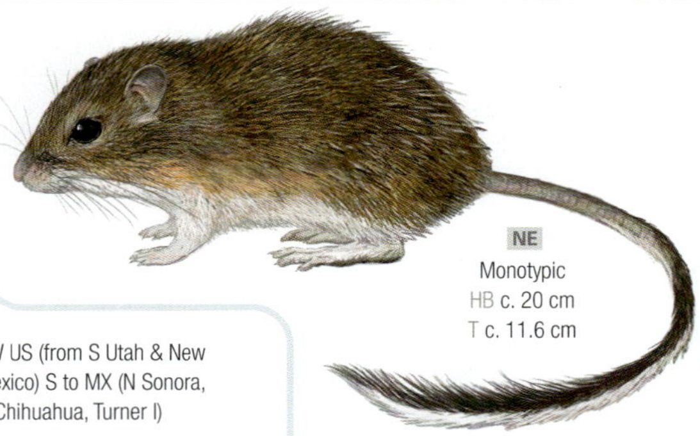

NE
Monotypic
HB c. 20 cm
T c. 11.6 cm

Durango Coarse-haired Pocket Mouse
Chaetodipus durangae

FR Souris-à-abajoues du Durango
DE Durango-Rauhaartaschenmaus
ES Ratón de abazones de Durango

Chihuahua (upper Presidio-Mezquital drainage), Durango (R Nazas)

Highland Coarse-haired Pocket Mouse
Chaetodipus collis

FR Souris-à-abajoues de Blair
DE Hochland-Rauhaartaschenmaus
ES Ratón de abazones de tierras altas

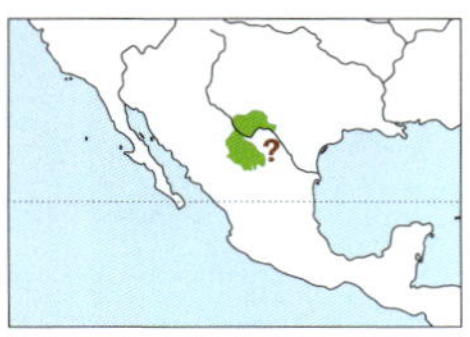

NE
2 sspp
HB 7.8–8.1 cm
T 8.8–11.5 cm
W 14–18 g

S US (extreme SW New Mexico, W Texas), N MX (Bolsón Mapimí)

Nelson's Coarse-haired Pocket Mouse
Chaetodipus nelsoni

FR Souris-à-abajoues de Nelson
DE Nelson-Rauhaartaschenmaus
ES Ratón de abazones de Nelson

LC
Monotypic
HB 7.8–8.1 cm
T 8.8–11.5 cm
W 14–18 g

Rocky and sandy soils from S Coahuila S to Jalisco

DD
Monotypic
HB c. 7.4 cm
T c. 9.2 cm
W c. 17 g

Lined Pocket Mouse
Chaetodipus lineatus

FR Souris-à-abajoues rayée
DE Streifen-Rauhaartaschenmaus
ES Ratón de abazones de líneas

1600–2400 m

Desert plains in SE Zacatecas & SW San Luis Potosí

NT
Monotypic
HB c. 8.3 cm T c. 10.7 cm
W 19–23 g

Goldman's Pocket Mouse
Chaetodipus goldmani

FR Souris-à-abajoues de Goldman
DE Goldman-Rauhaartaschenmaus
ES Ratón de abazones de Goldman

Sonoran Desert of S Sonora, W Chihuahua & N Sinaloa

Narrow-skulled Pocket Mouse
Chaetodipus artus

FR Souris-à-abajoues svelte
DE Schmalschädel-Rauhaartaschenmaus
ES Ratón de abazones de cráneo estrecho

LC
Monotypic
HB 8.6–9.2 cm
T 9.5–9.9 cm
W 13–27 g

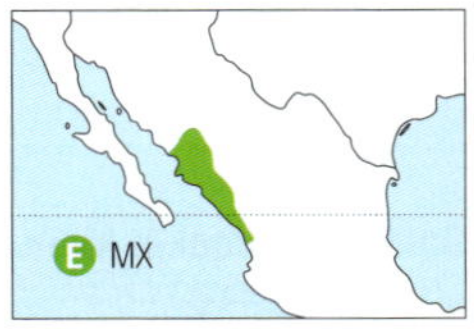

Pacific coast (S Sonora & SW Chihuahua to NW Durango, Sinaloa & N Nayarit)

Genus *MICRODIPODOPS*

(2 species)

Dark Kangaroo Mouse
Microdipodops megacephalus

FR Souris-kangourou foncée
DE Dunkle Kängurumaus
ES Ratón canguro oscuro

LC
12 sspp
HB c. 6.6 cm
T 6.7–10.3 cm
W 10–17 g

1190–2320 m

SE Oregon, NE & CE California, Nevada, extreme SW Idaho, W Utah

Pallid Kangaroo Mouse
Microdipodops pallidus

FR Souris-kangourou pâle
DE Fahle Kängurumaus
ES Ratón canguro pálido

LC
5 sspp
HB c. 6.6 cm
T 7.4–9.9 cm
W 10–17 g

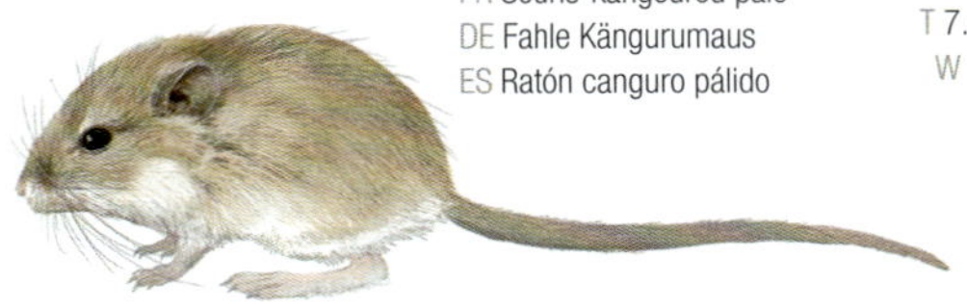

WC & SC Great Basin (Nevada), disjunctly in Deep Springs Valley (E California)

Genus *DIPODOMYS*

(20 species)

Desert Kangaroo Rat
Dipodomys deserti

FR Rat-kangourou du désert
DE Wüsten-Kängururatte
ES Rata canguro de desierto

LC
4 sspp
HB 13.4–15.5 cm
T 19.5–20.1 cm
W 83–148 g

60–1710 m

SW US (from E California & W Nevada) S to NW MX (NE Baja California, W Sonora)

Banner-tailed Kangaroo Rat
Dipodomys spectabilis

FR Rat-kangourou porte-étendard
DE Fahnenschwanz-Kängururatte
ES Rata canguro de cola de bandera

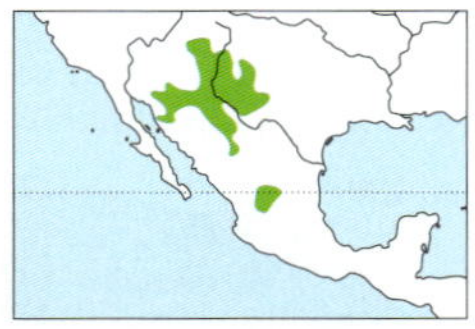

SW US (Arizona E to Texas), MX (N Sonora, Chihuahua; disjunctly in Mexican Plateau)

NT
6 sspp
HB 13–14.1 cm
T 18–21 cm
W 98–132 g

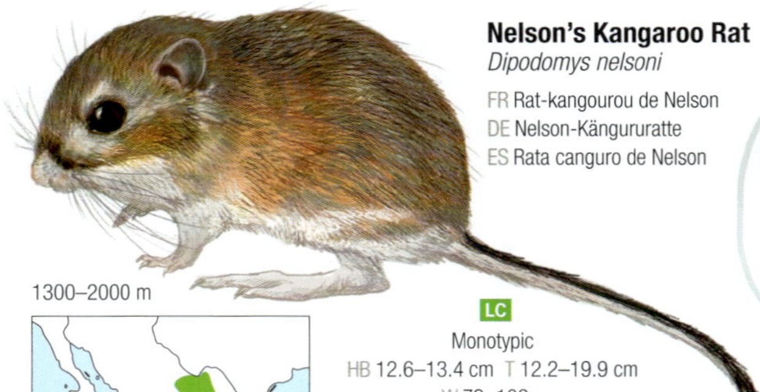

Nelson's Kangaroo Rat
Dipodomys nelsoni

FR Rat-kangourou de Nelson
DE Nelson-Kängururatte
ES Rata canguro de Nelson

1300–2000 m

SE Chihuahua & W Coahuila to N San Luis Potosí & S Nuevo León

LC
Monotypic
HB 12.6–13.4 cm T 12.2–19.9 cm
W 73–102 g

Texas Kangaroo Rat
Dipodomys elator

FR Rat-kangourou du Texas
DE Texas-Kängururatte
ES Rata canguro de Texas

VU
Monotypic
HB 9.9–14 cm
T 16.1–20 cm
W 65–90 g

Honey mesquite trees areas (NC Texas, adjacent SC Oklahoma)

Chihuahuan Desert of Mexican Plateau (SE Durango S to Querétaro)

Plateau Kangaroo Rat
Dipodomys ornatus

FR Rat-kangourou orné
DE Altiplano-Kängururatte
ES Rata canguro de altiplano

LC
Monotypic
HB 10.1–11 cm
T 15.1–19.2 cm
W c. 53 g

Valley of México of Distrito Federal, México & Hidalgo, S to N Oaxaca

Phillips's Kangaroo Rat
Dipodomys phillipsii

FR Rat-kangourou de Phillips
DE Phillips-Kängururatte
ES Rata canguro de Phillips

LC
3 sspp
HB 8.1–11.4 cm
T 14.9–19 cm
W c. 53 g

0–30 m

Coastal scrub plains in NW Baja California

San Quintin Kangaroo Rat
Dipodomys gravipes

FR Rat-kangourou de San Quintín
DE Baja-California-Känguruatte
ES Rata canguro de San Quintín

CR(PE)
Monotypic
HB 12.6–13.2 cm
T 16.8–18 cm W 79–91 g

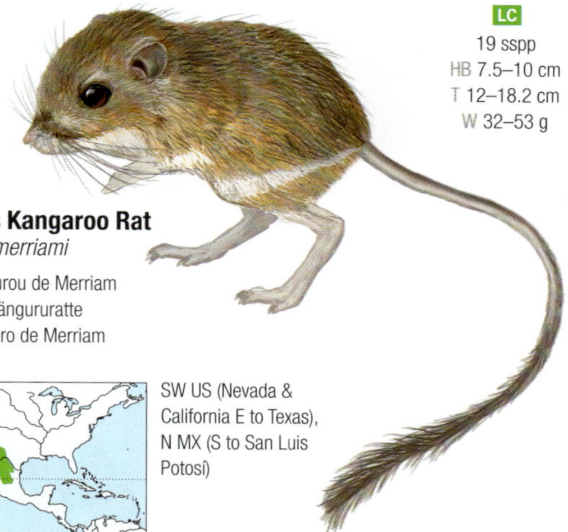

LC
19 sspp
HB 7.5–10 cm
T 12–18.2 cm
W 32–53 g

Merriam's Kangaroo Rat
Dipodomys merriami

FR Rat-kangourou de Merriam
DE Merriam-Känguruatte
ES Rata canguro de Merriam

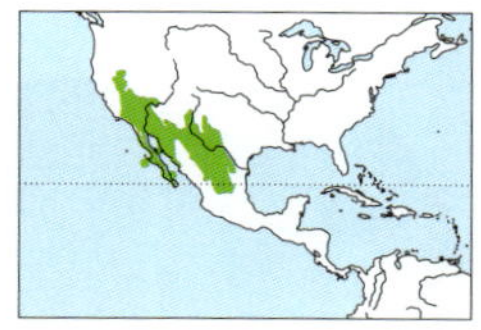

SW US (Nevada & California E to Texas), N MX (S to San Luis Potosí)

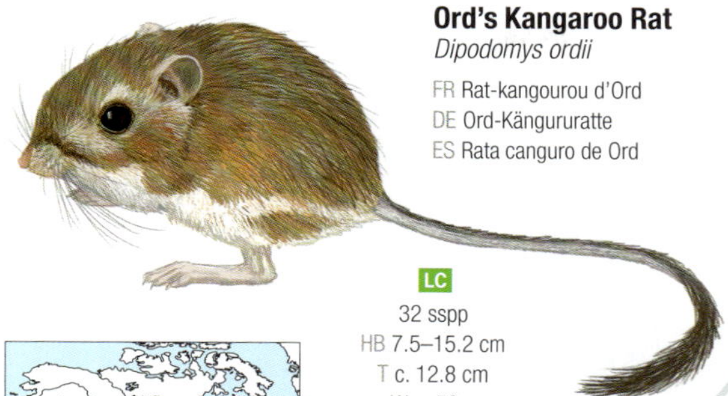

Ord's Kangaroo Rat
Dipodomys ordii

FR Rat-kangourou d'Ord
DE Ord-Känguruatte
ES Rata canguro de Ord

LC
32 sspp
HB 7.5–15.2 cm
T c. 12.8 cm
W c. 52 g

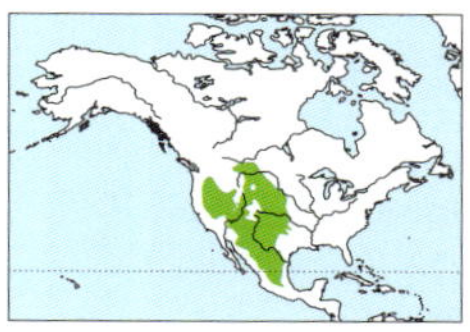

S CA (Alberta, Saskatchewan), W & S US (E to C Kansas) S to N MX (Hidalgo)

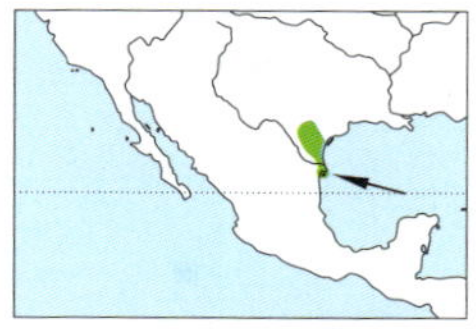

S US (SE Texas, Mustang & Padre Is), NE MX (barrier Is of N Tamaulipas)

ssp *sennetti*

Gulf Coast Kangaroo Rat
Dipodomys compactus

FR Rat-kangourou côtier
DE Golfküsten-Känguruatte
ES Rata canguro de la Costa del Golfo

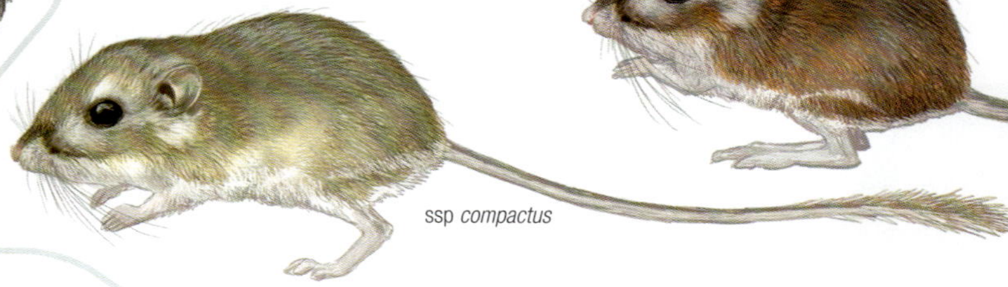

ssp *compactus*

LC
2 sspp
HB 9.9–12 cm
T 10.4–13.5 cm
W 44–60 g

up to 1300 m

Open areas in S Oregon & N California

California Kangaroo Rat
Dipodomys californicus

FR Rat-kangourou de Californie
DE Kalifornien-Känguruatte
ES Rata canguro de California

LC
3 sspp
HB 10.8–12.3 cm
T 15.2–22 cm
W 57–85 g

LC
3 sspp
HB 11.5–13 cm
T 17.5–20 cm
W 68–97 g

Narrow-faced Kangaroo Rat
Dipodomys venustus

FR Rat-kangourou des Santa Cruz
DE Santa-Cruz-Känguruatte
ES Rata canguro de cara estrecha

0–1770 m

Disjunct populations in WC California

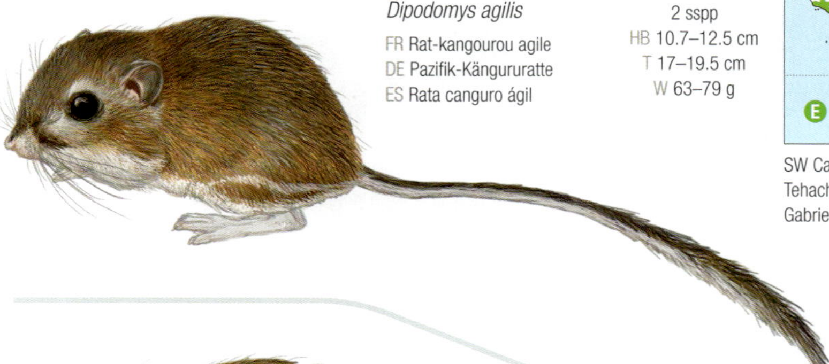

Agile Kangaroo Rat
Dipodomys agilis

FR Rat-kangourou agile
DE Pazifik-Känguruatte
ES Rata canguro ágil

LC
2 sspp
HB 10.7–12.5 cm
T 17–19.5 cm
W 63–79 g

SW California (Los Angeles Basin, Tehachapi, S Sierra Nevada, San Gabriel & San Bernardino Mts)

Giant Kangaroo Rat
Dipodomys ingens

FR Rat-kangourou géant
DE Riesenkänguruatte
ES Rata canguro gigante

200–760 m

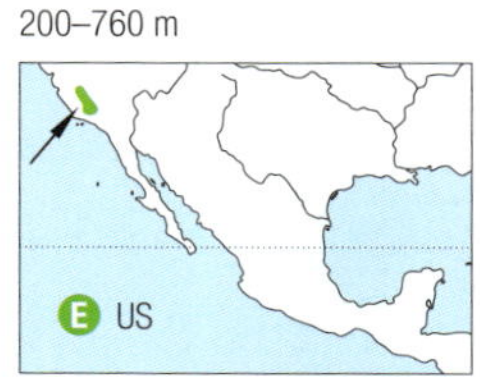

SC California (along SW edge of San Joaquin Valley)

EN
Monotypic
HB 13.7–14.9 cm
T 17.4–19.8 cm
W 93–195 g

Dulzura Kangaroo Rat
Dipodomys simulans

FR Rat-kangourou de Dulzura
DE San-Diego-Känguruatte
ES Rata canguro de Dulzura

LC
2 sspp
HB 11.1–12.1 cm
T 16–18.1 cm
W 55–70 g

0–2250 m

SW US (coastal basins of SW California), NW MX (Baja California Pen)

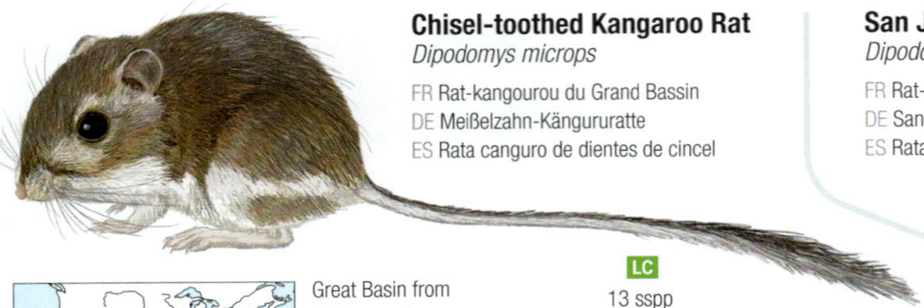

Chisel-toothed Kangaroo Rat
Dipodomys microps

FR Rat-kangourou du Grand Bassin
DE Meißelzahn-Känguruhratte
ES Rata canguro de dientes de cincel

Great Basin from SE Oregon & SW Idaho to S California & N Arizona

LC
13 sspp
HB 11–12 cm T 13.5–17.5 cm
W 40–70 g

San Joaquin Kangaroo Rat
Dipodomys nitratoides

FR Rat-kangourou du San Joaquin
DE San-Joaquin-Känguruhratte
ES Rata canguro de San Joaquín

VU
3 sspp
HB 7.4–11.2 cm
T 13.7–14.1 cm
W 40–53 g

SC California (San Joaquin, Panoche & Cuyama valleys, Carrizo Plain)

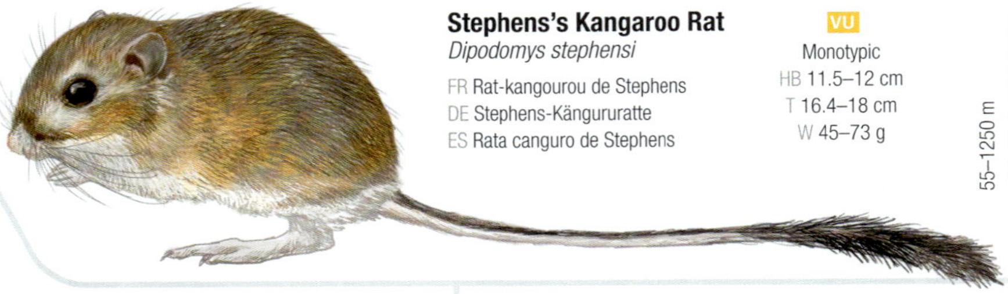

Stephens's Kangaroo Rat
Dipodomys stephensi

FR Rat-kangourou de Stephens
DE Stephens-Känguruhratte
ES Rata canguro de Stephens

VU
Monotypic
HB 11.5–12 cm
T 16.4–18 cm
W 45–73 g

55–1250 m

SW California (San Jacinto Valley and close lowlands, disjunctly near Warner Springs)

WC California (San Joaquin Valley & San Francisco Bay S to Morro Bay)

LC
9 sspp
HB 9–11.3 cm
T 16–20 cm
W 70–80 g

Heermann's Kangaroo Rat
Dipodomys heermanni

FR Rat-kangourou de Heermann
DE Heermann-Känguruhratte
ES Rata canguro de Heermann

LC
5 sspp
HB c. 11.9 cm
T c. 17 cm
W c. 72 g

Panamint Kangaroo Rat
Dipodomys panamintinus

FR Rat-kangourou des Panamint
DE Mojave-Känguruhratte
ES Rata canguro de Panamint

S along E of Sierra Nevada to Mojave Desert (EC & S California, WC Nevada)

1200–2670 m

Family GEOMYIDAE – Pocket Gophers

7 genera, 41 species. Near sea level to timberline in all habitats with friable, non-flooding soils. Nearctic and Neotropical Regions.

Mountain Pocket Gopher
Thomomys monticola

FR Gaufre de montagne
DE Sierra-Nevada-Taschenratte
ES Tuza de montaña

Genus *THOMOMYS*
(12 species)

S Cascade Range, C Sierra Nevada; disjunctly in S Yolla Bolly Mt (California, W Nevada)

LC
Monotypic
HB 12–16 cm
T 5–9 cm
W 70–110 g

Western Pocket Gopher
Thomomys mazama

FR Gaufre du Mazama
DE Mazama-Taschenratte
ES Tuza occidental

LC
14 extant sspp
HB 14–17 cm
T 5–8 cm
W 80–130 g

Multiple disjunct populations in N Whasington, W Oregon & NW California

Idaho Pocket Gopher
Thomomys idahoensis

FR Gaufre d'Idaho
DE Idaho-Taschenratte
ES Tuza de Idaho

Idaho and Montana form

Idaho, W Montana, SW Wyoming & extreme N Utah

LC
3 sspp
HB 12–15 cm
T 4–7 cm
W 45–90 g

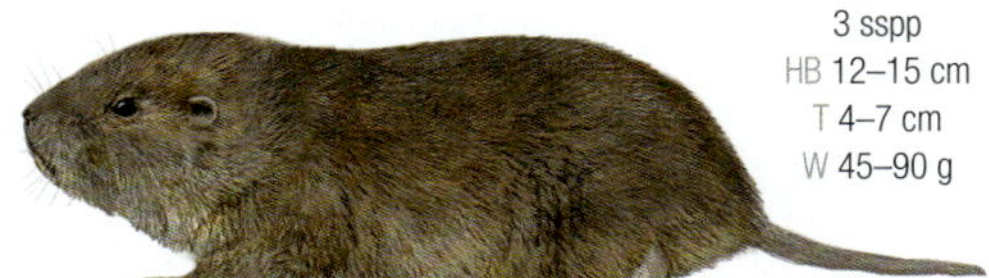

Wyoming form

Wyoming Pocket Gopher
Thomomys clusius

FR Gaufre du Wyoming
DE Wyoming-Taschenratte
ES Tuza de Wyoming

S Wyoming (2 sites in SE Sweetwater & SW Carbon Counties)

LC
Monotypic
HB 10–13 cm
T 5–7 cm
W 40–70 g

Northern Pocket Gopher
Thomomys talpoides

FR Gaufre gris
DE Nördliche Taschenratte
ES Tuza septentrional

LC
54 sspp
HB 11–19 cm
T 5–8 cm
W 65–209 g

900–3700 m

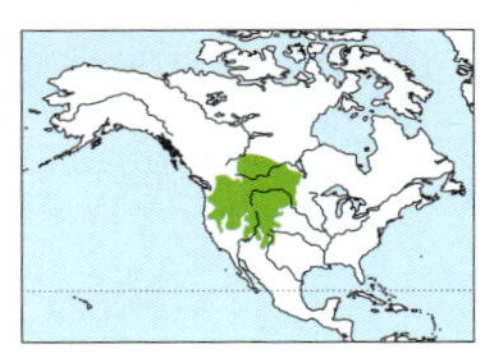

CA (from C Alberta) S to W US (S to California, N Arizona, N New Mexico)

Camas Pocket Gopher
Thomomys bulbivorus

FR Gaufre de la Willamette
DE Oregon-Taschenratte
ES Tuza de Camas

LC
Monotypic
HB 20–24 cm
T 8.5–9.5 cm
W 300–500 g

Willamette Valley (NW Oregon)

Townsend's Pocket Gopher
Thomomys townsendii

FR Gaufre de Townsend
DE Townsend-Taschenratte
ES Tuza de Townsend

LC
2 sspp
HB 14.5–23 cm
T 5.5–10 cm
W 120–420 g

up to 2000 m

SE Oregon, S Idaho, NE California, N Nevada (R Snake, R Humboldt & R Quinn, Honey Lake Valley)

Southern Pocket Gopher
Thomomys umbrinus

FR Gaufre méridional
DE Südliche Taschenratte
ES Tuza meridional

LC
4 sspp
HB 12–18 cm
T 5.5–8 cm
W 80–180 g

0–2400 m

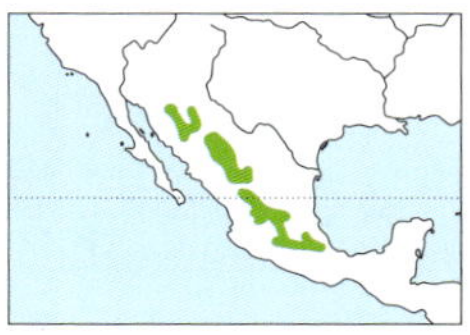

US (SE Arizona, SW New Mexico), MX (S to Trans-Mexican Volcanic Belt)

Sierra Madre Occidental Pocket Gopher
Thomomys sheldoni

FR Gaufre de Sheldon
DE Sierra Madre-Taschenratte
ES Tuza de Sierra Madre Occidental

NE
2 sspp
HB 12–18 cm
T 5.5–8 cm
W 75–130 g

above 2000 m

NC Chihuahua, W Durango, W Zacatecas, NE Nayarit

Black-and-brown Pocket Gopher
Thomomys atrovarius

FR Gaufre de Tatemeles
DE Schwarzbraune Taschenratte
ES Tuza negra y marrón

NE
2 sspp
HB 11–16.5 cm
T 5.5–8 cm
W 75–150 g

Sinaloa to S Nayarit, W Durango & NW Jalisco

NE
Monotypic
HB 11–14 cm
T 5–7.5 cm
W 75–110 g

Nayar Pocket Gopher
Thomomys nayarensis

FR Gaufre de la Nayar
DE Nayarit-Taschenratte
ES Tuza de Nayar

1300–2200 m

NE Nayarit (known from only 2 localities in Sierra del Nayar)

small form

large form

LC
128 sspp
HB 8–20 cm
T 5.5–9.5 cm
W 80–250 g

Botta's Pocket Gopher
Thomomys bottae

FR Gaufre de Botta
DE Gebirgstaschenratte
ES Tuza de Botta

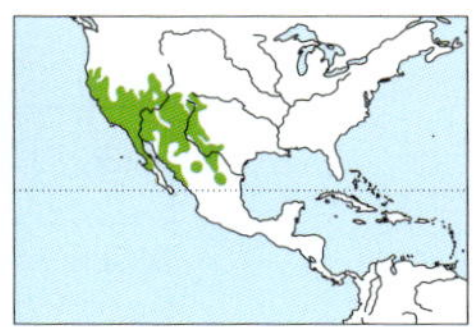

SW US (from S Oregon E through Great Basin) S to lowlands of N MX

Genus *GEOMYS*

(12 species)

Plains Pocket Gopher
Geomys bursarius

FR Gaufre brun
DE Flachland-Taschenratte
ES Tuza de llanura

LC
8 sspp
HB 13.5–23 cm
T 6–12 cm W 120–250 g

variants

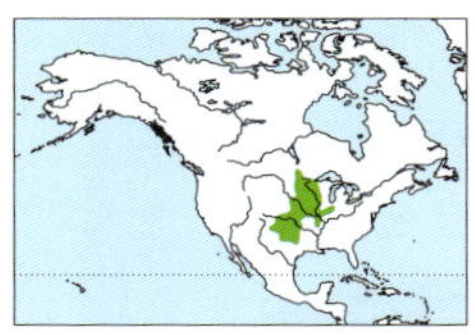

CA (extreme S Manitoba), S across C US Great Plains to New Mexico & Texas

Llano Pocket Gopher
Geomys texensis

FR Gaufre de Llano
DE Llano-Taschenratte
ES Tuza de Llano

LC
3 sspp
HB 12–21 cm T 5–8 cm
W 155–215 g

C & SC Texas

NE
2 sspp
HB 11.5–19 cm
T 5.5–10.5 cm
W 160–185 g

Hall's Pocket Gopher
Geomys jugossicularis

FR Gaufre de Hall
DE Hall-Taschenratte
ES Tuza de Hall

E Colorado, SW Nebraska, W Kansas

Sand Hills Pocket Gopher
Geomys lutescens

FR Gaufre des Sand Hills
DE Nebraska-Taschenratte
ES Tuza de Sand Hills

E Wyoming, W Nebraska, SW South Dakota, NE Colorado

NE
Monotypic
HB 11.5–19 cm
T 5.5–10.5 cm
W 160–185 g

Desert Pocket Gopher
Geomys arenarius

FR Gaufre du désert
DE Wüstentaschenratte
ES Tuza de desierto

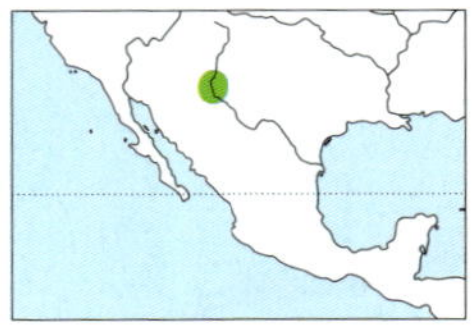

SC US (S New Mexico, W Texas), adjacent MX (NE Chihuahua)

NT
2 sspp
HB 13–22 cm
T 5–10.5 cm
W 165–255 g

Jones's Pocket Gopher
Geomys knoxjonesi

FR Gaufre de Jones
DE Jones-Taschenratte
ES Tuza de Jones

Scrubs in SE New Mexico & adjacent W Texas

LC
Monotypic
HB 11.5–19 cm
T 5.5–10.5 cm
W 160–185 g

Attwater's Pocket Gopher
Geomys attwateri

FR Gaufre d'Attwater
DE Attwater-Taschenratte
ES Tuza de Attwater

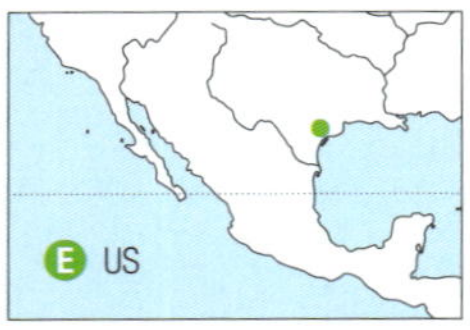

Grasses in gulf coast of SE Texas

LC
2 sspp
HB 13–17.5 cm
T 5–7 cm
W 130–160 g

Texas Pocket Gopher
Geomys personatus

FR Gaufre du Texas
DE Texas-Taschenratte
ES Tuza de Texas

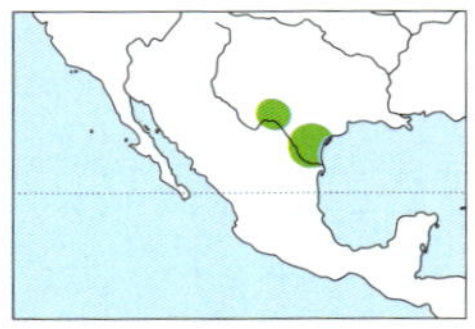

SE US (patchily in S Texas, with Mustang & Padre Is), NE MX (NE Tamaulipas)

LC
6 sspp
HB 14–28 cm
T 6–12 cm
W 165–400 g

Strecker's Pocket Gopher
Geomys streckeri

FR Gaufre de Strecker
DE Strecker-Taschenratte
ES Tuza de Strecker

SW Texas (N of Rio Grande in Dimmit & Zavala counties)

NE
Monotypic
HB 12–19 cm
T 7.5–10.5 cm
W 160–185 g

Tropical Pocket Gopher
Geomys tropicalis

FR Gaufre tropical
DE Tamaulipas-Taschenratte
ES Tuza de Tamaulipas

SE Tamaulipas (Coastal plains around Altamira and Tampico)

EN
Monotypic
HB 12–18 cm
T 6–9 cm
W 130–250 g

Baird's Pocket Gopher
Geomys breviceps

FR Gaufre de Baird
DE Baird-Taschenratte
ES Tuza de Baird

SE Oklahoma, SW Arkansas, E Texas (E of R Brazos), Louisiana

LC
2 sspp
HB 13–16 cm
T 5–7 cm
W 78–150 g

LC
6 extant sspp
HB 13–24 cm
T 5.5–12 cm
W 135–210 g

South-eastern Pocket Gopher
Geomys pinetis

FR Gaufre des pinèdes
DE Südöstliche Taschenratte
ES Tuza suroriental

C & S Alabama, C & S Georgia, N & C Florida

Genus *ZYGOGEOMYS*

(1 species)

EN
2 sspp
HB 17–24 cm
T 6–10 cm
W 190–580 g

Michoacan Pocket Gopher
Zygogeomys trichopus

FR Gaufre du Michoacan
DE Michoacán-Taschenratte
ES Tuza de Michoacán

2200–3000 m

NW & C Michoacán (Cerro Patambán & Cerro Tancítaro, vicinity of Pátzcuaro)

Genus *ORTHOGEOMYS*

(1 species)

Giant Pocket Gopher
Orthogeomys grandis

FR Grand Gaufre
DE Riesentaschenratte
ES Tuza gigante

lower elevation form

LC
16 sspp
HB 22–30 cm
T 9.5–14 cm
W 480–985 g

higher elevation form

0–2700 m

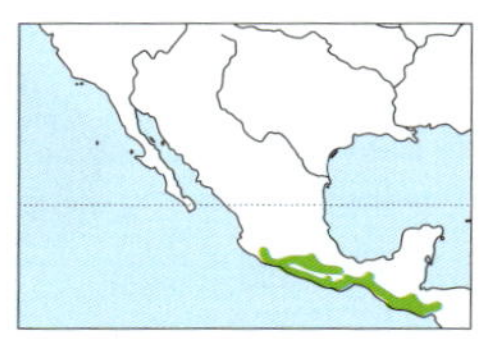

S MX (SE from S Colima), GT, SW HN, SV

Hispid Pocket Gopher
Heterogeomys hispidus

FR Gaufre hérissé
DE Borstige Taschenratte
ES Tuza hirsuta

LC
12 sspp
HB 21–27 cm
T 7.5–9 cm
W 450–600 g

0–2500 m

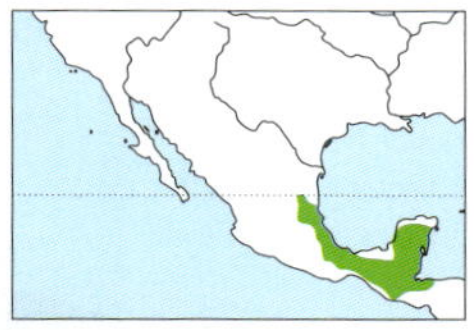

E & S MX (from SC Tamaulipas) S to BZ, GT & NW HN

Genus *HETEROGEOMYS*

(7 species)

Big Pocket Gopher
Heterogeomys lanius

FR Gaufre de Xuchil
DE Wollige Taschenratte
ES Tuza grande

CR
Monotypic
HB 25–29 cm
T 9–9.5 cm
W 850–900 g

above 2400 m

WC Veracruz (known only from 2 localities)

Chiriqui Pocket Gopher
Heterogeomys cavator

FR Gaufre du Chiriqui
DE Chiriquí-Taschenratte
ES Tuza de Chiriquí

LC
3 sspp
HB 22–27 cm
T 11–13 cm
W 500–910 g

0–3200 m

E CR, W PA

Darien Pocket Gopher
Heterogeomys dariensis

FR Gaufre du Darien
DE Darién-Taschenratte
ES Tuza de Darién

LC
2 sspp
HB 21–28 cm
T 12–14 cm
W 450–850 g

0–1200 m

E PA (Darién), NW CO (Chocó)

LC
4 sspp
HB 18–25 cm
T 8–10.5 cm
W 220–510 g

Variable Pocket Gopher
Heterogeomys heterodus

FR Gaufre du Costa Rica
DE Costa-Rica-Taschenratte
ES Tuza variable

LC
3 sspp
HB 22–27 cm
T 8.5–11.5 cm
W 400–915 g

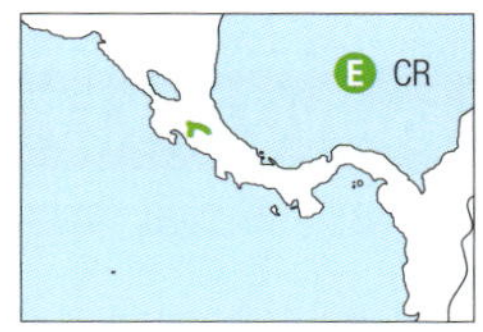

1000–2000 m

Central highlands in Alajuela, San José & Cartago prov

Cherrie's Pocket Gopher
Heterogeomys cherriei

FR Gaufre de Cherrie
DE Cherrie-Taschenratte
ES Tuza de Cherrie

0–1500 m

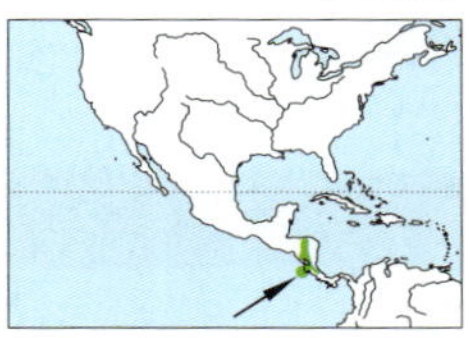

NC HN, NI, N & E CR (including parts of Nicoya Pen)

Underwood's Pocket Gopher
Heterogeomys underwoodi

FR Gaufre d'Underwood
DE Underwood-Taschenratte
ES Tuza de Underwood

LC
Monotypic
HB 18–21 cm
T 8.5–10.5 cm
W 230–330 g

0–1500 m

Pacific coast of CR & adjacent SW PA (Chiriquí)

Genus *PAPPOGEOMYS*

(1 species)

Buller's Pocket Gopher
Pappogeomys bulleri

FR Gaufre de Buller
DE Jalisco-Taschenratte
ES Tuza de Buller

LC
5 sspp
HB 12.5–18 cm
T 6–10 cm
W 100–250 g

up to 3000 m

From C Nayarit, trough Jalisco to Colima

Genus *CRATOGEOMYS*

(7 species)

Yellow-faced Pocket Gopher
Cratogeomys castanops

FR Gaufre à face jaune
DE Gelbgesicht-Taschenratte
ES Tuza de cara amarilla

LC
2 sspp
HB 15.5–23 cm
T 5–10.5 cm
W 185–410 g

10–2100 m

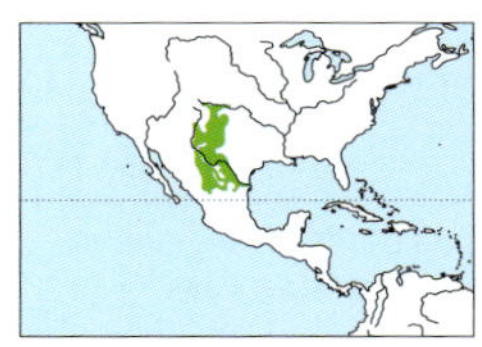

CS US (from SE Colorado & SW Kansas) S to MX (S to N Durango)

Goldman's Pocket Gopher
Cratogeomys goldmani

FR Gaufre de Goldman
DE Goldman-Taschenratte
ES Tuza de Goldman

LC
2 sspp
HB 14.5–21 cm
T 5.5–9.5 cm
W 130–350 g

750–2700 m

S Coahuila, NE Durango, Zacatecas, SW Nuevo León, San Luis Potosí

Oriental Basin Pocket Gopher
Cratogeomys fulvescens

FR Gaufre fauve
DE Puebla-Taschenratte
ES Tuza de la Cuenca Oriental

LC
Monotypic
HB 18–28 cm
T 8.5–11 cm
W 250–550 g

2300–2700 m

Oriental Basin of E Puebla, E Tlaxcala & extreme W Veracruz

Merriam's Pocket Gopher
Cratogeomys merriami
FR Gaufre de Merriam
DE Merriam-Taschenratte
ES Tuza de Merriam

slate-black morph

1800–4000 m

Patchily in S part of Valley of México E to W Puebla

LC
Monotypic
HB 20–28 cm
T 8–12.5 cm
W 450–700 g

chestnut-brown morph

Cofre de Perote Pocket Gopher
Cratogeomys perotensis
FR Gaufre du Perote
DE Veracruz-Taschenratte
ES Tuza de Cofre de Perote

LC
Monotypic
HB 20–25 cm
T 8–11.5 cm
W 280–600 g

2400–4000 m

S Hidalgo, N Tlaxcala, N Puebla, WC Veracruz (S to Cofre de Perote & Pico de Orizaba)

Volcan de Toluca Pocket Gopher
Cratogeomys planiceps
FR Gaufre du Toluca
DE Toluca-Taschenratte
ES Tuza del Volcán de Toluca

LC
Monotypic
HB 21–28 cm
T 7.5–11 cm
W 400–625 g

2500–3500 m

México State (N slopes of Volcán de Toluca, hills N & E of Valle de Bravo)

black morph

light-brown morph

Smoky Pocket Gopher
Cratogeomys fumosus
FR Gaufre fuligineux
DE Colima-Taschenratte
ES Tuza ahumada

LC
4 sspp
HB 15.5–28 cm
T 6.5–11 cm
W 250–1050 g

300–3000 m

Trans-Mexican Volcanic Belt (from Jalisco & Colima E to Distrito Federal)

Family ZENKERELLIDAE – Non-gliding Anomalure

1 genus, 1 species. Rainforests and semi-deciduous forests. Afrotropical Region.

Zenkerella
Zenkerella insignis
FR Anomalure aptère
DE Dornschwanzbilch
ES Anomaluro de Camerún

LC
Monotypic
HB 18–23 cm
T 15–18 cm
W 180–220 g

Genus *ZENKERELLA*

(1 species)

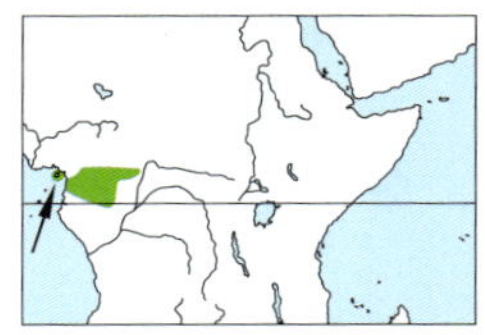

S CM, GQ (with Bioko I), N GA, N & W CD and SW CF; perhaps CG

Family ANOMALURIDAE – Gliding Anomalures

2 genera, 6 species. Montane, lowland rainforests, and woodlands. Afrotropical Region.

Beecroft's Anomalure
Anomalurus beecrofti
FR Anomalure de Beecroft
DE Beecroft-Dornschwanzhörnchen
ES Anomaluro de Beecroft

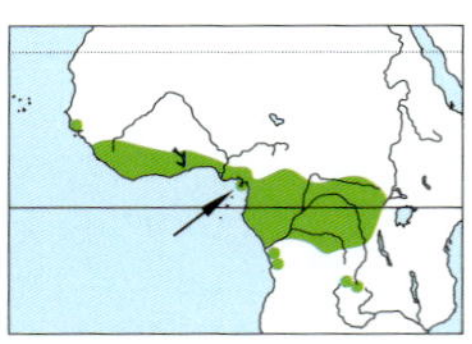

0–2500 m

From SN E to E CD, Bioko I; records in NW AO, NW ZM & W UG

LC
Monotypic
HB 25–31 cm
T 16–26 cm
W 640–660 g

ssp *auzembergeri*

Genus *ANOMALURUS*

(4 species)

ssp *peralbus*

LC
3 sspp
HB 40–46 cm
T 32–45 cm
W 1.3–1.8 kg

Pel's Anomalure
Anomalurus pelii
FR Anomalure de Pel
DE Pel-Dornschwanzhörnchen
ES Anomaluro de Pel

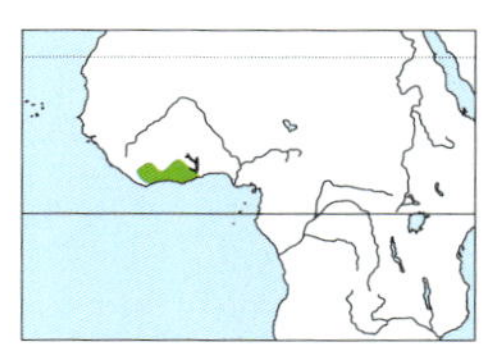

Primary forests in W Africa: E LR, S CI, S GH (W of R Volta)

Lesser Anomalure
Anomalurus pusillus

FR Anomalure pygmée
DE Zwerg-Dornschwanzhörnchen
ES Anomaluro enano

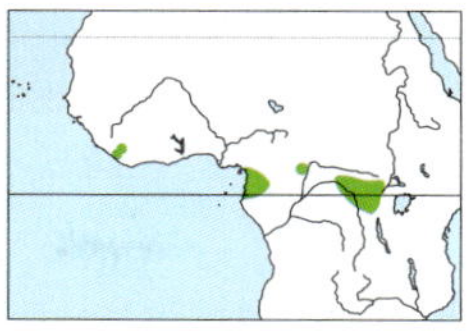

up to 2200 m

Discontinuously in LR, CM, GQ, NW GA, W CG, SW CF, N CD; perhaps W UG

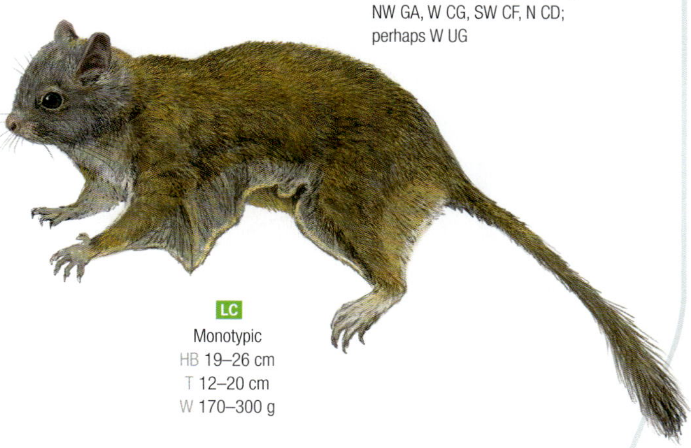

LC
Monotypic
HB 19–26 cm
T 12–20 cm
W 170–300 g

Lord Derby's Anomalure
Anomalurus derbianus

FR Anomalure de Derby
DE Lord-Derby-Dornschwanzhörnchen
ES Anomaluro de Lord Derby

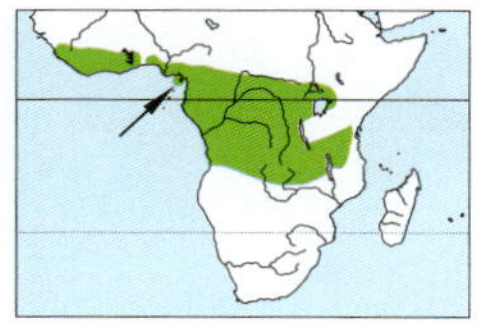

0–2400 m

From SL E to UG & SW KE, S to N AO, ZM, N MW, N MZ & TZ

variants

LC
Monotypic
HB 26–40 cm
T 22–33 cm
W 450–1100 g

Genus *IDIURUS*

(2 species)

Long-eared Pygmy Anomalure
Idiurus macrotis

FR Anomalure à oreilles longues
DE Großohr-Gleitbilch
ES Anomaluro de orejas grandes

LC
Monotypic
HB 7–12 cm
T 13–19 cm
W 23–35 g

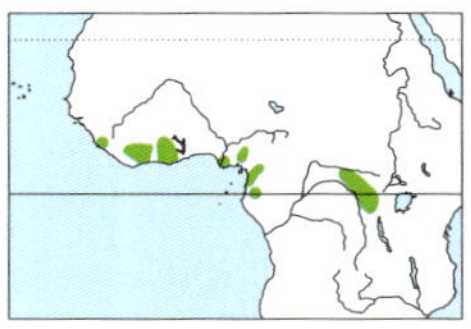

Disjunctly from SL to W GH, from SE NG to N GA, and NE CD (to Ituri Forest)

Zenker's Pygmy Anomalure
Idiurus zenkeri

FR Anomalure de Zenker
DE Zenker-Gleitbilch
ES Anomaluro pequeño

LC
Monotypic
HB 6–9 cm
T 7–13 cm
W 14–17 g

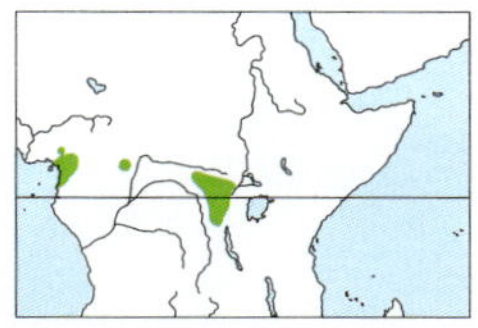

CM (SW & Mt Cameroon), GQ, CF, CD (Ngotto Forest & NE); perhaps W UG

Family PEDETIDAE – Springhares

1 genus, 2 species. Arid and semi-arid, flat and open savannas. Afrotropical Region.

Southern African Springhare
Pedetes capensis

FR Springhaas du Cap
DE Südlicher Springhase
ES Liebre saltadora de El Cabo

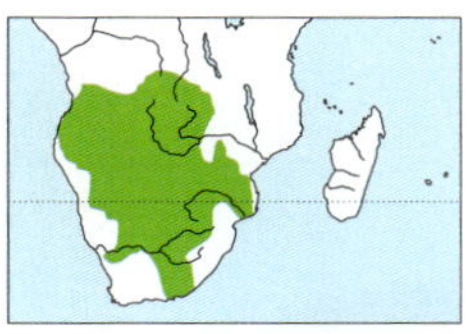

Widespread in S Africa from C AO & S CD to S MZ & ZA

LC
Monotypic
HB 34–46 cm
T 39–48 cm
W 2.5–3.5 kg

Genus *PEDETES*

(2 species)

East African Springhare
Pedetes surdaster

FR Springhaas du Kenya
DE Ostafrika-Springhase
ES Liebre saltadora oriental

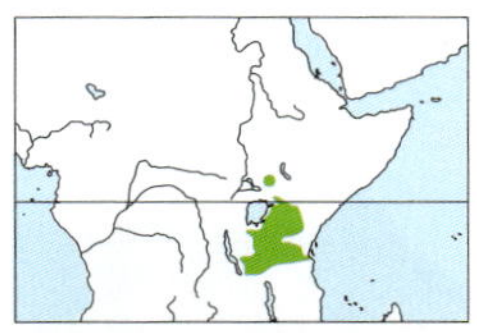

SW & SC KE, TZ (patchily from Nakuru S to L Rukwa); recorded in UG (Mt Moroto)

LC
Monotypic
HB 34–46 cm
T 39–48 cm
W 2.7–2.8 kg

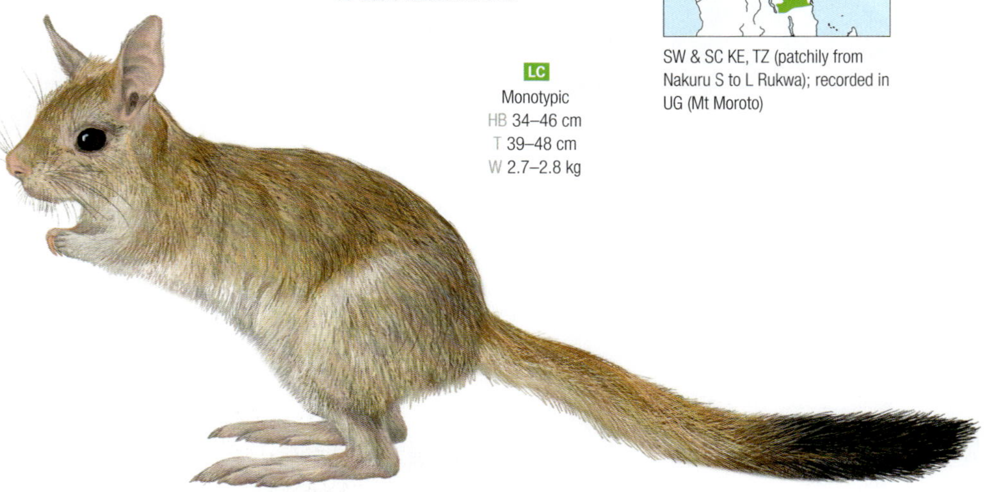

Family SMINTHIDAE – Birch Mice

1 genus, 19 species. Grasslands, steppes, semi-deserts, shrublands, subalpine meadows, and boreal forests. Palearctic Region.

Genus *SICISTA*

(19 species)

Chinese Birch Mouse
Sicista concolor

FR Siciste de Chine
DE China-Birkenmaus
ES Ratón listado de China

2140–4025 m

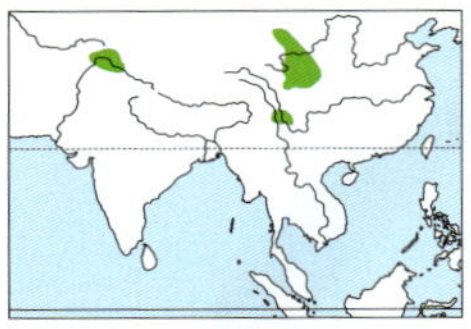

Kashmir (PK, IN), CN (Qinghai E to Shaanxi, S to Yunnan; probably SW Xinjiang)

LC
3 sspp
HB 5.1–7.6 cm
T 8.6–10.9 cm
W 5–8 g

Tian Shan Birch Mouse
Sicista tianschanica

FR Siciste des Tian Shan
DE Tienschan-Birkenmaus
ES Ratón listado de Tien Shan

LC
Monotypic
HB 5.7–7.3 cm
T 10.2–11.3 cm
W 7.2–13 g

400–3500 m

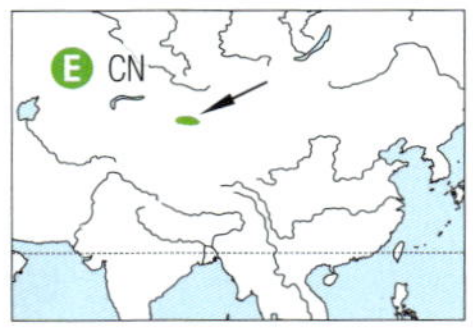

E Tian Shan, NE Narat range; exact limits unknown

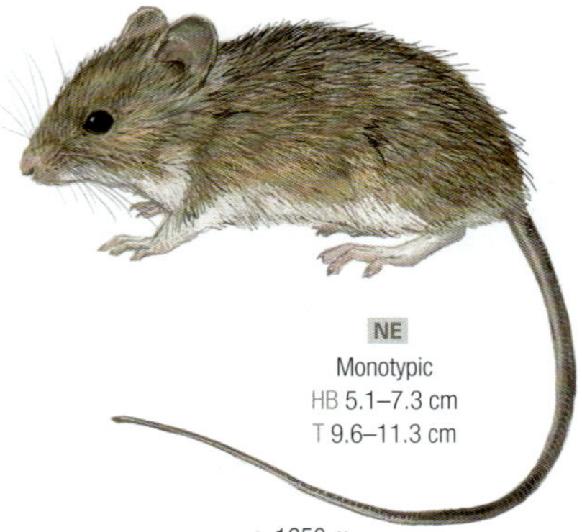

Talgar Birch Mouse
Sicista talgarica

FR Siciste de Talgar
DE Talgar-Birkenmaus
ES Ratón listado de Talgar

NE
Monotypic
HB 5.1–7.3 cm
T 9.6–11.3 cm

c. 1650 m

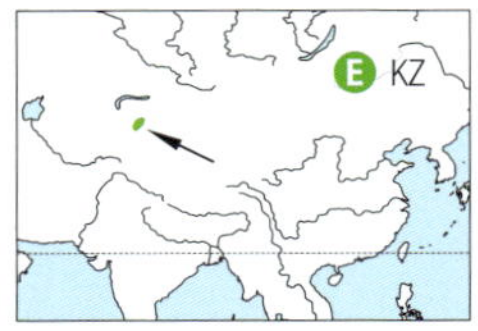

Trans-Ili Alatau–Talgar Mts, Malaya Almatinka Valley; E limit unknown

Terskey Birch Mouse
Sicista terskeica

FR Siciste des Terskey
DE Terskey-Birkenmaus
ES Ratón listado de Terskey

c. 2700 m

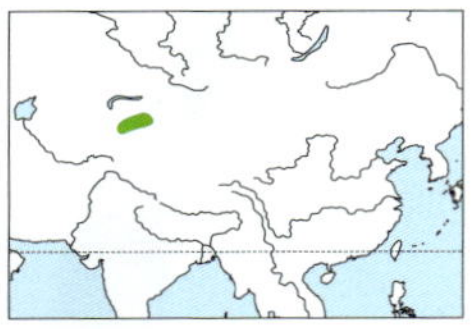

N Tian Shan (Terskey, Kungey & W Trans-Ili Alatau, Ketmen), C Tian Shan S to Atbashi (KZ, KG)

NE
Monotypic
HB 5.6–7 cm
T 8.6–10.3 cm
W c. 9.8 g

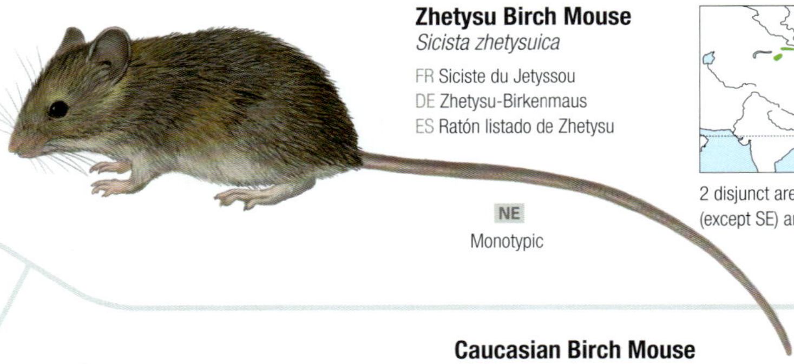

Zhetysu Birch Mouse
Sicista zhetysuica

FR Siciste du Jetyssou
DE Zhetysu-Birkenmaus
ES Ratón listado de Zhetysu

2 disjunct areas: Dzungar Alatau (except SE) and Tarbagatay & Saur

NE
Monotypic

Long-tailed Birch Mouse
Sicista caudata

FR Siciste à longue queue
DE Langschwanzbirkenmaus
ES Ratón listado de cola larga

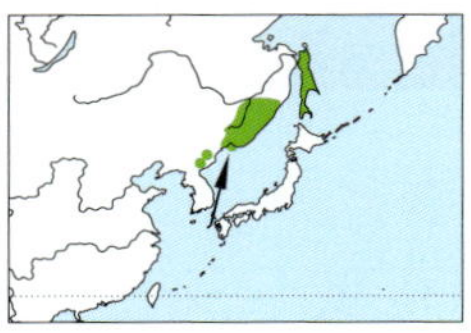

E RU (Sakhalin I, Sikhote-Alin Mts), NE CN (Heilongjiang, Jilin), NE KP

DD
Monotypic
HB 6.7–7.7 cm
T 9.9–11.8 cm
W 8.1–16.3 g

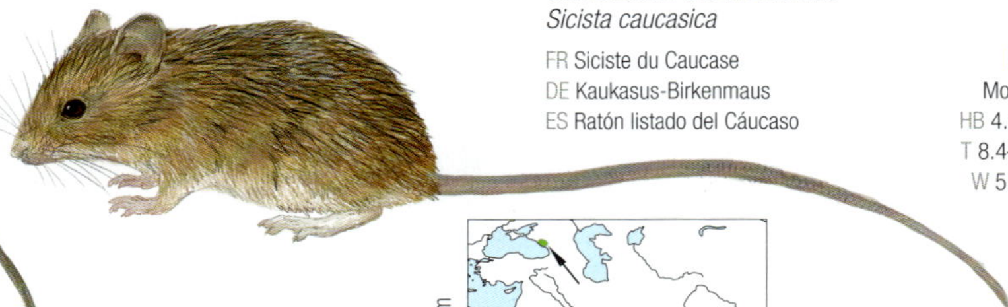

Caucasian Birch Mouse
Sicista caucasica

FR Siciste du Caucase
DE Kaukasus-Birkenmaus
ES Ratón listado del Cáucaso

NT
Monotypic
HB 4.4–6.9 cm
T 8.4–10.6 cm
W 5.8–7.2 g

1400–2700 m

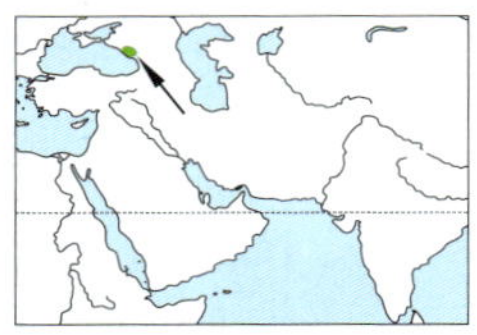

W Greater Caucasus on N from R Pshish E to N R Kizgich, on S in Abkhazia (RU, GE)

LC
Monotypic
HB 6.1–7.1 cm
T 8.6–10.7 cm
W 5.9–8.6 g

1500–2600 m

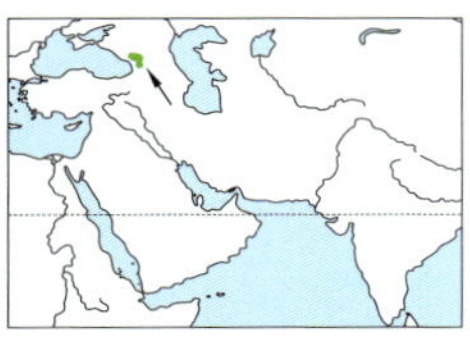

W Greater Caucasus (S RU, NW GE)

Kluchor Birch Mouse
Sicista kluchorica

FR Siciste du Klukhor
DE Kluchor-Birkenmaus
ES Ratón listado de Kluchor

EN
Monotypic
HB 6.2–6.7 cm T 9.4–10.9 cm
W 6.5–8 g

1850–2200 m

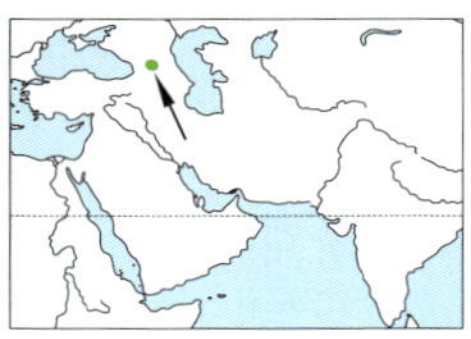

C Greater Caucasus in GE (Kazbegi) & RU (North Ossetia, perhaps Ingushetia)

Kazbeg Birch Mouse
Sicista kazbegica

FR Siciste du Kazbegi
DE Kazbeg-Birkenmaus
ES Ratón listado de Kazbegi

Armenian Birch Mouse
Sicista armenica

FR Siciste d'Arménie
DE Armenien-Birkenmaus
ES Ratón listado de Armenia

CR
Monotypic
HB c. 6.6 cm
T 9.8–10.1 cm
W 6.4–6.7 g

2000–2200 m

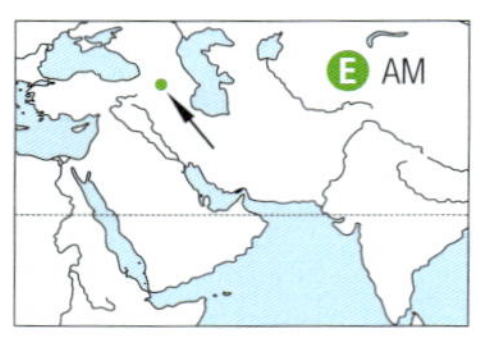

R Hrazdan Basin (Pambak and Tsakhkuniats ridges); perhaps Syunik Prov

Genus *EUCHOREUTES*

(1 species)

Long-eared Jerboa
Euchoreutes naso

FR Gerboise à longues oreilles
DE Riesenohr-Springmaus
ES Jerbo de orejas largas

NW CN (Xinjiang, Gansu, Qinghai, Inner Mongolia), S MN

LC
3 sspp
HB 9–10.5 cm
T 14.5–18 cm
W 23–45 g

Genus *ALLACTAGA*

(2 species)

Great Jerboa
Allactaga major

FR Grande Gerboise
DE Großer Pferdespringer
ES Jerbo grande

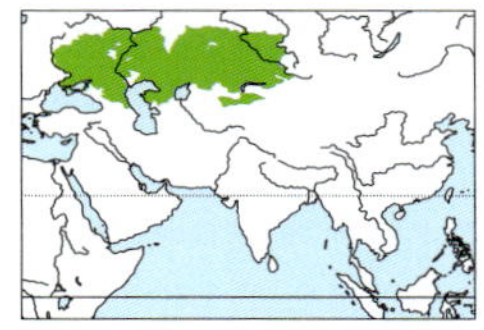

E UA & S RU E through KZ & N UZ to W Siberia & CN (W Xinjiang)

LC
3 sspp
HB 18.7–23 cm
T 20–30 cm
W 260–415 g

Severtzov's Jerboa
Allactaga severtzovi

FR Gerboise de Severtsov
DE Severtzov-Pferdespringer
ES Jerbo de Severtzov

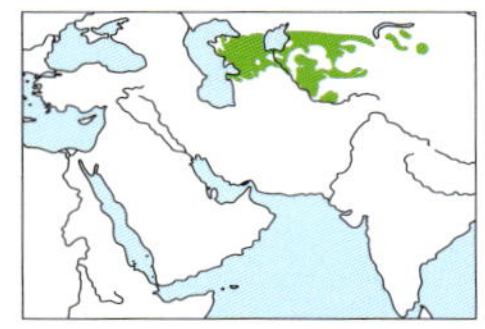

S KZ, UZ, NW & N TM, NW & SW TJ; probably adjacent CN (Xinjiang)

LC
2 sspp
HB 14.5–18.5 cm
T 20–25 cm W 122–220 g

Genus *ALLACTODIPUS*

(1 species)

Bobrinski's Jerboa
Allactodipus bobrinskii

FR Gerboise de Bobrinski
DE Bobrinski-Springmaus
ES Jerbo de Bobrinski

LC
Monotypic
HB 11–13.5 cm
T 17–20 cm
W 52–77 g

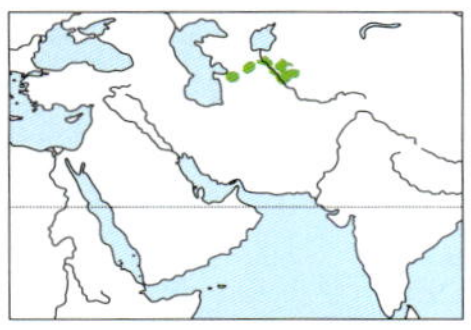

Karakum and Kyzylkum deserts in W & N TM and C & SW UZ

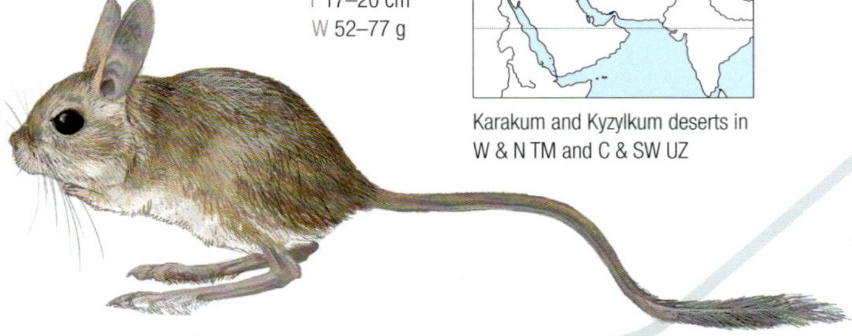

Genus *ORIENTALLACTAGA*

(3 species)

Siberian Jerboa
Orientallactaga sibirica

FR Gerboise de Sibérie
DE Sibirien-Pferdespringer
ES Jerbo de Siberia

Caspian Sea (KZ, N TM) E to RU (Tuva, Transbaikalia), MN & N CN (W Jilin)

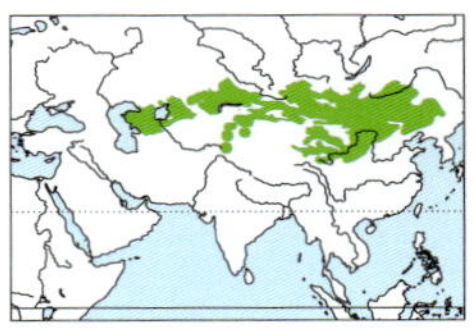

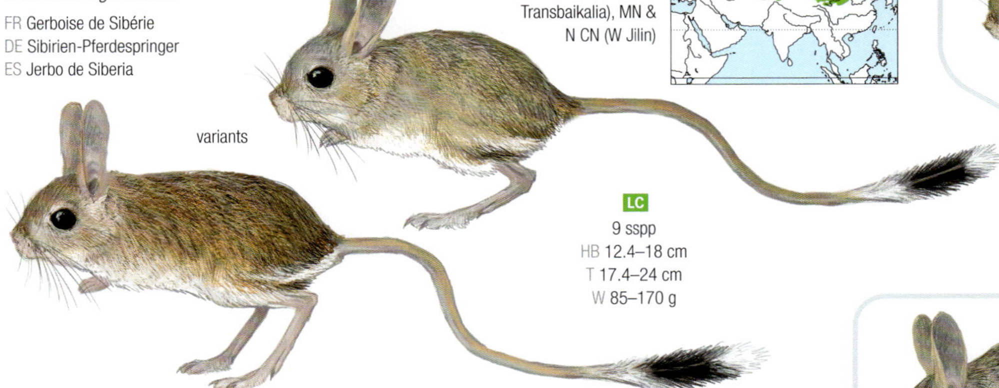

variants

LC
9 sspp
HB 12.4–18 cm
T 17.4–24 cm
W 85–170 g

Gobi Jerboa
Orientallactaga bullata

FR Gerboise de Gobi
DE Gobi-Pferdespringer
ES Jerbo del Gobi

LC
Monotypic
HB 12.5–15 cm
T 15.5–19.5 cm
W 70–108 g

NW & S MN, N CN (E Xinjiang, N Gansu, W Inner Mongolia, Ningxia)

LC
Monotypic
HB 12–13.5 cm
T 17–19.5 cm
W 63–88 g

Balikun Jerboa
Orientallactaga balikunica

FR Gerboise de Barkol
DE Balikun-Pferdespringer
ES Jerbo de Balikun

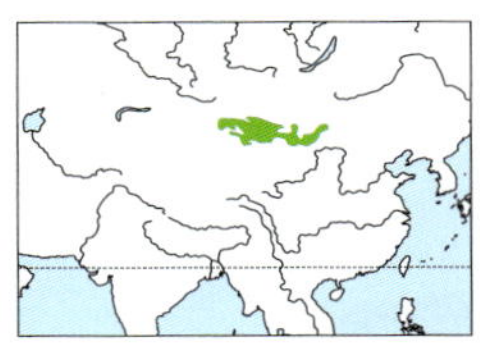

S MN, N CN (NE Xinjiang, NW Gansu, NW Inner Mongolia)

Genus *SCARTURUS*

(8 species)

Williams's Jerboa
Scarturus williamsi

FR Gerboise de Williams
DE Türkei-Pferdespringer
ES Jerbo de Williams

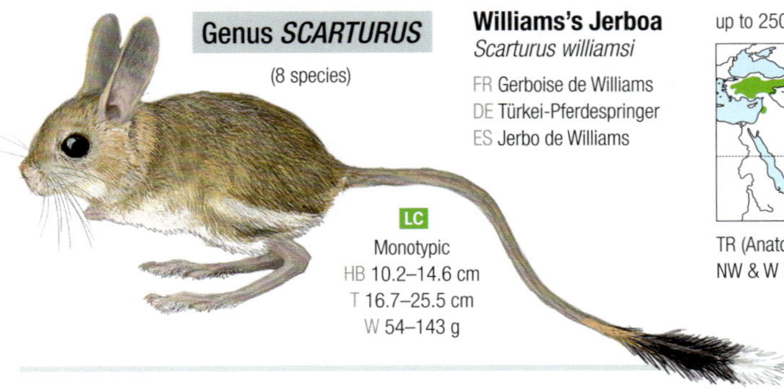

LC
Monotypic
HB 10.2–14.6 cm
T 16.7–25.5 cm
W 54–143 g

up to 2500 m

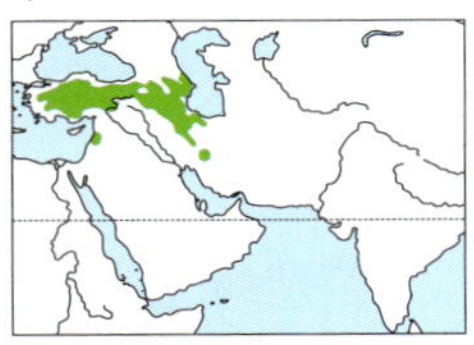

TR (Anatolia), AM, SE GE, AZ, NW & W IR, LB

Syrian Five-toed Jerboa
Scarturus aulacotis

FR Gerboise de Wagner
DE Syrien-Pferdespringer
ES Jerbo pentadáctilo de Siria

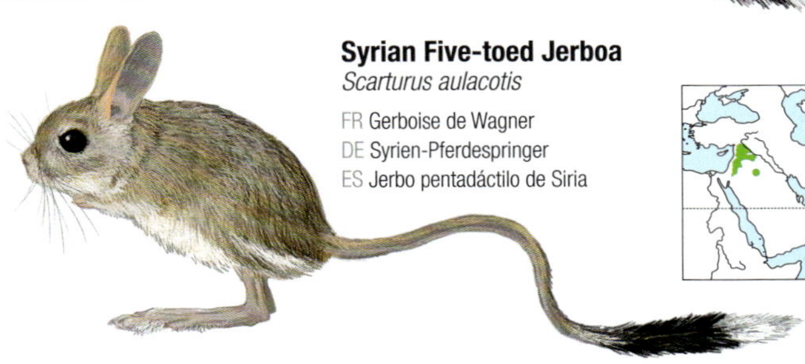

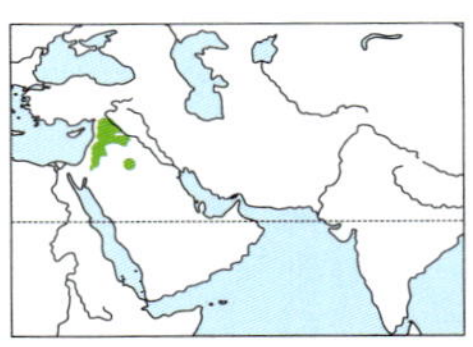

SC TR (Kilis), NW Arabian Pen W of R Euphrates (W SY, JO, NW SA)

NE
Monotypic
HB 9.6–12.8 cm
T 15.8–19.9 cm

Euphrates Jerboa
Scarturus euphraticus

FR Gerboise de l'Euphrate
DE Euphrat-Pferdespringer
ES Jerbo del Éufrates

LC
2 sspp
HB 7.7–14 cm
T 14.4–21 cm
W 48–92 g

400–3200 m

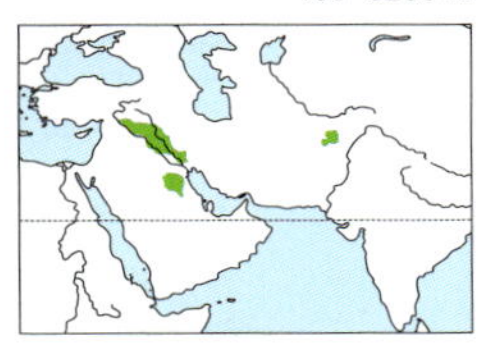

E TR, IQ E of R Euphrates, SW IR, AF, probably KW & N SA; limits uncertain

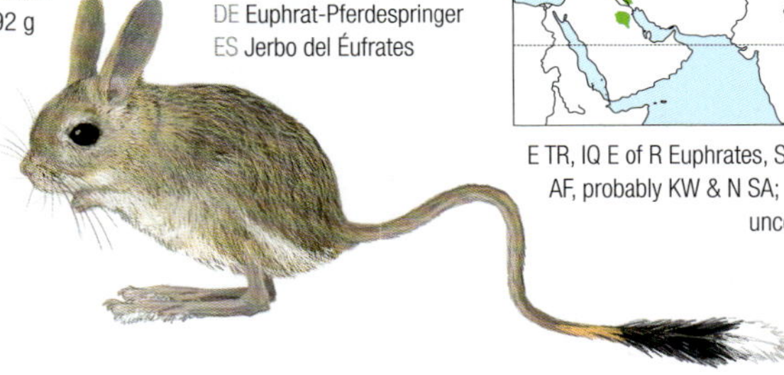

Hotson's Five-toed Jerboa
Scarturus hotsoni
FR Gerboise de Hotson
DE Hotson-Pferdespringer
ES Jerbo pentadáctilo de Hotson

LC
2 sspp
HB 9.7–16.1 cm
T 16.7–22 cm
W 48–92 g

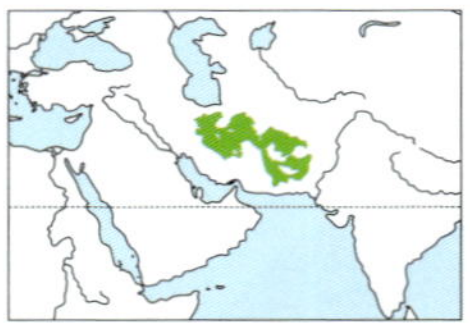

Gravel and clay-loess deserts in C & E IR, SW AF, SW PK

Small Five-toed Jerboa
Scarturus elater
FR Gerboise meneuse
DE Kleiner Pferdespringer
ES Jerbo pentadáctilo pequeño

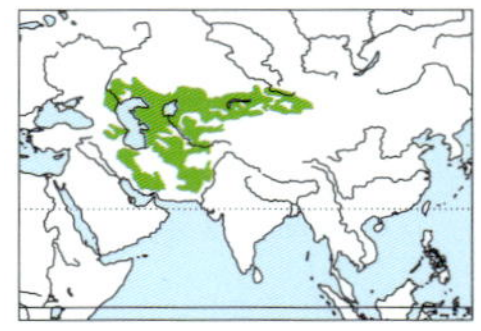

E TR & N Caucasus E through Central Asia to N CN, W MN, S to IR, AF & SW PK

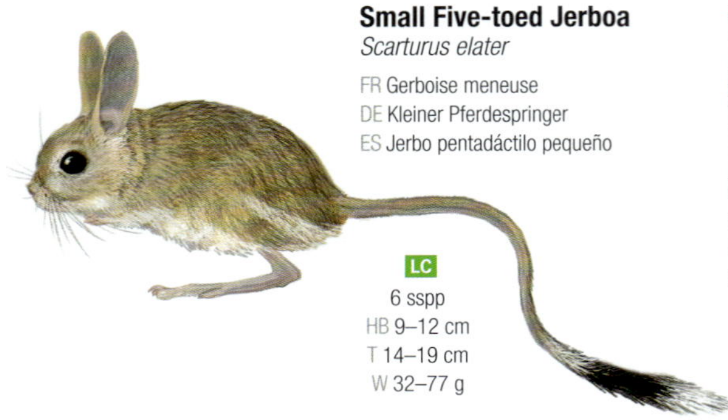

LC
6 sspp
HB 9–12 cm
T 14–19 cm
W 32–77 g

Tous Jerboa
Scarturus toussi
FR Gerboise de Tus
DE Touss-Pferdespringer
ES Jerbo de Tous

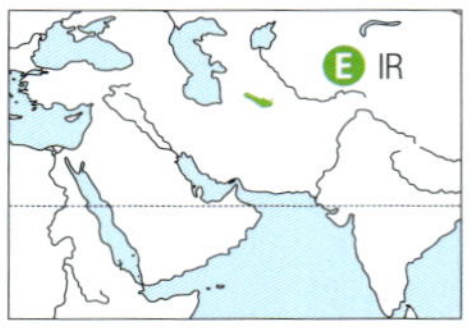

800–1200 m

Cheshmeh Gilas & Sarakhs in NE, areas between Zagros and Alborz Mts

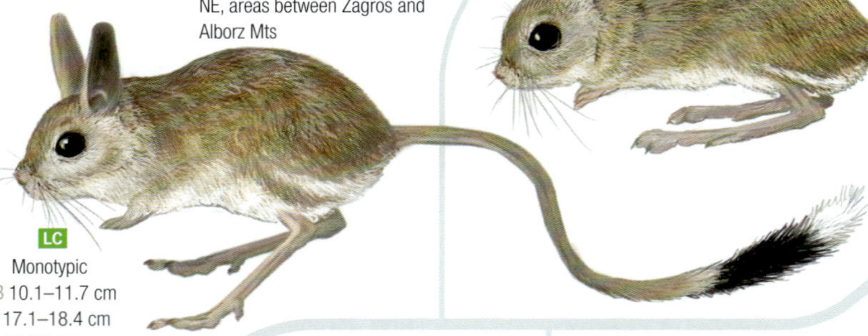

LC
Monotypic
HB 10.1–11.7 cm
T 17.1–18.4 cm
W 52–74 g

Vinogradov's Jerboa
Scarturus vinogradovi
FR Gerboise de Vinogradov
DE Vinogradov-Pferdespringer
ES Jerbo de Vinogradov

800–1200 m

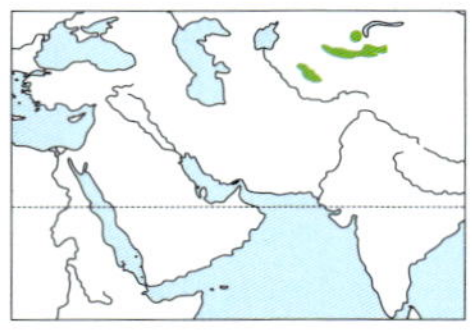

Semi-desert areas in S KZ, N KG, E UZ

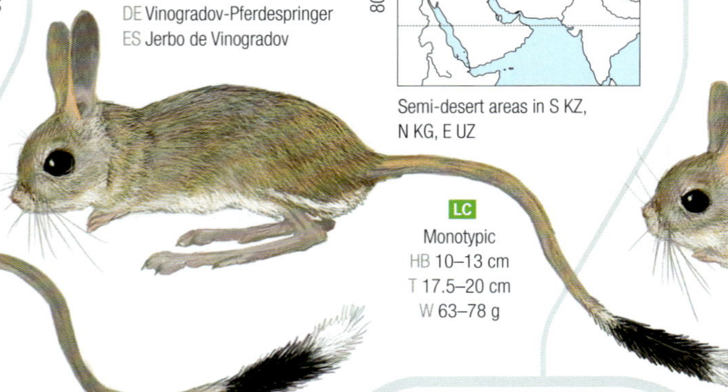

LC
Monotypic
HB 10–13 cm
T 17.5–20 cm
W 63–78 g

Four-toed Jerboa
Scarturus tetradactylus
FR Gerboise à quatre doigts
DE Vierzehen-Pferdespringer
ES Jerbo tetradáctilo

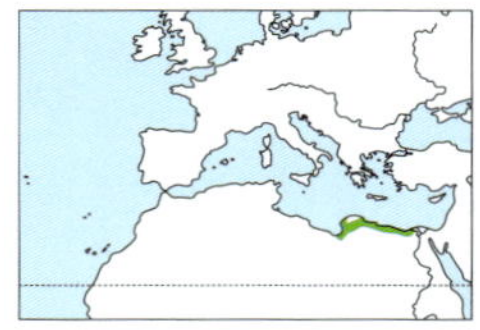

Coastal NE LY and NW EG

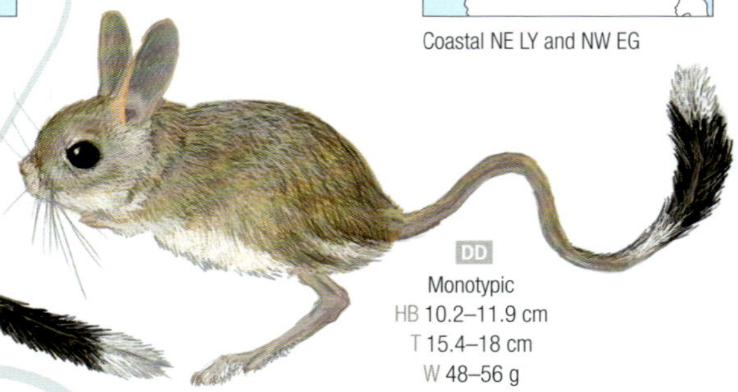

DD
Monotypic
HB 10.2–11.9 cm
T 15.4–18 cm
W 48–56 g

Genus *PYGERETMUS*

(3 species)

Greater Fat-tailed Jerboa
Pygeretmus shitkovi
FR Gerboise de Zhitkov
DE Große Fettschwanzspringmaus
ES Jerbo de cola grasa grande

L Balkhash, L Alakol, R Ili Valley, NE Betpak-Dala Desert

LC
Monotypic
HB 8.6–12.4 cm T 9.2–14.1 cm
W 40–86 g

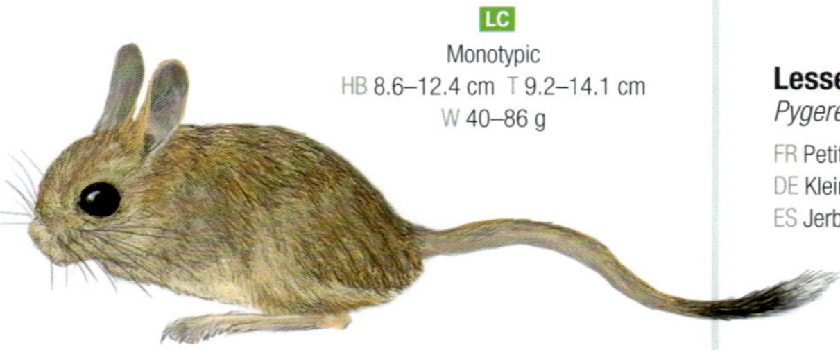

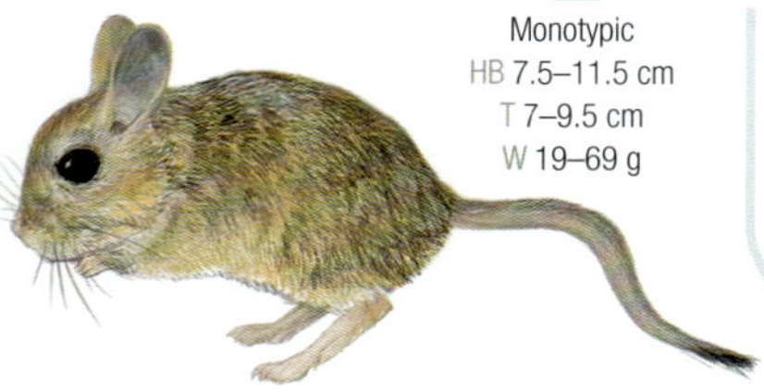

LC
Monotypic
HB 7.5–11.5 cm
T 7–9.5 cm
W 19–69 g

Lesser Fat-tailed Jerboa
Pygeretmus platyurus
FR Petite Gerboise
DE Kleine Fettschwanzspringmaus
ES Jerbo de cola grasa pequeño

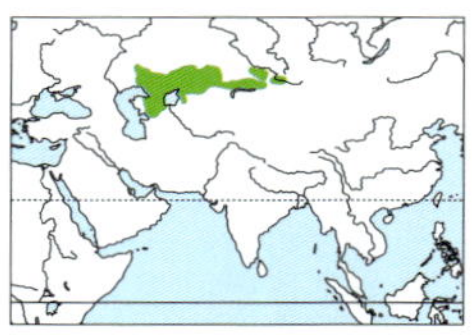

W, C & E KZ, NW TM (Balkhan Region), NW UZ (Karakalpakstan)

Dwarf Fat-tailed Jerboa
Pygeretmus pumilio
FR Gerboise lièvre
DE Zwerg-Fettschwanzspringmaus
ES Jerbo de cola grasa enano

LC
5 sspp
HB 9.5–13.5 cm
T 12.5–17.5 cm
W 30–78 g

From R Don (S RU) S to NE IR, E through KZ to S MN & N CN

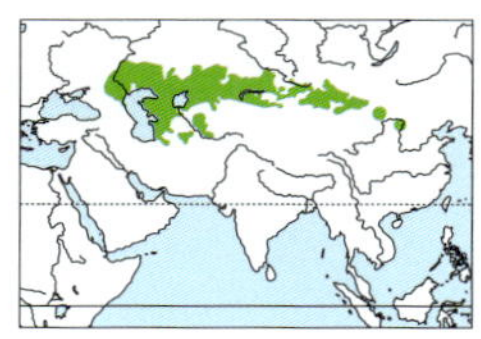

Genus *PARADIPUS*

(1 species)

Comb-toed Jerboa
Paradipus ctenodactylus
FR Gerboise à peignes
DE Kammzehen-Springmaus
ES Jerbo de pies en peine

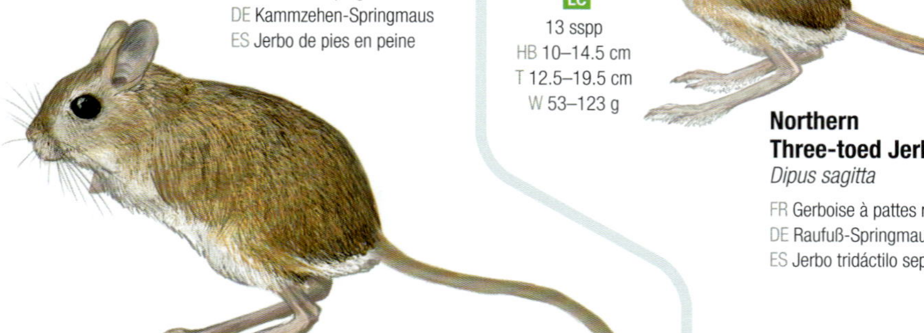

LC
Monotypic
HB 14–16.5 cm
T 18–22 cm
W 112–185 g

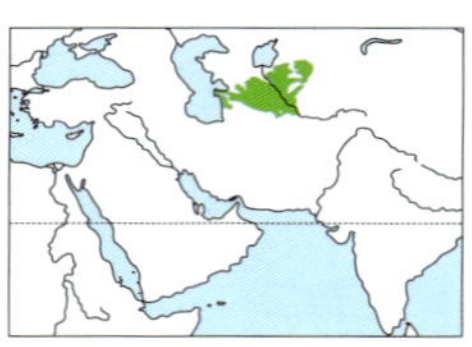

Karakum and Kyzylkum deserts (TM, UZ, S KZ)

Genus *DIPUS*

(2 species)

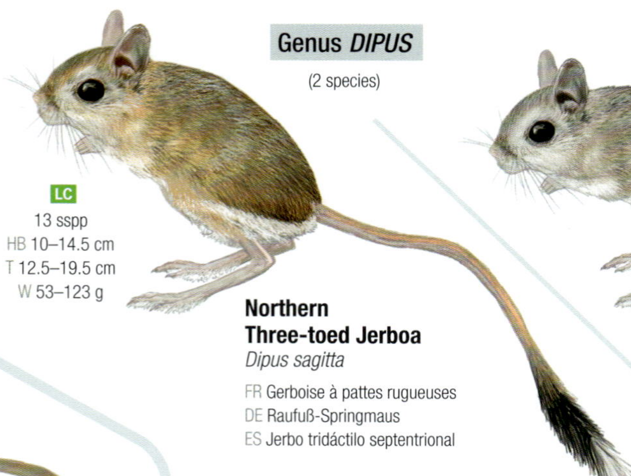

LC
13 sspp
HB 10–14.5 cm
T 12.5–19.5 cm
W 53–123 g

Northern Three-toed Jerboa
Dipus sagitta
FR Gerboise à pattes rugueuses
DE Raufuß-Springmaus
ES Jerbo tridáctilo septentrional

30–3175 m

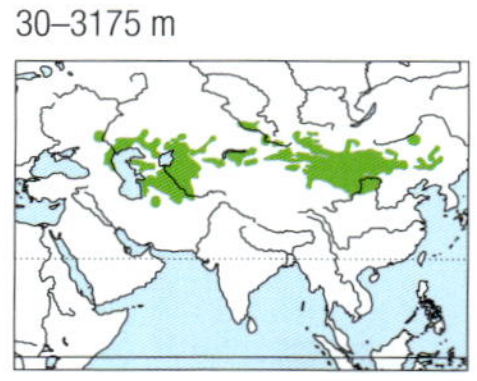

Deserts from R Don (RU), KJ, TM, UZ and N IR to Tuva, MN & N CN

Qaidam Three-toed Jerboa
Dipus deasyi
FR Gerboise de Deasy
DE Qaidam-Springmaus
ES Jerbo tridáctilo de Qaidam

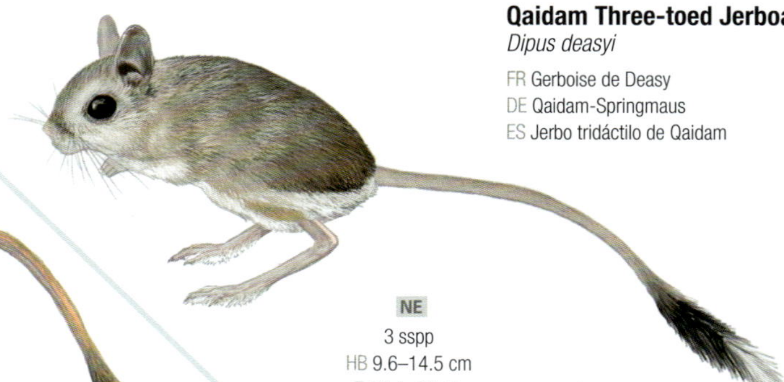

NE
3 sspp
HB 9.6–14.5 cm
T 14.4–20 cm
W 42–105 g

E Tian Shan & Tarim Basin (Xinjiang), S & W Qaidam Basin (Qinghai)

Genus *CHIMAERODIPUS*

(1 species)

Ningxia Three-toed Jerboa
Chimaerodipus auritus

FR Gerboise du Ningxia
DE Ningxia-Springmaus
ES Jerbo tridáctilo de Ningxia

NE

Monotypic
HB 9.8–11.8 cm
T 13.7–18.4 cm
W 36–68 g

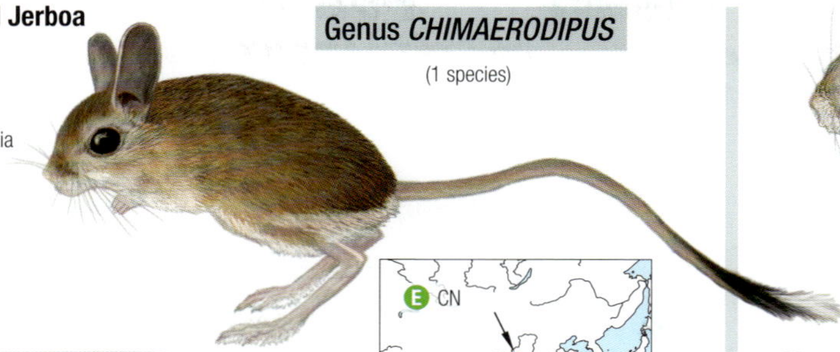

Known only from S Ningxia

Genus *STYLODIPUS*

(3 species)

MN, N CN (N Gansu, Inner Mongolia, Ningxia)

Mongolian Three-toed Jerboa
Stylodipus andrewsi

FR Gerboise d'Andrews
DE Östliche Dickschwanzspringmaus
ES Jerbo tridáctilo de Mongolia

LC

Monotypic
HB 12–13.5 cm
T 14–16.5 cm
W 60–95 g

Thick-tailed Three-toed Jerboa
Stylodipus telum

FR Gerboise javeline
DE Westliche Dickschwanzspringmaus
ES Jerbo tridáctilo de cola gruesa

LC

6 sspp
HB 11–13 cm
T 13–17 cm
W 45–73 g

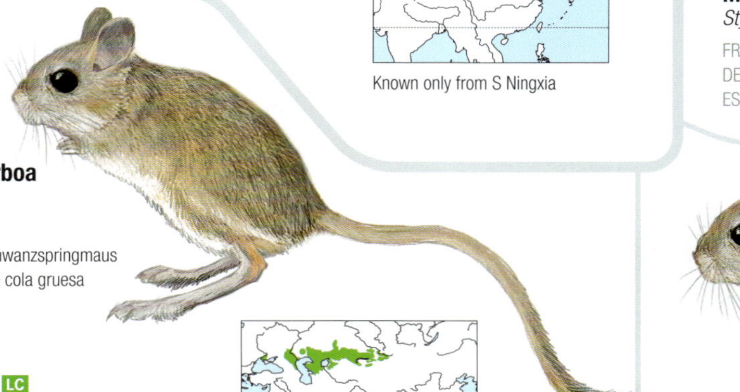

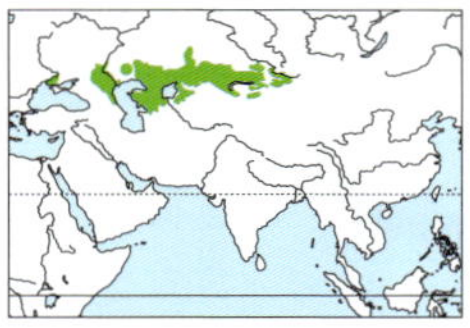

E UA & RU (N Caucasus) through C Asia steppes to N CN (N Xinjiang)

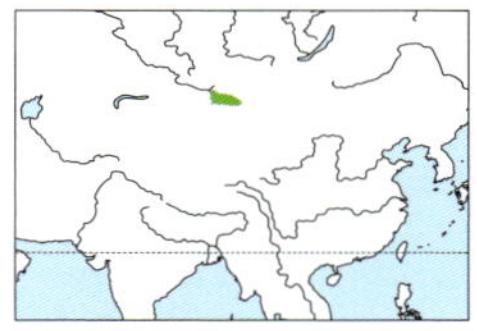

NW CN (NE Xinjiang), SW MN (Dzungarian Gobi Desert)

Dzungarian Three-toed Jerboa
Stylodipus sungorus

FR Gerboise de Dzoungarie
DE Mongolei-Dickschwanzspringmaus
ES Jerbo tridáctilo de Zungaria

LC

Monotypic
HB 12.5–13 cm
T 15–16.5 cm
W 80–95 g

Genus *EREMODIPUS*

(1 species)

Lichtenstein's Jerboa
Eremodipus lichtensteini

FR Gerboise de Lichtenstein
DE Lichtenstein-Springmaus
ES Jerbo tridáctilo de Lichtenstein

LC

3 sspp
HB 10.5–11.5 cm
T 13.5–16.5 cm
W 33–68 g

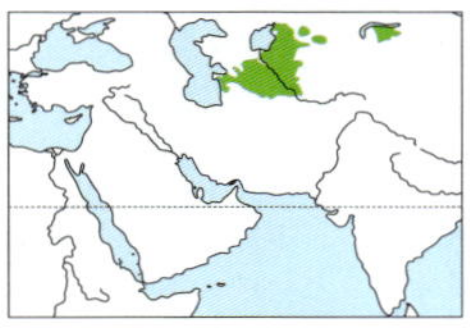

Karakum, Kyzylkum & Khazak deserts, S of L Balkhash (TM, UZ, S KZ)

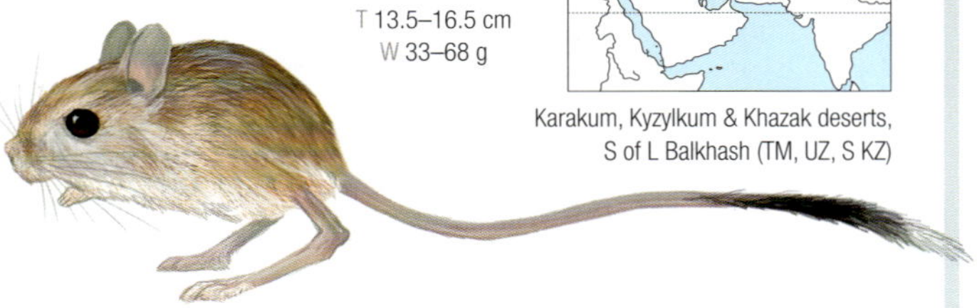

Genus *JACULUS*

(5 species)

Greater Egyptian Jerboa
Jaculus orientalis

FR Gerboise d'Erxleben
DE Große Wüstenspringmaus
ES Jerbo grande de Egipto

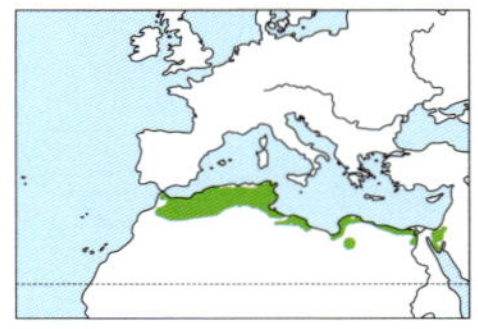

North Africa from NE MA E to EG and S IL (N Negev)

LC

3 sspp
HB 9.5–14.5 cm
T 17.2–21 cm
W 104–182 g

Blanford's Jerboa
Jaculus blanfordi

FR Gerboise de Blanford
DE Blanford-Springmaus
ES Jerbo de Blanford

LC

3 sspp
HB 13–15 cm
T 18.5–22 cm
W 79–141 g

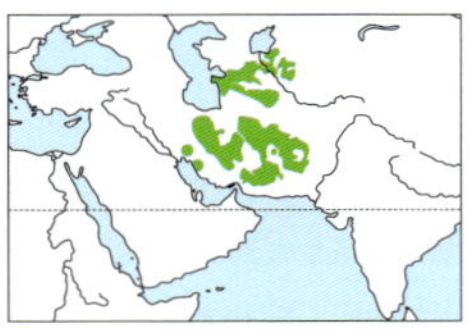

IR, W & S AF, SW PK, TM, C UZ, marginally S KZ (S Kyzylorda)

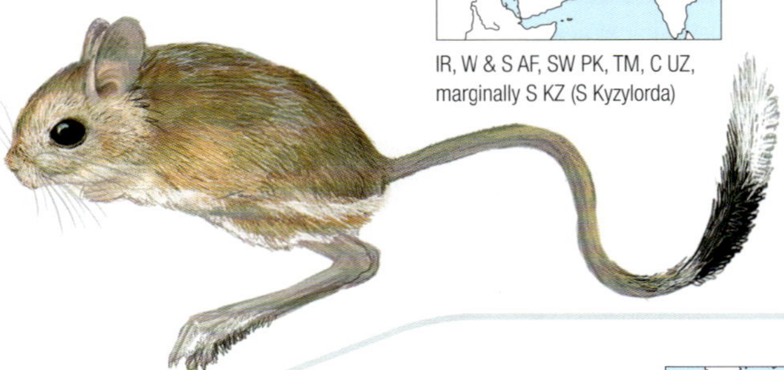

Lesser Egyptian Jerboa
Jaculus jaculus

FR Gerboise du désert
DE Kleine Wüstenspringmaus
ES Jerbo pequeño de Egipto

LC

2 sspp
HB 11.9–13 cm
T 19.1–19.7 cm
W 64–74 g

N Africa from MA E to EG & IL (probably W JO), S to ML, NE NG & SD

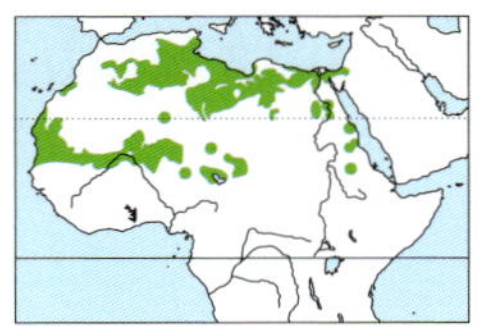

African Hammada Jerboa
Jaculus hirtipes

FR Gerboise de Saqqara
DE Hammada-Wüstenspringmaus
ES Jerbo africano de hamada

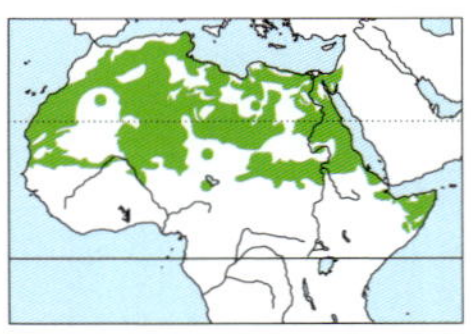

N Africa (S to NE NG & SD), Horn of Africa, W Middle East (IL, W JO)

NE

Monotypic
HB 9.5–14.5 cm
T 17.2–21 cm
W 43–91 g

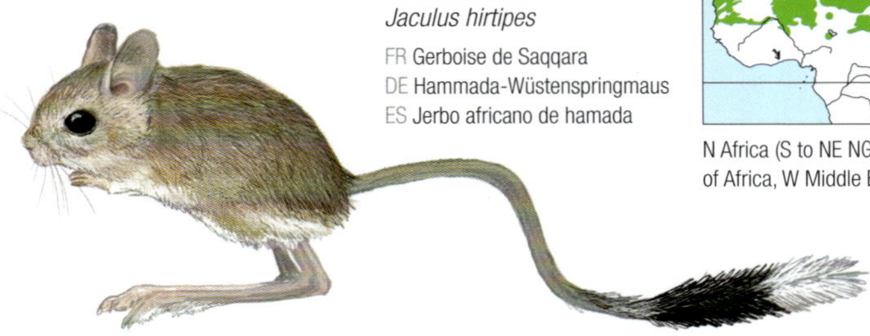

Arabian Jerboa
Jaculus loftusi

FR Gerboise de Loftus
DE Arabische Wüstenspringmaus
ES Jerbo de Arabia

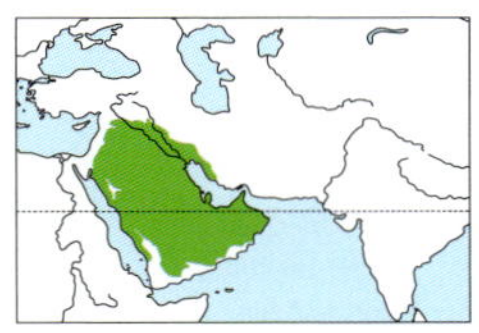

Deserts in SY, JO, IQ, SW IR, Arabian Pen

NE

2 sspp
HB 11.9–18 cm
T 20.7–26 cm
W 112–175 g

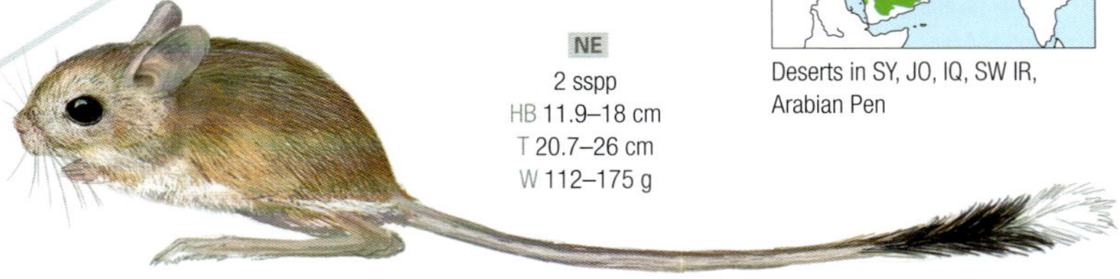

Family PLATACANTHOMYIDAE – Tree Mice

2 genera, 6 species. Tropical and subtropical montane forests. Palearctic and Indo-Malayan Regions.

Genus *PLATACANTHOMYS*

(1 species)

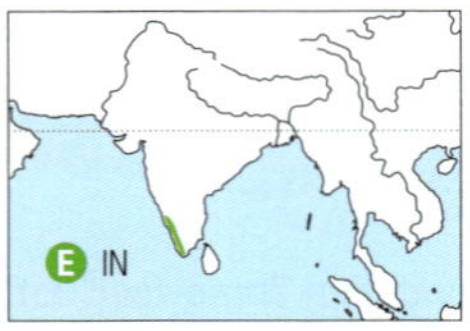

600–900 m

Western Ghats from 14° N to 8° N (Karnataka, Kerala, Tamil Nadu)

VU
Monotypic
HB 11.8–14 cm
T 7.6–10.6 cm

Malabar Spiny Tree Mouse
Platacanthomys lasiurus
FR Loir-épineux de Malabar
DE Südindien-Stachelbilch
ES Ratón arborícola de Malabar

Genus *TYPHLOMYS*

(5 species)

above 200 m

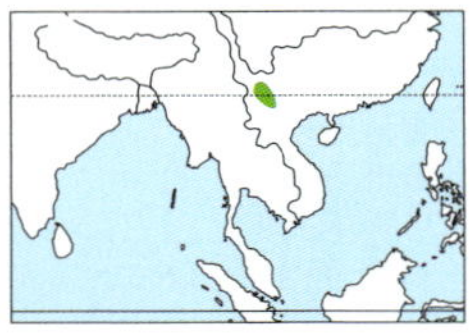

SC CN (W of R Red, SW Yunnan), NW VN (Lao Cai Prov)

Chapa Tree Mouse
Typhlomys chapensis
FR Loir-pygmée de Chapa
DE Vietnam-Zwergbilch
ES Ratón arborícola de Chapa

NE
Monotypic
HB 6.1–11.5 cm
T 8–12.6 cm
W 7.7–23 g

LC
2 sspp
HB 7.1–7.6 cm
T 9.6–10.2 cm
W 15–32 g

Soft-furred Tree Mouse
Typhlomys cinereus
FR Loir-pygmée de Chine
DE China-Zwergbilch
ES Ratón arborícola de pelaje suave

above 200 m

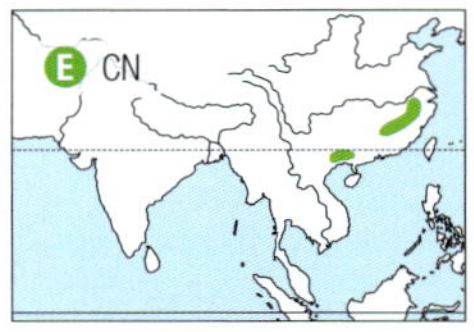

Montane forests in Guangxi, Anhui, Zhejiang, Jiangxi, Fujian

NE
Monotypic
HB c. 8.6 cm
T c. 9.8 cm
W c. 20 g

710–1300 m

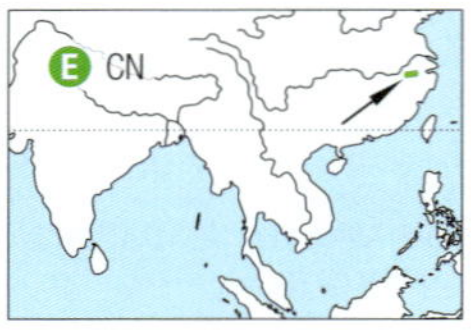

Huangshan & Qingliangfeng Mts (Anhui); perhaps more mts in S Anhui & W Zhejiang

Huangshan Tree Mouse
Typhlomys huangshanensis
FR Loir-pygmée des Huang Shan
DE Huangshan-Zwergbilch
ES Ratón arborícola de Huangshan

Dalou Mountains Tree Mouse
Typhlomys daloushanensis
FR Loir-pygmée des Dalou
DE Dalou Shan-Zwergbilch
ES Ratón arborícola de Dalou

above 200 m

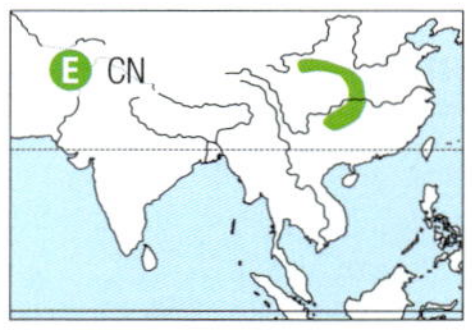

SE Gansu, S Shaanxi, E Sichuan, Chongqing, W Hubei, N Guizhou

NE
Monotypic
HB 7.2–10.5 cm
T 10.5–12.9 cm
W 15.4–31 g

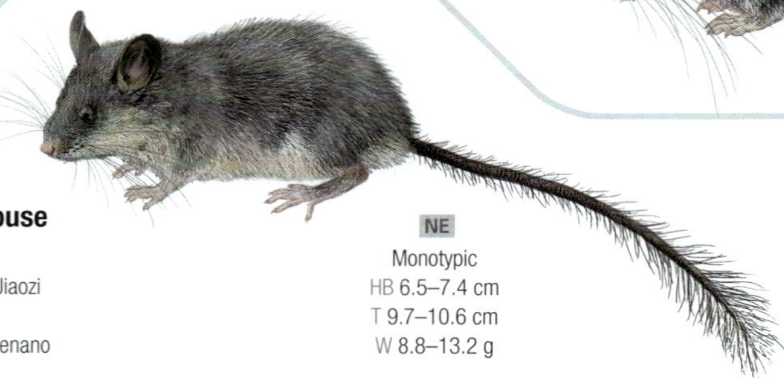

Dwarf Tree Mouse
Typhlomys nanus
FR Loir-pygmée du Jiaozi
DE Zwergbilch
ES Ratón arborícola enano

NE
Monotypic
HB 6.5–7.4 cm
T 9.7–10.6 cm
W 8.8–13.2 g

2000–3000 m

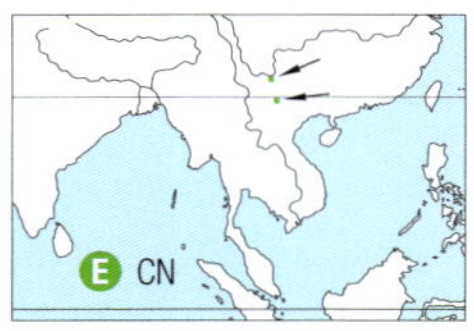

Known only from Mt Jiaozi & Mt Dawei (Yunnan)

Family SPALACIDAE – Muroid Mole-rats

7 genera, 29 species. Steppes, grasslands, meadows, forests, bamboo thickets, and farmlands. Palearctic, Afrotropical and Indo-Malayan Regions.

Genus *MYOSPALAX*

(5 species)

Altai Zokor
Myospalax myospalax
FR Zokor de l'Altaï
DE Sibirien-Blindmull
ES Zocor de Altái

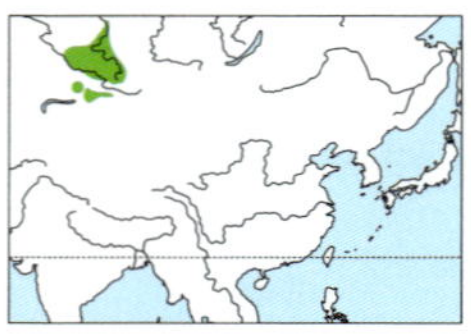

up to 2750 m

LC
3 sspp
HB 20–27 cm
T c. 4.8 cm
W 200–720 g

RU (Altai area N to Novosibirsk), E KZ (Altai Mts, Ayagoz, Tarbagatay)

Steppe Zokor
Myospalax aspalax
FR Zokor de Transbaïkalie
DE Daurien-Blindmull
ES Zocor de estepa

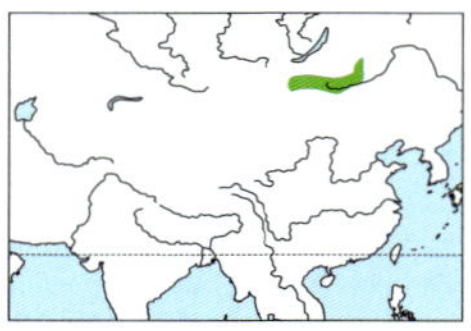

LC
Monotypic
HB 16.6–21 cm
T 3.4–6.5 cm
W c. 300 g

NE MN, RU (S Zabaykalsky Krai)

Armand's Zokor
Myospalax armandii
FR Zokor d'Armand
DE Armand-Blindmull
ES Zocor de Armand

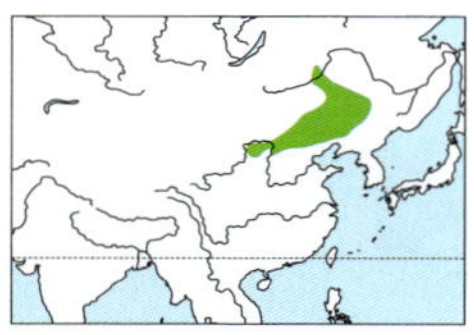

NE
Monotypic
HB 14–23 cm
T 4.8–6.9 cm
W 225–422 g

NE CN, RU (SE Zabaykalsky Krai)

Manchurian Zokor
Myospalax psilurus
FR Zokor mandchou
DE Mandschurei-Blindmull
ES Zocor de Manchuria

LC
Monotypic
HB 20–27 cm
T 3.5–5.5 cm
W 185–400 g

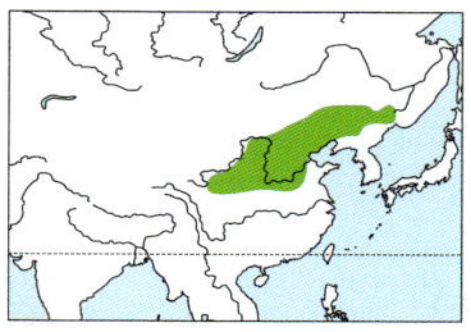

NE CN (S Gansu to S Heilongjiang), RU (SW Primorsky Krai)

Khingan Zokor
Myospalax epsilanus
FR Zokor des Khingan
DE Hinggan-Blindmull
ES Zocor de Khingan

NE
Monotypic
HB 19.8–22 cm
T c. 4.3 cm

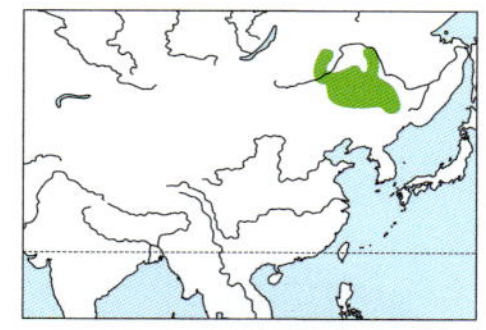

RU (S Zabaykalsky Krai), E MN, NE CN (NE Inner Mongolia, Heilongjiang)

Genus *EOSPALAX*

(7 species)

Fontanier's Zokor
Eospalax fontanierii
FR Zokor de Fontanier
DE China-Blindmull
ES Zocor de Fontanier

LC
Monotypic
HB 16.3–24 cm
T 3.8–6.9 cm
W 184–563 g

N Shaanxi, Shanxi, W Hebei, NE Inner Mongolia

Gansu Zokor
Eospalax cansus
FR Zokor du Gansu
DE Gansu-Blindmull
ES Zocor de Gansu

NE
Monotypic
HB 15.5–23 cm
T 3.3–6.6 cm
W 150–430 g

1500–2500 m

Gansu, Ningxia, Shaanxi, Sichuan, Hubei

Qinling Zokor
Eospalax rufescens
FR Zokor roux
DE Qin-Ling-Blindmull
ES Zocor de Qinling

NE
Monotypic
HB c. 18.6 cm
T c. 3.3 cm

above 2000 m

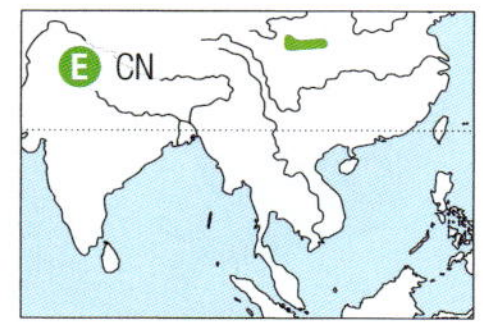

Qin Mts (S Gansu, S Shaanxi, N Sichuan)

LC
2 sspp
HB 14.9–17.2 cm
T 2.9–3.7 cm
W 160–440 g

Rothschild's Zokor
Eospalax rothschildi
FR Zokor de Rothschild
DE Rothschild-Blindmull
ES Zocor de Rothschild

1000–3000 m

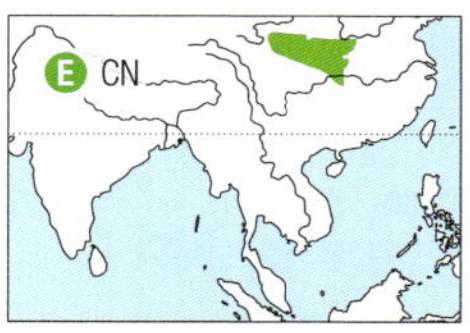

S Gansu, S & SE Shaanxi, N Sichuan, W Henan, W Hubei

NE
Monotypic
HB 14.5–20 cm
T 4.5–7 cm
W 118–280 g

Muli Zokor
Eospalax muliensis
FR Zokor du Muli
DE Muli-Blindmull
ES Zocor de Muli

c. 3700 m

Kangwu Ranch, Muli County (Sichuan); perhaps other sites in S Hengduan Mts

NE
Monotypic
HB 16–23.5 cm
T 3.3–6.1 cm
W 173–490 g

Plateau Zokor
Eospalax baileyi
FR Zokor des plateaux
DE Qinghai-Blindmull
ES Zocor de altiplano

2800–4500 m

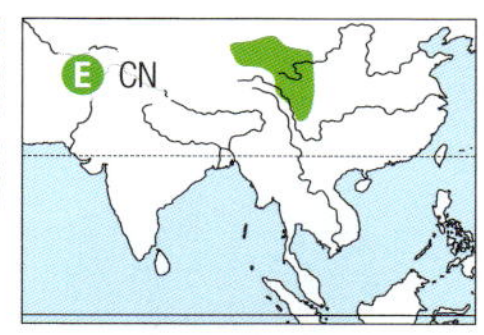

Qinghai, Gansu, W Sichuan

Smith's Zokor
Eospalax smithii
FR Zokor de Smith
DE Smith-Blindmull
ES Zocor de Smith

LC
Monotypic
HB 16.2–25 cm
T 3.4–3.9 cm
W 180–460 g

above 2000 m

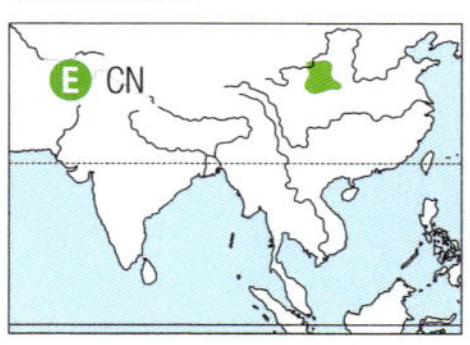

Gansu, Ningxia, N Sichuan, W Shaanxi

Genus *RHIZOMYS*

(3 species)

1220–3960 m

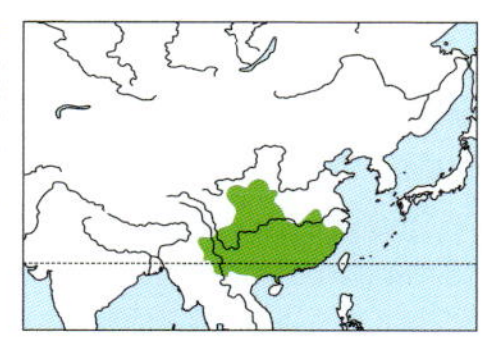

Bamboo thickets and pine forests in N MM, N VN, C & S CN

Chinese Bamboo Rat
Rhizomys sinensis
FR Rhizomys de Chine
DE China-Bambusratte
ES Rata de bambú de China

LC
6 sspp
HB 22–45 cm
T 5–9.6 cm
W c. 1.9 kg

Hoary Bamboo Rat
Rhizomys pruinosus
FR Rhizomys grisonnant
DE Graue Bambusratte
ES Rata de bambú canosa

LC
5 sspp
HB 24–35 cm
T 9–13 cm
W 1.5–2.5 kg

100–4000 m

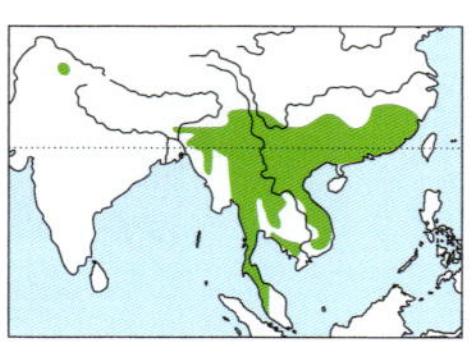

N & NE IN, SE Asia from S CN S to Perak (Thai-Malay Pen)

Indomalayan Bamboo Rat
Rhizomys sumatrensis
FR Rhizomys de Sumatra
DE Große Bambusratte
ES Rata de bambú de Indomalasia

SE Asia from E MM & SC CN (S Yunnan) S to ID (Sumatra)

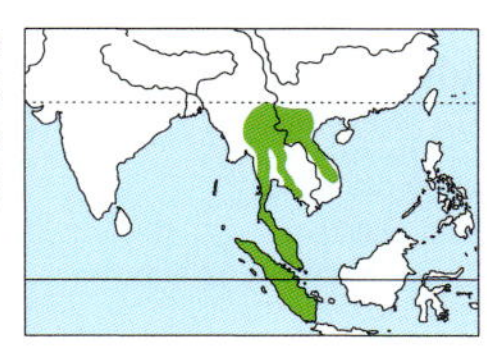

1000–4000 m

LC
4 sspp
HB 26–48 cm
T 10–20 cm
W 2.1–4 kg

Genus *CANNOMYS*

(1 species)

Lesser Bamboo Rat

Cannomys badius

FR Cannomys
DE Kleine Bambusratte
ES Rata de bambú pequeña

LC
Monotypic
HB 14.7–26 cm
T 4.7–9.6 cm
W 210–800 g

0–4000 m

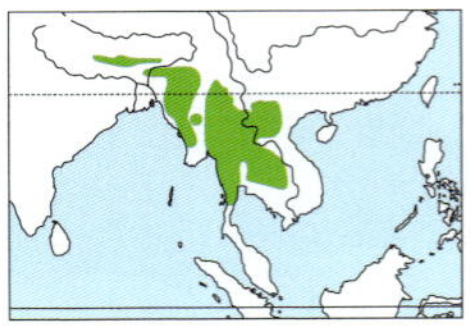

Himalaya (E from NP), SE BD and SE Asia E to NW VN & N KH

Genus *TACHYORYCTES*

(2 species)

African Root Rat

Tachyoryctes splendens

FR Rat-taupe est-africain
DE Ostafrika-Maulwurfsratte
ES Rata de raíces africana

LC
Monotypic
HB 15.9–27 cm
T 4.9–9.5 cm
W 140–330 g

variants

700–4150 m

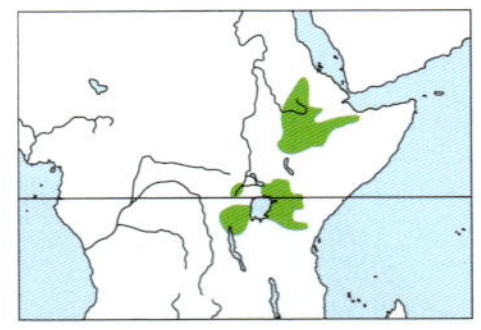

Disjunct in E Africa (ET, NW SO, S UG, SW KE, E CD, RW, BI, N TZ)

Giant Root Rat

Tachyoryctes macrocephalus

FR Rat-taupe géant
DE Riesenmaulwurfsratte
ES Rata de raíces gigante

EN
2 sspp
HB 22–31 cm
T 4–6.5 cm
W 330–930 g

3000–4150 m

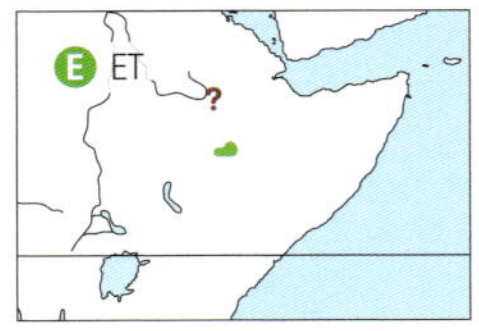

Alpine grassland in Bale Mts (Oromia Region); old records in N

Oltenia Blind Mole-rat

Spalax istricus

FR Spalax d'Olténie
DE Oltenien-Blindmaus
ES Rata topo ciega de Oltenia

CR(PE)
Monotypic
HB c. 24 cm
T not visible

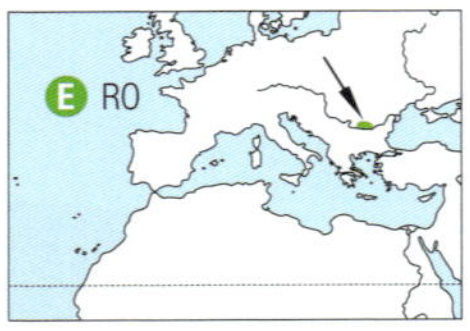

Between Carpathians and R Danube

Genus *SPALAX*

(8 species)

Bukovina Blind Mole-rat

Spalax graecus

FR Spalax de Bucovine
DE Bukowina-Blindmaus
ES Rata topo ciega de Bucovina

VU
Monotypic
HB 22–28 cm T not visible
W 415–700 g

40–850 m

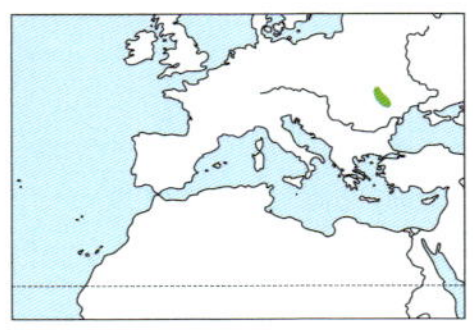

Bukovina region (SW UA, NE RO)

Podolsk Blind Mole-rat

Spalax zemni

FR Spalax de Podolie
DE Podolien-Blindmaus
ES Rata topo ciega de Podolia

VU
Monotypic
HB 20–31 cm T not visible
W 370–570 g

W of R Dnieper

EN
Monotypic

Mehely's Blind Mole-rat

Spalax antiquus

FR Spalax de Méhely
DE Mehely-Blindmaus
ES Rata topo ciega de Méhely

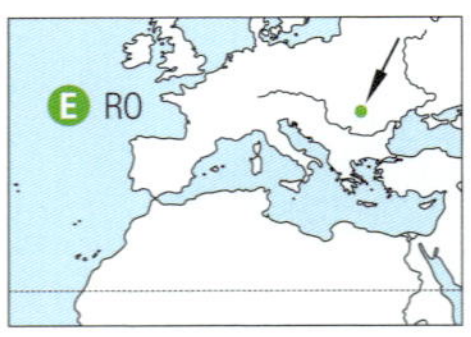

Carpathian Basin

Greater Blind Mole-rat

Spalax microphthalmus

FR Grand Spalax
DE Steppenblindmaus
ES Rata topo ciega grande

LC
2 sspp
HB 19–31 cm T not visible
W 120–818 g

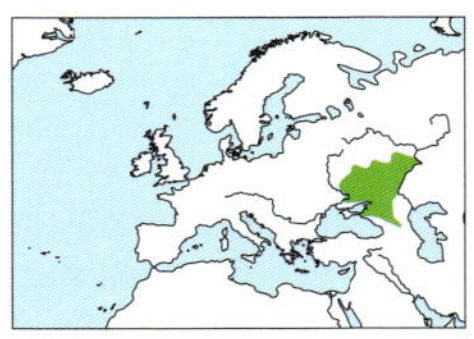

E UA & SW RU, between R Dnieper and R Volga

Giant Blind Mole-rat

Spalax giganteus

FR Spalax géant
DE Riesenblindmaus
ES Rata topo ciega gigante

LC
Monotypic
HB 25–35 cm T not visible
W ≤ 1 kg

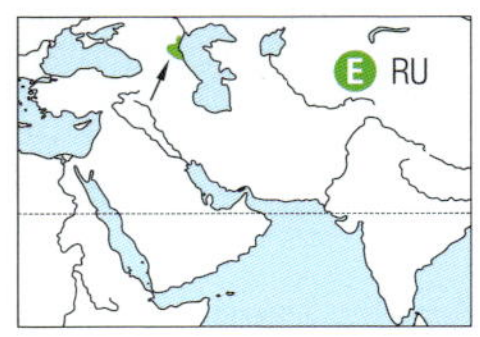

Semi-desert and steppes in Stavropol, Kalmykia, Chechnya, Dagestan

Sandy Blind Mole-rat

Spalax arenarius

FR Spalax des sables
DE Sandblindmaus
ES Rata topo ciega de los arenales

EN
Monotypic
HB 19–27 cm
T not visible
W 380–660 g

E bank of R Dnieper near its opening into the Black Sea

Kazakhstan Blind Mole-rat

Spalax uralensis

FR Spalax de l'Oural
DE Kasachstan-Blindmaus
ES Rata topo ciega de Kazajistán

NT
Monotypic
HB ≤ 31 cm
T not visible

Along floodplains of lower R Ural basin

DD
Monotypic
HB 15–24 cm T not visible
W 162–504 g

Lesser Blind Mole-rat
Nannospalax leucodon

FR Spalax de Hongrie
DE Westblindmaus
ES Rata topo ciega pequeña

0–2400 m

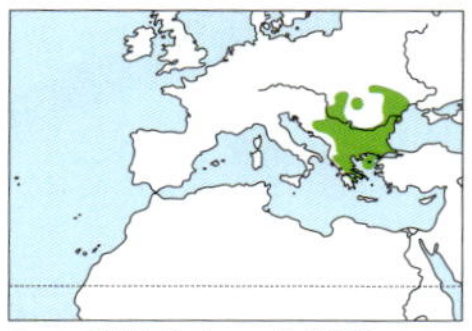

From E HU, Balkans & W TR through MD to SW UA

Genus *NANNOSPALAX*

(3 species)

Anatolian Blind Mole-rat
Nannospalax xanthodon

FR Spalax d'Anatolie
DE Anatolien-Blindmaus
ES Rata topo ciega de Anatolia

DD
Monotypic
HB 14.3–25 cm T not visible
W 130–522 g

50–2900 m

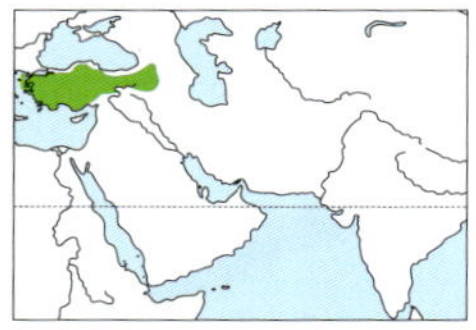

Asian TR, SW GE, NW AM, extreme NW IR, few E Aegean Is

DD
Monotypic
HB 13–22 cm
T not visible
W 73–252 g

Middle East Blind Mole-rat
Nannospalax ehrenbergi

FR Spalax d'Ehrenberg
DE Ehrenberg-Blindmaus
ES Rata topo ciega de Oriente Medio

0–2200 m

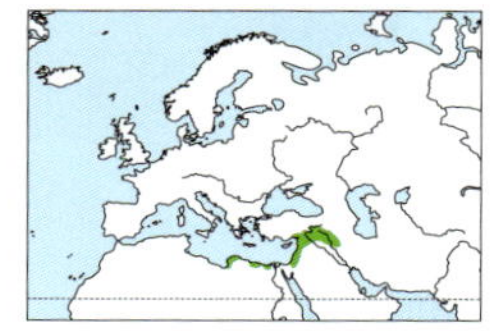

NE LY (N Cyrenaica), N EG, IL, W JO, LB, SY, SE TR, N IQ

Family CALOMYSCIDAE – Brush-tailed Mice

1 genus, 8 species. Dry rocky mountains, hillsides, and canyons. Palearctic Region.

Genus *CALOMYSCUS*

(8 species)

Zagros Mountains Brush-tailed Mouse
Calomyscus bailwardi

FR Calomysque de Bailward
DE Zagros-Maushamster
ES Ratón de cola de cepillo de Zagros

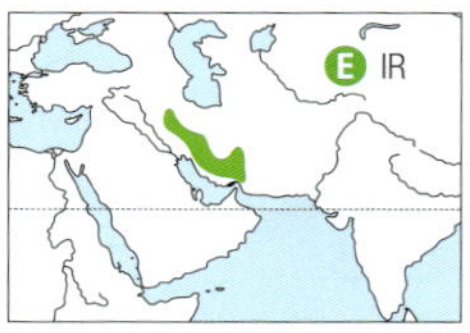

W & S IR (Zagros Mts, N Genu Mts); perhaps SE TR

LC
Monotypic
HB 8–9.2 cm T 8.1–9.2 cm
W 17–22 g

Hotson's Brush-tailed Mouse
Calomyscus hotsoni

FR Calomysque de Hotson
DE Hotson-Maushamster
ES Ratón de cola de cepillo de Hotson

65–1890 m

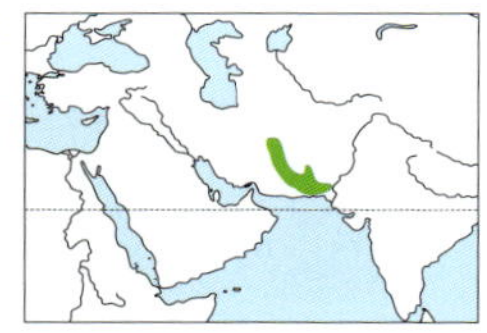

E & SE IR, SW & S PK

LC
Monotypic
HB 6.7–8.5 cm T 7.3–9 cm
W 14–25 g

Pakistan Brush-tailed Mouse
Calomyscus baluchi

FR Calomysque du Baloutchistan
DE Belutschistan-Maushamster
ES Ratón de cola de cepillo de Pakistán

LC
2 sspp
HB 7.3–9.7 cm
T 6.3–10.2 cm
W 15–30 g

400–3500 m

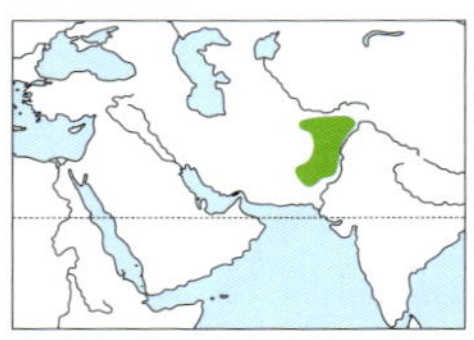

Rocky terrain in N & E AF, NW & W PK

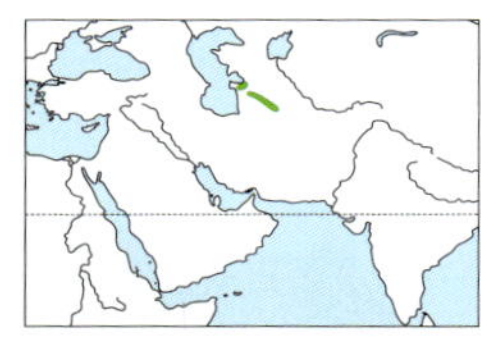

W & S TM, NE IR (Great Balkhan Mts, W & C Kopet Dag)

Great Balkhan Brush-tailed Mouse
Calomyscus mystax

FR Calomysque de Kashkarov
DE Turkmenistan-Maushamster
ES Ratón de cola de cepillo de Gran Balkán

LC
Monotypic
HB 7–9.8 cm
T 6.2–10.4 cm

Goodwin's Brush-tailed Mouse
Calomyscus elburzensis

FR Calomysque de l'Elburz
DE Elburs-Maushamster
ES Ratón de cola de cepillo de Goodwin

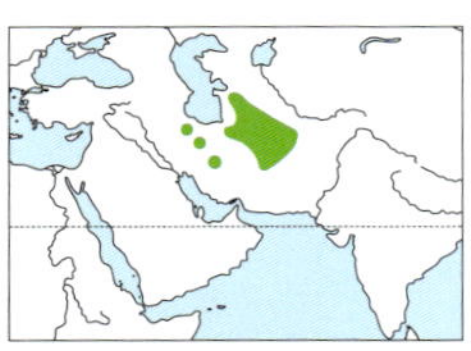

Mountains of N & NE IR (E of Alborz Mts), SW & S TM, NW AF

LC
3 sspp
HB 8.2–8.8 cm
T 8.4–9 cm

Noble Brush-tailed Mouse
Calomyscus grandis

FR Calomysque géant
DE Edler Maushamster
ES Ratón de cola de cepillo noble

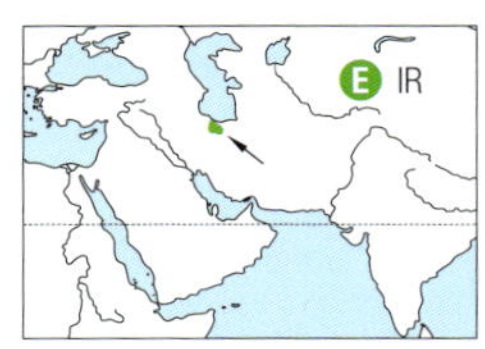

Mt Damavand, Central Alborz Mts; limits unknown

DD
Monotypic
HB 7.5–9.4 cm
T 8.8–9.9 cm

Urar Brush-tailed Mouse
Calomyscus urartensis

FR Calomysque d'Azerbaïdjan
DE Transkaukasus-Maushamster
ES Ratón de cola de cepillo de Urartu

LC
Monotypic
HB 7.8–9.1 cm
T 7.5–9.3 cm
W 15–39 g

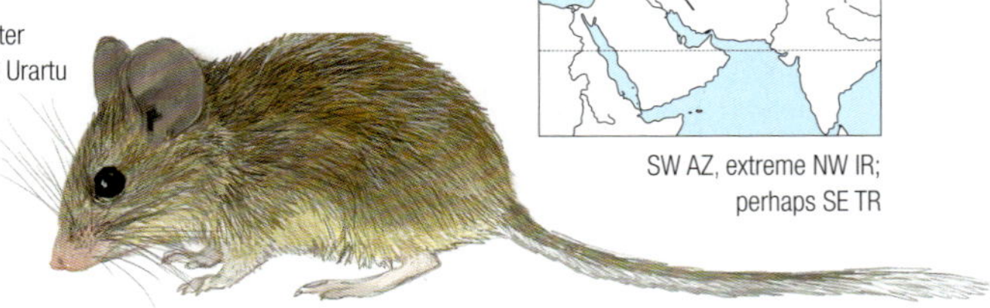

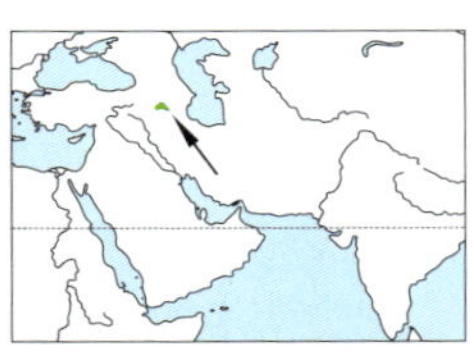

SW AZ, extreme NW IR; perhaps SE TR

Syrian Brush-tailed Mouse
Calomyscus tsolovi

FR Calomysque de Tsolov
DE Syrien-Maushamster
ES Ratón de cola de cepillo de Siria

DD
Monotypic
HB 6.6–7.5 cm
T 5.5–6.6 cm

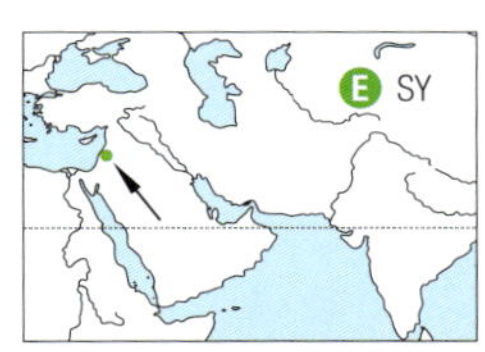

Der'a District (range based on only 14 specimens collected)

Family NESOMYIDAE – Pouched Rats, Climbing Mice and Fat Mice

22 genera, 71 species. Forests, marshes, savannas, and arid habitats. Afrotropical Region.

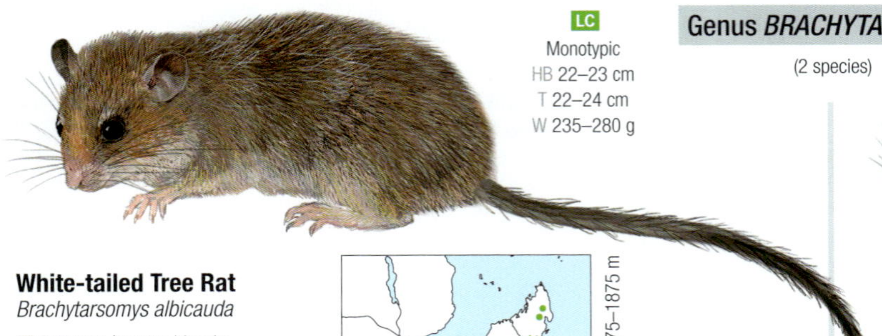

LC
Monotypic
HB 22–23 cm
T 22–24 cm
W 235–280 g

White-tailed Tree Rat
Brachytarsomys albicauda
FR Antsangy à queue blanche
DE Madagaskar-Weißschwanzratte
ES Rata arborícola de cola blanca

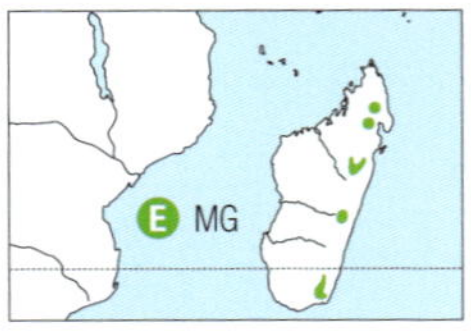

875–1875 m
Central Highlands

Genus *BRACHYTARSOMYS*
(2 species)

1200–2030 m

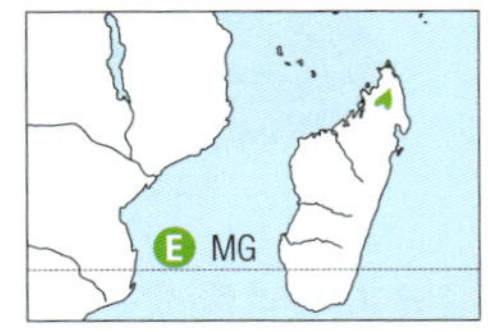

Northern Highlands

VU
Monotypic
HB 23–24 cm
T 26–27 cm
W 236–350 g

Hairy-tailed Tree Rat
Brachytarsomys villosus
FR Antsangy à queue touffue
DE Nördliche Madagaskar-Weißschwanzratte
ES Rata arborícola de cola peluda

Genus *GYMNUROMYS*
(1 species)

LC
Monotypic
HB 15.6–17.2 cm
T 14.9–19.9 cm
W 98–128 g

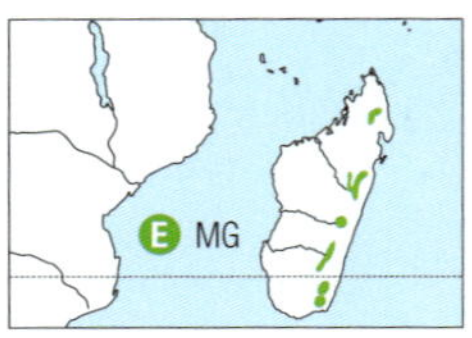

500–1625 m
Central Highlands

Sleek-furred Ground Rat
Gymnuromys roberti
FR Voalavoanala
DE Voalavoanala
ES Rata terrestre de pelaje liso

Genus *VOALAVO*
(2 species)

Northern Naked-tail Forest Mouse
Voalavo gymnocaudus
FR Voalavo à queue nue
DE Nördlicher Voalavo
ES Ratón de bosque de cola desnuda septentrional

LC
Monotypic
HB 8–9 cm
T 12–12.9 cm
W 17–25 g

1225–1950 m

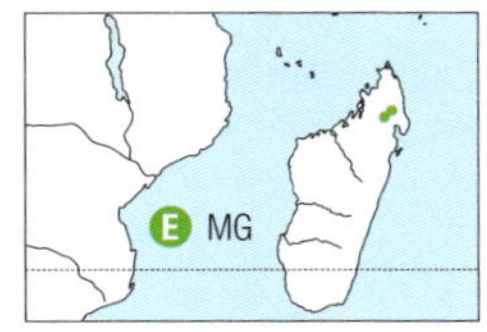

Northern Highlands (Anjanahari-be-Sud massif, Marojejy NP)

Anjozorobe Naked-tail Forest Mouse
Voalavo antsahabensis
FR Voalavo d'Antsahabe
DE Östlicher Voalavo
ES Ratón de bosque de cola desnuda de Anjozorobe

EN
Monotypic
HB 8.5–10 cm
T 10.2–12.3 cm
W 19–26 g

1250–1425 m

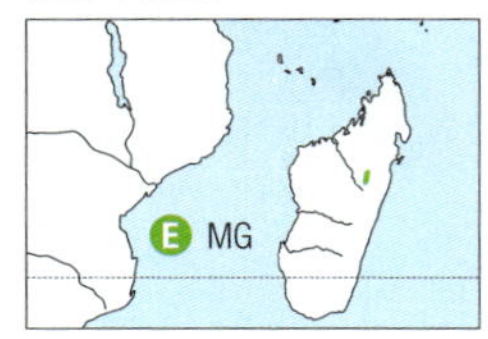

E edge of Central Highlands (Anjozorobe region)

Genus *ELIURUS*
(13 species)

LC
Monotypic
HB 11.1–16.4 cm
T 14.4–17.6 cm
W 42–62 g

Grandidier's Tufted-tail Rat
Eliurus grandidieri
FR Rat-loir de Grandidier
DE Grandidier-Bilchschwanz
ES Rata de cola de penacho de Grandidier

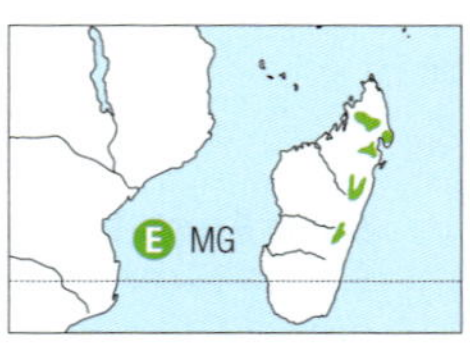

410–2050 m
Northern Highlands and E Central Highlands

NE
Monotypic
HB 14–16 cm
T 14–19.5 cm

100–320 m

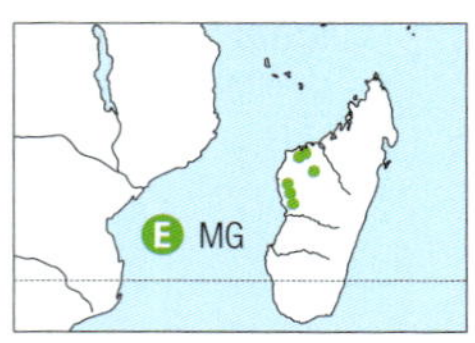

Western dry deciduous forests (Mahajanga Prov)

Limestone Tufted-tail Rat
Eliurus tsingimbato
FR Rat-loir des tsingys
DE Kalkstein-Bilchschwanz
ES Rata de cola de penacho de caliza

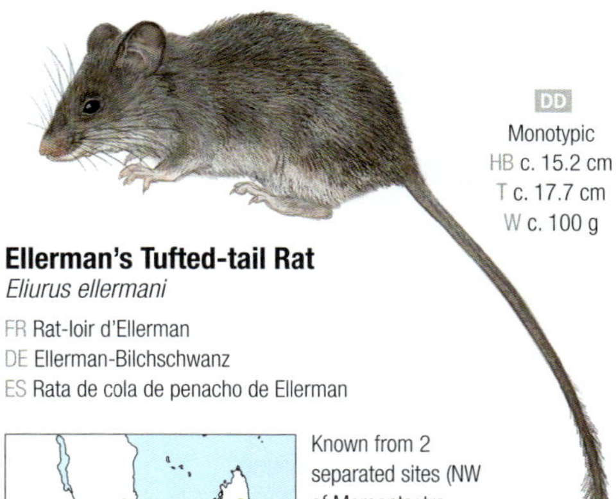

DD
Monotypic
HB c. 15.2 cm
T c. 17.7 cm
W c. 100 g

Ellerman's Tufted-tail Rat
Eliurus ellermani
FR Rat-loir d'Ellerman
DE Ellerman-Bilchschwanz
ES Rata de cola de penacho de Ellerman

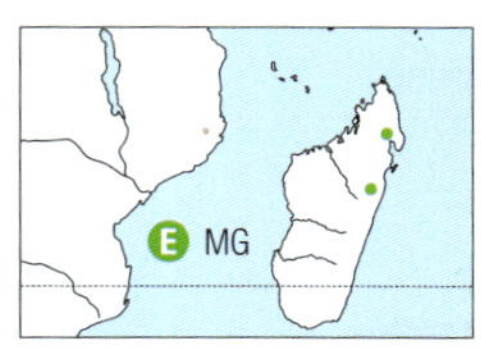

Known from 2 separated sites (NW of Maroantsetra, Lohariandava)

400–850 m

Tanala Tufted-tail Rat
Eliurus tanala
FR Rat-loir tanala
DE Tanala-Bilchschwanz
ES Rata de cola de penacho de los Tanala

400–1875 m

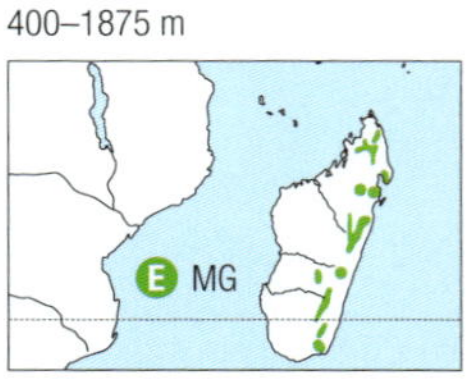

Humid forests from N Highlands S to Andohahela NP

LC
Monotypic
HB 14–15.9 cm
T 15.2–19.4 cm
W 66–97 g

LC
Monotypic
HB 11.7–13.6 cm
T 12.5–16.7 cm
W 51–75 g

Milne-Edwards's Tufted-tail Rat
Eliurus myoxinus
FR Rat-loir de Milne-Edwards
DE Milne-Edwards-Bilchschwanz
ES Rata de cola de penacho de Milne-Edwards

0–1250 m

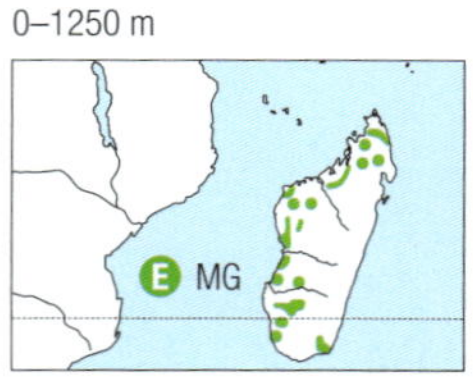

Humid forests in N to dry forests and spiny bush formations in W & S

Lesser Tufted-tail Rat
Eliurus minor
FR Petit Rat-loir
DE Kleiner Bilchschwanz
ES Rata de cola de penacho pequeña

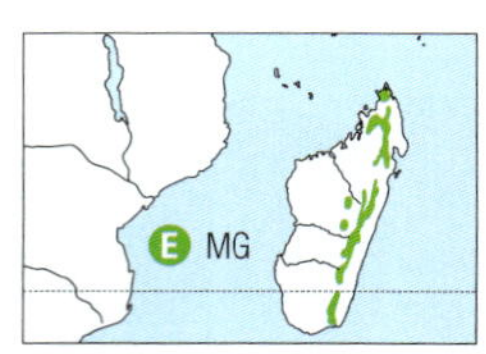

0–2030 m
From Montagne d'Ambre S to Anosyennes & Vohimena Mts

LC
Monotypic
HB 10.1–12.4 cm
T 11.9–13.7 cm
W 21–49 g

Webb's Tufted-tail Rat
Eliurus webbi

FR Rat-loir de Webb
DE Webb-Bilchschwanz
ES Rata de cola de penacho de Webb

LC
Monotypic
HB 14–15.9 cm
T 16.1–18.6 cm
W 66–97 g

0–1450 m

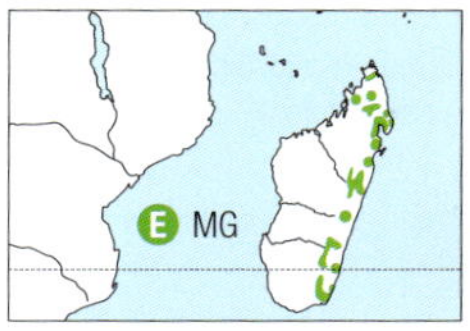

Humid forests of N and E

Petter's Tufted-tail Rat
Eliurus petteri

FR Rat-loir de Petter
DE Petter-Bilchschwanz
ES Rata de cola de penacho de Petter

EN
Monotypic
HB c. 13.3 cm
T c. 19.1 cm
W c. 74 g

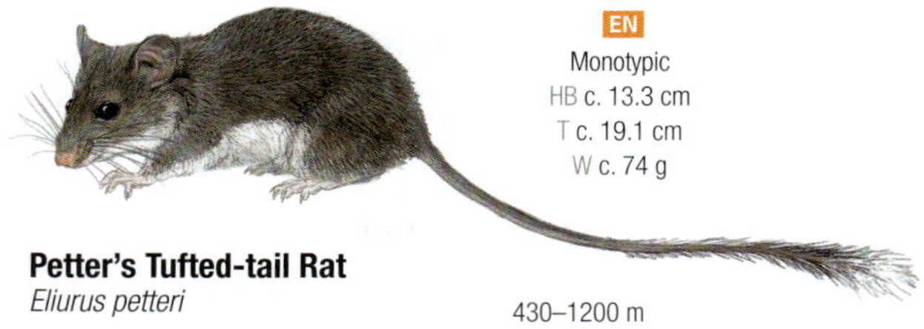

430–1200 m

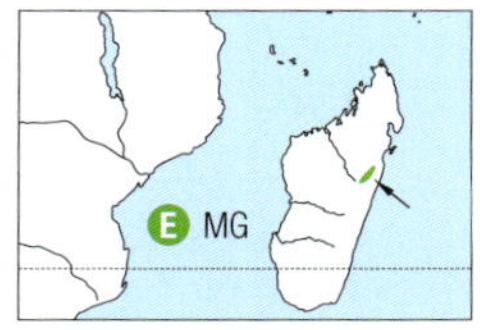

Known only from forests of Toamasina Prov

Antsingy Tufted-tail Rat
Eliurus antsingy

FR Rat-loir d'Antsingy
DE Antsingy-Bilchschwanz
ES Rata de cola de penacho de Antsingy

DD
Monotypic
HB 14.2–15.3 cm
T 15.3–19.5 cm
W 87–101 g

100–430 m

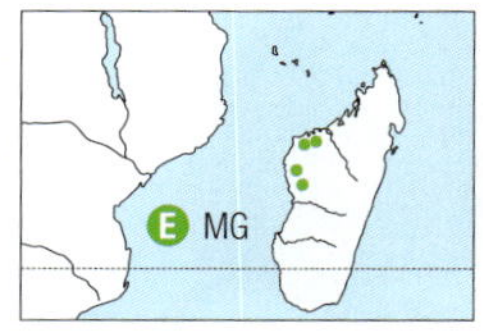

Limestone tsingy forests of CW (Kasijy region)

Major's Tufted-tail Rat
Eliurus majori

FR Rat-loir de Major
DE Major-Bilchschwanz
ES Rata de cola de penacho de Major

875–2400 m

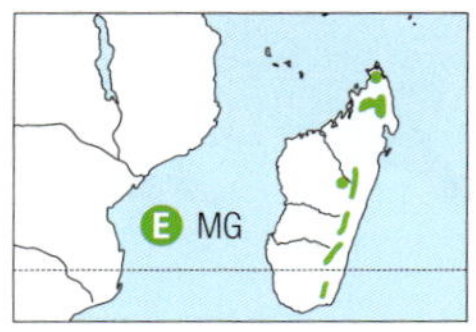

From Montagne d'Ambre S to Andohahela NP

LC
Monotypic
HB 13.8–16.4 cm T 15–19.2 cm
W 56–93 g

White-tailed Tufted-tail Rat
Eliurus penicillatus

FR Rat-loir à queue blanche
DE Büschel-Bilchschwanz
ES Rata de cola de penacho de punta blanca

1100–1670 m

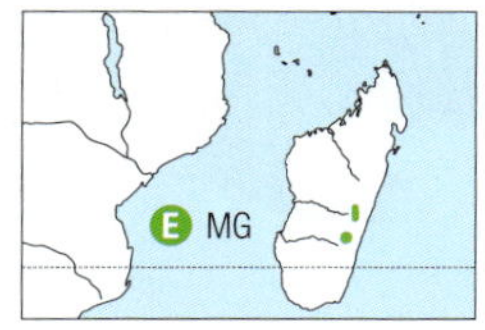

Ampitambe Forest, Ankerana

EN
Monotypic
HB c. 14.5 cm T c. 16.9 cm
W c. 70 g

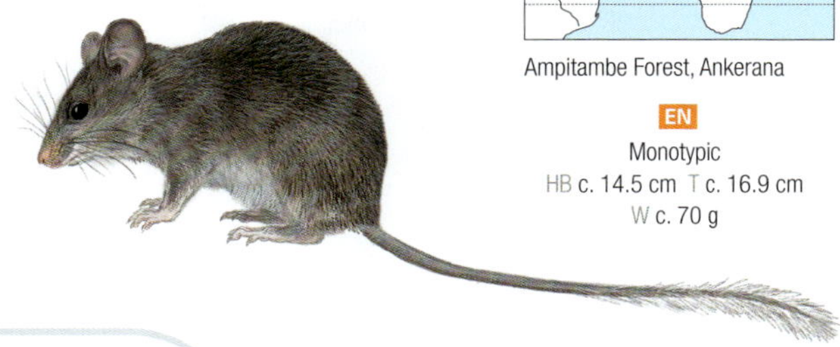

Carleton's Tufted-tail Rat
Eliurus carletoni

FR Rat-loir de Carleton
DE Carleton-Bilchschwanz
ES Rata de cola de penacho de Carleton

LC
Monotypic
HB 14.3–15 cm
T 16.4–18.3 cm
W 88–89 g

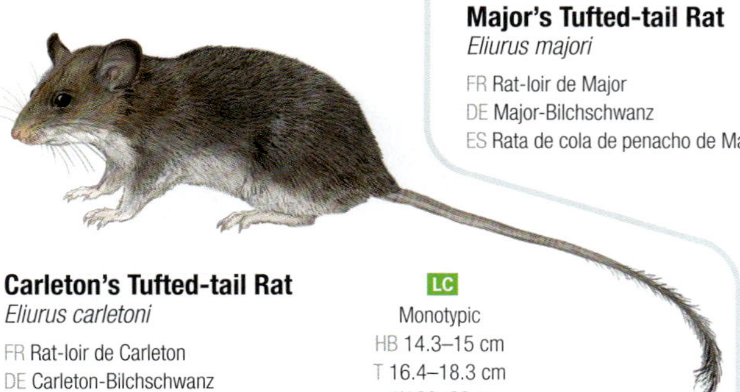

50–835 m

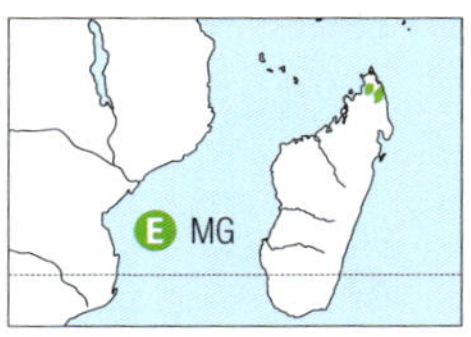

Dry deciduous forest in extreme N

Daniel's Tufted-tail Rat
Eliurus danieli

FR Rat-loir de Daniel
DE Daniel-Bilchschwanz
ES Rata de cola de penacho de Daniel

LC
Monotypic
HB 15–15.2 cm
T 17.9–19.5 cm
W 91–100 g

600–700 m

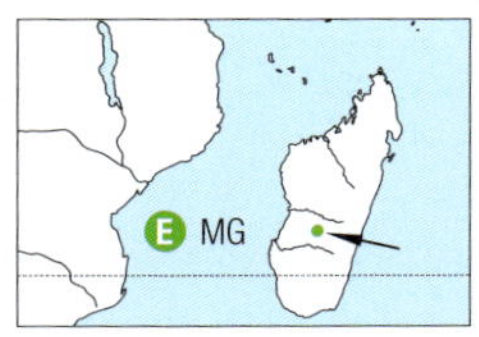

S Central Highlands (Isalo NP)

Genus *MACROTARSOMYS*

(3 species)

Bastard's Big-footed Mouse
Macrotarsomys bastardi

FR Kelibotra de Bastard
DE Bastard-Madagaskar-Großfußmaus
ES Ratón de pies grandes de Bastard

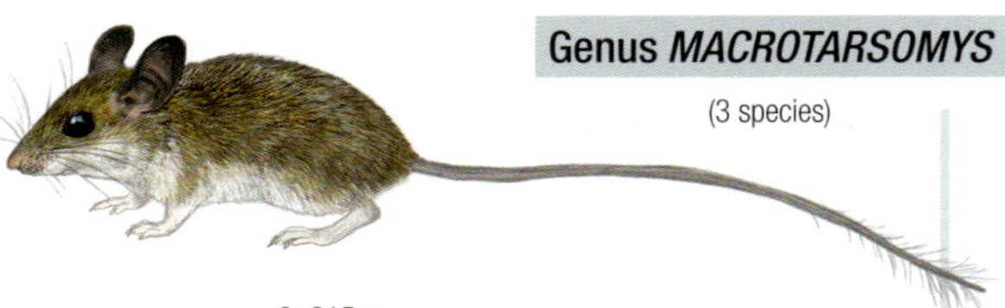

LC
2 sspp
HB 8.9–10.3 cm
T 12–14.5 cm
W 20–26 g

0–915 m

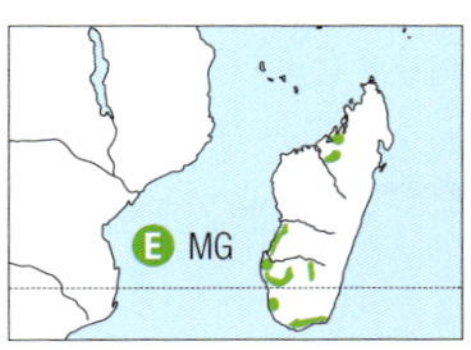

Dry deciduous forests and spiny bushes of E & S

Ankarafantsika Big-footed Mouse
Macrotarsomys ingens

FR Kelibotra d'Ankarafantsika
DE Ankarafantsika-Madagaskar-Großfußmaus
ES Ratón de pies grandes de Ankarafantsika

EN
Monotypic
HB 11.2–15 cm
T 18.3–24 cm
W 42–74 g

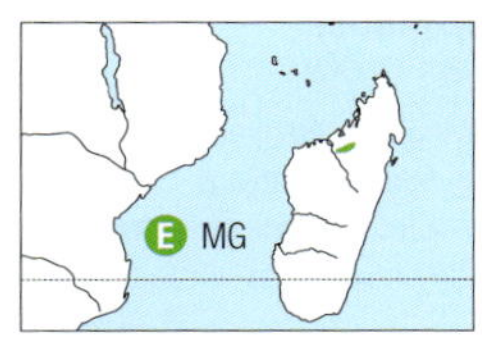

Ankarafantsika Forest

Petter's Big-footed Mouse
Macrotarsomys petteri

FR Kelibotra de Petter
DE Petter-Madagaskar-Großfußmaus
ES Ratón de pies grandes de Petter

DD
Monotypic
HB c. 16 cm
T c. 24 cm
W c. 105 g

c. 80 m

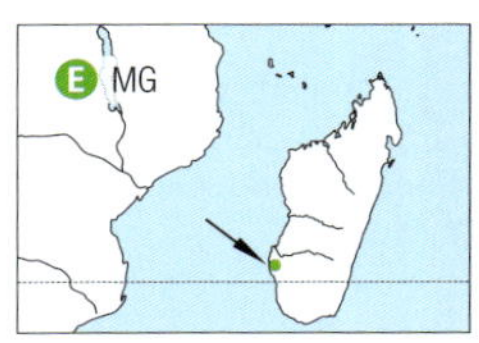

Andalomo Forest, Mikea Forest

Genus *MONTICOLOMYS*

(1 species)

Koopman's Forest Mouse
Monticolomys koopmani

FR Voalavo de Koopman
DE Madagaskar-Bergmaus
ES Ratón de bosque de Koopman

LC
Monotypic
HB 9.4–10.1 cm
T 13.4–14.3 cm
W 25–27 g

900–2030 m

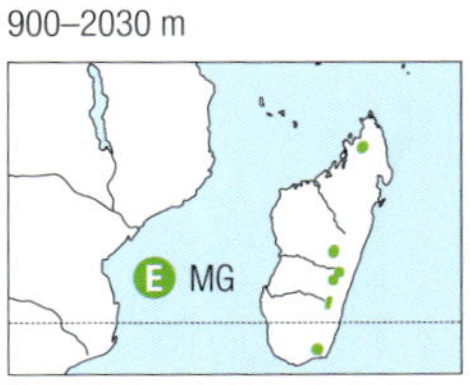

Humid forests dispersed in N, Central Highlands & E

Genus *HYPOGEOMYS*

(1 species)

Giant Jumping Rat
Hypogeomys antimena

FR Vositse géant
DE Madagaskar-Riesenratte
ES Rata saltadora gigante

40–100 m

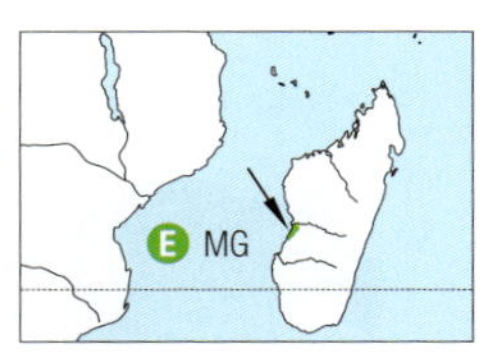

Coastal zone from R Tsiribihina S to N Menabe Antimena NP

CR
Monotypic
HB 30–34 cm
T 21–24 cm
W 1.1–1.3 kg

Genus *BRACHYUROMYS*

(2 species)

Small Short-tailed Rat
Brachyuromys betsileoensis

FR Petit Ramirohitra
DE Kleine Madagaskar-Kurzschwanzratte
ES Rata de cola corta pequeña

LC
Monotypic
HB 14–18.4 cm
T 7.7–9.5 cm
W 96–140 g

900–2660 m

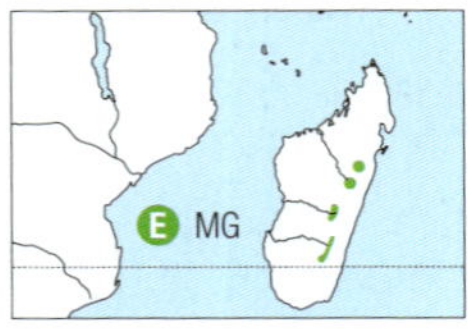

EC Highlands (from SE L Alaotra S to Andringitra Massif

Large Short-tailed Rat
Brachyuromys ramirohitra

FR Grand Ramirohitra
DE Große Madagaskar-Kurzschwanzratte
ES Rata de cola corta grande

LC
Monotypic
HB 14–16.5 cm
T 8.4–11 cm
W 64–117 g

1210–1960 m

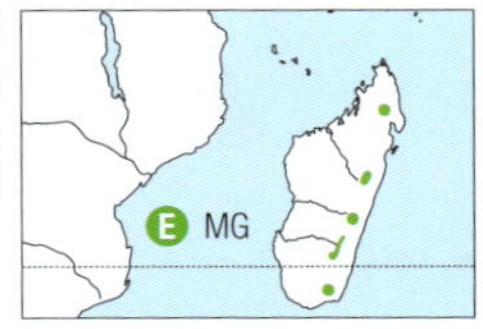

Eastern humid lowland to montane forests

Genus *NESOMYS*

(3 species)

Red Forest Rat
Nesomys rufus

FR Nésomys roux
DE Rote Inselratte
ES Rata de bosque roja

LC
Monotypic
HB 17–20 cm
T 16–18 cm
W 135–185 g

650–2000 m

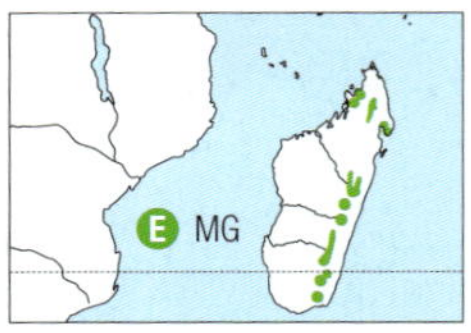

NE, C & E humid lowland to montane forests

LC
Monotypic
HB 19.5–20 cm
T 16.9–17.3 cm
W 193–239 g

Audebert's Forest Rat
Nesomys audeberti

FR Nésomys d'Audebert
DE Audebert-Inselratte
ES Rata de bosque de Audebert

0–1050 m

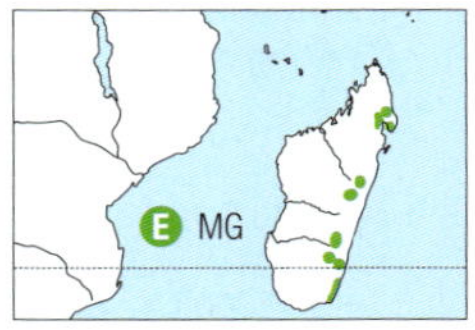

From Masoala Pen S to Manantantely (near Tolagnaro)

Lamberton's Forest Rat
Nesomys lambertoni

FR Nésomys de Lamberton
DE Lamberton-Inselratte
ES Rata de bosque de Lamberton

EN
Monotypic
HB 18.9–19.5 cm
T 18.3–19.1 cm
W 225–243 g

Limestone regions in Tsingy de Bemaraha area

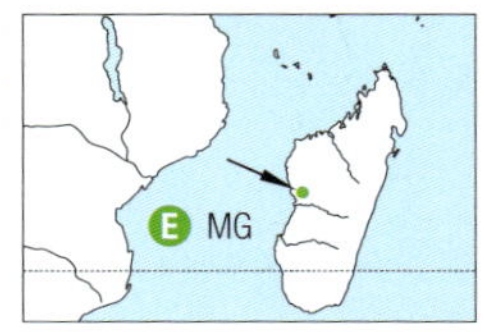

c. 100 m

Genus *DELANYMYS*

(1 species)

Delany's Swamp Mouse
Delanymys brooksi

FR Delanymys palustre
DE Delany-Sumpfklettermaus
ES Ratón de los pantanos de Delany

VU
Monotypic
HB 5–6.3 cm
T 8.7–11.1 cm
W 5–7 g

1700–3100 m

Albertine Rift E of L Kivu (E CD, W RW, SW UG)

Genus *MYSTROMYS*

(1 species)

VU
Monotypic
HB 11.6–19.9 cm
T 4.6–8.7 cm
W 49–111 g

African White-tailed Rat
Mystromys albicaudatus

FR Mystromys à queue blanche
DE Afrikanische Weißschwanzratte
ES Rata de cola blanca africana

above 1000 m

Grasslands and natural shrub-lands in SW & E ZA and LS

Genus *PETROMYSCUS*

(4 species)

Shortridge's Pygmy Rock Mouse
Petromyscus shortridgei

FR Petromysque de Shortridge
DE Shortridge-Zwergfelsenmaus
ES Ratón de roca enano de Shortridge

LC
Monotypic
HB 6.5–9.8 cm
T 7.4–10.4 cm
W 12–31 g

Rocky terrain in extreme SW AO & N NA

Short-eared Pygmy Rock Mouse
Petromyscus monticularis

FR Petromysque du Brukkaros
DE Brukkaros-Zwergfelsenmaus
ES Ratón de roca enano de orejas cortas

LC
Monotypic
HB 6.3–9.3 cm
T 6.6–8.4 cm

Rocky areas in S NA, NW ZA

Barbour's Pygmy Rock Mouse
Petromyscus barbouri

FR Petromysque de Barbour
DE Barbour-Zwergfelsenmaus
ES Ratón de roca enano de Barbour

LC
Monotypic
HB 7.1–7.5 cm
T 7.6–8.9 cm
W 14–16 g

Rocky outcrops with succulent vegetation

Common Pygmy Rock Mouse
Petromyscus collinus

FR Petromysque pygmée
DE Zwergfelsenmaus
ES Ratón de roca enano común

LC
Monotypic
HB 7.2–9.3 cm
T 7.8–12.2 cm
W 14–24 g

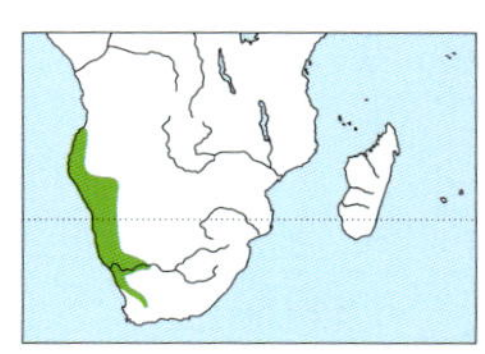

SW AO S through W NA to W ZA

Greater Long-tailed Pouched Rat
Beamys major

FR Grand Rat-hamster
DE Große Hamsterratte
ES Rata de abazones grande

500–1000 m

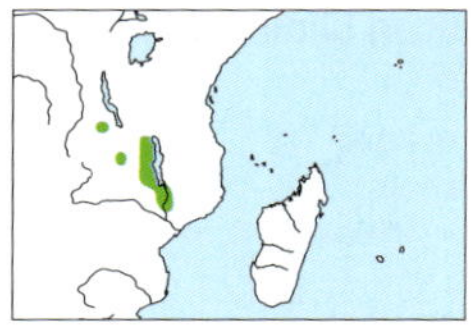

N & E ZM, MW & NC MZ (Mt Mabu)

NE
Monotypic
HB 13.6–17.3 cm
T 12.6–14.3 cm
W 95–102 g

Genus *BEAMYS*

(2 species)

Hinde's Long-tailed Pouched Rat
Beamys hindei

FR Petit Rat-hamster
DE Kleine Hamsterratte
ES Rata de abazones de Hinde

below 1200 m

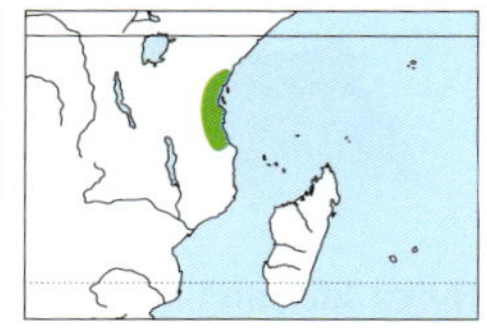

Coastal plains and montane areas of E KE & TZ

LC
Monotypic
HB 12.5–15.5 cm
T 11.4–12.7 cm
W 49–76 g

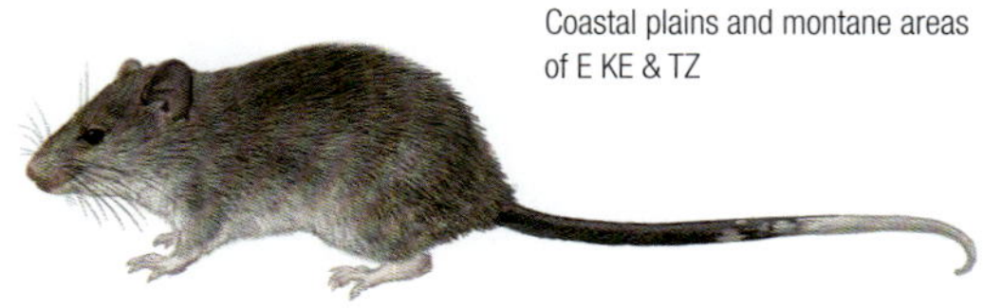

Genus *CRICETOMYS*

(4 species)

LC
Monotypic
HB 30–35 cm
T 32–43 cm
W 0.5–1.3 kg

Forest Giant Pouched Rat
Cricetomys emini

FR Cricétome d'Emin
DE Emin-Riesenhamsterratte
ES Rata de abazones gigante de bosque

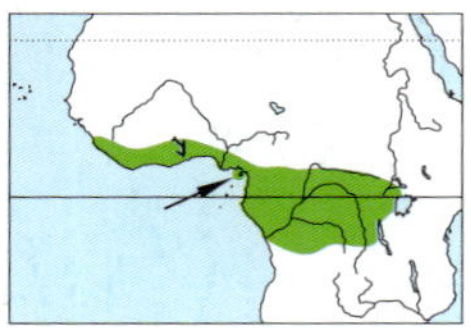

Tropical Africa from SL E to UG, RW & BI, S to CD & AO; Bioko

LC
Monotypic
HB 28–39 cm
T 25–40 cm
W 0.5–1.5 kg

Northern Giant Pouched Rat
Cricetomys gambianus

FR Cricétome de Gambie
DE Gambia-Riesenhamsterratte
ES Rata de abazones gigante septentrional

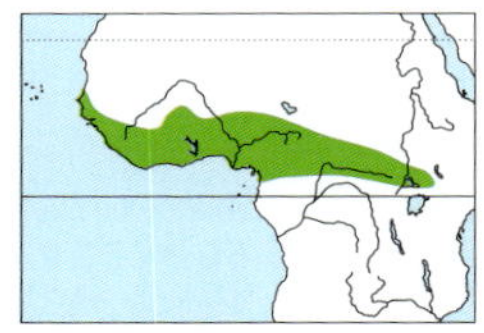

SN E to SS & UG, S to R Congo (CD)

Kivu Giant Pouched Rat
Cricetomys kivuensis

FR Cricétome du Kivu
DE Kivu-Riesenhamsterratte
ES Rata de abazones gigante de Kivu

NE
Monotypic
HB c. 32 cm
T c. 34 cm

Albertine Rift in E CD, SW UG, W RW & BI

Southern Giant Pouched Rat
Cricetomys ansorgei

FR Cricétome d'Ansorge
DE Südliche Riesenhamsterratte
ES Rata de abazones gigante meridional

LC
6 sspp
HB 29–41 cm
T 34–45 cm
W 0.9–2.8 kg

CD S of R Congo E to KE & TZ (with Unguja I), S to AO, ZM, ZW & NE ZA

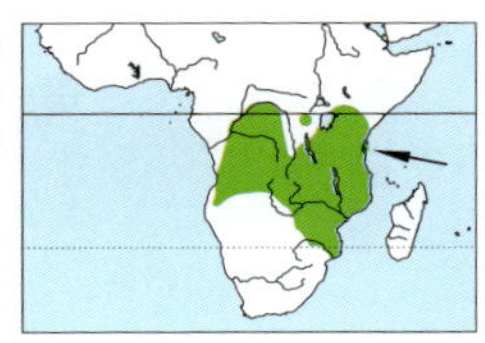

East African Pouched Mouse
Saccostomus mearnsi

FR Saccostome de Mearns
DE Ostafrika-Kurzschwanzhamsterratte
ES Ratón de abazones de África oriental

below 1500 m

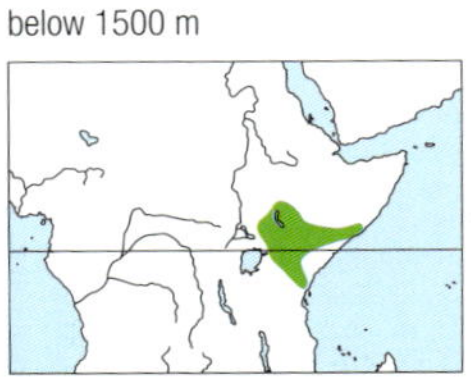

E UG, KE, SW ET & S SO

LC
Monotypic
HB 10.1–15 cm
T 3.8–7.4 cm
W 43–69 g

Genus *SACCOSTOMUS*

(3 species)

Gray-bellied Pouched Mouse
Saccostomus umbriventer

FR Saccostome à ventre gris
DE Graubauch-Kurzschwanzhamsterratte
ES Ratón de abazones de vientre gris

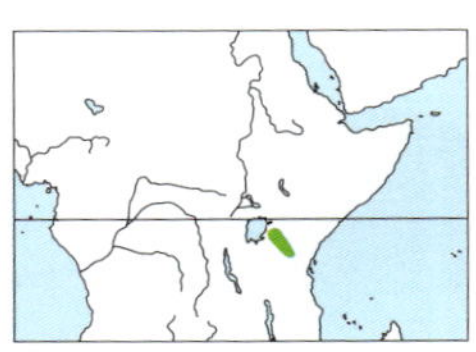

Masai Steppe savanna in SW KE & N TZ

NE
Monotypic
HB 13.3–13.6 cm
T 5.5–6 cm

Southern African Pouched Mouse
Saccostomus campestris

FR Saccostome du Cap
DE Südliche Kurzschwanzhamsterratte
ES Ratón de abazones de África meridional

below 1500 m

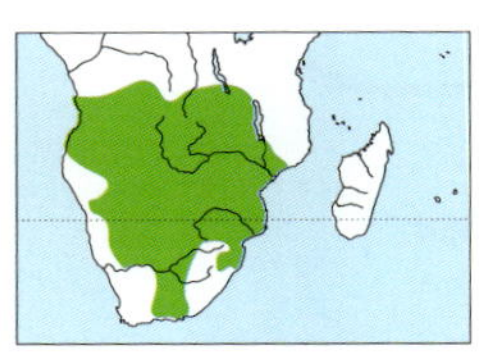

S Africa from AO, CD & TZ S to NA & ZA

LC
2 sspp
HB 9.5–17.3 cm
T 2.8–5.5 cm
W 33–81 g

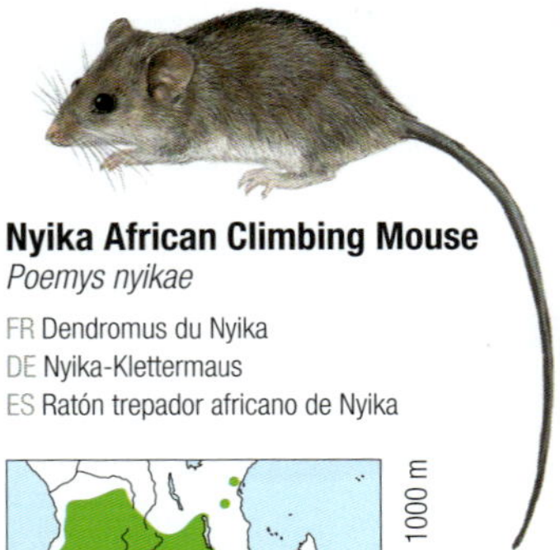

Nyika African Climbing Mouse
Poemys nyikae

FR Dendromus du Nyika
DE Nyika-Klettermaus
ES Ratón trepador africano de Nyika

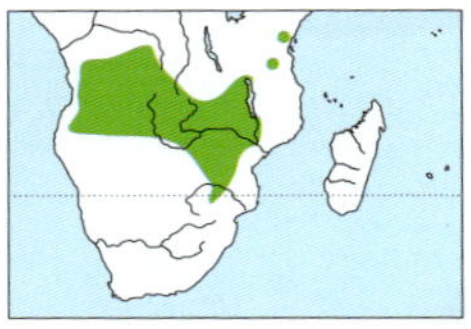

above 1000 m

AO E to TZ (extreme SW, Eastern Arc Mts) & C MZ, S to ZW & NE ZA

LC
Monotypic
HB 6.1–7.9 cm
T 8.5–9.7 cm
W 10–14 g

Genus *POEMYS*

(2 species)

NE
Monotypic
HB c. 7.2 cm
T c. 7.5 cm

Pecile's African Climbing Mouse
Poemys pecilei

FR Dendromus de Pecile
DE Pecile-Klettermaus
ES Ratón trepador africano del Congo

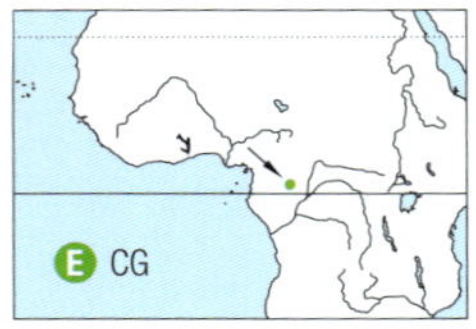

R Ogowe area

Genus *MEGADENDROMUS*

(1 species)

VU
Monotypic
HB 11.7–12.9 cm
T 9.2–10.6 cm
W 49–66 g

Nikolaus's African Climbing Mouse
Megadendromus nikolausi

FR Dendromus de Nikolaus
DE Riesenklettermaus
ES Ratón trepador africano de Nikolaus

3000–3800 m

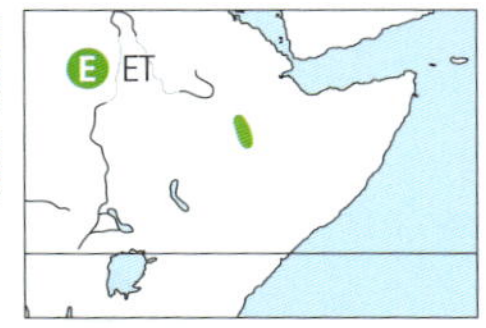

Highlands E of Rift Valley (Bale & Arsi Mts)

Genus *DENDROMUS*

(14 species)

Monard's African Climbing Mouse
Dendromus leucostomus

NE
Monotypic
HB 4.7–6.9 cm
T 5.7–8 cm

FR Dendromus de Monard
DE Monard-Klettermaus
ES Ratón trepador africano de Monard

Known from a single site (Caluquembe)

Gray African Climbing Mouse
Dendromus melanotis

FR Dendromus gris
DE Graue Klettermaus
ES Ratón trepador africano gris

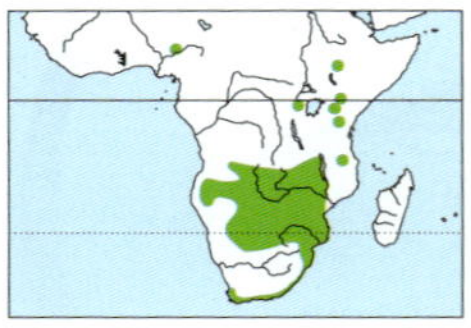

S Africa from SW AO E to MW, S to S ZA; disjunctly in NG, UG, SW ET, KE & TZ

LC
Monotypic
HB 5–7.7 cm
T 5.5–9.6 cm
W 6–15 g

Lachaise's African Climbing Mouse
Dendromus lachaisei

FR Dendromus de Lachaise
DE Lachaise-Klettermaus
ES Ratón trepador africano de Lachaise

50–600 m

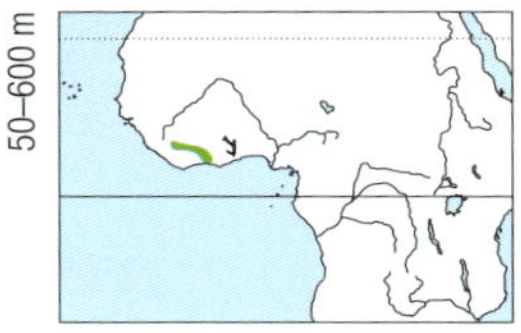

SE GN, LR, CI

DD
Monotypic
HB 5.5–7.8 cm
T 7.9–9.8 cm
W c. 11 g

Kumasi African Climbing Mouse
Dendromus kumasi

FR Dendromus de Kumasi
DE Kumasi-Klettermaus
ES Ratón trepador africano de Kumasi

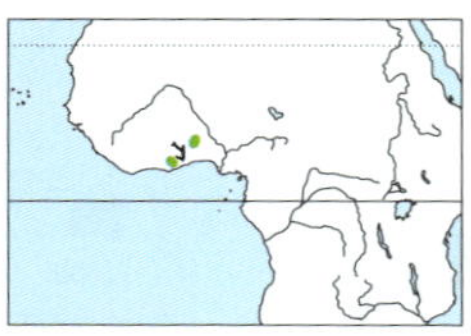

GH (Kumasi, Achimota), BJ

NE
Monotypic
HB c. 6 cm
T c. 9 cm

CR
Monotypic
HB 7.7–8.2 cm
T 12–13.2 cm
W 10–12 g

Mount Kahuzi African Climbing Mouse
Dendromus kahuziensis

FR Dendromus du Kahuzi
DE Kahuzi-Klettermaus
ES Ratón trepador africano de Kahuzi

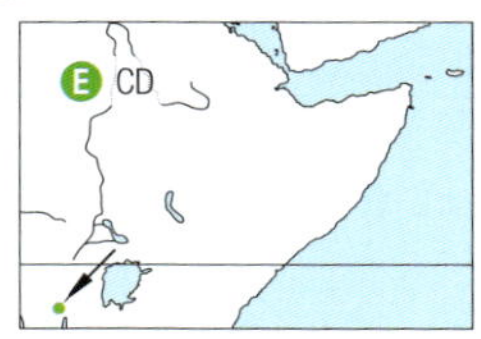

c. 2100 m

Known only from Mt Kahuzi

Lovat's African Climbing Mouse
Dendromus lovati

FR Dendromus de Lovat
DE Lovat-Klettermaus
ES Ratón trepador africano de Lovat

LC
Monotypic
HB 5.7–9.5 cm
T 5.7–8.7 cm
W 11–23 g

above 2500 m

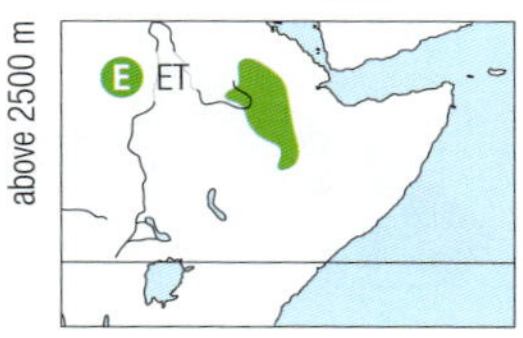

High-elevation regions

DD
Monotypic
HB 6.3–8.2 cm
T 9.1–11.5 cm
W 8–15 g

Rupp's African Climbing Mouse
Dendromus ruppi

FR Dendromus de Rupp
DE Rupp-Klettermaus
ES Ratón trepador africano de Rupp

1800–1900 m

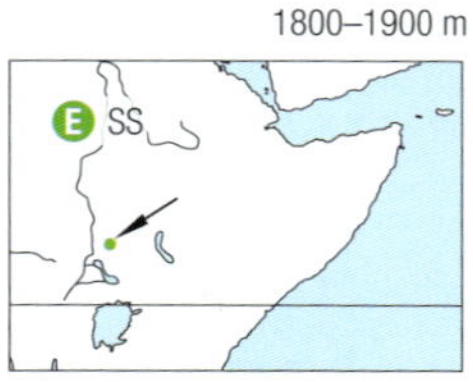

Known only from Imatong Mts

LC
Monotypic
HB 5.7–7 cm
T 7.8–9.5 cm
W 7–10 g

Banana African Climbing Mouse
Dendromus messorius

FR Dendromus des bananiers
DE Bananenklettermaus
ES Ratón trepador africano de bananera

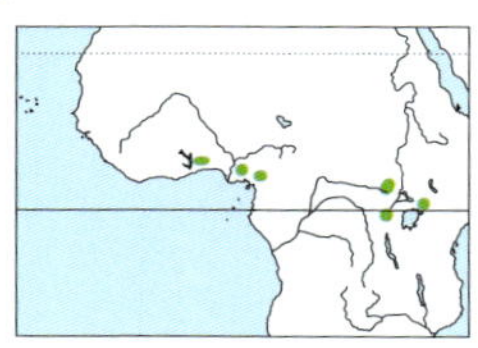

Disjunct in GH, TG, BJ, SE NG, CM, extreme S SS, NE CD, SE UG & W KE

Chestnut African Climbing Mouse
Dendromus mystacalis

FR Dendromus de Heuglin
DE Kastanienbraune Klettermaus
ES Ratón trepador africano castaño

below 1000 m

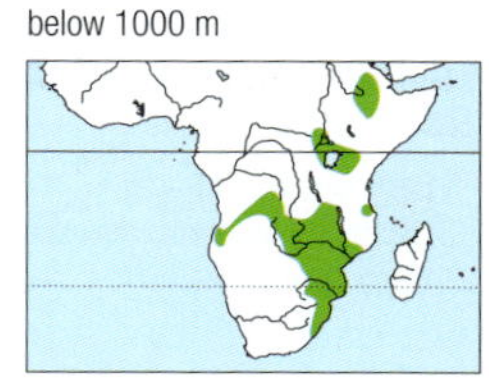

E & S Africa (ET, S SS, CD, UG & KE, from TZ S to E ZA, W to SW AO)

LC
Monotypic
HB 4.7–8 cm
T 6.4–10.1 cm
W 4–11 g

Kivu African Climbing Mouse
Dendromus nyasae

FR Dendromus du Kivu
DE Kivu-Klettermaus
ES Ratón trepador africano de Kivu

above 1300 m

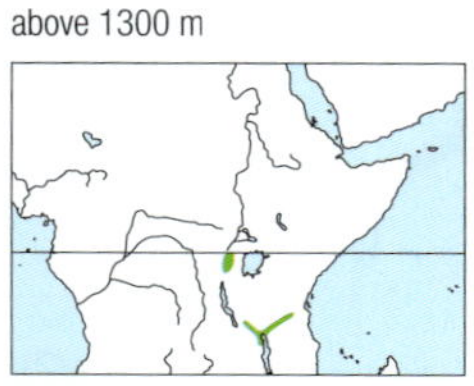

NE CD, SW UG & RW and S TZ, extreme NE ZM & N MW

NE
2 sspp
HB 6.4–8 cm T 8.4–10.5 cm
W 6–20 g

Montane African Climbing Mouse
Dendromus insignis

FR Dendromus des montagnes
DE Gebirgsklettermaus
ES Ratón trepador africano de montaña

LC
Monotypic
HB 7.8–9.2 cm T 8.2–11.1 cm
W 12–17 g

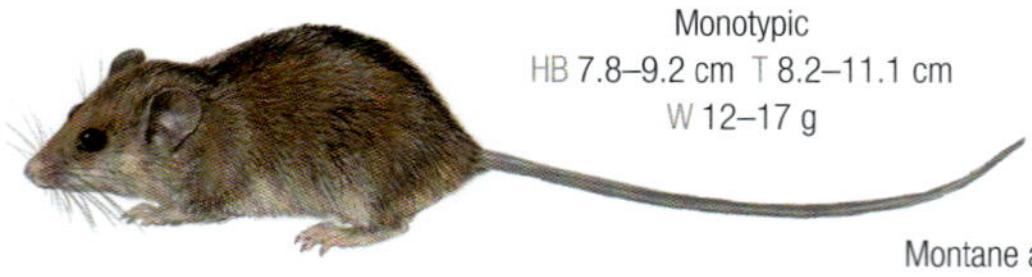

Montane areas of E Africa in E CD, W UG, RW, S KE & TZ (Eastern Arc Mts)

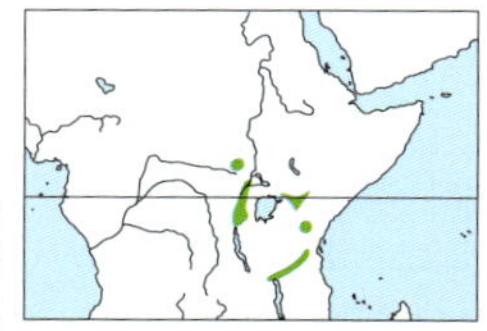

1500–4700 m

Brants's African Climbing Mouse
Dendromus mesomelas

FR Dendromus de Brants
DE Brants-Klettermaus
ES Ratón trepador africano de Brants

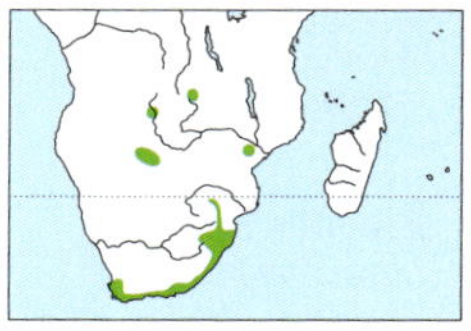

LC
Monotypic
HB 6.7–8.8 cm
T 7.7–12.1 cm
W 6–15 g

E & S ZA, SZ; disjunctly in S CD, NW ZM, N BW, NE NA (Caprivi) & C MZ

Vernay's African Climbing Mouse
Dendromus vernayi

FR Dendromus de Vernay
DE Vernay-Klettermaus
ES Ratón trepador africano de Vernay

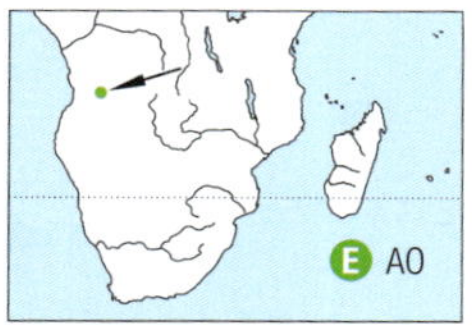

DD
Monotypic
HB 6.3–7.7 cm
T 8–8.6 cm

Known only from type locality (Chitau)

Cameroon African Climbing Mouse
Dendromus oreas

FR Dendromus du Cameroun
DE Kamerun-Klettermaus
ES Ratón trepador africano de Camerún

850–2900 m

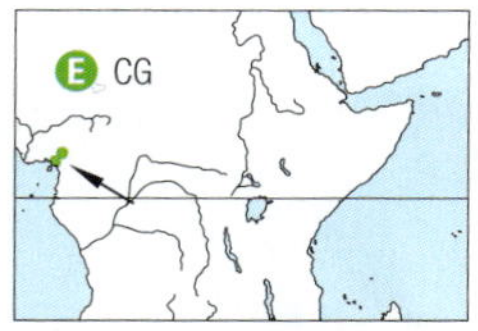

LC
Monotypic
HB 6–7.4 cm
T 8.9–10.4 cm
W 9–13 g

Manenguba, Kupé and Cameroon Mts

Genus *DENDROPRIONOMYS*

(1 species)

Velvet African Climbing Mouse
Dendroprionomys rousseloti

FR Dendromus de Brazzaville
DE Samtklettermaus
ES Ratón trepador africano de terciopelo

DD
Monotypic
HB c. 7.7 cm
T c. 10.8 cm

c. 300 m

Known only from type locality (Zoological Garden of Brazzaville)

Genus *PRIONOMYS*

(1 species)

DD
Monotypic
HB 7.9–8.5 cm
T 10.3–11.2 cm
W 15–21 g

100–300 m

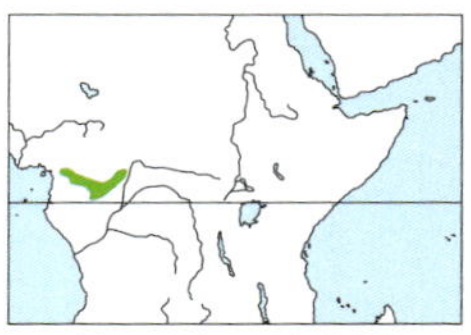

Low-lying rainforests in S CM, SW CF & N CD

Bates's African Climbing Mouse
Prionomys batesi

FR Dendromus de Bates
DE Dollman-Baummaus
ES Ratón trepador africano de Bates

Genus *STEATOMYS*

(8 species)

LC
Monotypic
HB 9.6–12.2 cm
T 3.5–5 cm
W 30–66 g

North-western Fat Mouse
Steatomys caurinus

FR Rat-adipeux du Nord-Ouest
DE Westafrikanische Fettmaus
ES Ratón grueso noroccidental

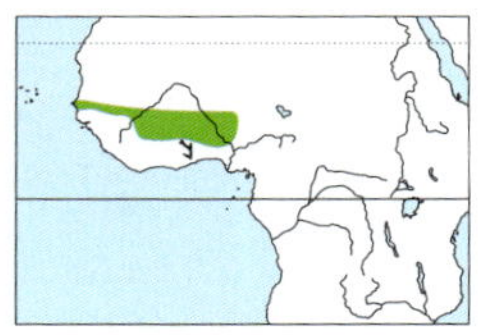

Continuously from SN E to ML, CI, BF, GH, TG, BJ & NW NG

LC
Monotypic
HB 7.8–9.3 cm
T 4.1–5 cm
W 14–24 g

Dainty Fat Mouse
Steatomys cuppedius

FR Rat-adipeux élégant
DE Zierliche Fettmaus
ES Ratón grueso delicado

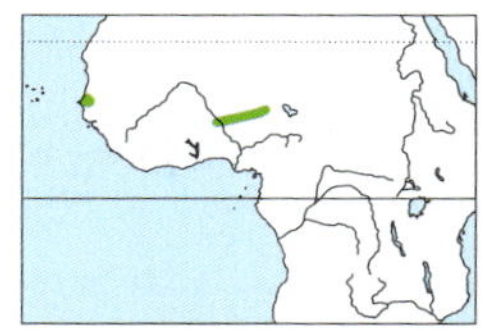

Arid savanna in SN and BJ, N NG & S NE

DD
Monotypic
HB c. 12 cm
T c. 5 cm

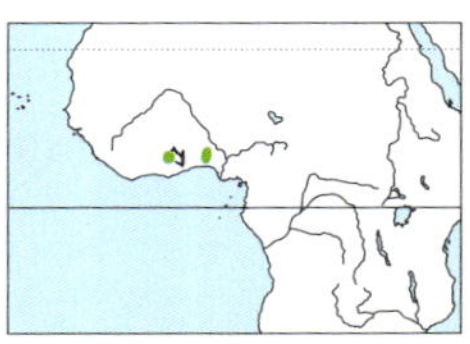

W GH, also reported from W NG

Jackson's Fat Mouse
Steatomys jacksoni

FR Rat-adipeux de Jackson
DE Jackson-Fettmaus
ES Ratón grueso de Jackson

LC
Monotypic
HB 11.5–12.5 cm
T 5.2–6.5 cm W 30–50 g

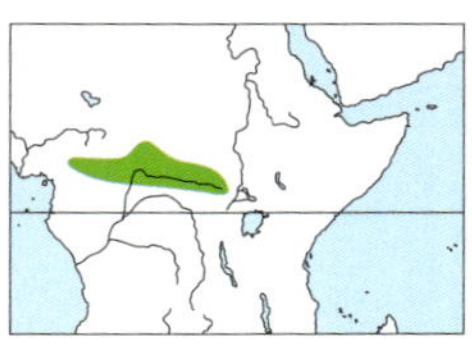

Narrowly from S CM E to CF, N CD & extreme SW SS

Pousargues's Fat Mouse
Steatomys opimus

FR Rat-adipeux de Pousargues
DE Pousargues-Fettmaus
ES Ratón grueso de Pousargues

LC
Monotypic
HB 9.7–12.4 cm
T 5.2–7.2 cm

above 500 m

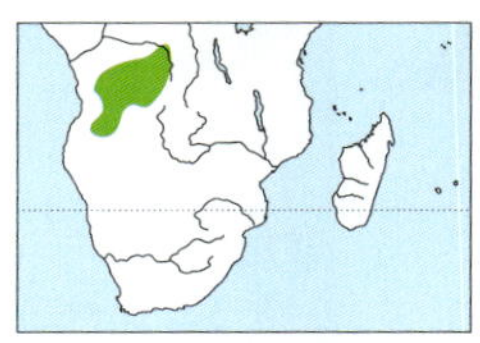

Moist savannas and woodlands in SW CD & AO

Bocage's Fat Mouse
Steatomys bocagei

FR Rat-adipeux de Bocage
DE Bocage-Fettmaus
ES Ratón grueso de Bocage

Tiny Fat Mouse
Steatomys parvus

FR Petit Rat-adipeux
DE Zwergfettmaus
ES Ratón grueso diminuto

LC
Monotypic
HB 5.5–8.3 cm
T 3.4–5.1 cm
W 8–15 g

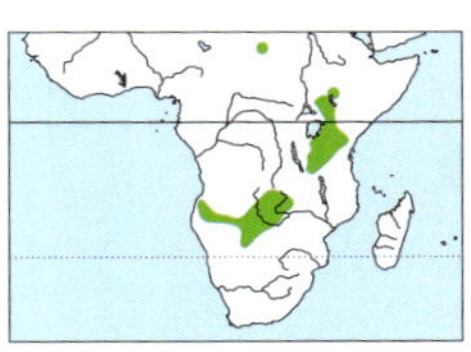

E Africa (SW ET to W TZ) & S Africa (S AO to W ZM); Jebel Marra (W SD)

Common Fat Mouse
Steatomys pratensis

FR Rat-adipeux de Peters
DE Gewöhnliche Fettmaus
ES Ratón grueso común

LC
Monotypic
HB 8.2–10.2 cm
T 4–5.6 cm
W 22–48 g

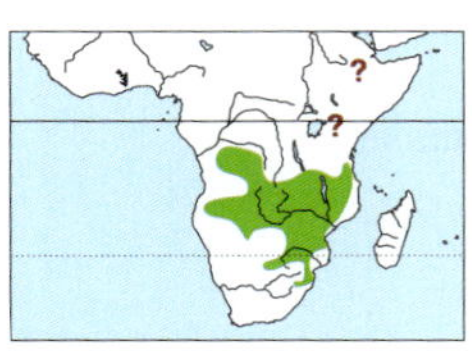

S Africa (S CD & S TZ, S to SE BW, N ZA & SZ); unconfirmed in ET & KE

Krebs's Fat Mouse
Steatomys krebsii

FR Rat-adipeux de Krebs
DE Krebs-Fettmaus
ES Ratón grueso de Krebs

LC
Monotypic
HB 7.3–10 cm
T 3.2–5.6 cm
W 15–42 g

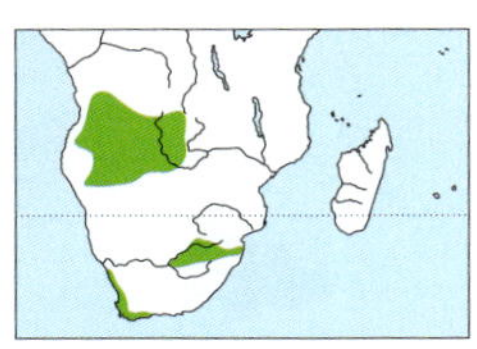

1 population in AO, W ZM, N NA & N BW, other 2 in NE & SW ZA, respectively

Genus *MALACOTHRIX*

(1 species)

Long-eared Desert Mouse
Malacothrix typica

FR Malacothrix à grandes oreilles
DE Afrikanische Langohrmaus
ES Ratón de desierto de orejas largas

LC
Monotypic
HB 6–8.7 cm
T 2.9–4.1 cm
W 15–23 g

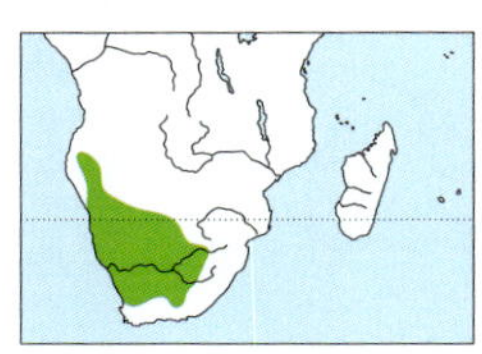

Extreme SW AO, NA, SW BW, W & C ZA

Family CRICETIDAE – True Hamsters, Voles, Lemmings and New World Rats and Mice

144 genera, 830 species. Virtually all terrestrial and some aquatic habitats. Holarctic and Neotropical Regions, marginally into Indo-Malayan Region.

Genus *MESOCRICETUS*

(4 species)

Golden Hamster
Mesocricetus auratus

FR Hamster doré
DE Goldhamster
ES Hámster de Siria

280–650 m

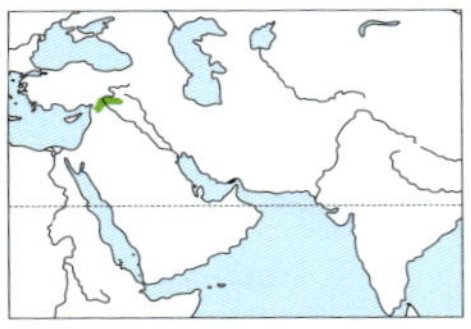

S TR & N SY (Aleppinian Plateau between Ebla and Şanlıurfa; possibly Hatay)

EN
Monotypic
HB 12.8–16.5 cm
T 1.3–2.1 cm
W 82–175 g

Ciscaucasian Hamster
Mesocricetus raddei

FR Hamster du Daghestan
DE Schwarzbrusthamster
ES Hámster del Cáucaso

200–2800 m

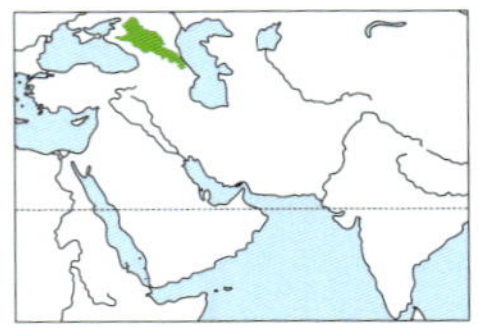

Between lower R Don and Dagestan Plateau (RU), recorded in NE GE

LC
2 sspp
HB 14.9–22 cm
T 1.2–2 cm
W 200–300 g

Brandt's Hamster
Mesocricetus brandti

NT Monotypic
HB 13.5–19.5 cm
T 1–3.9 cm
W 108–175 g

300–3000 m

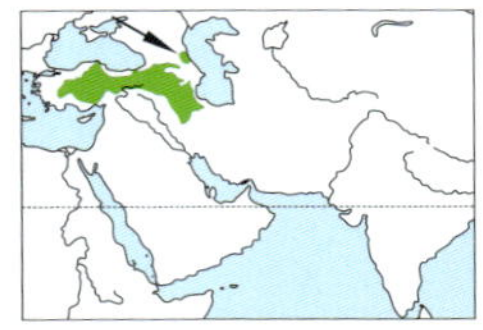

FR Hamster de Brandt
DE Brandt-Hamster
ES Hámster de Brandt

TR (C & E Anatolia), GE, AM, AZ, NW IR; disjunctly N of Caucasus (S RU)

Romanian Hamster
Mesocricetus newtoni

FR Hamster de Newton
DE Rumänischer Hamster
ES Hámster de Rumanía

up to 460 m

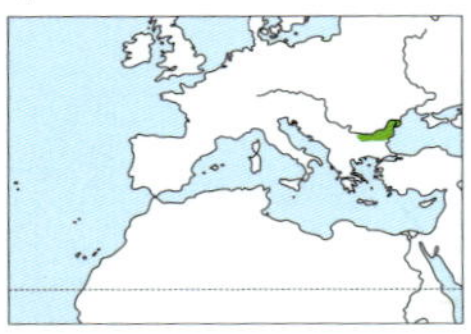

N BG & SE RO (lower Danube between Montana Prov and Black Sea)

NT
Monotypic
HB 13.5–16 cm
T 1.7–2.4 cm
W 60–130 g

Genus *PHODOPUS*

(3 species)

Roborovski's Desert Hamster
Phodopus roborovskii

LC
3 sspp
HB 7.7–7.9 cm
T 0.9–1.1 cm
W 19–20 g

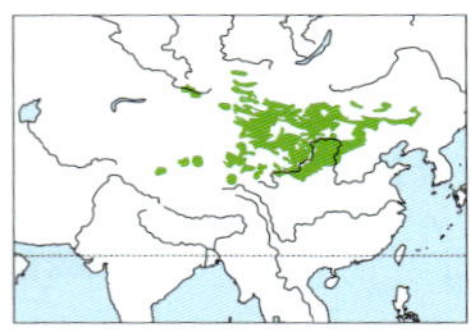

FR Hamster de Roborovski
DE Roborovski-Zwerghamster
ES Hámster enano de Roborovski

NE KZ, RU (S Tuva), NW & S MN, CN (E to NW Jilin & N Liaoning)

Campbell's Desert Hamster
Phodopus campbelli

LC
2 sspp
HB 7.9–10.3 cm
T 0.7–1.4 cm
W 23–28 g

up to 3500 m

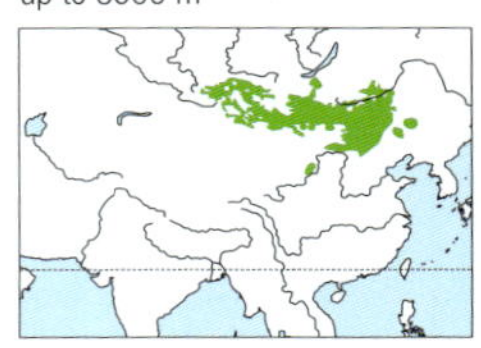

FR Hamster de Campbell
DE Campbell-Zwerghamster
ES Hámster enano de Campbell

RU (SE Siberia), MN, CN (W Ningxia, Inner Mongolia, N Hebei)

Striped Desert Hamster
Phodopus sungorus

FR Hamster de Dzoungarie
DE Dsungarischer Zwerghamster
ES Hámster enano de Zungaria

winter

summer

LC
Monotypic
HB 7–9 cm
T 0.5–1.5 cm
W 22–25 g

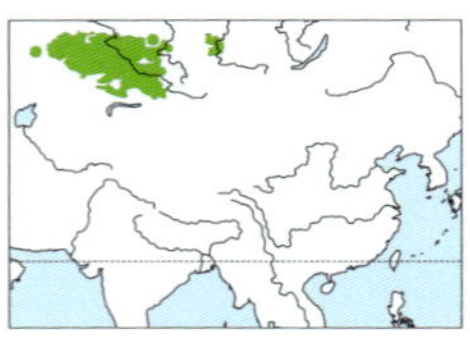

N, C & E KZ, RU (SW Siberia, isolated population in SC Siberia)

Genus *UROCRICETUS*

(2 species)

Ladakh Dwarf Hamster
Urocricetus alticola

FR Hamster du Ladakh
DE Ladakh-Zwerghamster
ES Hámster enano de Ladakh

3100–5200 m

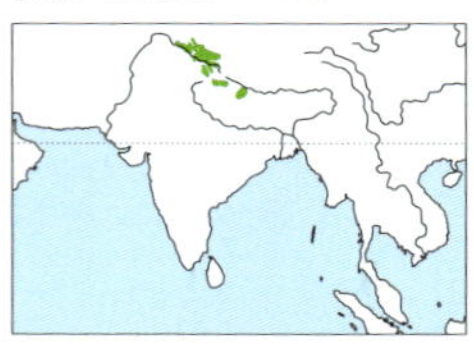

SW CN (SW Xinjiang, W Tibet), NW IN, W NP

LC
Monotypic
HB 8–10 cm
T 2.9–4.2 cm
W 22–48 g

Tibetan Dwarf Hamster
Urocricetus kamensis

LC
Monotypic
HB 8.8–11.2 cm
T 3–6.4 cm
W 20–40 g

3300–4100 m

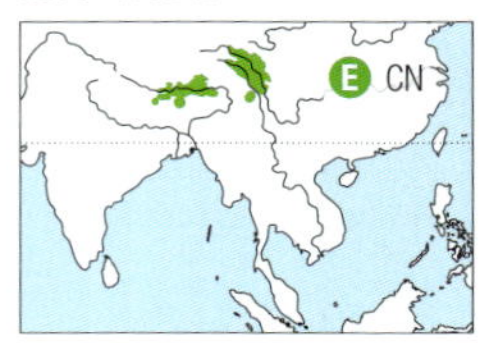

FR Hamster du Tibet
DE Tibet-Zwerghamster
ES Hámster enano de Tibet

Qinghai, Tibet, NW Sichuan

Genus *CANSUMYS*

(1 species)

Gansu Hamster
Cansumys canus

LC
Monotypic
HB 12.9–14 cm
T 9.5–10.9 cm
W 44–57 g

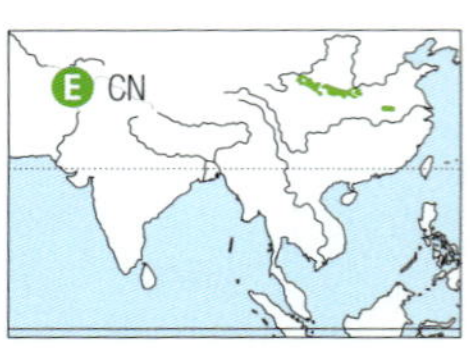

FR Hamster du Gansu
DE Gansu-Zwerghamster
ES Hámster de Gansu

Mountain forests in S Gansu, S Shaanxi & S Henan

Genus *TSCHERSKIA*

(1 species)

Greater Long-tailed Hamster
Tscherskia triton

LC
5 sspp
HB 14.2–22 cm
T 6.5–10.6 cm
W 92–241 g

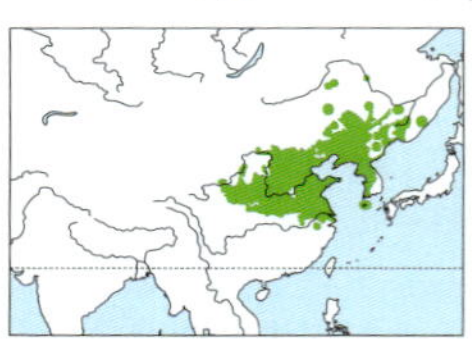

FR Hamster triton
DE Langschwanzhamster
ES Hámster grande de cola larga

CN, KR (with Cheju I), RU (Primorsky Krai)

Genus *ALLOCRICETULUS*

(2 species)

Mongolian Hamster
Allocricetulus curtatus

LC
Monotypic
HB 10–13 cm
T 1.6–2.6 cm
W 30–65 g

FR Hamster de Mongolie
DE Mongolischer Zwerghamster
ES Hámster de Mongolia

RU (S Tuva), MN, N CN (NE Xinjiang, Inner Mongolia, Gansu, Ningxia)

LC 2 sspp
HB 10.3–13.6 cm
T 2–3.1 cm
W 32–68 g

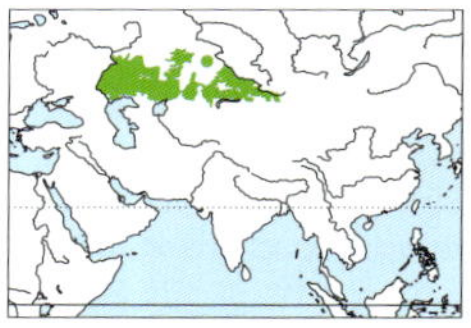

Eversmann's Hamster
Allocricetulus eversmanni

FR Hamster d'Eversmann
DE Eversmann-Zwerghamster
ES Hámster de Eversmann

RU (E of R Volga to Omsk Region), N KZ, NW CN (N Xinjiang)

Genus *CRICETUS*

(1 species)

CR
Monotypic
HB 16.5–32 cm
T 2.7–6.8 cm
W 146–860 g

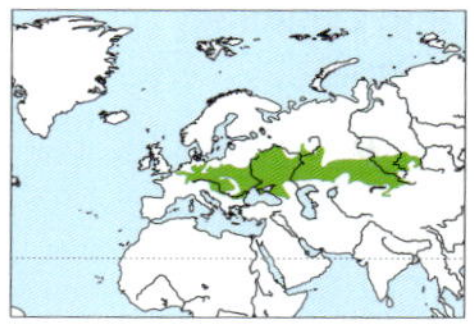

up to 2200 m

Common Hamster
Cricetus cricetus

FR Hamster d'Europe
DE Feldhamster
ES Hámster común

W Europe (E from BE) to R Yenisei (RU), KZ & NW CN (Xinjiang)

Genus *NOTHOCRICETULUS*

(1 species)

LC
2 sspp
HB 8.5–15 cm
T 1.9–3.9 cm
W 20–58 g

up to 3600 m

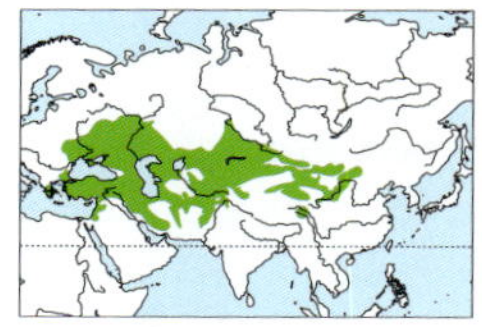

Gray Dwarf Hamster
Nothocricetulus migratorius

FR Hamster migrateur
DE Grauer Zwerghamster
ES Hámster enano gris

E Europe (E from GR & RO) to SW Siberia, MN & CN, S to IL, JO, IR & PK

LC
3 sspp
HB 8–13.5 cm
T 3.5–4.8 cm
W 15–50 g

up to 1900 m

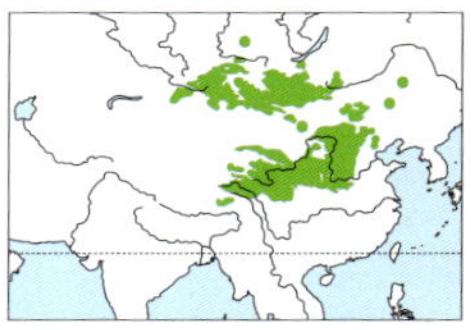

Long-tailed Dwarf Hamster
Cricetulus longicaudatus

FR Hamster à longue queue
DE Langschwanz-Zwerghamster
ES Hámster enano de cola larga

Extreme E KZ, RU (S Siberia), MN, N & C CN (S to NE Tibet)

Genus *CRICETULUS*

(3 species)

LC
5 sspp
HB 7.2–12.4 cm
T 1.5–3.3 cm
W 20–60 g

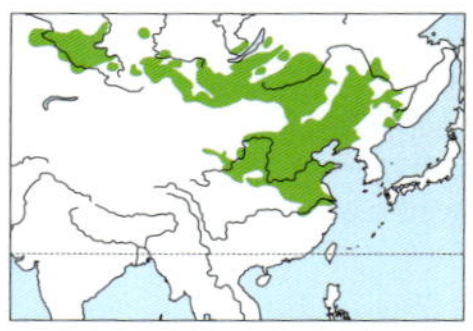

Striped Dwarf Hamster
Cricetulus barabensis

FR Hamster de Chine
DE Daurischer Zwerghamster
ES Hámster enano rayado

NE KZ, RU (S Siberia, Amur), N & E MN, NE CN, extreme N Korean Pen

LC
Monotypic
HB 7.7–11.4 cm
T 1.8–3.2 cm
W ≤ 60 g

Sokolov's Dwarf Hamster
Cricetulus sokolovi

FR Hamster de Sokolov
DE Sokolow-Zwerghamster
ES Hámster enano de Sokolov

Semi-deserts in MN & N CN (Inner Mongolia)

Genus *PROMETHEOMYS*

(1 species)

1500–2800 m

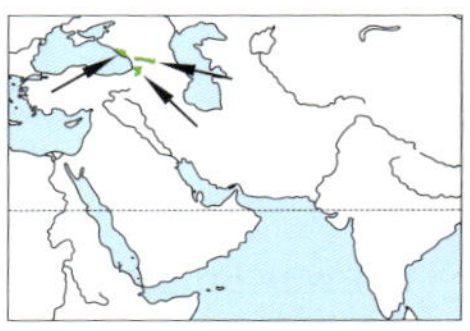

Mesic meadows in ciscaucasian RU, GE & NE TR

Long-clawed Mole Vole
Prometheomys schaposchnikowi

FR Lemming prométhée DE Prometheus-Maus
ES Topillo topo de uñas largas

LC
Monotypic
HB 12.6–16.5 cm
T 3.6–5.8 cm
W 59–88 g

Genus *NEOFIBER*

(1 species)

Round-tailed Muskrat
Neofiber alleni

FR Rat-musqué de Floride
DE Florida-Wasserratte
ES Rata almizclera de Florida

LC
4 sspp
HB 18.6–21 cm
T 9.9–16.8 cm
W 189–357 g

Submerged meadows in SE Georgia & Florida Pen

Genus *ONDATRA*

(1 species)

Common Muskrat
Ondatra zibethicus

FR Rat-musqué commun
DE Bisamratte
ES Rata almizclera

LC
15 sspp
HB 27–30 cm
T 20–25 cm
W 700–1800 g

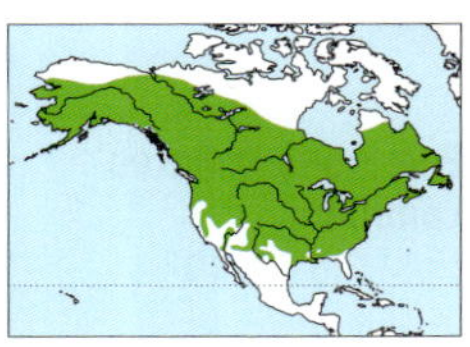

Alaska & N CA S through US (except arid regions of SW, S & Florida Pen)

Western Heather Vole
Phenacomys intermedius

FR Phénacomys des bruyères
DE Westliche Heidekraut-Wühlmaus
ES Topillo de brezo occidental

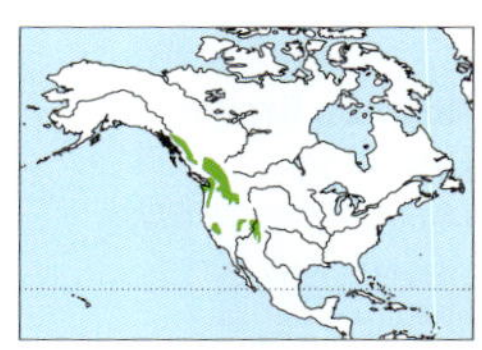

LC 5 sspp
HB 10.4–11.2 cm
T 2.6–4.1 cm
W c. 40 g

W CA (British Columbia, WC Alberta), W US (S to N New Mexico)

Genus *ARBORIMUS*

(3 species)

White-footed Vole
Arborimus albipes

FR Phénacomys à pieds blancs
DE Weißfuß-Baumwühlmaus
ES Topillo de pies blancos

0–1100 m

LC
Monotypic
HB 10.3–11 cm
T 6.2–7.1 cm

Oregon (R Columbia S along Coast Range, W Cascade Range) S to N California

Genus *PHENACOMYS*

(2 species)

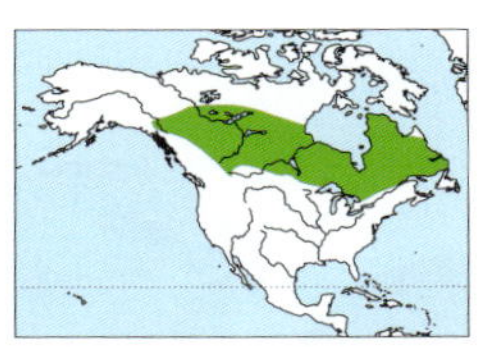

CA (from Yukon & British Columbia E to Labrador), US (NE Minnesota)

Eastern Heather Vole
Phenacomys ungava

FR Phénacomys d'Ungava
DE Östliche Heidekraut-Wühlmaus
ES Topillo de brezo oriental

LC
4 sspp
HB 8.8–11.2 cm
T 2.3–4 cm
W 15–47 g

NT
2 sspp
HB 9.8–10.4 cm
T 6–8.3 cm
W 25–50 g

up to 1600 m

Red Tree Vole
Arborimus longicaudus
FR Phénacomys arboricole
DE Rote Baumwühlmaus
ES Topillo arbóreo rojo

Oregon (Coast and W Cascade Range) S to Humboldt County (California)

LC
9 sspp
HB 9.3–11.3 cm
T 1.7–2.7 cm
W 27–35 g

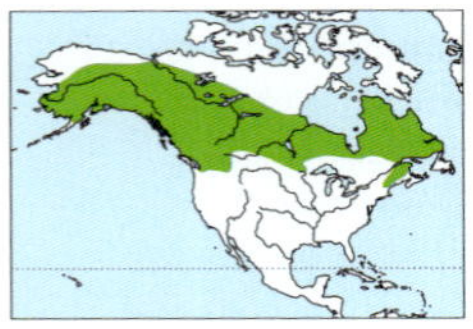

Northern Bog Lemming
Synaptomys borealis
FR Lemming boréal
DE Nördlicher Moorlemming
ES Lemming de pantano septentrional

Alaska, CA (E to Labrador) S to N US (Montana & New Hampshire)

Genus *SYNAPTOMYS*

(2 species)

LC
7 sspp
HB 10.5–13.1 cm
T 1.3–2.4 cm
W 21–50 g

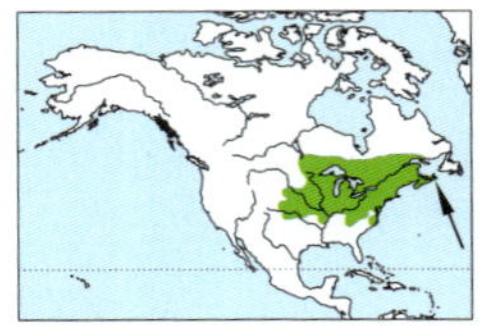

Southern Bog Lemming
Synaptomys cooperi
FR Lemming de Cooper
DE Südlicher Moorlemming
ES Lemming de pantano meridional

SE CA (SE Manitoba to Nova Scotia), NE US (S to NE Arkansas & NE Georgia)

NT
Monotypic
HB 9.5–10.3 cm
T 6.4–7.3 cm
W 20–45 g

Sonoma Tree Vole
Arborimus pomo
FR Phénacomys de Sonoma
DE Sonoma-Baumwühlmaus
ES Topillo arbóreo de Sonoma

N California (coastal forests from Klamath Mts S to Sonoma County)

Genus *MYOPUS*

(1 species)

LC Monotypic
HB 9–11.5 cm
T 1.2–2 cm
W 20–40 g

600–2450 m

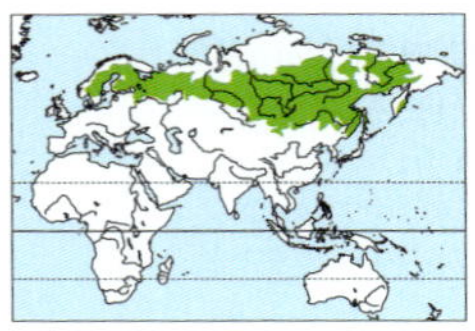

Wood Lemming
Myopus schisticolor
FR Lemming des bois
DE Waldlemming
ES Lemming de bosque

Fennoscandia E to Pacific coast of RU (and Sakhalin I), S to N MN & NE CN

LC
Monotypic
HB 8.6–11.9 cm
T 1–1.5 cm
W 16.2–51 g

300–1200 m

Amur Brown Lemming
Lemmus amurensis
FR Lemming de l'Amour
DE Amur-Lemming
ES Lemming del Amur

E Transbaikalia and R Amur E along Sea of Okhotsk to Kolyma Upland

Genus *LEMMUS*

(6 species)

NE
Monotypic
HB 8.6–11.9 cm
T 1–1.5 cm
W 16.2–51 g

300–1200 m

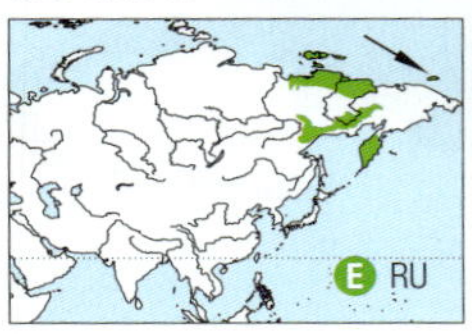

Eastern Siberian Brown Lemming
Lemmus paulus
FR Lemming d'Ognev
DE Ognev-Lemming
ES Lemming de Ognev

E Siberia between R Lena and R Kolyma, Kamchatka, New Siberian Is, Wrangel I

LC
Monotypic
HB 9.9–16 cm
T 0.8–1.8 cm
W 49–120 g

up to 800 m

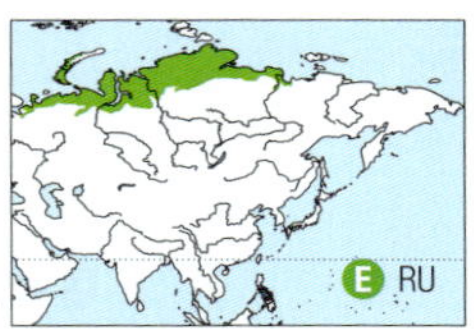

Western Siberian Brown Lemming
Lemmus sibiricus
FR Lemming de Sibérie
DE Sibirischer Lemming
ES Lemming de Siberia

NE European RU, Novaya Zemlya, W & C Siberia E to R Lena

LC Monotypic
HB 11.2–15 cm
T 1–1.8 cm
W 40–130 g

600–2450 m

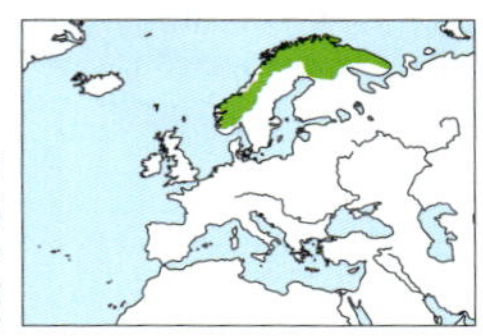

Norway Brown Lemming
Lemmus lemmus
FR Lemming de Norvège
DE Berglemming
ES Lemming de Noruega

NO, W & N SE, N FI, NW RU (Kola Pen); S limit fluctuates during cyclic peaks

LC
3 sspp
HB 11.5–13.3 cm
T 1.1–1.8 cm
W 51–90 g

up to 1500 m

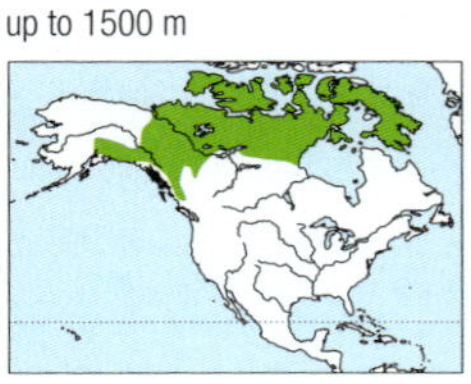

Nearctic Brown Lemming
Lemmus trimucronatus
FR Lemming brun
DE Brauner Lemming
ES Lemming de Canadá

E Alaska, CA (from Yukon to Nunavut, Canadian Arctic Arch, S to British Columbia & N Manitoba)

NE 7 sspp
HB 11.5–13.3 cm
T 1.1–1.8 cm
W 51–90 g

up to 1500 m

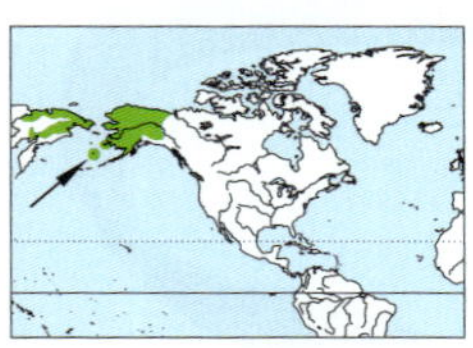

Beringian Brown Lemming
Lemmus nigripes
FR Lemming de Béringie
DE Bering-Lemming
ES Lemming de Bering

NE RU (E from R Kolyma, Kamchatka), US (Aleutian Is, Alaska), CA (W Yukon)

LC
3 sspp
HB 8.8–14 cm
T 1.1–2.1 cm
W 63–155 g

up to 1400 m

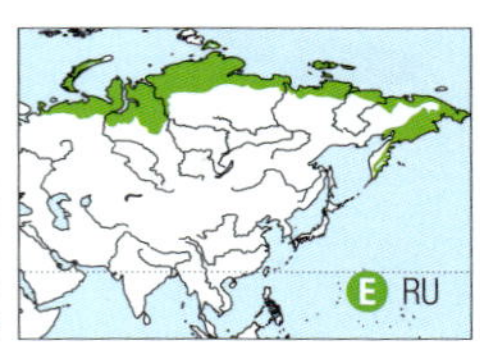

Palearctic Collared Lemming
Dicrostonyx torquatus
FR Lemming à collier
DE Echter Halsbandlemming
ES Lemming de collar paleártico

White Sea E to Chukchi & Kamchatka pen, Novaya Zemlya, New Siberian Is

Genus *DICROSTONYX*

(7 species)

Nearctic Collared Lemming
Dicrostonyx groenlandicus
FR Lemming du Groenland
DE Nördlicher Halsbandlemming
ES Lemming de collar de Groenlandia

LC
6 sspp
HB 11.3–15.1 cm
T 1.1–2.2 cm
W 52–114 g

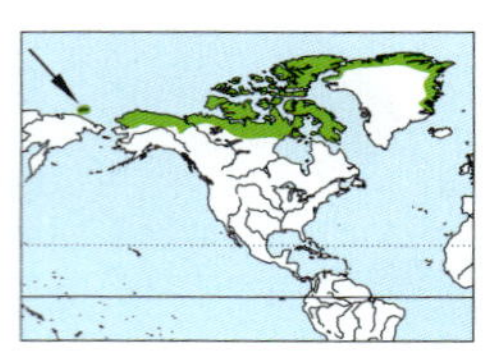

NE RU (Wrangel I), NW US (Alaska), N CA (N Yukon E to N Nunavut, Canadian Arctic Arch), GL

LC

Monotypic
HB 10.6–13.5 cm
T 0.9–1.5 cm
W 35–90 g

SE Northwest Territories to NE Manitoba, W of Hudson Bay

Richardson's Collared Lemming
Dicrostonyx richardsoni

FR Lemming de Richardson
DE Richardson-Halsbandlemming
ES Lemming de collar de Richardson

LC

Monotypic
HB 11.2–16 cm
T 1.2–2.2 cm
W 40–90 g

up to 1000 m

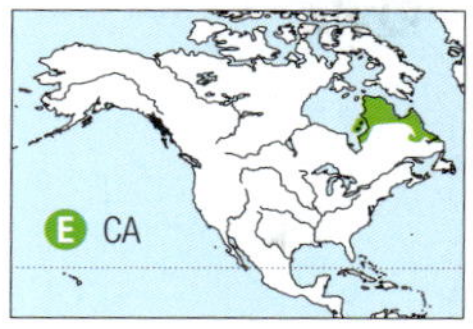

N Quebec & Labrador (with Belcher & King George Is)

Ungava Collared Lemming
Dicrostonyx hudsonius

FR Lemming d'Ungava
DE Ungava-Halsbandlemming
ES Lemming de collar de Ungava

LC

Monotypic
HB 11.7–11.8 cm
T 1.1–1.2 cm

1630–1710 m

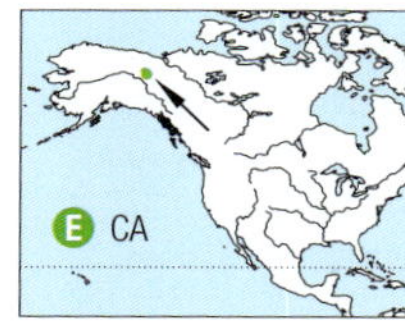

Rocky habitat in Ogilvie Mts (N Yukon)

Ogilvie Mountains Collared Lemming
Dicrostonyx nunatakensis

FR Lemming du Yukon
DE Ogilvie-Mountains-Halsbandlemming
ES Lemming de collar de Ogilvie

LC

Monotypic
HB 12.6–13.5 cm
T 0.9–1.3 cm
W 45–72 g

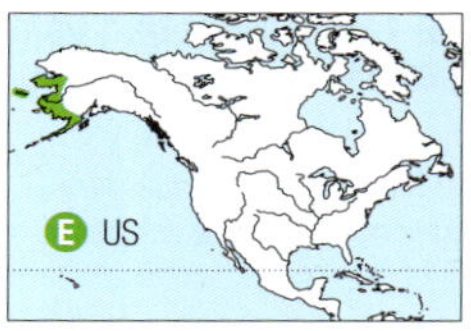

W Alaska, from Seward Pen S to Alaska Pen, St Lawrence I

Nelson's Collared Lemming
Dicrostonyx nelsoni

FR Lemming de Nelson
DE Nelson-Halsbandlemming
ES Lemming de collar de Nelson

DD

Monotypic

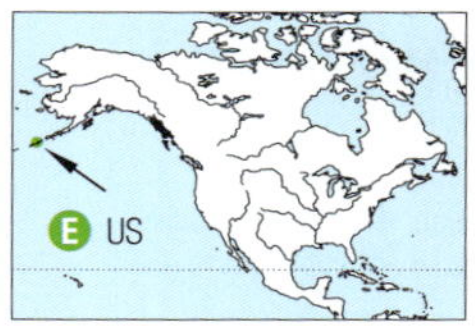

Unalaska I (Aleutian Is of Alaska)

Unalaska Collared Lemming
Dicrostonyx unalascensis

FR Lemming des Aléoutiennes
DE Unalaska-Halsbandlemming
ES Lemming de collar de Unalaska

LC

Monotypic
HB 10–14.2 cm
T 2.7–6.3 cm
W 27–52 g

up to 3000 m

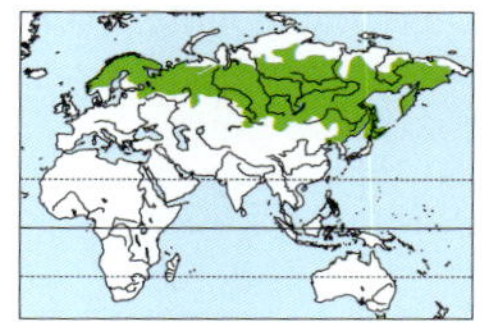

Fennoscandia E to Kamchatka Pen, S to S Ural, Altai Mts, N MN, N CN, N KP & N JP

Gray Red-backed Vole
Craseomys rufocanus

FR Campagnol de Sundevall
DE Graurötelmaus
ES Topillo rojo gris

Genus *CRASEOMYS*

(6 species)

LC

Monotypic
HB 9.3–10.6 cm
T 2.5–3.3 cm
W 23–28 g

Inner Mongolia, Shanxi, N Hebei, Beijing

Shanxi Red-backed Vole
Craseomys shanseius

FR Campagnol du Shanxi
DE Shanxi-Graurötelmaus
ES Topillo rojo de Shanxi

LC

Monotypic
HB 10–11.6 cm
T 3.7–5.1 cm
W 23–38 g

Korea Pen (except NE)

Korean Red-backed Vole
Craseomys regulus

FR Campagnol de Corée
DE Korea-Graurötelmaus
ES Topillo rojo de Corea

LC

Monotypic
HB 7.9–13.2 cm
T 4–7.9 cm
W 11–60 g

N & C Honshu and Kii Pen

Anderson's Red-backed Vole
Craseomys andersoni

FR Campagnol d'Anderson
DE Anderson-Graurötelmaus
ES Topillo rojo de Anderson

LC

Monotypic
HB 7–11.7 cm
T 3–5 cm
W 20–35 g

60–2400 m

C & S Honshu, Shikoku and Kyushu

Smith's Red-backed Vole
Craseomys smithii

FR Campagnol de Smith
DE Smith-Graurötelmaus
ES Topillo rojo de Smith

LC

Monotypic
HB 11.2–14.9 cm
T 4.4–6.8 cm
W 33–77 g

0–1900 m

E RU (C & S Sakhalin), JP (Hokkaido, S Kuril)

Hokkaido Red-backed Vole
Craseomys rex

FR Campagnol royal
DE Hokkaido-Graurötelmaus
ES Topillo rojo de Hokkaido

LC

Monotypic
HB 8.5–12.3 cm
T 2.1–5 cm
W 14.5–50 g

up to 2300 m

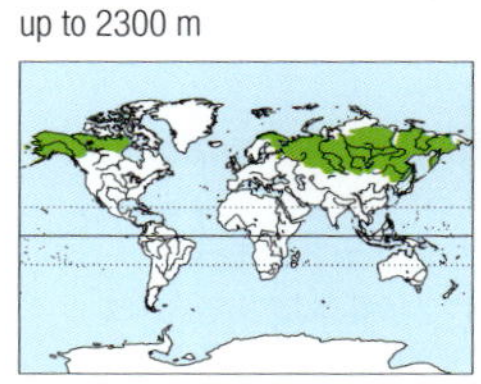

N Fennoscandia E across Siberia S to NE CN & N JP, Alaska, CA (E to Hudson Bay)

Northern Red-backed Vole
Clethrionomys rutilus

FR Campagnol boréal
DE Polarrötelmaus
ES Topillo rojo septentrional

Genus *CLETHRIONOMYS*

(5 species)

Western Red-backed Vole
Clethrionomys californicus

FR Campagnol de Merriam DE Westliche Rötelmaus
ES Topillo rojo de California

LC

3 sspp
HB 8.7–10.9 cm
T 3.4–5.6 cm
W 15–40 g

W Oregon (coast S from R Columbia) S to N Bay Area (NW California), E to Cascade Range

LC 29 sspp
HB 10–11.4 cm
T 3–4.4 cm
W 16–42 g

Southern Red-backed Vole
Clethrionomys gapperi

FR Campagnol de Gapper
DE Gapper-Rötelmaus
ES Topillo rojo de Gapper

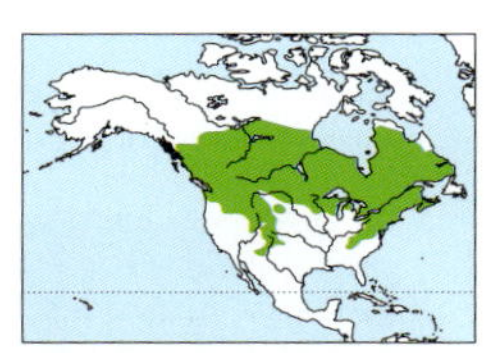

C & S CA, US (S in Rocky Mts to New Mexico, N in Great Plains & Appalachian Mts)

LC
Monotypic
HB 8–13.5 cm
T 3.2–7.2 cm
W 12–61 g

0–2350 m

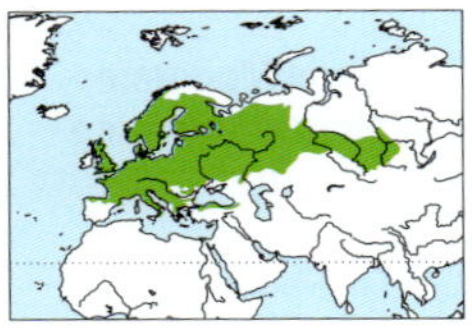

Bank Vole
Clethrionomys glareolus
FR Campagnol roussâtre
DE Rötelmaus
ES Topillo rojo europeo

W Europe E to R Yenisei, N to Arctic tundra, S to N TR & N KZ

LC 2 sspp
HB 10.4–12 cm
T 4.1–5.6 cm
W 24–33 g

1300–3500 m

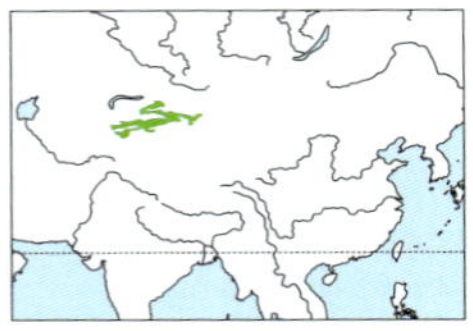

Tian Shan Red-backed Vole
Clethrionomys centralis
FR Campagnol d'Asie centrale
DE Tienschan-Rötelmaus
ES Topillo rojo de Tian Shan

Forests in SE KZ, KG, NW CN (NW Xinjiang)

LC 3 sspp
HB 9.3–11.5 cm
T 3.2–4.5 cm
W 24–39 g

1500–3000 m

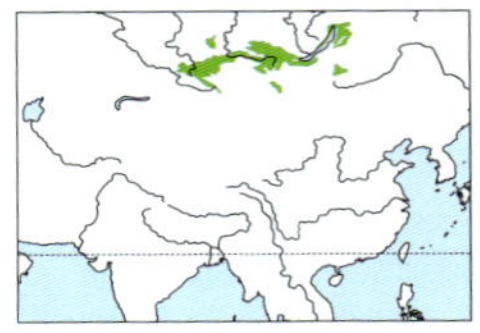

Large-eared Vole
Alticola macrotis
FR Campagnol à grandes oreilles
DE Großohrrötelmaus
ES Topillo de montaña de orejas grandes

Altai & Sayan Mts E to L Baikal and W Transbaikalia ranges (RU, MN & NW CN)

Genus *ALTICOLA*

(12 species)

LC Monotypic
HB 8.8–12 cm
T 1–2.8 cm
W 21–39 g

0–2000 m

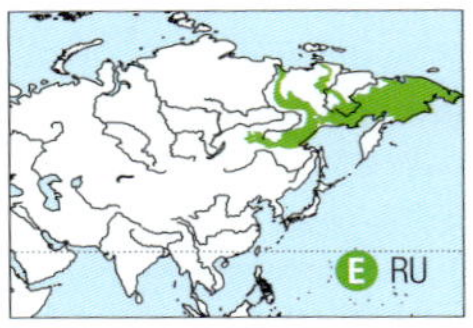

Lemming Mountain Vole
Alticola lemminus
FR Campagnol lemming
DE Lemming-Gebirgswühlmaus
ES Topillo de montaña lemming

E of R Lena and Yablonoi Mts to Chukchi Pen

LC 8 sspp
HB 8.7–11.9 cm
T 3.6–6.4 cm
W 21–55 g

500–4300 m

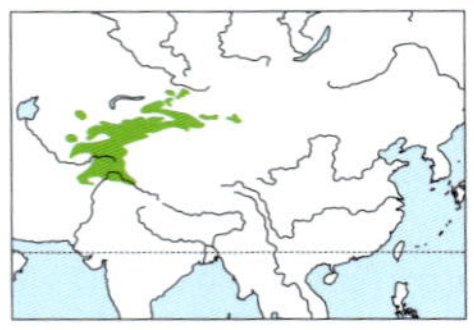

Silver Mountain Vole
Alticola argentatus
FR Campagnol argenté
DE Silberfell-Gebirgswühlmaus
ES Topillo de montaña plateado

Mts of Central Asia (E KZ S to N AF & Kashmir, E to W CN)

DD
Monotypic
HB 10.1–10.8 cm
T 2.8–3.2 cm

3600–4250 m

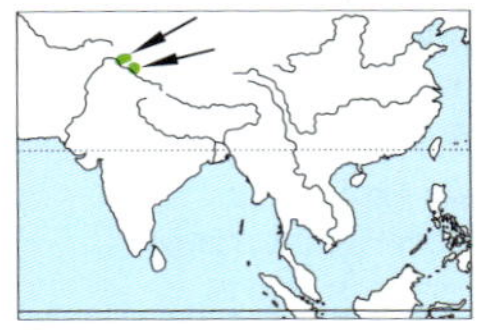

White-tailed Mountain Vole
Alticola albicauda
FR Campagnol à queue blanche
DE Weißschwanz-Gebirgswühlmaus
ES Topillo de montaña de cola blanca

Known from few specimens from Kashmir (NE PK, NW IN)

VU
Monotypic
HB 10.2–12.6 cm
T 4.1–6.5 cm

2600–4700 m

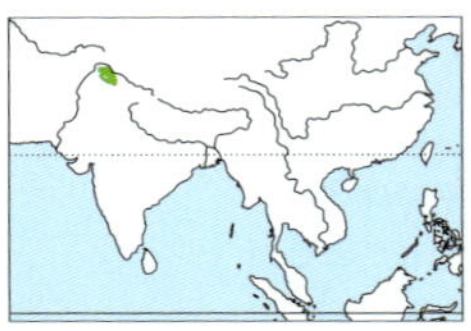

Kashmir Mountain Vole
Alticola montosus
FR Campagnol du Cachemire
DE Kaschmir-Gebirgswühlmaus
ES Topillo de montaña de Cachemira

Known from few localities in NE PK & NW IN

LC
3 sspp
HB 9.3–12.7 cm
T 1.3–2.7 cm
W 21–50 g

3340–6140 m

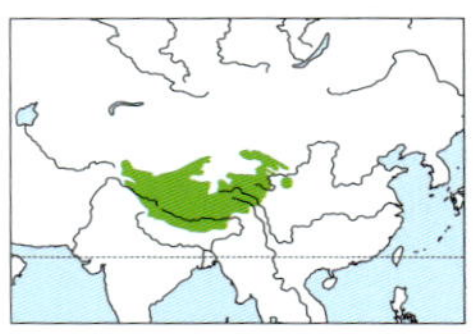

Stoliczka's Mountain Vole
Alticola stoliczkanus
FR Campagnol de Stoliczka
DE Stoliczka-Gebirgswühlmaus
ES Topillo de montaña de Stoliczka

Himalayas and Tibetan Plateau (N PK, N IN, NP, SW CN, possibly N BT)

LC
Monotypic
HB 9.5–12.9 cm
T 1.5–3 cm
W 28–63 g

up to 2500 m

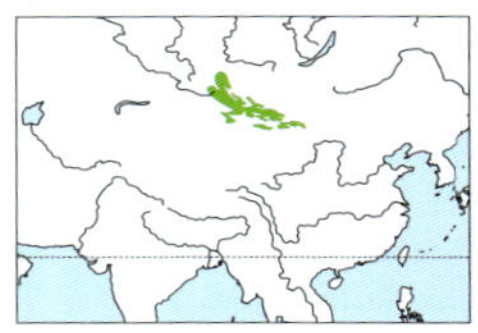

Gobi Altai Mountain Vole
Alticola barakshin
FR Campagnol de Bannikov
DE Gobi-Altai-Gebirgswühlmaus
ES Topillo de montaña de Altái

Altai Mts in RU (Tuva), MN & CN (Barköl County and nearby areas, NE Xinjiang)

LC 2 sspp
HB 10.1–13.5 cm
T 3.3–4.7 cm
W 34–58 g

400–3000 m

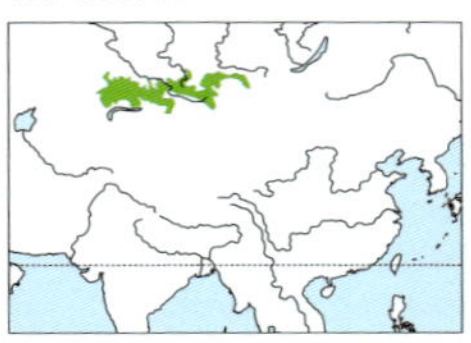

Strelzov's Mountain Vole
Alticola strelzovi
FR Campagnol de Streltzov
DE Flachkopf-Gebirgswühlmaus
ES Topillo de montaña de Strelzov

E KZ, RU (N to S Tuva), W MN & NW CN (Xinjiang)

LC 2 sspp
HB 9.3–13.5 cm
T 2–3.7 cm
W 24–69 g

Mongolian Mountain Vole
Alticola semicanus
FR Campagnol d'Allen
DE Mongolische Silberfell-Gebirgswühlmaus
ES Topillo de montaña de Mongolia

RU (extreme S Tuva), MN, NE CN (Inner Mongolia)

LC
Monotypic
HB 9.9–12.5 cm
T 3.6–4.9 cm
W 27–54 g

400–800 m

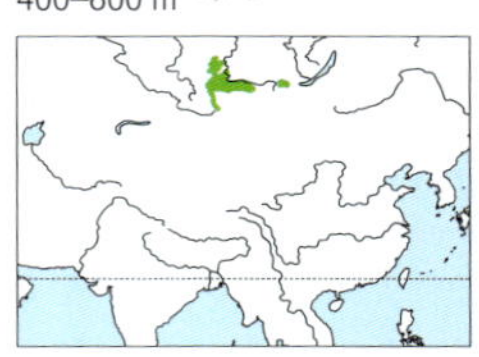

Tuva Mountain Vole
Alticola tuvinicus
FR Campagnol de Touva
DE Tuwa-Gebirgswühlmaus
ES Topillo de montaña de Tuva

S RU (Khakassia, Tuva), NW & N MN (Altai, disjunctly around L Khuvsgul)

Lake Baikal Mountain Vole
Alticola olchonensis
FR Campagnol du Baïkal DE Olchon-Gebirgswühlmaus
ES Topillo de montaña del Baikal

EN
Monotypic
HB 10.9–12.1 cm
T 3.3–3.7 cm
W 24–43 g

L Baikal (Olkhon & Ogoi Is and contiguous mainland in Maloye More)

NT Monotypic
HB 9.1–11.7 cm
T 2.9–4.8 cm

Royle's Mountain Vole
Alticola roylei
FR Campagnol de Royle
DE Royle-Gebirgswühlmaus
ES Topillo de montaña de Royle

2500–4300 m

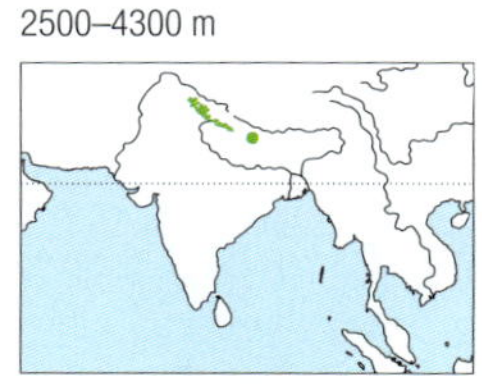

NW IN (Himachal Pradesh, Uttarakhand), W NP

Genus *CARYOMYS*

(2 species)

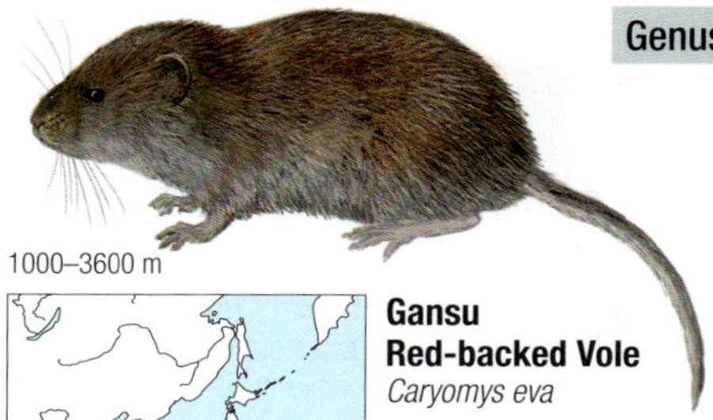

LC
2 sspp
HB 8.3–10.5 cm
T 4.6–6 cm
W 16–27 g

1000–3600 m

Gansu Red-backed Vole
Caryomys eva

FR Campagnol eva
DE Gansu-Rötelmaus
ES Topillo rojo de Gansu

Qinghai, Gansu, Ningxia, Shaanxi, Chongqing, Hubei, Sichuan

LC
2 sspp
HB 8.3–11 cm
T 2.2–4.6 cm
W 12–32 g

500–2150 m

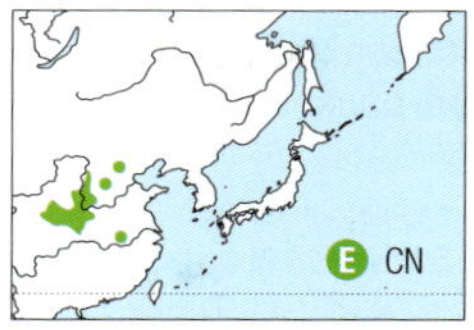

Kolan Red-backed Vole
Caryomys inez

FR Campagnol inez
DE Kolan-Rötelmaus
ES Topillo rojo de Kolan

Gansu, Ningxia, Shaanxi, Shanxi, Hebei, Henan, Anhui, Sichuan

NE
Monotypic
HB 8–14 cm
T 3.1–6.1 cm
W 22–60 g

1300–4250 m

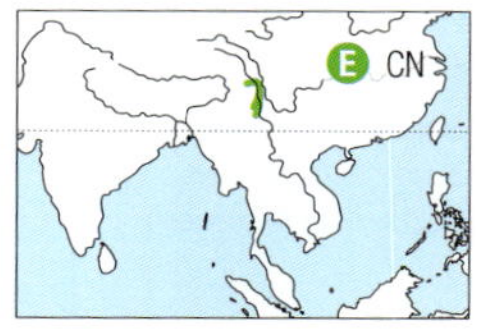

Loyal Red-backed Vole
Eothenomys fidelis

FR Campagnol fidèle
DE Lijiang-Rötelmaus
ES Topillo rojo leal

SW Sichuan, NW Yunnan

Genus *EOTHENOMYS*

(17 species)

NE Monotypic
HB 7–11.5 cm
T 2.1–5.5 cm
W 18–46 g

600–3000 m

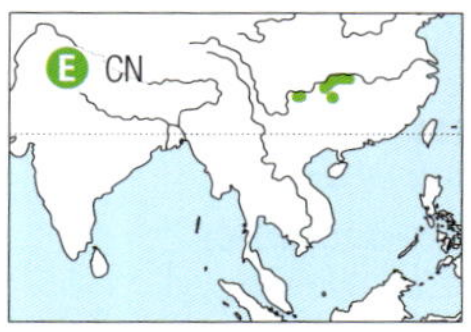

Chongqing Red-backed Vole
Eothenomys eleusis

FR Campagnol de Chongqing
DE Chongqing-Rötelmaus
ES Topillo rojo de Chongqing

N Guizhou, S & C Chongqing

LC
Monotypic
HB 8–14 cm
T 3.1–6.1 cm
W 22–60 g

1300–4250 m

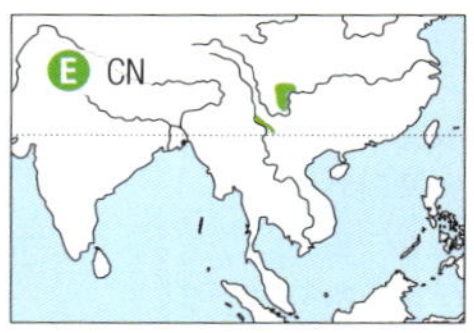

Yunnan Red-backed Vole
Eothenomys miletus

FR Campagnol du Yunnan
DE Yunnan-Rötelmaus
ES Topillo rojo de Yunnan

S Sichuan, WC Yunnan

NE
Monotypic

Shimian Red-backed Vole
Eothenomys shimianensis

FR Campagnol de Shimian
DE Shimian-Rötelmaus
ES Topillo rojo de Shimian

SC Sichuan (known from 2 localities in Shimian County)

LC Monotypic
HB 8–14 cm
T 3.1–6.1 cm
W 22–60 g

1300–4250 m

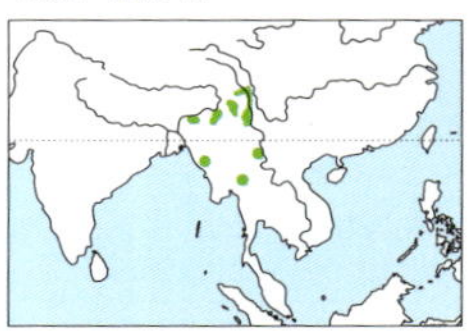

Kachin Red-backed Vole
Eothenomys cachinus

FR Campagnol kachin
DE Kachin-Rötelmaus
ES Topillo rojo de Kachin

N & WC MM, S CN (W Yunnan), scattered in NE IN & N TH; uncertain in N VN

LC
Monotypic
HB 7–11.5 cm
T 2.1–5.5 cm
W 18–46 g

600–3000 m

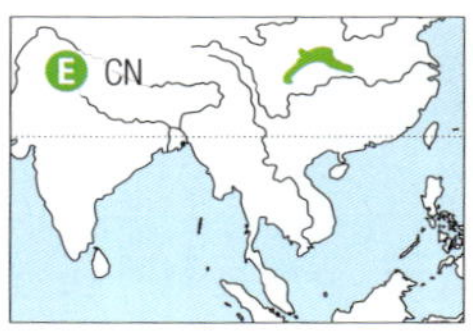

Pere David's Red-backed Vole
Eothenomys melanogaster

FR Campagnol à ventre noir
DE Père-David-Rötelmaus
ES Topillo rojo de Père David

SE Gansu, Sichuan, S Shanxi, N Chongqing, Guizhou, W Hubei

NE Monotypic
HB 7–11.5 cm
T 2.1–5.5 cm
W 18–46 g

600–3000 m

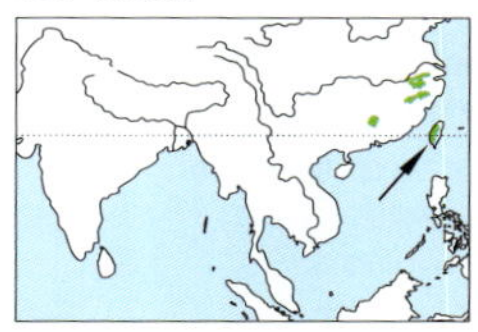

Zhejiang Red-backed Vole
Eothenomys colurnus

FR Campagnol du Zhejiang
DE Zhejiang-Rötelmaus
ES Topillo rojo de Zhejiang

E CN (from Anhui S to Guangdong), TW

NE
Monotypic

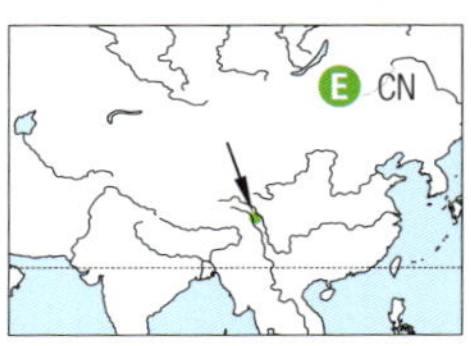

Jinyang Red-backed Vole
Eothenomys jinyangensis

FR Campagnol de Jinyang
DE Jinyang-Rötelmaus
ES Topillo rojo de Jinyang

S Sichuan (Jinyang & SW Zhaojue counties)

NE
Monotypic

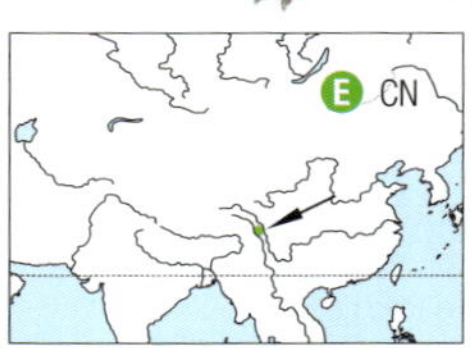

Luojishan Red-backed Vole
Eothenomys luojishanensis

FR Campagnol de Luojishan
DE Luojishan-Rötelmaus
ES Topillo rojo de Luojishan

S Sichuan (Luoji Mts at borders of Dechang, Puge, Ningnan & Miyi counties)

NE
Monotypic
HB 9.2–9.3 cm
T 5.1–5.9 cm

2400–3600 m

Hinton's Red-backed Vole
Eothenomys hintoni

FR Campagnol de Hinton
DE Hinton-Rötelmaus
ES Topillo rojo de Hinton

Mts in S Sichuan

NE
Monotypic

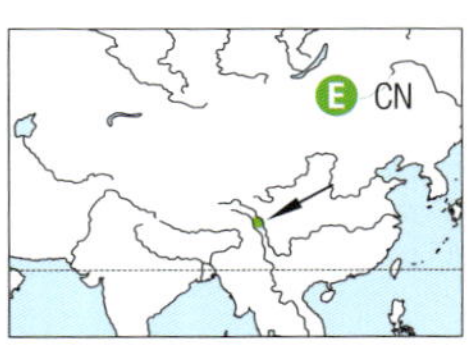

Meigu Red-backed Vole
Eothenomys meiguensis

FR Campagnol de Meigu
DE Meigu-Rötelmaus
ES Topillo rojo de Meigu

Mts in W Sichuan

NE Monotypic
HB 10.1–10.8 cm
T 6.8–7.6 cm

c. 3000 m

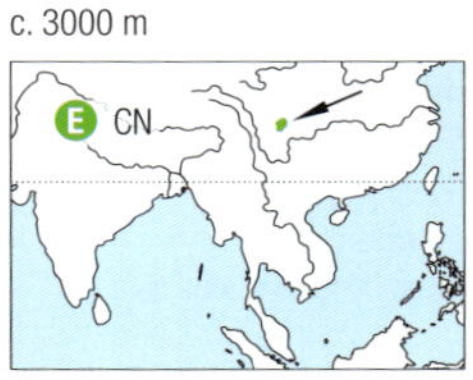

Tarquinius Red-backed Vole
Eothenomys tarquinius

FR Campagnol tarquin
DE Tarquinius-Rötelmaus
ES Topillo rojo tarquino

Mts in C Sichuan

LC 2 sspp
HB 8–9.2 cm
T 2.9–3.9 cm
W 13–15 g

1800–3350 m

Black-eared Red-backed Vole
Eothenomys olitor

FR Campagnol à oreilles noires
DE Schwarzohr-Rötelmaus
ES Topillo rojo de orejas negras

Mts in W & NE Yunnan

DD Monotypic
HB 10.5–11.5 cm
T 2.6–3.4 cm W 24–30 g

2700–4200 m

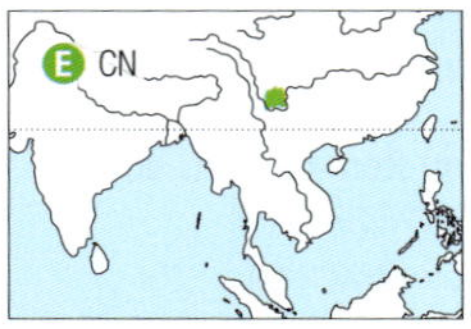

Yulongxuen Red-backed Vole
Eothenomys proditor
FR Campagnol du Yulong
DE Yulongchan-Rötelmaus
ES Topillo rojo de Yulongxuen

Mts between Sichuan and Yunnan

LC Monotypic
HB 10.2–13.4 cm
T 5.7–7.1 cm
W 38–57 g

1500–3400 m

Sichuan Red-backed Vole
Eothenomys chinensis
FR Campagnol de Chine
DE Sichuan-Rötelmaus
ES Topillo rojo de Sichuán

Sichuan (mts along R Dadu)

LC 4 sspp
HB 7.9–11 cm
T 3.2–5 cm
W 21–36 g

2500–3900 m

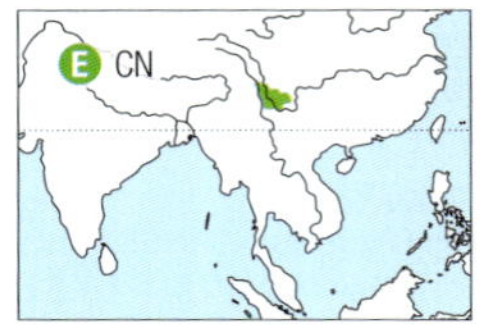

Southwest China Red-backed Vole
Eothenomys custos
FR Campagnol sentinelle
DE Südwestchina-Rötelmaus
ES Topillo rojo de China suroccidental

Yunnan & SW Sichuan

NT Monotypic
HB 9–10.8 cm
T 4.3–5.6 cm
W 22–39 g

2400–4250 m

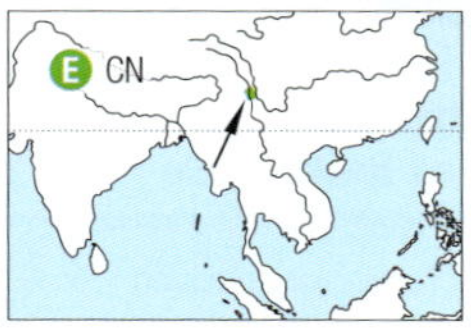

Ward's Red-backed Vole
Eothenomys wardi
FR Campagnol de Ward
DE Ward-Rötelmaus
ES Topillo rojo de Ward

NW Yunnan (Meili Snow Mts between R Salween and R Mekong valleys)

NT 3 sspp
HB 8.7–11.1 cm T 1.9–2.8 cm
W 21–23 g

2135–3655 m

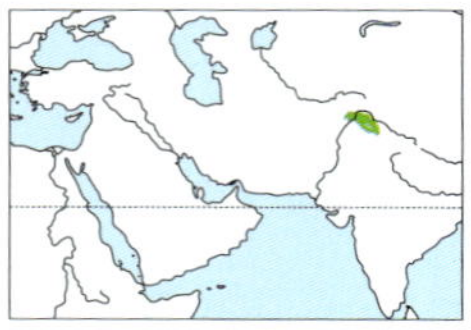

Burrowing Vole
Hyperacrius fertilis
FR Campagnol de True
DE Kaschmir-Wühlmaus
ES Topillo cavador

N PK (North-West Frontier Prov), NW IN (Jammu and Kashmir)

Genus *HYPERACRIUS*

(2 species)

LC
2 sspp
HB 10–13.8 cm
T 2.4–4.5 cm
W 55–60 g

1830–3290 m

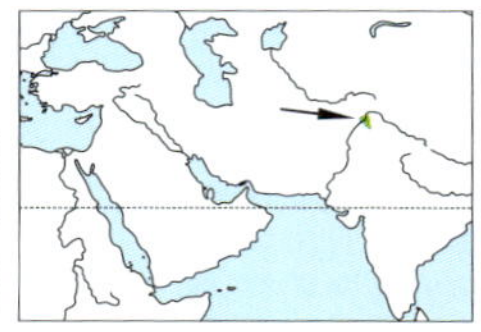

Murree Vole
Hyperacrius wynnei
FR Campagnol de Murree
DE Pakistan-Wühlmaus
ES Topillo de Murree

Himalayas of NW IN & N PK (including Murree Hills and Kaghan & Swat valleys)

Genus *DINAROMYS*

(1 species)

VU 3 sspp
HB 13.1–15.2 cm
T 8.3–10.7 cm
W 54–99 g

10–2200 m

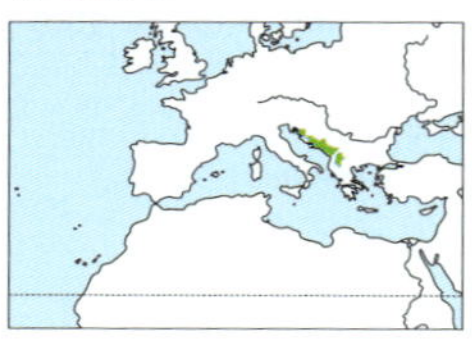

Balkan Snow Vole
Dinaromys bogdanovi
FR Campagnol de Martino
DE Bergmaus
ES Topillo nival de los Balcanes

Rocky habitats in Dinaric Mts (HR, BA, ME, XK, AL, W MK)

Northern Mole Vole
Ellobius talpinus
FR Campagnol taupe
DE Nördlicher Mull-Lemming
ES Topillo topo septentrional

LC 4 sspp
HB 8.4–11.5 cm
T 0.8–2.1 cm
W 24–56 g

variants

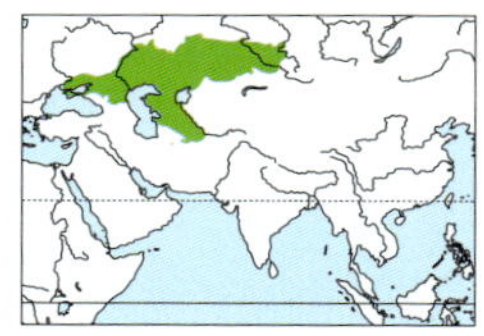

UA & S European RU E to SW Siberia & N KZ, S to NW UZ, TM & NE IR

Genus *ELLOBIUS*

(5 species)

LC 6 sspp
HB 9.6–12.5 cm
T 1–1.5 cm
W 38–60 g

up to 4000 m

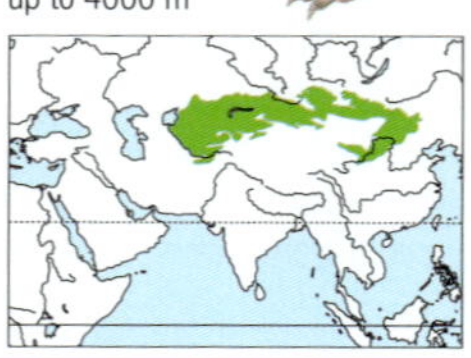

Eastern Mole Vole
Ellobius tancrei
FR Campagnol de Blasius
DE Zaisan-Mull-Lemming
ES Topillo topo oriental

Central Asia E to RU (Tuva), MN & N CN (S to Gansu & Shaanxi); probably AF

DD
Monotypic
HB 11–13.2 cm
T 0.8–1.7 cm
W 43–64 g

c. 3300 m

Alai Mole Vole
Ellobius alaicus
FR Campagnol des Alaï
DE Alai-Mull-Lemming
ES Topillo topo de Alay

Alai Mts

LC
Monotypic
HB 9.8–14.5 cm
T 0.9–2 cm
W 34–101 g

300–2600 m

Southern Mole Vole
Ellobius fuscocapillus
FR Campagnol du Baloutchistan
DE Südlicher Mull-Lemming
ES Topillo topo meridional

NE & SE IR, W, S & E TM, AF, marginally in W PK (Balochistan)

LC
Monotypic
HB 10–14 cm
T 0.7–1.9 cm
W 58–84 g

600–3000 m

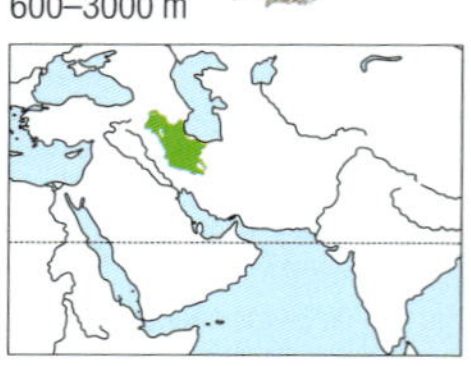

Transcaucasian Mole Vole
Ellobius lutescens
FR Campagnol fauve
DE Transkaukasischer Mull-Lemming
ES Topillo topo del Cáucaso

E TR (E Anatolia), AM, SW AZ (Nakhchivan), NW IR

Genus *EOLAGURUS*

(2 species)

LC
Monotypic
HB 13–16.2 cm
T 1.4–2.3 cm
W 60–140 g

up to 2100 m

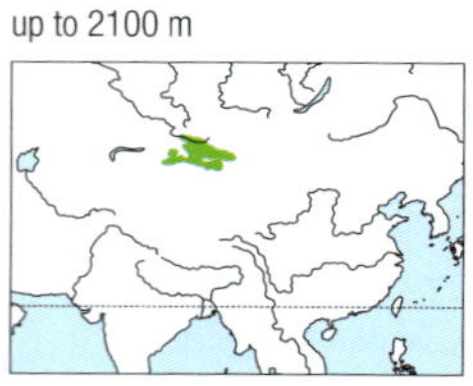

Yellow Steppe Lemming
Eolagurus luteus
FR Campagnol jaune
DE Gelblemming
ES Lemming de estepa dorado

E KZ (E Zaysan), NW CN (Dzungarian Basin), SW MN (Dzungarian Gobi)

LC
Monotypic
HB 12.5–13 cm
T 0.9–2 cm
W 58–80 g

up to 3100 m

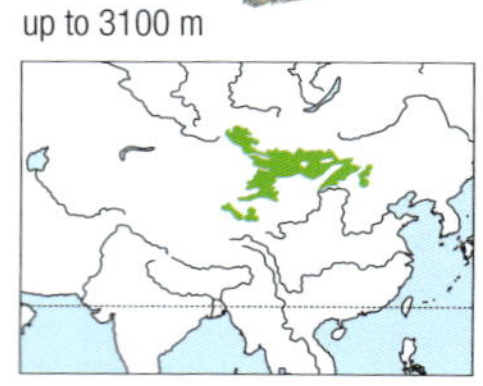

Przewalski's Steppe Lemming
Eolagurus przewalskii
FR Campagnol de Przewalski
DE Przewalski-Lemming
ES Lemming de estepa de Przewalski

NW & S MN, N CN (NW Gansu, Inner Mongolia, Qinghai)

Genus *LAGURUS*

(1 species)

Steppe Vole
Lagurus lagurus

FR Campagnol des steppes
DE Steppenlemming
ES Topillo de estepa

LC 5 sspp
HB 7.3–12.5 cm
T 0.6–1.6 cm W 17–46 g

up to 3000 m

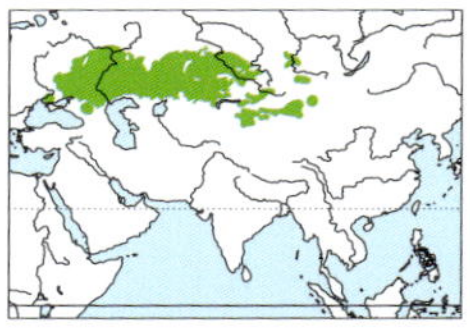

UA & S European RU E to S Siberia (Tuva), KZ, KG, NW CN (N Xinjiang) & MN

South-western Water Vole
Arvicola sapidus

FR Campagnol amphibie
DE Südwesteuropäische Schermaus
ES Rata de agua suroccidental

VU 2 sspp
HB 16.5–22 cm
T 10.7–13.5 cm
W 100–275 g

0–2300 m

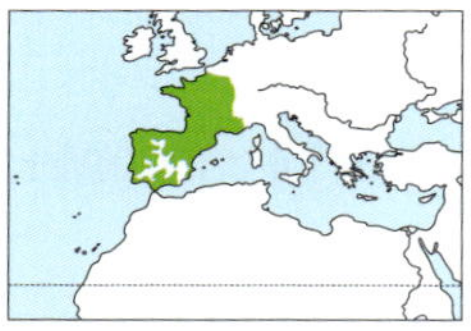

Aquatic habitats in FR, ES & PT

Genus *ARVICOLA*

(4 species)

aquatic morphotype

0–3210 m

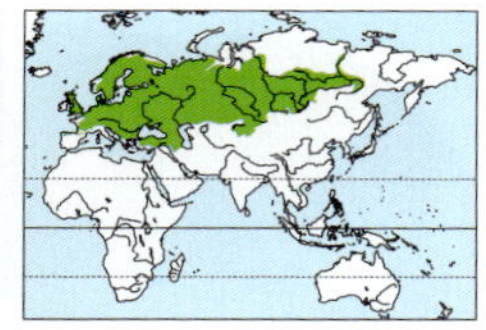

Europe E to Caucasus, Central Asia, S Siberia (E to R Lena), N MN & NW China

LC
Monotypic
HB 12–23 cm
T 5.4–13.9 cm
W 55–320 g

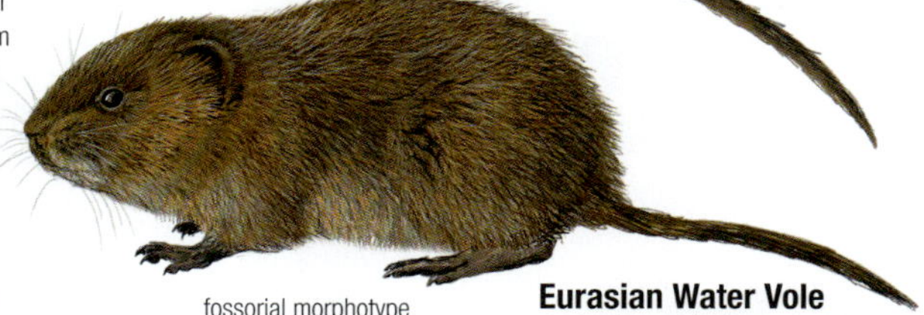

fossorial morphotype

Eurasian Water Vole
Arvicola amphibius

FR Grand Campagnol
DE Ostschermaus
ES Rata de agua euroasiática

NE
Monotypic
HB 16.2–19.1 cm
T 8.4–10.2 cm
W 83–139 g

up to 1600 m

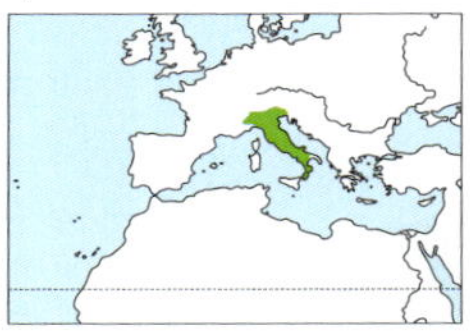

Italian Water Vole
Arvicola italicus

FR Campagnol d'Italie
DE Italien-Schermaus
ES Rata de agua de Italia

Humid habitats in IT, marginally reaching S CH (Ticino)

NE
Monotypic

75–2575 m

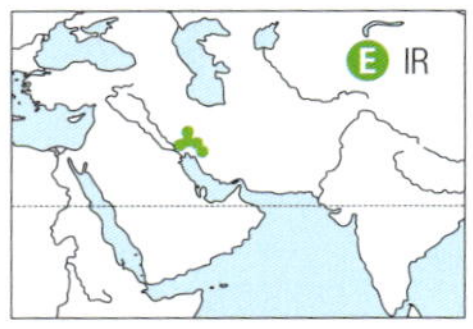

Persian Water Vole
Arvicola persicus

FR Campagnol de Sultanieh
DE Persien-Schermaus
ES Rata de agua de Persia

Alborz Mts (possibly further W)

Genus *CHIONOMYS*

(4 species)

Robert's Snow Vole
Chionomys roberti

FR Campagnol de Robert
DE Robert-Schneemaus
ES Topillo nival de Robert

LC
Monotypic

0–3200 m

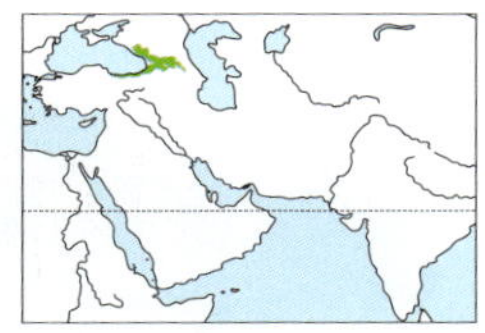

E Pontic Mts (NE TR), Caucasus (SW European RU, GE & N AZ)

HB 12.5–15.6 cm
T 8–10.8 cm
W 40–78 g

European Snow Vole
Chionomys syriacus

FR Campagnol des neiges
DE Schneemaus
ES Topillo nival europeo

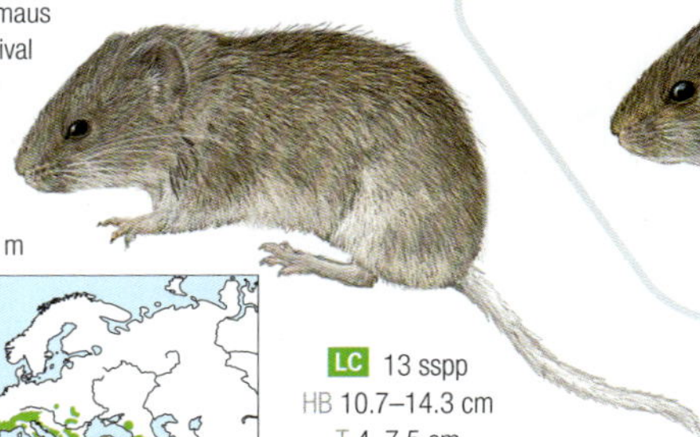

30–4700 m

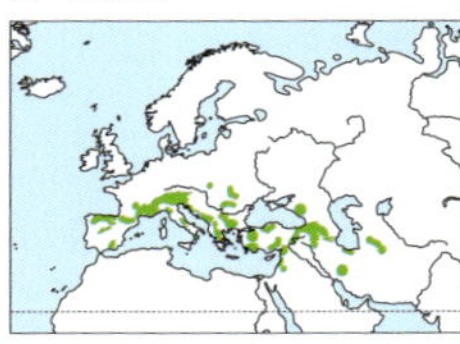

Iberian Pen E across S Europe, Anatolia & Caucasus to S TM, S to IL & S IR

LC 13 sspp
HB 10.7–14.3 cm
T 4–7.5 cm
W 29–62 g

Genus *LEMMISCUS*

(1 species)

LC
6 sspp
HB 9.2–11.4 cm
T 1.6–2.8 cm
W 17.5–25 g

300–3800 m

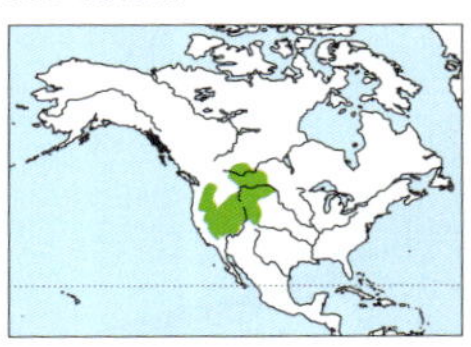

Sagebrush Vole
Lemmiscus curtatus

FR Campagnol des armoises
DE Beifuß-Wühlmaus
ES Topillo de artemisia

SW CA (S Alberta, S Saskatchewan), NW US (S to EC California & S Nevada)

Genus *VOLEMYS*

(2 species)

Sichuan Vole
Volemys millicens

FR Campagnol du Sichuan
DE Sichuan-Wühlmaus
ES Topillo de Sichuán

c. 4000 m

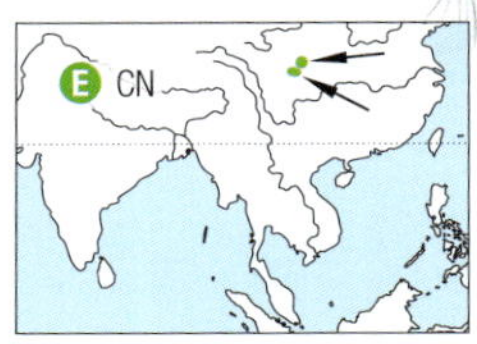

Forests in NC Sichuan (known from 3 sites)

NT
Monotypic
HB 8.3–9.5 cm
T 4.6–5.3 cm

Lazistan Snow Vole
Chionomys lasistanius

FR Campagnol du Lazistan
DE Lasistan-Schneemaus
ES Topillo nival de Lazistán

LC
3 sspp
HB 11.2–13 cm
T 5.7–7.9 cm
W 34–63 g

500–3000 m

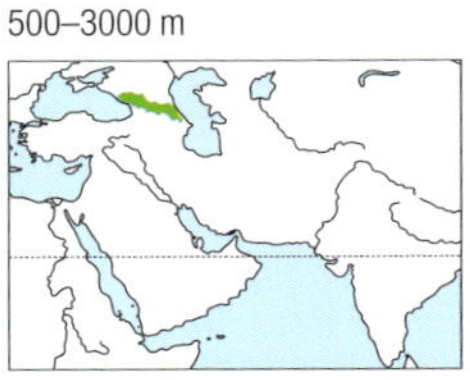

Gudaur Snow Vole
Chionomys gud

FR Campagnol du Goudaouri
DE Kaukasus-Schneemaus
ES Topillo nival del Cáucaso

Greater Caucasus (RU, GE, AZ)

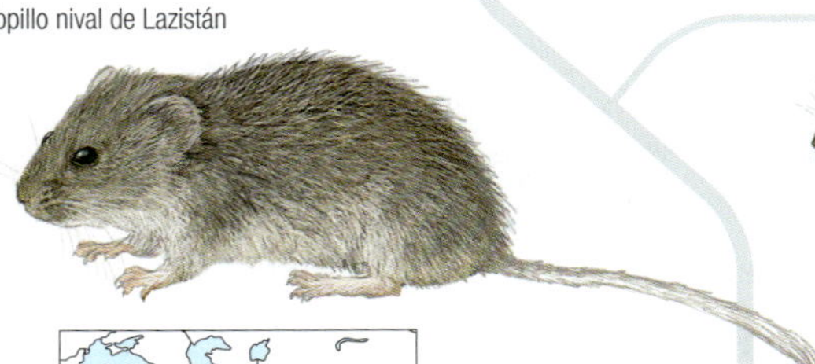

1550–2500 m

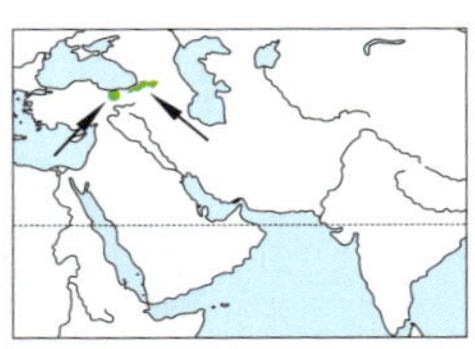

NE TR (Pontic Mts W to Giresun), S GE (Lesser Caucasus)

NE
Monotypic
HB 11.2–13 cm
T 5.7–8.2 cm
W 34–63 g

DD
Monotypic
HB 9–12.9 cm
T 4.7–7.2 cm

Marie's Vole
Volemys musseri

FR Campagnol de Musser
DE Marie-Wühlmaus
ES Topillo de Marie

Rocks and cliffs in NC Sichuan (Qionglai Shan)

VU
Monotypic
HB 9.9–12.9 cm
T 3.7–4.3 cm

2440–2550 m

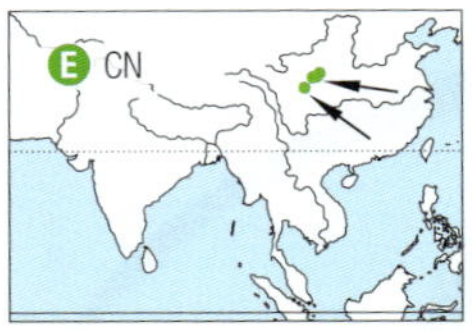

Duke of Bedford's Vole
Proedromys bedfordi

FR Campagnol de Bedford
DE Bedford-Wühlmaus
ES Topillo del Duque de Bedford

Moist montane forests in S Gansu & N Sichuan

Genus *PROEDROMYS*

(2 species)

DD
Monotypic
HB 10.2–13.1 cm
T 6.1–8.2 cm
W c. 60 g

2560–3100 m

Liangshan Vole
Proedromys liangshanensis

FR Campagnol du Liangshan
DE Liangshan-Wühlmaus
ES Topillo de Liangshan

Moist montane forests in SW Sichuan (Liangshan Mts)

LC
2 sspp
HB 10–14.8 cm
T 1.8–3 cm
W 23–55 g

130–2700 m

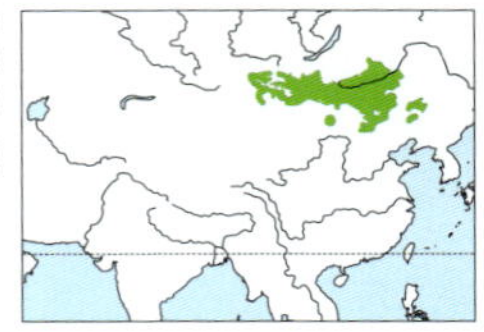

Brandt's Vole
Lasiopodomys brandtii

FR Campagnol de Brandt
DE Brandt-Wühlmaus
ES Topillo de Brandt

MN, RU (S Zabaykalsky Krai), NE CN (Inner Mongolia, N Hebei, Jilin)

LC
5 sspp
HB 9–11.4 cm
T 2–2.7 cm
W 18–50 g

up to 1700 m

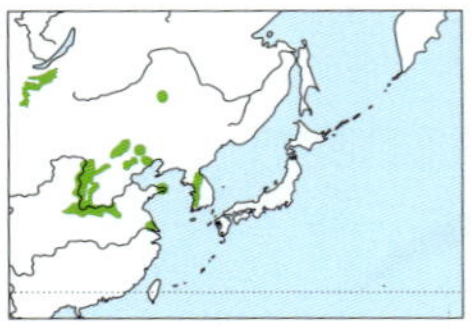

Mandarin Vole
Lasiopodomys mandarinus

FR Campagnol mandarin
DE Mandarin-Wühlmaus
ES Topillo mandarín

RU (S Buryatia), NC MN, NE & EC China, Korean Pen

LC
Monotypic
HB 8.1–14.8 cm
T 1.7–4.6 cm
W 18–68 g

up to 4200 m

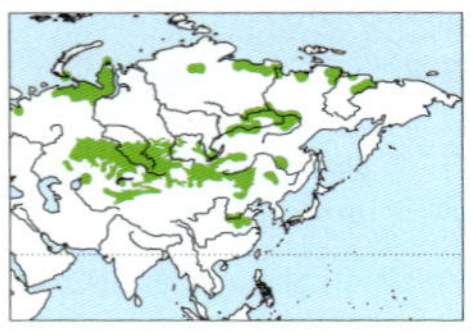

Narrow-headed Vole
Lasiopodomys gregalis

FR Campagnol grêle
DE Schmalkopf-Wühlmaus
ES Topillo de cabeza estrecha

Disjunct from White Sea to R Kolyma, Central Asia, SW Siberia, Yakutia, MN & N CN

Genus *LASIOPODOMYS*

(4 species)

Radde's Vole
Lasiopodomys raddei

FR Campagnol de Radde
DE Radde-Wühlmaus
ES Topillo de Radde

NE
Monotypic
HB 9.5–13 cm
T 2.1–3.7 cm
W 24–54 g

550–1150 m

RU (Zabaykalsky Krai, S of R Shilka & R Ingoda), NE MN

LC
Monotypic
HB 9.7–12.1 cm
T 3–5.2 cm
W 27–49 g

2100–4000 m

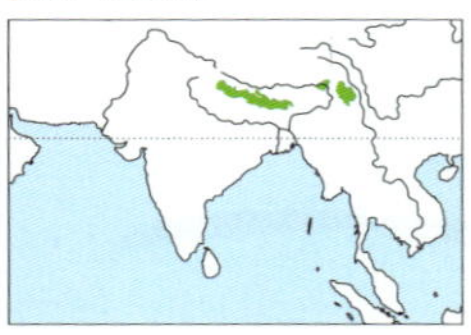

Sikkim Mountain Vole
Neodon sikimensis

FR Campagnol du Sikkim
DE Sikkim-Felswühlmaus
ES Topillo de montaña de Sikkim

Disjunct in E Himalayas from C & E NP through NE India to BT & SE Tibet

NE
Monotypic
HB 9.8–11.1 cm
T 3.3–3.6 cm

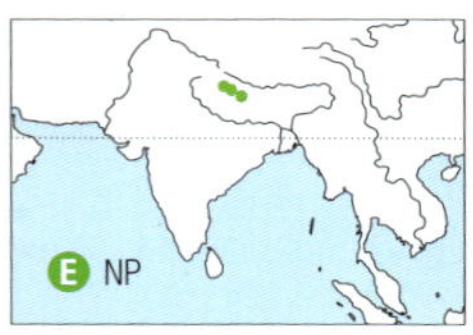

Nepalese Mountain Vole
Neodon nepalensis

FR Campagnol du Népal
DE Nepal-Felswühlmaus
ES Topillo de montaña de Nepal

West of R Trisuli

Genus *NEODON*

(10 species)

DD
Monotypic
HB 9.2–11.5 cm
T 2.4–3.9 cm
W 26–48 g

3640–3890 m

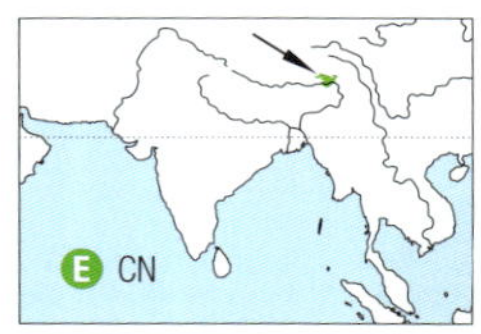

Linzhi Mountain Vole
Neodon linzhiensis

FR Campagnol de Linzhi
DE Linzhi-Felswühlmaus
ES Topillo de montaña de Linzhi

SE Tibet (known from 5 sites)

LC
Monotypic
HB 10.5–13.4 cm
T 6–7 cm

3400–4290 m

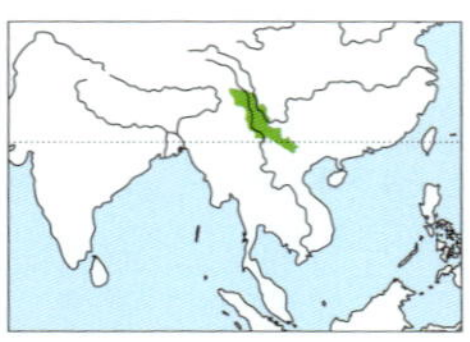

Clarke's Vole
Neodon clarkei

FR Campagnol de Clarke
DE Clarke-Felswühlmaus
ES Topillo de montaña de Clarke

SC CN (E Tibet, W & S Yunnan), NE MM, N VN

NE Monotypic
HB 8.9–11 cm
T 4.3–5.5 cm
W 26–41 g

Medog Mountain Vole
Neodon medogensis

FR Campagnol du Mêdog
DE Medog-Felswühlmaus
ES Topillo de montaña de Medog

Tibet (known from 2 sites in Mêdog County)

NE
Monotypic
HB 9.8–11.5 cm
T 3.8–5 cm
W 26–36 g

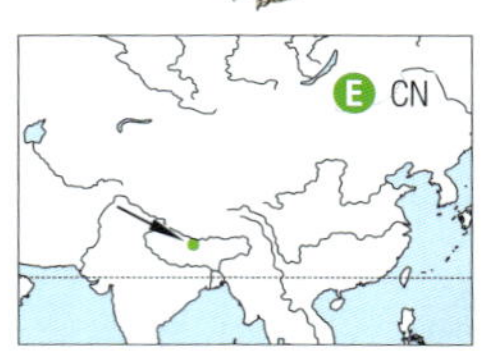

Nyalam Mountain Vole
Neodon nyalamensis

FR Campagnol du Nyalam
DE Nyalam-Felswühlmaus
ES Topillo de montaña de Nyalam

Secondary forests in Nyalam County (Tibet)

Irene Mountain Vole
Neodon irene

FR Campagnol irène DE Gansu-Felswühlmaus
ES Topillo de montaña de Irene

LC
Monotypic
HB 8–10.8 cm
T 2.2–4 cm
W 23–38 g

2900–3660 m

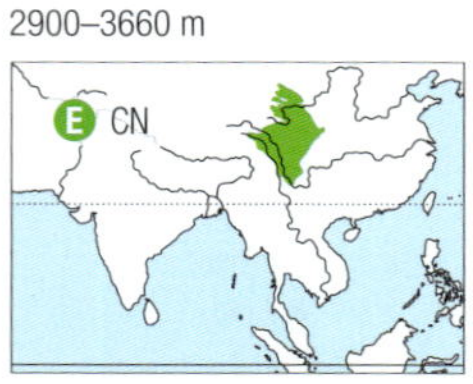

High mts in Qinghai, S Gansu, NE Tibet, W Sichuan & NW Yunnan

Forrest's Mountain Vole
Neodon forresti

FR Campagnol de Forrest DE Forrest-Felswühlmaus
ES Topillo de montaña de Forrest

DD
Monotypic
HB 10–13.4 cm
T 3.6–4.3 cm

3300–3650 m

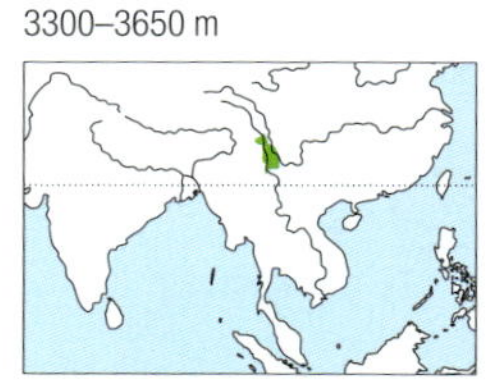

Rocky alpine meadows in N MM, SC CN (NW Yunnan)

LC Monotypic
HB 9.8–12.8 cm
T 2.6–3.5 cm
W 27–49 g

2950–4500 m

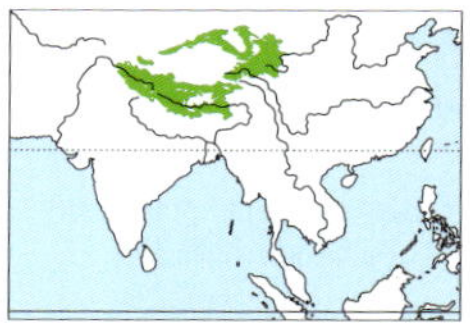

Blyth's Mountain Vole
Neodon leucurus
FR Campagnol de Blyth
DE Blyth-Felswühlmaus
ES Topillo de montaña de Blyth

Tibetan Plateau of W CN, NW IN & N NP

LC
Monotypic
HB 9.5–15 cm
T 2.2–4.2 cm
W 30–81 g

3700–4800 m

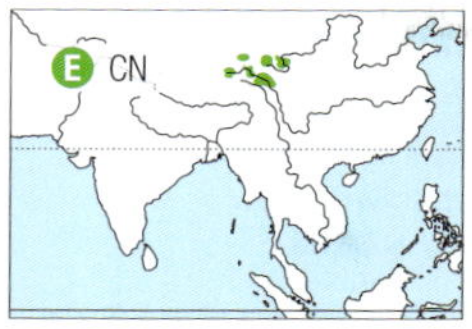

Smoky Mountain Vole
Neodon fuscus
FR Campagnol fuligineux
DE Sichuan-Felswühlmaus
ES Topillo de montaña ahumado

Moist meadows in S Qinghai

DD
Monotypic
HB 12.2–15.7 cm
T 4–6.5 cm
W 51–91 g

Evoron Vole
Alexandromys evoronensis
FR Campagnol de l'Evoron
DE Eworon-Wühlmaus
ES Topillo de Evoron

3 small isolated areas in lower R Amur Basin

Genus *ALEXANDROMYS*

(13 species)

DD
Monotypic
HB 11–15.8 cm
T 3.4–5.8 cm
W 41–97 g

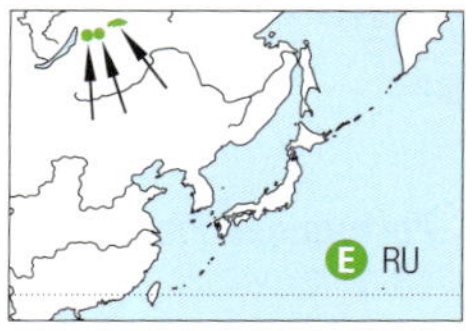

Muya Valley Vole
Alexandromys mujanensis
FR Campagnol de la Mouïa
DE Burjatien-Wühlmaus
ES Topillo de Muya

S Siberia (3 small isolated areas of NE Buryatia & N Zabaykalsky Krai)

LC 2 sspp
HB 11.4–15.2 cm
T 2.8–5.7 cm
W 25–95 g

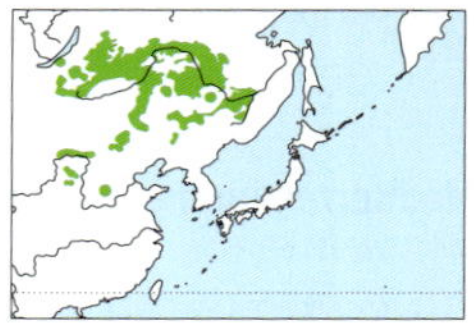

Maximowicz's Vole
Alexandromys maximowiczii
FR Campagnol de Maximowicz
DE Maxomowicz-Wühlmaus
ES Topillo de Maximowicz

E RU (SE Siberia), N & E MN, NE CN

LC
6 sspp
HB 11.5–17.6 cm
T 3.5–7.5 cm
W 36–117 g

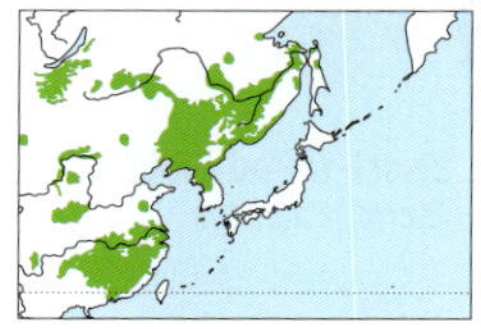

Reed Vole
Alexandromys fortis
FR Campagnol des roseaux
DE Schilfwühlmaus
ES Topillo de ribera

RU (Transbaikalia, Amur Basin), NE MN, E CN, Korean Pen

LC Monotypic
HB 10.8–17.6 cm
T 3.4–7.4 cm
W 30–149 g

Sakhalin Vole
Alexandromys sachalinensis
FR Campagnol de Sakhaline
DE Sachalin-Wühlmaus
ES Topillo de Sajalín

Forests in N & C Sakhalin I

LC Monotypic
HB 8.1–12.3 cm
T 2.1–3.9 cm
W 13–40 g

below 3000 m

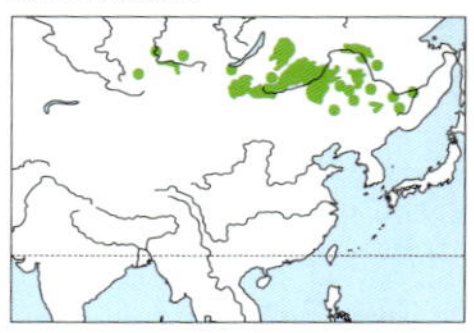

Mongolian Vole
Alexandromys mongolicus
FR Campagnol de Mongolie
DE Mongolische Wühlmaus
ES Topillo de Mongolia

S Siberia (E from Altai Republic), N & E MN, NE CN (Inner Mongolia to Jilin)

NE
Monotypic

above 2600 m

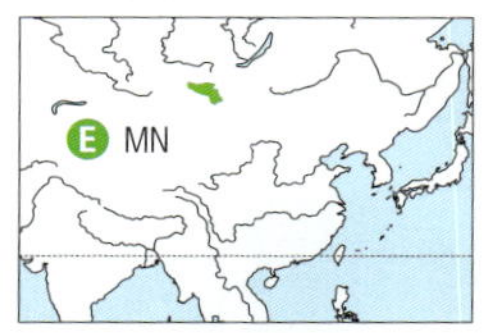

Alpine Vole
Alexandromys alpinus
FR Campagnol des Khangaï
DE Alpine Wühlmaus
ES Topillo montañoso de Mongolia

Khangai and adjacent areas

LC 3 sspp
HB 9.9–15 cm
T 2.1–4.2 cm
W 20–70 g

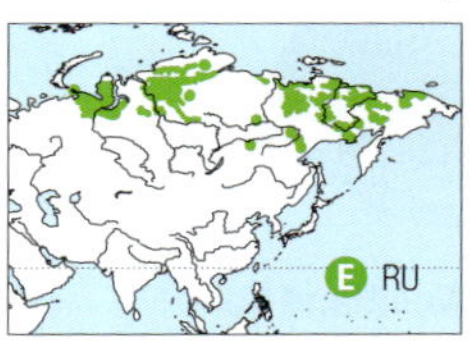

Middendorff's Vole
Alexandromys middendorffii
FR Campagnol de Middendorff
DE Middendorff-Wühlmaus
ES Topillo de Middendorff

Arctic and subarctic tundra from N Ural Mts to Chukotka

NE
Monotypic
HB 9.5–13 cm
T 2.8–4.4 cm
W 17–51 g

Shantar Vole
Alexandromys shantaricus
FR Campagnol de Gromov
DE Gromow-Wühlmaus
ES Topillo de Gromov

N & W Sea of Okhotsk coast (S Magadan, N & C Khabarovsk, SE Yakutia)

LC
2 sspp
HB 8.8–12 cm
T 2.9–4.9 cm
W 35–54 g

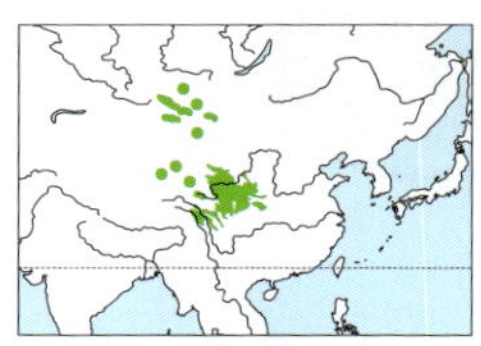

Lacustrine Vole
Alexandromys limnophilus
FR Campagnol lacustre
DE Seewühlmaus
ES Topillo lacustre

W MN, C CN (Qinghai, Gansu, Ningxia, Shaanxi, Sichuan); probably E Tibet

LC 25 sspp
HB 7.9–15 cm
T 2.8–6 cm
W 36–78 g

2500–2900 m

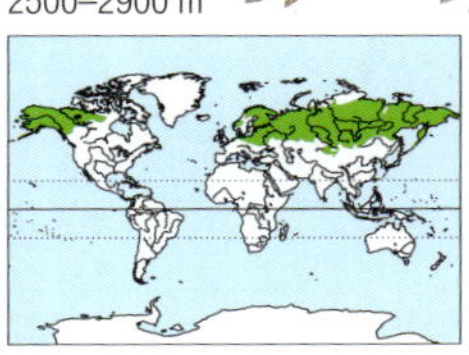

Root Vole
Alexandromys oeconomus
FR Campagnol nordique
DE Nordische Wühlmaus
ES Topillo de tundra

N Holarctic from NL & N Fennoscandia E to NW CA, S to HU & Tien Shan

LC
Monotypic
HB 10–14 cm
T 8–9.8 cm
W 37–43 g

1500–3670 m

Taiwan Vole
Alexandromys kikuchii
FR Campagnol de Taïwan
DE Taiwan-Wühlmaus
ES Topillo de Taiwán

Rugged mountainous ranges

LC
Monotypic
HB 9.5–13.6 cm
T 2.9–5 cm
W 22–62 g

Japanese Grass Vole
Alexandromys montebelli
FR Campagnol de Montebello
DE Japanische Wühlmaus
ES Topillo de Japón

Honshu, Kyushu, Sado & Noto-jima Is

Genus *MICROTUS*

(58 species)

LC
3 sspp
HB 6.5–12.5 cm
T 1.2–3.8 cm
W 16–61 g

200–3700 m

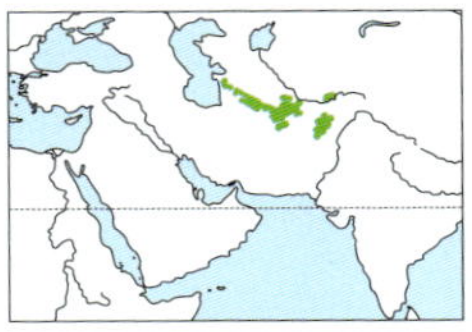

Afghan Vole
Microtus afghanus
FR Campagnol d'Afghanistan
DE Afghanistan-Wühlmaus
ES Topillo de Afganistán

W & S TM, NE IR, AF, SW TJ (Khatlon Region)

LC
2 sspp
HB ≤ 12.7 cm
T ≤ 3.4 cm
W ≤ 53 g

550–1850 m

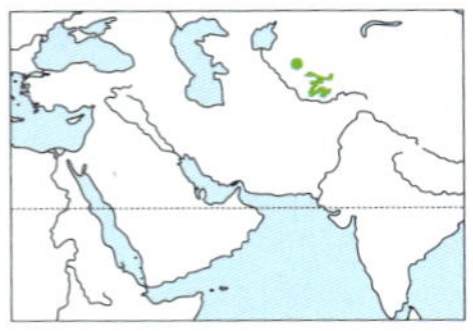

Bucharian Vole
Microtus bucharensis
FR Campagnol du Boukhara
DE Buchara-Wühlmaus
ES Topillo de Bujará

N Pamir-Alay ranges (UZ, TJ, extreme E TM)

LC 2 sspp
HB 8.3–11.5 cm
T 2.4–5 cm
W 21–33 g

1300–4500 m

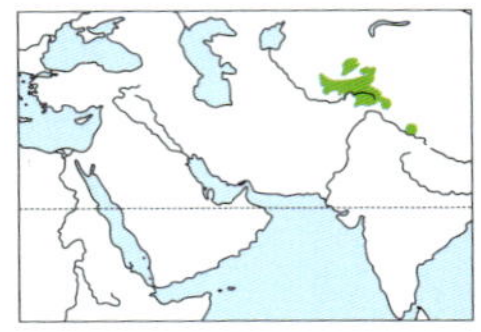

Juniper Vole
Microtus juldaschi
FR Campagnol des genévriers
DE Wacholder-Wühlmaus
ES Topillo de enebros

SE UZ, SW KG, TJ & NE AF; also NW PK & SW China (SW Xinjiang, NW Tibet)

LC Monotypic
HB 9.2–13.2 cm
T 3–5.3 cm
W 22–72 g

0–2140 m

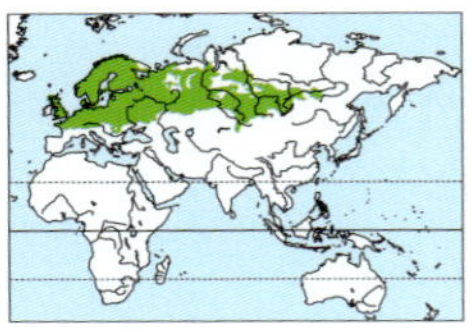

Short-tailed Field Vole
Microtus agrestis
FR Campagnol agreste
DE Erdmaus
ES Topillo agreste

UK E across Europe (with Carpathian Basin) to Siberia (R Lena), S to Altai Mts

NE
Monotypic
HB 9.7–14 cm
T 2.6–5.5 cm
W 20–59 g

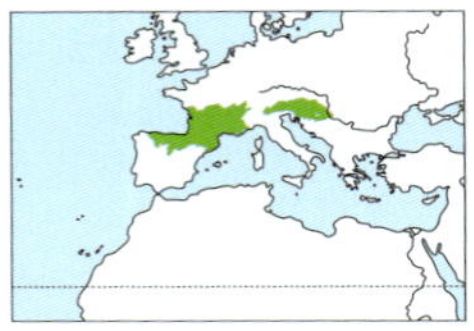

Mediterranean Field Vole
Microtus lavernedii
FR Campagnol de Lavernède
DE Südliche Erdmaus
ES Topillo de campo mediterráneo

Mediterranean Europe from N ES E to HU & RS

NE
Monotypic
HB c. 10 cm
T c. 3.9 cm

Portuguese Field Vole
Microtus rozianus
FR Campagnol de Bocage
DE Portugal-Erdmaus
ES Topillo de campo lusitano

NW Iberian Pen (NW ES, N PT)

Singing Vole
Microtus abbreviatus
FR Campagnol chanteur
DE Alaska-Wühlmaus
ES Topillo fonador

variants

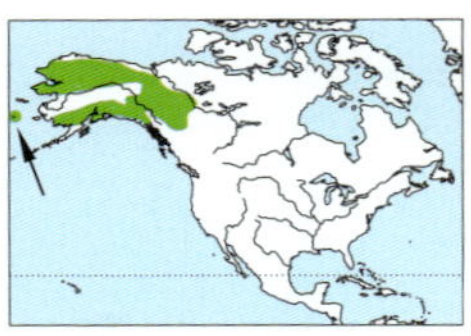

LC 7 sspp
HB 8.2–14.6 cm T 1.9–4.1 cm
W 22–79 g

US (Alaska, with St Matthew & Hall Is), CA (W Northwest Territories, Yukon)

Rock Vole
Microtus chrotorrhinus
FR Campagnol des rochers
DE Gelbnasen-Wühlmaus
ES Topillo de roca

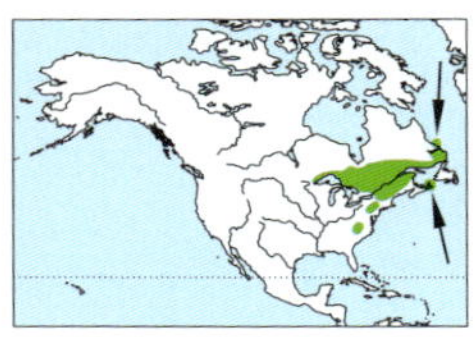

LC
3 sspp
HB 9.2–12.1 cm
T 4.2–6.4 cm
W 30–40 g

SE CA (S Ontario to E Labrador), NE US (NE Minnesota, Maine S to E Tennessee)

EN
Monotypic
HB 11.5–13.3 cm
T 4.9–6.5 cm
W c. 42 g

1800–3000 m

Zempoaltepec Vole
Microtus umbrosus
FR Campagnol du Zempoaltepec
DE Zempoaltepec-Wühlmaus
ES Topillo de Zempoaltépetl

NC Oaxaca (Mt Zempoaltépec slopes)

EN
Monotypic
HB 11.2–12.7 cm
T 3.1–4 cm
W 31–46 g

2100–2500 m

Tarabundi Vole
Microtus oaxacensis
FR Campagnol d'Oaxaca
DE Oaxaca-Wühlmaus
ES Topillo de Tarabundí

Oaxaca (mts bounded by R Santo Domingo & R Grande, Ixtlán District)

NT
Monotypic
HB 11.3–12.2 cm
T 3–3.6 cm
W 34–44 g

2500–3200 m

Guatemalan Vole
Microtus guatemalensis
FR Campagnol du Guatemala
DE Guatemala-Wühlmaus
ES Topillo de Guatemala

Cloud forests from C Chiapas (MX) SE to C GT

LC
7 sspp
HB 6.3–11 cm
T 1.7–2.5 cm
W 25–38 g

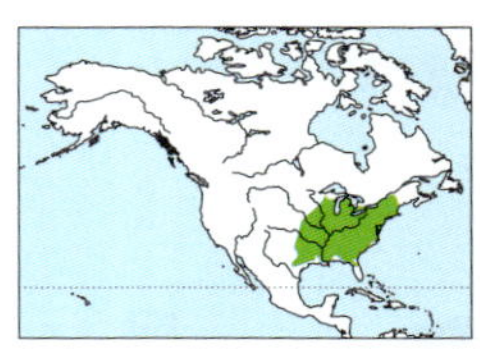

Woodland Vole
Microtus pinetorum
FR Campagnol sylvestre
DE Kiefernwühlmaus
ES Topillo de bosque

CA (extreme S Ontario, S Quebec), E US (except SE coast and Florida Pen)

Jalapan Vole
Microtus quasiater
FR Campagnol de Jalapa DE Jalapa-Wühlmaus
ES Topillo de Jalapa

500–2150 m

NT
Monotypic
HB 9.5–11 cm
T 1.7–2.5 cm
W 26–30 g

E slopes of Sierra Madre Oriental, from SE San Luis Potosí S to N Oaxaca

California Vole
Microtus californicus
FR Campagnol de Californie DE Kalifornien-Wühlmaus
ES Topillo de California

LC
17 sspp
HB 11.1–13.8 cm
T 3.8–5.8 cm
W 43–47 g

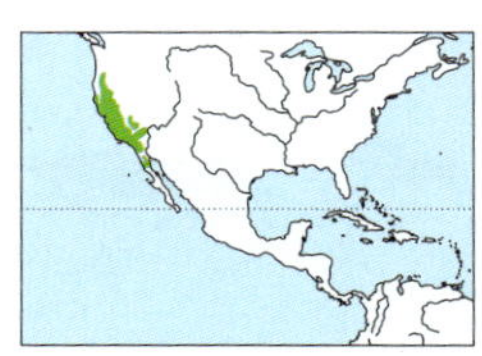

WC Oregon through C & W California (US) S to N Baja California (MX)

LC
9 sspp
HB 10.9–12.7 cm
T 2.4–3.5 cm
W 29–48 g

1200–4115 m

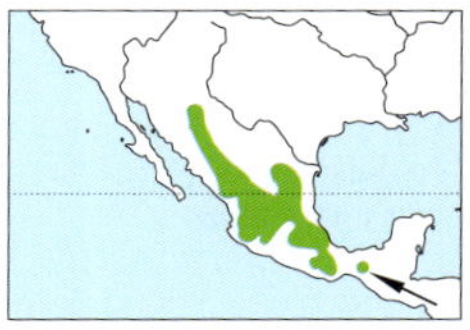

Mexican Vole
Microtus mexicanus
FR Campagnol du Mexique
DE Mexiko-Wühlmaus
ES Topillo de México

From S Utah & S Colorado (US) across MX (Sierra Madre Mts, S to W Chiapas)

NE
4 sspp
HB 10.9–12.7 cm
T 2.4–3.5 cm
W 25–43 g

Mogollon Vole
Microtus mogollonensis
FR Campagnol des Mogollon
DE Mogollon-Wühlmaus
ES Topillo de Mogollón

Extreme SW Colorado, Arizona, New Mexico, Texas (Guadalupe Mts)

LC
7 sspp
HB 10.6–13.1 cm
T 2.4–4.1 cm
W 37–48 g

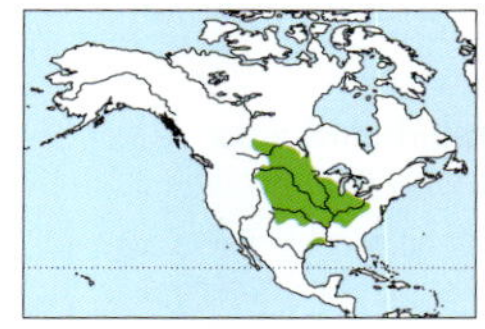

Prairie Vole
Microtus ochrogaster
FR Campagnol des Prairies
DE Prairiewühlmaus
ES Topillo de pradera

SC CA, C & CE US; relict in SW Louisiana & adjacent Texas

LC
Monotypic
HB 14.1–17.3 cm
T 4.5–5.3 cm
W 140–170 g

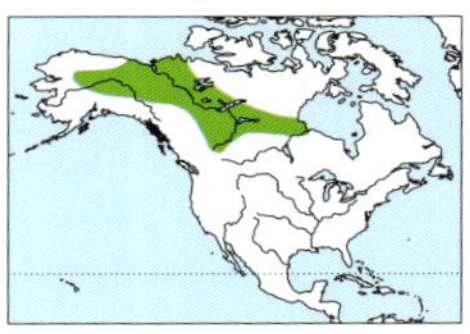

Taiga Vole
Microtus xanthognathus
FR Campagnol à joues jaunes
DE Gelbwangen-Wühlmaus
ES Topillo de taiga

C Alaska (US), W CA (S to C Alberta, E to E Hudson Bay)

NT
Monotypic
HB 10–13.5 cm
T 3–5.2 cm
W 30–78 g

0–1500 m

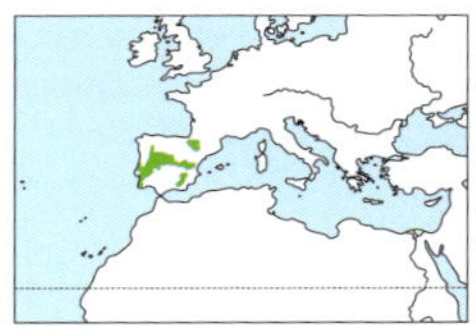

Cabrera's Vole
Microtus cabrerae
FR Campagnol de Cabrera
DE Cabrera-Wühlmaus
ES Topillo de Cabrera

Iberian Pen (ES, PT)

LC
4 sspp
HB 12.5–17.8 cm
T 6.6–9.8 cm
W 72–150 g

900–3000 m

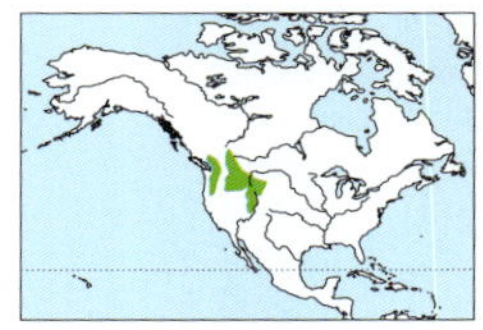

North American Water Vole
Microtus richardsoni
FR Campagnol de Richardson
DE Richardson-Wühlmaus
ES Topillo acuático norteamericano

SW CA, US (S along Cascade Range to Oregon and along Rocky Mts to C Utah)

LC
Monotypic
HB 10.8–12.3 cm
T 3.2–4.5 cm
W > 50 g

Gray-tailed Vole
Microtus canicaudus
FR Campagnol à queue grise
DE Grauschwanz-Wühlmaus
ES Topillo de cola gris

SE Washington (Clark County), NE Oregon (Willamette Valley)

LC
6 sspp
HB 12.1–16 cm
T 4.8–7.5 cm
W 47–83 g

0–1830 m

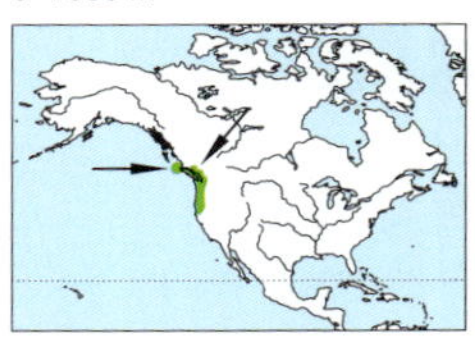

Townsend's Vole
Microtus townsendii
FR Campagnol de Townsend
DE Townsend-Wühlmaus
ES Topillo de Townsend

CA (SW British Columbia, Vancouver and some Is) S to NW California (US)

LC
15 sspp
HB 11.6–15.1 cm
T 2.4–6.9 cm
W 37–85 g

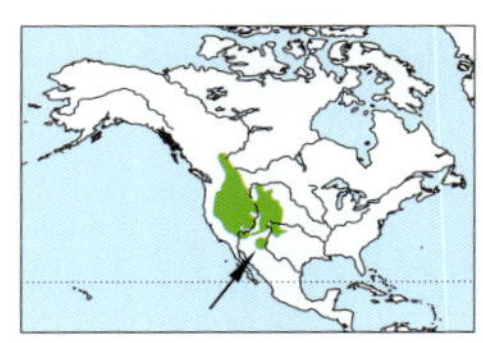

Montane Vole
Microtus montanus
FR Campagnol montagnard
DE Rocky-Mountains-Wühlmaus
ES Topillo de montaña

SW CA (SC British Columbia) S in montane areas of W US (S to New Mexico)

NE
Monotypic
HB 17.8–19.8 cm
T 4.3–5.5 cm
W 43–80 g

Florida Salt Marsh Vole
Microtus dukecampbelli
FR Campagnol de Floride
DE Florida-Wühlmaus
ES Topillo de Florida

WC Florida (salt marshes on Florida Gulf Coast)

Western Meadow Vole
Microtus drummondii
FR Campagnol de Drummond
DE Westliche Wiesenwühlmaus
ES Topillo de campiña occidental

NE
14 extant sspp
HB 10.7–13.1 cm
T 3.3–6.4 cm
W c. 44 g

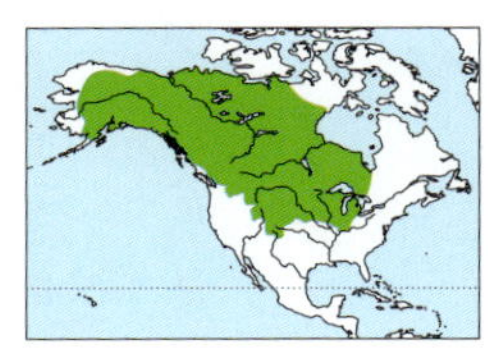

Alaska and most of W & C CA S through N & C US to New Mexico

LC
12 sspp
HB 10.7–13.1 cm
T 3.3–6.4 cm
W c. 44 g

Eastern Meadow Vole
Microtus pennsylvanicus
FR Campagnol des prés
DE Östliche Wiesenwühlmaus
ES Topillo de campiña oriental

ssp *breweri*

ssp *pennsylvanicus* variants

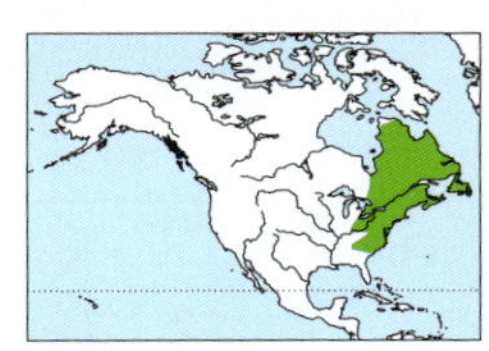

E CA (E Ontario to Newfoundland) S across NE US to South Carolina

LC 15 sspp
HB 10.6–14 cm
T 4.9–8.1 cm
W 37–59 g

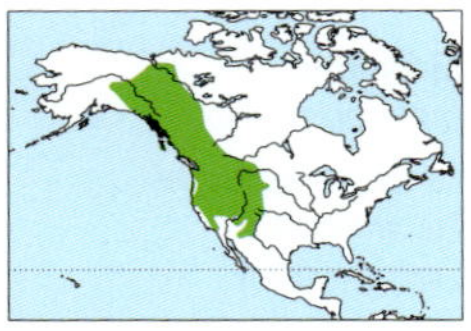

Long-tailed Vole
Microtus longicaudus

FR Campagnol à longue queue
DE Langschwanz-Wühlmaus
ES Topillo de cola larga

C Alaska S across W CA & W US to S California, Arizona & New Mexico

LC 4 sspp
HB 9.7–11.2 cm
T 3.2–4.2 cm
W c. 19 g

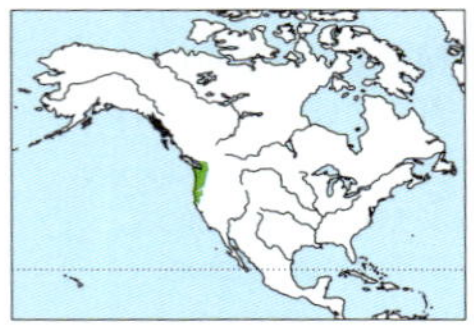

Creeping Vole
Microtus oregoni

FR Campagnol de l'Oregon
DE Oregon-Wühlmaus
ES Topillo de Oregón

CA (SW British Columbia), US (W Washington, W Oregon, NW California)

LC 3 sspp
HB 9–11.2 cm
T 3.2–4.9 cm
W 20–29 g

800–1400 m

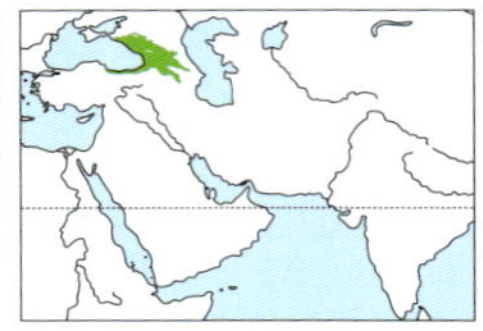

Major's Pine Vole
Microtus majori

FR Campagnol de Major
DE Östliche Kleinwühlmaus
ES Topillo de Major

SE shore of Black Sea (NE TR), N & S Greater Caucasus (RU, GE, AM, AZ)

LC
Monotypic
HB 7.9–11.5 cm
T 2–4.4 cm
W 10–26 g

0–2335 m

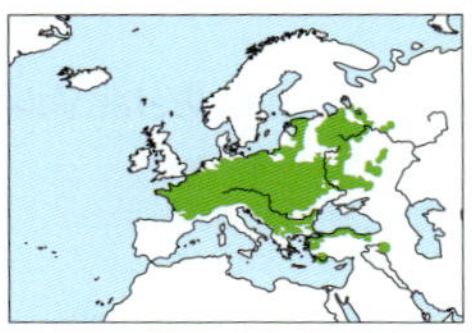

Common Pine Vole
Microtus subterraneus

FR Campagnol souterrain
DE Kurzohrmaus
ES Topillo europeo

W, C & E Europe, from FR to W European RU, SE to TR (W & N Anatolia)

LC Monotypic
HB 9.1–10.5 cm
T 3.3–4.2 cm
W 15–25 g

2000–2600 m

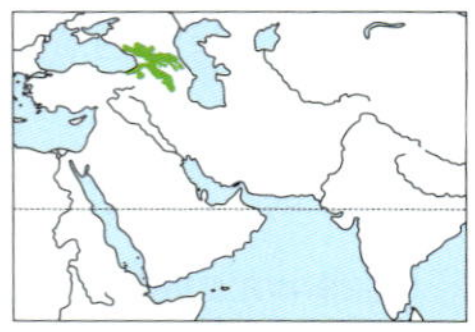

Caucasian Pine Vole
Microtus daghestanicus

FR Campagnol du Daghestan
DE Dagestan-Kleinwühlmaus
ES Topillo del Cáucaso

Caucasus (SW RU, GE, NE TR, AM, AZ, extreme NW IR)

LC
Monotypic
HB 9–11 cm
T 4–4.4 cm
W 19–29 g

up to 2800 m

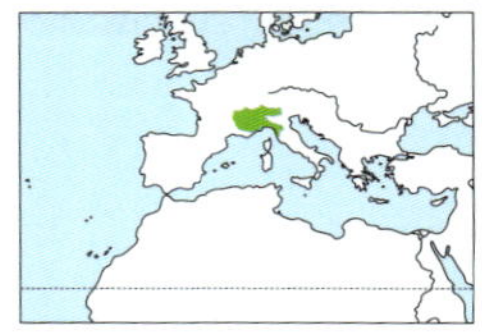

Alpine Pine Vole
Microtus multiplex

FR Campagnol de Fatio
DE Alpen-Kleinwühlmaus
ES Topillo de los Alpes

W Alps and NW Apennines (SE FR, S CH, NW IT)

LC
Monotypic
HB 8.8–11.5 cm
T 2.7–4 cm
W 14–32 g

0–1960 m

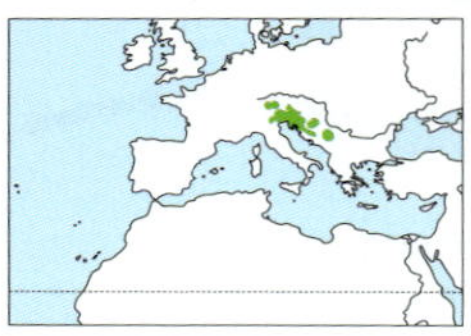

Liechtenstein's Pine Vole
Microtus liechtensteini

FR Campagnol de Liechtenstein
DE Illyrische Kleinwühlmaus
ES Topillo de Liechtenstein

E Alps and Dinaric Alps (from extreme S DE & NE IT E to W RS)

LC 2 sspp
HB 8.3–11.5 cm
T 3.1–4.7 cm
W 18–36 g

600–2340 m

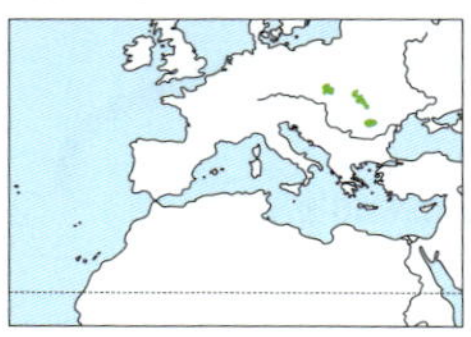

Tatra Pine Vole
Microtus tatricus

FR Campagnol des Tatras
DE Tatra-Kleinwühlmaus
ES Topillo de los Tatras

Carpathian Mts of S PL, N SK, SW UA & RO

LC
Monotypic
HB 8–11.1 cm
T 1.9–3.5 cm
W 19–32 g

0–3000 m

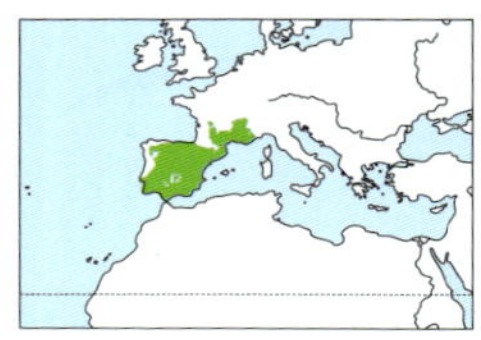

Mediterranean Pine Vole
Microtus duodecimcostatus

FR Campagnol provençal
DE Mittelmeer-Kleinwühlmaus
ES Topillo mediterráneo

Iberian Pen (except NW), S & SE FR

LC 2 sspp
HB 8–9.4 cm
T 2.3–3.1 cm
W 14–23 g

0–2000 m

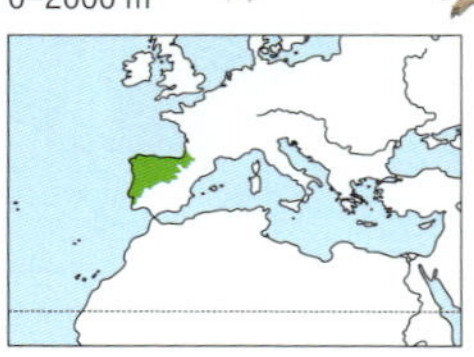

Lusitanian Pine Vole
Microtus lusitanicus

FR Campagnol basque
DE Iberien-Wühlmaus
ES Topillo lusitano

Extreme SW FR, NW ES, PT

LC Monotypic
HB 9.2–10.8 cm
T 2.2–3.7 cm
W 17–32 g

800–2035 m

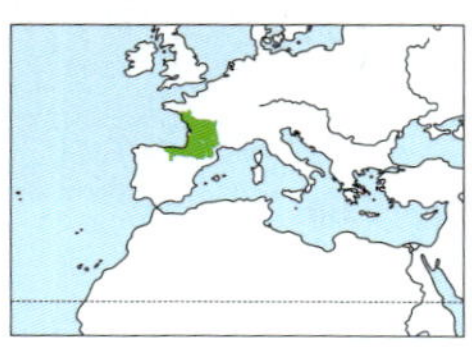

Pyrenean Pine Vole
Microtus gerbii

FR Campagnol de Gerbe
DE Pyrenäen-Kleinwühlmaus
ES Topillo pirenaico

FR (W of Massif Central, S of R Loire), AD, N ES (E Cantabria to E Pyrenees)

LC Monotypic
HB 8.3–10 cm
T 2.4–3 cm
W 16–24 g

up to 2800 m

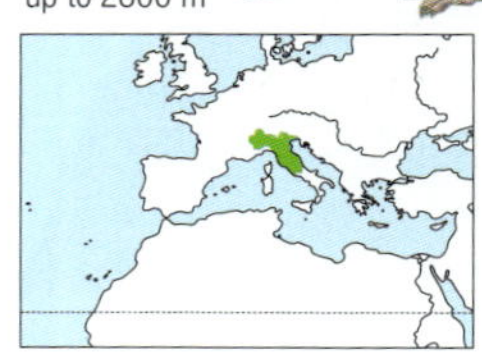

Savi's Pine Vole
Microtus savii

FR Campagnol de Savi
DE Italien-Kleinwühlmaus
ES Topillo de Savi

S CH (E Valais, Ticino), N & C IT

LC
Monotypic
HB 8.7–9.8 cm
T 2.3–2.7 cm
W 14–24 g

Calabria Pine Vole
Microtus brachycercus

FR Campagnol de Calabre
DE Kalabrien-Kleinwühlmaus
ES Topillo de Calabria

S from Abruzzo region

NE Monotypic
HB 9.6–10.4 cm
T 1.6–2.6 cm
W 18–21 g

175–1700 m

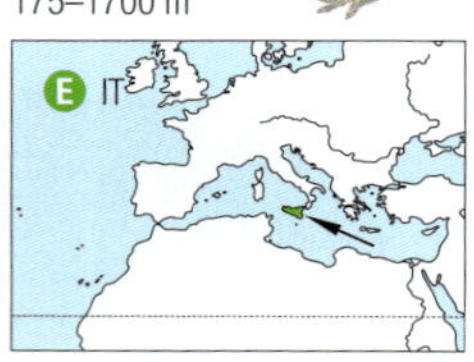

Sicilian Pine Vole
Microtus nebrodensis

FR Campagnol de Sicile
DE Sizilien-Kleinwühlmaus
ES Topillo de Sicilia

Sicily

LC
2 sspp
HB 7.1–11.9 cm
T 1.2–3 cm
W 20–43 g

0–2000 m

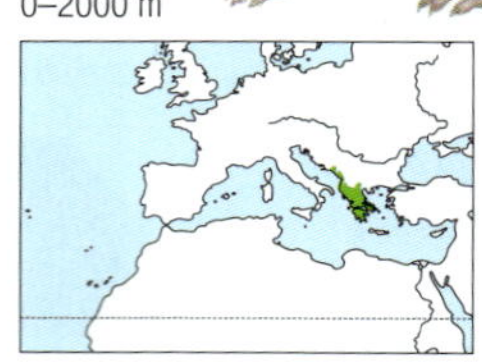

Thomas's Pine Vole
Microtus thomasi

FR Campagnol de Thomas
DE Balkan-Kleinwühlmaus
ES Topillo de Thomas

SE Balkans from S BA & ME to GR (mainland, Peloponnese, Euboea I)

LC Monotypic
HB 8.3–10.5 cm
T 2.3–3.9 cm
W 16–28 g

40–2050 m

Balkan Pine Vole
Microtus felteni

FR Campagnol de Felten
DE Felten-Kleinwühlmaus
ES Topillo de los Balcanes

Open habitats in S RS, XK, W MK, AL, NW GR

LC 2 sspp
HB 8.8–11.1 cm
T 1.8–2.5 cm
W 17–31 g

150–1400 m

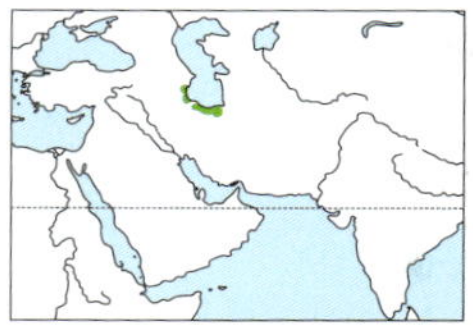

Schelkovnikov's Pine Vole
Microtus schelkovnikovi

FR Campagnol de Shelkovnikov
DE Schelkownikow-Kleinwühlmaus
ES Topillo de Schelkovnikov

Talysh Mts (SE AZ), Alborz Mts (NW IR)

NE Monotypic
HB 10.3–13.8 cm
T 2.2–3.5 cm
W 31–76 g

0–1500 m

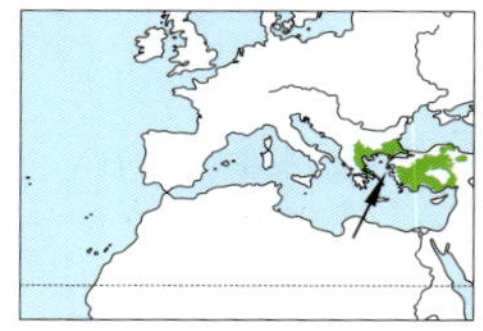

Harting's Vole
Microtus hartingi

FR Campagnol de Harting
DE Balkan-Wühlmaus
ES Topillo de Harting

RS, MK, SE BG, GR (with Lesbos I), TR (W & C Anatolia)

LC Monotypic
HB 10.9–13 cm
T 2.3–3.2 cm
W 26–47 g

up to 1700 m

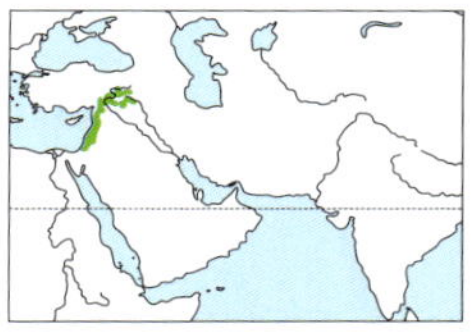

Levant Vole
Microtus guentheri

FR Campagnol de Günther
DE Levante-Wühlmaus
ES Topillo de Guenther

Meadows in SE TR, SY, LB, IL, NW JO

LC
2 sspp
HB 9.9–12.5 cm
T 1.8–3.3 cm
W 19–42 g

200–800 m

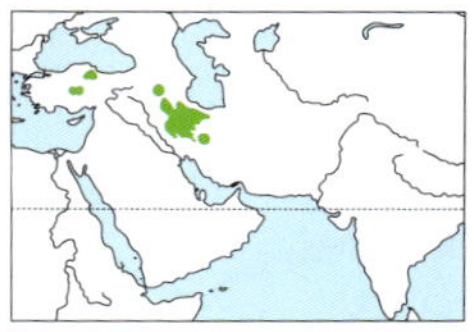

Dogramaci's Vole
Microtus dogramacii

FR Campagnol de Dogramaci
DE Dogramaci-Wühlmaus
ES Topillo de Dogramaci

Steppes in TR (C Anatolia) & NW IR

NE
Monotypic
HB 9.9–12.4 cm
T 2–3.1 cm

0–700 m

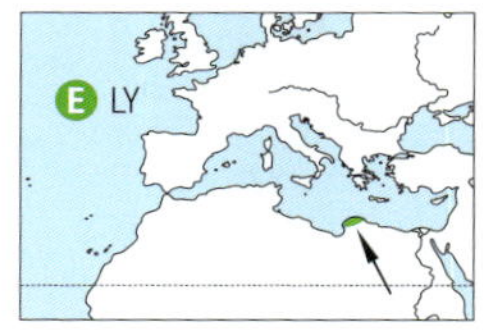

Cyrenaica Vole
Microtus mustersi

FR Campagnol de Musters
DE Cyrenaika-Wühlmaus
ES Topillo de Cirenaica

Coastal plain and inland plateau of Cyrenaica

LC
11 sspp
HB 9.3–11.6 cm
T 2.4–3.7 cm
W 20–38 g

up to 2000 m

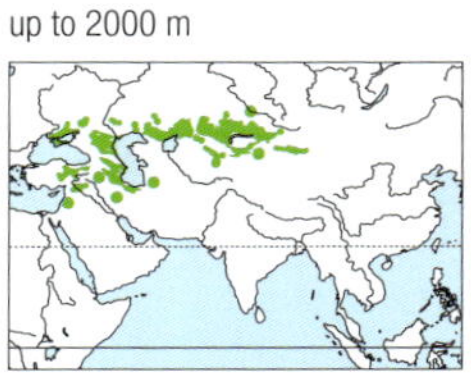

Social Vole
Microtus socialis

FR Campagnol social
DE Gesellige Wühlmaus
ES Topillo social

Fragmented fom R Dnieper & Crimea E to NW Xinjiang (CN), S to LB & NW IR

DD Monotypic
HB 10.5–12.5 cm
T 2.1–3.4 cm
W 23–53 g

530–980 m

Anatolian Vole
Microtus anatolicus

FR Campagnol d'Anatolie
DE Anatolien-Wühlmaus
ES Topillo de Anatolia

Dry open habitats in a small area in SE of C Anatolia

VU 3 sspp
HB 8.4–13 cm
T 2–3.6 cm
W 26–42 g

1000–2100 m

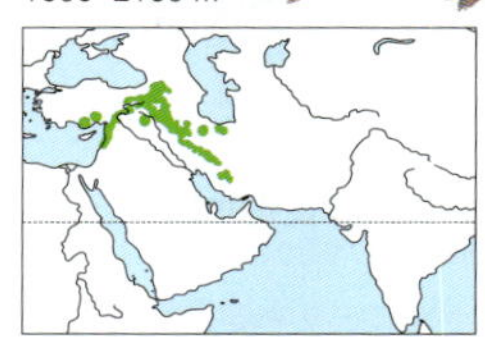

Iranian Vole
Microtus irani

FR Campagnol persan
DE Iran-Wühlmaus
ES Topillo de Irán

S & E TR and SC GE, S to LB, SE to W IR

DD
Monotypic
HB 9.7–12 cm
T 2–3.3 cm
W 23–36 g

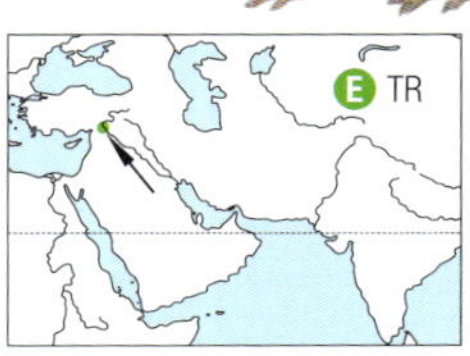

Elbeyli Vole
Microtus elbeyli

FR Campagnol d'Elbeyli
DE Elbeyli-Wühlmaus
ES Topillo de Elbeyli

Known only from Kilis and Elbey (Kilis Prov)

LC
Monotypic
HB 8.7–12.2 cm
T 2.1–3 cm
W 30–62 g

300–2500 m

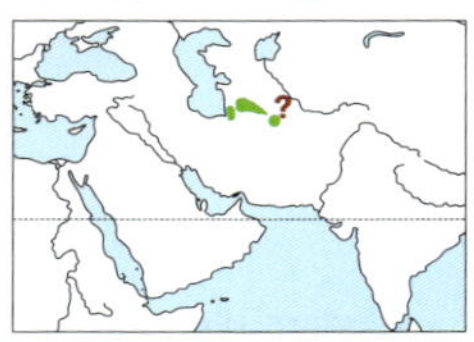

Kopet Dag Pine Vole
Microtus paradoxus

FR Campagnol paradoxal
DE Turkmenistan-Wühlmaus
ES Topillo de Kopet Dag

Kopet Dag (SW TM and adjacent IR); perhaps SE TM

LC
Monotypic
HB 9.3–12 cm
T 2.4–3.8 cm
W 18–40 g

up to 3000 m

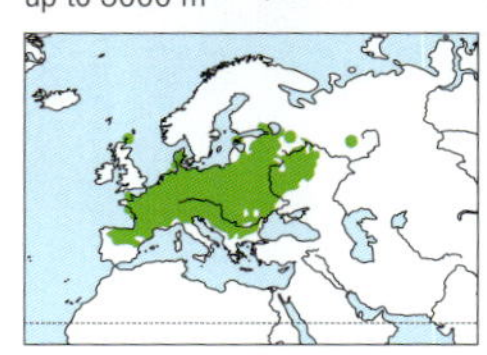

Common Vole
Microtus arvalis

FR Campagnol des champs
DE Feldmaus
ES Topillo campesino

N Iberian Pen E across most of Europe to SE FI & W European RU, SE to BG

NE Monotypic
HB 10.1–13.6 cm
T 3.2–5.1 cm
W 23–48 g

up to 3000 m

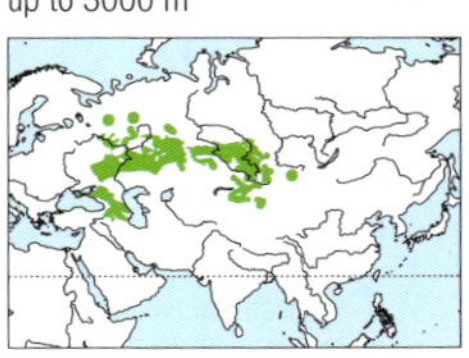

Altai Vole
Microtus obscurus

FR Campagnol obscur
DE Südostfeldmaus
ES Topillo de Altái

E UA, Crimea & Caucasus E across S Siberia to KZ, NW MN & NW CN (Xinjiang)

LC
Monotypic
HB 10.2–13 cm
T 3.3–5.5 cm
W 21–58 g

0–2480 m

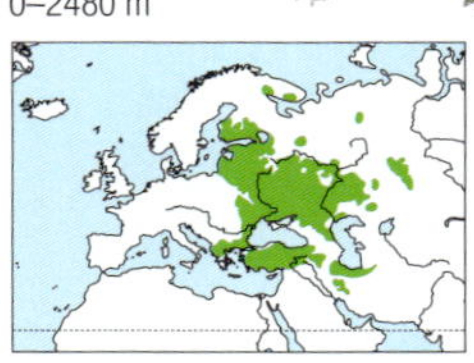

East European Vole
Microtus mystacinus

FR Campagnol d'Ondrias
DE Ostfeldmaus
ES Topillo europeo oriental

N & E Europe and Balkans E to R Tobol (W Siberia), S to TR and NW & N IR

DD
Monotypic
HB 11.4–15.2 cm
T 4.3–5.1 cm
W 41–57 g

c. 2700 m

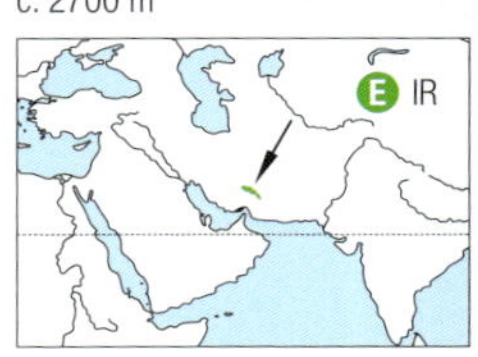

Kerman Vole
Microtus kermanensis

FR Campagnol de Kerman
DE Kerman-Wühlmaus
ES Topillo de Kermán

Known around type locality area (Zahrud-e Bala, Zarda, Deh Bakri)

Transcaspian Vole
Microtus transcaspicus

FR Campagnol transcaspien
DE Transkaspien-Feldmaus
ES Topillo del Caspio

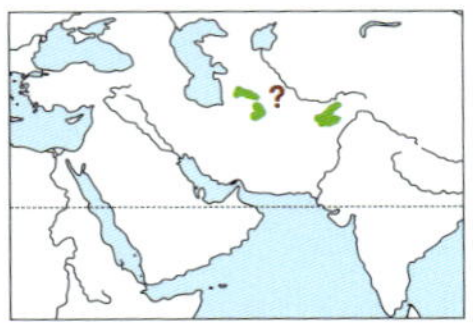

LC
Monotypic
HB 9.7–14 cm
T 3.8–5.5 cm
W 36–80 g

500–2800 m

W & C Kopet Dag (TM, IR), Binalud Mts (NE IR, E AF); perhaps SE TM

Tian Shan Vole
Microtus ilaeus

FR Campagnol du Kazakhstan
DE Tienschan-Feldmaus
ES Topillo de Tian Shan

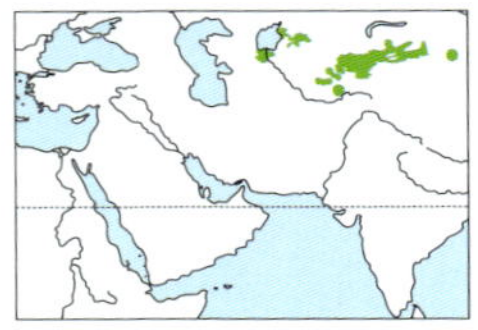

LC
2 sspp
HB 10.9–15 cm
T 3.5–5.6 cm
W 36–64 g

60–3200 m

Humid meadows in UZ, S KZ, KG, TJ & NW CN (Xinjiang)

Genus *NELSONIA*

(2 species)

Goldman's Diminutive Woodrat
Nelsonia goldmani

FR Néotoma des volcans
DE Goldman-Kleinstbuschratte
ES Rata de bosque pequeña de Goldman

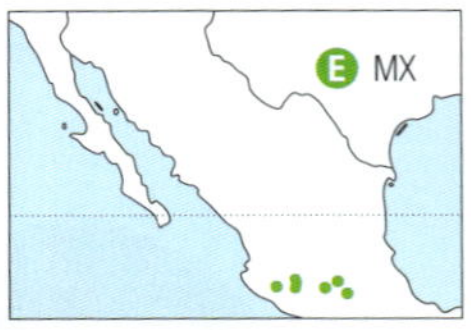

2000–3100 m

Disjunct in mts of S Jalisco, NE Colima, Michoacán, N México & N Morelos

EN 2 sspp
HB 11.7–13.3 cm
T 10.8–12.6 cm
W 43–57 g

Western Diminutive Woodrat
Nelsonia neotomodon

FR Néotoma de Nelson
DE Westliche Kleinstbuschratte
ES Rata de bosque pequeña occidental

LC
Monotypic
HB 11.8–12.1 cm
T 10.9–11.7 cm
W 43–55 g

above 2000 m

Sierra Madre Occidental (S Durango, S Zacatecas, Aguascalientes, N Jalisco)

Genus *HODOMYS*

(1 species)

0–1800 m

From S Sinaloa S to Guerrero & NW Oaxaca, E to Morelos & S Puebla

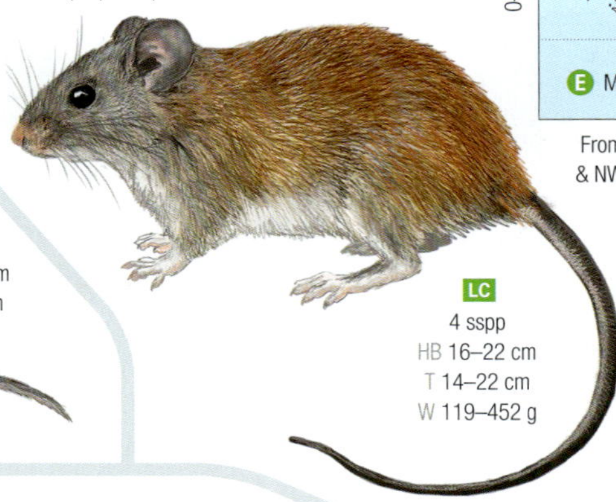

LC
4 sspp
HB 16–22 cm
T 14–22 cm
W 119–452 g

Allen's Woodrat
Hodomys alleni

FR Néotoma d'Allen
DE Allen-Buschratte
ES Rata de bosque de Allen

Genus *XENOMYS*

(1 species)

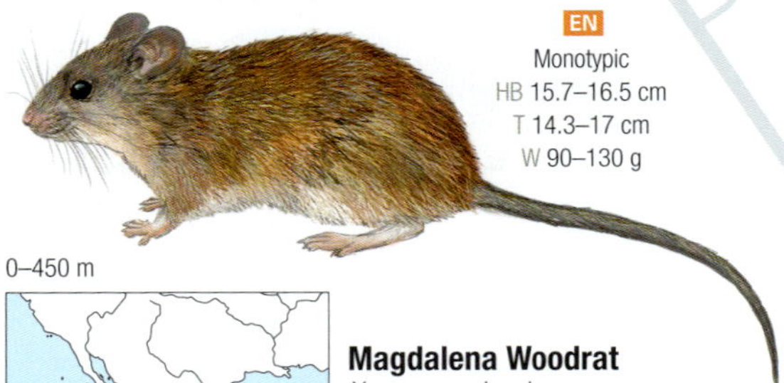

EN
Monotypic
HB 15.7–16.5 cm
T 14.3–17 cm
W 90–130 g

0–450 m

Coastal areas of Jalisco & Colima

Magdalena Woodrat
Xenomys nelsoni

FR Néotoma de Magdalena
DE Magdalena-Ratte
ES Rata de bosque de Magdalena

Bushy-tailed Woodrat
Neotoma cinerea

FR Néotoma à queue touffue
DE Buschschwanzratte
ES Rata de bosque de cola peluda

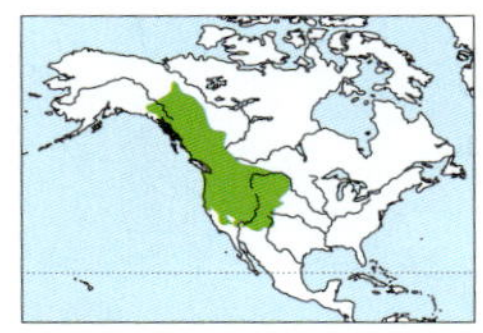

W CA (S from Northwest Territories), W US (S to Arizona & New Mexico, E to South Dakota)

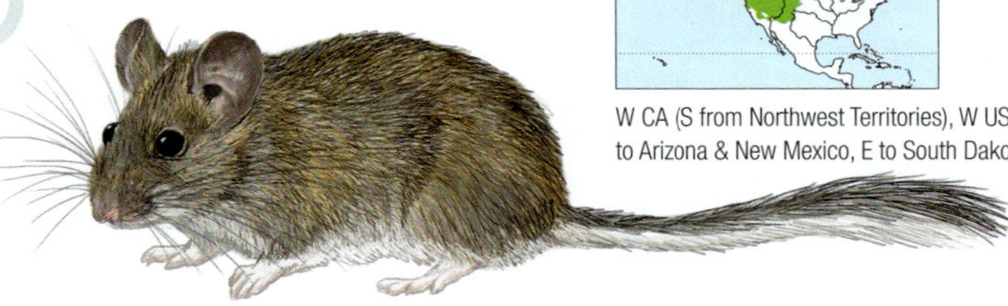

LC 13 sspp
HB 17.3–25 cm
T 12–22 cm
W 234–600 g

Genus *NEOTOMA*

(23 species)

Sonoran Woodrat
Neotoma phenax

FR Néotoma du Sonora
DE Sonora-Buschratte
ES Rata de bosque de Sonora

LC Monotypic
HB 18.1–21 cm
T 14.9–22 cm
W 18–279 g

0–150 m

Coastal SW Sonora & NW Sinaloa

Goldman's Woodrat
Neotoma goldmani

FR Néotoma de Goldman
DE Goldman-Buschratte
ES Rata de bosque de Goldman

LC
Monotypic
HB 14.9–15.2 cm
T 11.3–13.6 cm

1150–1830 m

N Mexican Plateau, from SE Chihuahua S to N Querétaro

NT
Monotypic
HB 16.4–24 cm
T 14.7–21 cm
W 282–403 g

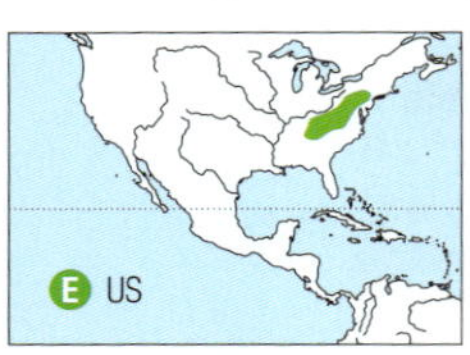

Appalachian region from extreme S New York SW to extreme NE Alabama

Allegheny Woodrat
Neotoma magister

FR Néotoma des Allegheny
DE Allegheny-Buschratte
ES Rata de bosque de Allegheny

Eastern Woodrat
Neotoma floridana

FR Néotoma de Floride
DE Florida-Buschratte
ES Rata de bosque oriental

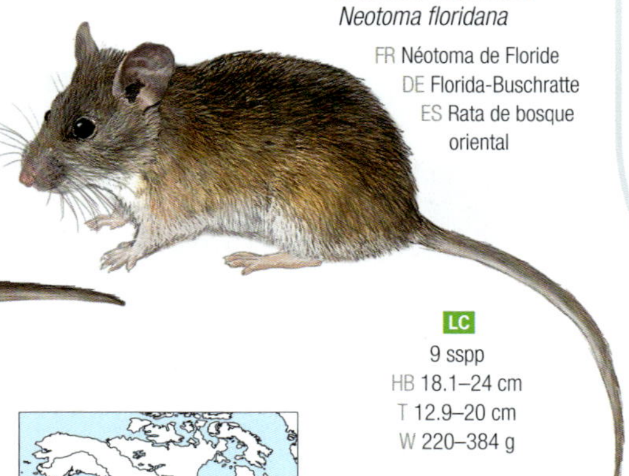

LC
9 sspp
HB 18.1–24 cm
T 12.9–20 cm
W 220–384 g

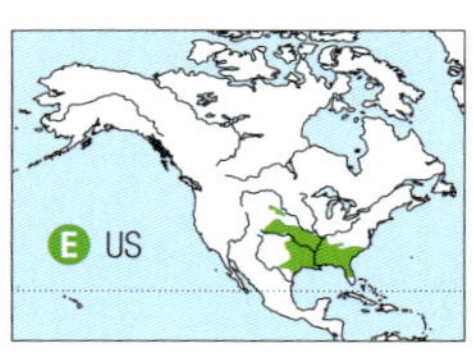

South Dakota & E Colorado E to North Carolina, S to Texas, Gulf coast & C Florida

LC
7 sspp
HB 20.6–21 cm
T 7.6–18.5 cm
W 145–350 g

0–2475 m

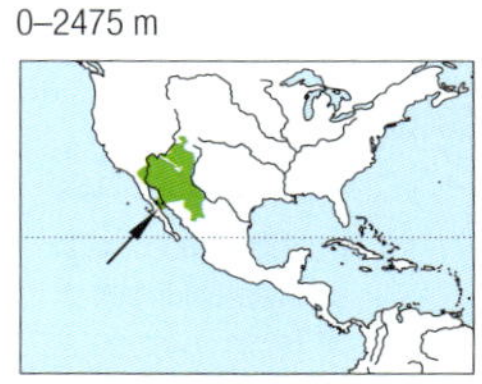

SW US (S from S Utah & SW Colorado) to NW MX (Sonora, C Chihuahua, Tiburon I)

White-throated Woodrat
Neotoma albigula

FR Néotoma à gorge blanche
DE Weißkehl-Buschratte
ES Rata de bosque de garganta blanca

Black-tailed Woodrat
Neotoma melanura

FR Néotoma à queue noire
DE Schwarzschwanz-Buschratte
ES Rata de bosque de cola negra

NE
Monotypic
HB 18.4–21 cm
T 14.9–18.5 cm
W 154–246 g

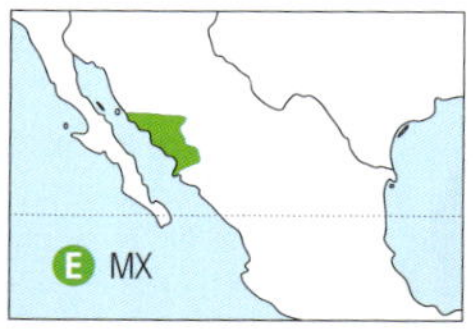

S Sonora & N Sinaloa

White-toothed Woodrat
Neotoma leucodon

FR Néotoma à dents blanches
DE Weißzahn-Buschratte
ES Rata de bosque de dientes blancos

LC
7 sspp

HB 20.7–21 cm
T 7.6–18.5 cm
W 145–200 g

1000–2700 m

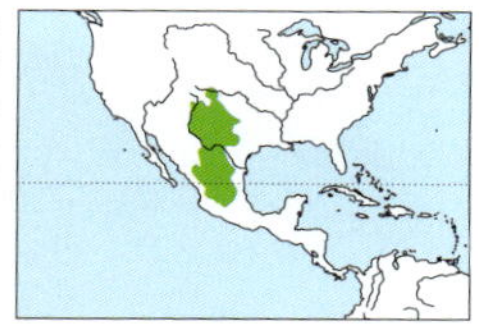

CS US (S from SE Colorado) S to C MX (Guanajuato & Queretaro)

Southern Plains Woodrat
Neotoma micropus

FR Néotoma à petites pattes
DE Kleinfuß-Buschratte
ES Rata de bosque meridional

0–1700 m

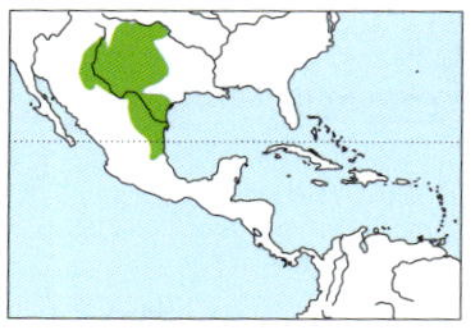

US (Great Plains S from SE Colorado & S Kansas) to MX (S to San Luis Potosi)

LC
5 sspp
HB 18–24 cm
T 13–17.5 cm
W 200–317 g

Nelson's Woodrat
Neotoma nelsoni

FR Néotoma de Perote
DE Nelson-Buschratte
ES Rata de bosque de Nelson

950–2300 m

E slope of Cofre de Perote Volcano and nearby area (Veracruz, Puebla)

CR
Monotypic
HB c. 19.5 cm
T c. 15.4 cm
W c. 198 g

Bolaños Woodrat
Neotoma palatina

FR Néotoma de Bolanos
DE Jalisco-Buschratte
ES Rata de bosque de Bolaños

VU
Monotypic
HB 18.2–22 cm
T 14.4–18 cm
W c. 198 g

940–1905 m

NE Jalisco & SE Zacatecas

Tamaulipan Woodrat
Neotoma angustapalata

FR Néotoma du Tamaulipas
DE Tamaulipas-Buschratte
ES Rata de bosque de Tamaulipas

NT
Monotypic
HB 18.5–22 cm
T 19.5–20 cm
W 180–240 g

0–1150 m

Sierra Madre Oriental (SW Tamaulipas and adacent San Luis Potosi)

Mexican Woodrat
Neotoma mexicana

FR Néotoma du Mexique
DE Mexiko-Buschratte
ES Rata de bosque de México

0–4045 m

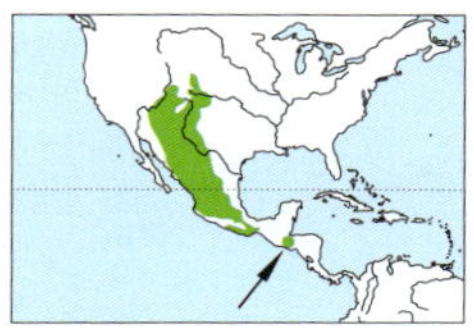

US (S from Utah & C Colorado), S through W & C MX to highlands of GT, SV & W HN

LC
19 sspp
HB 15.2–20 cm
T 13.3–22 cm
W 151–253 g

Painted Woodrat
Neotoma picta

FR Néotoma coloré
DE Guerrero-Buschratte
ES Rata de bosque pintada

NE
Monotypic
HB c. 17.3 cm
T 16.6–18.2 cm

up to 3048 m

CE Guerrero & WC Oaxaca

Guatemalan Woodrat
Neotoma ferruginea

FR Néotoma ferrugineux
DE Guatemala-Buschratte
ES Rata de bosque de Guatemala

NE 4 sspp
HB 18.9–19.2 cm
T 16.6–19.8 cm
W 150–300 g

MX (SE Oaxaca, Chiapas) SE to GT & SV; possibly W HN

Nicaraguan Woodrat
Neotoma chrysomelas

FR Néotoma du Nicaragua
DE Nicaragua-Buschratte
ES Rata de bosque de Nicaragua

LC
Monotypic
HB 18.4–22 cm
T 15–18.1 cm
W 152–187 g

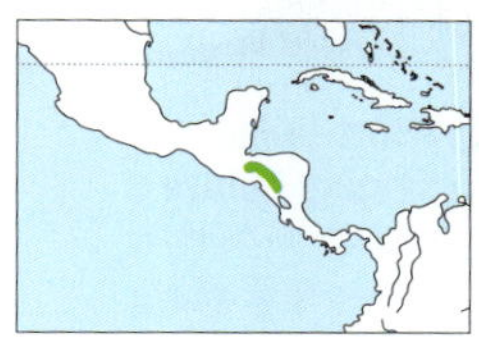

HN, NW NI

c. 1000 m

Stephens's Woodrat
Neotoma stephensi

FR Néotoma de Stephens
DE Stephens-Buschratte
ES Rata de bosque de Stephens

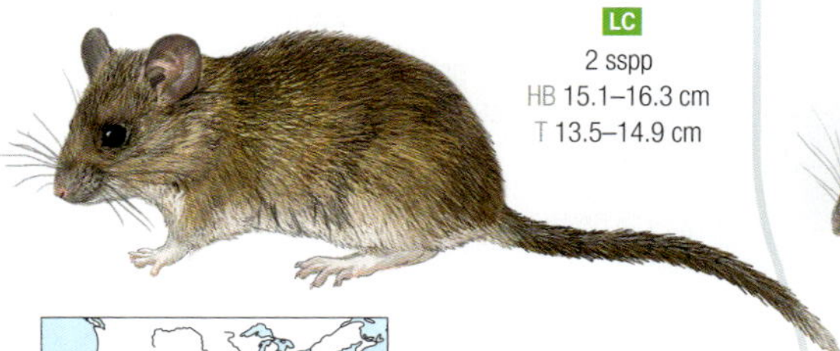

LC
2 sspp
HB 15.1–16.3 cm
T 13.5–14.9 cm

S Utah, N & C Arizona, W New Mexico

Dusky-footed Woodrat
Neotoma fuscipes

FR Néotoma à pattes noires
DE Dunkelfuß-Buschratte
ES Rata de bosque de pies oscuros

LC
6 sspp
HB 17.7–23 cm
T 15.8–24 cm
W 200–312 g

From R Columbia (W Oregon) S to Coastal Range & N Sierra Nevada (WC & EC California)

Big-eared Woodrat
Neotoma macrotis

FR Néotoma à oreilles longues
DE Großohr-Buschratte
ES Rata de bosque de orejas grandes

LC
5 sspp
HB 17.7–25 cm
T 15.8–21 cm
W 184–358 g

0–2550 m

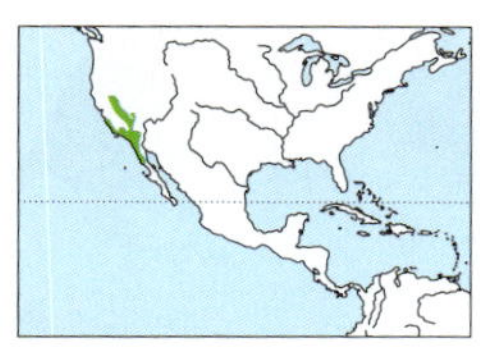

SW US (from C California) to NW MX (NW Baja California)

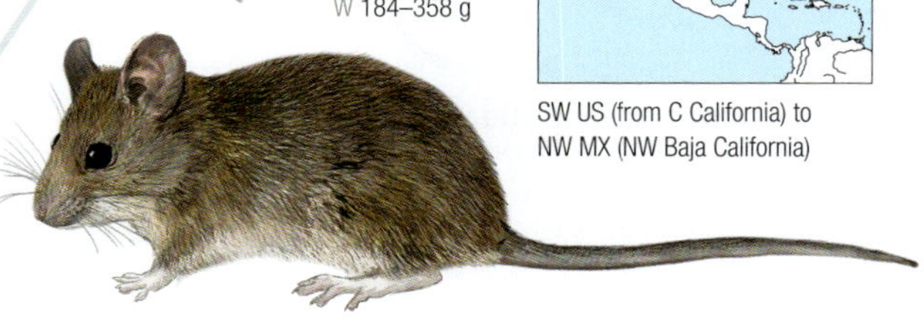

LC
3 extant sspp
HB 16.3–21 cm
T 12.3–18.7 cm
W 150–300 g

0–2620 m

Bryant's Woodrat
Neotoma bryanti
FR Néotoma de Bryant
DE Bryant-Buschratte
ES Rata de bosque de Bryant

SW US (from CW California) to NW MX (Baja California Pen, surrounding Is)

Angel de la Guarda Woodrat
Neotoma insularis
FR Néotoma d'Angel de la Guarda
DE Insel-Buschratte
ES Rata de bosque de Ángel de la Guarda

0–1300 m

DD
Monotypic
HB 17.4–17.8 cm
T 11.3–16.2 cm

Ángel de la Guarda I (Baja California)

Desert Woodrat
Neotoma lepida
FR Néotoma du désert
DE Wüsten-Buschratte
ES Rata de bosque de desierto

LC
3 sspp
HB 15.2–17.9 cm
T 10.1–17.3 cm
W 100–190 g

0–2800 m

W US (from SE Oregon, SW Idaho & NW Colorado) S to NW MX (E Baja California)

LC Monotypic
HB 14.7–17.6 cm
T 10.8–14.9 cm
W 100–199 g

0–1600 m

Arizona Woodrat
Neotoma devia
FR Néotoma d'Arizona
DE Arizona-Buschratte
ES Rata de bosque de Arizona

W Arizona (SW US) and extreme NW Sonora (NW MX)

Genus *OCHROTOMYS*

(1 species)

Golden Mouse
Ochrotomys nuttalli
FR Ochrotomys
DE Goldmaus
ES Ratón dorado

SE Missouri & Illinois E to E Virginia, S to E Texas, Gulf coast & C Florida

LC
5 sspp
HB 5.1–11.5 cm
T 5–9.7 cm
W 18–27 g

NE 7 sspp
HB 6.5–7.9 cm
T 3.5–5.6 cm
W 6–10 g

0–2000 m

Southern Pygmy Mouse
Baiomys brunneus
FR Baiomys austral
DE Südliche Amerikanische Zwergmaus
ES Ratón pigmeo oriental

S MX (Guerrero, Morelos, Puebla, Veracruz, Oaxaca, Chiapas), S GT, S HN, SV, NW NI

Genus *BAIOMYS*

(3 species)

LC
8 sspp
HB 5.3–7.6 cm
T 3.4–5.3 cm
W 6–9 g

0–2450 m

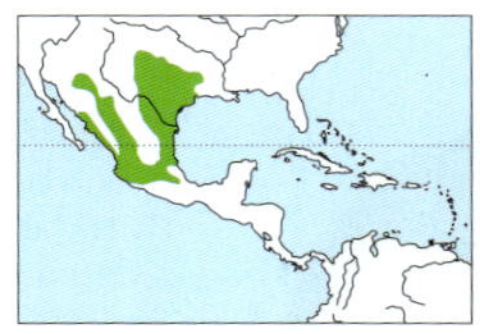

Northern Pygmy Mouse
Baiomys taylori
FR Baiomys de Taylor
DE Nördliche Amerikanische Zwergmaus
ES Ratón pigmeo de Taylor

S US (SE Arizona, SW New Mexico, SW Oklahoma, Texas) S to C MX (Michoacán, Puebla)

LC
Monotypic

Mexican Pygmy Mouse
Baiomys musculus
FR Baiomys de Colima
DE Mexikanische Zwergmaus
ES Ratón pigmeo occidental

S Nayarit, Jalisco, Colima, Michoacán

Genus *SCOTINOMYS*

(2 species)

LC
4 sspp
HB 6.9–8 cm
T 4.6–5.4 cm
W 10–13 g

1000–2940 m

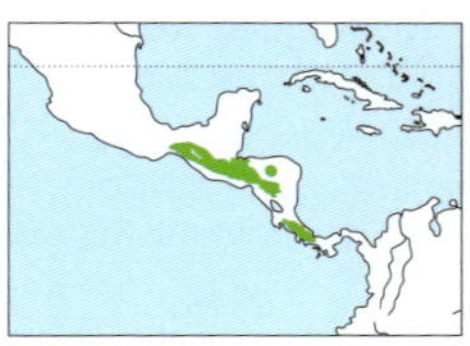

Short-tailed Singing Mouse
Scotinomys teguina
FR Scotinomys d'Alston
DE Kurzhaar-Braunmaus
ES Ratón fonador de cola corta

From SE MX (SE Oaxaca, Chiapas) E to C GT, HN, N SV, NW NI, CR & W PA

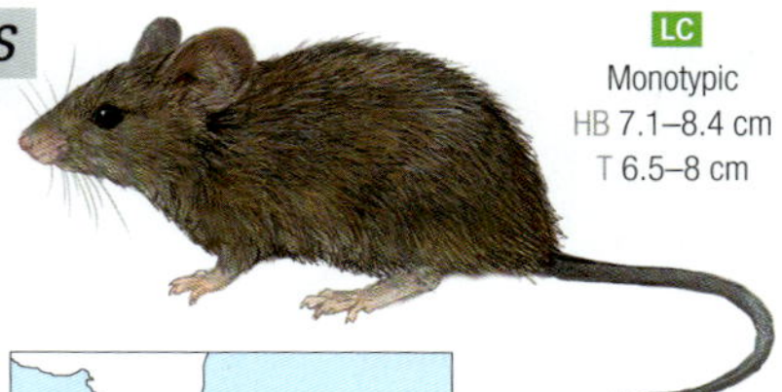

LC
Monotypic
HB 7.1–8.4 cm
T 6.5–8 cm

Long-tailed Singing Mouse
Scotinomys xerampelinus
FR Scotinomys du Chiriqui
DE Langhaar-Braunmaus
ES Ratón fonador de cola larga

C mountain regions of S CR & W PA

NT Monotypic
HB 16.5–17 cm
T 15.5–20 cm
W 43–57 g

915–1525 m

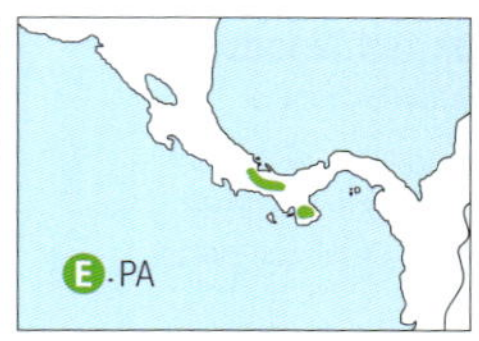

Yellow Deermouse
Isthmomys flavidus
FR Péromyscus jaune
DE Gelbe Isthmusratte
ES Ratón ciervo dorado

Chiriquí Volcano and surrounding areas, Azuero Pen

Genus *PODOMYS*

(1 species)

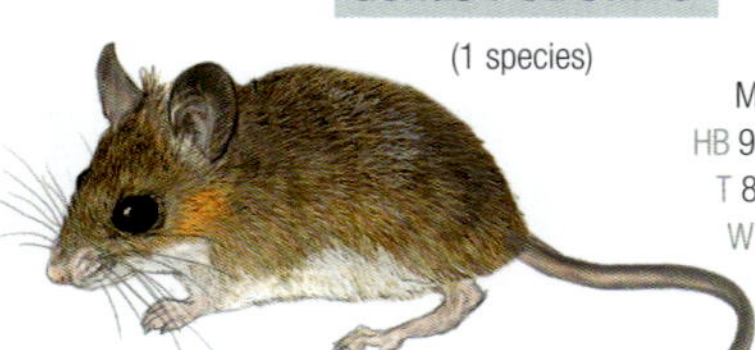

NT
Monotypic
HB 9.8–11.9 cm
T 8–10.1 cm
W 27–47 g

Florida Deermouse
Podomys floridanus
FR Péromyscus de Floride
DE Florida-Maus
ES Ratón ciervo de Florida

Florida Pen

Genus *NEOTOMODON*

(1 species)

LC
2 sspp
HB 9.6–12.8 cm
T 7.8–10.5 cm
W 40–60 g

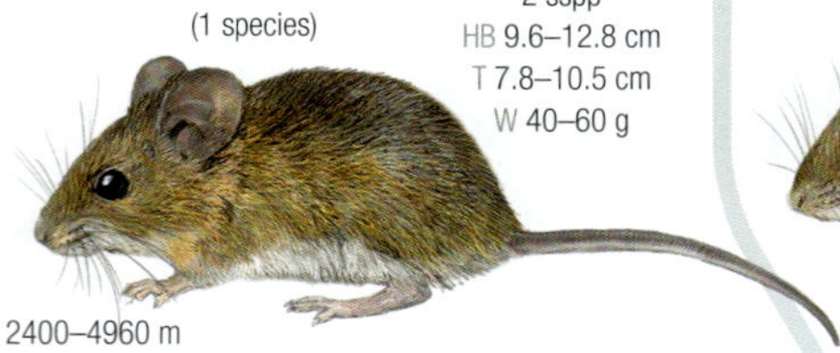

2400–4960 m

Volcano Deermouse
Neotomodon alstoni
FR Péromyscus d'Alston
DE Mexikanische Vulkanmaus
ES Ratón ciervo de volcanes

Trans-Mexican Volcanic Belt from WC Michoacán E to EC Veracruz

Genus *ISTHMOMYS*

(2 species)

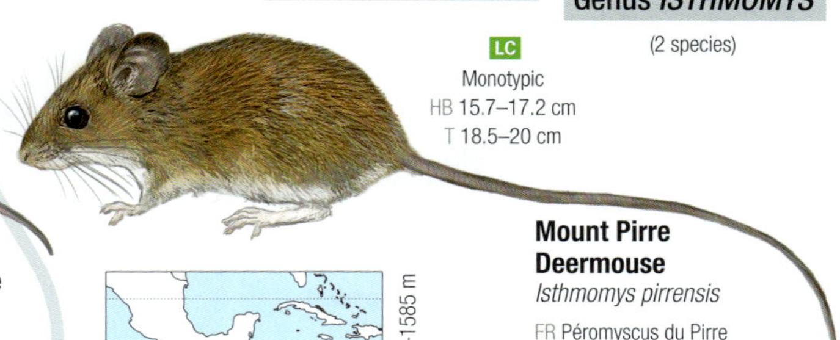

LC
Monotypic
HB 15.7–17.2 cm
T 18.5–20 cm

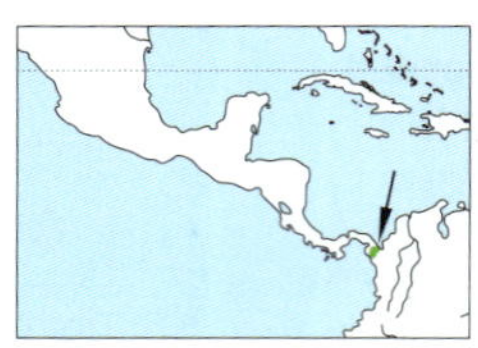

1065–1585 m

Mount Pirre Deermouse
Isthmomys pirrensis
FR Péromyscus du Pirre
DE Braune Isthmusratte
ES Ratón ciervo de Pirre

E PA, extreme NW CO

Genus *REITHRODONTOMYS*

(25 species)

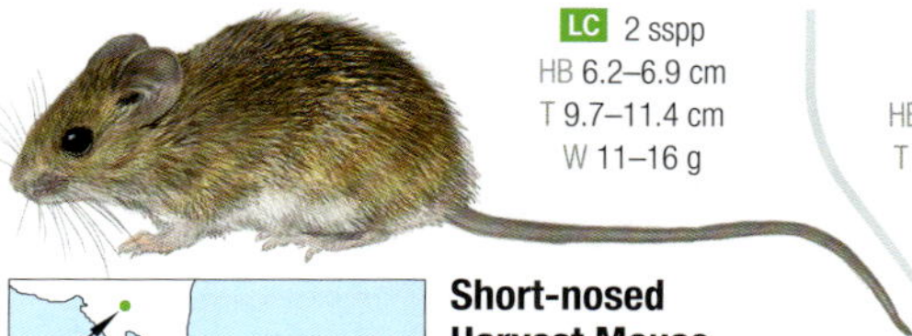

LC 2 sspp
HB 6.2–6.9 cm
T 9.7–11.4 cm
W 11–16 g

Short-nosed Harvest Mouse
Reithrodontomys brevirostris

FR Souris-moissonneuse à nez court
DE Kurznasen-Erntemaus
ES Ratón espiguero de hocico corto

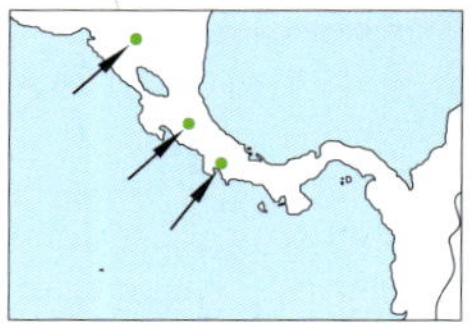

NC NI (Cordillera Dariense), NC & S CR (2 disjunct areas in Cordillera Central)

LC
Monotypic
HB 6.2–6.5 cm
T 10–11.2 cm
W c. 10 g

up to 900 m

Darien Harvest Mouse
Reithrodontomys darienensis

FR Souris-moissonneuse du Darien
DE Darién-Erntemaus
ES Ratón espiguero de Darién

E from Azuero Pen

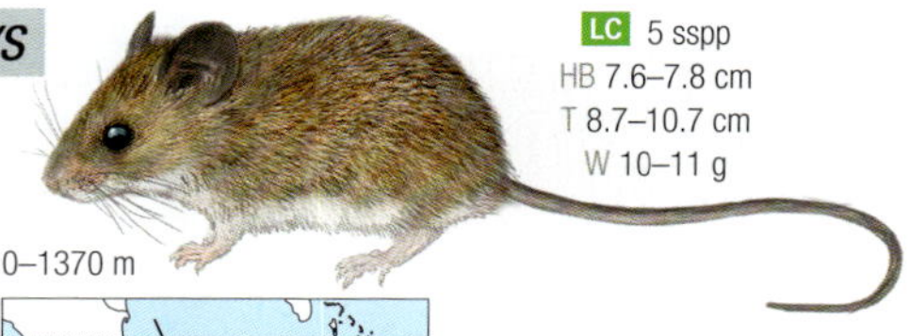

LC 5 sspp
HB 7.6–7.8 cm
T 8.7–10.7 cm
W 10–11 g

0–1370 m

Slender Harvest Mouse
Reithrodontomys gracilis

FR Souris-moissonneuse gracile
DE Zierliche Erntemaus
ES Ratón espiguero esbelto

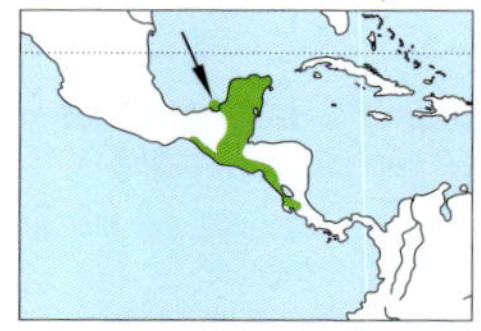

S MX (Yucatán Pen, coastal Chiapas) along Pacific coast to WC CR

LC 10 sspp
HB 6–8 cm
T 9–11.5 cm
W c. 19 g

90–1800 m

Mexican Harvest Mouse
Reithrodontomys mexicanus

FR Souris-moissonneuse du Mexique
DE Mexiko-Erntemaus
ES Ratón espiguero de México

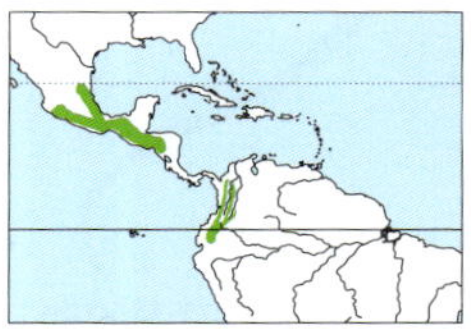

MX (from S Tamaulipas & WC Michoacán) across Middle America to CO & N EC

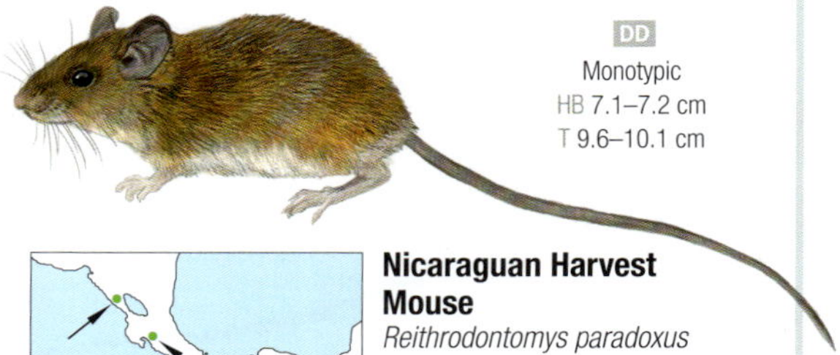

DD
Monotypic
HB 7.1–7.2 cm
T 9.6–10.1 cm

Nicaraguan Harvest Mouse
Reithrodontomys paradoxus

FR Souris-moissonneuse du Nicaragua
DE Nicaragua-Erntemaus
ES Ratón espiguero de Nicaragua

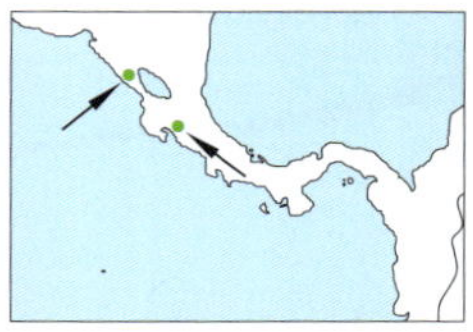

Isolated records in SW NI & WC CR

CR Monotypic
HB 8.4–8.9 cm
T 12.1–13.2 cm
W 18–21 g

Cozumel Harvest Mouse
Reithrodontomys spectabilis

FR Souris-moissonneuse de Cozumel
DE Cozumel-Erntemaus
ES Ratón espiguero de Cozumel

Cozumel I, off Yucatán Pen

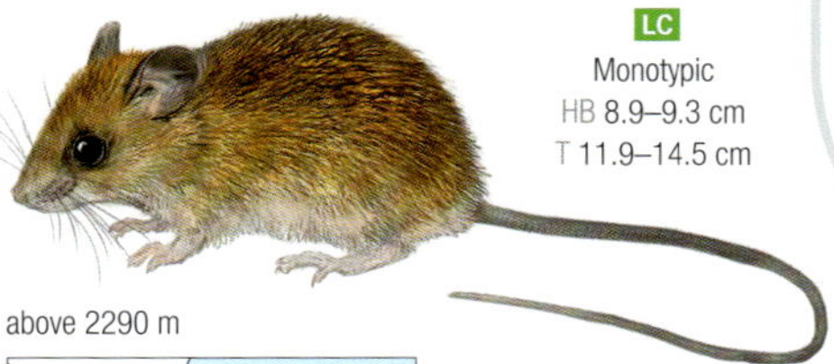

LC
Monotypic
HB 8.9–9.3 cm
T 11.9–14.5 cm

above 2290 m

Talamancan Harvest Mouse
Reithrodontomys creper

FR Souris-moissonneuse du Chiriqui
DE Talamanca-Erntemaus
ES Ratón espiguero de Talamanca

NC CR to WC PA

Rodriguez's Harvest Mouse
Reithrodontomys rodriguezi

FR Souris-moissonneuse de Rodriguez
DE Rodriguez-Erntemaus
ES Ratón espiguero de Rodríguez

Patchily recorded in Cordilleras Central & Talamanca

LC Monotypic
HB 7.7–8.3 cm
T 11.6–12.1 cm

EN
Monotypic
HB 8–12.4 cm
T 12–12.9 cm
W 18–23 g

2800–2910 m

Narrow-nosed Harvest Mouse
Reithrodontomys tenuirostris

FR Souris-moissonneuse à nez aigu
DE Schmalnasen-Erntemaus
ES Ratón espiguero de hocico estrecho

Extreme SE MX (Chiapas), SC GT

LC Monotypic
HB 6.8–7 cm
T 10.1–11.7 cm
W c. 20 g

2225–3050 m

Small-toothed Harvest Mouse
Reithrodontomys microdon

FR Souris-moissonneuse à petites dents
DE Kleinzahn-Erntemaus
ES Ratón espiguero de Guatemala

From SE MX (C Chiapas) S to C GT

Oaxacan Highlands Harvest Mouse
Reithrodontomys albilabris

FR Souris-moissonneuse d'Oaxaca
DE Oaxaca-Erntemaus
ES Ratón espiguero de Oaxaca

NE
Monotypic
HB 6.8–7 cm
T 10.1–11.7 cm
W c. 20 g

NC Oaxaca 2225–3050 m

Wagner's Harvest Mouse
Reithrodontomys wagneri

FR Souris-moissonneuse de Wagner
DE Wagner-Erntemaus
ES Ratón espiguero del Eje Neovolcánico

variants

NE
2 sspp
HB 6.8–7 cm
T 10.1–11.7 cm
W c. 20 g

2225–3050 m

From C Michoacán E to México, Sierra Madre del Sur in C Guerrero

NE
Monotypic
HB 6.7–8.5 cm
T 10.1–11.7 cm

915–2135 m

Costa Rican Harvest Mouse
Reithrodontomys cherrii

FR Souris-moissonneuse de Cherrie
DE Costa-Rica-Erntemaus
ES Ratón espiguero de Costa Rica

Highlands of Cordilleras Central & Talamanca

975–3350 m

Cordillera de Talamanca (C & SE CR, W PA)

Chiriquian Harvest Mouse
Reithrodontomys garichensis

FR Souris-moissonneuse du Gariché
DE Chiriquí-Erntemaus
ES Ratón espiguero de Chiriquí

NE
Monotypic
HB 7.5–7.9 cm
T 11–12 cm

c. 3385 m

DD Monotypic
HB c. 6.2 cm
T c. 9.9 cm

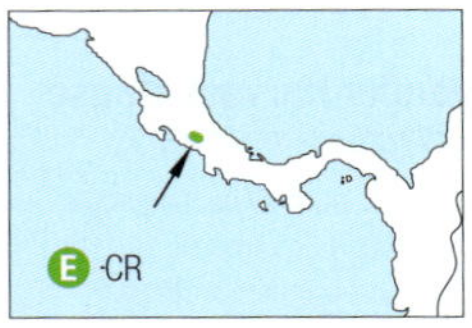

Cerro la Asunción (W Cordillera de Talamanca)

Musser's Harvest Mouse

Reithrodontomys musseri

FR Souris-moissonneuse de Musser
DE Musser-Erntemaus
ES Ratón espiguero de Musser

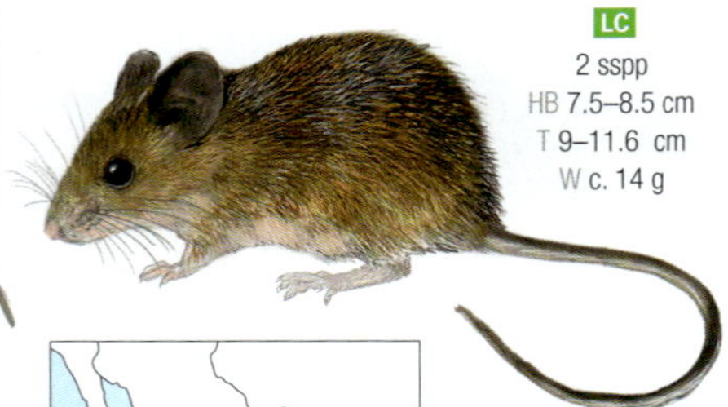

LC
2 sspp
HB 7.5–8.5 cm
T 9–11.6 cm
W c. 14 g

1830–4100 m

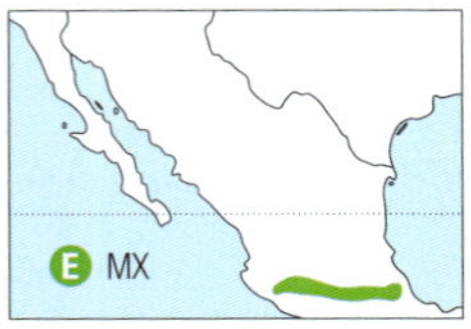

Trans-Mexican Volcanic Belt from SE Jalisco E to WC Veracruz

Volcano Harvest Mouse

Reithrodontomys chrysopsis

FR Souris-moissonneuse des volcans
DE Vulkan-Erntemaus
ES Ratón espiguero de los volcanes

Fulvous Harvest Mouse

Reithrodontomys fulvescens

FR Souris-moissonneuse fauve
DE Gelbliche Erntemaus
ES Ratón espiguero leonado

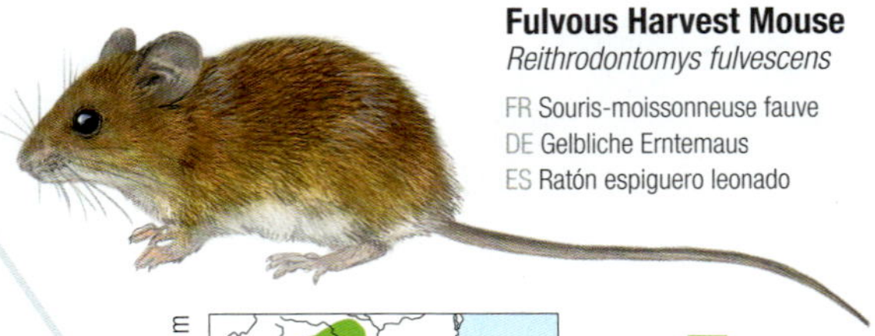

0–2600 m

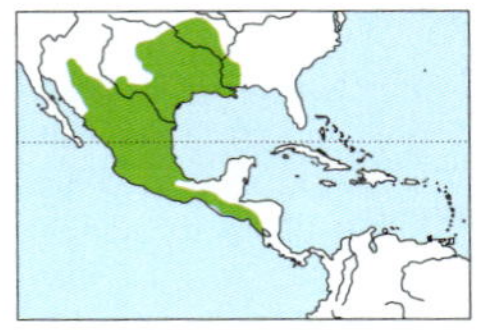

LC
17 sspp
HB 6.2–8.4 cm
T 7.2–11.6 cm
W 8–12 g

US (from S Arizona, SE Kansas & SW Missouri) S through MX to W NI

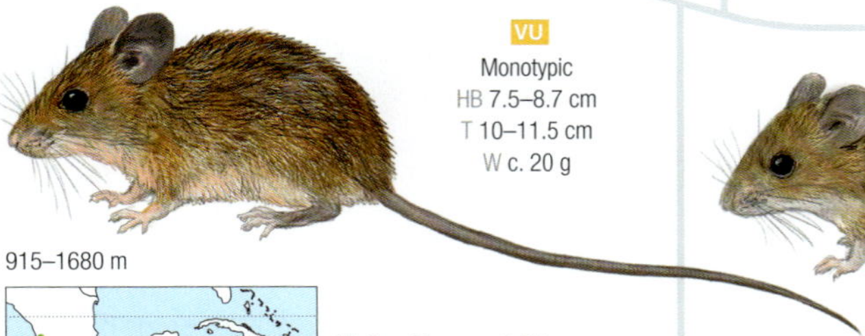

VU
Monotypic
HB 7.5–8.7 cm
T 10–11.5 cm
W c. 20 g

915–1680 m

SE Nayarit, NW Jalisco

Hairy Harvest Mouse

Reithrodontomys hirsutus

FR Souris-moissonneuse hirsute
DE Haarige Erntemaus
ES Ratón espiguero hirsuto

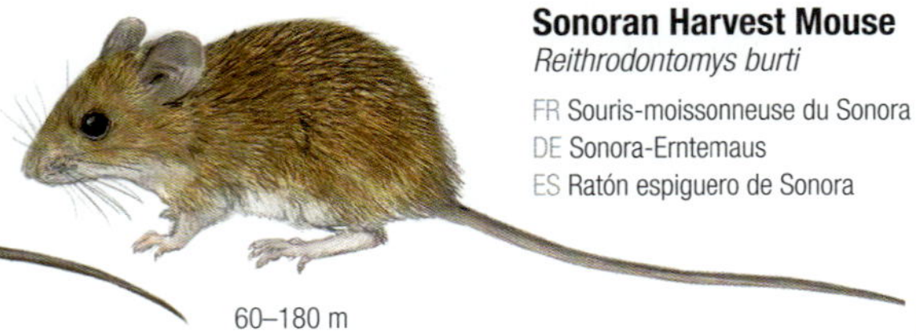

Sonoran Harvest Mouse

Reithrodontomys burti

FR Souris-moissonneuse du Sonora
DE Sonora-Erntemaus
ES Ratón espiguero de Sonora

60–180 m

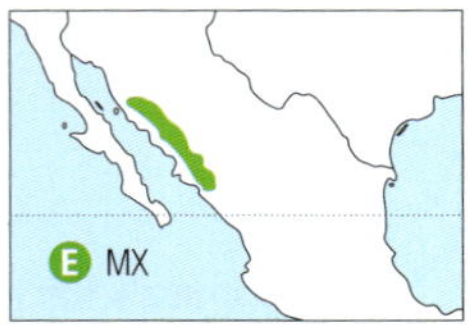

From WC Sonora to N Sinaloa

DD
Monotypic
HB 6.4–7 cm
T 5.2–6.6 cm

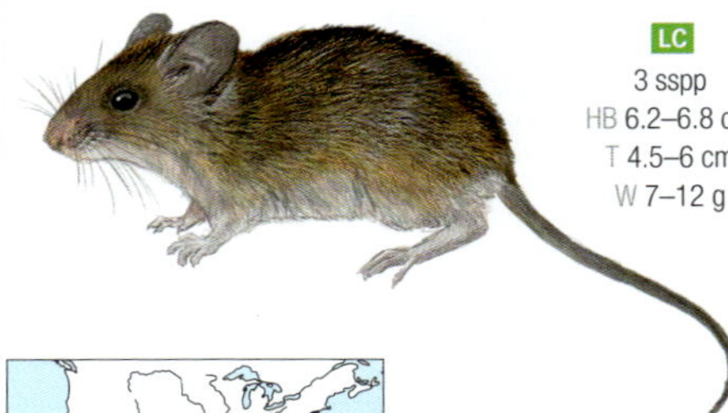

LC
3 sspp
HB 6.2–6.8 cm
T 4.5–6 cm
W 7–12 g

SE US (S from S Ohio & SW Maryland, W to Oklahoma & E Texas)

Eastern Harvest Mouse

Reithrodontomys humulis

FR Souris-moissonneuse de l'Est
DE Östliche Erntemaus
ES Ratón espiguero oriental

Western Harvest Mouse

Reithrodontomys megalotis

FR Souris-moissonneuse de l'Ouest
DE Westliche Erntemaus
ES Ratón espiguero occidental

-75–4000 m

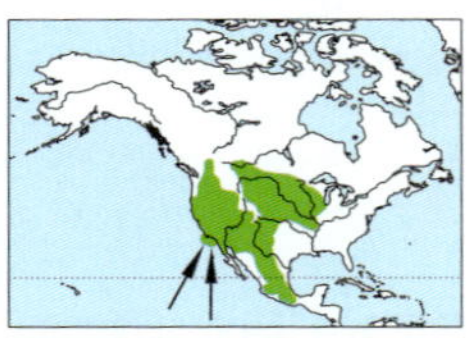

LC
15 sspp
HB 6.3–7.4 cm
T 5.5–9.6 cm
W 8–13 g

From SW CA through W & C US (with Channel Is, California) to S MX (Oaxaca)

LC 3 sspp
HB 5.5–6.5 cm
T 5.4–6.5 cm
W 7–9 g

1200–1920 m

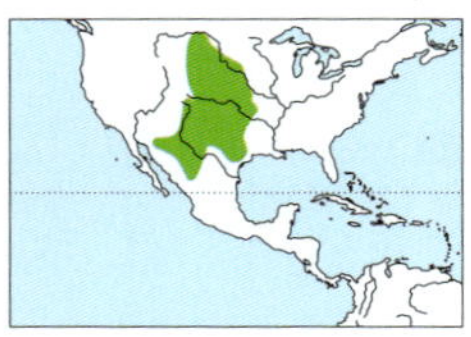

C US (from SE Montana & SW North Dakota) S to N MX (NC Durango)

Plains Harvest Mouse

Reithrodontomys montanus

FR Souris-moissonneuse des plateaux
DE Steppen-Erntemaus
ES Ratón espiguero de llanura

EN 2 sspp
HB 6.2–8 cm
T 5.6–9.5 cm
W 8–15 g

San Francisco Bay (California)

Salt-marsh Harvest Mouse

Reithrodontomys raviventris

FR Souris-moissonneuse à ventre gris
DE Salzsumpf-Erntemaus
ES Ratón espiguero de marisma

LC 7 sspp
HB 7.5–8.8 cm
T 8.1–11.3 cm
W 12–16 g

800–3200 m

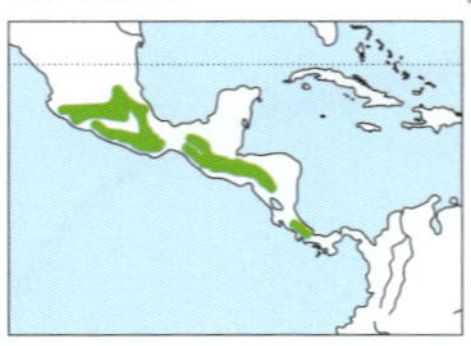

Highlands from C MX (SW Jalisco & Querétaro) to NC NI, C CR to W PA

Sumichrast's Harvest Mouse

Reithrodontomys sumichrasti

FR Souris-moissonneuse de Sumichrast
DE Sumichrast-Erntemaus
ES Ratón espiguero de Sumichrast

Zacatecan Harvest Mouse

Reithrodontomys zacatecae

FR Souris-moissonneuse du Zacatecas
DE Zacateca-Erntemaus
ES Ratón espiguero de Zacatecas

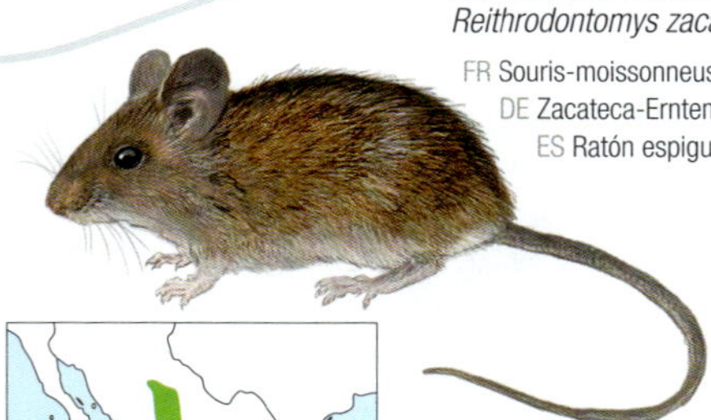

1830–2590 m

Sierra Madre Occidental from W Chihuahua S to WC Michoacán

LC Monotypic
HB 6.1–7 cm
T 6.7–8.7 cm
W 9–13 g

Southern Grasshopper Mouse

Onychomys torridus

FR Onychomys méridional
DE Südliche Grashüpfermaus
ES Ratón saltamontes meridional

0–1400 m

SW US (Nevada, S California, S Arizona), NW MX (N Baja California, Sonora, N Sinaloa)

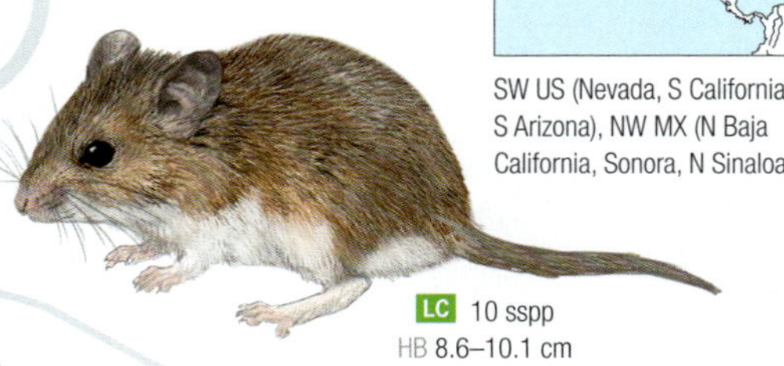

LC 10 sspp
HB 8.6–10.1 cm
T 3.3–6.2 cm

Genus *ONYCHOMYS*

(3 species)

1300–1580 m

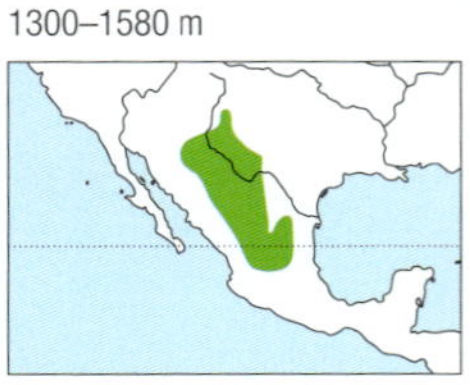

US (SE Arizona, SC New Mexico, W Texas), NC MX (S to Aguascalientes & San Luis Potosí)

LC 3 sspp
HB 8.5–10.4 cm
T 4.6–5.5 cm
W 21–29 g

Chihuahuan Grasshopper Mouse

Onychomys arenicola

FR Onychomys des sables
DE Chihuahua-Grashüpfermaus
ES Ratón saltamontes de Chihuahua

LC 13 sspp
HB 9–12.8 cm
T 2.9–6.2 cm
W 30–60 g

Northern Grasshopper Mouse

Onychomys leucogaster

FR Onychomys boréal
DE Nördliche Grashüpfermaus
ES Ratón saltamontes septentrional

SW CA (S Alberta, S Saskatchewan & S Manitoba) through W & C US to N MX (N Tamaulipas)

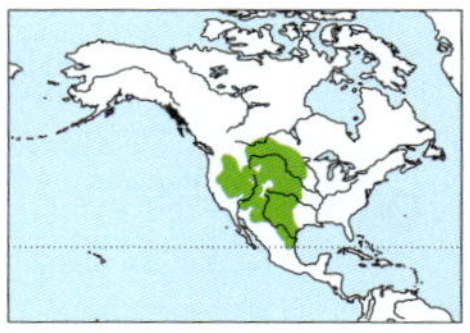

150–1400 m

Genus *OSGOODOMYS*

(1 species)

0–1400 m

Osgood's Deermouse
Osgoodomys banderanus

LC
2 sspp
HB 10.4–14.9 cm
T 9.6–12.5 cm
W 39–58 g

FR Péromyscus d'Osgood
DE Osgood-Hirschmaus
ES Ratón ciervo de Osgood

From Nayarit to E Michoacán and W Guerrero

Genus *PEROMYSCUS*

(80 species)

1155–2100 m

Hooper's Deermouse
Peromyscus hooperi

FR Péromyscus de Hooper
DE Hooper-Hirschmaus
ES Ratón ciervo de Hooper

LC
Monotypic
HB 8–8.6 cm
T 9.2–13.2 cm
W 20–24 g

From WC Coahuila S to NE Zacatecas & N San Luis Potosí

0–3230 m

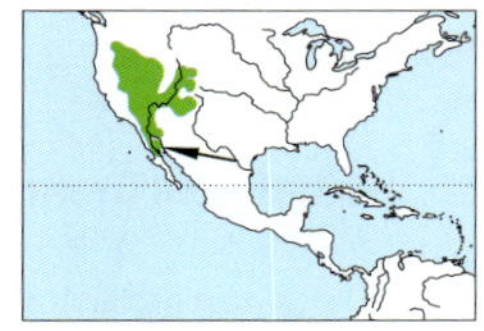

Canyon Deermouse
Peromyscus crinitus

LC 8 sspp
HB 7.3–8.5 cm
T 8.2–11.8 cm
W 17–24 g

FR Péromyscus des canyons
DE Canyon-Hirschmaus
ES Ratón ciervo de cañones

US (from Oregon, S Idaho & SW Wyoming) S to MX (N Baja California, NW Sonora)

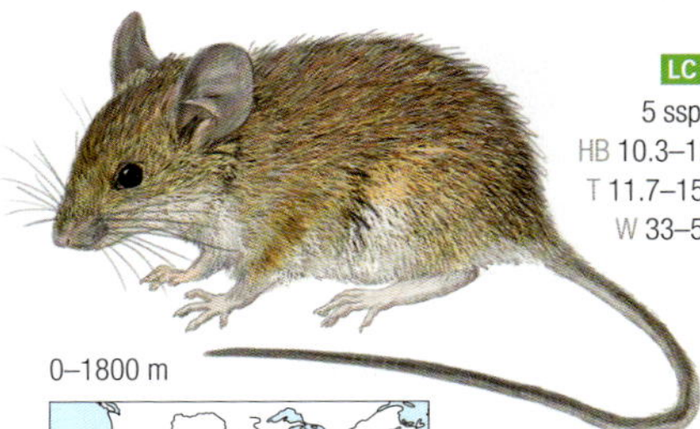

LC
5 sspp
HB 10.3–12.9 cm
T 11.7–15.6 cm
W 33–55 g

0–1800 m

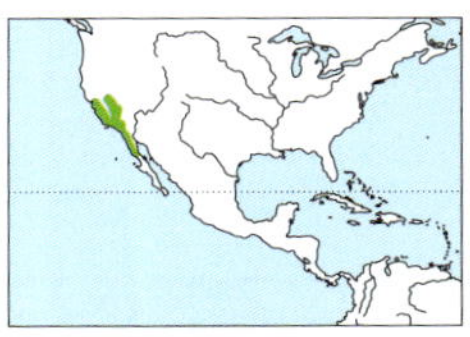

California Deermouse
Peromyscus californicus

FR Péromyscus de Californie
DE Kalifornien-Hirschmaus
ES Ratón ciervo de California

US (SW & C California), MX (NW Baja California)

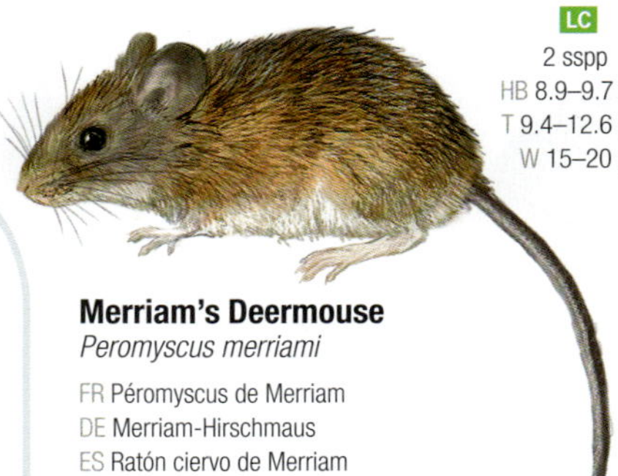

LC
2 sspp
HB 8.9–9.7 cm
T 9.4–12.6 cm
W 15–20 g

Merriam's Deermouse
Peromyscus merriami

FR Péromyscus de Merriam
DE Merriam-Hirschmaus
ES Ratón ciervo de Merriam

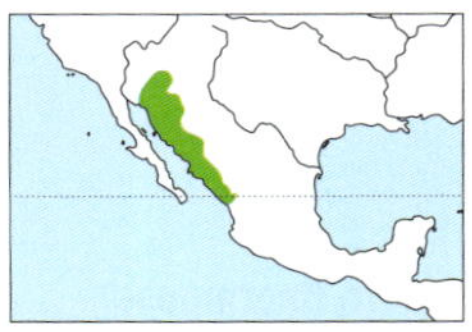

0–800 m

US (S Arizona), MX (W Sonora, N Sinaloa)

LC 17 sspp
HB 7.7–10 cm T 8.7–11.8 cm
W 13–22 g

0–2330 m

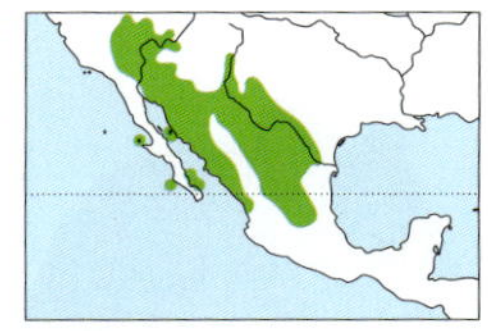

SW US, NC & NW MX (with several Is in Gulf of California)

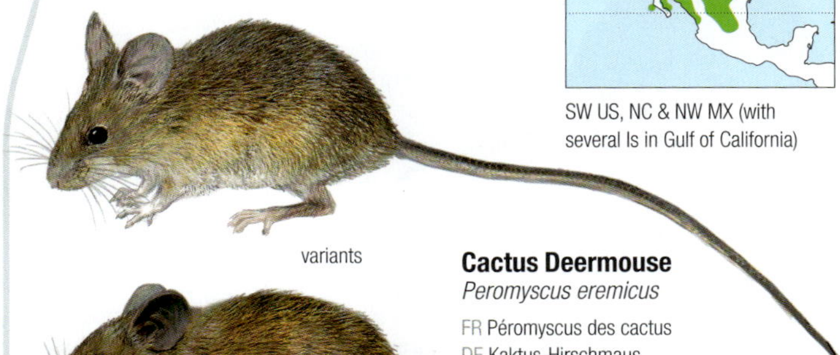

variants

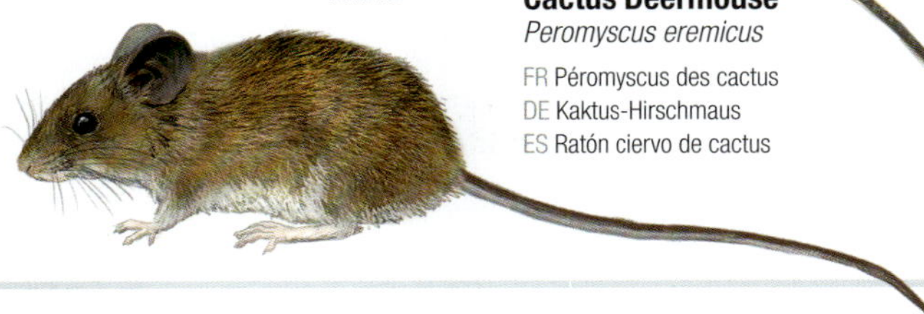

Cactus Deermouse
Peromyscus eremicus

FR Péromyscus des cactus
DE Kaktus-Hirschmaus
ES Ratón ciervo de cactus

LC 2 sspp
HB 8.4–9 cm T 10–12.8 cm
W 13–20 g

0–1650 m

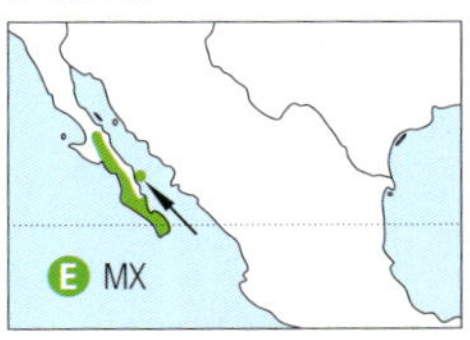

Southern Baja Deermouse
Peromyscus eva

FR Péromyscus eva
DE Südliche Baja-Hirschmaus
ES Ratón ciervo de Baja California meridional

Extreme S Baja California & Baja California Sur (with Carmen I)

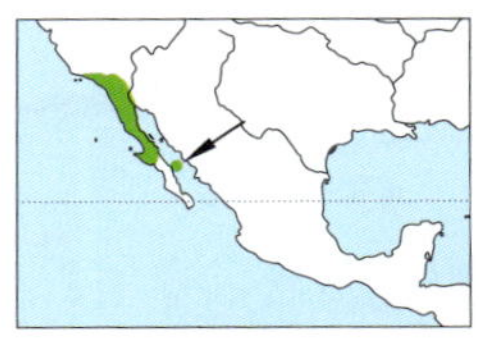

0–2300 m

US (SW California), MX (N & C Baja California Pen, with Coronado Is)

Northern Baja Deermouse
Peromyscus fraterculus

FR Péromyscus de San Diego
DE Nördliche Baja-Hirschmaus
ES Ratón ciervo de Baja California septentrional

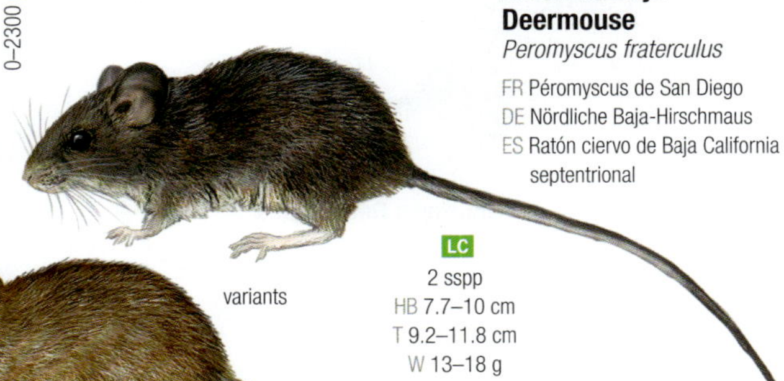

variants

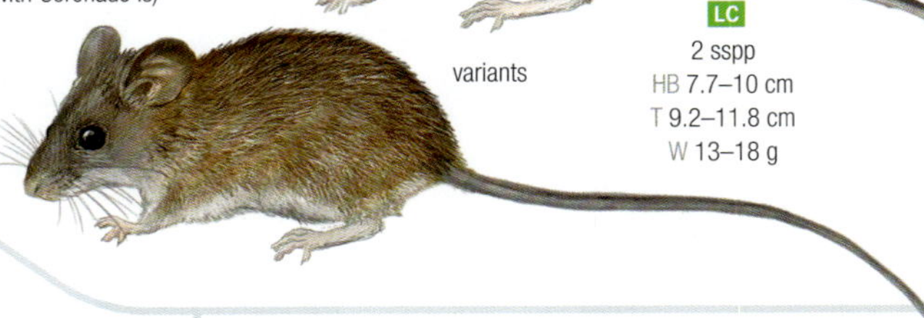

LC
2 sspp
HB 7.7–10 cm
T 9.2–11.8 cm
W 13–18 g

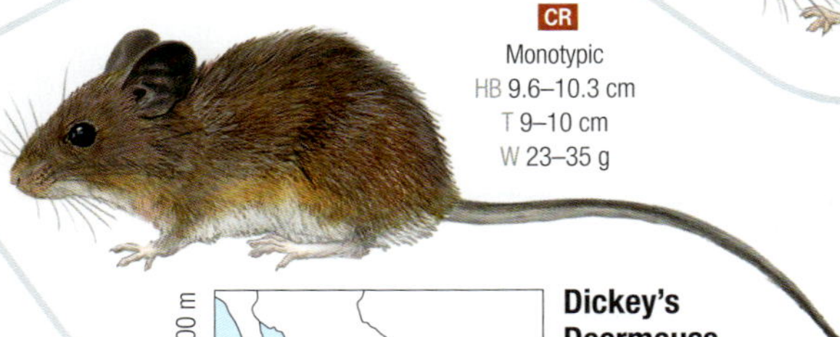

CR
Monotypic
HB 9.6–10.3 cm
T 9–10 cm
W 23–35 g

0–300 m

Dickey's Deermouse
Peromyscus dickeyi

FR Péromyscus de Dickey
DE Dickey-Hirschmaus
ES Ratón ciervo de Dickey

Tortuga I (Baja California Sur)

Monserrat Island Deermouse
Peromyscus caniceps

FR Péromyscus de Monserrate
DE Montserrat-Hirschmaus
ES Ratón ciervo de Montserrat

CR
Monotypic
HB 8–9.6 cm
T 10–12.4 cm
W 13–25 g

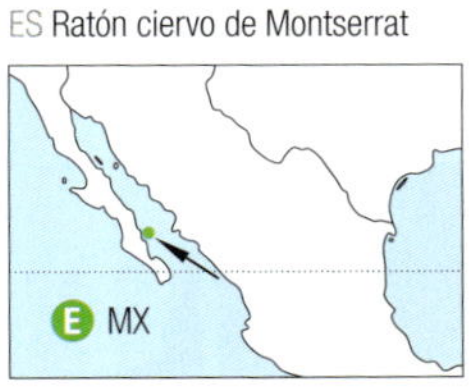

0–220 m

Montserrat I (Baja California Sur)

La Guarda Deermouse
Peromyscus guardia

FR Péromyscus d'Angel de la Guarda
DE Guarda-Hirschmaus
ES Ratón ciervo de La Guarda

CR(PE)
1 extant ssp
HB 9.6–10 cm
T 9.3–12.3 cm

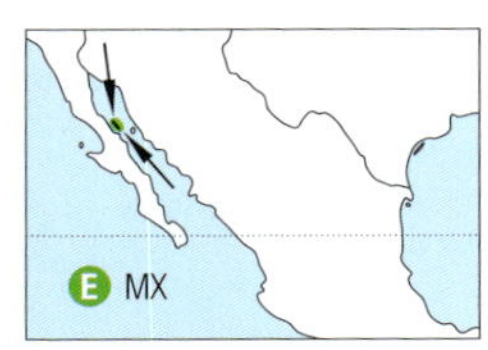

Ángel de la Guarda I and nearby Estanque I (Baja California)

SE Virginia S to Florida, W to Texas & Oklahoma, N to Missouri, Illinois & Kentucky

Cotton Deermouse
Peromyscus gossypinus

FR Péromyscus des cotonniers
DE Baumwoll-Hirschmaus
ES Ratón ciervo de algodón

LC
7 sspp
HB 7.1–11.6 cm
T 5.5–9.7 cm
W 17–46 g

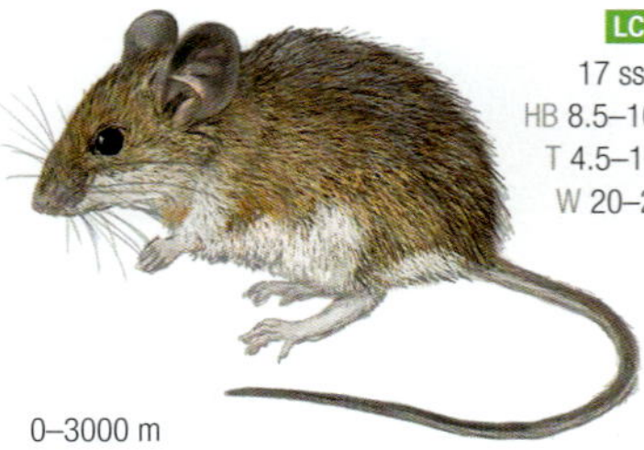

LC
17 sspp
HB 8.5–10.5 cm
T 4.5–10 cm
W 20–23 g

0–3000 m

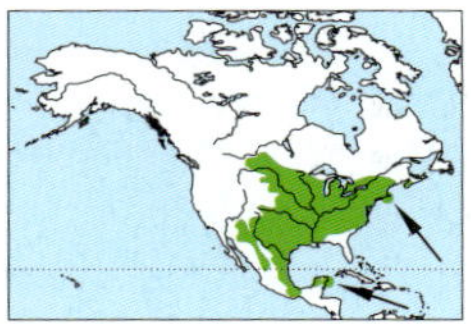

White-footed Deermouse
Peromyscus leucopus

FR Péromyscus à pattes blanches
DE Weißfuß-Hirschmaus
ES Ratón ciervo de pies blancos

SC & SE CA, C & E US (except SE coastal plains and Florida) S to S MX

Santa Cruz Deermouse
Peromyscus sejugis

FR Péromyscus de Santa Cruz
DE Santa-Cruz-Hirschmaus
ES Ratón ciervo de Santa Cruz

0–100 m

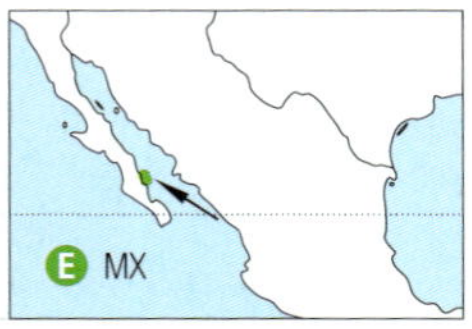

Santa Cruz I and nearby San Diego I (Baja California Sur)

EN
Monotypic
HB 9.5–10.3 cm
T 6.5–9.4 cm
W 14–25 g

Gambel's Deermouse
Peromyscus gambelii

FR Péromyscus de Gambel
DE Gambel-Hirschmaus
ES Ratón ciervo de Gambel

NE
Monotypic
W c. 20 g

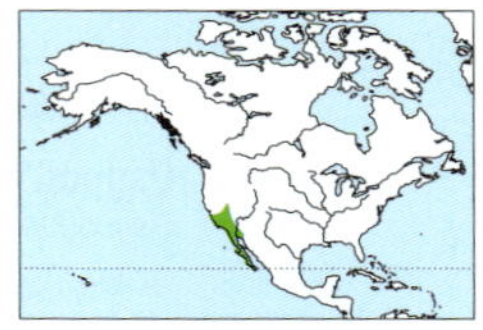

SW US (C & S California, Nevada, SW Arizona) S to MX (Baja California Pen)

North-western Deermouse
Peromyscus keeni

FR Péromyscus de Keen
DE Nordwestliche Hirschmaus
ES Ratón ciervo noroccidental

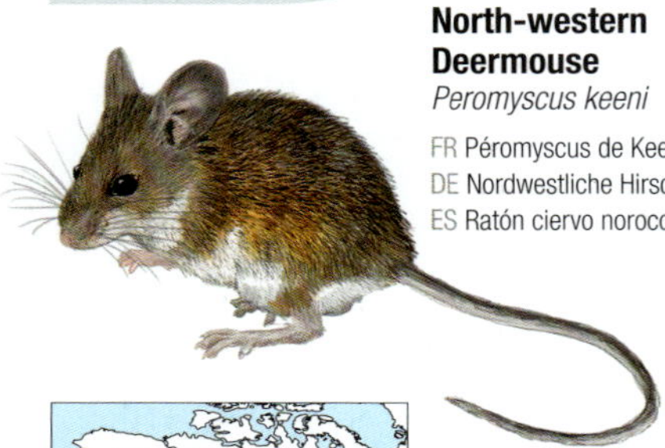

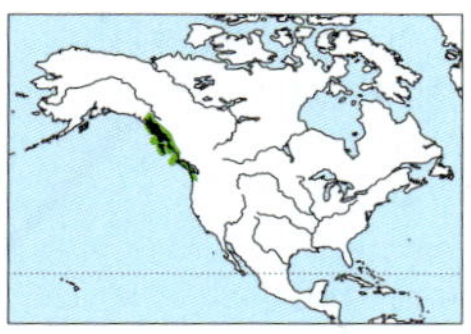

LC 18 sspp
HB 7–13.7 cm
T 7–12.6 cm
W 15–52 g

From S Alaska & SW Yukon across W CA (including coastal Is) to W Washington (US)

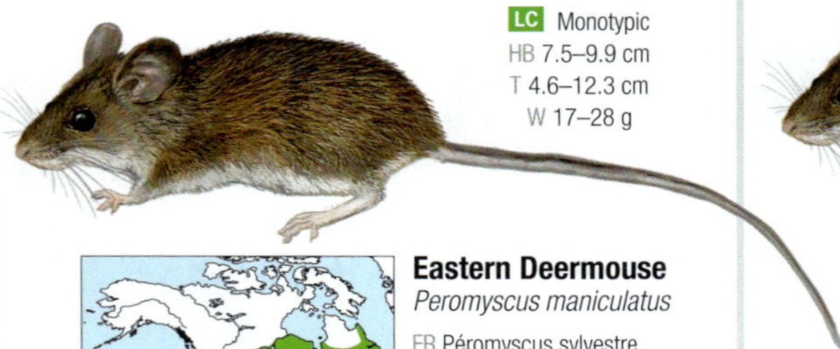

LC Monotypic
HB 7.5–9.9 cm
T 4.6–12.3 cm
W 17–28 g

0–3800 m

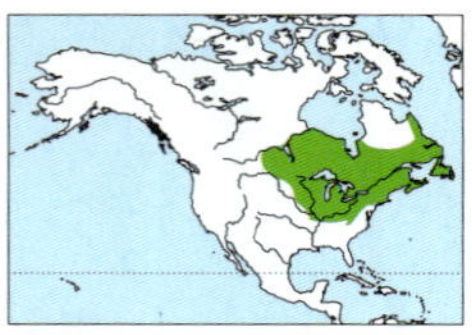

Eastern Deermouse
Peromyscus maniculatus

FR Péromyscus sylvestre
DE Östliche Hirschmaus
ES Ratón ciervo oriental

E CA (Manitoba E to Newfoundland), E US (S to Tennessee & North Carolina)

NE
Monotypic
HB 9.3–11.4 cm
T 5.9–8 cm

2400–3400 m

Surroundings of Nahuatzen (Michoacán)

Purepechus Deermouse
Peromyscus purepechus

FR Péromyscus purépecha
DE Purepechus-Hirschmaus
ES Ratón ciervo de Purépecha

Western Deermouse
Peromyscus sonoriensis

FR Péromyscus du Sonora
DE Westliche Hirschmaus
ES Ratón ciervo occidental

NE
Monotypic
HB 7.5–9.9 cm
T 4.6–12.3 cm
W 17–28 g

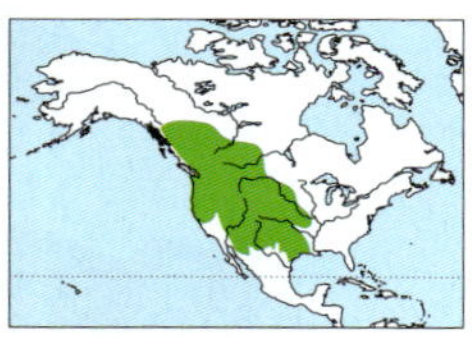

From Yukon (CA) S through W US (E to R Mississippi) to N Sonora (MX)

Yukon Deermouse
Peromyscus arcticus

FR Péromyscus du Yukon
DE Yukon-Hirschmaus
ES Ratón ciervo del Yukón

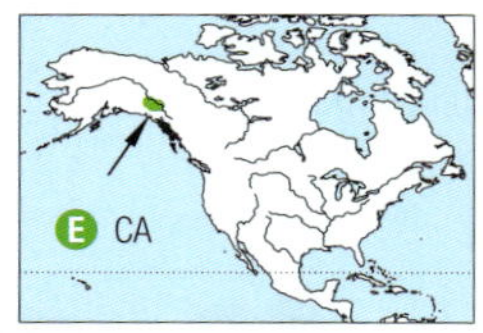

SW Yukon; likely W Northwest Territories

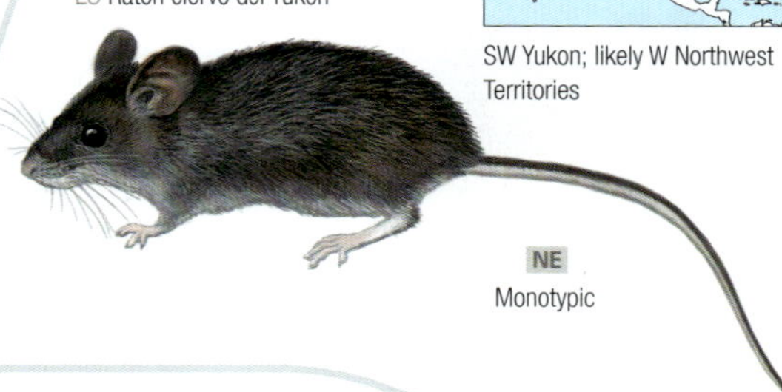

NE
Monotypic

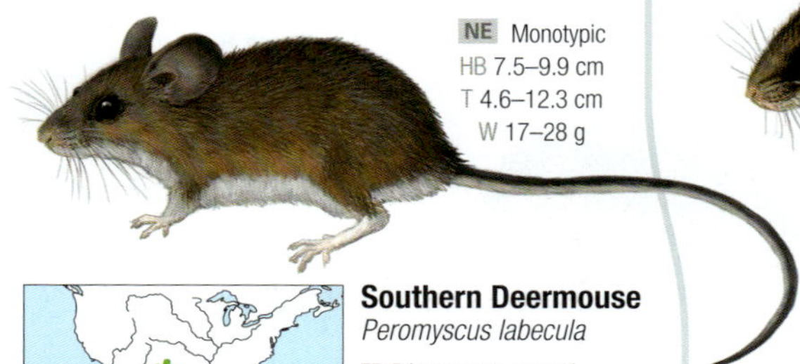

NE Monotypic
HB 7.5–9.9 cm
T 4.6–12.3 cm
W 17–28 g

Southern Deermouse
Peromyscus labecula

FR Péromyscus maculé
DE Südliche Hirschmaus
ES Ratón ciervo meridional

SC US (S New Mexico, W Texas) S to Oaxaca (MX)

EN
Monotypic
HB 11.1–13.3 cm
T 12.7–14.5 cm
W 40–60 g

Black-tailed Deermouse
Peromyscus melanurus

FR Péromyscus à queue noire
DE Schwarzschwanz-Hirschmaus
ES Ratón ciervo de cola negra

700–1900 m

Sierra Madre del Sur (Oaxaca)

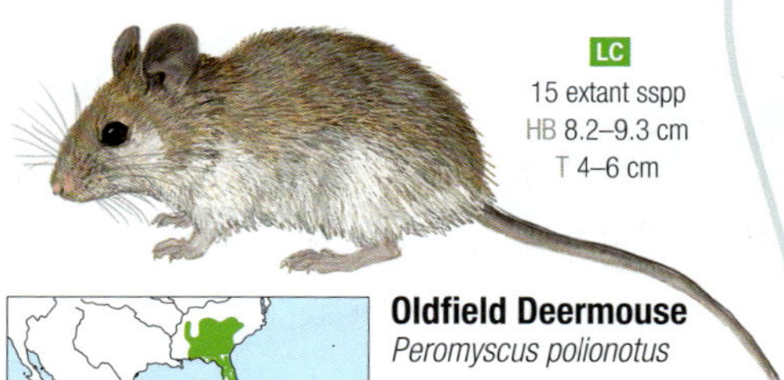

LC
15 extant sspp
HB 8.2–9.3 cm
T 4–6 cm

Oldfield Deermouse
Peromyscus polionotus

FR Péromyscus champêtre
DE Küsten-Hirschmaus
ES Ratón ciervo de erial

E Mississippi, Alabama, Georgia, S South Carolina, most of Florida Pen

LC Monotypic
HB 9.1–10.2 cm T 4.9–7.5 cm
W 17–28 g

1100–4300 m

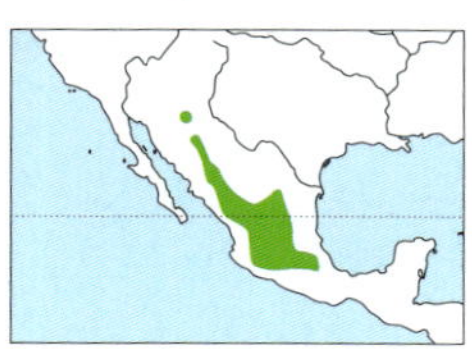

Black-eared Deermouse
Peromyscus melanotis

FR Péromyscus à oreilles noires
DE Schwarzohr-Hirschmaus
ES Ratón ciervo de orejas negras

US (SE Arizona), MX (Sierra Madre Occidental & Oriental and Mexican Plateau to Puebla)

1500–3000 m

Narrow band from S México, C & E Guerrero and Oaxaca

LC 2 sspp
HB 11.2–14.1 cm
T 11.6–15.1 cm
W 30–50 g

Broad-faced Deermouse
Peromyscus megalops

FR Péromyscus à tête large DE Breitgesicht-Hirschmaus
ES Ratón ciervo de cara ancha

Black-wristed Deermouse
Peromyscus melanocarpus

FR Péromyscus du Zempoaltepec
DE Schwarzhand-Hirschmaus
ES Ratón ciervo de muñeca negra

900–2800 m

EN
Monotypic
HB 10–13 cm
T 10–13.3 cm
W c. 59 g

Mt Zempoaltépec and Sierra Juárez (NC Oaxaca)

CR
Monotypic
HB 10.5–11.3 cm
T 9.7–10.9 cm

0–100 m

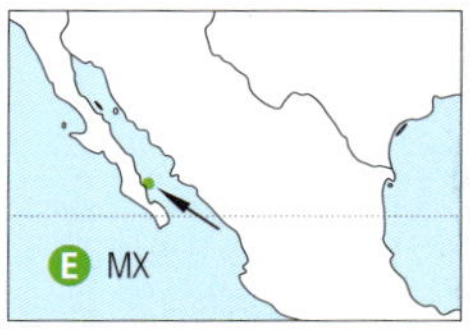

Catalina Deermouse
Peromyscus slevini
FR Péromyscus de Santa Catalina
DE Catalina-Hirschmaus
ES Ratón ciervo de Catalina

Santa Catalina I (Baja California Sur)

LC Monotypic
HB 9.2–12.9 cm
T 12.5–13.8 cm
W 25–40 g

0–1300 m

Tawny Deermouse
Peromyscus perfulvus
FR Péromyscus cannelle
DE Sumpf-Hirschmaus
ES Ratón ciervo de Guerrero

C & S Michoacán, W Guerrero

Jalisco Deermouse
Peromyscus chrysopus
FR Péromyscus du Jalisco
DE Jalisco-Hirschmaus
ES Ratón ciervo de Colima

NE
Monotypic
HB 7.4–13.6 cm
T 11–13.4 cm

Coastal Jalisco, Colima, extreme SW Michoacán

Plateau Deermouse
Peromyscus melanophrys
FR Péromyscus des plateaux
DE Plateau-Hirschmaus
ES Ratón ciervo de altiplano

50–2700 m

Morelos, Guerrero, Puebla, Oaxaca

LC Monotypic
HB 7.3–12 cm T 11.2–17.2 cm W 26–58 g

NE Monotypic
HB 7.3–12 cm
T 11.2–17.2 cm
W 26–58 g

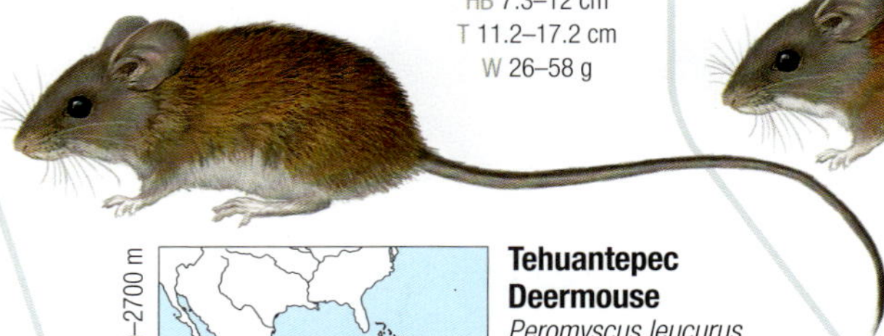

50–2700 m

Tehuantepec Deermouse
Peromyscus leucurus
FR Péromyscus de Tehuantepec
DE Tehuantepec-Hirschmaus
ES Ratón ciervo de Tehuantepec

River basins in Morelos, Puebla, Oaxaca & Chiapas

Small-footed Deermouse
Peromyscus micropus
FR Péromyscus à pattes courtes
DE Kleinfuß-Hirschmaus
ES Ratón ciervo de pies pequeños

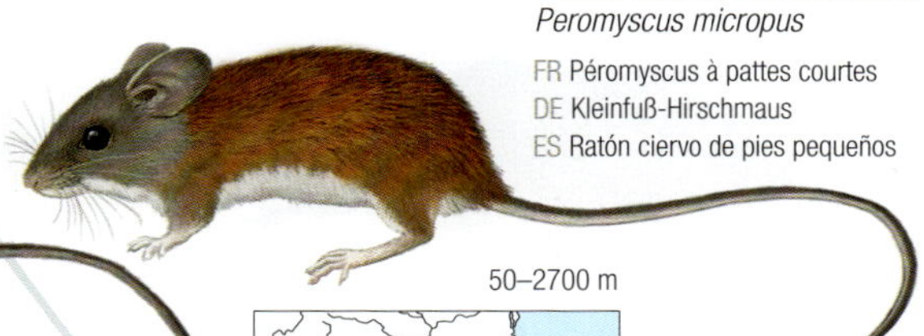

50–2700 m

Durango, Nayarit, Jalisco, Zacatecas, Aguascalientes

NE
Monotypic
HB 7.3–12 cm
T 11.2–17.2 cm
W 26–58 g

50–2700 m

Zamora Deermouse
Peromyscus zamorae
FR Péromyscus de Zamora
DE Zamora-Hirschmaus
ES Ratón ciervo de Zamora

S edge of Chihuahuan Desert and Mexican Plateau S to México & Hidalgo

NE Monotypic
HB 7.3–12 cm
T 11.2–17.2 cm
W 26–58 g

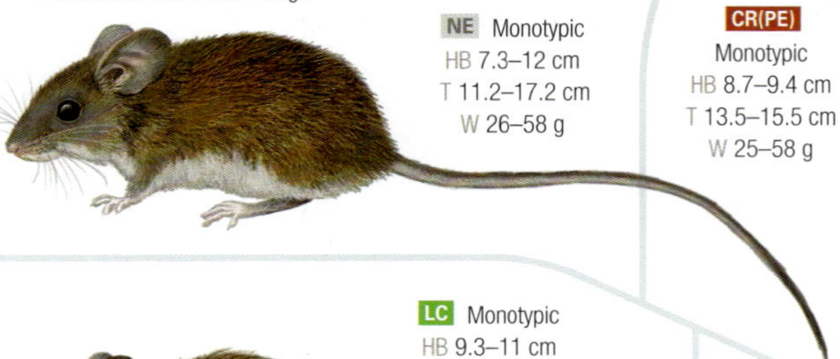

Puebla Deermouse
Peromyscus mekisturus
FR Péromyscus du Puebla
DE Puebla-Hirschmaus
ES Ratón ciervo de Puebla

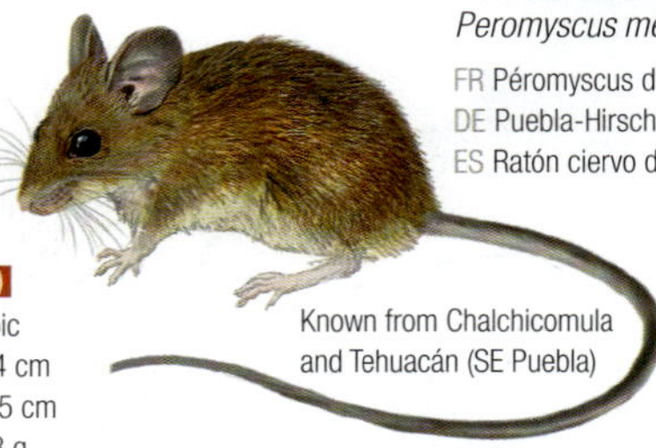

CR(PE)
Monotypic
HB 8.7–9.4 cm
T 13.5–15.5 cm
W 25–58 g

Known from Chalchicomula and Tehuacán (SE Puebla)

1700–2545 m

CR
Monotypic
HB 10.7–12.5 cm
T 10.2–12 cm

c. 2950 m

Known from Santa Eulalia (Huehuetenango)

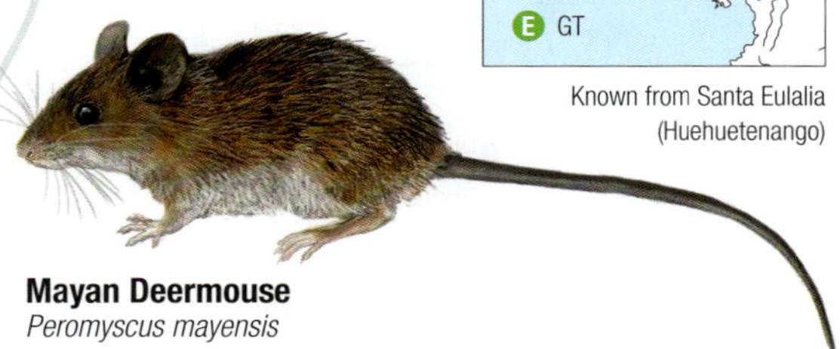

Mayan Deermouse
Peromyscus mayensis
FR Péromyscus maya DE Maya-Hirschmaus ES Ratón ciervo de los Maya

LC Monotypic
HB 9.3–11 cm
T 8.6–10.8 cm
W 28–31 g

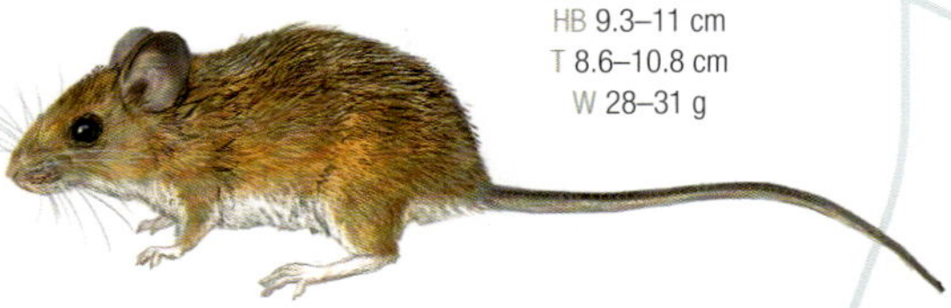

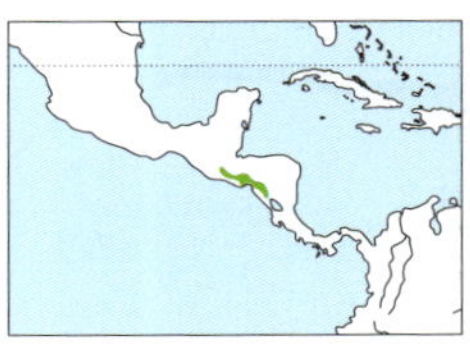

Stirton's Deermouse
Peromyscus stirtoni
FR Péromyscus de Stirton
DE Stirton-Hirschmaus
ES Ratón ciervo de Stirton

Pacific lowlands of SE GT, SV, S HN & NW NI

0–500 m

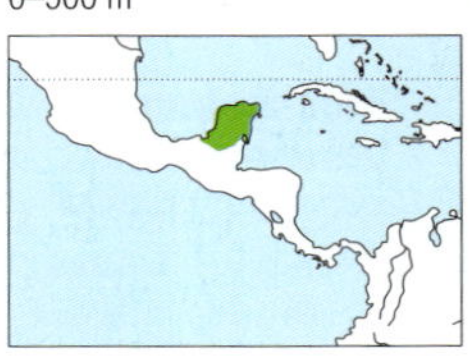

Yucatan Deermouse
Peromyscus yucatanicus
FR Péromyscus du Yucatan
DE Yucatán-Hirschmaus
ES Ratón ciervo de Yucatán

LC
2 sspp
HB 9.7–10.2 cm
T 8.4–11.7 cm
W 15–18 g

MX (Yucatán, Campeche, Quintana Roo), N GT

NE
Monotypic

Chimoxan Deermouse
Peromyscus tropicalis
FR Péromyscus de Chimoxan
DE Chimoxán-Hirschmaus
ES Ratón ciervo de Chimoxán

Low elevation mts in Caribbean slope

Talamancan Deermouse
Peromyscus nudipes
FR Péromyscus de la Talamanca
DE Talamanca-Hirschmaus
ES Ratón ciervo de Talamanca

NE
Monotypic
HB 11.1–13.7 cm
T 10.3–14 cm

Cordillera de Talamanca (S CR, NW PA)

above 2000 m

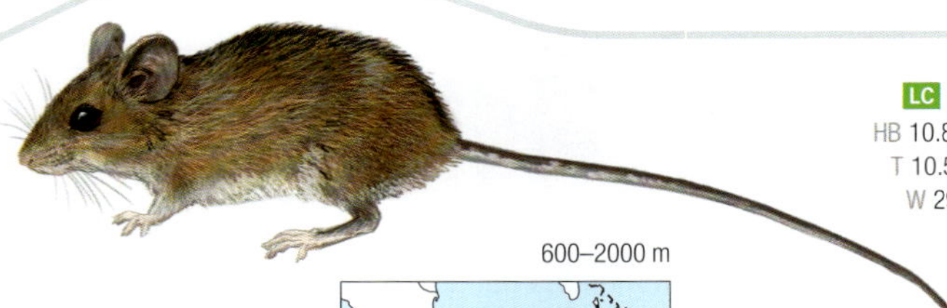

LC 7 sspp
HB 10.8–13.7 cm
T 10.5–14 cm
W 29–50 g

600–2000 m

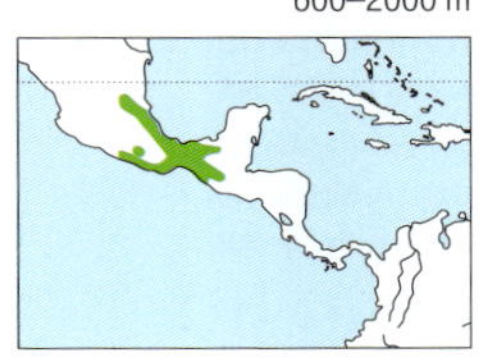

Mexican Deermouse
Peromyscus mexicanus
FR Péromyscus du Mexique
DE Mexiko-Hirschmaus
ES Ratón ciervo de México

MX (from S San Luis Potosí to Tabasco & from SE Guerrero to Chiapas), W GT

Naked-eared Deermouse
Peromyscus gymnotis

FR Péromyscus à oreilles nues
DE Nacktohr-Hirschmaus
ES Ratón ciervo de orejas desnudas

50–1175 m

LC
Monotypic
HB 11.2–13.3 cm
T 9–11 cm
W 31–50 g

MX (extreme SE Chiapas), S GT, S SV, S HN, W NI

VU Monotypic
HB 14.6–14.9 cm
T 15.7–17.8 cm
W 45–90 g

1400–2900 m

Chiapan Deermouse
Peromyscus zarhynchus

FR Péromyscus du Chiapas
DE Chiapas-Hirschmaus
ES Ratón ciervo de Chiapas

Forests in C Chiapas

NE
Monotypic
HB 14.5–14.7 cm
T 12.9–18 cm

1000–1700 m

Gardner's Deermouse
Peromyscus gardneri

FR Péromyscus de Gardner
DE Gardner-Hirschmaus
ES Ratón ciervo de Gardner

Along N slope of Sierra de los Cuchumatanes

NE
Monotypic

Nicaraguan Deermouse
Peromyscus nicaraguae

FR Péromyscus du Nicaragua
DE Nicaragua-Hirschmaus
ES Ratón ciervo de Nicaragua

Tropical deciduous forests in HN & NI

Salvadorean Deermouse
Peromyscus salvadorensis

FR Péromyscus du Salvador
DE El-Salvador-Hirschmaus
ES Ratón ciervo de El Salvador

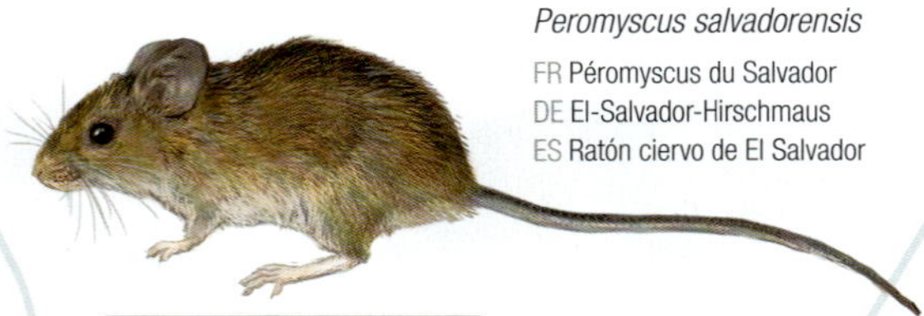

NE
Monotypic

Tropical deciduous forests in GT, SV & S HN

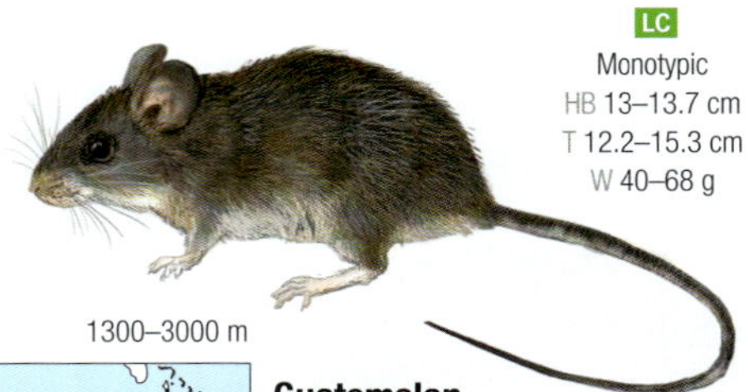

LC
Monotypic
HB 13–13.7 cm
T 12.2–15.3 cm
W 40–68 g

1300–3000 m

Guatemalan Deermouse
Peromyscus guatemalensis

FR Péromyscus du Guatemala
DE Guatemala-Hirschmaus
ES Ratón ciervo de Guatemala

MX (Sierra Madre of SE Chiapas), SW GT

NE
Monotypic

Baker's Deermouse
Peromyscus bakeri

FR Péromyscus de Baker
DE Baker-Hirschmaus
ES Ratón ciervo de Baker

Known only from Fuentes Georginas (Quetzaltenango)

NE
Monotypic

Carol Patton's Deermouse
Peromyscus carolpattonae

FR Péromyscus de Patton
DE Patton-Hirschmaus
ES Ratón ciervo de Patton

MX (Sierra Madre of S Chiapas), SW GT (W Central American Volcano Arc)

Large Deermouse
Peromyscus grandis

FR Grand Péromyscus DE Große Hirschmaus
ES Ratón ciervo grande

NT
Monotypic
HB 13.2–16.5 cm
T 14.2–16 cm

1200–2700 m

Between Tucurú and Purulhá & Sierra de las Minas

DD
Monotypic
HB 11.5–13.8 cm
T 11.4–16.2 cm
W 40–60 g

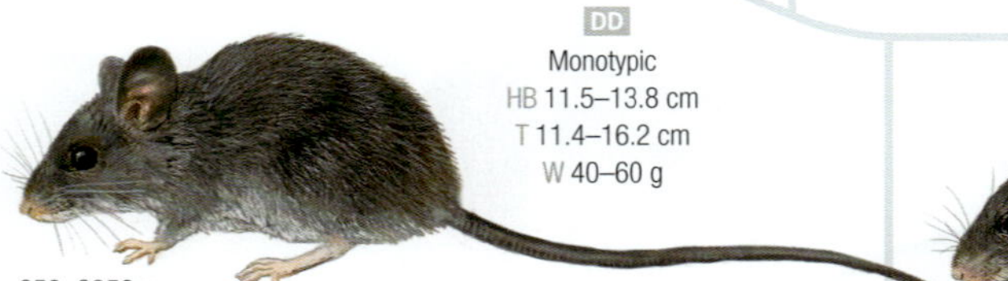

650–2950 m

Blackish Deermouse
Peromyscus furvus

FR Péromyscus noir
DE Schwärzliche Hirschmaus
ES Ratón ciervo negruzco

E slopes of Sierra Madre Oriental from Hidalgo S to N Oaxaca

NE
Monotypic

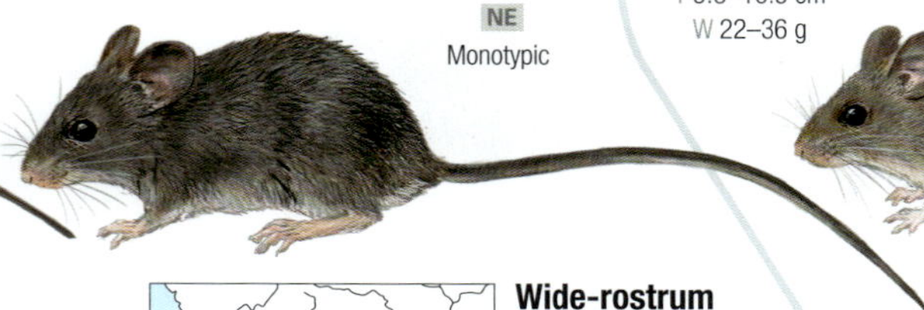

220–1400 m

Wide-rostrum Deermouse
Peromyscus latirostris

FR Péromyscus de Dalquest
DE Breitnasen-Hirschmaus
ES Ratón ciervo de rostro ancho

SE San Luis Potosí and NE Querétaro

DD
Monotypic
HB 9.2–10.1 cm
T 9.8–10.9 cm
W 22–36 g

Michoacan Deermouse
Peromyscus sagax

FR Péromyscus du Michoacan
DE Michoacán-Hirschmaus
ES Ratón ciervo de Michoacán

1300–1700 m

NW Michoacán (Los Reyes)

0–600 m

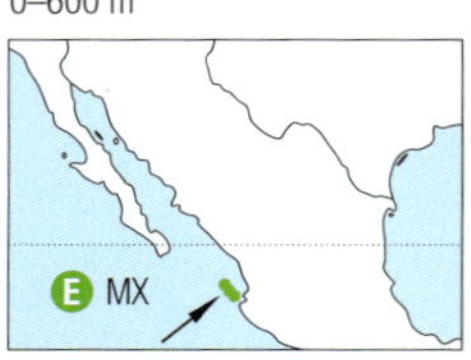

Tres Marias Deermouse
Peromyscus madrensis

FR Péromyscus des Marias
DE Tres-Marías-Hirschmaus
ES Ratón ciervo de Tres Marías

Marías Is (María Madre, María Cleofas, San Juanito), off Nayarit

EN
Monotypic
HB 10.4–12 cm
T 9.9–13 cm

CR
Monotypic
HB 8.3–9.5 cm
T 10–12.4 cm
W 16–25 g

San Esteban Deermouse
Peromyscus stephani

FR Péromyscus de San Esteban
DE San-Esteban-Hirschmaus
ES Ratón ciervo de San Esteban

0–540 m

San Esteban I, Gulf of California

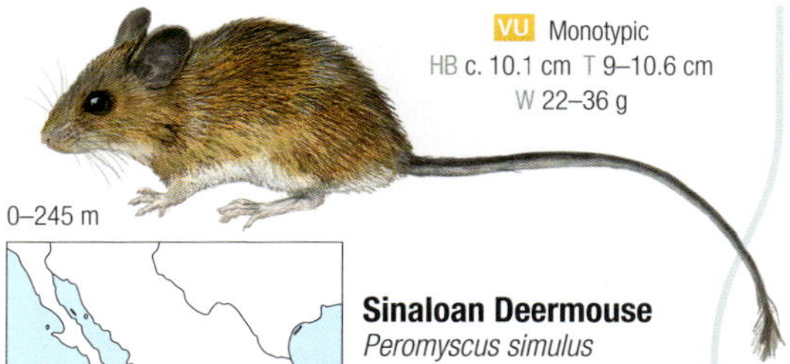

VU Monotypic
HB c. 10.1 cm T 9–10.6 cm
W 22–36 g

0–245 m

Sinaloan Deermouse
Peromyscus simulus
FR Péromyscus de San Blas
DE Sinaloa-Hirschmaus
ES Ratón ciervo de Sinaloa

Coastal areas of S Sinaloa and Nayarit

LC 4 sspp
HB 8.6–9.5 cm
T 8.9–11.5 cm
W 19–30 g

915–2590 m

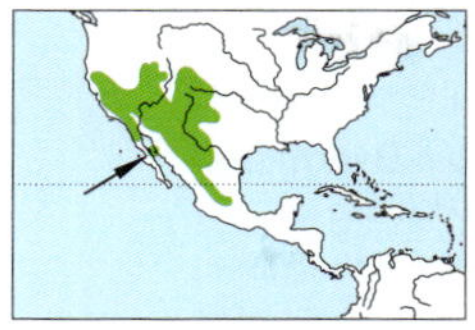

Brush Deermouse
Peromyscus boylii
FR Péromyscus de Boyle
DE Unterholz-Hirschmaus
ES Ratón ciervo cepillo

W US (S from N California) to C MX (Querétaro; also San Pedro Nolasco I)

Durango Deermouse
Peromyscus schmidlyi
FR Péromyscus de Schmidly
DE Durango-Hirschmaus
ES Ratón ciervo de Durango

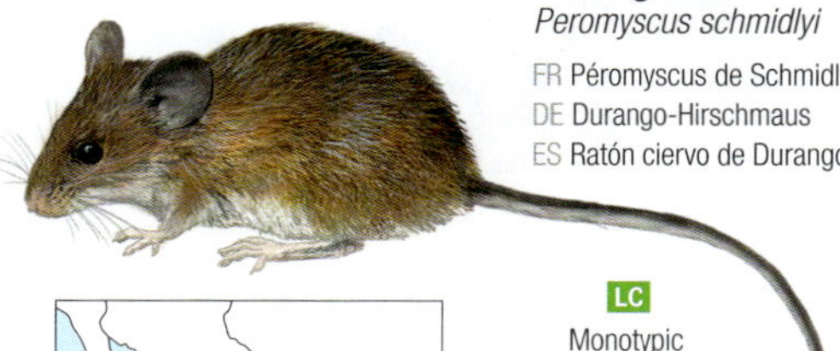

above 2000 m

LC
Monotypic
HB 9.5–11.9 cm
T 8.3–10.1 cm
W 19–30 g

Sierra Madre Occidental of E Sonora to E Sinaloa; possibly NE Nayarit

LC
2 sspp
HB 8.9–10.5 cm
T 9.1–11.5 cm
W 17–24 g

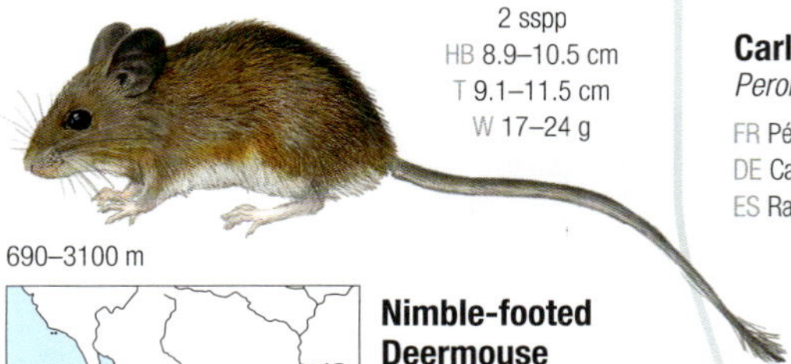

690–3100 m

Nimble-footed Deermouse
Peromyscus levipes
FR Péromyscus agile
DE Zartfuß-Hirschmaus
ES Ratón ciervo de pies ligeros

From extreme S Coahuila & S Tamaulipas S to N Puebla & C Morelos

Carleton's Deermouse
Peromyscus carletoni
FR Péromyscus de Carleton
DE Carleton-Hirschmaus
ES Ratón ciervo de Carleton

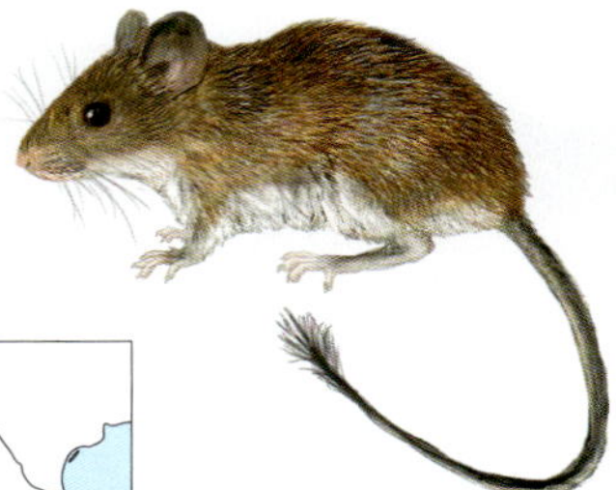

above 2000 m

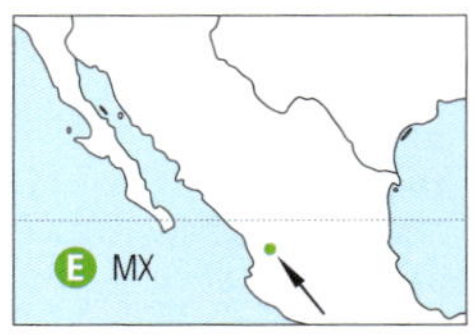

NE
Monotypic
HB 8.8–10.1 cm
T 8.7–11.4 cm

Sierra Madre Occidental of EC Nayarit; possibly W Zacatecas

Ensink's Deermouse
Peromyscus ensinki
FR Péromyscus d'Ensink
DE Ensink-Hirschmaus
ES Ratón ciervo de Ensink

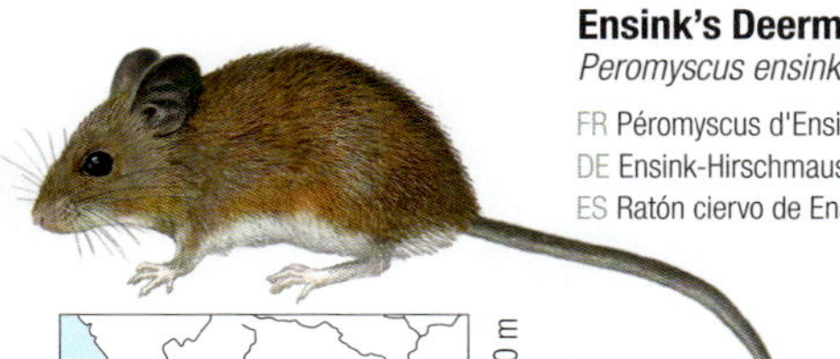

c. 2010 m

NE
Monotypic
HB c. 10.9 cm
T c. 10.8 cm

Mesic pine/oak forest near Zinapécuaro (NE Michoacán)

Greenbaum's Deermouse
Peromyscus greenbaumi
FR Péromyscus de Greenbaum
DE Greenbaum-Hirschmaus
ES Ratón ciervo de Greenbaum

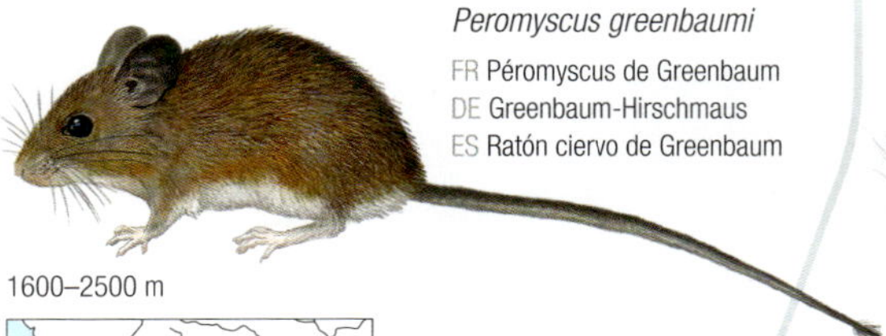

1600–2500 m

NE
Monotypic
HB c. 10.4 cm
T c. 10.3 cm

W Michoacán (Sierra de Coalcomán) N to along Colima–Jalisco border

Kilpatrick's Deermouse
Peromyscus kilpatricki
FR Péromyscus de Kilpatrick
DE Kilpatrick-Hirschmaus
ES Ratón ciervo de Kilpatrick

above 1600 m

NE
Monotypic
HB 8.5–10.2 cm
T 8.5–11.4 cm
W 24–38 g

Sierra Madre del Sur from C Michoacán E to Morelos; possibly México

LC
2 sspp
HB 8.8–11.8 cm
T 9–13.2 cm
W c. 26 g

1370–3810 m

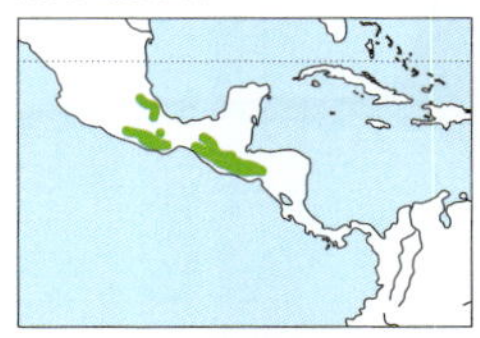

Orizaba Deermouse
Peromyscus beatae
FR Péromyscus d'Orizaba
DE Orizaba-Hirschmaus
ES Ratón ciervo de Orizaba

MX (SE Hidalgo, N Puebla, C Veracruz, Guerrero, Oaxaca, Chiapas), GT, W HN, SV

LC
Monotypic
HB c. 10.3 cm
T c. 10 cm
W 22–36 g

15–1980 m

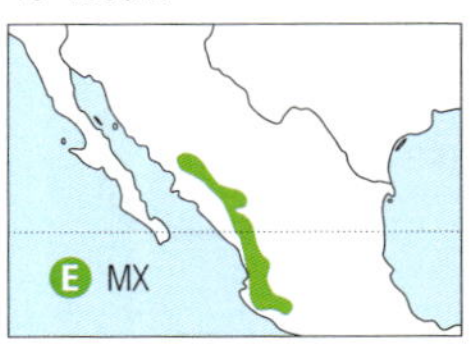

Gleaning Deermouse
Peromyscus spicilegus
FR Péromyscus glaneur
DE Sammelnde Hirschmaus
ES Ratón ciervo reluciente

SE Sinaloa & SW Chihuahua S along Sierra Madre Occidental to W Michoacán

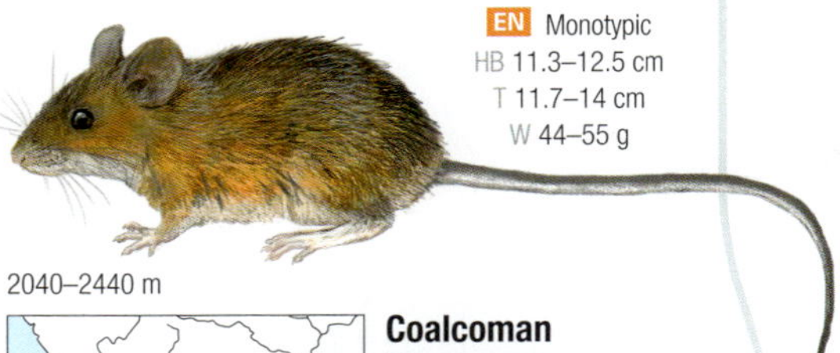

EN Monotypic
HB 11.3–12.5 cm
T 11.7–14 cm
W 44–55 g

2040–2440 m

Coalcoman Deermouse
Peromyscus winkelmanni
FR Péromyscus de Winkelmann
DE Winkelmann-Hirschmaus
ES Ratón ciervo de Coalcomán

SW Michoacán (Sierra Coalcomán), Guerrero (Sierra Madre del Sur)

NE 2 sspp
HB 11.1–13.5 cm
T 10.2–13.5 cm

Cordillera Deermouse
Peromyscus cordillerae
FR Péromyscus des cordillères
DE Oaxaca-Hirschmaus
ES Ratón ciervo de Oaxaca

From E Oaxaca (MX) SE to W HN, E SV

2300–2700 m

LC
Monotypic
HB 10.1–12.1 cm
T 10.2–11.7 cm
W 22–36 g

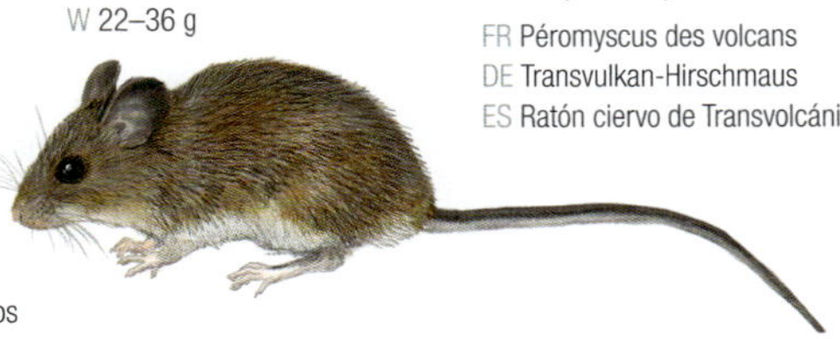

Transvolcanic Deermouse
Peromyscus hylocetes
FR Péromyscus des volcans
DE Transvulkan-Hirschmaus
ES Ratón ciervo de Transvolcánica

Volcanic Belt from N Colima & SW Jalisco E to Distrito Federal & Morelos

Aztec Deermouse
Peromyscus aztecus
FR Péromyscus aztèque
DE Azteken-Hirschmaus
ES Ratón ciervo de los Aztecas

500–3200 m

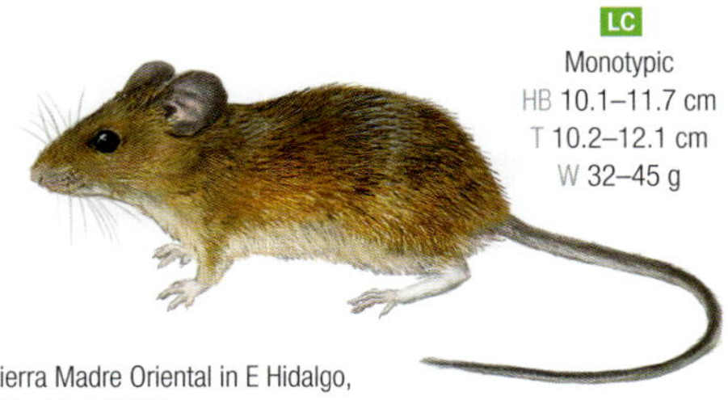

LC
Monotypic
HB 10.1–11.7 cm
T 10.2–12.1 cm
W 32–45 g

Sierra Madre Oriental in E Hidalgo, E Puebla & WC Veracruz

NE
Monotypic
HB 8.6–9.7 cm
T 8.9–10.9 cm
W 16–22 g

Northern White-ankled Deermouse
Peromyscus laceianus
FR Péromyscus de Lacey
DE Nördliche Weißfessel-Hirschmaus
ES Ratón ciervo de tobillo blanco septentrional

Extreme SC Oklahoma, extreme SE New Mexico, Texas

LC
2 sspp
HB 9.4–9.9 cm
T 8.1–12.7 cm
W 15–19 g

100–2500 m

From NC Chihuahua & N Coahuila S to Jalisco, Guanajuato & Hidalgo

Southern White-ankled Deermouse
Peromyscus pectoralis
FR Péromyscus à chevilles blanches
DE Südliche Weißfessel-Hirschmaus
ES Ratón ciervo de tobillo blanco meridional

NE
Monotypic

From NC Tamaulipas S to E Hidalgo

Tamaulipas White-ankled Deermouse
Peromyscus collinus
FR Péromyscus du Tamaulipas
DE Tamaulipas-Weißfessel-Hirschmaus
ES Ratón ciervo de tobillo blanco central

LC
4 sspp
HB 9.5–10.8 cm
T 7.6–12.3 cm
W 20–33 g

1710–2700 m

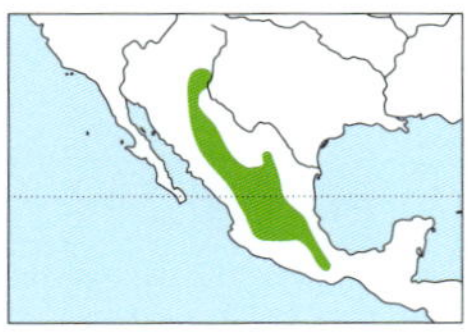
US (SW New Mexico), MX (from SE Coahuila & NW Chihuahua to Oaxaca)

Saxicoline Deermouse
Peromyscus gratus
FR Péromyscus gracieux
DE Felsliebende Hirschmaus
ES Ratón ciervo de roca

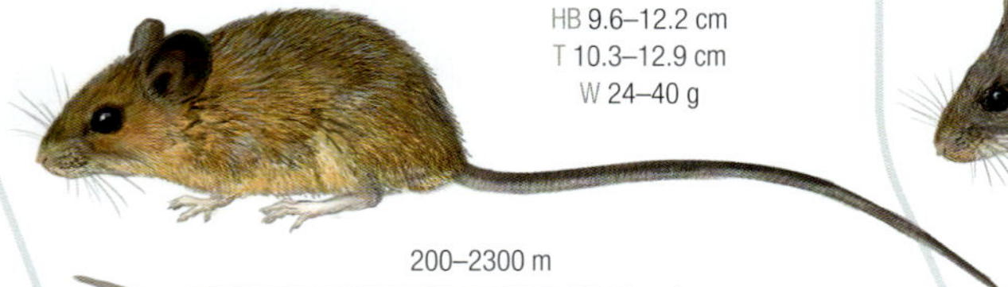

EN Monotypic
HB 9.6–12.2 cm
T 10.3–12.9 cm
W 24–40 g

200–2300 m

E of Sierra Madre Oriental in S Tamaulipas & NE San Luis Potosí

El Carrizo Deermouse
Peromyscus ochraventer
FR Péromyscus d'El Carrizo
DE El-Carrizo-Hirschmaus
ES Ratón ciervo de El Carrizo

Chihuahuan Deermouse
Peromyscus polius
FR Péromyscus du Chihuahua
DE Chihuahua-Hirschmaus
ES Ratón ciervo de Chihuahua

2110–2720 m

WC Chihuahua

NT
Monotypic
HB 9.9–11.4 cm
T 11.1–12 cm
W 22–36 g

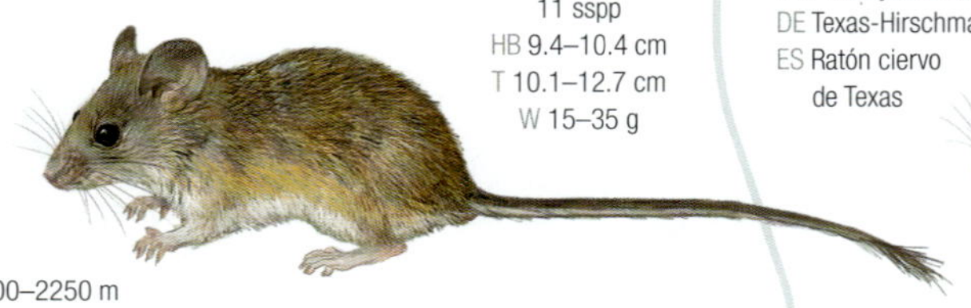

LC
11 sspp
HB 9.4–10.4 cm
T 10.1–12.7 cm
W 15–35 g

900–2250 m

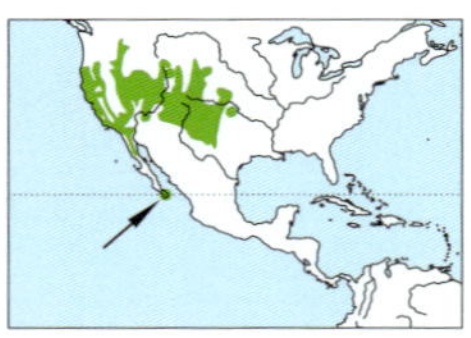
W & C US (S from Oregon, E to W Oklahoma), NW MX (N & S tip of Baja California Pen)

Pinyon Deermouse
Peromyscus truei
FR Péromyscus de True
DE Pinyon-Hirschmaus
ES Ratón ciervo de pinar

Texas Deermouse
Peromyscus attwateri
FR Péromyscus du Texas
DE Texas-Hirschmaus
ES Ratón ciervo de Texas

From SE Kansas & SW Missouri to SC Texas

LC
Monotypic
HB 9.1–10.6 cm
T 9.6–11.2 cm

LC 3 sspp
HB 9.9–10.7 cm
T 12.1–13.4 cm
W 25–35 g

2000–3115 m

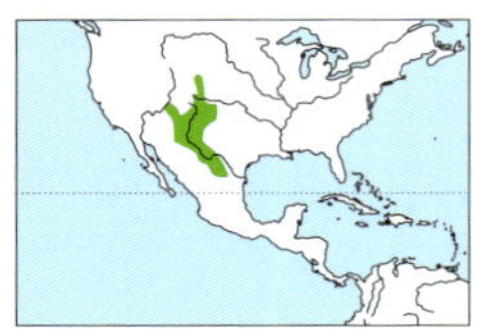
SC US (from NC Colorado & S Utah), N MX (N Chihuahua, Coahuila)

Northern Rock Deermouse
Peromyscus nasutus
FR Péromyscus du Colorado
DE Nördliche Felsen-Hirschmaus
ES Ratón ciervo de roca septentrional

Southern Rock Deermouse
Peromyscus difficilis
FR Péromyscus des rochers
DE Südliche Felsen-Hirschmaus
ES Ratón ciervo de Zacatecas

LC
2 sspp
HB 10–11.2 cm
T 11.2–13.8 cm
W 28–43 g

1200–3700 m

Sierra Madre Occidental (S to Guanajuato), N Sierra Madre Oriental (S to N San Luis Potosí)

Oaxacan Rock Deermouse
Peromyscus amplus
FR Péromyscus velu
DE Oaxaca-Felsen-Hirschmaus
ES Ratón ciervo de San Juan

NE
2 sspp
HB 10.7–11.5 cm
T 12.2–14.5 cm
W 24–32 g

N & E Valley of México (Hidalgo, Querétaro, W Veracruz, N México, Puebla, NW Oaxaca)

Fragmented range in mts of Valley of México & C Oaxaca

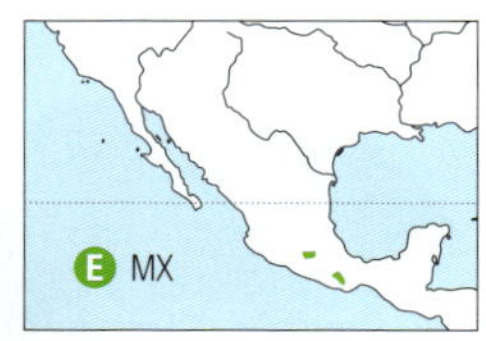

Felipe's Rock Deermouse
Peromyscus felipensis
FR Péromyscus de San Felipe
DE Felipe-Felsen-Hirschmaus
ES Ratón ciervo de San Felipe

NE
Monotypic
HB 10.7–11.6 cm
T 11.8–13.2 cm

c. 2375 m

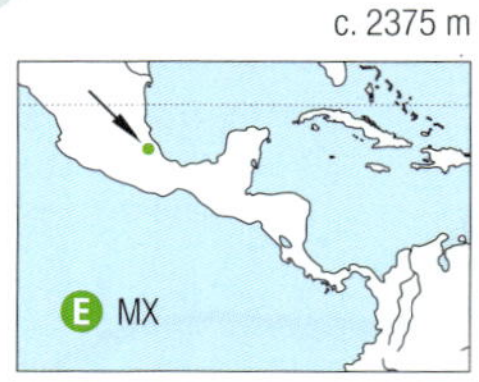

Known from a few sites in extreme E Puebla & WC Veracruz

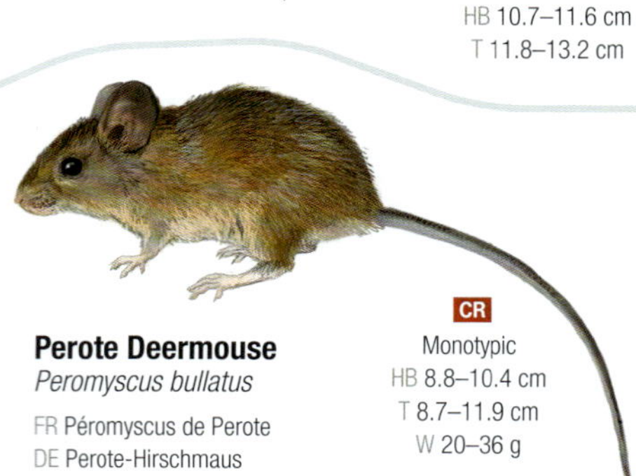

Perote Deermouse
Peromyscus bullatus
FR Péromyscus de Perote
DE Perote-Hirschmaus
ES Ratón ciervo de Perote

CR
Monotypic
HB 8.8–10.4 cm
T 8.7–11.9 cm
W 20–36 g

EN Monotypic
HB 14.5–16.6 cm
T 15–18.5 cm
W 62–113 g

3000–3500 m

Thomas's Big-toothed Deermouse
Megadontomys thomasi

FR Péromyscus de Thomas
DE Thomas-Großzahn-Hirschmaus
ES Ratón ciervo de dientes grandes de Thomas

C Guerrero (near Chilpancingo in Sierra Madre del Sur)

Genus *MEGADONTOMYS*

(3 species)

EN Monotypic
HB c. 14.5 cm T 15.5–18.5 cm
W 73–93 g

2400–3500 m

Oaxacan Big-toothed Deermouse
Megadontomys cryophilus

FR Péromyscus de Musser
DE Oaxaca-Großzahn-Hirschmaus
ES Ratón ciervo de dientes grandes de Oaxaca

NC Oaxaca (Sierra Juárez in Sierra Madre Oriental)

EN
Monotypic
HB 13.2–14.6 cm
T 17–17.2 cm
W c. 57 g

2000–3500 m

Sierra Madre Oriental (CW Veracruz, E Puebla, N Oaxaca)

Nelson's Big-toothed Deermouse
Megadontomys nelsoni

FR Péromyscus de Nelson DE Nelson-Großzahn-Hirschmaus
ES Ratón ciervo de dientes grandes de Nelson

Genus *HABROMYS*

(7 species)

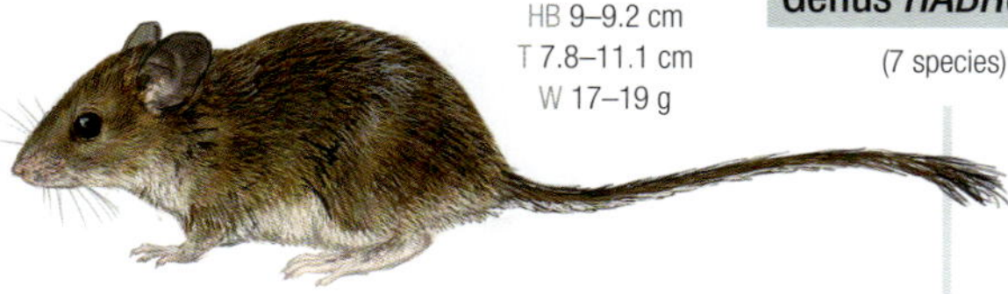

CR
Monotypic
HB 9–9.2 cm
T 7.8–11.1 cm
W 17–19 g

1830–2200 m

Xico Deermouse
Habromys simulatus

FR Habromys de Jico
DE Jico-Hirschmaus
ES Ratón ciervo de Xico

E slopes of Sierra Madre Oriental (Hidalgo, Veracruz)

EN Monotypic
HB 7.4–8.4 cm
T 7.4–8.1 cm
W 10–19 g

c. 2570 m

Delicate Deermouse
Habromys delicatulus

FR Habromys délicat
DE Zierliche Hirschmaus
ES Ratón ciervo delicado

NW México (Jilotepec), NE Michoacán (Cerro de Garnica)

CR Monotypic
HB 7.2–7.8 cm
T 7.2–8.9 cm
W 10–15 g

1800–2400 m

Schmidly's Deermouse
Habromys schmidlyi

FR Habromys de Schmidly
DE Schmidly-Hirschmaus
ES Ratón ciervo de Schmidly

Guerrero (N slopes of Sierra de Taxco)

CR Monotypic
HB 11–13.3 cm
T 10–14.7 cm
W 26–55 g

2500–3000 m

Ixtlan Deermouse
Habromys ixtlani

FR Habromys d'Ixtlan
DE Ixtlan-Hirschmaus
ES Ratón ciervo de Ixtlán

Oaxaca (N slopes of Sierra Juárez)

CR Monotypic
HB 9–9.3 cm
T 11–11.8 cm
W 17–22 g

2000–2650 m

Chinanteco Deermouse
Habromys chinanteco

FR Habromys chinantèque
DE Chinanteco-Hirschmaus
ES Ratón ciervo de los Chinanteco

Oaxaca (known only from Cerro Pelón, Sierra Juárez)

CR Monotypic
HB 10.4–11.6 cm
T 10.3–14.6 cm
W c. 32 g

2500–3000 m

Zempoaltepec Deermouse
Habromys lepturus

FR Habromys à queue fine
DE Zempoaltepec-Hirschmaus
ES Ratón ciervo de Zempoaltépetl

Oaxaca (known only from Mt Zempoaltépec)

NT Monotypic
HB 9.5–11.5 cm
T 9.2–11.5 cm
W 24–31 g

1950–3100 m

Crested-tailed Deermouse
Habromys lophurus

FR Habromys à queue hérissée
DE Kammschwanz-Hirschmaus
ES Ratón ciervo de cola crestada

SE MX (Sierra Madre of Chiapas) through GT to extreme NW SV

Genus *SIGMODON*

(14 species)

LC Monotypic
HB 10.6–15.2 cm
T 7.2–10.2 cm
W c. 74 g

below 1300 m

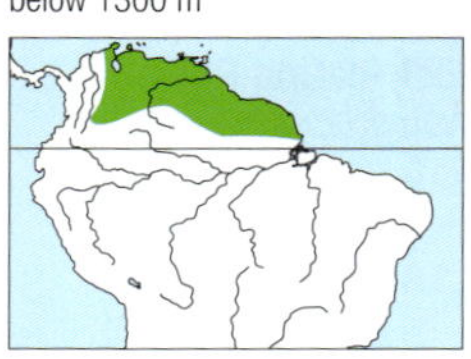

Groove-toothed Cotton Rat
Sigmodon alstoni

FR Sigmodon d'Alston
DE Furchenzahn-Baumwollratte
ES Rata de algodón de dientes estriados

Open habitats from NE CO to NE BR

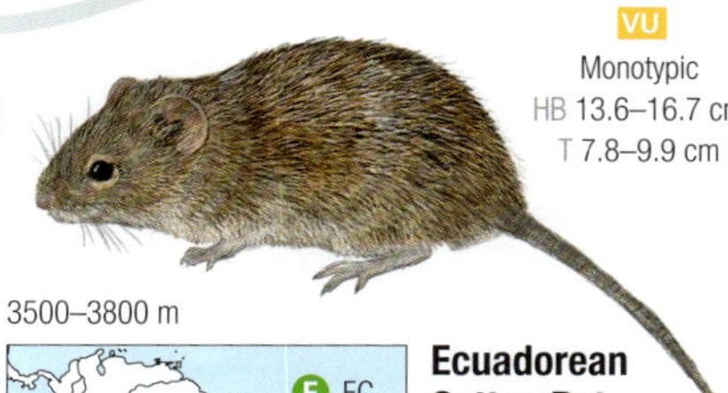

VU
Monotypic
HB 13.6–16.7 cm
T 7.8–9.9 cm

3500–3800 m

Ecuadorean Cotton Rat
Sigmodon inopinatus

FR Sigmodon andin
DE Ekuador-Baumwollratte
ES Rata de algodón de Ecuador

Known only from high-elevation Andes

0–1600 m

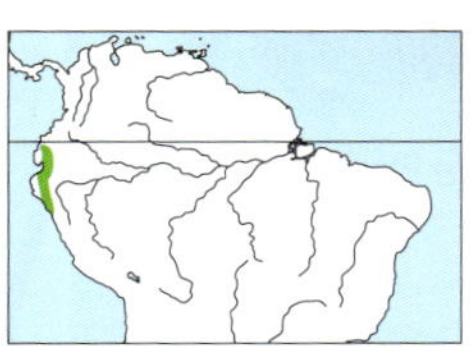

Pacific coastal plain and adjacent Andean foothills of W EC & NW PE

LC Monotypic
HB 12–17.2 cm
T 7.9–11.6 cm
W c. 72 g

Peruvian Cotton Rat
Sigmodon peruanus

FR Sigmodon du Pérou DE Peru-Baumwollratte
ES Rata de algodón de Perú

1800–3600 m

From SW Chihuahua and S Nuevo León S to C Oaxaca

White-eared Cotton Rat
Sigmodon leucotis

FR Sigmodon à oreilles blanches
DE Weißohr-Baumwollratte
ES Rata de algodón de orejas blancas

LC
2 sspp
HB 13.2–15.7 cm
T 8.4–10.5 cm
W ≤ 140 g

LC 5 sspp
HB 13.8–20 cm
T 10.1–11.8 cm
W 82–136 g

1290–2475 m

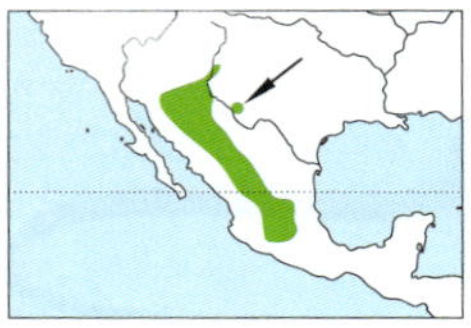

Tawny-bellied Cotton Rat
Sigmodon fulviventer
FR Sigmodon à ventre roux
DE Gelbbraune Baumwollratte
ES Rata de algodón de vientre leonado

US (SE Arizona, SW New Mexico, SW Texas), MX (S to W Michoacán)

LC Monotypic
HB 13.2–14.4 cm
T 8.5–11 cm
W 50–133 g

990–2375 m

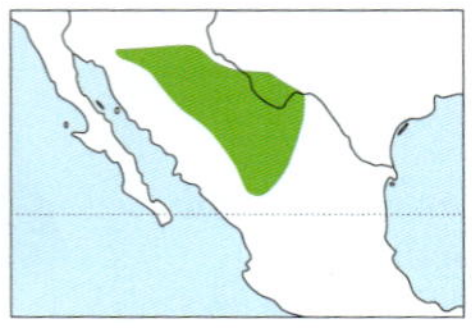

Yellow-nosed Cotton Rat
Sigmodon ochrognathus
FR Sigmodon à nez jaune
DE Gelbnasen-Baumwollratte
ES Rata de algodón de hocico dorado

US (SE Arizona, SW New Mexico, W Texas), MX (S to C Durango)

LC Monotypic
HB 12.5–23 cm
T 8.1–16.6 cm
W 100–225 g

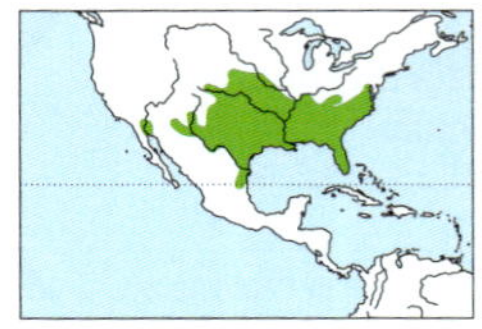

Hispid Cotton Rat
Sigmodon hispidus
FR Sigmodon hispide
DE Baumwollratte
ES Rata de algodón hirsuta

SC & SE US (from S Nebraska & Virginia) S to MX (Nuevo León & Tamaulipas); isolated in lower R Colorado

LC
Monotypic
HB 12.5–20 cm
T 7.5–13.8 cm
W 70–211 g

0–2500 m

West Mexican Cotton Rat
Sigmodon mascotensis
FR Sigmodon du Jalisco
DE Westmexiko-Baumwollratte
ES Rata de algodón de México occidental

From S Nayarit & S Durango S to NW Oaxaca, E to W Hidalgo & W Puebla

LC
3 extant sspp
HB 22–24 cm
T 9.5–15.5 cm
W 140–300 g

0–2000 m

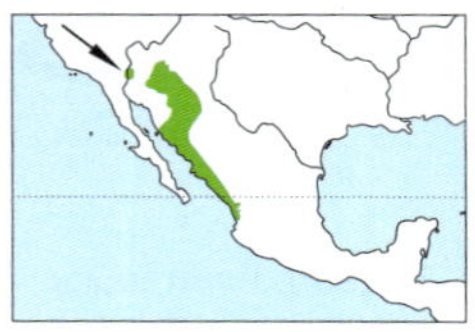

Arizona Cotton Rat
Sigmodon arizonae
FR Sigmodon d'Arizona
DE Arizona-Baumwollratte
ES Rata de algodón de Arizona

US (E California, W & S Arizona), MX (S to W Durango & Nayarit)

LC Monotypic
HB 13.8–23 cm
T 11.2–16.5 cm
W 70–133 g

0–1000 m

Toltec Cotton Rat
Sigmodon toltecus
FR Sigmodon toltèque
DE Tolteken-Baumwollratte
ES Rata de algodón de los Tolteca

E MX (from SE Tamaulipas S to Yucatán Pen & W Chiapas), BZ, N GT, NW HN

VU
Monotypic
HB 13.9–15.2 cm
T 8.8–12.7 cm
W ≤ 180 g

0–3050 m

Allen's Cotton Rat
Sigmodon alleni
FR Sigmodon d'Allen
DE Allen-Baumwollratte
ES Rata de algodón de Allen

Sierra Madre Occidental from S Sinaloa to Isthmus of Tehuantepec in E Oaxaca

LC
Monotypic
HB 12.5–23 cm
T 9.2–16.5 cm

500–1500 m

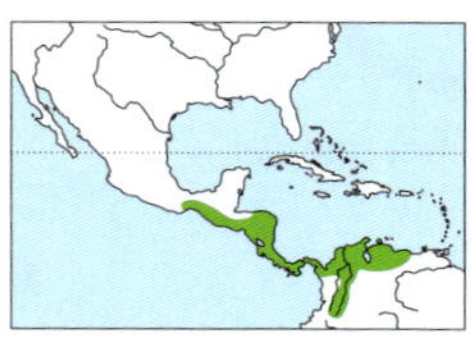

Burmeister's Cotton Rat
Sigmodon hirsutus
FR Sigmodon hirsute
DE Burmeister-Baumwollratte
ES Rata de algodón de Burmeister

SE MX (E Oaxaca & Chiapas) S to CO (Caribbean lowlands, Magdalena Valley), NW VE

NE
Monotypic
HB 14.6–16.3 cm
T 10.4–15.1 cm
W 62–113 g

up to 2700 m

Montane Cotton Rat
Sigmodon zanjonensis
FR Sigmodon de Zanjon
DE Berg-Baumwollratte
ES Rata de algodón de montaña

MX (Chiapas) E through GT to at least W HN

NE
Monotypic
HB c. 11.9 cm
T 8.8–11.1 cm

1000–2300 m

Miahuatlan Cotton Rat
Sigmodon planifrons
FR Sigmodon de Miahuatlan
DE Miahuatlán-Baumwollratte
ES Rata de algodón de Miahuatlán

S Oaxaca (pacific slope of Sierra de Miahuatlán)

Genus *NEUSTICOMYS*

(7 species)

VU
Monotypic
HB 9.4–11.8 cm
T 7.5–8.2 cm

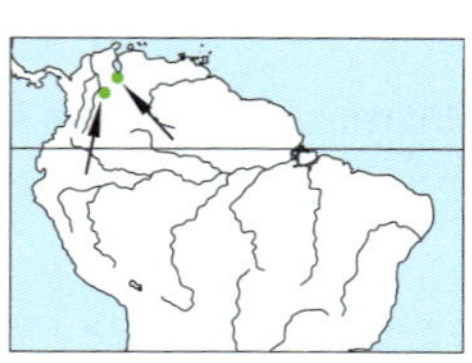

Musso's Fish-eating Rat
Neusticomys mussoi
FR Ichthyomys de Musso
DE Musso-Fischratte
ES Rata pescadora de Musso

Known from 3 localities in Andean N CO & W VE

LC
Monotypic
HB 9.5–21 cm
T 8.7–11.4 cm
W c. 75 g

1800–3800 m

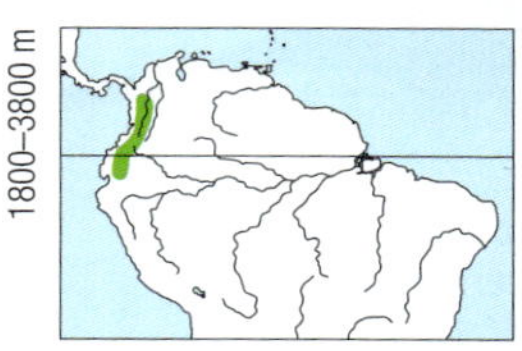

Montane Fish-eating Rat
Neusticomys monticolus
FR Ichthyomys de montagne
DE Gebirgs-Fischratte
ES Rata pescadora de montaña

W Andean slopes of CO and EC

1900–3750 m

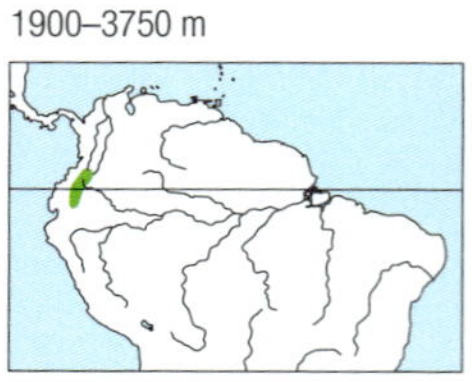

E slopes of Andes in S CO and N & C EC

NE Monotypic
HB 9.7–11.1 cm
T 9.2–10.8 cm
W c. 32 g

Voss's Fish-eating Rat
Neusticomys vossi
FR Ichthyomys de Voss
DE Voss-Fischratte
ES Rata pescadora de Voss

0–1400 m

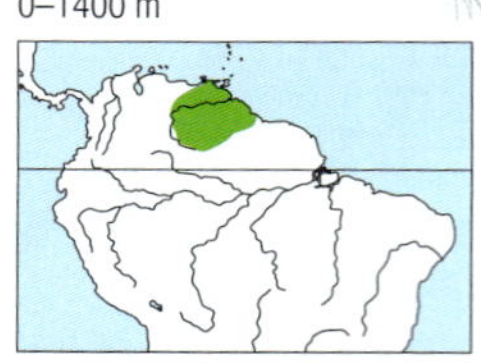

Known from a few localities in VE & GY

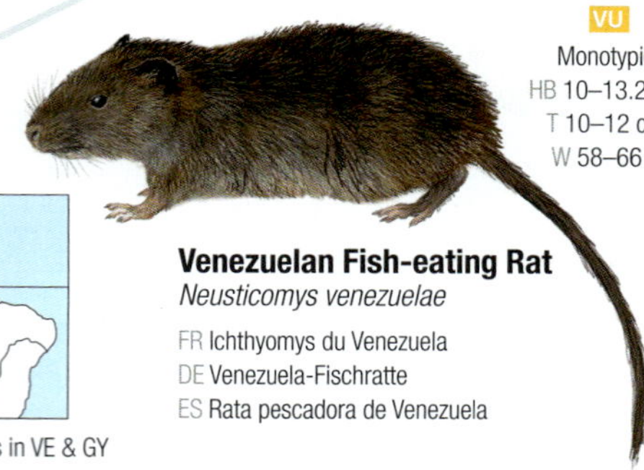

VU
Monotypic
HB 10–13.2 cm
T 10–12 cm
W 58–66 g

Venezuelan Fish-eating Rat
Neusticomys venezuelae
FR Ichthyomys du Venezuela
DE Venezuela-Fischratte
ES Rata pescadora de Venezuela

DD Monotypic
HB 9.6–11.5 cm
T 6.6–8.7 cm
W 21–50 g

200–450 m

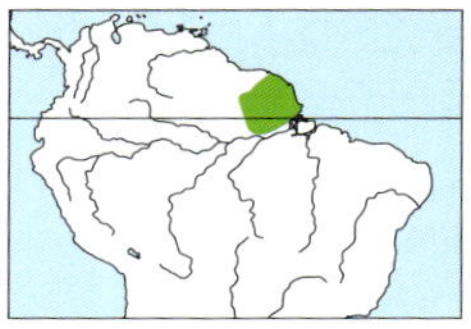

Guianan Fish-eating Rat
Neusticomys oyapocki
FR Ichthyomys de l'Oyapock
DE Guyana-Fischratte
ES Rata pescadora de Guayana

Scattered lowland sites in GF, S SR & BR (N of R Amazon)

DD Monotypic
HB 10.6–11.4 cm
T 7.5–9.5 cm
W 31–36 g

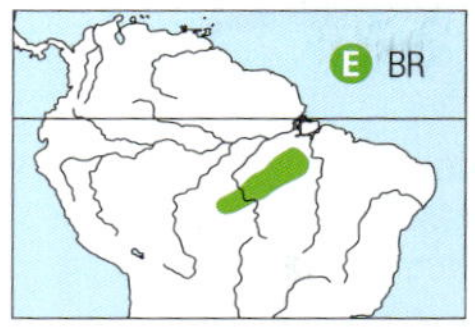

Ferreira's Fish-eating Rat
Neusticomys ferreirai
FR Ichthyomys de Ferreira
DE Ferreira-Fischratte
ES Rata pescadora de Ferreira

Scattered in Amazonian lowlands (S of R Amazon)

LC 2 sspp
HB 11.4–13.3 cm
T 9–11 cm
W c. 37 g

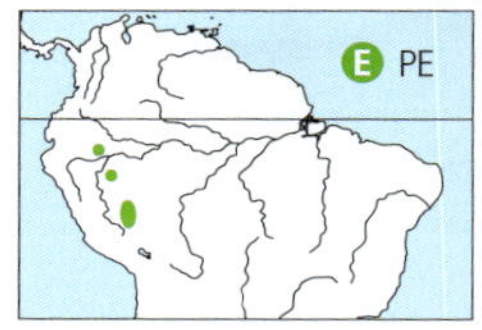

Peruvian Fish-eating Rat
Neusticomys peruviensis
FR Ichthyomys du Pérou
DE Peru-Fischratte
ES Rata pescadora de Perú

Known from a few localities in Amazonian lowlands

Genus *CHIBCHANOMYS*
(2 species)

DD Monotypic
HB 11.3–12.5 cm
T 11.5–13.3 cm
W 35–50 g

2400–2900 m

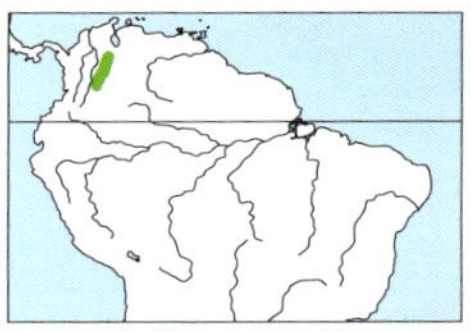

Northern Andean Water Mouse
Chibchanomys trichotis
FR Ichthyomys chibchan
DE Nördliche Andenwassermaus
ES Ratón de agua septentrional

E Andes Range in NC CO & extreme SW VE

DD Monotypic
HB 10.3–10.7 cm
T 10.8–12.2 cm
W 35–50 g

2400–3700 m

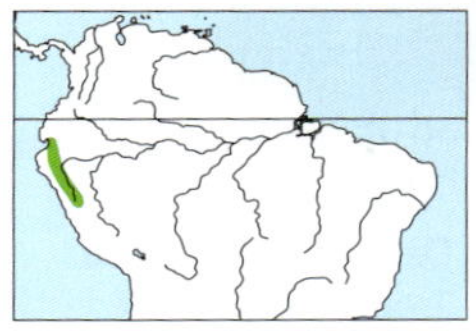

Orces's Andean Water Mouse
Chibchanomys orcesi
FR Ichthyomys d'Orces
DE Orces-Andenwassermaus
ES Ratón de agua de Orces

Páramo of S EC to C PE

Genus *ANOTOMYS*
(1 species)

EN Monotypic
HB 10.1–12.2 cm
T 12.5–15.3 cm W 40–60 g

2860–4000 m

Earless Water Mouse
Anotomys leander
FR Ichthyomys léandre
DE Ohrlose Wassermaus
ES Ratón de agua de Ecuador

A few Andean localities in W CO & NC EC

NT Monotypic
HB 9.5–17.5 cm
T 8–14.5 cm
W 38–147 g

700–1750 m

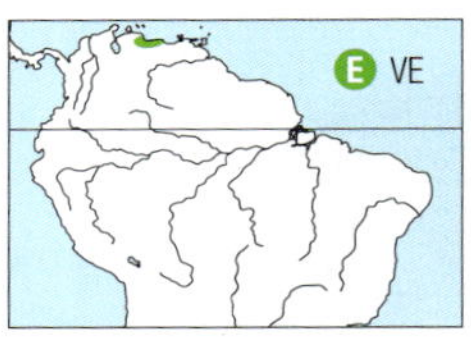

Pittier's Crab-eating Rat
Ichthyomys pittieri
FR Ichthyomys de Pittier
DE Pittier-Krabbenratte
ES Rata cangrejera de Pittier

Scattered in Cordillera Central

Genus *ICHTHYOMYS*
(5 species)

LC 3 sspp
HB 13.4–18.2 cm
T 12.5–15 cm

1000–2800 m

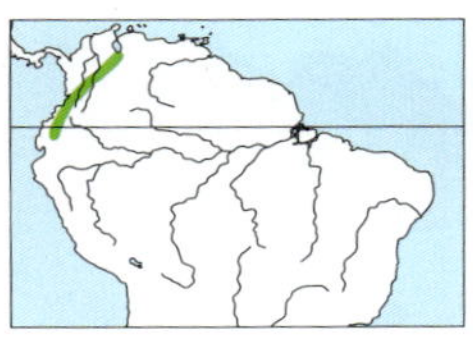

Common Crab-eating Rat
Ichthyomys hydrobates
FR Ichthyomys de Winge
DE Gemeine Krabbenratte
ES Rata cangrejera común

Andes forests of W VE, CO & NC EC

NE
Monotypic
HB c. 13.1 cm
T c. 13.7 cm

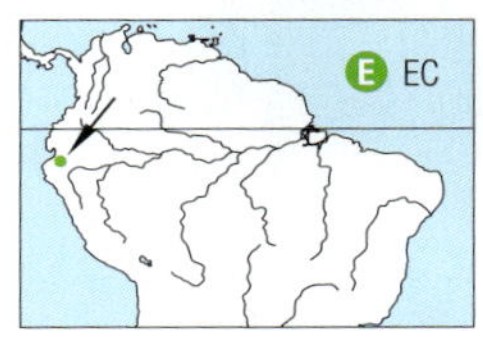

Pine's Crab-eating Rat
Ichthyomys pinei
FR Ichthyomys de Pine
DE Pine-Krabbenratte
ES Rata cangrejera de Pine

Known only from type locality (near Cantón Nabón, Azuay)

DD Monotypic
HB 14.3–19.7 cm
T 13.2–15.5 cm
W 123–155 g

0–1350 m

Tweedy's Crab-eating Rat
Ichthyomys tweedii
FR Ichthyomys de Tweedy
DE Tweedy-Krabbenratte
ES Rata cangrejera de Tweedy

C PA, lowlands and Andean foothills of W EC

DD Monotypic
HB c. 16 cm
T 17.7–18.5 cm
W c. 140 g

900–3400 m

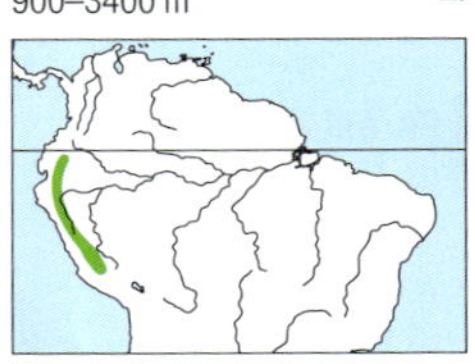

Stolzmann's Crab-eating Rat
Ichthyomys stolzmanni
FR Ichthyomys de Stolzmann
DE Stolzmann-Krabbenratte
ES Rata cangrejera de Stolzmann

Known from 3 sites in E Andean piedmont of EC and 4 sites in PE

LC
2 sspp
HB 9.2–13.3 cm
T 8.8–11.9 cm

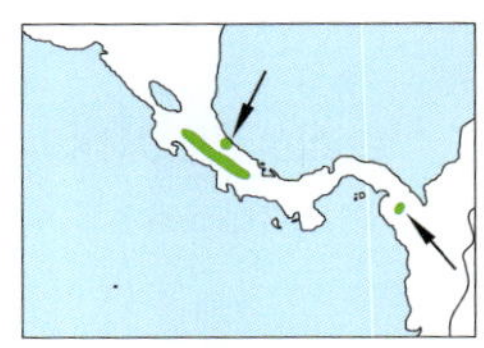

Goldman's Water Mouse
Rheomys raptor
FR Ichthyomys de Goldman
DE Goldman-Wassermaus
ES Ratón de agua de Goldman

Highlands in CR & W PA (mainly Pacific slope), E PA

Genus *RHEOMYS*
(4 species)

NT 2 sspp
HB 10.2–13.6 cm
T 10.9–14 cm

Thomas's Water Mouse
Rheomys thomasi
FR Ichthyomys de Thomas
DE Thomas-Wassermaus
ES Ratón de agua de Thomas

Highlands of SW MX, GT & SV; perhaps HN

Underwood's Water Mouse
Rheomys underwoodi
FR Ichthyomys d'Underwood
DE Underwood-Wassermaus
ES Ratón de agua de Underwood

1500–2000 m

Highlands of C CR & W PA

LC Monotypic
HB 13–14.2 cm
T 14.8–15.8 cm

Mexican Water Mouse
Rheomys mexicanus
FR Ichthyomys du Mexique
DE Mexiko-Wassermaus
ES Ratón de agua de México

EN
Monotypic
HB 10.5–14.8 cm
T 13.1–17.1 cm

0–2200 m

Pacific slope of C & SE Oaxaca

Genus *NEOMICROXUS*

(2 species)

LC Monotypic
HB 7.8–9 cm
T 6.2–7.5 cm

2600–3900 m

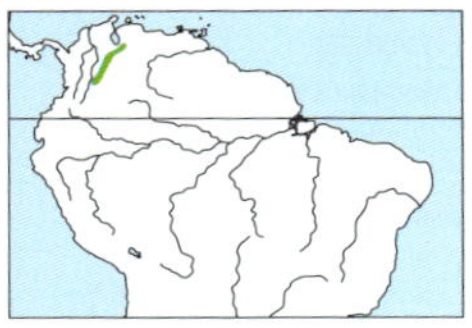

Bogota Grass Mouse
Neomicroxus bogotensis
FR Souris-d'alpage de Bogota
DE Bogotá-Grasmaus
ES Ratón de hierba de Bogotá

NE CO (E Andes), SW VE (Tamá Massif, Cordillera de Mérida)

EN
Monotypic
HB 7.4–10.3 cm
T 6–9.8 cm
W 15–19 g

above 3000 m

Ecuadorean Grass Mouse
Neomicroxus latebricola
FR Souris-d'alpage d'Équateur
DE Ekuador-Grasmaus
ES Ratón de hierba de Ecuador

Subalpine forests and páramo in NW & E Andes

Genus *CHINCHILLULA*

(1 species)

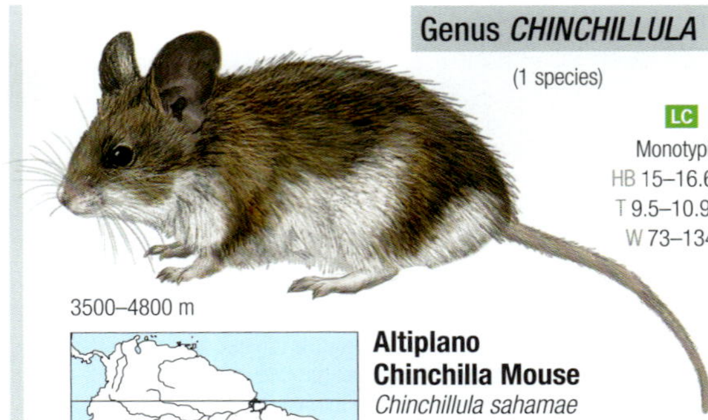

LC
Monotypic
HB 15–16.6 cm
T 9.5–10.9 cm
W 73–134 g

3500–4800 m

Altiplano Chinchilla Mouse
Chinchillula sahamae
FR Souris-chinchilla de l'Altiplano
DE Altiplano-Chinchillamaus
ES Ratón chinchilla de altiplano

Scattered in S PE, W BO & N CL

Genus *JULIOMYS*

(4 species)

Small Red-nosed Tree Mouse
Juliomys pictipes
FR Juliomys d'Osgood
DE Kleine Juliomaus
ES Ratón arborícola de hocico rojo pequeño

LC
Monotypic
HB 9–11 cm
T 8.2–11.2 cm
W 14–29 g

Atlantic Forest of SE & S BR, E PY & NE AR

NE Monotypic
HB 8.1–9.7 cm
T 10.2–11.6 cm
W 14–28 g

above 800 m

Delicate Red-nosed Tree Mouse
Juliomys ossitenuis
FR Juliomys délicat
DE Zierliche Juliomaus
ES Ratón arborícola de hocico rojo delicado

From Espírito Santo to Rio Grande do Sul

NT Monotypic
HB 8.5–9.3 cm
T 9.9–12.1 cm
W 14.5–23 g

Scattered montane localities in Minas Gerais, Rio de Janeiro & São Paulo

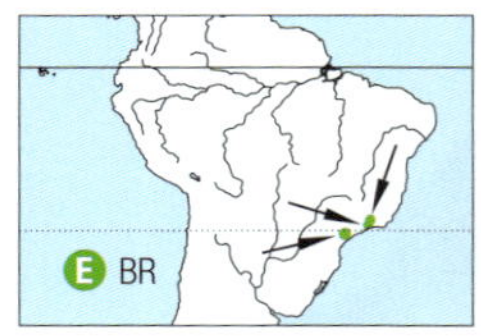

Montane Red-rumped Tree Mouse
Juliomys rimofrons
FR Juliomys montagnard
DE Berg-Juliomaus
ES Ratón arborícola de trasero rojo de montaña

Araucaria Forest Tree Mouse
Juliomys ximenezi
FR Juliomys de Christoff
DE Araukarien-Juliomaus
ES Ratón arborícola de Araucaria

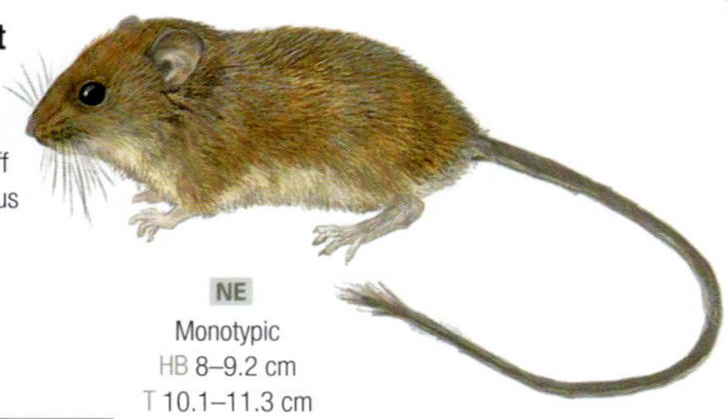

NE
Monotypic
HB 8–9.2 cm
T 10.1–11.3 cm

c. 800 m

Known only from Aparados da Serra NP (Rio Grande do Sul)

Genus *ABRAWAYAOMYS*

(1 species)

ssp *chebezi*

ssp *ruschii*

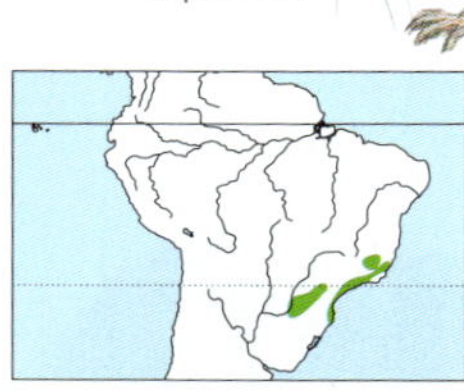

Scattered in BR (Minas Gerais & Espírito Santo to Santa Catarina), E PY & NE AR

Ruschi's Spiny Mouse
Abrawayaomys ruschii
FR Souris-barbée de Ruschi
DE Brasilien-Ruschi-Ratte
ES Ratón espinoso de Ruschi

LC
2 sspp
HB 8.9–13.5 cm
T 8.5–14.2 cm
W 25–63 g

Genus *DELOMYS*

(3 species)

Montane Atlantic Forest Rat
Delomys altimontanus
FR Délomys montagnard
DE Atlantische Bergwaldratte
ES Rata de bosque atlántico de montaña

above 1800 m

NE
Monotypic
HB 13–13.4 cm
T 12.3–12.6 cm
W 25–72 g

Restricted to Mantiqueira Mts (Caparaó & Itatiaia Mts)

Pallid Atlantic Forest Rat
Delomys sublineatus
FR Délomys pâle
DE Fahle Atlantische Waldratte
ES Rata de bosque atlántico pálida

LC
Monotypic
HB 12.6–14.6 cm
T 10.6–12.7 cm
W 41–60 g

0–1200 m

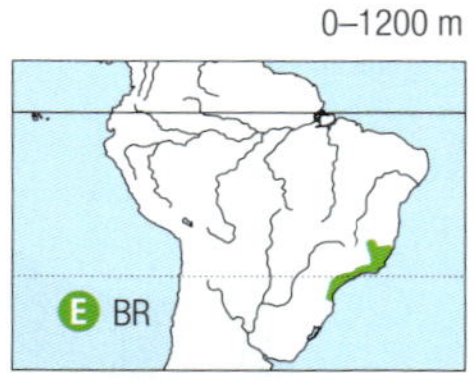

Atlantic Forest from Minas Gerais & Espírito Santo to Santa Catarina

LC
Monotypic
HB 10.4–13.5 cm
T 10–14.5 cm
W 20–56 g

0–2000 m

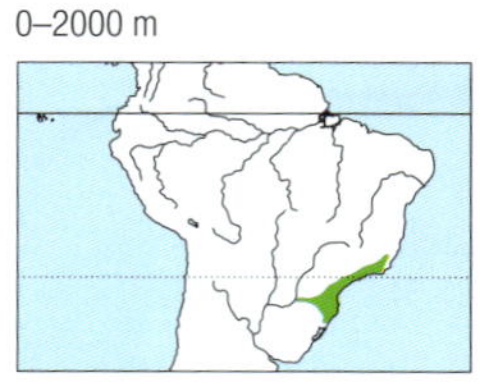

Striped Atlantic Forest Rat
Delomys dorsalis
FR Délomys sombre
DE Gestreifte Atlantische Waldratte
ES Rata de bosque atlántico rayada

Atlantic Forest of BR (Minas Gerais & Espírito Santo to Rio Grande do Sul) & NE AR

NE
Monotypic
HB 12.7–14.9 cm
T 9.4–9.9 cm
W 77–95 g

above 2000 m

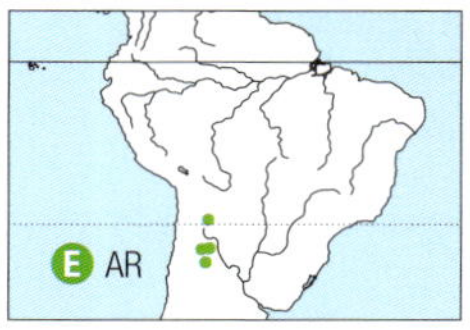

North-western Conyrat
Reithrodon caurinus
FR Reithrodon des Andes
DE Nordwestliche Kaninchenratte
ES Rata conejo noroccidental

Jujuy, Tucumán, Catamarca

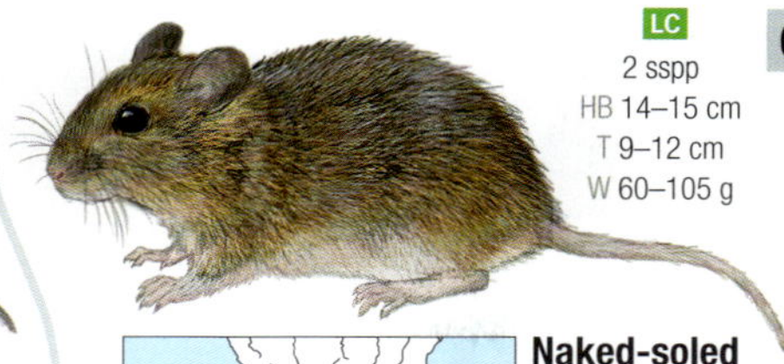

LC
2 sspp
HB 14–15 cm
T 9–12 cm
W 60–105 g

Naked-soled Conyrat
Reithrodon typicus
FR Reithrodon de l'Uruguay
DE Nacktsohlige Kaninchenratte
ES Rata conejo de plantas desnudas

BR (Rio Grande do Sul), UY, AR (Corrientes, Entre Ríos, probably NE Buenos Aires)

Genus *REITHRODON*
(3 species)

LC 3 sspp
HB 12.6–14.1 cm
T 7.9–9.5 cm
W 52–116 g

0–2200 m

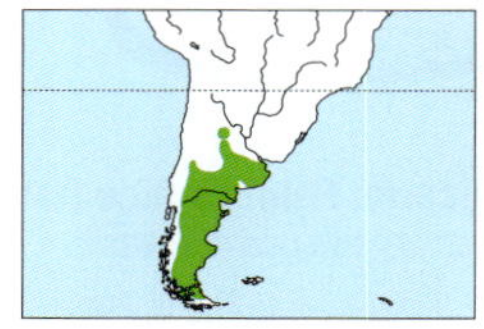

Hairy-soled Conyrat
Reithrodon auritus
FR Reithrodon de Patagonie
DE Haarsohlige Kaninchenratte
ES Rata conejo sin pelos

AR & adjacent CL S to Tierra del Fuego; unconfirmed in FK

Genus *ZYGODONTOMYS*
(2 species)

LC Monotypic
HB 14.1–16.1 cm
T 10.1–12 cm
W c. 57 g

below 1300 m

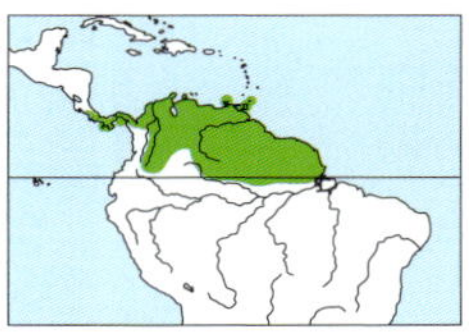

Short-tailed Cane Mouse
Zygodontomys brevicauda
FR Petite Souris-des-cannes
DE Kurzschwanz-Zuckermaus
ES Ratón de caña de cola corta

SW CR, PA (with Coiba, Cébaco & Pearl Is), CO, VE (with Margarita I), TT, Guianas, N BR

LC Monotypic
HB 13.5–15.2 cm
T 11–11.7 cm
W 45–81 g

350–1300 m

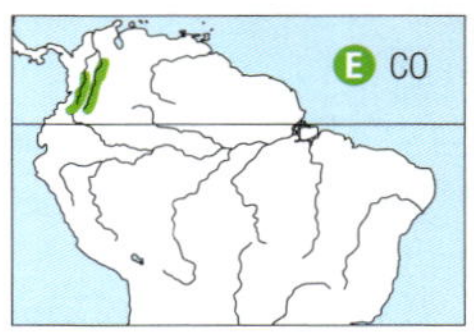

Colombian Cane Mouse
Zygodontomys brunneus
FR Grande Souris-des-cannes
DE Kolumbien-Zuckermaus
ES Ratón de caña de Colombia

Andean intermontane valleys in upper Cauca, Magdalena, Dagua & Patía rivers

LC
Monotypic
HB 8.8–11.4 cm
T 6–7.8 cm
W 20–34 g

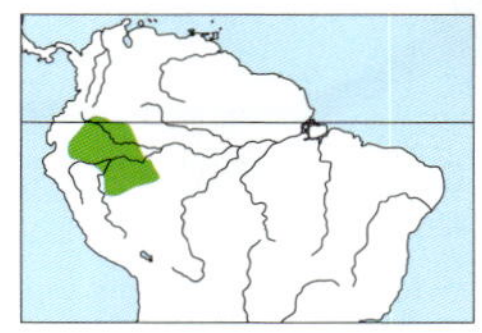

Ucayali South American Spiny Mouse
Scolomys ucayalensis
FR Scolomys de l'Ucayali
DE Ucayali-Neuweltstachelmaus
ES Ratón espinoso de Ucayali

Extreme S CO, E EC, NE PE, W BR

Genus *NEPHELOMYS*
(14 species)

LC Monotypic
HB 15.5–16.5 cm
T 18–19.5 cm

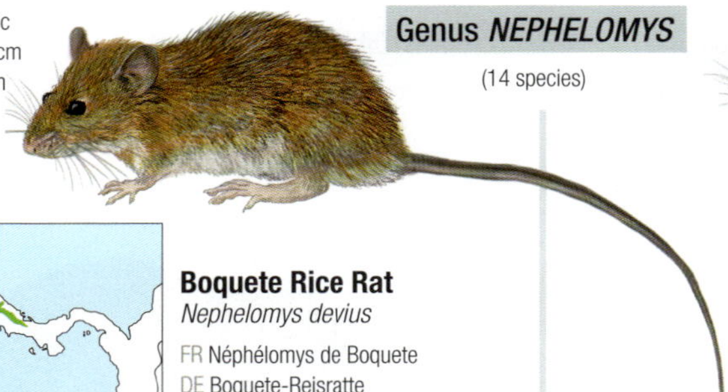

1200–2700 m

Boquete Rice Rat
Nephelomys devius
FR Néphélomys de Boquete
DE Boquete-Reisratte
ES Rata arrocera de Boquete

Highlands of CR and W PA

Santa Marta Rice Rat
Nephelomys maculiventer
FR Néphélomys de Santa Marta
DE Santa-Marta-Reisratte
ES Rata arrocera de Santa Marta

NE
Monotypic
HB 12.7–18 cm
T 14.8–19.4 cm

Slopes of Sierra Nevada de Santa Marta and Serranía de Perijá

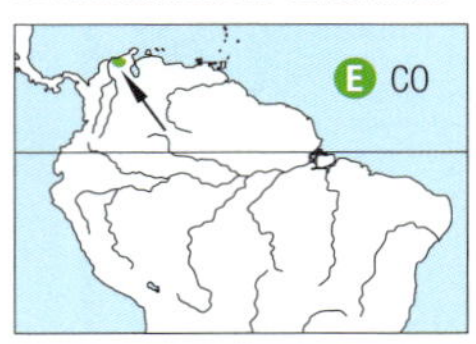

900–2600 m

Genus *SCOLOMYS*
(2 species)

LC Monotypic
HB 8.5–10.4 cm
T 5.5–7.6 cm
W 20–33 g

150–1200 m

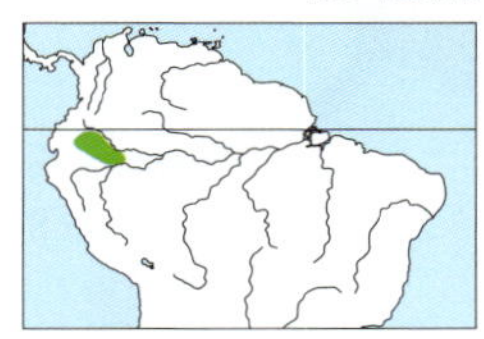

Gray South American Spiny Mouse
Scolomys melanops
FR Scolomys d'Anthony
DE Graue Neuweltstachelmaus
ES Ratón espinoso de Anthony

E EC & NE PE between N bank of R Napo, R Pastaza and N of R Amazon

NE
Monotypic
HB 15–15.5 cm
T 15.9–17 cm

1370–1585 m

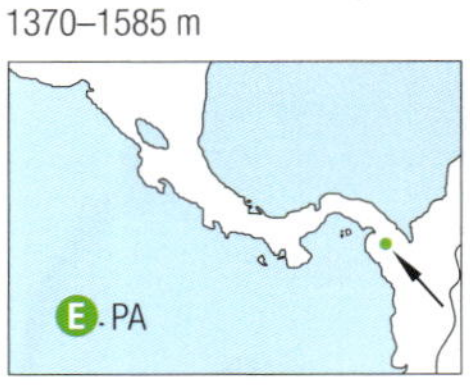

Mount Pirre Rice Rat
Nephelomys pirrensis
FR Néphélomys du Pirre
DE Mount-Pirre-Reisratte
ES Rata arrocera de Pirre

E PA; probably adjacent CO

NE Monotypic
HB 12.4–18 cm
T 11–19.6 cm
W 36–126 g

1200–3160 m

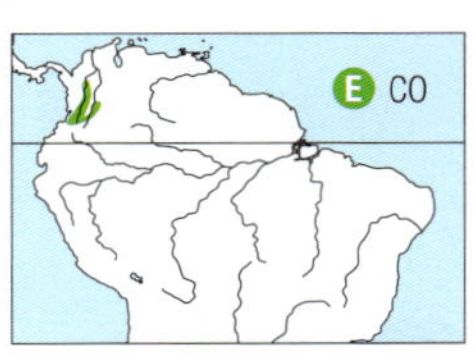

Western Colombian Rice Rat
Nephelomys pectoralis
FR Néphélomys de Colombie
DE Westkolumbianische Reisratte
ES Rata arrocera de Colombia occidental

All 3 Andean ranges

Coastal Cordilleran Rice Rat
Nephelomys caracolus
FR Néphélomys de Caracas
DE Küstenkordilleren-Reisratte
ES Rata arrocera de la Costa

1050–2300 m

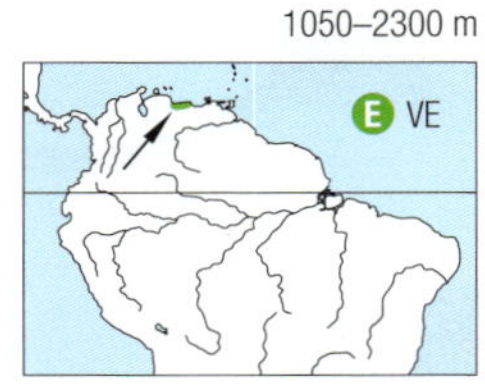

LC
Monotypic
HB c. 14.1 cm
T c. 16.2 cm
W c. 66 g

Cordillera de la Costa

1100–4000 m

Merida Rice Rat
Nephelomys meridensis
FR Néphélomys de Mérida
DE Mérida-Reisratte
ES Rata arrocera de Mérida

E Andes of NE CO and W VE

LC
Monotypic
HB 13.5–16 cm
T 17–17.2 cm
W 50–74 g

Child's Rice Rat
Nephelomys childi
FR Néphélomys de Child
DE Child-Reisratte
ES Rata arrocera de Child

NE
Monotypic
HB 12–18.4 cm
T 11–19.4 cm
W 43–86 g

1140–3400 m

All 3 Andean ranges

NE
Monotypic
HB 9.9–19.6 cm
T 15.4–19.1 cm
W 60–87 g

1900–3200 m

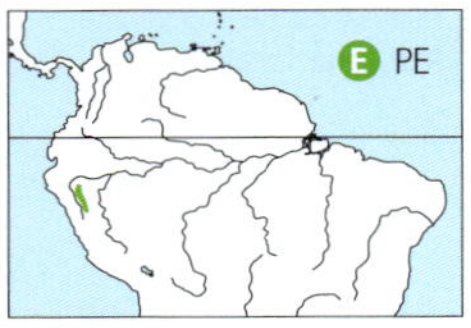

Ricardo Palma's Rice Rat
Nephelomys ricardopalmai

FR Néphélomys de Ricardo Palma
DE Ricardo-Palma-Reisratte
ES Rata arrocera de Ricardo Palma

Montane cloud forest E of R Marañón, W of R Huallaga (Amazonas, San Martín)

LC Monotypic
HB 10.2–16.2 cm
T 13.8–18.4 cm

White-throated Rice Rat
Nephelomys albigularis

FR Néphélomys de Tomes
DE Weißkehl-Reisratte
ES Rata arrocera de garganta blanca

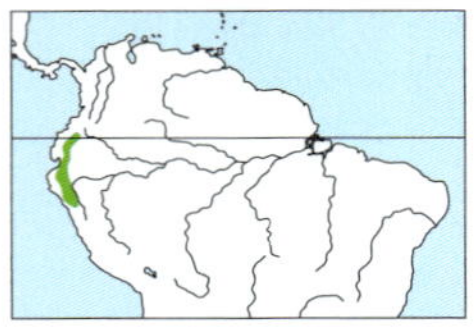

900–3100 m

NC EC (Carchi) S to NW PE (Huánuco)

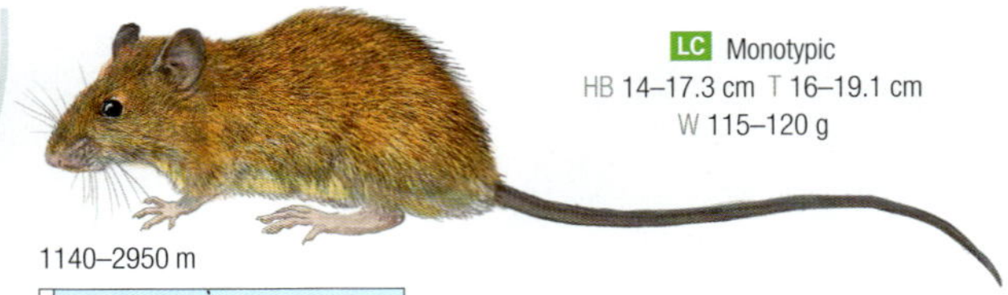

LC Monotypic
HB 14–17.3 cm T 16–19.1 cm
W 115–120 g

1140–2950 m

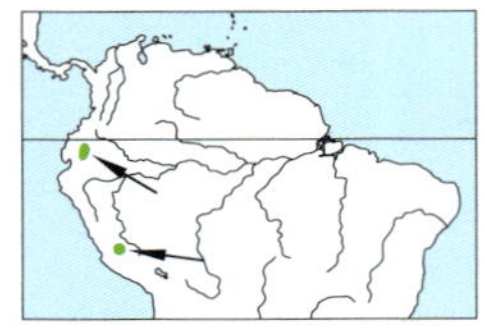

Greater Golden-bellied Rice Rat
Nephelomys auriventer

FR Néphélomys à ventre doré
DE Große Goldbauch-Reisratte
ES Rata arrocera de vientre dorado grande

Patchily in E Andes of C EC (upper R Pastaza) & SC PE (R Apurímac)

Keays's Rice Rat
Nephelomys keaysi

FR Néphélomys de Keays
DE Keays-Reisratte
ES Rata arrocera de Keays

1000–2600 m

LC
Monotypic
HB 12.5–16.5 cm
T 14.6–20.7 cm

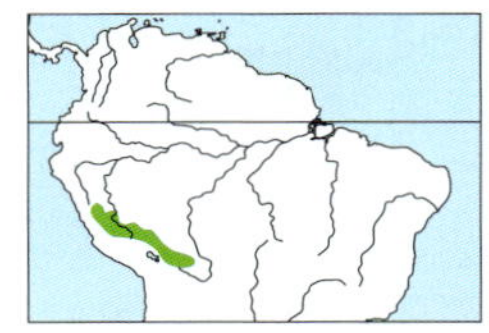

E Andean slopes from C PE to C BO

NE Monotypic
HB 11.4–16 cm
T 13.9–16.9 cm
W c. 42 g

1200–2800 m

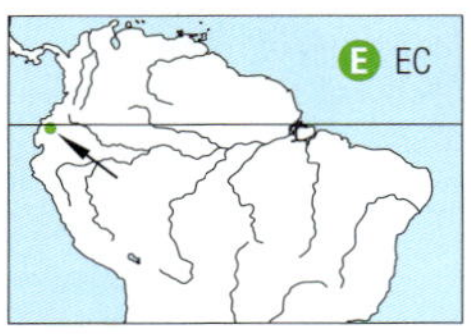

Gray-bellied Rice Rat
Nephelomys moerex

FR Néphélomys à ventre gris
DE Graubauch-Reisratte
ES Rata arrocera de vientre gris

NW & SW slopes of Volcán Pichincha and R Mindo

NE
Monotypic
HB 12.2–14.4 cm
T 14.7–17 cm
W c. 84 g

2000–2440 m

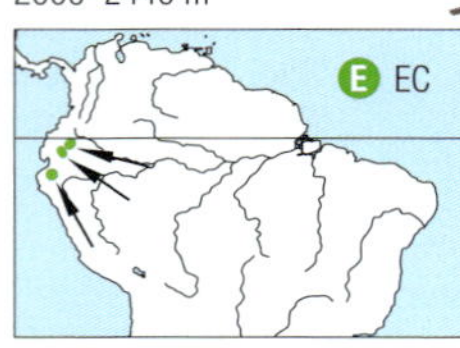

Lesser Golden-bellied Rice Rat
Nephelomys nimbosus

FR Néphélomys des brumes
DE Kleine Goldbauch-Reisratte
ES Rata arrocera de vientre dorado pequeña

Few localities in E Andes

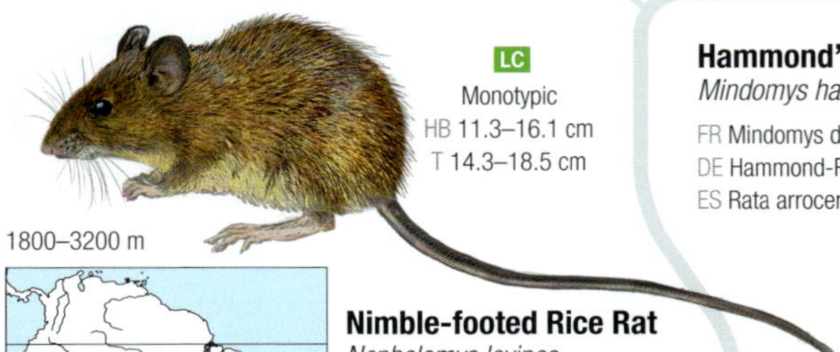

LC
Monotypic
HB 11.3–16.1 cm
T 14.3–18.5 cm

1800–3200 m

Nimble-footed Rice Rat
Nephelomys levipes

FR Néphélomys agile
DE Zartfüßige Reisratte
ES Rata arrocera de pies ligeros

E Andean slopes from SW PE to C BO

Hammond's Rice Rat
Mindomys hammondi

FR Mindomys de Hammond
DE Hammond-Reisratte
ES Rata arrocera de Hammond

1260–1330 m

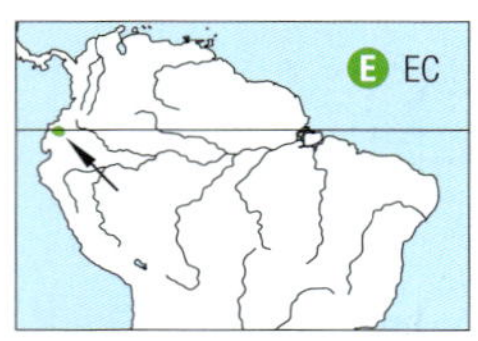

EN
Monotypic
HB 17.3–29 cm
T 22–25 cm

W Andean slope; uncorfirmed in Amazonia (Napo)

Genus *MINDOMYS*

(2 species)

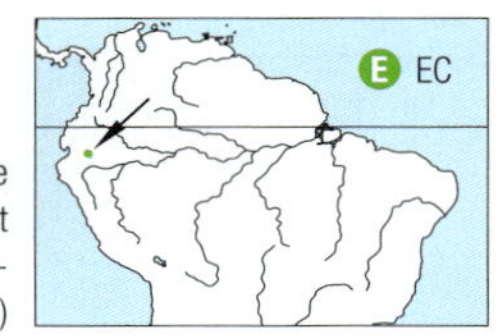

Known only from type locality (montane forest of Cordilleras Cóndor-Kutukú)

c. 1925 m

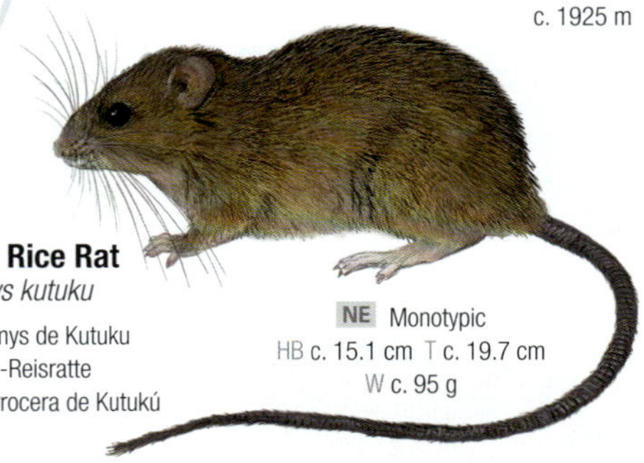

Kutuku Rice Rat
Mindomys kutuku

FR Mindomys de Kutuku
DE Kutuku-Reisratte
ES Rata arrocera de Kutukú

NE Monotypic
HB c. 15.1 cm T c. 19.7 cm
W c. 95 g

Ecominga Montane Rat
Pattonimus ecominga

FR Mindomys d'Ecominga
DE Ecominga-Bergreisratte
ES Rata montana de Ecominga

NE
Monotypic
HB 10.3–14.5 cm
T 13.5–18.8 cm
W 30–83 g

1660–2340 m

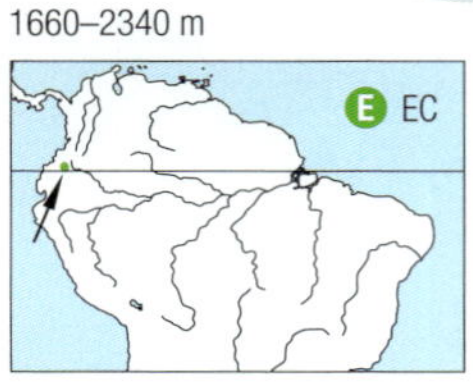

Known from several sites in Drácula R, W flank of Andes (Carchi)

Genus *PATTONIMUS*

(2 species)

c. 1200 m

Musser's Montane Rat
Pattonimus musseri

FR Mindomys de Musser
DE Musser-Bergreisratte
ES Rata montana de Musser

NE
Monotypic
HB 9.5–14 cm
T 12–19 cm
W 25–110 g

Know from type locality on W flank of Andes (Río Manduriacu R, Imbabura)

White-bellied Arboreal Rice Rat
Oecomys bicolor

FR Oecomys bicolore
DE Weißbauch-Baumreisratte
ES Rata arrocera arborícola de vientre blanco

0–1535 m

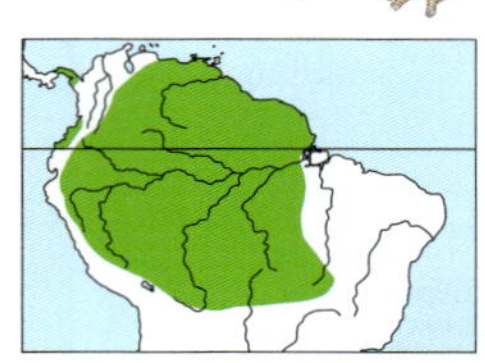

E PA, W CO & W EC, VE & Guianas through Amazon Basin to BO & C BR

LC
Monotypic
HB 9.4–10.7 cm
T 9.9–11.9 cm
W 25–37 g

Genus *OECOMYS*

(19 species)

LC Monotypic
HB 10.6–14 cm
T 13.2–15 cm
W 50–70 g

Yellow Arboreal Rice Rat
Oecomys flavicans

FR Oecomys jaune
DE Gelbe Baumreisratte
ES Rata arrocera arborícola dorada

0–2000 m

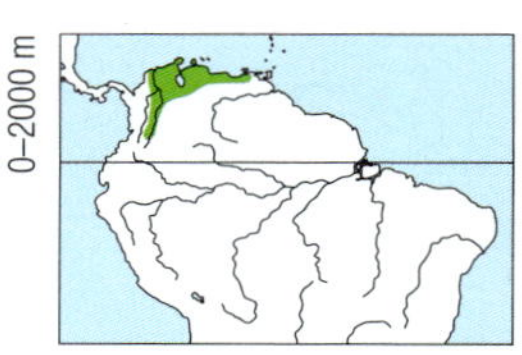

N & E CO and N VE (perhaps E to Delta Amacuro)

LC Monotypic
HB 10.7–12.7 cm T 12.6–15 cm
W 30–50 g

below 400 m

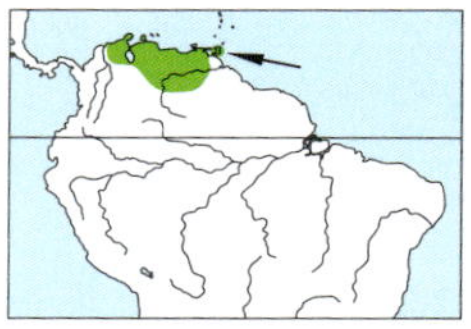

Savanna Arboreal Rice Rat
Oecomys speciosus

FR Oecomys superbe
DE Savannen-Baumreisratte
ES Rata arrocera arborícola de sabana

Extreme NE CO, N VE, TT (Trinidad I)

LC
Monotypic
HB 11.7–13.7 cm
T 13.5–16.5 cm
W 48–74 g

Long-furred Arboreal Rice Rat
Oecomys trinitatis

FR Oecomys de Trinidad
DE Langhaar-Baumreisratte
ES Rata arrocera arborícola de pelaje largo

0–2150 m

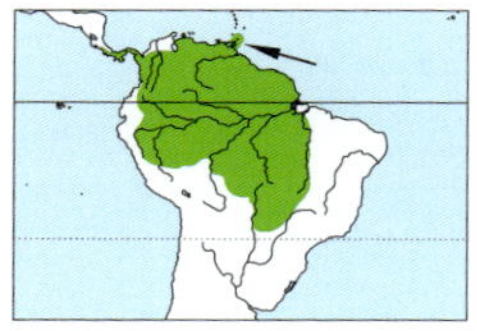

SW CR & PA E to TT, S to NE BO & C BR; exact limits unresolved

LC
Monotypic
HB 8.2–13 cm
T 10.1–15.2 cm
W 46–66 g

0–800 m

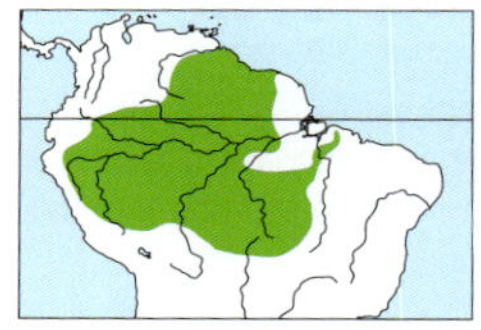

Robert's Arboreal Rice Rat
Oecomys roberti

FR Oecomys de Robert
DE Robert-Baumreisratte
ES Rata arrocera arborícola de Robert

SE CO, E & S VE, GY, SR, N & WC BR, E PE, extreme N & E BO

NE Monotypic
HB 8.9–13.9 cm
T 10.2–17.1 cm

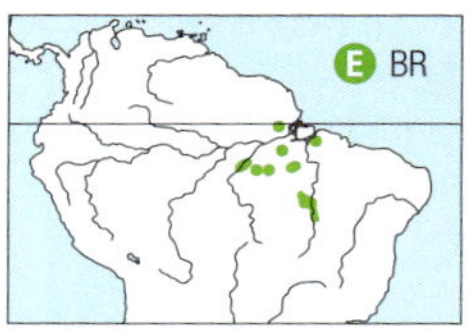

Tapajos Arboreal Rice Rat
Oecomys tapajinus

FR Oecomys du Tapajos
DE Tapajós-Baumreisratte
ES Rata arrocera arborícola de Tapajós

N (R Jari) and S (Tocantins, Araguaia, Xingu, Iriri & Tapajós rivers) of R Amazon

LC
Monotypic
HB 11.5–14.1 cm
T 13.7–16 cm
W 41–80 g

above 400 m

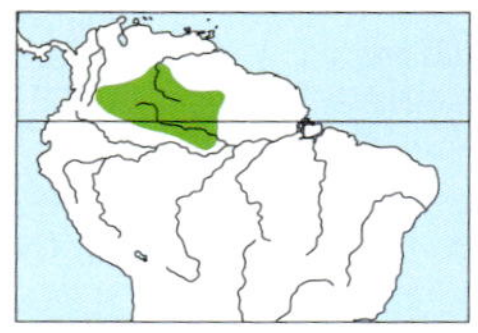

Unicolored Arboreal Rice Rat
Oecomys concolor

FR Oecomys de Wagner
DE Einfarb-Baumreisratte
ES Rata arrocera arborícola unicolor

E CO, S VE (S R Orinoco), NW BR (N of R Amazon)

LC
Monotypic
HB 10.1–12.1 cm
T 10.8–12.8 cm
W 30–48 g

0–1100 m

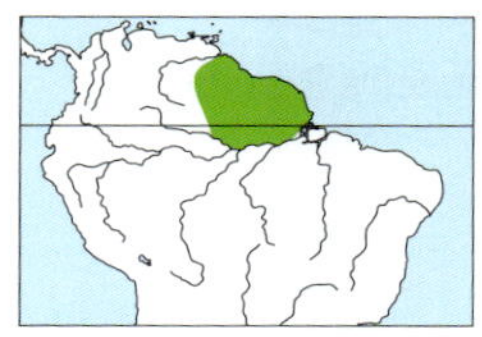

Guianan Arboreal Rice Rat
Oecomys auyantepui

FR Oecomys de l'Auyantepui
DE Guyana-Baumreisratte
ES Rata arrocera arborícola de Guayana

E VE, Guianas, NC BR (N of R Amazon)

LC Monotypic
HB 12.1–13.5 cm
T 14.2–16.8 cm
W 47–71 g

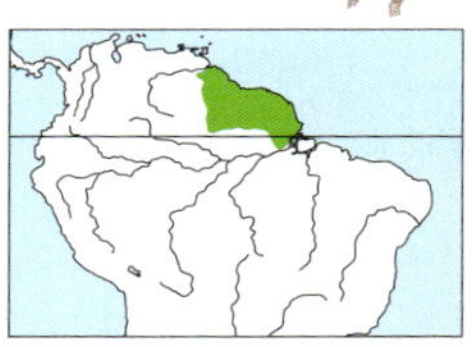

King Arboreal Rice Rat
Oecomys rex

FR Oecomys roi
DE Königsbaumreisratte
ES Rata arrocera arborícola real

E VE (Bolívar), Guianas, NC BR (N of R Amazon)

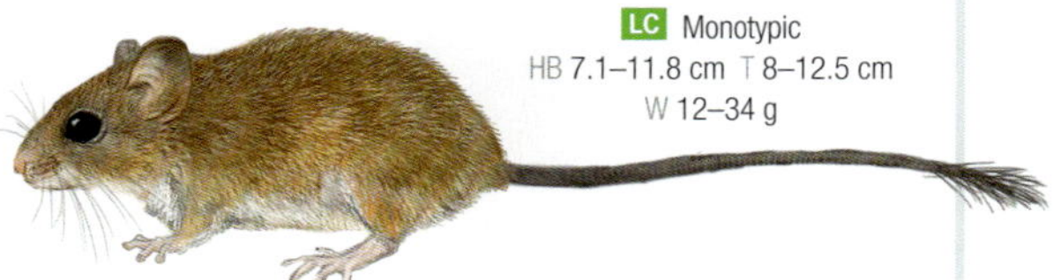

LC Monotypic
HB 7.1–11.8 cm T 8–12.5 cm
W 12–34 g

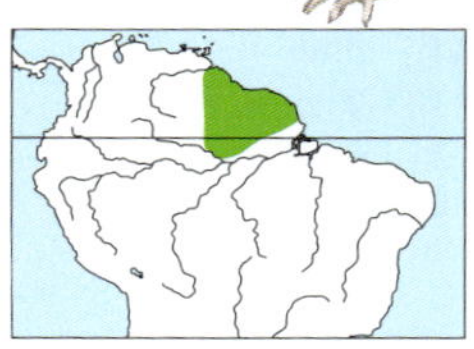

Red Arboreal Rice Rat
Oecomys rutilus

FR Oecomys roux
DE Rote Baumreisratte
ES Rata arrocera arborícola roja

E VE (Bolívar), Guianas, NC BR (N of R Amazon)

DD
Monotypic
HB 10.1–11.7 cm
T 11.7–13.9 cm
W c. 37 g

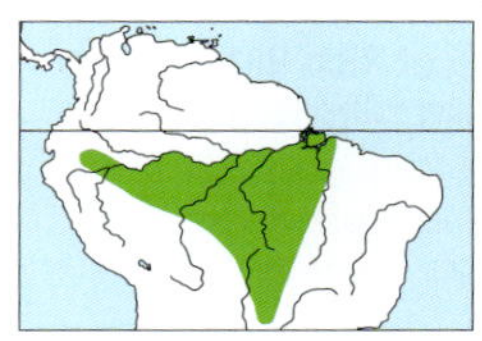

Brazilian Arboreal Rice Rat
Oecomys paricola

FR Oecomys du Para
DE Brasilien-Baumreisratte
ES Rata arrocera arborícola de Brasil

NE PE (Loreto), Amazonian BR (S of R Amazon); limits unresolved

LC Monotypic
HB 13.5–16.5 cm
T 15–18.4 cm
W 66–104 g

up to 1800 m

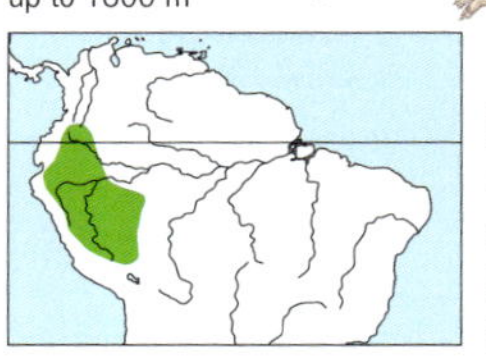

Foothill Arboreal Rice Rat
Oecomys superans

FR Oecomys des piémonts
DE Westamazonien-Baumreisratte
ES Rata arrocera arborícola de ladera

SC CO, E EC, E PE, extreme W BR

Dusky Arboreal Rice Rat
Oecomys phaeotis

FR Oecomys du Pérou
DE Dunkle Baumreisratte
ES Rata arrocera arborícola oscura

600–2100 m

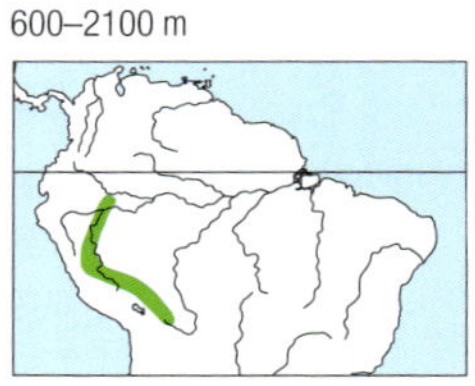

E Andean slopes from NE PE to C BO; limits unresolved

LC Monotypic
HB 10.2–11.8 cm T 10.7–12.3 cm
W 43–54 g

Franciscos' Arboreal Rice Rat
Oecomys franciscorum

FR Oecomys des Francisco
DE Francisco-Baumreisratte
ES Rata arrocera arborícola de los Franciscos

SC BR (Mato Grosso do Sul), NE AR (Formosa, Chaco); probably E PY

NE
Monotypic
HB 13.9–15.1 cm
T 16.6–17.6 cm
W 60–80 g

Mamore Arboreal Rice Rat
Oecomys mamorae

FR Oecomys du Mamoré
DE Mamoré-Baumreisratte
ES Rata arrocera arborícola de Mamoré

200–500 m

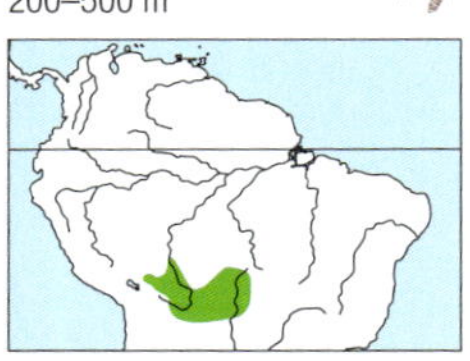

Much of BO, N PY, WC BR

LC
Monotypic
HB 12–17 cm
T 14.4–18 cm
W 48–120 g

Anderson's Arboreal Rice Rat
Oecomys sydandersoni

FR Oecomys d'Anderson
DE Anderson-Baumreisratte
ES Rata arrocera arborícola de Anderson

Woody islands in open savanna (Beni, Santa Cruz)

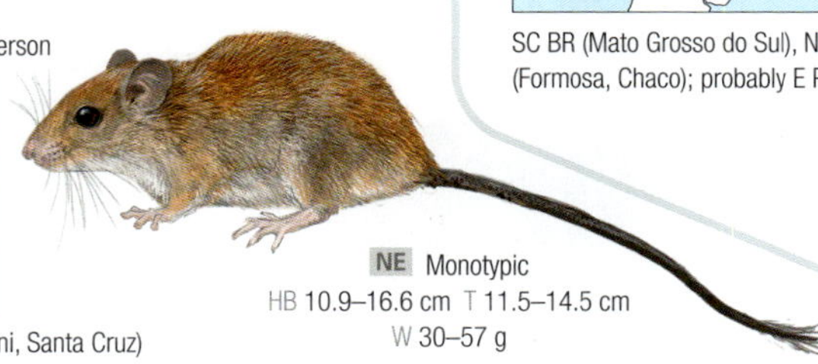

NE Monotypic
HB 10.9–16.6 cm T 11.5–14.5 cm
W 30–57 g

LC Monotypic
HB 12.1–13.9 cm
T 14.3–16.5 cm
W 47–71 g

0–770 m

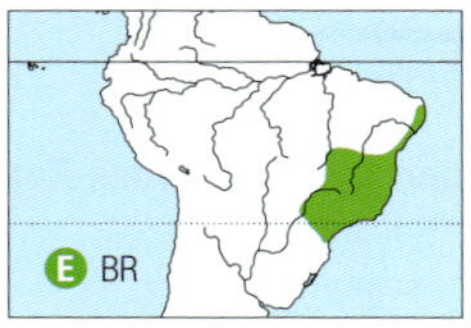

Atlantic Forest Arboreal Rice Rat
Oecomys catherinae
FR Oecomys de Santa Catarina
DE Atlantische Baumreisratte
ES Rata arrocera arborícola de bosque atlántico

Atlantic Forest and cerrado from Rio Grande do Norte S to Santa Catarina

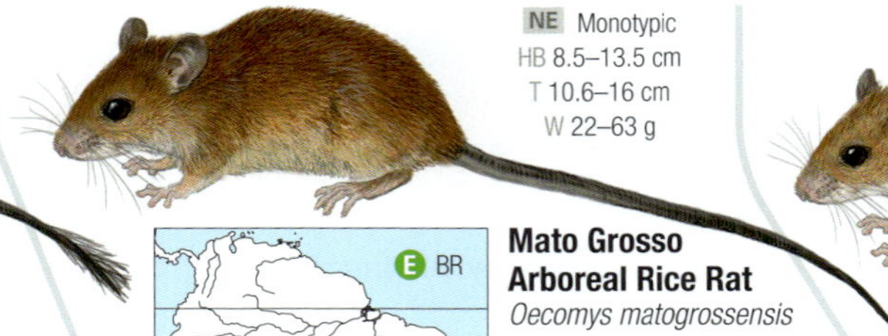

NE Monotypic
HB 8.5–13.5 cm
T 10.6–16 cm
W 22–63 g

Mato Grosso Arboreal Rice Rat
Oecomys matogrossensis
FR Oecomys du Mato Grosso
DE Mato-Grosso-Baumreisratte
ES Rata arrocera arborícola de Mato Grosso

Extreme SW Pará S to SW Mato Grosso (E to R Xingu), W to Vilhena (Rondônia)

Cleber's Arboreal Rice Rat
Oecomys cleberi
FR Oecomys de Cleber
DE Cleber-Baumreisratte
ES Rata arrocera arborícola de Cléber

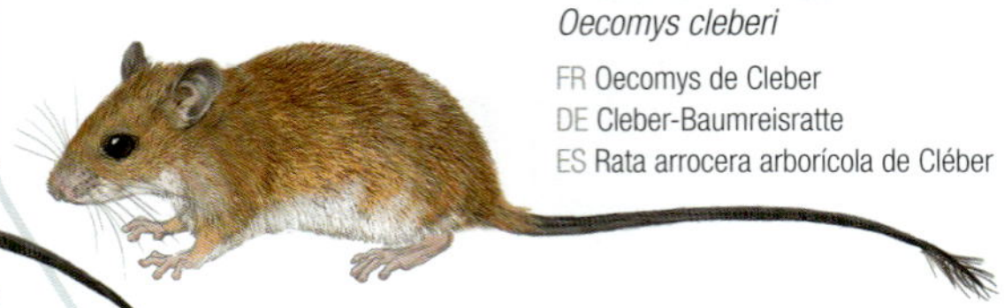

DD
Monotypic
HB 8.3–11.7 cm
T 8–12.3 cm

Riparian forests in S cerrado biome

Genus *EURYORYZOMYS*

(6 species)

LC
Monotypic
HB 13–16.6 cm
T 12.4–17.5 cm

0–1500 m

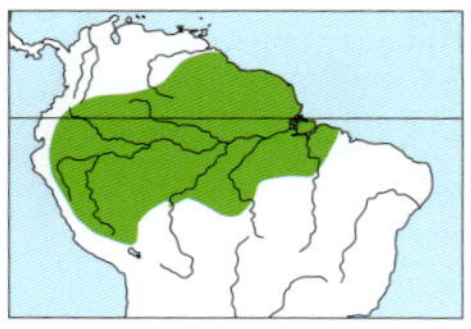

McConnell's Rice Rat
Euryoryzomys macconnelli
FR Oryzomys de McConnell
DE McConnell-Reisratte
ES Rata arrocera de McConnell

Mostly in Amazon Basin (S CO, S VE, Guianas, E EC, E PE, N BR)

DD
Monotypic
HB 12.8–14.2 cm
T 14.4–16 cm

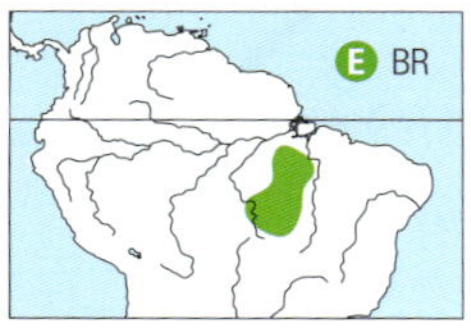

Emmons's Rice Rat
Euryoryzomys emmonsae
FR Oryzomys d'Emmons
DE Emmons-Reisratte
ES Rata arrocera de Emmons

Between R Xingu and R Tocantins (Pará, Mato Grosso)

LC Monotypic
HB 12.3–15.5 cm
T 13.4–16.2 cm

Big-headed Rice Rat
Euryoryzomys legatus
FR Oryzomys de Tarija
DE Großkopf-Reisratte
ES Rata arrocera de cabeza grande

SC BO (Chuquisaca, Santa Cruz, Tarija), NW AR (Jujuy, Salta)

LC Monotypic
HB 9.8–17.9 cm
T 9.9–18.1 cm
W 20–100 g

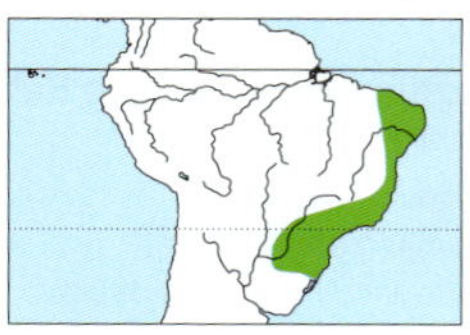

Russet Rice Rat
Euryoryzomys russatus
FR Oryzomys roux
DE Rostbraune Reisratte
ES Rata arrocera bermeja

E & S BR (from Ceará S to N Rio Grande do Sul), E PY, NE AR (Misiones)

VU
Monotypic
HB 14.1–16.1 cm
T 13.1–16 cm

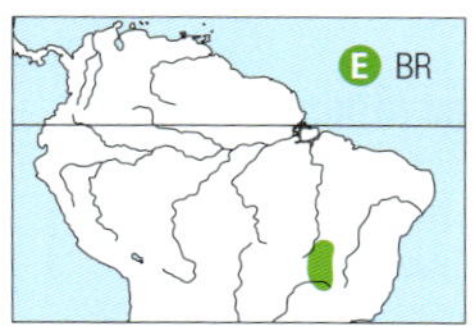

Buffy-sided Rice Rat
Euryoryzomys lamia
FR Oryzomys d'Araguari
DE Gelbbraune Reisratte
ES Rata arrocera de costados beige

Goiás, W Minas Gerais

LC
Monotypic
HB 10.8–16.3 cm
T 11.3–16 cm

50–2000 m

Elegant Rice Rat
Euryoryzomys nitidus
FR Oryzomys élégant
DE Elegante Reisratte
ES Rata arrocera elegante

SW Amazon Basin (E & SE PE, SW BR, N & E BO)

Genus *TRANSANDINOMYS*

(2 species)

LC
Monotypic
HB 10–14 cm
T 9–13 cm
W 39–75 g

0–1800 m

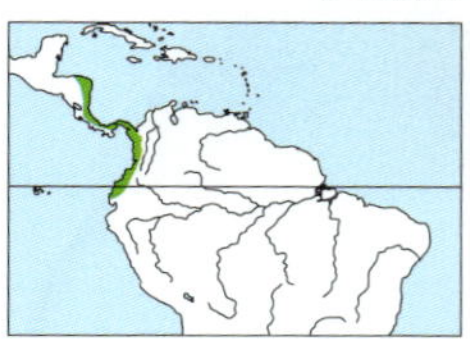

Long-whiskered Rice Rat
Transandinomys bolivaris
FR Oryzomys du Bolivar
DE Simón-Bolívar-Reisratte
ES Rata arrocera de vibrisas largas

Forests in E HN, E NI, CR, PA, W CO, W EC

LC
Monotypic
HB 12–15.1 cm
T 10.5–15.2 cm
W 47–74 g

0–1385 m

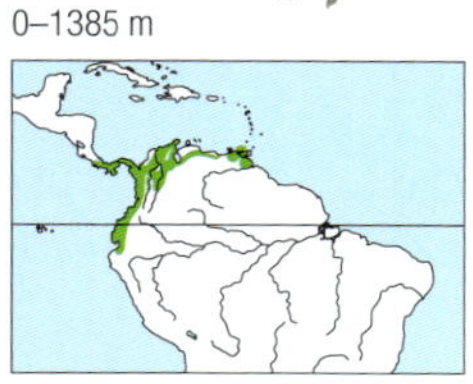

Talamancan Rice Rat
Transandinomys talamancae
FR Oryzomys de Talamanca
DE Talamanca-Reisratte
ES Rata arrocera de Talamanca

S CR, PA and N & W CO, E to N VE, S to W EC and extreme NW PE

LC Monotypic
HB 10–14 cm T 12–15 cm
W 30–55 g

0–1500 m

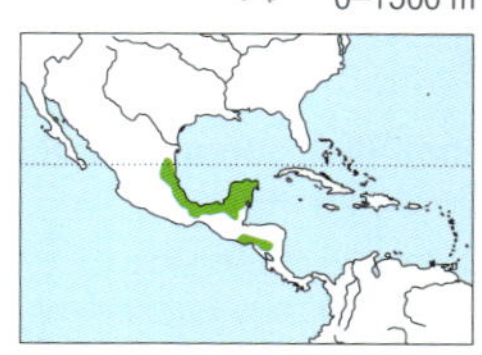

Long-nosed Rice Rat
Handleyomys rostratus
FR Oryzomys à long nez
DE Langnasen-Reisratte
ES Rata arrocera de hocico largo

E & S MX (with Yucatán Pen), N GT, SW HN, SV, NI

Genus *HANDLEYOMYS*

(9 species)

1550–2500 m

Sierra Madre Oriental and Oaxacan Highlands

VU Monotypic
HB 8–15.1 cm T 9.4–11.4 cm
W 20–44 g

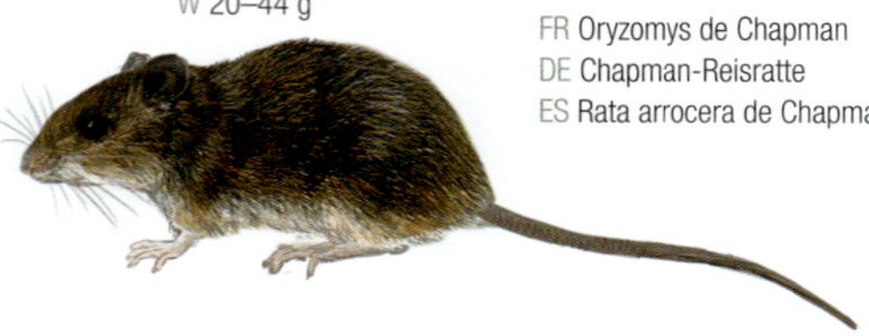

Chapman's Rice Rat
Handleyomys chapmani
FR Oryzomys de Chapman
DE Chapman-Reisratte
ES Rata arrocera de Chapman

LC Monotypic
HB 8.7–11.8 cm T 9.9–13.5 cm
W c. 25 g

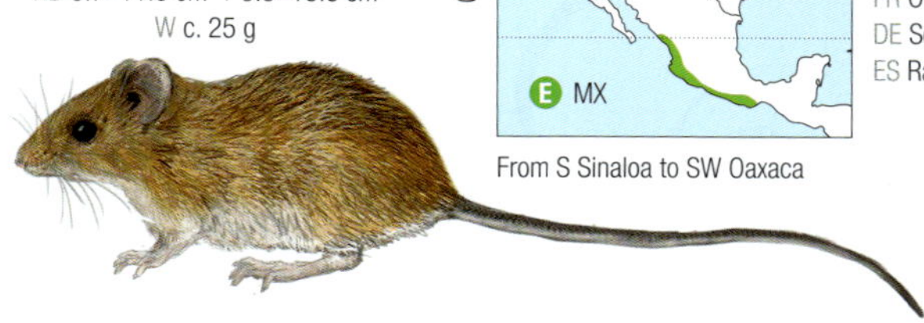

0–2000 m

From S Sinaloa to SW Oaxaca

Black-eared Rice Rat
Handleyomys melanotis
FR Oryzomys à oreilles noires
DE Schwarzohr-Reisratte
ES Rata arrocera de orejas negras

Guerrero Rice Rat
Handleyomys guerrerensis
FR Oryzomys du Guerrero
DE Guerrero-Reisratte
ES Rata arrocera de Guerrero

1000–2500 m

Sierra Madre del Sur

NE
Monotypic
HB 10.5–11 cm
T 11.6–11.8 cm

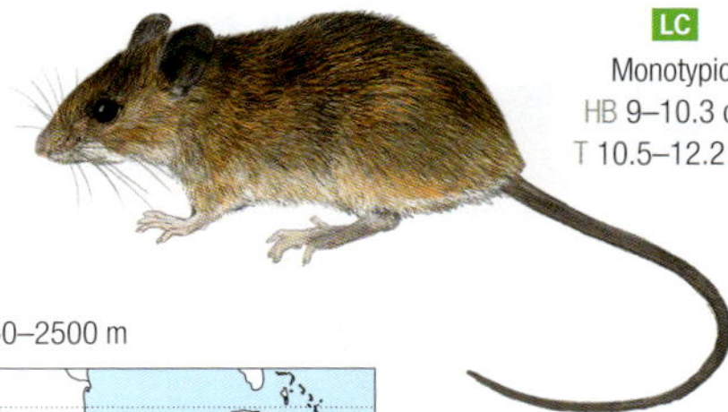

LC
Monotypic
HB 9–10.3 cm
T 10.5–12.2 cm

750–2500 m

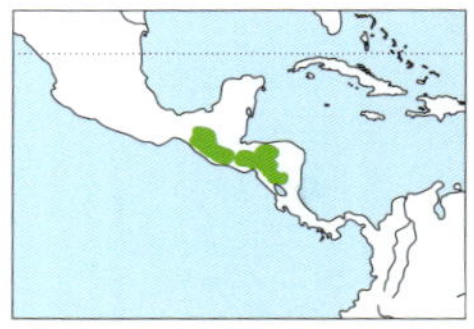

MX (Chiapas), HN, SV, NW NI

Cloud Forest Rice Rat
Handleyomys saturatior
FR Oryzomys sombre
DE Bergwald-Reisratte
ES Rata arrocera de bosque nuboso

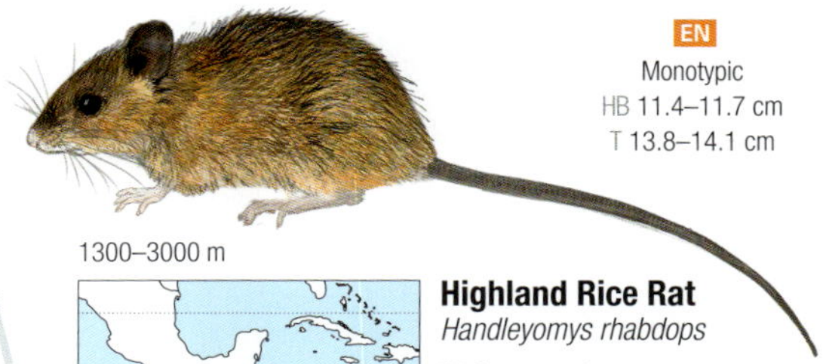

EN
Monotypic
HB 11.4–11.7 cm
T 13.8–14.1 cm

1300–3000 m

S MX (C Chiapas), WC & EC Guatemalan Highlands

Highland Rice Rat
Handleyomys rhabdops
FR Oryzomys de montagne
DE Hochland-Reisratte
ES Rata arrocera de tierras altas

LC Monotypic
HB 9–10.6 cm T 8.9–10.1 cm
W 20–32 g

0–2500 m

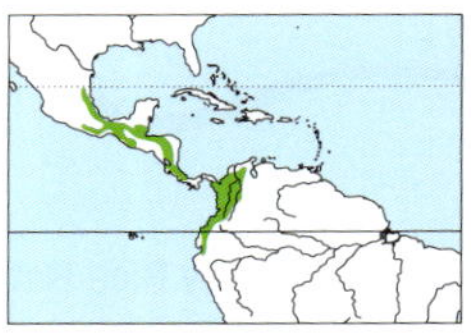

E & S MX through Middle America to NW CO & SW EC

Alfaro's Rice Rat
Handleyomys alfaroi
FR Oryzomys d'Alfaro
DE Alfaro-Reisratte
ES Rata arrocera de Alfaro

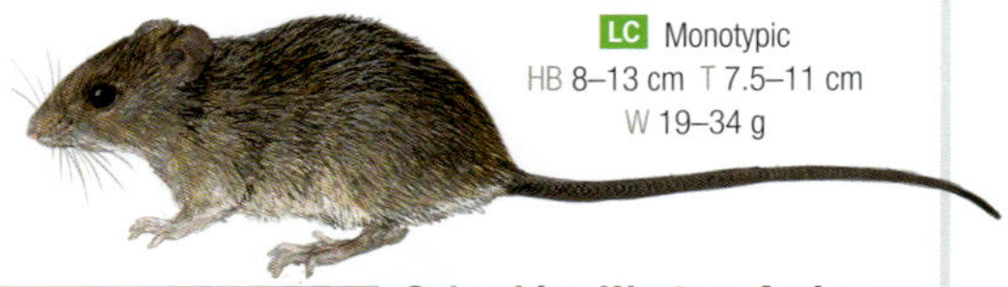

LC Monotypic
HB 8–13 cm T 7.5–11 cm
W 19–34 g

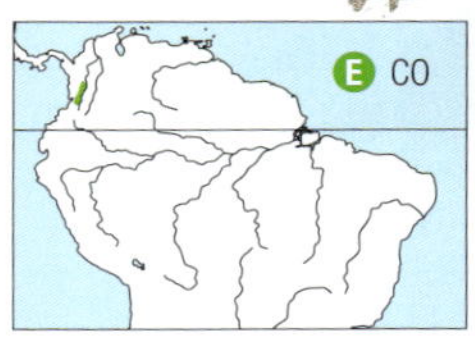

W Andes 1800–2530 m

Colombian Western Andes Cloud Forest Mouse
Handleyomys fuscatus
FR Oryzomys de Handley
DE Dunkelfuß-Reisratte
ES Ratón de bosque nuboso de Colombia occidental

LC Monotypic
HB 8.5–11.4 cm
T 7.5–10.6 cm
W 17–33 g

1700–2800 m

C Andes

Colombian Central Andes Cloud Forest Mouse
Handleyomys intectus
FR Oryzomys d'Antioquia
DE Weißfuß-Reisratte
ES Ratón de bosque nuboso de Colombia central

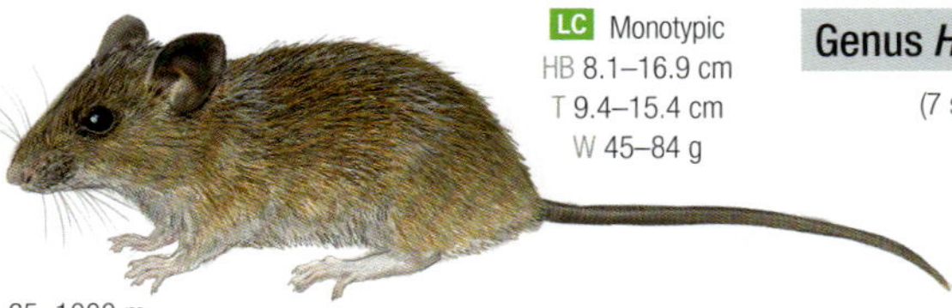

LC Monotypic
HB 8.1–16.9 cm
T 9.4–15.4 cm
W 45–84 g

65–1000 m

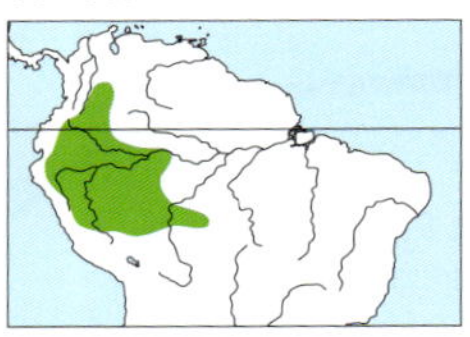

E Andes and W Amazon Basin (C & E CO, E EC, E PE, W BR, NW BO)

Western Amazonian Rice Rat
Hylaeamys perenensis
FR Oryzomys de Perené
DE Westamazonas-Reisratte
ES Rata arrocera del Amazonas occidental

Genus *HYLAEAMYS*
(7 species)

LC
Monotypic
HB 11.5–14.9 cm
T 10.7–12.8 cm
W 35–53 g

0–2000 m

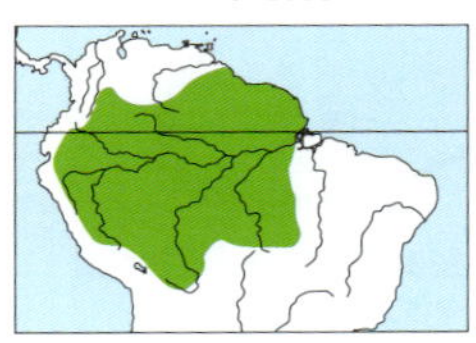

Amazon Basin (C & SE CO, S VE, Guianas, E EC, E PE and BR to C BO)

Yungas Rice Rat
Hylaeamys yunganus
FR Oryzomys des yungas
DE Yungas-Reisratte
ES Rata arrocera de Yungas

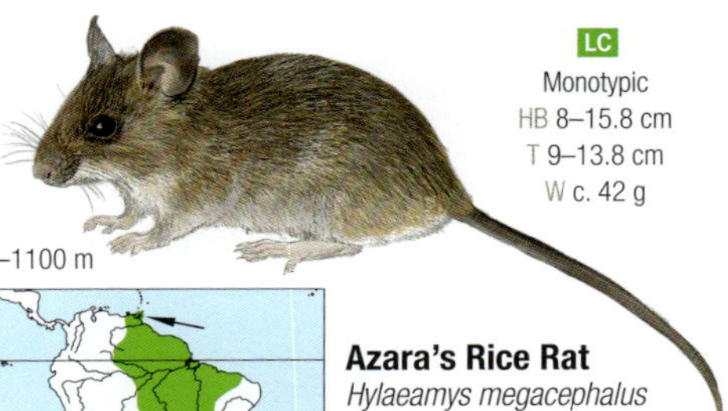

LC
Monotypic
HB 8–15.8 cm
T 9–13.8 cm
W c. 42 g

15–1100 m

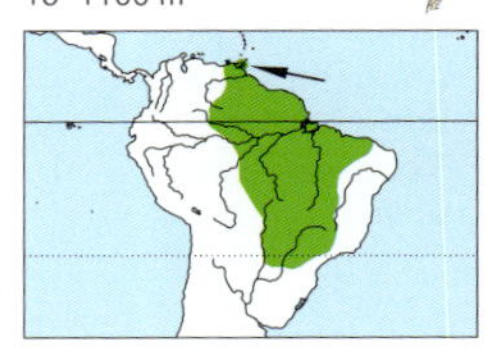

E & S VE, TT (Trinidad I), Guianas, N, C & SE BR, E PY

Azara's Rice Rat
Hylaeamys megacephalus
FR Oryzomys d'Azara
DE Azara-Reisratte
ES Rata arrocera de Azara

DD
Monotypic
HB 11.8–13.7 cm
T 12.7–18.9 cm

1160–1750 m

Few localities along upper R Pastaza (E Andean slope)

Tate's Rice Rat
Hylaeamys tatei
FR Oryzomys de Tate
DE Tate-Reisratte
ES Rata arrocera de Tate

Sowbug Rice Rat
Hylaeamys oniscus
FR Oryzomys du Pernambuco
DE Assel-Reisratte
ES Rata arrocera cochinita

NT
Monotypic
HB 14–16.2 cm
T 13.5–16.2 cm
W 57–95 g

0–400 m

N of R São Francisco (Paraíba, Pernambuco, Alagoas)

DD Monotypic
HB 13.1–15.2 cm
T 11–13 cm
W c. 57 g

c. 700 m

Restricted to E bank tributaries of R Iténez (Beni, Santa Cruz)

Bolivian Rice Rat
Hylaeamys acritus
FR Oryzomys de Bolivie
DE Bolivien-Reisratte
ES Rata arrocera de Bolivia

Atlantic Forest Rice Rat
Hylaeamys laticeps
FR Oryzomys côtier
DE Atlantische Reisratte
ES Rata arrocera de bosque atlántico

VU Monotypic
HB 12.2–17.4 cm
T 10–15.7 cm
W 51–97 g

0–300 m

E Atlantic Forest from S Bahia to N Rio de Janeiro

Genus *OLIGORYZOMYS*
(22 species)

LC 6 sspp
HB 7.5–8.9 cm
T 8.2–11.6 cm
W 10–17.5 g

0–1500 m

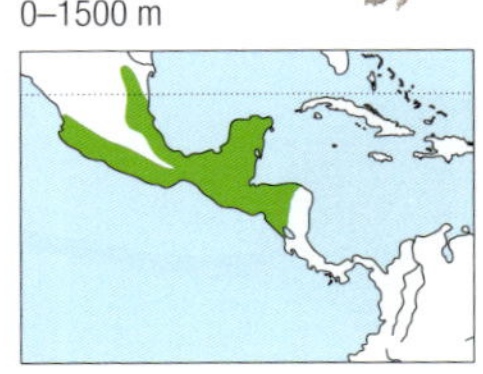

W, E & S MX (with Yucatán Pen) S to extreme NW NI

Fulvous Pygmy Rice Rat
Oligoryzomys fulvescens
FR Colilargo fauve
DE Gelbbraune Zwergreisratte
ES Rata arrocera pigmea leonada

Costa Rican Pygmy Rice Rat
Oligoryzomys costaricensis

FR Colilargo du Costa Rica
DE Costa-Rica-Zwergreisratte
ES Rata arrocera pigmea de Costa Rica

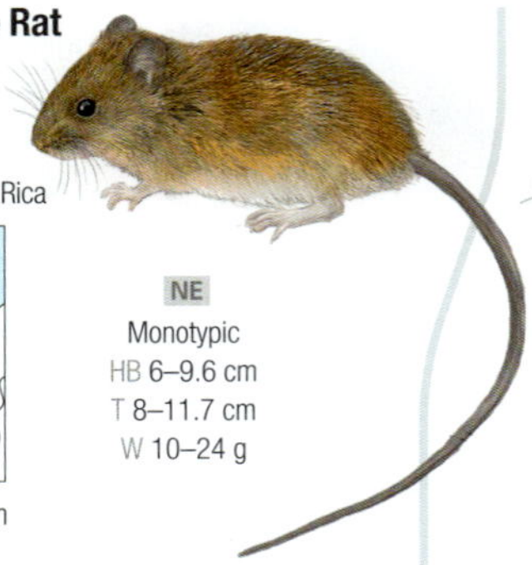

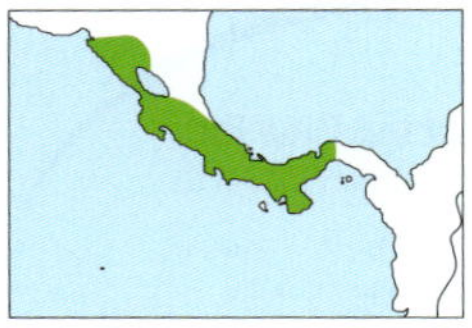

W NI, CR, W & C PA 0–1200 m

NE
Monotypic
HB 6–9.6 cm
T 8–11.7 cm
W 10–24 g

Sprightly Pygmy Rice Rat
Oligoryzomys vegetus

FR Colilargo vivace
DE Muntere Zwergreisratte
ES Rata arrocera pigmea briosa

LC
Monotypic
HB 7.5–10.5 cm
T 11–13.1 cm
W 11–16 g

840–3000 m

Highlands of CR and W PA

Delicate Pygmy Rice Rat
Oligoryzomys delicatus

NE Monotypic
HB 7.9–9.1 cm
T 9.5–11.9 cm
W 14–28 g

0–2000 m

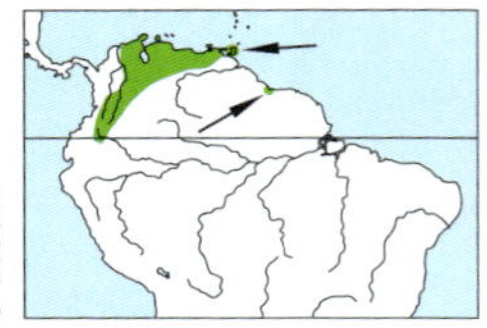

CO, extreme N EC, N VE, TT (Trinidad I), Guianas

FR Colilargo délicat
DE Zierliche Zwergreisratte
ES Rata arrocera pigmea delicada

Grayish Pygmy Rice Rat
Oligoryzomys griseolus

LC
Monotypic
HB 7.6–8.5 cm
T 9.6–11.4 cm

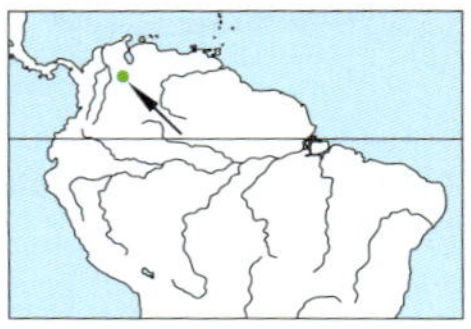

E CO (E Andes), extreme W VE (Táchira Andes)

FR Colilargo grisonnant
DE Graue Zwergreisratte
ES Rata arrocera pigmea gris

Hairy Pygmy Rice Rat
Oligoryzomys messorius

FR Colilargo glaneur
DE Haarige Zwergreisratte
ES Rata arrocera pigmea peluda

NE
Monotypic
HB 7.5–9 cm
T 8.5–10.5 cm
W 12–23 g

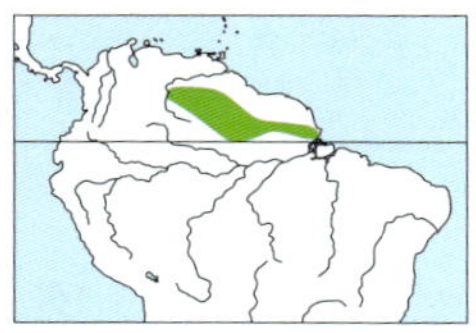

S VE, S GY, S SR, N BR (Roraima, Amapá)

Tschudi's Pygmy Rice Rat
Oligoryzomys destructor

LC
2 sspp
HB 9.1–9.6 cm
T 11.1–12.8 cm

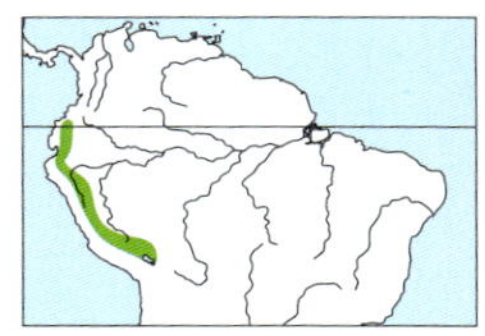

E Andean slopes of EC, PE & BO; possibly CO

FR Colilargo de Tschudi
DE Tschudi-Zwergreisratte
ES Rata arrocera pigmea de Tschudi

Sandy Pygmy Rice Rat
Oligoryzomys arenalis

LC
Monotypic
HB c. 8 cm
T c. 10 cm

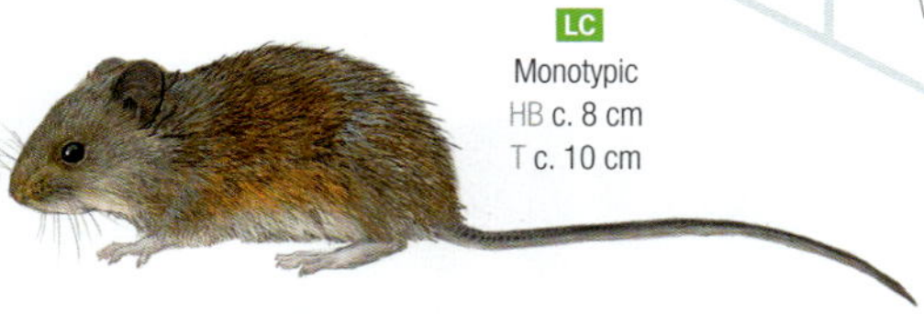

N coast and W slope of Andes; perhaps S EC

FR Colilargo des sables
DE Sand-Zwergreisratte
ES Rata arrocera pigmea de arena

Andean Pygmy Rice Rat
Oligoryzomys andinus

LC
Monotypic
HB c. 10.4 cm
T c. 14.9 cm

1700–4000 m

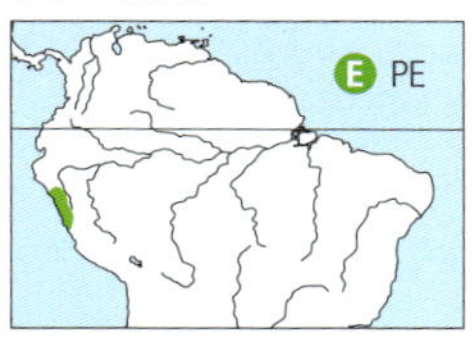

W Central Andean slopes

FR Colilargo des Andes
DE Anden-Zwergreisratte
ES Rata arrocera pigmea de los Andes

D'Elia's Pygmy Rice Rat
Oligoryzomys guille

NE
Monotypic
HB 6–12.3 cm
T 5.5–13.6 cm
W 6–31 g

0–4200 m

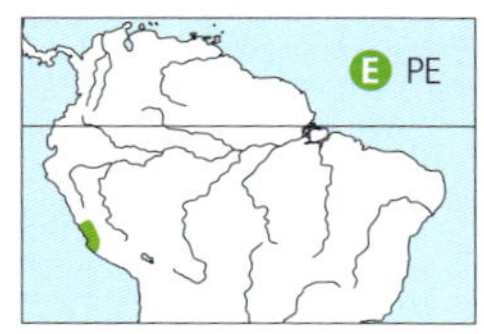

W slope of Andes from N Lima S to N Arequipa

FR Colilargo de Guillermo
DE D'Elía-Zwergreisratte
ES Rata arrocera pigmea de Guillermo D'Elía

Utiariti Pygmy Rice Rat
Oligoryzomys utiaritensis

LC
Monotypic
HB 7.8–10.4 cm
T 10.3–13.5 cm
W 14–32 g

110–570 m

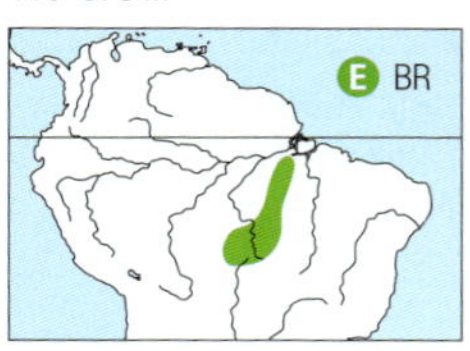

From SW Pará to NW Mato Grosso

FR Colilargo d'Utiariti
DE Utiariti-Zwergreisratte
ES Rata arrocera pigmea de Utiariti

Straw-colored Pygmy Rice Rat
Oligoryzomys stramineus

FR Colilargo jaune-paille DE Strohfarbene Zwergreisratte
ES Rata arrocera pigmea pajiza

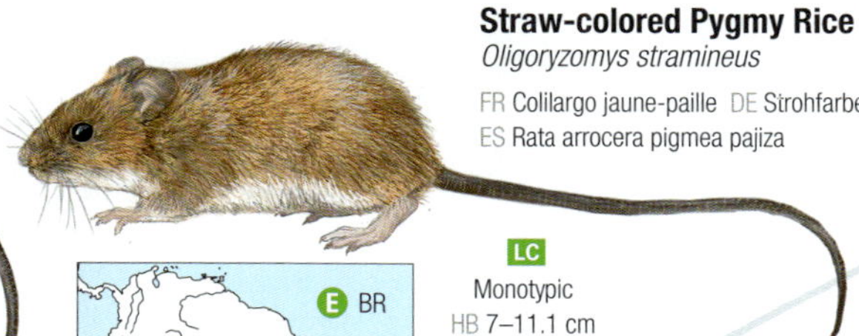

LC
Monotypic
HB 7–11.1 cm
T 9.5–13.4 cm

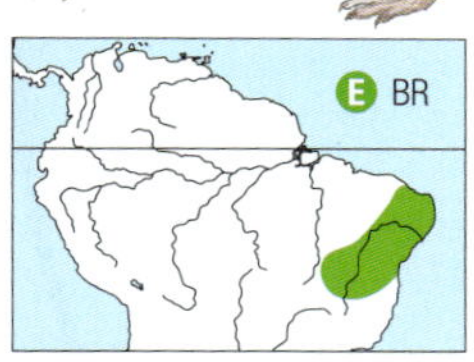

Caatinga and cerrado (Paraiba & Pernambuco S to Goiás & Minas Gerais)

Mato Grosso Pygmy Rice Rat
Oligoryzomys mattogrossae

FR Colilargo du Mato Grosso
DE Mato-Grosso-Zwergreisratte
ES Rata arrocera pigmea de Mato Grosso

NE
Monotypic
HB 9–9.5 cm
T 11.5–13 cm
W 16.4–18.8 g

Open areas in NE & SC BR and E PY

Highlands Pygmy Rice Rat
Oligoryzomys rupestris

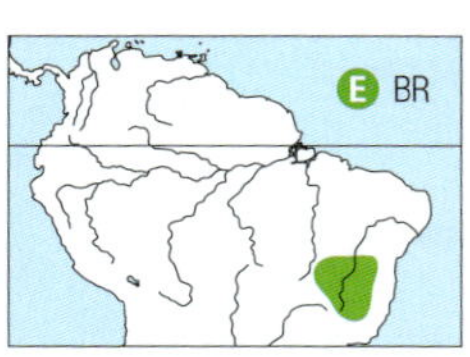

Known from few localities in Bahia, Goiás, Minas Gerais

FR Colilargo des rochers
DE Hochland-Zwergreisratte
ES Rata arrocera pigmea de tierras altas

DD
Monotypic
HB 7.6–9.9 cm
T 11.4–13.8 cm
W 10–20 g

Black-footed Pygmy Rice Rat
Oligoryzomys nigripes

FR Colilargo à pattes noires
DE Schwarzfuß-Zwergreisratte
ES Rata arrocera pigmea de pies negros

LC
Monotypic
HB 7.8–11.6 cm
T 10–14.9 cm
W 16–37 g

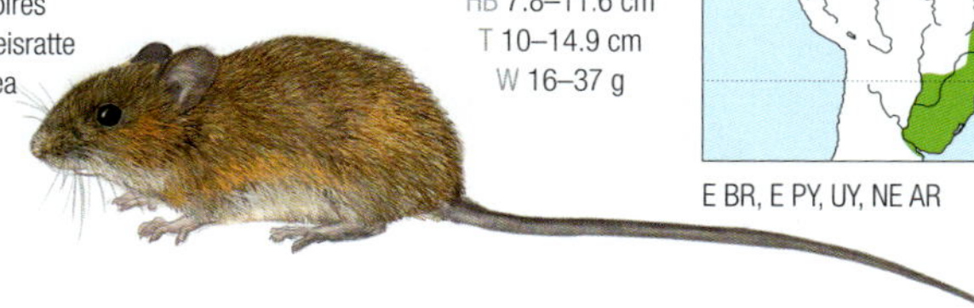

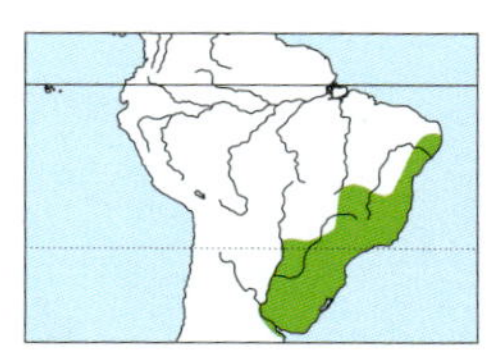

E BR, E PY, UY, NE AR

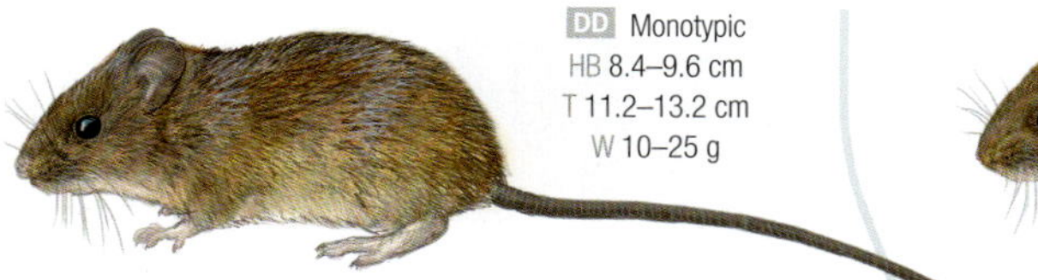

DD Monotypic
HB 8.4–9.6 cm
T 11.2–13.2 cm
W 10–25 g

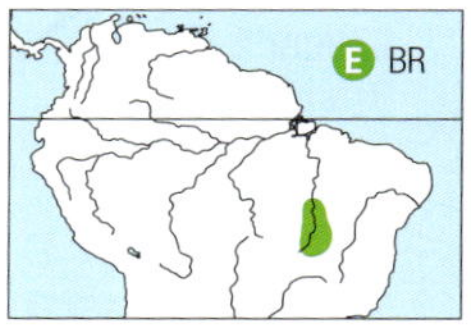

Cerrado of Tocantins and Goiás

Moojen's Pygmy Rice Rat
Oligoryzomys moojeni
FR Colilargo de Moojen
DE Moojen-Zwergreisratte
ES Rata arrocera pigmea de Moojen

DD
Monotypic
HB 7.8–10.1 cm
T 11.2–13.5 cm
W 12–34 g

700–2900 m

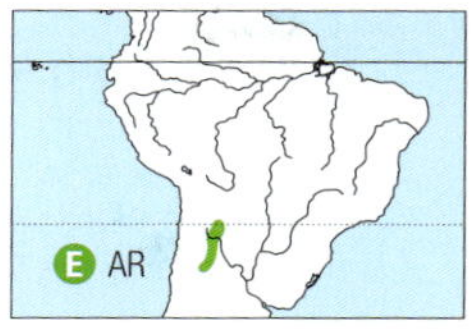

Highlands from Jujuy & Salta S to La Rioja; perhaps S BO

San Javier's Pygmy Rice Rat
Oligoryzomys brendae
FR Colilargo de Brenda
DE San-Javier-Zwergreisratte
ES Rata arrocera pigmea de San Javier

NE Monotypic
HB 8–10.8 cm
T 12–14.2 cm
W 21–33 g

1280–1960 m

Cloud mountain forests (Yungas) along E slope of Andes

Pacheco's Pygmy Rice Rat
Oligoryzomys pachecoi
FR Colilargo de Pacheco
DE Pacheco-Zwergreisratte
ES Rata arrocera pigmea de Pacheco

LC
Monotypic
HB c. 9.7 cm
T 10.5–15 cm
W c. 25 g

SE BO, SW BR, W & C PY, N AR

Chacoan Pygmy Rice Rat
Oligoryzomys chacoensis
FR Colilargo du Chaco
DE Chaco-Zwergreisratte
ES Rata arrocera pigmea del Chaco

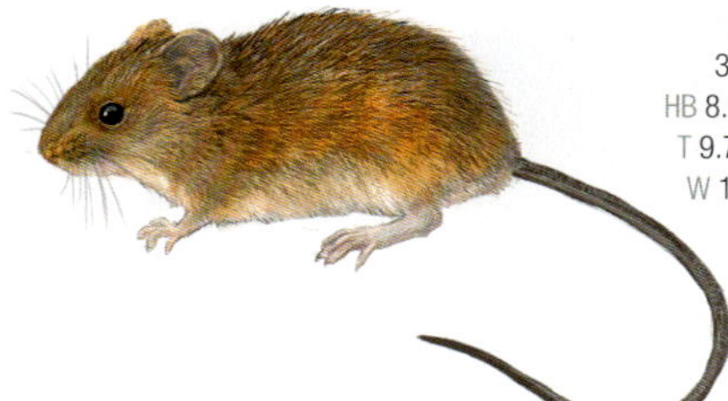

LC
3 sspp
HB 8.1–9.3 cm
T 9.7–13 cm
W 15–21 g

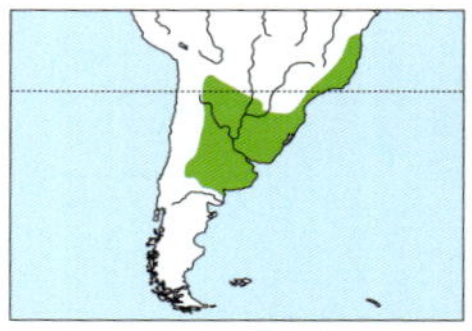

E BR and extreme SC BO (Tarija) S to N AR; limits uncertain

Yellow Pygmy Rice Rat
Oligoryzomys flavescens
FR Colilargo doré
DE Gelbe Zwergreisratte
ES Rata arrocera pigmea dorada

LC
Monotypic
HB 9.3–10.3 cm
T 10.9–12.7 cm
W 33–41 g

0–2000 m

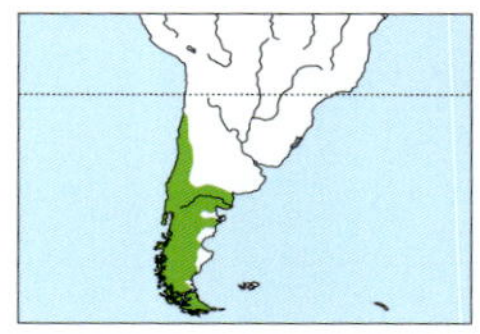

CL, W, SC & S AR

Long-tailed Pygmy Rice Rat
Oligoryzomys longicaudatus
FR Colilargo du Chili
DE Langschwanz-Zwergreisratte
ES Rata arrocera pigmea de cola larga

LC Monotypic
HB 7.8–9.8 cm
T 7.4–12.5 cm
W 14–19 g

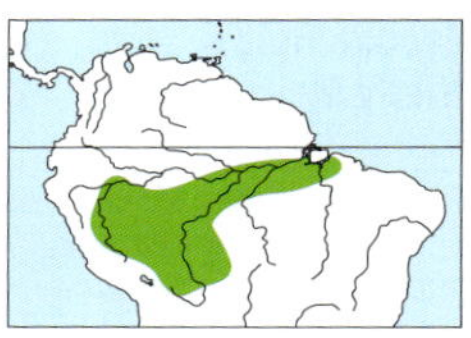

Amazon Basin of E PE, BR & N BO

Small-eared Pygmy Rice Rat
Oligoryzomys microtis
FR Colilargo à petites oreilles
DE Kleinohr-Zwergreisratte
ES Rata arrocera pigmea de orejas pequeñas

Genus *NEACOMYS*

(23 species)

LC
Monotypic
HB 7–8.3 cm
T 6.5–8.1 cm
W 10–19 g

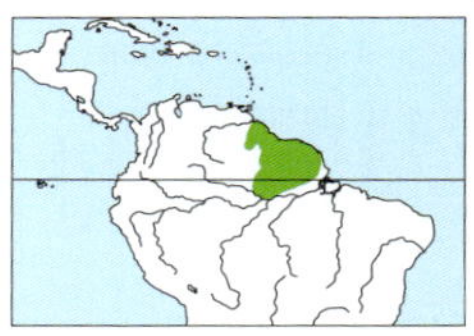

Extreme E VE, Guianas, N BR (Amazonas, Pará, Amapá)

Paracou Bristly Mouse
Neacomys paracou
FR Néacomys de Paracou
DE Paracou-Stachelreisratte
ES Ratón erizado de Paracou

NE Monotypic
HB c. 8 cm
T c. 10.2 cm
W c. 14 g

1595–2125 m

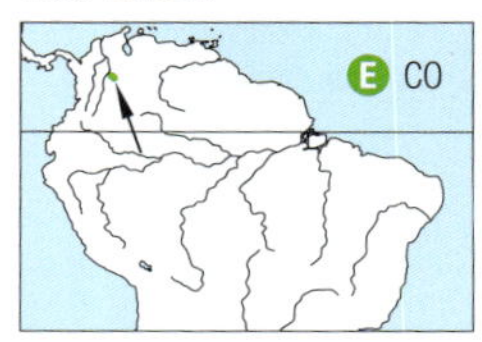

Known from 2 sites in Serranía de los Yariguíes (Santander); limits uncertain

Serrano Bristly Mouse
Neacomys serranensis
FR Néacomys des Yariguíes
DE Serrano-Stachelreisratte
ES Ratón erizado de Yariguíes

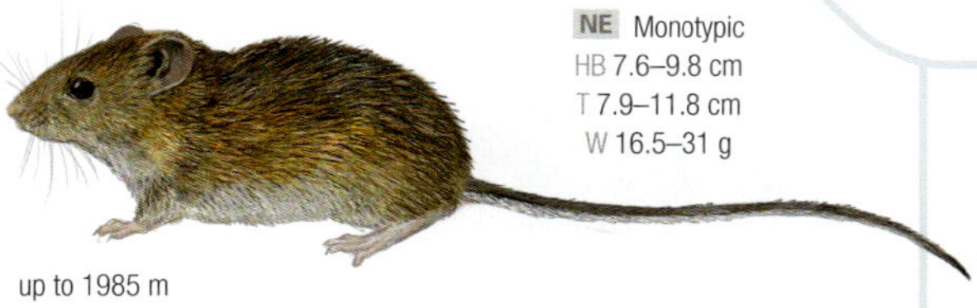

NE Monotypic
HB 7.6–9.8 cm
T 7.9–11.8 cm
W 16.5–31 g

up to 1985 m

S PE (Puno) to C BO (W Santa Cruz)

Vargas Llosa's Bristly Mouse
Neacomys vargasllosai
FR Néacomys de Vargas Llosa
DE Vargas-Llosa-Stachelreisratte
ES Ratón erizado de Vargas Llosa

LC Monotypic
HB 7.5–10.5 cm
T 8.8–12.2 cm
W 15–35 g

1125–2375 m

S R Marañón to N R Huallaga (Amazonas, San Martín, Huánuco)

Large Bristly Mouse
Neacomys spinosus
FR Néacomys épineux
DE Borsten-Stachelreisratte
ES Ratón erizado grande

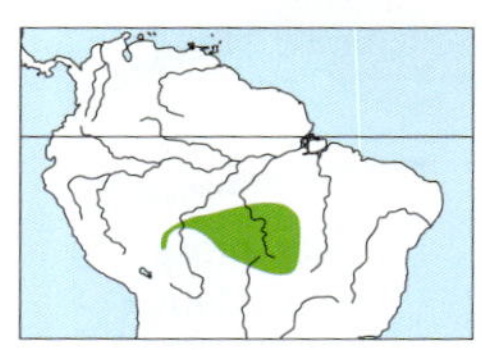

WC BR (Rondônia, Mato Grosso), N & E BO (Pando, La Paz, Santa Cruz)

Pleasant Bristly Mouse
Neacomys amoenus
FR Néacomys charmant
DE Schöne Stachelreisratte
ES Ratón erizado cordial

NE Monotypic
HB 6.5–7.9 cm
T 7.2–10.2 cm
W 20–30 g

Carcelen's Bristly Mouse
Neacomys carceleni
FR Néacomys de Carcelen
DE Carcelén-Stachelreisratte
ES Ratón erizado de Carcelén

200–1885 m

E EC (Napo S to Zamora Chinchipe) & E PE (Loreto S to Madre de Dios)

NE
Monotypic
HB 6.5–7.9 cm
T 7.2–10.2 cm
W 20–30 g

NE Monotypic
HB 6.4–7.5 cm T 8–11.9 cm
W 12.5–18 g

Golden-bellied Bristly Mouse
Neacomys auriventer
FR Néacomys à ventre doré
DE Goldbauch-Stachelreisratte
ES Ratón erizado de vientre dorado

1460–1885 m

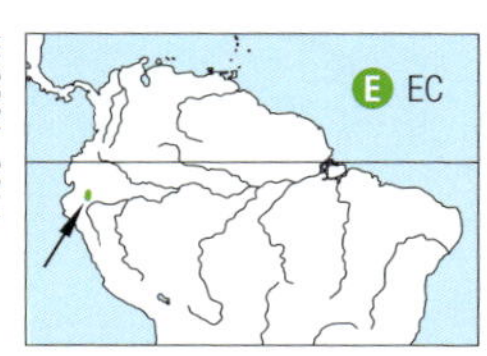

Known from 4 localities of Cordillera del Cóndor

LC Monotypic
HB 6.4–8.1 cm T 7.1–9.2 cm
W 11–19 g

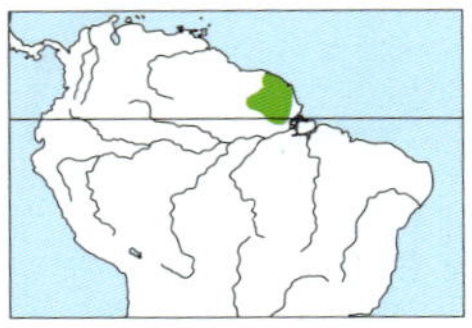

SE SR, GF, NE BR (Amapá, Pará)

Dubost's Bristly Mouse
Neacomys dubosti
FR Néacomys de Dubost
DE Dubost-Stachelreisratte
ES Ratón erizado de Dubost

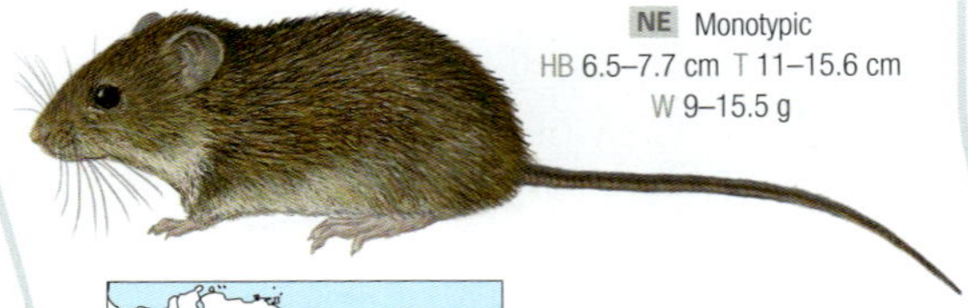

NE Monotypic
HB 6.5–7.7 cm T 11–15.6 cm
W 9–15.5 g

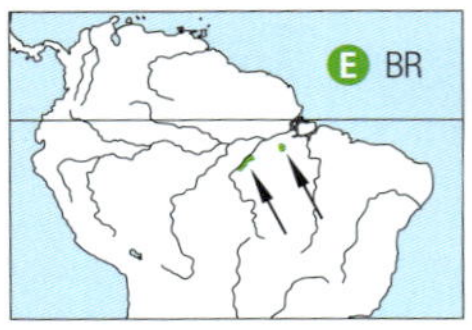

E upper & middle R Tapajós and W lower R Xingu

Voss's Bristly Mouse
Neacomys vossi
FR Néacomys de Voss
DE Voss-Stachelreisratte
ES Ratón erizado de Tapajó

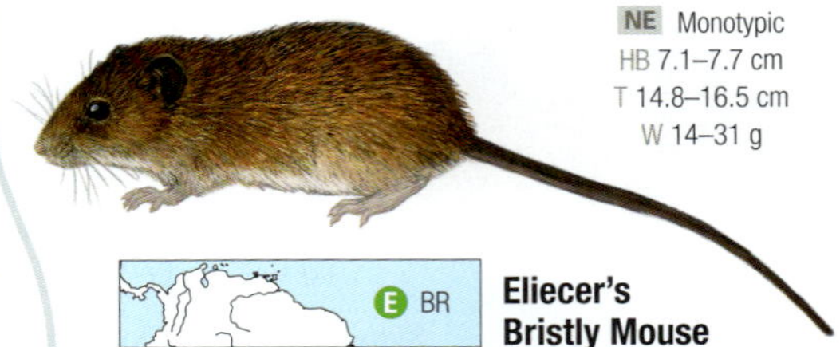

NE Monotypic
HB 7.1–7.7 cm
T 14.8–16.5 cm
W 14–31 g

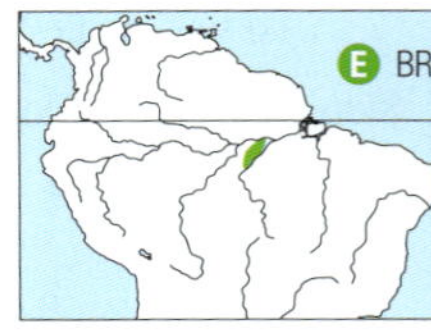

W bank of upper & medium R Tapajós (Pará)

Eliecer's Bristly Mouse
Neacomys elieceri
FR Néacomys d'Eliécer
DE Eliécer-Stachelreisratte
ES Ratón espinado de Eliécer

NE Monotypic
HB 6.8–7.5 cm
T 13.8–15.6 cm
W 11.5–18.5 g

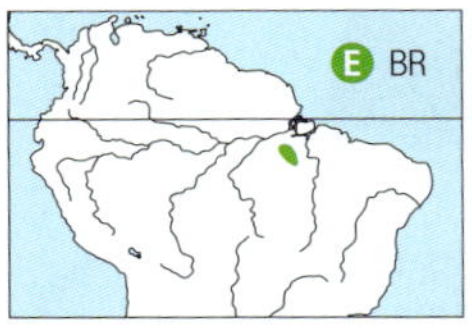

E lower R Xingu and Serra de Carajás region (SE Pará)

Xingu Bristly Mouse
Neacomys xingu
FR Néacomys du Xingu
DE Xingu-Stachelreisratte
ES Ratón erizado de Xingu

NE Monotypic
HB 6.2–7.7 cm
T 12.8–15.3 cm
W 9–16.5 g

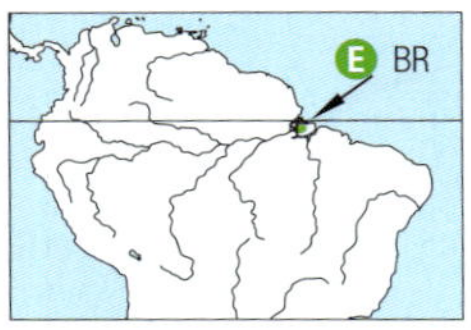

Known only from Marajó I (Pará)

Marajo Bristly Mouse
Neacomys marajoara
FR Néacomys de Marajo
DE Marajó-Stachelreisratte
ES Ratón erizado de Marajó

LC Monotypic
HB 6.4–8.5 cm
T 6.7–8.1 cm
W 10–16 g

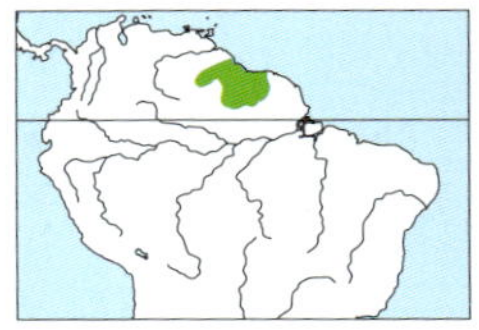

SE VE, GY, SR, probably adjacent NE BR

Guianan Bristly Mouse
Neacomys guianae
FR Néacomys du Guyana
DE Guyana-Stachelreisratte
ES Ratón erizado de Guayana

NE Monotypic
HB 6.9–10 cm
T 13–15.1 cm
W 13–16 g

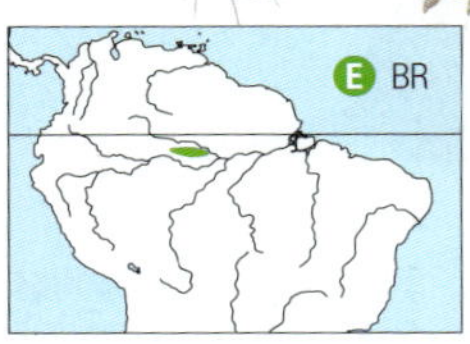

Right bank of R Jaú and left bank of R Japurá (Amazonas)

Jau Bristly Mouse
Neacomys jau
FR Néacomys du Jau
DE Jaú-Stachelreisratte
ES Ratón erizado de Jaú

LC
Monotypic
HB 6.5–7.9 cm
T 7–8.4 cm
W 10–15 g

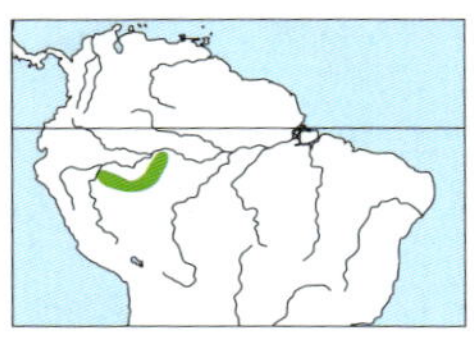

PE (R Gálvez Basin, Loreto), BR (C & lower R Juruá, Amazonas)

Minute Bristly Mouse
Neacomys minutus
FR Néacomys nain
DE Kleine Stachelreisratte
ES Ratón erizado ínfimo

NE
Monotypic
HB 6.5–8 cm
T 7.5–8.5 cm

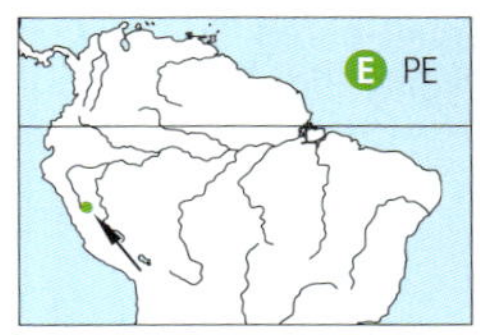

Known only from Tingo María NP (Huánuco)

Macedo Ruiz's Bristly Mouse
Neacomys macedoruizi
FR Néacomys de Macedo Ruiz
DE Macedo-Ruiz-Stachelreisratte
ES Ratón erizado de Macedo Ruiz

NE Monotypic
HB 6.7–7.9 cm
T 14.8–16.3 cm
W 11–15 g

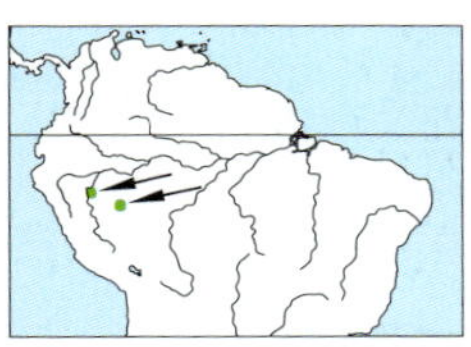

BR (both banks of R Juruá, Amazonas), PE (E bank of R Ucayali, Loreto)

Upper Jurua Bristly Mouse
Neacomys aletheia
FR Néacomys du Jurua
DE Juruá-Stachelreisratte
ES Ratón erizado verdadero

LC Monotypic
HB 6.5–7 cm T 6.8–9 cm
W 11–17 g

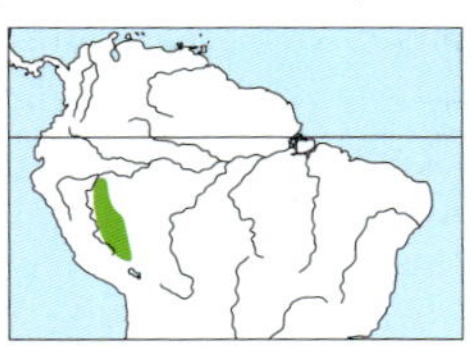

W Amazon Basin in E & SE PE and extreme W BR; probably adjacent BO

Musser's Bristly Mouse
Neacomys musseri
FR Néacomys de Musser
DE Musser-Stachelreisratte
ES Ratón erizado de Musser

NE
Monotypic
HB 6.2–9.9 cm
T 6.1–8.7 cm

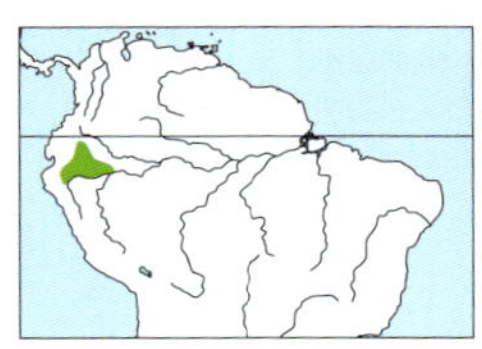

N of R Amazon in NE PE (Amazonas, Loreto) & E EC (Pastaza, Napo)

Rosalind's Bristly Mouse
Neacomys rosalindae
FR Néacomys de Rosalind
DE Rosalind-Stachelreisratte
ES Ratón erizado de Rosalind

Narrow-footed Bristly Mouse
Neacomys tenuipes
FR Néacomys à pattes fines
DE Schmalfuß-Stachelreisratte
ES Ratón erizado de pies estrechos

400–1750 m

Forests in NC & W CO, N VE and NW EC

LC Monotypic
HB 7.2–9.7 cm
T 7.4–10.8 cm
W 12–22 g

550–600 m

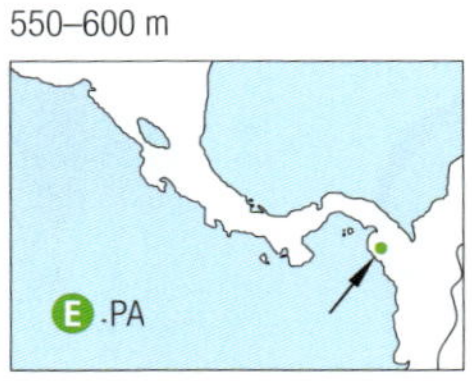

Known only from extreme E (Darién)

DD Monotypic
HB 6–8.8 cm T 7.4–8.7 cm

Painted Bristly Mouse
Neacomys pictus
FR Néacomys coloré
DE Bunte Stachelreisratte
ES Ratón erizado pintado

NE Monotypic
HB 6.7–8.2 cm T 9.2–11.8 cm
W 12.2–14.5 g

1050–1470 m

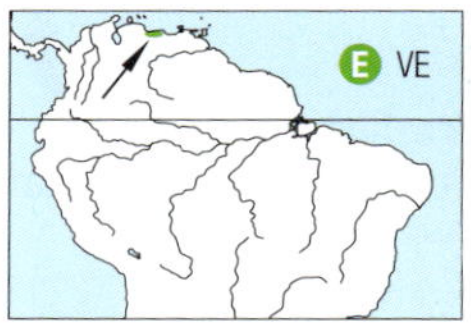

Leila Pessôa's Bristly Mouse
Neacomys leilae
FR Néacomys de Leila
DE Leila-Pessôa-Stachelreisratte
ES Ratón erizado de Leila

Central Coastal Cordillera (Aragua, Distrito Federal, Miranda)

Forest Small Rice Rat
Microryzomys minutus
FR Oryzomys nain
DE Kleinstreisratte
ES Rata arrocera pequeña de bosque

LC
Monotypic
HB 6.2–9.9 cm
T 11–12.6 cm
W 10–24 g

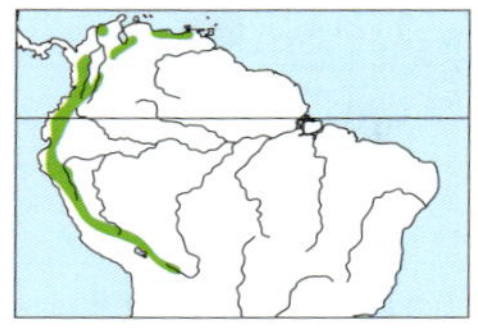

800–4265 m

VE (Coastal Ranges, Cordillera de Mérida), N & C Andes from CO to WC BO

Genus *MICRORYZOMYS*

(2 species)

LC
Monotypic
HB 6.2–9.9 cm
T 10.4–13.1 cm
W 10–24 g

Highland Small Rice Rat
Microryzomys altissimus
FR Oryzomys des sommets
DE Hochland-Kleinstreisratte
ES Rata arrocera pequeña de tierras altas

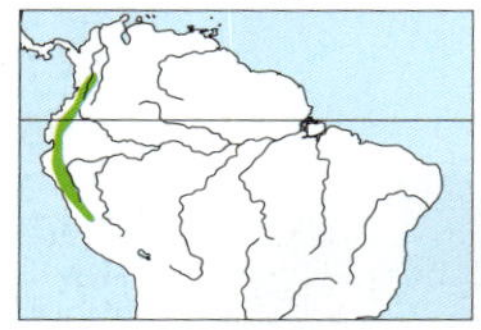

2500–4300 m

C Andes in CO and W & E Andes in EC and N PE

Floresta Nacional de Carajás (Pará)

NE
Monotypic
HB 6.8–8.1 cm
T 6.2–8.2 cm
W 10–15 g

655–745 m

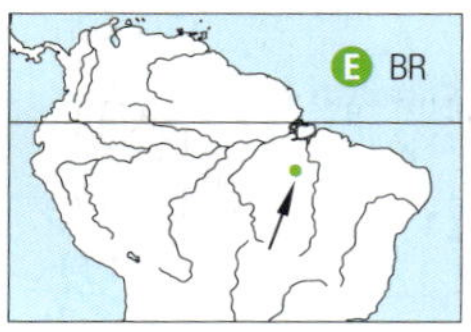

João Oliveira's Bristly Mouse
Neacomys oliveirai
FR Néacomys d'Oliveira
DE João-Oliveira-Stachelreisratte
ES Ratón erizado de Oliveira

Genus *DRYMOREOMYS*

(1 species)

White-throated Montane Forest Rat
Drymoreomys albimaculatus
FR Oryzomys à gorge blanche
DE Serra-do-Mar-Reisratte
ES Rata de bosque de montaña de garganta blanca

650–1200 m

Serra do Mar (from São Paulo to Santa Catarina)

NT
Monotypic
HB 11.5–14.9 cm
T 14.2–17.6 cm
W 46–58 g

Genus *OREORYZOMYS*

(1 species)

DD Monotypic
HB 7.5–9.4 cm
T 9.5–12 cm

Peruvian Rice Rat
Oreoryzomys balneator
FR Oryzomys de Banos
DE Peru-Reisratte
ES Rata arrocera de Perú

1500–2300 m

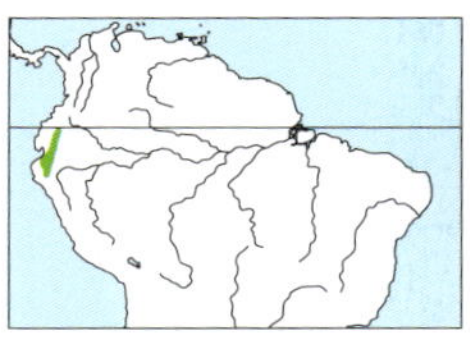

E Andes in EC and NE PE

Gray Rice Rat
Eremoryzomys polius
FR Oryzomys gris
DE Große Marañón-Reisratte
ES Rata arrocera gris

780–2205 m

Known from few localities in Amazonas, Cajamarca & La Libertad

DD
Monotypic
HB 14–17.2 cm
T 16.8–21 cm

Genus *EREMORYZOMYS*

(2 species)

Medium-tailed Rice Rat
Eremoryzomys mesocaudis
FR Oryzomys à queue médiane
DE Kleine Marañón-Reisratte
ES Rata arrocera de cola mediana

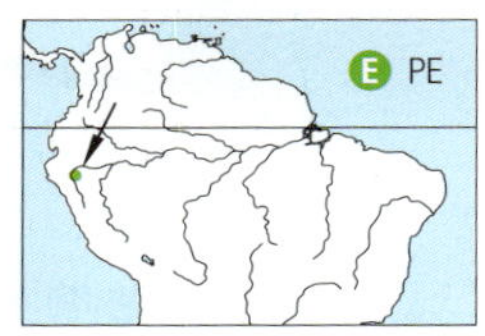

465–860 m

Known from few localities in Amazonas

NE Monotypic
HB 12–14 cm
T 11.3–14.5 cm

Genus *LUNDOMYS*

(1 species)

Lund's Water Rat
Lundomys molitor
FR Oryzomys de Lund
DE Lund-Wasserratte
ES Rata de agua de Lund

LC Monotypic
HB 17–24 cm
T 19–29 cm
W 250–490 g

S BR (Rio Grande do Sul), UY

Genus *SOORETAMYS*

(1 species)

SE & S BR, E PY, NE AR

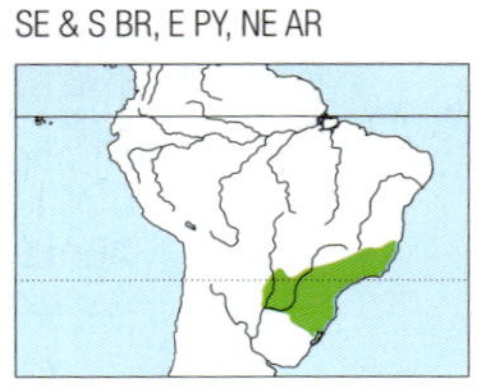

Paraguayan Rice Rat
Sooretamys angouya
FR Oryzomys du Paraguay
DE Paraguay-Reisratte
ES Rata arrocera de Paraguay

LC
Monotypic
HB 12.5–21 cm
T 16–24 cm
W 80–168 g

Genus *PSEUDORYZOMYS*

(1 species)

LC Monotypic
HB 9.9–12.7 cm
T 10.2–13.3 cm
W 32–46 g

below 1000 m

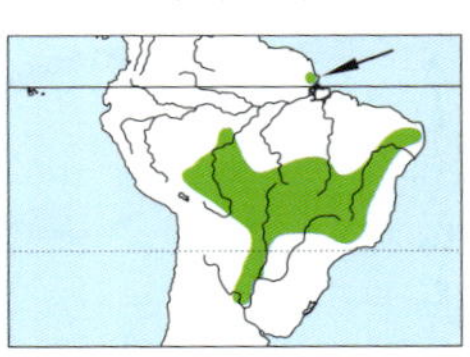

Brazilian False Rice Rat
Pseudoryzomys simplex
FR Oryzomys de Winge
DE Trug-Reisratte
ES Rata arrocera falsa de Brasil

BR, extreme SE PE, BO, PY, NE AR

Genus *HOLOCHILUS*

(7 species)

NE Monotypic
HB 13–23 cm
T 12.4–19.4 cm
W 131–132 g

Venezuelan Marsh Rat
Holochilus venezuelae
FR Oryzomys du Venezuela
DE Venezuela-Sumpfratte
ES Rata de marisma de Venezuela

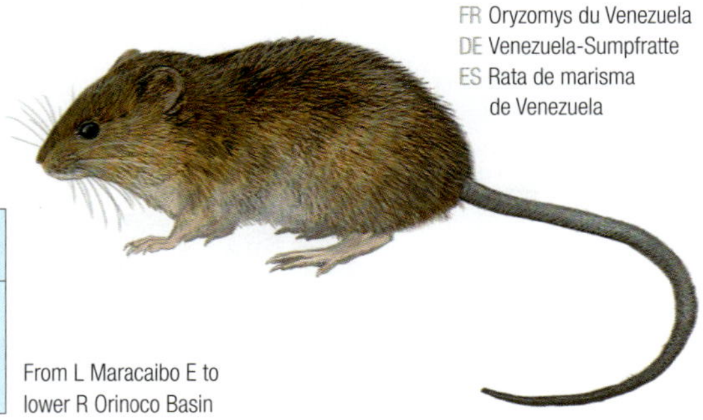

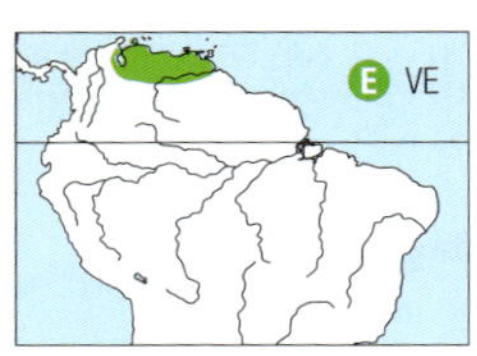

From L Maracaibo E to lower R Orinoco Basin

LC Monotypic
HB 13.6–17.1 cm
T 12.9–16.2 cm
W 80–167 g

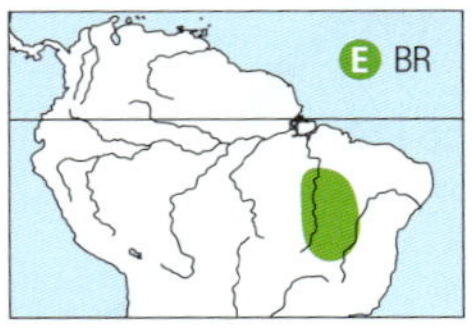

Cerrado Marsh Rat
Holochilus sciureus
FR Oryzomys écureuil
DE Cerrado-Sumpfratte
ES Rata de marisma del Cerrado

Cerrado biome in Goiás, Minas Gerais, Pará, Piauí & Tocantins

NE Monotypic
HB 13.4–20 cm T 9.5–19.4 cm
W 80–160 g

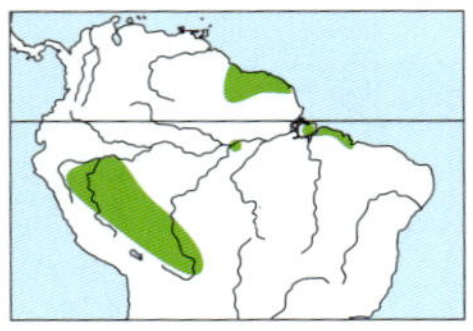

Amazonian Marsh Rat
Holochilus nanus
FR Oryzomys d'Amazonie
DE Amazonas-Sumpfratte
ES Rata de marisma pequeña

Lowlands of Amazon rainforest of Guianas, N BR, PE & BO

NE Monotypic
HB 13–20 cm
T 13–18 cm
W 65–255 g

Brazilian North-eastern Marsh Rat
Holochilus oxe
FR Oryzomys d'Alagoas
DE Nordöstliche Sumpfratte
ES Rata de marisma del noroeste

Atlantic Forest (Alagoas, Pernambuco), Mts of Crato (Ceará)

LC
Monotypic
HB 14–20 cm
T 14.8–18.3 cm
W 70–240 g

Chacoan Marsh Rat
Holochilus chacarius
FR Oryzomys du Chaco
DE Chaco-Sumpfratte
ES Rata de marisma del Chaco

S & E BO, PY, SW BR, N AR

LC
Monotypic
HB 14.1–24 cm
T 16–24 cm
W 130–370 g

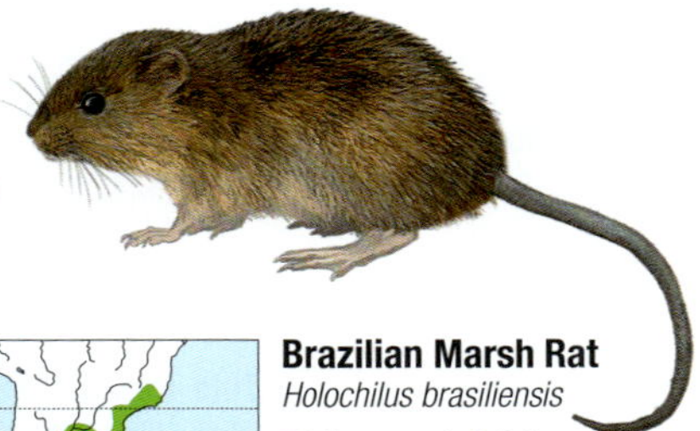

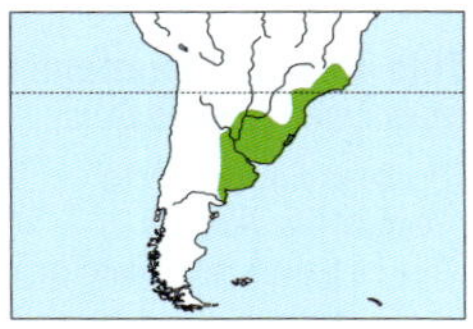

Brazilian Marsh Rat
Holochilus brasiliensis
FR Oryzomys du Brésil
DE Brasilien-Sumpfratte
ES Rata de marisma de Brasil

SE & S BR (S from Minas Gerais and Espíritu Santo), PY and UY to EC AR

Lagiglia's Marsh Rat
Holochilus lagigliai
FR Oryzomys de Lagiglia
DE Lagiglia-Sumpfratte
ES Rata de marisma de Atuel

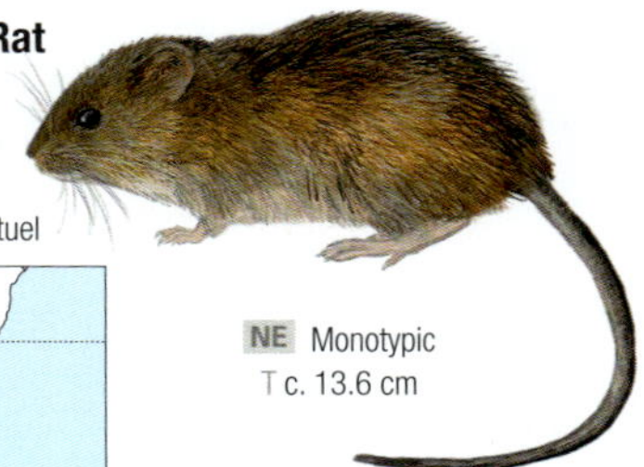

NE Monotypic
T c. 13.6 cm

Known only from Embalse El Nihuil (San Rafael, Mendoza)

Genus *CERRADOMYS*

(8 species)

NE Monotypic
HB 11.9–15.3 cm
T 14.4–19.5 cm

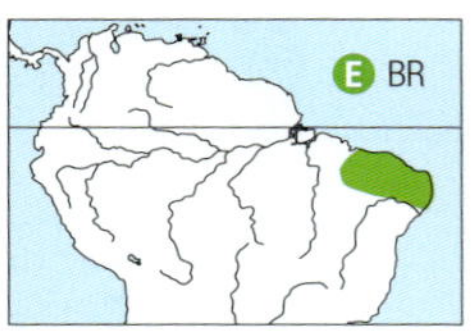

Langguth's Rice Rat
Cerradomys langguthi
FR Cerradomys de Langguth
DE Langguth-Cerradomaus
ES Rata arrocera de Langguth

From Maranhão to Pernambuco, N of R São Francisco

NE
Monotypic
HB 12.7–16 cm
T 15.3–20 cm

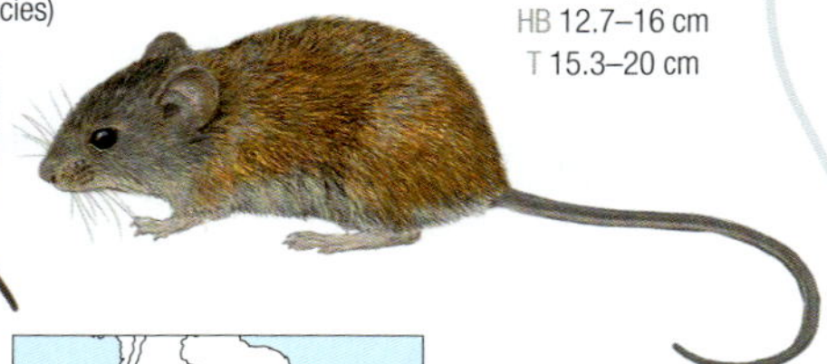

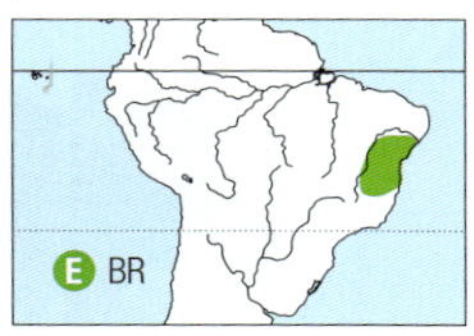

Vivo's Rice Rat
Cerradomys vivoi
FR Cerradomys de De Vivo
DE Vivo-Cerradomaus
ES Rata arrocera de Vivo

Bahia, Sergipe, Minas Gerais

LC Monotypic
HB 14–18.5 cm
T 17.1–23 cm

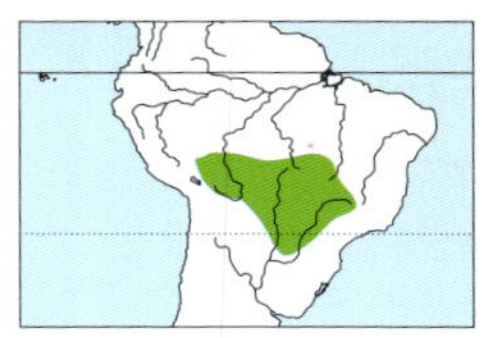

Maracaju Rice Rat
Cerradomys maracajuensis
FR Cerradomys de Maracaju
DE Maracaju-Cerradomaus
ES Rata arrocera de Maracaju

Extreme SE PE (Puno), BO, PY, SC BR

LC
Monotypic
HB 12.5–18.1 cm
T 14.1–19.8 cm

250–1180 m

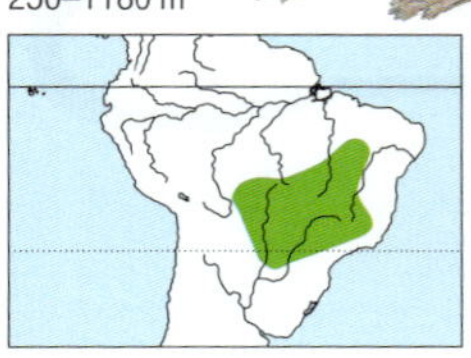

Lindbergh's Rice Rat
Cerradomys scotti
FR Cerradomys de Scott
DE Lindbergh-Cerradomaus
ES Rata arrocera de Lindbergh

E BO (Santa Cruz), C & E BR, Oriental Chaco of PY

NE
Monotypic
HB 11.1–14 cm
T 14.9–16.2 cm

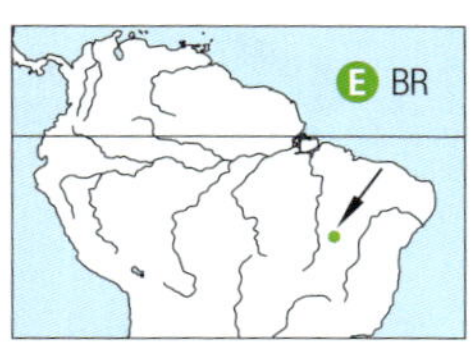

Akroa Rice Rat
Cerradomys akroai
FR Cerradomys des Akroas
DE Akroá-Cerradomaus
ES Rata arrocera de Akroá

Known only from Novo Jardim Municipality (Tocantins)

LC Monotypic
HB 15.3–17.9 cm
T 19.8–21 cm

Marinho's Rice Rat
Cerradomys marinhus
FR Cerradomys des Marinho
DE Marinho-Cerradomaus
ES Rata arrocera de Marinho

Tocantins, Bahia, Goiás, Minas Gerais

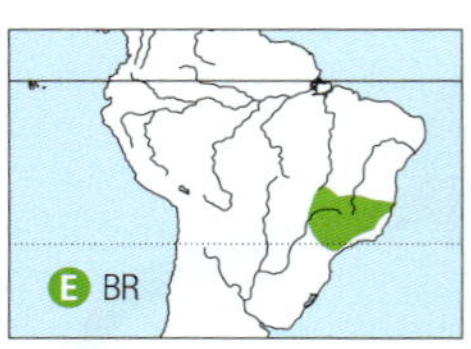

Bahia, Goiás, Minas Gerais, São Paulo

LC Monotypic
HB 12.5–17.9 cm
T 15–21 cm

Terraced Rice Rat
Cerradomys subflavus
FR Cerradomys ambré
DE Gelbliche Cerradomaus
ES Rata arrocera de terraza

Goytaca Rice Rat
Cerradomys goytaca
FR Cerradomys des Goytacas
DE Goytaca-Cerradomaus
ES Rata arrocera de Goytaca

Coast of Espírito Santo and Rio de Janeiro

NE
Monotypic
HB 11.6–16.6 cm
T 13–18.1 cm

Genus *AMPHINECTOMYS*

(1 species)

DD Monotypic
HB 18.7–19 cm
T 17.3–21 cm
W 214–225 g

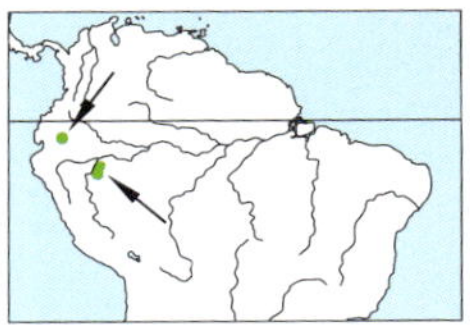

Ucayali Water Rat
Amphinectomys savamis

FR Nectomys de l'Ucayali
DE Ucayali-Wasserratte
ES Rata de agua de Ucayali

Known from a few sites in N & S bank tributaries of R Marañón (SE EC, NE PE)

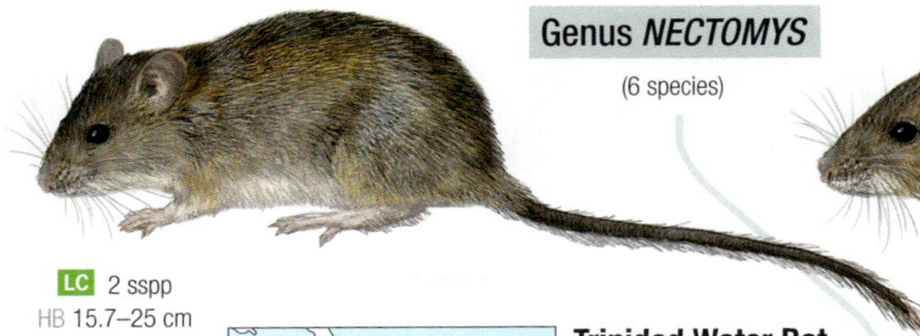

LC 2 sspp
HB 15.7–25 cm
T 14–23 cm
W 130–347 g

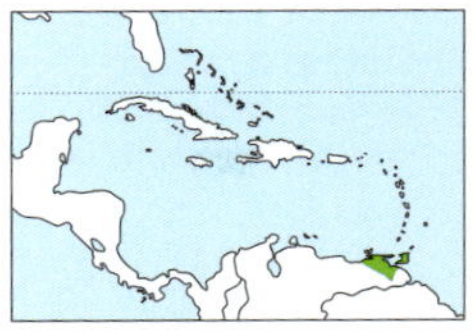

Trinidad Water Rat
Nectomys palmipes

FR Nectomys de Trinidad
DE Trinidad-Wasserratte
ES Rata de agua de Trinidad

NE VE (Paria Pen, Orinoco Delta), TT (Trinidad I)

Genus *NECTOMYS*

(6 species)

NE 2 sspp
HB 18.2–29 cm
T 19.1–27 cm

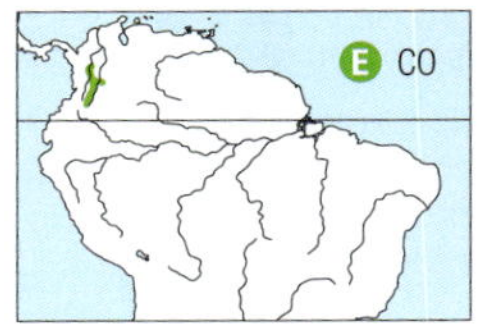

Magdalena Water Rat
Nectomys magdalenae

FR Nectomys du Magdalena
DE Magdalena-Wasserratte
ES Rata de agua de Magdalena

R Magdalena and R Cauca valleys

LC Monotypic
HB 14.5–27 cm
T 16.1–27 cm
W 141–430 g

105–3400 m

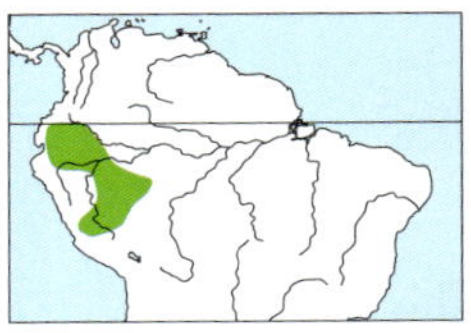

Western Amazonian Water Rat
Nectomys apicalis

FR Nectomys de Peters
DE Westamazonas-Wasserratte
ES Rata de agua del Amazonas occidental

E EC, E PE, W BR (Amazonas)

NE Monotypic
HB 20.5–21 cm
T 22–24 cm

c. 2225 m

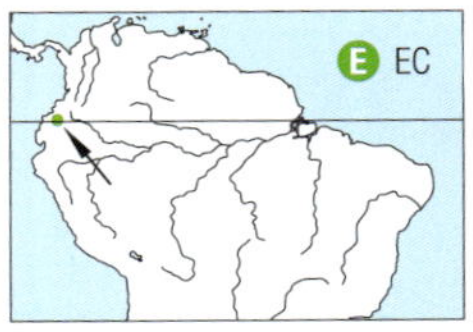

Ecuadorean Water Rat
Nectomys saturatus

FR Nectomys d'Équateur
DE Ecuador-Wasserratte
ES Rata de agua ecuatorial

Known only from type locality (Ibarra, Imbabura)

LC Monotypic
HB 12.7–29 cm T 12–25 cm
W 102–430 g

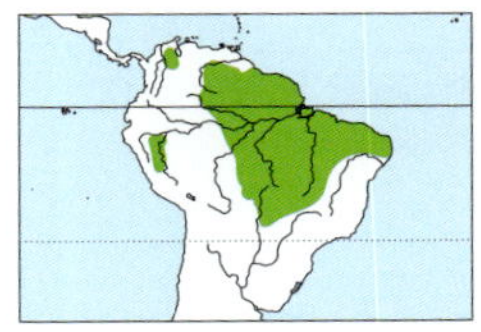

South American Water Rat
Nectomys rattus

FR Nectomys d'Amazonie
DE Südamerika-Wasserratte
ES Rata de agua suramericana

E CO, VE (Orinoco Basin), PE (Ucayali Basin), Guianas, BR (E to R Paraná and R São Francisco), PY

LC Monotypic
HB 11.9–26 cm
T 12.2–29 cm
W 112–380 g

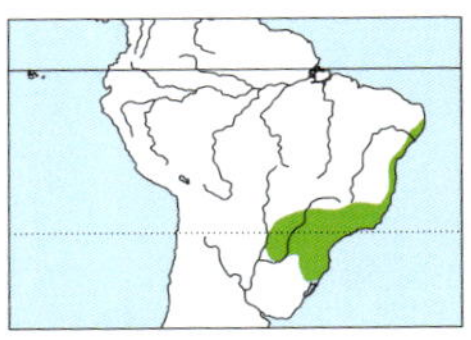

Atlantic Water Rat
Nectomys squamipes

FR Nectomys atlantique
DE Atlantische Wasserratte
ES Rata de agua atlántica

E BR (Atlantic Forest, S from Pernambuco), E PY, NE AR (Misiones)

Common Marsh Rice Rat
Oryzomys palustris

FR Oryzomys des marais
DE Gemeine Sumpfreisratte
ES Rata arrocera de marisma

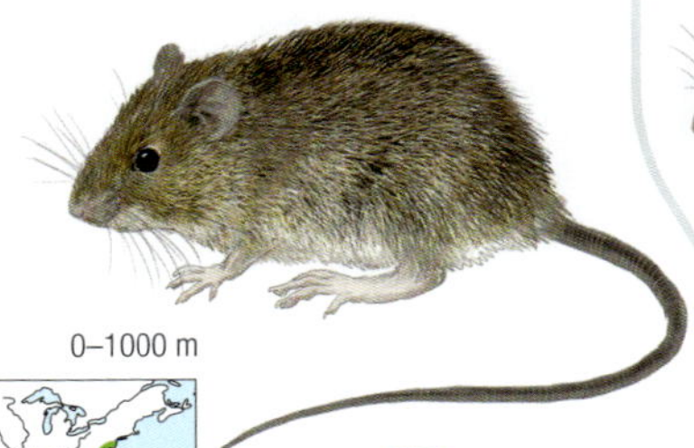

0–1000 m

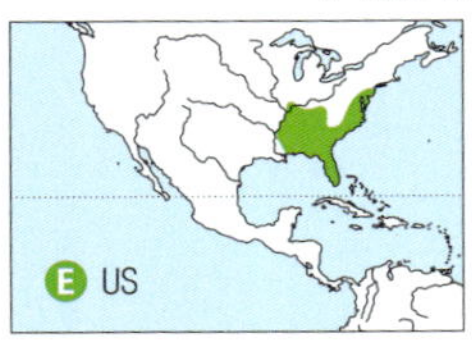

LC 5 sspp
HB 13.1–13.4 cm
T 12.1–14.3 cm
W 45–80 g

From S Illinois and S New Jersey to Florida (with lower Keys)

Genus *ORYZOMYS*

(6 species)

Texas Marsh Rice Rat
Oryzomys texensis

FR Oryzomys du Texas
DE Texas-Sumpfreisratte
ES Rata arrocera de marisma de Texas

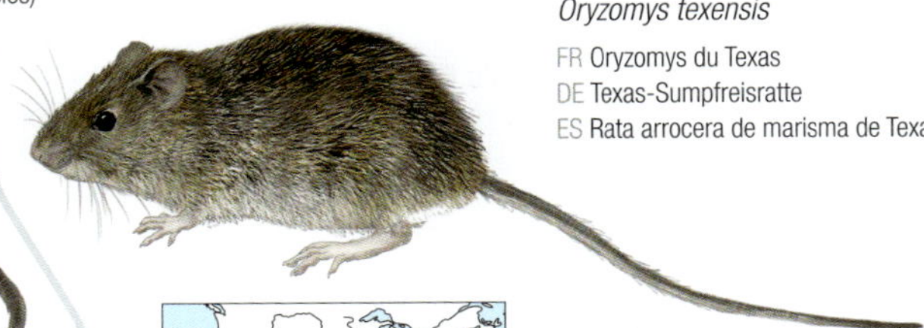

NE Monotypic
HB 12–14.6 cm
T 10–14 cm
W 40–74 g

US (SE Kansas & S Missouri, S to E Texas & W Mississippi), MX (NE Tamaulipas)

Coues's Marsh Rice Rat
Oryzomys couesi

FR Oryzomys de Coues
DE Coues-Sumpfreisratte
ES Rata arrocera de marisma de Coues

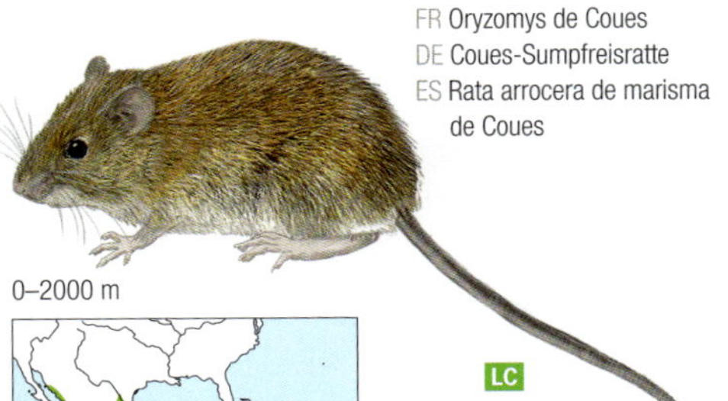

0–2000 m

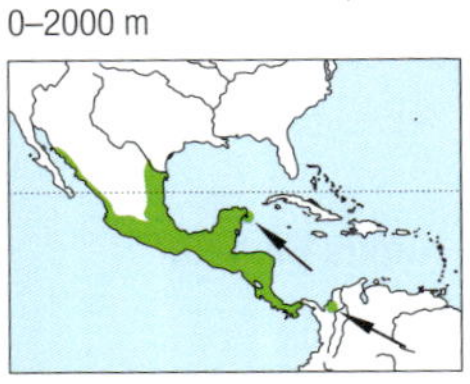

LC Monotypic
HB 9.1–14.4 cm
T 10.2–16 cm
W 43–82 g

US (extreme S Texas), MX (excluding NC plateau) to C PA, recorded in NW CO

NE Monotypic
HB 11.6–14.2 cm
T 12.9–17.3 cm

1200–1800 m

White-bellied Marsh Rice Rat
Oryzomys albiventer

FR Oryzomys à ventre blanc
DE Weißbauch-Sumpfreisratte
ES Rata arrocera de marisma de vientre blanco

Meseta de Anáhuac (C & E Jalisco, S Guanajuato, N Michoacán)

Nicaraguan Marsh Rice Rat
Oryzomys dimidiatus

FR Oryzomys du Nicaragua
DE Nicaragua-Sumpfreisratte
ES Rata arrocera de marisma de Nicaragua

DD Monotypic
HB 11.8–12.5 cm
T 11–11.5 cm
W c. 46 g

Known only from R Escondido area (near Rama)

EN
Monotypic
HB 9.4–15.2 cm
T 11.6–13.8 cm

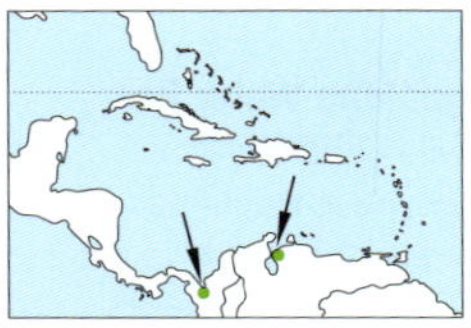

Gorgas's Marsh Rice Rat
Oryzomys gorgasi
FR Oryzomys de Gorgas
DE Gorgas-Sumpfreisratte
ES Rata arrocera de marisma de Gorgas

Known from 2 localities in coastal NW CO & NW VE; perhaps widespread

VU Monotypic
HB 11–19 cm T 10–16.3 cm
W 31–181 g

Santiago Galapagos Mouse
Nesoryzomys swarthi
FR Oryzomys de Santiago
DE Santiago-Galapagosratte
ES Ratón de Galápagos de Santiago

Restricted to Santiago I (Galápagos Is)

VU Monotypic
HB 10–12.1 cm T 7.7–10 cm
W 22–63 g

Small Fernandina Galapagos Mouse
Nesoryzomys fernandinae
FR Oryzomys de Fernandina
DE Kleine Fernandina-Galapagosratte
ES Ratón de Galápagos de Fernandina pequeño

Restricted to Fernandina I (Galápagos Is)

Genus *NESORYZOMYS*
(3 species)

VU Monotypic
HB 11–17.2 cm T 10–15.2 cm
W 77–136 g

Large Fernandina Galapagos Mouse
Nesoryzomys narboroughi
FR Oryzomys de Narborough
DE Große Fernandina-Galapagosratte
ES Ratón de Galápagos de Fernandina grande

Restricted to Fernandina I (Galápagos Is)

Genus *AEGIALOMYS*
(4 species)

VU
Monotypic
HB 10.6–15.2 cm
T 12–17.3 cm
W 62–103 g

Galapagos Rice Rat
Aegialomys galapagoensis
FR Oryzomys des Galapagos
DE Galapagos-Reisratte
ES Rata arrocera de Galápagos

Santa Fe I, maybe extinct on San Cristóbal & Santiago Is (Galápagos Is)

Yellowish Rice Rat
Aegialomys xanthaeolus
FR Oryzomys jaune
DE Gelbe Reisratte
ES Rata arrocera dorada

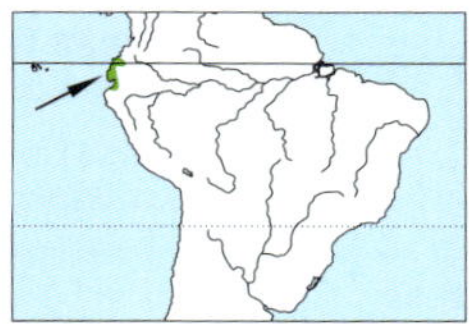

500–1800 m

W EC, extreme NW PE

LC
Monotypic
HB 10.5–11 cm
T 11.8–17.6 cm
W 50–73 g

NE
Monotypic
HB c. 13 cm
T 13.7–17.2 cm

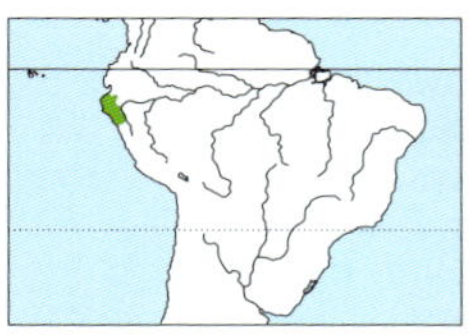

Baron's Rice Rat
Aegialomys baroni
FR Oryzomys de Baron
DE Baron-Reisratte
ES Rata arrocera de Baron

S EC to NC PE, from Pacific coast to W foothills of Andes

Ica Rice Rat
Aegialomys ica
FR Oryzomys d'Ica
DE Ica-Reisratte
ES Rata arrocera de Ica

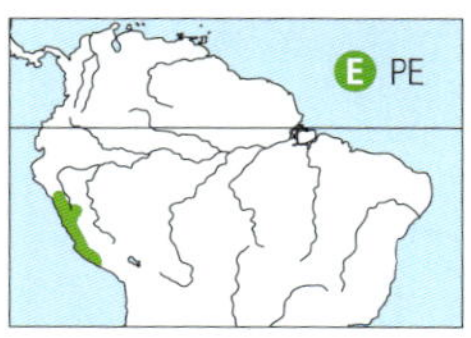

0–2200 m

From Ancash S to Arequipa

NE
Monotypic
HB 11–17.5 cm
T 11–20 cm
W 46–98 g

Genus *SIGMODONTOMYS*
(1 species)

LC
Monotypic
HB 12–17.2 cm
T 14.9–19 cm

0–1300 m

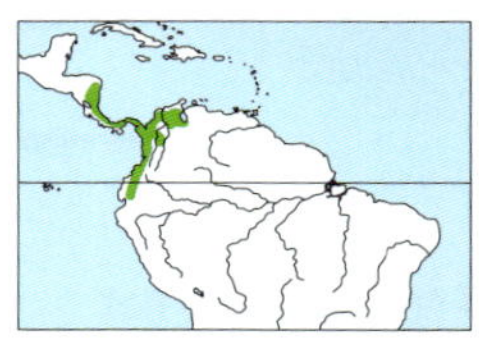

Alfaro's Water Rat
Sigmodontomys alfari
FR Mélanomys d'Alfaro
DE Alfaro-Wasserratte
ES Rata de agua de Alfaro

From extreme E HN to NW VE and W EC

1200–2500 m

Harris's Rice Water Rat
Tanyuromys aphrastus
FR Mélanomys à longue queue
DE Harris-Reiswasserratte
ES Rata de agua de Harris

SE CR, W PA; probably CO

Genus *TANYUROMYS*
(2 species)

DD Monotypic
HB 11.6–15.2 cm
T 17.6–23 cm
W c. 63 g

NE
Monotypic
HB 13.9–14.2 cm
T 20–21 cm

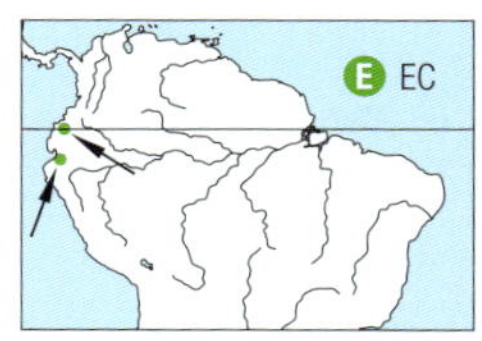

Carchi, Imbabura, Pichincha, Azuay; possibly in CO

Thomas Lee's Rice Water Rat
Tanyuromys thomasleei
FR Mélanomys de Lee
DE Thomas-Lee-Reiswasserratte
ES Rata de agua de Thomas Lee

Genus *MELANOMYS*
(6 species)

0–1000 m

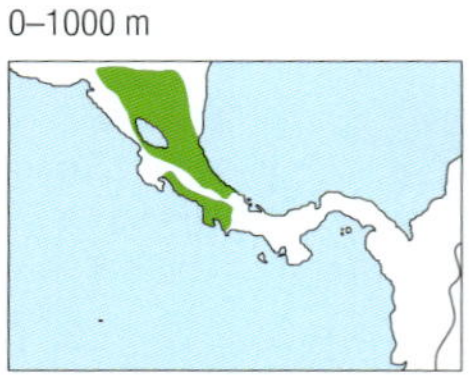

Black-and-yellow Rice Rat
Melanomys chrysomelas
FR Mélanomys noir et jaune
DE Schwarzgelbe Dunkelreisratte
ES Rata arrocera negra y dorada

SW HN, NI, CR; uncertain in PA

NE
Monotypic
HB 11–13 cm
T 8–10 cm

NE
Monotypic
HB 11.3–13 cm
T 7.6–9.9 cm

550–850 m

Heavily forested hill slopes; limits unresolved

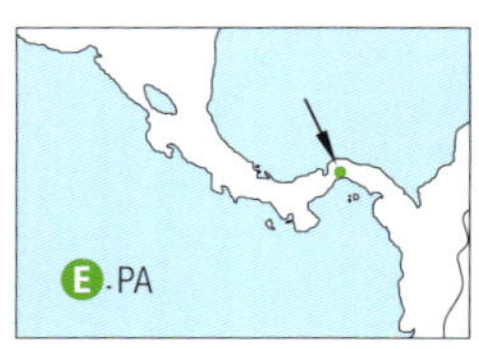

Cinnamon-rufous Rice Rat
Melanomys idoneus
FR Mélanomys roux
DE Zimtrote Dunkelreisratte
ES Rata arrocera canela

NE
Monotypic
HB 12.5–14.5 cm
T 8–10 cm
W 43–48 g

1000–2500 m

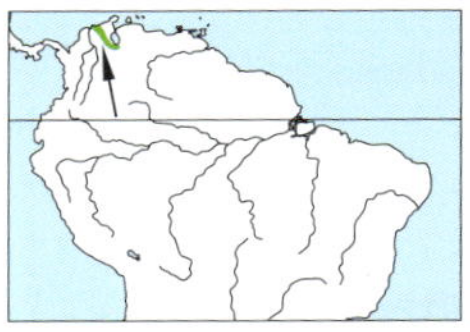

Colombian Rice Rat
Melanomys columbianus
FR Mélanomys de Colombie
DE Kolumbien-Dunkelreisratte
ES Rata arrocera de Colombia

N CO (Sierra Nevada de Santa Marta), extreme NW VE

LC
Monotypic
HB 10–14 cm
T 7–11.1 cm

0–2300 m

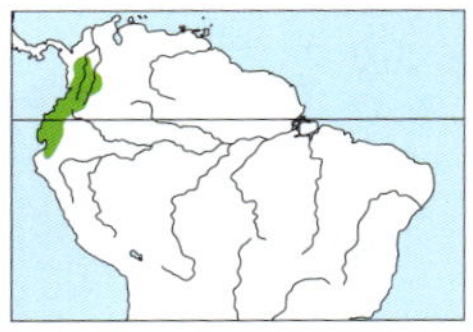

Dusky Rice Rat
Melanomys caliginosus
FR Mélanomys sombre
DE Nordwestliche Dunkelreisratte
ES Rata arrocera oscura

Andean & Pacific forests of WC CO to SW EC

LC Monotypic
HB c. 12.2 cm
T c. 9.2 cm

400–1200 m

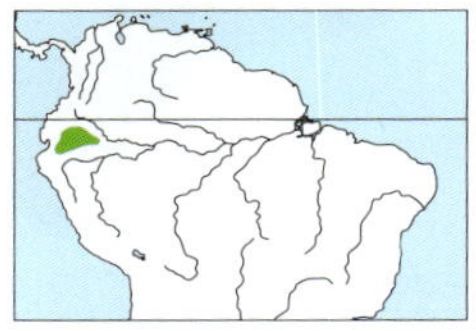

Robust Dark Rice Rat
Melanomys robustulus
FR Mélanomys robuste
DE Robuste Dunkelreisratte
ES Rata arrocera robusta

Lowland rainforest in E EC & N PE

CR(PE)
Monotypic
T 8.1–10.5 cm

600–800 m

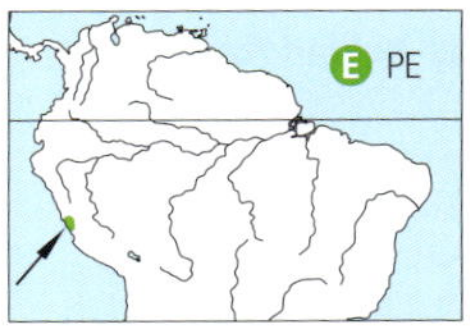

Zuniga's Dark Rice Rat
Melanomys zunigae
FR Mélanomys de Zuniga
DE Zúñiga-Dunkelreisratte
ES Rata arrocera de Zúñiga

Cerro San Jerónimo & Lomas de Atocongo

Genus *MICROAKODONTOMYS*

(1 species)

EN
Monotypic
HB c. 7 cm
T c. 9.3 cm

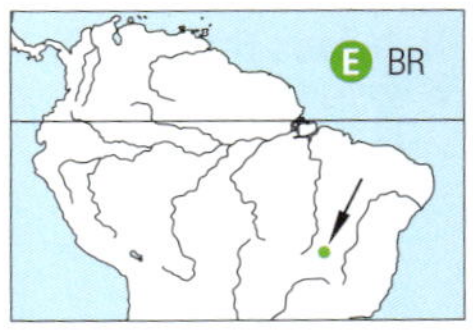

Intermediate Lesser Grass Mouse
Microakodontomys transitorius
FR Oryzomys intermédiaire
DE Brasilia-Ratte
ES Ratón de hierba mediano

Known only from type locality (Brasília NP, Distrito Federal)

Lagoa Santa Pink-lipped Mouse
Bibimys labiosus
FR Bibimys du Brésil
DE Lagoa-Santa-Pinknasenmaus
ES Ratón de labios rosados de Lagoa Santa

420–1000 m

LC
Monotypic
HB 7.6–9 cm
T 6.2–7.3 cm
W 23–26 g

From Minas Gerais S to Rio Grande do Sul

NE Monotypic
HB 11.4–13.5 cm
T 10.1–12.5 cm
W 50–91 g

530–1015 m

Brazilian Swamp Rat
Scapteromys meridionalis
FR Akodon méridional
DE Brasilien-Sumpflandratte
ES Rata de ciénaga de Brasil

From S Paraná to N Rio Grande do Sul

Genus *BIBIMYS*

(3 species)

LC Monotypic
HB 9.5–10 cm
T 7–7.5 cm
W 32–36 g

Chacoan Pink-lipped Mouse
Bibimys chacoensis
FR Bibimys du Chaco
DE Chaco-Pinknasenmaus
ES Ratón de labios rosados del Chaco

SE PY, NE AR (S Misiones, NW Corrientes)

VU
Monotypic
HB 9–11 cm
T 7.2–8 cm
W 32–42 g

Torres's Pink-lipped Mouse
Bibimys torresi
FR Bibimys de Torres
DE Torres-Pinknasenmaus
ES Ratón de labios rosados de Torres

Extreme S Entre Ríos, N Buenos Aires

Genus *SCAPTEROMYS*

(3 species)

Argentinean Swamp Rat
Scapteromys aquaticus
FR Akodon des marais
DE Argentinien-Sumpflandratte
ES Rata de ciénaga de Argentina

LC
Monotypic
HB 12.6–25 cm
T 10.8–14.7 cm
W 64–155 g

R Paraná & R Uruguay basins (PY, AR), marginally S BR & W UY

LC
Monotypic
HB 15–19.5 cm
T 13.3–17.6 cm
W 50–165 g

Uruguay Swamp Rat
Scapteromys tumidus
FR Akodon bouffi
DE Uruguay-Sumpflandratte
ES Rata de ciénaga de Uruguay

Extreme S BR (Rio Grande do Sul), most of UY

Genus *GYLDENSTOLPIA*

(2 species)

Cerrado Giant Rat
Gyldenstolpia planaltensis
FR Akodon de Planaltina
DE Cerrado-Riesenratte
ES Rata gigante del Cerrado

NE
Monotypic
HB 16–20 cm
T 7.5–11.8 cm
W 98–139 g

Mato Grosso, Distrito Federal

Fossorial Giant Rat
Gyldenstolpia fronto
FR Akodon du Chaco
DE Gyldenstolpe-Riesenratte
ES Rata gigante fosora

CR(PE)
Monotypic
HB c. 22 cm
T c. 11 cm

Known from a single locality in NE Chaco

Genus *KUNSIA*

(1 species)

Woolly Giant Rat
Kunsia tomentosus

FR Akodon laineux
DE Woll-Riesenratte
ES Rata gigante lanuda

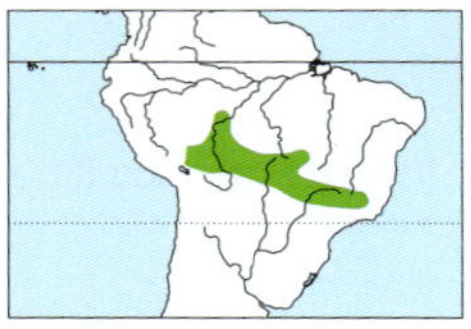

N & E BO, C BR; probably extreme SE PE

LC
Monotypic
HB 18.5–29 cm
T 14.7–19.6 cm
W 241–630 g

Genus *LENOXUS*

(1 species)

Andean Rat
Lenoxus apicalis

LC Monotypic
HB 11.8–16.5 cm
T 13.3–18.5 cm

FR Akodon des Andes
DE Andenratte
ES Rata de los Andes

1500–2500 m

E Andean slopes of SE PE & NW BO

Genus *BLARINOMYS*

(1 species)

LC
Monotypic
HB 9.1–12.1 cm
T 3–5.9 cm
W 12.5–49 g

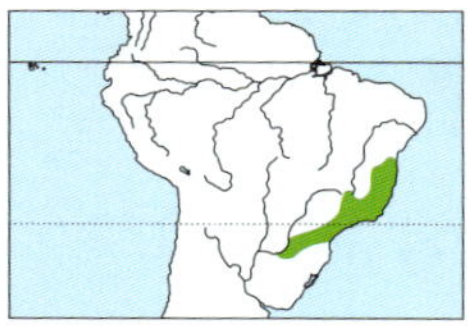

Atlantic Forest Burrowing Mouse
Blarinomys breviceps

FR Akodon musaraigne
DE Brasilianische Spitzmausratte
ES Ratón cavador de bosque atlántico

Atlantic Forest in E & SE BR and NE AR (Misiones)

Genus *BRUCEPATTERSONIUS*

(4 species)

NE Monotypic
HB 8.5–10.9 cm T 8.9–10.7 cm
W 20–35 g

Nebulous Brucie
Brucepattersonius nebulosus

FR Brucie nébuleux
DE Nebelwald-Brucie
ES Ratón hocicudo de bosque de niebla

Known from Serra da Bocaina, Serra dos Órgãos & Serra da Mantiqueira

DD
Monotypic
HB 9.3–10.9 cm
T 9.7–11.2 cm
W 20–27 g

1800–2400 m

Gray-bellied Brucie
Brucepattersonius griserufescens

FR Brucie à ventre gris
DE Graubauch-Brucie
ES Ratón hocicudo enano de vientre gris

Higher elevations of Caparaó, Mantiqueira & Serra dos Órgãos massifs

DD
Monotypic
HB 9.8–12.8 cm
T 8.2–9.9 cm
W 30–35 g

0–2000 m

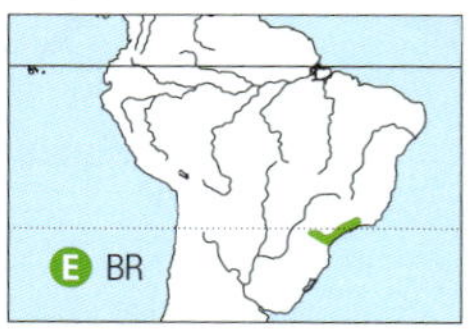

Short-tailed Brucie
Brucepattersonius soricinus

FR Brucie soricin
DE Kurzschwanz-Brucie
ES Ratón hocicudo enano de cola corta

Atlantic Forest from Paraná to S Rio de Janeiro

LC Monotypic
HB 8.9–11.3 cm
T 8.3–10.7 cm
W 40–45 g

0–1000 m

Ihering's Brucie
Brucepattersonius iheringi

FR Brucie d'Ihering
DE Ihering-Brucie
ES Ratón hocicudo enano de Ihering

S BR (Santa Catarina, Rio Grande do Sul), NE AR (Misiones)

Genus *JUSCELINOMYS*

(1 species)

DD
Monotypic
HB 13–19.2 cm
T 8.4–10 cm
W 61–99 g

200–700 m

Huanchaca Burrowing Mouse
Juscelinomys huanchacae

FR Akodon du Huanchaca
DE Huanchaca-Maulwurfsmaus
ES Ratón cavador de Huanchaca

Known from 4 close localities in Noel Kempff Mercado NP

NE Monotypic
HB 13.6–14.7 cm
T 8–9 cm
W 58–60 g

740–1050 m

Itapevi Hocicudo
Oxymycterus itapeby

FR Hocicudo d'Itapevi
DE Itapevi-Grabmaus
ES Ratón hocicudo de Itapevi

Paraná, São Paulo

Genus *OXYMYCTERUS*

(17 species)

LC Monotypic
HB 14–14.8 cm
T 8.6–9.5 cm

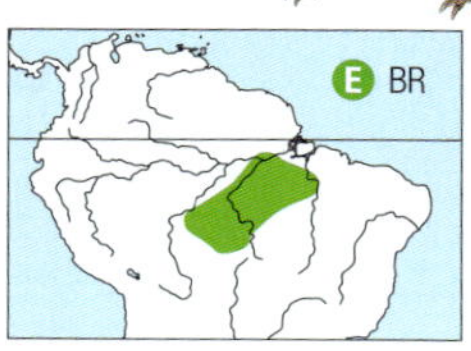

Amazonian Hocicudo
Oxymycterus amazonicus

FR Hocicudo d'Amazonie
DE Amazonas-Grabmaus
ES Ratón hocicudo del Amazonas

Along Madeira, Tapajós, Xingu & Tocantins river basins

LC Monotypic
HB 11.5–17.4 cm
T 7–12 cm

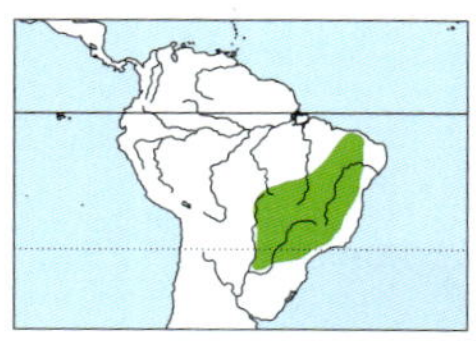

Spy Hocicudo
Oxymycterus delator

FR Hocicudo espion
DE Paraguay-Grabmaus
ES Ratón hocicudo delator

From NE BR S to E PY

Atlantic Forest Hocicudo
Oxymycterus dasytrichus

FR Hocicudo de Mucuri
DE Mata-Atlantica-Grabmaus
ES Ratón hocicudo de bosque atlántico

up to 1800 m

Atlantic Forest region

LC
Monotypic
HB 12.5–19.7 cm
T 9–15.6 cm
W 40–105 g

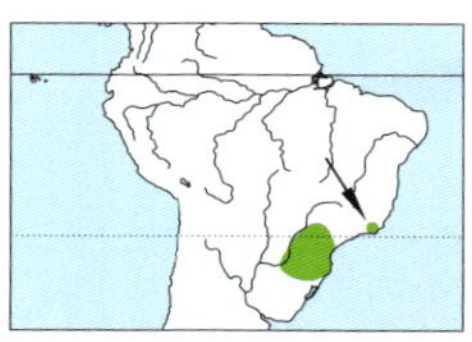

E PY, NE AR (Misiones), S & SE BR (from N Rio Grande do Sul to Rio de Janeiro)

LC Monotypic
HB 13.5–18 cm T 9.7–14.3 cm
W c. 100 g

Quaestor Hocicudo
Oxymycterus quaestor

FR Hocicudo questeur
DE Argentinien-Grabmaus
ES Ratón hocicudo cuestor

LC
2 sspp
HB 13.5–18.4 cm
T 8.5–12.2 cm

400–2500 m

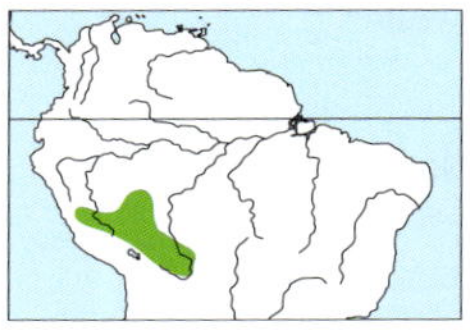

Inca Hocicudo
Oxymycterus inca
FR Hocicudo inca
DE Inka-Grabmaus
ES Ratón hocicudo de los Incas

E PE, W & C BO (S to Cochabamba, E to W Santa Cruz), adjacent W BR

LC Monotypic
HB 10.5–16 cm
T 7.5–12.4 cm
W 41–87 g

Andean Altiplano from S PE, BO & NW AR

Paramo Hocicudo
Oxymycterus paramensis
FR Hocicudo du paramo
DE Páramo-Grabmaus
ES Ratón hocicudo de páramo

700–4000 m

NE Monotypic
HB 11.8–13.6 cm
T 7.6–9.8 cm
W 50–75 g

3200– 4100 m

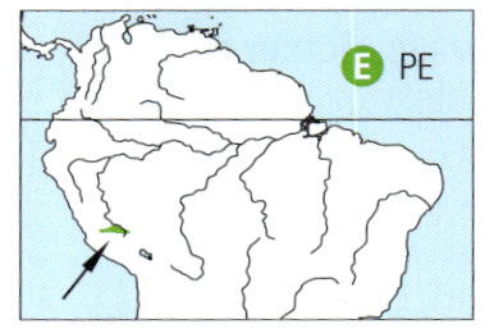

Cusco Hocicudo
Oxymycterus willkaurco
FR Hocicudo de Cusco
DE Cusco-Grabmaus
ES Ratón hocicudo de Vilcabamba

W Andes in S Cordillera de Vilcabamba & Cordillera de Vilcanota (Cusco)

Elfin Forest Hocicudo
Oxymycterus nigrifrons
FR Hocicudo à front noir
DE Elfenwald-Grabmaus
ES Ratón hocicudo de bosque delicado

2000–3350 m

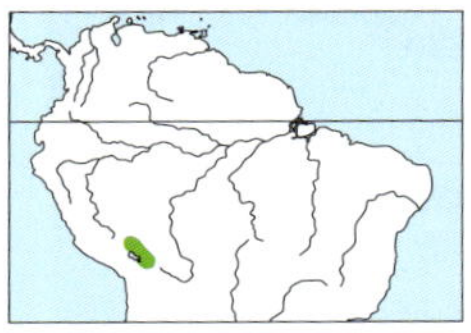

NE
Monotypic
HB 11.7–13.3 cm
T 7.9–9.6 cm

SE PE (S Cordillera Carabaya), WC BO (Cordillera Real)

NE Monotypic
HB 12.5–15.5 cm
T 9.5–14 cm

1000–2700 m

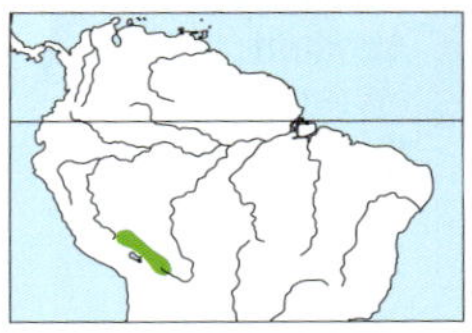

Upper Yungas Inca Hocicudo
Oxymycterus juliacae
FR Hocidudo de Juliaca
DE Nebelwald-Grabmaus
ES Ratón hocicudo de Yungas

SE PE (upper R Inambari), WC BO (upper R Chaparé)

LC Monotypic
HB 9.7–12.6 cm
T 7.7–9.1 cm
W 30–44 g

600–3500 m

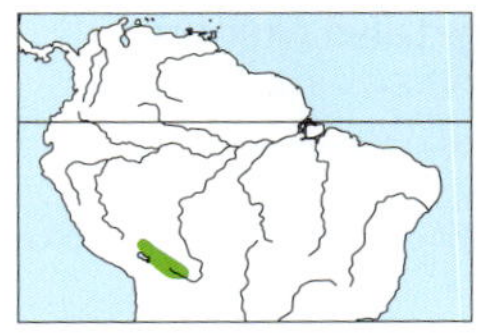

Small Yungas Hocicudo
Oxymycterus hiska
FR Petit Hocicudo
DE Kleine Yungas-Grabmaus
ES Ratón hocicudo pequeño de Yungas

Montane humid forest from SE PE to C BO

Quechuan Hocicudo
Oxymycterus hucucha
FR Hocicudo quechua
DE Quechua-Grabmaus
ES Ratón hocicudo de los Quechuas

2600–3000 m

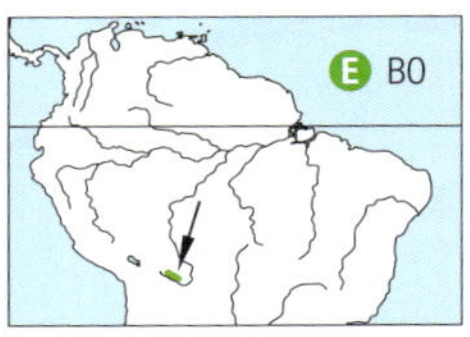

EN Monotypic
HB 9.9–10.9 cm
T 7.1–7.5 cm
W 26–36 g

Siberia Cloud Forest (Cochabamba, Santa Cruz)

LC
Monotypic
HB 11–13.9 cm
T 8–10.4 cm

above 2100 m

Mount Caparao Hocicudo
Oxymycterus caparoae
FR Hocicudo de Caparao
DE Caparaó-Grabmaus
ES Ratón hocicudo de Caparaó

Caparaó NP (Minas Gerais, Espírito Santo)

LC
Monotypic
HB 12.9–16.1 cm
T 6.9–11.8 cm
W 46–125 g

0–2000 m

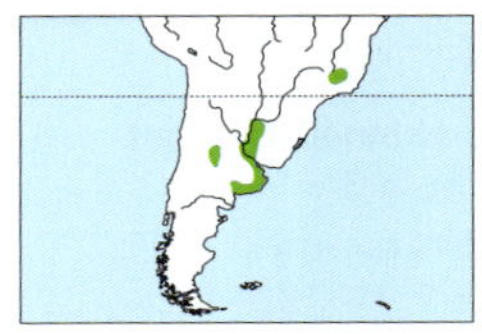

Red Hocicudo
Oxymycterus rufus
FR Hocicudo roux
DE Rote Grabmaus
ES Ratón hocicudo rojo

C & E AR, also SE BR (Minas Gerais)

LC Monotypic
HB 12.3–13.1 cm
T 8.7–9.7 cm
W 50–90 g

Darwin's Hocicudo
Oxymycterus nasutus
FR Hocicudo à nez long
DE Darwin-Grabmaus
ES Ratón hocicudo de Darwin

S BR (Paraná, Santa Catarina, Rio Grande do Sul), UY

VU Monotypic
HB 12.4–14.4 cm
T 7.7–9.7 cm
W 44–76 g

800–2200 m

Ravine Hocicudo
Oxymycterus wayku
FR Hocicudo des ravins
DE Schluchten-Grabmaus
ES Ratón hocicudo de barranco

E slopes of Sierra del Aconquija & Cumbres Calchaquíes (Tucumán)

NT Monotypic
HB 12.2–17.2 cm
T 7.5–10.4 cm
W 36–125 g

Cook's Hocicudo
Oxymycterus josei
FR Hocicudo de Cook
DE Cook-Grabmaus
ES Ratón hocicudo de Cook

S of R Negro in Soriano, Colonia, San José, Canelones & Maldonado

LC Monotypic
HB 9–10.9 cm
T 5.1–7 cm
W 26–40 g

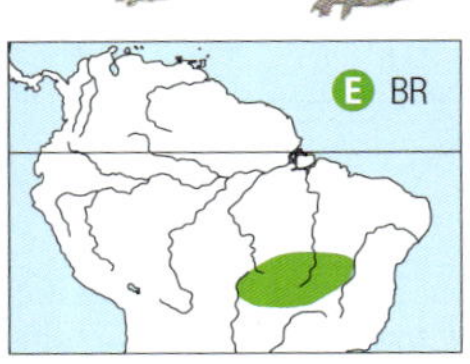

Common Cerrado Mouse
Thalpomys cerradensis
FR Akodon du Cerrado
DE Cerrado-Feldmaus
ES Ratón del Cerrado común

Mato Grosso, Goiás, Distrito Federal, Bahia

Genus *THALPOMYS*

(2 species)

LC Monotypic
HB c. 7.6 cm
T c. 4.6 cm
W c. 21 g

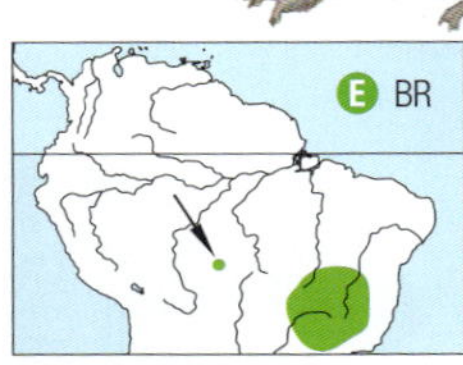

Hairy-eared Cerrado Mouse
Thalpomys lasiotis
FR Akodon à oreilles velues
DE Haarohr-Cerrado-Feldmaus
ES Ratón del Cerrado de oreja peluda

Bahia S to São Paulo, perhaps Mato Grosso; recorded in Rondônia

Genus *PODOXYMYS*

(1 species)

VU
Monotypic
HB 7.2–10.2 cm
T 7.8–10.2 cm

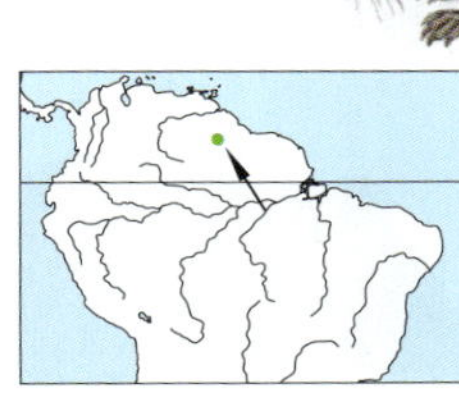

Roraima Mouse
Podoxymys roraimae
FR Akodon du Roraima
DE Roraima-Maus
ES Ratón de Roraima

Around Mt Roraima and Wei-Assipu-tepui (SE VE, W GY, N BR)

Genus *THAPTOMYS*

(1 species)

LC
2 sspp
HB 8.3–9.9 cm
T 3–5 cm
W 18–27 g

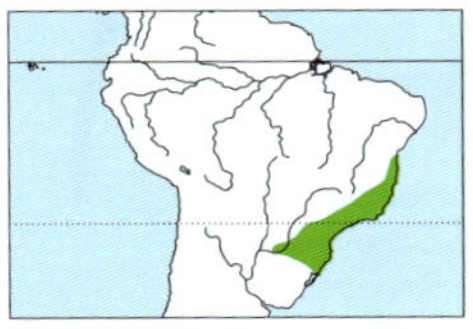

Blackish Grass Mouse
Thaptomys nigrita

FR Akodon noirâtre
DE Schwarze Grasmaus
ES Ratón de hierba negruzco

E & SE BR, SE PY, NE AR (Misiones)

DD
Monotypic
HB c. 13.2 cm
T c. 7.1 cm

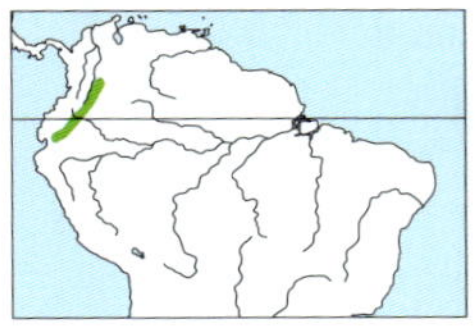

Ecuadorean Akodont
Necromys punctulatus

FR Akodon pointillé
DE Ekuador-Graslandmaus
ES Ratón cavador de Ecuador

Andes of C CO & W EC

Genus *NECROMYS*

(8 species)

LC Monotypic
HB 10.7–14.1 cm
T 6.9–10.2 cm
W 30–61 g

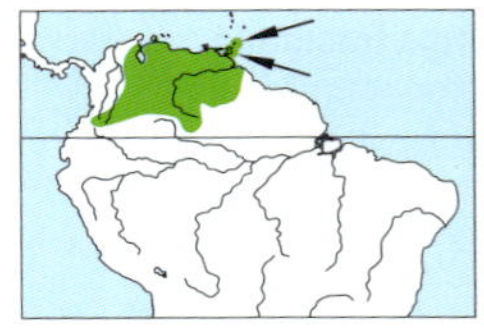

Northern Akodont
Necromys urichi

FR Akodon d'Urich
DE Nördliche Graslandmaus
ES Ratón cavador septentrional

CO, VE, TT, W GY; possibly extreme N BR

LC
Monotypic
HB 11.4–13.8 cm
T 5.9–9.6 cm
W 20–67.5 g

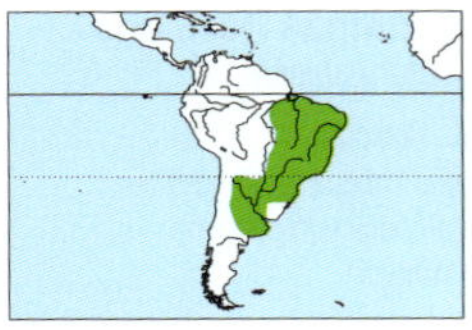

Hairy-tailed Akodont
Necromys lasiurus

FR Akodon à queue velue
DE Haarschwanz-Graslandmaus
ES Ratón cavador de cola peluda occidental

From S R Amazonas (BR) to PY & N AR, probably NW UY; limits unresolved

LC Monotypic
HB 11.4–13.8 cm
T 5.9–9.6 cm
W 20–67.5 g

Paraguayan Akodont
Necromys lenguarum

FR Akodon du Paraguay
DE Paraguay-Graslandmaus
ES Ratón cavador de cola peluda oriental

Extreme SE PE, BO, W PY, N AR, CW BR (Amazonas, Rondônia, Mato Grosso)

LC Monotypic
HB 9.5–10.3 cm
T 6.3–7.9 cm
W c. 25 g

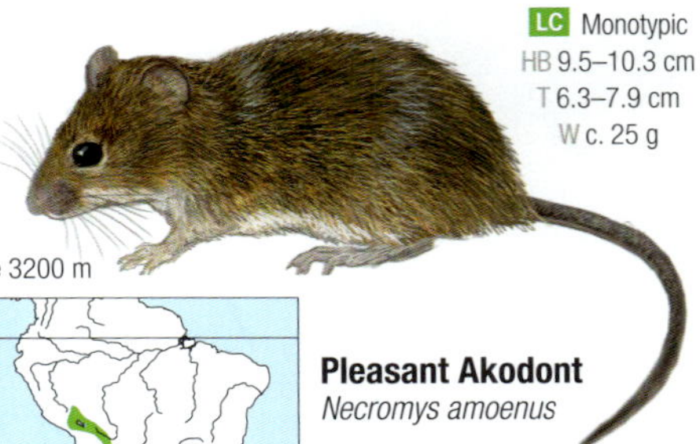

above 3200 m

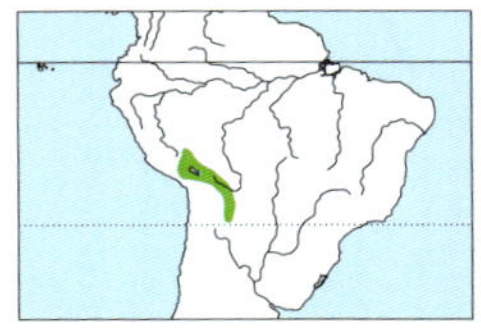

Pleasant Akodont
Necromys amoenus

FR Akodon aimable
DE Schöne Graslandmaus
ES Ratón cavador cordial

Altiplano & E Andean slopes in SE PE, W BO & NW AR

LC
Monotypic
HB 9.8–12.4 cm
T 6.3–8.9 cm
W 30–50 g

1400–4000 m

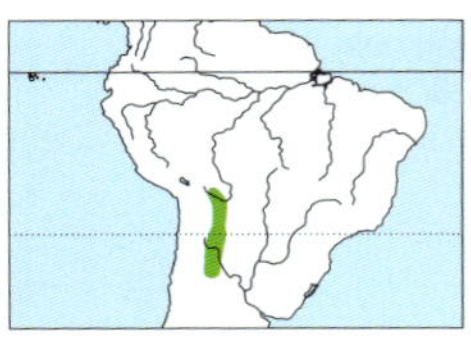

White-chinned Akodont
Necromys lactens

FR Akodon à menton blanc
DE Weißkehl-Graslandmaus
ES Ratón cavador de barbilla blanca

SC BO (Cochabamba S to Tarija), NW AR (Jujuy, Salta, Catamarca, Tucumán)

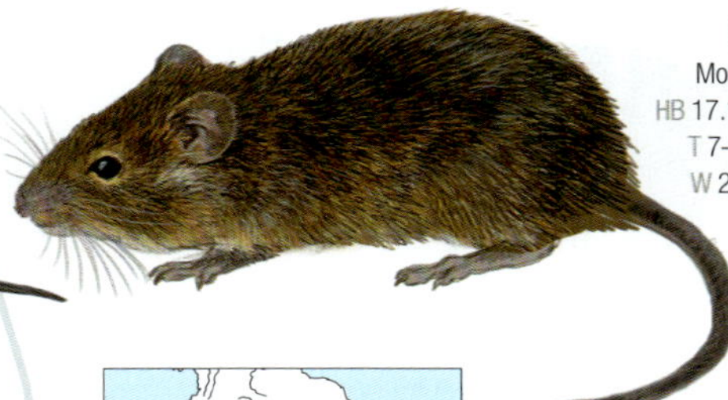

NE
Monotypic
HB 17.1–18.9 cm
T 7–7.7 cm
W 29–44 g

Lillo's Akodont
Necromys lilloi

FR Akodon de Lillo
DE Lillo-Graslandmaus
ES Ratón cavador de Lillo

Known from type locality and its vicinity (Burruyacu, Tucumán)

LC
2 sspp
HB 10.1–12.6 cm
T 5.7–7.8 cm
W 35–60 g

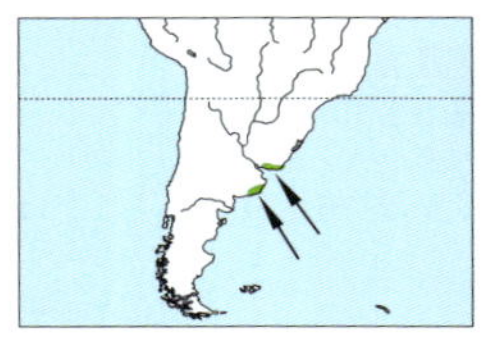

Dark-furred Akodont
Necromys obscurus

FR Akodon obscur
DE Dunkle Graslandmaus
ES Ratón cavador de pelaje oscuro

S UY, EC AR (SE Buenos Aires)

LC
Monotypic
HB c. 10 cm
T c. 8.3 cm
W c. 22 g

Kemp's Grass Mouse
Deltamys kempi

FR Akodon de Kemp
DE Kemp-Grasmaus
ES Ratón campestre de Kemp

S BR, UY, extreme CE AR

Genus *DELTAMYS*

(2 species)

NE
Monotypic
HB c. 8 cm
T c. 6.6 cm
W 10–22 g

c. 900 m

Araucaria Grass Mouse
Deltamys araucaria

FR Akodon des araucarias
DE Araukarien-Grasmaus
ES Ratón campestre de Araucaria

Known only from type locality (Rio Grande do Sul)

LC
Monotypic
HB 8.4–11.9 cm
T 7–10.1 cm
W 15–60 g

Genus *CASTORIA*

(1 species)

300–2100 m

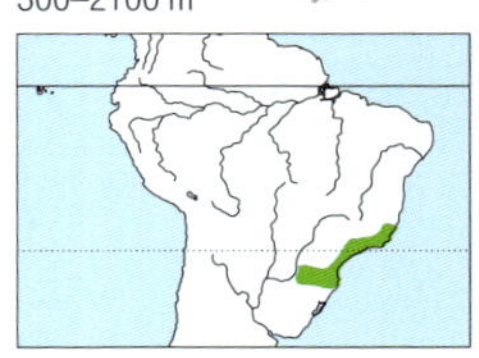

Serra do Mar Grass Mouse
Castoria angustidens

FR Akodon de la Serra do Mar
DE Serra-do-Mar-Grasmaus
ES Ratón campestre de la Sierra del Mar

SE & S BR (from Espírito Santo S to Rio Grande do Sul), NE AR (Misiones)

above 2000 m

Andes in EC & NW PE (isolated population in S); probably extreme S CO

LC 3 sspp
HB 8.1–11.3 cm T 5.5–9.2 cm
W 19–32 g

Soft-furred Grass Mouse
Akodon mollis

FR Akodon à poil doux
DE Weichhaar-Graslandmaus
ES Ratón campestre de pelaje suave

Genus *AKODON*

(41 species)

Colombian Grass Mouse
Akodon affinis

FR Akodon de Colombie DE Kolumbien-Graslandmaus
ES Ratón campestre de Colombia

LC Monotypic
HB c. 9.4 cm
T c. 6.5 cm

1800–2800 m

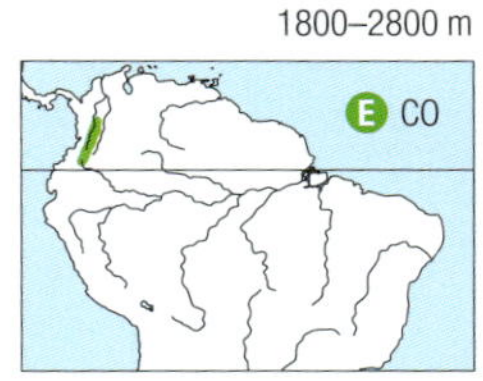

W Andes

LC
Monotypic
HB 9.6–11.9 cm
T 7–8.6 cm

1200–2400 m

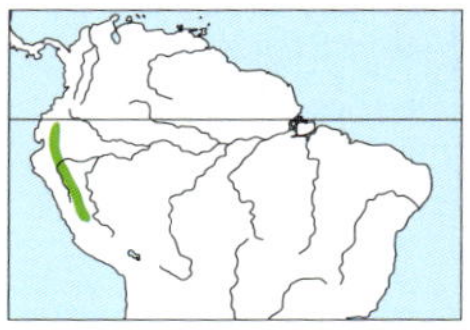

Yungas Grass Mouse
Akodon aerosus

FR Akodon cuivré
DE Yungas-Graslandmaus
ES Ratón campestre de Yungas septentrional

E Andean slopes from S EC to C PE

NE
Monotypic
HB 9.6–11.9 cm
T 7–8.6 cm

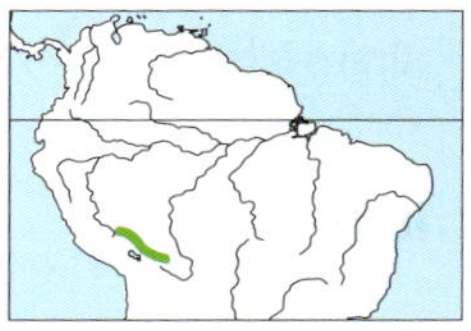

Dark Grass Mouse
Akodon baliolus

FR Akodon sombre
DE Kastanienbraune Graslandmaus
ES Ratón campestre de Yungas meridional

W Andes from SE PE (Cusco) to C BO (Cochabamba)

LC Monotypic
HB 9.5–11 cm T 7.4–10 cm

1900–2860 m

El Dorado Grass Mouse
Akodon orophilus

FR Akodon de l'Utcubamba
DE El-Dorado-Graslandmaus
ES Ratón de hierba de el Dorado

E Andean slopes E of R Marañón (Amazonas, N San Martín)

NE Monotypic
HB c. 10 cm
T c. 8.9 cm

2900–3800 m

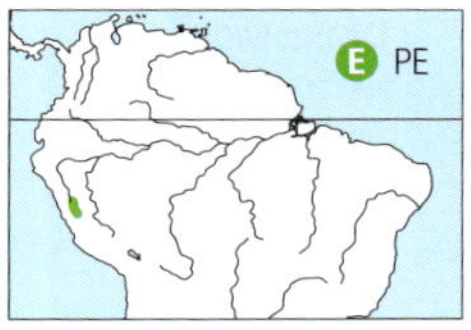

Arguedas's Grass Mouse
Akodon josemariarguedasi

FR Akodon d'Arguedas
DE Arguedas-Graslandmaus
ES Ratón campestre de Arguedas

S of R Huallaga in Huánuco

LC Monotypic
HB c. 9.7 cm
T c. 6.2 cm

2700–3800 m

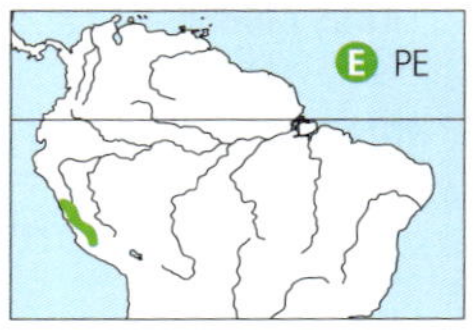

Junin Grass Mouse
Akodon juninensis

FR Akodon du Junin
DE Junín-Graslandmaus
ES Ratón campestre de Junín

W & E Andean slopes

LC 2 sspp
HB 8–9.2 cm
T 6.4–6.8 cm
W c. 21 g

1900–2500 m

Puno Grass Mouse
Akodon subfuscus

FR Akodon de Puno
DE Puno-Graslandmaus
ES Ratón campestre de Puna

Altiplano and Andean slopes from S PE to C BO

LC
Monotypic
HB c. 10.4 cm
T c. 9.1 cm

2000–3800 m

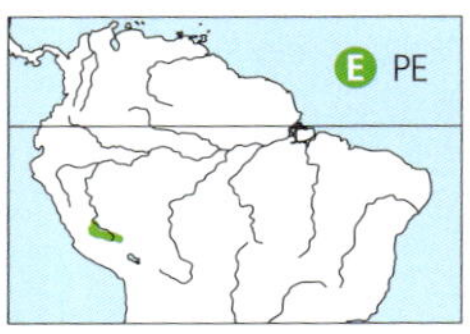

Cloud Forest Grass Mouse
Akodon torques

FR Akodon de Machu Picchu
DE Nebelwald-Graslandmaus
ES Ratón campestre de bosque nuboso

E Andean cloud forest

VU
Monotypic
HB c. 11 cm
T c. 8 cm

600–2000 m

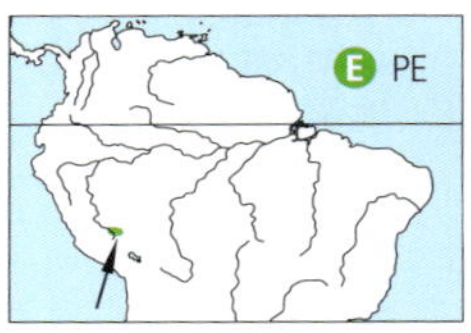

Silent Grass Mouse
Akodon surdus

FR Akodon sourd
DE Cusco-Graslandmaus
ES Ratón campestre silencioso

Machu Picchu area (Cusco)

NE
Monotypic
HB 9.8–12.2 cm
T 7.1–10 cm

above 2400 m

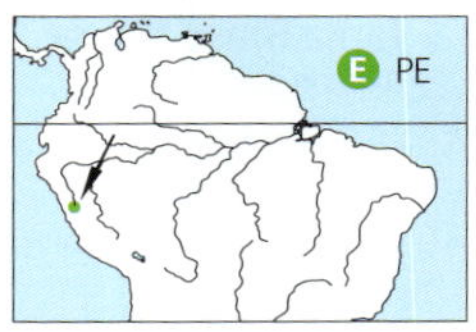

Kotosh Grass Mouse
Akodon kotosh

FR Akodon de Kotosh
DE Kotosh-Graslandmaus
ES Ratón campestre de Kotosh

Cordillera de Carpish (Huánuco)

LC Monotypic
HB 8.6–10.9 cm
T 6.9–7.7 cm
W 21–36 g

2350–4500 m

White-bellied Grass Mouse
Akodon albiventer

FR Akodon à ventre blanc
DE Weißbauch-Graslandmaus
ES Ratón campestre de vientre blanco

SE PE, SW BO, extreme NE CL, NW AR

LC
Monotypic
HB c. 9 cm
T c. 6.5 cm
W c. 23 g

3700–5000 m

Bolivian Grass Mouse
Akodon boliviensis

FR Akodon de Bolivie
DE Bolivien-Graslandmaus
ES Ratón campestre de Bolivia

SE PE through C BO to NW AR

DD Monotypic
HB 6.7–9.3 cm
T 4.9–7.4 cm
W 11–31 g

550–1500 m

Lindbergh's Grass Mouse
Akodon lindberghi

FR Akodon de Lindbergh
DE Lindbergh-Graslandmaus
ES Ratón campestre de Lindbergh

Cerrado (Goiás, Distrito Federal), Atlantic Forest (Minas Gerais, Rio de Janeiro)

LC Monotypic
HB 11.5–12 cm
T 9–9.3 cm
W 43–54 g

0–2000 m

Cursorial Grass Mouse
Akodon cursor

FR Akodon courseur
DE Eilige Graslandmaus
ES Ratón campestre corredor

From Paraíba S to Paraná

LC Monotypic
HB 10.3–11 cm
T 8.4–9 cm
W 23–29 g

above 800 m

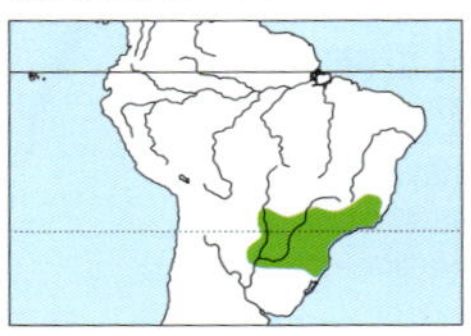

Montane Grass Mouse
Akodon montensis

FR Akodon des collines
DE Berg-Graslandmaus
ES Ratón campestre de monte

SE & S BR, C & E PY, NE AR

LC Monotypic
HB c. 8 cm
T c. 5.9 cm
W 10–20 g

1130–4560 m

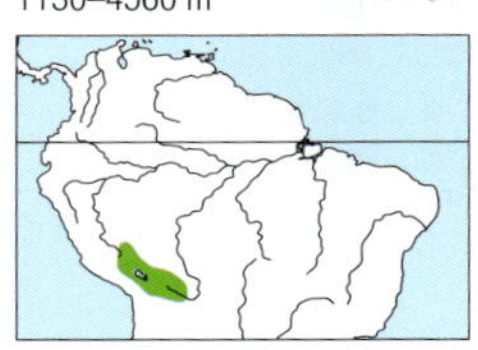

Altiplano Grass Mouse
Akodon lutescens

FR Akodon fauve
DE Altiplano-Graslandmaus
ES Ratón campestre de altiplano

E Andean slopes from SE PE to CW BO

LC Monotypic
HB c. 9.7 cm
T c. 9.4 cm

2000–3700 m

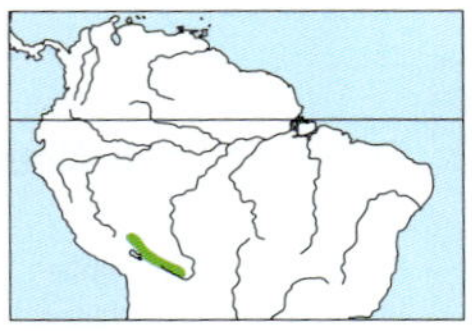

Thespian Grass Mouse
Akodon mimus
FR Akodon imitateur
DE Kleine Graslandmaus
ES Ratón campestre trágico

E Andean slopes from SE PE (Puno) to C BO (Santa Cruz)

LC
Monotypic
HB c. 9.6 cm
T c. 7.8 cm

1800–3700 m

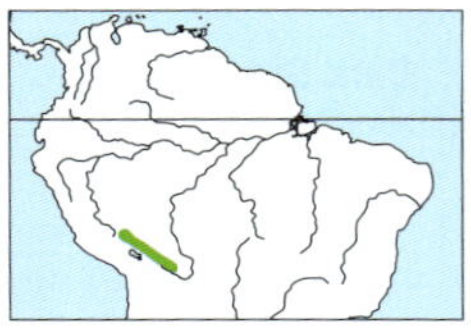

Koford's Grass Mouse
Akodon kofordi
FR Akodon de Koford
DE Koford-Graslandmaus
ES Ratón campestre de Koford

E Andean slopes from SE PE to CW BO

LC Monotypic
HB 9.5–10.8 cm
T 7.5–9.5 cm

700–3800 m

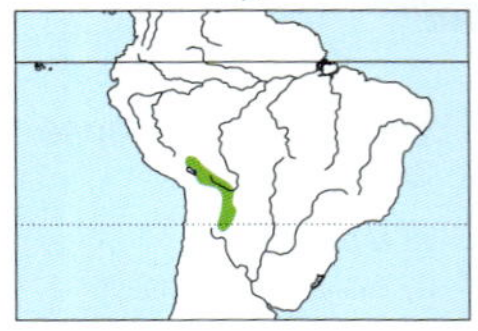

Smoky Grass Mouse
Akodon fumeus
FR Akodon fuligineux
DE Rauchgraue Graslandmaus
ES Ratón campestre ahumado

E Andean slopes in extreme SE PE (Puno), BO & NW AR (Salta, Jujuy)

LC Monotypic
HB 11–13.4 cm
T 7.5–8.4 cm
W 26–41 g

250–2450 m

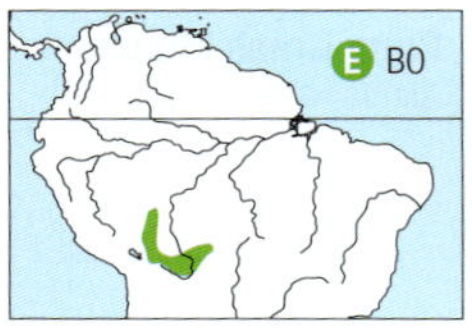

Day's Grass Mouse
Akodon dayi
FR Akodon de Day
DE Day-Graslandmaus
ES Ratón campestre de Day

Pando to WC Santa Cruz

NT Monotypic
HB 9.3–11.2 cm
T 7.7–10.1 cm
W 21–31 g

1800–3100 m

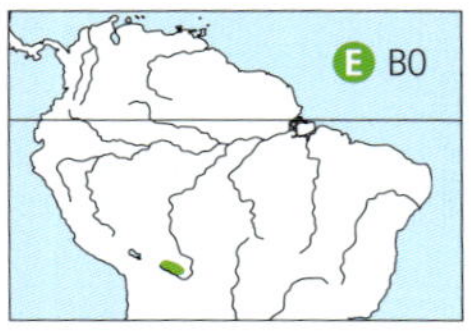

Cochabamba Grass Mouse
Akodon siberiae
FR Akodon du Cochabamba
DE Cochabamba-Graslandmaus
ES Ratón campestre de Cochabamba

Siberia Cloud Forest (Cochabamba, W Santa Cruz)

DD Monotypic
HB 7.8–9.4 cm
T 4.6–7.5 cm
W 10.5–27 g

400–3100 m

Unicolored Grass Mouse
Akodon caenosus
FR Akodon terreux
DE Einfarbige Graslandmaus
ES Ratón campestre unicolor

SC BO to NW AR

DD Monotypic
HB c. 11.6 cm
T c. 9.3 cm

900–2100 m

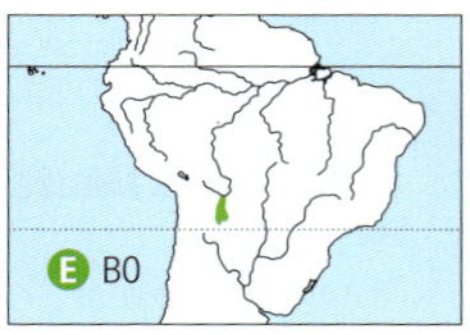

Tarija Grass Mouse
Akodon pervalens
FR Akodon du Tarija
DE Tarija-Graslandmaus
ES Ratón campestre de Tarija

Known from few localities in Chuquisaca & Tarija

LC Monotypic
HB 9.3–12.5 cm
T 6.1–9.1 cm
W 18.5–56 g

400–3000 m

Gray-bellied Grass Mouse
Akodon simulator
FR Akodon du Tucuman
DE Graubauch-Graslandmaus
ES Ratón campestre de vientre gris

S BO (Tarija), NW AR (Jujuy, Salta, Tucumán, Catamarca, La Rioja)

LC
Monotypic
HB 11.4–11.8 cm
T 5.2–8.5 cm
W 22–43 g

800–2800 m

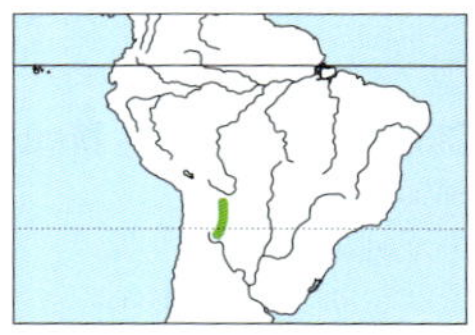

Budin's Grass Mouse
Akodon budini
FR Akodon de Budin
DE Budin-Graslandmaus
ES Ratón campestre de Budin

E Andean slopes in SC BO and NW AR

DD Monotypic
HB 10.5–15.8 cm
T 8–11 cm
W 18–32 g

365–3200 m

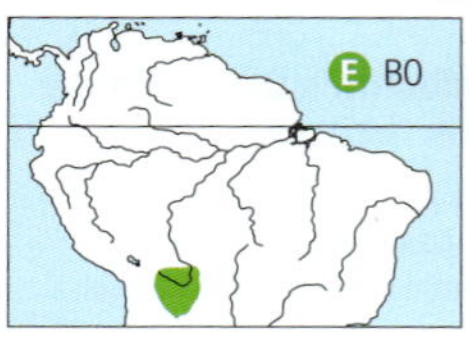

Variable Grass Mouse
Akodon varius
FR Akodon variable
DE Variable Graslandmaus
ES Ratón campestre variable

E Andean slopes

DD
Monotypic
HB 6.6–10.1 cm
T 4.8–7.6 cm

above 1800 m

Caparao Grass Mouse
Akodon mystax
FR Akodon de Caparao
DE Caparaó-Graslandmaus
ES Ratón campestre de Caparaó

Known only from type locality (W slope Mt Caparaó, Minas Gerais)

LC Monotypic
HB 8.8–12.5 cm
T 6–9.2 cm
W 39–42 g

900–1800 m

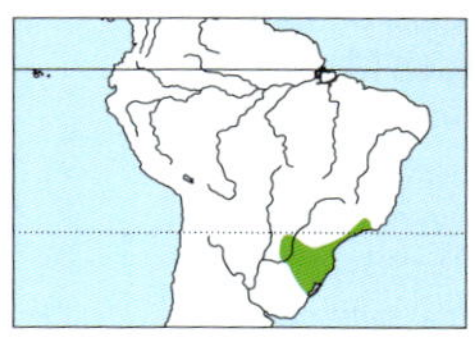

Parana Grass Mouse
Akodon paranaensis
FR Akodon du Parana
DE Paraná-Graslandmaus
ES Ratón campestre de Paraná

SE & S BR, E PY, NE AR (Misiones)

DD
Monotypic
HB 7.3–10 cm
T 5.8–7.5 cm

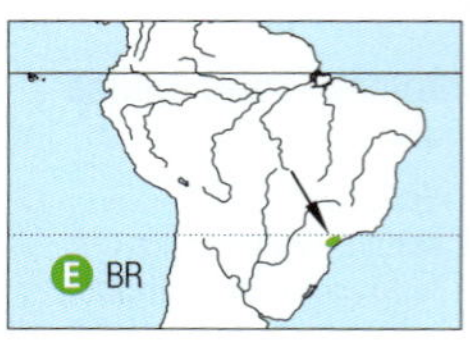

Sao Paulo Grass Mouse
Akodon sanctipaulensis
FR Akodon du Sao Paulo
DE São-Paulo-Graslandmaus
ES Ratón campestre de San Pablo

Known from few localities in Sierra do Mar (São Paulo)

LC Monotypic
HB 8.3–11.4 cm
T 6.8–8.9 cm
W 16–35 g

900–2400 m

Forest Grass Mouse
Akodon sylvanus
FR Akodon sylvestre
DE Wald-Graslandmaus
ES Ratón campestre de bosque

Known from few localities in Sierra de Santa Bárbara (SE Jujuy)

LC Monotypic
HB 6–10.3 cm T 4.6–8.3 cm
W 13–34 g

400–3500 m

Spegazzini's Grass Mouse
Akodon spegazzinii
FR Akodon de Spegazzini
DE Spegazzini-Graslandmaus
ES Ratón campestre de Spegazzini

E Andean slopes from Salta to La Rioja

NE Monotypic
HB 9.8–10.7 cm
T 6.2–7.6 cm
W 19.9–34 g

400–3500 m

NC & SC Mendoza

Monte Grass Mouse
Akodon oenos

FR Akodon de Mendoza
DE Montan-Graslandmaus
ES Ratón campestre de Mendoza

LC
Monotypic
HB 9.1–12.6 cm
T 6.7–8.5 cm
W 25–57 g

below 1000 m

SE BO, W PY, extreme SW BR, N AR

Toba Grass Mouse
Akodon toba

FR Akodon des Tobas
DE Toba-Graslandmaus
ES Ratón campestre del Chaco

LC Monotypic
HB 9.3–9.6 cm
T 6.8–8.4 cm
W 20–29 g

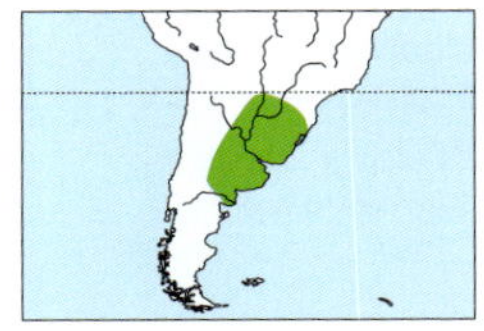

E & S PY, S BR, NE & EC AR, UY

Azara's Grass Mouse
Akodon azarae

FR Akodon d'Azara
DE Azara-Graslandmaus
ES Ratón campestre de Azara

DD
Monotypic
HB c. 9.2 cm
T c. 6 cm
W c. 19 g

Known from only 2 localities in S Misiones

Philip Myers's Grass Mouse
Akodon philipmyersi

FR Akodon de Myers
DE Myers-Graslandmaus
ES Ratón campestre de Philip Myers

NE
Monotypic
HB 8.4–9.7 cm
T 8–8.5 cm
W c. 20 g

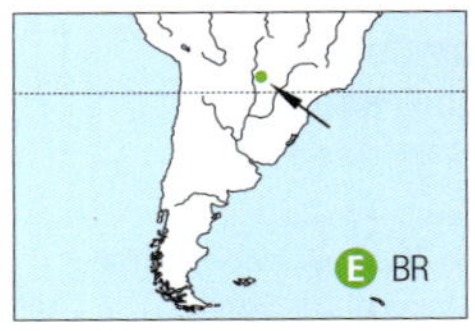

Known only from type locality (N Serra da Bodoquena, Mato Grosso do Sul)

Kadiweu Grass Mouse
Akodon kadiweu

FR Akodon des Caduveos
DE Kadiwéu-Graslandmaus
ES Ratón campestre de Kadiwéu

LC
Monotypic
HB 9.2–12.3 cm
T 8.8–9.7 cm
W 20–61 g

Extreme S BR (S Rio Grande do Sul), N & E UY

Reig's Grass Mouse
Akodon reigi

FR Akodon de Reig
DE Reig-Graslandmaus
ES Ratón campestre de Reig

LC
Monotypic
HB 7.9–10.3 cm
T 5.4–8.3 cm
W 22–48 g

1300–2250 m

Known from few localities in Córdoba & San Luis

Polop's Grass Mouse
Akodon polopi

FR Akodon de Polop
DE Polop-Graslandmaus
ES Ratón campestre de Polop

LC
Monotypic
HB 10.7–11.1 cm
T 7.4–7.6 cm
W 38–43 g

From Tucumán & Santiago del Estero S to NE Chubut

Dolores Grass Mouse
Akodon dolores

FR Akodon de Villa Dolores
DE Dolores-Graslandmaus
ES Ratón campestre de Dolores

LC Monotypic
HB c. 9.4 cm
T c. 5.9 cm
W c. 22 g

0–1000 m

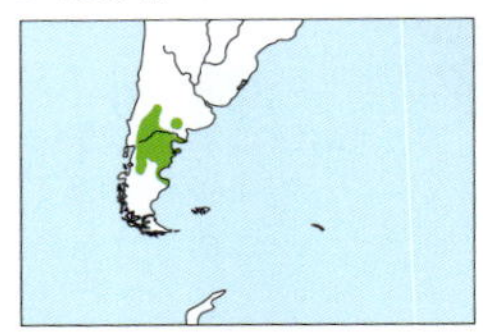

N Patagonia in SC & S AR and adjacent extreme E CL

Intelligent Grass Mouse
Akodon iniscatus

FR Akodon de Patagonie
DE Patagonien-Graslandmaus
ES Ratón campestre inteligente

LC
Monotypic
HB 11–11.9 cm
T 11.4–12.7 cm
W 30–44 g

1990–3500 m

Cordillera de Mérida

Merida Montane Mouse
Aepeomys lugens

FR Aepéomys du Mérida
DE Mérida-Andenmaus
ES Ratón de montaña de Mérida

Genus *AEPEOMYS*

(2 species)

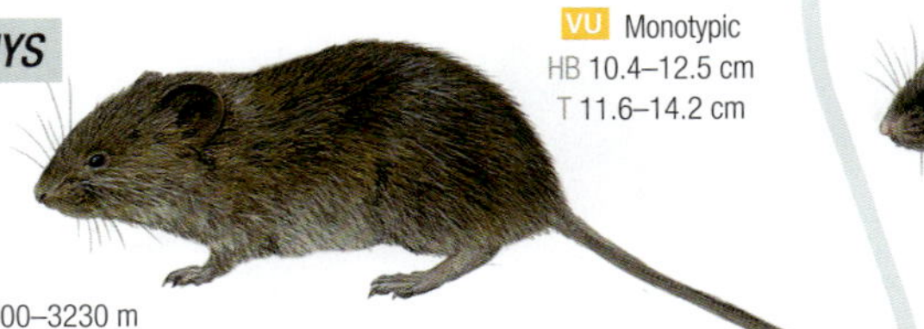

VU Monotypic
HB 10.4–12.5 cm
T 11.6–14.2 cm

1600–3230 m

Andes in Trujillo and Lara

Reig's Montane Mouse
Aepeomys reigi

FR Aepéomys de Reig
DE Reig-Andenmaus
ES Ratón de montaña de Reig

Smoky Forest Mouse
Chilomys fumeus

FR Chilomys sombre
DE Graue Kolumbien-Waldmaus
ES Ratón de bosque ahumado

1830–2700 m

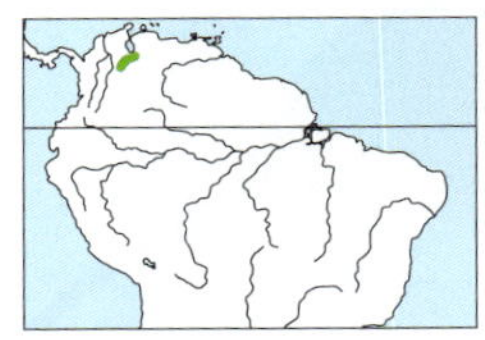

Andes of N CO and W VE

NE
Monotypic
HB 8.8–9.9 cm
T 12–13 cm
W 15–22 g

Long-tongued Arboreal Mouse
Rhagomys longilingua

FR Rhagomys du Pérou
DE Langzungen-Baummaus
ES Ratón arborícola de lengua larga

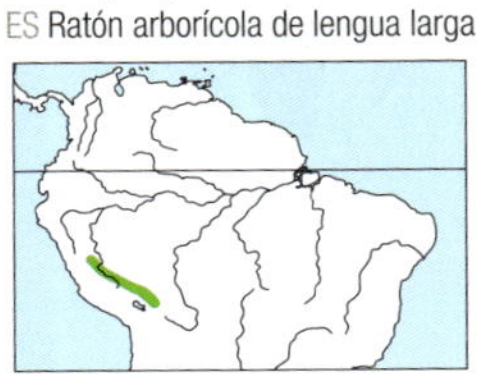

Genus *RHAGOMYS*

(3 species)

LC
Monotypic
HB 8.9–10.3 cm
T 9.3–10.4 cm
W 25–35 g

450–2100 m

Scattered localities in E Andean slopes of C & SE PE and CW BO

Genus *CHILOMYS*

(2 species)

1100–4000 m

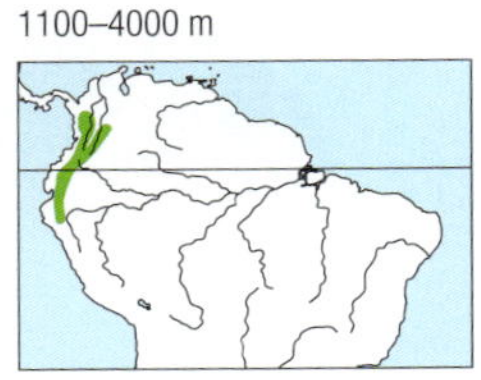

3 Andean ranges from WC CO to NW PE

LC
Monotypic
HB 8.6–9.9 cm
T 10.5–13 cm
W 15–26 g

Colombian Forest Mouse
Chilomys instans

FR Chilomys de Thomas
DE Subalpine Kolumbien-Waldmaus
ES Ratón de bosque de Colombia

NE
Monotypic
HB c. 9 cm
T c. 10 cm

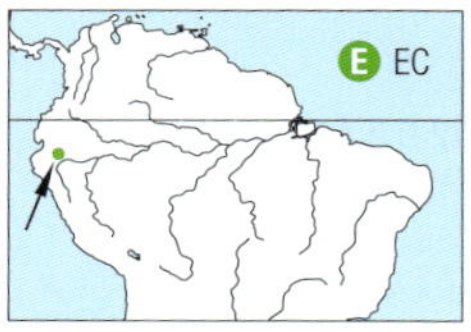

W foothills of Cordillera del Cóndor (Zamora Chinchipe)

Northern Arboreal Mouse
Rhagomys septentrionalis
FR Rhagomys d'Équateur
DE Nördliche Baummaus
ES Ratón arborícola septentrional

VU Monotypic
HB 8.2–9.8 cm T 8.5–10.7 cm
W 16–32 g

0–1700 m

Espíritu Santo, Minas Gerais, Rio de Janeiro, São Paulo, Santa Catarina

Brazilian Arboreal Mouse
Rhagomys rufescens
FR Rhagomys du Brésil
DE Brasilianische Baummaus
ES Ratón arborícola de Brasil

LC
Monotypic
HB 9.8–12.4 cm
T 13.5–18.4 cm
W 50–63 g

450–3000 m

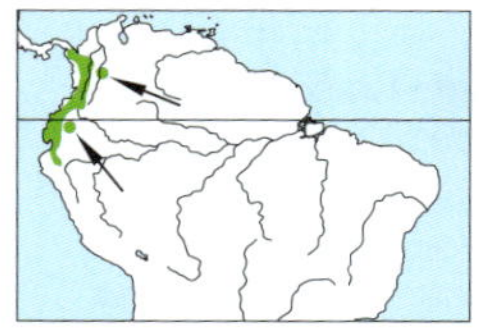

Extreme E PA, W CO, W EC, NW PE

Broad-footed Climbing Rat
Rhipidomys latimanus
FR Rhipidomys à pattes larges
DE Breitfuß-Neuweltklettermaus
ES Rata trepadora de pies anchos

Genus *RHIPIDOMYS*

(25 species)

Lesser Colombian Climbing Rat
Rhipidomys caucensis
FR Rhipidomys du Cauca
DE Kleine Cauca-Neuweltklettermaus
ES Rata trepadora de Colombia pequeña

1830–3500 m

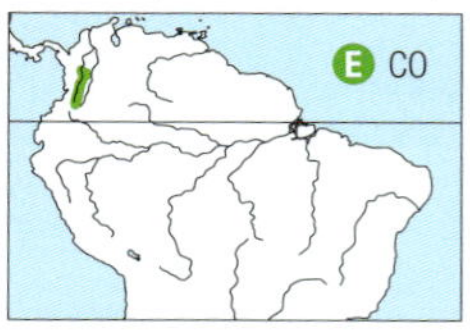

Andes in Cauca, Valle, Antioquia & Huila

DD
Monotypic
HB c. 10.3 cm
T c. 13.3 cm

NE Monotypic
HB 12.8–16.5 cm
T 18.4–19.2 cm

1830–2400 m

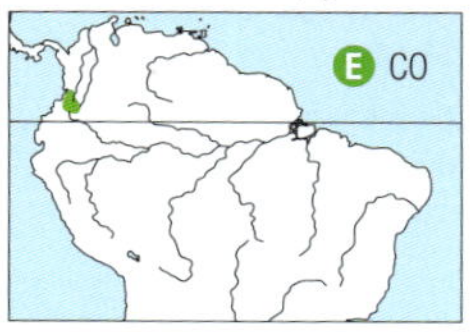

Andes in Cauca, Valle & Huila

Greater Colombian Climbing Rat
Rhipidomys similis
FR Rhipidomys de Colombie
DE Große Cauca-Neuweltklettermaus
ES Rata trepadora de Colombia grande

LC Monotypic
HB 8.8–11.2 cm
T 10.7–13.3 cm
W 25–46 g

1800–3100 m

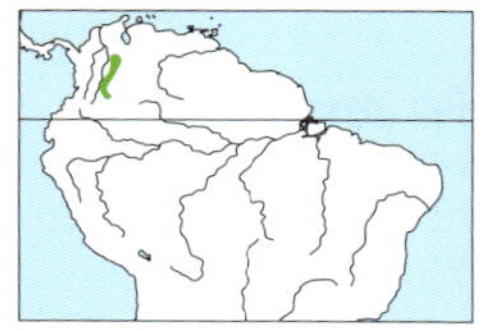

Extreme SW VE (Páramo de Tamá), E Andes in C CO

Tawny-bellied Climbing Rat
Rhipidomys fulviventer
FR Rhipidomys à ventre fauve
DE Ockerbauch-Neuweltklettermaus
ES Rata trepadora de vientre leonado

LC Monotypic
HB 15–21 cm T 17.1–19.5 cm
W 53–125 g

Coues's Climbing Rat
Rhipidomys couesi
FR Rhipidomys de Coues
DE Coues-Neuweltklettermaus
ES Rata de arroz de marisma de Coues

10–1500 m

C CO to N VE (with Margarita I), TT (Trinidad I)

Charming Climbing Rat
Rhipidomys venustus
FR Rhipidomys charmant
DE Schöne Neuweltklettermaus
ES Rata trepadora encantadora

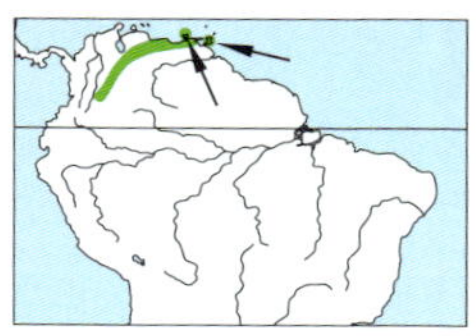

LC
Monotypic
HB 12.1–15 cm
T 12.3–16.5 cm
W 41–75 g

above 1000 m

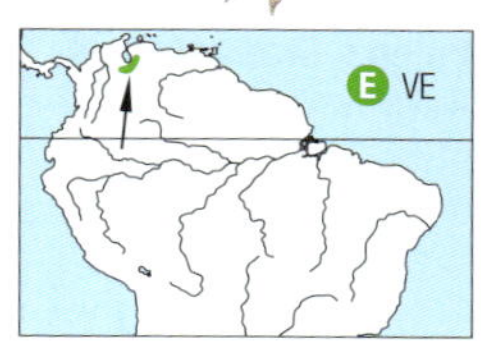

Cloud forests in Cordillera de Mérida

LC Monotypic
HB 12–15.2 cm T 13.4–18.8 cm
W 74–81 g

c. 710 m

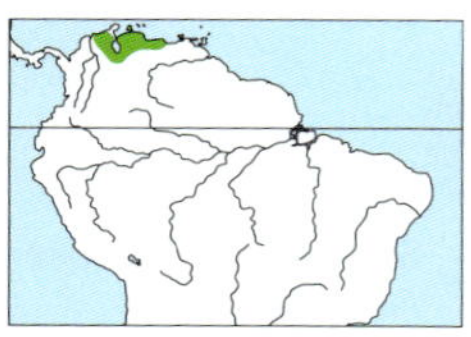

Moist and dry forests in N CO and NW & NC VE

Venezuelan Climbing Rat
Rhipidomys venezuelae
FR Rhipidomys du Venezuela
DE Venezuela-Neuweltklettermaus
ES Rata trepadora de Venezuela

NE
Monotypic
HB 10.9–12.8 cm
T 12.5–15.7 cm

945–2200 m

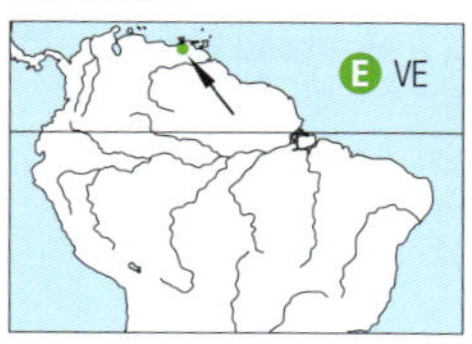

Serranía de Turimiquire

Turimiquire Climbing Rat
Rhipidomys tenuicauda
FR Rhipidomys du Turimiquire
DE Turimiquire-Neuweltklettermaus
ES Rata trepadora de Turimiquire

LC Monotypic
HB 6.3–10.3 cm
T 8.2–13 cm
W 20–39 g

1030–1850 m

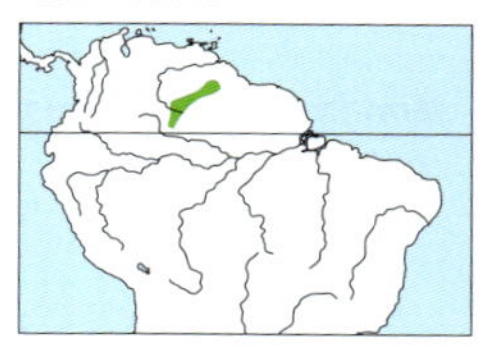

S VE (Amazonas, Bolívar), extreme W GY (Mt Roraima); perhaps N BR

Wetzel's Climbing Rat
Rhipidomys wetzeli
FR Rhipidomys de Wetzel
DE Wetzel-Neuweltklettermaus
ES Rata trepadora de Wetzel

Ochoa-Graterol's Climbing Rat
Rhipidomys ochoagrateroli
FR Rhipidomys d'Ochoa-Graterol
DE Ochoa-Graterol-Neuweltklettermaus
ES Rata trepadora de Ochoa-Graterol

NE
Monotypic
HB 11.1–19 cm
T 13.7–19 cm
W 37–81 g

1300–2225 m

Mts in Lara-Falcón, Cordillera de la Costa (Sierra de Aroa, Serranía del Litoral)

LC Monotypic
HB 17.7–21 cm T 22–24 cm
W 125–218 g

White-footed Climbing Rat
Rhipidomys leucodactylus
FR Rhipidomys à pattes blanches
DE Weißfuß-Neuweltklettermaus
ES Rata trepadora de pies blancos

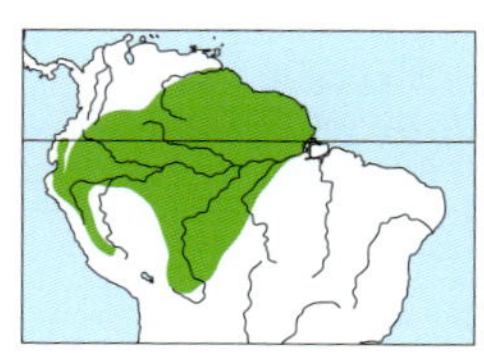

S CO, S VE and Guianas to EC, PE, C BR and C BO

DD
Monotypic
HB 11.8–13 cm
T 14.6–16.2 cm
W 46–62 g

1400–1800 m

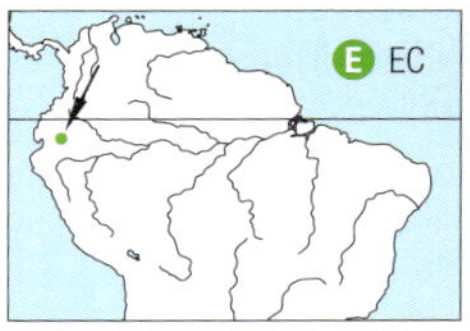

Albuja's Climbing Rat
Rhipidomys albujai
FR Rhipidomys d'Albuja
DE Albuja-Neuweltklettermaus
ES Rata trepadora de Albuja

Restricted to type locality in Andes (Sangay NP)

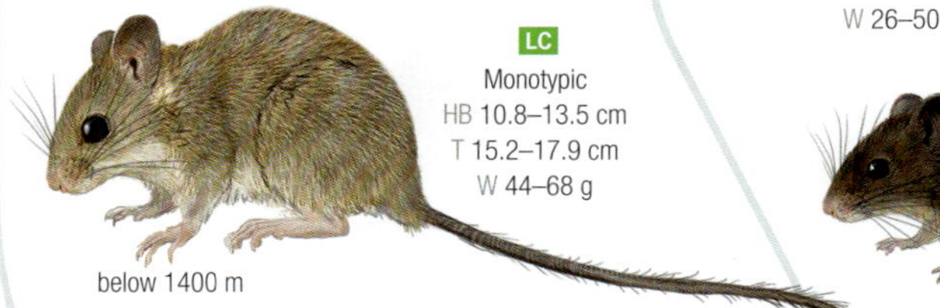

LC
Monotypic
HB 10.8–13.5 cm
T 15.2–17.9 cm
W 44–68 g

below 1400 m

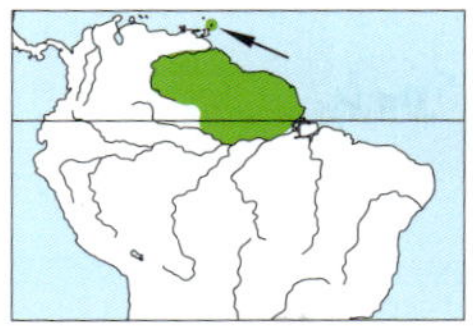

Splendid Climbing Rat
Rhipidomys nitela
FR Rhipidomys splendide
DE Pracht-Neuweltklettermaus
ES Rata trepadora magnífica

Amazonian lowlands in S VE, Guianas, NC BR & TT (Little Tobago I)

LC Monotypic
HB 10.2–12.3 cm T 14.8–18.1 cm
W 26–50 g

McConnell's Tepui Climbing Rat
Rhipidomys macconnelli
FR Rhipidomys de McConnell
DE Tepui-Neuweltklettermaus
ES Rata trepadora de McConnell

750–2600 m

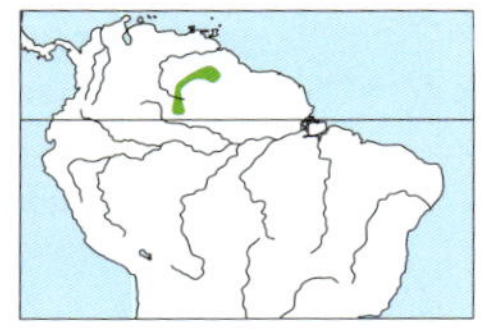

Guiana Shield tepuis in S VE and adjacent areas of GY & N BR

LC Monotypic
HB 11.6–15.6 cm
T 14.3–17.5 cm
W 50–87 g

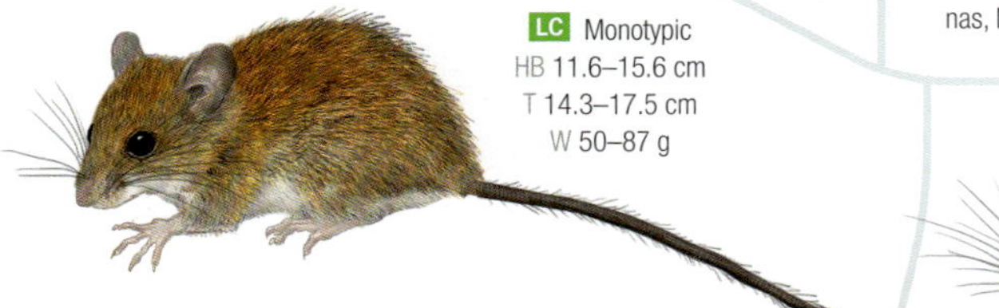

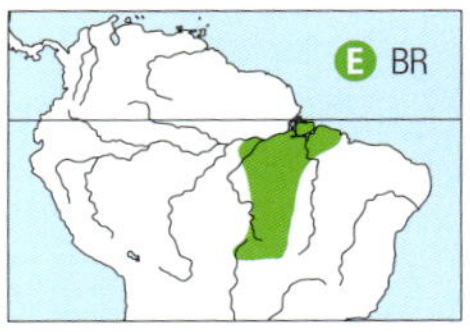

Snethlage's Climbing Rat
Rhipidomys emiliae
FR Rhipidomys de Snethlage
DE Snethlage-Neuweltklettermaus
ES Rata trepadora de Snethlage

S of R Amazon, from W bank of R Tapajós E to Maranhão, S to Mato Grosso

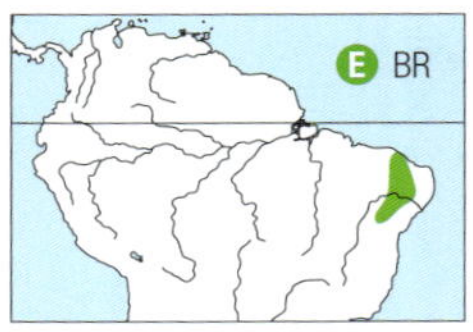

Cariri Climbing Rat
Rhipidomys cariri
FR Rhipidomys des Cariris
DE Cariri-Neuweltklettermaus
ES Rata trepadora de Cariri

Semiarid caatinga in Ceará, Pernambuco & N Bahia

DD Monotypic
HB 13–19 cm
T 14.3–27 cm

DD Monotypic
HB 9.9–14.1 cm T 11.3–16.5 cm
W 40–103 g

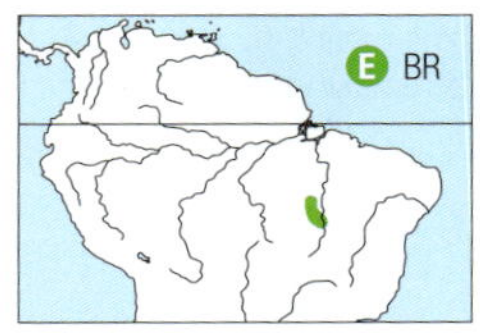

Ipuca Climbing Rat
Rhipidomys ipukensis
FR Rhipidomys des ipucas
DE Ipuca-Neuweltklettermaus
ES Rata trepadora de ipuca

Known only from 3 localities in Tocantins

LC Monotypic
HB 13–16.5 cm
T c. 17.5 cm

700–1800 m

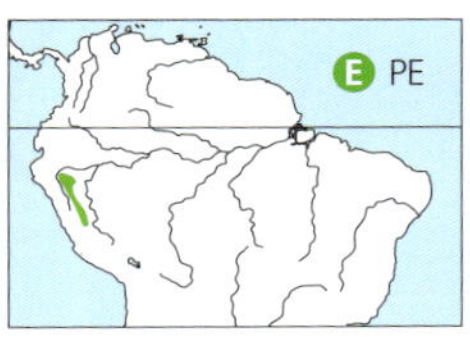

Lesser Peruvian Climbing Rat
Rhipidomys modicus
FR Rhipidomys du Pérou
DE Kleine Peru-Neuweltklettermaus
ES Rata trepadora de Perú pequeña

E Andean slopes S of R Marañón

LC Monotypic
HB 16.1–19 cm
T 17.7–19.4 cm
W 126–155 g

200–2500 m

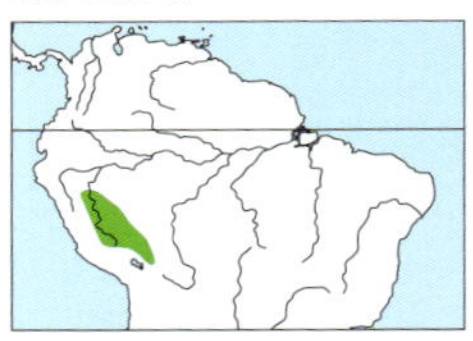

Gardner's Climbing Rat
Rhipidomys gardneri
FR Rhipidomys de Gardner
DE Gardner-Neuweltklettermaus
ES Rata trepadora de Gardner

E PE (from Loreto S to Madre de Dios), W BR (Acre), NW BO (La Paz)

DD
Monotypic
HB 15.2–15.4 cm
T 19.8–23 cm
W c. 129 g

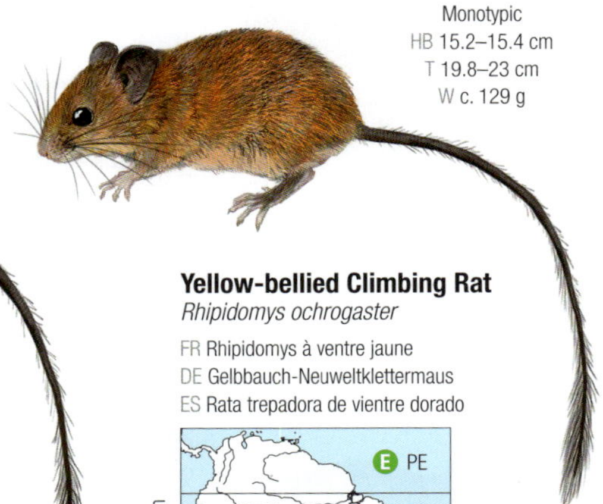

Yellow-bellied Climbing Rat
Rhipidomys ochrogaster
FR Rhipidomys à ventre jaune
DE Gelbbauch-Neuweltklettermaus
ES Rata trepadora de vientre dorado

1220–1940 m

E Andean slope in Puno

LC Monotypic
HB 11.3–15.7 cm
T 13.6–19 cm
W 61–112 g

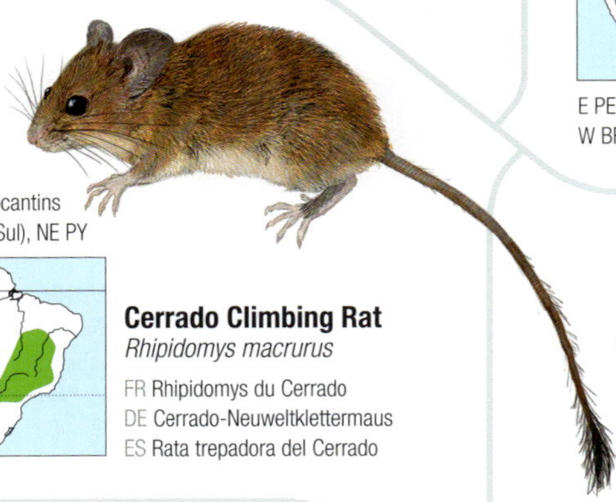

NE & C BR (Piauí & Tocantins S to Mato Grosso do Sul), NE PY

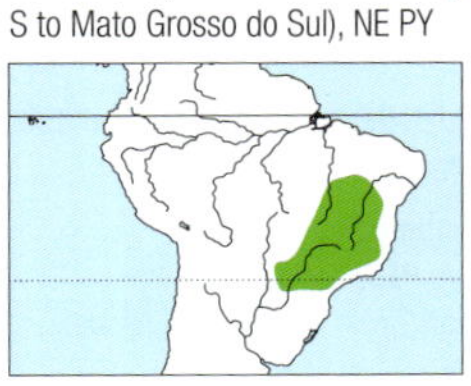

Cerrado Climbing Rat
Rhipidomys macrurus
FR Rhipidomys du Cerrado
DE Cerrado-Neuweltklettermaus
ES Rata trepadora del Cerrado

LC
Monotypic
HB 12.4–15.1 cm
T 14.7–16.7 cm
W 55–100 g

Long-tailed Climbing Rat
Rhipidomys mastacalis
FR Rhipidomys de Lund
DE Langschwanz-Neuweltklettermaus
ES Rata trepadora de cola larga

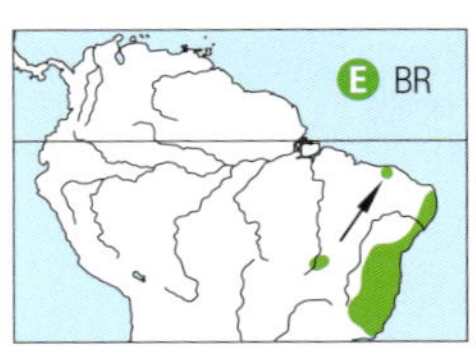

Paraíba S to Rio de Janeiro; populations of Ceará & Goiás perhaps of this species

DD
Monotypic
HB 9.6–13 cm
T 13.1–15 cm
W 36–68 g

Known from 3 localities in Minas Gerais & Espíritu Santo

up to 1300 m

Tribe's Climbing Rat
Rhipidomys tribei
FR Rhipidomys de Tribe
DE Tribe-Neuweltklettermaus
ES Rata trepadora de Tribe

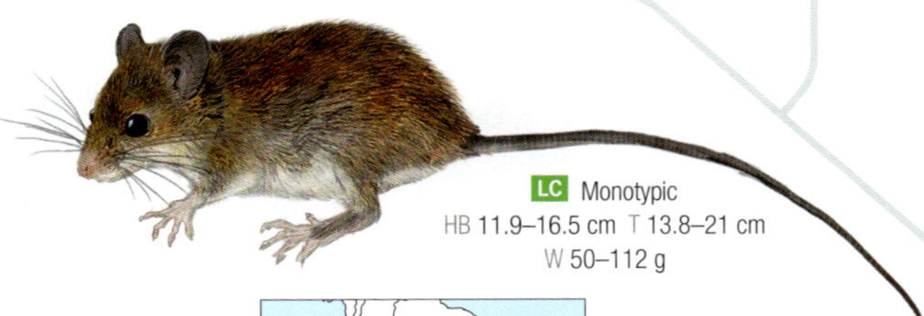

LC Monotypic
HB 11.9–16.5 cm T 13.8–21 cm
W 50–112 g

Rio de Janeiro and São Paulo including coastal islands

Sky Climbing Rat
Rhipidomys itoan
FR Rhipidomys céleste
DE Küstenwald-Neuweltklettermaus
ES Rata trepadora del cielo

LC
Monotypic
HB 12.5–13.7 cm
T 13.6–16.5 cm
W 51–70 g

360–2000 m

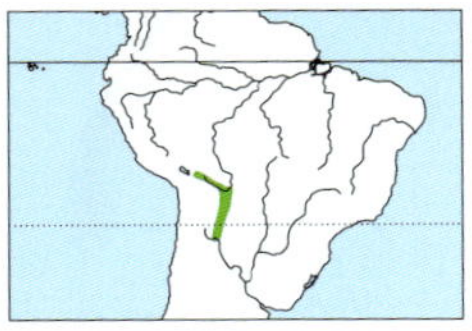

Southern Climbing Rat
Rhipidomys austrinus
FR Rhipidomys austral
DE Südliche Neuweltklettermaus
ES Rata trepadora meridional

Forested valleys of Yungas in WC BO S to N AR

Genus *THOMASOMYS*
(47 species)

VU
Monotypic
HB 10.5–12 cm
T c. 12.4 cm
W c. 36 g

2000–3600 m

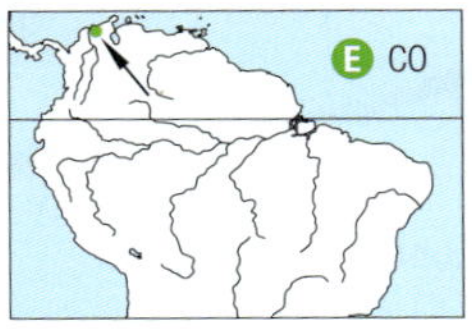

Unicolored Oldfield Mouse
Thomasomys monochromos
FR Thomasomys monochrome
DE Einfarbige Páramomaus
ES Ratón de erial unicolor

Sierra Nevada de Santa Marta (Magdalena)

DD
Monotypic
HB 10.4–12 cm

2090–3550 m

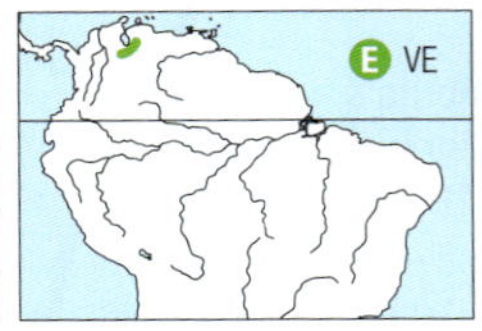

Venezuelan Oldfield Mouse
Thomasomys emeritus
FR Thomasomys du Venezuela
DE Venezuela-Páramomaus
ES Ratón de erial de Venezuela

Andes in Mérida and Trujillo

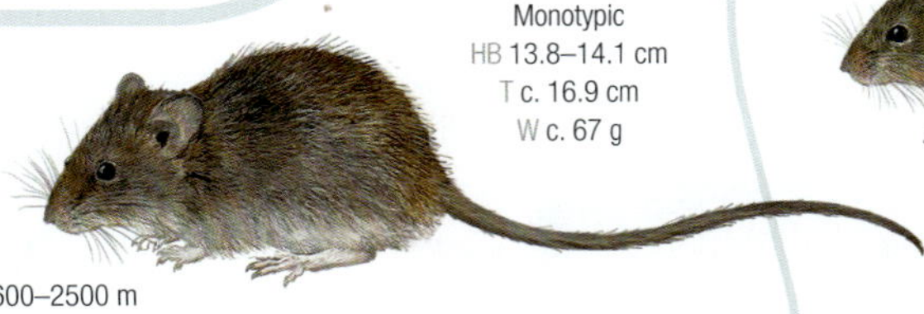

LC
Monotypic
HB 13.8–14.1 cm
T c. 16.9 cm
W c. 67 g

1600–2500 m

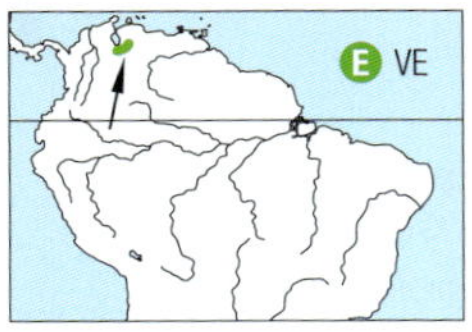

Dressy Oldfield Mouse
Thomasomys vestitus
FR Thomasomys velu
DE Schöne Páramomaus
ES Ratón de erial vistoso

Cordillera de Mérida

VU Monotypic
HB 10.7–12.6 cm T 12.4–15.5 cm
W 32–40 g

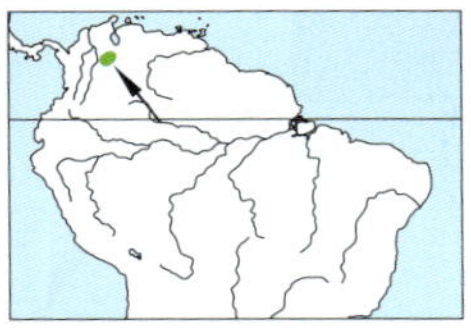

Woodland Oldfield Mouse
Thomasomys hylophilus
FR Thomasomys des bois
DE Waldland-Páramomaus
ES Ratón de erial de tierra arbolada

CO (E Andes), W VE (Cordillera de Mérida)

LC
Monotypic
HB 15.4–17.2 cm

1900–3810 m

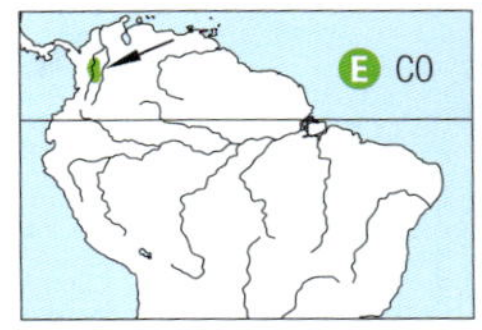

Niceforo Maria's Oldfield Mouse
Thomasomys nicefori
FR Thomasomys de Niceforo
DE Niceforo-Páramomaus
ES Ratón de erial de Niceforo María

Montane forests in N parts of W & C Andes

DD
Monotypic
HB 14.5–16.4 cm
T 20–22 cm

1830–3300 m

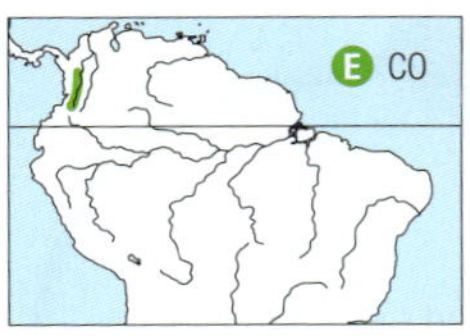

Popayan Oldfield Mouse
Thomasomys popayanus
FR Thomasomys de Popayan
DE Popayán-Páramomaus
ES Ratón de erial de Popayán

Montane forests of W Andes

Short-faced Oldfield Mouse
Thomasomys baeops
FR Thomasomys à tête courte
DE Kurzkopf-Páramomaus
ES Ratón de erial de cara corta

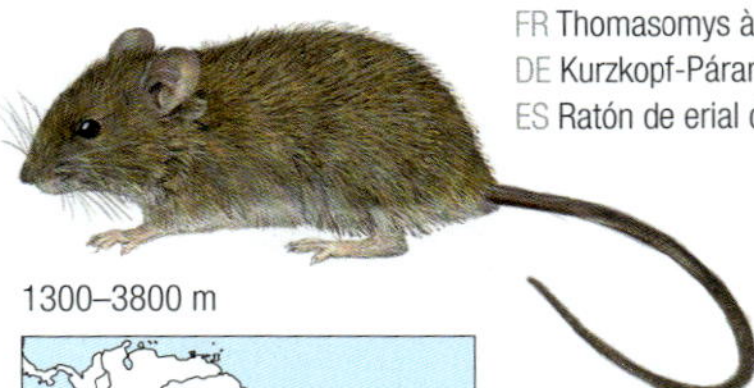

1300–3800 m

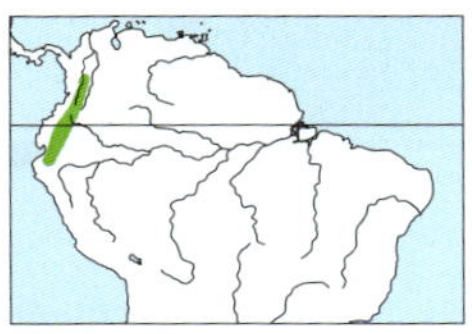

LC
Monotypic
HB 9.5–12 cm
T 10.6–14.7 cm
W 27–45 g

Andes of CO and EC

Snow-footed Oldfield Mouse
Thomasomys niveipes
FR Thomasomys à pattes blanches DE Schneefuß-Páramomaus
ES Ratón de erial de pies blancos

2550–3500 m

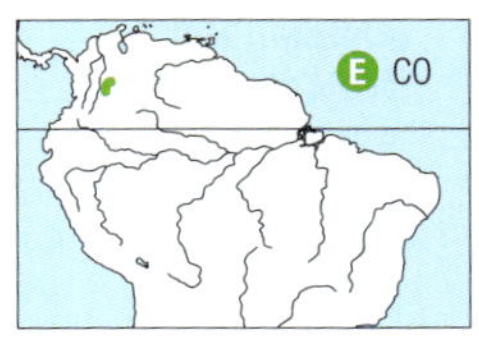

LC
Monotypic
HB 11–13.5 cm
T 11.5–13.3 cm
W 26–44 g

Páramo of E Andes

Principal Oldfield Mouse
Thomasomys princeps
FR Thomasomys cardinal
DE Kordilleren-Páramomaus
ES Ratón de erial principal

2650–3400 m

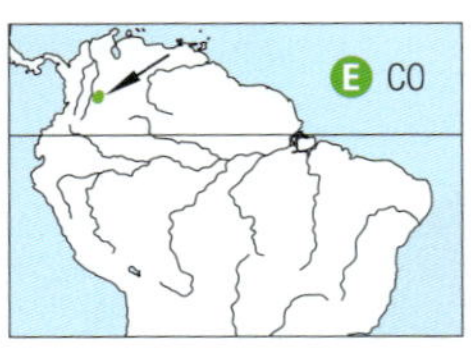

DD
Monotypic
HB 17.3–19.8 cm
T 21–25 cm
W c. 121 g

Montane forests of E Andes

VU
Monotypic
HB 11.2–13 cm
T c. 13.3 cm

2800–3800 m

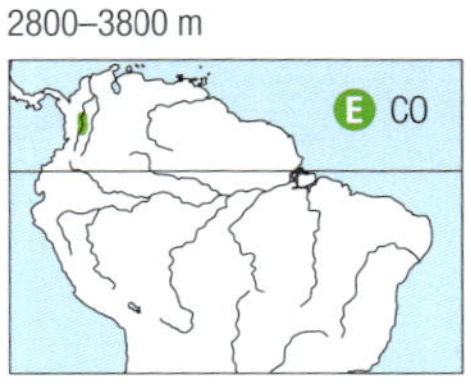

Silky Oldfield Mouse
Thomasomys bombycinus
FR Thomasomys soyeux
DE Seiden-Páramomaus
ES Ratón de erial sedoso

N part of W Andes

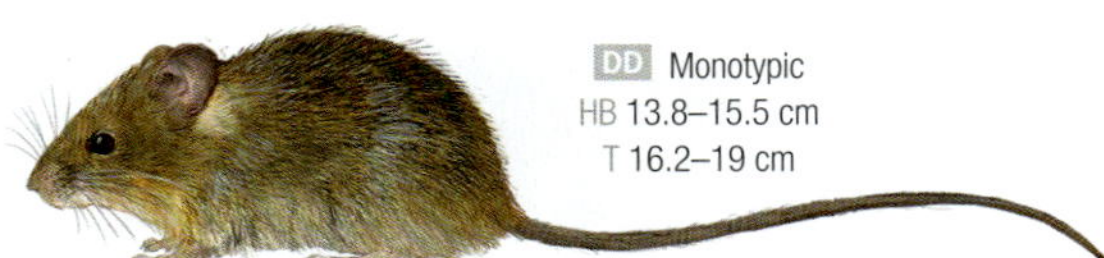

DD Monotypic
HB 13.8–15.5 cm
T 16.2–19 cm

2330–4000 m

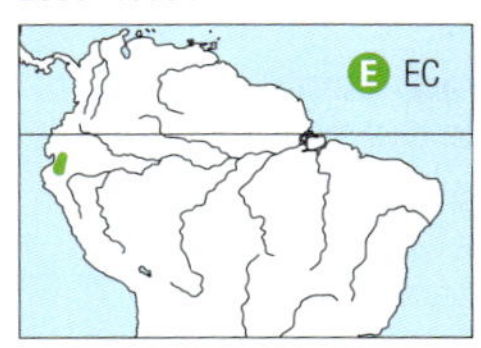

Red Andean Oldfield Mouse
Thomasomys auricularis
FR Thomasomys oreillard
DE Rote Páramomaus
ES Ratón de erial rojo de los Andes

W Andes in Cañar, Azuay, El Oro & Loja

2400–3800 m

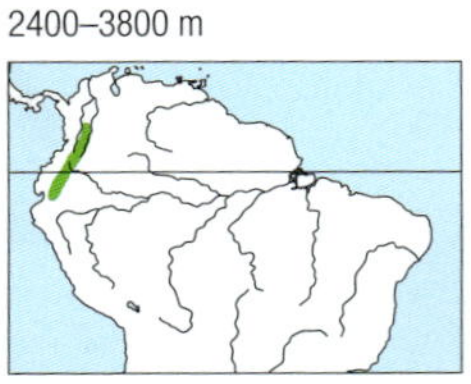

Andes of CO and EC

Cinnamon-colored Oldfield Mouse
Thomasomys cinnameus
FR Thomasomys cannelle
DE Zimtfarbene Páramomaus
ES Ratón de erial canela

LC
Monotypic
HB 8.2–9 cm
T 11–12.8 cm
W 14–19 g

Central Andes Oldfield Mouse
Thomasomys contradictus
FR Thomasomys à ventre jaune
DE Zentralanden-Páramomaus
ES Ratón de erial de los Andes centrales

2400–3870 m

Moist forests in C Andes

DD
Monotypic
HB 10.7–13.3 cm
T c. 15.3 cm

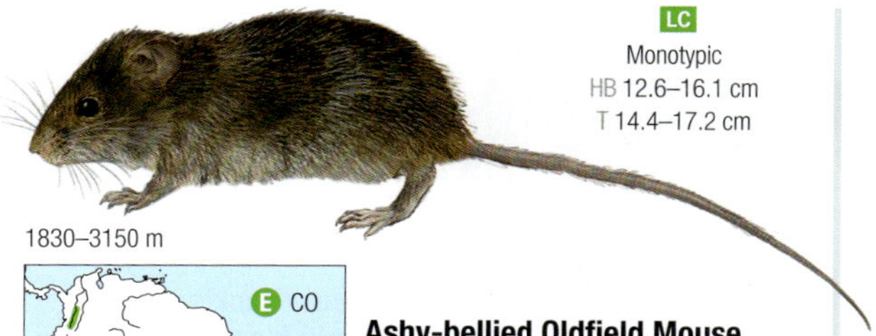

LC
Monotypic
HB 12.6–16.1 cm
T 14.4–17.2 cm

1830–3150 m

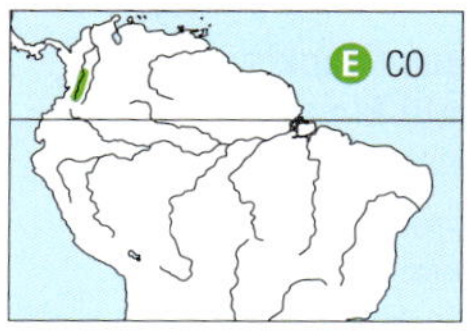

Ashy-bellied Oldfield Mouse
Thomasomys cinereiventer

FR Thomasomys à ventre gris
DE Graubauch-Páramomaus
ES Ratón de erial de vientre cenizo

W Andes

DD
Monotypic
HB 9.8–11.9 cm

2130–3310 m

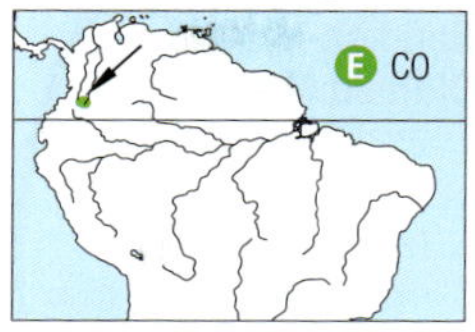

Colombian Oldfield Mouse
Thomasomys dispar

FR Thomasomys d'Andalucia
DE Kolumbien-Páramomaus
ES Ratón de erial de Colombia

S part of C & E Andes

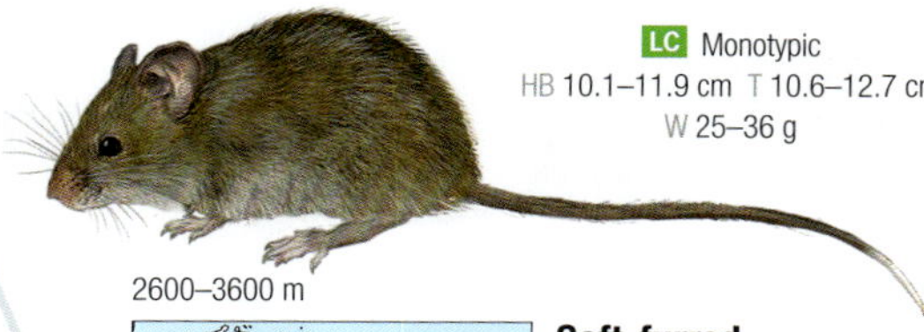

LC Monotypic
HB 10.1–11.9 cm T 10.6–12.7 cm
W 25–36 g

2600–3600 m

Soft-furred Oldfield Mouse
Thomasomys laniger

FR Thomasomys laineux
DE Weichfell-Páramomaus
ES Ratón de erial de pelaje suave

C & E Andes

LC Monotypic
HB 10.7–14.6 cm
T c. 14.7 cm
W c. 44 g

1200–3100 m

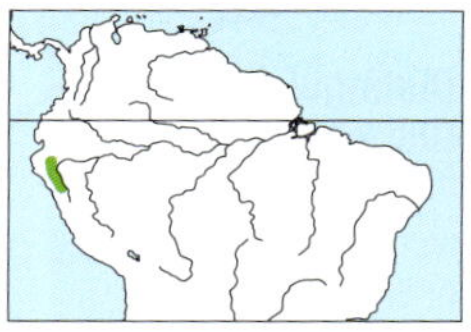

Ash-colored Oldfield Mouse
Thomasomys cinereus

FR Thomasomys cendré
DE Aschgraue Páramomaus
ES Ratón de erial cenizo

SW EC (Yacurí NP), NW PE
(W of R Marañón)

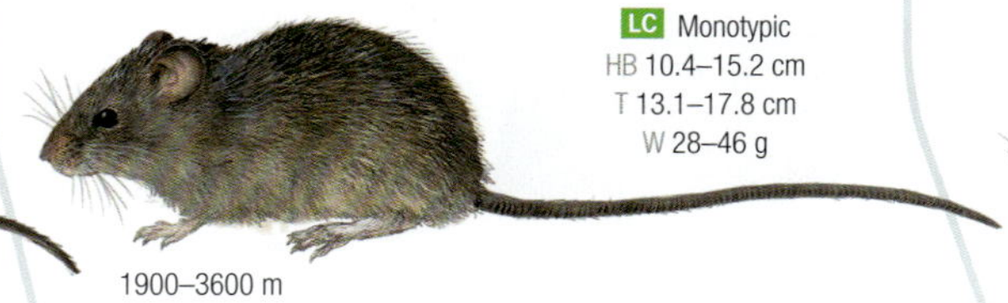

LC Monotypic
HB 10.4–15.2 cm
T 13.1–17.8 cm
W 28–46 g

1900–3600 m

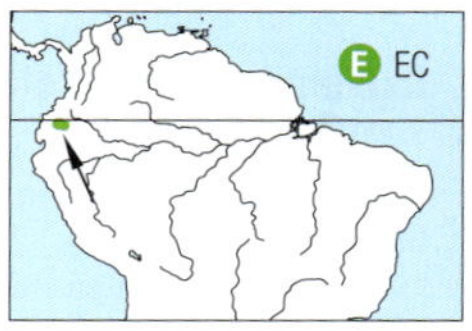

Wandering Oldfield Mouse
Thomasomys erro

FR Thomasomys du Sumaco
DE Napo-Páramomaus
ES Ratón de erial divagante

N part of E Andes (Napo)

Paramo Oldfield Mouse
Thomasomys paramorum

FR Thomasomys du paramo
DE Eigentliche Páramomaus
ES Ratón de erial de páramo

2000–4300 m

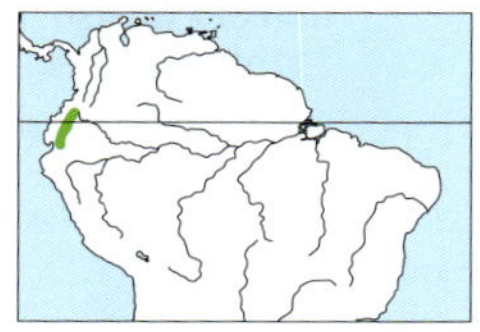

LC
Monotypic
HB 8.4–11.4 cm
T 11.3–13.3 cm
W 17–26 g

W Andes of S CO, Andes of EC
S to Huancabamba Depression

LC Monotypic
HB 9.5–12.7 cm
T 13.9–15.7 cm

1800–4500 m

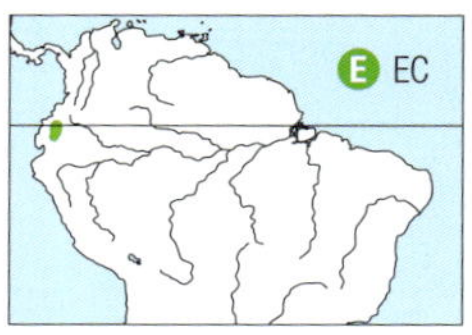

Forest Oldfield Mouse
Thomasomys silvestris

FR Thomasomys sylvestre
DE Wald-Páramomaus
ES Ratón de erial de bosque

W Andean slope; records from
E slope in Morona Santiago uncertain

DD
Monotypic
HB 10.1–12.5 cm
T 9.7–11.5 cm
W 31–50 g

2440–2850 m

Smoky Oldfield Mouse
Thomasomys fumeus

FR Thomasomys sombre
DE Rauchgraue Páramomaus
ES Ratón de erial ahumado

Known from 5 localities in E Andes
(Napo, Morona Santiago)

LC Monotypic
HB 11.6–12.6 cm
T 11.5–13.5 cm
W 32–57 g

1400–4500 m

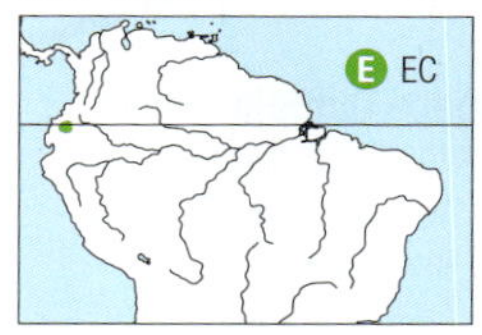

Pichincha Oldfield Mouse
Thomasomys vulcani

FR Thomasomys du Pichincha
DE Pichincha-Páramomaus
ES Ratón de erial de Pichincha

NC Andes (Pichincha, Carchi)

VU
Monotypic
HB 9.4–10.9 cm
T 12.2–15.1 cm
W 24–46 g

3380–3800 m

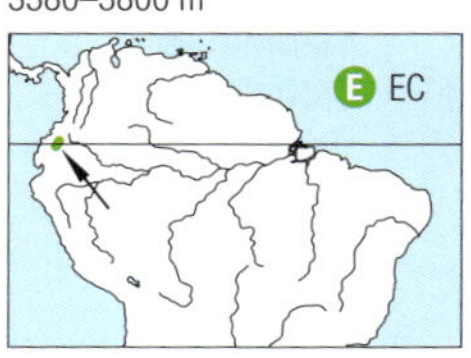

Ucucha Oldfield Mouse
Thomasomys ucucha

FR Thomasomys de Voss
DE Ucucha-Páramomaus
ES Ratón de erial de Ucucha

Known from crest of E Andes
(Carchi, Pichincha, Napo)

LC
Monotypic
HB c. 9.5 cm
T c. 13.8 cm

1150–3350 m

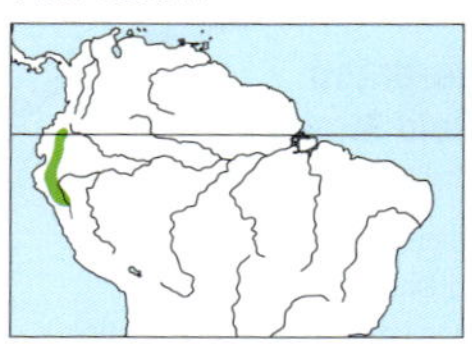

Taczanowski's Oldfield Mouse
Thomasomys taczanowskii

FR Thomasomys de Taczanowski
DE Taczanowski-Páramomaus
ES Ratón de erial de Taczanowski

Andes of EC & N PE (Piura, Lambayeque, Cajamarca, San Martín)

LC
Monotypic
HB 14.2–18 cm
T 17.2–25 cm
W 75–175 g

1460–3850 m

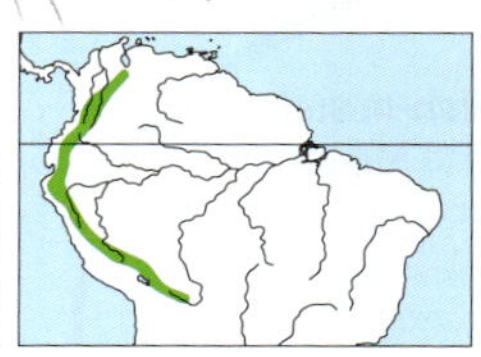

Golden Oldfield Mouse
Thomasomys aureus

FR Thomasomys doré
DE Gold-Páramomaus
ES Ratón de erial dorado

Andes from W VE, C & SW CO, EC
and PE to WC BO

LC
Monotypic
HB 9.9–12.5 cm
T 14.1–16.3 cm

2750–3350 m

White-tipped Oldfield Mouse
Thomasomys caudivarius

FR Thomasomys à pointe blanche
DE Weißschwanz-Páramomaus
ES Ratón de erial de cola blanca

W Andean slope from C EC to N PE;
records from NW EC unconfirmed

NE Monotypic
HB 13.7–14.5 cm
T 21–23 cm
W 63–68 g

1750–2215 m

Pardiñas's Oldfield Mouse
Thomasomys pardignasi

FR Thomasomys de Pardinas
DE Pardiñas-Páramomaus
ES Ratón de erial de Pardiñas

Cordillera del Cóndor (Morona
Santiago) and Kutukú (Zamora Chinchipe)

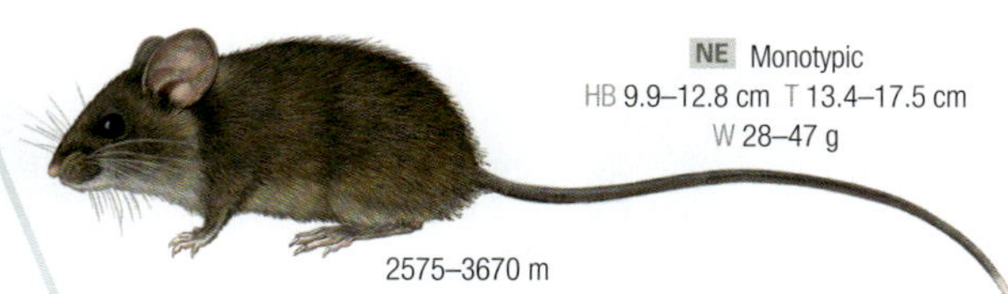

NE Monotypic
HB 9.9–12.8 cm T 13.4–17.5 cm
W 28–47 g

2575–3670 m

Salazar's Oldfield Mouse
Thomasomys salazari

FR Thomasomys de Salazar
DE Salazar-Páramomaus
ES Ratón de erial de Salazar

Known only from around type locality
(Sangay NP, Morona Santiago)

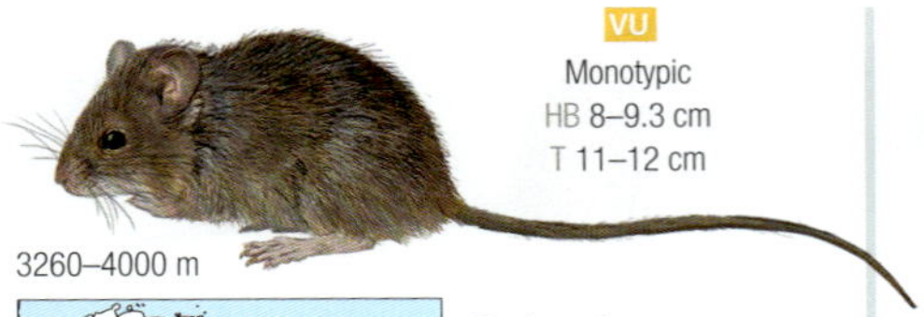

VU
Monotypic
HB 8–9.3 cm
T 11–12 cm

3260–4000 m

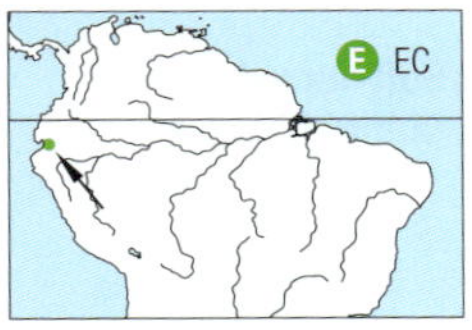

Hudson's Oldfield Mouse
Thomasomys hudsoni
FR Thomasomys de Hudson
DE Hudson-Páramomaus
ES Ratón de erial de Hudson

Known only from 2 localities in Bestión (Azuay)

DD Monotypic
HB 13.7–14.4 cm
T c. 19 cm

1200–3100 m

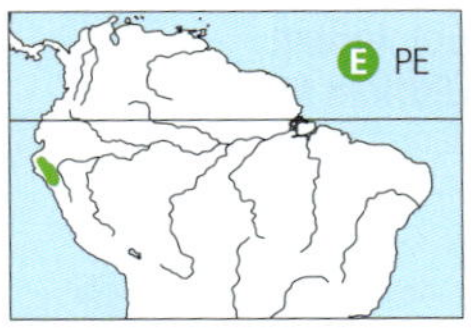

Reddish-backed Oldfield Mouse
Thomasomys pyrrhonotus
FR Thomasomys à dos roux
DE Rotrücken-Páramomaus
ES Ratón de erial de dorso rojizo

Andes in Piura and Cajamarca

NE
Monotypic
HB 18.4–18.8 cm
T 24–25 cm
W 70–154 g

2620–2720 m

Antonio Brack's Oldfield Mouse
Thomasomys antoniobracki
FR Thomasomys de Brack
DE Antonio-Brack-Páramomaus
ES Ratón de erial de Brack

Known only from type locality (Huancabamba Prov, Piura)

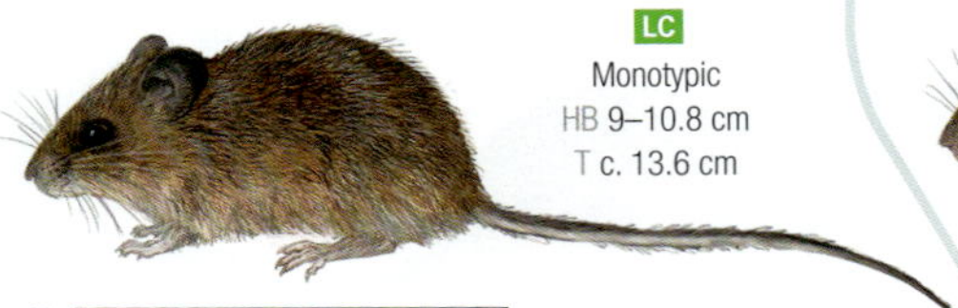

LC
Monotypic
HB 9–10.8 cm
T c. 13.6 cm

2460–3650 m

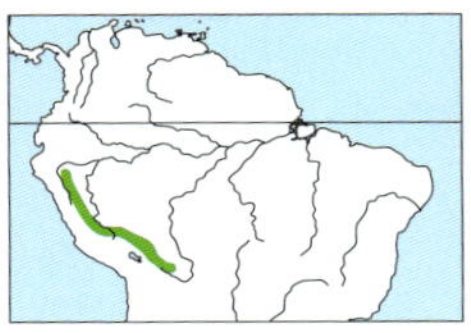

Montane Oldfield Mouse
Thomasomys oreas
FR Thomasomys de montagne
DE Bolivien-Páramomaus
ES Ratón de erial de montaña

Andes from N PE, S of R Marañón, to C BO

NT
Monotypic
HB 16.1–18.7 cm
T c. 18 cm

2050–4000 m

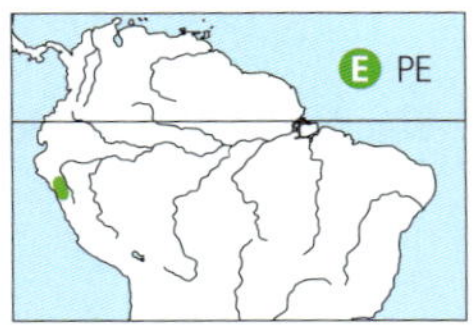

Cajamarca Oldfield Mouse
Thomasomys praetor
FR Thomasomys du Cajamarca
DE Cajamarca-Páramomaus
ES Ratón de erial de Cajamarca

W N Andes between Huancabamba Depression and R Marañón

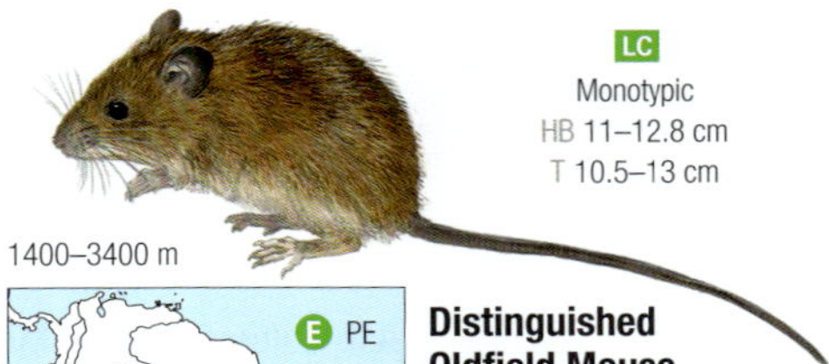

LC
Monotypic
HB 11–12.8 cm
T 10.5–13 cm

1400–3400 m

Distinguished Oldfield Mouse
Thomasomys notatus
FR Thomasomys à pattes sombres
DE Edle Páramomaus
ES Ratón de erial distinguido

E slopes of C Andes

LC Monotypic
HB c. 24 cm
T 28–33 cm
W 164–335 g

3200–3380 m

Apeco Oldfield Mouse
Thomasomys apeco
FR Thomasomys de l'Apeco
DE Apeco-Páramomaus
ES Ratón de erial de Apeco

Known from type locality in Valle de Los Chochos (Río Abiseo NP, San Martín)

VU
Monotypic
HB 12.6–14 cm
T 12.3–14.8 cm

3050–3660 m

Peruvian Oldfield Mouse
Thomasomys eleusis
FR Thomasomys du Pérou
DE Peru-Páramomaus
ES Ratón de erial de Perú

N part of E Andes, E of R Marañón

Strong-tailed Oldfield Mouse
Thomasomys ischyrus
FR Thomasomys à queue forte
DE Langschwanz-Páramomaus
ES Ratón de erial de cola fuerte

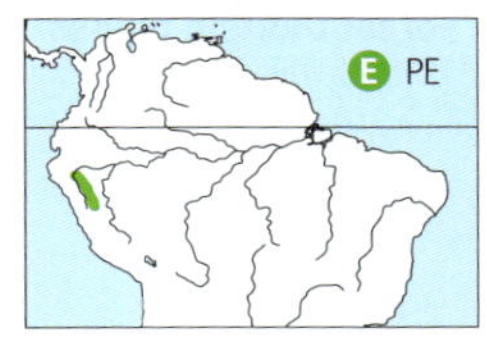

2280–3350 m

LC
Monotypic
HB 13–14 cm
T c. 15.1 cm

E slope of Andes (Amazonas, San Martín, Huánuco)

EN
Monotypic
HB c. 13.5 cm
T c. 17 cm

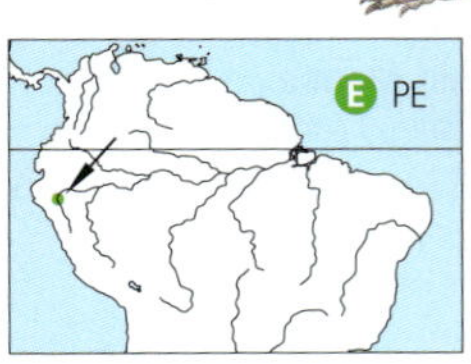

Reddish-nosed Oldfield Mouse
Thomasomys rosalinda
FR Thomasomys de Goncha
DE Rotnasen-Páramomaus
ES Ratón de erial de hocico rojizo

Known from type locality (Goncha, Amazonas)

LC
Monotypic
HB 15.3–16.8 cm
T 19.3–22 cm
W 140–166 g

3250–3380 m

Large-eared Oldfield Mouse
Thomasomys macrotis
FR Thomasomys à grandes oreilles
DE Großohr-Páramomaus
ES Ratón de erial de orejas grandes

Known from type locality and vicinity (Río Abiseo NP, San Martín)

LC Monotypic
HB 9.4–11.4 cm
T 14.4–15.9 cm

Ashaninka Oldfield Mouse
Thomasomys onkiro
FR Thomasomys de Luna
DE Asháninka-Páramomaus
ES Ratón de erial de los Asháninkas

Known from type locality in Cordillera de Vilcabamba (Junín); records from Tapichalaca R (EC) uncertain

LC
Monotypic
HB 11.1–13.4 cm
T c. 12.5 cm

2430–3850 m

Inca Oldfield Mouse
Thomasomys incanus
FR Thomasomys inca
DE Inka-Páramomaus
ES Ratón de erial de los Incas

E slope of Andes between San Martín and Junín

LC
Monotypic
HB 13.6–14.3 cm
T c. 15.5 cm

2050–3675 m

Kalinowski's Oldfield Mouse
Thomasomys kalinowskii
FR Thomasomys de Kalinowski
DE Kalinowski-Páramomaus
ES Ratón de erial de Kalinowski

Andes from Huánuco S to Ayacucho

NT
Monotypic
HB 8.3–10.1 cm
T c. 12 cm

275–4270 m

Slender Oldfield Mouse
Thomasomys gracilis
FR Thomasomys gracile
DE Zierliche Páramomaus
ES Ratón de erial esbelto

Montane forest of Yungas in Cusco

LC
Monotypic
HB 8.8–10.9 cm
T c. 13.3 cm

2000–3540 m

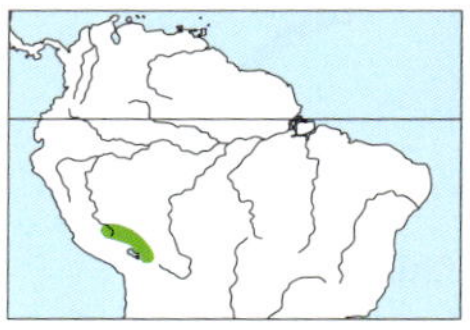

E Andean slopes from S PE to N BO

Daphne's Oldfield Mouse
Thomasomys daphne
FR Thomasomys daphné
DE Daphne-Páramomaus
ES Ratón de erial de Daphne

DD
Monotypic
HB 10.8–11 cm
T 12.2–12.8 cm
W 35–38 g

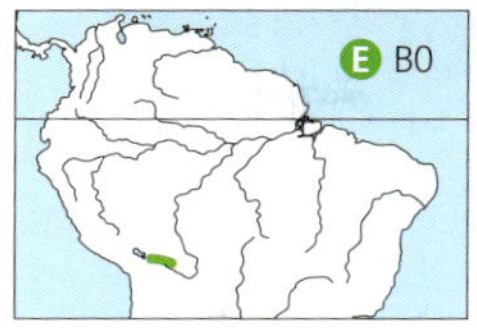

E Andean slopes in La Paz & Cochabamba

Anderson's Oldfield Mouse
Thomasomys andersoni
FR Thomasomys d'Anderson
DE Anderson-Páramomaus
ES Ratón de erial de Anderson

DD
Monotypic
HB 8.9–9.1 cm
T c. 13.8 cm

2350–2800 m

Andes in Cochabamba & Santa Cruz

Austral Oldfield Mouse
Thomasomys australis
FR Thomasomys austral
DE Südliche Páramomaus
ES Ratón de erial austral

Ladew's Oldfield Mouse
Thomasomys ladewi
FR Thomasomys de Ladew
DE Ladew-Páramomaus
ES Ratón de erial de Ladew

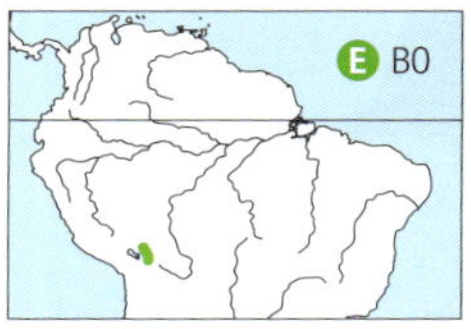

Montane forests in Andes of La Paz

2360–3300 m

LC
Monotypic
HB 12–13.8 cm
T c. 15.8 cm
W 42–64 g

Genus *PUNOMYS*

(2 species)

VU
Monotypic
HB 13.9–14.2 cm
T 4.6–6.1 cm

High W Andean slope in Puno, Arequipa, Moquegua & Tacna

Western Puna Mouse
Punomys lemminus
FR Punomys lemming
DE Westliche Puna-Maus
ES Ratón de puna occidental

NT
Monotypic
HB 12.6–16.3 cm
T 6.5–8.1 cm

4500–4770 m

E Andes in S PE (Puno) & NW BO (La Paz); specific affiliation of Bolivian record not fully resolved

Koford's Puna Mouse
Punomys kofordi
FR Punomys de Koford
DE Koford-Puna-Maus
ES Ratón de puna de Koford

Genus *ANDINOMYS*

(1 species)

Andes in S PE, extreme N CL, SW BO, NW AR

650–4500 m

LC Monotypic
HB 13.4–18.2 cm
T 10.7–16 cm
W 66–80 g

Andean Mouse
Andinomys edax
FR Andinomys vorace
DE Eigentliche Andenmaus
ES Ratón de los Andes

1950–4500 m

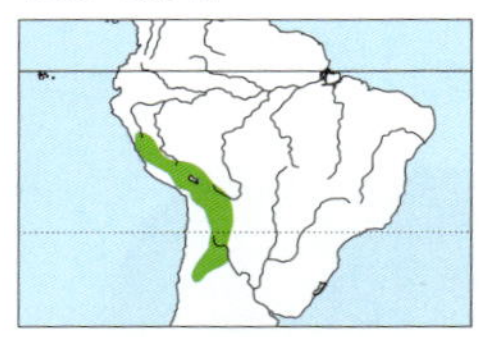

Highlands in C & S PE, extreme N CL, W BO and NW AR; probably S EC

LC
Monotypic
HB 9.7–14.9 cm
T 6.2–8.6 cm
W 63–69 g

Genus *NEOTOMYS*

(1 species)

Andean Swamp Rat
Neotomys ebriosus
FR Néotomys des Andes
DE Anden-Sumpfratte
ES Rata de ciénaga de los Andes

Salta Chinchilla Rat
Euneomys fossor
FR Eunéomys du Salta
DE Salta-Chinchillamaus
ES Rata chinchilla de Salta

WC AR (Mendoza, Neuquén, disjunct in Salta), adjacent CL

DD Monotypic
HB 14.5–14.9 cm
T 7.8–8.5 cm
W 78–86 g

Genus *EUNEOMYS*

(2 species)

C & S Patagonia, from WC AR and C CL S to Tierra del Fuego Is

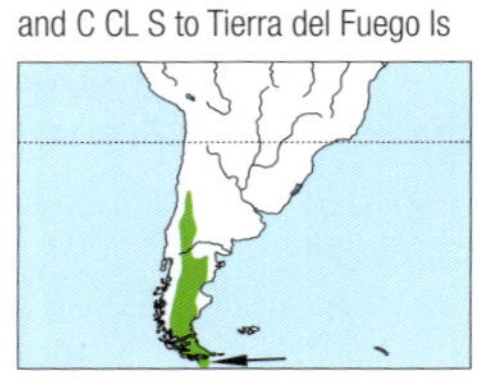

Patagonian Chinchilla Rat
Euneomys chinchilloides
FR Eunéomys chinchilla
DE Patagonien-Chinchillamaus
ES Rata chinchilla de Patagonia

DD Monotypic
HB 7–15.6 cm
T 4.6–8.1 cm
W 26–50 g

Genus *IRENOMYS*

(1 species)

LC
Monotypic
HB 10.8–13 cm
T 16.2–19.6 cm
W 30–67 g

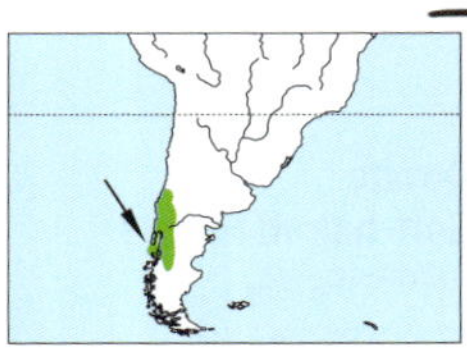

C & S CL and adjacent AR

Chilean Climbing Mouse
Irenomys tarsalis
FR Irenomys du Chili
DE Chile-Ratte
ES Ratón trepador de Chile

DD Monotypic
HB 10.3–10.7 cm
T 14.2–16.5 cm
W 35–37 g

Scattered in semiarid zone from Ceará S to SE Bahia

Cerrado Red-nosed Mouse
Wiedomys cerradensis
FR Wiedomys du Cerrado
DE Cerrado-Rotnasenmaus
ES Ratón de hocico rojo del Cerrado

Genus *WIEDOMYS*

(2 species)

Common Red-nosed Mouse
Wiedomys pyrrhorinos
FR Wiedomys à nez rouge
DE Gemeine Rotnasenmaus
ES Ratón de hocico rojo común

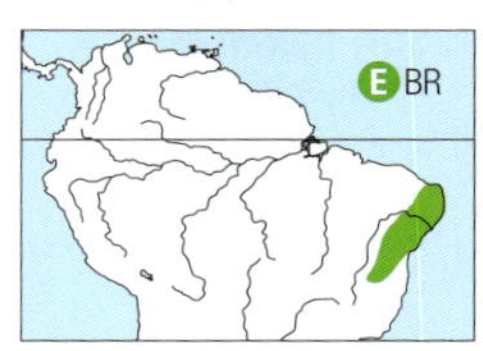

Scrub forest or caatinga from Ceará S to Minas Gerais

LC
Monotypic
HB 10.4–11.3 cm
T 11–18.5 cm
W 37–45 g

Genus *PHAENOMYS*

(1 species)

Rio de Janeiro Arboreal Rat
Phaenomys ferrugineus
FR Phaenomys ferrugineux
DE Rio-de-Janeiro-Ratte
ES Rata arborícola de Río de Janeiro

EN
Monotypic
HB 14.3–17 cm
T 18.7–19.9 cm
W 50–75 g

Atlantic Forest in Minas Gerais, Rio de Janeiro, São Paulo

900–1550 m

Genus *WILFREDOMYS*

(1 species)

EN Monotypic
HB 11.9–13 cm
T 16.1–21 cm
W 42–62 g

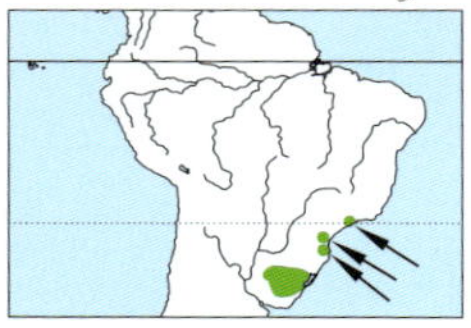

Red-nosed Tree Mouse
Wilfredomys oenax

FR Wilfredomys
DE Wilfred-Maus
ES Ratón arborícola de hocico rojo

Scattered in SE & S BR (from São Paulo S to Rio Grande do Sul), N UY

Genus *PAYNOMYS*

(1 species)

LC 2 sspp
HB 10.9–14.6 cm
T 4.7–6.5 cm
W 46–96 g

Andean Long-clawed Mouse
Paynomys macronyx

FR Paynomys des Andes
DE Anden-Langkrallenmaus
ES Ratón de uñas largas de los Andes

S Andes along CL–AR boundary, from Mendoza S to Strait of Magellan

Genus *CHELEMYS*

(1 species)

NT 2 sspp
HB 10.9–11.8 cm
T 5.5–6 cm
W 53–60 g

Coquimbo, Valparaíso, Metropolitana, Biobío, Araucanía, presumably Maule

Chilean Long-clawed Mouse
Chelemys megalonyx

FR Chélémys
DE Chile-Langkrallenmaus
ES Ratón de uñas largas de Chile

Genus *NOTIOMYS*

(1 species)

LC
Monotypic
HB 8–11.5 cm
T 3.5–4.2 cm
W 17–19 g

Edwards's Long-clawed Mouse
Notiomys edwardsii

FR Notiomys d'Edwards
DE Edwards-Langkrallenmaus
ES Ratón de uñas largas de Edwards

From S Río Negro to C Santa Cruz; a record from S CL needs confirmation

LC
Monotypic
HB 9.5–10.6 cm
T 3–4.5 cm
W 25–38 g

0–2000 m

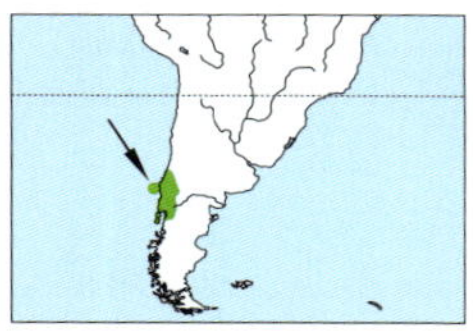

Valdivian Long-clawed Mouse
Geoxus valdivianus

FR Géoxus de Valdivia
DE Valdivia-Langkrallenmaus
ES Ratón topo de Valdivia

SW AR and CS CL, including Mocha & Chiloé Is

Genus *GEOXUS*

(4 species)

VU Monotypic
HB 10.8–12.6 cm
T 7.1–8.2 cm
W 42–61 g

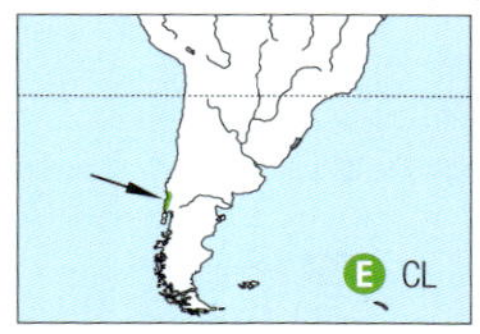

Large Long-clawed Mouse
Geoxus annectens

FR Géoxus de Patterson
DE Große Langkrallenmaus
ES Ratón topo grande

Known from few localities in a small portion of Coastal Range

NE Monotypic
HB c. 10.6 cm
T c. 6.7 cm

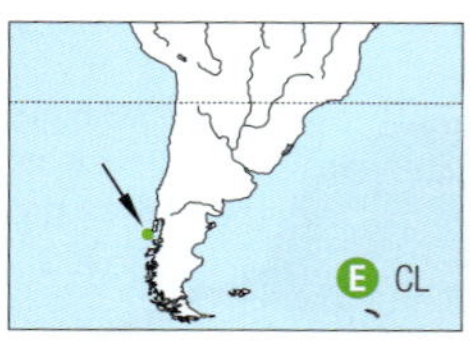

Guafo Island Long-clawed Mouse
Geoxus lafkenche

FR Géoxus de Guafo
DE Guafo-Langkrallenmaus
ES Ratón topo de Guafo

Guafo I (SW of Chiloé I)

NE Monotypic
HB c. 9.7 cm
T c. 4.5 cm

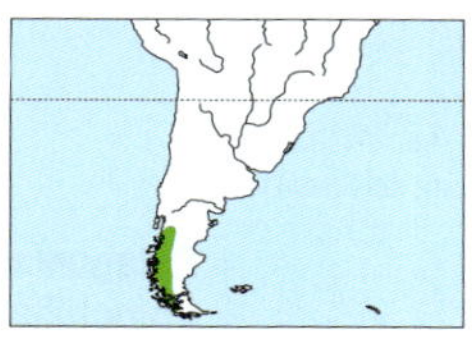

Michaelsen's Long-clawed Mouse
Geoxus michaelseni

FR Géoxus de Michaelsen
DE Michaelsen-Langkrallenmaus
ES Ratón topo de Michaelsen

S Andean continental AR and adjacent CL S to W Tierra del Fuego

LC
Monotypic
HB 13–13.8 cm
T 8.3–9.6 cm
W 45–65 g

Long-haired Soft-haired Mouse
Abrothrix longipilis

FR Abrothrix à long poil
DE Langhaar-Andenfeldmaus
ES Ratón de pelaje suave de pelo largo

Coastal to Andean piedmont in Central Chilean Mediterranean ecoregion

Genus *ABROTHRIX*

(9 species)

NE
Monotypic
HB 10–10.3 cm
T 7.3–7.7 cm
W 20–23 g

0–3500 m

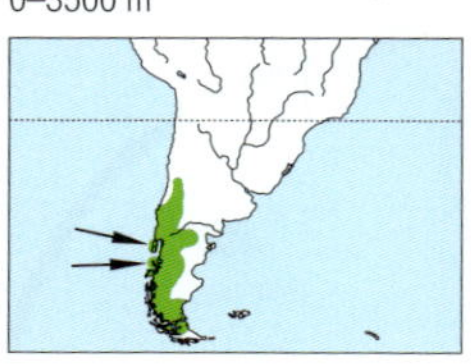

Hairy Soft-haired Mouse
Abrothrix hirta

FR Abrothrix velu
DE Haarige Andenfeldmaus
ES Ratón de pelaje suave peludo

SC CL and WC AR S to Tierra del Fuego

LC Monotypic
HB c. 10.7 cm T c. 8.3 cm
W c. 28 g

below 820 m

Mann's Soft-haired Mouse
Abrothrix manni

FR Abrothrix de Mann
DE Mann-Andenfeldmaus
ES Ratón de pelaje suave de Mann

SC CL (whith N Chiloé I), extreme WC AR

NT
Monotypic
HB 10–11.5 cm
T 6.9–8.5 cm

up to 800 m

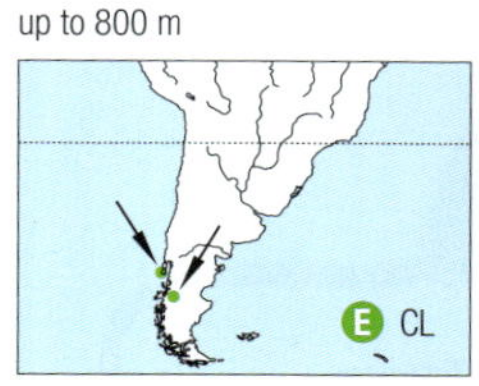

Sanborn's Soft-haired Mouse
Abrothrix sanborni

FR Abrothrix de Sanborn
DE Sanborn-Andenfeldmaus
ES Ratón de pelaje suave de Sanborn

C & S Chiloé I and EC Aysén

S CL, S AR (extreme EC Santa Cruz, S Tierra del Fuego)

LC
Monotypic
HB c. 9.6 cm
T 6.3–7 cm
W 21–22 g

Woolly Soft-haired Mouse
Abrothrix lanosa

FR Abrothrix laineux
DE Wollige Andenfeldmaus
ES Ratón de pelaje suave lanudo

above 2200 m

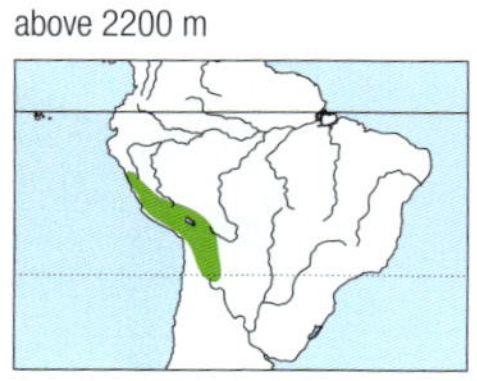

WC PE S through W BO to NW AR

Ornate Soft-haired Mouse
Abrothrix jelskii

FR Abrothrix de Jelski
DE Jelski-Andenfeldmaus
ES Ratón de pelaje suave de Jelski

LC Monotypic
HB 9.5–10.2 cm T 7.4–7.9 cm
W 28–32 g

LC Monotypic
HB 11–12.7 cm
T 8.2–8.8 cm
W 37–55 g

700–2200 m

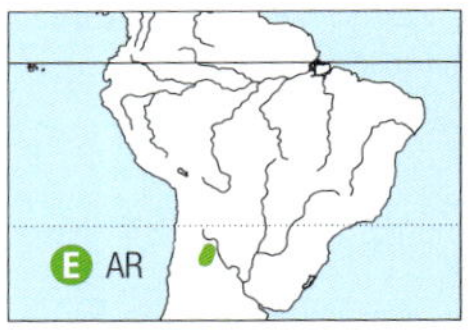

Gray Soft-haired Mouse
Abrothrix illutea

FR Abrothrix gris
DE Graue Andenfeldmaus
ES Ratón de pelaje suave gris

Catamarca, Tucumán

LC 2 sspp
HB 6.6–9.1 cm
T 4.8–6.6 cm
W 12–17 g

950–5000 m

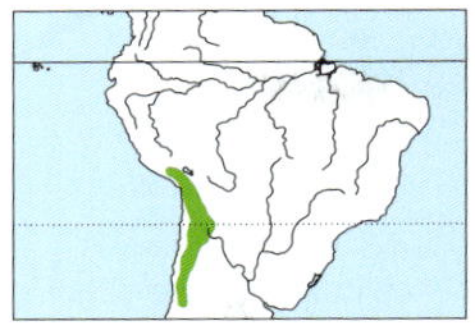

Andean Soft-haired Mouse
Abrothrix andina

FR Abrothrix des Andes
DE Eigentliche Andenfeldmaus
ES Ratón de pelaje suave de los Andes

S PE, CL (S to Santiago), W BO, NW & CW AR

LC Monotypic
HB 8–9.6 cm T 5.3–6 cm
W 13–25 g

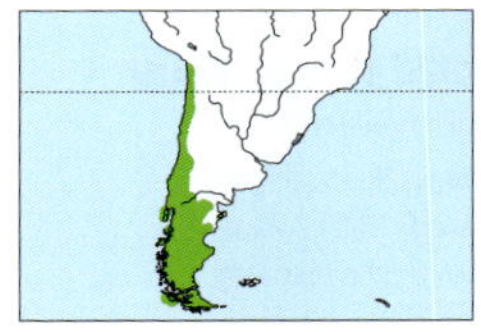

Olive Soft-haired Mouse
Abrothrix olivacea

FR Abrothrix olivâtre
DE Olivbraune Andenfeldmaus
ES Ratón de pelaje suave oliváceo

CL and WC AR S to Tierra del Fuego Is

Genus *CALASSOMYS*
(1 species)

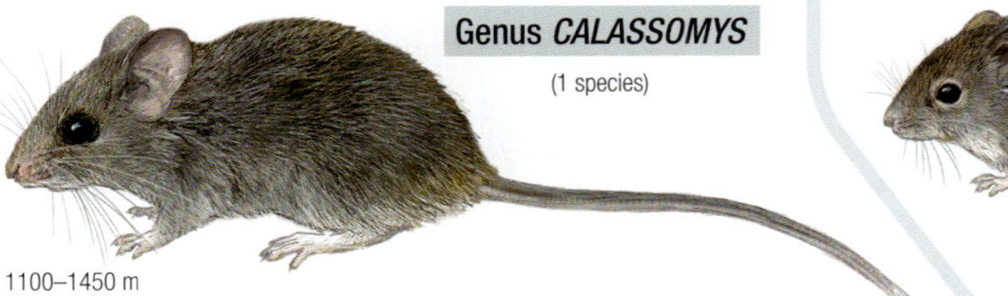

1100–1450 m

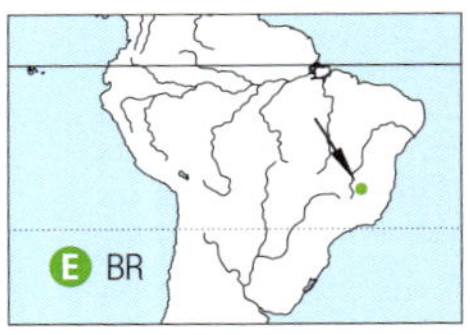

Calaça's White-tailed Mouse
Calassomys apicalis

FR Calassomys
DE Diamantina-Weißschwanzmaus
ES Ratón de cola blanca de Calaça

NE
Monotypic
HB 10.3–12.5 cm
T 14.9–15.8 cm
W 29–55 g

Known only from Sempre Vivas NP (Minas Gerais)

LC Monotypic
HB 7–8.9 cm T 4.3–6.7 cm
W 14.5–19 g

above 2500 m

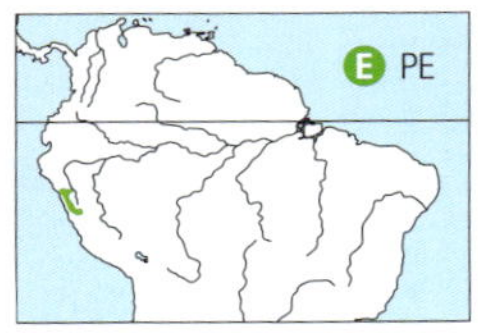

Peruvian Vesper Mouse
Calomys sorella

FR Calomys du Pérou
DE Peru-Vespermaus
ES Ratón laucha de Perú

Andes from La Libertad S to Lima & Pasco

Genus *CALOMYS*
(18 species)

Short-tailed Vesper Mouse
Calomys miurus

FR Calomys du Junin
DE Kurzschwanz-Vespermaus
ES Ratón laucha de cola corta

up to 4100 m

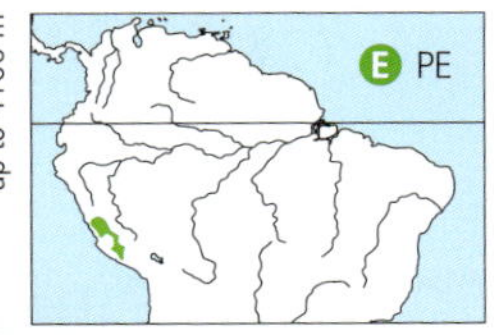

Andes from Lima & Junín S to N Arequipa

NE
Monotypic
HB 6.9–9.8 cm
T 4.2–8 cm
W 11.1–28 g

Coastal Vesper Mouse
Calomys chinchilico

FR Calomys farfadet
DE Küsten-Vespermaus
ES Ratón laucha de costa

280–980 m

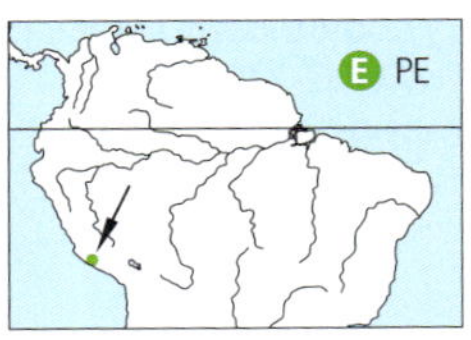

NE Monotypic
HB 6.2–8.9 cm
T 6–7.5 cm
W 11–19 g

Coastal desert in Lomas of Atiquipa & Taymara (Arequipa)

NE Monotypic
HB 6.5–9.9 cm T 5.9–8 cm
W 17–28 g

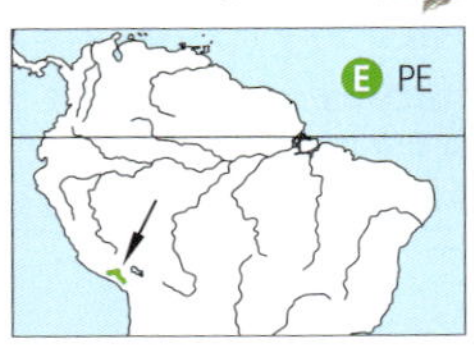

Arequipa Vesper Mouse
Calomys achaku

FR Calomys de l'Arequipa
DE Arequipa-Vespermaus
ES Ratón laucha de Arequipa

Known only from high Andes of Arequipa & SW Puno

NE
Monotypic
HB 7.6–10.3 cm
T 6.9–9 cm
W 13–26 g

above 2800 m

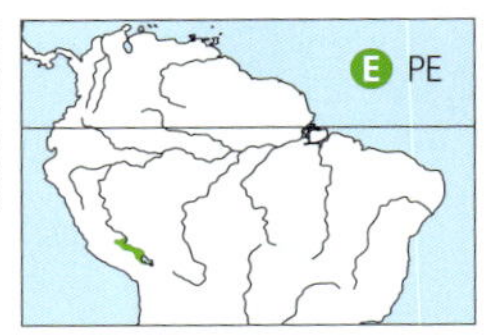

Long-tailed Vesper Mouse
Calomys frida

FR Calomys à longue queue
DE Langschwanz-Vespermaus
ES Ratón laucha de cola larga

Andean valleys of Cusco, Apurímac & Puno

LC
Monotypic
HB 6.3–6.5 cm
T 3–4.7 cm
W 10–16 g

2950–4850 m

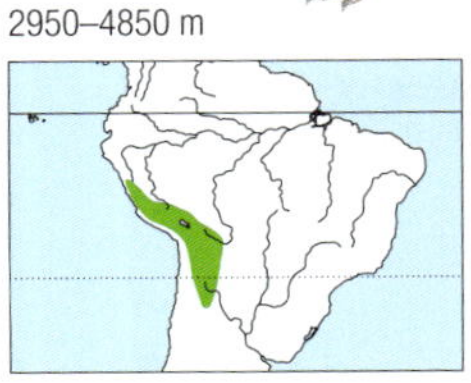

Andean Vesper Mouse
Calomys lepidus

FR Calomys des Andes
DE Anden-Vespermaus
ES Ratón laucha de los Andes

From C PE through extreme NE CL and BO to NW AR

LC
Monotypic
HB 8.2–9.3 cm
T 7.2–8.1 cm
W 15.1–24 g

0–3300 m

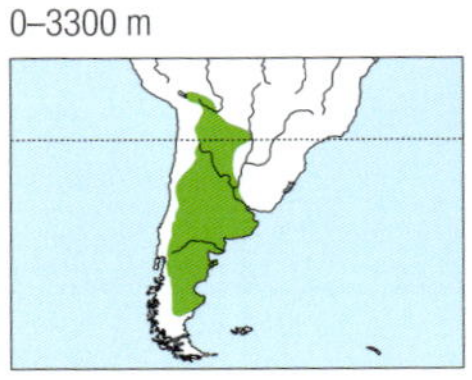

Drylands Vesper Mouse
Calomys musculinus

FR Calomys d'Argentine
DE Trockenland-Vespermaus
ES Ratón laucha de secano

CW & C BO and W PY to S AR

LC
Monotypic
HB 5.5–6.7 cm
T 4.5–5.4 cm
W 11–19 g

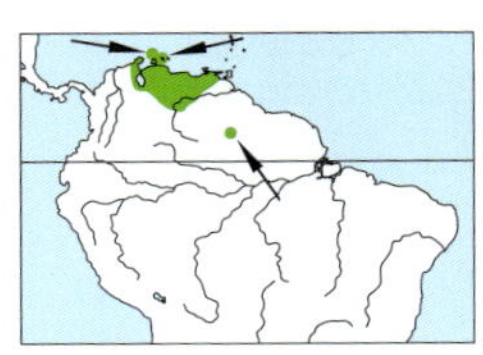

Hummelinck's Vesper Mouse
Calomys hummelincki

FR Calomys de Hummelinck
DE Hummelinck-Vespermaus
ES Ratón laucha de Hummelinck

NE CO (La Guajira), N & C VE, AW & CW; recorded in N BR (Roraima)

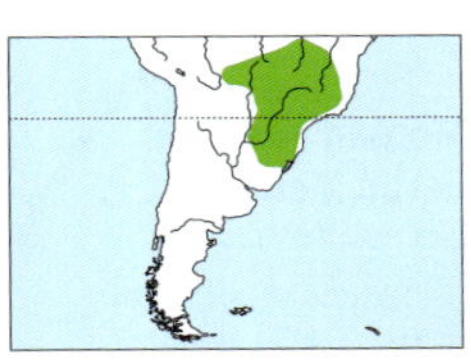

C, SE & S BR, E BO, E PA, NE AR

LC Monotypic
HB 7–8.5 cm T 4.7–7.4 cm
W 12.1–16.9 g

Delicate Vesper Mouse
Calomys tener

FR Calomys délicat
DE Zierliche Vespermaus
ES Ratón laucha delicado

LC Monotypic
HB 5.5–6.7 cm T 4.5–5.4 cm
W 11–19 g

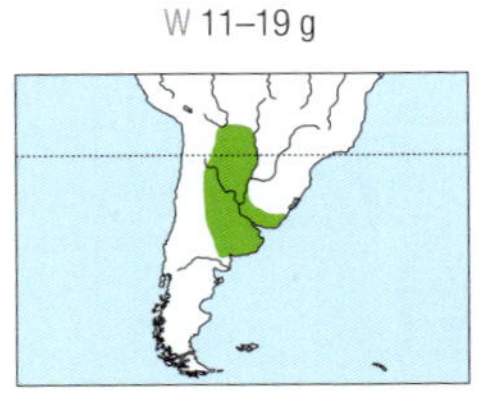

SE BO, W PY, N & C AR, UY, extreme S BR

Small Vesper Mouse
Calomys laucha

FR Calomys nain
DE Eigentliche Vespermaus
ES Ratón laucha pequeño

LC Monotypic
HB 8.6–11.3 cm T 6.2–8.2 cm
W 22–34 g

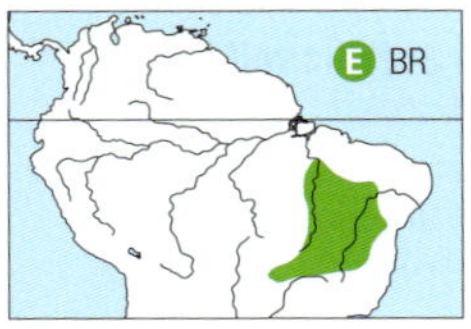

Rejected Vesper Mouse
Calomys expulsus

FR Calomys du Caatinga
DE Caatinga-Vespermaus
ES Ratón laucha excluyente

Caatinga and cerrado from Pernambuco S to Goiás

LC
Monotypic
HB 7.7–11.6 cm
T 6–7.5 cm
W 12.5–31 g

Tocantins Vesper Mouse
Calomys tocantinsi

FR Calomys du Tocantins
DE Tocantins-Vespermaus
ES Ratón laucha de Tocantins

Upper R Araguaia Basin (Tocantins, Mato Grosso, Goiás)

LC Monotypic
HB 9.9–12.2 cm
T 7.8–11 cm
W 30–52 g

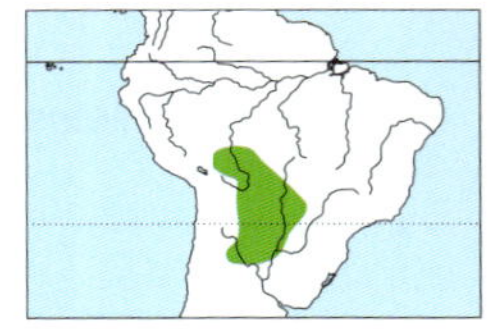

Large Vesper Mouse
Calomys callosus

FR Calomys de Rengger
DE Große Vespermaus
ES Ratón laucha grande

N & E BO, extreme W BR (Mato Grosso), PY, N AR

LC
Monotypic
HB 7.9–13.4 cm
T 5–8.7 cm
W 21–51 g

Crafty Vesper Mouse
Calomys callidus

FR Calomys rusé
DE Geschickte Vespermaus
ES Ratón laucha astuto

W BR (Rondônia, Mato Grosso, Mato Grosso do Sul), C PY, NE AR (along R Paraná)

LC
Monotypic
HB c. 10.6 cm
T c. 8.2 cm

c. 1200 m

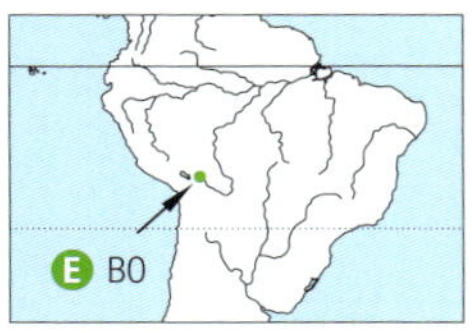

Bolivian Vesper Mouse
Calomys boliviae

FR Calomys de Bolivie
DE Bolivien-Vespermaus
ES Ratón laucha de Bolivia

Known only from type locality in Yungas (R Solocame, La Paz)

LC Monotypic
HB 9.3–12.1 cm
T 5.2–10 cm
W 60–85 g

600–2700 m

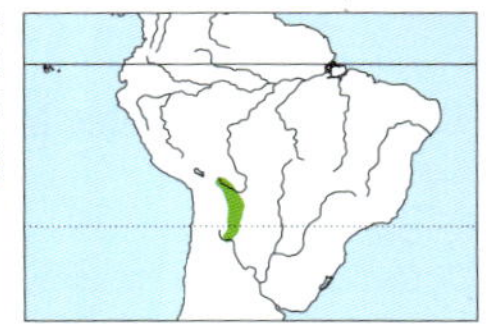

Fecund Vesper Mouse
Calomys fecundus

FR Calomys fécond
DE Fruchtbare Vespermaus
ES Ratón laucha fecundo

E Andean slopes from WC BO (Chuquisaca) to NW AR (Santiago del Estero)

LC Monotypic
HB 6.4–16.3 cm
T 4–11 cm
W 21–51 g

Cordoba Vesper Mouse
Calomys venustus

FR Calomys du Cordoba
DE Córdoba-Vespermaus
ES Ratón laucha de Córdoba

Santiago del Estero, Córdoba, San Luis

NE Monotypic
HB 8.6–11.2 cm T 7.5–8.5 cm
W 25–35 g

Cerqueira's Vesper Mouse
Calomys cerqueirai

FR Calomys de Cerqueira
DE Cerqueira-Vespermaus
ES Ratón laucha de Cerqueira

Known only from SE Minas Gerais

Genus *ELIGMODONTIA*

(7 species)

NE
Monotypic
HB 8.7–10.3 cm
T 8.2–10 cm
W 15–35 g

above 3000 m

Hairy-footed Gerbil Mouse
Eligmodontia hirtipes

FR Éligmodonte à pattes velues
DE Haarfüßige Hochlandwüstenmaus
ES Ratón colilargo de pies peludos

Andes in S PE, N CL & W BO

LC
Monotypic
HB 6.7–9.5 cm
T 5.9–9.2 cm
W 16–26 g

above 3500 m

Andean Gerbil Mouse
Eligmodontia puerulus

FR Éligmodonte des Andes
DE Anden-Hochlandwüstenmaus
ES Ratón colilargo de los Andes

NE CL, NW AR; uncertain in SW BO

NE Monotypic
HB 7.3–9.4 cm
T 7.4–10.5 cm
W c. 17.5 g

above 1800 m

Bolson Gerbil Mouse
Eligmodontia bolsonensis

FR Éligmodonte des bolsones
DE Bolsón-Hochlandwüstenmaus
ES Ratón colilargo de bolsón

Sierra de Ambato & Sierra del Manchao (Salta, Catamarca); limits unresolved

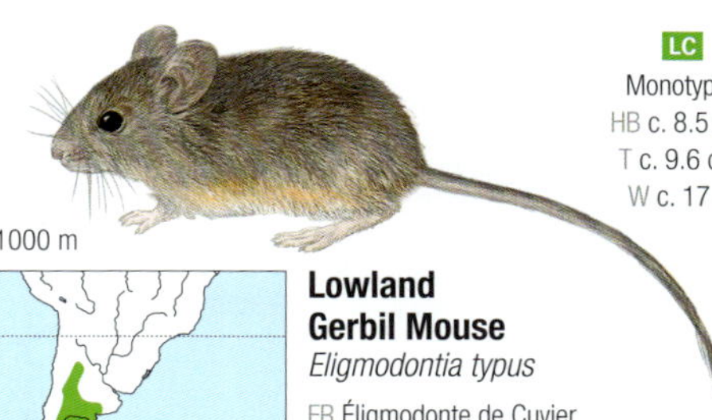

LC
Monotypic
HB c. 8.5 cm
T c. 9.6 cm
W c. 17 g

below 1000 m

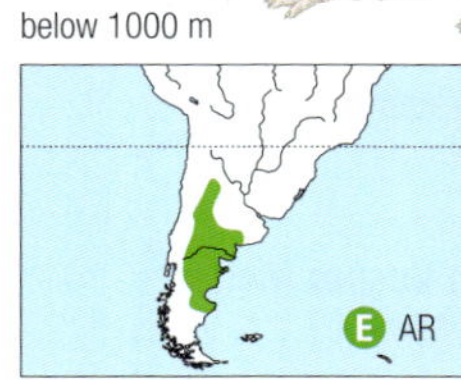

Lowland Gerbil Mouse
Eligmodontia typus

FR Éligmodonte de Cuvier
DE Eigentliche Hochlandwüstenmaus
ES Ratón colilargo de tierras bajas

Patagonia S to Santa Cruz

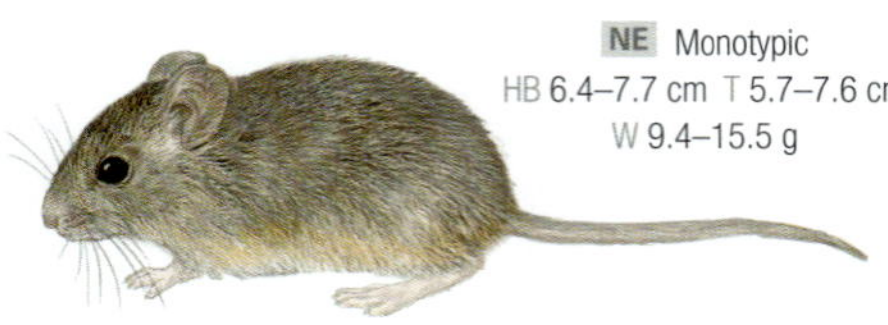

NE Monotypic
HB 6.4–7.7 cm T 5.7–7.6 cm
W 9.4–15.5 g

Dune Gerbil Mouse
Eligmodontia dunaris

FR Éligmodonte des dunes
DE Dünen-Hochlandwüstenmaus
ES Ratón colilargo de dunas

Restricted to Atacama Desert

LC
Monotypic
HB 6.9–8.9 cm
T 9.2–12.5 cm

500–3900 m

Monte Gerbil Mouse
Eligmodontia moreni

FR Éligmodonte de Moreno
DE Argentinien-Hochlandwüstenmaus
ES Ratón colilargo de montaña

E slopes of Andes from Salta S to Mendoza

LC
Monotypic
HB 8.3–9.5 cm
T 7–8.9 cm
W 12–24 g

0–3500 m

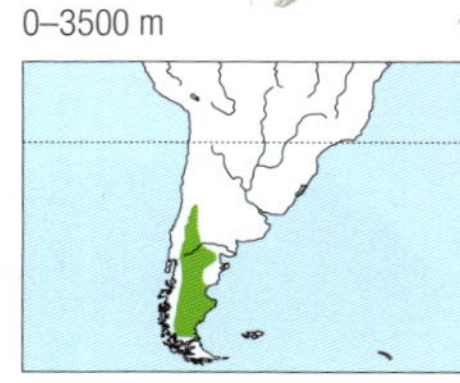

Morgan's Gerbil Mouse
Eligmodontia morgani

FR Éligmodonte de Morgan
DE Morgan-Hochlandwüstenmaus
ES Ratón colilargo de Morgan

WC & S continental AR and CL

Genus *GRAOMYS*

(4 species)

600–2700 m

Pale Pericote

Graomys domorum

FR Pericote pâle
DE Fahle Neuweltfeldratte
ES Pericote pálido

LC
3 sspp
HB 10.2–15 cm
T 14–17 cm
W 38–60 g

From C BO (Cochabamba) to NW AR (Tucumán)

DD Monotypic
HB c. 13.5 cm
T c. 15.5 cm
W c. 75 g

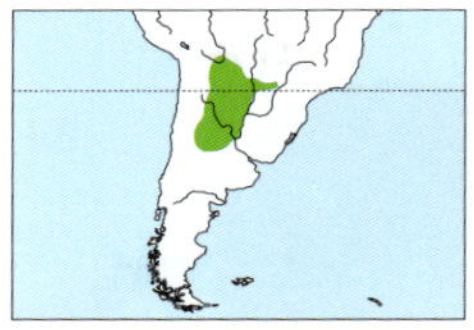

Chaco Pericote

Graomys chacoensis

FR Pericote du Chaco
DE Chaco-Neuweltfeldratte
ES Pericote del Chaco

SE BO, NC AR, PY, extreme CS BR (Mato Grosso do Sul)

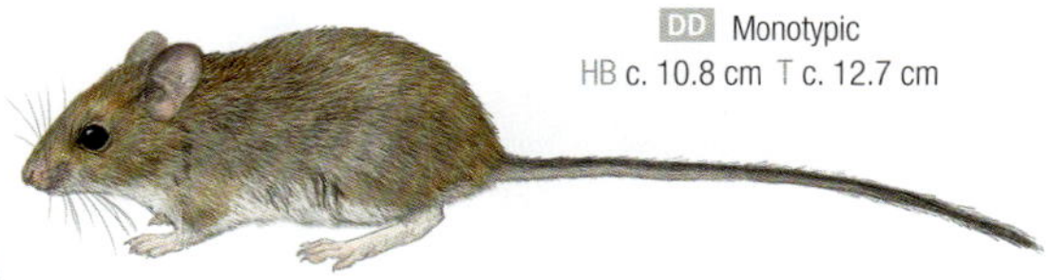

DD Monotypic
HB c. 10.8 cm T c. 12.7 cm

Otro Cerro Pericote

Graomys edithae

FR Pericote d'Otro Cerro
DE Edith-Neuweltfeldratte
ES Pericote de Cerro Otro

Restricted to type locality (S top of Sierra de Ambato)

Common Pericote

Graomys griseoflavus

FR Pericote gris
DE Gewöhnliche Neuweltfeldratte
ES Pericote dorado

LC
Monotypic
HB 11.8–16.9 cm
T 13.4–17.1 cm
W 30–73 g

From N Salta to NE Santa Cruz; probably SW BO

LC Monotypic
HB 9.9–11.5 cm
T 7.2–8.9 cm
W 29–30 g

2000–4500 m

Haggard's Leaf-eared Mouse

Phyllotis haggardi

FR Phyllotis de Haggard
DE Haggard-Blattohrmaus
ES Ratón orejudo de Haggard

W & E Andean slopes; exact limits unknown

LC
Monotypic
HB c. 10.7 cm
T 10.1–13.6 cm
W 22–55 g

870–4390 m

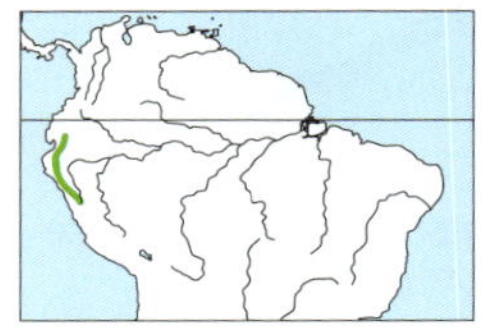

Andean Leaf-eared Mouse

Phyllotis andium

FR Phyllotis des Andes
DE Anden-Blattohrmaus
ES Ratón orejudo de los Andes

W & E Andean slopes from C EC to C PE

Genus *PHYLLOTIS*

(24 species)

LC Monotypic
HB 7.7–9.6 cm
T 6.2–9 cm
W c. 15.5 g

c. 50 m

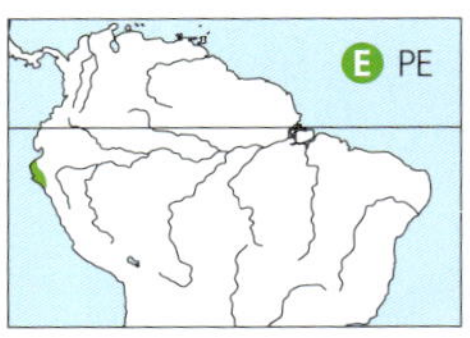

Peruvian Leaf-eared Mouse

Phyllotis gerbillus

FR Phyllotis gerbille
DE Peru-Blattohrmaus
ES Ratón orejudo de Perú

Sechura Desert

LC 3 sspp
HB 8–8.5 cm
T 10–10.5 cm
W 18–28 g

0–2800 m

Friendly Leaf-eared Mouse

Phyllotis amicus

FR Phyllotis familier
DE Freundliche Blattohrmaus
ES Ratón orejudo amigable

Coast & W Andean foothills from Piura S to Arequipa

NE
Monotypic
HB c. 10.6 cm
T 12.1–13.4 cm

1290–2425 m

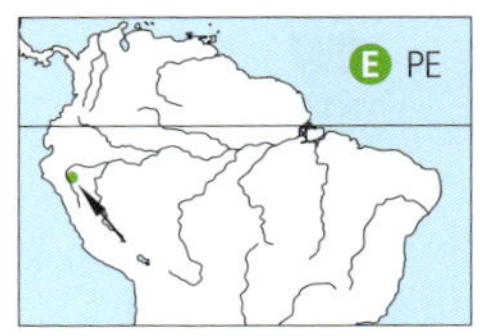

Narrow-nasal Leaf-eared Mouse

Phyllotis stenops

FR Phyllotis du Maranon
DE Schmalnasen-Blattohrmaus
ES Ratón orejudo de hocico estrecho

Known only from E R Marañón (Amazonas)

NE Monotypic
HB 11.3–13.4 cm
T 9.2–12.2 cm
W 41–68 g

3570–4270 m

Pearson's Leaf-eared Mouse

Phyllotis pearsoni

FR Phyllotis de Pearson
DE Pearson-Blattohrmaus
ES Ratón orejudo de Pearson

From S Cajamarca to N Ancash

Western Leaf-eared Mouse

Phyllotis occidens

FR Phyllotis occidental
DE Westliche Blattohrmaus
ES Ratón orejudo occidental

200–3800 m

From Ancash S to Lima

NE
Monotypic
HB c. 10.6 cm
T 11.5–12.5 cm
W 28–38 g

EN
Monotypic
HB c. 12.2 cm
T 10.1–13.6 cm
W 50–89 g

2600–3000 m

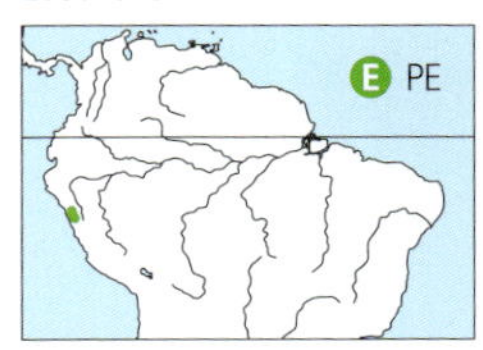

Ancash Leaf-eared Mouse

Phyllotis definitus

FR Phyllotis de l'Ancash
DE Ancash-Blattohrmaus
ES Ratón orejudo de Ancash

Restricted to Cordillera Negra (Ancash)

NE 2 sspp
HB 9–14.3 cm
T 8–14.1 cm
W 30–92 g

0–5570 m

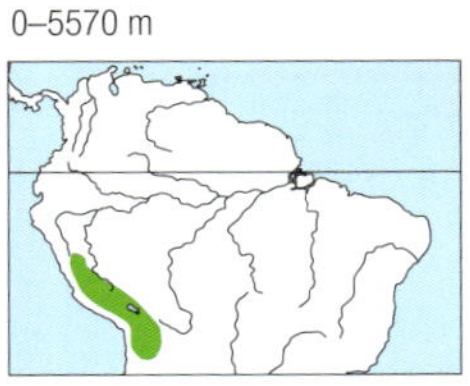

Both slopes of Andes in C & S PE, BO, W slope in N CL, E slope in AR

Rock-dwelling Leaf-eared Mouse

Phyllotis rupestris

FR Phyllotis des rochers
DE Felsen-Blattohrmaus
ES Ratón orejudo de trasero dorado

Lima Leaf-eared Mouse

Phyllotis limatus

FR Phyllotis de Lima
DE Lima-Blattohrmaus
ES Ratón orejudo de Lima

LC
Monotypic
HB c. 11.2 cm
T c. 12.8 cm
W 40–60 g

0–4000 m

CW PE S to N CL

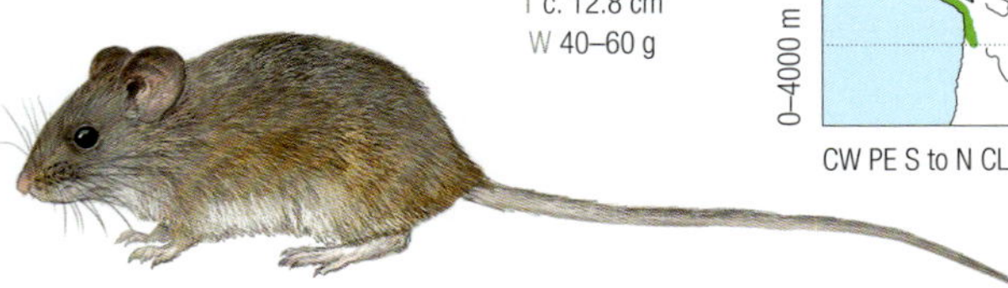

NE Monotypic
HB 4–12.2 cm T 8–15.2 cm
W 27–29 g

800–4000 m

Punta de Vacas Leaf-eared Mouse
Phyllotis vaccarum

FR Phyllotis de Punta de Vacas
DE Punta-de-Vacas-Blattohrmaus
ES Ratón orejudo de Punta de Vacas

Both sides of Andes in CL (Atacama S to Talca) & AR (Catamarca S to Neuquén)

2100–4500 m

Upper slopes of E Andes in S BO & NW AR

LC Monotypic
HB 10.2–14 cm
T 11.6–15.1 cm

Capricorn Leaf-eared Mouse
Phyllotis caprinus

FR Phyllotis du Capricorne
DE Ostandine Blattohrmaus
ES Ratón orejudo de Capricornio

LC Monotypic
HB 9–14.3 cm T 8–14.1 cm
W 30–92 g

Yellow-rumped Leaf-eared Mouse
Phyllotis xanthopygus

FR Phyllotis à croupe jaune
DE Gelbsteiß-Blattohrmaus
ES Ratón orejudo de dorso amarillento

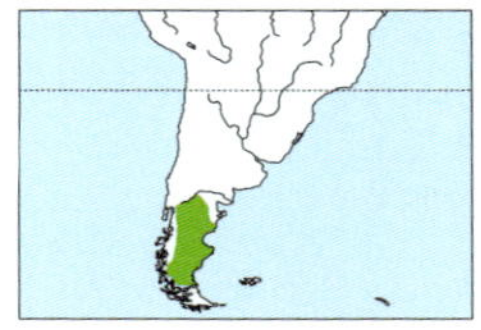

From NW Patagonia S to Strait of Magellan and Andes E to coast (CL, AR)

NE
Monotypic
HB 10–11.5 cm
T 11.1–14.4 cm
W 36–46 g

Camiari Leaf-eared Mouse
Phyllotis camiari

FR Phyllotis des sierras
DE Camiari-Blattohrmaus
ES Ratón orejudo serrano

Known from several localities in Córdoba & San Luis

500–1200 m

Known only from Sierra de la Ventana, SW Buenos Aires

NT
Monotypic
HB 12.7–15.1 cm
T 11–14.7 cm
W 62–110 g

Bonarian Leaf-eared Mouse
Phyllotis bonariensis

FR Phyllotis de Buenos Aires
DE Buenos-Aires-Blattohrmaus
ES Ratón orejudo de Buenos Aires

NE Monotypic
HB 9.5–14 cm T 6–14 cm
W 30–92 g

Pehuenche Leaf-eared Mouse
Phyllotis pehuenche

FR Phyllotis des Pehuenches
DE Pehuenche-Blattohrmaus
ES Ratón orejudo de los Pehuenches

1000–2300 m

Andes in SW Mendoza & W Neuquén

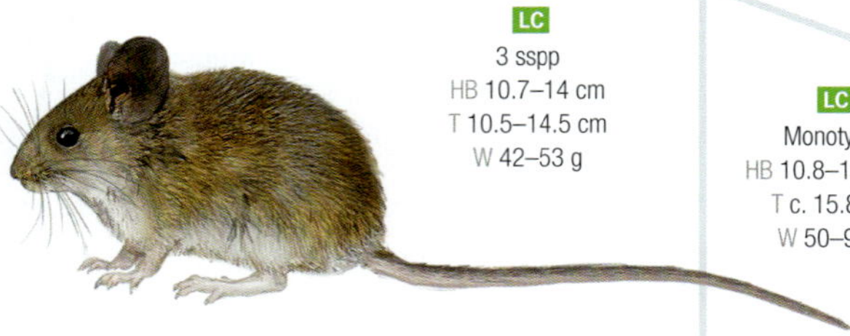

LC
3 sspp
HB 10.7–14 cm
T 10.5–14.5 cm
W 42–53 g

Darwin's Leaf-eared Mouse
Phyllotis darwinii

FR Phyllotis de Darwin
DE Darwin-Blattohrmaus
ES Ratón orejudo de Darwin

From Atacama S to Biobío

LC
Monotypic
HB 10.8–14.5 cm
T c. 15.8 cm
W 50–90 g

0–4000 m

Master Leaf-eared Mouse
Phyllotis magister

FR Phyllotis maître
DE Kommune Blattohrmaus
ES Ratón orejudo maestro

Andes of WC PE S to N CL

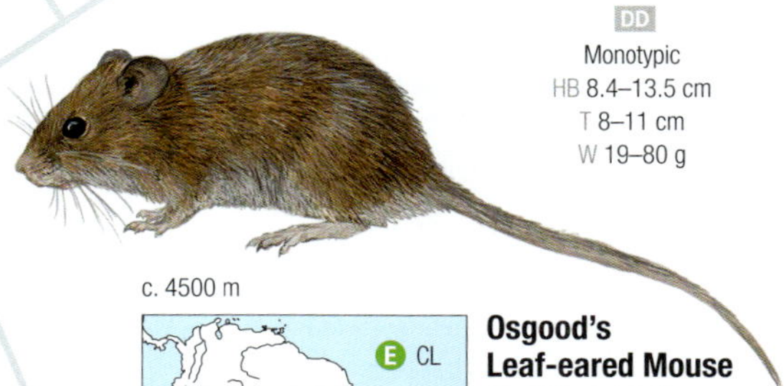

DD
Monotypic
HB 8.4–13.5 cm
T 8–11 cm
W 19–80 g

c. 4500 m

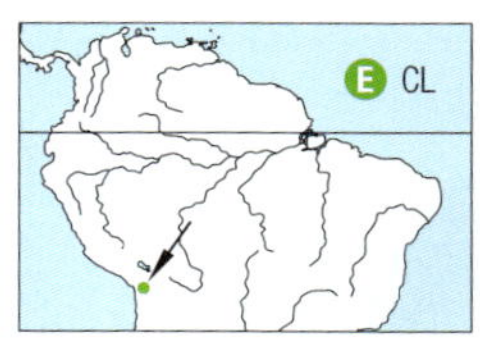

Osgood's Leaf-eared Mouse
Phyllotis osgoodi

FR Phyllotis d'Osgood
DE Osgood-Blattohrmaus
ES Ratón orejudo de Osgood

Known only from type locality around Parinacota (Arica y Parinacota)

LC
Monotypic
HB c. 11.2 cm
T c. 12.3 cm
W c. 57 g

2700–4300 m

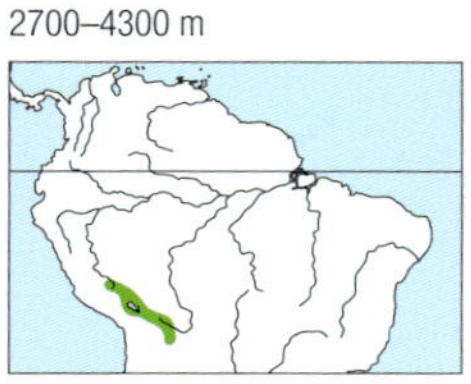

Bunch Grass Leaf-eared Mouse
Phyllotis osilae

FR Phyllotis d'Asillo
DE Büschelgras-Blattohrmaus
ES Ratón orejudo de tusoc

Andes of S PE and W & C BO

NE
Monotypic
HB 9.5–12.9 cm
T 10–13.2 cm
W 42–52 g

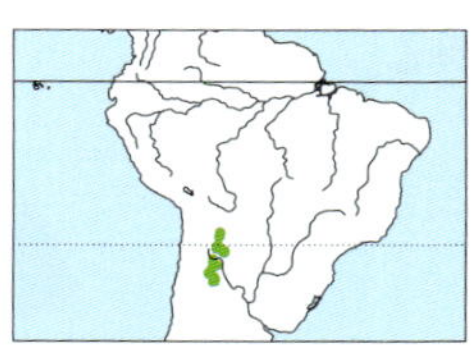

Tucuman Leaf-eared Mouse
Phyllotis tucumanus

FR Phyllotis du Tucuman
DE Tucumán-Blattohrmaus
ES Ratón orejudo de Tucumán

S BO, NW AR; limits uncertain

Walnut Leaf-eared Mouse
Phyllotis nogalaris

FR Phyllotis des noyers
DE Walnuss-Blattohrmaus
ES Ratón orejudo de nogal

NE
Monotypic
HB 11.8–13.7 cm
T 13–14.9 cm
W 43–82 g

1200–3100 m

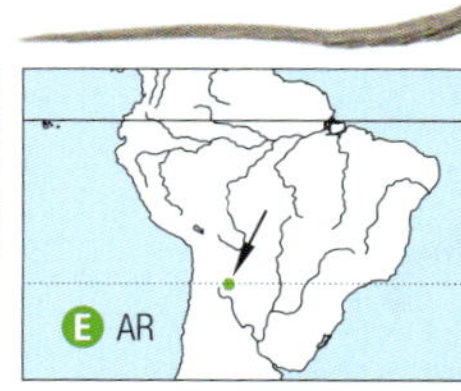

Jujuy, probably Salta

DD
Monotypic
HB 12.4–14.4 cm
T 12.1–13.9 cm
W 35–70 g

1235–2175 m

Los Alisos Leaf-eared Mouse
Phyllotis alisosiensis

FR Phyllotis de Los Alisos
DE Los-Alisos-Blattohrmaus
ES Ratón orejudo de los Alisos

Known only from 2 localities in Sierra del Aconquija (Tucumán)

Anita's Leaf-eared Mouse
Phyllotis anitae

FR Phyllotis d'Anita
DE Anita-Blattohrmaus
ES Ratón orejudo de Anita

DD
Monotypic
HB 8.2–12.3 cm
T 9.4–12.4 cm
W 20–50 g

1700–2800 m

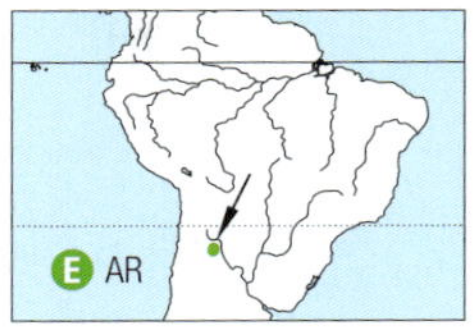

Known only from type locality (S of Hualinchay, Tucumán)

LC Monotypic
HB 11–14.5 cm
T 11.4–15.8 cm
W 41–70 g

2200–3500 m

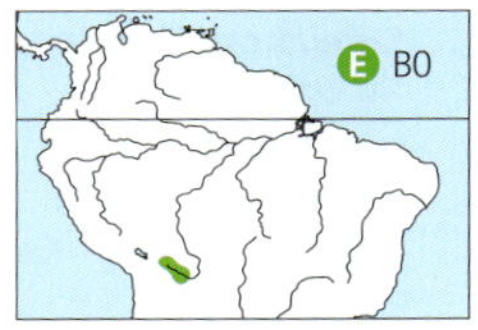

Wolffsohn's Leaf-eared Mouse
Tapecomys wolffsohni

FR Phyllotis de Wolffsohn
DE Wolffsohn-Blattohrmaus
ES Ratón orejudo de Wolffsohn

Cochabamba, Chuquisaca

Genus *TAPECOMYS*

(2 species)

LC
Monotypic
HB 12.4–15.5 cm
T 14.3–15.7 cm
W 49–95 g

1200–1500 m

Tapecua Leaf-eared Mouse
Tapecomys primus

FR Phyllotis de Tapecua
DE Tapecua-Blattohrmaus
ES Ratón orejudo de Tapecua

Lower Yungas forests in S BO (Tarija) & NW AR (Jujuy)

Genus *LOXODONTOMYS*

(1 species)

LC
Monotypic
HB 9.9–14.4 cm
T 8–11.2 cm
W 30–64 g

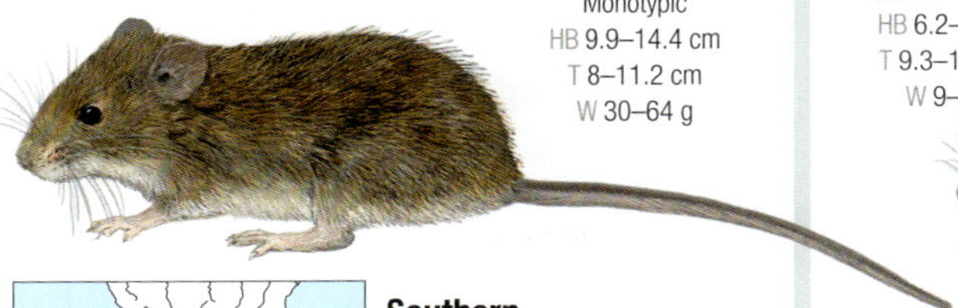

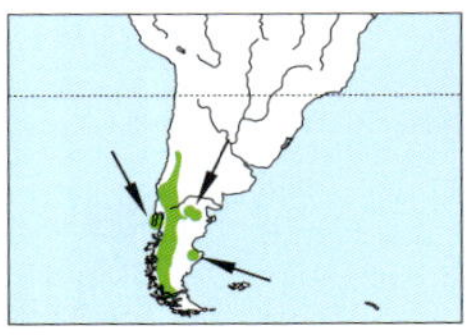

Southern Big-eared Mouse
Loxodontomys micropus

FR Phyllotis austral
DE Südliche Großohrmaus
ES Ratón orejudo meridional

S CL, SW AR

DD Monotypic
HB 6.2–8.1 cm
T 9.3–12.3 cm
W 9–14 g

Genus *SALINOMYS*

(1 species)

400–600 m

Delicate Salt Flat Mouse
Salinomys delicatus

FR Phyllotis des Salinas
DE Zierliche Salzmaus
ES Ratón de salinas

Catamarca, La Rioja, San Juan, Mendoza, San Luis

Genus *GALENOMYS*

(1 species)

DD
Monotypic
HB 10.5–13.2 cm
T 3–4.5 cm
W 59–60 g

3300–4650 m

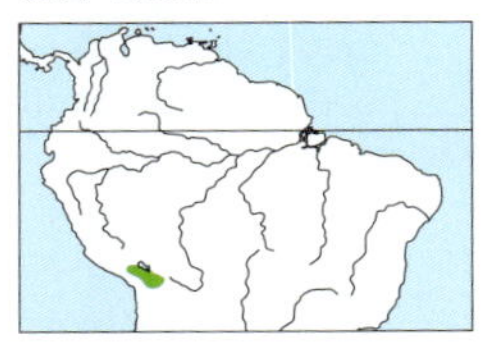

Garlepp's Mouse
Galenomys garleppii

FR Phyllotis de Garlepp
DE Garlepp-Maus
ES Ratón de Garlepp

SE PE (Puno), W BO (La Paz, Oruro); perhaps N CL

LC 2 sspp
HB 8.6–12.6 cm T 9.7–13.6 cm
W c. 45 g

Genus *ANDALGALOMYS*

(2 species)

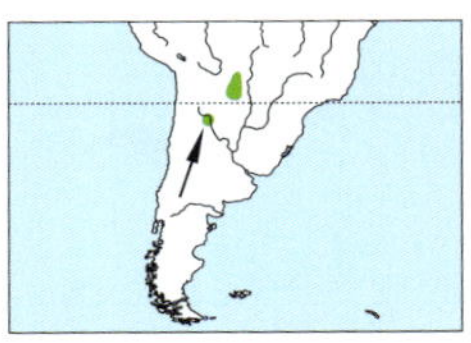

Pearson's Chaco Mouse
Andalgalomys pearsoni

FR Phyllotis du Paraguay
DE Pearson-Chacomaus
ES Ratón del Chaco de Pearson

SE BO (Santa Cruz), Chaco of W PY; supposedly NW AR (Salta)

LC 2 sspp
HB 7.5–11.3 cm
T 8.5–13.5 cm
W 25–27 g

Olrog's Chaco Mouse
Andalgalomys olrogi

FR Phyllotis d'Olrog
DE Olrog-Chacomaus
ES Ratón del Chaco de Olrog

Catamarca, La Rioja, Santiago del Estero, San Luis; perhaps Mendoza

Genus *AULISCOMYS*

(3 species)

LC Monotypic
HB 10–13.3 cm T 8.3–10.2 cm
W 44–60 g

2590–4820 m

Painted Big-eared Mouse
Auliscomys pictus

FR Phyllotis peint
DE Bunte Großohrmaus
ES Ratón de orejas grandes pintado

Andes of C & S PE and W BO

LC
2 sspp
HB c. 12.7 cm
T 6.8–9.5 cm
W c. 55 g

3500–5000 m

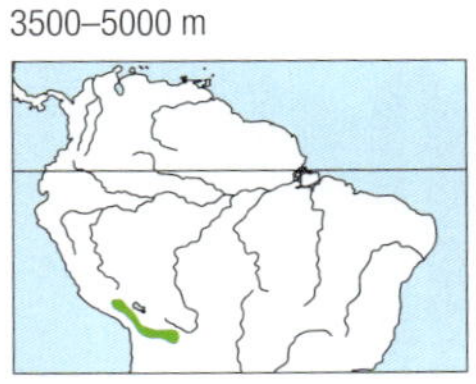

Bolivian Big-eared Mouse
Auliscomys boliviensis

FR Phyllotis de Bolivie
DE Bolivien-Großohrmaus
ES Ratón de orejas grandes de Bolivia

Altiplano of SW PE, N CL & WC BO

LC 2 sspp
HB 9.2–11.8 cm
T 4.9–6.4 cm
W 28–44 g

Andean Big-eared Mouse
Auliscomys sublimis

FR Phyllotis de l'Altiplano
DE Anden-Großohrmaus
ES Ratón de orejas grandes de los Andes

3200–5400 m

S PE, W BO, extreme NE CL, NW AR

Genus *NYCTOMYS*

(1 species)

0–2950 m

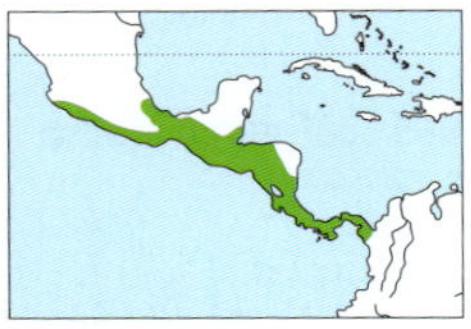

MX (from S Jalisco and S Veracruz) S to PA, excluding Yucatán Pen

LC
9 sspp
HB 12.8–14 cm
T 14.3–15.5 cm
W 40–55 g

Sumichrast's Vesper Rat
Nyctomys sumichrasti
FR Nyctomys de Sumichrast
DE Sumichrast-Vesperratte
ES Rata laucha de Sumichrast

Genus *OTONYCTOMYS*

(1 species)

MX (Yucatán Pen), NE GT, N BZ

Yucatan Vesper Rat
Otonyctomys hatti
FR Nyctomys du Yucatan
DE Yucatán-Kletterratte
ES Rata laucha de Yucatán

LC
Monotypic
HB 9–11.6 cm
T 6–12.7 cm
W 24–94 g

Genus *OTOTYLOMYS*

(2 species)

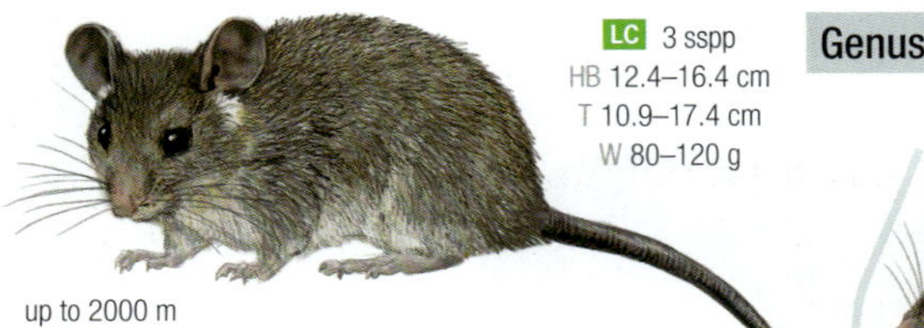

LC 3 sspp
HB 12.4–16.4 cm
T 10.9–17.4 cm
W 80–120 g

up to 2000 m

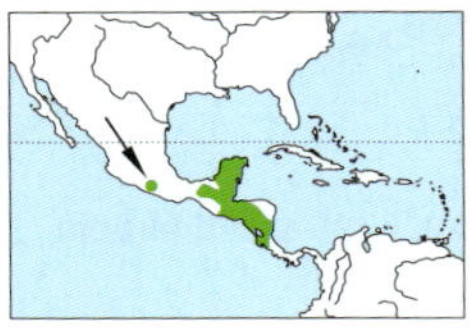

MX (S Tabasco, N Chiapas, Yucatán Pen, recorded in NC Guerrero) S to C CR

Big-eared Climbing Rat
Ototylomys phyllotis
FR Tylomys à grandes oreilles
DE Großohr-Kletterratte
ES Rata trepadora de orejas grandes

NE Monotypic
HB 13.9–19.4 cm
T 13.5–17.4 cm
W 55–165 g

700–1100 m

Known from 2 localities in Chiapas (El Ocote BR & La Pera CA)

La Pera Climbing Rat
Ototylomys chiapensis
FR Tylomys de La Pera
DE La-Pera-Kletterratte
ES Rata trepadora de La Pera

Genus *TYLOMYS*

(7 species)

LC 4 sspp
HB 18.8–27 cm
T 19.5–26 cm
W 280–326 g

0–1600 m

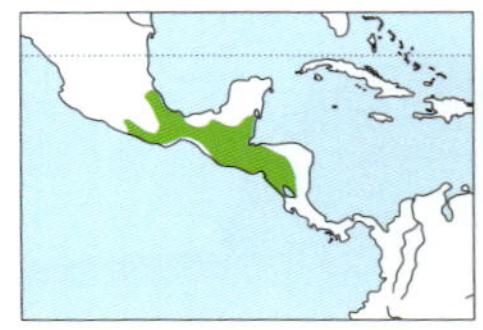

MX (from C Guerrero & C Veracruz) S to S NI, excluding N Yucatán Pen

Peters's Climbing Rat
Tylomys nudicaudus
FR Tylomys à queue nue
DE Nacktschwanz-Kletterratte
ES Rata trepadora de Peters

CR(PE)
Monotypic
HB c. 16.6 cm
T c. 15.8 cm
W c. 182 g

c. 550 m

Known only in Tuxtla Gutiérrez and surroundings (NW Chiapas)

Chiapan Climbing Rat
Tylomys bullaris
FR Tylomys de Tuxtla
DE Chiapas-Kletterratte
ES Rata trepadora de Chiapas

CR(PE)
Monotypic
HB c. 21 cm
T c. 23 cm
W c. 250 g

Known only from type locality (Tumbalá, N Chiapas)

Tumbala Climbing Rat
Tylomys tumbalensis
FR Tylomys de Tumbala
DE Tumbala-Kletterratte
ES Rata trepadora de Tumbalá

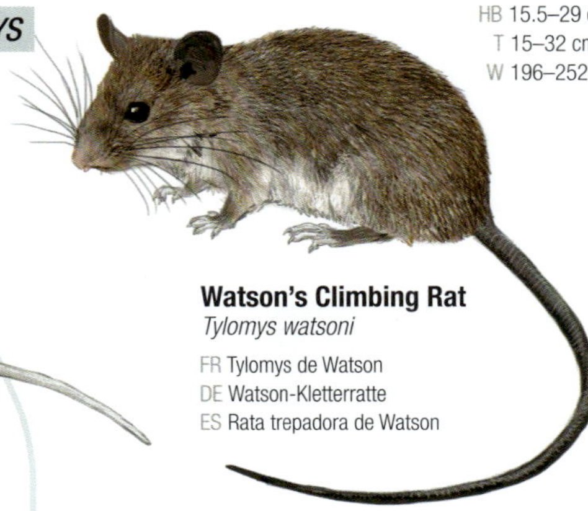

LC
Monotypic
HB 15.5–29 cm
T 15–32 cm
W 196–252 g

Watson's Climbing Rat
Tylomys watsoni
FR Tylomys de Watson
DE Watson-Kletterratte
ES Rata trepadora de Watson

0–2700 m

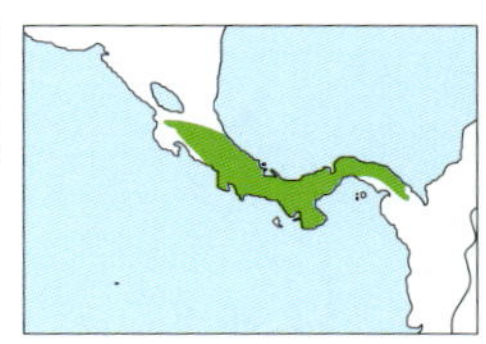

Caribbean & Pacific slopes and Central Highlands of CR and PA

DD
Monotypic
HB c. 22 cm
T c. 23 cm

Known only from a small area of Darién Gap

Fulvous-bellied Climbing Rat
Tylomys fulviventer
FR Tylomys à ventre fauve
DE Gelbbauch-Kletterratte
ES Rata trepadora de vientre bermejo

DD
Monotypic
HB c. 23 cm
T c. 19.9 cm

Panama Climbing Rat
Tylomys panamensis
FR Tylomys du Panama
DE Panama-Kletterratte
ES Rata trepadora de Panamá

Known localities are in Darién NP

LC 2 sspp
HB c. 23 cm T c. 27 cm
W 200–255 g

200–1300 m

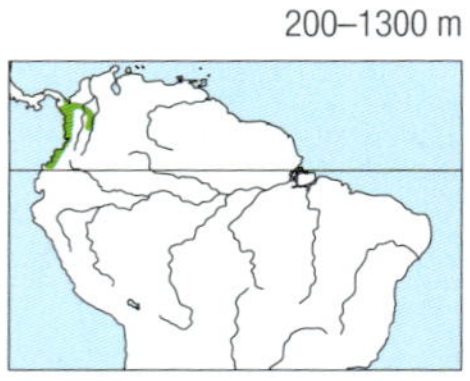

Tropical rain forest in W & C CO and NW EC

Mira Climbing Rat
Tylomys mirae
FR Tylomys du Mira
DE Mira-Kletterratte
ES Rata trepadora de Mira

Family MURIDAE – True Mice and Rats, Gerbils and relatives

161 genera, 843 species. All terrestrial and some aquatic habitats. Worldwide, except Antarctica.

Seurat's Spiny Mouse
Acomys seurati
FR Acomys de Seurat
DE Seurat-Stachelmaus
ES Ratón espinoso de Seurat

1000–2000 m

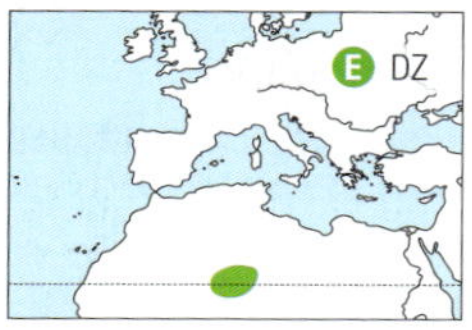

Mouydir, Tassili & Hoggar Mts in Sahara Desert

LC
Monotypic
HB 8.7–9.9 cm
T 10.5–12.5 cm

Genus *ACOMYS*

(21 species)

Northeast African Spiny Mouse
Acomys cahirinus
FR Acomys d'Égypte
DE Ägypten-Stachelmaus
ES Ratón espinoso de África nororiental

LC
Monotypic
HB 10–12 cm
T 9.2–13.7 cm
W 29–52 g

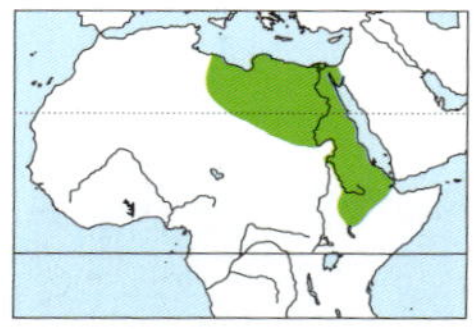

LY and EG S to ET and DJ, E to Sinai and extreme SE IL

Cyprus Spiny Mouse
Acomys nesiotes
FR Acomys de Chypre
DE Zypern-Stachelmaus
ES Ratón espinoso de Chipre

DD
Monotypic
HB 9.7–13.5 cm
T 9.1–11.8 cm
W 31–70 g

0–1200 m

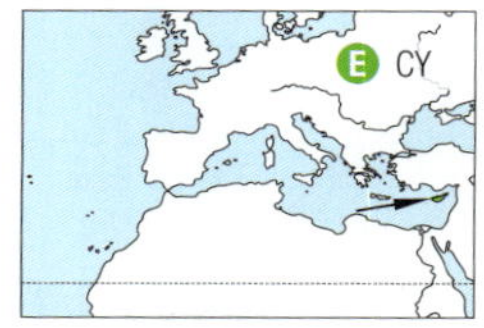

Known from 14 localities on N & S coasts; perhaps widespread

Crete Spiny Mouse
Acomys minous
FR Acomys de Crête
DE Kreta-Stachelmaus
ES Ratón espinoso de Creta

Crete

DD
Monotypic
HB 9–12.8 cm
T 8.9–12 cm
W 30–86 g

Turkish Spiny Mouse
Acomys cilicicus
FR Acomys de Silifke
DE Türkei-Stachelmaus
ES Ratón espinoso de Turquía

c. 10 m

Known from 3 sites close to Silifke, c. 15 km along coast; N limits unknown

DD
Monotypic
HB 9.1–12.1 cm
T 9.1–11.7 cm
W 29–63 g

Chudeau's Spiny Mouse
Acomys chudeaui
FR Acomys de Chudeau
DE Chudeau-Stachelmaus
ES Ratón espinoso de Chudeau

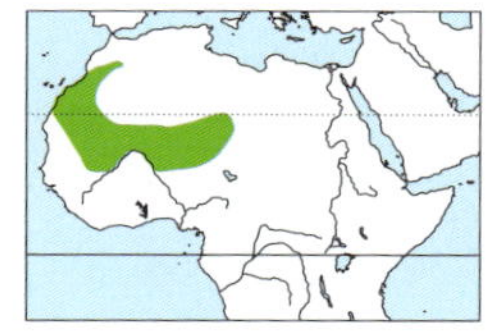

N & W Africa, N of R Niger

NE
Monotypic
HB 9.3–12 cm
T 7.2–12.5 cm
W 25–64 g

LC
Monotypic
HB 10.6–13.4 cm
T 9.5–11.7 cm

Mullah Spiny Mouse
Acomys mullah
FR Acomys mollah
DE Mullah-Stachelmaus
ES Ratón espinoso mulá

up to 1000 m

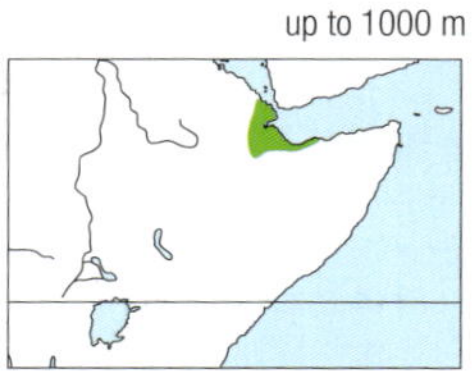

Horn of Africa (S ER, DJ, E ET & N SO)

LC
5 sspp
HB 8.5–12.9 cm
T 8.8–11.7 cm
W 30–45 g

Arabian Spiny Mouse
Acomys dimidiatus
FR Acomys d'Arabie
DE Sinai-Stachelmaus
ES Ratón espinoso de Arabia

up to 1900 m

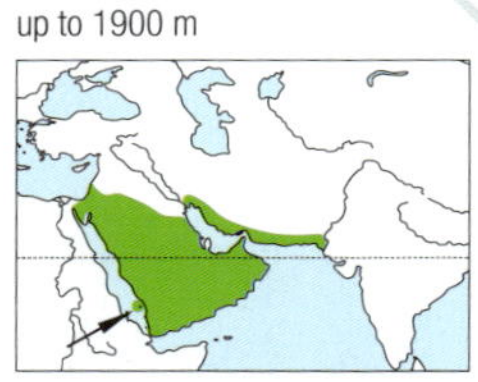

EG (Sinai), IL, JO, much of Arabian Pen and S IR to S PK

LC
Monotypic
HB 9.5–12.3 cm
T 8.5–11.5 cm
W 26–60 g

John's Spiny Mouse
Acomys johannis
FR Acomys de John
DE John-Stachelmaus
ES Ratón espinoso de John

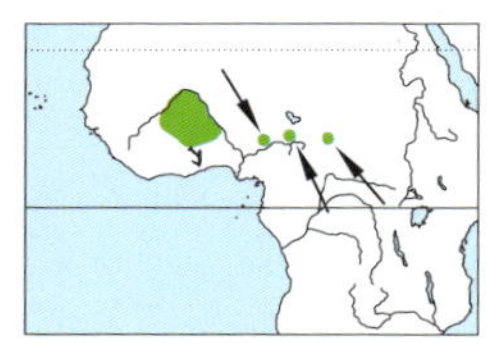

W Africa in Sudanian savannas S & W of R Niger; also N NG, N CM & S TD

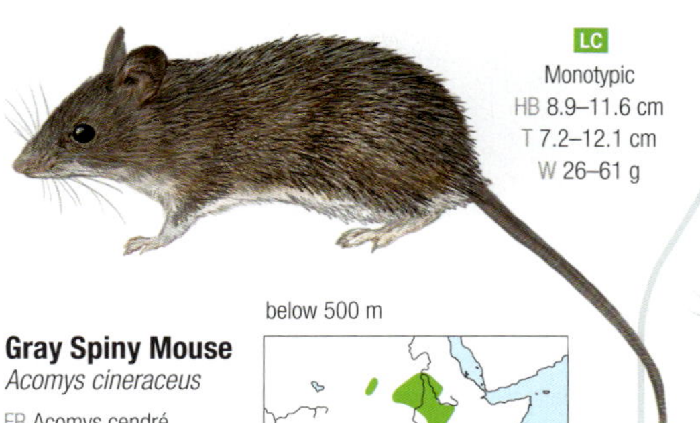

LC
Monotypic
HB 8.9–11.6 cm
T 7.2–12.1 cm
W 26–61 g

Gray Spiny Mouse
Acomys cineraceus
FR Acomys cendré
DE Graue Stachelmaus
ES Ratón espinoso gris

below 500 m

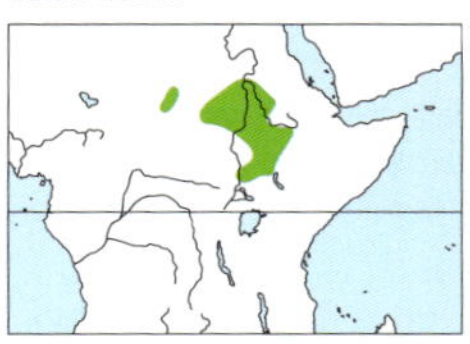

Woodland savanna in SD, SS, W ET; perhaps DJ

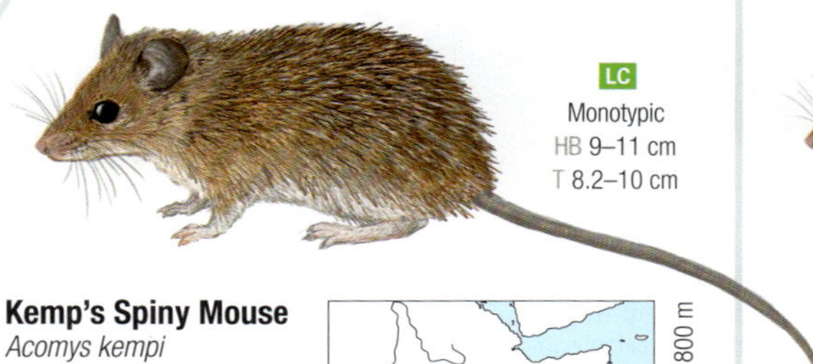

LC
Monotypic
HB 9–11 cm
T 8.2–10 cm

Kemp's Spiny Mouse
Acomys kempi
FR Acomys de Kemp
DE Kemp-Stachelmaus
ES Ratón espinoso de Kemp

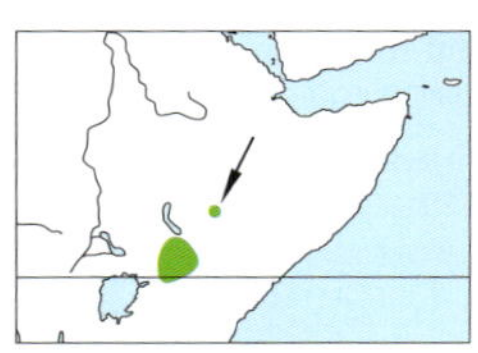

600–1800 m

S ET & KE, E of Rift Valley; perhaps S SO & NE TZ

LC
Monotypic
HB 9.2–11.4 cm
T 7–8.5 cm

Fiery Spiny Mouse
Acomys ignitus
FR Acomys de Dollman
DE Feuerrote Stachelmaus
ES Ratón espinoso fogoso

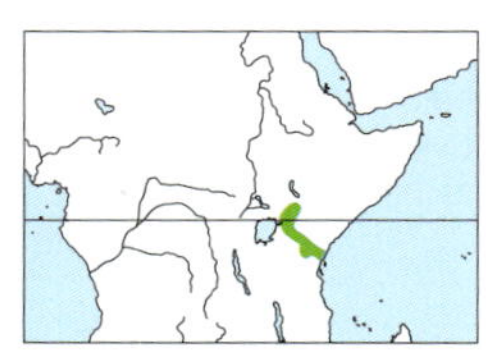

100–800 m

Somali-Masai bushland in E & S KE and NE TZ

LC
2 sspp
HB 7.9–9.5 cm
T 8.6–10.4 cm

Louise's Spiny Mouse
Acomys louisae

FR Acomys de Louise
DE Louise-Stachelmaus
ES Ratón espinoso de Louise

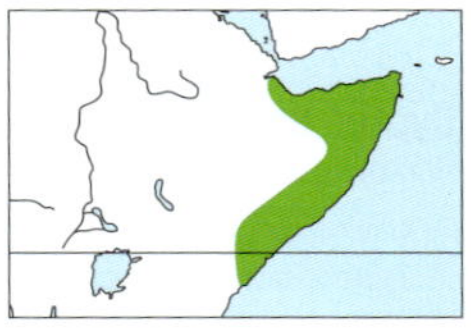

up to 1500 m

Somali-Masai bushland in DJ, SO and extreme E KE

LC
5 sspp
HB 8.9–11.4 cm
T 6.5–10 cm
W 22–47 g

Wilson's Spiny Mouse
Acomys wilsoni

FR Acomys de Wilson
DE Wilson-Stachelmaus
ES Ratón espinoso de Wilson

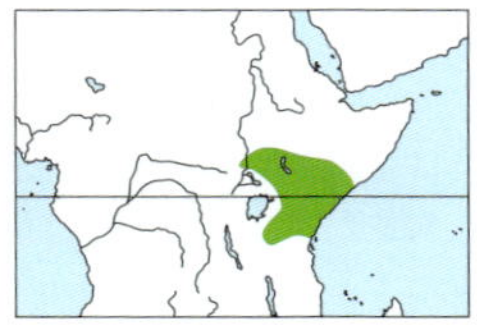

S SS, S ET, S SO, N UG, KE, N TZ

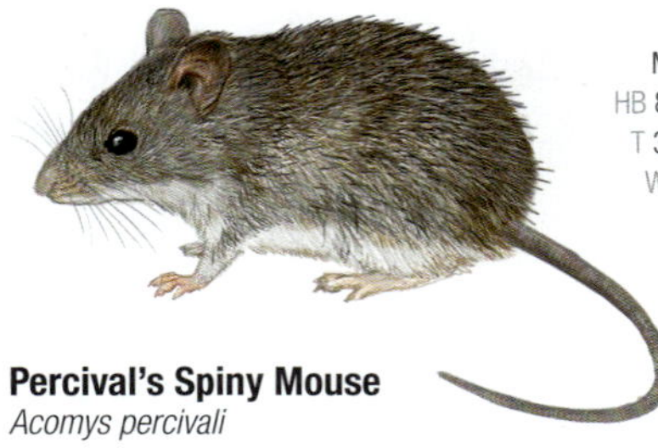

LC
Monotypic
HB 8.2–11.1 cm
T 3.9–9.2 cm
W 18–48 g

Percival's Spiny Mouse
Acomys percivali

FR Acomys de Percival
DE Percival-Stachelmaus
ES Ratón espinoso de Percival

Lowlands in SE SS, SW ET, NE UG and W KE

LC
2 sspp
HB 10–12 cm
T 6.8–7.8 cm
W 40–77 g

Golden Spiny Mouse
Acomys russatus

FR Acomys doré
DE Goldstachelmaus
ES Ratón espinoso dorado

up to 2285 m

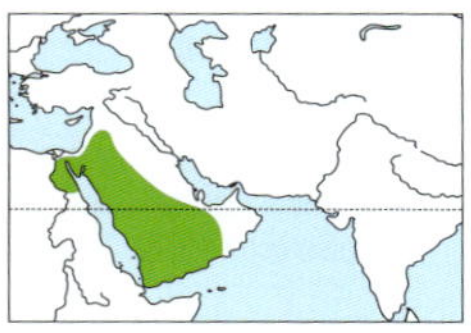

EG (Eastern Desert & Sinai), IL, PS (West Bank), SW SY, JO, Arabian Pen

LC
Monotypic
HB 7.5–9.7 cm
T 7.9–10.3 cm
W 14–24 g

Muze Spiny Mouse
Acomys muzei

FR Acomys de Muze
DE Muze-Stachelmaus
ES Ratón espinoso oriental

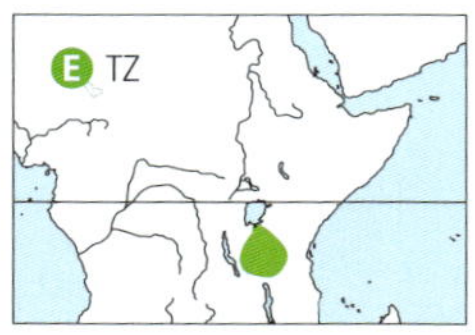

From L Victoria S to hills close to L Rukwa and W Eastern Arc Mts

LC
Monotypic
HB 7.9–9.7 cm
T 6.7–9 cm
W 20–31 g

Southern African Spiny Mouse
Acomys spinosissimus

FR Acomys de Peters
DE Zwergstachelmaus
ES Ratón espinoso de África meridional

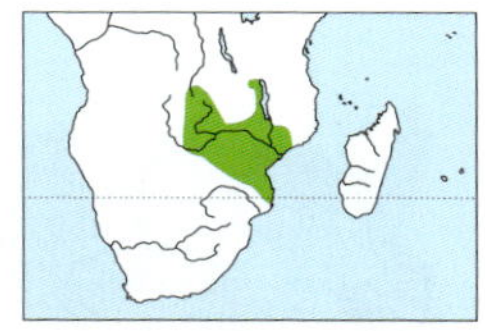

From S CD, NE ZM & N MW S to ZW & MZ, N of R Limpopo; S limits uncertain

LC
Monotypic
HB 7.8–9.4 cm
T 7.2–9.6 cm

Selous's Spiny Mouse
Acomys selousi

FR Acomys de Selous
DE Selous-Stachelmaus
ES Ratón espinoso de Selous

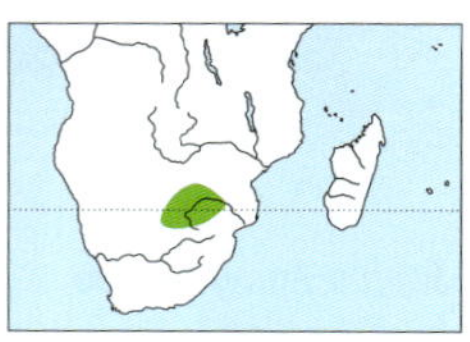

E BW, S ZW, N ZA; N limits uncertain

LC
Monotypic
HB 7.3–10.4 cm
T 6.2–9 cm
W 13–30 g

Nguru Spiny Mouse
Acomys ngurui

FR Acomys des Nguru
DE Nguru-Stachelmaus
ES Ratón espinoso de Nguru

up to 1100 m

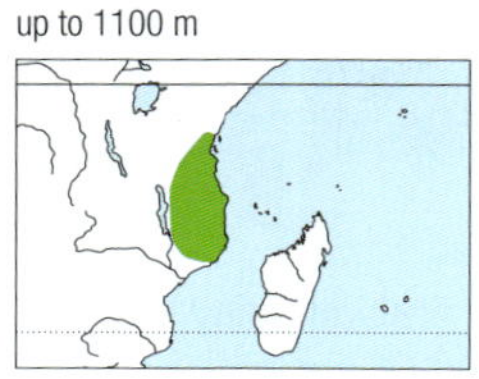

S TZ (E of Eastern Arc Mts), N MZ (N of R Zambezi), S MW

LC
Monotypic
HB 7.3–10.2 cm
T 6.5–8.9 cm
W 13–26 g

Cape Spiny Mouse
Acomys subspinosus

FR Acomys du Cap
DE Kap-Stachelmaus
ES Ratón espinoso de El Cabo

up to 1000 m

Rocky mountainous habitat in SW Western Cape

VU
Monotypic
HB 12–18 cm
T 3–9.9 cm
W 60–142 g

Black-clawed Brush-furred Rat
Lophuromys melanonyx

FR Rat-hérissé à griffes noires
DE Schwarzkrallen-Bürstenhaarmaus
ES Rata de pelaje de cepillo de uñas negras

above 3200 m

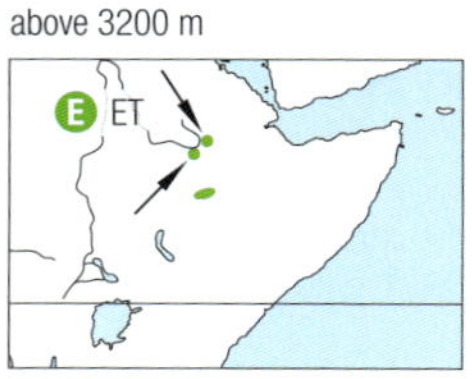

Rift Valley highlands

Genus *LOPHUROMYS*

(34 species)

LC
Monotypic
HB 12.5–12.9 cm
T 5.8–6.8 cm
W c. 69 g

Buff-spotted Brush-furred Rat
Lophuromys flavopunctatus

FR Rat-hérissé à taches jaunes
DE Gelbflecken-Bürstenhaarmaus
ES Rata de pelaje de cepillo moteado

1900–3000 m

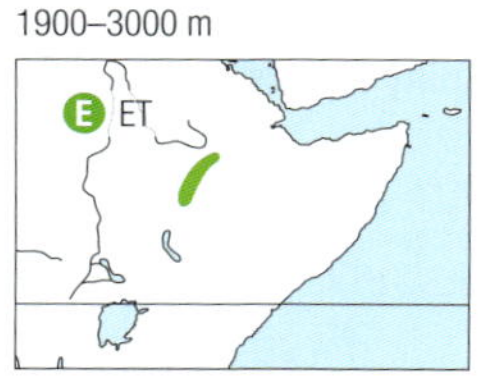

Several localities W of Rift Valley

North Western Rift Brush-furred Rat
Lophuromys menageshae

FR Rat-hérissé de Menagesha
DE Menagasha-Bürstenhaarmaus
ES Rata de pelaje de cepillo del Valle Noroccidental

DD
Monotypic
HB 12.9–14.3 cm
T 6–7.5 cm
W 72–77 g

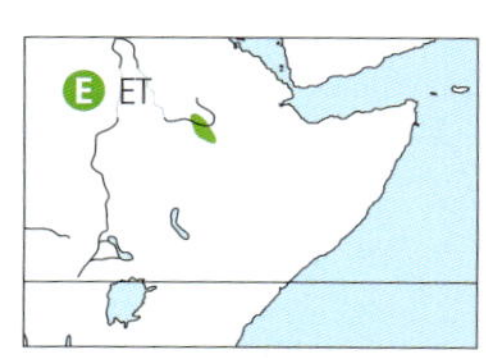

2100–2600 m

Mountain forests W of Rift Valley

NT
Monotypic
HB 9.6–13.4 cm
T 5–6.6 cm
W 29–60 g

above 2400 m

Short-tailed Brush-furred Rat
Lophuromys brevicaudus
FR Rat-hérissé à queue courte
DE Kurzschwanz-Bürstenhaarmaus
ES Rata de pelaje de cepillo de cola corta

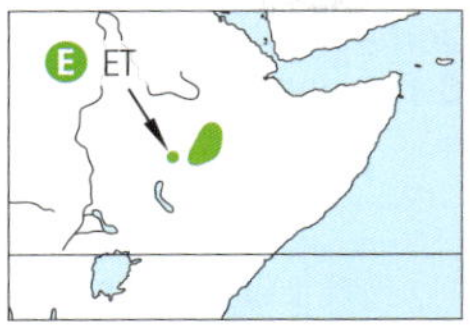

High plateaus W & E of Rift Valley

EN
Monotypic
HB 13–13.8 cm
T 7.8–8.6 cm
W 53–67 g

c. 1930 m

Sheko Forest Brush-furred Rat
Lophuromys pseudosikapusi
FR Rat-hérissé de Sheko
DE Sheko-Bürstenhaarmaus
ES Rata de pelaje de cepillo de Sheka

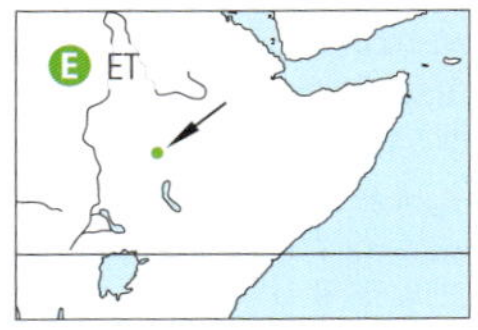

Sheko Forest

LC
Monotypic
HB 11.4–14.5 cm
T 5.8–8.5 cm
W 37–74 g

1600–3800 m

Simien Brush-furred Rat
Lophuromys simensis
FR Rat-hérissé des Simien
DE Simien-Bürstenhaarmaus
ES Rata de pelaje de cepillo de Simien

Simien Mts & L Tana region, W of Rift Valley

NE
Monotypic
HB 10.1–13.8 cm
T 5.5–8.4 cm
W 38–76 g

above 1800 m

Ethiopian Brush-furred Rat
Lophuromys brunneus
FR Rat-hérissé de Thomas
DE Äthiopien-Bürstenhaarmaus
ES Rata de pelaje de cepillo de Etiopía

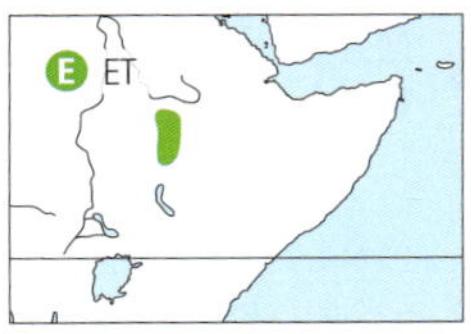

Mountain forests W of Rift Valley

LC
Monotypic
HB 9.4–11.4 cm
T 5.9–6.1 cm

Hutterer's Brush-furred Rat
Lophuromys huttereri
FR Rat-hérissé de Hutterer
DE Hutterer-Bürstenhaarmaus
ES Rata de pelaje de cepillo de Hutterer

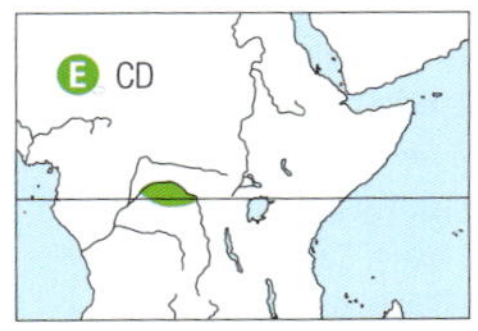

Few localities in Congo Basin on left bank of R Congo; limits uncertain

LC
2 sspp
HB 8.9–11.9 cm
T 4.7–7.4 cm W 29–52 g

From R Cross (NG) E to R Ubangui (CF), S to SW CG, Bioko I; recorded in right bank of R Congo

Naked-tailed Brush-furred Rat
Lophuromys nudicaudus
FR Rat-hérissé à queue nue
DE Nacktschwanz-Bürstenhaarmaus
ES Rata de pelaje de cepillo de cola desnuda

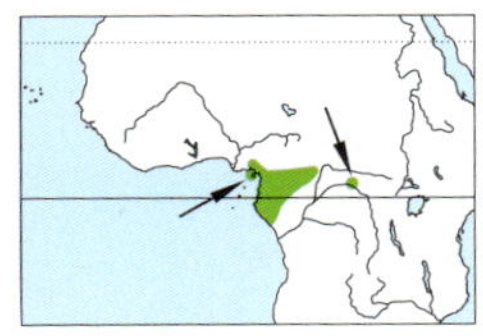

LC
Monotypic
HB 10.4–14.1 cm
T 5–7.8 cm
W 49–88 g

1200–3100 m

Mount Cameroon Brush-furred Rat
Lophuromys roseveari
FR Rat-hérissé de Rosevear
DE Kamerunberg-Bürsten-haarmaus
ES Rata de pelaje de cepillo de Camerún

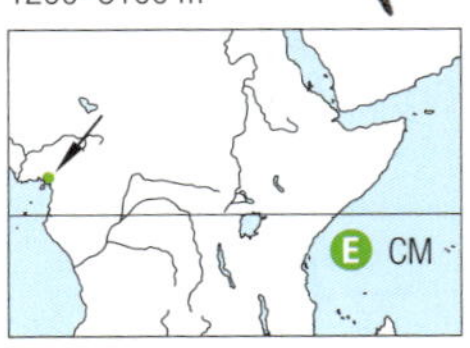

Mt Cameroon

NE
Monotypic
HB 11.5–15.4 cm
T 5.1–8.8 cm
W 43–71 g

c. 2600 m

Angolan Brush-furred Rat
Lophuromys angolensis
FR Rat-hérissé d'Angola
DE Angola-Bürstenhaarmaus
ES Rata de pelaje de cepillo de Angola

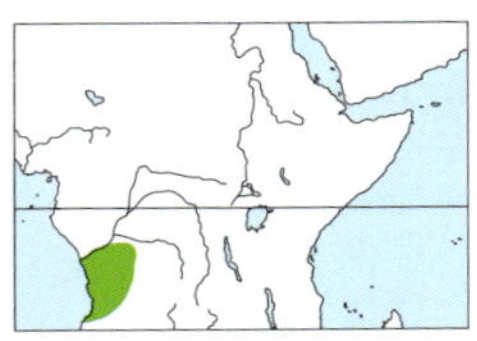
SW CD (E of R Congo), NW AO

LC
Monotypic
HB 11–14.5 cm
T 5.8–7.5 cm
W 34–87 g

below 1700 m

Western Brush-furred Rat
Lophuromys sikapusi
FR Rat-hérissé de Temminck
DE Rostbauch-Bürstenhaarmaus
ES Rata de pelaje de cepillo occidental

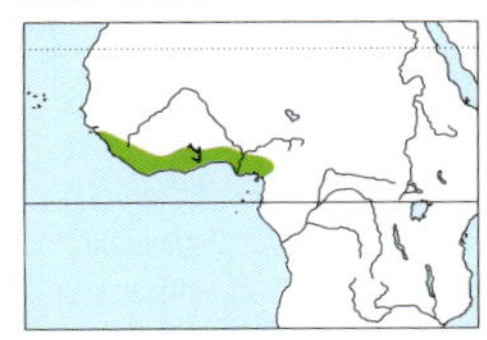
From GN and SL E to NG and CM, N of R Sanaga

NE
Monotypic
HB 10–16.5 cm
T 4.9–9.1 cm
W 45–111 g

Ansorge's Brush-furred Rat
Lophuromys ansorgei
FR Rat-hérissé d'Ansorge
DE Ansorge-Bürstenhaarmaus
ES Rata de pelaje de cepillo de Ansorge

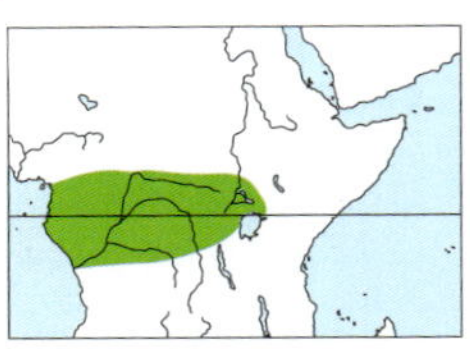
S CM, S of R Sanaga, E through Congo Basin to UG & KE

NT
Monotypic
HB 9.5–11.6 cm
T 4.8–5.6 cm
W 30–45 g

1900–2500 m

Rahm's Brush-furred Rat
Lophuromys rahmi
FR Rat-hérissé de Rahm
DE Rahm-Bürstenhaarmaus
ES Rata de pelaje de cepillo de Rahm

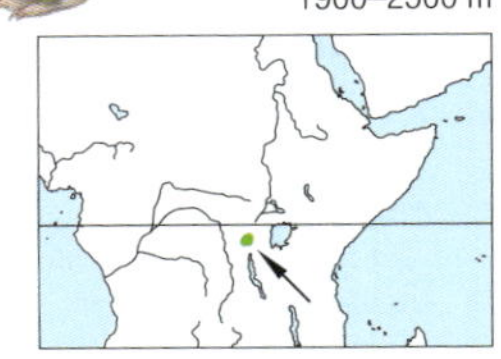
Albertine Rift around L Kivu (E CD, RW)

EN
Monotypic
HB 11.3–12.8 cm
T 7.4–7.7 cm
W 49–62 g

above 2000 m

Mount Oku Brush-furred Rat
Lophuromys dieterleni
FR Rat-hérissé de Dieterlen
DE Mount-Oku-Bürstenhaarmaus
ES Rata de pelaje de cepillo de Oku

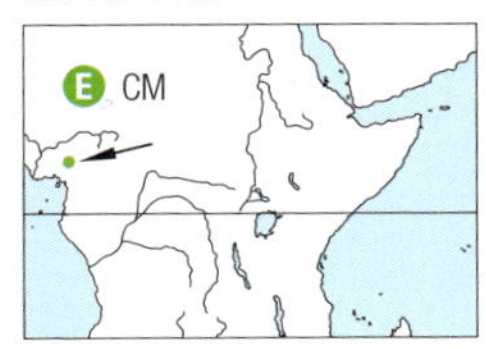

Mt Oku

CR
Monotypic
HB 8.8–9.4 cm
T 5.1–5.3 cm

Known only from Mt Lefo (Bamileke Plateau)

Mount Lefo Brush-furred Rat
Lophuromys eisentrauti
FR Rat-hérissé d'Eisentraut
DE Mount-Lefo-Bürstenhaarmaus
ES Rata de pelaje de cepillo de Lefu

1800–1900 m

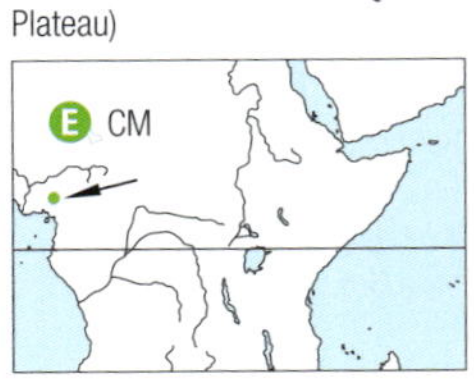

NE
Monotypic
HB 11.7–14.5 cm
T 6.6–8.4 cm
W 36–80 g

1700–3600 m

Dark-colored Brush-furred Rat
Lophuromys aquilus
FR Rat-hérissé du Kilimandjaro
DE Dunkle Bürstenhaarmaus
ES Rata de pelaje de cepillo de dorso oscuro

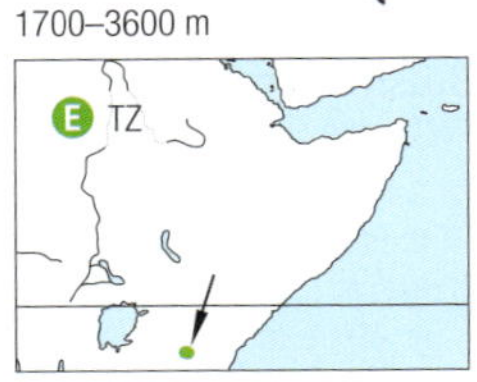

Mt Kilimanjaro

NE
Monotypic
HB 10.8–13.9 cm
T 4.2–8.8 cm
W 34–70 g

Zena Brush-furred Rat
Lophuromys zena
FR Rat-hérissé de Nyeri
DE Zena-Bürstenhaarmaus
ES Rata de pelaje de cepillo de Zena

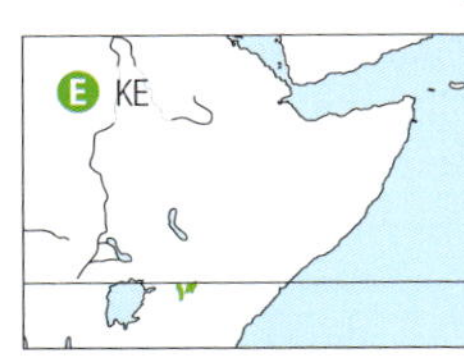

Aberdare Range & Mt Kenya

LC
Monotypic
HB 10–13 cm
T 7.2–8.7 cm W 32–59 g

Ethiopian Forest Brush-furred Rat
Lophuromys chrysopus

FR Rat-hérissé à pattes dorées
DE Goldfuß-Bürstenhaarmaus
ES Rata de pelaje de cepillo de bosque de Etiopía

1200–2760 m

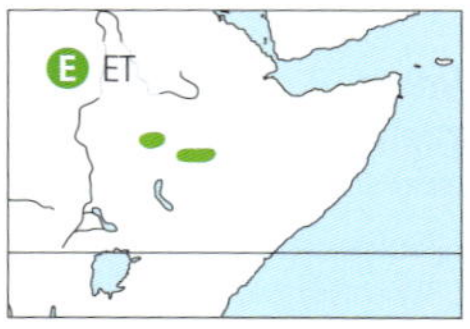

Mountain forests in both sides of Rift Valley

DD
Monotypic
HB 11.3–12.6 cm
T 4–8 cm
W 36–55 g

Rwenzori Brush-furred Rat
Lophuromys stanleyi

FR Rat-hérissé de Stanley
DE Ruwenzori-Bürstenhaarmaus
ES Rata de pelaje de cepillo de Rwenzori

c. 3700 m

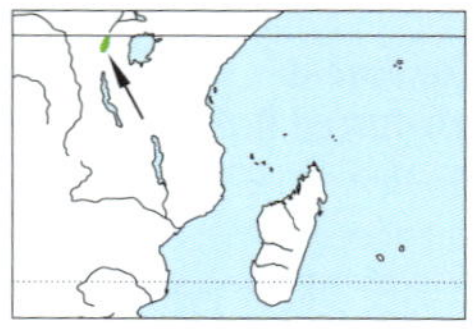

Rwenzori Mts (E CD, SW UG)

NE
Monotypic
HB 11.2–14 cm
T 4.3–8.1 cm
W 40–64 g

Verhagen's Brush-furred Rat
Lophuromys verhageni

FR Rat-hérissé de Verhagen
DE Verhagen-Bürstenhaarmaus
ES Rata de pelaje de cepillo de Verhagen

2300–3600 m

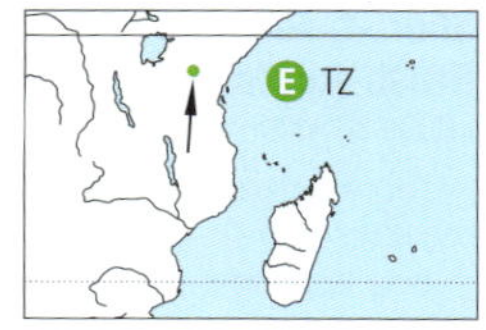

Mt Meru

DD
Monotypic
HB 11.2–14.5 cm
T 3–8 cm
W 38–87 g

Sabuni's Brush-furred Rat
Lophuromys sabunii

FR Rat-hérissé de Sabuni
DE Sabuni-Bürstenhaarmaus
ES Rata de pelaje de cepillo de Sabuni

c. 1750 m

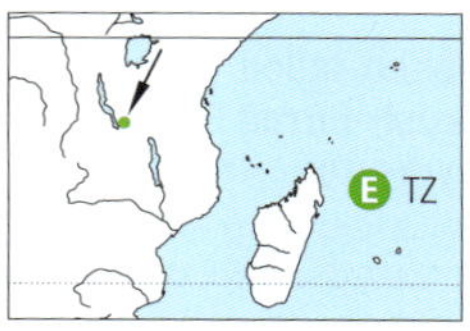

Ufipa Plateau

EN
Monotypic
HB 11.4–14.5 cm
T 6–6.5 cm
W 36–78 g

Mount Chercher Brush-furred Rat
Lophuromys chercherensis

FR Rat-hérissé du Chercher
DE Mount-Chercher-Bürstenhaarmaus
ES Rata de pelaje de cepillo de Chercher

2000–2700 m

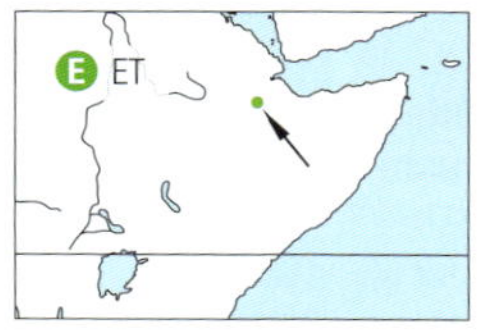

Chercher Mts on E side of Rift Valley

NE
Monotypic
HB c. 12 cm
T c. 8 cm

Margaret's Brush-furred Rat
Lophuromys margarettae

FR Rat-hérissé du Warges
DE Margaret-Bürstenhaarmaus
ES Rata de pelaje de cepillo de Margaret

above 2000 m

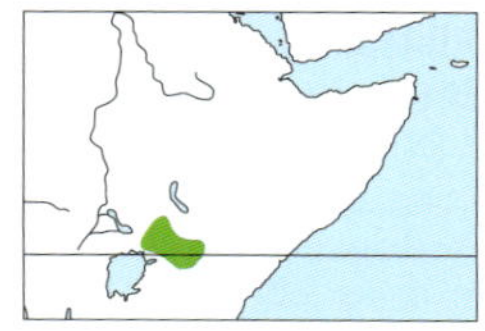

Highlands of W UG and KE

DD
Monotypic
HB 10.5–13.5 cm
T 2.4–8.1 cm
W 38–84 g

Machangu's Brush-furred Rat
Lophuromys machangui

FR Rat-hérissé de Machang'u
DE Machangu-Bürstenhaarmaus
ES Rata de pelaje de cepillo de Machangu

2000–2900 m

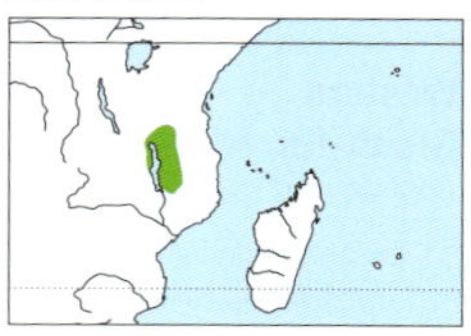

S TZ, N MW, N MZ

LC
Monotypic
HB 11.5–13.9 cm
T 5–9.4 cm
W 40–74 g

Kilonzo's Brush-furred Rat
Lophuromys kilonzoi

FR Rat-hérissé de Kilonzo
DE Kilonzo-Bürstenhaarmaus
ES Rata de pelaje de cepillo de Kilonzo

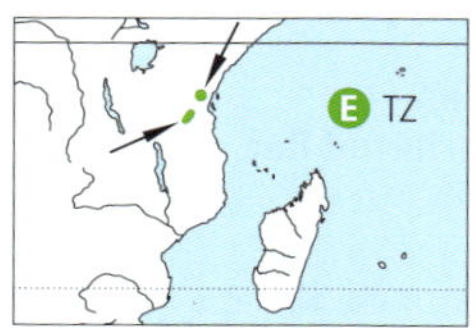

Usambara & Uluguru Mts

NE
Monotypic
HB 9.9–13.6 cm
T 5.4–7.6 cm
W 43–76 g

Albertine Rift Brush-furred Rat
Lophuromys laticeps

FR Rat-hérissé du Kivu
DE Breitkopf-Bürstenhaarmaus
ES Rata de pelaje de cepillo del Valle Albertino

2400–4000 m

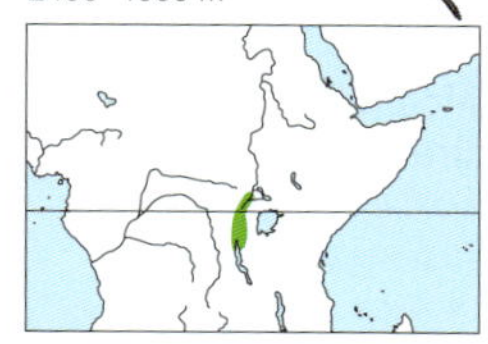

Albertine Rift in E CD, SW UG, RW and BI

DD
Monotypic
HB 11–12.9 cm
T 6.5–7.4 cm
W 45–56 g

Gray Brush-furred Rat
Lophuromys cinereus

FR Rat-hérissé du Kahuzi-Biega
DE Graue Bürstenhaarmaus
ES Rata de pelaje de cepillo gris

2275–2350 m

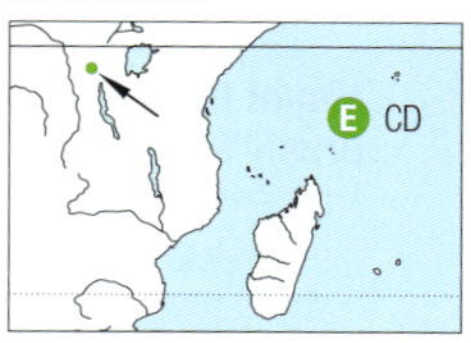

Known only from Kahuzi-Biéga NP

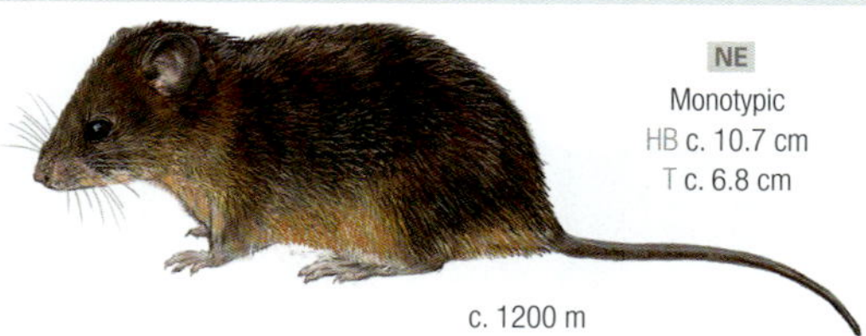

NE
Monotypic
HB c. 10.7 cm
T c. 6.8 cm

Congolese Brush-furred Rat
Lophuromys rita

FR Rat-hérissé du Congo
DE Kongo-Bürstenhaarmaus
ES Rata de pelaje de cepillo del Congo

c. 1200 m

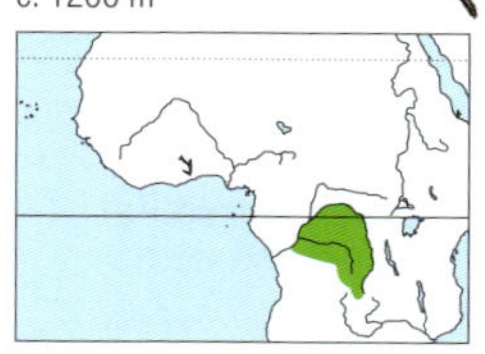

Left bank of R Congo (CD, NW ZM)

DD
Monotypic
HB 11.8–13.4 cm
T 5.9–8.5 cm
W 33–58 g

Makundi's Brush-furred Rat
Lophuromys makundii

FR Rat-hérissé de Makundi
DE Makundi-Bürstenhaarmaus
ES Rata de pelaje de cepillo de Makundi

above 2000 m

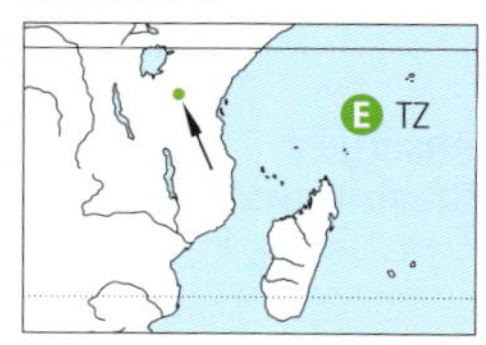

Mt Hanang

NE
Monotypic
HB 8.9–12.4 cm
T 5.1–7.2 cm

Dudu's Brush-furred Rat
Lophuromys dudui

FR Rat-hérissé de Dudu
DE Dudu-Bürstenhaarmaus
ES Rata de pelaje de cepillo de Dudu

below 1000 m

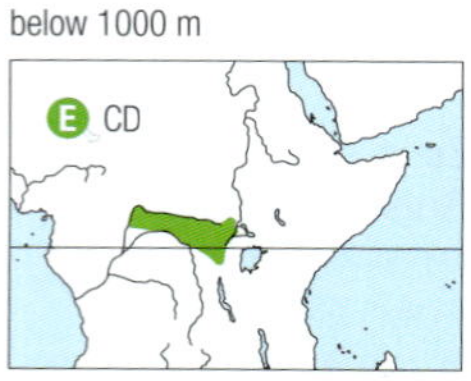

Mainly secondary lowland forest N of R Congo

LC
Monotypic
HB 9–11.3 cm
T 9–11.7 cm
W 28–41 g

Buff-bellied Brush-furred Rat
Lophuromys luteogaster

FR Rat-hérissé à ventre fauve
DE Gelbbauch-Bürstenhaarmaus
ES Rata de pelaje de cepillo de vientre leonado

700–1100 m

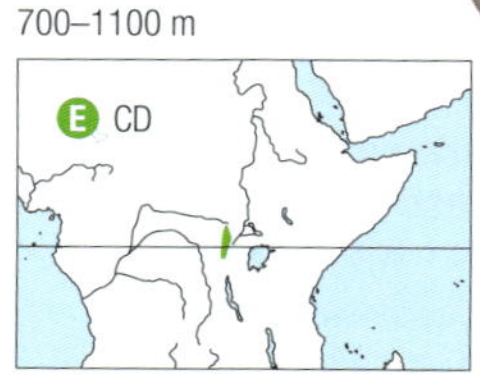

Lowland rainforest in NE; limits unknown

VU
Monotypic
HB 9.2–11.2 cm
T 7.3–9.5 cm
W 29–43 g

Western Rift Brush-furred Rat
Lophuromys medicaudatus

FR Rat-hérissé à longue queue
DE Kivu-Bürstenhaarmaus
ES Rata de pelaje de cepillo del Valle Occidental

1850–2500 m

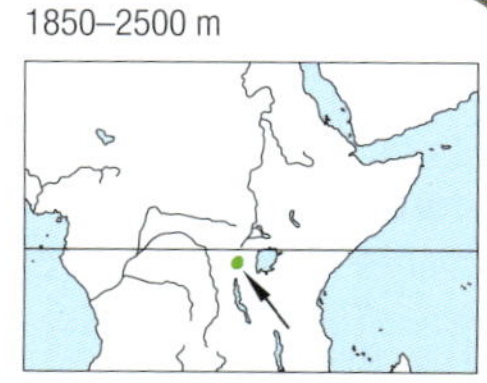

Albertine Rift around L Kivu in E CD & RW and SW UG (Bwindi)

Woosnam's Brush-furred Rat
Lophuromys woosnami

FR Rat-hérissé de Woosnam
DE Woosnam-Bürstenhaarmaus
ES Rata de pelaje de cepillo de Woosnam

LC
2 sspp
HB 10.6–13.5 cm
T 10.4–13.3 cm
W 38–64 g

1800–3880 m

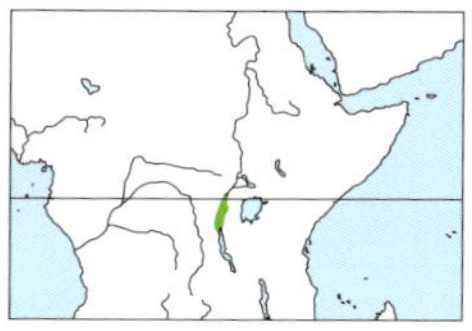

Rwenzori, Kigesi & Virunga Mts, Nyungwe & Kibira forests (E CD, SW UG, W RW, NW BI)

Genus *DEOMYS*

(1 species)

Congo Forest Rat
Deomys ferrugineus

FR Déomys roux
DE Kongo-Waldmaus
ES Rata de bosque de Congo

LC
2 sspp
HB 10.9–15 cm
T 15.9–20 cm
W 40–77 g

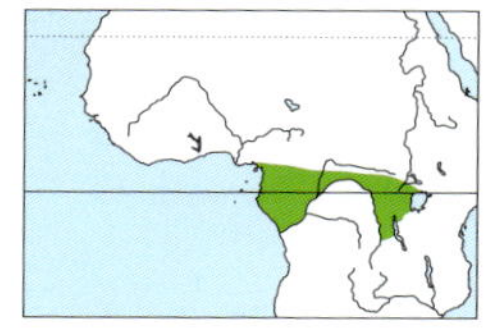

N Congo Basin from CM, GQ (with Bioko I) & CG E through CD to UG & TZ

Genus *URANOMYS*

(1 species)

Rudd's Bristle-furred Rat
Uranomys ruddi

FR Uranomys de Rudd
DE Weißbauch-Bürstenhaarmaus
ES Rata de pelaje erizado de Rudd

LC
Monotypic
HB 9.5–11.9 cm
T 6–7.3 cm W 26–43 g

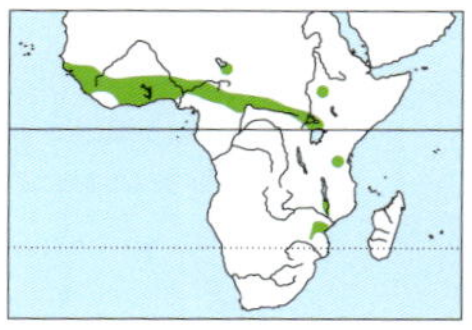

Disjunct in W, C & E Africa, from S SN E to SW ET, S to extreme E ZW & C MZ

Genus *TATERA*

(1 species)

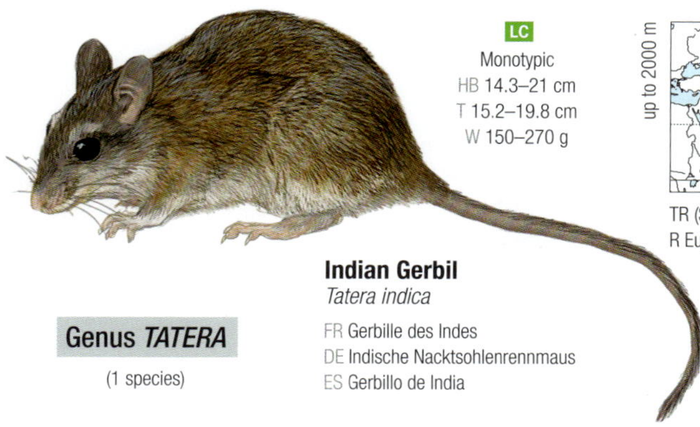

Indian Gerbil
Tatera indica

FR Gerbille des Indes
DE Indische Nacktsohlenrennmaus
ES Gerbillo de India

LC
Monotypic
HB 14.3–21 cm
T 15.2–19.8 cm
W 150–270 g

up to 2000 m

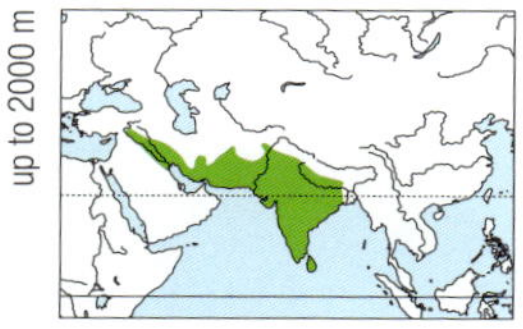

TR (SE Anatolia) & SY (E of R Euphrates) E to IN, NP & SL

Genus *GERBILLISCUS*

(18 species)

Fringe-tailed Gerbil
Gerbilliscus robustus

FR Gerbille robuste
DE Fransenschwanz-Nacktsohlenrennmaus
ES Gerbillo de cola de cepillo

LC
Monotypic
HB 12–19 cm
T 12.5–21 cm
W 80–152 g

up to 1600 m

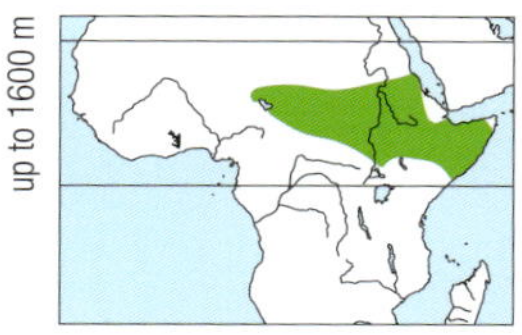

SE NE and TD E to SD, ER, ET and SO; possibly NE NG, N CM and N KE

Black-tailed Gerbil
Gerbilliscus nigricaudus

FR Gerbille à queue noire
DE Schwarzschwanz-Nacktsohlenrennmaus
ES Gerbillo de cola negra

LC
2 sspp
HB 13–19.3 cm
T 17–21 cm
W 80–195 g

up to 1500 m

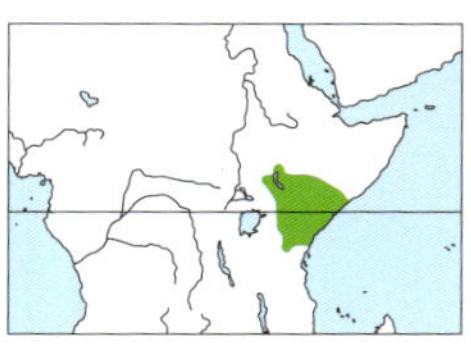

Somali-Masai savannas in SW ET, SW SO, KE and NE TZ

Phillips's Gerbil
Gerbilliscus phillipsi

FR Gerbille de Phillips
DE Phillips-Nacktsohlenrennmaus
ES Gerbillo de Phillips

LC
Monotypic
HB 11.6–14.5 cm
T 16.2–18.5 cm

Disjunct in SO, SW ET & WC KE

East African Gerbil
Gerbilliscus vicinus

FR Gerbille du Kitui
DE Ostafrika-Nacktsohlenrennmaus
ES Gerbillo de África oriental

NE
Monotypic
HB 10.9–15.7 cm
T 13.9–19.5 cm
W 54–104 g

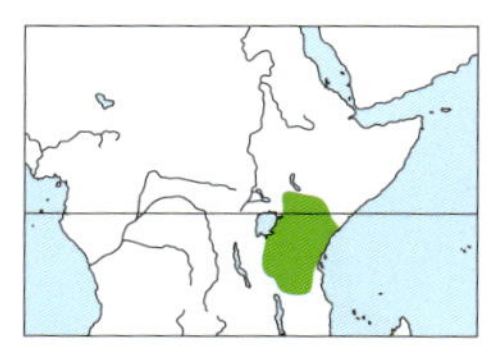

KE, TZ; N limits unknown

Paeba Hairy-footed Gerbil
Gerbilliscus paeba

FR Gerbille à pieds velus
DE Kalahari-Haarsohlenrennmaus
ES Gerbillo de pies peludos de Paeba

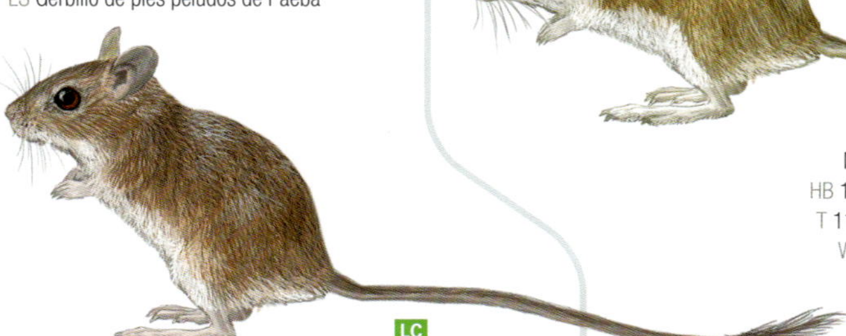

LC
5 sspp
HB 8.5–10.5 cm
T 10.2–12.5 cm
W 20–37 g

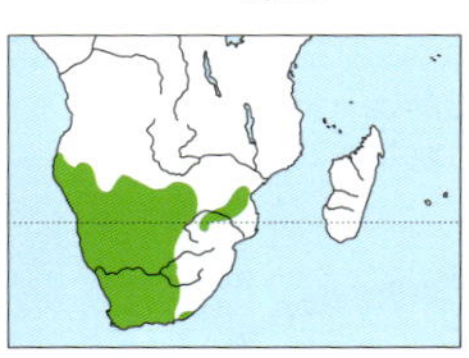

S Africa, from S AO E to MZ & ZW, S to ZA

LC
Monotypic
HB 10.4–11.8 cm
T 11.3–14.5 cm
W 30–43 g

Setzer's Hairy-footed Gerbil
Gerbilliscus setzeri

FR Gerbille de Setzer
DE Setzer-Haarsohlenrennmaus
ES Gerbillo de pies peludos de Setzer

Namib Desert in extreme SW AO & NW NA

Brush-tailed Hairy-footed Gerbil
Gerbilliscus vallinus

FR Gerbille des vallées
DE Pinselschwanz-Haarsohlenrennmaus
ES Gerbillo de pies peludos de cola de cepillo

S NA, ZA (Northern Cape)

LC
2 sspp
HB 9.6–11 cm
T 11.9–15.6 cm
W 30–43 g

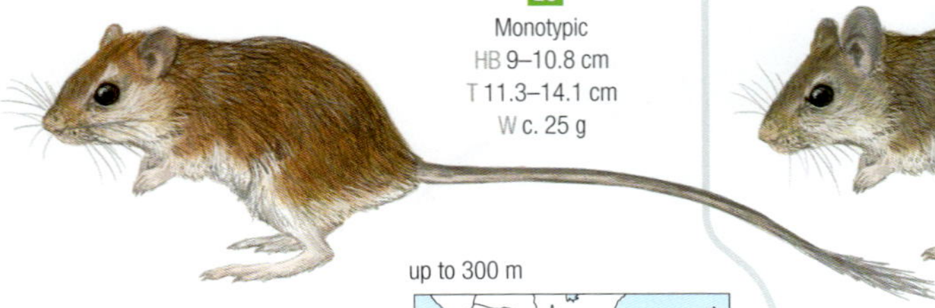

LC
Monotypic
HB 9–10.8 cm
T 11.3–14.1 cm
W c. 25 g

up to 300 m

From Swakopmund to Luderitz

Dune Hairy-footed Gerbil
Gerbilliscus tytonis
FR Gerbille des dunes
DE Dünen-Haarsohlenrennmaus
ES Gerbillo de pies peludos de duna

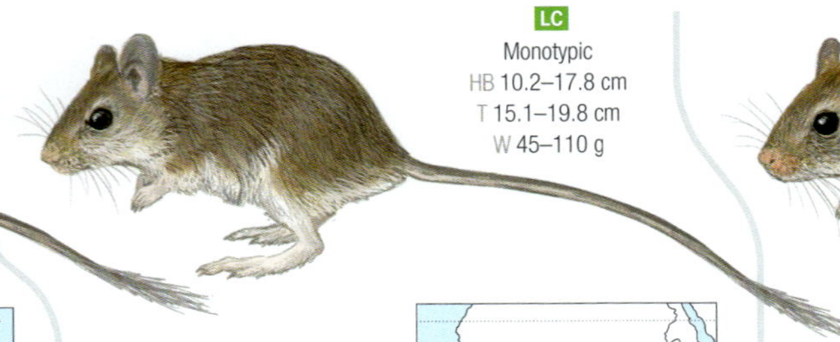

LC
Monotypic
HB 10.2–17.8 cm
T 15.1–19.8 cm
W 45–110 g

Guinean Gerbil
Gerbilliscus guineae
FR Gerbille de Guinée
DE Guinea-Nacktsohlenrennmaus
ES Gerbillo de Guinea

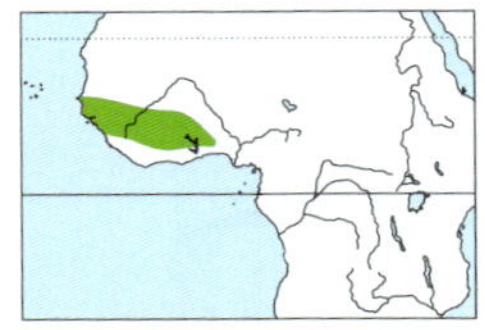

W Africa, from SN E to C BJ

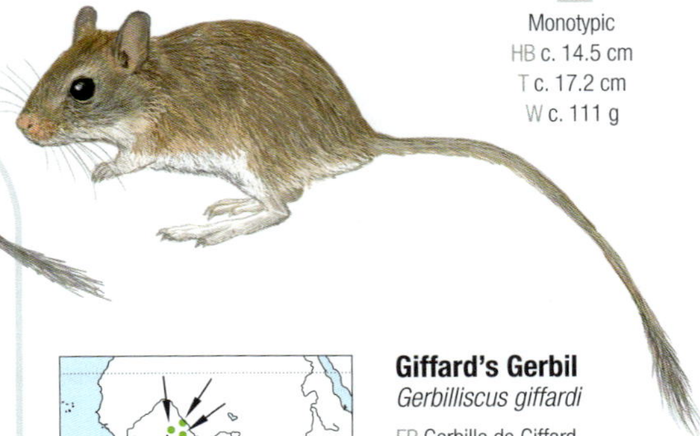

NE
Monotypic
HB c. 14.5 cm
T c. 17.2 cm
W c. 111 g

Giffard's Gerbil
Gerbilliscus giffardi
FR Gerbille de Giffard
DE Giffard-Nacktsohlenrennmaus
ES Gerbillo de Giffard

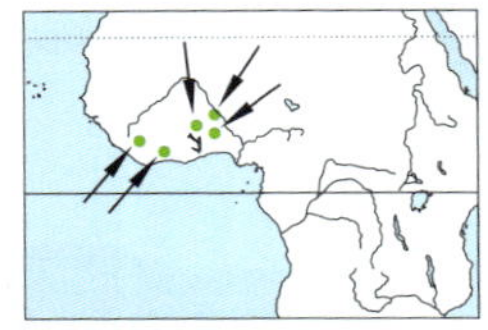

Scattered in E GN, CI, GH, SW NE & BJ; probably widespread

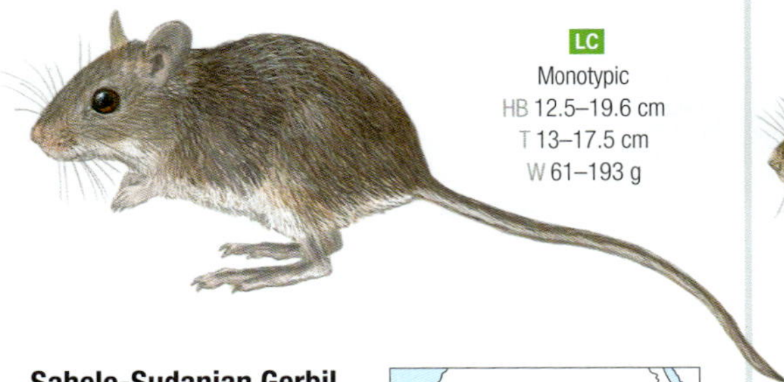

LC
Monotypic
HB 12.5–19.6 cm
T 13–17.5 cm
W 61–193 g

Sahelo-Sudanian Gerbil
Gerbilliscus gambianus
FR Gerbille de Gambie
DE Gambia-Nacktsohlenrennmaus
ES Gerbillo de Sahel de Sudán

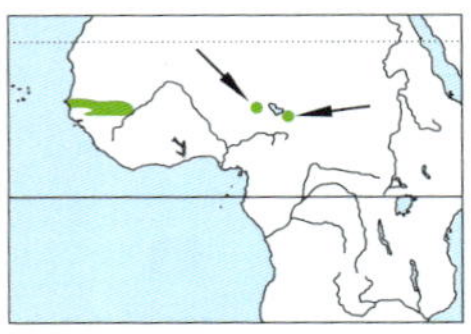

Sahelo-Sudanian savannas in W & S SN, SW ML, SE NE & extreme WC TD

Northern Savanna Gerbil
Gerbilliscus kempii
FR Gerbille de Kemp
DE Kemp-Nacktsohlenrennmaus
ES Gerbillo de sabana septentrional

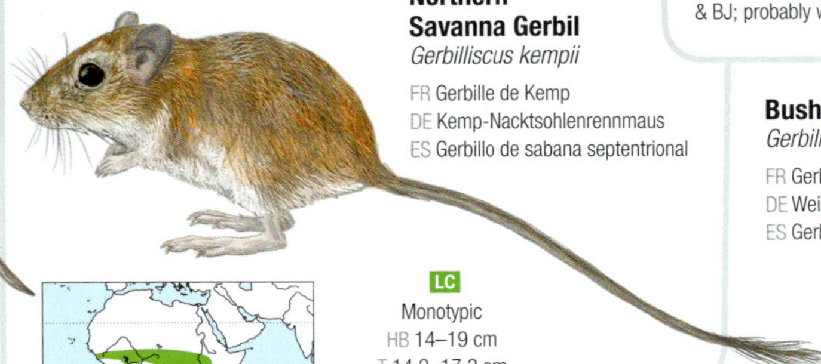

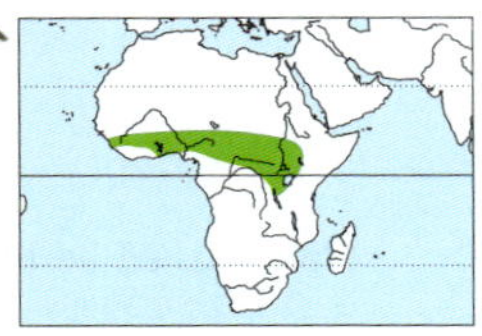

Sahelo-Sudanian & Guinean savannas from GN & SL E to SW ET, W KE & NW TZ

LC
Monotypic
HB 14–19 cm
T 14.2–17.3 cm
W 97–105 g

Bushveld Gerbil
Gerbilliscus leucogaster
FR Gerbille à ventre blanc
DE Weißbauch-Nacktsohlenrennmaus
ES Gerbillo de sabana arbolada

LC
Monotypic
HB 8.9–15.5 cm
T 12–17.5 cm
W 32–114 g

up to 1600 m

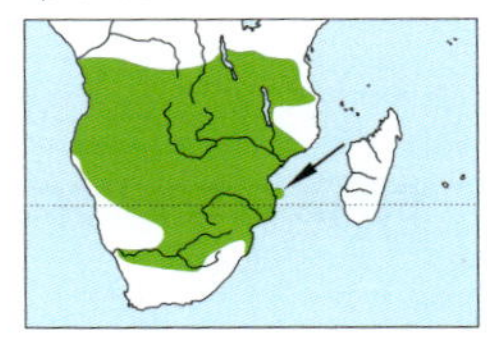

AO E through S CD to TZ, S to N ZA, SZ & MZ (with Bazaruto I)

Southern Savanna Gerbil
Gerbilliscus validus
FR Gerbille de savane
DE Savannen-Nacktsohlenrennmaus
ES Gerbillo de sabana meridional

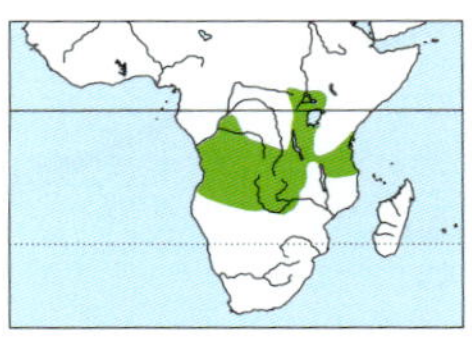

AO, CD, UG, RW, BI, SE KE, TZ, ZM & N ZW; possibly SD, SS & ET

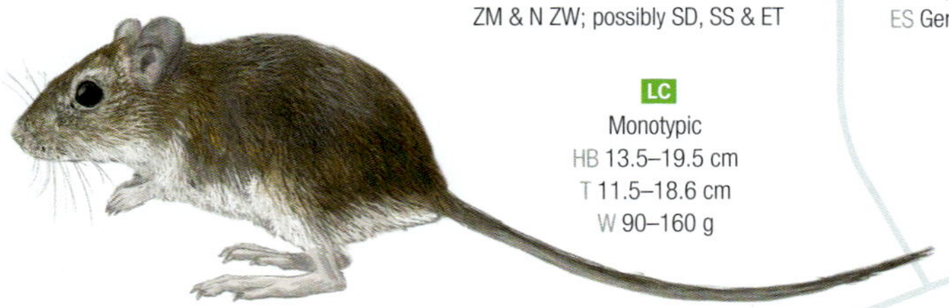

LC
Monotypic
HB 13.5–19.5 cm
T 11.5–18.6 cm
W 90–160 g

Cape Gerbil
Gerbilliscus afer
FR Gerbille du Cap
DE Kap-Nacktsohlenrennmaus
ES Gerbillo de El Cabo

Western Cape Province

LC
Monotypic
HB 12.4–15.7 cm
T 13.3–17.5 cm W 78–113 g

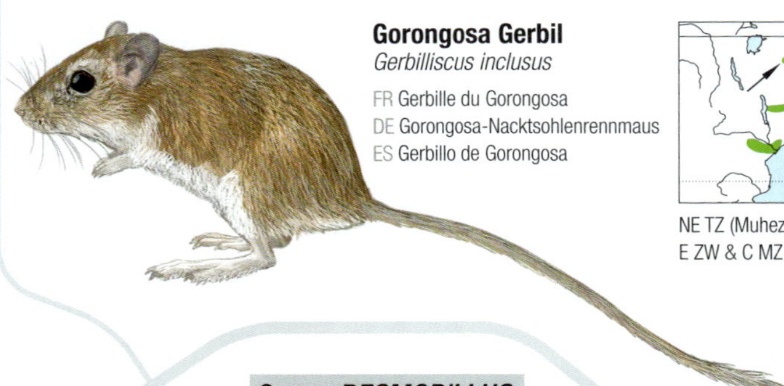

Gorongosa Gerbil
Gerbilliscus inclusus
FR Gerbille du Gorongosa
DE Gorongosa-Nacktsohlenrennmaus
ES Gerbillo de Gorongosa

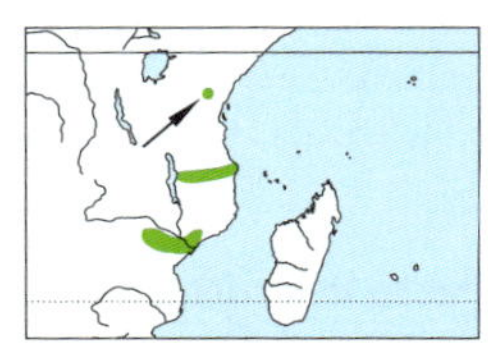

NE TZ (Muheza), S TZ & N MZ, E ZW & C MZ

LC
3 sspp
HB 15.2–16.1 cm
T 13.5–19.1 cm
W 99–154 g

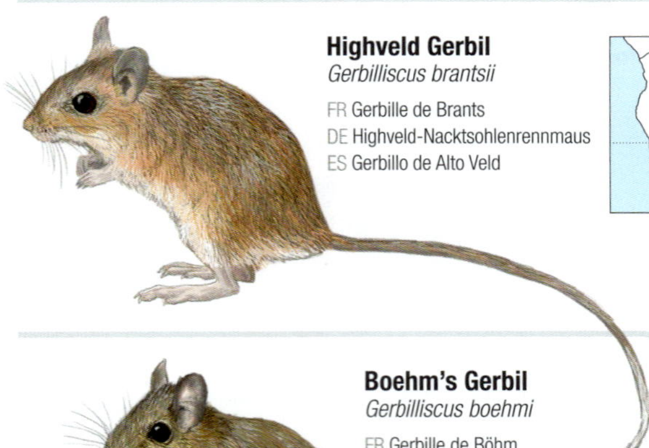

Highveld Gerbil
Gerbilliscus brantsii
FR Gerbille de Brants
DE Highveld-Nacktsohlenrennmaus
ES Gerbillo de Alto Veld

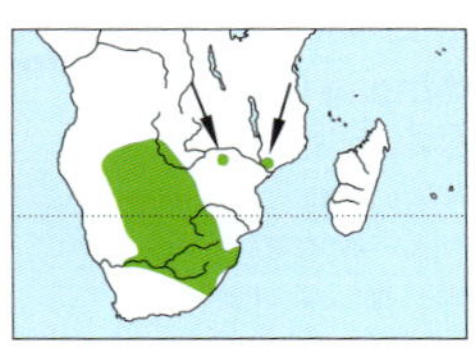

SE AO, SW ZM, E NA, BW, W ZW, ZA & LS; recorded in N ZW & N MZ

LC
2 sspp
HB 9.6–16.4 cm
T 10.3–18.6 cm
W 25–126 g

Boehm's Gerbil
Gerbilliscus boehmi
FR Gerbille de Böhm
DE Böhm-Nacktsohlenrennmaus
ES Gerbillo de Böhm

LC
Monotypic
HB 13.9–17.9 cm
T 19–23 cm
W c. 146 g

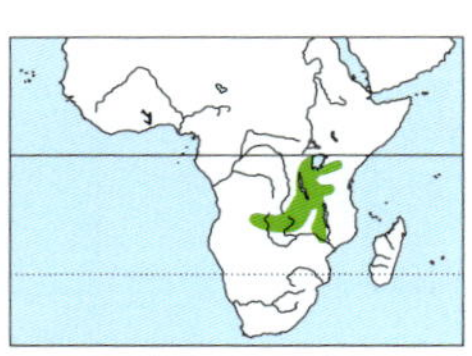

From S UG and SW KE S to E AO, N & E ZM and NW MZ

Genus *DESMODILLUS*

(1 species)

Cape Short-tailed Gerbil
Desmodillus auricularis
FR Gerbille à queue courte
DE Kurzschwanzrennmaus
ES Gerbillo de cola corta de El Cabo

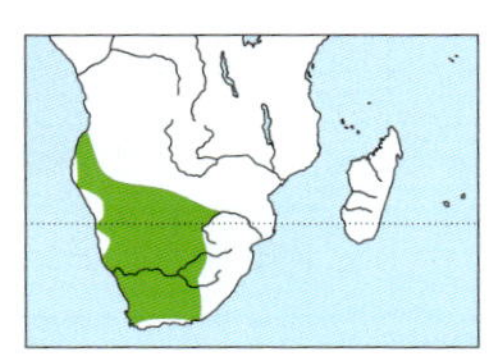

S Africa in SW AO, NA, BW & ZA; possibly S ZW

LC
Monotypic
HB 8.6–12.9 cm
T 7–9.8 cm
W 29–82 g

Genus *AMMODILLUS*

(1 species)

Ammodile
Ammodillus imbellis

FR Ammodile
DE Ammodill-Rennmaus
ES Ammodillo

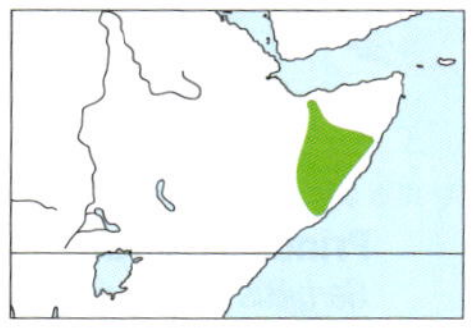

Few records in SO & E ET

DD
Monotypic
HB 8.4–10.5 cm
T 12.9–15 cm
W c. 31 g

Genus *TATERILLUS*

(8 species)

Congo Tateril
Taterillus congicus

FR Gerbille du Congo
DE Kongo-Taterillus
ES Gerbillo de Congo

LC
Monotypic
HB 9.5–14.2 cm
T 13.1–18.8 cm W 32–77 g

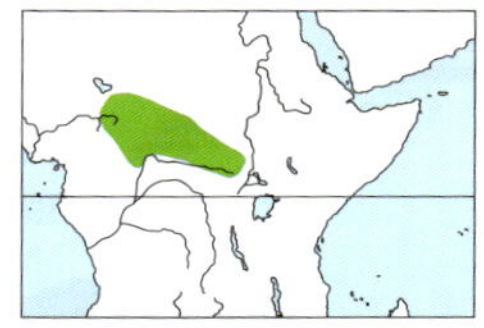

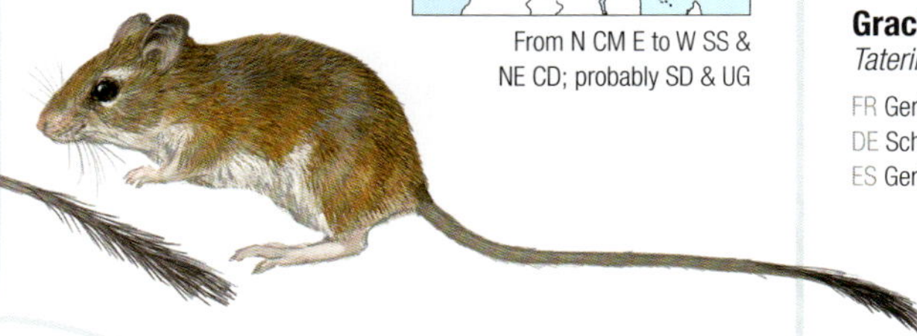

From N CM E to W SS & NE CD; probably SD & UG

Gracile Tateril
Taterillus gracilis

FR Gerbille gracile
DE Schlanker Taterillus
ES Gerbillo esbelto

LC
Monotypic
HB 9.6–13 cm
T 13.5–17.6 cm
W 45–81 g

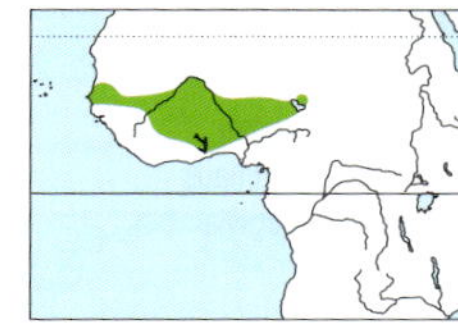

W Africa from SN E to NG & WC TD

Robbins's Tateril
Taterillus arenarius

FR Gerbille de Robbins
DE Mauretanien-Taterillus
ES Gerbillo de Robbins

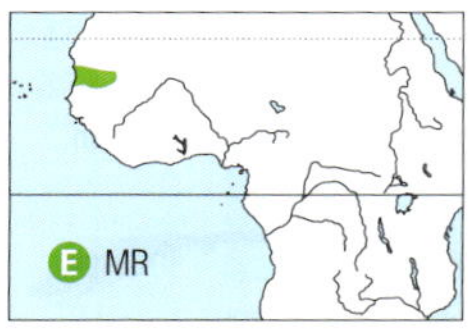

Arid zones of Sahelian region; possibly ML & NE

LC
Monotypic
HB 11.3–14.2 cm
T 14.1–15.5 cm
W 28–66 g

Senegal Tateril
Taterillus pygargus

FR Gerbille de Cuvier
DE Senegal-Taterillus
ES Gerbillo de Senegal

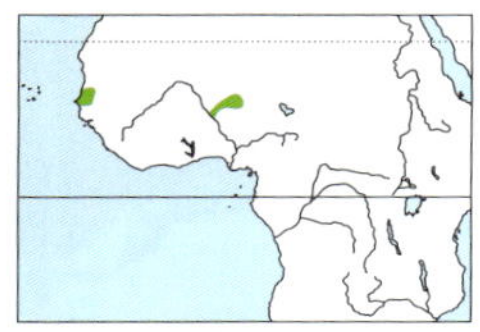

Disjunct in SN & SW NE; perhaps extreme S MR

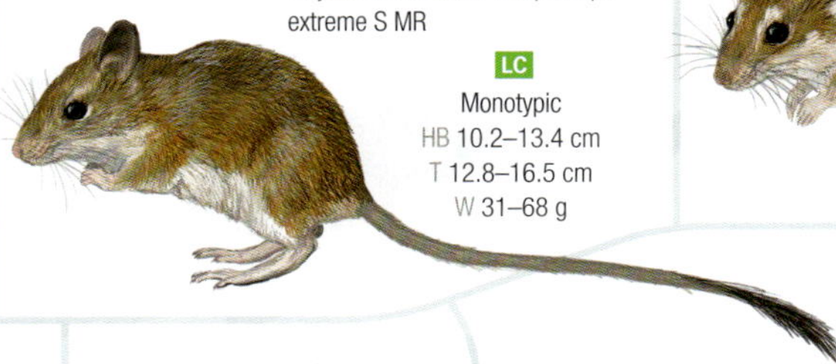

LC
Monotypic
HB 10.2–13.4 cm
T 12.8–16.5 cm
W 31–68 g

Petter's Tateril
Taterillus petteri

FR Gerbille de Petter
DE Petter-Taterillus
ES Gerbillo de Petter

LC
Monotypic
HB 10.1–13.5 cm
T 12.5–16.6 cm
W 30–64 g

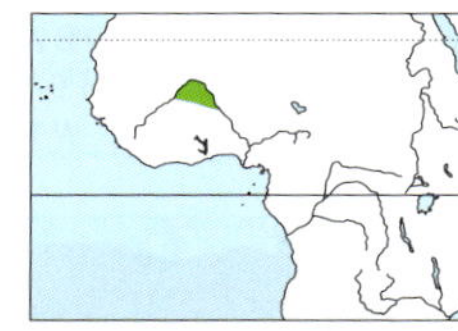

Sahelian savannas in SC ML, N BF & SW NE

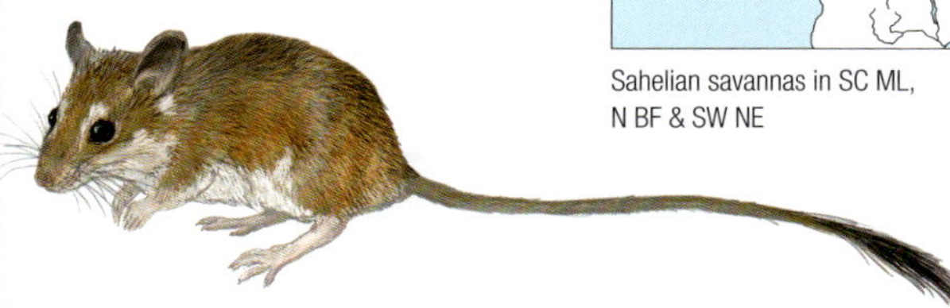

Lake Chad Tateril
Taterillus lacustris

FR Gerbille du Tchad
DE Tschadsee-Taterillus
ES Gerbillo del Chad

LC
Monotypic
HB 10.5–13.7 cm
T 13.4–16.2 cm
W 31–52 g

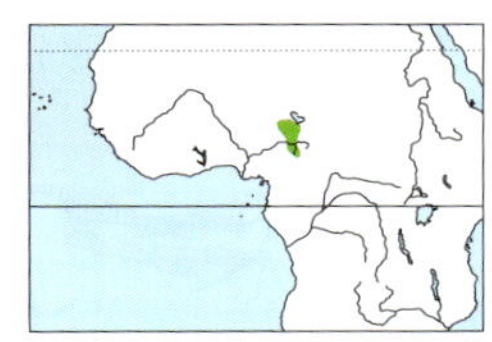

NE NG & N CM, S of L Chad

Emin's Tateril
Taterillus emini

FR Gerbille d'Emin
DE Emin-Taterillus
ES Gerbillo de Emin

LC
Monotypic
HB 10.5–11 cm
T 15–16 cm

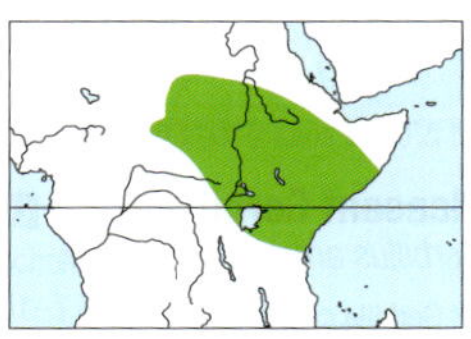

E & C Africa from SD, ET and C & S SO, S to NE CD and N TZ

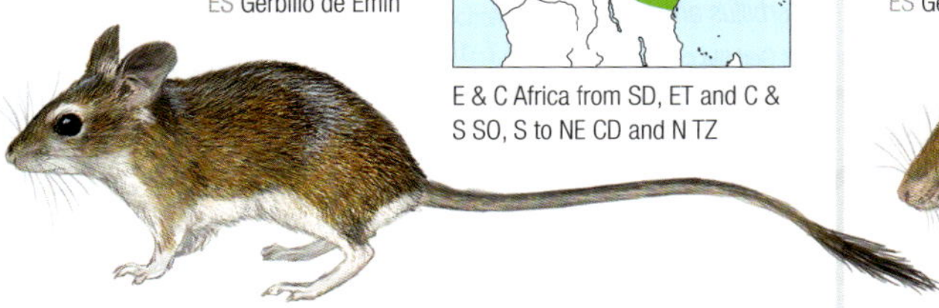

Tranier's Tateril
Taterillus tranieri

FR Gerbille de Tranier
DE Tranier-Taterillus
ES Gerbillo de Tranier

LC
Monotypic
HB 11.1–13.4 cm
T 15–18 cm
W 40–54 g

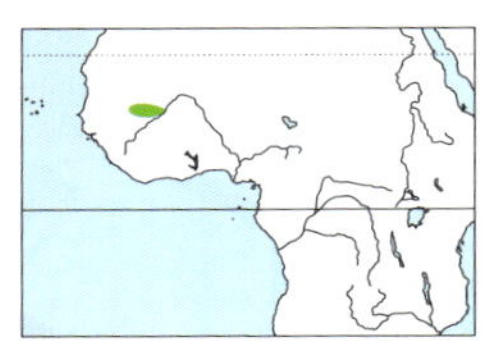

Cultivated and degraded savannas in S MR & W ML

Genus *GERBILLUS*

(48 species)

Large Aden Gerbil
Gerbillus poecilops

FR Gerbille d'Aden
DE Große Aden-Rennmaus
ES Gerbillo de Adén

LC
Monotypic
HB 15–22 cm
T 7.2–11.5 cm

Coastal mountain ranges in SW SA & W YE

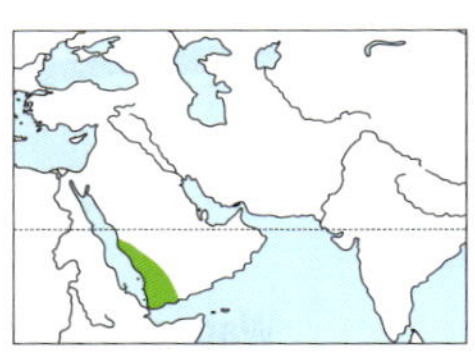

Pygmy Gerbil
Gerbillus henleyi

FR Gerbille de Henley
DE Henley-Rennmaus
ES Gerbillo pigmeo

LC
2 sspp
HB 5.5–7.3 cm
T 7.3–10.7 cm
W 8–14 g

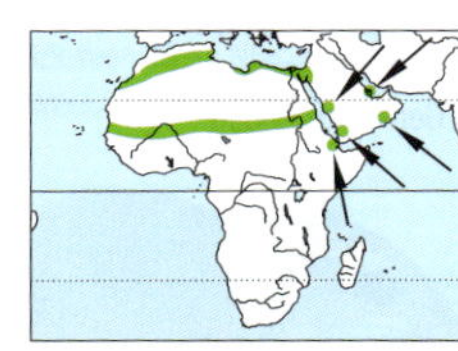

N Africa from EH & MA E to EG, S of Sahara from S MR & N SN E to N SD, Middle East, patchily in Arabian Pen & DJ

Sand Gerbil
Gerbillus syrticus

FR Gerbille de Syrte
DE Sandrennmaus
ES Gerbillo de arena

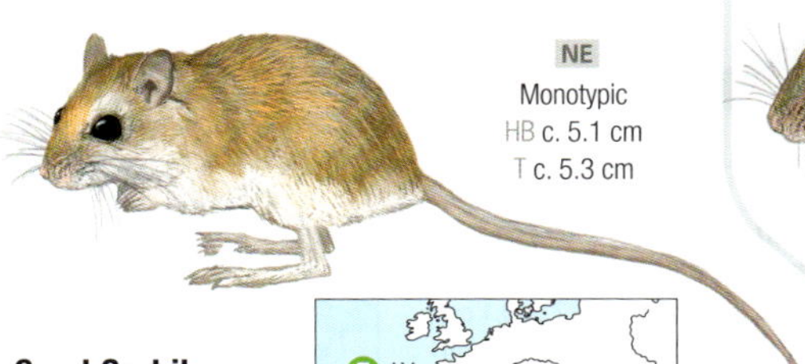

NE
Monotypic
HB c. 5.1 cm
T c. 5.3 cm

Known only from type locality near Nofilia

Balochistan Gerbil
Gerbillus nanus

FR Gerbille du Baloutchistan
DE Belutschistan-Rennmaus
ES Gerbillo de Baluchistán

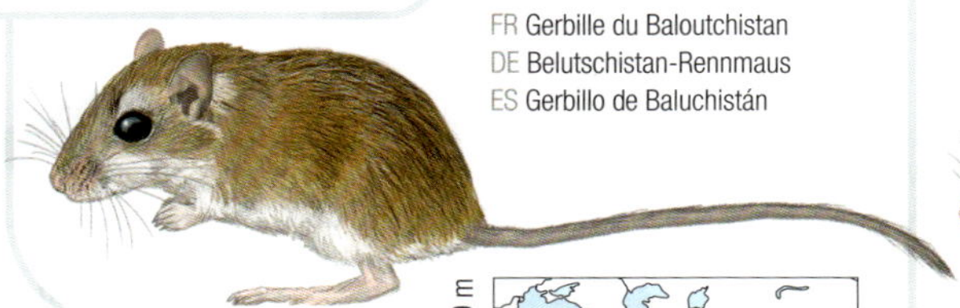

LC
4 sspp
HB 7.2–11 cm
T 8–14.5 cm
W 14–29 g

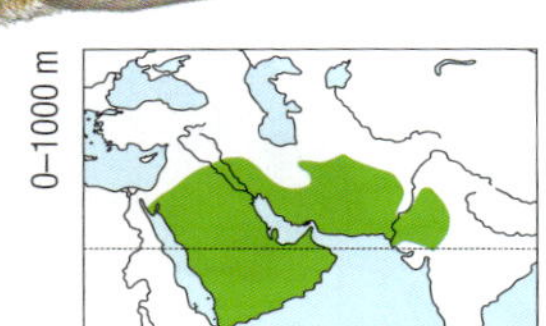

EG (Sinai), IL, JO, Arabian Pen, IQ E to NW IN

Least Gerbil
Gerbillus pusillus

FR Gerbille de Peters
DE Winzige Rennmaus
ES Gerbillo mínimo

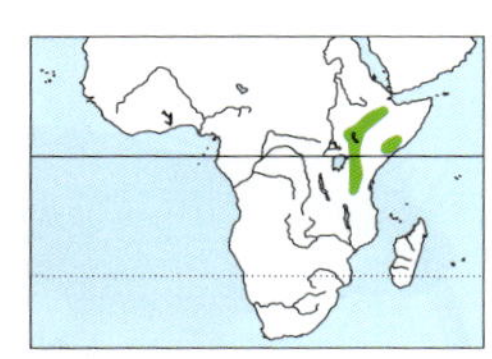

ET, SE SS, W KE, NC TZ, SW SO

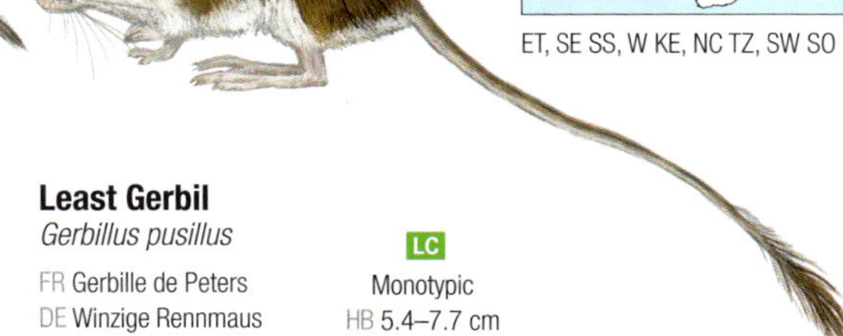

LC
Monotypic
HB 5.4–7.7 cm
T 8.2–11.5 cm

Lesser Egyptian Gerbil
Gerbillus gerbillus

FR Petite Gerbille
DE Kleine Ägypten-Rennmaus
ES Gerbillo de Egipto menor

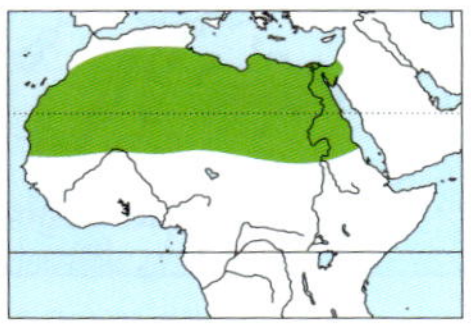

N Africa from MA & MR E to EG, N SD, SW IL, PS & SW JO; possibly N ER

LC
Monotypic
HB 7–10.5 cm
T 9–13.6 cm
W 14–35 g

Anderson's Gerbil
Gerbillus andersoni

FR Gerbille d'Anderson
DE Anderson-Rennmaus
ES Gerbillo de Anderson

LC
2 sspp
HB 7–11.5 cm
T 9–13.5 cm
W 16–38 g

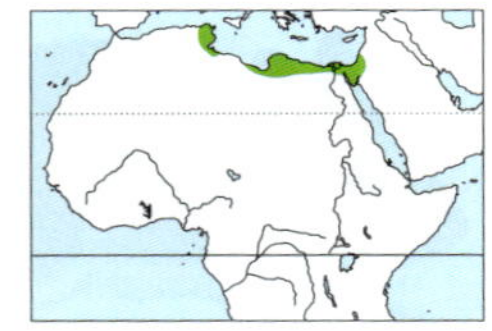

Coastal N Africa from TN E to EG, IL, PS & W JO

Lataste's Gerbil
Gerbillus latastei

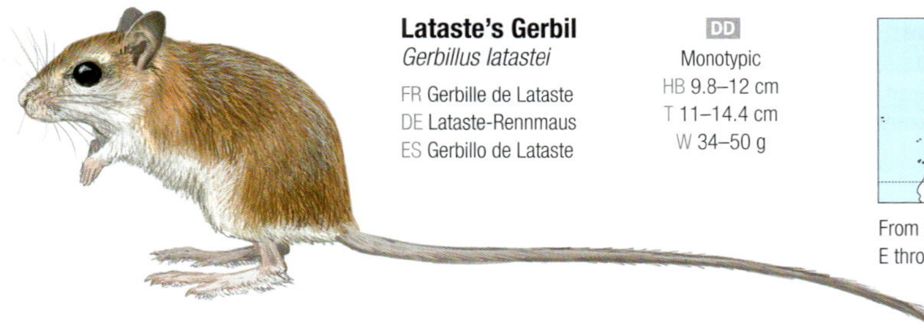

FR Gerbille de Lataste
DE Lataste-Rennmaus
ES Gerbillo de Lataste

DD
Monotypic
HB 9.8–12 cm
T 11–14.4 cm
W 34–50 g

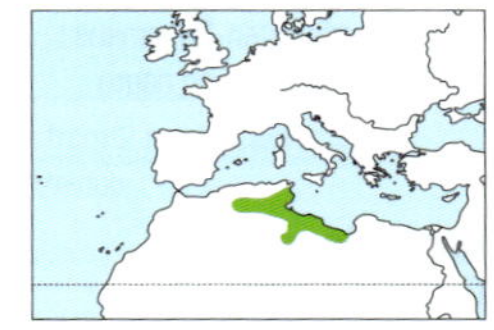

From Hadjeb El Djmel (NC DZ) E through TN to N LY

Nigerian Gerbil
Gerbillus nigeriae

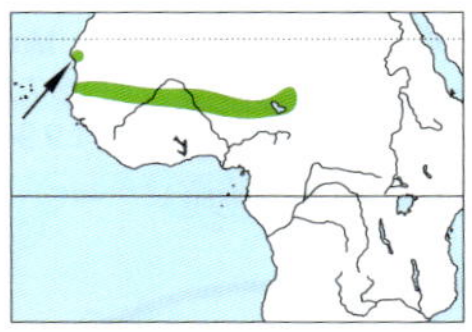

W MR, N SN, ML, NE, N BF, N NG, W TD

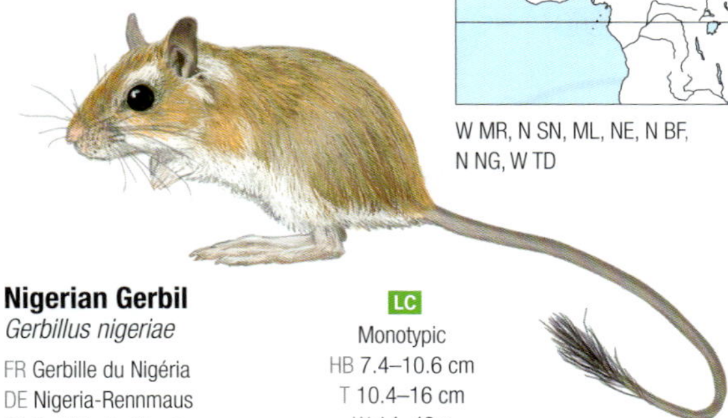

FR Gerbille du Nigéria
DE Nigeria-Rennmaus
ES Gerbillo de Nigeria

LC
Monotypic
HB 7.4–10.6 cm
T 10.4–16 cm
W 14–43 g

Moroccan Gerbil
Gerbillus hesperinus

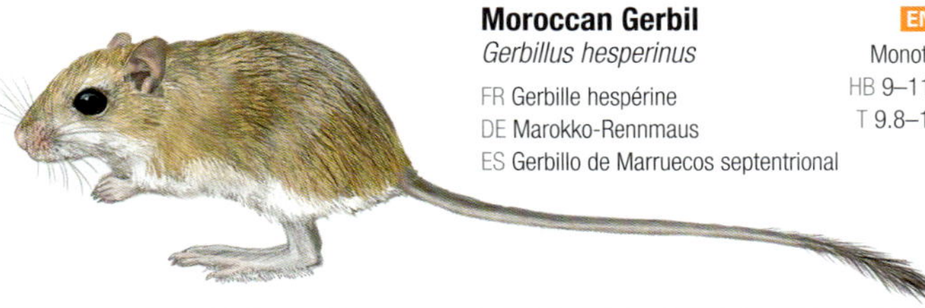

FR Gerbille hespérine
DE Marokko-Rennmaus
ES Gerbillo de Marruecos septentrional

EN
Monotypic
HB 9–11.4 cm
T 9.8–12 cm

Known from type locality (Essaouira)

Hoogstraal's Gerbil
Gerbillus hoogstraali

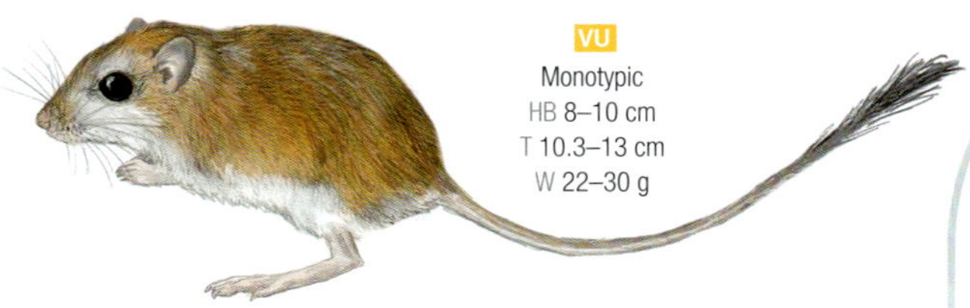

VU
Monotypic
HB 8–10 cm
T 10.3–13 cm
W 22–30 g

FR Gerbille de Hoogstraal
DE Hoogstraal-Rennmaus
ES Gerbillo de Hoogstraal

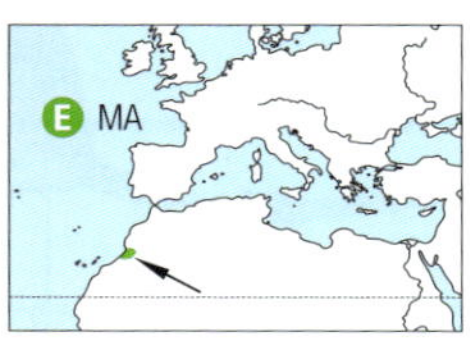

Sandy soils in Souss Valley

Flower's Gerbil
Gerbillus floweri

FR Gerbille de Flower
DE Flower-Rennmaus
ES Gerbillo de Flower

LC
Monotypic
HB 9.5–13 cm
T 13–15.8 cm
W 26–63 g

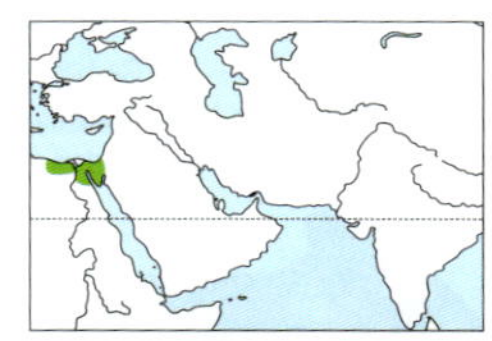

Both sides of Nile Delta & Sinai (N EG), Negev & coastal plains (S IL, PS)

Burton's Gerbil
Gerbillus burtoni

FR Gerbille de Burton
DE Burton-Rennmaus
ES Gerbillo de Burton

DD
Monotypic

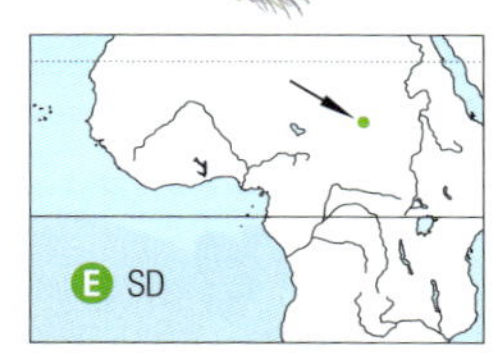

Dharfur area

Greater Egyptian Gerbil
Gerbillus pyramidum

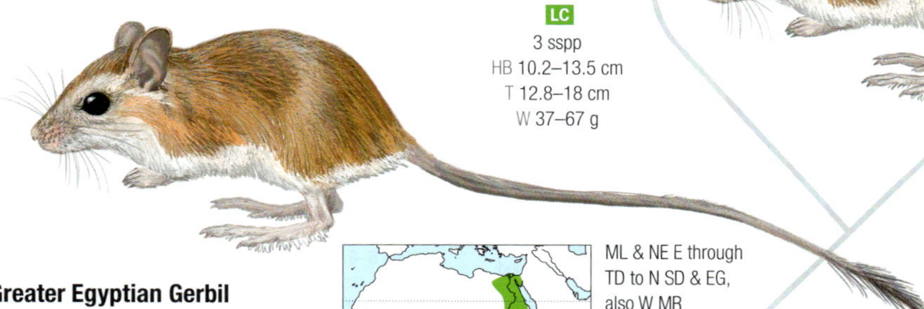

LC
3 sspp
HB 10.2–13.5 cm
T 12.8–18 cm
W 37–67 g

FR Grande Gerbille
DE Große Ägypten-Rennmaus
ES Gerbillo de Egipto grande

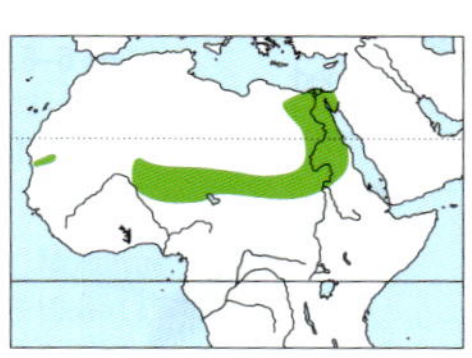

ML & NE E through TD to N SD & EG, also W MR

Berbera Gerbil
Gerbillus acticola

FR Gerbille de Berbera
DE Berbera-Rennmaus
ES Gerbillo de Berbera

DD
Monotypic
HB 9–10.4 cm
T 12.4–14.2 cm

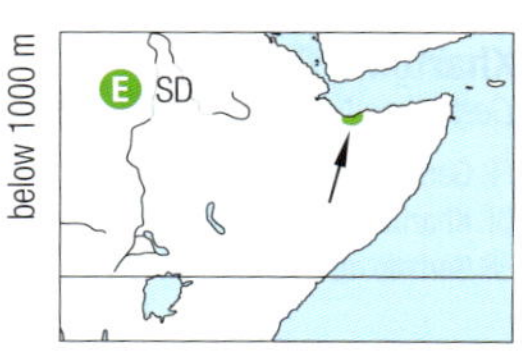

Arid NW coastal plains

Occidental Gerbil
Gerbillus occiduus

FR Gerbille occidentale
DE Westliche Rennmaus
ES Gerbillo de Marruecos meridional

Coast from 80 km SW Goulimine S of Anti-Atlas Mts to Tarfaya

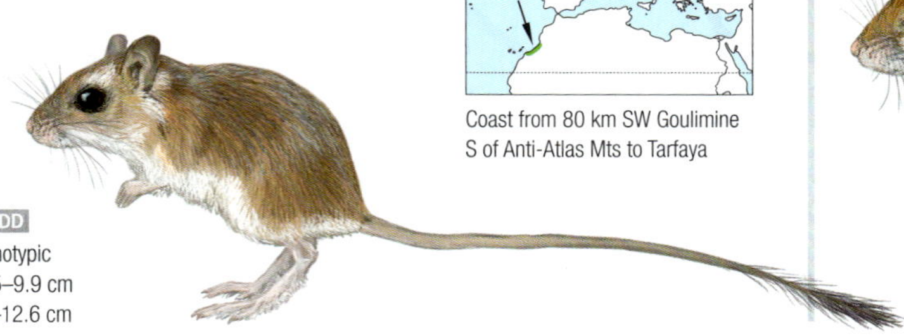

DD
Monotypic
HB 7.5–9.9 cm
T 9.8–12.6 cm

Tarabul Gerbil
Gerbillus tarabuli

FR Gerbille de Libye
DE Tarabul-Rennmaus
ES Gerbillo de Tarabul

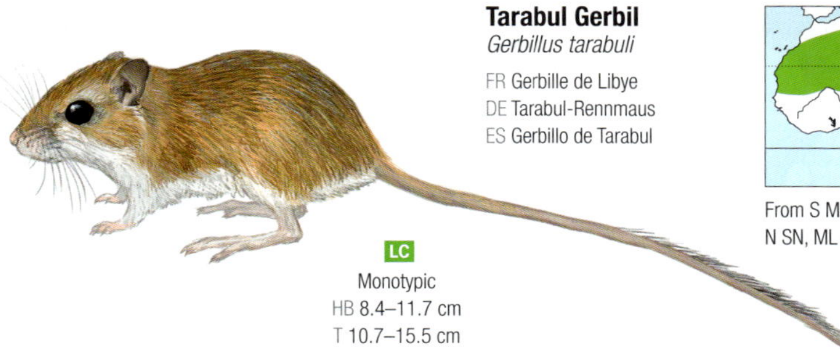

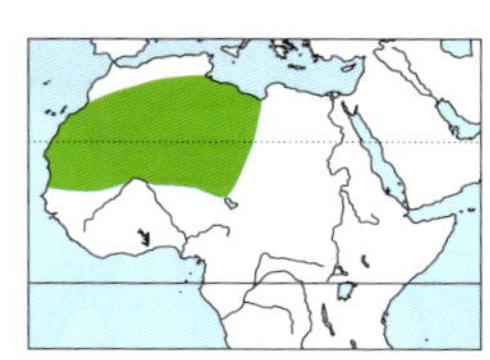

From S MA E to S TN & C LY, S to N SN, ML (N of R Niger) & NW TD

LC
Monotypic
HB 8.4–11.7 cm
T 10.7–15.5 cm
W 19–46 g

Agag Gerbil
Gerbillus agag
FR Gerbille d'Agageh
DE Agag-Rennmaus
ES Gerbillo de Agag

DD
3 sspp
HB 9–13.5 cm
T 10–14.3 cm
W 30–34 g

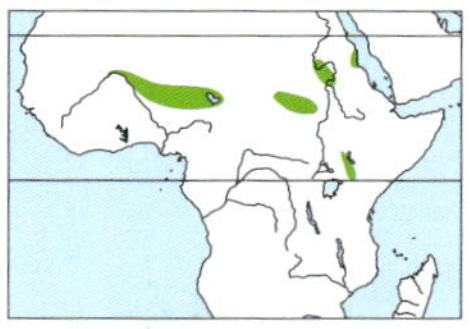

ML, S NE, N NG, extreme N CM, WC TD, SD, extreme SE SS, W KE

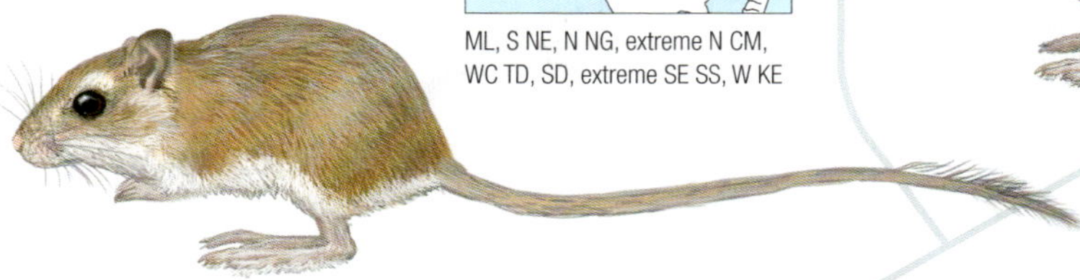

Rhoads's Gerbil
Gerbillus pulvinatus
FR Gerbille à coussinets
DE Dschibuti-Rennmaus
ES Gerbillo de Rhoads

LC
Monotypic
HB 8.6–10.5 cm
T 11.8–14.5 cm

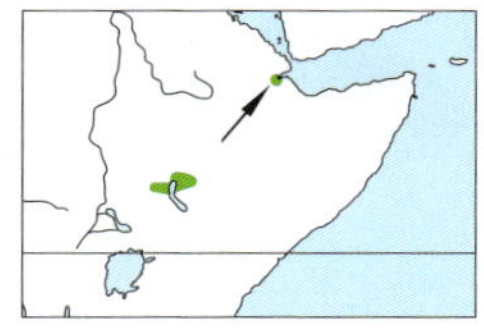

DJ, SW ET, NW KE; perhaps SE SS

LC
Monotypic
HB 8.7–10.6 cm
T 10.5–13 cm
W 20–35 g

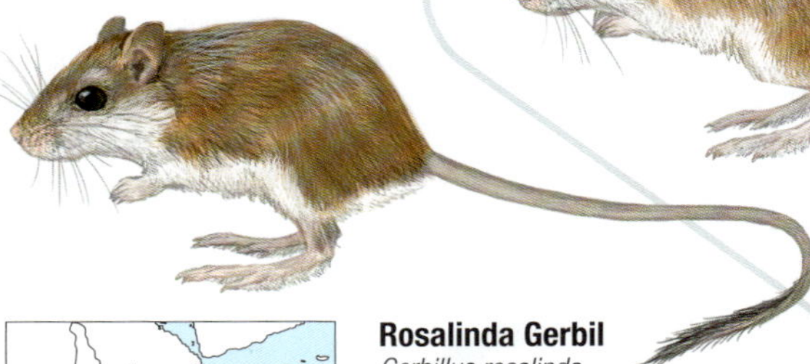

Indian Hairy-footed Gerbil
Gerbillus gleadowi
FR Gerbille de Gleadow
DE Indien-Haarfuß-Rennmaus
ES Gerbillo de pies peludos de India

LC
Monotypic
HB 7.5–10.2 cm
T 11.8–15 cm
W 20–28 g

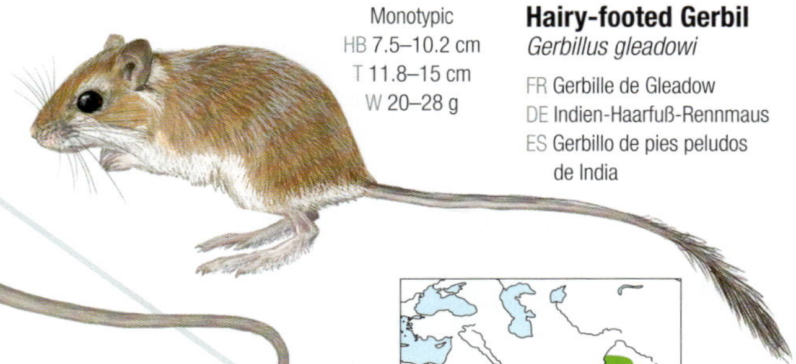

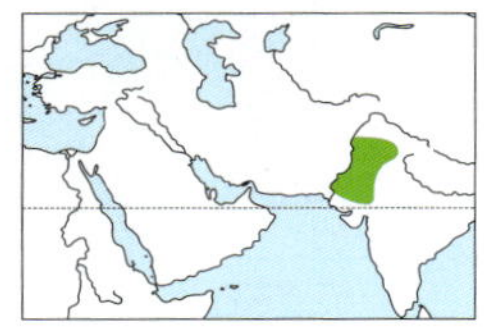

Indus Valley of PK (Punjab, Sind), NW IN (Rajasthan, Gujarat)

LC
Monotypic
HB 9.3–10.2 cm
T 13–14.3 cm
W 34–41 g

Dunn's Gerbil
Gerbillus dunni
FR Gerbille de Dunn
DE Dunn-Rennmaus
ES Gerbillo de Dunn

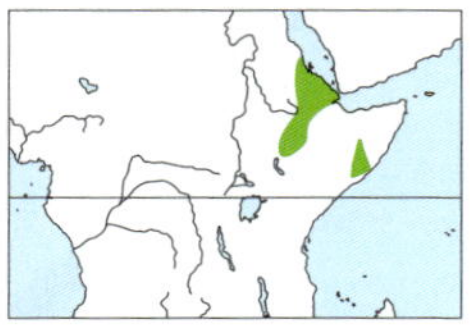

Steppe and dry savannas in ER, ET, DJ & SO

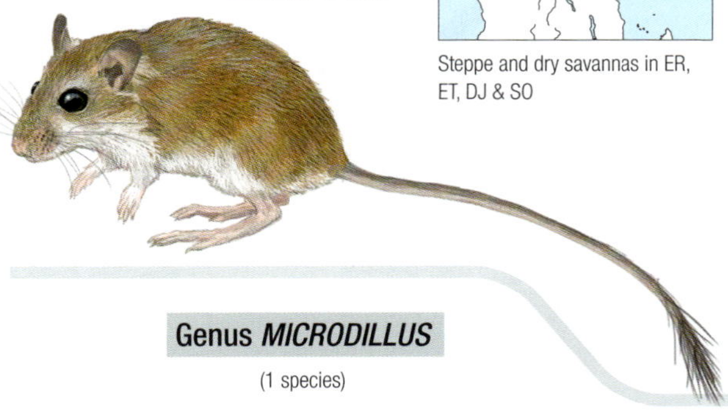

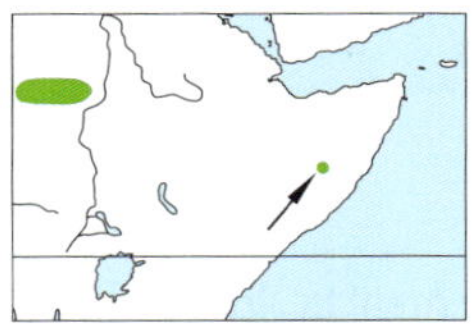

Rosalinda Gerbil
Gerbillus rosalinda
FR Gerbille rosalinda
DE Rosalinda-Rennmaus
ES Gerbillo rosalinda

Known from few localities in C SD (Kordofan region) & WC SO

Genus *MICRODILLUS*

(1 species)

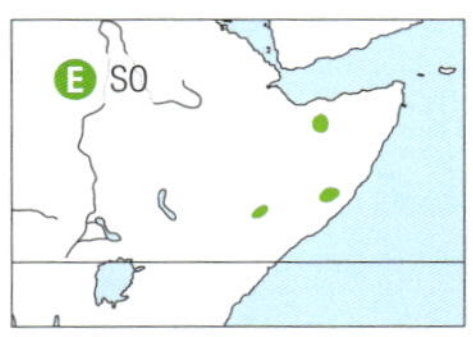

Known only from 3 localities

Somali Pygmy Gerbil
Microdillus peeli
FR Gerbille de Peel
DE Somalia-Zwergrennmaus
ES Gerbillo enano de Somalia

DD
Monotypic
HB 6.6–8 cm
T 5–6.5 cm
W 13–14 g

Genus *SEKEETAMYS*

(1 species)

LC
2 sspp
HB 9.8–13 cm
T 11.2–16.4 cm
W 26–50 g

Bushy-tailed Jird
Sekeetamys calurus
FR Gerbille à queue touffue
DE Buschschwanz-Rennmaus
ES Gerbillo de cola peluda

up to 600 m

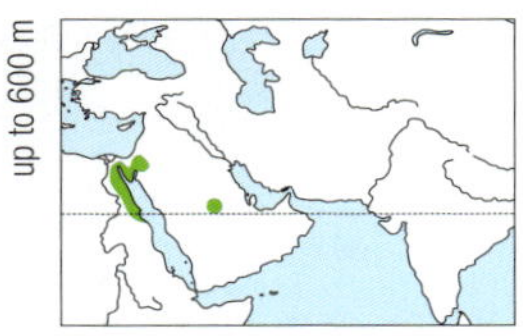

EG (Sinai, Eastern Desert), S IL, SW JO, C SA

Genus *MERIONES*

(17 species)

LC
Monotypic
HB 13.5–19 cm
T 11.5–15 cm
W 60–180 g

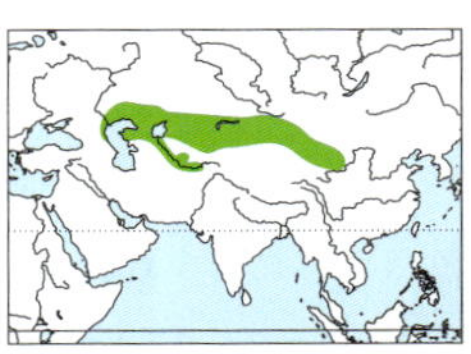

S European RU E to SW MN and NW CN (to N Gansu), S to N AF

Tamarisk Jird
Meriones tamariscinus
FR Mérione des tamaris
DE Tamarisken-Rennratte
ES Gerbillo de taraje

LC
6 sspp
HB 12–16.5 cm
T 12.5–19 cm
W 80–140 g

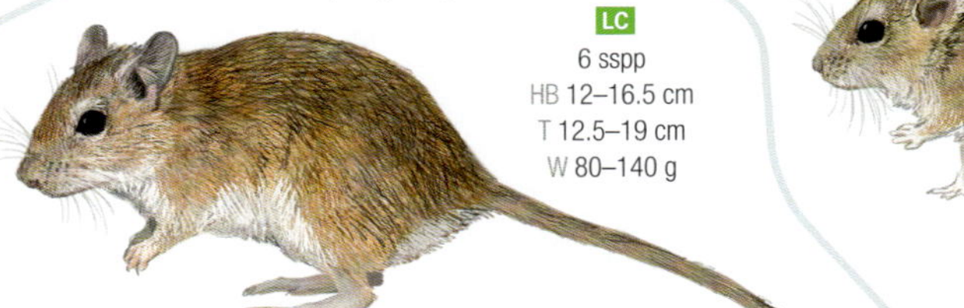

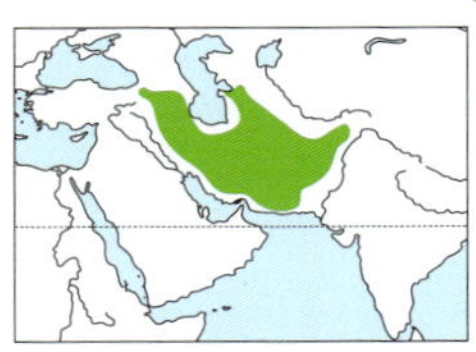

Caucasus (E TR, S AM, S AZ) E to NE IQ, IR, TM, AF & PK

Persian Jird
Meriones persicus
FR Mérione de Perse
DE Persien-Rennratte
ES Gerbillo de Persia

Central Midday Jird
Meriones meridianus
FR Mérione du midi
DE Zentrale Mittagsrennratte
ES Gerbillo de mediodía central

LC
3 sspp
HB 10.2–13.5 cm
T 7.9–13.4 cm
W 37–80 g

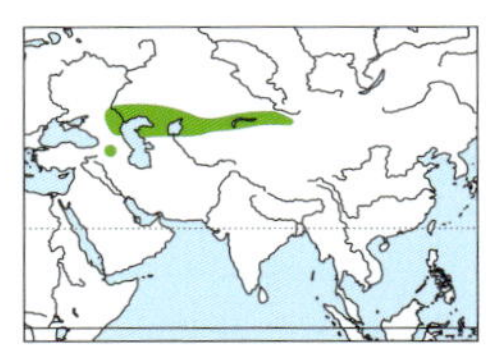

From N Caspian to MN (Dzungaria, up to Tahiyn-Shar-Nuruu Mts) & N Xinjiang (CN)

Eastern Midday Jird
Meriones psammophilus
FR Mérione psammophile
DE Östliche Mittagsrennratte
ES Gerbillo de mediodía oriental

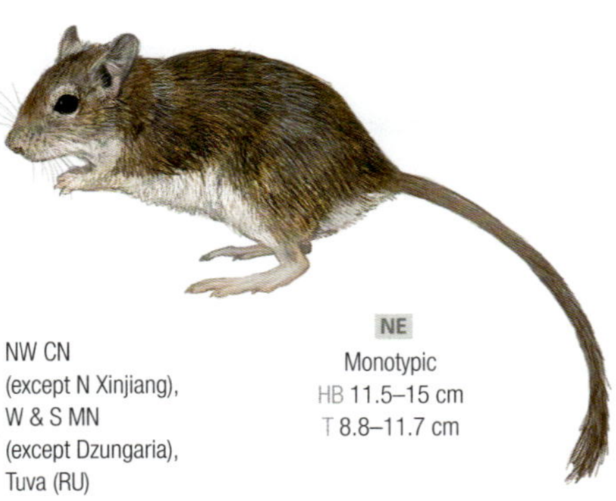

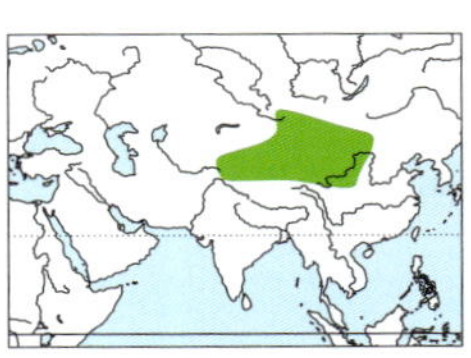

NW CN (except N Xinjiang), W & S MN (except Dzungaria), Tuva (RU)

NE
Monotypic
HB 11.5–15 cm
T 8.8–11.7 cm

Western Midday Jird
Meriones penicilliger
FR Mérione du Karakoum
DE Westliche Mittagsrennratte
ES Gerbillo de mediodía occidental

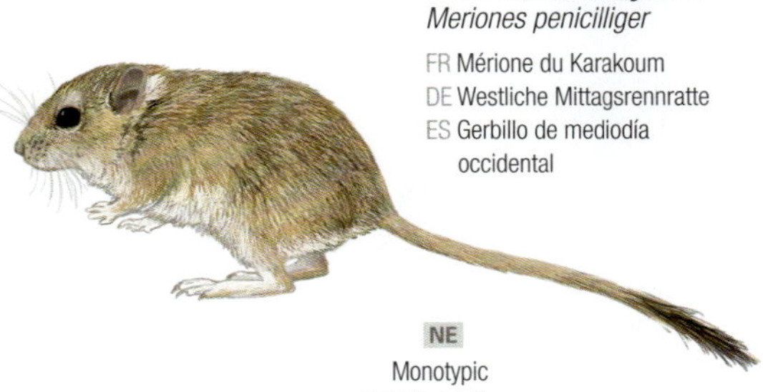

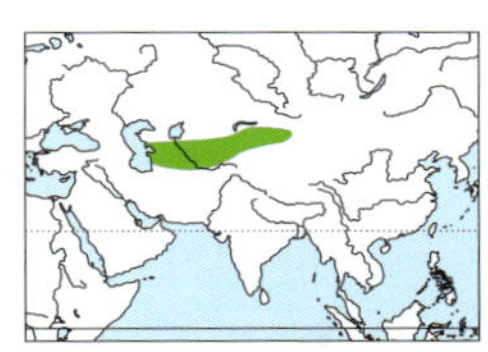

From E IR, TM & S KZ (N to Alakol Depression) E to NW CN (Xinjiang)

NE
Monotypic
HB 9.8–11.8 cm
T 10.3–12.4 cm

LC
Monotypic
HB 9.7–13.2 cm
T 8.5–10.6 cm
W 52–60 g

Mongolian Jird
Meriones unguiculatus
FR Mérione de Mongolie
DE Mongolische Rennratte
ES Gerbillo de Mongolia

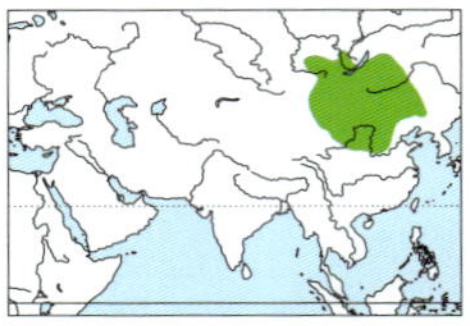

RU (SE Siberia, with Transbaikalia), C & E MN, NC & NE CN

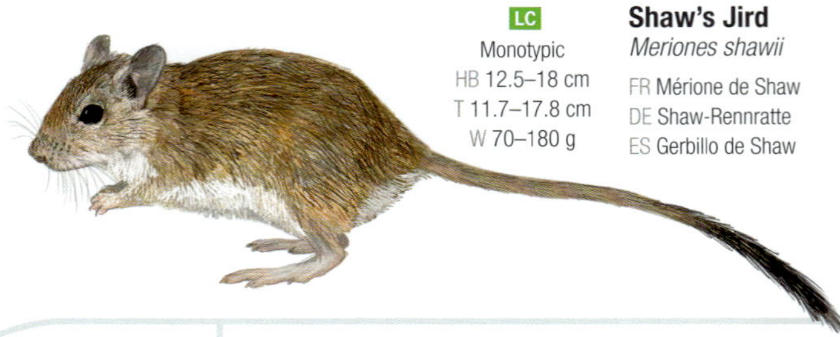

LC
Monotypic
HB 12.5–18 cm
T 11.7–17.8 cm
W 70–180 g

Shaw's Jird
Meriones shawii
FR Mérione de Shaw
DE Shaw-Rennratte
ES Gerbillo de Shaw

up to 2000 m

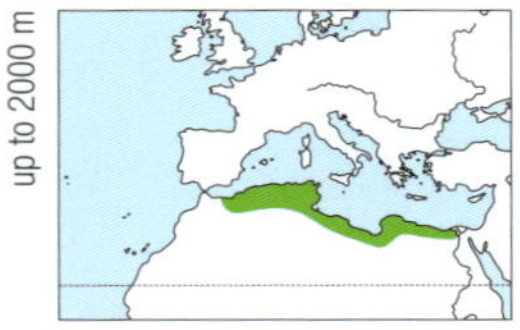

Coastal N Africa from E MA E to EG, W of R Nile

Moroccan Jird
Meriones grandis
FR Mérione du Maroc
DE Marokko-Rennratte
ES Gerbillo del Magreb

up to 2800 m

Up to Atlas Mts; E limit not clarified, possibly to DZ & TN

LC
Monotypic
HB 15–20 cm
T 16–18 cm
W 160–250 g

VU
Monotypic
HB 13–17 cm
T 12–18 cm
W 125–275 g

Buxton's Jird
Meriones sacramenti
FR Mérione de Buxton
DE Palästina-Rennratte
ES Gerbillo de Buxton

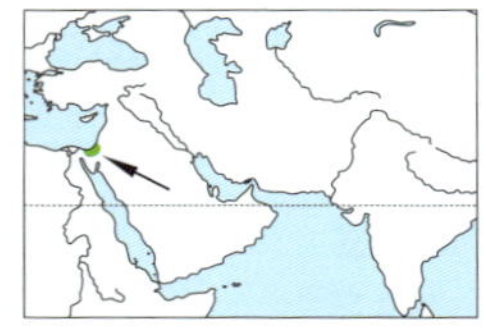

NE EG (coastal Sinai), S IL (Negev Desert), PS

Libyan Jird
Meriones libycus
FR Mérione de Libye
DE Libyen-Rennratte
ES Gerbillo de Libia

LC
3 sspp
HB 12.3–17 cm
T 11.5–19 cm
W 60–140 g

up to 2300 m

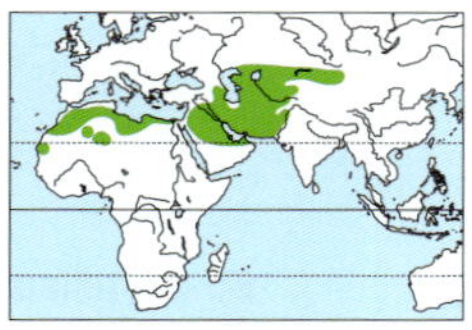

N Africa (from EH & MR to EG), Middle East & Transcaucasia E to CN & PK

Arabian Jird
Meriones arimalius
FR Mérione d'Arabie
DE Arabien-Rennratte
ES Gerbillo de Arabia

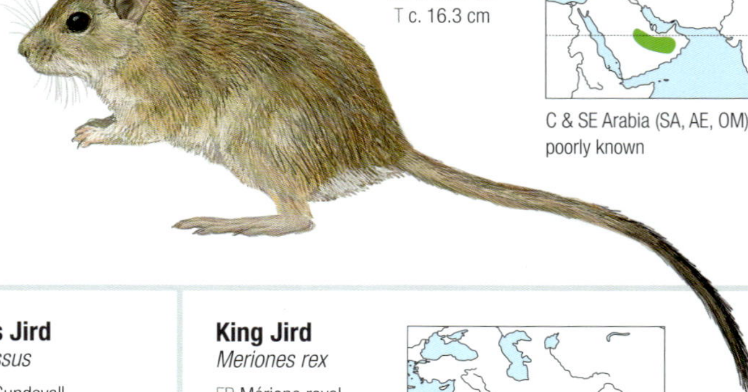

LC
Monotypic
HB c. 13.2 cm
T c. 16.3 cm

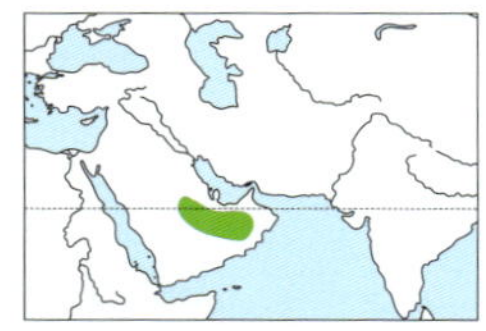

C & SE Arabia (SA, AE, OM); range poorly known

Sundevall's Jird
Meriones crassus
FR Mérione de Sundevall
DE Sundevall-Rennratte
ES Gerbillo de Sundevall

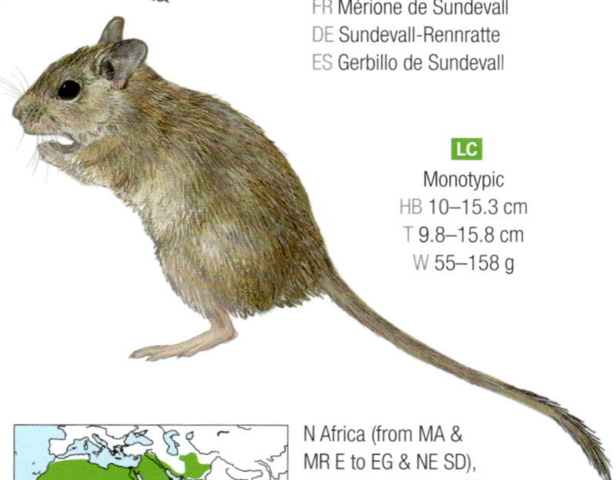

LC
Monotypic
HB 10–15.3 cm
T 9.8–15.8 cm
W 55–158 g

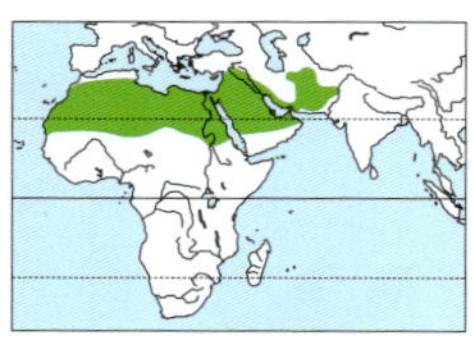

N Africa (from MA & MR E to EG & NE SD), Middle East, E to S TM & W PK

King Jird
Meriones rex
FR Mérione royal
DE Königsrennratte
ES Gerbillo real

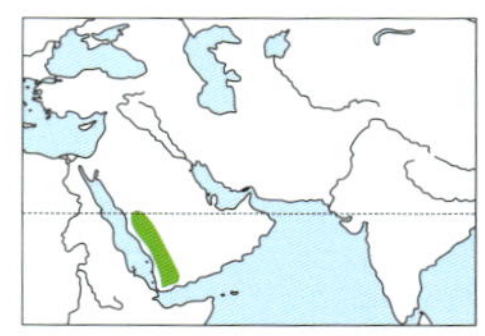

up to 2200 m

Highlands of SW Arabian Pen (SA, YE)

LC
Monotypic
HB 13.3–17.2 cm
T 10.5–19.6 cm

Tristram's Jird
Meriones tristrami
FR Mérione de Tristram
DE Tristram-Rennratte
ES Gerbillo de Tristram

1350–2000 m

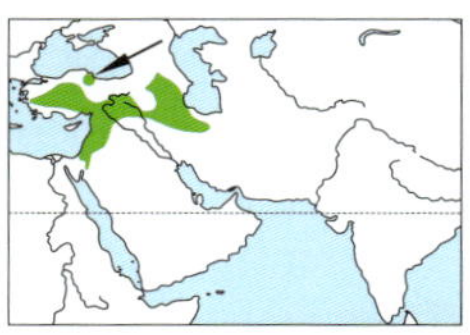

TR (Anatolia) & Caucasus S to IL & W Jordan, E to NW IR

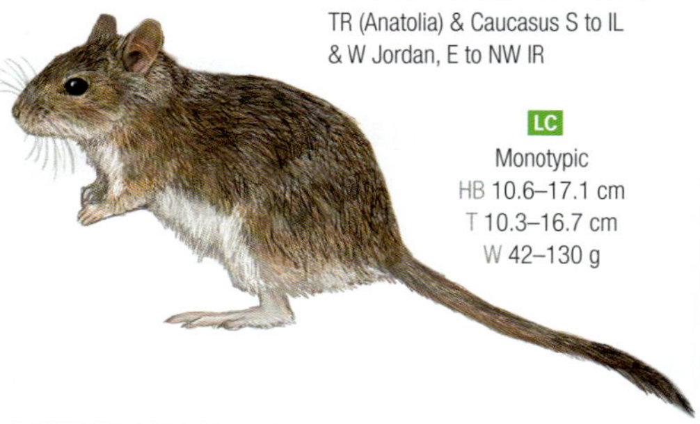

LC
Monotypic
HB 10.6–17.1 cm
T 10.3–16.7 cm
W 42–130 g

Vinogradov's Jird
Meriones vinogradovi
FR Mérione de Vinogradov
DE Vinogradow-Rennratte
ES Gerbillo de Vinográdov

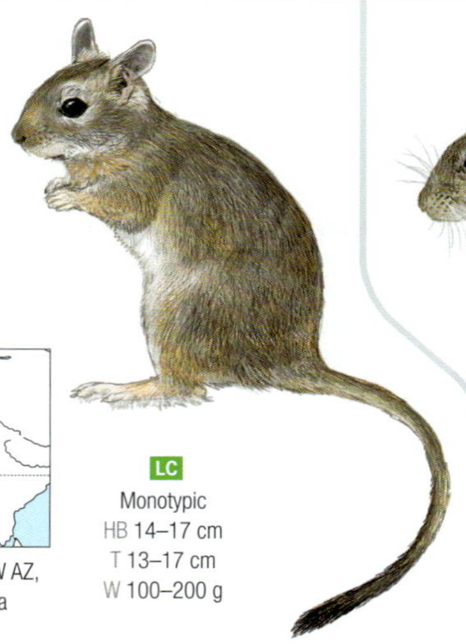

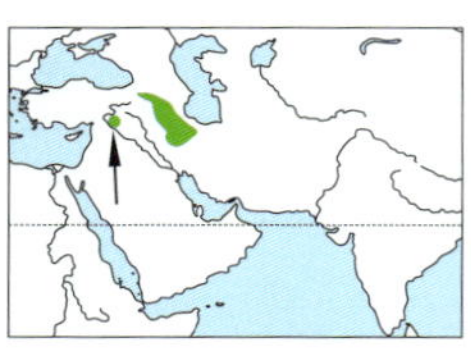

N SY, TR (E Anatolia), AM, SW AZ, NW IR; limits in Mesopotamia unresolved

LC
Monotypic
HB 14–17 cm
T 13–17 cm
W 100–200 g

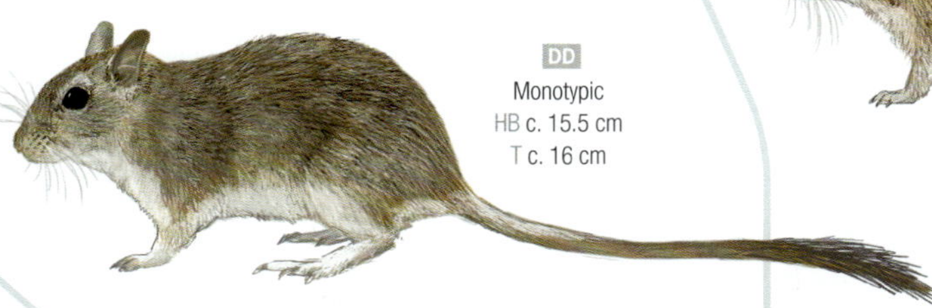

DD
Monotypic
HB c. 15.5 cm
T c. 16 cm

Zarudny's Jird
Meriones zarudnyi
FR Mérione de Zarudny
DE Zarudny-Rennratte
ES Gerbillo de Zarudny

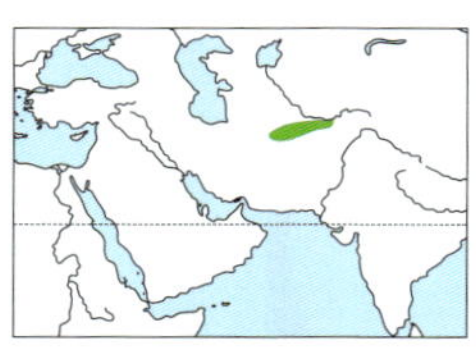

NE IR, S TM, N AF

LC
Monotypic
HB 11.5–14.3 cm
T 10–15.1 cm
W c. 70 g

Indian Desert Jird
Meriones hurrianae
FR Mérione de l'Haryana
DE Indische Wüstenrennratte
ES Gerbillo de desierto de India

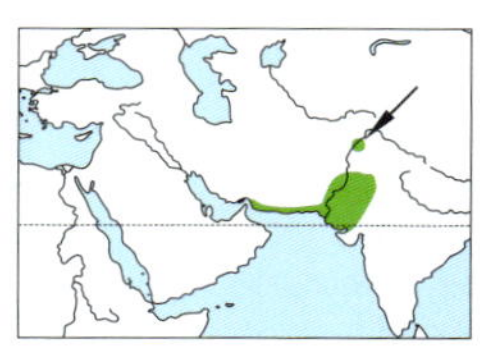

Coastal SE IR & PK, Indus Plain, Thar Desert (E PK, NW IN); perhaps AF

Fat Sand Rat
Psammomys obesus

FR Psammomys obèse
DE Fette Sandratte
ES Gerbillo de arena grueso

LC
3 sspp
HB 11.6–18.5 cm
T 8.8–14 cm
W 82–237 g

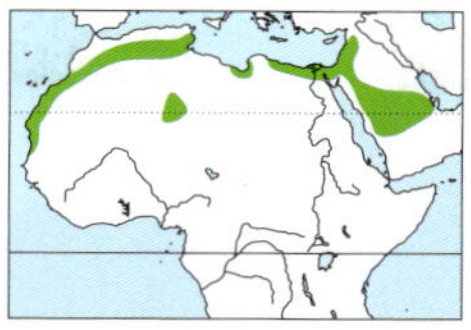

N Africa (from MA, EH & MR E to EG, NE SD), Middle East (SY S to C Arabia)

Genus *PSAMMOMYS*

(2 species)

DD
Monotypic
HB 11.5–13 cm
T 8–12 cm

Lesser Sand Rat
Psammomys vexillaris

FR Psammomys de Thomas
DE Schlanke Sandratte
ES Gerbillo de arena menor

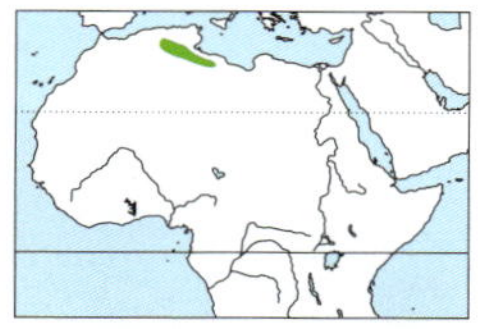

Arid regions of NE DZ, TN & NW LY

Genus *RHOMBOMYS*

(1 species)

LC
Monotypic
HB 15–18.5 cm
T 13–16 cm
W 169–275 g

Great Gerbil
Rhombomys opimus

FR Gerbille géante
DE Riesenrennmaus
ES Gerbillo grande

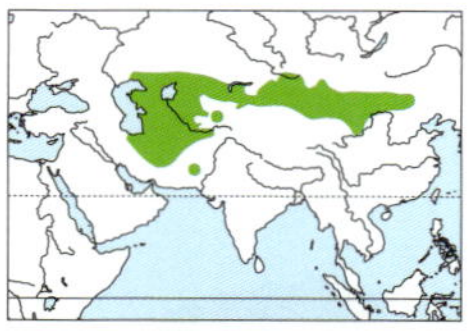

S KZ & IR E through Central Asia to CN & MN, also SW PK (Balochistan)

Genus *BRACHIONES*

(1 species)

LC
Monotypic
HB 6.7–10.3 cm
T 5.6–7.8 cm
W 12–42 g

Przewalski's Jird
Brachiones przewalskii

FR Gerbille de Przewalski
DE Przewalski-Rennratte
ES Jerbo de Przewalski

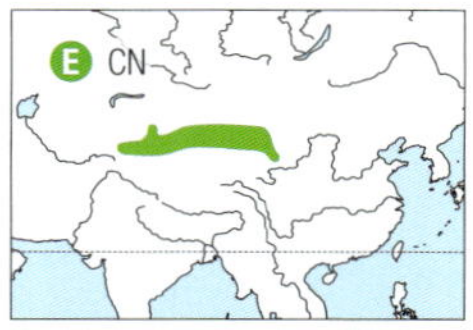

Xinjiang, Gansu, W Inner Mongolia

Genus *DESMODILLISCUS*

(1 species)

Pouched Gerbil
Desmodilliscus braueri

FR Gerbille de Brauer
DE Brauer-Zwergrennmaus
ES Gerbillo de abazones

LC
Monotypic
HB 5–7.3 cm
T 3.2–4.6 cm
W 4–13 g

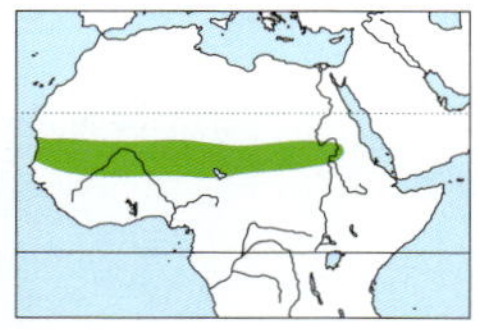

Sahelian savannas (18°–12° N) from MR & SN E to N SD

Fat-tailed Jird
Pachyuromys duprasi

FR Gerbille de Dupras
DE Fettschwanz-Rennmaus
ES Gerbillo de cola gruesa

Genus *PACHYUROMYS*

(1 species)

LC
Monotypic
HB 9.3–12 cm
T 5.5–6.5 cm
W 22–65 g

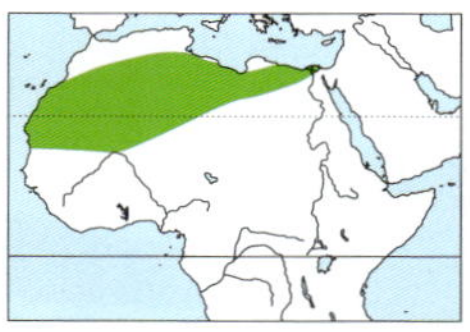

Arid regions of MA, EH, N MR, DZ, N ML, TN, LY & EG

Genus *LEIMACOMYS*

(1 species)

DD
Monotypic
HB c. 11.8 cm
T c. 3.7 cm

Büttner's African Forest Mouse
Leimacomys buettneri

FR Rat de Büttner
DE Furchenzahn-Waldmaus
ES Ratón de bosque de Büttner

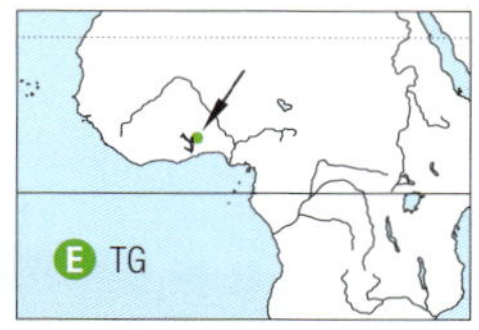

Atacora Mts

Greater Marmoset Rat
Hapalomys longicaudatus

FR Grand Hapalomys
DE Langschwanz-Seidenratte
ES Rata tití mayor

Genus *HAPALOMYS*

(3 species)

EN
Monotypic
HB 14–16.5 cm
T 17.6–20 cm

c. 500 m

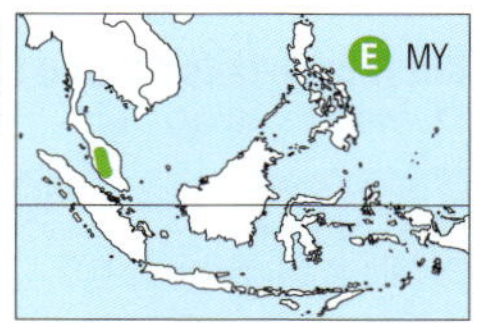

Scattered in Pen MY; presumably extinct in SE MM & TH

Maned Rat
Lophiomys imhausi

FR Rat-à-crinière
DE Mähnenratte
ES Rata de crin

Genus *LOPHIOMYS*

(1 species)

up to 3300 m

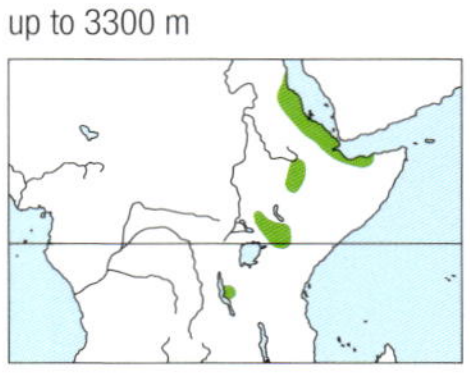

E SD, ER, ET, DJ, N SO, NE UG, W KE; recorded in W TZ

LC
Monotypic
HB 25–30 cm
T 14.5–20 cm
W 590–920 g

Lesser Marmoset Rat
Hapalomys delacouri

FR Petit Hapalomys
DE Delacour-Seidenratte
ES Rata tití menor

NT
Monotypic
HB 10.2–14.6 cm
T 13.5–17.1 cm

1200–1500 m

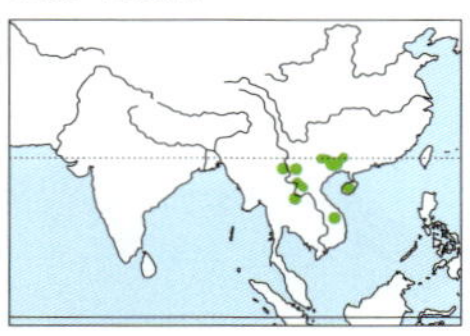

S CN (S Yunnan, S Guangxi, Hainan I), N LA, N TH (Loei Prov), C VN

Suntsov's Marmoset Rat
Hapalomys suntsovi

FR Hapalomys de Suntsov
DE Suntsow-Seidenratte
ES Rata tití de Suntsov

c. 540 m

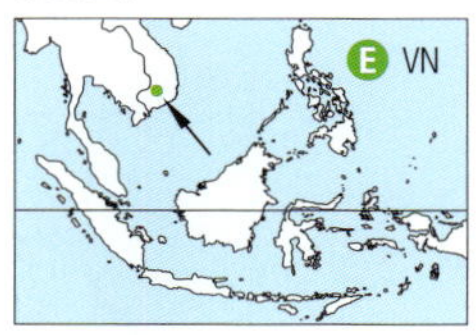

Known only from Bu Gia Map NP (Binh Phuoc Prov); perhaps adjacent E KH

NE
Monotypic
HB 12.7–14.6 cm
T 15–16.5 cm

Genus *PHLOEOMYS*

(2 species)

0–2300 m

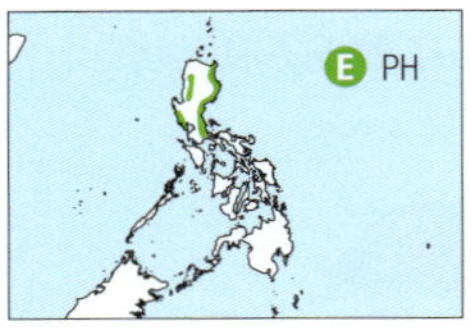

N & C Luzon I

Northern Luzon Giant Cloud Rat
Phloeomys pallidus

FR Rat-tacheté de Nehring
DE Nördliche Riesenborkenratte
ES Rata gigante de Luzón septentrional

LC
Monotypic
HB 39–42 cm
T 32–34 cm
W 2.2–2.7 kg

Southern Luzon Giant Cloud Rat
Phloeomys cumingi

FR Rat-tacheté de Cuming
DE Südliche Riesenborkenratte
ES Rata gigante de Luzón meridional

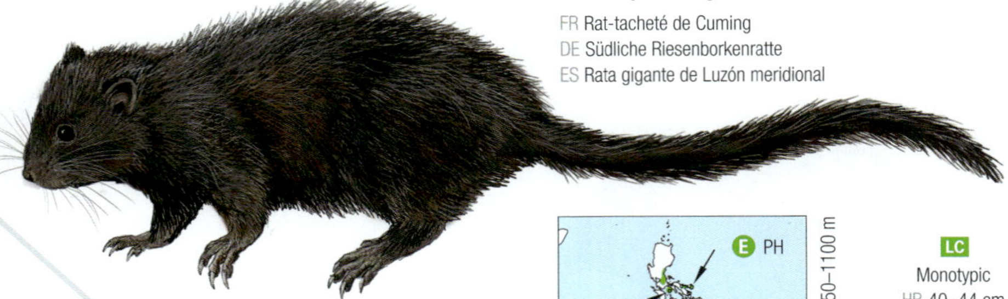

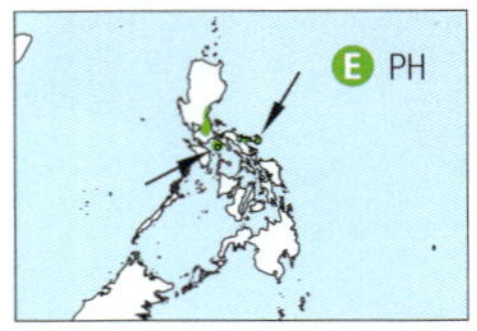

150–1100 m

C & S Luzon, Marinduque & Catanduanes Is

LC
Monotypic
HB 40–44 cm
T 27–31 cm
W 1.8–2.1 kg

Genus *CRATEROMYS*

(4 species)

Luzon Bushy-tailed Cloud Rat
Crateromys schadenbergi

FR Crateromys de Schadenberg
DE Schadenberg-Borkenratte
ES Rata de cola peluda de Schadenberg

EN
Monotypic
HB 30–37 cm
T 30–39 cm
W 1.3–1.5 kg

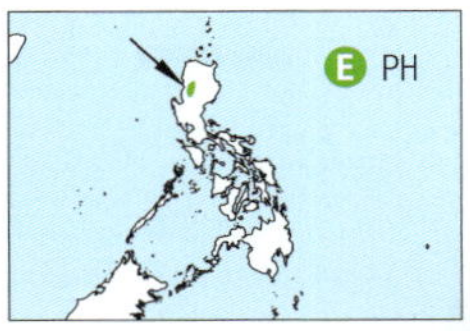

2000–2500 m

N Luzon I (Cordillera Central)

Panay Bushy-tailed Cloud Rat
Crateromys heaneyi

FR Crateromys de Heaney
DE Panay-Borkenratte
ES Rata de cola peluda de Panay

EN
Monotypic
HB c. 28 cm
T 30–34 cm
W 930–1042 g

up to 400 m

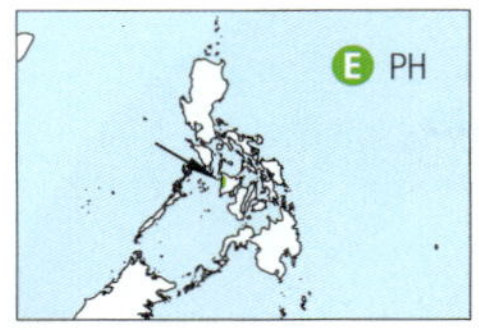

Panay I

Dinagat Bushy-tailed Cloud Rat
Crateromys australis

FR Crateromys de Dinagat
DE Dinagat-Borkenratte
ES Rata de cola peluda de Dinagat

DD
Monotypic
HB c. 26 cm T c. 28 cm

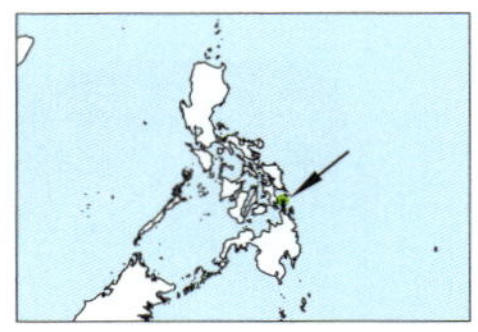

Dinagat I; possibly Siargao, Bucas Grande and some other small nearby Is

Ilin Bushy-tailed Cloud Rat
Crateromys paulus

FR Crateromys d'Ilin
DE Ilin-Borkenratte
ES Rata de cola peluda de Ilin

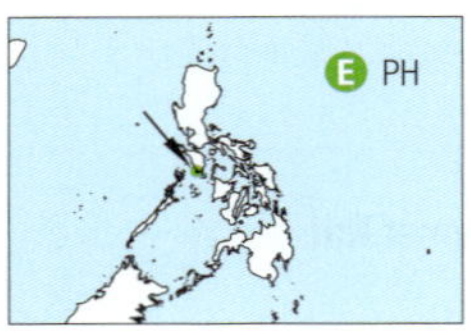

Ilin I

DD
Monotypic
HB c. 25 cm
T c. 21 cm

Genus *BATOMYS*

(6 species)

DD
Monotypic
HB 17.1–18.9 cm
T 11.1–12.5 cm
W 155–183 g

950–1130 m

Hamiguitan Hairy-tailed Rat
Batomys hamiguitan

FR Batomys de l'Hamiguitan
DE Hamiguitan-Haarschwanzratte
ES Rata de cola peluda de Hamiguitan

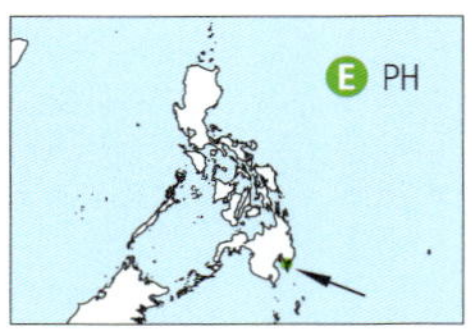

Mindanao I (Mt Hamiguitan)

Mindanao Hairy-tailed Rat
Batomys salomonseni

FR Batomys de Salomonsen
DE Mindanao-Haarschwanzratte
ES Rata de cola peluda de Mindanao

LC
Monotypic
HB 13.8–19.1 cm
T 13–16.7 cm
W 143–255 g

0–2400 m

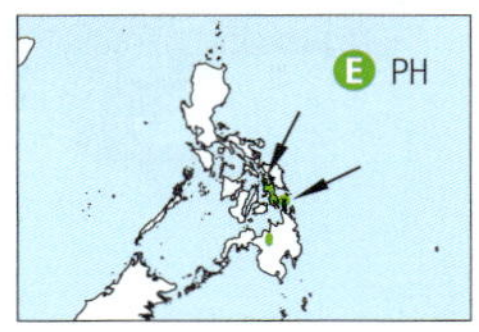

Biliran, Leyte, Dinagat & Mindanao Is

EN
Monotypic
HB 14–15.1 cm
T 10.2–11.8 cm
W c. 115 g

Russet Hairy-tailed Rat
Batomys russatus

FR Batomys roux
DE Dinagat-Haarschwanzratte
ES Rata de cola peluda rojiza

c. 350 m

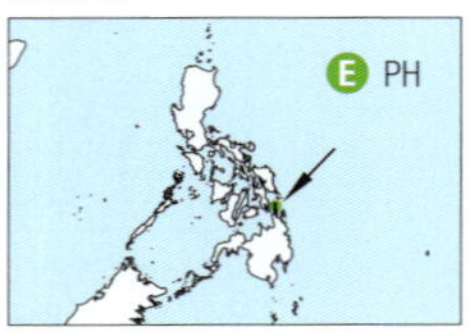

Dinagat I

Mount Isarog Hairy-tailed Rat
Batomys uragon

FR Batomys de l'Isarog
DE Mount-Isarog-Haarschwanzratte
ES Rata de cola peluda de Isarog

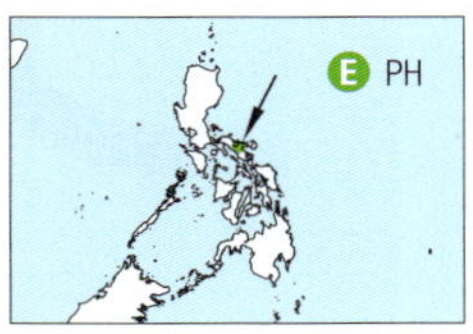

1350–1800 m

Luzon I (Mt Isarog)

LC
Monotypic
HB 17.7–21 cm
T 12.8–17.1 cm
W 135–200 g

Luzon Cordillera Hairy-tailed Rat
Batomys grantii

FR Batomys de Grant
DE Luzon-Haarschwanzratte
ES Rata de cola peluda de Luzón

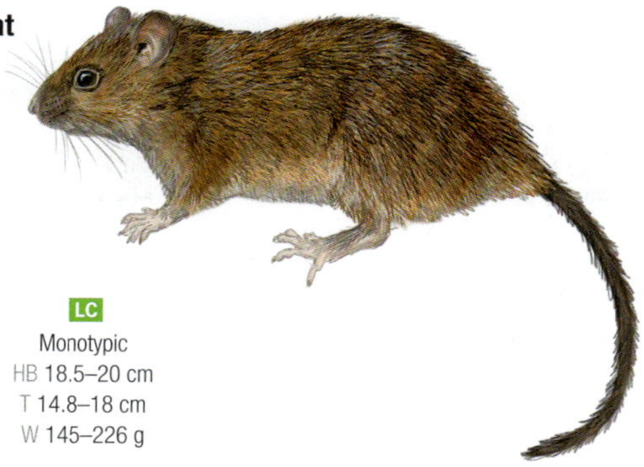

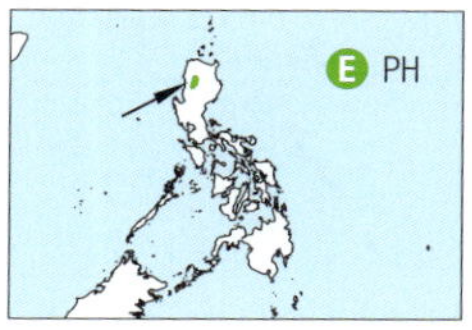

1600–2530 m

N Luzon I

LC
Monotypic
HB 18.5–20 cm
T 14.8–18 cm
W 145–226 g

Large-toothed Hairy-tailed Rat
Batomys dentatus

FR Batomys du Benguet
DE Großzähnige Haarschwanzratte
ES Rata de cola peluda de dientes grandes

DD
Monotypic
HB c. 19.5 cm
T c. 18.5 cm

c. 2135 m

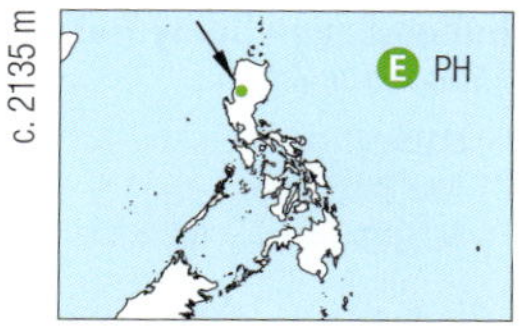

N Luzon I (Haights-in-the-Oaks, Benguet Prov)

Genus *CARPOMYS*

(2 species)

Black-tailed Dwarf Cloud Rat
Carpomys melanurus

FR Carpomys à queue noire
DE Schwarzschwanz-Zwergborkenratte
ES Rata enana de cola negra

2200–2300 m

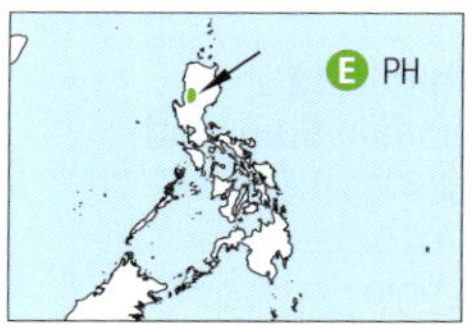

N Luzon I (Mt Data & Mt Pulag)

DD
Monotypic
HB 18–18.4 cm T 18–18.3 cm
W c. 165 g

Brown-tailed Dwarf Cloud Rat
Carpomys phaeurus

FR Carpomys à queue brune
DE Braunschwanz-Zwergborkenratte
ES Rata enana de cola marrón

LC
Monotypic
HB 16.6–17.5 cm
T 15.6–16.1 cm
W c. 123 g

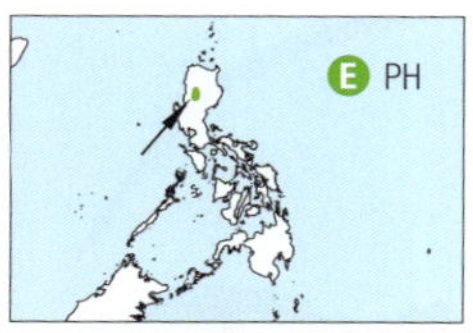

2150–2500 m

N Luzon I (Cordillera Central)

Genus *MUSSEROMYS*

(4 species)

Mount Anacuao Tree Mouse
Musseromys anacuao

FR Rat de l'Anacuao
DE Mount-Anacuao-Bergwaldmaus
ES Ratón arborícola de Anacuao

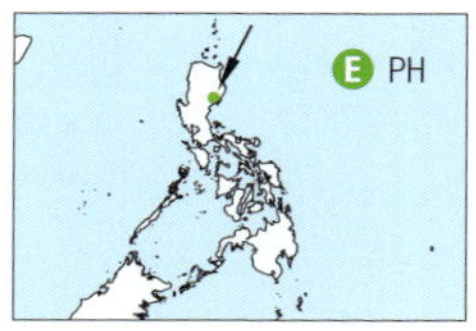

c. 1700 m

Luzon I (Mt Anacuao)

DD
Monotypic
HB 7.8–8.3 cm
T 8.2–8.6 cm
W 17–21 g

Mount Pulag Tree Mouse
Musseromys beneficus

FR Rat du Pulag
DE Mount-Pulag-Bergwaldmaus
ES Ratón arborícola de Pulag

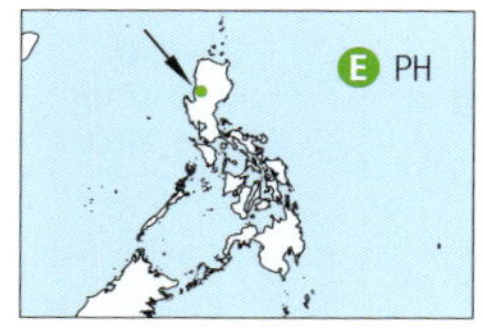

c. 2695 m

Luzon I (Mt Pulag)

DD
Monotypic
HB 7.5–8.1 cm
T c. 8.2 cm
W 18–22 g

Mount Amuyao Tree Mouse
Musseromys inopinatus

FR Rat de l'Amuyao
DE Mount-Amuyao-Bergwaldmaus
ES Ratón arborícola de Amuyao

DD
Monotypic
HB c. 7.8 cm
T 8.5–8.8 cm
W 17–19.5 g

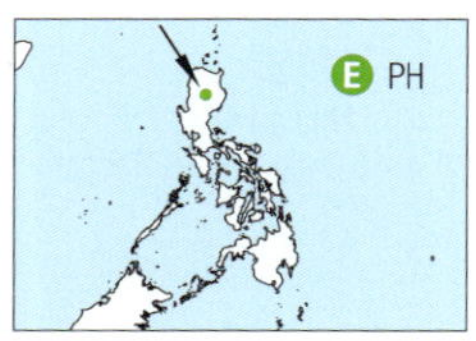

1650–2300 m

Luzon I (Mt Amuyao)

Mount Banahaw Tree Mouse
Musseromys gulantang

FR Rat du Banahaw
DE Mount-Banahaw-Bergwaldmaus
ES Ratón arborícola de Banahaw

c. 620 m

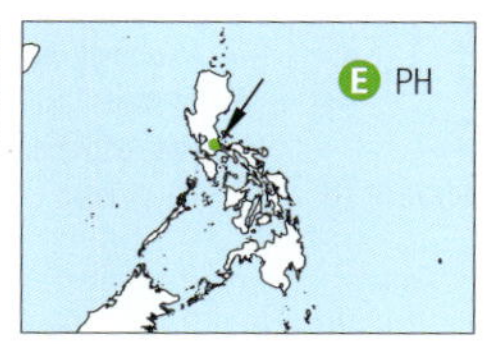

Luzon I (Mt Banahaw)

DD
Monotypic
HB c. 7.7 cm
T c. 10.1 cm
W c. 15.5 g

Genus *MICROMYS*

(2 species)

Eurasian Harvest Mouse
Micromys minutus

FR Rat des moissons
DE Eurasische Zwergmaus
ES Ratón espiguero

LC
Monotypic
HB 4.8–8 cm
T 3.9–7.2 cm
W 4–11 g

up to 1700 m

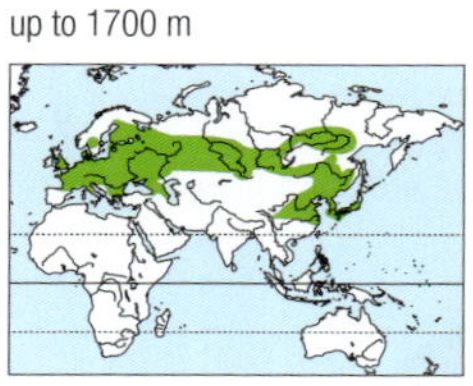

From N ES & GB E through Europe, RU & KZ to N MN, CN, KR & JP

Indochinese Harvest Mouse
Micromys erythrotis

FR Rat à oreilles rousses
DE Asiatische Zwergmaus
ES Ratón espiguero de Vietnam

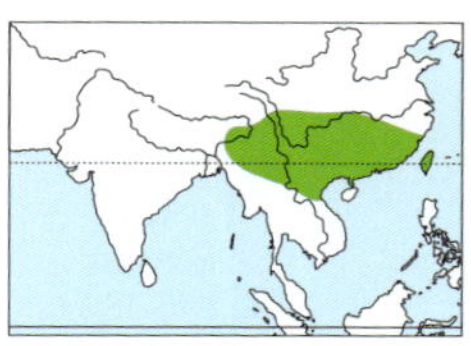

C & S CN (Yunnan & Sichuan to Fujian), NE IN, N MM, N LA, N VN; TW uncertain

NE
Monotypic
HB 4.6–6.8 cm
T 5.4–8.8 cm

Indochinese Mountain Spiny Rat
Maxomys moi

FR Maxomys du Vietnam
DE Indochina-Rajah-Ratte
ES Rata espinosa de montaña de Indochina

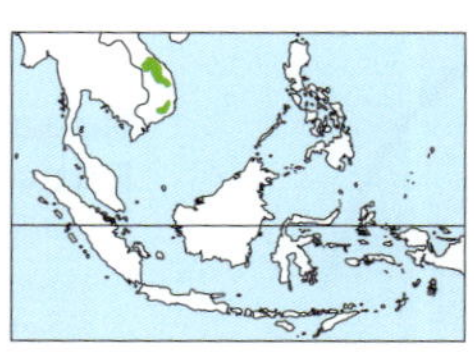

190–1500 m

S LA (possibly further N), S VN

LC
Monotypic
HB 14.2–21 cm
T 15.7–20 cm
W 80–220 g

Genus *MAXOMYS*

(18 species)

Bartels's Javan Spiny Rat
Maxomys bartelsii

FR Maxomys de Bartels
DE Bartels-Rajah-Ratte
ES Rata espinosa de Java de Bartels

LC
Monotypic
HB 12.7–17.8 cm
T 11.7–17 cm
W c. 93 g

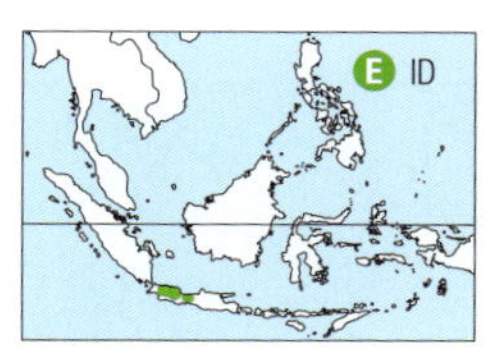

up to 1830 m

W & C Java

Indomalayan Spiny Rat
Maxomys surifer
FR Maxomys roux
DE Rote Rajah-Ratte
ES Rata espinosa de Indomalasia

LC
7 sspp
HB 15.5–23 cm
T 14.6–23 cm
W 90–285 g

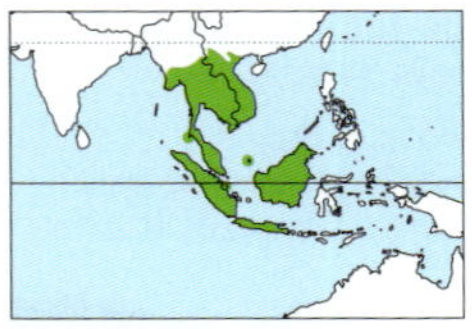

0–1680 m

S MM, CN (S Yunnan), LA & VN S through Thai-Malay Pen to Greater Sundas & many smaller Is

Palawan Spiny Rat
Maxomys panglima
FR Maxomys de Palawan
DE Palawan-Rajah-Ratte
ES Rata espinosa de Palawan

LC
Monotypic
HB 16.7–22 cm
T 18–23 cm
W 189–240 g

0–1950 m

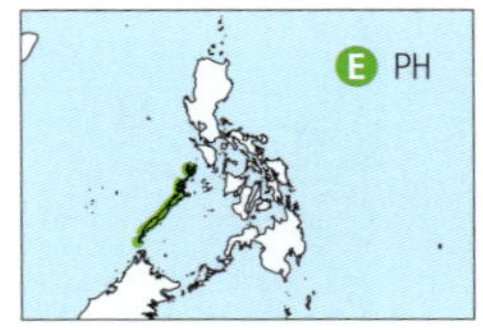

Palawan, Calauit, Busuanga, Culion & Balabac Is

Mentawai Archipelago Spiny Rat
Maxomys pagensis
FR Maxomys des Pagai
DE Mentawai-Rajah-Ratte
ES Rata espinosa de Mentawai

VU
Monotypic
HB 17–22 cm
T 16.3–20 cm

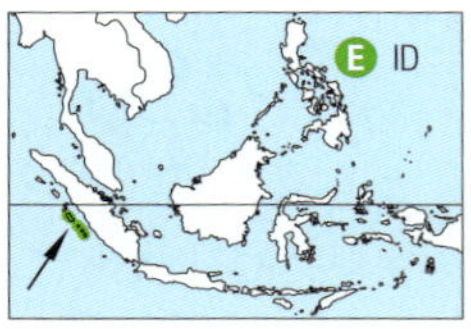

Mentawai Arch (Siberut, Sipora, North Pagai, South Pagai), off W Sumatra

Rajah Sundaic Spiny Rat
Maxomys rajah
FR Maxomys raja
DE Braune Rajah-Ratte
ES Rata espinosa de Sonda rajá

VU
Monotypic
HB 16.6–23 cm
T 16.2–21 cm
W 95–220 g

0–1100 m

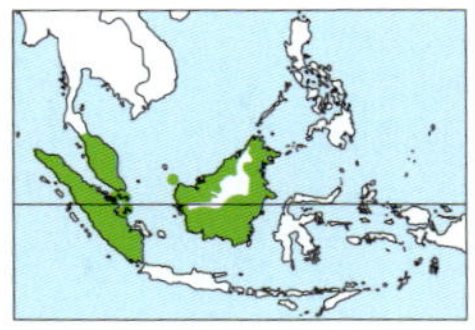

Thai-Malay Pen, Sumatra, Borneo & some offshore Is

VU
Monotypic
HB 10–12.6 cm
T 8.8–10.8 cm
W 35–80 g

up to 2100 m

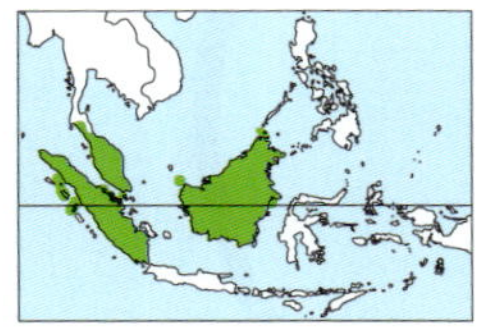

Thai-Malay Pen, Sumatra, Borneo & some offshore Is

Whitehead's Sundaic Spiny Rat
Maxomys whiteheadi
FR Maxomys de Whitehead
DE Whitehead-Rajah-Ratte
ES Rata espinosa de Sonda de Whitehead

VU
Monotypic
HB 16–19.3 cm
T 14.3–19 cm

Broad-nosed Sumatran Spiny Rat
Maxomys inflatus
FR Maxomys à nez large
DE Breitnasen-Rajah-Ratte
ES Rata espinosa de Sumatra de hocico ancho

900–1500 m

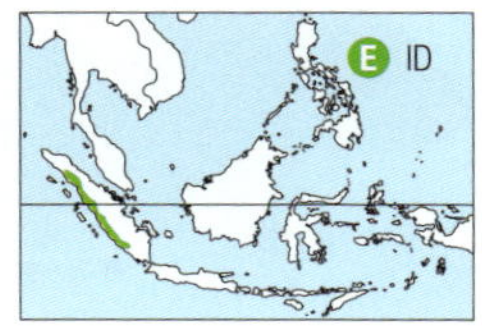

W Sumatra

LC
Monotypic
HB 12.4–16.3 cm
T 13.5–16.7 cm

above 900 m

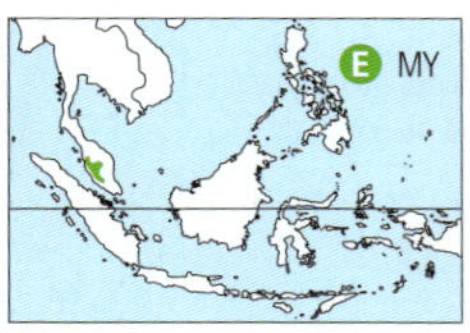

NW Pen MY; perhaps adjacent TH

Malayan Mountain Spiny Rat
Maxomys inas
FR Maxomys de l'Inas
DE Malaya-Rajah-Ratte
ES Rata espinosa de montaña de Malasia

Sumatran Mountain Spiny Rat
Maxomys hylomyoides
FR Maxomys du Korinchi
DE Sumatra-Rajah-Ratte
ES Rata espinosa de montaña de Sumatra

DD
Monotypic
HB 11.4–13.2 cm T 9.8–13.2 cm

Sumatra (C part of W mountain range)

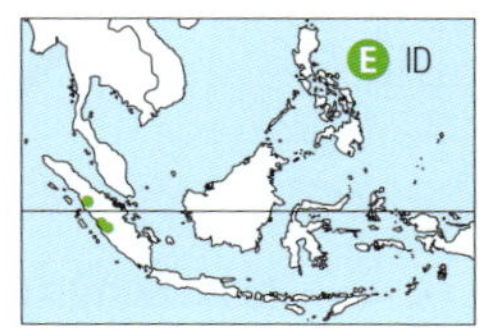

600–2225 m

Bornean Mountain Spiny Rat
Maxomys alticola
FR Maxomys de montagne
DE Gebirgs-Rajah-Ratte
ES Rata espinosa de montaña de Borneo

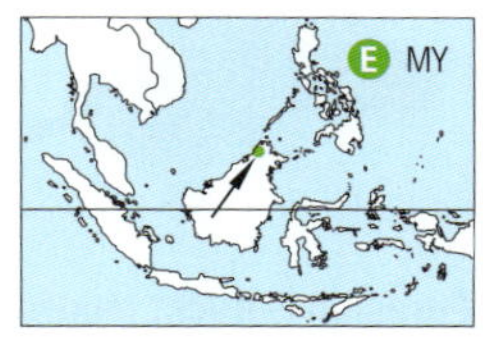

N Borneo: Sabah (Mt Kinabalu, Mt Trus Madi)

LC
Monotypic
HB 13.9–17.6 cm
T 12.8–17.6 cm

Tajuddin's Spiny Rat
Maxomys tajuddinii
FR Maxomys de Tajuddin
DE Tajuddin-Rajah-Ratte
ES Rata espinosa de Tajuddin

LC
Monotypic
HB 9.5–12.1 cm
T 10.6–12.2 cm W 50–70 g

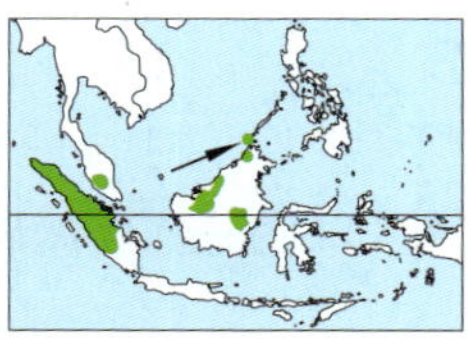

100–1000 m

SC Pen MY, Sumatra, N & C Borneo (with Balambangan I)

Small Bornean Spiny Rat
Maxomys baeodon
FR Petit Maxomys
DE Kleine Rajah-Ratte
ES Rata espinosa de Borneo pequeña

900–1400 m

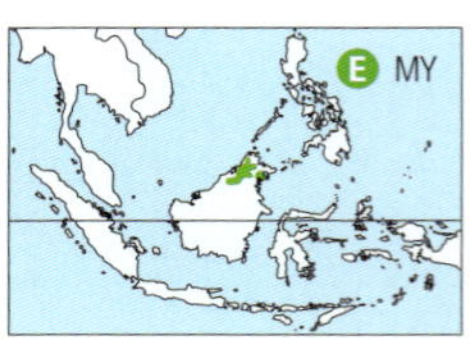

N Borneo: Sabah & Sarawak

DD
Monotypic
HB 12.6–14 cm
T 11.9–13.3 cm
W 53–67 g

LC
Monotypic
HB 11.5–16.3 cm
T 11.2–15.2 cm
W 95–240 g

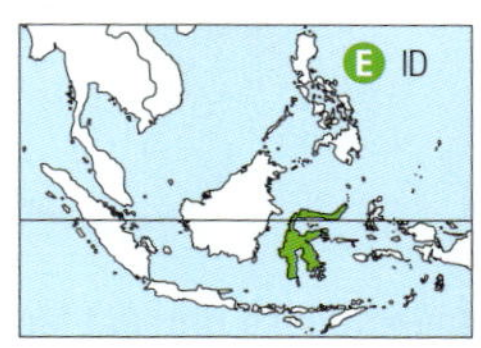

Sulawesi

Musschenbroek's Sulawesi Spiny Rat
Maxomys musschenbroekii
FR Maxomys de Musschenbroek
DE Musschenbroek-Rajah-Ratte
ES Rata espinosa de Célebes de Musschenbroek

Watts's Sulawesi Spiny Rat
Maxomys wattsi

EN
Monotypic
HB 16.4–18.5 cm
T 12.5–15.4 cm
W 110–130 g

FR Maxomys de Watts
DE Watts-Rajah-Ratte
ES Rata espinosa de Watts

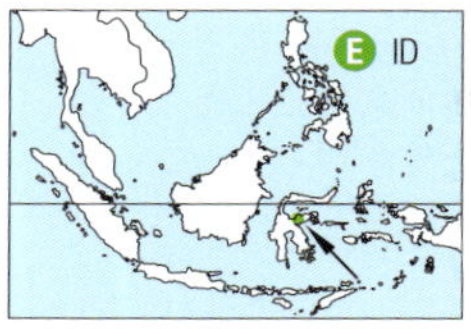

1430–1830 m

EC Sulawesi (Mt Tambusisi)

Ochraceous-bellied Bornean Spiny Rat
Maxomys ochraceiventer

FR Maxomys à ventre jaune
DE Ockerbauch-Rajah-Ratte
ES Rata espinosa de Borneo de vientre ocre

DD
Monotypic
HB 14–17.1 cm
T 13–17.5 cm

300–1700 m

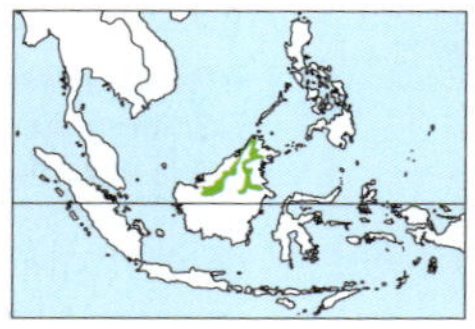

N & C Borneo: Sabah & Sarawak (MY), Kalimantan (ID)

DD
Monotypic
HB 14.5–21 cm
T 18–26 cm
W 165–210 g

Dollman's Sulawesi Spiny Rat
Maxomys dollmani

FR Maxomys de Dollman
DE Dollman-Rajah-Ratte
ES Rata espinosa de Célebes de Dollman

c. 1830 m

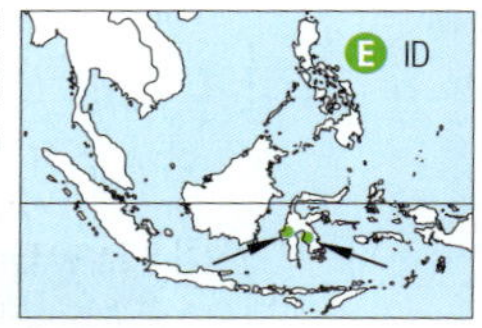

SC & SE Sulawesi (Quarles Range, Mekongga Mts)

Hellwald's Sulawesi Spiny Rat
Maxomys hellwaldii

LC
Monotypic
HB 18–22 cm
T 16–20 cm
W 150–270 g

FR Maxomys de Hellwald
DE Hellwald-Rajah-Ratte
ES Rata espinosa de Célebes de Hellwald

0–1005 m

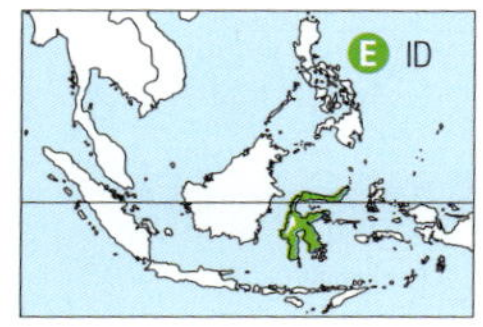

Sulawesi

Genus *CRUNOMYS*

(4 species)

Luzon Shrew Mouse
Crunomys fallax

DD
Monotypic
HB c. 10.5 cm
T c. 7.9 cm

FR Crunomys de Luçon
DE Luzon-Spitzmausratte
ES Ratón musaraña de Luzón

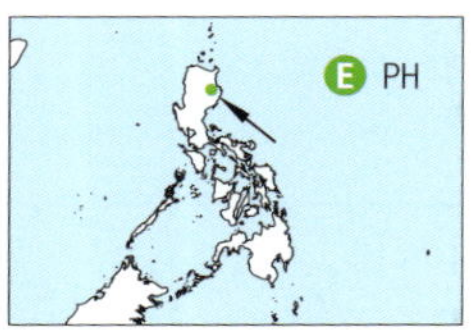

c. 300 m

NE Luzon I (Sierra Madre, Isabela Prov)

Mindanao Shrew Mouse
Crunomys melanius

LC
Monotypic
HB 9.8–13.3 cm
T 6.8–9.5 cm
W 58–71 g

FR Crunomys de Mindanao
DE Mindanao-Spitzmausratte
ES Ratón musaraña de Mindanao

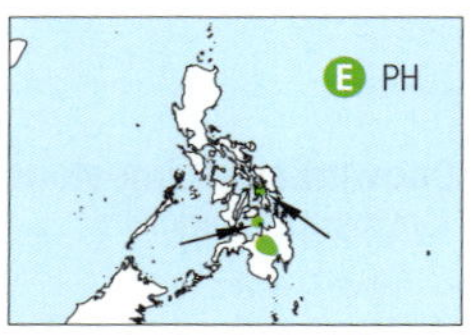

0–1500 m

Leyte, Camiguin & Mindanao Is

Katanglad Shrew Mouse
Crunomys suncoides

FR Crunomys du Katanglad
DE Katanglad-Spitzmausratte
ES Ratón musaraña de Katanglad

DD
Monotypic
HB c. 10.6 cm
T c. 7.9 cm

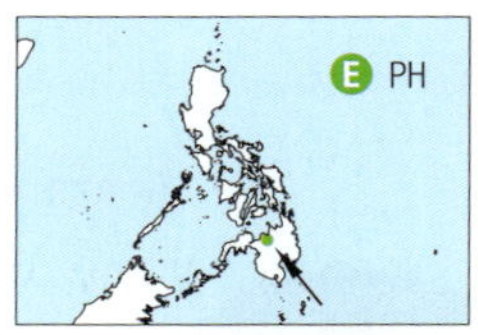

c. 2250 m

Mindanao I (Mt Kitanglad Range)

Sulawesi Shrew Mouse
Crunomys celebensis

FR Crunomys des Célèbes
DE Sulawesi-Spitzmausratte
ES Ratón musaraña de Célebes

DD
Monotypic
HB 11.5–12.7 cm
T 8–8.4 cm
W 35–55 g

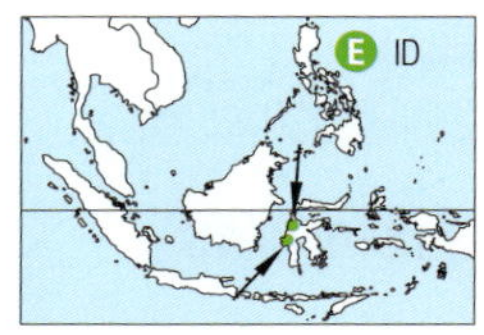

up to 1600 m

NC & WC Sulawesi (around L Lindu, Mt Gandangdewata)

Northern Sulawesi Spiny Rat
Echiothrix leucura

EN
Monotypic
HB 19.6–23 cm
T 21–27 cm
W 100–124 g

FR Grand Échiothrix
DE Nördliche Sulawesi-Stachelratte
ES Rata espinosa de Célebes septentrional

Genus *ECHIOTHRIX*

(2 species)

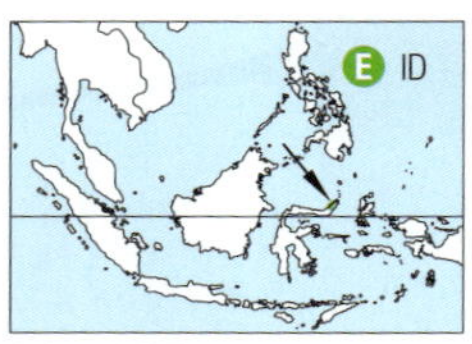

Sulawesi (NE end of Minahasa Pen)

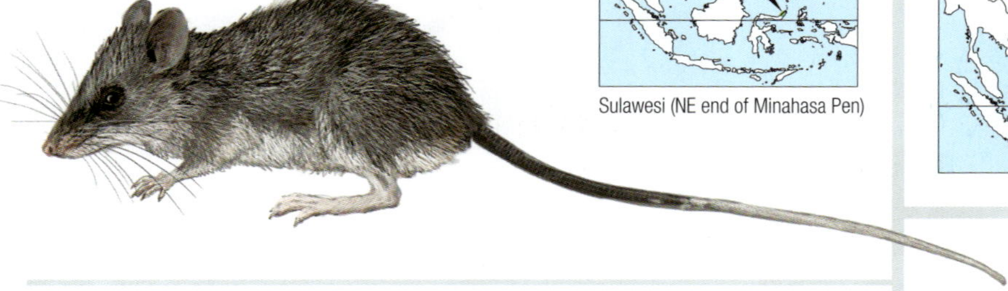

Edented Sulawesi Rat
Paucidentomys vermidax

FR Rat vermivore
DE Zahnarme Regenwurmratte
ES Rata desdentada de Célebes

Genus *PAUCIDENTOMYS*

(1 species)

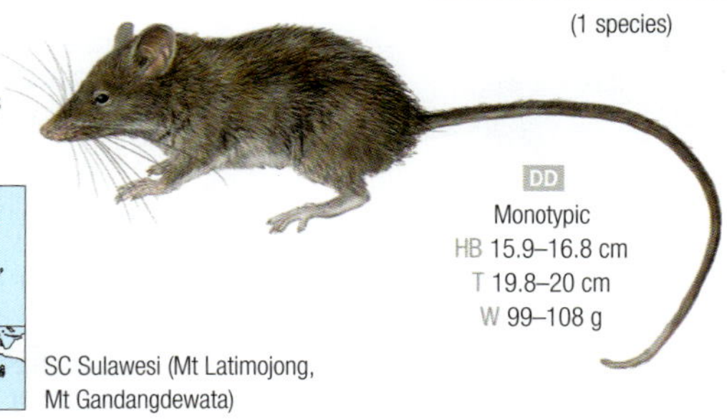

1570–2050 m

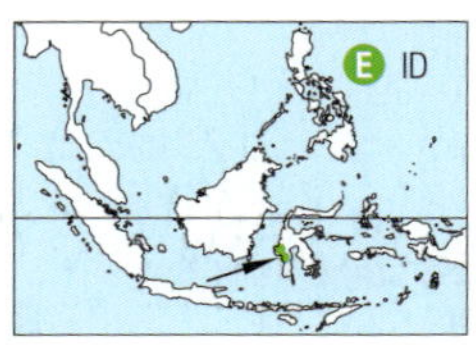

DD
Monotypic
HB 15.9–16.8 cm
T 19.8–20 cm
W 99–108 g

SC Sulawesi (Mt Latimojong, Mt Gandangdewata)

NT
Monotypic
HB 18.2–22 cm
T 23–26 cm
W 215–270 g

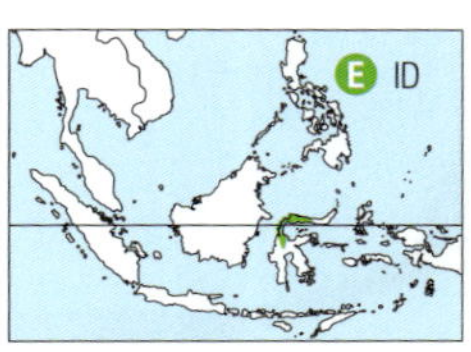

0–1100 m

Central Sulawesi Spiny Rat
Echiothrix centrosa

FR Petit Échiothrix
DE Zentral-Sulawesi-Stachelratte
ES Rata espinosa de Célebes central

NW Minahasa Pen & C Sulawesi

Genus *TATEOMYS*

(2 species)

Tate's Sulawesian Shrew Rat
Tateomys rhinogradoides

FR Rat de Tate
DE Tate-Sulawesi-Spitzmausratte
ES Rata musaraña de Célebes de Tate

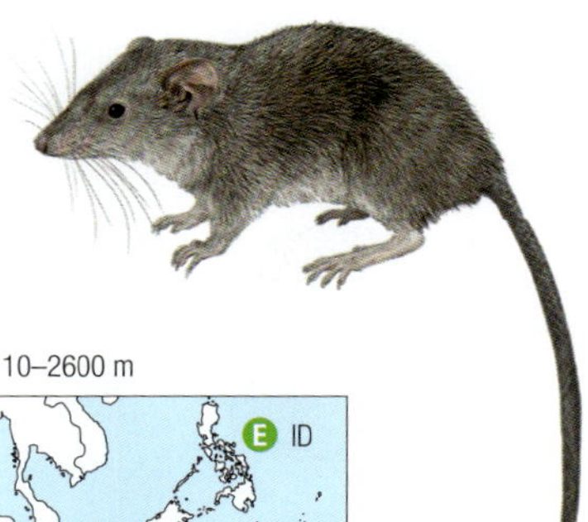

DD
Monotypic
HB 13.7–15.6 cm
T 15.2–17 cm
W 88–98 g

2110–2600 m

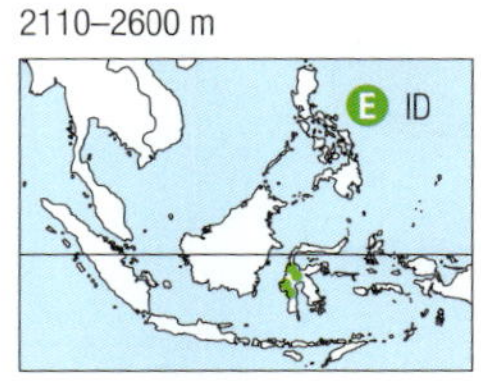

C Sulawesi (Nokilalaki, Rorekatimbo, Tokala, Gandangdewata & Latimojong Mts)

Long-tailed Sulawesian Shrew Rat
Tateomys macrocercus
FR Rat à longue queue
DE Langschwanz-Sulawesi-Spitzmausratte
ES Rata musaraña de Célebes de cola larga

DD
Monotypic
HB 11–12 cm
T 16–17.3 cm
W 43–55 g

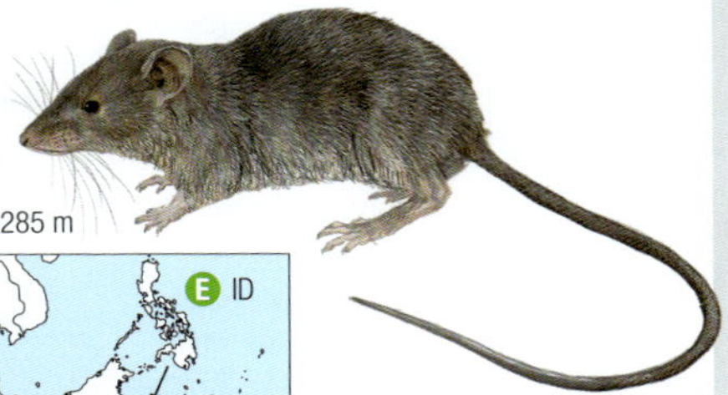

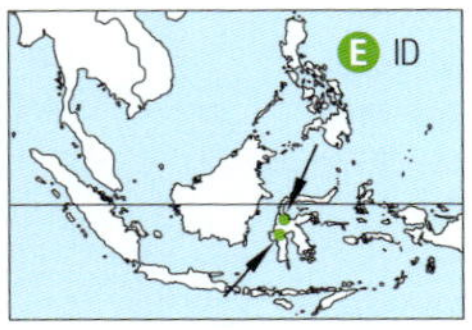

C Sulawesi (Mt Nokilalaki, Mt Gandangdewata)

Genus *MELASMOTHRIX*
(1 species)

DD
Monotypic
HB 11.5–12.1 cm
T 8.3–9.2 cm
W 42–52 g

Diurnal Sulawesian Shrew Rat
Melasmothrix naso
FR Rat diurne
DE Kleine Sulawesi-Spitzmausratte
ES Rata musaraña diurna de Célebes

Known only from 2 sites in C Sulawesi

Genus *HYORHINOMYS*
(1 species)

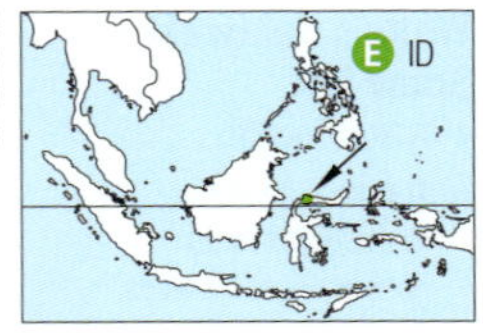

N Sulawesi (Mt Dako)

Hog-nosed Shrew Rat
Hyorhinomys stuempkei
FR Rat de Stümpke
DE Schweinsnasen-Spitzmausratte
ES Rata musaraña de hocico de cerdo

DD
Monotypic
HB 17.5–21 cm
T 18.5–22 cm
W 128–268 g

Genus *SOMMEROMYS*
(1 species)

DD
Monotypic
HB 9.5–10.8 cm
T 17.3–19.3 cm
W 24–30 g

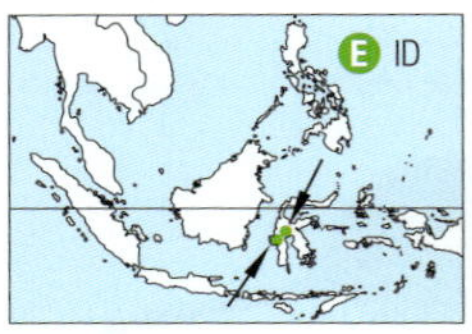

Known from few sites in C Sulawesi (Mt Tokala, Mt Gandangdewata)

Sommer's Sulawesi Rat
Sommeromys macrorhinos
FR Rat de Sommer
DE Sommer-Sulawesi-Langnasenratte
ES Rata de Célebes de Sommer

Genus *GRACILIMUS*
(1 species)

Sulawesi Slender Root Rat
Gracilimus radix
FR Rat des rhizomes
DE Sulawesi-Schlankratte
ES Rata de raíces de Célebes

DD
Monotypic
HB 10.8–12.5 cm
T 16.1–16.3 cm
W 34–45 g

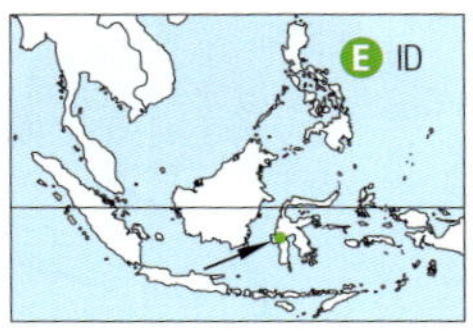

WC Sulawesi (Quarles Range)

Genus *WAIOMYS*
(1 species)

Mamasa Water Rat
Waiomys mamasae
FR Rat de Mamasa
DE Wasser-Spitzmausratte
ES Rata de agua de Mamasa

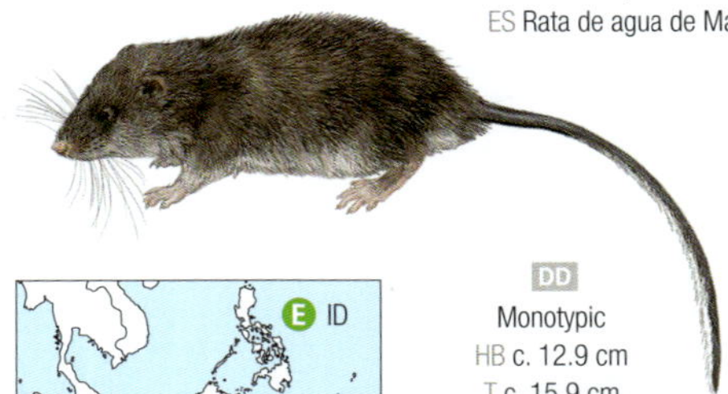

DD
Monotypic
HB c. 12.9 cm
T c. 15.9 cm
W c. 64 g

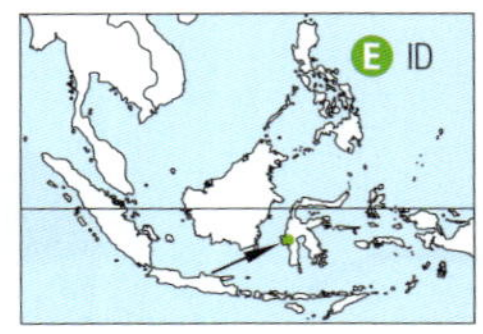

WC Sulawesi (known from 1 specimen from Mt Gandangdewata)

Genus *SAXATILOMYS*
(1 species)

Paulina's Limestone Rat
Saxatilomys paulinae
FR Rat de Jenkins
DE Paulina-Karstratte
ES Rata de caliza de Paulina

DD
Monotypic
HB 15.7–16.4 cm
T 18–20 cm
W 110–130 g

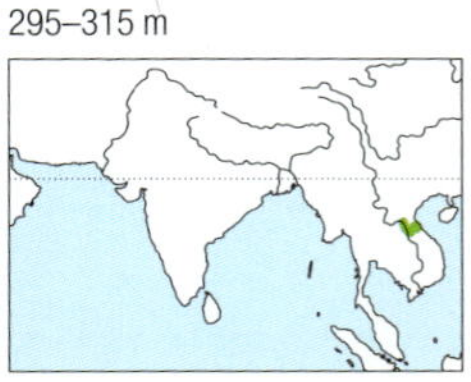

C LA (Khammouan Limestone NBCA), C VN (Quang Binh Prov)

Genus *TONKINOMYS*
(1 species)

Daovantien's Limestone Rat
Tonkinomys daovantieni
FR Rat de Dao Van Tien
DE Daovantien-Karstratte
ES Rata de caliza de Daovantien

DD
Monotypic
HB 18.4–22 cm
T 15.6–18.3 cm
W 140–205 g

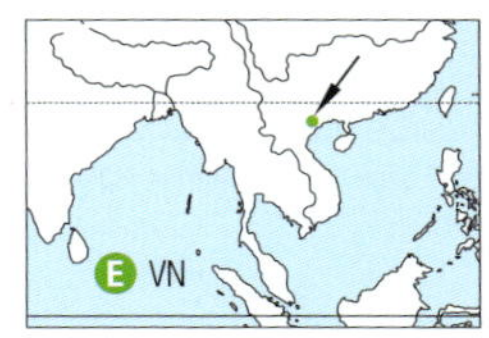

Huu Lien NR

Genus *MARGARETAMYS*
(4 species)

Lesser Margareta Rat
Margaretamys parvus
FR Rat des fûtaies
DE Kleine Margareta-Ratte
ES Rata margarita menor

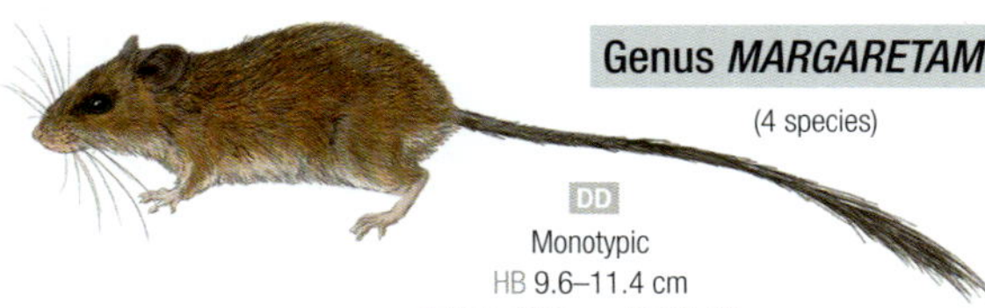

DD
Monotypic
HB 9.6–11.4 cm
T 15.4–18.4 cm W 23–40 g

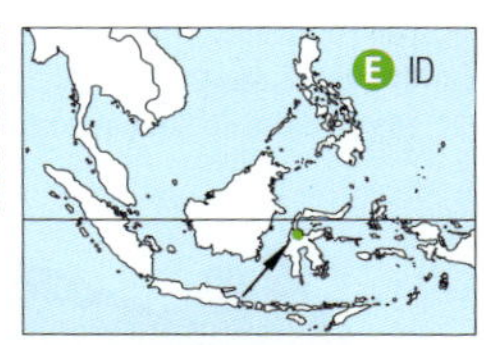

C Sulawesi (Mt Nokilalaki, Mt Kanino)

Christina's Margareta Rat
Margaretamys christinae
FR Rat de Christina
DE Christina-Margareta-Ratte
ES Rata margarita de Christina

EN
Monotypic
HB c. 11.1 cm
T c. 17.5 cm
W c. 49 g

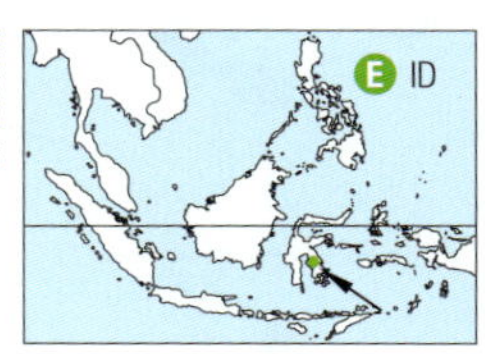

SE Sulawesi (Mt Moserosero)

DD
Monotypic
HB 11.7–14.7 cm
T 15–21 cm
W 50–85 g

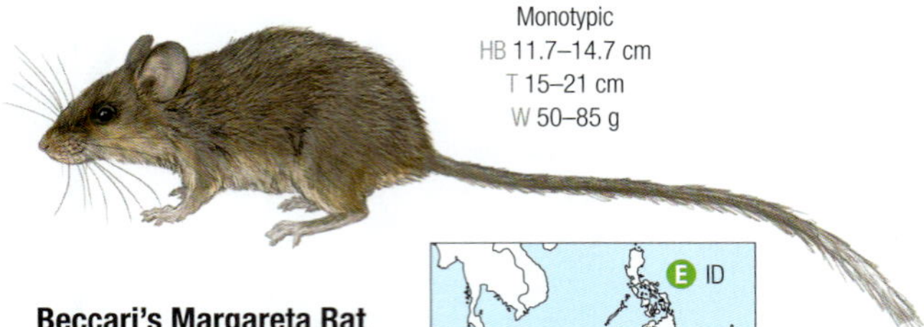

Beccari's Margareta Rat
Margaretamys beccarii
FR Rat de Beccari
DE Beccari-Margareta-Ratte
ES Rata margarita de Beccari

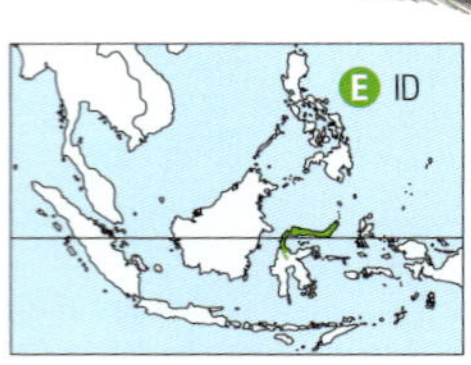

N & C Sulawesi

Elegant Margareta Rat
Margaretamys elegans
FR Rat élégant
DE Elegante Margareta-Ratte
ES Rata margarita elegante

NT
Monotypic
HB 16.5–19.7 cm
T 22–29 cm
W 85–150 g

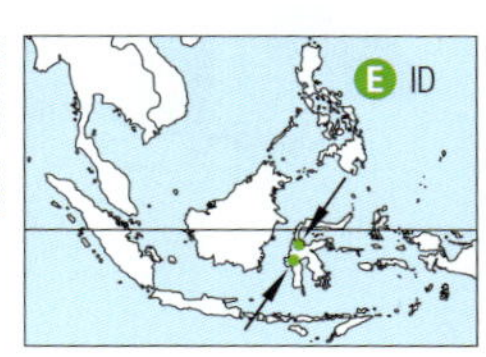

C Sulawesi (Mt Nokilalaki & S of Mamasa area; likely in other montane areas)

Genus *LENOTHRIX*

(1 species)

Gray Tree Rat

Lenothrix cana

FR Rat gris
DE Graue Haarbaumratte
ES Rata arborícola gris

LC
Monotypic
HB 14–21 cm
T 19.7–29 cm
W 81–220 g

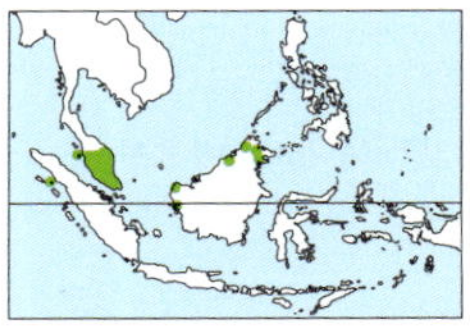

up to 550 m

Pen MY (with Penang I), Borneo, Tuangku I (off NW Sumatra)

Genus *DACNOMYS*

(1 species)

Millard's Rat

Dacnomys millardi

FR Rat de Millard
DE Millard-Ratte
ES Rata de Millard

DD
Monotypic
HB 21–27 cm
T 29–33 cm
W c. 300 g

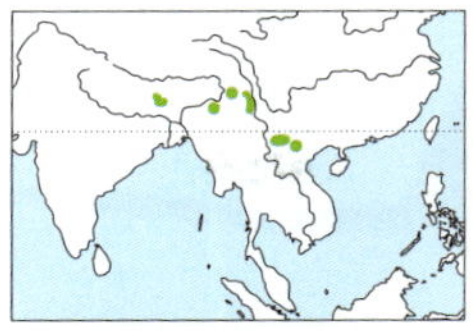

1050–3000 m

E NP, NE IN, SC CN (NW & S Yunnan), N LA, NW VN

Herbert's Long-tailed Giant Rat

Leopoldamys herberti

FR Rat de Herbert
DE Herbert-Langschwanz-Riesenratte
ES Rata gigante de cola larga de Herbert

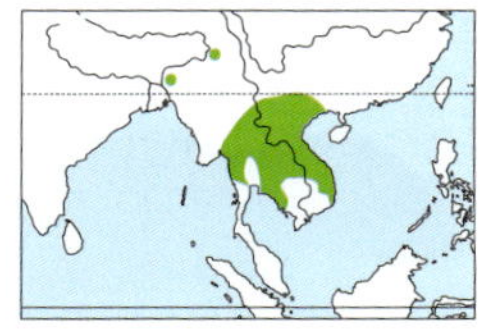

BD, NE IN, SC CN, SE MM, TH, LA, VN, N & W KH

NE
Monotypic
HB 23–28 cm
T 25–33 cm
W 203–276 g

Genus *LEOPOLDAMYS*

(9 species)

Millet's Long-tailed Giant Rat

Leopoldamys milleti

FR Rat de Millet
DE Millet-Langschwanz-Riesenratte
ES Rata gigante de cola larga de Millet

LC
Monotypic
HB 21–28 cm
T 29–36 cm

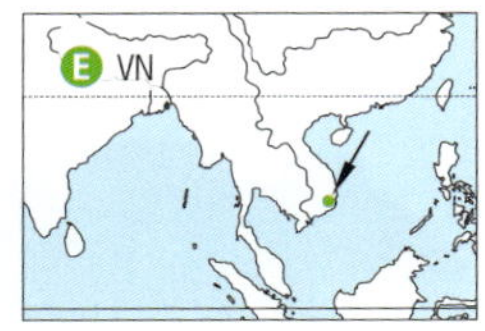

Langbian Plateau

Sundaic Mountain Long-tailed Giant Rat

Leopoldamys ciliatus

FR Rat cilié
DE Sunda-Langschwanz-Riesenratte
ES Rata gigante de cola larga de Sonda

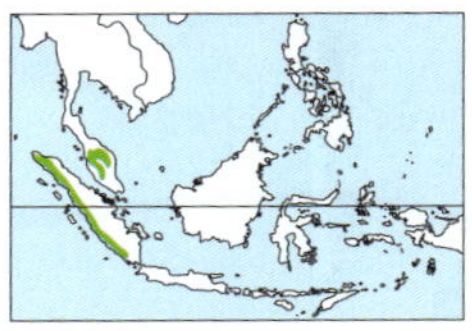

above 1000 m

Pen MY, W Sumatra (ID)

LC
Monotypic
HB 21–25 cm T 30–39 cm
W 204–425 g

Edwards's Long-tailed Giant Rat

Leopoldamys edwardsi

FR Rat de Milne-Edwards
DE Edwards-Langschwanz-Riesenratte
ES Rata gigante de cola larga de Edwards

LC
Monotypic
HB 21–29 cm T 26–31 cm
W 230–480 g

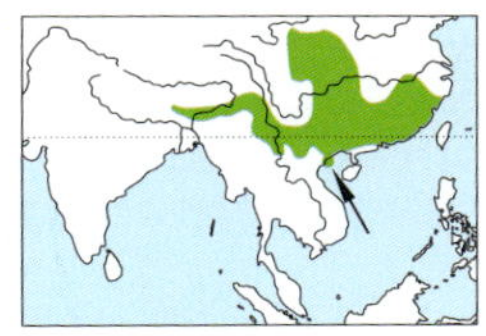

0–1400 m

NE IN, N MM, C & S CN, N VN (with Cat Ba I)

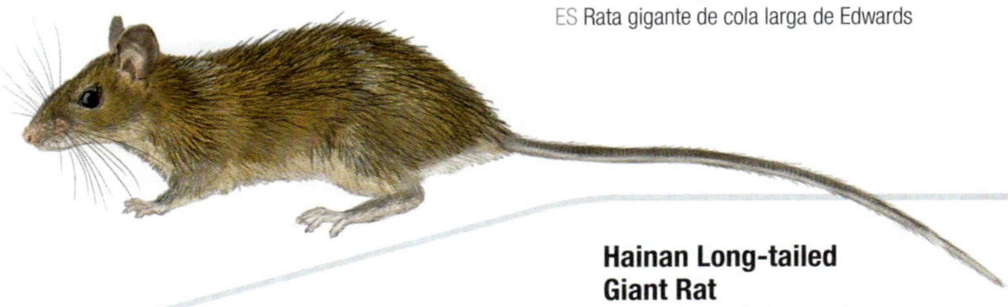

Neill's Long-tailed Giant Rat

Leopoldamys neilli

FR Rat de Neill
DE Neill-Langschwanz-Riesenratte
ES Rata gigante de cola larga de Neill

LC
Monotypic
HB 20–23 cm
T 24–30 cm
W c. 220 g

100–800 m

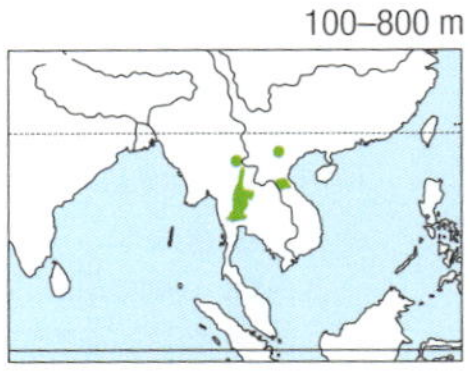

N & WC TH, C LA, N & WC VN

Hainan Long-tailed Giant Rat

Leopoldamys hainanensis

FR Rat de Xu
DE Hainan-Langschwanz-Riesenratte
ES Rata gigante de cola larga de Hainan

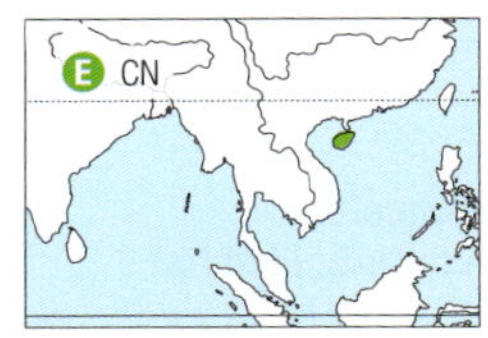

Hainan I

NE Monotypic
HB 18.5–25 cm T 23–29 cm
W 175–385 g

Indomalayan Long-tailed Giant Rat

Leopoldamys sabanus

FR Rat de Sabah
DE Südasiatische Langschwanz-Riesenratte
ES Rata gigante de cola larga de Indomalasia

LC
2 sspp
HB 20–27 cm
T 27–41 cm
W 250–500 g

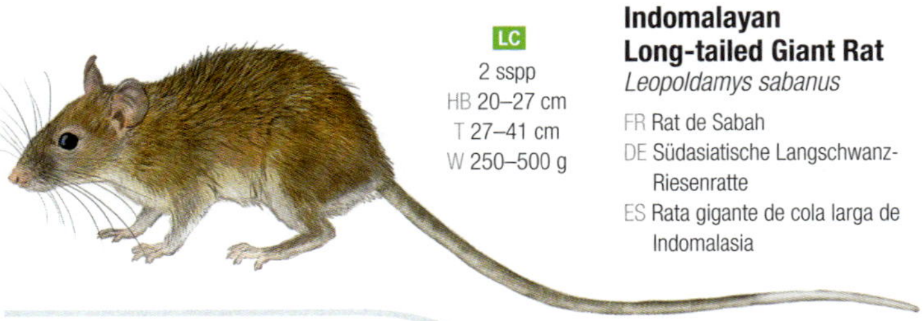

300–3100 m

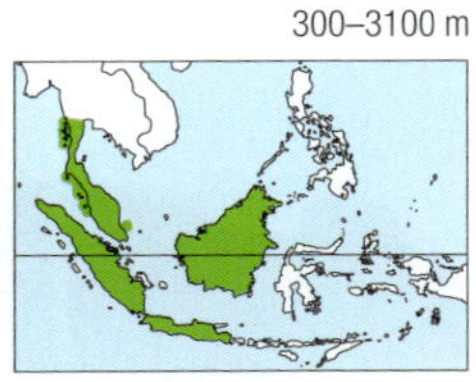

S MM, S TH, Thai-Malay Pen, Greater Sundas & some small offshore Is

Mentawai Archipelago Long-tailed Giant Rat

Leopoldamys siporanus

FR Rat de Sipora
DE Mentawai-Langschwanz-Riesenratte
ES Rata gigante de cola larga de Mentawai

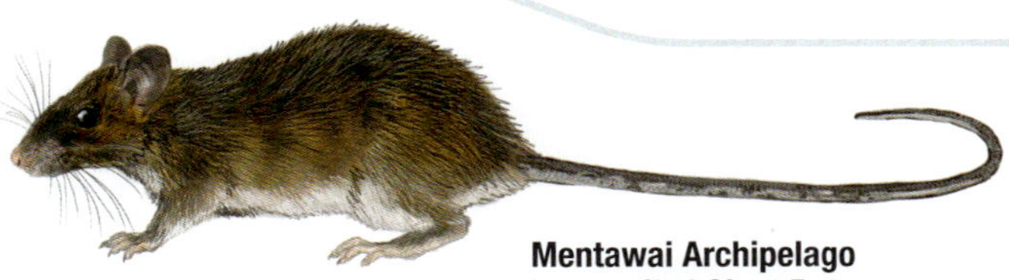

VU
Monotypic
HB 17.8–29 cm
T 22–33 cm

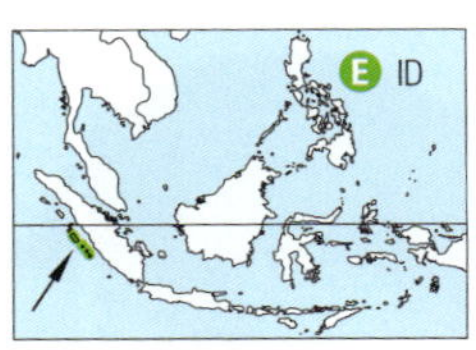

Mentawai Arch (Siberut, Sipora, North Pagai, South Pagai), off W Sumatra

Diwangkara's Long-tailed Giant Rat

Leopoldamys diwangkarai

FR Rat diwangkara
DE Diwangkara-Langschwanz-Riesenratte
ES Rata gigante de cola larga de Diwangkara

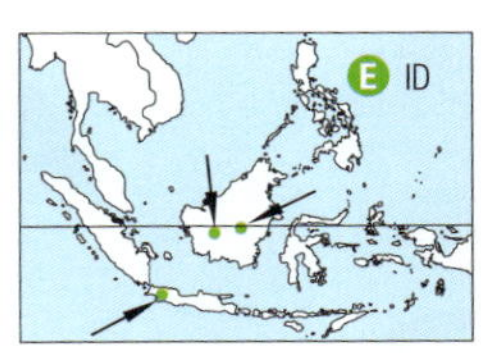

400–1400 m

Known only from 3 sites in C Borneo (Kalimantan) & W Java

DD
Monotypic
HB 19.7–22 cm
T 29–32 cm

LC
Monotypic
HB 13.8–16 cm
T 20–23 cm

Burmese Tree Rat
Chiromyscus chiropus
FR Rat de Fea
DE Burma-Fea-Baumratte
ES Rata arborícola de Birmania

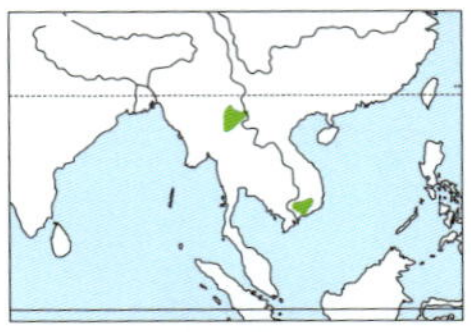

E MM, S VN; limits unknown

Genus *CHIROMYSCUS*

(3 species)

NE IN, MM, N & W TH, LA, VN, NE & SW KH, 1 record from Yunnan (SC CN)

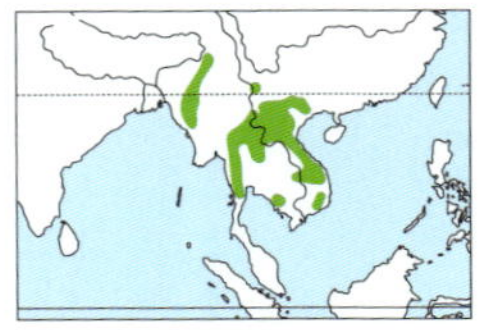

60–2800 m

LC
Monotypic
HB 13.5–15.5 cm
T 15.4–19.9 cm
W 58–98 g

Langbian Tree Rat
Chiromyscus langbianis
FR Rat du Lang Bian
DE Langbian-Fea-Baumratte
ES Rata arborícola de Langbian

NE
Monotypic
HB 14–18 cm
T 20–23 cm

Thomas's Tree Rat
Chiromyscus thomasi
FR Rat de Thomas
DE Thomas-Fea-Baumratte
ES Rata arborícola de Thomas

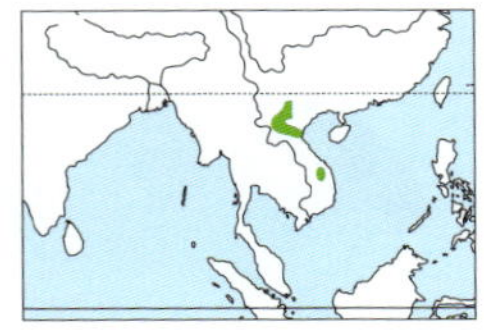

N & C VN, N LA; limits unknown

Genus *NIVIVENTER*

(24 species)

Sundaic Arboreal White-bellied Rat
Niviventer cremoriventer
FR Rat à ventre crème
DE Sunda-Weißbauchratte
ES Rata de vientre blanco arbórea de Sonda

LC
Monotypic
HB 12.5–14.8 cm
T 15.8–20 cm
W 51–58 g

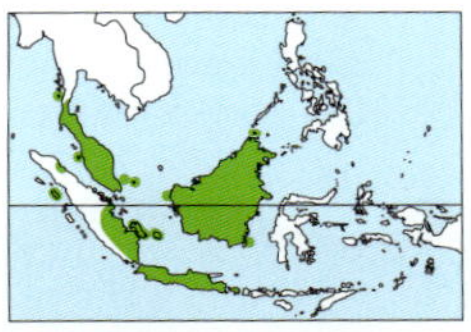

c. 1650 m

Extreme S MM (Mergui Arch), Thai-Malay Pen, Greater Sundas & offshore Is

Montane Bornean White-bellied Rat
Niviventer rapit
FR Rat rapit
DE Langschwanz-Weißbauchratte
ES Rata de vientre blanco de montaña de Borneo

600–3360 m

LC
Monotypic
HB c. 14.3 cm
T c. 19.7 cm
W c. 60 g

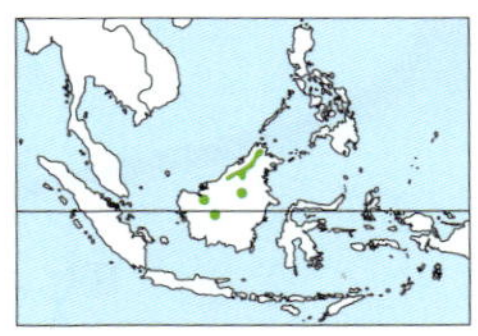

Highlands of Borneo (MY, ID)

Cameron Highlands White-bellied Rat
Niviventer cameroni
FR Rat de Cameron
DE Cameron-Hochland-Weißbauchratte
ES Rata de vientre blanco de Cameron

1520–2010 m

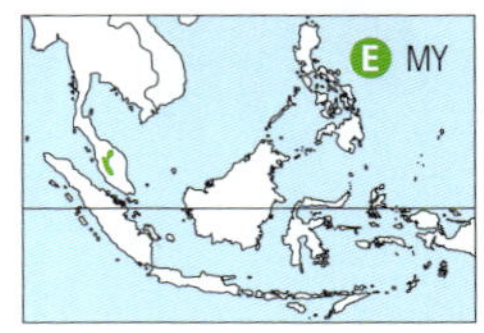

Cameron Highlands

VU
Monotypic
HB 13–17 cm
T 20–27 cm
W 58–129 g

Montane Sumatran White-bellied Rat
Niviventer fraternus
FR Rat jumeau
DE Sumatra-Weißbauchratte
ES Rata de vientre blanco de montaña de Sumatra

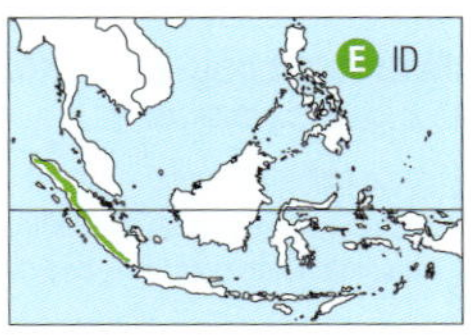

1250–3500 m

W Sumatra

LC
Monotypic
HB c. 16.2 cm T c. 23 cm

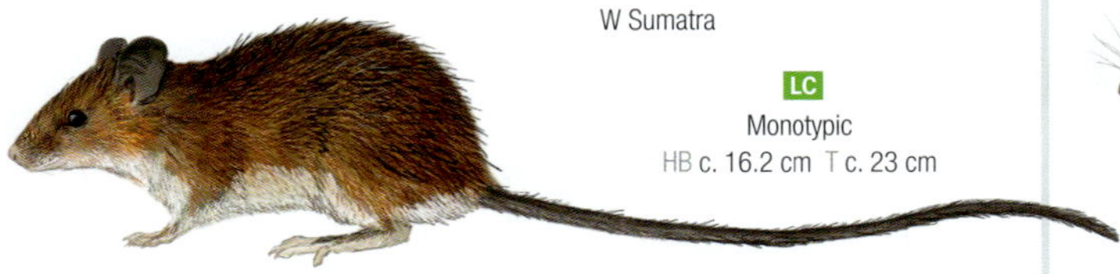

LC
Monotypic
HB c. 13.5 cm
T c. 17.8 cm

Montane Javan White-bellied Rat
Niviventer lepturus
FR Rat à queue fine
DE Java-Weißbauchratte
ES Rata de vientre blanco de montaña de Java

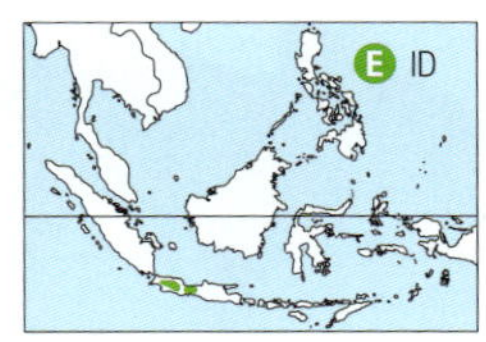

1000–2000 m

W & C Java

Limestone White-bellied Rat
Niviventer hinpoon
FR Rat du Korat
DE Karst-Weißbauchratte
ES Rata de vientre blanco de caliza

EN
Monotypic
HB 12–16 cm
T 12–16 cm
W 50–70 g

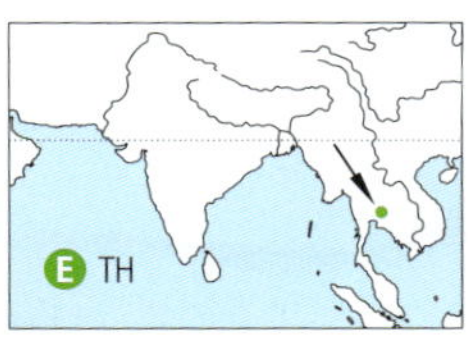

Limestone karst habitat

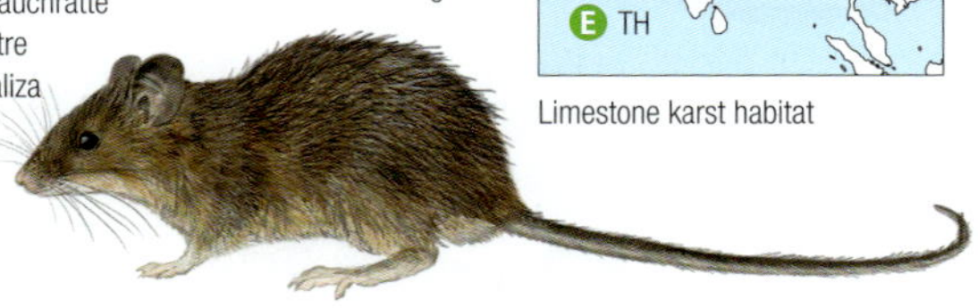

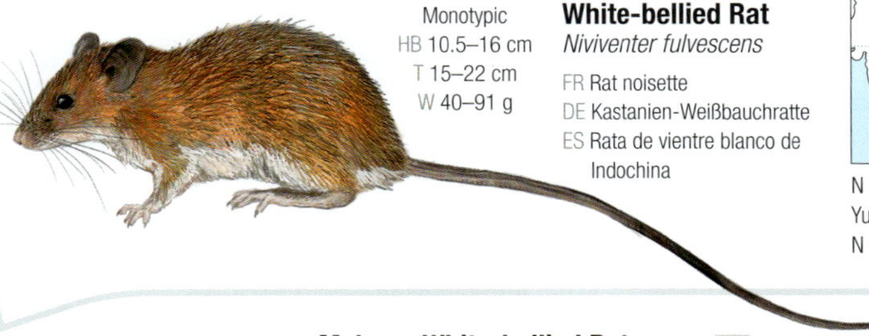

LC
Monotypic
HB 10.5–16 cm
T 15–22 cm
W 40–91 g

Indochinese White-bellied Rat
Niviventer fulvescens
FR Rat noisette
DE Kastanien-Weißbauchratte
ES Rata de vientre blanco de Indochina

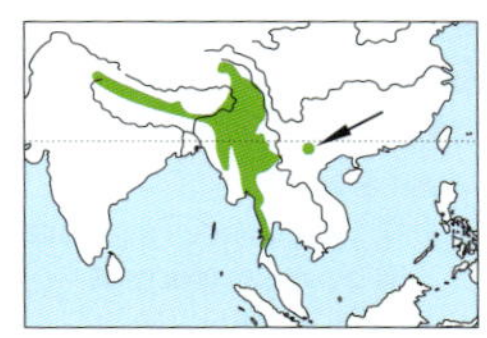

N IN & NP E to SC CN (E Tibet, Yunnan), MM, N VN; perhaps N PK & TH

Mekong White-bellied Rat
Niviventer mekongis
FR Rat du Mékong
DE Mekong-Weißbauchratte
ES Rata de vientre blanco de Mekong

NE
Monotypic
HB 10.5–16.3 cm
T 14.3–22 cm
W 46–126 g

50–2300 m

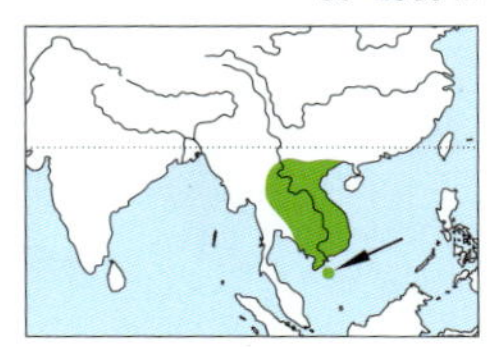

CN (S Yunnan, Guangxi), LA, VN

South China White-bellied Rat
Niviventer huang
FR Rat jaune
DE Fukien-Weißbauchratte
ES Rata de vientre blanco de China meridional

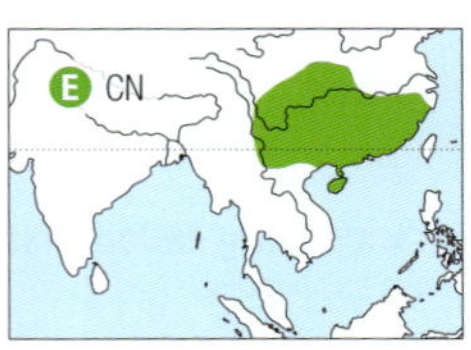

200–2300 m

E from Sichuan, Yunnan & Hubei, Hainan I

NE
Monotypic
HB 11.9–16.5 cm
T 14–20 cm
W 40–120 g

Feng's White-bellied Rat
Niviventer fengi
FR Rat de Feng
DE Feng-Weißbauchratte
ES Rata de vientre blanco de Feng

NE
Monotypic
HB 11–16 cm
T 12.7–18 cm
W 41–80 g

2900–3200 m

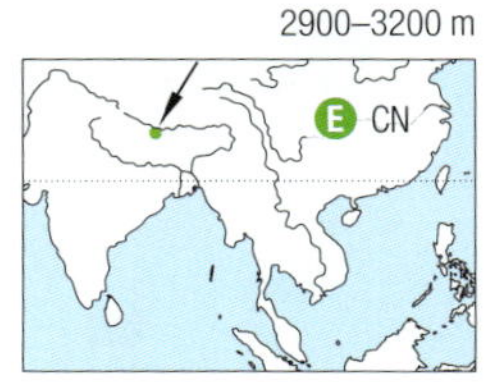

Jilong Valley (Tibet), near Nepalese border

Anderson's White-bellied Rat
Niviventer andersoni

FR Rat d'Anderson
DE Anderson-Weißbauchratte
ES Rata de vientre blanco de Anderson

LC
2 sspp
HB 13.6–20 cm
T 20–27 cm
W 88–182 g

2000–3000 m

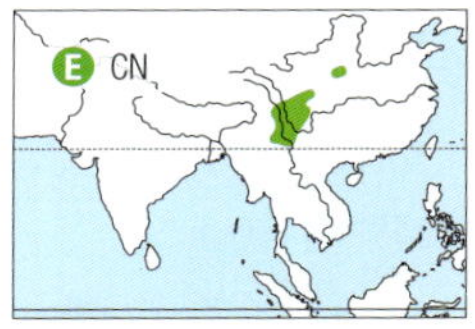

S Shaanxi, Sichuan, Yunnan

Sichuan White-bellied Rat
Niviventer excelsior

LC
Monotypic
HB 12.1–14.5 cm
T 18.5–22 cm
W 56–95 g

FR Rat du Sichuan
DE Sichuan-Weißbauchratte
ES Rata de vientre blanco de Sichuán

2300–3000 m

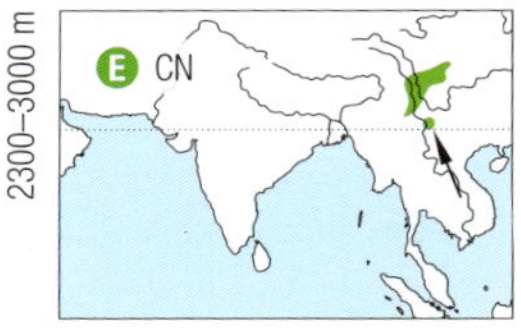

SW Sichuan, NW & C Yunnan

Soft-furred Taiwan White-bellied Rat
Niviventer culturatus

FR Rat de Taïwan
DE Taiwan-Weißbauchratte
ES Rata de vientre blanco de pelaje suave de Taiwán

LC
Monotypic
HB 13–15 cm
T 17–20 cm
W 62–213 g

2000–3000 m

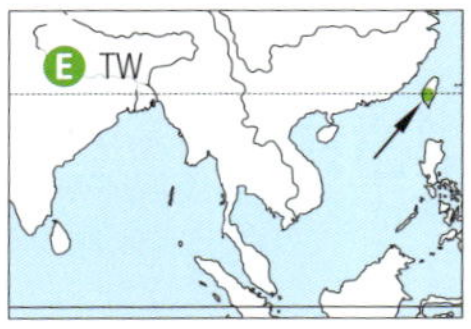

Highland forests

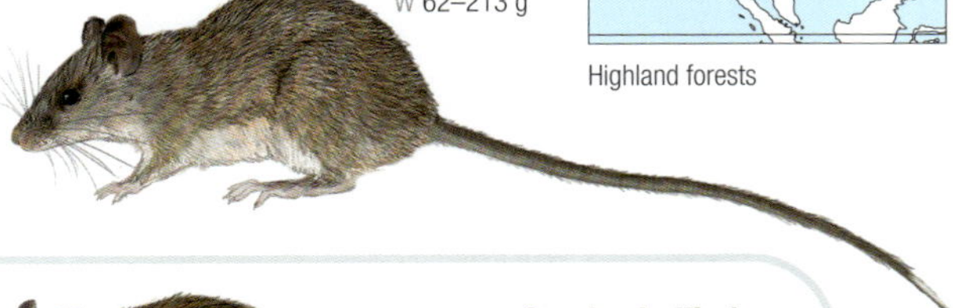

Brahman White-bellied Rat
Niviventer brahma

FR Rat brahmane
DE Brahma-Weißbauchratte
ES Rata de vientre blanco de Brahman

2000–2800 m

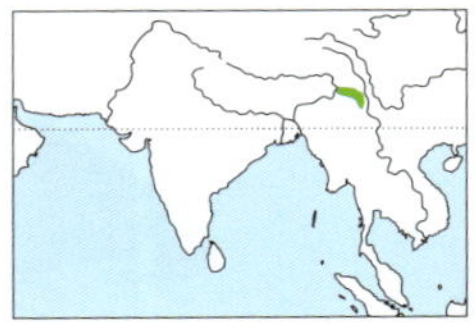

LC
Monotypic
HB 13.5–15.5 cm
T 20–24 cm

NE IN (Arunachal Pradesh), N MM (Kachin State), SC CN (Yunnan)

Himalayan White-bellied Rat
Niviventer niviventer

FR Rat à ventre blanc
DE Große Himalaya-Weißbauchratte
ES Rata de vientre blanco del Himalaya

LC
Monotypic
HB 13.3–16 cm
T c. 21 cm

235–2700 m

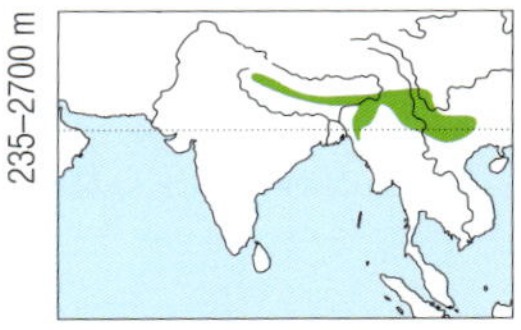

N & NE IN, NP, BT, N MM, N LA, N VN, SC CN

Smoke-bellied White-bellied Rat
Niviventer eha

FR Rat d'Aitken
DE Kleine Himalaya-Weißbauchratte
ES Rata de vientre blanco ahumado

LC
Monotypic
HB 11.2–13 cm
T 16.5–19.5 cm
W 21–63 g

2000–3700 m

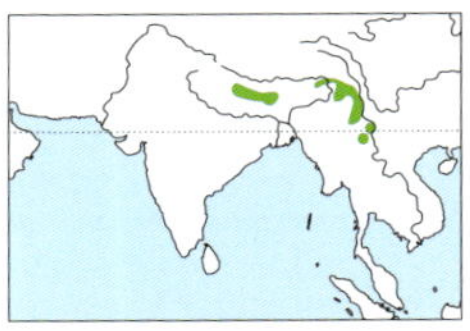

NP, IN (West Bengal, Sikkim, Arunachal Pradesh), N MM, SW & S CN; possibly BT

Spiny Taiwan White-bellied Rat
Niviventer coninga

FR Rat de Coxinga
DE Stachelige Taiwan-Weißbauchratte
ES Rata de vientre blanco espinosa de Taiwán

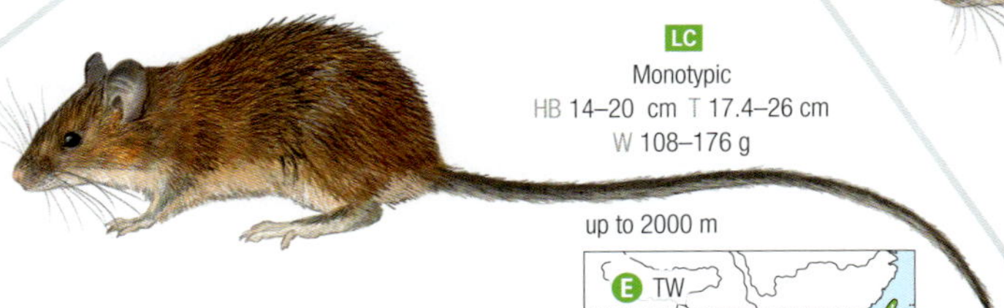

LC
Monotypic
HB 14–20 cm T 17.4–26 cm
W 108–176 g

up to 2000 m

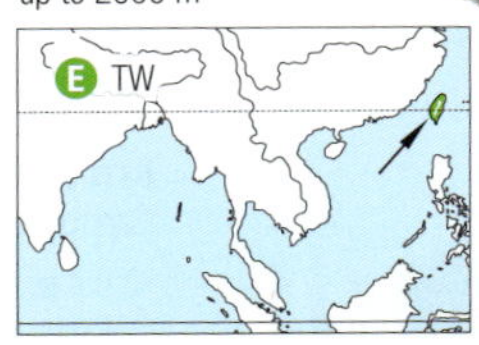

Lowland forests and scrubland

Hainan White-bellied Rat
Niviventer lotipes

NE
Monotypic
HB 10–18.3 cm
T 10.3–20 cm
W 32–118 g

FR Rat de Hainan
DE Hainan-Weißbauchratte
ES Rata de vientre blanco de Hainan

15–2300 m

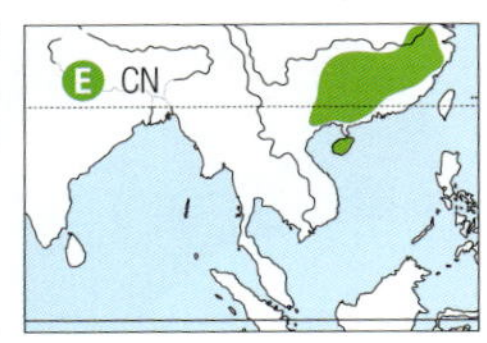

In SE, S of R Yangtze, Hainan I

Indochinese Mountain White-bellied Rat
Niviventer tenaster

FR Rat du Mulayit
DE Indochinesische Weißbauchratte
ES Rata de vientre blanco de montaña de Indochina

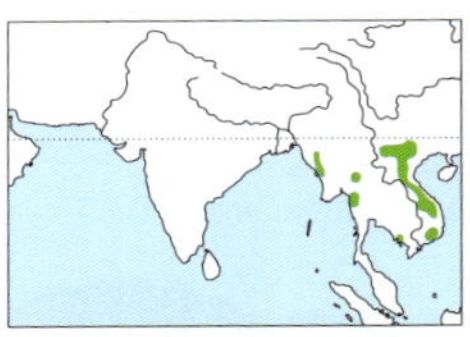

1300–3000 m

WC & S MM, NW TH (Chiang Mai Prov), LA, VN, S KH (Dâmrei Mts)

LC
Monotypic
HB 12–19 cm
T 17.5–23 cm
W 50–140 g

Least White-bellied Rat
Niviventer gladiusmaculus

FR Rat du Tibet
DE Tibet-Weißbauchratte
ES Rata de vientre blanco mínima

NE
Monotypic
HB 10.2–14.5 cm
T 12.3–18 cm W 32–82 g

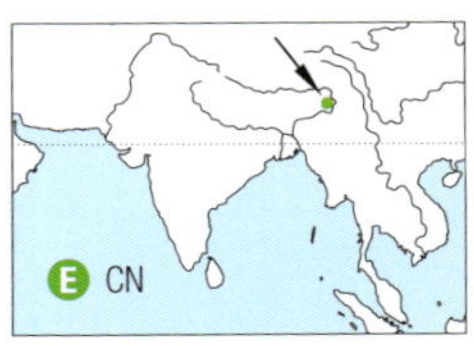

2980–4415 m

SE Tibet (known from Linzhi and its vicinity)

Pianma White-bellied Rat
Niviventer pianmaensis

FR Rat de Pianma
DE Pianma-Weißbauchratte
ES Rata de vientre blanco de Pianma

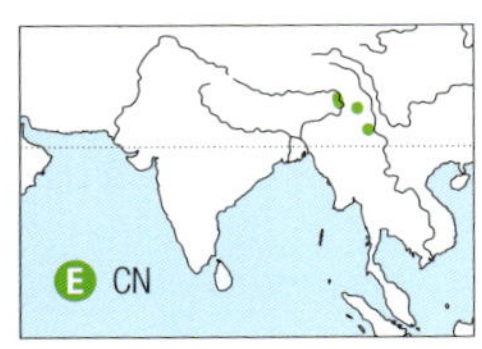

1160–4530 m

SE Tibet, SW Yunnan

NE
Monotypic
HB 12.8–17 cm T 17.9–23 cm
W 75–150 g

Confucian White-bellied Rat
Niviventer confucianus

FR Rat de Confusius
DE Chinesische Weißbauchratte
ES Rata de vientre blanco de Confucio

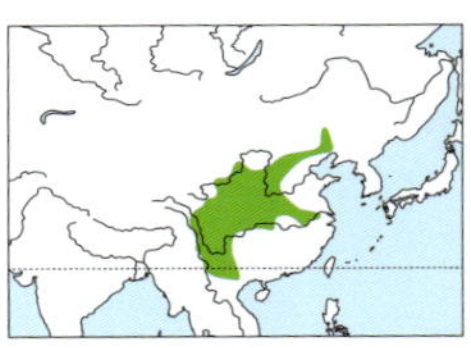

30–4475 m

C & N CN, N MM, N VN

LC
3 sspp
HB 11.3–21 cm
T 11.5–18.9 cm
W 43–124 g

Bukit White-bellied Rat
Niviventer bukit

NE
Monotypic
HB 12–15.7 cm T 14.4–18.5 cm
W 60–135 g

FR Rat de Bukit Besar
DE Bukit-Weißbauchratte
ES Rata de vientre blanco de Bukit

95–1400 m

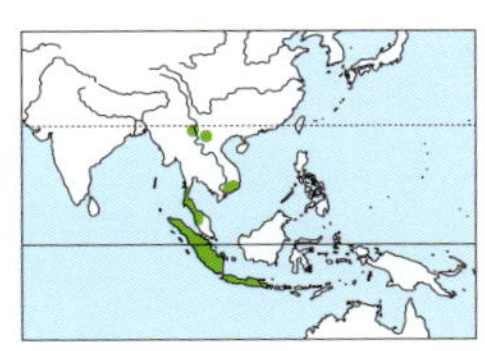

SC CN (Yunnan), N LA, S VN, Thai-Malay Pen, Sumatra, Java, Bali

Shandong White-bellied Rat
Niviventer sacer
FR Rat des temples
DE Shandong-Weißbauchratte
ES Rata de vientre blanco de Shandong

NE
Monotypic
HB 12.2–16.9 cm
T 12.1–17.9 cm

Mts & hills throughout Shandong

Ohia Rat
Srilankamys ohiensis
FR Rat d'Ohiya
DE Ohiya-Ratte
ES Rata de Ohiya

Genus *SRILANKAMYS*
(1 species)

VU
Monotypic
HB c. 14.5 cm
T c. 17.3 cm

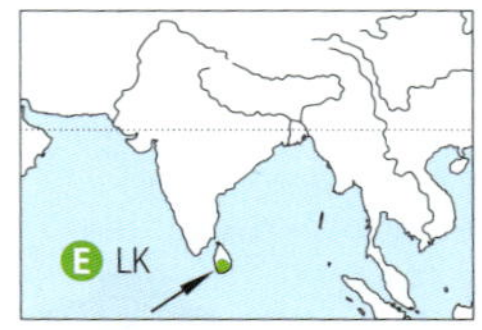

915–2310 m
Central Highlands

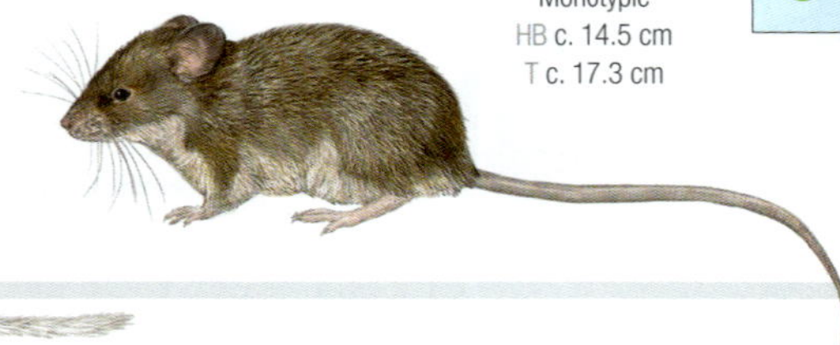

Genus *BERYLMYS*
(4 species)

Mackenzie's White-toothed Rat
Berylmys mackenziei
FR Rat de Mackenzie
DE Mackenzie-Weißzahnratte
ES Rata de dientes blancos de Mackenzie

DD
Monotypic
HB 15.5–24 cm
T 15–25 cm
W 120–140 g

1000–1850 m
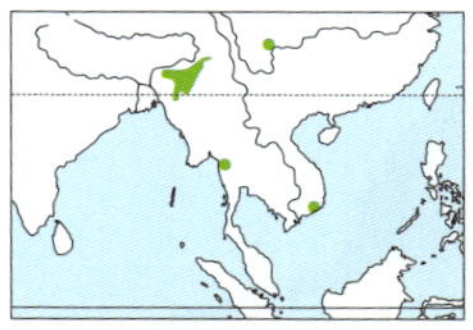
NE IN, NW & SE MM, CN (Sichuan), S VN

Manipur White-toothed Rat
Berylmys manipulus
FR Rat du Manipur
DE Manipur-Weißzahnratte
ES Rata de dientes blancos de Manipur

DD
Monotypic
HB 13.5–18.5 cm
T 14–18.7 cm
W c. 43 g

75–1525 m
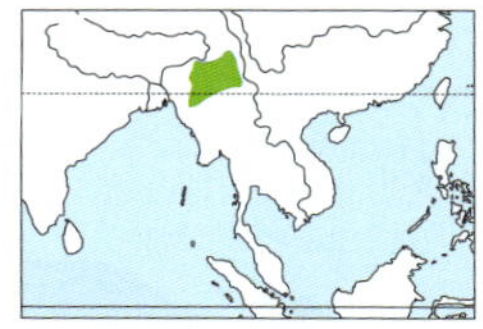
NE IN, N MM, SC CN (Yunnan)

Berdmore's White-toothed Rat
Berylmys berdmorei
FR Rat de Berdmore
DE Berdmore-Weißzahnratte
ES Rata de dientes blancos de Berdmore

LC
Monotypic
HB 19–25 cm
T 14.9–19.2 cm
W 158–300 g

0–1400 m
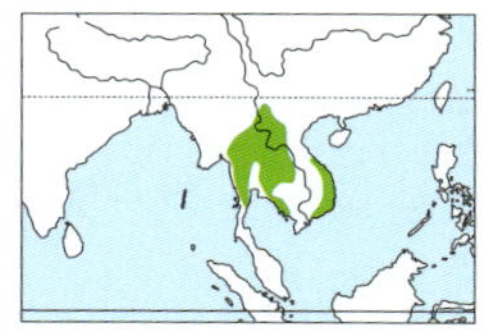
SC CN (Yunnan), SE MM, TH, N LA, VN

Bowers's White-toothed Rat
Berylmys bowersii
FR Rat de Bowers
DE Bowers-Weißzahnratte
ES Rata de dientes blancos de Bowers

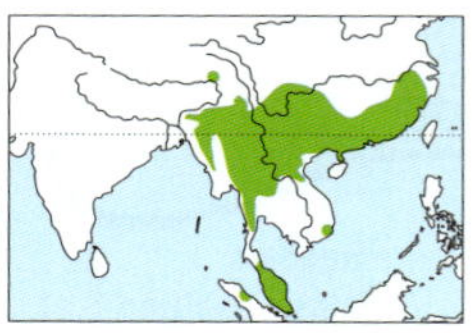
150–2620 m
NE IN E to SE CN, S to Thai-Malay Pen, NE Sumatra

LC
Monotypic
HB 23–30 cm
T 24–31 cm
W 270–650 g

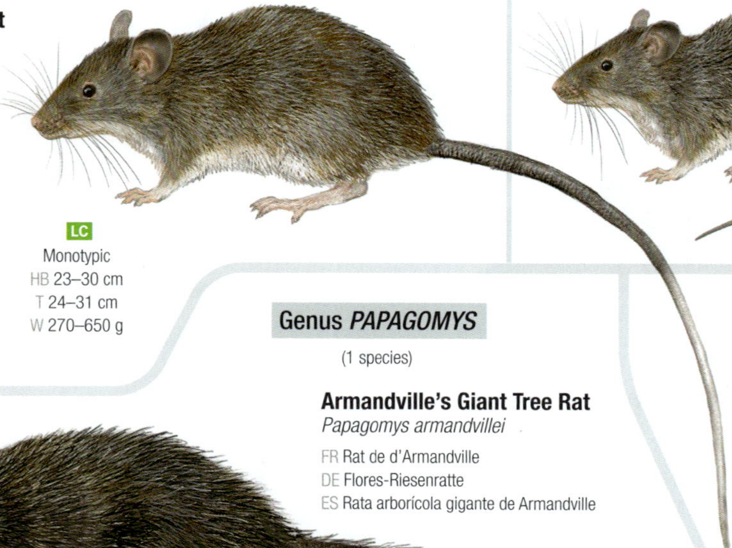

Genus *PAPAGOMYS*
(1 species)

Armandville's Giant Tree Rat
Papagomys armandvillei
FR Rat de d'Armandville
DE Flores-Riesenratte
ES Rata arborícola gigante de Armandville

NT
Monotypic
HB 27–43 cm
T 26–36 cm

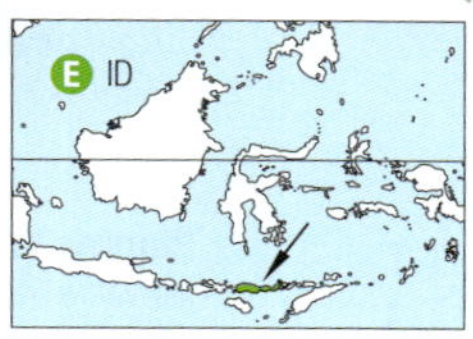

0–900 m
Flores I

Genus *PAULAMYS*
(1 species)

Paula's Long-nosed Rat
Paulamys naso
FR Rat de Paula
DE Flores-Langnasenratte
ES Rata de hocico largo de Paula

1000–2000 m
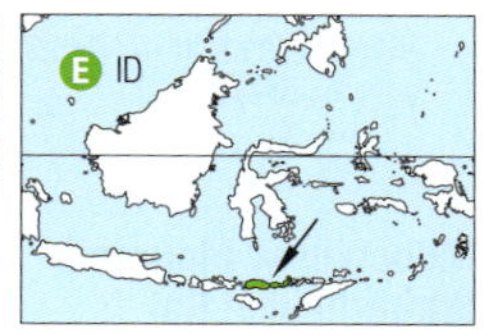

Flores I

EN
Monotypic
HB c. 16.4 cm
T c. 11.7 cm
W c. 122 g

Genus *KOMODOMYS*
(1 species)

VU
Monotypic
HB 12–18 cm
T 12–16.3 cm
W 59–155 g

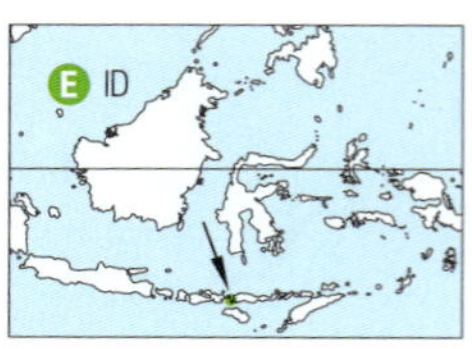

Rinca & Padar Is (Lesser Sunda Is); probably other Is (e.g. Komodo)

Komodo Rat
Komodomys rintjanus
FR Rat de Komodo
DE Komodo-Ratte
ES Rata de Komodo

Genus *SUNDAMYS*
(4 species)

Annandale's Sundaic Rat
Sundamys annandalei
FR Sundamys d'Annandale
DE Annandale-Sunda-Riesenratte
ES Rata de Sonda de Annandale

LC
Monotypic
HB 17.3–22 cm
T 22–26 cm
W 155–261 g

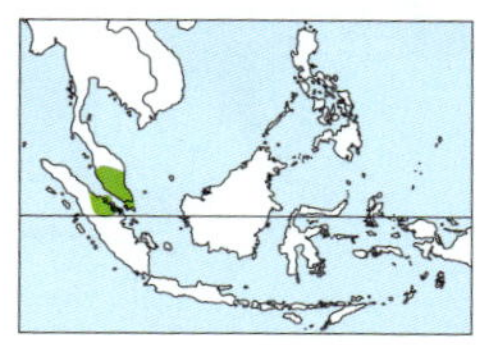
ID (E Sumatra, adjacent Rupat & Padang Is), Pen MY, SG

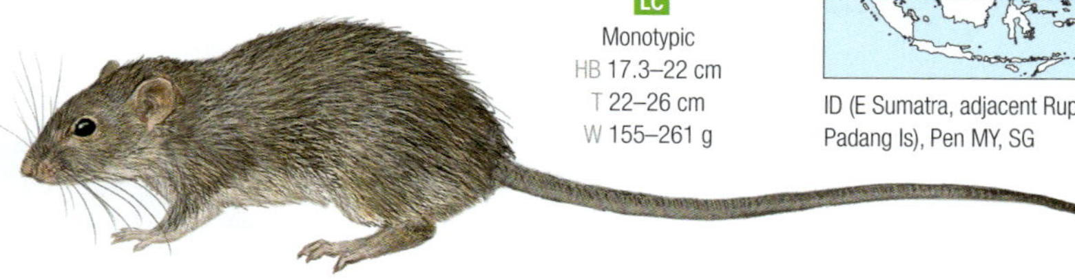

Mountain Giant Rat
Sundamys infraluteus
FR Sundamys des montagnes
DE Berg-Sunda-Riesenratte
ES Rata de Sonda de montaña

LC
2 sspp
HB 23–28 cm
T 29–34 cm
W 468–643 g

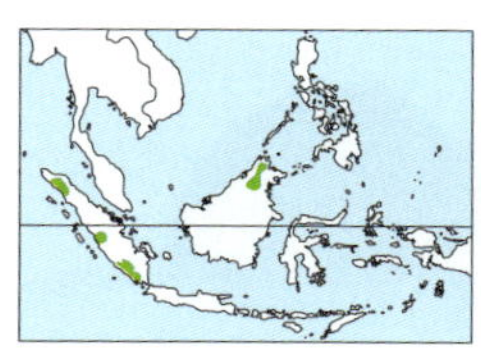
700–2930 m
W Sumatra & N Borneo (ID, MY, possibly BN)

Bartels's Rat
Sundamys maxi

FR Sundamys de Java
DE Bartels-Sunda-Riesenratte
ES Rata de Sonda de Bartels

VU
Monotypic
HB 22–27 cm
T 26–31 cm
W 521–643 g

900–1350 m

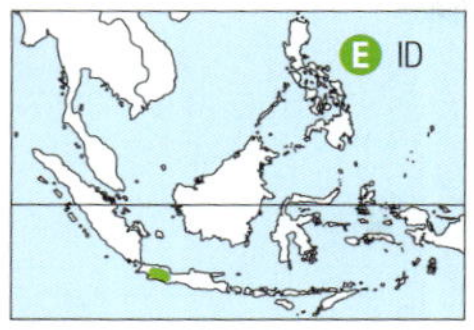

W Java

Müller's Rat
Sundamys muelleri

FR Sundamys de Müller
DE Müller-Sunda-Riesenratte
ES Rata de Sonda de Müller

LC
Monotypic
HB 18.5–30 cm
T 21–37 cm
W 196–292 g

0–1650 m

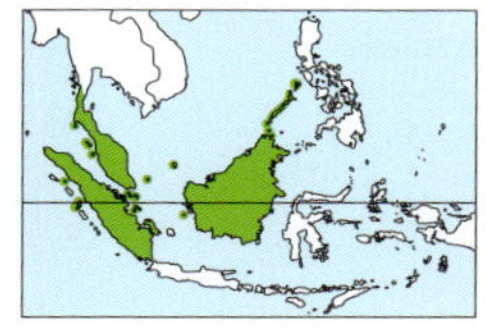

SW MM, Thai-Malay Pen, Sumatra, Borneo, Palawan & many offshore Is

Luzon Forest Rat
Bullimus luzonicus

FR Bullimus de Luçon
DE Luzon-Waldratte
ES Rata de bosque de Luzón

LC
Monotypic
HB 23–27 cm
T 19–23 cm W 351–520 g

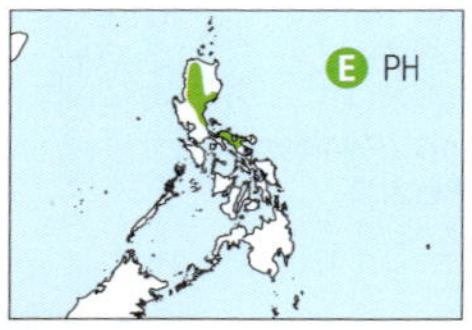

1000–1450 m

Luzon I

Genus *BULLIMUS*

(4 species)

Carleton's Forest Rat
Bullimus carletoni

FR Bullimus de Carleton
DE Carleton-Waldratte
ES Rata de bosque de Carleton

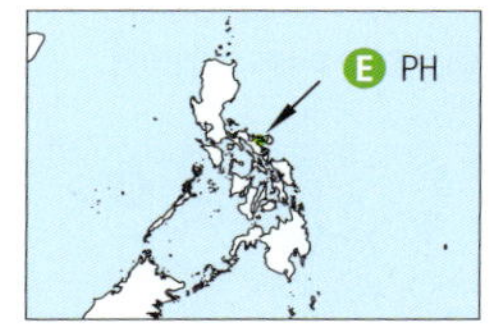

S Luzon I (E Caramoan Pen)

NE
Monotypic
HB 23–26 cm
T 18.7–22 cm
W 350–445 g

Mindanao Forest Rat
Bullimus bagobus

FR Bullimus de Mindanao
DE Mindanao-Waldratte
ES Rata de bosque de Mindanao

LC
Monotypic
HB 23–27 cm
T 17.5–19.6 cm
W 340–600 g

200–1800 m

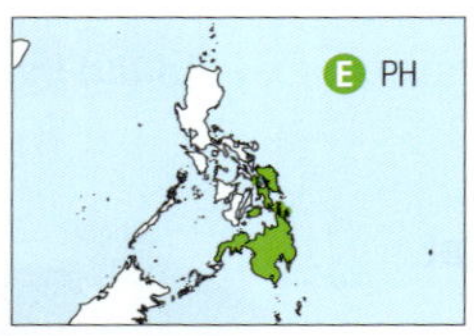

Samar, Maripipi, Leyte, Bohol, Dinagat, Siargao, Bucas Grande & Mindanao Is

Camiguin Forest Rat
Bullimus gamay

FR Bullimus de Camiguin
DE Camiguin-Waldratte
ES Rata de bosque de Camiguín

VU
Monotypic
HB 22–24 cm T 14.2–19.9
W 290–500 g

900–1475 m

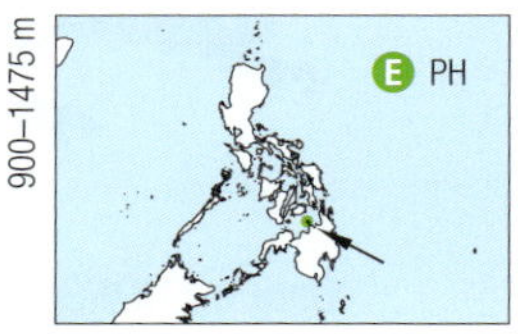

Camiguin I

Spiny Boki Mekot Rat
Halmaheramys bokimekot

FR Rat de Boki Mekot
DE Halmahera-Stachelratte
ES Rata espinosa de Boki Mekot

DD
Monotypic
HB 14.6–16.7 cm
T 11.9–13.2 cm
W 72–99 g

700–750 m

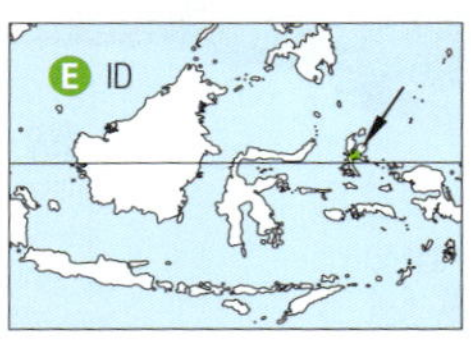

C Halmahera (Moluccas); subfossil records from Morotai I

Genus *HALMAHERAMYS*

(2 species)

Wallace's Large Spiny Rat
Halmaheramys wallacei

FR Rat de Wallace
DE Wallace-Stachelratte
ES Rata espinosa de Wallace

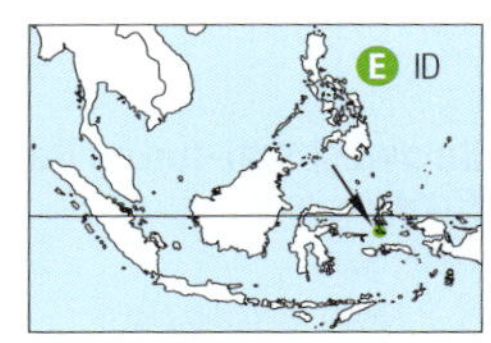

Bisa & Obi Is (Moluccas)

NE
Monotypic
HB 20–24 cm (F)
T 16.8–21 cm (F)
W 240–300 g (F)

Lovely-haired Rat
Taeromys callitrichus

FR Rat-laineux de Jentink
DE Schönhaar-Sulawesi-Ratte
ES Rata de pelo bonito

DD
Monotypic
HB 20–24 cm
T 22–26 cm
W 290–363 g

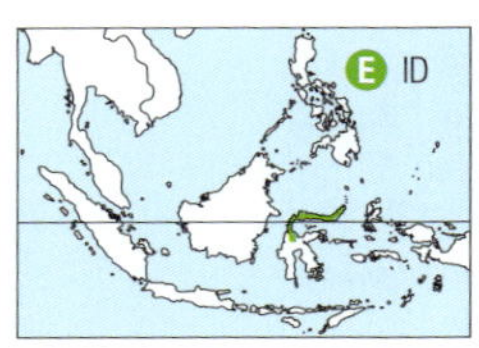

760–2260 m

Minahasa Pen & NE part of C Sulawesi

Genus *TAEROMYS*

(8 species)

Sulawesi Forest Rat
Taeromys punicans

FR Rat-laineux de Hollister
DE Rötliche Sulawesi-Ratte
ES Rata de bosque de Célebes

DD
Monotypic
HB 18.5–21 cm
T 15.6–18.5 cm

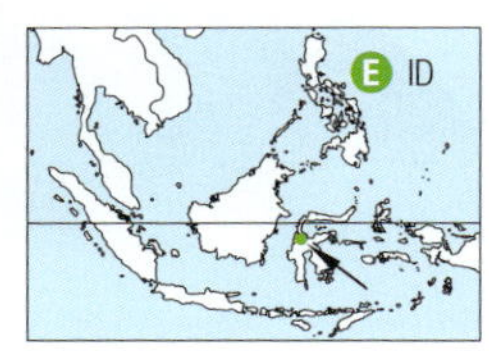

C Sulawesi (known only from Pinedapa area)

Tondano Rat
Taeromys taerae

FR Rat-laineux de Sody
DE Tondano-Sulawesi-Ratte
ES Rata de Tondano

600–800 m

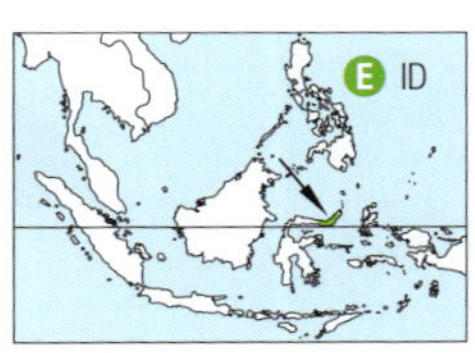

Sulawesi (Minahasa Pen)

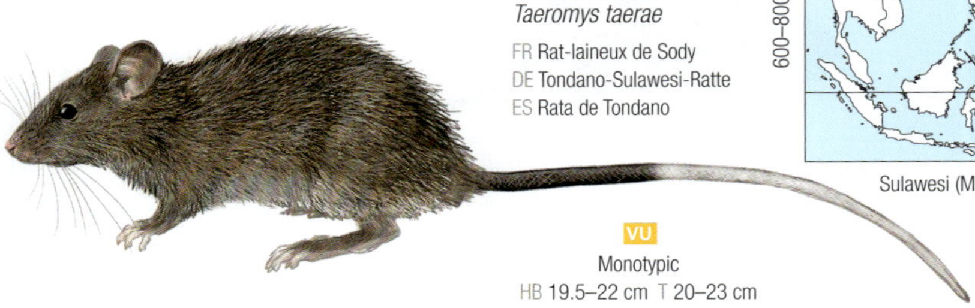

VU
Monotypic
HB 19.5–22 cm T 20–23 cm

Salokko Rat

Taeromys arcuatus

FR Rat-laineux du Salokko
DE Salokko-Sulawesi-Ratte
ES Rata de Salokko

LC
Monotypic
HB 20–21 cm
T 20–23 cm

1500–2000 m

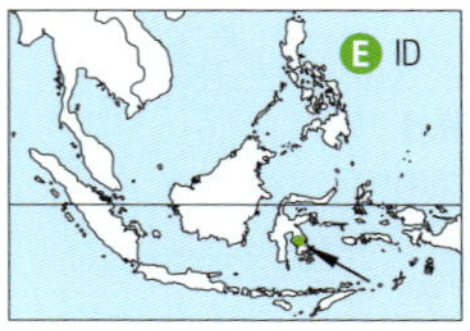

SE Pen of Sulawesi (Tanke Salokko)

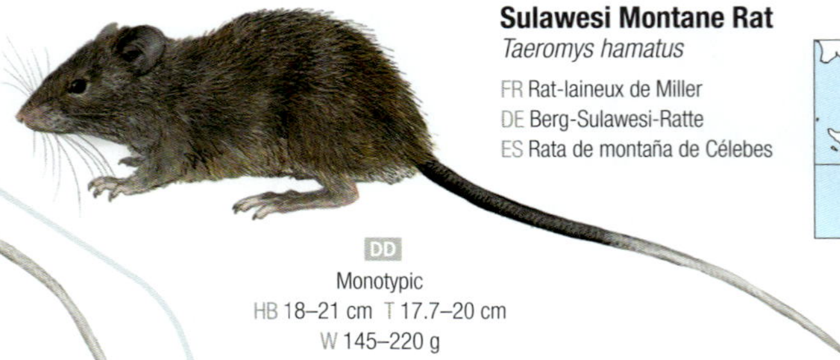

Sulawesi Montane Rat

Taeromys hamatus

FR Rat-laineux de Miller
DE Berg-Sulawesi-Ratte
ES Rata de montaña de Célebes

DD
Monotypic
HB 18–21 cm T 17.7–20 cm
W 145–220 g

1280–2285 m

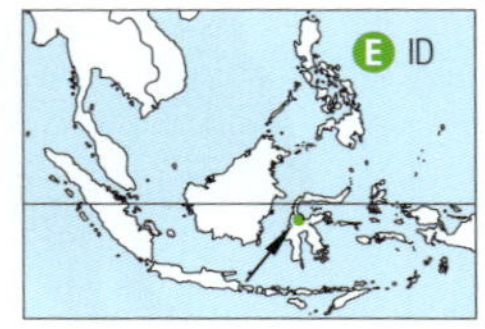

C Sulawesi

Small-eared Rat

Taeromys microbullatus

FR Rat-laineux à petites oreilles
DE Kleinohr-Sulawesi-Ratte
ES Rata de orejas pequeñas

DD
Monotypic
HB c. 20 cm
T c. 22 cm

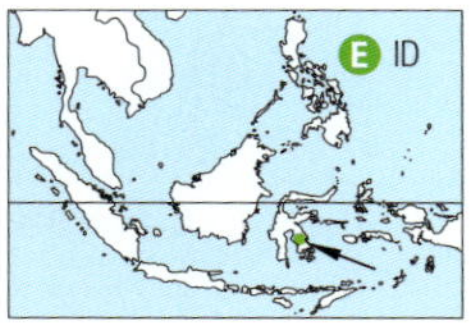

SE Pen of Sulawesi (Tanke Salokko)

c. 1500 m

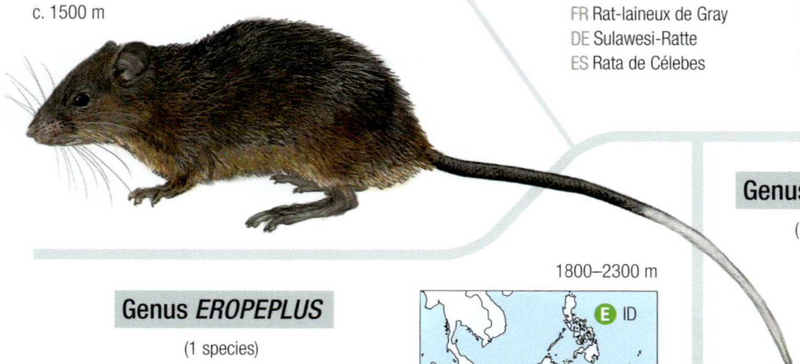

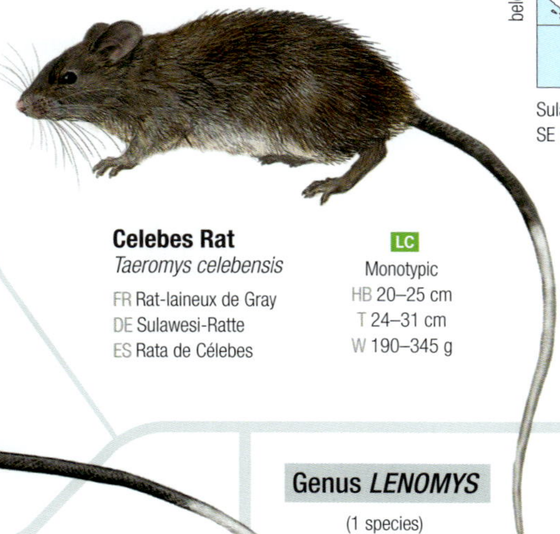

Celebes Rat

Taeromys celebensis

FR Rat-laineux de Gray
DE Sulawesi-Ratte
ES Rata de Célebes

LC
Monotypic
HB 20–25 cm
T 24–31 cm
W 190–345 g

below 1200 m

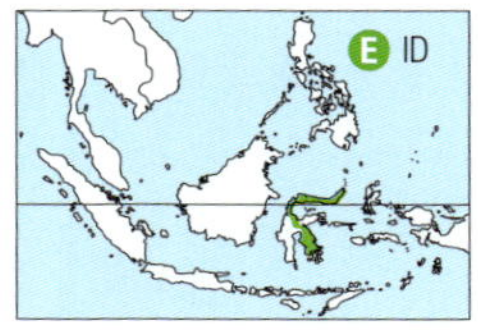

Sulawesi (Minahasa Pen, SE Pen & C)

Giant Sulawesi Rat

Taeromys dominator

FR Rat géant
DE Sulawesi-Riesenratte
ES Rata de Célebes gigante

LC
Monotypic
HB 20–28 cm
T 24–33 cm
W 350–500 g

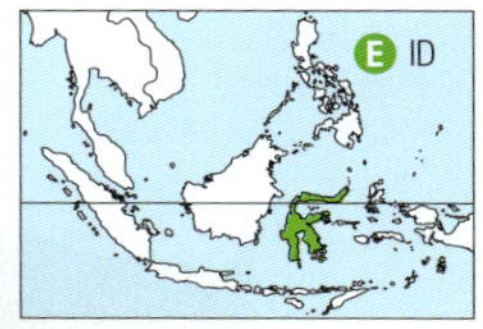

Forests of Sulawesi

Genus *EROPEPLUS*

(1 species)

1800–2300 m

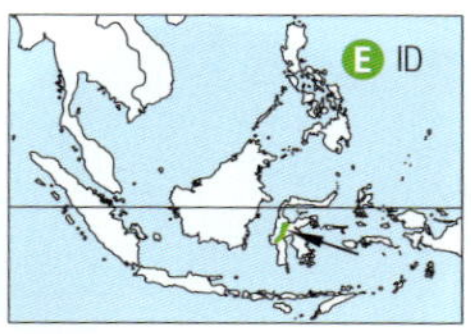

C Sulawesi

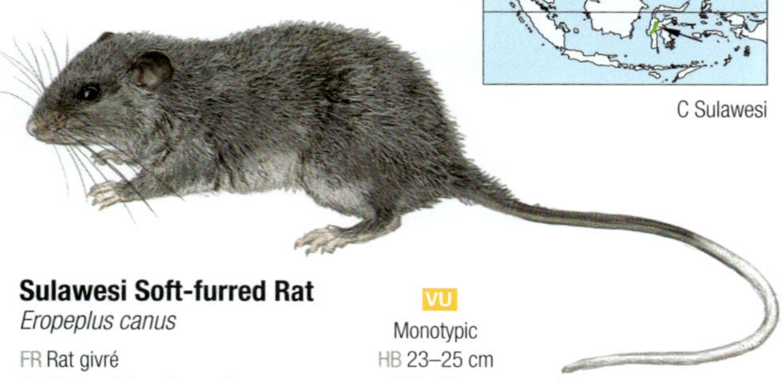

Sulawesi Soft-furred Rat

Eropeplus canus

FR Rat givré
DE Sulawesi-Langhaarratte
ES Rata de pelaje suave de Célebes

VU
Monotypic
HB 23–25 cm
T 27–30 cm
W 190–315 g

Genus *LENOMYS*

(1 species)

Meyer's Giant Rat

Lenomys meyeri

FR Rat de Meyer
DE Kleezahn-Riesenratte
ES Rata gigante de Meyer

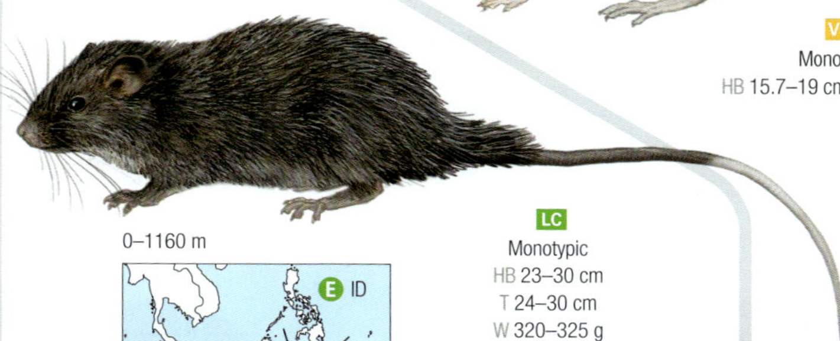

LC
Monotypic
HB 23–30 cm
T 24–30 cm
W 320–325 g

0–1160 m

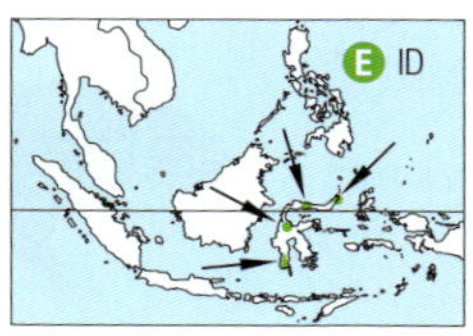

N, C & SW Sulawesi; possibly SE

Genus *FRATEROMYS*

(1 species)

North-eastern Peninsula Hill Rat

Frateromys fratrorum

FR Rat des Hose
DE Nordöstliche Sulawesi-Bergratte
ES Rata de colina nororiental

VU
Monotypic
HB 15.7–19 cm T 15–20 cm

up to 1980 m

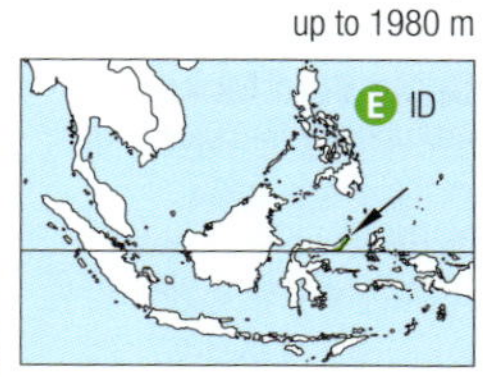

N Sulawesi (Minahasa Pen)

Karoko Hill Rat

Bunomys karokophilus

FR Bunomys des karokos
DE Karoko-Sulawesi-Bergratte
ES Rata de colina de Karoko

Genus *BUNOMYS*

(7 species)

DD
Monotypic
HB 15–19 cm
T 13.5–20 cm
W 95–175 g

825–1150 m

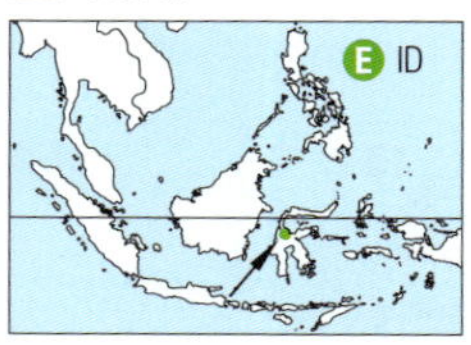

WC mountain block of C Sulawesi; perhaps widespread

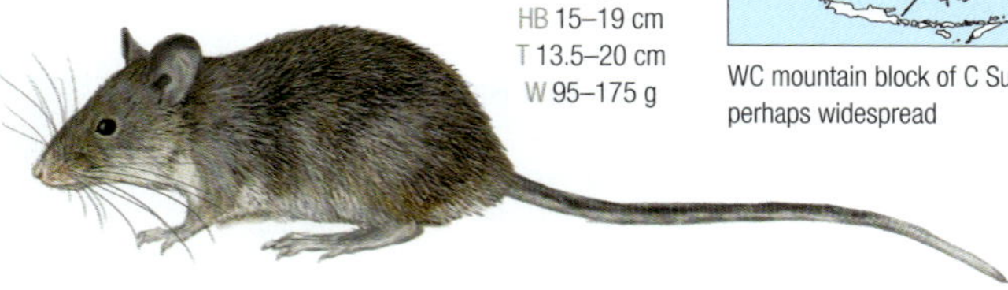

up to 1400 m

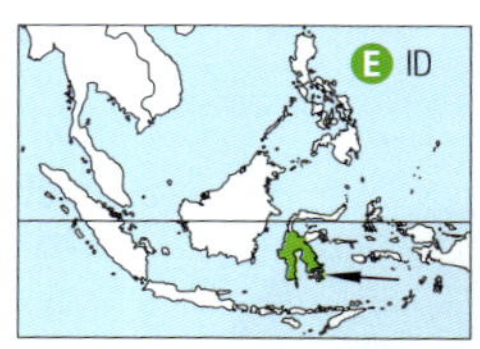

C & S Sulawesi, nearby Buton I

Andrews's Hill Rat

Bunomys andrewsi

FR Bunomys d'Andrews
DE Andrews-Sulawesi-Bergratte
ES Rata de colina de Andrews

LC
Monotypic
HB 13.7–19.5 cm
T 11–16.7 cm
W 95–222 g

1500–2000 m

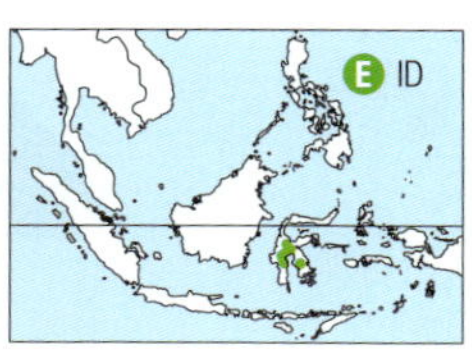

Sulawesi (WC core and SE Pen)

Montane Hill Rat

Bunomys penitus

FR Bunomys des montagnes
DE Zentral-Sulawesi-Bergratte
ES Rata de colina de montaña

LC
Monotypic
HB 15.5–24 cm
T 13.8–19 cm
W 95–170 g

Tana Toraja Hill Rat

Bunomys torajae

FR Bunomys des Torajas
DE Toraja-Sulawesi-Bergratte
ES Rata de colina de Toraja

DD
Monotypic
HB 15.6–17.5 cm
T 16–17 cm
W 98–128 g

2500–2600 m

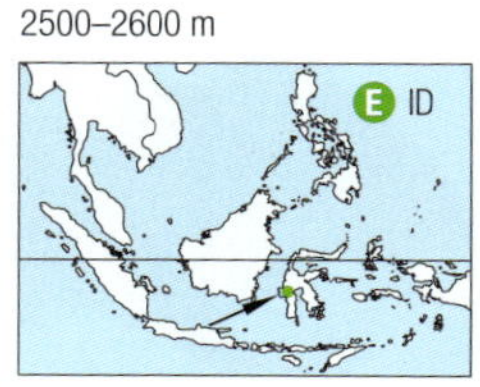

WC Sulawesi (Mt Gandangdewata, Quarles Range)

Tambusisi Hill Rat
Bunomys prolatus

FR Bunomys de Musser
DE Tambusisi-Sulawesi-Bergratte
ES Rata de colina de Tambusisi

EN
Monotypic
HB 15.6–17.9 cm
T 12.5–14.2 cm

c. 1830 m

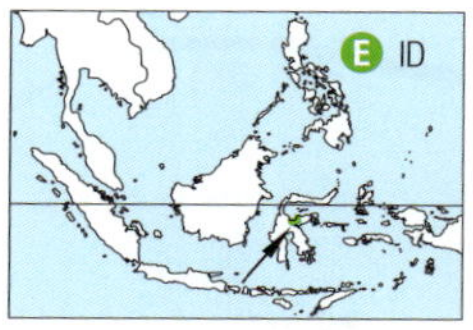

EC Sulawesi (Mt Tambusisi)

LC
Monotypic
HB 9.7–18 cm
T 9–18 cm
W 55–135 g

Common Hill Rat
Bunomys chrysocomus

FR Bunomys de Hoffmann
DE Gemeine Sulawesi-Bergratte
ES Rata de colina común

250–2200 m

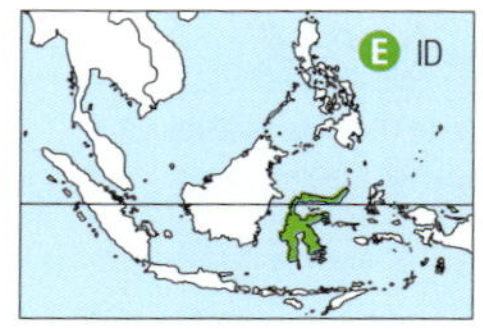

Throughout Sulawesi

Lompobatang Hill Rat
Bunomys coelestis

FR Bunomys du Lompobatang
DE Lompobatang-Sulawesi-Bergratte
ES Rata de colina de Lompobatang

EN
Monotypic
HB 14.7–17.9 cm
T 13.6–16.5 cm

1800–2500 m

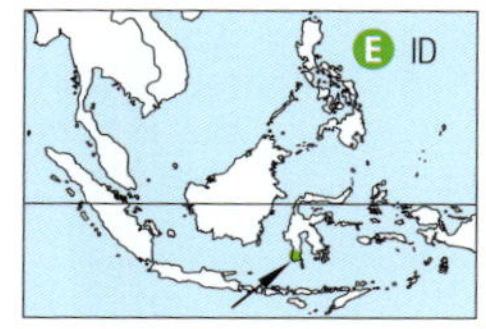

SW Sulawesi (Mt Lompobatang)

Short-tailed Bandicoot Rat
Nesokia indica

FR Nésokia à queue courte
DE Pestratte
ES Rata bandicut de cola corta

LC
Monotypic
HB 15–20 cm
T 9.1–15 cm
W 96–320 g

0–1600 m

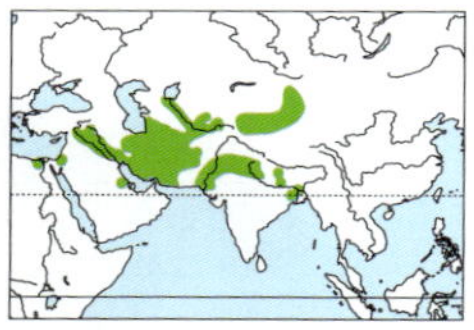

N EG (Nile Delta), Middle East, IR & S Central Asia E to NW CN & C BD

Genus *NESOKIA*

(2 species)

EN
Monotypic
HB 23–26 cm T 20–27 cm
W c. 519 g

Long-tailed Bandicoot Rat
Nesokia bunnii

FR Nésokia de Bunni
DE Rote Pestratte
ES Rata bandicut de cola larga

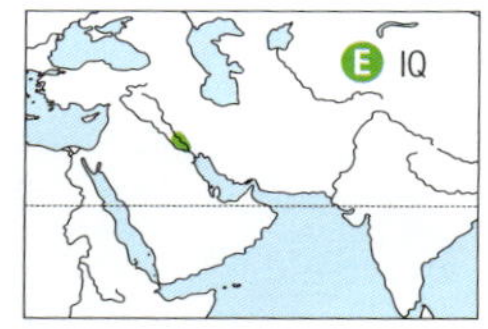

Tigris & Euphrates river valleys

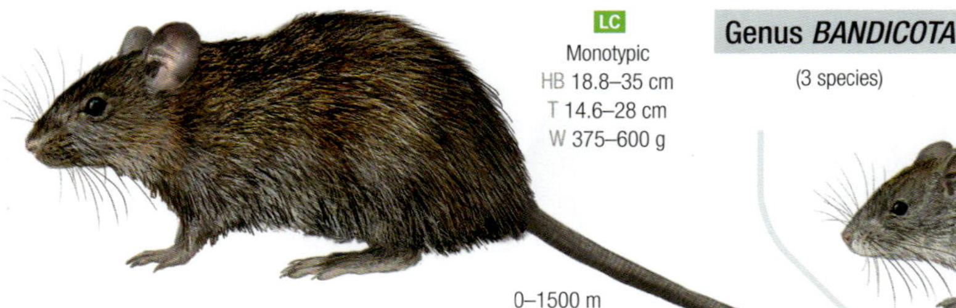

LC
Monotypic
HB 18.8–35 cm
T 14.6–28 cm
W 375–600 g

Genus *BANDICOTA*

(3 species)

Greater Bandicoot Rat
Bandicota indica

FR Grand Rat-bandicot
DE Große Bandikutratte
ES Rata bandicut mayor

0–1500 m

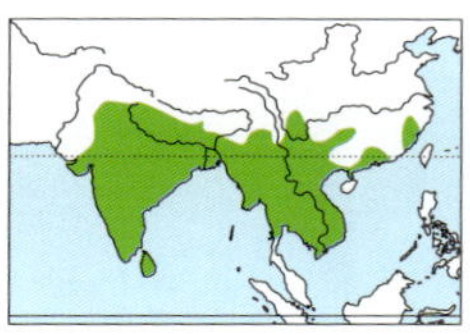

S & SE Asia, E from IN to CN (Fujian); with Cat Ba I (N VN); introduced to Pen MY, Java, perhaps TW

LC
Monotypic
HB 16.1–24 cm
T 11.2–18.8 cm

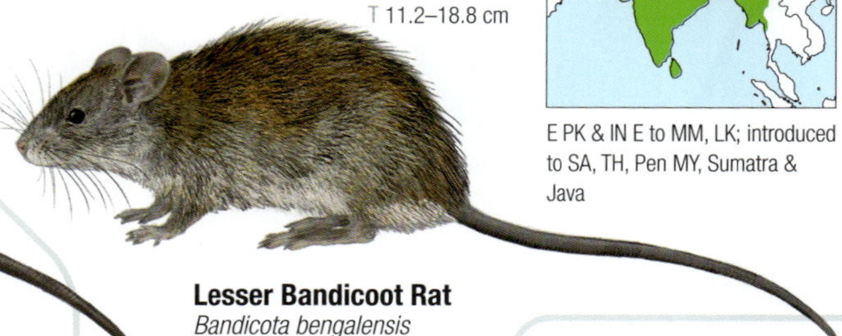

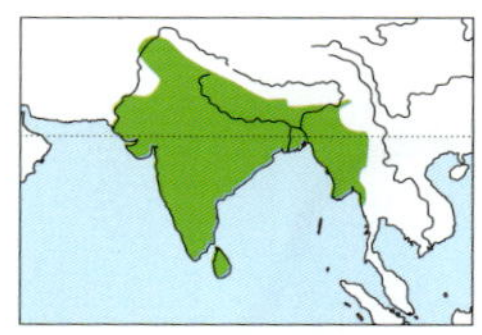

E PK & IN E to MM, LK; introduced to SA, TH, Pen MY, Sumatra & Java

Lesser Bandicoot Rat
Bandicota bengalensis

FR Petit Rat-bandicot
DE Kleine Bandikutratte
ES Rata bandicut menor

Savile's Bandicoot Rat
Bandicota savilei

FR Rat-bandicot de Savile
DE Savile-Bandikutratte
ES Rata bandicut de Savile

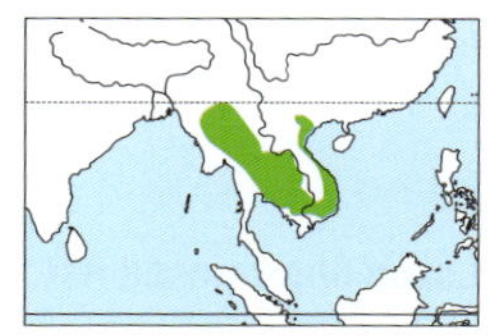

C MM, TH, KH, VN; perhaps lowlands of LA

LC
Monotypic
HB 14.5–22 cm
T 7.5–17.8 cm
W 140–220 g

Ryukyu Islands Tree Rat
Diplothrix legata

FR Rat des Ryukyu
DE Ryukyu-Ratte
ES Rata arborícola de Ryukyu

Genus *DIPLOTHRIX*

(1 species)

EN
Monotypic
HB c. 23 cm
T c. 25 cm
W 570–587 g

300–400 m

Ryukyu Is (Amami Ōshima, Tokunoshima, Okinawa)

Genus *NESOROMYS*

(1 species)

Seram Island Mountain Rat
Nesoromys ceramicus

FR Rat de Céram
DE Seram-Dschungelratte
ES Rata de montaña de Seram

EN
Monotypic
HB 11.8–13.5 cm
T 12.6–14 cm
W c. 66.5 g

1500–2000 m

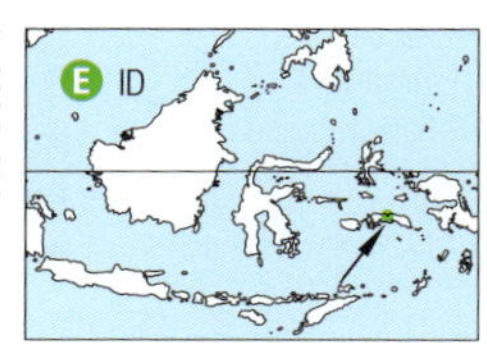

Seram I (Mt Manusela, perhaps other highlands), Moluccas

Genus *KADARSANOMYS*

(1 species)

Javan Bamboo Rat
Kadarsanomys sodyi

FR Rat de Kardasan
DE Java-Ratte
ES Rata de bambú de Java

EN
Monotypic
HB 16.2–21 cm
T 25–30 cm
W 175–230 g

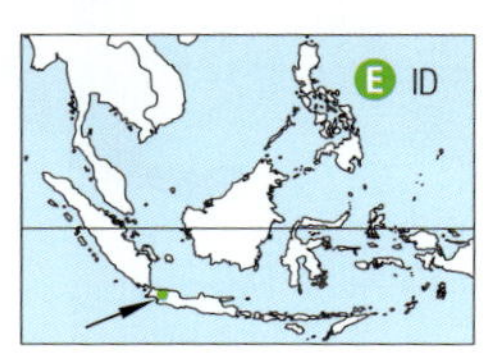

W Java (Mt Pangrango-Gede)

Genus *LIMNOMYS*

(2 species)

Gray-bellied Moss Mouse
Limnomys bryophilus
FR Rat bryophile
DE Graubauch-Moosratte
ES Ratón de musgo de vientre gris

LC
Monotypic
HB 12.1–14 cm
T 15.2–18.1 cm
W 48–90 g

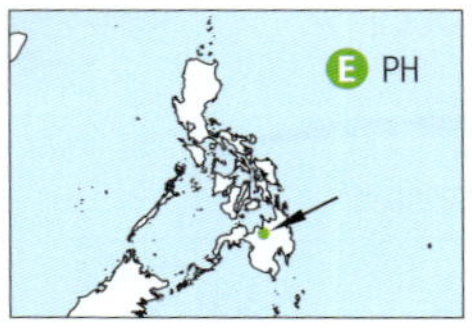

2250–2800 m

Mindanao I (Mt Kitanglad Range, perhaps other high peaks)

Long-tailed Moss Mouse
Limnomys sibuanus
FR Rat du Sibulan
DE Langschwanz-Moosratte
ES Ratón de musgo de cola larga

LC
Monotypic
HB 11.9–14.4 cm
T 14.7–17.4 cm
W 47–82 g

2000–2800 m

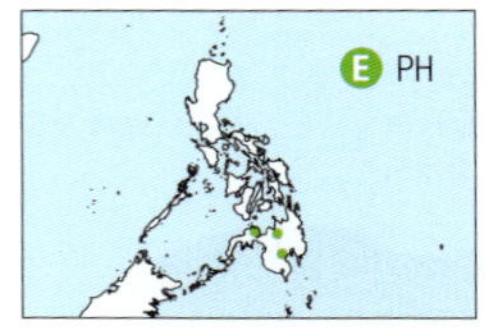

Mindanao I (Mt Malindang, Mt Kitanglad, Mt Apo)

Genus *TARSOMYS*

(2 species)

Dusky Long-footed Rat
Tarsomys apoensis
FR Rat de l'Apo
DE Dunkle Fersenratte
ES Rata de pies largos oscura

LC
Monotypic
HB 13.5–15.6 cm
T 11.6–12.6 cm
W 65–72 g

Mindanao I (including Mt Malindang, Mt Kitanglad & Mt Apo)

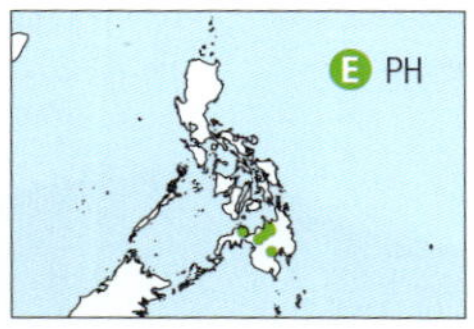

1550–2800 m

Spiny Long-footed Rat
Tarsomys echinatus
FR Rat du Matutum
DE Stachel-Fersenratte
ES Rata de pies largos espinosa

VU
Monotypic
HB 14.5–17.9 cm
T 12–15.8 cm

below 1300 m

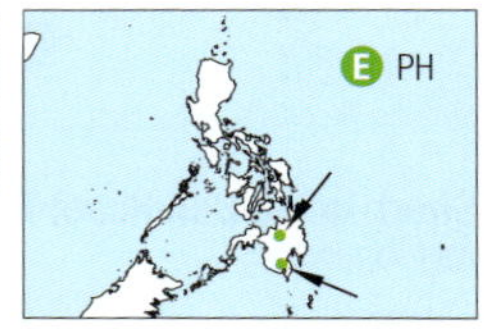

Mindanao I (lowlands around Mt Kitanglad & Mt Matutum)

Genus *PALAWANOMYS*

(1 species)

Palawan Mountain Rat
Palawanomys furvus
FR Rat de Palawan
DE Palawan-Ratte
ES Rata de montaña de Palawan

LC
Monotypic
HB 13.5–15.8 cm
T 14.9–16.2 cm
W 86–120 g

1500–1950 m

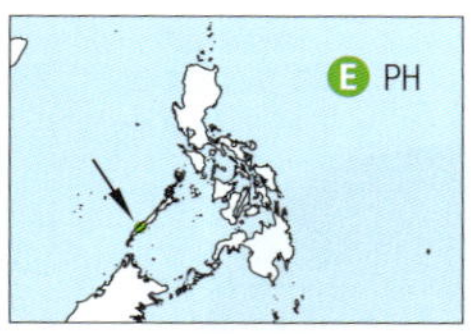

Palawan I (Mt Mantalingajan)

Genus *ABDITOMYS*

(1 species)

Luzon Broad-toothed Rat
Abditomys latidens
FR Rat à dents larges
DE Luzon-Breitzahnratte
ES Rata de dientes anchos de Luzón

DD
Monotypic
HB 27–23 cm
T 22–24 cm
W c. 269 g

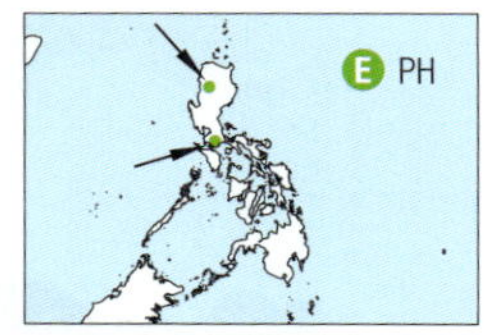

N & C Luzon I (known from 2 localities)

Genus *TRYPHOMYS*

(1 species)

Luzon Short-nosed Rat
Tryphomys adustus
FR Rat de Luçon
DE Mearns-Luzon-Ratte
ES Rata de pies largos de Luzón

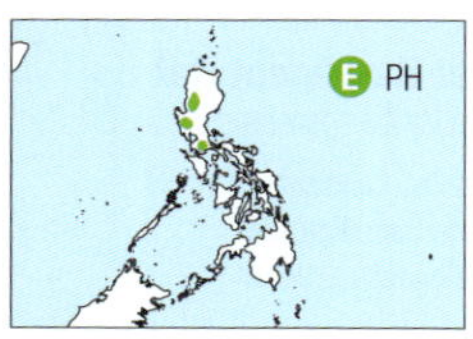

0–2400 m

Luzon I (Central Cordillera, Mt Makiling)

DD
Monotypic
HB 17.4–19.8 cm
T 15–18.1 cm
W c. 198 g

Genus *ANONYMOMYS*

(1 species)

Mindoro Climbing Rat
Anonymomys mindorensis
FR Rat-sauteur de Mindoro
DE Mindoro-Anonymus-Ratte
ES Rata trepadora de Mindoro

DD
Monotypic
HB 14.4–14.6 cm
T 19.6–22 cm
W 83–90 g

140–1650 m

N Mindoro I

Genus *RATTUS*

(66 species)

Timor Forest Rat
Rattus timorensis
FR Rat de Timor
DE Timor-Ratte
ES Rata de bosque de Timor

c. 1900 m

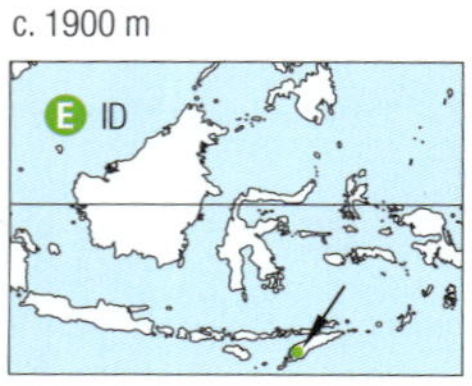

W Timor I (Mt Mutis)

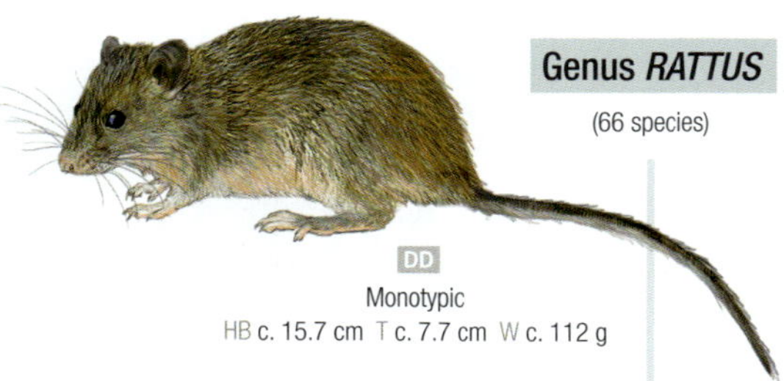

DD
Monotypic
HB c. 15.7 cm T c. 7.7 cm W c. 112 g

EN
Monotypic
HB c. 13.3 cm T c. 16.1 cm
W c. 81 g

Hainald's Flores Island Rat
Rattus hainaldi
FR Rat de Hainald
DE Hainald-Ratte
ES Rata de Flores de Hainald

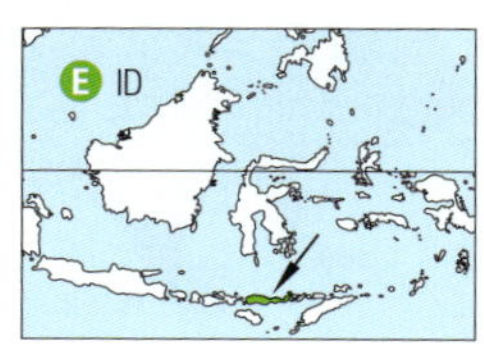

1000–2000 m

Flores I

LC
Monotypic
HB 21–24 cm
T 18.1–24 cm
W 320–390 g

up to 2400 m

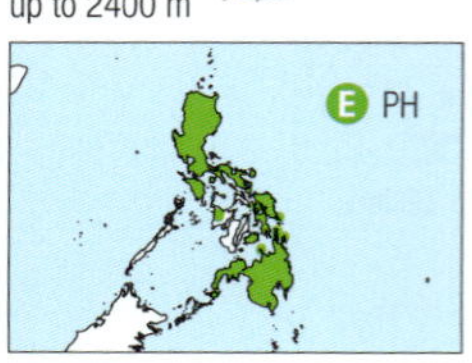

Widespread excluding Palawan & Batanes/ Babuyan groups

Philippine Forest Rat
Rattus everetti
FR Rat d'Everett
DE Philippinen-Ratte
ES Rata de bosque de Filipinas

North-eastern Xanthurus Rat
Rattus xanthurus
FR Rat à queue jaune
DE Nordöstliche Gelbschwanzratte
ES Rata de cola dorada nororiental

below 1000 m

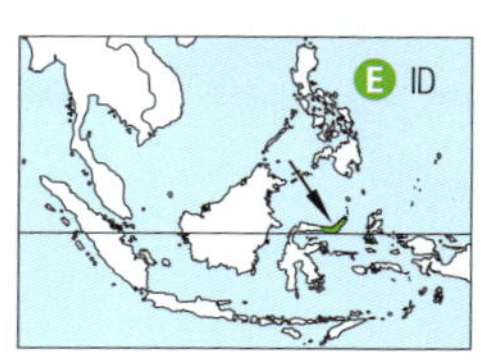

N Sulawesi (NE Minahasa Pen)

NT
Monotypic
HB 23–26 cm
T 29–33 cm

Peleng Island Xanthurus Rat
Rattus pelurus

FR Rat de Peleng
DE Peleng-Gelbschwanzratte
ES Rata de cola dorada de Peleng

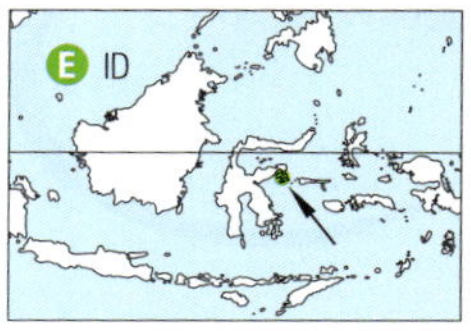

Peleng I & Banggai Is, off E Sulawesi

DD
Monotypic
HB 24–27 cm
T 24–28 cm

Lore Lindu Xanthurus Rat
Rattus facetus

FR Rat du Lindu
DE Lore-Lindu-Ratte
ES Rata de Lore Lindu

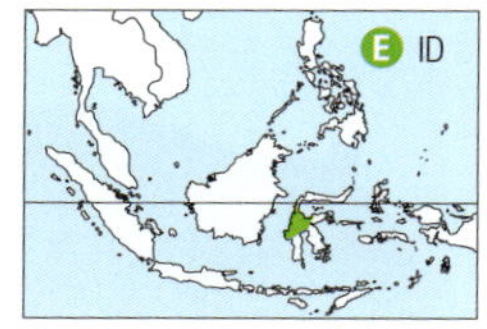

S part of Minahasa Pen & C Sulawesi

NE
Monotypic
HB 13.5–23 cm
T 17.5–26 cm

Marmoset Xanthurus Rat
Rattus marmosurus

FR Rat à queue de singe
DE Opossum-Gelbschwanzratte
ES Rata de cola dorada tití

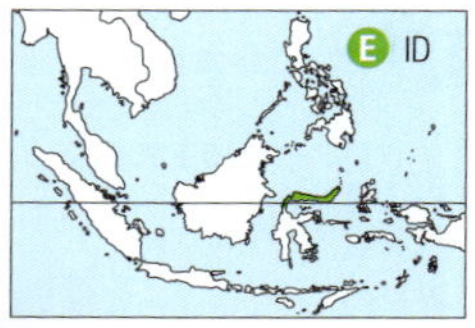

N Sulawesi (Minahasa Pen)

LC
Monotypic
HB 13.5–23 cm
T 17.5–26 cm

South-western Xanthurus Rat
Rattus bontanus

FR Rat du Bonthain
DE Südwestliche Gelbschwanzratte
ES Rata de cola dorada de Bonthain

DD
Monotypic
HB 18.7–24 cm
T 23–30 cm

SW Pen of Sulawesi (Mt Lompobatang and adjacent lowlands)

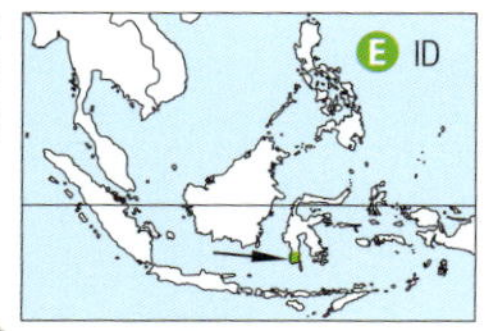

0–2500 m

South-eastern Xanthurus Rat
Rattus salocco

FR Rat du Salokko
DE Südöstliche Gelbschwanzratte
ES Rata de cola dorada suroriental

DD
Monotypic
HB c. 21 cm
T c. 26 cm

300–1500 m

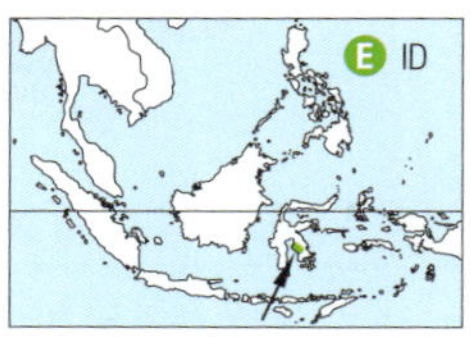

SE Pen of Sulawesi (Tanke Salokko, Mowewe lowlands)

Sula Archipelago Rat
Rattus elaphinus

FR Rat des Sula
DE Sula-Ratte
ES Rata de Soela

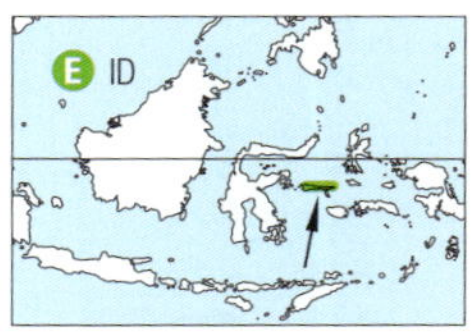

Sula Arch (Taliabu & Mangole Is)

NT
Monotypic
HB 12.8–21 cm
T 15.4–20 cm
W 123–197 g

Spiny Seram Island Rat
Rattus feliceus

FR Rat de Felix
DE Seram-Ratte
ES Rata espinosa de Seram

NT
Monotypic
HB 16.7–28 cm
T 16.5–18.5 cm
W 120–345 g

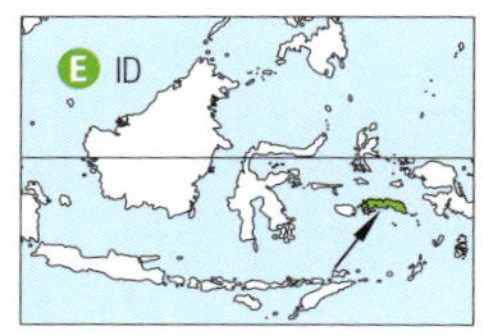

0–2000 m

Seram I (Moluccas)

Halmahara Rat
Rattus morotaiensis

FR Rat de Morotai
DE Molukken-Ratte
ES Rata de Halmahara

LC
Monotypic
HB 12–21 cm
T 16.5–22 cm
W 86–159 g

0–1000 m

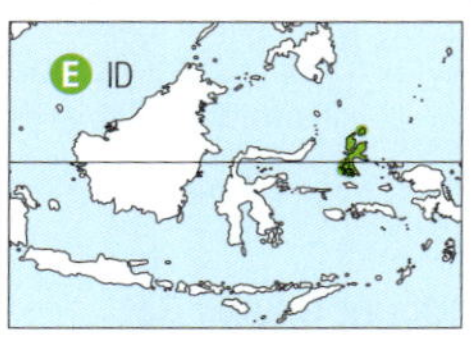

Morotai, Halmahera & Bacan Is (Moluccas); perhaps other Is (e.g. Ternate)

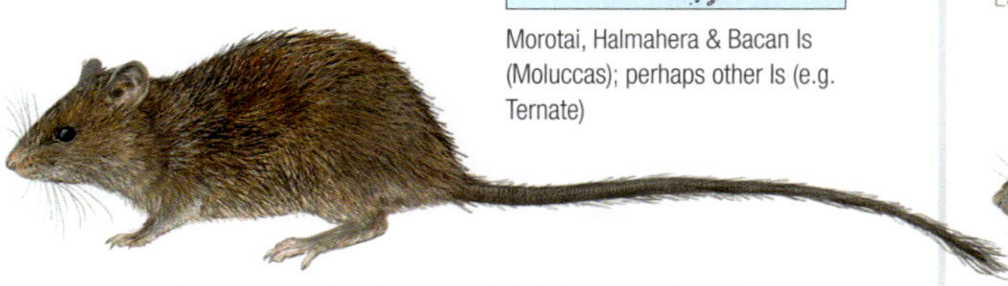

Canefield Rat
Rattus sordidus

FR Rat des cannes
DE Australische Zuckerrohrratte
ES Rata de cañaveral

LC
Monotypic
HB 11–21 cm
T 10–16 cm
W 50–260 g

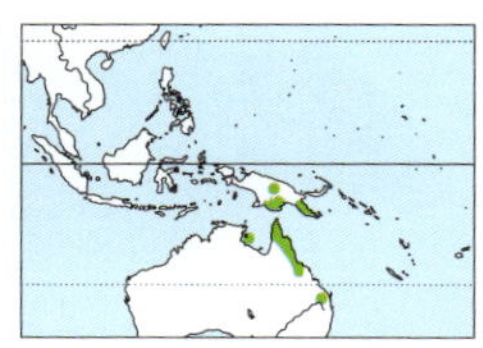

New Guinea (ID, PG), AU (N & E Qld, also South West I, off coast of NT, where perhaps extinct)

Australian Swamp Rat
Rattus lutreola

FR Rat des marais
DE Australische Riedgrasratte
ES Rata de ciénaga de Australia

LC
3 sspp
HB 12–20 cm
T 8–14.5 cm
W 55–160 g

NE Qld S to SE SA (with Fraser, North Stradbroke & Kangaroo Is), Tas & Bass Strait Is

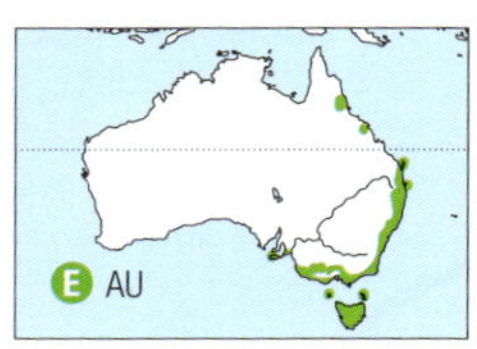

0–1600 m

Australian Pale Field Rat
Rattus tunneyi

FR Rat de Tunney
DE Blasse Australische Feldratte
ES Rata de campo de Australia

LC
2 sspp
HB 11.8–19.8 cm
T 7.8–19 cm
W 50–210 g

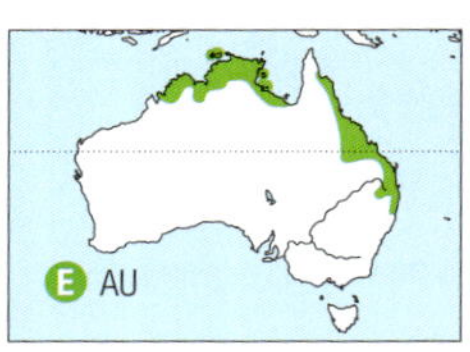

N, E & some Is (Bathurst, Melville, Groote Eylandt, Sir Edward Pellew Group)

Australian Bush Rat
Rattus fuscipes

FR Rat de brousse
DE Australische Schwarzfußratte
ES Rata peluda de Australia

LC
4 sspp
HB 10–20 cm
T 10–19.5 cm
W 50–225 g

0–2210 m

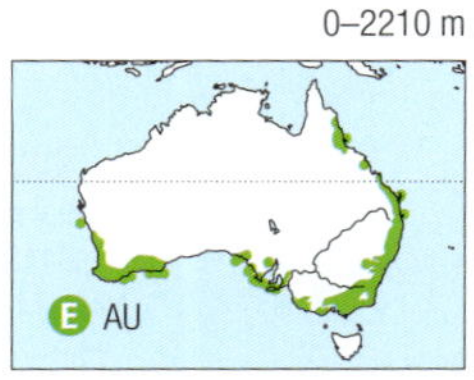

SW WA, SE Qld to SE SA, various offshore Is (including Kangaroo I)

Australian Long-haired Rat
Rattus villosissimus

FR Rat à poil long
DE Langhaarige Australische Ratte
ES Rata de pelo largo de Australia

LC
Monotypic
HB 13–22 cm
T 12–18 cm
W 60–280 g

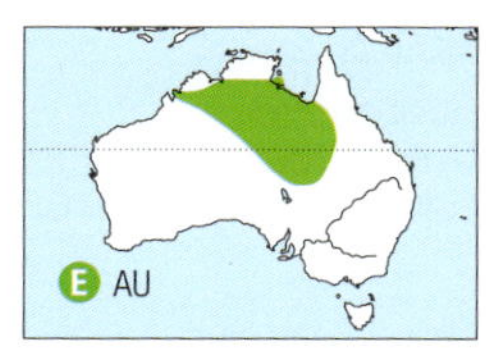

Core areas: Barkly L system (NT), Channel Country (SW Qld, NE SA), South West I

LC
Monotypic
HB 6.5–21 cm
T 7.8–15 cm
W 22–213 g

Australian Dusky Rat
Rattus colletti
FR Rat de Collett
DE Australische Dunkelratte
ES Rata oscura de Australia

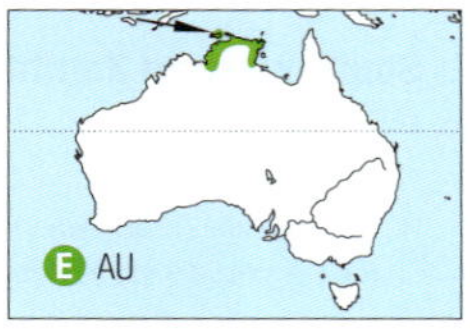

N NT & Melville I

Cape York Rat
Rattus leucopus
FR Rat de Cape York
DE Kap-York-Ratte
ES Rata de York

0–1200 m

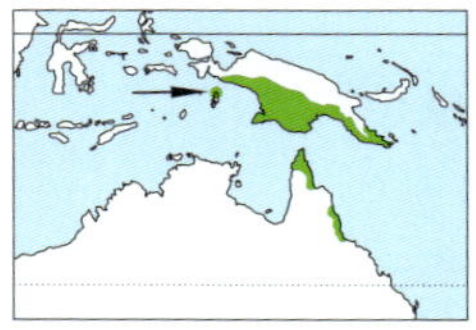

S New Guinea (ID, PG), AU (N & SE Cape York, Qld); probably Aru I

LC
5 sspp
HB 13.5–21 cm
T 14–21 cm
W 90–205 g

Yapen Island Rat
Rattus jobiensis
FR Rat de Yapen
DE Yapen-Ratte
ES Rata de Yapen

LC
Monotypic
HB 21–25 cm
T 18.6–23 cm
W 260–500 g

up to 600 m

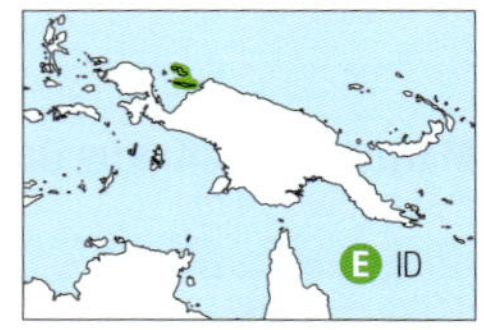

Supiori, Biak, Owi & Yapen Is (Cenderawasih Bay, off NW New Guinea)

LC
3 sspp
HB 12.3–16.8 cm
T 14.3–17.1 cm
W 88–133 g

New Guinea Slender Rat
Rattus verecundus
FR Rat discret
DE Schlanke Neuguinea-Ratte
ES Rata de Nueva Guinea esbelta

150–2700 m

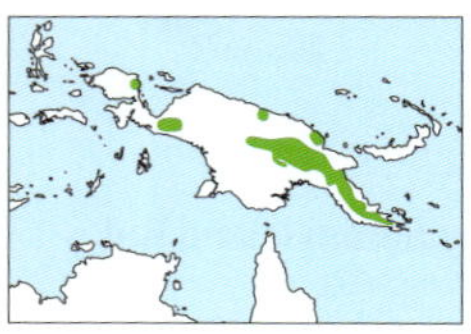

ID/PG: SE, C & E New Guinea

LC
Monotypic
HB 15.5–23 cm
T 13.5–18 cm
W 103–180 g

Papua New Guinea Rat
Rattus novaeguineae
FR Rat de Nouvelle-Guinée
DE Papua-Neuguinea-Ratte
ES Rata de Papúa Nueva Guinea

740–1520 m

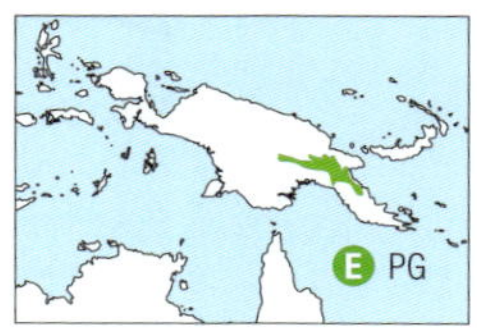

CE New Guinea

Large New Guinea Spiny Rat
Rattus praetor
FR Rat préteur
DE Große Neuguinea-Ratte
ES Rata espinosa de Nueva Guinea grande

0–2000 m

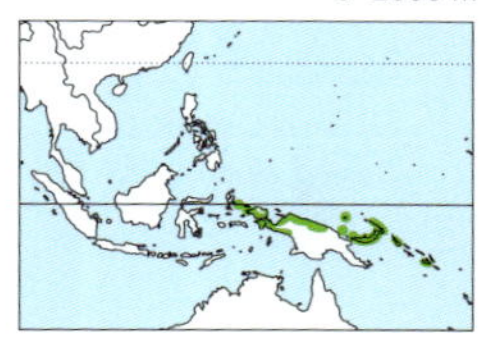

W & N New Guinea (perhaps Gebe I & Salawati I), Bismarck Arch, Solomon Is

LC
2 sspp
HB 15.7–24 cm
T 14.4–18.1 cm
W 164–228 g

Gag Island Rat
Rattus nikenii
FR Rat de Niken
DE Insel-Gag-Ratte
ES Rata de Gag

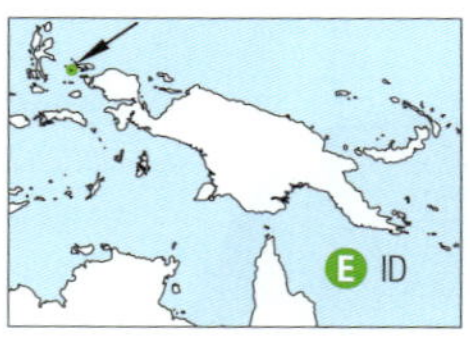

Gag I (Raja Ampat Is); perhaps other Is in region

VU
Monotypic
HB 13.5–30 cm
T 13.8–16.5 cm
W 113–175 g

Stein's New Guinea Rat
Rattus steini
FR Rat de Stein
DE Stein-Neuguinea-Ratte
ES Rata de Nueva Guinea de Stein

450–2800 m

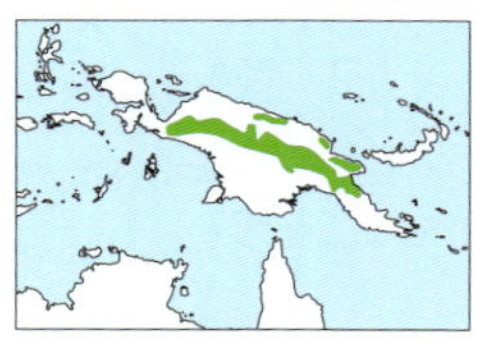

ID/PG: New Guinea (from Weyland Range E to Bowutu Mts)

LC
4 sspp
HB 14–19.3 cm
T 14–19.3 cm
W 110–220 g

Mount Giluwe Rat
Rattus giluwensis
FR Rat du Giluwe
DE Giluwe-Ratte
ES Rata de Giluwe

LC
Monotypic
HB 13.6–18.7 cm
T 8.4–10.8 cm
W 50–60 g

2195–3660 m

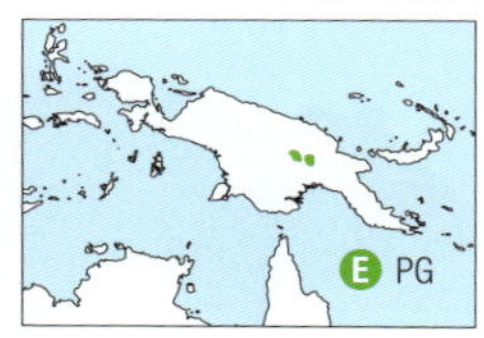

EC New Guinea (Mt Giluwe and its nearby peaks, Kaijende Highlands)

Pocock's New Guinea Highland Rat
Rattus pococki
FR Rat de Pocock
DE Pocock-Neuguinea-Ratte
ES Rata de altiplano de Nueva Guinea de Pocock

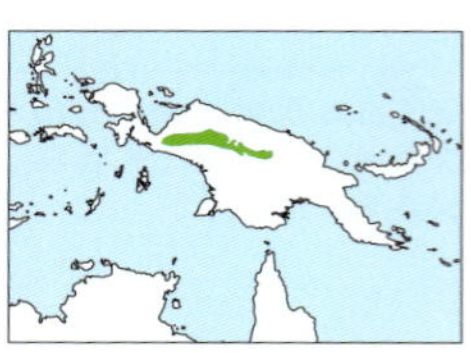

1500–2500 m

ID/PG: Central Range of W & C New Guinea

LC
Monotypic
HB c. 12.7 cm
T c. 12.6 cm

Richardson's New Guinea Mountain Rat
Rattus richardsoni
FR Rat de Richardson
DE Richardson-Neuguinea-Ratte
ES Rata de montaña de Nueva Guinea de Richardson

3225–4500 m

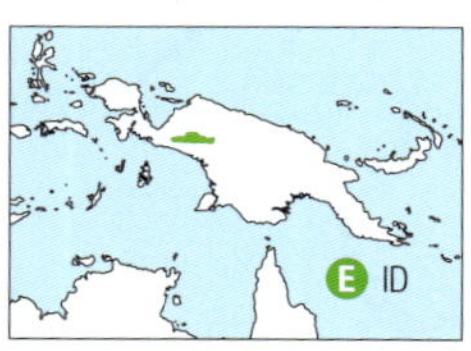

W New Guinea Highlands (including Mt Jaya, around L Habbema & Mt Trikora)

VU
Monotypic
HB 12.2–13.3 cm
T 12.8–14.2 cm
W 59–74 g

DD
Monotypic
HB 12.8–13.9 cm
T 9.6–10.1 cm W 69–78 g

c. 2950 m

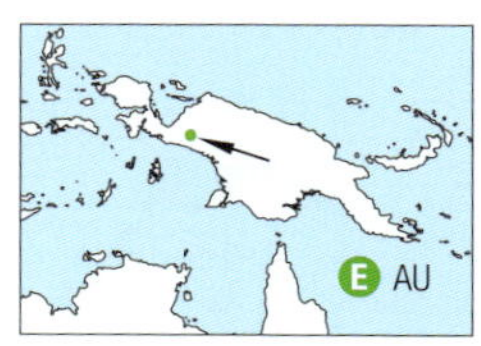

W New Guinea (W Snow Mts)

Arianus's New Guinea Mountain Rat
Rattus omichlodes
FR Rat des brumes
DE Misonne-Neuguinea-Ratte
ES Rata de montaña de Nueva Guinea de Arianus

EN
Monotypic
HB 12–16 cm
T 12.5–15.2 cm
W 50–88 g

Van Deusen's New Guinea Mountain Rat

Rattus vandeuseni

FR Rat de Van Deusen
DE Van-Deusen-Neuguinea-Ratte
ES Rata de montaña de Nueva Guinea de Van Deusen

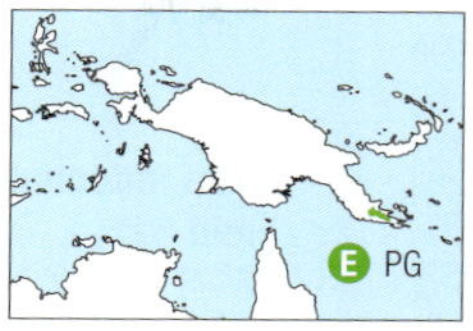

above 1300 m

New Guinea (extreme SE Papuan Pen)

Vogelkop Mountain Rat

Rattus arfakiensis

FR Rat des Arfak
DE Vogelkop-Ratte
ES Rata de montaña de Vogelkop

DD
Monotypic
HB c. 13.3 cm
T c. 13.5 cm

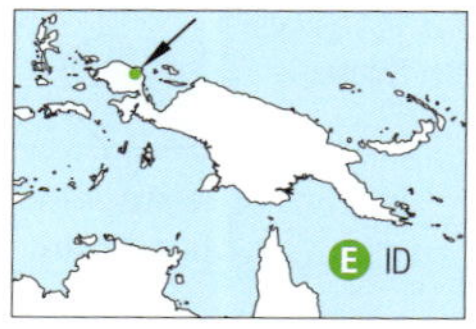

c. 2000 m

NW New Guinea (Arfak Mts in Vogelkop Pen)

Eastern New Guinea Mountain Rat

Rattus niobe

FR Rat niobé
DE Kleine Ostneuguinea-Ratte
ES Rata de montaña de Nueva Guinea oriental

760–4050 m

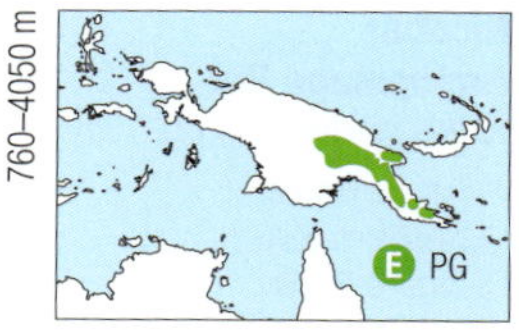

New Guinea (E Highlands & Huon Pen)

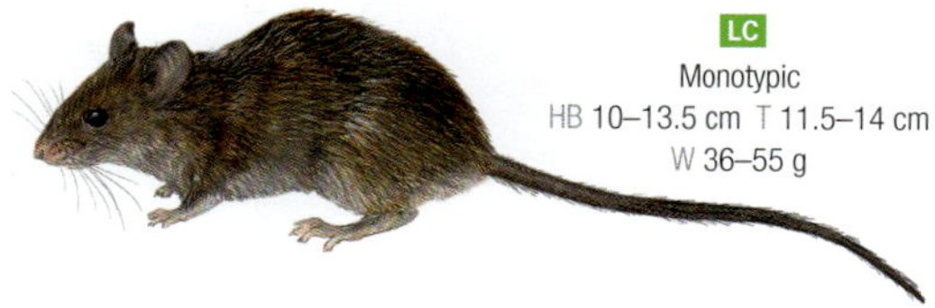

LC
Monotypic
HB 10–13.5 cm T 11.5–14 cm
W 36–55 g

Western New Guinea Mountain Rat

Rattus arrogans

FR Rat des Snow
DE Kleine Westneuguinea-Ratte
ES Rata de montaña de Nueva Guinea occidental

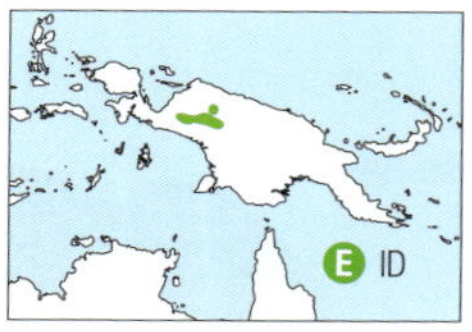

2200–4100 m

W New Guinea

LC
Monotypic
HB c. 13 cm
T c. 11.4 cm

Eastern New Guinea Rat

Rattus mordax

FR Rat kaisi
DE Ostneuguinea-Ratte
ES Rata de Nueva Guinea oriental

LC
2 sspp
HB 14.2–25 cm
T 11.5–20 cm
W 190–255 g

E New Guinea (Huon & Papuan Pen, D'Entrecasteaux Is, Woodlark I, Louisiade Arch)

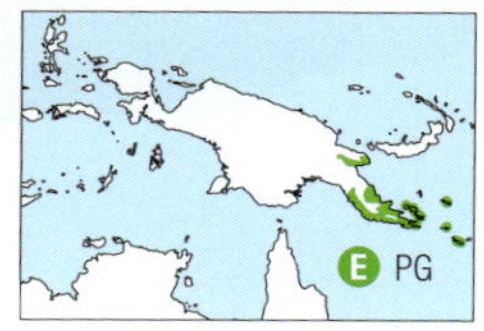

up to 2800 m

Sahyadris Forest Rat

Rattus satarae

FR Rat des Sahyadri
DE Sahyadris-Ratte
ES Rata de bosque de Sahyadris

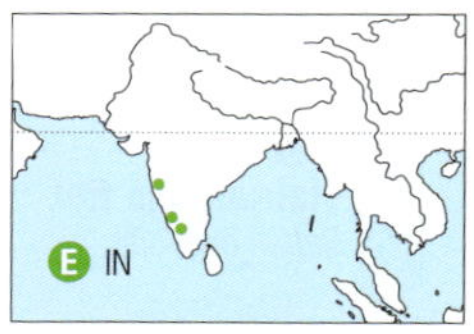

700–2150 m

Western Ghats (Maharashtra, Karnataka, Tamil Nadu)

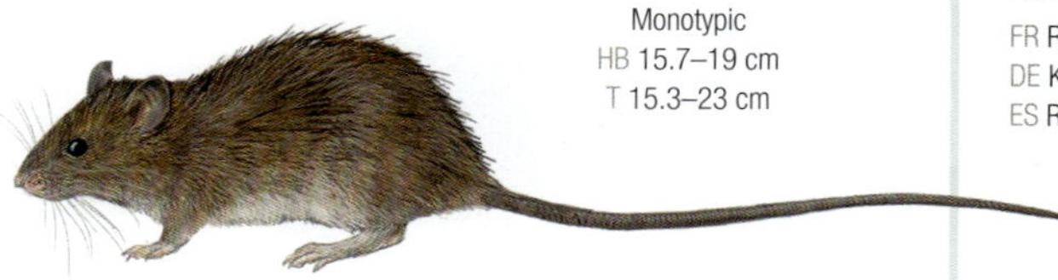

VU
Monotypic
HB 15.7–19 cm
T 15.3–23 cm

EN
Monotypic
HB 16.2–26 cm
T 18.7–23 cm

Ranjini's Rat

Rattus ranjiniae

FR Rat de Ranjini
DE Kerala-Ratte
ES Rata de Ranjini

up to 1000 m

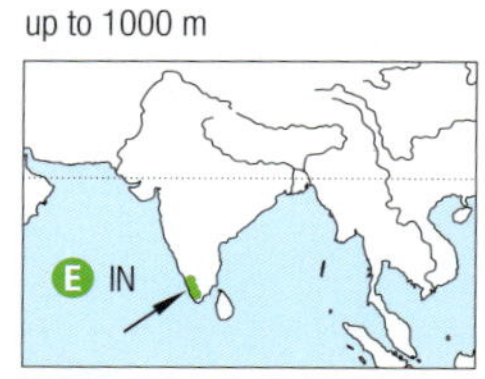

Scattered in S Kerala

DD
Monotypic
HB c. 26 cm
T c. 15 cm

Manus Island Spiny Rat

Rattus detentus

FR Rat de Manus
DE Manus-Ratte
ES Rata espinosa de Manus

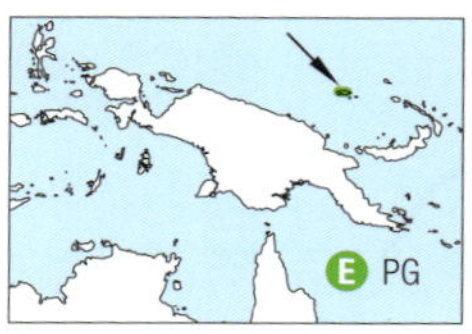

0–200 m

Manus I (Admiralty Is)

White-footed Indochinese Rat

Rattus nitidus

FR Rat soyeux
DE Himalaya-Ratte
ES Rata de Indochina de pies blancos

LC
Monotypic
HB 14.8–18 cm
T 13.5–21 cm
W 110–140 g

685 –2740 m

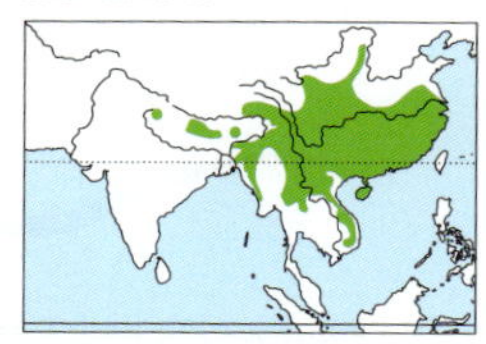

Himalayas E to N & E CN, S to MM, N TH & VN; introduced to PH, ID & PW

Sri Lankan Mountain Rat

Rattus montanus

FR Rat du Sri Lanka
DE Sri-Lanka-Ratte
ES Rata de montaña de Sri Lanka

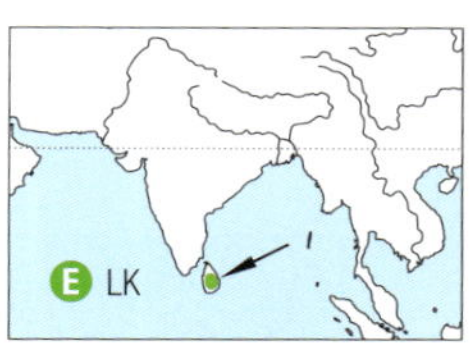

Fragmented locations in C Highlands

EN
Monotypic
HB 16–17 cm
T 21–23 cm

Brown Rat

Rattus norvegicus

FR Rat brun
DE Wanderratte
ES Rata parda

LC
Monotypic
HB 16–29 cm
T 12.2–25 cm
W 195–540 g

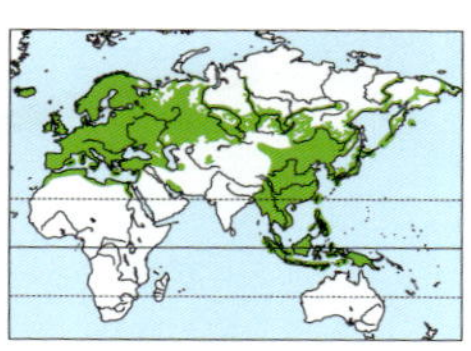

Native to N CN (possibly MN), SW Siberia & JP; introduced worldwide (modern places not mapped)

VU
Monotypic
HB 22–26 cm
T 19.2–21 cm

Andaman Archipelago Rat

Rattus stoicus

FR Rat des Andaman
DE Andamanen-Ratte
ES Rata de Andamán

0–200 m

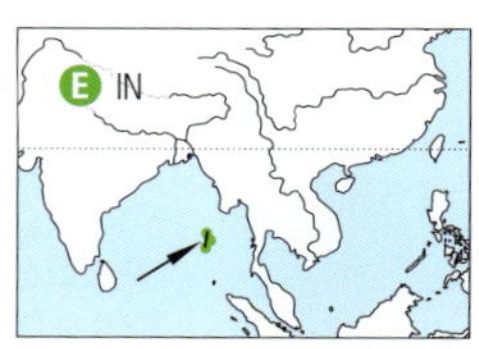

Andaman Is (Middle Andaman, South Andaman, Henry Lawrence)

Nicobar Archipelago Rat
Rattus burrus

FR Rat des Nicobar
DE Nikobaren-Ratte
ES Rata de Nicobar

EN

Monotypic
HB 17.4–21 cm
T 18.3–21 cm

up to 100 m

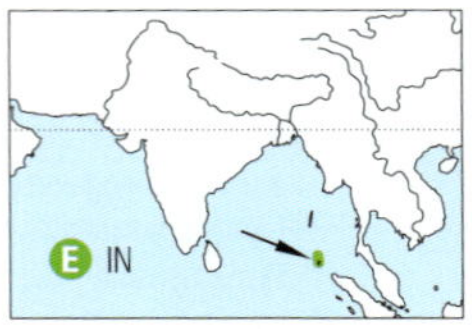

Nicobar Is (Trinkat, Little Nicobar, Great Nicobar)

VU

Monotypic
HB 22–24 cm
T 22–23 cm

Car Nicobar Rat
Rattus palmarum

FR Rat des palmiers
DE Car-Nicobar-Palmenratte
ES Rata de Car Nicobar

50–150 m

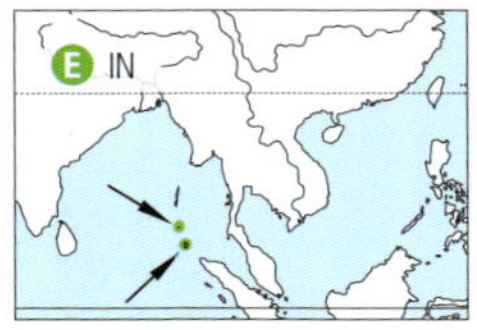

Nicobar Is (Car Nicobar, Great Nicobar)

NE

Monotypic
HB c. 15.4 cm
T c. 13.3 cm
W c. 84 g

Little Indochinese Field Rat
Rattus sakeratensis

FR Rat de Sakaerat
DE Thailand-Reisfeldratte
ES Rata de campo de Indochina pequeña

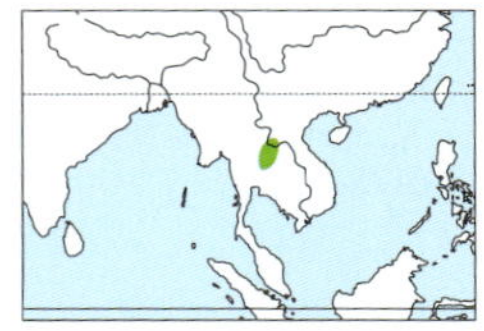

N & C TH, N LA (Vientiane plain); limits unknown

variants

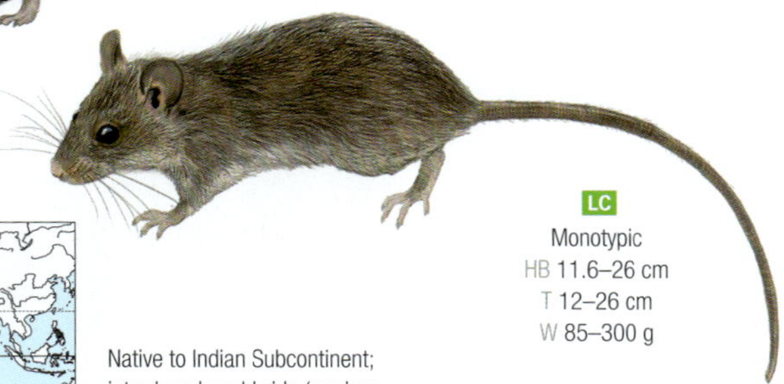

Roof Rat
Rattus rattus

FR Rat noir
DE Hausratte
ES Rata negra

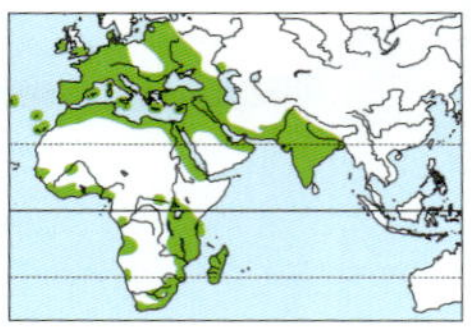

Native to Indian Subcontinent; introduced worldwide (modern places not mapped)

LC

Monotypic
HB 11.6–26 cm
T 12–26 cm
W 85–300 g

Oriental House Rat
Rattus tanezumi

FR Rat tanezumi
DE Asiatische Hausratte
ES Rata casera oriental

LC

Monotypic
HB 10.5–21 cm
T 12–23 cm
W 100–200 g

Native to E AF & N IN E to mainland SE Asia, perhaps TW & JP; widely introduced (modern places not mapped)

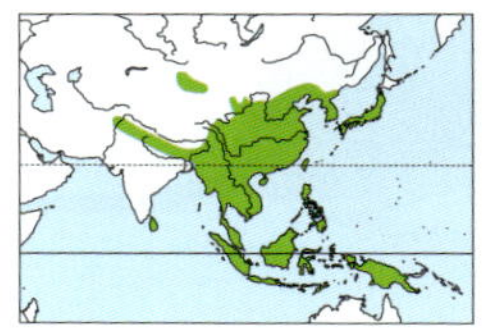

0–2000 m

Losea Rat
Rattus losea

FR Rat losea
DE Kleine Reisfeldratte
ES Rata de arrozal menor

LC

Monotypic
HB 12–18.5 cm
T 11–17.5 cm
W 22–92 g

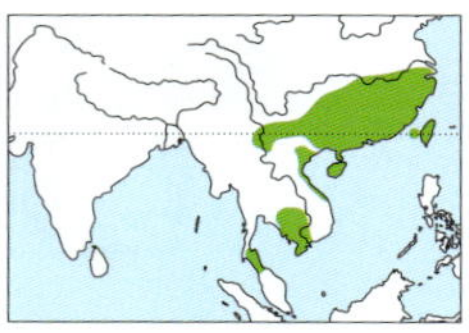

0–1000 m

SC & SE CN (with Hainan & Penghu Is), TW, TH, N LA, VN, KH, provisionally C Thai-Malay Pen

Osgood's Vietnamese Rat
Rattus osgoodi

FR Rat d'Osgood
DE Osgood-Ratte
ES Rata de Vietnam de Osgood

Langbian Plateau

900–2000 m

LC

Monotypic
HB 12.4–17.1 cm
T 14.7–16.1 cm

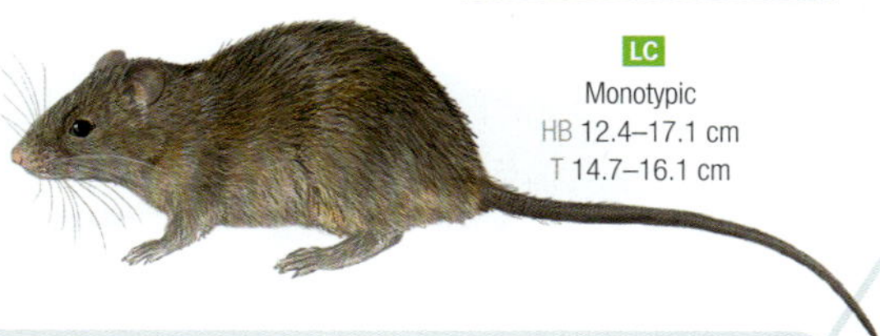

Malaysian Field Rat
Rattus tiomanicus

FR Rat de Tioman
DE Malaiische Feldratte
ES Rata de campo de Malasia

LC

Monotypic
HB 14–19 cm
T 15–20 cm
W 55–150 g

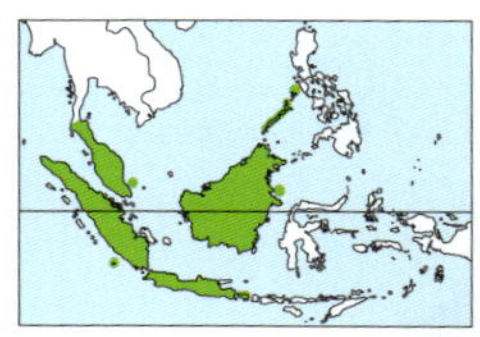

Thai-Malay Pen, Greater Sundas (MY/ID), Palawan (PH), many offshore Is

Tawitawi Forest Rat
Rattus tawitawiensis

FR Rat de Tawi-Tawi
DE Tawi-Tawi-Ratte
ES Rata de bosque de Tawitawi

DD

Monotypic
HB 16–21 cm
T 14.6–18 cm

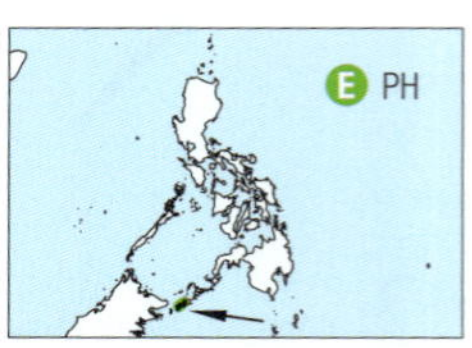

Sulu Arch (Tawi-Tawi I)

Kinabalu Rat
Rattus baluensis

FR Rat du Kinabalu
DE Kinabalu-Ratte
ES Rata de Kinabalu

LC

Monotypic
HB 15.8–18.8 cm
T 17.5–20 cm
W c. 108 g

1525–3810 m

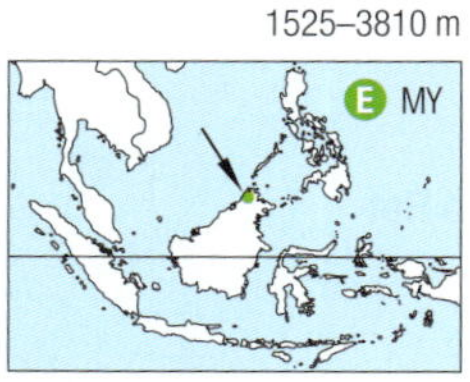

N Borneo (Mt Kinabalu)

Sumatran Mountain Rat
Rattus korinchi

FR Rat du Kerinci
DE Sumatra-Ratte
ES Rata de montaña de Sumatra

W Sumatra (Mt Talakmau, Mt Kerinci)

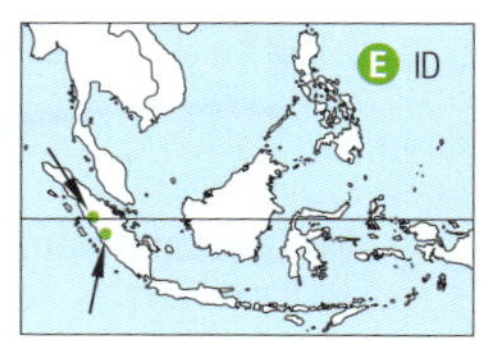

DD

Monotypic
HB 16.6–16.9 cm
T 21–22 cm

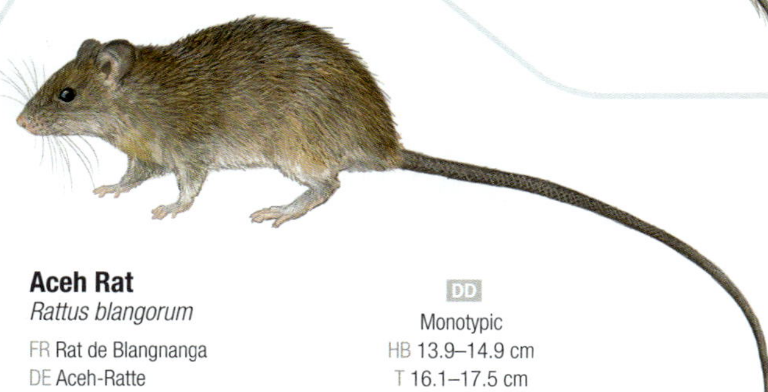

Aceh Rat
Rattus blangorum

FR Rat de Blangnanga
DE Aceh-Ratte
ES Rata de Aceh

DD

Monotypic
HB 13.9–14.9 cm
T 16.1–17.5 cm
W 74–91 g

c. 1100 m

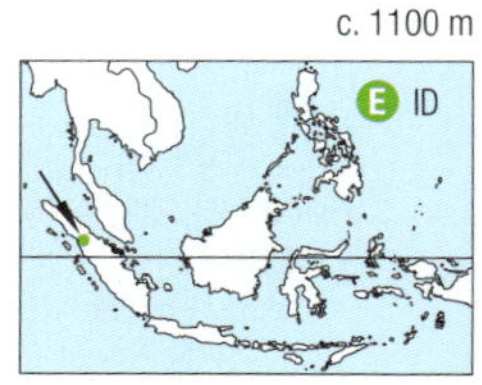

N Sumatra (foothills of Mt Leuser)

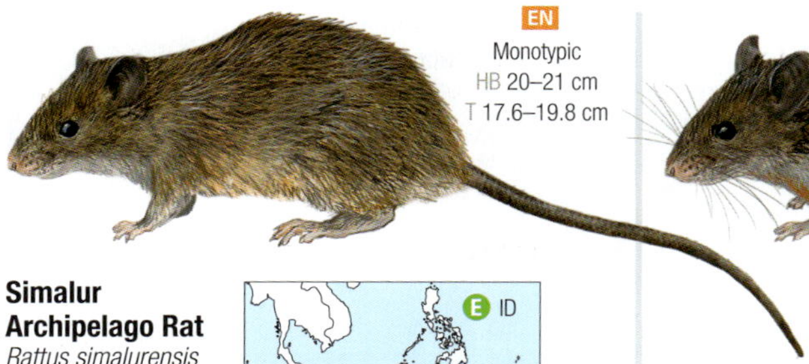

EN

Monotypic
HB 20–21 cm
T 17.6–19.8 cm

Simalur Archipelago Rat
Rattus simalurensis

FR Rat de Simeulue
DE Simalur-Ratte
ES Rata de Simalur

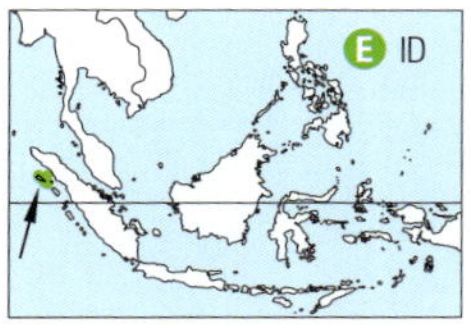

Simeulue, Siumat, Lasia & Babi Is (off NW Sumatra)

VU

Monotypic
HB c. 23 cm
T c. 21 cm

Mentawai Archipelago Rat
Rattus lugens

FR Rat des Mentawai
DE Mentawai-Ratte
ES Rata de Mentawai

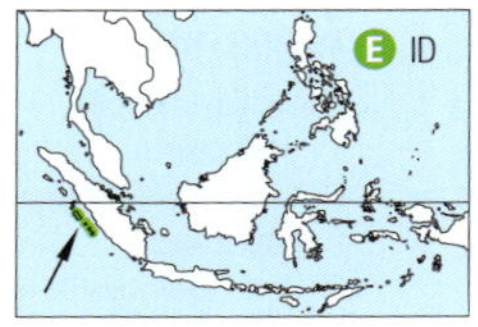

Mentawai Arch (Siberut, Sipora, North Pagai, South Pagai), off W Sumatra

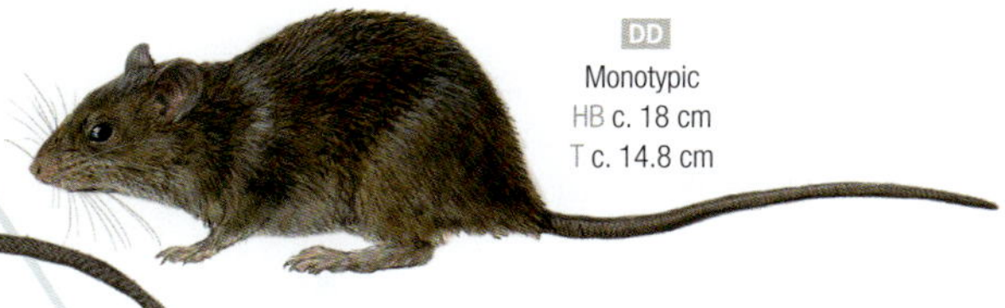

DD

Monotypic
HB c. 18 cm
T c. 14.8 cm

Burnished Enggano Rat
Rattus adustus

FR Rat basané
DE Indonesische Glattfellratte
ES Rata de Enggano pulida

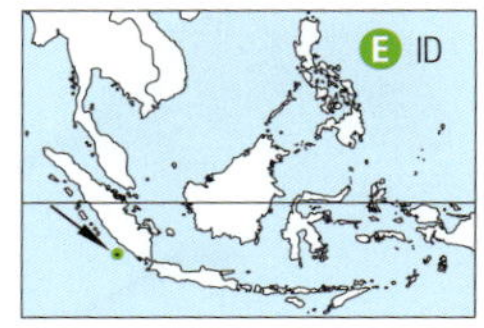

Enggano I, off SW Sumatra

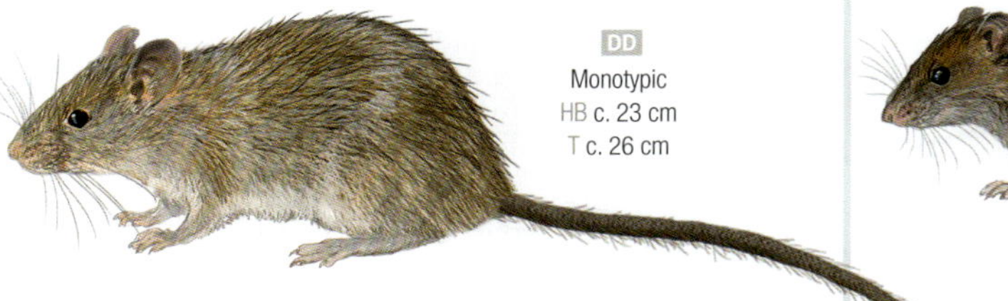

DD

Monotypic
HB c. 23 cm
T c. 26 cm

Enggano Island Rat
Rattus enganus

FR Rat d'Enggano
DE Enggano-Ratte
ES Rata de Enggano

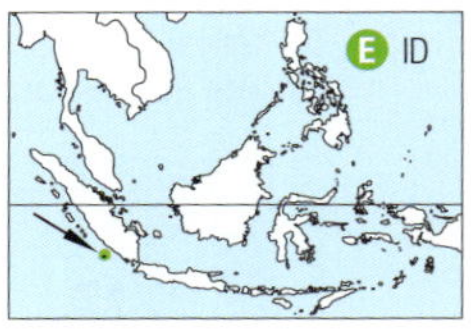

Enggano I, off SW Sumatra

LC

Monotypic
HB c. 19 cm
T c. 16.3 cm

Mindoro Mountain Rat
Rattus mindorensis

FR Rat de Mindoro
DE Schwarze Mindoro-Ratte
ES Rata de montaña de Mindoro

1000–1500 m

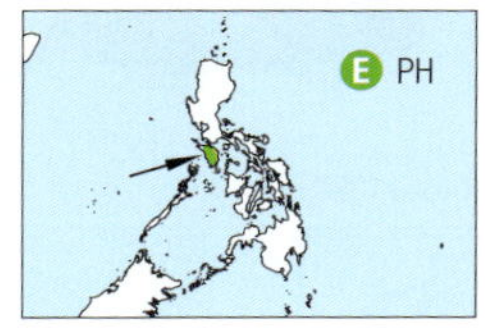

Mindoro I

Ricefield Rat
Rattus argentiventer

FR Rat des rizières
DE Reisfeldratte
ES Rata de arrozal

LC

Monotypic
HB 13.6–23 cm
T 14.9–21 cm
W 52–239 g

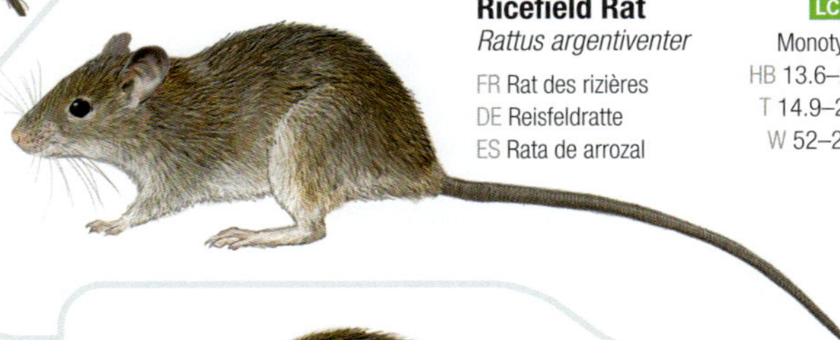

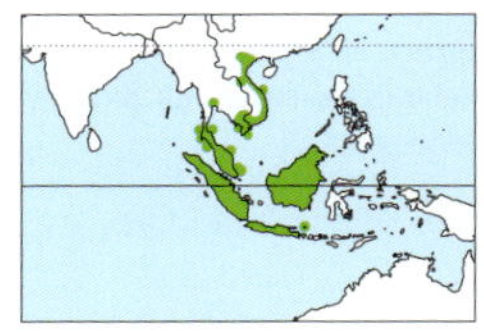

C LA, VN, KH, TH, Thai-Malay Pen, Greater Sundas, many offshore Is; introduced to PH & E ID

VU

Monotypic
HB 16.5–19.6 cm
T 21–26 cm
W 113–153 g

Hoogerwerf's Sumatran Rat
Rattus hoogerwerfi

FR Rat de Hoogerwerf
DE Hoogerwerf-Ratte
ES Rata de Sumatra de Hoogerwerf

2500–3000 m

N Sumatra (highlands of Mt Leuser)

Hoffmann's Sulawesi Rat
Rattus hoffmanni

FR Rat de Hoffmann
DE Hoffmann-Ratte
ES Rata de Célebes de Hoffmann

LC

Monotypic
HB 14.2–21 cm T 14.5–21 cm
W 90–250 g

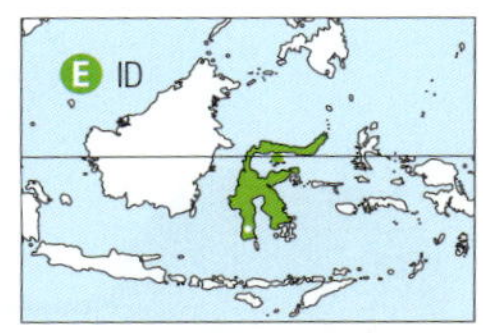

Sulawesi (except Mt Lompobatang on SW Pen), Togian Is (Malenge)

Lampobatang Sulawesi Rat
Rattus mollicomulus

FR Rat à poil doux
DE Lompobattang-Ratte
ES Rata de Lampobatang

LC

Monotypic
HB 15.3–15.7 cm
T 14.1–15.4 cm

1100–2000 m

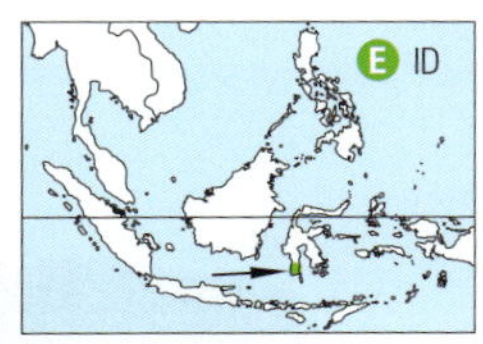

SW Sulawesi (Mt Lompobatang)

DD

Monotypic
HB c. 23 cm
T c. 21 cm

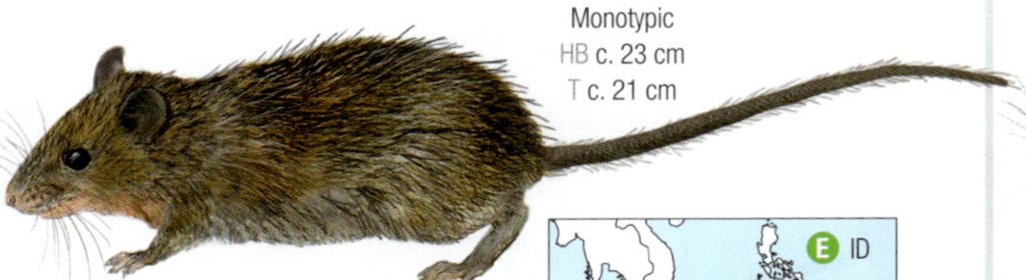

Koopman's Peleng Island Rat
Rattus koopmani

FR Rat de Koopman
DE Koopman-Ratte
ES Rata de Peleng de Koopman

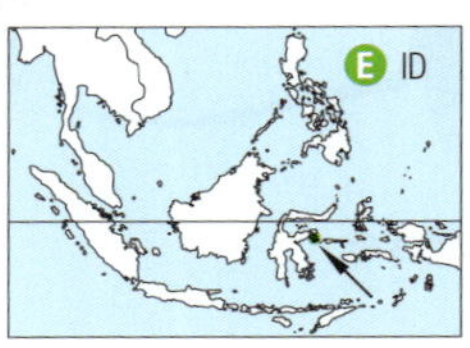

Banggai Is (Peleng)

LC

Monotypic
HB 15.5–20 cm
T 18.5–24 cm
W 100–150 g

Indochinese Forest Rat
Rattus andamanensis

FR Rat d'Indochine
DE Sikkim-Ratte
ES Rata de bosque de Indochina

up to 2000 m

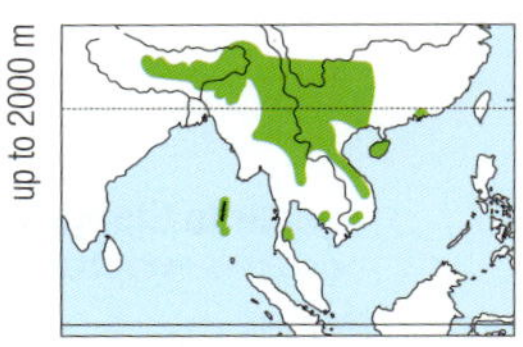

E NP E to S CN (with Hong Kong & Hainan I), S to Andaman & Nicobar, Is off E Pen TH, KH & S VN

Himalayan Rat
Rattus pyctoris

FR Rat pugiliste
DE Nepal-Ratte
ES Rata del Himalaya

LC

Monotypic
HB 14–21 cm
T 13.5–21 cm
W 100–200 g

1200–4250 m

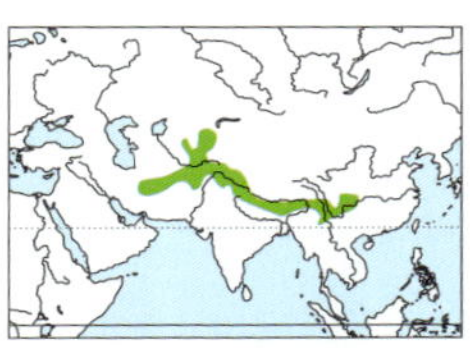

CE IR, AF, E UZ & SE KZ through Tian Shan, Pamir & Himalayas to S CN & N MM

Pacific Rat
Rattus exulans

FR Rat de Polynésie
DE Pazifik-Ratte
ES Rata del Pacífico

LC

Monotypic
HB 7.5–16.5 cm
T 10.2–19.7 cm W 23–60 g

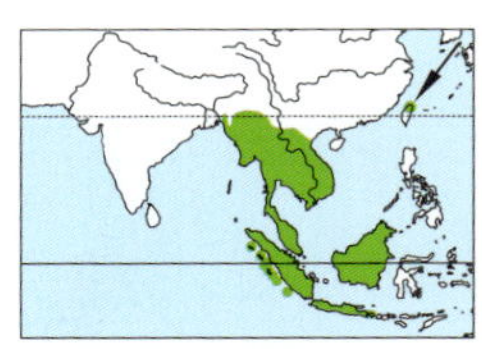

Native to mainland SE Asia, TW & Greater Sundas (including offshore Is); introduced to PH, E ID & Pacific Is

Genus *CHIROPODOMYS*

(6 species)

Indomalayan Pencil-tailed Tree Mouse
Chiropodomys gliroides

FR Souris-à-pinceau de Malaisie
DE Indomalayische Pinselschwanz-Baummaus
ES Ratón arborícola de cola de lápiz de Indomalasia

LC
Monotypic
HB 6.9–10.2 cm
T 9.4–14.3 cm
W 15–32 g

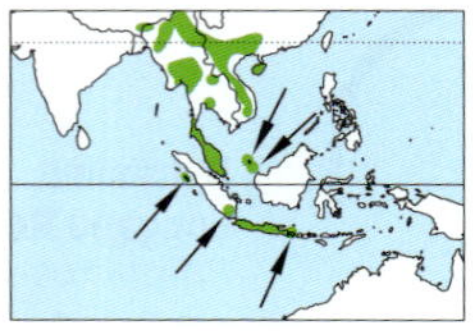

NE IN, SC CN, mainland SE Asia, ID (Nias I, S Sumatra, Tujuh Is, Natuna Is, Java, Bali)

Koopman's Pencil-tailed Tree Mouse
Chiropodomys karlkoopmani

FR Souris-à-pinceau de Koopman
DE Koopman-Pinselschwanz-Baummaus
ES Ratón arborícola de cola de lápiz de Koopman

VU
Monotypic
HB c. 10.7 cm
T c. 17.1 cm

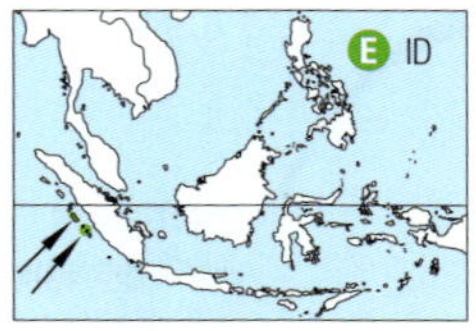

Mentawai Arch (Siberut, North Pagai), off W Sumatra

Lesser Pencil-tailed Tree Mouse
Chiropodomys pusillus

FR Petite Souris-à-pinceau
DE Kleine Pinselschwanz-Baummaus
ES Ratón arborícola de cola de lápiz menor

DD
Monotypic
HB 6.9–7.7 cm
T 8.1–9.6 cm

300–1220 m

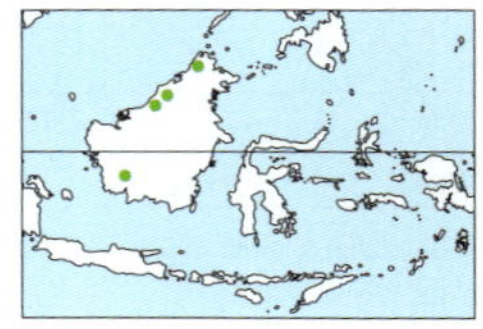

Borneo: Sabah & Sarawak (MY), S Kalimantan (ID); probably widespread

Gray-bellied Pencil-tailed Tree Mouse
Chiropodomys muroides

FR Souris-à-pinceau de Medway
DE Graubauch-Pinselschwanz-Baummaus
ES Ratón arborícola de cola de lápiz de vientre gris

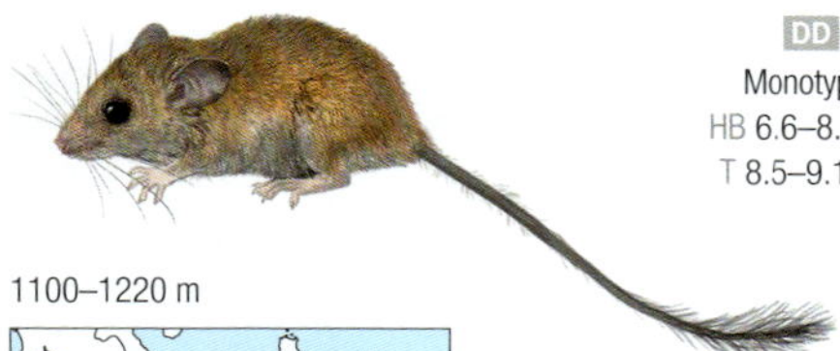

DD
Monotypic
HB 6.6–8.8 cm
T 8.5–9.1 cm

1100–1220 m

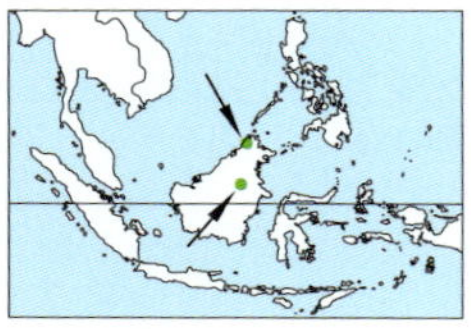

Borneo: Mt Kinabalu (MY), Long Petak (N Kalimantan, ID); probably widespread

Greater Pencil-tailed Tree Mouse
Chiropodomys major

FR Grande Souris-à-pinceau
DE Große Pinselschwanz-Baummaus
ES Ratón arborícola de cola de lápiz mayor

LC
Monotypic
HB 9.4–11.4 cm
T 10.9–14.4 cm
W 32–43 g

900–1500 m

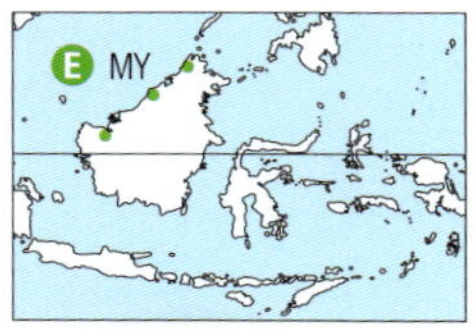

Borneo: Sarawak & Sabah (MY); probably Kalimantan (ID)

Palawan Pencil-tailed Tree Mouse
Chiropodomys calamianensis

FR Souris-à-pinceau de Palawan
DE Palawan-Pinselschwanz-Baummaus
ES Ratón arborícola de cola de lápiz de Palawan

DD
Monotypic
HB 10.9–12.2 cm
T 14–17.1 cm
W c. 40 g

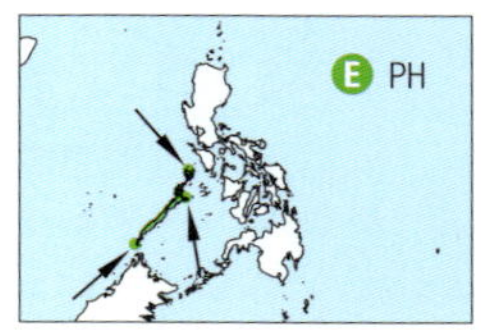

0–250 m

Palawan and nearby Calauit, Busuanga, Dumaran & Balabac Is

Genus *HAEROMYS*

(3 species)

Sundaic Ranee Mouse
Haeromys pusillus

FR Haeromys pygmée
DE Palawan-Zwergbaummaus
ES Ratón raní de Sonda

above 1000 m

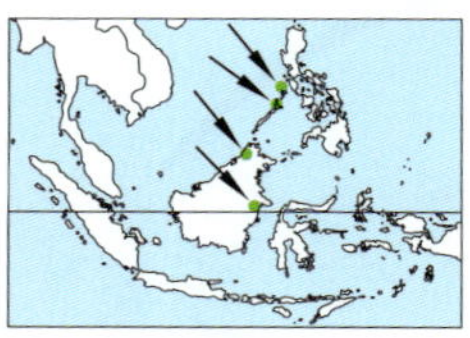

Few localities in N & E Borneo (MY, ID) and W PH (N Palawan, Calamian Is)

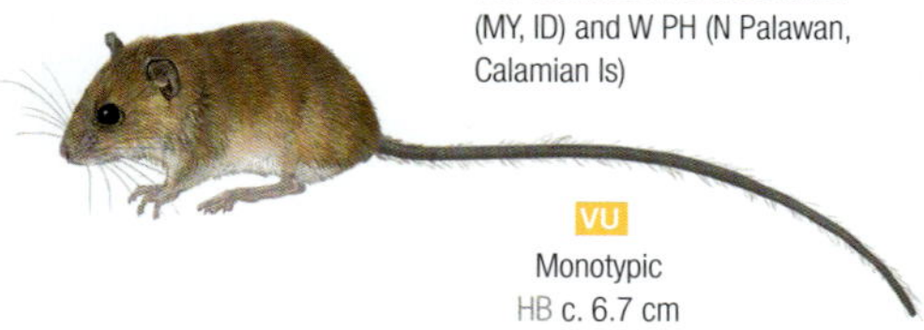

VU
Monotypic
HB c. 6.7 cm
T c. 12.3 cm

Margaret's Ranee Mouse
Haeromys margarettae

FR Haeromys de Margaret
DE Borneo-Zwergbaummaus
ES Ratón raní de Margaret

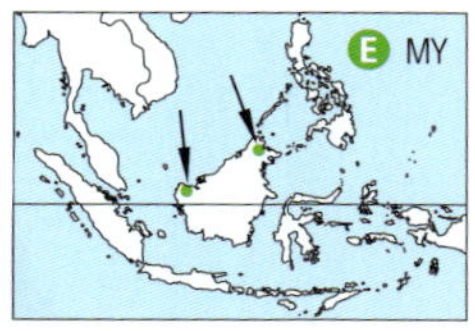

N & SW Borneo (known only from 2 localities in Sabah & Sarawak)

DD
Monotypic
HB c. 7.6 cm
T c. 14.4 cm

Lowland Sulawesi Ranee Mouse
Haeromys minahassae

FR Haeromys de Minahasa
DE Sulawesi-Zwergbaummaus
ES Ratón raní de Célebes

NT
Monotypic
HB 7–7.8 cm
T 12–13.1 cm
W 12–15 g

75–150 m

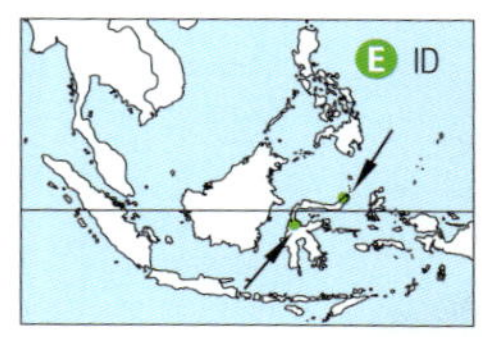

Known from 2 isolated regions in N & C Sulawesi

Genus *SORICOMYS*

(4 species)

Leonardo Shrew Mouse
Soricomys leonardocoi

FR Rat de Leonardo
DE Leonardo-Spitzkopfmaus
ES Ratón musaraña de Leonardo

DD
Monotypic
HB 9.8–11.5 cm
T 8.2–9.5 cm
W 26–36 g

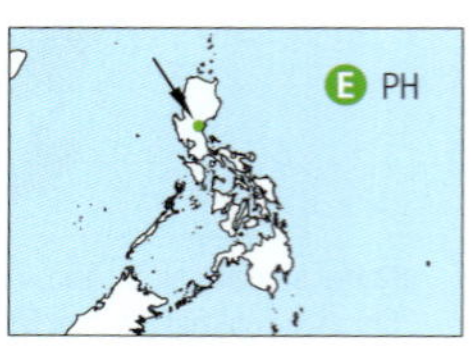

1475–1785 m

EC Luzon I (Mingan Mts)

Sierra Madre Shrew Mouse
Soricomys musseri

FR Rat de Musser
DE Musser-Spitzkopfmaus
ES Ratón musaraña de Sierra Madre

LC
Monotypic
HB 9.1–10.5 cm
T 8.9–9.3 cm
W 30–32 g

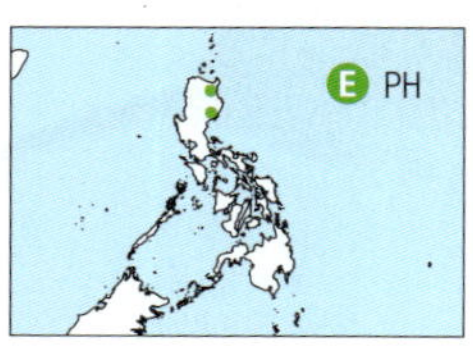

1450–1760 m

NE Luzon I (Mt Cetaceo, Mt Anacuao)

Kalinga Shrew Mouse
Soricomys kalinga

FR Rat kalinga
DE Kalinga-Spitzkopfmaus
ES Ratón musaraña de Bali

LC
Monotypic
HB 9.3–10.8 cm
T 8.5–10.1 cm
W 21–29 g

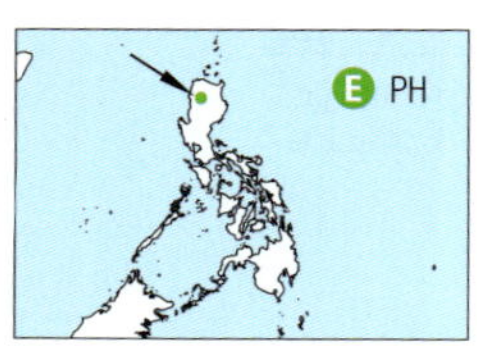

1600–2150 m

Luzon I (Mt Bali and surroundings)

Mountain Shrew Mouse
Soricomys montanus

FR Rat de Balete
DE Berg-Spitzkopfmaus
ES Ratón musaraña de montaña

LC
Monotypic
HB 9.3–10.3 cm
T 8.5–9.8 cm
W 23–31 g

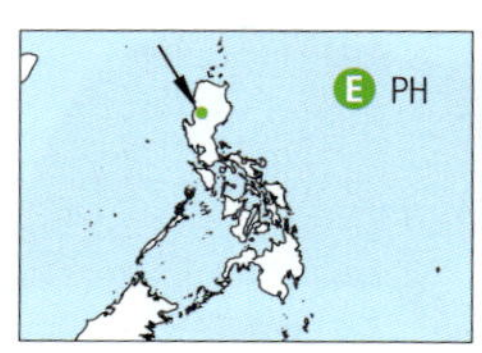

1950–2690 m

N Luzon I (S Cordillera Central)

Genus *CHROTOMYS*

(5 species)

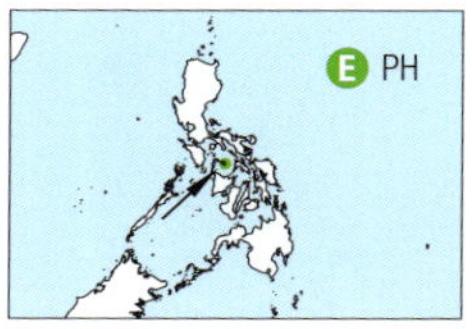

Sibuyan I

Sibuyan Striped Shrew Rat
Chrotomys sibuyanensis

FR Chrotomys de Sibuyan
DE Sibuyan-Streifenratte
ES Rata musaraña listada de Sibuyán

DD
Monotypic
HB c. 16 cm T c. 8.2 cm
W c. 104 g

Blazed Luzon Striped Shrew Rat
Chrotomys silaceus

FR Chrotomys orné
DE Ockergelbe Streifenratte
ES Rata musaraña listada de Luzón

LC
Monotypic
HB 13.6–17.3 cm
T 9.7–12.1 cm
W 71–160 g

1800–2800 m

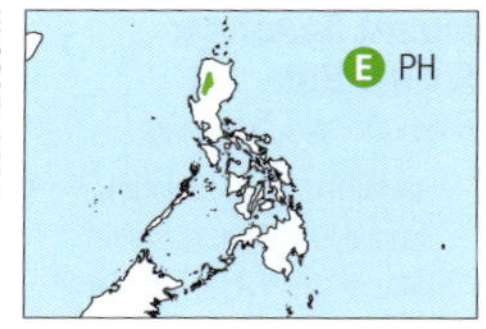

N Luzon I (Cordillera Central)

Isarog Striped Shrew Rat
Chrotomys gonzalesi

FR Chrotomys de l'Isarog
DE Isarog-Streifenratte
ES Rata musaraña listada de Isarog

NT
Monotypic
HB 16–19 cm
T 8.3–10.5 cm
W 70–190 g

1350–1750 m

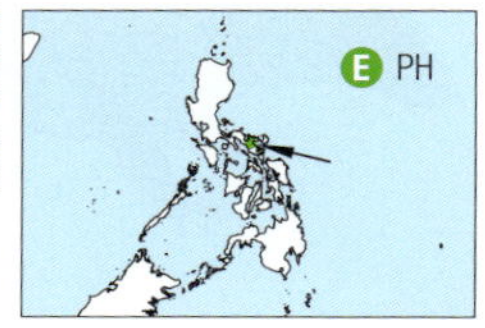

SE Luzon I (Mt Isarog, perhaps Saddle Peak & Mt Malinao)

Lowland Striped Shrew Rat
Chrotomys mindorensis

FR Chrotomys de Mindoro
DE Luzon-Flachlandstreifenratte
ES Rata musaraña listada de tierras bajas

LC
Monotypic
HB 15.5–18.6 cm
T 9.9–12.3 cm
W 152–199 g

below 2050 m

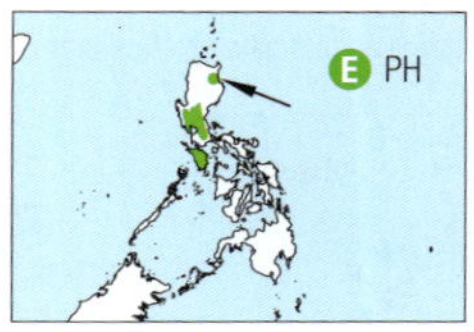

Luzon & Mindoro Is

Montane Striped Shrew Rat
Chrotomys whiteheadi

FR Chrotomys de Whitehead
DE Luzon-Bergstreifenratte
ES Rata musaraña listada de montaña

LC
Monotypic
HB 14.4–17.7 cm
T 9.5–13.3 cm
W 73–160 g

900–2700 m

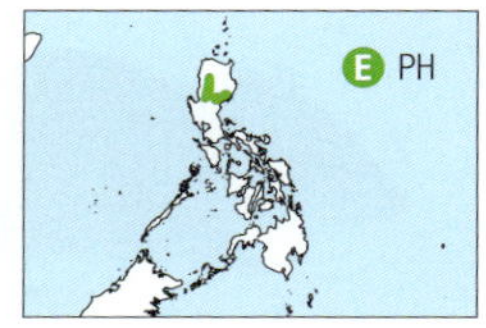

N Luzon I (Cordillera Central, Mt Palali in Caraballo Mts, Mt Anacuao in S Sierra Madre)

Genus *ARCHBOLDOMYS*

(2 species)

Isarog Shrew Mouse
Archboldomys luzonensis

FR Rat de l'Isarog
DE Isarog-Archboldmaus
ES Ratón musaraña de Isarog

VU
Monotypic
HB 10.1–11.1 cm
T 6.6–8 cm
W 31–36 g

1350–2000 m

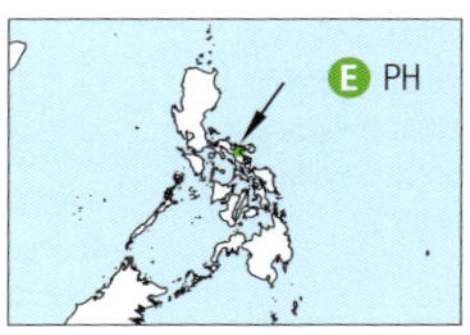

Luzon I (Mt Isarog)

Large Cordillera Shrew Mouse
Archboldomys maximus

FR Rat de la cordillère Centrale
DE Große Archboldmaus
ES Ratón musaraña de Cordillera

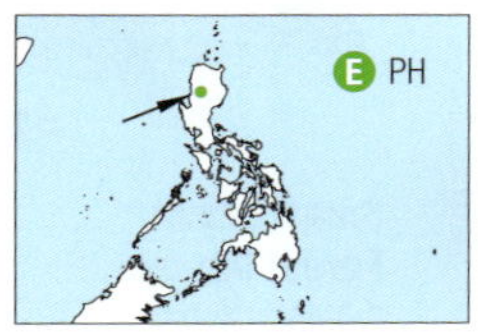

NC Luzon I (upper slopes of Mt Amuyao; probably widespread on nearby peaks)

DD
Monotypic
HB 10.2–11.6 cm
T 9–10.8 cm
W 40–55 g

Genus *RHYNCHOMYS*

(6 species)

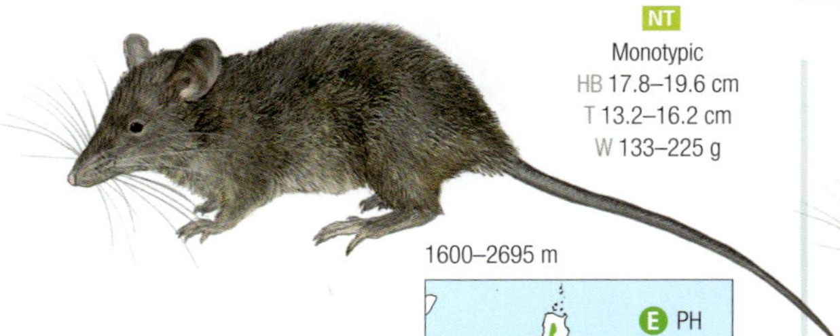

NT
Monotypic
HB 17.8–19.6 cm
T 13.2–16.2 cm
W 133–225 g

1600–2695 m

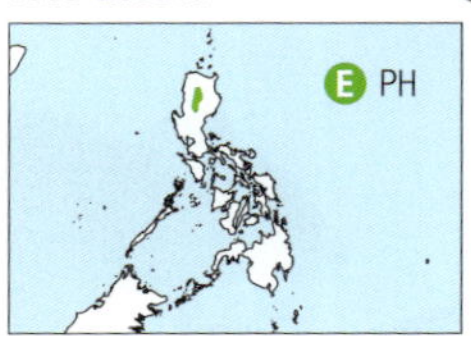

N Luzon I (Cordillera Central)

Mount Data Shrew Rat
Rhynchomys soricoides

FR Rhynchomys du Data
DE Mount-Data-Nasenratte
ES Rata musaraña de Data

NE
Monotypic
HB 17.5–19.9 cm
T 11.8–13.4 cm
W 132–176 g

1475–1785 m

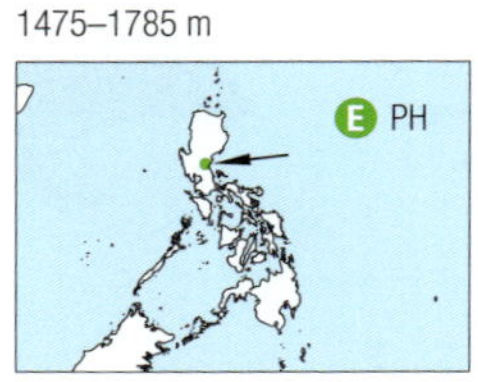

Luzon I (SW slope of Mt Mingan)

Mingan Shrew Rat
Rhynchomys mingan

FR Rhynchomys du Mingan
DE Mingan-Nasenratte
ES Rata musaraña de Mingan

DD
Monotypic
HB 16.4–18.8 cm
T 12–12.8 cm
W 129–156 g

Mount Tapulao Shrew Rat
Rhynchomys tapulao

FR Rhynchomys du Tapulao
DE Mount-Tapulao-Nasenratte
ES Rata musaraña de Tapulao

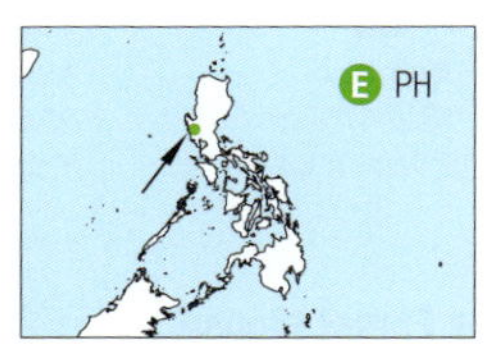

c. 2025 m

W Luzon I (Mt Tapulao)

VU
Monotypic
HB 17–18.7 cm
T 10.8–12.6 cm
W 110–156 g

Mount Isarog Shrew Rat
Rhynchomys isarogensis

FR Rhynchomys de l'Isarog
DE Mount-Isarog-Nasenratte
ES Rata musaraña de Isarog

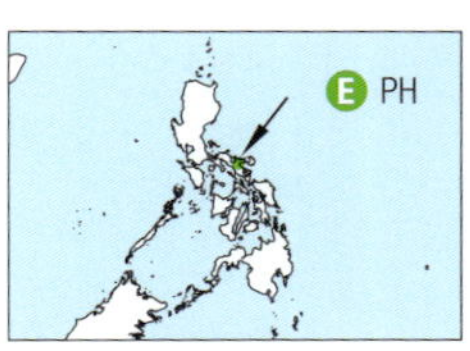

1125–1750 m

SE Luzon I (Mt Isarog)

NE
Monotypic
HB 15.7–19.2 cm
T 9.7–11.4 cm
W 134–182 g

Labo Shrew Rat
Rhynchomys labo

FR Rhynchomys du Labo
DE Labo-Nasenratte
ES Rata musaraña de Labo

1250–1415 m

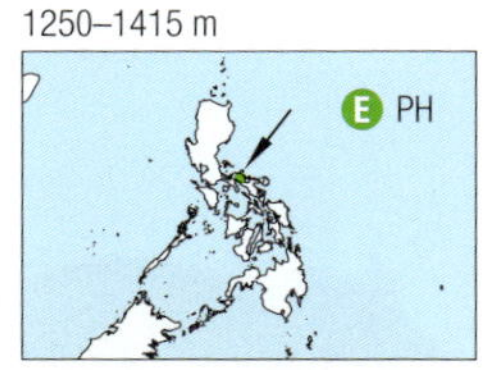

Luzon I (N slope of Mt Labo)

Mount Banahaw Shrew Rat
Rhynchomys banahao

FR Rhynchomys du Banahaw
DE Mount-Banahaw-Nasenratte
ES Rata musaraña de Banahaw

LC
Monotypic
HB 17.8–19 cm
T 12.6–13 cm
W 135–155 g

1250–1625 m

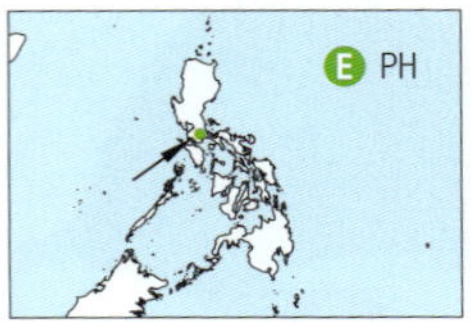

Luzon I (Mt Banahaw)

Genus *APOMYS*

(19 species)

Camiguin Forest Mouse
Apomys camiguinensis

FR Apomys de Camiguin
DE Camiguin-Philippinenwaldmaus
ES Ratón de bosque de Camiguín

VU
Monotypic
HB c. 10.6 cm
T 14–16 cm
W 33–48 g

1000–1400 m

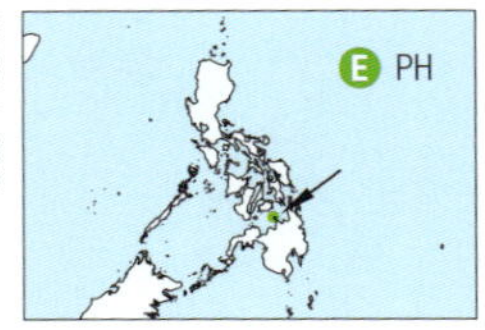

Camiguin I

Mindanao Mossy Forest Mouse
Apomys hylocoetes

FR Apomys des mousses
DE Nebelwald-Philippinenwaldmaus
ES Ratón de bosque de Mindanao

LC
Monotypic
HB 11.1–11.5 cm
T 12–15 cm
W 33–45 g

1800–2800 m

Mindanao I (Kitanglad Range, Mt Apo; probably Kamangkil–Lumut ranges)

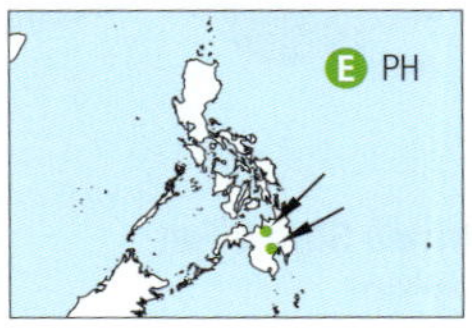

Mindanao Montane Forest Mouse
Apomys insignis

FR Apomys remarquable
DE Mindanao-Philippinenwaldmaus
ES Ratón de bosque de montaña de Mindanao

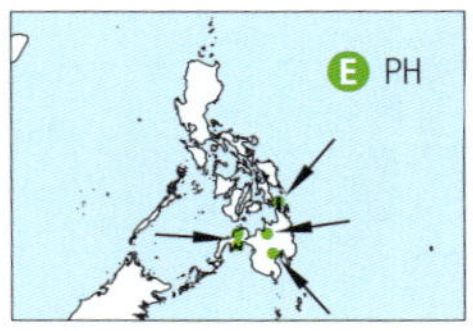

Mindanao & Dinagat Is

LC
Monotypic
HB 7.1–12.3 cm
T 13.4–17.6 cm
W 33–52 g

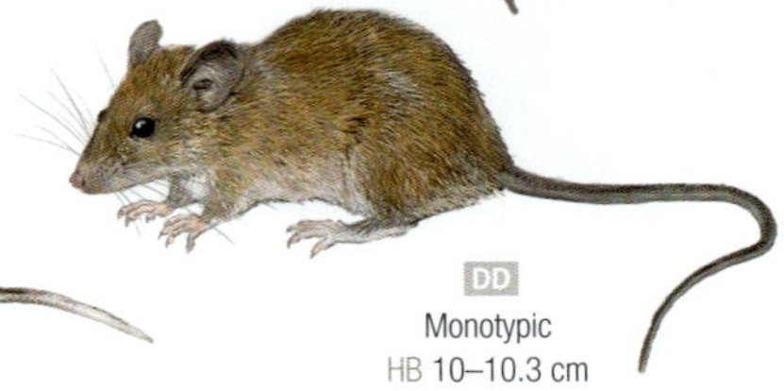

DD
Monotypic
HB 10–10.3 cm
T 12.2–12.6 cm

Mindanao Lowland Forest Mouse
Apomys littoralis

FR Apomys côtier
DE Küsten-Philippinenwaldmaus
ES Ratón de bosque de tierras bajas de Mindanao

15–1400 m

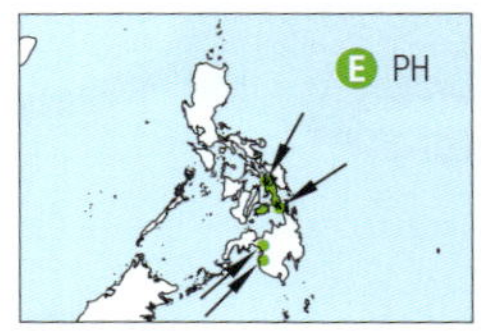

Mindanao I (Lanao del Sur & Maguindanao prov); perhaps Leyte, Biliran & Bohol Is

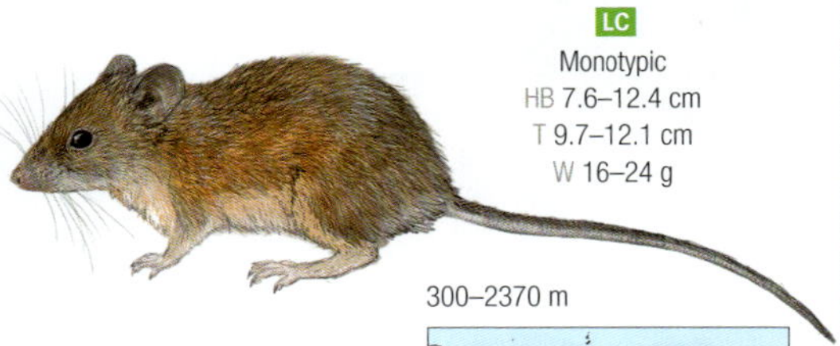

LC
Monotypic
HB 7.6–12.4 cm
T 9.7–12.1 cm
W 16–24 g

Least Philippine Forest Mouse
Apomys musculus

FR Petit Apomys
DE Zwerg-Philippinenwaldmaus
ES Ratón de bosque mínimo de Filipinas

300–2370 m

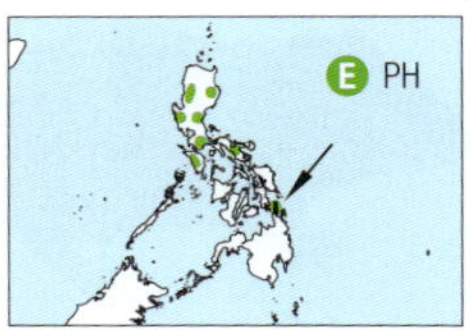

Luzon, Mindoro & Dinagat Is

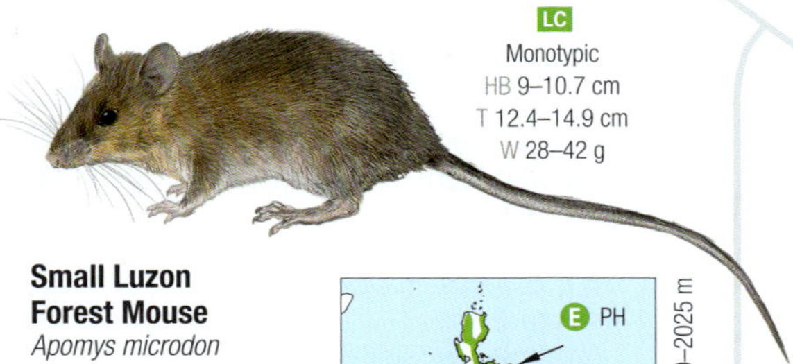

LC
Monotypic
HB 9–10.7 cm
T 12.4–14.9 cm
W 28–42 g

Small Luzon Forest Mouse
Apomys microdon

FR Apomys de Catanduanes
DE Kleinzahn-Philippinenwaldmaus
ES Ratón de bosque pequeño de Luzón

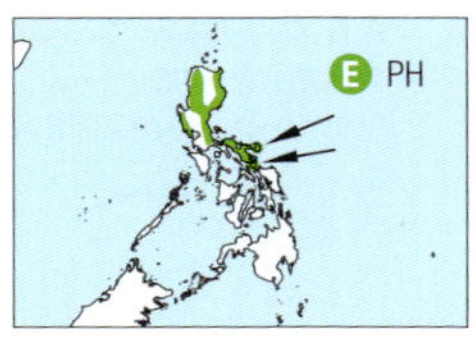

0–2025 m

Luzon & Catanduanes Is

Large Mindoro Forest Mouse
Apomys gracilirostris

FR Apomys de Mindoro
DE Schmalschnauzen-Philippinenwaldmaus
ES Ratón de bosque de Mindoro

1250–1950 m

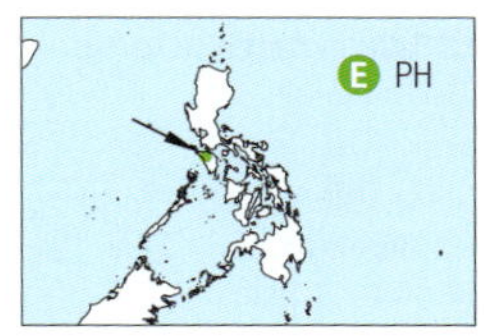

Mindoro I (recorded only on N ridge of Mt Halcon; likely widespread)

LC
Monotypic
HB 13.7–14.7 cm
T 13.5–18.5 cm
W 71–140 g

Sierra Madre Forest Mouse
Apomys sierrae

FR Apomys de la Sierra Madre
DE Sierra-Madre-Philippinenwaldmaus
ES Ratón de bosque de Sierra Madre

155–1700 m

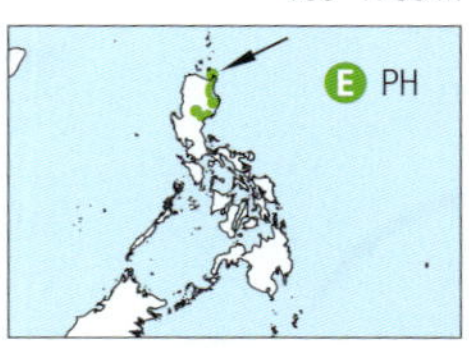

Luzon I (Sierra Madre, Mt Lataan, Mungiao Mts, Mt Palali), adjacent Palaui I

LC
Monotypic
HB 12.2–17.1 cm
T 11.8–15.4 cm W 64–111 g

Large Forest Mouse
Apomys magnus

FR Grand Apomys
DE Große Philippinenwaldmaus
ES Ratón de bosque grande

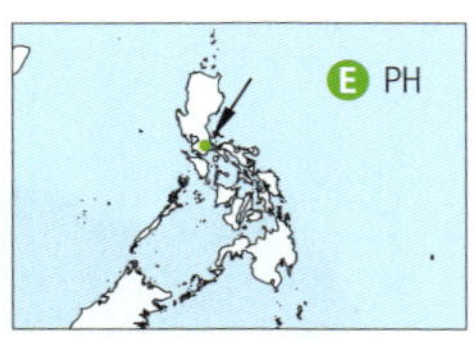

765–1465 m

Luzon I (Mt Banahaw; possibly on adjacent Mt San Cristobal)

LC
Monotypic
HB 13.7–16 cm
T 13.3–15.4 cm W 92–128 g

LC
Monotypic
HB 12.6–15.2 cm
T 12.9–15.3 cm
W 58–92 g

Aurora Forest Mouse
Apomys aurorae

FR Apomys d'Aurora
DE Aurora-Philippinenwaldmaus
ES Ratón de bosque de Aurora

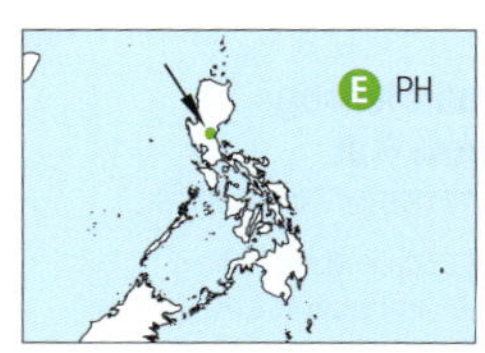

735–1675 m

E Luzon I (Mingan Mts of S Sierra Madre)

Zambales Forest Mouse
Apomys zambalensis

FR Apomys des Zambales
DE Zambales-Philippinenwaldmaus
ES Ratón de bosque de Zambales

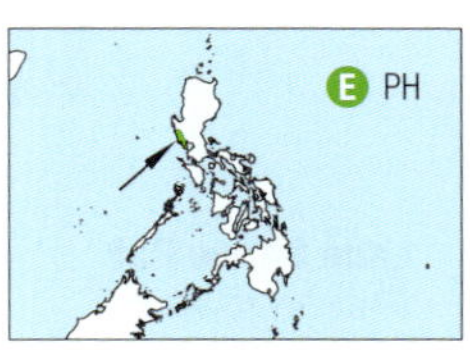

365–1690 m

W Luzon I (Zambales Mts)

LC
Monotypic
HB 12.7–16.3 cm
T 12.3–15.8 cm
W 67–112 g

Mount Irid Forest Mouse
Apomys iridensis

FR Apomys de l'Irid
DE Mount-Irid-Philippinenwaldmaus
ES Ratón de bosque de Irid

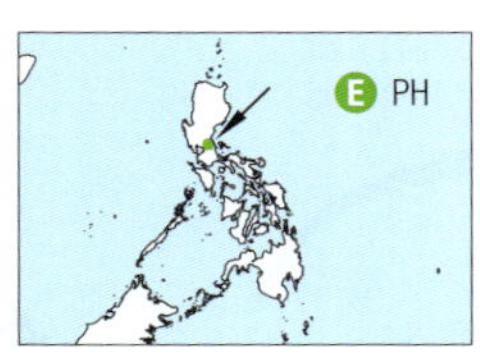

700–1320 m

E Luzon I (Mt Irid; possibly widespread in S Sierra Madre)

LC
Monotypic
HB 13.6–15.5 cm
T 13.2–15.7 cm
W 76–104 g

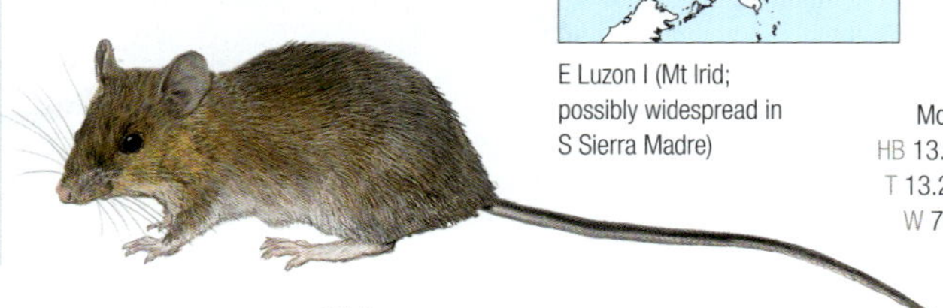

LC
Monotypic
HB 25–28 cm
T 11.6–13.8 cm
W 66–92 g

Mount Mingan Forest Mouse
Apomys minganensis

FR Apomys des Mingan
DE Mount-Mingan-Philippinenwaldmaus
ES Ratón de bosque de Mingan

1540–1785 m

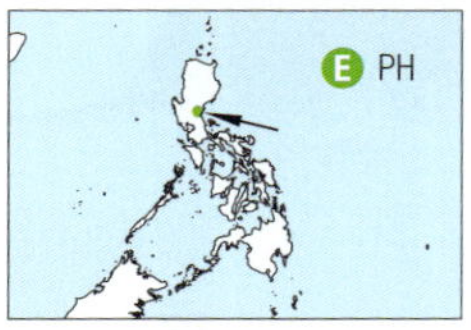

E Luzon I (Mingan Mts; perhaps on adjacent S Sierra Madre & Dingalan Mts)

DD
Monotypic
HB 12.3–14 cm
T 10.7–11.6 cm
W 60–84 g

Browns' Forest Mouse
Apomys brownorum

FR Apomys des Brown
DE Brown-Philippinenwaldmaus
ES Ratón de bosque de los Brown

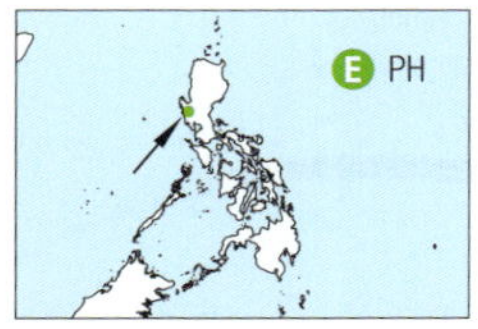

W Luzon I (Mt Tapulao; perhaps other high peaks in Zambales Mts)

LC
Monotypic
HB 12.5–15.4 cm
T 11.1–13.3 cm
W 71–92 g

Banahaw Forest Mouse
Apomys banahao

FR Apomys du Banahaw
DE Banahaw-Philippinenwaldmaus
ES Ratón de bosque de Banahaw

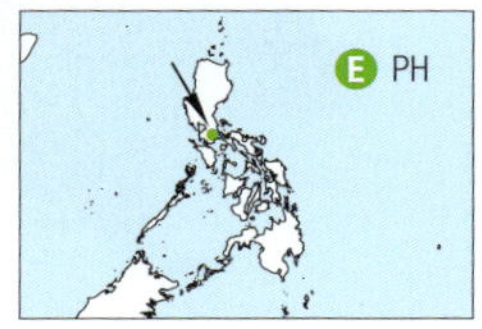

1465–1750 m

Luzon I (Mt Banahaw & Mt Banahaw de Lucban; likely Mt San Cristobal)

Long-nosed Luzon Forest Mouse
Apomys sacobianus

FR Apomys de la Sacobia
DE Langnasen-Philippinenwaldmaus
ES Ratón de bosque de Luzón de hocico largo

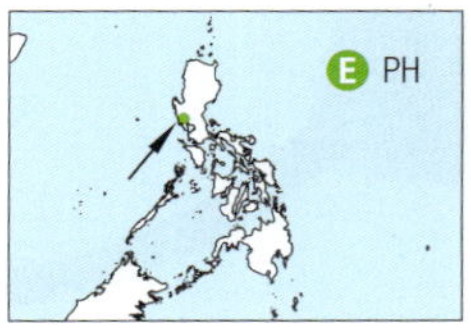

W Luzon I (Mt Pinatubo, Zambales Mts; perhaps widespread)

LC
Monotypic
HB 13.5–15.8 cm
T 11.5–15.9 cm W 71–105 g

LC
Monotypic
HB 14.3–16 cm
T 12.8–15.4 cm
W 102–128 g

Lubang Forest Mouse
Apomys lubangensis

FR Apomys de Lubang
DE Lubang-Philippinenwaldmaus
ES Ratón de bosque de Lubang

130–450 m

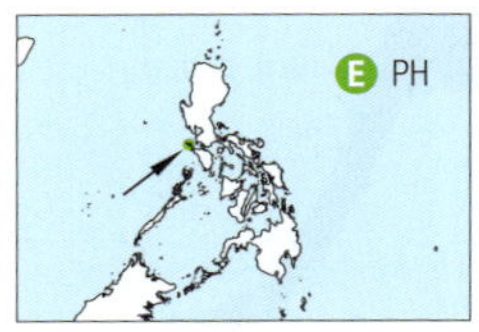

Lubang I (Mt Ambulong)

LC
Monotypic
HB 12.1–14.1 cm
T 12–14.4 cm
W 50–79 g

Luzon Cordillera Forest Mouse
Apomys abrae

FR Apomys d'Abra
DE Cordillera-Philippinenwaldmaus
ES Ratón de bosque de Luzón

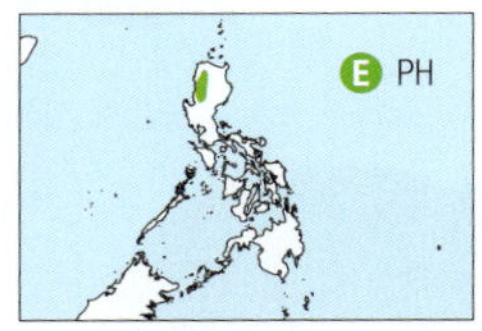

925–2200 m

NW Luzon I (Cordillera Central)

Northern Luzon Forest Mouse
Apomys datae

FR Apomys du Data
DE Mount-Data-Philippinenwaldmaus
ES Ratón de bosque de Luzón septentrional

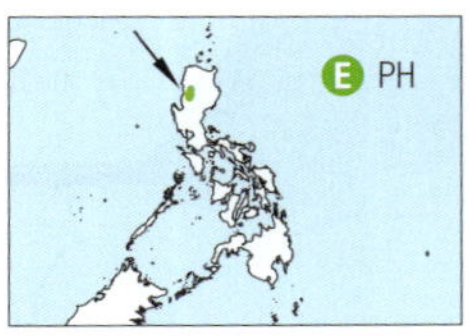

1500–2800 m

NW Luzon I (Cordillera Central)

LC
Monotypic
HB 12.5–15.5 cm
T 12.1–13.9 cm
W 63–105 g

Genus *ANISOMYS*

(1 species)

LC
Monotypic
HB 21–28 cm
T 27–36 cm
W 280–580 g

Uneven-toothed Rat
Anisomys imitator

FR Anisomys
DE Hörnchenzahnratte
ES Rata de dientes irregulares

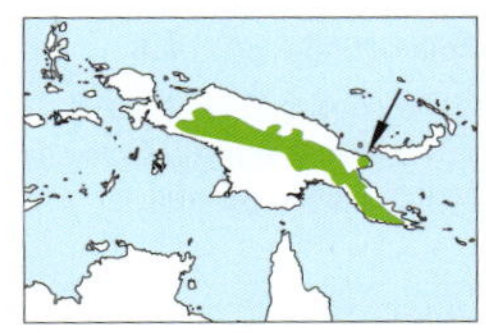

ID/PG: New Guinea (from Weyland Range E to E to Owen Stanley Range, Huon Pen)

Genus *LORENTZIMYS*

(1 species)

Long-footed Tree Mouse
Lorentzimys nouhuysii

FR Rat de Nouhuys
DE Neuguinea-Baumspringmaus
ES Ratón arborícola de pies largos

80–2700 m

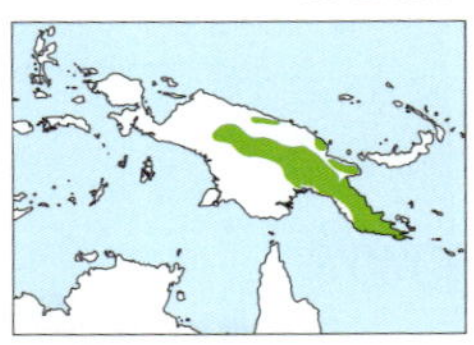

ID/PG: New Guinea (Central Range, Torricelli Mts, Adelbert Range, Huon Pen)

LC
Monotypic
HB 5.5–9 cm
T 11–12.5 cm
W 10–23 g

Genus *CHIRUROMYS*

(3 species)

Forbes's Tree Mouse
Chiruromys forbesi

FR Grand Chiruromys
DE Forbes-Breitkopf-Baummaus
ES Ratón arborícola de Forbes

LC
Monotypic
HB 13.4–16.1 cm
T 21–23 cm

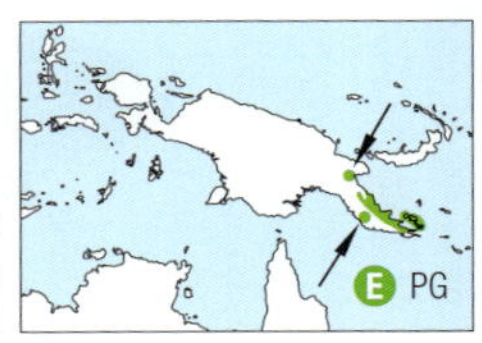

700–1300 m

E New Guinea (SE Pen, R Markham Valley, D'Entrecasteaux Is)

LC
Monotypic
HB 9.7–12 cm
T 14.7–17.2 cm
W 40–56 g

Broad-headed Tree Mouse
Chiruromys lamia

FR Chiruromys à tête large
DE Breitkopf-Baummaus
ES Ratón arborícola de cabeza ancha

1200–2300 m

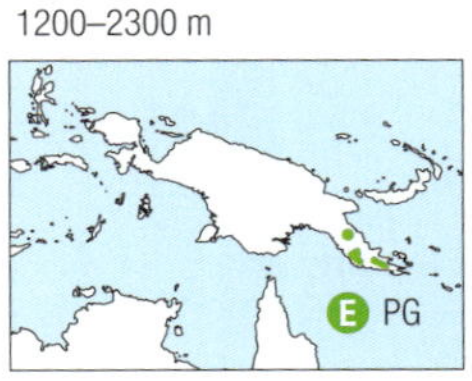

E New Guinea (SE Pen, with Astrolabe, Owen Stanley & Maneau ranges)

Lesser Tree Mouse
Chiruromys vates

FR Petit Chiruromys
DE Kleine Breitkopf-Baummaus
ES Ratón arborícola menor

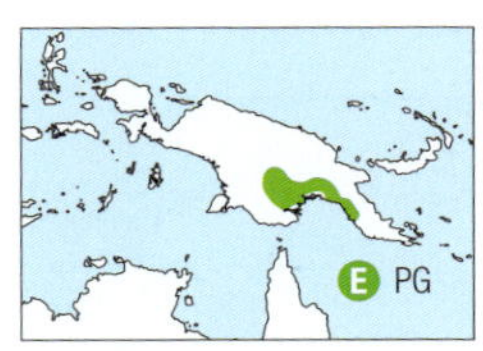

0–2200 m

SE New Guinea (from L Murray & upper R Fly E to Astrolabe Range & Kokoda Gap)

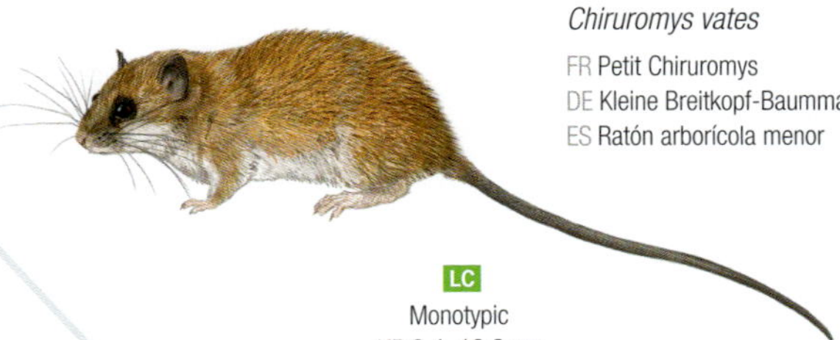

LC
Monotypic
HB 8.4–12.6 cm
T 12.8–18.3 cm
W 23–68 g

Genus *MACRUROMYS*

(2 species)

Greater Small-toothed Rat
Macruromys major

FR Grand Macruromys
DE Große Neuguinea-Kleinzahnratte
ES Rata de dientes pequeños mayor

LC
Monotypic
HB 22–26 cm
T 31–34 cm
W c. 350 g (F)

660–1900 m

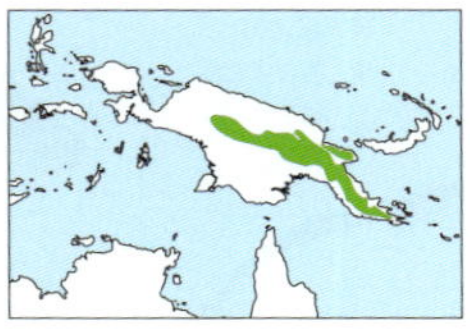

ID/PG: New Guinea Highlands (from Snow Mts E to Mt Simpson, Huon Pen)

Lesser Small-toothed Rat
Macruromys elegans

FR Petit Macruromys
DE Kleine Neuguinea-Kleinzahnratte
ES Rata de dientes pequeños menor

DD
Monotypic
HB 15.3–15.8 cm
T 21–22 cm

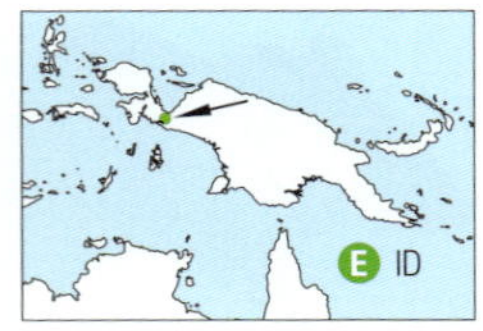

New Guinea (known from Mt Kunupi; possibly restricted to Weyland Range)

Genus *HYOMYS*

(2 species)

Western White-eared Giant Rat
Hyomys dammermani

FR Hyomys de Dammerman
DE Westliche Weißohr-Riesenratte
ES Rata gigante de orejas blancas occidental

1400–2800 m

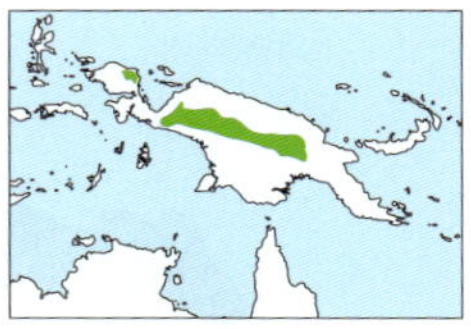

ID/PG: New Guinea (Arfak Mts, Central Range from Weyland Range E to Mt Hagen)

DD
Monotypic
HB 29–32 cm
T 24–32 cm
W 800–985 g

Eastern White-eared Giant Rat
Hyomys goliath

FR Hyomys à oreilles blanches
DE Östliche Weißohr-Riesenratte
ES Rata gigante de orejas blancas oriental

LC
2 sspp
HB 36–39 cm
T 30–38 cm
W 750–945 g

1400–2800 m

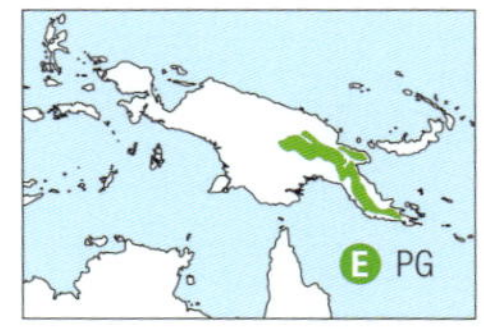

E New Guinea Highlands from Hagen area to extreme E Papuan Pen, Huon Pen

Genus *POGONOMYS*

(5 species)

Loria's Tree Mouse
Pogonomys loriae

FR Pogonomys de Loria
DE Loria-Greifschwanzratte
ES Ratón arborícola de Loria

LC
Monotypic
HB 12–16.8 cm
T 18–25 cm
W 53–128 g

100–3000 m

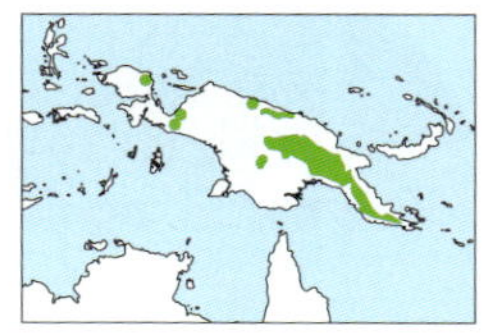

ID/PG: primarily E New Guinea, scattered in N & W as far as Arfak Mts

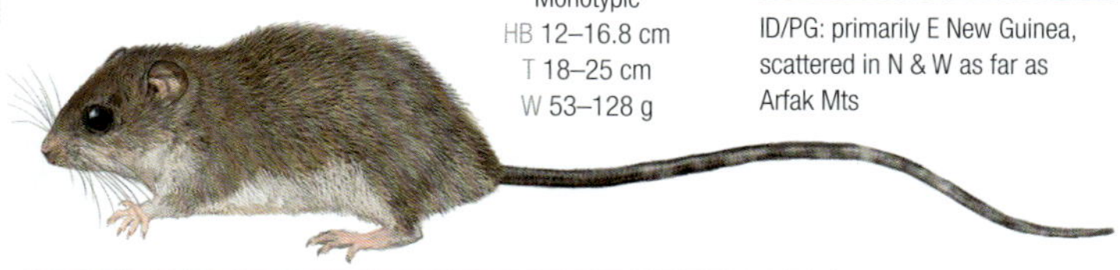

Chestnut Tree Mouse
Pogonomys macrourus

FR Pogonomys à longue queue
DE Kastanienbraune Greifschwanzratte
ES Ratón arborícola castaño

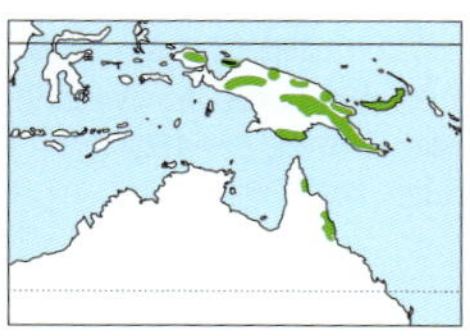

Mts of New Guinea, Yapen I, New Britain I (ID, PG), NE Qld (AU)

LC
Monotypic
HB 9.2–16.5 cm
T 12.6–21 cm
W 28–60 g

Gray-bellied Tree Mouse
Pogonomys sylvestris

FR Pogonomys sylvestre
DE Graubauch-Greifschwanzratte
ES Ratón arborícola de vientre gris

LC
Monotypic
HB 11–13 cm
T 13.5–17 cm
W 29–66 g

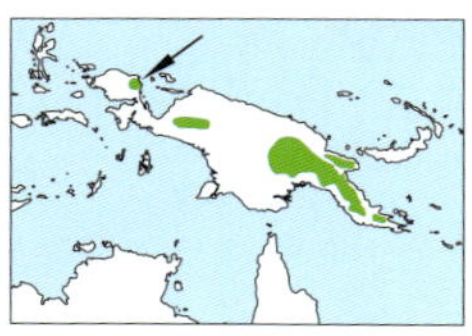

ID/PG: New Guinea (Arfak Mts, Snow Mts, Huon Pen, E Highlands)

Champion's Tree Mouse
Pogonomys championi

FR Pogonomys de Champion
DE Champion-Greifschwanzratte
ES Ratón arborícola de Champion

DD
Monotypic
HB 11.2–13.6 cm
T 14.9–17 cm
W 40–66 g

1400–2300 m

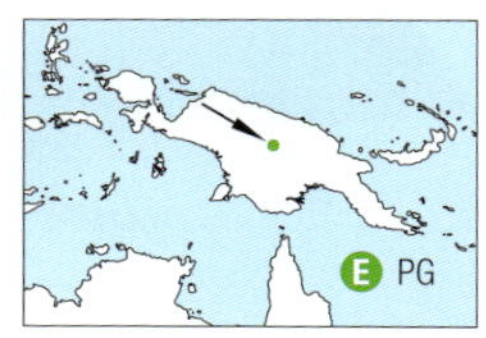

NC New Guinea (Telefomin & Tifalmin valleys in upper R Sepik drainage)

D'Entrecasteaux Archipelago Tree Mouse
Pogonomys fergussoniensis

FR Pogonomys de Fergusson
DE D'Entrecasteaux-Inseln-Greifschwanzratte
ES Ratón arborícola de D'Entrecasteaux

EN
Monotypic
HB 17–20 cm
T 25–26 cm
W 135–158 g

200–900 m

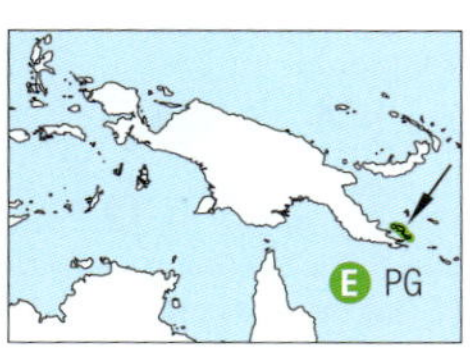

D'Entrecasteaux Is (Goodenough, Fergusson, Normanby)

Genus *COCCYMYS*

(3 species)

Rümmler's Brush Mouse
Coccymys ruemmleri

FR Coccymys de Rümmler
DE Rümmler-Bürstenschwanzmaus
ES Ratón de cepillo de Rümmler

LC
Monotypic
HB 9.3–11.4 cm
T 12.2–17.1 cm
W c. 34 g

2200–4050 m

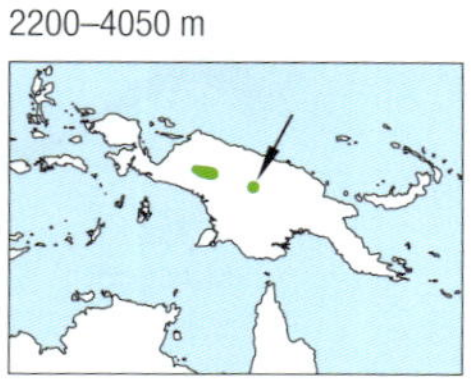

ID/PG: New Guinea Highlands (Snow & Star Mts)

Central Cordillera Brush Mouse
Coccymys shawmayeri

FR Coccymys de Shaw Mayer
DE Shaw-Mayer-Bürstenschwanzmaus
ES Ratón de cepillo de Cordillera Central

LC
Monotypic
HB 8.7–10.7 cm
T 13.2–17 cm
W 23–45 g

1590–3660 m

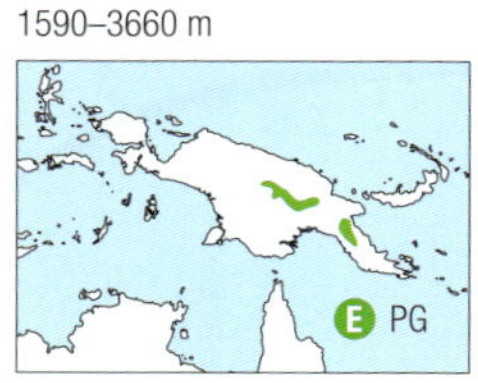

New Guinea Highlands (from Telefomin area E to Owen Stanley Range)

Tawny Brush Mouse
Coccymys kirrhos

FR Coccymys fauve
DE Gelbbraune Bürstenschwanzmaus
ES Ratón de cepillo leonado

DD
Monotypic
HB 8–10.5 cm
T 14.2–15.9 cm
W 22–28 g

1490–2500 m

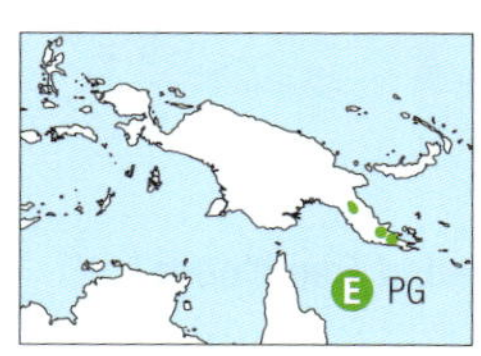

Highlands of Papuan Pen (with Mt Dayman & Mt Simpson in Maneau Range)

Genus *BRASSOMYS*

(1 species)

DD

Monotypic

HB 11.1–12.2 cm

T 14.4–16.2 cm

White-toothed Mouse

Brassomys albidens

FR Rat à dents blanches

DE Neuguinea-Weißzahnmaus

ES Ratón de dientes blancos

c. 3100 m

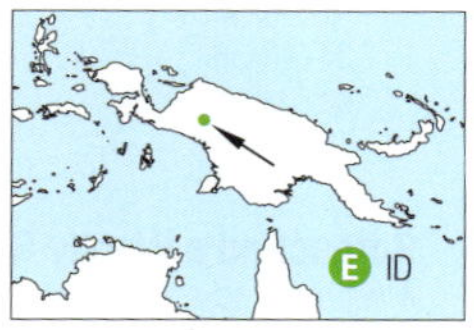

WC New Guinea (R Bele valley, near L Habbema; perhaps Mt Minni, Star Mts)

Genus *XENUROMYS*

(1 species)

LC

Monotypic

HB 27–34 cm

T 22–28 cm

W 900–1100 g

Rock-dwelling Giant Rat

Xenuromys barbatus

FR Rat barbu

DE Neuguinea-Weißschwanzratte

ES Rata gigante de roca

0–1600 m

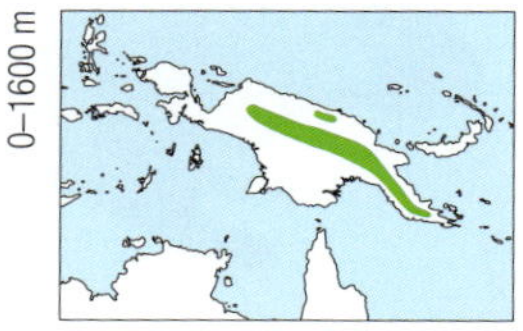

ID/PG: New Guinea (Central Range from R Taritatu E to Mt Dayman, Torricelli Mts)

Genus *MAMMELOMYS*

(2 species)

Highland Mosaic-tailed Rat

Mammelomys lanosus

FR Petit Mammelomys

DE Hochland-Mosaikschwanzratte

ES Rata de cola moteada de altiplano

LC

Monotypic

HB 13.2–17.1 cm

T 10.5–14.6 cm

W 100–123 g

1000–3200 m

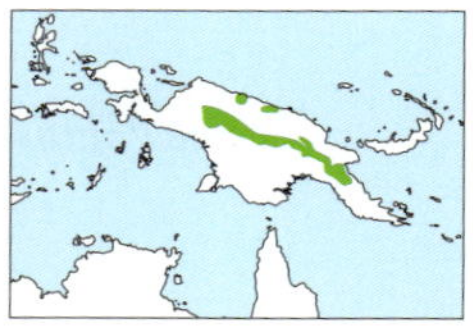

ID/PG: N, C & E New Guinea Mts

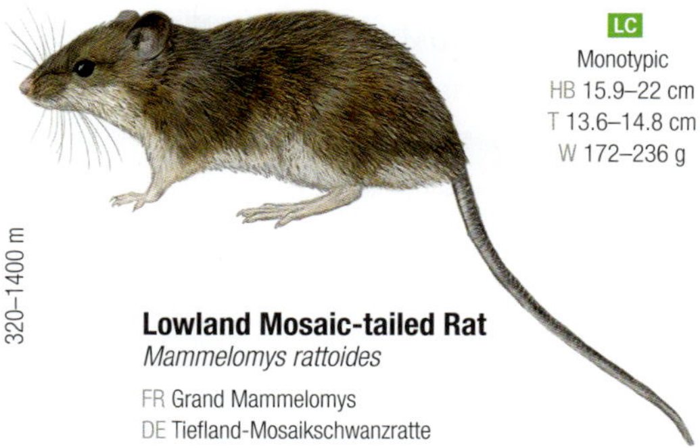

LC

Monotypic

HB 15.9–22 cm

T 13.6–14.8 cm

W 172–236 g

Lowland Mosaic-tailed Rat

Mammelomys rattoides

FR Grand Mammelomys

DE Tiefland-Mosaikschwanzratte

ES Rata de cola moteada de tierras bajas

320–1400 m

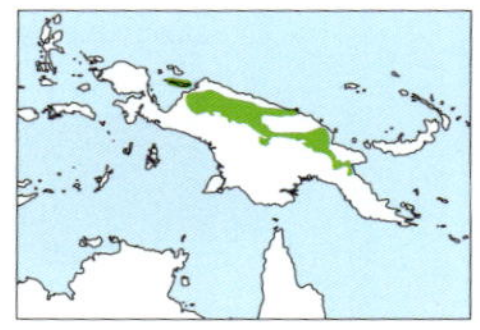

ID/PG: N New Guinea, Yapen I

Genus *POGONOMELOMYS*

(3 species)

LC

Monotypic

HB 17.3–18 cm

T 19–19.7 cm

Brass's Brush Mouse

Pogonomelomys brassi

FR Pogonomys de Brass

DE Brass-Mosaikschwanzratte

ES Ratón de cepillo de Brass

0–600 m

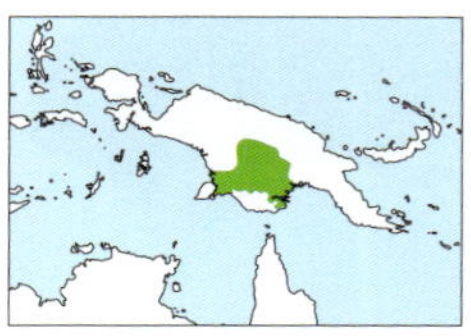

ID/PG: SC New Guinea lowlands

LC

Monotypic

HB c. 18 cm

T c. 19 cm

Bruijn's Brush Mouse

Pogonomelomys bruijnii

FR Pogonomys de Bruijn

DE Vogelkop-Mosaikschwanzratte

ES Ratón de cepillo de Brujin

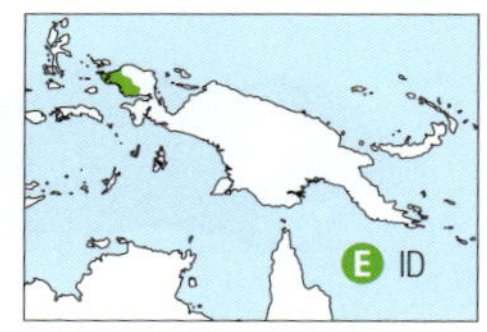

NW New Guinea (Salawati I, Vogelkop Pen)

LC

Monotypic

HB 13.4–15.2 cm

T 15.3–20 cm

W 75–144 g

Shaw Mayer's Brush Mouse

Pogonomelomys mayeri

FR Pogonomys de Shaw Mayer

DE Shaw-Mayer-Mosaikschwanzratte

ES Ratón de cepillo de Shaw Mayer

400–1500 m

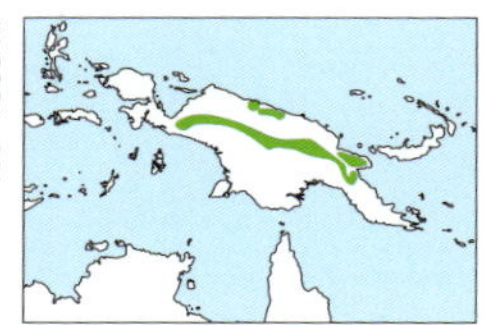

ID/PG: N & C New Guinea Mts

Genus *ABEOMELOMYS*

(1 species)

Papuan Highland Brush Mouse

Abeomelomys sevia

FR Rat de Sevia

DE Hochland-Bürstenmaus

ES Ratón de cepillo de altiplano de Papúa

LC

Monotypic

HB 11.4–13.8 cm

T 14–19.1 cm

W 40–65 g

1400–3560 m

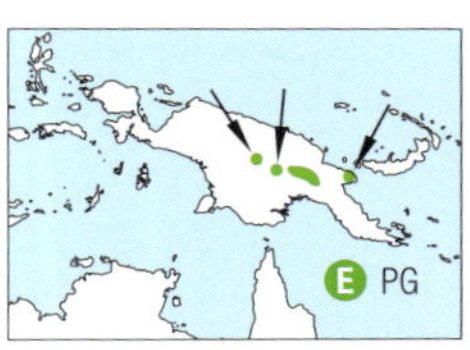

NE New Guinea (Mts from Victor Emanuel to Bismarck ranges, Huon Pen)

Genus *MALLOMYS*

(4 species)

LC

Monotypic

HB 40–43 cm

T 28–37 cm

W c. 1.9 kg

Subalpine Woolly Rat

Mallomys istapantap

FR Mallomys subalpin

DE Hochland-Riesenbaumratte

ES Rata lanuda subalpina

2450–3850 m

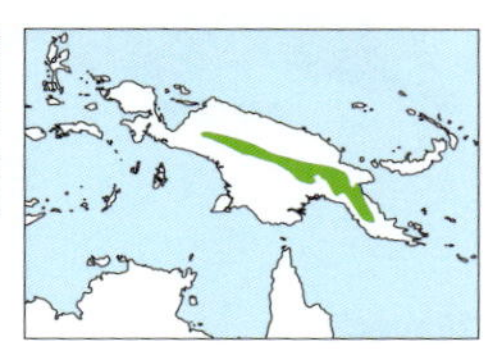

ID/PG: New Guinea Highlands

Alpine Woolly Rat

Mallomys gunung

FR Mallomys alpin

DE Berg-Riesenbaumratte

ES Rata lanuda alpina

3500–4050 m

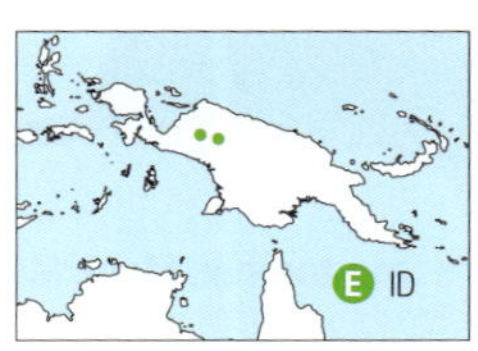

W New Guinea Highlands (known from 2 localities between Mt Jaya & Mt Trikor)

EN

Monotypic

HB 41–47 cm

T 35–37 cm

W c. 2 kg

LC
2 sspp
HB 34–41 cm
T 33–43 cm
W 1.4–2 kg

1100–2700 m

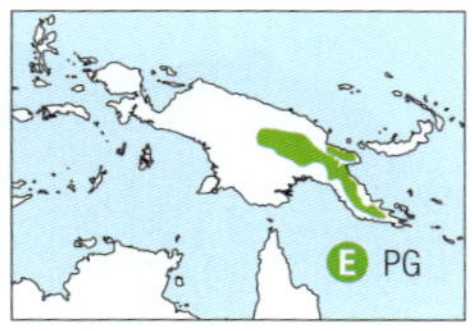

E New Guinea Highlands, Huon Pen

De Vis's Woolly Rat
Mallomys aroaensis
FR Mallomys de De Vis
DE De-Vis-Riesenbaumratte
ES Rata lanuda de De Vis

LC
2 sspp
HB 34–40 cm
T 34–42 cm
W 925–1500 g

Rothschild's Woolly Rat
Mallomys rothschildi
FR Mallomys de Rothschild
DE Rothschild-Riesenbaumratte
ES Rata lanuda de Rothschild

1200–3700 m

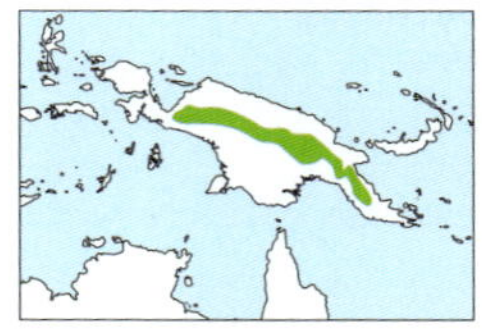

ID/PG: New Guinea Highlands

Genus *XEROMYS*

(1 species)

False Water Rat
Xeromys myoides
FR Rat des mangroves
DE Falsche Schwimmratte
ES Ratón de agua

VU
Monotypic
HB 7.2–13 cm
T 6.2–10 cm
W 32–64 g

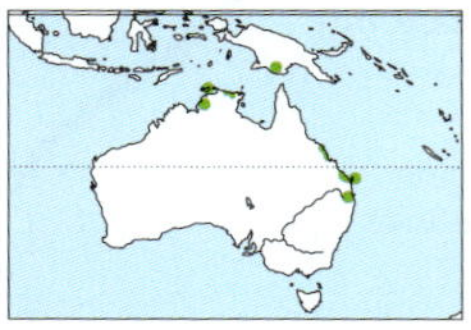

Patchy in N & E AU (from NW NT to SE Qld, some Is); recorded in S New Guinea

Eastern New Guinea Shrew Mouse
Pseudohydromys murinus
FR Souris-musaraigne orientale
DE Östliche Moosmaus
ES Ratón musaraña de Nueva Guinea oriental

1500–3400 m

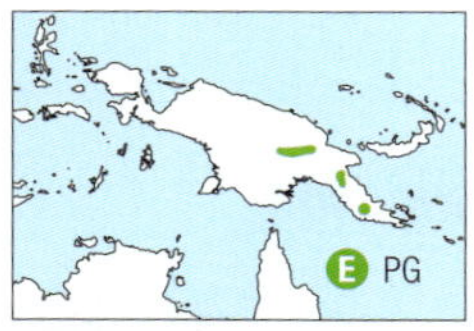

LC
Monotypic
HB 7–10.5 cm
T 8.1–10.2 cm
W 15–19.9 g

C & E New Guinea Highlands

Genus *PSEUDOHYDROMYS*

(12 species)

LC
Monotypic
HB 7–8 cm
T 7.9–8.8 cm

590–1570 m

Bishop Moss Mouse
Pseudohydromys berniceae
FR Souris-musaraigne de Bishop
DE Bishop-Moosmaus
ES Ratón de musgo de Bishop

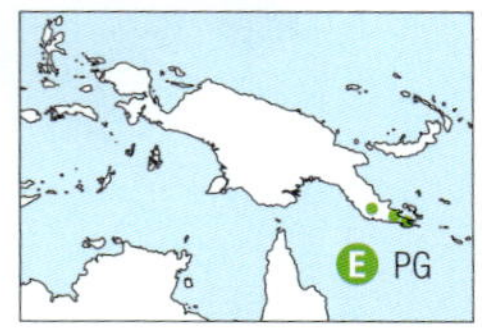

Mt Obree (Owen Stanley Range), Mt Simpson & Mt Pekopekowana (Maneau Range)

Laurie's Moss Mouse
Pseudohydromys eleanorae
FR Souris-musaraigne d'Eleanor
DE Eleonora-Moosmaus
ES Ratón de musgo de Laurie

2240–3050 m

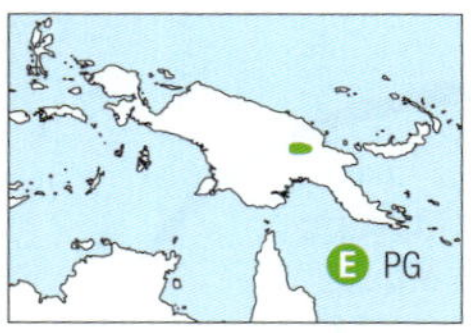

DD
Monotypic
HB 7–8 cm T 7.9–8.8 cm

New Guinea Central Highlands (Bismarck & Hagen ranges)

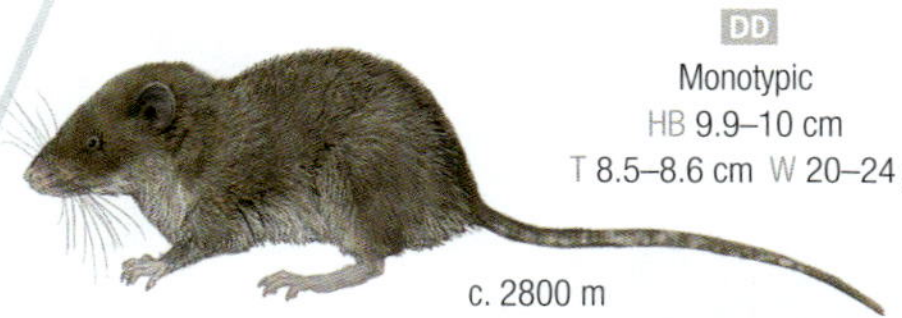

DD
Monotypic
HB 9.9–10 cm
T 8.5–8.6 cm W 20–24 g

c. 2800 m

Woolley's Moss Mouse
Pseudohydromys patriciae
FR Souris-musaraigne de Patricia
DE Woolley-Moosmaus
ES Ratón de musgo de Woolley

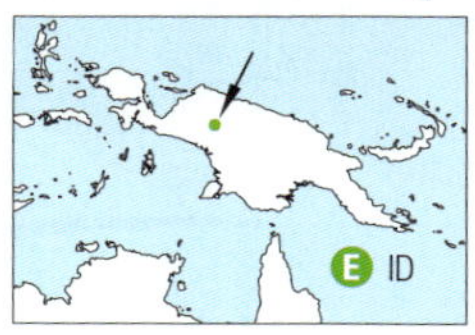

New Guinea (known only from vicinity of L Habbema, Snow Mts)

DD
Monotypic
HB 8.7–11.5 cm
T 8.5–9.5 cm

Western New Guinea Shrew Mouse
Pseudohydromys occidentalis
FR Souris-musaraigne occidentale
DE Westliche Moosmaus
ES Ratón musaraña de Nueva Guinea occidental

2300–3800 m

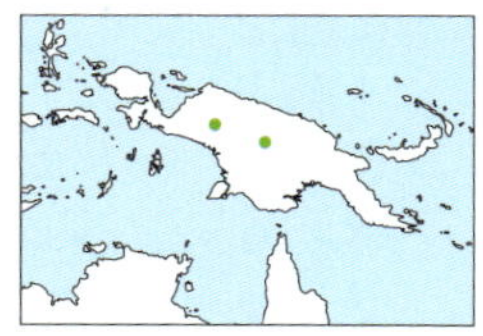

W & C New Guinea in L Habbema, Snow Mts (ID), Mt Trikora, Star Mts (PG)

DD
Monotypic
HB c. 10.8 cm
T c. 10.1 cm

Torricelli Mountains Shrew Mouse
Pseudohydromys musseri
FR Souris-musaraigne de Musser
DE Musser-Moosmaus
ES Ratón musaraña de Torricelli

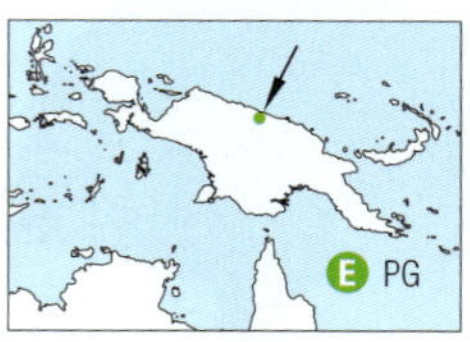

N New Guinea (known only from Mt Somoro, Torricelli Mts)

DD
Monotypic
HB c. 10.8 cm T c. 10.1 cm

White-bellied Moss Mouse
Pseudohydromys sandrae
FR Souris-musaraigne de Sandra
DE Weißbauch-Moosmaus
ES Ratón de musgo de vientre blanco

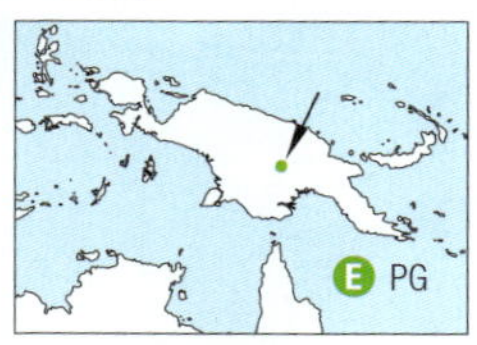

EC New Guinea (known only from Mt Sisa, R Kikori Basin)

Mottled-tailed Shrew Mouse
Pseudohydromys fuscus
FR Souris-musaraigne à queue tachetée
DE Fleckenschwanz-Moosmaus
ES Ratón musaraña de cola moteada

LC
Monotypic
HB 8–11 cm
T 7.2–9 cm
W 14.2–25 g

1600–3660 m

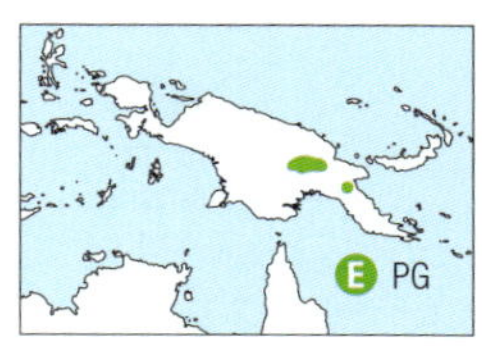

E New Guinea Highlands

LC
Monotypic
HB 8.5–10.3 cm
T 9–11.3 cm
W 13–21 g

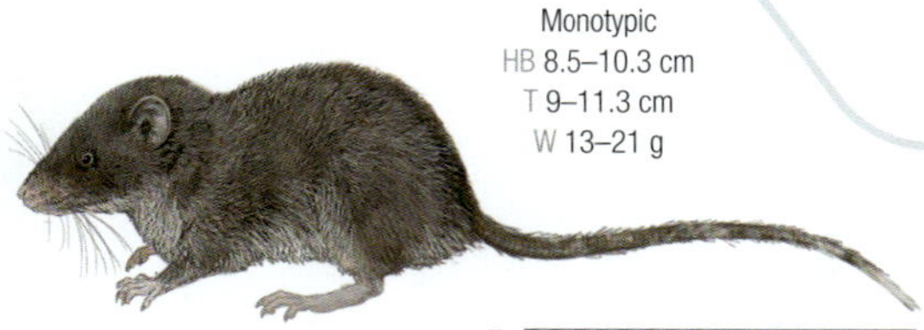

Shaw Mayer's Shrew Mouse
Pseudohydromys ellermani
FR Souris-musaraigne d'Ellerman
DE Shaw-Mayer-Moosmaus
ES Ratón musaraña de Shaw Mayer

1400–2800 m

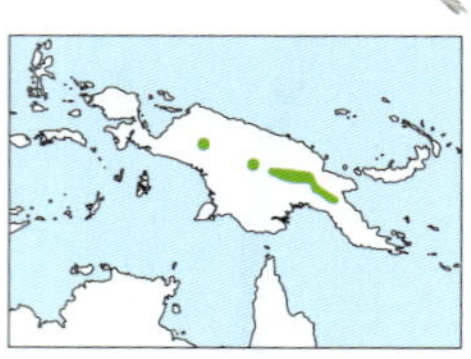

ID/PG: New Guinea Highlands (Snow Mts & from Star Mts to Mt Missim)

DD
Monotypic
HB 8.7–10.5 cm
T 9.5–10.3 cm
W 22–30 g

German's One-toothed Moss Mouse
Pseudohydromys germani
FR Souris-musaraigne de German
DE German-Moosmaus
ES Ratón de musgo de German

1300–1570 m

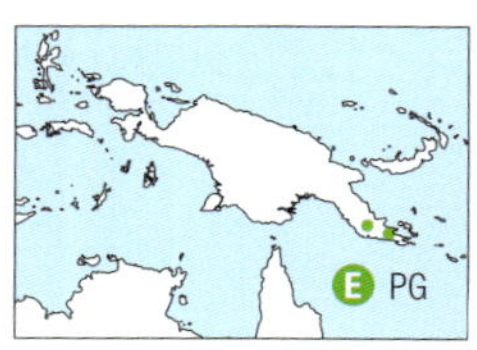

Mt Obree (Owen Stanley Range), Mt Dayman (Maneau Range)

Huon Small-toothed Moss Mouse
Pseudohydromys carlae

FR Souris-musaraigne de Carla
DE Huon-Halbinsel-Moosmaus
ES Ratón de musgo de Huon

DD
Monotypic
HB 8.9–10.3 cm
T 8.4–9.2 cm
W c. 15 g

2560–3000 m

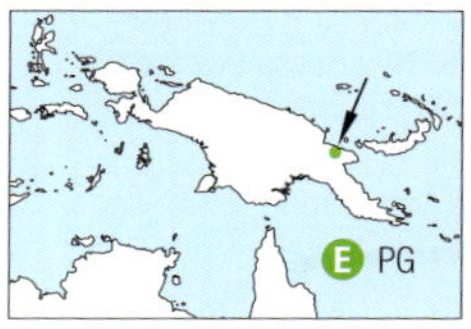

Huon Pen (2 localities in vicinity of Tep Tep, Finisterre Range)

Southern Small-toothed Moss Mouse
Pseudohydromys pumehanae

FR Souris-musaraigne de Pumehana
DE Südliche Moosmaus
ES Ratón de musgo de dientes pequeños meridional

DD
Monotypic
HB 9.6–10.1 cm
T 10–10.3 cm W 17–20 g

1550–2100 m

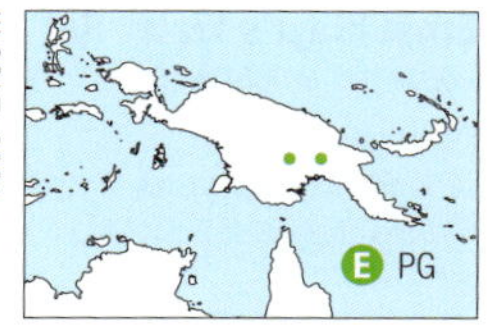

E New Guinea Highlands (upper R Strickland catchment & N Mt Karimui)

Genus *MIRZAMYS*

(2 species)

Mirza's Western Moss Rat
Mirzamys louiseae

FR Mirzamys de Louise
DE Westliche Moosratte
ES Rata de musgo de Mirza occidental

LC
Monotypic
HB 9.5–12 cm T 10–12.4 cm
W c. 25.6 g (F)

1900–3200 m

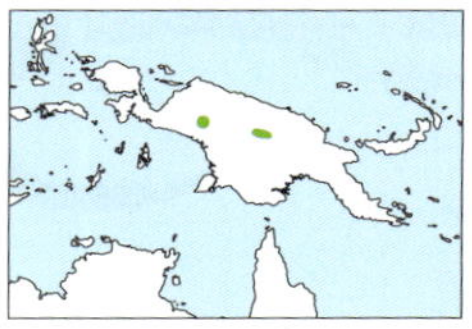

ID/PG: New Guinea (Star Mts, Hindenburg, Victor Emanuel & Blucher ranges; recorded in Snow Mts)

Mirza's Eastern Moss Rat
Mirzamys norahae

FR Mirzamys de Norah
DE Östliche Moosratte
ES Rata de musgo de Mirza oriental

DD
Monotypic
HB 10.5–11.5 cm
T 8.9–10 cm
W c. 30 g

above 2900 m

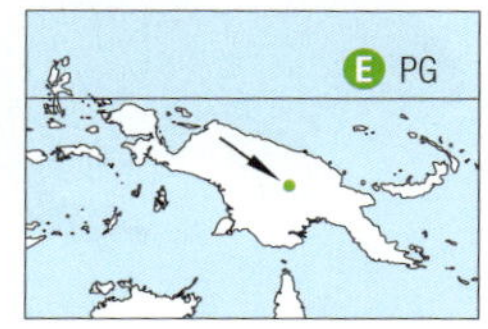

C New Guinea (Kaijende Highlands, Enga Prov)

Genus *MICROHYDROMYS*

(2 species)

DD
Monotypic
HB 7.8–8.6 cm
T 7.1–8.3 cm
W 12–13 g

Southern Groove-toothed Shrew Mouse
Microhydromys argenteus

FR Hydromys argenté
DE Südliche Kleinwassermaus
ES Ratón musaraña de dientes estriados meridional

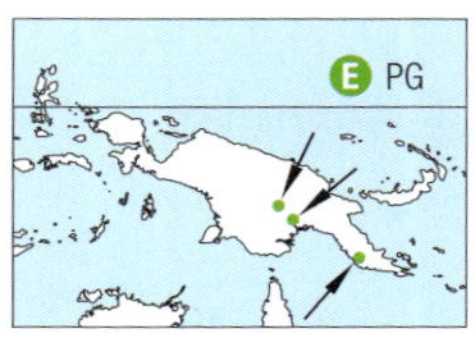

380–1450 m

E New Guinea Highlands (known from 3 localities in S foothills)

DD
Monotypic
HB 7.9–8.3 cm
T 8.4–9.2 cm W c. 11 g

Northern Groove-toothed Shrew Mouse
Microhydromys richardsoni

FR Hydromys de Richardson
DE Nördliche Kleinwassermaus
ES Ratón musaraña de dientes estriados septentrional

20–1500 m

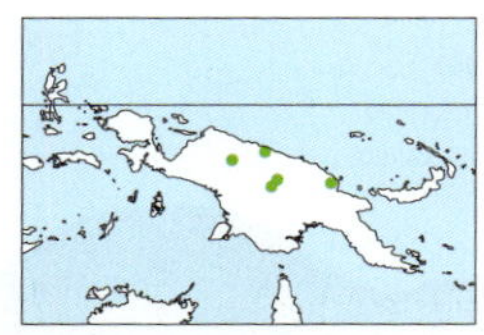

ID/PG: known from few localities in N catchments of New Guinea Highlands

Genus *PARALEPTOMYS*

(2 species)

EN
Monotypic
HB 11.8–13.5 cm
T 12.7–14.6 cm
W 54–58 g

Northern Water Rat
Paraleptomys rufilatus

FR Hydromys à flancs roux
DE Große Bergmoosratte
ES Rata de agua de Nueva Guinea septentrional

1200–1700 m

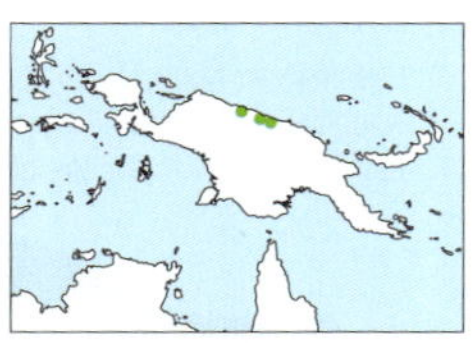

ID/PG: NC New Guinea (Cyclops, Princess Alexandra & Bewani Mts)

DD
Monotypic
HB 10.5–12.6 cm
T 12.5–13.5 cm W 31–35 g

Short-haired Water Rat
Paraleptomys wilhelmina

FR Hydromys du Wilhelmina
DE Kleine Bergmoosratte
ES Rata de agua de Nueva Guinea de pelo corto

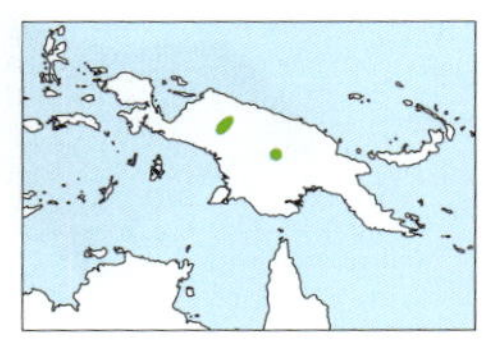

ID/PG: New Guinea Highlands (Mt Trikora, Tifalmin Valley, Star Mts)

Genus *LEPTOMYS*

(5 species)

Arfak Water Rat
Leptomys arfakensis

FR Leptomys des Arfak
DE Arfak-Neuguinea-Wasserratte
ES Rata de agua de Arfak

DD
Monotypic
HB 13.8–15.1 cm
T c. 14.6 cm

c. 1000 m

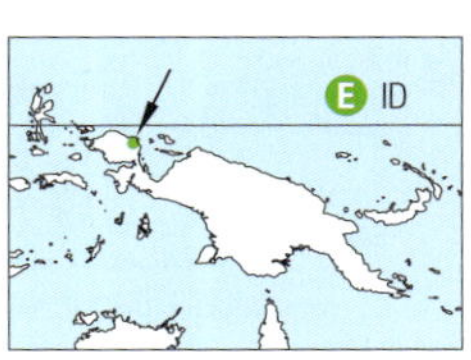

NW New Guinea (known from 2 specimens collected in 1925 in Arfak Mts)

Fly River Water Rat
Leptomys signatus

FR Leptomys à front blanc
DE Fly-Neuguinea-Wasserratte
ES Rata de agua de Fly

LC
Monotypic
HB 14–15.5 cm
T 14.3–15.5 cm
W c. 80 g

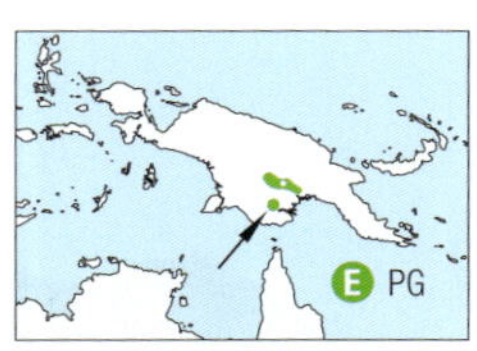

SC New Guinea (upper & lower R Fly, middle R Strickland, Mt Bosavi, Darai Plateau)

LC
Monotypic
HB 14.2–19.4 cm
T 12.5–16.4 cm
W 66–120 g

Elegant Water Rat
Leptomys elegans

FR Leptomys élégant
DE Elegante Neuguinea-Wasserratte
ES Rata de agua elegante

20–1500 m

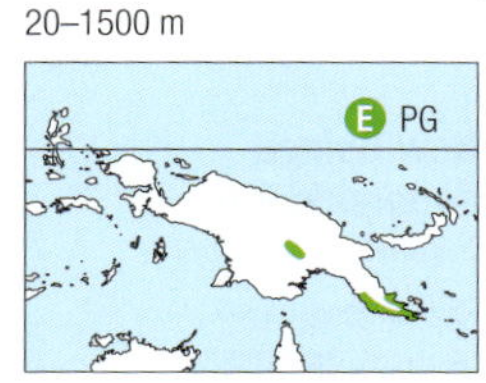

SE New Guinea (scattered in 2 areas: Kikori & Strickland rivers)

Ernst Mayr's Water Rat

Leptomys ernstmayri

FR Leptomys de Mayr
DE Ernst-Mayr-Neuguinea-Wasserratte
ES Rata de agua de Ernst Mayr

ID/PG: New Guinea (Foja Mts, Adelbert Range, Huon Pen, E Highlands)

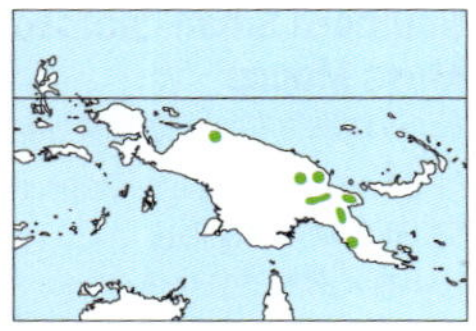

1150–2200 m

LC
Monotypic
HB 12.4–15.6 cm
T 13.2–17.1 cm
W 48–70 g

Small Water Rat

Leptomys paulus

FR Leptomys nain
DE Kleine Neuguinea-Wasserratte
ES Rata de agua pequeña

DD
Monotypic
HB 11.7–13.2 cm
T 13.8–16.3 cm
W 34–52 g

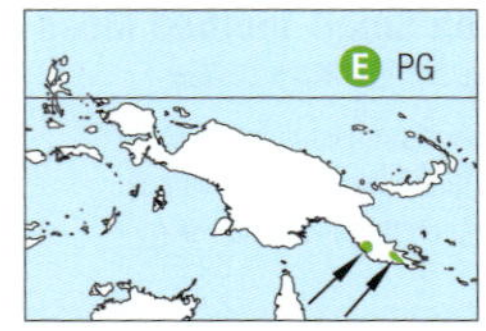

1240–1530 m

SE New Guinea (Astrolabe & Maneau ranges)

Genus *BAIYANKAMYS*

(2 species)

DD
Monotypic
HB 13.4–16 cm
T 15.4–18.7 cm

above 3200 m

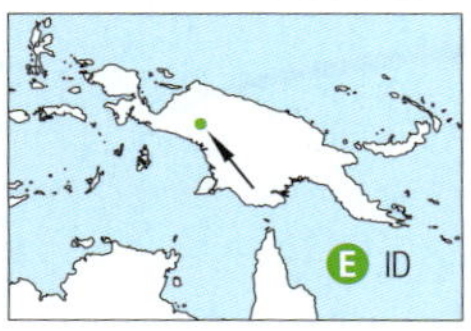

WC New Guinea (vicinity of L Habbema & close to Mt Trikora)

Mountain Water Rat

Baiyankamys habbema

FR Rat de l'Habbema
DE Habbema-Schwimmratte
ES Rata de agua montana

Shaw Mayer's Water Rat

Baiyankamys shawmayeri

FR Rat de Shaw Mayer
DE Shaw-Mayer-Schwimmratte
ES Rata de agua de Shaw Mayer

LC
Monotypic
HB 13–15.3 cm
T 14.8–18.5 cm
W 68–88 g

1980–3600 m

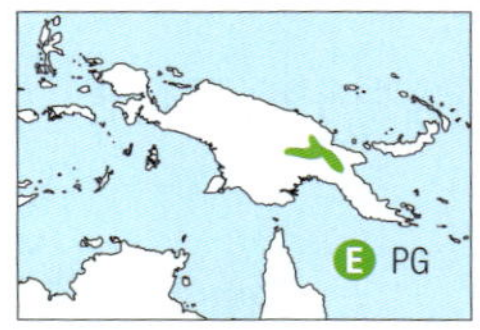

Mts of E New Guinea (from Doma Peaks E to vicinity of Mt Kaindi)

Genus *CROSSOMYS*

(1 species)

LC
Monotypic
HB 17.5–23 cm
T 21–27 cm
W c. 165 g

900–2700 m

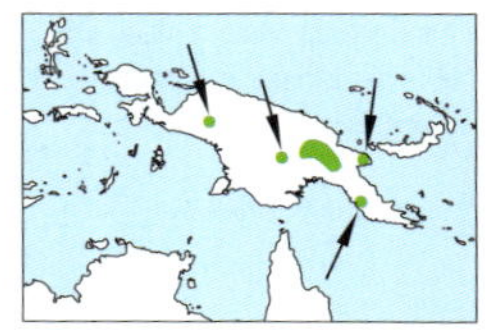

ID/PG: New Guinea (Swart Valley of Snow Mts, Mt Sisa, E Central Range, Huon Pen, Astrolabe Range)

Earless New Guinea Water Rat

Crossomys moncktoni

FR Crossomys de Monckton
DE Monckton-Schwimmratte
ES Rata de agua de Nueva Guinea de orejas pequeñas

Common Water Rat

Hydromys chrysogaster

FR Hydromys d'Australie
DE Goldbauch-Schwimmratte
ES Rata de agua de Australasia

LC
Monotypic
HB 19.5–39 cm
T 20–32 cm
W 210–1280 g

0–1900 m

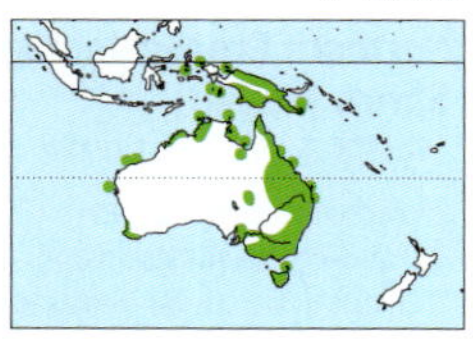

New Guinea (ID, PG), mainland AU, Tas, many Is & Archs

Genus *HYDROMYS*

(4 species)

DD
Monotypic
HB 12.8–16.7 cm
T 10.3–15.2 cm

Husson's Water Rat

Hydromys hussoni

FR Hydromys de Husson
DE Husson-Schwimmratte
ES Rata de agua de Husson

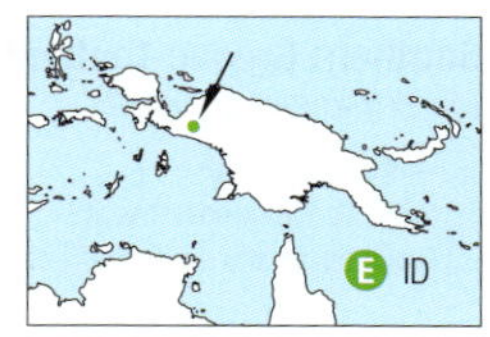

W New Guinea (catchment of L Paniai)

Ziegler's Water Rat

Hydromys ziegleri

FR Hydromys de Ziegler
DE Ziegler-Schwimmratte
ES Rata de agua de Ziegler

DD
Monotypic
HB c. 13.2 cm
T c. 11.8 cm

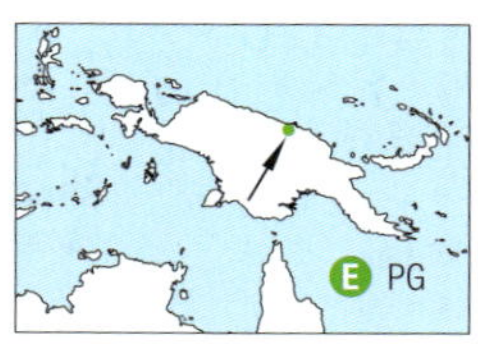

N New Guinea (foothills of Prince Alexander Range)

New Britain Water Rat

Hydromys neobritannicus

FR Hydromys de Nouvelle-Bretagne
DE Neubritannien-Schwimmratte
ES Rata de agua de Nueva Bretaña

DD
Monotypic
HB c. 29 cm
T c. 29 cm

0–1000 m

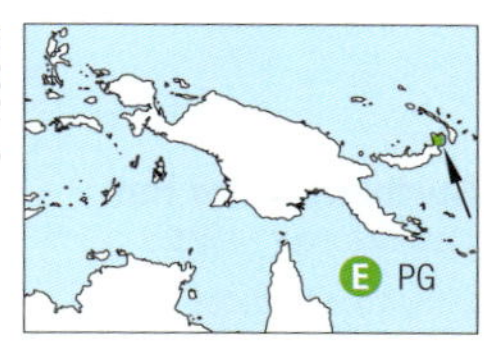

New Britain I (Bismarck Arch)

Genus *PARAHYDROMYS*

(1 species)

LC
Monotypic
HB 21–23 cm
T 23–27 cm
W 490–590 g

New Guinea Waterside Rat

Parahydromys asper

FR Hydromys rugueux
DE Bergschwimmratte
ES Rata de ribera de Nueva Guinea

700–2200 m

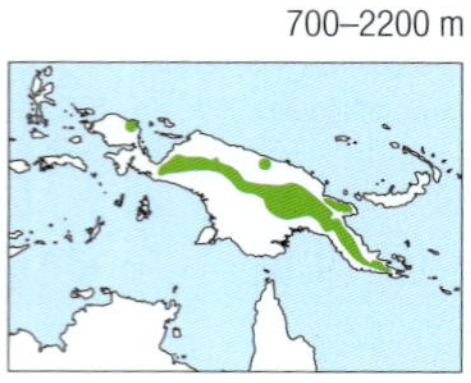

ID/PG: New Guinea (Vogelkop, N & C Highlands, Huon Pen)

Genus *UROMYS*

(11 species)

Great Key Island Giant Rat

Uromys siebersi

FR Uromys de Siebers
DE Kei-Inseln-Riesenratte
ES Rata gigante de Kai Besar

DD
Monotypic
HB 25–28 cm
T 22–23 cm

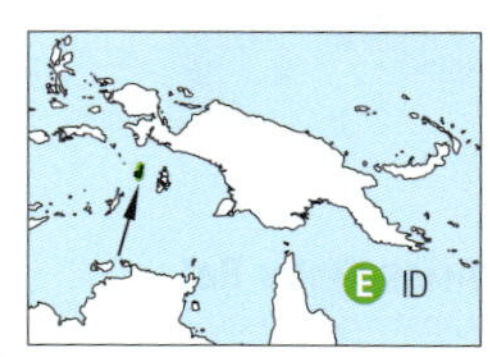

Kai Besar I (Kai Is)

NT
Monotypic
HB 25–30 cm
T 24–28 cm
W 570–730 g

New Britain Island Giant Rat

Uromys neobritannicus

FR Uromys de Nouvelle-Bretagne
DE Neubritannien-Riesenratte
ES Rata gigante de Nueva Bretaña

30–500 m

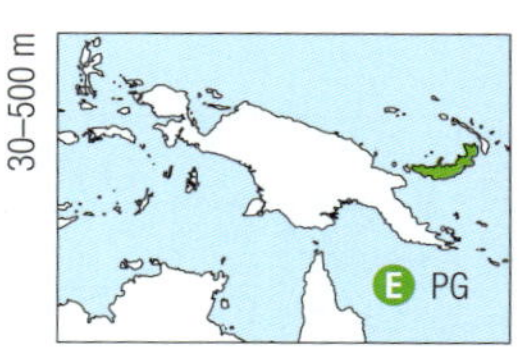

New Britain I (Bismarck Arch)

LC
3 sspp
HB 27–33 cm
T 29–40 cm
W 450–1020 g

Black-tailed Giant Rat
Uromys anak

FR Uromys à queue noire
DE Schwarzschwanz-Riesenratte
ES Rata gigante de cola negra

850–3000 m

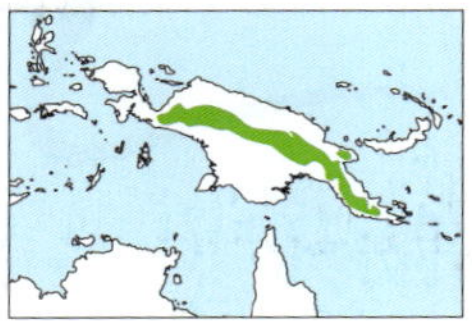

ID/PG: New Guinea Highlands, Huon Pen

CR
Monotypic
HB c. 25 cm
T c. 23 cm

Biak Island Giant Rat
Uromys boeadii

FR Uromys de Boeadi
DE Biak-Riesenratte
ES Rata gigante de Biak

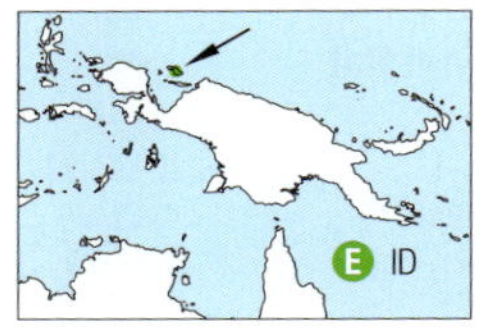

Biak I, off NW New Guinea; possibly adjacent Supiori I

CR(PE)
Monotypic
HB c. 23 cm
T c. 26 cm

Emma's Giant Rat
Uromys emmae

FR Uromys d'Emma
DE Emma-Riesenratte
ES Rata gigante de Emma

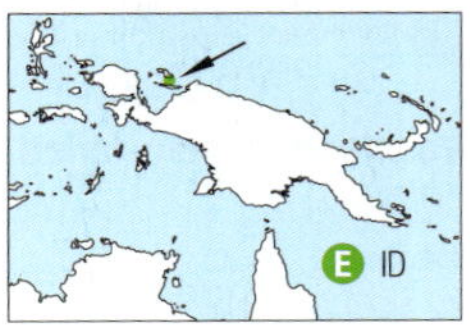

Padaido Is (Owi I), off NW New Guinea; possibly Supiori I

White-tailed Giant Rat
Uromys caudimaculatus

FR Uromys à queue nue
DE Weißschwanz-Riesenratte
ES Rata gigante de cola blanca

LC
3 sspp
HB 20–38 cm
T 21–36 cm
W 400–900 g

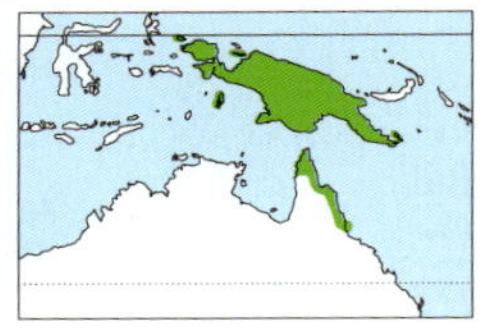

New Guinea (with Waigeo, Yapen, D'Entrecasteaux & Aru Is), AU (N & NE Qld)

0–1930 m

Masked White-tailed Giant Rat
Uromys hadrourus

FR Uromys du Thornton
DE Queensland-Riesenratte
ES Rata gigante de cola blanca enmascarada

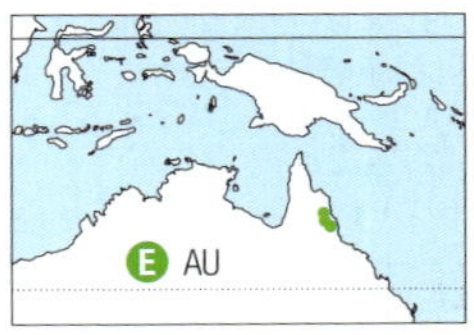

Rainforests in NE Qld

550–1240 m

NT
Monotypic
HB 17–18 cm
T 18–20 cm
W 170–220 g

CR
Monotypic
W 290–460 g

c. 150 m

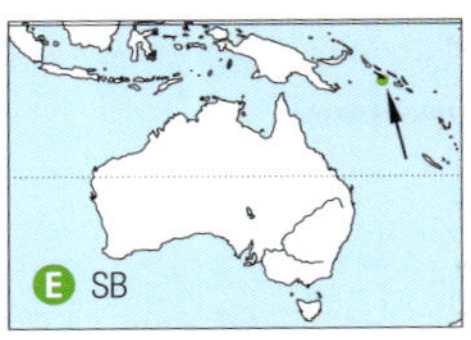

Vangunu Giant Rat
Uromys vika

FR Uromys de Vangunu
DE Vangunu-Riesenratte
ES Rata gigante de Vangunu

Vangunu I, possibly on nearby Is (Solomon Arch)

Emperor Giant Rat
Uromys imperator

FR Uromys empereur
DE Kaiserriesenratte
ES Rata gigante emperador

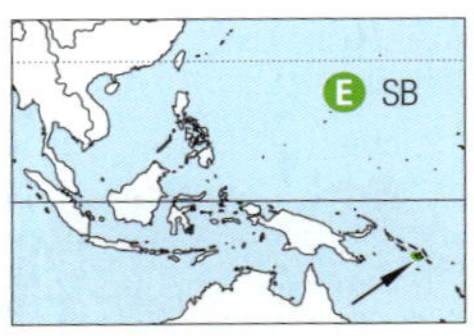

Guadalcanal I (Solomon Arch)

CR(PE)
Monotypic
HB 34–35 cm
T 25–26 cm

Guadalcanal Giant Rat
Uromys porculus

FR Uromys de Guadalcanal
DE Guadalcanal-Riesenratte
ES Rata gigante de Guadalcanal

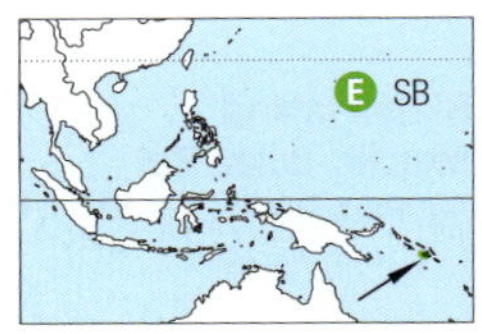

Guadalcanal I (Solomon Arch)

CR(PE)
Monotypic
HB c. 22 cm
T c. 13 cm

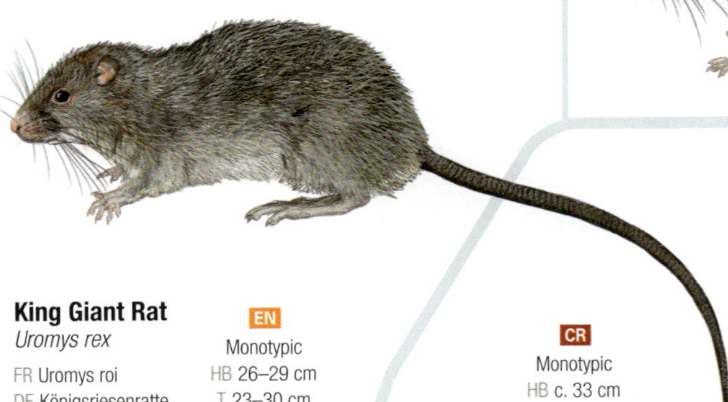

King Giant Rat
Uromys rex

FR Uromys roi
DE Königsriesenratte
ES Rata gigante real

EN
Monotypic
HB 26–29 cm
T 23–30 cm
W c. 420 g

20–600 m

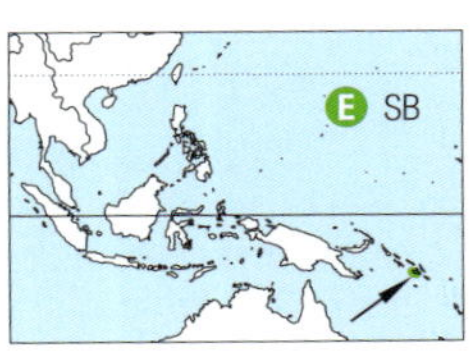

Guadalcanal I (Solomon Arch)

Genus *SOLOMYS*

(4 species)

Poncelet's Giant Rat
Solomys ponceleti

FR Solomys de Poncelet
DE Poncelet-Nacktschwanzratte
ES Rata gigante de Poncelet

CR
Monotypic
HB c. 33 cm
T 34–36 cm
W > 1000 g (F)

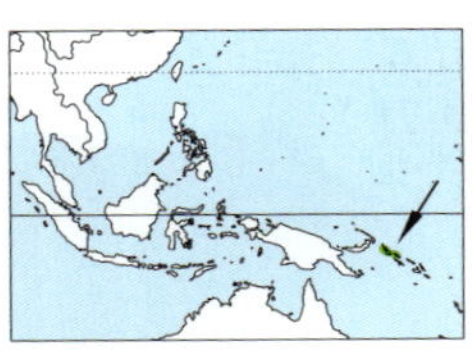

NW Solomon Arch: Bougainville I (PG) & Choiseul I (SB); fossils from Buka I

DD
Monotypic
HB c. 22 cm
T c. 22 cm

Ugi Island Giant Rat
Solomys salamonis

FR Solomys d'Ugi
DE Ugi-Insel-Nacktschwanzratte
ES Rata gigante de Ugi

Ugi I (Solomon Arch)

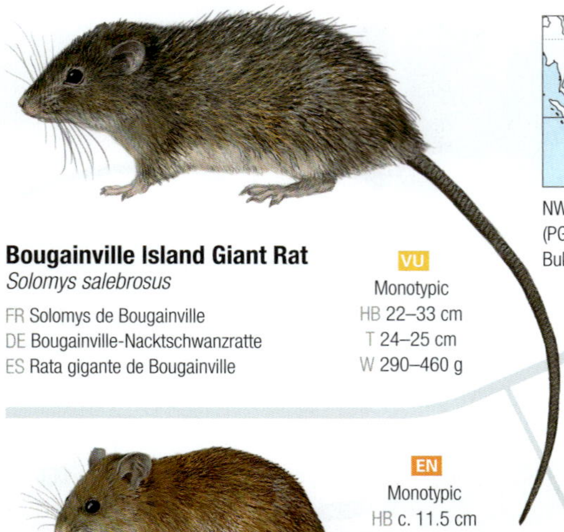

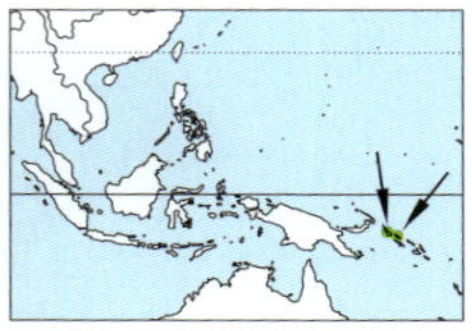

NW Solomon Arch: Bougainville I (PG), Choiseul I (SB); fossils from Buka I

Bougainville Island Giant Rat
Solomys salebrosus

VU
Monotypic
HB 22–33 cm
T 24–25 cm
W 290–460 g

FR Solomys de Bougainville
DE Bougainville-Nacktschwanzratte
ES Rata gigante de Bougainville

Isabel Island Giant Rat
Solomys sapientis

EN
Monotypic
HB 18.8–25 cm
T 19–26 cm

FR Solomys de Santa Isabel
DE Santa-Isabel-Nacktschwanzratte
ES Rata gigante de Isabel

c. 30 m

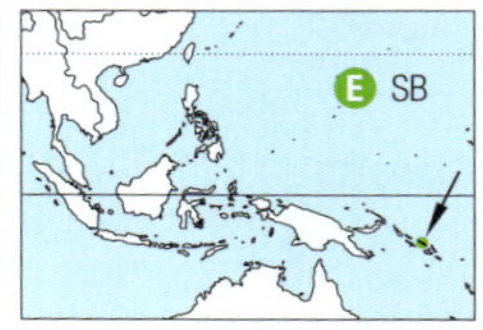

Santa Isabel I (Solomon Arch)

Genus *MELOMYS*

(21 species)

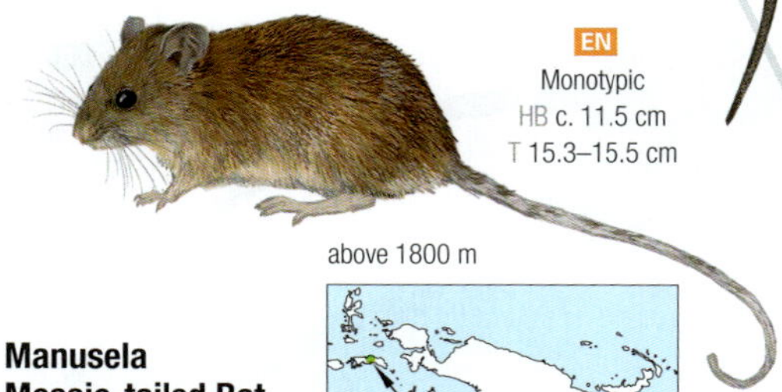

EN
Monotypic
HB c. 11.5 cm
T 15.3–15.5 cm

above 1800 m

Seram I (Mt Manusela; unconfirmed in other locations), Moluccas

Manusela Mosaic-tailed Rat
Melomys fraterculus

FR Mélomys du Manusela
DE Manusela-Mosaikschwanzratte
ES Rata de cola moteada de Manusela

LC
Monotypic

1200–1530 m

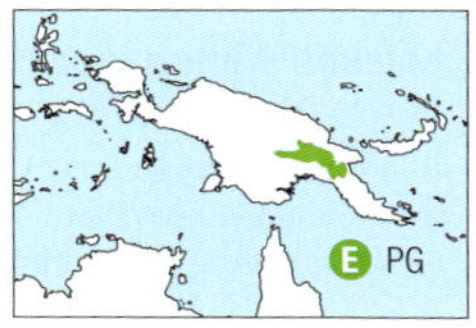

E New Guinea (Central Range; perhaps SE)

Dollman's Mosaic-tailed Rat
Melomys dollmani

FR Mélomys de Dollman
DE Dollman-Mosaikschwanzratte
ES Rata de cola moteada de Dollman

LC
Monotypic
HB 9–11.5 cm
T 11.3–12 cm

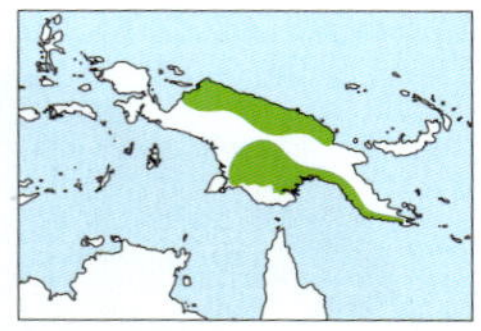

ID/PG: N, C & SE New Guinea

Papua Grassland Mosaic-tailed Rat
Melomys lutillus

FR Petit Mélomys
DE Neuguinea-Grasland-Mosaikschwanzratte
ES Rata de cola moteada de pradera de Papúa

DD
Monotypic
HB 11.1–11.2 cm
T 13.5–13.7 cm
W 61–73 g

Riama Island Mosaic-tailed Rat
Melomys howi

FR Mélomys de How
DE Riama-Mosaikschwanzratte
ES Rata de cola moteada de Riama

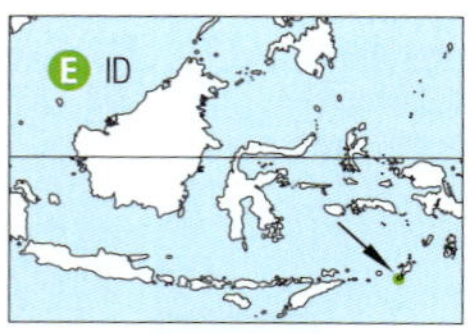

0–200 m

Riama I (S Moluccas)

Bannister's Mosaic-tailed Rat
Melomys bannisteri

FR Mélomys de Bannister
DE Bannister-Mosaikschwanzratte
ES Rata de cola moteada de Bannister

0–500 m

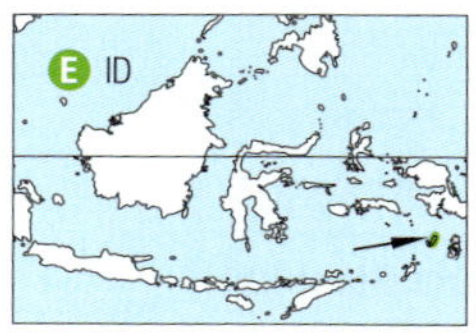

Kai Besar I (Kai Is)

EN
Monotypic
HB c. 13.8 cm
T c. 12.7 cm

Snow Mountains Grassland Mosaic-tailed Rat
Melomys frigicola

LC
Monotypic
HB 10.1–13.5 cm
T 12.2–13.6 cm

FR Mélomys des Snow
DE Schneegebirge-Mosaikschwanzratte
ES Rata de cola moteada de Sudirman

600–2200 m

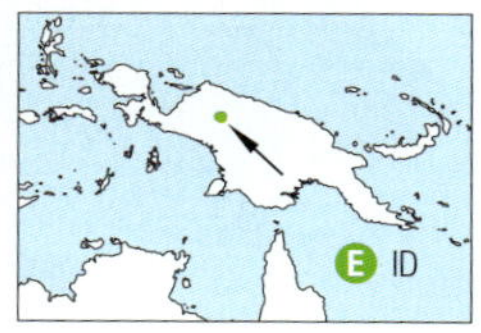

WC New Guinea (N slopes of Snow Mts and nearby R Baliem)

Grassland Mosaic-tailed Rat
Melomys burtoni

FR Mélomys de Burton
DE Grasland-Mosaikschwanzratte
ES Rata de cola moteada de pradera

LC
Monotypic
HB 12.2–14 cm
T 10.5–14 cm W 45–65 g

0–1200 m

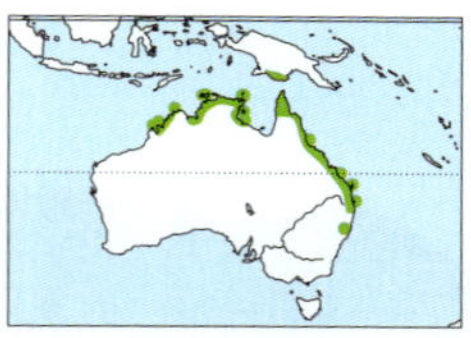

S New Guinea (Trans-Fly area), coastal N & E AU (with many offshore Is)

DD
Monotypic
HB c. 12.3 cm
T c. 12.8 cm
W c. 65 g

Pavel's Seram Mosaic-tailed Rat
Melomys paveli

FR Mélomys de Pavel
DE Pavel-Seram-Mosaikschwanzratte
ES Rata de cola moteada de Pavel

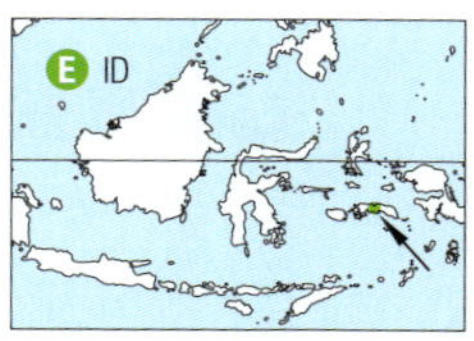

S coast of Seram I (Moluccas)

Manus Island Mosaic-tailed Rat
Melomys matambuai

EN
Monotypic
HB c. 15.1 cm T c. 14.1 cm
W c. 145 g

FR Mélomys de Manus
DE Manus-Mosaikschwanzratte
ES Rata de cola moteada de Manus

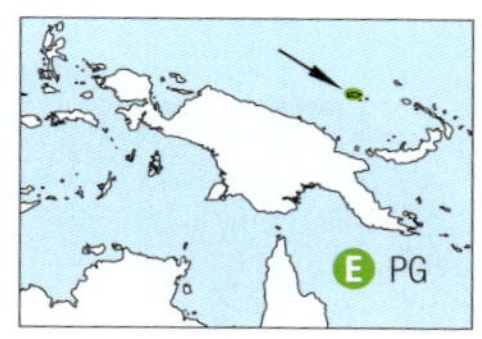

Manus I (Admiralty Is)

Bougainville Mosaic-tailed Rat
Melomys bougainville

DD
Monotypic
HB 14.9–17 cm
T 9.8–14 cm

FR Mélomys de Bougainville
DE Bougainville-Mosaikschwanzratte
ES Rata de cola moteada de Bougainville

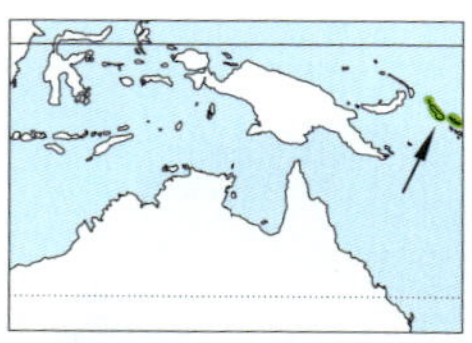

NW Solomon Arch: Bougainville I & Buka I (PG), Choiseul I (SB)

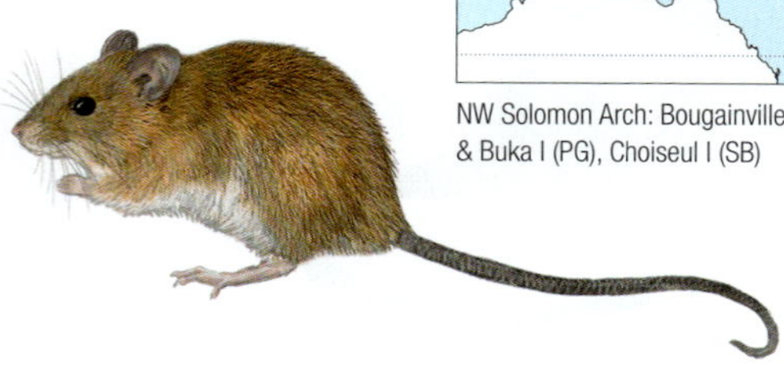

Dusky Seram Mosaic-tailed Rat
Melomys aerosus

EN
Monotypic
HB 15–16 cm T 12.5–13.8 cm
W 97–123 g

FR Mélomys cuivré
DE Dunkle Seram-Mosaikschwanzratte
ES Rata de cola moteada de Seram oscura

650–1830 m

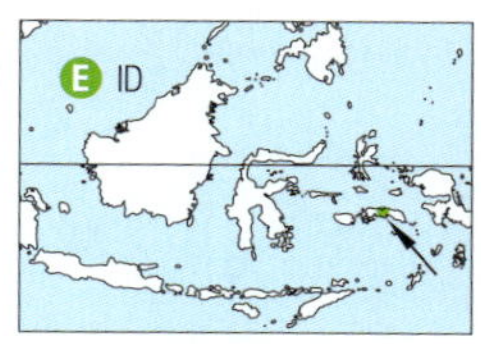

Seram I (Mt Manusela, Mt Hoaulu, near Kanikeh), Moluccas

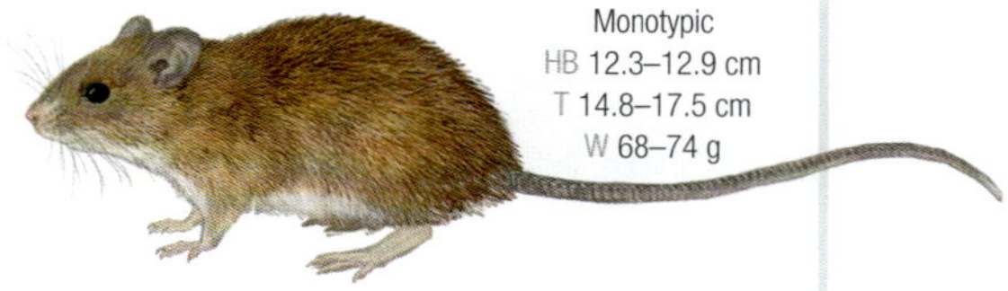

LC
Monotypic
HB 12.3–12.9 cm
T 14.8–17.5 cm
W 68–74 g

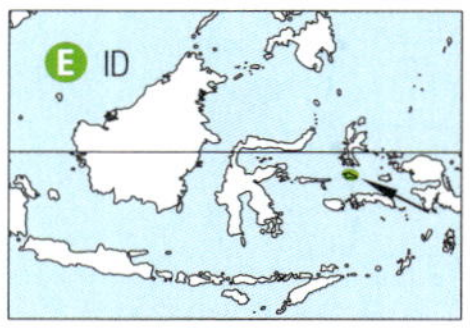

Obi I & Bisa I (S of Halmahera I, Moluccas)

Obi Island Mosaic-tailed Rat
Melomys obiensis
FR Mélomys d'Obi
DE Obi-Mosaikschwanzratte
ES Rata de cola moteada de Obi

Fawn-footed Mosaic-tailed Rat
Melomys cervinipes
FR Mélomys à pattes fauves
DE Kitzfüßige Mosaikschwanzratte
ES Rata de cola moteada de pies beige

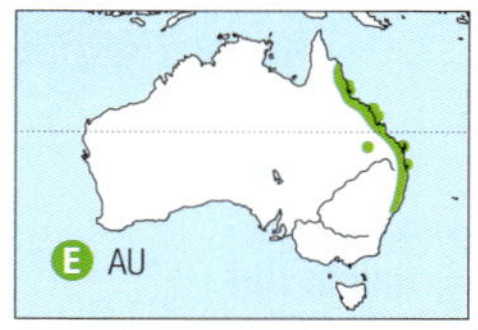

NE Qld S to C NSW (with Dunk, Hinchinbrook, Carlisle, Middle, Fraser & North Stradbroke Is)

LC
Monotypic
HB 9.5–20 cm
T 10.5–18 cm W 45–120 g

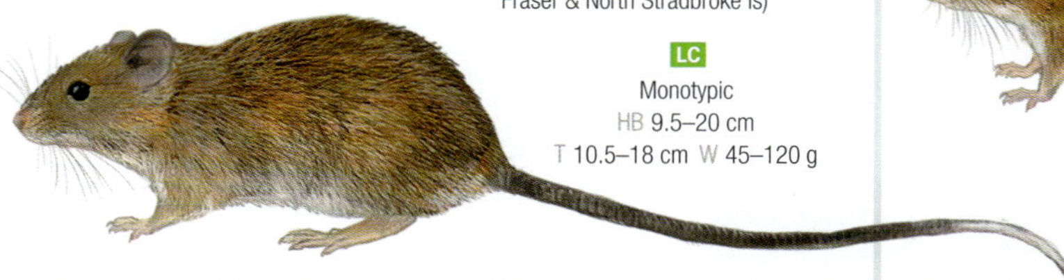

Cape York Mosaic-tailed Rat
Melomys capensis
FR Mélomys de Cape York
DE Kap-York-Mosaikschwanzratte
ES Rata de cola moteada de York

LC
Monotypic
HB 11.9–16.2 cm
T 12.9–17.2 cm
W 45–116 g

Cape York Pen (Qld), with some offshore Is

LC
7 sspp
HB 11.3–14.6 cm
T 10.8–17.5 cm
W 37–102 g

Black-tailed Mosaic-tailed Rat
Melomys rufescens
FR Mélomys roux
DE Schwarzschwanz-Mosaikschwanzratte
ES Rata de cola moteada y negra

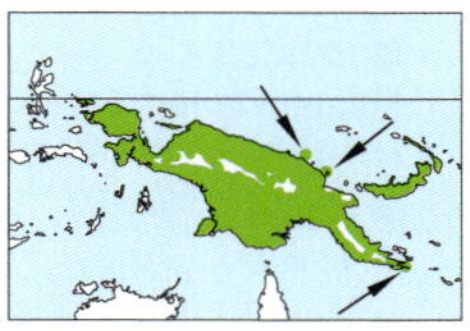

ID/PG: New Guinea (with Is off NW, N & SE), Bismarck Arch (New Britain, New Ireland)

Yamdena Island Mosaic-tailed Rat
Melomys cooperae
FR Mélomys de Cooper
DE Yamdena-Mosaikschwanzratte
ES Rata de cola moteada de Yamdena

DD
Monotypic
HB 11.2–14 cm
T 14–17 cm
W 60–96 g

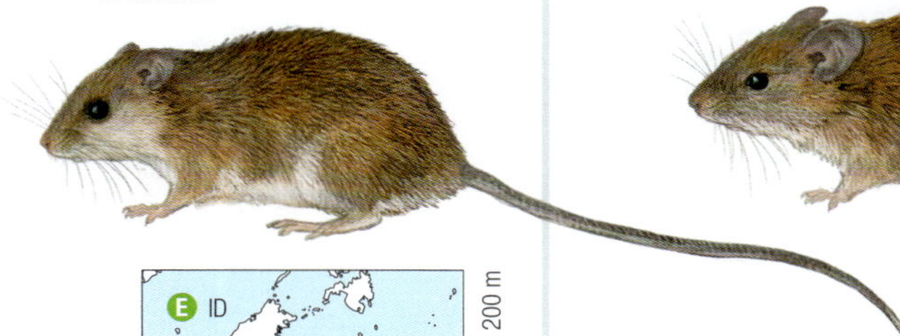

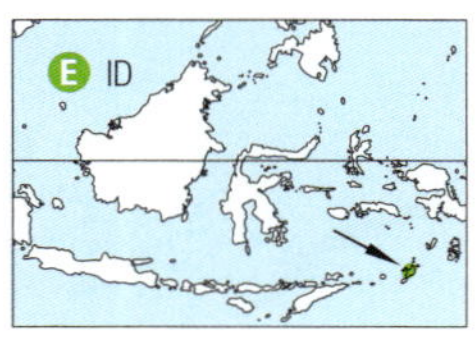

up to 200 m

Yamdena I (Tanimbar Is), S Moluccas

Short-tailed Talaud Mosaic-tailed Rat
Melomys caurinus
FR Mélomys à queue courte
DE Kurzschwanz-Talaud-Mosaikschwanzratte
ES Rata de cola moteada de Talaud de cola corta

EN
Monotypic
HB c. 17.6 cm
T 13.6–13.7 cm

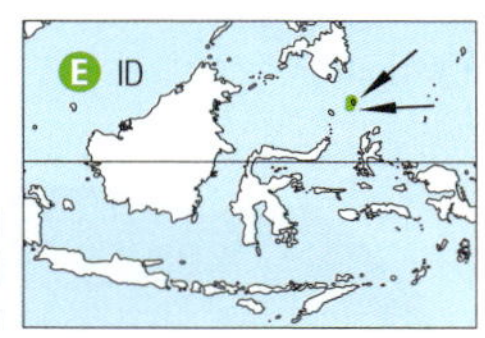

Karakelang I & Salebabu I (Talaud Is)

Long-tailed Talaud Mosaic-tailed Rat
Melomys talaudium
FR Mélomys des Talaud
DE Langschwanz-Talaud-Mosaikschwanzratte
ES Rata de cola moteada de Talaud de cola larga

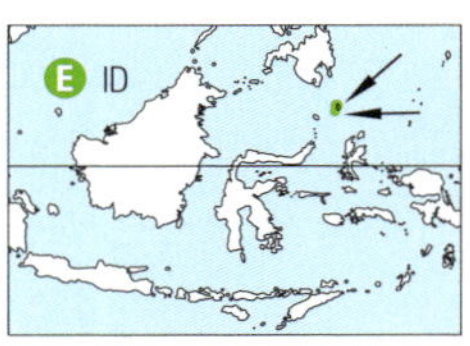

Karakelang I & Salebabu I (Talaud Is)

EN
Monotypic
HB 17.3–17.4 cm
T 15–19 cm

DD
Monotypic
HB c. 15 cm
T 20–21 cm

Seram Long-tailed Mosaic-tailed Rat
Melomys fulgens
FR Mélomys de Teluti
DE Langschwänzige Seram-Mosaikschwanzratte
ES Rata de cola moteada de Seram de cola larga

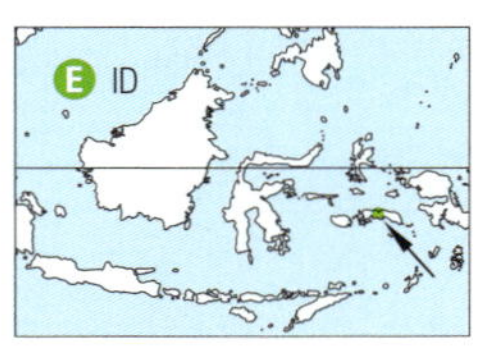

Seram I (Teloeti Bay, on S coast), Moluccas

White-bellied Mosaic-tailed Rat
Melomys leucogaster
FR Mélomys à ventre blanc
DE Weißbauch-Mosaikschwanzratte
ES Rata de cola moteada de vientre blanco

LC
Monotypic
HB ≤ 21 cm

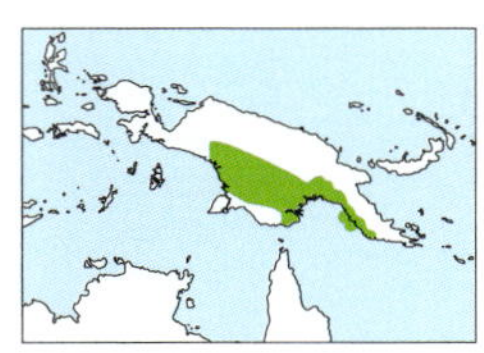

ID/PG: S New Guinea (from R Lorentz E to R Mori; some offshore Is)

Rossel Island Mosaic-tailed Rat
Melomys arcium
FR Mélomys de Rossel
DE Rossel-Mosaikschwanzratte
ES Rata de cola moteada de Rossel

DD
Monotypic
HB c. 13.8 cm
T c. 12.7 cm

50–700 m

Rossel I (Louisiade Arch)

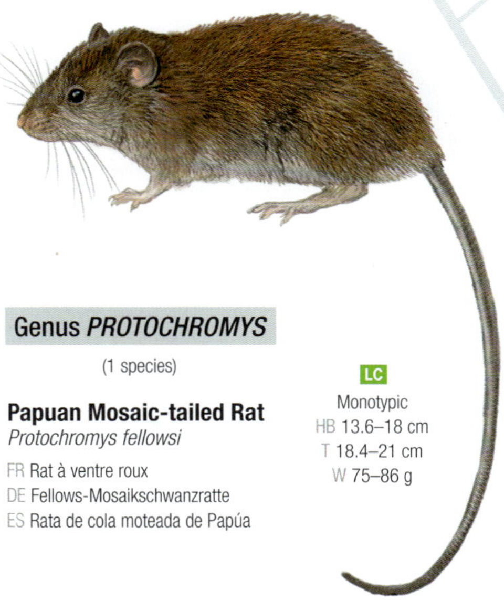

Genus *PROTOCHROMYS*

(1 species)

Papuan Mosaic-tailed Rat
Protochromys fellowsi
FR Rat à ventre roux
DE Fellows-Mosaikschwanzratte
ES Rata de cola moteada de Papúa

LC
Monotypic
HB 13.6–18 cm
T 18.4–21 cm
W 75–86 g

1800–2600 m

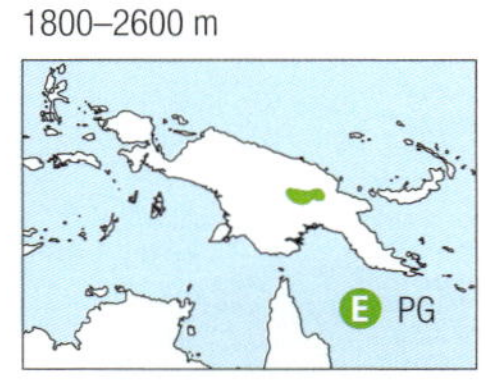

E New Guinea (from Porgera E to Hagen Range, Mt Wilhelm in Bismarck Range)

Genus *PARAMELOMYS*

(9 species)

EN
Monotypic
HB 13.5–16.2 cm

Gressitt's Mosaic-tailed Rat
Paramelomys gressitti

FR Mélomys de Gressitt
DE Gressitt-Mosaikschwanzratte
ES Rata de cola moteada de Gressitt

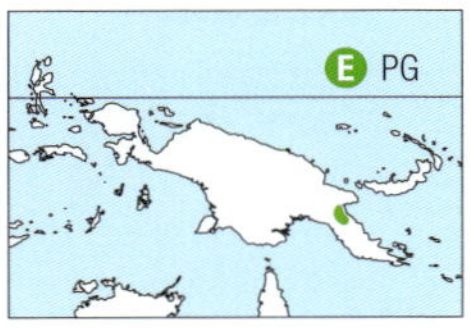

2300–2400 m

E New Guinea

Papuan Lowland Mosaic-tailed Rat
Paramelomys levipes

FR Mélomys à pattes fines
DE Papua-Tiefland-Mosaikschwanzratte
ES Rata de cola moteada de tierras bajas de Papúa

LC
Monotypic
HB 12.4–16.2 cm
T 11.2–14 cm W 72–117 g

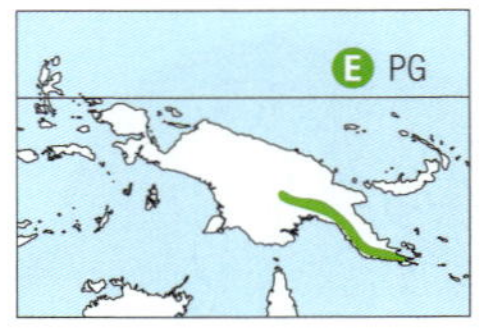

0–1200 m

E New Guinea (S side of Central Range)

LC
Monotypic
HB 13.5–19.5 cm
T 10.2–13.5 cm

0–1500 m

Lorentz's Mosaic-tailed Rat
Paramelomys lorentzii

FR Mélomys de Lorentz
DE Lorentz-Mosaikschwanzratte
ES Rata de cola moteada de Lorentz

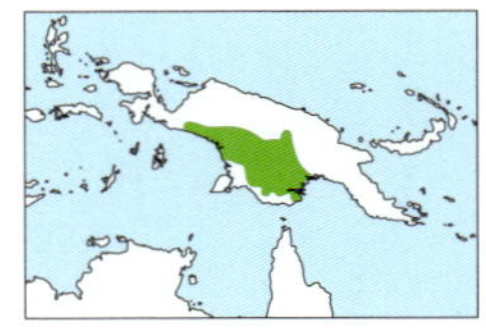

ID/PG: S New Guinea (from R Mimika E to Darai Plateau & Mt Bosavi area)

Long-nosed Mosaic-tailed Rat
Paramelomys naso

FR Mélomys à long nez
DE Langnasen-Mosaikschwanzratte
ES Rata de cola moteada de hocico largo

0–1000 m

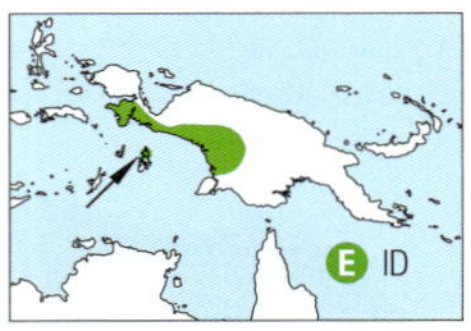

SW New Guinea, N Aru Is (Wokam I)

LC
Monotypic
HB 15.8–19.1 cm
T c. 13.2 cm

Montane Soft-furred Mosaic-tailed Rat
Paramelomys mollis

FR Mélomys doux
DE Weichfell-Mosaikschwanzratte
ES Rata de cola moteada de pelaje suave de montaña

1200–2200 m

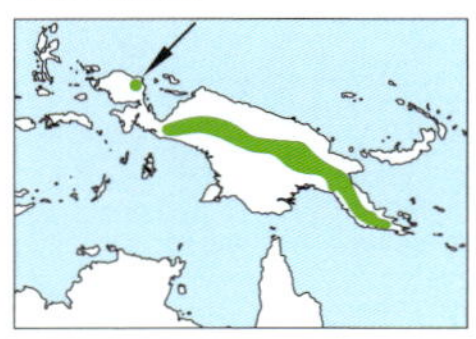

LC
Monotypic
HB 13–18.2 cm
T c. 14 cm

ID/PG: New Guinea Highlands

LC
Monotypic
HB 13.5–17 cm
T 11.5–14.5 cm W 78–118 g

0–700 m

Monckton's Mosaic-tailed Rat
Paramelomys moncktoni

FR Mélomys de Monckton
DE Monckton-Mosaikschwanzratte
ES Rata de cola moteada de Monckton

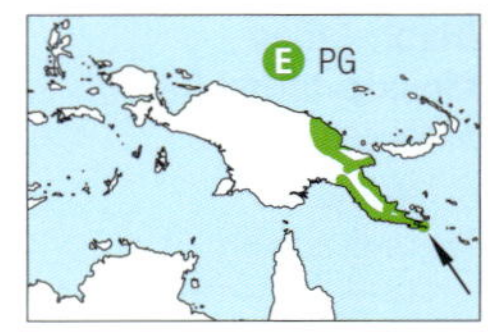

E New Guinea (NE & Papuan Pen, nearby Sideia I)

LC
Monotypic
HB 13–15.5 cm
T 10.5–12.5 cm
W 65–102 g

Common Lowland Mosaic-tailed Rat
Paramelomys platyops

FR Mélomys des plaines
DE Gemeine Tiefland-Mosaikschwanzratte
ES Rata de cola moteada de tierras bajas común

0–1200 m

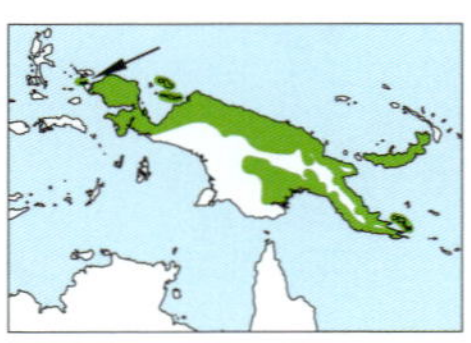

ID/PG: New Guinea (with Batanta, Yapen, Biak, New Britain & D'Entrecasteaux Is)

Stein's Mosaic-tailed Rat
Paramelomys steini

FR Mélomys de Stein
DE Stein-Mosaikschwanzratte
ES Rata de cola moteada de Stein

2000–2600 m

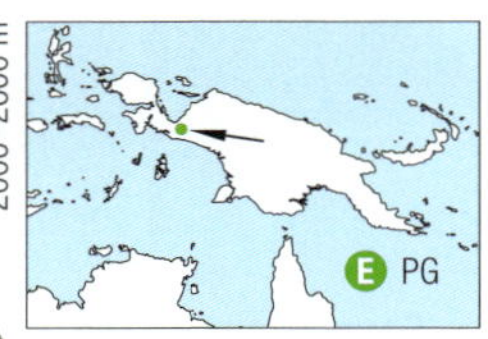

W New Guinea (Weyland Range)

DD
Monotypic
HB 12.4–13.1 cm T c. 12.8 cm

LC
Monotypic
HB 10–13 cm
T 9.6–14 cm
W 31–57 g

Mountain Mosaic-tailed Rat
Paramelomys rubex

FR Mélomys rubicond
DE Berg-Mosaikschwanzratte
ES Rata de cola moteada de montaña

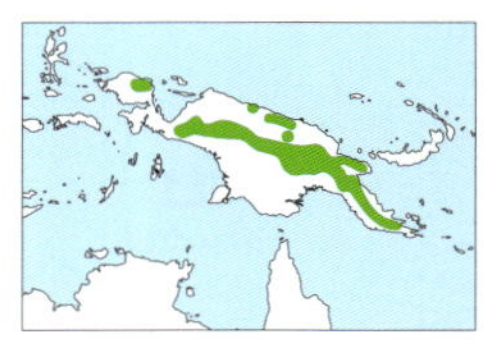

ID/PG: New Guinea (Central Range, Torricelli Mts, Huon Pen)

1200–2500 m

Genus *LEPORILLUS*

(1 species)

Greater Stick-nest Rat
Leporillus conditor

FR Grand Léporille
DE Große Häschenratte
ES Rata constructora mayor

NT
Monotypic
HB 17–26 cm
T 14.5–18 cm
W 180–450 g

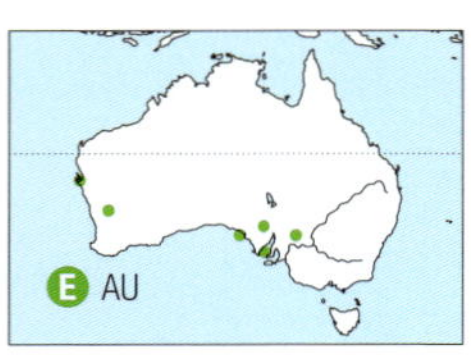

Native to Franklin Is (Nuyts Arch, SA); introduced to W WA, SA & NSW

Genus *CONILURUS*

(1 species)

VU
3 sspp
HB 15.2–19.5 cm
T 17–21 cm
W 110–170 g

Brush-tailed Rabbit Rat
Conilurus penicillatus

FR Conilure à pinceau
DE Bürstenschwanz-Kaninchenratte
ES Rata conejo de cola de cepillo

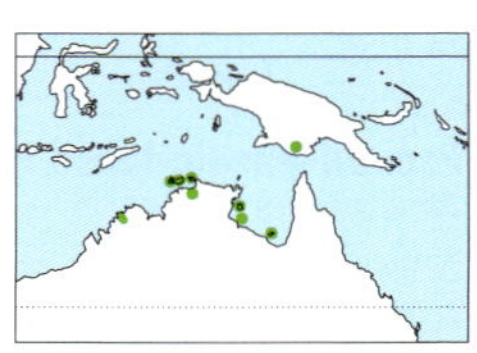

PG (Morehead, S New Guinea), AU (coast of WA, NT & NW Qld; some Is)

Genus *MESEMBRIOMYS*

(2 species)

VU
3 sspp
HB 25–30 cm
T 32–41 cm
W 580–880 g

Black-footed Tree Rat
Mesembriomys gouldii

FR Rat de Gould
DE Schwarzfuß-Baumratte
ES Rata arborícola de pies negros

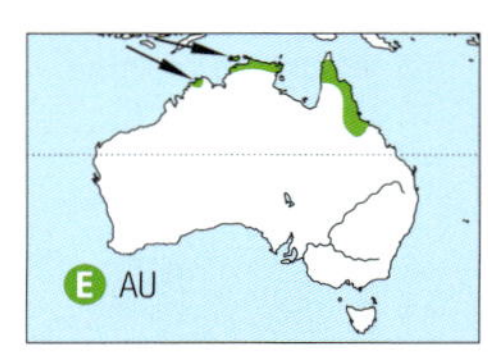

WA (Kimberley region), higher rainfall areas of NT (with Melville I), N Qld

Golden-backed Tree Rat

Mesembriomys macrurus

FR Rat à dos roux
DE Goldrücken-Baumratte
ES Rata arborícola de dorso dorado

NT
Monotypic
HB 18.8–24 cm
T 29–36 cm
W 205–330 g

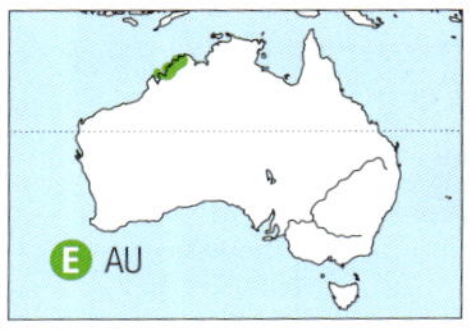

Rugged areas of N Kimberley region, with some Is (WA); formerly widespread

Kimberley Rock Rat

Zyzomys woodwardi

FR Zyzomys de Woodward
DE Kimberley-Dickschwanzratte
ES Rata de roca de Kimberley

LC
Monotypic
HB 12–16.6 cm
T 11–13 cm
W 80–190 g

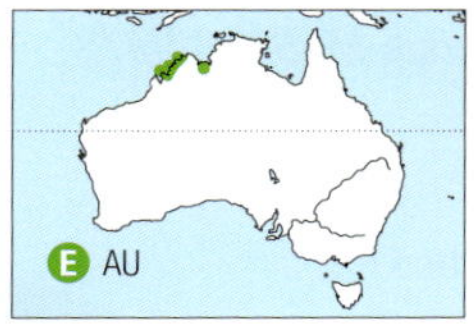

N Kimberley region, including at least 12 Is (WA)

Genus *ZYZOMYS*

(5 species)

Arnhem Land Rock Rat

Zyzomys maini

FR Zyzomys de Main
DE Arnhemland-Dickschwanzratte
ES Rata de roca de Tierra de Arnhem

VU
Monotypic
HB 11.5–17 cm
T 11–15 cm
W 65–180 g

Sandstone plateau & escarpment of W Arnhem Land (monsoonal NT)

Carpentarian Rock Rat

Zyzomys palatalis

FR Zyzomys de Wollogorang
DE Carpentaria-Dickschwanzratte
ES Rata de roca de Carpentaria

CR
Monotypic
HB 9–15 cm
T 10–15.2 cm
W 65–160 g

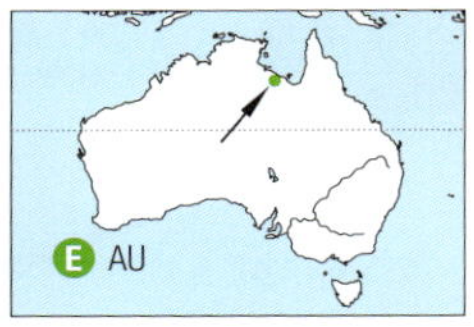

Restricted to a single locarion (Wollogorang Station) in NT

Common Australian Rock Rat

Zyzomys argurus

FR Zyzomys à queue d'argent
DE Gemeine Dickschwanzratte
ES Rata de roca de Australia

LC
Monotypic
HB 8.5–12.2 cm
T 9–13 cm
W 26–60 g

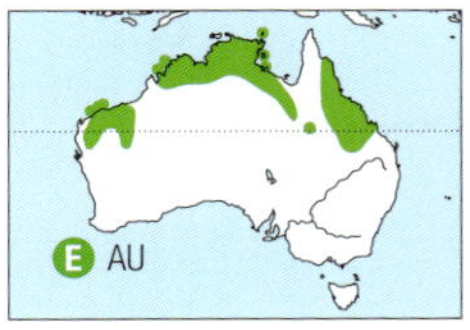

Fragmented range from Pilbara (WA) E to SE Qld, with many Is

Central Australian Rock Rat

Zyzomys pedunculatus

FR Zyzomys pédonculé
DE Zentralaustralische Dickschwanzratte
ES Rata de roca de Australia central

CR
Monotypic
HB 10.8–14 cm
T 11–14 cm
W 70–150 g

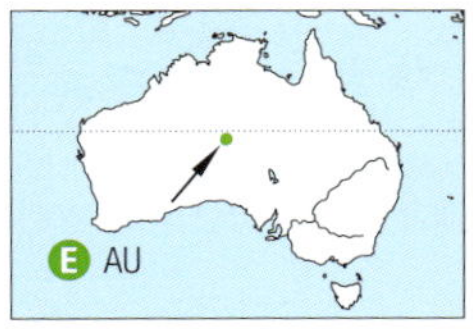

Chewings, Heavitree & W MacDonnell Ranges (NT)

Genus *LEGGADINA*

(2 species)

Central Short-tailed Mouse

Leggadina forresti

FR Rat de Forrest
DE Zentrale Australien-Zwergmaus
ES Ratón de cola corta de Australia central

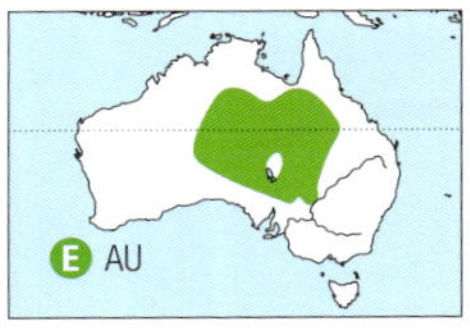

Low-rainfall areas in C & S NT, inland Qld, extreme EC WA, N SA, NW NSW

LC
Monotypic
HB 7–10 cm
T 5–7 cm
W 13–30 g

Northern Short-tailed Mouse

Leggadina lakedownensis

FR Rat de Lakeland
DE Nördliche Australien-Zwergmaus
ES Ratón de cola corta de Australia septentrional

LC
Monotypic
HB 5–9.2 cm
T 4–8 cm
W 15–30 g

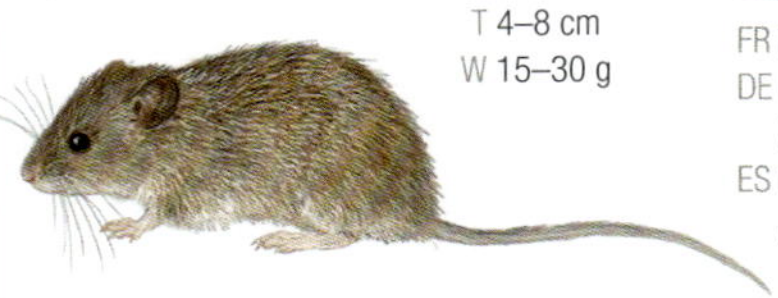

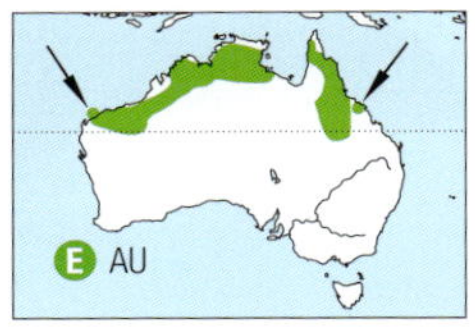

Monsoonal N mainland, also Thevenard I, translocated to nearby Serrurier I (NW WA)

Genus *NOTOMYS*

(5 species)

Fawn Hopping Mouse

Notomys cervinus

FR Notomys fauve
DE Rehbraune Australienhüpfmaus
ES Ratón saltador beige

NT
Monotypic
HB 9.5–12 cm
T 10.5–16 cm
W 30–50 g

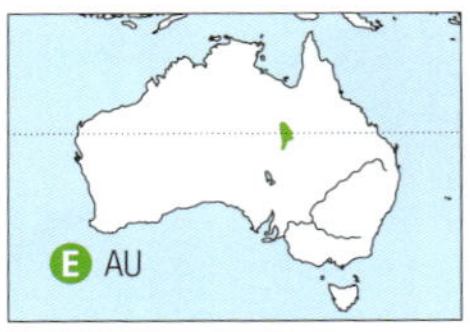

Restricted to Channel Country (SW Qld) & L Eyre Basin (NE SA)

Dusky Hopping Mouse

Notomys fuscus

FR Notomys brun
DE Dunkle Australienhüpfmaus
ES Ratón saltador oscuro

VU
Monotypic
HB 7.6–11.5 cm
T 11.5–15.5 cm
W 26–55 g

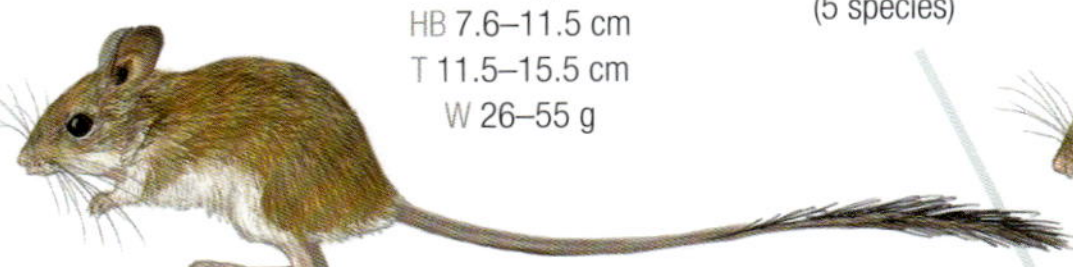

Arid inland, in E Simpson & S Strzelecki deserts (SW Qld, SA) & far NW NSW

Northern Hopping Mouse

Notomys aquilo

FR Notomys septentrional
DE Nördliche Australienhüpfmaus
ES Ratón saltador septentrional

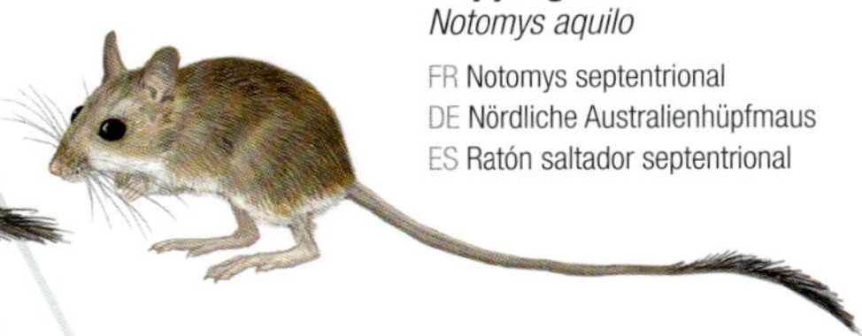

EN
Monotypic
HB 8.5–11.5 cm
T 15–17.3 cm
W 28–44 g

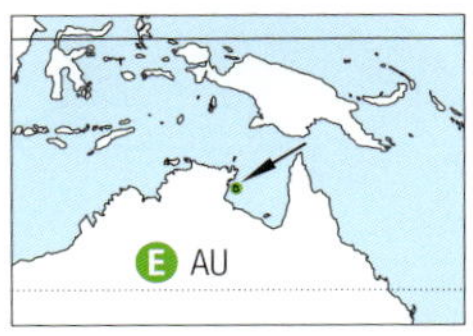

Groote Eylandt (NT) (all post-2000 confirmed records)

Mitchell's Hopping Mouse

Notomys mitchellii

FR Notomys de Mitchell
DE Mitchell-Australienhüpfmaus
ES Ratón saltador de Mitchell

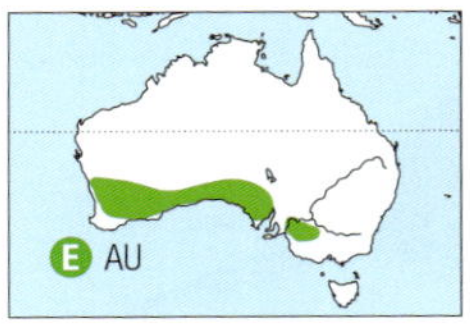

Discontinuous range in semiarid S mainland (WA, SA, W Vic)

LC
Monotypic
HB 10–12.5 cm
T 14–15.5 cm
W 40–60 g

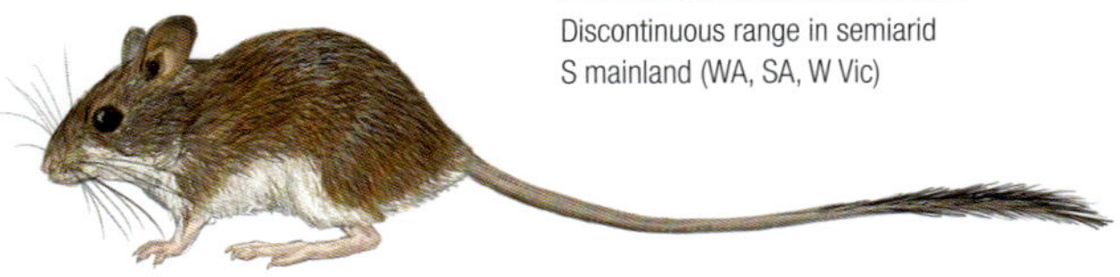

Spinifex Hopping Mouse

Notomys alexis

FR Notomys alexis
DE Spinifex-Australienhüpfmaus
ES Ratón saltador de spinifex

LC
3 sspp
HB 9–11.5 cm
T 11.5–15 cm
W 27–50 g

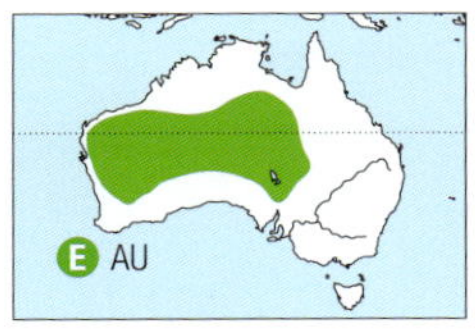

Inland WA, NT, SA & SW Qld

Genus *MASTACOMYS*

(1 species)

Australian Broad-toothed Rat
Mastacomys fuscus

FR Mastacomys sombre
DE Australische Breitzahnratte
ES Rata de dientes anchos

NT
2 sspp
HB 14.2–19.5 cm
T 10–13.5 cm W 97–180 g

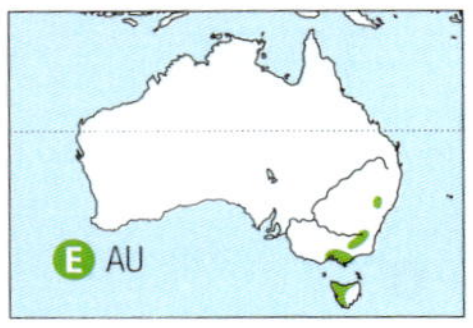

Vic (Otway & Dandenong ranges, Gippsland coast), NSW (Barrington Tops & near Brindabella Range), N & W Tas

Genus *PSEUDOMYS*

(20 species)

Bolam's Mouse
Pseudomys bolami

FR Pseudomys de Bolam
DE Bolam-Australienmaus
ES Ratón de Bolam

LC
Monotypic
HB 5–8 cm
T 7.1–10.3 cm
W 9–16 g

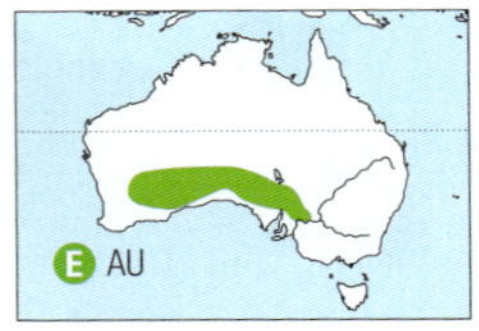

S WA, SA, extreme CW NSW

Little Native Mouse
Pseudomys delicatulus

FR Pseudomys délicat
DE Zierliche Australienmaus
ES Ratón nativo pequeño

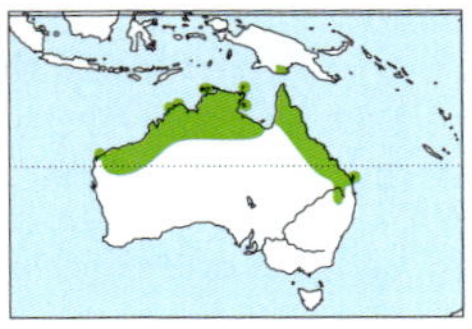

PG (Trans Fly plains, S New Guinea), N & E AU (from Pilbara to Qld, S to NE NSW, with some Is)

LC
2 sspp
HB 5.5–7.5 cm
T 5.5–8 cm
W 6–15 g

New Holland Mouse
Pseudomys novaehollandiae

FR Pseudomys de Nouvelle-Hollande
DE Neuholland-Australienmaus
ES Ratón castaño de Nueva Holanda

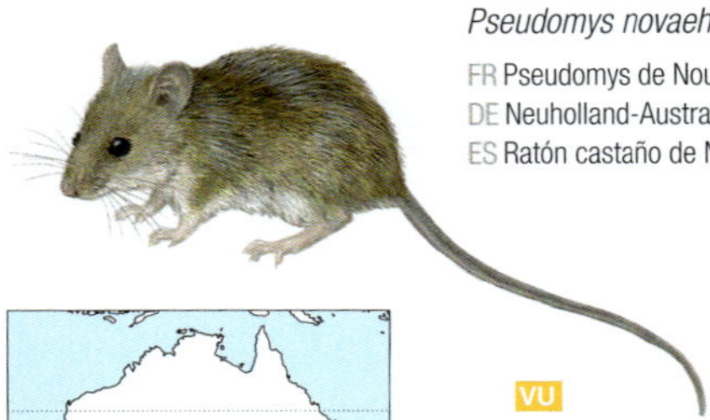

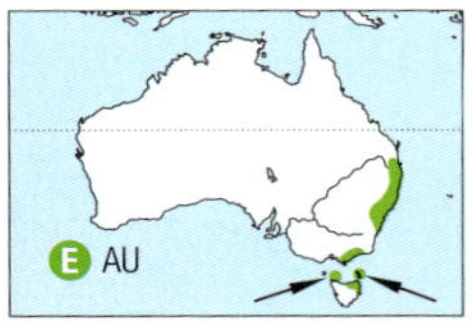

Coast in SE Qld, E NSW, SE Vic & Tas, Flinders I & Three Hummock I

VU
Monotypic
HB 6.5–9.5 cm
T 8.1–11 cm
W 12–25 g

Sandy Inland Mouse
Pseudomys hermannsburgensis

FR Pseudomys de Hermannsburg
DE Sand-Australienmaus
ES Ratón de interior de arena

LC
Monotypic
HB 6.5–8.5 cm
T 7–9 cm
W 8–15 g

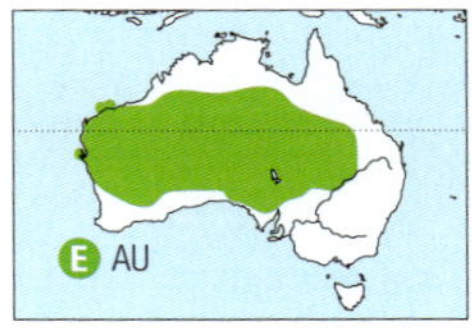

W, C & CE mainland, some offshore Is (Dirk Hartog, Rosemary, Dixon, Hope)

Gould's Mouse
Pseudomys gouldii

FR Pseudomys de Gould
DE Gould-Australienmaus
ES Ratón de Gould

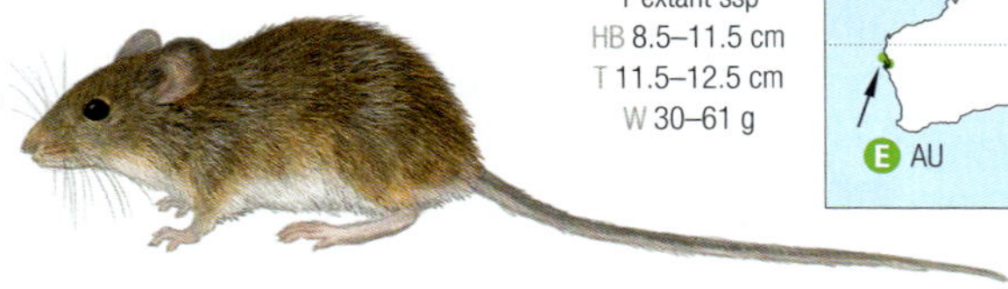

VU
1 extant ssp
HB 8.5–11.5 cm
T 11.5–12.5 cm
W 30–61 g

Shark Bay Is (Bernier, Faure); introduced to North West I, Montebello Is & Doole I (WA)

Long-tailed Mouse
Pseudomys higginsi

FR Pseudomys de Higgins
DE Langschwanz-Australienmaus
ES Ratón de cola larga

LC
Monotypic
HB 11.5–15 cm
T 14.5–20 cm
W 50–90 g

0–1600 m

Tas & Bruny I

Plains Mouse
Pseudomys australis

FR Pseudomys de Gray
DE Südliche Australienmaus
ES Ratón de llanura

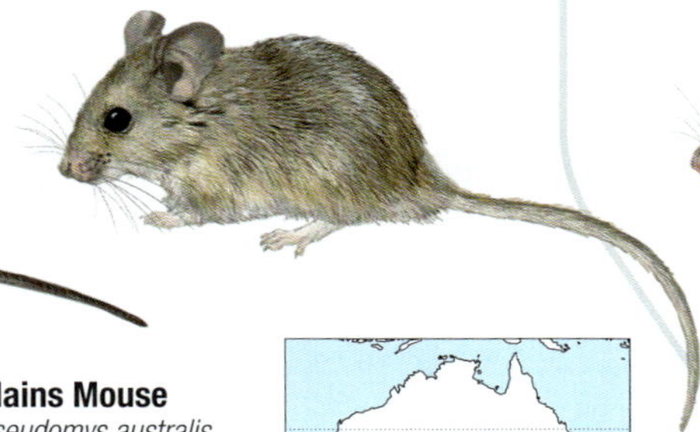

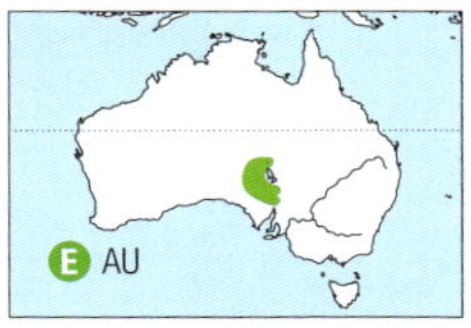

L Eyre Basin, from R Finke flood-out area (SE NT) to Pernatty Station (SA)

VU
Monotypic
HB 9–14.5 cm T 8.5–12.5 cm
W 30–65 g

Ash-gray Mouse
Pseudomys albocinereus

FR Pseudomys gris
DE Aschgraue Australienmaus
ES Ratón gris cenizo

LC
2 sspp
HB 6.3–11 cm
T 8.5–11.5 cm
W 14–40 g

Scrubby plains in W & S WA, Shark Bay Is (Bernier, Dorre, Dirk Hartog)

Desert Mouse
Pseudomys desertor

FR Pseudomys déserteur
DE Wüsten-Australienmaus
ES Ratón de desierto

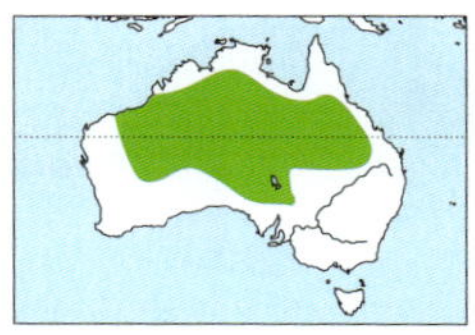

C arid zone and semiarid tropical N & C mainland

LC
Monotypic
HB 7–11 cm
T 8–10.3 cm W 11–35 g

LC
Monotypic
HB 6.5–9.5 cm
T 9–11.5 cm
W 16–22 g

Silky Mouse
Pseudomys apodemoides

FR Pseudomys soyeux
DE Seidige Australienmaus
ES Ratón sedoso

SE SA, W Vic

Australian Smoky Mouse
Pseudomys fumeus

FR Pseudomys fuligineux
DE Rauchgraue Australienmaus
ES Ratón ahumado de Australia

VU
Monotypic
HB 8.5–12.5 cm
T 11–15.5 cm
W 45–90 g

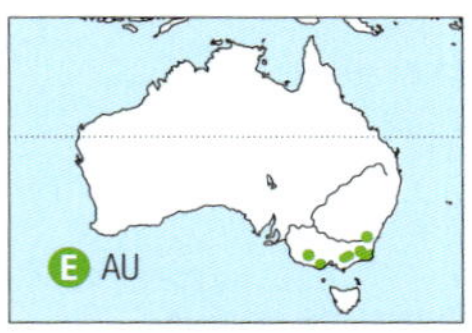

Patchily in SE mainland

Heath Mouse
Pseudomys shortridgei

FR Pseudomys de Shortridge
DE Heide-Australienmaus
ES Ratón de brezo

NT
Monotypic
HB 9–12 cm
T 8.5–10 cm
W 55–90 g

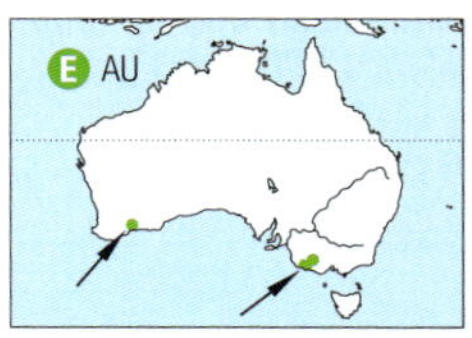

SW WA (Wheatbelt), SW Vic S to SE tip of SA

Hastings River Mouse
Pseudomys oralis

FR Pseudomys du Hastings
DE Hastings-River-Australienmaus
ES Ratón de Hastings

300–1250 m

SE Qld, NE NSW

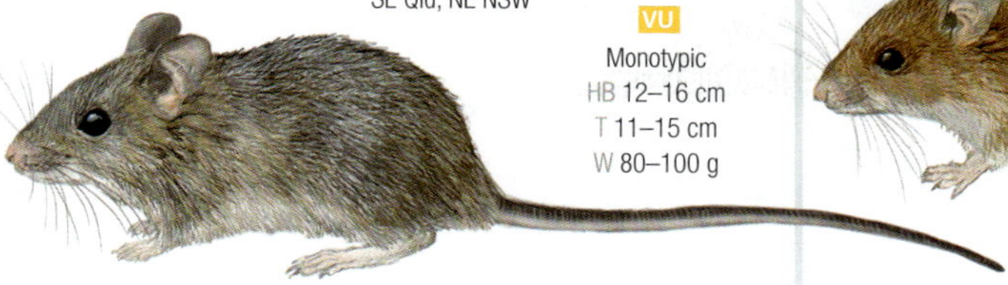

VU
Monotypic
HB 12–16 cm
T 11–15 cm
W 80–100 g

Eastern Chestnut Mouse
Pseudomys gracilicaudatus

FR Pseudomys à queue fine
DE Ostküsten-Australienmaus
ES Ratón castaño oriental

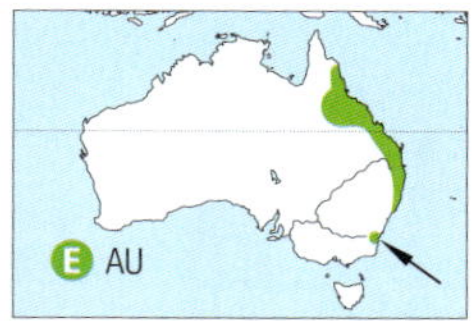

E coast from NE Qld S to EC NSW, disjunct in Jervis Bay (SE NSW)

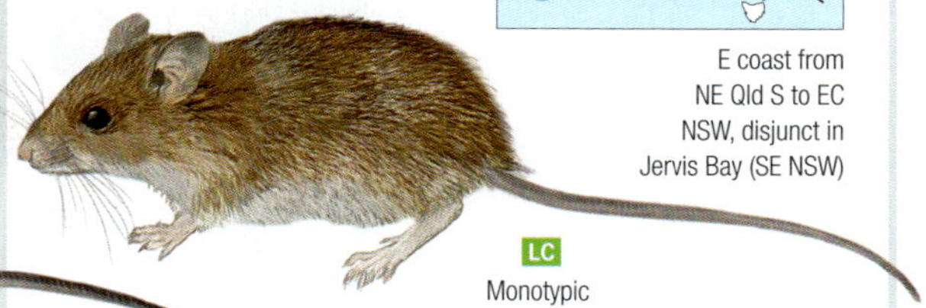

LC
Monotypic
HB 9–14.5 cm
T 7–12 cm W 45–118 g

LC
2 sspp
HB 7.1–13 cm
T 7.5–12.1 cm
W 17–56 g

Western Chestnut Mouse
Pseudomys nanus

FR Petit Pseudomys
DE Kleine Australienmaus
ES Ratón castaño occidental

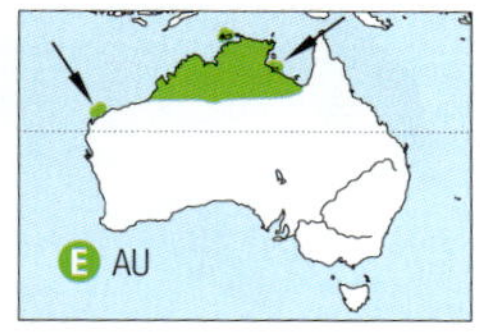

N mainland, some Is in WA (Barrow, Sholl, Potter) & NT (Melville & Bathurst, Sir Edward Pellew Group)

Kakadu Pebble Mouse
Pseudomys calabyi

FR Pseudomys de Calaby
DE Calaby-Australienmaus
ES Ratón de guijarros de Kakadu

VU
Monotypic
HB 6.8–9.5 cm
T 6.4–9.4 cm
W 12–20 g

Monsoon tropics of N NT, almost all records from Kakadu & Litchfield NPs

Central Pebble Mouse
Pseudomys johnsoni

FR Pseudomys de Johnson
DE Johnson-Australienmaus
ES Ratón de guijarros de Australia central

LC
Monotypic
HB 5.8–7.6 cm
T 6.3–9.5 cm
W 7–20 g

Monsoonal mainland through semiarid areas of mid-NT to NW Qld

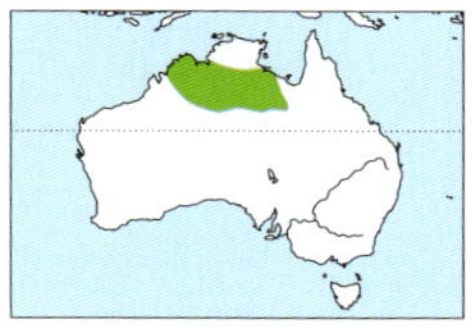

Eastern Pebble Mouse
Pseudomys patrius

FR Pseudomys de l'Inkerman
DE Queensland-Australienmaus
ES Ratón de guijarros oriental

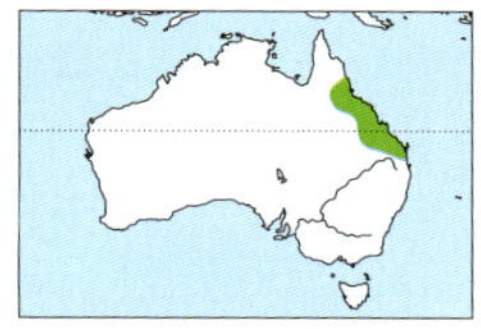

E Qld (from Paluma & Charters Towers S to Springsure & Gympie)

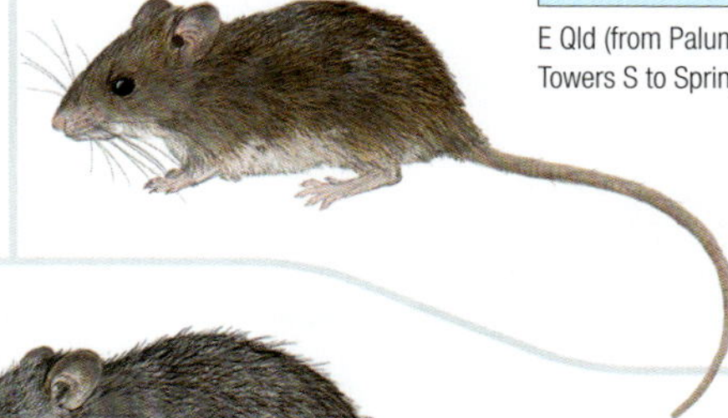

LC
Monotypic
HB 5.4–7.7 cm
T 5.5–8.5 cm
W 10–19 g

Western Pebble Mouse
Pseudomys chapmani

FR Pseudomys de Chapman
DE Westliche Australienmaus
ES Ratón de guijarros occidental

LC
Monotypic
HB 5.2–6.7 cm
T 7.3–7.9 cm
W 10–15 g

W & C parts of Pilbara region (WA)

Western Mouse
Pseudomys occidentalis

FR Pseudomys occidental
DE Stumpfschnauzige Australienmaus
ES Ratón de Australia occidental

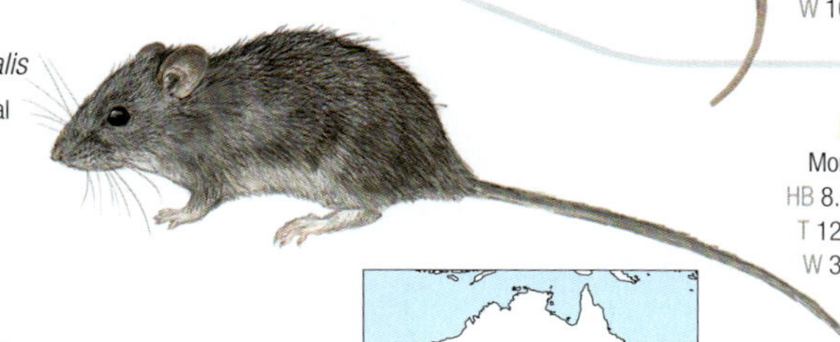

NT
Monotypic
HB 8.5–11 cm
T 12–14 cm
W 33–55 g

S Wheatbelt and coastal region of SW WA

Genus *DIOMYS*

(1 species)

Crump's Rat
Diomys crumpi

FR Rat de Crump
DE Crump-Maus
ES Rata de Crump

DD
Monotypic
HB 10.2–13.5 cm
T 10.9–13 cm

1000–2000 m

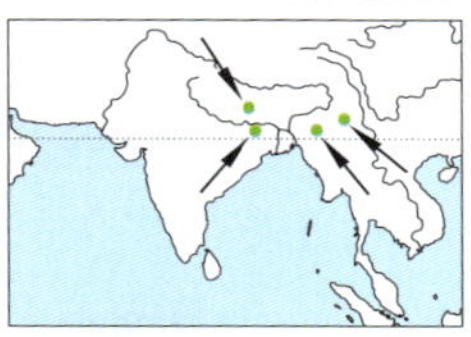

Known from single records in S NP, NE IN (Bihar, Manipur) & N MM

Genus *MADROMYS*

(1 species)

White-tailed Wood Rat
Madromys blanfordi

FR Rat de Blanford
DE Blanford-Ratte
ES Rata de bosque de cola blanca

LC
Monotypic
HB 15.1–18.2 cm
T 18.5–21 cm
W 98–110 g

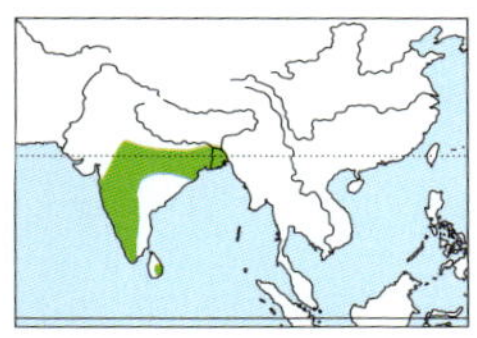

IN, SW BD, E & S LK

Genus *MILLARDIA*

(4 species)

Sand-colored Soft-furred Rat
Millardia gleadowi

FR Rat de Gleadow
DE Sandfarbene Weichfellratte
ES Rata de pelaje suave de color de arena

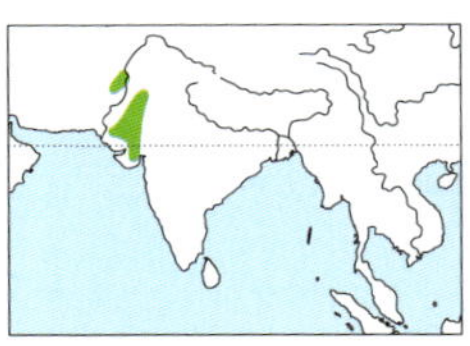

PK, NW IN (Rajasthan, Gujarat)

LC
Monotypic
HB 7.7–9.7 cm
T 6.7–9.3 cm

Kondana Soft-furred Rat
Millardia kondana

FR Rat de Singarth
DE Maharashtra-Weichfellratte
ES Rata de pelaje suave de Kondana

EN
Monotypic
HB 15–20 cm
T 11–18.6 cm

c. 1270 m

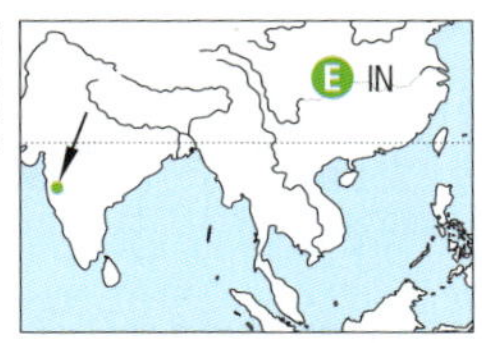

Known only from Sinhgarh Plateau area (W Maharashtra)

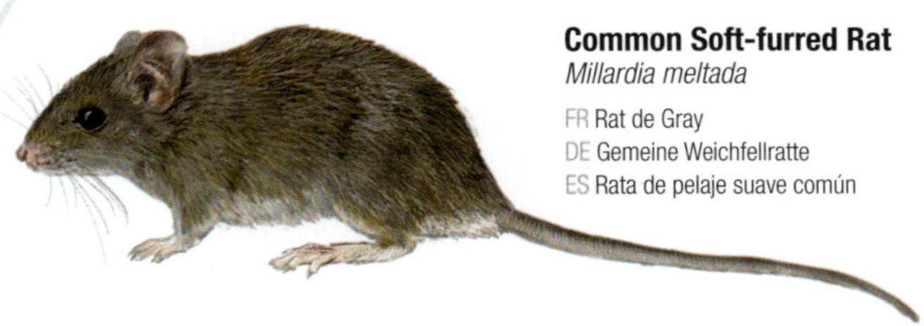

Common Soft-furred Rat
Millardia meltada

FR Rat de Gray
DE Gemeine Weichfellratte
ES Rata de pelaje suave común

LC
Monotypic
HB 10.7–15.4 cm
T 9.2–13.6 cm

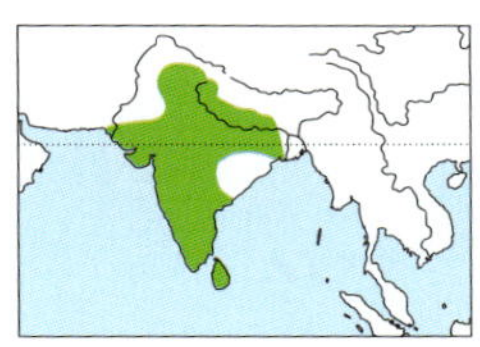

E PK, IN, S NP, LK

Burmese Soft-furred Rat

Millardia kathleenae

FR Rat de Ryley
DE Burma-Weichfellratte
ES Rata de pelaje suave de Birmania

LC
Monotypic
HB 13–16.5 cm
T 12–15.5 cm
W 98–110 g

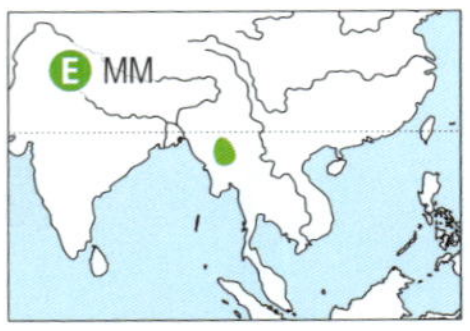

Sandy plains in C dry region

Cutch Rat

Cremnomys cutchicus

FR Rat du Cutch
DE Kutch-Indienfelsratte
ES Rata de Kutch

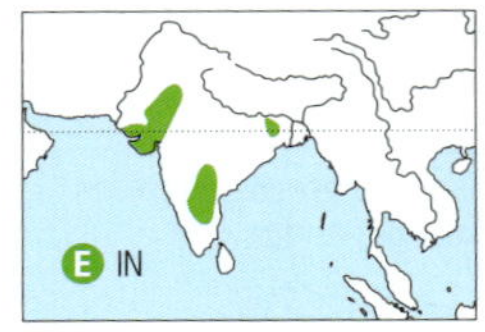

up to 1500 m

Rajasthan, Gujarat, Telangana, Andhra Pradesh, Karnataka, Tamil Nadu, Bihar, Jharkhand

LC
Monotypic
HB 10.4–14.9 cm
T 12–17.4 cm

Genus *CREMNOMYS*

(2 species)

Elvira Rat

Cremnomys elvira

FR Rat de Salem
DE Tamil-Nadu-Indienfelsratte
ES Rata de Elvira

CR
Monotypic
HB 12.6–14.9 cm
T 18–19.6 cm

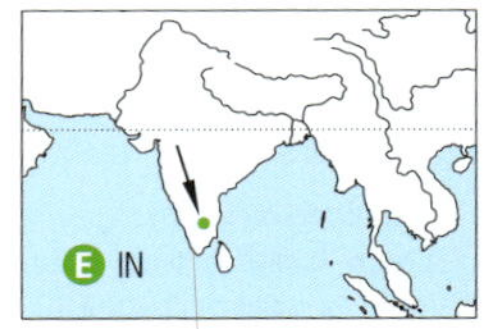

Known only from Kurumbapatti (Salem District, Eastern Ghats,Tamil Nadu)

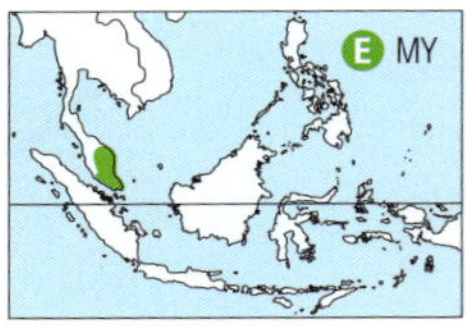

up to 1200 m

Lowland rainforests in Pen

Malayan Tree Rat

Pithecheir parvus

FR Pithécheir de Malaisie
DE Malaya-Affenfußratte
ES Rata arborícola de Malasia

DD
Monotypic
HB 12.2–18 cm
T 15.7–22 cm
W 49–146 g

Genus *PITHECHEIR*

(2 species)

Red Tree Rat

Pithecheir melanurus

FR Pithécheir de Java
DE Java-Affenfußratte
ES Rata arborícola roja

VU
Monotypic
HB 15.9–17.6 cm
T 19–21 cm

up to 1200 m

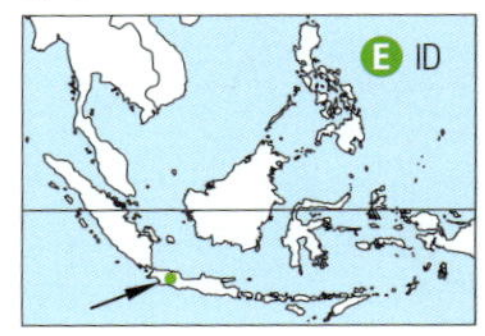

W Java

Genus *PITHECHEIROPS*

(1 species)

Bornean Tree Rat

Pithecheirops otion

FR Pithécheir de Bornéo
DE Borneo-Affenfußratte
ES Rata arborícola de Borneo

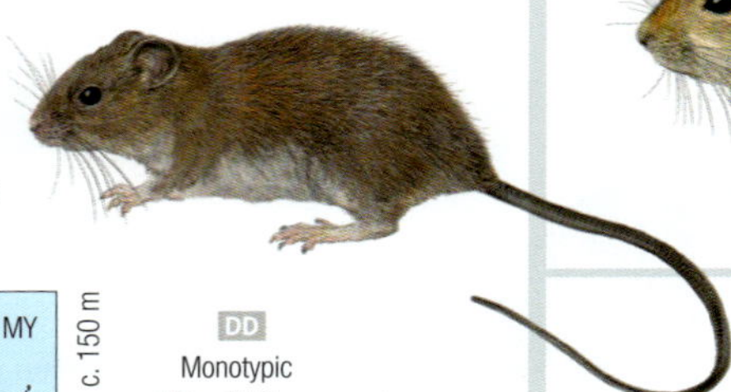

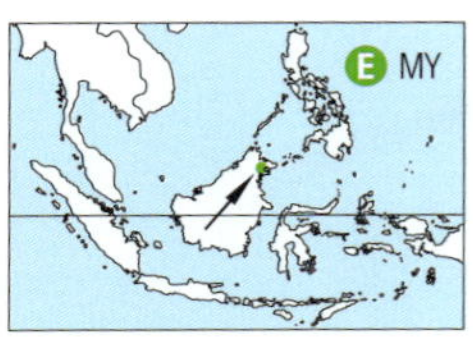

c. 150 m

N Borneo (known only from Danum Valley, Sabah)

DD
Monotypic
HB c. 11.3 cm
T c. 11.7 cm
W c. 36 g

LC
Monotypic
HB 14.1–15.8 cm
T 9.3–11.5 cm
W 75–111 g

Brants's Whistling Rat

Parotomys brantsii

FR Otomys de Brants
DE Brants-Pfeifratte
ES Rata silbadora de Brant

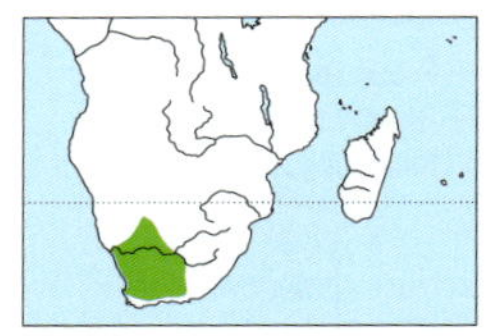

S & SE NA, extreme SW BW, W ZA (Northern Cape, Western Cape)

Genus *PAROTOMYS*

(2 species)

LC
Monotypic
HB 12.4–17 cm
T 8.2–11.2 cm
W 97–110 g

Littledale's Whistling Rat

Parotomys littledalei

FR Otomys de Littledale
DE Littledale-Pfeifratte
ES Rata silbadora de Littledale

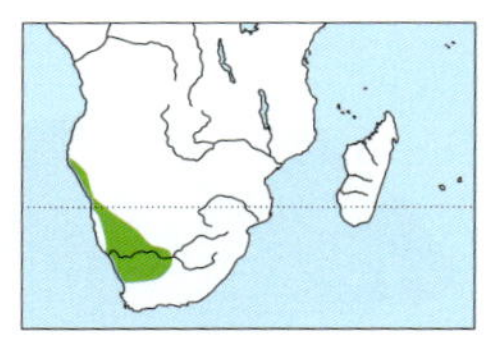

W NA along narrow coastal strip S to driest areas of NW ZA

Genus *OTOMYS*

(30 species)

Sloggett's Vlei Rat

Otomys sloggetti

FR Otomys de Sloggett
DE Sloggett-Lamellenzahnratte
ES Rata de laguna de Sloggett

LC
Monotypic
HB 12.4–17 cm
T 8.2–11.2 cm
W 97–110 g

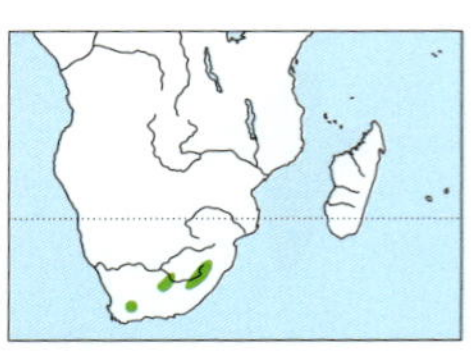

W NA along narrow coastal strip S to driest areas of NW ZA

NE
Monotypic
HB 11.1–12.4 cm
T 5.9–7.4 cm
W 75–83 g

Willan's Vlei Rat

Otomys willani

FR Otomys de Willan
DE Willan-Lamellenzahnratte
ES Rata de laguna de Willan

above 1000 m

S Great Escarpment, from Karoo NP (Western Cape) E to Giant's Castle (KwaZulu-Natal)

Fynbos Vlei Rat

Otomys karoensis

FR Otomys du Karoo
DE Fynbos-Lamellenzahnratte
ES Rata de laguna de fynbos

LC
Monotypic
HB 11–15.4 cm
T 8.6–11.1 cm
W 78–139 g

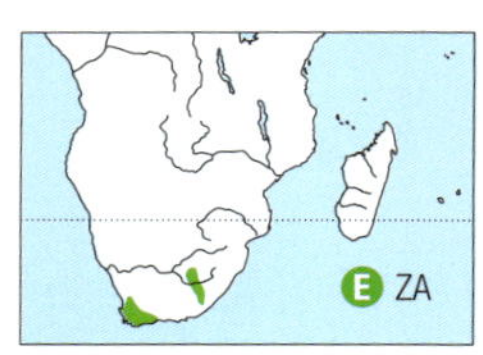

Western Cape (Cape Fold Belt Mts), S Free State & Eastern Cape N of 33° S

KwaZulu Natal Vlei Rat
Otomys laminatus

FR Otomys lamellé
DE KwaZulu-Lamellenzahnratte
ES Rata de laguna de Zululandia

NT
Monotypic
HB 15.8–21 cm
T 9.7–11.5 cm W 140–264 g

0–2000 m

Disjunct in Western Cape (Paarl Valley, Cape Pen) and Mpumalanga & KwaZulu-Natal (E slopes of Drakensberg)

LC
Monotypic
HB 11–20 cm
T 6.3–10.6 cm
W 61–156 g

Karoo Vlei Rat
Otomys unisulcatus

FR Otomys cafre
DE Karoo-Lamellenzahnratte
ES Rata de laguna de Karoo

Semiarid Succulent Karoo & Nama Karoo biomes (Northern Cape, Western Cape, Eastern Cape)

EN
Monotypic
HB 12.5–21 cm
T 5.7–9.6 cm
W 78–161 g

Barbour's Vlei Rat
Otomys barbouri

FR Otomys de Barbour
DE Barbour-Lamellenzahnratte
ES Rata de laguna de Barbour

above 3000 m

Mt Elgon (UG, KE)

LC
Monotypic
HB 13.9–16.5 cm
T 7.8–10.2 cm
W 67–106 g

Dent's Vlei Rat
Otomys denti

FR Otomys de Dent
DE Dent-Lamellenzahnratte
ES Rata de laguna de Dent

1950–3000 m

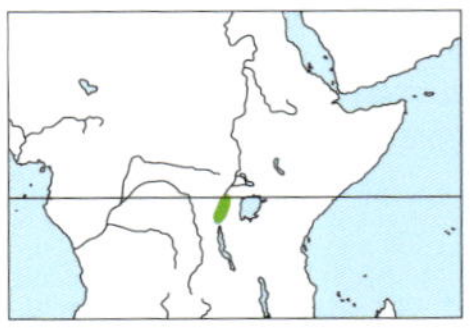

From Rwenzori Mts (UG, CD) S to Albertine Rift Mts (RW, BI, CD)

LC
Monotypic
HB 15.5–18.5 cm
T 9.3–10.8 cm

Tanzanian Vlei Rat
Otomys lacustris

FR Otomys de Tanzanie
DE Tansania-Lamellenzahnratte
ES Rata de laguna de Tanzania

1400–2300 m

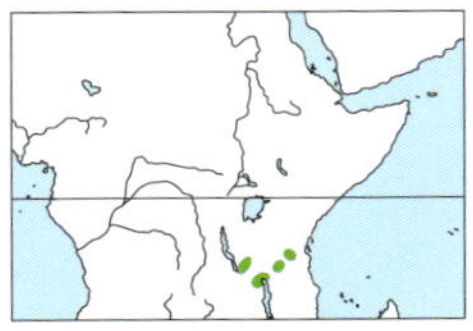

S Rift highlands of S TZ and neighboring ZM & MW

LC
Monotypic
HB 19.7–22 cm
T 8.7–12.7 cm

Angolan Vlei Rat
Otomys anchietae

FR Otomys d'Anchieta
DE Angola-Lamellenzahnratte
ES Rata de laguna de Angola

below 2000 m

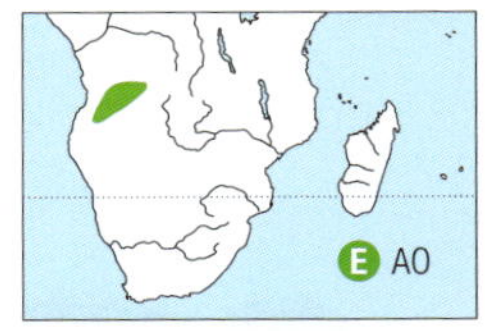

R Congo & R Okavango basins, recorded from source of R Cuito

NT
Monotypic
HB 13.3–20 cm
T 5.8–12 cm
W 63–211 g

South East African Vlei Rat
Otomys auratus

FR Otomys doré
DE Südostafrika-Lamellenzahnratte
ES Rata de laguna de África suroriental

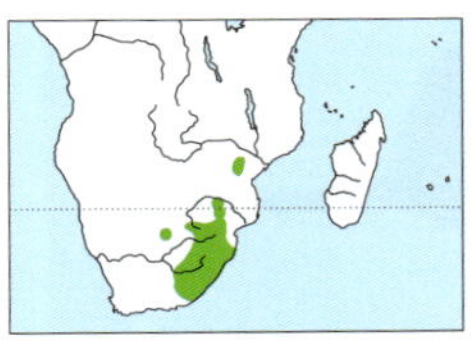

C plateau & Drakensberg (ZA, SZ, LS), isolated in Soutpansberg (N ZA) & Eastern Highlands (ZW, MZ)

Southern African Vlei Rat
Otomys irroratus

FR Otomys du fynbos
DE Südafrika-Lamellenzahnratte
ES Rata de laguna de África meridional

above 900 m

Western Cape, Eastern Cape

LC
Monotypic
HB 13–19 cm
T 8.5–11.4 cm
W 80–201 g

Cuanza Vlei Rat
Otomys cuanzensis

FR Otomys du Kwanza
DE Cuanza-Lamellenzahnratte
ES Rata de laguna de Cuanza

LC
Monotypic
HB 13–19 cm
T 8.5–11.4 cm
W 80–201 g

1000–2500 m

R Cuanza Basin and other rivers draining to Atlantic Ocean

Angoni Vlei Rat
Otomys angoniensis

FR Otomys angoni
DE Angoni-Lamellenzahnratte
ES Rata de laguna de Angoni

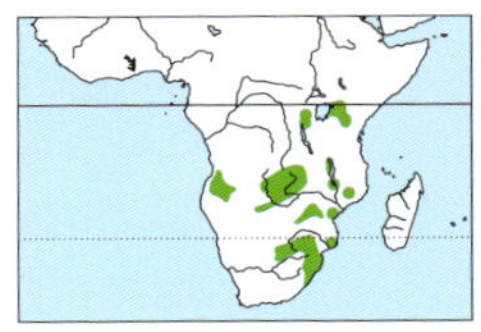

Disjunct from E CD, S UG & SW KE to SW AO, NA (Caprivi Strip) & NE ZA

LC
3 sspp
HB 12.4–19.2 cm
T 4–10 cm
W 60–182 g

VU
Monotypic
HB 14.5–18.2 cm
T 8.5–9.8 cm

Charada Vlei Rat
Otomys fortior

FR Otomys de Charada
DE Kaffa-Lamellenzahnratte
ES Rata de laguna de Charada

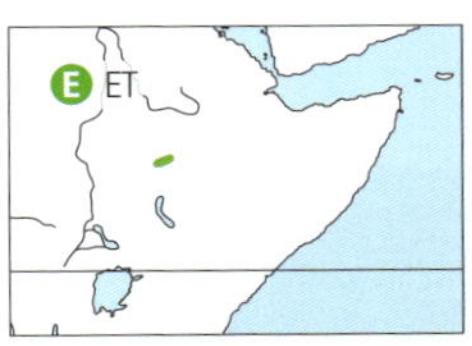

Known from few localities in SW (Beletta, Sheko & Charada forests)

Heller's Vlei Rat
Otomys helleri

FR Otomys de Heller
DE Heller-Lamellenzahnratte
ES Rata de laguna de Heller

LC
Monotypic
HB 12.3–18.5 cm
T 6.8–9.5 cm

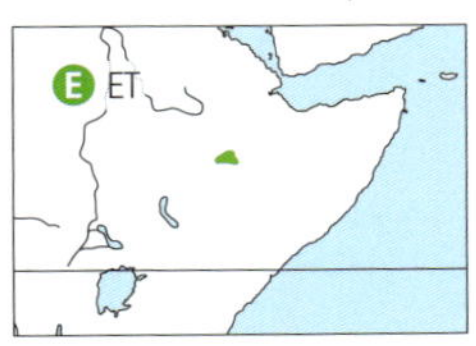

Bale Mts & Arussi Plateau (E Rift); possibly S to Melka Chireti (Somali Region)

LC
Monotypic
HB 15.5–17.8 cm
T 7.7–9.2 cm

Ethiopian Vlei Rat
Otomys typus

FR Otomys d'Éthiopie
DE Äthiopien-Lamellenzahnratte
ES Rata de laguna de Etiopía

1800–4000 m

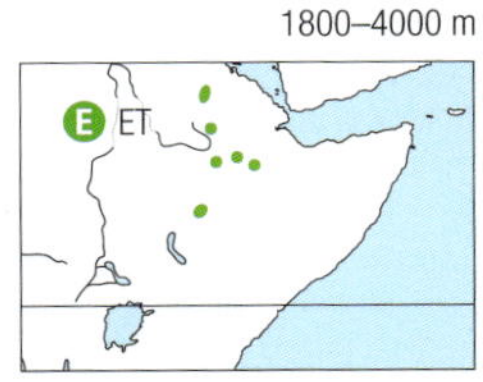

Widespread in W highlands, at least 1 locality (Hirna) E of Rift

Burton's Vlei Rat
Otomys burtoni

FR Otomys de Burton
DE Burton-Lamellenzahnratte
ES Rata de laguna de Burton

EN
Monotypic
HB 14.5–16.7 cm
T 7.5–8.4 cm
W 71–105 g

2000–4000 m

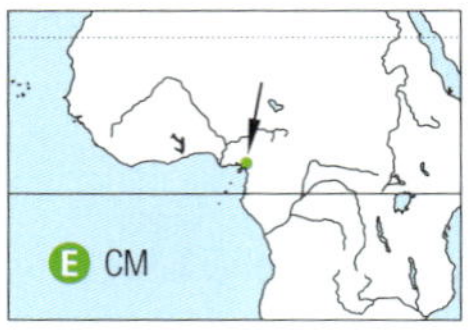

Mt Cameroon

Western Vlei Rat
Otomys occidentalis

FR Otomys occidental
DE Westliche Lamellenzahnratte
ES Rata de laguna occidental

VU
Monotypic
HB 13.1–16 cm
T 6.8–8.8 cm
W 69–88 g

1900–3000 m

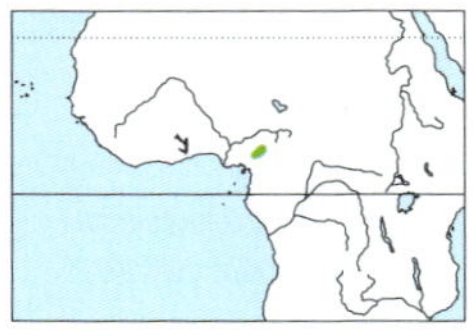

Gotel Mts & Mt Oku (NG, CM)

Eastern Arc Vlei Rat
Otomys sungae

FR Otomys des Usambara
DE Usambara-Lamellenzahnratte
ES Rata de laguna del Arco Oriental

NE
Monotypic
HB 15–18 cm
T 8–10 cm

1500–2100 m

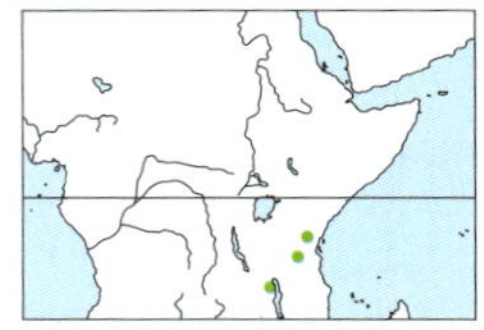

Usambara & Uluguru in Eastern Arc Mts (TZ), Nyika Plateau (MW, ZM)

Mount Elgon Vlei Rat
Otomys jacksoni

FR Otomys de Jackson
DE Mount-Elgon-Lamellenzahnratte
ES Rata de laguna de Elgon

NT
Monotypic
HB 12–17.1 cm
T 5.7–8.2 cm
W 70–121 g

3300–4200 m

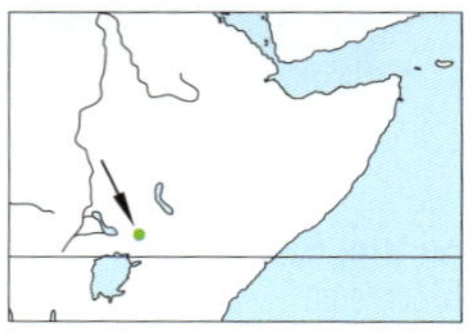

Mt Elgon (E UG, W KE)

Afroalpine Vlei Rat
Otomys orestes

FR Otomys afroalpin
DE Mount-Kenya-Lamellenzahnratte
ES Rata de laguna afroalpina

NE
Monotypic
HB 13.5–17.5 cm
T 6.1–9.3 cm

2700–4200 m

Mt Kenya & Aberdare Range

Thomas's Vlei Rat
Otomys thomasi

FR Otomys de Thomas
DE Thomas-Lamellenzahnratte
ES Rata de laguna de Thomas

VU
Monotypic
HB 16.2–18.2 cm
T 8.4–9.3 cm

2450–2700 m

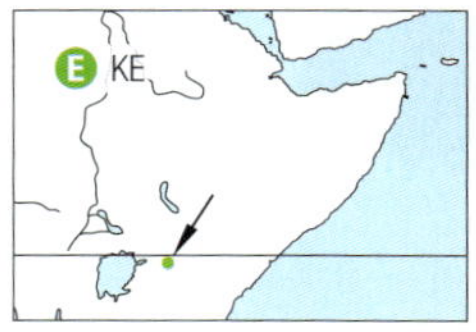

Mau Escarpment & Uasin Gishu Plateau (W rim of Rift Valley)

Udzungwa Vlei Rat
Otomys uzungwensis

FR Otomys des Udzungwa
DE Udzungwa-Lamellenzahnratte
ES Rata de laguna de Udzungwa

NT
Monotypic

1800–2750 m

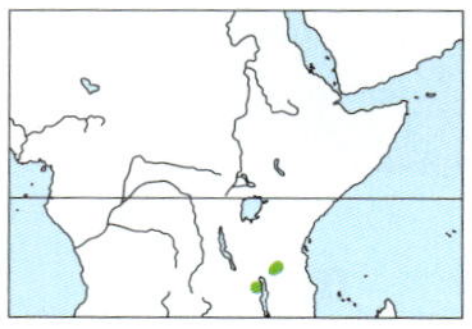

Udzungwa & Poroto Mts (S TZ), Nyika Plateau (MW, ZM)

East African Vlei Rat
Otomys tropicalis

FR Otomys tropical
DE Ostafrika-Lamellenzahnratte
ES Rata de laguna de África oriental

LC
Monotypic
HB 12.5–20 cm
T 5.2–10.2 cm
W 81–133 g

840–2250 m

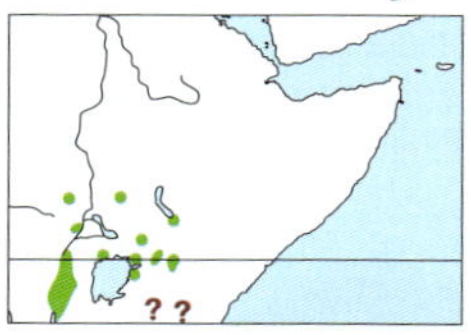

Albertine Rift (CD, UG, RW, BI), Imatong Mts (SS), Mt Elgon & Kenyan Rift (UG, KE); perhaps NE TZ

Dollman's Vlei Rat
Otomys dollmani

FR Otomys de Dollman
DE Dollman-Lamellenzahnratte
ES Rata de laguna de Dollman

NE
Monotypic
HB 13.8–15 cm
T 8.5–10.5 cm

Mt Gargues (Mathews Range)

Rwenzori Vlei Rat
Otomys dartmouthi

FR Otomys de Dartmouth
DE Ruwenzori-Lamellenzahnratte
ES Rata de laguna de Rwenzori

DD
Monotypic
HB c. 15 cm
T c. 9.3 cm

3300–4400 m

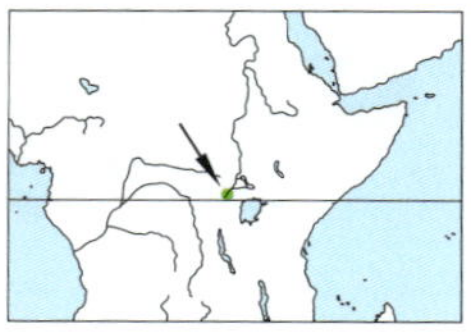

Rwenzori Mts (SW UG and neighboring CD)

Mount Kilimanjaro Vlei Rat
Otomys zinki

FR Otomys de Zink
DE Zink-Lamellenzahnratte
ES Rata de laguna de Zink

VU
Monotypic
HB 13.5–18.6 cm
T 6–10.8 cm

2350–4000 m

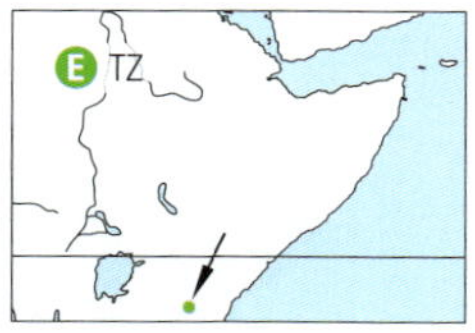

Mt Kilimanjaro

Simien Vlei Rat
Otomys simiensis

FR Otomys des Simien
DE Simien-Lamellenzahnratte
ES Rata de laguna de Simien

LC
Monotypic
HB 13.5–17.2 cm
T 7.7–8.8 cm

Simien Mts

Cheesman's Vlei Rat
Otomys cheesmani

FR Otomys de Cheesman
DE Cheesman-Lamellenzahnratte
ES Rata de laguna de Cheesman

CR(PE)
Monotypic
HB 16.5–21 cm
T 7.7–10.6 cm

2100–2500 m

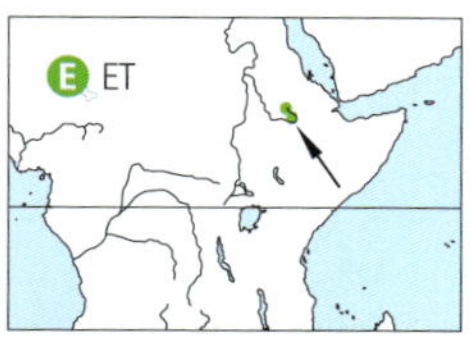

Known from 2 localities S of L Tana

Yalden's Vlei Rat
Otomys yaldeni

FR Otomys de Yalden
DE Yalden-Lamellenzahnratte
ES Rata de laguna de Yalden

VU
Monotypic
HB 12.5–15.5 cm
T 7–7.3 cm

2650–3800 m

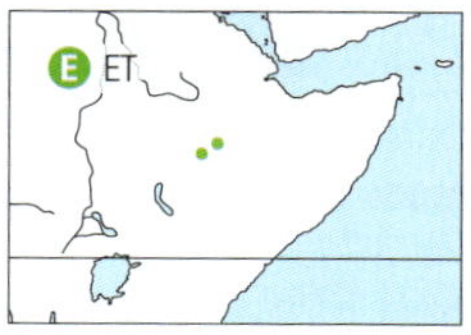

Bale Mts

Genus *GOLUNDA*

(1 species)

Indian Bush Rat

Golunda ellioti

FR Golunda indien
DE Kaffeeratte
ES Rata peluda de India

LC

2 sspp
HB 9.7–17 cm
T 7.4–13.1 cm
W 63–112 g

up to 2000 m

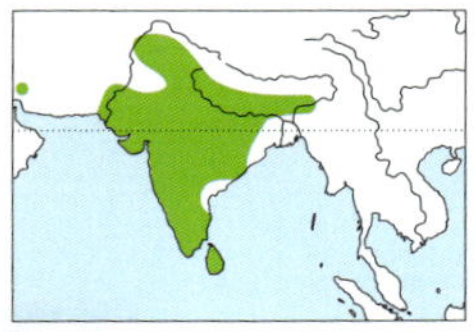

SE IR (Kerman Prov), PK, IN, S NP, LK

Genus *OENOMYS*

(2 species)

West African Rufous-nosed Rat

Oenomys ornatus

FR Oenomys du Ghana
DE Westafrika-Rotnasenratte
ES Rata de hocico rojo de África occidental

DD

Monotypic
HB 13–15.7 cm
T 17.5–20 cm
W 80–103 g

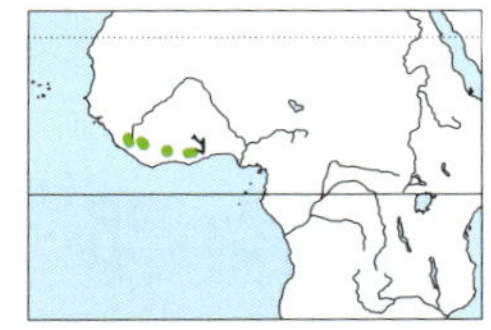

Upper Guinea Forest of W Africa, W of Dahomey Gap (from SL E to GH)

Common Rufous-nosed Rat

Oenomys hypoxanthus

FR Oenomys à museau roux
DE Gewöhnliche Rotnasenratte
ES Rata de hocico rojo común

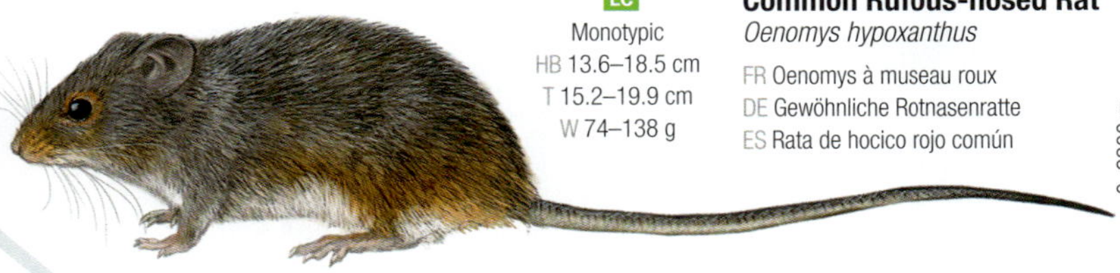

LC

Monotypic
HB 13.6–18.5 cm
T 15.2–19.9 cm
W 74–138 g

0–3300 m

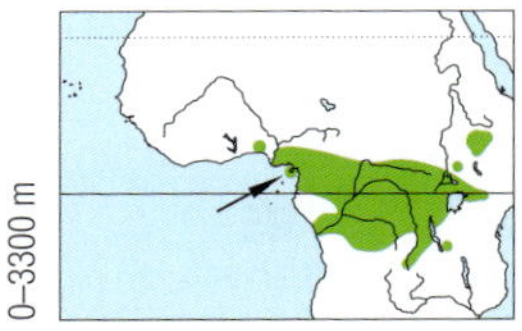

Tropical Africa from Dahomey Gap E to ET, S to N AO, S CD & SW TZ

Genus *THALLOMYS*

(4 species)

Loring's Thallomys

Thallomys loringi

FR Thallomys de Loring
DE Loring-Akazienratte
ES Rata de acacias de Loring

LC

Monotypic
HB 13–16 cm
T 12–18.3 cm
W c. 80 g

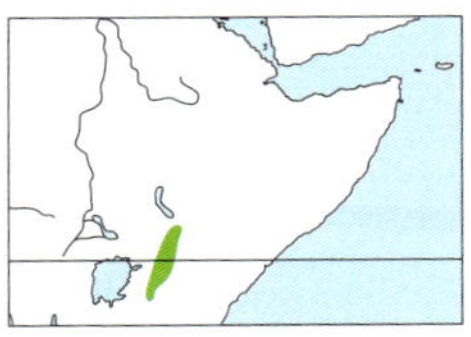

Narrowly distributed in W KE & extreme N TZ

Black-tailed Thallomys

Thallomys nigricauda

FR Thallomys à queue noire
DE Schwarzschwanz-Akazienratte
ES Rata de acacias de cola negra

LC

Monotypic
HB 11.2–16.5 cm
T 12.3–19.5 cm
W 38–85 g

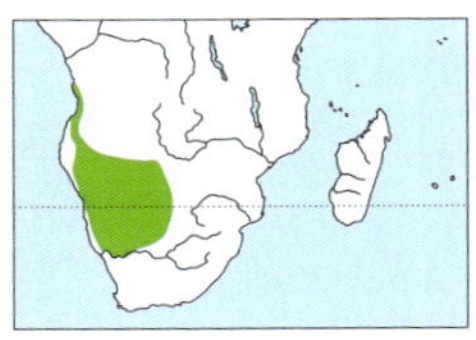

Coastal W AO, NA, W BW & NW ZA

Sundevall's Thallomys

Thallomys paedulcus

FR Thallomys de Sundevall
DE Sundevall-Akazienratte
ES Rata de acacias de Sundevall

LC

Monotypic
HB c. 13.1 cm
T 12–18.4 cm
W 32–116 g

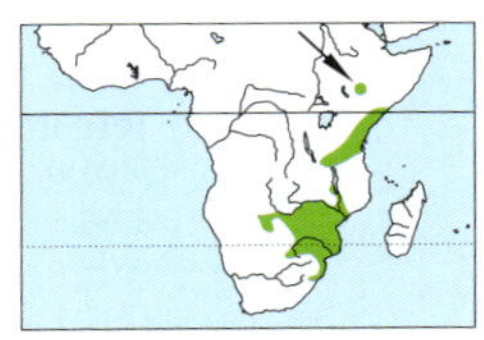

From S ET & S SO through E Africa to NE NA, BW & NE ZA

Shortridge's Thallomys

Thallomys shortridgei

FR Thallomys de Shortridge
DE Shortridge-Akazienratte
ES Rata de acacias de Shortridge

DD

Monotypic
HB 14–15 cm
T 18.3–21 cm

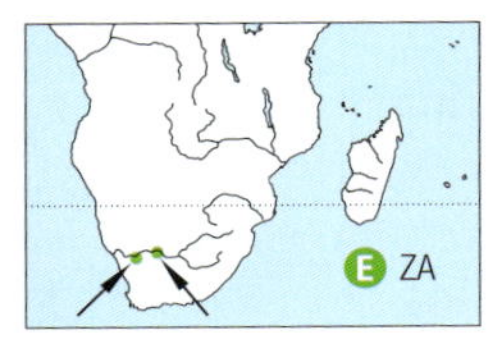

Known from 2 localities (Goodhouse, Upington) on S bank of R Orange

Genus *TYPOMYS*

(2 species)

Liberian Forest Striped Mouse

Typomys planifrons

FR Hybomys du Liberia
DE Liberia-Streifenmaus
ES Ratón listado de bosque de Liberia

100–1000 m

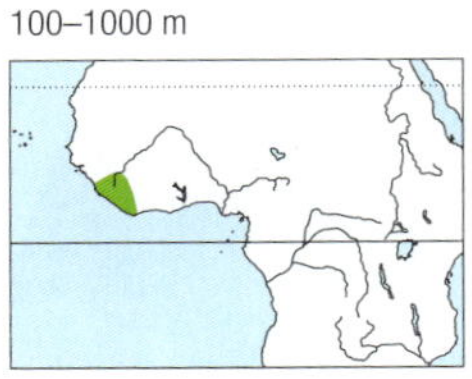

Upper Guinea Forest of W Africa (GN, SL, LR, SW CI, W of R Sassandra)

LC

Monotypic
HB 12–13 cm
T 9.1–10.4 cm
W 46–60 g

West African Striped Mouse

Typomys trivirgatus

FR Hybomys à trois bandes
DE Dreistreifenmaus
ES Ratón listado de África occidental

LC

Monotypic
HB 10.5–13 cm
T 8.5–11 cm
W 29–63 g

0–1200 m

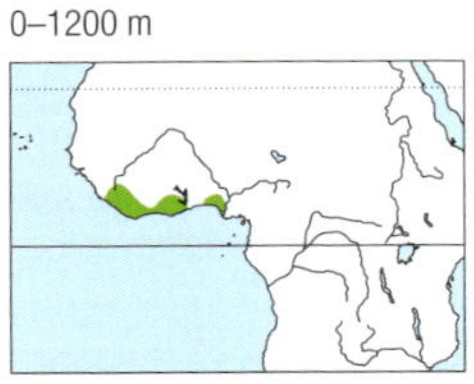

Widely in W Africa, from E SL E to SW NG (W of R Niger), absent from Dahomey Gap (SE GH, TG, BJ)

Genus *HYBOMYS*

(4 species)

Peters's Striped Mouse

Hybomys univittatus

FR Hybomys de Peters
DE Peters-Streifenmaus
ES Ratón listado de Peters

LC

Monotypic
HB 11.7–14 cm
T 10.5–12.1 cm
W 46–70 g

Congo Basin & Albertine Rift from R Niger E to L Victoria; isolated in NW ZM & neighboring CD

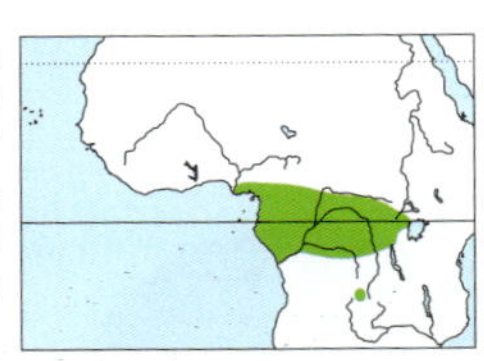

up to 1200 m

Moon Mountains Striped Mouse

Hybomys lunaris

FR Hybomys du Rwenzori
DE Ruwenzori-Streifenmaus
ES Ratón listado de Montes Luna

VU

Monotypic
HB c. 10.8 cm
T c. 11.5 cm

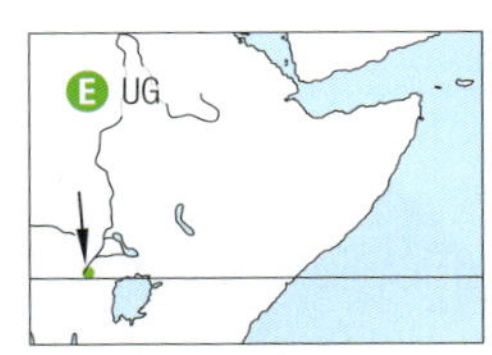

Rwenzori Mts; possibly Kanyaware (Kibale Forest) 60 km of known range

Cameroon Highland Striped Mouse

Hybomys rufocanus

FR Hybomys du Cameroun
DE Kamerun-Streifenmaus
ES Ratón listado de Bakossi

NE

Monotypic
HB 10.5–13.7 cm
T 10.8–12.1 cm W 55–75 g

up to 2000 m

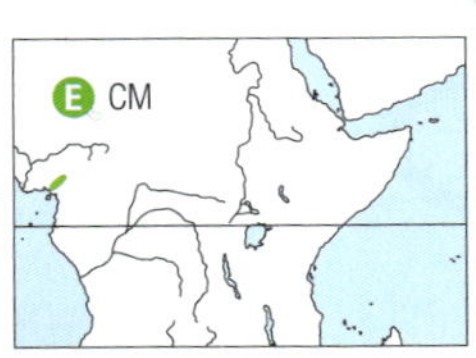

Highlands of W

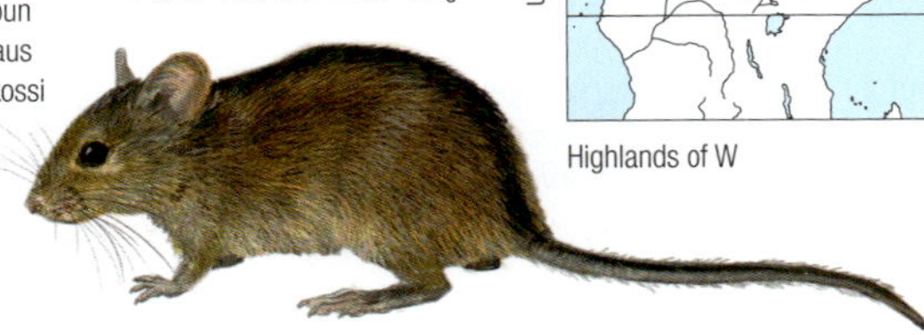

Father Basilio's Striped Mouse

Hybomys basilii

FR Hybomys de Bioko
DE Bioko-Streifenmaus
ES Ratón listado de Basilio

450–2000 m

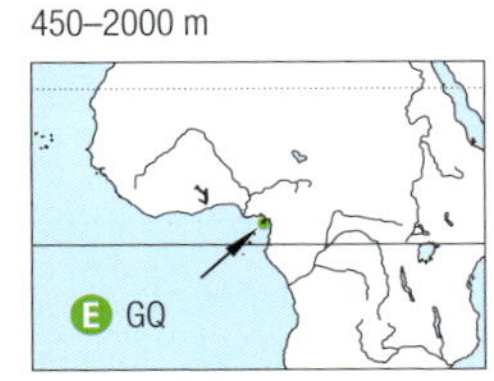

Bioko I

EN

Monotypic
HB 11.8–16 cm
T 9–13.2 cm
W 70–95 g

Genus *STOCHOMYS*

(1 species)

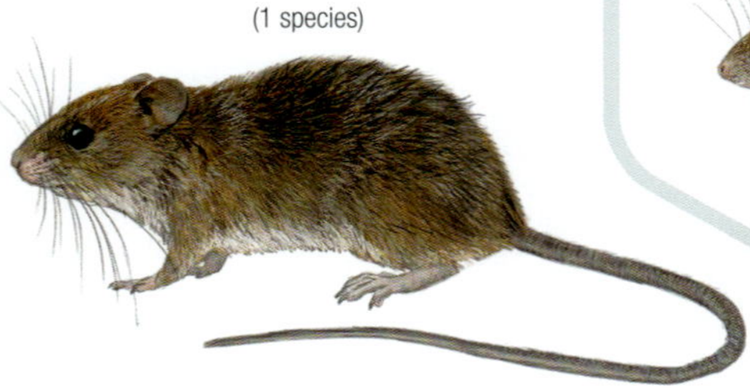

Target Rat

Stochomys longicaudatus

FR Rat fléché
DE Zielratte
ES Rata objetivo

LC
Monotypic
HB 10–16.5 cm
T 18.5–23 cm
W 70–102 g

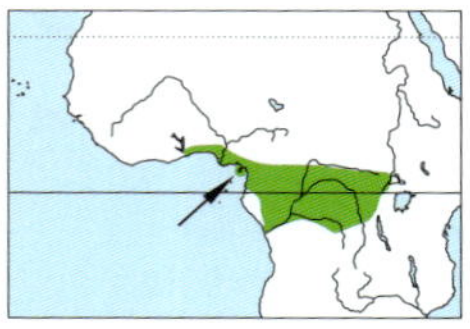

From R Volta (GH) E through Congo Basin to Albertine Rift (UG), S to NW AO (Cabinda) & SC CD; Bioko I

Genus *DEPHOMYS*

(2 species)

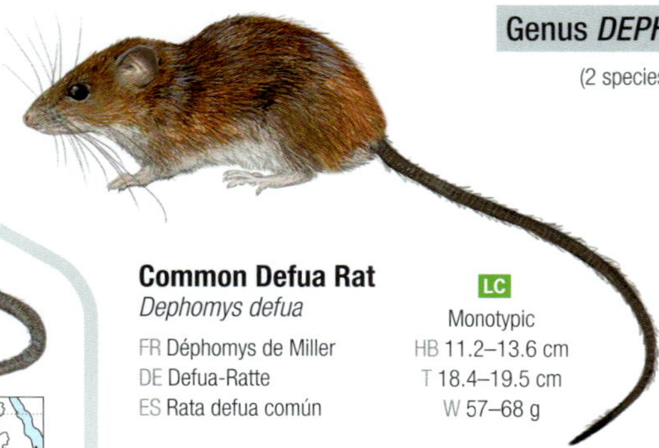

Common Defua Rat

Dephomys defua

FR Déphomys de Miller
DE Defua-Ratte
ES Rata defua común

LC
Monotypic
HB 11.2–13.6 cm
T 18.4–19.5 cm
W 57–68 g

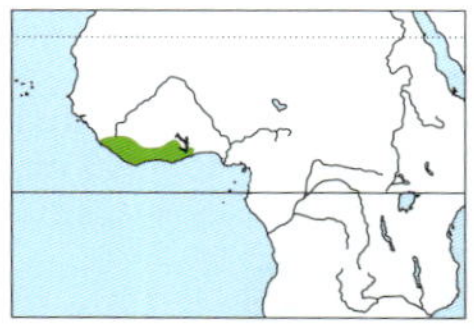

Rainforests in W Africa, from SL to TG

Ivory Coast Defua Rat

Dephomys eburneae

FR Déphomys de Côte d'Ivoire
DE Liberianische Defua-Ratte
ES Rata defua de Costa de Marfil

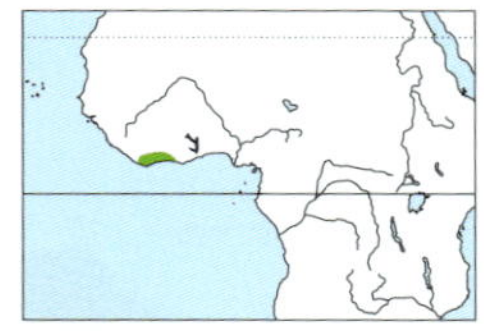

E LR, S CI

NE
Monotypic
HB 12–14.2 cm
T 18.2–20 cm
W 47–59 g

West African Rock Rat

Aethomys stannarius

FR Aethomys occidental
DE Westafrika-Buschlandratte
ES Rata de roca de África occidental

DD
Monotypic
HB 12.6–16.9 cm
T 12.4–20 cm
W 60–120 g

N NG, extending marginally to N CM

Genus *MICAELAMYS*

(2 species)

Namaqua Rock Rat

Micaelamys namaquensis

FR Rat namaquois
DE Namaqua-Buschlandratte
ES Rata de roca de Namaqua

LC
Monotypic
HB 8–14.7 cm
T 10.7–19.7 cm
W 33–89 g

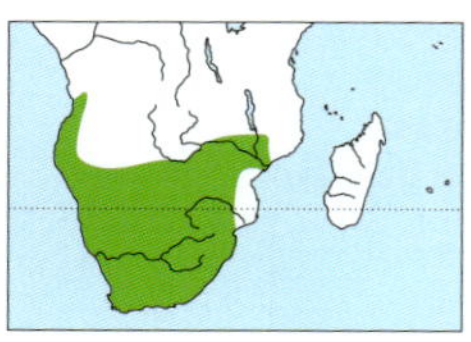

S Africa in W AO, NA, BW, extreme S ZM, ZW, S MW, C MZ, SZ, LS & ZA

LC
Monotypic
HB 9.3–12.5 cm
T 9.6–13.8 cm
W 19–65 g

Grant's Rock Rat

Micaelamys granti

FR Rat de Grant
DE Grant-Buschlandratte
ES Rata de roca de Grant

Parts of Northern Cape, Western Cape & Eastern Cape

Genus *AETHOMYS*

(9 species)

Hinde's Rock Rat

Aethomys hindei

FR Aethomys de Hinde
DE Hinde-Buschlandratte
ES Rata de roca de Hinde

LC
Monotypic
HB 12–17.5 cm
T 12–18.8 cm
W 87–158 g

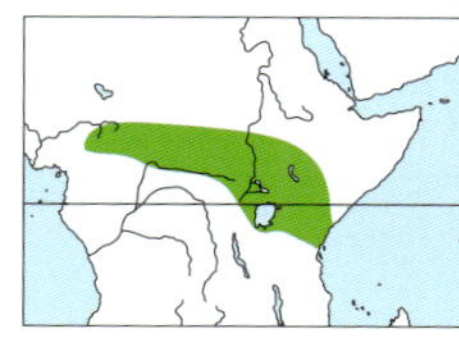

below 2000 m

C CM E to S ET, KE & N TZ

Red Rock Rat

Aethomys chrysophilus

FR Aethomys roux
DE Rote Buschlandratte
ES Rata de roca roja

LC
Monotypic
HB 12–16.9 cm
T 13.5–19 cm W 63–112 g

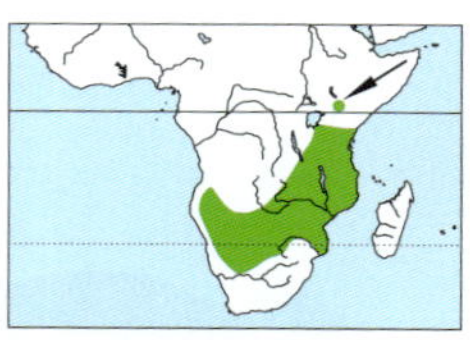

E & S Africa from EC & SW KE to SW AO, NA, N ZA & MZ

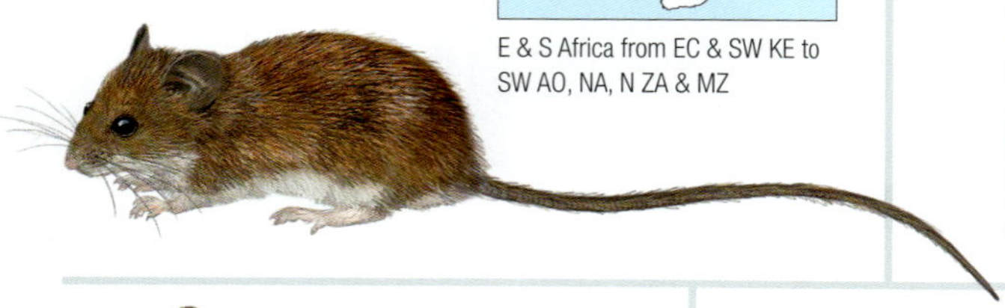

Tete Rock Rat

Aethomys ineptus

FR Aethomys de Tete
DE Mosambik-Buschlandratte
ES Rata de roca de Tete

LC
Monotypic
HB 14–15.5 cm
T 14.4–17 cm
W 65–107 g

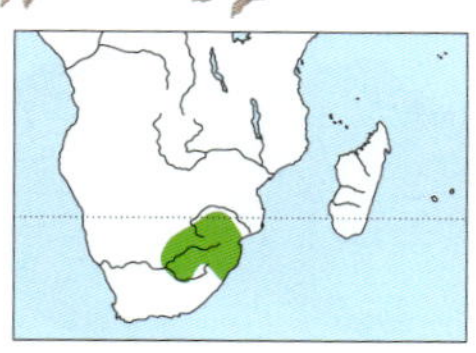

Savannas in ZA, extreme SE BW, SZ & nearby S MZ (apparently S of R Limpopo)

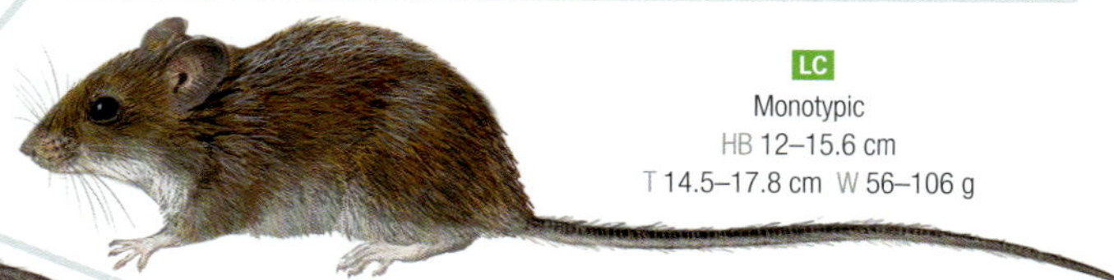

LC
Monotypic
HB 12–15.6 cm
T 14.5–17.8 cm W 56–106 g

Nyika Rock Rat

Aethomys nyikae

FR Aethomys du Nyika
DE Nyika-Buschlandratte
ES Rata de roca de Nyika

800–2300 m

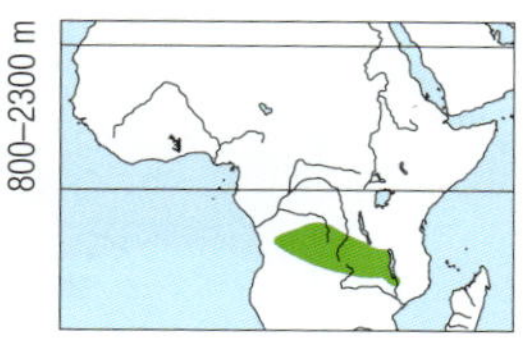

NE AO, S CD, N ZM, MW

Kaiser's Rock Rat

Aethomys kaiseri

FR Aethomys de Kaiser
DE Kaiser-Buschlandratte
ES Rata de roca de Kaiser

LC
Monotypic
HB 13.9–19.6 cm
T 12.4–16.7 cm
W 78–144 g

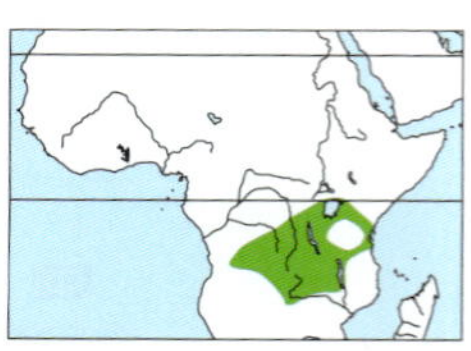

From N shore of L Victoria (S UG & S KE), S to NC AO, N ZM, MW & NW MZ

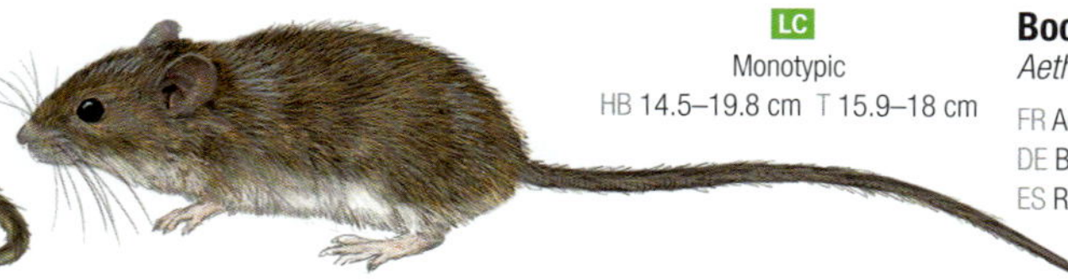

LC
Monotypic
HB 14.5–19.8 cm T 15.9–18 cm

Bocage's Rock Rat

Aethomys bocagei

FR Aethomys de Bocage
DE Bocage-Buschlandratte
ES Rata de roca de Bocage

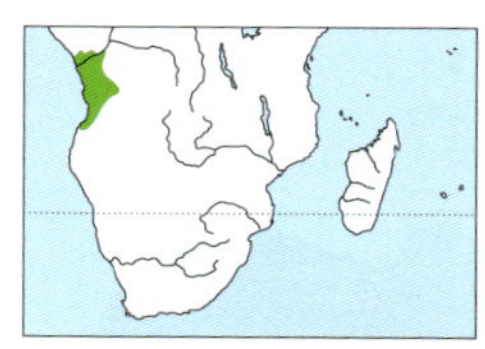

W AO & neighboring extreme SW CD

DD
Monotypic
HB 15.5–20 cm
T 16.6–19.4 cm

Eastern Highlands of E ZW & W MZ

Selinda Rock Rat

Aethomys silindensis

FR Aethomys du Selinda
DE Selinda-Buschlandratte
ES Rata de roca de Selinda

LC
Monotypic
HB 14–16.2 cm
T 11.2–13.5 cm
W 60–120 g

Thomas's Rock Rat

Aethomys thomasi

FR Aethomys de Thomas
DE Thomas-Buschlandratte
ES Rata de roca de Thomas

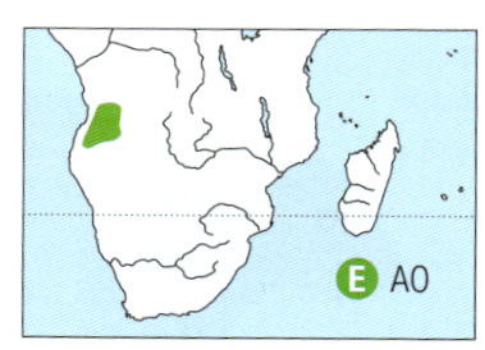

Central Plateau

Genus *THAMNOMYS*

(4 species)

Kemp's Thicket Rat
Thamnomys kempi

FR Thamnomys de Kemp
DE Kemp-Dickichtratte
ES Rata de acacias de Kemp

VU

Monotypic
HB 13–17.3 cm
T 17–23 cm
W 62–108 g

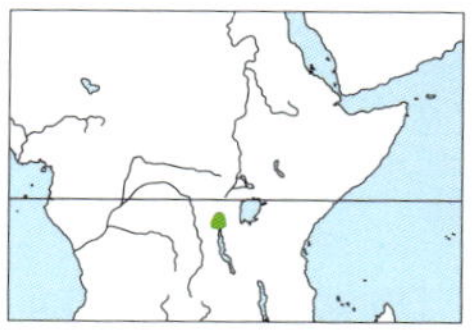

L Kivu region of Albertine Rift (E CD, RW, BI)

Hatt's Thicket Rat
Thamnomys major

FR Thamnomys de Hatt
DE Große Dickichtratte
ES Rata de acacias de Hatt

NE

Monotypic
HB c. 15.6 cm
T c. 20 cm

c. 3600 m

Known only from N slope Mt Karisimbi (Kivu Volcanoes)

Schouteden's Thicket Rat
Thamnomys schoutedeni

FR Thamnomys de Schouteden
DE Schouteden-Dickichtratte
ES Rata de acacias de Schouteden

DD

Monotypic
HB 11.4–12.2 cm
T 17.1–20 cm
W 40–50 g

800–1450 m

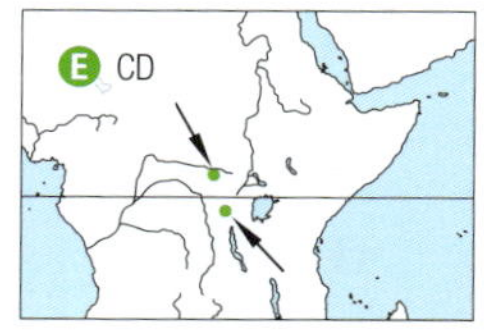

Known only from Medje (Ituri Forest) & Irangi

Thomas's Thicket Rat
Thamnomys venustus

FR Thamnomys de Thomas
DE Thomas-Dickichtratte
ES Rata de acacias de Thomas

1800–3000 m

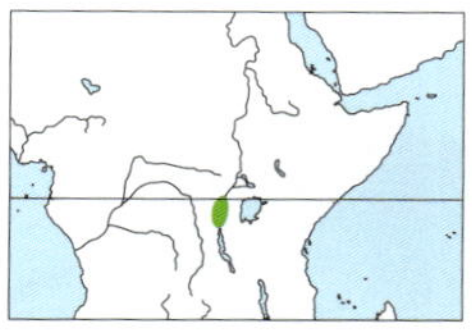

Few localities near L Albert & L Kivu (Albertine Rift) in E CD & UG

LC

Monotypic
HB 14–16 cm
T 18.3–21 cm
W 59–70 g

Genus *GRAMMOMYS*

(14 species)

Western Rainforest Thicket Rat
Grammomys poensis

NE

Monotypic
HB 11.2–14.7 cm
T 16.3–20 cm
W 55–92 g

FR Grammomys de Fernando Po
DE Westliche Akazienmaus
ES Rata de matorral de bosque lluvioso occidental

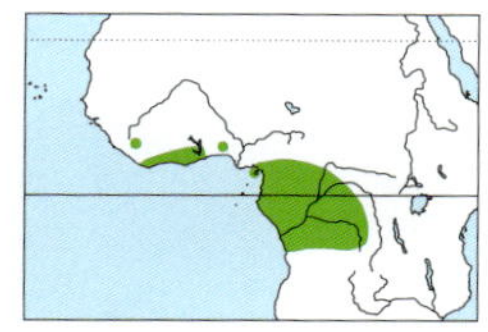

Mt Nimba (SW GN), from SE LR to Togo, SW NG, from CM through Congo Basin S to N AO, Bioko I

Eastern Rainforest Thicket Rat
Grammomys kuru

FR Grammomys du Congo
DE Zentralafrika-Akazienmaus
ES Rata de matorral de bosque lluvioso oriental

LC

Monotypic
HB 10.9–14.5 cm
T 14–17.2 cm
W 31–60 g

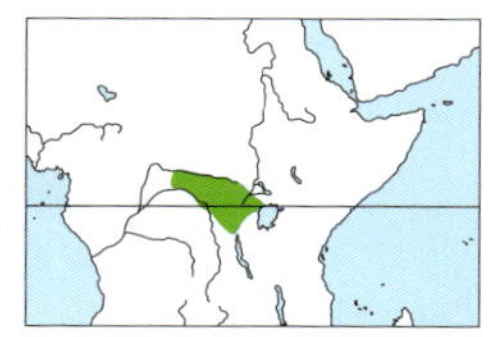

NE part of Congo Basin (NE CD) to W UG, RW & BI

Selous's Thicket Rat
Grammomys selousi

FR Grammomys de Selous
DE Selous-Akazienmaus
ES Rata de matorral de Selous

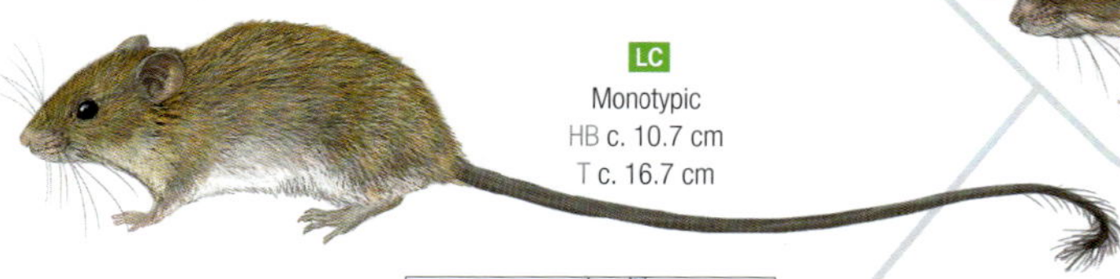

LC

Monotypic
HB c. 10.7 cm
T c. 16.7 cm

Known only from Kichi Forest

Mozambique Thicket Rat
Grammomys cometes

FR Grammomys du Mozambique
DE Mosambik-Akazienmaus
ES Rata de matorral de Mozambique

LC

Monotypic
HB 11.2–12.4 cm
T 14.2–19.4 cm
W c. 52 g

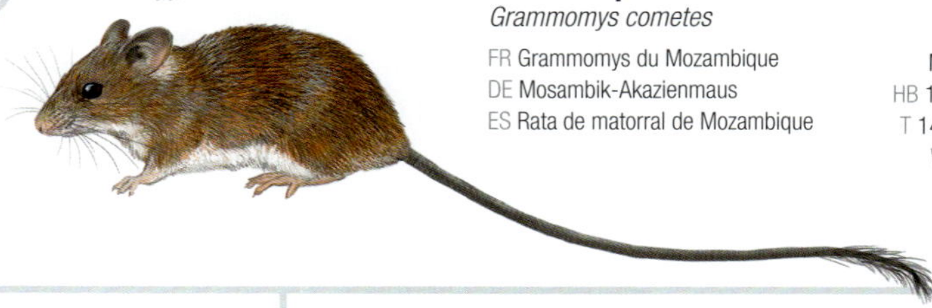

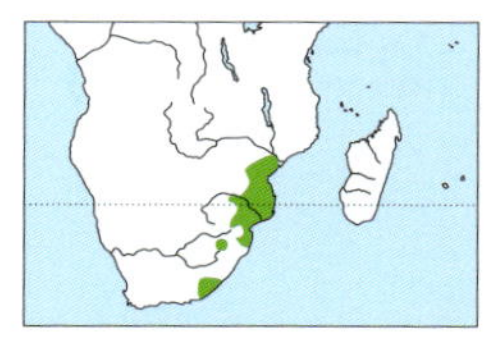

E parts of S Africa from MZ just S of R Zambezi S to Eastern Cape (ZA)

Woodland Thicket Rat
Grammomys dolichurus

FR Grammomys du Cap
DE Wald-Akazienmaus
ES Rata de matorral de bosque

LC

Monotypic
HB 9–13.6 cm
T 12.9–18.5 cm
W 20–54 g

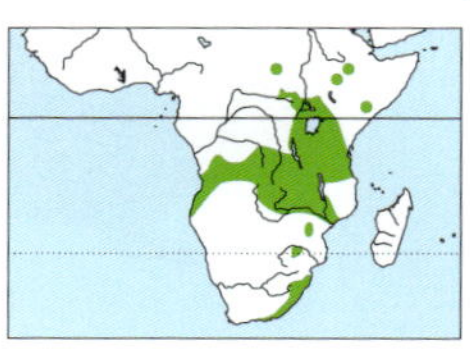

E & S Africa, from NE CD & UG, W to AO, S to coastal SE ZA; isolated in WC SS, S ET & SW SO

Albertine Rift Thicket Rat
Grammomys dryas

FR Grammomys du Rwenzori
DE Dryas-Akazienmaus
ES Rata de matorral del Valle Albertino

LC

Monotypic
HB 10–13 cm T 14.2–17.7 cm
W 30–59 g

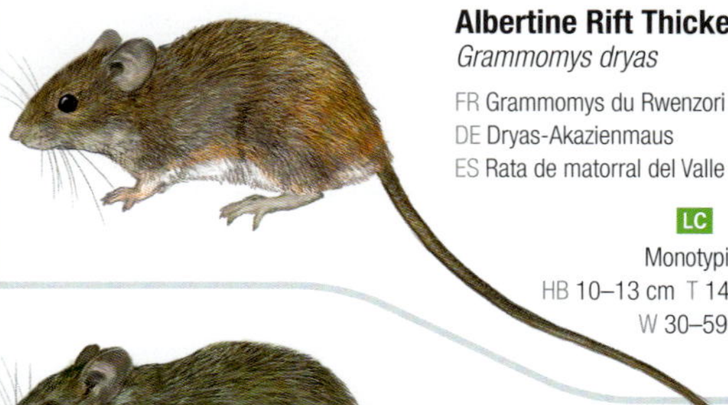

above 1000 m

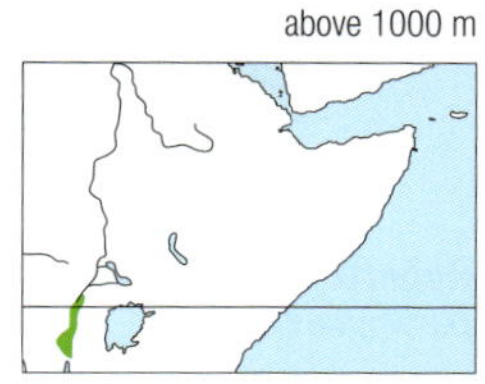

Albertine Rift Mts (E CD, UG, BI; possibly RW)

East African Thicket Rat
Grammomys ibeanus

FR Grammomys d'Osgood
DE Östliche Akazienmaus
ES Rata de matorral de África oriental

LC

Monotypic
HB 11.2–13 cm
T 12.9–22 cm
W 27–60 g

1900–3600 m

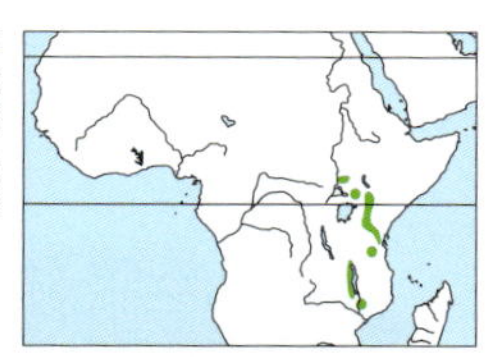

Rift Valley system & Eastern Arc Mts (S SS, E UG, KE, TZ, MW & MZ)

Bunting's Thicket Rat
Grammomys buntingi

FR Grammomys de Bunting
DE Bunting-Akazienmaus
ES Rata de matorral de Bunting

DD

Monotypic
HB 10–12.5 cm
T 16.5–18 cm

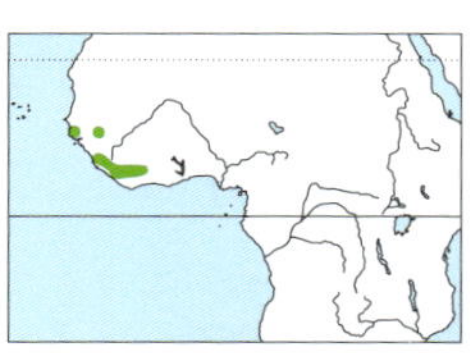

W Africa from S GN E to C CI, 2 localities in S SN

Mount Kenya Thicket Rat
Grammomys gigas

FR Grammomys géant
DE Mount-Kenya-Akazienmaus
ES Rata de matorral del Monte Kenia

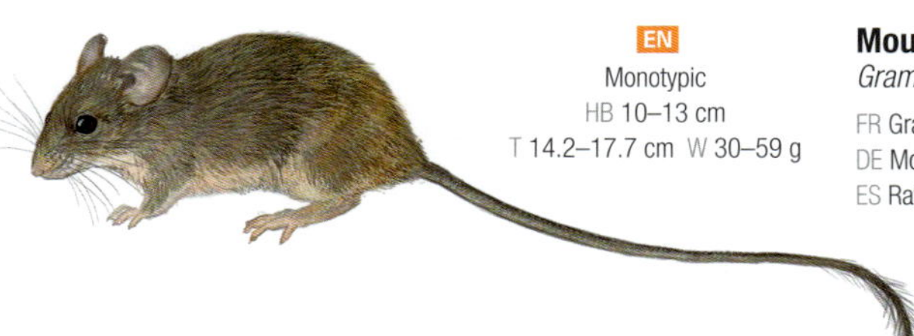

EN

Monotypic
HB 10–13 cm
T 14.2–17.7 cm W 30–59 g

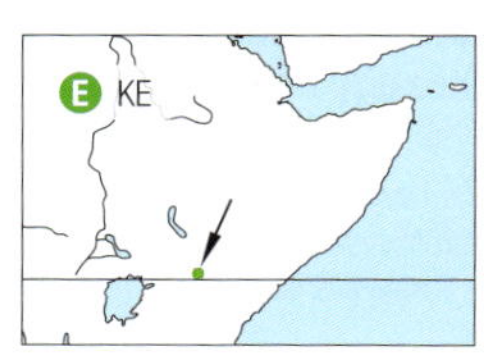

Mt Kenya

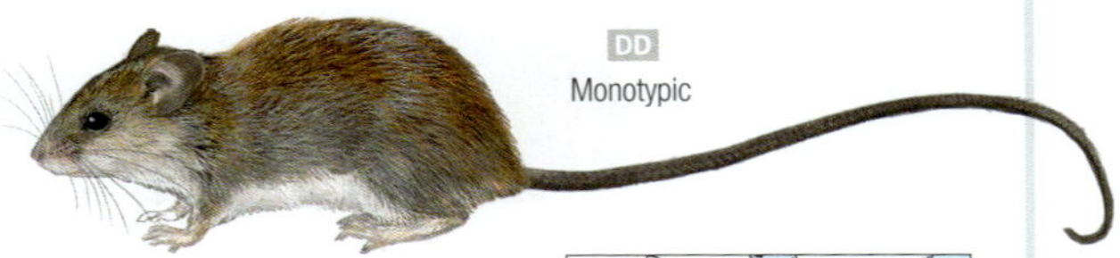

Short-snouted Thicket Rat
Grammomys brevirostris

FR Grammomys à museau court
DE Kurzschnauzen-Akazienmaus
ES Rata de matorral de hocico corto

DD
Monotypic

Known only from Lemesikio (Loita Plains on KE–TZ border)

Macmillan's Thicket Rat
Grammomys macmillani

FR Grammomys de Macmillan
DE Macmillan-Akazienmaus
ES Rata de matorral de Macmillan

LC
Monotypic
HB 9.8–11 cm T 14.9–18.6 cm
W 31–60 g

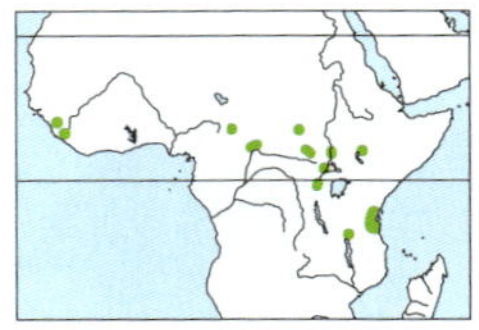

Savannas from SL E to SS & extreme S ET, S to KE & TZ; perhaps farther S

Ethiopian Thicket Rat
Grammomys minnae

FR Grammomys de Minna
DE Äthiopien-Akazienmaus
ES Rata de matorral de Etiopía

VU
Monotypic
HB c. 11 cm T c. 17.3 cm
W c. 36 g

c. 1800 m

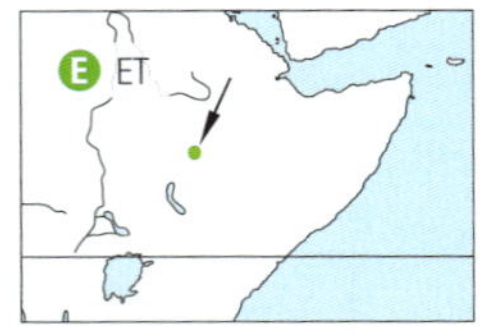

Known from 2 montane localities in Rift Valley

Arid Woodland Thicket Rat
Grammomys aridulus

FR Grammomys du Soudan
DE Darfur-Akazienmaus
ES Rata de matorral de bosque árido

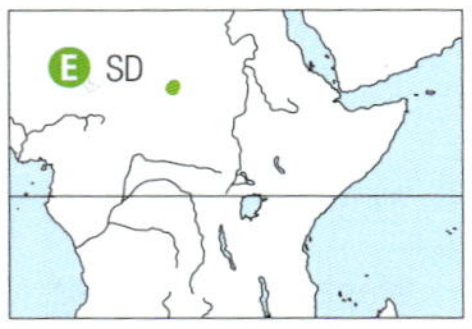

Jebel Marra

above 1000 m

DD
Monotypic
HB 11–11.6 cm
T 18.1–18.8 cm
W 41–46 g

Gray-headed Thicket Rat
Grammomys caniceps

FR Grammomys à tête grise
DE Graukopf-Akazienmaus
ES Rata de matorral de cabeza gris

DD
Monotypic
HB 8.6–10.5 cm T 14.7–15.5 cm
W 18–40 g

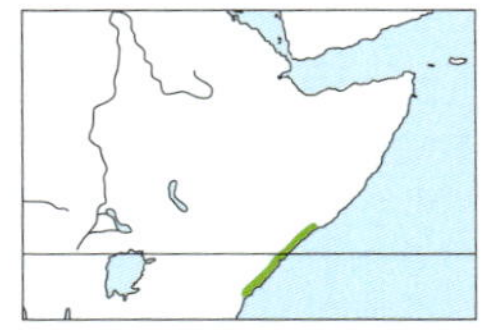

Narrow coastal strip from S SO S to Malindi (E KE)

Genus *DASYMYS*

(14 species)

NE
Monotypic
HB c. 15.3 cm
T c. 13.2 cm
W c. 111 g

Cape Shaggy Rat
Dasymys capensis

FR Dasymys du Cap
DE Kap-Wollhaarratte
ES Rata peluda de El Cabo

Few isolated localities in Western Cape

LC
Monotypic
HB 13.7–19.2 cm
T 11.9–16 cm
W 103–218 g

African Shaggy Rat
Dasymys incomtus

FR Dasymys d'Afrique
DE Afrikanische Wollhaarratte
ES Rata peluda de África

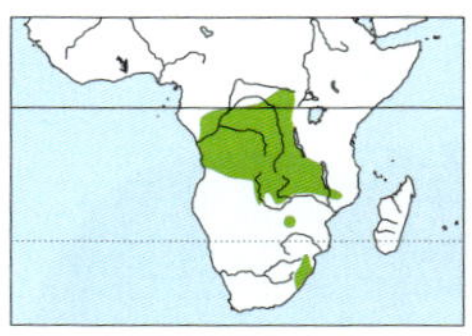

Much of C Africa (GA, CG, CD, AO, ZM, MW, MZ) and SW ZW, E ZA & SZ

DD
Monotypic
HB c. 15.2 cm
T c. 13.4 cm W c. 117 g

Fox's Shaggy Rat
Dasymys foxi

FR Dasymys de Fox
DE Fox-Wollhaarratte
ES Rata peluda de Fox

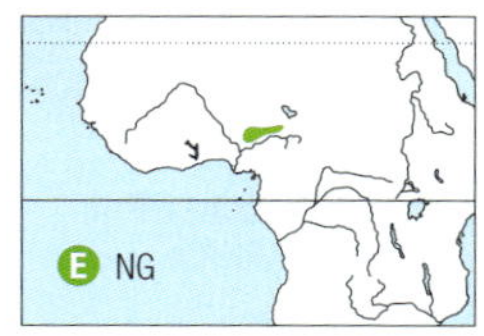

C Jos Plateau

NE
Monotypic
HB c. 14.9 cm
T c. 12.8 cm
W c. 90 g

Roberts's Shaggy Rat
Dasymys robertsii

FR Dasymys de Roberts
DE Roberts-Wollhaarratte
ES Rata peluda de Roberts

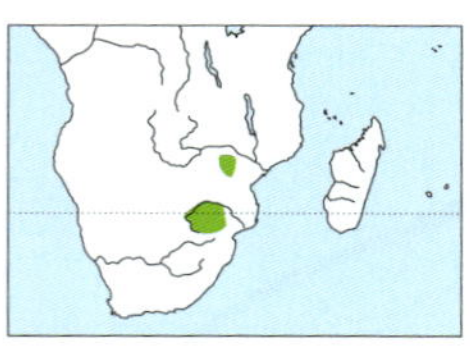

NE ZW, extreme E BW, NE ZA

NE
Monotypic
HB c. 15 cm
T c. 13.4 cm
W c. 107 g

Middle Shaggy Rat
Dasymys medius

FR Dasymys intermédiaire
DE Mittlere Wollhaarratte
ES Rata peluda de altitud media

up to 2600 m

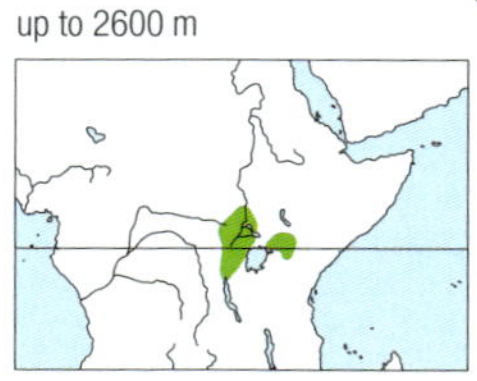

Mts of Albertine Rift & E African Rift

West African Shaggy Rat
Dasymys rufulus

FR Dasymys roux
DE Rote Wollhaarratte
ES Rata peluda de África occidental

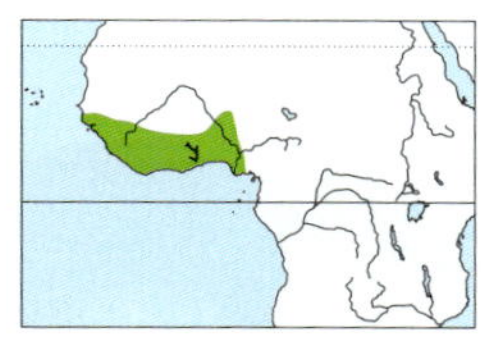

W Africa from SN E to NG

LC
Monotypic
HB 13–17 cm
T 12–16 cm
W 45–125 g

NE
Monotypic
HB 11.2–17.9 cm
T 10.3–14 cm
W 48–135 g

Glover Allen's Shaggy Rat
Dasymys alleni

FR Dasymys d'Allen
DE Allen-Wollhaarratte
ES Rata peluda de Glover Allen

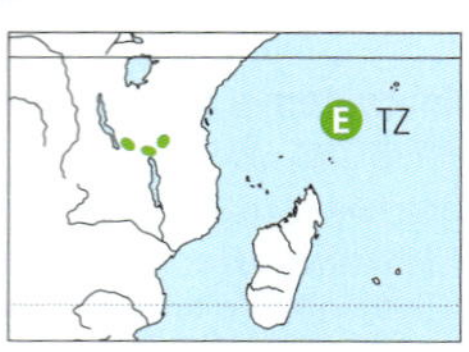

W rim of Albertine Rift & Mts of Southern Rift

NE
Monotypic
HB c. 12.2 cm
T c. 11.1 cm
W c. 77 g

Rwandan Shaggy Rat
Dasymys rwandae

FR Dasymys du Rwanda
DE Ruanda-Wollhaarratte
ES Rata peluda de Ruanda

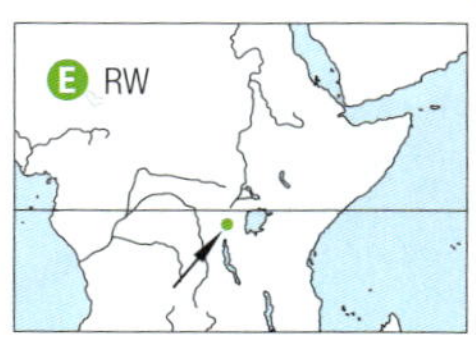

Kinigi and neighboring areas (Virunga Volcanoes)

NE
Monotypic
HB c. 14.1 cm
T c. 12.8 cm
W c. 78 g

Tanzanian Shaggy Rat
Dasymys sua

FR Dasymys de Tanzanie
DE Tansania-Wollhaarratte
ES Rata peluda de Tanzania

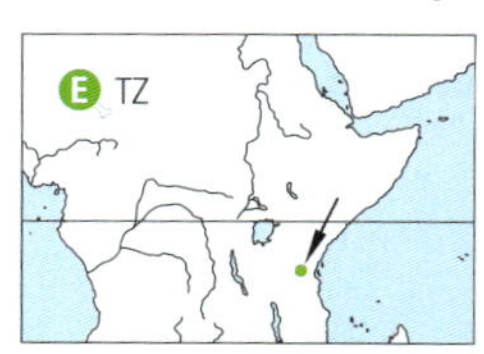

Few localities in Uluguru Mts

Grayish-brown Shaggy Rat
Dasymys griseifrons
FR Dasymys d'Éthiopie
DE Graubraune Wollhaarratte
ES Rata peluda gris parda

NE
Monotypic
HB c. 16.6 cm
T c. 13.5 cm

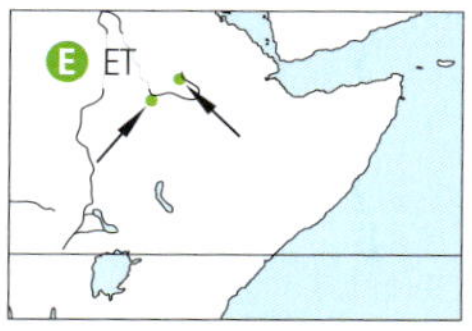

Near lakes Tana & Jigga

EN
Monotypic
HB 13.2–15.6 cm
T 9.3–11.5 cm
W 69–100 g

Rwenzori Shaggy Rat
Dasymys montanus
FR Dasymys de montagne
DE Berg-Wollhaarratte
ES Rata peluda de Rwenzori

above 2600 m

Rwenzori Mts

DD
Monotypic
HB 15.2–20 cm
T 14–18.6 cm
W 102–164 g

Angolan Shaggy Rat
Dasymys nudipes
FR Dasymys d'Angola
DE Angola-Wollhaarratte
ES Rata peluda de Angola

Central Highlands

NE
Monotypic
HB c. 16.3 cm
T c. 15.3 cm
W c. 119 g

Crawford-Cabral's Shaggy Rat
Dasymys cabrali
FR Dasymys de Cabral
DE Cabral-Wollhaarratte
ES Rata peluda de Crawford-Cabral

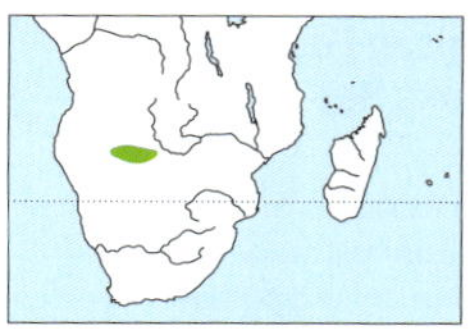
SE AO, NE NA, NW BW

Mount Cameroon Shaggy Rat
Dasymys longipilosus
FR Dasymys du Cameroun
DE Kamerun-Wollhaarratte
ES Rata peluda de Camerún

NE
Monotypic
HB c. 15.8 cm
T c. 13 cm

above 2000 m

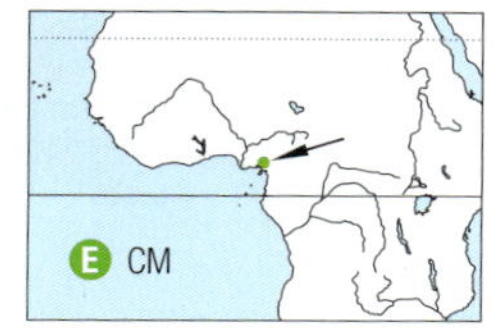

Mt Cameroon

Genus *DESMOMYS*

(2 species)

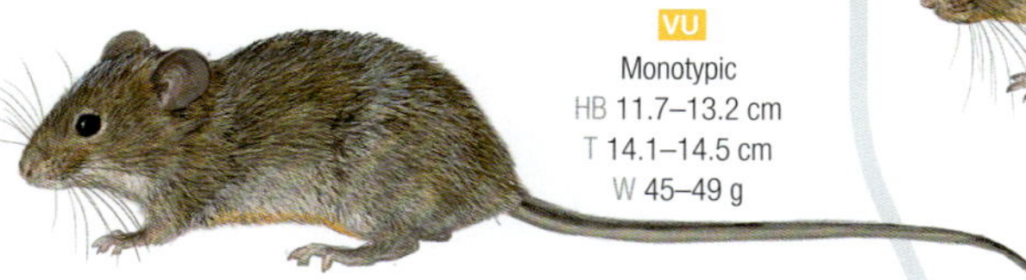

VU
Monotypic
HB 11.7–13.2 cm
T 14.1–14.5 cm
W 45–49 g

Yalden's Rat
Desmomys yaldeni
FR Rat de Yalden
DE Yalden-Strohratte
ES Rata de Yalden

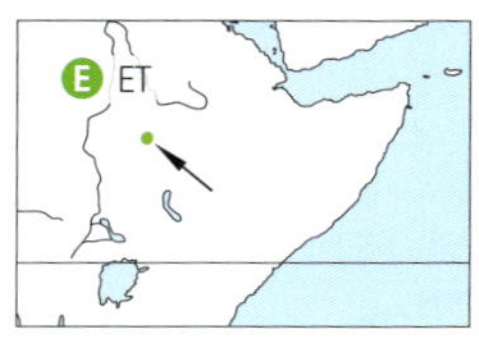

Sheko Forest

Genus *LAMOTTEMYS*

(1 species)

EN
Monotypic
HB 13–14.9 cm
T 12.3–14 cm
W 62–85 g

2100–2900 m

Mount Oku Rat
Lamottemys okuensis
FR Rat de Lamotte
DE Oku-Ratte
ES Rata de Oku

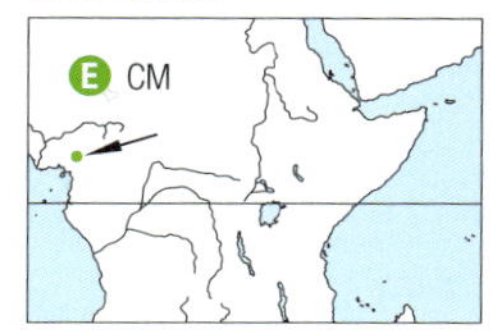

Mt Oku

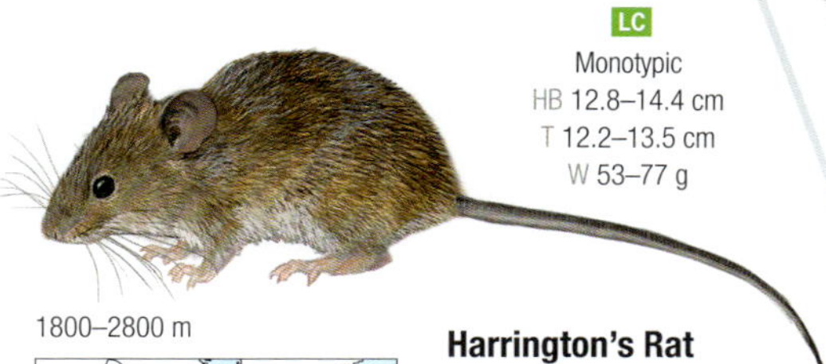

LC
Monotypic
HB 12.8–14.4 cm
T 12.2–13.5 cm
W 53–77 g

1800–2800 m

Harrington's Rat
Desmomys harringtoni
FR Rat de Harrington
DE Harrington-Strohratte
ES Rata de Harrington

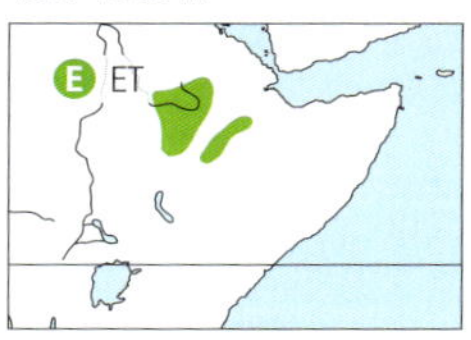

Widespread in highlands, particularly W of Rift Valley

Genus *PELOMYS*

(5 species)

LC
Monotypic
HB 13–17 cm
T 10.5–16.9 cm
W 100–170 g

East African Groove-toothed Swamp Rat
Pelomys fallax
FR Pélomys du Mozambique
DE Ostafrika-Furchenzahn-Bachratte
ES Rata de ciénaga de dientes estriados de África oriental

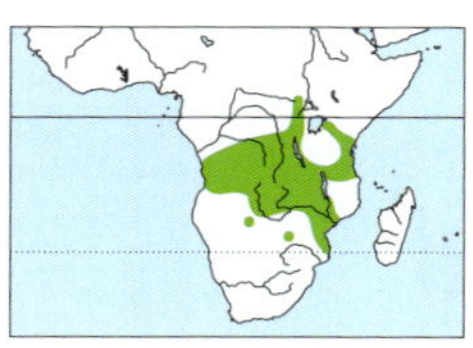
From Albertine Rift, S KE & E TZ W to AO, S to Okavango Delta & S MZ

DD
Monotypic
HB 11.6–12.7 cm
T 13.5–16.6 cm W 41–68 g

Hopkins's Groove-toothed Swamp Rat
Pelomys hopkinsi
FR Pélomys de Hopkins
DE Hopkins-Furchenzahn-Bachratte
ES Rata de ciénaga de dientes estriados de Hopkins

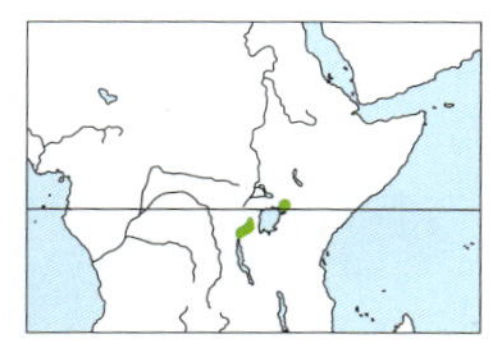
Few localities in SW UG, RW, BI, NW TZ & SW KE

Angolan Groove-toothed Swamp Rat
Pelomys campanae
FR Pélomys d'Angola
DE Angola-Furchenzahn-Bachratte
ES Rata de ciénaga de dientes estriados de Angola

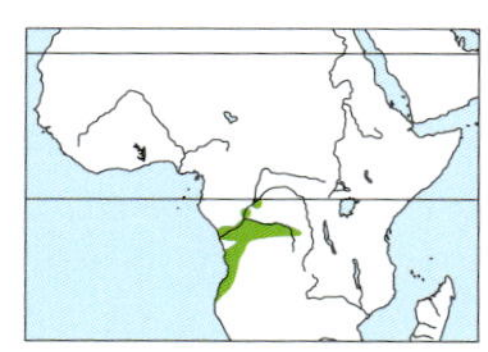
N & W AO, W & SC CD

LC
Monotypic
HB 12–17 cm T 13.8–14.6 cm

NT
Monotypic
HB 13.2–14.9 cm
T 14.8–15.6 cm
W 60–83 g

Lake Victoria Groove-toothed Swamp Rat
Pelomys isseli
FR Pélomys d'Issel
DE Viktoriasee-Furchenzahn-Bachratte
ES Rata de ciénaga de dientes estriados del Victoria

Ssese Is (Bugala, Bunyama, Kome) in L Victoria (UG), Nyabera Marsh (SW KE)

LC
Monotypic
HB 10.1–16.1 cm
T 10–13.1 cm
W 42–56 g

Least Groove-toothed Swamp Rat
Pelomys minor
FR Pélomys nain
DE Kleine Furchenzahn-Bachratte
ES Rata de ciénaga de dientes estriados mínima

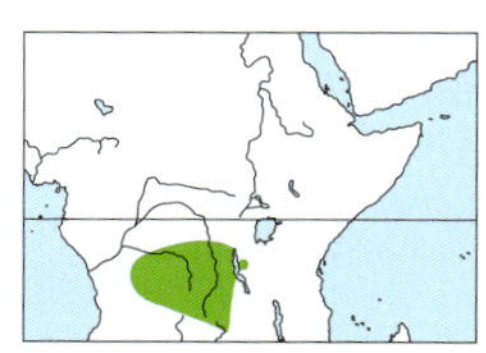
S CD and neighboring NE AO, N ZM & W TZ

Ethiopian Three-toed Grass Rat
Mylomys rex
FR Mylomys roi
DE Äthiopien-Furchenzahnratte
ES Rata de hierba de tres dedos de Etiopía

DD
Monotypic
HB c. 21 cm
T c. 17.5 cm

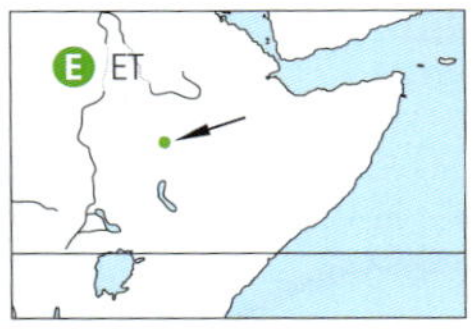

Charada Forest (Kaffa)

Genus *MYLOMYS*

(2 species)

up to 2300 m

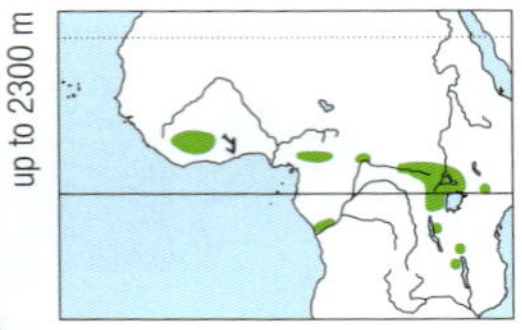

Tropical Africa from SE GN (Mt Nimba) E to C KE, S to N MW

Dybowski's Three-toed Grass Rat
Mylomys dybowskii
FR Mylomys de Dybowski
DE Dybowski-Furchenzahnratte
ES Rata de hierba de tres dedos de Dybowski

LC
Monotypic
HB 12.2–19.4 cm
T 10.4–18 cm
W 46–190 g

LC
Monotypic
HB 9.1–12.6 cm
T 7.2–9 cm
W 30–47 g

up to 2700 m

Mesic Four-striped Grass Rat
Rhabdomys dilectus
FR Rhabdomys des montagnes
DE Schöne Vierstreifengrasmaus
ES Rata de hierba de cuatro listas mésica

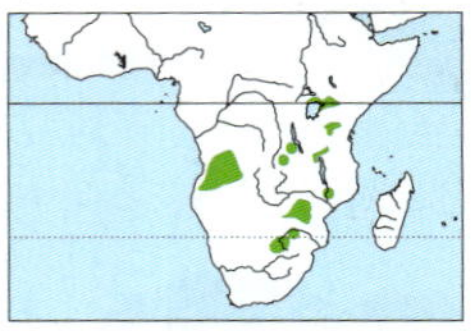

Montane habitats from SE UG & C KE S to NE ZA; isolated on plateau of AO

Genus *RHABDOMYS*

(5 species)

NE
Monotypic
HB 7.5–14 cm T 6.7–10.2 cm
W 14–70 g

KwaZulu Natal Four-striped Grass Rat
Rhabdomys chakae
FR Rhabdomys de Chaka
DE KwaZulu-Vierstreifengrasmaus
ES Rata de hierba de cuatro listas de KwaZulu

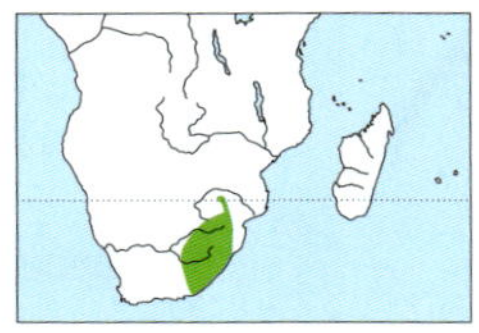

Wetter E ZA, from Soutpansberg S through SZ & LS to Eastern Cape

LC
Monotypic
HB c. 10.7 cm
T 9.5–12 cm
W 32–55 g

West-central South African Four-striped Grass Rat
Rhabdomys bechuanae
FR Rhabdomys du Botswana
DE Betschuanaland-Vierstreifengrasmaus
ES Rata de hierba de cuatro listas de África suroccidental

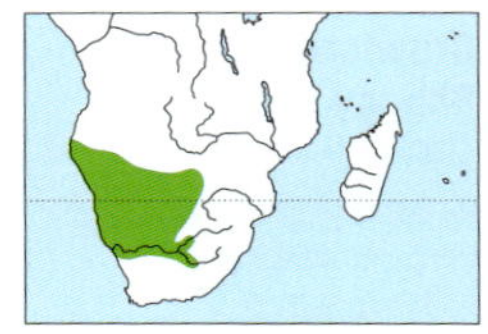

Drier W parts of S Africa (SW AO, NA, BW, NW & C ZA)

Karoo Four-striped Grass Rat
Rhabdomys intermedius
FR Rhabdomys du Karoo
DE Karoo-Vierstreifengrasmaus
ES Rata de hierba de cuatro listas de Karoo

LC
Monotypic
HB 9.6–13.5 cm
T 9–12.1 cm W 28–64 g

Narrow band through semiarid Nama Karoo

LC
Monotypic
HB 9.3–13.4 cm
T 9.9–12.8 cm W 24–54 g

Xeric Four-striped Grass Rat
Rhabdomys pumilio
FR Rhabdomys côtier
DE Eigentliche Vierstreifengrasmaus
ES Rata de hierba de cuatro listas xérica

Narrow coastal strip from R Orange mouth (NW Northern Cape) to Port Elizabeth (S Eastern Cape)

LC
Monotypic
HB 10.1–12.1 cm
T 11–14.5 cm
W 28–56 g

Genus *LEMNISCOMYS*

(11 species)

Barbary Striped Grass Mouse
Lemniscomys barbarus
FR Rat-rayé de Barbarie
DE Berber-Streifengrasmaus
ES Ratón de hierba listado de Berbería

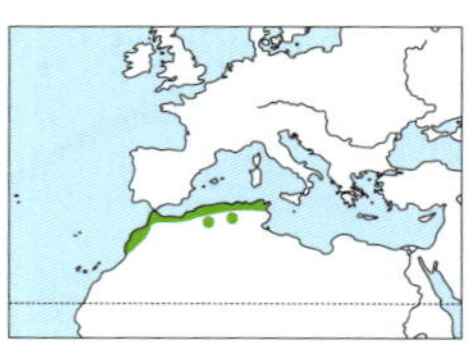

Narrow coastal strip in MA, DZ & TN, mostly N of Atlas Mts

LC
Monotypic
HB 9.6–16.1 cm
T 8.8–16 cm
W 33–82 g

Single-striped Grass Mouse
Lemniscomys rosalia
FR Rat-rayé de Tanzanie
DE Einstreifengrasmaus
ES Ratón de hierba de una lista

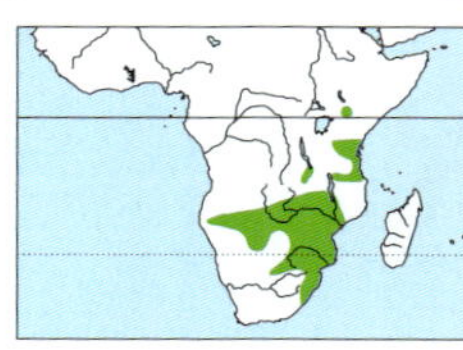

E & S Africa, from KE S to NE ZA, W to S AO & N NA

DD
Monotypic
HB 11.3–13.7 cm
T 12.3–15.3 cm

Rosevear's Striped Grass Mouse
Lemniscomys roseveari
FR Rat-rayé de Rosevear
DE Rosevear-Streifengrasmaus
ES Ratón de hierba listado de Rosevear

Known from 2 localities (Zambezi, Solwezi); possibly widespread

LC
Monotypic
HB 11.6–13.8 cm
T 11.8–14.9 cm

Griselda Striped Grass Mouse
Lemniscomys griselda
FR Rat-rayé d'Angola
DE Griselda-Streifengrasmaus
ES Ratón de hierba listado de Griselda

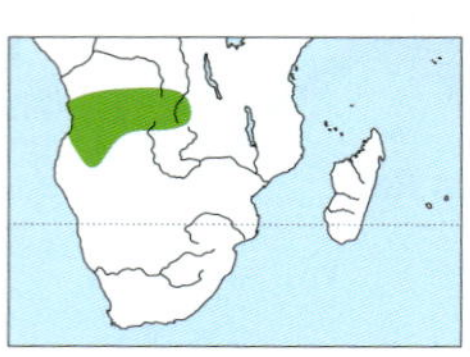

AO and neighboring parts of CD & NW ZM

LC
Monotypic
HB 9.6–11.9 cm
T 11–13.3 cm
W 24–60 g

Heuglin's Striped Grass Mouse
Lemniscomys zebra
FR Rat-rayé de Heuglin
DE Heuglin-Streifengrasmaus
ES Ratón de hierba listado de Heuglin

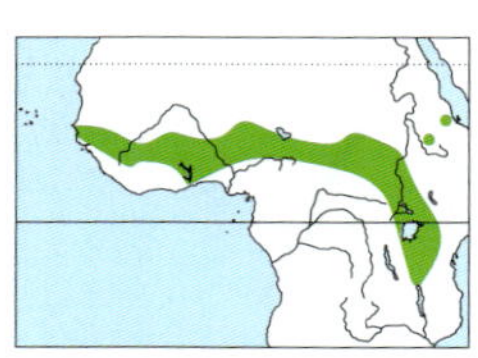

Arid Sahelo-Sudanian savannas from SN E to ER, S to S TZ

LC
Monotypic
HB 9.7–13.9 cm
T 9.6–15.3 cm
W 28–67 g

Typical Striped Grass Mouse
Lemniscomys striatus
FR Rat-rayé d'Afrique
DE Eigentliche Streifengrasmaus
ES Ratón de hierba listado típico

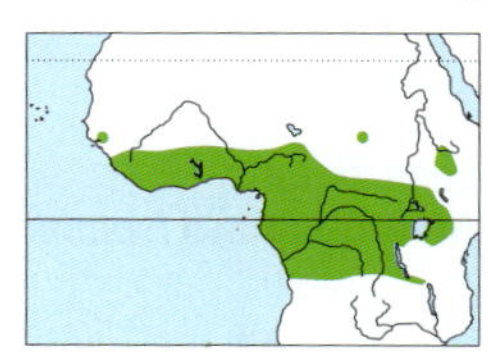

Tropical Africa from SL E to W KE, S to N AO, S CD, NE ZM & N MW; isolated in GW, SW SD & SW ET

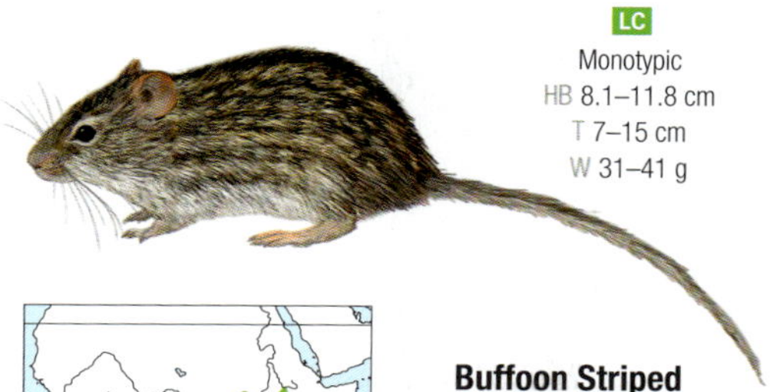

LC

Monotypic
HB 8.1–11.8 cm
T 7–15 cm
W 31–41 g

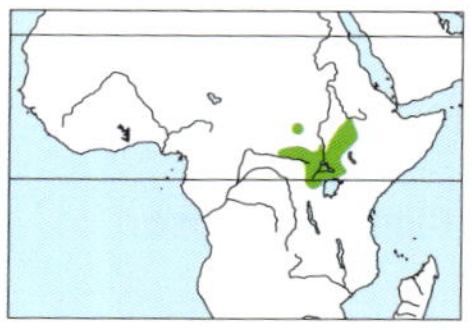

E Africa in NE CD, S SS, UG, SW ET, W KE, extreme NW TZ & N RW

Buffoon Striped Grass Mouse
Lemniscomys macculus

FR Rat-rayé d'Ouganda
DE Bajazzo-Streifengrasmaus
ES Ratón de hierba listado bufón

LC

Monotypic
HB 9.1–12.7 cm
T 9.4–13.4 cm
W 26–46 g

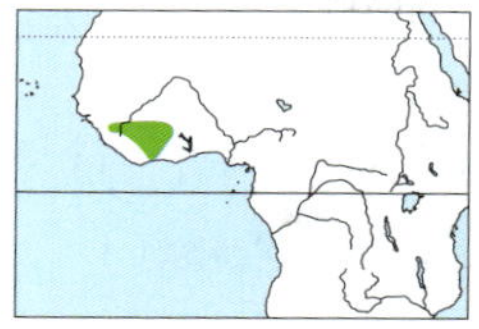

W Africa W of Dahomey Gap (from GN & SL to extreme S BF & NW GH)

Bellier's Striped Grass Mouse
Lemniscomys bellieri

FR Rat-rayé de Bellier
DE Bellier-Streifengrasmaus
ES Ratón de hierba listado de Bellier

LC

Monotypic
HB 9.3–12.6 cm
T 9.5–13.5 cm
W 37–41 g

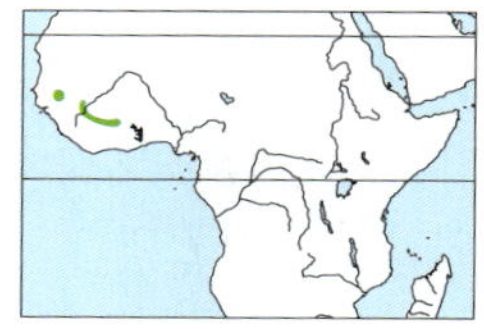

W Africa W of Dahomey Gap (SN, ML, GN, CI)

Senegal Striped Grass Mouse
Lemniscomys linulus

FR Rat-rayé du Sénégal
DE Senegal-Streifengrasmaus
ES Ratón de hierba listado de Senegal

Mittendorf's Striped Grass Mouse
Lemniscomys mittendorfi

FR Rat-rayé de Mittendorf
DE Mittendorf-Streifengrasmaus
ES Ratón de hierba listado de Mittendorf

LC

Monotypic
HB 8.5–9.8 cm
T 7.5–8.4 cm
W 22–34 g

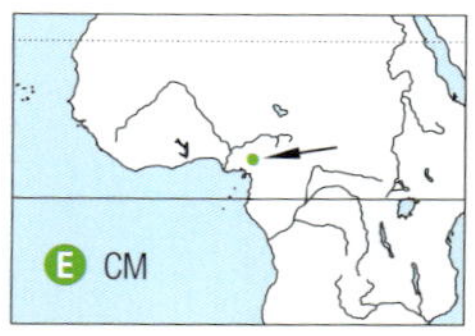

c. 2300 m

Mt Oku

Hoogstraal's Striped Grass Mouse
Lemniscomys hoogstraali

FR Rat-rayé de Hoogstraal
DE Hoogstraal-Streifengrasmaus
ES Ratón de hierba listado de Hoogstraal

DD

Monotypic
HB c. 12.7 cm
T c. 12.8 cm

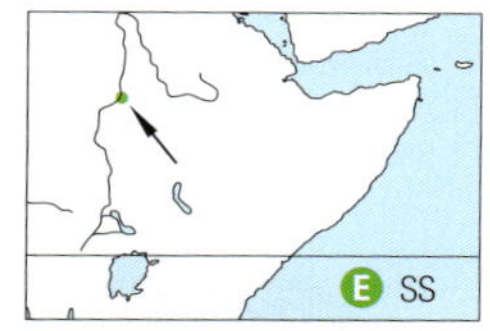

Known only from Paloich, Upper Nile

Genus *ARVICANTHIS*

(7 species)

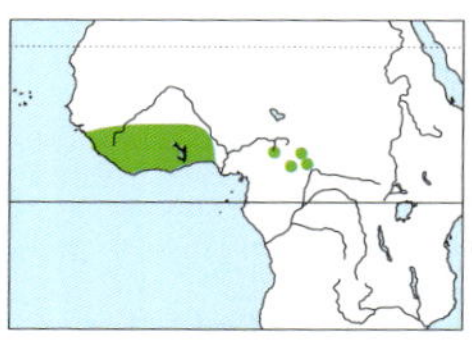

GN E to NG, isolated records from N CM & CF

Guinean Grass Rat
Arvicanthis rufinus

FR Rat-roussard de Temminck
DE Rote Grasratte
ES Rata de hierba de Guinea

LC

Monotypic
HB 11.5–17.7 cm
T 12.7–14.6 cm W 59–168 g

Sudanian Grass Rat
Arvicanthis ansorgei

FR Rat-roussard d'Ansorge
DE Ansorge-Grasratte
ES Rata de hierba de Sudán

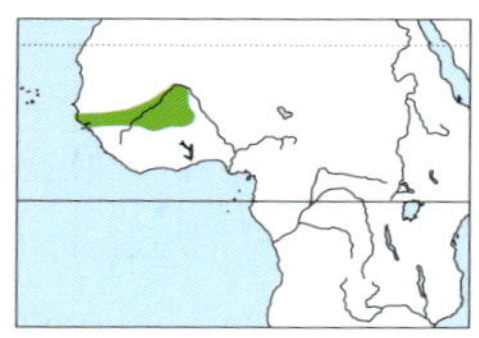

W Africa from S SN & GW E through S ML to BF

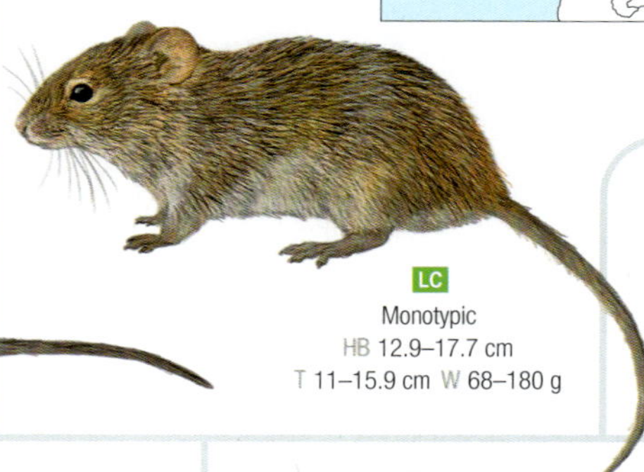

LC

Monotypic
HB 12.9–17.7 cm
T 11–15.9 cm W 68–180 g

Neumann's Grass Rat
Arvicanthis neumanni

FR Rat-roussard de Neumann
DE Neumann-Grasratte
ES Rata de hierba de Neumann

LC

Monotypic
HB 11.5–16 cm
T 8.8–11.8 cm
W 48–72 g

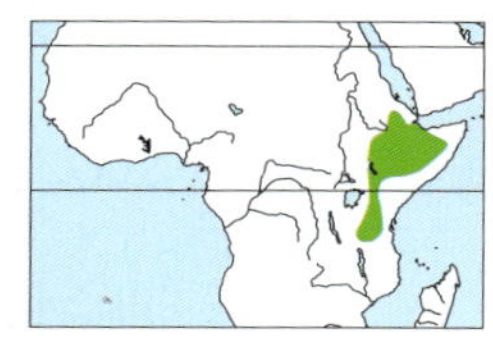

400–1500 m

E Africa from N ET & N SO S to C TZ

Nairobi Grass Rat
Arvicanthis nairobae

FR Rat-roussard de Nairobi
DE Nairobi-Grasratte
ES Rata de hierba de Nairobi

LC

Monotypic
HB 11.8–16.7 cm
T 9.4–12.9 cm
W 51–135 g

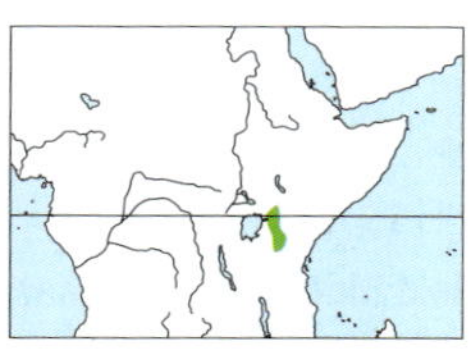

Narrow band from C & W KE to N TZ; perhaps S ET

Abyssinian Grass Rat
Arvicanthis abyssinicus

FR Rat-roussard d'Abyssinie
DE Äthiopien-Grasratte
ES Rata de hierba de Abisinia

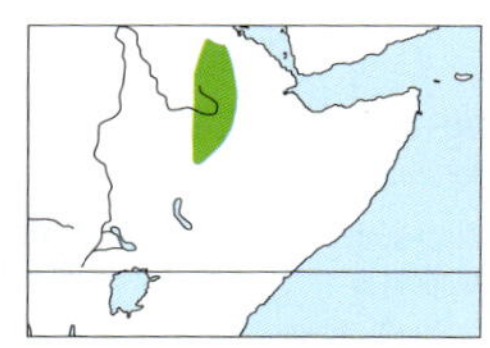

2000–3400 m

Ethiopian highlands on both sides of Rift Valley (ER, ET)

LC

Monotypic
HB 12.5–15 cm T 9.8–10.9 cm W 72–125 g

African Grass Rat
Arvicanthis niloticus

FR Rat-roussard du Nil
DE Nil-Grasratte
ES Rata de hierba de África

LC

Monotypic
HB 12.7–18.8 cm
T 9.2–15.5 cm
W 89–160 g

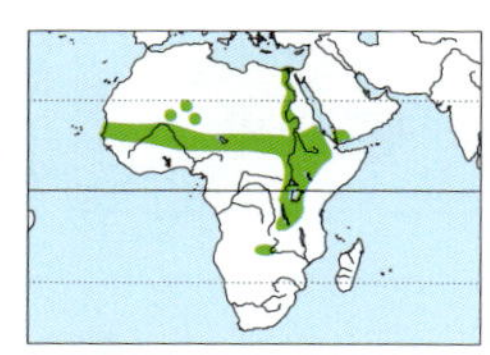

200–1200 m

From MR & SN E to ET, S to TZ & SW ZM, N along R Nile to Nile Delta (EG), isolated in SW Arabia (YE)

Blick's Grass Rat
Arvicanthis blicki

FR Rat-roussard de Blick
DE Blick-Grasratte
ES Rata de hierba de Blick

NT

Monotypic
HB 16–17.5 cm
T 10.2–11.2 cm
W 140–150 g

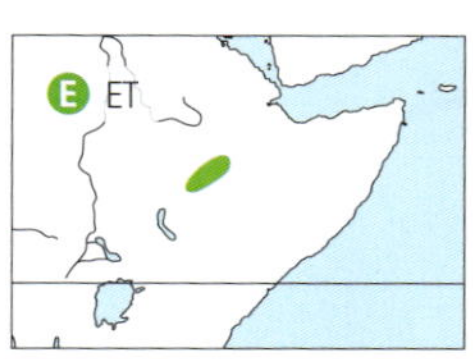

2750–4050 m

Chilalo & Bale Mts (E of Rift Valley)

EN

Monotypic
HB 9.8–14 cm
T 12–13.8 cm

Genus *HADROMYS*

(2 species)

Hume's Hadromys
Hadromys humei

FR Rat de Hume
DE Manipur-Starkratte
ES Rata de Hume

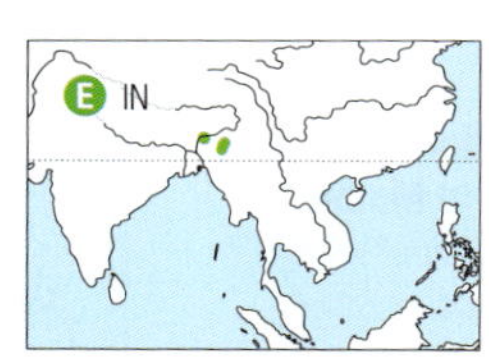

900–1300 m

Known from 3 locations in Assam & Manipur

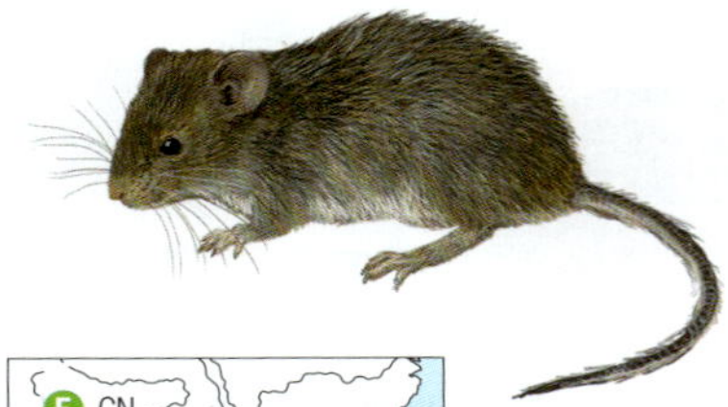

DD
Monotypic
HB 12.3–13.6 cm
T 11.4–13.2 cm
W 41–77 g

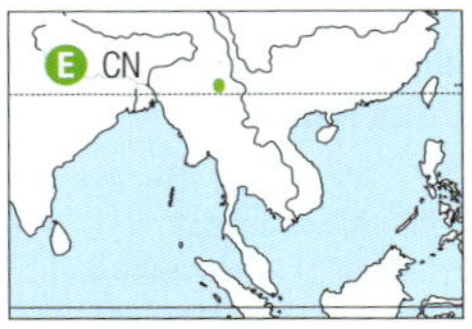

Ruili County (W Yunnan)

970–1300 m

Yunnan Hadromys
Hadromys yunnanensis
FR Rat du Yunnan
DE Yunnan-Starkratte
ES Rata de Yunnan

Genus *VERNAYA*

(1 species)

Vernay's Climbing Mouse
Vernaya fulva
FR Vernaya rousse
DE China-Klettermaus
ES Ratón trepador de Vernay

above 2100 m

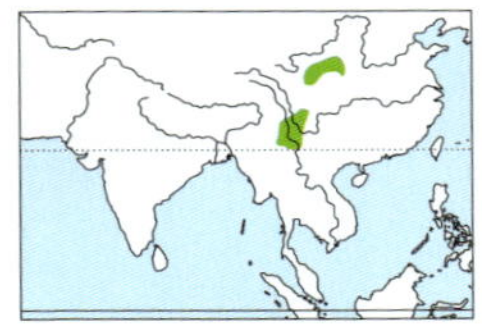

C & SC CN (S Gansu, SW Shaanxi, Sichuan, Yunnan), NE MM

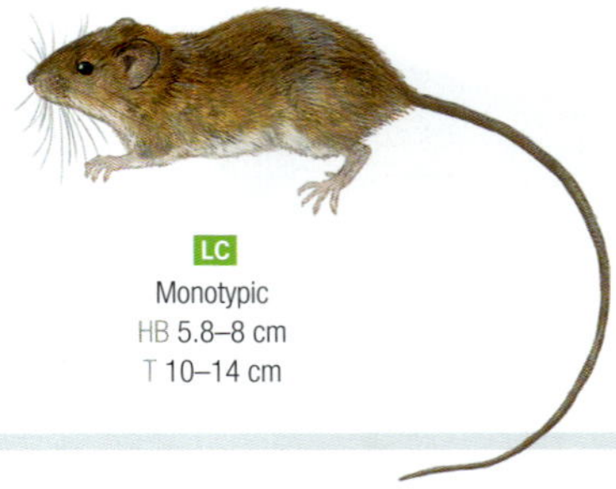

LC
Monotypic
HB 5.8–8 cm
T 10–14 cm

Genus *VANDELEURIA*

(3 species)

Indomalayan Long-tailed Climbing Mouse
Vandeleuria oleraceus
FR Vandeleuria d'Asie
DE Asiatische Langschwanzklettermaus
ES Ratón trepador de cola larga de Indomalasia

LC
Monotypic
HB c. 6.8 cm
T c. 10.5 cm
W c. 10 g

150–1500 m

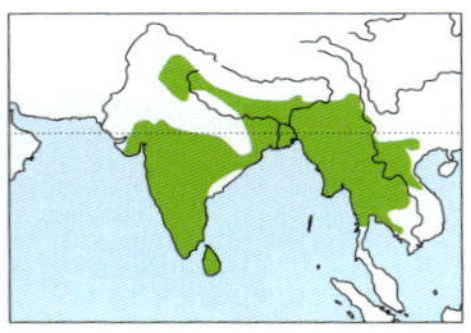

Most of Indian subcontinent, S CN (W & S Yunnan), mainland SE Asia N of Isthmus of Kra

Nilgiri Long-tailed Climbing Mouse
Vandeleuria nilagiricus
FR Vandeleuria des Nilgiri
DE Nilgiri-Langschwanzklettermaus
ES Ratón trepador de cola larga de Nilgiri

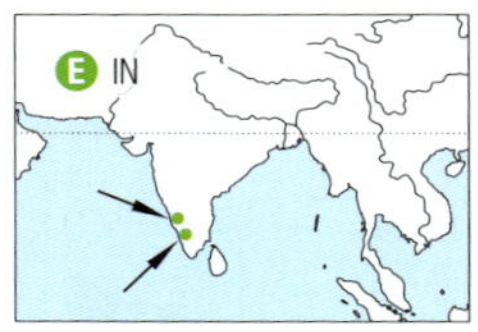

S Western Ghats; perhaps N

EN
Monotypic
HB c. 10.5 cm
T c. 14.2 cm

Sri Lankan Long-tailed Climbing Mouse
Vandeleuria nolthenii
FR Vandeleuria de Nolthenius
DE Sri-Lanka-Langschwanzklettermaus
ES Ratón trepador de cola larga de Sri Lanka

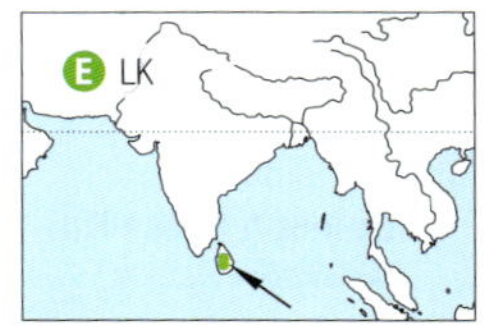

C highlands

1320–2130 m

EN
Monotypic
HB 12–12.5 cm
T 12–12.5 cm

Genus *APODEMUS*

(22 species)

2200–3600 m

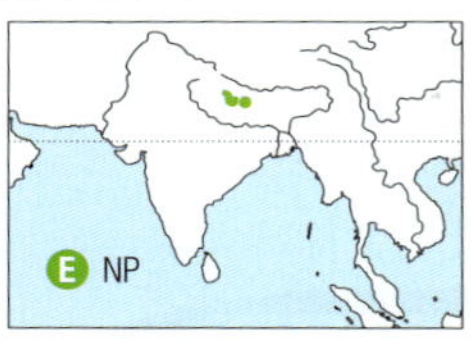

Rhododendron and coniferous mountain forests

Nepalese Field Mouse
Apodemus gurkha
FR Mulot du Népal
DE Nepal-Waldmaus
ES Ratón de campo de Nepal

LC
Monotypic
HB 8.5–11.6 cm
T 9.7–13 cm W 23–40 g

LC
Monotypic
HB 6.5–10 cm
T 7–11 cm
W 10–20 g

0–2500 m

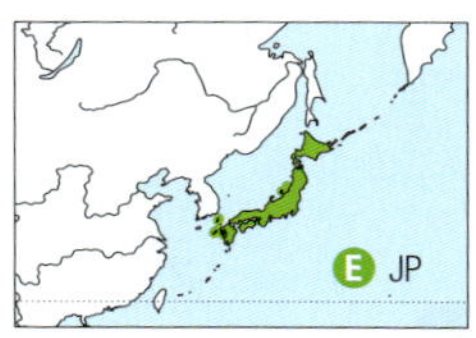

Hokkaido, Honshu, Shikoku, Kyushu, adjacent Is

Small Japanese Field Mouse
Apodemus argenteus
FR Mulot geisha
DE Kleine Japan-Waldmaus
ES Ratón de campo pequeño de Japón

Steppe Field Mouse
Apodemus witherbyi
FR Mulot de Witherby
DE Steppenwaldmaus
ES Ratón de campo de estepa

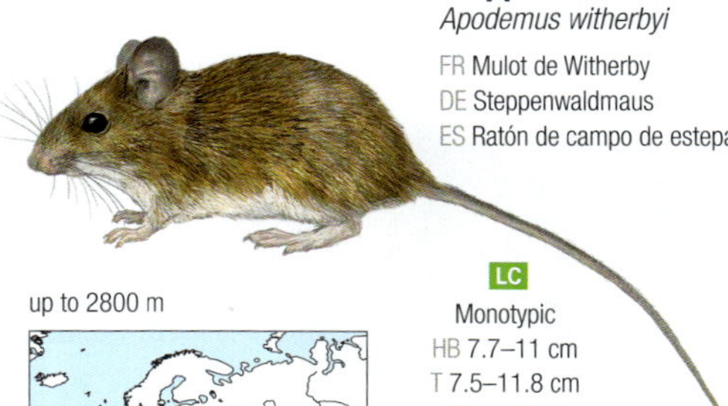

up to 2800 m

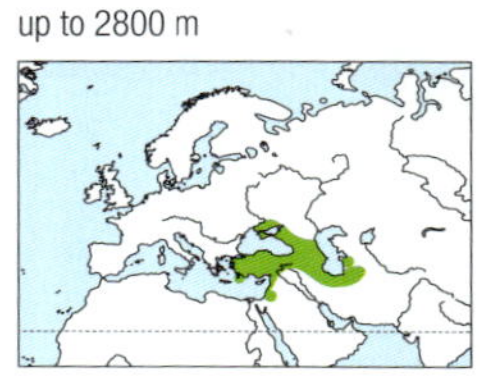

LC
Monotypic
HB 7.7–11 cm
T 7.5–11.8 cm
W 16–34 g

SE UA (E of R Dnieper) & E Aegean Is (GR) E to N IR & TM, S to IL & NW JO

Long-tailed Field Mouse
Apodemus sylvaticus
FR Mulot sylvestre
DE Waldmaus
ES Ratón de campo de cola larga

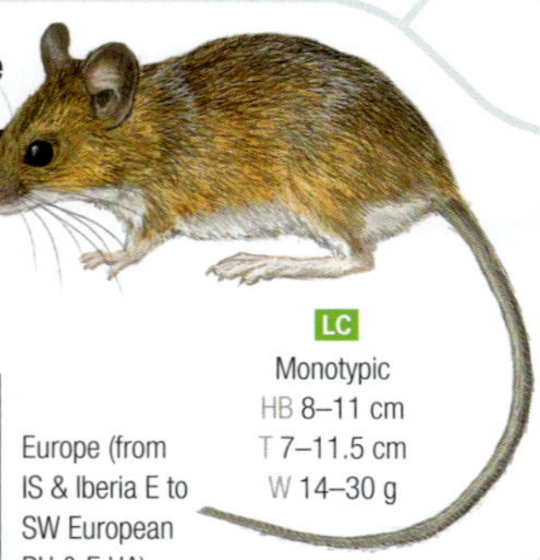

0–2000 m

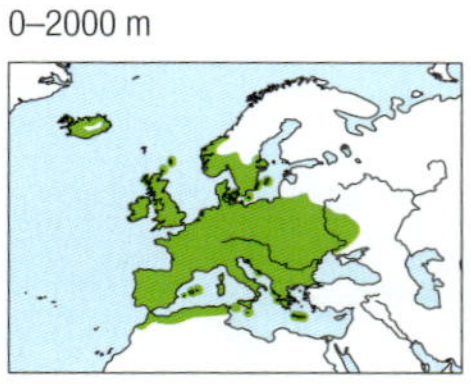

Europe (from IS & Iberia E to SW European RU & E UA), MA, DZ & TN

LC
Monotypic
HB 8–11 cm
T 7–11.5 cm
W 14–30 g

up to 2000 m

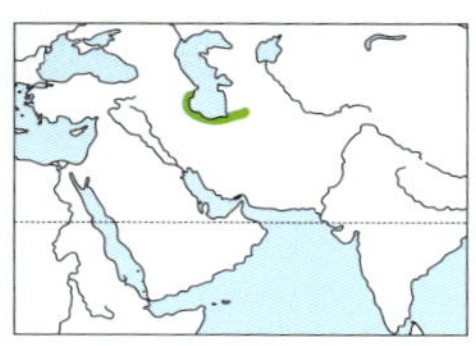

SE AZ & N IR along coast of Caspian Sea; possibly SW TM

Hyrcanian Field Mouse
Apodemus hyrcanicus
FR Mulot d'Hyrcanie
DE Talysh-Waldmaus
ES Ratón de campo de Hircania

NT
Monotypic
HB 9.5–10.5 cm
T 9.4–10.7 cm
W 22–35 g

Caucasus Field Mouse
Apodemus ponticus
FR Mulot de la mer Noire
DE Kaukasus-Waldmaus
ES Ratón de campo del Cáucaso

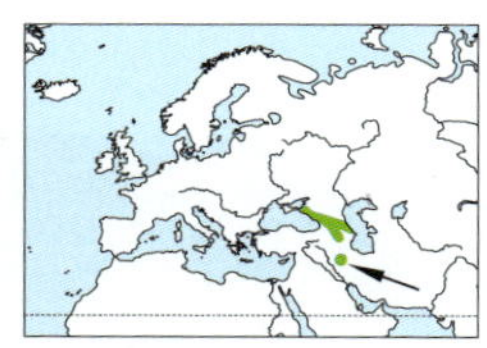

Caucasus (S European RU, GE, AM, AZ) & NW IR (Zagros Mts)

LC
Monotypic
HB 8.3–10.3 cm
T 9.7–10.4 cm

LC
Monotypic
HB 8.1–13 cm
T 8.6–13.5 cm
W 22–56 g

Yellow-necked Field Mouse
Apodemus flavicollis
FR Mulot à collier
DE Gelbhalsmaus
ES Ratón de campo de cuello amarillo

0–2500 m

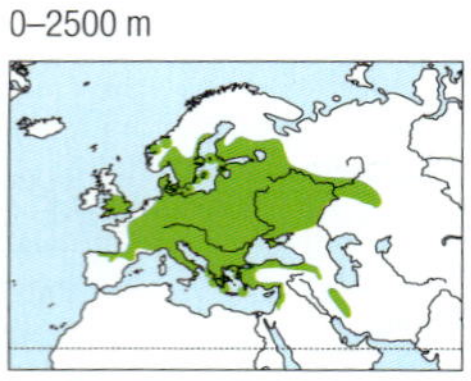

Europe, from Britain & N ES E to Ural Mts, SE to N IL & W IR (Zagros Mts)

Kashmir Field Mouse
Apodemus rusiges
FR Mulot du Cachemire
DE Kaschmir-Waldmaus
ES Ratón de campo de Cachemira

LC
Monotypic
HB 7.7–11.2 cm
T 7.8–12.2 cm
W c. 37 g

1980–3350 m

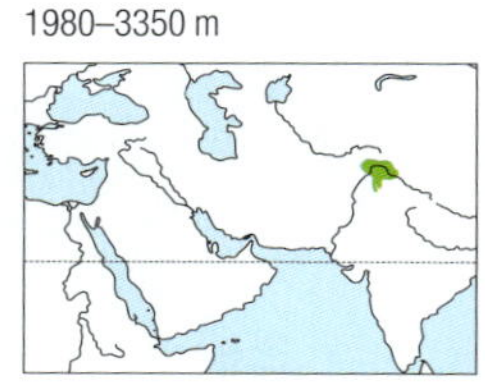

N PK, adjacent NW IN

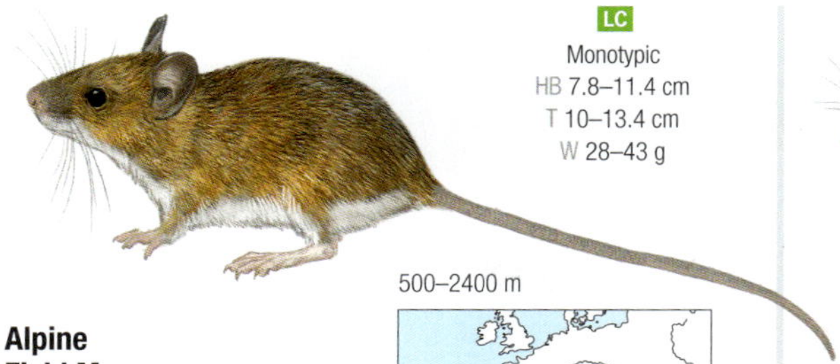

LC
Monotypic
HB 7.8–11.4 cm
T 10–13.4 cm
W 28–43 g

Alpine Field Mouse
Apodemus alpicola

FR Mulot alpestre
DE Alpenwaldmaus
ES Ratón de campo de los Alpes

500–2400 m

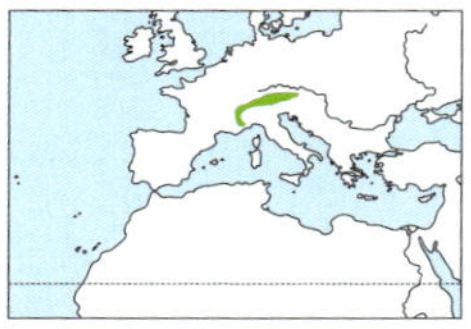

Alps of SE FR, CH, S DE, LI, AT & N IT

LC
Monotypic
HB 7–10.5 cm
T 6.5–11 cm
W 13–28 g

Herb Field Mouse
Apodemus uralensis

FR Mulot pygmée
DE Zwergwaldmaus
ES Ratón de campo de los Urales

0–1400 m

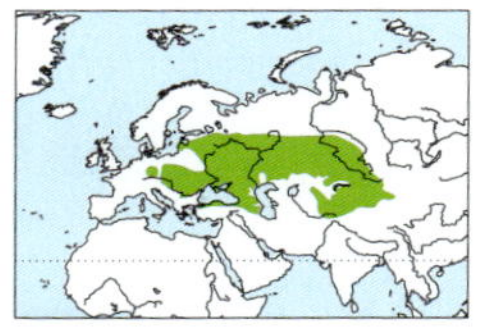

From C & E Europe E through RU, N TR (Anatolia) & Caucasus to W CN (Xinjiang)

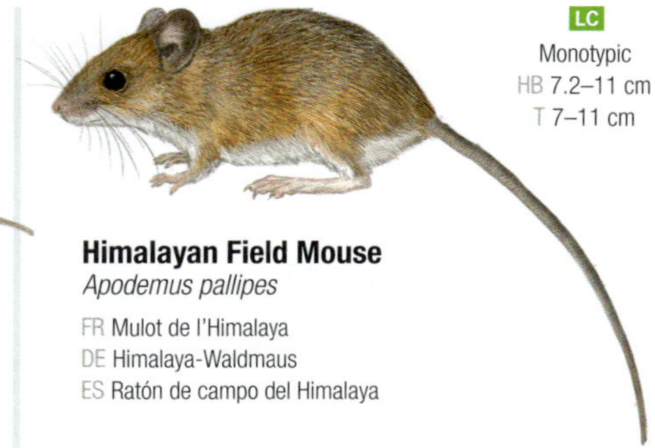

LC
Monotypic
HB 7.2–11 cm
T 7–11 cm

Himalayan Field Mouse
Apodemus pallipes

FR Mulot de l'Himalaya
DE Himalaya-Waldmaus
ES Ratón de campo del Himalaya

1460–3970 m

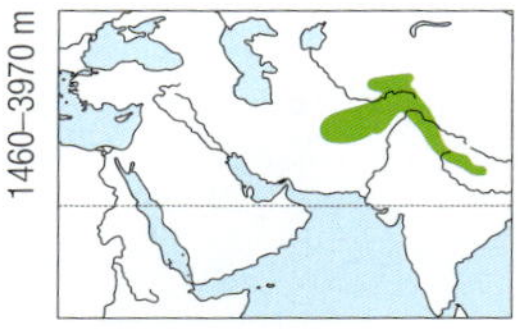

AF, S KG, TJ, N PK, extreme SW CN (Tibet), NW IN, W NP

LC
Monotypic
HB 9.8–13.2 cm
T 9.8–14.5 cm
W 25–57 g

0–2700 m

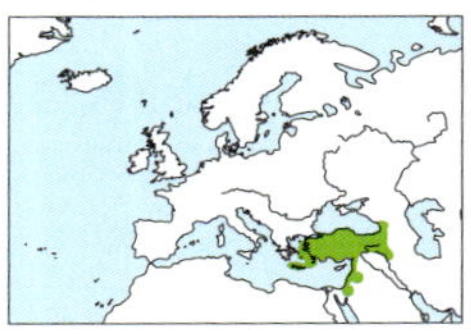

Eastern Broad-toothed Field Mouse
Apodemus mystacinus

FR Mulot rupestre
DE Orient-Felsenmaus
ES Ratón de campo de dientes anchos oriental

GR (Crete & Aegean Is), TR (Anatolia), W GE, W SY, LB, N IL, PS, JO, N IQ

LC
Monotypic
HB 10–15 cm
T 10.2–14.6 cm
W 28–56 g

0–1620 m

Western Broad-toothed Field Mouse
Apodemus epimelas

FR Mulot du Parnasse
DE Europäische Felsenmaus
ES Ratón de campo de dientes anchos occidental

HR (with Korçula & Mljet Is), throug Balkans to Ionian & Aegean Is (GR)

LC
Monotypic
HB 8–14 cm
T 7–13 cm
W 20–60 g

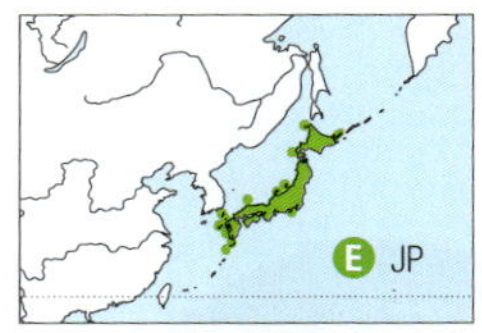

Large Japanese Field Mouse
Apodemus speciosus

FR Mulot du Japon
DE Große Japan-Waldmaus
ES Ratón de campo grande de Japón

Hokkaido, Honshu, Shikoku, Kyushu, many smaller Is

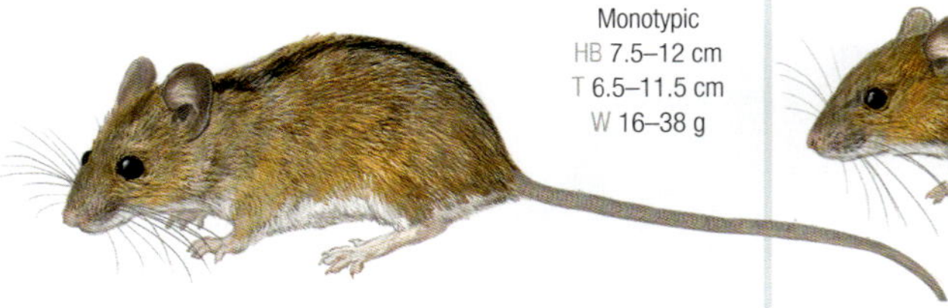

LC
Monotypic
HB 7.5–12 cm
T 6.5–11.5 cm
W 16–38 g

Striped Field Mouse
Apodemus agrarius

FR Mulot rayé
DE Brandmaus
ES Ratón de campo listado

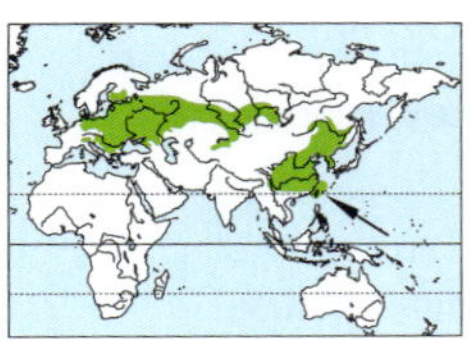

C Europe E to SC Siberia & NW CN, disjunct in Russian Far East & E MN S to N MM, S CN, TW & JP (Senkaku Is)

LC
Monotypic
HB 8.8–11 cm
T 8.3–10.5 cm

Chevrier's Field Mouse
Apodemus chevrieri

FR Mulot de Chevrier
DE Chevrier-Waldmaus
ES Ratón de campo de Chevrier

300–1800 m

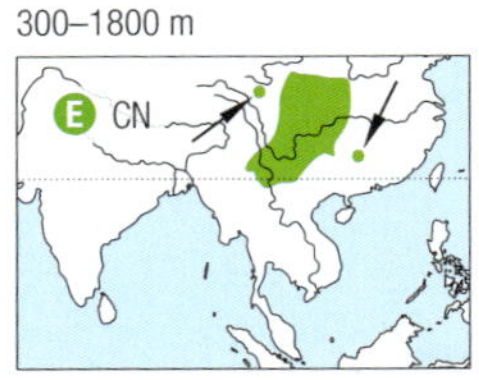

S Gansu, S Shaanxi, Sichuan, Chongqing, W Hubei, Yunnan, Guizhou, Hunan

LC
Monotypic
HB 9.2–10.7 cm
T 10–12 cm

Large-eared Field Mouse
Apodemus latronum

FR Mulot du Sichuan
DE Sichuan-Waldmaus
ES Ratón de campo de orejas grandes

2700–4000 m

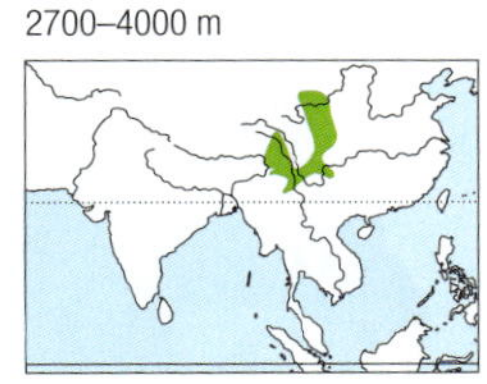

NE IN, N MM, CN (E Qinghai, E Tibet, Sichuan, N Yunnan)

LC
Monotypic
HB 8–11.8 cm
T 7.5–10.3 cm
W 19–27 g

Korean Field Mouse
Apodemus peninsulae

FR Mulot de Corée
DE Korea-Waldmaus
ES Ratón de campo de Corea

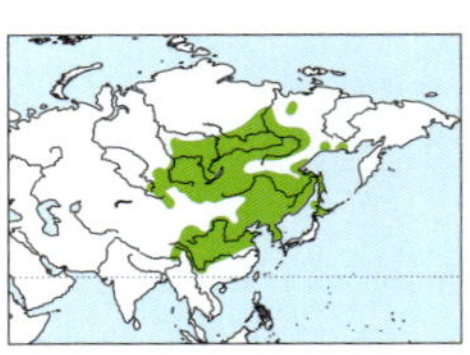

S Siberia & Russian Far East, MN, NE & C CN, Korea Pen, N JP (Hokkaido)

NE
Monotypic
HB 9.2–10.7 cm
T 9.3–11 cm
W 20–33 g

Black Field Mouse
Apodemus nigrus

FR Mulot noir
DE Schwarze Waldmaus
ES Ratón de campo negro

Highlands of Fanjing Mt NR (Guizhou) & Jinfo Mt NR (Chongqing)

LC
Monotypic
HB 8.2–10 cm
T 10–11.9 cm
W 18.5–39 g

Taiwan Field Mouse
Apodemus semotus

FR Mulot de Taïwan
DE Taiwan-Waldmaus
ES Ratón de campo de Taiwán

1400–3700 m

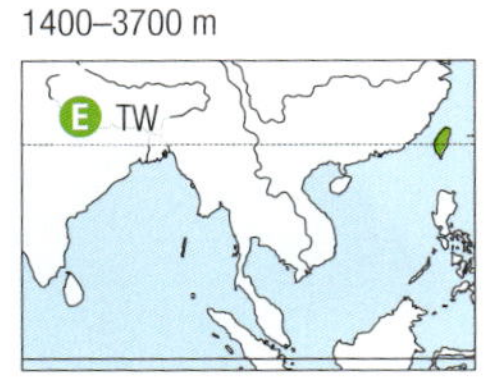

Primarily in montane areas

Yunnan Field Mouse
Apodemus ilex

FR Mulot du Yunnan
DE Yunnan-Waldmaus
ES Ratón de campo de Yunnan

NE
Monotypic

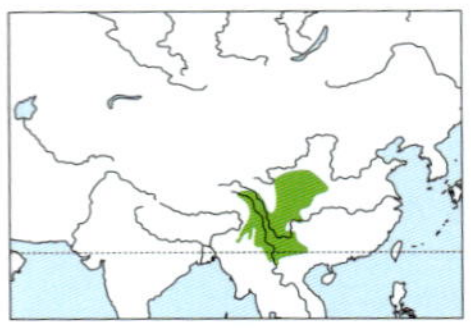

SC CN (E Tibet to Sichuan & Yunnan), NE IN, N & NC MM

South China Field Mouse
Apodemus draco

FR Mulot dragon
DE Südchina-Waldmaus
ES Ratón de campo de China meridional

LC
Monotypic

Guizhou & Chongqing E to Hebei & Fujian

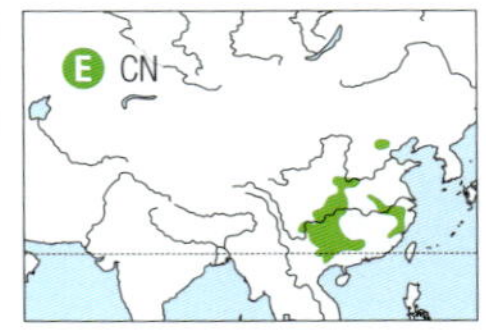

up to 3820 m

Genus *TOKUDAIA*

(3 species)

EN
Monotypic
HB 10.3–16 cm
T 8.3–13.5 cm

Amami Spiny Rat
Tokudaia osimensis

FR Mulot d'Amami
DE Amami-Inselstachelratte
ES Rata espinosa de Amami

Amami Ōshima (Ryukyu Is)

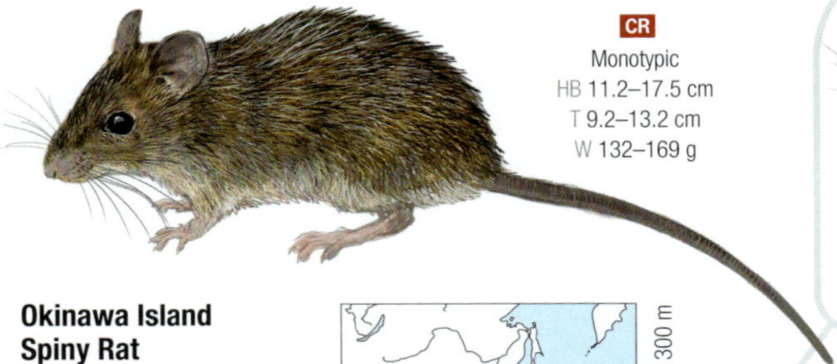

CR
Monotypic
HB 11.2–17.5 cm
T 9.2–13.2 cm
W 132–169 g

Okinawa Island Spiny Rat
Tokudaia muenninki

FR Mulot d'Okinawa
DE Okinawa-Inselstachelratte
ES Rata espinosa de Okinawa

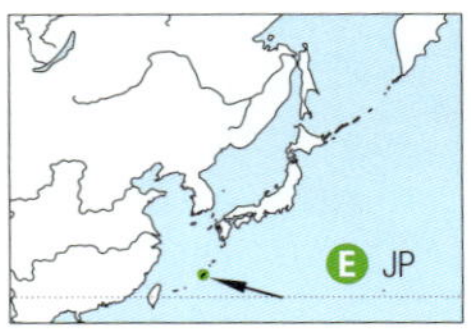

Okinawa I (Ryukyu Is)

above 300 m

EN
Monotypic
HB 10.3–16 cm
T 8.3–13.5 cm

Tokunoshima Spiny Rat
Tokudaia tokunoshimensis

FR Mulot de Tokunoshima
DE Tokunoshima-Inselstachelratte
ES Rata espinosa de Tokunoshima

Tokunoshima (Ryukyu Is)

LC
Monotypic
HB 13.5–17.5 cm
T 17–20 cm
W 94–120 g

Common Swamp Rat
Malacomys longipes

FR Malacomys du Gabon
DE Gemeine Großohr-Sumpfratte
ES Rata de ciénaga común

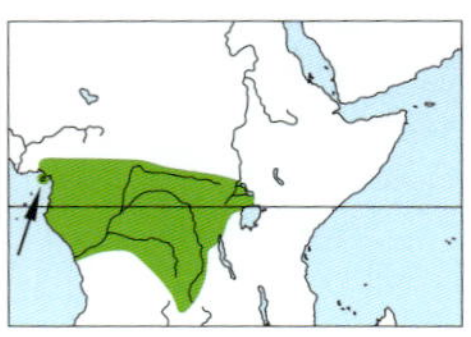

From extreme SE NG E to L Victoria, S to extreme NE AO & NW ZM; Bioko I

Genus *MALACOMYS*

(3 species)

Cansdale's Swamp Rat
Malacomys cansdalei

FR Malacomys de Cansdale
DE Cansdale-Großohr-Sumpfratte
ES Rata de ciénaga de Cansdale

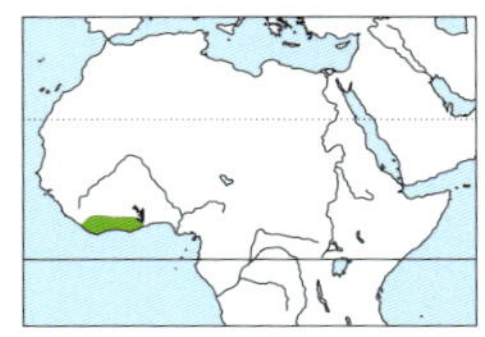

Upper Guinea Forest of W Africa, W of R Volta (E LR, CI, GH)

LC
Monotypic
HB 13.3–16.2 cm
T 17.8–21 cm
W 63–101 g

Edwards's Swamp Rat
Malacomys edwardsi

FR Malacomys d'Edwards
DE Edwards-Großohr-Sumpfratte
ES Rata de ciénaga de Edwards

LC
Monotypic
HB 12.6–16 cm
T 15.5–18 cm
W 46–80 g

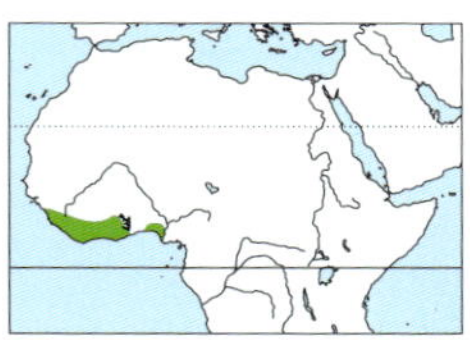

Upper Guinea Forest of W Africa, W of R Niger (from GN E to SW NG)

Genus *MUS*

(42 species)

LC
3 sspp
HB 7.1–11.2 cm
T 5.3–9.6 cm
W 18–22 g

Saxicolous Mouse
Mus saxicola

FR Souris des roches
DE Felsen-Stachelzwergmaus
ES Ratón de roca

up to 1000 m

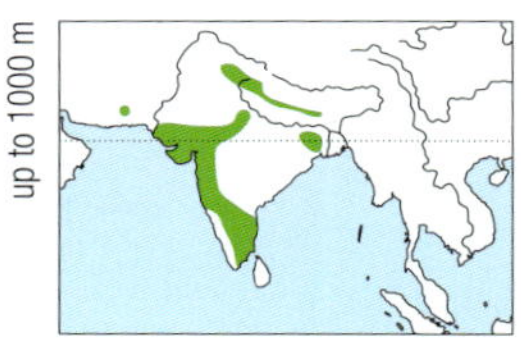

SW & SE PK (Balochistan, Sind), widely in IN, S NP

Flat-haired Mouse
Mus platythrix

FR Souris à poil plat
DE Flachkopf-Stachelzwergmaus
ES Ratón de pelaje plano

up to 2000 m

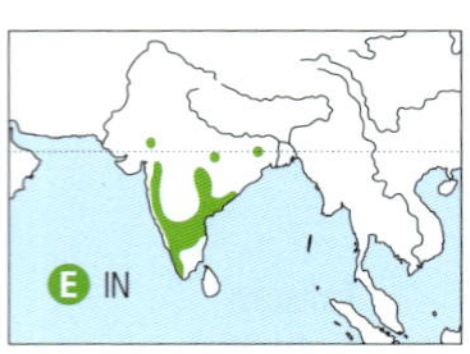

Rajasthan, Madhya Pradesh, West Bengal, Maharashtra, Andhra Pradesh, Karnataka, Tamil Nadu, Kerala

LC
Monotypic
HB 7.7–12.3 cm
T 5.1–8.1 cm
W c. 18 g

LC
Monotypic
HB 6.2–8 cm
T 5–6.2 cm

Phillips's Mouse
Mus phillipsi

FR Souris de Phillips
DE Phillips-Stachelzwergmaus
ES Ratón de Phillips

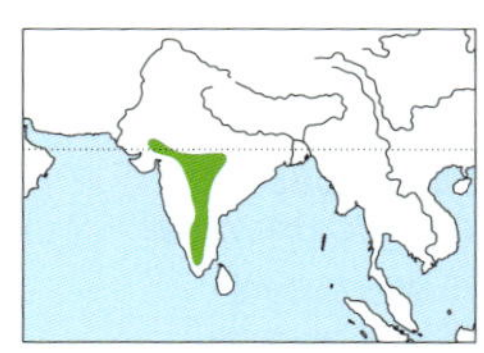

Rajasthan, Gujarat, Madhya Pradesh, Maharashtra, Telangana, Andhra Pradesh, Karnataka, Tamil Nadu

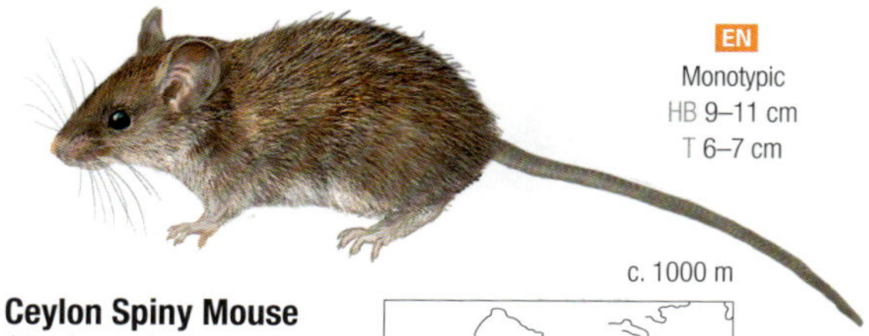

EN
Monotypic
HB 9–11 cm
T 6–7 cm

Ceylon Spiny Mouse

Mus fernandoni

FR Souris du Sri Lanka
DE Ceylon-Stachelzwergmaus
ES Ratón espinoso de Sri Lanka

c. 1000 m

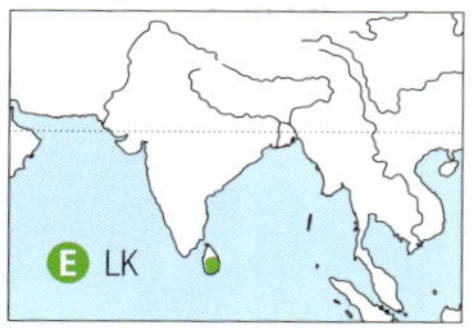

Found only at few locations

LC
Monotypic
HB 9.5–12 cm
T 6–9 cm
W c. 34 g

Shortridge's Mouse

Mus shortridgei

FR Souris de Shortridge
DE Shortridge-Stachelzwergmaus
ES Ratón de Shortridge

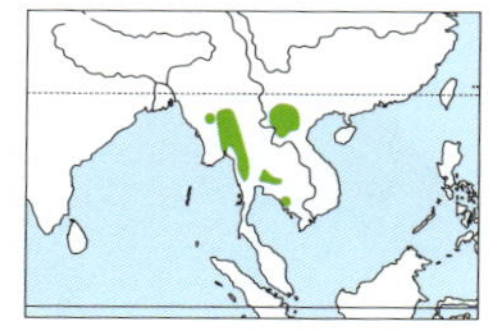

MM, TH, LA, NW VN, SW KH

Sumatran Shrew-like Mouse

Mus crociduroides

FR Souris de Sumatra
DE Sumatra-Zwergmaus
ES Ratón musaraña de Sumatra

DD
Monotypic
HB 7.7–9.7 cm
T 11.1–12.9 cm

2200–3000 m

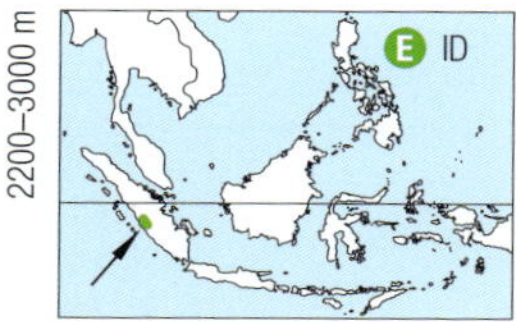

High mountain forest in W Sumatra

Indochinese Shrew-like Mouse

Mus pahari

FR Souris du Sikkim
DE Indochina-Maus
ES Ratón musaraña de Indochina

LC
Monotypic
HB 7.5–10.5 cm
T 7–10 cm
W 20–28 g

up to 2000 m

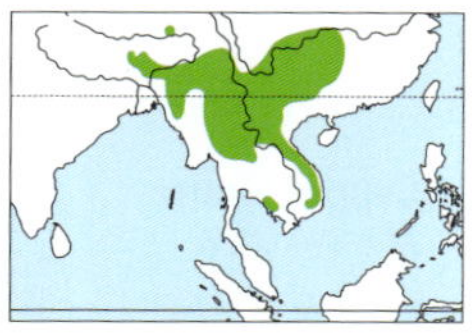

NE IN, BH, MM, S CN, TH, LA, VN, SW KH

VU
Monotypic
HB c. 9.8 cm
T c. 10.2 cm

Mayor's Mouse

Mus mayori

FR Souris de Mayor
DE Mayor-Maus
ES Ratón de Mayor

165–2310 m

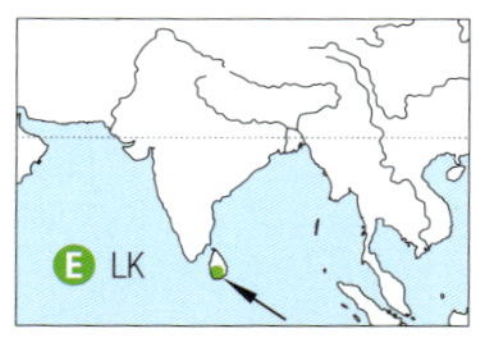

Scattered localities in rainforests of C & SW

LC
Monotypic
HB 8–9.2 cm
T 8.4–9.2 cm

Javan Shrew-like Mouse

Mus vulcani

FR Souris de Java
DE Java-Zwergmaus
ES Ratón musaraña de Java

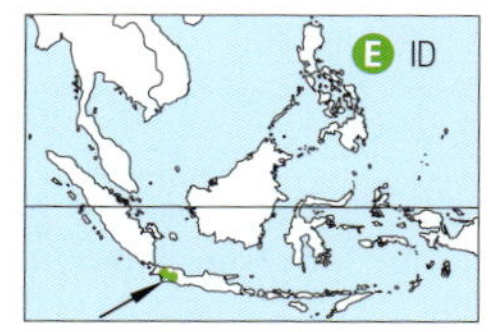

Montane forests of W Java

LC
Monotypic
HB 7–9.5 cm
T 5–7 cm
W 8–17 g

Fawn-colored Mouse

Mus cervicolor

FR Souris fauve
DE Falbmaus
ES Ratón beige

up to 2000 m

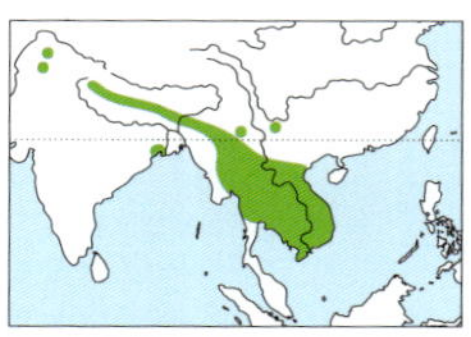

NE PK, N IN, NP, MM, S CN (Yunnan), TH, LA, VN, KH; likely introduced to Sumatra & Java (ID)

LC
Monotypic
HB 7.7–9.6 cm
T 8.3–9.1 cm
W 16.5–23 g

Cook's Mouse

Mus cookii

FR Souris de Cook
DE Cook-Maus
ES Ratón de Cook

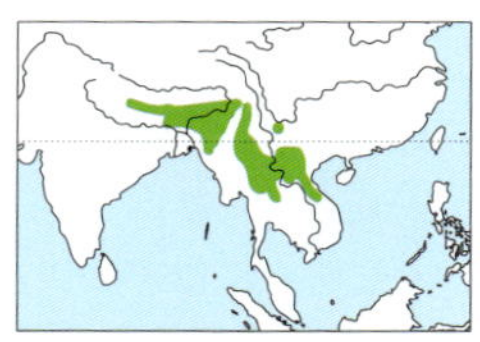

NP, NE IN, BH, BD, MM, S CN (Yunnan), TH, LA, VN

Ryukyu Mouse

Mus caroli

FR Souris de Rothschild
DE Reisfeldmaus
ES Ratón de Ryukyu

LC
Monotypic
HB 6.2–9.5 cm
T 6.5–9.5 cm
W 11.5–19.5 g

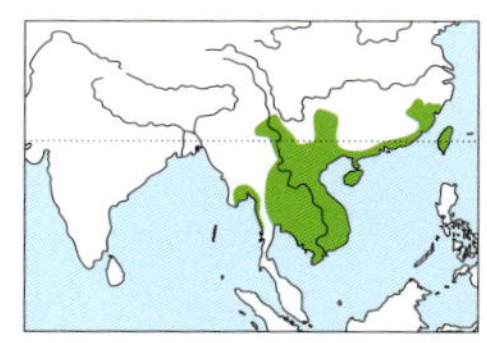

S CN, TW, S MM, TH, LA, VN, KH; origin uncertain in Pen MY, W ID & JP (Okinawa I); introduced to Flores I (ID)

LC
Monotypic
HB 5.3–7 cm
T 5.1–7 cm

Earth-colored Mouse

Mus terricolor

FR Souris terreuse
DE Erdfarbene Maus
ES Ratón terrizo

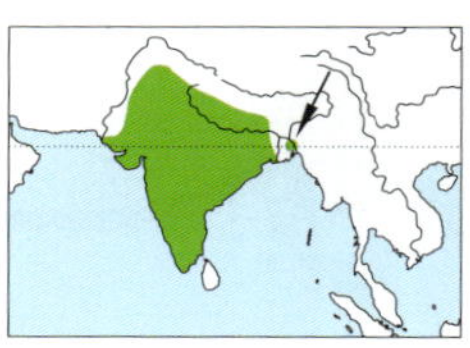

E PK, IN, NP, BD; introduced to NE Sumatra (ID)

EN
Monotypic
HB 9–10 cm T 7.1–9.4 cm W 23–30 g

Servant Mouse

Mus famulus

FR Souris des Nilgiri
DE Südindische Maus
ES Ratón servil

Fragmented locations in Western Ghats of Kerala & Tamil Nadu

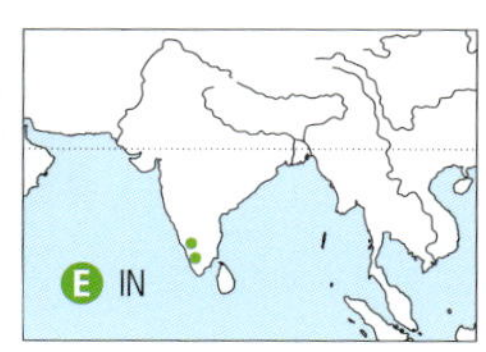

LC
Monotypic
HB 6.6–7 cm T 5.6–6.7 cm
W 9–15 g

Sheath-tailed Mouse

Mus fragilicauda

FR Souris à queue fragile
DE Laos-Maus
ES Ratón de cola frágil

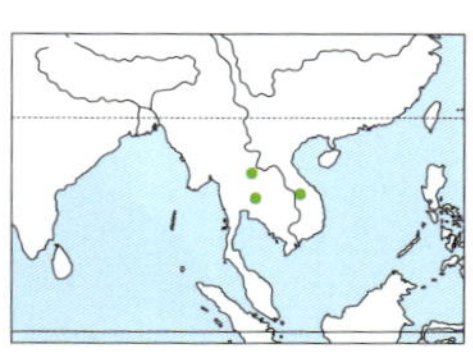

Scattered localities in NC & SC TH & SE LA

Little Indian Field Mouse

Mus booduga

FR Souris indienne
DE Kleine Indien-Maus
ES Ratón de India pequeño

LC
Monotypic
HB 5.6–6.3 cm
T 5.5–7.5 cm
W 10–15 g

0–4000 m

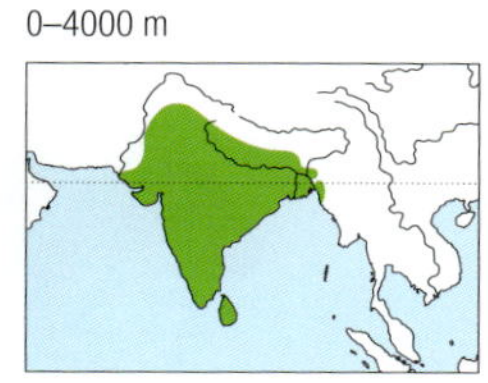

E PK, IN, S NP, LK, BD

NE
Monotypic
HB 5.3–9.1 cm
T 5.2–7.3 cm
W 16–18 g

Blyth's Mouse
Mus nitidulus

FR Souris de Blyth
DE Blyth-Maus
ES Ratón de Blyth

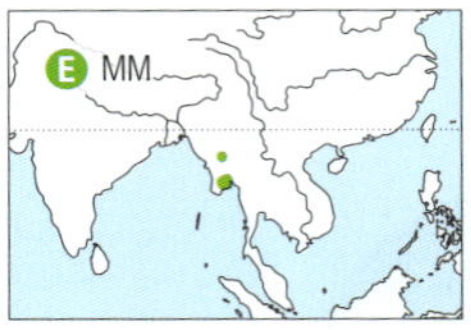

Locally abundant in ricefield habitat of Central Basin

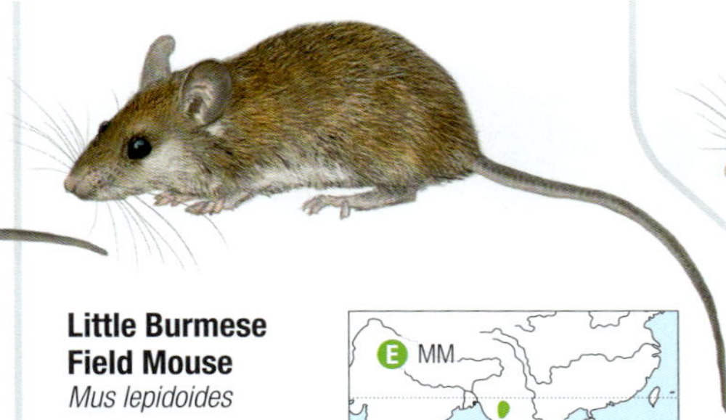

Little Burmese Field Mouse
Mus lepidoides

FR Souris de Birmanie
DE Kleine Burma-Maus
ES Ratón de Birmania pequeño

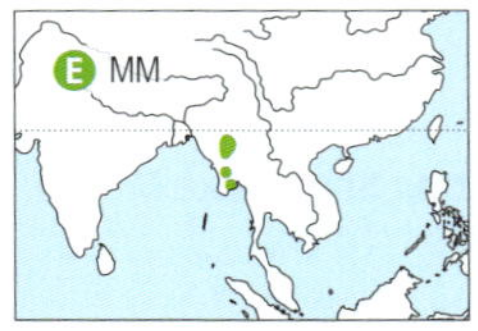

Central Dry Zone and along R Irrawaddy

NE
Monotypic
HB 5.2–7.7 cm
T 4.4–5.8 cm
W 6–8 g

House Mouse
Mus musculus

FR Souris domestique
DE Hausmaus
ES Ratón casero

LC
4 sspp
HB 7–10.3 cm
T 6.7–10.4 cm
W 12–39 g

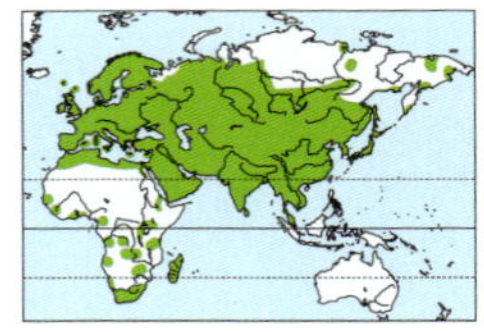

Europe, Asia & Africa; introduced worldwide (modern range not shaded)

LC
Monotypic
HB 7.9–9.3 cm
T 4.7–7.3 cm
W 7–19 g

Western Mediterranean Mouse
Mus spretus

FR Souris sauvage
DE Heckenhausmaus
ES Ratón moruno

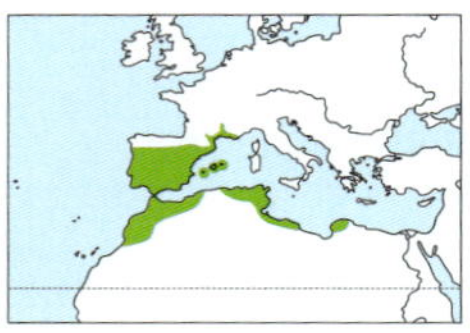

PT, ES (with Balearic Is), S FR, MA, N DZ, TN, N LY

Macedonian Mouse
Mus macedonicus

FR Souris de Macédoine
DE Balkan-Hausmaus
ES Ratón de Macedonia

LC
Monotypic
HB 6.5–9.8 cm
T 5.6–8.2 cm
W 11–30 g

up to 1500 m

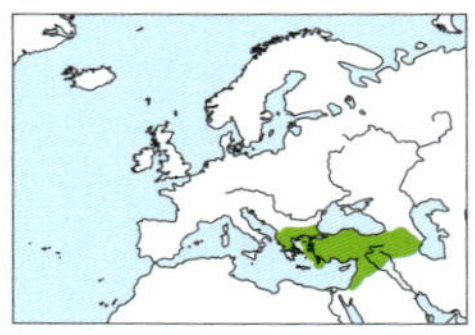

E Balkans, some Aegean Is, TR, Caucasus, Levant, NW IR; pehaps IQ

Cypriot Mouse
Mus cypriacus

FR Souris de Chypre
DE Zypern-Hausmaus
ES Ratón de Chipre

LC
Monotypic
HB 7–9.5 cm
T 6.3–9.2 cm
W 11.5–24 g

0–1605 m

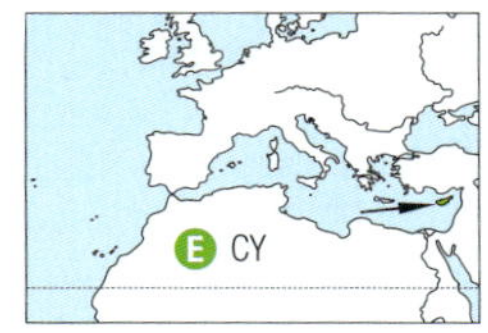

Quite widespread in upland areas

LC
Monotypic
HB 5.5–9.3 cm
T 5–7 cm
W 9–20 g

Mound-building Mouse
Mus spicilegus

FR Souris glaneuse
DE Ährenmaus
ES Ratón de estepa

0–200 m

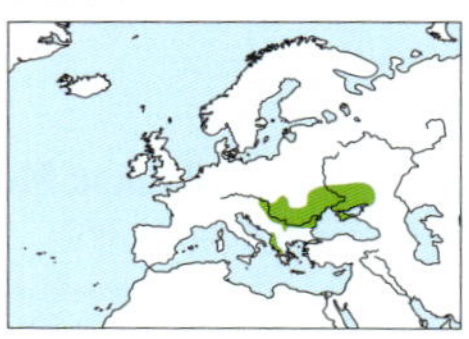

SE Europe in W Balkans and from AT E to BG, RO, UA & RU

LC
Monotypic
HB 7–8 cm
T 4.6–5.2 cm
W c. 16 g

Ethiopian Striped Mouse
Mus imberbis

FR Souris à bande dorsale
DE Äthiopische Streifenmaus
ES Ratón listado de Etiopía

1900–3400 m

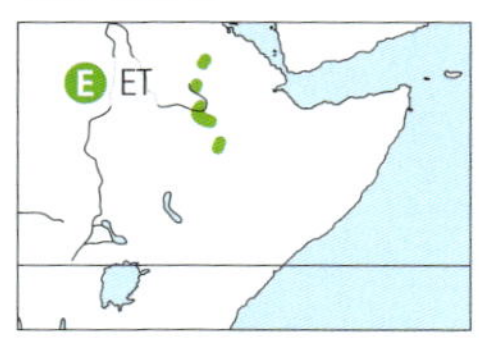

High plateau on both sides of Rift Valley

LC
Monotypic
HB 5.6–8.4 cm
T 4.2–6.5 cm
W 8–16 g

Gray-bellied Mouse
Mus triton

FR Souris triton
DE Triton-Zwergmaus
ES Ratón de vientre gris

1000–3500 m

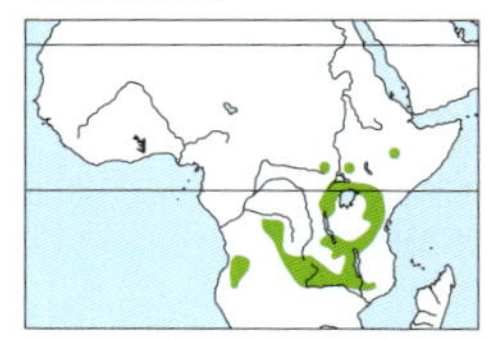

Extreme S SS, S ET (Bale Mts), NE CD, from UG & KE S to AO, ZM & WC MZ

Harenna Mouse
Mus harennensis

FR Souris de l'Harenna
DE Harenna-Zwergmaus
ES Ratón de Harenna

NE
Monotypic
HB 6.9–8.4 cm
T 5.7–6.5 cm

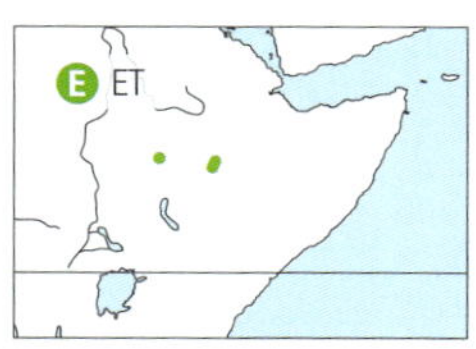

1950–3730 m

Bale Mts (most records from Harenna Forest) & Chingawa Forest

Callewaert's Mouse
Mus callewaerti

FR Souris de Callewaert
DE Callewaert-Zwergmaus
ES Ratón de Callewaert

DD
Monotypic
HB 8–8.9 cm
T 4.3–5.7 cm
W 14–18 g

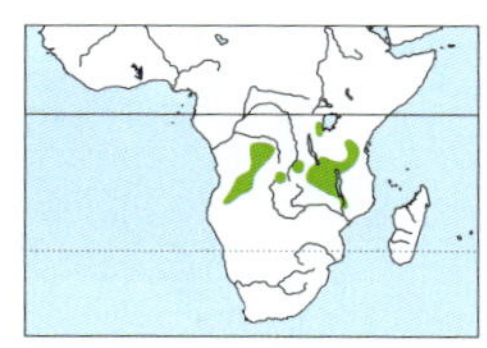

C & NE AO, S & SE CD, TZ (NW, Eastern Arc Mts), much of NE ZM (recorded in extreme NW), MW

Baoule Mouse
Mus baoulei

FR Souris des Baoulés
DE Baoule-Zwergmaus
ES Ratón de Baoule

LC
Monotypic
HB 5.6–7 cm
T 3.8–4.5 cm
W 6–8 g

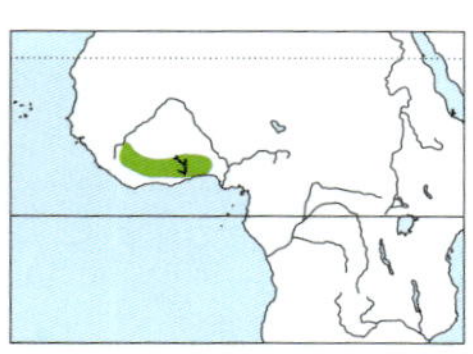

Few localities in W Africa (E GN, CI, GH, TG, BJ, W NG)

Thomas's Mouse
Mus sorella

FR Souris de Thomas
DE Thomas-Zwergmaus
ES Ratón de Thomas

LC
Monotypic
HB 5.1–7.3 cm
T 3.4–4.6 cm

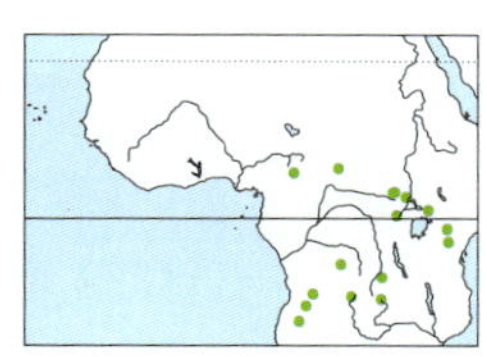

Scattered in N CM, N CF, NE & S CD, UG, KE, N TZ, AO & N ZM

Gounda Mouse
Mus goundae

FR Souris de la Gounda
DE Gounda-Zwergmaus
ES Ratón de Gounda

DD
Monotypic
HB 5.5–6.2 cm
T 3–3.2 cm

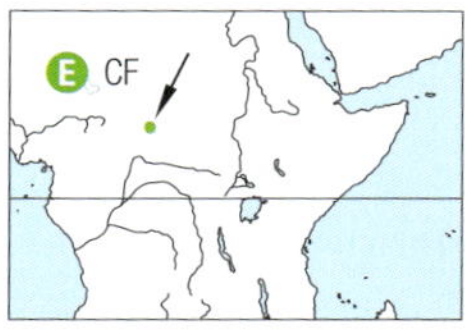

Known only from type locality (close to R Gounda)

Ubangui Mouse
Mus oubanguii

DD
Monotypic
HB 5–6.9 cm
T 2.6–4.1 cm

FR Souris de l'Oubangui
DE Ubangi-Zwergmaus
ES Ratón de Ubangui

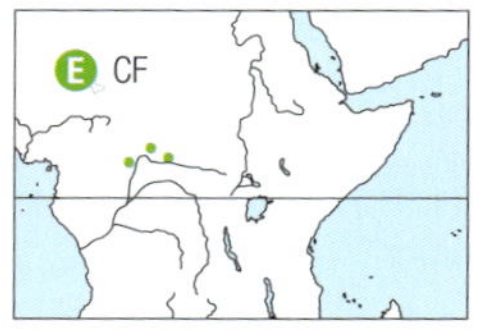

Known from just 3 localities, N of R Oubangui

Neave's Mouse
Mus neavei

FR Souris de Neave
DE Neave-Zwergmaus
ES Ratón de Neave

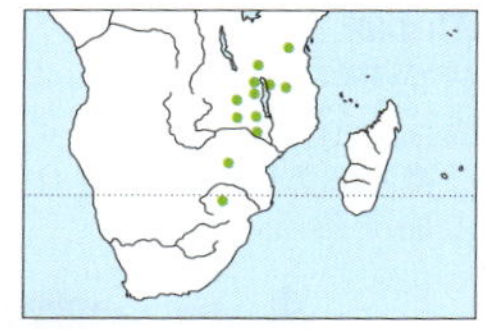

E ZM, MW, TZ; records from N MZ, ZW & N ZA lack genetic verification

DD
Monotypic
HB 5.8–10.6 cm
T 3.3–4.8 cm

Mahomet Mouse
Mus mahomet

FR Souris de Mahomet
DE Mahomet-Zwergmaus
ES Ratón de Mahomet

LC
Monotypic
HB 6.3–7.3 cm
T 4.6–6 cm
W 6–13 g

1500–3200 m

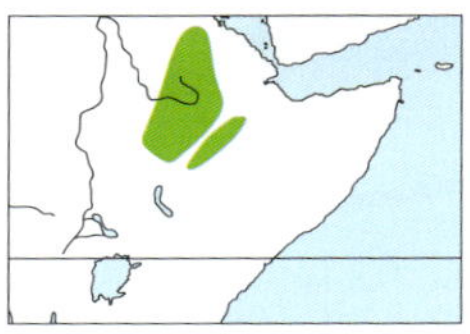

Highlands of ER & ET

Toad Mouse
Mus bufo

FR Souris crapaud
DE Krötenzwergmaus
ES Ratón sapo

LC
Monotypic
HB 6.9–7.7 cm
T 6–6.6 cm

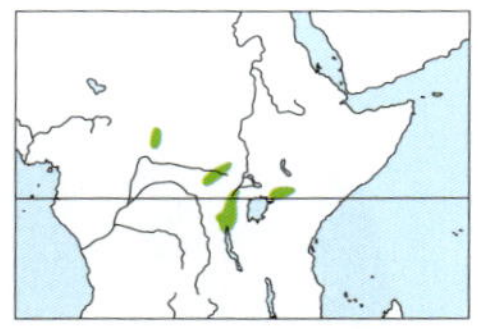

Albertine Rift (E CD, SW UG, RW, BI), C CF, N CD, W KE

Peters's Mouse
Mus setulosus

LC
Monotypic
HB 5.2–8.4 cm
T 4.3–6.3 cm
W 8–23 g

FR Souris hérissée
DE Peters-Zwergmaus
ES Ratón de Peters

500–2300 m

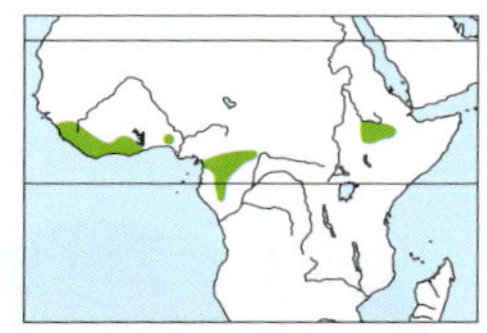

W, C & E Africa, from GN E to GH, W NG, CM, CF, GA, W CG, ET

Hausa Mouse
Mus haussa

LC
Monotypic
HB 4.1–5.8 cm
T 3.1–4.4 cm

FR Souris des Haoussas
DE Hausa-Zwergmaus
ES Ratón de los hausa

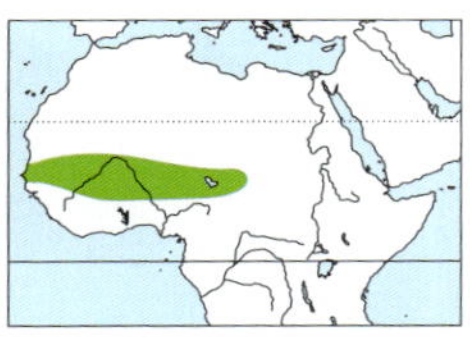

Sahelo-Sudanian savannas of W Africa from S MR & SN E to TD

Matthey's Mouse
Mus mattheyi

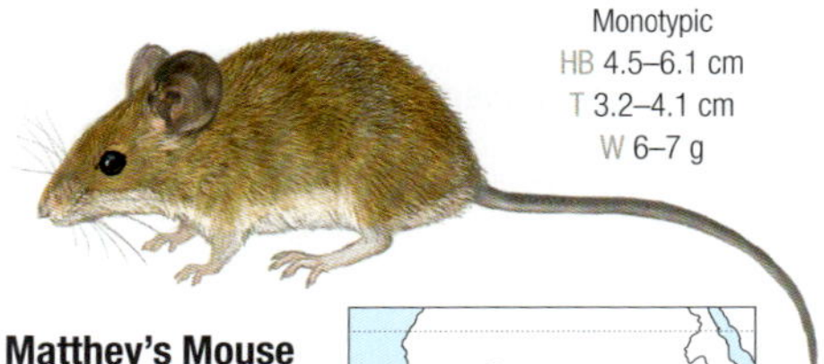

LC
Monotypic
HB 4.5–6.1 cm
T 3.2–4.1 cm
W 6–7 g

FR Souris de Matthey
DE Matthey-Zwergmaus
ES Ratón de Matthey

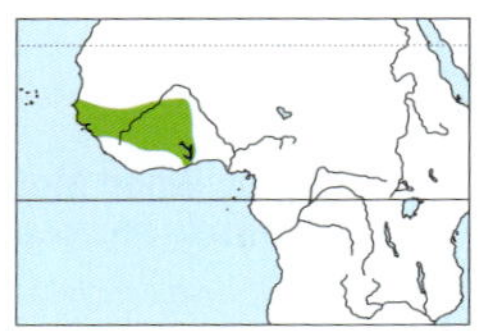

Savannas of W Africa from SN E to BF, GH & TG

Desert Pygmy Mouse
Mus indutus

LC
Monotypic
HB 4.5–6.5 cm
T 3–5.2 cm
W 3–8 g

FR Souris du désert
DE Wüsten-Zwergmaus
ES Ratón pigmeo de desierto

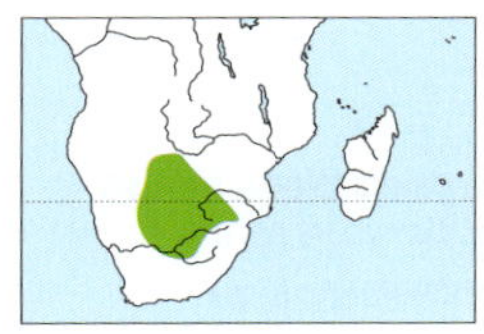

BW & N ZA; perhaps S AO, SW ZM, S MW, NA, W ZW & S ZA

West African Pygmy Mouse
Mus musculoides

FR Souris de Temminck
DE Temminck-Zwergmaus
ES Ratón pigmeo de África occidental

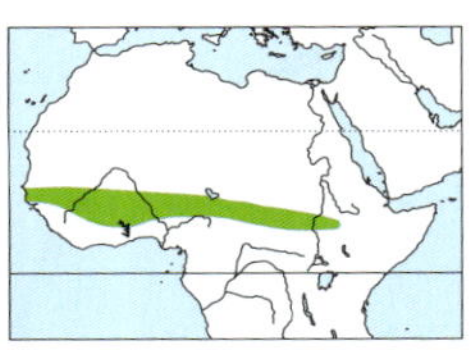

Restricted to Sudanian savanna belt from SN to W ET

LC
Monotypic
HB 5.8–7.5 cm
T 4.7–5.3 cm
W 6–9 g

Delicate Mouse
Mus tenellus

FR Souris délicate
DE Zierliche Zwergmaus
ES Ratón delicado

600–2000 m

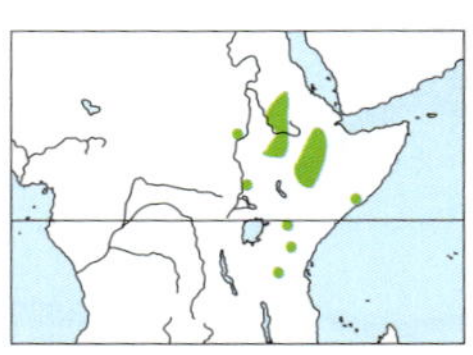

ET and very patchy in SD, SS, SO, KE & TZ

LC
Monotypic
HB 4.4–6.2 cm
T 3.4–4.3 cm

Sub-Saharan Pygmy Mouse
Mus minutoides

FR Souris naine
DE Eigentliche Afrika-Zwergmaus
ES Ratón pigmeo de Sahel

LC
Monotypic
HB 4–7.2 cm
T 2.7–5.7 cm
W 3–12 g

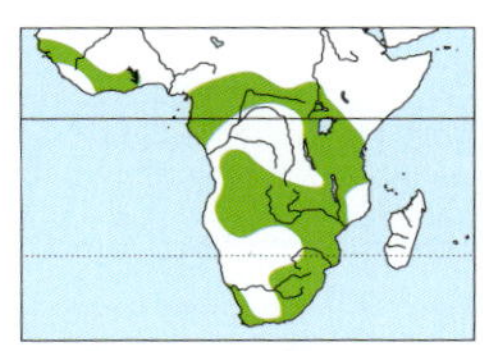

Very widespread across sub-Saharan Africa from W, C & E to S Africa

Setzer's Mouse
Mus setzeri

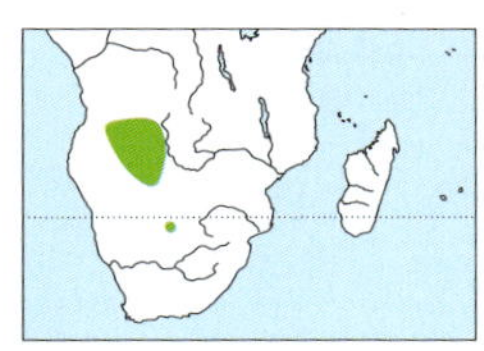

SE AO, W ZM, NE NA, NW BW, recorded in S BW

FR Souris de Setzer
DE Setzer-Zwergmaus
ES Ratón de Setzer

LC
Monotypic
HB c. 5.4 cm
T 3.1–4.8 cm
W 5–9 g

Genus *HEIMYSCUS*

(1 species)

African Smoky Mouse
Heimyscus fumosus

FR Rat de De Balsac
DE Afrikanische Rauchmaus
ES Ratón ahumado de África

LC
Monotypic
HB 8–9.8 cm
T 8.6–11.1 cm
W 16–31 g

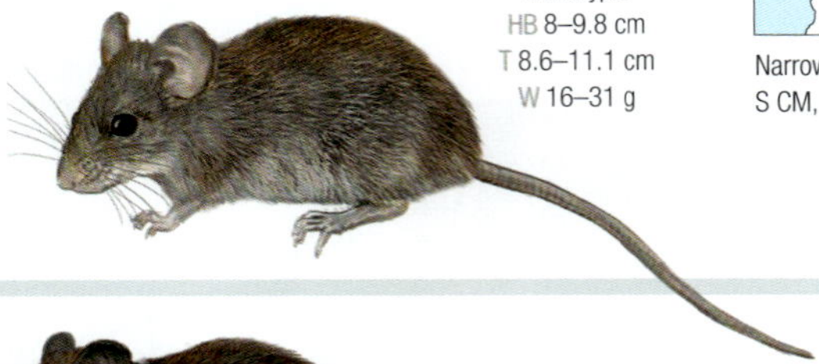

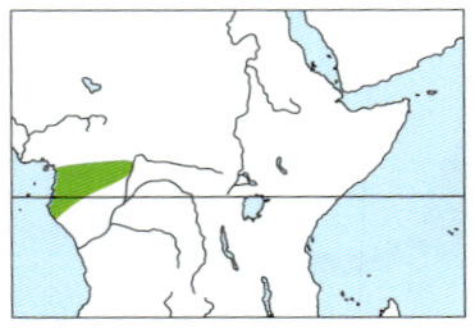

Narrowly in C African rainforests of S CM, SW CF, GQ, GA & NW CG

Genus *HYLOMYSCUS*

(22 species)

Mount Kenya Wood Mouse
Hylomyscus endorobae

FR Hylomysque du Kenya
DE Mount-Kenya-Afrikawaldmaus
ES Ratón de bosque del Monte Kenia

LC
Monotypic
HB 9.2–12 cm
T 12.3–17.2 cm

above 1900 m

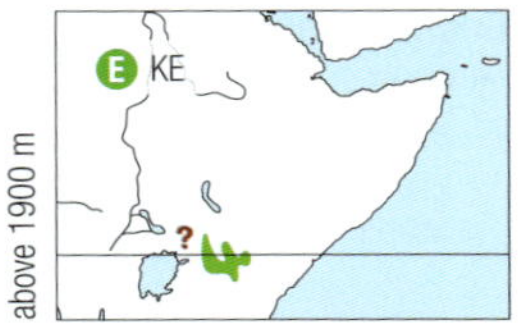

Highlands (Mau Escarpment, Aberdare Mts, Mt Kenya; possibly Mt Elgon)

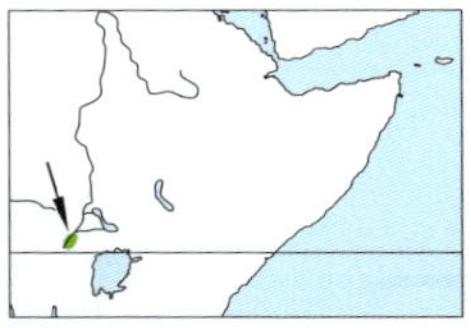

Higher slopes of Rwenzori Mts (NE CD, SW UG)

Montane Wood Mouse
Hylomyscus denniae

FR Hylomysque du Rwenzori
DE Berg-Afrikawaldmaus
ES Ratón de bosque de montaña

LC
Monotypic
HB 9–11.5 cm
T 12.2–15.7 cm

LC
Monotypic
HB 8.1–10.7 cm
T 11.7–15.5 cm
W 16–30 g

Albertine Rift Wood Mouse
Hylomyscus vulcanorum

FR Hylomysque des volcans
DE Vulkan-Afrikawaldmaus
ES Ratón de bosque del Valle Albertino

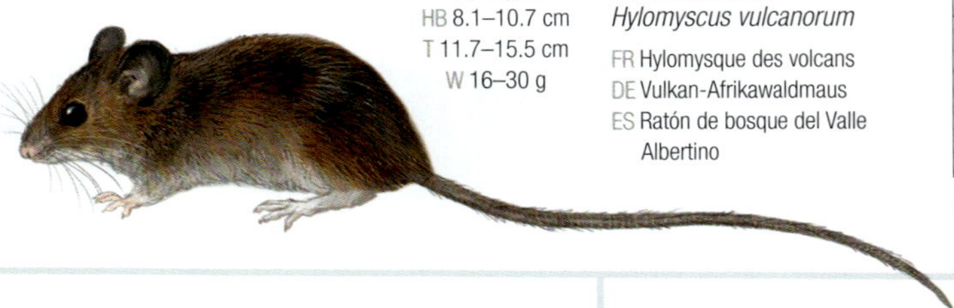

above 1670 m

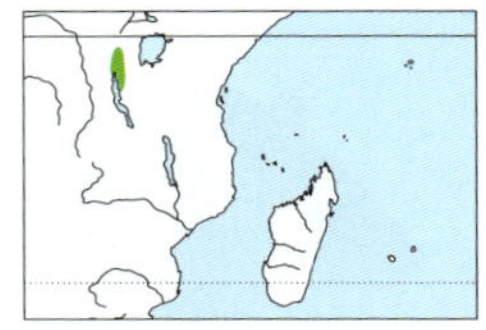

C Albertine Rift in E CD, RW & BI

EN
Monotypic
HB 9.8–11.5 cm
T 11.1–14.1 cm
W 25–36 g

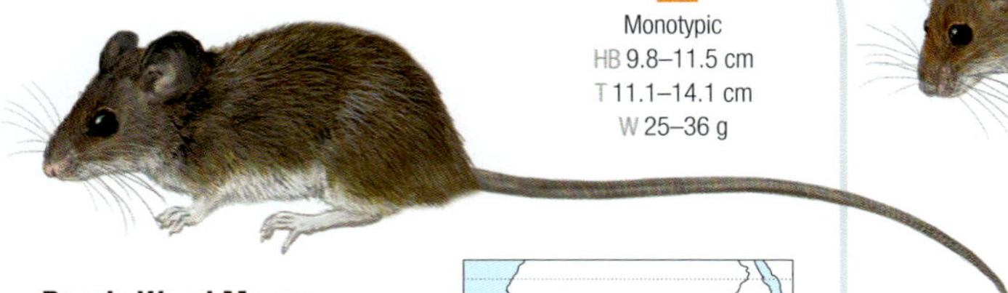

Baer's Wood Mouse
Hylomyscus baeri

FR Hylomysque de Baer
DE Baer-Afrikawaldmaus
ES Ratón de bosque de Baer

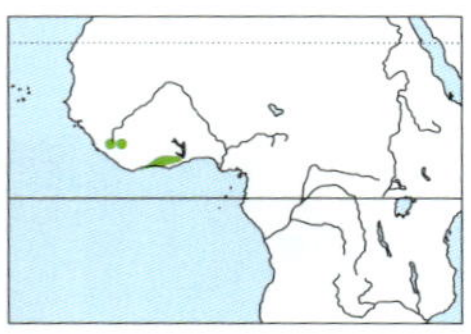

Known from few localities in SL, GN, CI & GH

Beaded Wood Mouse
Hylomyscus aeta

FR Hylomysque de Thomas
DE Perlen-Afrikawaldmaus
ES Ratón de bosque de borde

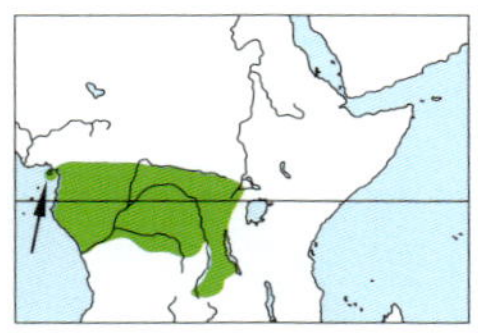

From S CM, GQ (with Bioko I) & GA E through Congo Basin to Albertine Rift

EN
Monotypic
HB 8.9–11 cm
T 13–15 cm
W 25–41 g

Mount Oku Wood Mouse
Hylomyscus grandis

FR Hylomysque de l'Oku
DE Mount-Oku-Afrikawaldmaus
ES Ratón de bosque de Oku

c. 2100 m

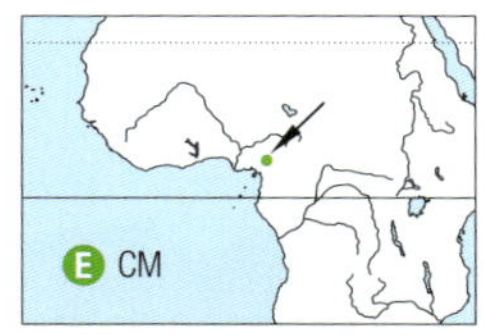

Known from 4 specimens of upper slopes of Mt Oku

Lesser Wood Mouse
Hylomyscus parvus

FR Petit Hylomysque
DE Kleine Afrikawaldmaus
ES Ratón de bosque menor

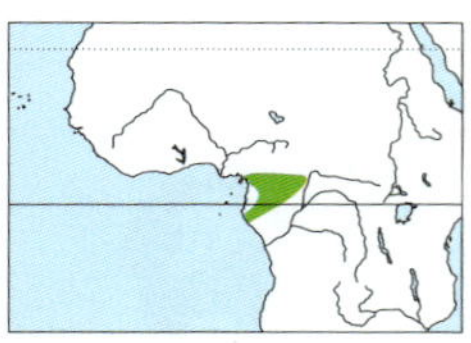

Scattered localities in C Africa (W & S CM, SW CF, GA, N CG, N CD)

LC
Monotypic
HB 5.6–7.7 cm
T 8.7–12.4 cm
W 10–16 g

Pygmy Wood Mouse
Hylomyscus pygmaeus

FR Hylomysque pygmée
DE Zwerg-Afrikawaldmaus
ES Ratón de bosque pigmeo

c. 360 m

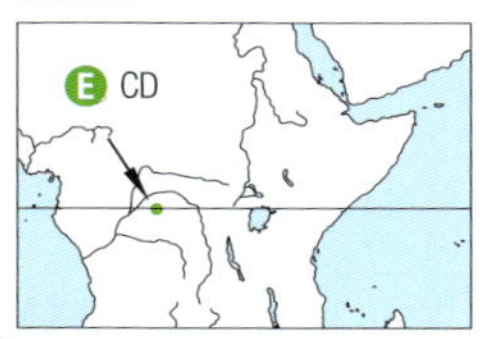

Known only from type locality (Baliko, Tshuapa Prov)

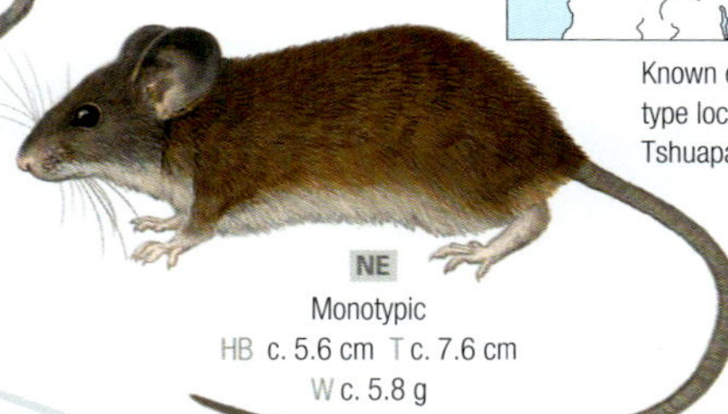

NE
Monotypic
HB c. 5.6 cm T c. 7.6 cm
W c. 5.8 g

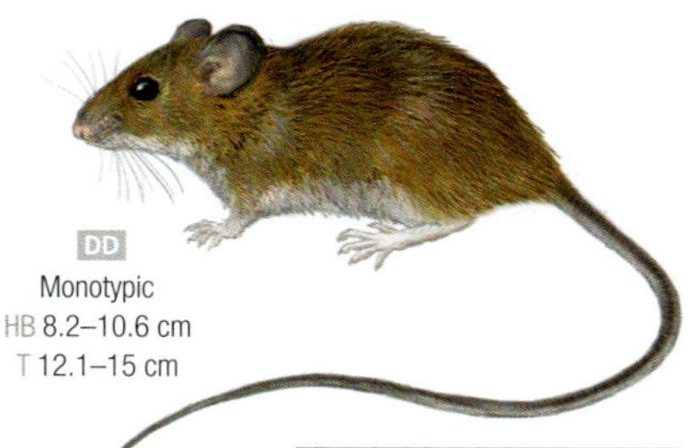

DD
Monotypic
HB 8.2–10.6 cm
T 12.1–15 cm

Heinrichs' Wood Mouse
Hylomyscus heinrichorum

FR Hylomysque des Heinrich
DE Heinrich-Afrikawaldmaus
ES Ratón de bosque de los Heinrich

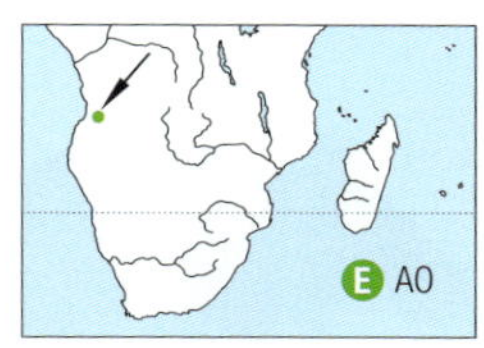

W highlands (Mt Moco, Mt Soque)

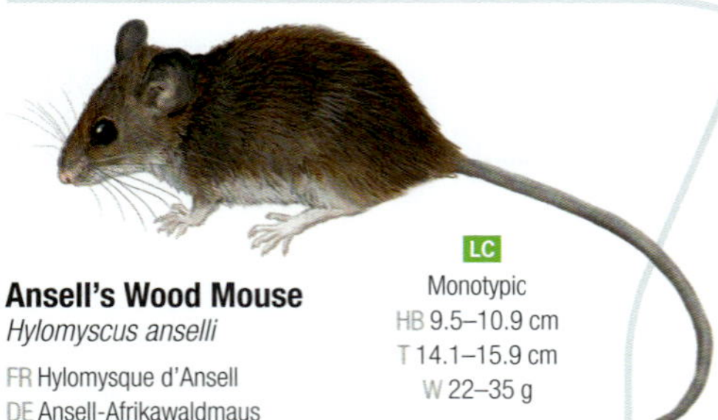

Ansell's Wood Mouse
Hylomyscus anselli

FR Hylomysque d'Ansell
DE Ansell-Afrikawaldmaus
ES Ratón de bosque de Ansell

LC
Monotypic
HB 9.5–10.9 cm
T 14.1–15.9 cm
W 22–35 g

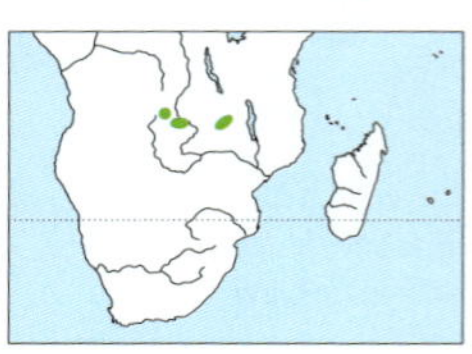

1220–2300 m

Ufipa Plateau (SW TZ) & highlands in N ZM; perhaps neighboring CD & AO

NE
Monotypic
HB c. 10.2 cm
T c. 14.6 cm
W c. 27 g

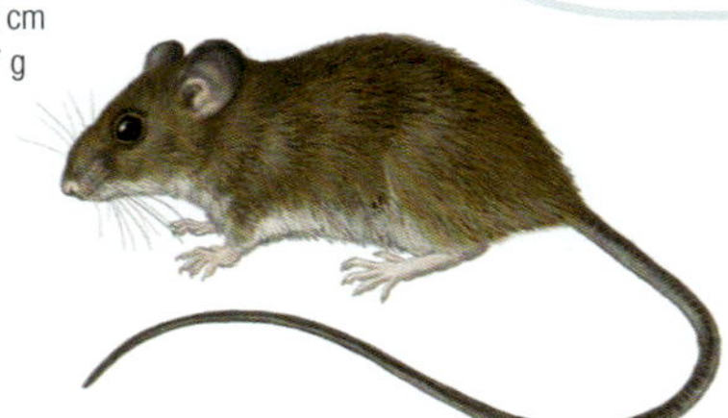

Stanley's Wood Mouse
Hylomyscus stanleyi

FR Hylomysque de Stanley
DE Stanley-Afrikawaldmaus
ES Ratón de bosque de Stanley

c. 2200 m

Known only from Mbizi Forest R

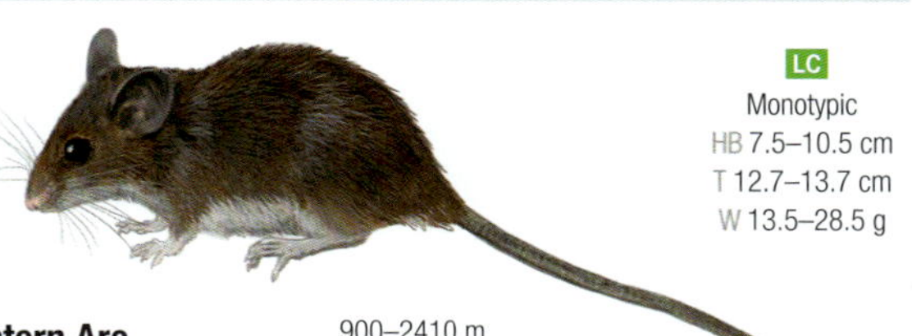

LC
Monotypic
HB 7.5–10.5 cm
T 12.7–13.7 cm
W 13.5–28.5 g

Eastern Arc Wood Mouse
Hylomyscus arcimontensis

FR Hylomysque de l'Arc
DE Östliche Afrikawaldmaus
ES Ratón de bosque del Arco Oriental

900–2410 m

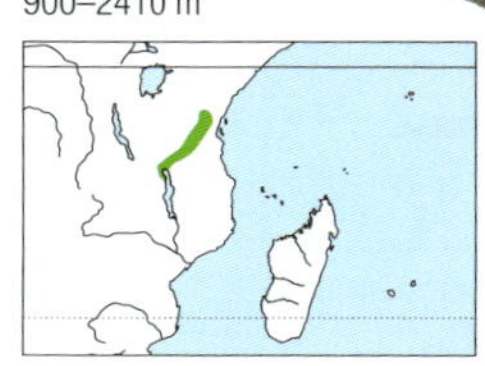

TZ (Eastern Arc Mts from South Pare Mts S to Mt Rungwe), N MW (Misuku Mts)

Mahale Wood Mouse
Hylomyscus mpungamachagorum

FR Hylomysque des Mahale
DE Mahale-Afrikawaldmaus
ES Ratón de bosque de Mahale

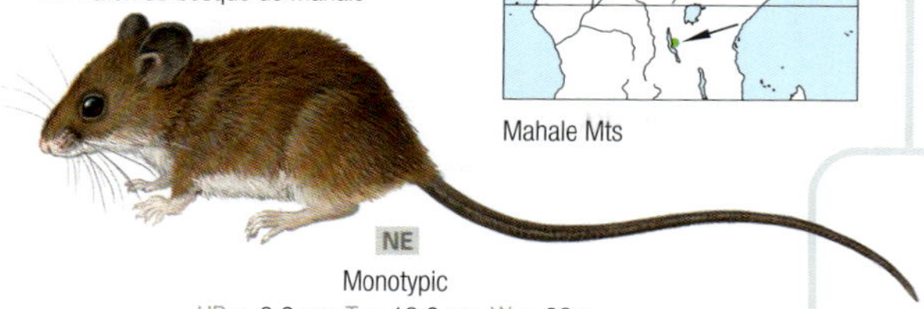

c. 2100 m

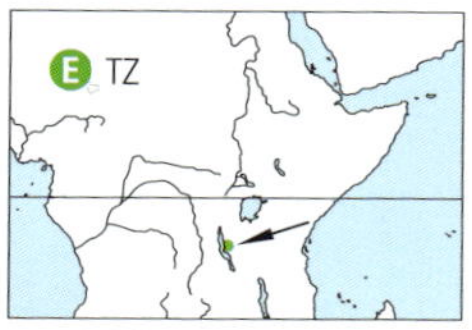

Mahale Mts

NE
Monotypic
HB c. 9.8 cm T c. 13.8 cm W c. 22 g

Mother Ellen's Wood Mouse
Hylomyscus thornesmithae

FR Hylomysque de Thorne Smith
DE Ellen-Afrikawaldmaus
ES Ratón de bosque de Ellen

c. 320 m

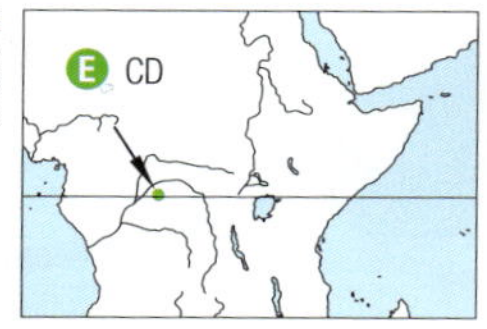

Known from 2 locations, c. 250 km S of R Congo, off right bank of R Tshuapa

NE
Monotypic
HB c. 8.3 cm
T c. 12.1 cm W c. 13.9 g

Kerbis-Peterhans's Wood Mouse
Hylomyscus kerbispeterhansi

FR Hylomysque de Kerbis-Peterhans
DE Kerbis-Peterhans-Afrikawaldmaus
ES Ratón de bosque de Kerbis-Peterhans

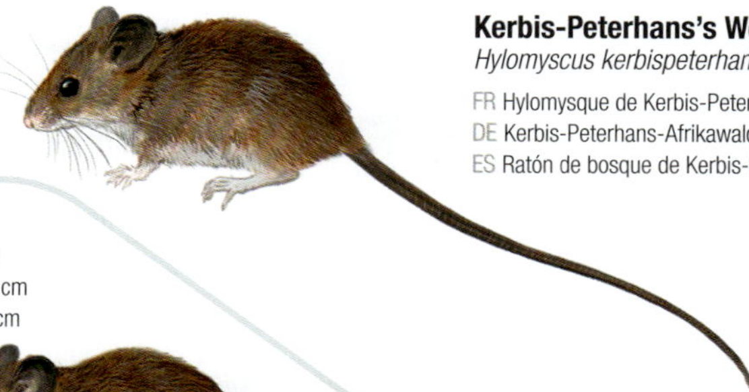

above 2300 m

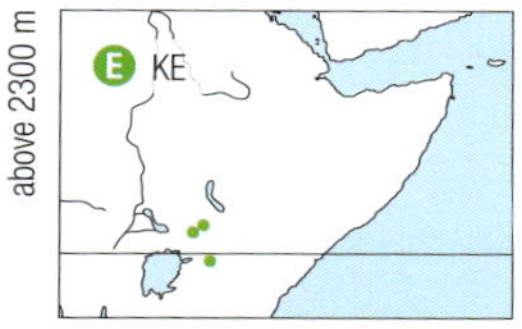

Mt Elgon, Cherangani Hills & Mau Escarpment, W of Rift Valley; possibly UG (W Mt Elgon)

LC
Monotypic
HB 7.6–10 cm
T 11.9–13.8 cm
W 18–26 g

Allen's Wood Mouse
Hylomyscus alleni

FR Hylomysque d'Allen
DE Allen-Afrikawaldmaus
ES Ratón de bosque de Allen

LC
Monotypic
HB 6.7–9.6 cm
T 10.1–15.2 cm
W 10–30 g

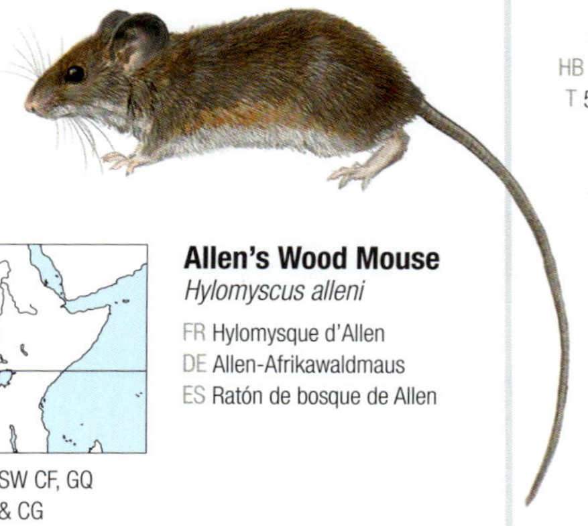

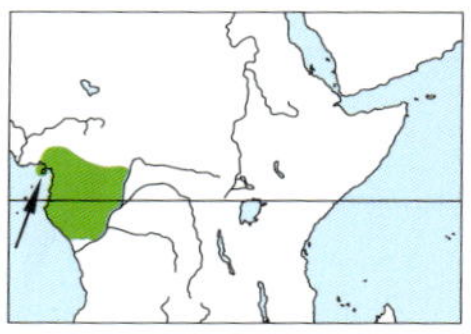

C Africa in S CM, SW CF, GQ (with Bioko I), GA & CG

Dahomey Gap Wood Mouse
Hylomyscus pamfi

FR Hylomysque du Dahomey
DE Dahomey-Afrikawaldmaus
ES Ratón de bosque de Dahomey

DD
Monotypic
HB 5.6–10.5 cm
T 5.9–12.1 cm

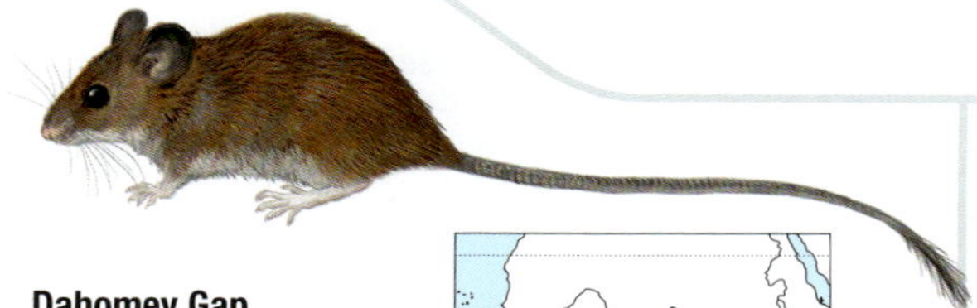

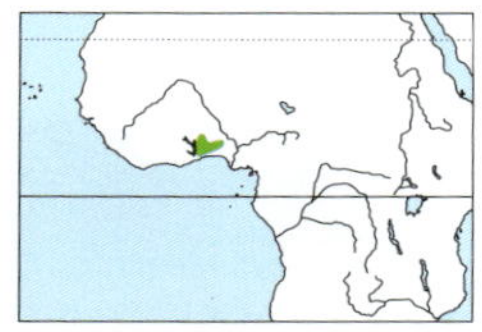

Dahomey Gap in TG, BJ & W NG; presumably extreme E GH, E of R Volta

West African Wood Mouse
Hylomyscus simus

FR Hylomysque de Brosset
DE Westliche Afrikawaldmaus
ES Ratón de bosque de África occidental

NE
Monotypic
HB 6.6–12.8 cm
T 5.2–14.3 cm
W 12–22 g

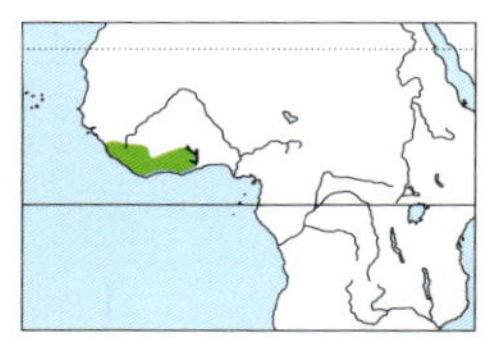

SW W Africa from SL & GN E to GH

Walter Verheyen's Wood Mouse
Hylomyscus walterverheyeni

FR Hylomysque de Verheyen
DE Walter-Verheyen-Afrikawaldmaus
ES Ratón de bosque de Walter Verheyen

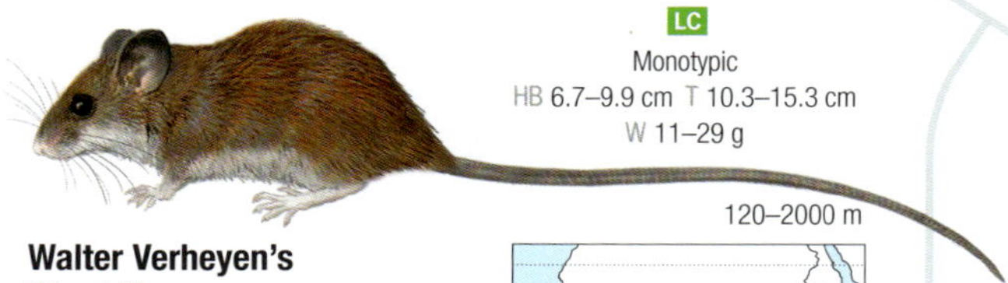

LC
Monotypic
HB 6.7–9.9 cm T 10.3–15.3 cm
W 11–29 g

120–2000 m

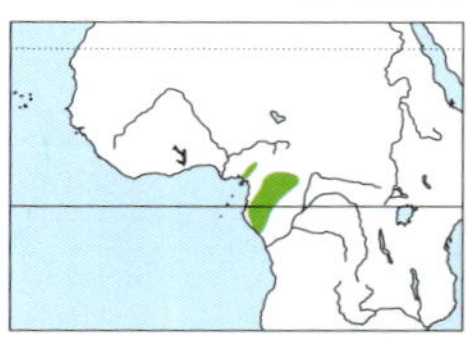

C Africa in Cameroon volcanic line (SW CM, S CF, GA, W CG; perhaps E NG)

Stella Wood Mouse
Hylomyscus stella

FR Hylomysque stella
DE Stella-Afrikawaldmaus
ES Ratón de bosque de estela

800–2250 m

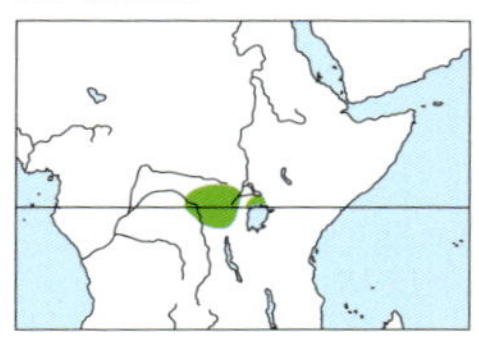

E Congo Basin in CD, RW & BI, extending to N L Victoria (S UG)

LC
Monotypic
HB 7.2–10.4 cm
T 11.1–14 cm
W 14–24 g

Tanzanian Wood Mouse
Hylomyscus kaimosae

FR Hylomysque de Kaimosi
DE Tansania-Afrikawaldmaus
ES Ratón de bosque de Tanzania

NE
Monotypic
HB c. 8.6 cm
T c. 13.5 cm

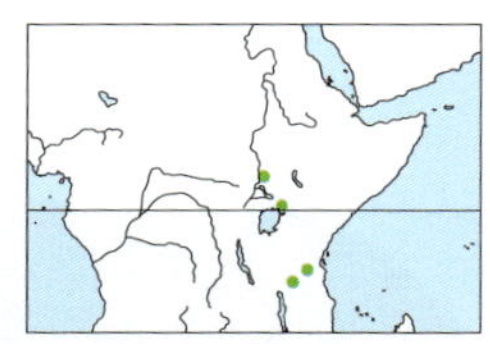

Scattered in E Africa: Imoteng Mts (SS), Kakamega Forest (KE), Mbete & Mufindi (TZ)

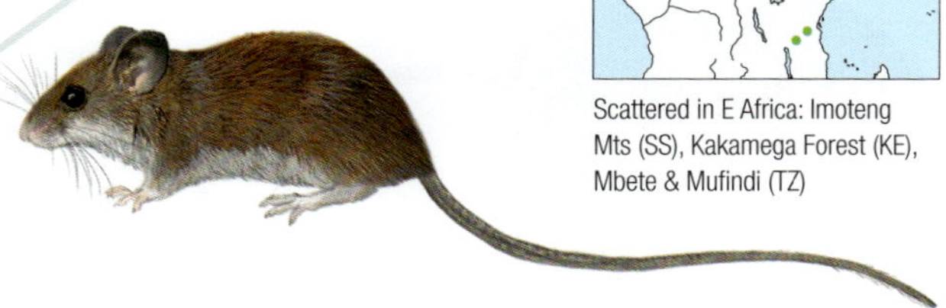

Angolan Wood Mouse
Hylomyscus carillus

FR Hylomysque d'Angola
DE Angola-Afrikawaldmaus
ES Ratón de bosque de Angola

LC
Monotypic
HB 8.7–9.7 cm
T 11.8–14.6 cm

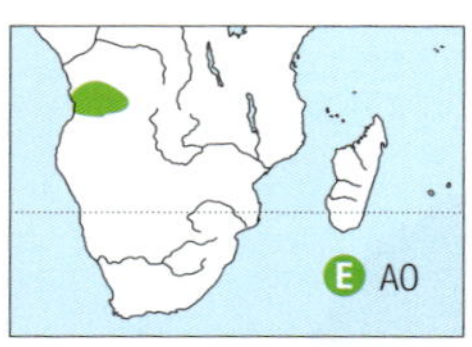

W & C; perhaps neighboring CD

Genus *SERENGETIMYS*

(1 species)

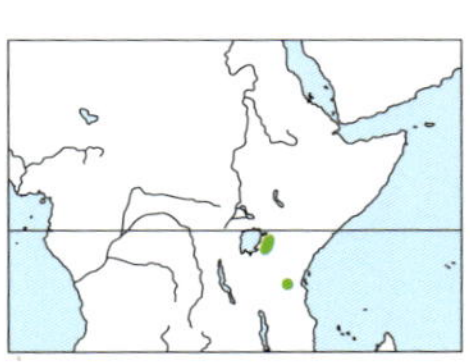

Mostly Serengeti Plains on KE–TZ border, recorded in EC TZ; perhaps RW

Dwarf Serengeti Mouse
Serengetimys pernanus

FR Rat du Serengeti
DE Serengeti-Zwergmaus
ES Ratón multimamas enano

DD
Monotypic
HB 7.3–8.8 cm
T 6–7.8 cm

Genus *MONTEMYS*

(1 species)

Delicate Montane Mouse
Montemys delectorum

FR Rat délicat
DE Zierliche Montanmaus
ES Ratón de pelaje suave delicado

1000–2400 m

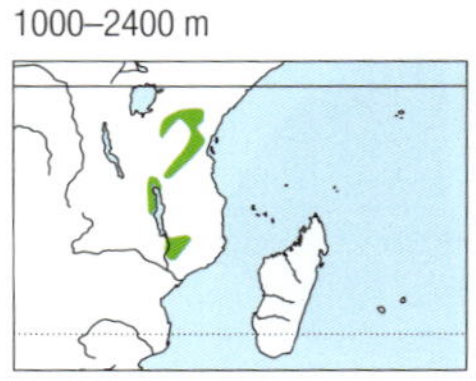

LC
Monotypic
HB 9–12 cm
T 11–14.2 cm
W 20–38 g

Eastern Arc Mts and adjacent highlands in KE & TZ, S to MW & MZ

Genus *CHINGAWAEMYS*

(1 species)

Chingawa Forest Rat
Chingawaemys rarus

FR Rat de Chingawa
DE Chingawa-Waldratte
ES Ratón de Chingawa

c. 2340 m

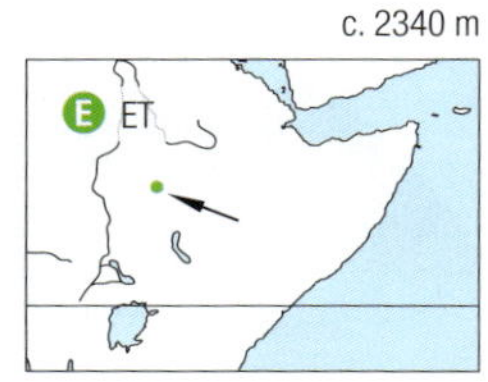

Known only from Chingawa Forest (25 km N of Tepi)

NE
Monotypic
HB c. 10.2 cm T c. 13.6 cm
W c. 27 g

LC
Monotypic
HB 9.2–12.4 cm T 11.5–15.1 cm
W 28–64 g

above 1800 m

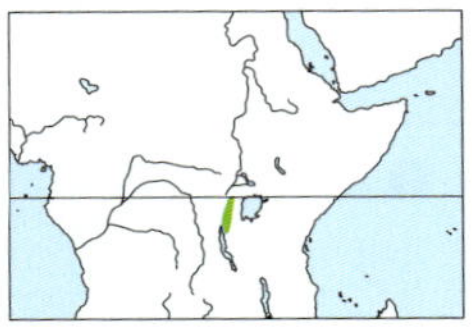

Albertine Rift from SW UG to SW BI

De Graaff's Soft-furred Mouse
Praomys degraaffi

FR Praomys de De Graaff
DE De-Graaff-Weichhaarratte
ES Ratón de pelaje suave de De Graaff

Genus *PRAOMYS*

(13 species)

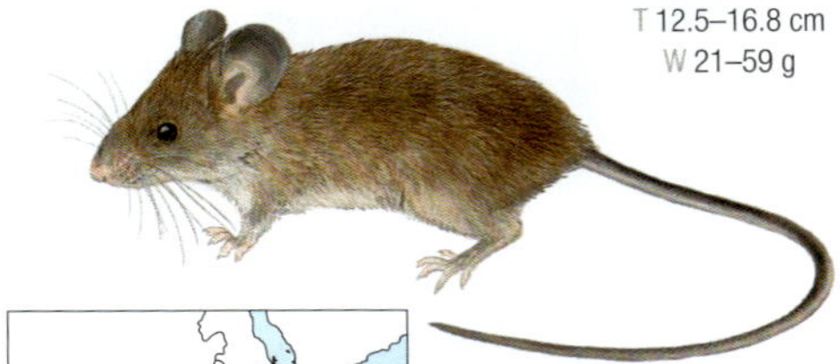

LC
Monotypic
HB 9.4–13.1 cm
T 12.5–16.8 cm
W 21–59 g

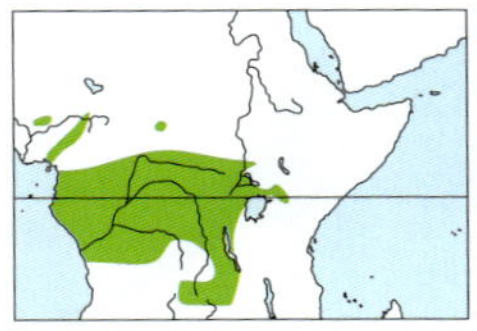

Tropical Africa from NG E to KE, S to N AO & ZM

Jackson's Soft-furred Mouse
Praomys jacksoni

FR Praomys de Jackson
DE Jackson-Weichhaarratte
ES Ratón de pelaje suave de Jackson

LC
Monotypic
HB 8–11.8 cm
T 10.2–14.5 cm
W c. 26 g

Least Soft-furred Mouse
Praomys minor

FR Praomys mineur
DE Zwerg-Weichhaarratte
ES Ratón de pelaje suave mínimo

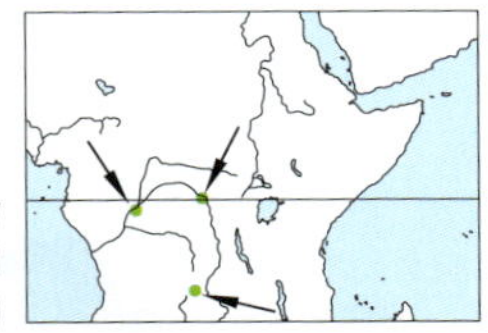

CD (scattered S & W of R Congo), NW ZM (Mwinilunga District)

NT
Monotypic
HB 11–13.9 cm T 11.3–17.1 cm
W 31–57 g

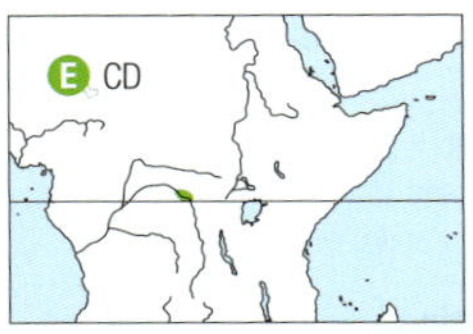

Along both banks of middle-upper reaches of R Congo

Riverine Soft-furred Mouse
Praomys mutoni

FR Praomys de Masako
DE Muton-Weichhaarratte
ES Ratón de pelaje suave de la ribera del Congo

Dalton's Soft-furred Mouse
Praomys daltoni

FR Praomys de Dalton
DE Dalton-Weichhaarratte
ES Ratón de pelaje suave de Dalton

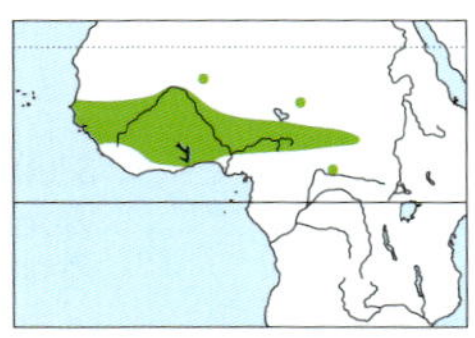

W Africa, from SN E to NW SS; outlying records in NE ML, C TD & S CF

LC
Monotypic
HB 9.8–13.3 cm
T 10.7–14.2 cm
W 24–66 g

Tullberg's Soft-furred Mouse
Praomys tullbergi

FR Praomys de Tullberg
DE Tullberg-Weichhaarratte
ES Ratón de pelaje suave de Tullberg

c. 1200 m

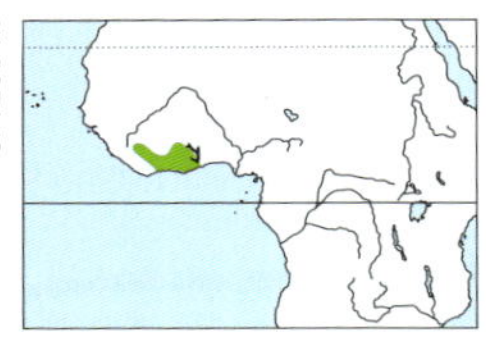

Upper Guinean forest from SE GN E to GH, W of R Volta

LC
Monotypic
HB 10.8–13.1 cm
T 11.5–15.3 cm
W 28–54 g

Misonne's Soft-furred Mouse
Praomys misonnei

FR Praomys de Misonne
DE Misonne-Weichhaarratte
ES Ratón de pelaje suave de Misonne

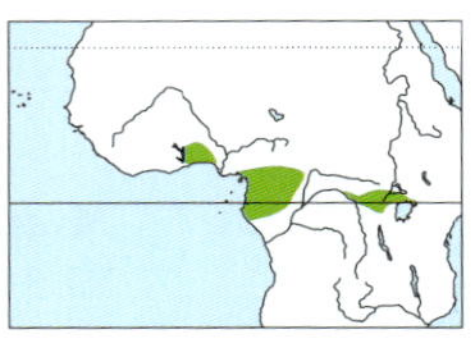

Tropical Africa from GH (R Volta) S to GA, E to SW KE (Kakamega Forest)

LC
Monotypic
HB 8.9–12.3 cm
T 11.3–16.3 cm

LC
Monotypic
HB 10.9–14.7 cm
T 12.4–17.1 cm
W 31–73 g

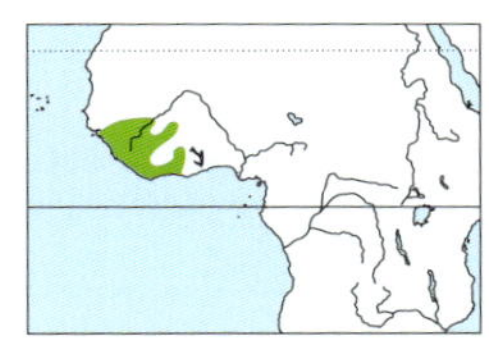

up to 1300 m

Tropical W Africa from SN (Basse Casamance) to CI–GH border

West African Soft-furred Mouse
Praomys rostratus

FR Praomys de Miller
DE Westafrika-Weichhaarratte
ES Ratón de pelaje suave de África occidental

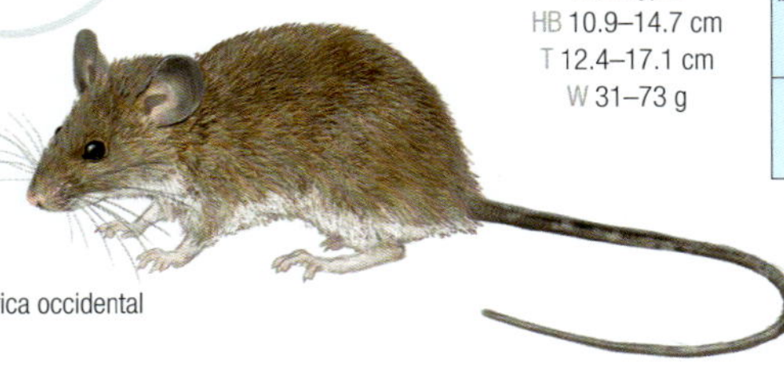

Coetzee's Soft-furred Mouse
Praomys coetzeei

FR Praomys de Coetzee
DE Coetzee-Weichhaarratte
ES Ratón de pelaje suave de Coetzee

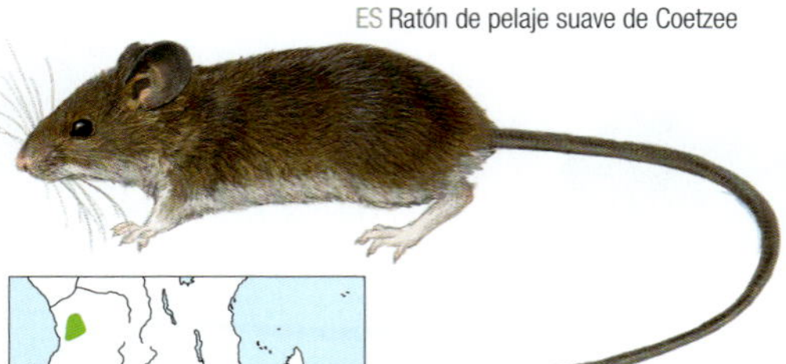

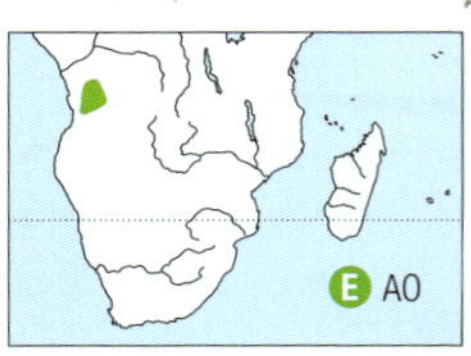

In NW, perhaps NE

DD
Monotypic
HB 9–14 cm
T 12.3–16.3 cm

Petter's Soft-furred Mouse
Praomys petteri

FR Praomys de Petter
DE Petter-Weichhaarratte
ES Ratón de pelaje suave de Petter

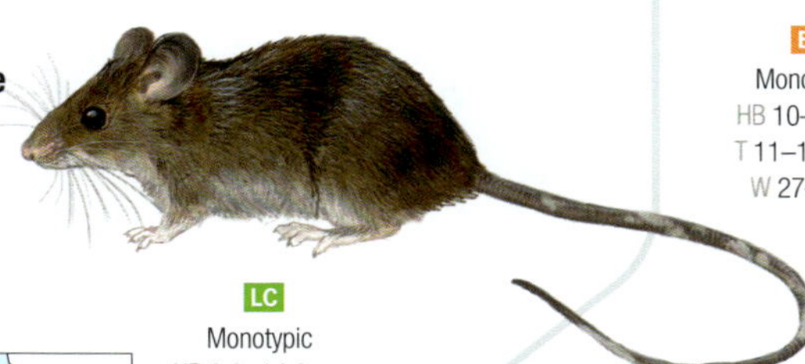

LC
Monotypic
HB 9.9–14.3 cm
T 10–16.5 cm
W 40–61 g

below 500 m

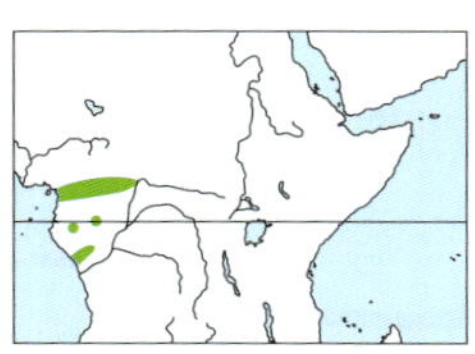

C Africa from R Sanaga (CM) E to R Ubangi (CF), S to SW CG

EN
Monotypic
HB 10–13 cm
T 11–14.7 cm
W 27–62 g

Cameroon Soft-furred Mouse
Praomys morio

FR Praomys du Cameroun
DE Kamerun-Weichhaarratte
ES Ratón de pelaje suave de Camerún

above 1000 m

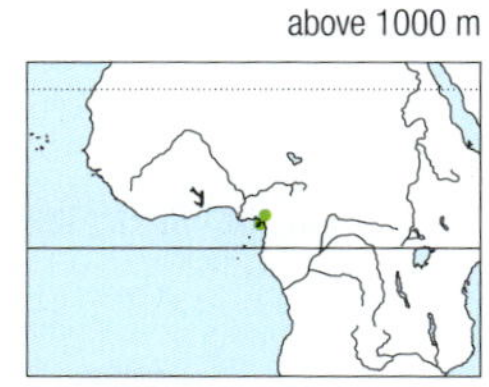

Mt Cameroon (W CM), Bioko I (GQ)

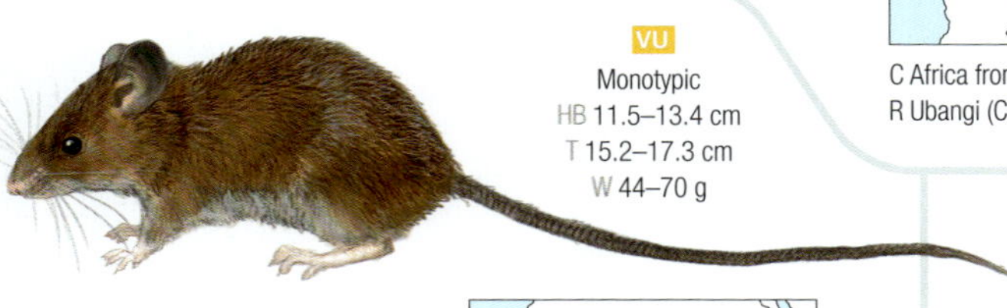

VU
Monotypic
HB 11.5–13.4 cm
T 15.2–17.3 cm
W 44–70 g

Hartwig's Soft-furred Mouse
Praomys hartwigi

FR Praomys de Hartwig
DE Hartwig-Weichhaarratte
ES Ratón de pelaje suave de Hartwig

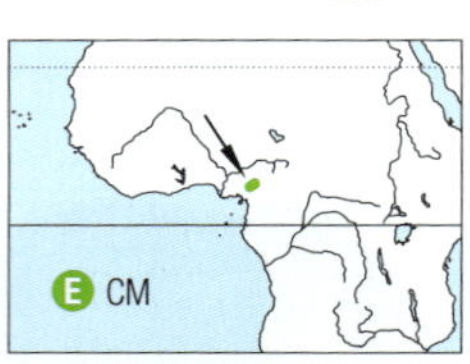

Bamenda Highlands (Mt Oku, Mt Manengouba)

Gotel Mountain Soft-furred Mouse
Praomys obscurus

FR Praomys des Gotel
DE Gotel-Weichhaarratte
ES Ratón de pelaje suave de Gotel

EN
Monotypic
HB 10.8–13.7 cm
T 14.9–17.4 cm
W 34–60 g

1900–2300 m

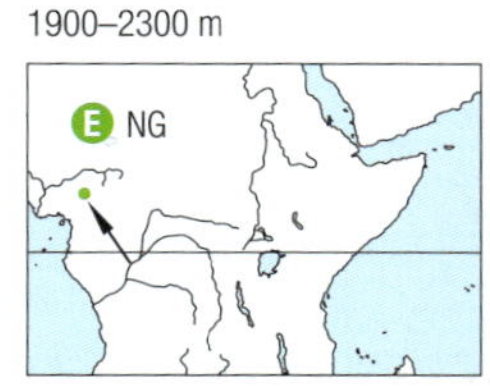

Gotel Mts

Genus *CONGOMYS*

(2 species)

Verschuren's Swamp Mouse
Congomys verschureni

FR Congomys de Verschuren
DE Verschuren-Weichhaarratte
ES Ratón de pelaje suave de Verschuren

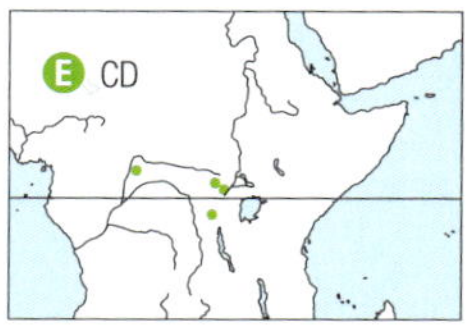

up to 1800 m

Scattered localities N & E of R Congo

DD
Monotypic
HB 10.9–13.5 cm
T 12.3–16.2 cm
W 40–65 g

Lukolela Swamp Mouse
Congomys lukolelae

FR Congomys de Lukolela
DE Lukolela-Weichhaarratte
ES Ratón de pelaje suave de Lukolela

LC
Monotypic
HB 10.9–14.4 cm
T 11.6–15.9 cm
W 27–57 g

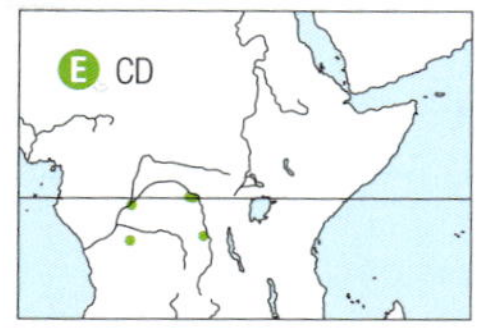

Scattered localities S & W of R Congo

Genus *MYOMYSCUS*

(1 species)

Verreaux's Meadow Mouse
Myomyscus verreauxii

FR Myomyscus de Verreaux
DE Verreaux-Wiesenratte
ES Ratón de campiña de Verreaux

LC
Monotypic
HB 9–11.3 cm
T 13–14.8 cm
W 22–52 g

Western Cape (from R Olifants E to Plettenberg Bay & Knysna)

Genus *COLOMYS*

(4 species)

East African Wading Rat
Colomys goslingi

FR Rat de Gosling
DE Afrikanische Waldbachmaus
ES Ratón vadeador de Gosling

LC
3 sspp
HB 10.9–15.1 cm
T 14.3–18.2 cm
W 41–72 g

470–2740 m

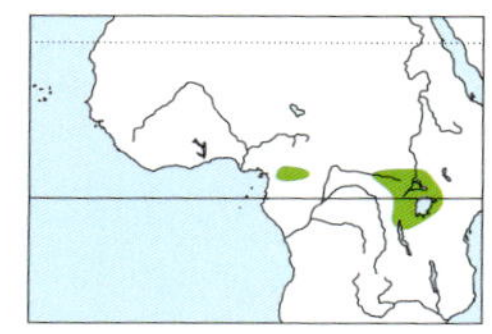

Disjunct in S CM and E CD, S SS, UG, W KE, RW & BI

Cameroonean Wading Rat
Colomys eisentrauti

FR Rat du Cameroun
DE Eisentraut-Waldbachmaus
ES Ratón vadeador de Camerún

NE
Monotypic
HB 11.6–13.7 cm
T 16.3–18.9 cm
W 62–82 g

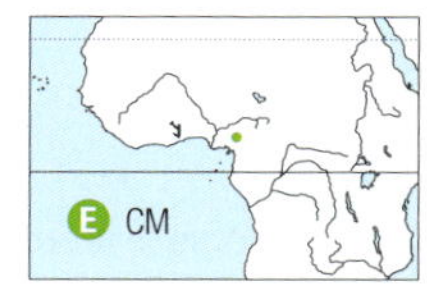

SW Bamenda Highlands

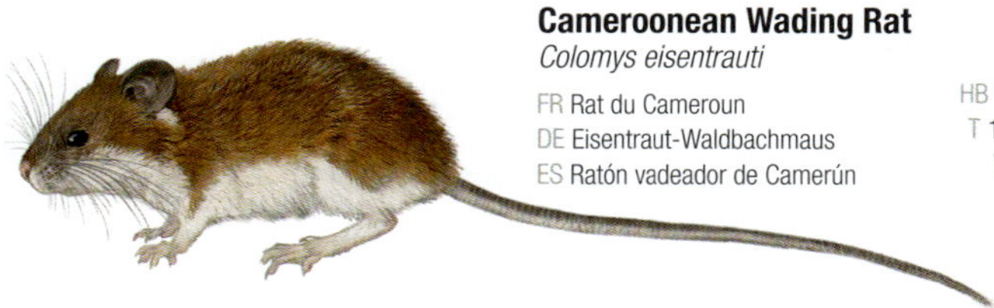

Congo Wading Rat
Colomys lumumbai

FR Rat du Congo
DE Kongo-Waldbachmaus
ES Ratón vadeador de Lumumba

NE
Monotypic
HB 11.2–14.6 cm
T 14.2–18.2 cm

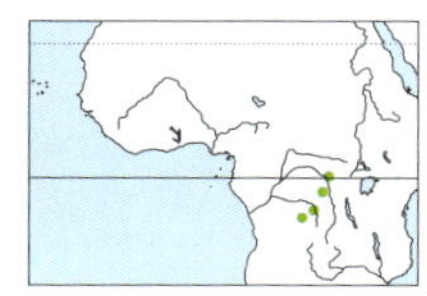

up to 800 m

Known from just 5 localities in CD & AO; probably widespread

Wologizi Wading Rat
Colomys wologizi

FR Rat du Wologizi
DE Wologizi-Waldbachmaus
ES Ratón vadeador de Wologizi

NE
Monotypic
HB 11.6–13 cm
T 13.4–15 cm
W 47–48 g

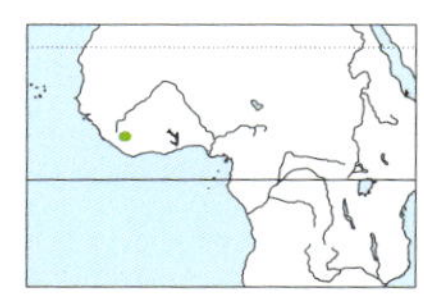

500–600 m

Upper St Paul & Lofa rivers, in Ziama–Wonegizi–Wologizi Mts (LB, GN)

Genus *ZELOTOMYS*

(2 species)

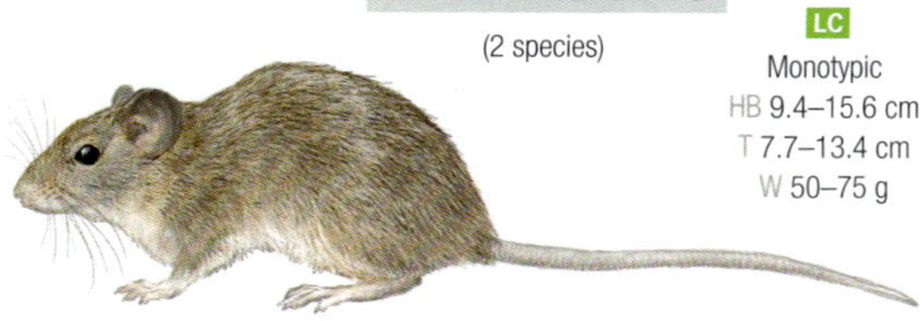

LC
Monotypic
HB 9.4–15.6 cm
T 7.7–13.4 cm
W 50–75 g

Woosnam's Broad-headed Mouse
Zelotomys woosnami

FR Zélotomys de Woosnam
DE Woosnam-Breitkopfratte
ES Ratón de cabeza ancha de Woosnam

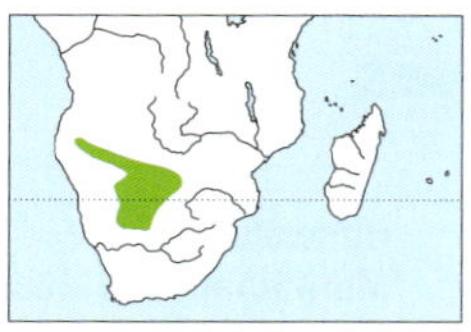

S AO through N & E NA and BW to NW ZA

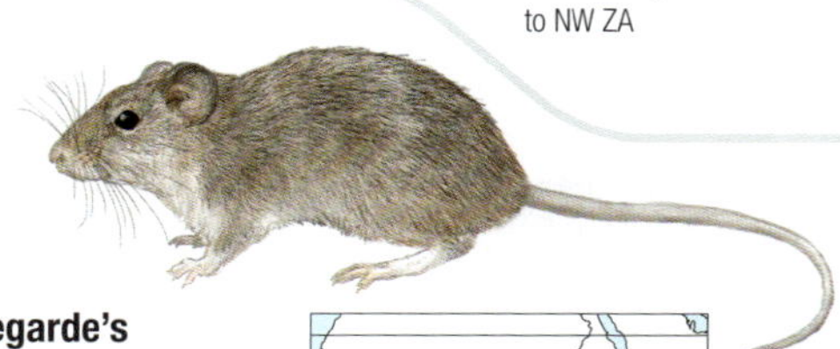

Hildegarde's Broad-headed Mouse
Zelotomys hildegardeae

FR Zélotomys de Hildegarde
DE Hildegarde-Breitkopfratte
ES Ratón de cabeza ancha de Hildegarde

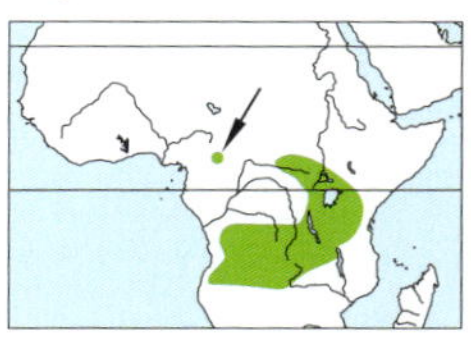

AO & ZM to E Africa; isolated in W CF

LC
Monotypic
HB 10–12.2 cm
T 8–10.3 cm
W 38–74 g

Genus *NILOPEGAMYS*

(1 species)

Ethiopian Amphibious Rat
Nilopegamys plumbeus

FR Rat d'Éthiopie
DE Äthiopische Waldbachmaus
ES Rata de agua de Etiopía

c. 2800 m

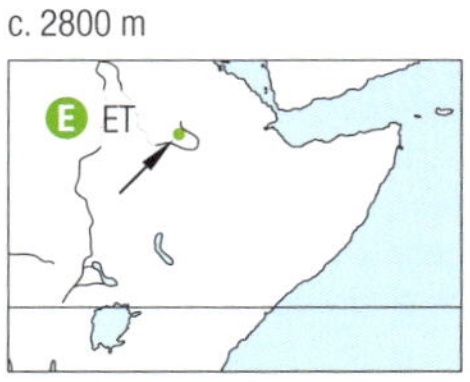

Known from type locality (Gojjam, Ethiopian highlands)

CR(PE)
Monotypic
HB c. 14.8 cm
T c. 18 cm
W c. 80 g

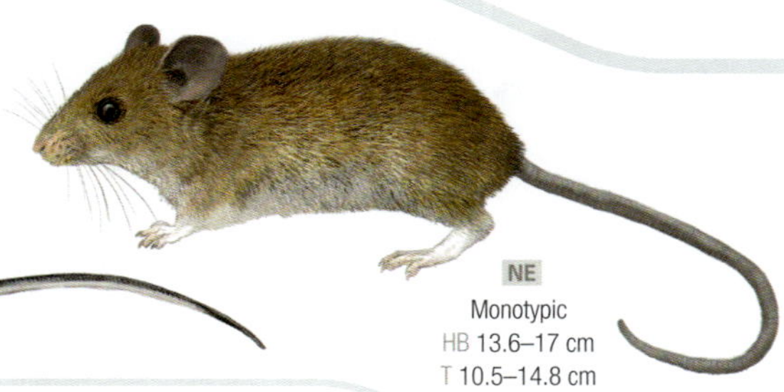

NE
Monotypic
HB 13.6–17 cm
T 10.5–14.8 cm

Zima's Ethiopian Rat
Stenocephalemys zimai

FR Rat de Zima
DE Zima-Hochlandratte
ES Rata de Etiopía de Zima

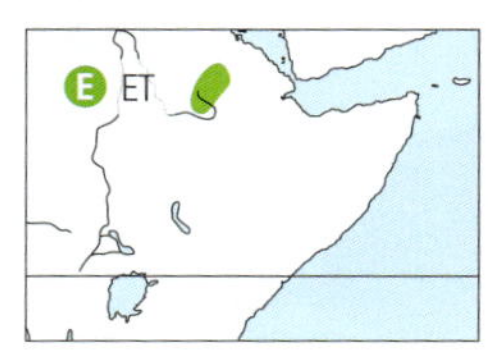

Found in 4 mountain areas (Semien, Guna, Choke & Abohoy Gara Mts)

Genus *STENOCEPHALEMYS*

(6 species)

White-footed Ethiopian Rat
Stenocephalemys albipes

FR Rat à pattes blanches
DE Weißfuß-Hochlandratte
ES Rata de Etiopía de pies blancos

LC
Monotypic
HB 10.7–17.1 cm T 13.4–19.5 cm
W 40–79 g

820–3300 m

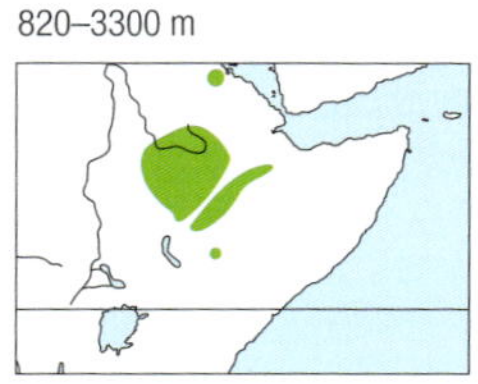

ET, ER

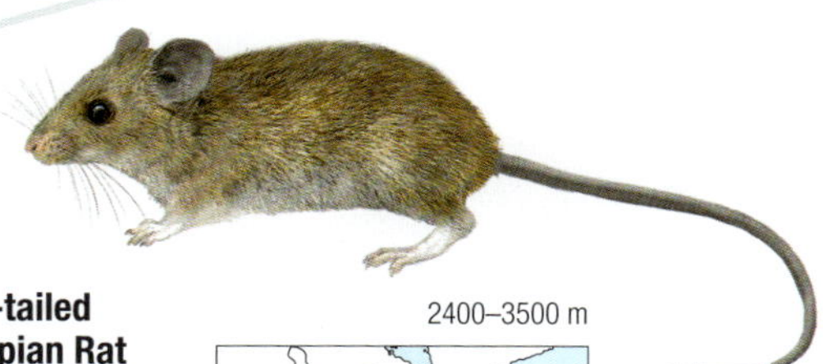

Gray-tailed Ethiopian Rat
Stenocephalemys griseicauda

FR Rat à queue grise
DE Grauschwanz-Hochlandratte
ES Rata de Etiopía de cola gris

2400–3500 m

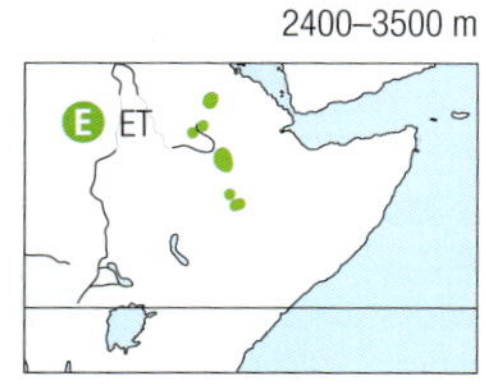

W & E of Rift Valley

LC
Monotypic
HB 13–17.5 cm
T 10.5–16 cm
W 70–136 g

NE
Monotypic
HB 12.9–15.4 cm
T 11.4–16.1 cm

Sokolov's Ethiopian Rat
Stenocephalemys sokolovi

FR Rat de Sokolov
DE Sokolov-Hochlandratte
ES Rata de Etiopía de Sokolov

3200–3500 m

Regions of Debre Sina, Ankober, Guwassa & Borena Saynt NP

DD
Monotypic
HB 12.4–13.8 cm
T 14.3–17.9 cm

Rupp's Ethiopian Rat
Stenocephalemys ruppi

FR Rat de Rupp
DE Rupp-Hochlandratte
ES Rata de Etiopía de Rupp

2700–3200 m

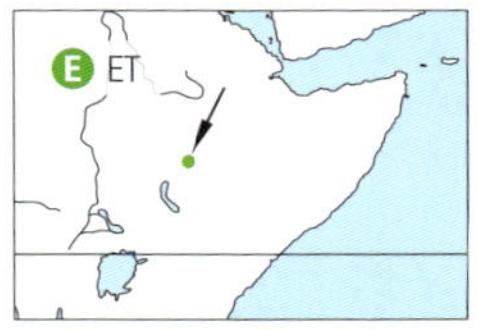

Gughe Mts, W of Rift Valley

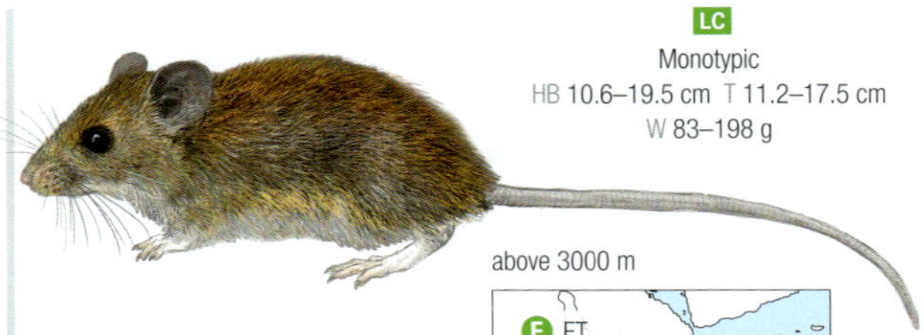

LC
Monotypic
HB 10.6–19.5 cm T 11.2–17.5 cm
W 83–198 g

White-tailed Ethiopian Rat
Stenocephalemys albocaudatus

FR Rat à queue blanche
DE Weißschwanz-Hochlandratte
ES Rata de Etiopía de cola blanca

above 3000 m

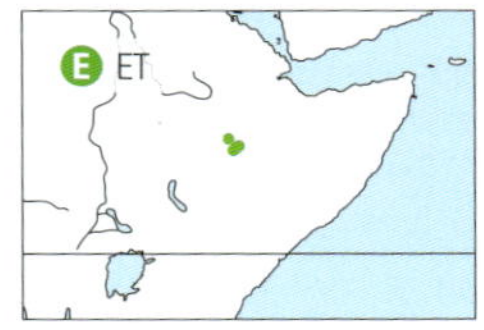

E of Rift Valley

Yemen White-bellied Rocky Mouse
Ochromyscus yemeni

FR Rat du Yémen
DE Jemen-Wiesenratte
ES Ratón de campiña de Yemen

DD
Monotypic
HB 7.8–12 cm
T 15.1–18 cm

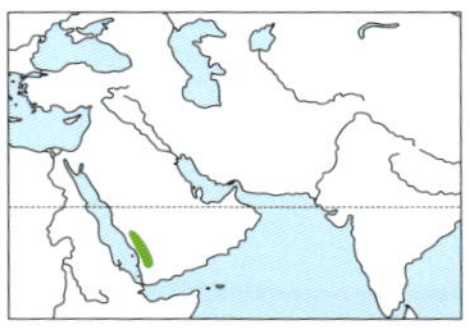

Narrow region of SW SA & NW YE

Genus *OCHROMYSCUS*
(2 species)

Brockman's White-bellied Rocky Mouse
Ochromyscus brockmani

FR Rat de Brockman
DE Brockman-Wiesenratte
ES Ratón de campiña de Brockman

LC
Monotypic
HB 9.3–12.5 cm
T 11.5–17.5 cm
W 22–39 g

1000–3000 m

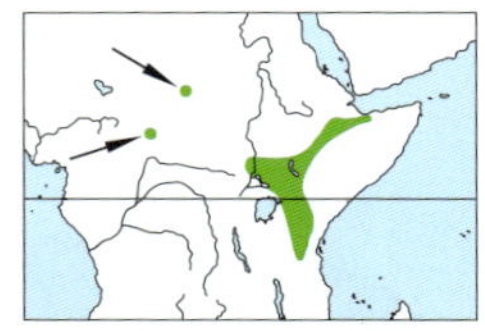

ET, NW SO, S SS, N UG, KE, N TZ; isolated in SW SD & NC CF

Angolan Multimammate Mouse
Mastomys angolensis

FR Mastomys d'Angola
DE Angola-Wiesenratte
ES Ratón de campiña de Angola

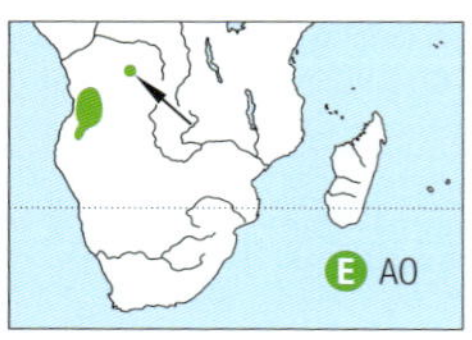

above 1000 m

Primarily on Angolan Plateau

LC
Monotypic
HB 10.9–12.7 cm
T 10.3–12.8 cm

Kollmannsperger's Multimammate Mouse
Mastomys kollmannspergeri

FR Mastomys de Kollmannsperger
DE Kollmannsperger-Vielzitzenmaus
ES Ratón multimamas de Kollmannsperger

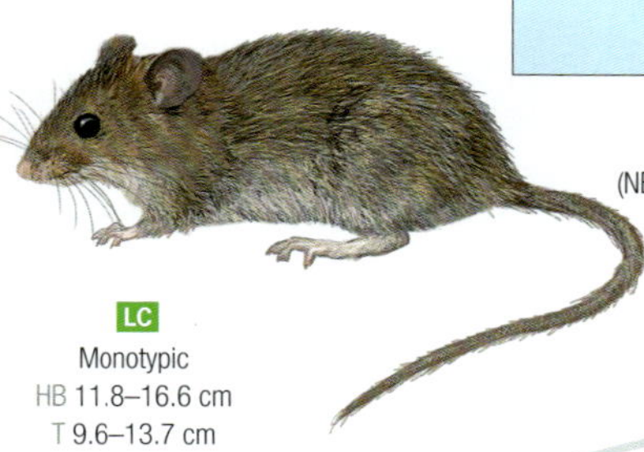

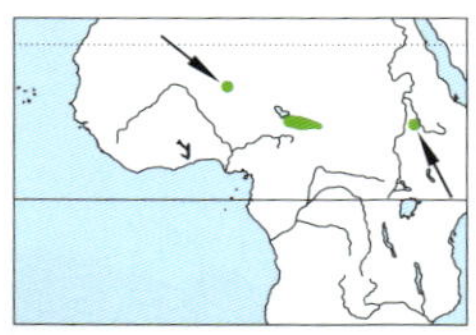

Scattered localities in W NE, near L Chad (NE NG, SW TD) & SE SD

LC
Monotypic
HB 11.8–16.6 cm
T 9.6–13.7 cm
W 54–160 g

Genus *MASTOMYS*
(8 species)

Hubert's Multimammate Mouse
Mastomys huberti

FR Mastomys de Hubert
DE Hubert-Vielzitzenmaus
ES Ratón multimamas de Hubert

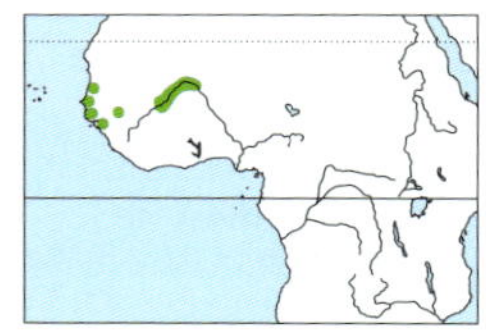

Mainly R Senegal Delta (SN, S MR) & Inner Niger Delta (ML), patchily in W & S SN and W GN

LC
Monotypic
HB 7.4–14.6 cm T 6.2–12 cm
W 10–91 g

Southern African Multimammate Mouse
Mastomys coucha

FR Mastomys austral
DE Südliche Vielzitzenmaus
ES Ratón multimamas de África meridional

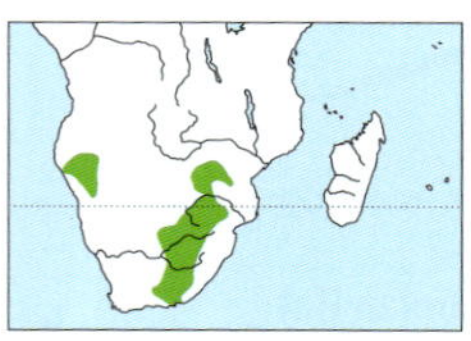

S Africa in NW NA & S of R Zambezi from C ZW S to SC ZA; probably E & S BW

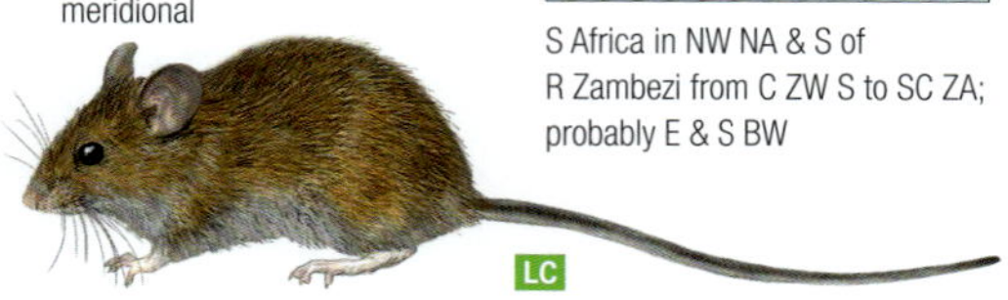

LC
Monotypic
HB 8.2–11.9 cm T 7–10.7 cm
W 22–40 g

Awash Multimammate Mouse
Mastomys awashensis

FR Mastomys de l'Awash
DE Awash-Vielzitzenmaus
ES Ratón multimamas de Awash

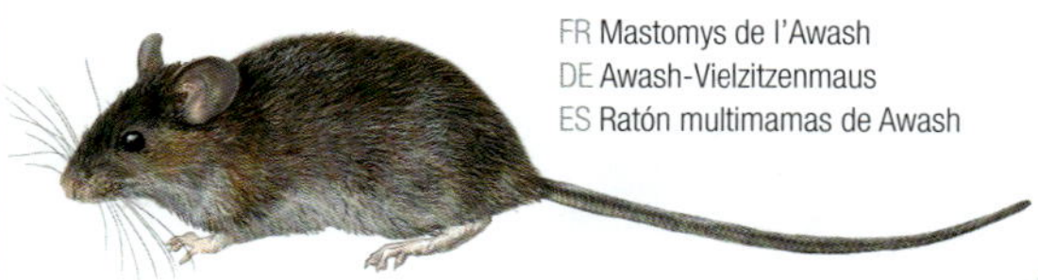

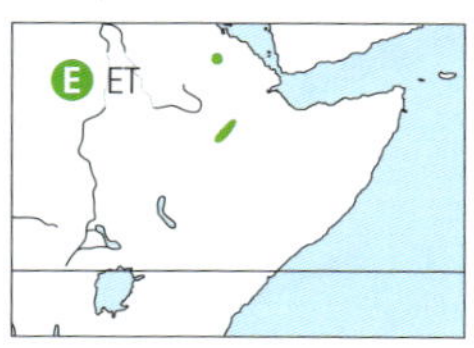

Known from few localities, mostly in Rift Valley

LC
Monotypic
HB 11.6–12.7 cm
T 11.1–12.3 cm

Natal Multimammate Mouse
Mastomys natalensis

FR Mastomys du Natal
DE Natal-Vielzitzenmaus
ES Ratón multimamas de Natal

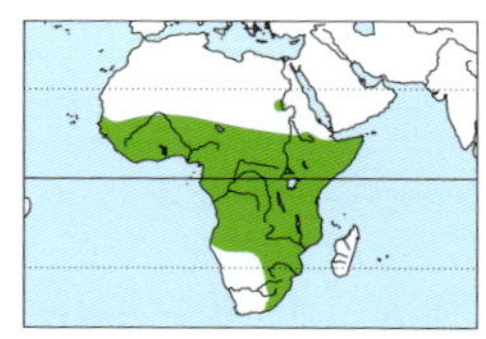

Sub-Saharan Africa from SN E to SO, S to N NA, N BW & E ZA; isolated in N SD

LC
Monotypic
HB 7.6–15.5 cm T 7–17.4 cm
W 15–98 g

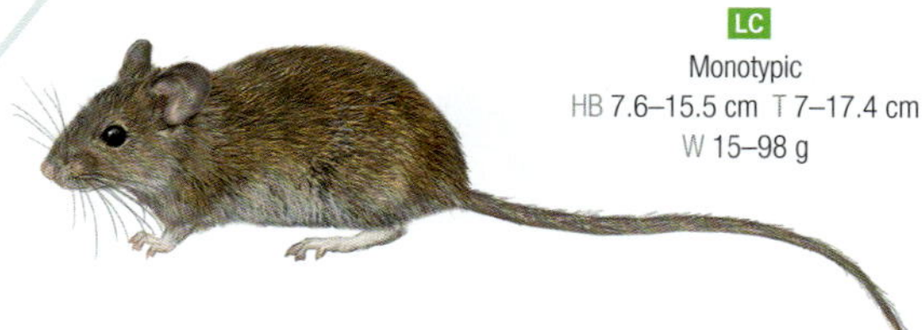

LC
Monotypic
HB 9.2–17.8 cm
T 8.1–15.1 cm
W 32–107 g

Reddish-white Multimammate Mouse
Mastomys erythroleucus

FR Mastomys roux-blanc
DE Guinea-Vielzitzenmaus
ES Ratón multimamas blanquirrojo

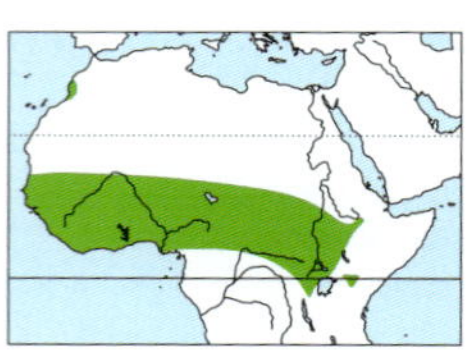

N savannas from SN E to C ET, S to CM, NE CD, BI & C KE; isolated in WC MA

Shortridge's Multimammate Mouse
Mastomys shortridgei

FR Mastomys de Shortridge
DE Shortridge-Vielzitzenmaus
ES Ratón multimamas de Shortridge

LC
Monotypic
HB 10.3–13.7 cm
T 8.6–11.8 cm
W 35–74 g

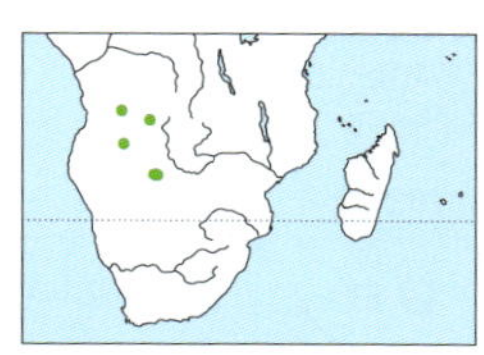

Scattered in S & E AO and along R Okavango in Caprivi Strip (NA & NW BW)

Family CTENODACTYLIDAE – Gundis

4 genera, 5 species. Essentially desert or semi-desert terrains. Palearctic and Afrotropical Regions.

Genus *PECTINATOR*

(1 species)

LC
Monotypic
HB 15.5–19 cm
T 6–8 cm
W ≤ 194 g

Speke's Gundi
Pectinator spekei
FR Goundi de Speke
DE Buschschwanz-Gundi
ES Gundi de Speke

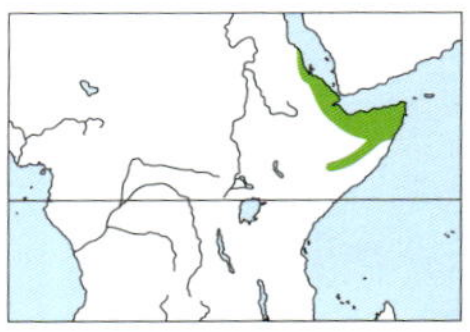

Horn of Africa (ER, DJ, ET, SO)

Genus *MASSOUTIERA*

(1 species)

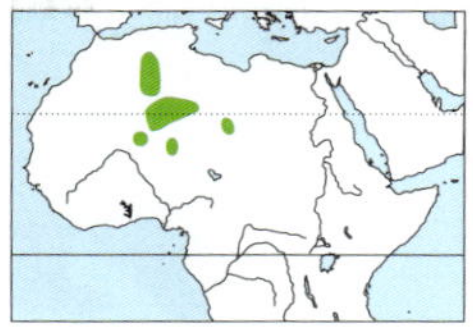

C & SE DZ, E ML, W NE, NW TD; perhaps SW LY

Mzab Gundi
Massoutiera mzabi
FR Goundi du Mzab
DE Langhaar-Gundi ES Gundi de Mzab

LC Monotypic
HB 12.5–21 cm
T 3.3–8.5 cm
W ≤ 234 g

Genus *FELOVIA*

(1 species)

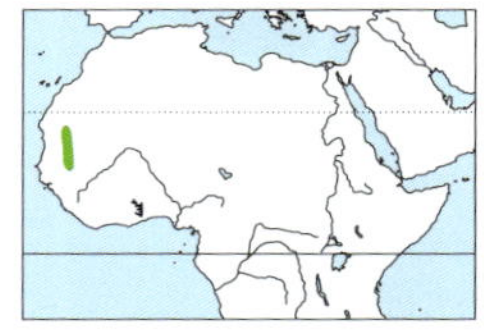

NW Africa (C MR, E SN, W ML)

Felou Gundi
Felovia vae
FR Goundi de Félou
DE Senegal-Gundi ES Gundi de Félou

LC Monotypic
HB 16.9–19 cm
T 6.7–8 cm
W ≤ 188 g

Genus *CTENODACTYLUS*

(2 species)

up to 2900 m

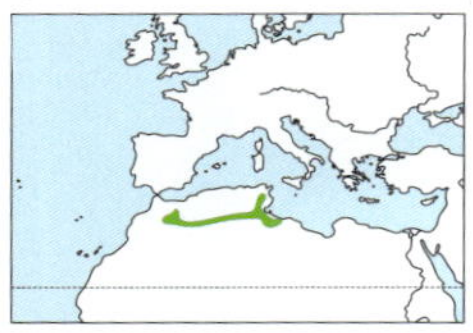

NW Africa (NE MA, N DZ, TN, NW LY)

Common Gundi
Ctenodactylus gundi
FR Goundi de l'Atlas
DE Eigentlicher Gundi ES Gundi común

LC Monotypic
HB 15–23 cm
T 2–4.5 cm
W ≤ 396 g

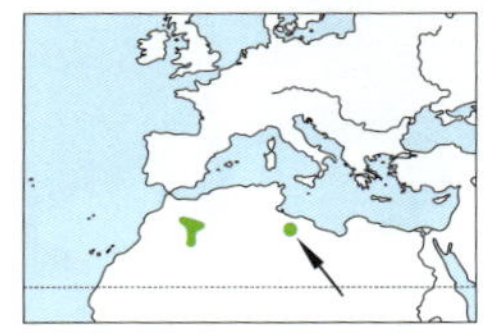

NW Africa (NE MA, NW DZ, NW LY)

Desert Gundi
Ctenodactylus vali
FR Goundi du désert
DE Wüstengundi ES Gundi del desierto

DD Monotypic
HB 12.4–18.5 cm
T 2.7–4.3 cm
W ≤ 185 g

Family DIATOMYIDAE – Kha-nyou

1 genus, 1 species. Limestone karst in tropical deciduous, semi-evergreen, and evergreen forests. Indo-Malayan Region.

Genus *LAONASTES*

(1 species)

Kha-nyou
Laonastes aenigmamus
FR Kanyou
DE Laotische Felsenratte
ES Rata de roca de Laos

LC
Monotypic
HB 21–30 cm
T 12.3–16.1 cm
W 309–414 g

C LA (Nam Kading CA & Phou Hin Poun CA), C VN (Phong Nha-Ke Bang NP)

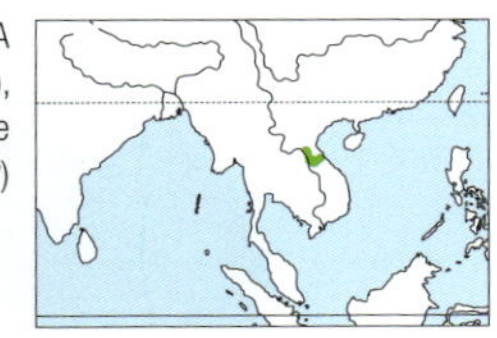

160–400 m

Family HYSTRICIDAE – Old World Porcupines

3 genera, 11 species. Mediterranean coastal shrublands, warm and cold savanna grasslands, dry semi-deserts, lowland and montane tropical forests, forested hills and steppes, acacia and miombo woodlands, and agricultural and urbanized habitats. Afrotropical, Palearctic and Indo-Malayan Regions.

Genus *TRICHYS*

(1 species)

LC
Monotypic
HB 35–48 cm
T 17.5–23 cm
W 1.5–2.2 kg

up to 1220 m

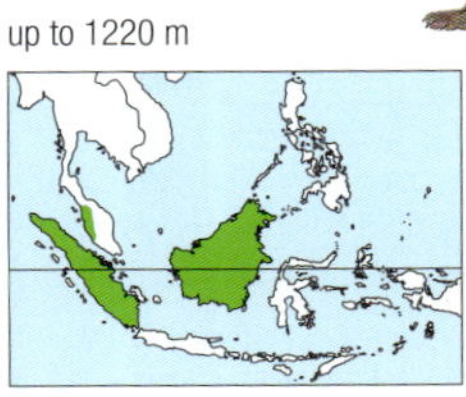

W Pen MY, Sumatra & Borneo (MY, ID, BN)

Asian Long-tailed Porcupine
Trichys fasciculata
FR Porc-épic à longue queue
DE Pinselstachler
ES Puercoespín de cola larga

LC Monotypic
HB 36–60 cm
T 10–26 cm
W 1.5–4.3 kg

0–3000 m

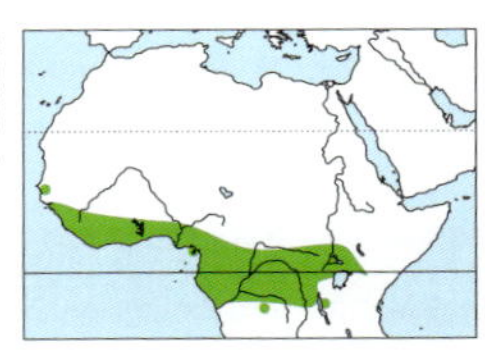

African Brush-tailed Porcupine
Atherurus africanus
FR Athérure africain
DE Afrikanischer Quastenstachler
ES Puercoespín africano de cola de cepillo

GN E to S SS, UG & E KE S to AO (Cabinda, Carumbo L) & E TZ (L Tanganyika), Bioko I

Genus *ATHERURUS*

(2 species)

NE IN E to C & S CN (with Hainan I) S to Pen MY (with some Is); perhaps BD

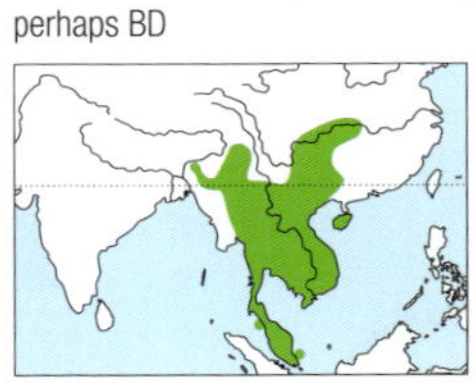

Asian Brush-tailed Porcupine
Atherurus macrourus
FR Athérure asiatique
DE Asiatischer Quastenstachler
ES Puercoespín asiático de cola de cepillo

LC Monotypic
HB 38–57 cm
T 16–19.5 cm
W 1.5–4.3 kg

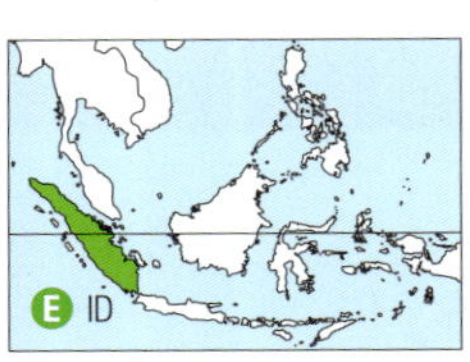

Sumatra

0–300 m

LC Monotypic
HB 45–56 cm T 9–11 cm
W 3.8–5.4 kg

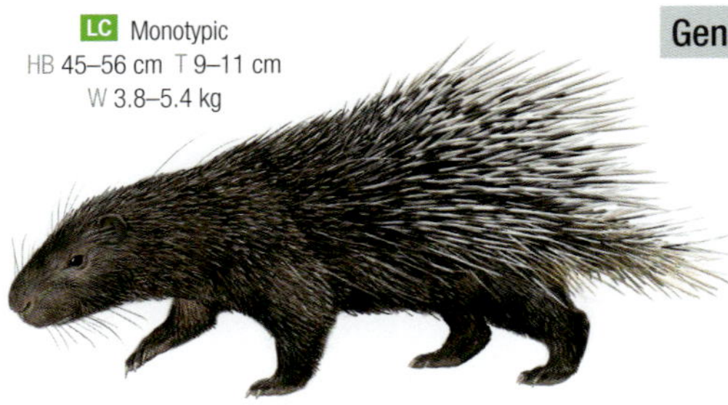

Sumatran Porcupine
Hystrix sumatrae
FR Porc-épic de Sumatra DE Sumatra-Stachelschwein ES Puercoespín de Sumatra

Genus *HYSTRIX*

(8 species)

Thick-spined Porcupine
Hystrix crassispinis
FR Porc-épic de Bornéo DE Borneo-Stachelschwein ES Puercoespín de espinas gruesas

up to 1200 m

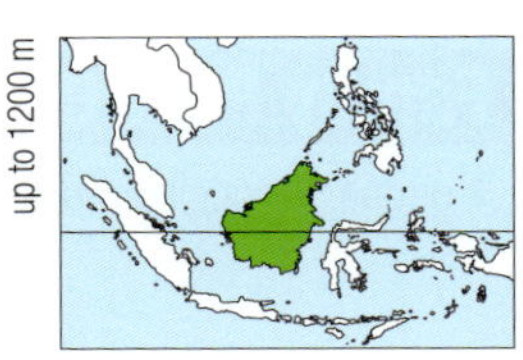

Borneo (MY, ID, BN)

LC Monotypic
HB 45–66 cm T 6.5–19 cm
W 3.8–5.4 kg

VU
Monotypic
HB 45–66 cm
T 6.4–19 cm
W 3.8–5.4 kg

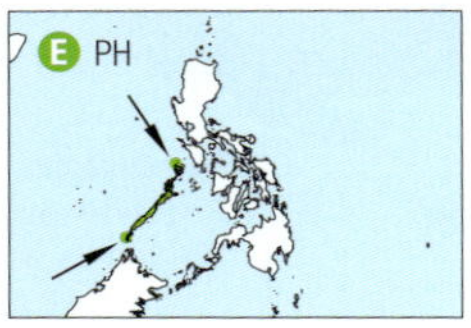

Palawan, adjacent Busuanga & Balabac Is

Philippine Porcupine
Hystrix pumila
FR Porc-épic des Philippines
DE Palawan-Stachelschwein
ES Puercoespín de Filipinas

Malayan Porcupine
Hystrix brachyura
FR Porc-épic de Malaisie DE Malaiisches Stachelschwein
ES Puercoespín de Malasia

LC
3 sspp
HB 45–93 cm
T 6–17 cm
W 8–27 kg

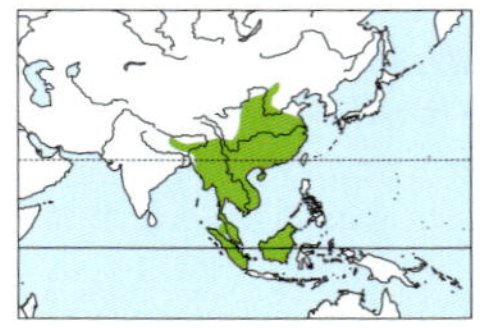

S Himalayas E to C & S CN (with Hainan I), S to Pen MY, Sumatra & Borneo

LC
Monotypic
HB 45–66 cm
T 6–17 cm
W 8–27 kg

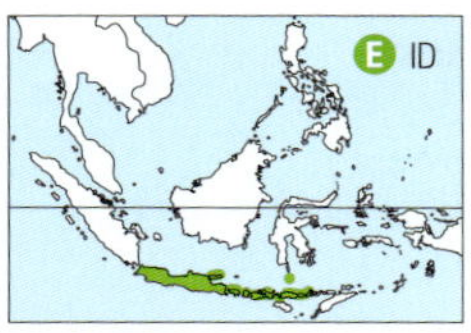

Java (with Madura I), Lesser Sunda Is (Bali, Lombok, Sumbawa, Flores), Selayar Is (Rinca, Tanah Jampea)

Sunda Porcupine
Hystrix javanica
FR Porc-épic de Java DE Sunda-Stachelschwein
ES Puercoespín de la Sonda

LC Monotypic
HB 45–93 cm
T 6–17 cm
W 8–27 kg

up to 3500 m

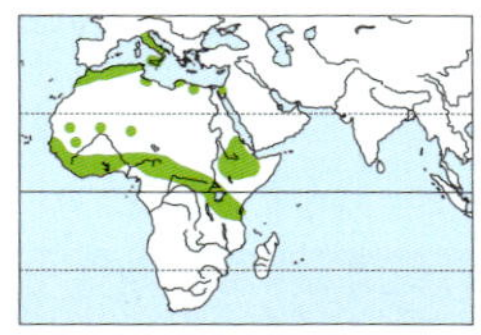

Crested Porcupine
Hystrix cristata
FR Porc-épic à crête DE Gewöhnliches Stachelschwein ES Puercoespín crestado

IT, Sicily, Africa (MA to EG, S SD & SS to SO, SN to TZ, isolated in MR, ML & NE)

LC Monotypic
HB 63–80 cm
T 10.5–13 cm
W 10–24 kg

0–2000 m

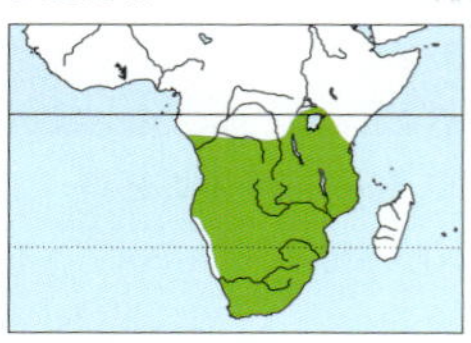

From S CG & S CD E to SW KE & TZ, S to ZA, excluding Namibian coast

Cape Porcupine
Hystrix africaeaustralis
FR Porc-épic du Cap
DE Südafrikanisches Stachelschwein
ES Puercoespín de El Cabo

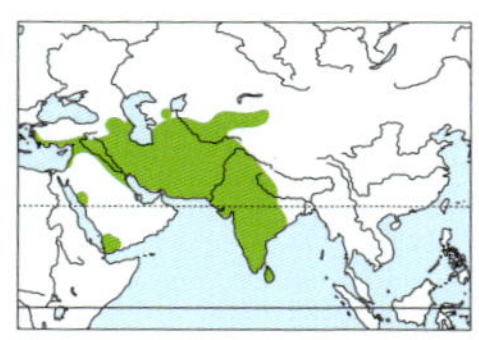

LC Monotypic
HB 45–93 cm
T 6–17 cm
W 8–27 kg

S TR, Levant & Arabian Pen, NE to Transcaucasia & S KZ, SE to IN & LK

Indian Crested Porcupine
Hystrix indica
FR Porc-épic indien
DE Indisches Stachelschwein ES Puercoespín de India

Family THRYONOMYIDAE – Cane Rats

1 genus, 2 species. Grassland habitats. Afrotropical Region.

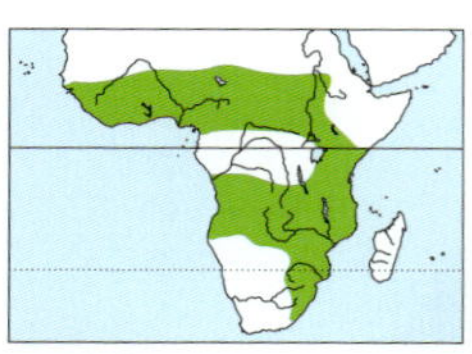

SN E to SD, S to E ZA; absent from Horn of Africa, WC Africa & arid SW

LC
Monotypic
HB 63–77 cm
T 16–19.5 cm
W 3.2–9 kg

Greater Cane Rat
Thryonomys swinderianus
FR Grand Aulacode
DE Große Rohrratte
ES Rata de cañaveral grande

Genus *THRYONOMYS*
(2 species)

LC
Monotypic
HB 41–57 cm
T 11–17.5 cm
W 1.4–2.4 kg

Lesser Cane Rat
Thryonomys gregorianus
FR Petit Aulacode
DE Kleine Rohrratte
ES Rata de cañaveral pequeña

up to 2600 m

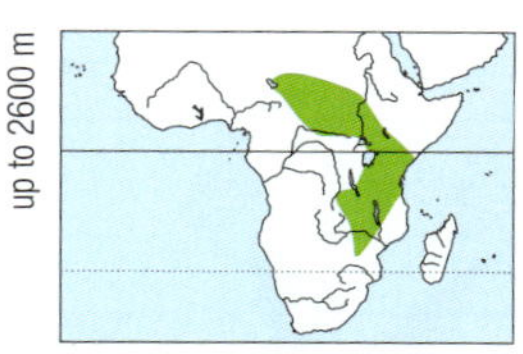

Tropical C & E Africa in N CM & S TD E to W ET & KE S to ZW & NW MZ

Family PETROMURIDAE – Noki

1 genus, 1 species. Rocky arid habitats. Afrotropical Region.

Genus *PETROMUS*
(1 species)

Noki
Petromus typicus
FR Noki
DE Felsenratte
ES Rata damán

LC
Monotypic
HB 13.5–22 cm T 11.6–17.5 cm
W 88–285 g

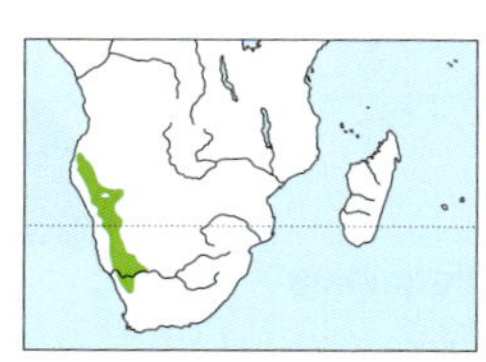

From SW AO S through W NA to NW ZA

Family HETEROCEPHALIDAE – Naked Mole-rat

1 genus, 1 species. Subterranean, arid bushland, thicket, and semi-desert scrub in the Horn of Africa. Afrotropical Region.

Genus *HETEROCEPHALUS*

(1 species)

Naked Mole-rat
Heterocephalus glaber
FR Hétérocéphale
DE Nacktmull
ES Rata topo lampiña

LC
Monotypic
HB 7–11 cm
T 3–5 cm
W 15–70 g

300–1500 m

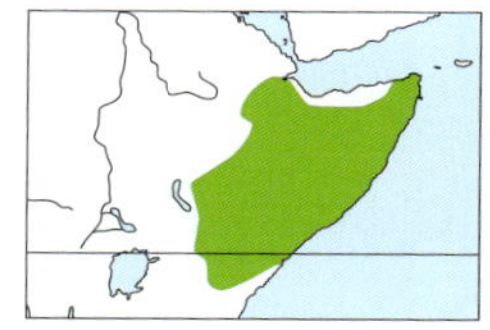

DJ, E & S ET, SO, N & E KE

Family BATHYERGIDAE – African Mole-rats

5 genera, 25 species. Subterranean in open forests to savannas and semi-deserts under mesic to xeric conditions. Afrotropical Region.

Genus *HELIOPHOBIUS*

(1 species)

LC
5 sspp
HB 13.1–19.5 cm
T 0.6–1.7 cm
W 51–331 g

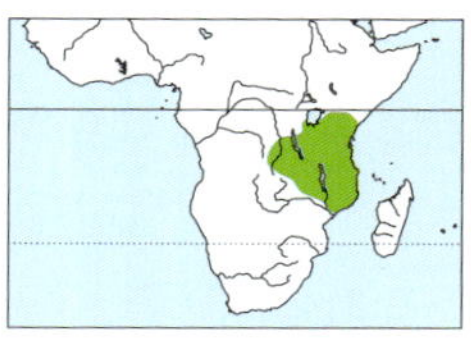

E CD, S KE, TZ, E ZM, MW, MZ
(N of R Zambezi)

Silvery Mole-rat
Heliophobius argenteocinereus
FR Bathyergue argenté
DE Silbermull
ES Rata topo plateada

Genus *BATHYERGUS*

(2 species)

LC Monotypic
HB 17–23 cm T 4–5.2 cm
W 195–710 g

Namaqua Dune Mole-rat
Bathyergus janetta
FR Bathyergue namaquois
DE Namaqua-Strandgräber
ES Rata topo de dunas de Namaqua

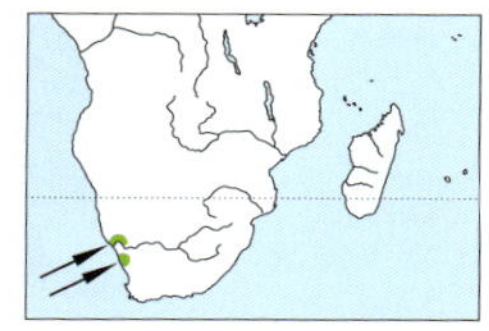

S NA, W ZA (W Northern Cape)

LC Monotypic
HB 27–35 cm T 11–17.5 cm
W 0.57–1.3 kg

Cape Dune Mole-rat
Bathyergus suillus
FR Bathyergue des dunes
DE Kap-Strandgräber
ES Rata topo de dunas de El Cabo

Western Cape from Knysna
to Lamberts Bay

Genus *GEORYCHUS*

(1 species)

LC
Monotypic
HB 8.5–23 cm
T 1.7–2.4 cm
W 88–272 g

Western Cape, Eastern Cape,
isolated in Mpumalanga & KwaZulu-Natal

Cape Mole-rat
Georychus capensis
FR Bathyergue du Cap
DE Kap-Blessmull
ES Rata topo de El Cabo

Genus *CRYPTOMYS*

(5 species)

DD
Monotypic
HB 13.5–14.2 cm (F)
T 1.7–2.3 cm (F)

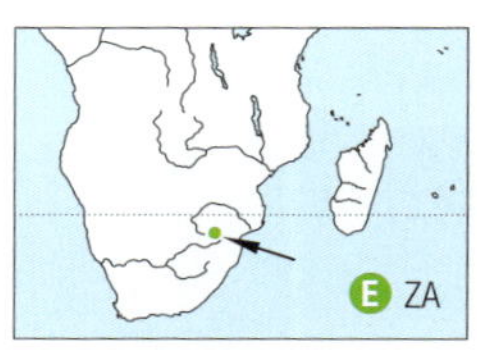

Described from Patryshoek
(Gauteng); range unknown

Mahali Mole-rat
Cryptomys mahali
FR Bathyergue mahali
DE Mahali-Graumull
ES Rata topo de Mahali

NE
Monotypic
HB c. 14.7 cm

S ZW, extreme E BW, N ZA
(N Limpopo)

Matabeleland Mole-rat
Cryptomys nimrodi
FR Bathyergue du Zimbabwe
DE Matabeleland-Graumull
ES Rata topo de Matabelelandia

LC
Monotypic

Northern Cape, Western Cape,
Eastern Cape, Free State

Common Mole-rat
Cryptomys hottentotus
FR Bathyergue hottentot
DE Hottentotten-Graumull
ES Rata topo común

LC
Monotypic
HB 15–17.7 cm
T 2–2.5 cm

Extreme S MZ, E ZA (Mpumalanga, KwaZulu-Natal), SZ, LS

Natal Mole-rat
Cryptomys natalensis
FR Bathyergue du Natal
DE Natal-Graumull
ES Rata topo de Natal

LC
Monotypic
HB c. 14 cm
T c. 2 cm

S Limpopo, E North West, Gauteng, Mpumalanga

Highveld Mole-rat
Cryptomys pretoriae
FR Bathyergue de Pretoria
DE Highveld-Graumull
ES Rata topo de Pretoria

Genus *FUKOMYS*

(16 species)

DD

Monotypic

HB c. 11.8 cm

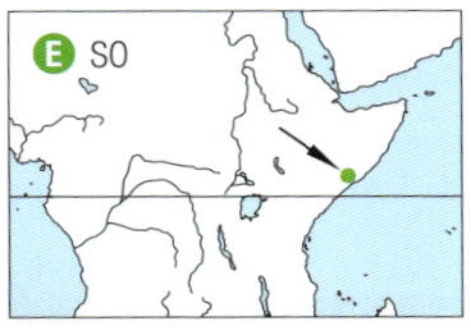

Somali Striped Mole-rat

Fukomys ilariae

FR Bathyergue d'Ilaria

DE Somalia-Graumull

ES Rata topo de Somalia

Known only from type specimen probably collected near Mogadishu

LC

Monotypic

HB 12–23 cm

T c. 1.5 cm

W 51–283 g

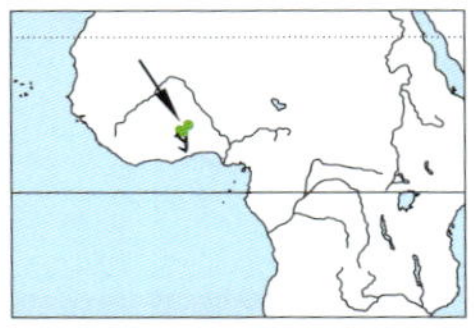

Ghana Mole-rat

Fukomys zechi

FR Bathyergue de Zech

DE Togo-Graumull

ES Rata topo de Ghana

EC GH, N TG (between R Oti & R Volta)

DD

Monotypic

HB 16.1–19.1 cm

T 16–25 cm

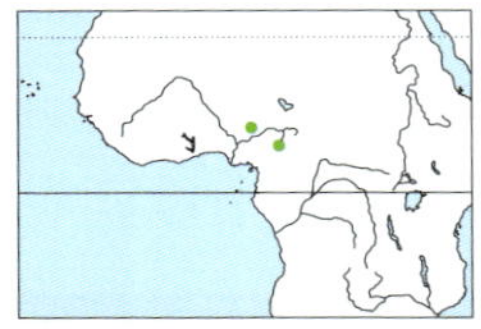

Nigerian Mole-rat

Fukomys foxi

FR Bathyergue du Nigeria

DE Nigeria-Graumull

ES Rata topo de Nigeria

N NG (Jos Plateau), C CM (Ngaoundéré)

LC

Monotypic

HB 15.7–20 cm

T 1.4–2.7 cm

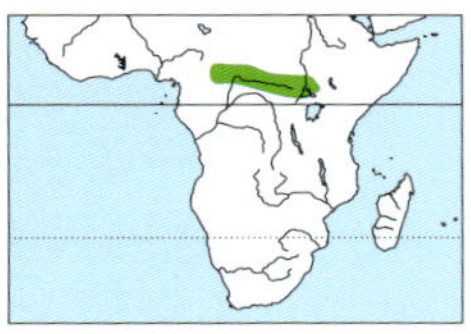

Ochre Mole-rat

Fukomys ochraceocinereus

FR Bathyergue ocre

DE Zentralafrika-Graumull

ES Rata topo ocre

CF, N CD, SS, NW UG; perhaps E CM & NW KE

LC Monotypic

HB 14.1–16.5 cm

T 6–15 cm

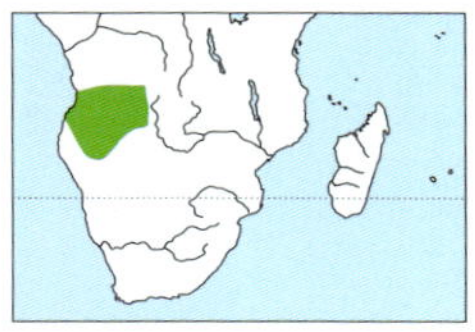

Bocage's Mole-rat

Fukomys bocagei

FR Bathyergue de Bocage

DE Bocage-Graumull

ES Rata topo de Bocage

C & S AO, N NA (Ovamboland); perhaps extreme S CD & W ZM

DD

Monotypic

HB 8.7–15 cm

T 1.4–2.1 cm

W 64–134 g

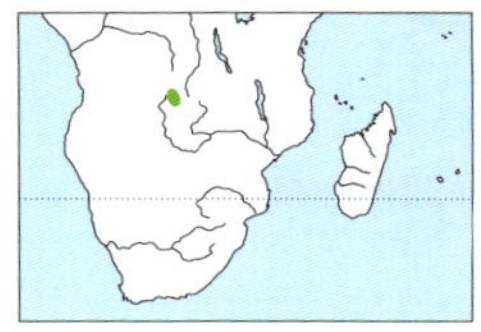

Caroline's Mole-rat

Fukomys vandewoestijneae

FR Bathyergue de Caroline

DE Katanga-Graumull

ES Rata topo de Caroline

Border between S CD and NW ZM; probably E AO

LC

Monotypic

HB 13.5–26 cm

T 2.3–3.4 cm

W 218–440 g

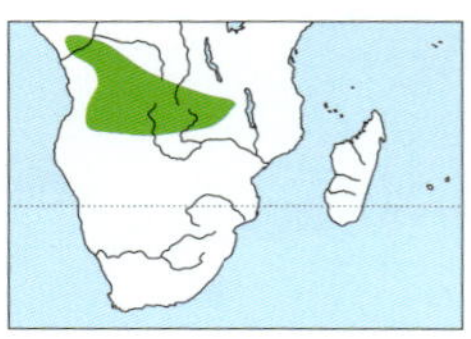

Mechow's Mole-rat

Fukomys mechowii

FR Bathyergue de Mechow

DE Riesengraumull

ES Rata topo de Mechow

S & E CD, E AO, N ZM; unconfirmed in SW TZ (Ufipa) & N MW (Rumphi)

DD Monotypic

HB c. 11.5 cm T c. 0.9 cm

W 38–80 g

2600–2625 m

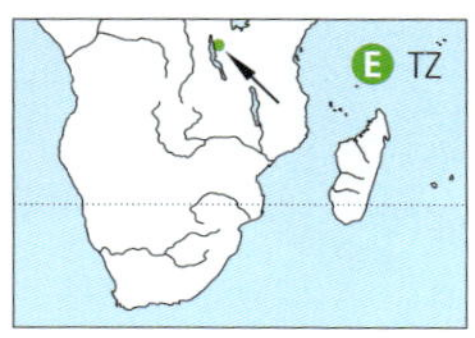

Livingstone's Mole-rat

Fukomys livingstoni

FR Bathyergue de Livingstone

DE Livingstone-Graumull

ES Rata topo de Livingstone

Known only from vicinity of Ujiji (near L Tanganika)

1855–2200 m

EN Monotypic

HB c. 11.1 cm T c. 1.2 cm

W 35–140 g

Hanang Mole-rat

Fukomys hanangensis

FR Bathyergue du Hanang

DE Hanang-Graumull

ES Rata topo de Hanang

Manyara Region

NE

Monotypic

HB c. 10.5 cm

T c. 1 cm

W 52–71 g

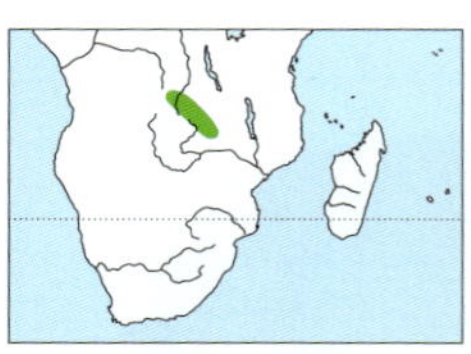

Zambian Mole-rat

Fukomys amatus

FR Bathyergue de Zambie

DE Sambia-Graumull

ES Rata topo de Zambia

SE CD, C ZM

NE

Monotypic

HB 13–15.7 cm

T 1.5–2 cm

W 111–156 g

500–1760 m

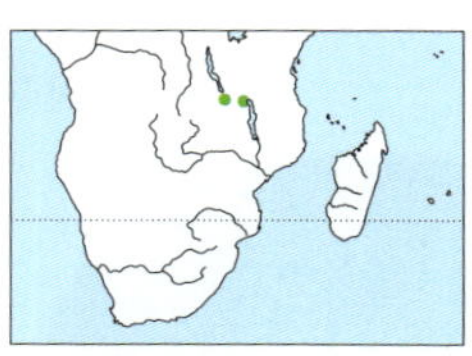

Whyte's Mole-rat

Fukomys whytei

FR Bathyergue de Whyte

DE Malawi-Graumull

ES Rata topo de Whyte

N MW, NE ZM, SE TZ

LC

Monotypic

HB 12.5–16.5 cm

T 8–13 cm

W 54–92 g

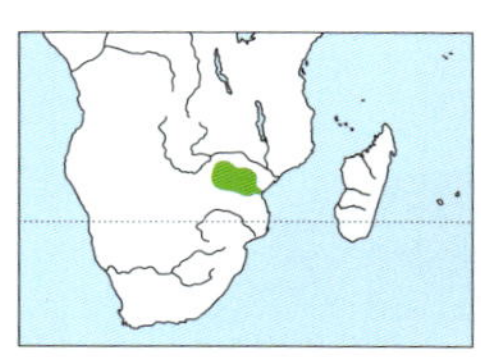

Mashona Mole-rat

Fukomys darlingi

FR Bathyergue de Darling

DE Mashona-Graumull

ES Rata topo de Mashona

N & E ZW, C MZ

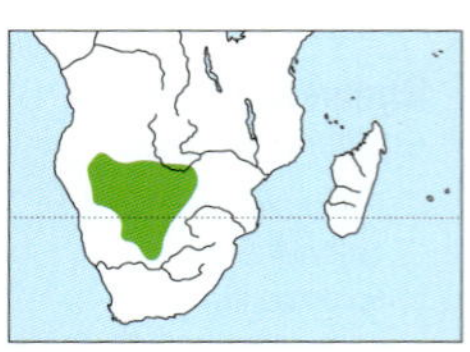

S AO, S ZM, E NA, most of BW, W ZW & NW ZA (Northern Cape)

LC

Monotypic

HB 15.5–17.9 cm

T 2.1–2.4 cm

W 114–210 g

Damara Mole-rat

Fukomys damarensis

FR Bathyergue damara DE Damara-Graumull ES Rata topo de Damara

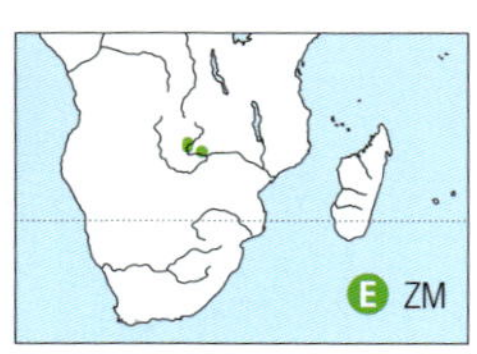

Along R Kafue

LC

Monotypic

HB 10.8–13.5 cm

T 1.4–2.3 cm

W 65–126 g

Ansell's Mole-rat

Fukomys anselli

FR Bathyergue d'Ansell DE Ansell-Graumull ES Rata topo de Ansell

Along R Zambezi

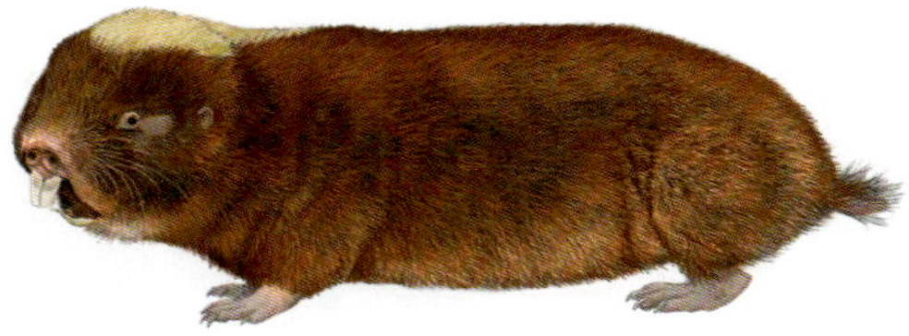

Micklem's Mole-rat
Fukomys micklemi

NE Monotypic
HB 16.5–17 cm T 1.8–2 cm

FR Bathyergue de Micklem DE Micklem-Graumull ES Rata topo de Micklem

Kafue NP

VU
Monotypic
HB 9.6–12.9 cm
T 1.3–2 cm
W 61–139 g

Kafue Mole-rat
Fukomys kafuensis

FR Bathyergue du Kafue DE Kafue-Graumull ES Rata topo de Kafue

Family ERETHIZONTIDAE – New World Porcupines

3 genera, 18 species. Tundra and boreal forests in the north, temperate deciduous forests, deserts, grasslands, and lowland and montane tropical forests. Nearctic and Neotropical Regions.

Genus *CHAETOMYS*

(1 species)

VU
Monotypic
HB 36–45 cm
T 26–27 cm
W 1–2 kg

From Sergipe S to N Rio de Janeiro

Broomstraw-spined Porcupine
Chaetomys subspinosus

FR Coendou subépineux
DE Borstenbaumstachler
ES Puercoespín espinoso

Genus *ERETHIZON*

(1 species)

North American Porcupine
Erethizon dorsatum

FR Porc-épic d'Amérique
DE Nordamerikanischer Baumstachler
ES Puercoespín de Norteamérica

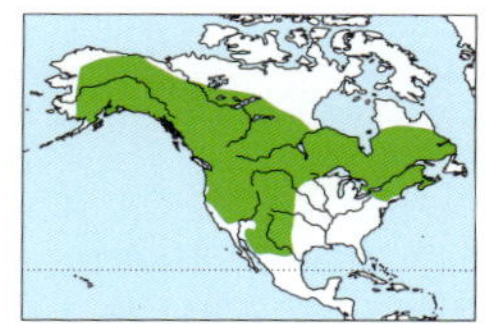

Alaska, S half of CA, N & W US (scattered in E), N MX

LC
7 sspp
HB 60–130 cm
T 16–25 cm
W 5–12 kg

Genus *COENDOU*

(16 species)

LC
Monotypic
HB 29–48 cm
T 31–43 cm
W 2.3–2.9 kg

N Atlantic Forest S to right bank of R São Francisco (Alagoas)

Yellow Quill-tipped Porcupine
Coendou prehensilis

FR Coendou du Brésil
DE Eigentlicher Greifstachler
ES Puercoespín arborícola

E Amazon (Pará, Maranhão), montane forests in Ceará

Baturite Porcupine
Coendou baturitensis

FR Coendou de Baturité
DE Baturité-Greifstachler
ES Puercoespín de Baturité

DD
Monotypic
HB 46–55 cm
T 32–47 cm
W 3.4–3.5 kg

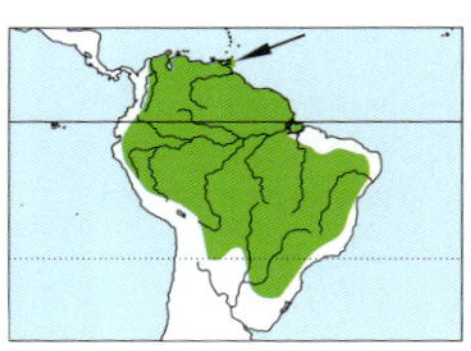

Widespread in Amazon Forest E of Andes S to NW AR & PY; TT (Trinidad I)

Amazonian Long-tailed Porcupine
Coendou longicaudatus

FR Coendou à longue queue
DE Amazonas-Greifstachler
ES Puercoespín arborícola de cola larga

ssp *longicaudatus*

ssp *boliviensis*

NE
2 sspp
HB 45–57 cm
T 46–60 cm
W 2.3–5.6 kg

0–2300 m

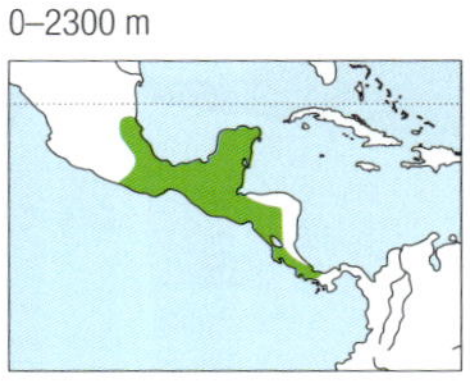

MX (from San Luis Potosí & Guerrero) S to W PA

Mexican Hairy Porcupine
Coendou mexicanus

FR Coendou du Mexique
DE Mexiko-Greifstachler
ES Puercoespín de México

LC
Monotypic
HB 35–46 cm
T 20–36 cm
W 1.4–2.6 kg

800–3500 m

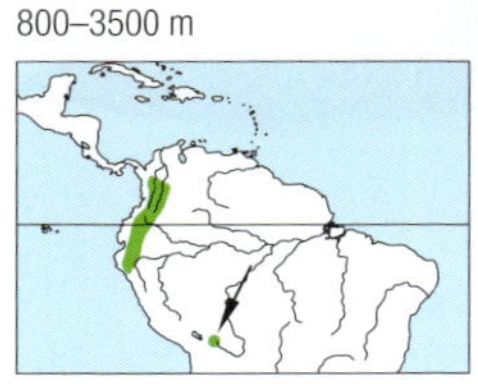

All 3 Andean ranges in C & S CO, S in both slopes to N PE (Lambayeque), also N BO (Cochabamba)

LC
Monotypic
HB 31–37 cm
T 10.5–15 cm

Stump-tailed Porcupine
Coendou rufescens

FR Coendou rougeâtre
DE Kurzschwanz-Greifstachler
ES Puercoespín de cola corta

DD

Monotypic
HB 33–44 cm
T 26–41 cm
W 2–3 kg

0–3300 m

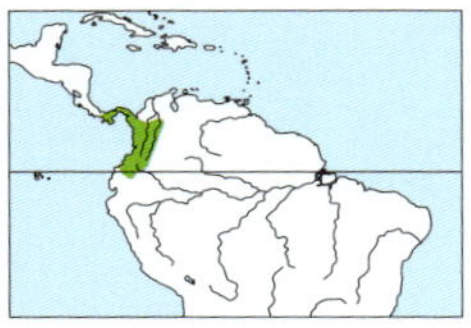

E PA, trans-Andean CO, NW EC

Quichua Porcupine
Coendou quichua

FR Coendou des Andes DE Rothschild-Greifstachler
ES Puercoespín quichua

E Andes from N PE (San Martín) S to W BO; W Andes in N PE (Cajamarca)

LC

Monotypic
HB 38–50 cm
T 33–54 cm
W 3.4–4.7 kg

Bicolor-spined Porcupine
Coendou bicolor

FR Coendou bicolore DE Südamerika-Greifstachler ES Puercoespín bicolor

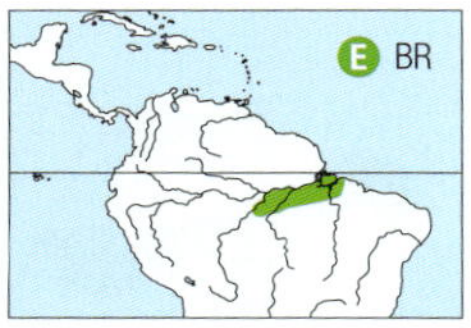

S of R Amazon, E of R Madeira to Marajó I (Amazonas, Pará, perhaps Maranhão)

DD

Monotypic
HB 29–38 cm
T 28–37 cm
W c. 950 g

Eastern Amazonian Dwarf Porcupine
Coendou nycthemera

FR Coendou noir
DE Koopman-Greifstachler
ES Puercoespín negro

Pernambuco, Alagoas

EN

Monotypic
HB 33–44 cm
T 29–32 cm
W 1.4–1.6 kg

Pernambuco Dwarf Porcupine
Coendou speratus

FR Coendou du Pernambuco DE Pernambuco-Greifstachler
ES Puercoespín de Pernambuco

0–1150 m

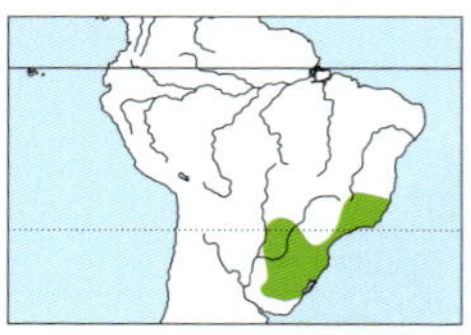

SE BR (S from Minas Gerais & Espírito Santo), E PY (R Tebicuary), N UY, NE AR (Misiones)

LC

Monotypic
HB 24–55 cm
T 20–38 cm
W c. 1 kg

Paraguayan Hairy Dwarf Porcupine
Coendou spinosus

FR Coendou épineux
DE Paraguay-Greifstachler
ES Puercoespín de Paraguay

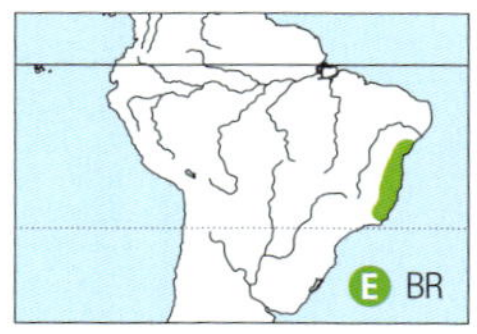

Likely from S Sergipe to extreme E Minas Gerais & N Rio de Janeiro

LC

Monotypic
HB 29–35 cm
T 18–22 cm

Bahian Hairy Dwarf Porcupine
Coendou insidiosus

FR Coendou de Bahia DE Bahia-Greifstachler ES Puercoespín de Bahía

DD

Monotypic
HB 29–37 cm
T 17–19 cm

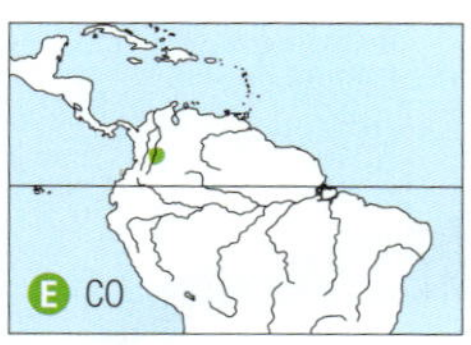

W foothills (possibly also lower slopes) of E Andes (Cundinamarca)

Blackish Hairy Dwarf Porcupine
Coendou vestitus

FR Coendou brun
DE Schwarzbrauner Greifstachler
ES Puercoespín marrón

Frosted Porcupine
Coendou pruinosus

FR Coendou givré
DE Raureif-Greifstachler
ES Puercoespín escarchado

LC

Monotypic
HB 32–38 cm
T c. 19 cm

55–2600 m

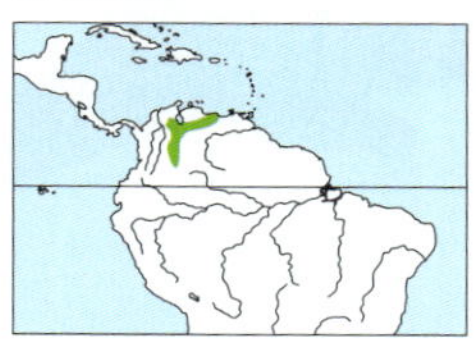

NC CO (E Andes, Serranía de la Macarena), W & N VE (Serranía de Perijá E to Cordillera de la Costa)

DD

Monotypic
HB 26–29 cm
T 21–25 cm

below 500 m

Lowlands of E EC & NE PE; perhaps N of R Amazon E to R Negro

Western Amazonian Dwarf Porcupine
Coendou ichillus

FR Coendou nain
DE Streifen-Greifstachler
ES Puercoespín manchado

DD

Monotypic
HB c. 29 cm
T c. 26 cm
W c. 600 g

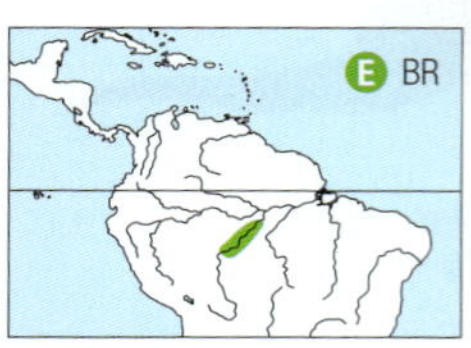

Both sides of R Madeira (Amazonas, Rondônia); perhaps widespread

Roosmalens' Porcupine
Coendou roosmalenorum

FR Coendou des Roosmalen
DE Roosmalen-Greifstachler
ES Puercoespín de los Roosmalen

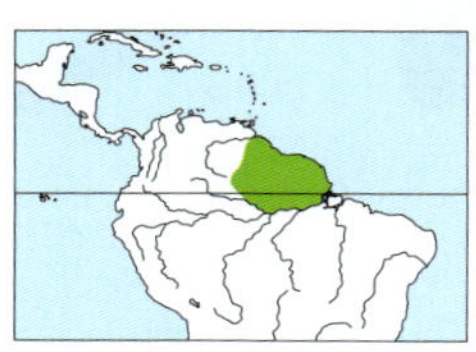

Lowlands N of R Amazon, E of R Orinoco & R Negro (E VE, Guianas, N BR)

Black-tailed Porcupine
Coendou melanurus

FR Coendou à queue noire
DE Schwarzschwanz-Greifstachler
ES Puercoespín de cola negra

LC

Monotypic
HB 28–38 cm
T 22–36 cm
W 1.5–2.4 kg

Family CUNICULIDAE – Pacas

1 genus, 2 species. Lowland and highland forests, and tropical alpine tundra habitats, generally near water. Neotropical Region.

Genus *CUNICULUS*

(2 species)

LC 5 sspp
HB 50–77 cm
T 1.3–3.5 cm
W 5–14 kg

0–2000 m

E & S MX S to BO, PY, NE AR & UR; introduced to Cuba & Lesser Antilles

Lowland Paca
Cuniculus paca
FR Paca des plaines DE Tieflandpaka ES Paca de llanura

NT Monotypic
HB 50–80 cm
T 2.1–3 cm
W 3.2–5.2 kg

1700–4000 m

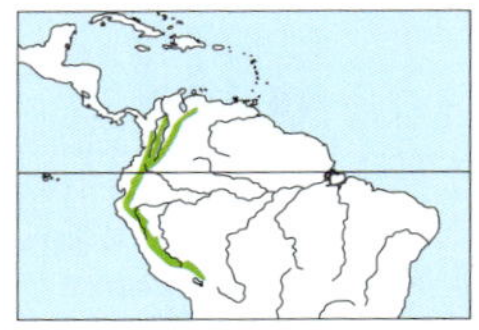

Andes of NW VE, CO, EC, PE & N BO (La Paz Dep)

Mountain Paca
Cuniculus taczanowskii
FR Paca des montagnes DE Hochlandpaka ES Paca de montaña

Family CAVIIDAE – Cavies, Capybaras and Maras

7 genera, 23 species. Arid and semi-arid steppes, dry grasslands, thornscrub, savannas, mesic grasslands, and wetlands. Neotropical Region.

Genus *CAVIA*

(6 species)

LC 7 sspp
HB 21–39 cm
W 500–790 g

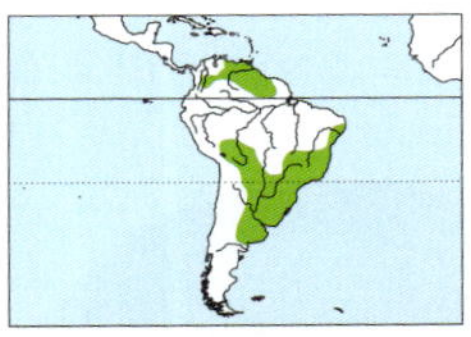

CO, VE, Guianas, BR, BO, PY, N AR, UR; recorded in SE PE

Brazilian Guinea Pig
Cavia aperea
FR Cobaye du Brésil
DE Gemeines Meerschweinchen
ES Cobaya común

LC
8 sspp
HB 22–27 cm
W 295–390 g

0–4500 m

W PE, highlands of BO, NW AR & N CL

Montane Guinea Pig
Cavia tschudii
FR Cobaye de Tschudi
DE Tschudi-Meerschweinchen
ES Cobaya de montaña

LC
Monotypic
HB 22–27 cm

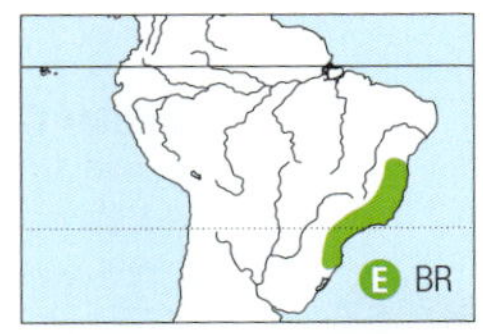

From Minas Gerais S to Santa Catarina

Shiny Guinea Pig
Cavia fulgida
FR Cobaye luisant
DE Glanzmeerschweinchen
ES Cobaya brillante

DD Monotypic
HB 28–29 cm
W 700–725 g

3000–3800 m

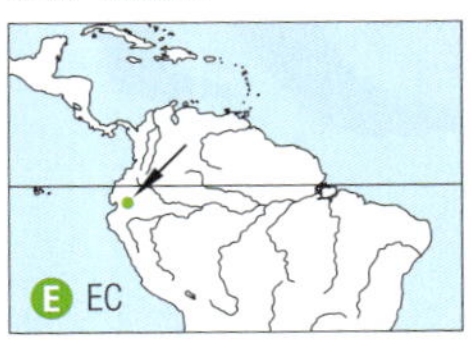

Mt Chimborazo; probably beyond type locality

Sacha Guinea Pig
Cavia patzelti
FR Cobaye de Patzelt
DE Patzelt-Meerschweinchen
ES Cobaya de Patzelt

LC
Monotypic
HB 22–34 cm
W 440–840 g

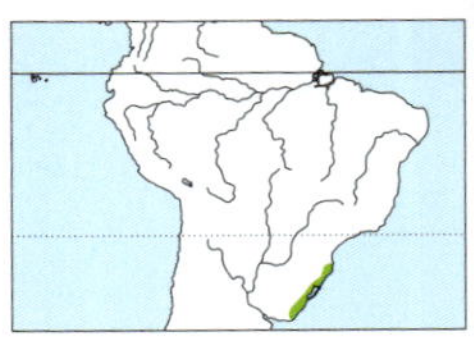

SE BR (Santa Catarina, Rio Grande do Sul), E UY (Rocha)

Greater Guinea Pig
Cavia magna
FR Grand Cobaye
DE Riesenmeerschweinchen
ES Cobaya grande

CR
Monotypic
HB 27–31 cm
W 495–680 g

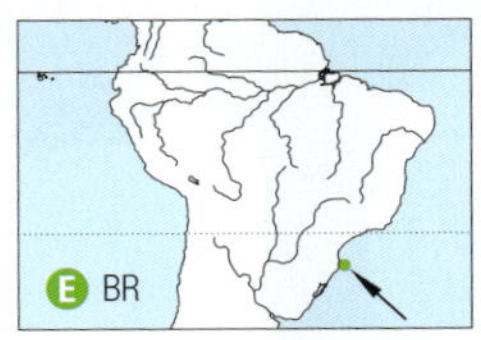

Moleques do Sul Is (Santa Catarina)

Santa Catarina's Guinea Pig
Cavia intermedia
FR Cobaye du Santa Catarina
DE Santa-Catarina-Meerschweinchen
ES Cobaya de Santa Catarina

LC
3 sspp
HB 22–23 cm
W c. 400 g

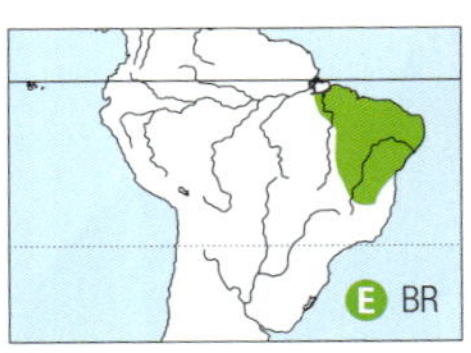

S to NW Minas Gerais, W limits undefined

Spix's Yellow-toothed Cavy
Galea spixii
FR Cobaye de Spix
DE Spix-Wieselmeerschweinchen
ES Cuy de Spix

LC
Monotypic
HB 20–23 cm
W 150–330 g

Chapada dos Veadeiros NP (Goiás); limits poorly defined

Eastern Yellow-toothed Cavy
Galea flavidens
FR Cobaye à dents jaunes
DE Östliches Wieselmeerschweinchen
ES Cuy de dientes amarillos oriental

Genus *GALEA*

(5 species)

DD
3 sspp

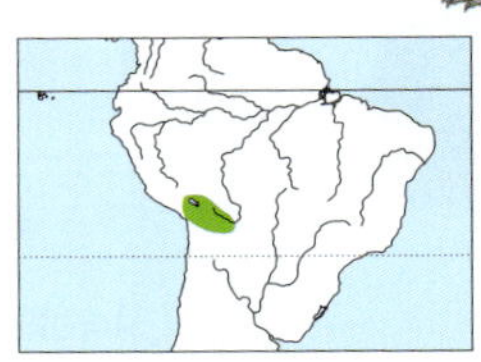

Andean Altiplano in SE PE, BO & N CL

Highland Yellow-toothed Cavy
Galea musteloides
FR Cobaye belette
DE Wieselmeerschweinchen
ES Cuy de montaña

Andes of S BO (Tarija) & NW AR (Jujuy)

DD
Monotypic
HB c. 24 cm

Southern Highland Yellow-toothed Cavy
Galea comes
FR Cobaye du Jujuy DE Südliches Wieselmeerschweinchen
ES Cuy de dientes amarillos meridional

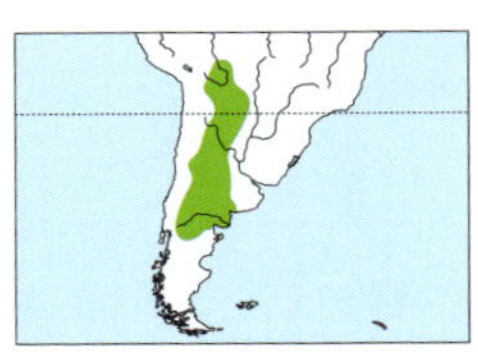

C & S BO, W PY, AR (S to coastal Buenos Aires & N Chubut)

LC
3 sspp
HB 19.8–23 cm
W 180–280 g

Lowland Yellow-toothed Cavy
Galea leucoblephara
FR Cobaye à lunettes DE Tiefland-Wieselmeerschweinchen
ES Cuy de dientes amarillos de tierras bajas

Genus *MICROCAVIA*

(6 species)

LC
Monotypic

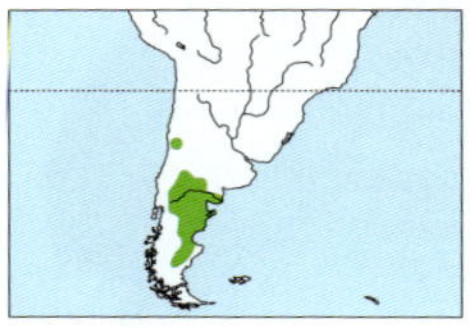

W & S AR (from Mendoza & S Buenos Aires S to Santa Cruz), W CL (Aysén)

Southern Mountain Cavy
Microcavia australis
FR Cobaye austral
DE Südliches Zwergmeerschweinchen
ES Cuy de Patagonia

NE
Monotypic

From S Salta S to N Mendoza & N San Luis, E to Santiago del Estero & Córdoba

Thomas's Mountain Cavy
Microcavia maenas
FR Cobaye de La Rioja
DE Thomas-Zwergmeerschweinchen
ES Cuy de Thomas

NE
Monotypic

Santiago del Estero; perhaps Salta, Chaco, Córdoba & Santa Fe

Jayat's Mountain Cavy
Microcavia jayat
FR Cobaye de Jayat
DE Jayat-Zwergmeerschweinchen
ES Cuy de Jayat

LC
2 sspp
HB 19–20 cm
W c. 380 g

3700–4000 m

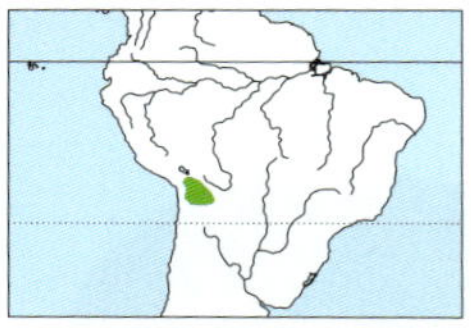

Altiplano of SW BO (SW La Paz, Oruro, N Potosí) & NE CL (Tarapacá)

Northern Mountain Cavy
Microcavia niata
FR Cobaye de Bolivie
DE Nördliches Zwergmeerschweinchen
ES Cuy de los Andes

LC
Monotypic
HB c. 18.6 cm
W 150–220 g

3000–4500 m

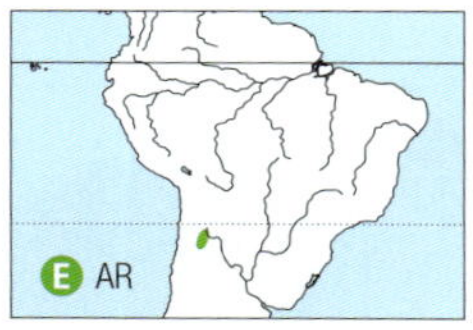

Salta, Catamarca, Tucumán

Shipton's Mountain Cavy
Microcavia shiptoni
FR Cobaye de Shipton
DE Shipton-Zwergmeerschweinchen
ES Cuy de Shipton

NE
Monotypic
HB 19–19.5 cm

c. 5000 m

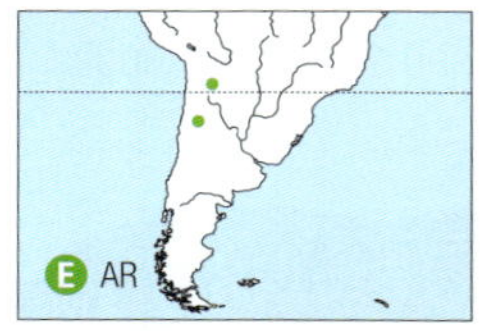

Known only from Chorrillos (Salta) & Quebrada de Santo Domingo (La Rioja)

Sorojchi Mountain Cavy
Microcavia sorojchi
FR Cobaye sorojchi
DE Sorojchi-Zwergmeerschweinchen
ES Cuy soroche

Genus *HYDROCHOERUS*

(2 species)

LC Monotypic
HB 107–134 cm T 1–2 cm
W 35–65 kg

Greater Capybara
Hydrochoerus hydrochaeris
FR Grand Capybara
DE Wasserschwein
ES Capibara

E of Andes in CO, VE, Guianas, BR, EC, PE, BO, PY, UY and NW & E AR

Lesser Capybara
Hydrochoerus isthmius
FR Capybara du Panama
DE Kleines Wasserschwein
ES Capibara pequeña

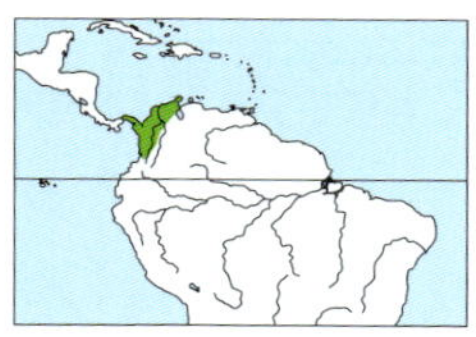

E PA, W CO, NW VE (Zulia)

DD Monotypic
HB c. 102 cm

Genus *KERODON*

(2 species)

LC
Monotypic
HB c. 30 cm
W c. 612 g

Rock Cavy
Kerodon rupestris
FR Cobaye des rochers
DE Felsenmeerschweinchen
ES Mocó de roca

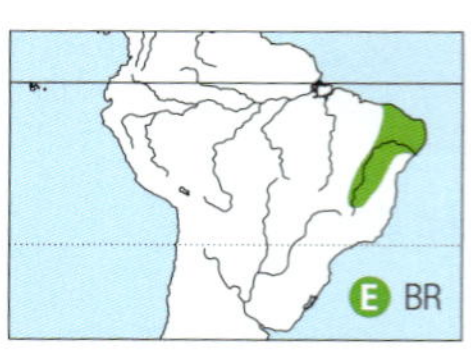

Ceará S to N Minas Gerais

DD
Monotypic
HB c. 38 cm
W c. 1 kg

Acrobatic Cavy
Kerodon acrobata
FR Cobaye acrobate
DE Großes Felsenmeerschweinchen
ES Mocó acróbata

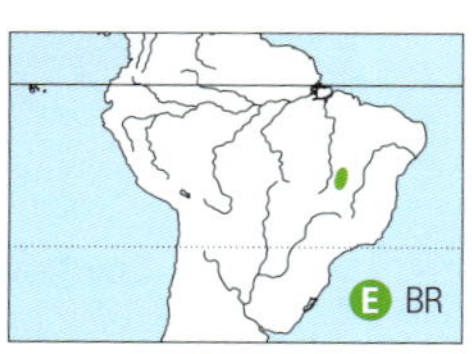

NE Goiás, adjacent Tocantins (W of Serra Geral de Goiás)

Genus *DOLICHOTIS*

(1 species)

NT
2 sspp
HB 60–80 cm
T 2.5–4 cm
W 7–9 kg

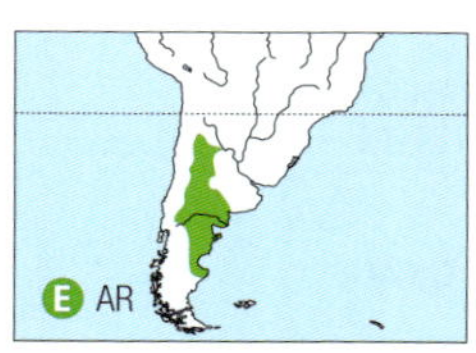

Catamarca S to Santa Cruz, E to Buenos Aires; perhaps S CL

Patagonian Mara
Dolichotis patagonum
FR Mara de Patagonie
DE Großer Pampashase
ES Mara de Patagonia

Genus *PEDIOLAGUS*

(1 species)

Chacoan Mara
Pediolagus salinicola
FR Mara du Chaco
DE Kleiner Pampashase
ES Mara del Chaco

LC
Monotypic
HB 42–48 cm
T 1.9–2 cm
W 1.8–2.3 kg

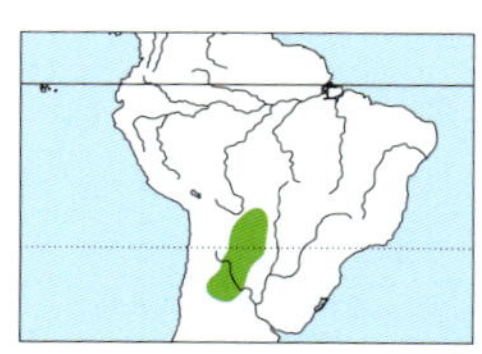

Extreme SE BO, W PY, NW AR (S to Córdoba)

Family DASYPROCTIDAE – Agoutis and Acouchis

2 genera, 15 species. Tropical forested areas, brush lands, dry forests, and savannas. Neotropical Region.

Genus *DASYPROCTA*

(13 species)

CR
Monotypic
HB 44–56 cm
T 2–3 cm
W 2–4 kg

50–600 m

Lowlands of Veracruz, N Oaxaca, NW Chiapas & W Tabasco; introduced to Cuba

Mexican Agouti
Dasyprocta mexicana
FR Agouti du Mexique
DE Mexiko-Aguti
ES Agutí de México

LC
Monotypic
HB 48–60 cm
T 2–5.5 cm
W 3.2–4.2 kg

up to 2400 m

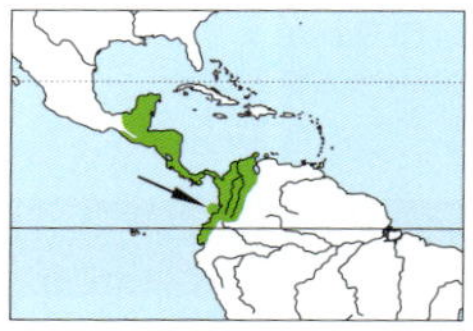

From S MX to CO (with Gorgona I), W VE & NW EC (S to Guayas); introduced to Cayman Is

Central American Agouti
Dasyprocta punctata
FR Agouti ponctué
DE Mittelamerika-Aguti
ES Agutí de Centroamérica

NT
Monotypic
HB 43–45 cm
T 3–4 cm

Coiba I

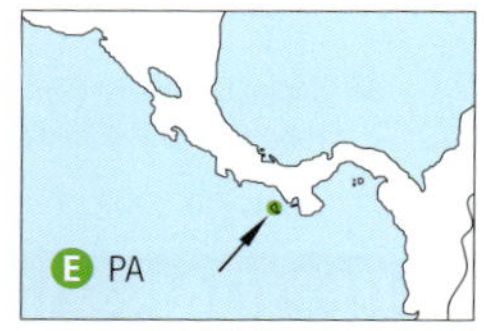

Coiban Agouti
Dasyprocta coibae
FR Agouti de Coiba
DE Coiba-Aguti
ES Agutí de Coiba

EN
Monotypic
HB c. 43 cm
W 0.6–4 kg (F)

Roatán I

Roatan Island Agouti
Dasyprocta ruatanica
FR Agouti de Roatan
DE Roatán-Aguti
ES Agutí de Roatán

LC
Monotypic
HB 54–76 cm
T 2–4 cm
W 3.5–6 kg

up to 1000 m

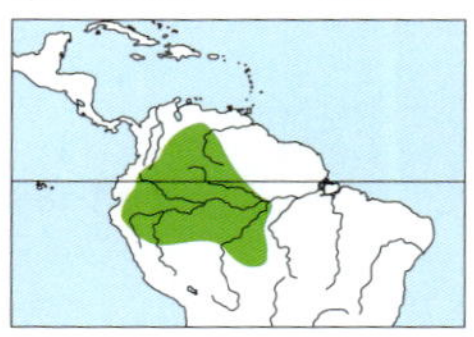

SE CO, S VE, W BR (E to R Negro & R Madeira), E EC, E PE (N of Junín)

Black Agouti
Dasyprocta fuliginosa
FR Agouti cendré
DE Mohrenaguti
ES Agutí negro

LC
Monotypic
HB 47–65 cm
T 1–3 cm
W 2.1–5.9 kg

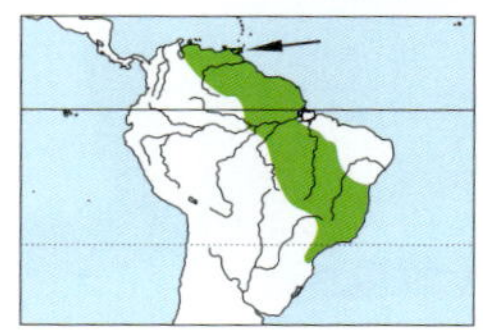

N VE (E of L Maracaibo), TT (Trinidad I), Guianas, BR (E of R Negro, SE to C coast); introduced to Lesser Antilles

Common Red-rumped Agouti
Dasyprocta leporina
FR Agouti doré DE Goldaguti
ES Agutí de dorso rojo

NT
Monotypic
HB 47–56 cm
T 2–2.3 cm
W 3–4.4 kg

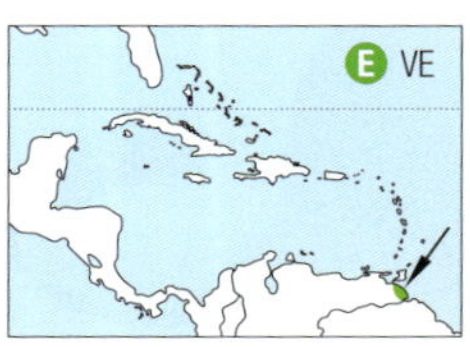

Known from 2 localities in Orinoco Delta

Guamara Agouti
Dasyprocta guamara
FR Agouti de l'Orénoque
DE Orinoco-Aguti
ES Agutí del Orinoco

DD Monotypic
HB 46–56 cm
T 1–2.5 cm
W 2.2–2.8 kg

S of R Amazon, from R Tapajós to W of R Tocantins, Marajó & Mexiana Is

0–200 m

Orange Agouti
Dasyprocta croconota
FR Agouti à croupe orange
DE Orangefarbenes Aguti
ES Agutí naranja

LC
Monotypic
HB 45–52 cm
T 1.8–3 cm

0–900 m

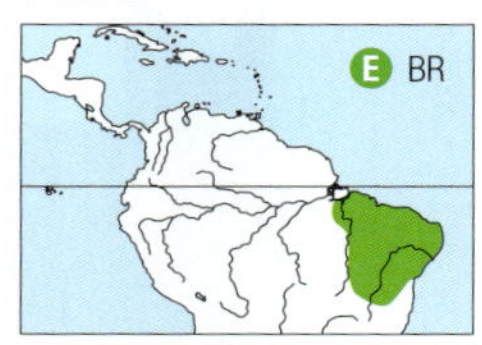

From E Pará along coast to EC Bahia, inland to E Tocantins & N Minas Gerais

Black-rumped Agouti
Dasyprocta prymnolopha
FR Agouti à croupe noire
DE Schwarzsteißaguti
ES Agutí de dorso negro

DD
Monotypic
HB 44–50 cm
W 2.3–3.8 kg

Littoral Paraíba & Pernambuco

Iack's Red-rumped Agouti
Dasyprocta iacki
FR Agouti d'Iack
DE Pernambuco-Aguti
ES Agutí de Iack

DD
Monotypic
HB 44–54 cm
T 1.1–3.8 cm
W 3–5.2 kg

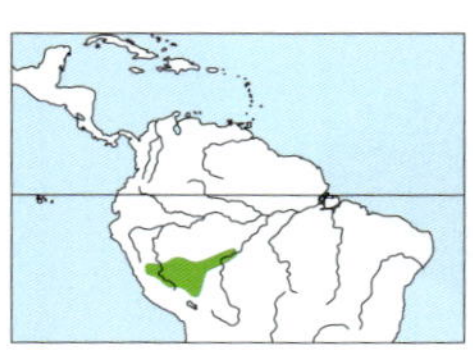

PE (E from Junín), W BR (upper R Purus & R Madeira), N BO (Pando, Beni)

Brown Agouti
Dasyprocta variegata
FR Agouti brun DE Vielfarbiges Aguti
ES Agutí marrón

DD
Monotypic
HB c. 63 cm
T c. 2.4 cm

1000–2000 m

Urubamba & Marcapata basins in E slope of Andes

Kalinowski's Agouti
Dasyprocta kalinowskii
FR Agouti de Kalinowski
DE Kalinowski-Aguti
ES Agutí de Kalinowski

DD
Monotypic
HB 43–57 cm
T 1–3.5 cm
W 2.4–3.2 kg

0–700 m

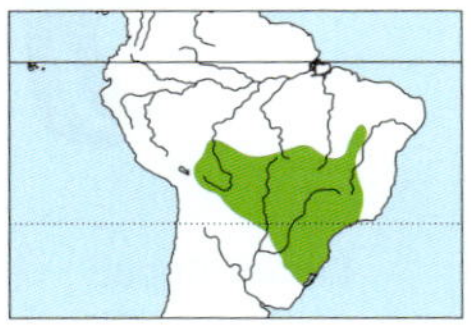

Azara's Agouti
Dasyprocta azarae
FR Agouti d'Azara
DE Azara-Aguti
ES Agutí de Azara

BR (S from Bahia), BO (S of R Guaporé, E of R Beni), PY, AR (Formosa, Corrientes, Misiones)

Genus *MYOPROCTA*

(2 species)

LC
Monotypic
HB 33–39 cm
T 5.1–7.8 cm
W 1–1.4 kg

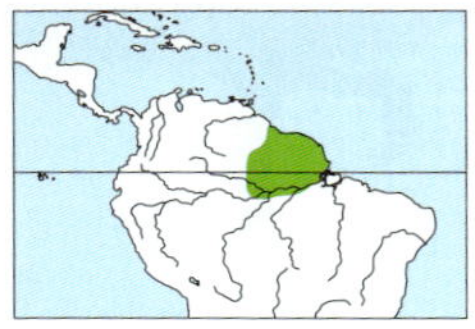

Red Acouchi
Myoprocta acouchy
FR Agouti rouge
DE Rotes Acouchi
ES Acuchí rojo

Guianas, N BR (E of R Branco, N of R Amazon); reported S of R Amazon

LC
Monotypic
HB 30–38 cm
T 4–5.8 cm
W 0.8–1.2 kg

50–1200 m

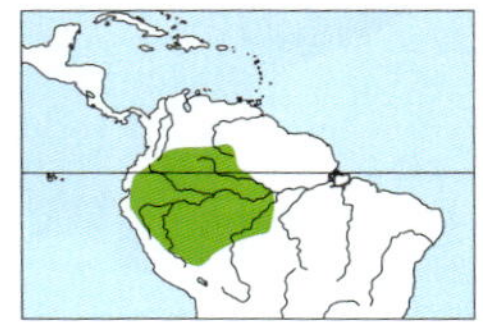

Green Acouchi
Myoprocta pratti
FR Agouti vert
DE Grünes Acouchi
ES Acuchí verde

From Andes E to lower R Negro and beyond of R Madeira (SE CO, S VE, W BR, E EC, E PE, N BO)

Family CHINCHILLIDAE – Viscachas and Chinchillas

3 genera, 6 species. Semi-arid and arid plains to grasslands, low shrublands, and high-elevation steppes in western and southern South America from sea level to 6000 m. Neotropical Region.

Genus *LAGOSTOMUS*

(1 species)

LC
3 sspp
HB 39–61 cm
T 13.5–20 cm
W 3.5–8.8 kg

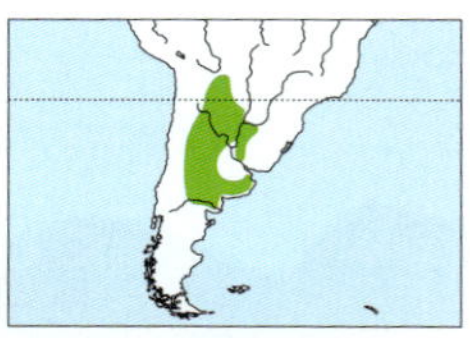

Common Plains Viscacha
Lagostomus maximus
FR Viscache des plaines
DE Viscacha
ES Vizcacha de llanura

SE BO, W & S PY, AR (S to Río Negro)

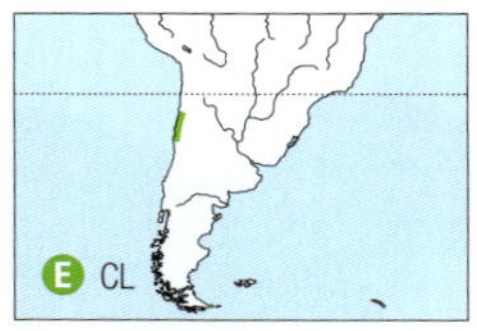

Coquimbo, Atacama

Chilean Chinchilla
Chinchilla lanigera
FR Chinchilla à longue queue
DE Langschwanz-Chinchilla
ES Chinchilla de cola larga

EN
Monotypic
HB 22–24 cm
T 14–17 cm
W 369–493 g

Genus *CHINCHILLA*

(2 species)

CL (Antofagasta S to Atacama); relict populations seem to persist in BO & AR

Short-tailed Chinchilla
Chinchilla chinchilla
FR Chinchilla à queue courte
DE Kurzschwanz-Chinchilla
ES Chinchilla de cola corta

EN
Monotypic
HB 30–38 cm
T ≤ 10 cm
W 500–850 g

Genus *LAGIDIUM*

(3 species)

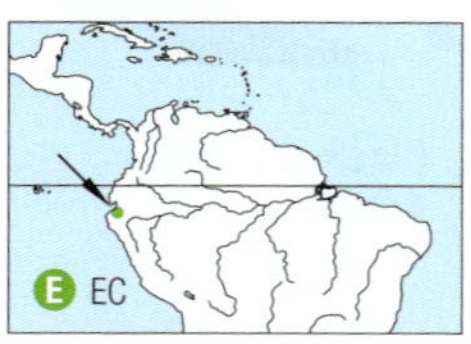

2000–2480 m

Known only from Cerro El Ahuaca (Loja)

DD
Monotypic
HB c. 40 cm
T c. 40 cm
W c. 2 kg

Ecuadorean Mountain Viscacha
Lagidium ahuacaense
FR Viscache d'Équateur DE Ecuador-Bergviscacha ES Vizcacha del Ahuaca

up to 4000 m

SW AR (W Santa Cruz), adjacent CL (Magallanes, reported from Aysén)

DD
Monotypic
HB 47–48 cm
T 30–31 cm
W c. 2 kg

Wolffsohn's Mountain Viscacha
Lagidium wolffsohni
FR Viscache de Wolffsohn DE Südliches Bergviscacha ES Vizcacha de Wolffsohn

variants

LC
Monotypic
HB 29–46 cm
T 21–38 cm
W 0.75–2.1 kg

700–4800 m

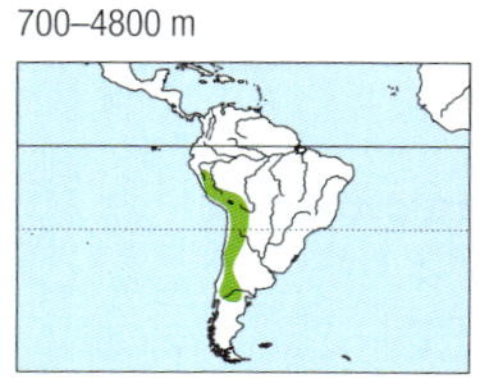

Common Mountain Viscacha
Lagidium viscacia
FR Viscache de montagne
DE Eigentliches Bergviscacha
ES Vizcacha de montaña

Andes from C PE to BO, N & C CL and W AR (S to Chubut)

Family DINOMYIDAE – Pacarana

1 genus, 1 species. Low tropical forests at elevations of about 240 m to high montane wet forests of up to 3400 m. Neotropical Region.

Genus *DINOMYS*

(1 species)

240–3400 m

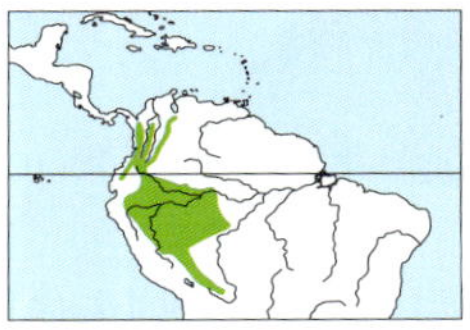

W VE, CO (3 Andean ranges), EC (N Andes, Amazon Basin), E PE, W BR, BO

Pacarana
Dinomys branickii
FR Pacarana
DE Pakarana
ES Pacarana

LC
Monotypic
HB 73–79 cm
T 14–23 cm
W 7.3–15 kg

Family ABROCOMIDAE – Inca Rats and Chinchilla Rats

2 genera, 10 species. Arid temperate montane, lowland Mediterranean chaparral, and tropical cloud forests. Neotropical Region.

Genus *CUSCOMYS*

(2 species)

DD Monotypic
HB c. 35 cm T c. 26 cm
W c. 910 g (F)

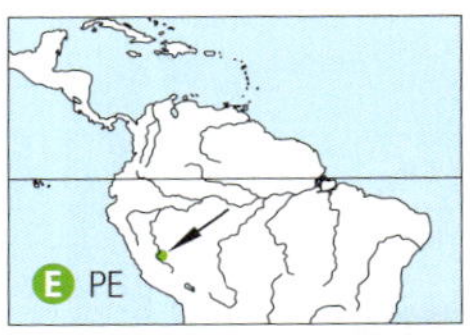

c. 3370 m

Headwaters of R Pomureni, W of crest of R Apurímac–R Ene divide (Junín)

Ashaninka Inca Rat
Cuscomys ashaninka
FR Abrocome ashaninka DE Asháninka-Inkaratte ES Rata inca de los Asháninka

c. 2700 m

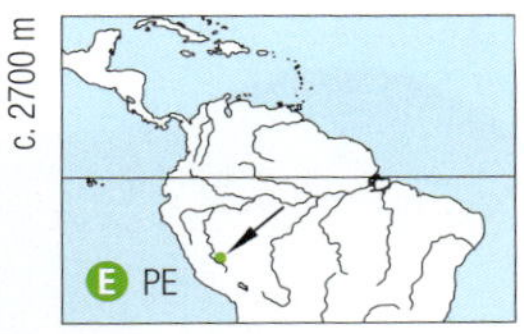

Vicinity of Machu Picchu (Cusco)

DD
Monotypic

Machu Picchu Inca Rat
Cuscomys oblativus
FR Abrocome des Incas DE Machu-Picchu-Inkaratte
ES Rata inca de Machu Picchu

Genus *ABROCOMA*

(8 species)

CR
Monotypic
HB 17–18 cm
T 13.2–15 cm

1815–2500 m

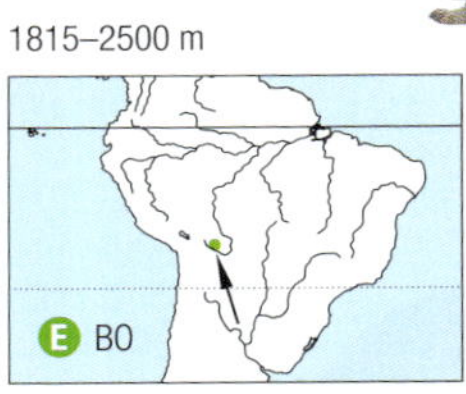

Comarapa (Santa Cruz)

Bolivian Chinchilla Rat
Abrocoma boliviensis
FR Abrocome de Bolivie
DE Bolivien-Chinchillaratte
ES Rata chinchilla de Bolivia

LC
2 sspp
HB 17–23 cm
T 13–18 cm
W 150–307 g

Antofagasta S to O'Higgins

0–2000 m

Bennett's Chinchilla Rat
Abrocoma bennettii
FR Abrocome de Bennett
DE Chile-Chinchillaratte
ES Rata chinchilla de Bennett

LC Monotypic
HB 15.7–19.6 cm
T 5–9.6 cm
W 90–105 g

3700–5000 m

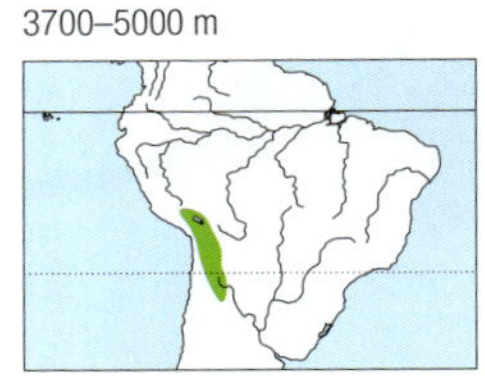

Ashy Chinchilla Rat
Abrocoma cinerea
FR Abrocome cendré
DE Graue Chinchillaratte
ES Rata chinchilla cinérea

PE (S from Arequipa), SW BO, NW AR (S to Catamarca), CL (S to Antofagasta)

c. 3000 m

S end of Sierra de Ambato (Catamarca)

DD
Monotypic
HB 19.7–20 cm
T 13–14.4 cm

Budin's Chinchilla Rat
Abrocoma budini
FR Abrocome de Budin DE Catamarca-Chinchillaratte
ES Rata chinchilla de Budin

Famatina Range (La Rioja)

DD Monotypic
HB 16.4–18.2 cm
T 10.8–11.7 cm

Famatina Chinchilla Rat
Abrocoma famatina
FR Abrocome du Famatina DE Famatina-Chinchillaratte
ES Rata chinchilla de Famatina

LC
Monotypic
HB 16–19.6 cm
T 9.6–12 cm
W 164–180 g

2115–2885 m

Sierra del Tontal (San Juan)

Sierra del Tontal Chinchilla Rat
Abrocoma schistacea
FR Abrocome du Tontal
DE Sierra-del-Tontal-Chinchillaratte
ES Rata chinchilla de Tontal

DD Monotypic
HB c. 17.6 cm
T 11.2–16.4 cm
W 141–174 g

above 1880 m

Sierra de Uspallata (Mendoza)

Uspallata Chinchilla Rat
Abrocoma uspallata
FR Abrocome d'Uspallata
DE Uspallata-Chinchillaratte
ES Rata chinchilla de Uspallata

DD Monotypic
HB 16.5–19.1 cm
T c. 9.4 cm

c. 3000 m

Know from Punta de Vacas (Mendoza)

Mendozan Chinchilla Rat
Abrocoma vaccarum
FR Abrocome du Mendoza
DE Mendoza-Chinchillaratte
ES Rata chinchilla de Mendoza

Family CTENOMYIDAE – Tuco-tucos

1 genus, 67 species. Sand fields, dunes, and forests. Neotropical Region.

Genus *CTENOMYS*

(67 species)

CR
Monotypic
HB 16.8–25 cm
T 6.7–8 cm
W 180–234 g

Colonial Tuco-tuco
Ctenomys sociabilis

FR Tuco-tuco sociable
DE Gesellige Kammratte
ES Tuco tuco colonial

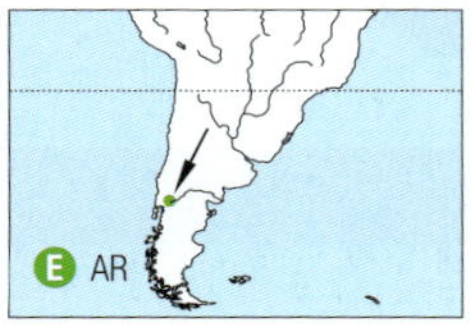

Narrow area between R Traful and L Nahuel Huapi, W of R Limay (S Neuquén)

NE Monotypic
HB c. 25 cm (F) T c. 7.7 cm (F)
W c. 140 g (F)

c. 800 m

Plebiscite Tuco-tuco
Ctenomys plebiscitum

FR Tuco-tuco d'Esquel
DE Plebiszit-Kammratte
ES Tuco tuco de Esquel

Known from 2 populations in vicinity of Esquel (Chubut)

LC
Monotypic
HB c. 21 cm
T c. 9.1 cm

up to 800 m

Emilio's Tuco-tuco
Ctenomys emilianus

FR Tuco-tuco d'Emilio
DE Emilio-Budin-Kammratte
ES Tuco tuco de Budin

Known only from Chos Malal (Neuquén)

DD Monotypic
W c. 600 g

2 disjunct populations in Sierra del Aconquija (Tucumán)

260–3000 m

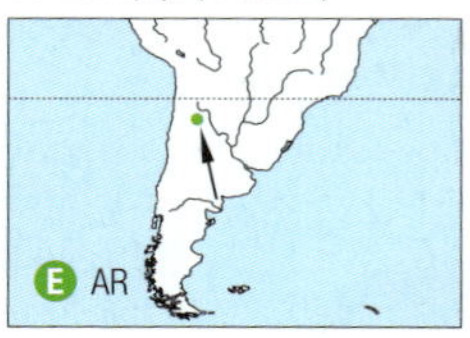

Robust Tuco-tuco
Ctenomys tuconax

FR Tuco-tuco robuste
DE Robuste Kammratte
ES Tuco tuco robusto

LC
Monotypic
HB 23–25 cm

W Santa Cruz

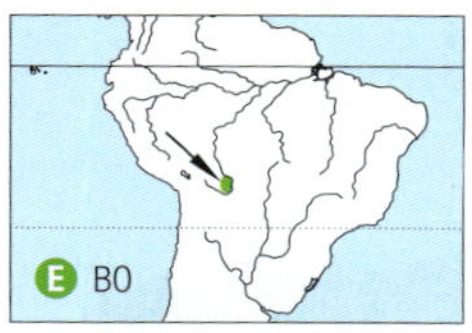

Steinbach's Tuco-tuco
Ctenomys steinbachi

FR Tuco-tuco de Steinbach
DE Steinbach-Kammratte
ES Tuco tuco de Steinbach

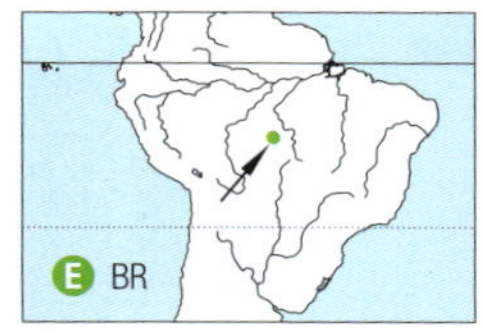

Bicolored Tuco-tuco
Ctenomys bicolor

FR Tuco-tuco bicolore
DE Zweifarbige Kammratte
ES Tuco tuco bicolor

Known from José Bonifácio Indian Village & Pimenta Bueno (Rondônia)

NE
Monotypic
HB c. 31 cm
T c. 8.9 cm
W c. 418 g

black morph

brown morph

NE
Monotypic
HB c. 23 cm
T c. 8 cm

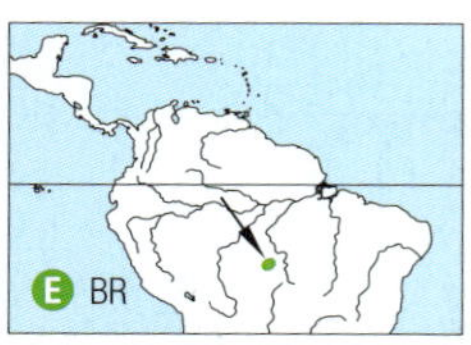

Rondon's Tuco-tuco
Ctenomys rondoni

FR Tuco-tuco de Rondon
DE Rondônia-Kammratte
ES Tuco tuco de Rondonia

Known from Vilhena (Rondônia) & Juruena (Mato Grosso)

NE
Monotypic
HB c. 24 cm
T c. 7.1 cm

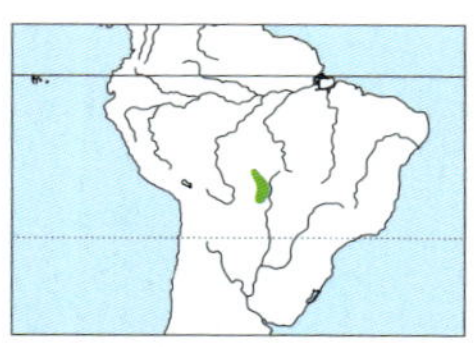

Natterer's Tuco-tuco
Ctenomys nattereri

FR Tuco-tuco de Natterer
DE Natterer-Kammratte
ES Tuco tuco de Natterer

SW BR (Rondônia, Mato Grosso), E BO (Santa Cruz)

LC
Monotypic
HB 18.6–28 cm
T 7.1–9.5 cm
W 420–650 g

Santa Cruz

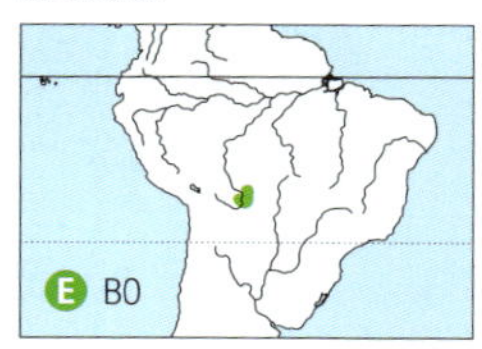

Bolivian Tuco-tuco
Ctenomys boliviensis

FR Tuco-tuco de Bolivie
DE Bolivien-Kammratte
ES Tuco tuco de Bolivia

NE
Monotypic
HB 15.3–22 cm
T 3.5–9 cm
W 83–360 g

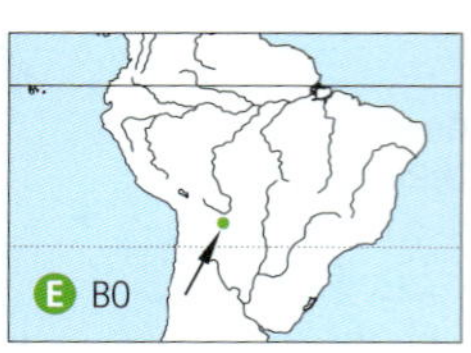

Anderson's Tuco-tuco
Ctenomys andersoni

FR Tuco-tuco d'Anderson
DE Anderson-Kammratte
ES Tuco tuco de Anderson

Cerro Itahuaticua

NE
Monotypic
HB 15.9–23 cm
T 2.5–9.6 cm
W 82–390 g

800–1800 m

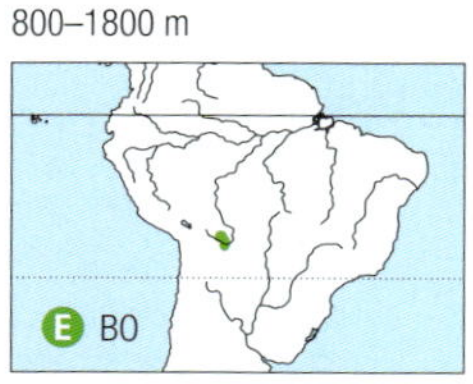

Erika's Tuco-tuco
Ctenomys erikacuellarae

FR Tuco-tuco d'Erika
DE Erika-Kammratte
ES Tuco tuco de Erika

E slope of Andes in Santa Cruz & Chuquisaca

NE
Monotypic
HB 14.1–15.7 cm
T 5.8–6.3 cm
W 79–105 g

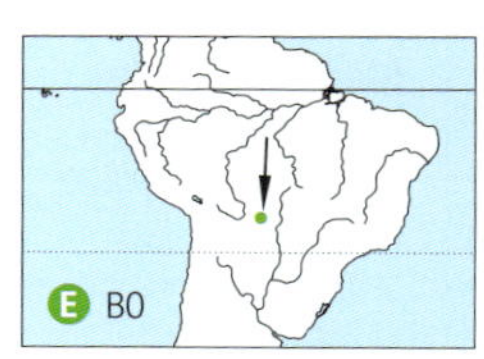

Yates's Tuco-tuco
Ctenomys yatesi

FR Tuco-tuco de Yates
DE Yates-Kammratte
ES Tuco tuco de Yates

Known from near Roboré (Santa Cruz)

DD
Monotypic
HB c. 15.6 cm
T c. 4.6 cm

Northern Chaco; probably in adjacent BO & BR

Black-backed Tuco-tuco
Ctenomys dorsalis
FR Tuco-tuco à dos noir
DE Streifenkammratte
ES Tuco tuco de dorso negro

DD
Monotypic
W 176–341 g

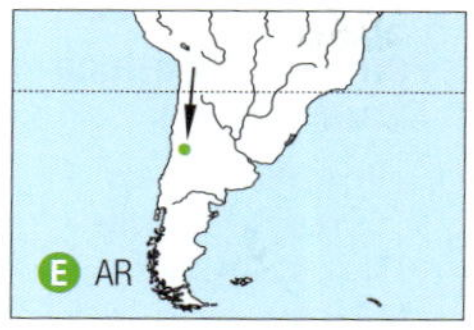

Médanos del Borbollón area (Mendoza)

Guaymallen Tuco-tuco
Ctenomys validus
FR Tuco-tuco du Guaymallen
DE Guaymallén-Kammratte
ES Tuco tuco de Guaymallén

DD
Monotypic
HB c. 20 cm
T c. 9.7 cm

c. 1000 m

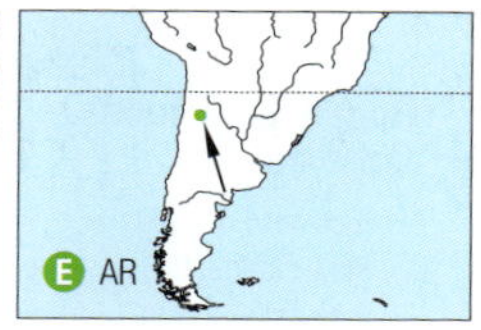

Known only from La Puntilla, near Tinogasta (Catamarca)

Puntilla Tuco-tuco
Ctenomys coludo
FR Tuco-tuco de La Puntilla
DE Puntilla-Kammratte
ES Tuco tuco de La Puntilla

DD
Monotypic
HB c. 16 cm
T c. 7.4 cm

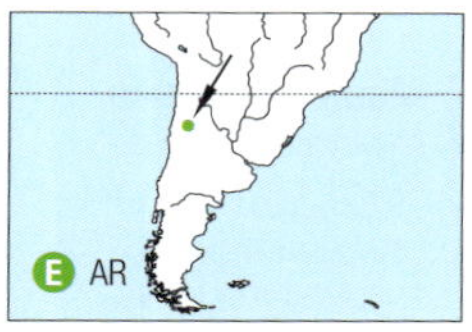

Sierra de Famatina (La Rioja)

Famatina Tuco-tuco
Ctenomys famosus
FR Tuco-tuco de la Famatina
DE Famatina-Kammratte
ES Tuco tuco de Famatina

DD
Monotypic
HB c. 20 cm
T c. 9.7 cm

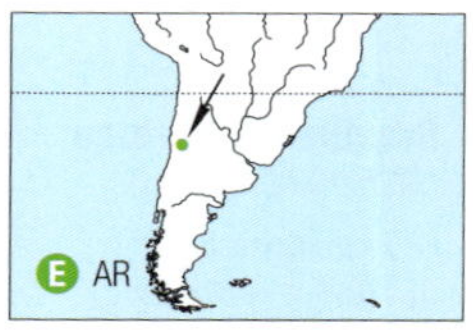

Known from Cañada Honda (San Juan)

San Juan Tuco-tuco
Ctenomys johannis
FR Tuco-tuco du San Juan
DE San-Juan-Kammratte
ES Tuco tuco de San Juan

DD
Monotypic
HB c. 19 cm
T c. 6.9 cm

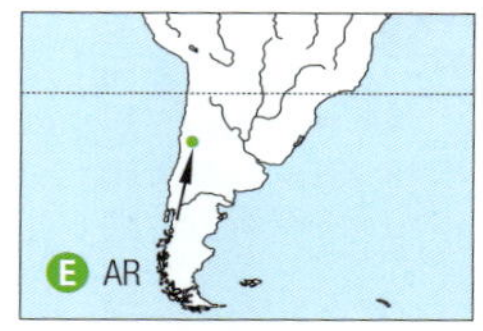

Sierra del Tontal (San Juan)

Sierra Tontal Tuco-tuco
Ctenomys tulduco
FR Tuco-tuco du Tontal
DE Sierra-Tontal-Kammratte
ES Tuco tuco de Tontal

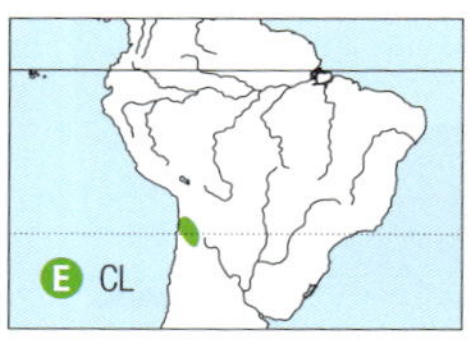

Oasis de Pica (Tarapacá desert), E Antofogasta

DD
2 sspp

Long-tailed Tuco-tuco
Ctenomys fulvus
FR Tuco-tuco coloré
DE Langschwanzkammratte
ES Tuco tuco de cola larga

LC
3 sspp
HB 19.3–25 cm
T 7.5–9.4 cm

up to 5000 m

Andes in S PE (S from Moquegua), N CL (Arica y Parinacota), SW BO & NW AR

Andean Tuco-tuco
Ctenomys opimus
FR Tuco-tuco de la puna
DE Anden-Kammratte
ES Tuco tuco de los Andes

DD
Monotypic
HB c. 19.5 cm
T c. 8.5 cm
W c. 270 g

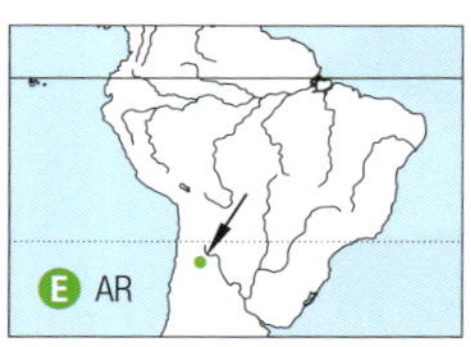

Known only from Los Cardones (Tucumán)

Scaglia's Tuco-tuco
Ctenomys scagliai
FR Tuco-tuco de Scaglia
DE Scaglia-Kammratte
ES Tuco tuco de Scaglia

DD
Monotypic
HB c. 20 cm
T c. 9 cm

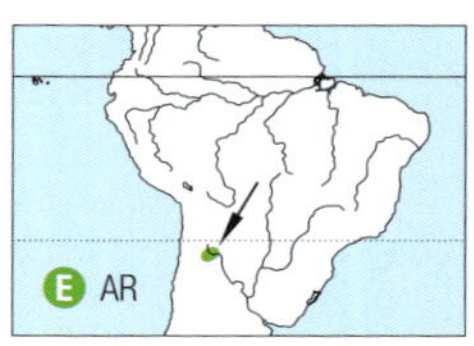

Known from few localities in Yungas & Chaco Seco ecoregions (Salta)

Salta Tuco-tuco
Ctenomys saltarius
FR Tuco-tuco de Salta
DE Salta-Kammratte
ES Tuco tuco de Salta

LC
Monotypic
HB 15.2–23 cm
T 5.5–8.5 cm

SE BR (S Rio Grande do Sul), N & C UY

Collared Tuco-tuco
Ctenomys torquatus
FR Tuco-tuco à collier
DE Halsbandkammratte
ES Tuco tuco de collar

VU
Monotypic
HB 16.4–22 cm
T 6.7–9.2 cm
W 170–307 g

Region known as "Coxilha da Lombas" (Rio Grande do Sul)

Lami Tuco-tuco
Ctenomys lami
FR Tuco-tuco de Lami
DE Lami-Kammratte
ES Tuco tuco de Lami

DD
Monotypic
T c. 7.4 cm
W c. 193 g

From Laguna Beach (Santa Catarina) S to São José do Norte (Rio Grande do Sul)

Minute Tuco-tuco
Ctenomys minutus
FR Petit Tuco-tuco
DE Kleine Kammratte
ES Tuco tuco diminuto

DD
Monotypic
HB c. 15.9 cm T c. 7.5 cm
W c. 200 g

c. 200 m

Known from few sites in Manoel Viana & Maçambará (Rio Grande do Sul)

Ibicui Tuco-tuco
Ctenomys ibicuiensis
FR Tuco-tuco de l'Ibicui
DE Ibicui-Kammratte
ES Tuco tuco de Ibicuí

CR
Monotypic
HB 17.6–20 cm
T 7.6–9.9 cm
W 175–279 g

Roig's Tuco-tuco
Ctenomys roigi
FR Tuco-tuco de Roig
DE Roig-Kammratte
ES Tuco tuco de Roig

Small area along R Paraná (Corrientes)

LC
Monotypic
HB c. 20 cm
T c. 6.7 cm

Perrens's Tuco-tuco
Ctenomys perrensi
FR Tuco-tuco de Perrens
DE Corrientes-Kammratte
ES Tuco tuco de Perrens

W Corrientes

NT
Monotypic
HB 17.3–19.5 cm
T 7.2–8.2 cm
W 165–300 g

Pearson's Tuco-tuco
Ctenomys pearsoni
FR Tuco-tuco de Pearson
DE Pearson-Kammratte
ES Tuco tuco de Pearson

NE AR (E Entre Ríos), S UY (E to Rocha)

NT
Monotypic
HB 19–22 cm T 8.6–9.9 cm
W 269–376 g

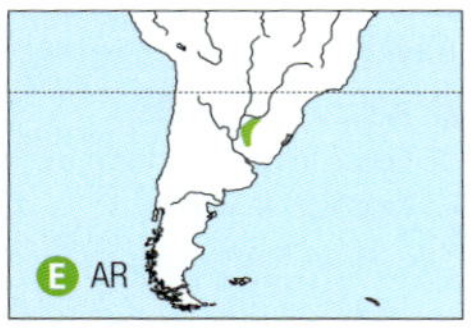

D'Orbigny's Tuco-tuco
Ctenomys dorbignyi
FR Tuco-tuco de d'Orbigny
DE D'Orbigny-Kammratte
ES Tuco tuco de D'Orbigny

Patchily in Esteros del Iberá & Pampa regions (Corrientes, Entre Ríos)

DD
Monotypic

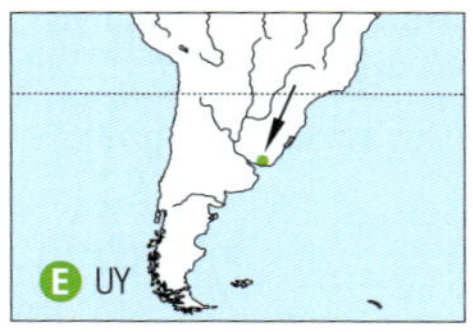

Brazilian Tuco-tuco
Ctenomys brasiliensis
FR Tuco-tuco du Brésil
DE Brasilien-Kammratte
ES Tuco tuco de Brasil

Known only from Minas (Lavalleja)

LC
2 sspp

900–2000 m

Maule Tuco-tuco
Ctenomys maulinus
FR Tuco-tuco du Maule
DE Maule-Kammratte
ES Tuco tuco de Maule

C CL (Maule, Ñuble, La Araucanía), AR (Neuquén, Río Negro)

EN Monotypic
HB c. 21 cm
T 9.6–10.3 cm
W 349–366 g

Dune Tuco-tuco
Ctenomys australis
FR Tuco-tuco austral
DE Dünenkammratte
ES Tuco tuco austral

Along Atlantic coast in S Buenos Aires

LC
Monotypic
HB 16.5–20 cm
T 6.8–8.9 cm

up to 3400 m

Mendoza Tuco-tuco
Ctenomys mendocinus
FR Tuco-tuco de Mendoza
DE Mendoza-Kammratte
ES Tuco tuco de Mendoza

San Juan, Mendoza, Córdoba, San Luis, La Pampa, W Buenos Aires

NE
Monotypic
HB 15.9–17.7 cm

N Córdoba

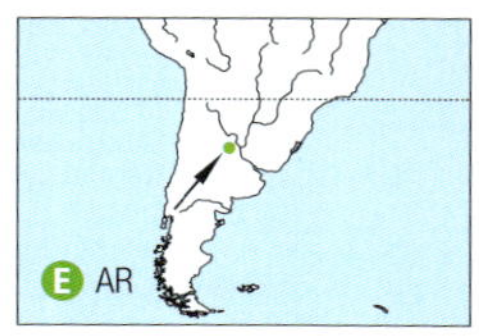

Rosendo Pascual's Tuco-tuco
Ctenomys rosendopascuali
FR Tuco-tuco de Pascual
DE Pascual-Kammratte
ES Tuco tuco de Rosendo Pascual

EN
Monotypic
HB c. 25 cm
T c. 7.4 cm

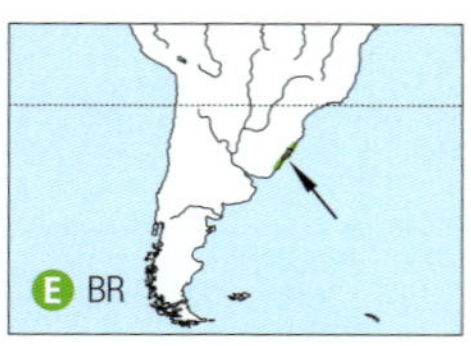

Flamarion's Tuco-tuco
Ctenomys flamarioni
FR Tuco-tuco de Flamarion
DE Flamarion-Kammratte
ES Tuco tuco de Flamarion

From Arroio do Sal S to R Chuí (Rio Grande do Sul, Santa Catarina); perhaps N UR

Rio Negro Tuco-tuco
Ctenomys rionegrensis
FR Tuco-tuco du Rio Negro
DE Rio-Negro-Kammratte
ES Tuco tuco de Río Negro

SW UR (Río Negro), NE AR (Entre Ríos)

EN
Monotypic
HB 16.2–19 cm
T 6.5–8.3 cm

variants

EN
Monotypic
HB c. 17 cm
T c. 4.3 cm

Pundt's Tuco-tuco
Ctenomys pundti
FR Tuco-tuco de Pundt
DE Pundt-Kammratte
ES Tuco tuco de Pundt

Córdoba, San Luis

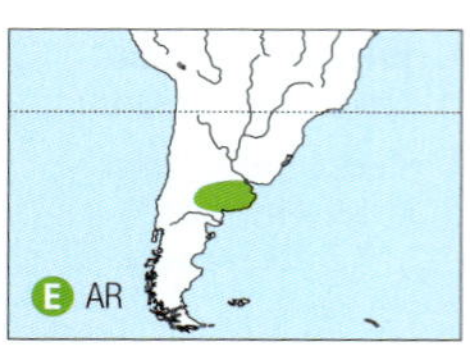

LC
3 sspp
HB 15.6–17.9 cm
T 5.6–7.5 cm

Talas Tuco-tuco
Ctenomys talarum
FR Tuco-tuco des talas
DE La-Plata-Kammratte
ES Tuco tuco de Las Talas

Patchily in coast and inland (Buenos Aires, La Pampa)

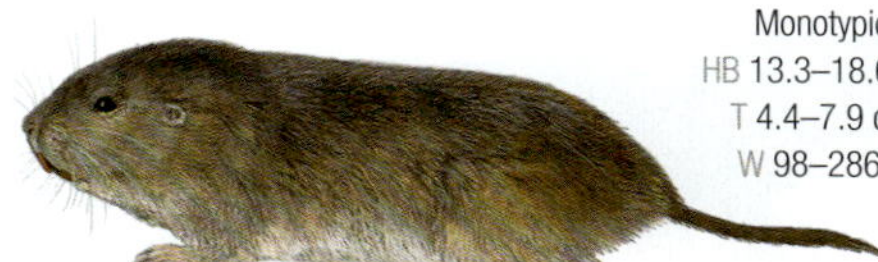

NE
Monotypic
HB 13.3–18.6 cm
T 4.4–7.9 cm
W 98–286 g

2500–2750 m

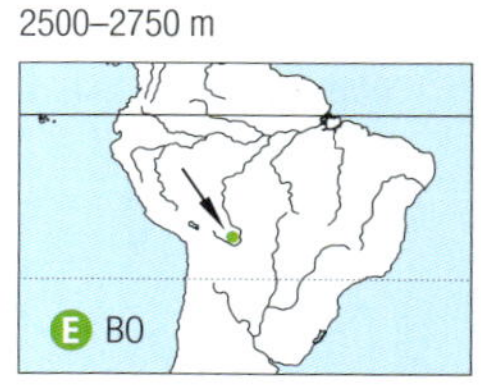

Lessa's Tuco-tuco
Ctenomys lessai
FR Tuco-tuco de Lessa
DE Lessa-Kammratte
ES Tuco tuco de Lessa

Known from Lluthu Pampa (Cochabamba)

c. 2000 m

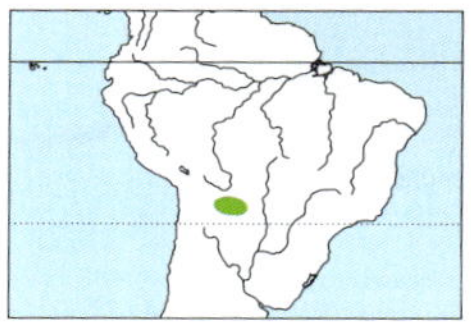

S BO (Santa Cruz, Chuquisaca, Tarija), NW PY (Boquerón)

Chacoan Tuco-tuco
Ctenomys conoveri

FR Tuco-tuco de Conover
DE Chaco-Kammratte
ES Tuco tuco del Chaco

LC
Monotypic
HB ≤ 56 cm
T ≤ 12.2 cm
W ≤ 1.2 kg (M)

LC
Monotypic
HB 20–22 cm
T 6.8–7.1 cm

3450–4000 m

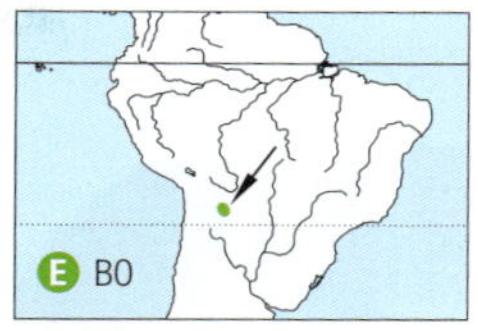

NW Tarija

Lewis's Tuco-tuco
Ctenomys lewisi

FR Tuco-tuco de Lewis
DE Lewis-Kammratte
ES Tuco tuco de Lewis

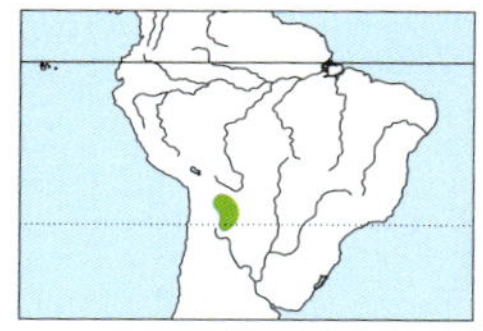

S Bolivia (E Potosí, S Tarija), NW AR (Jujuy, Salta)

Forest Tuco-tuco
Ctenomys frater

FR Tuco-tuco des forêts
DE Waldkammratte
ES Tuco tuco de bosque

LC
5 sspp
HB 17.3–19.6 cm
T 5–8.5 cm

EN
Monotypic
HB 13.8–15.1 cm

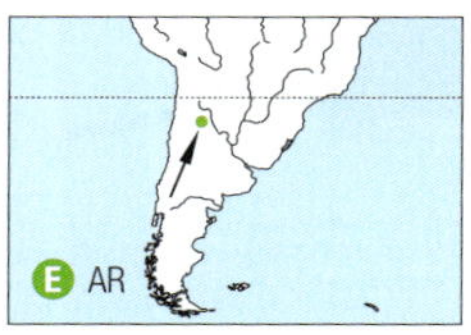

Tucumán

Furtive Tuco-tuco
Ctenomys occultus

FR Tuco-tuco furtif
DE Heimliche Kammratte
ES Tuco tuco furtivo

NT
Monotypic
HB 17.6–17.9 cm
T 7.8–8.4 cm

Formosa, Chaco, Santiago del Estero, Santa Fe

Argentine Tuco-tuco
Ctenomys argentinus

FR Tuco-tuco d'Argentine
DE Argentinien-Kammratte
ES Tuco tuco de Argentina

EN
Monotypic
HB 16.1–17.2 cm

Tucumán

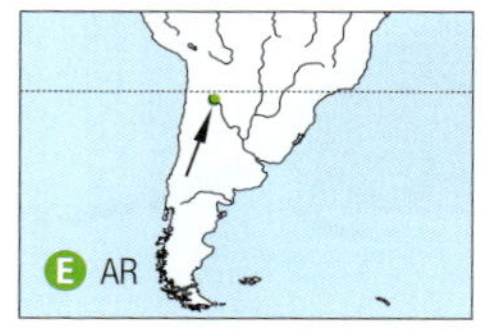

Mottled Tuco-tuco
Ctenomys latro

FR Tuco-tuco marbré
DE Gelbbraune Kammratte
ES Tuco tuco moteado

LC
Monotypic
HB 11.5–17.2 cm
T 7.9–8.5 cm

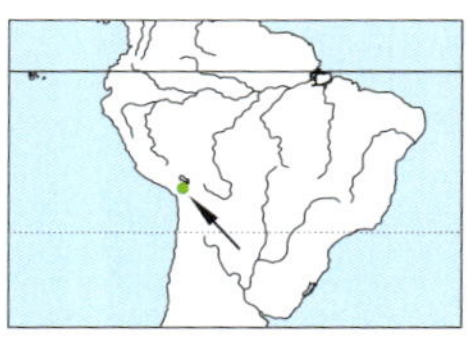

Andean altiplano in S PE (Puno) & W BO (La Paz)

White-toothed Tuco-tuco
Ctenomys leucodon

FR Tuco-tuco à dents blanches
DE Weißzahnkammratte
ES Tuco tuco de dientes blancos

DD
Monotypic
HB c. 17.2 cm
T c. 7.1 cm

Tucumán

Tucuman Tuco-tuco
Ctenomys tucumanus

FR Tuco-tuco de Tucuman
DE Tucumán-Kammratte
ES Tuco tuco de Tucumán

EN
Monotypic
HB 16.9–21 cm

SW Ñeembucú

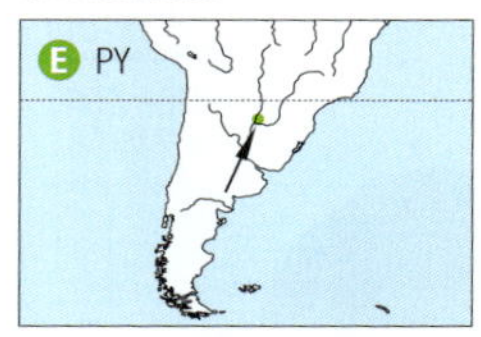

Pilar Tuco-tuco
Ctenomys pilarensis

FR Tuco-tuco de Pilar
DE Pilar-Kammratte
ES Tuco tuco de Pilar

NE
Monotypic
HB 15.2–17.9 cm
T 5.3–7.3 cm
W 105–165 g

Near coastal areas of Valdés Pen (Chubut)

Bidau's Tuco-tuco
Ctenomys bidaui

FR Tuco-tuco de Bidau
DE Bidau-Kammratte
ES Tuco tuco de Bidau

LC
Monotypic

C & S AR, S CL (with Riesco I), Isla Grande de Tierra del Fuego

Magellanic Tuco-tuco
Ctenomys magellanicus

FR Tuco-tuco de Magellan
DE Magellan-Kammratte
ES Tuco tuco de Magallanes

LC
Monotypic
HB 17.2–19.4 cm T 8.1–8.8 cm
W 152–229 g

Neuquén, Río Negro, W Chubut

Patagonian Tuco-tuco
Ctenomys haigi

FR Tuco-tuco de Haig
DE Patagonien-Kammratte
ES Tuco tuco de Patagonia

c. 575 m

SW Chubut

Lago Blanco Tuco-tuco
Ctenomys fodax

FR Tuco-tuco du Lago Blanco
DE Lago-Blanco-Kammratte
ES Tuco tuco del Lago Blanco

DD
Monotypic
HB c. 26 cm
T c. 9.8 cm

DD
Monotypic
T 5.1–8.1 cm

AR (from SW Chubut S to C Santa Cruz), adjacent CL (Aysén)

Silky Tuco-tuco
Ctenomys sericeus

FR Tuco-tuco soyeux
DE Seidenkammratte
ES Tuco tuco sedoso

Thales's Tuco-tuco
Ctenomys thalesi

FR Tuco-tuco de Thales
DE Thales-Kammratte
ES Tuco tuco de Thales

NE Chubut (known from 2 localities close to Atlantic coast, S of R Chubut)

NE
Monotypic
HB 12.6–15.5 cm
T 5.7–6.6 cm
W 43–96 g

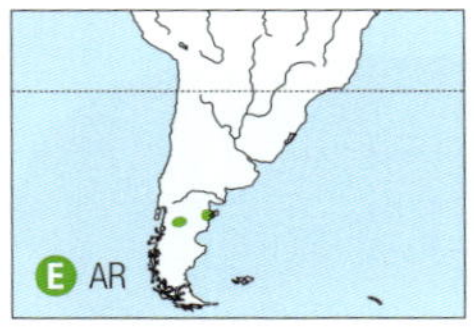

Contreras's Tuco-tuco
Ctenomys contrerasi

FR Tuco-tuco de Contreras
DE Contreras-Kammratte
ES Tuco tuco de Contreras

Disjunct in WC Chubut (S of R Chubut) and close to Atlantic coast (from R Chubut N to Ameghino Isthmus)

NE
2 sspp
HB 11.6–16.4 cm
T 4.9–6.8 cm
W 43–146 g

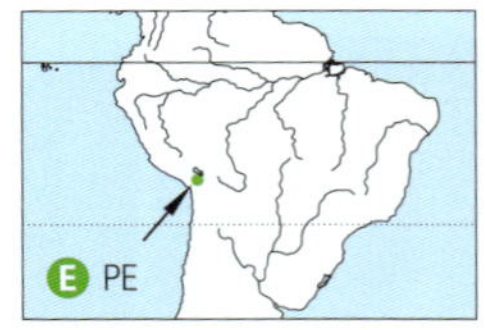

Peruvian Tuco-tuco
Ctenomys peruanus

FR Tuco-tuco du Pérou
DE Peru-Kammratte
ES Tuco tuco de Perú

Puno

LC
Monotypic
HB c. 22 cm
T c. 8.8 cm

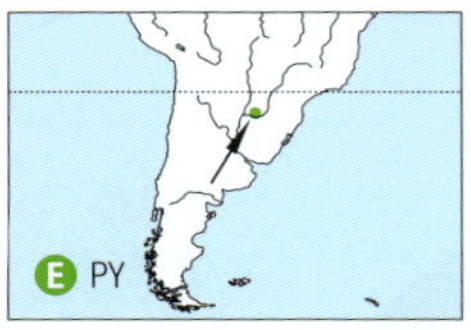

Paraguayan Tuco-tuco
Ctenomys paraguayensis

FR Tuco-tuco du Paraguay
DE Paraguay-Kammratte
ES Tuco tuco de Paraguay

Known from W Ayolas (Misiones)

NE
Monotypic
HB 16.6–17.4 cm
T 7.3–8 cm
W 146–187 g

c. 1000 m

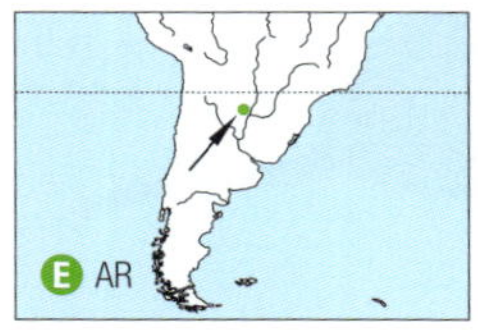

Bonetto's Tuco-tuco
Ctenomys bonettoi

FR Tuco-tuco de Bonetto
DE Bonetto-Kammratte
ES Tuco tuco de Bonetto

Few sites around Capitán Solari (SE Chaco)

EN
Monotypic
HB 17.1–18.3 cm
T 6.1–7.7 cm
W 185–220 g

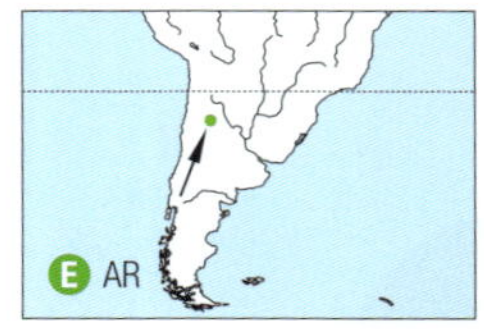

Foch's Tuco-tuco
Ctenomys fochi

FR Tuco-tuco de Foch
DE Foch-Kammratte
ES Tuco tuco de Foch

Vicinity of Chumbicha (Catamarca)

DD
Monotypic
HB c. 16.2 cm
T c. 7.6 cm

Jujuy Tuco-tuco
Ctenomys juris

FR Tuco-tuco du Jujuy
DE Jujuy-Kammratte
ES Tuco tuco de Jujuy

Known from El Chaguaral (Jujuy)

DD
Monotypic
HB c. 17.7 cm
T c. 7.2 cm

770–2025 m

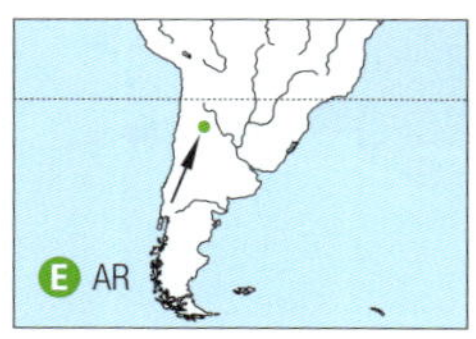

Catamarca Tuco-tuco
Ctenomys knighti

FR Tuco-tuco de Knight
DE Catamarca-Kammratte
ES Tuco tuco de Catamarca

Catamarca, La Rioja

DD
Monotypic
HB c. 20 cm
T c. 8.2 cm

DD
Monotypic
HB c. 21 cm
T c. 7.6 cm

320–2500 m

Monte Tuco-tuco
Ctenomys viperinus

FR Tuco-tuco de Vipos
DE Trancas-Kammratte
ES Tuco tuco de Vipos

In Yungas around Vipos (N Tucumán)

DD
Monotypic

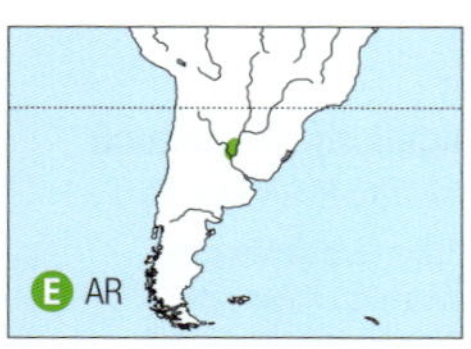

E Santa Fe (along R Paraná & R San Javier)

Santa Fe Tuco-tuco
Ctenomys yolandae

FR Tuco-tuco de Yolanda
DE Santa-Fe-Kammratte
ES Tuco tuco de Yolanda

c. 1000 m

NW Córdoba (small area along R Cruz del Eje)

EN
Monotypic
HB 13.2–16.3 cm
T c. 7.4 cm

Cordoba Tuco-tuco
Ctenomys bergi

FR Tuco-tuco du Cordoba DE Córdoba-Kammratte
ES Tuco tuco de Córdoba

CR
Monotypic
HB 23–25 cm (F)
W 176–279 g

Sierras Grandes de Los Gigantes (Córdoba)

Osvaldo Reig's Tuco-tuco
Ctenomys osvaldoreigi

FR Tuco-tuco d'Osvaldo Reig
DE Osvaldo-Reig-Kammratte
ES Tuco tuco de Osvaldo Reig

DD
Monotypic
HB c. 18.3 cm
T c. 7.7 cm

Brown Tuco-tuco
Ctenomys pontifex

FR Tuco-tuco marron
DE Mittlere Kammratte
ES Tuco tuco marrón

E of Andes in Mendoza

Family OCTODONTIDAE – Viscacha Rats, Degus, Rock Rats and Coruro

7 genera, 15 species. Coastal areas in central Chile, pre-Andean and Andean regions, desert and semi-desert scrublands, rainforests, and Mediterranean habitats. Neotropical Region.

Genus *TYMPANOCTOMYS*

(3 species)

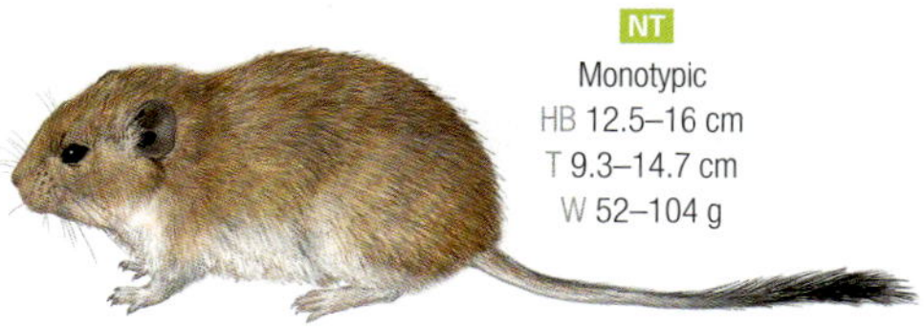

NT
Monotypic
HB 12.5–16 cm
T 9.3–14.7 cm
W 52–104 g

300–1400 m

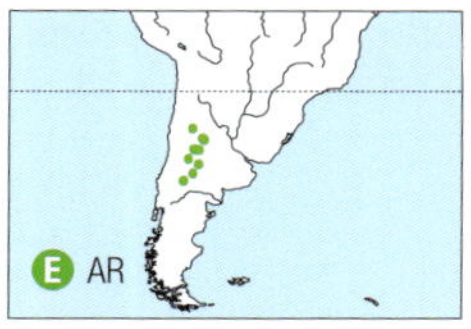

Red Viscacha Rat
Tympanoctomys barrerae
FR Octodon de Barrera
DE Rote Viscacharatte
ES Rata vizcacha colorada

Patchily in arid sites in San Juan, Mendoza, La Pampa & Neuquén

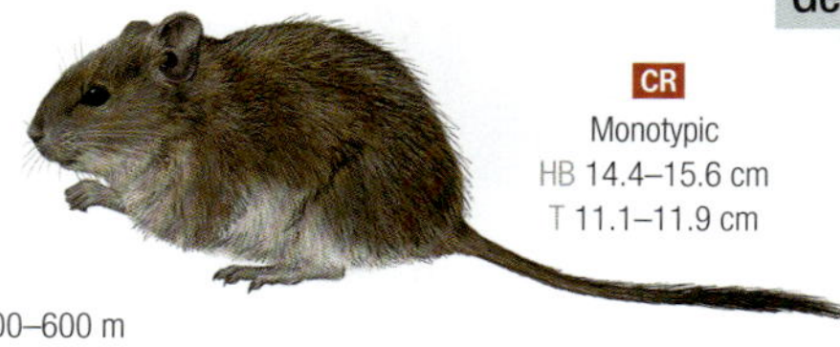

CR
Monotypic
HB 14.4–15.6 cm
T 11.1–11.9 cm

500–600 m

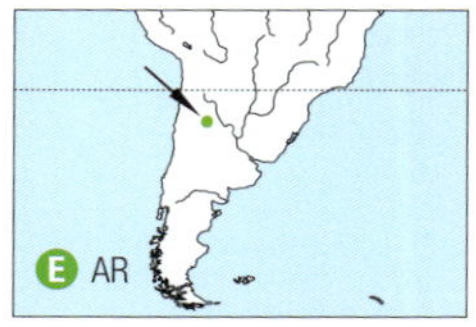

Chalchalero Viscacha Rat
Tympanoctomys loschalchalerosorum
FR Octodon des Chalchaleros
DE Chalchaleros-Viscacharatte
ES Rata vizcacha de Los Chalchaleros

Known from SW Quimilo (La Rioja)

DD
Monotypic
HB 11.1–13.6 cm
T 11.1–12 cm
W 50–80 g

c. 500 m

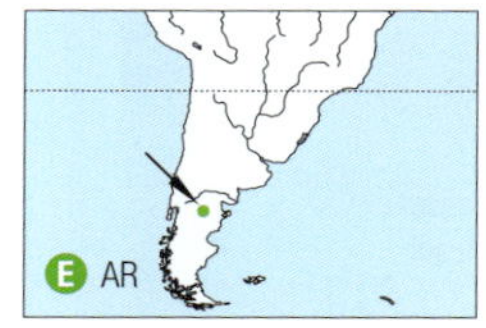

Kirchners' Viscacha Rat
Tympanoctomys kirchnerorum
FR Octodon des Kirchner
DE Kirchner-Viscacharatte
ES Rata vizcacha de Patagonia

Known from near Los Adobes (N Chubut)

Genus *PIPANACOCTOMYS*

(1 species)

CR
Monotypic
HB 16.6–17.8 cm
T 12.7–14.5 cm

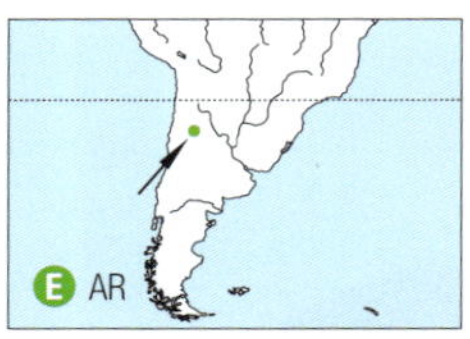

Golden Viscacha Rat
Pipanacoctomys aureus
FR Octodon doré
DE Goldene Viscacharatte
ES Rata vizcacha dorada

Salar de Pipanaco (Catamarca)

Genus *OCTOMYS*

(1 species)

up to 800 m

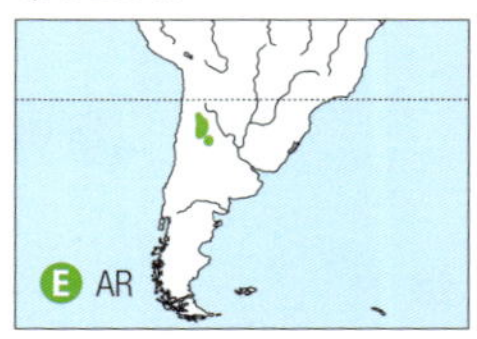

LC
Monotypic
HB 14–33 cm
T 10.5–17.9 cm
W c. 96 g

Catamarca, La Rioja, San Juan, San Luis, Mendoza

Common Viscacha Rat
Octomys mimax
FR Octodon vizcacha DE Viscacharatte ES Rata vizcacha del Monte

Genus *OCTODONTOMYS*

(1 species)

LC
Monotypic
HB 16.1–18.2 cm
T 10–19 cm
W 100–200 g

2000–5000 m

Mountain Degu
Octodontomys gliroides
FR Octodon à longue queue
DE Pinselschwanzratte
ES Degú del Monte

Andean & sub-Andean zones of SW BO, N CL & NW AR

Genus *OCTODON*

(5 species)

LC
Monotypic
HB 16.9–21 cm
T 8.1–13.8 cm
W 200–300 g

0–2000 m

Common Degu
Octodon degus
FR Octodon du Chili
DE Gewöhnlicher Degu
ES Degú común

Lower slopes of Andes between Atacama and Valparaíso

Mocha I (Biobío)

c. 390 m

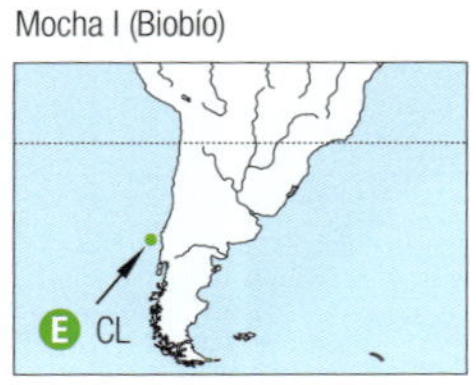

Pacific Degu
Octodon pacificus
FR Octodon de Mocha
DE Pazifikdegu
ES Degú de Isla Mocha

CR
Monotypic
HB 12.5–19.5 cm
T 16.5–17 cm
W c. 290 g

NE
Monotypic
HB c. 19.2 cm
T c. 13.4 cm
W c. 195 g

500–1100 m

Ricardo Ojeda's Degu
Octodon ricardojeda
FR Octodon d'Ojeda
DE Ricardo-Ojeda-Degu
ES Degú de Ojeda

AR (W Neuquén), CL (Malalcauello R, NE La Araucanía)

VU
Monotypic
HB 15–20 cm
T 10.2–16.7 cm
W c. 92 g

0–1200 m

Bridges's Degu
Octodon bridgesii
FR Octodon de Bridges
DE Walddegu
ES Degú de Bridges

CL (Coast Range & Andes from O'Higgins to La Araucanía), AR (Neuquén)

NT
Monotypic
HB 16.7–22 cm
T 15.2–16.1 cm
W c. 233 g

Coastal Degu
Octodon lunatus
FR Octodon à dents de lune
DE Küstendegu ES Degú de costa

0–1200 m

Coast Range from Bosque Fray Jorge NP (Coquimbo) to Quilpué (Valparaíso)

DD
Monotypic
HB 14.8–19.2 cm
T 6.8–8.5 cm
W 105–160 g

900–2000 m

Porter's Rock Rat
Aconaemys porteri
FR Tunduco de Porter
DE Porter-Felsenratte
ES Tunduco de Porter

CL (from Villarica to Puyehue volcanoes, Los Lagos), AR (Lanín to Nahuel Huapi NPs, Neuquén)

Genus *ACONAEMYS*
(3 species)

DD
Monotypic
HB 14–15.9 cm
T 5.8–6.8 cm
W 83–110 g

Sage's Rock Rat
Aconaemys sagei
FR Tunduco de Sage
DE Sage-Felsenratte
ES Tunduco de Sage

C CL (La Araucanía), WC AR (L Quillén & L Hui Hui, Neuquén)

LC
Monotypic
HB 15–17 cm
T 5.5–7.8 cm
W 80–230 g

1000–4000 m

Chilean Rock Rat
Aconaemys fuscus
FR Tunduco du Chili
DE Südamerikanische Felsenratte
ES Tunduco de Chile

Andes of C CL (from Curicó, Maule to Temuco, La Araucanía), CW AR (Mendoza)

Genus *SPALACOPUS*
(1 species)

0–3400 m

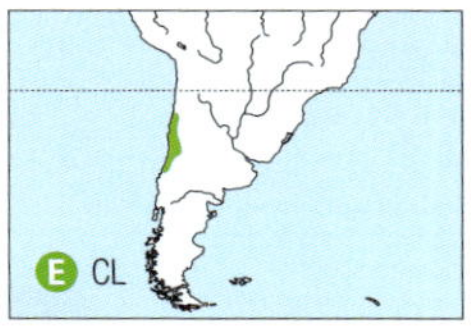

Coruro
Spalacopus cyanus
FR Cururo
DE Coruro
ES Cururo

Coast & Andes from Caldera (Atacama) S to Quirihue (Biobío)

LC
3 sspp
HB 11.5–16.5 cm
T 4–5.7 cm
W 80–120 g

Family ECHIMYIDAE – Hutias, Coypu and South American Spiny-rats

28 genera, 102 species. Lowland and highland forested habitats, lowland open dry habitats, and wetlands. Neotropical Region.

Genus *PLAGIODONTIA*
(1 species)

EN
3 sspp
HB 31–40 cm
T 12.5–15.3 cm
W ≤ 1.8 kg

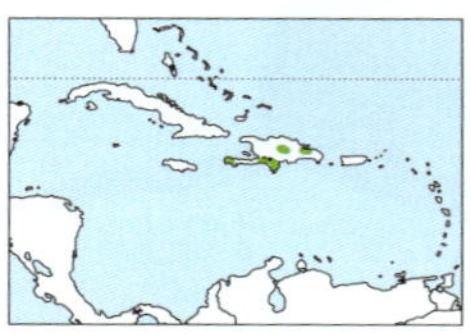

Hispaniolan Hutia
Plagiodontia aedium
FR Hutia de Saint-Domingue
DE Zaguti
ES Jutía de la Española

SW & SE HT (La Hotte & La Selle massifs), N & SW DO

Genus *GEOCAPROMYS*
(2 species)

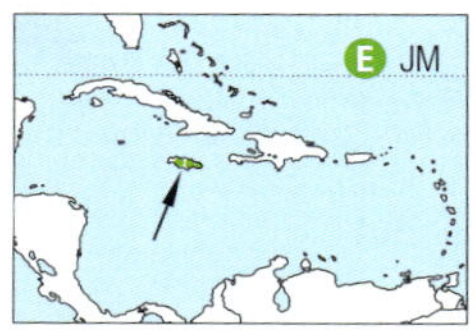

Severely fragmented range

Jamaican Hutia
Geocapromys brownii
FR Hutia de Jamaïque
DE Jamaika-Ferkelratte
ES Jutía de Jamaica

EN
Monotypic
HB 37–45 cm
T 4–6.4 cm
W ≤ 2 kg

VU
Monotypic
HB 28–32 cm
T c. 5.5 cm

Ingraham's Hutia
Geocapromys ingrahami
FR Hutia des Bahamas
DE Bahama-Ferkelratte
ES Jutía de Bahamas

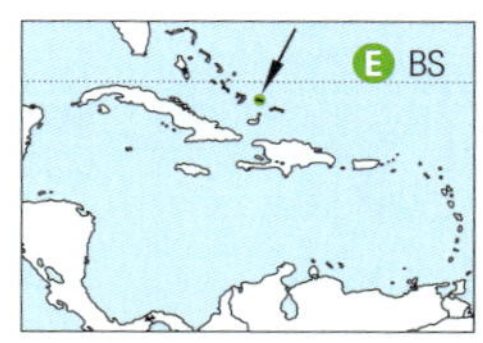

Known only from East Plana Cay; introduced to Little Wax & Warderick Wells cays

Genus *CAPROMYS*
(2 species)

LC
5 extant sspp
HB 30–62 cm
T 13.1–31 cm
W ≤ 7 kg

Mainland, Isla de la Juventud, Jardines de la Reina Arch, Sabana Arch

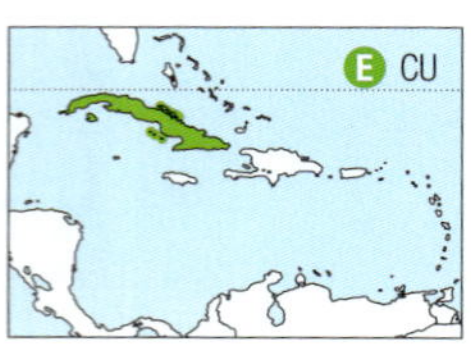

Desmarest's Hutia
Capromys pilorides
FR Hutia de Desmarest
DE Hutiaconga
ES Jutía de Desmarest

NE
Monotypic

Known from 1 specimen from an islet near Cayo Largo (Los Canarreos Arch)

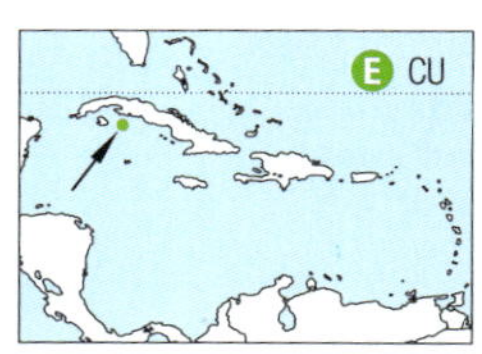

Garrido's Hutia
Capromys garridoi
FR Hutia de Garrido
DE Garrido-Hutia
ES Jutía de Garrido

Genus *MESOCAPROMYS*

(5 species)

CR
Monotypic
HB c. 25 cm
T c. 17.6 cm
W ≤ 483 g

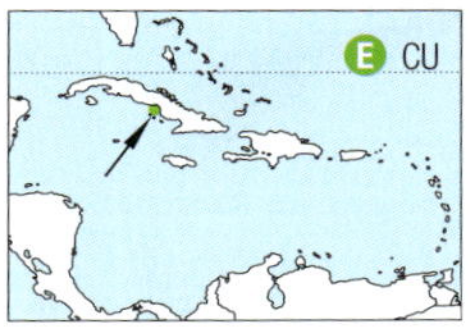

Cayos Salinas in Cayos Ana María; introduced to Cayo La Loma

Cabrera's Hutia
Mesocapromys angelcabrerai

FR Hutia de Cabrera
DE Cabrera-Zwerghutia
ES Jutía de Cabrera

EN
Monotypic
HB c. 29 cm
T c. 19.8 cm
W ≤ 1 kg

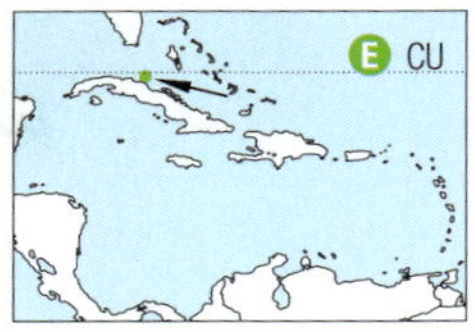

Cayo Fragoso (Sabana Arch); introduced to Cayos Pasaje & Sagra

Eared Hutia
Mesocapromys auritus

FR Hutia oreillard
DE Langohr-Zwerghutia
ES Jutía de orejas grandes

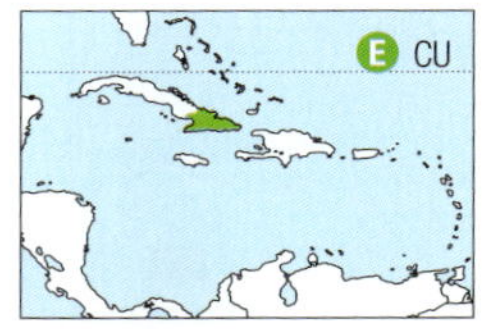

W to Malagueta Bay & Cauto Estuary; limits poorly known

Black-tailed Hutia
Mesocapromys melanurus

FR Hutia à queue noire
DE Schwarzschwanz-Zwerghutia
ES Jutía de cola negra

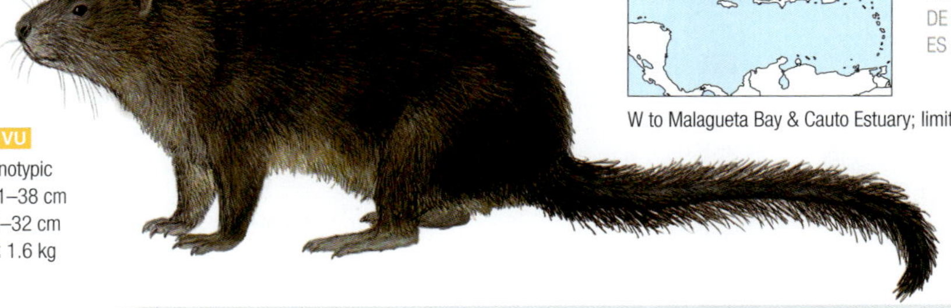

VU
Monotypic
HB 31–38 cm
T 21–32 cm
W ≤ 1.6 kg

CR(PE)
Monotypic

Cayo Juan García

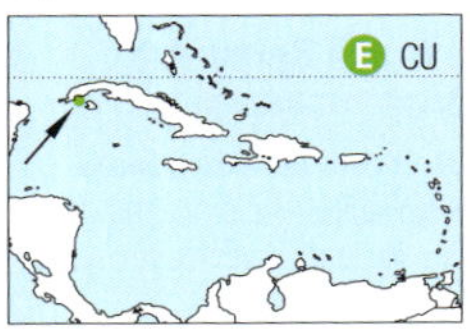

San Felipe Hutia
Mesocapromys sanfelipensis

FR Hutia de San Felipe
DE San-Felipe-Zwerghutia
ES Jutía de San Felipe

Ciénaga de Zapata (Matanzas)

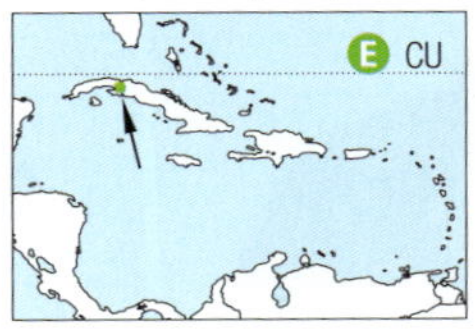

Dwarf Hutia
Mesocapromys nana

FR Hutia nain
DE Zwerghutia
ES Jutía enana

CR(PE)
Monotypic
HB c. 20 cm
T c. 16.4 cm
W 400–450 g

Genus *MYSATELES*

(1 species)

NT
3 sspp
HB 33–45 cm
T 25–35 cm
W ≤ 2.8 kg

Mainland & Isla de la Juventud

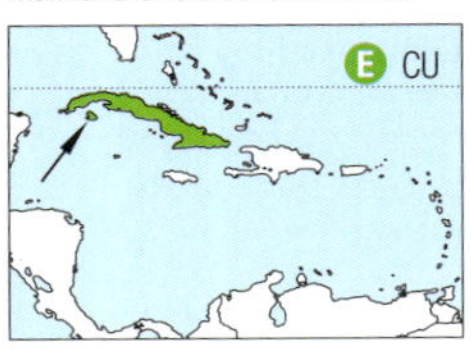

Prehensile-tailed Hutia
Mysateles prehensilis

FR Hutia à queue préhensile
DE Langschwanzhutia
ES Jutía de cola prensil

Genus *CARTERODON*

(1 species)

250–1100 m

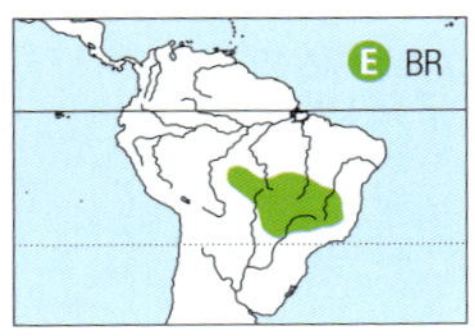

Minas Gerais, Goiás, Distrito Federal, Mato Grosso, Mato Grosso do Sul

DD
Monotypic
HB 13.5–25 cm
T 6.5–9.2 cm
W 92–195 g

Groove-toothed Spiny-rat
Carterodon sulcidens

FR Rat-épineux du Cerrado DE Furchenzahn-Stachelratte
ES Rata espinosa de dientes con surcos

Genus *CLYOMYS*

(1 species)

C BR (Bahia & Minas Gerais S to Mato Grosso do Sul & São Paulo), E PY

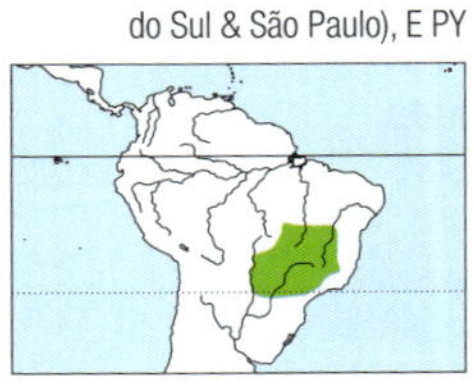

Broad-headed Spiny-rat
Clyomys laticeps

FR Rat-épineux à tête large
DE Breitkopf-Stachelratte
ES Rata espinosa de cabeza ancha

LC
Monotypic
HB 15–29 cm
T 4.8–8.9 cm
W 100–350 g

Genus *EURYZYGOMATOMYS*

(1 species)

0–2300 m

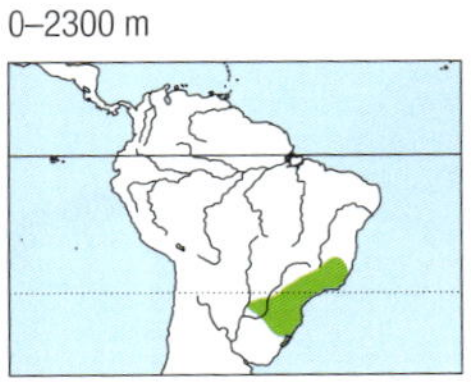

SE & S BR (Espírito Santo S to Rio Grande do Sul), NE AR (Misiones), S PY

Guiara
Euryzygomatomys spinosus

FR Rat-épineux guira
DE Guiara
ES Rata guira

LC
Monotypic
HB 16.3–20 cm
T 6–6.4 cm
W 180–210 g

Genus *TRINOMYS*

(10 species)

800–1100 m

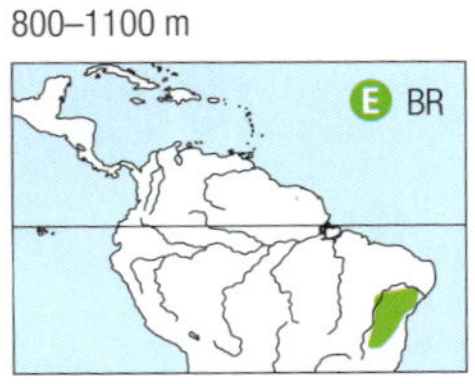

Bahia, Minas Gerais, Sergipe

LC
2 sspp
HB 15.3–19 cm
T 14.8–17.5 cm
W 120–230 g

White-spined Spiny-rat
Trinomys albispinus

FR Rat-épineux à poils blancs
DE Weißstachel-Atlantikstachelratte
ES Rata espinosa de espinas blancas

NT
Monotypic
HB 17–24 cm
T 15–22 cm
W 100–290 g

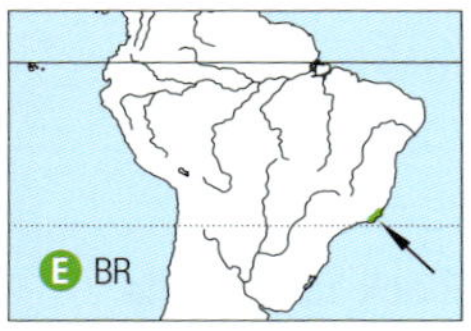

NE & E Rio de Janeiro State

Elias's Spiny-rat
Trinomys eliasi
FR Rat-épineux d'Elias
DE Elias-Atlantikstachelratte
ES Rata espinosa de Elias

DD
Monotypic
HB 18.7–27 cm
T 17.1–23 cm
W 195–350 g

below 400 m

Few sites in S Espírito Santo & EC Minas Gerais

Rigid-spined Atlantic Spiny-rat
Trinomys paratus
FR Rat-épineux hérissé
DE Lanzen-Atlantikstachelratte
ES Rata espinosa de espinas duras

LC
2 sspp
HB 18–23 cm
T 16.8–23 cm
W 113–235 g

From coastal Bahia & Espírito Santo S to E Minas Gerais & N Rio de Janeiro

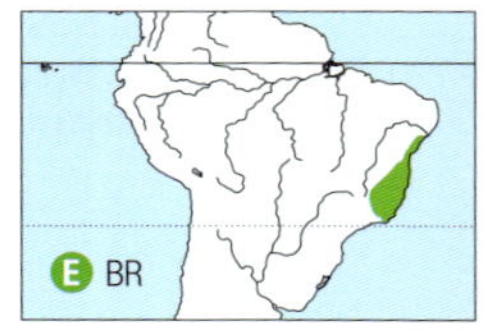

Elegant-spined Atlantic Spiny-rat
Trinomys setosus
FR Rat-épineux velu
DE Haarige Atlantikstachelratte
ES Rata espinosa peluda

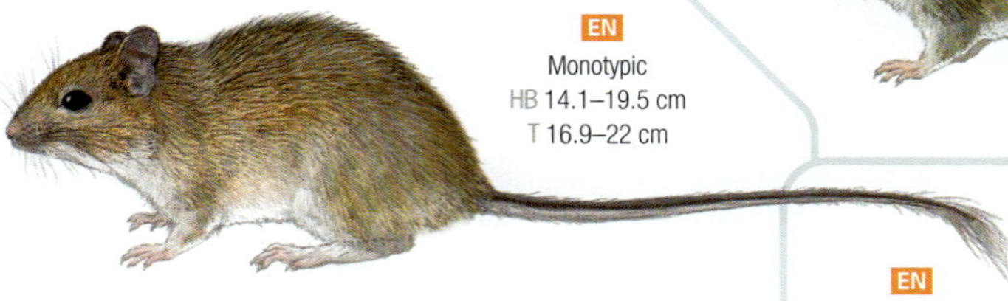

EN
Monotypic
HB 14.1–19.5 cm
T 16.9–22 cm

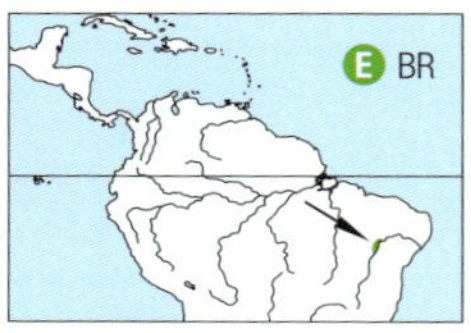

Left bank of middle R São Francisco (NW Bahia)

Yonenaga's Spiny-rat
Trinomys yonenagae
FR Rat-épineux d'Yonenaga
DE Yonenaga-Atlantikstachelratte
ES Rata espinosa de Yonenaga

EN
Monotypic
HB 15–18 cm
T 14.4–17.2 cm

Moojen's Spiny-rat
Trinomys moojeni
FR Rat-épineux de Moojen
DE Moojen-Atlantikstachelratte
ES Rata espinosa de Moojen

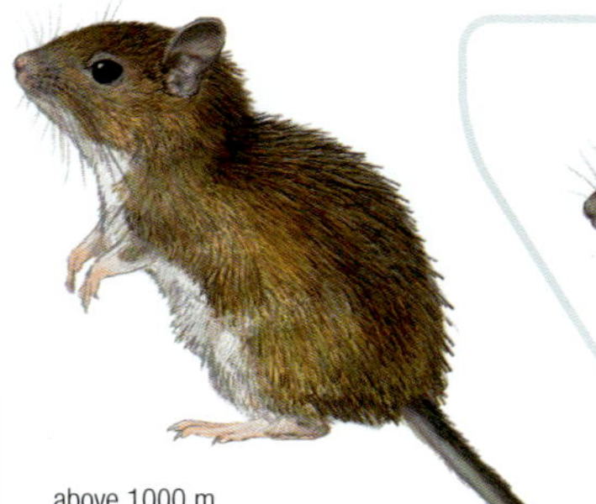

above 1000 m

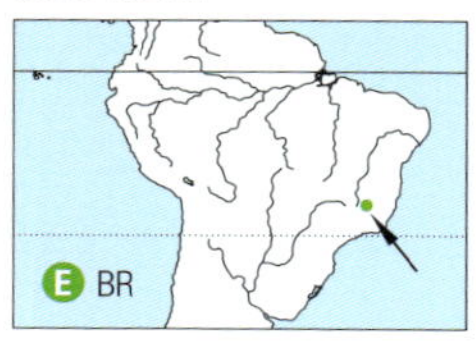

S Espinhaço Mts (EC Minas Gerais)

S coast of Bahia

20–40 m

DD
Monotypic
HB c. 22 cm
T 18.5–20 cm
W c. 259 g

Pau Brasil Spiny-rat
Trinomys mirapitanga
FR Rat-épineux des mirapitangas
DE Mirapitanga-Atlantikstachelratte
ES Rata espinosa de dorso oscuro

0–1000 m

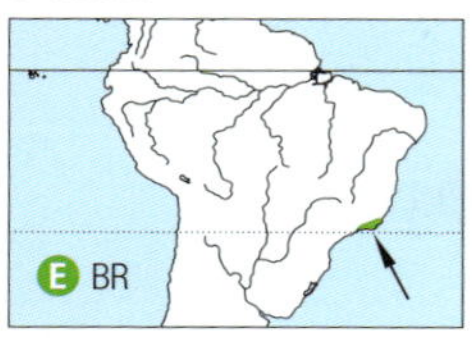

Rio de Janeiro and N coastal border of São Paulo

Rio de Janeiro Spiny-rat
Trinomys dimidiatus
FR Rat-épineux de Rio de Janeiro
DE Rio-de-Janeiro-Atlantikstachelratte
ES Rata espinosa atlántica

LC
Monotypic
HB 17–24 cm
T 15–22 cm
W 100–290 g

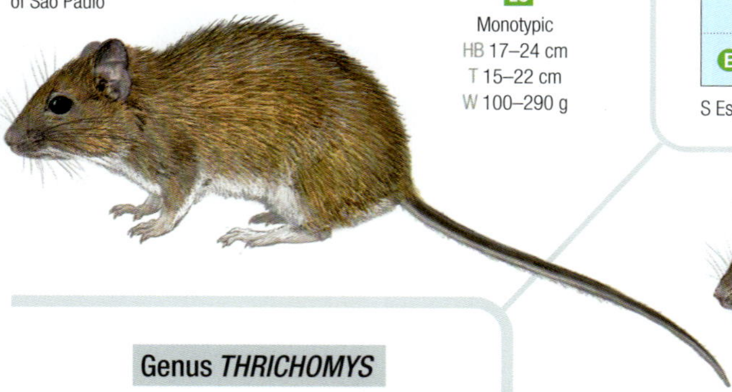

LC
2 sspp
HB 16.3–23 cm T 15–22 cm
W 162–270 g

Gracile Atlantic Spiny-rat
Trinomys gratiosus
FR Rat-épineux gracieux
DE Zierliche Atlantikstachelratte
ES Rata espinosa grácil

above 600 m

Serra do Mar & Mantiqueira Mts (Espírito Santo, Rio de Janeiro, Minas Gerais)

LC Monotypic
HB 18.5–22 cm
T 17–22 cm
W 139–233 g

above 600 m

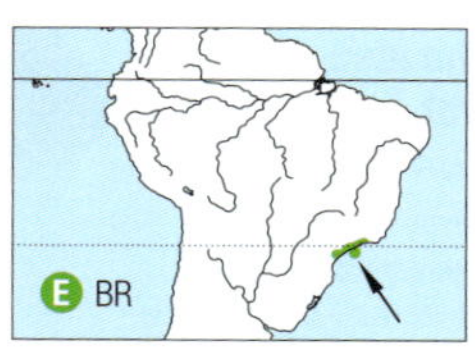

SW Rio de Janeiro (with Ilha Grande off S coast), São Paulo

Sao Paulo Spiny-rat
Trinomys iheringi
FR Rat-épineux d'Ihering
DE Ihering-Atlantikstachelratte
ES Rata espinosa de Ihering

Genus *THRICHOMYS*

(5 species)

DD
Monotypic
HB 12–24 cm T 13.5–21 cm
W 120–365 g

15–800 m

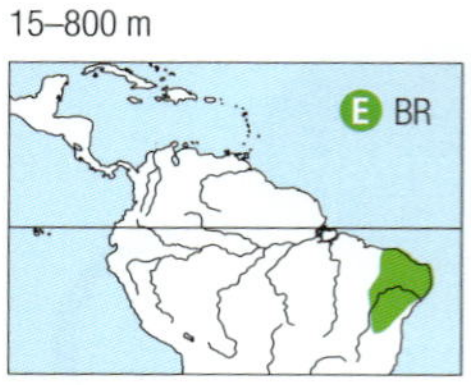

From Ceará & Rio Grande do Norte S to C Bahia

Sao Lourenço Punare
Thrichomys laurentius
FR Punaré de Sao Lourenço
DE Laurentius-Punaré
ES Punaré de San Lorenzo

LC Monotypic
HB 18–22 cm
T 11–20 cm
W c. 350 g

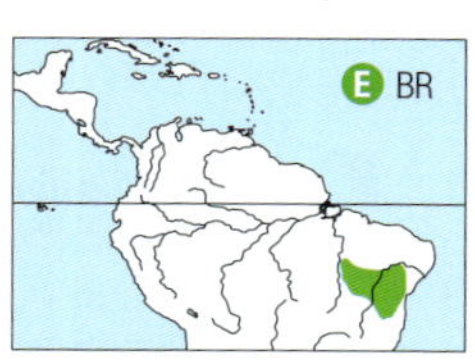

From C Bahia W to C Tocantins

260–1030 m

Jacobina Punare
Thrichomys inermis
FR Punaré des plateaux
DE Jacobina-Punaré ES Punaré de Jacobina

LC
Monotypic
HB 19.7–21 cm
T 16.4–17.9 cm
W 210–300 g

Lagoa Santa Punare
Thrichomys apereoides
FR Punaré commun
DE Lagoa-Santa-Punaré
ES Punaré de Lagoa Santa

890–1110 m

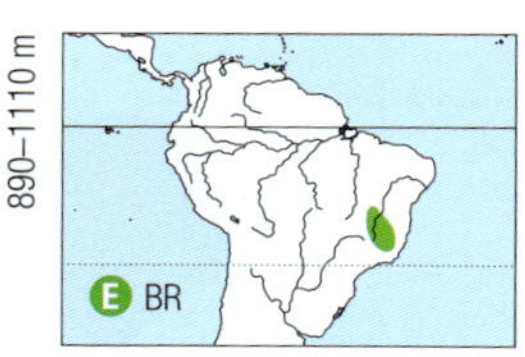

Bahia, Minas Gerais

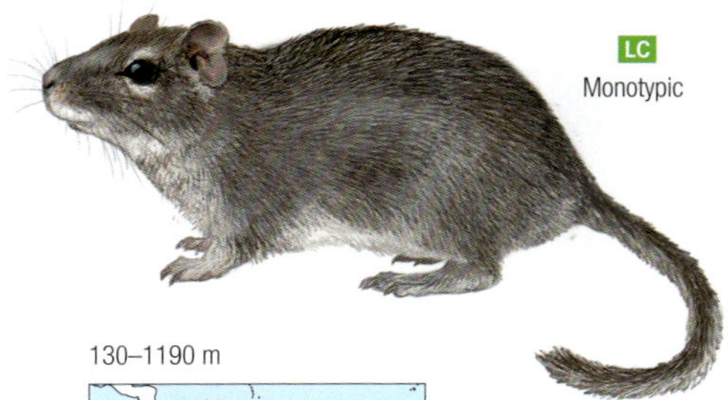

LC
Monotypic

130–1190 m

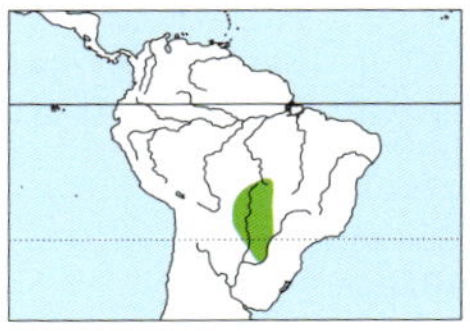

CW BR (Mato Grosso, Mato Grosso do Sul), extreme E BO (Santa Cruz), PY

Pantanal Punare
Thrichomys pachyurus
FR Punaré du Paraguay
DE Pantanal-Punaré
ES Punaré de Paraguay

NE
Monotypic
HB 11.9–22 cm
T 11.1–20 cm
W c. 350 g

90–370 m

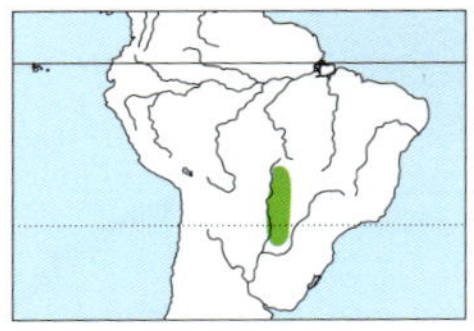

CW BR (Mato Grosso, Mato Grosso do Sul), PY

Foster's Punare
Thrichomys fosteri
FR Punaré de Foster
DE Foster-Punaré
ES Punaré de Foster

Genus *CALLISTOMYS*

(1 species)

EN
Monotypic
HB 25–29 cm
T 27–32 cm
W ≤ 480 g

Coastal Atlantic Forest near Ilhéus & Itabuna (Bahia)

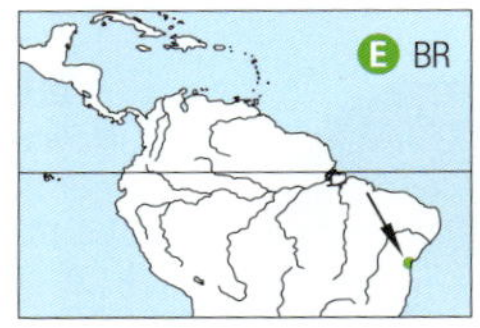

Painted Tree-rat
Callistomys pictus
FR Rat-épineux coloré
DE Schwarzweiße Baumratte
ES Rata arborícola pintada

Genus *MYOCASTOR*

(1 species)

up to 1200 m

BO, SE BR, PY, N AR, UY, CL (with Chiloé I); introduced widely

Coypu
Myocastor coypus
FR Ragondin
DE Nutria
ES Coipú

LC
4 sspp
HB 47–57 cm
T 34–40 cm
W ≤ 6.7 kg

Genus *HOPLOMYS*

(1 species)

0–1250 m

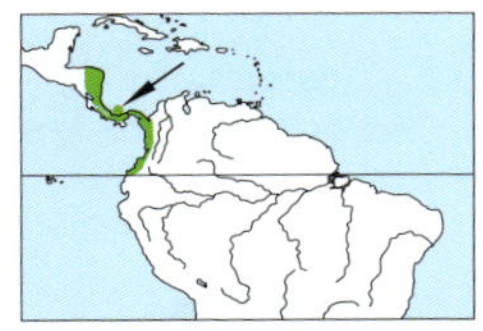

E HN, E NI, CR, PA (with Escudo de Veraguas I), W CO, NW EC

Armored Rat
Hoplomys gymnurus
FR Rat-épineux cuirassé
DE Lanzenratte
ES Rata acorazada

LC
4 sspp
HB 21–32 cm
T 11.4–24 cm
W 220–820 g

0–1700 m

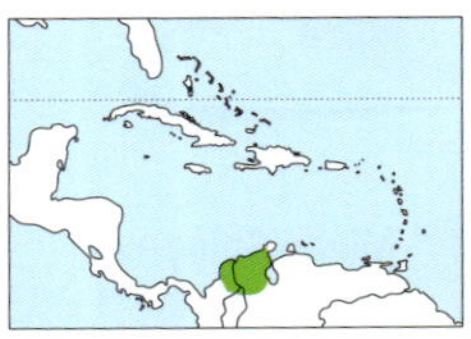

From lower R Magdalena Valley (NE CO) E to L Maracaibo (NW VE)

LC
Monotypic
HB c. 22 cm T c. 16.8 cm

Colombian Spiny-rat
Proechimys canicollis
FR Rat-épineux de Colombie
DE Kolumbien-Kurzstachelratte
ES Rata espinosa de Colombia

Genus *PROECHIMYS*

(22 species)

NT
Monotypic
HB 26–30 cm
T c. 19.6 cm

0–800 m

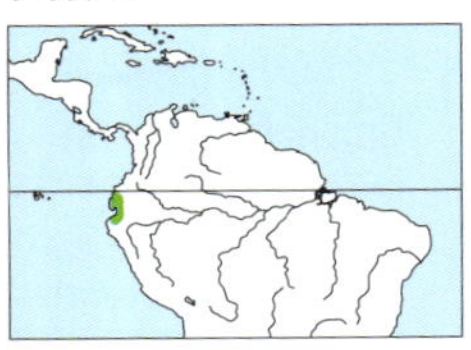

Pacific coastal lowlands of EC & NW PE

Pacific Spiny-rat
Proechimys decumanus
FR Rat-épineux côtier
DE Pazifik-Kurzstachelratte
ES Rata espinosa del Pacífico

LC
Monotypic
HB 16.5–27 cm
T 11.8–23 cm
W 160–400 g

50–2000 m

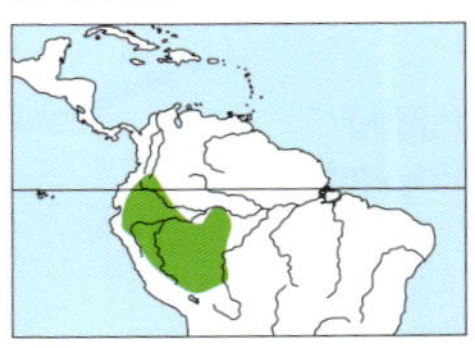

E Andean slopes & Amazon Basin of S CO, E EC, E PE, N BO & W BR

Simons's Spiny-rat
Proechimys simonsi
FR Rat-épineux de Simons
DE Simons-Kurzstachelratte
ES Rata espinosa de Simons

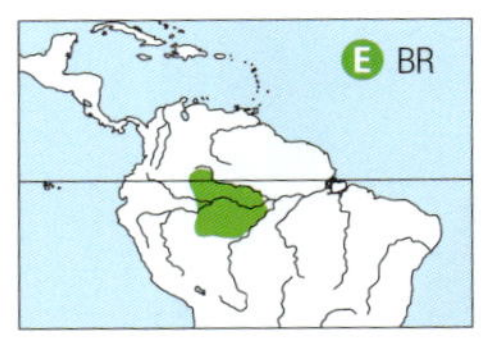

Both sides of R Amazon, W of R Negro & R Madeira; possibly SE CO

Stiff-spined Spiny-rat
Proechimys echinothrix
FR Rat-épineux du Jurua
DE Igel-Kurzstachelratte
ES Rata espinosa de espinas rígidas

LC
Monotypic
HB 14.1–24 cm
T 10.6–21 cm
W 240–280 g

LC
10 sspp
HB c. 29 cm
T c. 24 cm
W 0.26–1 kg

0–500 m

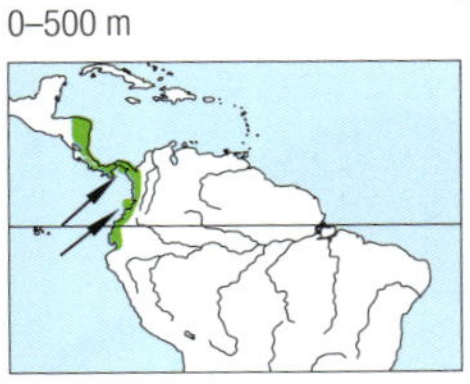

E HN, E & C NI, CR, PA (with Pearl Is), W CO (with Gorgona I), W EC

Tomes's Spiny-rat
Proechimys semispinosus
FR Rat-épineux de Tomes
DE Tomes-Kurzstachelratte
ES Rata espinosa de Tomes

DD
Monotypic
HB c. 25 cm
T c. 17.5 cm

170–700 m

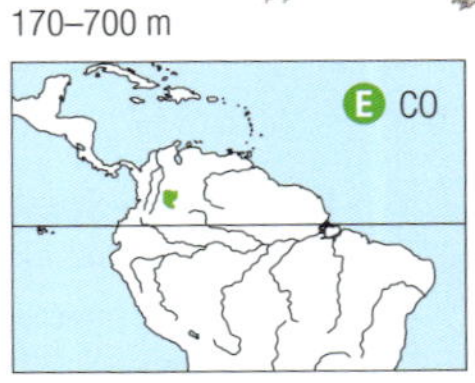

E of E Andes in headwaters of R Meta & R Guaviare

O'Connell's Spiny-rat
Proechimys oconnelli
FR Rat-épineux d'O'Connell
DE O'Connell-Kurzstachelratte
ES Rata espinosa de O'Connell

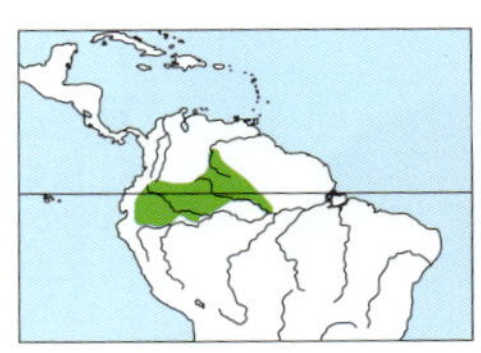

S & E CO, S VE, NW BR, E EC & N PE, largely N of R Amazon, E to R Negro

Napo Spiny-rat
Proechimys quadruplicatus
FR Rat-épineux du Napo
DE Napo-Kurzstachelratte
ES Rata espinosa de Napo

LC
Monotypic
HB 21–29 cm
T 12.5–20 cm
W 300–700 g

50–400 m

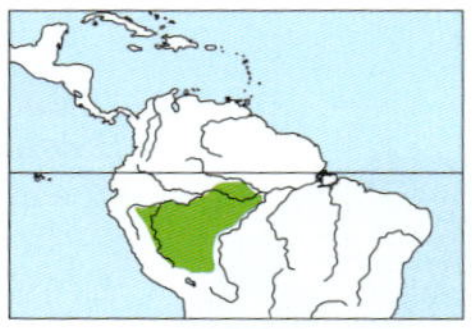

Steere's Spiny-rat
Proechimys steerei

FR Rat-épineux de Steere
DE Steere-Kurzstachelratte
ES Rata espinosa de Steere

E PE, NW BO & W BR, largely S of R Amazon, N of R Amazon just W of lower R Negro

LC
Monotypic
HB 21–49 cm
T 12–21 cm
W 300–800 g

0–300 m

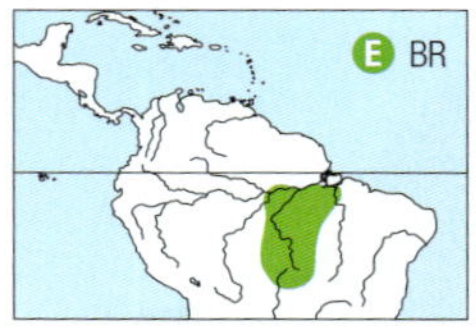

E BR

Goeldi's Spiny-rat
Proechimys goeldii

FR Rat-épineux de Goeldi
DE Goeldi-Kurzstachelratte
ES Rata espinosa de Goeldi

E Amazon Basin mostly S of R Amazon & E of R Madeira, S to Mato Grosso

LC
Monotypic
HB 18.8–27 cm
T 10–17.1 cm

LC
Monotypic
HB 18–23 cm
T 11–18.6 cm

0–500 m

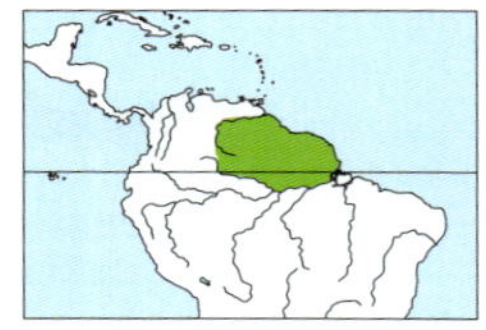

Cayenne Spiny-rat
Proechimys guyannensis

FR Rat-épineux de Cayenne
DE Cayenne-Kurzstachelratte
ES Rata espinosa de Cayena

Amazon Basin in S & E VE, Guianas & NE BR (E of R Negro, N of R Amazon)

LC
Monotypic
HB 20–23 cm
T c. 16 cm
W 170–240 g

0–1000 m

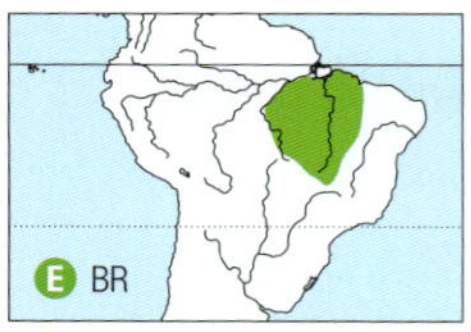

E BR

Robert's Spiny-rat
Proechimys roberti

FR Rat-épineux de Robert
DE Robert-Kurzstachelratte
ES Rata espinosa de Robert

Amazonian lowlands, S of R Amazon, S to cerrado (Goiás, Mato Grosso)

LC
Monotypic
HB 14.6–19.3 cm
T 10.6–14.1 cm
W 100–120 g

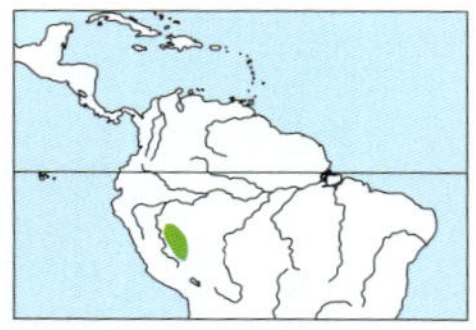

Patton's Spiny-rat
Proechimys pattoni

FR Rat-épineux de Patton
DE Patton-Kurzstachelratte
ES Rata espinosa de Patton

Headwaters of R Juruá (W BR, E & SE PE)

DD
Monotypic
HB 14.4–19 cm
T 9.5–14 cm
W 130–170 g

120–180 m

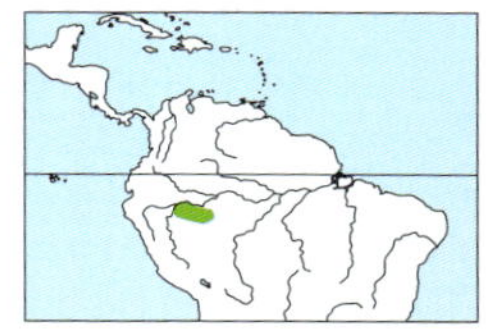

Kulina Spiny-rat
Proechimys kulinae

FR Rat-épineux des Kulinas
DE Kulina-Kurzstachelratte
ES Rata espinosa de Kulina

NE Peru (S of R Amazon), W Brazil (middle R Juruá)

DD Monotypic
HB 15.4–21 cm
T 8.8–15.2 cm
W 110–180 g

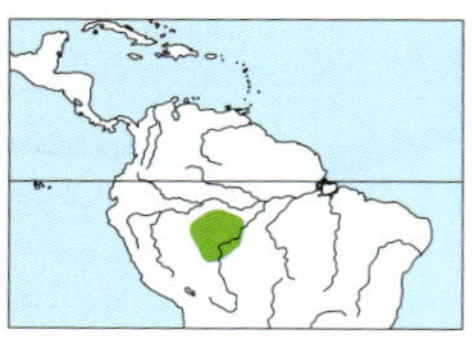

Gardner's Spiny-rat
Proechimys gardneri

FR Rat-épineux de Gardner
DE Gardner-Kurzstachelratte
ES Rata espinosa de Gardner

W BR between R Madeira and R Juruá, extreme N BO

DD Monotypic
HB 21–22 cm
T 14.5–16 cm
W c. 370 g

50–900 m

E CO

Boyaca Spiny-rat
Proechimys chrysaeolus

FR Rat-épineux du Boyaca
DE Boyacá-Kurzstachelratte
ES Rata espinosa de Boyacá

Lower Cauca & Magdalena valleys W of E Andes

DD
Monotypic
HB 22–23 cm
T 20–21 cm

0–500 m

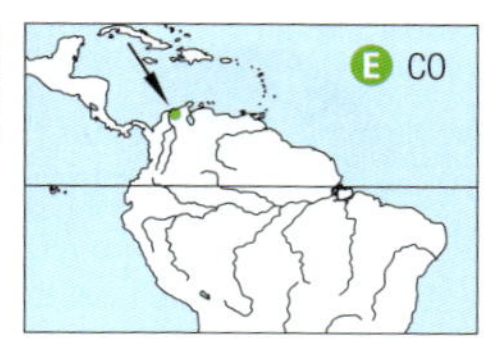

E CO

Minca Spiny-rat
Proechimys mincae

FR Rat-épineux de Minca
DE Minca-Kurzstachelratte
ES Rata espinosa de Minca

Lower Magdalena Valley in Sierra Nevada de Santa Marta

LC 3 sspp
HB 21–24 cm
T 17–20 cm
W 340–400 g

0–800 m

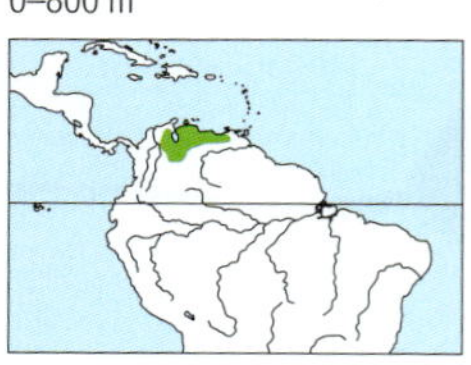

Guaira Spiny-rat
Proechimys guairae

FR Rat-épineux de La Guaira
DE Guaira-Kurzstachelratte
ES Rata espinosa de La Guaira

NE CO, W & N VE (E to Cordillera Central)

DD
Monotypic
HB 26–27 cm
T 20–21 cm
W c. 460 g

0–1300 m

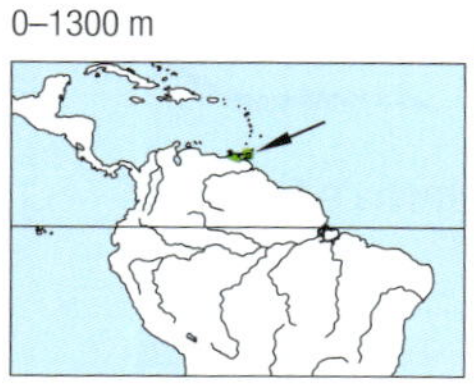

Trinidad Spiny-rat
Proechimys trinitatis

FR Rat-épineux de Trinidad
DE Trinidad-Kurzstachelratte
ES Rata espinosa de Trinidad

TT (Trinidad I), adjacent coastal lowlands of NE VE

DD
Monotypic
HB c. 21 cm
T c. 15.4 cm

130–2300 m

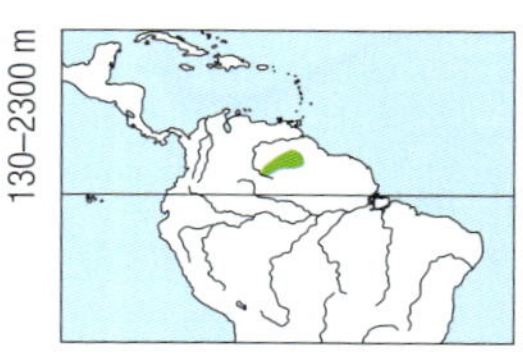

Guianan Spiny-rat
Proechimys hoplomyoides

FR Rat-épineux du Roraima
DE Guyana-Kurzstachelratte
ES Rata espinosa de Guayana

S & SE VE (Amazonia & Guianan Shield tepuis), adjacent W GY & N BR

LC Monotypic
HB 18.9–25 cm
T 13.7–16.7 cm
W 240–490 g

200–1800 m

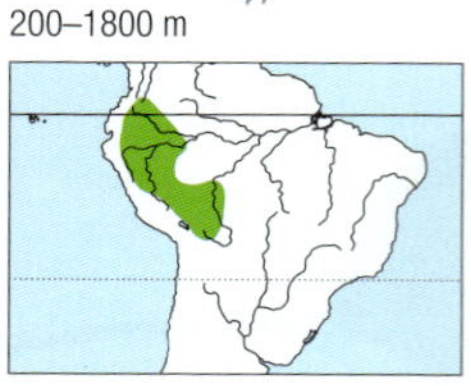

Short-tailed Spiny-rat
Proechimys brevicauda

FR Rat-épineux à queue courte
DE Kurzschwanz-Kurzstachelratte
ES Rata espinosa de cola corta

S CO, E EC, E PE, NW BO & W BR (Acre, perhaps Amazonas)

LC
Monotypic
HB 22–25 cm
T 13.1–15 cm

50–1000 m

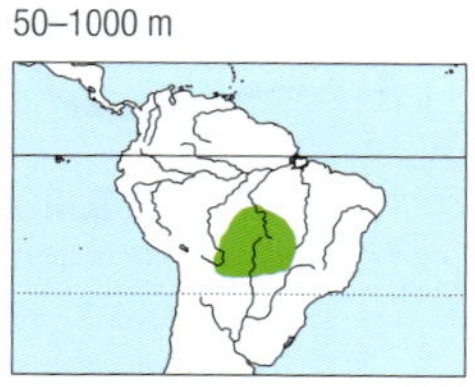

Long-tailed Spiny-rat
Proechimys longicaudatus

FR Rat-épineux longicaude
DE Langschwanz-Kurzstachelratte
ES Rata espinosa de cola larga

SC BR, E BO, N PY

0–500 m

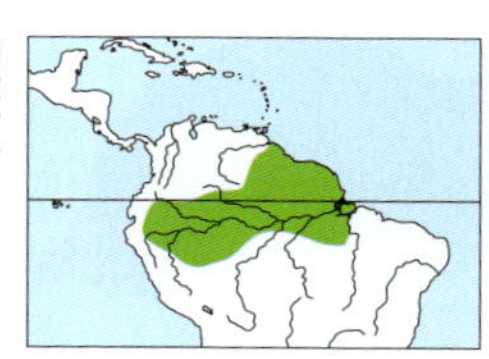

Cuvier's Spiny-rat
Proechimys cuvieri

FR Rat-épineux de Cuvier
DE Cuvier-Kurzstachelratte
ES Rata espinosa de Cuvier

E EC, NE PE, SE CO & W BR E to R Amazon mouth, N to E VE & Guianas

LC Monotypic
HB 21–23 cm
T 14.4–16.8 cm
W 260–420 g

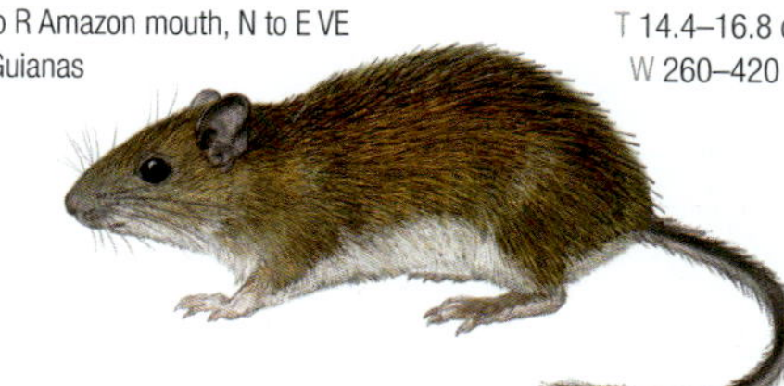

Genus *ISOTHRIX*

(6 species)

Barbara Brown's Brush-tailed Rat
Isothrix barbarabrownae

c. 1850 m

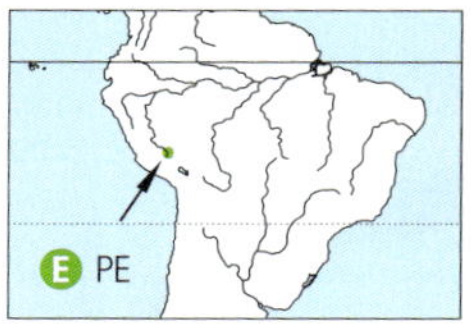

Cultural Zone of Manu NP & BR

FR Rat-épineux de Brown
DE Barbara Brown-Borstenschwanzratte
ES Rata de cola de cepillo de Bárbara Brown

DD
Monotypic

DD
Monotypic
HB 21–26 cm
T 16.8–28 cm

Orinoco Brush-tailed Rat
Isothrix orinoci

100–1600 m

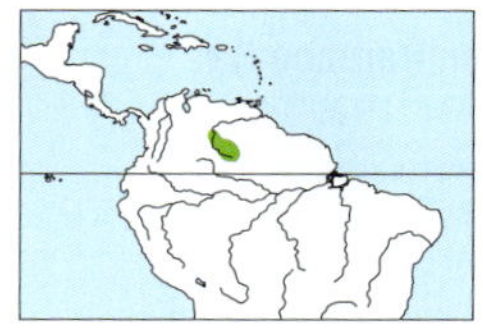

Upper Orinoco & Casiquiare rivers of extreme E CO & S VE; perhaps N BR

FR Rat-épineux de l'Orénoque
DE Orinoco-Borstenschwanzratte
ES Rata de cola de cepillo del Orinoco

Rio Negro Brush-tailed Rat
Isothrix negrensis

0–200 m

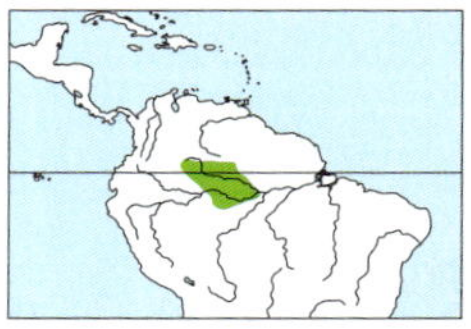

E CO & NW BR through R Negro Basin, S of R Amazon from lower R Juruá E to near R Madeira

LC
Monotypic
HB 20–29 cm T 18.2–27 cm

FR Rat-épineux du Rio Negro DE Rio Negro-Borstenschwanzratte ES Rata de cola de cepillo de Río Negro

Yellow-crowned Brush-tailed Rat
Isothrix bistriata

100–250 m

W Amazon Basin in E EC, E PE, W BR, N BO; probably SE CO

FR Rat-épineux couronné
DE Zweistreifen-Borstenschwanzratte
ES Rata de cola de cepillo de corona amarilla

LC
Monotypic
HB 21–27 cm T 21–27 cm
W 300–530 g

LC
Monotypic
HB 17–23 cm
T 17–23 cm

Plain Brush-tailed Rat
Isothrix pagurus

0–115 m

From left bank of lower R Negro to mouth of R Pitinga & from lower R Madeira to lower R Tapajós

FR Rat-épineux à queue de brosse
DE Schlichte Borstenschwanzratte
ES Rata de cola de cepillo lisa

Sinnamary Brush-tailed Rat
Isothrix sinnamariensis

0–270 m

GF SW to S SR & GY

FR Rat-épineux du Sinnamary
DE Sinnamary-Borstenschwanzratte
ES Rata de cola de cepillo de Sinnamary

LC Monotypic
HB c. 21 cm
T c. 26 cm

Genus *LONCHOTHRIX*

(1 species)

LC
Monotypic
HB 15.5–22 cm
T 15–23 cm
W ≤ 300 g

Tuft-tailed Spiny Tree-rat
Lonchothrix emiliae

below 100 m

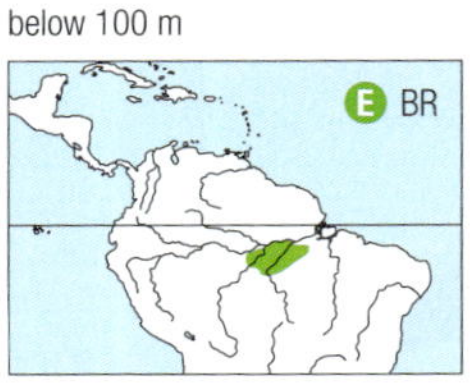

S of R Amazon, from lower reaches of R Madeira, R Tapajós & R Xingu

FR Rat-épineux d'Emilie
DE Büschelschwanz-Stachelratte
ES Rata arborícola de cola de penacho

LC
Monotypic
HB 17.1–18.6 cm
T 16–19.2 cm
W ≤ 160 g

Ferreira's Spiny Tree-rat
Mesomys hispidus

FR Rat-épineux arboricole
DE Amazonas-Stachelratte
ES Rata arborícola de Ferreira

0–1000 m

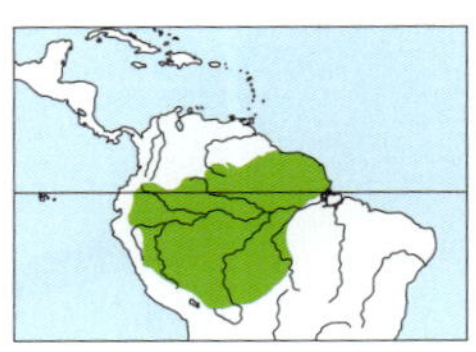

S & E CO, S VE, E EC, E PE, N BO, E through BR to R Tapajós (S of R Amazon), N to Amapá & Guianas

Genus *MESOMYS*

(4 species)

DD
Monotypic
HB c. 18 cm
T c. 21 cm

1555–1980 m

E Andean slopes in Amazonas & San Martín

Long-haired Spiny Tree-rat
Mesomys leniceps

FR Rat-épineux du Pérou DE Langhaar-Amazonas-Stachelratte
ES Rata arborícola de pelo largo

Hidden Spiny Tree-rat
Mesomys occultus

FR Rat-épineux furtif
DE Büschelschwanz-Amazonas-Stachelratte
ES Rata arborícola oculta

LC
Monotypic
HB c. 16.7 cm
T c. 18.3 cm

S of R Amazon in lower R Juruá & R Urucu

LC
Monotypic
HB 15.4–19.6 cm
T 12.2–19.9 cm
W ≤ 139 g

Para Spiny Tree-rat
Mesomys stimulax

0–250 m

S of R Amazon from R Tapajós to R Tocantins

FR Rat-épineux du Para
DE Pará-Amazonas-Stachelratte
ES Rata arborícola de Pará

Genus *KANNABATEOMYS*

(1 species)

LC
2 sspp
HB 23–35 cm T 30–42 cm
W ≤ 570 g

Atlantic Bamboo Rat
Kannabateomys amblyonyx

0–900 m

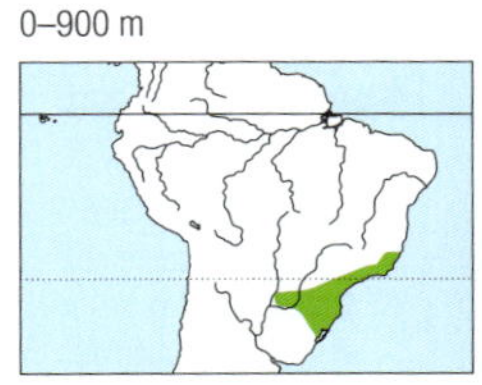

SE BR (from Rio de Janeiro S to Santa Catarina), E PY (Paraná Basin), NE AR (Misiones)

FR Rat-épineux des bambous
DE Bambus-Fingerratte
ES Rata de bambú atlántica

Genus *DACTYLOMYS*

(3 species)

LC
3 sspp
HB c. 31 cm
T c. 39 cm
W ≤ 325 g

0–300 m

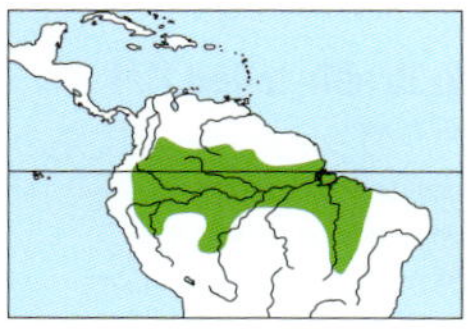

Amazon Basin of SE CO, S VE, E EC, NE PE, NW BO & BR

Amazon Bamboo Rat
Dactylomys dactylinus
FR Coro-coro d'Amazonie
DE Amazonas-Fingerratte
ES Rata de bambú del Amazonas

LC
Monotypic
HB 27–29 cm T c. 41 cm
W ≤ 600 g

200–1000 m

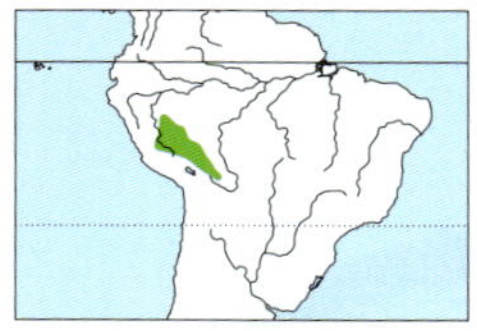

SW Amazon Basin, from C PE S to C BO, E to W BR (C R Juruá)

Bolivian Bamboo Rat
Dactylomys boliviensis
FR Coro-coro de Bolivie
DE Bolivien-Fingerratte
ES Rata de bambú de Bolivia

DD
Monotypic
HB c. 24 cm
T c. 32 cm

1000–3300 m

E Andean slopes from C PE to W BO

Montane Bamboo Rat
Dactylomys peruanus
FR Coro-coro du Pérou
DE Peru-Fingerratte
ES Rata de bambú de Perú

Genus *OLALLAMYS*

(2 species)

2300–3300 m

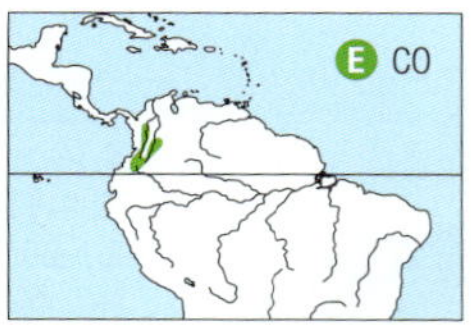

E & C Andes

DD 2 sspp
HB 15–18 cm
T 25–26 cm

White-tailed Olalla Rat
Olallamys albicauda
FR Rat-épineux à queue blanche DE Olalla-Ratte ES Rata olalla de cola blanca

c. 2800 m

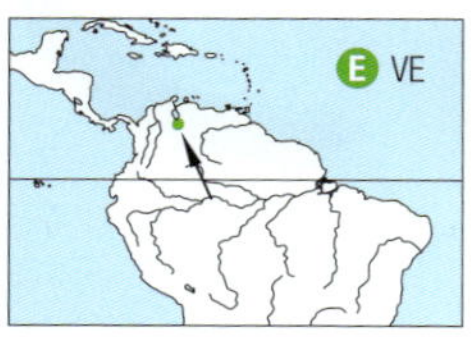

C Cordillera de Mérida

DD
Monotypic
HB c. 22 cm
T c. 34 cm

Greedy Olalla Rat
Olallamys edax
FR Rat-épineux de Mérida DE Große Olalla-Ratte ES Rata olalla glotona

Genus *DIPLOMYS*

(2 species)

LC
2 sspp
HB c. 34 cm
T c. 20 cm
W ≤ 492 g

300–1500 m

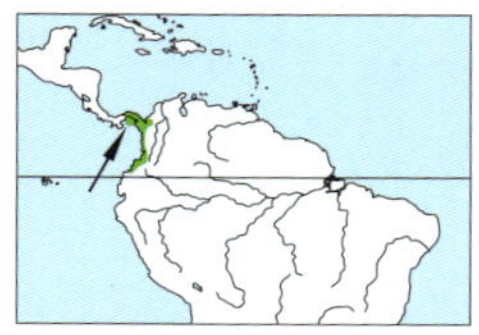

From C PA (with Pearl Is) to W CO (Chocó, Antioquia) & NW EC

Central American Rufous Tree-rat
Diplomys labilis
FR Rat-épineux du Panama
DE Panama-Baumstachelratte
ES Rata arborícola de Centroamérica

DD
Monotypic
HB 21–39 cm
T 17.8–27 cm

Cauca Valley near Medellín; perhaps Serranía de San Lucas (Bolívar)

Colombian Rufous Tree-rat
Diplomys caniceps
FR Rat-épineux à tête grise
DE Graukopf-Baumstachelratte
ES Rata arborícola de Bolivia

Genus *SANTAMARTAMYS*

(1 species)

CR
Monotypic
HB c. 19 cm
T c. 27 cm
W c. 492 g

680–2000 m

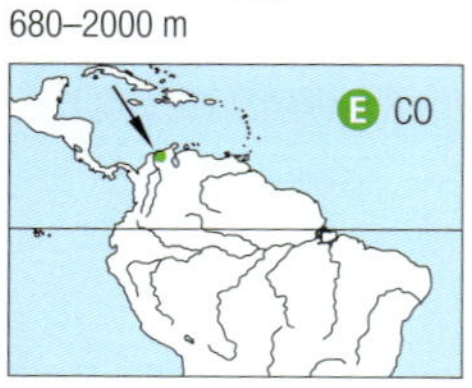

Sierra Nevada de Santa Marta; perhaps NW VE

Red Crested Tree-rat
Santamartamys rufodorsalis
FR Rat-épineux à crête rouge
DE Rotschopf-Baumratte
ES Rata arborícola de dorso rojo

Genus *PATTONOMYS*

(4 species)

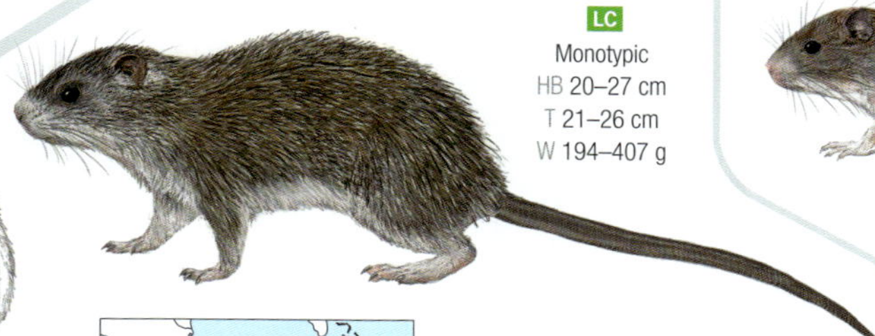

LC
Monotypic
HB 20–27 cm
T 21–26 cm
W 194–407 g

0–600 m

From lower R Magdalena Basin to Caribbean coast

Colombian Speckled Tree-rat
Pattonomys semivillosus
FR Rat-épineux moucheté
DE Kolumbien-Tüpfelbaumratte
ES Rata arborícola de Colombia

0–700 m

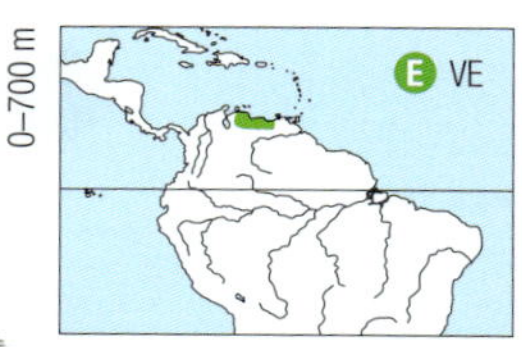

Cordillera de la Costa & C Los Llanos (from Falcón & Lara to Guárico)

DD
Monotypic
HB c. 22 cm T c. 22 cm

Carriker's Speckled Tree-rat
Pattonomys carrikeri
FR Rat-épineux de Carriker DE Carriker-Tüpfelbaumratte
ES Rata arborícola de Carriker

DD
Monotypic
HB c. 25 cm
T c. 25 cm

Yellow Speckled Tree-rat
Pattonomys flavidus
FR Rat-épineux jaunâtre
DE Gelbe Tüpfelbaumratte
ES Rata arborícola de manchas amarillas

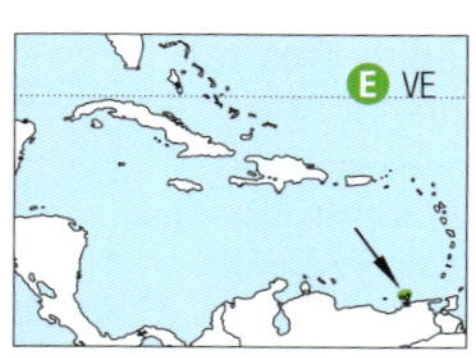

Margarita I, off NC coast

0–100 m

Middle R Orinoco Basin

DD
Monotypic
HB c. 27 cm
T c. 23 cm

Orinocan Speckled Tree-rat
Pattonomys punctatus
FR Rat-épineux ponctué DE Orinoco-Tüpfelbaumratte
ES Rata arborícola del Orinoco

Genus *LEIUROMYS*

(1 species)

DD
Monotypic
HB c. 22 cm
T c. 22 cm

120–575 m

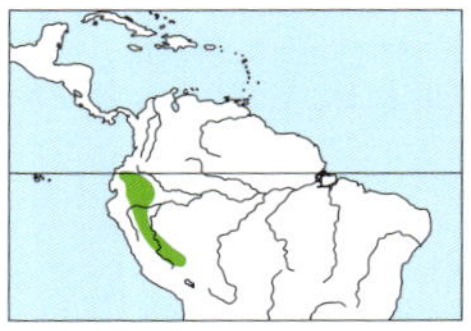

Bare-tailed Tree-rat
Leiuromys occasius

FR Rat-épineux à queue nue
DE Nacktschwanz-Tüpfelbaumratte
ES Rata arborícola de cola desnuda

W Amazon Basin of E EC & E PE

Genus *TOROMYS*

(3 species)

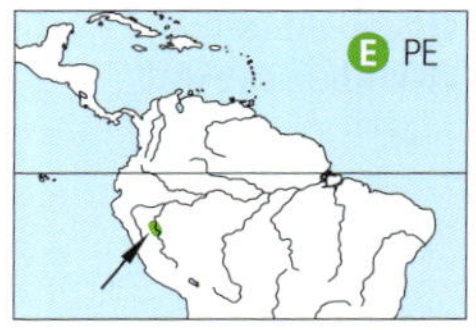

Known from 2 localities along R Ucayali

White-bellied Toro
Toromys albiventris

FR Toromys à ventre blanc
DE Weißbauch-Tororatte
ES Rata arborícola de vientre blanco

NE
Monotypic

90–160 m

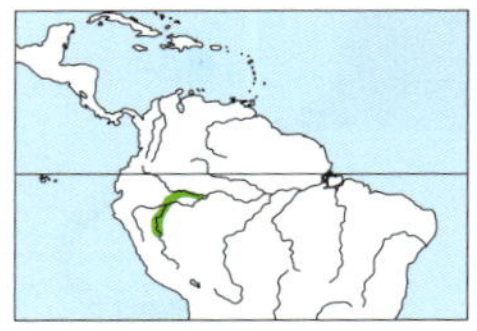

W Amazonia in NE PE & extreme SE CO (Puerto Nariño); unconfirmed in EC

Peruvian Toro
Toromys rhipidurus

FR Toromys du Pérou
DE Peru-Tororatte
ES Rata arborícola de Perú

DD
Monotypic
HB 21–26 cm
T 18–21 cm
W c. 315 g

LC
Monotypic
HB c. 30 cm
T c. 28 cm

Both sides of lower R Amazon, from R Negro to its mouth at Caviana I

Black Toro
Toromys grandis

FR Toromys géant
DE Schwarze Tororatte
ES Rata arborícola grande

Genus *MAKALATA*

(3 species)

LC
Monotypic
HB 21–28 cm T 17–24 cm
W 340–400 g

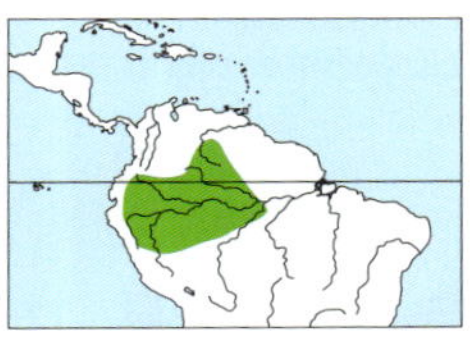

SE CO, E EC, NC & EC PE, S VE & BR (to left bank of lower R Negro & lower R Madeira)

Long-tailed Armored Tree-rat
Makalata macrura

FR Makalata à longue queue
DE Langschwanz-Breitstachelratte
ES Rata arborícola de cola larga

Red-nosed Armored Tree-rat
Makalata didelphoides

LC Monotypic
HB 15–25 cm
T 15.8–23 cm
W ≤ 405 g

FR Makalata à nez roux
DE Rotnasen-Breitstachelratte
ES Rata arborícola de Brasil

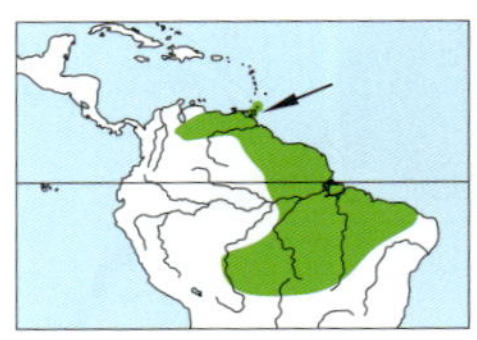

N VE, TT, Guianas, BR (E of R Negro & R Madeira, E to Ceará), SW to NE BO

Presumably E Amazonia (Pará, Maranhão) but range uncertain

DD
Monotypic

Dark Armored Tree-rat
Makalata obscura

FR Makalata obscur
DE Dunkle Breitstachelratte
ES Rata arborícola oscura

Genus *ECHIMYS*

(3 species)

0–200 m

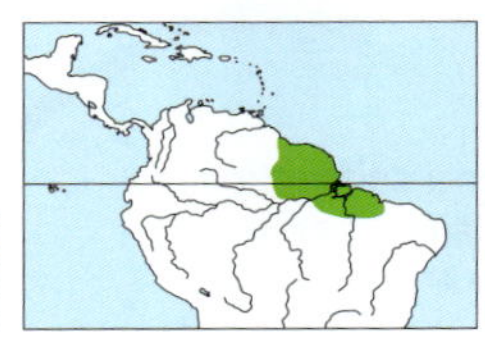

Guianas, NE BR (Guianan region & E Amazon Basin to E of R Xingu)

White-faced Tree-rat
Echimys chrysurus

FR Rat-épineux à face blanche
DE Weißnasen-Stachelratte
ES Rata arborícola de cara blanca

LC
Monotypic
HB 24–33 cm
T 25–39 cm
W ≤ 760 g

DD Monotypic
HB 28–33 cm T 29–38 cm

230–1005 m

Dark Tree-rat
Echimys saturnus

FR Rat-épineux noir
DE Dunkle Stachelratte
ES Rata arborícola espinosa oscura

E Andean foothills & Amazonian lowlands in E EC & N PE

DD
Monotypic
HB c. 24 cm
T c. 34 cm

S of R Amazon between R Madeira & R Tapajós

Vieira's Tree-rat
Echimys vieirai

FR Rat-épineux de Vieira
DE Vieira-Stachelratte
ES Rata arborícola de Vieira

Genus *PHYLLOMYS*

(14 species)

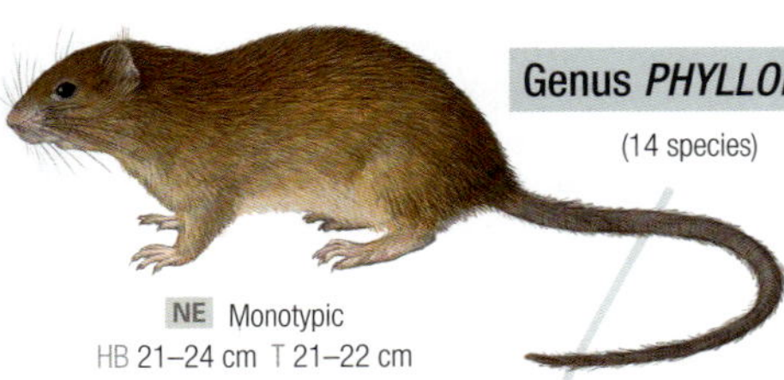

NE Monotypic
HB 21–24 cm T 21–22 cm
W 250–340 g

Few localities in Minas Gerais & Distrito Federal

Central Atlantic Tree-rat
Phyllomys centralis

FR Rat-épineux de Brasilia
DE Zentralbrasilianische Küstenbaumratte
ES Rata arborícola atlántica central

LC
Monotypic
HB 17.5–22 cm
T 19.2–27 cm
W 150–320 g

Golden Atlantic Tree-rat
Phyllomys blainvilii

FR Rat-épineux de Blainville
DE Goldfarbige Küstenbaumratte
ES Rata arborícola atlántica dorada

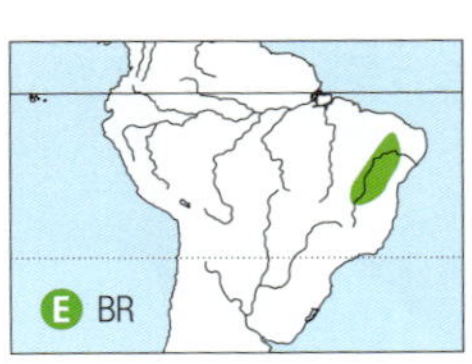

From S Ceará to N Minas Gerais

500–800 m

R Paraopeba & R das Velhas (Minas Gerais)

Brazilian Atlantic Tree-rat
Phyllomys brasiliensis

FR Rat-épineux du Brésil
DE Brasilien-Küstenbaumratte
ES Rata arborícola atlántica de Brasil

EN
Monotypic
HB c. 21 cm
T c. 17 cm
W c. 240 g

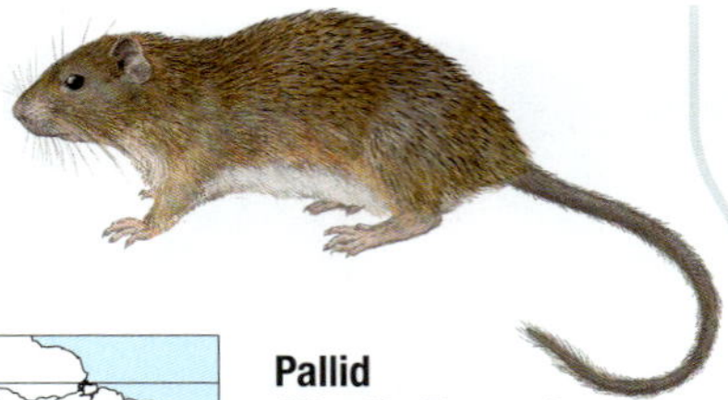

DD
Monotypic
HB 18–23 cm
T 18–23 cm
W 135–290 g

NE Bahia through NE Minas Gerais to N Espírito Santo

Pallid Atlantic Tree-rat
Phyllomys lamarum
FR Rat-épineux pâle
DE Helle Küstenbaumratte
ES Rata arborícola atlántica pálida

0–1000 m

São Paulo, Paraná, Santa Catarina

Black-spined Atlantic Tree-rat
Phyllomys nigrispinus
FR Rat-épineux à dos noir
DE Schwarzstachel-Küstenbaumratte
ES Rata arborícola atlántica de dorso negro

LC
Monotypic
HB 19.5–26 cm
T 19.8–26 cm
W 250–400 g

DD
Monotypic
HB 20–21 cm
T 16–25 cm
W c. 190 g

0–850 m

São Paulo, Paraná, Santa Catarina, Rio Grande do Sul

Southern Atlantic Tree-rat
Phyllomys sulinus
FR Rat-épineux austral
DE Südliche Küstenbaumratte
ES Rata arborícola atlántica meridional

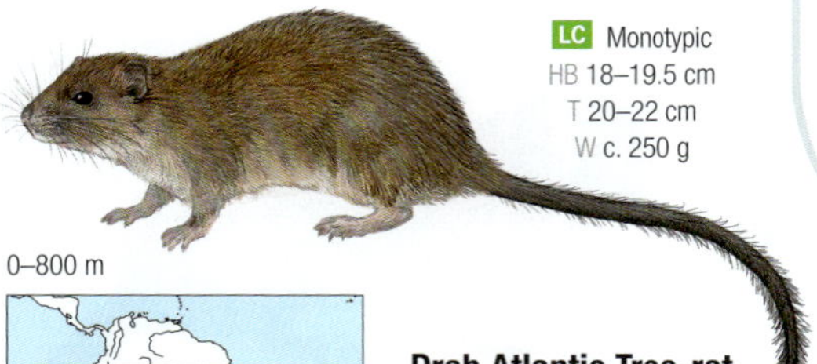

LC Monotypic
HB 18–19.5 cm
T 20–22 cm
W c. 250 g

0–800 m

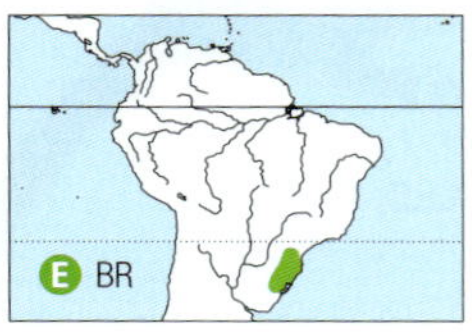

Paraná, Santa Catarina, Rio Grande do Sul

Drab Atlantic Tree-rat
Phyllomys dasythrix
FR Rat-épineux bistre
DE Düstere Küstenbaumratte
ES Rata arborícola atlántica apagada

DD
Monotypic
HB 19–23 cm
T 20–22 cm
W 200–250 g

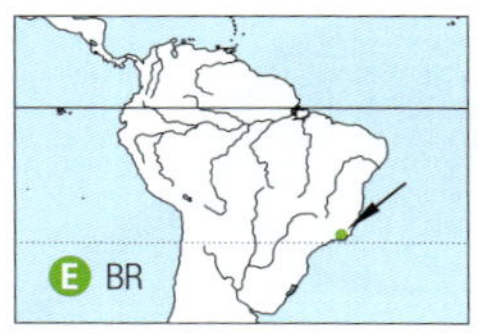

Vicinity of Ubatuba (extreme NE São Paulo)

Kerr's Atlantic Tree-rat
Phyllomys kerri
FR Rat-épineux de Kerr
DE Kerr-Küstenbaumratte
ES Rata arborícola atlántica de Kerr

EN Monotypic
HB 18.4–21 cm
T c. 20 cm
W c. 145 g

SE Minas Gerais, N Rio de Janeiro

Lund's Atlantic Tree-rat
Phyllomys lundi
FR Rat-épineux de Lund
DE Lund-Küstenbaumratte
ES Rata arborícola atlántica de Lund

CR
Monotypic
HB c. 22 cm
T c. 22 cm
W c. 207 g

c. 1850 m

Mantiqueira Mts (Minas Gerais)

Serra da Mantiqueira Atlantic Tree-rat
Phyllomys mantiqueirensis
FR Rat-épineux de la Mantiqueira
DE Serra-da-Mantiqueira-Küstenbaumratte
ES Rata arborícola atlántica de Mantiqueira

LC
Monotypic
HB 21–24 cm
T 19–22 cm
W 150–380 g

Along coast from Paraíba to NE São Paulo

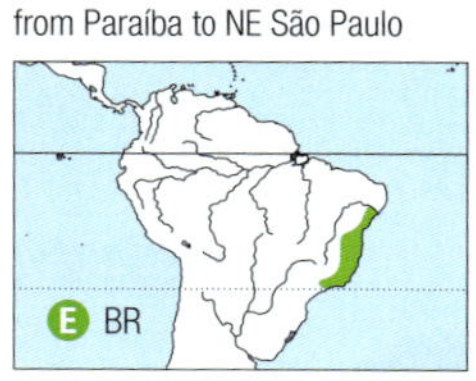

0–1000 m

Patton's Atlantic Tree-rat
Phyllomys pattoni
FR Rat-épineux à flancs roux
DE Patton-Küstenbaumratte
ES Rata arborícola atlántica de Patton

CR
Monotypic
HB c. 28 cm
T c. 20 cm

Known only from Helvécia (extreme S Bahia)

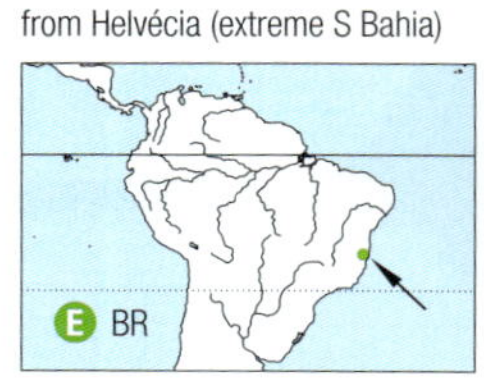

Unicolored Tree-rat
Phyllomys unicolor
FR Rat-épineux unicolore
DE Einfarb-Küstenbaumratte
ES Rata arborícola atlántica unicolor

LC
Monotypic
HB 22–23 cm T 22–26 cm
W 248–386 g

0–1000 m

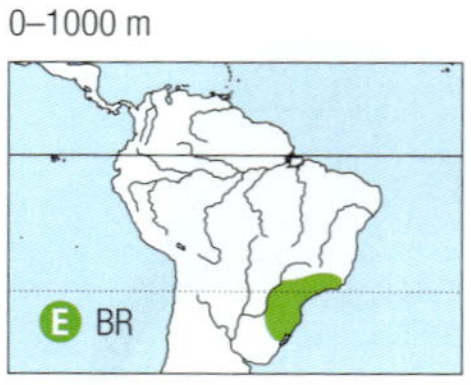

Minas Gerais & Rio de Janeiro S to Rio Grande do Sul

Long-furred Atlantic Tree-rat
Phyllomys medius
FR Rat-épineux à long poil
DE Langhaar-Küstenbaumratte
ES Rata arborícola atlántica de pelo largo

EN
Monotypic
HB 27–30 cm T 27–34 cm
W c. 432 g

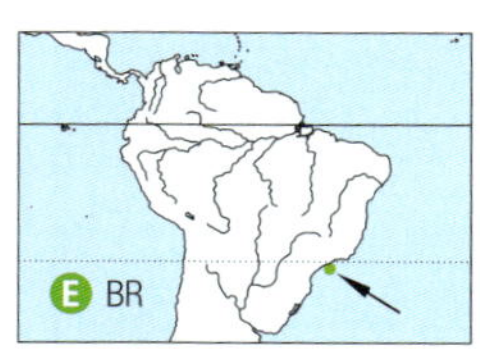

0–1375 m

São Sebastião I, off São Paulo

Giant Atlantic Tree-rat
Phyllomys thomasi
FR Rat-épineux de Sao Sebastiao
DE Riesen-Küstenbaumratte
ES Rata arborícola atlántica gigante

Family APLODONTIIDAE – Mountain Beaver

1 genus, 1 species. Moist forested temperate habitats from sea level to timberline, particularly near streams in second growth, where undergrowth is thick and tangled. Nearctic Region.

Genus *APLODONTIA*

(1 species)

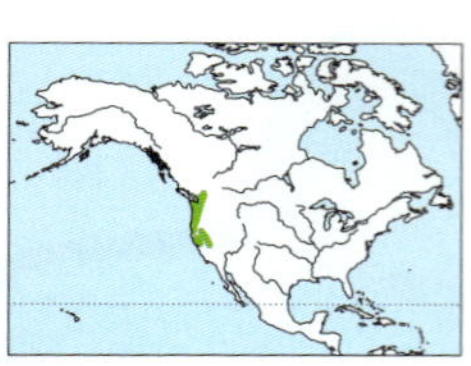

SW CA (SW British Columbia), NW US (S to C California)

Mountain Beaver
Aplodontia rufa
FR Aplodonte
DE Stummelschwanzhörnchen
ES Castor de montaña

LC
7 sspp
HB 23–43 cm
T 2–5.5 cm
W 0.6–1.4 kg

Family SCIURIDAE – Tree, Flying and Ground Squirrels, Chipmunks, Prairie Dogs and Marmots

63 genera, 309 species. From deserts through grasslands and forests to mountains and tundra, from cold temperate zones to tropics. Holarctic, Neotropical, Afrotropical and Indo-Malayan Regions.

Genus *SCIURILLUS*

(1 species)

LC
3 sspp
HB 8.9–11.5 cm
T 8.9–12 cm
W 33–45 g

Disjunct in PE (Upper Amazon), SE CO, CN BR, Guianas & NE BR; perhaps S VE

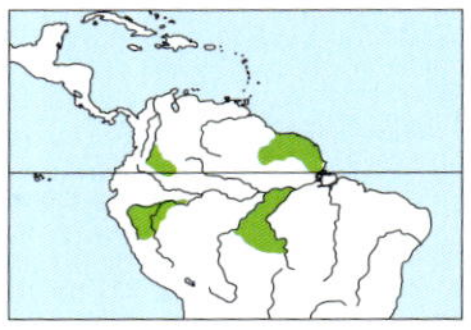

Neotropical Pygmy Squirrel
Sciurillus pusillus

FR Écureuil pygmée
DE Neuwelt-Zwerghörnchen
ES Ardilla pigmea neotropical

ssp *melanochra*

ssp *dandolena*

NT
3 sspp
HB 32–40 cm
T 35–36 cm
W c. 1.6 kg

Sri Lankan Giant Squirrel
Ratufa macroura

FR Écureuil du Sri Lanka
DE Sri-Lanka-Riesenhörnchen
ES Ardilla gigante de Sri Lanka

below 500 m

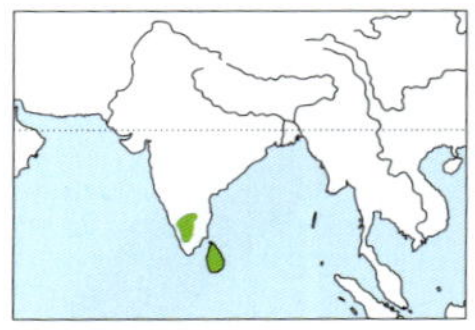

S IN (Eastern & Western Ghats), LK

ssp *macroura*

Genus *RATUFA*

(4 species)

Pale Giant Squirrel
Ratufa affinis

FR Écureuil de Singapour
DE Fahles Riesenhörnchen
ES Ardilla gigante pálida

NT
9 sspp
HB 33–34 cm
T 41–42 cm
W 1–1.2 kg

ssp *affinis*

ssp *baramensis*

ssp *cothurnata*

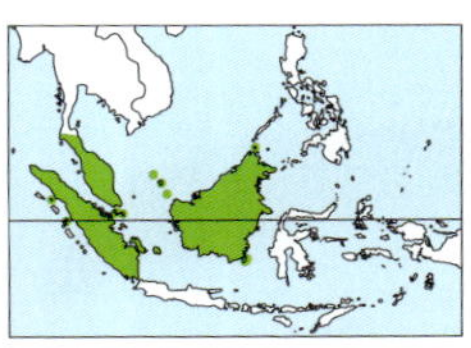

S TH, Pen MY, SG and Sumatra, Borneo, Natuna Is & offshore Is (ID, MY, BN)

LC
4 sspp
HB 34–45 cm
T 38–49 cm
W 0.9–2.3 kg

ssp *indica*

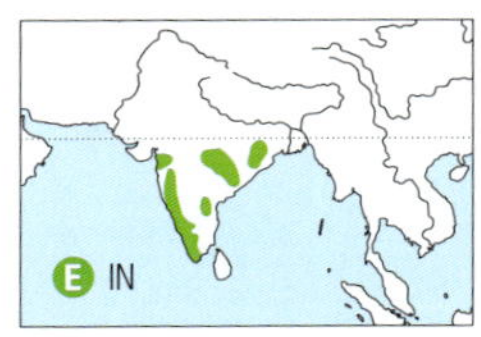

Fragmented in Western & Eastern Ghats, C highlands & Dangs region

Indian Giant Squirrel
Ratufa indica

FR Écureuil indien
DE Indisches Riesenhörnchen
ES Ardilla gigante de India

ssp *centralis*

ssp *dealbata*

ssp *bicolor*

ssp *gigantea*

NT
11 sspp
HB 33–36 cm
T 41–42 cm
W 1–1.2 kg

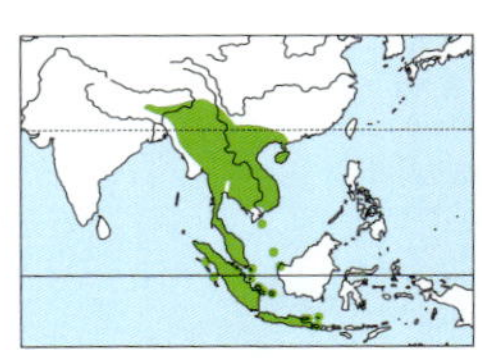

From E NP & NE IN E to SC CN, S to Pen MY, ID (Sumatra, Bali, Java, Natuna Is), offshore Is

Black Giant Squirrel
Ratufa bicolor

FR Écureuil bicolore
DE Schwarzes Riesenhörnchen
ES Ardilla gigante negra

Genus *FUNAMBULUS*

(6 species)

LC 2 sspp
HB 13.4–15.5 cm
T 13–13.5 cm
W 95–103 g

0–4000 m

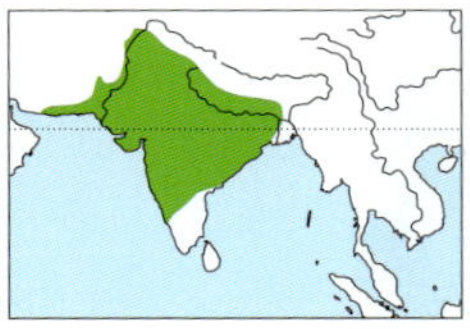

Extreme SE IR, E PK, IN, S NP, W BD; introduced to AU, IL & Andaman Is

Northern Palm Squirrel
Funambulus pennantii

FR Écureuil de Pennant
DE Fünfstreifen-Palmenhörnchen
ES Ardilla de palma septentrional

VU Monotypic
HB 11–13 cm
T 9.8–13.5 cm
W c. 42 g

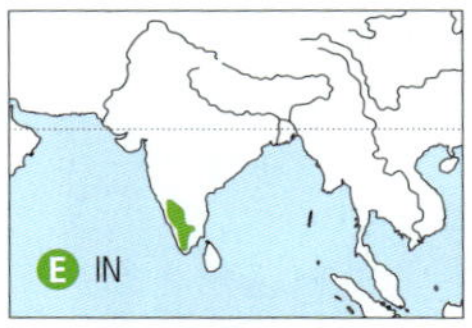

S Western Ghats

Dusky Palm Squirrel
Funambulus sublineatus

FR Écureuil sombre
DE Dunkles Palmenhörnchen
ES Ardilla de palma oscura

VU
Monotypic
HB 10–13.1 cm
T 7.9–12.2 cm
W c. 70 g

0–2135 m

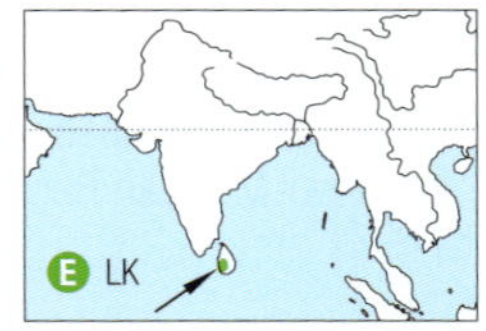

Fragmented in C & SW forests

Sri Lankan Palm Squirrel
Funambulus obscurus

FR Écureuil de Pelzeln DE Sri-Lanka-Palmenhörnchen
ES Ardilla de palma de Sri Lanka

VU
Monotypic
HB 14.4–15.4 cm
T 15.8–16.8 cm
W c. 168 g

Layard's Palm Squirrel
Funambulus layardi

FR Écureuil de Layard DE Layard-Palmenhörnchen
ES Ardilla de palma de Layard

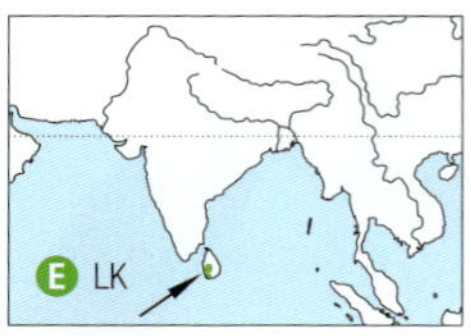

Fragmented in C & SW forests

LC 3 sspp
HB 14.7–14.9 cm
T 14.8–15.8 cm
W 99–117 g

up to 1370 m

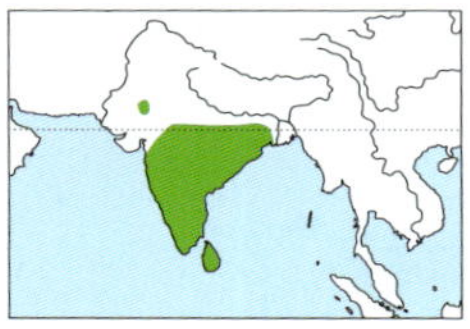

IN (up to c. 20–24° N), LK

Indian Palm Squirrel
Funambulus palmarum

FR Écureuil palmiste
DE Dreistreifen-Palmenhörnchen
ES Ardilla de palma de India

LC 2 sspp
HB 15.9–16 cm
T 13.9–14.3 cm
W c. 139 g

700–2100 m

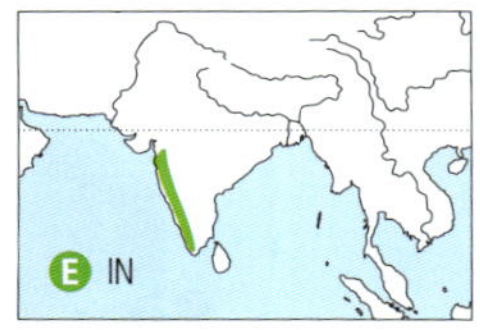

W coast (up to 20° N)

Jungle Palm Squirrel
Funambulus tristriatus

FR Écureuil à trois bandes
DE Dschungel-Palmenhörnchen
ES Ardilla de palma de jungla

Genus *EXILISCIURUS*

(3 species)

DD
Monotypic
HB 7.2–7.3 cm
T 4–5.1 cm
W 16.5–17 g

up to 1700 m

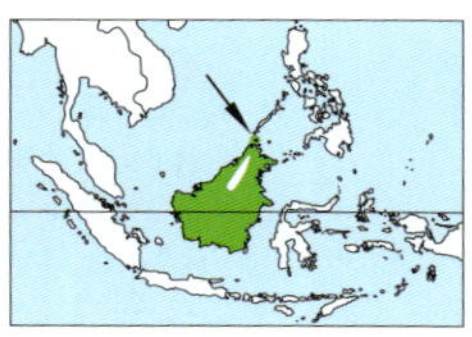

Borneo and Banggi I (MY, ID, BN)

Least Pygmy Squirrel
Exilisciurus exilis

FR Écureuil menu
DE Borneo-Kleinsthörnchen
ES Ardilla pigmea de llanura

LC
Monotypic
HB 8.5–8.6 cm
T 6.3–6.5 cm
W 22–24 g

600–3000 m

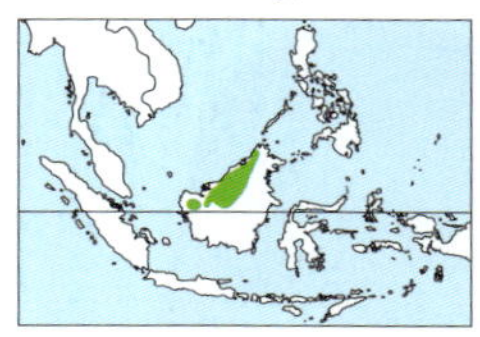

N & W Borneo Mts in MY (Sabah, Sarawak), ID (Kalimantan) & BN

Tufted Pygmy Squirrel
Exilisciurus whiteheadi

FR Écureuil de Whitehead
DE Quasten-Zwerghörnchen
ES Ardilla pigmea de penachos

LC Monotypic
HB c. 8.7 cm T 6.2–6.8 cm
W c. 28 g

0–2000 m

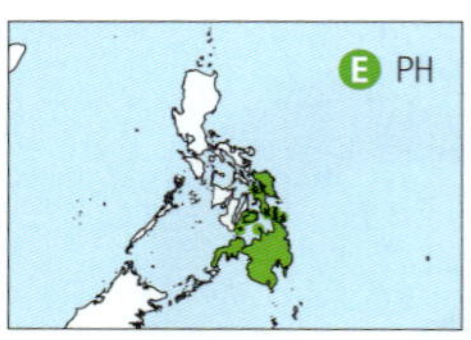

Samar, Biliran, Leyte & Bohol S to Mindanao & Basilan, with small Is

Philippine Pygmy Squirrel
Exilisciurus concinnus

FR Écureuil élégant
DE Philippinen-Zwerghörnchen
ES Ardilla pigmea de Filipinas

LC
4 sspp
HB 11.9–12.1 cm
T 10.1–10.2 cm
W 54–56 g

2000–3000 m

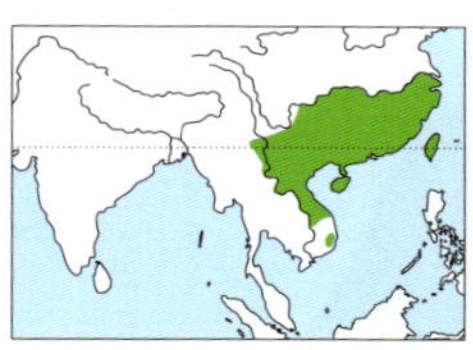

SE CN (with Hainan I), TW, VN, LA

Maritime Striped Squirrel
Tamiops maritimus

FR Écureuil maritime
DE Küsten-Zwergstreifenhörnchen
ES Ardilla rayada marítima

LC 4 sspp
HB c. 13 cm T c. 10 cm
W c. 88 g

2135–3050 m

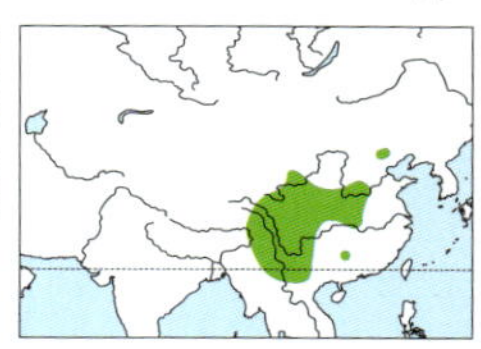

S & E CN, N & NE MM, N LA, N VN

Swinhoe's Striped Squirrel
Tamiops swinhoei

FR Écureuil de Swinhoe
DE Swinhoe-Zwergstreifenhörnchen
ES Ardilla rayada de Swinhoe

LC
2 sspp
HB 11.7–11.8 cm
T 10.8–11.1 cm
W c. 56 g

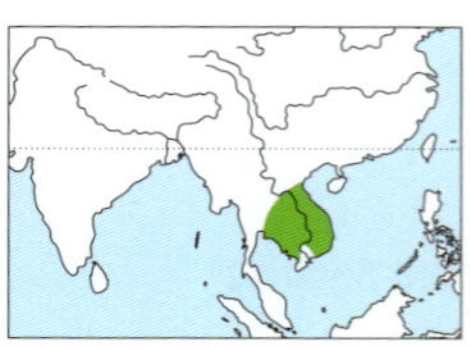

E TH, C & S LA, VN, KH

Cambodian Striped Squirrel
Tamiops rodolphii

FR Écureuil du Cambodge
DE Kambodscha-Zwergstreifenhörnchen
ES Ardilla rayada de Camboya

LC
6 sspp
HB 11.3–11.4 cm
T 10.8–10.9 cm
W 49–52 g

above 700 m

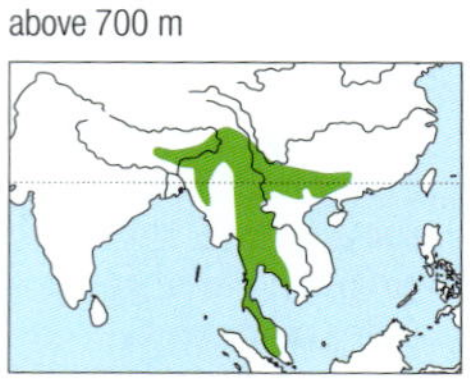

E NP, BT, NE IN, MM, S CN, NW LA, N VN, W KH, TH, Pen MY

Himalayan Striped Squirrel
Tamiops mcclellandii

FR Écureuil de l'Himalaya
DE Himalaya-Zwergstreifenhörnchen
ES Ardilla rayada del Himalaya

Genus *TAMIOPS*

(5 species)

2530–3005 m

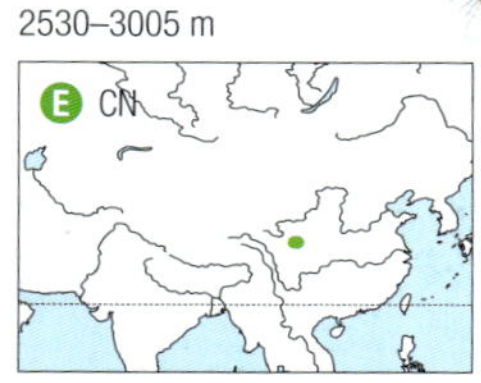

Known only from Wanglang NR (Sichuan)

Minshan Striped Squirrel
Tamiops minshanicus

FR Écureuil des Min
DE Minshan-Zwergstreifenhörnchen
ES Ardilla rayada de Sichuan

NE
Monotypic
HB 12–14 cm
T 9.4–10.5 cm
W 73–95 g

Genus *DREMOMYS*

(6 species)

2000–3500 m

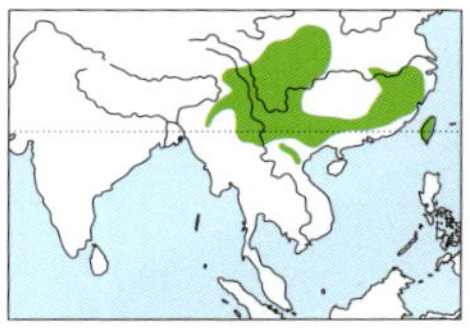

NE IN, C, S & E CN, N MM, N VN, TW

LC
8 sspp
HB 17–20 cm
T 13.8–18 cm
W 160–225 g

Perny's Long-nosed Squirrel
Dremomys pernyi
FR Écureuil de Perny DE Perny-Rotwangenhörnchen ES Ardilla de hocico largo de Perny

LC
6 sspp
HB 16.5–20 cm T 13.5–22 cm
W 150–200 g

900–3000 m

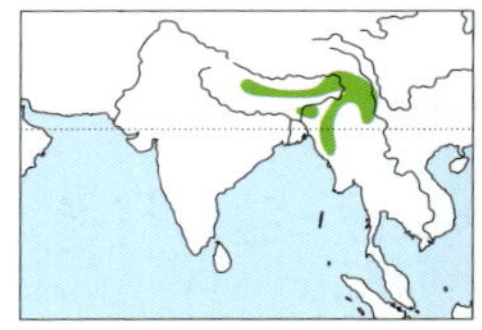

E NP, BT, NE IN, extreme NE BD, extreme S CN, N & W MM

Orange-bellied Himalayan Squirrel
Dremomys lokriah
FR Écureuil à ventre roux
DE Orangebauch-Himalayahörnchen
ES Ardilla del Himalaya de vientre naranja

C & S CN (with Hainan I), N VN

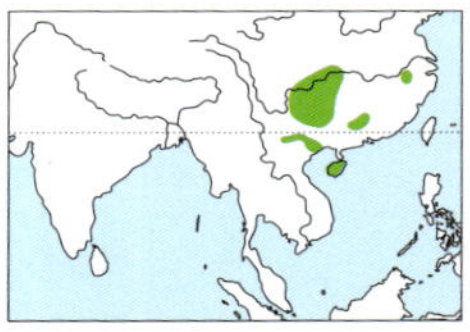

LC
2 sspp
HB 19.4–21 cm
T 14–16.2 cm
W c. 245 g

ssp *pyrrhomerus*

ssp *riudonensis*

Red-hipped Squirrel
Dremomys pyrrhomerus
FR Écureuil à hanches rousses DE Rothüftenhörnchen ES Ardilla de flancos rojos

Asian Red-cheeked Squirrel
Dremomys rufigenis
FR Écureuil à joues rousses
DE Echtes Rotwangenhörnchen
ES Ardilla de mejillas rojas

LC
4 sspp
HB 18.5–20 cm
T 14.5–16.5 cm

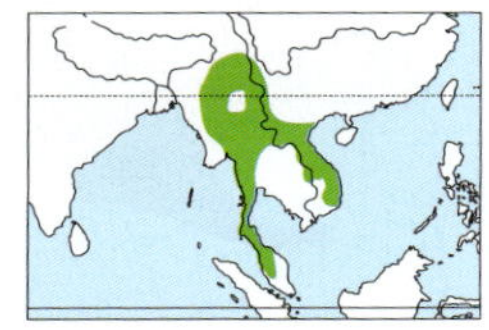

S CN (W Yunnan), MM, TH, LA, C & S VN S to Pen MY

DD
Monotypic
HB c. 22 cm
T c. 16.8 cm

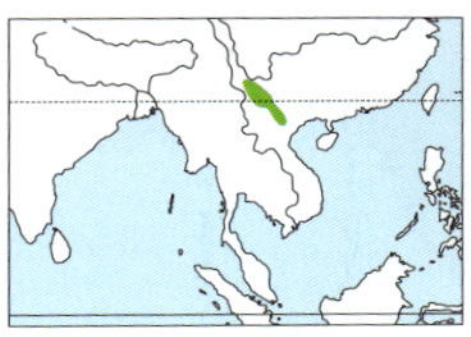

Red River Valley, from SC Yunnan (CN) to N VN

Red-throated Squirrel
Dremomys gularis
FR Écureuil à gorge rousse
DE Ockerkehlhörnchen
ES Ardilla de garganta roja

NE
Monotypic

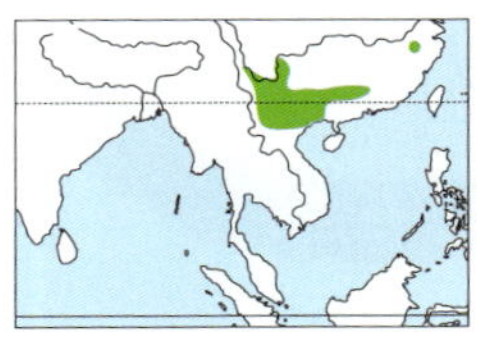

E & S CN (Anhui, Hunan, Yunnan, SW Guangxi, likely S Guizhou & N Guandong), N VN

Chinese Red-cheeked Squirrel
Dremomys ornatus
FR Écureuil du Yunnan DE China-Rotwangenhörnchen ES Ardilla de Annam

Genus *RUBRISCIURUS*

(1 species)

0–1510 m

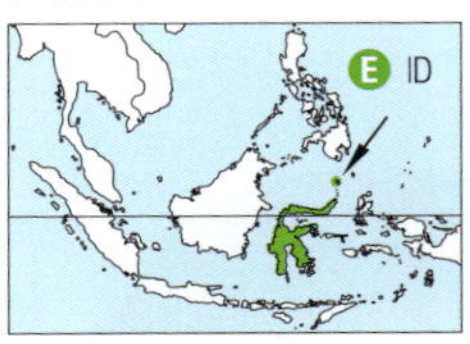

Sulawesi, Sangihe Is

Sulawesi Giant Squirrel
Rubrisciurus rubriventer
FR Écureuil des Célèbes
DE Sulawesi-Riesenhörnchen
ES Ardilla gigante de Célebes

VU
Monotypic
HB 15–30 cm
T 18–25 cm
W 500–860 g

LC
2 sspp
HB 16.5–18.8 cm T 14–19 cm

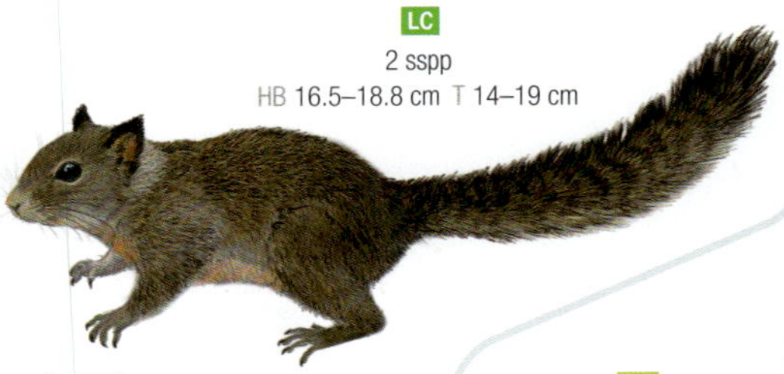

up to 1700 m

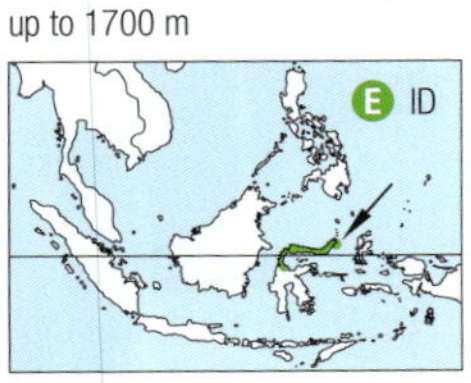

N Sulawesi, Lembeh I

Whitish Dwarf Squirrel
Prosciurillus leucomus
FR Écureuil givré
DE Weißliches Sulawesi-Zwerghörnchen
ES Ardilla enana blanquecina

Alston's Squirrel
Prosciurillus alstoni
FR Écureuil d'Alston
DE Alston-Zwerghörnchen
ES Ardilla de Alston

NT
Monotypic
HB 15.7–19.5 cm
T 13.5–18 cm
W 135–210 g

30–1200 m

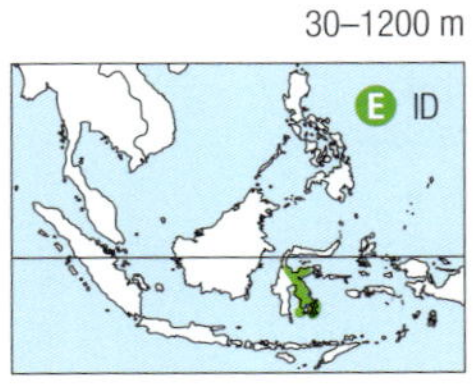

C, E & SE Sulawesi, Kabaena & Buton Is

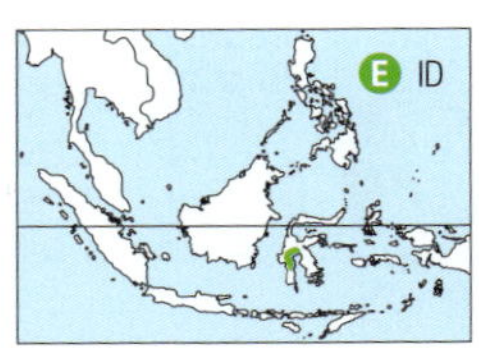

C Sulawesi (coastal NW Gulf of Boni); limits poorly known

Weber's Dwarf Squirrel
Prosciurillus weberi
FR Écureuil de Weber DE Weber-Zwerghörnchen
ES Ardilla enana de Weber

EN
Monotypic
HB c. 18.7 cm
T c. 14.2 cm

Genus *PROSCIURILLUS*

(7 species)

350–2800 m

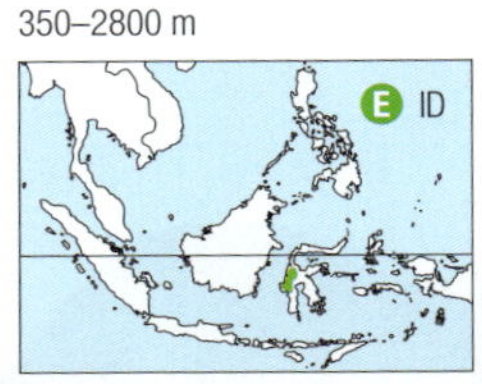

C Sulawesi (highlands from L Lindu to Latimojong Mts)

Mount Topapu Squirrel
Prosciurillus topapuensis
FR Écureuil du Topapu DE Topapu-Zwerghörnchen
ES Ardilla de Topapu

NT
Monotypic
HB 15.5–19 cm
T 12–16.5 cm
W 150–210 g

LC Monotypic
HB c. 19 cm
T c. 18 cm

Sangihe Is

up to 1000 m

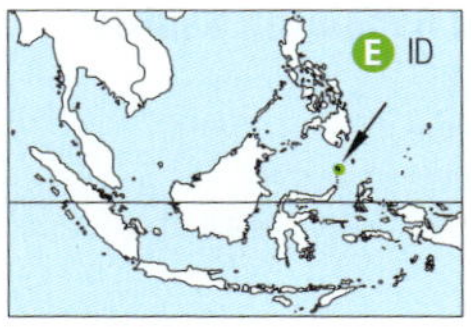

Sanghir Squirrel
Prosciurillus rosenbergii
FR Écureuil de Rosenberg
DE Sangihe-Zwerghörnchen
ES Ardilla de Sangihe

LC
Monotypic
HB 10.2–15 cm T 5.5–12 cm
W 42–110 g

up to 2200 m

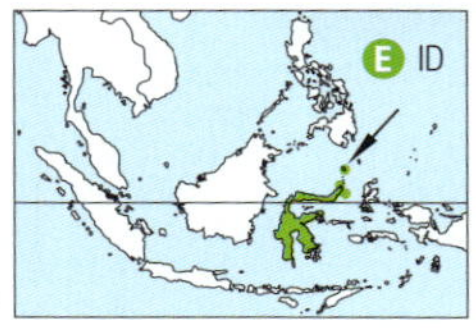

Sulawesi (with Talisei & Lembeh Is and Sangihe Is)

Celebes Dwarf Squirrel
Prosciurillus murinus
FR Écureuil murin
DE Sulawesi-Zwerghörnchen
ES Ardilla enana de Célebes

DD
Monotypic
HB 11.5–14.8 cm
T 7.2–13 cm

1500–2000 m

SE Sulawesi (Mekongga Mts)

Secretive Dwarf Squirrel
Prosciurillus abstrusus
FR Écureuil discret
DE Mekongga-Zwerghörnchen
ES Ardilla enana sigilosa

Genus *HYOSCIURUS*

(2 species)

WC Sulawesi Mts

LC
Monotypic
HB 19.5–24 cm
T 6.5–12 cm
W 228–370 g

1280–2285 m

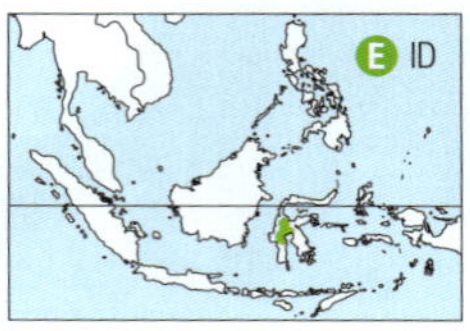

Montane Long-nosed Squirrel
Hyosciurus heinrichi
FR Écureuil de Heinrich
DE Bergland-Ferkelhörnchen
ES Ardilla de hocico largo de montaña

VU
Monotypic
HB 21–25 cm
T 7–12.5 cm
W 293–520 g

170–1700 m

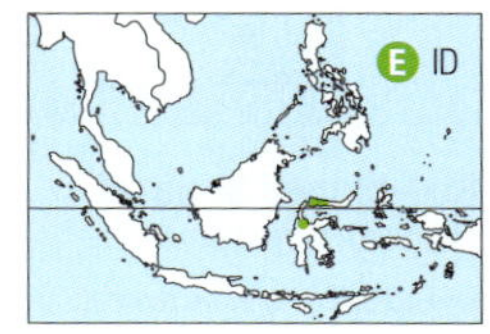

N & C Sulawesi

Lowland Long-nosed Squirrel
Hyosciurus ileile
FR Écureuil de l'Ile-ile
DE Flachland-Ferkelhörnchen
ES Ardilla de hocico largo de tierras bajas

Genus *NANNOSCIURUS*

(1 species)

LC
4 sspp
HB c. 7.9 cm T 6.2–7 cm
W c. 17 g

Black-eared Squirrel
Nannosciurus melanotis
FR Écureuil à oreilles noires
DE Schwarzohr-Zwerghörnchen
ES Ardilla de orejas negras

up to 1070 m

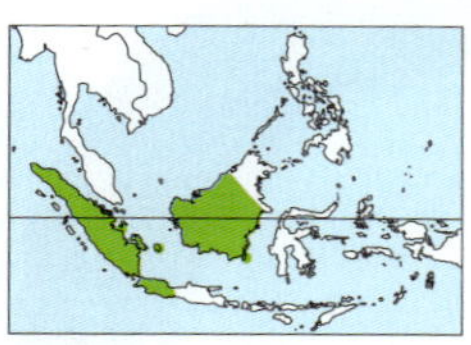

Sumatra (with Singkep I), Bangka & Belitung Is, W Java, Borneo (with Laut I)

LC Monotypic
HB 15.5–17.5 cm T 9.8–11.1 cm
W c. 130 g

Bornean Mountain Ground Squirrel
Sundasciurus everetti
FR Écureuil d'Everett
DE Everett-Hörnchen
ES Ardilla de montaña de Borneo

above 975 m

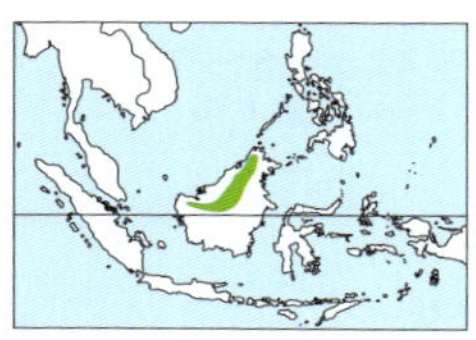

MY, ID: N, C & W Borneo Mts

Genus *SUNDASCIURUS*

(20 species)

NT 7 sspp
HB 23–25 cm T 23–25 cm
W 430–433 g

Horse-tailed Squirrel
Sundasciurus hippurus
FR Écureuil à queue de cheval
DE Pferdeschwanz-Hörnchen
ES Ardilla de cola de caballo

up to 1525 m

S VN (Ca Mau Prov), Thai-Malay Pen, Sumatra & Borneo (MY, ID, BN)

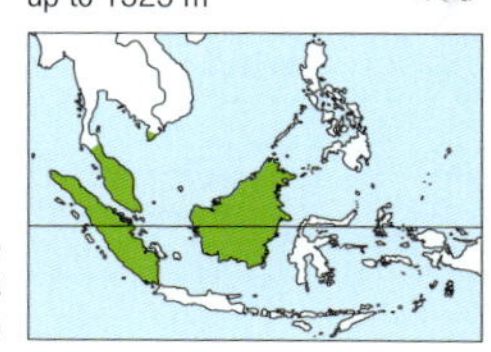

LC
Monotypic
HB c. 19 cm
T 17.2–19.2 cm
W c. 244 g

up to 2100 m

Samar, Biliran, Leyte, Bohol, Dinagat, Siargao, Mindanao, Basilan

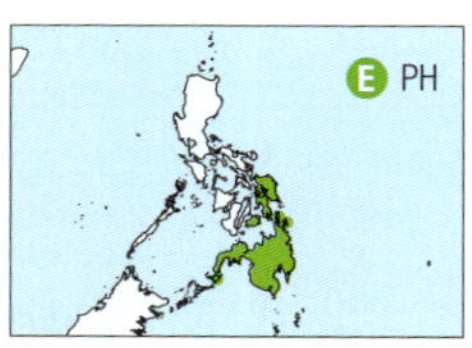

Philippine Tree Squirrel
Sundasciurus philippinensis
FR Écureuil des Philippines
DE Philippinen-Hörnchen
ES Ardilla arborícola de Filipinas

LC
Monotypic
HB c. 20 cm T c. 19 cm
W c. 285 g

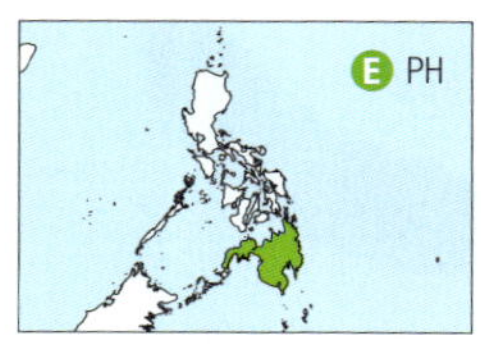

Mindanao and nearby adjacent small Is

Mindanao Squirrel
Sundasciurus mindanensis
FR Écureuil de Mindanao
DE Mindanao-Hörnchen
ES Ardilla de Mindanao

Davao Squirrel
Sundasciurus davensis
FR Écureuil du Davao
DE Davao-Hörnchen
ES Ardilla de Dávao

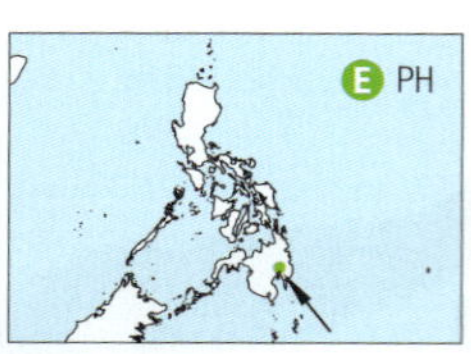

DD
Monotypic
HB c. 19. 8 cm
T c. 18.2 cm (M)

Known only from Tagum (Davao Prov, Mindanao)

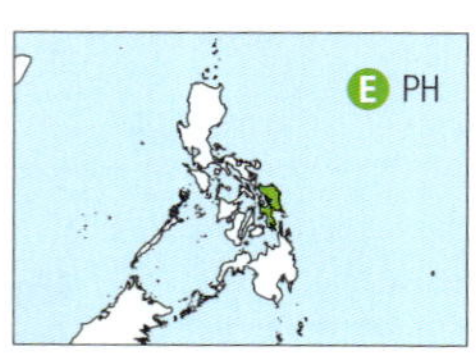

Samar, Leyte

Samar Squirrel
Sundasciurus samarensis
FR Écureuil de Samar
DE Samar-Hörnchen
ES Ardilla de Samar

LC
Monotypic
HB 18.5–19.1 cm
T 15.8–16.1 cm
W 222–243 g

Palawan Montane Squirrel
Sundasciurus rabori
FR Écureuil de Rabor
DE Palawan-Berghörnchen
ES Ardilla de montaña de Palawan

above 800 m

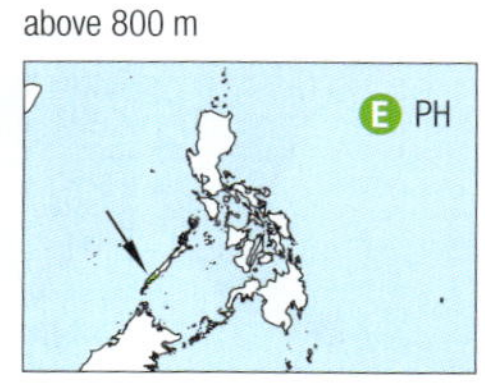

DD
Monotypic
HB 16.3–18.3 cm
T 13.5–14 cm
W c. 163 g

Palawan

LC Monotypic
HB c. 20 cm T c. 16 cm
W 222–257 g

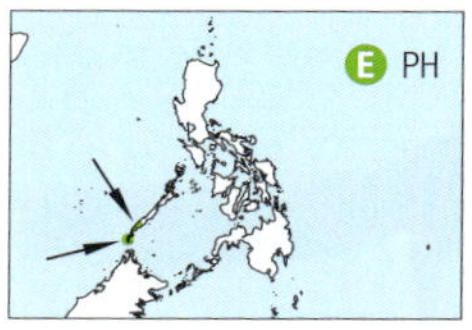

S Palawan, Bancalan, Matangule, Balabac

Southern Palawan Tree Squirrel
Sundasciurus steerii

FR Écureuil de Steere
DE Südliches Palawan-Hörnchen
ES Ardilla meridional de Palawan

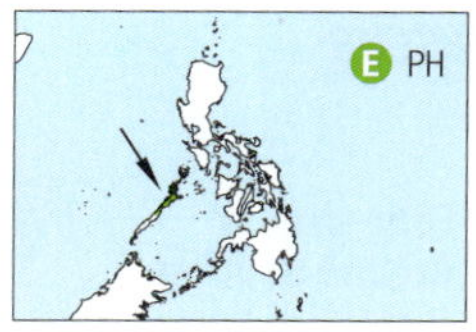

N & C Palawan

Northern Palawan Tree Squirrel
Sundasciurus juvencus

FR Écureuil de Palawan
DE Nördliches Palawan-Hörnchen
ES Ardilla septentrional de Palawan

LC Monotypic
HB 19.5–20 cm T 15.5–17.2 cm
W 244–284 g

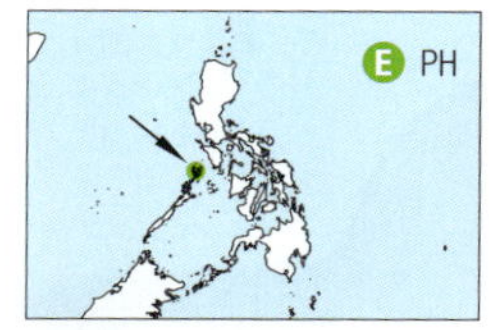

Calamian Is (except Busuanga I)

Culion Tree Squirrel
Sundasciurus moellendorffi

FR Écureuil de Möllendorff
DE Culion-Hörnchen
ES Ardilla de Culión

NT
Monotypic
HB c. 20 cm T c. 19 cm
W c. 190 g (F)

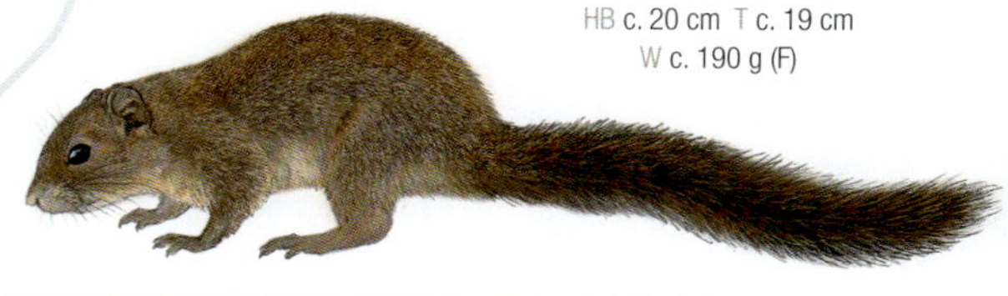

Busuanga I (Calamian Is)

LC
Monotypic
HB c. 20 cm T 16–16.5 cm

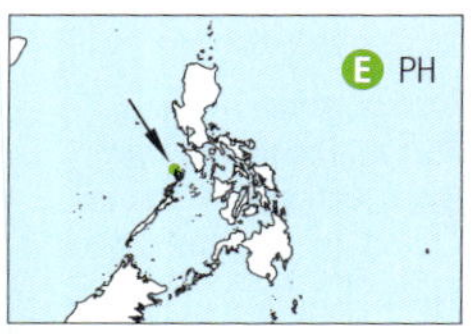

Busuanga Squirrel
Sundasciurus hoogstraali

FR Écureuil de Busuanga
DE Busuanga-Hörnchen
ES Ardilla de Busuanga

LC 5 sspp
HB c. 14 cm T 11.3–11.5 cm
W 81–85 g

above 1650 m

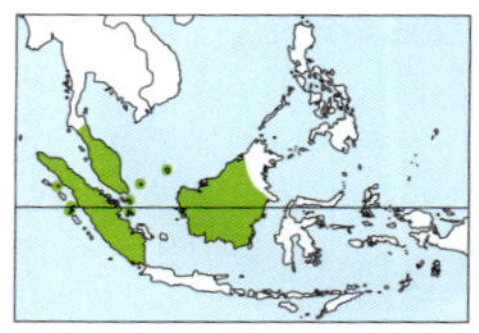

Thai-Malay Pen, Sumatra (with Banyak & Batu Is), Borneo, Anambas, Natunas & N Lingga arch

Slender Squirrel
Sundasciurus tenuis

FR Écureuil grêle
DE Sunda-Schlankhörnchen
ES Ardilla esbelta

NE
Monotypic
HB c. 15 cm
T c. 11.5 cm

490–2530 m

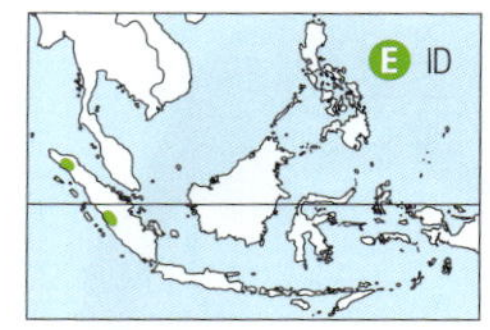

W Sumatra (Mt Leuser & Mt Kerinci in Barisan Mts)

Sumatran Mountain Squirrel
Sundasciurus altitudinis

FR Écureuil de Sumatra
DE Sumatra-Berghörnchen
ES Ardilla de montaña de Sumatra

NE
Monotypic
HB c. 15.5 cm
T c. 10.4 cm

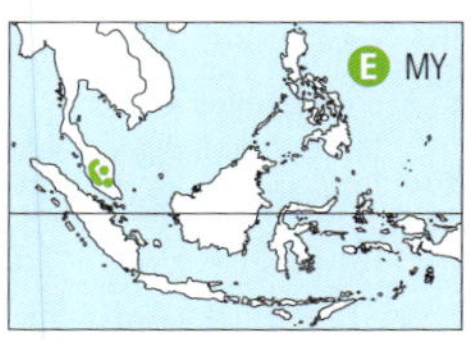

Mt Tahan and other peaks on Peninsula

Upland Squirrel
Sundasciurus tahan

FR Écureuil du Tahan
DE Tahan-Hörnchen
ES Ardilla de tierras altas

LC Monotypic
HB 12.6–13 cm
T 11.2–12.5 cm
W 55–60 g

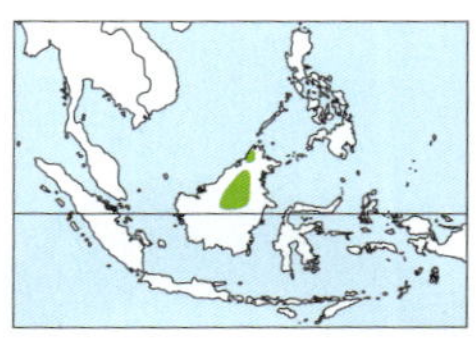

N & C Borneo Mts in Sabah & Sarawak (MY) & NW Kalimantan (ID)

Jentink's Squirrel
Sundasciurus jentinki

FR Écureuil de Jentink
DE Jentink-Hörnchen
ES Ardilla de Jentink

LC Monotypic
HB 15.2–15.4 cm
T 11.2–12.2 cm
W 124–132 g

600–1500 m

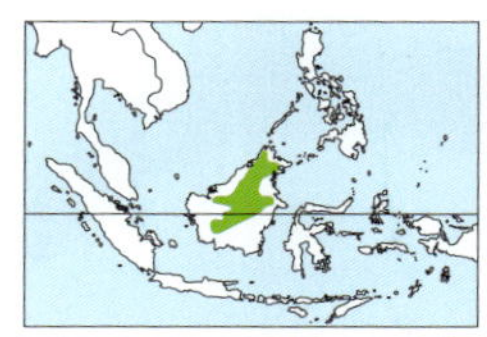

N, C & W Borneo Mts in Sabah & Sarawak (MY) & NW Kalimantan (ID)

Brooke's Squirrel
Sundasciurus brookei

FR Écureuil de Brooke
DE Brooke-Hörnchen
ES Ardilla de Brooke

LC
3 sspp
HB 12.7–16.9 cm
T 7.4–10.9 cm
W 60–125 g

up to 1400 m

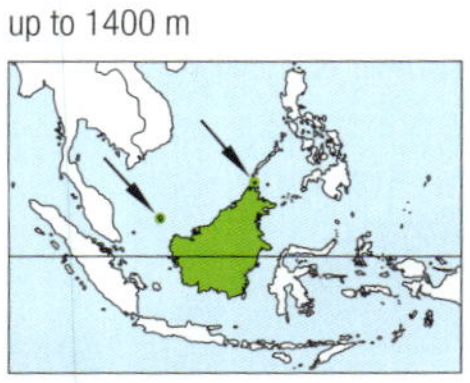

Borneo, Banggi & Balambangan Is (possibly Malawali I), N Natuna Is (Bunguran, Lingung, Laut)

Low's Squirrel
Sundasciurus lowii

FR Écureuil de Low
DE Low-Hörnchen
ES Ardilla de Low

NE Monotypic
HB 13.3–15.2 cm
T 8.2–8.9 cm

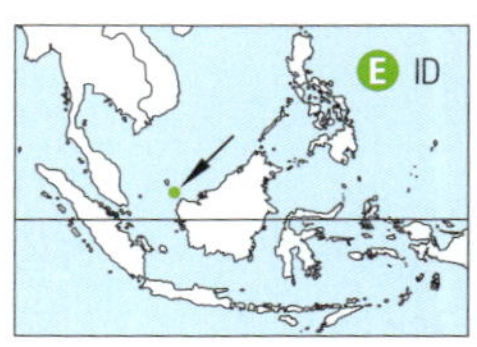

S Natuna Is (only recorded in Sirhassen I); possibly W Borneo

Natuna Squirrel
Sundasciurus natunensis

FR Écureuil des Natuna
DE Natuna-Hörnchen
ES Ardilla de Natuna

NE 3 sspp
HB 11.2–14 cm
T 74–103 cm
W 49–77 g

0–1370 m

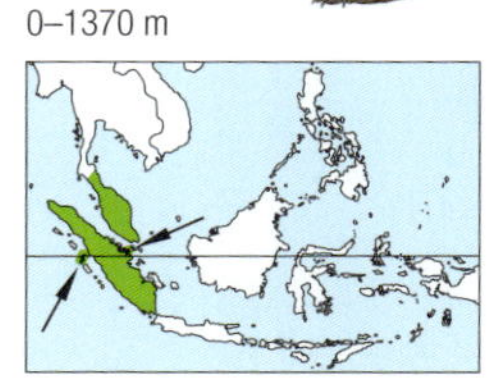

Extreme S TH, Pen MY, ID (Sumatra, Batu & Riau archs)

Robinson's Squirrel
Sundasciurus robinsoni

FR Écureuil de Robinson
DE Robinson-Hörnchen
ES Ardilla de Robinson

VU
Monotypic
HB 11.6–11.9 cm
T 7.5–7.8 cm

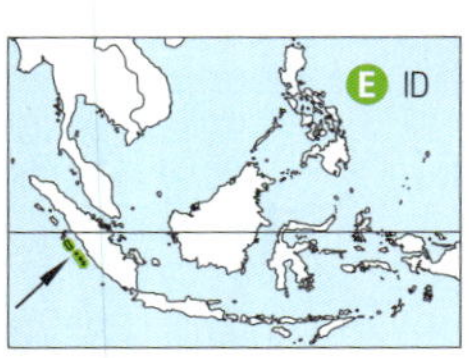

Mentawai Arch (Siberut, Sipora, North Pagai, South Pagai), off W Sumatra

Fraternal Squirrel
Sundasciurus fraterculus

FR Écureuil frère
DE Mentawi-Hörnchen
ES Ardilla fraternal

Genus *MENETES*

(1 species)

LC 7 sspp
HB c. 18.2 cm
T 13.3–13.8 cm
W 172–176 g

E, C & S MM, TH (to S of Isthmus of Kra), S CN (S Yunnan), LA, S VN, KH

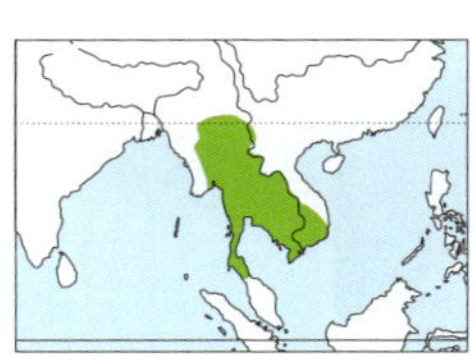

Indochinese Ground Squirrel
Menetes berdmorei

FR Écureuil de Berdmore
DE Berdmore-Hörnchen
ES Ardilla de Berdmore

Genus *RHINOSCIURUS*

(1 species)

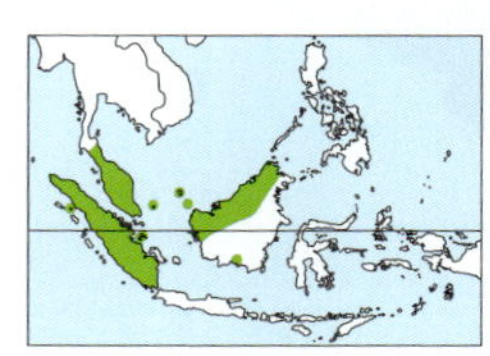

Thai-Malay Pen, Sumatra (with Tuangku I), N, W & S Borneo, Anambas, Natuna & Lingga arch

Shrew-faced Squirrel
Rhinosciurus laticaudatus

FR Écureuil à long nez
DE Langnasenhörnchen
ES Ardilla de cara de musaraña

NT
3 sspp
HB 19.5–21 cm
T 11.5–14 cm

LC 5 sspp
HB 18.7–19.4 cm
T 10–10.4 cm
W 175–182 g

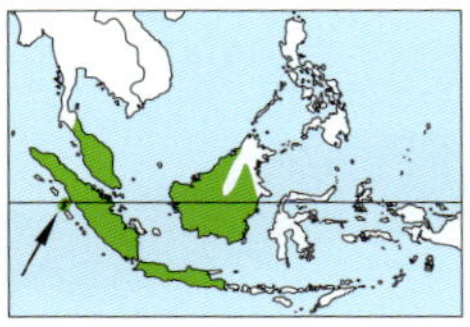

Thai-Malay Pen, Sumatra (with Batu Is), Java, Borneo

Three-striped Ground Squirrel
Lariscus insignis
FR Écureuil à trois raies
DE Dreistreifenhörnchen
ES Ardilla terrestre de tres rayas

Genus *LARISCUS*

(4 species)

up to 1530 m

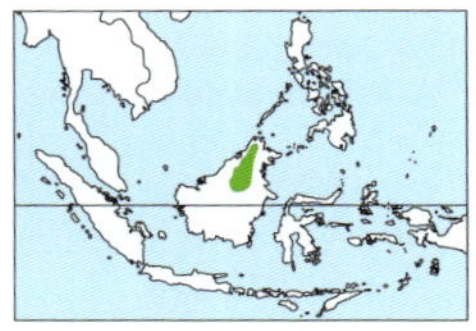

N & NC Borneo, from Mt Kinabalu to Kelabit Highlands (MY, ID)

Four-striped Ground Squirrel
Lariscus hosei
FR Écureuil à quatre raies
DE Vierstreifenhörnchen
ES Ardilla terrestre de cuatro rayas

LC Monotypic
HB c. 18.9 cm
T c. 8.6 cm
W c. 215 g (F)

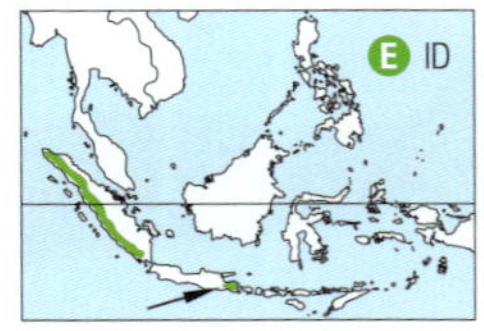

W Sumatra (Barisan Range), E Java (Ijen Mts)

Niobe Ground Squirrel
Lariscus niobe
FR Écureuil niobé
DE Niobe-Dreistreifenhörnchen
ES Ardilla terrestre de Niobe

DD
2 sspp
HB 18.9–19.4 cm
T 8.5–9 cm

NT
3 sspp
HB c. 20 cm
T 8.6–8.8 cm

Mentawai Arch (Siberut, Sipora, North Pagai, South Pagai),off W Sumatra

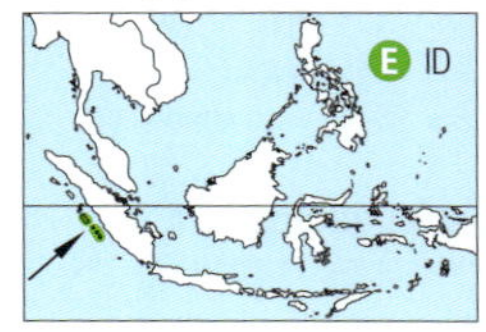

Mentawai Three-striped Squirrel
Lariscus obscurus
FR Écureuil obscur
DE Mentawi-Dreistreifenhörnchen
ES Ardilla terrestre de Mentawai

Genus *GLYPHOTES*

(1 species)

Sculptor Squirrel
Glyphotes simus
FR Écureuil sculpteur
DE Borneo-Zwerghörnchen
ES Ardilla pigmea de Borneo

N Borneo (Mts of Sabah & Sarawak)

DD
Monotypic
HB 9.4–14.4 cm
T 9.5–10.6 cm

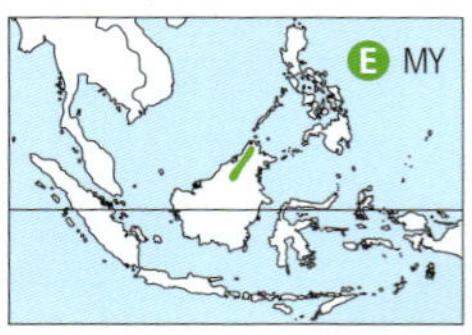

285–1800 m

Genus *CALLOSCIURUS*

(15 species)

LC Monotypic
HB 14.8–15.3 cm
T 13.9–15.7 cm
W c. 278 g

1100–1700 m

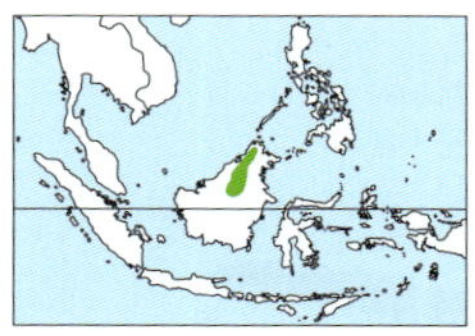

N & WC Borneo in Sabah & Sarawak (MY) & NW Kalimantan (ID)

Borneo Black-banded Squirrel
Callosciurus orestes
FR Écureuil de Bornéo
DE Borneo-Schönhörnchen
ES Ardilla de bandas negras de Borneo

LC
Monotypic
HB 23–24 cm T 24–25 cm
W c. 370 g

300–1800 m

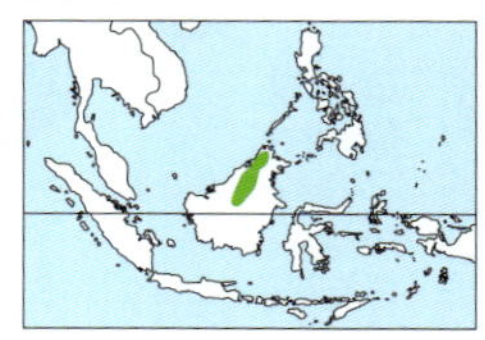

Kinabalu Squirrel
Callosciurus baluensis
FR Écureuil du Kinabalu
DE Kinabalu-Schönhörnchen
ES Ardilla de Kinabalu

MY/ID: N & WC Borneo in Sabah, Sarawak & NW Kalimantan

NT
Monotypic
HB 15–17 cm
T 14–16 cm
W 134–150 g

up to 900 m

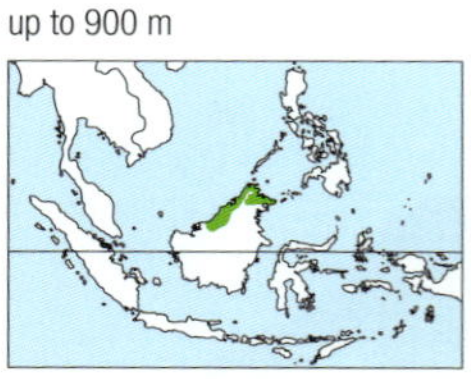

MY/BN: N & W Borneo

Ear-spot Squirrel
Callosciurus adamsi
FR Écureuil à oreilles tachetées
DE Ohrfleckschönhörnchen
ES Ardilla de orejas moteadas

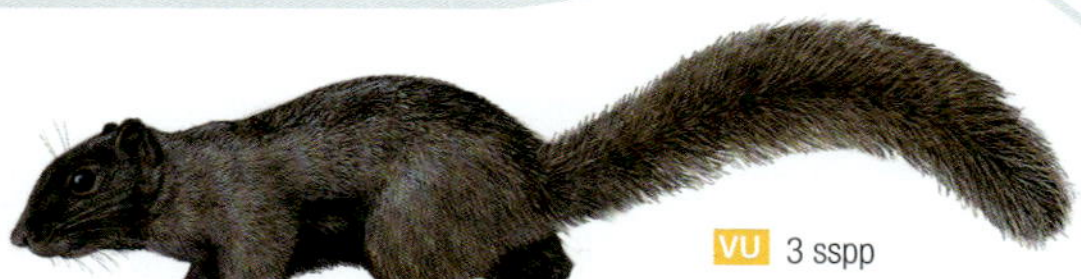

VU 3 sspp
HB c. 21 cm T c. 18 cm
W c. 292 g

Mentawai Arch (Siberut, Sipora, North Pagai, South Pagai), off W Sumatra

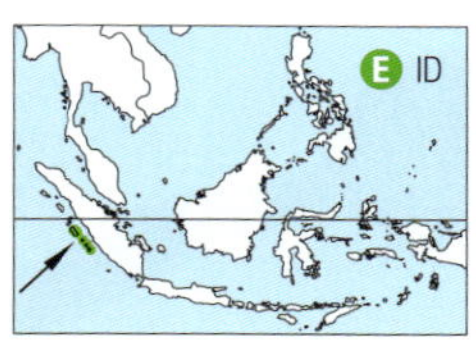

Mentawai Squirrel
Callosciurus melanogaster
FR Écureuil des Mentawai
DE Mentawi-Schönhörnchen
ES Ardilla de Mentawai

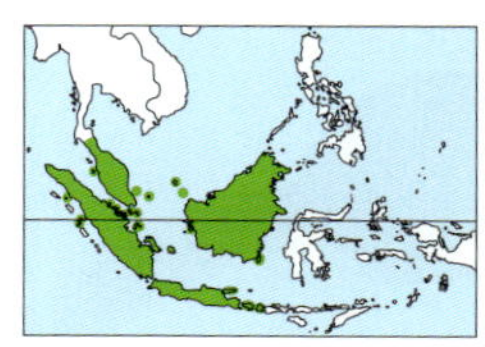

Plantain Squirrel
Callosciurus notatus
FR Écureuil tricolore
DE Bananenschönhörnchen
ES Ardilla de platanero

Thai-Malay Pen, Sumatra, Java, Borneo, Bali, Lombok, many offshore Is

LC
6 sspp
HB 20–24 cm
T 17.5–18.6 cm
W 219–234 g

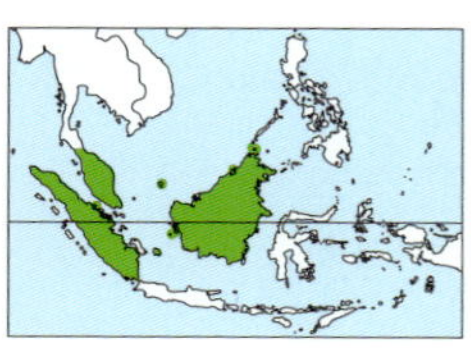

Thai-Malay Pen, Sumatra (with Rupat I), Borneo (with offshore Is), Natuna & Belitung Is

up to 1200 m

ssp *caedis*

LC
17 sspp
HB c. 24 cm
T c. 23 cm
W 353–403 g

Prevost's Squirrel
Callosciurus prevostii
FR Écureuil de Prévost DE Prevost-Schönhörnchen
ES Ardilla de Prevost

ssp *prevostii*

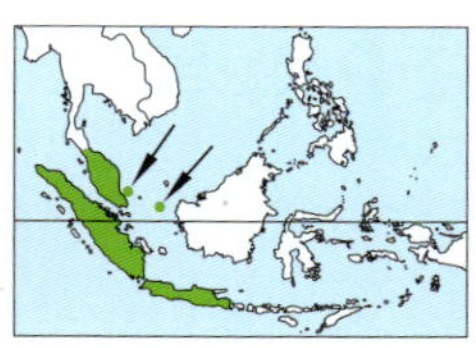

Thai-Malay Pen (with Tioman I), Sumatra, Java, Tambelan Arch (Saddle I)

Black-striped Squirrel
Callosciurus nigrovittatus
FR Écureuil à bandes noires
DE Schwarzstreifen-Schönhörnchen
ES Ardilla de rayas negras

LC 4 sspp
HB 18.4–19.9 cm T 15.9–18.2 cm
W 202–239 g

E NP, NE IN, BT, BD, MM, S CN (Tibet, small population in S Yunnan)

LC
7 sspp
HB 18.7–23 cm
T 11–22 cm
W 230–300 g

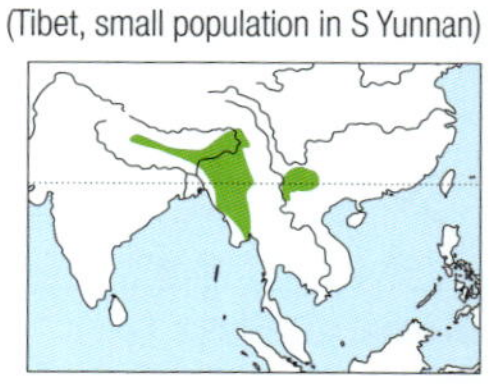

Irrawaddy Squirrel
Callosciurus pygerythrus
FR Écureuil de l'Irrawaddy DE Irawadi-Schönhörnchen ES Ardilla de Irrawaddy

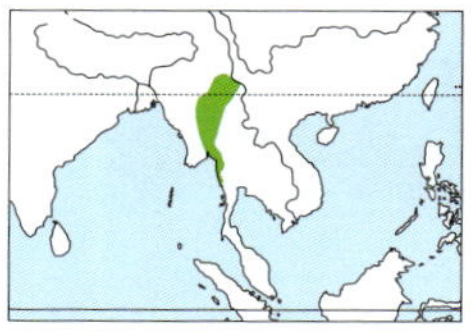

S CN (Yunnan), E & S MM (from upper R Irrawaddy & R Sittaung E to R Salween)

LC
Monotypic
HB 21–24 cm
T 20–25 cm
W 258–377 g

Phayre's Squirrel
Callosciurus phayrei
FR Écureuil de Phayre
DE Phayre-Schönhörnchen
ES Ardilla de Phayre

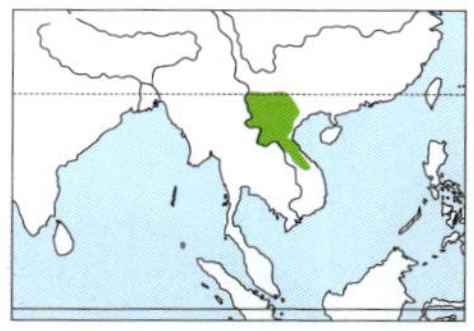

S CN (S Yunnan), N & E LA, N & W VN

Inornate Squirrel
Callosciurus inornatus
FR Écureuil terne
DE Einfarbiges Schönhörnchen
ES Ardilla inornada

LC
Monotypic
HB c. 29 cm
T 17.6–21 cm
W c. 325 g

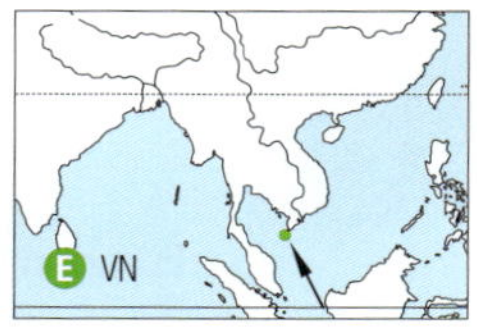

Hon Khoai I

Hon Khoai Squirrel
Callosciurus honkhoaiensis
FR Écureuil de Hon Khoai
DE Hon-Khoai-Schönhörnchen
ES Ardilla de Hon Khoai

NE Monotypic
HB c. 17.5 cm T c. 18 cm
W c. 152 g

c. 1000 m

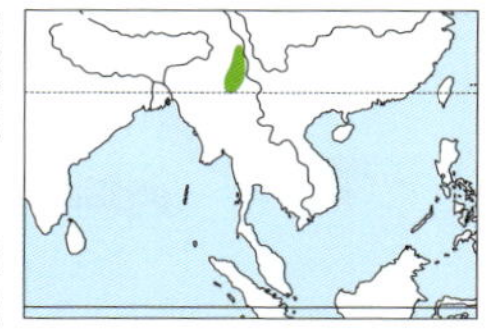

N MM, CN (NW Yunnan)

Anderson's Squirrel
Callosciurus quinquestriatus
FR Écureuil à cinq raies
DE Anderson-Schönhörnchen
ES Ardilla de Anderson

LC 2 sspp
HB c. 20 cm T 18–21 cm
W 258–315 g

LC
8 sspp
HB 21–22 cm
T 22–24 cm
W 266–313 g

up to 2500 m

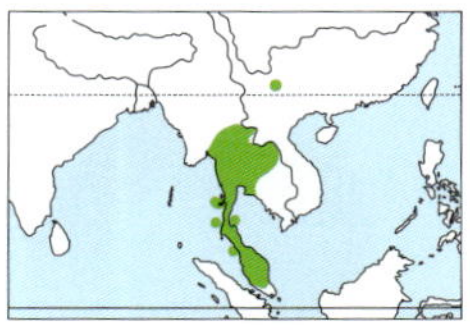

MM & TH S to Pen MY, with Mergui Arch & other Is; perhaps CN (Yunnan)

Gray-bellied Squirrel
Callosciurus caniceps
FR Écureuil à ventre gris
DE Graubauchschönhörnchen
ES Ardilla de vientre gris

LC
24 sspp
HB 21–23 cm T 17.6–27 cm
W 286–375 g

Pallas's Squirrel
Callosciurus erythraeus
FR Écureuil de Pallas
DE Pallas-Schönhörnchen
ES Ardilla de Pallas

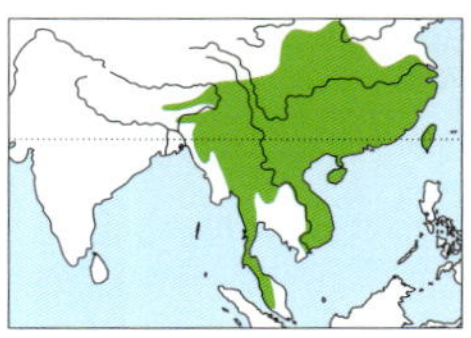

NE IN & BT E to MM, SE CN (with Hainan I) & TW, S to VN, TH & Pen MY; introduced to JP, HK, Europe & AR

ssp *boonsongi*

variants

ssp *annellatus*

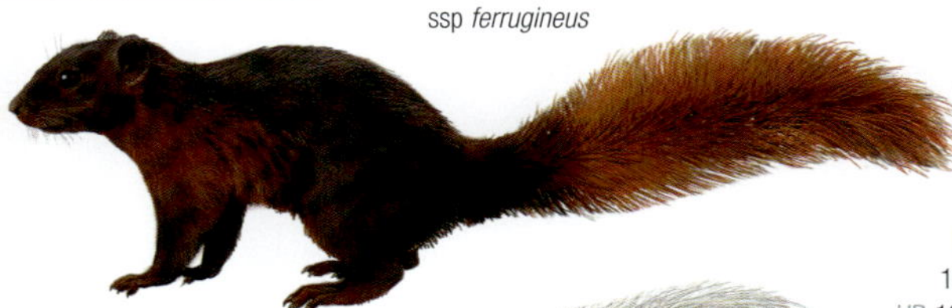

ssp *ferrugineus*

LC
16 sspp
HB 19.1–21 cm
T 17.3–22 cm
W c. 278 g

ssp *finlaysonii*

ssp *bocourti*

variants

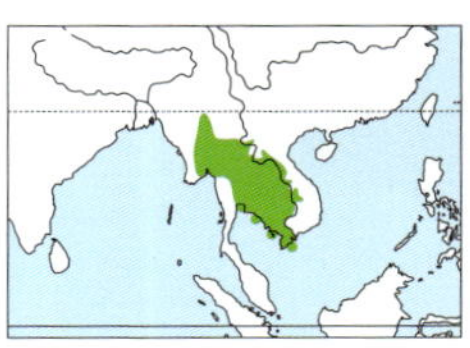

SC MM, TH, LA, KH, S VN (Mekong Delta), offshore Is; introduced to SG, JP & N IT

Finlayson's Squirrel
Callosciurus finlaysonii
FR Écureuil de Finlayson
DE Finlayson-Schönhörnchen
ES Ardilla de Finlayson

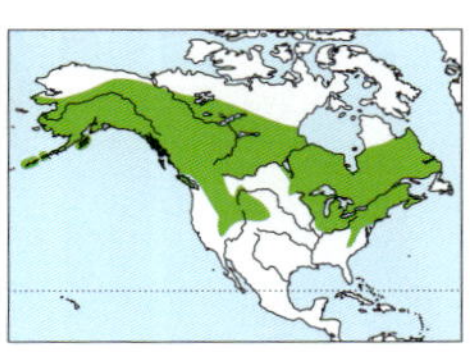

Widely across CA & US (Alaska, S to Arizona in W and to N Georgia in E)

LC
20 sspp
HB 18.7–19 cm
T c. 12.4 cm
W 194–213 g

ssp *hudsonicus*

North American Red Squirrel
Tamiasciurus hudsonicus
FR Écureuil roux DE Rothörnchen
ES Ardilla roja de Norteamérica

ssp *petulans*

Genus *TAMIASCIURUS*

(3 species)

NE
4 sspp
HB 19.4–22 cm
T 10.4–14.6 cm
W 231–223 g

SC Wyoming, SE Utah, Colorado, Arizona, New Mexico, W Texas

Fremont's Squirrel
Tamiasciurus fremonti
FR Écureuil de Frémont
DE Südwestliches Rothörnchen
ES Ardilla de Frémont

Douglas's Squirrel
Tamiasciurus douglasii
FR Écureuil de Douglas
DE Douglas-Rothörnchen
ES Ardilla de Douglas

ssp *douglasii*

LC
3 sspp
HB 18.1–18.5 cm
T 11.9–14.5 cm
W 199–207 g

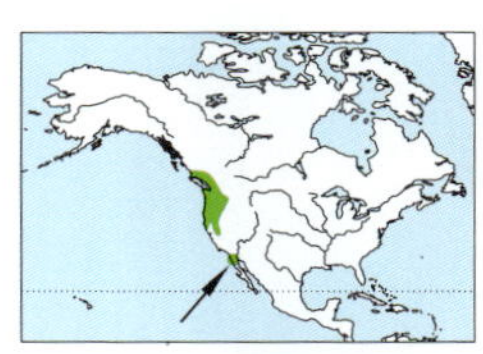

CA (SW British Columbia), US (Washington to California), NW MX (Baja California)

ssp *mearnsi*

Genus *SCIURUS*

(29 species)

ssp *chiliensis*

ssp *alpinus*

LC

22 sspp

HB 21–25 cm

T 15–20 cm

W 235–480 g

0–3100 m

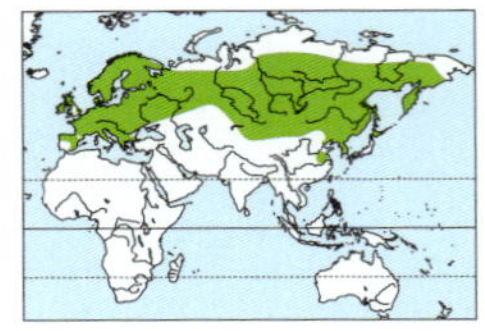

From IE, UK & Iberian Pen E through Eurasia to Pacific coast & JP; introduced S of its native range & KN

ssp *exalbidus*

Eurasian Red Squirrel

Sciurus vulgaris

FR Écureuil d'Eurasie
DE Eurasisches Eichhörnchen
ES Ardilla roja de Eurasia

ssp *leucourus*

ssp *vulgaris*

NE

Monotypic

W 380–486 g

Calabria, Basilicata

600–1600 m

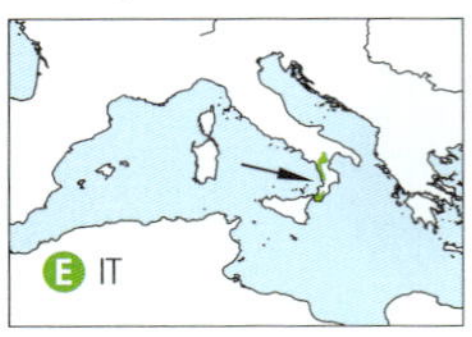

Calabrian Black Squirrel

Sciurus meridionalis

FR Écureuil de Calabre
DE Kalabrisches Eichhörnchen
ES Ardilla de Calabria

LC

Monotypic

HB 16–22 cm

T 13–17 cm

W 250–310 g

Honshu, Shikoku; formerly Kyushu & Awaji (no recent records)

Japanese Squirrel

Sciurus lis

FR Écureuil du Japon
DE Japan-Eichhörnchen
ES Ardilla de Japón

Caucasian Squirrel

Sciurus anomalus

FR Écureuil du Caucase
DE Kaukasus-Eichhörnchen
ES Ardilla del Cáucaso

LC

3 sspp

HB 19.2–25 cm

T 12–16.2 cm

W 250–410 g

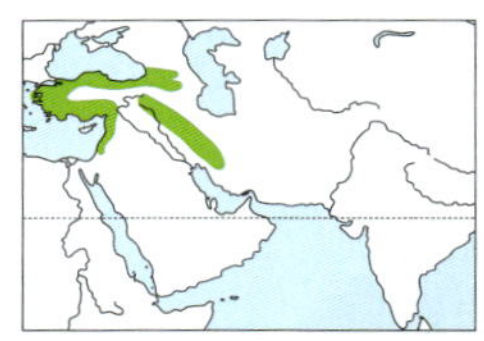

GR (Lesbos I), TR, Transcaucasia, Levant, N IQ, W & S IR (Zagros Mts)

ssp *ferreus*

LC

6 sspp

HB 26–27 cm

T 21–22 cm

W 594–619 g

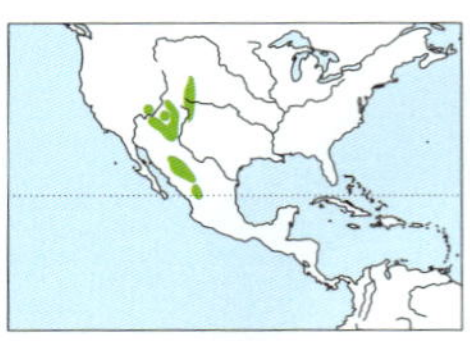

Abert's Squirrel

Sciurus aberti

FR Écureuil d'Abert
DE Abert-Hörnchen
ES Ardilla de Abert

ssp *aberti*

US (from S Wyoming & Utah) S to NW MX (Sierra Madre Occidental S to Durango); introduced to other US sites

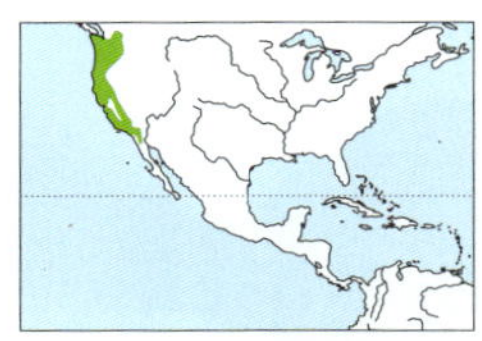

W US (Washington, Oregon, California) S to MX (extreme N Baja California)

Western Gray Squirrel

Sciurus griseus

FR Écureuil de Californie
DE Westliches Grauhörnchen
ES Ardilla gris occidental

LC

3 sspp

HB 26–32 cm

T 24–31 cm

W 520–942 g

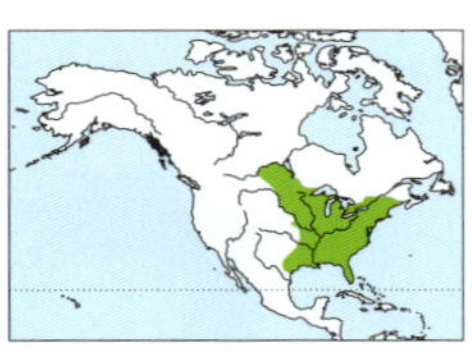

Eastern Gray Squirrel

Sciurus carolinensis

FR Écureuil gris
DE Östliches Grauhörnchen
ES Ardilla gris oriental

CA (Saskatchewan E to Quebec), E US (S to E Texas & Florida); introduced to Europe, ZA, W US & CA

LC

5 sspp

HB 20–31 cm

T 15–25 cm

W 300–710 g

LC

10 sspp

HB 26–37 cm T 20–33 cm

W 0.5–1.4 kg

melanistic morph

ssp *niger*

ssp *shermani*

Eastern Fox Squirrel

Sciurus niger

FR Écureuil fauve
DE Fuchshörnchen
ES Ardilla zorro oriental

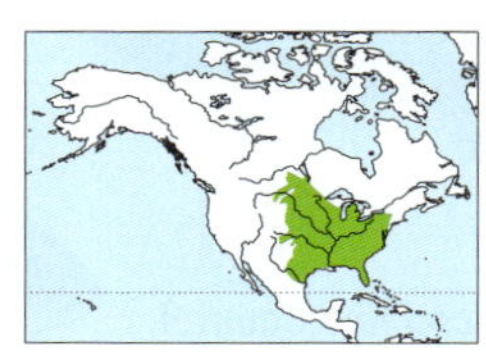

CA (S Saskatchewan, S Manitoba) S across E US to NE MX (along R Grande); introduced to other US & CA sites

LC 3 sspp
HB 25–26 cm
T c. 25 cm
W 669–761 g

1500–2700 m

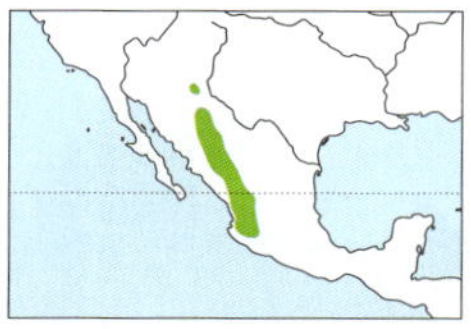

Mexican Fox Squirrel
Sciurus nayaritensis

FR Écureuil du Nayarit
DE Mexiko-Fuchshörnchen
ES Ardilla zorro de México

US (Chiricahua Mts in SE Arizona), W MX (Sierra Madre Occidental from E Sonora S to N Colima)

DD
Monotypic
HB 25–26 cm
T c. 25 cm
W 667–736 g

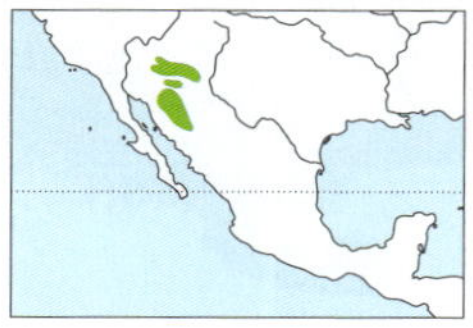

Arizona Gray Squirrel
Sciurus arizonensis

FR Écureuil d'Arizona
DE Arizona-Grauhörnchen
ES Ardilla gris de Arizona

US (S Arizona, W New Mexico), NW MX (NC Sonora)

Peters's Squirrel
Sciurus oculatus

FR Écureuil de Peters
DE Peters-Hörnchen
ES Ardilla de Peters

Forests in arid mountains of S Mexican Plateau

LC 3 sspp
HB 51–56 cm T 26–27 cm
W 550–750 g

LC
Monotypic
HB 22–25 cm
T 22–25 cm
W 290–510 g

Allen's Squirrel
Sciurus alleni

FR Écureuil d'Allen
DE Allen-Hörnchen
ES Ardilla de Allen

E Coahuila, S Nuevo León, W Tamaulipas, N San Luis Potosí

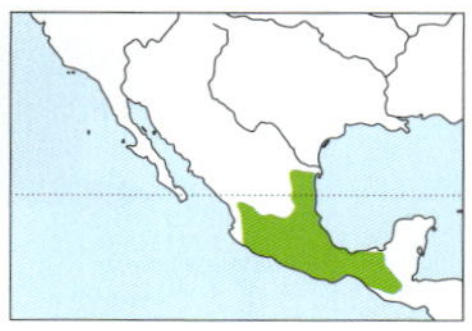

E, C & S MX (S from Nayarit & Tamaulipas) to SW & S GT

LC
2 sspp
HB 23–31 cm T 21–28 cm
W 375–690 g

Red-bellied Squirrel
Sciurus aureogaster

FR Écureuil du Mexique
DE Rotbauchhörnchen
ES Ardilla de vientre rojo

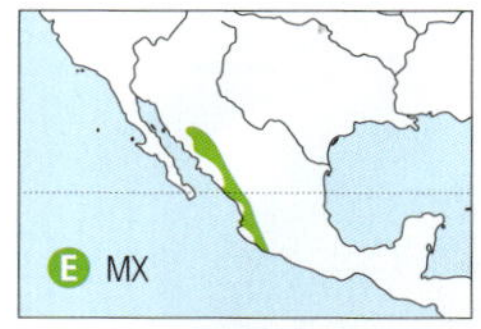

below 2000 m

LC 4 sspp
HB 24–25 cm T 24–26 cm
W 335–441 g

Along Sierra Madre Occidental from Sonora & SW Chihuahua S to Colima

Collie's Squirrel
Sciurus colliaei

FR Écureuil de Collie
DE Collie-Hörnchen
ES Ardilla de Collie

Variegated Squirrel
Sciurus variegatoides

FR Écureuil multicolore
DE Bunthörnchen
ES Ardilla variable

LC
15 sspp
HB c. 26 cm
T 26–28 cm
W 469–537 g

ssp *variegatoides*

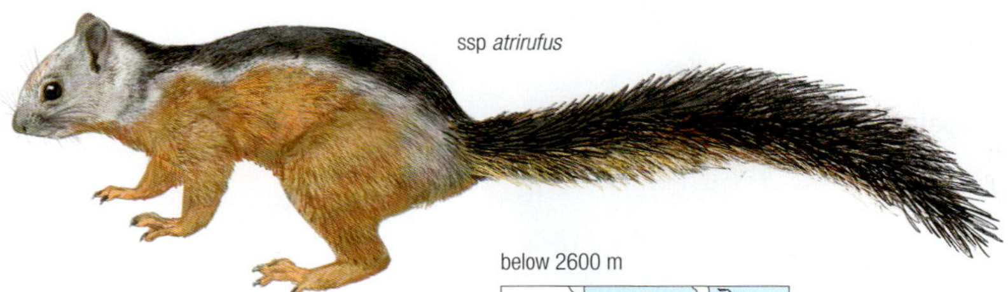

ssp *atrirufus*

below 2600 m

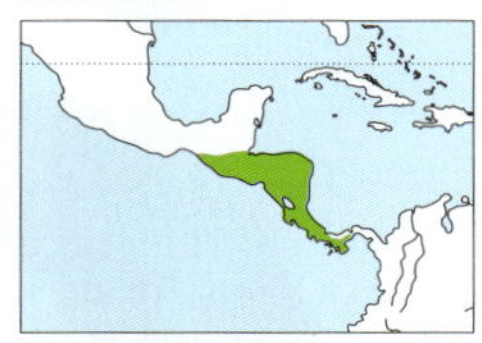

From extreme S MX (SE Chiapas) & S GT to W PA

ssp *managuensis*

ssp *dorsalis*

ssp *melania*

ssp *goldmani*

Richmond's Squirrel
Sciurus richmondi

FR Écureuil de Richmond
DE Richmond-Hörnchen
ES Ardilla de Richmond

NT
Monotypic
HB 16–22 cm T 13–18.4 cm
W 235–268 g

Rainforests of Central & Caribbean lowlands

LC
3 sspp
HB 20–32 cm
T 19.4–27 cm
W 320–540 g

Yucatan Squirrel
Sciurus yucatanensis

FR Écureuil du Yucatan
DE Yucatán-Hörnchen
ES Ardilla de Yucatán

Yucatán Pen (MX, BZ, N GT)

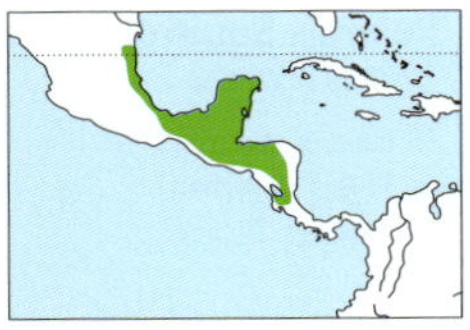

From E MX (S Tamaulipas) S to GT, BZ, SV, HN, NI & N CR

LC
5 sspp
HB c. 21 cm
T 17–18 cm
W 268–287 g

Deppe's Squirrel
Sciurus deppei
FR Écureuil de Deppe
DE Deppe-Hörnchen
ES Ardilla de Deppe

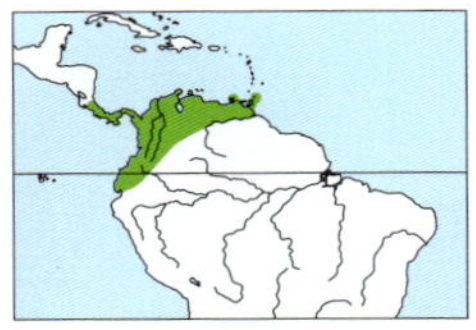

CR, PA, W & C CO, EC, W & N VE (with Margarita I), TT

Red-tailed Squirrel
Sciurus granatensis
FR Écureuil à queue rousse
DE Rotschwanzhörnchen
ES Ardilla de cola roja

LC
32 sspp
HB 20–28 cm
T 14–28 cm
W 212–520 g

2200–3500 m

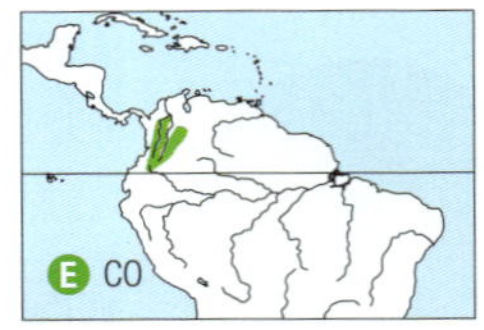

E CO

All 3 Andean ranges

Andean Squirrel
Sciurus pucheranii
FR Écureuil de Colombie
DE Anden-Hörnchen
ES Ardilla de los Andes

DD
3 sspp
HB 14–18.4 cm T 11.9–16 cm
W 100–146 g

600–2700 m

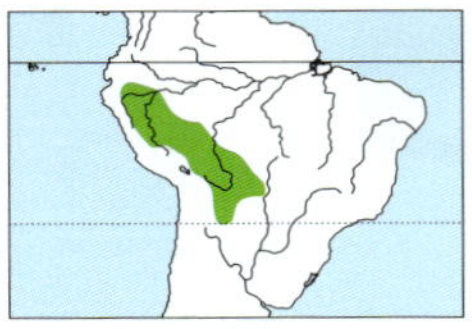

E of Andes in PE, BO, NW AR & SW BR

Bolivian Squirrel
Sciurus ignitus
FR Écureuil de Bolivie
DE Bolivien-Hörnchen
ES Ardilla de Bolivia

LC
5 sspp
HB 14–22 cm
T 15–23 cm
W 183–242 g

up to 400 m

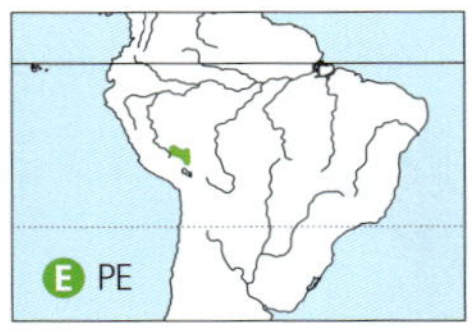

E PE

Madre de Dios

Sanborn's Squirrel
Sciurus sanborni
FR Écureuil de Sanborn
DE Sanborn-Hörnchen
ES Ardilla de Sanborn

DD Monotypic
HB 15.2–17.5 cm T 16.1–18.4 cm

LC 11 sspp
HB 16–18.6 cm
T 16.3–25 cm
W 160–380 g

SE CO, VE, Guianas, most of BR, NE AR

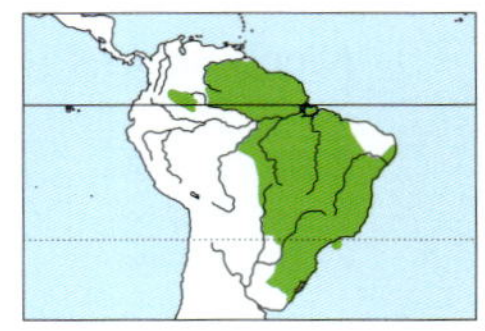

Guianan Squirrel
Sciurus aestuans
FR Écureuil de Guyane
DE Guayana-Hörnchen
ES Ardilla de Guayana

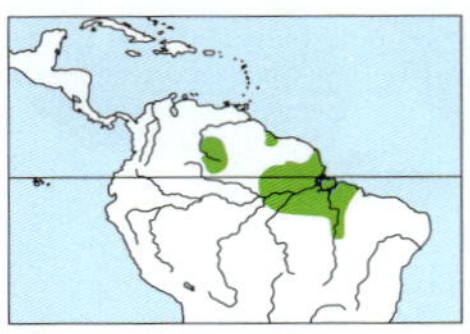

Disjunct in S VE, NE GY & N BR; range poorly known

DD
2 sspp
HB 15.5–17.7 cm
T 16.5–19.5 cm
W 150–165 g

Yellow-throated Squirrel
Sciurus gilvigularis
FR Écureuil du Brésil
DE Gelbkehlhörnchen
ES Ardilla de garganta amarilla

0–2000 m

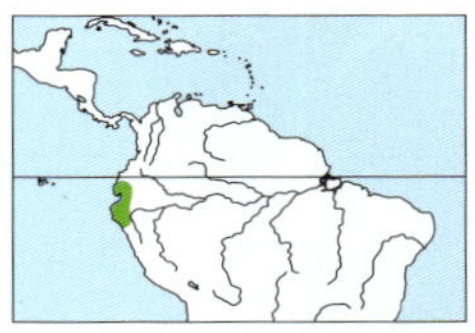

SW EC, extreme NW PE; introduced to Lima city (PE)

LC Monotypic
HB 18–32 cm
T 25–33 cm
W 460–495 g

Guayaquil Squirrel
Sciurus stramineus
FR Écureuil du Guayaquil DE Guayaquil-Hörnchen
ES Ardilla de Guayaquil

600–2500 m

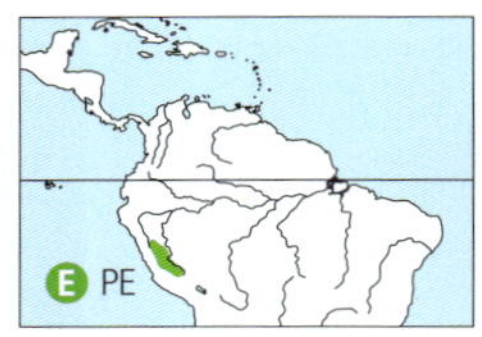

E PE

E slopes of Andes; perhaps EC & BO

DD
Monotypic
HB 24–25 cm
T 21–25 cm

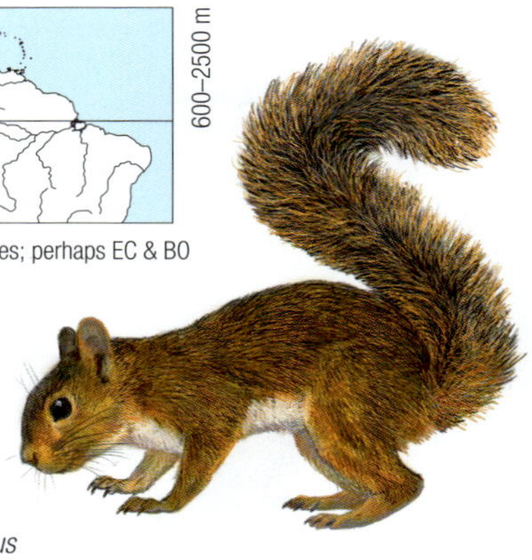

Junin Red Squirrel
Sciurus pyrrhinus
FR Écureuil du Junin DE Junín-Hörnchen ES Ardilla de Junín

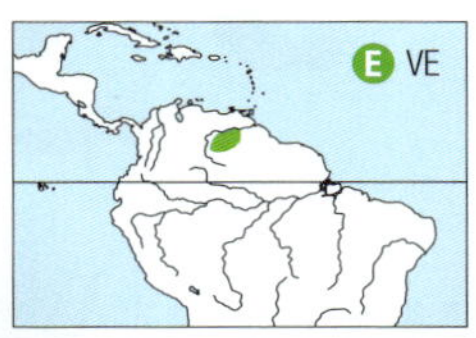

E VE

S of R Orinoco, from CO border E to Ciudad Bolívar

DD
Monotypic
HB 27–30 cm
T 24–31 cm
W 550–625 g

Fiery Squirrel
Sciurus flammifer
FR Écureuil du Venezuela
DE Venezuela-Hörnchen
ES Ardilla fogosa

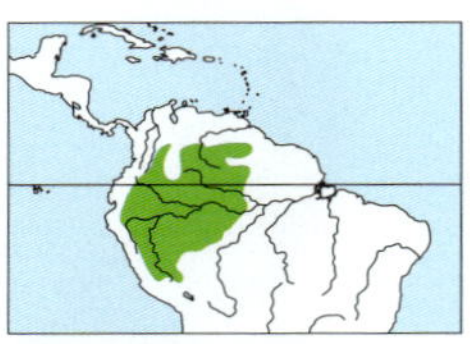

E CO, S VE & W BR S to E EC & NE PE, E of Andes; likely extreme SW GY

LC
2 sspp
HB 24–29 cm
T 24–29 cm
W 500–900 g

Northern Amazon Red Squirrel
Sciurus igniventris
FR Écureuil de Wagner
DE Nordamazonas-Rothörnchen
ES Ardilla del Amazonas septentrional

Southern Amazon Red Squirrel
Sciurus spadiceus
FR Écureuil d'Olfers
DE Südamazonas-Rothörnchen
ES Ardilla del Amazonas meridional

LC 3 sspp
HB 24–29 cm
T 23–34 cm
W 570–660 g

E of Andes in S CO, E EC, E PE, BO & W BR; limits poorly known

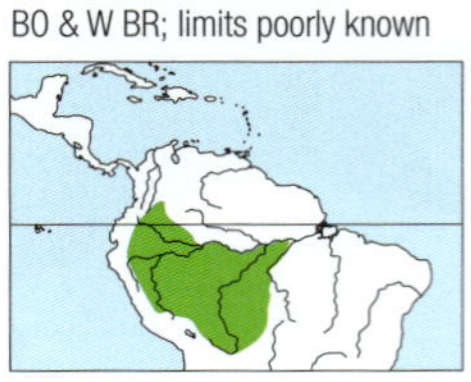

Genus *MICROSCIURUS*

(4 species)

Central American Dwarf Squirrel
Microsciurus alfari
FR Écureuil d'Alfaro
DE Mittelamerika-Zwerghörnchen
ES Ardilla enana de Centroamérica

LC
6 sspp
HB 10.8–14.6 cm
T 8–13 cm
W 72–105 g

up to 2600 m

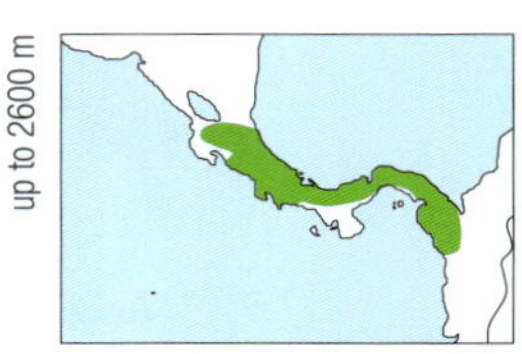

CR, PA, adjacent W CO; marginally in S NI

PA, W CO, NW EC

LC
3 sspp
HB 13.5–14.8 cm T 9.4–11.6 cm
W c. 120 g

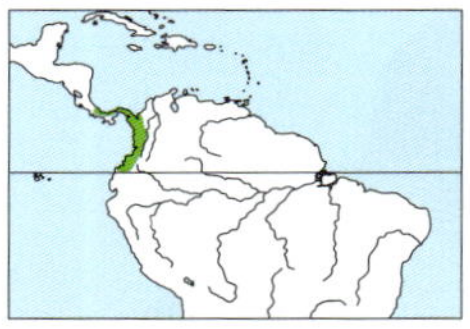

Western Dwarf Squirrel
Microsciurus mimulus
FR Écureuil d'Esmeraldas
DE Westliches Zwerghörnchen
ES Ardilla enana occidental

Santander Dwarf Squirrel
Microsciurus santanderensis
FR Écureuil du Santander DE Santander-Zwerghörnchen
ES Ardilla enana de Santander

100–3800 m

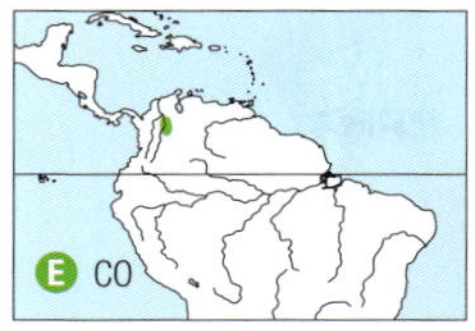

Between R Magdalena and E Andes, W to upper W Andes (Santander)

DD
Monotypic
HB 13.3–16 cm
T 13.6–15.2 cm

below 2000 m

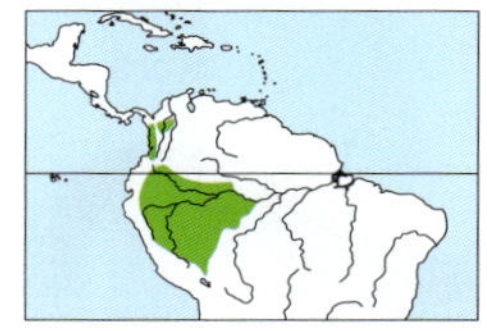

CO, W BR, EC, PE

LC
8 sspp
HB 12–16 cm
T 9.6–15 cm
W 60–128 g

Amazon Dwarf Squirrel
Microsciurus flaviventer
FR Écureuil d'Amazonie
DE Amazonas-Zwerghörnchen
ES Ardilla enana del Amazonas

Genus *SYNTHEOSCIURUS*

(1 species)

1900–2300 m

Montane forests of CR & W PA

DD
Monotypic
HB 15–18.5 cm
T 12–15.2 cm
W 143–170 g

Bangs's Mountain Squirrel
Syntheosciurus brochus
FR Écureuil du Panama
DE Mittelamerika-Berghörnchen
ES Ardilla de montaña de Bangs

Genus *RHEITHROSCIURUS*

(1 species)

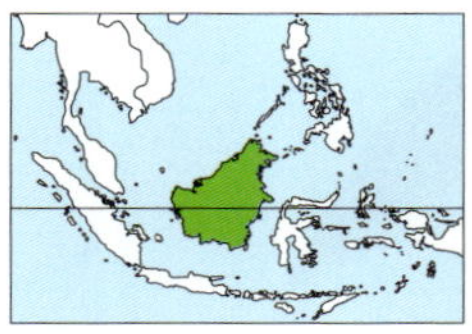

below 1100 m

Borneo (MY, ID, BN)

Tufted Ground Squirrel
Rheithrosciurus macrotis
FR Écureuil à pinceaux
DE Borneo-Hörnchen
ES Ardilla terrestre de penachos

VU
Monotypic
HB 33–35 cm
T 30–34 cm
W 1.2–1.3 kg

Genus *EOGLAUCOMYS*

(1 species)

LC
2 sspp
HB 23–30 cm
T 25–33 cm
W 300–734 g

900–4000 m

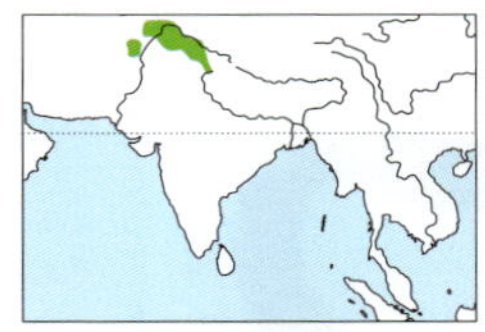

NE AF, N PK, NW IN (Jammu and Kashmir E to C Uttarakhand)

Kashmir Flying Squirrel
Eoglaucomys fimbriatus
FR Polatouche du Cachemire
DE Kaschmir-Gleithörnchen
ES Ardilla voladora de Cachemira

1200–3000 m

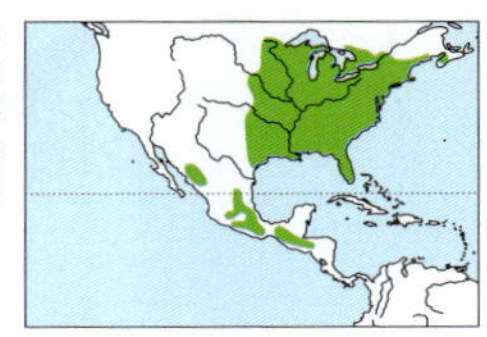

SE CA (S Ontario E to Nova Scotia), E half of US, mts of MX, S GT & HN

Southern Flying Squirrel
Glaucomys volans
FR Petit Polatouche
DE Südliches Gleithörnchen
ES Ardilla voladora meridional

LC 11 sspp
HB 11.7–13.8 cm T 8–12 cm
W 42–141 g

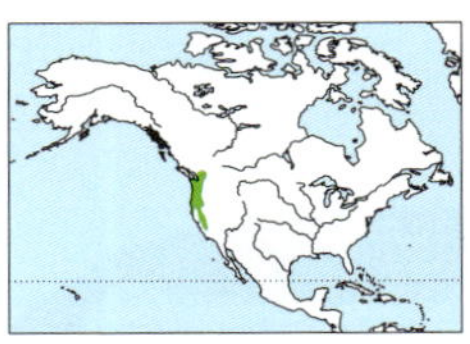

SW CA (SW British Columbia; perhaps further N), US (W Washington S to S California)

Humboldt's Flying Squirrel
Glaucomys oregonensis
FR Polatouche d'Oregon
DE Oregon-Gleithörnchen
ES Ardilla voladora de Oregón

NE
Monotypic
HB c. 16.9 cm
T c. 14.3 cm

Genus *GLAUCOMYS*

(3 species)

LC
18 sspp
HB 11.5–19 cm
T 12.2–16 cm
W 110–185 g

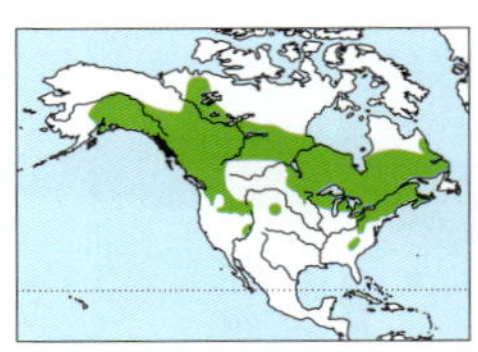

Northern Flying Squirrel
Glaucomys sabrinus
FR Grand Polatouche
DE Nördliches Gleithörnchen
ES Ardilla voladora septentrional

Alaska, most of CA, disjunct in N US (reaching S Utah, South Dakota, E Tennessee)

LC
4 sspp
HB 16.5–23 cm
T 15.9–21 cm
W 120–231 g

up to 1800 m

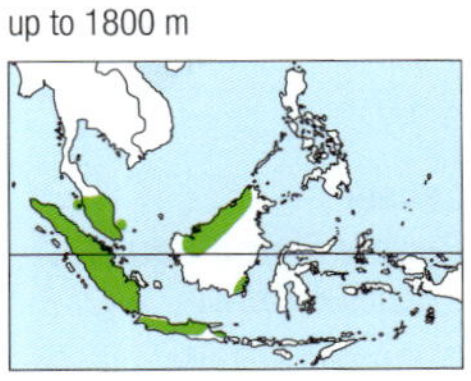

Thai-Malay Pen (with Tioman I & Penang I), Sumatra, Java, Borneo

Javanese Flying Squirrel
Iomys horsfieldii
FR Polatouche de Horsfield
DE Horsfield-Gleithörnchen
ES Ardilla voladora de Java

Genus *IOMYS*

(2 species)

EN
Monotypic
HB 17.3–20 cm
T 17.5–19.2 cm

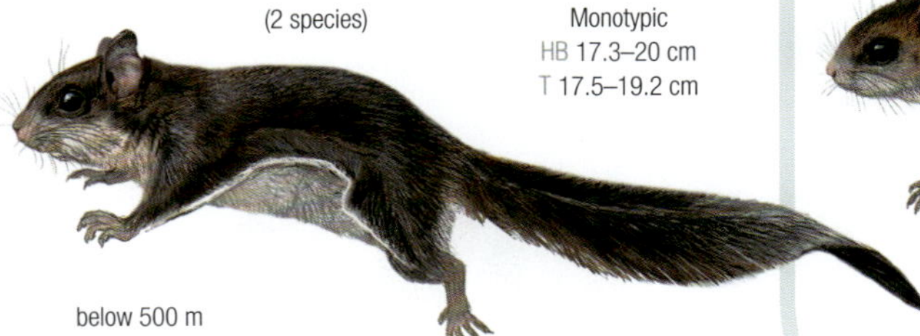

below 500 m

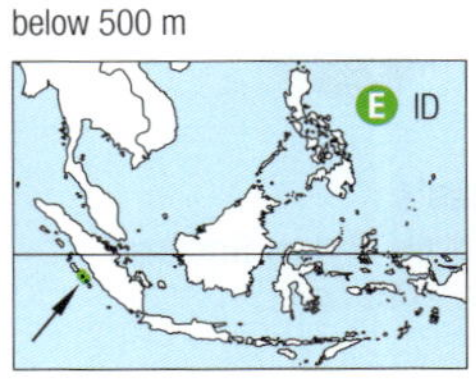

Mentawai Arch (Sipora, North Pagai), off W Sumatra

Mentawai Flying Squirrel
Iomys sipora
FR Polatouche des Mentawai
DE Mentawi-Gleithörnchen
ES Ardilla voladora de Mentawai

Genus *PRIAPOMYS*

(1 species)

NE
Monotypic
HB 18–21 cm
T 12.5–21 cm
W 150–204 g

1400–2440 m

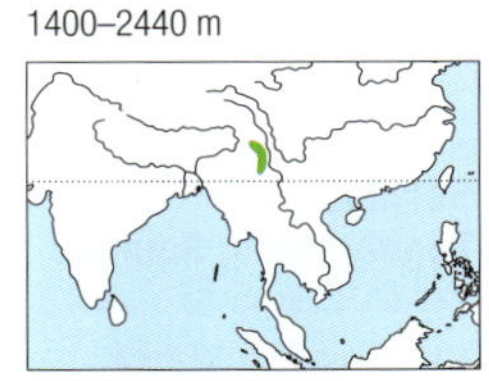

N MM (Kachin), S CN (Mt Gaoligong & upper R Mekong Valley, Yunnan)

Himalayan Large-eared Flying Squirrel
Priapomys leonardi
FR Polatouche de Leonard
DE Himalaya-Großohrhörnchen
ES Ardilla voladora de Yunnan

Genus *HYLOPETES*

(10 species)

c. 1200 m

DD

Monotypic HB c. 14.2 cm
T c. 14.3 cm

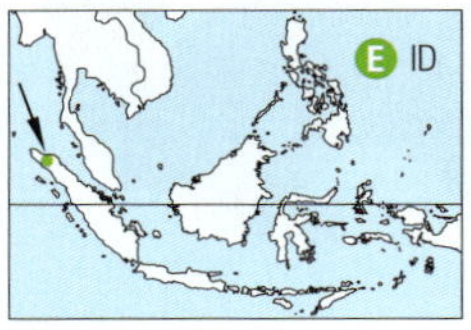

N Sumatra (Aceh)

Sumatran Flying Squirrel
Hylopetes winstoni

FR Polatouche de Winston
DE Sumatra-Gleithörnchen
ES Ardilla voladora de Sumatra

NT 2 sspp
HB 25–33 cm T 25–31 cm
W c. 534 g

Palawan & Bancalan

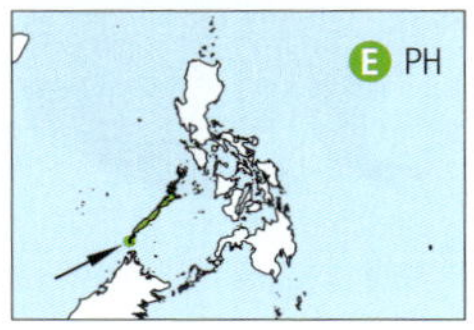

Palawan Flying Squirrel
Hylopetes nigripes

FR Polatouche de Palawan
DE Palawan-Gleithörnchen
ES Ardilla voladora de Palawan

Red-cheeked Flying Squirrel
Hylopetes spadiceus

FR Polatouche à joues rouges
DE Rotwangen-Gleithörnchen
ES Ardilla voladora de mejillas rojas

LC
3 sspp
HB 13.5–18.4 cm
T 10.2–16.6 cm
W 70–157 g

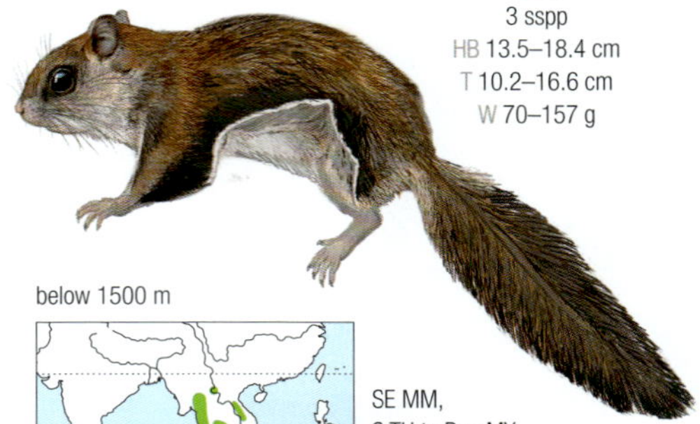

below 1500 m

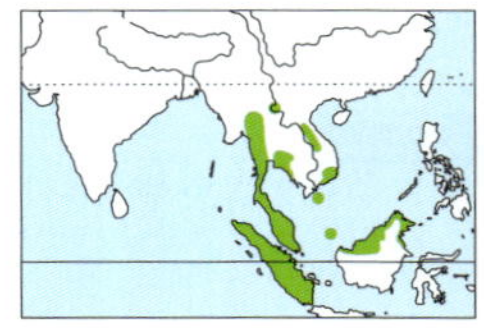

SE MM,
S TH to Pen MY,
NW & C LA, W KH,
S VN, Sumatra,
Borneo, Natuna Is

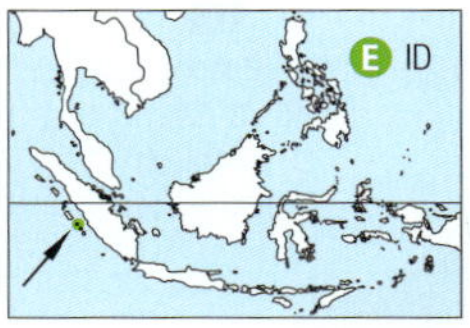

Sipora Flying Squirrel
Hylopetes sipora

FR Polatouche de Sipora
DE Sipora-Gleithörnchen
ES Ardilla voladora de Sipora

Mentawai Arch (Sipora), off W Sumatra

EN
Monotypic
HB c. 14 cm
T c. 16 cm
W c. 89 g

DD
2 sspp
HB 11.5–12.7 cm
T 9.1–12.3 cm
W 39–43 g

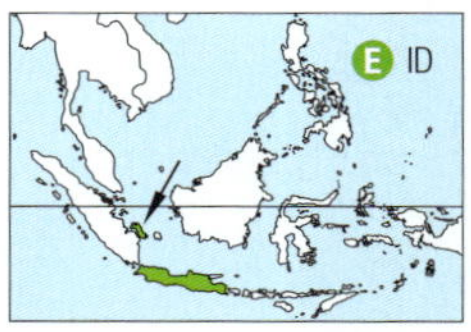

Bangka I (off NE Sumatra), Java

Gray-cheeked Flying Squirrel
Hylopetes sagitta

FR Polatouche à joues grises
DE Grauwangen-Gleithörnchen
ES Ardilla voladora de mejillas grises

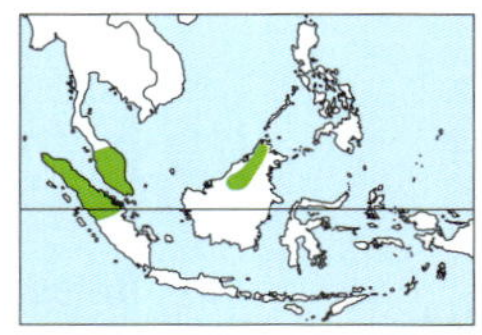

Thai-Malay Pen, N Sumatra, N Borneo

Jentink's Flying Squirrel
Hylopetes platyurus

FR Polatouche de Jentink
DE Jentink-Gleithörnchen
ES Ardilla voladora de Jentink

DD
Monotypic
HB 11–15.3 cm
T 10–13.6 cm
W 67–99 g

DD
Monotypic
HB 13.3–19.7 cm
T 11.9–12 cm

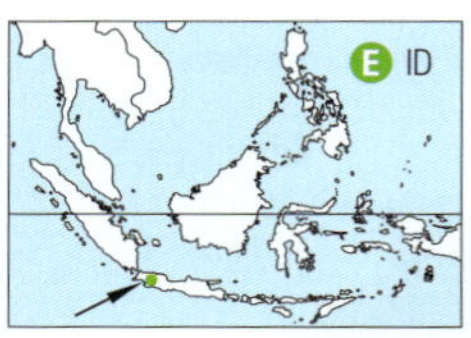

Known only from Mt Pangrango, W Java

Bartels's Flying Squirrel
Hylopetes bartelsi

FR Polatouche de Bartels
DE Bartels-Gleithörnchen
ES Ardilla voladora de Bartels

NE
Monotypic
HB c. 17.2 cm
T 15.9 cm

Hainan I

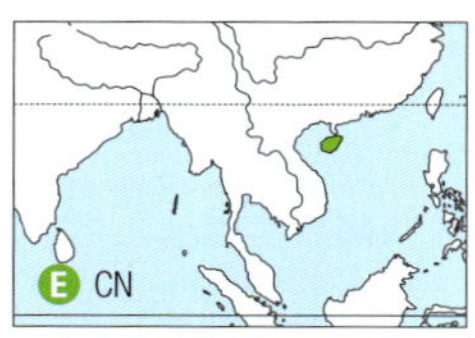

Hainan Flying Squirrel
Hylopetes electilis

FR Polatouche de Hainan
DE Hainan-Gleithörnchen
ES Ardilla voladora de Hainan

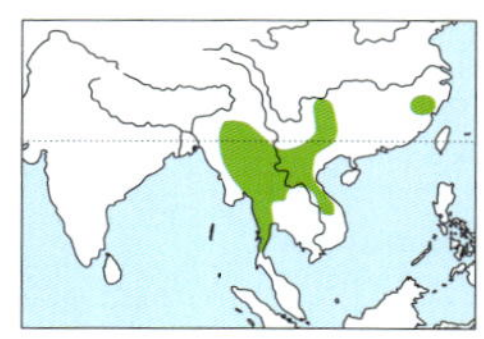

NE IN (E Nagaland, Manipur), SE CN (Guizhou, Guangxi, Fujian), MM, TH, LA, VN

LC
Monotypic
HB 14.4–19.7 cm
T 12.8–17.4 cm
W 113–189 g

Indochinese Flying Squirrel
Hylopetes phayrei

FR Polatouche de Phayre
DE Phayre-Gleithörnchen
ES Ardilla voladora de Indochina

1500–4500 m

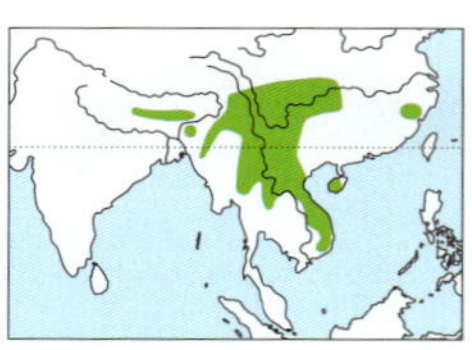

Particolored Flying Squirrel
Hylopetes alboniger

FR Polatouche bicolore
DE Schwarzweiß-Gleithörnchen
ES Ardilla voladora bicolor

E NP, NE IN, BT, S CN (E to Zhejiang, Hainan I), MM, N TH, LA, NE KH, VN

LC 3 sspp
HB 17.5–25 cm
T 18–23 cm
W 236–270 g

Genus *PETAURILLUS*

(3 species)

DD
Monotypic
HB 8.5–9.5 cm T 8–9.8 cm
W 17–31 g

Peninsula (Selangor)

Selangor Pygmy Flying Squirrel
Petaurillus kinlochii

FR Polatouche du Selangor
DE Selangor-Kleinstgleithörnchen
ES Ardilla voladora pigmea de Selangor

DD
Monotypic
HB 7.5–9 cm T 8–9.8 cm
W c. 21 g

MY/BN: N Borneo (Sepilok, Tasek Merimbun and Sarawak S to Niah Caves

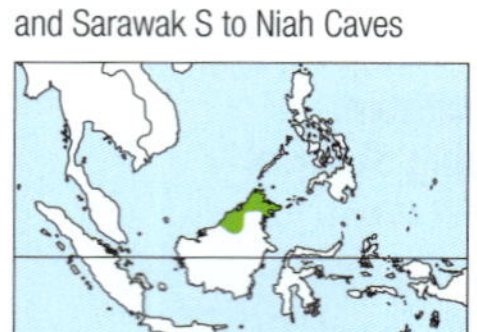

Hose's Pygmy Flying Squirrel
Petaurillus hosei

FR Polatouche de Hose
DE Hose-Kleinstgleithörnchen
ES Ardilla voladora pigmea de Hose

DD
Monotypic
HB 6.8–7.2 cm
T 6.2–6.7 cm
W c. 13.5 g

Known only from Baram (E Sarawak), NW Borneo

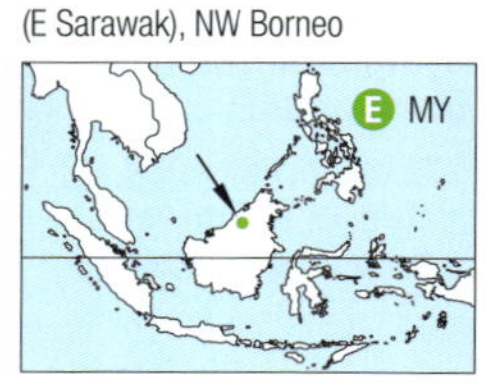

Lesser Pygmy Flying Squirrel
Petaurillus emiliae

FR Polatouche d'Emilia
DE Emilia-Kleinstgleithörnchen
ES Ardilla voladora pigmea menor

Genus *PETINOMYS*

(8 species)

VU
Monotypic
HB 9.2–12.8 cm
T 8.5–11.8 cm
W 26–58 g

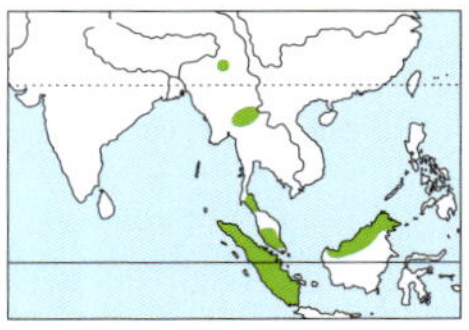

Temminck's Flying Squirrel
Petinomys setosus

FR Polatouche à ventre blanc
DE Temminck-Gleithörnchen
ES Ardilla voladora de Temminck

N & E MM, NW TH, Thai-Malay Pen, Sumatra, N third of Borneo

LC
Monotypic
HB 18.7–34 cm
T 24–29 cm
W 300–712 g

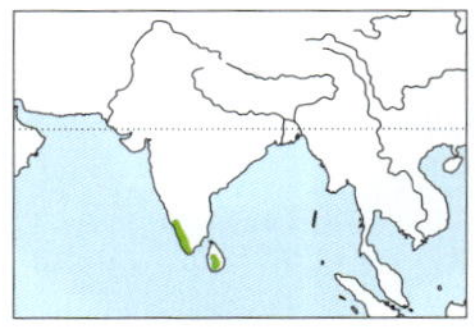

Travancore Flying Squirrel
Petinomys fuscocapillus

FR Polatouche du Travancore
DE Travancore-Gleithörnchen
ES Ardilla voladora de Travancore

S IN (Western Ghats), C & S LK

LC
Monotypic
HB c. 31 cm
T c. 26 cm
W c. 1.1 kg

Dinagat, Siargao, Mindanao, Basilan

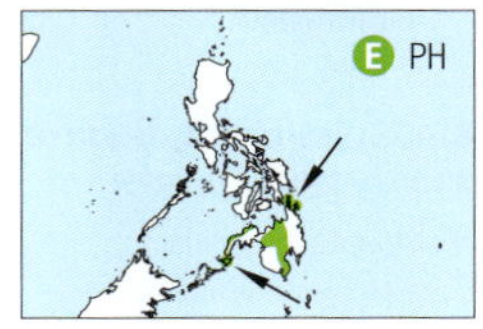

Basilan Flying Squirrel
Petinomys crinitus

FR Polatouche de Basilan
DE Basilan-Gleithörnchen
ES Ardilla voladora de Basilán

Mindanao Flying Squirrel
Petinomys mindanensis

FR Polatouche de Mindanao
DE Mindanao-Gleithörnchen
ES Ardilla voladora de Mindanao

LC
Monotypic
HB 32–37 cm
T 34–46 cm

500–1600 m

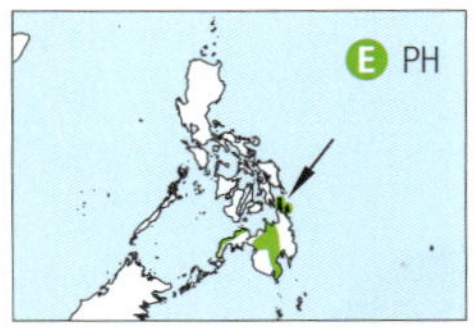

Dinagat, Siargao, Mindanao

VU
Monotypic
HB 14.2–19.5 cm
T 15.5–19.2 cm
W 52–110 g

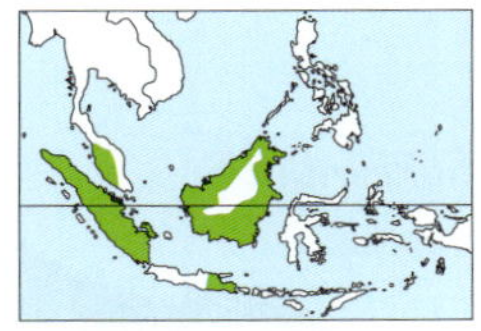

Pen MY, Sumatra, Bangka I, Borneo, E Java

Whiskered Flying Squirrel
Petinomys genibarbis

FR Polatouche à moustaches
DE Schnurrbart-Gleithörnchen
ES Ardilla voladora bigotuda

VU
Monotypic
HB 23–28 cm
T 21–23 cm
W c. 433 g

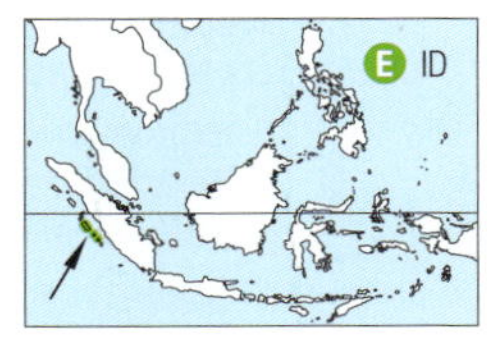

Mentawai Arch (Siberut, Sipora, North Pagai), off W Sumatra

Siberut Flying Squirrel
Petinomys lugens

FR Polatouche sombre
DE Siberut-Gleithörnchen
ES Ardilla voladora de Siberut

VU Monotypic
HB 9.2–12 cm
T 8.9–12 cm
W 22–52 g

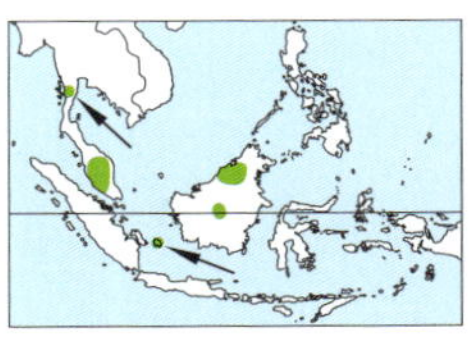

Extreme S MM, Pen MY, Borneo, Belitung I (off SE Sumatra)

Vordermann's Flying Squirrel
Petinomys vordermanni

FR Polatouche de Vordermann
DE Vordermann-Gleithörnchen
ES Ardilla voladora de Vordermann

DD
Monotypic
HB 23–28 cm T 23–25 cm
W c. 388 g

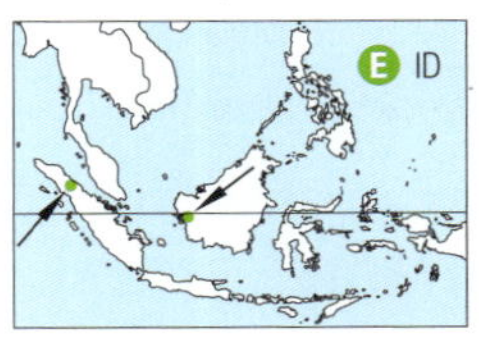

N Sumatra, SW Borneo (Kalimantan)

Hagen's Flying Squirrel
Petinomys hageni

FR Polatouche de Hagen
DE Hagen-Gleithörnchen
ES Ardilla voladora de Hagen

Genus *PTEROMYS*

(2 species)

Siberian Flying Squirrel
Pteromys volans

FR Polatouche de Sibérie
DE Europäisches Gleithörnchen
ES Ardilla voladora de Siberia

LC
4 sspp
HB 12–23 cm
T 9–14.9 cm
W 95–200 g

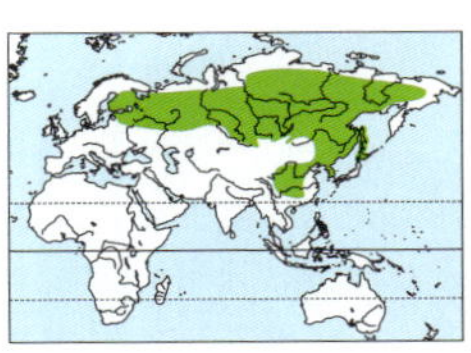

N Europe E to Russian Far East, N MN, N & C CN, Korean Pen, JP (Hokkaido)

LC Monotypic
HB 14–20 cm T 9.5–14 cm
W c. 152 g

Honshu, Shikoku, Kyushu

Japanese Flying Squirrel
Pteromys momonga

FR Polatouche du Japon DE Japan-Gleithörnchen ES Ardilla voladora de Japón

Genus *AEROMYS*

(2 species)

DD
2 sspp
HB 37–38 cm
T 39–44 cm
W 1.1–1.2 kg

Black Flying Squirrel
Aeromys tephromelas

FR Polatouche noir
DE Schwarzes Gleithörnchen
ES Ardilla voladora negra

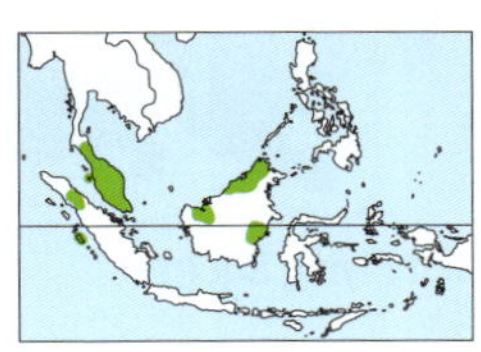

Pen MY (with Penang I), N Sumatra (with Siberut I), scattered in N, W & E Borneo; perhaps N TH

LC Monotypic
HB 30–36 cm T 37–41 cm
W 1.1–1.4 kg

Thomas's Flying Squirrel
Aeromys thomasi

FR Polatouche de Thomas
DE Thomas-Gleithörnchen
ES Ardilla voladora de Thomas

up to 1600 m

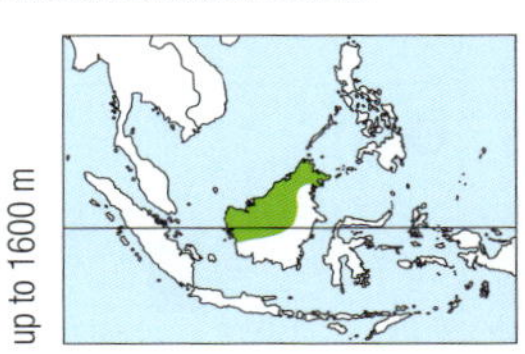

Borneo, except S (ID, MY, BN)

Genus *BISWAMOYOPTERUS*

(3 species)

CR
Monotypic
HB c. 40 cm
T c. 60 cm

100–350 m

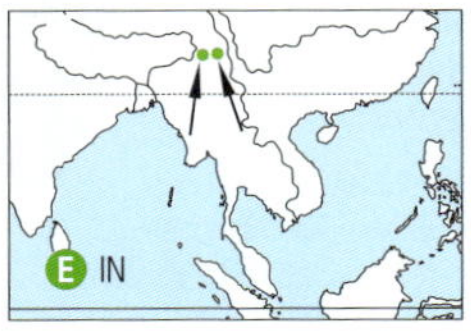

Namdapha Flying Squirrel
Biswamoyopterus biswasi

FR Polatouche de Namdapha
DE Biswas-Riesengleithörnchen
ES Ardilla voladora de Namdapha

Known only from Namdapha NP (Arunachal Pradesh); perhaps N MM

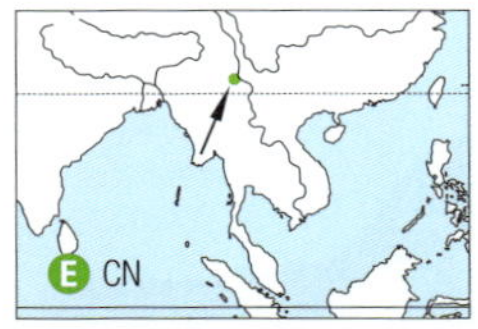

Known only from Mt Gaoligong (Yunnan)

NE
Monotypic
HB c. 44 cm
T c. 52 cm
W c. 1.4 kg

Mount Gaoligong Flying Squirrel
Biswamoyopterus gaoligongensis

FR Polatouche du Gaoligong DE Gaoligong-Riesengleithörnchen
ES Ardilla voladora de Gaoligong

DD
Monotypic
HB c. 46 cm
T c. 62 cm
W c. 1.8 kg

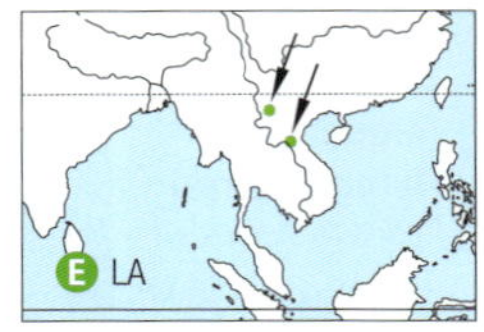

Laotian Flying Squirrel
Biswamoyopterus laoensis

FR Polatouche du Laos
DE Laos-Riesengleithörnchen
ES Ardilla voladora de Laos

Range unknown; 2 specimens collected in markets in N & C

Genus *EUPETAURUS*

(3 species)

EN
Monotypic
HB 42–61 cm
T 38–54 cm
W 1.4–2.5 kg

2400–4800 m

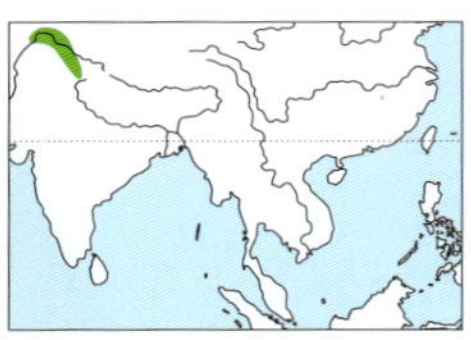

Western Woolly Flying Squirrel
Eupetaurus cinereus

FR Polatouche laineux
DE Westliches Felsgleithörnchen
ES Ardilla voladora lanuda occidental

W Himalayas in N PK & NW IN (Kashmir); possibly W NP

NE
Monotypic
HB 41–44 cm
T 40–44 cm
W 1.4–1.9 kg

3400–4450 m

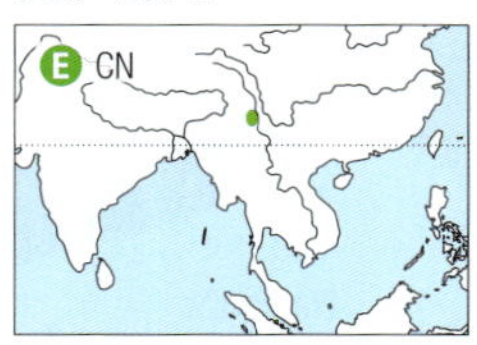

Yunnan Woolly Flying Squirrel
Eupetaurus nivamons

FR Polatouche du Yunnan
DE Yunnan-Felsgleithörnchen
ES Ardilla voladora lanuda de Yunnan

Mt Gaoligong & Biluo Snow Mt; perhaps adjacent MM

NE
Monotypic
HB c. 47 cm

2700–4000 m

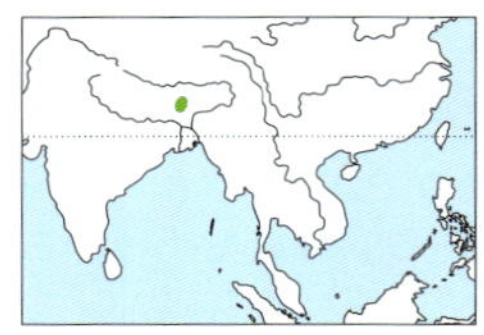

Tibetan Woolly Flying Squirrel
Eupetaurus tibetensis

FR Polatouche du Tibet
DE Tibet-Felsgleithörnchen
ES Ardilla voladora lanuda de Tibet

SC Tibet (CN), N slopes of Himalayas in Sikkim (IN) & BT

Genus *AERETES*

(1 species)

NT
2 sspp
HB 27–35 cm
T 27–36 cm

2500–3000 m

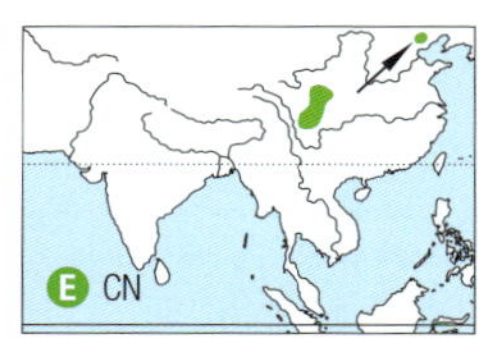

2 isolated populations: Gansu & Sichuan and Hebei

Northern Chinese Flying Squirrel
Aeretes melanopterus

FR Polatouche de Chine DE Furchenzahn-Gleithörnchen
ES Ardilla voladora de China septentrional

Genus *PTEROMYSCUS*

(1 species)

below 3000 m

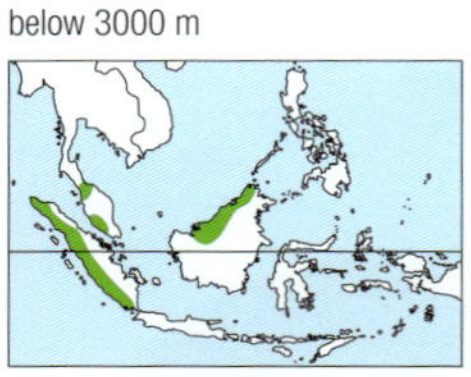

Smoky Flying Squirrel
Pteromyscus pulverulentus

FR Polatouche fuligineux
DE Rauchgraues Gleithörnchen
ES Ardilla voladora ahumada

S TH, Pen MY, W Sumatra, NW Borneo (MY, ID, BN)

EN 2 sspp
HB 20–29 cm T 17.7–23 cm
W 134–315 g

Genus *BELOMYS*

(1 species)

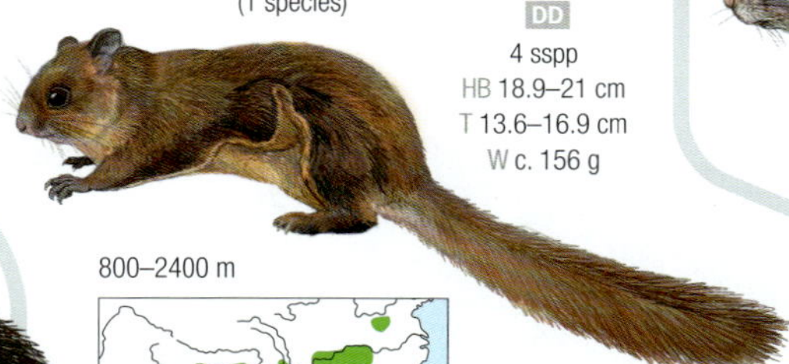

DD
4 sspp
HB 18.9–21 cm
T 13.6–16.9 cm
W c. 156 g

800–2400 m

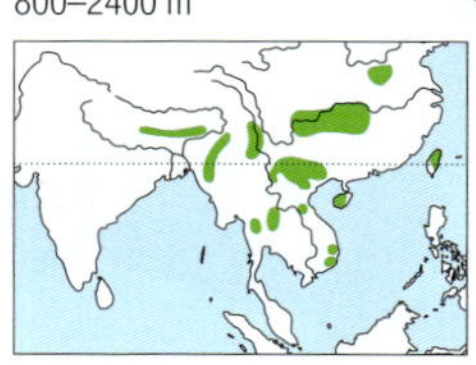

Hairy-footed Flying Squirrel
Belomys pearsonii

FR Polatouche à pieds velus
DE Haarfuß-Gleithörnchen
ES Ardilla voladora de patas peludas

Patchily in NP, NE IN, BT, CN (with Hainan I), TW, N MM, TH, LA, VN

Genus *TROGOPTERUS*

(1 species)

From E Qinghai to Hebei, S to N Yunnan & Guizhou, isolated in S Liaoning, E Henan, S Qinghai & S Tibet

1360–2750 m

NT
Monotypic
HB 20–33 cm
T 26–30 cm

Complex-toothed Flying Squirrel
Trogopterus xanthipes

FR Polatouche à pattes jaunes
DE Komplexzahn-Gleithörnchen
ES Ardilla voladora de dientes con crestas

Genus *PETAURISTA*

(19 species)

Honshu, Shikoku, Kyushu

Japanese Giant Flying Squirrel
Petaurista leucogenys

FR Pétauriste à joues blanches
DE Japan-Riesengleithörnchen
ES Ardilla voladora gigante de Japón

LC
3 sspp
HB 27–48 cm
T 28–41 cm
W 0.49–1.2 kg

LC
Monotypic
HB 32–44 cm
T 29–38 cm
W 0.73–1.2 kg

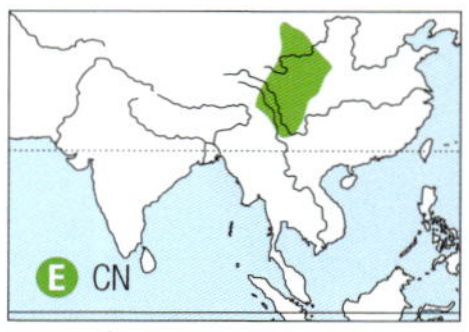

2000–3400 m

Qinghai, E Tibet, Gansu, Sichuan, NW Yunnan

Chinese Giant Flying Squirrel
Petaurista xanthotis
FR Pétauriste de Chine DE China-Riesengleithörnchen
ES Ardilla voladora gigante de China

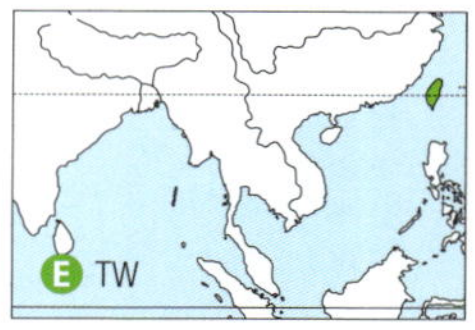

1200–3750 m

Restricted to hardwood and coniferous forests

NE
Monotypic
HB 35–51 cm
T 36–52 cm
W 0.88–1.9 kg

Taiwan Giant Flying Squirrel
Petaurista lena
FR Pétauriste de Taïwan
DE Taiwan-Riesengleithörnchen
ES Ardilla voladora gigante de Taiwán

NE
Monotypic
HB 30–37 cm
T 36–40 cm

Gray-headed Giant Flying Squirrel
Petaurista caniceps
FR Pétauriste à tête grise
DE Graukopf-Riesengleithörnchen
ES Ardilla voladora gigante de cabeza gris

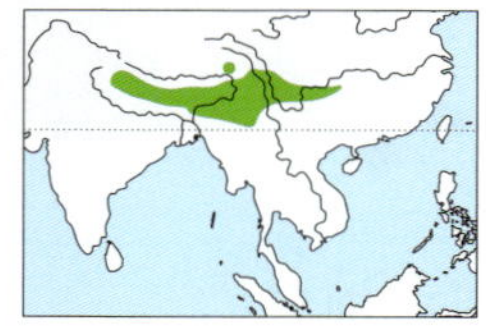

NP, NE IN (N West Bengal E to Arunachal Pradesh & Nagaland), BT, S CN (S Tibet, Sichuan, Guizhou, Yunnan)

Formosan Giant Flying Squirrel
Petaurista grandis
FR Pétauriste de Swinhoe
DE Formosa-Riesengleithörnchen
ES Ardilla voladora gigante de Formosa

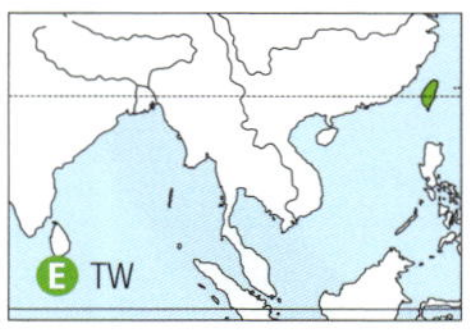

NE
Monotypic

Hardwood forests

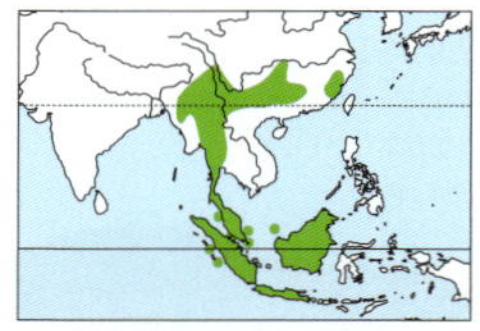

SE CN & MM through Thai-Malay Pen to Sumatra, Java, Borneo & offshore Is

LC
15 sspp
HB 28–52 cm
T 34–63 cm
W 0.99–2.9 kg

Red Giant Flying Squirrel
Petaurista petaurista
FR Pétauriste roux DE Rotes Riesengleithörnchen
ES Ardilla voladora gigante roja

Spotted Giant Flying Squirrel
Petaurista elegans
FR Pétauriste tacheté DE Geflecktes Riesengleithörnchen
ES Ardilla voladora gigante moteada

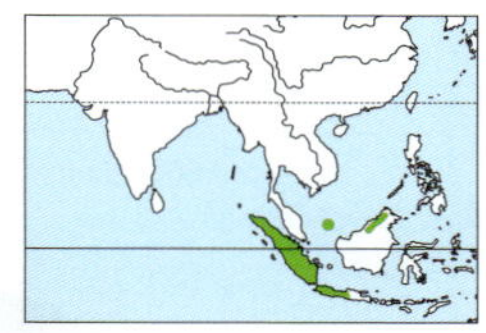

Sumatra, Java (with Kembang I), N Borneo Mts, Natuna Is

LC
3 sspp

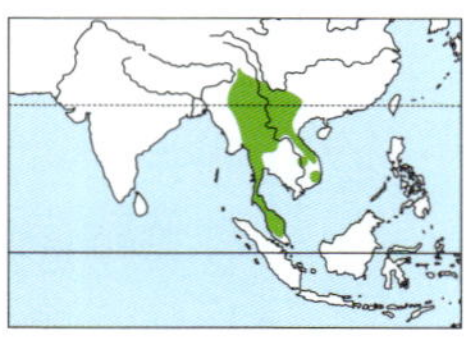

200–4000 m

NE IN, N & E MM, S CN (Yunnan, Guangxi), LA, VN, TH, Pen MY

NE
Monotypic
HB c. 36 cm
T c. 38 cm

Indochinese Giant Flying Squirrel
Petaurista marica
FR Pétauriste marica
DE Indochina-Riesengleithörnchen
ES Ardilla voladora gigante de Indochina

White-bellied Giant Flying Squirrel
Petaurista albiventer
FR Pétauriste à ventre blanc
DE Weißbauch-Riesengleithörnchen
ES Ardilla voladora gigante de vientre blanco

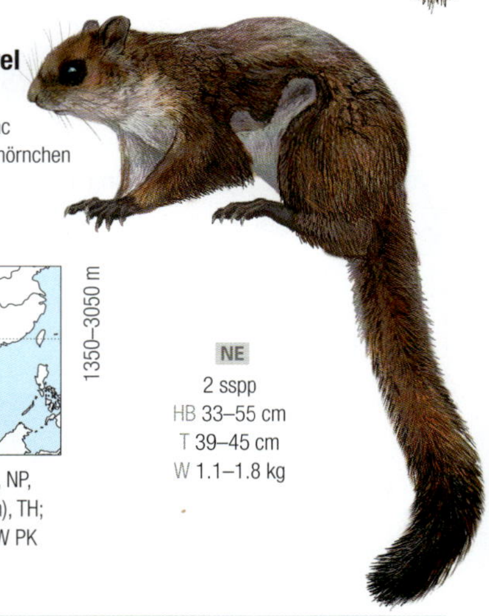

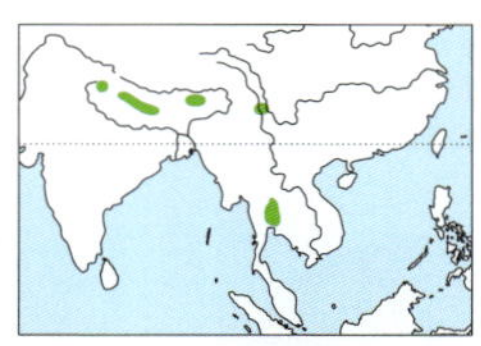

1350–3050 m

N IN (Punjab, Uttarakhand), NP, S CN (SE Tibet, NW Yunnan), TH; unconfirmed in NE AF & NW PK

NE
2 sspp
HB 33–55 cm
T 39–45 cm
W 1.1–1.8 kg

Chindwin Giant Flying Squirrel
Petaurista sybilla
FR Pétauriste de Sybil DE Chindwin-Riesengleithörnchen
ES Ardilla voladora gigante de Chindwin

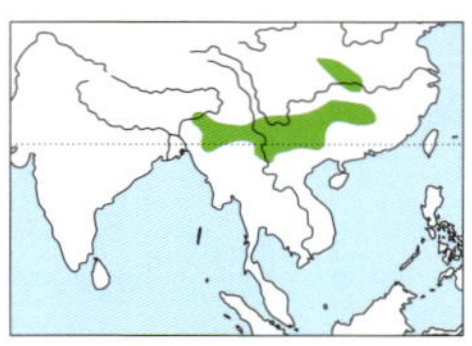

NE IN (S Assam, Meghalaya, Manipur), N MM, C & S CN (E to Hubei & Hunan)

NE
Monotypic
HB c. 34 cm
T c. 34 cm

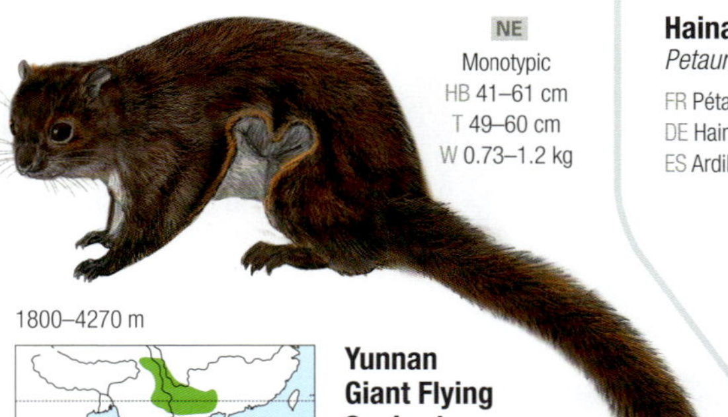

NE
Monotypic
HB 41–61 cm
T 49–60 cm
W 0.73–1.2 kg

1800–4270 m

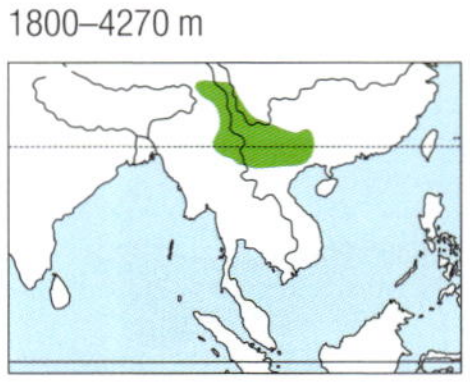

CN (Tibet, Sichuan, Yunnan, Guangxi) extending to N MM, N LA & N VN

Yunnan Giant Flying Squirrel
Petaurista yunanensis
FR Pétauriste du Yunnan
DE Yunnan-Riesengleithörnchen
ES Ardilla voladora gigante de Yunnan

Hainan Giant Flying Squirrel
Petaurista hainana
FR Pétauriste de Hainan
DE Hainan-Riesengleithörnchen
ES Ardilla voladora gigante de Hainan

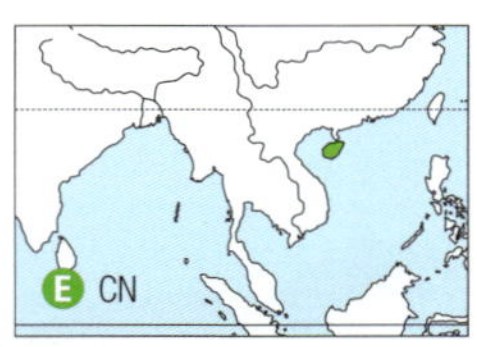

Hainan I

NE
Monotypic
HB 40–54 cm
T 45–59 cm

LC
5 sspp
HB 35–58 cm
T 40–61 cm
W 1.2–1.9 kg

800–3500 m

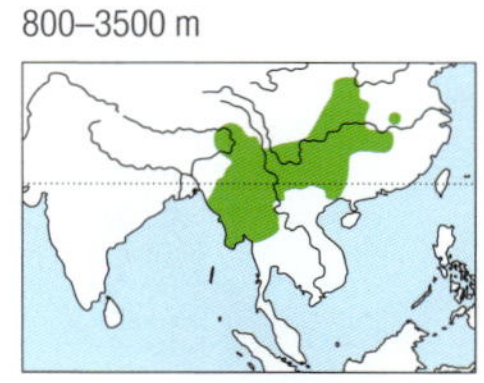

NE IN, C CN (N to Shaanxi, E to Hubei), MM, N TH

Red-and-white Giant Flying Squirrel
Petaurista alborufus
FR Pétauriste blanc et roux
DE Rotweißes Riesengleithörnchen
ES Ardilla voladora gigante roja y blanca

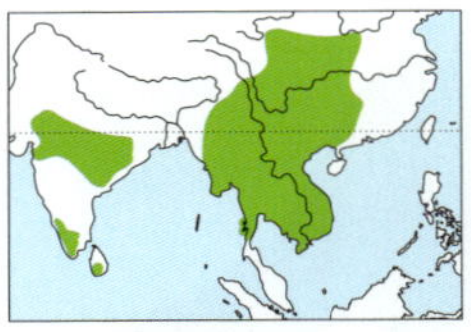

Fragmented in C, S & NE IN, LK and CN (S from E Gansu & Shaanxi) to most of mainland SE Asia, Mergui Arch

Indian Giant Flying Squirrel

Petaurista philippensis

FR Pétauriste indien
DE Indien-Riesengleithörnchen
ES Ardilla voladora gigante de India

LC
5 sspp
HB 38–61 cm
T 48–69 cm
W 1.3–2.3 kg

c. 3000 m

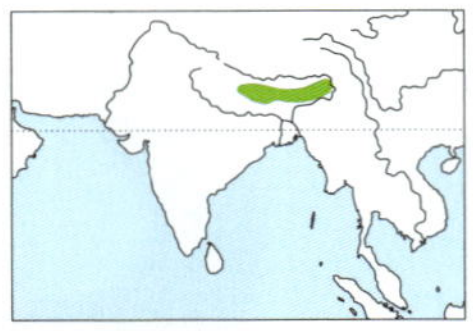

NP, NE IN (Sikkim, Arunachal Pradesh), BT, CN (S Tibet)

LC
Monotypic
HB 36–42 cm
T 41–55 cm
W 1.3–1.8 kg

Hodgson's Giant Flying Squirrel

Petaurista magnificus

FR Pétauriste magnifique
DE Hodgson-Riesengleithörnchen
ES Ardilla voladora gigante de Hodgson

Bhutan Giant Flying Squirrel

Petaurista nobilis

FR Pétauriste noble
DE Bhutan-Riesengleithörnchen
ES Ardilla voladora gigante de Bután

NT
2 sspp
HB 35–49 cm
T 38–49 cm
W 2.3–3 kg

1500–3000 m

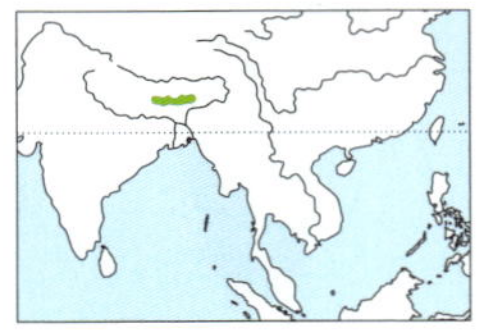

NP, NE India (Sikkim, N West Bengal), BT

Mechuka Giant Flying Squirrel

Petaurista mechukaensis

FR Pétauriste de Mechuka DE Mechuka-Riesengleithörnchen
ES Ardilla voladora gigante de Mechuka

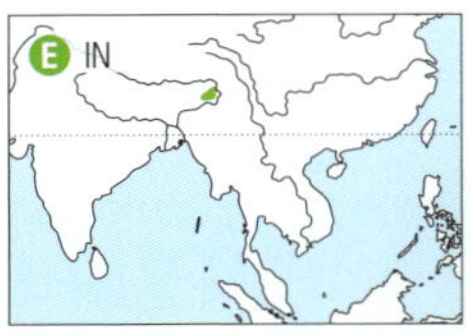

Mechuka Valley (Arunachal Pradesh)

1800–2000 m

DD Monotypic
HB 46–53 cm
T 52–77 cm

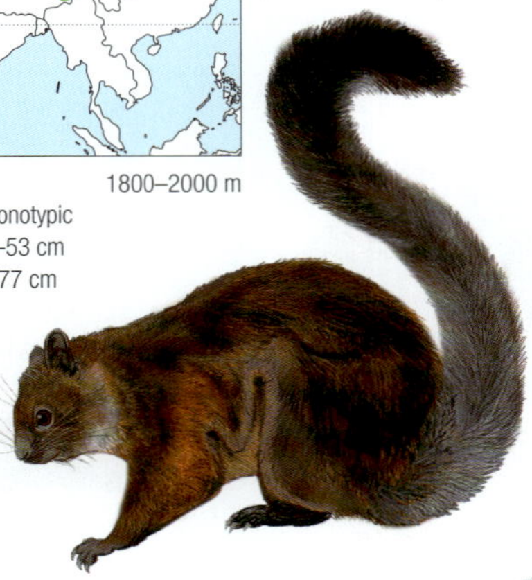

NT
Monotypic
HB 40–59 cm
T 57–60 cm

Mishmi Hills (Upper Dibang Valley & Anjaw districts, Arunachal Pradesh)

Mishmi Giant Flying Squirrel

Petaurista mishmiensis

FR Pétauriste des Mishmi
DE Mishmi-Riesengleithörnchen
ES Ardilla voladora gigante de Mishmi

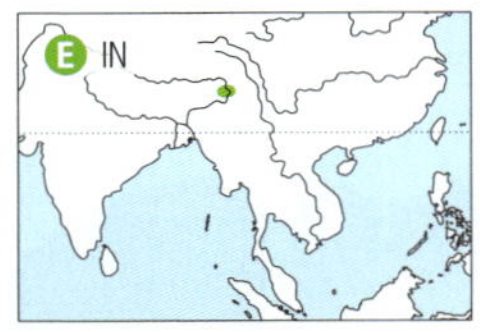

Mebo Giant Flying Squirrel

Petaurista siangensis

FR Pétauriste du Siang
DE Siang-Riesengleithörnchen
ES Ardilla voladora gigante de Mebo

Assumed to be between R Siang and R Dibang (Arunachal Pradesh)

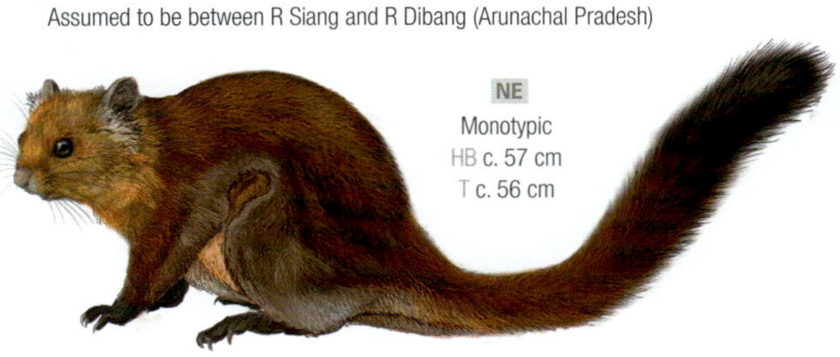

NE
Monotypic
HB c. 57 cm
T c. 56 cm

Genus *SPERMOPHILOPSIS*

(1 species)

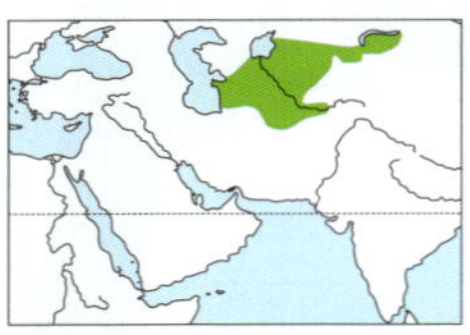

NE IR (Khorasan Prov), N AF, TM, UZ, SW TJ, S KZ

Long-clawed Ground Squirrel

Spermophilopsis leptodactyla

FR Écureuil leptodactyle
DE Langkrallenziesel
ES Ardilla terrestre de uñas largas

LC
3 sspp
HB 23–27 cm
T 2.7–8.1 cm
W 548–620 g

Genus *ATLANTOXERUS*

(1 species)

LC
Monotypic
HB c. 17.4 cm
T c. 13.3 cm
W c. 250 g

150–4000 m

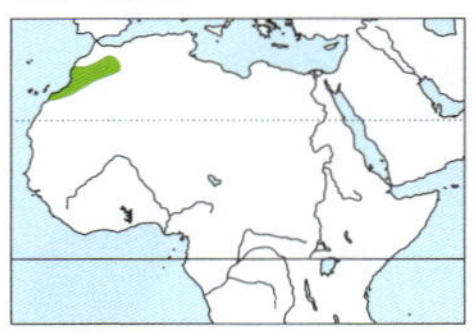

Atlas Mts in NW Africa; introduced to Canary Is (ES)

Barbary Ground Squirrel

Atlantoxerus getulus

FR Écureuil de Barbarie
DE Atlashörnchen
ES Ardilla terrestre de Berbería

Genus *XERUS*

(1 species)

0–2000 m

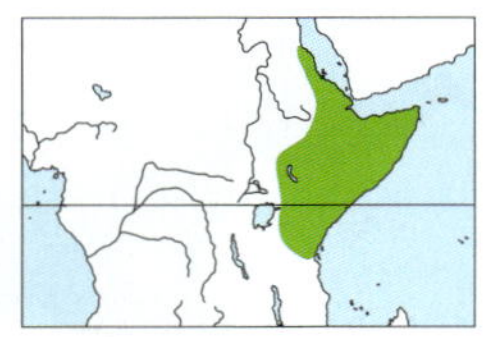

NE SD & Horn of Africa S to SE SS, UG, KE & NE TZ

LC
8 sspp
HB 23–27 cm
T 18.2–20 cm
W 252–369 g

Unstriped Ground Squirrel

Xerus rutilus

FR Écureuil de Cretzschmar
DE Schlichtborstenhörnchen
ES Ardilla terreste sin rayas

Genus *GEOSCIURUS*

(2 species)

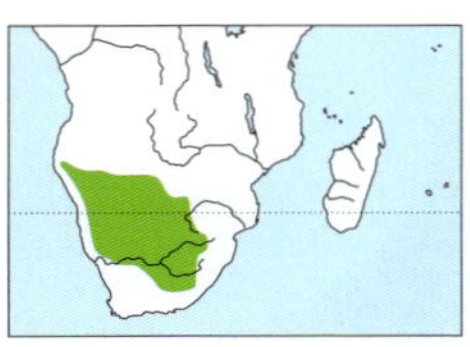

NA, BW, N & C ZA, W LS

South African Ground Squirrel

Geosciurus inauris

FR Écureuil du Cap
DE Kap-Borstenhörnchen
ES Ardilla terrestre de Sudáfrica

LC Monotypic
HB 24–25 cm T 19.4–21 cm
W 423–649 g

LC
Monotypic
HB 22–29 cm
T 21–28 cm
W c. 640 g

SW AO through W NA to near ZA border

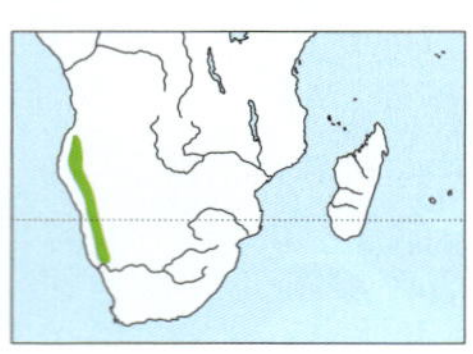

Damara Ground Squirrel

Geosciurus princeps

FR Écureuil du Kaokoveld
DE Damara-Borstenhörnchen
ES Ardilla terrestre de los Damara

Genus *EUXERUS*

(1 species)

Striped Ground Squirrel

Euxerus erythropus

FR Écureuil de Geoffroy
DE Gestreiftes Borstenhörnchen
ES Ardilla terrestre rayada

LC
6 sspp
HB 22–29 cm
T 18.5–26 cm
W 500–950 g

600–1800 m

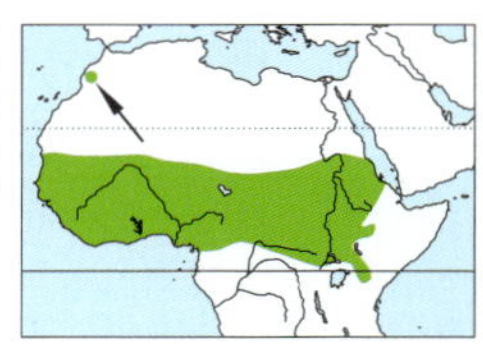

W & C Africa, from S MR & SN E to SD, ER, ET & KE; disjunt in C MA

Genus *MYOSCIURUS*

(1 species)

African Pygmy Squirrel
Myosciurus pumilio

FR Écureuil du Gabon
DE Afrika-Zwerghörnchen
ES Ardilla pigmea de África

LC
Monotypic
HB 7.2–7.5 cm T 5–5.5 cm
W c. 16.1 g

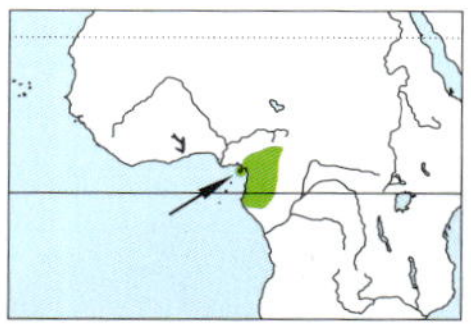

CM S through GQ (with Bioko I), GA & NW CG

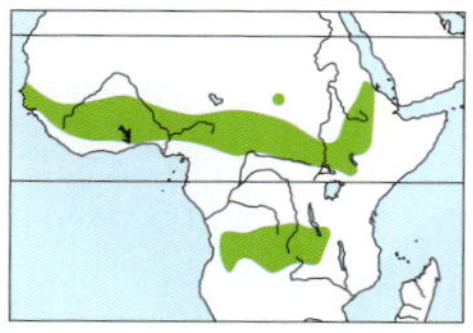

From SN & GM E to SS (isolated in Jebel Marra), ER, ET, UG & NW KE; also C AO, ZM, S CD & SW TZ

Gambian Sun Squirrel
Heliosciurus gambianus

FR Écureuil de Gambie
DE Graufußsonnenhörnchen
ES Ardilla sol de Gambia

LC
16 sspp
HB 20–22 cm
T 23–24 cm
W 245–329 g

Genus *HELIOSCIURUS*

(6 species)

LC
21 sspp
HB 22–23 cm
T 24–25 cm
W 361–380 g

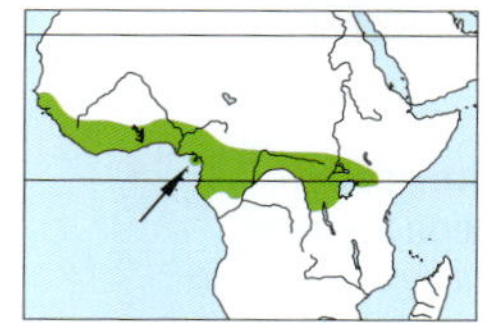

W & C Africa from SN & GM E to KE & NW TZ, S to GA, CG & EC CD, Bioko I

Red-legged Sun Squirrel
Heliosciurus rufobrachium

FR Écureuil de Waterhouse
DE Rotfußsonnenhörnchen
ES Ardilla sol de patas rojas

Ruwenzori Sun Squirrel
Heliosciurus ruwenzorii

FR Écureuil du Rwenzori
DE Ruwenzori-Sonnenhörnchen
ES Ardilla sol de Rwenzori

ssp *schoutedeni*

LC
4 sspp
HB 21–22 cm
T c. 25 cm
W c. 300 g

1600–2700 m

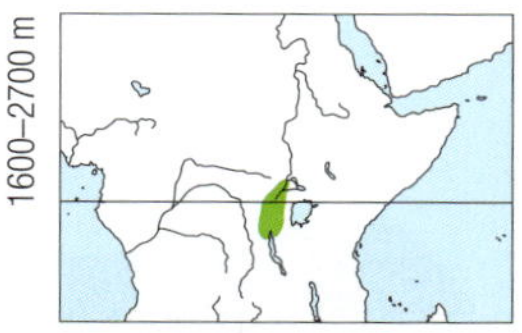

Albertine Rift in E CD, W UG, RW & BI

ssp *ruwenzorii*

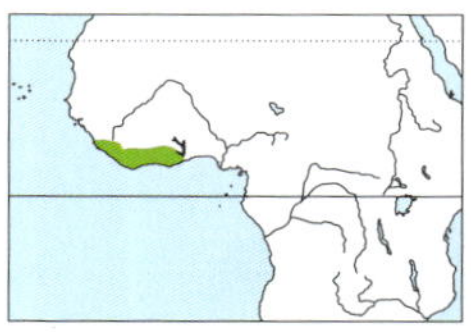

SL & extreme SW GN E to S GH (W of R Volta)

Small Sun Squirrel
Heliosciurus punctatus

FR Écureuil pointillé DE Kleines Sonnenhörnchen ES Ardilla sol pequeña

DD
2 sspp
HB 18.2–18.5 cm
T 20–21 cm
W 166–169 g

Mutable Sun Squirrel
Heliosciurus mutabilis

FR Écureuil variable
DE Variables Sonnenhörnchen
ES Ardilla sol mutable

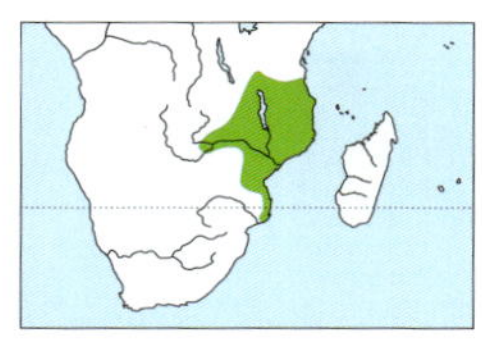

S TZ, S & E ZM, MW, MZ, E ZW (Chirinda Forest)

LC
5 sspp
HB 22–23 cm T 23–27 cm
W 333–382 g

DD
Monotypic
HB 23–24 cm
T 26–27 cm
W c. 315 g

Zanj Sun Squirrel
Heliosciurus undulatus

FR Écureuil zanj
DE Zanj-Sonnenhörnchen
ES Ardilla sol Zanj

up to 2000 m

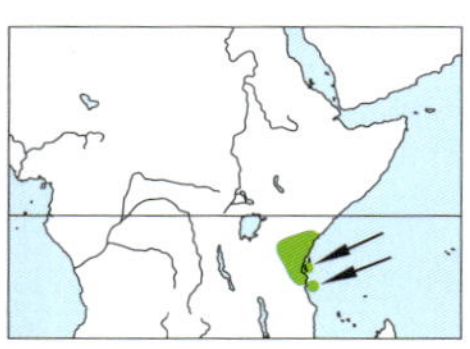

SE KE, NE TZ (with Unguja & Mafia Is)

Genus *PROTOXERUS*

(2 species)

LC
12 sspp
HB 28–30 cm
T 30–31 cm
W 538–761 g

NT
2 sspp
HB 24–25 cm
T c. 30 cm
W 415–525 g

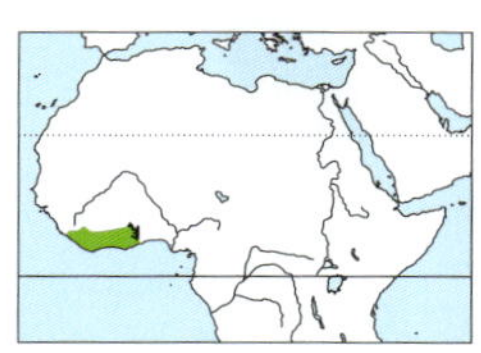

W Africa in S GN, S SL, LR, S CI & SW GH

Slender-tailed Squirrel
Protoxerus aubinnii

FR Écureuil à queue fine
DE Schlankschwanzhörnchen
ES Ardilla de cola esbelta

Forest Giant Squirrel
Protoxerus stangeri

FR Écureuil de Stanger DE Ölpalmenhörnchen
ES Ardilla gigante de bosque

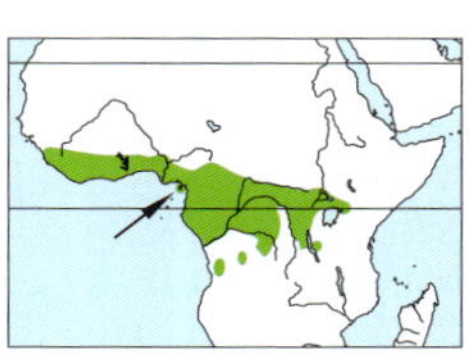

W & C Africa from SL E to UG & W KE, S to N AO, S CD & W TZ, Bioko I

Genus *EPIXERUS*

(1 species)

up to 1000 m

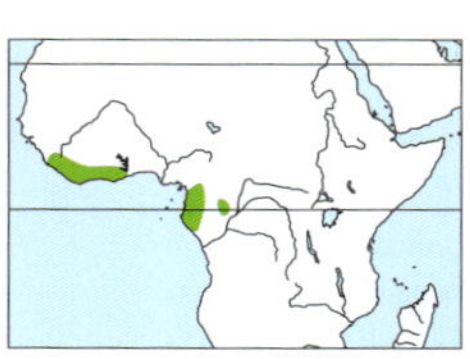

Disjunct in SL & S GN E to GH; SW CM, GQ, W GA & CG

LC 3 sspp
HB 28–29 cm T c. 28 cm
W 388–652 g

African Palm Squirrel
Epixerus ebii

FR Écureuil ébien
DE Afrika-Palmenhörnchen
ES Ardilla de palma occidental

Genus *FUNISCIURUS*

(10 species)

LC
9 sspp
HB 19.1–19.3 cm
T 14.5–15.1 cm
W 225–240 g

0–1650 m

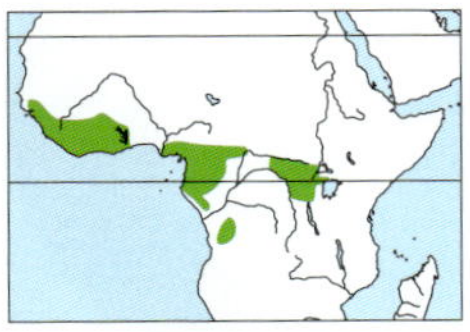

Disjunct in 4 main areas of W & C Africa, from GM E to UG & RW, S to AO

Fire-footed Rope Squirrel
Funisciurus pyrropus

FR Écureuil de Cuvier
DE Feuerfußhörnchen
ES Ardilla listada africana de pies de fuego

Kintampo Rope Squirrel
Funisciurus substriatus

FR Écureuil de Kintampo
DE Kintampo-Rotschenkelhörnchen
ES Ardilla listada africana de Kintampo

DD Monotypic
HB 16.1–16.5 cm T c. 15.5 cm
W c. 186 g

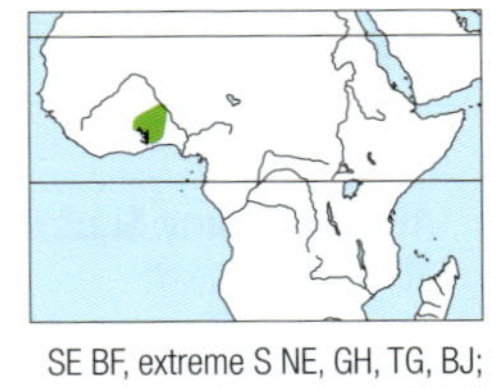

SE BF, extreme S NE, GH, TG, BJ; perhaps extreme W NG

Red-cheeked Rope Squirrel
Funisciurus leucogenys

FR Écureuil à tête orange
DE Rotwangenhörnchen
ES Ardilla listada africana de cuello rojo

LC
3 sspp
HB 20–21 cm
T 14.7–14.9 cm
W 252–271 g

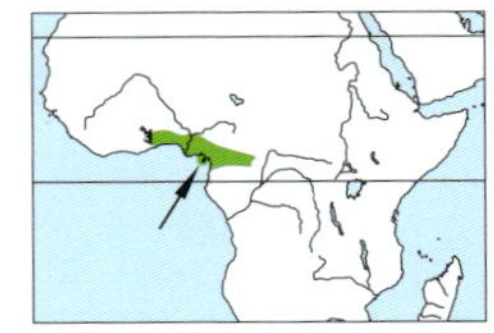

From SE GH (E of Volta Basin) to SW CF (upper R Sanga) & GQ (with Bioko I)

LC 4 sspp
HB 17.2–17.7 cm T 16.6–16.7 cm
W c. 218 g

Thomas's Rope Squirrel
Funisciurus anerythrus

FR Écureuil de Thomas
DE Thomas-Rotschenkelhörnchen
ES Ardilla listada africana de Thomas

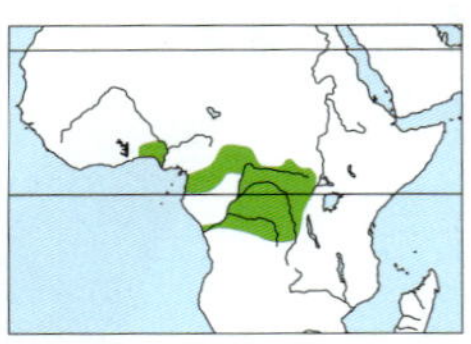

W Africa in BJ & NG, C Africa from CM & S TD E to UG, S to AO (Cabinda) & much of Congo Basin (CD)

LC
Monotypic
HB 16.1–16.5 cm
T 14.9–16.1 cm
W c. 107 g

Lady Burton's Rope Squirrel
Funisciurus isabella

FR Écureuil de Lady Burton
DE Lady-Burton-Rotschenkelhörnchen
ES Ardilla listada africana de Lady Burton

up to 2100 m

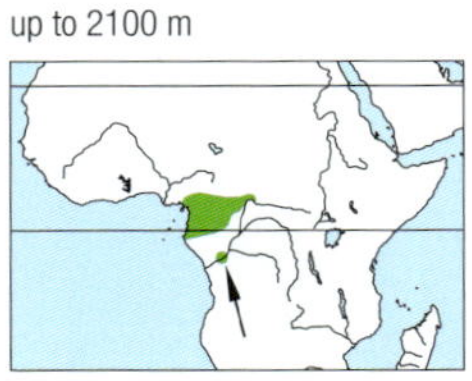

WC Africa from W CM Highlands to W CF, S to C GA & W CG (recorded in E)

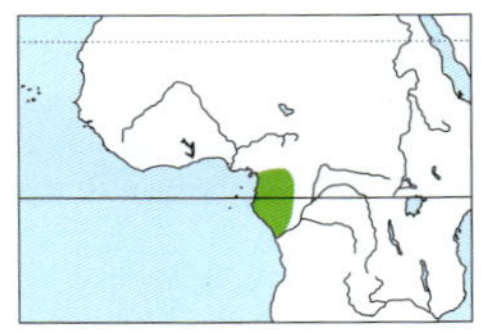

S CM, GA, W CG, extreme SW CD, NW AO (Cabinda)

Ribboned Rope Squirrel
Funisciurus lemniscatus

FR Écureuil rubané
DE Gebändertes Rotschenkelhörnchen
ES Ardilla listada africana huesuda

LC 2 sspp
HB 16.8–17.1 cm T 13.5–13.6 cm
W c. 141 g

LC
Monotypic
HB 16–17.4 cm
T 16.1–16.2 cm
W c. 111 g

Congo Rope Squirrel
Funisciurus congicus

FR Écureuil de Kuhl
DE Kongo-Rotschenkelhörnchen
ES Ardilla listada africana de Congo

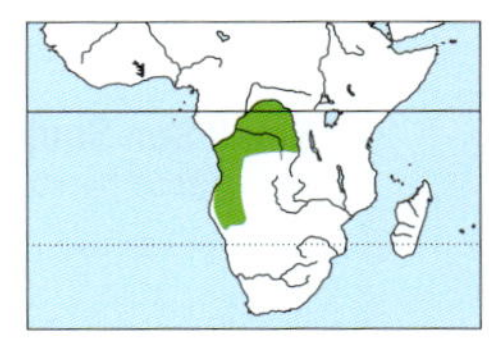

CD (S of R Congo), W AO, NW NA

LC
4 sspp
HB 21–22 cm
T 18.9–19 cm
W c. 268 g

above 1500 m

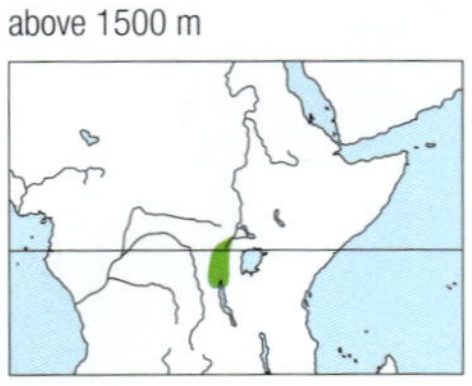

Albertine Rift Mts (CD, UG, RW, BI)

Carruther's Mountain Squirrel
Funisciurus carruthersi

FR Écureuil de Carruthers
DE Carruthers-Rotschenkelhörnchen
ES Ardilla listada africana de Carruthers

Between Ogooué and Ngounié rivers

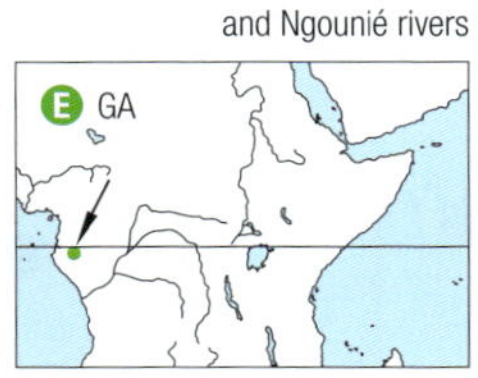

DD
Monotypic
HB 18.5–21 cm
T 19–23 cm
W 195–205 g

Du Chaillu's Rope Squirrel
Funisciurus duchaillui

FR Écureuil de Du Chaillu
DE Du-Chaillu-Rotschenkelhörnchen
ES Ardilla listada africana de Du Chaillu

DD
Monotypic
HB 18.4–25 cm
T 19.9–20 cm
W c. 135 g

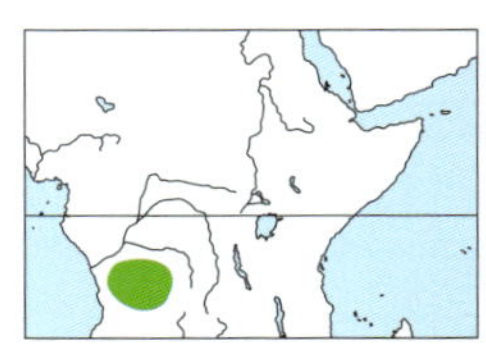

C Africa, from S CD to NE AO

Lunda Rope Squirrel
Funisciurus bayonii

FR Écureuil de Bocage
DE Lunda-Rotschenkelhörnchen
ES Ardilla listada africana de Lunda

Genus *PARAXERUS*

(11 species)

LC
Monotypic
HB 15.4–15.5 cm
T 15.9–16.6 cm
W 125–144 g

Green Bush Squirrel
Paraxerus poensis

FR Écureuil de Fernando Po
DE Grünes Buschhörnchen
ES Ardilla de matorral verde

up to 1600 m

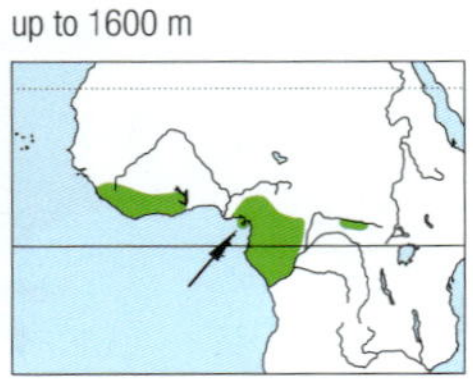

Disjunct from SL to E GH (W of Volta Basin), from SE NG to E CG, S to NW AO (Cabinda) & N CD, Bioko I

DD
Monotypic
HB 19.5–20 cm
T c. 19 cm
W c. 250 g

Cooper's Mountain Squirrel
Paraxerus cooperi

FR Écureuil de Cooper
DE Cooper-Buschhörnchen
ES Ardilla de matorral de Cooper

above 1400 m

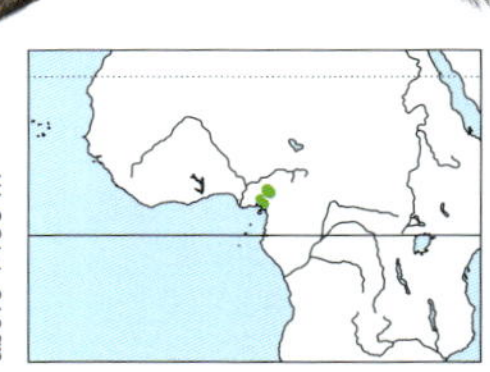

N of R Sanaga in SE NG & W CM

LC 4 sspp
HB c. 13 cm
T 13.8–14.6 cm
W 70–75 g

up to 2300 m

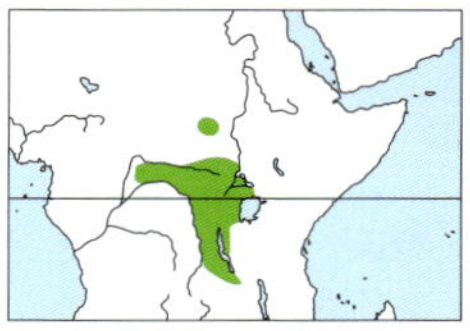

Boehm's Bush Squirrel
Paraxerus boehmi
FR Écureuil de Böhm
DE Böhm-Buschhörnchen
ES Ardilla de matorral de Böhm

C Africa from SS through CD, UG, RW & BI S to NW TZ & N ZM

below 1500 m

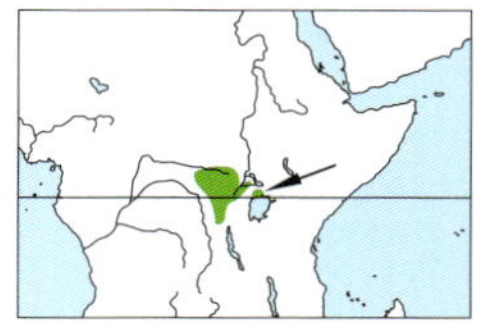

Alexander's Bush Squirrel
Paraxerus alexandri
FR Écureuil d'Alexander
DE Alexander-Buschhörnchen
ES Ardilla de matorral de Alexander

NE CD, W & S UG

LC
Monotypic
HB 13–15 cm T 10.3–11 cm
W 40–45 g

0–2500 m

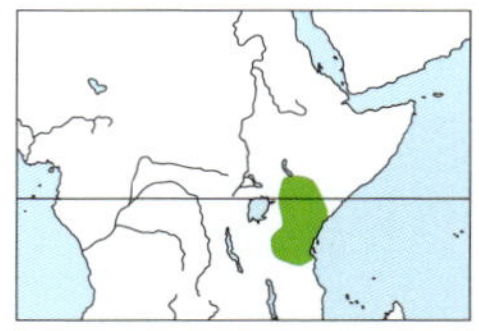

KE, TZ; possibly extending to SS & S ET

LC
8 sspp
HB 16–16.5 cm
T 16.4–16.6 cm
W 125–137 g

Ochre Bush Squirrel
Paraxerus ochraceus
FR Écureuil ocré DE Ockerbuschhörnchen ES Ardilla de matorral ocre

LC 7 sspp
HB c. 21 cm T 20–21 cm
W 307–312 g

up to 2000 m

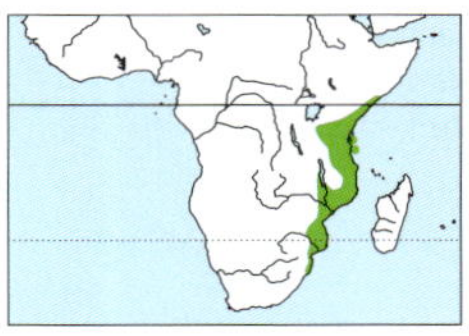

Red Bush Squirrel
Paraxerus palliatus
FR Écureuil mantelé
DE Rotschwanzbuschhörnchen
ES Ardilla de matorral roja

S SO & E KE S to TZ (with Unguja & Mafia Is), MW, MZ, SE ZW (Chirinda Forest) & NE ZA

LC
4 sspp
HB 16.8–17.1 cm
T 15.7–16.3 cm

SE KE, E TZ, SE MW, NE MZ

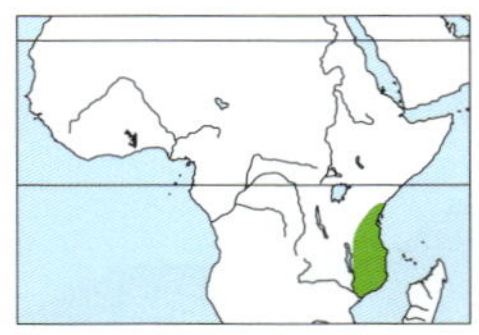

Striped Bush Squirrel
Paraxerus flavovittis
FR Écureuil à raies jaunes
DE Gestreiftes Buschhörnchen
ES Ardilla de matorral rayada

Mt Kilimanjaro, Eastern Arc Mts (Usambara, Uluguru, Udzungwa)

LC
2 sspp
HB 21–23 cm
T 18–18.4 cm
W c. 243 g

Swynnerton's Bush Squirrel
Paraxerus vexillarius
FR Écureuil de Kershaw
DE Swynnerton-Buschhörnchen
ES Ardilla de matorral de Swynnerton

Smith's Bush Squirrel
Paraxerus cepapi
FR Écureuil de Smith DE Smith-Buschhörnchen
ES Ardilla de matorral de Smith

LC
10 sspp
HB c. 23 cm
T c. 18 cm
W 180–186 g

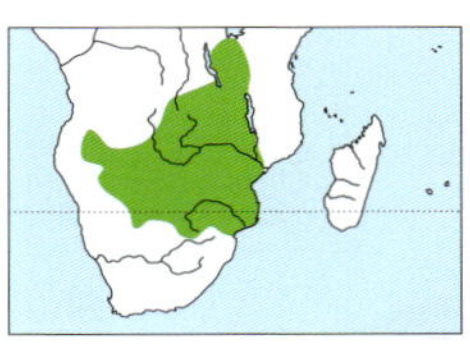

W TZ S to NE ZA and from S AO & NE NA E to MZ

DD
Monotypic
HB c. 23 cm
T 19.7–19.8 cm
W 680–685 g

Black-and-red Bush Squirrel
Paraxerus lucifer
FR Écureuil rouge et noir
DE Schwarzrotes Buschhörnchen
ES Ardilla de matorral negra y roja

above 2000 m

SW TZ, N MW; perhaps NE ZM

EN
Monotypic
HB 21–21.5 cm
T c. 21 cm

Mt Namuli

1200–1850 m

Vincent's Bush Squirrel
Paraxerus vincenti
FR Écureuil de Vincent
DE Vincent-Buschhörnchen
ES Ardilla de matorral de Vincent

Genus *MARMOTA*

(15 species)

Hoary Marmot
Marmota caligata
FR Marmotte des Rocheuses
DE Eisgraues Murmeltier
ES Marmota canosa

LC
3 sspp
HB 45–60 cm
T 17–25 cm
W 3.6–9 kg

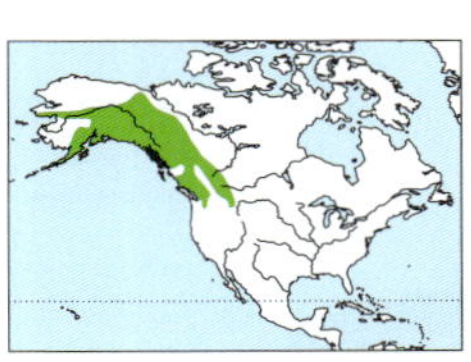

W CA (Yukon, Northwest Territories, British Columbia, Alberta), US (S Alaska, Washington, Montana, Idaho)

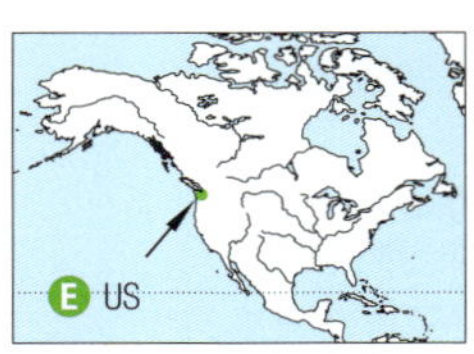

Olympic Mts (W Washington)

Olympic Marmot
Marmota olympus
FR Marmotte des Olympic
DE Olympic-Mountains-Murmeltier
ES Marmota de las Olímpicas

LC
Monotypic
HB 67–75 cm T 18–24 cm
W 2.7–11 kg

CR
Monotypic
HB 66–69 cm
T 19.3–22 cm
W 3–7 kg

1000–1500 m

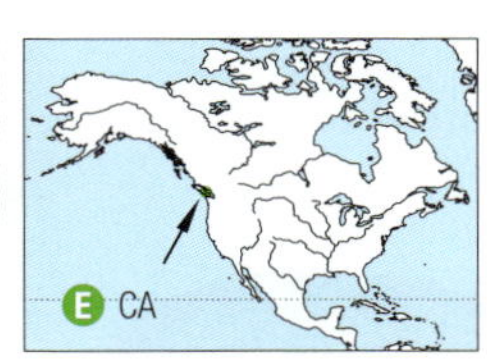

Vancouver I

Vancouver Island Marmot
Marmota vancouverensis
FR Marmotte de l'île Vancouver
DE Vancouver-Island-Murmeltier
ES Marmota de la Isla de Vancouver

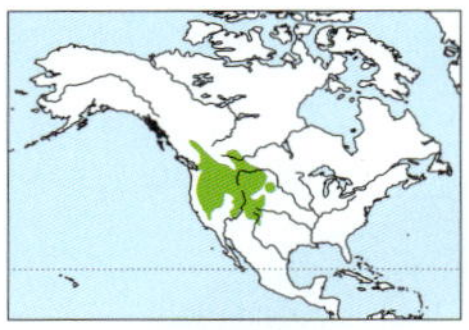

Yellow-bellied Marmot

Marmota flaviventris

FR Marmotte à ventre jaune
DE Gelbbauchmurmeltier
ES Marmota de vientre amarillo

SW CA (S British Columbia, S Alberta, SW Saskatchewan), W US (S to S California & N New Mexico)

LC
10 sspp
HB 47–70 cm
T 13–22 cm
W 1.6–5.2 kg

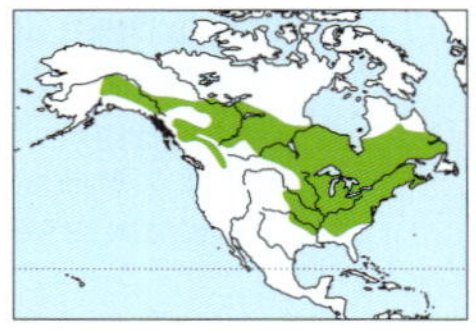

Woodchuck

Marmota monax

FR Marmotte commune
DE Waldmurmeltier
ES Marmota de Canadá

S CA (S of tree line), US (Alaska, S to N Idaho in W, to Arkansas & Alabama in E)

LC
9 sspp
HB 42–66 cm
T 10–15.5 cm
W 3.4–4.1 kg

Alaska Marmot

Marmota broweri

FR Marmotte d'Alaska
DE Alaska-Murmeltier
ES Marmota de Alaska

LC
Monotypic
HB 54–65 cm
T 13–18 cm
W 2.5–4 kg

N Alaska (Brooks Range from Chukchi Sea coast to Alaska/Yukon border); perhaps CA

1200–3000 m

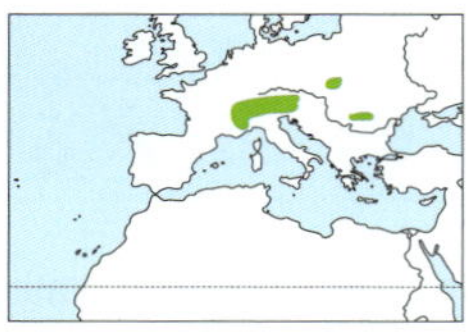

Alpine Marmot

Marmota marmota

FR Marmotte des Alpes
DE Alpen-Murmeltier
ES Marmota alpina

Alps, High Tatra Mts; reintroduced to Carpathian Mts, introduced to Pyrenees, Massif Central & Apennine Mts

LC
2 sspp
HB 50–60 cm
T 14–16.8 cm
W 2.5–5 kg

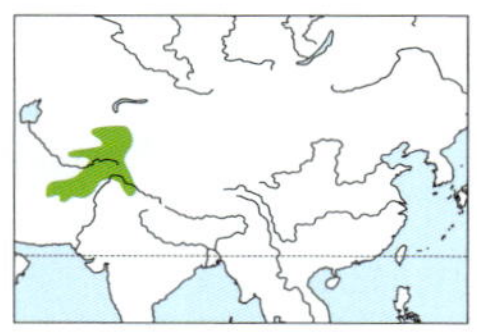

Long-tailed Marmot

Marmota caudata

FR Marmotte à longue queue
DE Langschwanzmurmeltier
ES Marmota de cola larga

SE KZ, E UZ, KG, TJ, W CN (W Xinjiang), AF, N PK, N IN (Kashmir)

LC
3 sspp
HB c. 50 cm
T c. 18.2 cm
W 3–5 kg

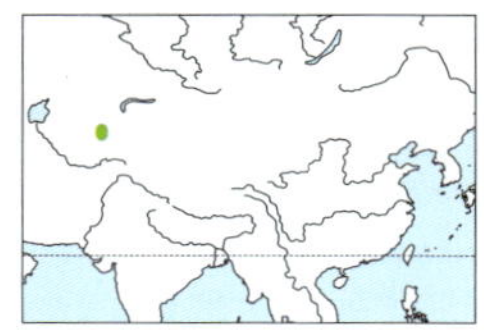

Menzbier's Marmot

Marmota menzbieri

FR Marmotte de Menzbier
DE Menzbier-Murmeltier
ES Marmota de Menzbier

W Tian Shan (KZ, KG, UZ)

VU
2 sspp
HB c. 49 cm
T c. 12 cm
W 3–4 kg

up to 4000 m

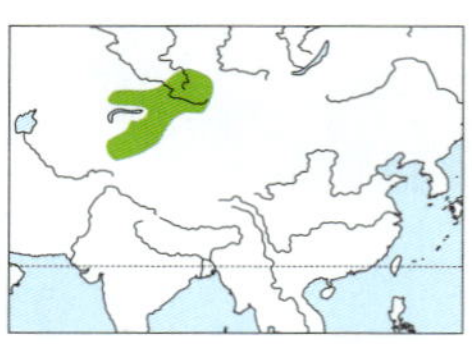

Gray Marmot

Marmota baibacina

FR Marmotte des steppes
DE Graues Murmeltier
ES Marmota gris

Altai & Tian Shan in RU, W MN, E KZ, KG, W CN (Xinjiang)

LC
2 sspp
HB 46–65 cm
T 13–15.4 cm
W 4.2–6.5 kg

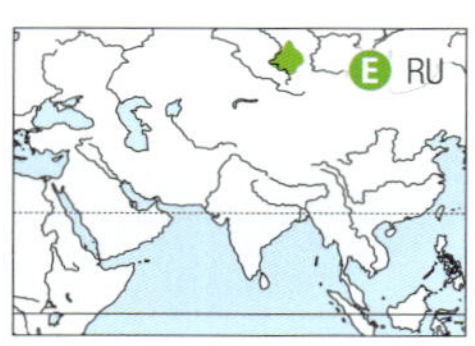

Forest Steppe Marmot

Marmota kastschenkoi

FR Marmotte de Kachtchenko
DE Waldsteppenmurmeltier
ES Marmota de bosque

S of Novosibirsk, Kemerovo & Tomsk in SC Siberia

NE
Monotypic
HB 49–64 cm
T 12.4–20 cm
W 3–9 kg

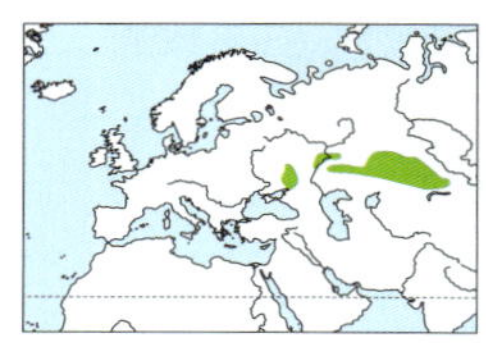

Bobak Marmot

Marmota bobak

FR Marmotte bobak
DE Steppenmurmeltier
ES Marmota de estepa

Disjunct from E UA across S RU to KZ; introduced to Caucasus Mts (Dagestan, RU)

LC
3 sspp
HB 49–57 cm
T 10.6–13 cm
W c. 5.7 kg

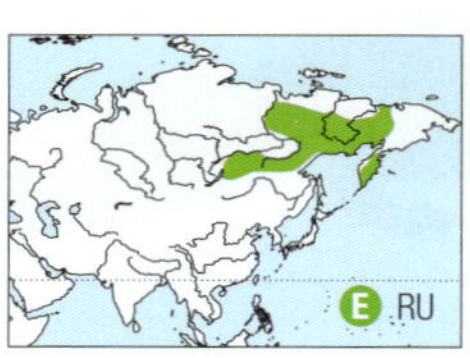

Black-capped Marmot

Marmota camtschatica

FR Marmotte du Kamchatka
DE Schwarzkappenmurmeltier
ES Marmota de caperuza negra

From L Baikal & Yakutia E to Chukotka, Kamchatka Pen & Sea of Okhotsk

LC
3 sspp
HB 46–53 cm
T c. 16.5 cm
W 2.9–5 kg

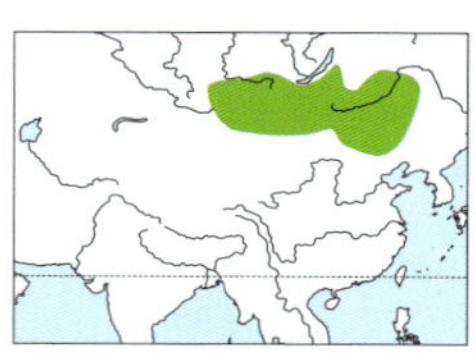

Tarbagan Marmot

Marmota sibirica

FR Marmotte de Mongolie
DE Sibirien-Murmeltier
ES Marmota de Siberia

MN, NE CN (Inner Mongolia, Heilongjiang), adjacent S Siberia (RU)

EN
2 sspp
HB 36–49 cm
T 11.2–12 cm
W 5–8 kg

LC 2 sspp
HB 46–65 cm T 12.5–15 cm
W 4–9.2 kg

up to 5670 m

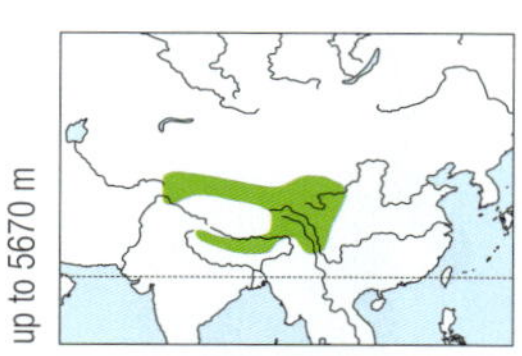

Himalayas & Tibetan Plateau in SE TJ, N PK, N IN, NP, BT & CN; limits uncertain

Himalayan Marmot

Marmota himalayana

FR Marmotte de l'Himalaya
DE Himalaya-Murmeltier
ES Marmota del Himalaya

Genus *OTOSPERMOPHILUS*

(3 species)

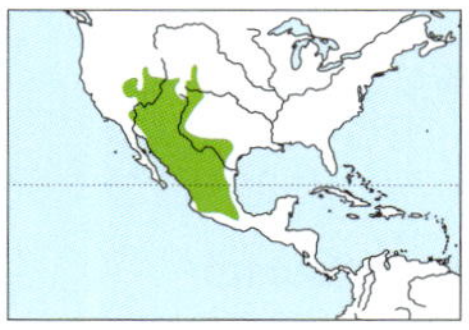

SW US (S from E Nevada, Utah & Colorado) to S MX (Puebla & N Oaxaca)

LC
8 sspp
HB 43–54 cm
T 17.4–26 cm
W 450–875 g

Common Rock Squirrel
Otospermophilus variegatus

FR Spermophile des rochers DE Felsenziesel ES Ardilla de roca

California Ground Squirrel
Otospermophilus beecheyi

FR Spermophile de Californie
DE Kalifornien-Ziesel
ES Ardilla de roca de California

ssp *atricapillus*

ssp *beecheyi*

LC 2 sspp
HB c. 28 cm T 12.7 cm

California in Sierra Nevada (N to L Almanor) & Central Valley S to Baja California Pen (MX); Channel Is

E US

S Washington, W Oregon, N California (S to R San Joaquin Delta)

NE Monotypic
HB c. 34 cm T c. 14.6 cm

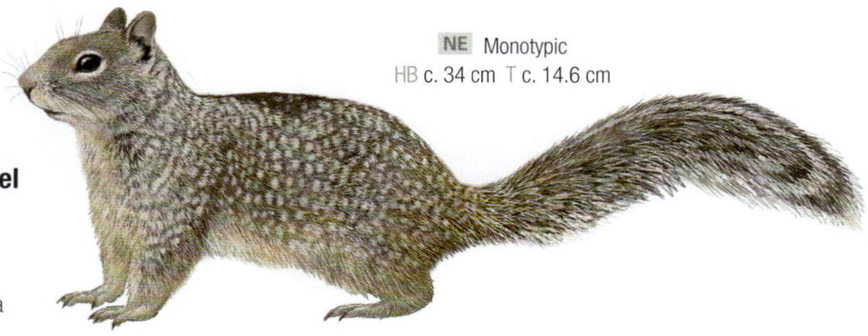

Douglas's Ground Squirrel
Otospermophilus douglasii

FR Spermophile de Douglas
DE Douglas-Kalifornien-Ziesel
ES Ardilla de roca de Baja California

Genus *CALLOSPERMOPHILUS*

(3 species)

LC
Monotypic
HB 28–31 cm
T 9.2–11.8 cm
W 200–350 g

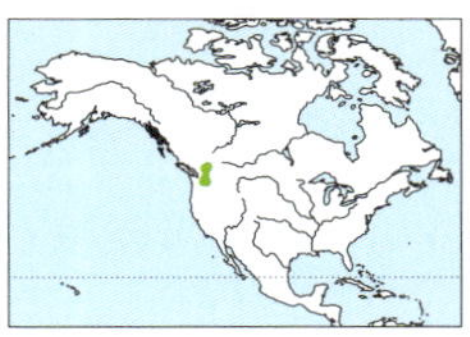

Cascade Range of SW British Columbia (CA) & W Washington (US)

Cascade Golden-mantled Ground Squirrel
Callospermophilus saturatus

FR Spermophile des Cascades
DE Kaskaden-Goldmantelziesel
ES Ardilla terrestre de manto dorado de las Cascadas

LC 13 sspp
HB c. 18 cm T 8.3–8.7 cm
W 160–178 g

above 1220 m

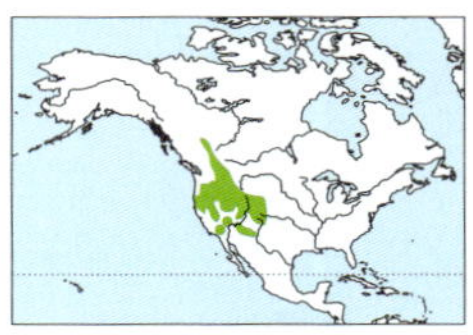

Common Golden-mantled Ground Squirrel
Callospermophilus lateralis

FR Spermophile à mante dorée
DE Goldmantelziesel
ES Ardilla terrestre de manto dorado

From CA (E British Columbia, W Alberta), S through W US (to California, Arizona & New Mexico)

NT
Monotypic
HB 16.8–16.9 cm T 6.4–6.5 cm
W 151–152 g

3000–3750 m

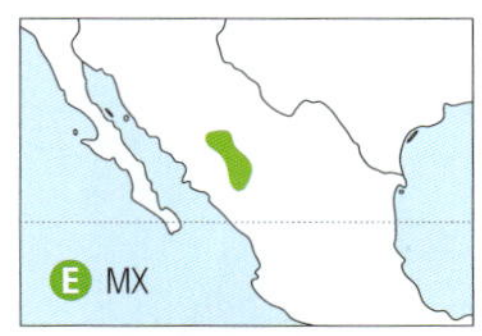

E MX

N Sierra Madre Occidental (extreme W Sonora, SW Chihuahua, NW Durango)

Sierra Madre Ground Squirrel
Callospermophilus madrensis

FR Spermophile de la Sierra Madre
DE Sierra-Madre-Goldmantelziesel
ES Ardilla terrestre de manto dorado de Sierra Madre

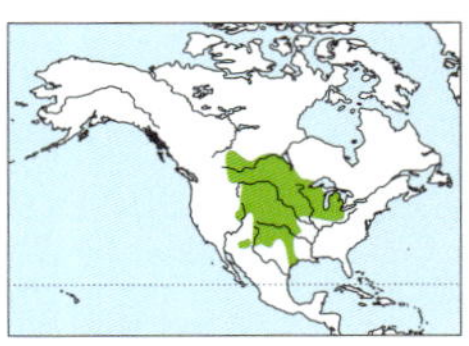

Thirteen-lined Ground Squirrel
Ictidomys tridecemlineatus

FR Spermophile rayé
DE Dreizehnstreifenziesel
ES Ardilla terrestre de trece franjas

SC CA (Alberta, Saskatchewan, Manitoba) S through C US to Arizona & Texas Gulf Coast

LC
10 sspp
HB 17–31 cm
T 6–13.2 cm
W 110–270 g

Genus *ICTIDOMYS*

(3 species)

NE
Monotypic
HB 22–32 cm T 8–13 cm
W 90–210 g

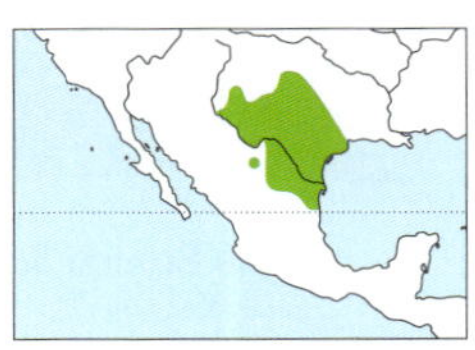

Rio Grande Ground Squirrel
Ictidomys parvidens

FR Spermophile du Rio Grande
DE Rio-Grande-Ziesel
ES Ardilla terrestre de Río Grande

US (S New Mexico, W Texas), NE MX (S to Zacatecas & S Tamaulipas, isolated in NE Durango)

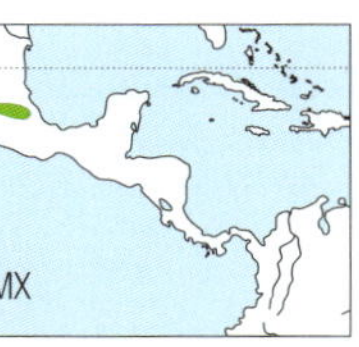

E MX

Central Plateau from E Jalisco to WC Veracruz

Mexican Ground Squirrel
Ictidomys mexicanus

FR Spermophile du Mexique
DE Mexiko-Ziesel
ES Ardilla terrestre de México

LC
Monotypic
HB 32–38 cm
T 12.4–16.6 cm
W 217–398 g

Genus *POLIOCITELLUS*

(1 species)

Franklin's Ground Squirrel
Poliocitellus franklinii

FR Spermophile de Franklin
DE Franklin-Ziesel
ES Ardilla terrestre de Franklin

LC
Monotypic
HB 23–24 cm T c. 12.8 cm
W 424–461 g

N Great Plains (SC CA, NC US)

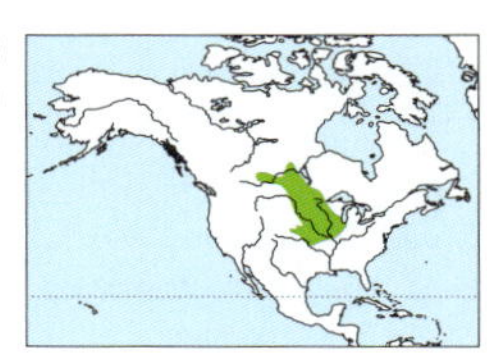

Genus *XEROSPERMOPHILUS*

(4 species)

Mohave Ground Squirrel
Xerospermophilus mohavensis

FR Spermophile du Mohave
DE Mojave-Ziesel
ES Ardilla terreste de Mojave

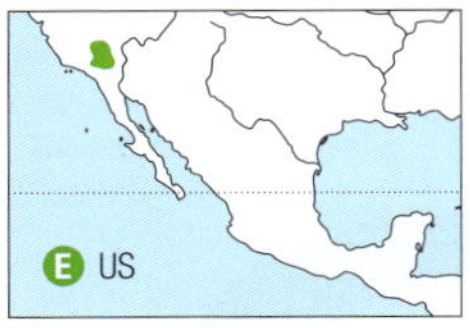

NW Mohave Desert & Owens Valley (S California)

NT
Monotypic
HB 21–23 cm
T 5.7–7.2 cm
W 70–300 g

Round-tailed Ground Squirrel
Xerospermophilus tereticaudus

FR Spermophile à queue ronde
DE Rundschwanzziesel
ES Ardilla terrestre de cola redonda

S US (S Nevada, S California, W & S Arizona), MX (N Baja California, Sonora, extreme NW Sinaloa)

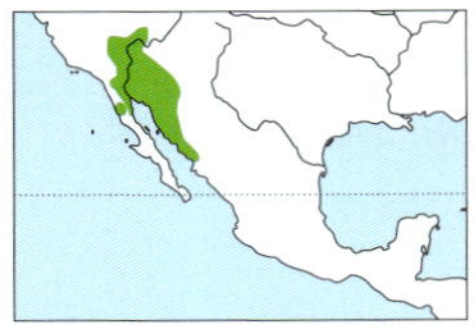

LC
4 sspp
HB 20–28 cm
T 6–11.2 cm
W 110–170 g

Spotted Ground Squirrel
Xerospermophilus spilosoma

FR Spermophile moucheté
DE Fleckenziesel
ES Ardilla terrestre moteada

US (S from SE Wyoming & S South Dakota) to MX (S to Jalisco & Guanajuato)

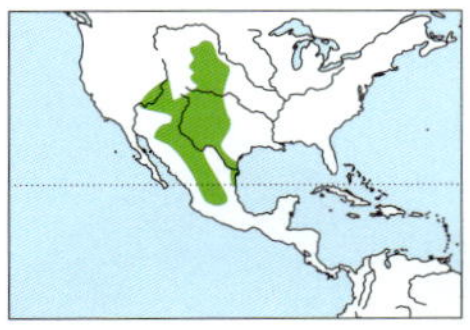

LC
13 sspp
HB 18.5–25 cm
T 5.5–9.2 cm
W 100–200 g

Perote Ground Squirrel
Xerospermophilus perotensis

FR Spermophile de Perote
DE Perote-Ziesel
ES Ardilla terrestre de Perote

2200–2700 m

NE Puebla, extreme W Veracruz

EN
Monotypic
HB 24–26 cm
T 5.7–7.8 cm

Genus *CYNOMYS*

(5 species)

Utah Prairie Dog
Cynomys parvidens

FR Chien-de-prairie de l'Utah
DE Utah-Präriehund
ES Perrito de la pradera de Utah

EN
Monotypic
HB 29–37 cm
T 4.7–6.2 cm
W 0.78–1.1 kg

SC Utah

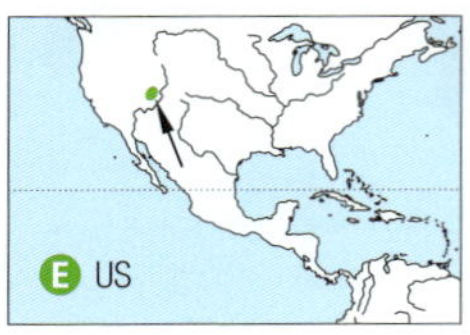

Gunnison's Prairie Dog
Cynomys gunnisoni

FR Chien-de-prairie de Gunnison
DE Gunnison-Präriehund
ES Perrito de la pradera de Gunnison

SE Utah, Colorado, C & E Arizona, N & NW New Mexico

LC
2 sspp
HB 31–39 cm T 4–6.1 cm
W 0.46–1.3 kg

White-tailed Prairie Dog
Cynomys leucurus

FR Chien-de-prairie à queue blanche
DE Weißschwanz-Präriehund
ES Perrito de la pradera de cola blanca

LC
Monotypic
HB 35–37 cm
T 5.5–5.7 cm
W 0.92–1.1 kg

SC Montana, W & C Wyoming, NE Utah, W Colorado

Black-tailed Prairie Dog
Cynomys ludovicianus

FR Chien-de-prairie à queue noire
DE Schwarzschwanz-Präriehund
ES Perrito de la pradera de cola negra

Extreme S CA (S Saskatchewan) across C USA to NW MX (extreme NE Sonora, NW Chihuahua)

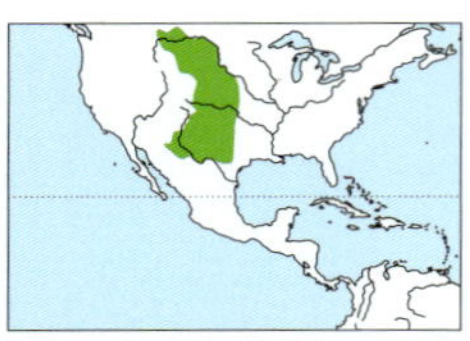

LC
2 sspp
HB c. 37 cm
T 8.4–8.7 cm
W 819–905 g

Mexican Prairie Dog
Cynomys mexicanus

FR Chien-de-prairie du Mexique
DE Mexiko-Präriehund
ES Perrito de la pradera de México

EN
Monotypic
HB 38–44 cm
T 10–10.5 cm
W 0.93–1.1 kg

SE Coahuila, SW Nuevo León, N San Luis Potosí

Genus *UROCITELLUS*

(12 species)

Idaho Ground Squirrel
Urocitellus brunneus

FR Spermophile d'Idaho
DE Idaho-Ziesel
ES Ardilla terrestre de Idaho

EN
2 sspp
HB 17.4–18 cm
T c. 55 cm
W 117–119 g

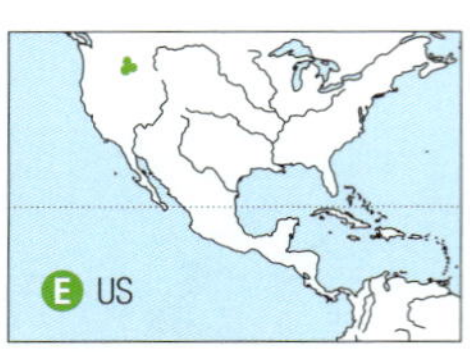

Idaho (Valley, Adams, Washington, Payette & Gem counties)

Townsend's Ground Squirrel
Urocitellus townsendii

FR Spermophile de Townsend
DE Townsend-Ziesel
ES Ardilla terrestre de Townsend

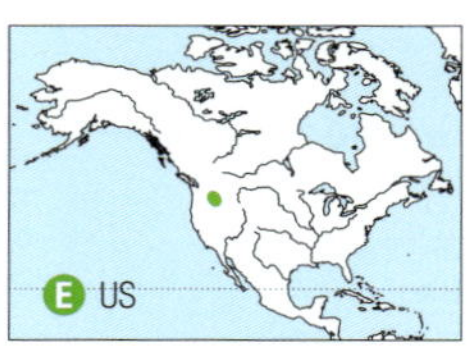

Washington (R Yakima Valley)

VU
2 sspp
HB 20–23 cm
T 3.2–5.4 cm
W 125–174 g

Piute Ground Squirrel
Urocitellus mollis

FR Spermophile à poil doux
DE Piute-Ziesel
ES Ardilla terrestre de Piute

LC
3 sspp
HB 20–23 cm
T 4.4–6.1 cm
W 82–205 g

SE Oregon, Idaho, E California, Nevada, Utah

Merriam's Ground Squirrel
Urocitellus canus

FR Spermophile de Merriam
DE Merriam-Ziesel
ES Ardilla terrestre de Merriam

E Oregon, SW Idaho, extreme NE California, NW Nevada

LC
2 sspp
HB 15.3–16 cm
T 3.8–4.1 cm
W c. 154 g

NT
Monotypic
HB 18.5–24 cm
T 3.2–6.5 cm
W c. 210 g

Washington Ground Squirrel
Urocitellus washingtoni

FR Spermophile du Washington
DE Washington-Ziesel
ES Ardilla terrestre de Washington

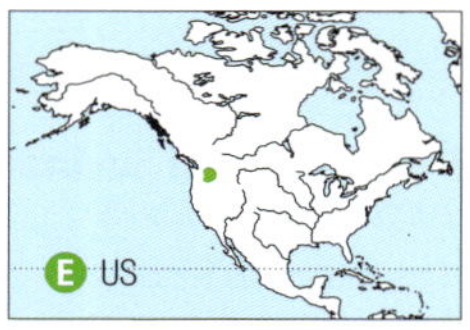

Columbia Plateau (E Washington, NE Oregon)

E Oregon, SW Idaho, NE & NC California, N Nevada

LC
3 sspp
HB 20–21 cm
T 6.3–6.5 cm
W 229–265 g

Belding's Ground Squirrel
Urocitellus beldingi

FR Spermophile de Belding
DE Belding-Ziesel
ES Ardilla terrestre de Belding

LC
Monotypic
HB c. 22 cm
T 6.9–7.2 cm
W 347–394 g

Uinta Ground Squirrel
Urocitellus armatus

FR Spermophile des Uinta
DE Uinta-Ziesel
ES Ardilla terrestre de Uinta

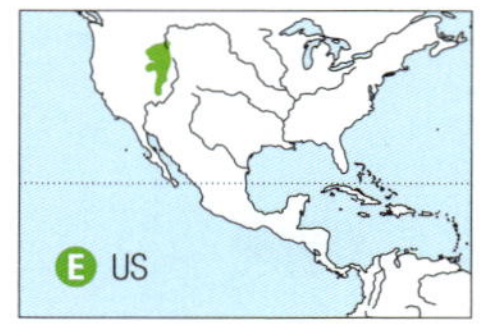

SW Montana, SE Idaho & W Wyoming to SC Utah

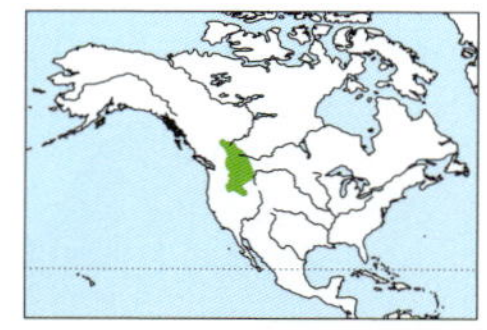

CA (SE British Columbia, W Alberta), US (E Washington, N & C Idaho, W Montana, NE Oregon)

LC
2 sspp
HB 25–26 cm
T 8.4–10.1 cm
W 442–490 g

Columbian Ground Squirrel
Urocitellus columbianus

FR Spermophile du Columbia
DE Columbia-Ziesel
ES Ardilla terrestre de Columbia

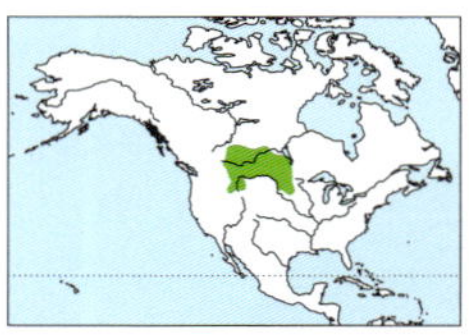

N Great Plains in CA (S Alberta E to S Manitoba) & N US (S to Montana, NE South Dakota & extreme NW Iowa)

Richardson's Ground Squirrel
Urocitellus richardsonii

FR Spermophile de Richardson
DE Richardson-Ziesel
ES Ardilla terrestre de Richardson

LC
Monotypic
HB 26–34 cm
T 5.5–8.8 cm
W 120–745 g

LC
3 sspp
HB c. 20 cm
T c. 7.3 cm
W 284–330 g

Wyoming Ground Squirrel
Urocitellus elegans

FR Spermophile élégant
DE Wyoming-Ziesel
ES Ardilla terrestre de Wyoming

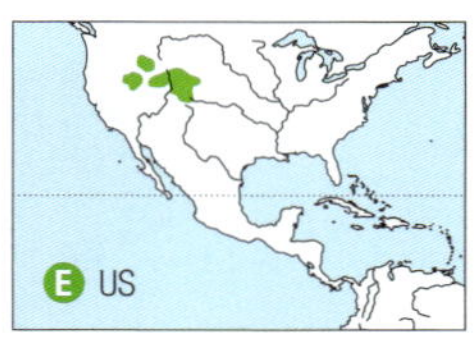

Disjunct in SW Montana, SE Oregon, Idaho, S Wyoming, N Nevada, NE Utah, N & W Colorado, extreme W Nebraska

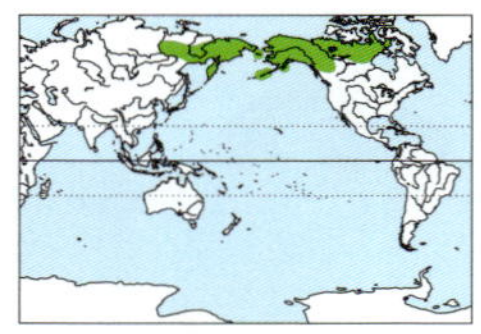

RU (NE Siberia), US (Alaska, St Lawrence, Shumagin & Kodiak Is), CA (E to Hudson Bay, S to N British Columbia)

Arctic Ground Squirrel
Urocitellus parryii

FR Spermophile arctique
DE Arktis-Ziesel
ES Ardilla terrestre del Ártico

LC
10 sspp
HB 26–27 cm
T 8.8–11 cm
W 524–673 g

LC
6 sspp
HB 21–31 cm
T 10–14 cm
W 250–580 g

Long-tailed Ground Squirrel
Urocitellus undulatus

FR Spermophile de Sibérie
DE Langschwanzziesel
ES Ardilla terrestre de cola larga

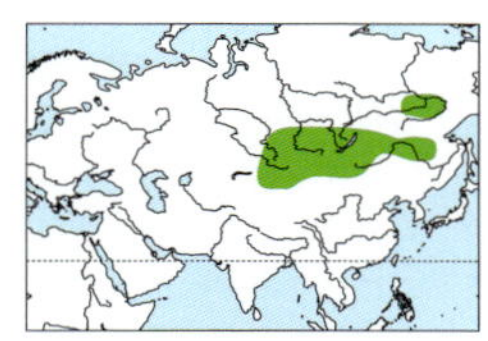

Tian Shan & Altai Mts, Transbaikal, NE Siberia, R Amur (E KZ, RU, MN, NW & NE CN)

Genus *SPERMOPHILUS*

(15 species)

EN Monotypic
HB 17.4–23 cm T 3.1–9 cm
W 125–380 g

European Ground Squirrel
Spermophilus citellus

FR Spermophile d'Europe
DE Europäischer Ziesel
ES Ardilla terrestre de Europa

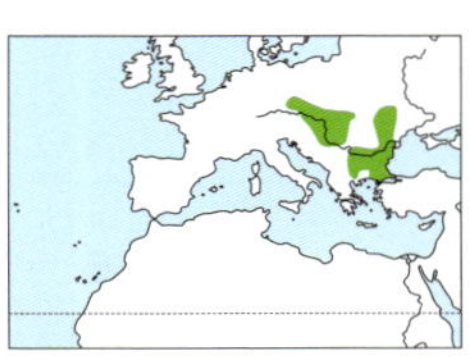

0–2500 m

2 disjunct areas: Pannonian (CZ E to W RO, S to N RS), Balkan (UA S to N GR & NW TR); reintroduced to PL

above 1000 m

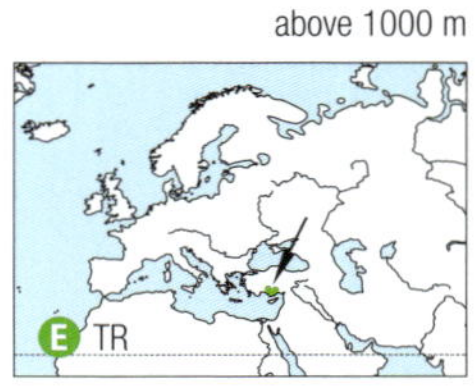

Taurus Mts (Erenkaya to Morca Yaylasi and Salamut Plateau to Mut)

Tauren Ground Squirrel
Spermophilus taurensis

FR Spermophile des Taurus
DE Taurus-Ziesel
ES Ardilla terrestre de Tauro

LC
Monotypic
HB c. 20 cm
T c. 6.1 cm
W c. 201 g (F)

NT
Monotypic
HB 14.5–23 cm
T 3.3–3.5 cm
W 172–312 g

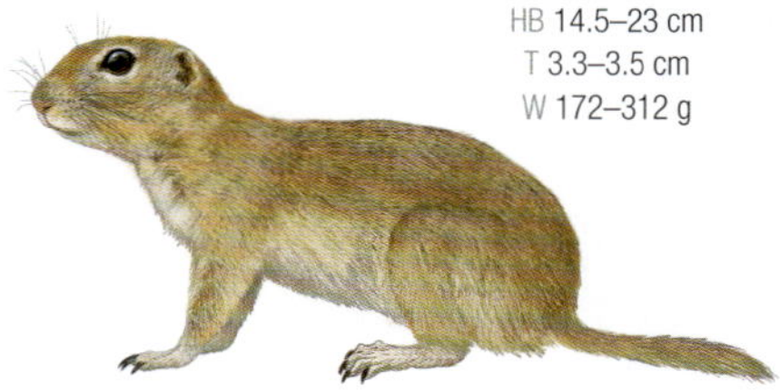

Asia Minor Ground Squirrel
Spermophilus xanthoprymnus

FR Spermophile de Bennett
DE Kleinasien-Ziesel
ES Ardilla terrestre de Asia Menor

800–2900 m

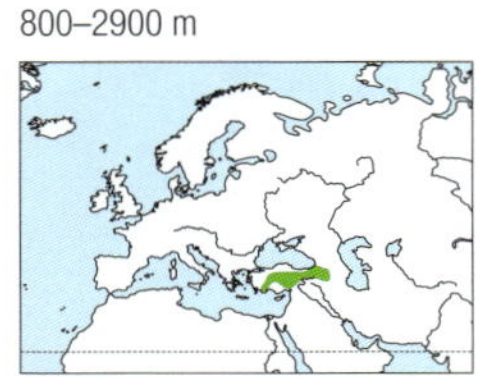

TR (C lowlands, E highlands, Teke Pen), W AM (Alagez) & NW IR (Maku)

below 500 m

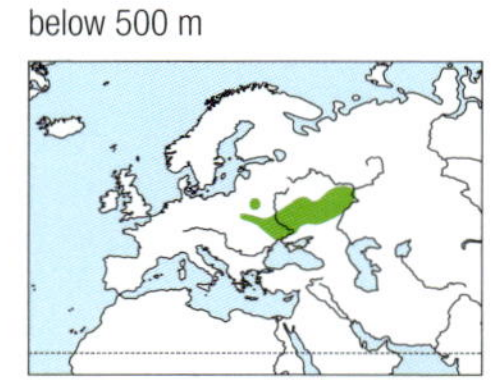

E PL, MD, UA, RU (E to R Volga), isolated in W BY

Speckled Ground Squirrel
Spermophilus suslicus

FR Spermophile tacheté
DE Perlziesel
ES Ardilla terrestre manchada

NT
3 sspp
HB 18–26 cm
T 3.6–5.7 cm
W 180–220 g

LC
Monotypic
HB c. 19 cm
T 5.5–6.3 cm
W c. 224 g

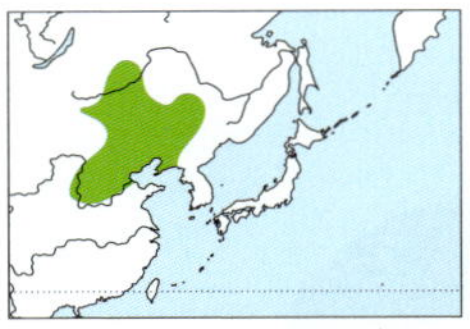

SE Transbaikal in SE RU, E MN & NE CN (S to Shaanxi & Henan)

Daurian Ground Squirrel
Spermophilus dauricus

FR Spermophile de Daourie
DE Daurischer Ziesel
ES Ardilla terrestre de Dauria

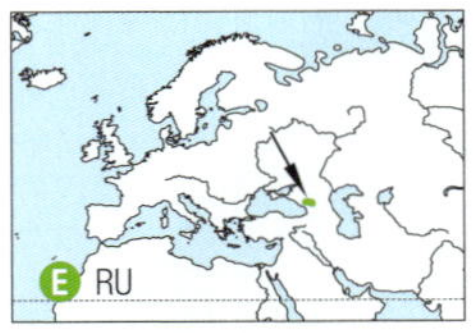

N slopes of C Greater Caucasus Range

Caucasian Mountain Ground Squirrel
Spermophilus musicus

FR Spermophile du Caucase
DE Kaukasus-Ziesel
ES Ardilla terrestre del Cáucaso

NT Monotypic
HB 20–24 cm T 3.4–5 cm

LC
4 sspp
HB 17.5–26 cm
T 2.5–5 cm
W c. 235 g

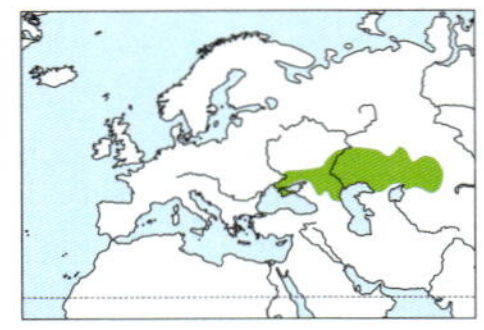

SE UA, Ciscaucasia & lower R Volga region (RU) to W KZ

Little Ground Squirrel
Spermophilus pygmaeus

FR Spermophile nain
DE Kleinziesel
ES Ardilla terrestre pigmea

Russet Ground Squirrel
Spermophilus major

FR Spermophile roussâtre
DE Rotgelber Ziesel
ES Ardilla terrestre bermeja

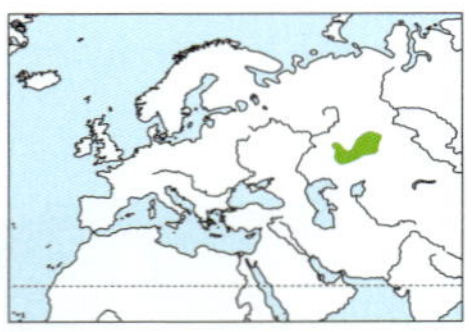

SW RU, N KZ; introduced to N Caucasus Range

LC Monotypic
HB 25–32 cm T 7.3–10.5 cm
W 500–570 g

up to 2100 m

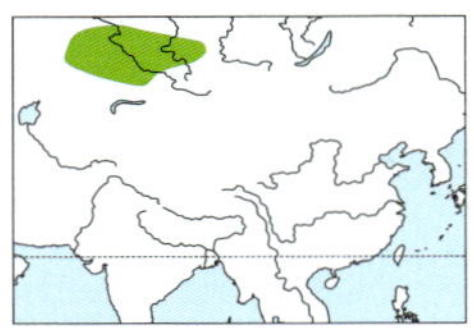

SW Siberia (RU), NE KZ

Red-cheeked Ground Squirrel
Spermophilus erythrogenys

FR Spermophile à joues rouges
DE Rotwangenziesel
ES Ardilla terrestre de mejillas rojas

LC
Monotypic
HB 18.8–19.3 cm
T 4.1–4.6 cm
W c. 355 g

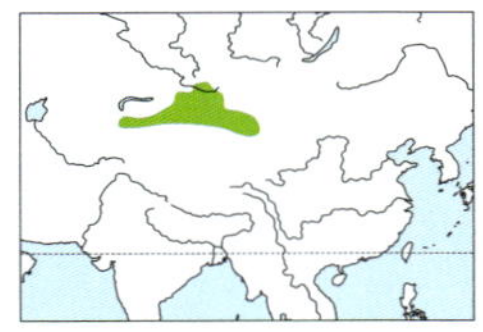

E KZ, N KG, extreme W MN, NW CN (N Xinjiang)

Brandt's Ground Squirrel
Spermophilus brevicauda

FR Spermophile de Brandt
DE Kurzschwanz-Ziesel
ES Ardilla terrestre de Brandt

LC
3 sspp
HB c. 28 cm
T c. 5.1 cm
W 143–436 g

LC
3 sspp
HB 22–28 cm
T 7.1–8.5 cm
W 0.27–1.1 kg

Yellow Ground Squirrel
Spermophilus fulvus

FR Spermophile jaune
DE Gelbziesel
ES Ardilla de tierra amarilla

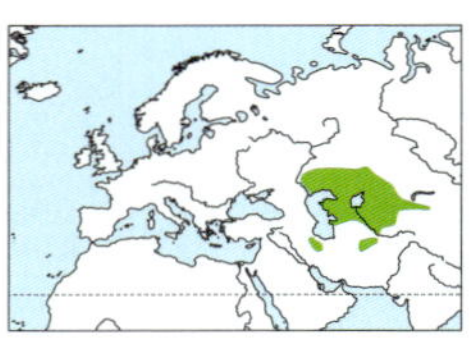

RU (E of R Volga) E to C KZ, S to NW & NE IR, NW AF, W TJ; limits uncertain

Relict Ground Squirrel
Spermophilus relictus

FR Spermophile relique
DE Westliches Tienschan-Ziesel
ES Ardilla terrestre relicta

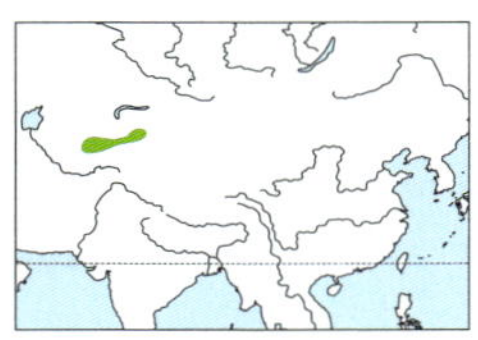

600–2800 m

Tian Shan (SE KZ, E UZ, KG, NW TJ)

LC
Monotypic
HB 23–27 cm
T 5.8–7.6 cm

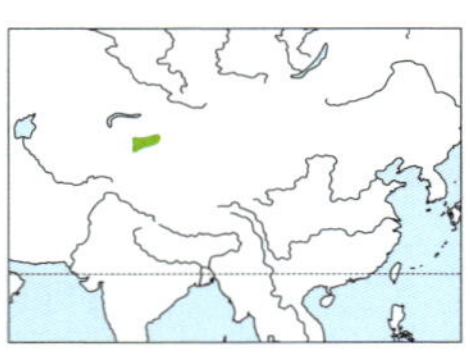

E Tian Shan (SE KZ, NE KG, W CN in W Xinjiang)

Tian Shan Ground Squirrel
Spermophilus nilkaensis

FR Spermophile de Rall
DE Östliches Tienschan-Ziesel
ES Ardilla terrestre de Tien Shan

LC
Monotypic
HB 20–24 cm T 6–7.5 cm
W 290–405 g

Alashan Ground Squirrel
Spermophilus alashanicus

FR Spermophile de l'Alashan
DE Alashan-Ziesel
ES Ardilla terrestre de Alashan

up to 3200 m

SC MN, NC CN (Inner Mongolia, Ningxia, Gansu, Qinghai, Shaanxi, Shanxi)

LC
Monotypic
HB c. 22 cm
T c. 7.6 cm
W 192–224 g

LC
Monotypic
HB 19.8–23 cm
T 3.5–5.3 cm

Pallid Ground Squirrel
Spermophilus pallidicauda

FR Spermophile pâle DE Fahlziesel
ES Ardilla terrestre pálida

S MN, N CN (extreme E Xinjiang, N Gansu, Inner Mongolia)

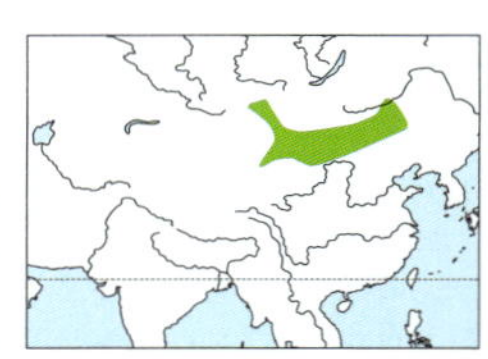

Genus *NOTOCITELLUS*

(2 species)

Tropical Ground Squirrel

Notocitellus adocetus

FR Spermophile des Tropiques
DE Tropischer Ziesel
ES Ardilla terrestre tropical

LC
2 sspp
HB 16.8–17.5 cm
T 13.2–15 cm
W 163–250 g

c. 1200 m

From SW Jalisco across Michoacán to N Guerrero & S Distrito Federal

Ring-tailed Ground Squirrel

Notocitellus annulatus

FR Spermophile à queue annelée
DE Ringelschwanzziesel
ES Ardilla terrestre de cola anillada

LC
2 sspp
HB c. 22 cm
T c. 21 cm
W 361–387 g

SW Nayarit across Jalisco to W Guerrero

Genus *AMMOSPERMOPHILUS*

(4 species)

Harris's Antelope Squirrel

Ammospermophilus harrisii

FR Spermophile de Harris
DE Harris-Antilopenziesel
ES Ardilla antílope de Harris

LC
2 sspp
HB 27–28 cm
T 6.7–9.2 cm
W 113–150 g

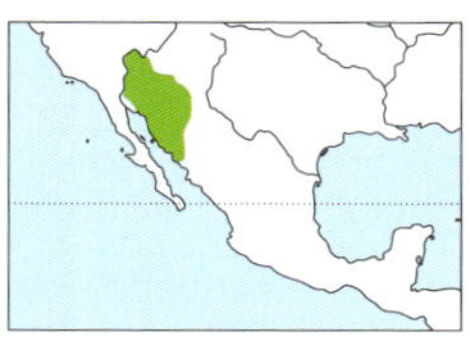

US (Arizona, SW New Mexico) to MX (N Sonora)

Nelson's Antelope Squirrel

Ammospermophilus nelsoni

FR Spermophile de Nelson
DE Nelson-Antilopenziesel
ES Ardilla antílope de Nelson

San Joaquin Valley (S California)

EN
Monotypic
HB 23–27 cm
T 6.7–7.8 cm
W 142–179 g

Texas Antelope Squirrel

Ammospermophilus interpres

FR Spermophile du Texas
DE Texas-Antilopenziesel
ES Ardilla antílope de Texas

LC
Monotypic
HB 22–23 cm
T 6.8–8.4 cm
W 99–122 g

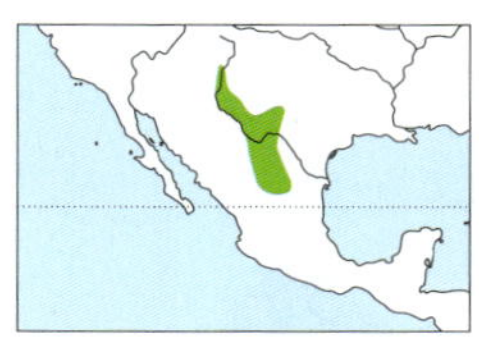

US (New Mexico, W Texas), MX (E Chihuahua, Coahuila, NE Durango, extreme N Zacatecas)

White-tailed Antelope Squirrel

Ammospermophilus leucurus

FR Spermophile à queue blanche
DE Weißschwanz-Antilopenziesel
ES Ardilla antílope de cola blanca

LC
10 sspp
HB 18.8–22 cm T 4.2–7.1 cm
W 96–117 g

US (S from SE Oregon & SW Idaho) to MX (Baja California Pen, with Espíritu Santo & Partida Is)

Genus *EUTAMIAS*

(1 species)

Siberian Chipmunk

Eutamias sibiricus

FR Tamia de Sibérie
DE Burunduk
ES Ardilla listada de Siberia

LC
3 sspp
HB 14.7–15 cm
T 10.6–11.6 cm
W 93–99 g

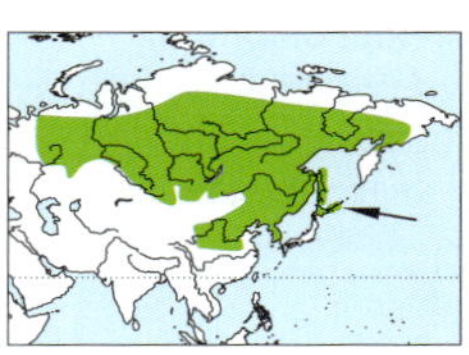

RU (to E Siberia), NE KZ, N MN, CN, Korean Pen, N JP; introduced to W Europe

Genus *TAMIAS*

(1 species)

Eastern Chipmunk

Tamias striatus

FR Tamia rayé
DE Streifen-Backenhörnchen
ES Ardilla listada oriental

LC
11 sspp
HB 14.6–15 cm
T 8.8–9.3 cm
W 94–101 g

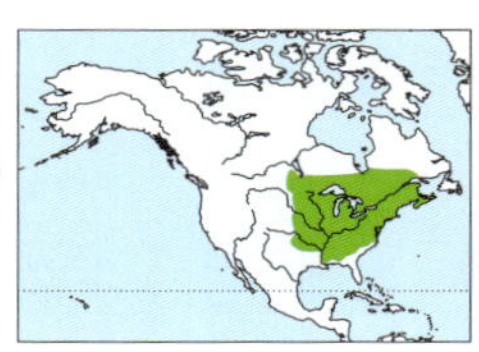

S CA (SE Saskatchewan to Nova Scotia), E half US (S to Louisiana)

Genus *NEOTAMIAS*

(24 species)

Least Chipmunk

Neotamias minimus

FR Tamia mineur
DE Kleines Backenhörnchen
ES Ardilla listada mínima

LC
21 sspp
HB 10.6–11.4 cm
T 8–8.5 cm
W 44–50 g

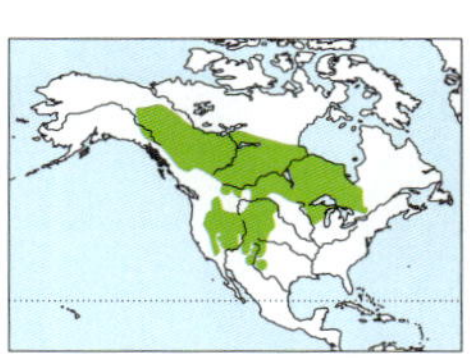

CA (from Yukon E to W Quebec), W US (S to California, Arizona & New Mexico)

Alpine Chipmunk

Neotamias alpinus

FR Tamia de montagne
DE Sierra-Nevada-Backenhörnchen
ES Ardilla listada alpina

LC
Monotypic
HB 10.5–10.6 cm
T 6.9–7.5 cm
W c. 36 g

2300–3900 m

Sierra Nevada, from Tuolumne to Tulare counties (SC California)

Panamint Chipmunk

Neotamias panamintinus

FR Tamia des Panamint
DE Panamint-Backenhörnchen
ES Ardilla listada de los Panamint

1230–3180 m

Mts of SW Nevada & SE California

LC 2 sspp
HB 10.7–11.9 cm
T 8.7–9.2 cm
W 53–54 g

LC 4 sspp
HB 12.2–12.7 cm T 8.7–9.5 cm
W 57–63 g

1500–3300 m

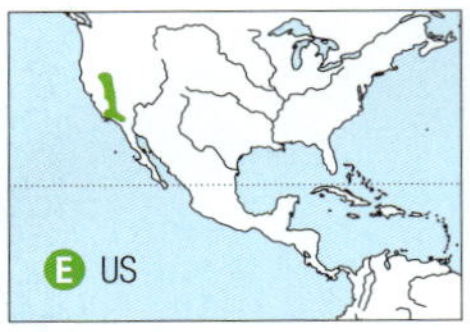

Lodgepole Chipmunk
Neotamias speciosus
FR Tamia de Californie
DE Küstenkiefer-Backenhörnchen
ES Ardilla listada de Lodgepole

Sierra Nevada S to San Bernardino & San Gabriel Mts, Mt Pinos (E California, extreme W Nevada)

Sierra Nevada (EC California, WC Nevada)

Long-eared Chipmunk
Neotamias quadrimaculatus
FR Tamia à grandes oreilles
DE Langohr-Backenhörnchen
ES Ardilla listada de orejas largas

LC Monotypic
HB 13.6–13.8 cm
T 9.4–10.3 cm
W 78–87 g

LC
2 sspp
HB 12.2–12.7 cm
T 9.8–10.9 cm
W 57–67 g

720–2400 m

Red-tailed Chipmunk
Neotamias ruficaudus
FR Tamia à queue rousse
DE Rotschwanz-Backenhörnchen
ES Ardilla listada de cola roja

CA (SE British Columbia, SW Alberta), US (E Washington, NE Idaho, W Montana)

EN
Monotypic
HB 12.5–12.6 cm T 8–9.8 cm
W 52–60 g

Palmer's Chipmunk
Neotamias palmeri
FR Tamia de Palmer DE Palmer-Backenhörnchen
ES Ardilla listada de Palmer

Spring Mts in Clark County (S Nevada)

2100–3600 m

LC
7 sspp
HB 12.2–12.5 cm
T 8.9–11.9 cm
W 51–74 g

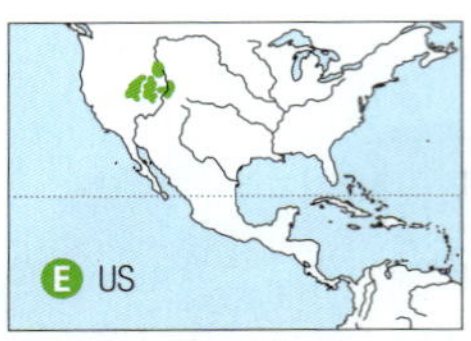

Uinta Chipmunk
Neotamias umbrinus
FR Tamia des Uinta
DE Uinta-Backenhörnchen
ES Ardilla listada de Uinta

Extreme S Montana S to E California, N Arizona & N Colorado

SW US (from S Idaho & S Wyoming) to N MX (S to NE Sinaloa & NW Durango; Sierra del Carmen in Coahuila)

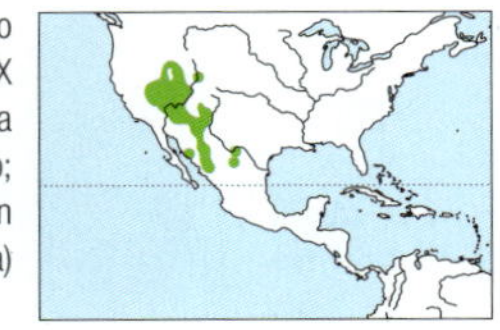

LC
6 sspp
HB 12.3–12.8 cm T 9.4–10.4 cm
W 65–74 g

Cliff Chipmunk
Neotamias dorsalis
FR Tamia des falaises DE Felsen-Backenhörnchen ES Ardilla listada de acantilado

Gray-collared Chipmunk
Neotamias cinereicollis
FR Tamia à col gris
DE Grauhals-Backenhörnchen
ES Ardilla listada de collar gris

1950–3440 m

C & E Arizona, W New Mexico

LC
2 sspp
HB 12.9–13.7 cm
T 9.8–10.8 cm
W c. 70 g

LC
3 sspp
HB 12.5–13.2 cm
T c. 10 cm
W 45–70 g

Colorado Chipmunk
Neotamias quadrivittatus
FR Tamia du Colorado
DE Colorado-Backenhörnchen
ES Ardilla listada de Colorado

E Utah, Colorado, N Arizona, New Mexico, extreme W Oklahoma

LC
Monotypic
HB 12–12.4 cm
T 9.1–9.5 cm
W 53–58 g

E Utah, W Colorado, NE Arizona

Hopi Chipmunk
Neotamias rufus
FR Tamia roux
DE Hopi-Backenhörnchen
ES Ardilla listada de los Hopi

2100–2400 m

Buller's Chipmunk
Neotamias bulleri
FR Tamia de Buller
DE Buller-Backenhörnchen
ES Ardilla listada de Buller

Sierra Madre Occidental (border between Durango, Zacatecas and Jalisco)

VU
Monotypic
HB 13–13.4 cm
T 8.5–10.5 cm
W 66–75 g

1600–3600 m

Gray-footed Chipmunk
Neotamias canipes
FR Tamia à pattes grises
DE Graufuß-Backenhörnchen
ES Ardilla listada de patas grises

New Mexico (Capitan, Jicarilla, Gallinas & Sacramento Mts), W Texas (Sierra Diablo & Guadalupe Mts)

LC
2 sspp
HB 12.9–13.2 cm
T 9.8–10.8 cm
W c. 70 g

2550–2850 m

Sierra del Carmen Chipmunk
Neotamias solivagus
FR Tamia du Coahuila
DE Sierra-del-Carmen-Backenhörnchen
ES Ardilla listada de Sierra del Carmen

Sierra del Carmen in Sierra Madre Oriental (SE Coahuila, WC Nuevo León)

NE
Monotypic
HB 11.9–14.4 cm
T 8.5–11.2 cm
W c. 83 g

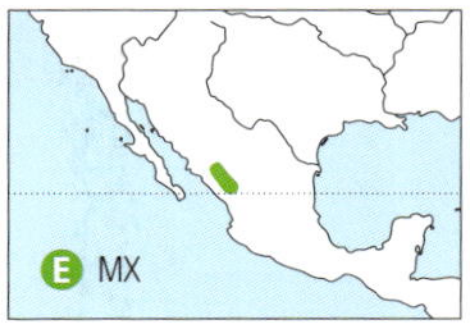

Sierra Madre Occidental from extreme SW Chihuahua S to W Durango

1950–2550 m

LC Monotypic
HB 12.5–15.6 cm
T 8–11 cm
W c. 83 g

Durango Chipmunk
Neotamias durangae

FR Tamia du Durango DE Durango-Backenhörnchen
ES Ardilla listada de Durango

LC
3 sspp
HB 12.4–12.8 cm
T 10.3–11.7 cm
W c. 69 g

US (from S California) S to NW MX (Baja California Pen)

1200–3000 m

California Chipmunk
Neotamias obscurus

FR Tamia du chaparral
DE Kalifornien-Backenhörnchen
ES Ardilla listada de California

LC 3 sspp
HB 13.2–13.5 cm
T 10.6–11.6 cm
W 68–71 g

0–2940 m

Merriam's Chipmunk
Neotamias merriami

FR Tamia de Merriam
DE Merriam-Backenhörnchen
ES Ardilla listada de Merriam

US (C & S California) S to NW MX (extreme N Baja California)

LC 2 sspp
HB 13.3–13.8 cm
T 10.6–11.2 cm
W c. 70 g

below 1800 m

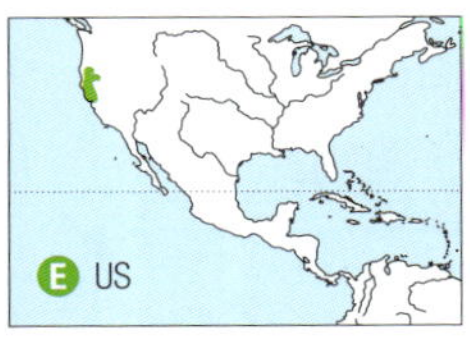

Sonoma Chipmunk
Neotamias sonomae

FR Tamia de Sonoma
DE Sonoma-Backenhörnchen
ES Ardilla listada de Sonoma

NW California (Siskiyou County to San Francisco Bay)

LC 2 sspp
HB 14.3–14.7 cm T 10.3–10.7 cm
W 86–94 g

1200–1800 m

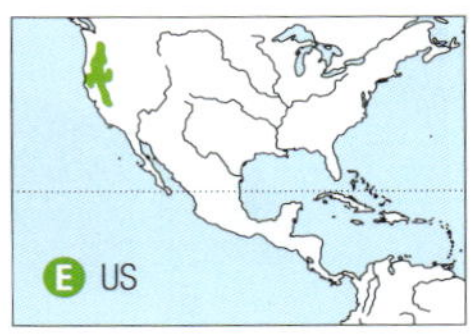

Shadow Chipmunk
Neotamias senex

FR Tamia d'Allen
DE Allen-Backenhörnchen
ES Ardilla listada de Allen

Sierra Nevada, Cascade & other ranges (C Oregon, N California, W Nevada)

Townsend's Chipmunk
Neotamias townsendii

FR Tamia de Townsend
DE Townsend-Backenhörnchen
ES Ardilla listada de Townsend

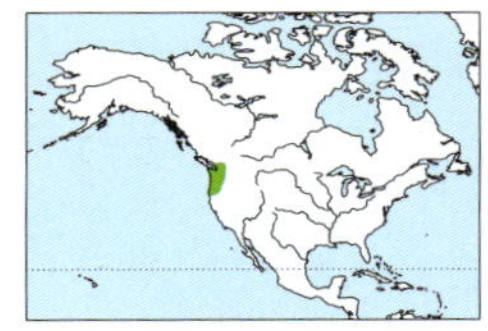

CA (SW British Columbia), US (Washington, Oregon)

LC 2 sspp
HB 13.9–14.6 cm T 11–11.6 cm
W 70–76 g

0–1280 m

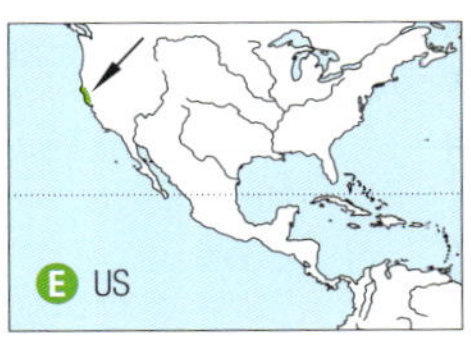

Yellow-cheeked Chipmunk
Neotamias ochrogenys

FR Tamia des séquoias
DE Gelbwangen-Backenhörnchen
ES Ardilla listada de mejillas amarillas

N coast of California, from R Van Duzen to S Sonoma County

LC Monotypic
HB 14.8–15.2 cm
T 10.9–11.5 cm
W 80–180 g

LC
14 sspp
HB 11.9–12.3 cm
T 8.6–9.5 cm
W 43–58 g

Yellow-pine Chipmunk
Neotamias amoenus

FR Tamia amène
DE Sumpfkiefer-Backenhörnchen
ES Ardilla listada de pino amarillo

up to 3000 m

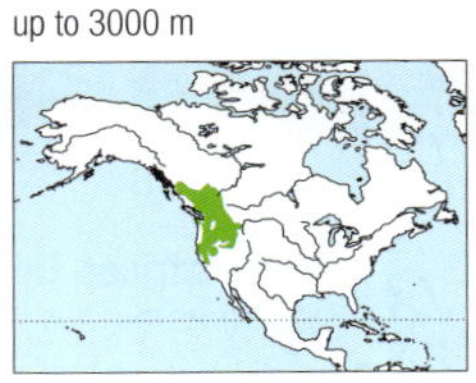

CA (C British Columbia, SW Alberta), US (S to California, N Nevada & NW Utah)

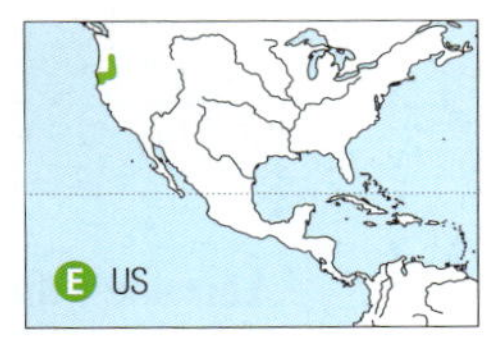

Siskiyou Mts (SW Oregon, NW California)

LC
2 sspp
HB 14.4–14.7 cm
T 10.5–10.7 cm
W c. 75 g

Siskiyou Chipmunk
Neotamias siskiyou

FR Tamia des Siskiyou
DE Siskiyou-Backenhörnchen
ES Ardilla listada de Siskiyou

Genus *SCIUROTAMIAS*

(2 species)

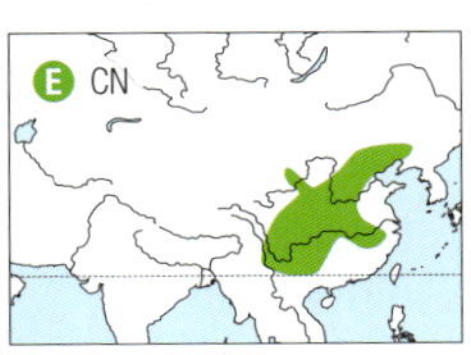

Pere David's Rock Squirrel
Sciurotamias davidianus

FR Tamia de David
DE Pater-David-Rothörnchen
ES Ardilla de Père David

Widespread in rocky habitats

LC
3 sspp
HB 20–21 cm
T 14.1–14.3 cm
W c. 260 g

Forrest's Rock Squirrel
Sciurotamias forresti

FR Tamia de Forrest
DE Forrest-Rothörnchen
ES Ardilla de Forrest

LC
Monotypic
HB c. 22 cm (F)
T c. 16 cm (F)

c. 3000 m

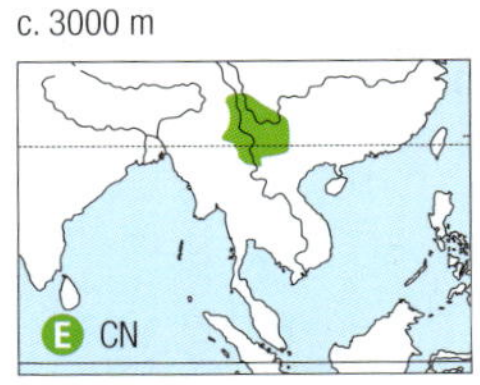

S Sichuan, Yunnan; perhaps N MM, LA & VN

Family GLIRIDAE – Dormice

9 genera, 30 species. Temperate and tropical forests, savannas, steppes, and deserts. Palearctic and Afrotropical Regions.

Genus *GRAPHIURUS*

(15 species)

LC
Monotypic
HB 12–15.5 cm
T 6.2–12.2 cm
W 79–104 g

0–650 m

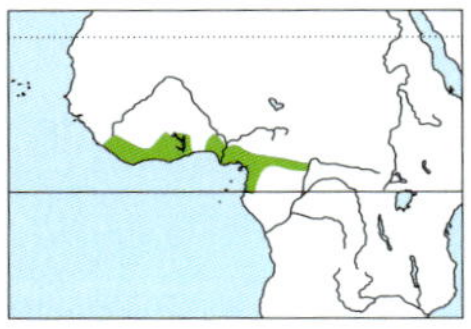

Nagtglas's African Dormouse
Graphiurus nagtglasii

FR Loir de Nagtglas
DE Großer Afrikanischer Bilch
ES Lirón de Nagtglas

W & C Africa, from SE SL to E CM (except BJ) & CF, S to GA; S limit uncertain

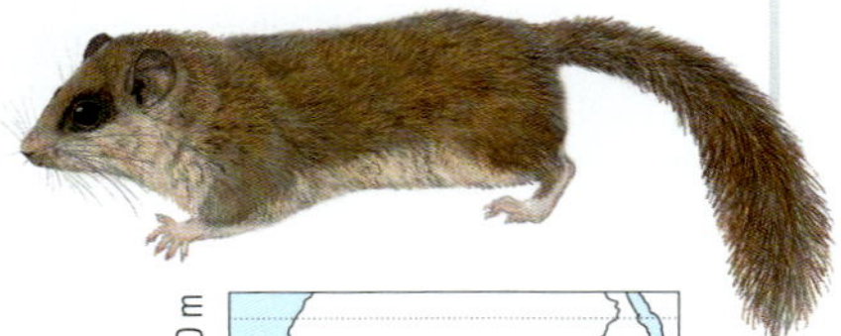

DD
Monotypic
HB 8.3–9.8 cm
T 5.5–7 cm
W 20–29 g

Thick-tailed African Dormouse
Graphiurus crassicaudatus

FR Loir de Jentink
DE Jentink-Bilch
ES Lirón de cola gruesa

0–600 m

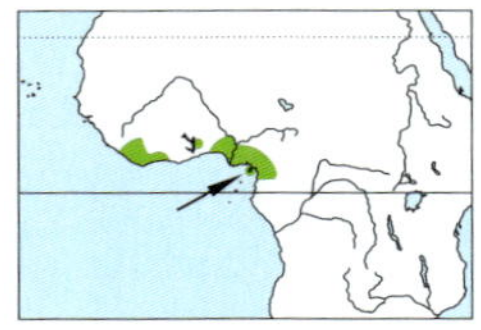

W & C Africa from LR & SW GN E to SW CM (except BJ), Bioko I (GQ)

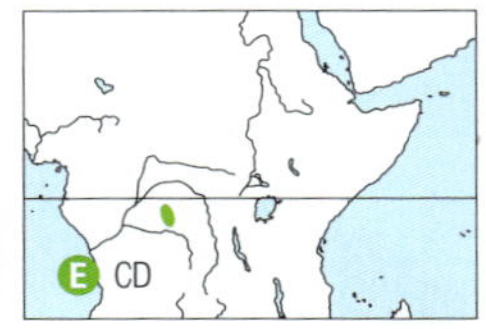

Walter Verheyen's African Dormouse
Graphiurus walterverheyeni

FR Loir de Verheyen
DE Verheyen-Bilch
ES Lirón de Walter Verheyen

Known from R Lukenie & Wafania, near left bank of R Luilaka; limits unknown

DD
Monotypic
HB c. 7.2 cm
T c. 6.4 cm

1000–2000 m

Angolan African Dormouse
Graphiurus angolensis

FR Loir d'Angola
DE Angola-Bilch
ES Lirón de Angola

C & SC AO, NW ZM

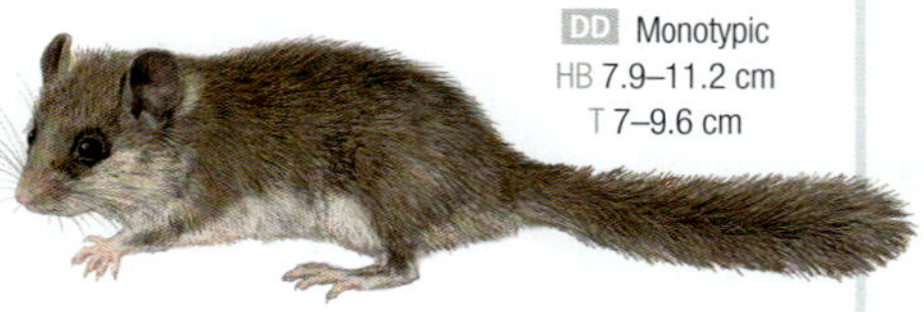

DD Monotypic
HB 7.9–11.2 cm
T 7–9.6 cm

400–1000 m

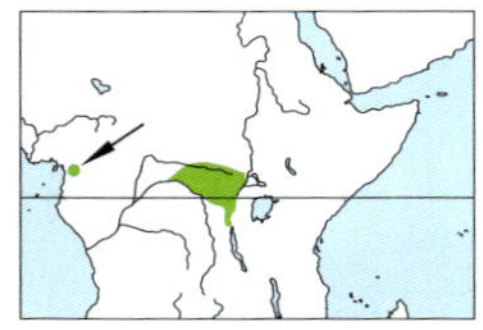

NE CD (N & E of R Congo & R Lualaba), 1 locality in SW CM (Lolodorf)

LC
Monotypic
HB 8.6–10.7 cm
T 7.3–9.5 cm

Christy's African Dormouse
Graphiurus christyi

FR Loir de Christy DE Christy-Bilch ES Lirón de Christy

DD Monotypic
HB 6.9–8.4 cm
T 6.5–7.5 cm

up to 1500 m

S of country, but limits unknown

Johnston's African Dormouse
Graphiurus johnstoni

FR Loir de Johnston
DE Johnston-Bilch
ES Lirón de Johnston

LC Monotypic
HB 7.5–9.2 cm
T 5.4–8.1 cm

0–1525 m

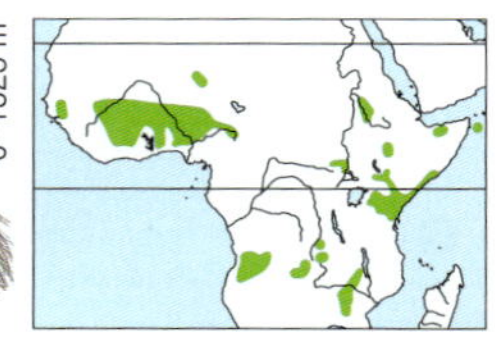

Patchily in SN, GM & GW E to SE SD, SS & SO, S to S AO & SW ZW; perhaps ZA

Kellen's African Dormouse
Graphiurus kelleni

FR Loir de Kellen
DE Kellen-Bilch
ES Lirón de Kellen

LC
Monotypic
HB 7.2–9.3 cm T 5.4–7.4 cm
W 12–24 g

0–2600 m

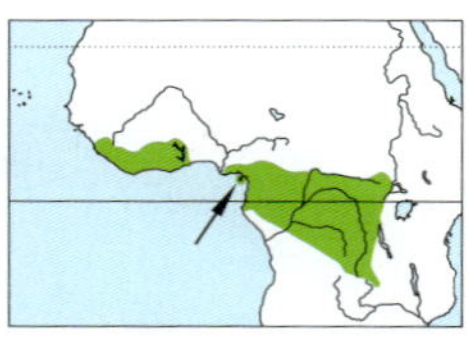

Lorraine's African Dormouse
Graphiurus lorraineus

FR Loir africain
DE Lorraine-Bilch
ES Lirón de Lorraine

W & C Africa from S GN & SL to GH and from SE NG (E of R Niger) E to E CD, S to NE AO & NC ZM, Bioko I

0–1400 m

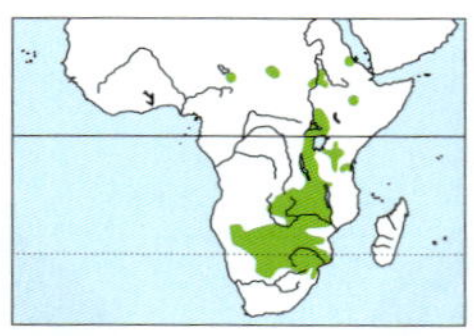

Sub-Saharan Africa in N CM, SW TD, W & E SD, ER & C ET, S through E Africa to E AO, NE NA, N ZA & SZ

Noack's African Dormouse
Graphiurus microtis

FR Loir à oreilles courtes
DE Kleinohrbilch ES Lirón de Noack

LC
Monotypic
HB 7.5–11.5 cm
T 6.2–8.6 cm
W 17.6–42 g

DD
Monotypic
HB c. 16 cm
T c. 13 cm

975–1300 m

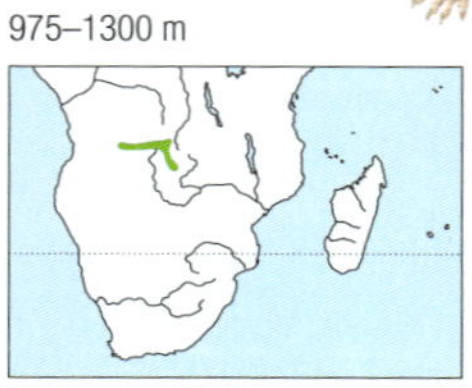

Monard's African Dormouse
Graphiurus monardi

FR Loir de Monard
DE Monard-Bilch
ES Lirón de Monard

C Africa in E AO, SW CD & NW ZM

LC
Monotypic
HB 8.7–11.7 cm
T 7.2–8.9 cm
W 24–34 g

0–4100 m

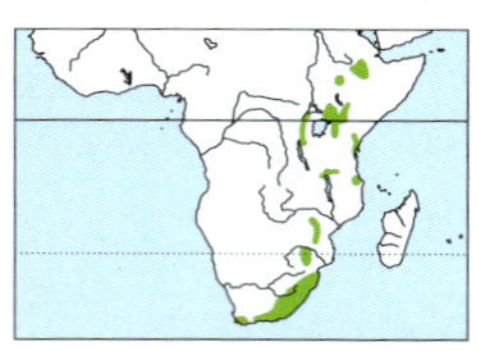

Forest African Dormouse
Graphiurus murinus

FR Loir murin
DE Afrikanischer Bilch
ES Lirón de bosque

E & S Africa patchily in ET, KE, UG, RW, BI, TZ, MW, ZM, ZW, MZ, ZA, SZ & LS

LC
Monotypic
HB 11.7–14.5 cm
T 10.3–15 cm
W 72–85 g

0–1585 m

Northern Cape, Western Cape, Eastern Cape

Spectacled African Dormouse
Graphiurus ocularis

FR Loir du Cap
DE Brillenbilch
ES Lirón de anteojos

LC
Monotypic
HB 9.5–12.2 cm
T 6.5–9.8 cm
W 30–53 g

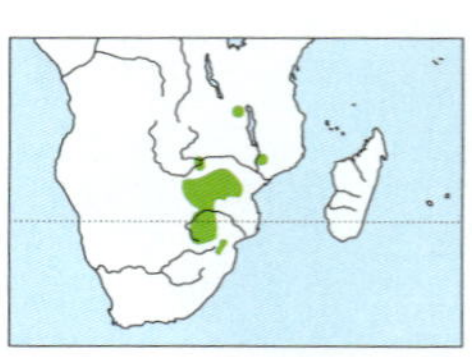

Flat-headed African Dormouse
Graphiurus platyops

FR Loir de Thomas
DE Südafrikanischer Felsenbilch
ES Lirón de cabeza plana

SE Africa in NE & S ZM, S MW, ZW, WC MZ, E BW, NE ZA & SZ

LC
Monotypic
HB 10.5–11.9 cm T 9.6–11.8 cm

400–1650 m

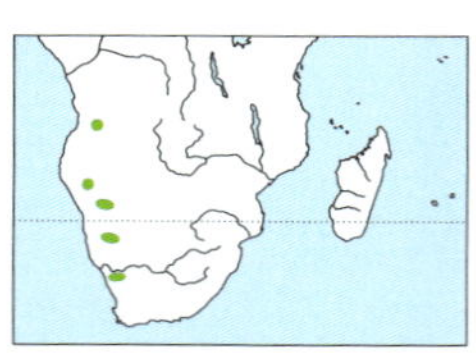

Rupicolous African Dormouse
Graphiurus rupicola

FR Loir des rochers
DE Namaqualand-Felsenbilch
ES Lirón de roca

SW Africa in WC AO, C NA & NW ZA (NW Northern Cape)

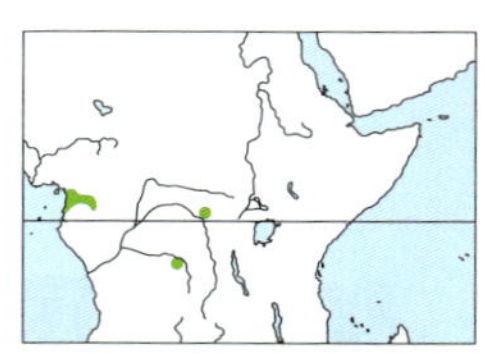

Short-eared African Dormouse
Graphiurus surdus

FR Loir sourd
DE Kurzohrbilch
ES Lirón de orejas cortas

C Africa in S CM, GQ, N GA and NE & SC CD (Masako, Inkongo); limits unknown

DD Monotypic
HB 8.7–11 cm
T 6.5–8.2 cm
W 18–34 g

Genus *GLIRULUS*

(1 species)

Japanese Dormouse
Glirulus japonicus

FR Loir du Japon
DE Japan-Bilch
ES Lirón de Japón

400–1880 m

LC
Monotypic
HB 6.6–9.3 cm
T 3.8–5.9 cm
W 14–45 g

Honshu, Shikoku, Kyushu, Oki Is

Genus *GLIS*

(2 species)

LC
Monotypic
HB 13.1–18.5 cm
T 10–17.5 cm
W 79–228 g

0–2000 m

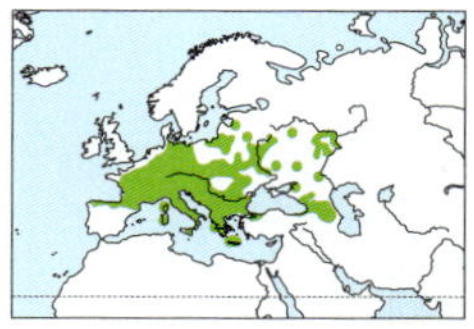

Europe, from N ES E to middle R Volga (RU) & Caucasus; introduced to UK

European Fat Dormouse
Glis glis

FR Loir gris
DE Europäischer Siebenschläfer
ES Lirón gris

Iranian Fat Dormouse
Glis persicus

FR Loir de Perse DE Iranischer Siebenschläfer ES Lirón de Irán

SE AZ (S R Kura & R Aras), IR (E to E Golestan); recorded in TM

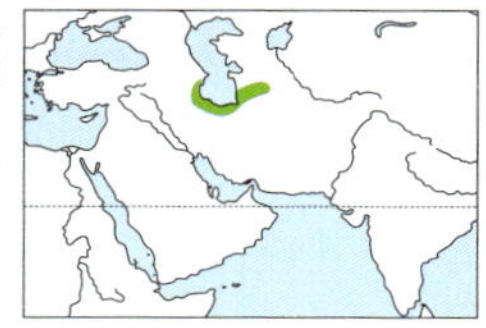

NE Monotypic
HB 14.8–25 cm T 11.5–20 cm

Genus *MUSCARDINUS*

(1 species)

Hazel Dormouse
Muscardinus avellanarius

FR Loir muscardin
DE Haselmaus
ES Muscardino

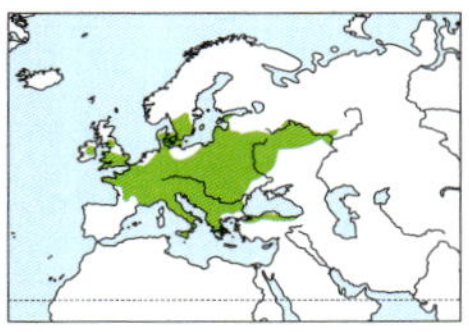

up to 2000 m

LC Monotypic
HB 6.5–9.1 cm
T 5.7–8.6 cm
W 17–30 g

From British Is & FR through most of Europe to RU (R Volga), S to Sicily & TR

NE Mediterranean in SE BG & W TR (E Thrace, W Anatolia); perhaps NE GR

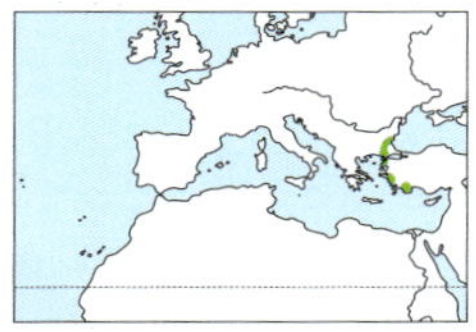

VU Monotypic
HB 8.6–13.6 cm
T 6.5–9.4 cm
W 21–70 g

Roach's Mouse-tailed Dormouse
Myomimus roachi

FR Loir de Roach DE Mausschläfer ES Lirón de Roach

Genus *MYOMIMUS*

(3 species)

DD Monotypic
HB 7.5–8.8 cm
T 6–6.7 cm

1800–2800 m

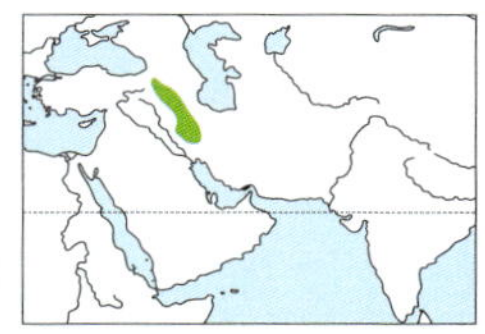

Extreme E TR, NW IR (Zagros Mts); limits unknown

Setzer's Mouse-tailed Dormouse
Myomimus setzeri

FR Loir de Setzer DE Setzer-Mausschläfer ES Lirón de Setzer

SW TM (Kopet Dag & Malyy Balkhan Mts), NE IR; historic remains from WC AF

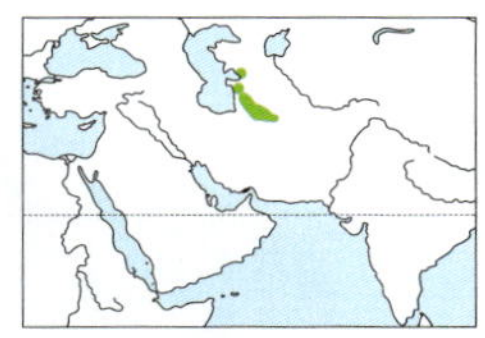

Ognev's Mouse-tailed Dormouse
Myomimus personatus

FR Loir d'Ognev DE Ognev-Mausschläfer
ES Lirón de Ognev

700–2400 m

DD Monotypic
HB 7.3–7.7 cm T 5.3–6.9 cm
W c. 11.6 g

Genus *SELEVINIA*

(1 species)

Desert Dormouse
Selevinia betpakdalaensis

FR Loir du désert DE Salzkrautbilch
ES Lirón de desierto

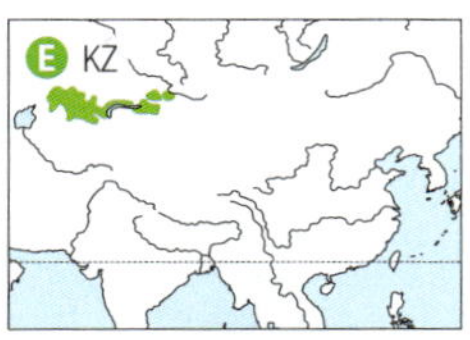

Dar'yalyktakyr Desert E to Zaysan & Alakol basins; perhaps NW CN, N limits uncertain

DD Monotypic
HB 7.5–9.5 cm T 5.8–7.7 cm
W 19–21 g

Genus *CHAETOCAUDA*

(1 species)

2320–4890 m

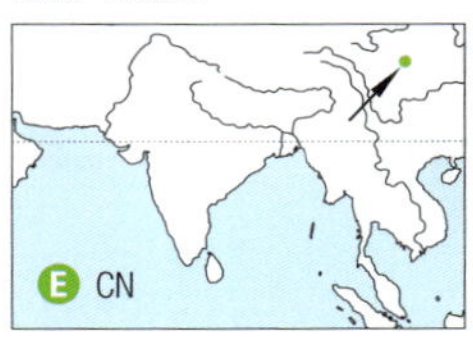

Known only from Wanglang NR (N Sichuan)

DD Monotypic
HB 9–9.1 cm T 9.2–10.2 cm
W 24–36 g

Sichuan Dormouse
Chaetocauda sichuanensis

FR Lérotin du Sichuan DE Sichuan-Bilch ES Lirón de Sichuán

Eurasian Forest Dormouse
Dryomys nitedula

FR Lérotin de forêt
DE Baumschläfer
ES Lirón euroasiático

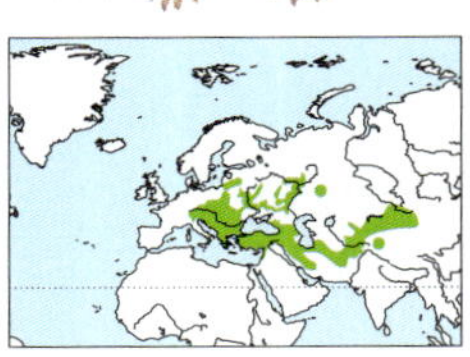

From E CH through Europe (disjunct in S IT), Anatolia & Caucasus to C RU, W MN & W CN

LC
Monotypic
HB 7.9–10.3 cm
T 6.6–10.1 cm

DD
Monotypic
HB 8.3–9.6 cm
T 4.8–7.6 cm
W 17–32 g

Woolly Forest Dormouse
Dryomys laniger

FR Lérotin laineux
DE Felsenschläfer
ES Lirón lanudo

1620–2950 m

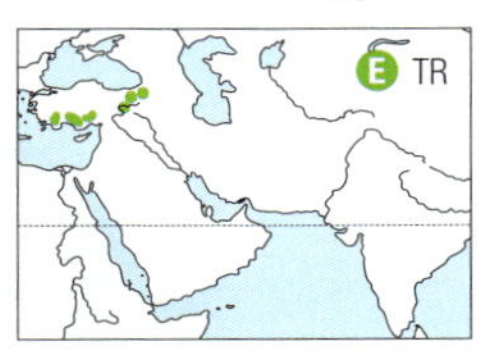

W Taurus Mts & E Anatolia

Genus *DRYOMYS*

(3 species)

VU
Monotypic
HB 9.9–10.3 cm
T c. 9.3 cm
W 33–38 g

1980–2590 m

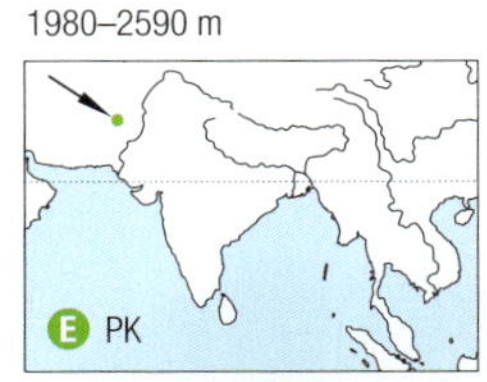

Know from NE of Quetta, Urak Valley & Ziarat (Balochistan Prov)

Niethammer's Forest Dormouse
Dryomys niethammeri

FR Lérotin de Niethammer
DE Belutschistan-Baumschläfer
ES Lirón de Niethammer

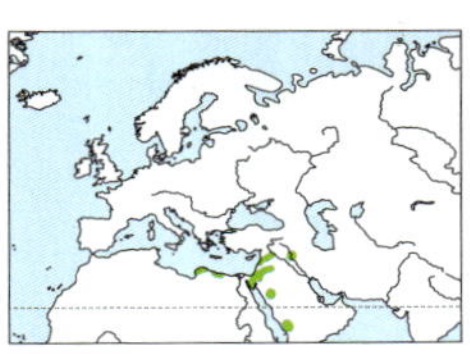

0–2850 m

From E Libya, EG & Sinai Pen to N SY, N IQ & SW SA; old records from S TR

Genus *ELIOMYS*

(3 species)

Black-tailed Garden Dormouse
Eliomys melanurus

FR Lérot d'Asie DE Löffelbilch
ES Lirón careto de cola negra

LC Monotypic
HB 11.1–14.4 cm T 10–13.6 cm
W 38–63 g

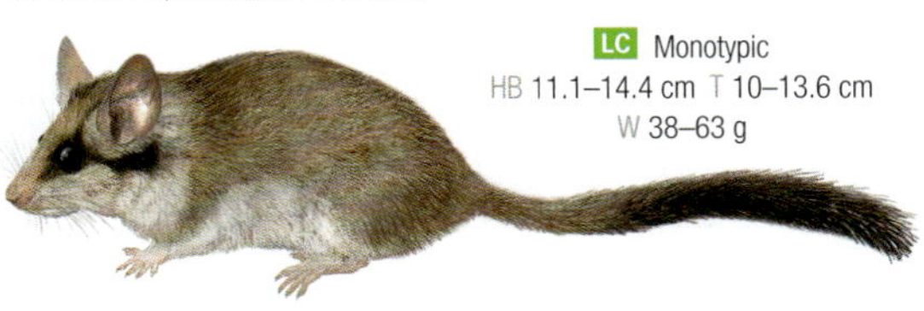

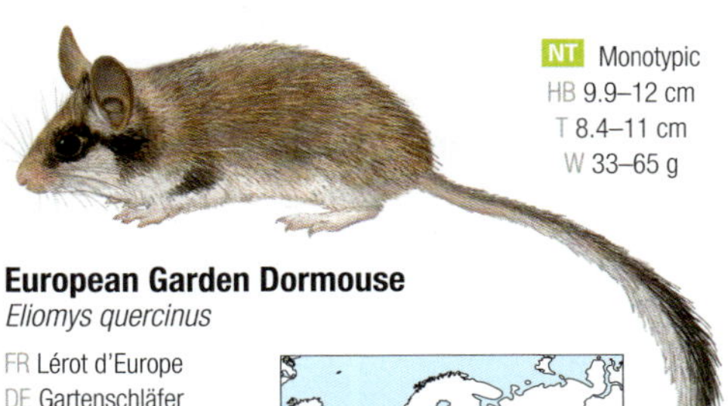

NT Monotypic
HB 9.9–12 cm
T 8.4–11 cm
W 33–65 g

European Garden Dormouse
Eliomys quercinus

FR Lérot d'Europe
DE Gartenschläfer
ES Lirón careto europeo

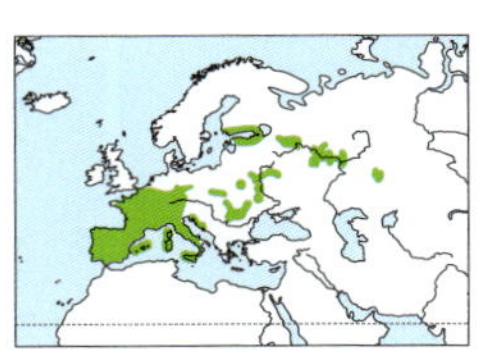

From Iberian Pen, FR & IT fragmented through Europe to E Ural Mts (RU)

0–2300 m

Maghreb Garden Dormouse
Eliomys munbyanus

FR Lérot du Maghreb
DE Maghreb-Gartenschläfer
ES Lirón careto del Magreb

0–3800 m

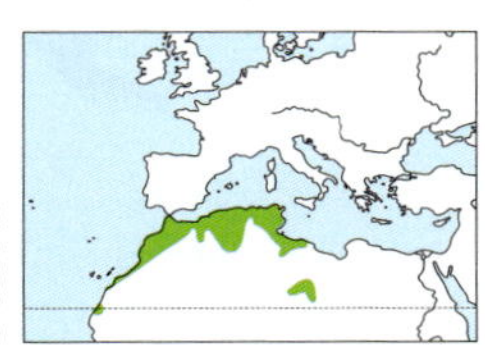

N Africa from SW EH, MA, N DZ & TN to W LY (E to Murqub, S to Fezzan)

LC
Monotypic
HB 10–14 cm T 9.6–11.8 cm
W 42–62 g

Order EULIPOTYPHLA (4 families)

Family ERINACEIDAE – Hedgehogs and Gymnures

10 genera, 26 species. Deserts, steppes, grasslands, temperate forests, and urban environments (hedgehogs) or humid, tropical forests (gymnures). Palearctic, Afrotropical and Indo-Malayan Regions.

Genus *ERINACEUS*

(4 species)

West European Hedgehog
Erinaceus europaeus

FR Hérisson commun
DE Braunbrustigel
ES Erizo europeo

up to 2000 m

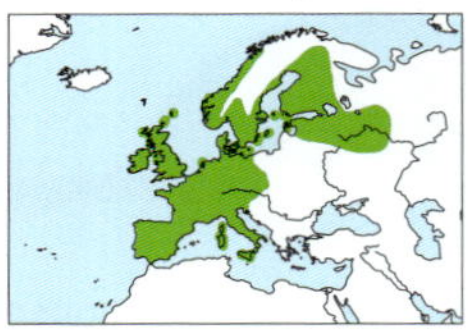

LC Monotypic
HB 21–24 cm
T 2.4–3.2 cm
W 0.5–1.2 kg

W Europe N to Fennoscandia, E to NW RU, many Is; introduced to NZ & Azores

Northern White-breasted Hedgehog
Erinaceus roumanicus

FR Hérisson des Balkans
DE Nördlicher Weißbrustigel
ES Erizo de los Balcanes

at least 1400 m

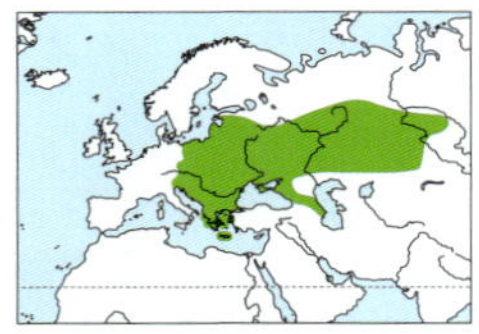

LC 5 sspp
HB 21–31 cm
T 2.4–3.1 cm
W 0.6–1.2 kg

From C & E Europe, Baltic & Balkans E to W Siberia, S to Creta & N Caucasus

Southern White-breasted Hedgehog
Erinaceus concolor

FR Hérisson de Turquie
DE Südlicher Weißbrustigel
ES Erizo oriental de vientre blanco

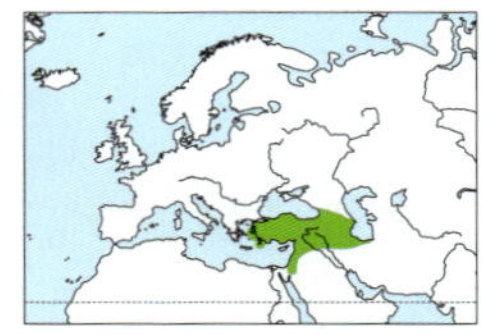

LC Monotypic
HB 20–26 cm
T 2–3 cm
W 0.5–1.5 kg

Aegean Is & TR E to S Caucasus, N IQ & NW IR, S to Levant (LB, SY, IL, JO)

Amur Hedgehog
Erinaceus amurensis

FR Hérisson de Mandchourie
DE Amur-Igel
ES Erizo de Manchuria

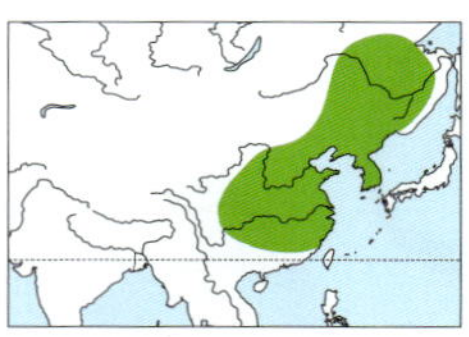

LC
Monotypic
HB 15.8–30 cm
T 1.7–4.2 cm
W 0.8–1.4 kg

E RU (R Amur basin, E from R Zeya), Korean Pen, C & E CN (S to E Sichuan & N Guangdong)

Genus *ATELERIX*

(4 species)

North African Hedgehog
Atelerix algirus

FR Hérisson d'Algérie
DE Nordafrikanischer Igel
ES Erizo moruno

up to 2000 m

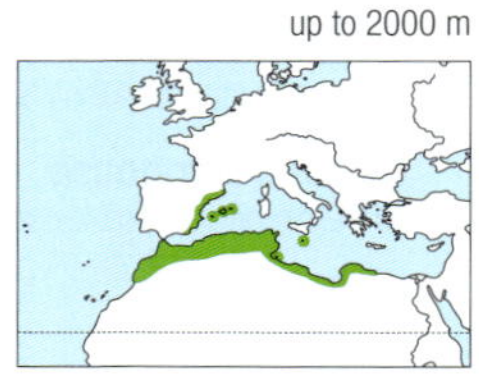

LC
2 sspp
HB 20–27 cm
T 1.5–3 cm
W 500–865 g

Native from MA E to NE LY; old introductions to E ES, Balearic Is & Malta; modern (Canary Is) not shaded

Four-toed Hedgehog
Atelerix albiventris

FR Hérisson à ventre blanc
DE Weißbauchigel
ES Erizo africano de vientre blanco

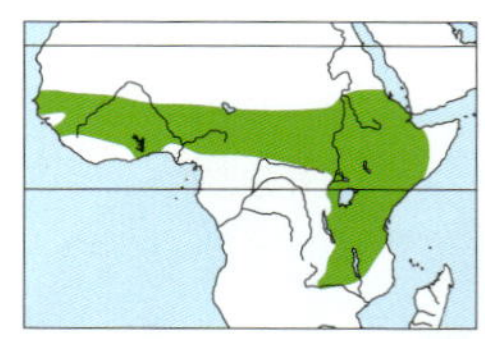

LC
Monotypic
HB 14–26 cm
T 0.9–2.5 cm
W 250–680 g

SN E to ER, DJ & SO, S through E Africa to R Zambezi

Somali Hedgehog
Atelerix sclateri

FR Hérisson de Somalie
DE Somalischer Igel
ES Erizo de Somalia

LC
Monotypic
HB 21–26 cm
T 1.4–2 cm

Maritime plain near hills in arid scrublands

Southern African Hedgehog
Atelerix frontalis

FR Hérisson d'Afrique du Sud
DE Kap-Igel
ES Erizo de Sudáfrica

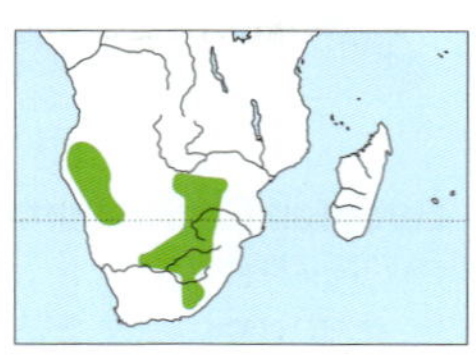

LC
2 sspp
HB 17–21 cm
T 1.9–3 cm
W 291–479 g

SW AO, N & C NA and E BW, W & C ZW, ZA

Genus *PARAECHINUS*

(4 species)

Desert Hedgehog
Paraechinus aethiopicus

FR Hérisson du désert
DE Äthiopischer Igel
ES Erizo del desierto

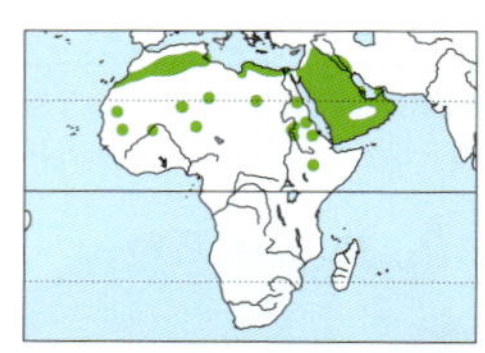

LC
Monotypic
HB 13–24 cm T 1–3 cm
W 285–700 g

Sahara from MR E to ET, deserts in Middle East, Djerba (TN), BH, Tunb Is (S IR)

Brandt's Hedgehog
Paraechinus hypomelas

FR Hérisson de Brandt
DE Brandt-Igel
ES Erizo de Brandt

LC

5 sspp
HB 13.6–29 cm
T 1.8–3.8 cm

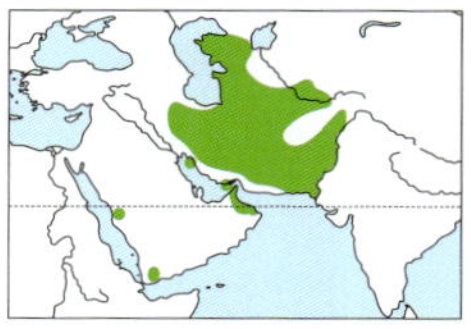

Arabian Pen (SA, YE, OM) & IR (with Kharg & Tunb Is) N to SW KZ & SW TJ, E to PK

Indian Hedgehog
Paraechinus micropus

FR Hérisson indien
DE Indischer Igel
ES Erizo de India

LC

Monotypic
HB 13–23 cm T 1–4 cm
W 300–600 g

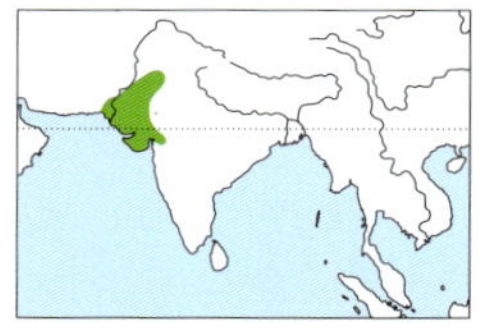

SE PK (SE Balochistan, Punjab, Sind), NW IN (W Rajasthan, Gujarat, NW Maharashtra)

Bare-bellied Hedgehog
Paraechinus nudiventris

FR Hérisson à ventre nu
DE Nacktbauchigel
ES Erizo de vientre desnudo

LC

Monotypic

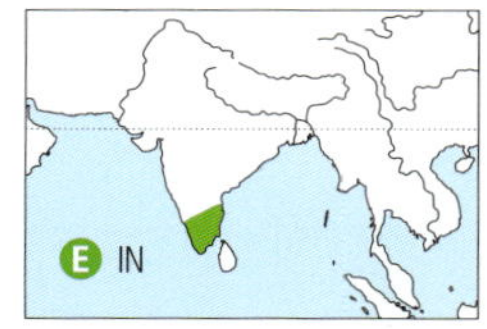

Andhra Pradesh, Tamil Nadu, Kerala

Genus *HEMIECHINUS*

(2 species)

Common Long-eared Hedgehog
Hemiechinus auritus

FR Hérisson à oreilles longues
DE Eigentlicher Langohrigel
ES Erizo de orejas largas dorado

LC

3 sspp
HB 16–28 cm T 1–5.5 cm
W 0.2–1.1 kg

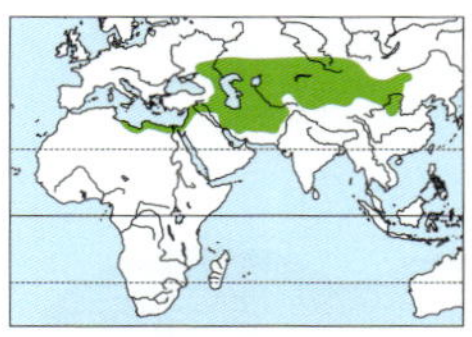

From E UA to MN & C CN, NE Africa (N LY, N EG), Middle East to W PK; introduced to CY

Indian Long-eared Hedgehog
Hemiechinus collaris

FR Hérisson à collier
DE Indischer Langohrigel
ES Erizo de orejas largas de India

LC

Monotypic
HB 14–18 cm T c. 2.3 cm
W 257–306 g

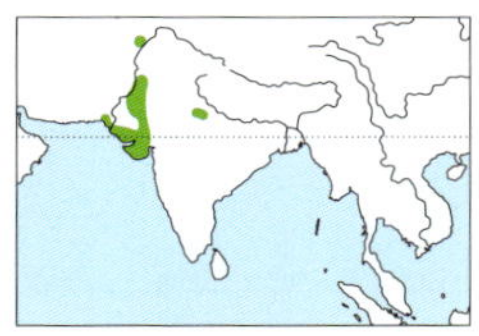

PK (KPK, Punjab, Sind), IN (Gujarat, Rajasthan, Uttar Pradesh)

Genus *MESECHINUS*

(4 species)

Wang's Forest Hedgehog
Mesechinus wangi

FR Hérisson de Wang
DE Gaoligong-Igel
ES Erizo de Wang

NE

Monotypic
HB 17.7–24 cm T 1.4.–1.8 cm
W 336–451 g

2100–2680 m

Gaoligongshan National NR (Yunnan)

Small-toothed Forest Hedgehog
Mesechinus miodon

FR Hérisson du Shaanxi
DE Steppenigel
ES Erizo de dientes pequeños

NE

Monotypic
HB 12–22 cm T 2.5–4.6 cm
W 230–750 g

E Ningxia, N Shaanxi

Daurian Hedgehog
Mesechinus dauuricus

FR Hérisson de Daourie
DE Daurischer Igel
ES Erizo del Gobi

LC

Monotypic
HB 24–27 cm
T 2.6–3.4 cm
W 1–1.1 kg

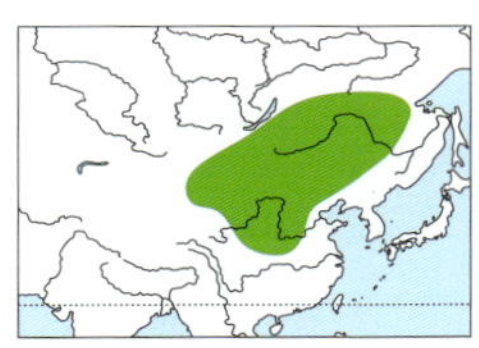

RU (Transbaikalia E to upper R Amur Basin), C & E MN, N & C CN

Hugh's Hedgehog
Mesechinus hughi

FR Hérisson de Hugh
DE Hugh-Igel
ES Erizo de Hugh

LC

Monotypic
HB 14.8–23 cm
T 1.2–2.4 cm
W 112–750 g

S Gansu, S Shaanxi, Shanxi, S Henan, NE & SE Sichuan

Genus *ECHINOSOREX*

(1 species)

Moonrat
Echinosorex gymnurus

FR Grand Gymnure
DE Großer Rattenigel
ES Gimnuro

LC

2 sspp
HB 25–46 cm
T 16.5–30 cm
W 0.4–2 kg

Thai-Malay Pen, Sumatra (with Tebingtinggi I), Borneo (with Labuan I)

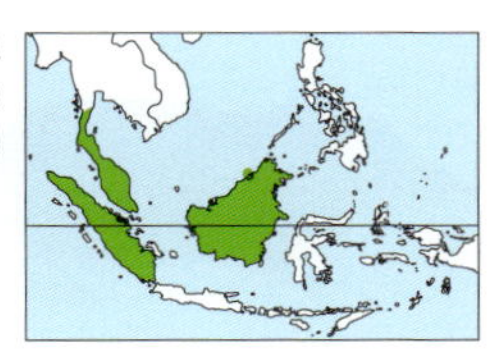

Mindanao Gymnure
Podogymnura truei

FR Gymnure de Mindanao
DE Mindanao-Rattenigel
ES Gimnuro de Mindanao

LC
Monotypic
HB 20–22 cm
T 4–7 cm
W 52–82 g

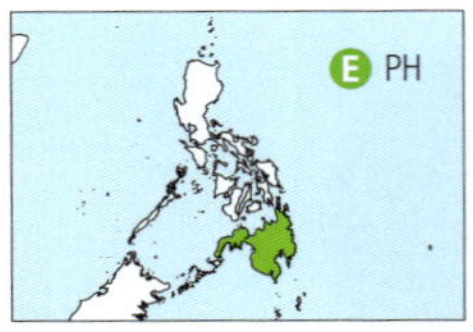

1300–2900 m

Mindanao I

Genus *PODOGYMNURA*

(2 species)

Dinagat Gymnure
Podogymnura aureospinula

FR Gymnure de Dinagat
DE Dinagat-Rattenigel
ES Gimnuro de Dinagat

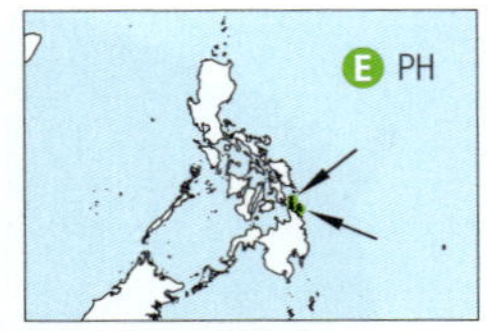

Dinagat, Bucas Grande

EN
Monotypic
HB 19–21 cm
T 5.9–7.3 cm

Genus *HYLOMYS*

(3 species)

DD
Monotypic
HB 11.5–13.5 cm
T 8.2–9.4 cm

Long-eared Gymnure
Hylomys megalotis

FR Gymnure oreillard
DE Langohr-Rattenigel
ES Gimnuro de orejas largas

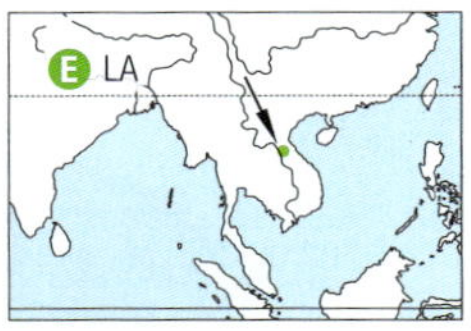

Khammouan Prov

VU
Monotypic
HB 10–12.7 cm
T 1.5–3.2 cm

2000–3000 m

Dwarf Gymnure
Hylomys parvus

FR Gymnure nain
DE Zwergrattenigel
ES Gimnuro enano

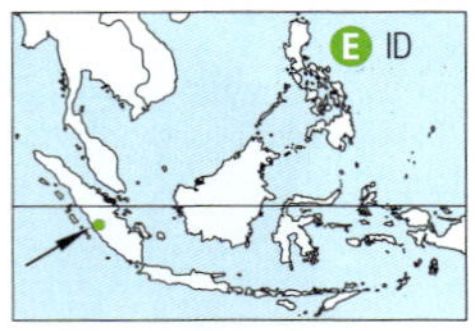

WC Sumatra (Mt Kerinci)

LC
7 sspp
HB 9–15.6 cm
T 1–3 cm
W 45–80 g

Short-tailed Gymnure
Hylomys suillus

FR Petit Gymnure
DE Kurzschwanz-Rattenigel
ES Gimnuro de cola corta

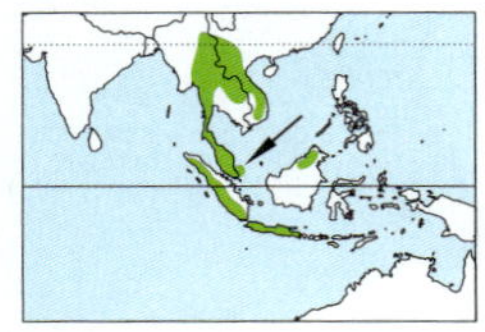

90–3400 m

From S CN (Yunnan), E MM, LA, VN S to Thai-Malay Pen (with Tioman I), W Sumatra, Java, N Borneo

Shrew Gymnure
Neotetracus sinensis

FR Gymnure de Chine
DE Spitzmausigel
ES Gimnuro de China

Genus *NEOTETRACUS*

(1 species)

1500–2700 m

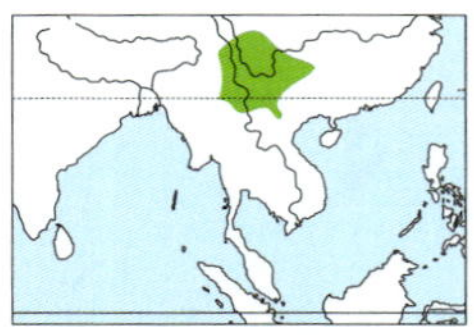

SC CN (Sichuan, Guizhou, Yunnan), adjacent NE MM & N VN

LC
Monotypic
HB 10.8–11 cm
T 6.2–6.5 cm
W 36–38 g

Hainan Gymnure
Neohylomys hainanensis

FR Gymnure de Hainan
DE Hainan-Rattenigel
ES Gimnuro de Hainan

Genus *NEOHYLOMYS*

(1 species)

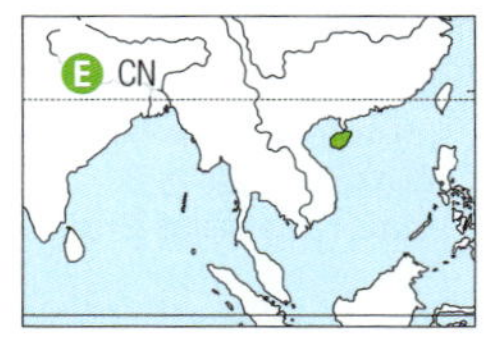

Hainan I

EN
Monotypic
HB 12–14.7 cm
T 3.6–4.3 cm
W 50–70 g

Family SORICIDAE – Shrews

29 genera, 476 species. Typically moist areas with dense ground cover, from sea level to elevations of about 4400 m. Holarctic, Indo-Malayan, Afrotropical and Neotropical Regions.

Genus *SOREX*

(86 species)

Alpine Shrew
Sorex alpinus

FR Musaraigne alpine
DE Alpenspitzmaus
ES Musaraña alpina

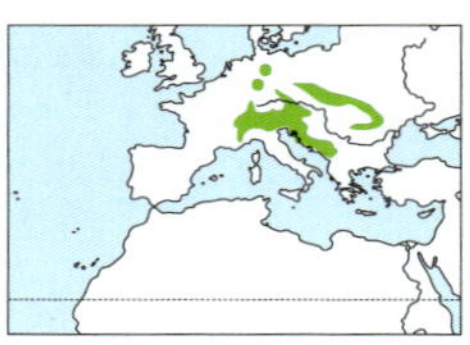

Fragmented in Alps, Balkans, Carpathians, Tatra Mts, some isolated mts in DE & CZ

NT
3 sspp
HB 6.2–7.7 cm
T 6–7.5 cm
W 5.5–11.5 g

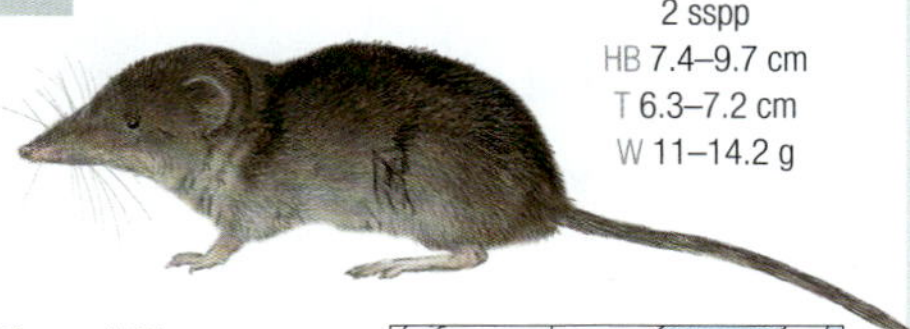

DD
2 sspp
HB 7.4–9.7 cm
T 6.3–7.2 cm
W 11–14.2 g

Ussuri Shrew
Sorex mirabilis

FR Musaraigne de l'Oussouri
DE Ussuri-Spitzmaus
ES Musaraña de Ussuri

E RU (Primorsky Krai), Korean Pen, NE CN (Heilongjiang, Jilin)

LC
Monotypic
HB 6–6.8 cm
T 5–6.1 cm
W 5–10 g

Chinese Highland Shrew
Sorex excelsus

FR Musaraigne des sommets
DE Chinesische Hochland-Spitzmaus
ES Musaraña de tierras altas de China

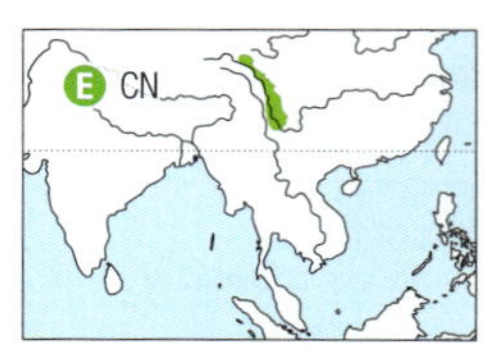

2000–4000 m

SE Qinghai, SE Tibet, W Sichuan (W of R Yalong), NW Yunnan; perhaps NP

LC
Monotypic
HB 5.4–7.7 cm
T 5.5–6.2 cm

c. 3000 m

Greater Striped Shrew
Sorex cylindricauda

FR Musaraigne du Sichuan
DE Rückenstreifen-Spitzmaus
ES Musaraña listada mayor

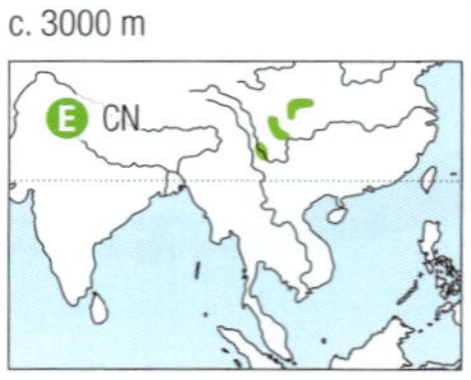

Gansu, Shaanxi, Sichuan, Yunnan

LC
4 sspp
HB 5.5–7.7 cm
T 4.2–6.6 cm

2000–4400 m

Lesser Striped Shrew
Sorex bedfordiae

FR Musaraigne de Bedford
DE Kleine Streifenspitzmaus
ES Musaraña listada menor

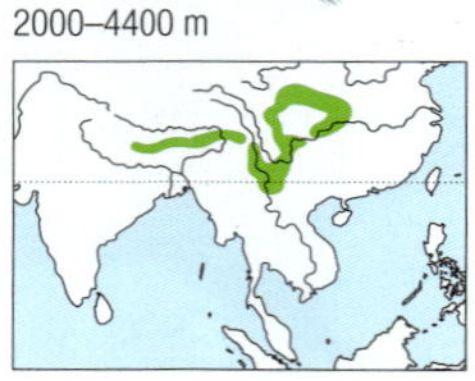

NP, CN (Gansu & Shaanxi S to S Tibet, W Sichuan, Yunnan), adjacent MM

LC
Monotypic
HB 7–8 cm
T 4.3–5.4 cm
W 9.2–13 g

450–2400 m

Radde's Shrew
Sorex raddei

FR Musaraigne de Radde
DE Radde-Spitzmaus
ES Musaraña de Radde

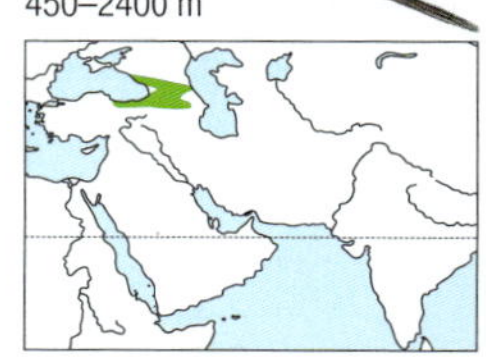

Caucasus (S European RU, GE, AM, AZ), NE TR

LC
5 sspp
HB 5.8–8.6 cm T 3–4.8 cm
W 7–15.5 g

Flat-skulled Shrew
Sorex roboratus
FR Musaraigne forte
DE Flachkopf-Rotzahnspitzmaus
ES Musaraña de cráneo aplastado

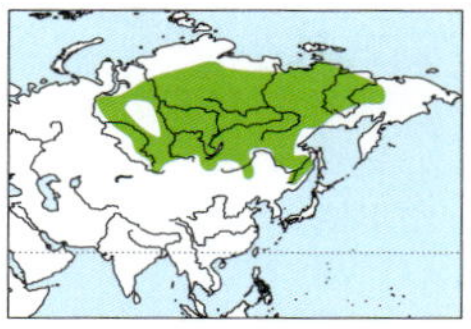

RU (from Urals E through Siberia to R Kolyma & Primorsky Krai), N MN, NE CN

LC
11 sspp
HB 3–5.3 cm
T 1.7–3.3 cm
W 1.4–4 g

Eurasian Least Shrew
Sorex minutissimus
FR Musaraigne minuscule
DE Knirpsspitzmaus
ES Musaraña mínima de Eurasia

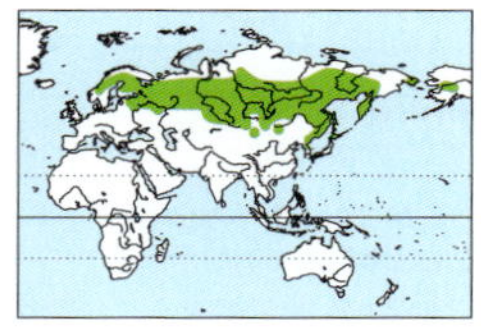

From Fennoscandia through RU to Pacific coast, also W KZ, MN, NE CN, Korean Pen, JP & Alaska (US)

LC
Monotypic
HB 5.5–5.8 cm
T 4.2–5.5 cm
W 2.7–5.5 g

Azumi Shrew
Sorex hosonoi
FR Musaraigne d'Hosono
DE Azumi-Spitzmaus
ES Musaraña de Azumi

900–3000 m

Middle and high mts in C Honshu

Slender Shrew
Sorex gracillimus
FR Musaraigne gracile
DE Schlanke Spitzmaus
ES Musaraña esbelta

LC
3 sspp
HB 4.5–6.6 cm T 3.6–4.9 cm
W 2.5–5.3 g

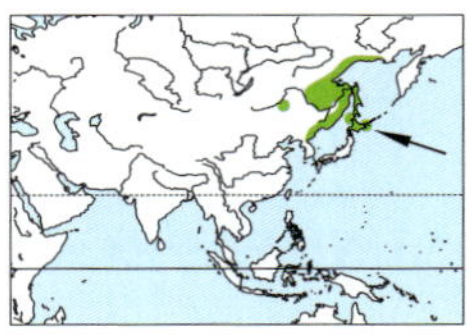

E RU (from S Magadan S to Primorsky Krai, Sakhalin), NE CN, KP, N JP (Hokkaido), S Kuril Is

Laxmann's Shrew
Sorex caecutiens
FR Musaraigne lapone
DE Maskenspitzmaus
ES Musaraña de Laxmann

LC
8 sspp
HB 5–7.3 cm T 2.8–4.5 cm
W 4–12.5 g

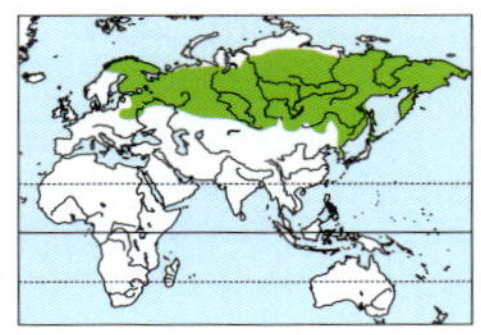

From Fennoscandia & PL E through RU to Pacific coast, S to KZ, MN, NW & NE CN, JP

Shinto Shrew
Sorex shinto
FR Musaraigne shinto
DE Shinto-Spitzmaus
ES Musaraña sintoista

LC
3 sspp
HB 5.1–7.2 cm T 4.4–5.7 cm
W 4.1–6.2 g

0–1200 m

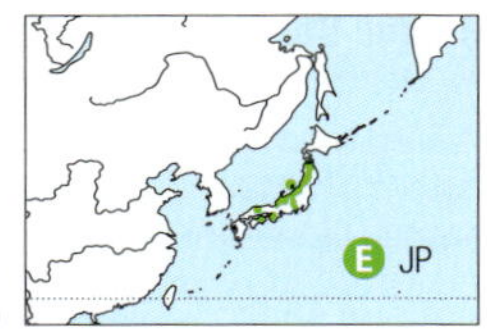

Honshu, Shikoku (Shizuchi & Tsurugi Mts), N Sado

LC
6 sspp
HB 5.4–8.6 cm T 3.7–5.5 cm
W 6.1–16.3 g

Taiga Shrew
Sorex isodon
FR Musaraigne de la taïga
DE Taigaspitzmaus
ES Musaraña de taiga

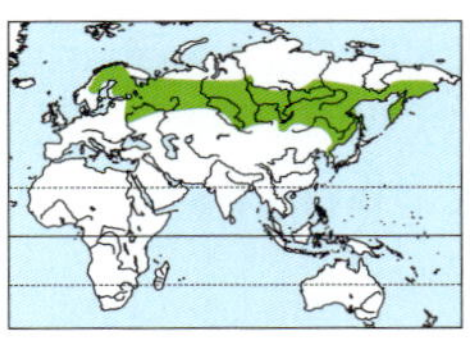

From Fennoscandia & NE BY through RU to Pacific coast, S to N MN, NE CN & Korean Pen

LC
Monotypic
HB 7–9.1 cm
T 4–5.5 cm
W 6.9–15 g

Long-clawed Shrew
Sorex unguiculatus
FR Musaraigne à longues griffes
DE Amur-Langkrallen-Spitzmaus
ES Musaraña de uñas largas

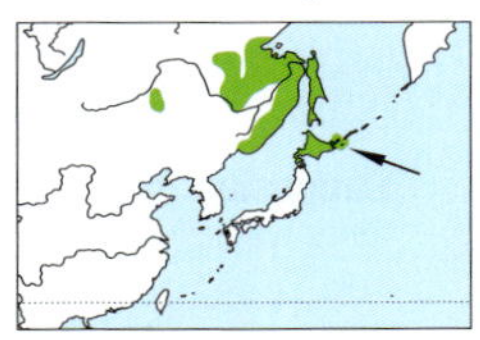

RU (S Far East, Sakhalin I), NE CN, N JP (Hokkaido), S Kuril Is; perhaps KP

DD
Monotypic
HB 6.4–8.5 cm
T 4.9–6.8 cm

2700–3000 m

Chinese Shrew
Sorex sinalis
FR Musaraigne de Chine
DE Chinesische Spitzmaus
ES Musaraña de China

Mts of S Gansu, S Shaanxi & W Sichuan; perhaps Tibet

LC
2 sspp
HB 5.6–8.2 cm
T 3.7–5.2 cm
W 5–14.5 g

Common Shrew
Sorex araneus
FR Musaraigne carrelet
DE Waldspitzmaus
ES Musaraña bicolor

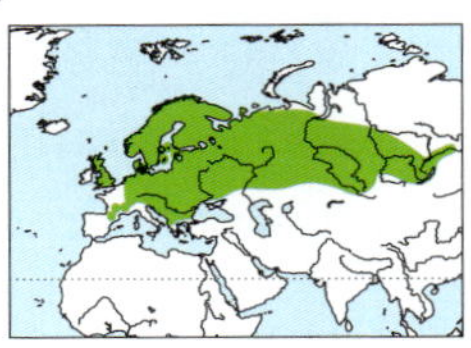

From Great Britain E through Europe & RU to SW Yakutia & N MN; isolated in Pyrenees & C FR

LC
Monotypic
HB 6.3–7.2 cm
T 3.8–4.4 cm
W 4.5–8 g

0–2000 m

Iberian Shrew
Sorex granarius
FR Musaraigne ibérique
DE Iberische Waldspitzmaus
ES Musaraña ibérica

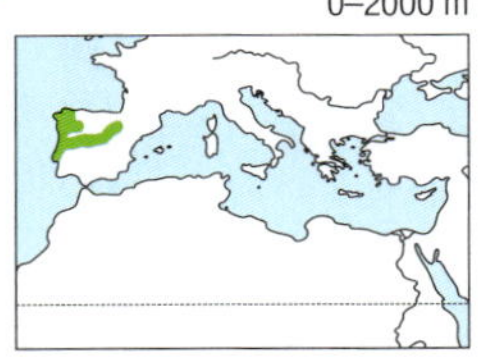

N PT (N of R Tagus), ES (Galicia, W León, NW Zamora, Central System range)

LC
Monotypic
HB 5.5–8.2 cm
T 4–5.5 cm
W 5.7–13 g

Valais Shrew
Sorex antinorii
FR Musaraigne du Valais
DE Valais-Spitzmaus
ES Musaraña de Valais

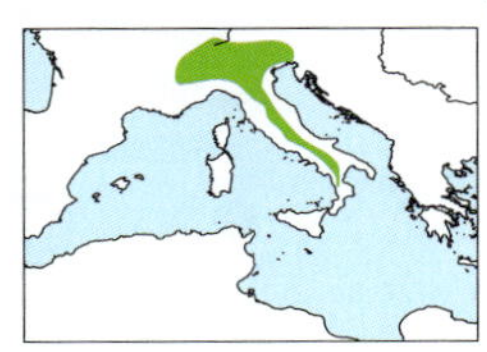

Alps & Apennines of CH, SE FR & IT

LC
Monotypic
HB 6.8–8 cm
T 3.7–4.6 cm
W 6.5–11.8 g

Crowned Shrew
Sorex coronatus
FR Musaraigne couronnée
DE Schabrackenspitzmaus
ES Musaraña de Millet

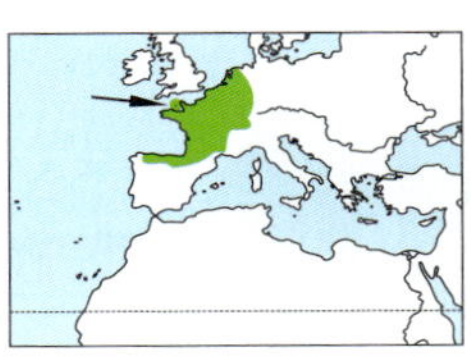

N ES, FR & CH, N to DE, Jersey I

Caucasian Shrew
Sorex satunini
FR Musaraigne de Satunin
DE Kaukasus-Spitzmaus
ES Musaraña del Cáucaso

LC
2 sspp
HB 5.7–8.1 cm T 3.8–4.4 cm
W 5.5–11.5 g

0–2600 m

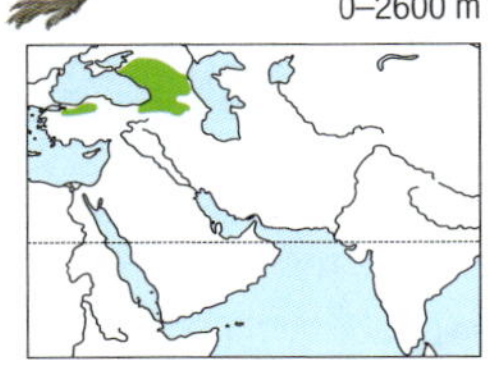

N TR, Caucasus (S European RU, GE, AM, AZ)

LC
4 sspp
HB 5.4–7.8 cm
T 2.5–3.9 cm
W 6–10.3 g

Siberian Large-toothed Shrew
Sorex daphaenodon
FR Musaraigne à dents foncées
DE Sibirische Großzahnspitzmaus
ES Musaraña de dientes grandes de Siberia

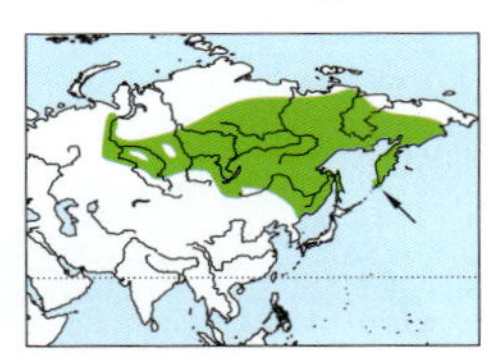

RU (from Urals E to Pacific coast, Sakhalin, Paramushir I), N MN, NE CN, NE KP

DD
Monotypic
HB 6.2–6.4 cm
T 3.8–4.3 cm

c. 3000 m

Gansu Shrew
Sorex cansulus

FR Musaraigne du Gansu
DE Gansu-Spitzmaus
ES Musaraña de Gansu

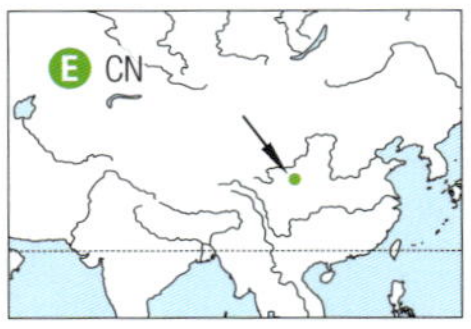

Known only by 3 specimens from 2 sites in S Gansu; perhaps N Sichuan

LC
10 sspp
HB 6.1–8.6 cm
T 2–3.8 cm
W 8–11.5 g

Tundra Shrew
Sorex tundrensis

FR Musaraigne de la toundra
DE Tundra-Rotzahnspitzmaus
ES Musaraña de tundra

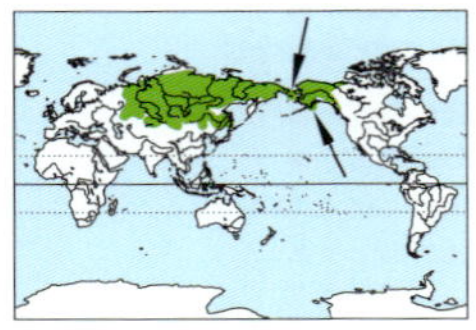

From W Urals (RU) to Alaska (US) & NW CA; S to N & E KZ, MN, NW & NE CN

LC
Monotypic
HB 6.5–7.7 cm
T 3.4–4.5 cm
W 6–10.3 g

Tian Shan Shrew
Sorex asper

FR Musaraigne des Tian Shan
DE Tian-Shan-Spitzmaus
ES Musaraña de Tian Shan

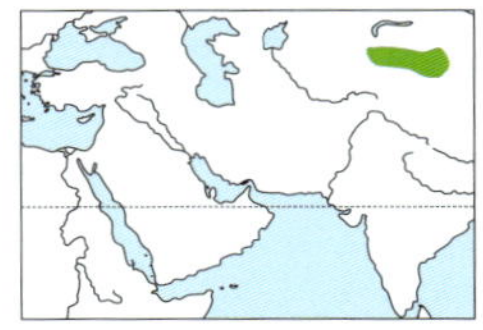

up to 3000 m

Tian Shan in SE KZ, KG & NW CN (Xinjiang)

LC
Monotypic
HB 6.8–7.8 cm
T 3.3–4.5 cm
W 6.5–10 g

up to 1160 m

Apennine Shrew
Sorex samniticus

FR Musaraigne des Apennins
DE Italienische Waldspitzmaus
ES Musaraña de los Apeninos

Low mts along Peninsula S to Calabria

LC
2 sspp
HB 6.8–8.9 cm
T 3.6–4.8 cm
W 5–13 g

Arctic Shrew
Sorex arcticus

FR Musaraigne nordique
DE Arktische Spitzmaus
ES Musaraña del Ártico

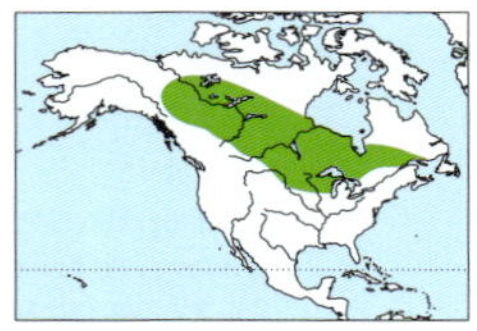

CA (from C Yukon E to E Quebec), N US (S to North Dakota E to NW Michigan)

LC
Monotypic
HB 6–7.2 cm
T 3.7–4.2 cm
W 7–8 g

Maritime Shrew
Sorex maritimensis

FR Musaraigne des Maritimes
DE Nova-Scotia-Spitzmaus
ES Musaraña marítima

New Brunswick, Nova Scotia

LC
6 sspp
HB 4–6.4 cm T 3.3–4.5 cm
W 2.6–7 g

Eurasian Pygmy Shrew
Sorex minutus

FR Musaraigne pygmée
DE Zwergspitzmaus
ES Musaraña enana de Eurasia

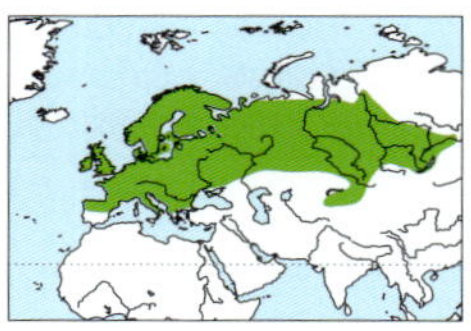

From British Is & Iberian Pen E through Europe & Siberia to SW Yakutia, S to Tian Shan

LC
2 sspp
HB 4.3–7.4 cm T 3.3–5 cm
W 2–9 g

Caucasian Pygmy Shrew
Sorex volnuchini

FR Musaraigne de Volnuchin
DE Kaukasus-Zwergspitzmaus
ES Musaraña enana del Cáucaso

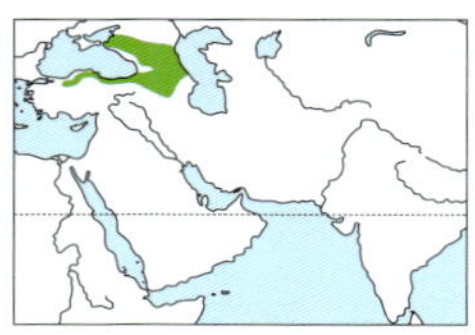

N TR, Caucasus (S European RU, GE, AM, AZ), NW IR (Talysh Mts)

DD
Monotypic
HB 5.5–6.9 cm
T 4.2–5 cm
W 3.6–8.2 g

Buchara Shrew
Sorex buchariensis

FR Musaraigne du Pamir
DE Buchara-Spitzmaus
ES Musaraña de Buchara

Pamir-Alay Mts in UZ & TJ

DD
Monotypic
HB 4.1–6.4 cm
T 3.2–5.4 cm

2000–4000 m

Tibetan Shrew
Sorex thibetanus

FR Musaraigne du Tibet
DE Tibet-Spitzmaus
ES Musaraña de Tibet

C CN (Tibetan Plateau, mainly highlands of Qinghai & S Gansu); perhaps W NP

DD
Monotypic
HB 4.1–6.4 cm
T 3.2–5.4 cm

2000–4000 m

Kozlov's Shrew
Sorex kozlovi

FR Musaraigne de Kozlov
DE Kozlov-Spitzmaus
ES Musaraña de Kozlov

Highlands of SE Qinghai, E Tibet & N Sichuan

LC
Monotypic
HB 5.7–7.4 cm
T 4–4.7 cm

Kashmir Shrew
Sorex planiceps

FR Musaraigne à tête plate
DE Kaschmir-Zwergspitzmaus
ES Musaraña de Cachemira

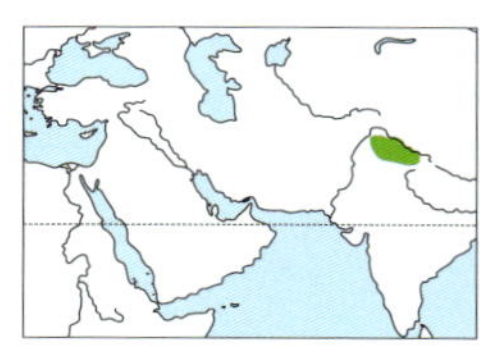

2280–3970 m

NE PK, NW IN, CN (SW Tibet)

Trowbridge's Shrew
Sorex trowbridgii

FR Musaraigne de Trowbridge
DE Trowbridge-Spitzmaus
ES Musaraña de Trowbridge

LC
5 sspp
HB 5.7–7.5 cm
T 5–6.4 cm
W 4–6 g

up to 1820 m

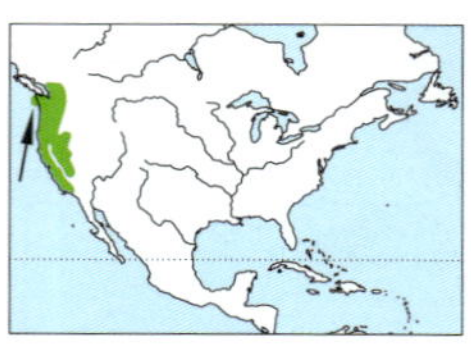

CA (SW British Columbia), W US (S to California & WC Nevada, Destruction I)

Arizona Shrew
Sorex arizonae

FR Musaraigne d'Arizona
DE Arizona-Spitzmaus
ES Musaraña de Arizona

LC
Monotypic
HB 5–7.3 cm T 3.7–4.6 cm
W 2–5 g

1500–2600 m

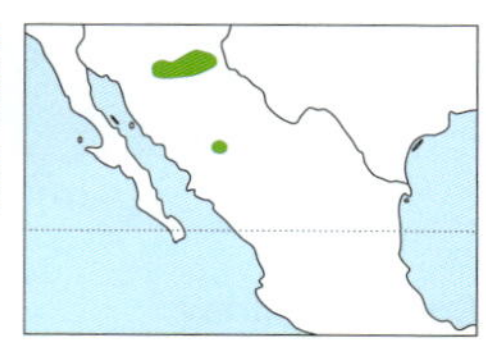

SW US (SE Arizona, SW New Mexico), NW MX (W Chihuahua)

Merriam's Shrew
Sorex merriami

FR Musaraigne de Merriam
DE Merriam-Spitzmaus
ES Musaraña de Merriam

LC
Monotypic
HB 4.8–6.9 cm
T 3–4.2 cm
W 3–7 g

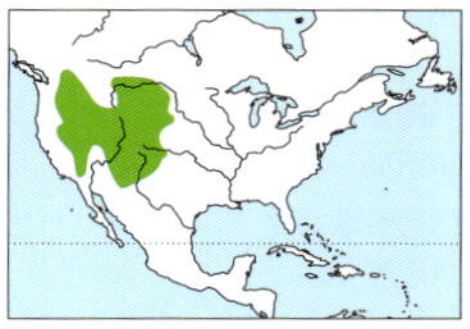

SW CA (extreme S British Columbia), W US (S to S California, Arizona & New Mexico)

Alto Shrew
Sorex altoensis

FR Musaraigne de Carraway
DE Alto-Spitzmaus
ES Musaraña de altitud

NE
Monotypic
HB 6–7.6 cm
T 3.8–5.4 cm
W c. 6 g

2100–3650 m

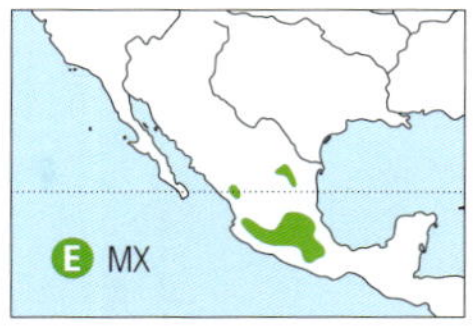

Sierra Madre Occidental & Sierra Madre Oriental, connecting in S

Jalisco Shrew
Sorex mediopua

FR Musaraigne du Jalisco
DE Jalisco-Spitzmaus
ES Musaraña de Jalisco

LC
Monotypic
HB 6.1–7.5 cm
T 3.8–4.9 cm
W c. 6 g

1875–3050 m

Jalisco, Michoacán, México, Guerrero

Saussure's Shrew
Sorex saussurei

FR Musaraigne de De Saussure
DE Saussure-Spitzmaus
ES Musaraña de Saussure

LC
Monotypic
HB 6.5–7.7 cm
T 4.3–6 cm
W c. 6 g

2100–3650 m

Jalisco, Colima, Michoacán, México, Distrito Federal, Morelos, Puebla

McCarthy's Shrew
Sorex mccarthyi

FR Musaraigne de McCarthy
DE McCarthy-Spitzmaus
ES Musaraña de McCarthy

NE
Monotypic
HB c. 5.6 cm
T c. 5.1 cm
W c. 4.2 g

c. 2560 m

Known only from Celaque NP (Lempira)

Salvin's Shrew
Sorex salvini

FR Musaraigne de Salvin
DE Salvin-Spitzmaus
ES Musaraña de Salvin

NE
2 sspp
HB 6–6.9 cm
T 4.5–4.9 cm
W 6–7 g

1900–2950 m

SE MX (Chiapas), GT (Huehuetenango, Quetzaltenango, Totonicapán, Alta Verapaz)

Sclater's Shrew
Sorex sclateri

FR Musaraigne de Sclater
DE Sclater-Spitzmaus
ES Musaraña de Sclater

CR
Monotypic
HB 6.8–7.3 cm
T 5.2–5.3 cm
W c. 7 g

c. 1700 m

Known from 2 localities in N & E Chiapas

Pale-toothed Shrew
Sorex stizodon

FR Musaraigne à dents pâles
DE Hellzahnspitzmaus
ES Musaraña de dientes pálidos

CR
Monotypic
HB c. 6.6 cm
T c. 4.1 cm
W c. 7 g

c. 2700 m

Reserva Ecológica Huitepec (C Chiapas)

Chiapan Shrew
Sorex chiapensis

FR Musaraigne du Chiapas
DE Chiapas-Spitzmaus
ES Musaraña de Chiapas

NE
Monotypic
HB 6.1–7.5 cm
T 4.4–5.1 cm
W c. 6 g

3030–3160 m

Highlands of SE MX (E Chiapas) & W GT

Ibarra's Shrew
Sorex ibarrai

FR Musaraigne d'Ibarra
DE Ibarra-Spitzmaus
ES Musaraña de Ibarra

NE
Monotypic
HB 7.1–7.9 cm
T 3.8–6.6 cm
W 6–13.5 g

1475–2800 m

Alta Verapaz, Baja Verapaz, El Progreso

Cruz's Long-tailed Shrew
Sorex cruzi

FR Musaraigne de Cruz
DE Cruz-Spitzmaus
ES Musaraña de Cruz

NE
Monotypic
HB 5.4–5.5 cm
T 5.2–5.3 cm

c. 1835 m

Cusuco NP (Sierra del Merendón)

Mexican Large-toothed Shrew
Sorex macrodon

FR Musaraigne à grandes dents
DE Mexikanische Großzahn-Spitzmaus
ES Musaraña de dientes grandes de México

VU
Monotypic
HB 7.9–8.4 cm T 4.9–5.2 cm
W 9.8–11.3 g

1200–2600 m

NE Puebla, CW Veracruz, N Oaxaca

NE
Monotypic
HB 6.1–7.8 cm
T 4.3–5.7 cm
W c. 7 g

Sierra Shrew
Sorex madrensis

FR Musaraigne du Guatemala
DE Sierra-Madre-Spitzmaus
ES Musaraña de Sierra Madre

2300–3100 m

Sierra Madre in highlands of SW

NE
Monotypic
HB 6.1–7.6 cm
T 4.8–6.1 cm
W c. 6 g

Mutable Shrew
Sorex mutabilis

FR Musaraigne variable
DE Variable Spitzmaus
ES Musaraña mudable

700–3000 m

Guerrero, Oaxaca

LC
Monotypic
HB 6.7–7.8 cm
T 4.5–5.7 cm
W c. 6 g

Verapaz Shrew
Sorex veraepacis

FR Musaraigne de Verapaz
DE Verapaz-Spitzmaus
ES Musaraña de Verapaz

2680–3350 m

Sierra de los Cuchumatanes (Huehuetenango), E to around Cobán (Alta Verapaz)

Rock Shrew
Sorex dispar

FR Musaraigne à longue queue
DE Langschwanzspitzmaus
ES Musaraña de roca

LC
3 sspp
HB 4.8–7.9 cm
T 4.6–6.7 cm W 2–8 g

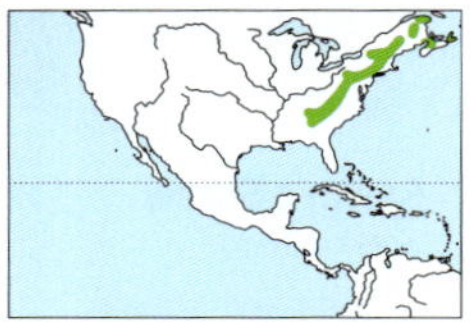

Appalachian Mts of E CA (E Quebec, Nova Scotia, New Brunswick) & E US (S to Tennessee & NE Georgia)

Smoky Shrew
Sorex fumeus

FR Musaraigne fuligineuse
DE Rauchfarbene Spitzmaus
ES Musaraña ahumada

LC
2 sspp
HB 5.9–7.5 cm
T 4.3–5.3 cm W 5–9 g

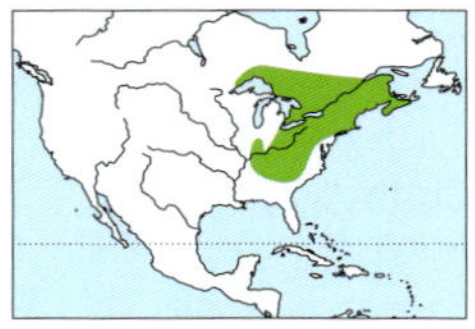

SE CA (S Ontario E to Nova Scotia), NE US (S to NE Georgia & NW South Carolina)

Inyo Shrew
Sorex tenellus

FR Musaraigne d'Inyo
DE Inyo-Spitzmaus
ES Musaraña de Inyo

LC
Monotypic
HB 4.8–7.4 cm
T 3.6–4.8 cm W 3–4 g

Great Basin in EC California & SW Nevada; isolated record in WC Utah

Dwarf Shrew
Sorex nanus

FR Musaraigne naine
DE Westamerikanische Zwergspitzmaus
ES Musaraña enana

LC
Monotypic
HB 5–6 cm
T 2.7–4.5 cm W 2–4 g

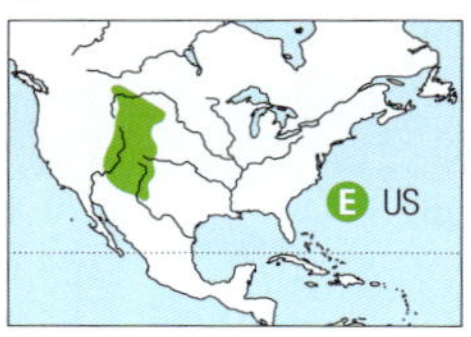

From Montana & SW South Dakota S to Arizona & New Mexico

LC
Monotypic
HB 5.3–6.6 cm
T 3.5–4.6 cm W c. 5 g

Mexican Long-tailed Shrew
Sorex oreopolus

FR Musaraigne fantôme
DE Vulkan-Langschwanzspitzmaus
ES Musaraña de cola larga de México

c. 2500 m

Jalisco, México, Distrito Federal, Morelos, Tlaxcala, Puebla

LC
Monotypic
HB 5.4–7.1 cm
T 3.3–4.3 cm
W c. 6 g

Orizaba Long-tailed Shrew
Sorex orizabae

FR Musaraigne d'Orizaba
DE Orizaba-Langschwanzspitzmaus
ES Musaraña de cola larga de Orizaba

2060–4175 m

Michoacán, México, Distrito Federal, Morelos, Tlaxcala, Puebla, Veracruz

Chestnut-bellied Shrew
Sorex ventralis

FR Musaraigne à ventre brun
DE Kastanienbauch-Spitzmaus
ES Musaraña de vientre castaño

LC
Monotypic
HB 6.1–6.9 cm
T 3.6–4.3 cm W c. 5 g

2760–3000 m

Mexico E through Distrito Federal & Tlaxcala to Puebla, SW to Oaxaca

Veracruz Shrew
Sorex veraecrucis

FR Musaraigne du Veracruz
DE Veracruz-Spitzmaus
ES Musaraña de Veracruz

LC
2 sspp
HB 4.6–7.2 cm
T 4.5–7.5 cm W c. 6 g

1600–3000 m

WC & S Veracruz, E Puebla, Oaxaca

DD
Monotypic
HB 6.6–7.7 cm
T 4.5–6.7 cm
W c. 6 g

Ixtlan Shrew
Sorex ixtlanensis

FR Musaraigne d'Ixtlan
DE Ixtlan-Spitzmaus
ES Musaraña de Ixtlán

1900–3000 m

Guerrero, Oaxaca

Olympic Shrew
Sorex rohweri

FR Musaraigne de Rohwer
DE Rohwer-Spitzmaus
ES Musaraña de Olympic

LC
Monotypic
HB 4.7–6.9 cm
T 3.7–4.9 cm

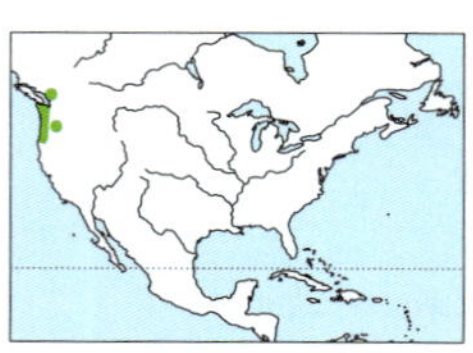

0–700 m

SW CA (SW British Columbia), NW US (W Washington, NW Oregon)

South-eastern Shrew
Sorex longirostris

FR Musaraigne à long nez
DE Langschnauzen-Spitzmaus
ES Musaraña suroriental

LC
3 sspp
HB 4.4–5.7 cm
T 3.9–4.4 cm
W 2–6 g

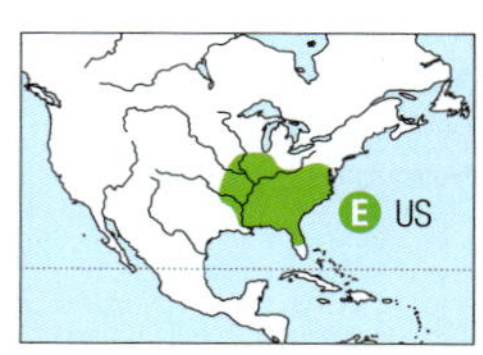

0–760 m

From N Illinois, N Indiana & N Maryland S to C Florida, W to Missouri & E Oklahoma

Masked Shrew
Sorex cinereus

FR Musaraigne cendrée
DE Amerikanische Maskenspitzmaus
ES Musaraña ceniza

LC

7 sspp
HB 4.2–6.7 cm
T 2.9–5 cm W 2.5–6 g

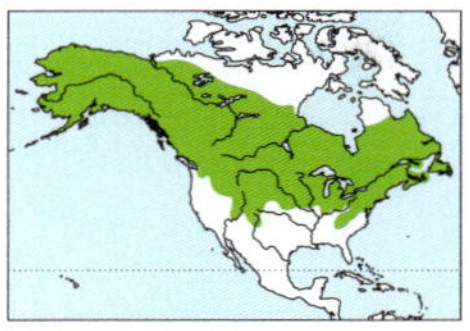

CA (from Yukon E to Labrador), US (Alaska, S to Utah, New Mexico & N Georgia)

Maryland Shrew
Sorex fontinalis

FR Musaraigne du Maryland
DE Maryland-Spitzmaus
ES Musaraña de Maryland

NE

Monotypic
HB 5.1–5.8 cm
T 2.7–3.9 cm W 2.3–4.5 g

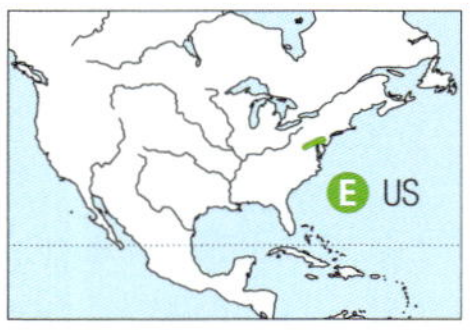

SE Pennsylvania, W New Jersey, NE West Virginia, N Maryland, N Delaware

Mount Lyell Shrew
Sorex lyelli

FR Musaraigne du Lyell
DE Mount-Lyell-Spitzmaus
ES Musaraña de Monte Lyell

LC

Monotypic
HB 5–6.8 cm
T 3.9–4.4 cm W 4–6 g

2100–3630 m

EC Sierra Nevada (EC California)

Zacatecas Shrew
Sorex emarginatus

FR Musaraigne de la Sierra Madre
DE Zacatecas-Spitzmaus
ES Musaraña de Zacatecas

LC

Monotypic
HB 5–6.3 cm
T 3.8–4.4 cm W 5–6.5 g

1830–3660 m

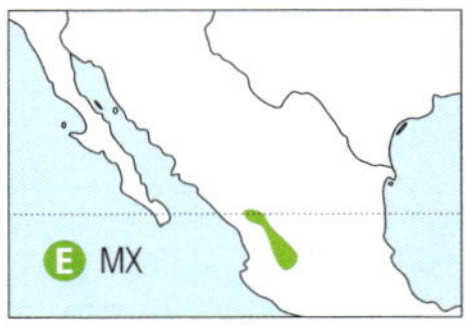

Durango, Zacatecas, Jalisco

Carmen Mountain Shrew
Sorex milleri

FR Musaraigne de Miller
DE Miller-Spitzmaus
ES Musaraña de Carmen

VU

Monotypic
HB 5.1–5.8 cm
T 3.7–4.4 cm W 2.9–4.8 g

2400–3700 m

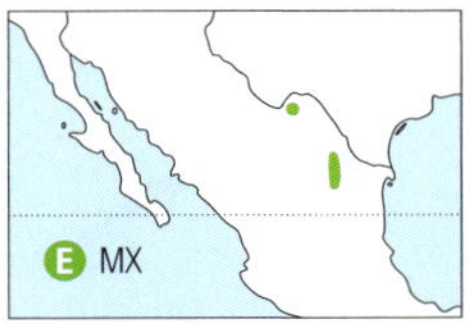

SE Coahuila & W Nuevo León, disjunct in NW Coahuila

Preble's Shrew
Sorex preblei

FR Musaraigne de Preble
DE Preble-Spitzmaus
ES Musaraña de Preble

LC

Monotypic
HB 4.4–5.7 cm
T 2.8–3.8 cm W 2–4 g

1280–2750 m

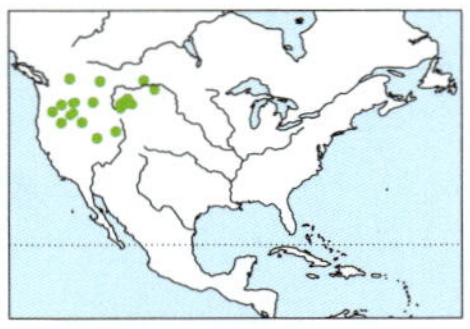

Scattered in Columbia Plateau & N Great Plains of CA (SC British Columbia) & NW US (S to N Utah)

Prairie Shrew
Sorex haydeni

FR Musaraigne des steppes
DE Präriespitzmaus
ES Musaraña de pradera

LC

Monotypic
HB 4.7–6.7 cm
T 3–4.1 cm W 2–5 g

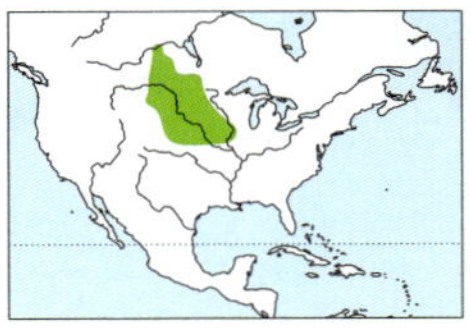

SC CA (C & S Saskatchewan, S Manitoba), NC US (S to N Kansas & N Missouri)

Pribilof Island Shrew
Sorex pribilofensis

FR Musaraigne des Pribilof
DE Pribilof-Rotzahnspitzmaus
ES Musaraña de Pribilof

EN

Monotypic
HB 5.9–6.7 cm
T 3.1–3.2 cm W 4–6 g

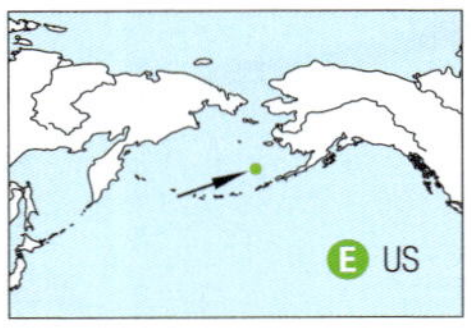

St Paul I (Pribilof Is), off W Alaska

Barren Ground Shrew
Sorex ugyunak

FR Musaraigne de Beaufort
DE Kanadische Tundraspitzmaus
ES Musaraña de tierras áridas

LC

Monotypic
HB 4.7–7 cm
T 3.2–3.7 cm W 4–5 g

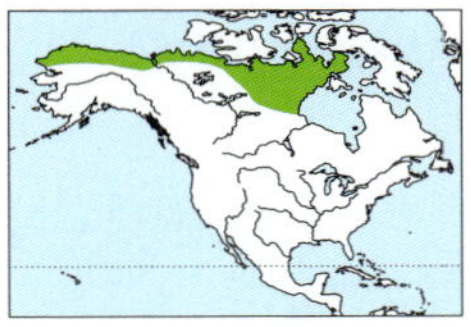

US (N Alaska), CA (N Yukon E to extreme NE Saskatchewan & NW Manitoba)

Portenko's Shrew
Sorex portenkoi

FR Musaraigne de Portenko
DE Portenko-Rotzahnspitzmaus
ES Musaraña de Portenko

DD

Monotypic
HB 5.3–6 cm
T 2.5–3.6 cm W c. 3.7 g

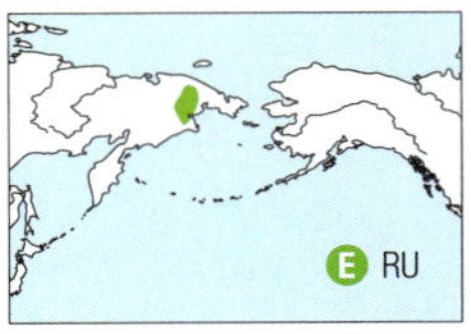

Chukchi Pen

LC

Monotypic
HB 6.2–7.1 cm
T 3.2–3.7 cm
W 4–5 g

Saint Lawrence Island Shrew
Sorex jacksoni

FR Musaraigne de Saint Lawrence
DE St. Lawrence-Rotzahnspitzmaus
ES Musaraña de San Lorenzo

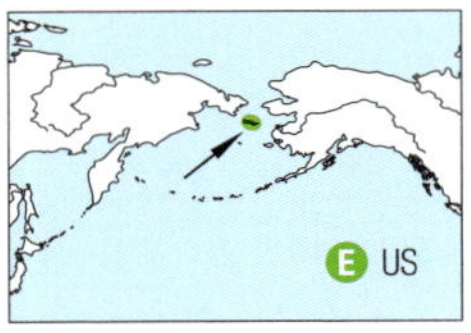

St Lawrence I, off W Alaska

LC

Monotypic
HB 5–6.8 cm
T 4.3–5.4 cm
W 3.7–6 g

Kamchatka Shrew
Sorex camtschaticus

FR Musaraigne du Kamtchatka
DE Kamschatka-Spitzmaus
ES Musaraña de Kamchatka

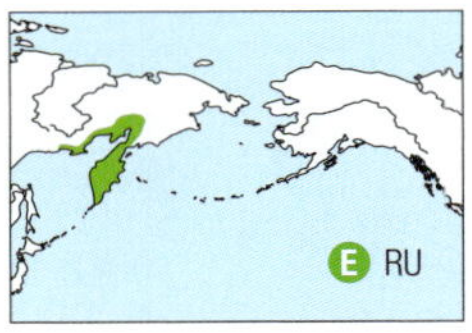

Far East (Magadan Region, Kamchatka Krai)

Paramushir Shrew
Sorex leucogaster

FR Musaraigne de Paramouchir
DE Paramuschir-Spitzmaus
ES Musaraña de Paramushir

DD

Monotypic
HB 4.9–6.5 cm
T 3.8–4.6 cm

Paramushir I (N Kuril Is)

Eastern Pygmy Shrew
Sorex hoyi

FR Musaraigne de Hoy
DE Östliche Amerikanische Zwergspitzmaus
ES Musaraña enana de América oriental

LC

Monotypic
HB 5.2–7 cm
T 2.8–3.7 cm
W 1–7 g

0–1640 m

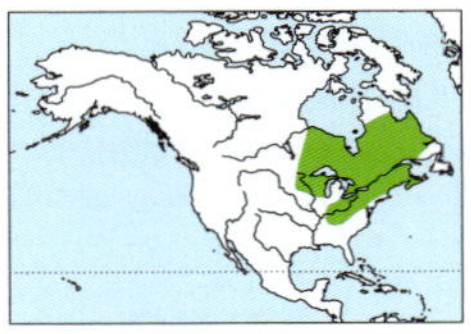

E CA (from E Manitoba to Nova Scotia), NE US (S along Appalachians to N Georgia)

Western Pygmy Shrew

Sorex eximius

0–1640 m

FR Musaraigne d'Alaska
DE Westliche Amerikanische Zwergspitzmaus
ES Musaraña enana de América occidental

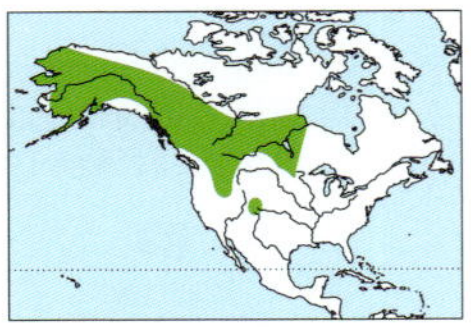

NE
2 sspp
HB 5.2–7 cm
T 2.8–3.7 cm W 1–7 g

W & C CA (E to Manitoba), W US (Alaska, S to SE Wyoming & NC Colorado)

Vagrant Shrew

Sorex vagrans

0–2100 m

FR Musaraigne errante
DE Wanderspitzmaus
ES Musaraña errante

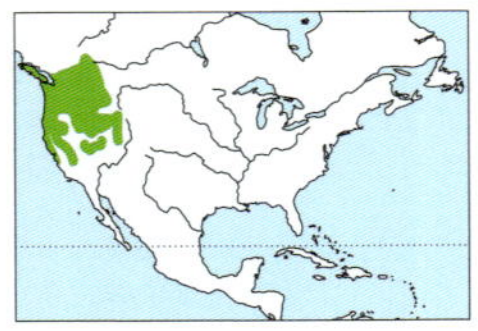

LC
3 sspp
HB 5.6–7.1 cm
T 3.2–5 cm W 3–9 g

SW CA (S British Columbia), W US (S to C California, E to W Montana, W Wyomying & N Utah)

LC
9 sspp
HB 4.5–7 cm
T 3.1–4 cm
W 3–7 g

Ornate Shrew

Sorex ornatus

0–2000 m

FR Musaraigne ornée
DE Schönspitzmaus
ES Musaraña ornada

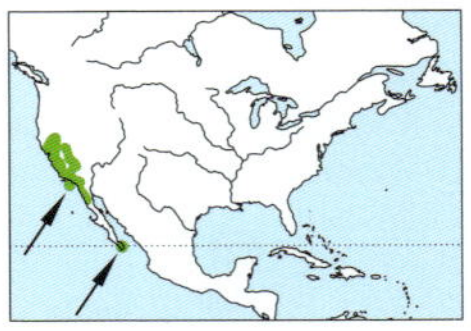

SW US (C & S California, with Santa Catalina I), NW MX (Baja California Pen)

LC
2 sspp
HB 6.2–9.8 cm
T 5.8–7.1 cm
W 8.7–20 g

American Water Shrew

Sorex palustris

FR Musaraigne palustre
DE Amerikanische Wasserspitzmaus
ES Musgaño americano

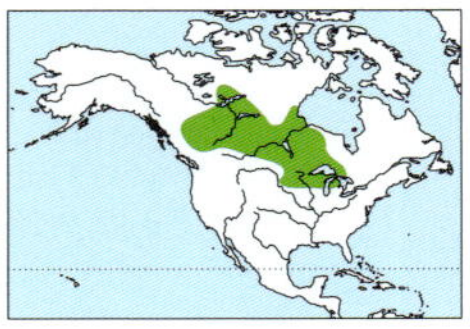

CA (E British Columbia E to C Ontario), US (S to South Dakota, Minnesota & Michigan)

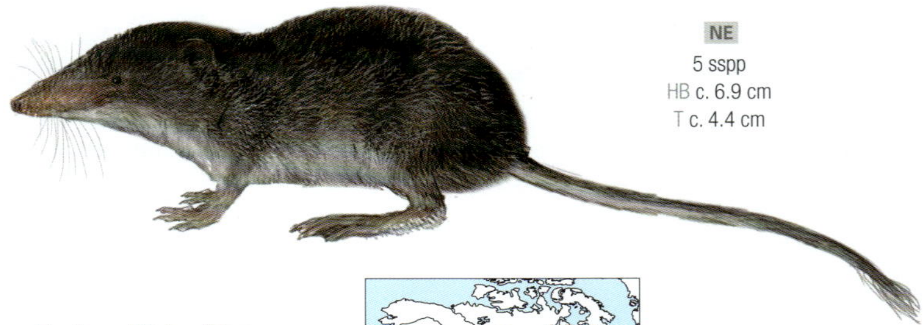

NE
5 sspp
HB c. 6.9 cm
T c. 4.4 cm

Eastern Water Shrew

Sorex albibarbis

FR Musaraigne de Cope
DE Östliche Wasserspitzmaus
ES Musgaño oriental

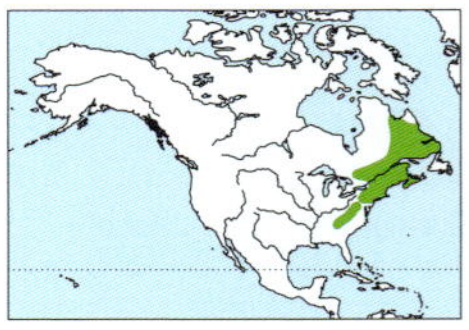

SE CA (Quebec & Labrador to Nova Scotia), NE US (S along Appalachians to Tennessee)

Western Water Shrew

Sorex navigator

FR Musaraigne de Washington
DE Westliche Wasserspitzmaus
ES Musgaño occidental

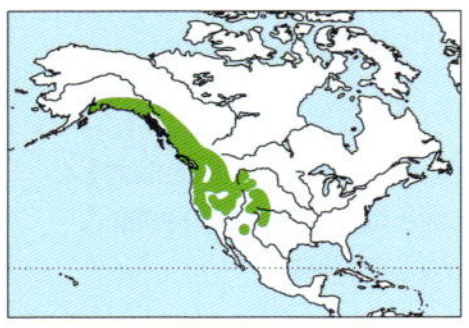

North American Cordillera & W coastal areas (S Alaska, W CA & W US); limits uncertain

NE
2 sspp
HB 5.6–8.8 cm
T 6.2–7.9 cm
W 6.5–21.5 g

LC
3 sspp
HB 7–10.1 cm
T 6–8 cm
W 10–21 g

Marsh Shrew

Sorex bendirii

FR Musaraigne de Bendire
DE Pazifische Sumpfspitzmaus
ES Musaraña de marisma

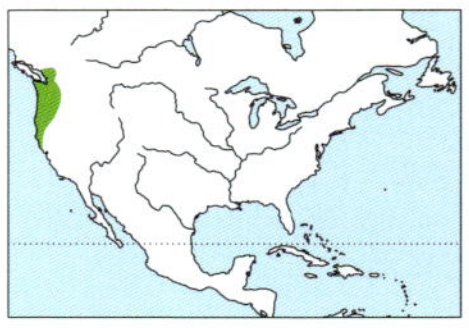

SW CA (SW British Columbia), NW US (W Washington, W Oregon, NW California)

Pacific Shrew

Sorex pacificus

FR Musaraigne du Pacifique
DE Pazifische Spitzmaus
ES Musaraña del Pacífico

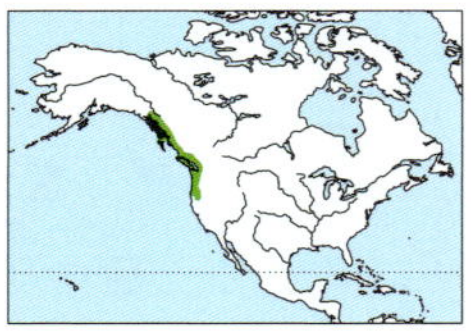

SW CA (W British Columbia), NW US (SE Alaska, W Washington, Oregon, N California)

LC
12 sspp
HB 5.8–8.1 cm
T 4.5–6.8 cm
W 5–16 g

LC
2 sspp
HB 6.7–8.9 cm
T 4.9–6.8 cm
W 7–14 g

Fog Shrew

Sorex sonomae

FR Musaraigne de Sonoma
DE Nebelspitzmaus
ES Musaraña de Sonoma

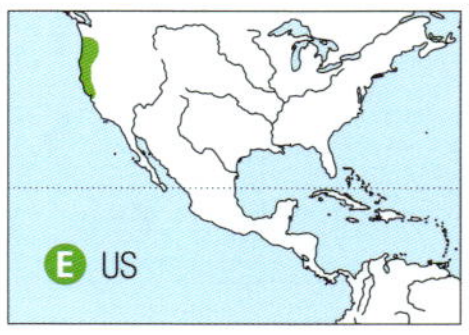

Pacific coast in Oregon & California

Northern Montane Shrew

Sorex obscurus

FR Musaraigne sombre
DE Nördliche Dunkle Rotzahnspitzmaus
ES Musaraña de montaña septentrional

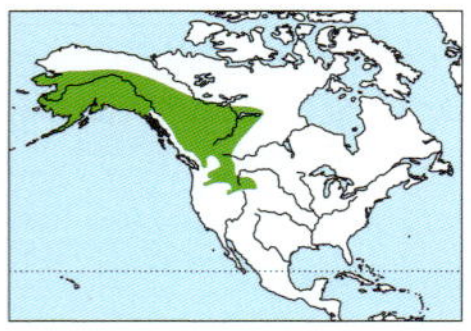

W CA (excluding coast, E to Manitoba), US (Alaska, W Oregon, Idaho, NW Wyoming)

NE
4 sspp
HB 5.5–7.7 cm
T 3.7–5.5 cm
W 4–10 g

ssp *neomexicanus*

LC
2 sspp
HB 5.4–7 cm
T 3.7–5.1 cm
W 4–10 g

Southern Montane Shrew

Sorex monticola

FR Musaraigne de montagne
DE Südliche Dunkle Rotzahnspitzmaus
ES Musaraña de montaña meridional

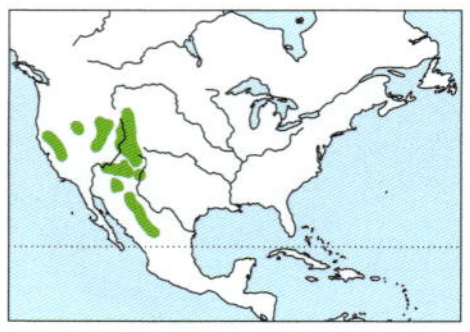

Mts of W US (S from EC California, Utah & Wyoming) to N MX (S to S Durango)

Genus *CRYPTOTIS*

(52 species)

Nelson's Small-eared Shrew
Cryptotis nelsoni

FR Musaraigne de Nelson
DE Nelson-Kleinohrspitzmaus
ES Musaraña de orejas pequeñas de Nelson

CR
Monotypic
HB 7.2–7.9 cm
T 2.2–3.3 cm
W 3–8 g

1460–1525 m

Volcán San Martín (Veracruz)

Big Mexican Small-eared Shrew
Cryptotis magnus

FR Musaraigne d'Oaxaca
DE Große Mexikanische Kleinohrspitzmaus
ES Musaraña de orejas pequeñas de México grande

1300–3000 m

NC Oaxaca

VU
Monotypic
HB 8.6–8.8 cm
T 3.7–5.3 cm
W c. 7 g

Small Mexican Small-eared Shrew
Cryptotis mexicanus

FR Musaraigne du Mexique
DE Kleine Mexikanische Kleinohrspitzmaus
ES Musaraña de orejas pequeñas de México pequeña

1040–2500 m

SW Tamaulipas, SE San Luis Potosí, NE Querétaro, Hidalgo, WC Veracruz, WC México

LC
Monotypic
HB 5.6–7.5 cm
T 2–3.2 cm
W 3–10 g

LC
Monotypic
HB 5.6–7.5 cm
T 2–3.2 cm
W 3–10 g

Grizzled Small-eared Shrew
Cryptotis obscurus

FR Musaraigne obscure
DE Grauhaarige Kleinohrspitzmaus
ES Musaraña de orejas pequeñas canosa

1040–2500 m

SW Tamaulipas, SE San Luis Potosí, NE Querétaro, Hidalgo, WC Veracruz, WC México

VU
Monotypic
HB 5.8–7.8 cm
T 2.3–3.6 cm
W 6.5–10 g

Phillips's Small-eared Shrew
Cryptotis phillipsii

FR Musaraigne de Phillips
DE Phillips-Kleinohrspitzmaus
ES Musaraña de orejas pequeñas de Phillips

1060–2600 m

Sierra de Miahuatlán (SC Oaxaca)

DD
Monotypic

Eastern Cordillera Small-eared Shrew
Cryptotis brachyonyx

FR Musaraigne à griffes courtes
DE Östliche Kordilleren-Kleinohrspitzmaus
ES Musaraña de orejas pequeñas de la Cordillera oriental

1300–2715 m

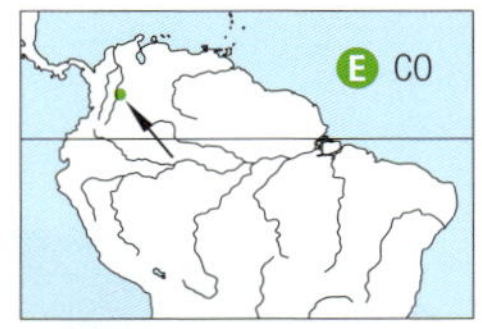

W slopes of C part of E Andes

LC
Monotypic
HB 6–7.6 cm
T 2.7–3.4 cm

Colombian Small-eared Shrew
Cryptotis colombianus

FR Musaraigne de Colombie
DE Kolumbien-Kleinohrspitzmaus
ES Musaraña de orejas pequeñas de Colombia

1750–2800 m

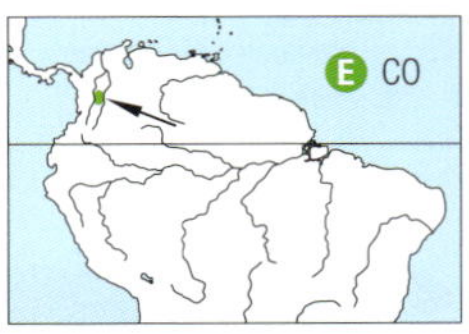

C Andes in Antioquia & Caldas

DD
Monotypic
HB 5.5–6.4 cm
T 2.7–2.9 cm

Honduran Small-eared Shrew
Cryptotis hondurensis

FR Musaraigne du Honduras
DE Honduras-Kleinohrspitzmaus
ES Musaraña de orejas pequeñas de Honduras

c. 1700 m

Known from 3 sites in Francisco Morazán

EN
Monotypic
HB 6.6–7.3 cm
T 2.4–3.1 cm

Darien Small-eared Shrew
Cryptotis merus

FR Musaraigne du Darien
DE Darién-Kleinohrspitzmaus
ES Musaraña de orejas pequeñas de Darién

1400–1500 m

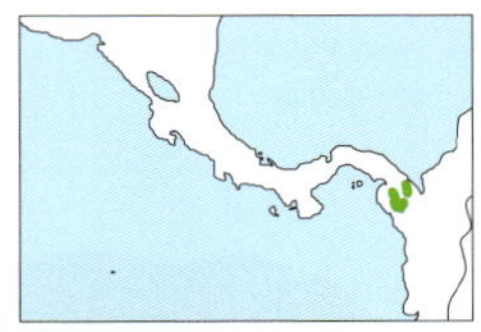

Border of E PA & NW CO

LC
Monotypic
HB 5.5–8.3 cm
T 1.9–3.7 cm
W 5–8 g

Blackish Small-eared Shrew
Cryptotis nigrescens

FR Musaraigne noire
DE Schwärzliche Kleinohrspitzmaus
ES Musaraña de orejas pequeñas negruzca

800–2900 m

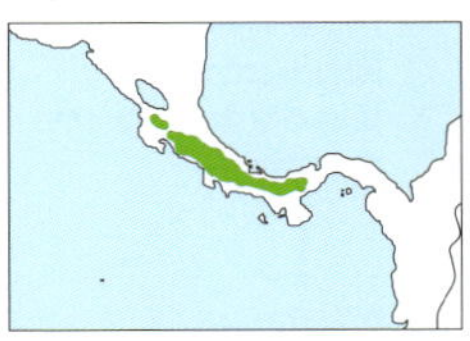

Highlands of CR & W PA

NE
Monotypic
HB 7.8–8.1 cm
T 3.3–3.5 cm W c. 6 g

Lacandona Small-eared Shrew
Cryptotis lacandonensis

FR Musaraigne de Lacandona
DE Lacandona-Kleinohrspitzmaus
ES Musaraña de orejas pequeñas de Lacandona

c. 90 m

Lacandona Rainforest (Chiapas); perhaps adjacent GT

LC
Monotypic
HB 6.1–9 cm
T 2.4–3.3 cm W 4–6 g

Yucatan Small-eared Shrew
Cryptotis mayensis

FR Musaraigne des Mayas
DE Yucatán-Kleinohrspitzmaus
ES Musaraña de orejas pequeñas de Yucatán

0–650 m

Yucatán Pen (SE MX, NE GT, N BZ); collected in owl pellets in Guerrero (SC MX)

Merriam's Small-eared Shrew
Cryptotis merriami

FR Musaraigne de Choate
DE Merriam-Kleinohrspitzmaus
ES Musaraña de orejas pequeñas de Merriam

LC
Monotypic
HB 6.5–7.6 cm
T 2.4–3.3 cm

Highlands of SE MX (Chiapas), GT, HN, SV, NW NI, NW CR

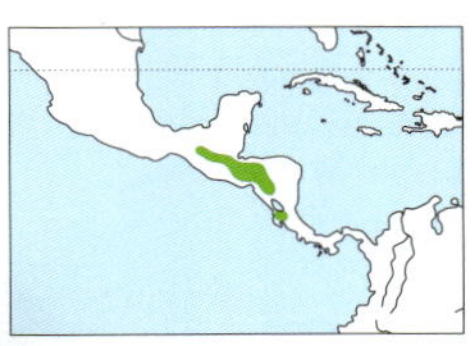

600–1720 m

Berlandier's Least Shrew
Cryptotis berlandieri

FR Musaraigne de Berlandier
DE Berlandier-Kleinohrspitzmaus
ES Musaraña mínima de Berlandier

NE
Monotypic
HB 5.3–7 cm
T c. 1.7 cm

C US (S Texas), MX (S to Nayarit, Jalisco & Michoacán)

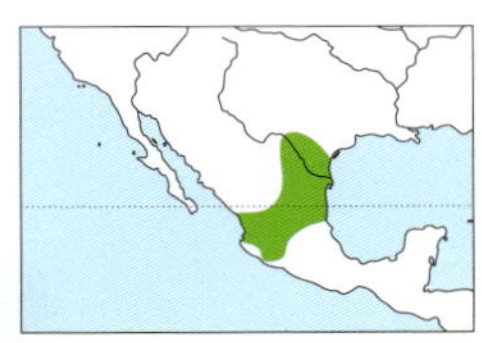

0–2750 m

DD
Monotypic
HB 4.8–7.7 cm
T 1.7–2.4 cm
W 3–10 g

Central American Least Shrew
Cryptotis orophilus

FR Musaraigne d'Irazu
DE Mittelamerika-Kleinohrspitzmaus
ES Musaraña mínima de Centroamérica

1150–1985 m

W HN, extreme NE SV, NI S to C CR

DD
Monotypic
HB 7–7.1 cm
T 2.6–2.9 cm
W 6–7 g

Tropical Least Shrew
Cryptotis tropicalis

FR Musaraigne des Tropiques
DE Tropische Kleinohrspitzmaus
ES Musaraña mínima tropical

1110–1400 m

Highlands of SE MX (E Chiapas), GT, BZ, W HN & NW SV

LC
2 sspp
HB 5.2–6.8 cm
T 1.3–2.3 cm
W 3–10 g

North American Least Shrew
Cryptotis parvus

FR Petite Musaraigne
DE Nordamerika-Kleinohrspitzmaus
ES Musaraña mínima de Norteamérica

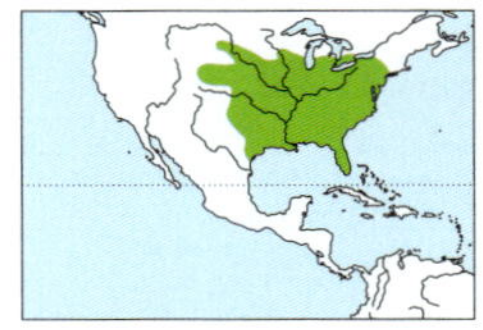

0–905 m

SE CA (SE Ontario), E US (E from SW Wyoming & C Colorado, S to Texas & Florida); limits uncertain

NE
Monotypic
HB 5.1–7.7 cm
T c. 2.2 cm

Puebla Least Shrew
Cryptotis pueblensis

FR Musaraigne du Puebla
DE Puebla-Kleinohrspitzmaus
ES Musaraña mínima de Puebla

30–2300 m

San Luis Potosí, Veracruz, Hidalgo, Puebla, Oaxaca, Tabasco, Chiapas

NE
Monotypic
HB 6.5–6.9 cm
T c. 2.3 cm

Mexican Least Shrew
Cryptotis soricinus

FR Musaraigne soricine
DE Mexiko-Kleinohrspitzmaus
ES Musaraña mínima de México

1250–2350 m

Distrito Federal, México

DD
Monotypic
HB 6.9–8.7 cm
T 2.4–3.6 cm
W 8–16 g

Popocatepetl Broad-clawed Shrew
Cryptotis alticola

FR Musaraigne alticole
DE Popocatepetl-Kleinohrspitzmaus
ES Musaraña de uñas anchas de Popocatépetl

2460–4400 m

C Trans-Mexican Volcanic Belt from Jalisco & Colima E to Puebla

LC
2 sspp
HB 6.3–8.4 cm
T 2.6–3.6 cm W 5–14 g

Goldman's Broad-clawed Shrew
Cryptotis goldmani

FR Musaraigne de Goldman
DE Goldman-Kleinohrspitzmaus
ES Musaraña de uñas anchas de Goldman

1550–3200 m

Omiltemi area of Guerrero, Oaxaca; probably Chiapas

DD
Monotypic
HB 6.6–7.8 cm
T 2.4–3.3 cm
W 3–5 g

Oaxacan Broad-clawed Shrew
Cryptotis peregrinus

FR Musaraigne pèlerine
DE Oaxaca-Kleinohrspitzmaus
ES Musaraña de uñas anchas de Oaxaca

c. 2860 m

Sierra de Cuatro Venados & Sierra Yucuyagua (W & C Oaxaca)

LC
Monotypic
HB 6–7.9 cm
T 3–4.1 cm
W 5–10 g

Talamancan Broad-clawed Shrew
Cryptotis gracilis

FR Musaraigne de Talamanca
DE Talamancan-Kleinohrspitzmaus
ES Musaraña de uñas anchas de Talamanca

1800–3400 m

Highlands from NC CR to W PA

DD
Monotypic
T c. 3.2 cm

Santa Barbara Broad-clawed Shrew
Cryptotis cavatorculus

FR Musaraigne de Santa Barbara
DE Santa-Barbara-Kleinohrspitzmaus
ES Musaraña de uñas anchas de Santa Bárbara

c. 1900 m

Montaña de Santa Bárbara (Santa Bárbara)

DD
Monotypic
HB 6.7–8 cm
T 2.4–3 cm
W 7–11 g

Celaque Broad-clawed Shrew
Cryptotis celaque

FR Musaraigne du Celaque
DE Celaque-Kleinohrspitzmaus
ES Musaraña de uñas anchas de Celaque

1430–2560 m

Celaque NP (Lempira)

EN
Monotypic
HB 7.3–8.5 cm
T 2.7–3.2 cm

Dark Mexican Broad-clawed Shrew
Cryptotis griseoventris

FR Musaraigne à ventre gris
DE Dunkle Mexiko-Kleinohrspitzmaus
ES Musaraña de uñas anchas de México

above 2600 m

Highlands of Los Altos de Chiapas (C Chiapas); possibly Las Margaritas (E Chiapas)

Omoa Broad-clawed Shrew
Cryptotis mccarthyi

FR Musaraigne d'Omoa
DE Omoa-Kleinohrspitzmaus
ES Musaraña de uñas anchas de Omoa

Sierra de Omoa (Cortés); perhaps W to Santa Bárbara

1500–1620 m

DD
Monotypic
HB 7.2–8.4 cm
T 2.1–2.7 cm W 7–13 g

Muscular Broad-clawed Shrew
Cryptotis lacertosus

FR Musaraigne armée
DE Robuste Kleinohrspitzmaus
ES Musaraña de uñas anchas musculosa

Highlands of Sierra de los Cuchumatanes around San Mateo Ixtatán (Huehuetenango)

2680–3110 m

DD
Monotypic
HB 7.5–8.7 cm
T 2.4–3 cm W 10–17 g

NE

Monotypic
HB c. 8 cm
T c. 2.5 cm

c. 1730 m

Honduran Broad-clawed Shrew
Cryptotis magnimanus

FR Musaraigne à grandes pattes
DE Großpfoten-Kleinohrspitzmaus
ES Musaraña de uñas anchas de Honduras

Cordillera de Montecillos (Comayagua)

DD

Monotypic
HB 6.4–8.1 cm
T 2.2–3.2 cm
W 7–11 g

2740–3350 m

Mam Broad-clawed Shrew
Cryptotis mam

FR Musaraigne mam
DE Mam-Kleinohrspitzmaus
ES Musaraña de uñas anchas maya

Sierra de los Cuchumatanes (Huehuetenango); perhaps adjacent HN (Quiché)

LC

Monotypic
HB 7.5–9.4 cm
T 2.5–3.4 cm
W 16–19 g

1200–3350 m

Goodwin's Broad-clawed Shrew
Cryptotis goodwini

FR Musaraigne de Goodwin
DE Goodwin-Kleinohrspitzmaus
ES Musaraña de uñas anchas de Goodwin

W Sierra Madre

NE

Monotypic
HB c. 8.6 cm
T c. 2.7 cm

Mataquescuintla Broad-clawed Shrew
Cryptotis matsoni

FR Musaraigne de Matson
DE Matson-Kleinohrspitzmaus
ES Musaraña de orejas pequeñas de Matson

E Sierra Madre

NE

Monotypic
HB 8.8–9.6 cm
T 3–3.1 cm W 15–17 g

Montecristo Broad-clawed Shrew
Cryptotis montecristo

FR Musaraigne de Montecristo
DE Montecristo-Kleinohrspitzmaus
ES Musaraña de orejas pequeñas de Montecristo

Cerro Montecristo (SE GT, NW SV, W HN)

NE

Monotypic
HB 7.1–7.7 cm
T 2.6–3 cm
W 8–10 g

Cucurucho Broad-clawed Shrew
Cryptotis eckerlini

FR Musaraigne d'Eckerlin
DE Eckerlin-Kleinohrspitzmaus
ES Musaraña de orejas pequeñas de Eckerlin

Cerro Cucurucho (E Sierra Madre)

DD

Monotypic
HB 7.7–8.6 cm
T 2.2–3.4 cm
W 10–15 g

Highland Broad-clawed Shrew
Cryptotis oreoryctes

FR Musaraigne de Chelema
DE Alta-Verapaz-Kleinohrspitzmaus
ES Musaraña de uñas anchas de tierras altas

above 2000 m

Chelemhá Cloud Forest R

NE

Monotypic
HB c. 8 cm
T c. 4.6 cm

c. 1550 m

Monteverde Small-eared Shrew
Cryptotis monteverdensis

FR Musaraigne de Monteverde
DE Monteverde-Kleinohrspitzmaus
ES Musaraña de orejas pequeñas de Monteverde

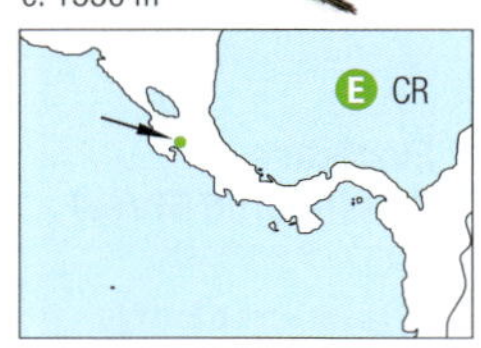

Near Monteverde in Tilarán Highlands

EN

Monotypic
HB c. 8.1 cm
T c. 4.9 cm

1200–1850 m

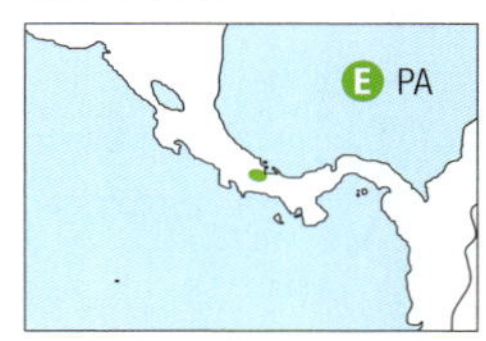

W Highlands

Enders's Small-eared Shrew
Cryptotis endersi

FR Musaraigne d'Enders
DE Enders-Kleinohrspitzmaus
ES Musaraña de orejas pequeñas de Enders

EN

Monotypic
HB 7.7–8.1 cm
T 3.4–4 cm
W 10–12 g

c. 1730 m

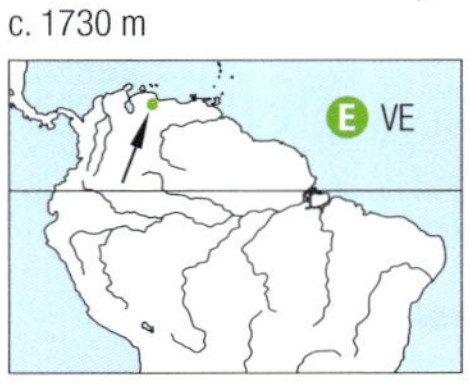

Highlands of Sierra de Aroa

Aroa Small-eared Shrew
Cryptotis aroensis

FR Musaraigne d'Aroa
DE Aroa-Kleinohrspitzmaus
ES Musaraña de orejas pequeñas de Aroa

LC

Monotypic
HB 7.4–9.6 cm
T 2–2.7 cm
W c. 10 g

2700–3500 m

Thomas's Small-eared Shrew
Cryptotis thomasi

FR Musaraigne de Thomas
DE Thomas-Kleinohrspitzmaus
ES Musaraña de orejas pequeñas de Thomas

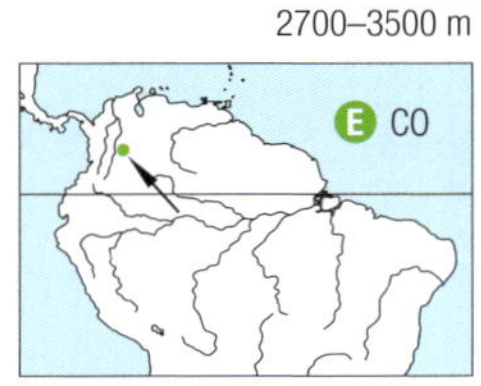

C part of E Andes

DD

Monotypic
HB 7.1–8.3 cm
T 3.2–4 cm
W 10–12 g

2430–3100 m

Dinira Small-eared Shrew
Cryptotis dinirensis

FR Musaraigne de Dinira
DE Dinira-Kleinohrspitzmaus
ES Musaraña de orejas pequeñas de Dinira

N part of E Andes

DD

Monotypic
HB 7.2–8.4 cm
T 3.3–3.4 cm
W 9–11 g

Venezuelan Small-eared Shrew
Cryptotis venezuelensis

FR Musaraigne du Venezuela
DE Venezuela-Kleinohrspitzmaus
ES Musaraña de orejas pequeñas de Venezuela

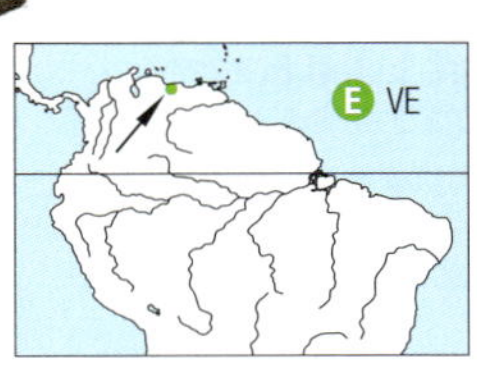

Serranía del Litoral in Cordillera de la Costa

VU

Monotypic
HB 6.7–10.2 cm
T 2.5–4.1 cm
W 10–15 g

1640–3950 m

Merida Small-eared Shrew
Cryptotis meridensis

FR Musaraigne de Mérida
DE Mérida-Kleinohrspitzmaus
ES Musaraña de orejas pequeñas de Mérida

C & S Cordillera de Mérida; probably N part of Cordillera

LC

Monotypic
HB 7.9–8.7 cm
T 3.4–3.8 cm
W 10–15 g

2385–3330 m

Tama Small-eared Shrew
Cryptotis tamensis

FR Musaraigne de Tama
DE Tamá-Kleinohrspitzmaus
ES Musaraña de orejas pequeñas de Tamá

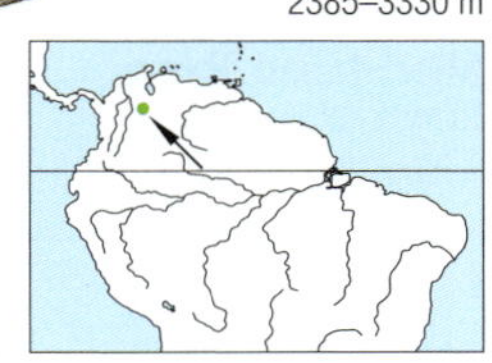

NE CO (Tamá highlands), NW VE (W Táchira); perhaps more extensive

DD
Monotypic
HB c. 6.8 cm
T c. 3.6 cm

2000–3200 m

Perija Small-eared Shrew
Cryptotis perijensis

FR Musaraigne de la Perija
DE Perijá-Kleinohrspitzmaus
ES Musaraña de orejas pequeñas de Perijá

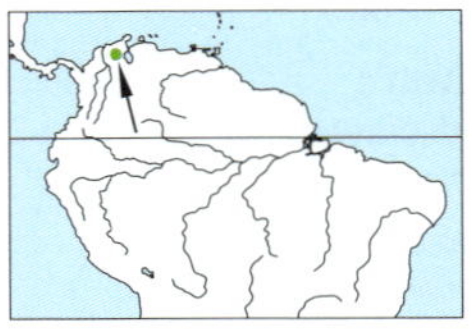

N half of Sierra de Perijá (NE CO, NW VE)

LC
Monotypic
HB 7.7–9.8 cm
T 2.9–4.6 cm
W 15.5–17 g

Medellin Small-eared Shrew
Cryptotis medellinius

FR Musaraigne de Medellin
DE Medellín-Kleinohrspitzmaus
ES Musaraña de orejas pequeñas de Medellín

2000–3800 m

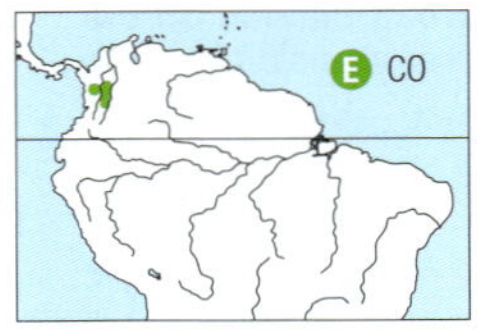

N tip of W Andes, N half of C Andes

LC
Monotypic
HB c. 8.6 cm
T c. 4.2 cm

1500–3375 m

Western Colombian Small-eared Shrew
Cryptotis squamipes

FR Musaraigne à écailles
DE Westliche Kolumbien-Kleinohrspitzmaus
ES Musaraña de orejas pequeñas de Colombia occidental

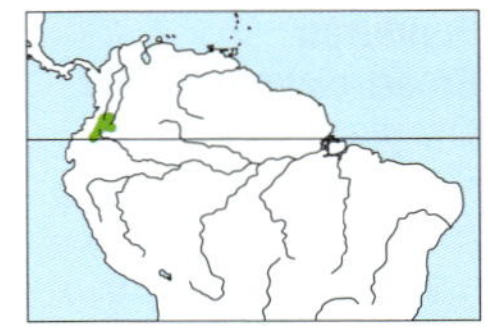

W & C Andes of SW CO & NC EC

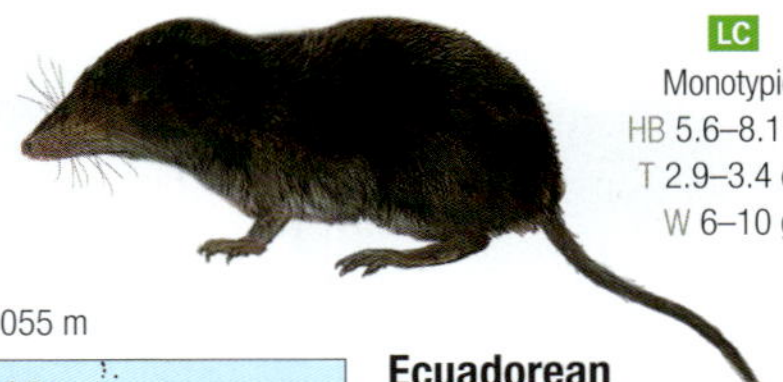

LC
Monotypic
HB 5.6–8.1 cm
T 2.9–3.4 cm
W 6–10 g

1675–4055 m

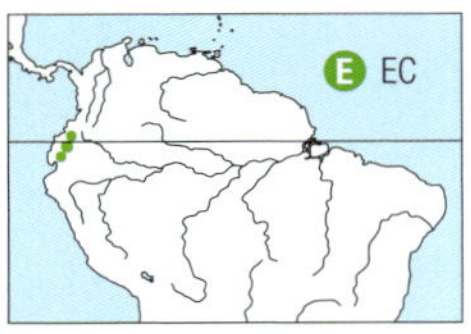

W foothills of N & C Andes

Ecuadorean Small-eared Shrew
Cryptotis equatoris

FR Musaraigne d'Équateur
DE Ecuador-Kleinohrspitzmaus
ES Musaraña de orejas pequeñas de Ecuador

NE
Monotypic
HB 5.4–8.3 cm
T 2.4–3.5 cm
W 6–13 g

Osgood's Small-eared Shrew
Cryptotis osgoodi

FR Musaraigne d'Osgood
DE Osgood-Kleinohrspitzmaus
ES Musaraña de orejas pequeñas de Osgood

1700–3700 m

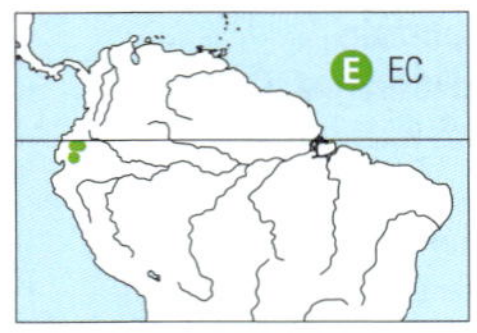

NC Andes

LC
Monotypic
HB 6.5–9.9 cm
T 2.2–3.8 cm W 9–16.5 g

Wandering Small-eared Shrew
Cryptotis montivagus

FR Musaraigne d'Anthony
DE Wandernde Kleinohrspitzmaus
ES Musaraña de orejas pequeñas errante

2500–3800 m

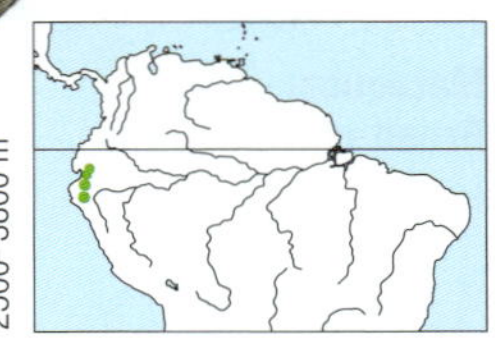

Andes of C & S EC (Chimborazo, Azuay, Loja) & NW PE (Piura)

DD
Monotypic
HB 5.4–9.3 cm
T 3–4.1 cm
W 12–17 g

2800–3900 m

Blind Small-eared Shrew
Cryptotis niausa

FR Musaraigne aveugle
DE Blinde Kleinohrspitzmaus
ES Musaraña de orejas pequeñas ciega

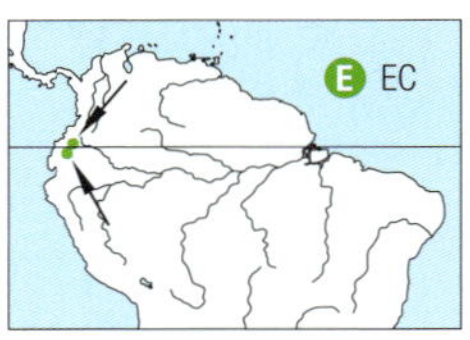

NC Andes

NE
Monotypic
HB 7–9.6 cm
T 2.9–3.7 cm
W 11–16 g

2700–3280 m

Evaristo's Small-eared Shrew
Cryptotis evaristoi

FR Musaraigne d'Evaristo
DE Evaristo-Kleinohrspitzmaus
ES Musaraña de orejas pequeñas de Evaristo

Tabaconas Namballe National Sanctuary

DD
Monotypic
HB 6.3–7.3 cm
T 3.1–3.5 cm
W c. 9 g

Peruvian Small-eared Shrew
Cryptotis peruviensis

FR Musaraigne du Pérou
DE Peru-Kleinohrspitzmaus
ES Musaraña de orejas pequeñas de Perú

2050–3150 m

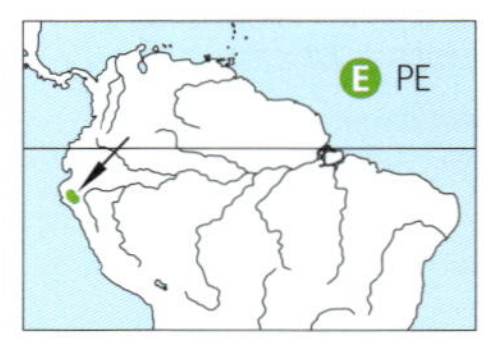

W & E Andes slopes in N

Genus *BLARINA*

(5 species)

Elliot's Short-tailed Shrew
Blarina hylophaga

FR Musaraigne d'Elliot
DE Elliot-Kurzschwanzspitzmaus
ES Musaraña colicorta de Elliot

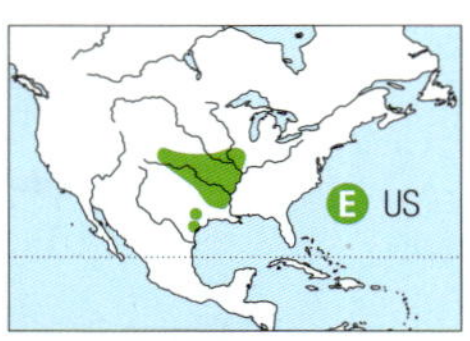

From NE Colorado, S Nebraska & Missouri S to N Louisiana & SE Texas; limits uncertain

LC
2 sspp
HB 7.2–9.6 cm
T 1.9–2.7 cm
W 8–16 g

Northern Short-tailed Shrew
Blarina brevicauda

FR Grande Musaraigne
DE Nördliche Kurzschwanzspitzmaus
ES Musaraña colicorta septentrional

LC
11 sspp
HB 9–11.4 cm
T 1.7–3.2 cm
W 11–30 g

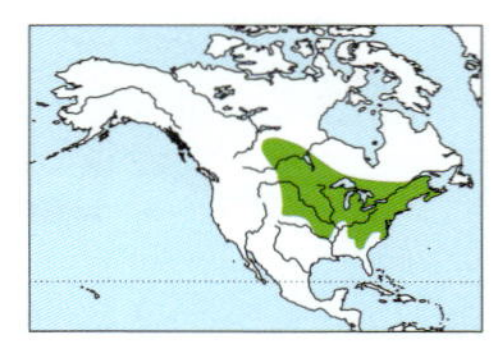

SE CA (E from C Saskatchewan), C & NE US (S to Arkansas & Georgia)

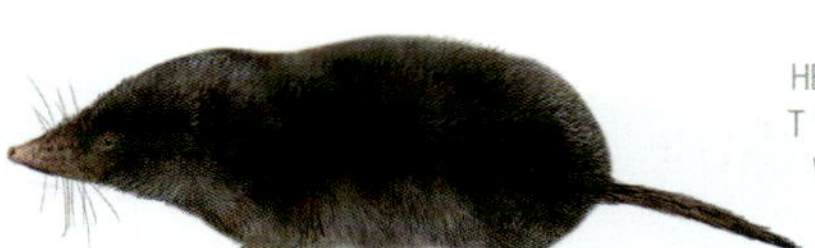

LC
2 sspp
HB 6.9–8 cm
T 1.2–2.6 cm
W 5–13 g

Southern Short-tailed Shrew
Blarina carolinensis

FR Musaraigne de Caroline
DE Südliche Kurzschwanzspitzmaus
ES Musaraña colicorta meridional

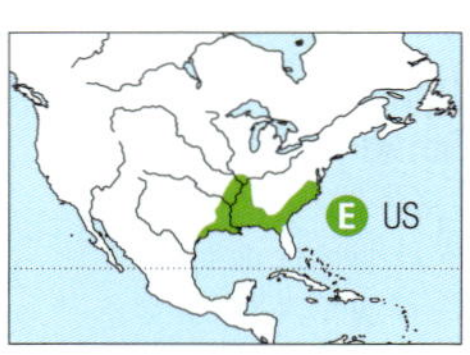

From S Illinois & Virginia S to E Texas & N Florida

NE
Monotypic
HB 6.6–7.6 cm
T 1.8–2.6 cm
W 5–13 g

Everglades Short-tailed Shrew
Blarina peninsulae

FR Musaraigne de Floride
DE Everglades-Kurzschwanzspitzmaus
ES Musaraña colicorta de Everglades

NE, C & S Florida (except C Everglades region); possibly extreme SE Georgia

NE
Monotypic
HB 7.8–9.1 cm
T 2.2–2.5 cm
W 11.1–17 g

Sherman's Short-tailed Shrew
Blarina shermani

FR Musaraigne de Sherman
DE Sherman-Kurzschwanzspitzmaus
ES Musaraña colicorta de Sherman

SW coast of Florida (just N of Fort Myers S to near Royal Palm)

Genus *BLARINELLA*

(2 species)

Sichuan Short-tailed Shrew
Blarinella quadraticauda

FR Musaraigne du Moupin
DE Sichuan-Kurzschwanzspitzmaus
ES Musaraña colicorta de Sichuán

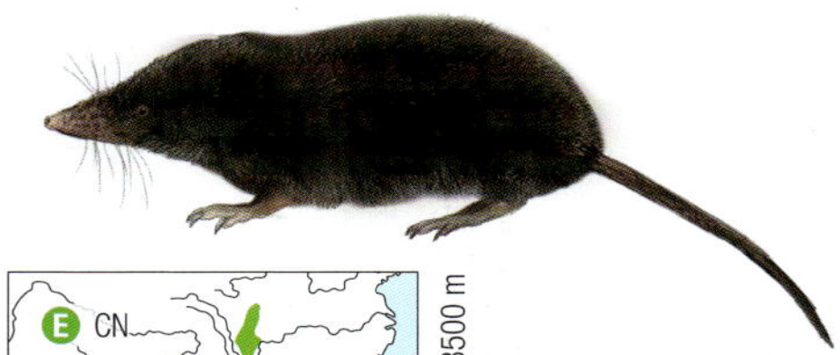

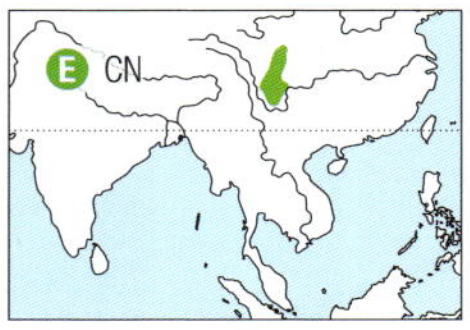

1600–3500 m

W Sichuan Mts

NT
Monotypic
HB 6.5–8.1 cm
T 4–6 cm

Burmese Short-tailed Shrew
Blarinella wardi

FR Musaraigne de Ward
DE Burma-Kurzschwanzspitzmaus
ES Musaraña colicorta de Birmania

1600–3000 m

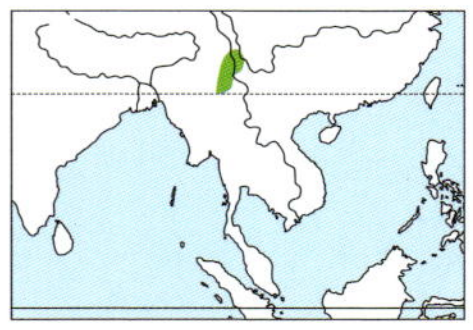

NE MM, SC CN (NW & W Yunnan)

LC
Monotypic
HB 6–6.9 cm
T 3.2–4.3 cm

Genus *PARABLARINELLA*

(1 species)

LC
Monotypic
HB 5.2–7.4 cm
T 3.1–4.2 cm
W c. 8 g

Indochinese Short-tailed Shrew
Parablarinella griselda

FR Musaraigne griselda
DE Indochina-Kurzschwanzspitzmaus
ES Musaraña colicorta de Indochina

1300–3300 m

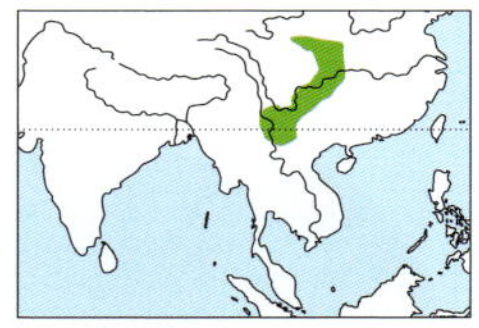

C & SC CN (Gansu, Shaanxi, Sichuan, Chongqing, Hubei, Yunnan, Guizhou), N VN

Genus *ANOUROSOREX*

(4 species)

Chinese Mole Shrew
Anourosorex squamipes

LC
Monotypic
HB 7.4–11 cm
T 0.8–1.9 cm

FR Musaraigne taupe
DE Maulwurfspitzmaus
ES Musaraña topo de China

300–4000 m

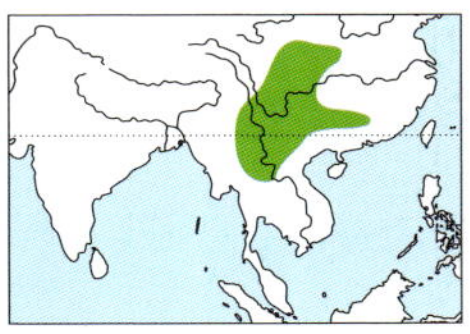

C & S CN (from S Gansu S to Guangdong), N & E MM, N LA, N VN, N TH

LC
Monotypic
HB 5.1–9.8 cm
T 0.7–1.3 cm

Taiwanese Mole Shrew
Anourosorex yamashinai

FR Musaraigne de Yamashina
DE Taiwan-Maulwurfspitzmaus
ES Musaraña topo de Taiwán

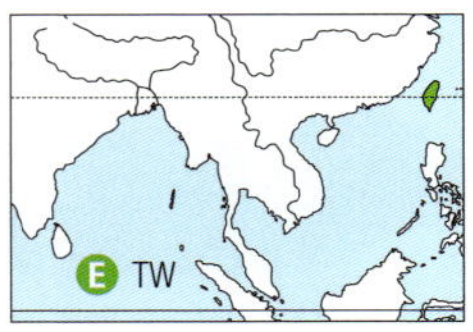

300–3000 m

Widespread

Assam Mole Shrew
Anourosorex assamensis

LC
Monotypic
HB 8.5–11.9 cm
T 1–1.5 cm

FR Musaraigne d'Assam
DE Assam-Maulwurfspitzmaus
ES Musaraña topo de Assam

1500–3100 m

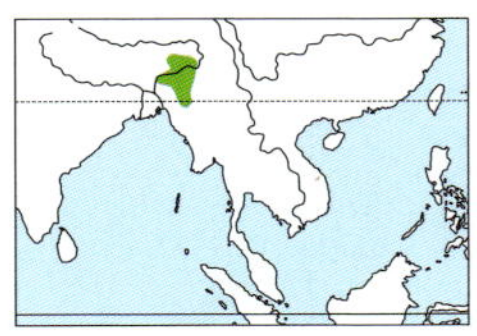

SC CN (Shannan in extreme SE Tibet), NE IN, adjacent WC MM (Tedim)

Giant Mole Shrew
Anourosorex schmidi

DD
Monotypic

FR Musaraigne de Schmid
DE Riesenmaulwurfspitzmaus
ES Musaraña topo gigante

1500–3100 m

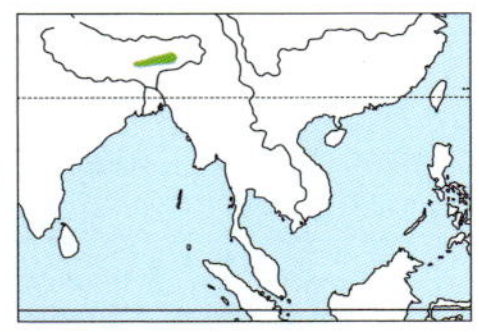

BT (Gomchu), NE IN (Sikkim, Arunachal Pradesh)

Genus *NOTIOSOREX*

(5 species)

LC
Monotypic
HB 5.3–6.8 cm
T 2.5–3.5 cm
W 3–6.3 g

Desert Gray Shrew
Notiosorex crawfordi

FR Musaraigne de Crawford
DE Graue Wüstenspitzmaus
ES Musaraña gris de desierto

0–2600 m

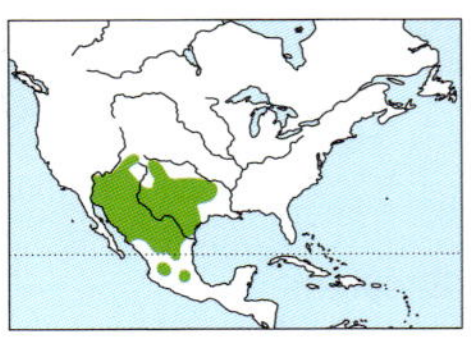

SC US (from S Colorado) S to N & C MX (S to E Jalisco & Hidalgo)

Cockrum's Gray Shrew
Notiosorex cockrumi

FR Musaraigne de Cockrum
DE Cockrum-Wüstenspitzmaus
ES Musaraña gris de Cockrum

675–1195 m

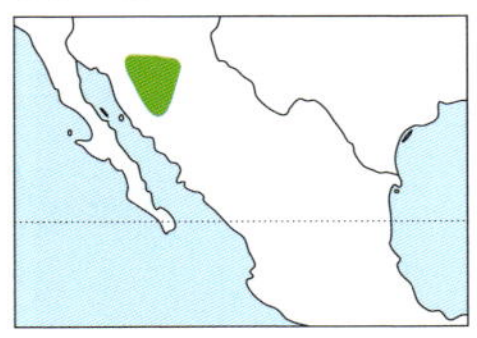

SE Arizona (SW US) S to C Sonora (NW MX)

LC
Monotypic
HB 5.2–6.1 cm
T 1.9–2.7 cm
W 3–6.3 g

NE
3 sspp
W 3–6 g

Ticul's Gray Shrew
Notiosorex tataticuli

FR Musaraigne d'Alvarez
DE Ticul-Wüstenspitzmaus
ES Musaraña gris de Ticul

0–2200 m

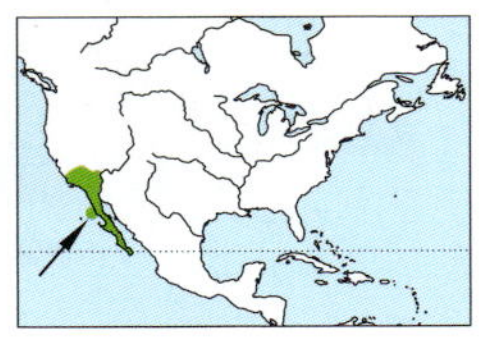

US (S California & extreme S Nevada), MX (Baja California Pen, with San Martín I)

LC
Monotypic
HB 5.4–7.3 cm
T 2.3–3.2 cm
W 5–6.3 g

Large-eared Gray Shrew
Notiosorex evotis

FR Musaraigne à oreilles longues
DE Großohr-Wüstenspitzmaus
ES Musaraña gris de orejas largas

0–550 m

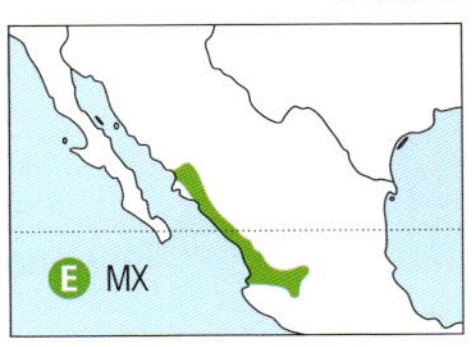

Sinaloa, Nayarit, Jalisco, Michoacán

VU
Monotypic
HB 5.9–6.2 cm
T 2.8–3.1 cm
W 5–6.3 g

Villa's Gray Shrew
Notiosorex villai

FR Musaraigne de Villa
DE Villa-Wüstenspitzmaus
ES Musaraña gris de Villa

580–1340 m

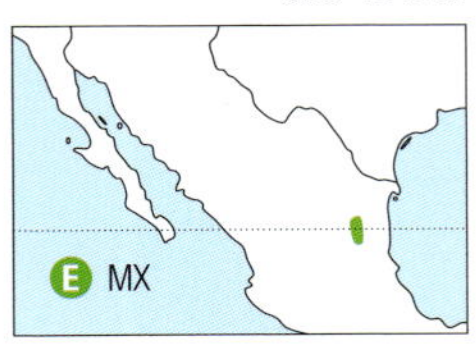

Known from 3 localities in SW Tamaulipas; potentially wider

Genus *MEGASOREX*

(1 species)

LC
Monotypic
HB 6.9–8.4 cm
T 3.2–4.9 cm
W 9–12 g

Mexican Shrew
Megasorex gigas

FR Musaraigne géante
DE Mexikanische Wüstenspitzmaus
ES Musaraña gigante de México

80–1800 m

N Nayarit S to W Oaxaca

Genus *PSEUDOSORICULUS*

(1 species)

Taiwanese Brown-toothed Shrew

Pseudosoriculus fumidus

FR Musaraigne de Taïwan
DE Taiwan-Braunzahnspitzmaus
ES Musaraña de dientes marrones de Taiwán

LC
Monotypic
HB 5.3–7.1 cm
T 3.7–5.2 cm
W 5.7–7.8 g

1000–3600 m

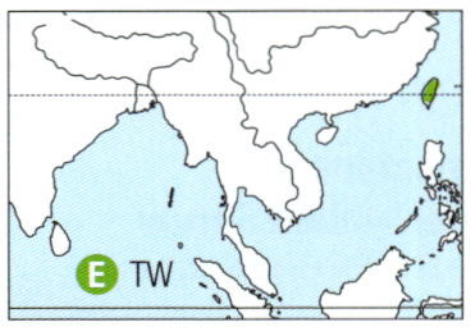

Montane habitats

Arboreal Brown-toothed Shrew

Episoriculus macrurus

FR Musaraigne de Blanford
DE Kleine Braunzahnspitzmaus
ES Musaraña de dientes marrones arbórea

LC
2 sspp
HB 4.8–7.3 cm
T 7.6–10.1 cm

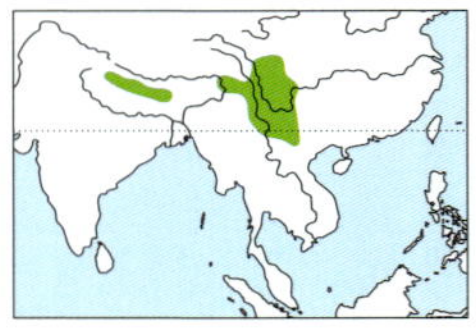

N NP, NE IN, S CN (S & SE Tibet, Sichuan, Yunnan), N MM, NW VN

Genus *EPISORICULUS*

(7 species)

Bailey's Brown-toothed Shrew

Episoriculus baileyi

FR Musaraigne de Bailey
DE Bailey-Braunzahnspitzmaus
ES Musaraña de dientes marrones de Bailey

NE
Monotypic
HB 6.3–8.1 cm
T 6–7.6 cm

1565–2680 m

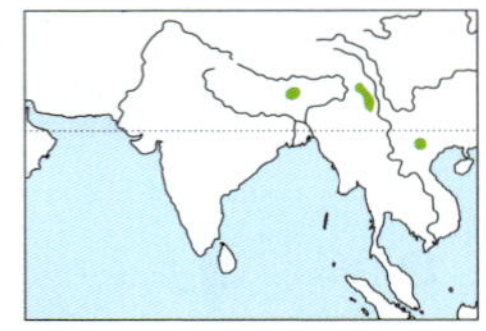

SW CN (S Tibet), NE IN, N MM, N VN

Long-tailed Brown-toothed Shrew

Episoriculus leucops

FR Musaraigne du Népal
DE Langschwanz-Braunzahnspitzmaus
ES Musaraña de dientes marrones de cola larga

LC
Monotypic
HB 5.3–7.6 cm
T 5.8–8.2 cm

1800–3500 m

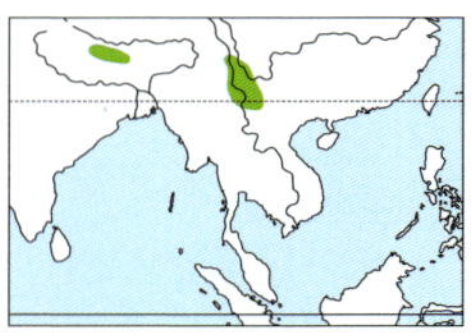

NP, S CN (S Tibet, W Yunnan, perhaps SW Sichuan)

Hodgson's Brown-toothed Shrew

Episoriculus caudatus

FR Musaraigne de Hodgson
DE Hodgson-Braunzahnspitzmaus
ES Musaraña de dientes marrones de Hodgson

LC
Monotypic
HB 5.2–7 cm
T 4.6–6.5 cm

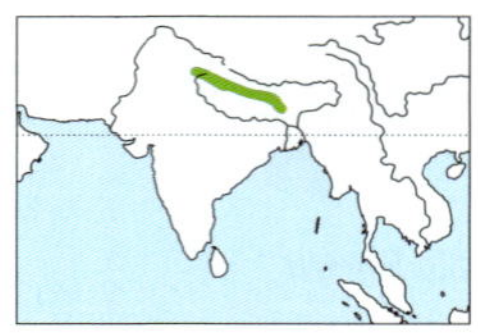

N IN, NP, SW CN (S Tibet)

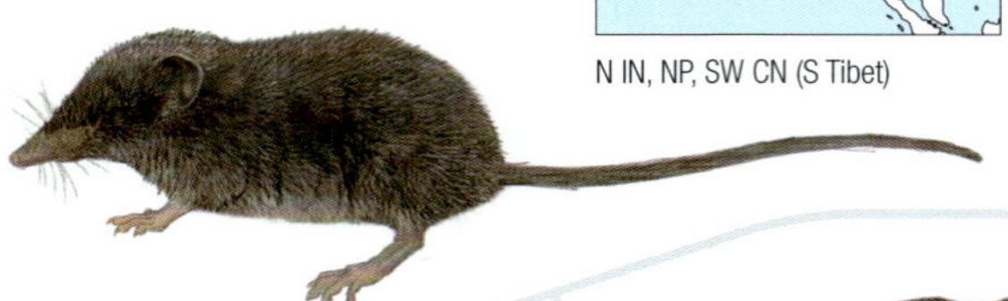

Sichuan Brown-toothed Shrew

Episoriculus sacratus

FR Musaraigne de l'Emei
DE Sichuan-Braunzahnspitzmaus
ES Musaraña de dientes marrones de Sichuán

1700–3500 m

W Sichuan Mts

NE
Monotypic
HB 5.8–7.4 cm
T 4.8–6.9 cm

Hidden Brown-toothed Shrew

Episoriculus umbrinus

FR Musaraigne ombrée
DE Saluen-Braunzahnspitzmaus
ES Musaraña de dientes marrones ocultadiza

NE
Monotypic
HB 4.4–7.4 cm
T 4.2–5.8 cm

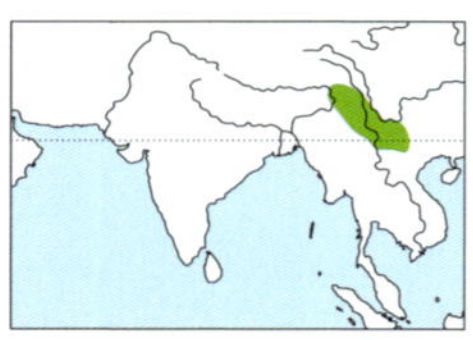

1800–3500 m

SC CN (SE Tibet, Yunnan), NE IN, N MM, extreme N VN

Nepalese Brown-toothed Shrew

Episoriculus soluensis

FR Musaraigne du Solu
DE Nepal-Braunzahnspitzmaus
ES Musaraña de dientes marrones de Nepal

NE
Monotypic
HB 6.2–6.6 cm
T 4.9–5.6 cm

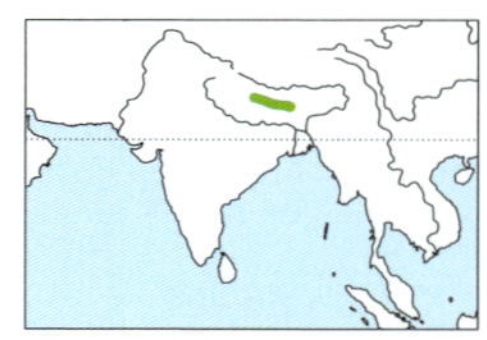

NP, NE IN (Sikkim); limits unclear

Genus *CHODSIGOA*

(10 species)

LC
2 sspp
HB 6.2–8.6 cm
T 5.6–7.3 cm

1200–3500 m

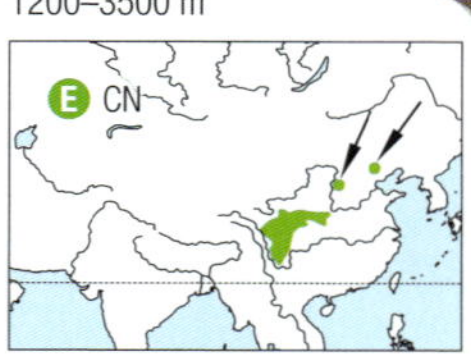

De Winton's Brown-toothed Shrew

Chodsigoa hypsibia

FR Musaraigne de De Winton
DE De Winton-Braunzahnspitzmaus
ES Musaraña de dientes marrones de De Winton

SE Qinghai, S Gansu, S Shaanxi, E Tibet, Sichuan, N Yunnan, disjunct in Hebei & S Anhui

DD
Monotypic
HB 4.7–6.4 cm
T 4.1–5.2 cm

1800–3500 m

Pygmy Brown-toothed Shrew

Chodsigoa parva

FR Musaraigne de Lijiang
DE Yunnan-Braunzahnspitzmaus
ES Musaraña de dientes marrones pigmea

W Sichuan, Yunnan; limits unclear

NE
Monotypic
HB 6.2–7.3 cm
T 5.4–6.4 cm
W 4.7–5.8 g

Dabieshan Brown-toothed Shrew

Chodsigoa dabieshanensis

FR Musaraigne des Dabie
DE Dabieshan-Braunzahnspitzmaus
ES Musaraña de dientes marrones de Dabie

750–1250 m

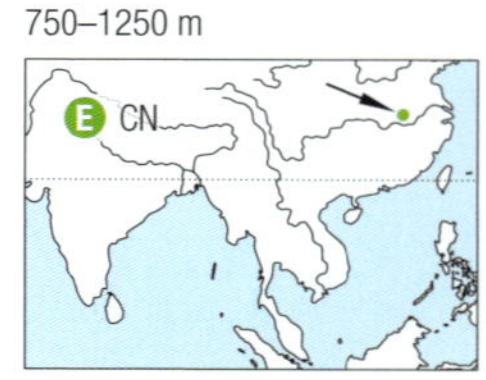

Dabie Mts (Yaoleping, Bancang & Foziling NRs) in Anhui

Smith's Brown-toothed Shrew

Chodsigoa smithii

NT
Monotypic
HB 7.6–8.4 cm
T 9.3–11 cm

FR Musaraigne de Smith
DE Smith-Braunzahnspitzmaus
ES Musaraña de dientes marrones de Smith

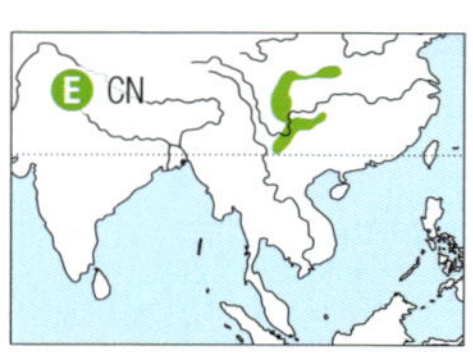

900–3000 m

S Shaanxi, Sichuan, E Yunnan, Guizhou

Salenski's Brown-toothed Shrew

Chodsigoa salenskii

FR Musaraigne de Salenski
DE Salenski-Braunzahnspitzmaus
ES Musaraña de dientes marrones de Salenski

DD
Monotypic
HB c. 7.8 cm
T c. 11 cm

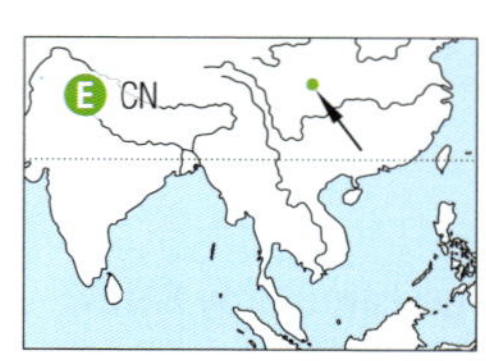

c. 2750 m

Known only from type locality in N Sichuan

NE
Monotypic
HB 6.9–7.5 cm
T 8.4–8.7 cm

above 2000 m

Dusky Brown-toothed Shrew
Chodsigoa furva
FR Musaraigne de Birmanie
DE Dunkle Braunzahnspitzmaus
ES Musaraña de dientes marrones oscura

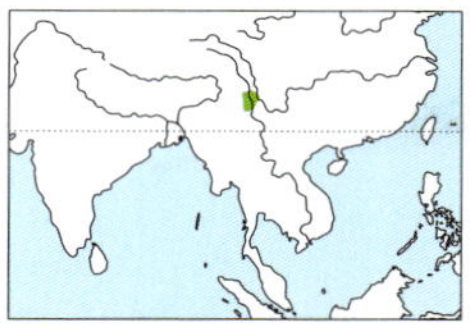

Few localities in SC CN (NW Yunnan) & adjacent N MM; limits unclear

Lesser Taiwanese Brown-toothed Shrew
Chodsigoa sodalis
FR Musaraigne sociable
DE Kleine Taiwan-Braunzahnspitzmaus
ES Musaraña de dientes marrones de Taiwán menor

DD
Monotypic
HB 6.5–7.1 cm
T 6.4–7.3 cm
W 4.2–5.6 g

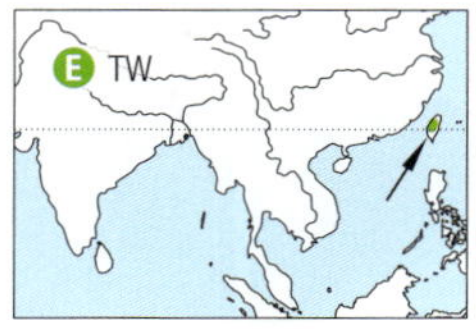

600–2500 m

Montane broad-leaved forests

DD
Monotypic
HB 5.8–7.4 cm
T 5.1–8.3 cm

Van Sung's Brown-toothed Shrew
Chodsigoa caovansunga
FR Musaraigne de Cao Van Sung
DE Van Sung-Braunzahnspitzmaus
ES Musaraña de dientes marrones de Van Sung

350–2000 m

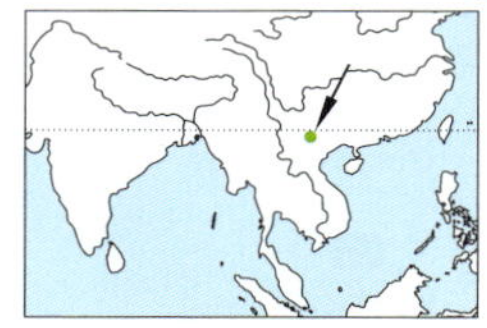

Known from 2 sites in SC CN (S Yunnan) & N VN (Ha Giang Prov)

Hoffmann's Brown-toothed Shrew
Chodsigoa hoffmanni
FR Musaraigne de Hoffmann
DE Hoffmann-Braunzahnspitzmaus
ES Musaraña de dientes marrones de Hoffmann

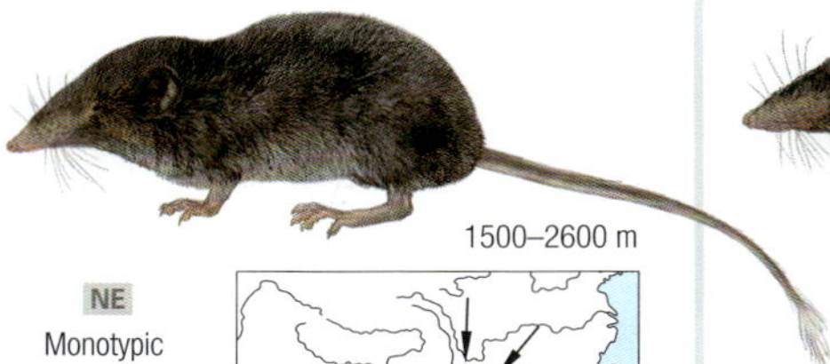

1500–2600 m

NE
Monotypic
HB 5.8–7.5 cm
T 7.4–8.8 cm

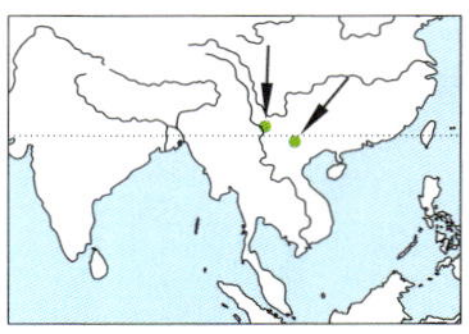

SC CN (Wuliang Shan & Ailao Shan, Yunnan), N VN (Mt Tay Con Linh II)

LC
2 sspp
HB 6.2–7.7 cm
T 7.7–9.9 cm

Lowe's Brown-toothed Shrew
Chodsigoa parca
FR Musaraigne étriquée
DE Lowe-Braunzahnspitzmaus
ES Musaraña de dientes marrones de Lowe

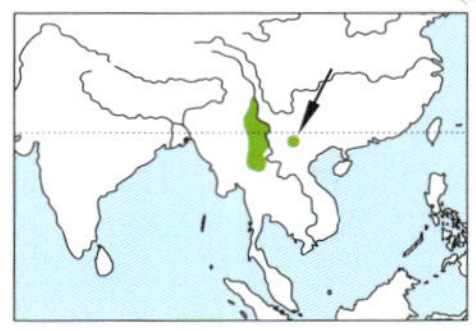

SC CN (Yunnan), NE & E MM, N TH, NW VN (Sa Pa, Lao Cai Prov)

Genus *SORICULUS*

(1 species)

LC
2 sspp
HB 7–9.4 cm
T 3.2–5 cm

Himalayan Shrew
Soriculus nigrescens
FR Musaraigne de l'Himalaya
DE Sikkim-Großklauenspitzmaus
ES Musaraña del Himalaya

1000–4300 m

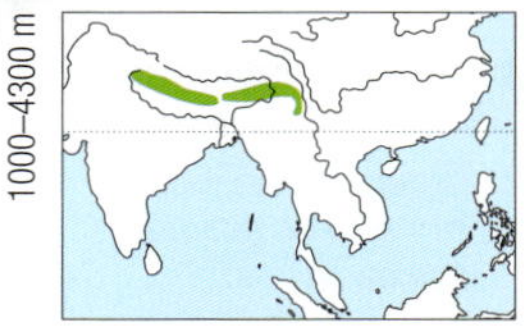

NW & NE IN, NP, BT, SC CN (S Tibet, NW Yunnan, W of R Salween), N MM

Iberian Water Shrew
Neomys anomalus
FR Crossope de Cabrera
DE Iberische Sumpfspitzmaus
ES Musgaño de Cabrera

LC
Monotypic
HB 7.3–1 cm
T 5.2–6.1 cm

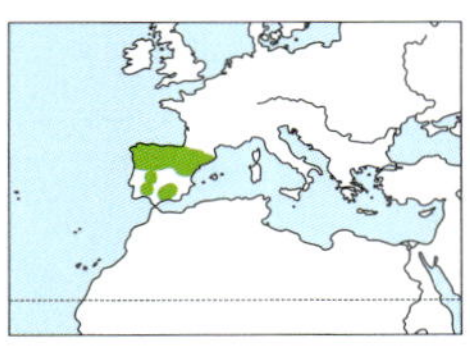

up to 2100 m

Iberian Pen (PT, ES)

Mediterranean Water Shrew
Neomys milleri
FR Crossope de Miller
DE Sumpfspitzmaus
ES Musgaño de Miller

NE
Monotypic
HB 7.1–8.7 cm
T 4.5–5.9 cm

up to 2100 m

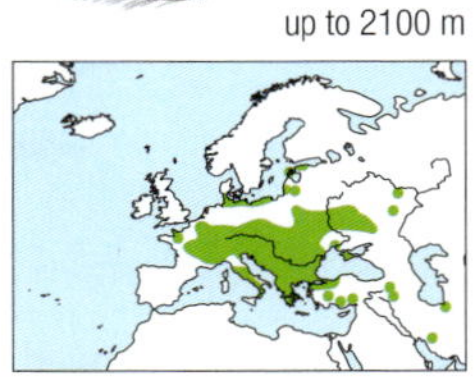

From FR to European RU, S through Mediterranean; sparse in N Europe, TR & IR

Genus *NEOMYS*

(4 species)

Transcaucasian Water Shrew
Neomys teres
FR Crossope du Caucase
DE Kaukasische Bachspitzmaus
ES Musgaño del Cáucaso

up to 2500 m

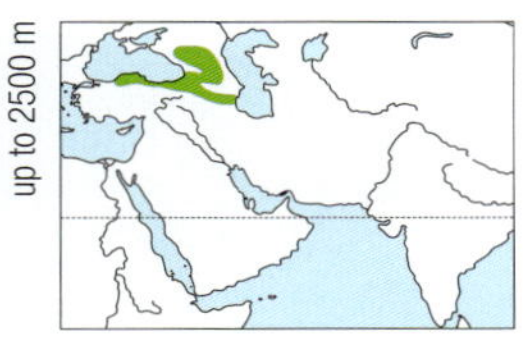

NE TR, Caucasus

LC
4 sspp
HB 8.3–10.1 cm T 5.8–7.3 cm
W 11–28 g

Eurasian Water Shrew
Neomys fodiens
FR Crossope aquatique
DE Wasserspitzmaus
ES Musgaño patiblanco

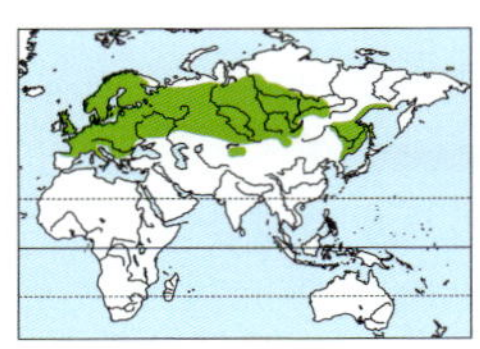

Great Britain & N ES through Europe to C Siberia, E to Pacific coast

LC
4 sspp
HB 7.5–10.3 cm
T 5.8–7.3 cm
W 8.5–25 g

Genus *CROSSOGALE*

(3 species)

Bornean Water Shrew
Crossogale phaeura
FR Chimarrogale de Bornéo
DE Borneo-Wasserspitzmaus
ES Musgaño de Borneo

EN
Monotypic
HB 9.1–11 cm
T 8.1–9.5 cm

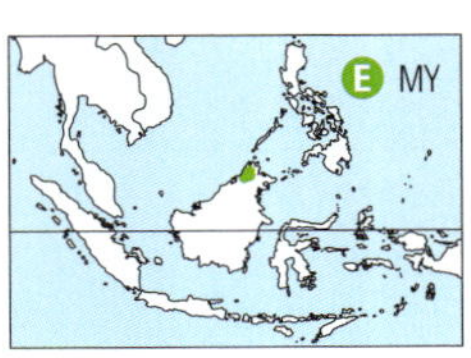

N Borneo in 3 mts of Sabah (Kinabalu, Crocker Range, Trus Madi); perhaps S Borneo

DD
Monotypic
HB 11.2–11.5 cm
T 9–9.5 cm

Sumatran Water Shrew
Crossogale sumatrana
FR Chimarrogale de Sumatra
DE Sumatra-Wasserspitzmaus
ES Musgaño de Sumatra

500–1600 m

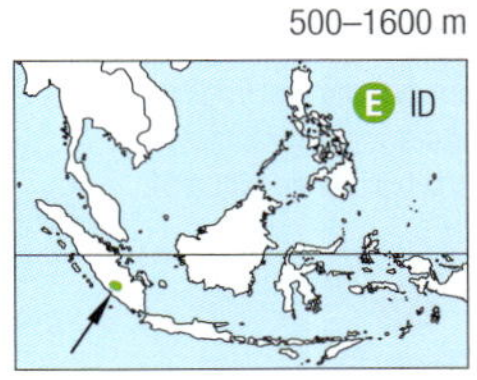

S Sumatra (known only from Padang Highlands)

NT
Monotypic
HB 8.5–12.1 cm
T 8.6–10.2 cm W 29–36 g

Malayan Water Shrew
Crossogale hantu
FR Chimarrogale de Malaisie
DE Malaiische Wasserspitzmaus
ES Musgaño de Malasia

300–1675 m

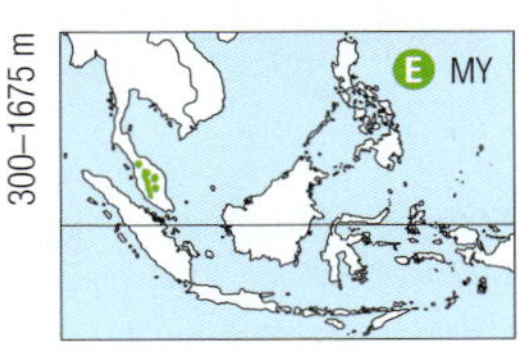

Peninsula in Kedah, Perak, Kelantan, Pahang & Selangor

Genus *CHIMARROGALE*

(5 species)

Chinese Water Shrew
Chimarrogale styani

FR Chimarrogale de Chine
DE Chinesische Wasserspitzmaus
ES Musgaño de China

LC
Monotypic
HB 9.6–10.8 cm
T 6.1–8.5 cm

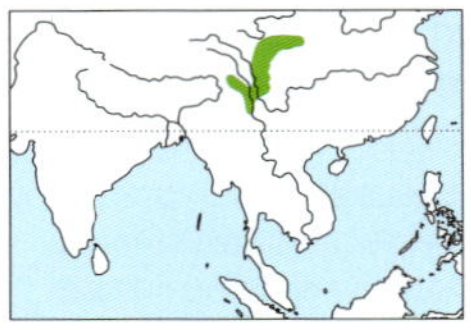

1700–3500 m

SC CN (SE Qinghai, S Gansu, SW Shaanxi, SE Tibet, W Sichuan, NW Yunnan), N MM

Himalayan Water Shrew
Chimarrogale himalayica

LC
Monotypic
HB 9.6–12.9 cm
T 6.2–9.8 cm

FR Chimarrogale de l'Himalaya
DE Himalaya-Wasserspitzmaus
ES Musgaño del Himalaya

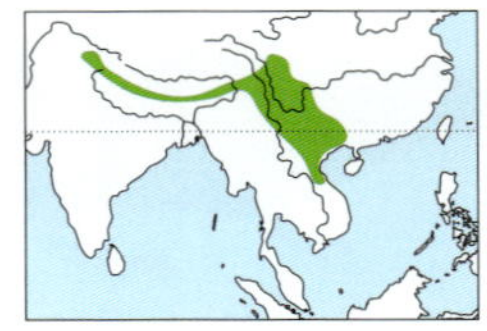

610–3045 m

S Himalaya from NW IN E to SC CN (SE Tibet, Sichuan, Yunnan), N MM, LA, N & C VN

Vietnamese Water Shrew
Chimarrogale varennei

NE
Monotypic
T c. 6.2 cm

FR Chimarrogale du Vietnam
DE Vietnam-Wasserspitzmaus
ES Musgaño del sur de Vietnam

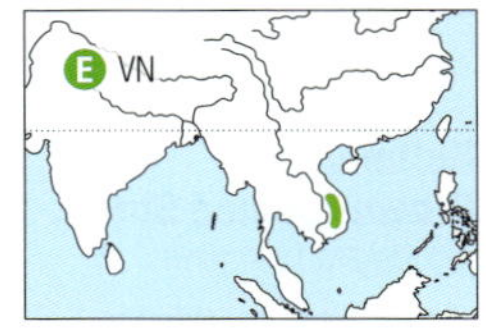

Kon Tum, Dak Lak & Lam Dong prov

NE
Monotypic
HB 8–13 cm
T 8.1–10.1 cm

Leander's Water Shrew
Chimarrogale leander

FR Chimarrogale de Taïwan
DE Taiwan-Wasserspitzmaus
ES Musgaño de Taiwán

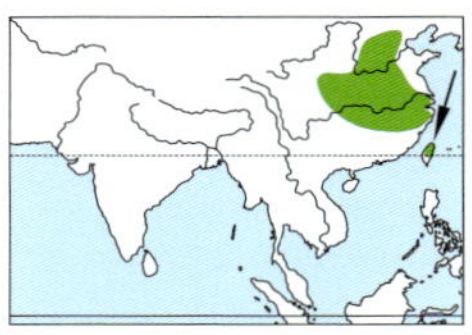

300–2200 m

C & E CN, TW

Japanese Water Shrew
Chimarrogale platycephala

FR Chimarrogale du Japon
DE Japanische Wasserspitzmaus
ES Musgaño de Japón

LC
Monotypic
HB 11.2–13.5 cm
T 8.2–11 cm
W 37–55 g

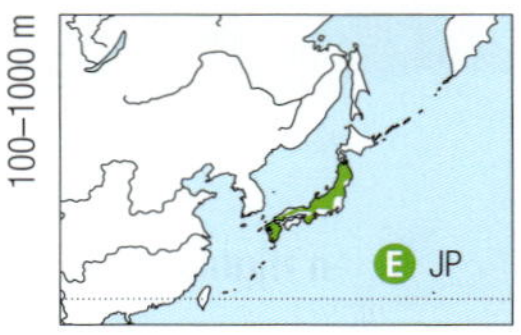

100–1000 m

Honshu, Kyushu; fossil records on Shikoku

Genus *NECTOGALE*

(1 species)

Elegant Water Shrew
Nectogale elegans

FR Nectogale élégante
DE Gebirgsbachspitzmaus
ES Musgaño elegante

LC
Monotypic
HB 9.9–12.5 cm
T 9.6–11.5 cm

1500–3900 m

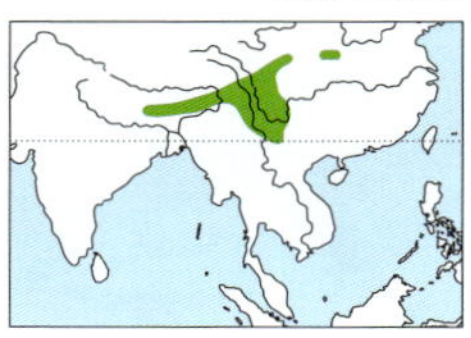

Himalayas in E NP, NE IN, BT, C & SC CN (Shaanxi, Sichuan, SE Tibet, Yunnan), N MM

Genus *PALAWANOSOREX*

(1 species)

Palawan Moss Shrew
Palawanosorex muscorum

FR Pachyure des mousses
DE Palawan-Moosspitzmaus
ES Musaraña de musgo de Palawan

NE
Monotypic
HB 8.4–9.9 cm
T 5.3–6.8 cm
W 16–24 g

1550–1950 m

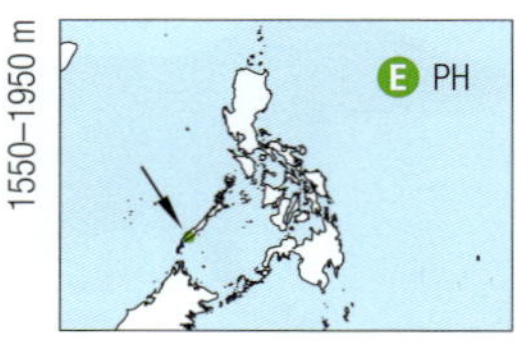

Palawan (higher elevations of Mt Mantalingahan)

Genus *SOLISOREX*

(1 species)

950–2310 m

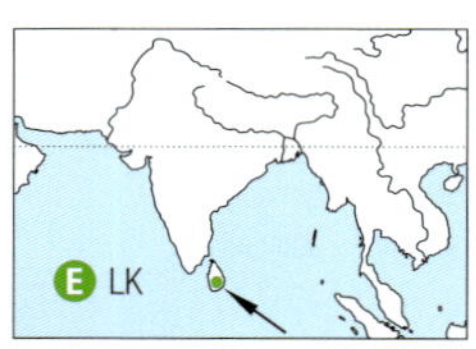

C highlands

Pearson's Long-clawed Shrew
Solisorex pearsoni

FR Pachyure de Pearson
DE Pearson-Langkrallenspitzmaus
ES Musaraña de uñas largas de Pearson

EN
Monotypic
HB 12–13 cm
T 6–7 cm

Genus *FEROCULUS*

(1 species)

Kelaart's Long-clawed Shrew
Feroculus feroculus

FR Pachyure de Kelaart
DE Kelaart-Langkrallenspitzmaus
ES Musaraña de uñas largas de Kelaart

EN
Monotypic
HB 9.8–11.8 cm
T 6.5–7.3 cm

1850–2400 m

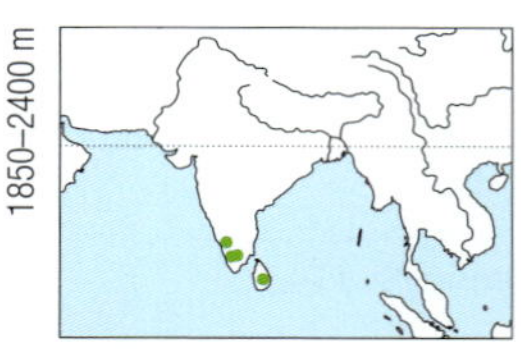

S IN (Western Ghats), C LK

Genus *SUNCUS*

(19 species)

Etruscan Shrew
Suncus etruscus

FR Pachyure étrusque
DE Wimperspitzmaus
ES Musarañita

LC
7 sspp
HB 3.3–5 cm
T 2.1–3 cm
W 1.2–2.7 g

S Europe, N Africa, Middle East, C & S Asia, mainland SE Asia, MG

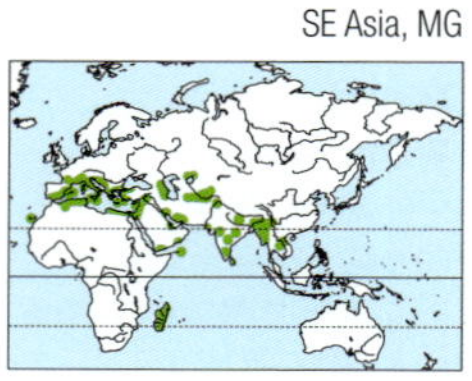

Sri Lankan Shrew
Suncus fellowesgordoni

FR Pachyure de Fellowes-Gordon
DE Sri-Lanka-Wimperspitzmaus
ES Musaraña pigmea de Sri Lanka

EN
Monotypic
HB 4.4–5.8 cm
T 3.3–3.8 cm

1000–1830 m

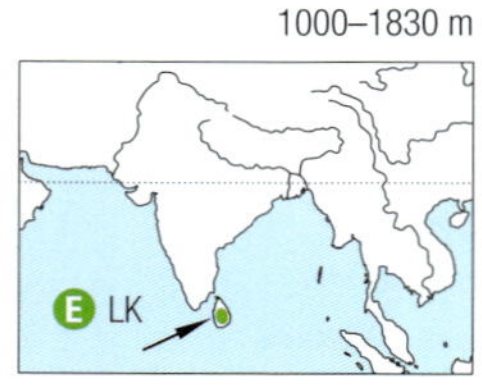

C highlands

Malayan Pygmy Shrew
Suncus malayanus

FR Pachyure de Malaisie
DE Malaya-Wimperspitzmaus
ES Musaraña pigmea de Malasia

LC
Monotypic
HB 3.7–5.1 cm
T 2.2–3.4 cm W 1.1–2.4 g

up to 1465 m

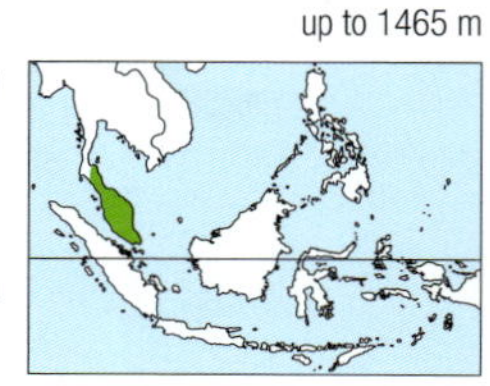

Thai-Malay Pen

Bornean Pygmy Shrew

Suncus hosei

FR Pachyure de Hose
DE Borneo-Wimperspitzmaus
ES Musaraña pigmea de Borneo

DD
Monotypic
HB 5.3–5.9 cm
T 2–2.8 cm
W 2–5 g

0–800 m

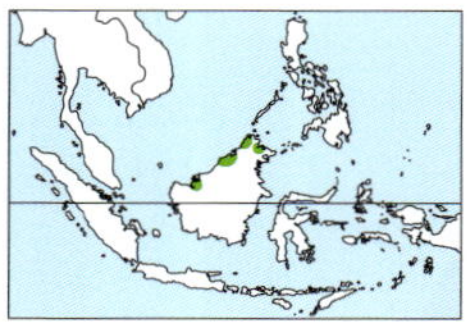

N Borneo, scattered in MY (Sabah & Sarawak) & BN (Belalong); perhaps Kalimantan (ID)

Jungle Shrew

Suncus zeylanicus

FR Pachyure du Sri Lanka
DE Ceylon-Moschusspitzmaus
ES Musaraña de jungla

EN
Monotypic
HB 10.8–11.4 cm
T 9.2–9.7 cm

150–1065 m

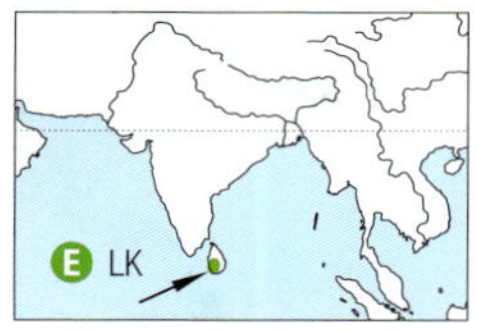

Known from 3 sites in Central & Sabaragamuwa prov; perhaps Western Prov

Asian House Shrew

Suncus murinus

FR Pachyure musquée
DE Moschusspitzmaus
ES Musaraña casera de Asia

LC
Monotypic
HB 9–16 cm
T 4.5–11 cm
W 23–147 g

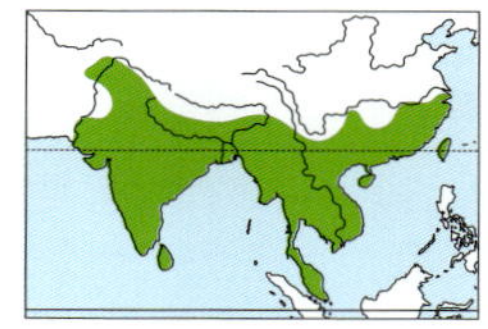

Indo-Malayan Region, S CN (with Hainan), LK & TW; introduced to E Africa, Arabia, ID, PH, JP & many other Is

Sri Lankan Highland Shrew

Suncus montanus

FR Pachyure des montagnes
DE Hochland-Moschusspitzmaus
ES Musaraña de tierras altas de Sri Lanka

VU
Monotypic
HB 7.2–10.6 cm
T 4.6–7.1 cm

150–2520 m

Few sites in Central & Sabaragamuwa prov, including Sinharaja Forest

Nilgiri Highland Shrew

Suncus niger

FR Pachyure de Horsfield
DE Nilgiri-Moschusspitzmaus
ES Musaraña de tierras altas de Nilgiri

NE
Monotypic
HB 7.4–11.4 cm
T 5.4–8.6 cm W 13–24 g

1050–2440 m

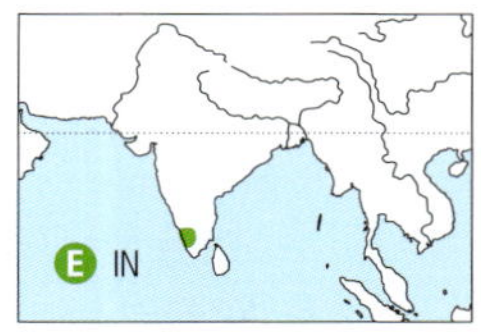

Few disjunct sites in Karnataka, Kerala & Tamil Nadu (Nilgiri & Palni hills)

Anderson's Shrew

Suncus stoliczkanus

FR Pachyure d'Anderson
DE Stoliczka-Wimperspitzmaus
ES Musaraña de Anderson

LC
Monotypic
HB 5–8.5 cm
T 3.2–5.5 cm
W 8–15 g

up to 500 m

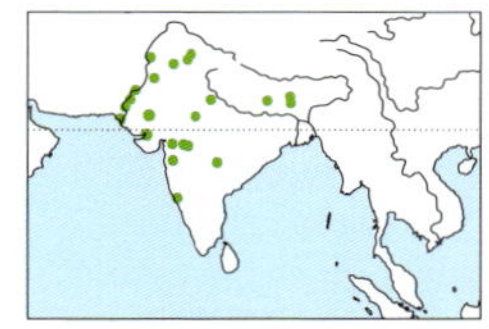

Scattered in PK, IN & NP

Flores Shrew

Suncus mertensi

FR Pachyure de Florès
DE Flores-Wimperspitzmaus
ES Musaraña de Flores

EN
Monotypic
HB 6.2–6.8 cm
T 5.4–6.2 cm

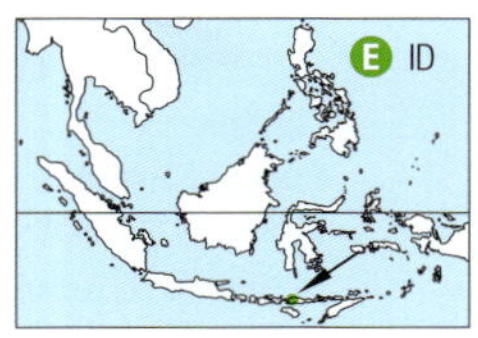

W Flores (known only from near L Rana Mese)

Black Shrew

Suncus ater

FR Pachyure noire
DE Kinabalu-Moschusspitzmaus
ES Musaraña negra

DD
Monotypic
HB c. 7.5 cm
T c. 5.7 cm

c. 1675 m

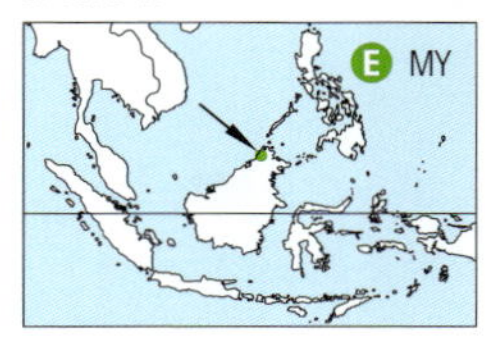

N Borneo (known from Mt Kinabalu; perhaps more widely in highlands)

Day's Shrew

Suncus dayi

FR Pachyure de Day
DE Day-Wimperspitzmaus
ES Musaraña de Day

EN
Monotypic
HB 7–7.8 cm
T 6–8.8 cm
W 8–10 g

1500–2500 m

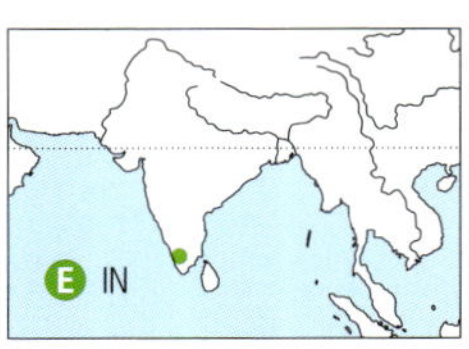

Tamil Nadu (Nilgiri & Palni hills), Kerala (Thrissur, Eravikulam NP)

Taita Shrew

Suncus aequatorius

FR Pachyure des Taita
DE Taita-Wimperspitzmaus
ES Musaraña de Taita

EN
Monotypic
HB 5.3–8.5 cm
T 5.1–6.7 cm
W 7.5–11 g

1500–1600 m

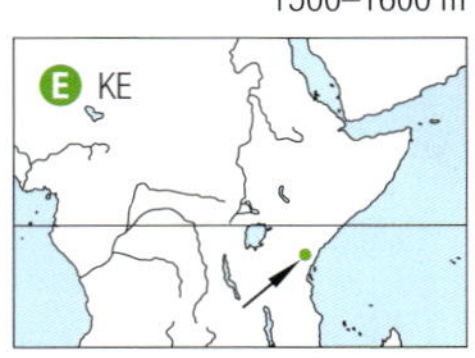

Taita Hills (Mt Sagalla & Chawia Forest); perhaps NE TZ

Greater Dwarf Shrew

Suncus lixa

FR Pachyure cantinière
DE Mittelgroße Wimperspitzmaus
ES Musaraña enana mayor

LC
Monotypic
HB 5.2–8.1 cm
T 3.6–5.1 cm
W c. 4.9 g

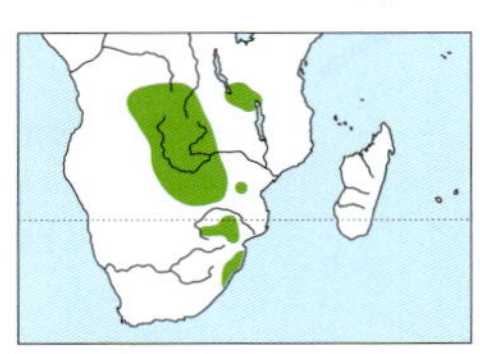

S Africa (S CD, E AO, ZM, SW TZ, N MW, ZW, NE NA, N BW, S MZ, SZ, E ZA)

Hutu-Tutsi Dwarf Shrew

Suncus hututsi

FR Pachyure des Hutus-Tutsis
DE Hutu-Tutsi-Wimperspitzmaus
ES Musaraña enana de Hutu-Tutsi

DD
Monotypic
HB 4.4–5.8 cm
T 2.8–3.3 cm
W 1.7–2.4 g

1170–2350 m

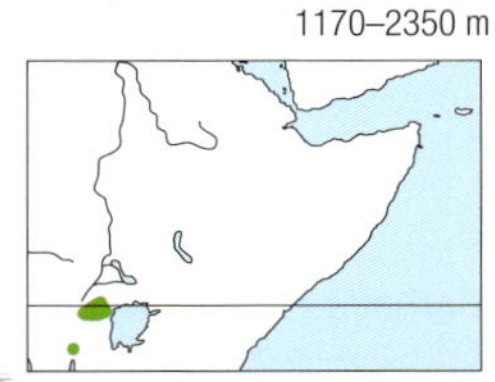

S UG, BI; perhaps E CD & RW

Least Dwarf Shrew

Suncus infinitesimus

FR Pachyure de Heller
DE Zwergwimperspitzmaus
ES Musaraña enana mínima

LC
Monotypic
HB 4.2–6.2 cm
T 2.3–3.1 cm
W 2.8–3.1 g

1700–2100 m

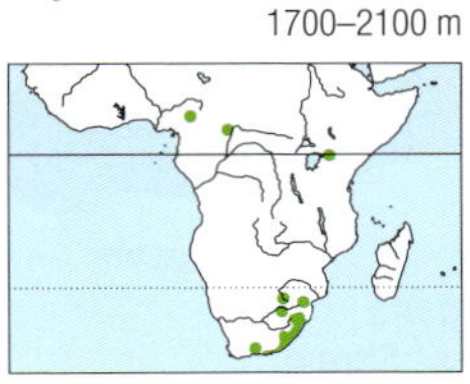

S & E ZA, SZ, disjunct in C CM, SW CF (both from single records) & W KE; probably widespread

Remy's Pygmy Shrew

Suncus remyi

FR Pachyure de Rémy
DE Remy-Wimperspitzmaus
ES Musaraña enana de Remy

LC
Monotypic
HB 4.2–4.9 cm
T 1.7–2.1 cm
W 1.5–1.8 g

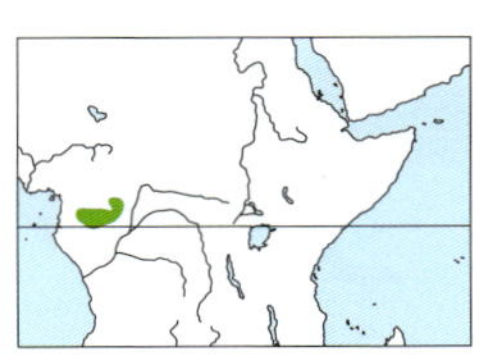

WC Africa in NE GA, NW CG & SW CF; perhaps SW CM

Lesser Dwarf Shrew

Suncus varilla

FR Pachyure des termitières
DE Kleine Wimperspitzmaus
ES Musaraña enana menor

LC
Monotypic
HB 4.4–6.8 cm
T 2.5–4.5 cm
W c. 6.5 g

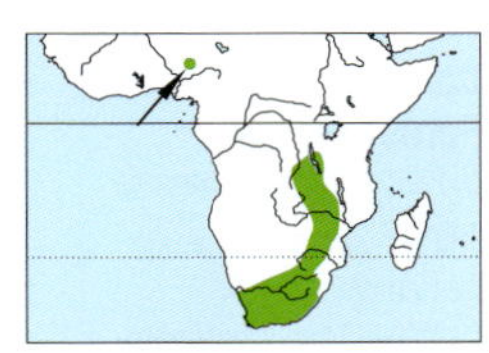

SE CD & W TZ S through E Africa to ZA & LS; 1 record from C NG

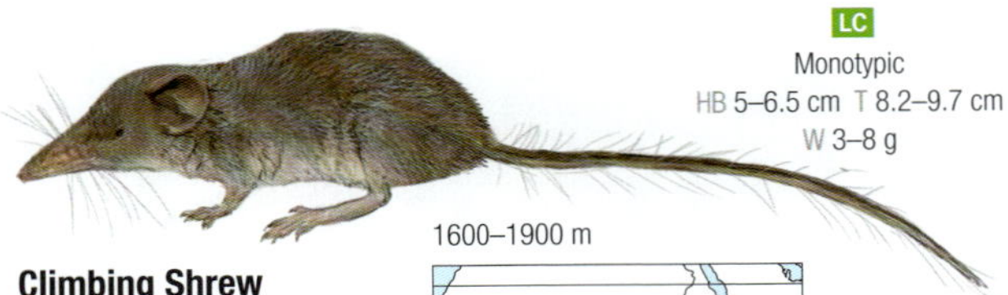

LC
Monotypic
HB 5–6.5 cm T 8.2–9.7 cm
W 3–8 g

Climbing Shrew
Suncus megalurus

FR Pachyure grimpeuse
DE Langschwanz-Wimperspitzmaus
ES Musaraña trepadora

1600–1900 m

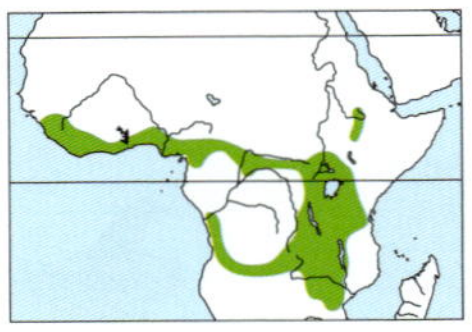

Sub-Saharan Africa from GN E to W KE, S to C AO & C MZ, isolated in W ET

Genus *RUWENZORISOREX*

(1 species)

Rwenzori Shrew
Ruwenzorisorex suncoides

FR Pachyure du Rwenzori
DE Ruwenzori-Spitzmaus
ES Musaraña de Rwenzori

VU
Monotypic
HB 8.4–10.2 cm
T 5.5–7 cm W 18–23 g

1800–2350 m

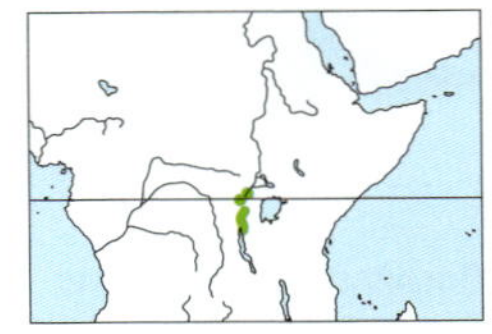

Albertine Rift (E CD, SW UG, RW, BI)

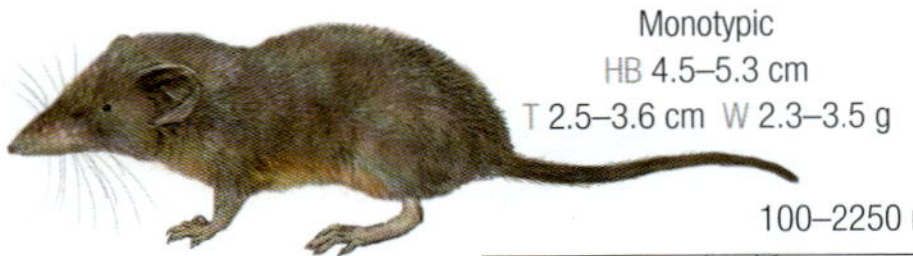

LC
Monotypic
HB 4.5–5.3 cm
T 2.5–3.6 cm W 2.3–3.5 g

Johnston's Forest Shrew
Sylvisorex johnstoni

FR Pachyure de Johnston
DE Johnston-Waldmoschusspitzmaus
ES Musaraña de bosque de Johnston

100–2250 m

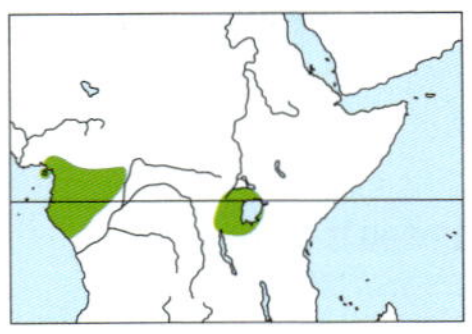

WC Africa (S CM, SW CF, GQ, GA, CG), EC Africa (EC CD, S UG, RW, BI, NW TZ)

Genus *SYLVISOREX*

(15 species)

Akaibe's Forest Shrew
Sylvisorex akaibei

FR Pachyure d'Akaibe
DE Akaibe-Waldmoschusspitzmaus
ES Musaraña de bosque de Akaibe

390–495 m

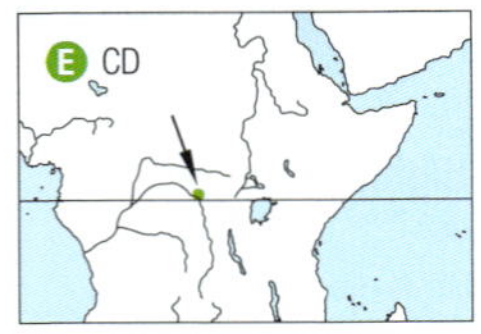

Right bank of R Congo

DD
Monotypic
HB 6–7.3 cm T 3–3.4 cm W 4.5–6.5 g

NT
Monotypic
HB 7–9.3 cm T 4.5–5.6 cm
W 9–20 g

Moon Forest Shrew
Sylvisorex lunaris

FR Pachyure de la Lune
DE Mondberg-Waldmoschusspitzmaus
ES Musaraña de bosque de las Montañas de la Luna

1600–3300 m

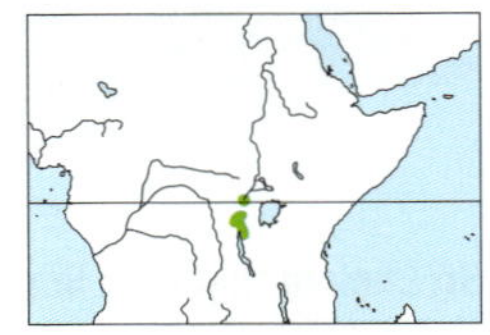

EC CD (with Idjwi I in L Kivu), SW UG, W RW, W BI

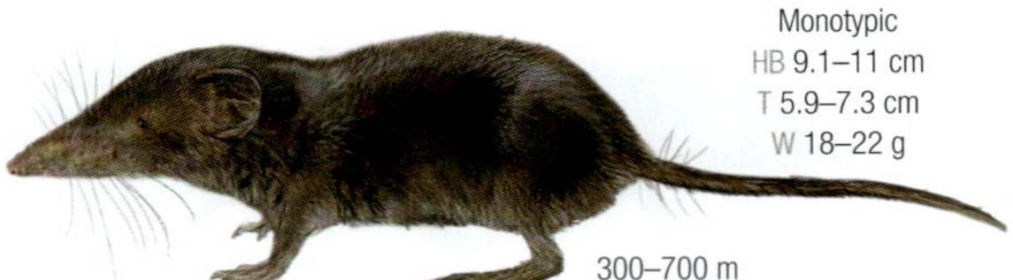

LC
Monotypic
HB 9.1–11 cm
T 5.9–7.3 cm
W 18–22 g

Greater Forest Shrew
Sylvisorex ollula

FR Pachyure de Thomas
DE Große Waldmoschusspitzmaus
ES Musaraña de bosque mayor

300–700 m

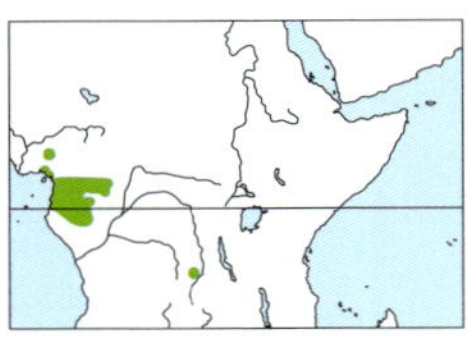

SE NG (Obudu Plateau), S CM, SW CF, GQ, GA, SE CD (1 record); perhaps W CG

Lesser Forest Shrew
Sylvisorex oriundus

FR Pachyure bondissante
DE Kleinere Waldmoschusspitzmaus
ES Musaraña de bosque menor

DD
Monotypic
HB c. 7.2 cm
T c. 6.2 cm

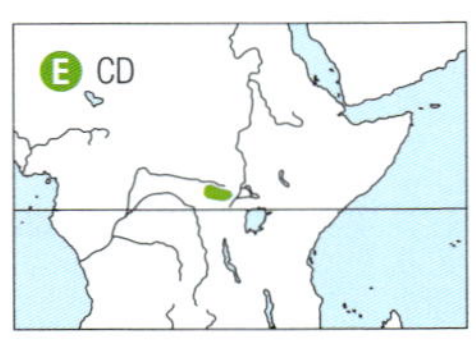

NE lowland tropical moist forest (Medje, Ituri)

2000–2400 m

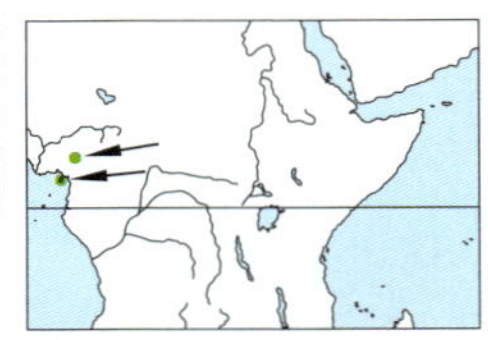

N Bioko (GQ), SW CM

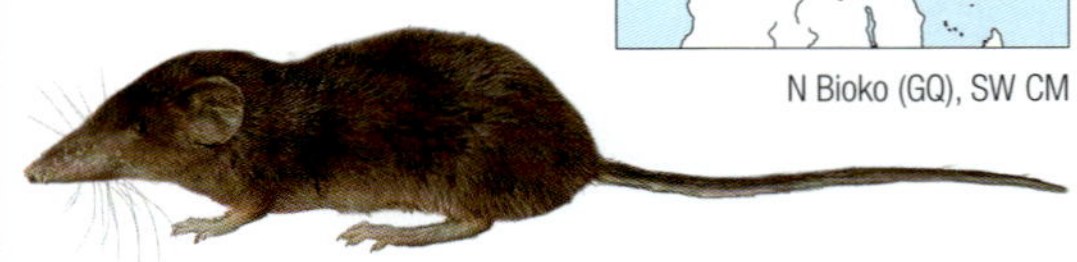

Bioko Forest Shrew
Sylvisorex isabellae

FR Pachyure de Santa Isabel
DE Bioko-Waldmoschusspitzmaus
ES Musaraña de bosque de Bioko

VU
Monotypic
HB 5.4–6.3 cm
T 5.4–5.9 cm
W 5.5–9 g

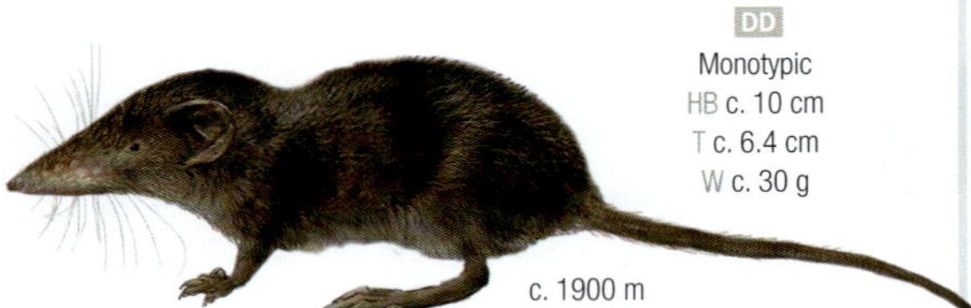

DD
Monotypic
HB c. 10 cm
T c. 6.4 cm
W c. 30 g

Corbet's Forest Shrew
Sylvisorex corbeti

FR Pachyure de Corbet
DE Corbet-Waldmoschusspitzmaus
ES Musaraña de bosque de Corbet

c. 1900 m

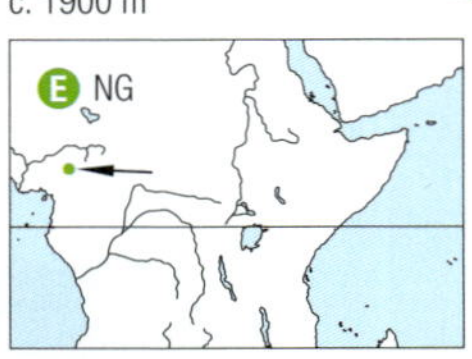

Gotel Mts

DD
Monotypic
HB c. 5.2 cm
T c. 4.4 cm

Bamenda Forest Shrew
Sylvisorex silvanorum

FR Pachyure de Bamenda
DE Bamenda-Waldmoschusspitzmaus
ES Musaraña de bosque de Bamenda

1800–2400 m

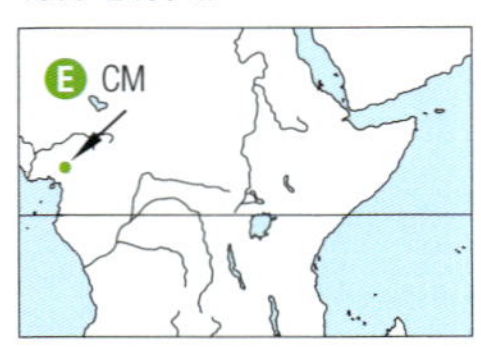

Bamenda Highlands

DD
Monotypic
HB c. 7.4 cm
T c. 6.7 cm
W c. 5 g

Rainforest Shrew
Sylvisorex pluvialis

FR Pachyure des pluies
DE Korup-Waldmoschusspitzmaus
ES Musaraña de bosque lluvioso

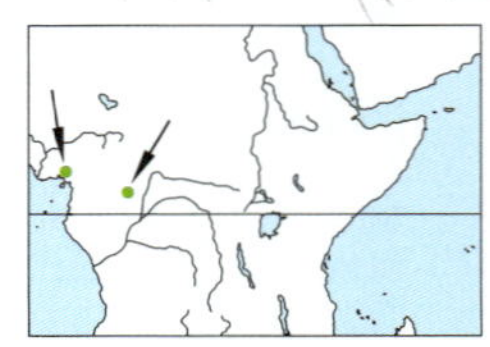

SW CM (Korup NP), SW CF (Kongana)

VU
Monotypic
HB 5.5–6 cm T 5.5–6.1 cm
W 4–5 g

Cameroonian Forest Shrew
Sylvisorex camerunensis

FR Pachyure du Cameroun
DE Kamerun-Waldmoschusspitzmaus
ES Musaraña de bosque de Camerún

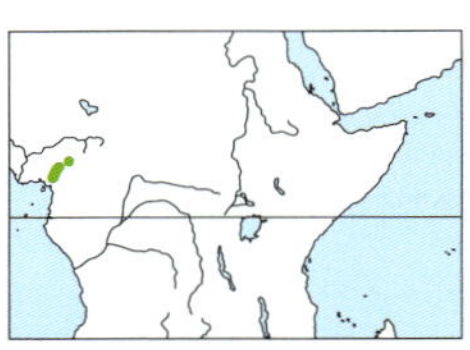

2000–2300 m

SE NG (Gotel Mts), mts of W CM

EN
Monotypic
HB 6.1–7.1 cm
T 5–5.9 cm W c. 8 g

Mount Cameroon Forest Shrew
Sylvisorex morio

FR Pachyure morion
DE Kamerunberg-Waldmoschusspitzmaus
ES Musaraña de bosque del Monte Camerún

1000–2100 m

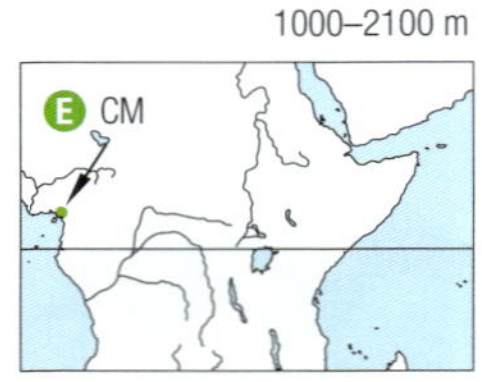

Mt Cameroon

DD
Monotypic
HB c. 6.4 cm
T c. 3.6 cm W c. 3 g

Kongana Forest Shrew
Sylvisorex konganensis

FR Pachyure de la Kongana
DE Kongana-Waldmoschusspitzmaus
ES Musaraña de bosque de Kongana

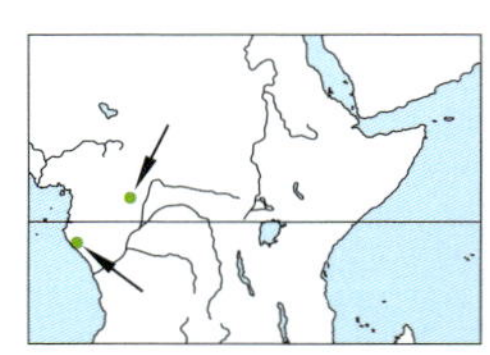

SW CF, N CG, SW GA

Volcano Forest Shrew
Sylvisorex vulcanorum

FR Pachyure des volcans
DE Vulkan-Waldmoschusspitzmaus
ES Musaraña de bosque de los Volcanes

1780–3100 m

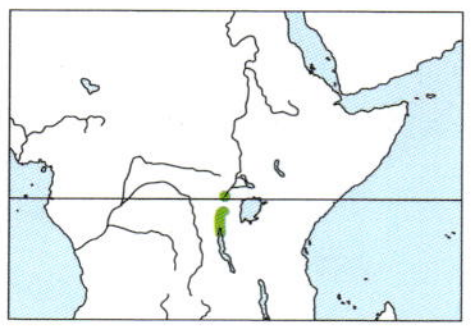

Albertine Rift (EC CD, SW UG, RW, BI)

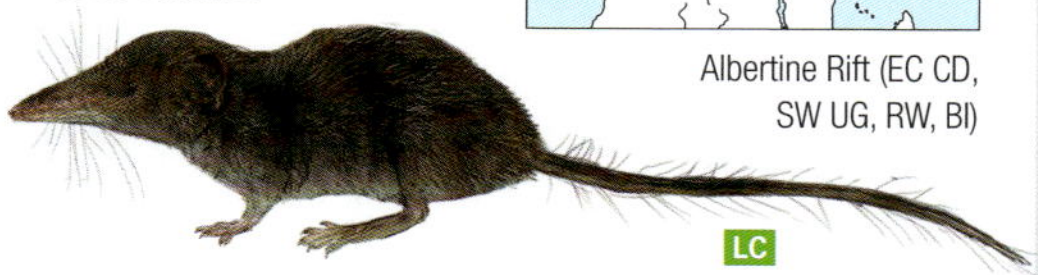

LC
Monotypic
HB 4.3–5.6 cm T 4.7–5.1 cm
W c. 3.5 g

Grant's Forest Shrew
Sylvisorex granti

FR Pachyure de Grant
DE Grant-Waldmoschusspitzmaus
ES Musaraña de bosque de Grant

LC
2 sspp
HB 4.5–6.5 cm T 4.7–6.3 cm
W 3–6 g

above 1500 m

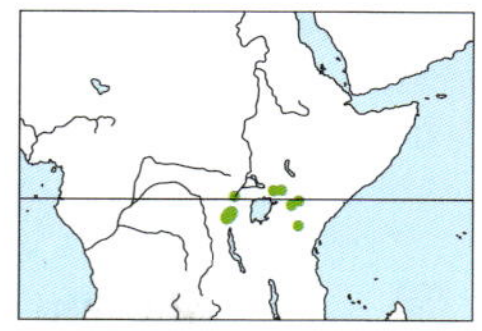

Albertine Rift (EC CD, SW UG, W RW, NW BI), EC UG, W & SW KE, NE TZ

Howell's Forest Shrew
Sylvisorex howelli

LC
2 sspp
HB 5.3–7.2 cm
T 3.6–4.8 cm
W 2.5–5 g

FR Pachyure de Howell
DE Howell-Waldmoschusspitzmaus
ES Musaraña de bosque de Howell

above 900 m

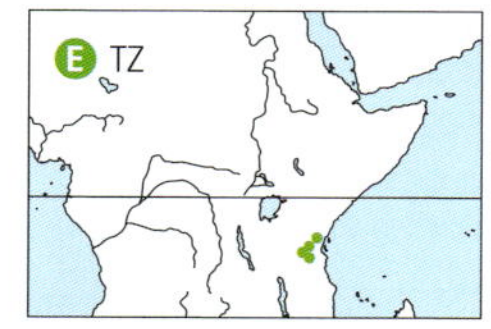

Uluguru & Usambara Mts

Armored Hero Shrew
Scutisorex somereni

FR Pachyure de Someren
DE Panzerspitzmaus
ES Musaraña acorazada

LC
Monotypic
HB 12–15.3 cm
T 7.6–9.9 cm
W 42–92 g

400–2500 m

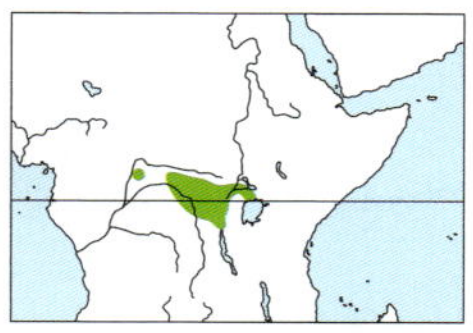

NC & NE CD, C & SW UG, W RW, NW BI

Genus *SCUTISOREX*

(2 species)

Thor's Hero Shrew
Scutisorex thori

FR Pachyure de Thor
DE Thor-Panzerspitzmaus
ES Musaraña de Thor

DD
Monotypic
HB c. 14.6 cm
T c. 10.5 cm
W c. 47 g

c. 355 m

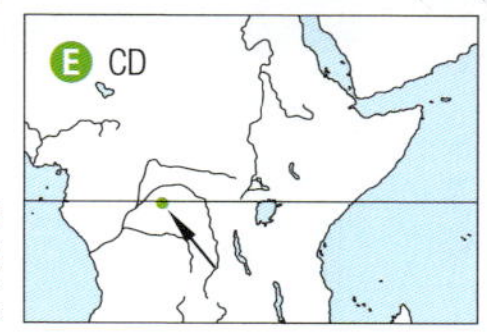

Known only from Baleko near R Tshuapa (Tshuapa Prov)

LC
Monotypic
HB 8–9.7 cm
T 3.7–4.8 cm
W 16–21 g

Greater Large-headed Shrew
Paracrocidura maxima

FR Crocidure à tête large
DE Große Großkopfspitzmaus
ES Musaraña de cabeza grande mayor

1600–2680 m

EC CD, SW UG, W RW, NW BI

Genus *PARACROCIDURA*

(3 species)

LC
Monotypic
HB 6.5–9 cm
T 3.3–3.8 cm
W 7–10 g

Lesser Large-headed Shrew
Paracrocidura schoutedeni

FR Crocidure de Schouteden
DE Kleine Großkopfspitzmaus
ES Musaraña de cabeza grande menor

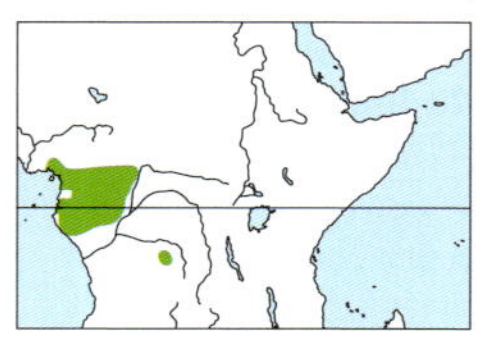

S CM, SW CF, GA, CG, SC CD; possibly GQ

DD
Monotypic
HB c. 8.3 cm
T c. 4.6 cm

Grauer's Large-headed Shrew
Paracrocidura graueri

FR Crocidure de Grauer
DE Grauer-Großkopfspitzmaus
ES Musaraña de cabeza grande de Grauer

c. 2000 m

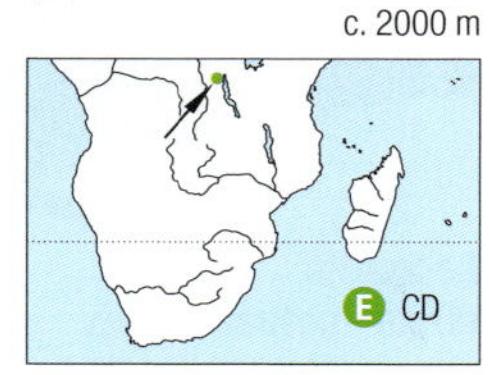

Known from type locality in Itombwe Mts

Genus *DIPLOMESODON*

(1 species)

Piebald Shrew
Diplomesodon pulchellus

FR Crocidure tachetée
DE Turkestan-Wüstenspitzmaus
ES Musaraña moteada

LC
Monotypic
HB 5.9–7.4 cm T 2.5–3.3 cm
W 6–12 g

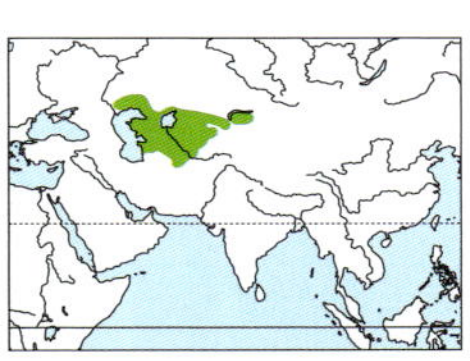

Extreme SE European RU, W & S KZ, UZ, TM

Malayan White-toothed Shrew
Crocidura malayana

FR Crocidure de Malaisie
DE Malaya-Weißzahnspitzmaus
ES Musaraña de Malasia

Genus *CROCIDURA*

(219 species)

LC
Monotypic
HB 7.7–10.2 cm
T 5.6–7.8 cm

50–1500 m

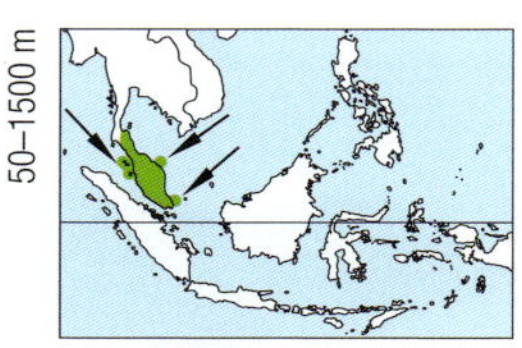

Thay-Malay Pen, with several offshore Is

Peninsular White-toothed Shrew
Crocidura negligens

FR Crocidure péninsulaire
DE Halbinsel-Weißzahnspitzmaus
ES Musaraña de Malaca

LC
Monotypic
HB 7.5–9.2 cm
T 5.5–6.9 cm
W 9.5–12.5 g

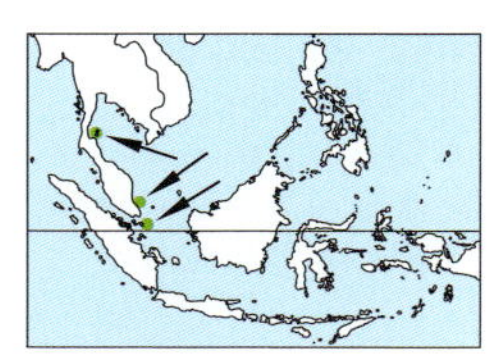

Is off Pen TH (Ko Samui) & Pen MY (Tioman, Mapur); doubtful record from Mt Tahan (MY)

Black-footed White-toothed Shrew
Crocidura nigripes

FR Crocidure à pattes noires
DE Schwarzfuß-Weißzahnspitzmaus
ES Musaraña de pies negros

LC
2 sspp
HB 6.4–8.5 cm
T 4.2–7.1 cm
W 7.9–16 g

30–2200 m

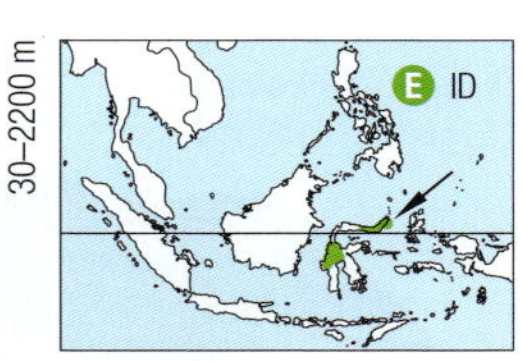

Sulawesi (E Minahasa Pen & C, Lembeh I)

Bornean White-toothed Shrew
Crocidura foetida

FR Crocidure de Bornéo
DE Borneo-Weißzahnspitzmaus
ES Musaraña de Borneo

LC
3 sspp
HB 6.4–10 cm
T 5.4–8.3 cm
W 9.2–16 g

0–1900 m

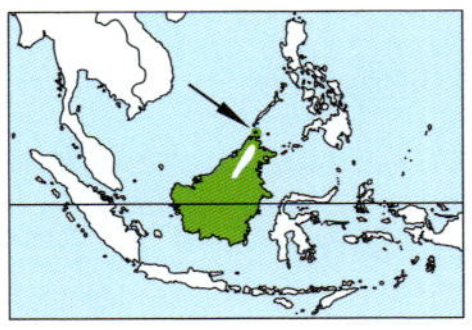

Borneo in MY (Sabah, Sarawak), ID & BN; perhaps Balambangan I

Kinabalu White-toothed Shrew
Crocidura baluensis

FR Crocidure du Kinabalu
DE Kinabalu-Weißzahnspitzmaus
ES Musaraña de Kinabalu

LC
Monotypic
HB 9–11 cm T 6.2–8.5 cm
W 13–19 g

1675–3655 m

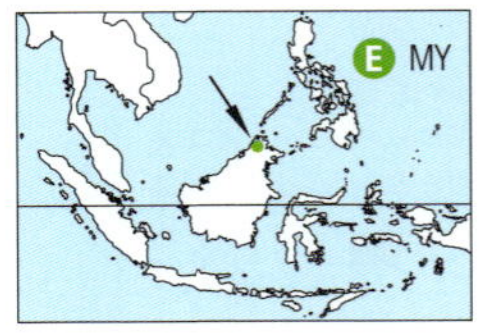

N Borneo in Sabah (upper slopes of Mt Kinabalu)

LC
2 sspp
HB 7.2–10.4 cm
T 4.2–6.7 cm
W 13–17 g

Thick-tailed White-toothed Shrew
Crocidura brunnea

FR Crocidure brune
DE Dickschwanz- Weißzahnspitzmaus
ES Musaraña de cola gruesa de Java

0–1600 m

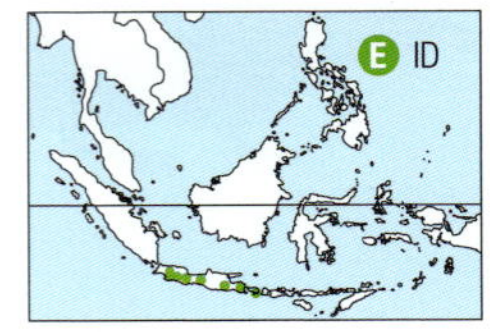

Java, Bali

LC
2 sspp
HB 7.3–10 cm
T 6.2–8.6 cm
W 9.5–18.2 g

Oriental White-toothed Shrew
Crocidura orientalis

FR Crocidure de Java
DE Orientalische Weißzahnspitzmaus
ES Musaraña oriental

1500–2700 m

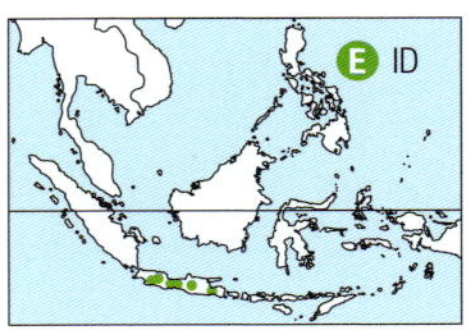

Java

Bangka White-toothed Shrew
Crocidura vosmaeri

FR Crocidure de Vosmaer
DE Bangka-Weißzahnspitzmaus
ES Musaraña de Bangka

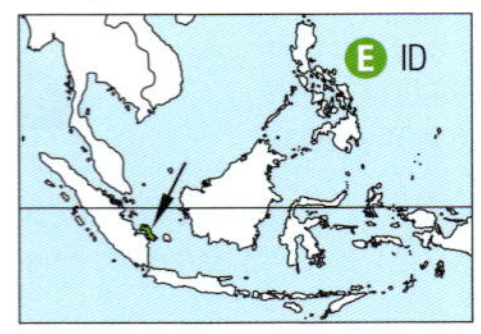

Bangka I, off SE Sumatra; perhaps opposite mainland of Sumatra

DD
Monotypic
HB 6.1–6.5 cm
T 3.7–5.3 cm W 5.8–8.8 g

LC
Monotypic
HB 7.2–8.3 cm
T 5.1–6.1 cm
W 10–12 g

Hutan White-toothed Shrew
Crocidura hutanis

FR Crocidure des forêts
DE Hutan-Weißzahnspitzmaus
ES Musaraña de bosque

300–600 m

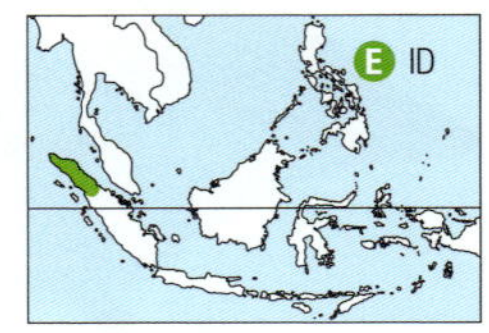

N Sumatra; perhaps more widely across Is

LC
Monotypic
HB 5.6–7.3 cm
T 5.1–6.2 cm W 5–8 g

Beccari's White-toothed Shrew
Crocidura beccarii

FR Crocidure de Beccari
DE Beccari-Weißzahnspitzmaus
ES Musaraña de Beccari

1600–2100 m

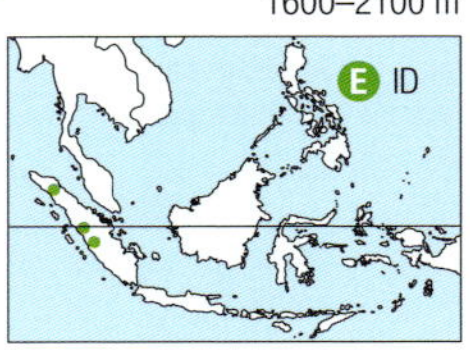

N & W Sumatra

Sumatran Giant White-toothed Shrew
Crocidura lepidura

FR Crocidure de Lyon
DE Sumatra-Weißzahnspitzmaus
ES Musaraña gigante de Sumatra

LC
Monotypic
HB 9–11.2 cm
T 6–8.4 cm W 13–21 g

0–2000 m

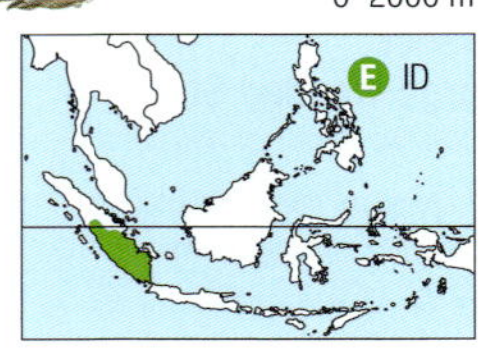

Upper montane rainforest in W Sumatra to hillsides in E & S Sumatra

DD
Monotypic
HB c. 10 cm
T c. 5.9 cm

Mount Malindang White-toothed Shrew
Crocidura grandis

FR Crocidure du Malindang
DE Malindang-Weißzahnspitzmaus
ES Musaraña de Malindang

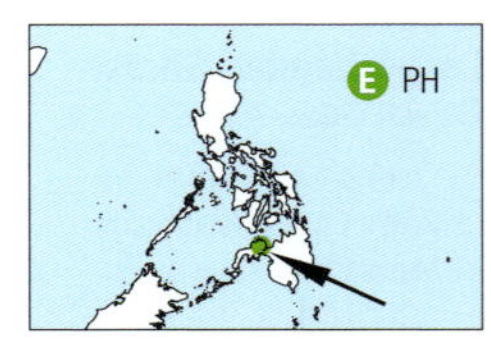

Mindanao (Mt Malindang)

LC
Monotypic
HB c. 7.7 cm T 5.2–7.5 cm
W 9–13 g

Mindanao White-toothed Shrew
Crocidura beata

FR Crocidure de Mindanao
DE Mindanao-Weißzahnspitzmaus
ES Musaraña de Mindanao

0–2800 m

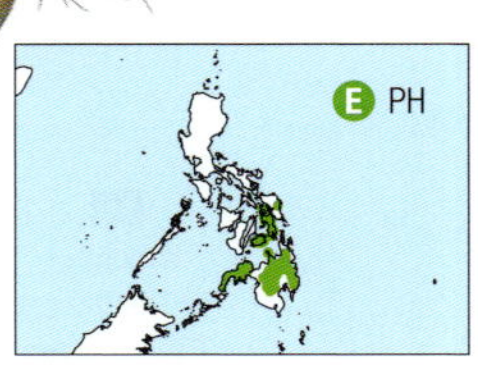

Samar, Leyte, Maripipi, Biliran, Bohol, Mindanao, Camiguin

DD
Monotypic
HB c. 9.1 cm
T 7.3–7.8 cm
W c. 13.5 g

Mindoro White-toothed Shrew
Crocidura mindorus

FR Crocidure de Mindoro
DE Mindoro-Weißzahnspitzmaus
ES Musaraña de Mindoro

1550–1920 m

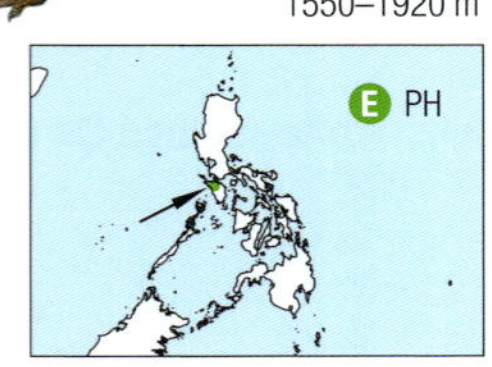

Mindoro (high elevations of Mt Halcon)

Panay White-toothed Shrew
Crocidura panayensis

FR Crocidure de Panay
DE Panay-Weißzahnspitzmaus
ES Musaraña de Panay

DD
Monotypic
HB 6.5–7.4 cm
T 5.8–6.4 cm
W 10–12.5 g

450–900 m

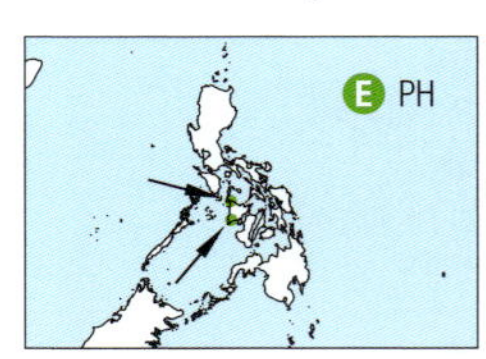

Panay (NW peninsula & SW)

Negros White-toothed Shrew
Crocidura negrina

FR Crocidure de Negros
DE Negros-Weißzahnspitzmaus
ES Musaraña de Negros

EN
Monotypic
HB c. 8.8 cm
T 6.3–8.3 cm
W 11–15.5 g

500–1450 m

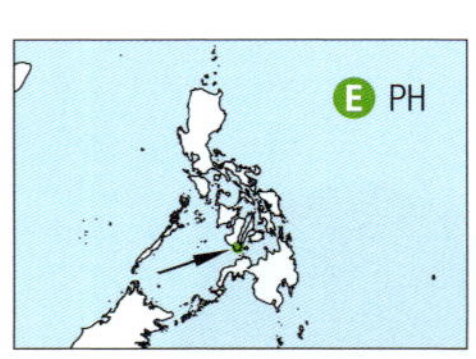

S Negros (known by 6 specimens from Cuernos de Negros volcanoes)

LC
Monotypic
HB 7.8–8.5 cm
T 4.7–6.4 cm W 8.5–10 g

Luzon White-toothed Shrew
Crocidura grayi

FR Crocidure de Luçon
DE Luzon-Weißzahnspitzmaus
ES Musaraña de Luzón

250–2400 m

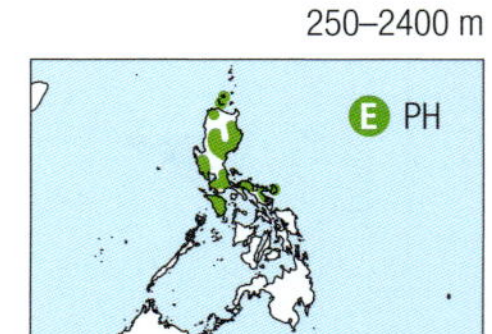

Luzon, Calayan (Babuyan Group), Mindoro, Catanduanes

DD
Monotypic
HB 8.4–9.7 cm
T 6.6–7.5 cm
W 9.5–13.5 g

320–1330 m

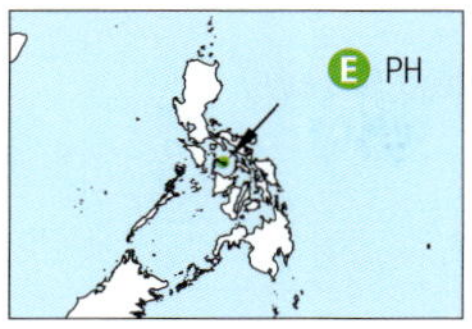

Sibuyan (Mt Guiting-Guiting)

Sibuyan White-toothed Shrew
Crocidura ninoyi
FR Crocidure de Ninoy
DE Sibuyan-Weißzahnspitzmaus
ES Musaraña de Sibuyán

LC
Monotypic
HB 7–8.6 cm T 7.6–9.6 cm
W 8–19.5 g

Palawan White-toothed Shrew
Crocidura palawanensis
FR Crocidure de Palawan
DE Palawan-Weißzahnspitzmaus
ES Musaraña de Palawan

900–1950 m

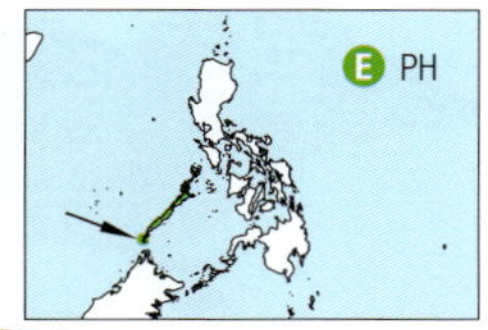

Palawan, Balabac I

Javan Long-tailed White-toothed Shrew
Crocidura abscondita
FR Crocidure cachée
DE Java-Langschwanz-Weißzahnspitzmaus
ES Musaraña de cola larga de Java

c. 1600 m

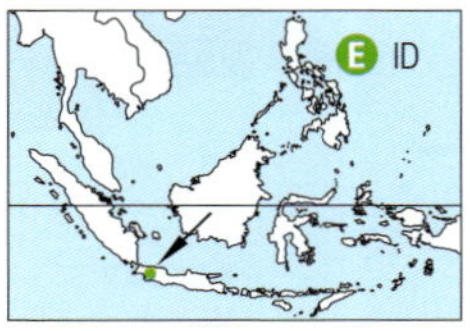

W Java (Mt Gede-Pangrango area)

DD
Monotypic
HB c. 7.3 cm
T c. 8.8 cm
W c. 7.5 g

Gathorne's White-toothed Shrew
Crocidura gathornei
FR Crocidure de Gathorne
DE Gathorne-Weißzahnspitzmaus
ES Musaraña de Gathorne

DD
Monotypic
HB 7.7–9.1 cm
T 5.7–7 cm

2745–3355 m

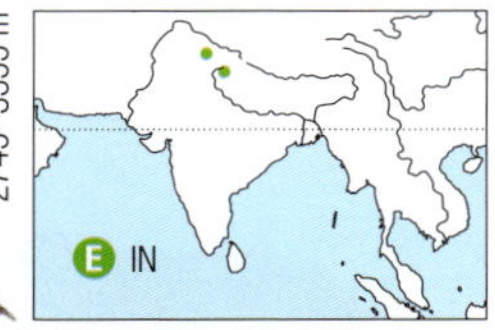

Uttarakhand (Kumaon), Himachal Pradesh (Kulu Valley)

Horsfield's White-toothed Shrew
Crocidura horsfieldii
FR Crocidure de Horsfield
DE Horsfield-Weißzahnspitzmaus
ES Musaraña de Horsfield

DD
Monotypic
HB 5.8–6.8 cm T 4.4–4.9 cm

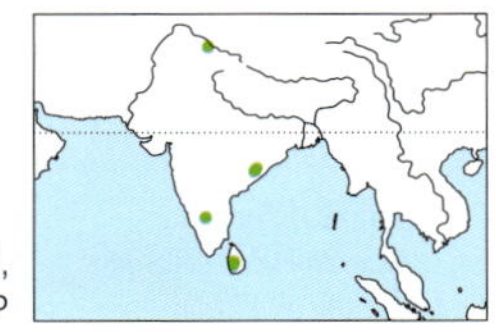

IN (Ladakh, Araku Valley, Mysore), C & SW LK; unlikely NP

EN
Monotypic
HB 6.4–7.6 cm
T 7.1–7.8 cm W 6.5–9 g

460–1040 m

Sinharaja White-toothed Shrew
Crocidura hikmiya
FR Crocidure de Sinharâja
DE Sinharaja-Weißzahnspitzmaus
ES Musaraña de Sinharaja

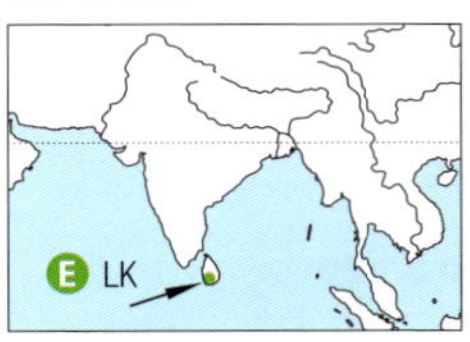

Known only from 2 localities in Sinharaja Forest

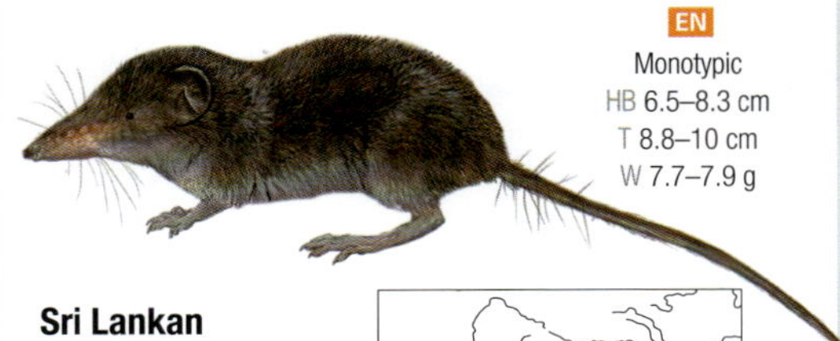

EN
Monotypic
HB 6.5–8.3 cm
T 8.8–10 cm
W 7.7–7.9 g

Sri Lankan White-toothed Shrew
Crocidura miya
FR Crocidure du Sri Lanka
DE Sri-Lanka-Weißzahnspitzmaus
ES Musaraña de Sri Lanka

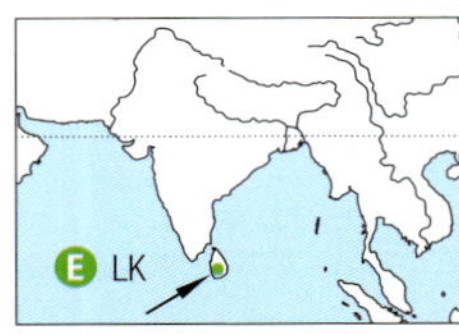

900–1985 m

Several sites in highlands of Central & Uva prov

LC
Monotypic
HB 5.4–7.4 cm
T 4.7–6 cm

Lesser Ryukyu White-toothed Shrew
Crocidura watasei
FR Crocidure de Watase
DE Kleine Ryukyu-Weißzahnspitzmaus
ES Musaraña de Ryukyu menor

0–400 m

Ryukyu Is, including 15 small Is of Amami & Okinawa Is

LC
Monotypic
HB 7.5–9.5 cm
T 6.5–8.4 cm

Southeast Asian White-toothed Shrew
Crocidura fuliginosa
FR Crocidure fuligineuse
DE Südostasien-Weißzahnspitzmaus
ES Musaraña de Asia suroriental

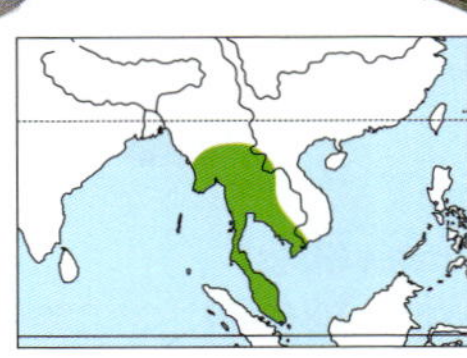

MM, LA, TH, KH, S VN, Pen MY, with offshore Is

Large White-toothed Shrew
Crocidura dracula
FR Crocidure dracula
DE Große Weißzahnspitzmaus
ES Musaraña grande

NE
2 sspp
HB 7.8–9.5 cm T 6.5–8 cm

C & SC CN (Sichuan, Yunnan, Chongqing, Guizhou, Guangxi), N MM, N LA, N VN; limits unclear

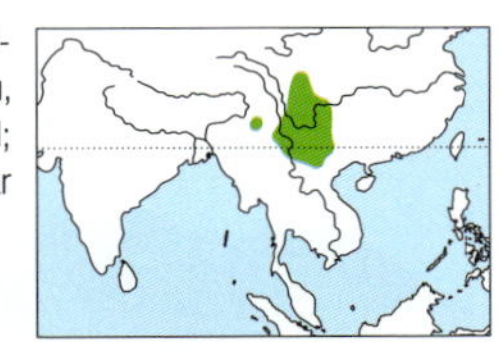

up to 3000 m

LC
Monotypic
HB 6.6–7.7 cm
T 8–10.5 cm

Sumatran Long-tailed White-toothed Shrew
Crocidura paradoxura
FR Crocidure paradoxale
DE Sumatra-Langschwanz-Weißzahnspitzmaus
ES Musaraña de Sumatra

1000–2400 m

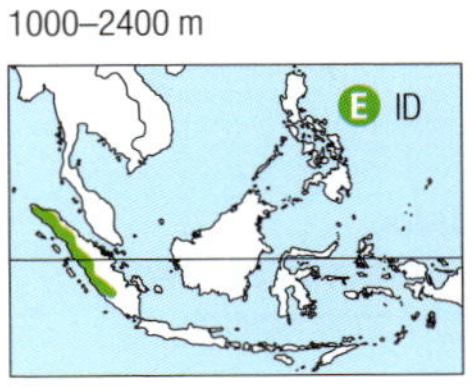

N & W Sumatra

Dsinezumi White-toothed Shrew
Crocidura dsinezumi
FR Crocidure du Japon
DE Dsinezumi-Weißzahnspitzmaus
ES Musaraña de Dsinezumi

Honshu, Shikoku, Kyushu, many small Is; introduced to Hokkaido & Jeju I (KR)

LC
Monotypic
HB 6.1–8.4 cm
T 3.9–5.4 cm

LC
Monotypic
HB 6.2–9.1 cm
T 4.1–6.1 cm

Asian Gray White-toothed Shrew
Crocidura attenuata
FR Crocidure mince
DE Graue Weißzahnspitzmaus
ES Musaraña gris de Asia

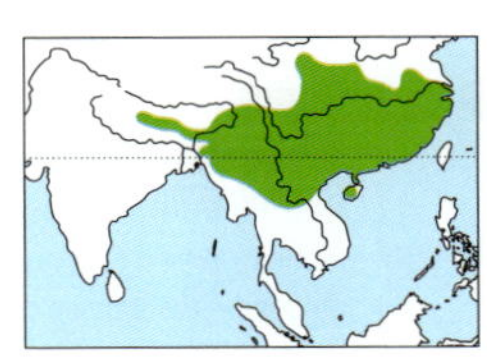

NP, NE IN, C & S CN (from SE Tibet & S Gansu E to Fujian, Hainan I), MM, TH, LA, VN

NE
Monotypic
HB 6.1–7.8 cm
T 4.8–5.9 cm
W 10.5–13.5 g

Anhui White-toothed Shrew
Crocidura anhuiensis

FR Crocidure d'Anhui
DE Anhui-Weißzahnspitzmaus
ES Musaraña de Anhui

c. 500 m

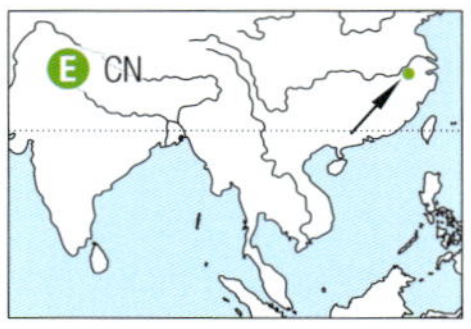

Anhui (known from Wild Monkey Valley, Mt Huang)

NE
Monotypic
T 4.8–6.2 cm
W 2.2–7.8 g

Dongjiangyuan White-toothed Shrew
Crocidura dongyangjiangensis

FR Crocidure de Dongyang
DE Dongjiangyuan-Weißzahnspitzmaus
ES Musaraña de Dongyang

up to 615 m

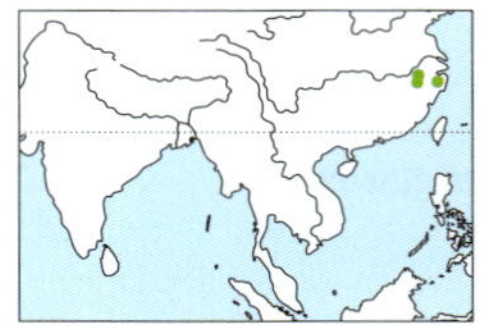

Few sites in Anhui (Mt Huang) & Zhejiang (Moxinjian NR in Chun'an, Dong Jiang Yuan NR in Dongyang)

LC
Monotypic
HB 6.9–8.6 cm T 4.7–6.3 cm

Taiwanese Gray White-toothed Shrew
Crocidura tanakae

FR Crocidure de Tanaka
DE Taiwan-Weißzahnspitzmaus
ES Musaraña de Taiwán

0–2200 m

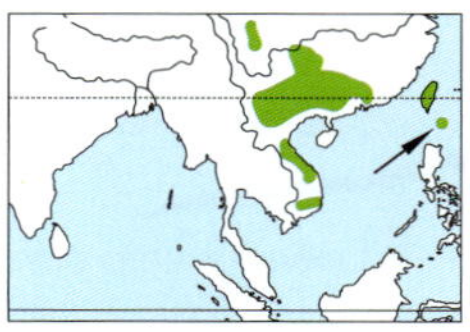

C & S CN (S from Sichuan), TW, LA, VN, N PH (Batan & Sabtang Is); perhaps KH

DD
Monotypic
HB 5.6–7 cm
T 3.8–4.7 cm

Chinese White-toothed Shrew
Crocidura rapax

FR Crocidure rapace
DE China-Weißzahnspitzmaus
ES Musaraña de dientes blancos de China

0–2000 m

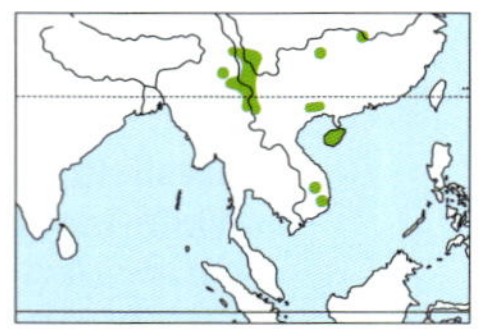

NE IN (Meghalaya), N MM, S CN (Sichuan, Yunnan, Guizhou, Hunan, Guangxi, Hainan I), S VN

NE
3 sspp
HB 5.3–7.3 cm
T 3.4–5.8 cm

Kuroda's White-toothed Shrew
Crocidura tadae

FR Crocidure de Taïwan
DE Kuroda-Weißzahnspitzmaus
ES Musaraña de Kuroda

0–2000 m

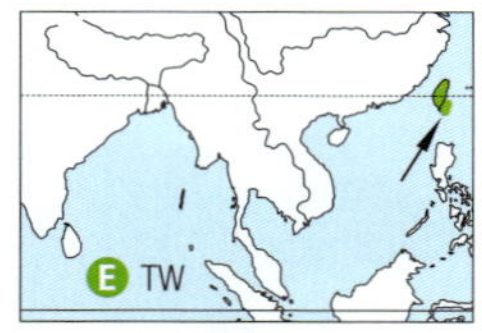

Mainland & off SE Is: Orchid I (Lanyu), Green I (Lutao)

Ussuri White-toothed Shrew
Crocidura lasiura

LC
Monotypic
HB 6.6–10 cm
T 2.9–4.1 cm
W 8.8–27 g

FR Crocidure de l'Oussouri
DE Ussuri-Weißzahnspitzmaus
ES Musaraña de dientes blancos de Ussuri

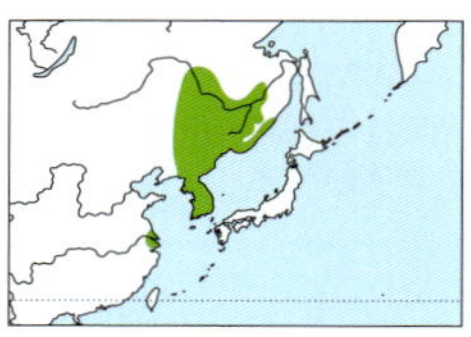

E RU (SE Far East), NE & E CN, Korean Pen

Zaitsev's White-toothed Shrew
Crocidura zaitsevi

DD
Monotypic
HB 4.8–5.8 cm
T 3.3–4.1 cm

FR Crocidure de Zaïtsev
DE Zaitsev-Weißzahnspitzmaus
ES Musaraña de Zaitsev

1650–2300 m

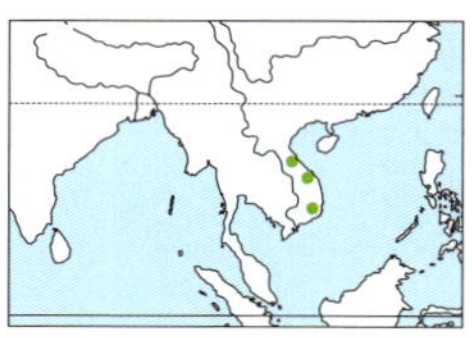

C & S VN (Ngoc Linh Mt, Phong Nha-Ke Bàng NP, Bi Doup Núi Bà NP, Hòn Bà Mt & Chu Yang Sin NP)

Indochinese White-toothed Shrew
Crocidura indochinensis

FR Crocidure d'Indochine
DE Indochina-Weißzahnspitzmaus
ES Musaraña de Indochina

LC
Monotypic
HB 5.3–6.1 cm
T 4–5 cm

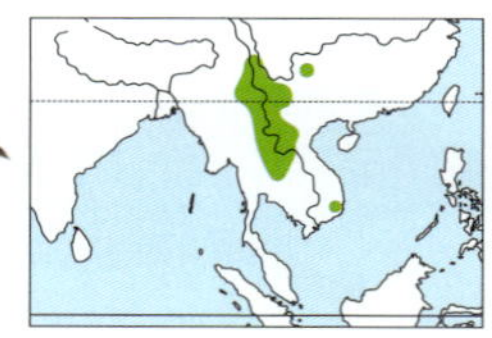

1200–2400 m

SW CN (Yunnan, Guizhou), N & W MM, LA, TH, S VN

Hainan White-toothed Shrew
Crocidura wuchihensis

FR Crocidure du Wuzhi
DE Hainan-Weißzahnspitzmaus
ES Musaraña de Hainan

DD
Monotypic
HB 5.9–6.5 cm
T 3.7–4.2 cm
W 3.5–5 g

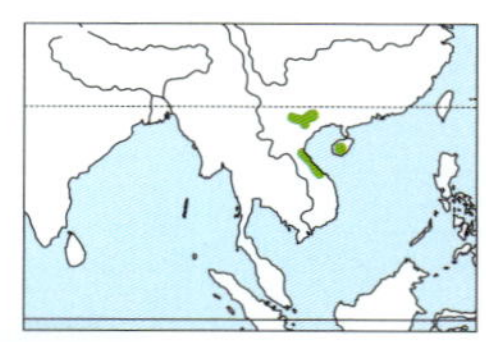

1000–1300 m

S CN (Guangxi, Hainan I), N & C VN; perhaps widespread

DD
Monotypic
HB 7.8–8.4 cm
T 4.6–5 cm

Timor White-toothed Shrew
Crocidura tenuis

FR Crocidure de Timor
DE Timor-Weißzahnspitzmaus
ES Musaraña de Timor

200–3000 m

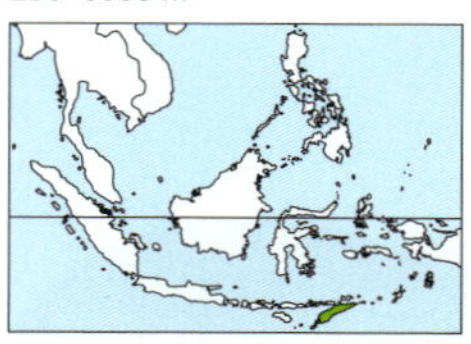

Timor I (ID, TL)

Javan White-toothed Shrew
Crocidura maxi

FR Crocidure de Max
DE Java-Weißzahnspitzmaus
ES Musaraña de Java

LC
Monotypic
HB 5.4–7.5 cm
T 4.4–5.3 cm W 4.3–6.9 g

0–1440 m

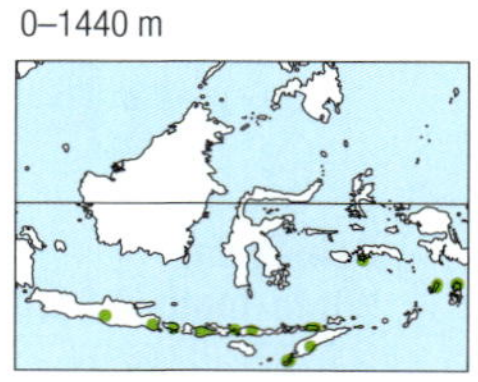

C & E Java, Lesser Sundas (Bali E to Alor, Timor, Roti), Ambon (Moluccas), Kai Besar (Kai), Wokam (Aru)

Javan Ghost White-toothed Shrew
Crocidura umbra

NT
Monotypic
HB 6.3–7.1 cm
T 4.7–5.8 cm
W 5–7.2 g

FR Crocidure fantôme
DE Gede-Weißzahnspitzmaus
ES Musaraña fantasma de Java

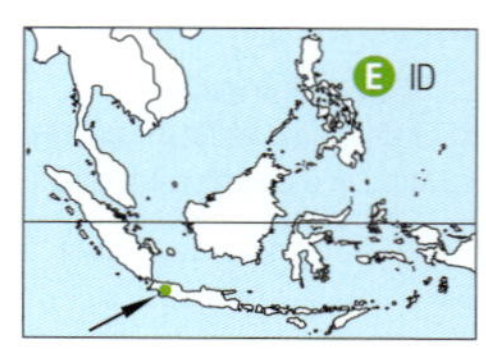

W Java (known from small area on Mt Gede-Pangrango)

LC
Monotypic
HB 5.3–7 cm
T 3.5–5.9 cm
W 3.4–6.3 g

1055–2045 m

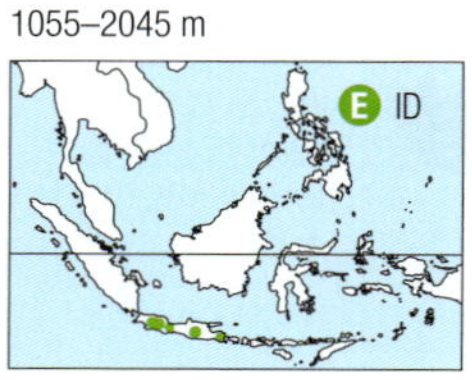

Java

Sunda White-toothed Shrew
Crocidura monticola

FR Crocidure du Sunda
DE Sunda-Weißzahnspitzmaus
ES Musaraña de Sonda

NE
Monotypic
HB 5.4–5.5 cm
T 3.4–4.1 cm

Neglected White-toothed Shrew
Crocidura neglecta

FR Crocidure de Sumatra
DE Vergessene Weißzahnspitzmaus
ES Musaraña descuidada

Known from N & W Sumatra (Mt Tujuh, Bukit Lawan); perhaps distributed by Sundaland

DD
Monotypic
HB 6.8–7.2 cm
T 4.9–5.9 cm

Phu Quoc White-toothed Shrew
Crocidura phuquocensis

FR Crocidure de Phu Quoc
DE Phu-Quoc-Weißzahnspitzmaus
ES Musaraña de Phu Quoc

25–40 m

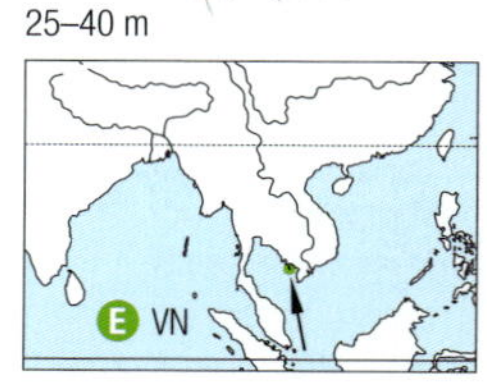

Phu Quoc I (Kien Giang Prov)

LC
Monotypic
HB 5.4–6.6 cm
T 4–4.8 cm

50–480 m

Phan Luong White-toothed Shrew
Crocidura phanluongi
FR Crocidure de Phan Luong
DE Phan-Luong-Weißzahnspitzmaus
ES Musaraña de Phan Luong

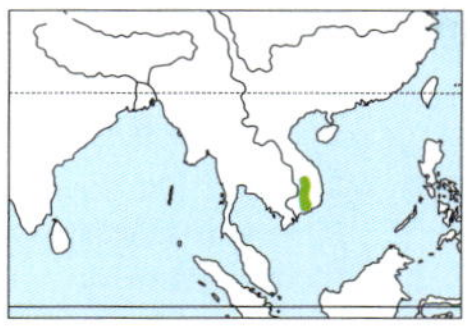

E KH, S VN

DD
Monotypic
HB 7–7.8 cm
T 6.5–6.8 cm

Sokolov White-toothed Shrew
Crocidura sokolovi
FR Crocidure de Sokolov
DE Sokolov-Weißzahnspitzmaus
ES Musaraña de Sokolov

2300–2400 m

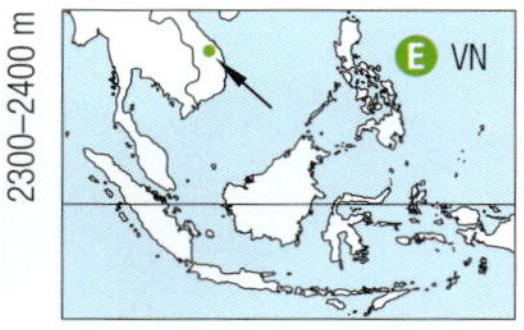

S Annamite Range (Ngoc Linh Mt)

Cranbrook's White-toothed Shrew
Crocidura cranbrooki
FR Crocidure de Cranbrook
DE Cranbrook-Weißzahnspitzmaus
ES Musaraña de Cranbrook

DD
Monotypic
HB 6.5–8.6 cm
T 6.5–8.8 cm

915–1220 m

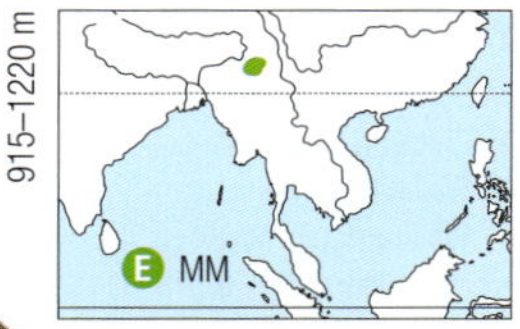

High river valleys and hills near N borders

DD
Monotypic
T c. 4.5 cm

150–550 m

Hill's White-toothed Shrew
Crocidura hilliana
FR Crocidure de Hill
DE Hill-Weißzahnspitzmaus
ES Musaraña de Hill

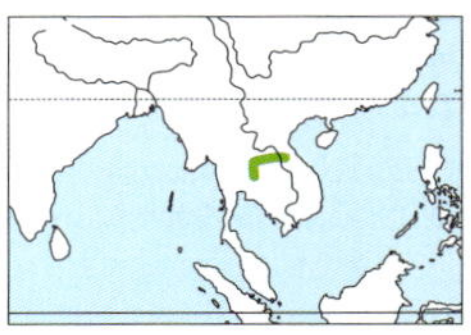

Few localities in C & NE TH and LA

Voracious White-toothed Shrew
Crocidura vorax
LC
Monotypic
HB 5.4–9 cm
T 4.1–5.1 cm
FR Crocidure vorace
DE Gefräßige Weißzahnspitzmaus
ES Musaraña voraz

c. 3600 m

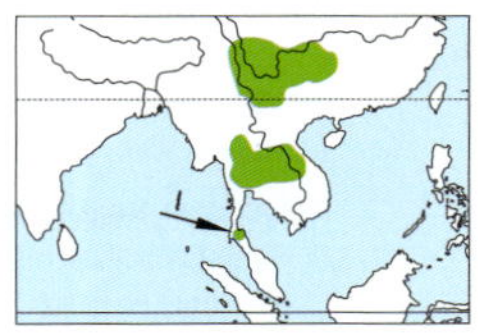

S CN (S Sichuan, Yunnan, Guizhou, Hunan), TH, LA; uncertain in NE IN & Hainan I (CN)

DD
Monotypic
HB 5–6.5 cm T 3.7–4.7 cm

Sa Pa White-toothed Shrew
Crocidura sapaensis
FR Crocidure de Sa Pa
DE Sa-Pa-Weißzahnspitzmaus
ES Musaraña de Sa Pa

1930–2200 m

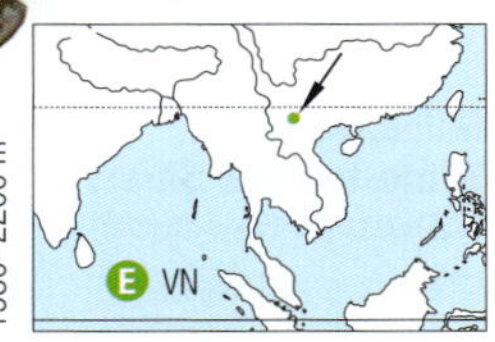

Lao Cai Prov (Sa Pa District, possibly other adjacent mts)

DD
Monotypic
HB 4.9–5.8 cm T 3–3.3 cm

Annamite White-toothed Shrew
Crocidura annamitensis
FR Crocidure annamite
DE Annam-Weißzahnspitzmaus
ES Musaraña de Annamita

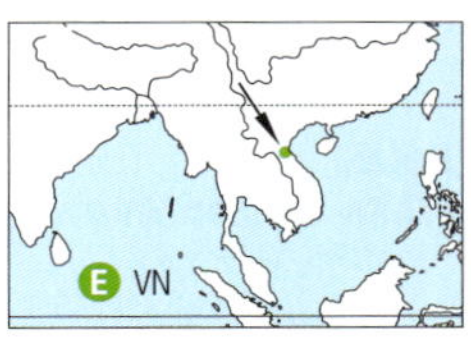

Huong Son District (Ha Tinh Prov); likely N Annamite Range & bordering LA

DD
Monotypic
HB 4.7–5.3 cm
T 3.4–3.7 cm

Vietnamese White-toothed Shrew
Crocidura guy
FR Crocidure de Guy
DE Vietnam-Weißzahnspitzmaus
ES Musaraña de Vietnam

300–800 m

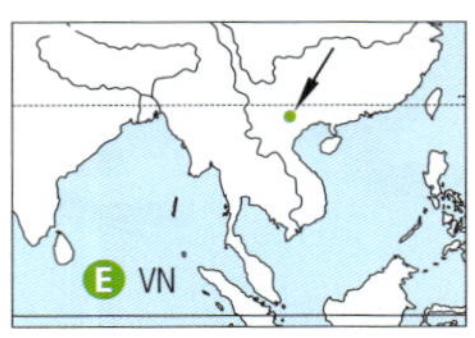

Known only from Na Hang NR (Tuyen Quang Prov)

NE
Monotypic
HB c. 4.8 cm
T c. 2.7 cm

Ke Go White-toothed Shrew
Crocidura kegoensis
FR Crocidure de Ke Go
DE Ke-Go-Weißzahnspitzmaus
ES Musaraña de Ke Go

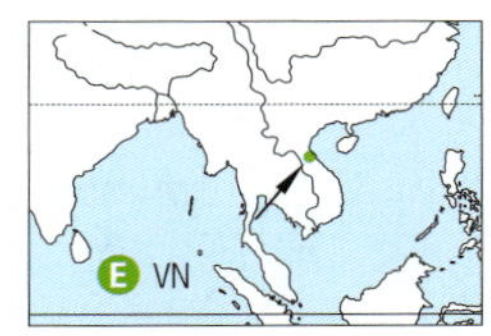

Known only from Ke Go NR (Ha Tinh Prov)

VU
Monotypic
HB c. 8.5 cm
T c. 10.3 cm

Andaman Spiny White-toothed Shrew
Crocidura hispida
FR Crocidure épineuse
DE Andamanen-Stachelweißzahnspitzmaus
ES Musaraña espinosa de Andamán

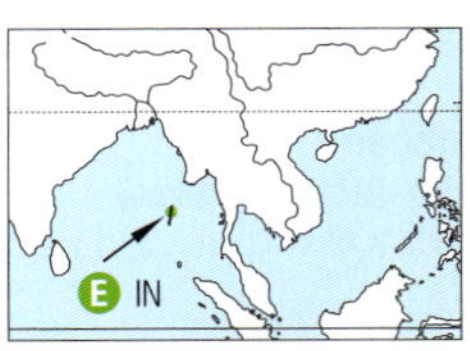

up to 100 m

N Middle Andaman I (Andaman Is)

Andaman White-toothed Shrew
Crocidura andamanensis
FR Crocidure des Andaman
DE Andamanen-Weißzahnspitzmaus
ES Musaraña de Andamán

CR
Monotypic
HB 7.6–11.4 cm
T 6.9–8.6 cm

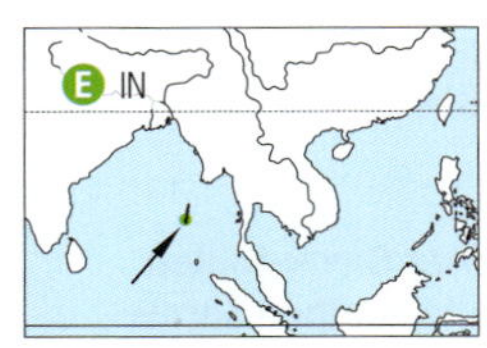

up to 380 m

South Andaman I in MacPherson Strait & Mt Harriet NP (Andaman Is)

CR
Monotypic
HB 10–10.7 cm
T 9.2–9.6 cm

Jenkins's White-toothed Shrew
Crocidura jenkinsi
FR Crocidure de Jenkins
DE Jenkins-Weißzahnspitzmaus
ES Musaraña de Jenkins

up to 385 m

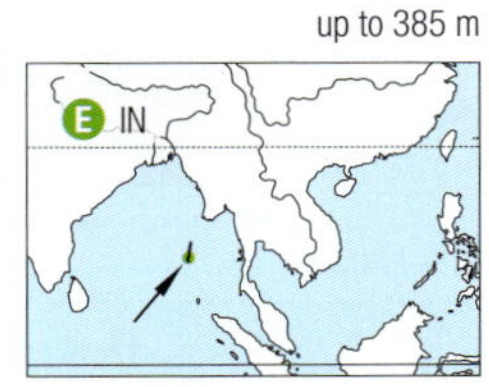

South Andaman I in Wright Myo & Mt Harriet NP (Andaman Is)

up to 100 m

Nicobar White-toothed Shrew
Crocidura nicobarica
FR Crocidure des Nicobar
DE Nikobar-Weißzahnspitzmaus
ES Musaraña de Nicobar

CR
Monotypic
HB c. 12 cm
T c. 9 cm

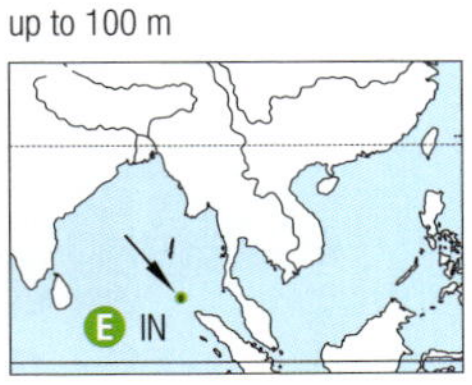

S Great Nicobar I (Nicobar Is)

Narcondam White-toothed Shrew
Crocidura narcondamica
FR Crocidure de Narcondam
DE Narcondom-Weißzahnspitzmaus
ES Musaraña de Narcondam

NE
Monotypic
HB 6.3–6.7 cm T 5.8–5.6 cm

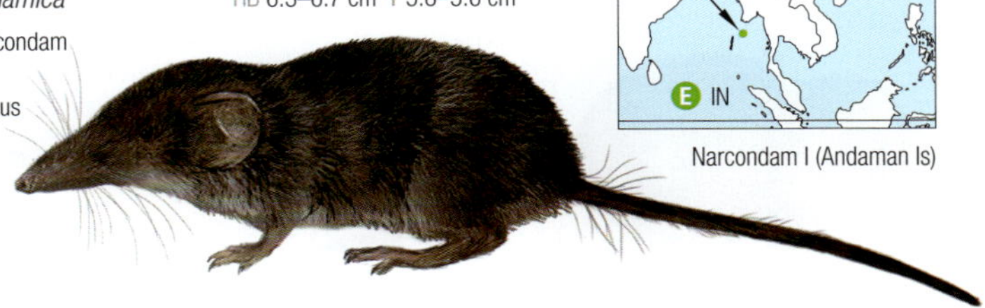

c. 11 m

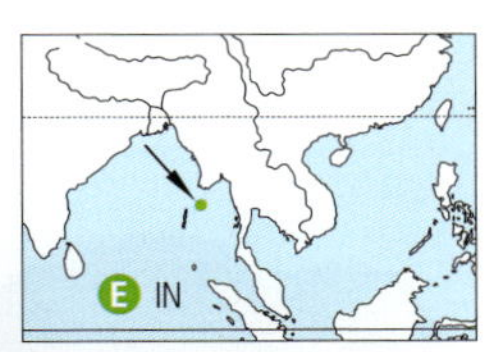

Narcondam I (Andaman Is)

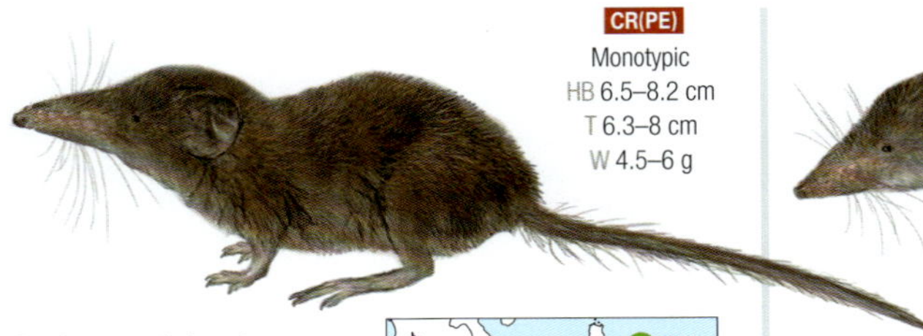

CR(PE)
Monotypic
HB 6.5–8.2 cm
T 6.3–8 cm
W 4.5–6 g

Christmas Island White-toothed Shrew
Crocidura trichura
FR Crocidure de Christmas
DE Weihnachtsinsel-Weißzahnspitzmaus
ES Musaraña de Isla de Navidad

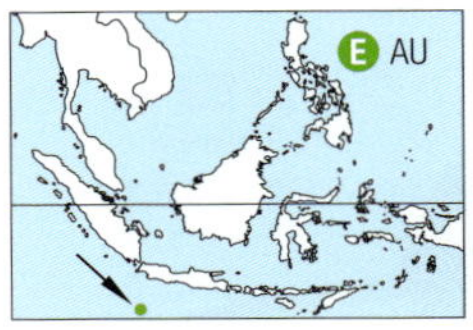

Christmas I

LC
Monotypic
HB 7.7–8.6 cm
T 3.1–3.8 cm
W 7.5–9.4 g

Siberian White-toothed Shrew
Crocidura sibirica
FR Crocidure de Sibérie
DE Sibirien-Weißzahnspitzmaus
ES Musaraña de Siberia

SC RU (S regions of W & C Siberia), extreme NE KZ

LC
3 sspp
HB 5.1–7.4 cm
T 2.6–4 cm
W 6.2–8.2 g

Shantung White-toothed Shrew
Crocidura shantungensis
FR Crocidure du Shandong
DE Asiatische Gartenspitzmaus
ES Musaraña de Asia menor

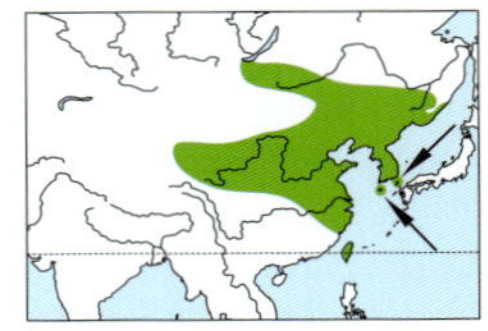

SE RU (Transbaikalia, Far East), N & E MN, C & E CN, Korean Pen (with Jeju I), TW, JP (Tsushima I)

NE
4 sspp
HB 7.7–8.6 cm
T 4.1–5.6 cm
W 8.5–13.3 g

Güldenstädt's White-toothed Shrew
Crocidura gueldenstaedtii
FR Crocidure du Caucase
DE Güldenstädt-Weißzahnspitzmaus
ES Musaraña de campo

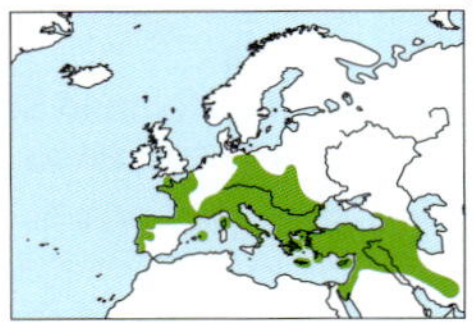

C & S Europe E to Caucasus & Iran, S to Levant & EG (Sinai), Minorca, Corsica, Crete & Sark Is; perhaps SA

LC
3 sspp
HB 4.7–8 cm
T 2.5–4 cm
W 6.5–9.4 g

Lesser White-toothed Shrew
Crocidura suaveolens
FR Crocidure des jardins
DE Gartenspitzmaus
ES Musaraña de Pallas

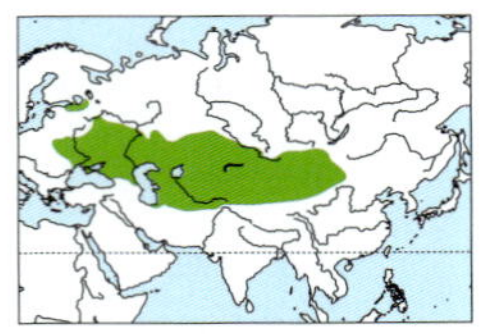

From S BY, UA & S RU through Central Asia to NW CN and W & C MN, S to N IR & Pamir; probably N AF

LC
Monotypic
HB 5.6–6.5 cm
T 3.9–4.6 cm
W 4–6.5 g

Cyrenaica White-toothed Shrew
Crocidura aleksandrisi
FR Crocidure d'Aleksandrs
DE Cyrenaika-Weißzahnspitzmaus
ES Musaraña de Cirenaica

N part of Cyrenaica

LC
Monotypic
HB 5.5–6.2 cm
T 4.3–4.8 cm

500–3000 m

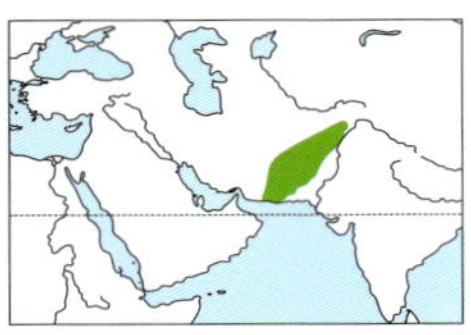

Zarudny's White-toothed Shrew
Crocidura zarudnyi
FR Crocidure de Zarudny
DE Zarudny-Weißzahnspitzmaus
ES Musaraña de Zarudny

SE IR, S & E AF, SW PK (Balochistan)

EN
Monotypic
HB 6.5–9 cm
T 4.1–5.1 cm

0–300 m

Greater Ryukyu White-toothed Shrew
Crocidura orii
FR Crocidure d'Orii
DE Große Ryukyu-Weißzahnspitzmaus
ES Musaraña de Ryukyu mayor

Amami Is (Amami Ōshima, Kakeromajima, Tokunoshima), C Ryukyu Is

DD
Monotypic
HB 6.6–7.2 cm
T 4–5.5 cm
W 4.5–7.1 g

200–800 m

Batak White-toothed Shrew
Crocidura batakorum
FR Crocidure des Batak
DE Batak-Weißzahnspitzmaus
ES Musaraña de los Batak

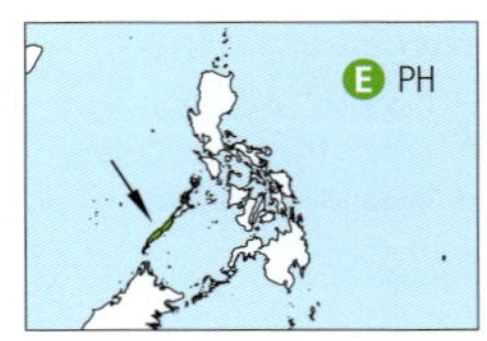

C & S Palawan

Lesser Black-footed White-toothed Shrew
Crocidura lea
FR Crocidure lionne
DE Kleine Schwarzfuß-Weißzahnspitzmaus
ES Musaraña de pies negros pequeña

LC
Monotypic
HB 5.6–6.9 cm
T 4.4–5.5 cm
W 3.4–6.6 g

400–1600 m

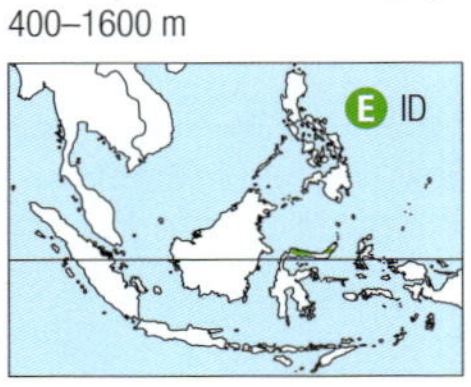

N Sulawesi, probably restricted to Minahasa Pen (Mt Dako, Mt Buliohuto, Temboan)

LC
Monotypic
HB 9–9.2 cm
T 10.2–13.7 cm
W 9.5–17.7 g

Northern Elongated White-toothed Shrew
Crocidura elongata
FR Crocidure élancée
DE Schlanke Weißzahnspitzmaus
ES Musaraña alargada

500–1600 m

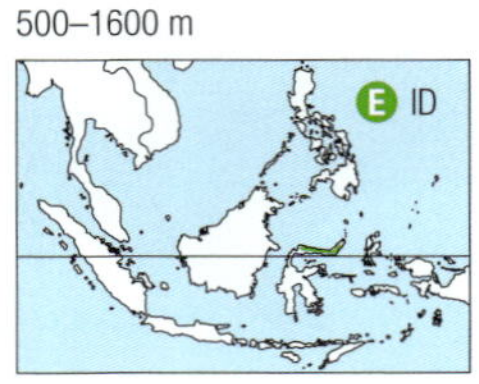

N Sulawesi (apparently restricted to Minahasa Pen)

LC
Monotypic
HB 7.4–10.5 cm
T 5.7–7.7 cm
W 11.5–22 g

650–1700 m

Temboan White-toothed Shrew
Crocidura rhoditis
FR Crocidure rosée
DE Temboan-Weißzahnspitzmaus
ES Musaraña rosada

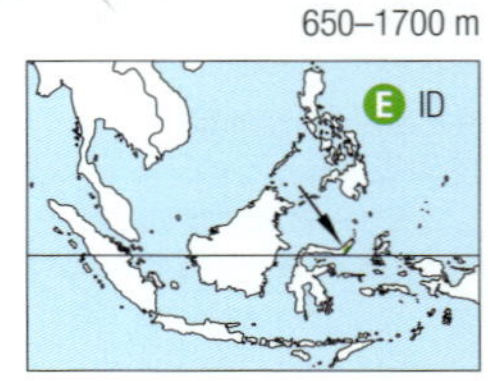

N Sulawesi in E Minahasa Pen (Temboan, Mt Ambang)

North Peninsula White-toothed Shrew
Crocidura pseudorhoditis
FR Crocidure sosie
DE Nordhalbinsel-Weißzahnspitzmaus
ES Musaraña pseudorosada

NE
Monotypic
HB 7.5–9.6 cm
T 5.9–8.4 cm
W 10.3–18.5 g

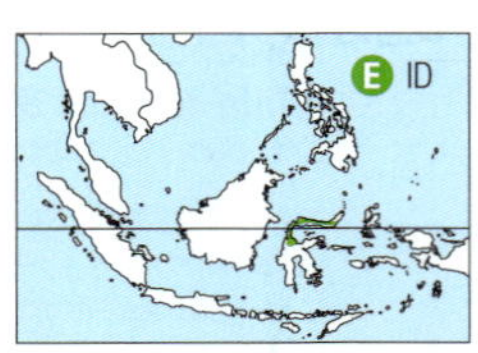

500–2200 m

N & NE Sulawesi (Torompupu, Rorekatimbo, Dako, Buliohuto & Ambang Mts)

Mossy Forest White-toothed Shrew
Crocidura musseri
FR Crocidure de Musser
DE Musser-Weißzahnspitzmaus
ES Musaraña de bosque de musgo

1800–2300 m

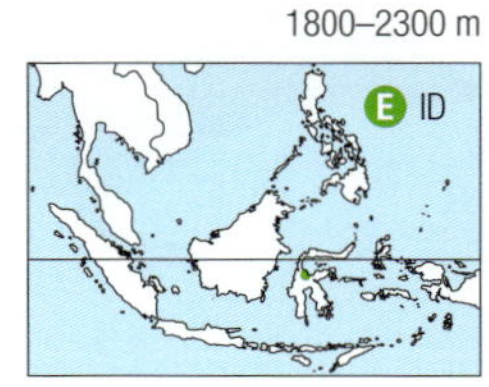

C Sulawesi (Mt Rorekatimbo, Mt Nokilalaki)

DD
Monotypic
HB 11.2–14.4 cm
T 5.1–6.7 cm
W 6.2–12 g

NE
Monotypic
HB 7.1–8.8 cm
T 9.4–12.1 cm
W 7.5–14.2 g

Small Elongated White-toothed Shrew
Crocidura microelongata

FR Crocidure menue
DE Kleine Langschwanz-Weißzahnspitzmaus
ES Musaraña alargada pequeña

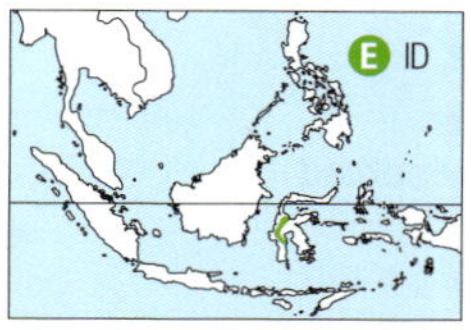

700–2600 m

WC Sulawesi (Latimojong, Gandang Dewata, Torompupu & Rorekatimbo Mts)

NE
Monotypic
HB 7.3–8.7 cm
T 4.9–7.3 cm
W 7.4–16 g

Pallid-footed White-toothed Shrew
Crocidura pallida

FR Crocidure à pattes pâles
DE Weißfuß-Weißzahnspitzmaus
ES Musaraña rosada de pies pálidos

100–2500 m

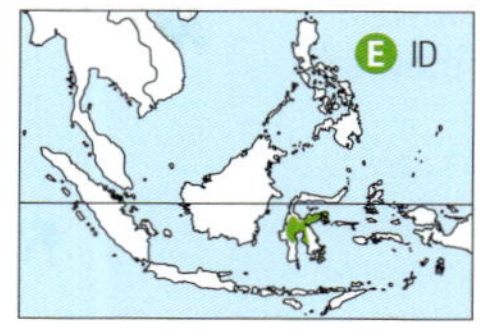

C & SE Sulawesi (from Gandang Dewata E to Tompotika, from Torompupu S to Latimojong & Mekongga Mts)

NE
Monotypic
HB 8–8.6 cm
T 5–6.4 cm
W 13.5–18.5 g

Thick-tailed Sulawesi White-toothed Shrew
Crocidura caudicrassa

FR Crocidure à queue épaisse
DE Dickschwänzige Weißzahnspitzmaus
ES Musaraña de cola gruesa de Célebes

2120–2600 m

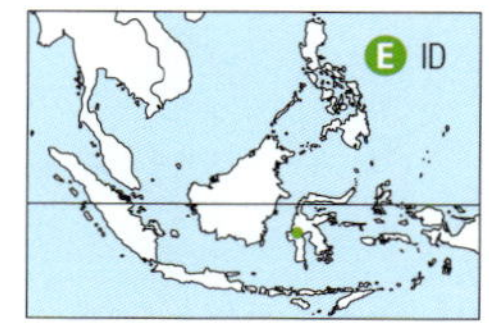

WC Sulawesi (Mt Gandang Dewata; 1 record 40 km ESE in Rindingallo)

NE
Monotypic
HB 7.6–8.6 cm
T 5.4–6.4 cm
W 10.5–15.2 g

Short-tailed White-toothed Shrew
Crocidura brevicauda

FR Crocidure à queue courte
DE Kurzschwänzige Weißzahnspitzmaus
ES Musaraña de cola corta

2300–2500 m

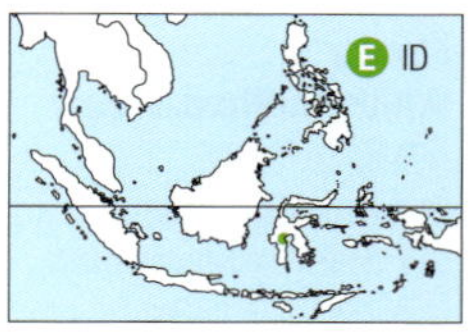

WC Sulawesi (known only from Mt Latimojong)

NE
Monotypic
HB 8.5–9.6 cm
T 8.6–14.4 cm
W 8.6–19.5 g

Southern Elongated White-toothed Shrew
Crocidura quasielongata

FR Crocidure svelte
DE Südliche Langschwanz-Weißzahnspitzmaus
ES Musaraña cuasialargada

200–1800 m

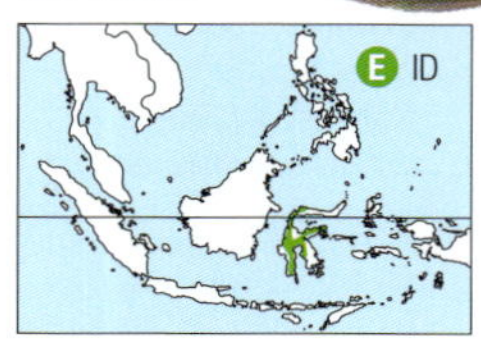

Widespread on Sulawesi (except SE & NE)

NE
Monotypic
HB 7.7–8.1 cm
T 5.9–6.7 cm
W 8.5–12.7 g

South-west Peninsula White-toothed Shrew
Crocidura australis

FR Crocidure australe
DE Südwest-Halbinsel-Weißzahnspitzmaus
ES Musaraña rosada meridional

1660–2550 m

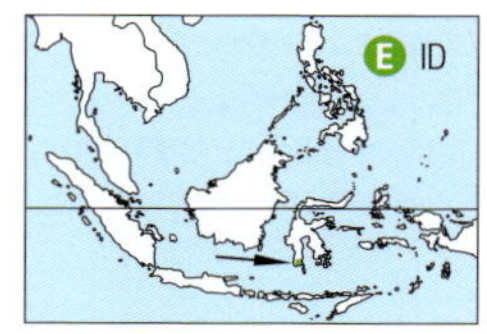

SW Sulawesi (Mt Bawakaraeng)

NE
Monotypic
HB 7–8.7 cm
T 5.2–7.6 cm W 5–12.8 g

West Sulawesi White-toothed Shrew
Crocidura ordinaria

FR Crocidure intermédiaire
DE West-Sulawesi-Weißzahnspitzmaus
ES Musaraña ordinaria

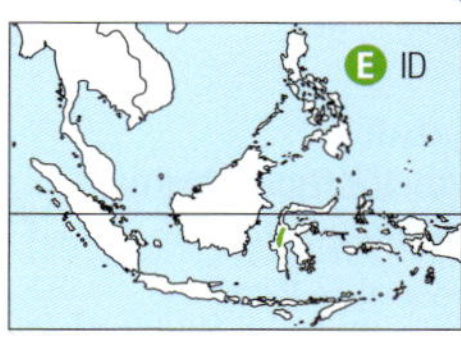

200–2600 m

WC Sulawesi (Mt Gandang Dewata, Mt Torompupu)

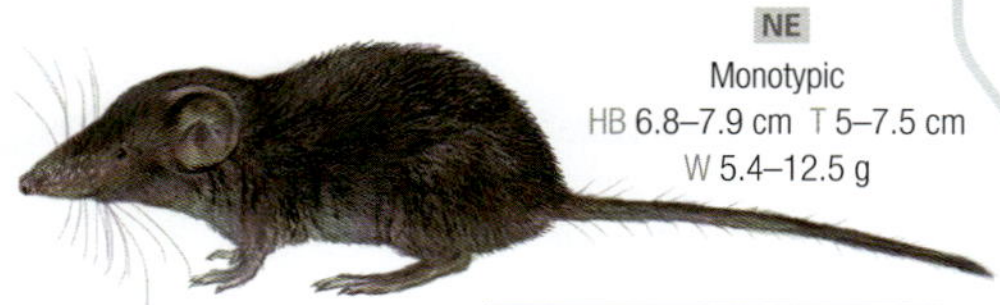

NE
Monotypic
HB 6.8–7.9 cm T 5–7.5 cm
W 5.4–12.5 g

West-central Sulawesi White-toothed Shrew
Crocidura solita

FR Crocidure familière
DE West-Zentral-Sulawesi-Weißzahnspitzmaus
ES Musaraña corriente

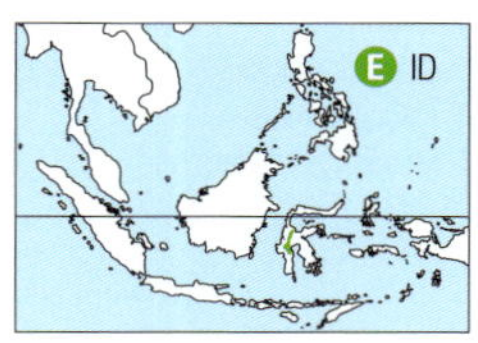

700–2600 m

WC Sulawesi (Latimojong, Rorekatimbo & Gandang Dewata Mts)

LC
Monotypic
HB 5.8–6.5 cm
T 2.7–4.4 cm
W 3–7 g

Sulawesi Tiny White-toothed Shrew
Crocidura levicula

FR Crocidure légère
DE Kleine Sulawesi-Weißzahnspitzmaus
ES Musaraña de Pinedapa

400–2000 m

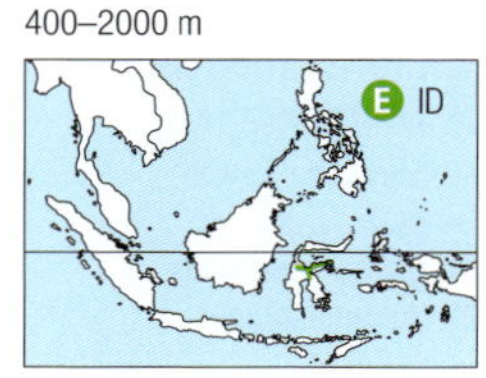

C & SE Sulawesi

NE
Monotypic
HB 6.6–8.5 cm
T 7.9–9.8 cm
W 5.8–12 g

Hairy-tailed White-toothed Shrew
Crocidura caudipilosa

FR Crocidure à queue velue
DE Sulawesi-Haarschwanz-Weißzahnspitzmaus
ES Musaraña de cola peluda

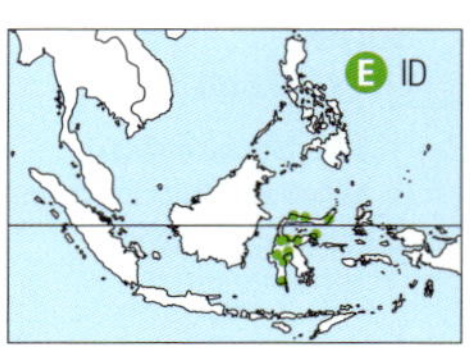

500–2600 m

Widespread on Sulawesi

NE
Monotypic
HB 6.4–7.5 cm
T 5.6–6.5 cm
W 4.3–8.2 g

Typical Sulawesi White-toothed Shrew
Crocidura normalis

FR Crocidure modèle
DE Gemeine Sulawesi-Weißzahnspitzmaus
ES Musaraña normal

1400–2600 m

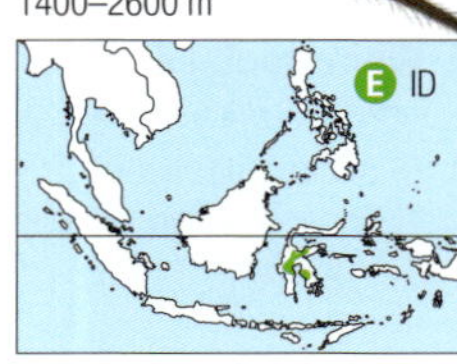

WC, EC & SE Sulawesi (Gandang Dewata, Latimojong, Rorekatimbo, Katopasa & Mekongga Mts)

NE
Monotypic
HB 5.9–6.5 cm
T 2.9–4.7 cm
W 3.2–5.9 g

Tiny Sulawesi White-toothed Shrew
Crocidura parva

FR Petite Crocidure
DE Winzige Sulawesi-Weißzahnspitzmaus
ES Musaraña diminuta de Célebes

1700–2500 m

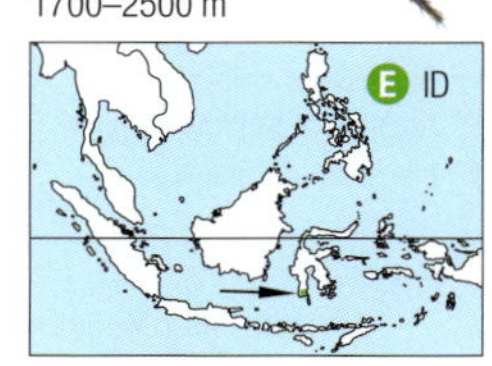

SW Sulawesi (Mt Bawakaraeng)

Central Sulawesi White-toothed Shrew
Crocidura mediocris

FR Crocidure morne
DE Zentral-Sulawesi-Weißzahnspitzmaus
ES Musaraña regular

NE
Monotypic
HB 5.2–6.7 cm
T 3.7–5 cm
W 2.9–5.9 g

200–1900 m

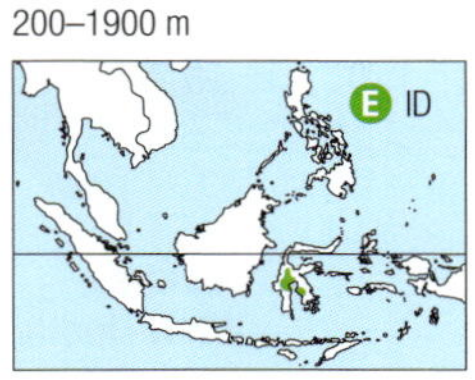

WC & SE Sulawesi (Torompupu, Balease, Latimojong & Mekongga Mts, Salu Tiwo)

NE
Monotypic
HB 6–6.7 cm T 3.6–4.4 cm
W 4.3–6.6 g

Dark White-toothed Shrew
Crocidura tenebrosa

FR Crocidure ombrée
DE Düstere Sulawesi-Weißzahnspitzmaus
ES Musaraña oscura

1400–1500 m

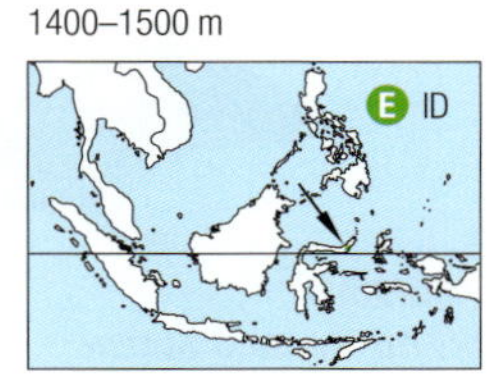

N Sulawesi in E Minahasa Pen (recorded only in Mt Ambang)

NE

Monotypic
HB 6–6.8 cm
T 4.2–5.3 cm
W 4.3–6.4 g

1400–1600 m

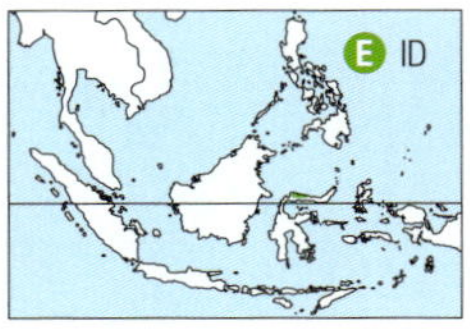

N Sulawesi, probably endemic to Minahasa Pen (Mt Dako, Mt Buliohuto)

Balete's White-toothed Shrew
Crocidura baletei

FR Crocidure de Balete
DE Balete-Weißzahnspitzmaus
ES Musaraña de Balete

LC

Monotypic
HB 6.2–7.8 cm
T 3–4 cm

North African White-toothed Shrew
Crocidura pachyura

FR Crocidure de Sardaigne
DE Nordafrikanische Hausspitzmaus
ES Musaraña norteafricana

0–1000 m

ES (Ibiza I), IT (Sardinia & Pantelleria I), range uncertain in N Africa (NE DZ, TN); possibly native to N Africa

LC

2 sspp
HB 4.4–8.6 cm
T 2.4–4.7 cm
W 5–16 g

Greater White-toothed Shrew
Crocidura russula

FR Crocidure musette
DE Hausspitzmaus
ES Musaraña mayor

below 1000 m

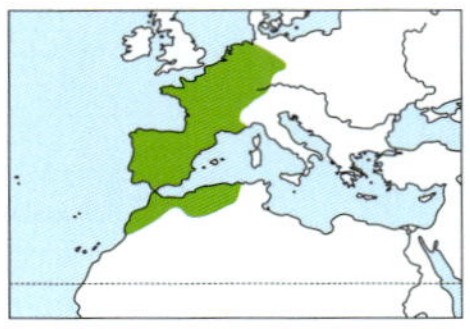

W Europe (DE, BE, NL, LU, CH, FR, ES, PT & many Is), NW Africa (N MA, N DZ); introduced to Canary Is & IE

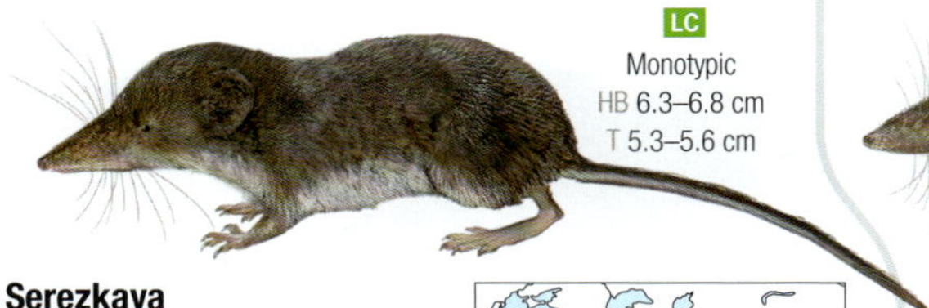

LC

Monotypic
HB 6.3–6.8 cm
T 5.3–5.6 cm

Serezkaya White-toothed Shrew
Crocidura serezkyensis

FR Crocidure du Sarez
DE Kleine Felsen-Weißzahnspitzmaus
ES Musaraña de rocas menor

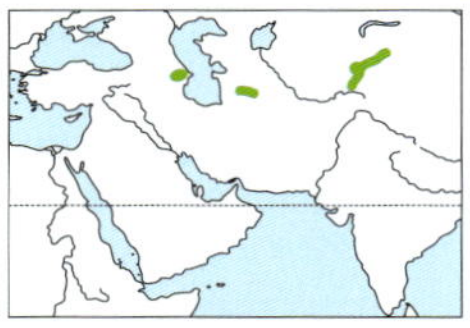

KZ, KG, TJ, TM, disjunct in AZ

LC

Monotypic
HB 5.6–6.4 cm T 2.8–3.9 cm
W 5–7 g

Whitaker's White-toothed Shrew
Crocidura whitakeri

FR Crocidure de Whitaker
DE Whitaker-Weißzahnspitzmaus
ES Musaraña de Whitaker

0–1800 m

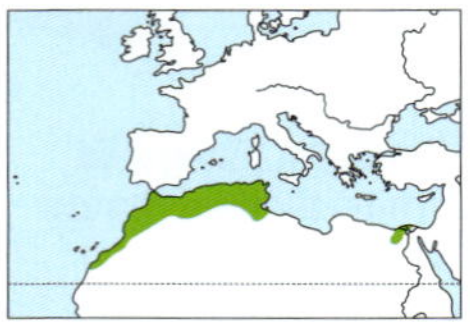

NW Africa from NW EH E to TN, N EG around Nile Delta, perhaps continuously from MA to EG; possibly LY

DD

Monotypic
HB 5.7–7.1 cm
T 5.5–5.8 cm

Flower's White-toothed Shrew
Crocidura floweri

FR Crocidure de Flower
DE Flower-Weißzahnspitzmaus
ES Musaraña de Flower

Nile Delta, mummified remains have been identified in Thebes

DD

Monotypic
HB 4.8–6.2 cm
T 2.8–4 cm

Egyptian Pygmy White-toothed Shrew
Crocidura religiosa

FR Crocidure d'Égypte
DE Kleine Ägyptische Weißzahnspitzmaus
ES Musaraña pigmea de Egipto

Nile Delta & 1 record S on Nile; perhaps between Giza and Thebes (where mummified specimens found)

LC

6 sspp
HB 5.9–7.2 cm
T 3.1–4.1 cm
W 5.9–11.1 g

Bicolored White-toothed Shrew
Crocidura leucodon

FR Crocidure bicolore
DE Feldspitzmaus
ES Musaraña de Hermann

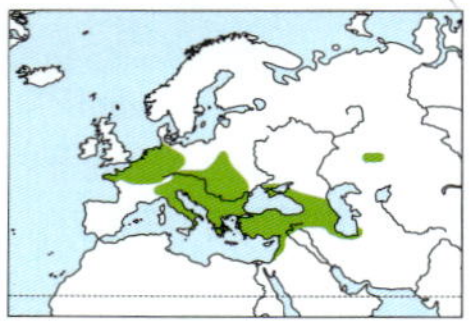

Europe from FR E to RU (Urals) S to TR, Caucasus, Levant (S to IL), N IR, SW TM

DD

Monotypic
HB 5.6–6.9 cm
T 3.3–4.2 cm
W 6–7 g

Saharan White-toothed Shrew
Crocidura tarfayensis

FR Crocidure de Tarfaya
DE Sahara-Weißzahnspitzmaus
ES Musaraña del Sáhara

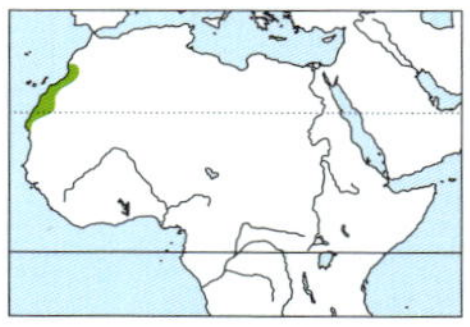

NW Africa in SW MA, EH & NW MR

Arabian White-toothed Shrew
Crocidura arabica

FR Crocidure d'Arabie
DE Arabien-Weißzahnspitzmaus
ES Musaraña de Arabia

LC

Monotypic
HB 5.1–5.4 cm
T 3.6–4.1 cm W c. 5.9 g

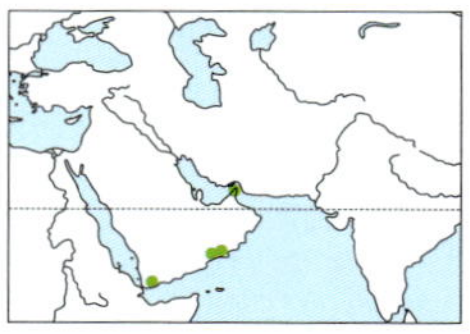

S Arabian Pen in YE (Aden) & OM (Dhofar, Musandam)

DD

Monotypic
HB 6.8–7.2 cm
T 4.7–5.1 cm
W 5.2–6.9 g

Dhofar White-toothed Shrew
Crocidura dhofarensis

FR Crocidure du Dhofar
DE Dhofar-Weißzahnspitzmaus
ES Musaraña de Dhofar

410–735 m

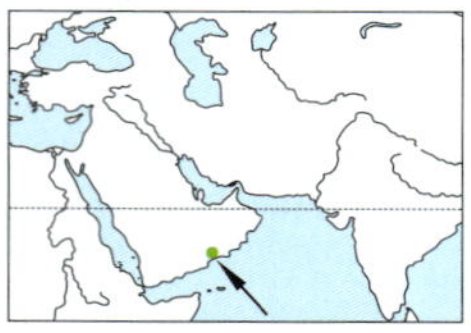

Coastal plateaus of YE (Al Mahrah) & OM (Dhofar)

LC

Monotypic
HB 5.8–6.3 cm
T 4.2–4.3 cm
W 4–5 g

Negev White-toothed Shrew
Crocidura ramona

FR Crocidure du Ramon
DE Negev-Weißzahnspitzmaus
ES Musaraña de Negev

200–950 m

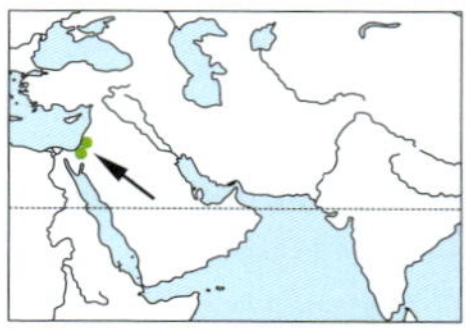

IL & PS in Negev Desert (Mitzpe Ramon, Sede Boquer) & Judean Desert (Sartaber)

Katinka's White-toothed Shrew
Crocidura katinka

FR Crocidure de Katinka
DE Katinka-Weißzahnspitzmaus
ES Musaraña de Katinka

DD

Monotypic

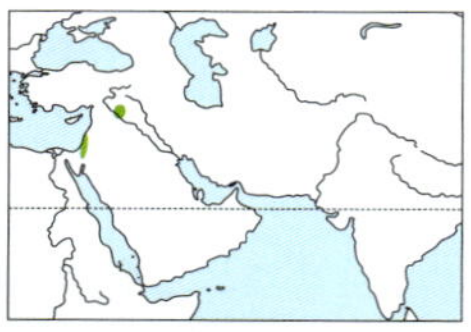

SY (Near Halabiya at Euphrates & Qal'at Sukkara); possibly extinct from IL, PS & LB; perhaps SW IR

Jackass White-toothed Shrew
Crocidura arispa

FR Crocidure des Taurus
DE Taurus-Weißzahnspitzmaus
ES Musaraña de Jackass

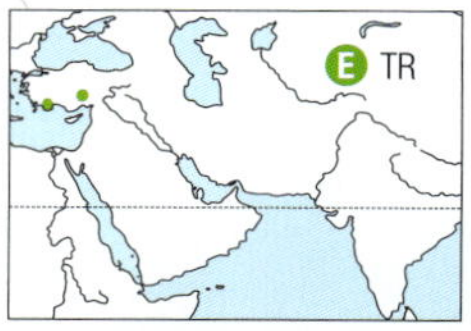

Known from 2 separated sites in Taurus Mts

LC

Monotypic
HB 5.7–7.5 cm
T 4.3–5.4 cm
W c. 9 g

DD

Monotypic

Armenian White-toothed Shrew
Crocidura armenica

FR Crocidure d'Arménie
DE Armenien-Weißzahnspitzmaus
ES Musaraña de Armenia

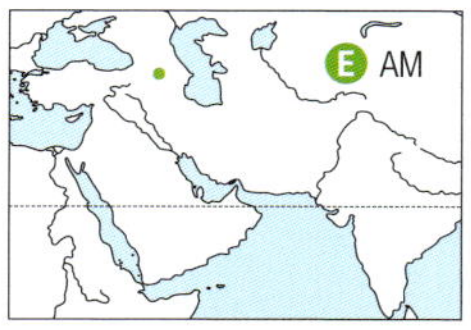

Known from type locality (14 km down river from Garni); perhaps AZ

DD

Monotypic
HB 6.7–8.3 cm
T 4–5.1 cm
W 9.8–14 g

Caspian White-toothed Shrew
Crocidura caspica

FR Crocidure de la Caspienne
DE Kaspien-Weißzahnspitzmaus
ES Musaraña del Caspio

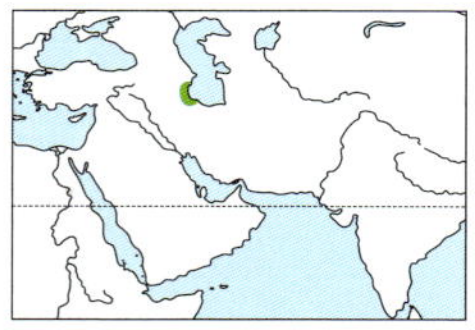

Lankaran Lowland & Talysh Mts (SE AZ, NW IR)

DD

Monotypic
HB 7.8–8.5 cm
T 5.6–6.3 cm

100–300 m

Iranian White-toothed Shrew
Crocidura susiana

FR Crocidure de Susiane
DE Iran-Weißzahnspitzmaus
ES Musaraña de Irán

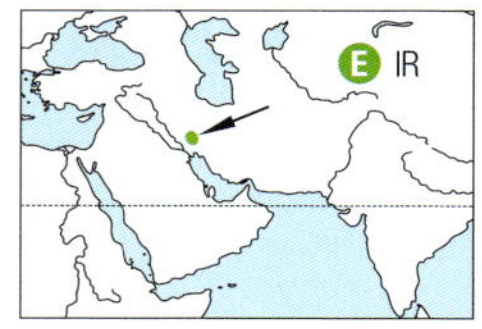

Known from vicinity of Dezful (Khuzestan Prov); perhaps wider range

Pale Gray White-toothed Shrew
Crocidura pergrisea

FR Crocidure gris-pâle
DE Hellgraue Weißzahnspitzmaus
ES Musaraña gris pálida

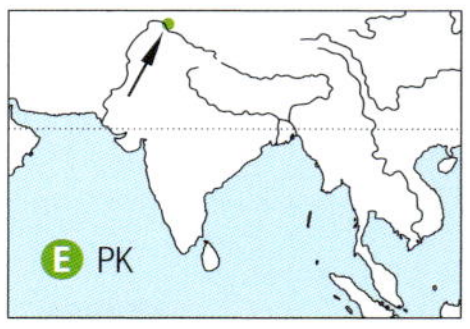

Known from Skoro Loomba (Baltistan, Kashmir); specimens from other sites might belong to *C. zarudnyi*

DD

Monotypic
HB 7.2–7.5 cm
T 5.3–5.4 cm

DD

Monotypic
HB 6–9 cm
T 3.9–5.3 cm

Kashmir White-toothed Shrew
Crocidura pullata

FR Crocidure sombre
DE Kashmir-Weißzahnspitzmaus
ES Musaraña de dientes blancos de Cachemira

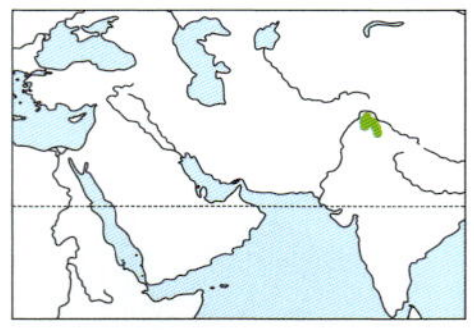

1830–2650 m

W Himalaya in Jammu and Kashmir (IN, PK)

EN

Monotypic
HB 6.5–7.8 cm
T 3.5–4.2 cm
W 7–8 g

140–1400 m

Cretan White-toothed Shrew
Crocidura zimmermanni

FR Crocidure de Zimmermann
DE Kreta-Spitzmaus
ES Musaraña de Creta

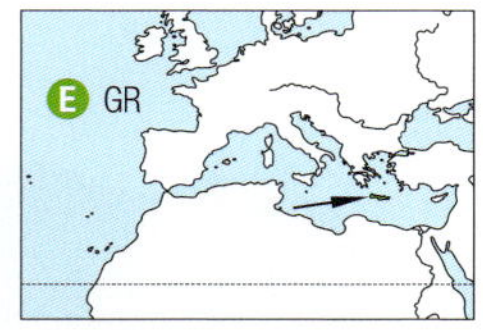

Crete I (mainly C mountain areas, including Nida & Omalos plateaus)

LC

3 sspp
HB 5–7.9 cm T 2.8–4.5 cm
W 4–10.5 g

0–1000 m

Sicilian White-toothed Shrew
Crocidura sicula

FR Crocidure de Sicile
DE Sizilien-Spitzmaus
ES Musaraña de Sicilia

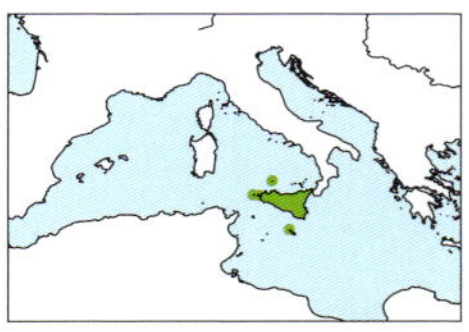

IT (Sicily, Ustica, Gozo & Egadi Is; possibly Comino Is), MT (probably extinct)

EN

Monotypic
HB 5.4–7.4 cm
T 3.1–4.8 cm
W 4–12 g

Canarian White-toothed Shrew
Crocidura canariensis

FR Crocidure des Canaries
DE Kanaren-Spitzmaus
ES Musaraña de Canarias

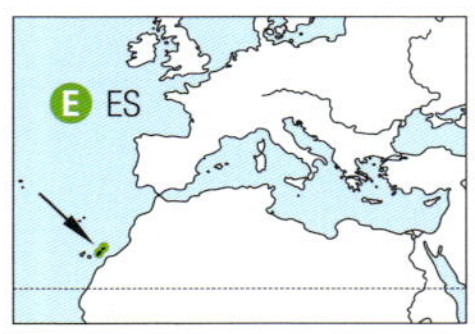

E Canary Is (Lanzarote, Chinijo Arch in Lobos & Montaña Clara, Fuerteventura)

West African Pygmy White-toothed Shrew
Crocidura obscurior

FR Crocidure obscure
DE Westafrika-Weißzahnspitzmaus
ES Musaraña pigmea de África occidental

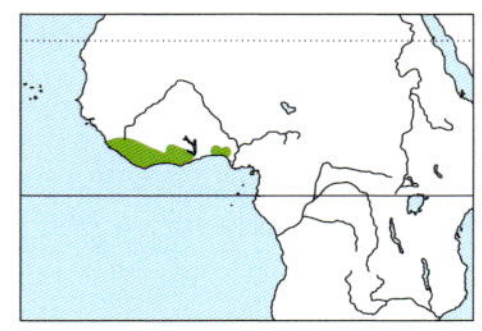

up to 600 m

W Africa in SE GN, E LR, S CI & W GH; perhaps SL (where it could be *C. eburnea*) & S NG

LC

Monotypic
HB 4.5–6.1 cm
T 3–3.8 cm
W 2.5–4.3 g

LC

Monotypic
HB 5.2–6 cm T 2.3–3.1 cm
W 3.1–3.8 g

up to 600 m

Ivory Coast White-toothed Shrew
Crocidura eburnea

FR Crocidure de Côte d'Ivoire
DE Elfenbeinküsten-Weißzahnspitzmaus
ES Musaraña de Costa de Marfil

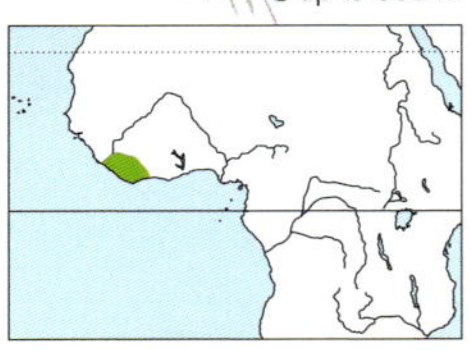

W Africa in SE GN, E LR & SW CI; perhaps SL (where it could be *C. obscurior*)

LC

Monotypic
HB 6–10.7 cm
T 3.1–6.9 cm

above 1000 m

Moonshine White-toothed Shrew
Crocidura luna

FR Crocidure séléné
DE Mondberg-Weißzahnspitzmaus
ES Musaraña luna

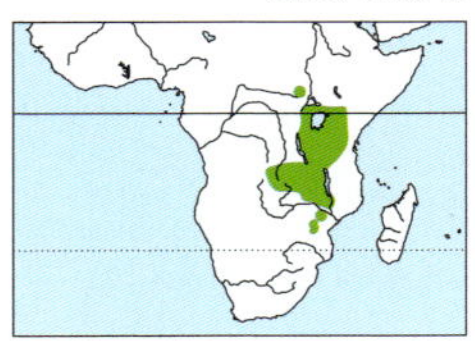

NE CD, S UG & SW KE S to ZM & WC MZ, isolated in E ZW (Inyanga Highlands)

LC

Monotypic
HB c. 7.1 cm
T c. 5.1 cm

Greenwood's White-toothed Shrew
Crocidura greenwoodae

FR Crocidure de Greenwood
DE Greenwood-Weißzahnspitzmaus
ES Musaraña de Greenwood

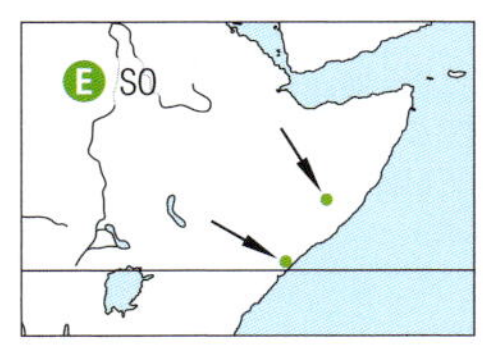

Recorded only from 2 localities

DD

Monotypic
HB c. 4 cm
T c. 3 cm

Somali Dwarf White-toothed Shrew
Crocidura nana

FR Crocidure naine
DE Kleine Somalia-Weißzahnspitzmaus
ES Musaraña enana de Somalia

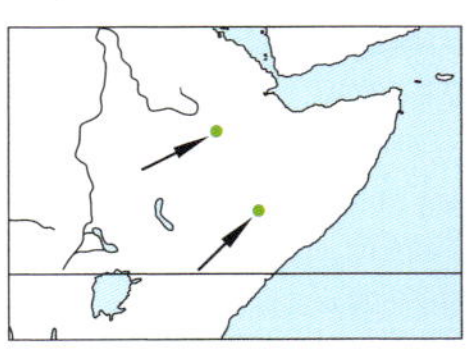

C ET (near Addis Ababa), SW SO (known from 2 localities in Dollo)

Bottego's White-toothed Shrew
Crocidura bottegi

FR Crocidure de Bottego
DE Bottego-Weißzahnspitzmaus
ES Musaraña de Bottego

up to 1750 m

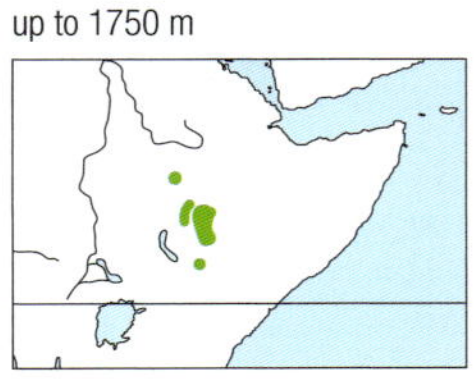

SW ET, NC KE (1 record from Marsabit)

DD

Monotypic
HB 4.6–5.1 cm
T 2.9–3 cm
W c. 4 g

DD

Monotypic
HB 7.8–8.5 cm
T 4.3–4.8 cm

c. 4050 m

Bekele's White-toothed Shrew
Crocidura afeworkbekelei

FR Crocidure de Bekele
DE Bekele-Weißzahnspitzmaus
ES Musaraña de Bekele

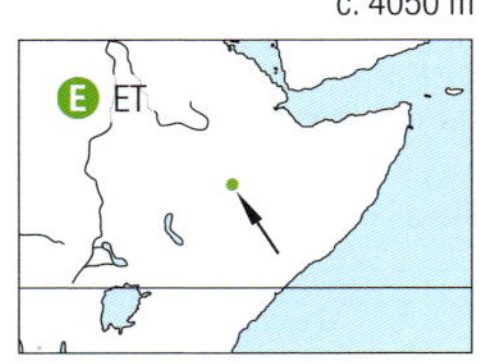

Bale Mts in Sanetti Plateau (Konteh Area, perhaps other parts of Plateau)

DD

Monotypic

HB 8.4–9.9 cm

T 6–7.3 cm

c. 1900 m

Yalden's White-toothed Shrew

Crocidura yaldeni

FR Crocidure de Yalden

DE Yalden-Weißzahnspitzmaus

ES Musaraña de Yalden

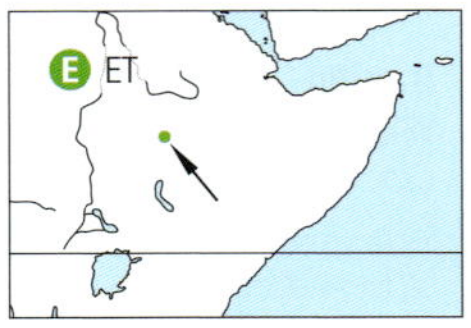

Known from Beletta Forest; perhaps widespread

NT

Monotypic

HB 6.5–8.5 cm

T 4.3–5.8 cm

W 7–14 g

Glass's White-toothed Shrew

Crocidura glassi

FR Crocidure de Glass

DE Glass-Weißzahnspitzmaus

ES Musaraña de Glass

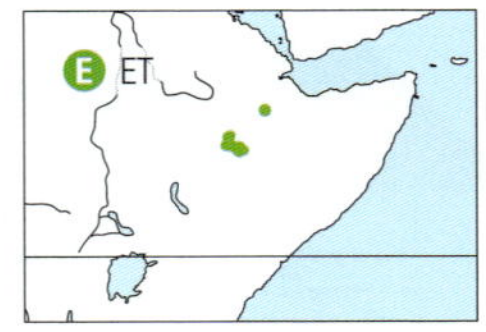

2700–4050 m

E of Rift Valley in Bale Mts, Mt Badda & Chercher Highlands

Thalia's White-toothed Shrew

Crocidura thalia

FR Crocidure thalie

DE Thalia-Weißzahnspitzmaus

ES Musaraña de Thalia

LC

Monotypic

HB 7.7–10 cm T 5.5–6.9 cm

W 9–26 g

515–3300 m

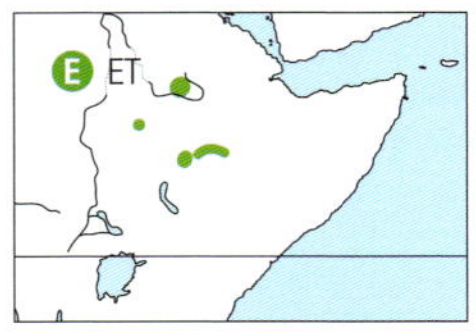

W & E of Rift Valley in 3 disjunct regions

Bale White-toothed Shrew

Crocidura bottegoides

FR Crocidure du Balé

DE Bale-Weißzahnspitzmaus

ES Musaraña de Bale

EN

Monotypic

HB 4.6–5.7 cm

T 4.3–4.7 cm

W 2.5–3.5 g

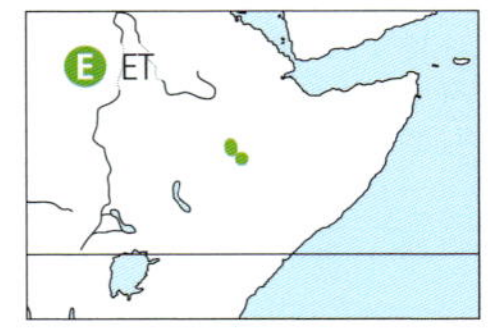

2400–3280 m

Bale Mts & Mt Albasso

LC

Monotypic

HB 7.2–9.3 cm

T 4.1–4.6 cm

Bailey's White-toothed Shrew

Crocidura baileyi

FR Crocidure de Bailey

DE Bailey-Weißzahnspitzmaus

ES Musaraña de Bailey

2700–3550 m

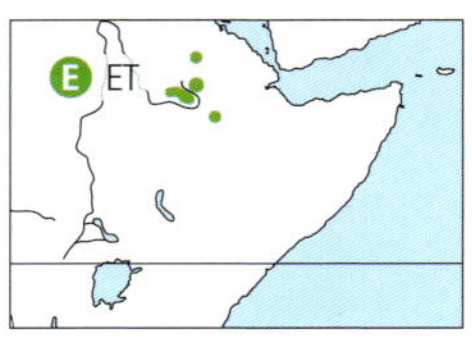

Highlands W of Rift Valley (Simien Mts, Amedamit Mts, Enjiabara, Debra Sina, Ankober)

NE

Monotypic

HB 6.8–8.3 cm

T 3.1–5 cm

W 8.5–15 g

Makeda White-toothed Shrew

Crocidura makeda

FR Crocidure de Makeda

DE Makeda-Weißzahnspitzmaus

ES Crocidura de Makeda

3120–3785 m

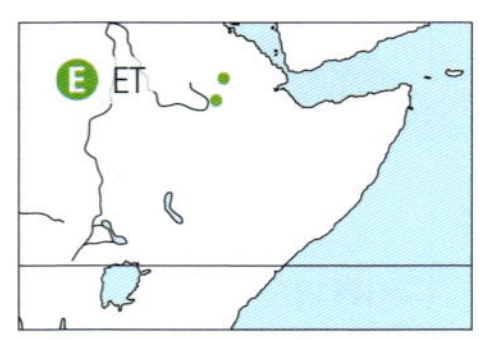

Borena Saynt NP & Abohoy Gara CA (part of Mt Abuna Yosef)

VU

Monotypic

HB 7.3–9.2 cm

T 4.7–5.5 cm

Lucina's White-toothed Shrew

Crocidura lucina

FR Crocidure lucine

DE Lucina-Weißzahnspitzmaus

ES Musaraña de Lucina

3000–4050 m

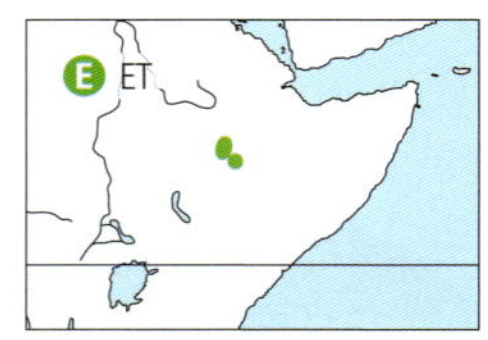

Mt Badda & Bale Mts E of Rift Valley

EN

Monotypic

HB 6–8 cm

T 4.8–5.7 cm

Guramba White-toothed Shrew

Crocidura phaeura

FR Crocidure du Guramba

DE Guramba-Weißzahnspitzmaus

ES Musaraña de Guramba

1100–2400 m

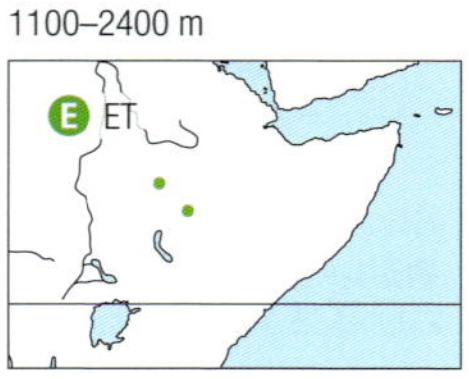

Mt Guramba & Nechisar NP, W of Rift Valley

CR

Monotypic

HB 6.5–7.6 cm T 4.4–4.8 cm

W 7–9.5 g

2400–2630 m

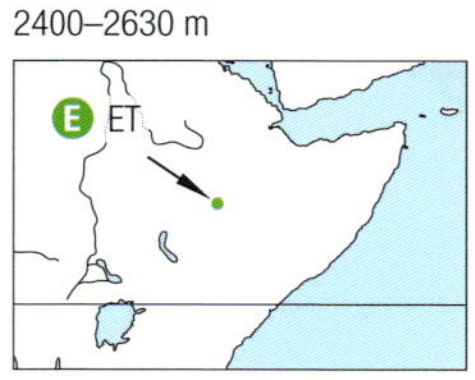

Harenna Forest in Bale Mts

Harenna White-toothed Shrew

Crocidura harenna

FR Crocidure de l'Harenna

DE Harenna-Weißzahnspitzmaus

ES Musaraña de Harenna

Macmillan's White-toothed Shrew

Crocidura macmillani

FR Crocidure de Macmillan

DE Macmillan-Weißzahnspitzmaus

ES Musaraña de Macmillan

1220–1930 m

Koteke, Middle Godjeb Valley, Sheko Forest

NT

Monotypic

HB 7.3–8.7 cm

T 5.2–5.7 cm

W c. 8.8 g

LC

Monotypic

HB 6.5–9.2 cm

T 3.9–6.3 cm

Smoky White-toothed Shrew

Crocidura fumosa

FR Crocidure enfumée

DE Rauchgraue Weißzahnspitzmaus

ES Musaraña de dientes blancos ahumada

800–4000 m

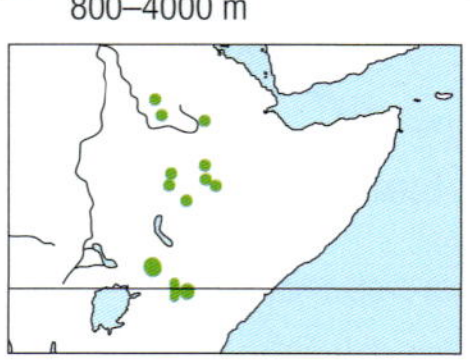

WC KE (Mt Kenya, Aberdare Range, Cherangani Hills), ET (research is needed)

DD

Monotypic

HB c. 10.5 cm

T c. 6 cm

Zaphiro's White-toothed Shrew

Crocidura zaphiri

FR Crocidure de Zaphiro

DE Zaphiro-Weißzahnspitzmaus

ES Musaraña de Zaphiro

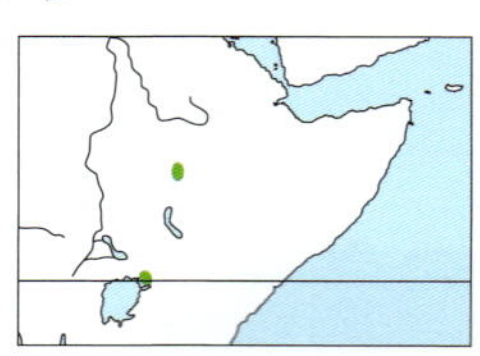

SW ET, WC KE (Kaimosi in Kisumu County)

DD

Monotypic

HB 7.7–10 cm

T 5.5–6.8 cm

W 11–15 g

Ugandan Lowland White-toothed Shrew

Crocidura selina

FR Crocidure d'Ouganda

DE Uganda-Tiefland-Weißzahnspitzmaus

ES Musaraña de tierras bajas de Uganda

up to 1800 m

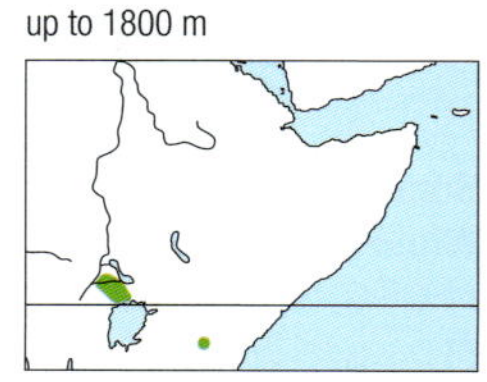

C UG (Mabira, Kibanda & Mbanga forests), S KE (Chyulu Hills)

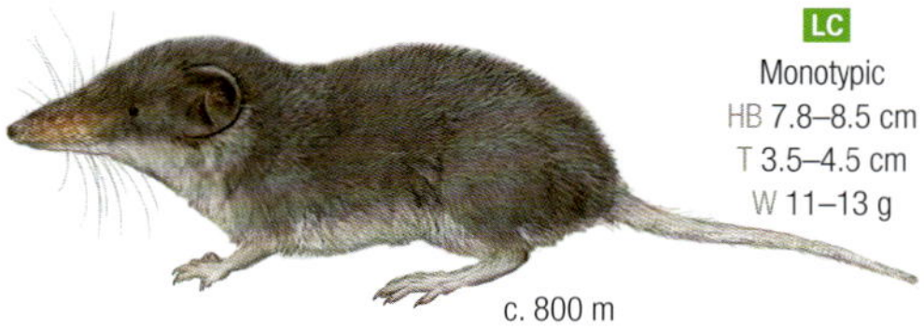

LC

Monotypic
HB 7.8–8.5 cm
T 3.5–4.5 cm
W 11–13 g

MacArthur's White-toothed Shrew
Crocidura macarthuri

FR Crocidure de MacArthur
DE MacArthur-Weißzahnspitzmaus
ES Musaraña de MacArthur

c. 800 m

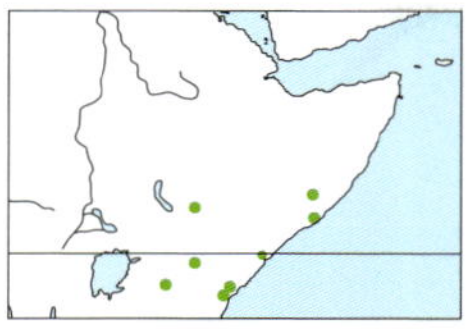

Scattered throughout KE and C & S SO

DD

Monotypic
HB 6.8–7.1 cm
T 5.7–5.8 cm

Nyiro White-toothed Shrew
Crocidura macowi

FR Crocidure du Ng'iro
DE Nyiro-Weißzahnspitzmaus
ES Musaraña de Nyiro

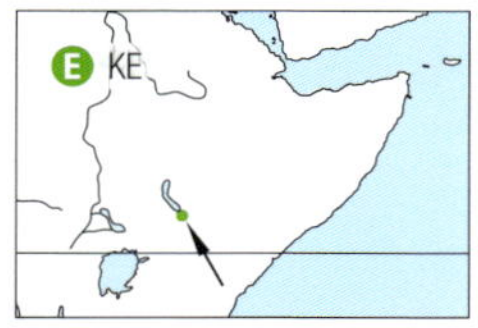

Known only from type locality at Mt Nyiro

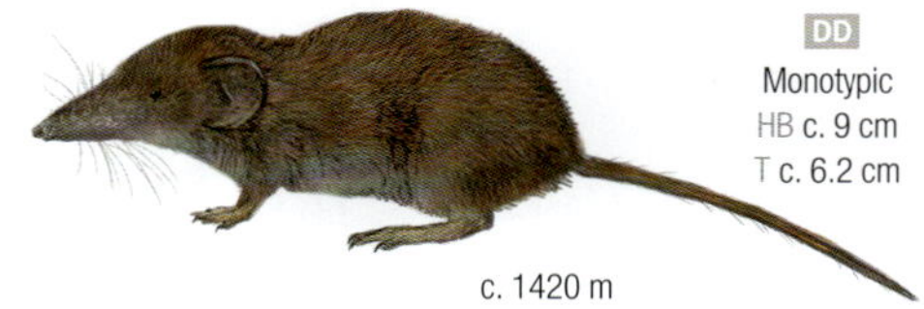

DD

Monotypic
HB c. 9 cm
T c. 6.2 cm

Ultimate White-toothed Shrew
Crocidura ultima

FR Crocidure de Nyeri
DE Jombeni-Weißzahnspitzmaus
ES Musaraña máxima

c. 1420 m

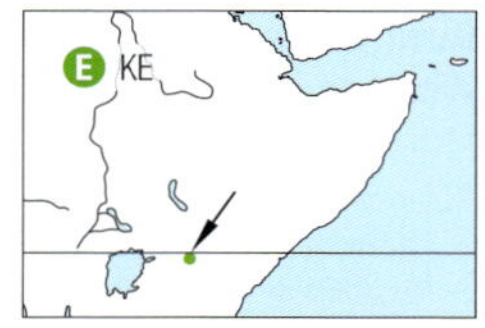

Jombeni Range near Nyeri; potentially Aberdare Range

DD

Monotypic
HB 9–9.2 cm
T 5.9–6.5 cm

Rainey's White-toothed Shrew
Crocidura raineyi

FR Crocidure de Rainey
DE Rainey-Weißzahnspitzmaus
ES Musaraña de Rainey

Mt Gargues (Mathews Range)

LC

Monotypic
HB 5.5–6.4 cm
T 4.5–5.5 cm
W 4–8 g

East African Highland White-toothed Shrew
Crocidura allex

FR Crocidure d'Osgood
DE Ostafrikanische Hochland-Weißzahnspitzmaus
ES Musaraña de tierras altas de África oriental

2000–4000 m

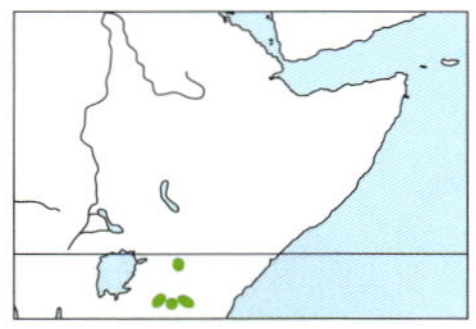

SW KE, NE TZ (Ngorogoro, Mt Meru, Mt Kilimanjaro)

LC

Monotypic
HB 8.9–9.8 cm T 5.5–6.7 cm
W 12–14 g

Xanthippe's White-toothed Shrew
Crocidura xantippe

FR Crocidure de Xanthippe
DE Xanthippe-Weißzahnspitzmaus
ES Musaraña de Xanthippe

1100–1300 m

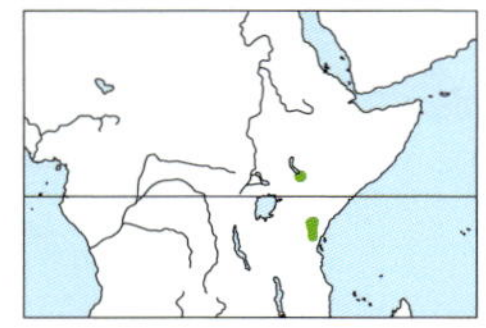

SE KE, NE TZ; disjunct record from NW KE needs to be confirmed

LC

Monotypic
HB 5.5–6.8 cm
T 3.3–4 cm
W 2–4.6 g

Elgon White-toothed Shrew
Crocidura elgonius

FR Crocidure de l'Elgon
DE Elgon-Weißzahnspitzmaus
ES Musaraña de Elgon

below 1000 m

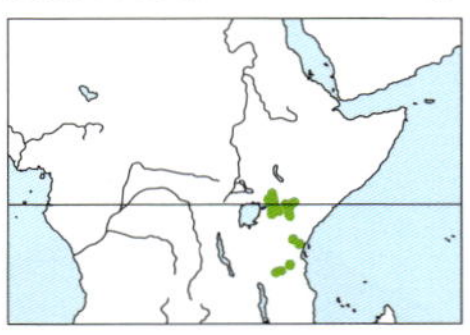

W & S KE, E TZ (East & West Usambara, Uluguru & Udzungwa Mts); perhaps EC UG

DD

Monotypic
HB 6.5–9.7 cm T 4.1–5.7 cm
W 11–15 g

Fischer's White-toothed Shrew
Crocidura fischeri

FR Crocidure de Fischer
DE Fischer-Weißzahnspitzmaus
ES Musaraña de Fischer

c. 800 m

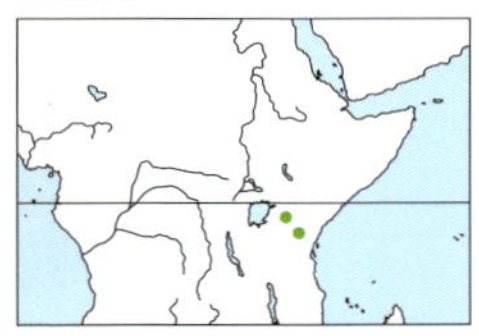

SC KE (N of L Natron), NE TZ (Himo)

DD

Monotypic
HB c. 6.5 cm
T c. 5.2 cm

Peters's White-toothed Shrew
Crocidura gracilipes

FR Crocidure du Kilimandjaro
DE Peters-Weißzahnspitzmaus
ES Musaraña de Peters

Type locality accepted to be Mt Kilimanjaro; reports across W & E Africa unconfirmed

Desperate White-toothed Shrew
Crocidura desperata

FR Crocidure furieuse
DE Verzweifelte Weißzahnspitzmaus
ES Musaraña desesperada

VU

Monotypic
HB 8–10.1 cm T c. 6.7 cm
W 16–27 g

above 1500 m

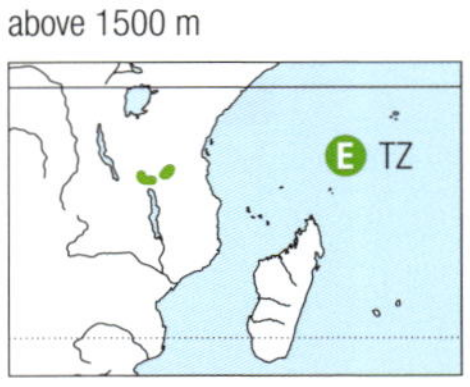

Relict forest patches in Rungwe & Udzungwa Mts

VU

Monotypic
HB 10.4–10.8 cm
T 7–8.3 cm
W 16.5–22 g

Telford's White-toothed Shrew
Crocidura telfordi

FR Crocidure de Telford
DE Telford-Weißzahnspitzmaus
ES Musaraña de Telford

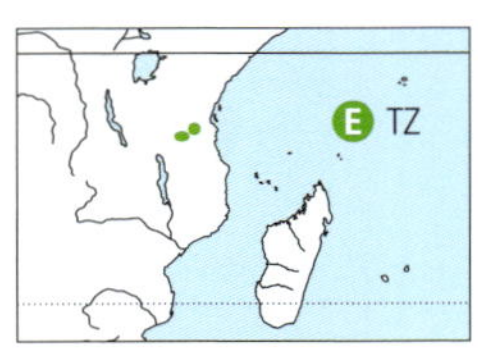

Eastern Arc Mts (Uluguru & Udzungwa Mts)

DD

Monotypic
HB 8.3–10.1 cm
T 5.5–7.1 cm
W 9.2–17 g

Kilimanjaro White-toothed Shrew
Crocidura monax

FR Crocidure moine
DE Kilimandscharo-Weißzahnspitzmaus
ES Musaraña del Kilimanjaro

1500–3000 m

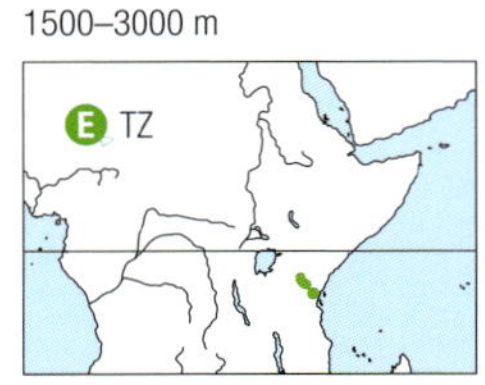

Mt Kilimanjaro & North Pare Mts

Mduma's White-toothed Shrew
Crocidura mdumai

FR Crocidure de Mduma
DE Mduma-Weißzahnspitzmaus
ES Musaraña de Mduma

EN

Monotypic
HB 7.6–9.1 cm
T 5.2–6.5 cm
W 7.3–9.6 g

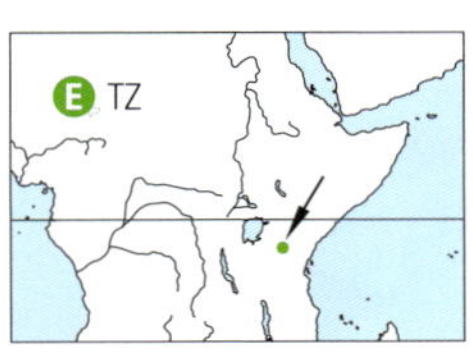

2000–2370 m

Ngorongoro Crater

Montane White-toothed Shrew
Crocidura montis

FR Crocidure de montagne
DE Montane Weißzahnspitzmaus
ES Musaraña de dientes blancos de montaña

LC

Monotypic
HB 6.3–9.5 cm
T 4.6–6.9 cm

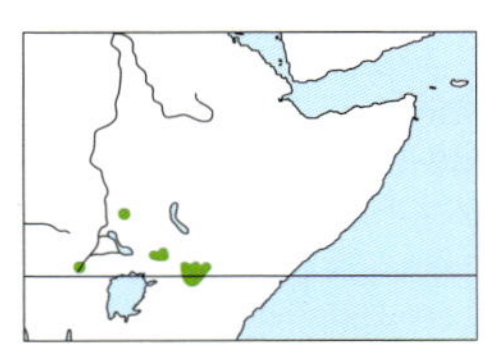

2130–4200 m

NE CD, SW & EC UG, SE SS, W & C KE

NT
Monotypic
HB 6.5–8.5 cm
T 4.5–6 cm
W 3.5–11 g

Newmark's White-toothed Shrew
Crocidura newmarki
FR Crocidure de Newmark
DE Newmark-Weißzahnspitzmaus
ES Musaraña de Newmark

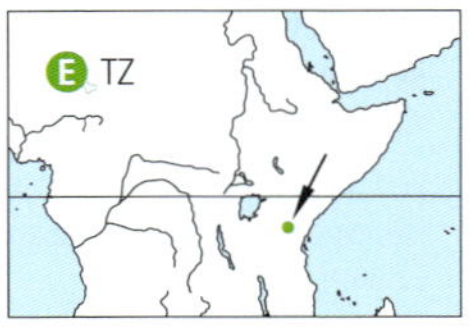

1950–3500 m

Mt Meru

Munissi's White-toothed Shrew
Crocidura munissii
FR Crocidure de Munissi
DE Munissi-Weißzahnspitzmaus
ES Musaraña de Munissi

LC
Monotypic
HB 7.5–10.6 cm
T 6.6–9.5 cm
W 9.5–19.5 g

S Eastern Arc Mts (Ukaguru, Uluguru, Rubeho & Udzungwa Mts)

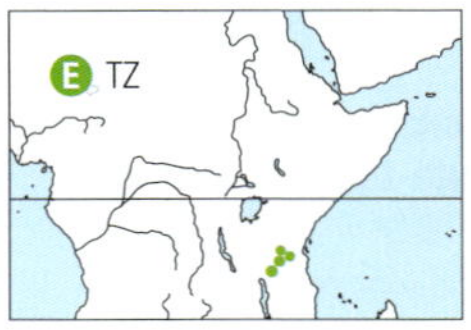

above 1450 m

Tanzanian White-toothed Shrew
Crocidura tansaniana
FR Crocidure de Tanzanie
DE Tansania-Weißzahnspitzmaus
ES Musaraña de Tanzania

NT
Monotypic
HB 8–11 cm
T 5.8–7.6 cm
W 9.5–20 g

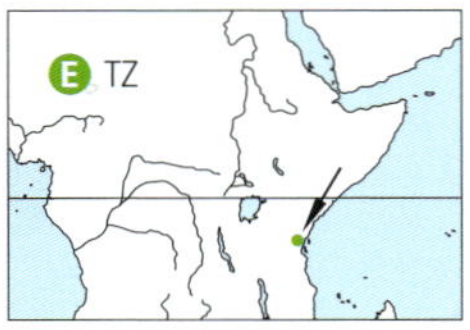

West & East Usambara Mts

Usambara White-toothed Shrew
Crocidura usambarae
FR Crocidure des Usambara
DE Usambara-Weißzahnspitzmaus
ES Musaraña de Usambara

VU
Monotypic
HB 7.5–9.3 cm
T 5.4–6.2 cm
W 8.4–10.5 g

1100–1830 m

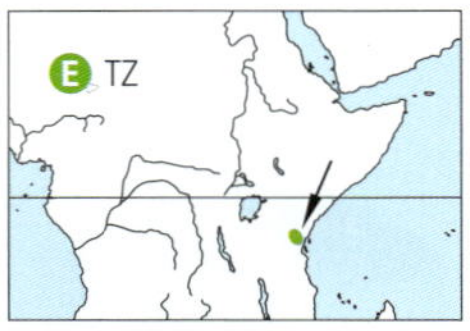

Magamba, Shume, Chome, Mazumbai

LC
Monotypic
HB 6.5–8.1 cm
T 5.6–6.6 cm
W 5.5–10 g

Niobe's White-toothed Shrew
Crocidura niobe
FR Crocidure niobé
DE Niobe-Weißzahnspitzmaus
ES Musaraña de Niobe

1600–2250 m

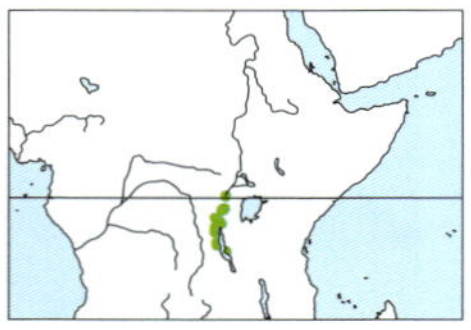

E CD, SW UG, W RW, NW BI, WC TZ

LC
2 sspp
HB 6.4–8.5 cm
T 3.1–5.1 cm

Desert White-toothed Shrew
Crocidura smithii
FR Crocidure de Smith
DE Wüsten-Weißzahnspitzmaus
ES Musaraña de desierto

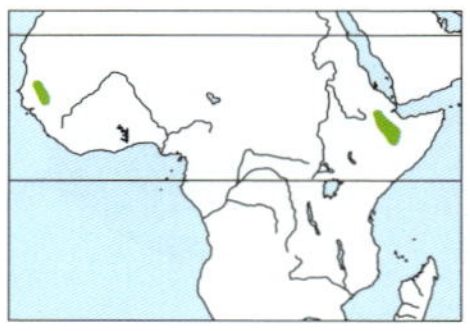

E SN and extremely far in E ET

NT
Monotypic
HB 6.4–8.8 cm
T 6–7.5 cm
W 5–13 g

Kivu White-toothed Shrew
Crocidura kivuana
FR Crocidure du Kivu
DE Kivu-Weißzahnspitzmaus
ES Musaraña de Kivu

1700–3300 m

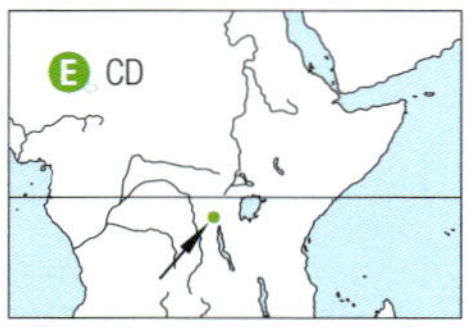

Kahuzi Mts (small area W of L Kivu between Lwiro and Mt Kahuzi); probably widespread

LC
Monotypic
HB 4.2–5 cm
T 2.9–3.8 cm

Sahelian Tiny White-toothed Shrew
Crocidura pasha
FR Crocidure pacha
DE Kleine Sahel-Weißzahnspitzmaus
ES Musaraña diminuta de Sahel

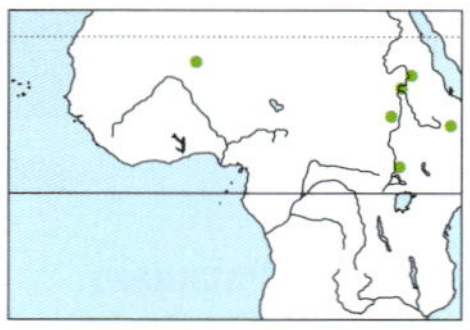

NE ML, EC & SE SD, SE SS, NE ET; status uncertain in EG (a mummified individual reported from Thebes)

Roosevelt's White-toothed Shrew
Crocidura roosevelti
FR Crocidure de Roosevelt
DE Roosevelt-Weißzahnspitzmaus
ES Musaraña de Roosevelt

LC
Monotypic
HB 6.2–8.2 cm
T 5.9–6.9 cm
W c. 9 g

below 1200 m

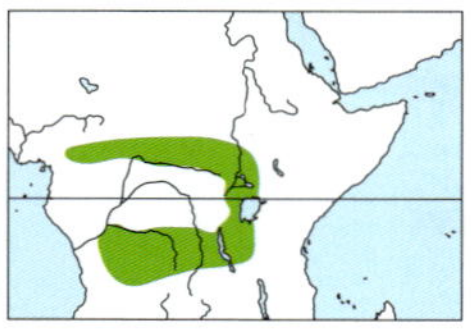

Forest-savanna margin of C Africa in CM, CF, CD, UG, RW, TZ & NE AO; perhaps SS

Long-footed White-toothed Shrew
Crocidura crenata
FR Crocidure sauteuse
DE Langfuß-Weißzahnspitzmaus
ES Musaraña de pies largos

LC
Monotypic
HB 6.2–7.9 cm
T 7.2–9.1 cm
W 7.8–9.1 g

100–700 m

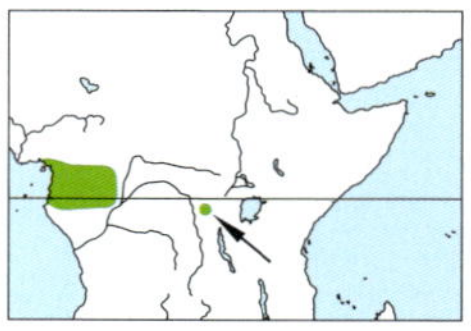

S CM, SW CF, GQ (Mbini), N & C GA, N CG, 1 record from EC CD

Grasse's White-toothed Shrew
Crocidura grassei
FR Crocidure de Grassé
DE Grasse-Weißzahnspitzmaus
ES Musaraña de Grasse

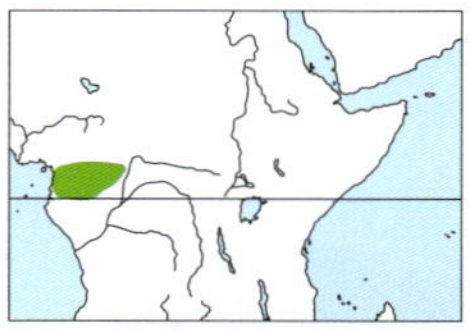

S CM, SW CF, GQ (Mbini), N GA, N CG

LC
Monotypic
HB 7.6–8.6 cm
T 6.9–8 cm
W 11–13 g

Bicolored African White-toothed Shrew
Crocidura fuscomurina
FR Crocidure de Heuglin
DE Zweifarb-Weißzahnspitzmaus
ES Musaraña de África bicolor

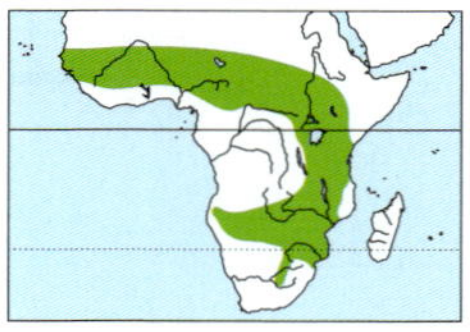

Sub-Saharan Africa from SN E to W ET, S to S AO, N NA & NE ZA, avoiding densely forested W & C Africa

LC
Monotypic
HB 4.8–7.5 cm
T 2.7–4.8 cm
W 2–5 g

Flat-headed White-toothed Shrew
Crocidura planiceps
FR Crocidure à tête plate
DE Flachkopf-Weißzahnspitzmaus
ES Musaraña de cabeza plana

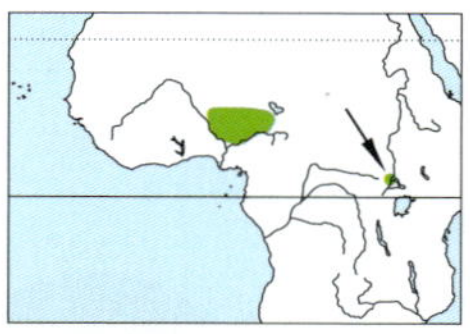

N NG, NW UG; perhaps SD, ET & CD

DD
Monotypic
HB 5–6.6 cm
T 3.9–4.8 cm
W 2–3 g

LC
Monotypic
HB 5.4–8 cm
T 4.5–5.7 cm
W 4.5–8 g

Crosse's White-toothed Shrew
Crocidura crossei

FR Crocidure de Crosse
DE Crosse-Weißzahnspitzmaus
ES Musaraña de Crosse

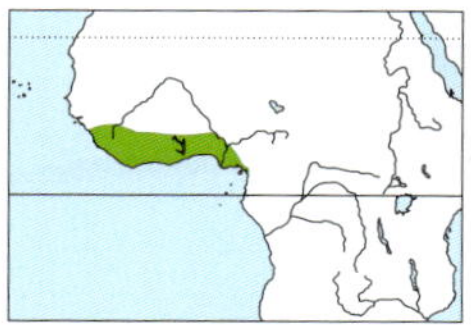

W Africa from GN E to SW CM

LC
Monotypic
HB 6.4–8.6 cm T 4.7–6.3 cm
W 7.5–12 g

Jouvenet's White-toothed Shrew
Crocidura jouvenetae

FR Crocidure de Jouvenet
DE Jouvenet-Weißzahnspitzmaus
ES Musaraña de Jouvenet

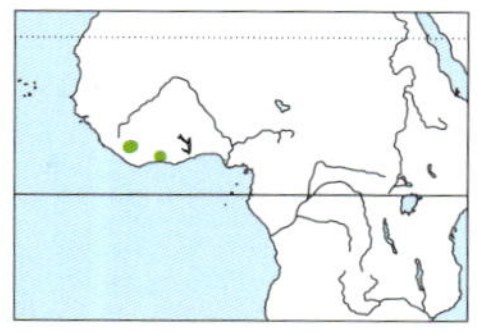

S GN, LR, CI; perhaps wider because limits with *C. crossei* uncertain

LC
Monotypic
HB 4.5–6.8 cm
T 2.9–3.7 cm
W 3–5.1 g

Mauritanian White-toothed Shrew
Crocidura lusitania

FR Crocidure de Mauritanie
DE Mauretanien- Weißzahnspitzmaus
ES Musaraña de Mauritania

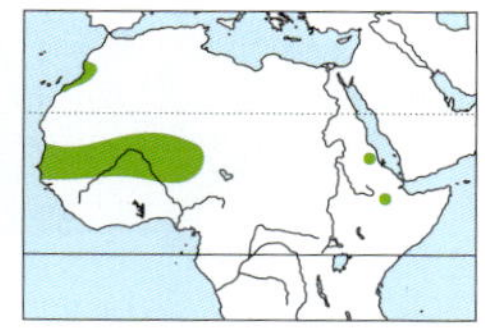

S MA, N EH, S MR & SN, E to W NE and NW NG, 2 isolated records in NW ER & NC ET

Naked-tailed White-toothed Shrew
Crocidura littoralis

FR Crocidure des rives
DE Nacktschwanz-Weißzahnspitzmaus
ES Musaraña de cola desnuda

LC
Monotypic
HB 8.5–10 cm
T 6.4–7.7 cm
W 20–24 g

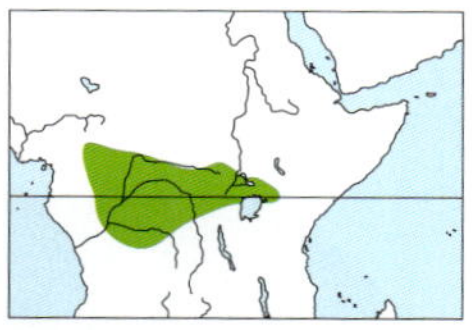

E CM & CG E through SW CF & CD to UG & WC KE

Gracile White-toothed Shrew
Crocidura maurisca

FR Crocidure d'Entebbe
DE Grazile Nacktschwanz-Weißzahnspitzmaus
ES Musaraña grácil

LC
Monotypic
HB 7.5–9.7 cm
T 6–7.2 cm
W 8.3–18 g

1500–2400 m

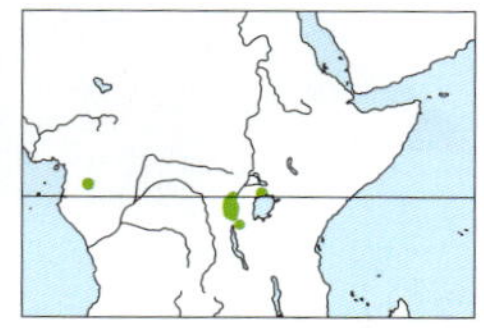

E CD, RW, BI, S UG, isolated record in N GA

LC
Monotypic
HB 6.7–9.3 cm
T 3.3–6.9 cm

Reddish-gray White-toothed Shrew
Crocidura cyanea

FR Crocidure ardoisée
DE Rotgraue Weißzahnspitzmaus
ES Musaraña gris rojiza

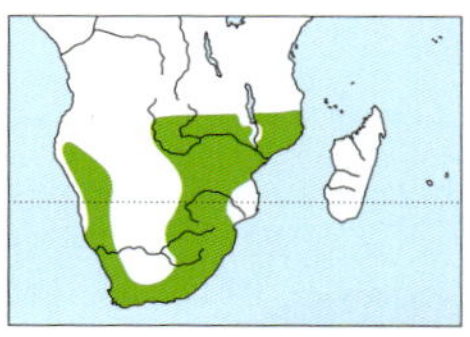

S Africa in ZM, ZW, MZ & E BW (in E), extreme SW AO & NA (in W), S to ZA, SZ & LS

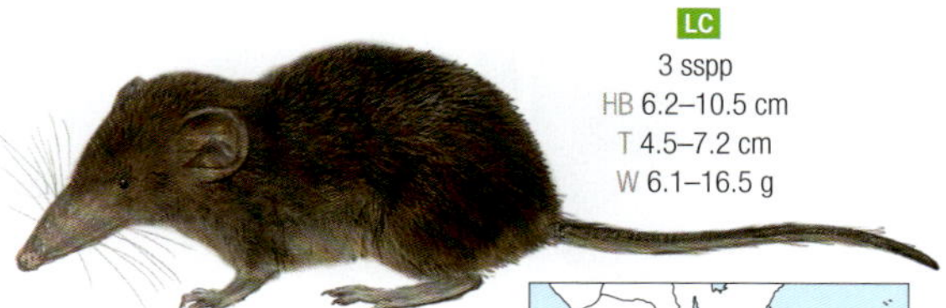

LC
3 sspp
HB 6.2–10.5 cm
T 4.5–7.2 cm
W 6.1–16.5 g

Swamp White-toothed Shrew
Crocidura mariquensis

FR Crocidure des marais
DE Sumpf-Weißzahnspitzmaus
ES Musaraña de ciénaga

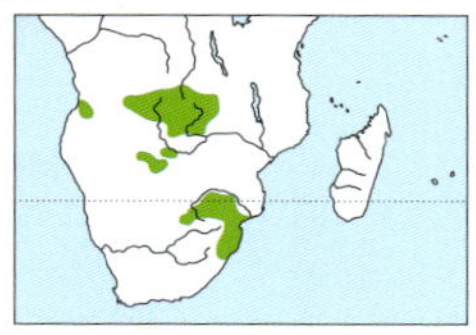

WC AO, S CD, ZM, NE NA, N BW, extreme NW ZW, E ZA, SZ, S MZ

LC
Monotypic
HB 6–8 cm
T 4.3–5.6 cm
W 6–8 g

Lesser Gray-brown White-toothed Shrew
Crocidura silacea

FR Crocidure du Mpumalanga
DE Kleine graubraune Weißzahnspitzmaus
ES Musaraña gris marrón menor

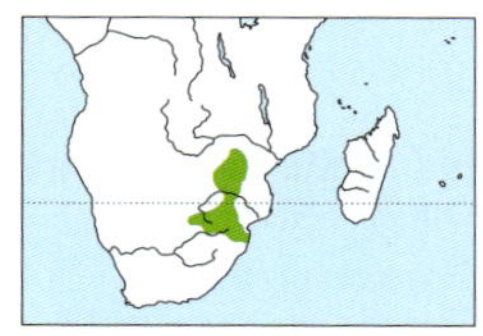

ZW, NE ZA, SZ, extreme S MZ

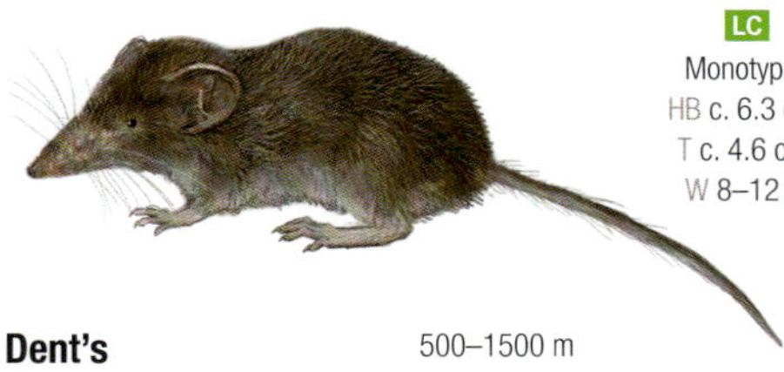

LC
Monotypic
HB c. 6.3 cm
T c. 4.6 cm
W 8–12 g

Dent's White-toothed Shrew
Crocidura denti

FR Crocidure de Dent
DE Dent-Weißzahnspitzmaus
ES Musaraña de Dent

500–1500 m

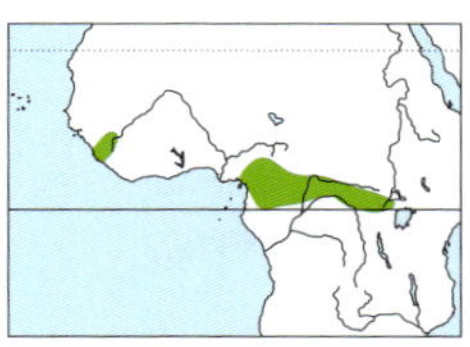

SE NG, CM & N GA through S CF, N CG & N CD to WC UG, disjunct in GN & SL

LC
Monotypic
HB 6.4–7.7 cm
T 3.9–5.6 cm
W 4.6–7.7 g

Hildegarde's White-toothed Shrew
Crocidura hildegardeae

FR Crocidure de Hildegarde
DE Hildegarde-Weißzahnspitzmaus
ES Musaraña de Hildegarde

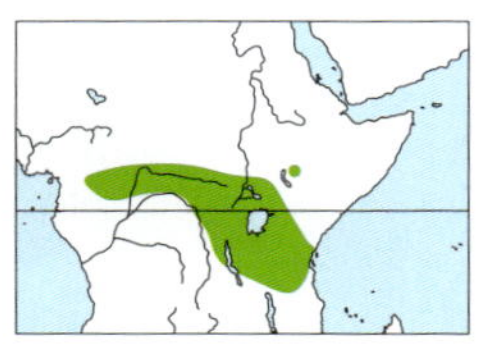

From SE CM, S CF & N CG through N & E CD to W KE & TZ, records in S SS & SW ET; limits uncertain

LC
Monotypic
HB c. 6.1 cm
T c. 4 cm
W c. 7 g

Mamfe White-toothed Shrew
Crocidura virgata

FR Crocidure à bande dorsale
DE Mamfe-Weißzahnspitzmaus
ES Musaraña de Mamfe

c. 850 m

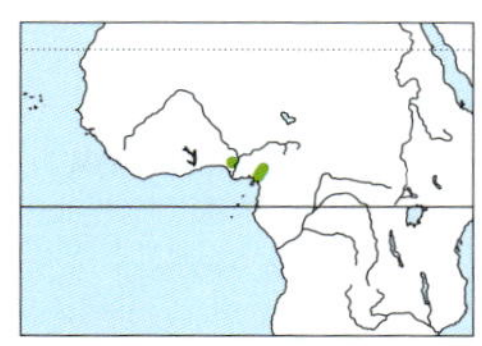

S NG, W CM

Sao Tome White-toothed Shrew
Crocidura thomensis

FR Crocidure de Sao Tomé
DE São-Tomé-Weißzahnspitzmaus
ES Musaraña de Santo Tomé

EN
Monotypic
HB 8.3–10 cm
T 7–8.6 cm

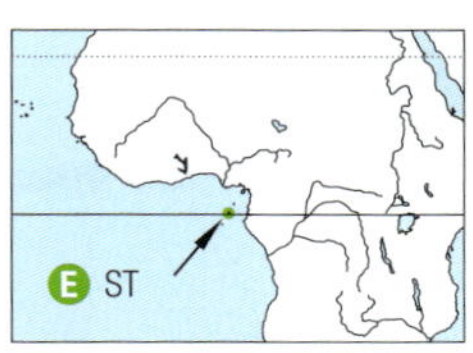

São Tomé I

Fingui White-toothed Shrew
Crocidura fingui

FR Crocidure de Principe
DE Fingui-Weißzahnspitzmaus
ES Musaraña fingui

DD
Monotypic
HB 7.5–10.7 cm
T 7–8.6 cm
W 10–13 g

130–310 m

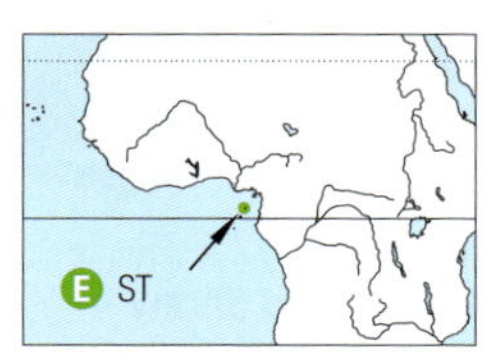

N part of Príncipe I; probably throughout island

Fraser's White-toothed Shrew
Crocidura poensis

FR Crocidure de Fernando Po
DE Fraser-Weißzahnspitzmaus
ES Musaraña de Fraser

LC
Monotypic
HB 7.1–9.8 cm
T 4.6–6.6 cm
W 8–20 g

up to 1000 m

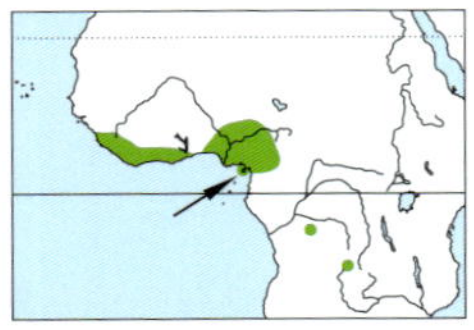

From GN & SL E to SW CF, W CG, GA & GQ (Mbini, Bioko I), isolated records in SW CD & N ZM

Turbo White-toothed Shrew
Crocidura turba

FR Crocidure turbulente
DE Haarschwanz-Weißzahnspitzmaus
ES Musaraña tumultuosa

LC
Monotypic
HB 9.9–10.2 cm
T 5.5–5.9 cm
W 17–21 g

up to 1500 m

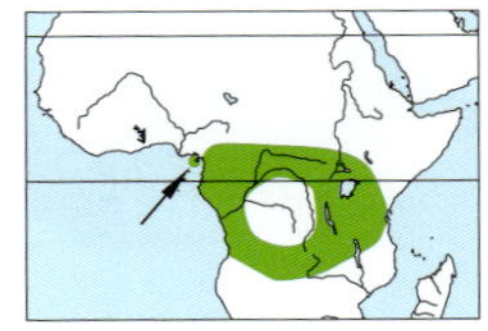

C Africa except C Congo Basin, from S CM E to W KE, S to AO & ZM, Bioko I (GQ)

Medabo White-toothed Shrew
Crocidura similiturba

FR Crocidure de Medabo
DE Medabo-Weißzahnspitzmaus
ES Musaraña semitumultuosa

NE
Monotypic
HB 8.6–10.2 cm T 5.7–6.7 cm
W 13–21 g

Known only from Medabo (Kafa zone); perhaps widespread in SW

Nigerian White-toothed Shrew
Crocidura nigeriae

FR Crocidure du Nigeria
DE Nigeria-Weißzahnspitzmaus
ES Musaraña de Nigeria

LC
Monotypic
HB 9–10.9 cm
T 5.3–6.4 cm
W 11–24 g

up to 600 m

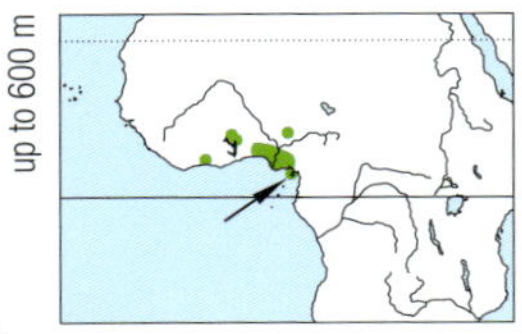

SE CI, NE GH, C & S NG, WC CM, Bioko I (GQ)

Hun White-toothed Shrew
Crocidura attila

FR Crocidure d'Attila
DE Hunnen-Weißzahnspitzmaus
ES Musaraña de Hun

LC
Monotypic
HB 7.5–9 cm
T 4.5–5.5 cm
W c. 5 g

up to 1900 m

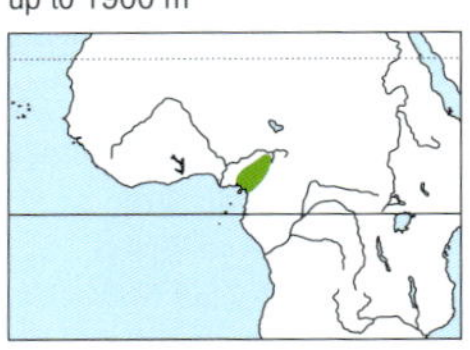

SE NG, W CM; perhaps CF & NE CD

Fox's White-toothed Shrew
Crocidura foxi

FR Crocidure de Fox
DE Fox-Weißzahnspitzmaus
ES Musaraña de Fox

LC
Monotypic
HB 9.1–11.4 cm
T 5.3–7.2 cm
W 13–22 g

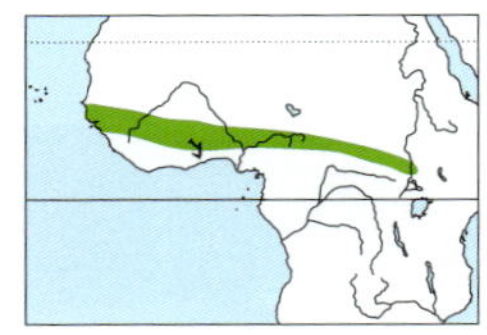

SN E through Sudan Savanna Zone to S SS

Büttikofer's White-toothed Shrew
Crocidura buettikoferi

FR Crocidure de Büttikofer
DE Büttikofer-Weißzahnspitzmaus
ES Musaraña de Buettikofer

LC
Monotypic
HB 7.2–10 cm
T 5.1–6.2 cm

Scattered through W Africa in SE GN, LR, SW CI, S GH & S NG

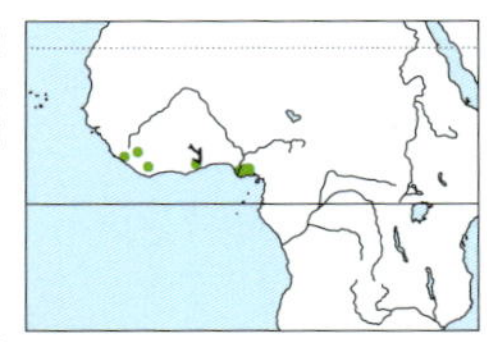

Therese's White-toothed Shrew
Crocidura theresae

FR Crocidure de Thérèse
DE Thérèse-Weißzahnspitzmaus
ES Musaraña de Therese

LC
Monotypic
HB 7.8–10.1 cm
T 3.4–5.5 cm
W 7–18.4 g

c. 1600 m

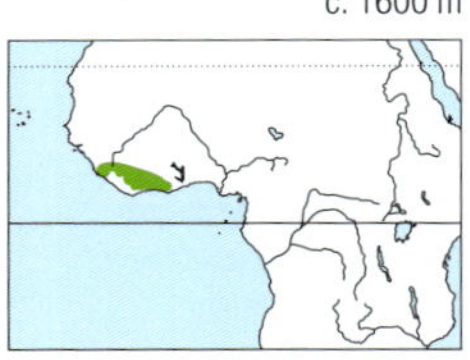

W Africa in SL, SE GN, CI & SW GH; perhaps LR

Large-headed White-toothed Shrew
Crocidura grandiceps

FR Crocidure à tête longue
DE Großkopf-Weißzahnspitzmaus
ES Musaraña de cabeza grande

NT
Monotypic
HB 9.8–11.9 cm T 5.5–7.4 cm
W 19–27 g

c. 1800 m

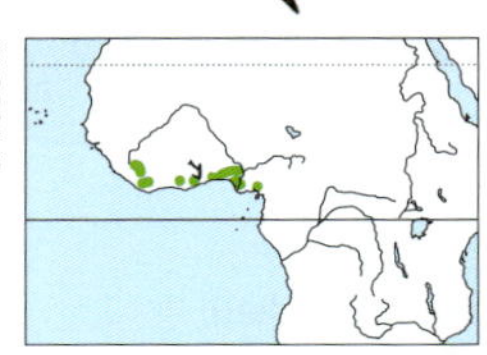

Scattered in W Africa in SW GN, LR, W CI, S GH, S BJ, S NG, possibly SW CM

Wimmer's White-toothed Shrew
Crocidura wimmeri

FR Crocidure de Wimmer
DE Wimmer-Weißzahnspitzmaus
ES Musaraña de Wimmer

CR
Monotypic
HB 8–9 cm T c. 5.7 cm
W 22–28 g

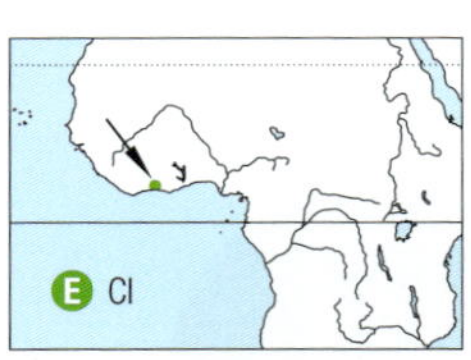

Known only from Adiopodoumé & Banco NP

African Black White-toothed Shrew
Crocidura nigrofusca

FR Crocidure brun-noir
DE Schwarze Afrika-Weißzahnspitzmaus
ES Musaraña negra de África

LC
Monotypic
HB 6.5–9.5 cm
T 5.5–7.2 cm
W 13–23 g

up to 2000 m

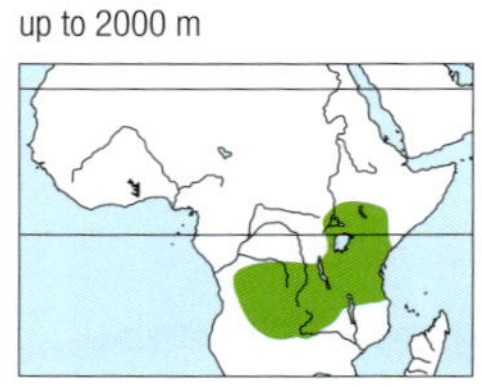

S SS & S ET S through E Africa to N & E AO & ZM; unconfirmed in CM

Savanna Dwarf White-toothed Shrew
Crocidura nanilla

FR Crocidure pygmée
DE Kleine Savannen-Weißzahnspitzmaus
ES Musaraña enana de sabana

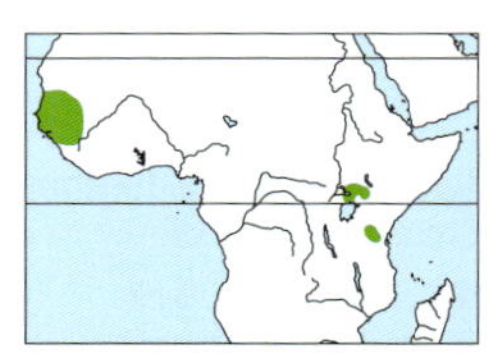

W Africa (SW MR, SN, GM, SW ML, GW, GN, N SL), E Africa (UG, NW KE, NE TZ); perhaps CI, GH & NG

LC
Monotypic
HB 4.1–5.8 cm
T 3.1–4.1 cm

NT
Monotypic
HB 7.5–9 cm
T 5–5.7 cm
W 8–19 g

Nimba White-toothed Shrew
Crocidura nimbae

FR Crocidure du Nimba
DE Nimba-Weißzahnspitzmaus
ES Musaraña de Nimba

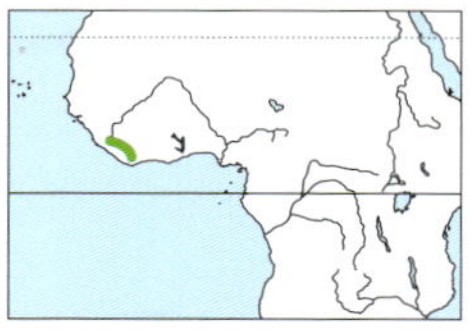

W Africa in S SL, SE GN, N LR & SW CI

Nimba Giant White-toothed Shrew
Crocidura nimbasilvanus

FR Crocidure sylvaine
DE Nimba-Riesenweißzahnspitzmaus
ES Musaraña de Nimba gigante

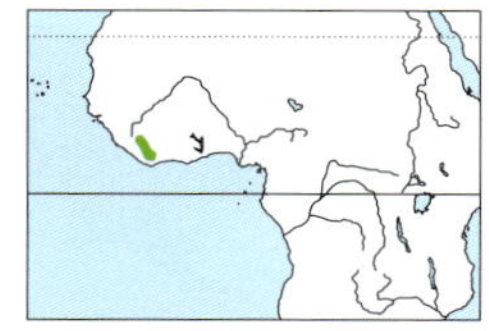

W Africa in SE GN to SW CI

LC
Monotypic
HB 12.5–15.1 cm
T 9.8–12.5 cm
W 47–67 g

LC
Monotypic
HB 5–5.5 cm
T 4–4.5 cm
W 4–5 g

Doucet's White-toothed Shrew
Crocidura douceti

FR Crocidure de Doucet
DE Doucet-Weißzahnspitzmaus
ES Musaraña de Doucet

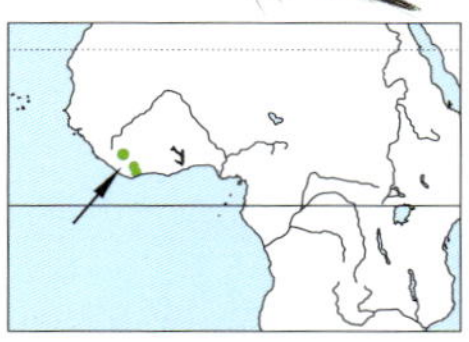

W Africa in SE GN & SW CI; unconfirmed in NG & GH (R Mamang R)

LC
Monotypic
HB 5.6–6.6 cm
T 6.2–9.5 cm
W 3–8 g

West African White-toothed Shrew
Crocidura muricauda

FR Crocidure à queue de souris
DE Westafrika-Langschwanz-Weißzahnspitzmaus
ES Musaraña de África occidental

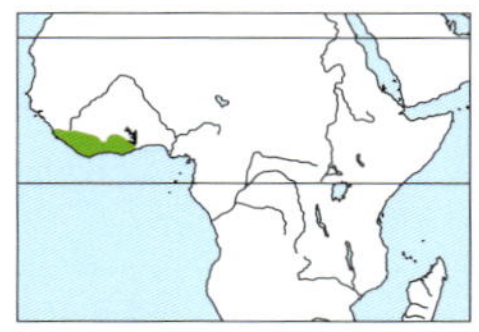

W Africa in SL, SE GN, LR, CI & S GH

LC
Monotypic
HB 8.5–9 cm
T 3.5–5.5 cm
W 18–23 g

Lamotte's White-toothed Shrew
Crocidura lamottei

FR Crocidure de Lamotte
DE Lamotte-Weißzahnspitzmaus
ES Musaraña de Lamotte

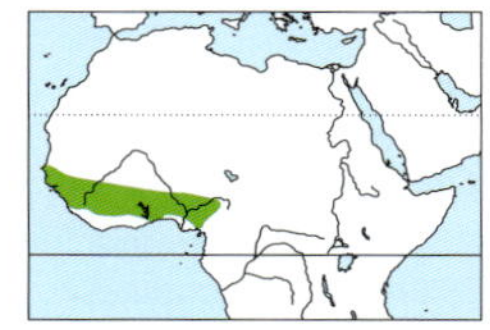

W Africa from SN E to NG & extreme SW CM

LC
Monotypic
HB 8.1–11.7 cm
T 3.6–5.8 cm
W 20–40 g

up to 1500 m

Greater Red White-toothed Shrew
Crocidura flavescens

FR Crocidure blonde
DE Südafrika-Riesenweißzahnspitzmaus
ES Musaraña roja mayor

Coastal & E ZA, LS, SZ, S MZ

LC
Monotypic
HB 6.9–9.6 cm
T 3.9–5.9 cm
W 8–24 g

Lesser Red White-toothed Shrew
Crocidura hirta

FR Crocidure roussâtre
DE Ockerfarbene Weißzahnspitzmaus
ES Musaraña roja menor

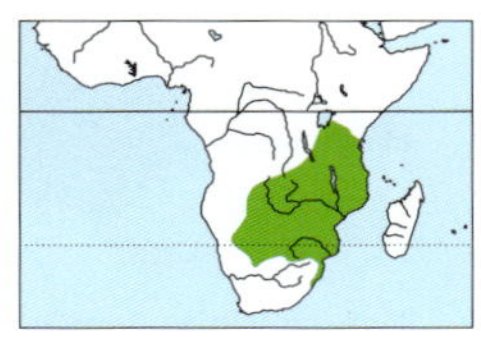

TZ, SE CD, E AO, ZM, MW, MZ, E NA, BW, ZW, N & E ZA, E SZ

DD
Monotypic
HB c. 9.6 cm
T c. 5.4 cm

c. 1095 m

Heather White-toothed Shrew
Crocidura erica

FR Crocidure d'Angola
DE Graue Angola-Weißzahnspitzmaus
ES Musaraña de brezo

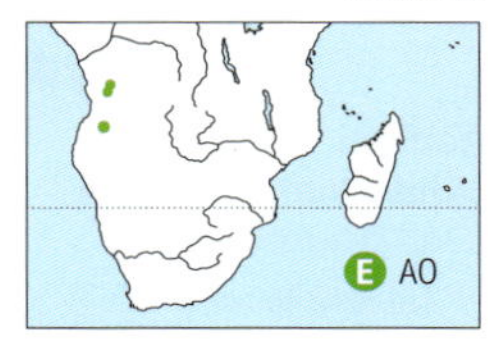

Known only from 3 localities in W

LC
Monotypic
HB 6.2–8.9 cm
T 4.6–6.3 cm

Blackish White-toothed Shrew
Crocidura nigricans

FR Crocidure noire
DE Schwarze Angola-Weißzahnspitzmaus
ES Musaraña negruzca

Central Plateau

LC
Monotypic
HB 15.5–17.5 cm
T 10–13.2 cm
W 51–76 g

Goliath White-toothed Shrew
Crocidura goliath

FR Crocidure géante
DE Goliath-Riesenweißzahnspitzmaus
ES Musaraña goliat

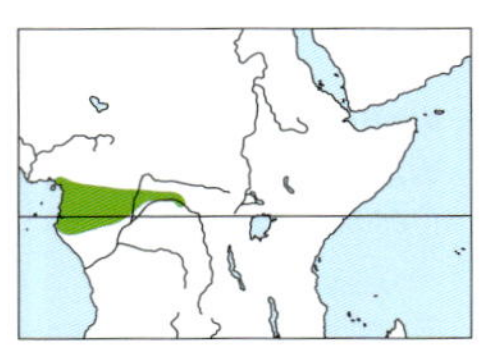

WC Africa in S CM, SW CF, GQ (Mbini), GA, N CG & NW CD

Somali White-toothed Shrew
Crocidura somalica

FR Crocidure de Somalie
DE Somali-Weißzahnspitzmaus
ES Musaraña de Somalia

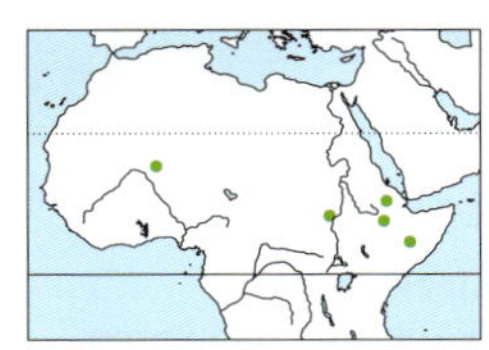

EC ML (Adrar de Iforas), S SD (Angolo), several sites in N, C & SE ET

LC
Monotypic
HB 6.3–7.7 cm
T 4–4.6 cm W c. 8 g

LC
Monotypic
HB 11–14 cm T 7–10 cm
W 33–65 g

African Giant White-toothed Shrew
Crocidura olivieri

FR Crocidure d'Olivier
DE Afrikanische Riesenweißzahnspitzmaus
ES Musaraña gigante de África

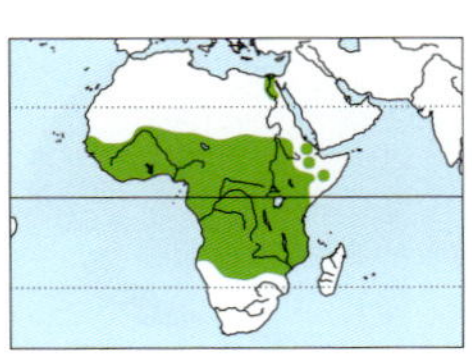

Sub-Saharan Africa from SN E to W ET, S to BW, scattered in C & E ET, disjunct along R Nile (EG)

LC
Monotypic
HB 7.5–10 cm
T 4.2–6.3 cm

Savanna Path White-toothed Shrew
Crocidura viaria

FR Crocidure des chemins
DE Große Savannen-Weißzahnspitzmaus
ES Musaraña de sendero de sabana

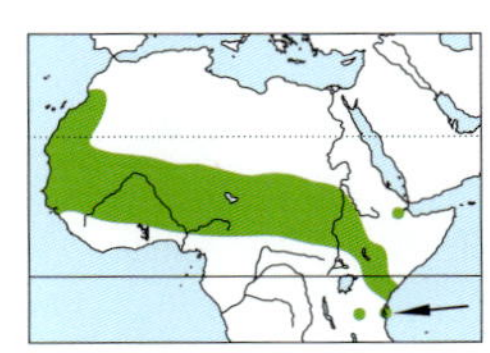

S MA & extreme W DZ S to GN, E to ET & KE, Unguja I (where perhaps extinct); isolated in N ET & C TZ

Savanna White-toothed Shrew
Crocidura fulvastra

FR Crocidure jaunâtre
DE Savannen-Weißzahnspitzmaus
ES Musaraña de sabana

LC
Monotypic
HB 8.6–12 cm
T 4.9–6.4 cm
W 10–29 g

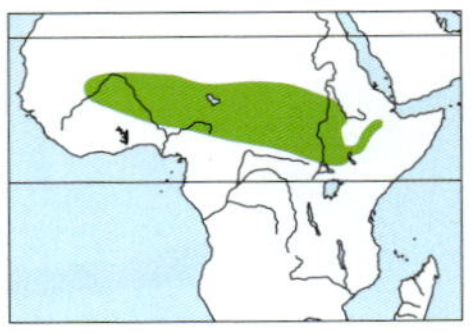

Sahelian region from ML E to W & C ET; mummified specimen from Thebes (EG)

DD
Monotypic
HB c. 11.5 cm T c. 6.4 cm

Ugandan White-toothed Shrew
Crocidura mutesae

FR Crocidure de Kampala
DE Uganda-Weißzahnspitzmaus
ES Musaraña de Uganda

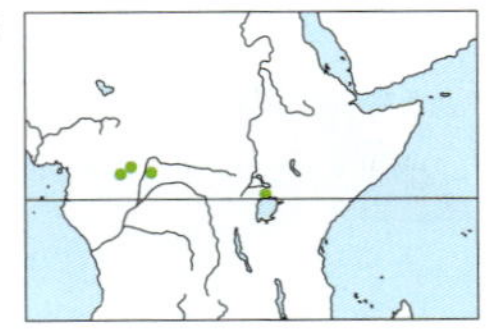

SW CF, NW CD, SE UG

LC
Monotypic
HB 6.8–8.5 cm
T 3.2–5.2 cm
W 8–12 g

Small-footed White-toothed Shrew
Crocidura parvipes

FR Crocidure à pattes courtes
DE Kleinfuß-Weißzahnspitzmaus
ES Musaraña de pies pequeños

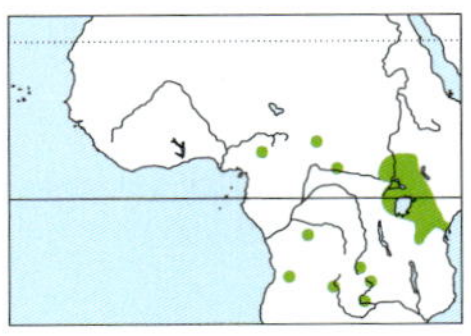

E Africa from S SS to N TZ, scattered in CM, CF, AO & ZM; rather unknown

LC
Monotypic
HB 5–6.7 cm
T 4.5–4.9 cm
W c. 5.6 g

Cinderella White-toothed Shrew
Crocidura cinderella

FR Crocidure cendrillon
DE Cinderella-Weißzahnspitzmaus
ES Musaraña cenicienta

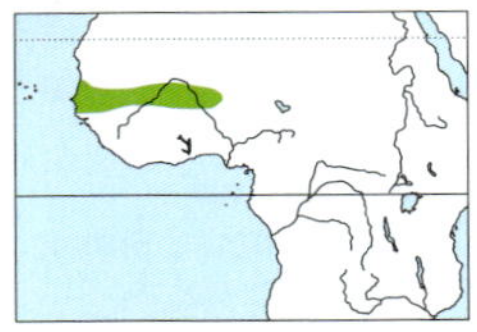

WC Africa in S MR, SN, GM, ML, N BF & W NE

LC
Monotypic
HB 7.5–8.6 cm
T 3.4–5.3 cm
W c. 13.4 g

Voi White-toothed Shrew
Crocidura voi

FR Crocidure de Voi
DE Voi-Weißzahnspitzmaus
ES Musaraña de Voi

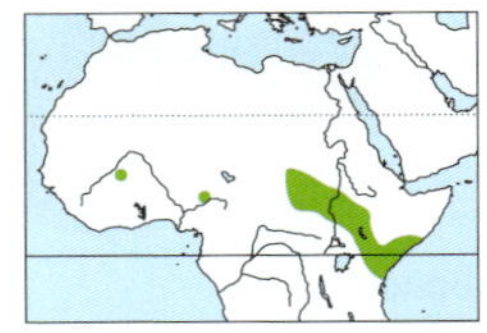

S SD, SS, SW ET, KE, S SO, 2 isolated records from SC ML & E NG

LC
Monotypic
HB 5.2–6.6 cm
T 3.4–4.1 cm
W c. 4.5 g

Yankari White-toothed Shrew
Crocidura yankariensis

FR Crocidure de Yankari
DE Yankari-Weißzahnspitzmaus
ES Musaraña de Yankari

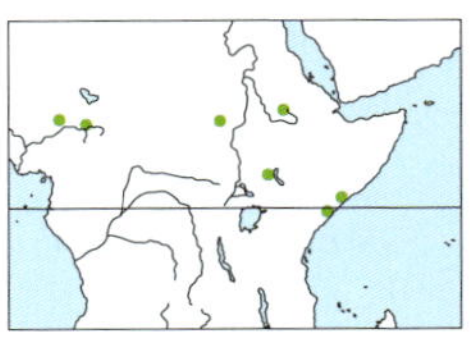

Scattered in NE NG, N CM, S SD, NW ET, NW KE & S SO

1500–2380 m

Montane swamps in E CD (Mt Kahuzi) & SW UG (Echuya Forest, Bwindi Impenetrable NP)

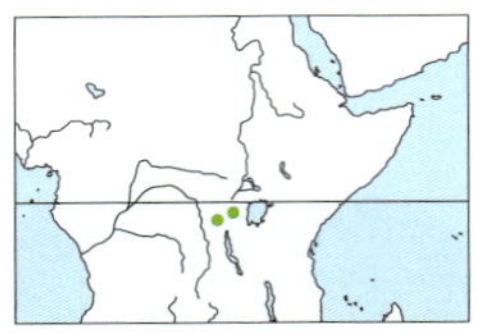

EN
Monotypic
HB 8.2–10.6 cm
T 6.1–8.1 cm
W 16–35 g

Kahuzi White-toothed Swamp Shrew
Crocidura stenocephala

FR Crocidure du Kahuzi
DE Kahuzi-Weißzahnspitzmaus
ES Musaraña de Kahuzi

EN
Monotypic
HB c. 9.1 cm
T c. 5.7 cm

CE Africa in SC & SW UG, EC CD & NW RW

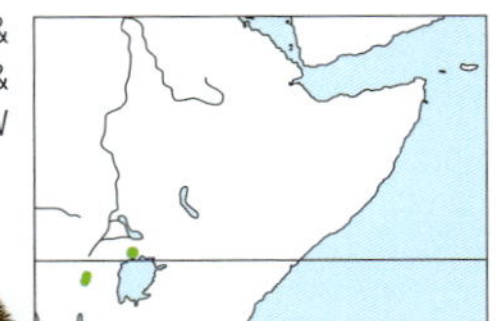

Tarella White-toothed Shrew
Crocidura tarella

FR Crocidure de Rutshuru
DE Tarella-Weißzahnspitzmaus
ES Musaraña de Tarella

LC
Monotypic
HB 6.3–6.9 cm
T 5.3–6 cm
W c. 6.5 g

Dramatic White-toothed Shrew
Crocidura ludia

FR Crocidure belluaire
DE Drama-Weißzahnspitzmaus
ES Musaraña cuidadora

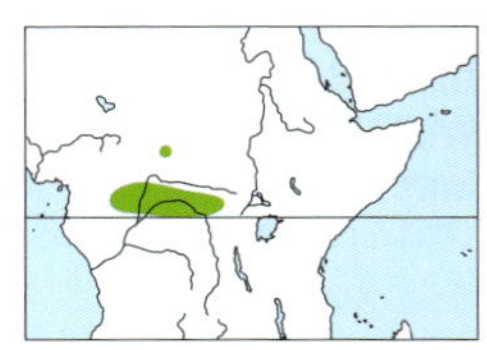

Restricted to lowland rainforests of CD & CF; perhaps CG & CM

LC
Monotypic
HB 5.3–8.5 cm
T 4.8–6.9 cm
W 6–12 g

600–2200 m

Jackson's White-toothed Shrew
Crocidura jacksoni

FR Crocidure de Jackson
DE Jackson-Weißzahnspitzmaus
ES Musaraña de Jackson

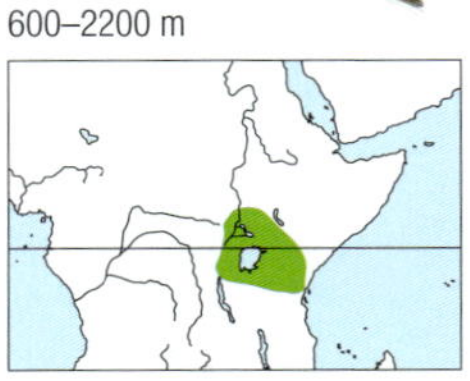

E Africa in E CD, S SS, UG, KE, RW, BI & N TZ

VU
Monotypic
HB 8.7–10.5 cm
T 6.9–8.5 cm
W 18–34 g

Kivu Long-haired White-toothed Shrew
Crocidura lanosa

FR Crocidure laineuse
DE Langfell-Weißzahnspitzmaus
ES Musaraña de pelo largo de Kivu

1850–2450 m

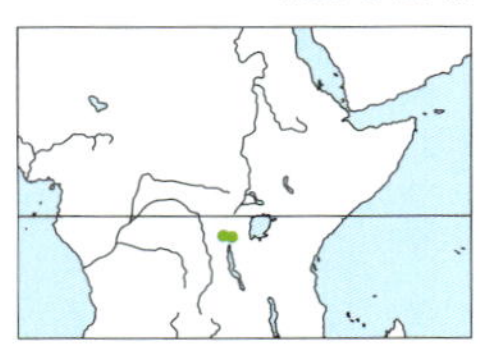

Montane areas of EC CD (Kivu, Lemera, Irangi, Kahuzi-Biéga NP, Tshibati) & RW (Uinka & Nyungwe Forest)

S NG, S CM, S CF, GQ (Mbini, Bioko I), N GA, CG, CD, W RW, W BI

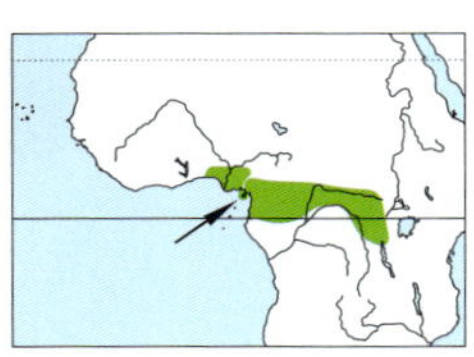

Long-tailed White-toothed Shrew
Crocidura dolichura

FR Crocidure à longue queue
DE Langschwanz-Weißzahnspitzmaus
ES Musaraña de cola larga

LC
Monotypic
HB 6–7 cm
T 7.3–7.5 cm
W 4–7 g

Savanna White-toothed Swamp Shrew
Crocidura longipes

FR Crocidure à pattes longues
DE Feuchtsavannen-Weißzahnspitzmaus
ES Musaraña de ciénaga y sabana

DD
Monotypic
HB 9.5–11.1 cm
T 5.6–6.9 cm
W 15–24 g

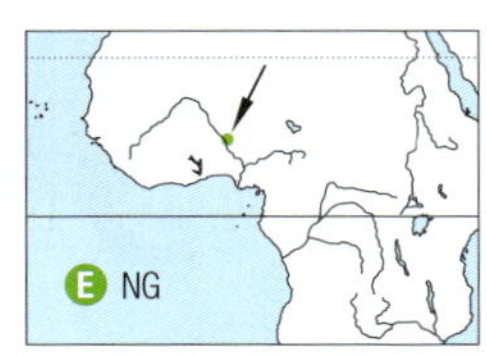

Known from 2 swamp areas near Kainji NP (Dada & Iella)

EN

Monotypic
HB 7.5–7.6 cm
T 4.3–4.7 cm
W 12–13 g

Cameroon White-toothed Shrew
Crocidura picea
FR Crocidure bitumée
DE Kamerun-Weißzahnspitzmaus
ES Musaraña de Camerún

1200–1800 m

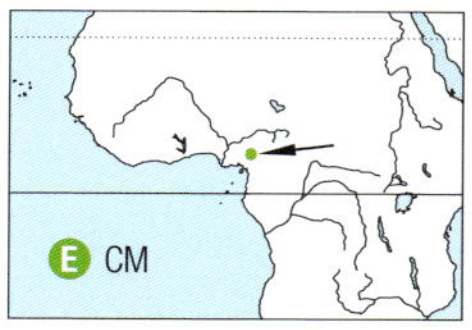

Bamenda Highlands & Mt Oku

VU

Monotypic
HB 7.7–8.7 cm
T 6.1–6.2 cm
W 13–15 g

Manenguba White-toothed Shrew
Crocidura manengubae
FR Crocidure du Manengouba
DE Manenguba-Weißzahnspitzmaus
ES Musaraña de Manenguba

1000–1800 m

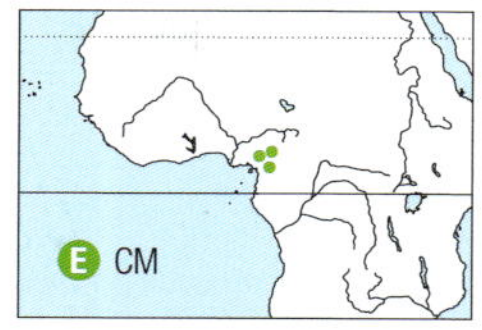

Bamenda (L Manenguba) & Adamaoua highlands; perhaps Yaoundé

VU

Monotypic
HB 6.8–7.8 cm
T 5.3–6.2 cm
W 8–11 g

Eisentraut's White-toothed Shrew
Crocidura eisentrauti
FR Crocidure d'Eisentraut
DE Eisentraut-Weißzahnspitzmaus
ES Musaraña de Eisentraut

1850–3000 m

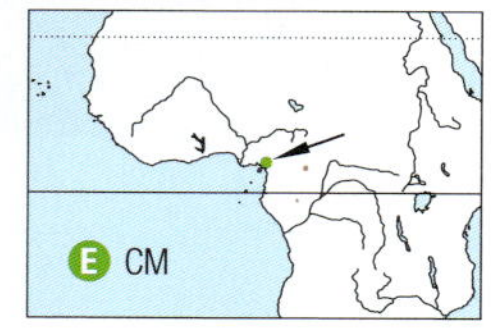

Restricted to top of Mt Cameroon

DD

Monotypic
HB 11.2–12.3 cm
T 4.8–5.6 cm

Upemba White-toothed Shrew
Crocidura zimmeri
FR Crocidure de Zimmer
DE Upemba-Weißzahnspitzmaus
ES Musaraña de Upemba

1750–1800 m

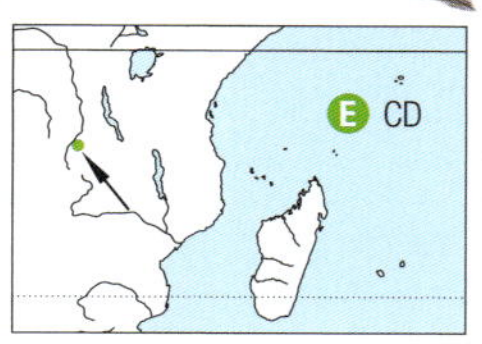

Upemba NP

DD

Monotypic
HB c. 6.5 cm T c. 4.5 cm
W c. 5.6 g

Misotshi-Kabogo White-toothed Shrew
Crocidura lwiroensis
FR Crocidure de Misotchi-Kabogo
DE Lwiro-Weißzahnspitzmaus
ES Musaraña de Misotshi-Kabogo

c. 1250 m

Known from type locality in W bank of L Tanganyika

LC

Monotypic
HB 6.5–7.4 cm
T 4.5–5.2 cm
W 7–9.6 g

African Dusky White-toothed Shrew
Crocidura caliginea
FR Crocidure de Medje
DE Dunkle Kongo-Weißzahnspitzmaus
ES Musaraña de África oscura

c. 750 m

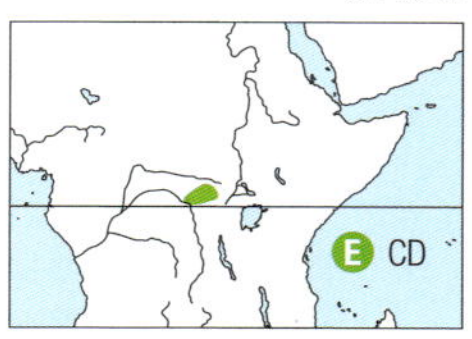

Masako Forest, Epulu Forest, Kisangani (with Kungulu I)

Congo White-toothed Shrew
Crocidura congobelgica
FR Crocidure du Congo
DE Kongo-Weißzahnspitzmaus
ES Musaraña del Congo

LC

Monotypic
HB 7.4–7.9 cm
T c. 5.9 cm W c. 17 g

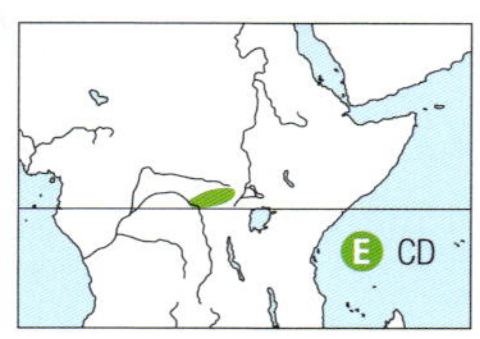

Ituri Forest, Medje, Okapi Wildlife R, Masako Forest

Latona's White-toothed Shrew
Crocidura latona
FR Crocidure léto
DE Latona-Weißzahnspitzmaus
ES Musaraña de Latona

LC

Monotypic
HB 6.5–7.6 cm
T c. 5.9 cm
W c. 6.6 g

up to 500 m

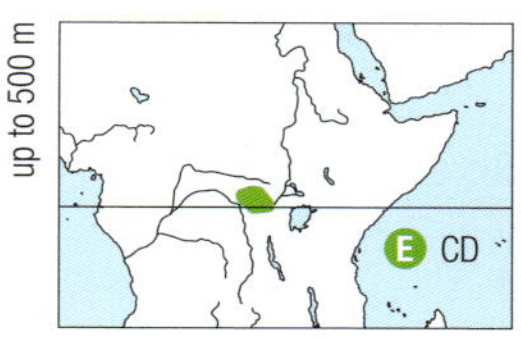

Medje, Okapi Wildlife R, Masako Forest, Kisangani (with Mbiye Is)

Polia's White-toothed Shrew
Crocidura polia
FR Crocidure grise
DE Polia-Weißzahnspitzmaus
ES Musaraña de Polia

DD

Monotypic
HB c. 5.8 cm
T c. 7.2 cm

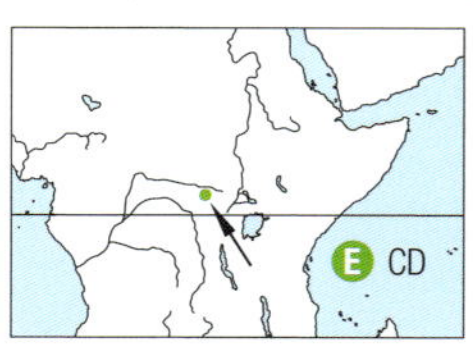

Known only from Medje

Ansells' White-toothed Shrew
Crocidura ansellorum
FR Crocidure des Ansell
DE Ansell-Weißzahnspitzmaus
ES Musaraña de los Ansell

EN

Monotypic
HB 5.6–5.8 cm T 4.1–5.1 cm
W 4.2–5 g

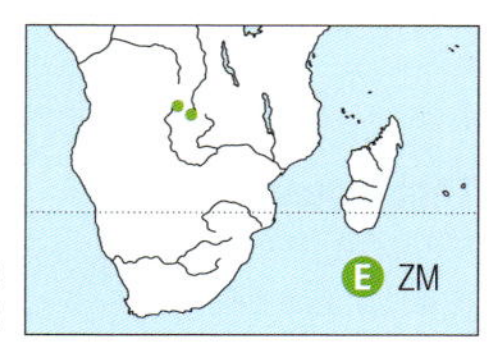

Restricted to Ikelenge Pedicle; perhaps adjacent CD & E AO

Pitman's White-toothed Shrew
Crocidura pitmani
FR Crocidure de Pitman
DE Pitman-Weißzahnspitzmaus
ES Musaraña de Pitman

DD

Monotypic
HB c. 5 cm
T c. 3.6 cm

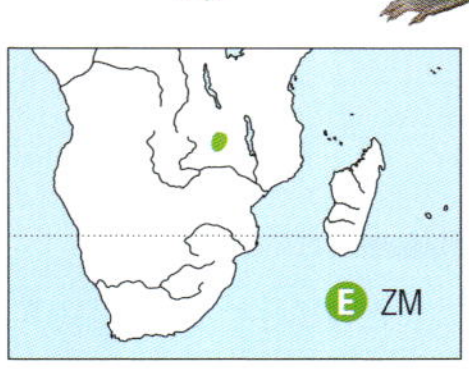

Known from type locality (Maluwe-Sarengi District); perhaps across N & C

LC

Monotypic
HB 5.4–7.4 cm
T 4.3–4.6 cm

Makwassie White-toothed Shrew
Crocidura maquassiensis
FR Crocidure de Maquassi
DE Makwassie-Weißzahnspitzmaus
ES Musaraña de Makwassie

up to 1500 m

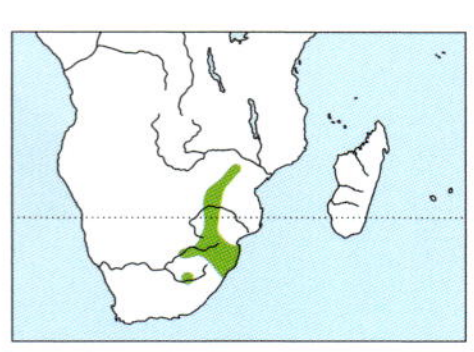

E & S ZW, E ZA, SZ

Genus *SURDISOREX*

(3 species)

Aberdare Mole Shrew
Surdisorex norae
FR Musaraigne de Nora
DE Aberdare-Wühlspitzmaus
ES Musaraña topo de Aberdare

2700–3350 m

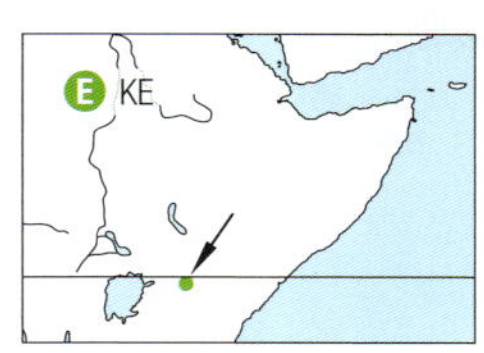

Aberdare Range

LC

Monotypic
HB 8.2–11 cm
T 2–3.5 cm
W 22–27 g

Mount Kenya Mole Shrew
Surdisorex polulus

FR Musaraigne du Kenya
DE Mount-Kenya-Wühlspitzmaus
ES Musaraña topo de Monte Kenya

DD
Monotypic
HB 8.5–10 cm T 2.4–3.2 cm
W 16–21 g

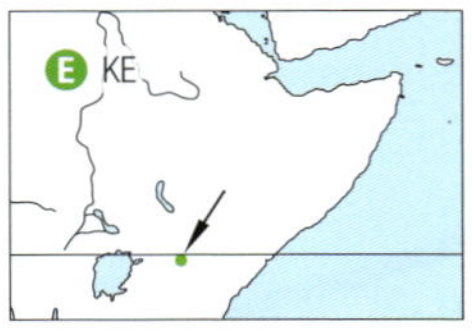

c. 3260 m

Mt Kenya

Schlitter's Mole Shrew
Surdisorex schlitteri

DD
Monotypic
HB c. 8.4 cm T c. 2.9 cm

FR Musaraigne de Schlitter
DE Mount-Elgon-Wühlspitzmaus
ES Musaraña topo de Schlitter

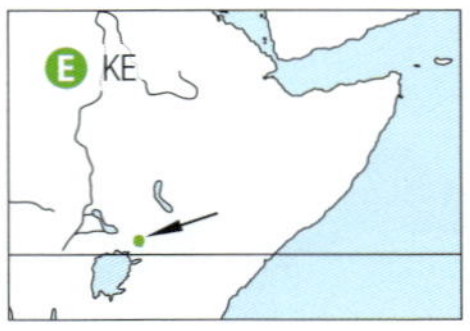

c. 3150 m

Mt Elgon

Genus *CONGOSOREX*
(3 species)

CR
Monotypic
HB 6.8–7.6 cm
T 3.5–4 cm
W 7.4–8.3 g

Phillips' Congo Shrew
Congosorex phillipsorum

FR Musaraigne des Udzungwa
DE Phillips-Kongo-Spitzmaus
ES Musaraña del Congo de los Phillips

1500–2085 m

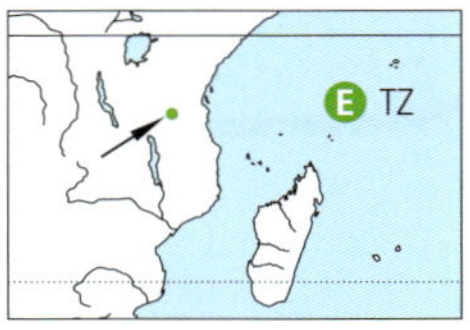

Udzungwa Mts (Ndundulu & Nyumbanitu forests)

DD
Monotypic
HB c. 6 cm
T c. 2.4 cm

Greater Congo Shrew
Congosorex polli

FR Musaraigne de Poll
DE Große Kongo-Spitzmaus
ES Musaraña del Congo mayor

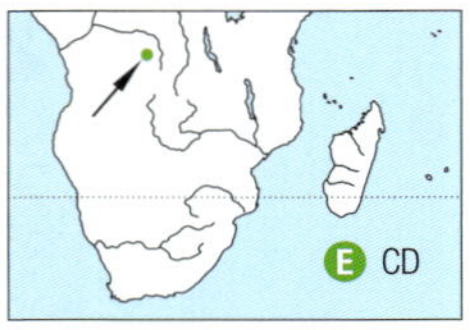

Known from type locality (Kasaï-Central Prov)

LC
Monotypic
HB 5.3–9.5 cm
T 1.9–2.6 cm
W 4–12 g

Lesser Congo Shrew
Congosorex verheyeni

FR Musaraigne de Verheyen
DE Kleine Kongo-Spitzmaus
ES Musaraña del Congo menor

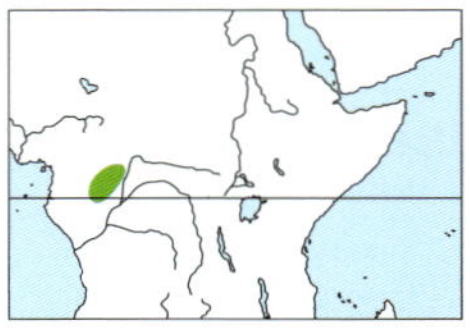

CG (Odzala-Kokoua NP), SW CF (Ngotto Forest)

Genus *MYOSOREX*
(19 species)

LC Monotypic
HB 7.5–9.4 cm
T 3.4–4.5 cm
W 6.9–20 g

1120–1700 m

Meester's Mouse Shrew
Myosorex meesteri

FR Musaraigne de Meester
DE Meester-Mausspitzmaus
ES Musaraña ratón de Meester

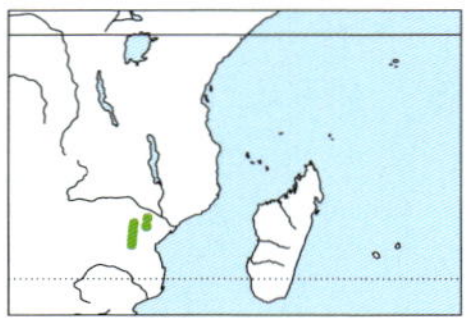

ZW (E highlands), C MZ (Mt Gorongosa)

VU
Monotypic
HB 7.8–10.5 cm T 3.6–5.1 cm
W 9–17 g

Dark-footed Mouse Shrew
Myosorex cafer

FR Musaraigne à pattes sombres
DE Dunkelfuß-Mausspitzmaus
ES Musaraña ratón de pies oscuros

up to 2000 m

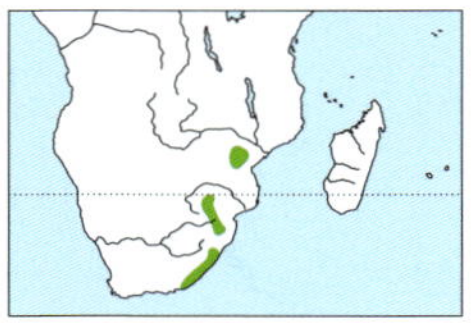

EC ZW, WC MZ, E ZA & SZ

VU
Monotypic
HB 8.1–11.4 cm
T 4.6–5.7 cm

Sclater's Mouse Shrew
Myosorex sclateri

FR Musaraigne du Kwazulu
DE Sclater-Mausspitzmaus
ES Musaraña ratón de Sclater

KwaZulu-Natal

Thin Mouse Shrew
Myosorex tenuis

FR Musaraigne fine
DE Zierliche Mausspitzmaus
ES Musaraña ratón fina

EN
Monotypic
HB 7.5–9.2 cm T 3–4.5 cm
W 7.1–20 g

up to 1600 m

Limpopo, Mpumalanga

South African Mouse Shrew
Myosorex varius

FR Musaraigne sud-africaine
DE Südafrika-Mausspitzmaus
ES Musaraña ratón de Suráfrica

LC
Monotypic
HB 7.6–8.8 cm T 3.8–4.8 cm
W 6–11 g

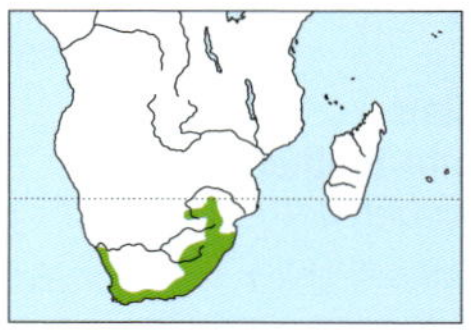

0–3000 m

ZA, LS, W SZ

EN
2 sspp
HB 7.3–9.3 cm
T 4.6–7.9 cm
W 9–21 g

240–3600 m

Knysna Forest, Boosmansbos Forest, Langeberg Mts

Long-tailed Mouse Shrew
Myosorex longicaudatus

FR Musaraigne de Knysna
DE Langschwanz-Mausspitzmaus
ES Musaraña ratón de cola larga

Kilimanjaro Mouse Shrew
Myosorex zinki

FR Musaraigne de Zink
DE Kilimanjaro-Mausspitzmaus
ES Musaraña ratón del Kilimanjaro

LC
Monotypic
HB 8.4–10 cm
T 3.3–3.7 cm
W 14.5–19.5 g

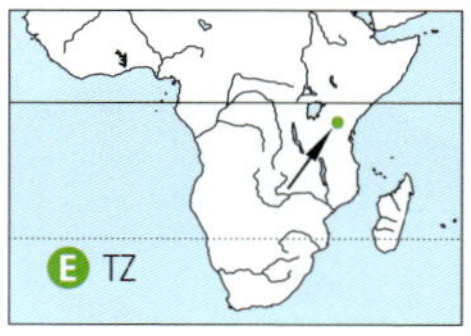

2470–4000 m

Mt Kilimanjaro

EN
Monotypic
HB 6.8–7.5 cm
T 4–4.3 cm
W 7.6–10.5 g

600–2250 m

Geata Mouse Shrew
Myosorex geata

FR Musaraigne des Uluguru
DE Uluguru-Mausspitzmaus
ES Musaraña ratón de Geata

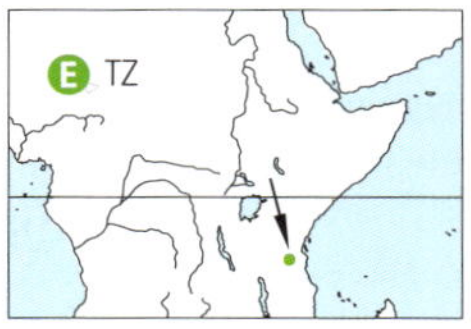

Uluguru Mts

EN
Monotypic
HB 7–8.4 cm
T 3.6–4.6 cm
W 8–12.5 g

900–2000 m

Kihaule's Mouse Shrew
Myosorex kihaulei

FR Musaraigne de Kihaule
DE Kihaule-Mausspitzmaus
ES Musaraña ratón de Kihaule

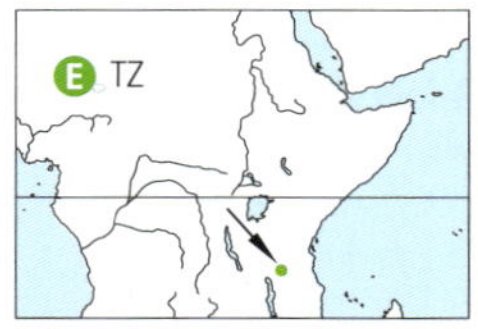

Udzungwa Mts

EN
Monotypic
HB c. 7.5 cm
T c. 4.1 cm
W c. 8 g

c. 2285 m

Nyika Mouse Shrew
Myosorex gnoskei

FR Musaraigne de Gnoske
DE Gnoske-Mausspitzmaus
ES Musaraña ratón de Nyika

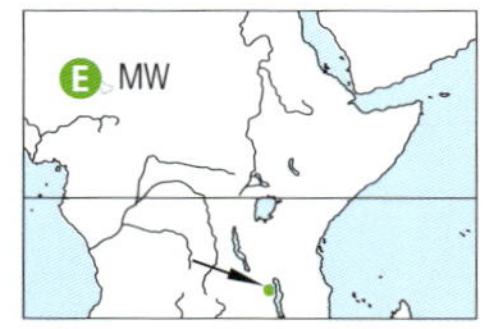

Nyika NP

DD
Monotypic
HB 5–5.5 cm
T c. 4.4 cm

Schaller's Mouse Shrew
Myosorex schalleri

FR Musaraigne de Schaller
DE Schaller-Mausspitzmaus
ES Musaraña ratón de Schaller

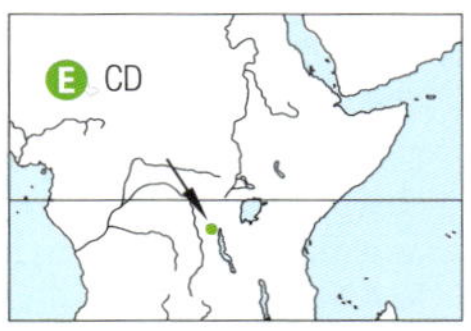

Itombwe Mts

DD
Monotypic
HB c. 8.9 cm
T c. 2.8 cm
W c. 12 g

1950–2000 m

Kabogo Mouse Shrew
Myosorex kabogoensis

FR Musaraigne de Misotchi-Kabogo
DE Kabogo-Mausspitzmaus
ES Musaraña ratón de Kabogo

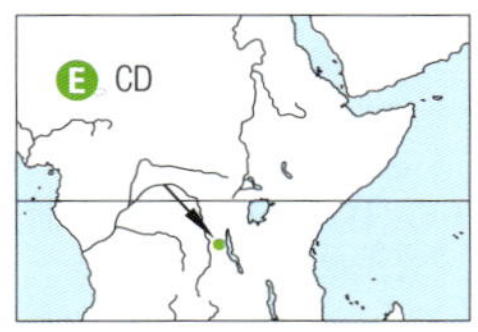

Mt Misotshi

LC
Monotypic
HB 7.9–9.1 cm
T 3.4–4 cm
W 12–22 g

1850–2300 m

Babault's Mouse Shrew
Myosorex babaulti

FR Musaraigne de Babault
DE Babault-Mausspitzmaus
ES Musaraña ratón de Babault

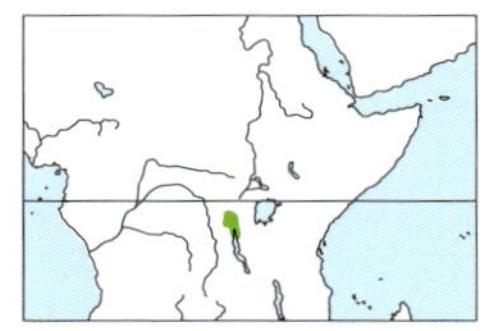

EC CD (with Idjwi I in L Kivu), RW, BI, SW UG (Bwindi Impenetrable NP)

EN
Monotypic
HB 8.1–9.3 cm
T 3.1–4 cm
W 13.5–18 g

1800–4000 m

Montane Mouse Shrew
Myosorex blarina

FR Musaraigne du Rwenzori
DE Ruwenzori-Mausspitzmaus
ES Musaraña ratón de montaña

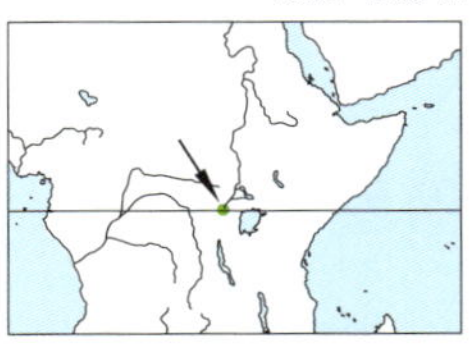

Rwenzori Mts (NE CD, SW UG); perhaps SW UG (Kibale)

VU
Monotypic
HB c. 8.8 cm
T c. 2.6 cm
W c. 14 g

c. 1880 m

Bururi Mouse Shrew
Myosorex bururiensis

FR Musaraigne du Bururi
DE Bururi-Mausspitzmaus
ES Musaraña ratón de Bururi

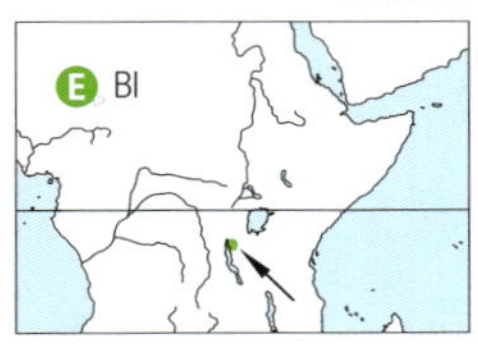

Bururi Forest R

VU
Monotypic
HB 5.3–7.8 cm
T 4.2–4.9 cm
W 6–8.5 g

2200–3300 m

Kahuzi Mouse Shrew
Myosorex jejei

FR Musaraigne de Jeje
DE Kahuzi-Mausspitzmaus
ES Musaraña ratón de Kahuzi

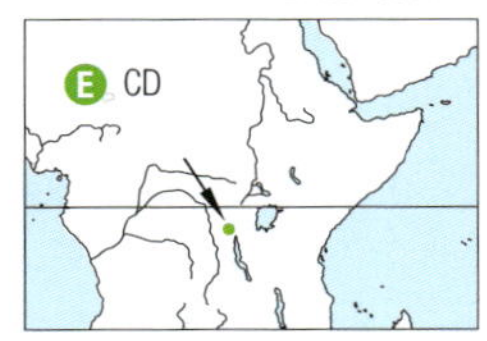

Kahuzi-Biéga NP

CR
Monotypic
HB 7.1–8.2 cm
T 3.7–4.2 cm
W 12–15 g

1400–3000 m

Eisentraut's Mouse Shrew
Myosorex eisentrauti

FR Musaraigne d'Eisentraut
DE Eisentraut-Mausspitzmaus
ES Musaraña ratón de Eisentraut

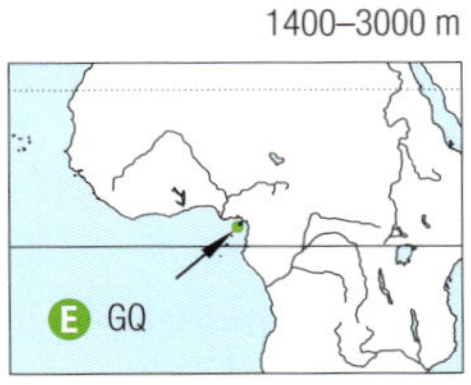

Bioko I

EN
Monotypic
HB c. 8.4 cm
T c. 3.5 cm
W c. 15 g

c. 1100 m

Rumpi Mouse Shrew
Myosorex rumpii

FR Musaraigne des Rumpi
DE Rumpi-Mausspitzmaus
ES Musaraña ratón de Rumpi

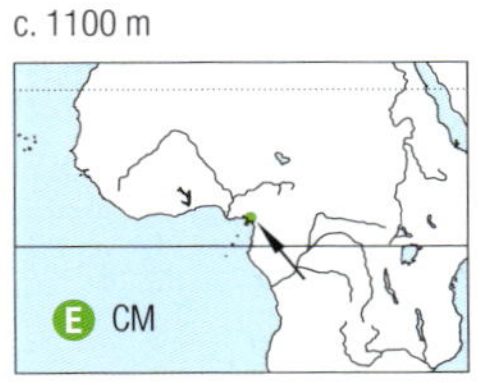

Rumpi Hills

VU
Monotypic
HB 7.2–7.3 cm
T 4–4.1 cm
W 11–12 g

1800–2300 m

Oku Mouse Shrew
Myosorex okuensis

FR Musaraigne de l'Oku
DE Oku-Mausspitzmaus
ES Musaraña ratón de Oku

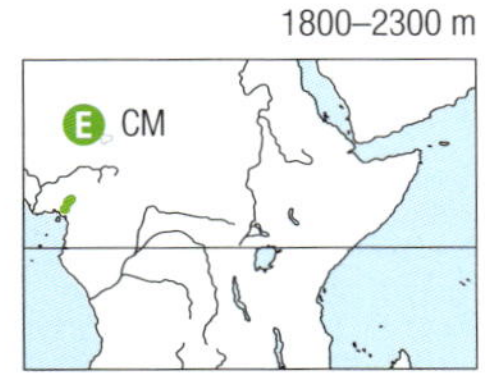

Bamenda Highlands

Family TALPIDAE – Moles, Desmans, Star-nosed Mole and Shrew Moles

19 genera, 60 species. Various terrestrial and aquatic biomes from boreal zones to tropics. Holarctic and Indo-Malayan Regions.

Genus *UROPSILUS*

(8 species)

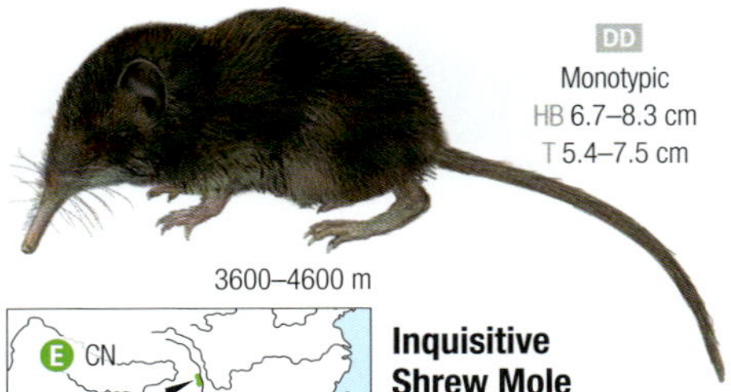

DD
Monotypic
HB 6.7–8.3 cm
T 5.4–7.5 cm

3600–4600 m

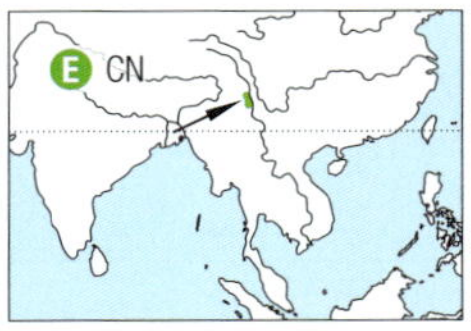

Inquisitive Shrew Mole

Uropsilus investigator

FR Taupe du Yunnan
DE Yunnan-Spitzmausmaulwurf
ES Topo musaraña de Yunnan

NW Yunnan; perhaps NE MM

Snow Mountain Shrew Mole

Uropsilus nivatus

FR Taupe du Xue Shan
DE Yulong-Spitzmausmaulwurf
ES Topo musaraña de alta montaña

NE
Monotypic
HB 6.8–7.4 cm T 5.7–6.5 cm
W 5.9–7.5 g

3000–3600 m

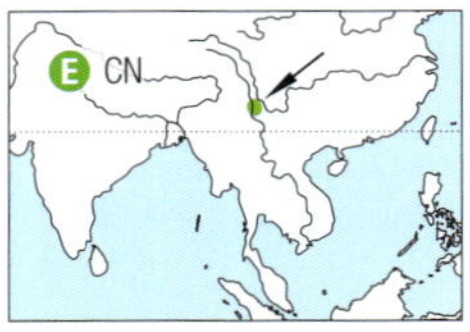

NW Yunnan (E of R Salween, near to Lijiang and N of Dali); perhaps N MM

NE
Monotypic
HB 7.2–8.2 cm
T 6.7–7.3 cm
W c. 11 g

Equivalent-teeth Shrew Mole

Uropsilus aequodonenia

FR Taupe équidonte
DE Gleichzahn-Spitzmausmaulwurf
ES Topo musaraña de Sichuán

2430–3100 m

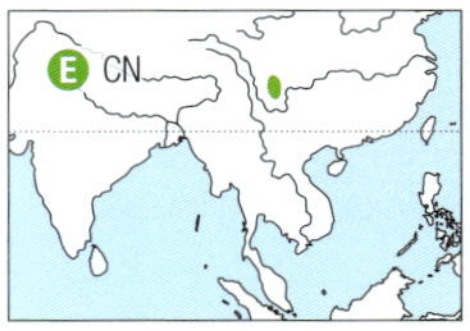

SW Sichuan (small area surrounded by Dadu, Jinsha & Yalu rivers)

DD
Monotypic
HB 6.5–8.3 cm
T 5.9–7.2 cm
W 7.4–11.2 g

Anderson's Shrew Mole

Uropsilus andersoni

FR Taupe d'Anderson
DE Anderson-Spitzmausmaulwurf
ES Topo musaraña de Anderson

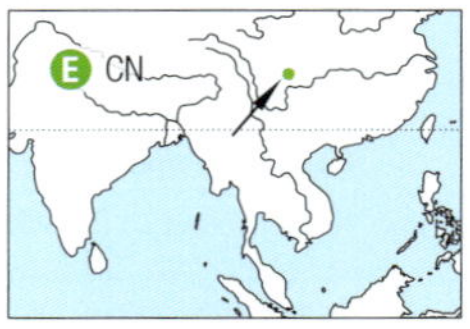

C Sichuan (Mt Emei and adjacent area)

LC
Monotypic
HB c. 6.6 cm (F)
T c. 5.5 cm (F)

Gracile Shrew Mole

Uropsilus gracilis

FR Taupe gracile
DE China-Spitzmausmaulwurf
ES Topo musaraña esbelto

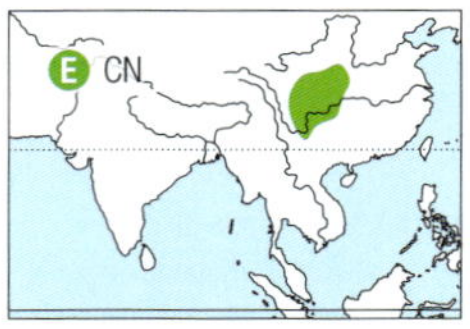

Shaanxi, Sichuan, Chongqing, Yunnan

NE
Monotypic
HB c. 6.7 cm (F)
T c. 5.7 cm (F)

Black-backed Shrew Mole

Uropsilus atronates

FR Taupe à croupe noire
DE Schwarzrücken-Spitzmausmaulwurf
ES Topo musaraña de dorso negro

c. 2135 m

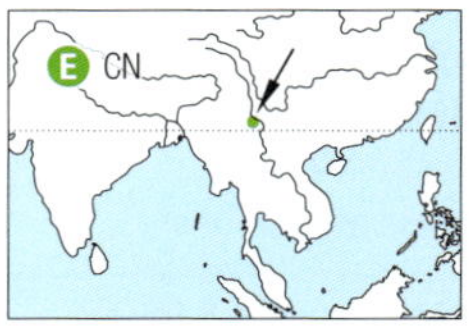

C Yunnan (region S of Dali, between Salween and Mekong rivers)

Dabie Mountains Shrew Mole

Uropsilus dabieshanensis

FR Taupe des Dabie
DE Dabieshan-Spitzmausmaulwurf
ES Topo musaraña de Dabie

NE
Monotypic
HB 6.6–8.2 cm
T 5.2–5.4 cm
W 6.2–8.9 g

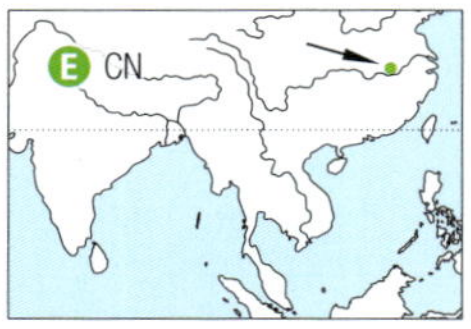

1065–1275 m

Anhui (Yaoluoping National NR, Huoshan Foziling Provincial NR)

Chinese Shrew Mole

Uropsilus soricipes

FR Taupe du Sichuan
DE Sichuan-Spitzmausmaulwurf
ES Topo musaraña de China

LC
Monotypic
HB 6.6–8 cm T 5–6.9 cm

below 2200 m

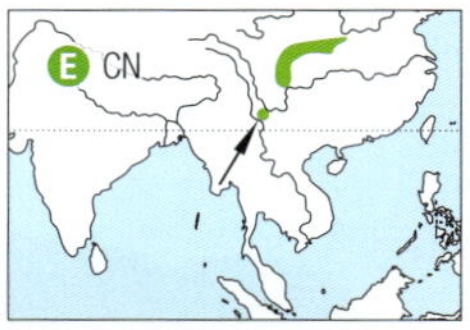

Gansu, Shaanxi, Sichuan, Yunnan

Genus *SCAPANULUS*

(1 species)

Gansu Mole

Scapanulus oweni

FR Taupe du Gansu
DE Gansu-Maulwurf
ES Topo de Gansu

LC
Monotypic
HB 8–9.1 cm
T 3.5–4.9 cm

1500–3000 m

Qinghai, Gansu, Shaanxi, Sichuan, Chongqing, Hubei

Genus *ALPISCAPTULUS*

(1 species)

Medog Mole

Alpiscaptulus medogensis

FR Taupe du Mêdog
DE Medog-Maulwurf
ES Topo alpino de Medog

NE
Monotypic
HB 8.9.–11.1 cm
T 4–4.2 cm
W 26–35 g

2400–3700 m

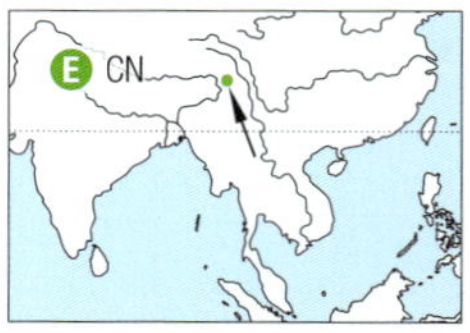

Known only from Mt Namjagbarwa (Medog County, SE Tibet)

Genus *PARASCALOPS*

(1 species)

LC
Monotypic
HB 12.5–14 cm
T 2.6–3.3 cm
W 41–63 g

Hairy-tailed Mole

Parascalops breweri

FR Taupe à queue velue
DE Haarschwanzmaulwurf
ES Topo de cola peluda

0–1950 m

SE CA (S Ontario, S Quebec), NE US (S to E Tennessee & extreme NE Georgia)

Genus *SCAPANUS*

(5 species)

Coast Mole
Scapanus orarius

FR Taupe du Pacifique
DE Pazifischer Maulwurf
ES Topo costero del Pacífico

LC
2 sspp
HB 10.6–16.8 cm
T 2.1–4.6 cm
W 61–91 g

up to 350 m

SW CA (SW British Columbia), NW US (Washington, Oregon, NW California, marginally WC Idaho)

Townsend's Mole
Scapanus townsendii

FR Taupe de Townsend
DE Townsend-Maulwurf
ES Topo de Townsend

LC
2 sspp
HB 16.1–18.6 cm
T 3.4–5.1 cm
W 64–171 g

0–1675 m

SW CA (SW British Columbia), US (W Washington, W Oregon, NW California)

LC
4 sspp
HB 13.1–15.3 cm
T 2.3–3.9 cm
W 56–74 g

Northern Broad-footed Mole
Scapanus latimanus

FR Taupe de Californie
DE Nördlicher Kalifornien-Maulwurf
ES Topo de pies anchos septentrional

up to 3000 m

From SC Oregon S to Santa Barbara & Yosemite Valley (California)

NE
Monotypic
HB 11.1–13.6 cm
T 2.2–3.4 cm
W 50–59 g

Southern Broad-footed Mole
Scapanus occultus

FR Taupe de Grinnell
DE Südlicher Kalifornien-Maulwurf
ES Topo de pies anchos meridional

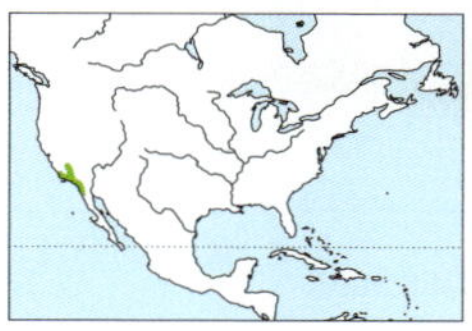

US (California S from Yosemite Valley & Santa Barbara), MX (Baja California S to Laguna Hanson)

Mexican Mole
Scapanus anthonyi

FR Taupe d'Anthony
DE Baja-California-Maulwurf
ES Topo de México

NE
Monotypic
HB 11–11.4 cm
T 2.1–2.5 cm
W 30–35 g

1150–2000 m

Baja California (known from 4 sites in Sierra de San Pedro Mártir)

Eastern Mole
Scalopus aquaticus

FR Taupe à queue glabre
DE Ostamerikanischer Maulwurf
ES Topo de cola lampiña

Genus *SCALOPUS*

(1 species)

LC
16 sspp
HB 12.9–15.4 cm
T 1.5–3 cm
W 40–64 g

0–1635 m

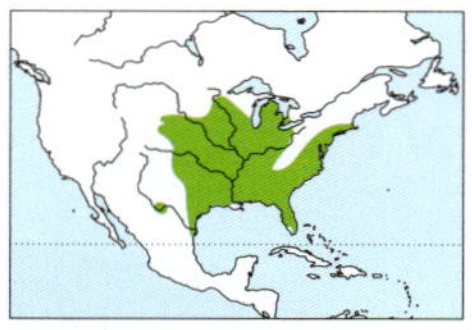

CA (S tip of Ontario), C & E US (S to Texas & Florida), NE MX (N Coahuila, NE Tamaulipas)

Genus *SCAPTONYX*

(1 species)

LC
2 sspp
HB 7.2–9 cm
T 2.6–4.5 cm
W 9.8–13.8 g

Long-tailed Mole
Scaptonyx fusicauda

FR Taupe à longue queue
DE Langschwanzmaulwurf
ES Topo de cola larga

2000–4500 m

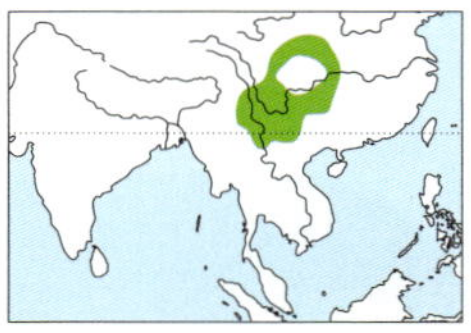

C CN (Shaanxi, Sichuan, Chongqing, SE Qinghai, N Guizhou, Yunnan), N MM, N VN

Genus *UROTRICHUS*

(1 species)

Japanese Shrew Mole
Urotrichus talpoides

FR Taupe du Japon
DE Japan-Spitzmull
ES Topo musaraña de Japón

Honshu, Shikoku, Kyushu, adjacent Is (including Tsushima)

LC
2 sspp
HB 8.9–10.4 cm
T 2.7–3.8 cm
W 14.5–28 g

Genus *DYMECODON*

(1 species)

True's Shrew Mole
Dymecodon pilirostris

FR Taupe de True
DE True-Spitzmull
ES Topo musaraña de True

above 1000 m

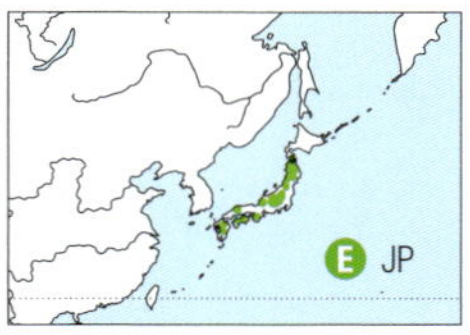

Honshu, Shikoku, Kyushu

LC
Monotypic
HB 7–8.4 cm T 3.2–4.4 cm
W 8–14.5 g

American Shrew Mole
Neurotrichus gibbsii

FR Taupe naine
DE Amerika-Spitzmull
ES Topo musaraña americano

Genus *NEUROTRICHUS*

(1 species)

LC
3 sspp
HB 6.3–7.1 cm
T 2.9–5 cm
W 6–14.5 g

0–2440 m

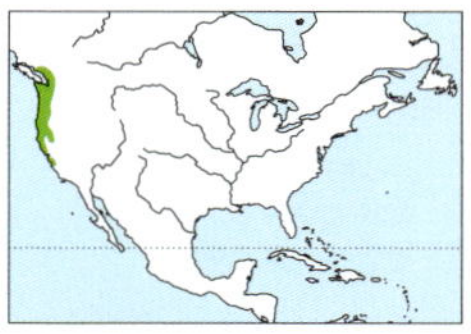

CA (SW British Columbia), NW US (W Washington to W California S to San Benito County, Destruction I)

Genus *CONDYLURA*

(1 species)

LC
2 sspp
HB 9.6–12.8 cm
T 6.5–8.3 cm
W 27–52 g

Star-nosed Mole
Condylura cristata
FR Taupe à nez étoilé
DE Sternmull
ES Topo de nariz estrellada

0–1675 m

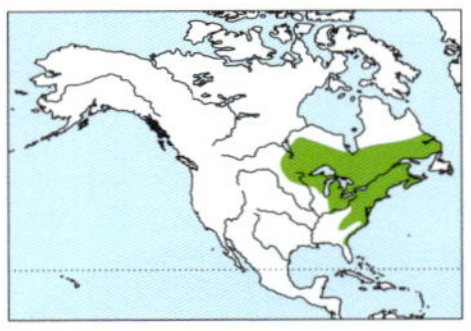

SE CA (from S Manitoba E to Newfoundland), NE US (S to South Carolina, SE Georgia & extreme N Florida)

Genus *DESMANA*

(1 species)

EN
Monotypic
HB 19.3–24 cm
T 18–21 cm
W 370–450 g

Russian Desman
Desmana moschata
FR Desman de Russie
DE Russischer Desman
ES Desmán ruso

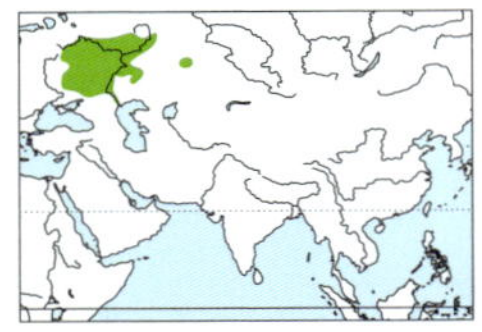

Dnieper, Don, Volga, Kama & Ural basins (RU, E UA, NW KZ); introduced to upper R Ob (W Siberia)

Genus *GALEMYS*

(1 species)

EN
Monotypic
HB 10.8–14.5 cm
T 12.3–16 cm
W 44–80 g

Pyrenean Desman
Galemys pyrenaicus
FR Desman des Pyrénées
DE Pyrenäen-Desman
ES Desmán ibérico

15–2700 m

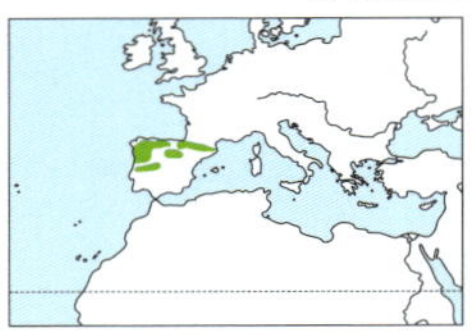

N Iberian Pen in N & C ES, N PT (N of upper R Mondego) & S FR (Pyrenees)

Genus *TALPA*

(13 species)

Altai Mole
Talpa altaica
FR Taupe de Sibérie
DE Altai-Maulwurf
ES Topo de Altái

LC
Monotypic
HB 12.4–23 cm
T 2–3.4 cm
W ≤ 225 g

up to 3500 m

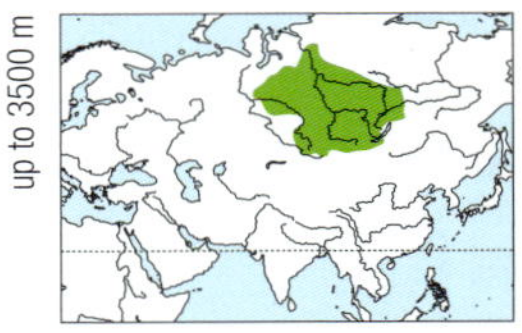

Siberian RU (from R Irtysh E to L Baikal, from R Yenisei S to Altai Mts), NE KZ, NW MN

Pere David's Mole
Talpa davidiana
FR Taupe de David
DE Pater-David-Maulwurf
ES Topo del Padre David

1575–2240 m

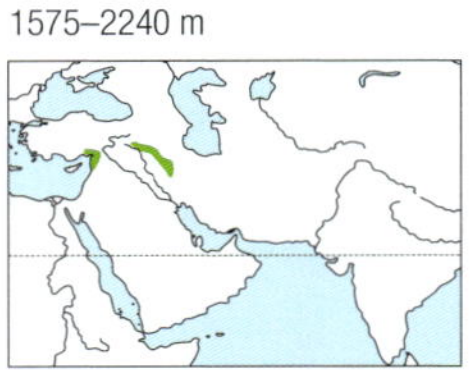

DD
Monotypic
HB 11.5–13.4 cm
T 1.8–3 cm
W 42–80 g

SC & SE TR (Adana, Gaziantep, Bitlis & Hakkari prov), NW IR (Kordestan); perhaps W SY

Talysch Mole
Talpa talyschensis
FR Taupe des Talysh
DE Talysch-Maulwurf
ES Topo de Talysch

0–300 m

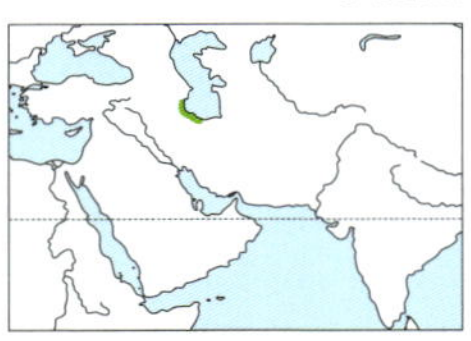

NE
Monotypic
HB 10.4–11.4 cm
T 2–2.5 cm
W 31–49 g

SW Caspian coast between Lenkoran (AZ) and Chalus in W Mazandaran (IR)

Ognev's Mole
Talpa ognevi
FR Taupe d'Ognev
DE Ognev-Maulwurf
ES Topo de Ognev

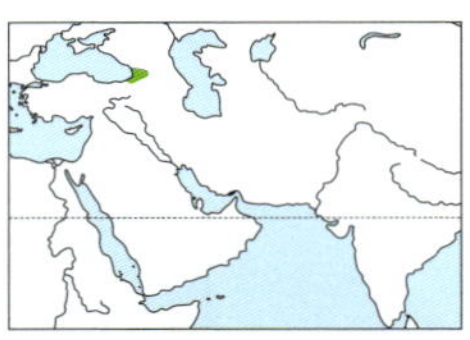

NE
Monotypic
HB 13.4–14.2 cm
T 2–2.6 cm
W 62–91 g

SE Black Sea coast in NE TR (Hopa District) & adjacent GE (inland to upper R Kura); range uncertain

Caucasian Mole
Talpa caucasica
FR Taupe du Caucase
DE Kaukasus-Maulwurf
ES Topo del Cáucaso

up to 2800 m

LC
Monotypic
HB 9.9–14.2 cm
T 2.2–4.2 cm
W 38–102 g

SW Ciscaucasia (from Sea of Azov N to R Yeya) & Greater Caucasus N of line Nalchik–Sukhumi

Levant Mole
Talpa levantis
FR Taupe du Levant
DE Schwarzmeer-Maulwurf
ES Topo oriental

0–2400 m

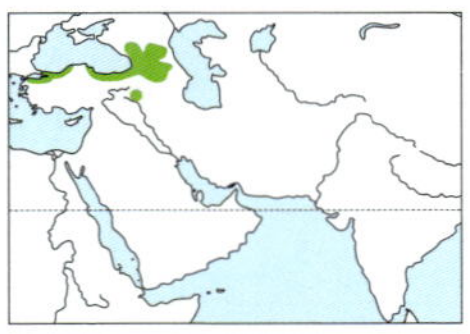

LC
2 sspp
HB 8.2–12.5 cm
T 2–3.7 cm
W 21–67 g

TR (N Anatolia, isolated around L Van), S European RU (S of Kuban & Terek rivers), GE, AM, W AZ

Balkan Mole
Talpa stankovici
FR Taupe des Balkans
DE Balkan-Maulwurf
ES Topo de los Balcanes

0–2300 m

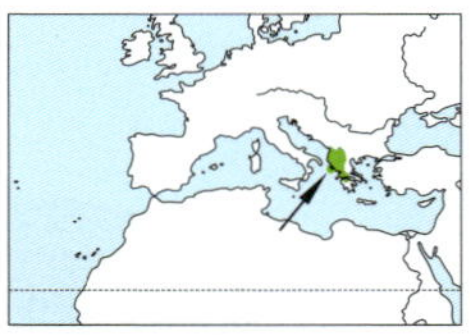

LC
2 sspp
HB 12.1–14.3 cm
T 2.6–4.2 cm
W 48–96 g

W MK, W GR (with Corfu I), coastal lowlands of S ME & adjacent AL

Blind Mole
Talpa caeca
FR Taupe aveugle
DE Blindmaulwurf
ES Topo ciego

0–2500 m

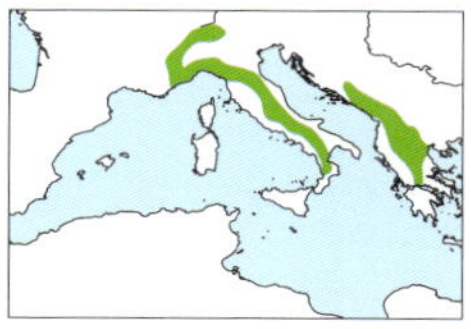

LC
Monotypic
HB 10.3–12.4 cm
T 2.6–4.3 cm
W 38–67 g

W Alps (CH, NW IT, SE FR), Apennine Mts S to Sila Massif (IT), Balkans (from SE BA to W GR)

Roman Mole
Talpa romana
FR Taupe romaine
DE Römischer Maulwurf
ES Topo romano

0–2000 m

LC
Monotypic
HB 9.5–16 cm T 2.2–3.7 cm
W 61–134 g

S from Umbria; presumably extinct in Sicily (last recorded in 1885)

Iberian Mole
Talpa occidentalis
FR Taupe ibérique
DE Spanischer Maulwurf
ES Topo ibérico

0–2300 m

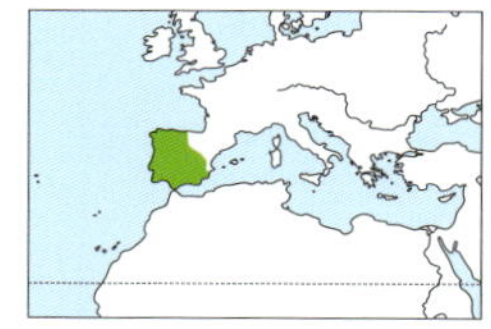

LC
Monotypic
HB 9.6–13 cm T 1.9–3.5 cm
W 34–66 g

Iberian Pen (PT, ES), widespread (but absent in most of E, Pyrenees & Ebro Basin)

Martinos' Mole
Talpa martinorum
FR Taupe des Martino
DE Martino-Maulwurf
ES Topo de los Martino

0–445 m

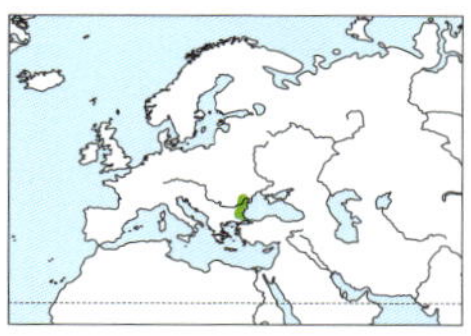

NE
Monotypic
HB 12.4–13.4 cm
T 2.4–3.2 cm
W 45–65 g

SW Black Sea coast from Burgas (BG) to Istanbul (TR)

European Mole
Talpa europaea
FR Taupe d'Europe
DE Maulwurf
ES Topo europeo

0–2700 m

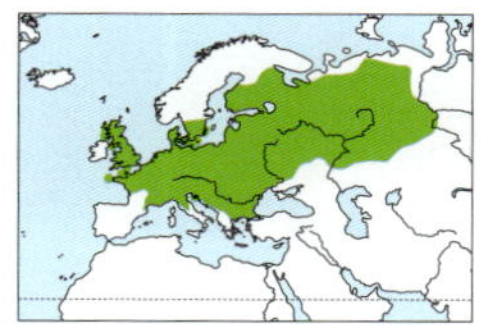

LC
Monotypic
HB 11.3–15.9 cm
T 2.5–4 cm
W 72–128 g

From British Is & FR E through much of continental Europe to W Asian RU (Irtysh & Ob rivers)

Aquitanian Mole
Talpa aquitania
FR Taupe d'Aquitaine
DE Aquitanien-Maulwurf
ES Topo de Aquitania

0–2000 m

NE
Monotypic
HB 11.6–20 cm
T 1.9–3.8 cm W 45–120 g

FR (W & S of R Loire), N & NE Iberian Pen (E Cantabrian coast E to N Catalonia)

Genus *OREOSCAPTOR*

(1 species)

LC
Monotypic
HB 8.2–11.1 cm
T 2–2.6 cm
W 20–32 g

Japanese Mountain Mole
Oreoscaptor mizura
FR Taupe de montagne
DE Japan-Bergmaulwurf
ES Topo de montaña de Japón

500–2400 m

Honshu

Ussuri Mole
Mogera robusta
FR Taupe de l'Oussouri
DE Ussuri-Maulwurf
ES Topo de Ussuri

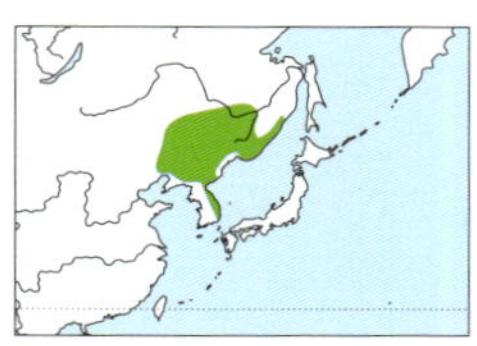

NE CN (Inner Mongolia, Heilongjiang, Jilin, Liaoning), RU (SE Far East), Korean Pen

LC
Monotypic
HB 14.7–16.5 cm
T 1.6–2 cm
W 96–127 g

Genus *MOGERA*

(10 species)

LC
Monotypic
HB 12.3–18 cm
T 1.1–2.9 cm
W 63–178 g

Large Japanese Mole
Mogera wogura
FR Taupe de Temminck
DE Japan-Maulwurf
ES Topo de Japón grande

W Honshu, Shikoku, Kyushu, adjacent Is

Small Japanese Mole
Mogera imaizumii
FR Taupe d'Imaizumi
DE Kleiner Japan-Maulwurf
ES Topo de Japón pequeño

0–2000 m

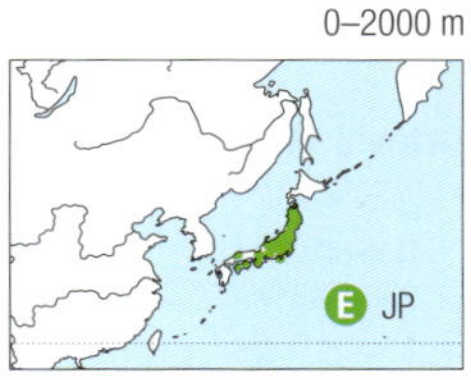

LC
Monotypic
HB 10.2–15.4 cm
T 0.8–2.2 cm
W 36–109 g

Mainly E Honshu & adjacent Is, isolated populations in W Honshu & Shikoku

Sado Mole
Mogera tokudae
FR Taupe de Sado
DE Sado-Maulwurf
ES Topo de Sado

NT
Monotypic
HB 13.1–16.3 cm
T 2.3–2.7 cm
W 82–120 g

Sado I

EN
Monotypic
HB 15.7–17 cm
T 2.3–2.9 cm
W 95–161 g

Echigo Mole
Mogera etigo
FR Taupe d'Echigo
DE Echigo-Maulwurf
ES Topo de Echigo

Honshu (Echigo Plain in Niigata Prefecture)

Senkaku Mole
Mogera uchidai

FR Taupe des Senkaku
DE Ryukyu-Maulwurf
ES Topo de Senkaku

Senkaku Is (Uotsuri-jima)

VU

Monotypic
HB c. 12.3 cm
T c. 1.2 cm
W c. 43 g

Formosan Mole
Mogera insularis

FR Taupe insulaire
DE Formosa-Inselmaulwurf
ES Topo de Taiwán

0–2500 m

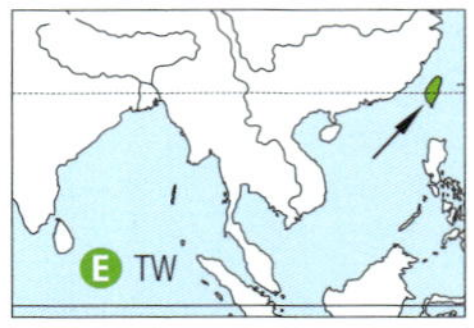

Cultivated farms, riverbanks, forests

LC

Monotypic
HB 11.2–13.4 cm
T 0.6–1.1 cm
W 42–72 g

Hainan Mole
Mogera hainana

FR Taupe de Hainan
DE Hainan-Inselmaulwurf
ES Topo de Hainan

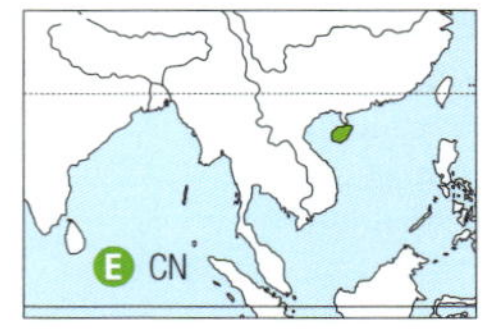

Hainan I

NE

Monotypic
HB 12–13.4 cm
T 0.9–1.2 cm

Kano's Mole
Mogera kanoana

FR Taupe de Kano
DE Kano-Maulwurf
ES Topo de Kano

up to 2800 m

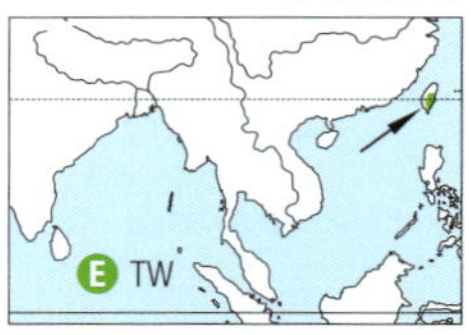

Mainly broadleaf forests

NE

Monotypic
HB 11.3–13.3 cm
T 0.8–1.3 cm W 23–59 g

La Touche's Mole
Mogera latouchei

FR Taupe de La Touche
DE La Touche-Maulwurf
ES Topo de La Touche

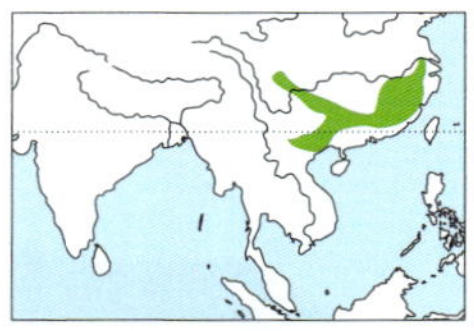

Farmland and possibly forests in S CN & N VN

NE

Monotypic
HB 11.6–13 cm
T 1.2–1.6 cm W 33–52 g

Genus *EUROSCAPTOR*

(10 species)

Himalayan Mole
Euroscaptor micrurus

FR Taupe de l'Himalaya
DE Himalaya-Maulwurf
ES Topo del Himalaya

up to 2700 m

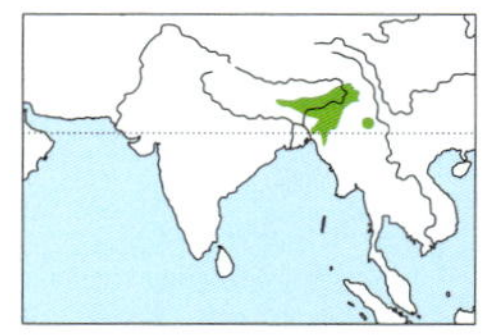

E NP, BT, NE IN, SC CN (W Yunnan)

LC

Monotypic
HB 12.8–13.5 cm
T 0.5–0.9 cm
W 44–72 g

Greater Chinese Mole
Euroscaptor grandis

FR Taupe de l'Emei
DE Großer China-Maulwurf
ES Topo de China grande

C Sichuan, W Yunnan; perhaps adjacent MM

LC

Monotypic
HB c. 15 cm
T c. 1 cm

Malaysian Mole
Euroscaptor malayanus

FR Taupe de Malaisie
DE Malaysia-Maulwurf
ES Topo de Malasia

1300–1600 m

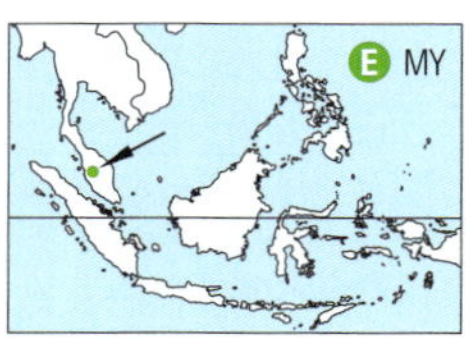

Cameron Highlands on Peninsula

NE

Monotypic
HB 12.8–13.4 cm
T 0.4–0.8 cm
W 44–71 g

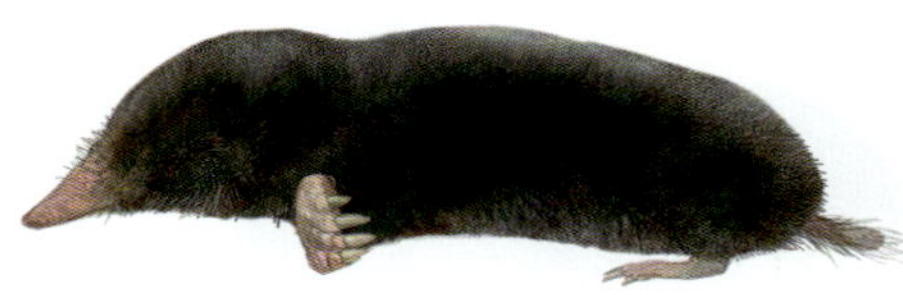

Long-nosed Mole
Euroscaptor longirostris

FR Taupe à long nez
DE Langnasenmaulwurf
ES Topo de hocico largo

1800–2900 m

Gansu, Shaanxi, Sichuan, Chongqing, Guizhou, Hubei, Hunan, Guangxi, Jiangxi, Fujian

LC

Monotypic
HB 9–14.5 cm
T 1.1–2.5 cm

Kloss's Mole
Euroscaptor klossi

FR Taupe de Kloss
DE Kloss-Maulwurf
ES Topo de Kloss

1200–2550 m

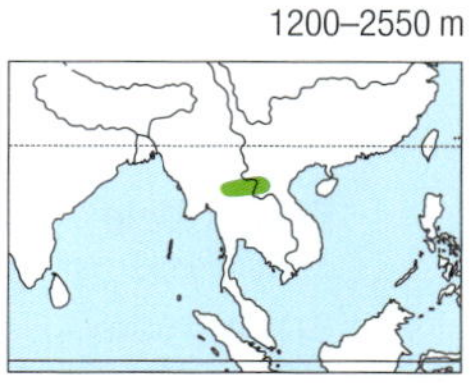

Mountain forests in N TH & NW LA

LC

Monotypic
HB 11.5–13.5 cm
T 0.9–1.3 cm
W 48–62 g

Kuznetsov's Mole
Euroscaptor kuznetsovi

FR Taupe de Kuznetsov
DE Kuznetsov-Maulwurf
ES Topo de Kuznetsov

750–950 m

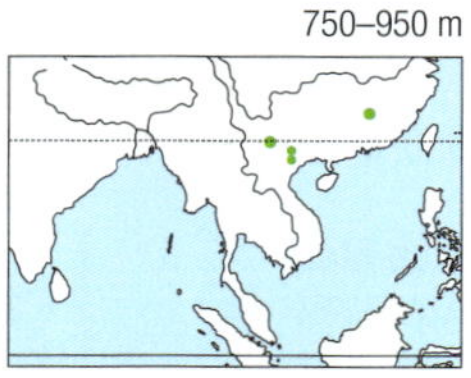

NE VN (Cao Bang & Vinh Phuc prov), S CN (Yunnan, Jiangxi)

NE

Monotypic
HB 11.7–13.6 cm
T 1.4–1.7 cm
W 42–77 g

Orlov's Mole
Euroscaptor orlovi

FR Taupe d'Orlov
DE Orlov-Maulwurf
ES Topo de Orlov

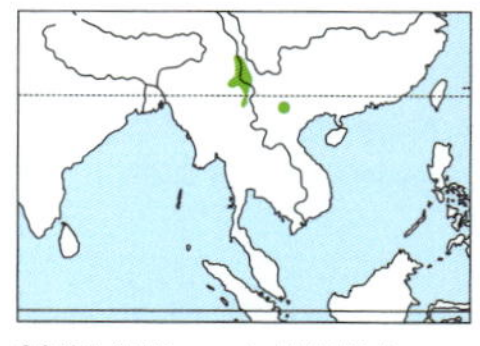

SC CN (W Yunnan), NW VN (Lao Cai Prov); perhaps N LA E to N VN, W of R Red

NE

Monotypic
HB 10.6–12.9 cm
T 1.5–1.7 cm
W 26–48 g

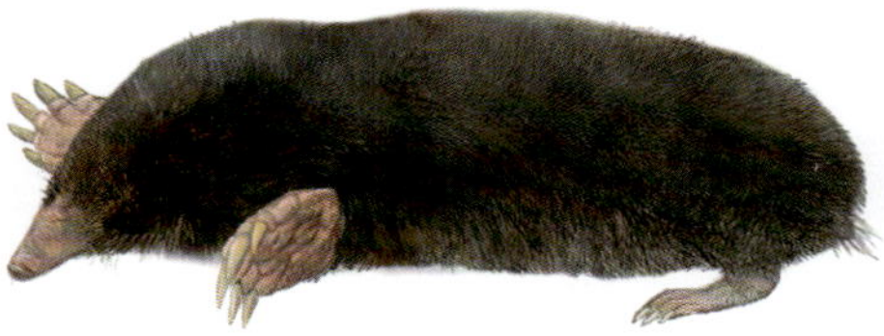

Vietnamese Mole
Euroscaptor subanura
FR Taupe anoure
DE Vietnam-Maulwurf
ES Topo de Vietnam

DD
Monotypic
HB 11.3–13.1 cm
T 0.2–0.5 cm
W 33–50 g

200–300 m

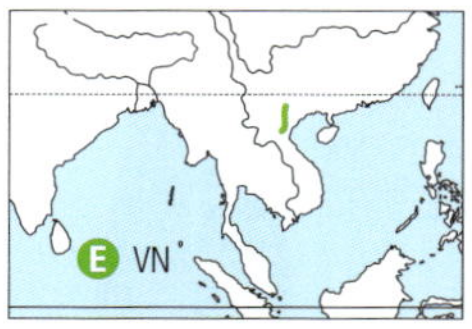

Semideciduous forests in limestone mts

Small-toothed Mole
Euroscaptor parvidens
FR Taupe à petites dents
DE Pakho-Maulwurf
ES Topo de dientes pequeños

DD
Monotypic
HB 13.5–15 cm
T 0.5–1 cm
W 37–72 g

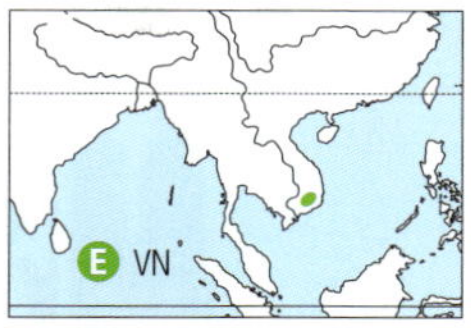

Dalat Plateau

Ngoc Linh Mole
Euroscaptor ngoclinhensis
FR Taupe du Ngoc Linh
DE Ngoc-Linh-Maulwurf
ES Topo de Ngoc Linh

NE
Monotypic
HB 12–12.3 cm
T 0.6–0.7 cm

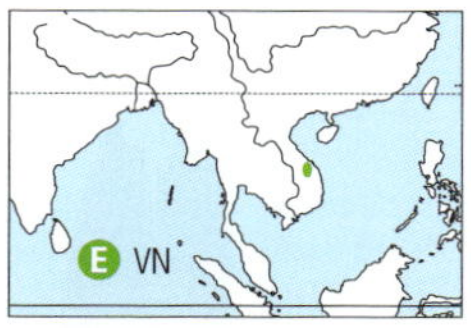

C highlands (Quang Nam, Kon Tum & perhaps Gia Lai prov)

Genus *PARASCAPTOR*

(1 species)

LC
Monotypic
HB 11.5–12.6 cm
T 1–1.4 cm
W 36–47 g

White-tailed Mole
Parascaptor leucurus
FR Taupe à queue blanche
DE Weißschwanzmaulwurf
ES Topo de cola blanca

400–3000 m

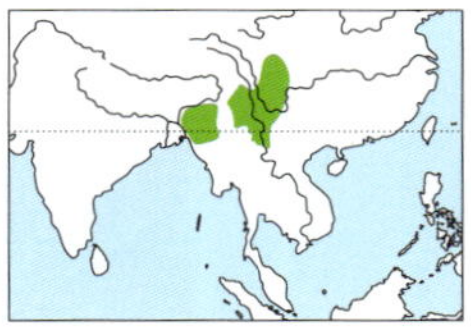

NE IN (S of R Brahmaputra, possibly N in Arunachal Pradesh), BD, MM, CN (Sichuan, Yunnan); perhaps N LA

Genus *SCAPTOCHIRUS*

(1 species)

LC
Monotypic
HB 10–12.6 cm
T 1.4–2.3 cm

Short-faced Mole
Scaptochirus moschatus
FR Taupe de Mongolie
DE Kurzgesichtmaulwurf
ES Topo de cara corta

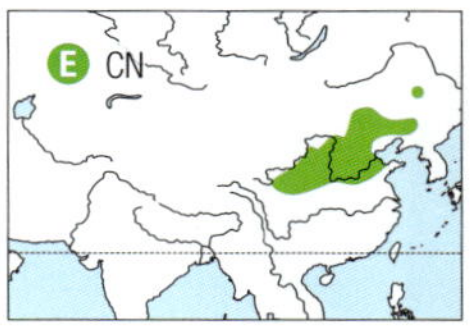

NE & C CN (Heilongjiang S to Gansu, Henan & Jiangsu); probably MN

Family SOLENODONTIDAE – Solenodons

2 genera, 2 species. Dry, moist, and wet tropical pine forests, and human-modified landscapes. Neotropical Region.

Genus *ATOPOGALE*

(1 species)

Cuban Solenodon
Atopogale cubana
FR Solénodonte de Cuba
DE Kuba-Schlitzrüssler
ES Almiquí

EN
Monotypic
HB 20–36 cm
T 13–19 cm
W 455–950 g

400–900 m

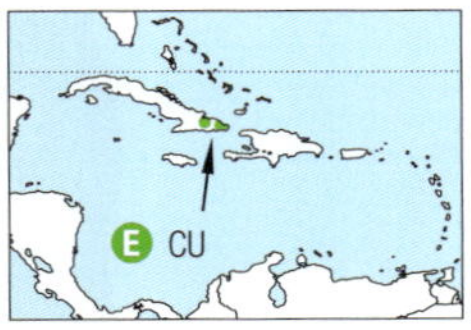

Now restricted to Nipe-Sagua-Baracoa Massif in E

Genus *SOLENODON*

(1 species)

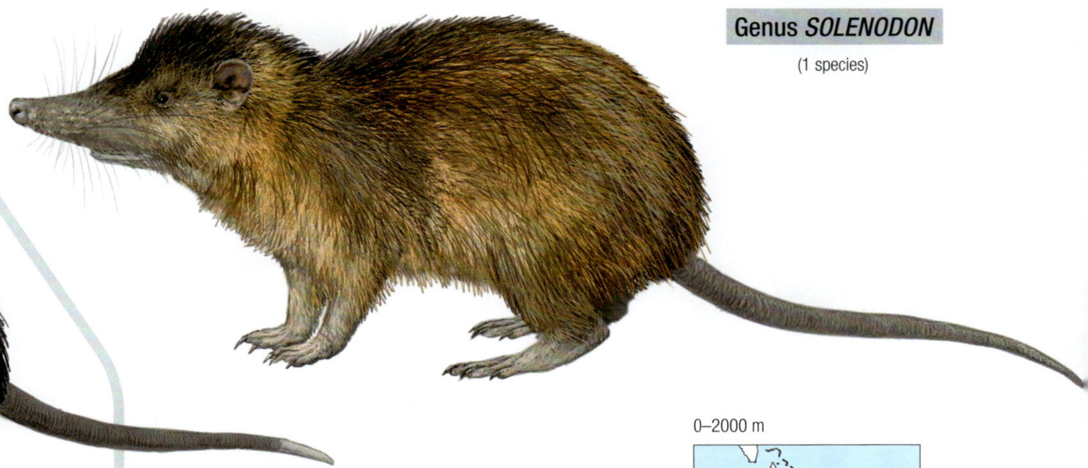

Hispaniolan Solenodon
Solenodon paradoxus
FR Solénodonte d'Haïti
DE Haiti-Schlitzrüssler
ES Solenodonte de la Española

LC
3 sspp
HB 27–49 cm
T 20–25 cm
W 620–1166 g

0–2000 m

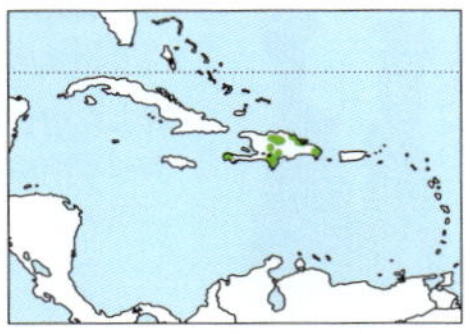

N & E DO, HT (Massif de la Hotte, Massif de la Selle)

Order CHIROPTERA (21 families)

Family PTEROPODIDAE – Old World Fruit Bats

47 genera, 191 species. From montane and lowland tropical rainforests to open woodlands and deserts. Palearctic, Afrotropical, Indo-Malayan and Australasian Regions.

Genus *MEGAEROPS*

(4 species)

Temminck's Tailless Fruit Bat
Megaerops ecaudatus

LC
Monotypic
HB 6.5–8.4 cm
T tailless
FA 5–5.8 cm
W 22–28 g

FR Cynoptère écaudé
DE Temminck-Altwelt-Fruchtfledermaus
ES Megaerop de Temminck

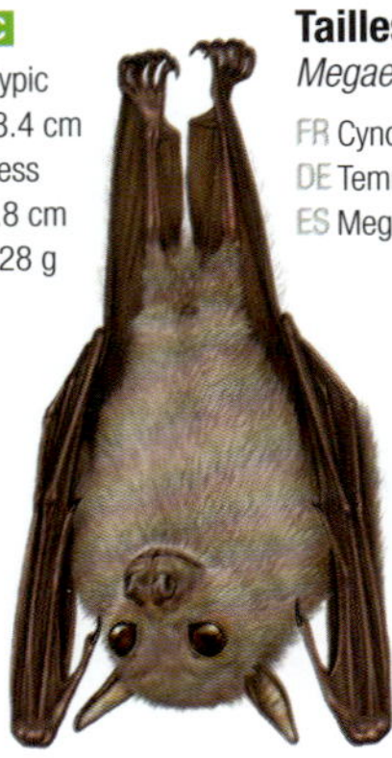

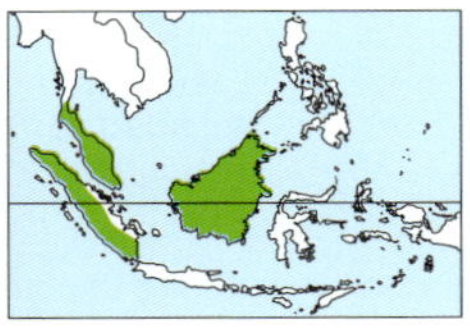

Thai-Malay Pen, N & W Sumatra, Borneo; perhaps NE IN (Arunachal Pradesh)

Javan Tailless Fruit Bat
Megaerops kusnotoi

VU
Monotypic
HB 5.8–7.5 cm
T tailless
FA 4.9–5.4 cm
W c. 19.9 g

FR Cynoptère de Kusnoto
DE Java-Altwelt-Fruchtfledermaus
ES Megaerop de Kusnoto

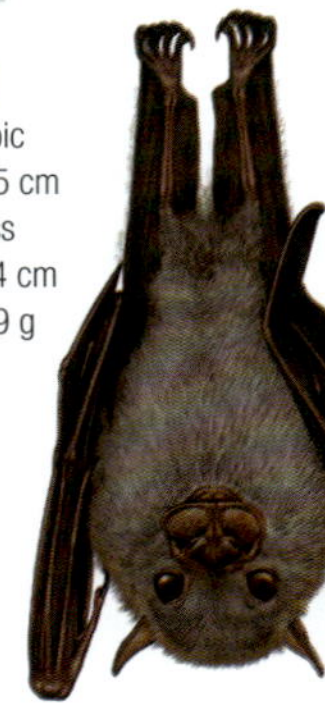

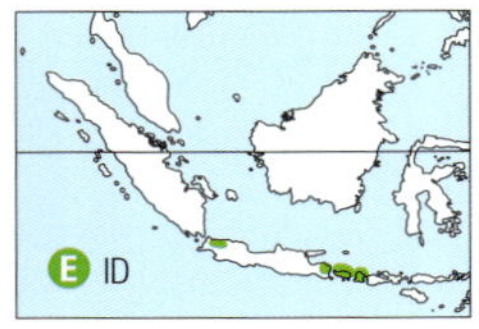

W & E Java, Bali, Lombok

Ratanaworabhan's Tailless Fruit Bat
Megaerops niphanae

LC
Monotypic
HB 7.7–9 cm
T tailless
FA 5.2–6.3 cm
W 22–28 g

FR Cynoptère de Niphan
DE Ratanaworabhan-Altwelt-Fruchtfledermaus
ES Megaerop de Niphan

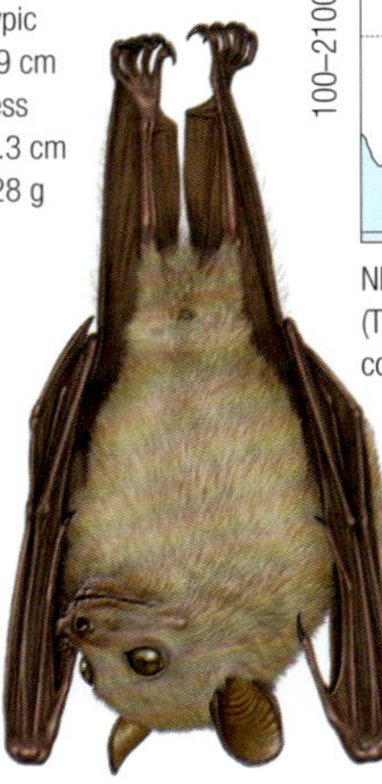

100–2100 m

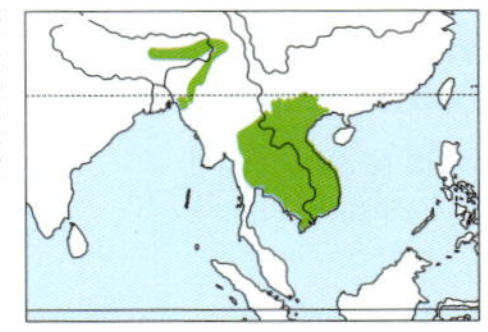

NE IN, BD (Chittagong), SE Asia (TH, LA, VN, KH); probably MM to connect disjunct areas

White-collared Fruit Bat
Megaerops albicollis

NE
Monotypic
HB 4.6–5.1 cm
T 0–0.4 cm
FA 4.6–5.1 cm
W 16–26 g

FR Cynoptère à col blanc
DE Weißkragen-Altwelt-Fruchtfledermaus
ES Megaerop de cuello blanco

below 110 m

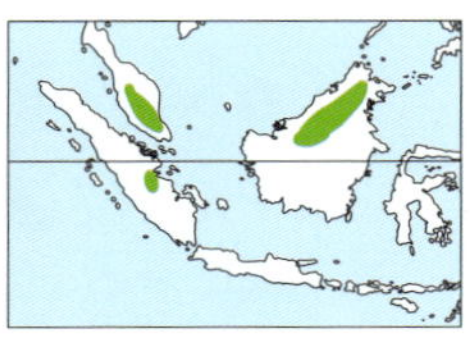

Few localities in Pen MY, C Sumatra (Jambi, Riau), Borneo

Genus *CYNOPTERUS*

(7 species)

Greater Short-nosed Fruit Bat
Cynopterus sphinx

LC
6 sspp
HB 8.9–10.9 cm
T 1.3–1.8 cm
FA 6.7–7.6 cm
W 28–70 g

FR Cynoptère à nez court
DE Großer Kurznasenflughund
ES Cinóptero de nariz corta

0–400 m

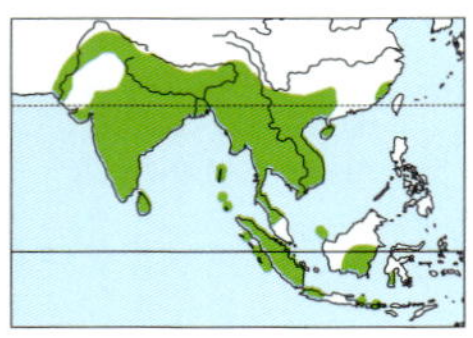

PK & IN (excluding Thar Desert) E to S CN, S to Greater Sundas, W Sumbawa, S Sulawesi, many Is

Lesser Short-nosed Fruit Bat
Cynopterus brachyotis

LC
8 sspp
HB 7.2–9.6 cm
T 1–1.6 cm
FA 5.5–6.6 cm
W 27–45 g

FR Cynoptère de Müller
DE Kleiner-Kurznasenflughund
ES Cinóptero de Müller

0–1600 m

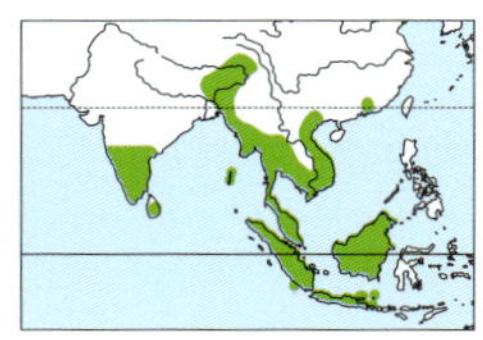

S & NE IN, LK, BT, BD & N MM to VN, S to Greater Sundas, Andaman Is, Enggano I; uncertain in S CN

Horsfield's Short-nosed Fruit Bat
Cynopterus horsfieldii

LC
4 sspp
HB 8.5–12 cm
T 1.2–1.9 cm
FA 6.4–8.9 cm
W 43–63 g

FR Cynoptère de Horsfield
DE Horsfield-Kurznasenflughund
ES Cinóptero de Horsfield

0–1460 m

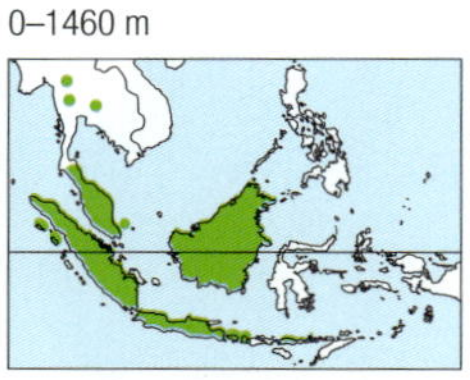

W & C TH (scattered), Thai-Malay Pen (with Tioman I), Greater Sundas, Lombok, Flores; perhaps Sumbawa

Peters's Short-nosed Fruit Bat
Cynopterus luzoniensis

LC
Monotypic
HB 9.5–10.8 cm
T 0.9–1.4 cm
FA 5.7–6.8 cm
W 26–45 g

FR Cynoptère de Luçon
DE Peters-Kurznasenflughund
ES Cinóptero de Luzón

0–1600 m

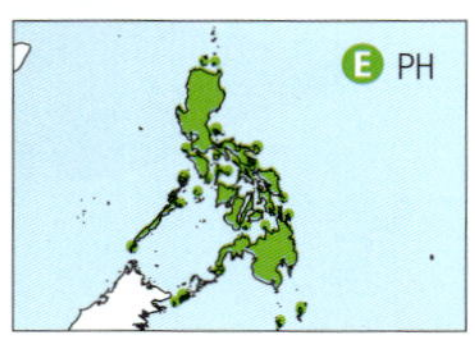

Palawan, Sulu Arch (Sibutu, Sanga-Sanga, Bongao)

Minute Short-nosed Fruit Bat
Cynopterus minutus

LC
Monotypic
HB 7.6–8.3 cm
T 1–1.2 cm
FA 5.4–5.9 cm
W 19–24 g

FR Petit Cynoptère
DE Zwerg-Kurznasenflughund
ES Cinóptero pequeño

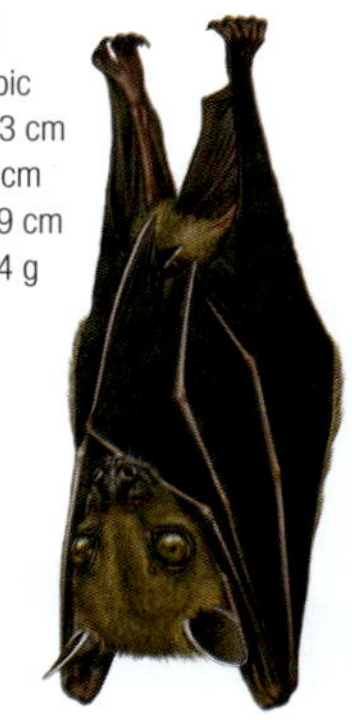

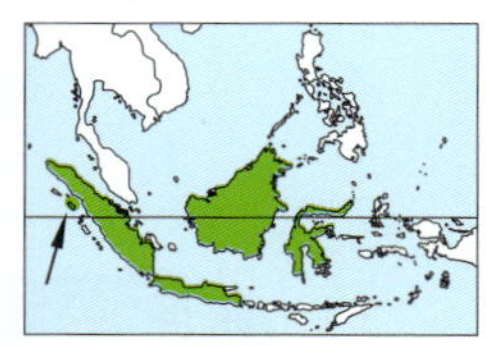

Sumatra (with Nias I), Borneo, Java, Sulawesi

Nusatenggara Short-nosed Fruit Bat
Cynopterus nusatenggara

LC
3 sspp
HB c. 8.8 cm
T c. 0.5 cm
FA 5.5–6.5 cm
W 24–37 g

FR Cynoptère de Nusa Tenggara
DE Nusa-Tenggara-Kurznasenflughund
ES Cinóptero de Nusa Tenggara

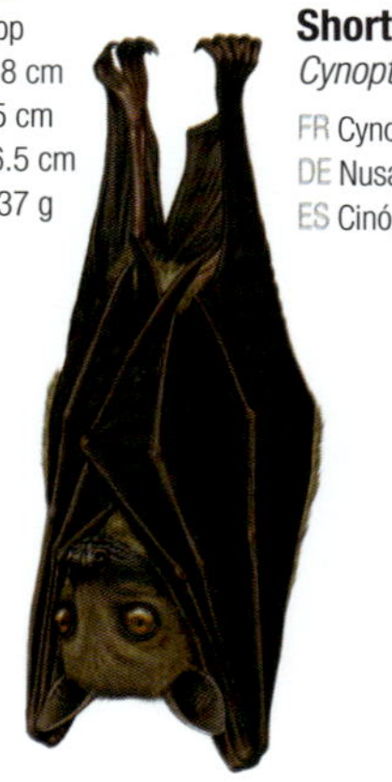

up to 100 m

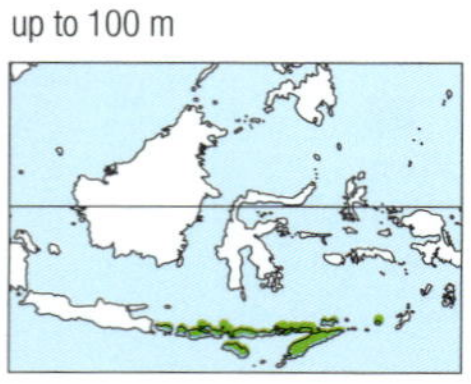

Lesser Sunda Is (from Bali to Wetar, Timor & Babar)

Indonesian Short-nosed Fruit Bat
Cynopterus titthaecheilus

LC
3 sspp
HB 11.4–12.7 cm
T 0.9–1.2 mm
FA 7.4–8.4 cm
W 76–84 g

FR Cynoptère de Temminck
DE Indonesien-Kurznasenflughund
ES Cinóptero de Indonesia

0–800 m

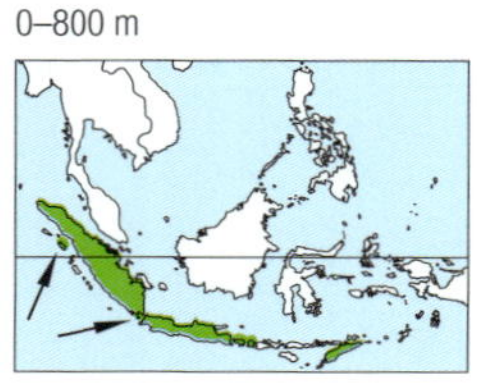

Sumatra (with Nias, Krakatau & Sebesi Is), Java, Bali, Lombok, Timor

Genus *PTENOCHIRUS*

(3 species)

Lesser Musky Fruit Bat
Ptenochirus minor

LC
Monotypic
HB 10.4–11.3 cm
T 0.7–1.2 cm
FA 7–7.8 cm
W c. 50 g

FR Cynoptère musqué
DE Kleiner Moschusflughund
ES Ptenoquiro pequeño

0–1600 m

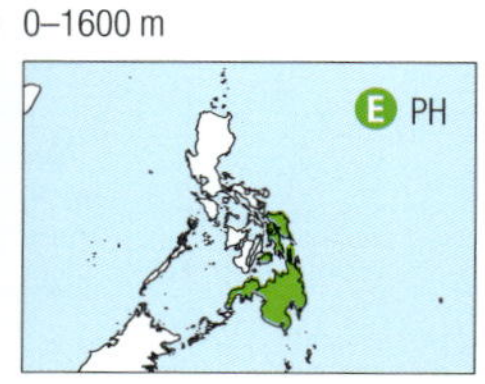

C & E Visayan Is (Samar, Biliran, Leyte, Bohol), Dinagat, Mindanao

Greater Musky Fruit Bat
Ptenochirus jagorii

FR Cynoptère de Jagor
DE Großer Moschusflughund
ES Ptenoquiro grande

LC Monotypic
HB 11.4–12.7 cm
T 0.6–1.8 cm
FA 7.6–9.1 cm
W 62–97 g

0–1950 m

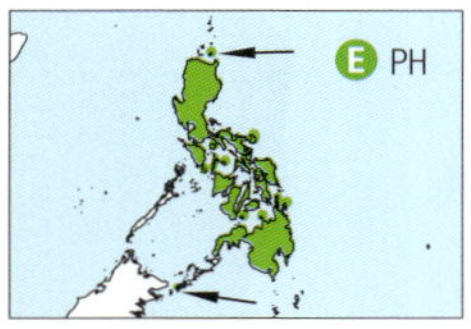

Widespread (except Palawan region), including Babuyan Is (Camiguin) & Sulu Arch (Bongao, Sanga-Sanga)

VU
Monotypic
HB 7.1–7.6 cm
T 0.3–0.5 cm
FA 4.7–5.2 cm
W 16–21 g

Mindanao Fruit Bat
Ptenochirus wetmorei

FR Cynoptère de Wetmore
DE Mindanao-Moschusflughund
ES Ptenoquiro de Wetmore

800–1200 m

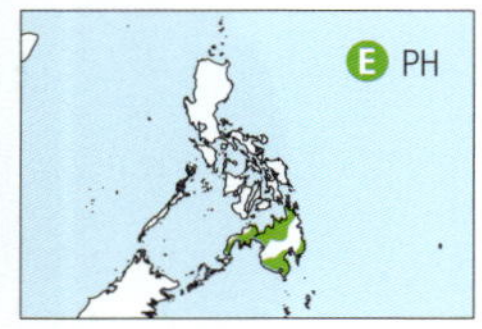

Mindanao

VU
Monotypic
HB 11.6–12 cm
T 1.7–2 cm
FA 8.1–8.3 cm
W 86–91 g

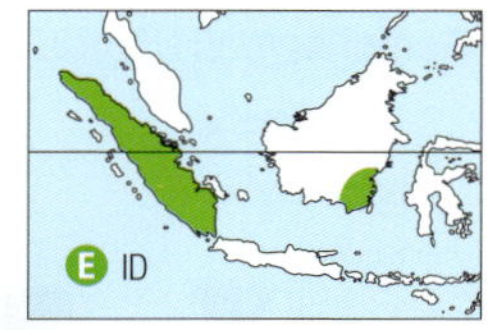

Sumatra, SE Borneo (Kalimantan)

Brooks's Dayak Fruit Bat
Dyacopterus brooksi

FR Cynoptère de Brooks
DE Brooks-Dayak-Flughund
ES Diacóptero de Brooks

Genus *DYACOPTERUS*

(3 species)

NT Monotypic
HB 10.6–13 cm T 1.6–2.6 cm
FA 7.4–8.1 cm W 70–100 g

Common Dayak Fruit Bat
Dyacopterus spadiceus

FR Cynoptère dayak
DE Gewöhnlicher Dayak-Flughund
ES Diacóptero dayak

0–1190 m

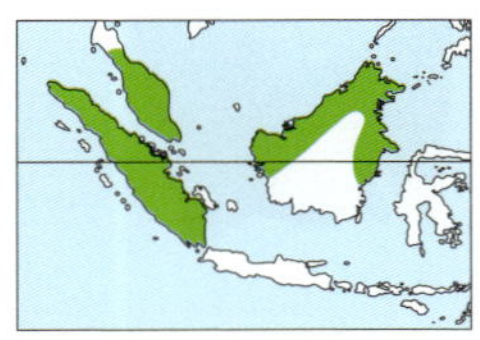

Thai-Malay Pen, Sumatra, W, N & E Borneo

EN
Monotypic
HB 13–14.4 cm
T 1.8–2.9 cm
FA 9.1–9.6 cm
W 138–148 g

260–1680 m

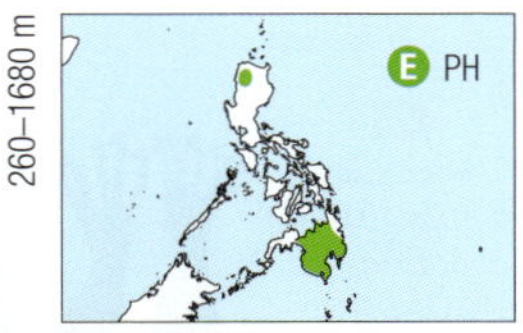

Few records from Luzon & Mindanao; suspected to be widespread

Philippine Large-headed Fruit Bat
Dyacopterus rickarti

FR Cynoptère de Rickart
DE Philippinen-Dayak-Flughund
ES Diacóptero de Rickart

Genus *SPHAERIAS*

(1 species)

Blanford's Fruit Bat
Sphaerias blanfordi

FR Cynoptère de Blanford
DE Blanford-Flughund
ES Cinóptero de Blanford

LC Monotypic
HB 7.5–9 cm T tailless
FA 5.2–6.1 cm
W c. 31 g

300–2700 m

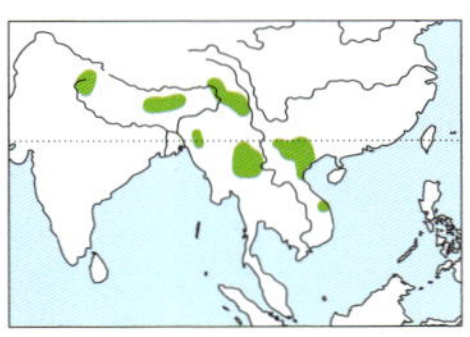

N IN, E NP, BT, N MM, SC CN (SE Tibet, W & S Yunnan), N TH, NW LA, N & C VN; perhaps widespread

LC Monotypic
HB 5–7.4 cm
T tailless
FA 3.8–4.3 cm
W 11–14 g

Bornean Spotted-winged Fruit Bat
Balionycteris maculata

FR Cynoptère tacheté
DE Borneo-Fleckenkurznasenflughund
ES Balionícteris maculado

up to 1000 m

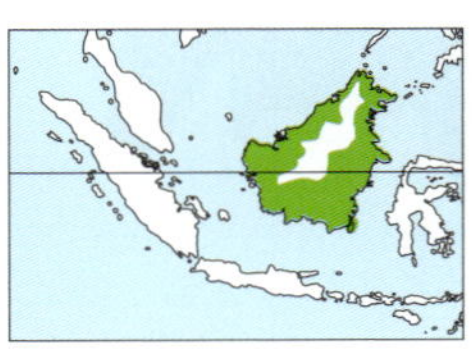

Borneo (MY, ID, BN)

LC Monotypic
HB 5.5–7 cm
T tailless
FA 4–4.8 cm
W 14–18 g

Bornean Pygmy Fruit Bat
Aethalops aequalis

FR Cynoptère de Bornéo
DE Borneo-Zwergflughund
ES Aethalops de Borneo

550–2700 m

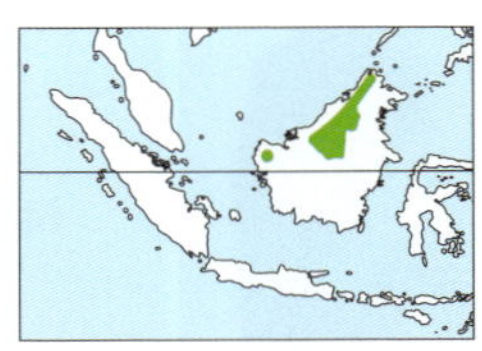

C Mts of Borneo in MY (Sabah, Sarawak) & ID (Kalimantan)

LC
Monotypic
HB 9.4–11.1 cm
T c. 0.6 cm
FA 7–8.2 cm
W 67–99 g

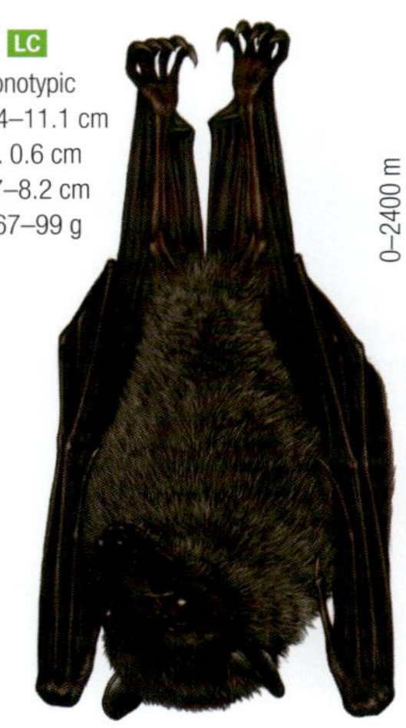

0–2400 m

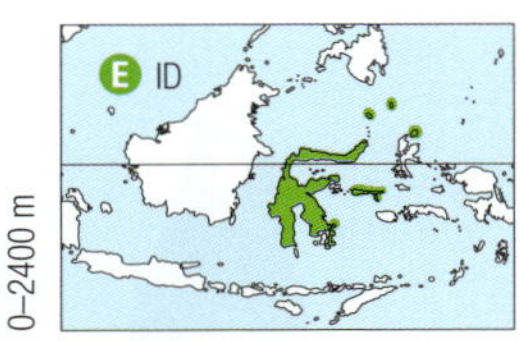

Talaud Is (Karakelong), Sangihe Is (Sangihe), N Moluccas (Morotai), Sulawesi (with Wowoni & Buton), Sula Is

Common Swift Fruit Bat
Thoopterus nigrescens

FR Cynoptère sombre
DE Schwarzflügelflughund
ES Toptérido cenizo

Genus *BALIONYCTERIS*

(2 species)

LC Monotypic
HB 5.6–6.6 cm
T tailless
FA 4–4.2 cm
W 9–14 g

Malayan Spotted-winged Fruit Bat
Balionycteris seimundi

FR Cynoptère de Seimund
DE Malaya-Fleckenkurznasenflughund
ES Balionícteris de Seimund

up to 1500 m

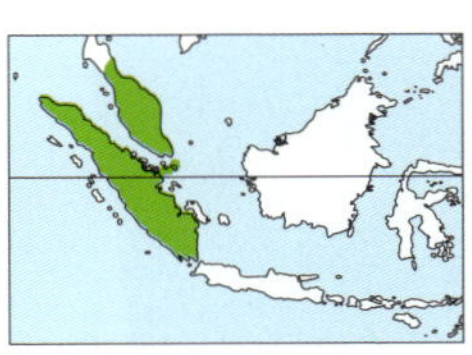

Thai-Malay Pen, Sumatra, SW Riau Arch (Bintan, Durian, Galang)

Genus *AETHALOPS*

(2 species)

LC 3 sspp
HB 6.3–7.5 cm
T tailless
FA 4.3–4.8 cm
W 17–22 g

Common Pygmy Fruit Bat
Aethalops alecto

FR Cynoptère pygmée
DE Gewöhnlicher Zwergflughund
ES Aethalops pigmeo

1000–2700 m

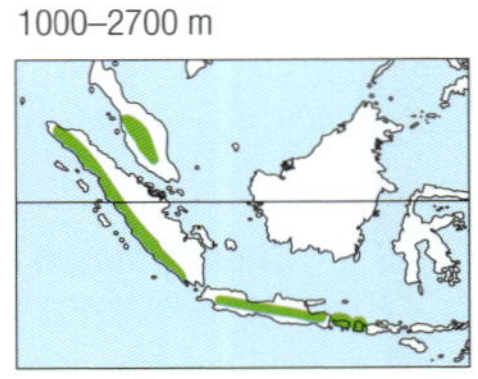

W Pen MY, ID (W Sumatra, Java, Bali, Lombok)

Genus *THOOPTERUS*

(2 species)

Suhaniah's Swift Fruit Bat
Thoopterus suhaniahae

FR Cynoptère de Suhaniah
DE Suhaniah-Schwarzflügelflughund
ES Toptérido de Suhaniah

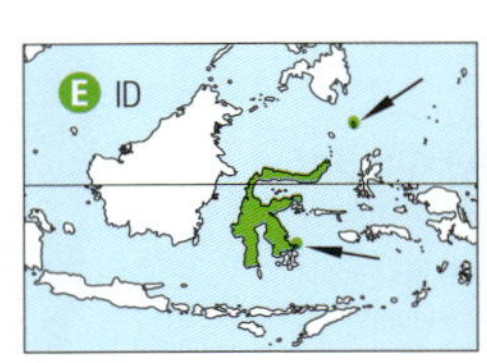

60–2100 m

Talaud Is, Sulawesi (with Wowoni I)

LC
Monotypic
HB 8.9–10.9 cm
T vestigial
FA 7.4–8.2 cm
W 52–100 g

Genus *ALIONYCTERIS*

(1 species)

LC
Monotypic
HB 6.6–7.2 cm
T tailless
FA 4.4–5 cm
W 14–18 g

1500–2250 m

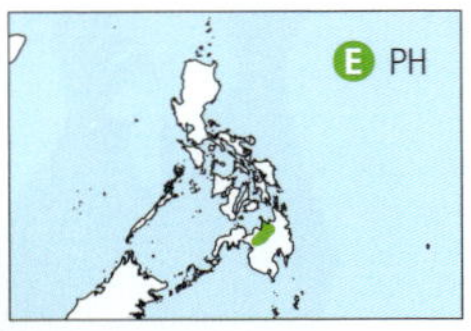

C Mindanao (Mt Kitanglad, neighboring Kalatungan Mt Range)

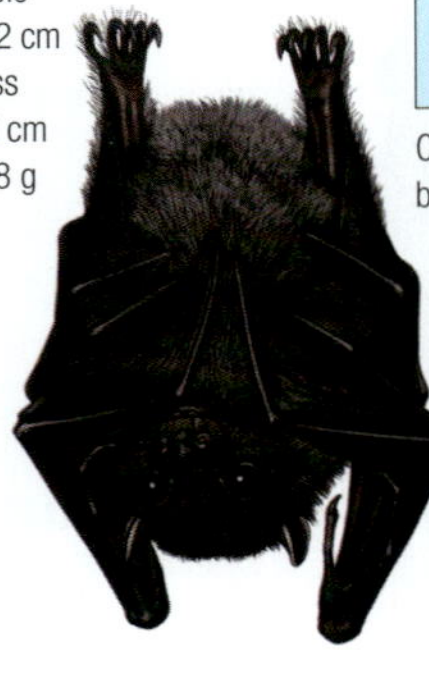

Mindanao Pygmy Fruit Bat
Alionycteris paucidentata
FR Cynoptère de Mindanao
DE Mindanao-Kleinflughund
ES Alionícteris de Mindanao

Genus *HAPLONYCTERIS*

(1 species)

LC
Monotypic
HB 6.9–7.8 cm
T tailless
FA 4.6–5.2 cm
W 16–21 g

150–2250 m

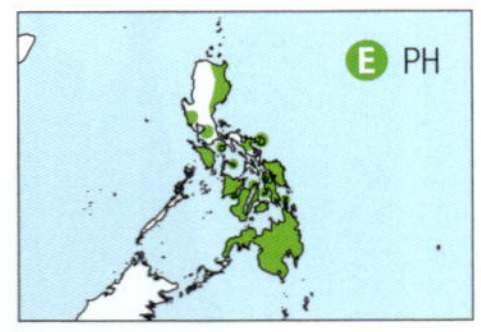

Widespread (excluding Palawan, Sulu Arch & Batanes/Babuyan Is)

Fischer's Pygmy Fruit Bat
Haplonycteris fischeri
FR Cynoptère de Fischer
DE Fischer-Kleinflughund
ES Haplonícteris de Fischer

Genus *OTOPTEROPUS*

(1 species)

LC
Monotypic
HB 6.9–7.4 cm
T tailless
FA 4.4–4.7 cm
W 12–15 g

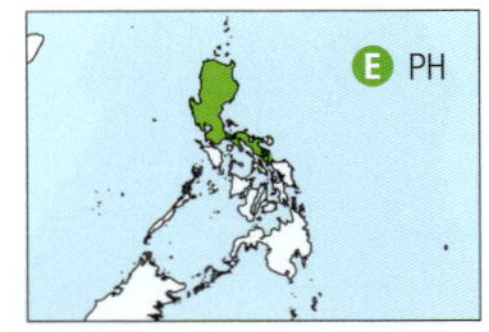

Luzon 200–2530 m

Luzon Pygmy Fruit Bat
Otopteropus cartilagonodus
FR Cynoptère à poil long
DE Luzon-Kleinflughund
ES Otoptérido de Luzón

Genus *LATIDENS*

(1 species)

EN
Monotypic
HB 10.2–10.9 cm
T tailless
FA 6.6–6.9 cm
W c. 69 g

800–1100 m

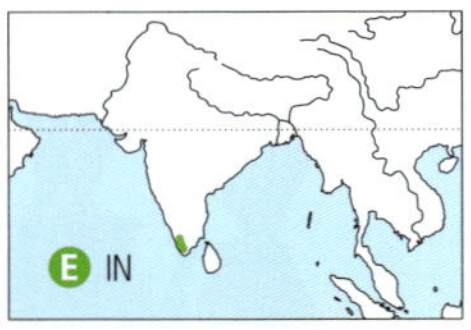

S Western Ghats in Kerala (Periyar WS, Agasthiyar Hill, High Wavy Mts) & Tamil Nadu (Anaimalai Hills)

Salim Ali's Fruit Bat
Latidens salimalii
FR Cynoptère de Salim Ali
DE Salim-Ali-Fruchtfledermaus
ES Latideno de Salim Ali

LC
2 sspp
HB 5.5–7.2 cm
T tailless
FA 4.4–4.8 cm
W 16–20 g

above 600 m

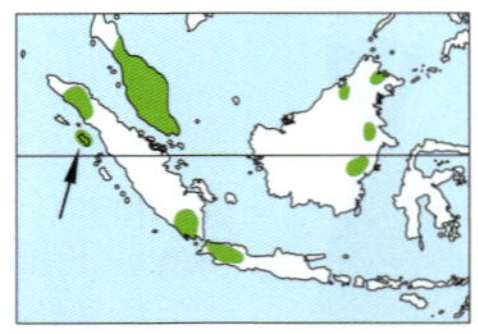

Thai-Malay Pen, N & S Sumatra (with Nias I), W Java, Borneo

Sundaic Black-capped Fruit Bat
Chironax melanocephalus
FR Cynoptère à tête noire
DE Sunda-Schwarzkappenflughund
ES Chironax de Sunda

Genus *CHIRONAX*

(2 species)

NE
Monotypic
HB c. 5.6 cm
T tailless
FA 4.3–4.8 cm
W 12–17 g

up to 960 m

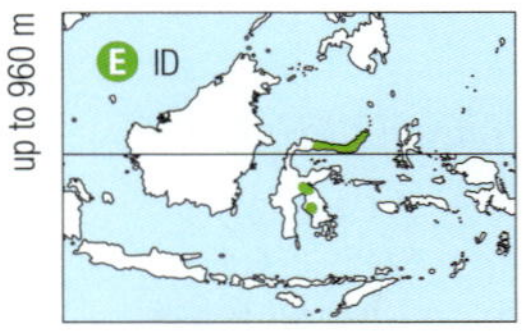

NE, CE & SE Sulawesi

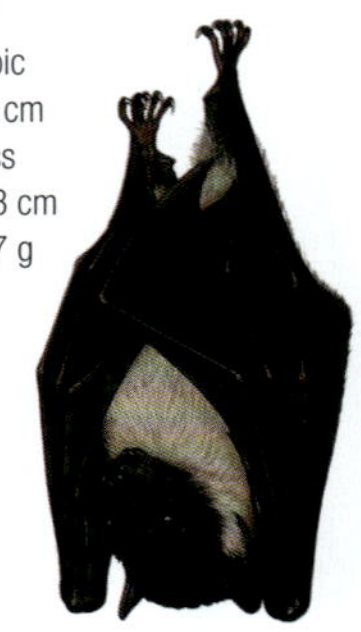

Sulawesi Black-capped Fruit Bat
Chironax tumulus
FR Cynoptère de Sulawesi
DE Sulawesi-Schwarzkappenflughund
ES Chironax de Sulawesi

Genus *SCOTONYCTERIS*

(3 species)

NT
2 sspp
HB 6.5–8.5 cm
T tailless
FA 4.7–5.5 cm
W 16–24 g

0–1100 m

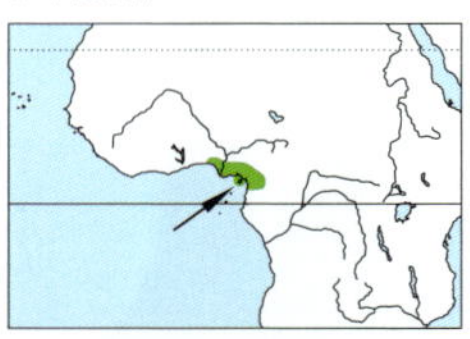

W Africa in CM & Bioko I (GQ); unconfirmed in SE NG

Zenker's Fruit Bat
Scotonycteris zenkeri
FR Scotonyctère de Zenker
DE Zenker-Harlekinflughund
ES Escotonícteris de Zenker

0–530 m

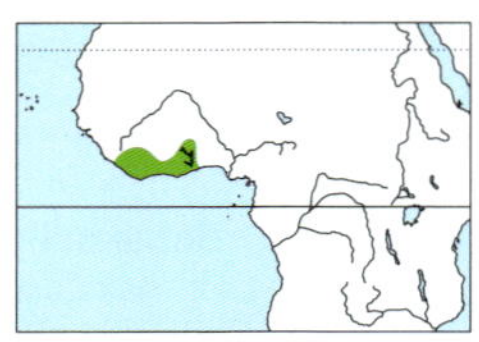

W Africa in GN, LR, CI & GH

LC Monotypic
HB 6.5–8.5 cm T tailless
FA 4.7–5.5 cm W 16–24 g

Hayman's Fruit Bat
Scotonycteris occidentalis
FR Scotonyctère de Hayman
DE Hayman-Harlekinflughund
ES Escotonícteris de Hayman

Genus *PENTHETOR*

(1 species)

LC
2 sspp
HB 7.8–10.3 cm
T 0.8–1.3 cm
FA 5.4–6.6 cm
W 34–55 g

0–600 m

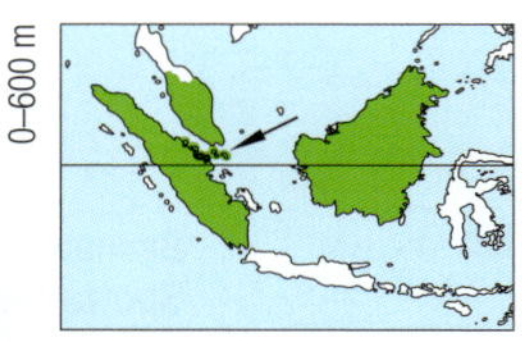

Pen MY, Sumatra, Riau Arch, Borneo

Lucas's Short-nosed Fruit Bat
Penthetor lucasii
FR Cynoptère de Lucas
DE Lucas-Kurznasenflughund
ES Penthetor de Lucas

LC 2 sspp
HB 6.5–8.5 cm T tailless
FA 4.7–5.5 cm
W 16–24 g

310–1100 m

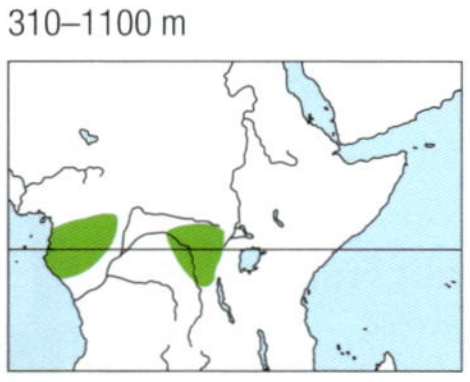

C Africa in S CM, SW CF, GQ, GA, CG & E CD; perhaps RW, BI & NW TZ

Bergmans's Fruit Bat
Scotonycteris bergmansi
FR Scotonyctère de Bergmans
DE Bergmans-Harlekinflughund
ES Escotonícteris de Bergmans

335–750 m

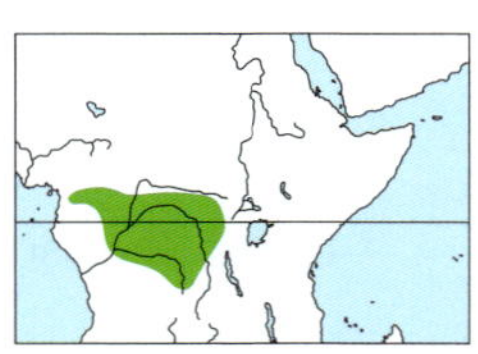

C Africa in S CM, SW CF, N CG & CD

LC Monotypic
HB 7–10 cm T tailless
FA 5–6.2 cm
W 24–41 g

Short-palated Fruit Bat
Casinycteris argynnis
FR Casinyctère doré
DE Goldfarbener Kurzschnauzenflughund
ES Casinícteris dorado

Genus *CASINYCTERIS*

(3 species)

NE
Monotypic
HB 11.5–12.2 cm
T tailless
FA 7.3–8.1 cm
W 60–95 g

120–1200 m

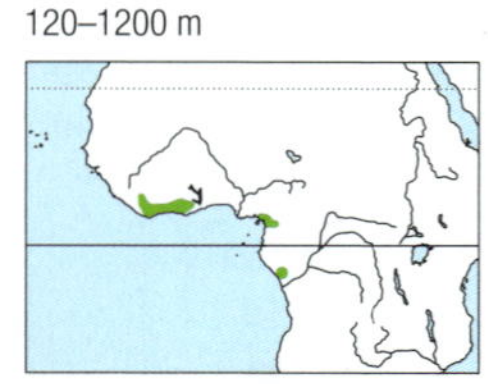

Disjunctly in LR, CI, GH, CM & CG

Pohle's Fruit Bat
Casinycteris ophiodon
FR Casinyctère de Pohle
DE Pohle-Kurzschnauzenflughund
ES Casinícteris de Pohle

Campo-Ma'an Fruit Bat
Casinycteris campomaanensis

FR Casinyctère de Campo-Ma'an
DE Campo-Ma'an-Kurzschnauzenflughund
ES Casinícteris de Campo Ma'an

DD
Monotypic
T tailless
FA c. 7.3 cm
W c. 49 g

0–500 m

Known from Campo-Ma'an area (South Region); perhaps NG & CG

Genus *EONYCTERIS*

(3 species)

Lesser Dawn Bat
Eonycteris spelaea

FR Éonyctère des cavernes
DE Höhlen-Langzungenflughund
ES Eonícteris pequeño

LC
4 sspp
HB 7.9–13 cm
T 1.1–2.5 cm
FA 6.1–8 cm
W 40–79 g

0–1000 m

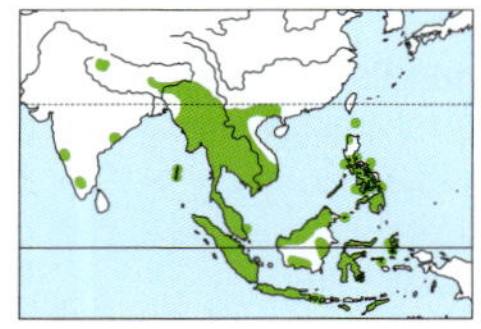

IN & NP E to S CN, mainland SE Asia, Andaman Is, Greater & Lesser Sundas, Sulawesi, N Moluccas, PH

Greater Dawn Bat
Eonycteris major

FR Grand Éonyctère
DE Borneo-Langzungenflughund
ES Eonícteris grande

NT Monotypic
HB 12.4–13 cm T 1.8–2.7 cm
FA 6.1–8.5 cm W 91–98 g

0–1100 m

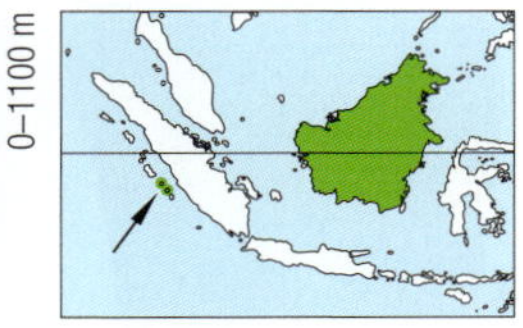

Borneo, Mentawai Is (Sipora, North Pagai)

Philippine Dawn Bat
Eonycteris robusta

FR Éonyctère des Philippines
DE Philippinen-Langzungenflughund
ES Eonícteris de las Filipinas

VU
Monotypic
HB 10.7–12.7 cm
T 2–2.8 cm
FA 6.7–8.2 cm
W 56–80 g

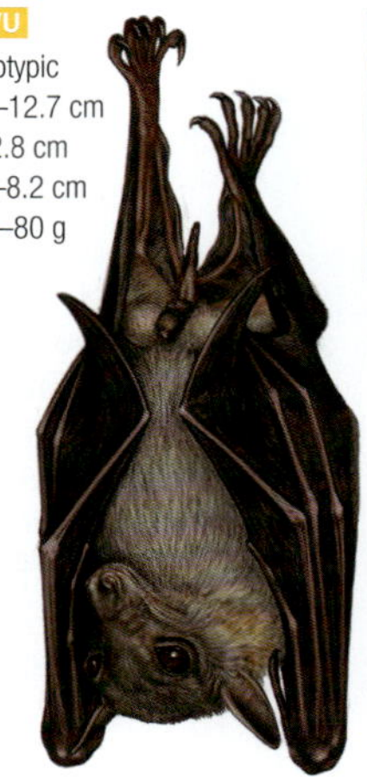

0–1100 m

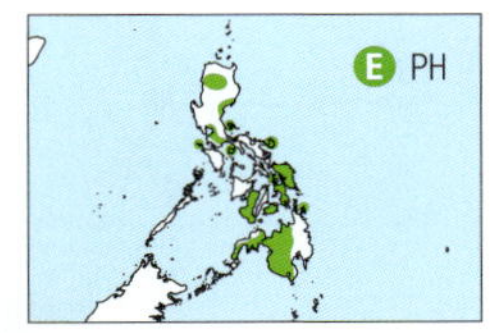

Most of large Is (excluding Palawan, Sulu Arch & Batanes/Babuyan Is)

Genus *ROUSETTUS*

(7 species)

Geoffroy's Rousette
Rousettus amplexicaudatus

FR Roussette de Geoffroy
DE Geoffroy-Flughund
ES Rosetus de Geoffroy

LC 3 sspp
HB 12.8–15.4 cm
T 1.3–2.4 cm
FA 7–9.2 cm
W 64–106 g

0–2200 m

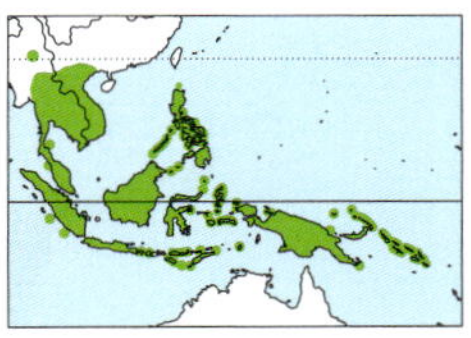

Mainland SE Asia (S from C MM & SC CN), ID, PH, New Guinea (with nearby Is), Solomon Is

Bare-backed Rousette
Rousettus spinalatus

FR Roussette à dos nu
DE Nacktrücken-Flughund
ES Rosetus de espalda desnuda

VU
Monotypic
HB c. 10.5 cm
T 0.9–1.8 cm
FA 8.1–8.9 cm
W 66–94 g

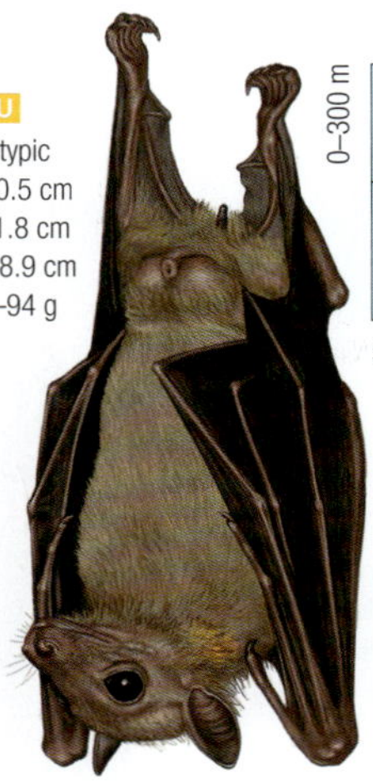

0–300 m

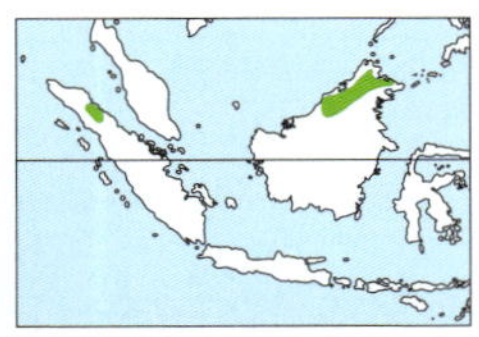

Sumatra (ID), N Borneo (MY, BN, ID)

Leschenault's Rousette
Rousettus leschenaultii

FR Roussette de Leschenault
DE Leschenault-Flughund
ES Rosetus de Leschenault

NT
3 sspp
HB 8–12.5 cm
T 1.3–1.7 cm
FA 7.7–9.6 cm
W 40–92 g

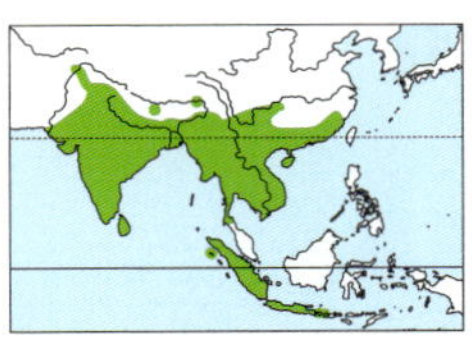

PK, IN, LK E to S CN (with Hainan), mainland SE Asia, Sumatra (with Simeulue I), Java, Bali & Lombok

Linduan Rousette
Rousettus linduensis

FR Roussette du Lindu
DE Lindu-Flughund
ES Rosetus de Lindu

DD Monotypic
HB 10.8–11.5 cm T 2.6–3.2 cm
FA 7.6–7.7 cm W 80–99 g

900–1200 m

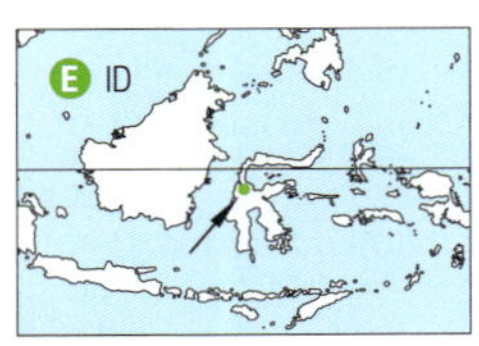

C Sulawesi (Lore Lindu NP)

Egyptian Rousette
Rousettus aegyptiacus

FR Roussette d'Égypte
DE Nil-Flughund
ES Rosetus de Egipto

LC
6 sspp
HB 13.8–19.2 cm
T 0.7–2.5 cm
FA 8.2–10.6 cm
W 81–171 g

0–4000 m

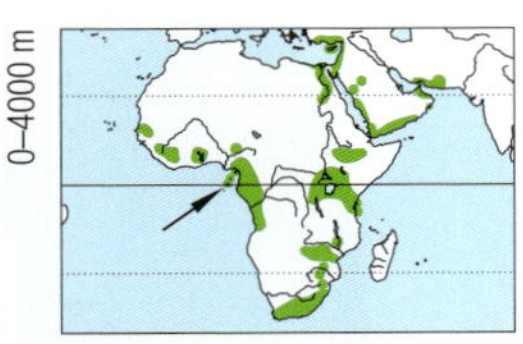

Middle East (from S TR & CY E to SW PK), Africa (from SN to WC AO, from ET to ZA, Bioko I, ST)

Malagasy Rousette
Rousettus madagascariensis

FR Roussette de Madagascar
DE Madagaskar-Flughund
ES Rosetus de Madagascar

VU
Monotypic
HB 11.9–14.3 cm
T 0.6–1.3 cm
FA 6.6–8 cm
W 31–81 g

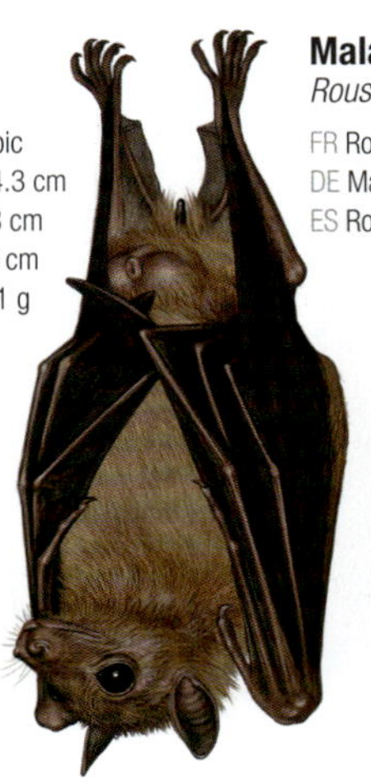

0–1150 m

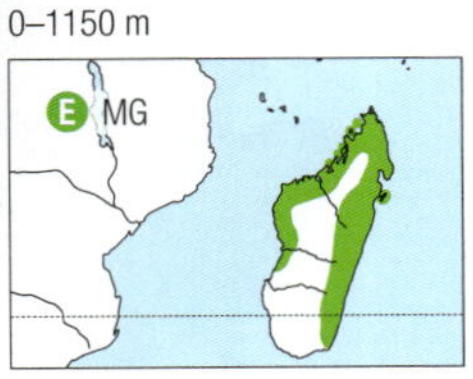

Except C highlands & arid SW region

Comoro Rousette
Rousettus obliviosus

FR Roussette des Comores
DE Komoren-Flughund
ES Rosetus de las Comoras

VU
Monotypic
HB 12.1–14.2 cm
T 1.4–2.5 cm
FA 7–7.8 cm
W 42–73 g

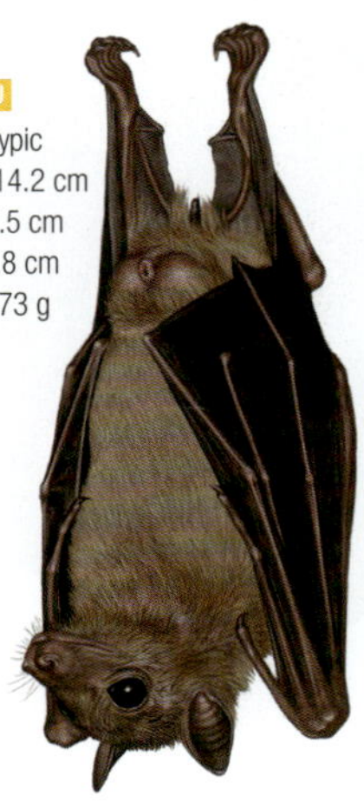

0–1750 m

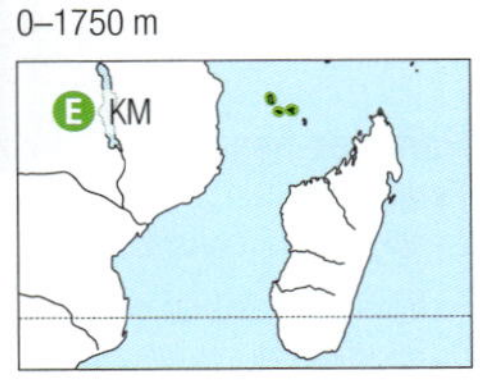

Grande Comore, Mohéli, Anjouan

Genus *PILONYCTERIS*

(1 species)

Sulawesi Fruit Bat
Pilonycteris celebensis

FR Roussette de Sulawesi
DE Sulawesi-Flughund
ES Rosetus de Sulawesi

VU
Monotypic
HB 8.8–10.9 cm
T 2.3–3 cm
FA 7.2–8 cm
W 64–91 g

0–1400 m

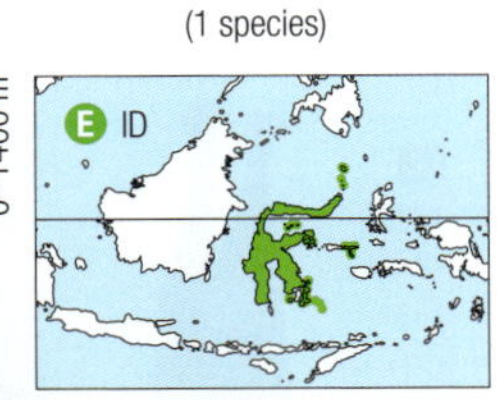

Sulawesi (with nearby Is), Sangihe Is (Sangihe, Siau, Tagulandang), Sula Is (Mangole, Sanana)

Genus *STENONYCTERIS*

(1 species)

Long-haired Fruit Bat
Stenonycteris lanosus

FR Roussette à poil long
DE Langhaariger Flughund
ES Estenonícteris piloso

LC
Monotypic
HB 11.4–17.3 cm
T 0.9–2.5 cm
FA 8.5–9.5 cm
W 94–162 g

500–4000 m

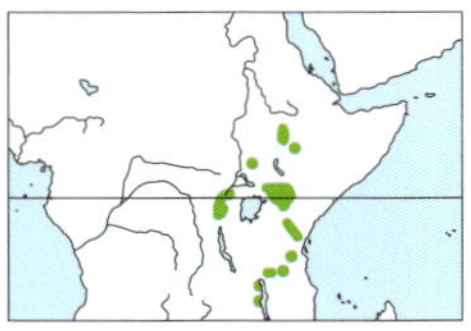

E Africa, fragmented in highlands of W & S ET, S SS, E CD, UG, KE, RW, TZ & MW; likely BI

Genus *MEGALOGLOSSUS*

(2 species)

Western Woermann's Fruit Bat
Megaloglossus azagnyi

FR Mégaloglosse d'Azagny
DE Westlicher Afrika-Langzungenflughund
ES Megaloglosus de Azagny

LC
Monotypic
HB 6.4–8.2 cm
T tailless
FA 3.8–4.2 cm
W 9–14 g

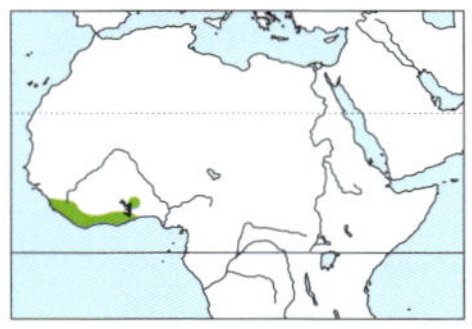

0–760 m

W Africa from GN E to TG

Eastern Woermann's Fruit Bat
Megaloglossus woermanni

FR Mégaloglosse de Woermann
DE Östlicher Afrika-Langzungenflughund
ES Megaloglosus de Woermann

LC
Monotypic
HB 6–7.9 cm
T tailless
FA 3.8–4.9 cm
W 10–25 g

0–1950 m

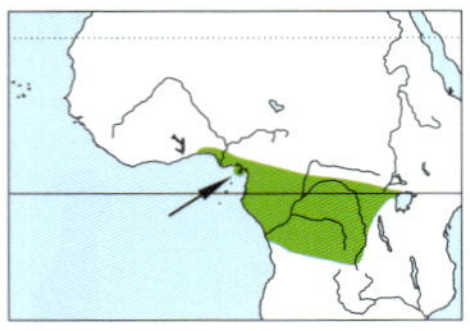

From S BJ & NG E to CD & UG, S to N AO, also Bioko I (GQ)

Genus *LISSONYCTERIS*

(1 species)

Angolan Soft-furred Fruit Bat
Lissonycteris angolensis

FR Lissonyctère d'Angola
DE Angola-Samtfellflughund
ES Lisonícteris de Angola

LC
5 sspp
HB 9–14 cm
T 0.6–2.1 cm
FA 6.8–9 cm
W 60–87 g

0–4000 m

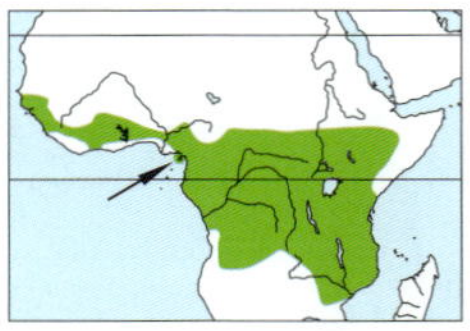

W & C Africa, from SN & GM E to ET (Highlands), S to W AO, N ZM, E ZW & C MZ, Bioko I (GQ)

Genus *MYONYCTERIS*

(4 species)

Little Collared Fruit Bat
Myonycteris torquata

FR Myonyctère à collier
DE Kleiner Halskrausenflughund
ES Mionícteris pequeño

LC
Monotypic
HB 8.6–13 cm
T 0–1.2 cm
FA 5.5–6.6 cm
W 28–43 g (F)

0–1935 m

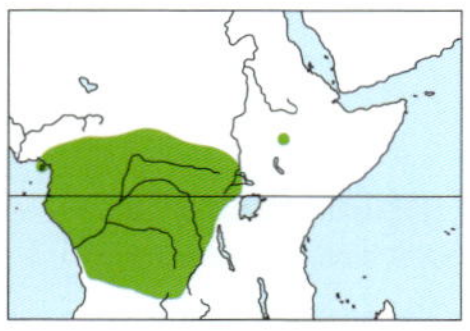

From S CM E to SS & UG, S to N AO & extreme NW ZM; recorded in Beletta Forest (ET)

Sao Tome Collared Fruit Bat
Myonycteris brachycephala

FR Myonyctère de Sao Tomé
DE São-Tomé-Halskrausenflughund
ES Mionícteris de Sao Tomé

EN
Monotypic
HB 9.4–10.6 cm
T c. 0.1 cm
FA 6.2–6.4 cm

300–1300 m

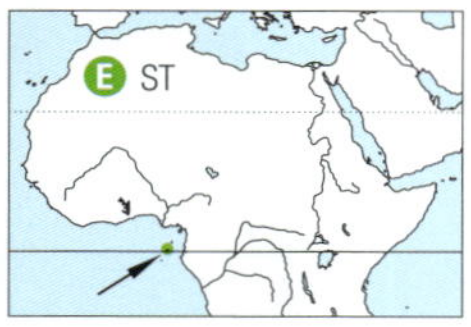

São Tomé I

Sierra Leone Collared Fruit Bat
Myonycteris leptodon

FR Myonyctère ouest-africain
DE Sierra-Leone-Halskrausenflughund
ES Mionícteris de Sierra Leona

30–1350 m

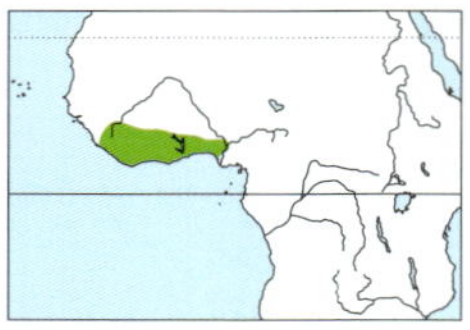

W Africa from SL E to SW NG (Niger Delta); E limits uncertain

LC Monotypic
HB 8.8–12 cm
T 0–1.3 cm
FA 5.6–6.7 cm
W 27–54 g

Bergmans's Collared Fruit Bat
Myonycteris relicta

FR Myonyctère est-africain
DE Relikt-Halskrausenflughund
ES Mionícteris de Bergmans

LC
Monotypic
HB 9–11.5 cm
T 0.5–1.4 cm
FA 6.5–7.5 cm
W 48–54 g

150–1300 m

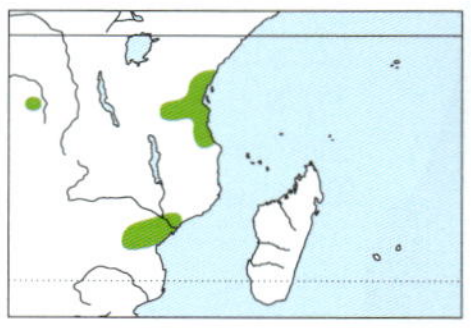

KE (Shimba Hills), TZ (Nguru & Usambara Mts), CD, C MZ (Chinizuia), E ZW (conflux of R Haroni & R Rusitu)

Genus *HYPSIGNATHUS*

(1 species)

0–1800 m

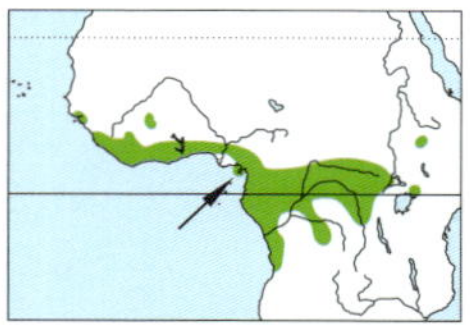

W & C Africa from GN E to W ET, UG & W KE, S to NW AO, Bioko I (GQ)

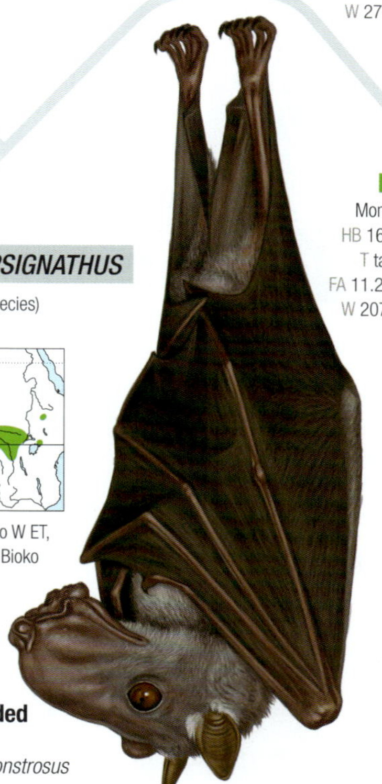

LC
Monotypic
HB 16–30 cm
T tailless
FA 11.2–13.9 cm
W 207–419 g

Hammer-headed Fruit Bat
Hypsignathus monstrosus

FR Hypsignathe monstrueux
DE Hammerkopf
ES Hipsignatus cabezudo

Genus *PLEROTES*

(1 species)

Broad-faced Fruit Bat
Plerotes anchietae

FR Plérote d'Anchieta
DE Anchieta-Breitgesichtflughund
ES Plerotes de Anchieta

LC
Monotypic
HB 7–9.6 cm
T tailless
FA 4.5–5.2 cm
W c. 20 g

1000–2000 m

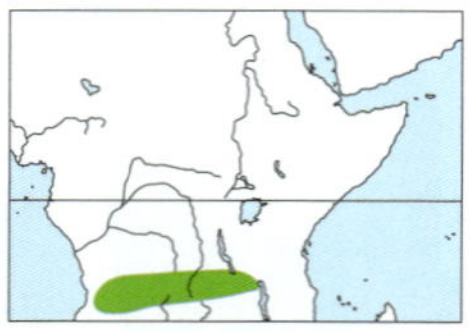

C Africa, from C AO E through S CD & NE ZM to N MW

Genus *EPOMOPS*

(2 species)

Franquet's Epauletted Fruit Bat
Epomops franqueti

FR Épomophore de Franquet
DE Franquet-Epaulettenflughund
ES Epomops de Franquet

LC
Monotypic
HB 11–18 cm
T 0–0.1 cm
FA 7.7–10.1 cm
W 61–172 g

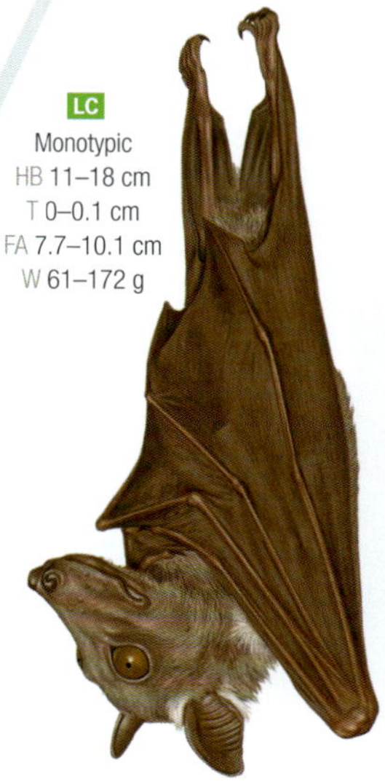

0–1850 m

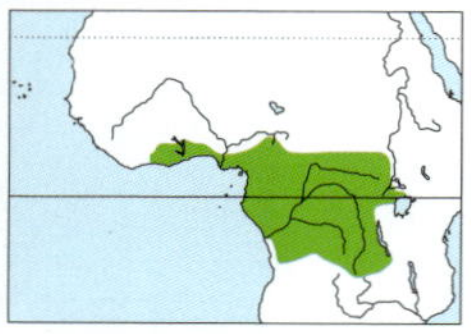

W & C Africa from SE CI E to SS & UG, S to N AO & N ZM

Genus *DOBSONIA*

(13 species)

Lesser Naked-backed Fruit Bat
Dobsonia minor

FR Petite Roussette
DE Kleiner Nacktrückenflughund
ES Dobsonia pequeño

LC
Monotypic
HB 10.2–13.5 cm
T 0.5–2.8 cm
FA 7.8–9.2 cm
W 63–90 g

0–600 m

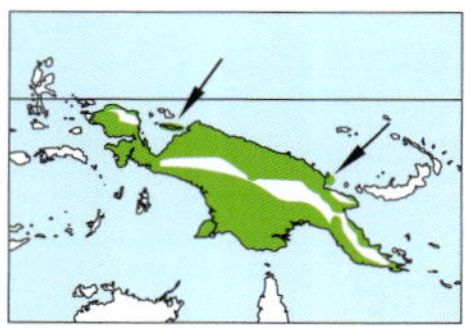

ID/PG: New Guinea (with Yapen & Bagabag Is)

Moluccan Naked-backed Fruit Bat
Dobsonia moluccensis

FR Roussette des Moluques
DE Molukken-Nacktrückenflughund
ES Dobsonia de las Molucas

LC
2 sspp
HB 18.5–25 cm
T 2.1–4.2 cm
FA 12–16.2 cm
W 325–600 g

0–2700 m

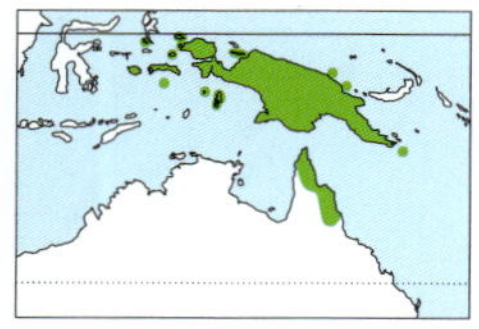

Bacan & S Moluccas, New Guinea (with Raja Ampat, Yapen, Kariru, Karkar & Sideia Is), NE AU (N Qld)

Western Naked-backed Fruit Bat
Dobsonia peronii

FR Roussette de Péron
DE Westlicher Nacktrückenflughund
ES Dobsonia de Péron

LC 2 sspp
HB 14.1–15.8 cm
T 2.9–3.2 cm
FA 11.4–12.1 cm
W 162–235 g

0–880 m

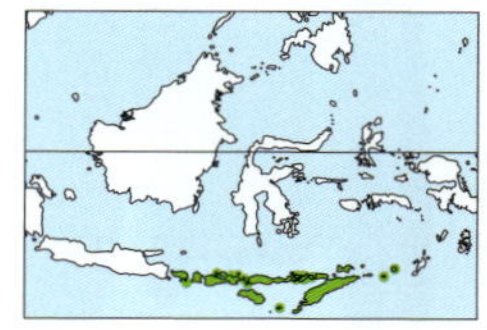

Lesser Sundas (from Bali E to Timor, Sermata & Babar)

Andersen's Naked-backed Fruit Bat
Dobsonia anderseni

FR Roussette d'Andersen
DE Andersen-Nacktrückenflughund
ES Dobsonia de Andersen

LC
Monotypic
HB 13–21.2 cm
T 1.5–3.2 cm
FA 10.9–13.5 cm
W 156–325 g

0–1500 m

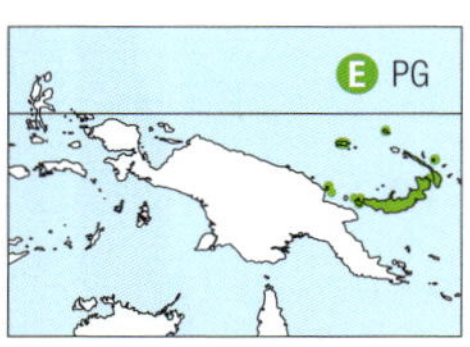

Bismarck Arch, Manus, Mussau & several offshore Is (Karkar, Bagabag, Tolokiwa, Sakar, Umboi)

Panniet Naked-backed Fruit Bat
Dobsonia pannietensis

FR Roussette de Panaeati
DE Panniet-Nacktrückenflughund
ES Dobsonia de Panniet

NT
2 sspp
HB 15–20 cm
T 2.1–3.6 cm
FA 10.8–12.5 cm
W 180–304 g

0–1800 m

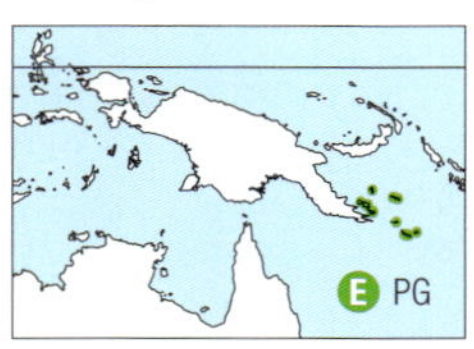

D'Entrecasteaux Is, Louisiade Arch, Trobriand Is (Kiriwina), Woodlark Is

Philippine Naked-backed Fruit Bat
Dobsonia chapmani

FR Roussette de Chapman
DE Chapman-Nacktrückenflughund
ES Dobsonia de Chapman

CR
Monotypic
T 2.3–2.6 cm
FA 12.3–13.1 cm
W 138–143 g

0–860 m

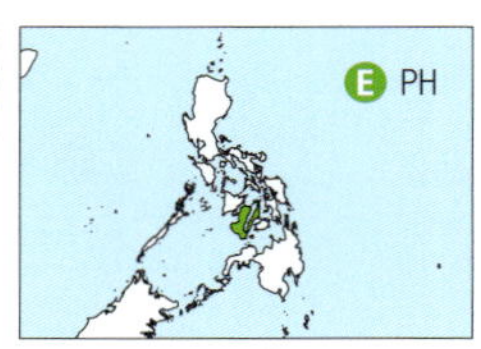

Cebu, Negros

Biak Naked-backed Fruit Bat
Dobsonia emersa

FR Roussette de Biak
DE Biak-Nacktrückenflughund
ES Dobsonia de Biak

VU
Monotypic
HB 15.8–17 cm
T 2.4–2.7 cm
FA 11.2–11.9 cm
W 178–218 g

0–30 m

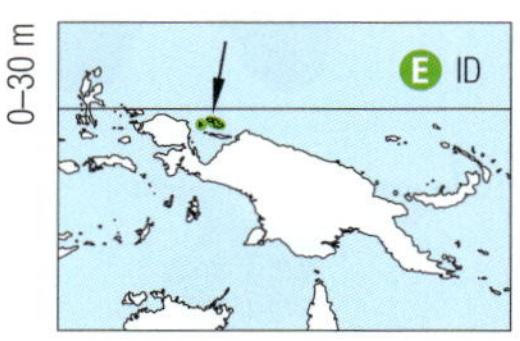

Schouten Is (Numfor, Biak-Supiori, Owi), off NW New Guinea

Sulawesi Naked-backed Fruit Bat
Dobsonia exoleta

FR Roussette de Minahasa
DE Sulawesi-Nacktrückenflughund
ES Dobsonia de Célebes

LC
Monotypic
HB 15–18.2 cm
T 2.3–3.4 cm
FA 10.5–12.4 cm
W 162–275 g

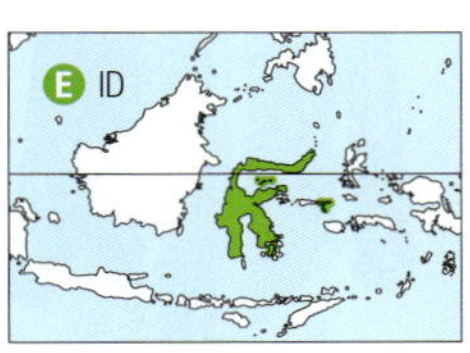

Sulawesi (with Togian & Muna Is), Sula Is (Mangole, Sanana); probably all Togian Is, possibly Buton I

Greenish Naked-backed Fruit Bat
Dobsonia viridis

FR Roussette verdâtre
DE Grüner Nacktrückenflughund
ES Dobsonia verdoso

LC
Monotypic
HB 15.4–17.6 cm
T 2.5–3.9 cm
FA 10.9–11.8 cm
W 173–229 g

0–1000 m

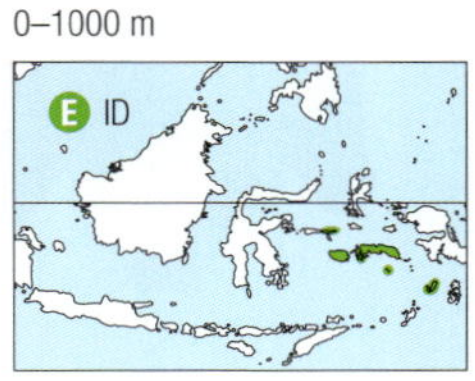

Sula Is (Mangole), C Moluccas (Buru, Seram, Ambon), Banda Is (Banda Neira), Kai Is; perhaps Tanimbar Is

Beaufort's Naked-backed Fruit Bat
Dobsonia beauforti

FR Roussette de Beaufort
DE Waigeo-Nacktrückenflughund
ES Dobsonia de Beaufort

LC
Monotypic
HB 12–15.5 cm
T 2–2.8 cm
FA 10–11.4 cm
W 116–163 g

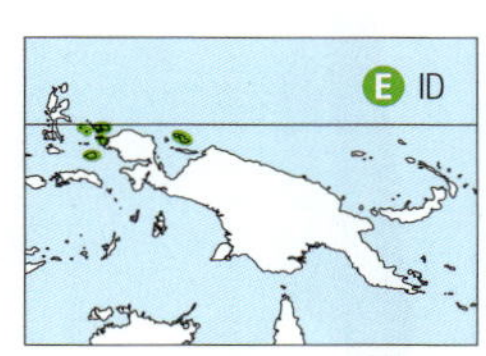

Raja Ampat Is (Gebe, Gag, Waigeo, Batanta, perhaps Misool), Schouten Is (Biak-Supior, Owi)

Halmahera Naked-backed Fruit Bat
Dobsonia crenulata

FR Roussette crénelée
DE Halmahera-Nacktrückenflughund
ES Dobsonia de Halmahera

LC Monotypic
HB 16.5–20 cm
T 2.3–3.8 cm
FA 11.9–13.6 cm
W 220–320 g

0–1000 m

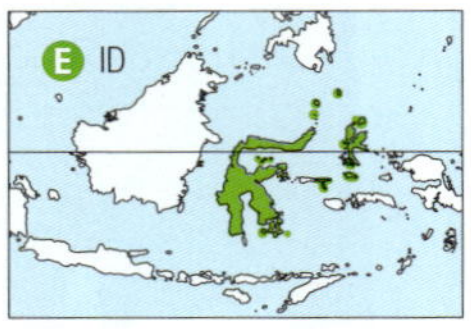

Sulawesi (with offshore Is), Talaud Is, Sangihe Is, Sula Is (Mangole, Sanana), C Moluccas

Solomons Naked-backed Fruit Bat
Dobsonia inermis

FR Roussette des Salomon
DE Salomonen-Nacktrückenflughund
ES Dobsonia de las Salomón

LC 2 sspp
HB 11.8–18.8 cm
T 1.3–3.2 cm
FA 9.1–11.6 cm
W 125–215 g

0–1300 m

Solomon Is (PG, SB)

New Britain Naked-backed Fruit Bat
Dobsonia praedatrix

FR Roussette de Nouvelle-Bretagne
DE Neubritannien-Nacktrückenflughund
ES Dobsonia de Nueva Bretaña

LC Monotypic
HB 11.6–18 cm
T 2.1–3.7 cm
FA 10.1–12.5 cm
W 141–260 g

0–300 m

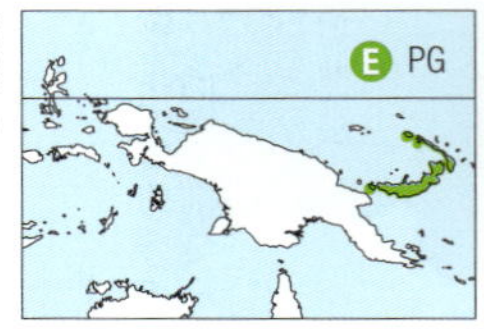

Bismarck Arch (with Umboi I and nearby Is)

Genus *EIDOLON*

(2 species)

African Straw-colored Fruit Bat
Eidolon helvum

FR Roussette-paillée d'Afrique
DE Palmenflughund
ES Eidolon de África

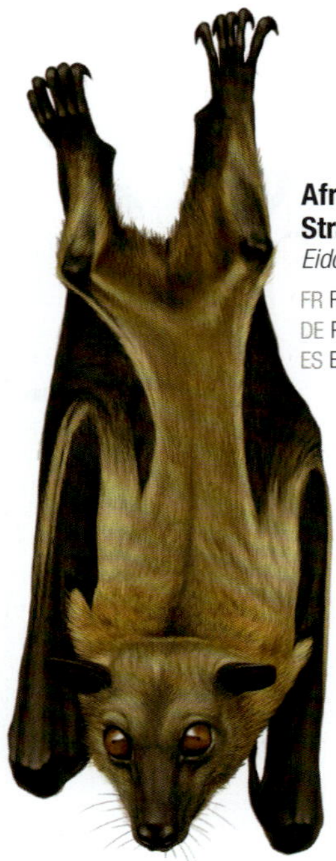

NT
3 sspp
HB 15–19.5 cm
T 1–1.5 cm
FA 11.7–13.2 cm
W 230–350 g

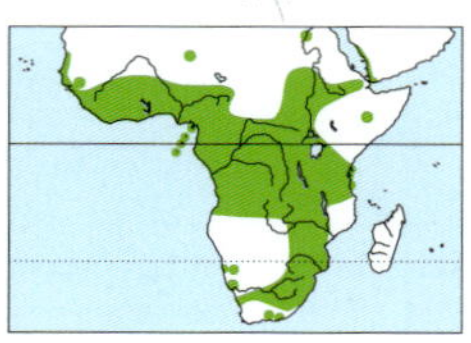

0–2000 m

Across sub-Saharan & equatorial Africa, including Gulf of Guinea Is & Zanzibar Arch, SW Arabian Pen

Madagascan Straw-colored Fruit Bat
Eidolon dupreanum

FR Roussette-paillée de Madagascar
DE Madagaskar-Palmenflughund
ES Eidolon de Madagascar

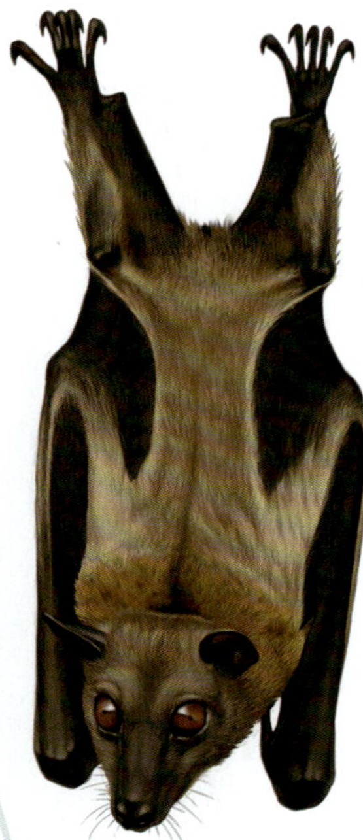

VU
Monotypic
HB 19–21 cm
T 1.4–2.2 cm
FA 11.7–13.4 cm
W 285–300 g

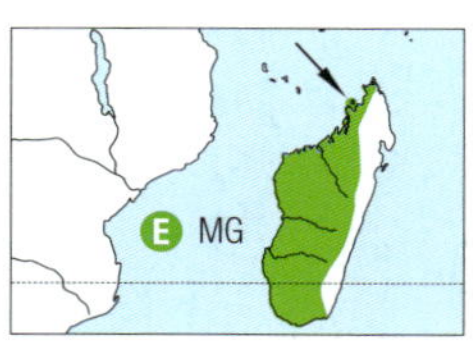

10–1200 m

W coast & Central High Plateau, Nosy Be

Unstriped Tube-nosed Fruit Bat
Paranyctimene raptor

FR Nyctimène rapace
DE Ungestreifter Röhrennasenflughund
ES Paraníctimo rapaz

LC Monotypic
HB 6.3–9.3 cm T 1.6–2.1 cm
FA 4.8–5.4 cm W 22–33 g

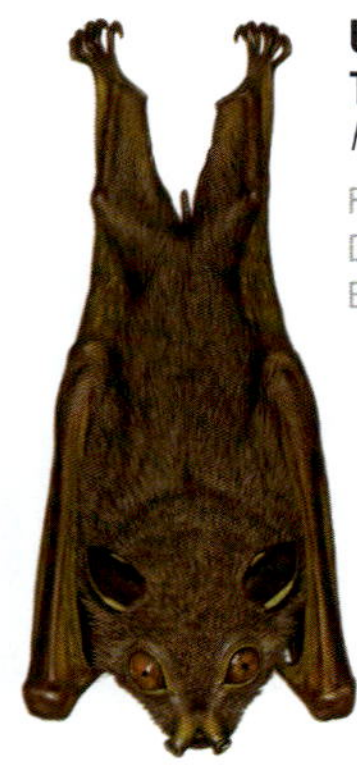

0–1350 m

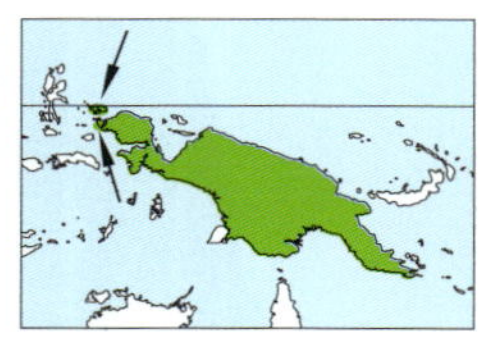

ID/PG: New Guinea (possibly widespread in N lowlands), Raja Ampat Is (Waigeo, Salawati)

Genus *PARANYCTIMENE*

(2 species)

Steadfast Tube-nosed Fruit Bat
Paranyctimene tenax

FR Nyctimène de Nijhoff
DE Standhafter Röhrennasenflughund
ES Paraníctimo de Nijhoff

0–1350 m

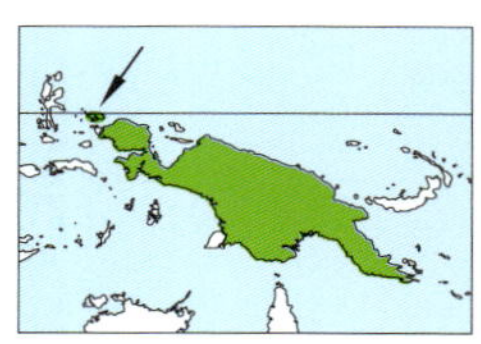

ID/PG: New Guinea (primarily S lowlands), Raja Ampat Is (Waigeo)

LC 2 sspp
T 1.5–2.1 cm
FA 5.1–5.5 cm
W c. 25 g

Broad-striped Tube-nosed Fruit Bat
Nyctimene aello

FR Grand Nyctimène
DE Breitstreifen-Röhrennasenflughund
ES Níctimo grande

LC
Monotypic
HB 10.3–14.4 cm
T 1.4–3.1 cm
FA 7.7–8.8 cm
W 70–93 g

0–1000 m

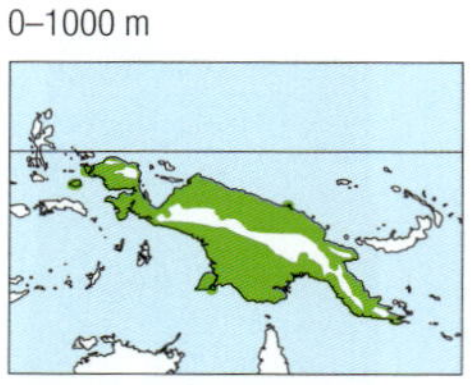

ID/PG: lowland New Guinea, Raja Ampat Is (Waigeo, Misool), Kairiru I & Admosin I, off N coast

Genus *NYCTIMENE*

(15 species)

Round-eared Tube-nosed Fruit Bat
Nyctimene cyclotis

FR Nyctimène à oreilles rondes
DE Rundohr-Röhrennasenflughund
ES Níctimo de orejas redondas

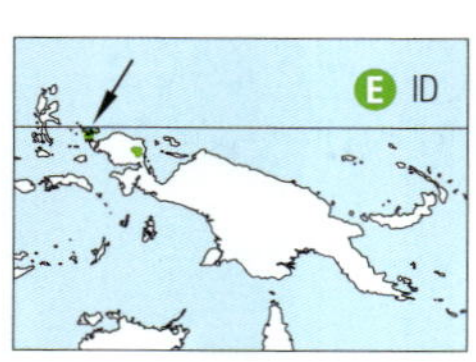

NW New Guinea (Arfak Mts); tentatively Raja Ampat Is (Waigeo, Mansuar)

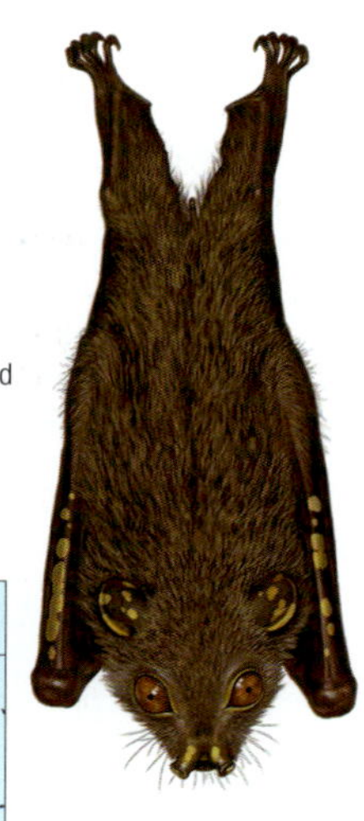

DD
Monotypic
FA 5.4–5.6 cm

Mountain Tube-nosed Fruit Bat
Nyctimene certans

FR Nyctimène de montagne
DE Gebirgs-Röhrennasenflughund
ES Níctimo montano

LC
Monotypic
HB 8.1–11.8 cm
T 1.2–3 cm
FA 5.5–6.9 cm
W 33–48 g

700–3000 m

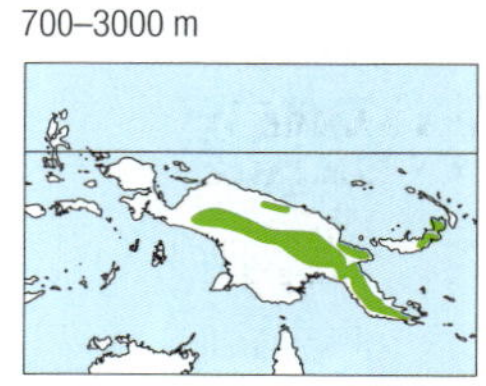

ID/PG: New Guinea Mts; tentatively Bismarck Arch (E New Britain)

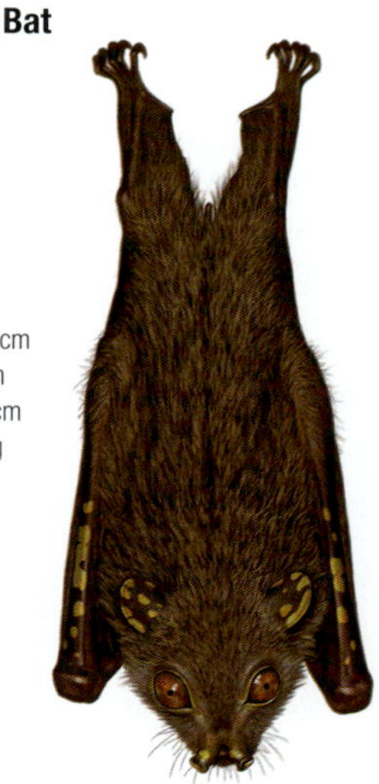

Happy Tube-nosed Fruit Bat
Nyctimene wrightae

FR Nyctimène de Wright
DE Wright-Röhrennasenflughund
ES Níctimo de Wright

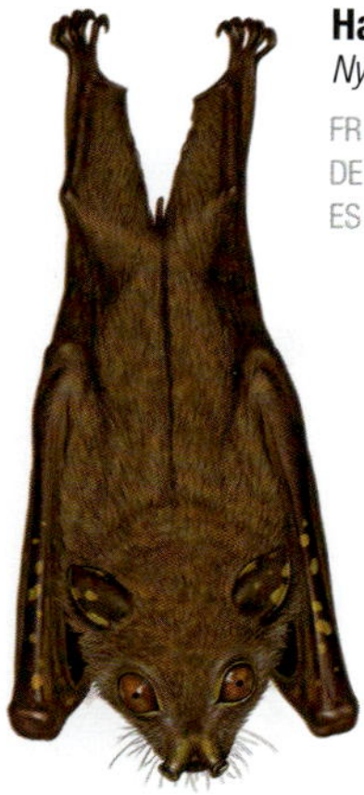

LC Monotypic
HB 8.4–8.6 cm T 1.9–2.2 cm
FA 5.2–6.3 cm W 34–36 g

40–770 m

ID/PG: New Guinea in lowland to montane rainforests

Lesser Tube-nosed Fruit Bat
Nyctimene varius

FR Petit Nyctimène
DE Kleiner Röhrennasenflughund
ES Níctimo pequeño

700–900 m

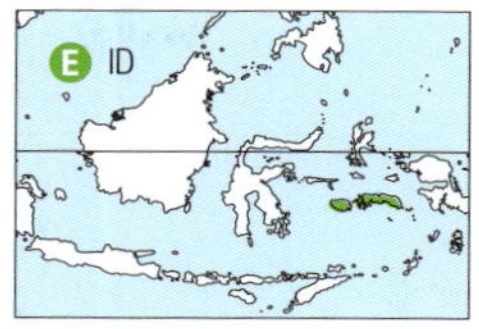

C Moluccas (Buru, Seram)

VU Monotypic
FA c. 5.5 cm

Common Tube-nosed Fruit Bat
Nyctimene albiventer

FR Nyctimène à ventre blanc
DE Gemeiner Röhrennasenflughund
ES Níctimo de vientre blanco

LC 2 sspp
HB 7.4–11.5 cm T 1.8–2.5 cm
FA 5.5–6.7 cm W 27–55 g

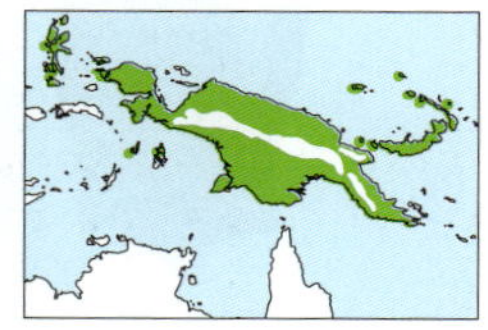

N Moluccas, Raja Ampat Is, Kai Is, N Aru Is, lowland New Guinea, Admiralty Is, Bismarck Arch

Dragon Tube-nosed Fruit Bat
Nyctimene draconilla

FR Nyctimène dragon
DE Drachen-Röhrennasenflughund
ES Níctimo dragón

DD
Monotypic
HB 7–8.8 cm
T 1.5–2.2 cm
FA 4.8–5.4 cm
W 26–31 g

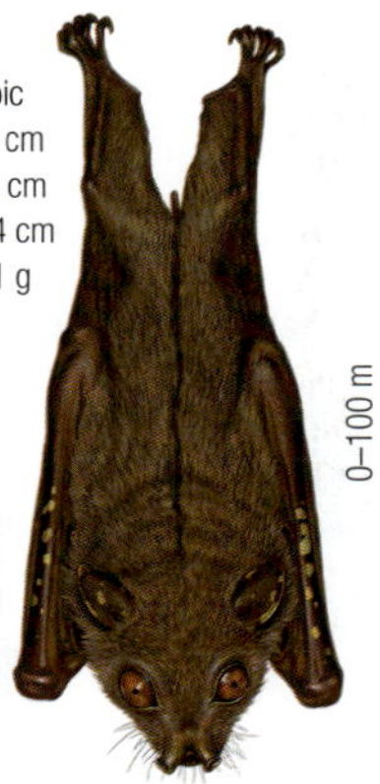

0–100 m

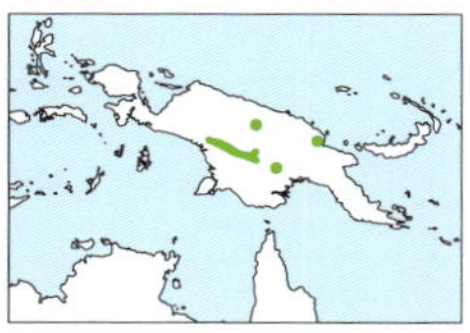

ID/PG: mainly lowlands S of Central Range of New Guinea

Demonic Tube-nosed Fruit Bat
Nyctimene masalai

FR Nyctimène de Nouvelle-Irlande
DE Waldgeist-Röhrennasenflughund
ES Níctimo de Nueva Irlanda

DD Monotypic
HB 9.4–10.3 cm T 2.1–2.2 cm
FA 6.3–6.7 cm W 45–52 g

Bismarck Arch (New Ireland, perhaps other Is)

Malaita Tube-nosed Fruit Bat
Nyctimene malaitensis

FR Nyctimène de Malaita
DE Malaita-Röhrennasenflughund
ES Níctimo de Malaita

LC Monotypic
HB c. 11.8 cm T c. 2.3 cm
FA c. 6.5 cm

0–1100 m

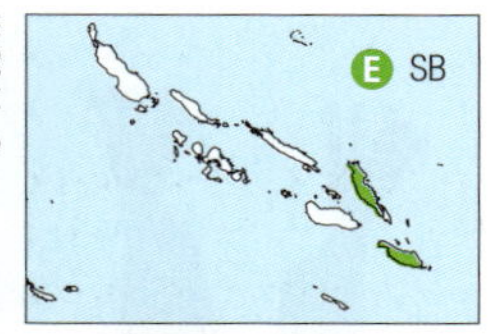

Malaita, Makira; perhaps smaller satellite Is (such as Ugi)

Umboi Tube-nosed Fruit Bat
Nyctimene vizcaccia

FR Nyctimène d'Umboi
DE Umboi-Röhrennasenflughund
ES Níctimo de Umboi

LC
2 sspp
HB 7.4–11.5 cm
T 1.5–2.5 cm
FA 5.2–6.7 cm
W 27–55 g

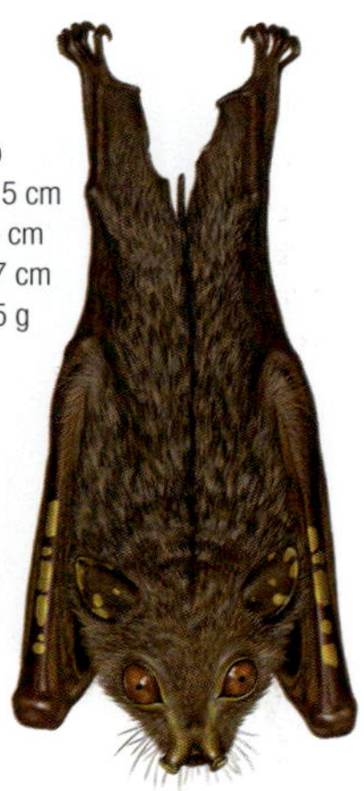

0–1800 m

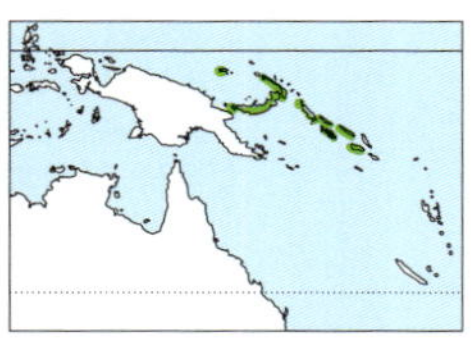

Admiralty Is (Manus), Bismarck Arch (New Ireland, New Britain, Umboi), Solomon Is (Buka S to Guadalcanal)

Island Tube-nosed Fruit Bat
Nyctimene major

FR Nyctimène géant
DE Großer Röhrennasenflughund
ES Níctimo gigante

LC
5 sspp
HB 9–13.8 cm
T 1.8–3.3 cm
FA 6.5–8.7 cm
W 52–100 g

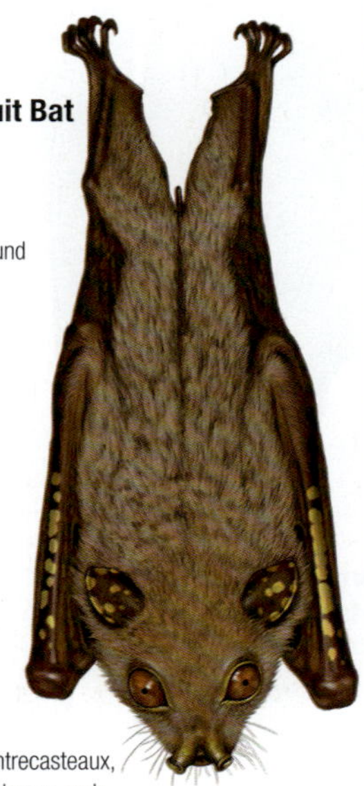

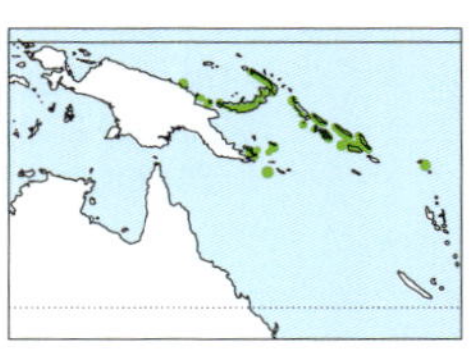

0–900 m

Bismarck, D'Entrecasteaux, Louisiade & Solomon arch, several other Is (E to Nendö)

Philippine Tube-nosed Fruit Bat
Nyctimene rabori

FR Nyctimène des Philippines
DE Philippinen-Röhrennasenflughund
ES Níctimo de las Filipinas

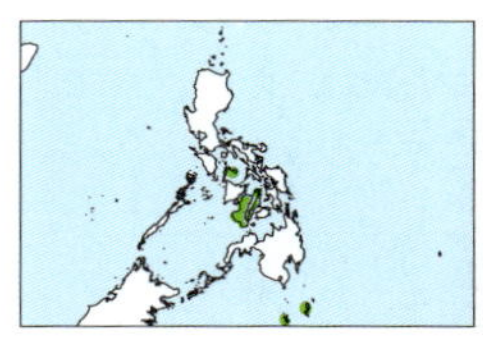

0–1300 m

PH (Sibuyan, Cebu, Negros, probably Panay), tentatively Talaud & Sangihe (ID)

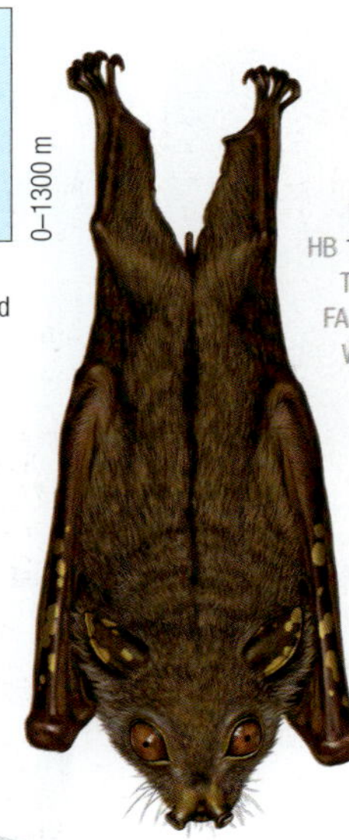

EN
Monotypic
HB 11.3–12.4 cm
T 1.9–3 cm
FA 7.3–8.1 cm
W 60–74 g

Pallas's Tube-nosed Fruit Bat
Nyctimene cephalotes

FR Nyctimène à grosse tête
DE Pallas-Röhrennasenflughund
ES Níctimo de Pallas

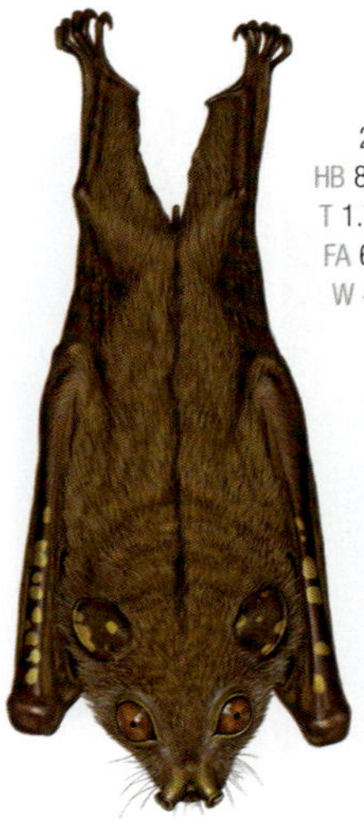

LC
2 sspp
HB 8.7–10 cm
T 1.7–2.5 cm
FA 6.2–7 cm
W 40–49 g

0–1800 m

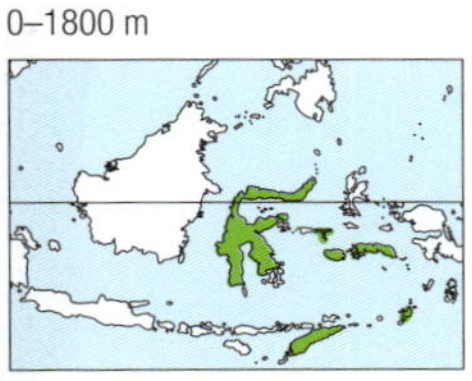

Sulawesi, Sula Is (Mangole, Sanana), C Moluccas, Tanimbar Is (Yamdena), Timor (where perhaps extinct)

Keast's Tube-nosed Fruit Bat
Nyctimene keasti

FR Nyctimène de Keast
DE Keast-Röhrennasenflughund
ES Níctimo de Keast

NT
3 sspp
HB 6.9–8.9 cm
T 1.8–2.5 cm
FA 4.1–4.6 cm
W 29–30 g

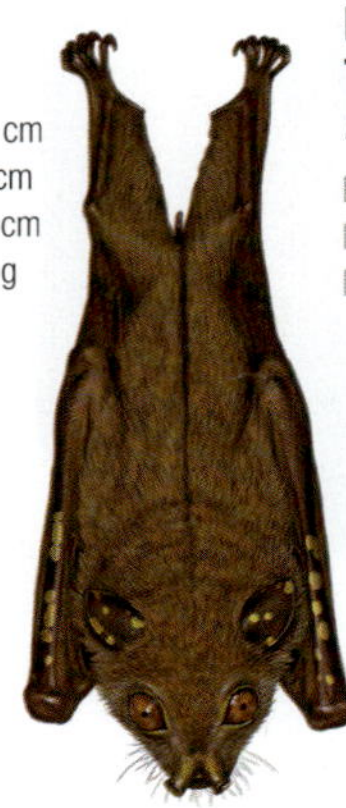

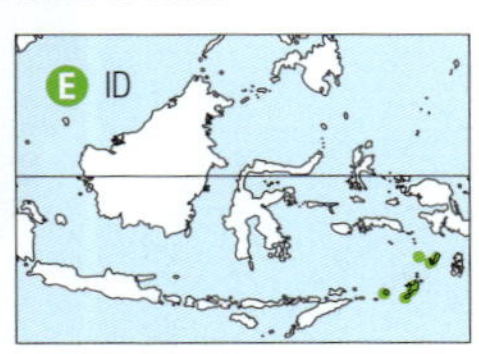

0–1000 m

Babar I, Kai Is, Tanimbar Is (Larat, Yamdena, Salaru); unconfirmed in Flores, Timor & Gag Is

Queensland Tube-nosed Fruit Bat
Nyctimene robinsoni

FR Nyctimène du Queensland
DE Queensland-Röhrennasenflughund
ES Níctimo de Queensland

LC
Monotypic
HB 8.3–9.3 cm
T 2.1–2.4 cm
FA 6.5–6.9 cm
W 42–59 g

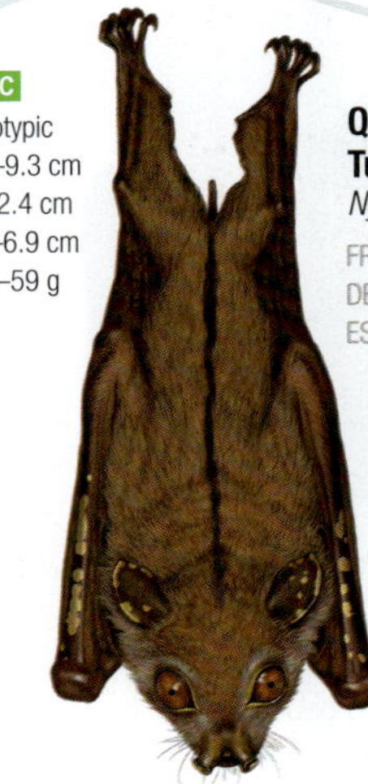

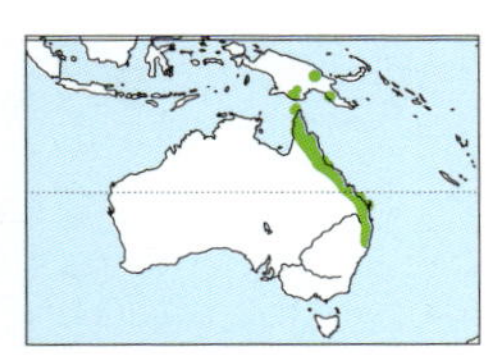

NE & E AU (E Qld & NE NSW, with several Is); tentatively PG

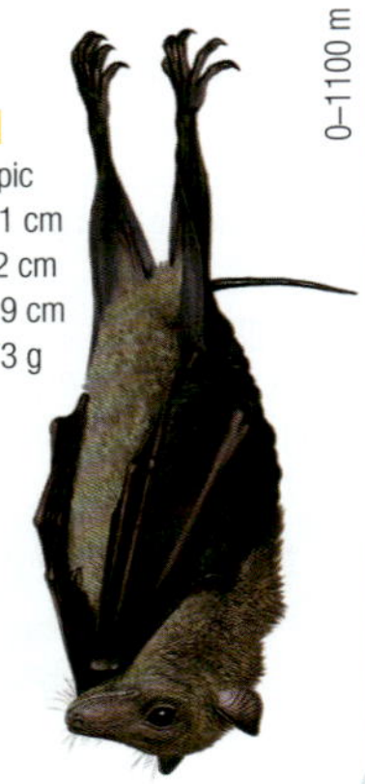

VU

Monotypic
HB 9.5–11 cm
T 5.7–6.2 cm
FA 6.5–6.9 cm
W 56–73 g

0–1100 m

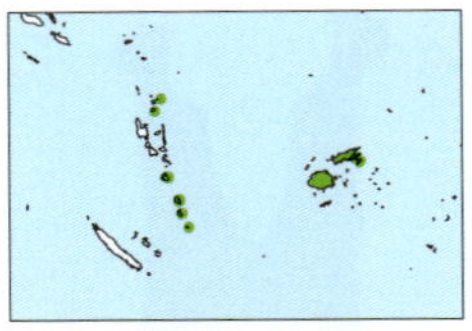

VU (Mota Lava, Santa Maria, Efate S to Anatom), FJ (Vanua Levu, Taveuni, Viti Levu)

Fijian Long-tailed Fruit Bat

Notopteris macdonaldii

FR Notoptère des Fidji
DE Fidschi-Langschwanzflughund
ES Notóptero de Fiyi

Genus *NOTOPTERIS*

(2 species)

New Caledonia Long-tailed Fruit Bat

Notopteris neocaledonicus

FR Notoptère de Nouvelle-Calédonie
DE Neukaledonien-Langschwanzflughund
ES Notóptero de Nueva Caledonia

EN

Monotypic
HB 9.3–10 cm
T 4.3–5.4 cm
FA 5.8–6 cm

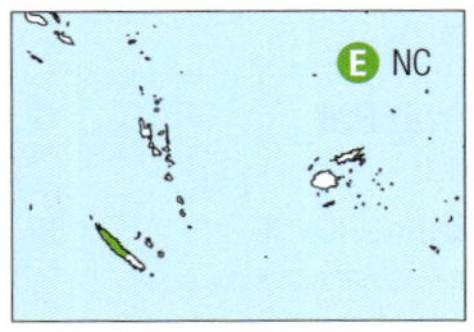

Known with certainty from only 2 caves in N

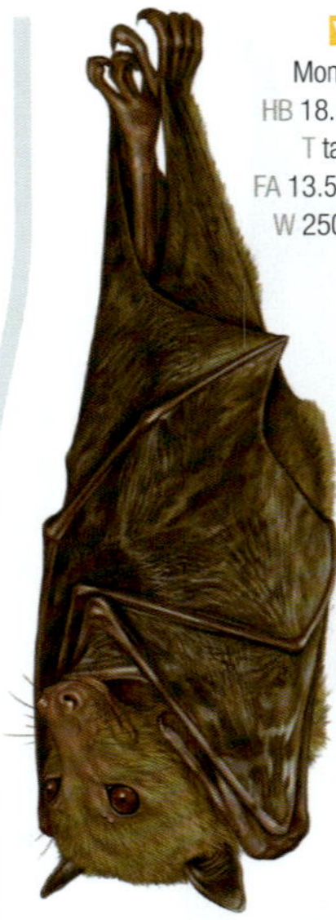

VU

Monotypic
HB 18.5–24 cm
T tailless
FA 13.5–14.5 cm
W 250–375 g

0–1200 m

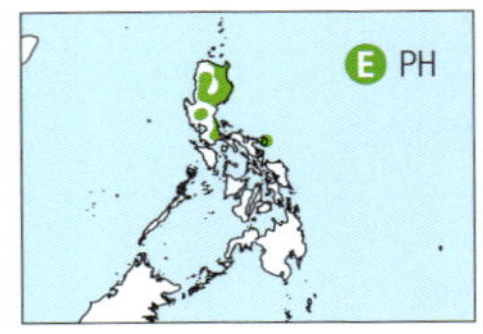

Luzon, Catanduanes

White-winged Flying Fox

Desmalopex leucopterus

FR Roussette leucoptère
DE Weißflügelflughund
ES Desmalopex de alas blancas

Genus *DESMALOPEX*

(2 species)

Mindoro Pallid Flying Fox

Desmalopex microleucopterus

FR Roussette microptère
DE Mindoro-Weißflügelflughund
ES Desmalopex de Mindoro

EN

Monotypic
HB 13.3–15.5 cm
T tailless
FA 9.7–10.3 cm
W 129–156 g

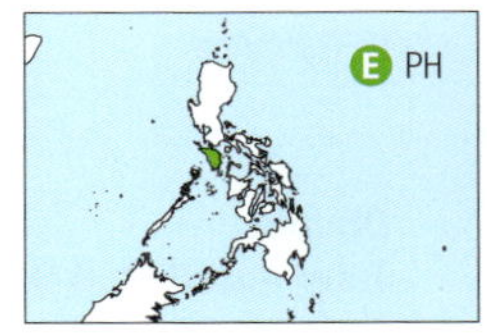

Mindoro 0–750 m

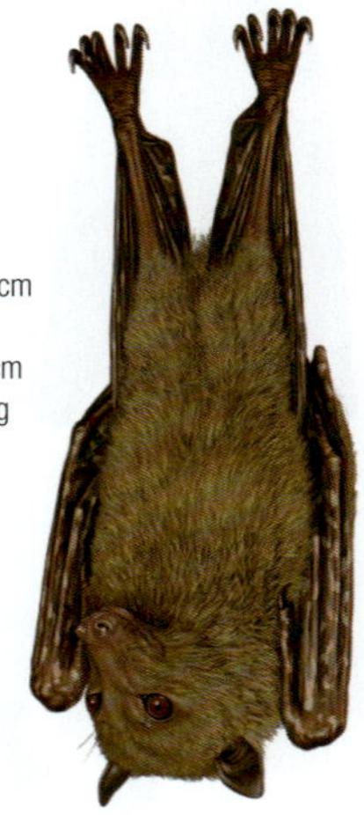

Genus *MIRIMIRI*

(1 species)

Fijian Monkey-faced Fruit Bat

Mirimiri acrodonta

FR Roussette des Fidji
DE Fidschi-Flughund
ES Mirimiri de Fiyi

CR

Monotypic
HB 17–20 cm
T tailless
FA 11.1–12.7 cm
W 222–264 g

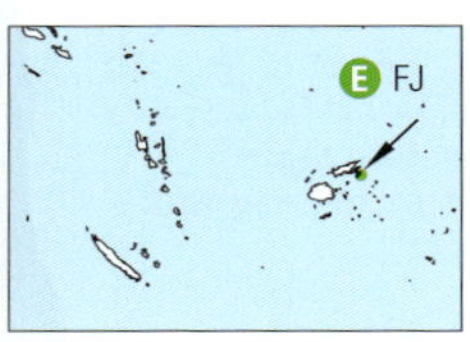

Taveuni above 800 m

EN

Monotypic
HB 23–28 cm
T tailless
FA 14.1–16 cm
W 600–700 g

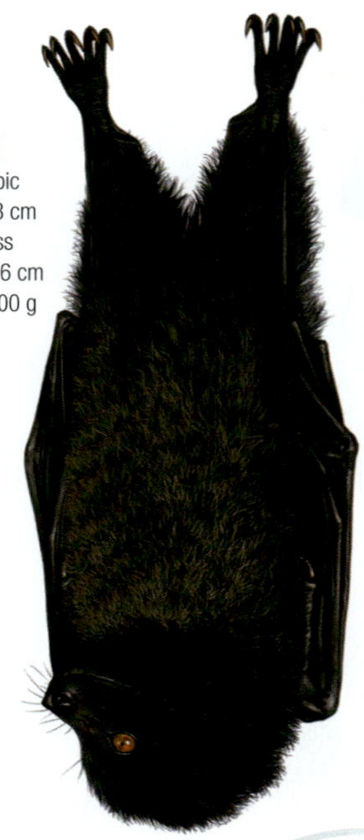

Genus *PTERALOPEX*

(5 species)

0–1900 m

PG/SB: Solomon Is (Bougainville, Choiseul); perhaps Santa Isabel

Bougainville Monkey-faced Fruit Bat

Pteralopex anceps

FR Roussette de Bougainville
DE Bougainville-Affengesichtflughund
ES Pteralopex de Bougainville

Guadalcanal Monkey-faced Fruit Bat

Pteralopex atrata

FR Roussette de Guadalcanal
DE Guadalcanal-Affengesichtflughund
ES Pteralopex de Guadalcanal

0–400 m

New Georgia, Guadalcanal

EN

Monotypic
HB 19.6–24 cm
T tailless
FA 12.9–14.7 cm
W 438–506 g

Montane Monkey-faced Fruit Bat

Pteralopex pulchra

FR Roussette de montagne
DE Berg-Affengesichtflughund
ES Pteralopex montano

CR

Monotypic
HB c. 16.1 cm
T tailless
FA c. 11.8 cm
W c. 280 g

1200–2500 m

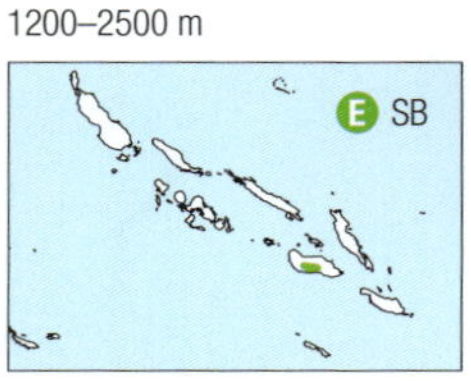

Known only from holotype in montane Guadalcanal

New Georgia Monkey-faced Fruit Bat

Pteralopex taki

FR Roussette de Nouvelle-Géorgie
DE New-Georgia-Affengesichtflughund
ES Pteralopex de Nueva Georgia

VU

Monotypic
HB c. 19 cm
T tailless
FA 11.2–12.3 cm
W 225–351 g

0–400 m

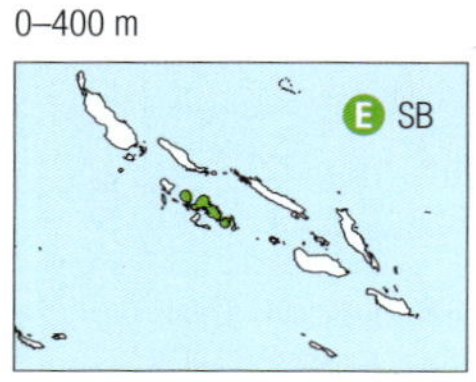

Kolombangara, New Georgia, Vangunu

Greater Monkey-faced Fruit Bat

Pteralopex flanneryi

FR Roussette de Flannery
DE Flannery-Affengesichtflughund
ES Pteralopex de Flannery

CR

Monotypic
HB 25–28 cm
T tailless
FA 15.9–16.9 cm
W c. 790 g

0–200 m

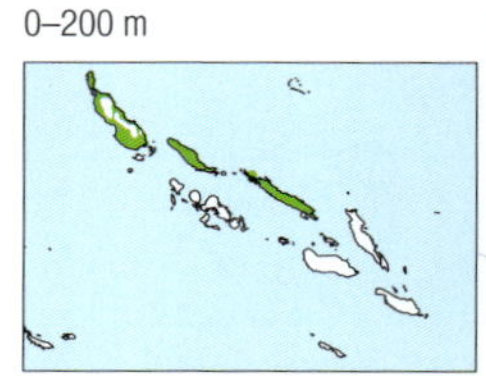

PG/SB: Solomon Is (Buka, Bougainville, Choiseul, Barora Fa, Santa Isabel)

Genus *MELONYCTERIS*

(1 species)

Black-bellied Blossom Bat
Melonycteris melanops

FR Mélonyctère à ventre noir
DE Schwarzbauch-Blütenflughund
ES Melonícteris de vientre cenizo

LC
Monotypic
HB 7.8–11 cm
T tailless
FA 5–6.8 cm
W 31–63 g

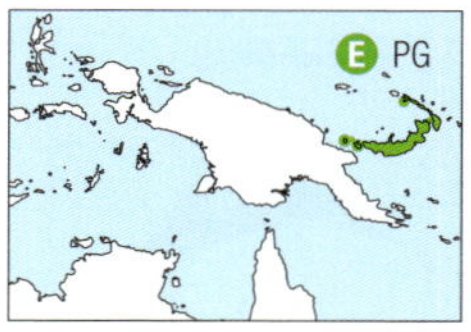

Bismarck Arch 0–1600 m

Fardoulis's Blossom Bat
Nesonycteris fardoulisi

FR Mélonyctère de Fardoulis
DE Fardoulis-Blütenflughund
ES Melonícteris de Fardoulis

ssp *fardoulisi*

ssp *maccoyi*

NT
4 sspp
HB 8.7–9.6 cm
T tailless
FA 5.8–6.1 cm
W 39–54 g

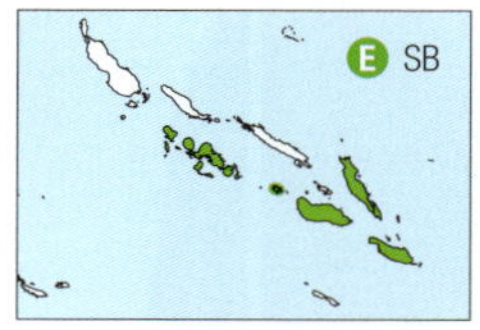

Vella Lavella, Kolombangara, New Georgia, Vangunu, Russell, Malaita, Guadalcanal, Makira

Genus *NESONYCTERIS*

(2 species)

Woodford's Blossom Bat
Nesonycteris woodfordi

FR Mélonyctère de Woodford
DE Woodford-Blütenflughund
ES Melonícteris de Woodford

0–1100 m

PG/SB: Solomon Is (Buka, Bougainville, Shortland Is, Fauro, Choiseul, Santa Isabel, Nggela)

LC
2 sspp
HB 8–10.6 cm
T tailless
FA 4.8–5.9 cm
W 25–50 g

Genus *STYLOCTENIUM*

(2 species)

Wallace's Stripe-faced Fruit Bat
Styloctenium wallacei

FR Roussette de Wallace
DE Sulawesi-Streifengesichtflughund
ES Estilo de Wallace

NT Monotypic
HB 14.5–20 cm
T tailless
FA 9–10.3 cm
W 174–218 g

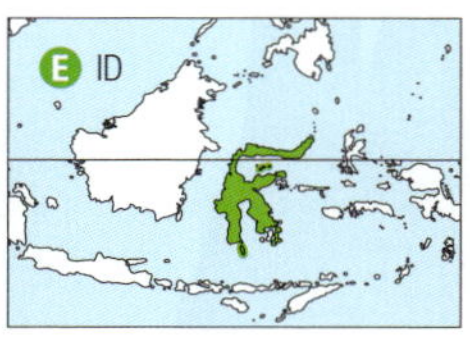

0–1100 m

Sulawesi, nearby Togian, Buton & Selayar Is

Mindoro Stripe-faced Fruit Bat
Styloctenium mindorensis

FR Roussette de Mindoro
DE Mindoro-Streifengesichtflughund
ES Estilo de Mindoro

EN
Monotypic
HB 14.8–18 cm
T tailless
FA 10.1–11.4 cm
W 149–212 g

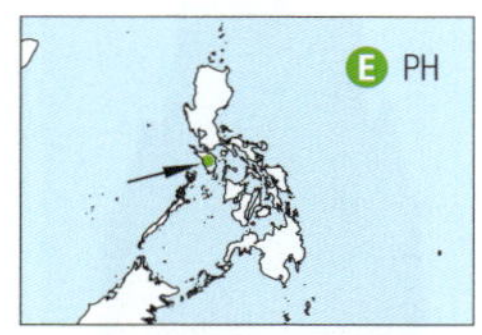

Mindoro (known only from Mt Siburan)

Genus *NEOPTERYX*

(1 species)

Small-toothed Fruit Bat
Neopteryx frosti

FR Roussette de Frost
DE Kleinzahnflughund
ES Neóptero de Frost

EN
Monotypic
HB c. 15.5 cm
T tailless
FA 10.5–11.1 cm
W 190–250 g

225–1005 m

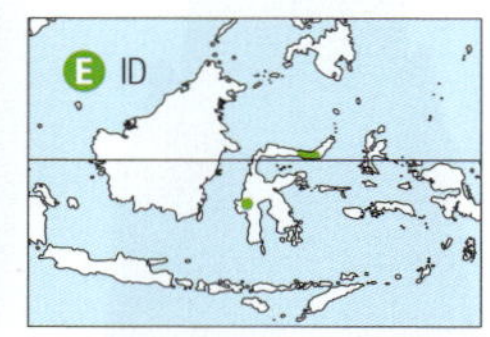

Known from 2 sites in Sulawesi (Tamalanti & Bogani Nani Wartabone NP)

Genus *ACERODON*

(5 species)

Sulawesi Flying Fox
Acerodon celebensis

FR Acérodon de Sulawesi
DE Großer Sulawesi-Flughund
ES Zorro volador de Célebes

0–1500 m

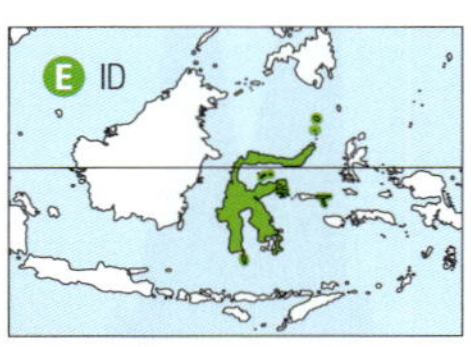

Sulawesi & nearby Is, Sula Is (Mangole, Sanana)

VU
Monotypic
HB 19–22 cm
T tailless
FA 13–14.5 cm
W c. 374 g

Talaud Flying Fox
Acerodon humilis

FR Acérodon des Talaud
DE Talaud-Flughund
ES Zorro volador de Talaud

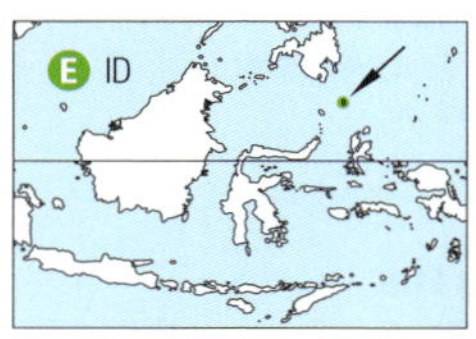

Talaud Is (Salebabu, Karakelong)

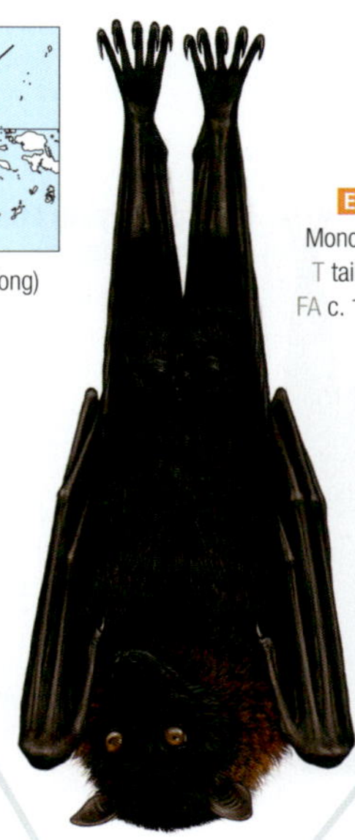

EN
Monotypic
T tailless
FA c. 14 cm

Golden-capped Flying Fox
Acerodon jubatus

FR Acérodon couronné
DE Goldkronenflughund
ES Zorro volador coronado

0–1100 m

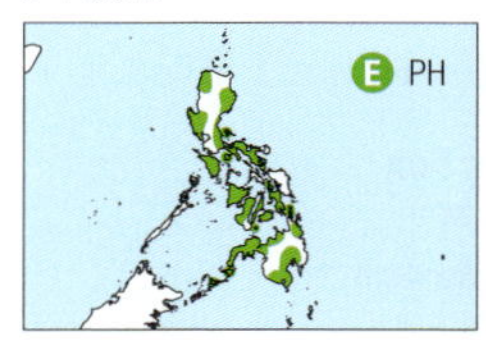

Widespread (except Batanes, Babuyan & Palawan regions; probably exctinct in Panay)

EN
2 extant sspp
HB 26–30 cm
T tailless
FA 18.2–20.5 cm
W 0.8–1.4 kg

Palawan Flying Fox
Acerodon leucotis

FR Acérodon de Palawan
DE Palawan-Flughund
ES Zorro volador de Palawan

VU
2 sspp
T tailless
FA 13–16.5 cm

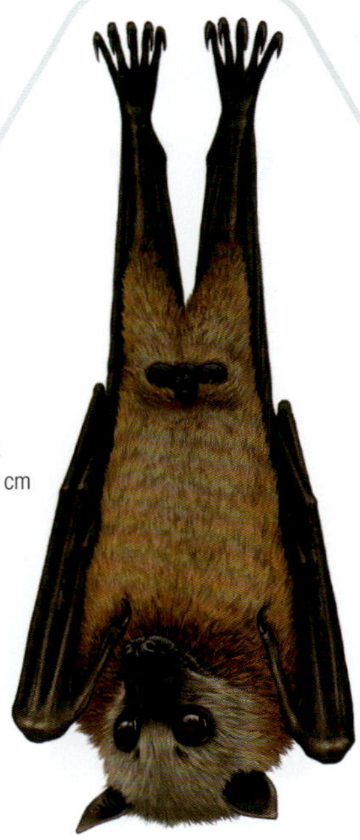

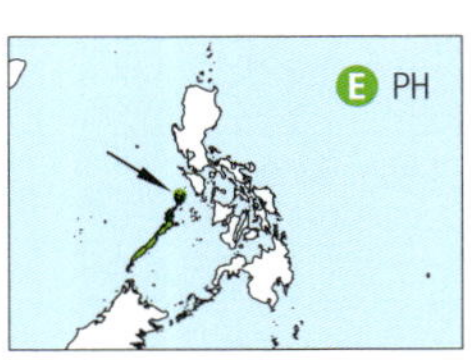

Calamian Is (Busuanga), Palawan & Balabac Is

Sunda Flying Fox
Acerodon mackloti

FR Acérodon de Macklot
DE Sunda-Flughund
ES Zorro volador de Macklot

VU
5 sspp
HB c. 23 cm
T tailless
FA 13.9–14.8 cm
W c. 518 g

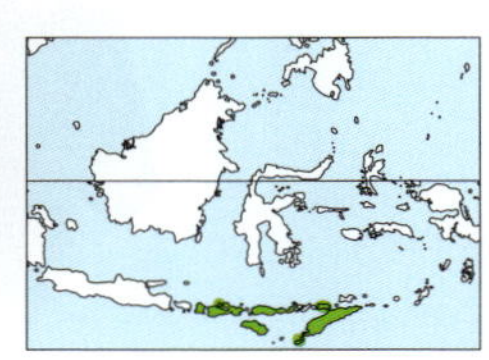

0–450 m

Lesser Sundas (Lombok, Sumbawa, Moyo, Sumba, Flores, Alor, Roti, Timor)

Genus *PTEROPUS*

(57 species)

Moluccan Masked Flying Fox
Pteropus personatus

FR Roussette masquée
DE Maskenflughund
ES Zorro volador enmascarado

LC
2 sspp
HB 12.4–14.9 cm
T tailless
FA 8.6–9.8 cm
W 92–153 g

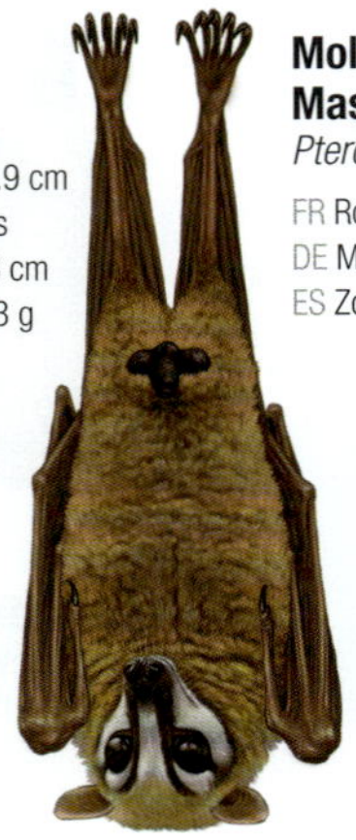

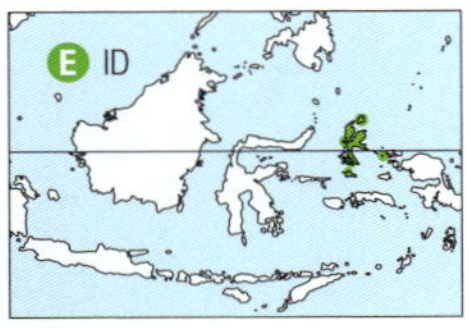

N Moluccas (Halmahera & Obi groups), Raja Ampat Is (Gag)

Lombok Flying Fox
Pteropus lombocensis

FR Roussette de Lombok
DE Lombok-Flughund
ES Zorro volador de Lombok

up to 880 m

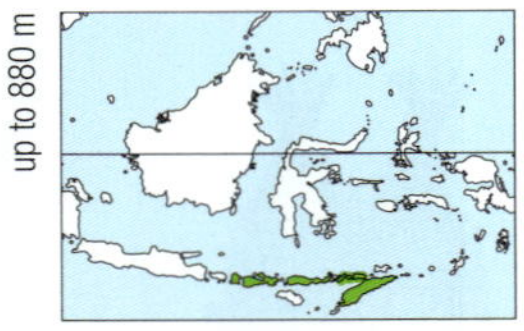

W Lesser Sundas (Lombok, Sumbawa, Flores, Lembata, Pantar, Alor, Timor)

DD
3 sspp
HB 15–18 cm
T tailless
FA 11–12.2 cm
W 215–360 g

Little Red Flying Fox
Pteropus scapulatus

FR Roussette à collier
DE Kleiner Roter Flughund
ES Zorro volador de Queensland

LC
Monotypic
HB 12.2–20 cm
T tailless
FA 11.6–14.3 cm
W 258–600 g

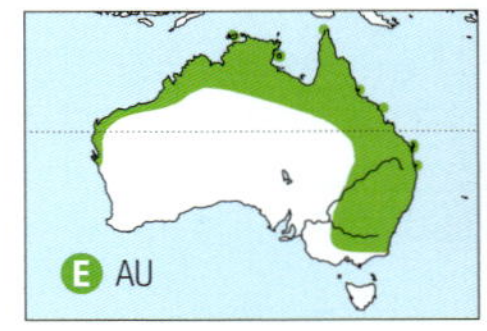

Inland to subcoastal areas of N & E; also some NT, Torres Strait & Qld Is

Big-eared Flying Fox
Pteropus macrotis

FR Roussette à grandes oreilles
DE Großohrflughund
ES Zorro volador orejudo

LC 2 sspp
HB 17.9–24 cm
T tailless
FA 11.2–15.7 cm
W 214–480 g

0–500 m

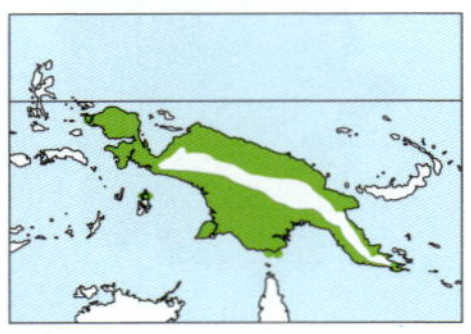

New Guinea, Raja Ampat Is (Salawati), Aru Is; uncertain in Boigu Is (AU)

Gilliards' Flying Fox
Pteropus gilliardorum

FR Roussette des Gilliard
DE Gilliard-Flughund
ES Zorro volador de los Gilliard

VU Monotypic
HB 15.4–17.8 cm T tailless
FA 11.4–12 cm W c. 400 g (M)

200–2300 m

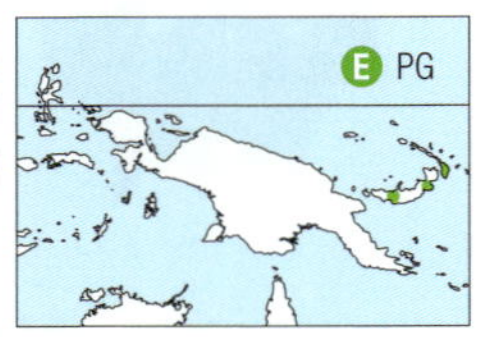

Bismarck Arch (New Britain, E New Ireland)

Sanborn's Flying Fox
Pteropus mahaganus

FR Roussette de Sanborn
DE Sanborn-Flughund
ES Zorro volador de Sanborn

up to 600 m

PG/SB: Buka, Bougainville, Choiseul, Santa Isabel, Tanabuli

LC
Monotypic
HB 19–23 cm
T tailless
FA 12.6–15.3 cm
W 325–520 g

Dwarf Flying Fox
Pteropus woodfordi

FR Roussette de Woodford
DE Zwergflughund
ES Zorro volador de Woodford

LC Monotypic
HB 9.2–14.5 cm
T tailless
FA 8–9.6 cm
W c. 150 g

up to 1200 m

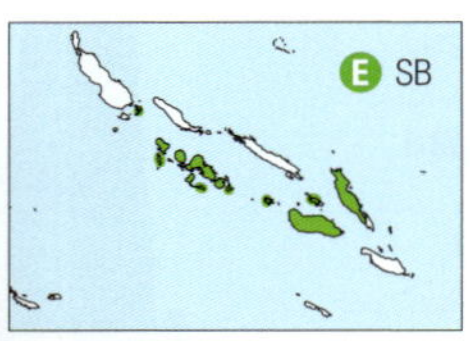

Fauro, New Georgia Is, Russell, Nggela, Malaita, Guadalcanal

Pohnpei Flying Fox
Pteropus molossinus

FR Roussette de Ponape
DE Pohnpei-Flughund
ES Zorro volador de Ponape

EN
Monotypic
HB 14–14.8 cm
T tailless
FA 9.4–9.8 cm
W c. 129 g

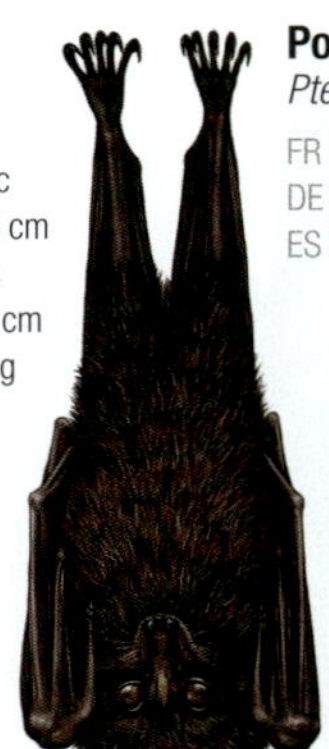

0–770 m

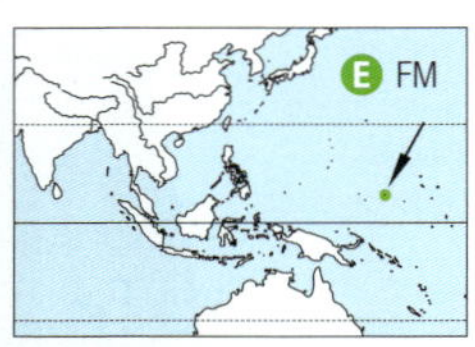

Pakin Atoll, Ant Atoll, Pohnpei I; unconfirmed in Mortlock Is

Chuuk Flying Fox
Pteropus pelagicus

FR Roussette de Chuuk
DE Mortlock-Flughund
ES Zorro volador de Chuuk

EN
2 sspp
HB 15–18.6 cm
T tailless
FA 10.1–10.5 cm
W 142–203 g

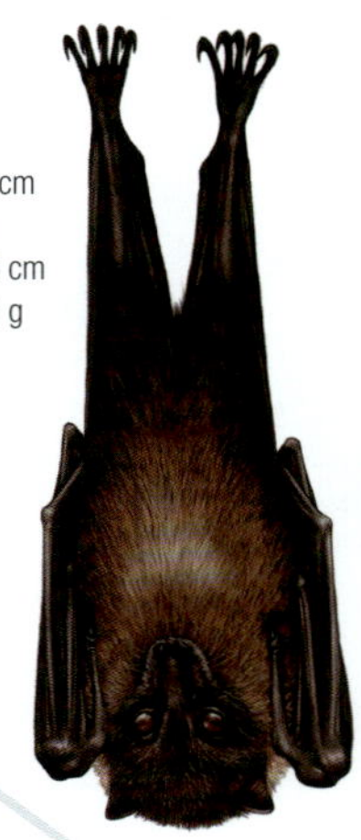

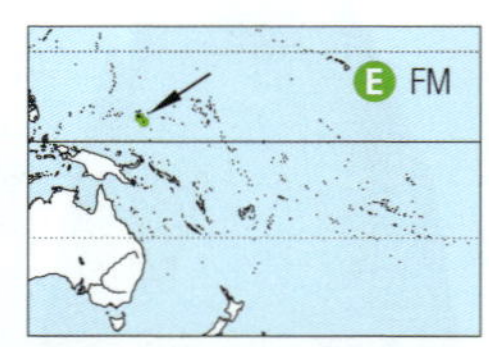

C & S Mortlock Is (perhaps extirpated from Losap Atoll in N Mortlocks), Chuuk Lagoon, Namonuito Atoll

Ornate Flying Fox
Pteropus ornatus

FR Roussette ornée
DE Schmuckflughund
ES Zorro volador de las islas de la Lealtad

VU
2 sspp
HB 18–18.7 cm
T tailless
FA 14.2–16.5 cm
W 285–325 g

0–1100 m

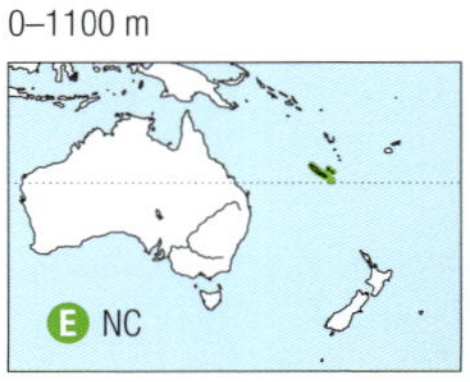

Grande Terre, Île des Pins, Loyalty Is (Lifou I, Maré)

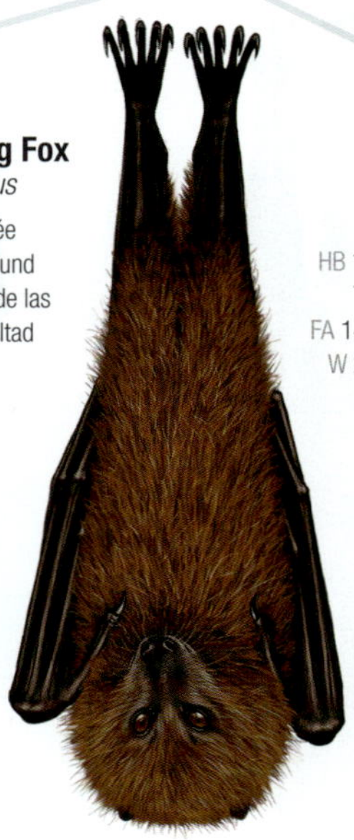

Seram Flying Fox
Pteropus ocularis

FR Roussette de Céram
DE Seram-Flughund
ES Zorro volador de Ceram

VU
Monotypic
HB c. 19.5 cm
T tailless
FA 13.2–14.5 cm
W 200–400 g

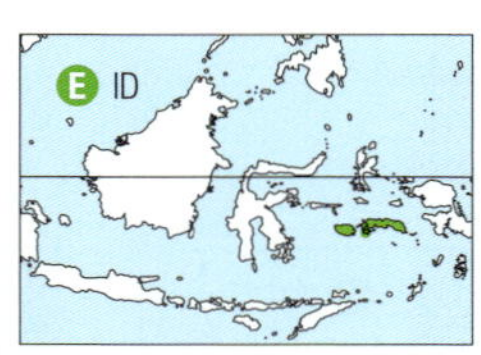

C Moluccas (Buru, Ambon, Seram)

Great Flying Fox
Pteropus neohibernicus

FR Roussette papoue
DE Großer Flughund
ES Zorro volador de Bismarck

0–1400 m

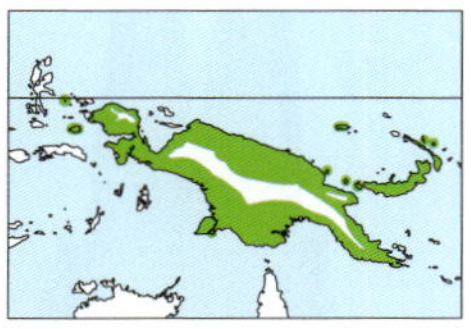

Raja Ampat Is (Gebe, Misool), New Guinea, Bismarck Arch, Admiralty Is (Manus)

LC
2 sspp
HB 23–37 cm
T tailless
FA 16.5–22 cm
W 1–1.5 kg

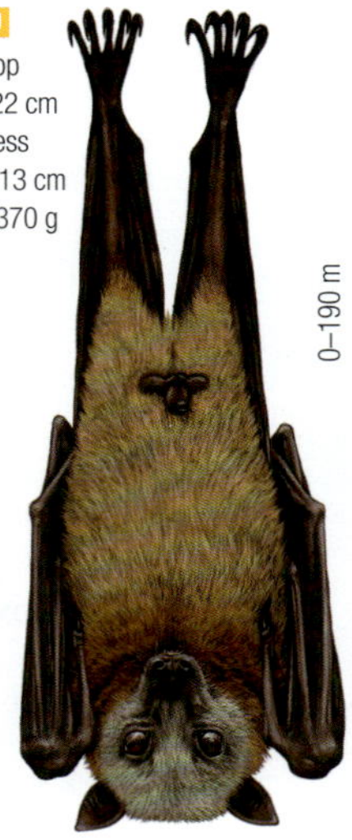

Wallacean Gray Flying Fox
Pteropus griseus

VU
3 sspp
HB 17–22 cm
T tailless
FA 11.3–13 cm
W 230–370 g

FR Roussette grise
DE Grauer Wallacea-Flughund
ES Zorro volador gris

0–190 m

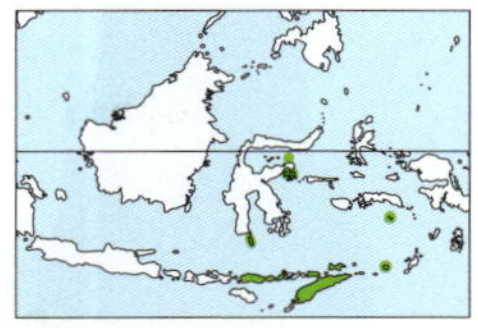

E Lesser Sundas (Flores, Alor, Wetar, Timor, Babar), Mantalu Daka, Peleng, Selayar & Banda Is

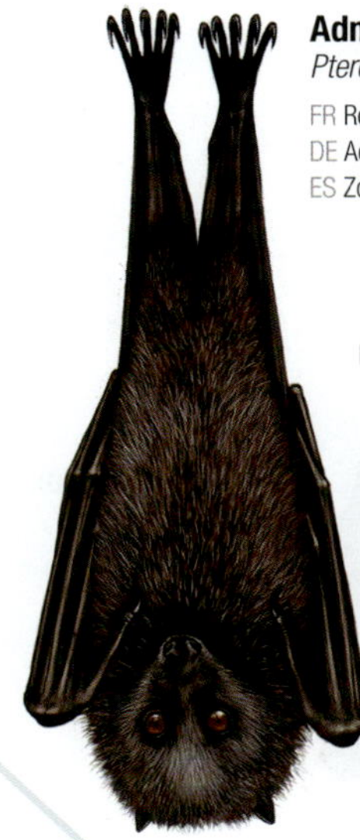

Admiralty Flying Fox
Pteropus admiralitatum

FR Roussette de l'Amirauté
DE Admiralitätsinseln-Flughund
ES Zorro volador de Almirantazgo

LC Monotypic
HB 17.1–23 cm T tailless
FA 11.8–13 cm W 205–334 g

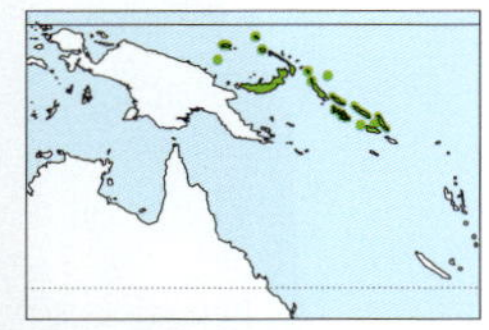

Admiralty, Purdy, St Mattias, Bismarck Arch, Nissan, Carteret & Solomon Is (S to Malaita & Guadalcanal)

Island Flying Fox
Pteropus hypomelanus

FR Roussette hypomélane
DE Inselflughund
ES Zorro volador pequeño

NT
16 sspp
HB 19.4–22 cm
T tailless
FA 12.5–14.5 cm
W 300–420 g

0–900 m

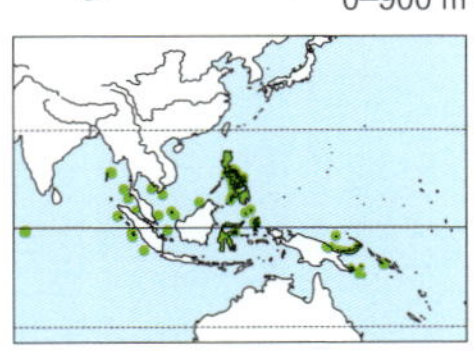

Many offshore Is & coastal lowlands from MV E through SE Asia to Melanesia

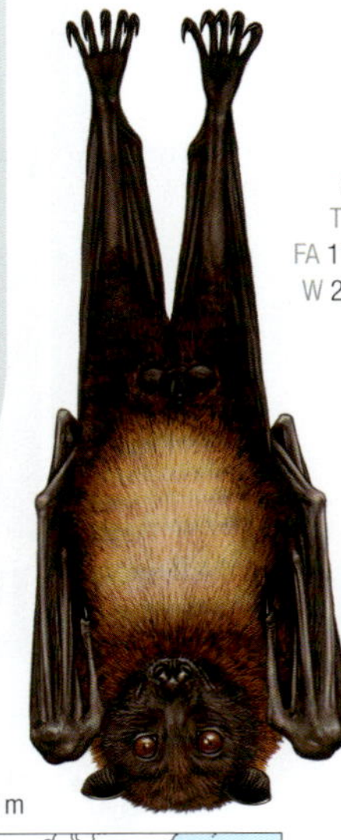

Black-eared Flying Fox
Pteropus melanotus

FR Roussette à oreilles noires
DE Schwarzohrflughund
ES Zorro volador de Andamán

VU
5 sspp
T tailless
FA 11–16.5 cm
W 220–500 g

0–900 m

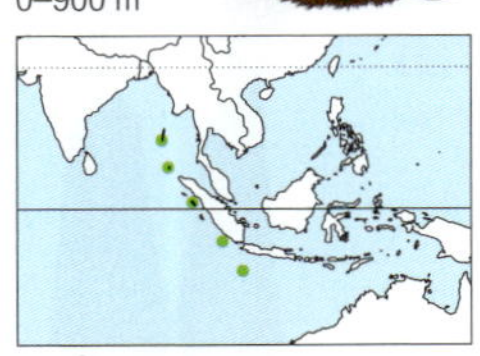

Andaman Is (South Andaman, Rutland), Nicobar Is, Enggano & Nias Is (off W Sumatra), Christmas I (AU)

Black Flying Fox
Pteropus alecto

0–1000 m

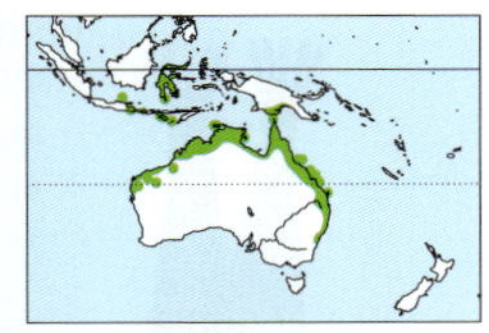

Lesser Sundas (Java E to Savu), Sulawesi (with Selayar I), S New Guinea, N & E AU (with several Is)

FR Roussette alecto
DE Schwarzer Flughund
ES Zorro volador negro

LC
4 sspp
HB 18.6–28 cm
T tailless
FA 15.3–19.1 cm
W 590–880 g

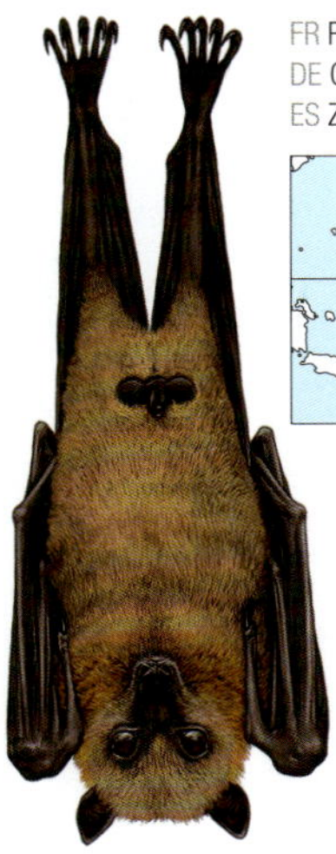

Philippine Gray Flying Fox
Pteropus speciosus

FR Roussette des Philippines
DE Grauer Philippinen-Flughund
ES Zorro volador de Filipinas

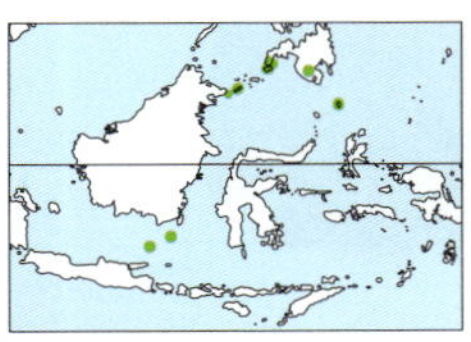

PH (S Mindanao, Sulu Arch), ID (Karakelong, Masalembu & Mata Siri Is)

DD
Monotypic
T tailless
FA 12–12.5 cm
W c. 400 g

Spectacled Flying Fox
Pteropus conspicillatus

FR Roussette à lunettes
DE Brillenflughund
ES Zorro volador de anteojos

0–200 m

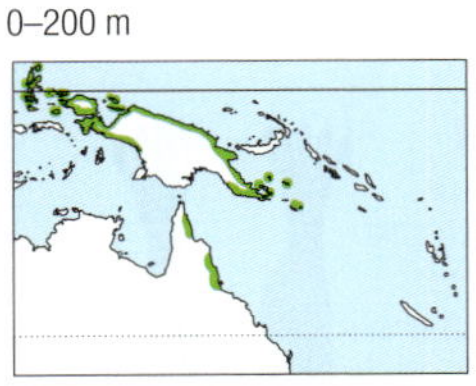

N Moluccas, coastal New Guinea & nearby Is, NE AU (NE Qld)

EN
2 sspp
HB 23–29 cm
T tailless
FA 15–18.9 cm
W 0.5–1 kg

Geelvink Bay Flying Fox
Pteropus pohlei

FR Roussette de Pohle
DE Pohle-Flughund
ES Zorro volador de Yapen

VU
Monotypic
HB 17.6–21 cm
T tailless
FA 12.6–13.5 cm
W 315–360 g

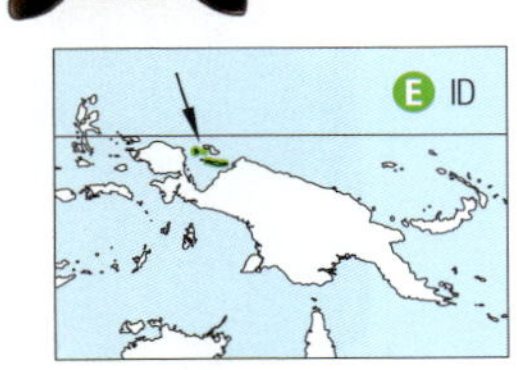

Geelvink Bay Is (Rani, Numfor, Yapen), off NW New Guinea

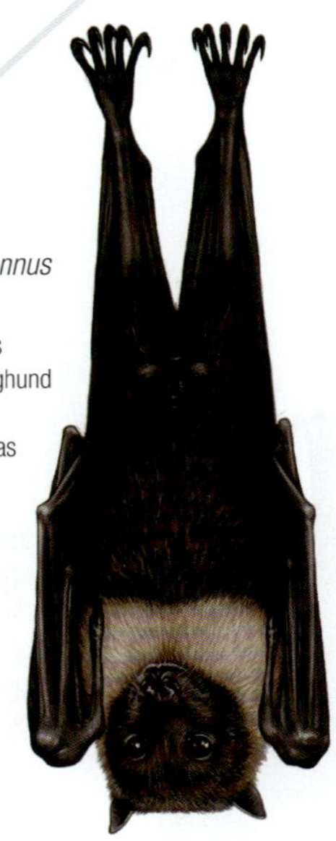

Marianas Flying Fox
Pteropus mariannus

FR Roussette des Mariannes
DE Marianen-Flughund
ES Zorro volador de las Marianas

EN
3 sspp
HB 19.5–25 cm
T tailless
FA 13.4–15.4 cm
W 270–577 g

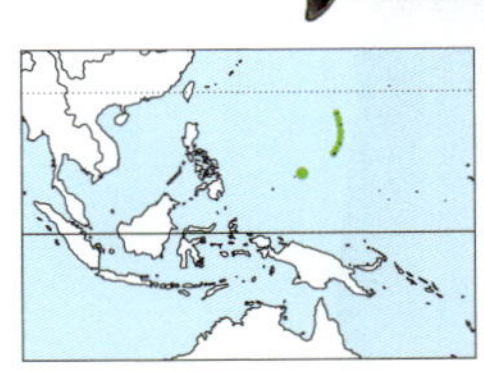

Mariana Is (Farallon de Pajaros S to Guam), W Caroline Is (Ulithi Atoll)

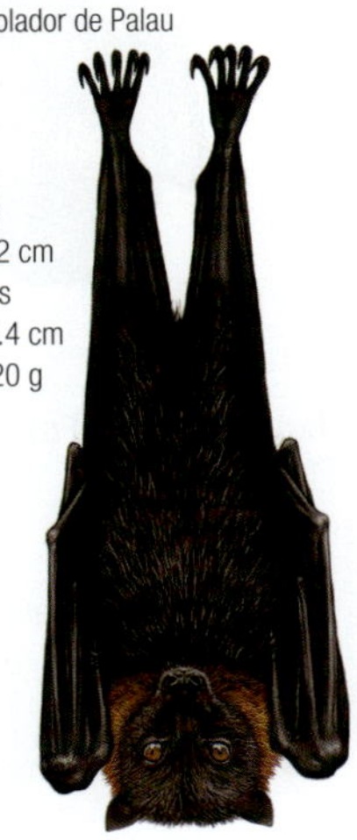

Palau Flying Fox
Pteropus pelewensis

FR Roussette de Palau
DE Palau-Flughund
ES Zorro volador de Palau

VU
2 sspp
HB 14.8–32 cm
T tailless
FA 10.1–14.4 cm
W 130–420 g

0–215 m

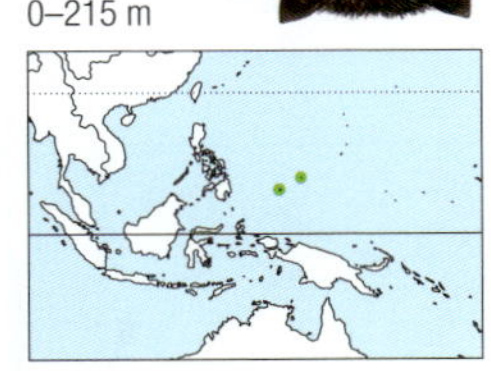

Palau Arch (with Kayangel Atoll, Sonsorol & Fanna Is), W Caroline Is (Yap)

Pacific Flying Fox
Pteropus tonganus

FR Roussette du Pacifique
DE Tonga-Flughund
ES Zorro volador del Pacífico

LC
3 sspp
HB 14.5–27 cm
T tailless
FA 11.4–17.5 cm
W 420–550 g

c. 0 m

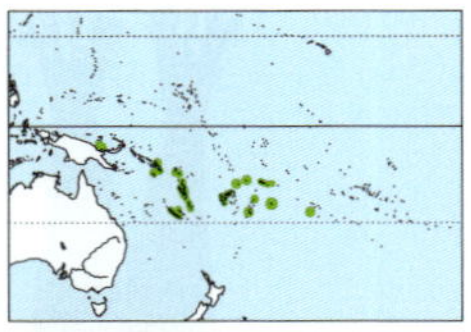

From Koil & Karkar Is (NE New Guinea) E to Cook Is (Mangaia, Rarotonga)

Kosrae Flying Fox
Pteropus ualanus

FR Roussette de Kosrae
DE Kosrae-Flughund
ES Zorro volador de Kosrae

EN
Monotypic
HB 18.7–23 cm
T tailless
FA 12.7–13.3 cm

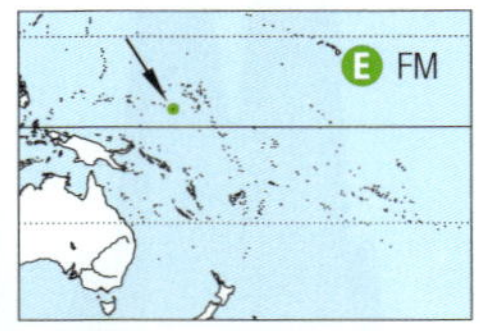

Kosrae I 0–635 m

Nicobar Flying Fox
Pteropus faunulus

FR Roussette des Nicobar
DE Nikobaren-Flughund
ES Zorro volador de Nicobar

EN
Monotypic
HB c. 17 cm
T tailless
FA 11–13 cm
W 112–233 g

0–200 m

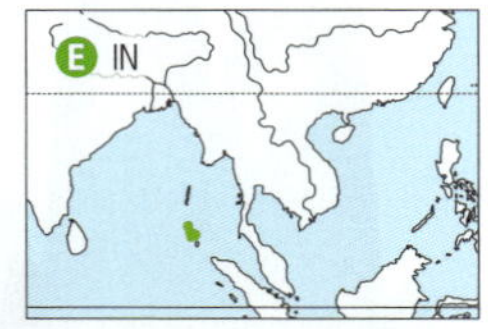

Nicobar Is (Teressa, Bompoka, Camorta, Trinkat, Nankowry, Katchall); probably extinct on Car Nicobar

Ontong Java Flying Fox
Pteropus howensis

FR Roussette d'Ontong Java
DE Ontong-Java-Flughund
ES Zorro volador de Ontong Java

CR Monotypic
HB 17.6–19.6 cm T tailless
FA 11.8–12.2 cm

Ontong Java Atoll; possibly Nukumanu I (PG)

New Britain Masked Flying Fox
Pteropus capistratus

FR Roussette bridée
DE Neubritannien-Flughund
ES Zorro volador de Nueva Bretaña

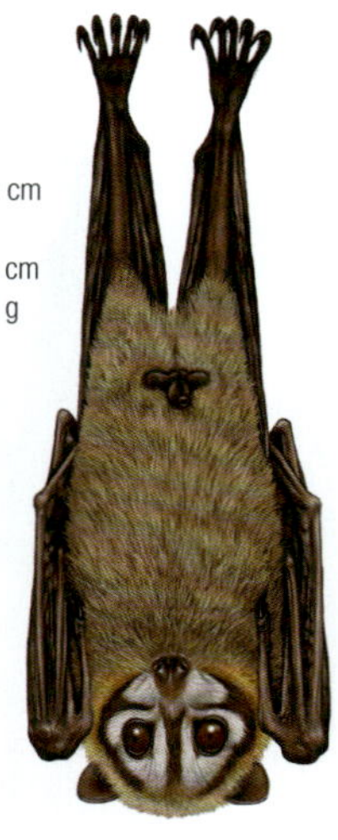

VU
Monotypic
HB 14.5–19.2 cm
T tailless
FA 10.2–12.2 cm
W 165–290 g

up to 1200 m

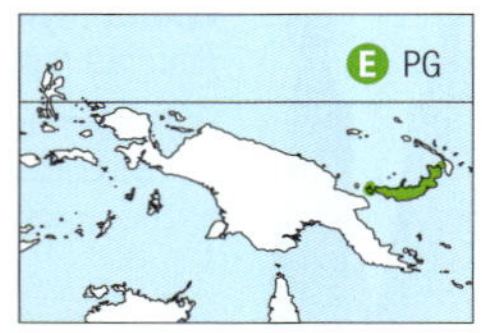

Bismarck Arch (Duke of York, Mioko, New Britain, Sakar, Umboi)

New Ireland Masked Flying Fox
Pteropus ennisae

FR Roussette de Nouvelle-Irlande
DE Neuirland-Flughund
ES Zorro volador de Nueva Irlanda

Bismarck Arch (New Ireland)

VU Monotypic
HB 15.5–19 cm T tailless
FA 10.9–11.8 cm
W 170–250 g

New Caledonia Flying Fox
Pteropus vetula

FR Roussette de Nouvelle-Calédonie
DE Neukaledonien-Flughund
ES Zorro volador de Nueva Caledonia

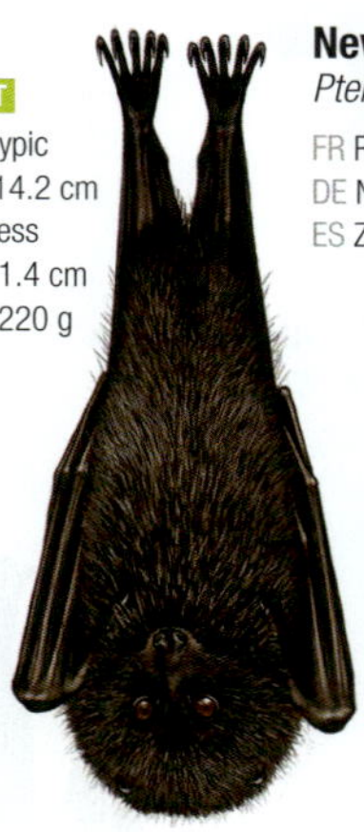

NT
Monotypic
HB 12.2–14.2 cm
T tailless
FA 9.2–11.4 cm
W 120–220 g

New Caledonia (Grande Terre, Île des Pins)

Common Samoan Flying Fox
Pteropus samoensis

FR Roussette des Samoa
DE Gewöhnlicher Samoa-Flughund
ES Zorro volador común de Samoa

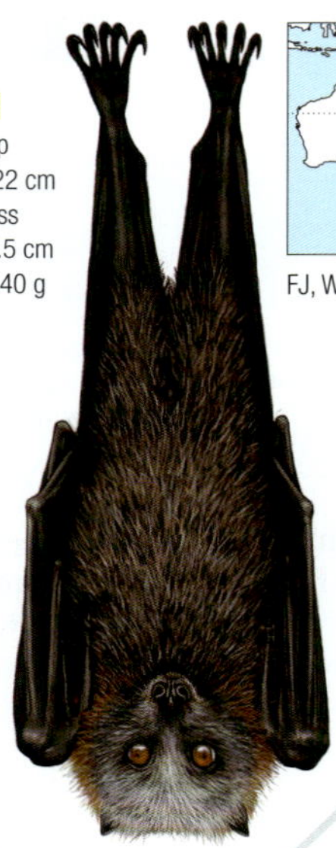

NT
2 sspp
HB 17.5–22 cm
T tailless
FA 12–15.5 cm
W 220–440 g

FJ, WS, AS

Banks Flying Fox
Pteropus fundatus

FR Roussette des Banks
DE Banks-Flughund
ES Zorro volador de las Islas de Banks

EN
Monotypic
HB 13.7–15.2 cm
T tailless
FA 9.9–10.6 cm
W 184–190 g

Banks Is (Mota, Vanua Lava)

Vanuatu Flying Fox
Pteropus anetianus

FR Roussette du Vanuatu
DE Vanuatu-Flughund
ES Zorro volador de Vanuatu

VU 7 sspp
HB 15.5–20 cm T tailless
FA 11.7–13.5 cm
W c. 346 g

From Banks Is S to Aneityum

Vanikoro Flying Fox
Pteropus tuberculatus

FR Roussette de Vanikoro
DE Vanikoro-Flughund
ES Zorro volador de Vanikoro

EN
Monotypic
HB 13.6–17 cm
T tailless
FA 11.2–13 cm
W c. 230 g

c. 0 m

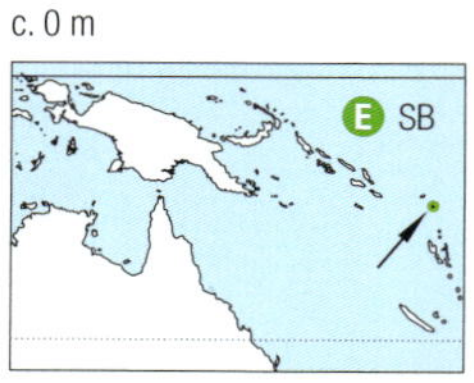

Santa Cruz Is (Vanikoro, Teanu)

Temotu Flying Fox
Pteropus nitendiensis

FR Roussette de Nendo
DE Temotu-Flughund
ES Zorro volador de la Isla de Ndeni

EN
Monotypic
HB 17.1–17.5 cm
T tailless
FA 11.6–12.1 cm
W 210–320 g

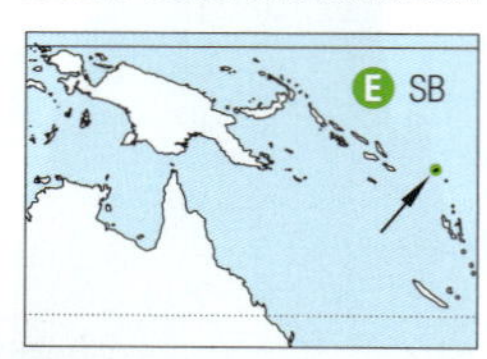

Santa Cruz Is (Tinakula, Tõmotu Neo, Nendö)

Solomons Flying Fox
Pteropus rayneri

FR Roussette de Rayner
DE Salomonen-Flughund
ES Zorro volador de las Salomón

0–700 m

PG/SB: Solomon Is (from Buka S to Guadalcanal & Malaita)

NT
5 sspp
HB 23–27 cm
T tailless
FA 13.5–18.7 cm
W 550–870 g

VU
Monotypic
HB 16.6–18 cm
T tailless
FA 12.1–12.9 cm
W c. 288 g

0–900 m

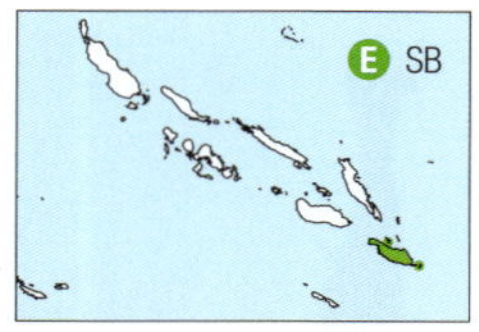

Makira, Ugi, Santa Ana

Makira Flying Fox
Pteropus cognatus

FR Roussette de Makira
DE Makira-Flughund
ES Zorro volador de Makira

EN
Monotypic
HB 14.5–18.5 cm
T tailless
FA 9.5–12.8 cm
W c. 179 g

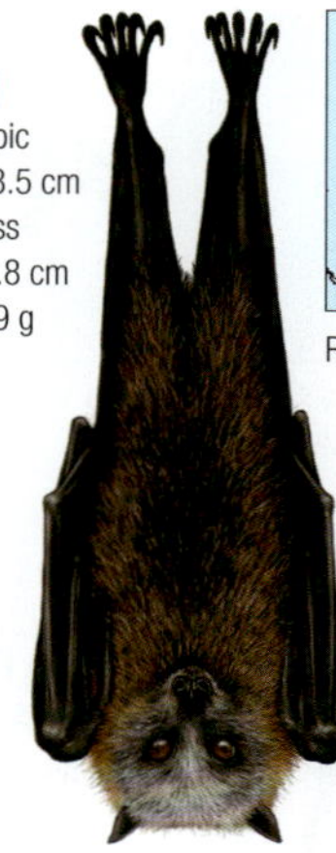

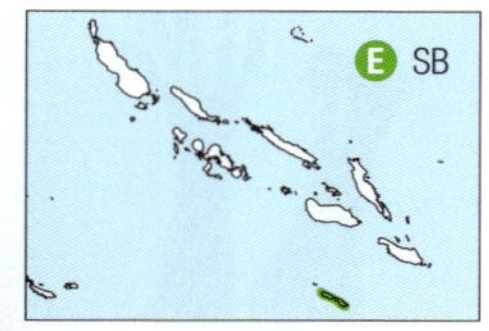

Rennell

Rennell Flying Fox
Pteropus rennelli

FR Roussette de Rennell
DE Rennell-Flughund
ES Zorro volador de Rennell

VU
Monotypic
HB 22–28 cm
T tailless
FA 15.1–17.7 cm
W 0.4–1.3 kg

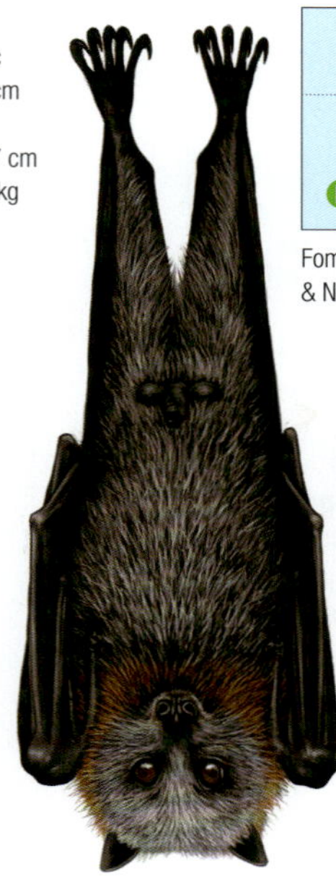

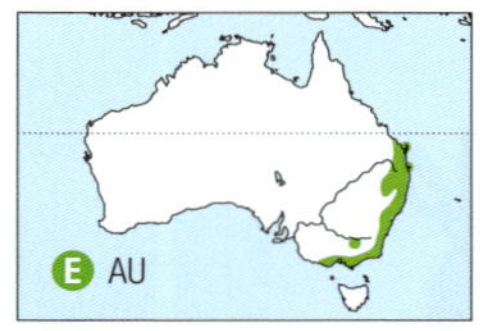

Fom SE Qld (with Fraser, Moreton & North Stradbrooke Is) S to Vic

Gray-headed Flying Fox
Pteropus poliocephalus

FR Roussette à tête grise
DE Graukopf-Flughund
ES Zorro volador de cabeza gris

Livingstone's Flying Fox
Pteropus livingstonii

FR Roussette de Livingstone
DE Livingstone-Flughund
ES Zorro volador de Livingstone

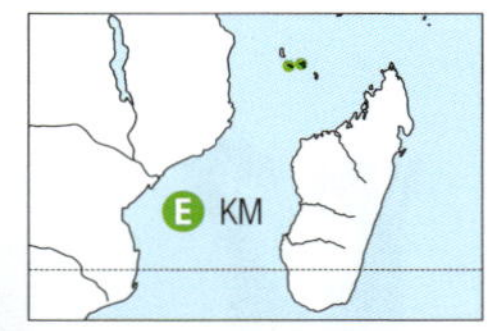

Anjouan, Mohéli

CR
Monotypic
HB c. 34 cm
T tailless
FA 16.1–17.2 cm
W 576–601 g

VU
Monotypic
HB 15.1–17 cm
T tailless
FA 9.4–10.8 cm
W 145–155 g

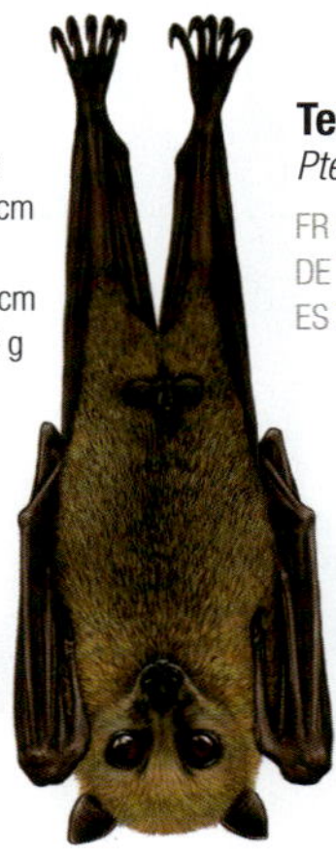

Temminck's Flying Fox
Pteropus temminckii

FR Roussette de Temminck
DE Temminck-Flughund
ES Zorro volador de Temminck

0–800 m

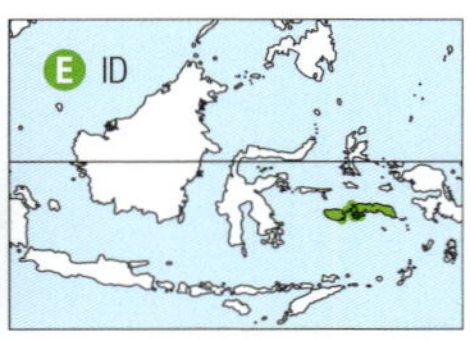

C Moluccas (Buru, Ambon, Seram, nearby small Is)

0–45 m

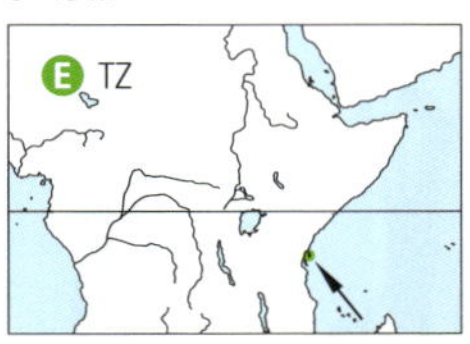

Pemba I

Pemba Flying Fox
Pteropus voeltzkowi

FR Roussette de Pemba
DE Pemba-Flughund
ES Zorro volador de Pemba

VU
Monotypic
HB 22–26 cm
T tailless
FA 14.5–16.3 cm
W 430–610 g

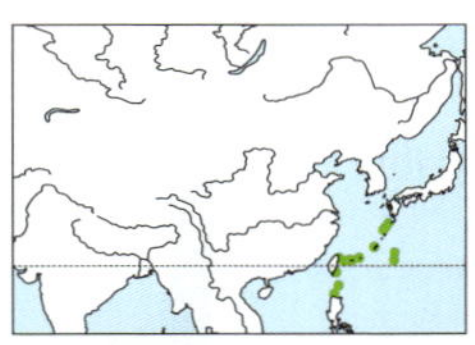

JP (Ryukyu Is), TW (Lutao, Lanyu & Guishan Dao Is, probably extinct on mainland), N PH

Ryukyu Flying Fox
Pteropus dasymallus

FR Roussette des Ryukyu
DE Ryukyu-Flughund
ES Zorro volador de Ryukyu

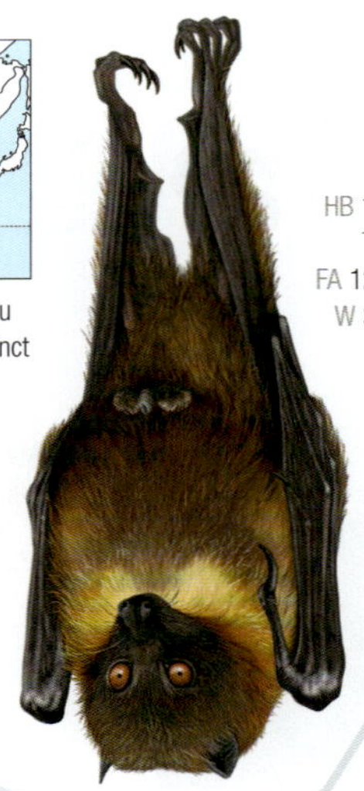

VU
5 sspp
HB 18.5–23 cm
T tailless
FA 12.4–14.9 cm
W 318–662 g

EN
Monotypic
T tailless
FA 13.1–14.5 cm
W 353–616 g

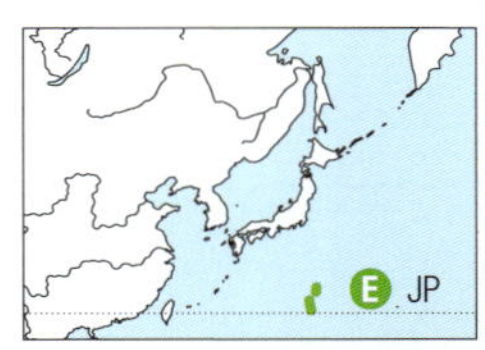

Ogasawara Is (Chichijima, Hahajima), Iwo Is (Kita-Iwoto, Iwoto, Minami-Iwoto); perhaps other Is

Bonin Flying Fox
Pteropus pselaphon

FR Roussette des Bonin
DE Bonin-Flughund
ES Zorro volador de Bonin

0–1250 m

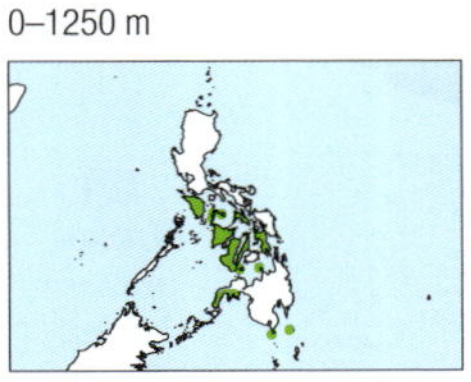

C & S PH (Mindoro S to NW Mindanao & Balut Is), ID (Miangas I, off North Sulawesi)

Little Golden-mantled Flying Fox
Pteropus pumilus

FR Roussette naine
DE Goldmantel-Flughund
ES Zorro volador dorado

NT
Monotypic
HB 15.5–18 cm
T tailless
FA 9.4–11.3 cm
W 145–200 g

Lyle's Flying Fox
Pteropus lylei

FR Roussette de Lyle
DE Lyle-Flughund
ES Zorro volador de Lyle

VU
Monotypic
HB c. 20 cm
T tailless
FA 14.5–16 cm
W 390–480 g

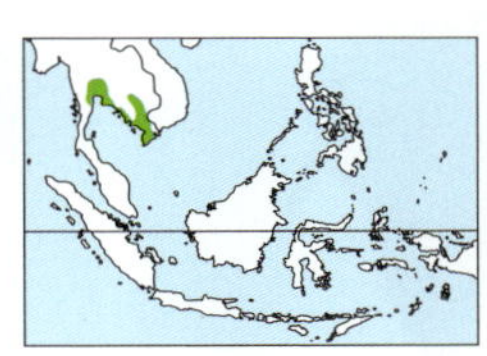

S TH, C & S KH, S VN; 1 record from Yunnan (CN) considered alien

0–2000 m

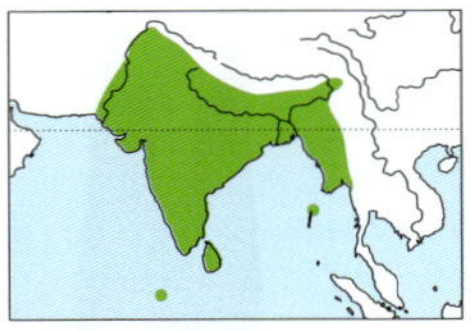

W PK, IN, NP, BT, BD, W MM, LK, MV, Andaman Is

Indian Flying Fox
Pteropus medius
FR Roussette indienne
DE Indischer Flughund
ES Zorro volador de la India

LC
3 sspp
HB 22–25 cm
T tailless
FA 16.3–17.6 cm
W 1–1.2 kg

0–400 m

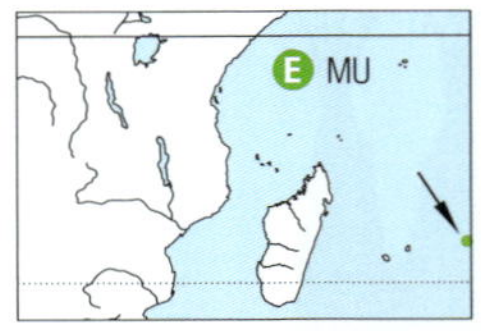

Rodrigues I; specimen from Round I might be a vagrant

Rodrigues Flying Fox
Pteropus rodricensis
FR Roussette de Rodrigues
DE Rodrigues-Flughund
ES Zorro volador de Rodrigues

EN Monotypic
HB 15–20 cm T tailless
FA 12.1–12.8 cm
W c. 278 g

EN
6 sspp
HB 22–29 cm
T tailless
FA 17.5–23 cm
W 0.7–1.4 kg

0–1370 m

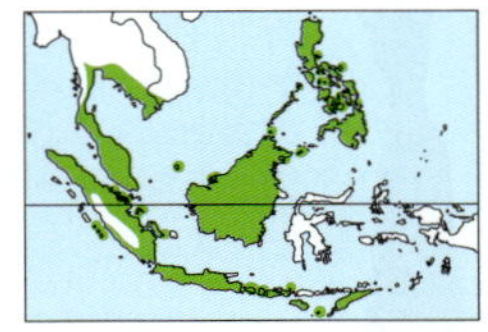

S mainland SE Asia, Greater & Lesser Sundas (with Lingga Arch, Bangka, S Mentawai & Natuna Is), PH

Large Flying Fox
Pteropus vampyrus
FR Roussette de Malaisie
DE Kalong-Flughund
ES Zorro volador de cuello rojizo

0–1200 m

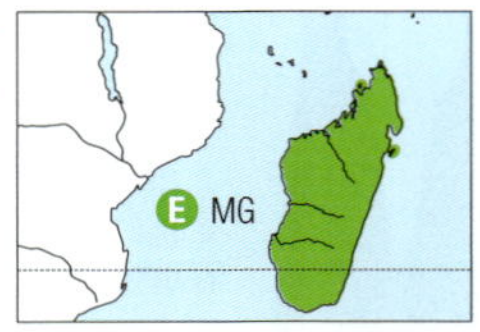

Mainland and Nosy Be, Nosy Komba, Nosy Tanikely & Nosy Boraha

Malagasy Flying Fox
Pteropus rufus
FR Roussette marron
DE Großer Madagaskar-Flughund
ES Zorro volador de Madagascar

VU
Monotypic
HB 23–27 cm
T tailless
FA 15.4–17.3 cm
W 550–750 g

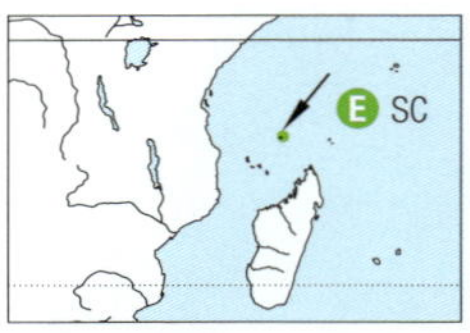

Aldabra Atoll

Aldabra Flying Fox
Pteropus aldabrensis
FR Roussette d'Aldabra
DE Aldabra-Flughund
ES Zorro volador de Aldabra

EN
Monotypic
HB 18.4–18.9 cm
T tailless
FA 12.8–14.1 cm
W 257–395 g

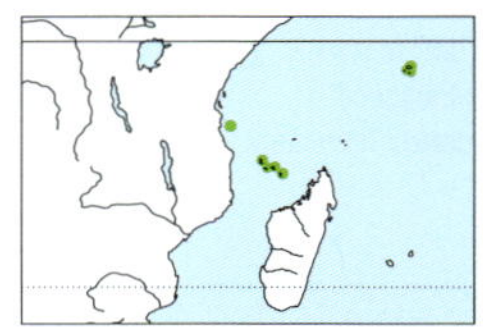

0–1000 m

E granitic SC (Mahé, Praslin, La Digue, Silhouette), Mafia I (TZ), Comoro Is

Seychelles Flying Fox
Pteropus seychellensis
FR Roussette des Seychelles
DE Seychellen-Flughund
ES Zorro volador de las Seychelles

LC
2 sspp
HB c. 23 cm
T tailless
FA 14.7–15.8 cm
W 370–390 g

EN
Monotypic
HB c. 23 cm
T tailless
FA 14.3–17.1 cm
W 380–540 g (M)

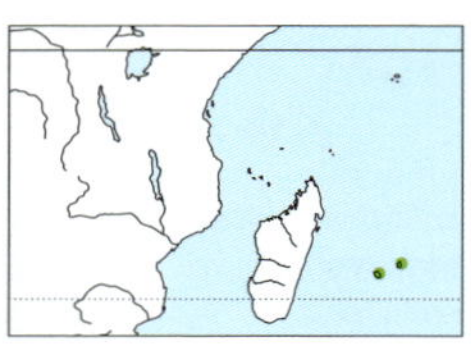

Mauritius (MU), Réunion (FR)

Greater Mascarene Flying Fox
Pteropus niger
FR Roussette des Mascareignes
DE Maskarenen-Flughund
ES Zorro volador mayor de las Islas Mascareñas

CR(PE)
Monotypic
HB c. 29 cm
T tailless
FA c. 19 cm

Aru Is

Aru Flying Fox
Pteropus aruensis
FR Roussette des Aru
DE Aru-Flughund
ES Zorro volador de Aru

Kai Flying Fox
Pteropus keyensis
FR Roussette des Kei
DE Kei-Flughund
ES Zorro volador de Kei

DD
Monotypic
HB c. 29 cm
T tailless
FA 17.9–18.8 cm

Kai Is

Black-bearded Flying Fox
Pteropus melanopogon
FR Roussette à barbe noire
DE Schwarzbart-Flughund
ES Zorro volador de barba negra

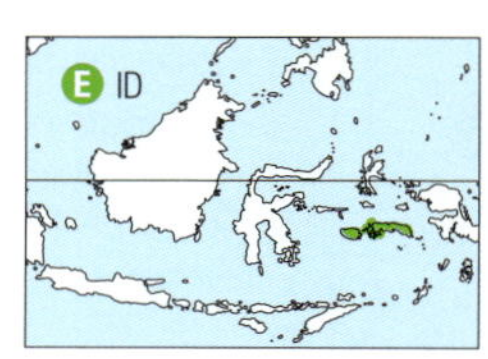

C Moluccas (Buru, Ambon, Seram, and adjacent Is)

EN
Monotypic
HB c. 29 cm
T tailless
FA 18.8–20 cm
W 860–900 g

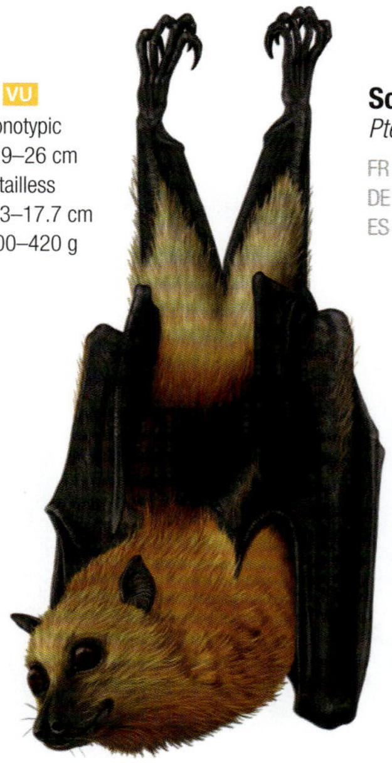

VU
Monotypic
HB 19–26 cm
T tailless
FA 16.3–17.7 cm
W 300–420 g

South Moluccan Flying Fox
Pteropus chrysoproctus
FR Roussette d'Ambon
DE Südmolukken-Flughund
ES Zorro volador de Ambon

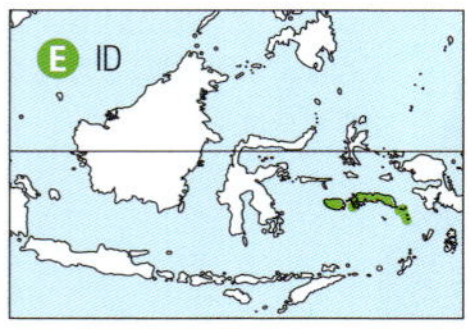

C Moluccas (Buru, Ambon, Seram), Watubela Arch

VU
Monotypic
HB c. 21 cm
T tailless
FA 18.8–20 cm
W 860–900 g

North Moluccan Flying Fox
Pteropus caniceps
FR Roussette de Bacan
DE Nordmolukken-Flughund
ES Zorro volador de Molucas

0–1500 m

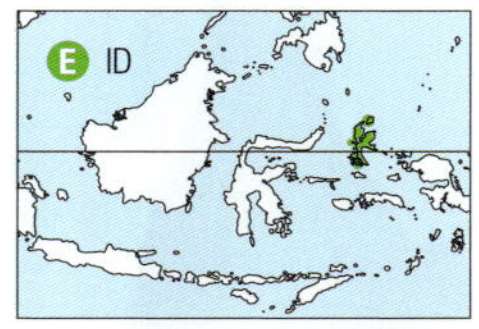

N Moluccas (Halmahera, satellite Is)

Family RHINOPOMATIDAE – Mouse-tailed Bats

1 genus, 6 species. Arid and semi-arid habitats, roosting in caves, rocky niches, and man-made structures. Palearctic, Afrotropical and Indo-Malayan Regions.

Genus *RHINOPOMA*

(6 species)

Greater Mouse-tailed Bat
Rhinopoma microphyllum
FR Grand Rhinopome
DE Große Mausschwanzfledermaus
ES Rinopoma grande

LC
Monotypic
HB 7.8–8.1 cm
T 4.8–8.5 cm
FA 5.7–7.4 cm
W 14–40 g

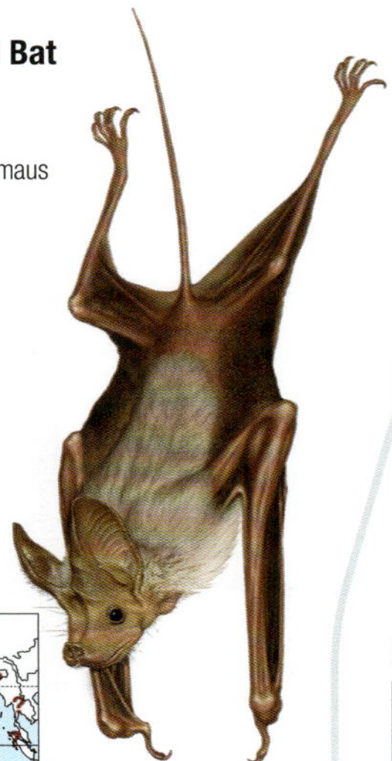

0–1200 m

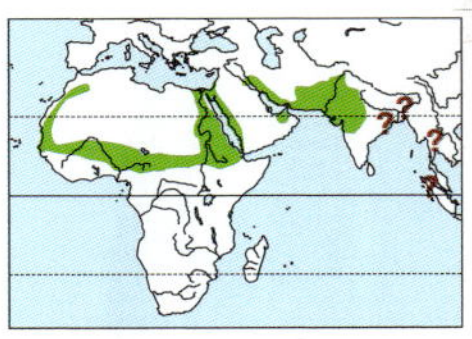
NW Africa, Sahel & Nile Valley E through Middle East to NW & C IN; perhaps E IN, BD, MM, TH & Sumatra

Muscat Mouse-tailed Bat
Rhinopoma muscatellum
FR Rhinopome de Muscat
DE Muscat-Mausschwanzfledermaus
ES Rinopoma de Mascate

LC
2 sspp
HB 5.3–6.4 cm
T 5.6–7.5 cm
FA 4.8–5.5 cm
W 4.6–14 g

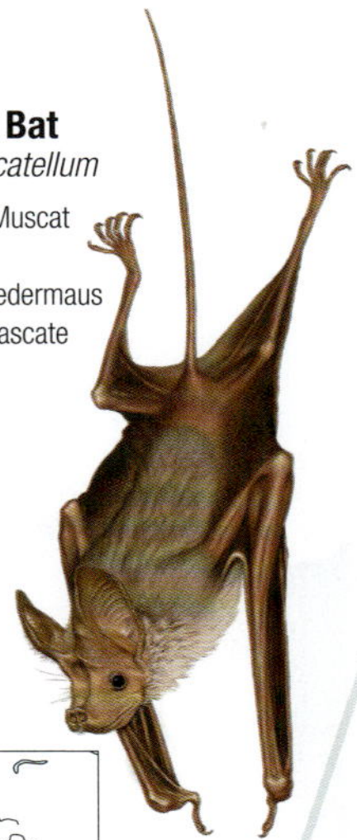

0–1190 m

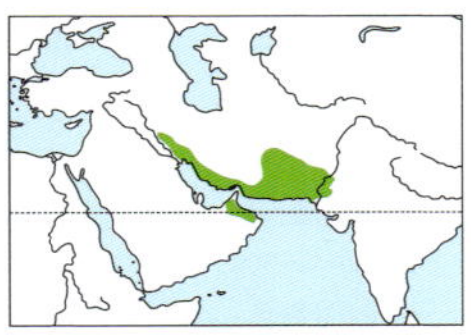
SE Arabian Pen (Al-Hajar Mts), W & S IR, S AF, S PK; perhaps IN

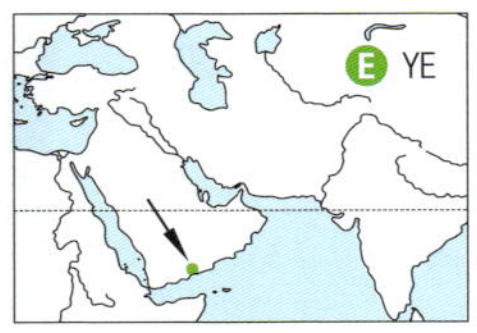

Known only from Ash Sheher (Hadramaut Prov)

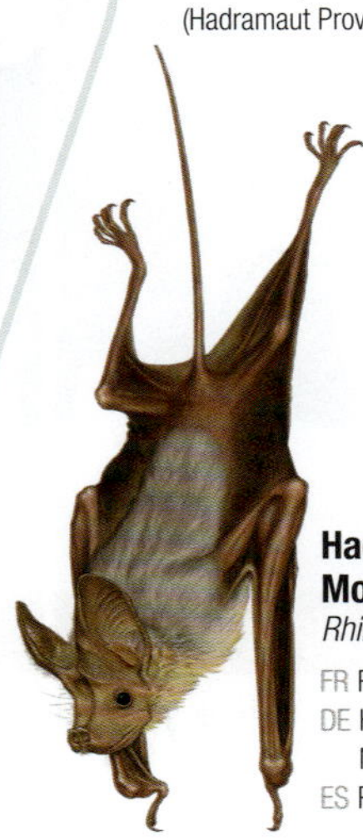

EN
Monotypic
HB 5.8–6.8 cm
T 5.4–6.2 cm
FA 5.2–5.6 cm
W 7–11.2 g

Hadramaut Mouse-tailed Bat
Rhinopoma hadramauticum
FR Rhinopome du Yémen
DE Hadramaut-Mausschwanzfledermaus
ES Rinopoma de Yemen

up to 1100 m

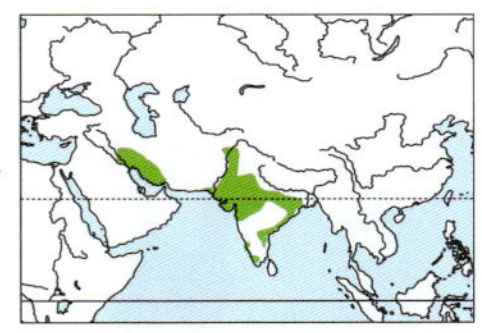
From SE IQ to SW IR (W of Strait of Hormuz), from E AF and N & E PK to most of IN; E limit unclear

LC
Monotypic
HB 5.3–6.5 cm
T 5.4–7.8 cm
FA 4.6–6.5 cm
W 6.5–12 g

Lesser Mouse-tailed Bat
Rhinopoma hardwickii
FR Rhinopome de Hardwicke
DE Kleine Mausschwanzfledermaus
ES Rinopoma pequeño

up to 1100 m

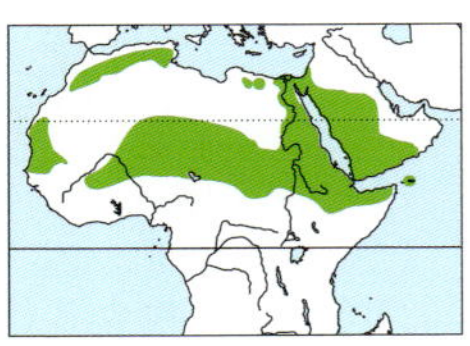
Arid N Africa, Levant & W Arabian Pen, Socotra I

LC
2 sspp
HB 5.2–7.2 cm
T 5.4–8.2 cm
FA 5–6.7 cm
W 6.5–17 g

Arabian Mouse-tailed Bat
Rhinopoma cystops
FR Rhinopome d'Égypte
DE Arabische Mausschwanzfledermaus
ES Rinopoma de Egipto

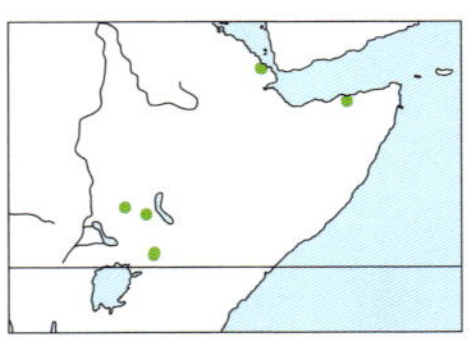
S ER (Assab), N SO (Galgala Oasis), W KE, extreme NE UG

DD
Monotypic
T 5.5–7.6 cm
FA 4.5–5.3 cm
W 5–11 g

MacInnes's Mouse-tailed Bat
Rhinopoma macinnesi
FR Rhinopome de MacInnes
DE MacInnes-Mausschwanzfledermaus
ES Rinopoma de MacInnes

Family CRASEONYCTERIDAE – Hog-nosed Bat

1 genus, 1 species. Limestone caves in mixed or dry deciduous forests. Indo-Malayan Region.

Genus *CRASEONYCTERIS*

(1 species)

up to 500 m

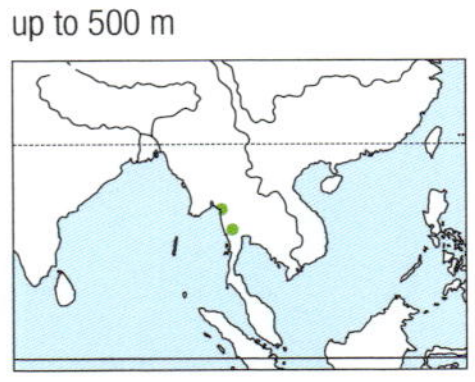
SE MM (Kayin & Mon states), WC TH (Kanchanaburi Prov)

Kitti's Hog-nosed Bat
Craseonycteris thonglongyai
FR Craséonyctère
DE Hummelfledermaus
ES Murciélago abejorro

NT
Monotypic
HB 2.8–3.4 cm
T tailless
FA 2.2–2.8 cm
W 2–3.2 g

Family MEGADERMATIDAE – False-vampire Bats

6 genera, 6 species. Tropical and subtropical rainforests, dry forests, and savannas. Afrotropical, Indo-Malayan, Australasian and marginally Palearctic Regions.

Genus *LAVIA*

(1 species)

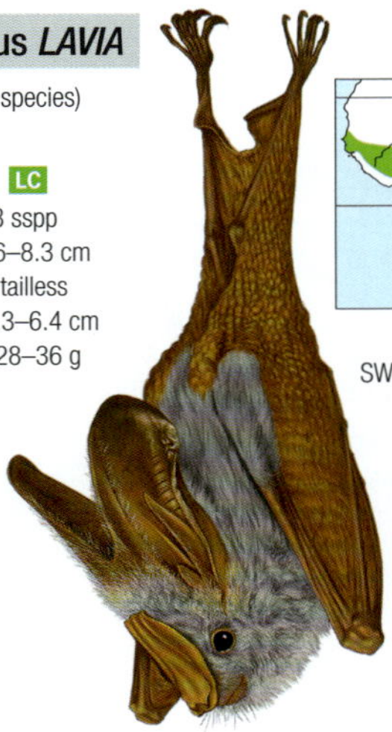

LC
3 sspp
HB 6–8.3 cm
T tailless
FA 5.3–6.4 cm
W 28–36 g

0–2000 m

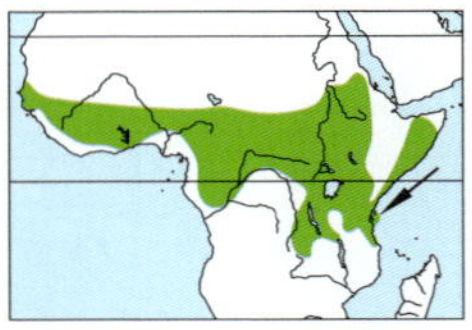

SN & GM E to SD & SO, S to SW CG, N ZM, N MW & TZ (with Unguja I)

Yellow-winged False-vampire
Lavia frons

FR Mégaderme à ailes orangées
DE Gelbflügelfledermaus
ES Megaderma de alas amarillas

Genus *CARDIODERMA*

(1 species)

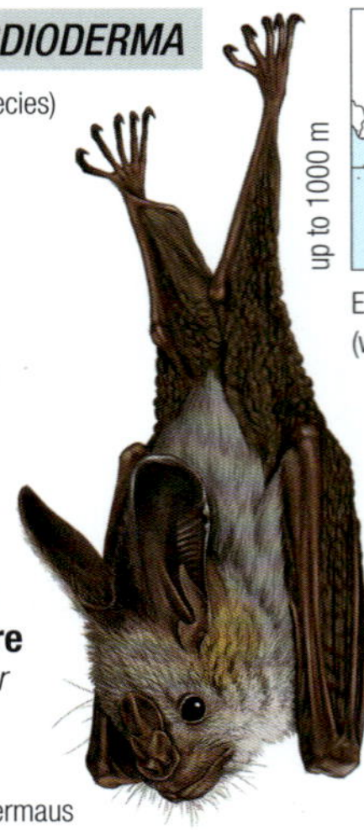

LC
Monotypic
HB 7–7.7 cm
T tailless
FA 5.3–5.9 cm
W 21–35 g

up to 1000 m

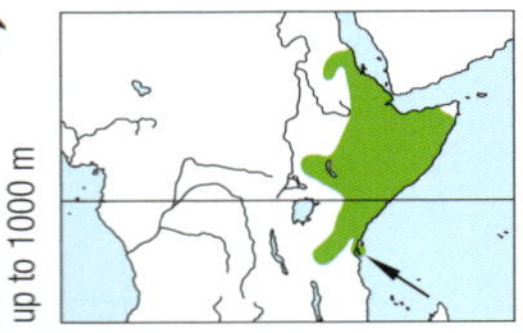

E Africa, from E SD & ER S to C TZ (with Unguja I)

Heart-nosed False-vampire
Cardioderma cor

FR Mégaderme nez-en-cœur
DE Herznasenfledermaus
ES Megaderma cor

Genus *MEGADERMA*

(1 species)

Lesser Asian False-vampire
Megaderma spasma

FR Mégaderme de Seba
DE Kleiner Falscher Vampir
ES Megaderma de Seba

LC
17 sspp
HB 5.4–8.1 cm
T tailless
FA 5.3–6.2 cm
W 23–28 g

up to 1000 m

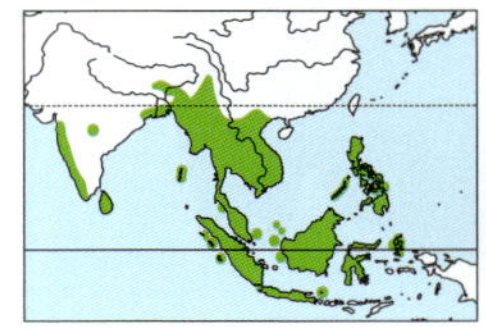

IN, LK, BD, SC CN (Yunnan), entire mainland and most of insular SE Asia E to PH & Moluccas

Genus *LYRODERMA*

(1 species)

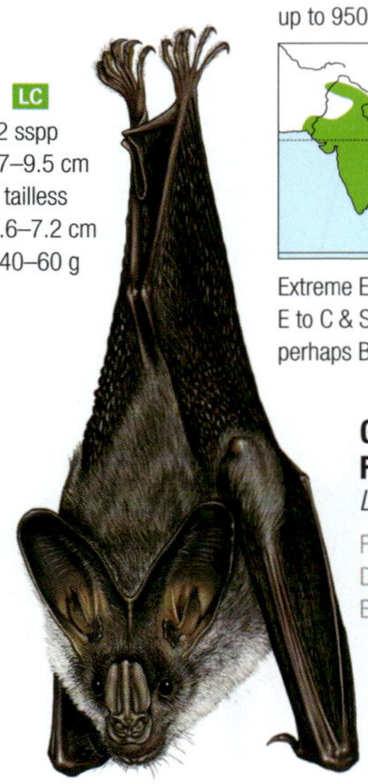

LC
2 sspp
HB 7–9.5 cm
T tailless
FA 5.6–7.2 cm
W 40–60 g

up to 950 m

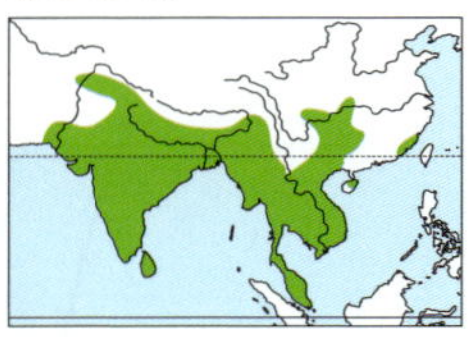

Extreme E AF, PK, IN, LK, NP E to C & S CN, mainland SE Asia; perhaps BT

Greater Asian False-vampire
Lyroderma lyra

FR Mégaderme lyre
DE Großer Falscher Vampir
ES Megaderma lyra

Genus *EUDISCODERMA*

(1 species)

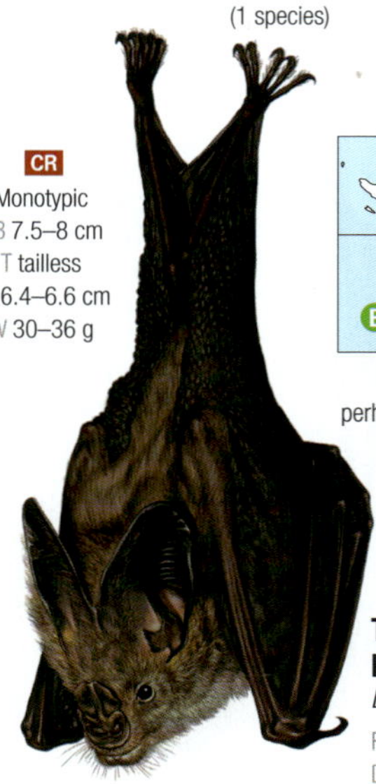

CR
Monotypic
HB 7.5–8 cm
T tailless
FA 6.4–6.6 cm
W 30–36 g

Known only from Bala Forest; perhaps in similar forested habitat in Pen MY

Thongaree's False-vampire
Eudiscoderma thongareeae

FR Mégaderme de Thongaree
DE Thongarees Falscher Vampir
ES Megaderma de Thongaree

Genus *MACRODERMA*

(1 species)

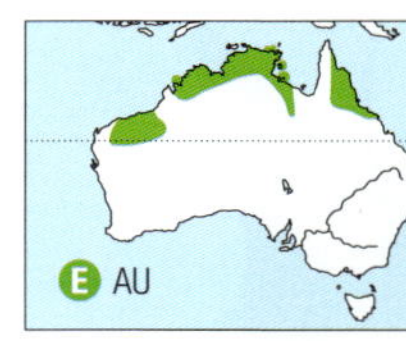

Disjunct in N WA, N NT, NE Qld & offshore Is (Koolan, Milingimbi, Elcho, Groote Eylandt, Pellew)

Ghost False-vampire
Macroderma gigas

FR Macroderme d'Australie
DE Australische Gespenstfledermaus
ES Megaderma de Australia

VU
Monotypic
HB 10–13 cm
T tailless
FA 9.6–11.3 cm
W 130–170 g

Family RHINONYCTERIDAE – Trident Bats

4 genera, 9 species. From dry to forest savannas and forests in tropics, marginally in subtropics. Afrotropical, Australasian and marginally Palearctic Regions.

Genus *RHINONICTERIS*

(1 species)

Orange Diamond-faced Bat
Rhinonicteris aurantia

FR Rhinonyctère doré
DE Orangefarbene Australien-Blattnase
ES Rinonicterio dorado

up to 800 m

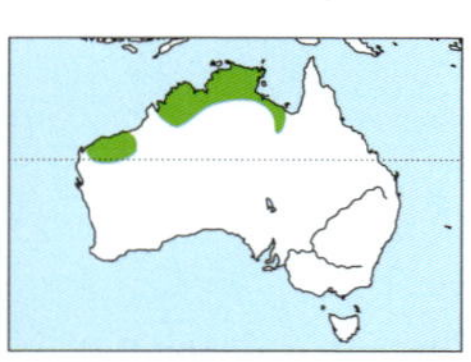

W Pilbara region (WA), disjunct in NE WA (Kimberley region), N NT & NW Qld

LC
Monotypic
HB 4–5.6 cm
T 2.3–2.9 cm
FA 4.5–5.1 cm
W 6–12 g

Persian Trident Bat
Triaenops persicus

FR Triaenops de Perse
DE Persische Dreizahnblattnase
ES Rinonicterio persa

LC
Monotypic
HB 5.6–7.2 cm
T 2.7–3.8 cm
FA 4.8–5.8 cm
W 6–18 g

0–1550 m

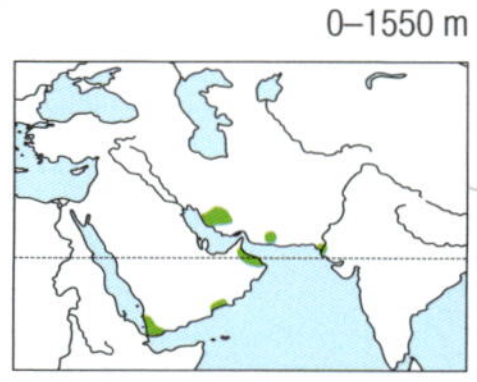

Along W & S coast of YE, SW & NE OM, E AE, S IR & S PK

Genus *TRIAENOPS*

(4 species)

Yemeni Trident Bat
Triaenops parvus

FR Triaenops du Yémen
DE Jemen-Dreizahnblattnase
ES Rinonicterio del Yemen

0–650 m

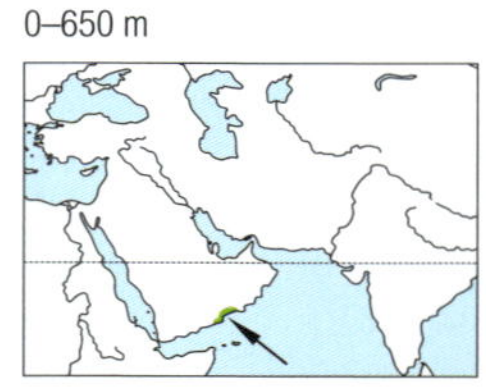

Belt of c. 300 km along coast of S Arabian Pen in E YE & SW OM

DD
Monotypic
HB 5.2–6.1 cm
T 3–3.4 cm
FA 4.4–4.8 cm
W 4–9 g

LC

Monotypic
HB 5.5–7.2 cm
T 2.5–3.8 cm
FA 4.8–6 cm
W 8–17 g

African Trident Bat
Triaenops afer

FR Triaenops d'Afrique
DE Afrika-Dreizahnblattnase
ES Rinonicterio africano

0–1700 m

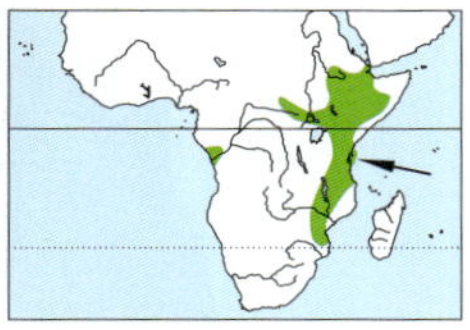

From ET, DJ & SO S to C MZ, W to E CF & NE CD (with Zanzibar Arch), disjunct in SW CG & NW AO

Rufous Trident Bat
Triaenops menamena

FR Triaenops roux
DE Madagaskar-Dreizahnblattnase
ES Rinonicterio pardo

LC

Monotypic
HB 5.6–6.6 cm
T 2.8–3.5 cm
FA 4.6–5.6 cm
W 6–16 g

0–1450 m

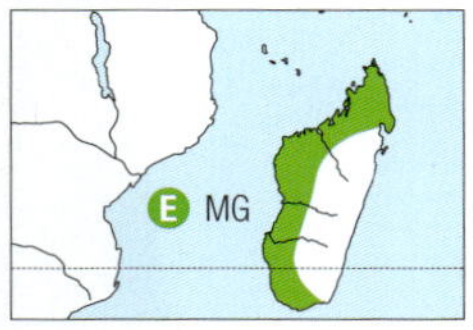

Mostly in lower & drier W, occasionally reported from Highlands

Genus *CLOEOTIS*

(1 species)

0–1540 m

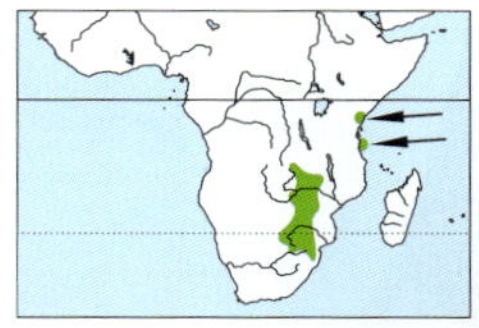

SE KE, Mafia I (TZ), SE CD, ZM, MZ, ZW, SE BW, NE ZA, SZ

LC

Monotypic
HB 3–4.3 cm
T 1.8–3.6 cm
FA 3.1–3.9 cm
W 3–6 g

Percival's Trident Bat
Cloeotis percivali

FR Cléothe de Percival
DE Percival-Kleinohr-Dreizahnblattnase
ES Rinonicterio de Percival

Genus *PARATRIAENOPS*

(3 species)

Grandidier's Trident Bat
Paratriaenops auritus

FR Triaenops de Grandidier
DE Grandidier-Dreizahnblattnase
ES Rinonicterio de Grandidier

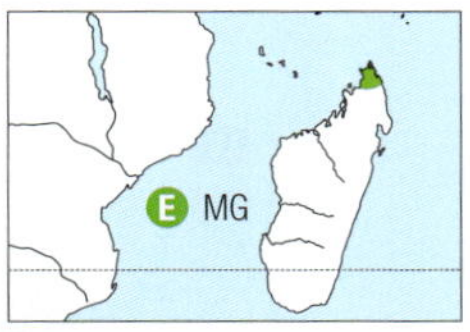

Most extreme N, S to c. 14° S (between Ambilobe and Antsohihy)

VU

Monotypic
HB 4.7–6.4 cm
T 1.9–2.8 cm
FA 4.3–5.1 cm
W 5–8 g

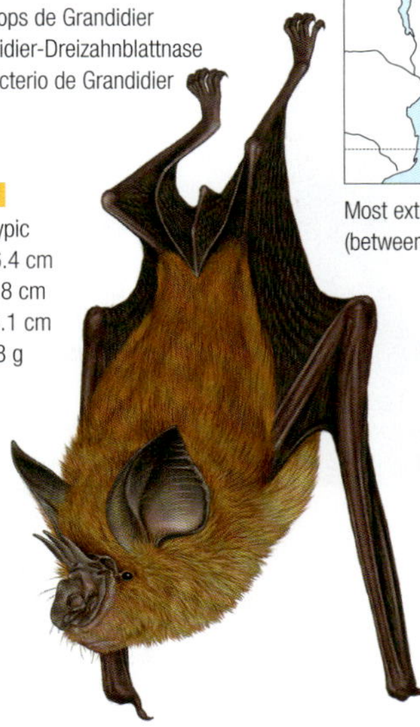

Trouessart's Trident Bat
Paratriaenops furcula

FR Triaenops de Trouessart
DE Trouessart-Dreizahnblattnase
ES Rinonicterio de Trouessart

LC

Monotypic
HB 5–6 cm
T 1.9–2.7 cm
FA 4.2–4.9 cm
W 3–9 g

0–140 m

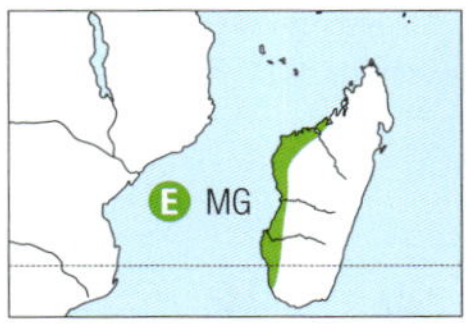

Narrow belt along most of W coast

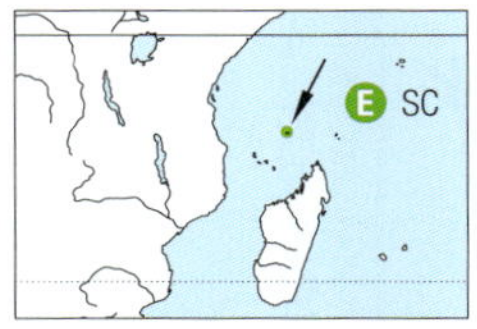

Picard I (Aldabra Atoll)

DD

Monotypic
T 1.7–2.1 cm
FA 4.1–4.5 cm
W c. 6.5 g

Paulian's Trident Bat
Paratriaenops pauliani

FR Triaenops de Paulian
DE Paulian-Dreizahnblattnase
ES Rinonicterio de Paulian

Family HIPPOSIDERIDAE – Old World Leaf-nosed Bats

7 genera, 90 species. Deserts, arid grasslands, savannas, and tropical forests, from sea level to mid-elevations. Palearctic, Afrotropical, Indo-Malayan and Australasian Regions.

Arabian Trident Leaf-nosed Bat
Asellia arabica

FR Asellia d'Arabie
DE Südarabische Dreizackblattnase
ES Aselia de Arabia

DD

Monotypic
HB 4.1–6 cm
T 1.6–2.9 cm
FA 4.3–4.6 cm
W 6–13 g

Genus *ASELLIA*

(4 species)

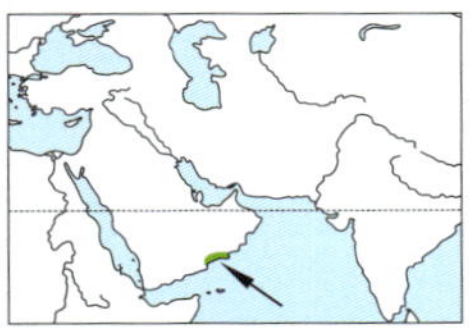

Narrow coastal strip of S Arabian Pen in SE YE & SW OM

Somalian Trident Leaf-nosed Bat
Asellia italosomalica

FR Asellia de Somalie
DE Somalische Dreizackblattnase
ES Aselia de Somalia

DD

Monotypic
HB 4.1–6 cm
T 1.6–2.9 cm
FA 4.3–4.8 cm
W 6–13 g

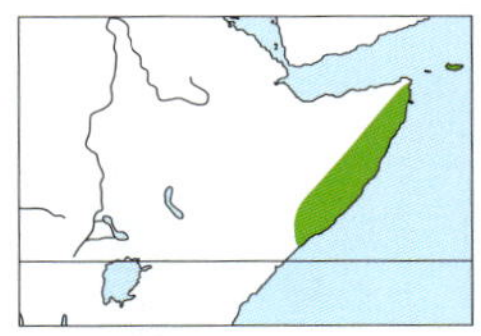

SO, Socotra I (YE)

Patrizi's Trident Leaf-nosed Bat
Asellia patrizii

FR Asellia de Patrizi
DE Patrizi-Dreizackblattnase
ES Aselia de Patrizi

LC

Monotypic
HB 4.2–4.6 cm
T 1.3–2 cm
FA 3.8–4.3 cm
W 3–4 g

up to 1000 m

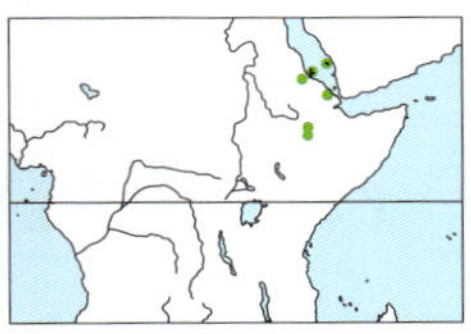

Few localities in ER, ET & offshore Is in Red Sea

Geoffroy's Trident Leaf-nosed Bat
Asellia tridens

FR Asellia trident
DE Nordafrikanische Dreizackblattnase
ES Aselia tridente

LC

2 sspp
HB 4.1–6 cm
T 1.6–2.9 cm
FA 4.5–5.5 cm
W 6–13 g

up to 2000 m

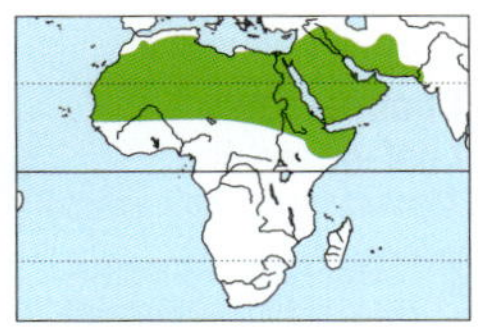

N Africa, from MA, MR & SN E to EG & Horn of Africa, Middle East E to W AF & S PK

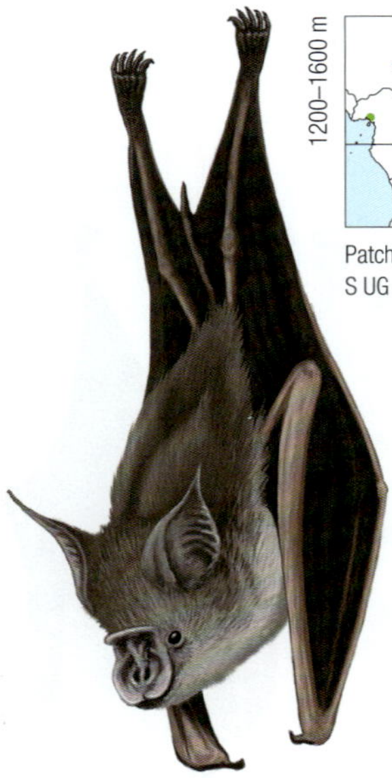

1200–1600 m

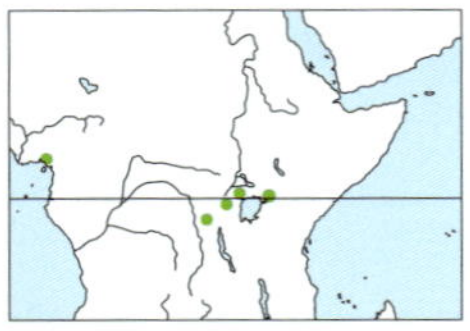

Patchily from SW CM, EC CD, S UG & W KE

DD
Monotypic
HB 9.2–9.7 cm
T 2.3–4.5 cm
FA 7.4–8 cm
W 39–53 g

Greater Leaf-nosed Bat
Doryrhina camerunensis

FR Phyllorhine du Cameroun
DE Kamerun-Rundblattnase
ES Doryrina de Camerún

Genus *DORYRHINA*

(8 species)

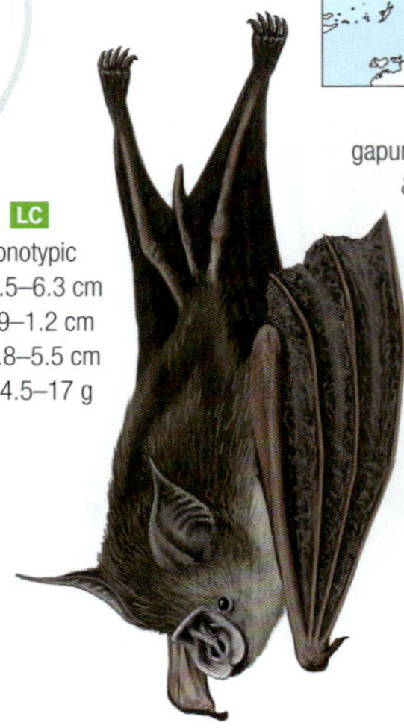

LC
Monotypic
HB 5.5–6.3 cm
T 0.9–1.2 cm
FA 4.8–5.5 cm
W 14.5–17 g

0–2700 m

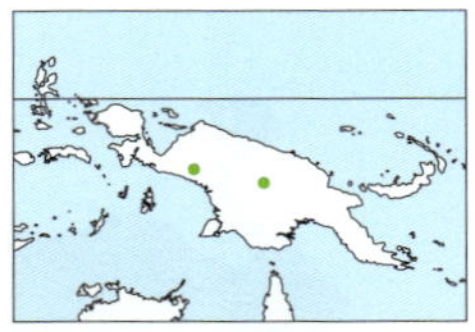

New Guinea in Tembagapura (ID) and around Tifalmin & Telefomin (PG); probably across Central Range

Telefomin Leaf-nosed Bat
Doryrhina corynophyllus

FR Phyllorhine de Telefomin
DE Telefomin-Rundblattnase
ES Doryrina de Telefomin

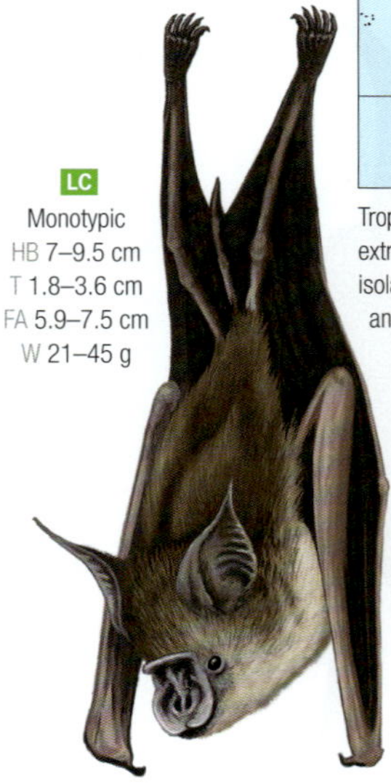

LC
Monotypic
HB 7–9.5 cm
T 1.8–3.6 cm
FA 5.9–7.5 cm
W 21–45 g

up to 2000 m

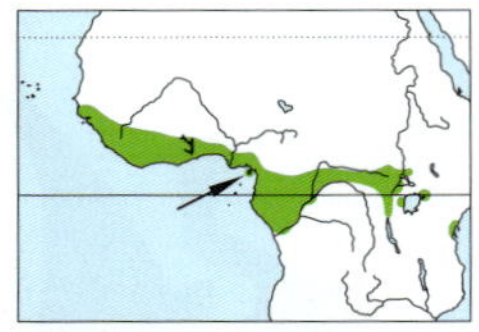

Tropical Africa from SN E to extreme S SS, W UG, RW & BI, isolated in W & SE KE and NE TZ, Bioko I

Cyclops Leaf-nosed Bat
Doryrhina cyclops

FR Phyllorhine cyclope
DE Zyklopen-Rundblattnase
ES Doryrina cíclope

VU
Monotypic
HB c. 5.1 cm
T 1.1–1.5 cm
FA 4.9–5.1 cm
W c. 11 g

200–300 m

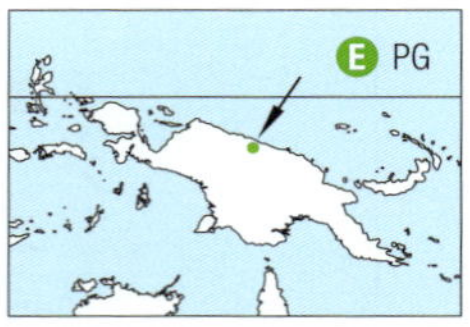

Known only from 3 zones of Bewani Mts (North Coastal Range)

Hill's Leaf-nosed Bat
Doryrhina edwardshilli

FR Phyllorhine d'Edwards Hill
DE Hill-Rundblattnase
ES Doryrina de Edwards Hill

LC
Monotypic
HB 4.4–5.2 cm
T c. 2.3 cm
FA 4.5–4.7 cm

0–2400 m

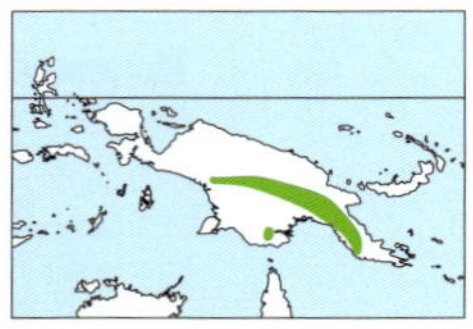

ID/PG: New Guinea, mostly along Central Range

Fly River Leaf-nosed Bat
Doryrhina muscinus

FR Phyllorhine de Fly River
DE Fly-River-Rundblattnase
ES Doryrina de Fly River

LC
Monotypic
HB 4–5 cm
T 2.2–2.8 cm
FA 3.8–4.2 cm
W 6–10 g

0–1400 m

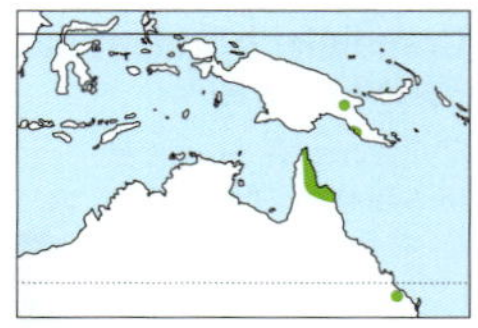

PG (E New Guinea), NE AU (coastal Qld from E Cape York to S of Cooktown); S limits unclear

Semon's Leaf-nosed Bat
Doryrhina semoni

FR Phyllorhine de Semon
DE Semon-Rundblattnase
ES Doryrina de Semon

VU
Monotypic
HB 4–4.6 cm
T 2.6–2.7 cm
FA 4.2–4.6 cm
W 4.6–6.4 g

Northern Leaf-nosed Bat
Doryrhina stenotis

FR Phyllorhine à oreilles fines
DE Nördliche Rundblattnase
ES Doryrina norteña

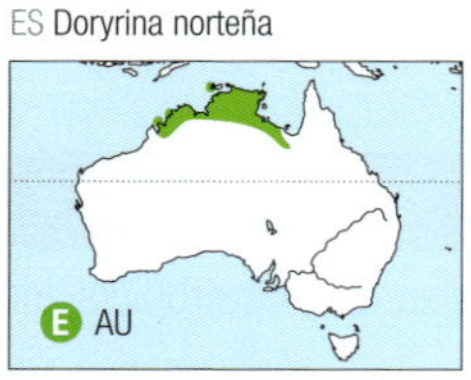

NE WA, N NT, NW Qld (Mt Isa), offshore Is (Boongaree, Koolan, Bathurst)

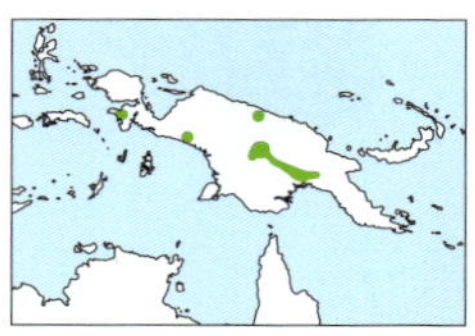

ID/PG: New Guinea in Fakfak, North Coastal & Central Ranges (mostly reported near Tifalmin & Telefomin)

30–2440 m

LC
3 sspp
HB 4.1–4.5 cm
T c. 2.7 cm
FA c. 4.4 cm
W c. 8 g

Wollaston's Leaf-nosed Bat
Doryrhina wollastoni

FR Phyllorhine de Wollaston
DE Wollaston-Rundblattnase
ES Doryrina de Wollaston

up to 1500 m

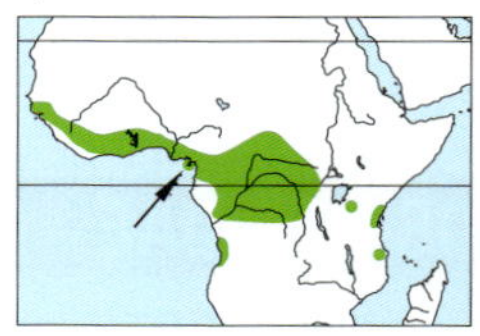

Tropical Africa from SN E to extreme W UG & E CD, Bioko I (GQ), isolated in W AO, SE KE & TZ

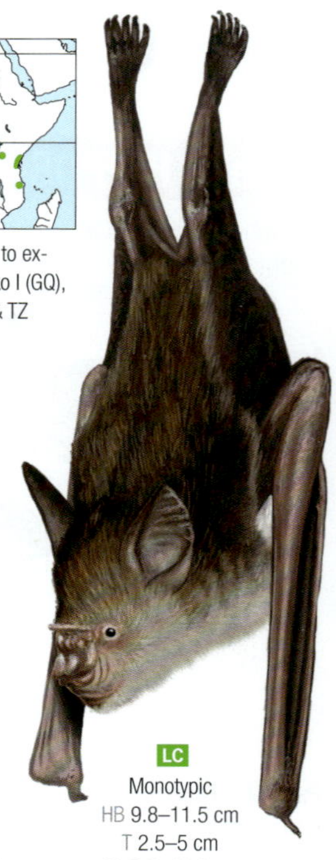

LC
Monotypic
HB 9.8–11.5 cm
T 2.5–5 cm
FA 9.5–12.4 cm
W 85–138 g

Giant Leaf-nosed Bat
Macronycteris gigas

FR Phyllorhine géante
DE Riesenrundblattnase
ES Macronícteris gigante

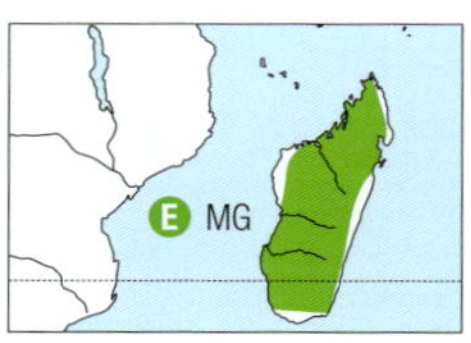

up to 1350 m

Widespread on variety of forested and wooded habitats

Commerson's Leaf-nosed Bat
Macronycteris commersonii

FR Phyllorhine de Commerson
DE Commerson-Rundblattnase
ES Macronícteris de Commerson

NT
Monotypic
HB 10.4–11 cm
T 3.1–4.5 cm
FA 8–10.3 cm
W 39–98 g

Genus *MACRONYCTERIS*

(5 species)

Madagascar Cryptic Leaf-nosed Bat
Macronycteris cryptovalorona

FR Phyllorhine cryptique
DE Madagaskar-Rundblattnase
ES Macronícteris de Madagascar

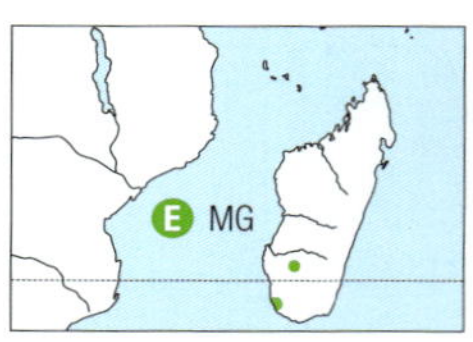

Known from Isalo NP & Androimpano Cave (near Itampolo)

NE
Monotypic
HB 8.9–9.2 cm
T 3–3.4 cm
FA 8–8.1 cm
W 26–42 g

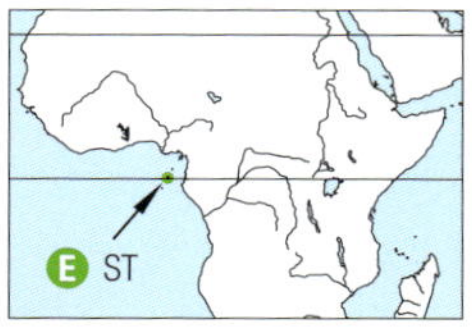

São Tomé I

LC
Monotypic
HB 10.2–10.6 cm
T 2.9–3 cm

Sao Tome Leaf-nosed Bat
Macronycteris thomensis

FR Phyllorhine de Sao Tomé
DE São-Tomé-Rundblattnase
ES Macronícteris de Sao Tomé

Striped Leaf-nosed Bat
Macronycteris vittatus

FR Phyllorhine rayée
DE Gestreifte Rundblattnase
ES Macronícteris de rayas

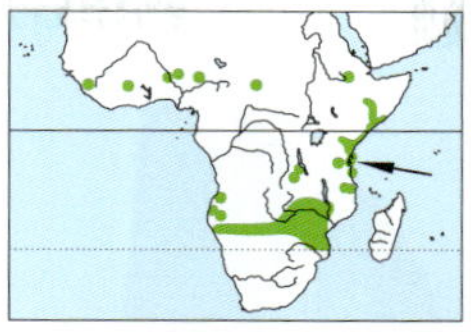

Savannas of Africa, patchily from SL E to ET, S to NA, BW, ZW & NE ZA, also Zanzibar Arch

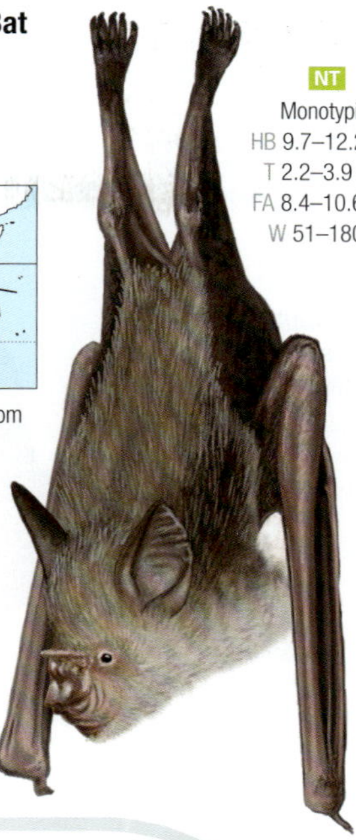

NT
Monotypic
HB 9.7–12.2 cm
T 2.2–3.9 cm
FA 8.4–10.6 cm
W 51–180 g

Dong Bac Trident Bat
Aselliscus dongbacanus

FR Phyllorhine du Dong Bac
DE Dong-Bac-Dreizackblattnase
ES Aselisco de Dong Bac

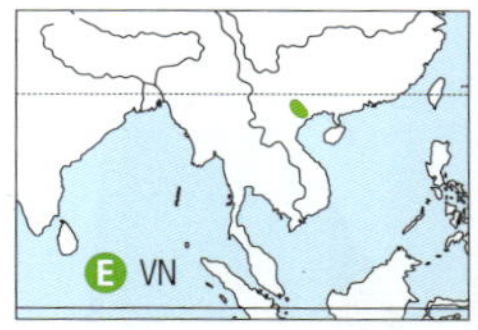

0–500 m

Karst and mountainous areas, including some offshore Is

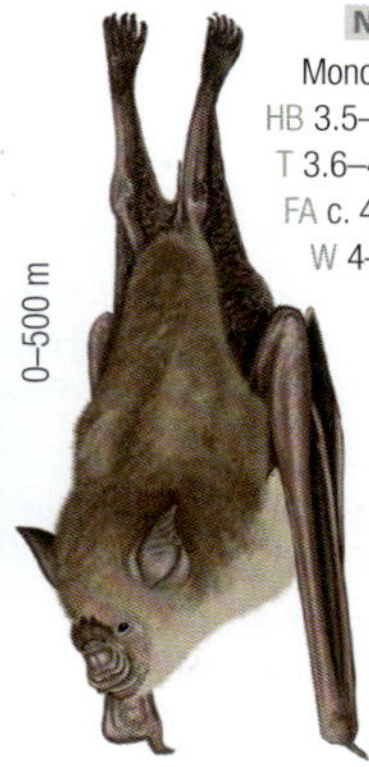

NE
Monotypic
HB 3.5–4.7 cm
T 3.6–4.2 cm
FA c. 4.3 cm
W 4–8 g

Genus *ASELLISCUS*

(3 species)

Temminck's Trident Bat
Aselliscus tricuspidatus

LC 4 sspp
HB 3.6–4.9 cm
T 1.7–2.7 cm
FA 3.6–4.5 cm
W c. 4.6 g

FR Phyllorhine de Temminck
DE Temminck-Dreizackblattnase
ES Aselisco de Temminck

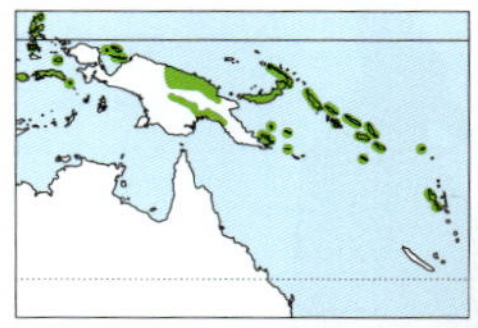

0–900 m

Moluccas, New Guinea (E mainland & nearest Is), Bismarck Arch, Solomon Is, N VU

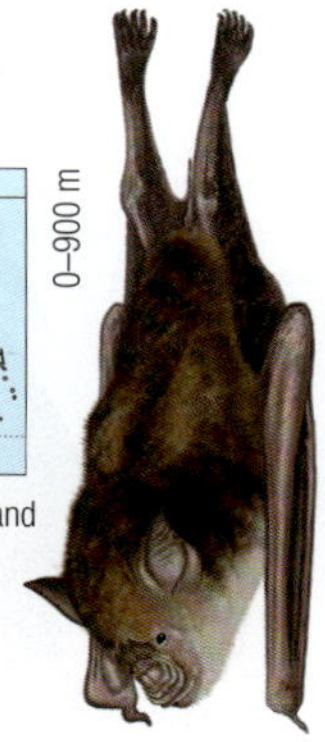

East Asian Tailless Leaf-nosed Bat
Coelops frithii

FR Phyllorhine de Frith
DE Ostasiatische Schwanzlose Blattnase
ES Coelops de Frith

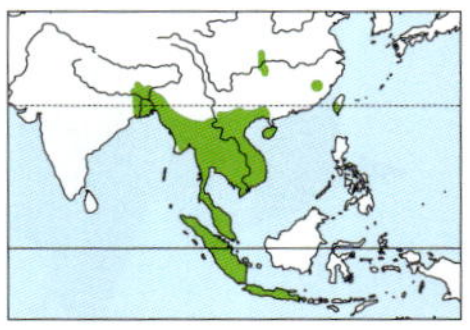

0–1370 m

NE IN & BD E to C & S CN (with Hainan I), TW, mainland SE Asia S to Sumatra, Java & Bali

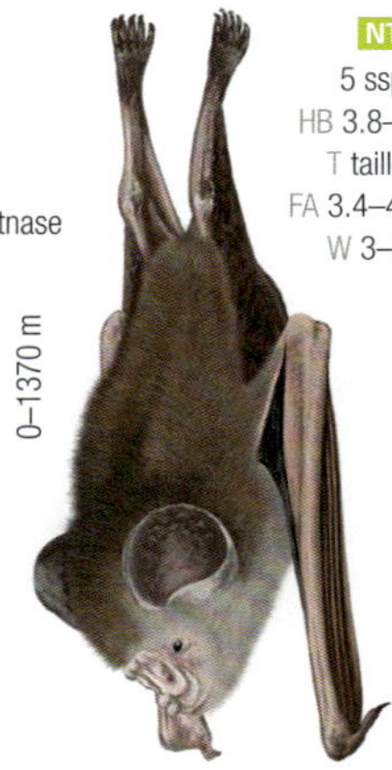

NT
5 sspp
HB 3.8–5 cm
T tailless
FA 3.4–4.4 cm
W 3–7 g

Stoliczka's Trident Bat
Aselliscus stoliczkanus

FR Phyllorhine de Stoliczka
DE Stoliczka-Dreizackblattnase
ES Aselisco de Stoliczka

LC
Monotypic
HB 4–5 cm
T 3.3–4.4 cm
FA 3.9–4.5 cm
W c. 6.9 g

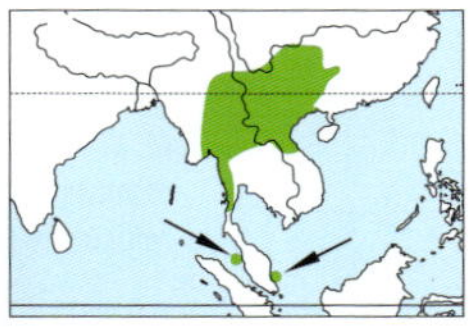

NE MM & SC CN S to S TH, C LA & C VN, also Penang & Tioman Is

Genus *COELOPS*

(3 species)

Malayan Tailless Leaf-nosed Bat
Coelops robinsoni

FR Phyllorhine de Robinson
DE Malaiische Schwanzlose Blattnase
ES Coelops de Robinson

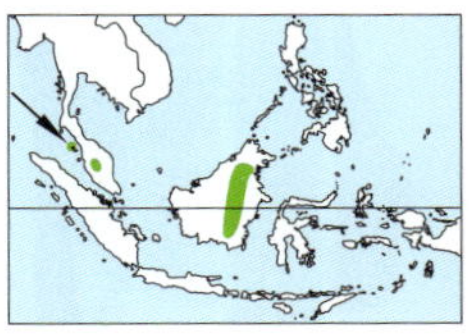

Thai-Malay Pen (with Tarutao I), Borneo (MY, ID)

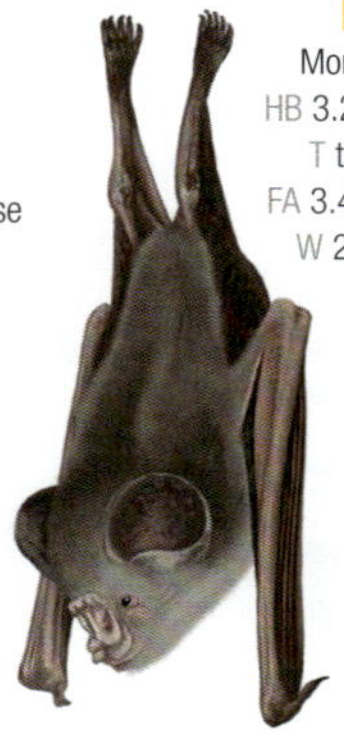

VU
Monotypic
HB 3.2–3.6 cm
T tailless
FA 3.4–3.7 cm
W 2.5–4 g

Philippine Tailless Leaf-nosed Bat
Coelops hirsutus

FR Phyllorhine hirsute
DE Philippinische Schwanzlose Blattnase
ES Coelops de Filipinas

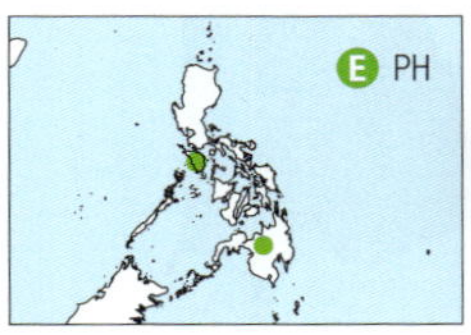

Mindoro & Mindanao (Bukidnon Prov)

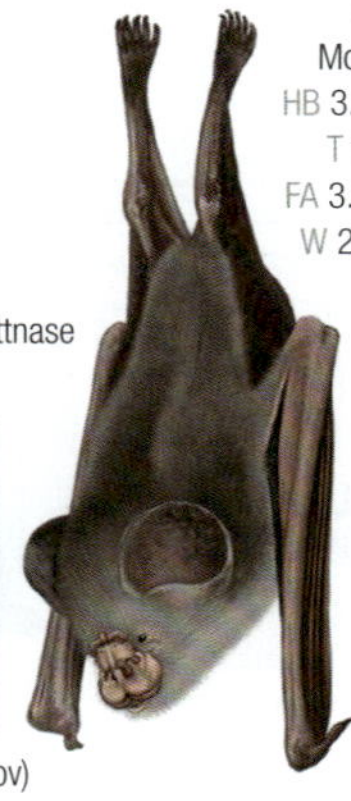

DD
Monotypic
HB 3.5–4.2 cm
T tailless
FA 3.3–3.6 cm
W 2.4–3.6 g

VU
Monotypic
HB 4.7–6.1 cm
T 0.3–0.6 cm
FA 4.8–5.3 cm
W c. 8 g

Genus *ANTHOPS*

(1 species)

0–200 m

Solomon Is (Buka, Bougainville, Choiseul, Santa Isabel, Florida, Guadalcanal)

Solomons Leaf-nosed Bat
Anthops ornatus

FR Anthopse des Salomon
DE Blumennasenfledermaus
ES Anthops de las Salomón

Large-eared Leaf-nosed Bat
Hipposideros megalotis

FR Phyllorhine à grandes oreilles
DE Großohr-Rundblattnase
ES Hiposidérido de grandes orejas

LC
Monotypic
HB c. 3.5 cm
T 2.1–2.7 cm
FA 3.4–3.9 cm
W c. 5 g

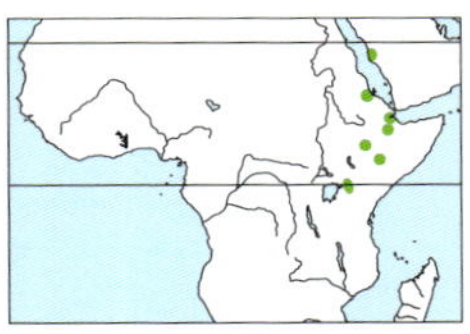

Patchily in W coast of SA, ER, DJ, ET & KE

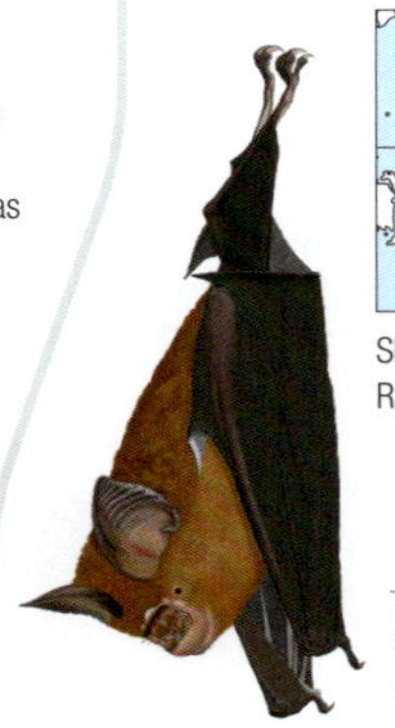

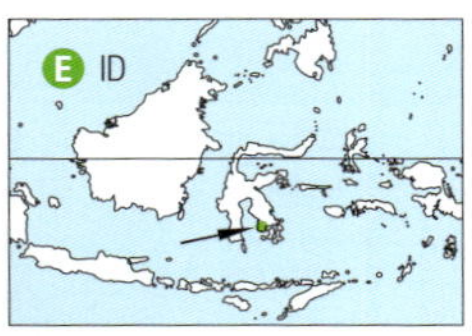

SE Sulawesi (known only from Rawa Aopa Watumohai NP)

DD
Monotypic
T 1.6–2.1 cm
FA 4–4.3 cm
W 6.8–8.5 g

Boeadi's Leaf-nosed Bat
Hipposideros boeadii

FR Phyllorhine de Boeadi
DE Boeadi-Rundblattnase
ES Hiposidérido de Boeadi

Genus *HIPPOSIDEROS*

(66 species)

Ha Long Leaf-nosed Bat
Hipposideros alongensis

FR Phyllorhine d'Halong
DE Halong-Rundblattnase
ES Hiposidérido de Halong

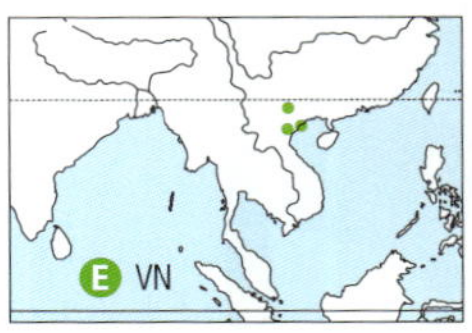

Scattered in mainland (Na Hang NR, Ba Be & Cuc Phuong NPs), Cat Ba I

VU
2 sspp
FA 6.8–7.6 cm
W 22–35 g

Great Himalayan Leaf-nosed Bat
Hipposideros armiger

FR Phyllorhine de l'Himalaya
DE Große Himalaya-Rundblattnase
ES Hiposidérido de Himalaya

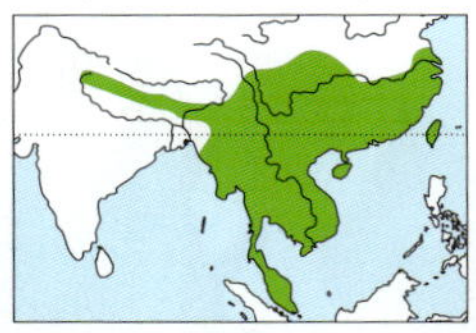

N IN, NP, C, S & SE CN (with Hainan I), TW, mainland SE Asia

LC
4 sspp
HB 8–11 cm
T 4.8–7 cm
FA 8.5–10.3 cm
W 44–67 g

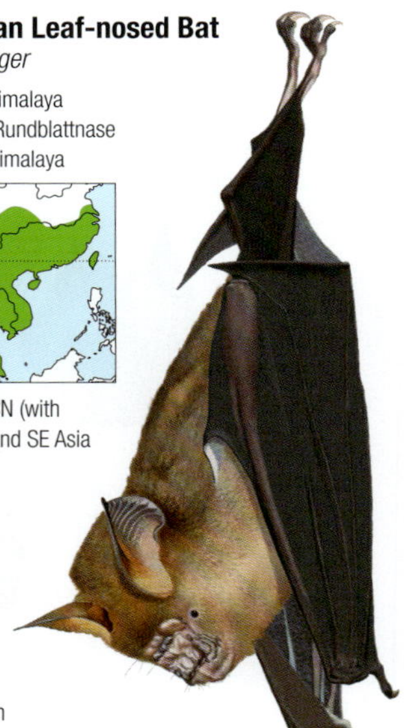

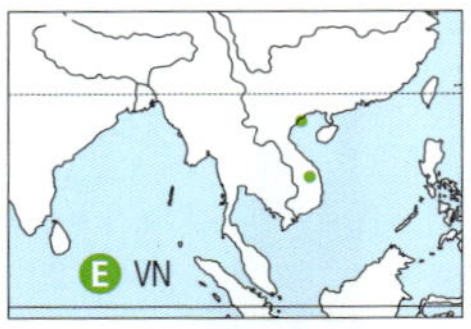

Known from Cat Ba I (Ha Long Bay) & Chu Mom Ray NP (Kon Tum Prov)

Griffin's Leaf-nosed Bat
Hipposideros griffini

FR Phyllorhine de Griffin
DE Griffin-Rundblattnase
ES Hiposidérido de Griffin

NT
Monotypic
FA 8.3–9 cm
W c. 44 g

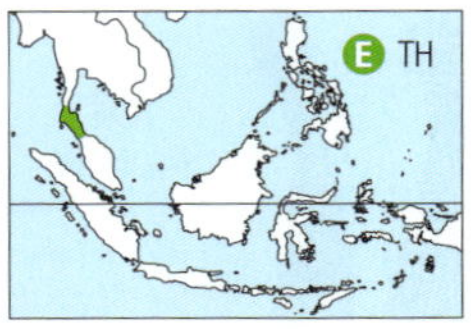

Thai-Malay Pen

Pendlebury's Leaf-nosed Bat
Hipposideros pendleburyi

FR Phyllorhine de Pendlebury
DE Pendlebury-Rundblattnase
ES Hiposidérido de Pendlebury

VU
Monotypic
T 4.8–7 cm
FA 7.5–8.1 cm
W 44–67 g

Yaeyama Is (Yonaguni, Iriomote, Ishigaki, Hateruma)

Lesser Leaf-nosed Bat
Hipposideros turpis

FR Phyllorhine des Ryukyu
DE Ryukyu-Rundblattnase
ES Hiposidérido de Ryukyu

EN
Monotypic
HB 6.9–8.8 cm
T 4.2–5.1 cm
FA 6.4–7.3 cm
W 25–33 g

Grand Leaf-nosed Bat
Hipposideros grandis

FR Grande Phyllorhine
DE Große Rundblattnase
ES Hiposidérido grande

LC
Monotypic
HB 6.9–8 cm
T 3–4.5 cm
FA 5.9–6.2 cm
W 17–18 g

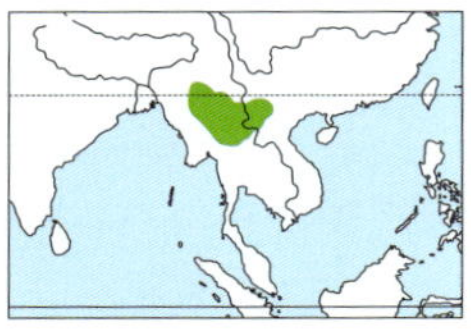

SC CN (Yunnan), MM, NW TH, N LA & NW VN

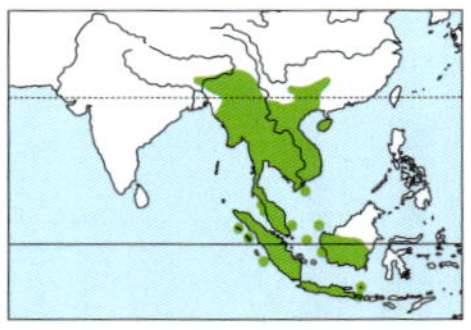

NE IN, NE BD, S CN (with Hainan I), mainland SE Asia, Greater Sundas (with many offshore Is)

Horsfield's Leaf-nosed Bat
Hipposideros larvatus

FR Phyllorhine de Horsfield
DE Horsfield-Rundblattnase
ES Hiposidérido de Horsfield

LC
Monotypic
HB 5.2–7.5 cm
T 3.2–3.5 cm
FA 5–6.7 cm
W 15–23 g

below 1000 m

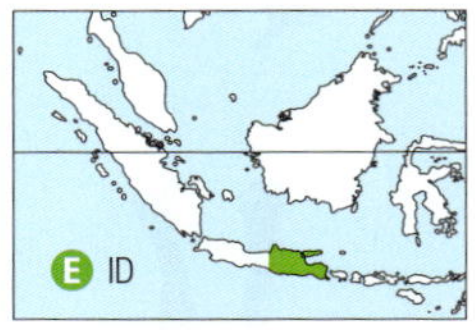

E Java & Madura I

Maduran Leaf-nosed Bat
Hipposideros madurae

FR Phyllorhine de Madura
DE Madura-Rundblattnase
ES Hiposidérido de Madura

NT
2 sspp
T 2.4–3.1 cm
FA 5.3–5.8 cm

up to 1000 m

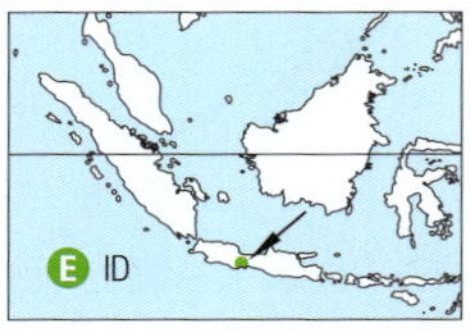

WC Java (known only from type locality in Pangandaran)

Sorensen's Leaf-nosed Bat
Hipposideros sorenseni

FR Phyllorhine de Sorensen
DE Sorensen-Rundblattnase
ES Hiposidérido de Sorensen

EN
Monotypic
HB 5–6 cm
T 3–3.6 cm
FA 5.5–6 cm

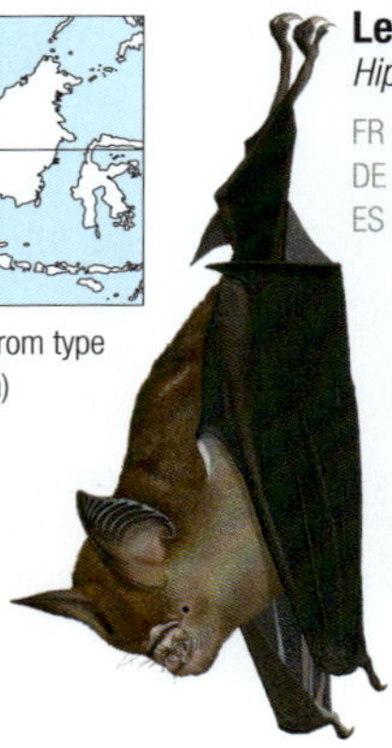

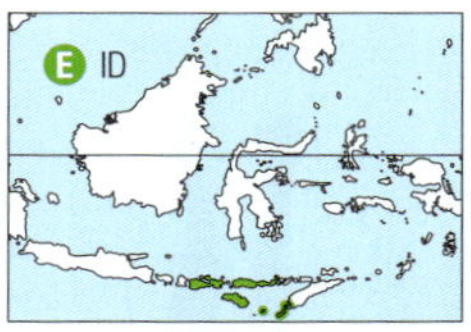

below 1000 m

Lesser Sundas (Sumbawa, Flores, Sumba, Savu, Roti, Semau, W Timor)

Sumban Leaf-nosed Bat
Hipposideros sumbae

FR Phyllorhine de Sumba
DE Sumba-Rundblattnase
ES Hiposidérido de Sumba

LC
3 sspp
T 2.4–3.2 cm
FA 4.8–5.7 cm

Shield-faced Leaf-nosed Bat
Hipposideros lylei

FR Phyllorhine de Lyle
DE Lyle-Rundblattnase
ES Hiposidérido de Lyle

LC
Monotypic
HB 7.2–9.5 cm
T 4.8–5.5 cm
FA 7.3–8.4 cm

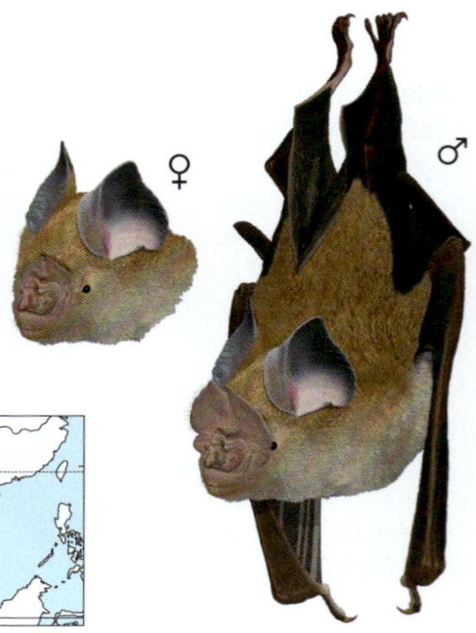

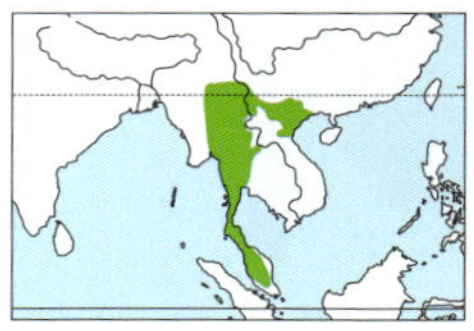

E & S MM, SC CN (Yunnan), W & S TH, N VN, Thai-Malay Pen

Swinhoe's Leaf-nosed Bat
Hipposideros swinhoei

FR Phyllorhine de Swinhoe
DE Pratt-Rundblattnase
ES Hiposidérido de Swinhoe

LC
Monotypic
HB 9.1–11 cm
T 5–6.2 cm
FA 7.9–8.9 cm

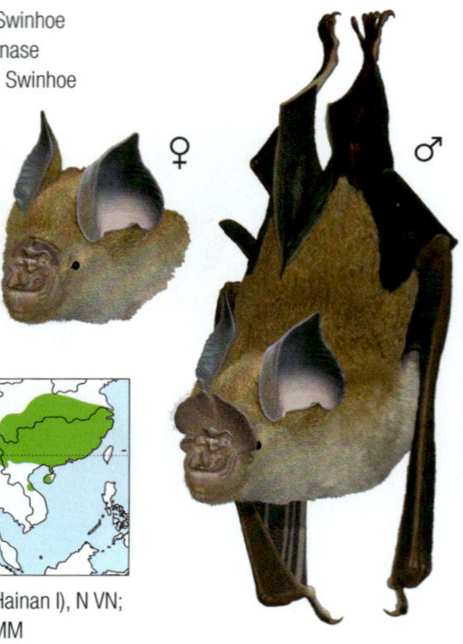

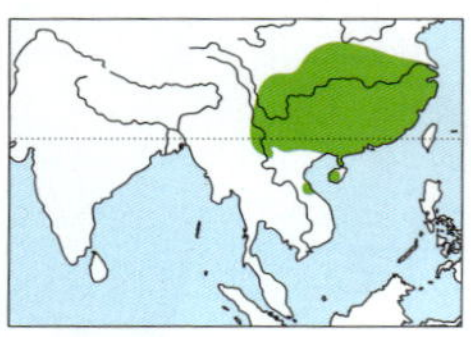

C, E & S CN (with Hainan I), N VN; unconfirmed in N MM

Shield-nosed Leaf-nosed Bat
Hipposideros scutinares

FR Phyllorhine à blason
DE Nasenschild-Rundblattnase
ES Hiposidérido de escudo

VU
Monotypic
T 5–5.4 cm
FA 7.8–8.3 cm
W c. 42 g

up to 160 m

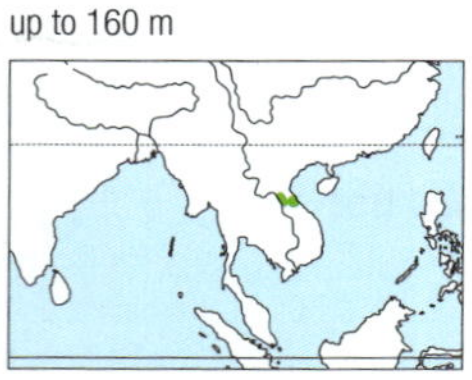

C LA, adjacent C VN

Makira Leaf-nosed Bat
Hipposideros demissus

FR Phyllorhine de Makira
DE Makira-Rundblattnase
ES Hiposidérido de Makira

EN
Monotypic
HB 6.5–7 cm
T 3.4–4.3 cm
FA c. 6.7 cm
W c. 27 g

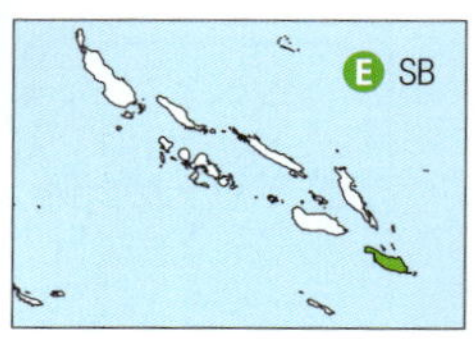

Makira

LC
Monotypic
HB 6.6–10 cm
T 3.2–5.1 cm
FA 5.8–9.6 cm
W 22–57 g

0–1210 m

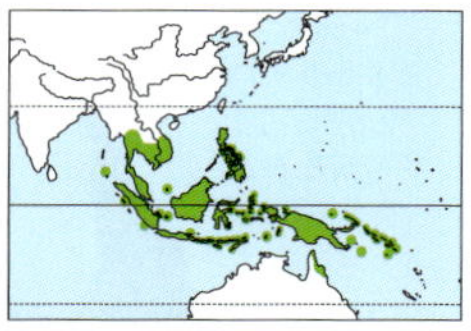

Mainland SE Asia, N Nicobar Is, PH, ID, PG, SB, NE AU

Diadem Leaf-nosed Bat
Hipposideros diadema
FR Phyllorhine à diadème
DE Diadem-Rundblattnase
ES Hiposidérido de diadema

0–400 m

Solomon Is (Bougainville, New Georgia, Santa Isabel, San Jorge, Malaita, Guadalcanal)

Fierce Leaf-nosed Bat
Hipposideros dinops
FR Phyllorhine féroce
DE Salomonen-Rundblattnase
ES Hiposidérido de las Salomón

VU
Monotypic
HB 8.8–10.5 cm
T 5.4–6.3 cm
FA 8.6–9.9 cm
W c. 78 g

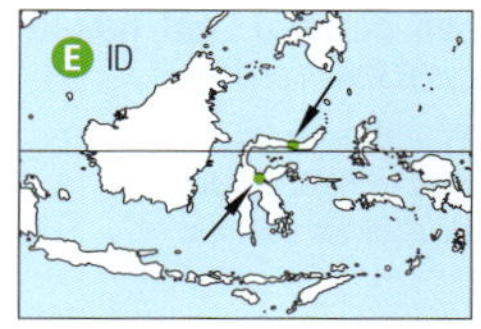

N & C Sulawesi

Crested Leaf-nosed Bat
Hipposideros inexpectatus
FR Phyllorhine à crête
DE Sulawesi-Rundblattnase
ES Hiposidérido crestado

DD
Monotypic
FA c. 10.1 cm

VU
Monotypic
HB 6.6–7.7 cm
T 3.2–4.6 cm
FA 6.8–7.4 cm
W 22–35 g

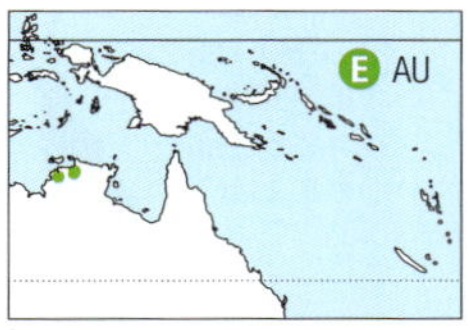

Small area in N NT (few sites in Kakadu NP area & Litchfield NP)

Arnhem Leaf-nosed Bat
Hipposideros inornatus
FR Phyllorhine de l'Arnhem
DE Arnhem-Rundblattnase
ES Hiposidérido de Arnhem

up to 1000 m

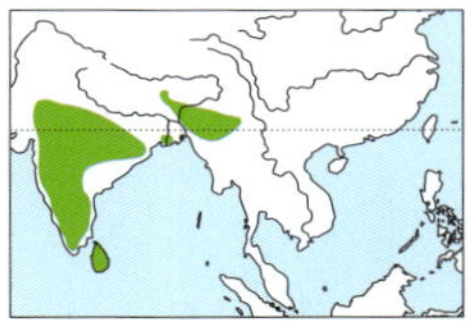

IN (Peninsula & NE), BD, LK, N MM

LC
3 sspp
T 3.5–5.8 cm
FA 7.5–9.9 cm

Indian Leaf-nosed Bat
Hipposideros lankadiva
FR Phyllorhine indienne
DE Indien-Rundblattnase
ES Hiposidérido indio

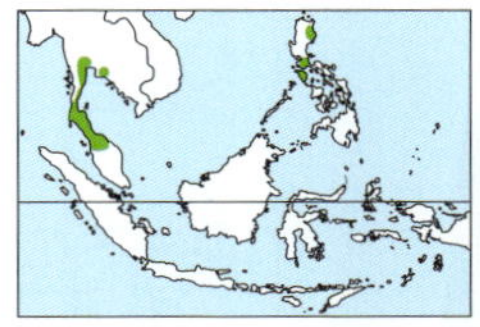

S TH, Pen MY, PH (Luzon, Mindoro)

Large Asian Leaf-nosed Bat
Hipposideros lekaguli
FR Phyllorhine de Lekagul
DE Lekagul-Rundblattnase
ES Hiposidérido de Lekagul

NT
Monotypic
FA 6.4–7.9 cm

Peleng Leaf-nosed Bat
Hipposideros pelingensis
FR Phyllorhine de Peleng
DE Peleng-Rundblattnase
ES Hiposidérido de Peleng

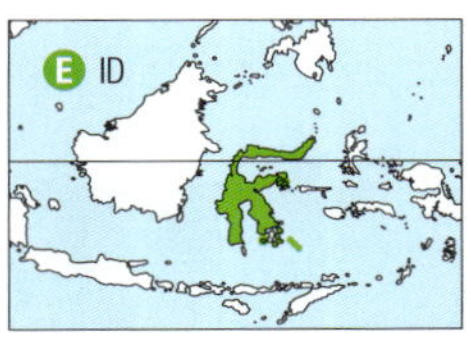

0–1000 m

Sulawesi (with Peleng, Buton & Kabaena Is), Tukangbesi Is (Wangi-wangi, Kaledupa, Tomea)

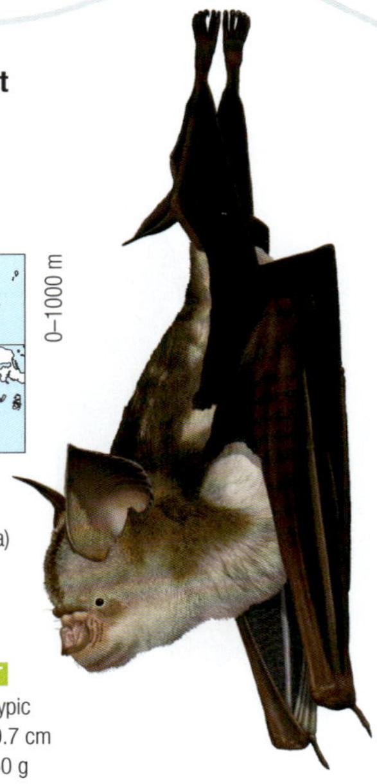

NT
Monotypic
FA 9.3–9.7 cm
W c. 50 g

EN
Monotypic
HB 5–5.5 cm
T 1.8–2.3 cm
FA 4.2–4.7 cm

Short-tailed Leaf-nosed Bat
Hipposideros curtus
FR Phyllorhine à queue courte
DE Kurzschwanz-Rundblattnase
ES Hiposidérido de cola corta

0–500 m

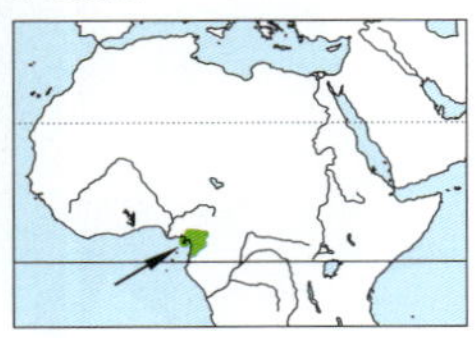

Patchily in CM & GQ (with Bioko I)

NT
Monotypic
HB 5–5.5 cm
T 1.7–2.7 cm
FA 4.4–5 cm
W 4–8 g

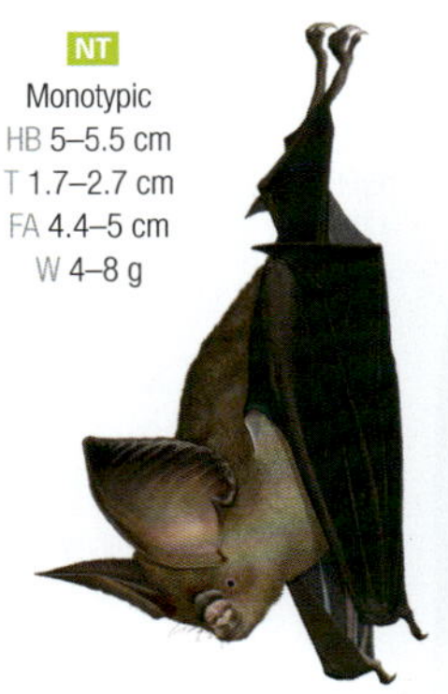

Jones's Leaf-nosed Bat
Hipposideros jonesi
FR Phyllorhine de Jones
DE Jones-Rundblattnase
ES Hiposidérido de Jones

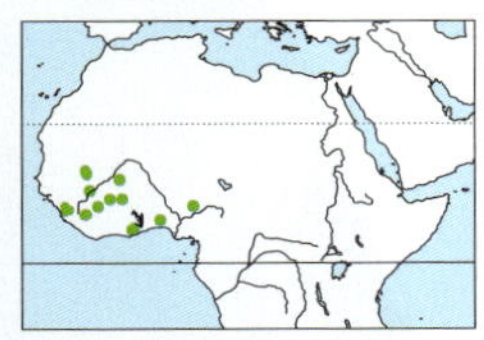

Patchily in W Africa (GN, SL, LR, ML, BF, CI, GH, NG)

VU
Monotypic
HB 4.1–4.7 cm
T 1.9–2.1 cm
FA 3.8–4.2 cm
W c. 5 g

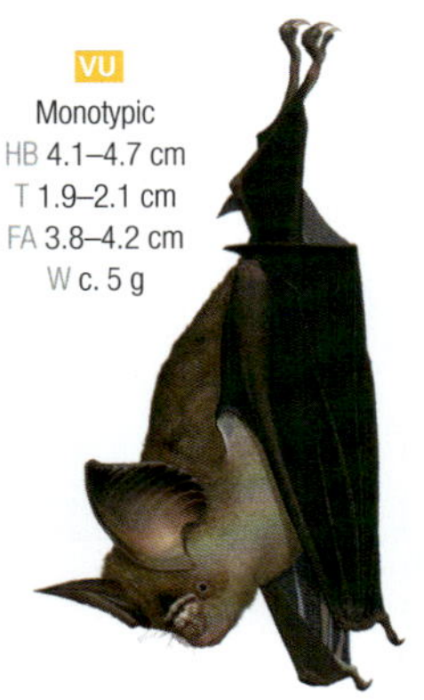

Aellen's Leaf-nosed Bat
Hipposideros marisae
FR Phyllorhine d'Aellen
DE Aellen-Rundblattnase
ES Hiposidérido de Aellen

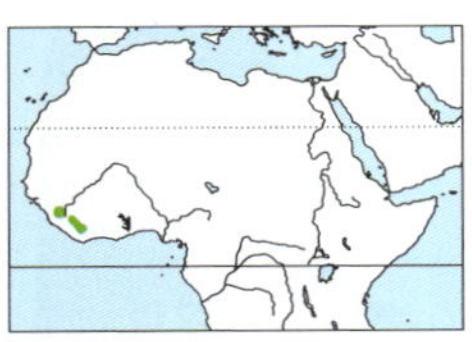

Very patchily in W Africa (GN, SL, LR, CI)

Schneider's Leaf-nosed Bat
Hipposideros speoris
FR Phyllorhine de Schneider
DE Schneider-Rundblattnase
ES Hiposidérido de Schneider

up to 1385 m

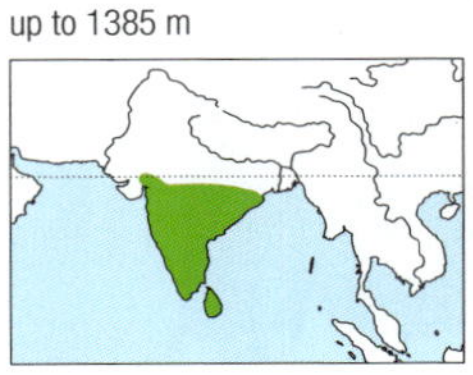

IN, LK

LC
Monotypic
HB 4.6–6.2 cm
FA 4.5–5.4 cm

Spurred Leaf-nosed Bat
Hipposideros calcaratus
FR Phyllorhine à éperons
DE Sporn-Rundblattnase
ES Hiposidérido de espolones

0–1000 m

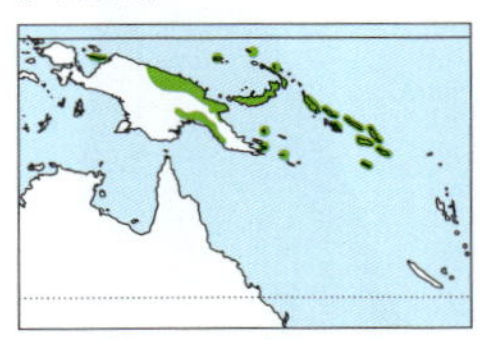

New Guinea (mainland, Yapen, Kiriwina, Normanby, Misima & Manus Is), Bismarck Arch, Solomon Is

LC
2 sspp
HB 4.9–7 cm
T 3.1–4.4 cm
FA 4.6–5.6 cm
W c. 11.5 g

Fawn-colored Leaf-nosed Bat
Hipposideros cervinus

LC
2 sspp
HB 5–5.5 cm
T 2.1–3.1 cm
FA 4.4–5.3 cm
W 6–9 g

FR Phyllorhine de Gould
DE Rehbraune Rundblattnase
ES Hiposidérido de Gould

0–1400 m

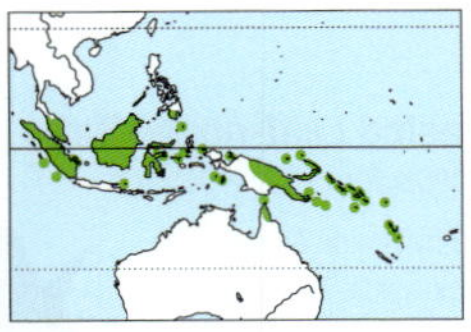

Pen MY, PH (Palawan, Mindanao), across ID & PG E to SB, VU & NE AU

Cox's Leaf-nosed Bat
Hipposideros coxi

EN
Monotypic
FA 5.3–5.5 cm

FR Phyllorhine de Cox
DE Cox-Rundblattnase
ES Hiposidérido de Cox

0–1200 m

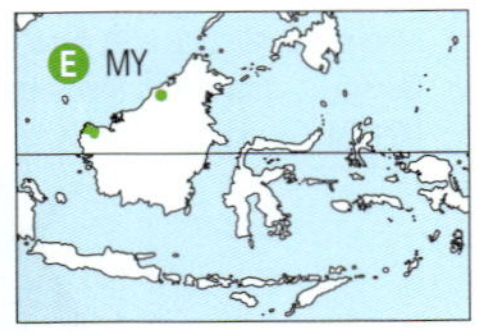

Few sites in W Borneo (N & SW Sarawak); unconfirmed in Kalimantan (ID)

Cantor's Leaf-nosed Bat
Hipposideros galeritus

LC
4 sspp
HB 4.5–5.9 cm
T 2.9–3.7 cm
FA 4.5–5.1 cm
W c. 8.6 g

FR Phyllorhine de Cantor
DE Cantor-Rundblattnase
ES Hiposidérido de Cantor

0–1100 m

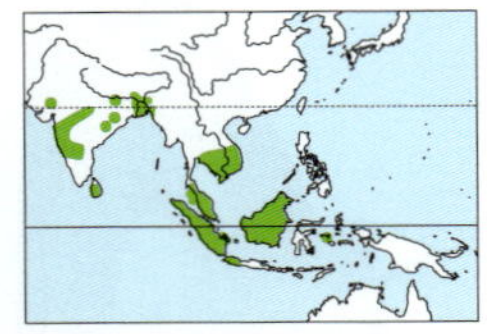

IN, BD, LK, S mainland SE Asia S to Sumatra, Bangka, Borneo, W & C Java & Sula Is (Sanana)

Aba Leaf-nosed Bat
Hipposideros abae

LC
Monotypic
HB 6–7 cm
T 2.8–4 cm
FA 5.4–6.5 cm
W 17–19 g

FR Phyllorhine d'Aba
DE Aba-Rundblattnase
ES Hiposidérido de Aba

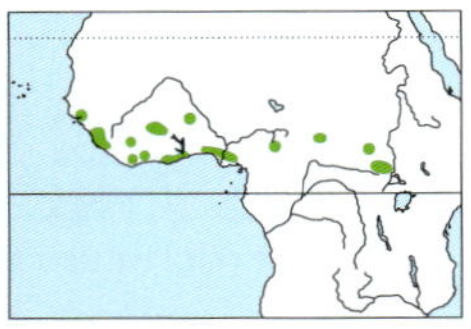

Patchily in Africa N of equator from GW & GN E to SS & extreme NW UG

Benito Leaf-nosed Bat
Hipposideros beatus

LC
2 sspp
HB 4.3–5 cm
T 2–3.1 cm
FA 3.9–4.8 cm
W 6–9 g

FR Phyllorhine du Benito
DE Benito-Rundblattnase
ES Hiposidérido de Benito

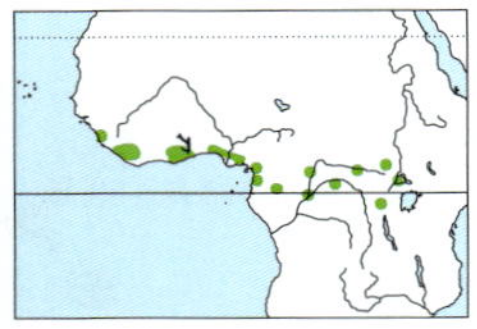

Patchily in Africa from SL & LR E to GA, CG, CD & extreme NW UG

Sundevall's Leaf-nosed Bat
Hipposideros caffer

LC
Monotypic
HB 4.5–6 cm
T 2.5–3.8 cm
FA 4.2–5.2 cm
W 5–11 g

FR Phyllorhine de Cafrerie
DE Gewöhnliche Rundblattnase
ES Hiposidérido de Cafrería

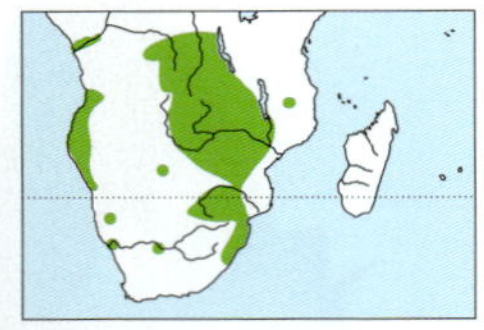

Across most of sub-Saharan Africa, but N & W boundaries unknown

Sooty Leaf-nosed Bat
Hipposideros fuliginosus

LC
Monotypic
HB 8–10 cm
T 2.9–4 cm
FA 5.1–6 cm
W 18–20 g

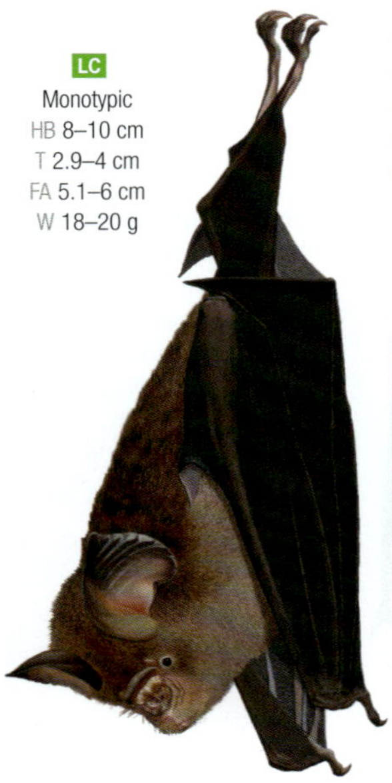

FR Phyllorhine fuligineuse
DE Temminck-Rundblattnase
ES Hiposidérido tiznado

up to 1300 m

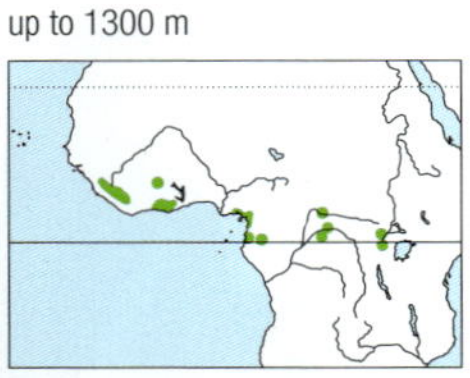

Patchily across tropical Africa from SL & GN E to N CD & SW UG

Lamotte's Leaf-nosed Bat
Hipposideros lamottei

CR
Monotypic
HB 9.6–10.3 cm
T 3.5–4.1 cm
FA 5.5–5.7 cm
W 9–10 g

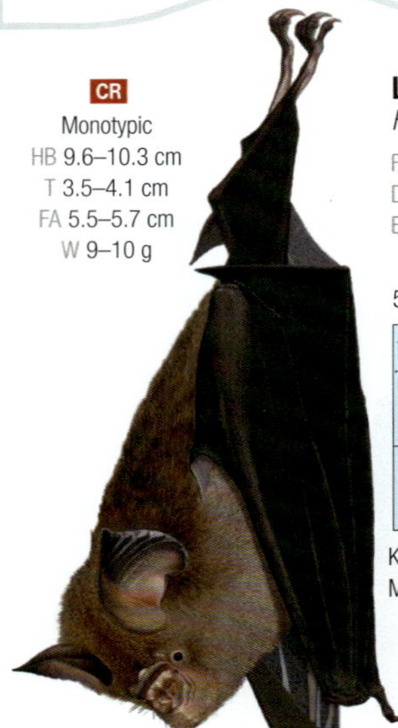

FR Phyllorhine de Lamotte
DE Lamotte-Rundblattnase
ES Hiposidérido de Lamotte

500–1400 m

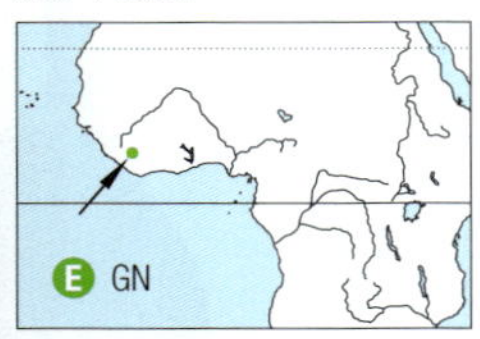

Known only from Guinean side of Mt Nimba

Noack's Leaf-nosed Bat
Hipposideros ruber

LC
Monotypic
HB 5.2–6.2 cm
T 3–4.1 cm
FA 4.7–5.5 cm
W 9–12 g

FR Phyllorhine de Noack
DE Noack-Rundblattnase
ES Hiposidérido de Noack

up to 1000 m

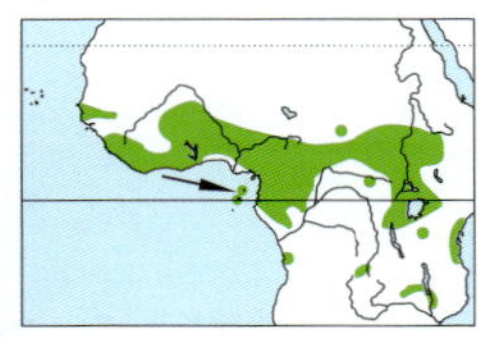

Tropical Africa, from SN & GM E to ET, S to CG, CD & TZ, isolated in NW AO, ZM, MW & MZ; also ST

Maghreb Leaf-nosed Bat
Hipposideros tephrus

FR Phyllorhine cendrée
DE Maghreb-Rundblattnase
ES Hiposidérido de Magreb

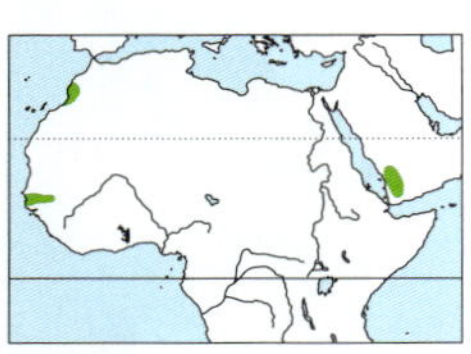

Certainty in MA, SN, SA & YE; perhaps continuously from MR & SN E to SS, ET & ER

LC
Monotypic
HB 4.5–5 cm
T 2.7–3.5 cm

Common Dusky Leaf-nosed Bat
Hipposideros ater

FR Phyllorhine sombre
DE Dunkle Rundblattnase
ES Hiposidérido sombreado común

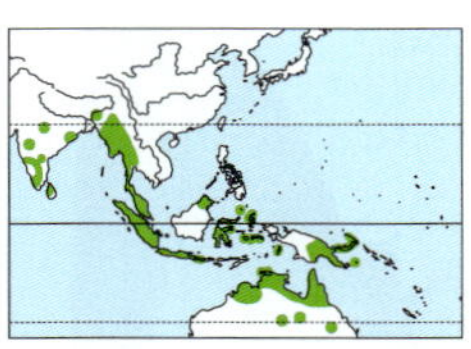

IN, LK & MM S to Pen MY & Sumatra, E to New Guinea, Bismarck Arch & N AU

LC
6 sspp
HB 3.5–4.5 cm
T 1.7–2.8 cm
FA 1.7–4.3 cm
W 4.5–10 g

Philippine Dusky Leaf-nosed Bat
Hipposideros antricola

NE
Monotypic
HB 4.4–5 cm
T 2.7–3.6 cm
FA 3.8–4.3 cm
W 5–7 g

FR Phyllorhine des antres
DE Philippinische Rundblattnase
ES Hiposidérido sombreado de Filipinas

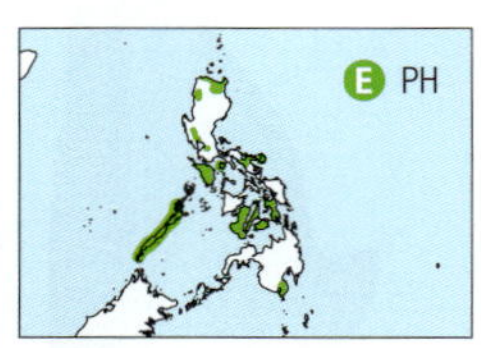

From Luzon S to Mindanao, Palawan

LC
7 sspp
FA 4.3–4.8 cm

Bicolored Leaf-nosed Bat
Hipposideros bicolor

FR Phyllorhine bicolore
DE Zweifarb-Rundblattnase
ES Hiposidérido bicolor

up to 600 m

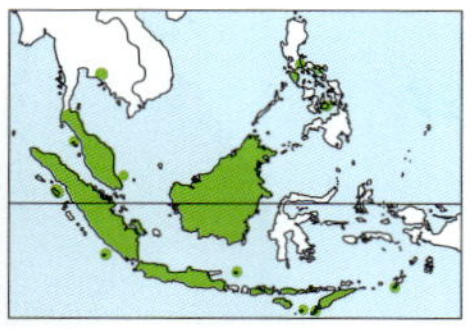

S TH, Thai-Malay Pen, Sumatra, Bangka, Borneo, Java E to Timor & Tanimbar Is, PH

DD
Monotypic
HB c. 4.2 cm
T c. 2.3 cm
FA c. 4.3 cm

Short-headed Leaf-nosed Bat
Hipposideros breviceps

FR Phyllorhine des Mentawai
DE Kurzkopf-Rundblattnase
ES Hiposidérido de Mentawai

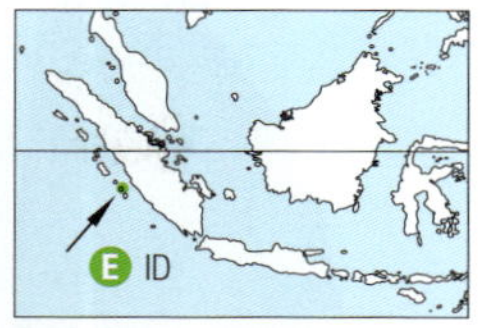

Mentawai Is (North Pagai), off WC Sumatra

LC
2 sspp
FA 3.2–3.6 cm

Least Leaf-nosed Bat
Hipposideros cineraceus

FR Petite Phyllorhine
DE Aschgraue Rundblattnase
ES Hiposidérido pequeño

0–1480 m

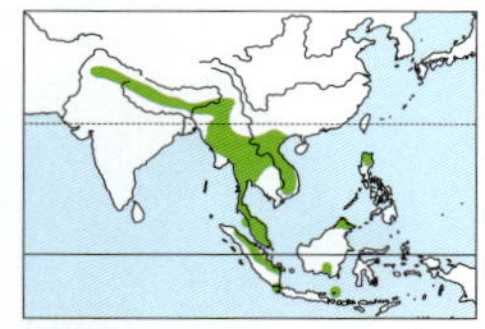

N PK, N IN, mainland SE Asia to Pen MY (with Langkawi I), Sumatra, Krakatau, Borneo, Kangean, PH (Luzon)

DD
Monotypic
HB 8.1–8.3 cm
T 2.7–2.9 cm
FA c. 4.9 cm
W c. 9.5 g

Large Mindanao Leaf-nosed Bat
Hipposideros coronatus

FR Phyllorhine de Mindanao
DE Große Mindanao-Rundblattnase
ES Hiposidérido de Mindanao

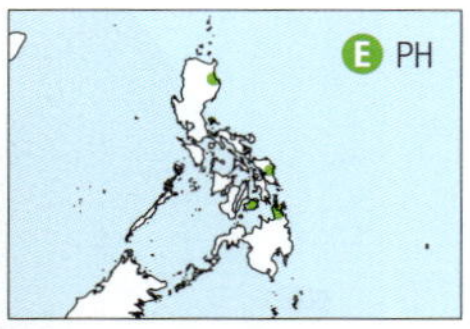

Scattered records on Luzon, Polillo, Bohol, Samar & Mindanao

DD
Monotypic

Timor Leaf-nosed Bat
Hipposideros crumeniferus

FR Phyllorhine de Timor
DE Timor-Rundblattnase
ES Hiposidérido de Timor

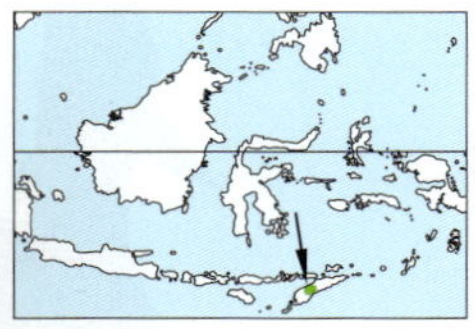

Timor I (ID, TL)

NT
Monotypic
FA 3.4–3.7 cm

Bornean Leaf-nosed Bat
Hipposideros doriae

FR Phyllorhine de Doria
DE Borneo-Rundblattnase
ES Hiposidérido de Doria

up to 1500 m

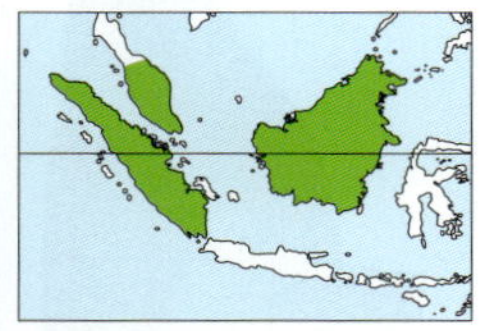

Pen MY, Sumatra, Borneo

VU
Monotypic
HB 3.6–4.1 cm
T 2.1–2.4 cm
FA 3.4–3.7 cm

Durga Das's Leaf-nosed Bat
Hipposideros durgadasi

FR Phyllorhine de Durga Das
DE Durga-Das-Rundblattnase
ES Hiposidérido de Durga Das

345–1100 m

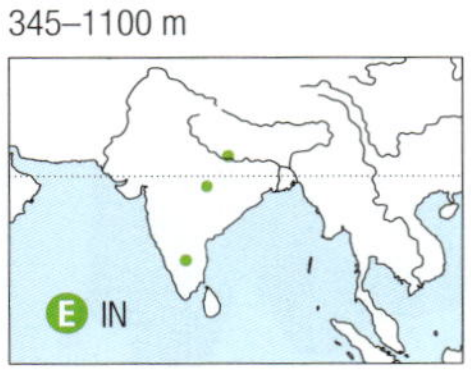

Few sites in Mirzapur (Uttar Pradesh), Jabalpur (Madhya Pradesh) & Kolar (Karnataka)

Dayak Leaf-nosed Bat
Hipposideros dyacorum

FR Phyllorhine dayak
DE Dayak-Rundblattnase
ES Hiposidérido dayak

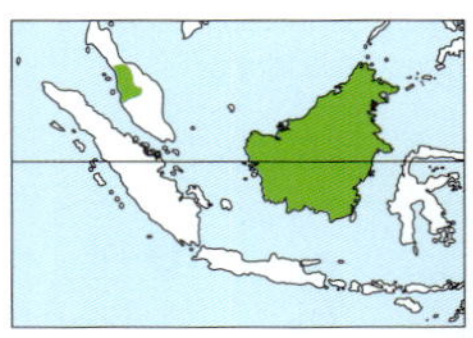

Pen MY, Borneo

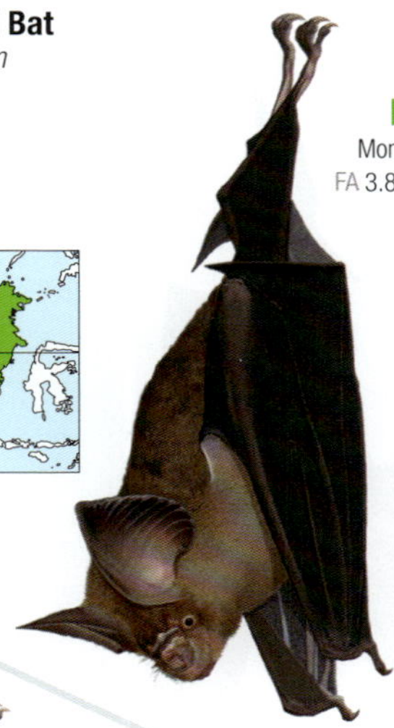

LC
Monotypic
FA 3.8–4.3 cm

DD
Monotypic
FA 3.9–4 cm

House-dwelling Leaf-nosed Bat
Hipposideros einnaythu

FR Phyllorhine des maisons
DE Myanmar-Rundblattnase
ES Hiposidérido de Myanmar

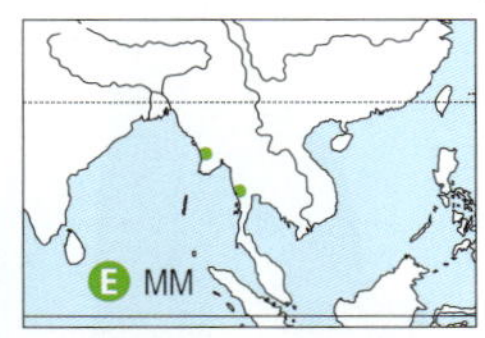

Known from Kan Thar Yar Beach (Rakhine State) & Tanintharyi Region

LC
2 sspp
HB c. 3.6 cm
T 2.8–3.5 cm
FA 3.8–4.3 cm

Andersen's Leaf-nosed Bat
Hipposideros gentilis

FR Phyllorhine d'Andersen
DE Andersen-Rundblattnase
ES Hiposidérido de Andersen

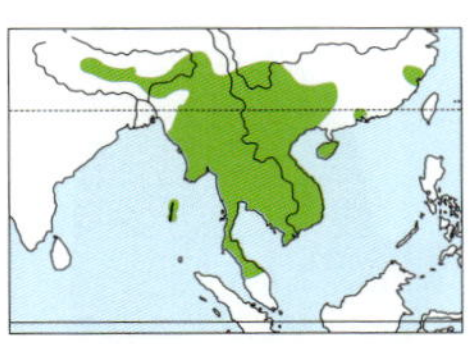

NP, NE IN, BD, SC & SE CN (with Hainan I), S through mainland SE Asia to Pen MY, Andaman Is

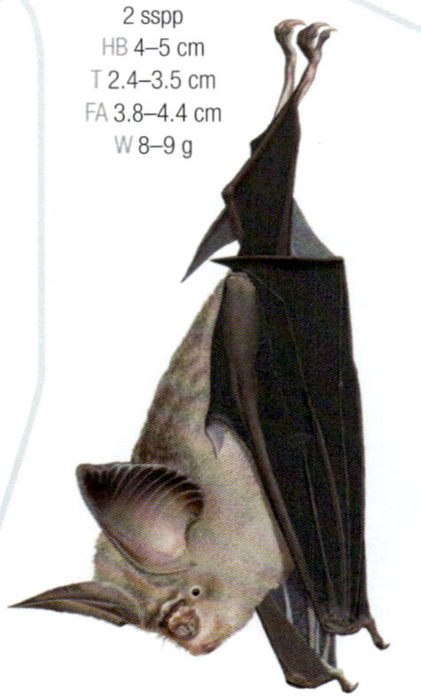

LC
2 sspp
HB 4–5 cm
T 2.4–3.5 cm
FA 3.8–4.4 cm
W 8–9 g

Fulvus Leaf-nosed Bat
Hipposideros fulvus

FR Phyllorhine fauve
DE Gelbrote Rundblattnase
ES Hiposidérido leonado

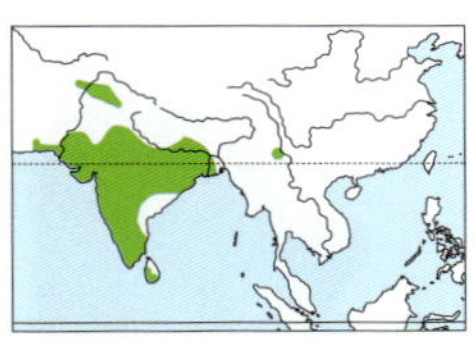

NE AF, PK, IN, NP, BD, LK, recorded in SC CN (W Yunnan)

VU
Monotypic
FA 3.6–3.9 cm

Thailand Leaf-nosed Bat
Hipposideros halophyllus

FR Phyllorhine de Thaïlande
DE Thailand-Rundblattnase
ES Hiposidérido de Tailandia

0–480 m

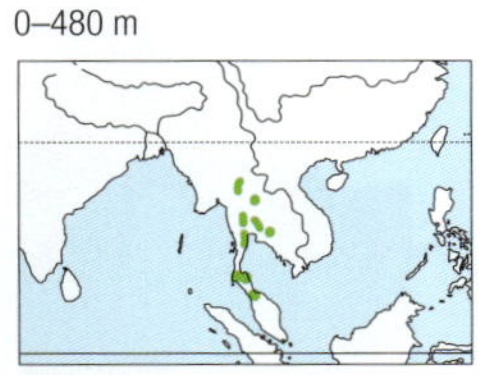

Scattered records in N, C & S TH, N Pen MY (Perlis State)

CR
Monotypic
HB 4.1–4.2 cm
T 2.3–2.4 cm
FA 3.8–4.9 cm

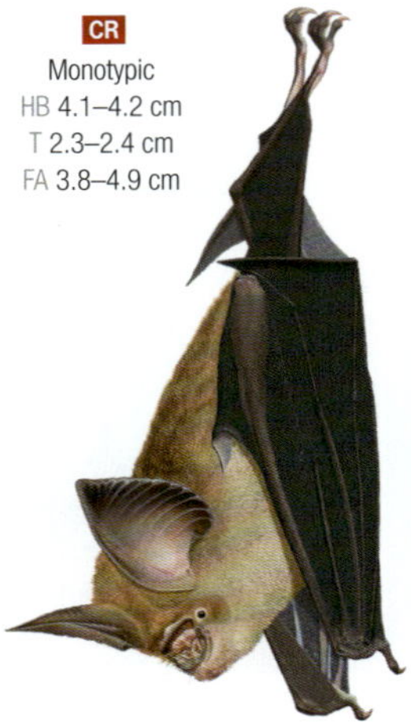

Kolar Leaf-nosed Bat
Hipposideros hypophyllus

FR Phyllorhine du Kolar
DE Kolar-Rundblattnase
ES Hiposidérido de Kolar

up to 570 m

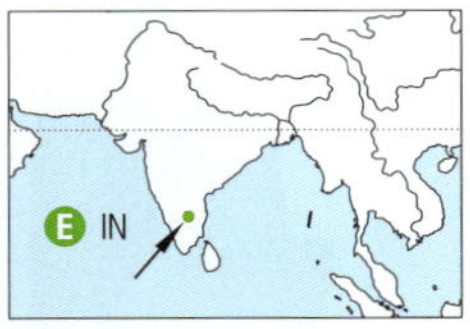

Known from Hanumanahalli (Kolar District); previously on nearby Therahalli, where now absent

VU
Monotypic
FA 4.6–4.9 cm

Phou Khao Khouay Leaf-nosed Bat
Hipposideros khaokhouayensis

FR Phyllorhine du Khao Khouay
DE Phou-Khao-Khouay-Rundblattnase
ES Hiposidérido de Khao Khouay

180–400 m

Known from Tak Leuk waterfall (Phou Khao Khouay CA) & Ban Nampe (Vientiane Prov)

NE
Monotypic
FA 3.8–4.5 cm

Kunz's Leaf-nosed Bat
Hipposideros kunzi

FR Phyllorhine de Kunz
DE Kunz-Rundblattnase
ES Hiposidérido de Kunz

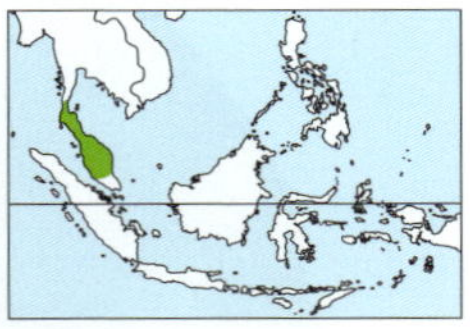

Thai-Malay Pen

DD
Monotypic
FA 4–4.2 cm

Big-eared Leaf-nosed Bat
Hipposideros macrobullatus

FR Phyllorhine de Tate
DE Heinrich-Rundblattnase
ES Hiposidérido orejudo

c. 300 m

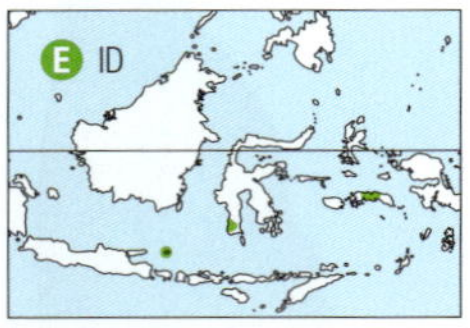

Known from Kangean, SW Sulawesi & Seram

LC
2 sspp
HB 5.7–8 cm
T 3–4.4 cm
FA 5–6.7 cm
W 13–23 g

Maggie Taylor's Leaf-nosed Bat
Hipposideros maggietaylorae

FR Phyllorhine de Taylor
DE Maggie-Taylor-Rundblattnase
ES Hiposidérido de Maggie Taylor

0–380 m

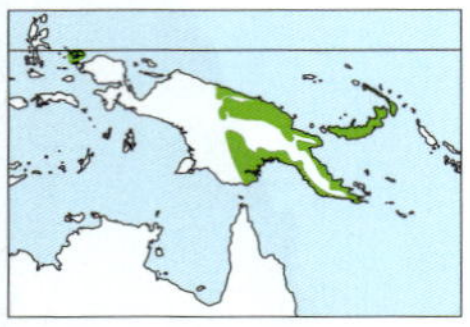

ID/PG: Raja Ampat Is (Waigeo, Batanta), New Guinea (perhaps more widespread in W), Bismarck Arch

DD
Monotypic
FA c. 4.6 cm

Malayan Leaf-nosed Bat
Hipposideros nequam

FR Phyllorhine de Malaisie
DE Malaya-Rundblattnase
ES Hiposidérido malayo

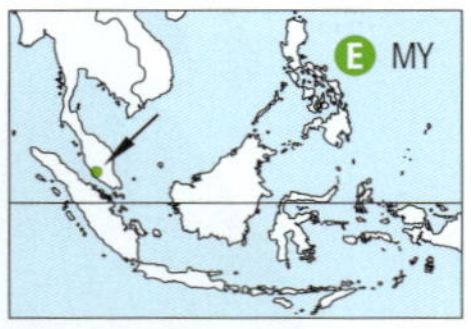

Peninsula (known only from Klang, Selangor)

EN
Monotypic
FA 3.8–4.2 cm

Nicobar Leaf-nosed Bat
Hipposideros nicobarulae

FR Phyllorhine des Nicobar
DE Nikobaren-Rundblattnase
ES Hiposidérido de Nicobar

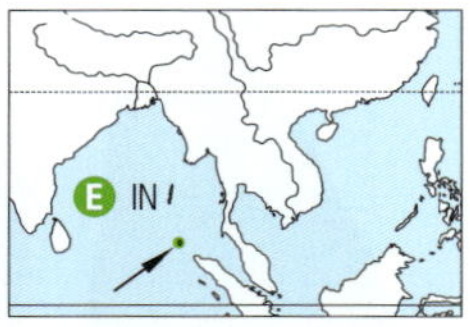

Nicobar Is

LC
Monotypic
HB 5.3–5.8 cm
T 1.9–2.4 cm
FA 4.2–4.8 cm
W 7–12 g

Philippine Forest Leaf-nosed Bat
Hipposideros obscurus

FR Phyllorhine obscure
DE Philippinen-Rundblattnase
ES Hiposidérido filipino

up to 1100 m

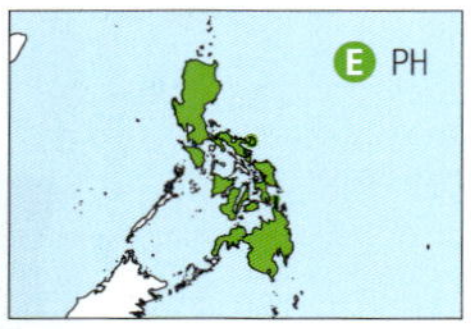

Widespread across primary and disturbed forests

VU
Monotypic
T 2.6–3.4 cm
FA 4.6–4.9 cm
W 9.3–10.3 g

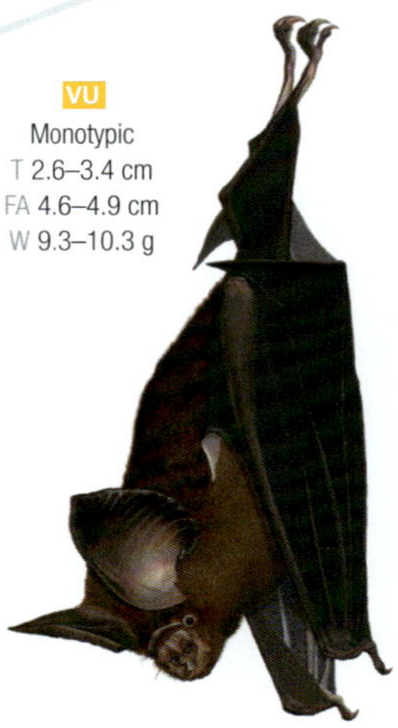

Orbiculus Leaf-nosed Bat
Hipposideros orbiculus

FR Phyllorhine à petit disque
DE Kleinblatt-Rundblattnase
ES Hiposidérido de disco pequeño

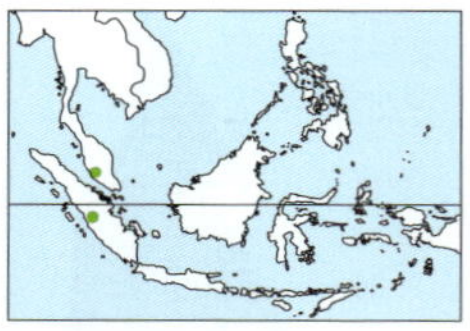

Pen MY, Sumatra (ID)

LC
Monotypic
HB 4.9–5.9 cm
T 2.2–3.5 cm
FA 4.9–5.3 cm
W 9.3–10.1 g

Biak Leaf-nosed Bat
Hipposideros papua

FR Phyllorhine papoue
DE Biak-Rundblattnase
ES Hiposidérido de Biak

100–300 m

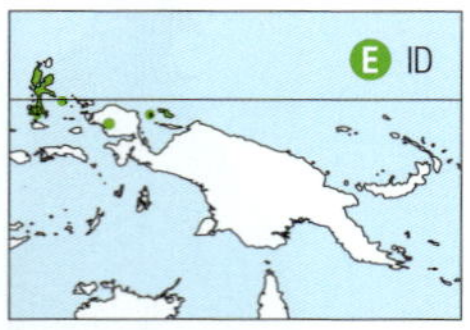

Moluccas (Halmahera, Bacan), Gebe I, W New Guinea (Vogelkop), Geelvink Bay Is (Numfor, Supiori, Biak)

EN
Monotypic
FA 3.8–3.9 cm

Pomona Leaf-nosed Bat
Hipposideros pomona

FR Phyllorhine pomone
DE Pomona-Rundblattnase
ES Hiposidérido pomona

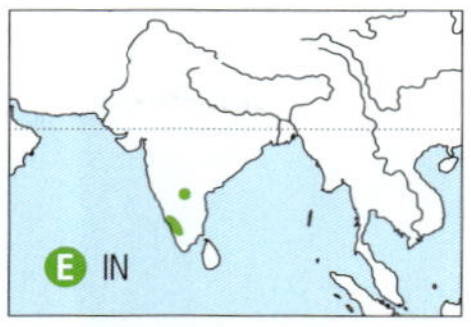

Andhra Pradesh, Karnataka, Kerala, Tamil Nadu

LC
Monotypic
HB 5.7–7.3 cm
T 1.8–2.7 cm
FA 3.7–4.1 cm
W c. 4 g

Philippine Pygmy Leaf-nosed Bat
Hipposideros pygmaeus

FR Phyllorhine pygmée
DE Kleine Philippinen-Rundblattnase
ES Hiposidérido filipino pequeño

0–200 m

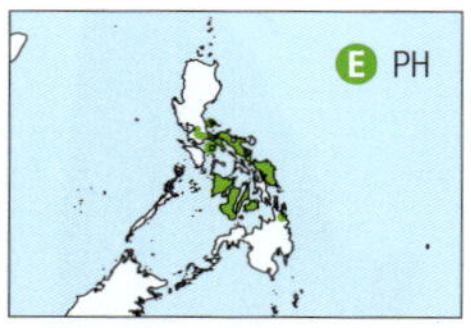

S Luzon, Polillo, Marinduque, Samar, Panay, Negros, Cebu, Bohol, N Mindanao

Ridley's Leaf-nosed Bat
Hipposideros ridleyi

FR Phyllorhine de Ridley
DE Ridley-Rundblattnase
ES Hiposidérido de Ridley

VU
Monotypic
T 2.5–2.9 cm
FA 4.7–5.1 cm
W 7.5–13.5 g

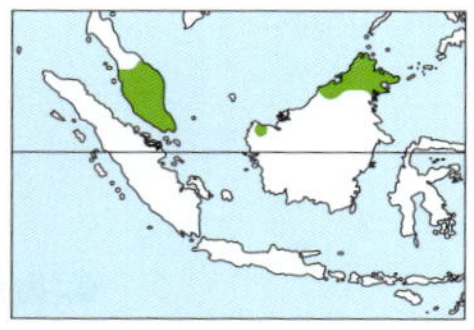

Pen MY, N & SW Borneo (MY, BN); probably Kalimantan (ID) & Pen TH

Laotian Leaf-nosed Bat
Hipposideros rotalis

FR Phyllorhine du Laos
DE Laos-Rundblattnase
ES Hiposidérido de Laos

LC
Monotypic
T 3.1–3.9 cm
FA 4.5–4.9 cm
W c. 11 g

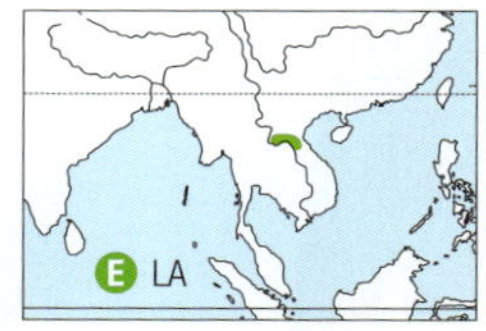

Reported from few sites

Family RHINOLOPHIDAE – Horseshoe Bats

1 genus, 107 species. Deserts, forests, and mountainous areas, mainly tropical but extending into temperate regions to c. 52° N. Palearctic, Afrotropical, Indo-Malayan and Australasian Regions.

Genus *RHINOLOPHUS*

(107 species)

Adam's Horseshoe Bat
Rhinolophus adami

FR Rhinolophe du Congo
DE Adam-Hufeisennase
ES Herradura del Congo

DD
Monotypic
T 2.5–2.8 cm
FA 4.6–5 cm

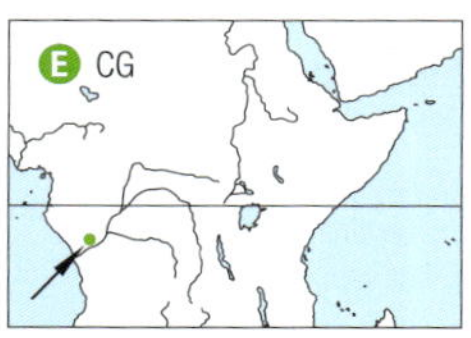

Kimanika Cave (Kouilou)

Maendeleo Horseshoe Bat
Rhinolophus maendeleo

FR Rhinolophe de Tanga
DE Tanga-Hufeisennase
ES Herradura de Tanga

DD
Monotypic
HB 4.6–4.9 cm
T 2.3–2.6 cm
FA 4.7–4.9 cm
W 6–15.5 g

0–1900 m

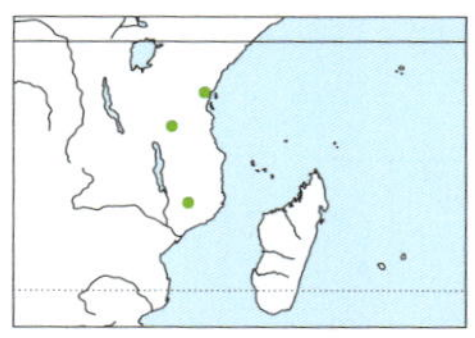

NE & C TZ (Amboni Cave Forest, Mazumbai Forest & Iringa region), N MZ (Mt Namuli)

Halcyon Horseshoe Bat
Rhinolophus alcyone

FR Rhinolophe alcyone
DE Alcyon-Hufeisennase
ES Herradura alción

LC
Monotypic
HB 5.7–6.8 cm
T 1.8–3.2 cm
FA 4.8–5.6 cm
W 14–23 g

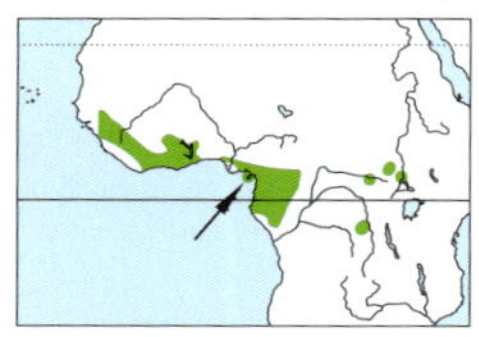

W & C Africa from SN to TG and from S NG E to SW CF, S to CG; patchily in Congo Basin, SW SS & W UG

Lander's Horseshoe Bat
Rhinolophus landeri

FR Rhinolophe de Lander
DE Lander-Hufeisennase
ES Herradura de Lander

LC
Monotypic
HB 5.2–6 cm
T c. 2.7 cm
FA 3.5–4.9 cm
W 5–11 g

0–1800 m

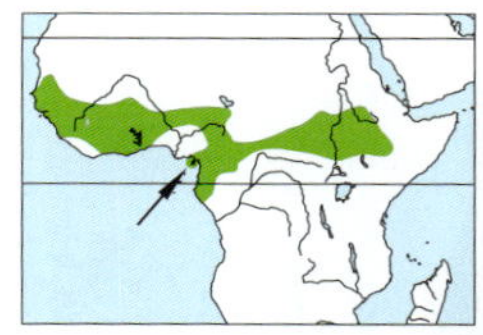

S MR & SN through most of W & C Africa to S SD & ET, Bioko I; limits uncertain

orange morph

Guinean Horseshoe Bat
Rhinolophus guineensis

FR Rhinolophe de Guinée
DE Guinea-Hufeisennase
ES Herradura de Guinea

EN
Monotypic
HB 4.9–5.7 cm
T 2.3–3 cm
FA 4.4–5 cm
W 8–11 g

up to 1600 m

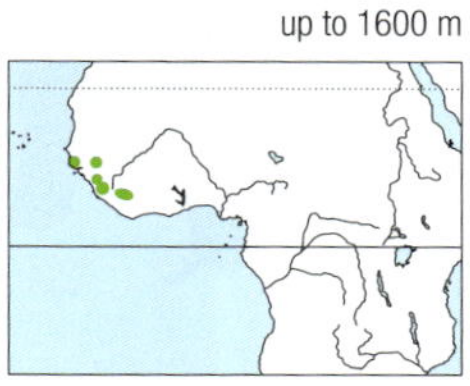

Scattered records in S SN, GN, C SL, N LR & EC CI; perhaps GM

Blasius's Horseshoe Bat
Rhinolophus blasii

FR Rhinolophe de Blasius
DE Blasius-Hufeisennase
ES Herradura de Blasius

LC
4 sspp
HB 4.6–5.4 cm
T 2–3.1 cm
FA 4.1–5.1 cm
W 7–15 g

500–2300 m

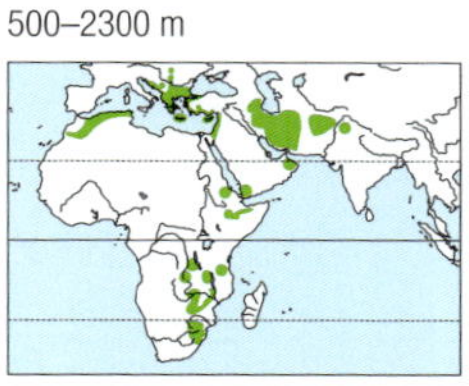

SE Europe (E from HR), SW Asia (E to N PK), NW Africa (N MA, N DZ, N TN), E Africa (from W ER S to E ZA)

Peters's Horseshoe Bat
Rhinolophus lobatus

FR Rhinolophe du Zambèze
DE Peters-Hufeisennase
ES Herradura de Peters

NE
Monotypic
HB 5.1–5.6 cm
T 2.1–3.1 cm
FA 4.3–5 cm
W 5–10 g

0–1980 m

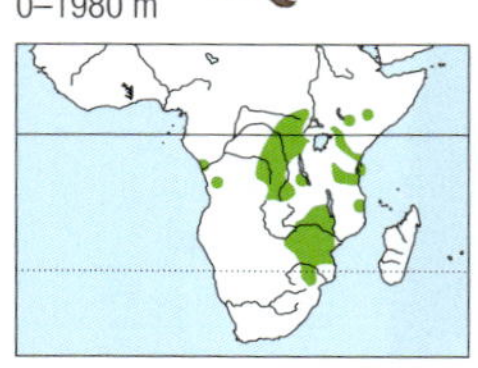

Certainty in MZ, tentatively in Congo Basin (W to N AO), S SS, KE & TZ (with Unguja I) S to NE ZA

Swinny's Horseshoe Bat
Rhinolophus swinnyi

FR Rhinolophe de Swinny
DE Swinny-Hufeisennase
ES Herradura de Swinny

LC
Monotypic
HB 4.2–4.7 cm
T 1.8–2.7 cm
FA 4.1–4.5 cm

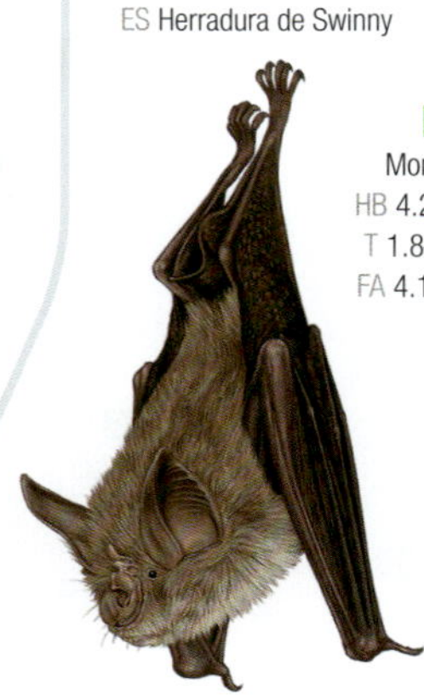

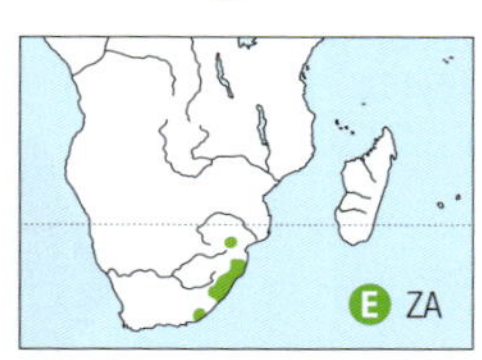

KwaZulu-Natal, Eastern Cape; specimens from Mpumalanga tentatively included here

Cape Horseshoe Bat
Rhinolophus capensis

FR Rhinolophe du Cap
DE Kap-Hufeisennase
ES Herradura de El Cabo

LC
Monotypic
HB 5.6–6.6 cm
T 2.4–3.2 cm
FA 4.7–5.1 cm
W 9–14.6 g

Coastal Northern, Western & Eastern Cape; unconfirmed in coastal SW NA

Bushveld Horseshoe Bat
Rhinolophus simulator

FR Rhinolophe de brousse
DE Bushveld-Hufeisennase
ES Herradura de Bushveld

LC
3 sspp
HB 3.9–6.5 cm
T 1.6–3 cm
FA 3.8–4.9 cm
W 4.5–11.2 g

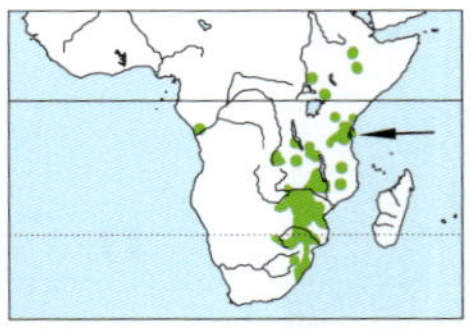

Scattered mainly in E & SE Africa from C ET, S SS, SW & SE CD S to ZW, SE BW & E ZA

Dent's Horseshoe Bat
Rhinolophus denti

FR Rhinolophe de Dent
DE Dent-Hufeisennase
ES Herradura de Dent

LC
2 sspp
HB 4.4–6 cm
T 1.7–2.4 cm
FA 3.7–4.4 cm
W 4–9 g

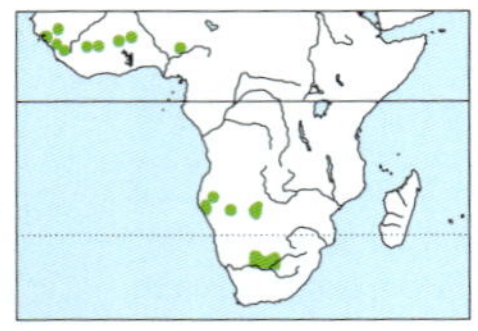

W Africa (from SE SN E to S BF & C NG), SW Africa (SW AO, N NA, BW & NC ZA); perhaps CG & SS

Cameroon Horseshoe Bat
Rhinolophus alticolus

FR Rhinolophe du Cameroun
DE Kamerun-Hufeisennase
ES Herradura de Camerún

NE
Monotypic
HB 5.5–5.6 cm
T 2.2–2.6 cm
FA c. 4.9 cm
W c. 12.5 g

130–2300 m

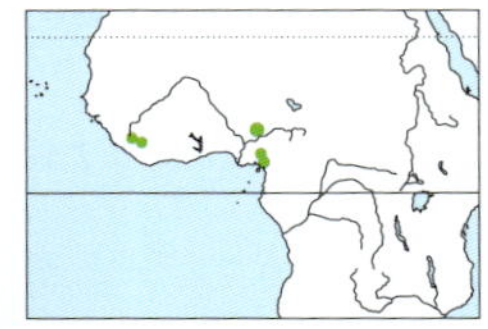

W Africa in N LR, SE GN, C NG & WC CM

Mehely's Horseshoe Bat
Rhinolophus mehelyi

FR Rhinolophe de Méhelÿ
DE Mehely-Hufeisennase
ES Herradura mediana

VU
2 sspp
HB 4.2–6.4 cm
T 2.1–3.7 cm
FA 4.8–5.6 cm
W 10–23 g

0–2000 m

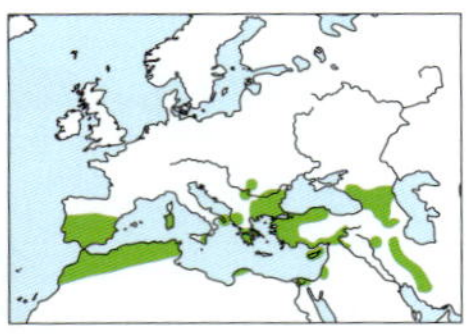

N Africa (MA E to TN, LY & EG), S Europe, SW Asia (TR, Caucasus, Levant & IR); perhaps FR & AF

Mediterranean Horseshoe Bat
Rhinolophus euryale

FR Rhinolophe euryale
DE Mittelmeer-Hufeisennase
ES Herradura mediterránea

NT
Monotypic
HB 4.3–5.8 cm
T 2.2–3 cm
FA 4.3–5.1 cm
W 7.5–17.5 g

0–1360 m

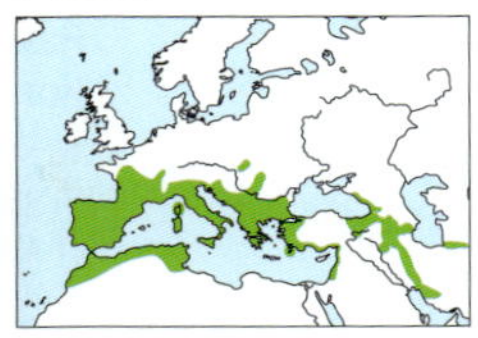

S Europe, N Africa (N MA, N DZ, TN), SW Asia (TR, Caucasus, Levant, IR, N IQ); perhaps CY & TM

Hildebrandt's Horseshoe Bat
Rhinolophus hildebrandtii

FR Rhinolophe de Hildebrandt
DE Hildebrandt-Hufeisennase
ES Herradura de Hildebrandt

LC
4 sspp
HB 7.3–8.4 cm
T 3.5–4.1 cm
FA 6.1–6.9 cm
W 22–31 g

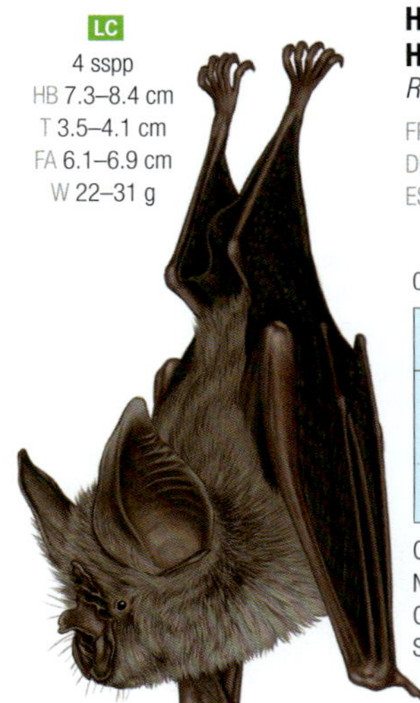

0–1300 m

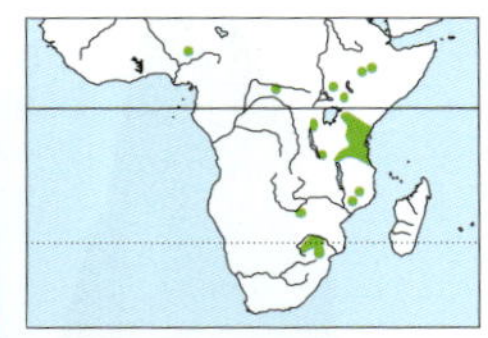

Certainty in KE, TZ, RW, BI, N MZ, NW ZW & NE ZA, tentatively in C NG, NC CD, S SS & ET; perhaps S CD

Mozambican Horseshoe Bat
Rhinolophus mossambicus

FR Rhinolophe du Mozambique
DE Mosambik-Hufeisennase
ES Herradura de Mozambique

60–1000 m

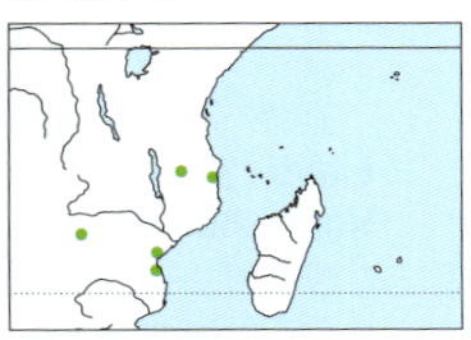

MZ (5 sites), NW ZW (1 site); possibly in much of ZA, MW, MZ & ZW

LC
Monotypic
FA 6–6.5 cm

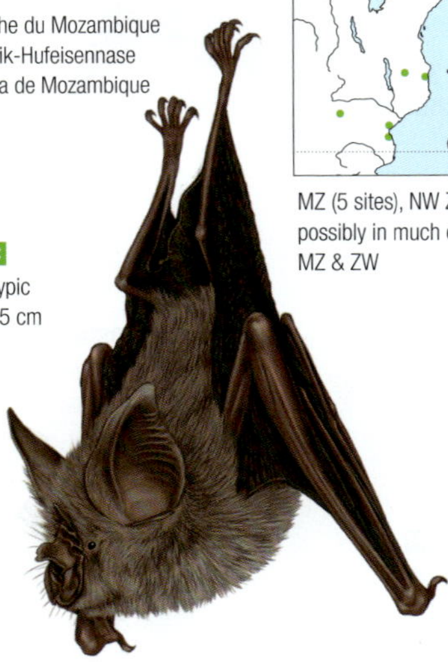

Rüppell's Horseshoe Bat
Rhinolophus fumigatus

FR Rhinolophe de Rüppell
DE Rüppell-Hufeisennase
ES Herradura de Rüppell

LC
6 sspp
HB 4–7.4 cm
T 2–3.9 cm
FA 4.7–6 cm
W 11–24 g

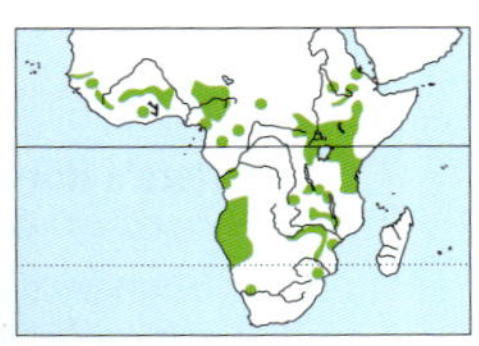

Widespread but scattered in sub-Saharan Africa

Eloquent Horseshoe Bat
Rhinolophus eloquens

FR Rhinolophe éloquent
DE Lautfreudige Hufeisennase
ES Herradura elocuente

LC
2 sspp
HB 6.8–7.2 cm
T 2–4.5 cm
FA 5.3–6.3 cm
W 13–40 g

up to 1500 m

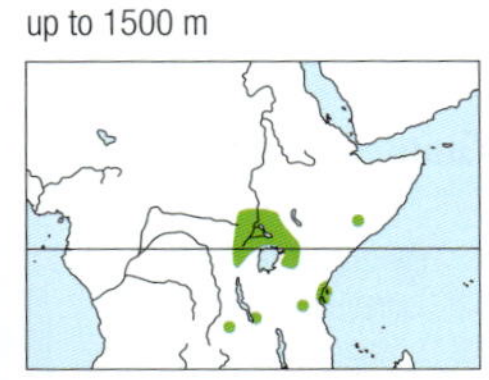

S SS, NE & SE CD, S SO, UG, RW, KE & TZ (with Pemba & Unguja Is); perhaps S ET S to MW & MZ

Darling's Horseshoe Bat
Rhinolophus darlingi

FR Rhinolophe de Darling
DE Darling-Hufeisennase
ES Herradura de Darling

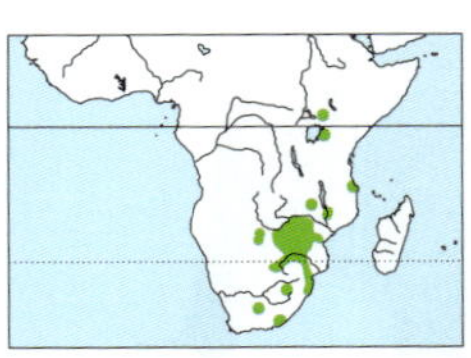

SE Africa (SE ZM, S MW, W MZ, ZW, N BW, SZ, C & E ZA), isolated in UG & TZ

LC
Monotypic
HB 5–6.2 cm
T 2–3.7 cm
FA 4.2–5.1 cm
W 6–13 g

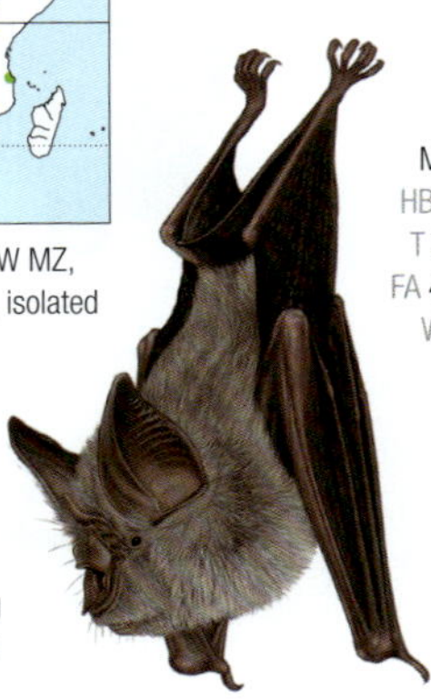

Damara Horseshoe Bat
Rhinolophus damarensis

FR Rhinolophe damara
DE Damara-Hufeisennase
ES Herradura de Damara

LC
Monotypic
HB 6.2–6.4 cm
T 2.1–3.5 cm
FA 4.4–5.3 cm
W 8–14 g

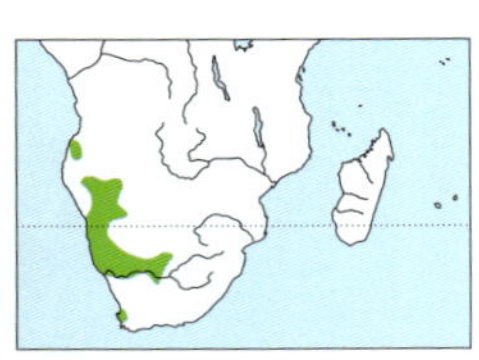

SW Africa (SW AO, NA, W & C ZA)

Decken's Horseshoe Bat
Rhinolophus deckenii

FR Rhinolophe de Decken
DE Decken-Hufeisennase
ES Herradura de Decken

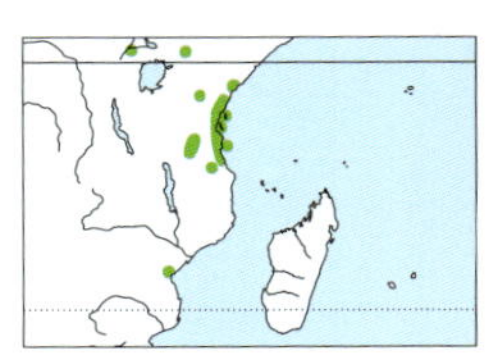

E Africa in WC UG, WC & SE KE, E TZ (with Pemba, Unguja & Mafia Is), apparently C MZ (Chinizuia Forest)

NT
Monotypic
HB 5.1–6.7 cm
T 2.5–3.3 cm
FA 4.8–5.6 cm

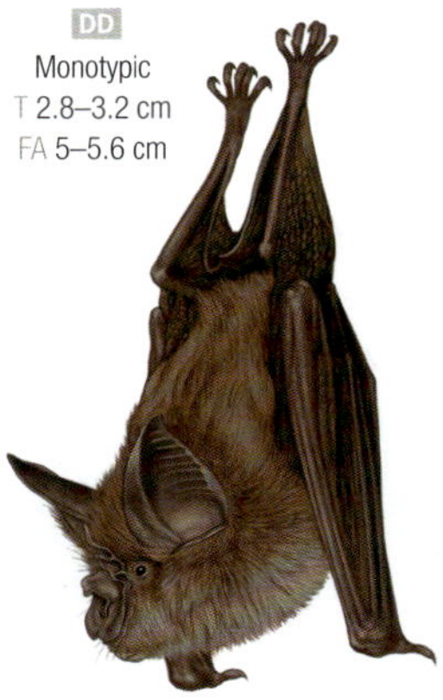

DD
Monotypic
T 2.8–3.2 cm
FA 5–5.6 cm

African Forest Horseshoe Bat
Rhinolophus silvestris

FR Rhinolophe sylvestre
DE Waldhufeisennase
ES Herradura forestal

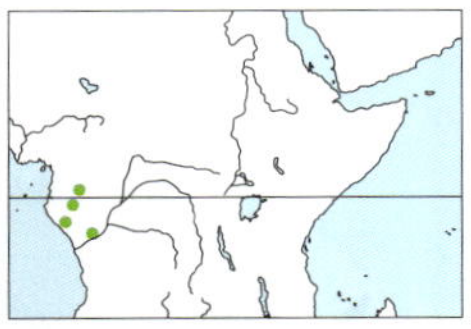

Known from 3 sites in GA (Benga, Belinga & Dumbu caves, Latoursville) & 1 site in CG (Meya-Nzouari)

VU
Monotypic
HB 6.4–7.1 cm
T 3–4.1 cm
FA 5.2–5.7 cm
W 16.5–25 g

Upland Horseshoe Bat
Rhinolophus hillorum

FR Rhinolophe de Lofa
DE Hochland-Hufeisennase
ES Herradura montana

1400–1950 m

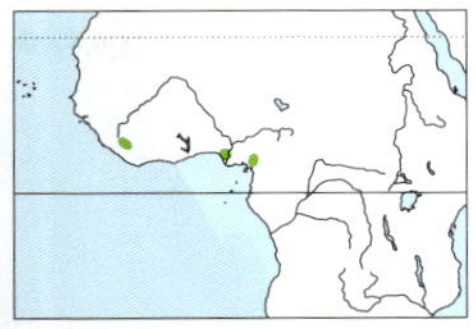

Scattered in W Africa (certainty in SE GN, N LR, S NG & WC CM)

DD
Monotypic
HB 5.7–5.8 cm
T 2.9–3.1 cm
FA 5.3–5.5 cm
W 19–24 g

Sakeji Horseshoe Bat
Rhinolophus sakejiensis

FR Rhinolophe de la Sakeji
DE Sakeji-Hufeisennase
ES Herradura de Sakeji

c. 1390 m

Known only from type locality (Ikelenge Pedicle)

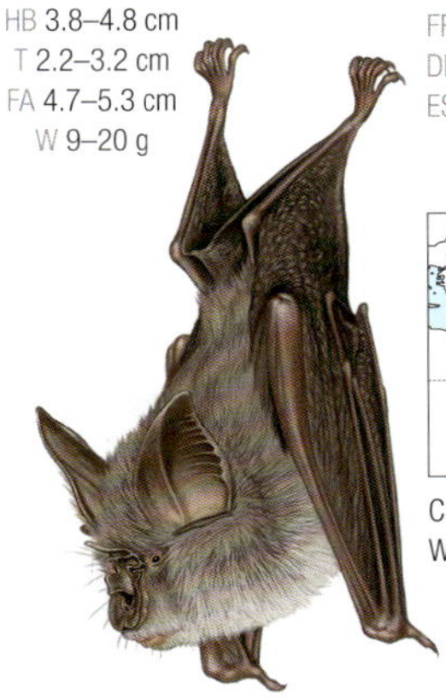

LC
Monotypic
HB 3.8–4.8 cm
T 2.2–3.2 cm
FA 4.7–5.3 cm
W 9–20 g

Bokhara Horseshoe Bat
Rhinolophus bocharicus

FR Rhinolophe du Boukhara
DE Buchara-Hufeisennase
ES Herradura de Bujará

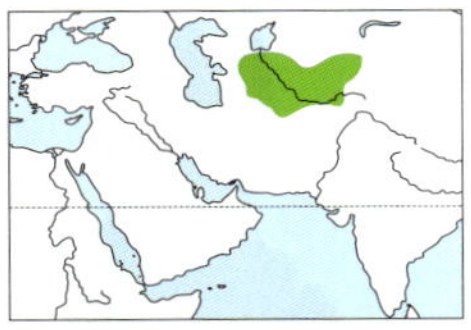

C Asia in SC KZ, UZ, W KG, TM, W & C TJ and N AF

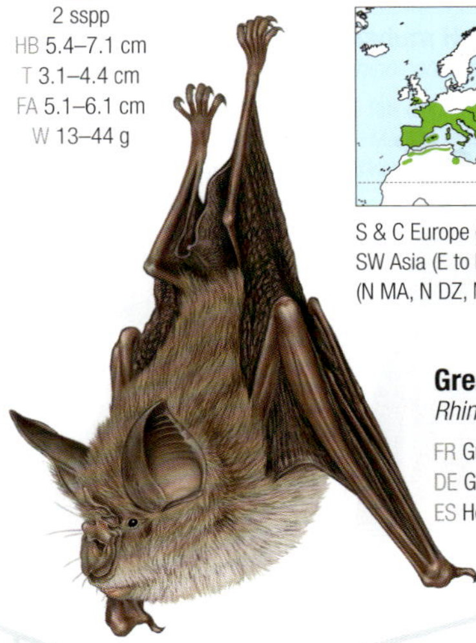

LC
2 sspp
HB 5.4–7.1 cm
T 3.1–4.4 cm
FA 5.1–6.1 cm
W 13–44 g

0–3500 m

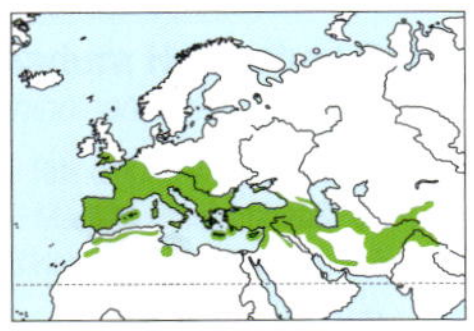

S & C Europe (with Mediterranean Is), SW Asia (E to KG & N IN), NW Africa (N MA, N DZ, N TN, NW LY)

Greater Horseshoe Bat
Rhinolophus ferrumequinum

FR Grand Rhinolophe
DE Große Hufeisennase
ES Herradura grande

LC
10 sspp
HB 4.6–5 cm
T 2.2–4 cm
FA 4.2–5.9 cm
W 10–25 g

Geoffroy's Horseshoe Bat
Rhinolophus clivosus

FR Rhinolophe de Cretzschmar
DE Geoffroy-Hufeisennase
ES Herradura de Geoffroy

0–2300 m

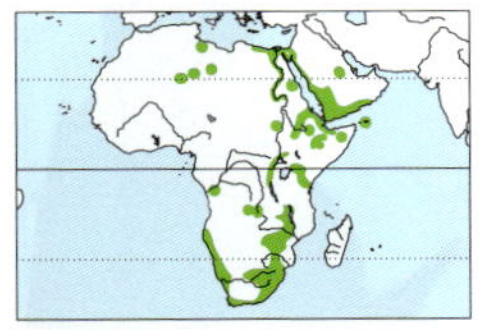

Patchily in N, E & S Africa, SW Asia (S Levant, W & SE Arabian Pen)

NE
2 sspp
HB 5.5–7.9 cm
T 2.5–4.4 cm
FA 4.9–6.4 cm

Greater Japanese Horseshoe Bat
Rhinolophus nippon

FR Rhinolophe nippon
DE Große Japan-Hufeisennase
ES Herradura de Japón

0–3500 m

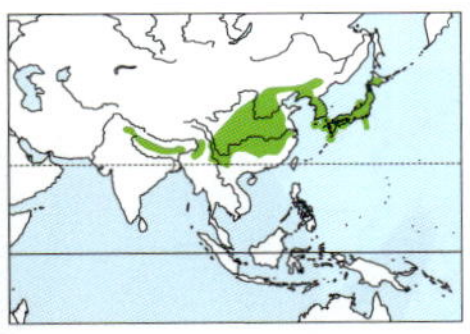

N IN, NP, BT, N BD, S, C & E CN, Korean Pen, JP (with some offshore Is)

NE
Monotypic
HB c. 6 cm
T c. 3.4 cm
FA 4.8–5 cm

N Cyrenaica 0–660 m

Horacek's Horseshoe Bat
Rhinolophus horaceki

FR Rhinolophe de Horacek
DE Horacek-Hufeisennase
ES Herradura de Horacek

Maclaud's Horseshoe Bat
Rhinolophus maclaudi

FR Rhinolophe de Maclaud
DE Maclaud-Hufeisennase
ES Herradura de Maclaud

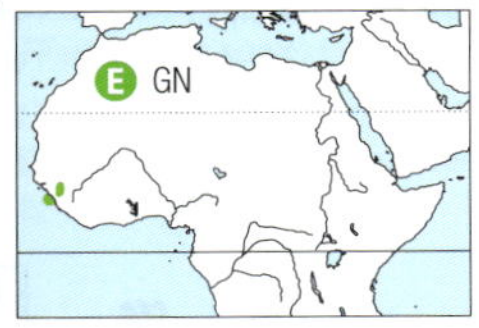

Fouta Djallon Highlands N to R Gessorewoul, S to SL border, Conakry I

EN
Monotypic
HB 7.3–9.4 cm
T 3.8–4.3 cm
FA 6.4–6.9 cm
W 30–33 g

EN
Monotypic
HB 7.4–7.6 cm
T 3.5–3.7 cm
FA c. 6 cm
W c. 20 g

c. 600 m

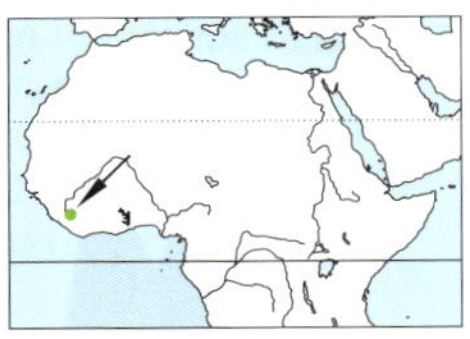

SE GN (Ziama Massif), NW LR (Wonegizi Mts)

Ziama Horseshoe Bat
Rhinolophus ziama

FR Rhinolophe du Ziama
DE Ziama-Hufeisennase
ES Herradura de Ziama

CR
Monotypic
HB c. 6.3 cm
T c. 2.9 cm
FA c. 5.4 cm
W c. 16.5 g

Hill's Horseshoe Bat
Rhinolophus hilli

FR Rhinolophe de Hill
DE Hill-Hufeisennase
ES Herradura de Hill

1750–2515 m

Uwinka & Ruta Bansugera in Nyungwe Forest NP

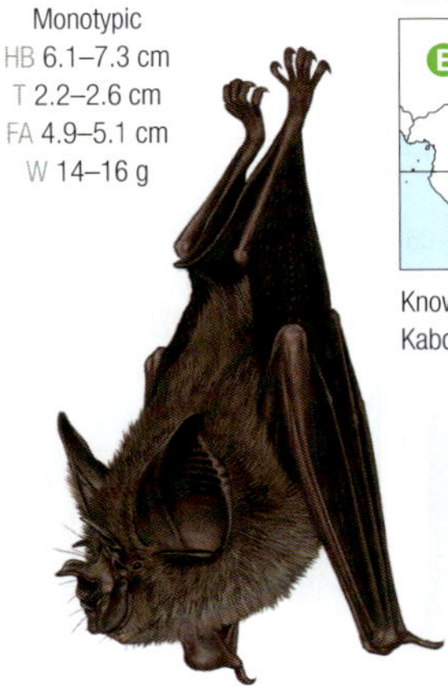

EN
Monotypic
HB 6.1–7.3 cm
T 2.2–2.6 cm
FA 4.9–5.1 cm
W 14–16 g

1880–1950 m

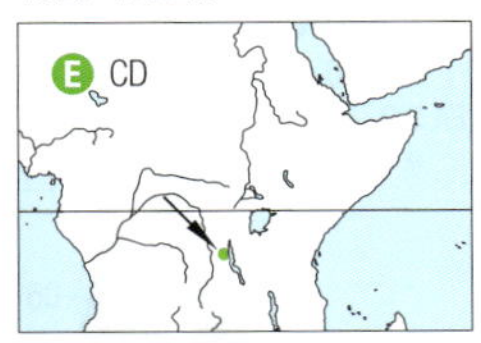

Known only from Misotschi-Kabogo Highlands

Willard's Horseshoe Bat
Rhinolophus willardi

FR Rhinolophe de Willard
DE Willard-Hufeisennase
ES Herradura de Willard

NE
2 sspp
HB 3.6–5.6 cm
T 1.5–2.9 cm
FA 3.7–4.2 cm

Little Japanese Horseshoe Bat
Rhinolophus cornutus

FR Rhinolophe cornu
DE Kleine Japan-Hufeisennase
ES Herradura pequeño de Japón

From Hokkaido S to Amami Is; dubious record from Guangxi (S CN)

NE
Monotypic
HB 3.6–4.3 cm
FA 3.8–4 cm

Okinawa Horseshoe Bat
Rhinolophus pumilus

FR Rhinolophe d'Okinawa
DE Okinawa-Hufeisennase
ES Herradura de Okinawa

Okinawa Is (Iheyajima, Kumejima), Kerama Is (Tokashikijima), Miyako Is (Miyakojima, Irabu)

EN
Monotypic
HB 3.9–4.8 cm
T 1.8–2.6 cm
FA 4–4.3 cm

Yaeyama Horseshoe Bat
Rhinolophus perditus

FR Rhinolophe des Yaeyama
DE Yaeyama-Hufeisennase
ES Herradura de Yaeyama

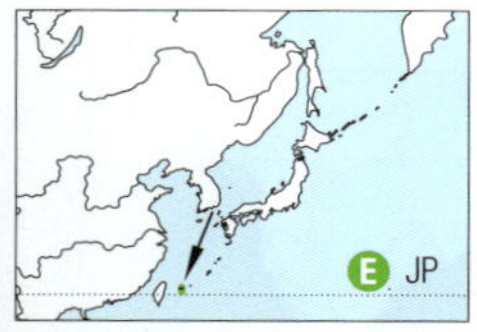

Yaeyama Is (Ishigakijima, Taketomi)

NE
Monotypic
HB 3.9–4.8 cm
T 1.8–2.6 cm
FA 4–4.3 cm

Imaizumi's Horseshoe Bat
Rhinolophus imaizumii

FR Rhinolophe d'Imaizumi
DE Imaizumi-Hufeisennase
ES Herradura de Imaizumi

Yaeyama Is (Iriomotejima, Kohamajima)

NE
Monotypic
HB 4.2–4.8 cm
T 1.9–2.6 cm
FA 4.2–4.4 cm
W 6.9–8.6 g

Mountain Horseshoe Bat
Rhinolophus monticolus

FR Rhinolophe de montagne
DE Gebirgs-Hufeisennase
ES Herradura montano

650–1320 m

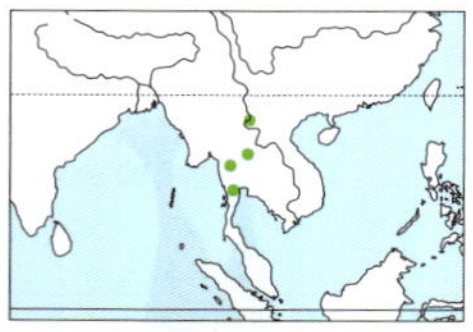

NW LA (Luang Namtha), N & W TH (Pu Nam Rom Stream, Phu Soun Sai & Mae Wong NPs)

DD
Monotypic
HB 4.1–5.9 cm
T 1.6–2.9 cm
FA 3.8–4.3 cm
W 5.5–8.8 g

Shortridge's Horseshoe Bat
Rhinolophus shortridgei

FR Rhinolophe de Shortridge
DE Shortridge-Hufeisennase
ES Herradura de Shortridge

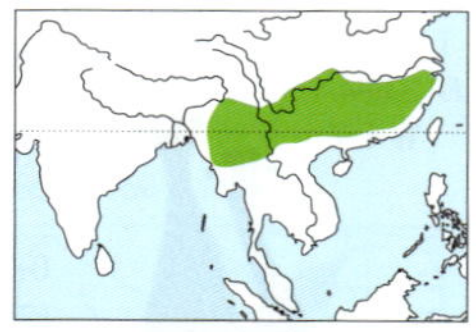

NE IN (Falta, West Bengal), N & C MM, SC & SE CN

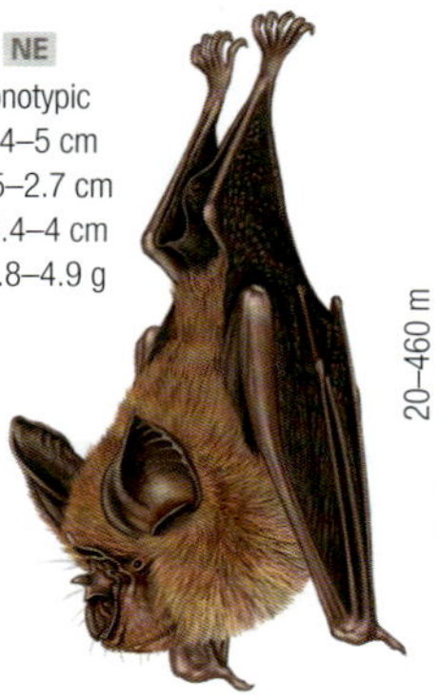

NE
Monotypic
HB 4–5 cm
T 1.5–2.7 cm
FA 3.4–4 cm
W 4.8–4.9 g

Formosan Horseshoe Bat
Rhinolophus monoceros

FR Rhinolophe unicorne
DE Taiwan-Hufeisennase
ES Herradura unicornio

20–460 m

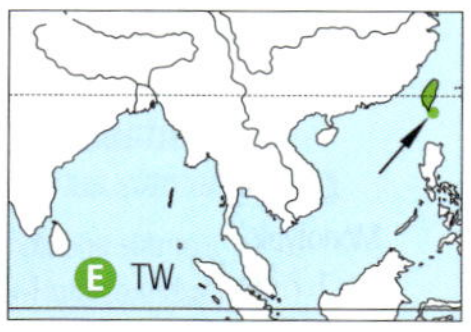

Lowlands of mainland & Orchid I; dubious record from Guizhou (CN)

LC
9 sspp
HB 3–4.2 cm
T 1.3–2.6 cm
FA 3.3–4.2 cm
W 3.3–8 g

Least Horseshoe Bat
Rhinolophus pusillus

FR Rhinolophe pusille
DE Kleinste Hufeisennase
ES Herradura de nariz pequeña

200–1370 m

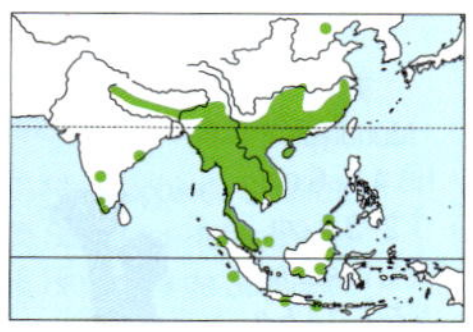

N & S IN E to CN (N to Beijing area), mainland SE Asia S to Greater Sundas

LC
Monotypic
HB 3.5–3.7 cm
T 1.6–1.9 cm
FA 3.3–3.8 cm

Little Nepalese Horseshoe Bat
Rhinolophus subbadius

FR Rhinolophe du Népal
DE Kleine Nepal-Hufeisennase
ES Herradura de Nepal

c. 1230 m

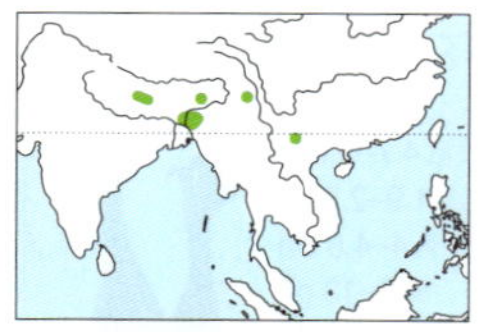

C & E NP, BT, NE IN, N BD, N MM & SC CN (Yunnan); dubious in VN

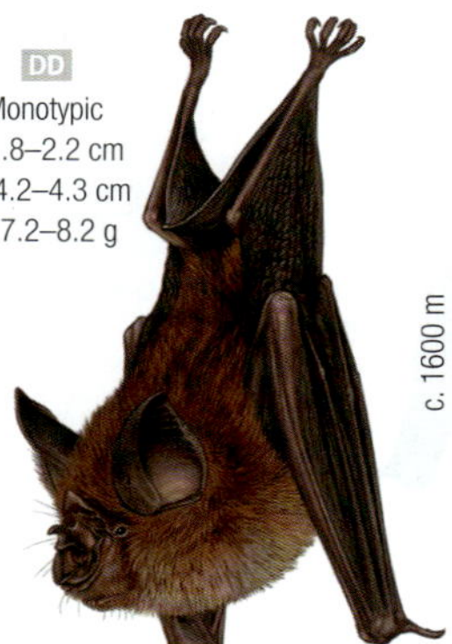

DD
Monotypic
T 1.8–2.2 cm
FA 4.2–4.3 cm
W 7.2–8.2 g

Convex Horseshoe Bat
Rhinolophus convexus

FR Rhinolophe convexe
DE Cameron-Highlands-Hufeisennase
ES Herradura convexo

c. 1600 m

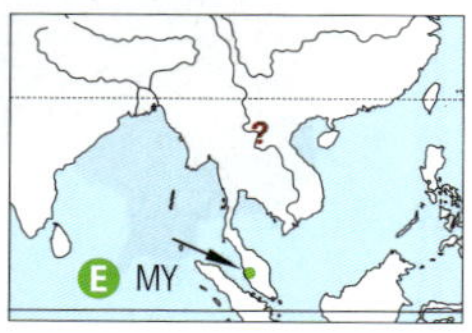

Peninsula (Selangor & Pahang states); uncertain in N LA

EN
2 sspp
HB c. 3.7 cm
T 1.3–2.1 cm
FA 3.7–4.1 cm
W c. 9.9 g

Andaman Horseshoe Bat
Rhinolophus cognatus

FR Rhinolophe de Giglioli
DE Andamanen-Hufeisennase
ES Herradura de Andamán

up to 600 m

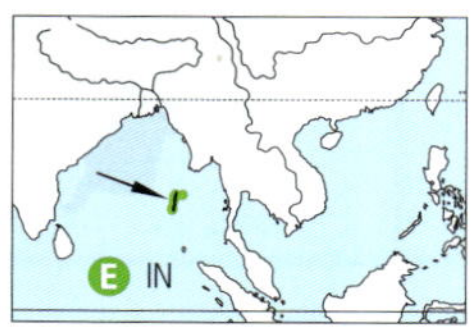

Andaman Is (North, South, Point, Interview, Baratang, Narcondam)

NE
2 sspp
HB 3.7–5.1 cm
T 1.4–2.2 cm
FA 3.6–4.3 cm
W 4–7.2 g

Glossy Horseshoe Bat
Rhinolophus refulgens

FR Rhinolophe irisé
DE Glänzende Hufeisennase
ES Herradura iriscente

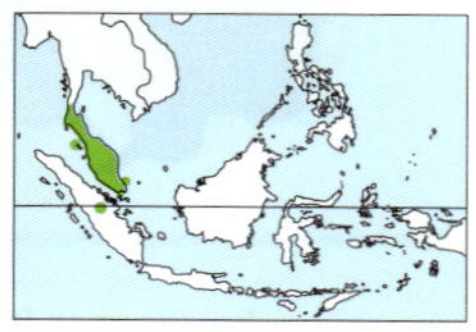

Thai-Malay Pen & SG (with Tarutao & Tioman Is), EC Sumatra (ID)

LC
3 sspp
HB 3.3–5.4 cm
T 1.3–2.6 cm
FA 3.7–4.3 cm
W 3.7–8.2 g

Blyth's Horseshoe Bat
Rhinolophus lepidus

FR Rhinolophe de Blyth
DE Blyth-Hufeisennase
ES Herradura de Blyth

100–2340 m

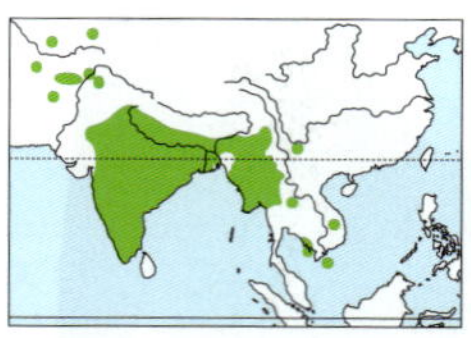

SE UZ, SW KG, AF, N PK, most of IN & BD, NP, SC CN (Yunnan), mainland SE Asia (with several Is)

LC
2 sspp
HB c. 5.1 cm
T 1.1–1.6 cm
FA 4.6–5.4 cm
W 10.5–13.5 g

Creagh's Horseshoe Bat
Rhinolophus creaghi

FR Rhinolophe de Creagh
DE Creagh-Hufeisennase
ES Herradura de Creagh

0–1500 m

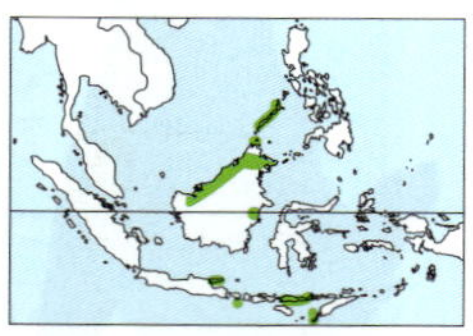

PH (Palawan), Borneo (with Banggi I), Lesser Sundas (Madura, Nusa Penida, Flores, Semau, Roti)

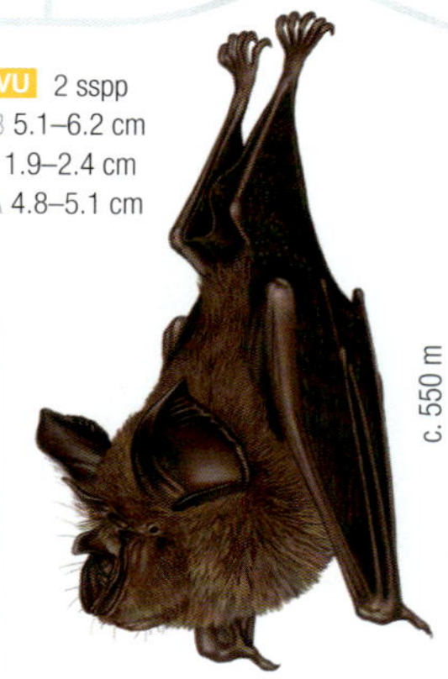

VU 2 sspp
HB 5.1–6.2 cm
T 1.9–2.4 cm
FA 4.8–5.1 cm

Knud's Horseshoe Bat
Rhinolophus canuti

FR Rhinolophe de Knud
DE Knud-Andersen-Hufeisennase
ES Herradura de Knud

c. 550 m

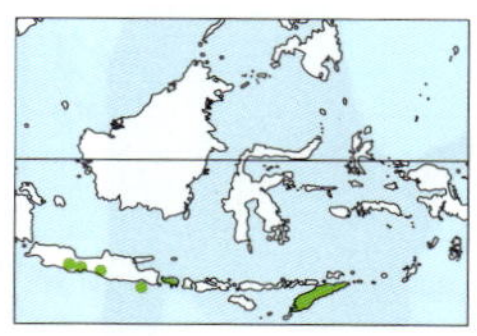

Java (3 localities), Nusa Barong I, Bali, Timor

McIntyre's Horseshoe Bat
Rhinolophus mcintyrei

DD
Monotypic
HB 4.8–6.4 cm
T 1.6–2.3 cm
FA 4.7–5.4 cm
W 9–15 g

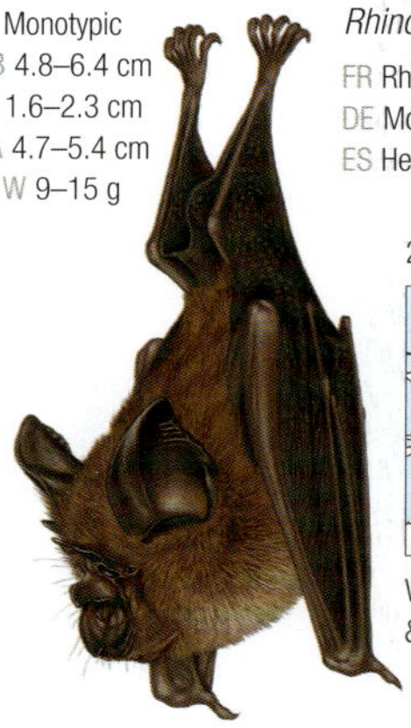

FR Rhinolophe de McIntyre
DE McIntyre-Hufeisennase
ES Herradura de McIntyre

270–1600 m

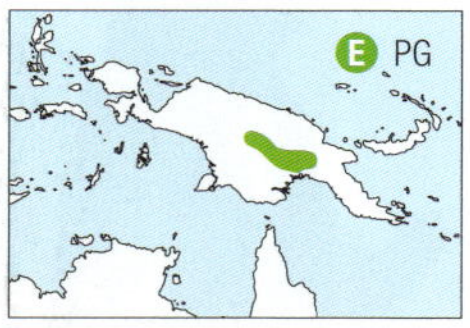

Western, Sandaun, Southern & Eastern Highlands prov

New Guinea Broad-eared Horseshoe Bat
Rhinolophus euryotis

LC
5 sspp
HB 5.4–7.3 cm
T 1.6–2.8 cm
FA 5–6 cm
W 14.1–27 g

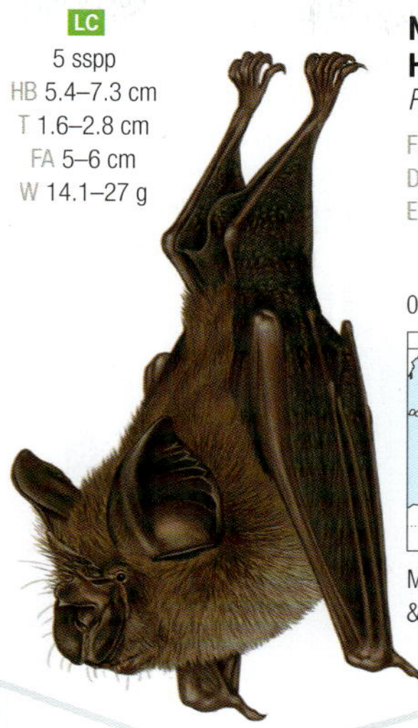

FR Rhinolophe euryote
DE Neuguinea-Hufeisennase
ES Herradura de Nueva Guinea

0–2050 m

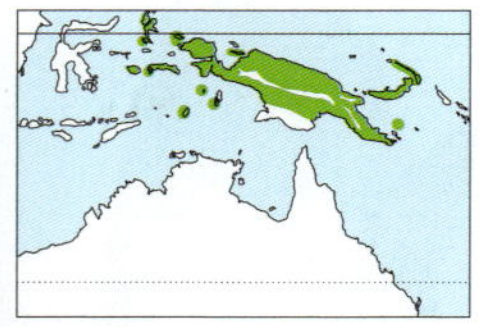

Moluccas (Halmahera S to Tanimbar & Aru), New Guinea, Bismarck Arch

Sulawesi Broad-eared Horseshoe Bat
Rhinolophus tatar

LC
Monotypic
HB 5.1–6.2 cm
T 1.6–2.2 cm
FA 4.7–5.2 cm
W 9–13 g

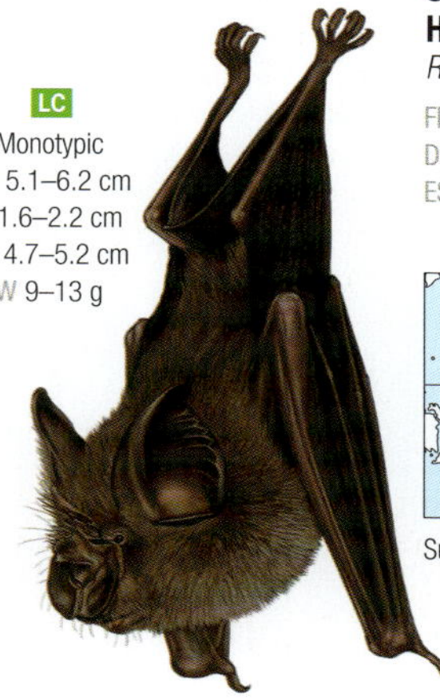

FR Rhinolophe d'Archbold
DE Tate-Archbold-Hufeisennase
ES Herradura de Archbold

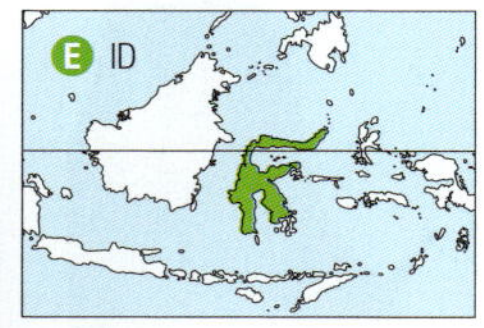

Sulawesi

Poso Horseshoe Bat
Rhinolophus belligerator

EN
Monotypic
HB 5.2–6.1 cm
T 1.5–2.1 cm
FA 4.7–5.2 cm
W 10–12.5 g

FR Rhinolophe de Poso
DE Zentralsulawesi-Hufeisennase
ES Herradura de Poso

c. 0 m

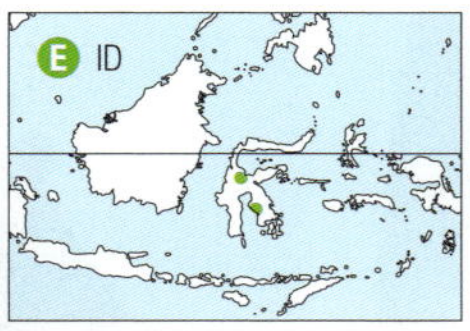

C & SE Sulawesi

Bornean Woolly Horseshoe Bat
Rhinolophus proconsulis

EN
Monotypic
FA 4.6–4.8 cm

FR Rhinolophe proconsul
DE Wollhaarige Borneo-Hufeisennase
ES Herradura lanoso de Borneo

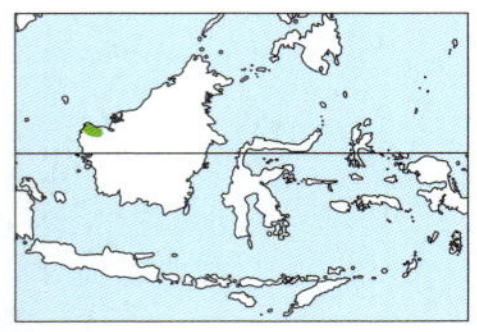

WC Borneo (SW Sarawak, apparently C & E Kalimantan, but exact localities uncertain)

Arcuate Horseshoe Bat
Rhinolophus arcuatus

DD
5 sspp
HB 4.9–6.2 cm
T 1.5–2.1 cm
FA 4–5 cm
W 6–9 g

FR Rhinolophe arqué
DE Bogen-Hufeisennase
ES Herradura arqueado

0–2500 m

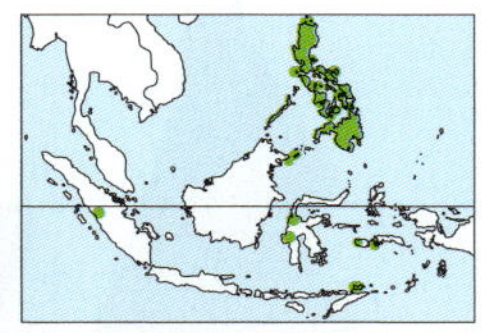

PH, ID (WC Sumatra, Wetar, Buru, Ambon, undescribed taxon in C Sulawesi)

Philippine Forest Horseshoe Bat
Rhinolophus inops

LC
Monotypic
HB 5.5–7 cm
T 1.7–2.7 cm
FA 4.9–5.7 cm
W 10–19.6 g

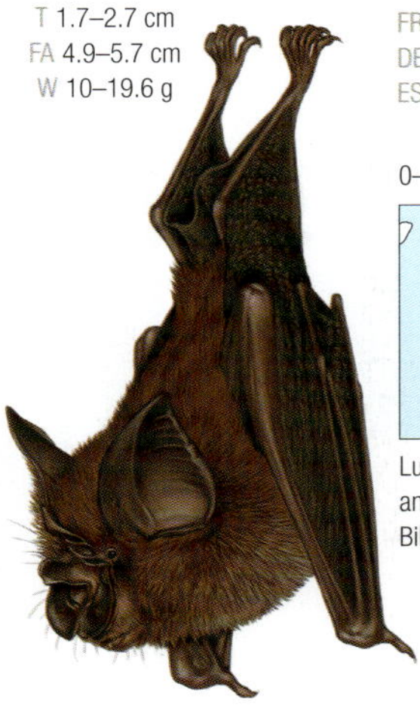

FR Rhinolophe de Mindanao
DE Philippinen-Hufeisennase
ES Herradura de Mindanao

0–2250 m

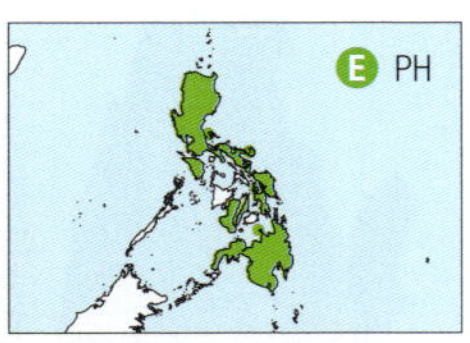

Luzon, Polillo, Mindoro, Catanduanes, Samar, Negros, Cebu, Leyte, Biliran, Camiguin, Mindanao

Small Rufous Horseshoe Bat
Rhinolophus subrufus

0–1000 m

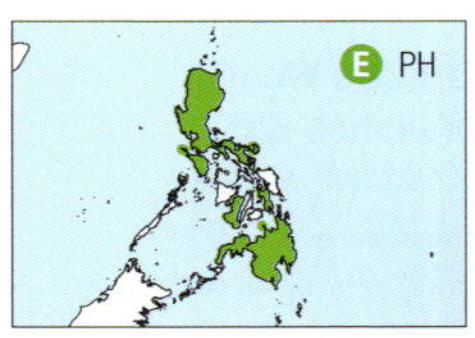

Luzon, Lubang, Mindoro, Catanduanes, Negros, Leyte, Camiguin, Mindanao

DD
2 sspp
HB 6–6.3 cm
T 2.2–2.9 cm
FA 5.3–5.7 cm
W 12–19 g

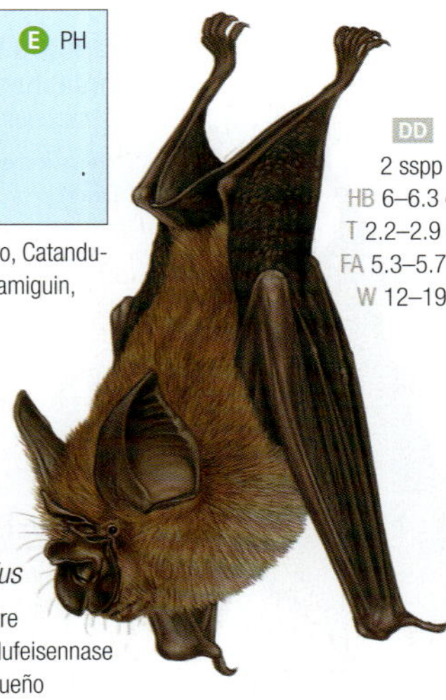

FR Rhinolophe roussâtre
DE Kleine Rotbraune Hufeisennase
ES Herradura rufo pequeño

Large Rufous Horseshoe Bat
Rhinolophus rufus

0–600 m

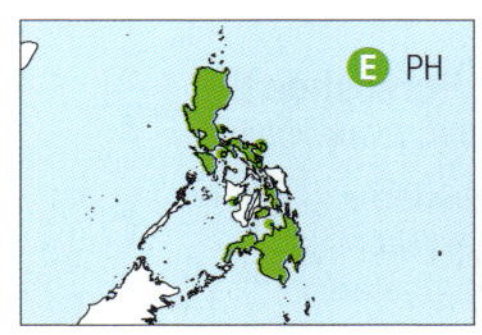

Luzon, Polillo, Mindoro, Marinduque, Catanduanes, Guimaras, Bohol, Leyte, Camiguin, Mindanao

NT
Monotypic
HB 8–9.7 cm
T 3–3.3 cm
FA 6.8–7.3 cm
W 26–34 g

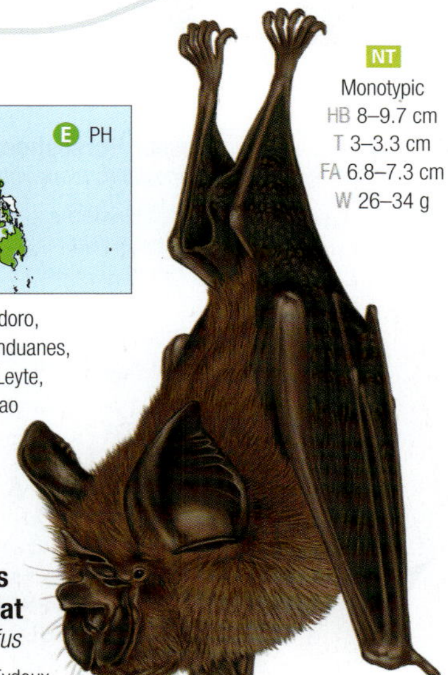

FR Rhinolophe d'Eydoux
DE Große Rotbraune Hufeisennase
ES Herradura rufo grande

Intermediate Horseshoe Bat
Rhinolophus affinis

LC
6 sspp
HB 4.6–6.8 cm
T 1.7–3.5 cm
FA 4.6–5.7 cm
W 9.9–19 g

FR Rhinolophe de Horsfield
DE Mittelgroße Asiatische Hufeisennase
ES Herradura de Horsfield

200–2000 m

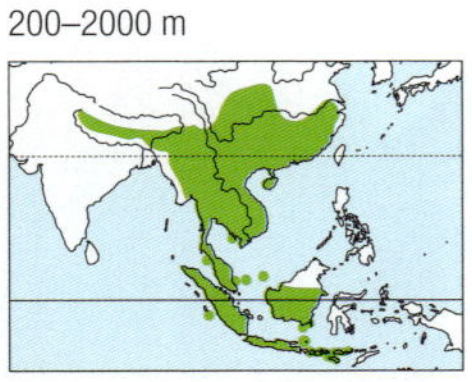

N IN & NP E to C & SC CN, mainland SE Asia, Anambas & Natuna Is, Greater & Lesser Sundas E to Flores

Homfray's Horseshoe Bat
Rhinolophus andamanensis

NE
Monotypic
HB c. 5.8 cm
T 1.8–3.1 cm
FA 4.7–5.7 cm

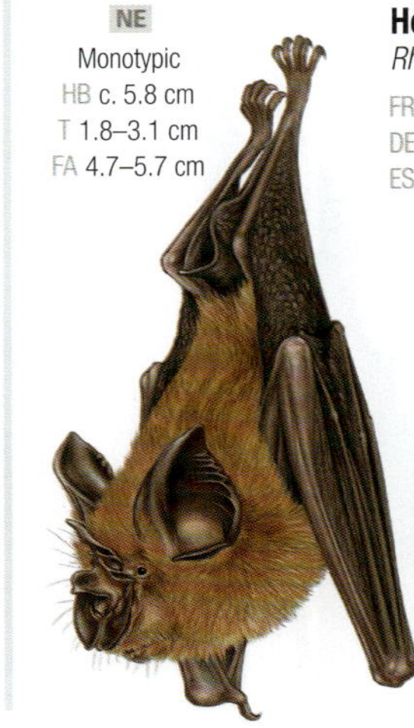

FR Rhinolophe de Homfray
DE Dobson-Andamanen-Hufeisennase
ES Herradura de Homfray

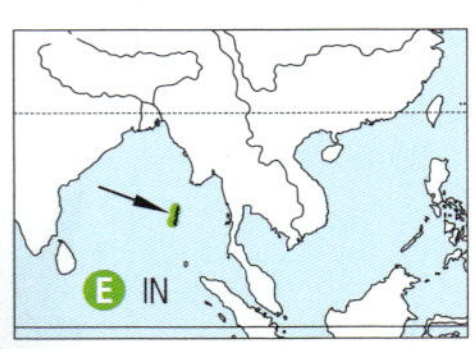

Andaman Is (North, Interview, Middle, Baratang, South, Little)

Croslet Horseshoe Bat
Rhinolophus coelophyllus

LC Monotypic
HB 3.6–5 cm T 1.6–2.4 cm
FA 4.2–4.8 cm W 6.2–8.6 g

FR Rhinolophe de Peters
DE Saluen-Hufeisennase
ES Herradura thai

0–1500 m

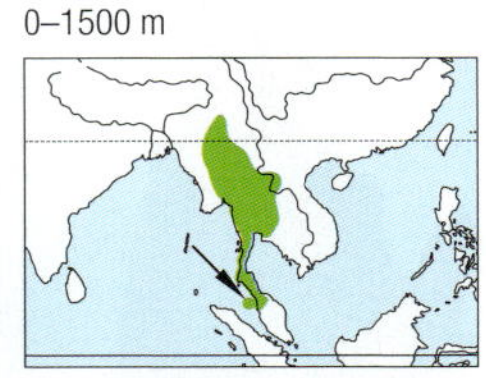

NC, C & SE MM, TH (with Tarutao I), C LA, NW Pen MY (with Langkawi I)

LC
Monotypic
HB 4.7–4.8 cm
T 1.7–2.3 cm
FA 4.3–4.8 cm
W 7.5–12.5 g

Shamel's Horseshoe Bat
Rhinolophus shameli
FR Rhinolophe de Shamel
DE Shamel-Hufeisennase
ES Herradura de Shamel

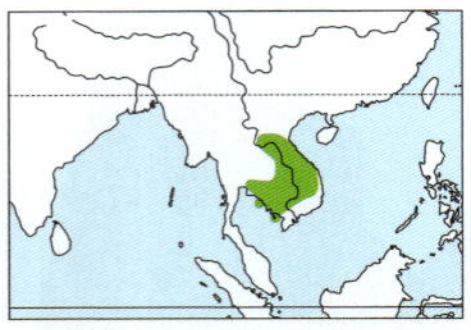

E & SE TH (with Koh Chang I), C & S LA, SC VN (with Phu Quoc I), KH

LC
Monotypic
HB 3.4–6 cm
T 1.4–2.2 cm
FA 4.1–4.6 cm
W 5–9 g

Vietnamese Brown Horseshoe Bat
Rhinolophus microglobosus
FR Rhinolophe à petit nez
DE Braune Vietnam-Hufeisennase
ES Herradura de Vietnam

0–1080 m

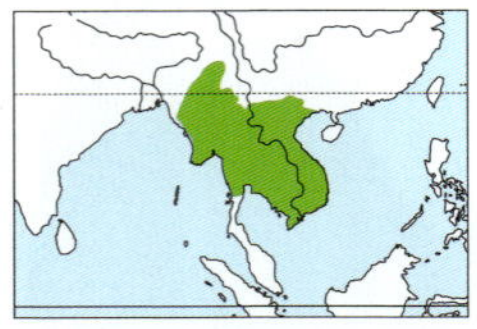

MM, SC CN (Yunnan), TH, LA, KH, at least N & C VN; limits uncertain

LC
Monotypic
HB 4.5–5.7 cm
T 1.3–2.1 cm
FA 4.3–4.8 cm
W 6.9–10 g

Lesser Brown Horseshoe Bat
Rhinolophus stheno
FR Rhinolophe sthéno
DE Stheno-Hufeisennase
ES Herradura de Stheno

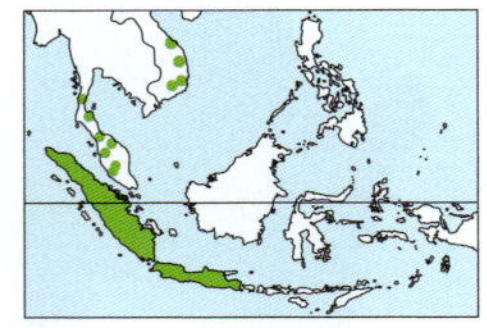

C & S VN, Thai-Malay Pen, Sumatra, Java; limits uncertain

Indian Rufous Horseshoe Bat
Rhinolophus rouxii
FR Rhinolophe de Roux
DE Rotbraune Indien-Hufeisennase
ES Herradura de Roux

summer

winter

0–1370 m

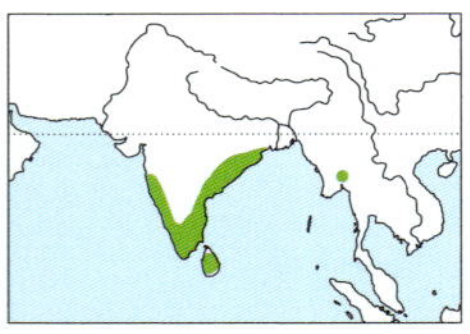

W, E & S IN, LK, SC MM; perhaps SC CN (Yunnan)

LC
2 sspp
HB 4.2–6.6 cm
T 2.1–3.3 cm
FA 4.4–5.2 cm
W 12.2–14.8 g

Chinese Rufous Horseshoe Bat
Rhinolophus sinicus
FR Rhinolophe de Chine
DE Rotbraune China-Hufeisennase
ES Herradura rufo de China

500–2770 m

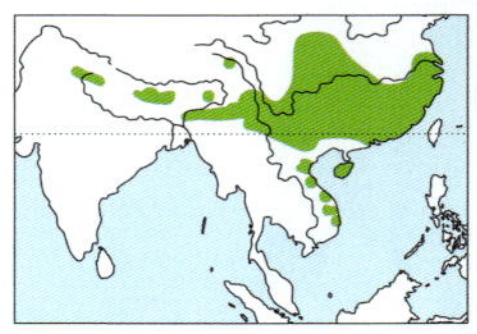

N IN, NP, N MM, C, SC & SE CN (with Hainan I), N & C VN

LC
2 sspp
HB 4.3–5.3 cm
T 2.1–3 cm
FA 4.3–5.6 cm
W 8.9–10.9 g

400–1100 m

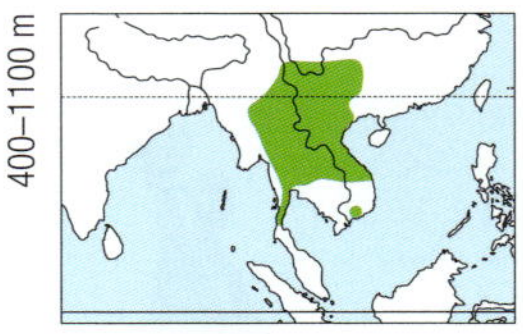

SC CN, E MM, TH, LA, VN

LC
Monotypic
HB 4.8–5 cm
T 1.8–2.8 cm
FA 4–4.8 cm
W 7–11 g

Thomas's Horseshoe Bat
Rhinolophus thomasi
FR Rhinolophe de Thomas
DE Thomas-Hufeisennase
ES Herradura de Thomas

LC
6 sspp
HB 3.5–4.8 cm
T 2.1–3.2 cm
FA 3.5–4.2 cm
W 4.7–9 g

Lesser Horseshoe Bat
Rhinolophus hipposideros
FR Petit Rhinolophe
DE Kleine Hufeisennase
ES Herradura pequeño

0–2100 m

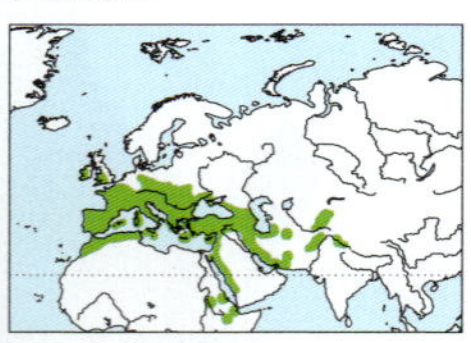

C & S Europe, Middle East to N IN, N Africa (N MA, N DZ, N TN), E Africa (SW SD, C ER, DJ, C ET)

NE
2 sspp
HB c. 6.3 cm
T 2.9–3.8 cm
FA 5.3–5.5 cm
W 16–18 g

Francis's Woolly Horseshoe Bat
Rhinolophus francisi
FR Rhinolophe de Francis
DE Francis-Wollhaar-Hufeisennase
ES Herradura de Francis

430–1600 m

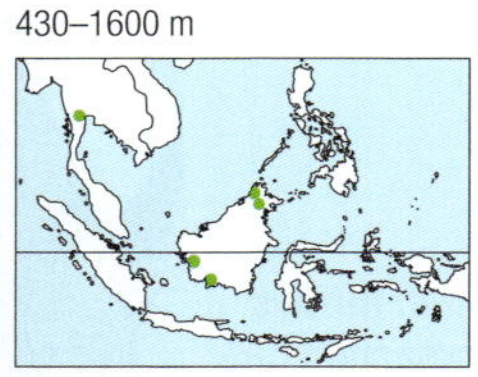

SW TH (known from type locality), N, SW & SC Borneo; perhaps VN

NT
Monotypic
HB 4.7–5.1 cm
T 1.9–3 cm
FA 3.7–4.8 cm
W 6.8–11 g

Lesser Woolly Horseshoe Bat
Rhinolophus sedulus
FR Rhinolophe diligent
DE Kleine Wollhaar-Hufeisennase
ES Herradura lanoso pequeño

0–1500 m

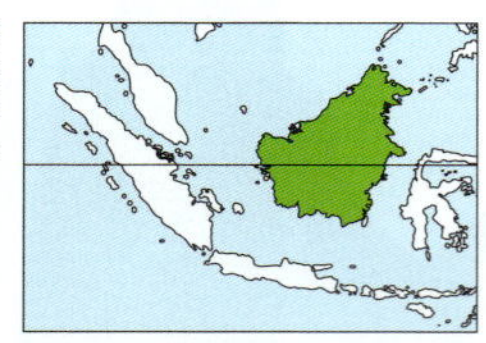

Borneo (MY, ID, BN)

NE Monotypic
HB 4.7–5.1 cm
T 1.9–3 cm
FA 3.9–4.2 cm
W 6.3–7.1 g

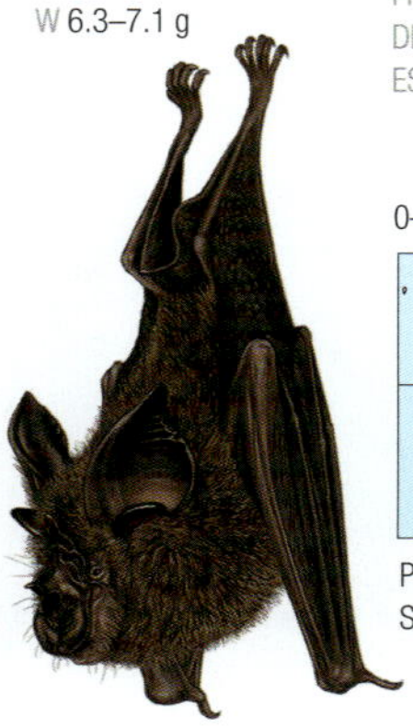

Yong Hoi Sen's Woolly Horseshoe Bat
Rhinolophus yonghoiseni
FR Rhinolophe de Yong
DE Yong-Hoi-Sen-Wollhaar-Hufeisennase
ES Herradura de Yong

0–1500 m

Peninsula (Pahang, Negeri Sembilan, Selangor, Johor, Kedah, Perak)

0–1800 m

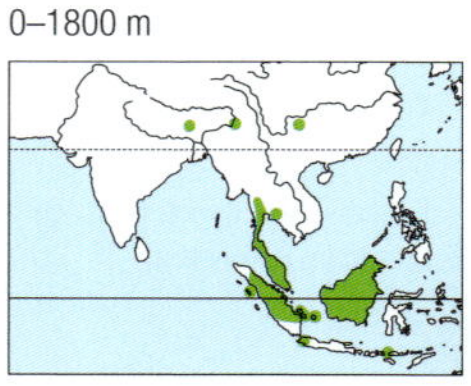

NE IN, SC CN (Guizhou), S MM, C & S TH, Pen MY, SG, Sumatra, Borneo, W Java, Banta

NT
4 sspp
HB 5.1–6.5 cm
T 2.5–3.8 cm
FA 4.5–5.7 cm
W 10–20 g

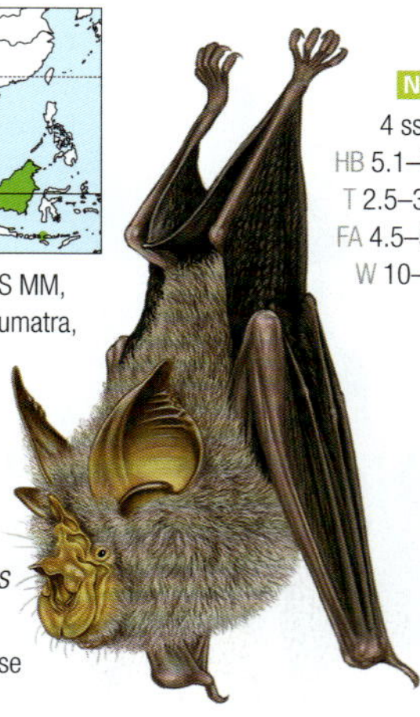

Trefoil Horseshoe Bat
Rhinolophus trifoliatus
FR Rhinolophe à trèfle
DE Kleeblatt-Hufeisennase
ES Herradura de trébol

NE
3 sspp
HB 7.5–9.9 cm
T 3.6–6.1 cm
FA 6.8–8.1 cm
W 27–45 g

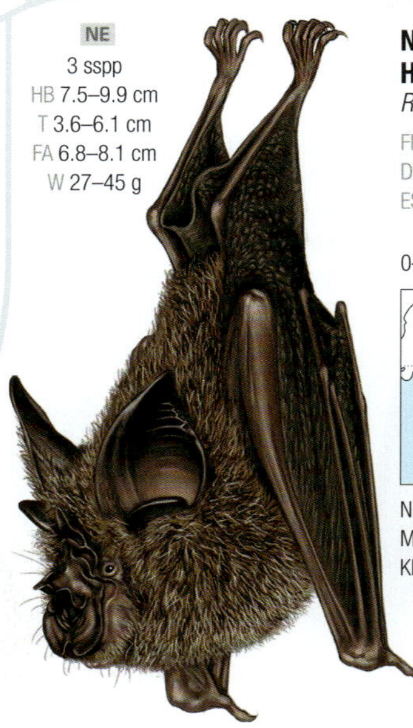

Northern Woolly Horseshoe Bat
Rhinolophus perniger
FR Rhinolophe noir
DE Nördliche Wollhaar-Hufeisennase
ES Herradura lanoso norteño

0–925 m

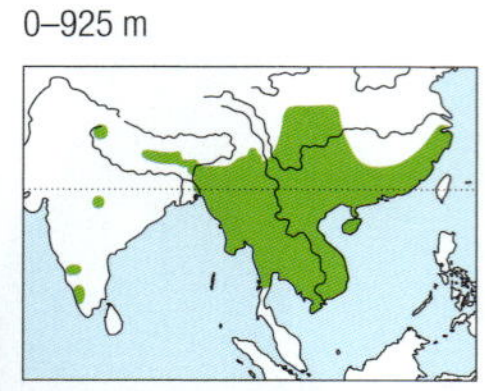

N, C & SW IN, C & E NP, BT, BD, MM, SC, C & SE CN, N TH, LA, VN, KH; possibly Pen TH

NE

Monotypic
FA 5.9–6.5 cm
W 22–32 g

Selangor Woolly Horseshoe Bat
Rhinolophus luctoides

FR Rhinolophe du Selangor
DE Selangor-Wollhaar-Hufeisennase
ES Herradura de Selangor

above 600 m

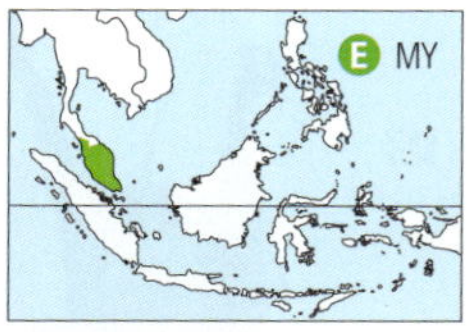

Montane regions; perhaps more extensive in continental SE Asia

NE

Monotypic
HB 7–8.5 cm
T 4–5.2 cm
FA 6–6.9 cm
W 25–41 g

Malaysian Woolly Horseshoe Bat
Rhinolophus morio

FR Rhinolophe de Singapour
DE Malaya-Wollhaar-Hufeisennase
ES Herradura de Singapur

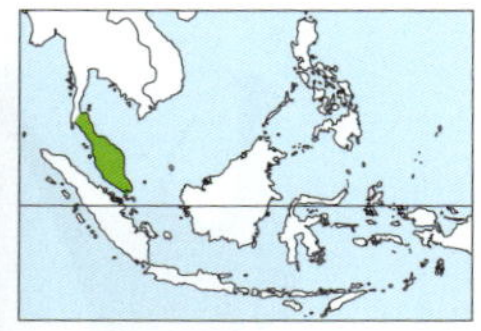

Lowlands of Thai-Malay Pen; limits uncertain

NE

Monotypic
HB c. 8.9 cm
T 3.8–5.2 cm
FA 6.3–7.1 cm
W 27–37 g

up to 1600 m

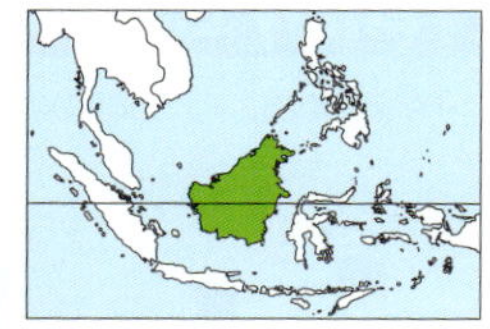

Borneo (MY, ID, BN)

Andersen's Woolly Horseshoe Bat
Rhinolophus foetidus

FR Rhinolophe fétide
DE Andersen-Wollhaar-Hufeisennase
ES Herradura lanoso de Andersen

LC

Monotypic
HB c. 8.9 cm
T 3.8–5.2 cm
FA 6.3–7.1 cm
W 27–37 g

up to 1600 m

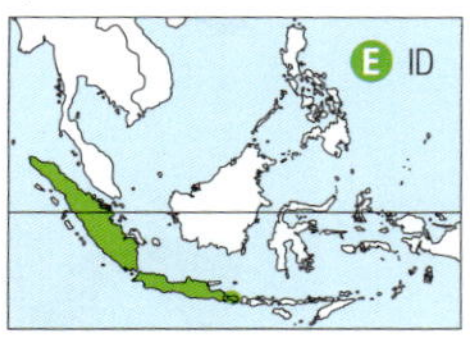

Sumatra, Java, Bali

Great Woolly Horseshoe Bat
Rhinolophus luctus

FR Rhinolophe deuil
DE Große Wollhaar-Hufeisennase
ES Herradura lanoso grande

LC

2 sspp
HB 6.5–7.5 cm
T 3.7–4.8 cm
FA 5.5–6.4 cm

Beddome's Woolly Horseshoe Bat
Rhinolophus beddomei

FR Rhinolophe de Beddome
DE Beddome-Wollhaar-Hufeisennase
ES Herradura de Beddome

0–1070 m

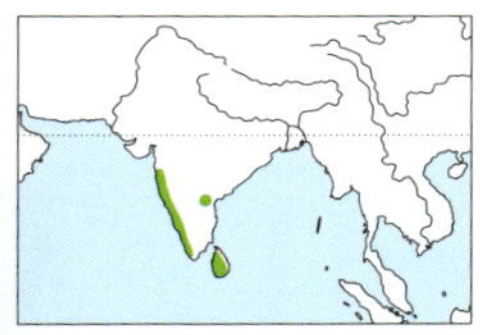

W & EC IN (Maharashtra, Karnataka, Kerala, Andhra Pradesh), LK

LC

Monotypic
HB 6.3–9 cm
T 2.8–4.1 cm
FA 5.4–6.2 cm
W 11.9–25 g

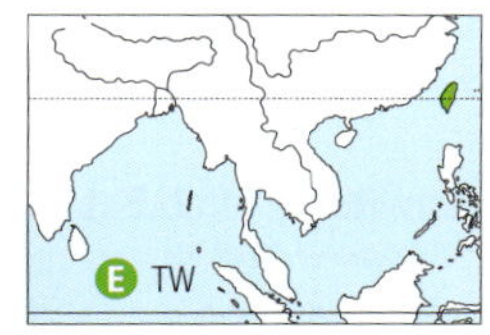

Primary and secondary forests

Formosan Woolly Horseshoe Bat
Rhinolophus formosae

FR Rhinolophe de Formose
DE Formosa-Wollhaar-Hufeisennase
ES Herradura de Formosa

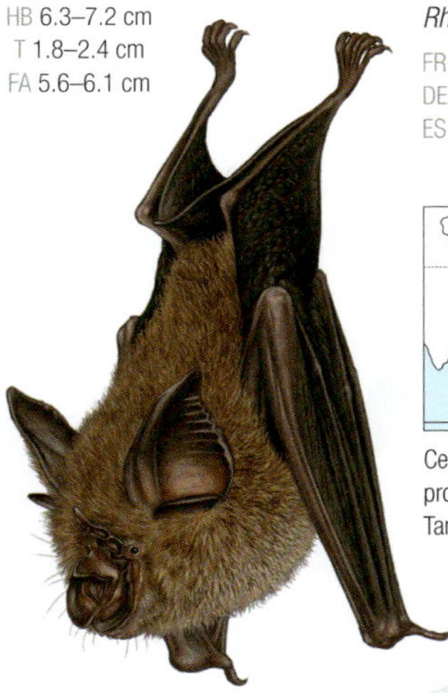

LC

Monotypic
HB 6.3–7.2 cm
T 1.8–2.4 cm
FA 5.6–6.1 cm

Thailand Horseshoe Bat
Rhinolophus thailandensis

FR Rhinolophe de Thaïlande
DE Thailand-Hufeisennase
ES Herradura de Tailandia

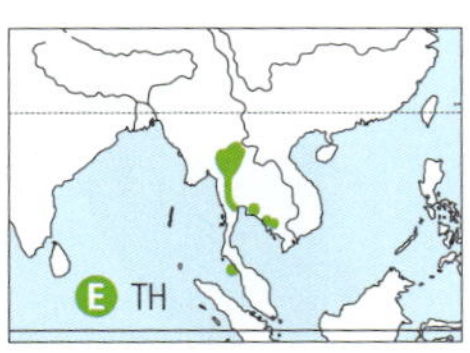

Certainty in Chiang Mai Prov, probably in NW, W & SE (with Tarutao I); perhaps KH

LC

Monotypic
HB 5.2–7.2 cm
T 1.8–2.8 cm
FA 5.5–5.9 cm
W 20–22 g

600–1300 m

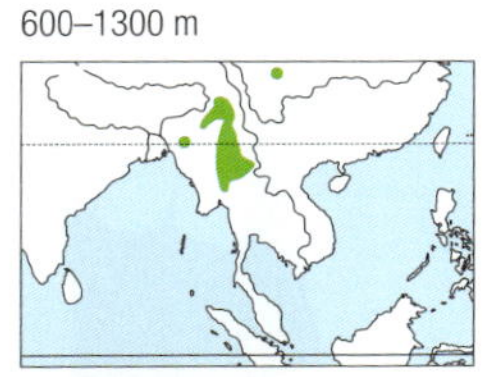

SC CN (Sichuan, Yunnan, perhaps Guizhou), tentatively NE IN & MM; perhaps LA

Dobson's Horseshoe Bat
Rhinolophus yunanensis

FR Rhinolophe du Yunnan
DE Dobson-Hufeisennase
ES Herradura de Yunnan

DD

Monotypic
HB 5.1–6.5 cm
T 1.5–1.9 cm
FA 5.2–5.6 cm

Chiew Kwee's Horseshoe Bat
Rhinolophus chiewkweeae

FR Rhinolophe de Chiew Kwee
DE Chiew-Kwee-Hufeisennase
ES Herradura de Chiew Kwee

0–1275 m

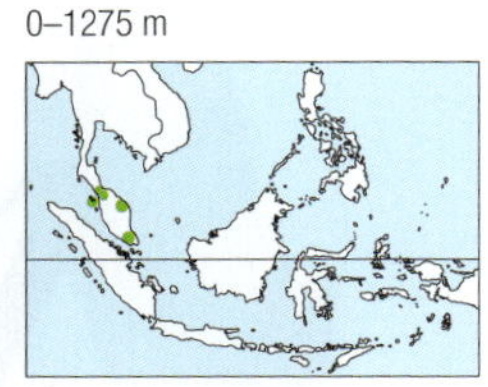

Thai-Malay Pen (with Langkawi I)

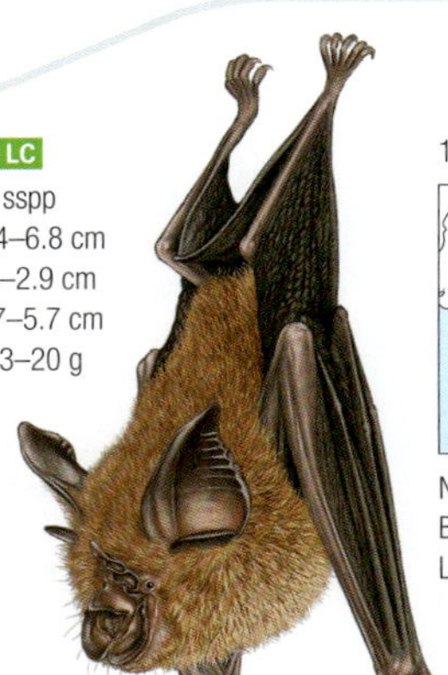

LC

2 sspp
HB 5.4–6.8 cm
T 1.6–2.9 cm
FA 4.7–5.7 cm
W 13–20 g

Pearson's Horseshoe Bat
Rhinolophus pearsonii

FR Rhinolophe de Pearson
DE Pearson-Hufeisennase
ES Herradura de Pearson

160–3380 m

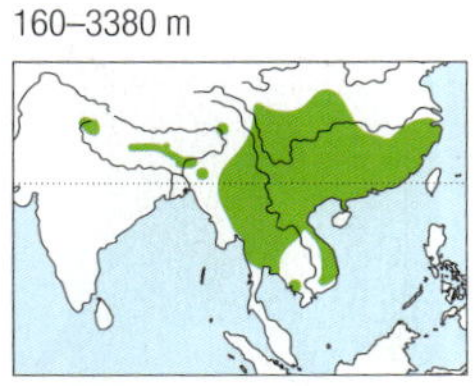

N IN (E from Uttarakhand), NP, BT, BD, C, S & SE CN, MM, TH, LA, VN, KH

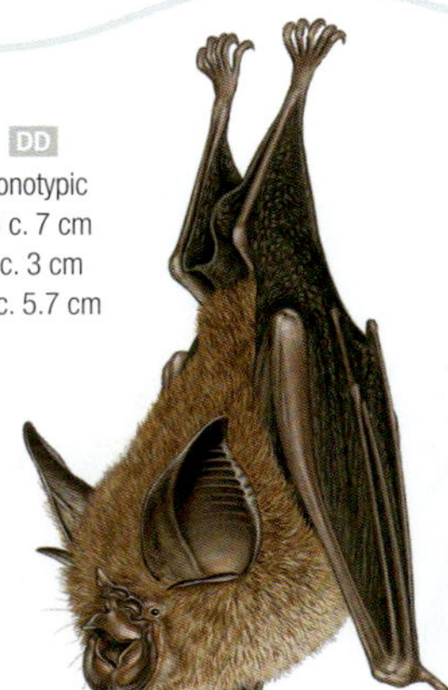

DD

Monotypic
HB c. 7 cm
T c. 3 cm
FA c. 5.7 cm

Mitred Horseshoe Bat
Rhinolophus mitratus

FR Rhinolophe mitré
DE Mitra-Hufeisennase
ES Herradura de mitra

up to 300 m

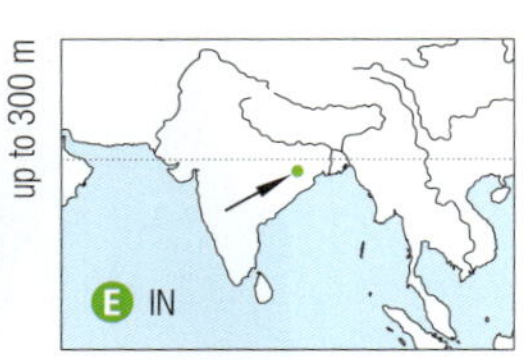

Known only from Chaibasa (Jharkhand)

Family EMBALLONURIDAE – Sheath-tailed Bats

14 genera, 55 species. Closed and open forests to savannas and semi-deserts. Afrotropical, Indo-Malayan, Australasian, Neotropical and marginally Holarctic Regions.

Naked-rumped Tomb Bat
Taphozous nudiventris

LC
4 sspp
HB 8.3–10.5 cm
T 2.2–4.6 cm
FA 7–8 cm
W 30–35 g

FR Taphien à ventre nu
DE Nacktbauch-Grabfledermaus
ES Tafozo de vientre desnudo

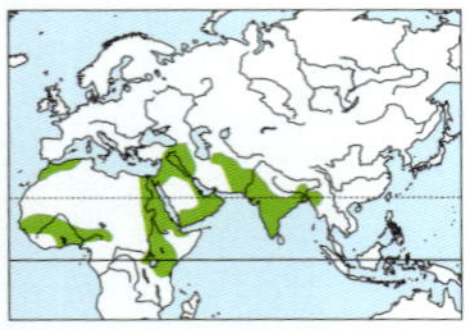

N, W, C & E Africa, Middle East, S Caucasus, S Asia from TM to MM

Egyptian Tomb Bat
Taphozous perforatus

LC 4 sspp
HB 7.1–8.5 cm T 2.2–3 cm
FA 5.4–6.7 cm W 23–33 g

FR Taphien perforé
DE Ägyptische Grabfledermaus
ES Tafozo egipcio

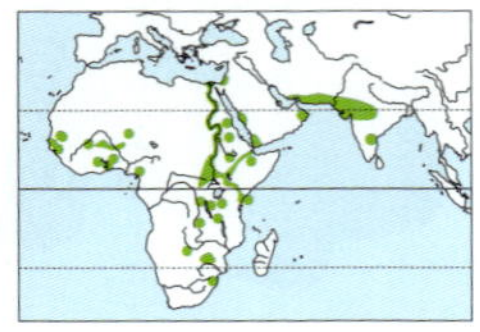

Nile Valley, scattered in W & E Africa, Middle East (IL, Arabian Pen, S IR) E to IN

Genus *TAPHOZOUS*
(14 species)

Mauritian Tomb Bat
Taphozous mauritianus

LC
Monotypic
HB 7.6–8.8 cm
T 1.5–2.8 cm
FA 5.8–6.5 cm
W 20–36 g

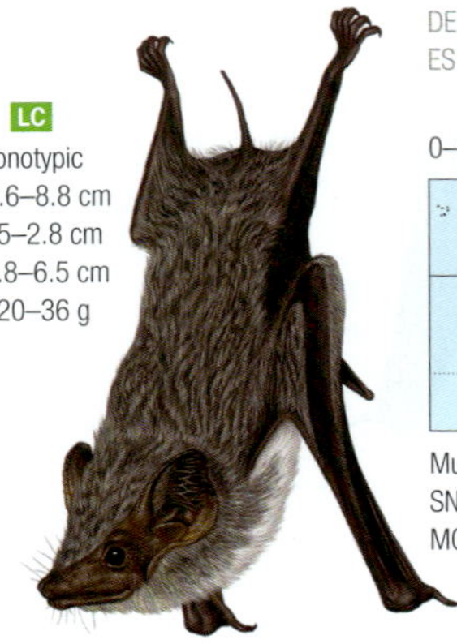

FR Taphien de Maurice
DE Mauritius-Grabfledermaus
ES Tafozo de Mauricio

0–900 m

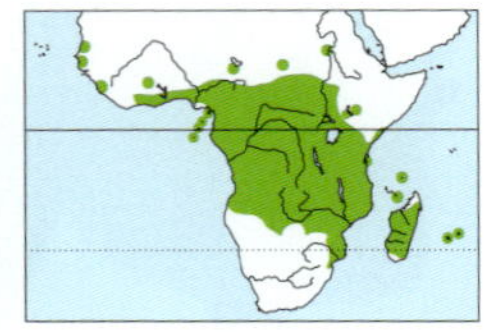

Much of sub-Saharan Africa, from SN E to S SO, S to NE NA & E ZA, MG, Gulf of Guinea & E African Is

Hamilton's Tomb Bat
Taphozous hamiltoni

FR Taphien de Hamilton
DE Hamilton-Grabfledermaus
ES Tafozo de Hamilton

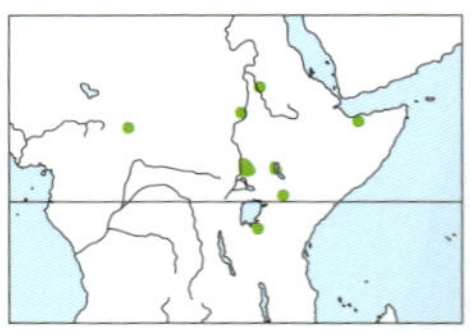

Few records in SD, SO (Lower Sheikh), SS, UG, KE, TZ (Shanwa) & TD (Sarh); limits poorly known

DD
Monotypic
HB 8–8.4 cm
T 2–3.5 cm
FA 6.1–7.1 cm
W 19.4–22 g

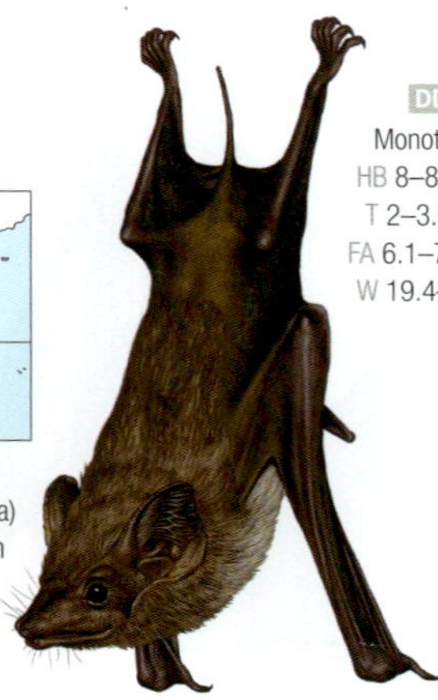

Hildegarde's Tomb Bat
Taphozous hildegardeae

FR Taphien de Hildegarde
DE Hildegarde-Grabfledermaus
ES Tafozo de Hildegarde

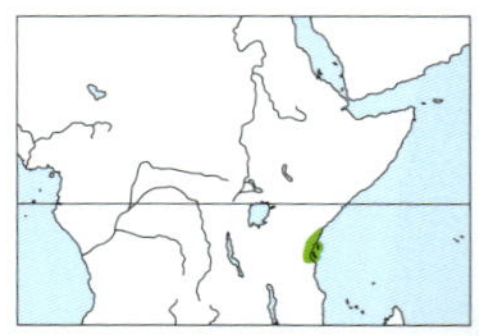

0–215 m

Coast of SE KE (perhaps C) & NE TZ (from Lower R Tana S to Dar es Salaam), Pemba & Unguja Is

EN
Monotypic
HB 7.9–8.2 cm
T 2.5–2.9 cm
FA 6.3–6.6 cm
W 24–36 g

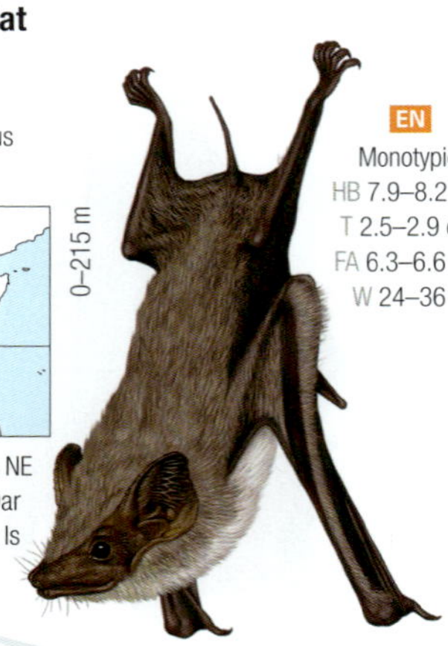

Long-winged Sheath-tailed Bat
Taphozous longimanus

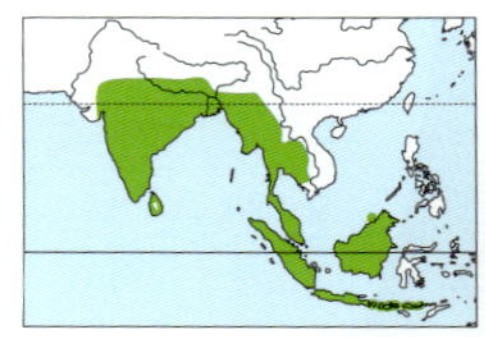

0–1200 m

Indian subcontinent (NP, IN, BD, LK), SE Asia (MM, TH, KH) S to Greater & Lesser Sundas (E to Flores)

FR Taphien à longs bras
DE Langflügel-Grabfledermaus
ES Tafozo alilargo

LC
4 sspp
HB 7.3–8.6 cm
T 2–3 cm
FA 5.4–6.3 cm

Black-bearded Tomb Bat
Taphozous melanopogon

LC
4 sspp
HB 6.7–8.6 cm
T 1.2–2.6 cm
FA 6–6.8 cm
W 21–26 g

FR Taphien à barbe noire
DE Schwarzbart-Grabfledermaus
ES Tafozo de barba negra

0–800 m

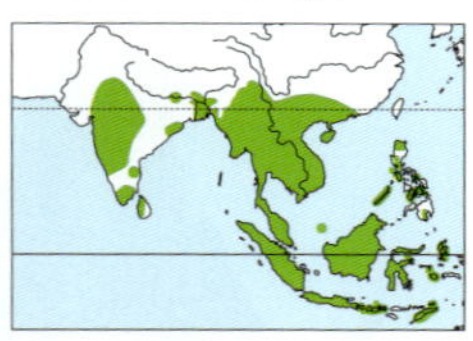

IN, LK, S CN & SE Asia S to Greater & Lesser Sundas, Sulawesi, Moluccas, PH

Theobald's Tomb Bat
Taphozous theobaldi

LC 2 sspp
HB 8.8–9.5 cm T 2.5–3 cm
FA 7–7.6 cm W 40–50 g

FR Taphien de Theobald
DE Theobald-Grabfledermaus
ES Tafozo de Theobald

0–1200 m

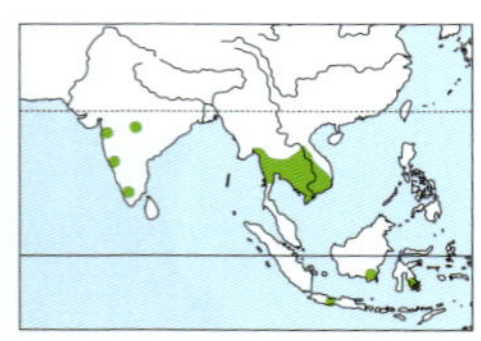

Patchily from C & SW IN, SC CN (Yunnan, not mapped) & SE Asia S to Borneo, Java & Sulawesi

Indonesian Tomb Bat
Taphozous achates

DD 2 sspp
HB 7.4–7.8 cm T 2.5–2.7 cm
FA 5.9–6.4 cm W 21–23 g

FR Taphien d'Indonésie
DE Indonesien-Grabfledermaus
ES Tafozo indonesio

0–1000 m

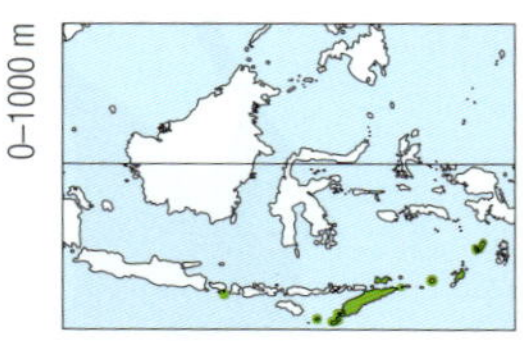

Lesser Sundas (Nusa Penida, Savu, Semau, Roti, Timor, Wetar, Leti, Babar), Tanimbar, Kai Is

Arnhem Sheath-tailed Bat
Taphozous kapalgensis

FR Taphien de Kapalga
DE Arnhemland-Grabfledermaus
ES Tafozo de Kapalga

WA (extreme E Kimberley region), NT (W Top End, perhaps farther E)

LC
Monotypic
HB 7.5–8.5 cm
T 2–2.3 cm
FA 5.7–6.3 cm
W 26–29 g

Coastal Sheath-tailed Bat
Taphozous australis

FR Taphien de Gould
DE Küsten-Grabfledermaus
ES Tafozo de Gould

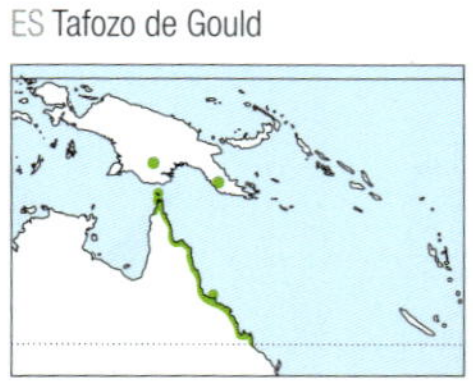

0–300 m

PG (lower R Strickland & Port Moresby, perhaps along S coast), NE AU (S to Keppel Bay), Torres Strait Is

NT Monotypic
HB 7.8–9 cm T 2–2.9 cm
FA 6.3–6.7 cm W 20–30 g

Common Sheath-tailed Bat
Taphozous georgianus

WA (Pilbara & Kimberley), NT, Qld (N Mt Isa Inlier & W Gulf Plains), offshore Is

FR Taphien de King George
DE Gewöhnliche Grabfledermaus
ES Tafozo australiano

LC
Monotypic
HB 6.2–8 cm
T 2.1–3.9 cm
FA 6.5–7.5 cm
W 19–51 g

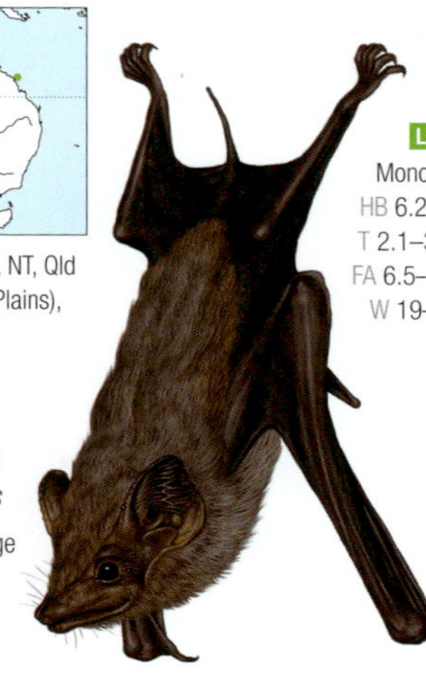

Hill's Sheath-tailed Bat
Taphozous hilli

FR Taphien de Hill
DE Hill-Grabfledermaus
ES Tafozo de Hill

LC
Monotypic
HB 6.3–8.1 cm
T 2.6–3.5 cm
FA 6.3–7.2 cm
W 20–25 g

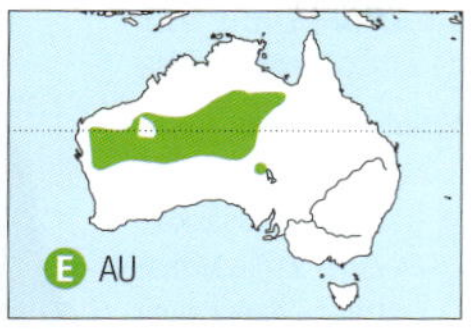

WA (semiarid Gibson Desert, Pilbara, Murchison), NT, extreme NW Qld, SA

Troughton's Sheath-tailed Bat
Taphozous troughtoni

FR Taphien de Troughton
DE Troughton-Grabfledermaus
ES Tafozo de Troughton

LC
Monotypic
HB 7.9–8.6 cm
T 3.1–3.7 cm
FA 7.3–7.6 cm
W 20–29 g

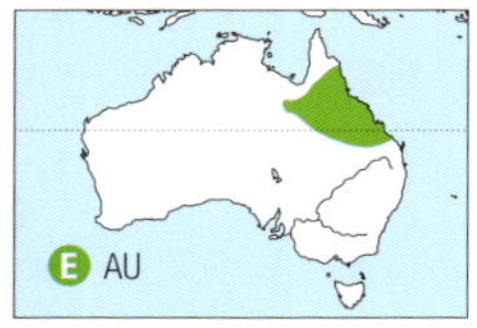

WC, C & E Qld

Genus *SACCOLAIMUS*

(4 species)

Pel's Pouched Bat
Saccolaimus peli

FR Taphien de Pel
DE Pel-Glattnasenfreischwanz
ES Tafozo de Pel

LC Monotypic
HB 11–13.5 cm
T 2.7–3.6 cm
FA 8.4–9.5 cm
W 77–114 g

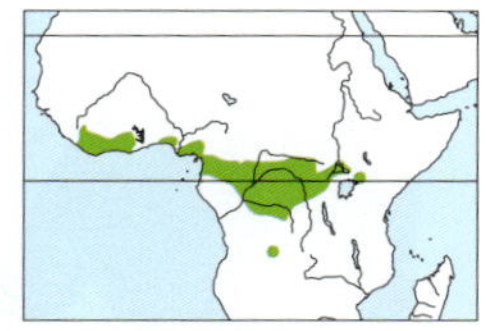

Forest belt of W & C Africa, from S GN & LR E to CD & W UG, recorded in W KE & E AO

Bare-rumped Sheath-tailed Bat
Saccolaimus saccolaimus

FR Taphien de Temminck
DE Nacktbauch-Glattnasenfreischwanz
ES Tafozo de Temminck

LC
5 sspp
HB 8.8–10 cm
T 2.2–3.5 cm
FA 6.8–7.8 cm
W 40–50 g

0–1200 m

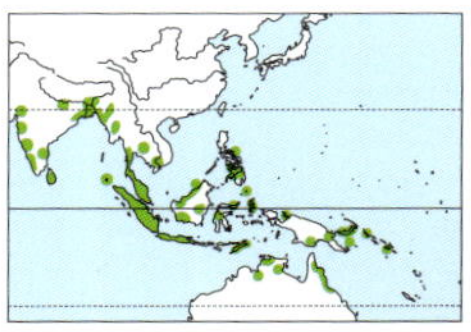

S & SE Asia from IN (with Great Nicobar I) & LK to PH, New Guinea, Solomon Is & N AU

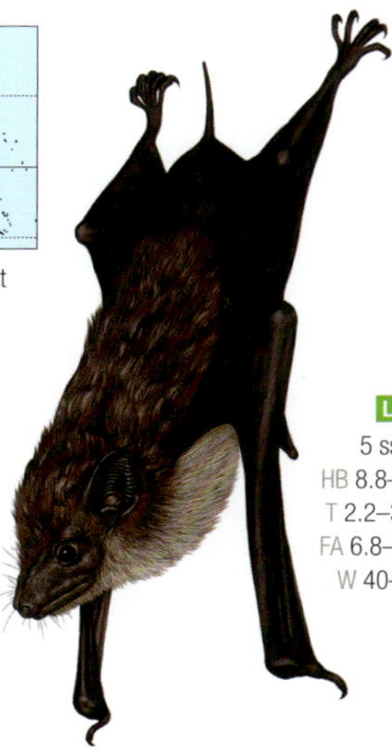

Papuan Sheath-tailed Bat
Saccolaimus mixtus

FR Taphien papou
DE Papua-Glattnasenfreischwanz
ES Tafozo de Papúa

NT
Monotypic
HB 7.2–7.9 cm
T 2.2–2.9 cm
FA 6.1–6.8 cm
W 21–27 g

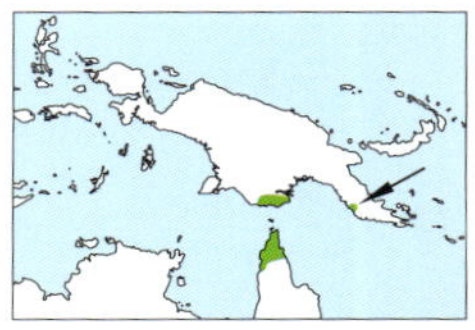

PG (S coast in Western Prov & Capital District), AU (N Cape York Pen); possibly Torres Strait Is

Yellow-bellied Sheath-tailed Bat
Saccolaimus flaviventris

FR Taphien à ventre jaune
DE Gelbbauch-Glattnasenfreischwanz
ES Tafozo de vientre amarillo

LC
Monotypic
HB 7.5–9 cm
T 2.5–3.5 cm
FA 7.2–8.7 cm
W 30–60 g

up to 600 m

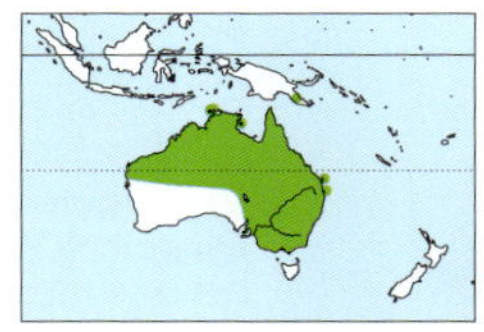

Much of mainland AU, with Tiwi, Groote Eylandt, Fraser & Moreton Is, PG (2 old records)

Genus *EMBALLONURA*

(8 species)

Lesser Sheath-tailed Bat
Emballonura monticola

FR Petite Emballonure
DE Kleine Freischwanzfledermaus
ES Embalonuro pequeño

LC
Monotypic
HB 4–4.7 cm
T 11–1.4 cm
FA 3.8–4.5 cm
W 4.5–7 g

0–1800 m

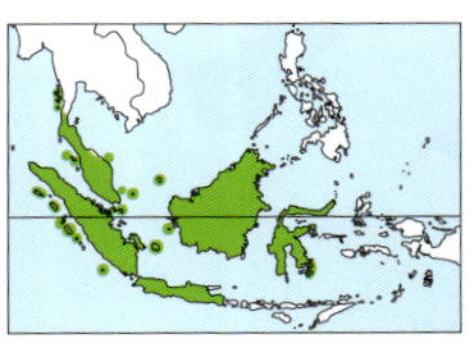

Thai-Malay Pen, Sumatra, Borneo, Java, Sulawesi, Anambas, Natuna Is & several offshore Is

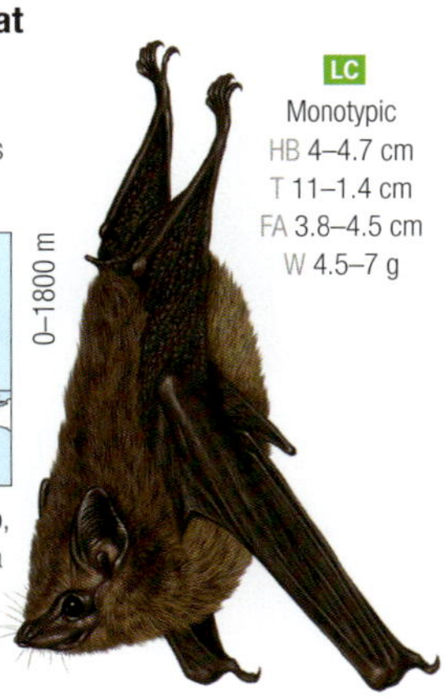

Small Asian Sheath-tailed Bat
Emballonura alecto

FR Emballonure des Philippines
DE Philippinen-Freischwanzfledermaus
ES Embalonuro de Filipinas

LC
4 sspp
HB 4.6–4.8 cm
T 1.2–1.7 cm
FA 4.3–4.8 cm
W 4.5–7 g

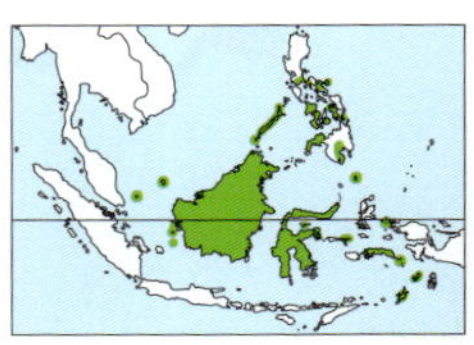

PH, Anambas, Natuna Besar I, Borneo, Talaud Is, Sulawesi, Moluccas, Raja Ampat Is (Gag)

Raffray's Sheath-tailed Bat
Emballonura raffrayana

FR Emballonure de Raffray
DE Raffray-Freischwanzfledermaus
ES Embalonuro de Raffray

LC
3 sspp
HB 3.7–5.8 cm
T 1–1.4 cm
FA 3.7–4.7 cm
W 4.2–10.2 g

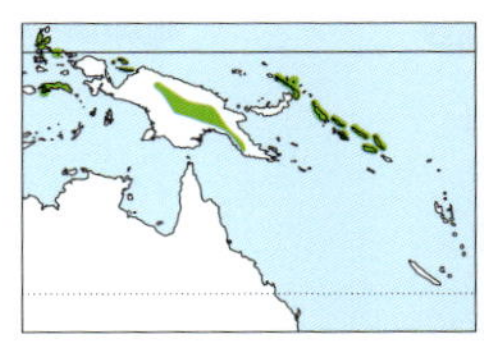

Moluccas (Halmahera, Seram, Ambon, Gebe), New Guinea (with Geelwink Bay Is), Bismarck Arch, Solomon Is

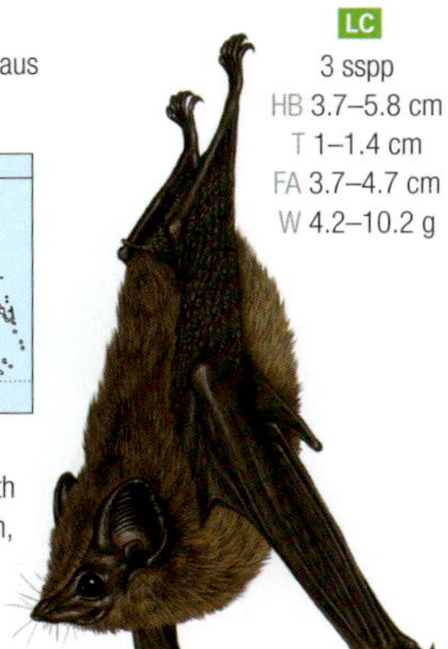

New Guinea Sheath-tailed Bat
Emballonura furax

FR Grande Emballonure
DE Neuguinea-Freischwanzfledermaus
ES Embalonuro grande

LC
Monotypic
HB 5.4–6.1 cm
T 1–1.9 cm
FA 4.8–5.3 cm
W 9.5–14 g

0–1500 m

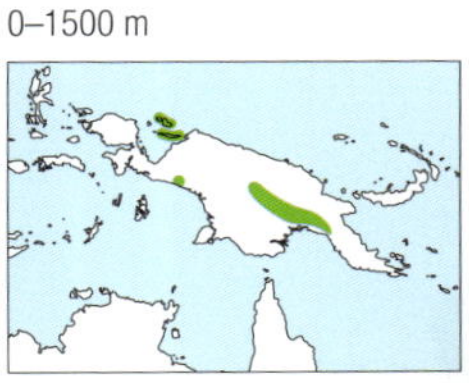

ID/PG: New Guinea (Biak & Yapen Is, Fakfak District, E Central Range)

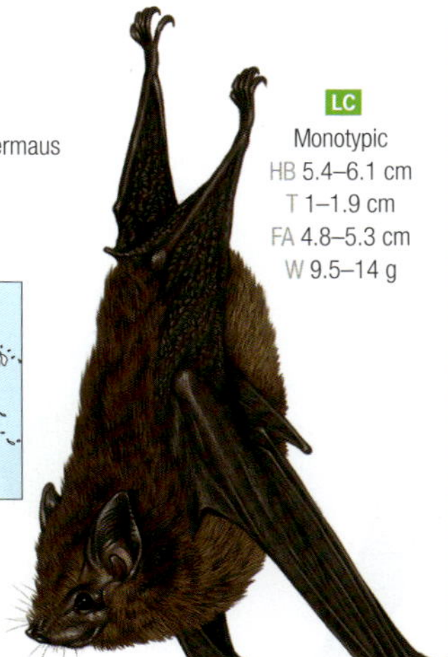

Beccari's Sheath-tailed Bat
Emballonura beccarii

FR Emballonure de Beccari
DE Beccari-Freischwanzfledermaus
ES Embalonuro de Beccari

LC
3 sspp
HB 3.4–5 cm
T 1–2.2 cm
FA 3.7–4.4 cm
W 3.5–5.3 g

0–1500 m

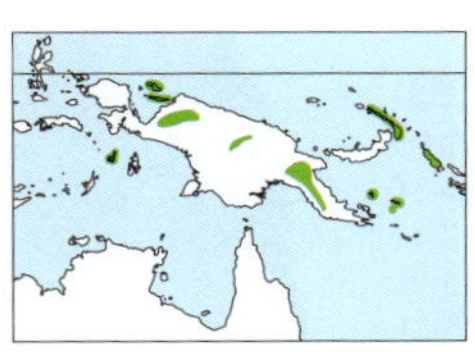

ID/PG: Kai Is, New Guinea, Biak, Yapen, New Ireland, Bougainville & Trobriand Is, Louisiade Arch

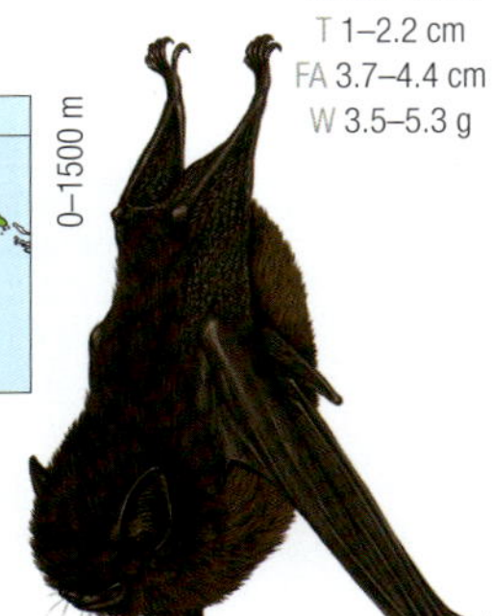

Large-eared Sheath-tailed Bat
Emballonura dianae

FR Emballonure de Diana
DE Diana-Freischwanzfledermaus
ES Embalonuro de Diana

LC
3 sspp
HB 4.8–6.4 cm
T 1.1–1.6 cm
FA 4.2–4.8 cm
W 5–10.5 g

0–1400 m

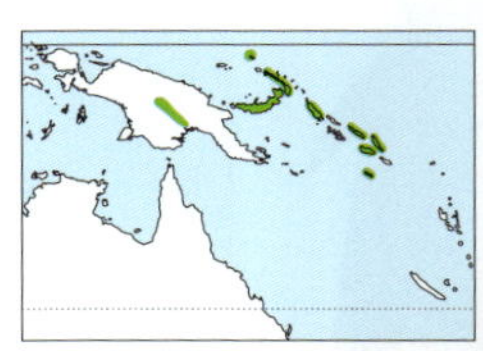

PG/SB: New Guinea (Western, Southern Highlands & Gulf prov), Bismarck Arch, Solomon Is

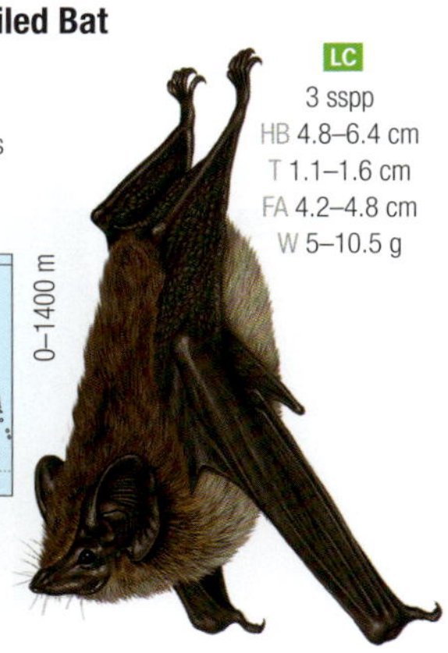

Seri's Sheath-tailed Bat

Emballonura serii

FR Emballonure de Seri
DE Seri-Freischwanzfledermaus
ES Embalonuro de Seri

VU
Monotypic
HB 4.4–6.7 cm
T 1–1.9 cm
FA 4.4–5.3 cm
W 3.9–7 g

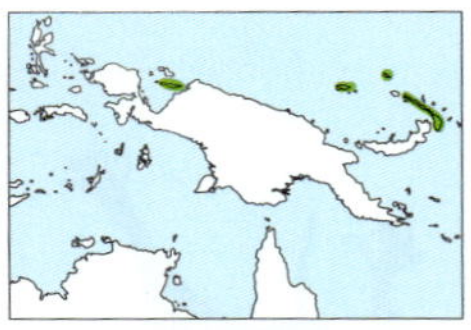

0–300 m

ID (Yapen I in Geelvink Bay), PG (Bismarck Arch in Manus, Los Negros, Mussau & New Ireland)

Pacific Sheath-tailed Bat

Emballonura semicaudata

FR Emballonure du Pacifique
DE Pazifik-Freischwanzfledermaus
ES Embalonuro del Pacífico

EN
4 sspp
HB 4.1–4.9 cm
FA 4.4–4.8 cm
W 5.5–8 g

0–300 m

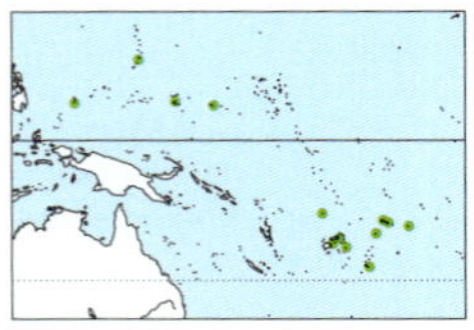

GU & MP (currently only in Aguigan I), PW, Caroline Is (Chuuk, Pohnpei), FJ, TO, WS, AS

Western Sheath-tailed Bat

Paremballonura tiavato

FR Emballonure des rochers
DE Westliche Madagaskar-Freischwanzfledermaus
ES Embalonuro occidental

LC
Monotypic
HB 4–4.6 cm
T 1.5–1.8 cm
FA 3.5–4.1 cm
W 2.7–3.8 g

0–350 m

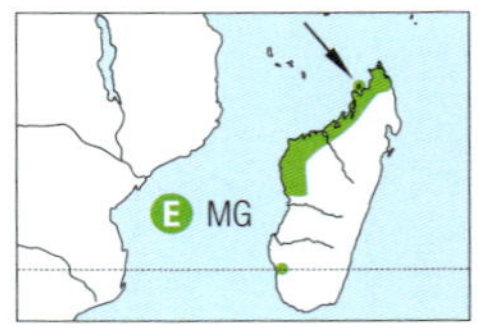

From Daraina region to Tsingy de Bemaraha NP (with Nosy Be & Nosy Komba), recorded in Toliara

Genus *PAREMBALLONURA*

(2 species)

Peters's Sheath-tailed Bat

Paremballonura atrata

FR Emballonure de Madagascar
DE Östliche Madagaskar-Freischwanzfledermaus
ES Embalonuro de Peters

LC
Monotypic
HB 4.5–4.9 cm
T 1.8–2 cm
FA 3.7–4.1 cm
W 5–7.1 g

0–900 m

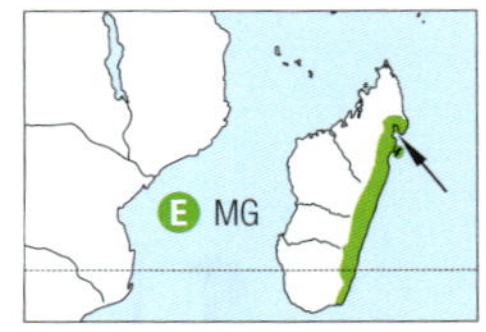

From Maroantsetra S to Tôlanaro (with Nosy Mangabé & Nosy Boraha)

Genus *COLEURA*

(4 species)

African Sheath-tailed Bat

Coleura afra

FR Emballonure d'Afrique
DE Afrika-Freischwanzfledermaus
ES Embalonuro africano

LC
Monotypic
HB 5.5–6.5 cm
T 1–1.1 cm
FA 4.9–5.5 cm
W 10–12 g

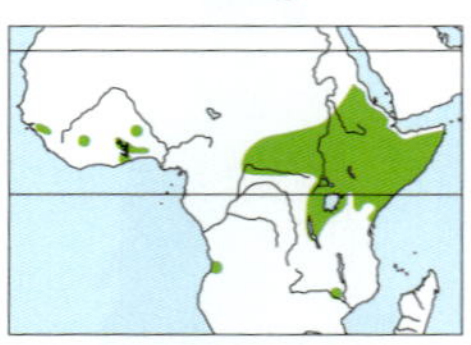

GW & GN E to W NG, also from CF & CD E to SO & KE, recorded in AO (Benguela) & N MZ

Arabian Sheath-tailed Bat

Coleura gallarum

FR Emballonure de Somalie
DE Arabische Freischwanzfledermaus
ES Embalonuro de Somalia

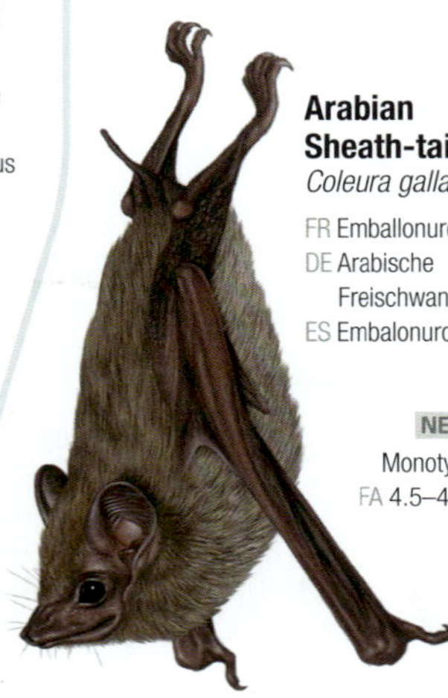

NE
Monotypic
FA 4.5–4.8 cm

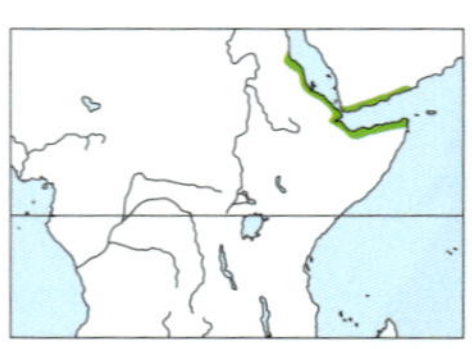

Coastal SD, ER, ET, DJ, N SO, S YE & SW OM; limits uncertain

Seychelles Sheath-tailed Bat

Coleura seychellensis

FR Emballonure des Seychelles
DE Seychellen-Freischwanzfledermaus
ES Embalonuro de las Seychelles

CR
2 sspp
HB 5.5–6.5 cm
FA 5.2–5.6 cm
W 10.2–11.1 g

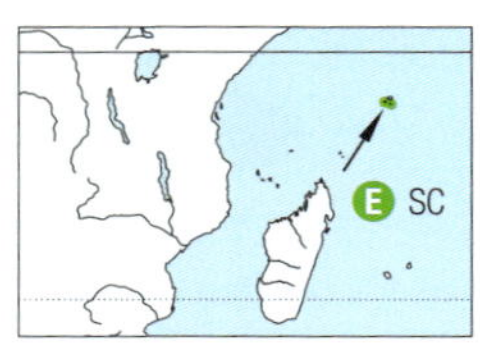

NW Mahé, Silhouette; extinct on Praslin & La Digue

Madagascar Sheath-tailed Bat

Coleura kibomalandy

FR Emballonure à ventre blanc
DE Madagaskar-Freischwanzfledermaus
ES Embalonuro de Madagascar

DD
Monotypic
HB 6.3–6.4 cm
T 1.1–1.8 cm
FA 4.8–5.2 cm
W 8.4–12.5 g

15–115 m

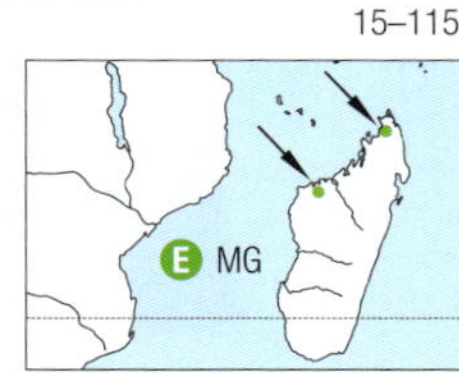

Known only from Ankarana SR & Tsingy de Namoroka NP

Genus *MOSIA*

(1 species)

Dark Sheath-tailed Bat

Mosia nigrescens

FR Emballonure sombre
DE Dunkle Freischwanzfledermaus
ES Embalonuro oscuro

LC
3 sspp
HB 3–4.5 cm
T 0.7–1.7 cm
FA 3–3.8 cm
W 2.5–5.7 g

0–1600 m

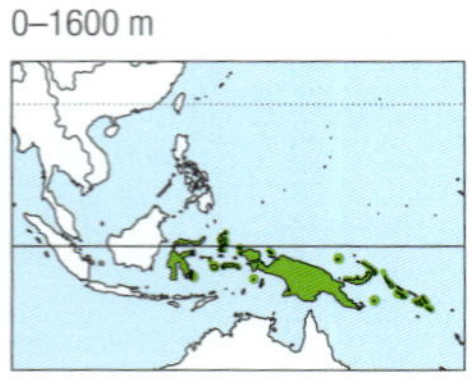

Sulawesi, Moluccas, New Guinea (with Raja Ampat & Geelvink Bay Is), Bismarck Arch, Solomon Is

Genus *CORMURA*

(1 species)

Chestnut Sac-winged Bat

Cormura brevirostris

FR Cormure des souches
DE Kastanienbraune Sackflügelfledermaus
ES Cormura castaño

LC
Monotypic
HB 4.6–5.8 cm
T 1–1.6 cm
FA 4.1–5 cm
W 7–11 g

0–1000 m

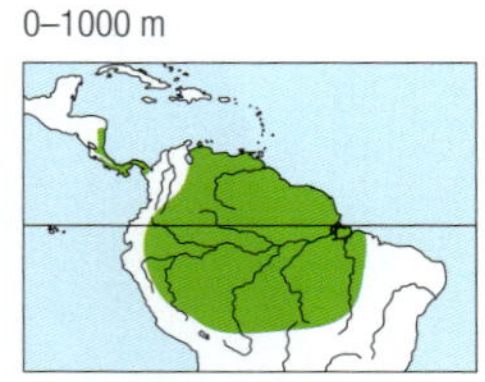

From E NI to PA, W of Andes to NE EC, E of Andes from VE & Guianas to Amazonian PE, C BR & N BO

Gray Sac-winged Bat
Balantiopteryx plicata

FR Emballonure de Peters
DE Peters-Sackflügelfledermaus
ES Embalonuro de sacos de Peters

LC
2 sspp
HB 4.7–5.3 cm
T 1.3–2.3 cm
FA 3.7–4.6 cm
W 5–6 g

0–1500 m

MX (S from Sonora & tip of Baja California Sur) to NI & NW CR

Genus *BALANTIOPTERYX*

(3 species)

Thomas's Sac-winged Bat
Balantiopteryx io

FR Emballonure de Thomas
DE Thomas-Sackflügelfledermaus
ES Embalonuro de sacos de Thomas

VU
Monotypic
HB 3.8–4.4 cm
T 0.8–1.8 cm
FA 3.6–4.1 cm
W 3–4 g

0–500 m

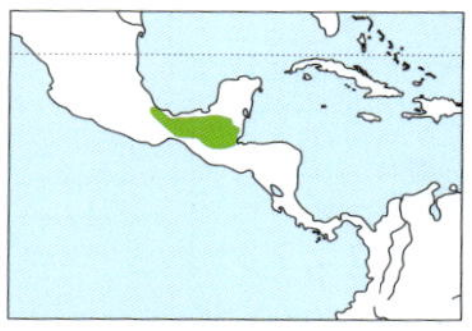

Atlantic slope of MX (from S Veracruz & Oaxaca) S to EC GT & BZ

Ecuadorian Sac-winged Bat
Balantiopteryx infusca

FR Emballonure d'Équateur
DE Ecuador-Sackflügelfledermaus
ES Embalonuro de sacos ecuatoriano

VU
Monotypic
HB 3.7–4.2 cm
T 1.2–1.8 cm
FA 3.8–4.2 cm
W 4–5 g

150–1200 m

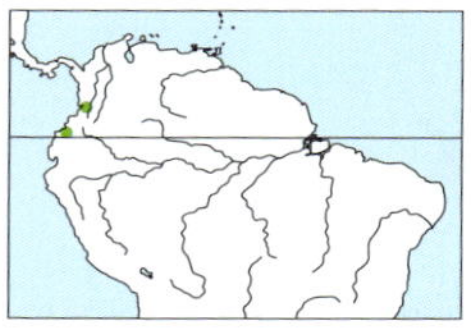

Few sites on Pacific slope of SW CO (Río Chanco in Valle del Cauca) & NW EC (Esmeraldas, Imbabura)

Genus *CYTTAROPS*

(1 species)

Short-eared Bat
Cyttarops alecto

FR Cytarope furieux
DE Dunkelgraue Kurzohrfledermaus
ES Embalonuro orejicorto

LC
Monotypic
HB 4.7–5.5 cm
T 2–2.5 cm
FA 4.5–4.7 cm
W 6–7 g

0–500 m

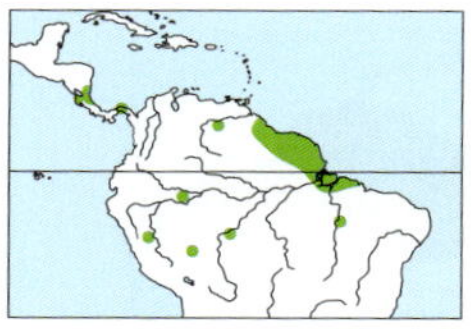

From NI to PA, VE (Bolívar) E to N BR (Pará), scattered in Amazon Basin (S CO, E PE, W BR & N BO)

Northern Ghost Bat
Diclidurus albus

FR Diclidure fantôme
DE Weiße Gespenstfledermaus
ES Dicliduro norteño

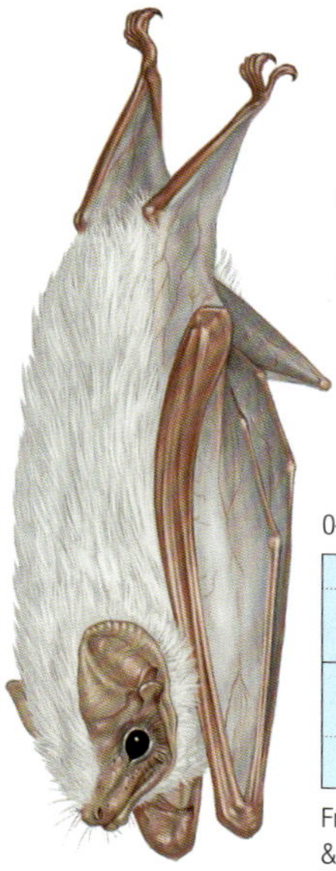

LC
2 sspp
HB 6.8–8.2 cm
T 1.8–2.2 cm
FA 6.3–6.9 cm
W 17–24 g

0–1700 m

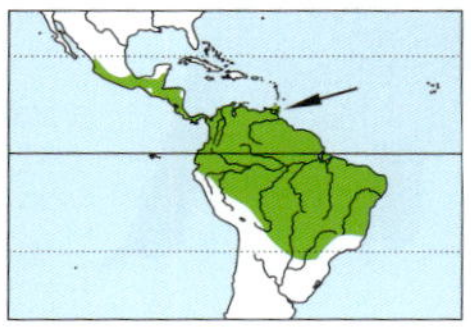

From SW & S MX S to EC, E PE, & most of BR, TT (Trinidad)

Genus *DICLIDURUS*

(4 species)

Greater Ghost Bat
Diclidurus ingens

FR Grande Diclidure
DE Große Gespenstfledermaus
ES Dicliduro grande

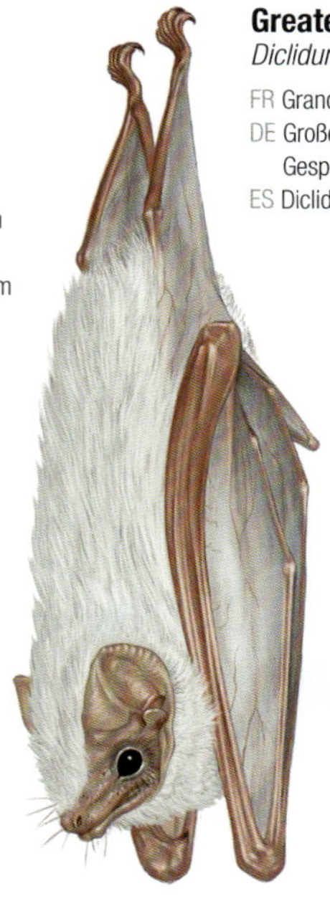

DD
Monotypic
HB c. 8.3 cm
T c. 2.2 cm
FA 6.8–7.2 cm
W 20–26 g

0–200 m

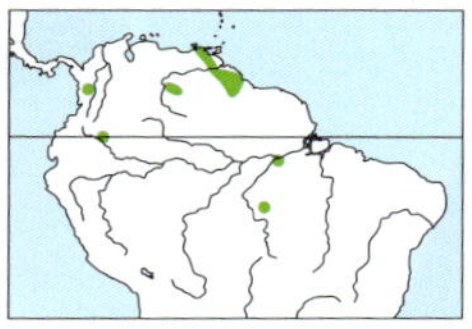

Scattered in CO (Chocó, Putumayo), VE, GY, N & C BR (Pará, Mato Grosso); range unresolved

Lesser Ghost Bat
Diclidurus scutatus

FR Diclidure des cyclanthes
DE Kleine Gespenstfledermaus
ES Dicliduro pequeño

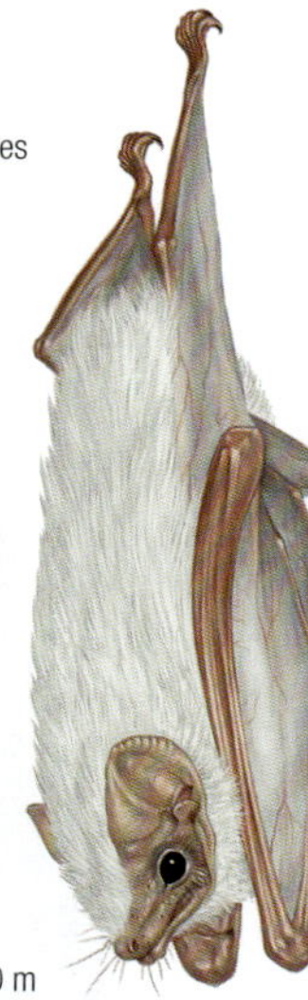

LC
Monotypic
HB 5.7–6.8 cm
T 1.7–1.9 cm
FA 5.1–5.9 cm
W c. 13 g

0–1000 m

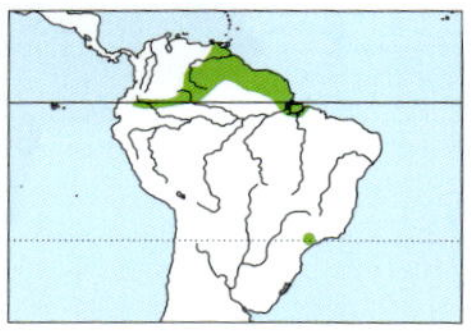

NE EC (Orellana), NE PE (Loreto), E CO (Vaupés), S & E VE, Guianas, N BR (1 record in São Paulo)

Isabelline Ghost Bat
Diclidurus isabella

FR Diclidure isabelle
DE Isabellfarbene Gespenstfledermaus
ES Dicliduro isabelino

LC
Monotypic
HB c. 6.8 cm
T 1.5–2.4 cm
FA c. 5.4 cm
W 13–15 g

0–200 m

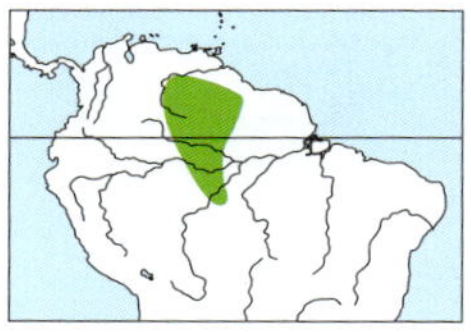

From S VE (Amazonas) & GY to NW BR (Amazonas)

Genus *PEROPTERYX*

(5 species)

Lesser Dog-like Bat
Peropteryx macrotis

FR Péroptère des cavernes
DE Kleine Hundskopffledermaus
ES Peróptero menor

LC
Monotypic
HB 4.2–5.3 cm
T 1.1–1.5 cm
FA 3.8–4.8 cm
W 4–7 g

0–1000 m

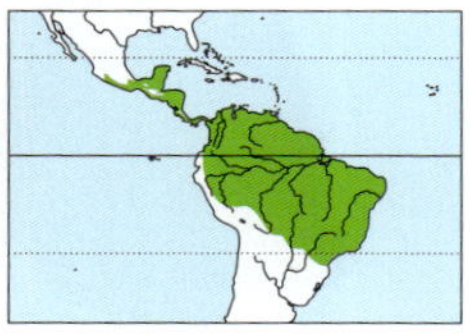

S MX (from Michoacán & S Veracruz) S to E PE, N & E BO, N PY & S BR

Greater Dog-like Bat
Peropteryx kappleri

FR Grand Péroptère
DE Große Hundskopffledermaus
ES Peróptero mayor

LC
2 sspp
HB 6.3–7.5 cm
T 1.1–2 cm
FA 4.5–5.4 cm
W 7–13 g

0–850 m

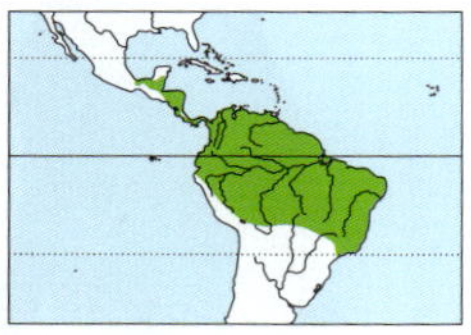

S MX (from S Veracruz) S to CO, EC, PE & N BO, E to Atlantic Forest of E BR

Trinidad Dog-like Bat
Peropteryx trinitatis

FR Péroptère de Trinidad
DE Trinidad-Hundskopffledermaus
ES Peróptero de Trinidad

DD
2 sspp
HB 4.5–4.6 cm
T 1.1–1.8 cm
FA 3.6–4.3 cm
W 3.7–8 g

0–400 m

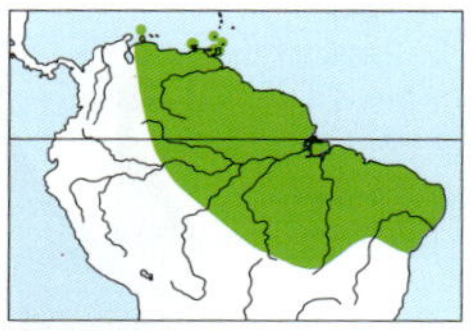

Most of VE, Guianas, N, NE & C BR, several Caribbean Is (AW, Margarita, GD, TT)

White-winged Dog-like Bat
Peropteryx leucoptera

FR Péroptère à ailes blanches
DE Weißflügel-Hundskopffledermaus
ES Peróptero aliblanco

LC
2 sspp
HB 4.7–5.5 cm
T 1.1–2 cm
FA 4.1–5.2 cm
W 5.5–6.4 g

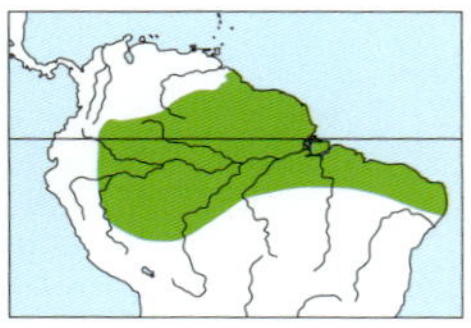

SE CO, S VE & Guianas through Amazon Basin to E PE, N BO & NE BR

Pale-winged Dog-like Bat
Peropteryx pallidoptera

FR Péroptère à ailes claires
DE Blassflügel-Hundskopffledermaus
ES Peróptero aliclaro

DD
Monotypic
HB 4.6–5.3 cm
T 1.1–1.4 cm
FA 3.9–4.3 cm
W 4.1–6 g

150–400 m

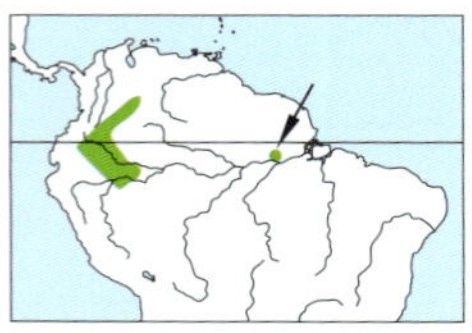

CO (including Llanos), E EC, NE PE, N BR

Genus *CENTRONYCTERIS*

(2 species)

Thomas's Shaggy Bat
Centronycteris centralis

FR Centronyctère de Thomas
DE Thomas-Spornscheidenschwanz
ES Centronícteris de Thomas

LC
Monotypic
HB 4.2–5.9 cm
T 1.8–4 cm
FA 4.3–4.9 cm
W 4–6 g

up to 1450 m

S MX (from Gulf of Mexico) through Central America (both slopes in S CR & PA) to SE PE & NW BO

Common Shaggy Bat
Centronycteris maximiliani

FR Centronyctère de Maximilian
DE Maximilian-Spornscheidenschwanz
ES Centronícteris de Maximilian

LC
Monotypic
HB 4.3–6.1 cm
T 2–2.6 cm
FA 4.1–4.8 cm
W 4.5–9 g

0–200 m

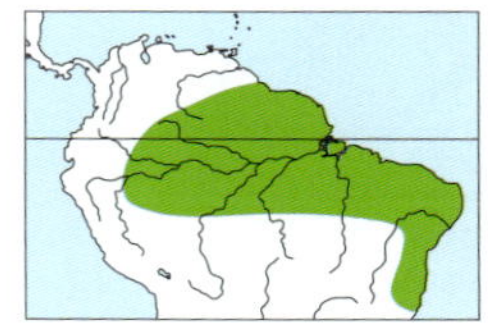

SE CO, S VE, Guianas, NE PE, N & E BR S to Vitória (Espírito Santo)

Genus *RHYNCHONYCTERIS*

(1 species)

Proboscis Bat
Rhynchonycteris naso

FR Nasin des rivières
DE Nasenfledermaus
ES Murciélago narigón

LC
Monotypic
HB 3.6–4.8 cm
T 1.1–1.7 cm
FA 3.6–4 cm
W 3–6 g

0–1500 m

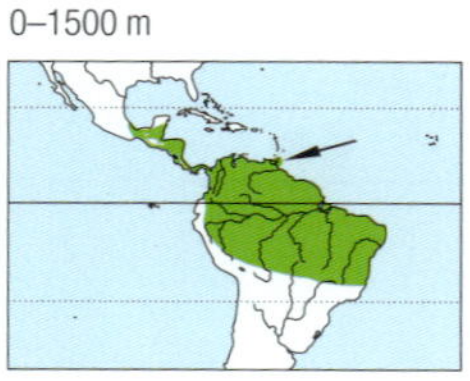

From S MX (Veracruz, Oaxaca) S to E EC, E PE & N BO, E to C & E BR, TT (Trinidad)

Genus *SACCOPTERYX*

(5 species)

Greater Sac-winged Bat
Saccopteryx bilineata

FR Saccoptère à deux bandes
DE Große Sackflügelfledermaus
ES Sacóptero mayor

LC
Monotypic
HB 4.7–5.6 cm
T 1.6–2.3 cm
FA 4.4–4.8 cm
W 6–9.3 g

up to 1000 m

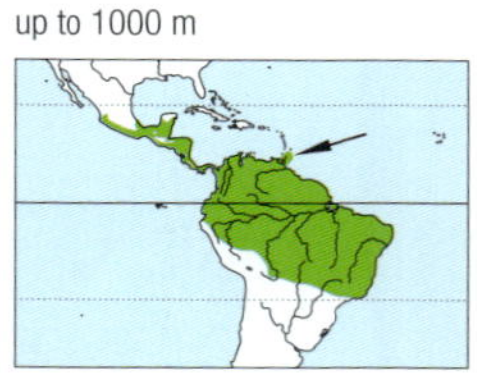

From C MX (Jalisco, Veracruz) S to EC, E PE & N BO, E to E BR (S to Rio de Janeiro), TT

Lesser Sac-winged Bat
Saccopteryx leptura

FR Saccoptère commun
DE Kleine Sackflügelfledermaus
ES Sacóptero menor

LC
Monotypic
HB 3.8–5.1 cm
T 0.9–1.9 cm
FA 3.7–4.3 cm
W 3–6 g

0–915 m

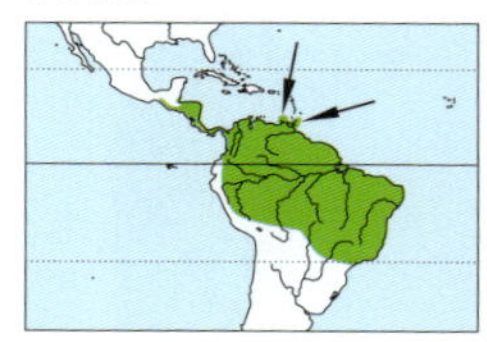

From S MX (Chiapas) S to E EC, E PE, N BO, E to E & SE BR (S to São Paulo), also Margarita I (VE) & TT

Frosted Sac-winged Bat
Saccopteryx canescens

FR Saccoptère givré
DE Weißgraue Sackflügelfledermaus
ES Sacóptero escarchado

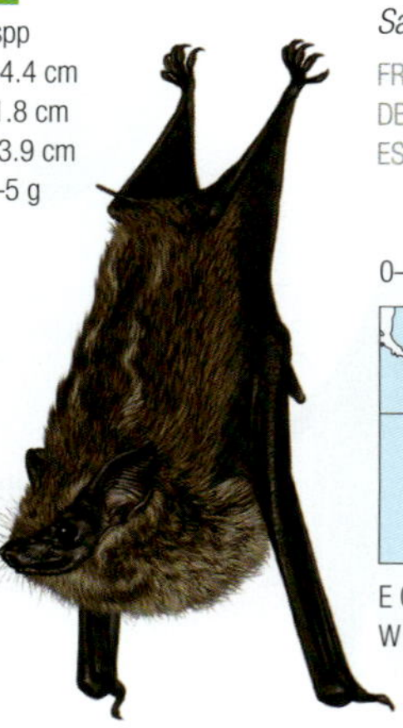

LC
2 sspp
HB 3.8–4.4 cm
T 1.6–1.8 cm
FA 3.5–3.9 cm
W 3–5 g

0–500 m

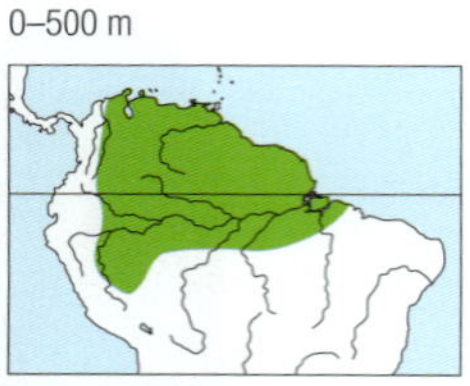

E CO, most of VE, Guianas, E PE, W & N BR; limits unclear

Antioquian Sac-winged Bat
Saccopteryx antioquensis

FR Saccoptère d'Antioquia
DE Antioquia-Sackflügelfledermaus
ES Sacóptero de Antioquia

EN
Monotypic
HB 4–4.4 cm
T 1–1.2 cm
FA 3.6–3.8 cm
W c. 5 g

650–1200 m

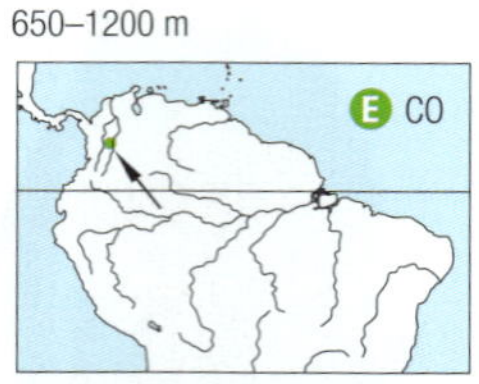

SE Antioquia, along E slope of C Andes

Amazonian Sac-winged Bat
Saccopteryx gymnura

FR Saccoptère à queue nue
DE Amazonas-Sackflügelfledermaus
ES Sacóptero amazónico

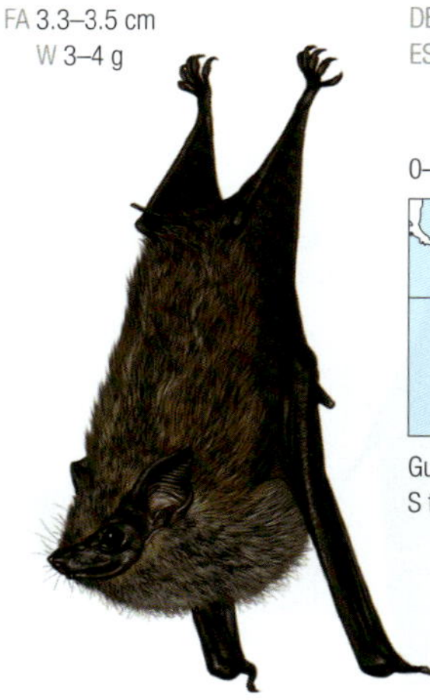

DD
Monotypic
HB 3.9–4 cm
T 1.3–1.4 cm
FA 3.3–3.5 cm
W 3–4 g

0–120 m

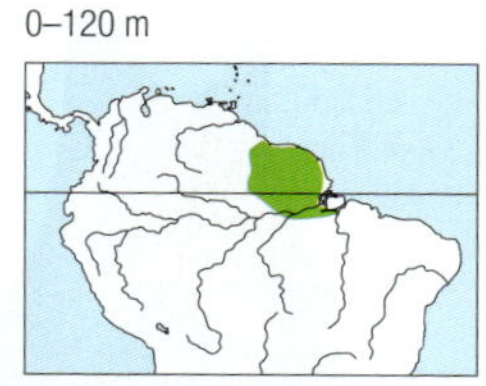

Guianas, N Amazonian BR (Pará) S to Santarém, E to Belém

Family NYCTERIDAE – Slit-faced Bats

1 genus, 14 species. Deserts, arid grasslands, savannas, and tropical forests, from sea level to mid-elevations on mountains. Afrotropical and Indo-Malayan Regions.

Genus *NYCTERIS*

(14 species)

Bates's Slit-faced Bat
Nycteris arge

FR Nyctère de Bates
DE Bates-Schlitznase
ES Nicteris de Bates

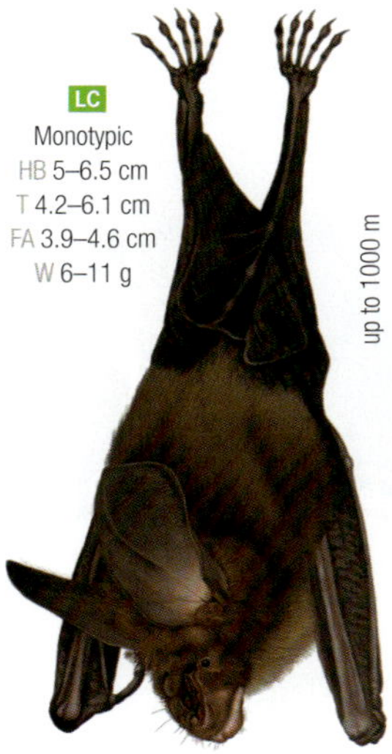

LC
Monotypic
HB 5–6.5 cm
T 4.2–6.1 cm
FA 3.9–4.6 cm
W 6–11 g

up to 1000 m

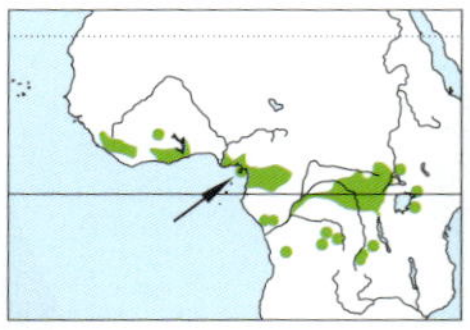

Patchily in tropical forests of W & C Africa, from SL E to SW KE & NW TZ, S to N AO, Bioko I

Intermediate Slit-faced Bat
Nycteris intermedia

FR Nyctère d'Aellen
DE Intermediäre Schlitznase
ES Nícteris intermedio

LC
Monotypic
HB 4–5 cm
T 4.1–5.3 cm
FA 3.6–3.8 cm
W 6–9 g

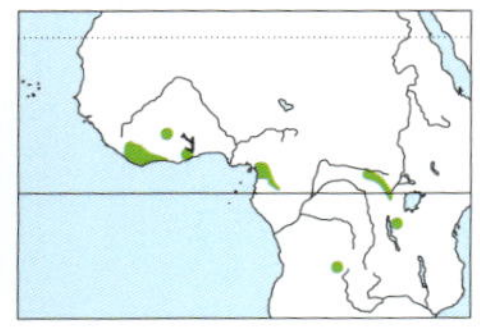

Very patchily in tropical forests of W & C Africa in GN, LR, CI, GH, CM, GA, NE CD, NW TZ & NE AO

Dwarf Slit-faced Bat
Nycteris nana

FR Nyctère nain
DE Kleinste Schlitznase
ES Nícteris enano

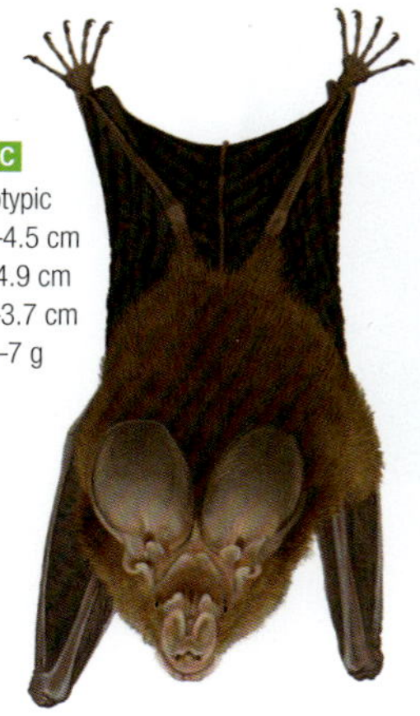

LC
Monotypic
HB 3.8–4.5 cm
T 3.7–4.9 cm
FA 3.2–3.7 cm
W 4–7 g

up to 2100 m

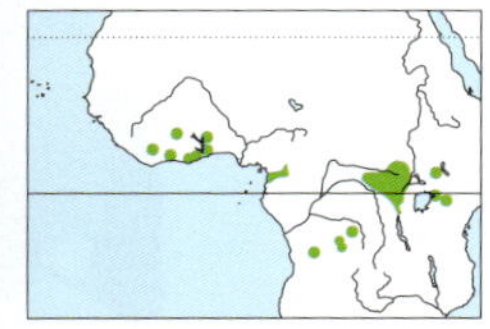

Patchily in tropical forests of CI, GH, CM, GQ, CD, SS, W UG, W KE, W BI & N AO

Dja Slit-faced Bat
Nycteris major

FR Nyctère du Dja
DE Dja-Schlitznase
ES Nicteris de Dja

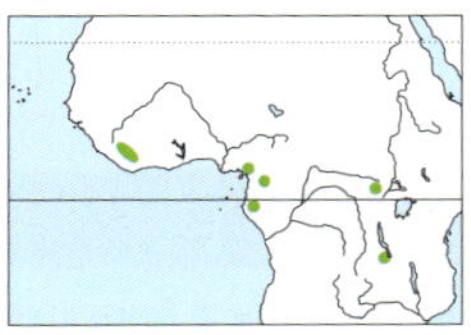

Very patchily in tropical forests of GN, LR, CI, CM, GA, NE CD & NE ZM

DD
Monotypic
HB 5.1–7.3 cm
T 5.5–6.4 cm
FA 4.5–4.9 cm
W 8–16 g

Malayan Slit-faced Bat
Nycteris tragata

FR Nyctère de Malaisie
DE Malaiische Schlitznase
ES Nícteris malayo

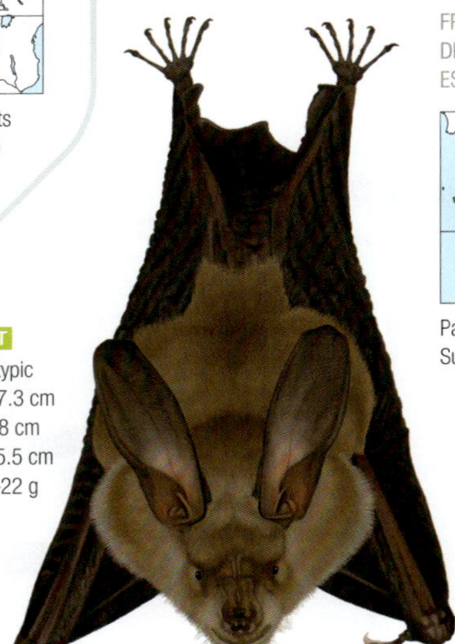

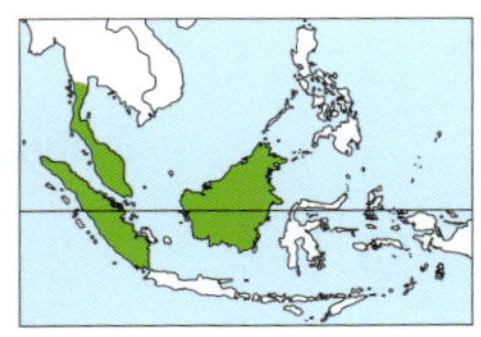

Patchily in Thai-Malay Pen, Sumatra & Borneo

NT
Monotypic
HB 6.8–7.3 cm
T 7–7.8 cm
FA 4.6–5.5 cm
W 13–22 g

Javan Slit-faced Bat
Nycteris javanica

FR Nyctère de Java
DE Java-Schlitznase
ES Nícteris de Java

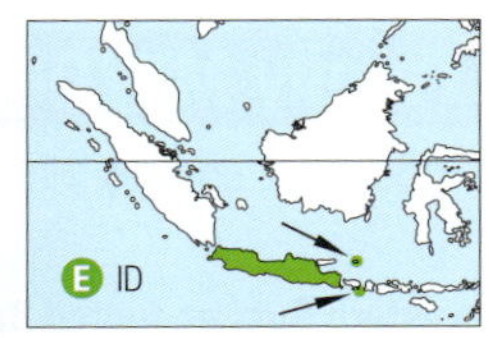

Java, Kangean Is & Nusa Penida I; very doubtful in West Timor

VU
Monotypic
FA 4.3–5 cm

Hairy Slit-faced Bat
Nycteris hispida

FR Nyctère hérissé
DE Gemeine Schlitznase
ES Nícteris erizado

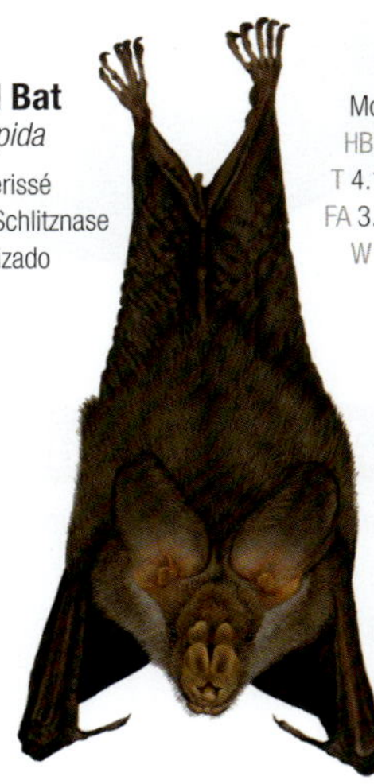

LC
Monotypic
HB 4–5 cm
T 4.1–5.6 cm
FA 3.4–4.4 cm
W 6–12 g

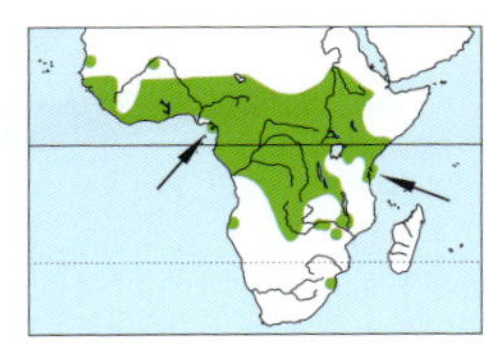

Tropical Africa from MR & SN E to ET, KE & SO, S to AO, BW, ZW & MZ, apparently isolated in E ZA

Andersen's Slit-faced Bat
Nycteris aurita

FR Nyctère à oreilles longues
DE Andersen-Schlitznase
ES Nícteris de Andersen

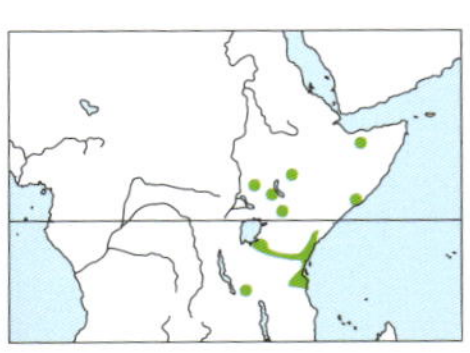

Patchily in savannas of E Africa (SO, ET, SS, KE & TZ)

LC
Monotypic
HB 4–5 cm
T 4.5–5.6 cm
FA 3.7–4.5 cm
W 6–12 g

Large Slit-faced Bat
Nycteris grandis

FR Grand Nyctère
DE Große Schlitznase
ES Nícteris grande

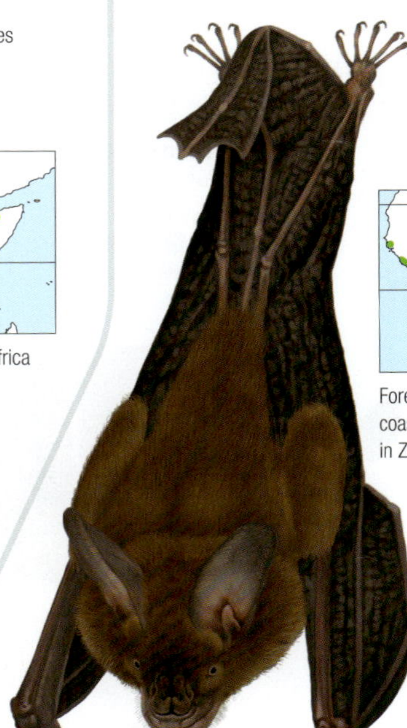

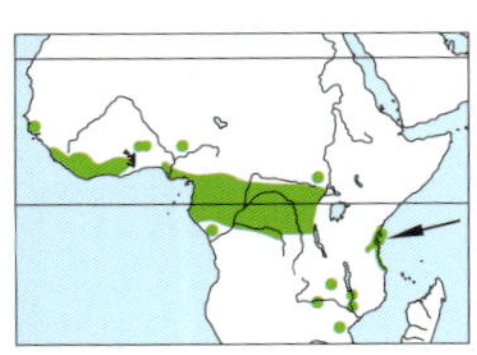

Forest belt from SN E to E CD, coastal KE, TZ & NE MZ, scattered in ZM, S MW, N ZW & WC MZ

LC
Monotypic
HB 7–9 cm
T 6–8.4 cm
FA 5.2–6.5 cm
W 21–43 g

Large-eared Slit-faced Bat
Nycteris macrotis

FR Nyctère de Dobson
DE Großohrige Schlitznase
ES Nícteris de orejas grandes

0–2200 m

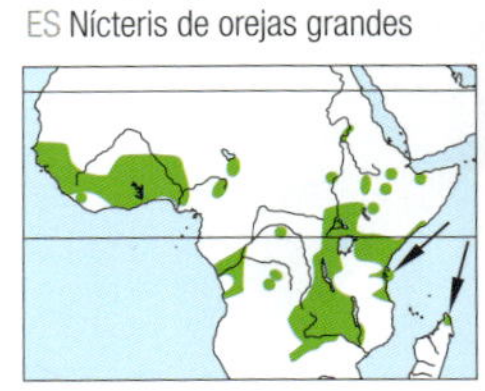

Tropical Africa from SN E to SD, ET, SO, KE & TZ, S to NW AO, S CD, NE BW, N ZW & C MZ, N MG

LC Monotypic
HB 4.5–7 cm T 4.5–6.5 cm
FA 4–5.5 cm W 15–18 g

Parisi's Slit-faced Bat
Nycteris parisii
FR Nyctère de Parisi
DE Parisi-Schlitznase
ES Nícteris de Parisi

DD
Monotypic
HB 4.6–4.7 cm
T 4.4–5 cm
FA 3.8–4.1 cm
W c. 7 g

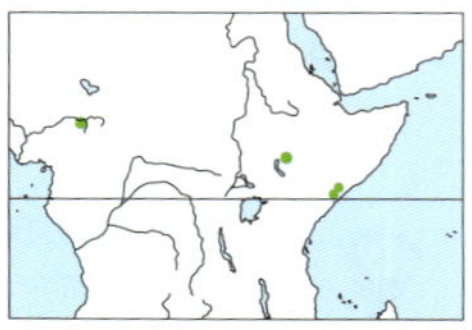

Known from just 4 sites in N CM, S ET & S SO

Wood's Slit-faced Bat
Nycteris woodi
FR Nyctère de Wood
DE Wood-Schlitznase
ES Nicteris de Wood

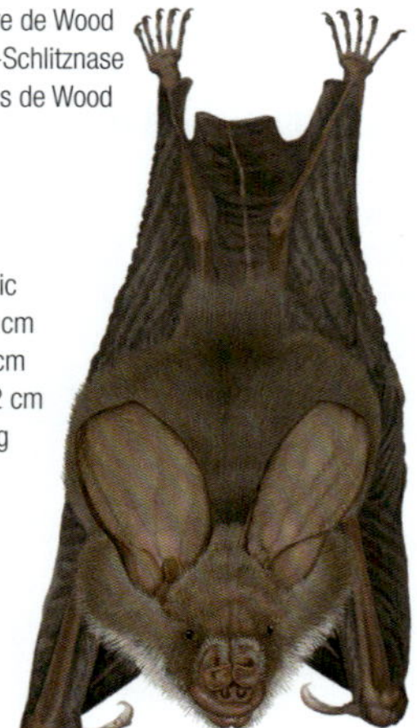

LC
Monotypic
HB 4–4.8 cm
T 4–4.5 cm
FA 3.5–4.2 cm
W 7–9 g

up to 1250 m

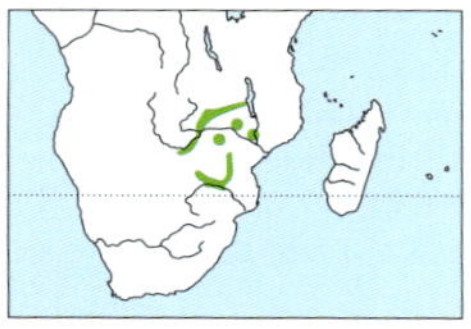

ZM, MW, ZW, C MZ, extreme NE ZA (Limpopo)

Gambian Slit-faced Bat
Nycteris gambiensis
FR Nyctère de Gambie
DE Gambia-Schlitznase
ES Nícteris gambiano

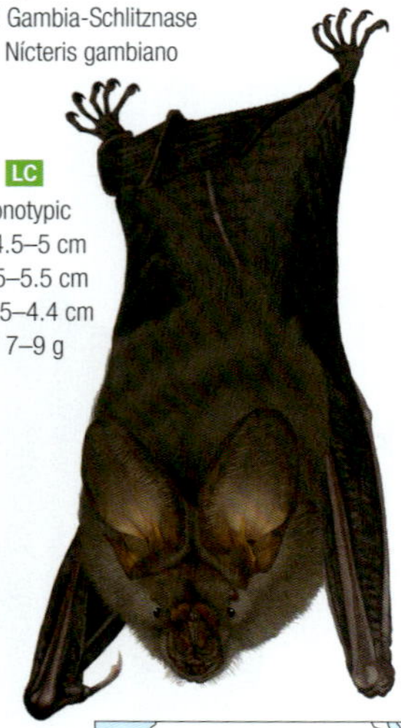

LC
Monotypic
HB 4.5–5 cm
T 4.5–5.5 cm
FA 3.5–4.4 cm
W 7–9 g

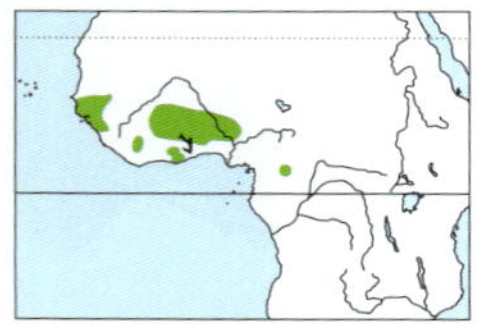

W Africa from SN E to NG, recorded in SE CM

Egyptian Slit-faced Bat
Nycteris thebaica
FR Nyctère de Thébaïde
DE Ägyptische Schlitznase
ES Nicteris egipcio

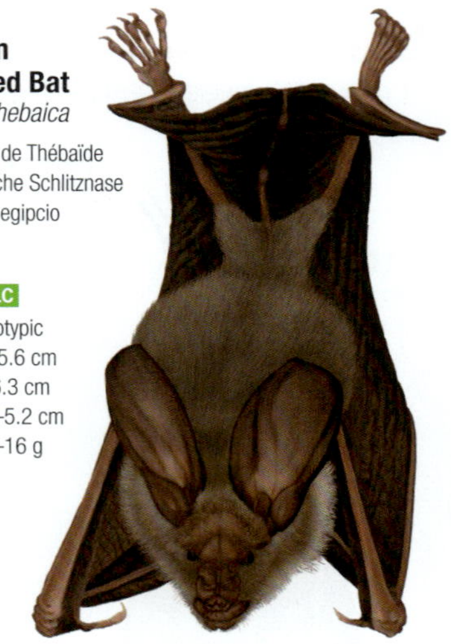

LC
Monotypic
HB 4–5.6 cm
T 4–6.3 cm
FA 3.4–5.2 cm
W 6–16 g

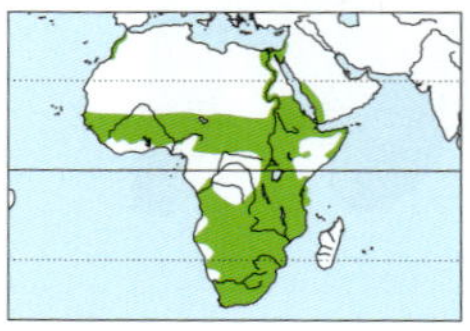

0–1500 m

From SN E to SO, S to ZA (absent in most of Congo Basin), N along R Nile to Levant, also MA, W SA, W YE

Family MYZOPODIDAE – Madagascar Sucker-footed Bats

1 genus, 2 species. Humid and dry deciduous forests, forest edges, open areas with palms, and often marshy areas. Afrotropical Region (Madagascar).

Genus *MYZOPODA*

(2 species)

Western Sucker-footed Bat
Myzopoda schliemanni
FR Myzopode de Schliemann
DE Schliemann-Haftscheibenfledermaus
ES Mizopoda de Schliemann

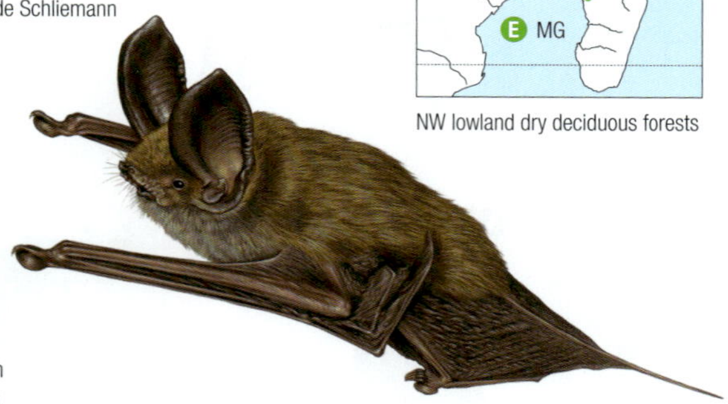

LC
Monotypic
HB 4.8–6 cm
T 4.4–4.7 cm
FA 4.5–4.9 cm
W 7.8–10.3 g

0–200 m

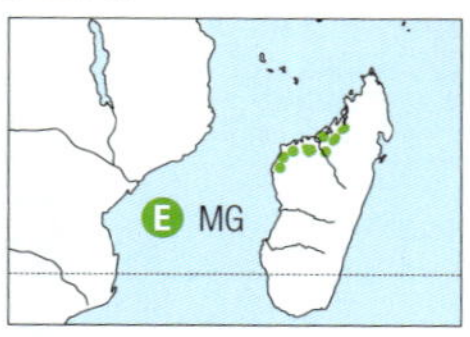

NW lowland dry deciduous forests

Eastern Sucker-footed Bat
Myzopoda aurita
FR Myzopode de Madagascar
DE Madagaskar-Haftscheibenfledermaus
ES Mizopoda de Madagascar

LC
Monotypic
HB 6.1–6.5 cm
T 4.0–5.3 cm
FA 4.6–5.2 cm
W 7.2–10.5 g

0–1000 m

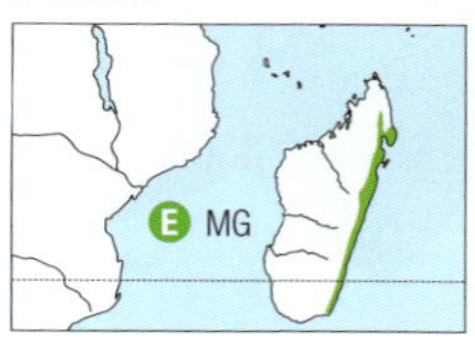

E lowland and littoral forests

Family MYSTACINIDAE – New Zealand Short-tailed Bats

1 genus, 2 species. Primarily interiors of indigenous New Zealand forests from sea level to high elevations. Australasian Region (New Zealand).

Genus *MYSTACINA*

(2 species)

Lesser New Zealand Short-tailed Bat
Mystacina tuberculata
FR Petite Mystacine
DE Kleine Neuseelandfledermaus
ES Murciélago neozelandés pequeño

VU
3 sspp
HB 6–7 cm
T 0.6–0.8 cm
FA 3.7–4.7 cm
W 10–22 g

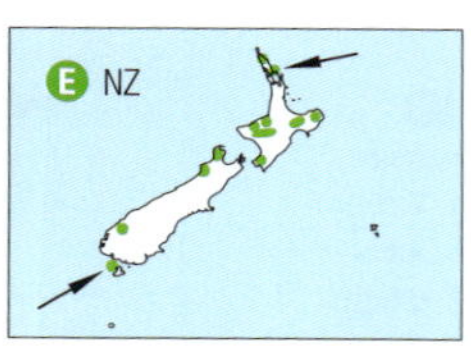

North Island (with Little Barrier I), South Island, Codfish I

Greater New Zealand Short-tailed Bat
Mystacina robusta
FR Grande Mystacine
DE Große Neuseelandfledermaus
ES Murciélago neozelandés grande

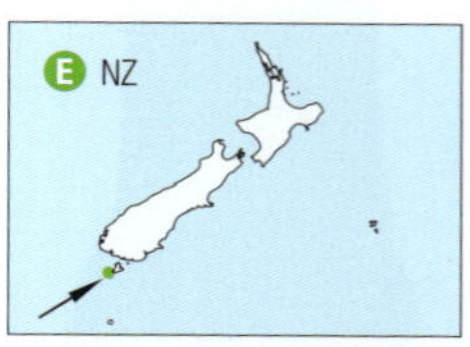

Might still occur on Big South Cape I & nearby Putauhina I or other Is off Stewart I; perhaps extinct

CR(PE)
Monotypic
HB 7–9 cm
T c. 1.5 cm
FA 4.5–4.7 cm
W c. 24 g

Family NOCTILIONIDAE – Bulldog Bats

1 genus, 2 species. Forests and savannas, as long as there are some trees for roosting and water bodies for foraging. Neotropical and marginally Nearctic Regions.

Genus *NOCTILIO*

(2 species)

Greater Bulldog Bat
Noctilio leporinus
FR Grand Noctilion
DE Großes Hasenmaul
ES Noctilio grande

0–500 m

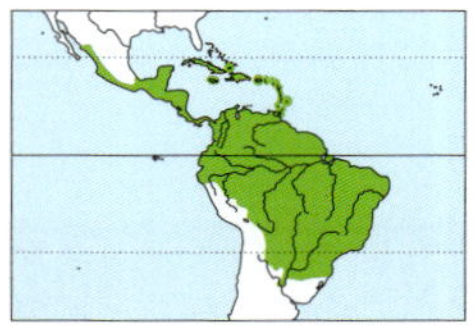

West Indies, W & S MX S to W EC, NW PE, E BO, PY, N AR & S BR

LC
3 sspp
HB 8.2–10 cm
T 2.3–2.8 cm
FA 8.1–8.7 cm
W 50–90 g

Lesser Bulldog Bat
Noctilio albiventris
FR Petit Noctilion
DE Kleines Hasenmaul
ES Noctilio pequeño

0–1100 m

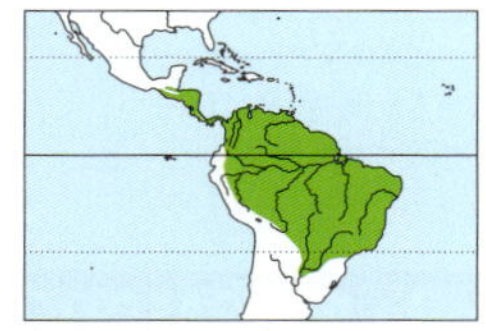

Extreme S MX (Chiapas) S to E EC, E PE, BO, PY, NE AR & S BR

LC
4 sspp
HB 6.5–6.8 cm
T 1.3–1.6 cm
FA 5.4–7 cm
W 22–42 g

Family FURIPTERIDAE – Smoky Bat and Thumbless Bat

2 genera, 2 species. Deserts into tropical deciduous and evergreen forests. Neotropical Region.

Genus *AMORPHOCHILUS*

(1 species)

Smoky Bat
Amorphochilus schnablii
FR Furie de Schnabl
DE Felsenküsten-Stummeldaumen
ES Murciélago ahumado

0–2680 m

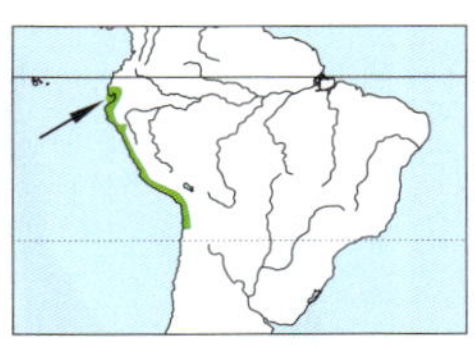

Narrow strip W of Andes in C & S EC (with Puná I), PE & N CL

VU
Monotypic
HB 3.9–4.7 cm
T 2.7–3.4 cm
FA 3.4–3.8 cm
W 3–10 g

Genus *FURIPTERUS*

(1 species)

Thumbless Bat
Furipterus horrens
FR Furiptère hérissé
DE Gemeiner Stummeldaumen
ES Murciélago sin pulgar

LC
Monotypic
HB 3.3–4.1 cm
T 2–3.8 cm
FA 3.5–3.8 cm
W 3.1–4.2 g

250–1200 m

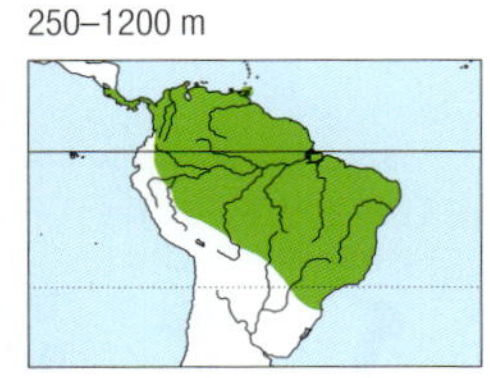

From SW NI (Refugio Bartola) S to E EC, E PE, N BO & BR, TT (Trinidad)

Family THYROPTERIDAE – Disk-winged Bats

1 genus, 5 species. Lowland tropical forests. Neotropical Region.

Genus *THYROPTERA*

(5 species)

DD
Monotypic
HB 3.8–4.6 cm
T 2.5–2.9 cm
FA 3.6–3.8 cm
W c. 5 g

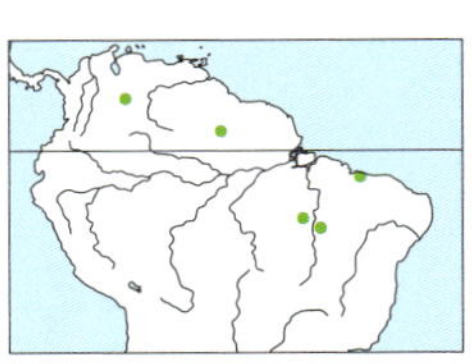

Scattered in CO (Llanos), SW GY & NE BR; probably wider range

De Vivo's Disk-winged Bat
Thyroptera devivoi
FR Thyroptère de De Vivo
DE De-Vivo-Haftscheibenfledermaus
ES Tiróptero de De Vivo

LC
Monotypic
HB 4.1–4.7 cm
T 2.7–3.5 cm
FA 3.2–3.4 cm

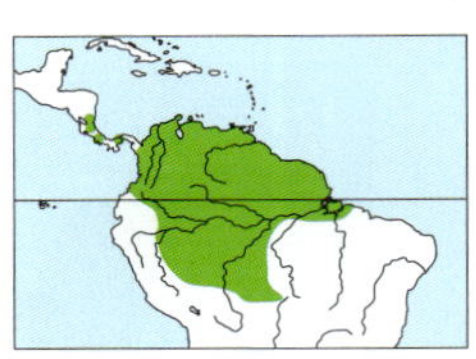

Scattered in S NI, NE & SE CR & C PA, South America from CO E & S to N & WC BR, E PE & N BO

Peters's Disk-winged Bat
Thyroptera discifera
FR Thyroptère de Peters
DE Peters-Haftscheibenfledermaus
ES Tiróptero de Peters

DD
Monotypic
HB 4.8–5.6 cm
T 3–3.1 cm
FA 3.7–4.1 cm
W 4–6 g

up to 700 m

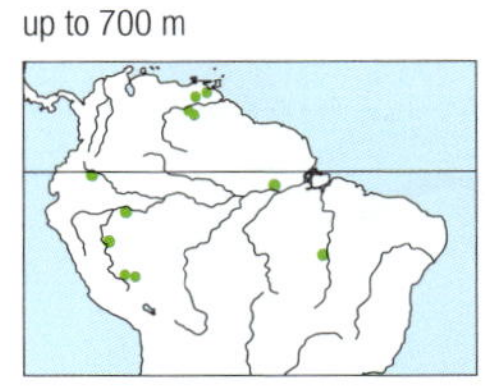

Scattered in E VE, E EC (Orellana), E PE & BR (Pará, Tocantins)

LaVal's Disk-winged Bat
Thyroptera lavali
FR Thyroptère de LaVal
DE LaVal-Haftscheibenfledermaus
ES Tiróptero de LaVal

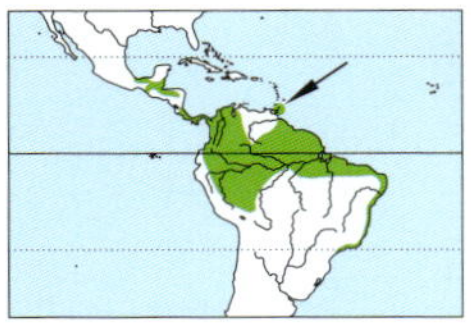

Disjunct from S MX to PA, widely from CO E to E & SE BR, S to E PE & N BO, TT (Trinidad)

LC
Monotypic
HB 4.2–4.9 cm
T 2.5–3 cm
FA 3.3–4 cm
W 3.4–5.1 g

Spix's Disk-winged Bat
Thyroptera tricolor
FR Thyroptère tricolore
DE Spix-Haftscheibenfledermaus
ES Tiróptero tricolor

up to 300 m

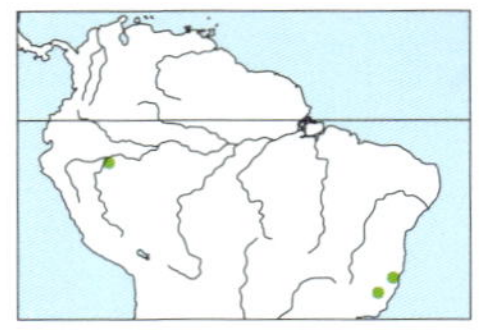

Known from 3 sites in NE PE (Yavarí & Ucayali interfluve) & SE BR (Rio Doce State Park, Sooretama R)

DD
Monotypic
HB 3.8–4.2 cm
T 2.6–2.7 cm
FA 3.3–3.4 cm

Patricia's Disk-winged Bat
Thyroptera wynneae
FR Thyroptère de Wynne
DE Patricia-Haftscheibenfledermaus
ES Tiróptero de Wynne

Family MORMOOPIDAE – Ghost-faced Bats, Naked-backed Bats and Mustached Bats

2 genera, 18 species. Arid and semi-arid regions to coastal lowlands, tropical rainforests, and montane forests up to elevations of 2700 m. Nearctic and Neotropical Regions.

Genus *MORMOOPS*

(2 species)

LC Monotypic
HB 5.2–5.6 cm
T 2.8–3 cm
FA 4.3–5 cm
W 6–11 g

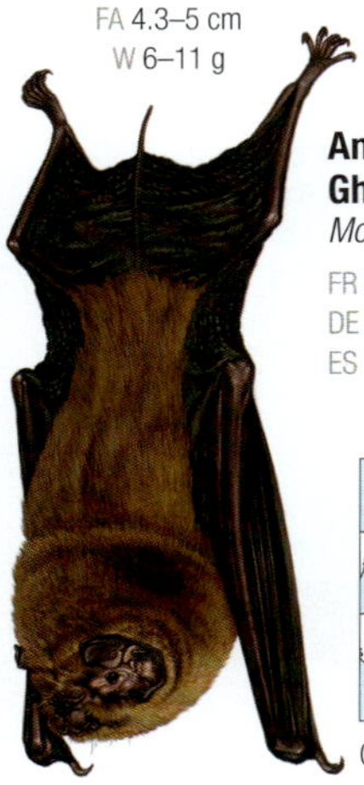

Antillean Ghost-faced Bat
Mormoops blainvillei
FR Mormoops de Blainville
DE Antillen-Kinnblattfledermaus
ES Mormópido de Blainville

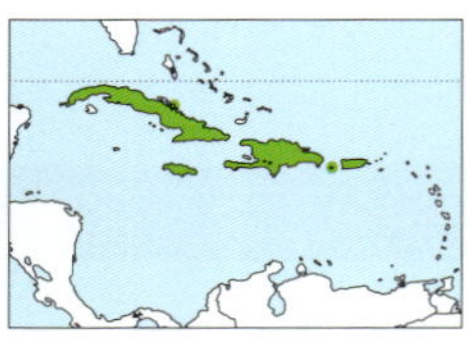

CU, JM, HT, DO, PR (with Mona I)

LC
4 sspp
HB 5.6–7.3 cm
T 1.8–3.1 cm
FA 5–6.1 cm
W 11–22 g

Peters's Ghost-faced Bat
Mormoops megalophylla
FR Mormoops de Peters
DE Peters-Kinnblattfledermaus
ES Mormópido de Peters

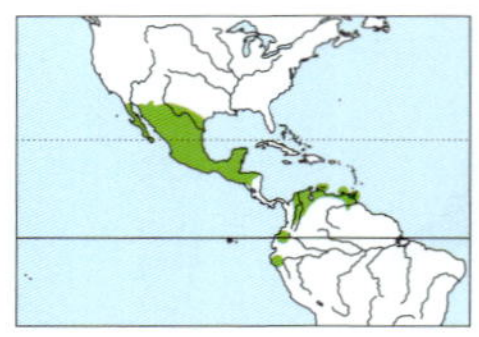

S US (SW Arizona, Texas) & Baja California (MX) S to NI, CO, EC, NW PE & VE, also AW to TT (Trinidad)

LC
2 sspp
HB 5–5.7 cm
T 1.8–2.5 cm
FA 4.6–5 cm
W 8–11 g

Davy's Naked-backed Bat
Pteronotus davyi
FR Ptéronote de Davy
DE Davy-Nacktrückenfledermaus
ES Pteronotus de Davy

up to 2700 m

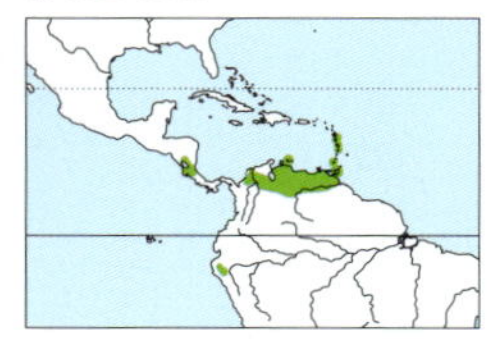

NI, CR, N CO, N VE, NW PE, Lesser Antilles (Marie-Galante S to Trinidad, CW)

Genus *PTERONOTUS*

(16 species)

NE Monotypic
HB 4.8–5.2 cm T 2–2.3 cm
FA 4–4.7 cm W 7–8.7 g

Thomas's Naked-backed Bat
Pteronotus fulvus
FR Ptéronote fauve
DE Thomas-Nacktrückenfledermaus
ES Pteronotus de Thomas

below 600 m

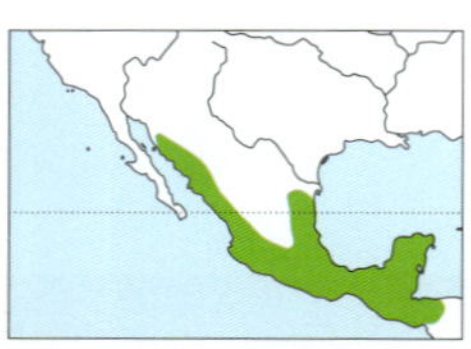

MX (from Sonora & Tamaulipas) S to SV & W HN

up to 1000 m

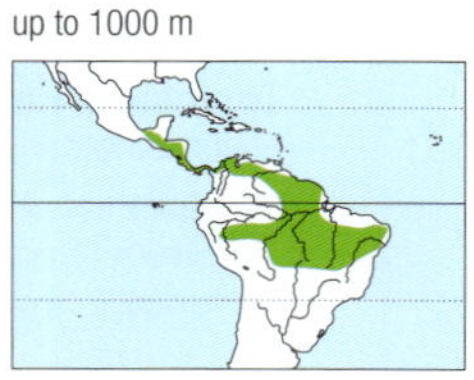

S MX S along Central America, Caribbean coast of South America & Guianas to C & NE BR, E PE & NE BO

Big Naked-backed Bat
Pteronotus gymnonotus
FR Ptéronote à dos nu
DE Große Nacktrückenfledermaus
ES Pteronotus de espalda desnuda

LC
Monotypic
HB 5.1–6.9 cm
T 1.7–2.7 cm
FA 5–5.6 cm
W 12–18 g

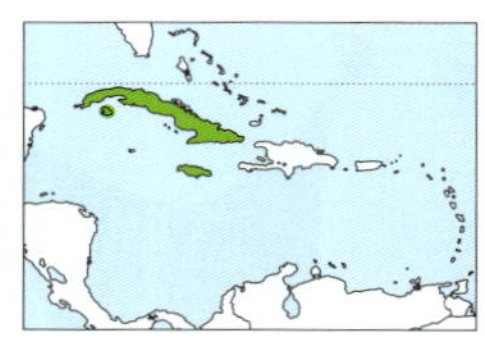

CU (with Isla de la Juventud), JM

Macleay's Mustached Bat
Pteronotus macleayii
FR Ptéronote de Macleay
DE Macleay-Schnurrbartfledermaus
ES Pteronotus de Macleay

LC
2 sspp
HB 4.1–5.3 cm
T 2–2.5 cm
FA 4.1–4.5 cm
W 4–8 g

LC 2 sspp
HB 4–4.4 cm T 1.6–2.2 cm
FA 3.5–4 cm W 3–6 g

Sooty Mustached Bat
Pteronotus quadridens
FR Ptéronote fuligineux
DE Rußfarbene Schnurrbartfledermaus
ES Pteronotus tiznado

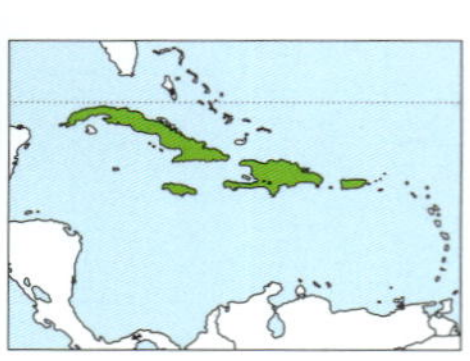

CU, JM, HT, DO, PR

LC
Monotypic
HB 4.5–5.5 cm
T 1.6–2.1 cm
FA 4.3–4.8 cm
W 6.5–10 g

Wagner's Lesser Mustached Bat
Pteronotus personatus
FR Ptéronote masqué
DE Kleine Wagner-Schnurrbartfledermaus
ES Pteronotus de Wagner

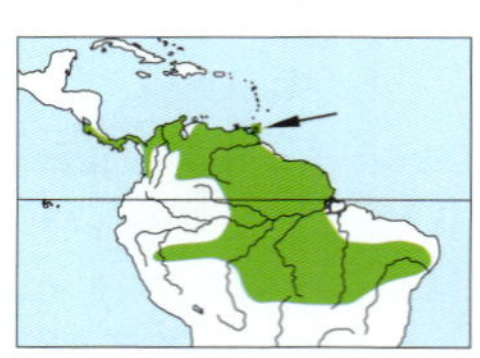

SW NI S through CR & PA to NE & C BR, E PE & NE BO, TT (Trinidad)

NE
Monotypic
HB 4.5–5.4 cm
T 1.5–1.8 cm
FA 4.1–4.5 cm
W 6.5–7.4 g

Dobson's Lesser Mustached Bat
Pteronotus psilotis
FR Ptéronote à oreilles nues
DE Kleine Dobson-Schnurrbartfledermaus
ES Pteronotus de Dobson

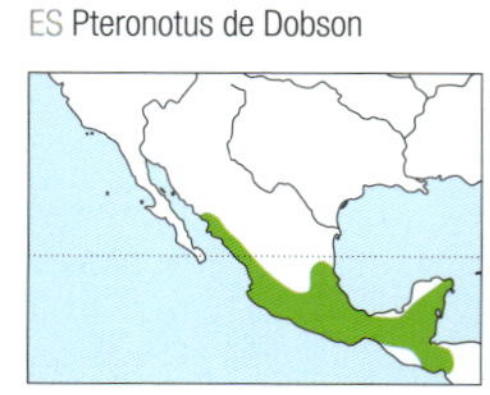

MX (from Sonora & Tamaulipas) S to SV & W HN

Parnell's Common Mustached Bat
Pteronotus parnellii

FR Ptéronote de Parnell
DE Parnell-Schnurrbartfledermaus
ES Pteronotus de Parnell

LC
Monotypic
HB c. 5.9 cm
T 1.8–2.2 cm
FA 5–5.4 cm
W 10.5–16 g

0–1300 m

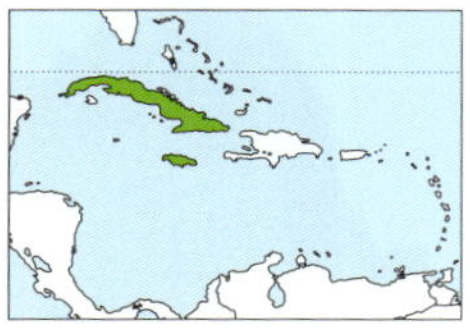

CU, JM

Hispaniolan Common Mustached Bat
Pteronotus pusillus

FR Ptéronote d'Hispaniola
DE Hispaniola-Schnurrbartfledermaus
ES Pteronotus de Hispaniola

NE
Monotypic
HB c. 5.3 cm
T c. 2 cm
FA 4.9–5.3 cm

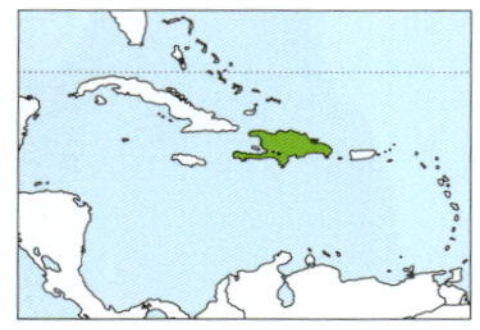

HT, DO

Puerto Rican Common Mustached Bat
Pteronotus portoricensis

FR Ptéronote de Porto Rico
DE Puerto-Rico-Schnurrbartfledermaus
ES Pteronotus de Puerto Rico

NE
Monotypic
HB 5.8–6.9 cm
T 1.3–2.2 cm
FA 5–5.4 cm
W 11.3–14.3 g

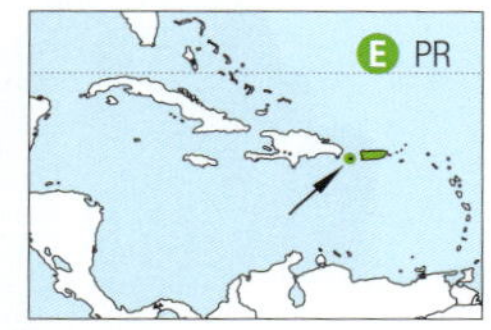

Including Mona I

Mexican Common Mustached Bat
Pteronotus mexicanus

FR Ptéronote du Mexique
DE Mexiko-Schnurrbartfledermaus
ES Pteronotus mexicano

NE
Monotypic
HB 6.4–7 cm
T 1.9–2.4 cm
FA 5.3–5.8 cm
W c. 13.5 g

up to 1600 m

Lowlands along Pacific coast from Sonora to Guerrero, E to Puebla

Mesoamerican Common Mustached Bat
Pteronotus mesoamericanus

FR Ptéronote d'Amérique Centrale
DE Mittelamerika-Schnurrbartfledermaus
ES Pteronotus de América Central

LC
Monotypic
HB 7–7.2 cm
T 2.3–2.7 cm
FA 5.5–6.3 cm
W 16–19.5 g

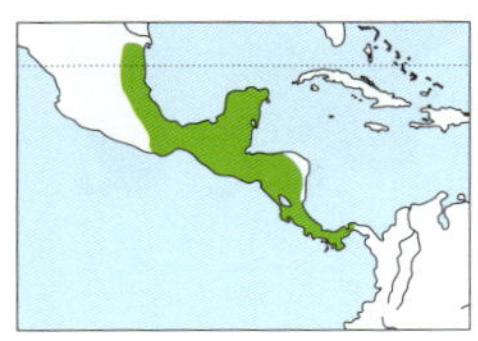

MX (from along Gulf Coast & Isthmus of Tehuantepec) S to Canal Zone (PA)

Allen's Common Mustached Bat
Pteronotus fuscus

FR Ptéronote des Caraïbes
DE Allen-Schnurrbartfledermaus
ES Pteronotus de Allen

NE
Monotypic
HB 6.3–7.2 cm
T 2–2.7 cm
FA 5.6–6.2 cm
W 13.4–22 g

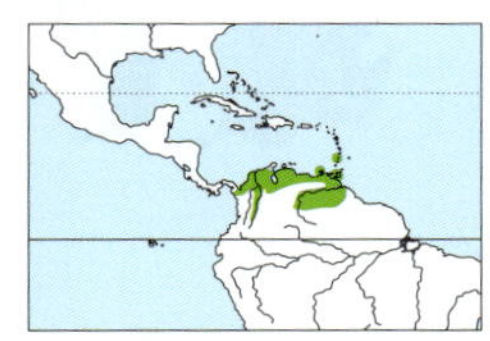

CO, VE, NW GY, Lesser Antilles (St Vincent, Trinidad, Margarita)

Paraguana Common Mustached Bat
Pteronotus paraguanensis

FR Ptéronote de Paraguana
DE Paraguaná-Schnurrbartfledermaus
ES Pteronotus de Paraguaná

EN
Monotypic
HB 6.1–6.4 cm
T 1.7–2.6 cm
FA 5.3–5.6 cm
W 12–17 g

0–120 m

Paraguaná Pen (Falcón)

Wagner's Common Mustached Bat
Pteronotus rubiginosus

FR Ptéronote rubigineux
DE Wagner-Schnurrbartfledermaus
ES Pteronotus rufo

LC
Monotypic
HB 6.3–7.9 cm
T 2.4–3.1 cm
FA 6.1–6.7 cm
W 20–27 g

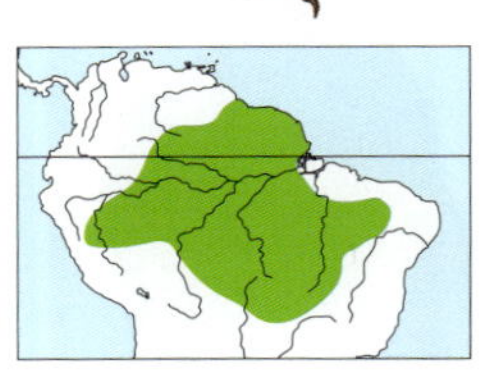

S VE & Guianas S to C & NE BR, E PE & NE BO

Amazonian Common Mustached Bat
Pteronotus alitonus

FR Ptéronote alitone
DE Amazonas-Schnurrbartfledermaus
ES Pteronotus bigotudo amazónico

NE
Monotypic
HB 6.5–7 cm
T 2.4–3.1 cm
FA 5.9–6.4 cm
W 18–26 g

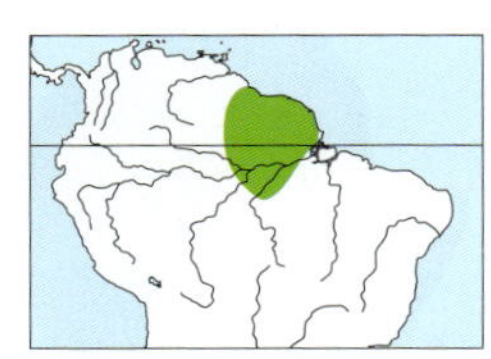

Guianas, N BR

Family PHYLLOSTOMIDAE – New World Leaf-nosed Bats

62 genera, 226 species. Tropical rainforests, tropical dry forests, subtropical cloud forests, transformed landscapes, and deserts. Nearctic and Neotropical Regions.

Genus *MACROTUS*

(2 species)

Californian Leaf-nosed Bat

Macrotus californicus

LC
Monotypic
HB 8.2–10.2 cm
T 2.7–4.4 cm
FA 4.4–5.3 cm
W 11–15 g

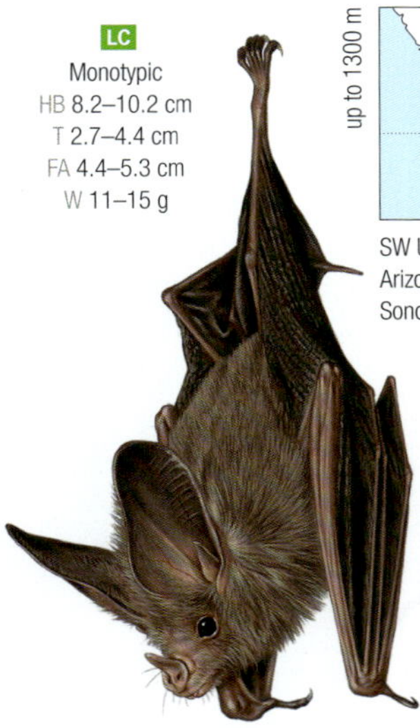

up to 1300 m

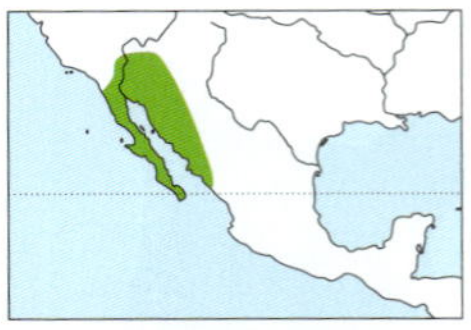

SW US (S California, S Nevada, Arizona), NW MX (Baja California, Sonora, SW Chihuahua, N Sinaloa)

FR Macrotus de Californie
DE Kalifornien-Großohrblattnase
ES Macroto de California

Waterhouse's Leaf-nosed Bat

Macrotus waterhousii

LC
6 sspp
HB 8.5–10.8 cm
T 2.8–4.7 cm
FA 4.7–5.6 cm
W 12–18 g

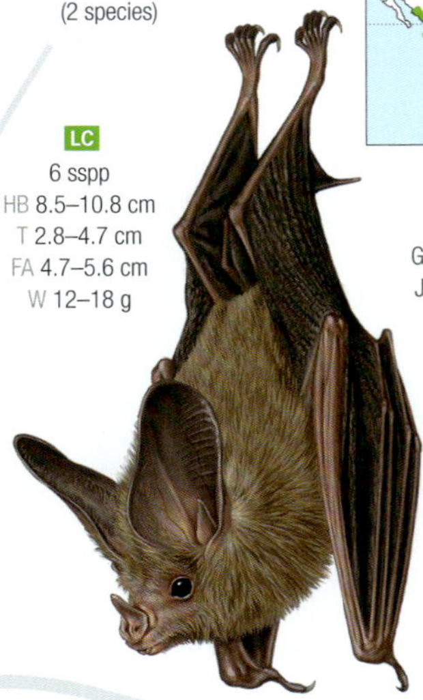

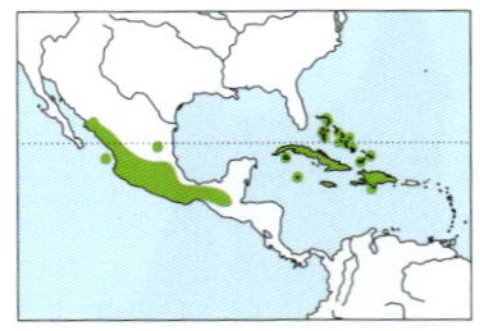

up to 1300 m

NW MX (with Marías I) S to GT, BS, TC, CU (with Isla de la Juventud), JM, KY, Hispaniola (with Beata I)

FR Macrotus de Waterhouse
DE Waterhouse-Großohrblattnase
ES Macroto de Waterhouse

Genus *LAMPRONYCTERIS*

(1 species)

Orange-throated Bat

Lampronycteris brachyotis

LC
Monotypic
HB 5–6.1 cm
T 0.7–1.4 cm
FA 3.9–4.3 cm
W 9–15 g

below 525 m

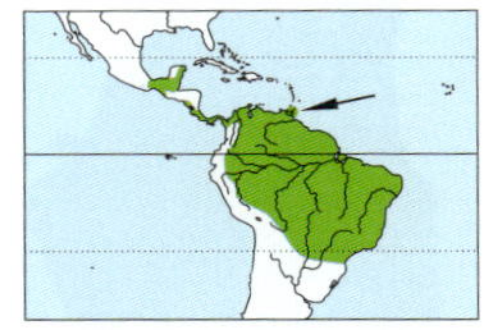

S MX (from S Veracruz & Oaxaca) S to E EC, E PE, BO & BR (to São Paulo), TT (Trinidad)

FR Lampronyctère à oreilles courtes
DE Gelbbauch-Blattnase
ES Lampronicteris de garganta naranja

Genus *MICRONYCTERIS*

(13 species)

Tiny Big-eared Bat

Micronycteris minuta

LC
Monotypic
HB 4.2–5.8 cm
T 0.7–1.5 cm
FA 3.1–4 cm
W 5–8.5 g

FR Micronyctère menu
DE Zwerggroßohrblattnase
ES Micronícteris menudo

below 1600 m

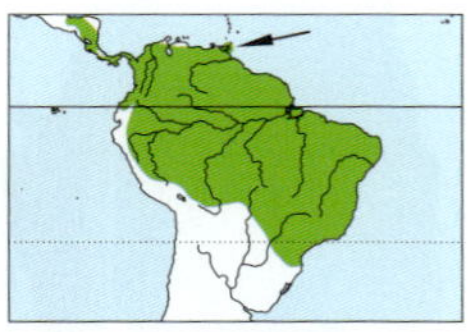

From extreme E GT (Sierra del Caral) to EC, E PE, BO & BR, TT (Trinidad)

Simmons's Big-eared Bat

Micronycteris simmonsae

NE
Monotypic
HB 4.6–5.8 cm
T 1.1–1.2 cm
W c. 5.9 g

FR Micronyctère de Simmons
DE Simmons-Großohrblattnase
ES Micronícteris de Simmons

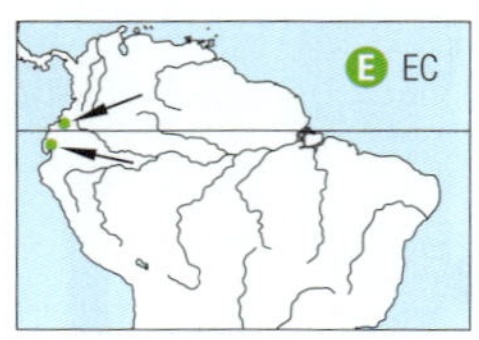

W Andes (Guayas, Esmeraldas)

Sanborn's Big-eared Bat

Micronycteris sanborni

LC
Monotypic
HB 3.7–4.9 cm
T 1.1–1.5 cm
FA 3.2–3.7 cm
W 5–8 g

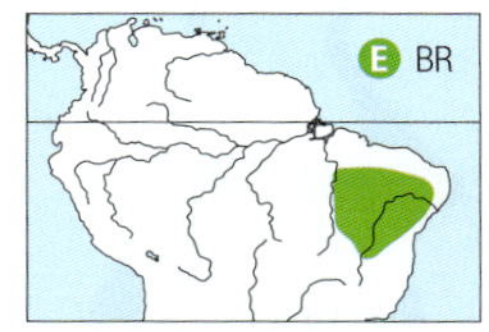

Maranhão, Piauí, Ceará, Paraíba, Pernambuco, Alagoas, Sergipe, Tocantins, Bahia, Minas Gerais

FR Micronyctère de Sanborn
DE Sanborn-Großohrblattnase
ES Micronícteris de Sanborn

Yates's Big-eared Bat

Micronycteris yatesi

DD
Monotypic
HB 4.8–6 cm
T 0.7–0.9 cm
FA 3.4–3.7 cm

FR Micronyctère de Yates
DE Yates-Großohrblattnase
ES Micronícteris de Yates

Zurima (Chuquisaca), Refugio Los Volcanes & Estancia Patuju (Santa Cruz); perhaps W BR

Northern Big-eared Bat

Micronycteris tresamici

NE
Monotypic
HB 4.2–5.3 cm
T 9–10.5 cm
W c. 5.4 g

FR Micronyctère du Honduras
DE Nördliche Großohrblattnase
ES Micronícteris de Honduras

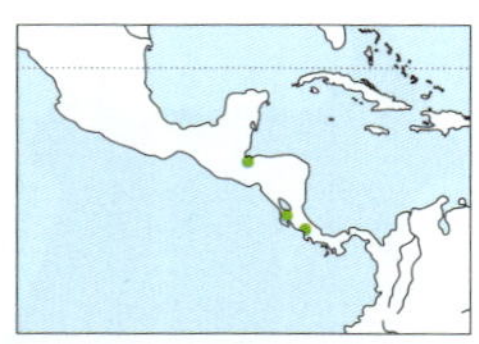

HN (Atlántida), CR (Guanacaste, Alajuela)

Hairy Big-eared Bat

Micronycteris hirsuta

FR Grand Micronyctère
DE Haarige Großohrblattnase
ES Micronícteris de orejas peludas

below 1500 m

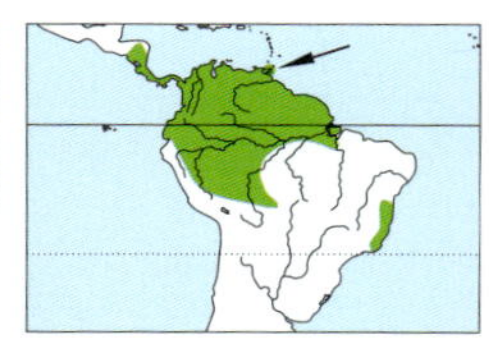

E HN S to EC, E PE, N BO, W & N BR, apparently disjunct in SE BR, TT (Trinidad)

LC
Monotypic
HB 5.4–7.5 cm
T 1.1–2 cm
FA 4.1–4.9 cm
W 10.3–18.4 g

Schmidts' Big-eared Bat

Micronycteris schmidtorum

LC
Monotypic
HB 5.3–6.7 cm
T 1–1.7 cm
FA 3.2–3.5 cm
W 5–10 g

FR Micronyctère des Schmidt
DE Schmidt-Großohrblattnase
ES Micronícteris de los Schmidt

10–160 m

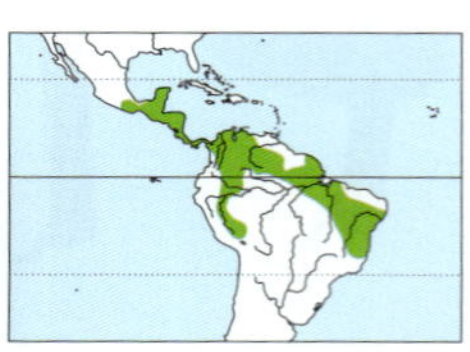

SE MX (Yucatán Pen, Chiapas) S to CO, VE, GF, E PE & BR; dubious in Cozumel I (MX)

Brosset's Big-eared Bat

Micronycteris brosseti

DD
Monotypic
HB 5.1–6.1 cm
T 1–1.4 cm
FA 3.1–3.4 cm
W 4.3–5 g

FR Micronyctère de Brosset
DE Brosset-Großohrblattnase
ES Micronícteris de Brosset

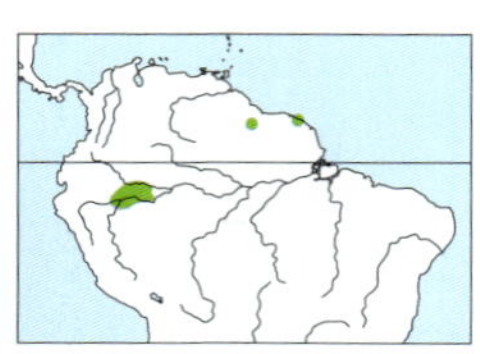

NE PE (Loreto), GY, GF, 1 possible record in SE BR (São Paulo); possibly widespread

Giovanni's Big-eared Bat

Micronycteris giovanniae

DD
Monotypic
HB c. 5.5 cm
T c. 1.6 cm
FA c. 3.7 cm
W c. 8.6 g

FR Micronyctère de Giovanni
DE Giovanni-Großohrblattnase
ES Micronícteris de Giovanni

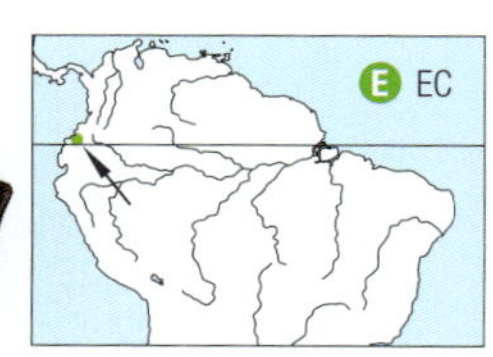

Known only from type locality in N Esmeraldas

Matses Big-eared Bat
Micronycteris matses

FR Micronyctère des Matses
DE Matses-Großohrblattnase
ES Micronícteris de Matses

DD
Monotypic
HB 5–5.5 cm
T 1.3–1.7 cm
FA 3.7–3.9 cm
W 8.6–14 g

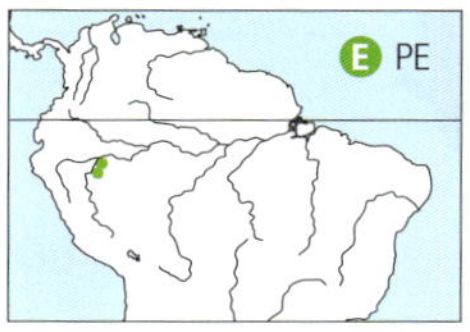

150–170 m

Known from Nuevo San Juan & Sierra del Divisor (Loreto); perhaps throughout Amazon Basin

Little Big-eared Bat
Micronycteris megalotis

FR Micronyctère oreillard
DE Kleine Großohrblattnase
ES Micronícteris común de Gray

LC
Monotypic
HB 3.8–5.9 cm
T 1–1.6 cm
FA 3.1–3.6 cm
W 5.5–6.3 g

up to 3000 m

N South America S to SE PE, N BO & SE BR, Lesser Antilles (Margarita, Grenada, Trinidad, Tobago)

Common Big-eared Bat
Micronycteris microtis

FR Micronyctère modeste
DE Gewöhnliche Großohrblattnase
ES Micronícteris común de Miller

LC
2 sspp
HB 4.4–4.8 cm
T 0.8–1.7 cm
FA 3–3.7 cm
W 5.5–10 g

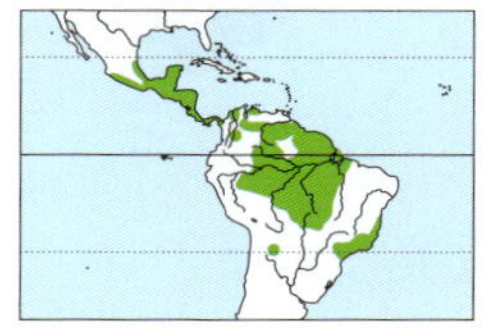

MX (S from Jalisco & Tamaulipas) to CO, VE, Guianas, E BO & BR, recorded in NW AR (Salta)

Saint Vincent Big-eared Bat
Micronycteris buriri

FR Micronyctère de Saint-Vincent
DE St.-Vincent-Großohrblattnase
ES Micronícteris de Saint Vincent

DD
Monotypic
HB 6.6–7.2 cm
T 1.2–1.8 cm
FA 3.6–3.9 cm

0–645 m

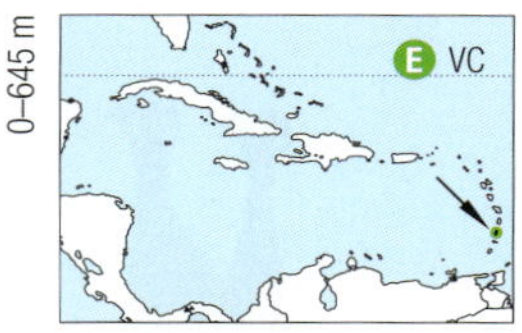

St Vincent I

Genus *DESMODUS*

(1 species)

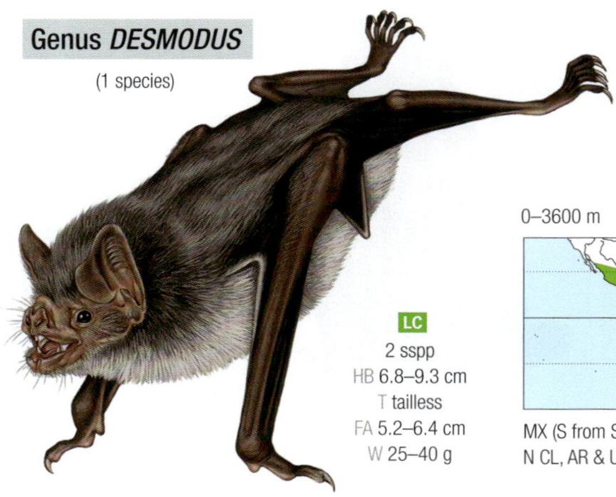

Common Vampire Bat
Desmodus rotundus

FR Vampire commun
DE Gewöhnlicher Vampir
ES Vampiro común

LC
2 sspp
HB 6.8–9.3 cm
T tailless
FA 5.2–6.4 cm
W 25–40 g

0–3600 m

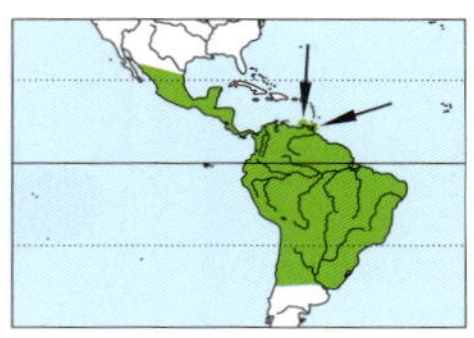

MX (S from Sonora & Tamaulipas) to N CL, AR & UY, Margarita & Trinidad Is

White-winged Vampire Bat
Diaemus youngii

Genus *DIAEMUS*

(1 species)

FR Vampire des oiseaux
DE Weißflügelvampir
ES Vampiro de alas blancas

LC
Monotypic
HB 8.2–8.5 cm
T tailless
FA 5.1–5.4 cm
W 31–48 g

0–500 m

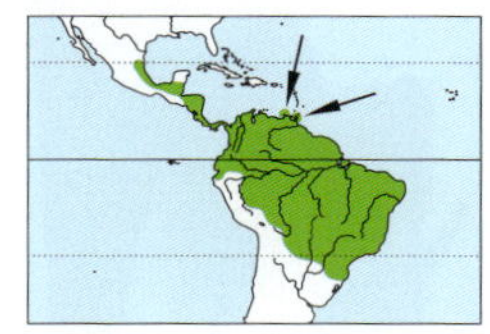

E MX (S from Tamaulipas) to E PE, BO, PY, S BR & N AR, Margarita & Trinidad Is

Genus *DIPHYLLA*

(1 species)

Hairy-legged Vampire Bat
Diphylla ecaudata

FR Vampire à pattes velues
DE Kammzahnvampir
ES Vampiro de patas peludas

LC
2 sspp
HB 6.5–9.3 cm
T tailless
FA 4.9–5.6 cm
W 20–43 g

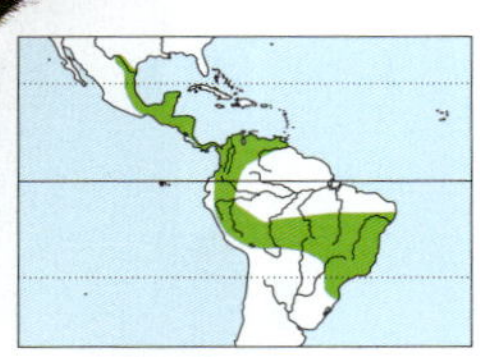

From S Texas (US) through E MX to EC, PE, N BO & BR (except C Amazon Basin)

LC
2 sspp
HB 5.3–6.7 cm
T 4.2–6.5 cm
FA 4.7–5.7 cm
W 12–17 g

Genus *LONCHORHINA*

(6 species)

Common Sword-nosed Bat
Lonchorhina aurita

FR Lonchorhine de Tomes
DE Eigentliche Schwertnase
ES Loncorino de Tomes

0–1500 m

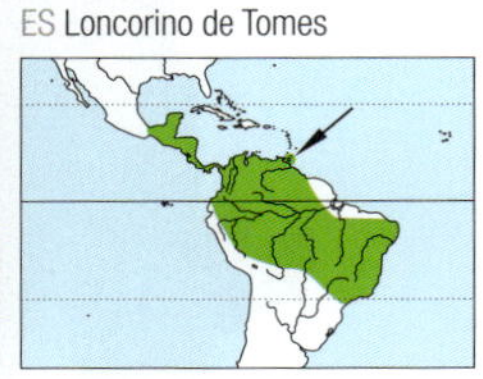

MX (S from Veracruz, Oaxaca) to EC, E PE, N BO & SE BR, TT (Trinidad)

EN
Monotypic
HB 5.3–5.7 cm
T 4.1–4.8 cm
FA 4.1–4.4 cm
W 9.7–10 g

70–90 m

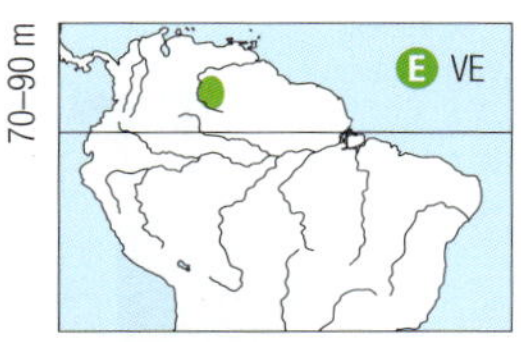

Known from vicinity of type locality along R Orinoco (Bolívar, Amazonas)

Fernandez's Sword-nosed Bat
Lonchorhina fernandezi

FR Lonchorhine de Fernandez
DE Fernandez-Schwertnase
ES Loncorino de Fernández

DD
Monotypic
HB 5.7–7.2 cm
T 5.4–6.7 cm
FA 5.2–5.7 cm
W 14–19 g

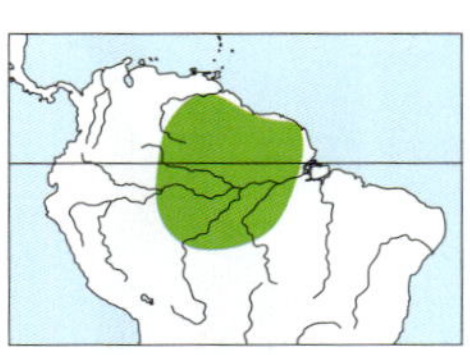

S VE, Guianas, N BR

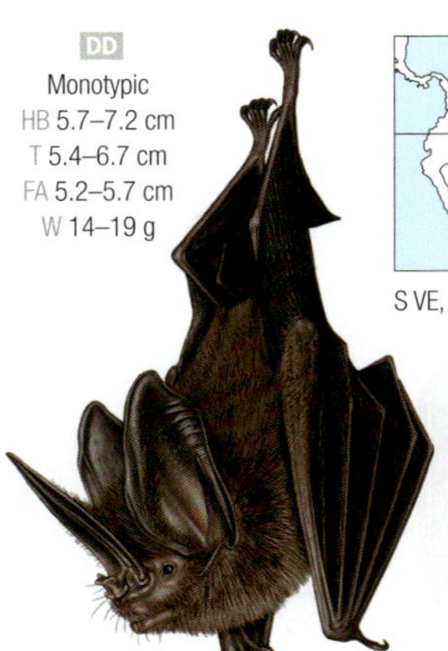

Uncommon Sword-nosed Bat
Lonchorhina inusitata

FR Lonchorhine de Guyane
DE Nördliche Schwertnase
ES Loncorino de Guayana Francesa

VU
Monotypic
HB 4.7–5.4 cm
T 4.7–5.7 cm
FA 4–4.4 cm
W 8–12 g

Orinoco Sword-nosed Bat
Lonchorhina orinocensis

FR Lonchorhine de l'Orénoque
DE Orinoco-Schwertnase
ES Loncorino de Orinoco

below 300 m

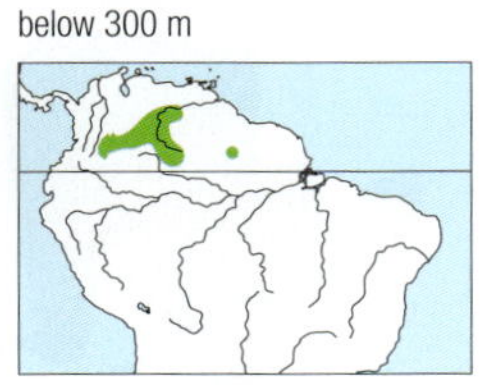

E CO, S VE, SW GY

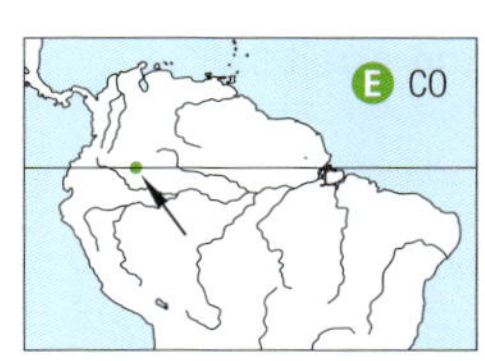

Known from type locality in SE Chiribiquete NP

NE
Monotypic
HB 7.4–7.6 cm
T 5.7–6.9 cm
FA 6.3–6.5 cm
W c. 33 g

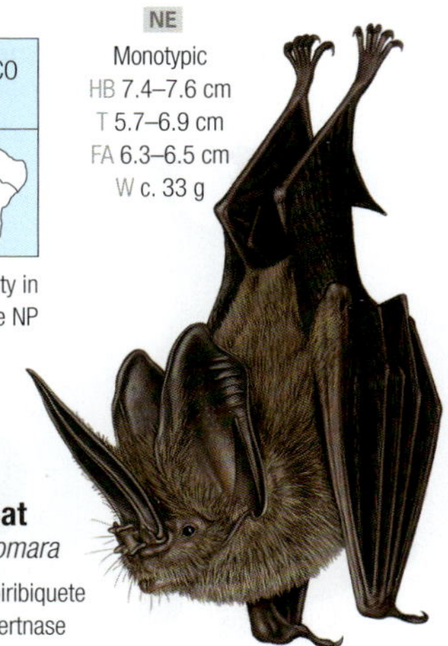

Chiribiquete Sword-nosed Bat
Lonchorhina mankomara

FR Lonchorhine de Chiribiquete
DE Chiribiquete-Schwertnase
ES Loncorino de Chiribiquete

Marinkelle's Sword-nosed Bat
Lonchorhina marinkellei

VU
Monotypic
HB 6.4–7.3 cm
T 5.9–7 cm
FA 5.9–6.3 cm

FR Lonchorhine de Marinkelle
DE Marinkelle-Schwertnase
ES Loncorino de Marinkelle

180–240 m

Guiana Shield (Guaviare, Vaupés)

Genus *MACROPHYLLUM*

(1 species)

Long-legged Bat
Macrophyllum macrophyllum

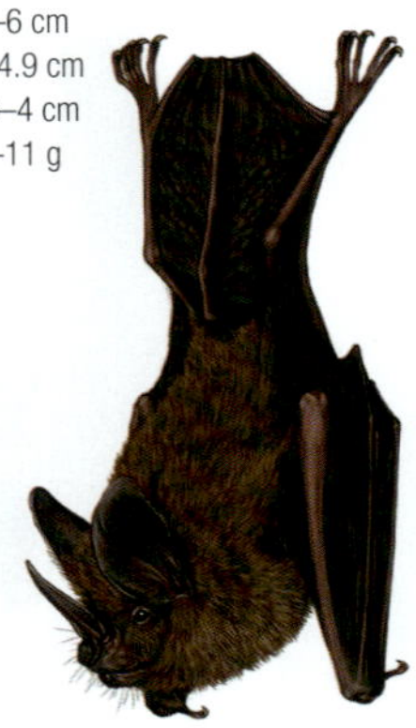

LC
Monotypic
HB 4–6 cm
T 3.8–4.9 cm
FA 3.3–4 cm
W 6–11 g

FR Macrophylle pêcheur
DE Langbeinfledermaus
ES Macrófilo de patas largas

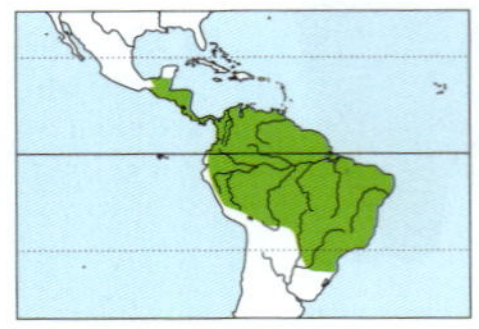

Extreme S MX (Tabasco, E Chiapas) S to E PE, N & E BO, PY, extreme NE AR & BR

Atlantic Forest Fringe-lipped Bat
Trachops ehrhardti

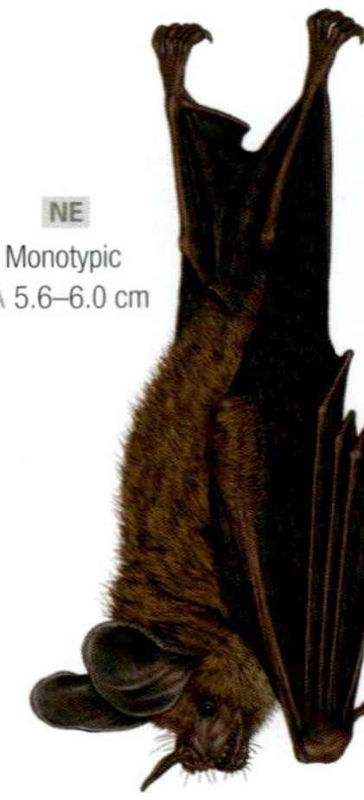

NE
Monotypic
FA 5.6–6.0 cm

0–500 m

Atlantic Forest of SE

FR Trachope d'Ehrhardt
DE Atlantische Fransenlippenfledermaus
ES Tracops de Ehrhardt

Genus *GARDNERYCTERIS*

(3 species)

Striped Hairy-nosed Bat
Gardnerycteris crenulatum

LC
3 sspp
HB 8–9.1 cm
T 2.1–2.7 cm
FA 4.8–5.2 cm
W 12–16 g

FR Gardnéryctère crénelé
DE Streifen-Haarblattnase
ES Gardnericterio de nariz peluda

E & S CO, VE, TT (Trinidad), Guianas, BR, E EC, E PE, N BO

Keenan's Hairy-nosed Bat
Gardnerycteris keenani

NE
Monotypic
HB 4–6.6 cm
T 2.3–2.4 cm
FA 4.8–5.2 cm
W 12–16 g

FR Gardnéryctère de Keenan
DE Keenan-Haarblattnase
ES Gardnericterio de Keenan

0–600 m

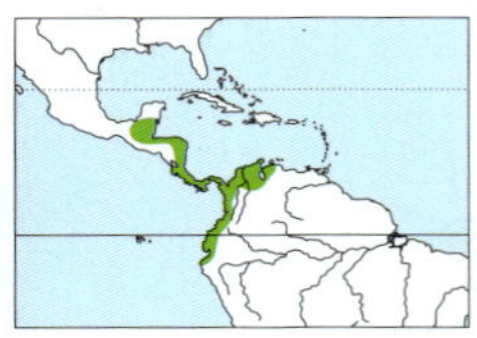

From S MX through Caribbean slope to PA & W of Andes in CO, NW VE, W EC & extreme NW PE

Genus *TRACHOPS*

(2 species)

Common Fringe-lipped Bat
Trachops cirrhosus

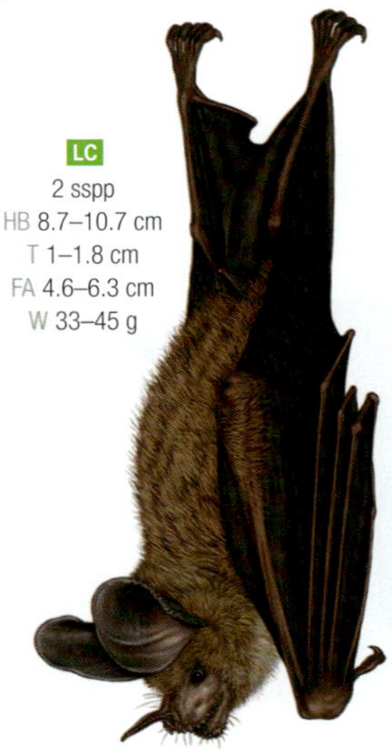

LC
2 sspp
HB 8.7–10.7 cm
T 1–1.8 cm
FA 4.6–6.3 cm
W 33–45 g

FR Trachope verruqueux
DE Fransenlippenfledermaus
ES Tracops de labios verrugosos

0–500 m

MX (from S Veracruz & Oaxaca) S to EC, PE, BO & BR, TT (Trinidad)

Koepcke's Hairy-nosed Bat
Gardnerycteris koepckeae

DD
Monotypic
HB c. 6.4 cm
T c. 1.6 cm
FA 4.5–5.1 cm
W 11–15 g

FR Gardnéryctère de Koepcke
DE Koepcke-Haarblattnase
ES Gardnericterio de Koepcke

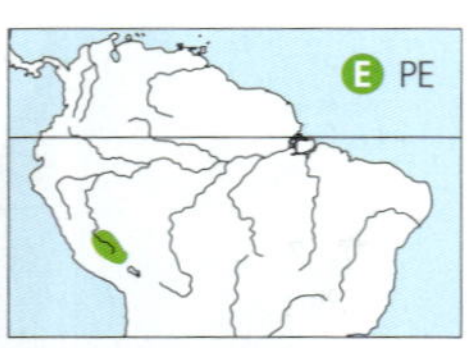

E Andes slope forests in Ayacucho & Junín

Genus *LOPHOSTOMA*

(7 species)

Kalko's Round-eared Bat
Lophostoma kalkoae

DD
Monotypic
T 0.8–0.9 cm
FA 4.4–4.6 cm

FR Lophostome de Kalko
DE Kalko-Rundohrblattnase
ES Lofostoma de Kalko

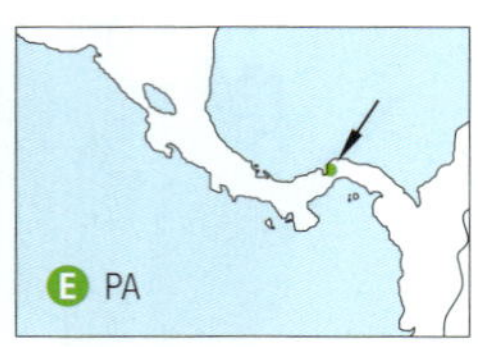

Known only from type locality in Soberanía NP (Colón)

Pygmy Round-eared Bat
Lophostoma brasiliense

LC
Monotypic
HB 4.2–6.1 cm
T 0.8–1.3 cm
FA 3.4–4 cm
W 9.9–13.8 g

FR Lophostome des marais
DE Kleine Rundohrblattnase
ES Lofostoma pigmeo

below 500 m

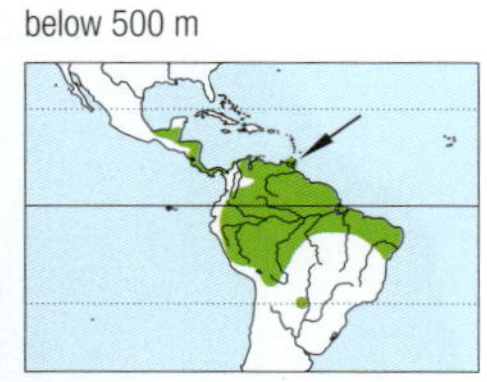

S MX (Veracruz, S Yucatán Pen) S to CO, VE, TT (Trinidad), Guianas, EC, PE, BR, BO, PY (isolated record)

Carriker's Round-eared Bat
Lophostoma carrikeri

LC
Monotypic
HB 6.6–7.6 cm
T 1–1.5 cm
FA 4.2–4.8 cm
W 18–25 g

FR Lophostome à ventre blanc
DE Carriker-Rundohrblattnase
ES Lofostoma de vientre blanco

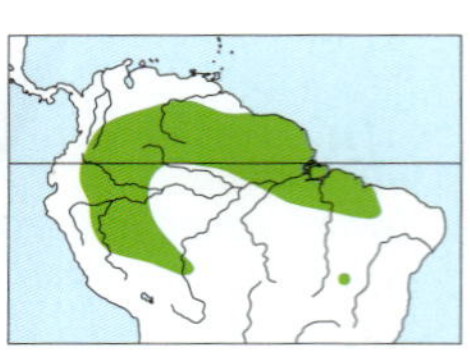

E CO, S VE, Guianas, NE EC, E PE, N BR, N BO

Schulz's Round-eared Bat
Lophostoma schulzi

LC
Monotypic
HB 6.3–6.9 cm
T 1.2–1.3 cm
FA 4.2–4.5 cm
W c. 15 g

FR Lophostome de Schulz
DE Schulz-Rundohrblattnase
ES Lofostoma de Schulz

above 1500 m

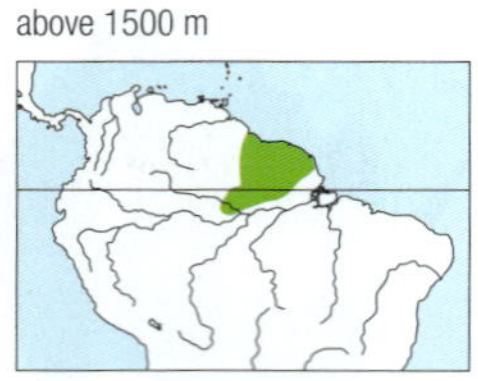

Guianas, N BR (Amazonas, Pará, Amapá)

Western Round-eared Bat
Lophostoma occidentale

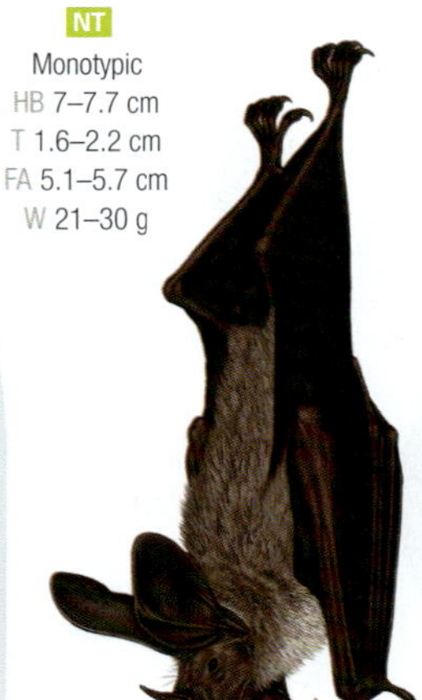

NT
Monotypic
HB 7–7.7 cm
T 1.6–2.2 cm
FA 5.1–5.7 cm
W 21–30 g

FR Lophostome occidental
DE Westliche Rundohrblattnase
ES Lofostoma occidental

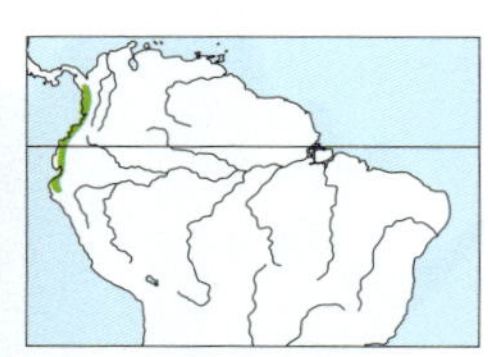

W CO (Chocó, Valle del Cauca, Cauca), W lowlands & foothills of EC & NW PE (Tumbes, Piura)

LC
Monotypic
HB 4.9–7 cm
T 1.1–1.8 cm
FA 4.7–5.3 cm
W 14–23 g

Davis's Round-eared Bat
Lophostoma evotis

FR Lophostome de Davis
DE Davis-Rundohrblattnase
ES Lofostoma de Davis

below 100 m

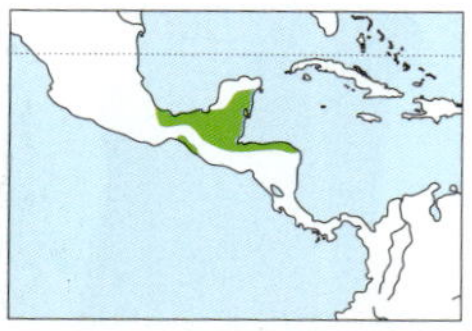

S MX (S Veracruz, NE Oaxaca, S Yucatán Pen, Chiapas), BZ, N GT, N HN

LC
3 sspp
HB 6.6–8.9 cm
T 1–2.2 cm
FA 5–5.9 cm
W 25–39 g

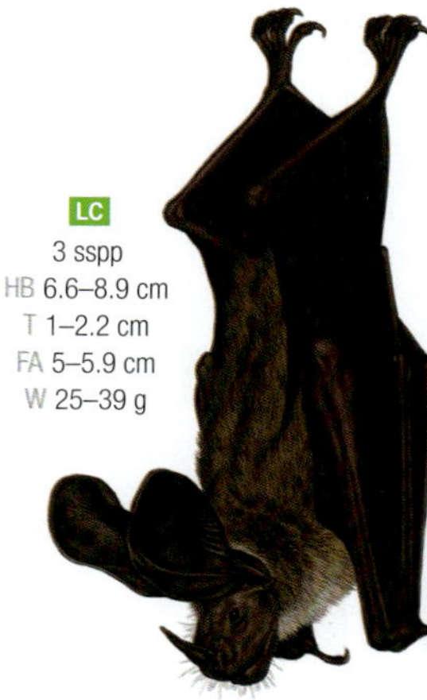

White-throated Round-eared Bat
Lophostoma silvicola

FR Grand Lophostome
DE Weißkehl-Rundohrblattnase
ES Lofostoma de garganta blanca

up to 600 m

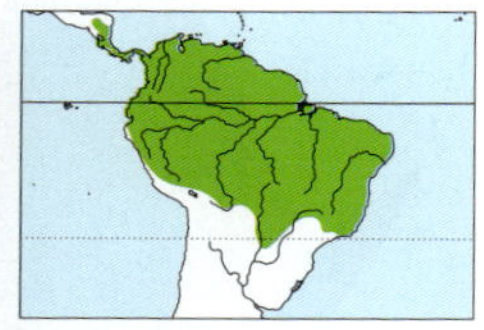

From HN & NI S to EC, PE, BO, PY & BR

DD
Monotypic
HB 6.9–8.1 cm
T 1.6–1.9 cm
FA 4.8–5.9 cm
W 27–34 g

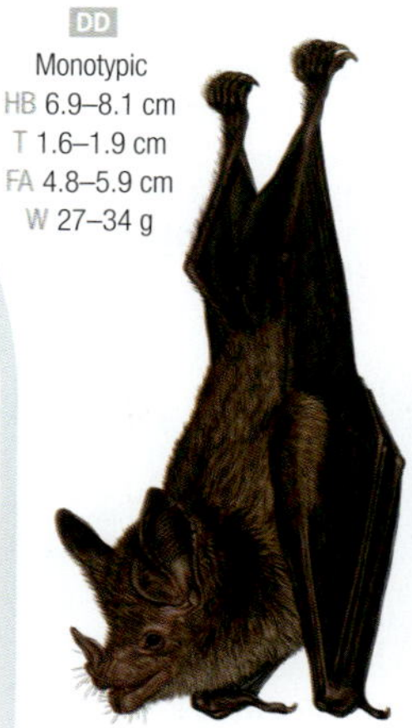

Greater Round-eared Bat
Tonatia bidens

FR Tonatia de Spix
DE Große Rundohrblattnase
ES Tonatia grande

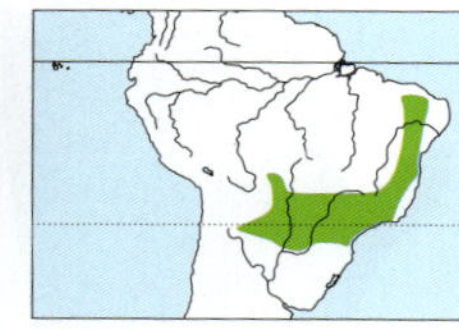

E BR, E BO, PY, N AR

Genus *PHYLLODERMA*

(1 species)

LC 3 sspp
HB 8.3–10.9 cm T 1.4–2.1 cm
FA 6.6–8.1 cm W 25–39 g

Pale-faced Bat
Phylloderma stenops

FR Phylloderme de Peters
DE Peters-Lanzennase
ES Filoderma pálido

0–1750 m

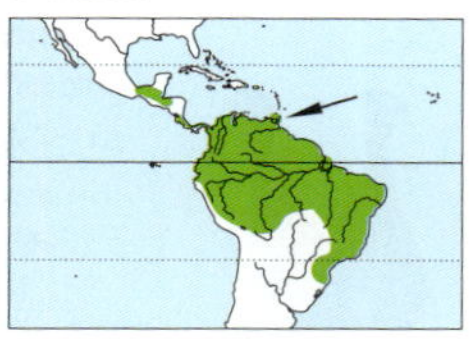

From S MX (Veracruz, Oaxaca) to EC, PE, BO, N & E BR, TT (Trinidad)

Genus *TONATIA*

(3 species)

NE Monotypic
HB c. 6.9 cm T 0.8–2.1 cm
FA 5.7–6.3 cm

Northern Stripe-headed Round-eared Bat
Tonatia bakeri

FR Tonatia de Baker
DE Nördliche Rundohrblattnase
ES Tonatia de Baker

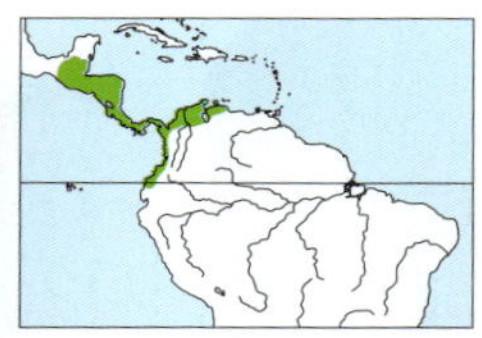

Extreme SE MX (Lacandon Forest, Chiapas) S to N CO, NW EC, NW VE (N of Cordillera de Mérida)

Southern Stripe-headed Round-eared Bat
Tonatia maresi

FR Tonatia de Mares
DE Südliche Rundohrblattnase
ES Tonatia de Mares

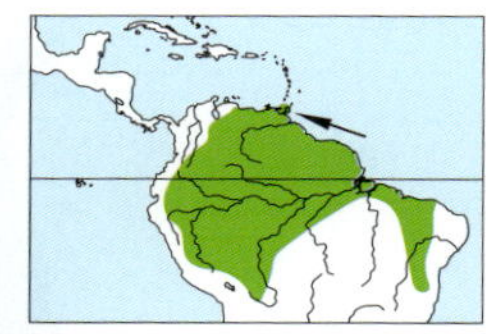

VE (E & S of Cordillera de Mérida), TT (Trinidad), Guianas, E CO, EC, PE, BO, N & E BR

NE Monotypic
HB c. 7.5 cm
T 0.4–2.5 cm
FA 5.3–6 cm

Genus *PHYLLOSTOMUS*

(4 species)

LC 2 sspp
HB 6.6–9.7 cm
T 1.2–1.7 cm
FA 6–6.8 cm
W 26–51 g

Pale Spear-nosed Bat
Phyllostomus discolor

FR Phyllostome des fleurs
DE Helle Lanzennase
ES Filostoma de nariz pálida

up to 610 m

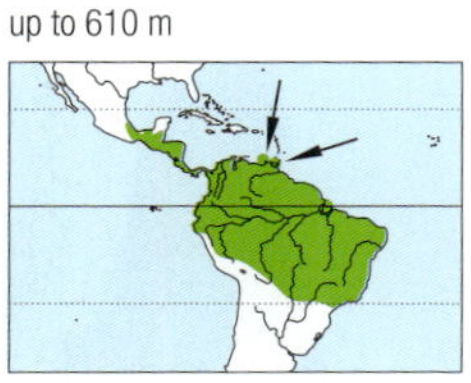

S MX (from Veracruz & Oaxaca) S to PE, BO, NW PY & BR, Margarita & Trinidad Is; perhaps NW AR

LC
Monotypic
HB 6.5–10 cm
T 1.5–2.4 cm
FA 5.8–7.1 cm
W 30–57 g

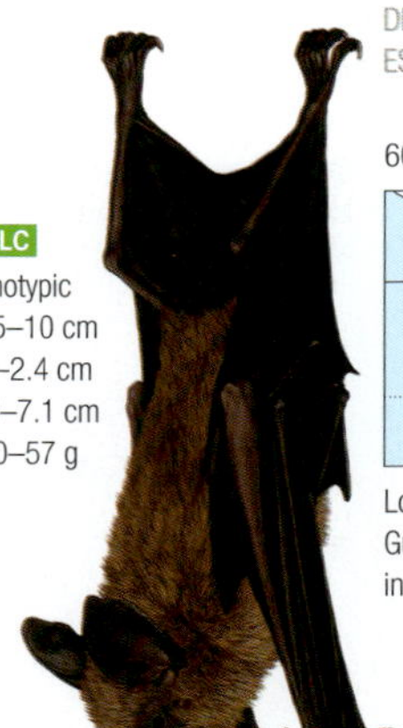

Lesser Spear-nosed Bat
Phyllostomus elongatus

FR Phyllostome des sous-bois
DE Kleine Lanzennase
ES Filostoma pequeño

60–1100 m

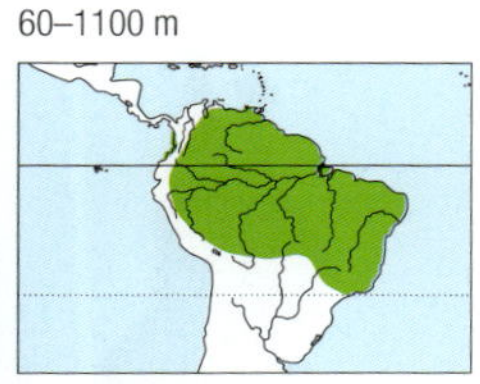

Lowlands E of Andes in CO, VE, Guianas, PE, BO & BR, isolated in W CO & NW EC

LC
3 sspp
HB 9–12.4 cm
T 1.2–3.1 cm
FA 7.7–9.4 cm
W 55–140 g

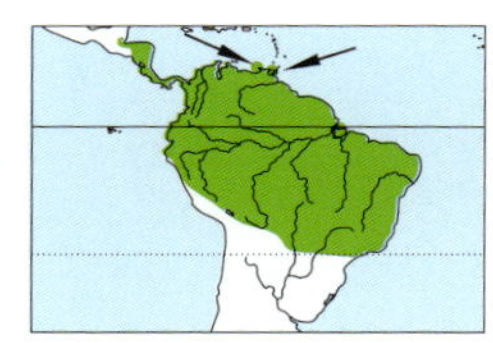

0–1800 m

S BZ & E GT S through both coasts to E PE, N & E BO, N PY & SE BR, Margarita & Trinidad Is

Greater Spear-nosed Bat
Phyllostomus hastatus

FR Grand Phyllostome
DE Große Lanzennase
ES Filostoma grande

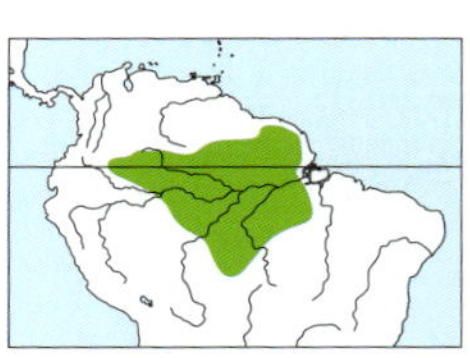

SE CO, Guianas, N BR (E to Pará, S to Mato Grosso)

Guianan Spear-nosed Bat
Phyllostomus latifolius

FR Phyllostome des cavernes
DE Guayana-Lanzennase
ES Filostoma de las Guayanas

LC
Monotypic
HB 7.8–8.2 cm
T 1.3–1.7 cm
FA 5.5–6 cm
W 27–31 g

Genus *CHROTOPTERUS*

(1 species)

Woolly False Vampire Bat
Chrotopterus auritus

FR Chrotoptère oreillard
DE Großohr-Wolffledermaus
ES Crotóptero lanudo

0–2200 m

S MX (from S Veracruz & Oaxaca) S to E PE, BO, PY, N AR & S BR

LC
Monotypic
HB 10.3–12.2 cm
T 0.7–1.7 cm
FA 7.8–8.6 cm
W 66–96 g

Genus *MIMON*

(2 species)

Southern Golden Bat
Mimon bennettii

FR Mimon doré
DE Südliche Haarnasenfledermaus
ES Mimón dorado

LC
Monotypic
HB 6.5–7.4 cm
T 1.5–2.6 cm
FA 5.5–6.1 cm
W 19–25 g

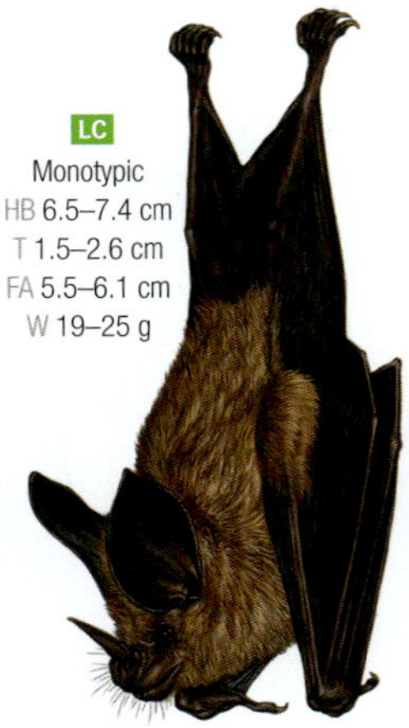

up to 1000 m

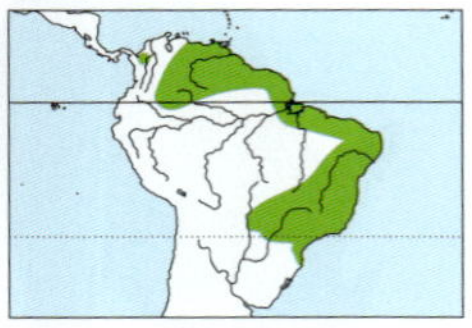

N & E CO, VE, Guianas, N, E & CW BR

Cozumelan Golden Bat
Mimon cozumelae

FR Mimon de Cozumel
DE Cozumel-Haarnasenfledermaus
ES Mimón de Cozumel

LC
Monotypic
HB 8.5–9.5 cm
T 2–2.5 cm
FA 5.3–5.9 cm
W 21–23 g

below 600 m

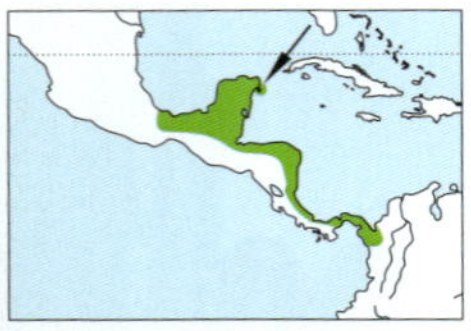

From S MX (S Veracruz, Oaxaca, Yucatán Pen, Cozumel I) S across E Caribbean slope to NW CO

Genus *VAMPYRUM*

(1 species)

Spectral Bat
Vampyrum spectrum

FR Faux-vampire
DE Große Spießblattnase
ES Vampiro falso

NT
Monotypic
HB 12.5–15.8 cm
T tailless
FA 9.8–11 cm
W 126–190 g

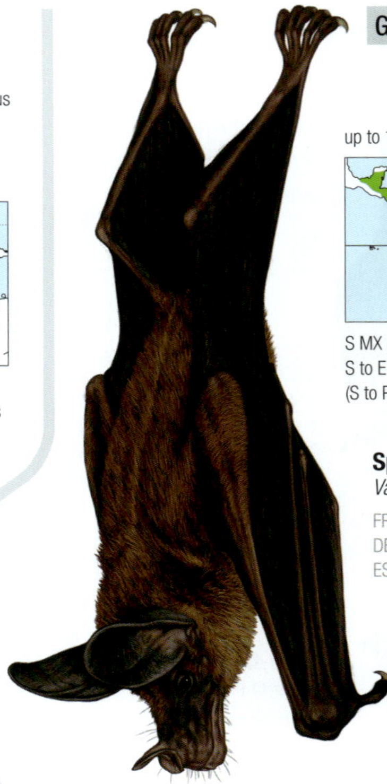

up to 1650 m

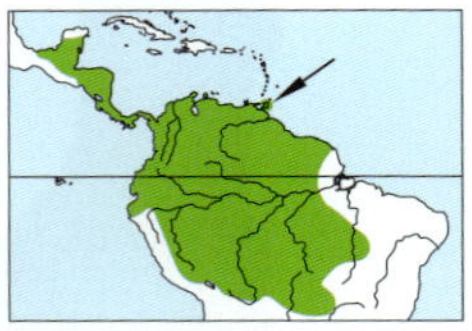

S MX (from S Veracruz & Oaxaca) S to E PE, N BO & Amazonian BR (S to Pantanal), TT (Trinidad)

Genus *MONOPHYLLUS*

(2 species)

Insular Single-leaf Bat
Monophyllus plethodon

FR Monophylle des Petites Antilles
DE Kleine-Antillen-Blattnase
ES Monófilo de las Pequeñas Antillas

LC
2 sspp
HB 6.7–8.4 cm
T 0.8–1.6 cm
FA 3.9–4.6 cm
W 12.5–17.2 g

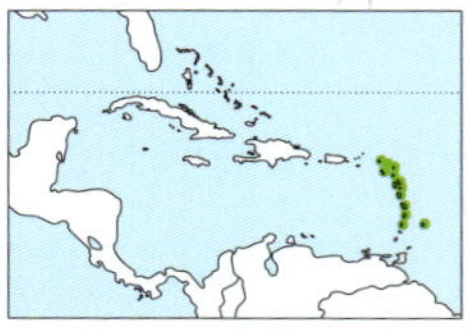

Lesser Antilles, from AI S to VC (St Vincent) & BB

Leach's Single-leaf Bat
Monophyllus redmani

FR Monophylle de Redman
DE Große-Antillen-Blattnase
ES Monófilo de Redman

LC
3 sspp
HB 5.8–8 cm
T 0.7–1.1 cm
FA 3.5–4.3 cm
W 6–13.4 g

-40–1515 m

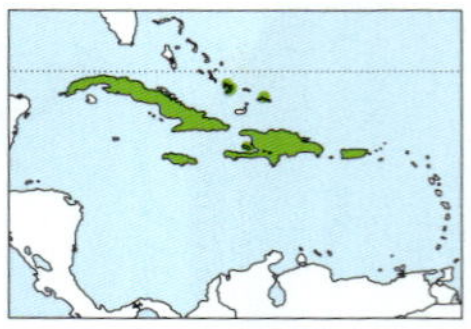

S BS (Acklins & Crooked Is), TC, CU, JM, HT (with Gonâve I), DO, PR

Genus *GLOSSOPHAGA*

(9 species)

Commissaris's Long-tongued Bat
Glossophaga commissarisi

FR Glossophage de Commissaris
DE Commissaris-Blütenfledermaus
ES Glosófago de Commissaris

LC
2 sspp
HB 4.2–6.1 cm
T 0.4–1.1 cm
FA 3.2–3.5 cm
W 8–11 g

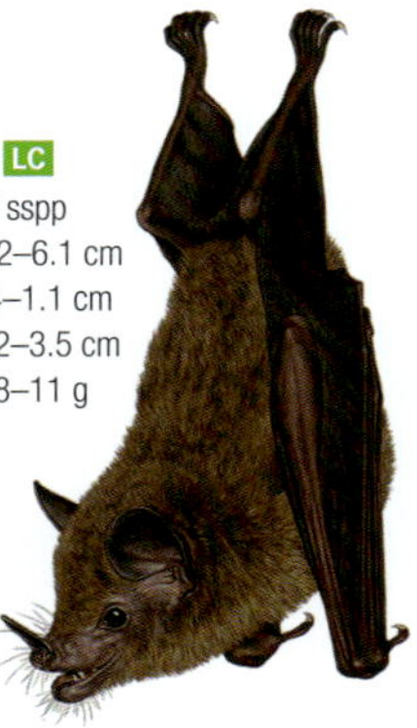

0–2000 m

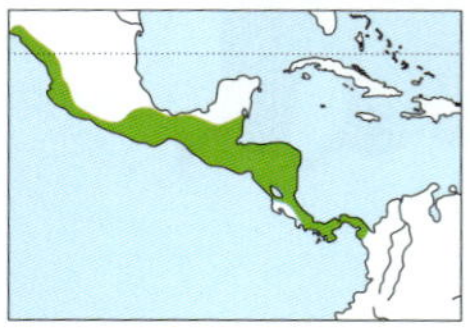

W & S MX (S from Sinaloa & Veracruz) through Central America to E PA; perhaps W CO

Baker's Long-tongued Bat
Glossophaga bakeri

FR Glossophage de Baker
DE Baker-Blütenfledermaus
ES Glosófago de Baker

NE
Monotypic
HB 5.6–6.1 cm
T 0.5–0.7 cm
FA 3.2–3.5 cm
W 6.8–9.2 g

up to 300 m

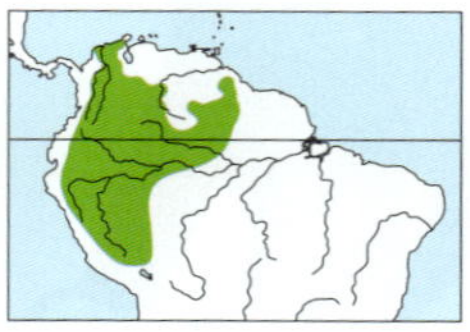

W Amazonia in CO, WC GY, EC, PE & W BR

Western Long-tongued Bat
Glossophaga morenoi

FR Glossophage de Moreno
DE Moreno-Blütenfledermaus
ES Glosófago de Moreno

LC
2 sspp
HB 5.7–7.5 cm
T 0.5–1.1 cm
FA 3.2–3.7 cm
W 7–10 g

0–1500 m

Colima & Michoacán E to Tlaxcala & Puebla, SE to Chiapas

Gray's Long-tongued Bat
Glossophaga leachii

FR Glossophage de Leach
DE Leach-Blütenfledermaus
ES Glosófago de Leach

LC
Monotypic
HB 4.9–6.1 cm
T 0.5–1.1 cm
FA 3.5–3.9 cm
W 9–11 g

0–2380 m

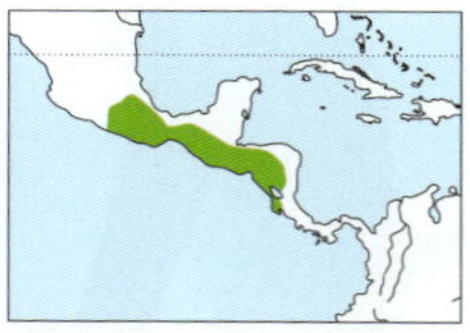

C MX (Jalisco, Colima, Michoacán, Morelos, Tlaxcala) S along Pacific slope to NW CR

Miller's Long-tongued Bat
Glossophaga longirostris

FR Glossophage de Miller
DE Miller-Blütenfledermaus
ES Glosófago de Miller

LC
4 sspp
HB 5.1–7.5 cm
T 0.4–1.1 cm
FA 3.5–4.1 cm
W 10–17 g

0–2000 m

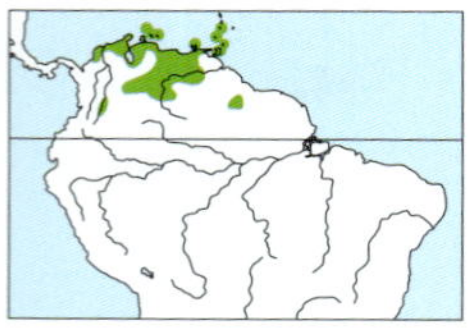

N & E CO, N VE, SW GY, N BR (Roraima), S Lesser Antilles

Pallas's Long-tongued Bat
Glossophaga soricina

FR Glossophage murin
DE Pallas-Blütenfledermaus
ES Glosófago de Pallas

LC
Monotypic
HB 4.5–5.9 cm
T 0.5–1.2 cm
FA 3.2–3.8 cm
W 7–12 g

up to 2500 m

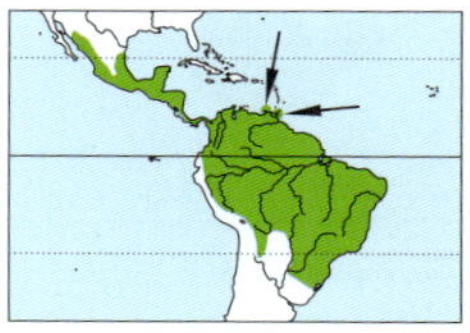

W & E MX lowlands through Central America to E EC, E PE, BO, PY, S BR & N AR, Margarita & Trinidad Is

NE
Monotypic
HB 5.1–5.6 cm
T 0.7–0.9 cm
FA 3.6–3.9 cm

Merriam's Long-tongued Bat
Glossophaga mutica
FR Glossophage des Marias
DE Merriam-Blütenfledermaus
ES Glosófago de Merriam

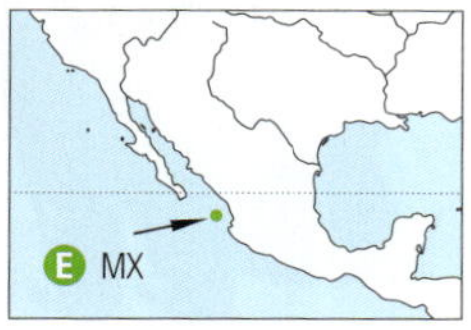

Marías Is, off Nayarit

NE
Monotypic
HB 5–5.7 cm
T 0.6–1 cm
FA 3.5–4 cm

Ecuadorian Long-tongued Bat
Glossophaga valens
FR Glossophage d'Équateur
DE Ekuador-Blütenfledermaus
ES Glosófago de Ecuador

W of Andes in W EC & W PE

NE
Monotypic
HB 5.4–5.5 cm
T 0.6–0.8 cm
FA 3.7–3.9 cm

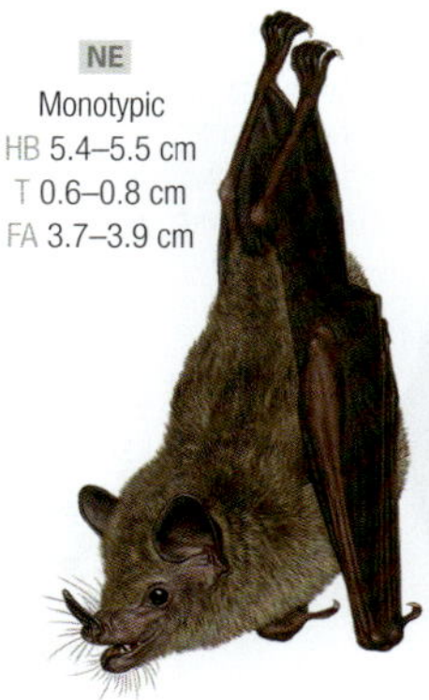

Jamaican Long-tongued Bat
Glossophaga antillarum
FR Glossophage de Jamaïque
DE Jamaica-Blütenfledermaus
ES Glosófago de Jamaica

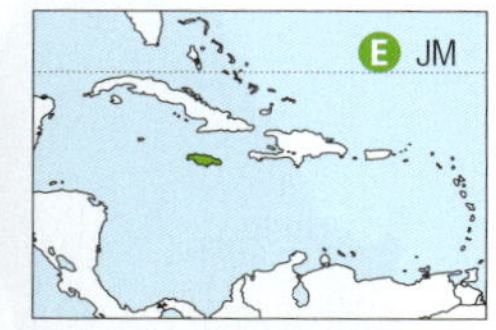

Wide variety of subtropical and tropical habitats

Genus *LEPTONYCTERIS*

(3 species)

VU
Monotypic
HB 7.5–8.4 cm
T tailless
FA 5.1–5.5 cm
W 21–26 g

Southern Long-nosed Bat
Leptonycteris curasoae
FR Leptonyctère de Curaçao
DE Curaçao-Blütenfledermaus
ES Leptonícteris de Curaçao

up to 1500 m

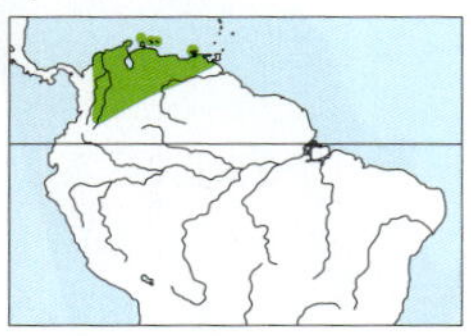

Dry tropical areas of N & E CO and N & W VE (with Margarita I), AW, CW, BQ (Bonaire)

EN
Monotypic
HB 7.6–8.5 cm
T tailless
FA 5.3–5.9 cm
W 18–30 g

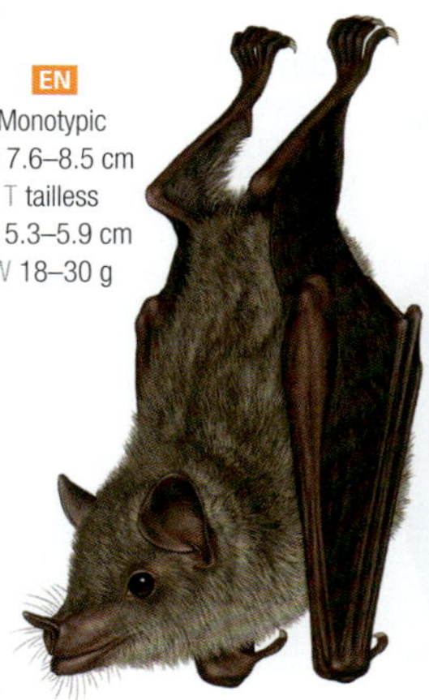

Greater Long-nosed Bat
Leptonycteris nivalis
FR Grand Leptonyctère
DE Große Mexiko-Blütenfledermaus
ES Leptonícteris grande

800–3000 m

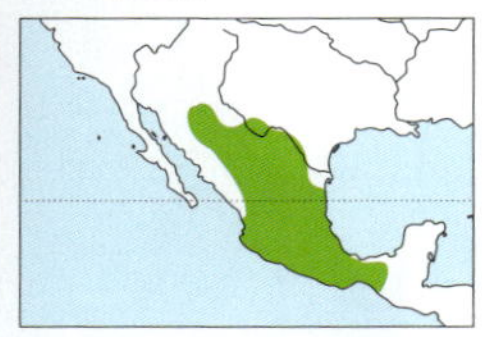

SW US (S New Mexico & SW Texas) S through N & C MX (S to Puebla & Oaxaca)

NT
Monotypic
HB 7.4–9 cm
T tailless
FA 5.1–5.5 cm
W 15–28 g

Lesser Long-nosed Bat
Leptonycteris yerbabuenae
FR Petit Leptonyctère
DE Kleine Mexiko-Blütenfledermaus
ES Leptonícteris pequeño

up to 2500 m

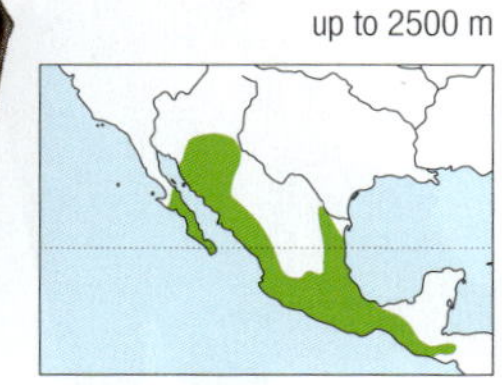

SW US (S Arizona, SW New Mexico), MX, GT, SV, W HN

Genus *BRACHYPHYLLA*

(2 species)

LC
3 sspp
HB 8.5–9.4 cm
T tailless
FA 6.1–6.8 cm
W 36–53 g

Antillean Fruit-eating Bat
Brachyphylla cavernarum
FR Brachyphylle des cavernes
DE Großer Antillen-Fruchtvampir
ES Bracófilo de las cavernas

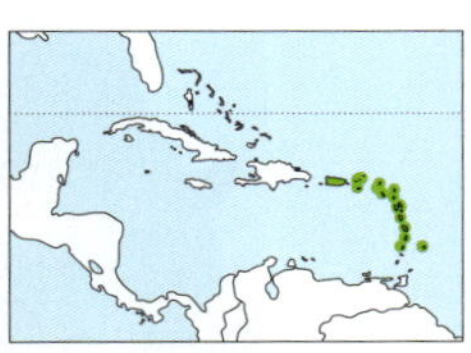

PR, Lesser Antilles from VG (Norman I) & VI S to VC (St Vincent) & BB

LC
Monotypic
HB 7.6–9 cm
T tailless
FA 5.4–6.4 cm
W 27–41 g

Cuban Fruit-eating Bat
Brachyphylla nana
FR Brachyphylle de Cuba
DE Kleiner Antillen-Fruchtvampir
ES Bracófilo de Cuba

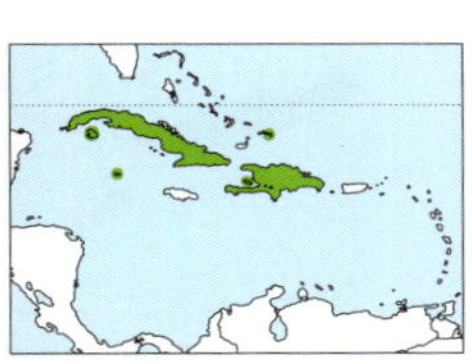

CU (with Isla de la Juventud), KY (Grand Cayman), TC (Middle Caicos), HT (with Gonâve I), DO

Genus *PHYLLONYCTERIS*

(2 species)

CR
Monotypic
HB 7.2–7.6 cm
T 0.7–1 cm
FA 4.4–4.8 cm
W 14–15 g

Jamaican Flower Bat
Phyllonycteris aphylla
FR Phyllonyctère de Jamaïque
DE Jamaika-Blütenfledermaus
ES Filonícteris de Jamaica

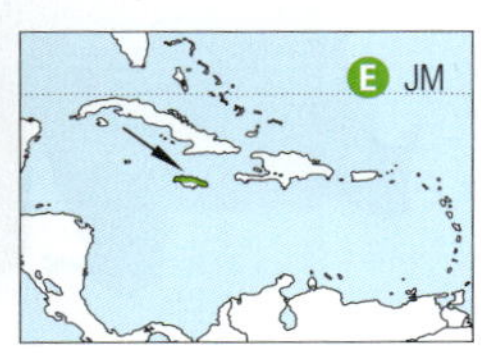

Currently only known in just 1 cave (Portland)

LC
2 sspp
HB 7.5–8.7 cm
T 0.6–1.8 cm
FA 4.3–5.1 cm
W 15–29 g

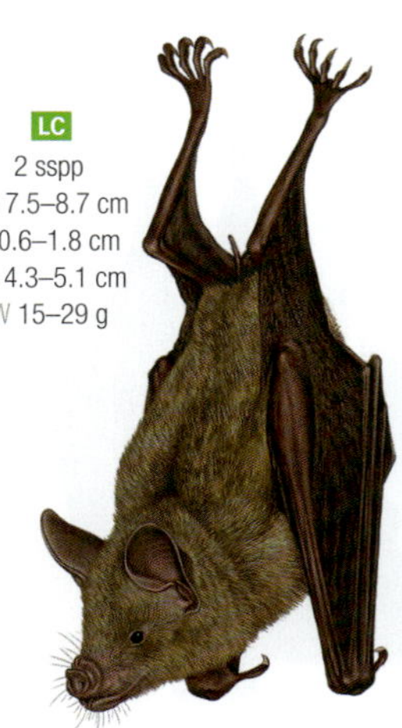

Cuban Flower Bat
Phyllonycteris poeyi
FR Phyllonyctère de Cuba
DE Kuba-Blütenfledermaus
ES Filonícteris de Cuba

0–1700 m

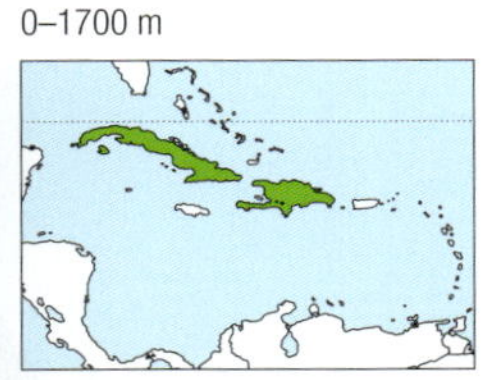

CU (with Isla de la Juventud), HT, DO; 2 probably vagrants in Florida Keys (US)

Genus *EROPHYLLA*

(2 species)

LC
2 sspp
HB 8–8.8 cm
T 1.3–1.7 cm
FA 4.7–5.1 cm
W 16–21 g

Brown Flower Bat
Erophylla bombifrons
FR Érophylle de Porto Rico
DE Braunstirn-Blütenfledermaus
ES Erófilo de Puerto Rico

HT, DO, PR

LC
4 sspp
HB 6.9–8.3 cm
T 1–1.9 cm
FA 4.3–5.1 cm
W 13–21 g

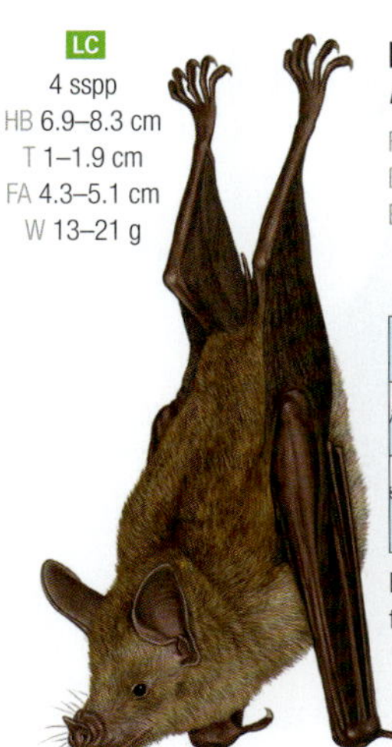

Buffy Flower Bat
Erophylla sezekorni
FR Érophylle de Sezekorn
DE Gelbbraune Blütenfledermaus
ES Erófilo de Sezekorn

BS, TC, CU (with Isla de la Juventud), KY, JM

Genus *ANOURA*

(10 species)

Tailed Tailless Bat

Anoura caudifer

FR Anoura des tunnels
DE Geschwänzte Langnasenfledermaus
ES Anoura de cola corta

LC
Monotypic
HB 4.6–6.4 cm
T 0.4–0.5 cm
FA 3.3–3.9 cm
W 8–13 g

0–1500 m

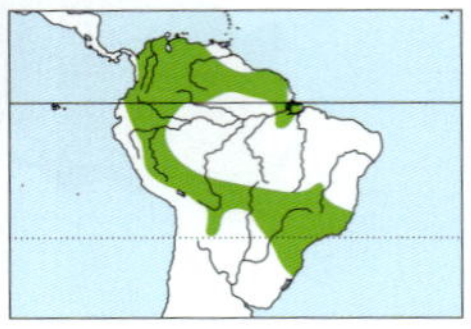

CO, VE, Guianas, BR (surrounding C Amazon Basin), EC, PE, BO, NW AR; doubtful in PY

Ecuadorian Tailless Bat

Anoura aequatoris

FR Anoura d'Équateur
DE Ecuador-Langnasenfledermaus
ES Anoura de Ecuador

LC Monotypic
HB 5.1–5.9 cm T 0–0.7 cm
FA 3.2–3.8 cm W 8–9 g

1000–1500 m

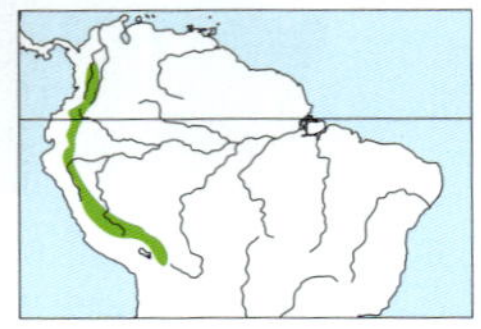

W & C Andes from NW CO (Antioquia) S to WC Bolivia (La Paz); E limits unknown

Cadena's Tailless Bat

Anoura cadenai

FR Anoura de Cadena
DE Cadena-Langnasenfledermaus
ES Anoura de Cadena

DD Monotypic
HB 5.9–6.1 cm T tailless
FA 3.6–3.7 cm

800–1560 m

Known from W slope of SW Andes; perhaps adjacent N EC

Tube-lipped Tailless Bat

Anoura fistulata

FR Anoura à longues lèvres
DE Röhrenlippen-Langnasenfledermaus
ES Anoura de labio largo

DD
Monotypic
HB 5.7–6.7 cm
T 0.4–0.7 cm
FA 3.5–4 cm
W 9–17 g

1300–2275 m

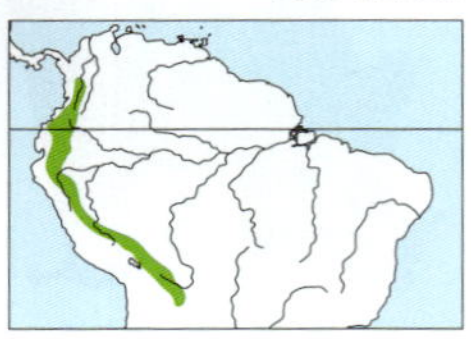

E slopes of Andes in CO, PE & BO, both Andean slopes in EC

Luis Manuel's Tailless Bat

Anoura luismanueli

FR Anoura de Luis Manuel
DE Luis-Manuel-Langnasenfledermaus
ES Anoura de Luis Manuel

LC
Monotypic
HB 5.5–5.9 cm
T 0.3–0.5 cm
FA 3.4–3.7 cm
W 7.5–10 g

1100–2400 m

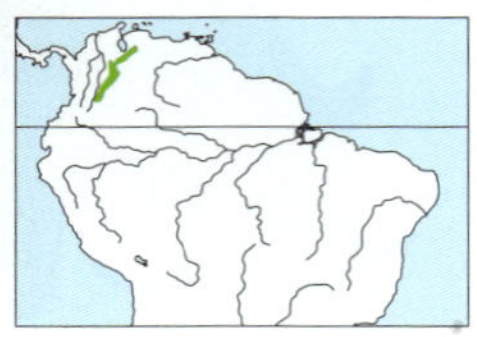

E Andes in CO, Andes of W VE (Cordillera de Mérida)

Peruvian Tailless Bat

Anoura javieri

FR Anoura de Javier
DE Peru-Langnasenfledermaus
ES Anoura de Javier

NE
Monotypic
HB 6.1–6.9 cm
T 0–0.8 cm
FA 3.7–3.8 cm
W 10–11 g

1900–3450 m

Upper montane forests of Manu NP & BR (Cusco); perhaps BO (Yungas)

Handley's Tailless Bat

Anoura cultrata

FR Anoura de Handley
DE Handley-Langnasenfledermaus
ES Anoura de Handley

LC
Monotypic
HB 5.4–7.9 cm
T 0.3–0.4 cm
FA 3.4–4.3 cm
W 9–23 g

50–2600 m

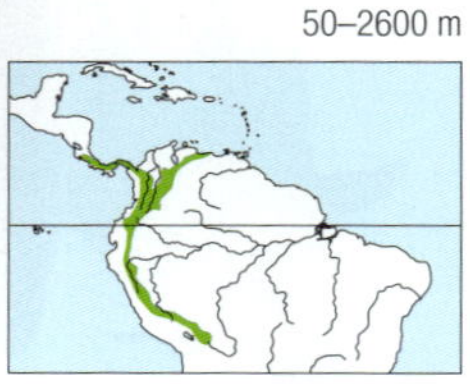

CR, PA, CO, N & W VE, EC, PE, WC BO (S to Cochabamba)

Broad-toothed Tailless Bat

Anoura latidens

FR Anoura à dents larges
DE Breitzahn-Langnasenfledermaus
ES Anoura de dientes largos

LC
Monotypic
HB 5.9–7.7 cm
T tailless
FA 4–4.6 cm
W 15–16 g

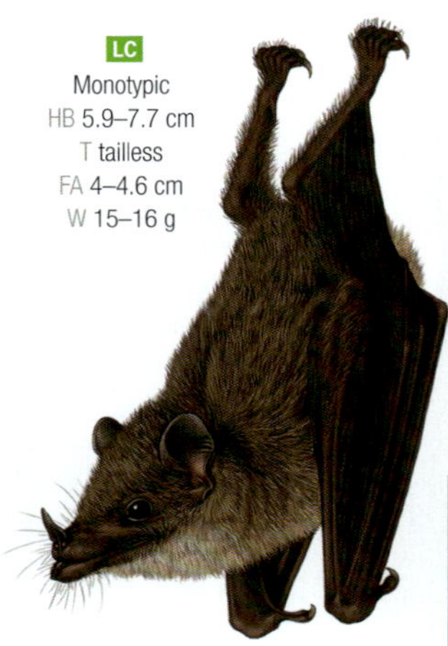

up to 2600 m

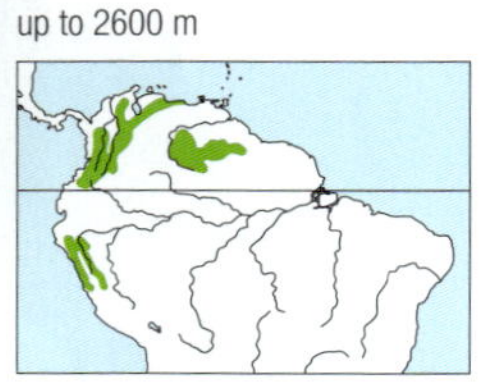

Discontinously in Andean CO & PE, VE, WC GY

Geoffroy's Tailless Bat

Anoura geoffroyi

FR Anoura de Geoffroy
DE Geoffroy-Langnasenfledermaus
ES Anoura de Geoffroy

LC
2 sspp
HB 5.8–7.3 cm
T tailless
FA 4–4.5 cm
W 13–18 g

up to 3500 m

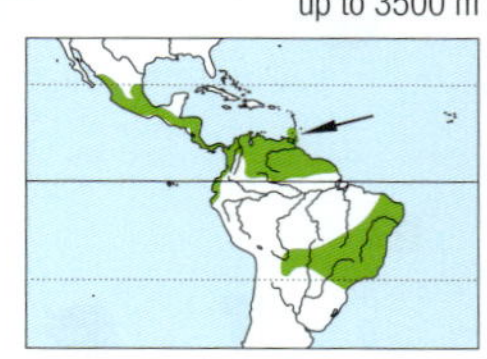

MX (from S Sinaloa & S Tamaulipas) S to NW PE & N BR, disjunct in E BR to E BO, TT (Trinidad), GD

Tschudi's Tailless Bat

Anoura peruana

FR Anoura du Pérou
DE Tschudi-Langnasenfledermaus
ES Anoura de Perú

LC
Monotypic
HB 5.9–8.1 cm
T tailless
FA 4.3–4.9 cm
W 13–18 g

10–3800 m

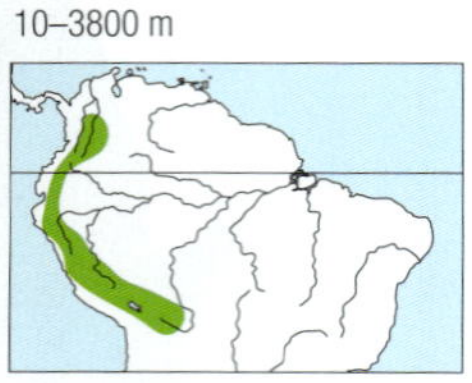

Andean zone of W CO, W & E EC, W & E PE, W & C BO

Genus *HYLONYCTERIS*

(1 species)

Underwood's Long-tongued Bat

Hylonycteris underwoodi

FR Hylonyctère d'Underwood
DE Underwood-Blattnase
ES Hilonícteris de Underwood

LC
2 sspp
HB 4.8–6 cm
T 0.3–1 cm
FA 3.1–3.6 cm
W 6.5–8.5 g

0–2640 m

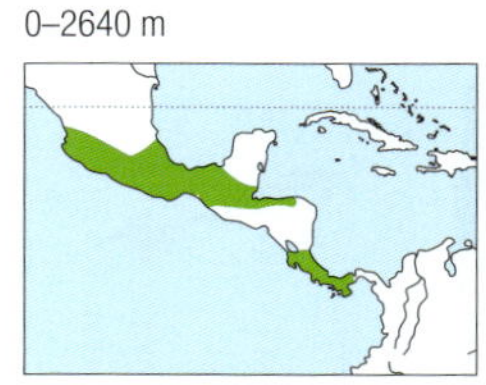

S MX (fom Nayarit & Veracruz) S to N HN and from SE NI to W PA; gap in NI might just reflect lack of data

Genus *CHOERONISCUS*

(3 species)

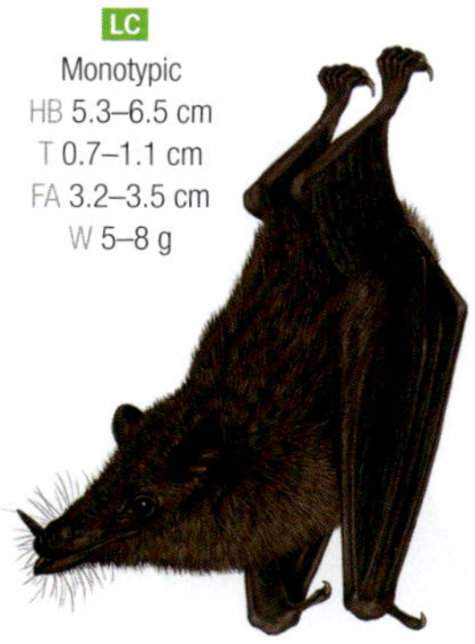

LC
Monotypic
HB 5.3–6.5 cm
T 0.7–1.1 cm
FA 3.2–3.5 cm
W 5–8 g

Godman's Long-tailed Bat
Choeroniscus godmani

FR Oxyrhine de Godman
DE Godman-Langschwanz-Blumenfledermaus
ES Coeronisco de Godman

0–1400 m

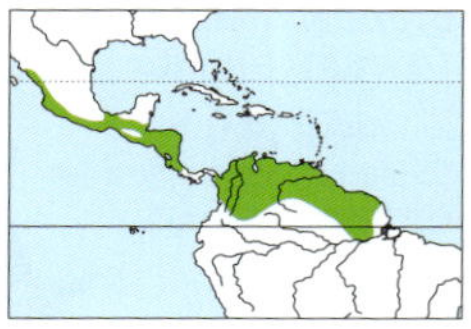

W & S MX (from Sinaloa & Veracruz) S to C CO, VE, Guianas & N BR (W Pará)

LC
Monotypic
HB 5.2–6 cm
T 0.4–1.1 cm
FA 2.6–3.8 cm
W 7–12 g

Lesser Long-tailed Bat
Choeroniscus minor

FR Petite Oxyrhine
DE Kleine Langschwanz-Blumenfledermaus
ES Coeronisco pequeño

150–1030 m

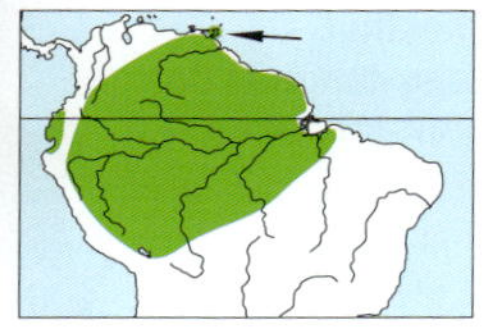

E CO, E VE, TT (Trinidad), Guianas, both Andean slopes of EC & PE, Amazonian Basin in BR & BO

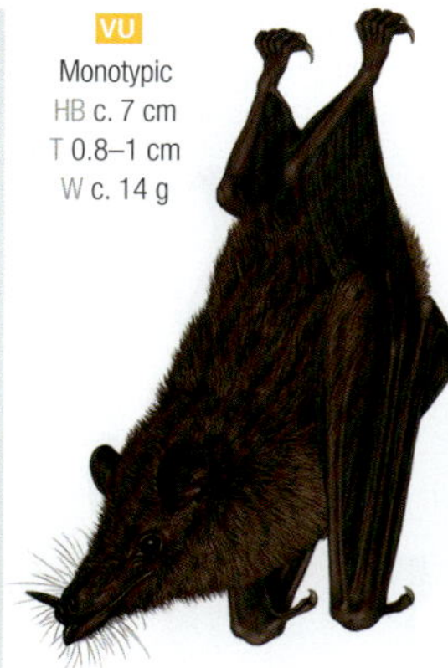

VU
Monotypic
HB c. 7 cm
T 0.8–1 cm
W c. 14 g

Greater Long-tailed Bat
Choeroniscus periosus

FR Grande Oxyrhine
DE Große Langschwanz-Blumenfledermaus
ES Coeronisco grande

0–400 m

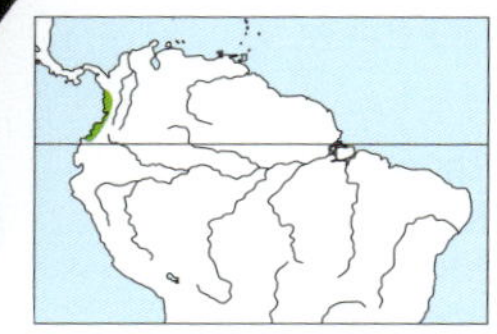

Pacific lowlands of W CO (Chocó S to Nariño), NW EC (Esmeraldas)

Genus *CHOERONYCTERIS*

(1 species)

Mexican Long-tongued Bat
Choeronycteris mexicana

FR Choéronyctère du Mexique
DE Langschnauzenfledermaus
ES Coeronícteris de México

NT
Monotypic
HB 8.1–10.3 cm
T 0.6–1.2 cm
FA 3.9–4.5 cm
W c. 20 g

300–3600 m

SW US (S California E to extreme S Texas) S throughout most of MX (with Marías Is) to SV & HN

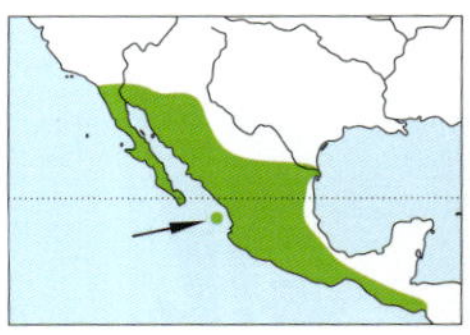

Genus *DRYADONYCTERIS*

(1 species)

Capixaba Nectar-feeding Bat
Dryadonycteris capixaba

FR Dryadonyctère d'Espírito Santo
DE Capixaba-Dryades-Fledermaus
ES Driadonícteris de Capixaba

DD
Monotypic
HB 5–5.6 cm
T 0.4–0.6 cm
FA 2.9–3.2 cm
W 4.1–6 g

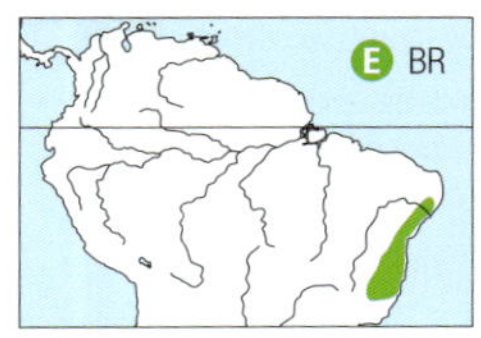

Pernambuco, Alagoas, Sergipe, Minas Gerais, Espírito Santo; probably Bahia

Genus *MUSONYCTERIS*

(1 species)

VU
Monotypic
HB 8.1–8.9 cm
T 0.8–1.2 cm
FA 4.1–4.4 cm
W 10–13.5 g

Banana Bat
Musonycteris harrisoni

FR Musonyctère des bananiers
DE Bananenfledermaus
ES Musonícteris del bananero

up to 1700 m

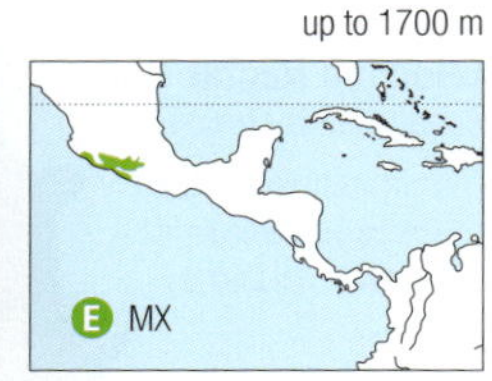

Coast from Jalisco to Guerrero, reaching inland to Morelos, México & Puebla

Genus *LICHONYCTERIS*

(2 species)

LC
Monotypic
HB 4.6–5.5 cm
T 0.6–1 cm
FA 3.3–3.5 cm
W 6–10 g

Pale Brown Long-tongued Bat
Lichonycteris degener

FR Lichonyctère du Brésil
DE Hellbraune Langzungenfledermaus
ES Liconícteris de Brasil

up to 600 m

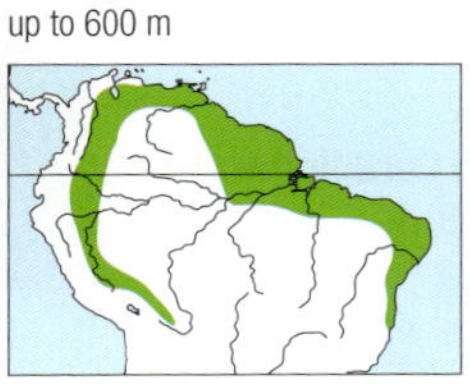

E CO, N VE, GY, GF, N & E BR, E EC, E PE, C BO; perhaps SR

LC
Monotypic
HB 4.6–5.5 cm
T 0.6–0.9 cm
FA 3.1–3.5 cm
W 6–8.8 g

Dark Long-tongued Bat
Lichonycteris obscura

FR Lichonyctère sombre
DE Dunkle Langzungenfledermaus
ES Liconícteris oscuro

up to 1000 m

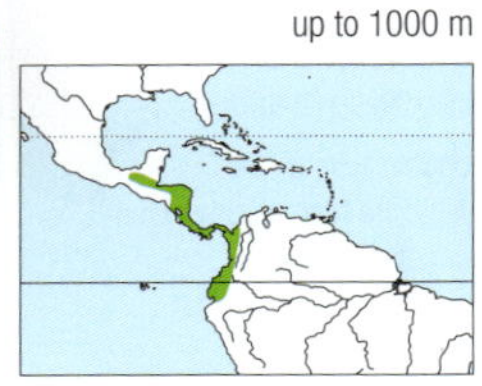

From S MX (Tabasco, Chiapas) to W Andes in CO & EC

Genus *SCLERONYCTERIS*

(1 species)

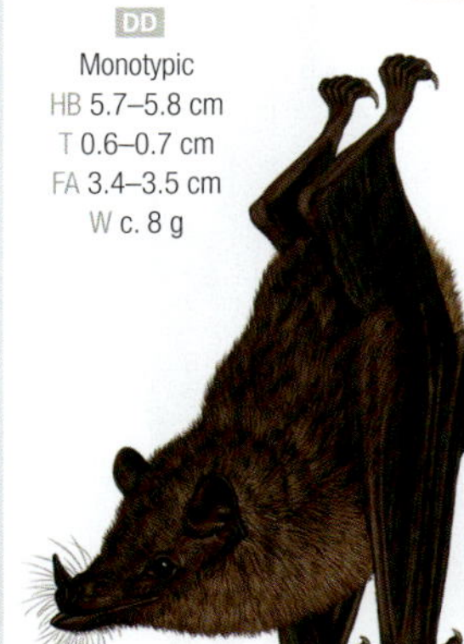

DD
Monotypic
HB 5.7–5.8 cm
T 0.6–0.7 cm
FA 3.4–3.5 cm
W c. 8 g

Ega Long-tongued Bat
Scleronycteris ega

FR Scléronyctère d'Ega
DE Ega-Langzungenfledermaus
ES Escleronícteris de Ega

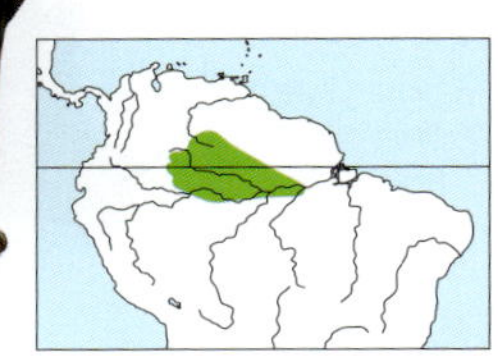

Few verified records from S VE (Amazonas) & N BR (Amazonas, Pará)

Genus *HSUNYCTERIS*

(4 species)

DD
Monotypic
HB 5.2–5.8 cm
T 0.7–1 cm
FA 3.1–3.3 cm
W c. 7 g

Cadena's Nectar Bat
Hsunycteris cadenai

FR Lonchophylle de Cadena
DE Cadena-Nektarfledermaus
ES Sunícteris de Cadena

up to 2400 m

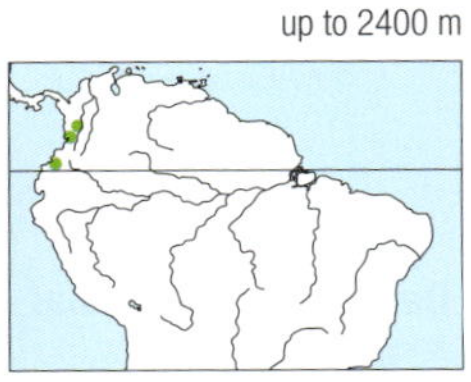

Known from few localities in W CO (Risaralda, Valle del Cauca) & NW EC (Esmeraldas)

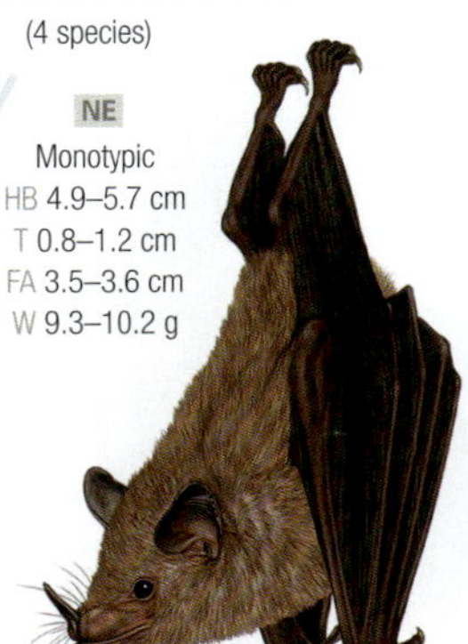

NE
Monotypic
HB 4.9–5.7 cm
T 0.8–1.2 cm
FA 3.5–3.6 cm
W 9.3–10.2 g

Dashe's Nectar Bat
Hsunycteris dashe

FR Lonchophylle de Dashe
DE Dashe-Nektarfledermaus
ES Sunícteris de Dashe

below 200 m

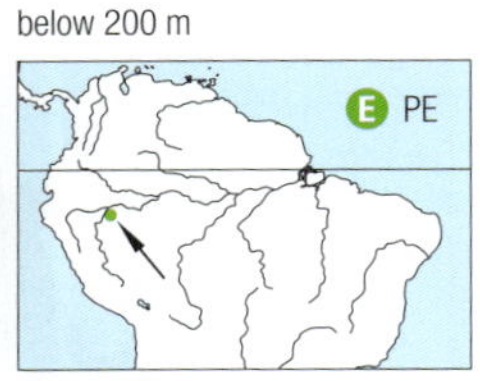

Known only from type locality in Yavarí-Ucayali interfluve

DD
Monotypic
HB 5.1–5.9 cm
T 0.5–1 cm
FA 3.1–3.4 cm
W 5–12 g

Patton's Nectar Bat
Hsunycteris pattoni

FR Lonchophylle de Patton
DE Patton-Nektarfledermaus
ES Sunícteris de Patton

up to 1000 m

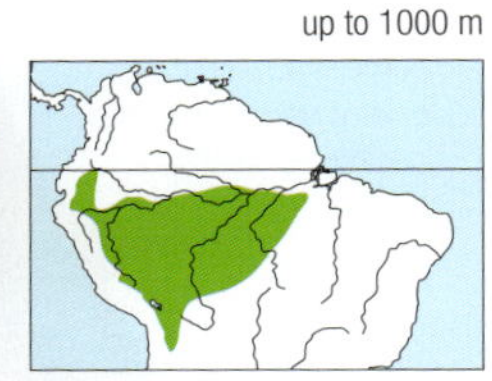

Scattered sites in Amazon Basin of SE CO, W & C BR (Amazonas, Pará), E EC, E PE & NE BO

up to 1300 m

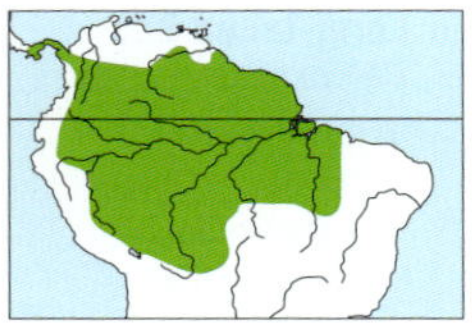

C & E PA, CO, VE (mostly S of R Orinoco), Guianas, Amazonian BR, E EC, E PE, BO

LC Monotypic
HB 4.5–5.6 cm T 0.4–1.1 cm
FA 2.9–3.4 cm W 5–10 g

Thomas's Nectar Bat
Hsunycteris thomasi

FR Lonchophylle de Thomas
DE Thomas-Nektarfledermaus
ES Sunícteris de Thomas

Genus *LIONYCTERIS*
(1 species)

up to 1400 m

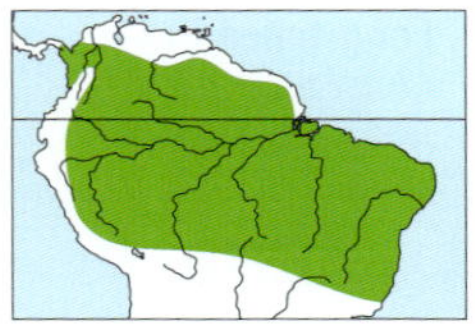

E PA, CO, S VE, Guianas, BR, E EC, E PE, N BO

LC
Monotypic
HB 4–6.3 cm
T 0.5–1 cm
FA 3.2–3.7 cm
W 6–13 g

Chestnut Long-tongued Bat
Lionycteris spurrelli

FR Lionyctère des cavernes
DE Spurrell-Langzungenfledermaus
ES Lionícteris de Spurrell

up to 500 m

E BR

Alagoas, Sergipe, NE Bahia

NT
Monotypic
HB 5–5.6 cm
T 0.8–1.2 cm
FA 3.4–3.7 cm
W 7.5–11 g

Brazilian Nectar Bat
Lonchophylla mordax

FR Lonchophylle du Brésil
DE Brasilien-Nektarfledermaus
ES Loncófilo de Brasil

Genus *LONCHOPHYLLA*
(13 species)

500–1000 m

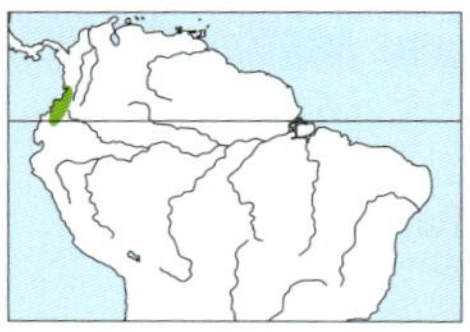

W Andes in SW CO (Chocó, Valle del Cauca, Nariño) & NW EC (Esmeraldas)

DD
Monotypic
HB 6.7–7.4 cm
T 0.7–1.1 cm
FA 4.2–4.8 cm
W 19–23 g

Chocoan Nectar Bat
Lonchophylla chocoana

FR Lonchophylle du Choco
DE Chocó-Nektarfledermaus
ES Loncófilo del Chocó

10–1500 m

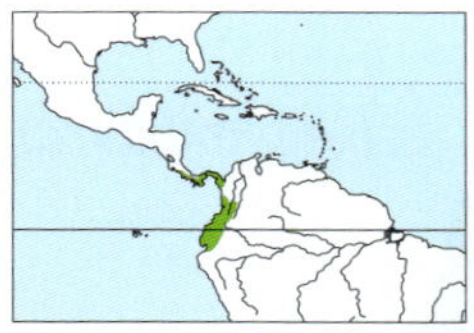

S CR S to Pacific lowlands of PA, W CO and coastal & Andean NW EC

LC
Monotypic
HB 5.4–6.5 cm
T 0.5–1.2 cm
FA 3.2–3.5 cm
W 7–10 g

Goldman's Nectar Bat
Lonchophylla concava

FR Lonchophylle de Goldman
DE Goldman-Nektarfledermaus
ES Loncófilo de Goldman

75–510 m

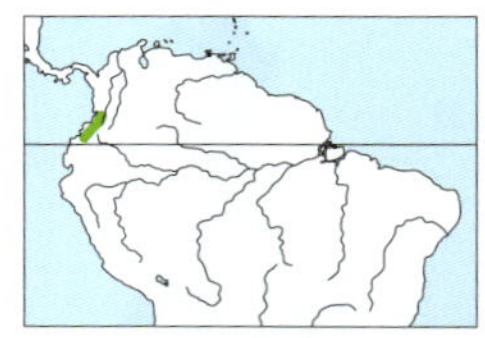

Known from few sites in coastal plains of SW CO (Valle del Cauca, Nariño) & NW EC (Imbabura)

DD
Monotypic
HB 5.2–6.2 cm
T 0.7–1.2 cm
FA 3.3–3.5 cm
W c. 8.7 g

Pacific Forest Long-tongued Bat
Lonchophylla fornicata

FR Lonchophylle de Woodman
DE Woodman-Nektarfledermaus
ES Loncófilo de Woodman

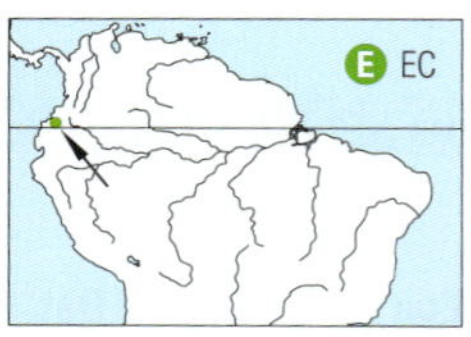

E EC

Known from type locality in Cordillera de Toisán (Esmeraldas)

DD
Monotypic
HB c. 6.1 cm
T c. 1.1 cm
FA c. 4.7 cm
W c. 22 g

Orces's Long-tongued Bat
Lonchophylla orcesi

FR Lonchophylle d'Orces
DE Orces-Nektarfledermaus
ES Loncófilo de Orces

0–2000 m

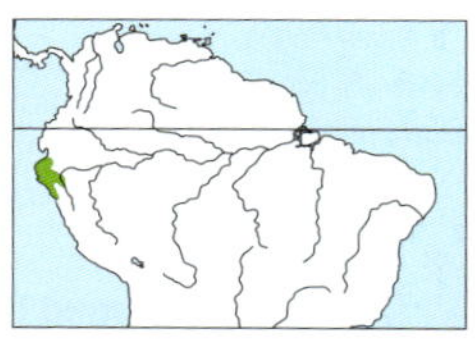

SW EC, NW PE

NT
Monotypic
HB 5.9–6.8 cm
T 0.7–1.3 cm
FA 3.5–4.1 cm
W c. 10 g

Western Nectar Bat
Lonchophylla hesperia

FR Lonchophylle d'Allen
DE Westliche Nektarfledermaus
ES Loncófilo de Allen

75–2015 m

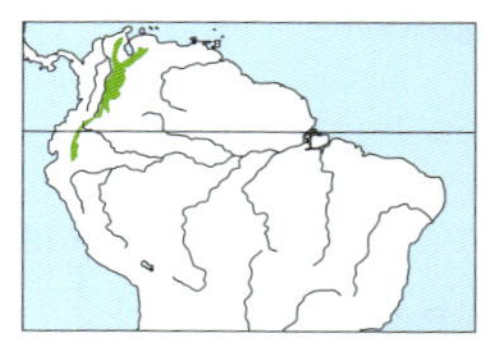

NW VE (Serranía del Perijá, Cordillera de Mérida), CO (E Andes, N Serranía de la Macarena), EC (E Andes)

DD
Monotypic
HB 6.1–7.1 cm
T 0.6–1.1 cm
FA 4–4.7 cm
W 11–16 g

Eastern Cordilleran Nectar Bat
Lonchophylla orienticollina

FR Lonchophylle de la Macarena
DE Ostkordilleren-Nektarfledermaus
ES Loncófilo de la Macarena

500–1200 m

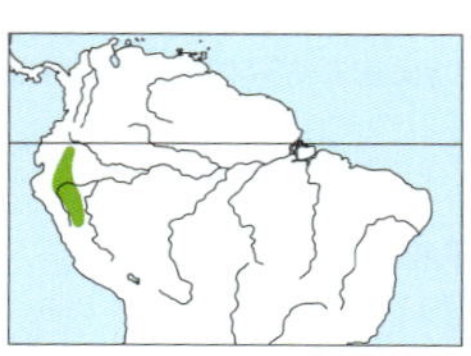

E EC, E PE

LC
Monotypic
HB 6.6–8.3 cm
T 0.5–0.6 cm
FA 4.3–4.8 cm
W 16–21 g

Handley's Nectar Bat
Lonchophylla handleyi

FR Lonchophylle de Handley
DE Handley-Nektarfledermaus
ES Loncófilo de Handley

up to 1500 m

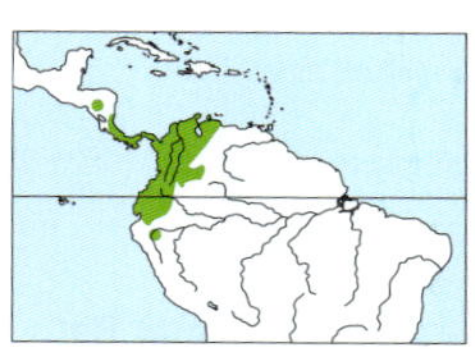

From NW & SE NI to W & C CO, NW VE, EC and NW PE

LC
Monotypic
HB 5.6–7.5 cm
T 0.6–1.2 cm
FA 3.9–4.6 cm
W 13–16 g

Orange Nectar Bat
Lonchophylla robusta

FR Lonchophylle du Panama
DE Robuste Nektarfledermaus
ES Loncófilo de Panamá

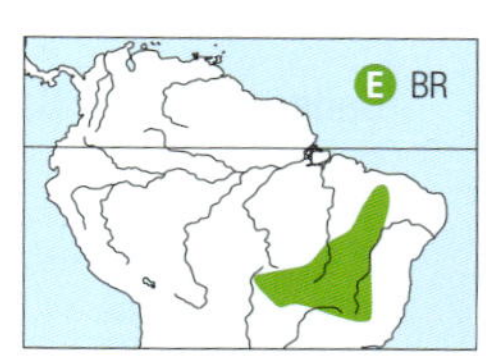

E BR

Cerrado (Tocantins S to Mato Grosso do Sul), transitional cerrado/caatinga (Piauí)

EN
Monotypic
HB 4.8–6.3 cm
T 0.6–0.8 cm
FA 3.4–3.8 cm
W 9.8–11.5 g

Dekeyser's Nectar Bat
Lonchophylla dekeyseri

FR Lonchophylle de Dekeyser
DE Dekeyser-Nektarfledermaus
ES Loncófilo de Dekeyser

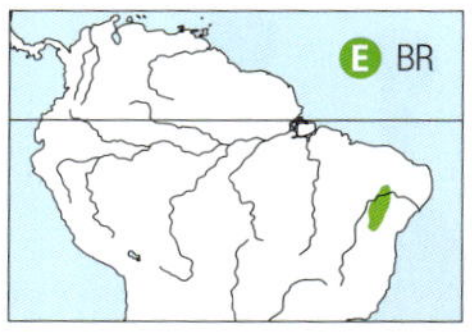

Pernambuco, Bahia

NE
Monotypic
FA 3.2–3.6 cm
W 7–9.5 g

Pale-bellied Nectar Bat
Lonchophylla inexpectata
FR Lonchophylle du Caatinga
DE Hellbäuchige Nektarfledermaus
ES Loncófilo de Caatinga

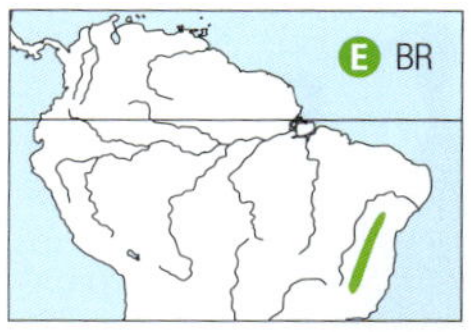

720–1130 m

Cerrado biome of Bahia & Minas Gerais

EN
Monotypic
HB 5.5–6.3 cm
T 1–1.2 cm
FA 3.6–4.1 cm
W 10–12 g

Bokermann's Nectar Bat
Lonchophylla bokermanni
FR Lonchophylle de Bokermann
DE Bokermann-Nektarfledermaus
ES Loncófilo de Bokermann

0–900 m

Atlantic Forest in Bahia, Rio de Janeiro & São Paulo; probably Espírito Santo

LC
Monotypic
FA 3.4–3.7 cm

Peracchi's Nectar Bat
Lonchophylla peracchii
FR Lonchophylle de Peracchi
DE Peracchi-Nektarfledermaus
ES Loncófilo de Peracchi

Genus *PLATALINA*

(1 species)

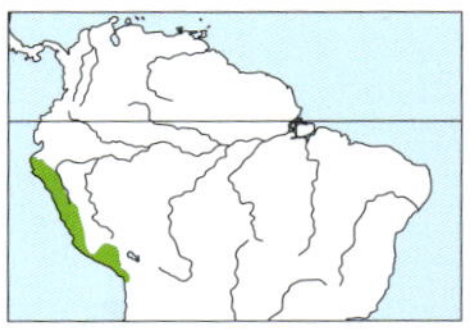

0–2600 m

W PE (from Piura to Tacna), extreme NW CL

NT
Monotypic
HB 6.7–8 cm
T 0.6–0.8 cm
FA 4.6–5.3 cm
W 8–16 g

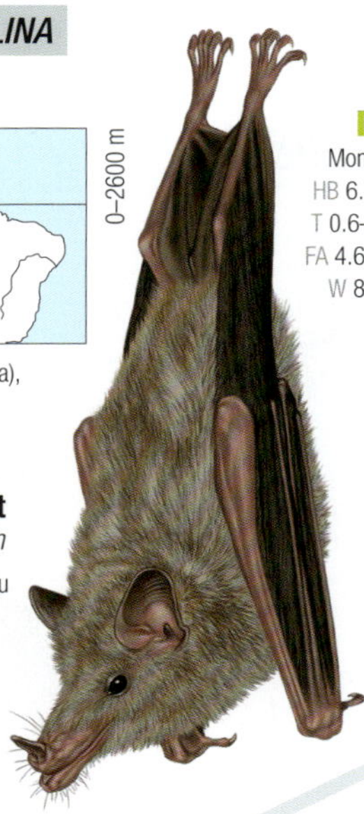

Long-snouted Bat
Platalina genovensium
FR Lonchophylle du Pérou
DE Langmaulfledermaus
ES Platalino de Perú

Genus *XERONYCTERIS*

(1 species)

From Rio Grande do Norte S to N Minas Gerais

DD
Monotypic
T 0.6–0.8 cm
FA 3.5–3.9 cm
W 8–16 g

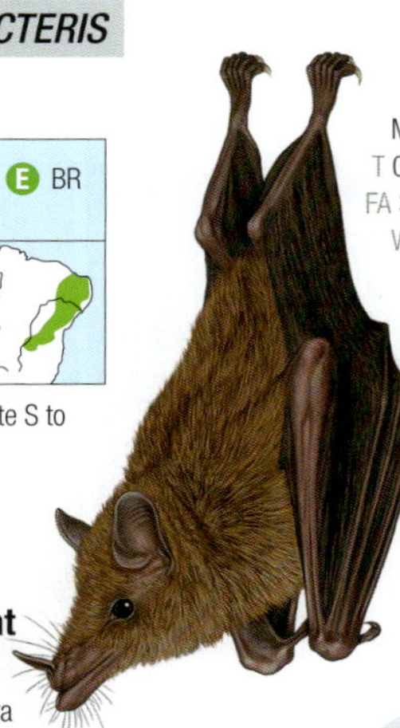

Vieira's Long-tongued Bat
Xeronycteris vieirai
FR Lonchophylle de Vieira
DE Vieira-Langzungenfledermaus
ES Xeronícteris de Vieira

up to 1000 m

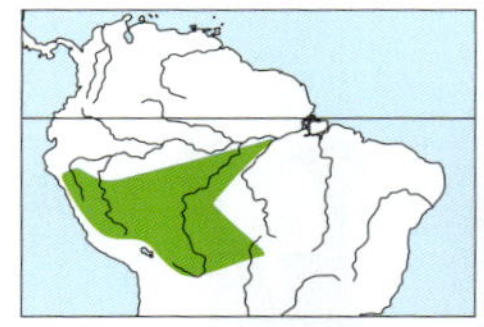

Lowland forests of W BR, E PE & N BO, S of R Amazon

LC
Monotypic
HB 5.2–5.8 cm
T 0.5–1.4 cm
FA 3.4–3.7 cm
W 11–16 g

Ben Keith's Short-tailed Bat
Carollia benkeithi
FR Carollia de Ben Keith
DE Keith-Kurzschwanzblattnase
ES Carolia de Ben Keith

Genus *CAROLLIA*

(8 species)

up to 1000 m

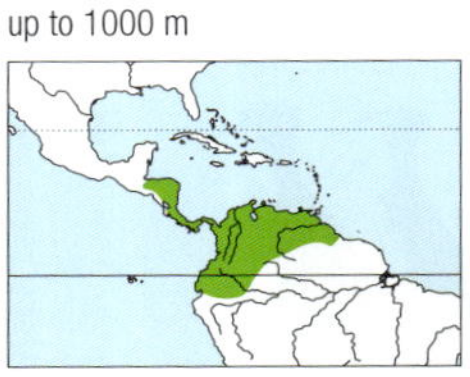

From W HN S to CO, VE, GY, EC & N PE, N of R Amazon

LC
Monotypic
HB 4.8–6 cm
T 0.7–1.4 cm
FA 3.4–3.8 cm
W 11–16 g

Chestnut Short-tailed Bat
Carollia castanea
FR Carollia d'Allen
DE Kastanienbraune Kurzschwanzblattnase
ES Carolia de Allen

0–1200 m

Pacific slope from Jalisco (W MX) to NW CR

LC
Monotypic
HB 5.6–7.8 cm
T 0.5–1.5 cm
FA 3.4–4 cm
W c. 18 g

Gray Short-tailed Bat
Carollia subrufa
FR Carollia de Hahn
DE Hahn-Kurzschwanzblattnase
ES Carolia de Hahn

1300–2250 m

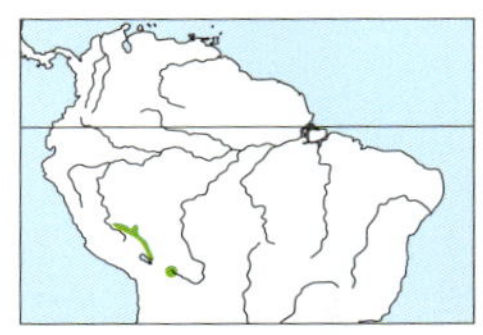

SE PE (Madre de Dios, Cusco, Puno), W & C BO (La Paz)

LC
Monotypic
HB 5.9–6.6 cm
T 0.7–1 cm
FA 4.1–4.4 cm
W 17–21 g

Manu Short-tailed Bat
Carollia manu
FR Carollia du Manu
DE Manu-Kurzschwanzblattnase
ES Carolia de Manu

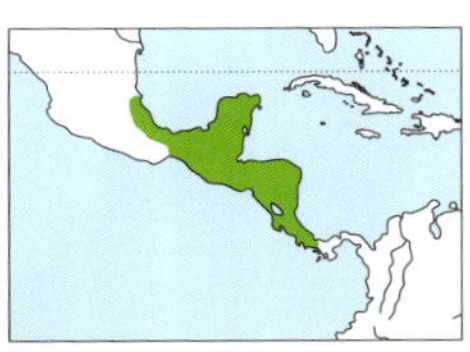

0–2400 m

MX (Gulf slope from San Luis Potosí & N Veracruz to Yucatán Pen, Oaxaca, Chiapas) S to W PA

LC
Monotypic
HB 6–7.9 cm
T 0.7–1.4 cm
FA 3.7–4.2 cm
W 12–21 g

Sowell's Short-tailed Bat
Carollia sowelli
FR Carollia de Sowell
DE Sowell-Kurzschwanzblattnase
ES Carolia de Sowell

up to 2400 m

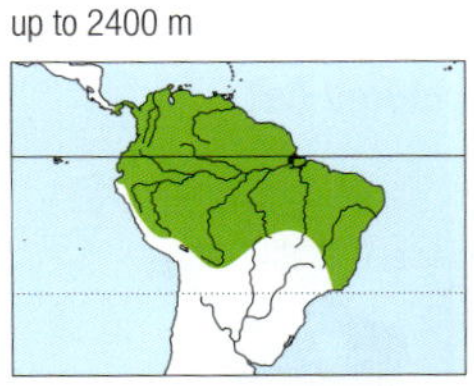

C PA (Altos de Campana NP) S to PE, N BO & SE BR

LC
Monotypic
HB 4.8–5.9 cm
T 0.6–1.2 cm
FA 3.7–4.1 cm
W 11–15 g

Silky Short-tailed Bat
Carollia brevicaudum
FR Carollia à queue courte
DE Seidige Kurzschwanzblattnase
ES Carolia de Schinz

DD
Monotypic
HB c. 5 cm
T c. 1.1 cm
FA c. 4 cm

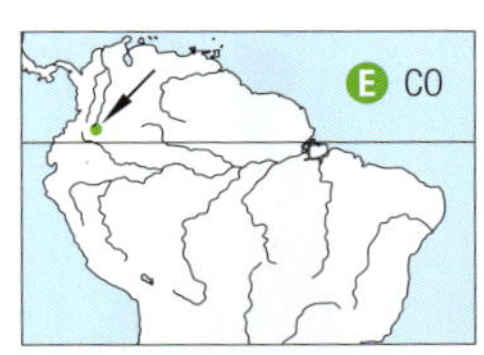

Known only from type locality in Florencia (Caquetá)

Hernandez's Short-tailed Bat
Carollia monohernandezi
FR Carollia d'Hernandez
DE Hernandez-Kurzschwanzblattnase
ES Carolia de Hernández

up to 3000 m

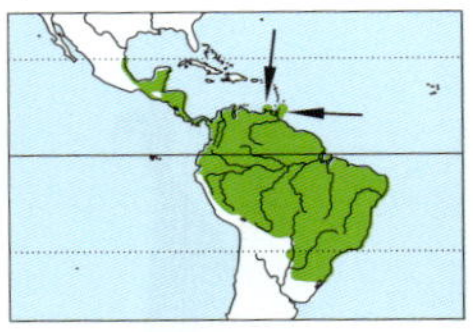

From NE MX (S Tamaulipas) S to PE, BO, PY, N AR & S BR, Margarita I (VE), TT; dubious record from GD

LC
3 sspp
HB 4.8–7 cm
T 0.8–1.6 cm
FA 4.1–4.5 cm
W 15–25 g

Seba's Short-tailed Bat
Carollia perspicillata
FR Carollia commune
DE Brillenblattnase
ES Carolia comuna

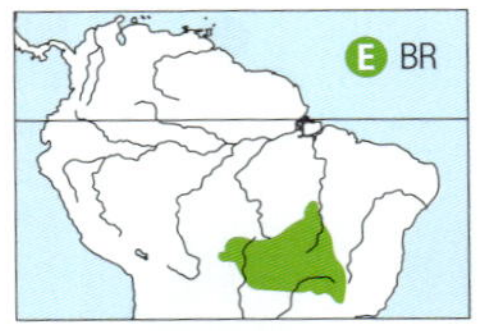

Tocantins, S Mato Grosso, SW Goiás, Minas Gerais

DD
Monotypic
T c. 0.9 cm
FA 4.5–4.7 cm

Behn's Big-eared Bat
Glyphonycteris behnii
FR Glyphonyctère de Behn
DE Behn-Großohrblattnase
ES Glifonícteris de Behn

180–1030 m

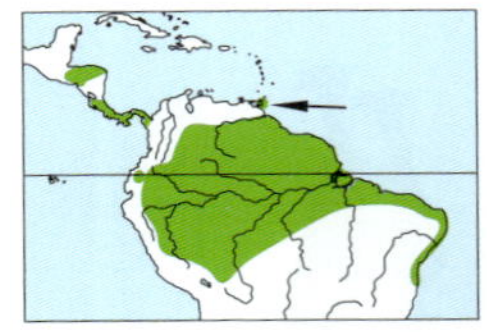

HN, CR, PA, E CO, VE, TT (Trinidad), Guianas, BR, EC, E PE, N BO; expected in NI

LC
Monotypic
HB 6.1–8.4 cm
T 0.5–1.1 cm
FA 5.1–5.9 cm
W 18–30 g

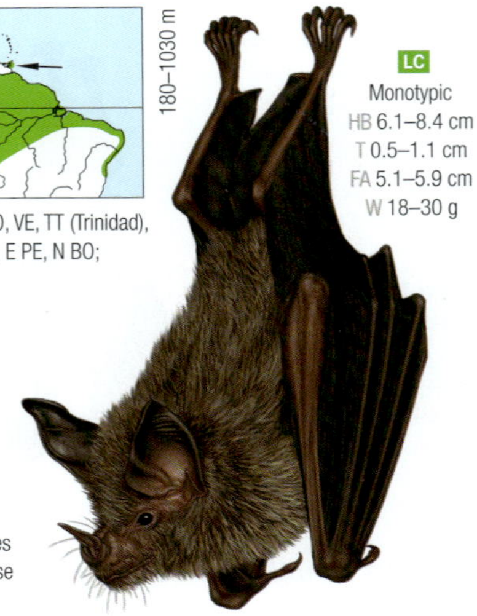

Davies's Big-eared Bat
Glyphonycteris daviesi
FR Glyphonyctère de Davies
DE Davies-Großohrblattnase
ES Glifonícteris de Davies

Genus *TRINYCTERIS*

(1 species)

25–460 m

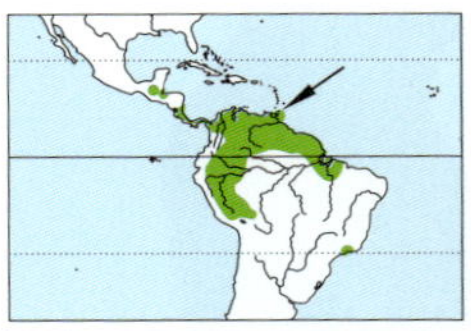

From S MX (Chiapas) S to E EC, E PE, NW BO & SE BR, TT (Trinidad)

LC
Monotypic
HB 5.5–6.8 cm
T 0.9–1.3 cm
FA 3.6–4.1 cm
W 7–8.9 g

Niceforo's Big-eared Bat
Trinycteris nicefori
FR Trinyctère de Niceforo
DE Niceforo-Großohrblattnase
ES Trinícteris de Niceforo

Genus *NEONYCTERIS*

(1 species)

DD
Monotypic
FA c. 3.4 cm

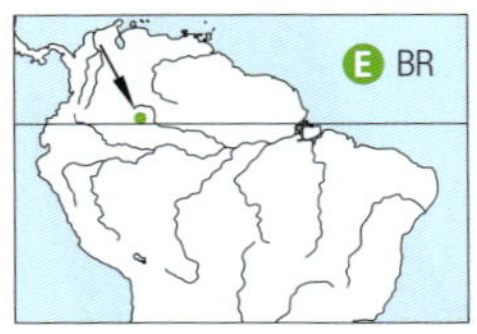

Known only from Tahuapunta (R Uaupés) in W Amazonia

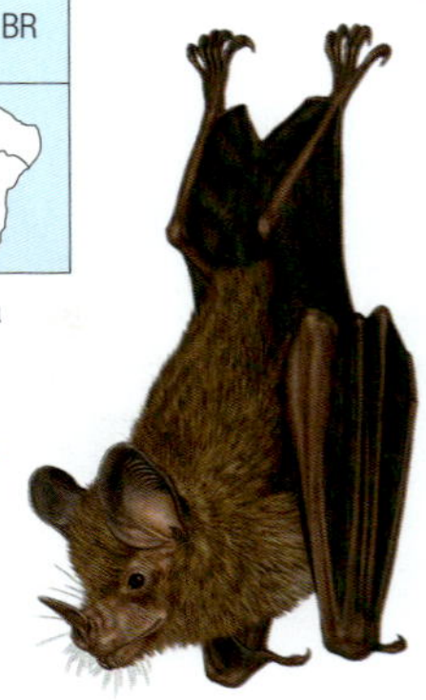

Least Big-eared Bat
Neonycteris pusilla
FR Néonyctère nain
DE Kleinst-Großohrblattnase
ES Neonícteris pequeño

Genus *GLYPHONYCTERIS*

(3 species)

LC Monotypic
HB 4.8–5.6 cm T 0.9–1.2 cm
FA 3.8–4.4 cm W 7–11 g

15–1300 m

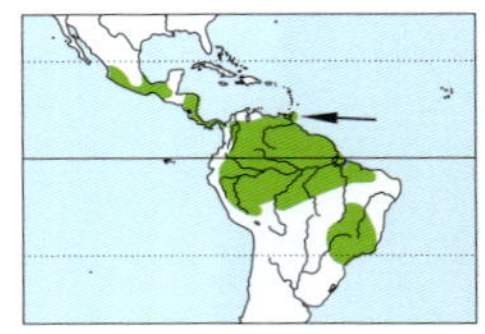

From C MX (Nayarit & Veracruz) S to E EC, E PE & E BR, TT (Trinidad)

Tricolored Big-eared Bat
Glyphonycteris sylvestris
FR Glyphonyctère sylvestre
DE Dreifarbige Großohrblattnase
ES Glifonícteris silvestre

Genus *RHINOPHYLLA*

(3 species)

0–1700 m

Pacific slope of CO (Chocó S to Nariño), EC (Esmeraldas S to Chimborazo)

NT
Monotypic
HB 4.7–6 cm
T tailless
FA 3.5–3.7 cm
W 12–16 g

Hairy Little Fruit Bat
Rhinophylla alethina
FR Rhinophylle de Handley
DE Wollhaarige Kleinfruchtfledermaus
ES Rinófilo de Handley

LC
Monotypic
HB 4.4–4.8 cm
T tailless
FA 2.9–3.2 cm
W 6.5–7 g

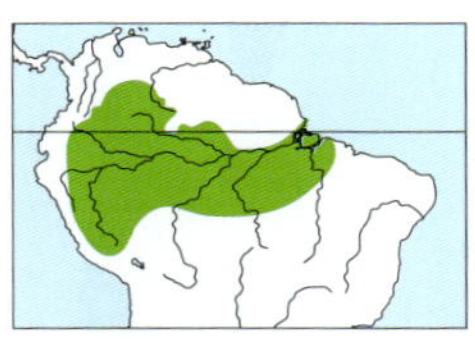

Amazon Basin of CO, VE, BR, EC & PE

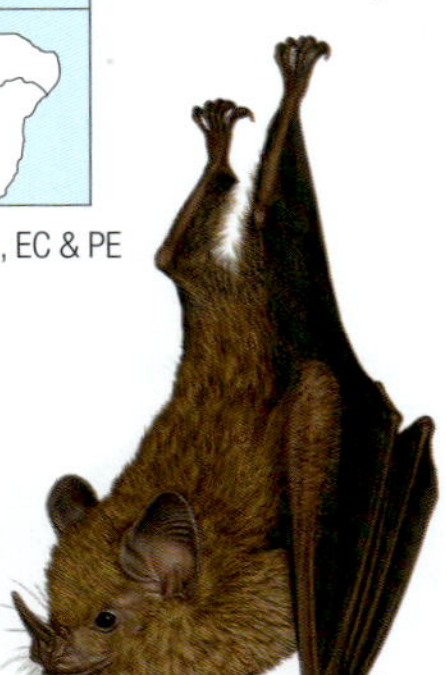

Fischer's Little Fruit Bat
Rhinophylla fischerae
FR Rhinophylle de Fischer
DE Fischer-Kleinfruchtfledermaus
ES Rinófilo de Fischer

0–1400 m

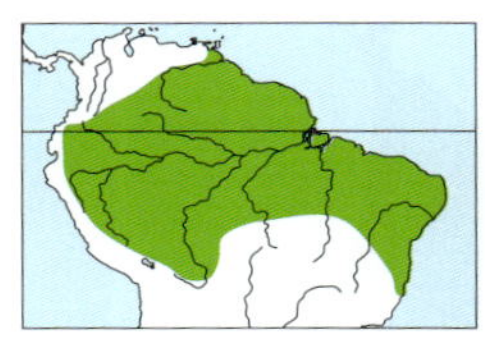

E of Andes in CO, EC, PE & BO, E to VE, Guianas & BR (Amazon & Atlantic coast S to N Espírito Santo)

LC
Monotypic
HB 4–5.7 cm
T tailless
FA 2.9–3.7 cm
W 8–13.5 g

Dwarf Little Fruit Bat
Rhinophylla pumilio
FR Rhinophylle naine
DE Zwerg-Kleinfruchtfledermaus
ES Rinófilo enano

Genus *STURNIRA*

(24 species)

1700–3000 m

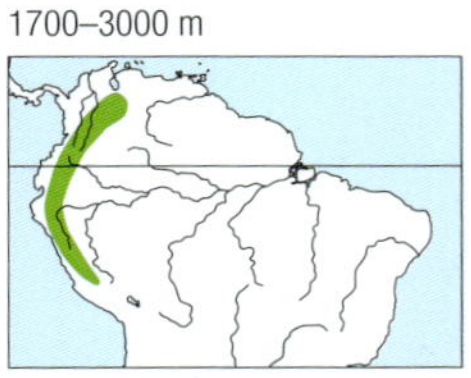

Andean regions of W VE, CO, EC & PE

LC Monotypic
HB 6.2–6.7 cm T tailless
FA 3.9–4.3 cm W 14–21 g

Bidentate Yellow-shouldered Bat
Sturnira bidens
FR Sturnire bidentée
DE Gelbschulterfledermaus
ES Esturniro bidente

Lesser Yellow-shouldered Bat
Sturnira nana
FR Sturnire naine
DE Kleine Gelbschulterfledermaus
ES Esturniro enano

1250–1670 m

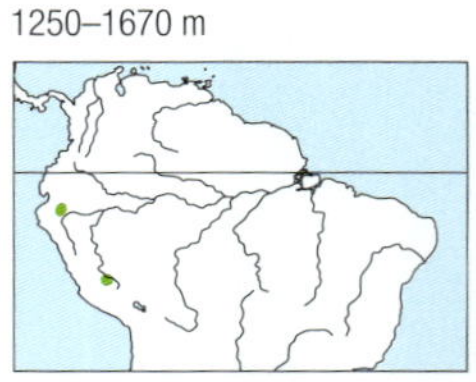

Known only from SE EC (Cordillera del Cóndor) & SC PE (Huanhuachayo, Ayacucho)

EN
Monotypic
HB 4.7–5 cm
T tailless
FA 3.3–3.6 cm
W 10–12 g

1650–3165 m

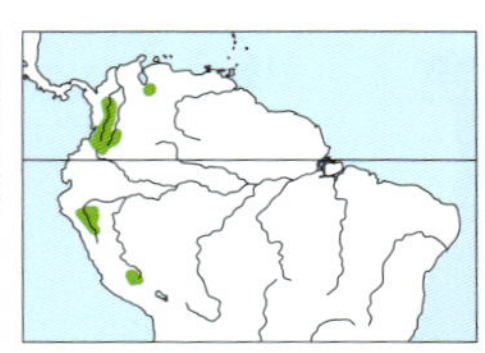

W CO, W VE, EC & PE; records from EC do not have precise localities

LC
Monotypic
HB 8.8–10.1 cm
T tailless
FA 5.5–6.1 cm
W 47–54 g

Arata-Thomas Yellow-shouldered Bat
Sturnira aratathomasi
FR Sturnire Arata-Thomas
DE Arata-Thomas-Gelbschulterfledermaus
ES Esturniro de Arata-Thomas

Honduran Yellow-shouldered Bat
Sturnira hondurensis

LC
Monotypic
HB 6.6–7 cm
T tailless
FA 4.1–4.5 cm
W 17–23 g

FR Sturnire du Honduras
DE Honduras-Gelbschulterfledermaus
ES Esturniro de Honduras

above 800 m

N MX (S from S Sinaloa & S Tamaulipas), GT, SV, HN, NI

Burton's Yellow-shouldered Bat
Sturnira burtonlimi

DD
Monotypic
HB 7–7.2 cm
T tailless
FA c. 4.4 cm
W c. 19 g

FR Sturnire de Burton Lim
DE Burton-Gelbschulterfledermaus
ES Esturniro de Burton Lim

1290–1500 m

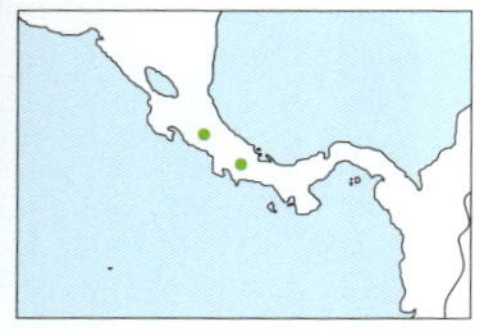

Known only from W PA (Santa Clara, Chiriquí) & C CR (Cartago)

Highland Yellow-shouldered Bat
Sturnira ludovici

LC
Monotypic
HB 6.6–7.7 cm
T tailless
FA 4.4–5 cm
W 22–30 g

FR Sturnire d'Anthony
DE Hochland-Gelbschulterfledermaus
ES Esturniro de Anthony

30–2900 m

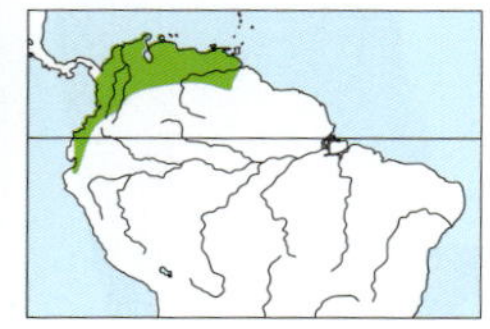

CO, VE & EC

Adriana's Yellow-shouldered Bat
Sturnira adrianae

NE
2 sspp
HB 5.5–8.5 cm
T tailless
FA 4.4–5.2 cm
W 19–35 g

FR Sturnire d'Adriana
DE Adriana-Gelbschulterfledermaus
ES Esturniro de Adriana

60–2400 m

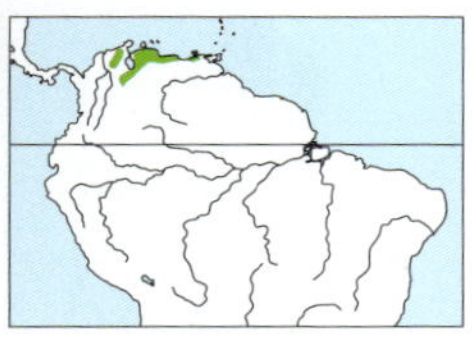

All Andean and coastal mts of VE, adjacent E Andes in CO

Tschudi's Yellow-shouldered Bat
Sturnira oporaphilum

FR Sturnire de Tschudi
DE Tschudi-Gelbschulterfledermaus
ES Esturniro de Tschudi

LC Monotypic
HB 6.6–7.5 cm T tailless
FA 4.4–4.8 cm W 17–23 g

300–2950 m

E EC, E PE, BO & NW AR

Mistratoan Yellow-shouldered Bat
Sturnira mistratensis

DD
Monotypic
HB c. 6.8 cm
T tailless
FA c. 4.3 cm
W c. 24 g

FR Sturnire de Mistrato
DE Mistrato-Gelbschulterfledermaus
ES Esturniro de Mistrató

c. 1000 m

Known from type locality (Mistrató, Risaralda) in W Andes

Soriano's Yellow-shouldered Bat
Sturnira sorianoi

DD
Monotypic
HB 5.7–6.9 cm
T tailless
FA 4.3–4.4 cm
W 17–21 g

FR Sturnire de Soriano
DE Soriano-Gelbschulterfledermaus
ES Esturniro de Soriano

up to 3000 m

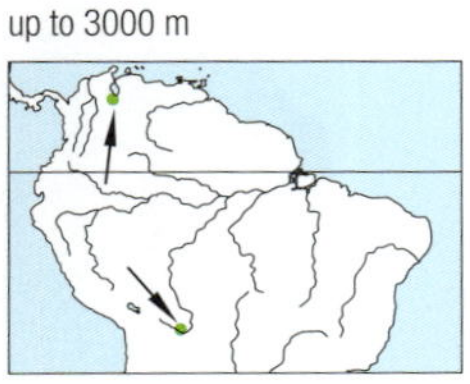

Known from isolated sites in NW VE (Mérida) & E BO (Santa Cruz)

Choco Yellow-shouldered Bat
Sturnira koopmanhilli

DD Monotypic
HB 7.2–8 cm T tailless
FA 4.8–5.3 cm
W 25–36 g

FR Sturnire du Choco
DE Chocó-Gelbschulterfledermaus
ES Esturniro del Chocó

300–2000 m

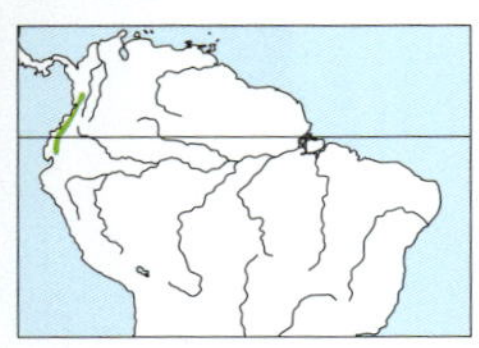

Lowlands & adjacent Andean slopes from W CO (Chocó) S to NW EC (Chimborazo)

Talamancan Yellow-shouldered Bat
Sturnira mordax

LC
Monotypic
HB 7–7.5 cm
T tailless
FA 4.3–4.9 cm
W 24–31 g

FR Sturnire de la Talamanca
DE Talamanca-Gelbschulterfledermaus
ES Esturniro de Talamanca

600–2900 m

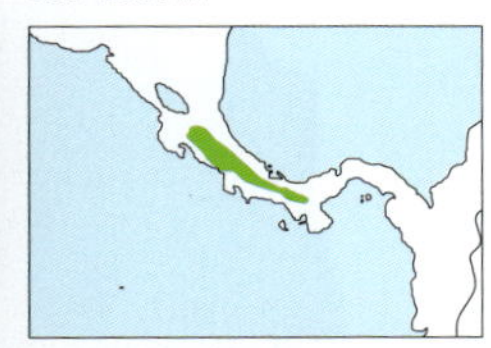

NW CR (Cordillera de Guanacaste) to W PA (Cordillera Central)

Perla Yellow-shouldered Bat
Sturnira perla

DD
Monotypic
HB 5.6–6.6 cm
T tailless
FA 4.1–4.5 cm
W c. 18 g

FR Sturnire de La Perla
DE Perla-Gelbschulterfledermaus
ES Esturniro de La Perla

35–700 m

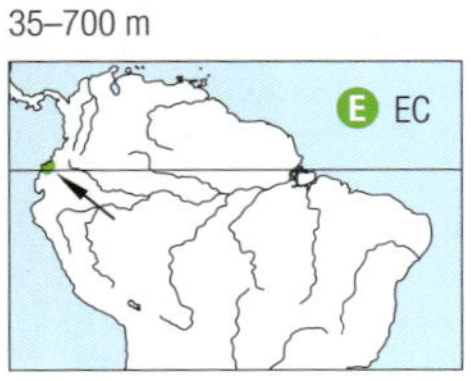

Tropical rainforest lowlands of Chocó forest; probably SW CO

Tilda's Yellow-shouldered Bat
Sturnira tildae

LC
Monotypic
HB 6.6–7 cm
T tailless
FA 4.3–5.5 cm
W 23–30 g

FR Sturnire des forêts
DE Tilda-Gelbschulterfledermaus
ES Esturniro de Tilda

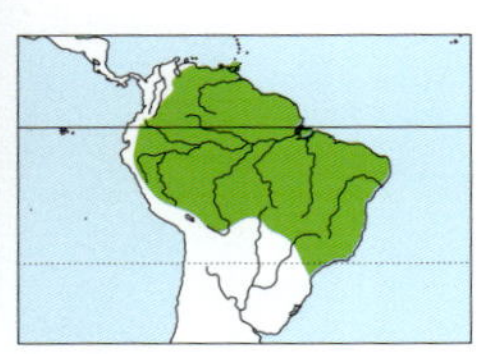

E Andean slopes & lowlands of CO, VE, TT (Trinidad), Guianas, BR, EC, PE & N BO

Hairy Yellow-shouldered Bat
Sturnira erythromos

LC Monotypic
HB 5.3–5.9 cm
T tailless
FA 3.8–4.4 cm
W 12–17 g

FR Sturnire des Andes
DE Haarige Gelbschulterfledermaus
ES Esturniro de los Andes

500–3740 m

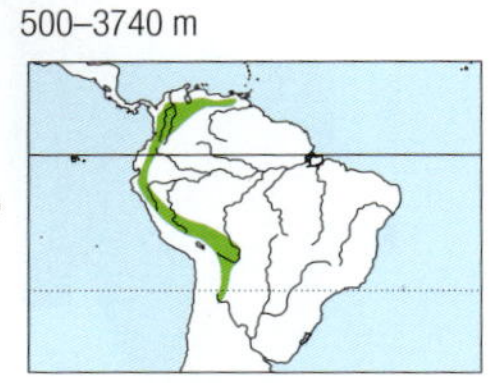

Andes of VE, CO, EC, PE, BO & N AR

LC
Monotypic
HB 6.5–6.8 cm
T tailless
FA 4.2–4.5 cm
W 19–20 g

Bogota Yellow-shouldered Bat
Sturnira bogotensis

FR Sturnire de Bogota
DE Bogotá-Gelbschulterfledermaus
ES Esturniro de Bogotá

1200–3340 m

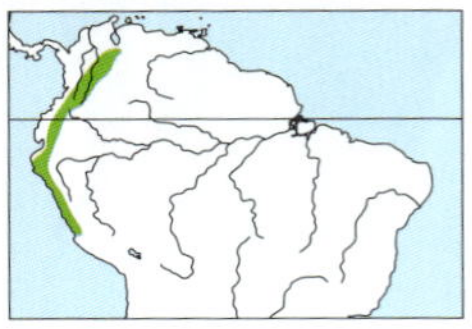

W VE, CO, EC & PE

LC Monotypic
HB 8.4–9.3 cm
T tailless
FA 5.6–6.1 cm
W 41–59 g

Greater Yellow-shouldered Bat
Sturnira magna

FR Grande Sturnire
DE Große Gelbschulterfledermaus
ES Esturniro gigante

up to 2300 m

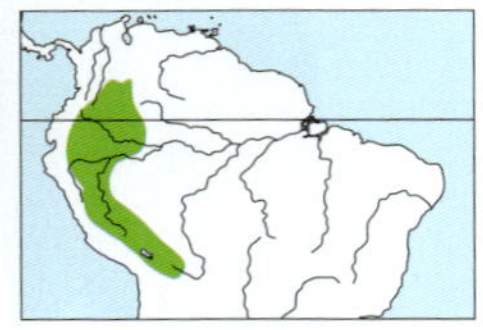

CO (W Amazon Basin) S along E of Andes in E EC, PE, N BO (La Paz, Cochabamba) & BR (Acre)

LC
Monotypic
HB 5.4–6.5 cm
T tailless
FA 3.7–4.2 cm
W 13–18 g

Northern Yellow-shouldered Bat
Sturnira parvidens

FR Sturnire du Mexique
DE Nördliche Gelbschulterfledermaus
ES Esturniro de México

up to 1600 m

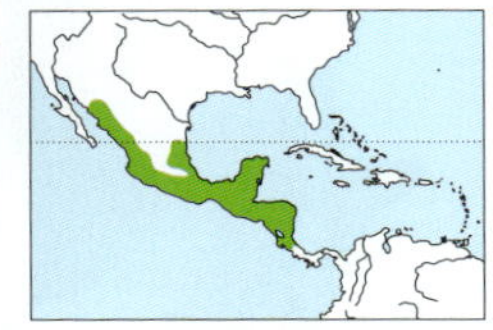

MX (from S Sonora & C Tamaulipas) S to N CR (S to Cordillera de Talamanca)

LC
Monotypic
HB 6.4–7.7 cm
T tailless
FA 4.1–4.7 cm
W 18.7–26 g

Baker's Yellow-shouldered Bat
Sturnira bakeri

FR Sturnire de Baker
DE Baker-Gelbschulterfledermaus
ES Esturniro de Baker

400–2000 m

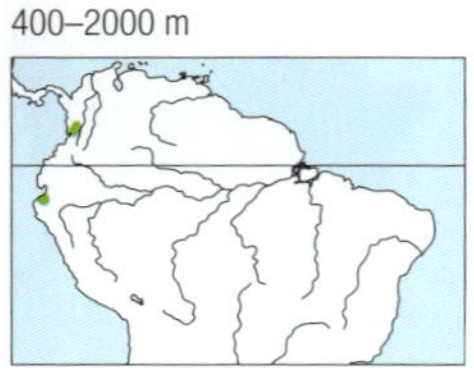

Pacific coast of CO (Chocó, Valle del Cauca), SW EC (El Oro), NW PE (Tumbes)

LC
Monotypic
HB 5–7.4 cm
T tailless
FA 3.9–4.6 cm
W 14–26 g

Little Yellow-shouldered Bat
Sturnira lilium

FR Sturnire fleur-de-lys
DE Kleinste Gelbschulterfledermaus
ES Esturniro pequeño

up to 1500 m

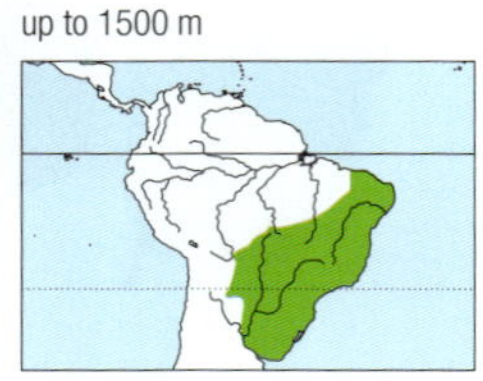

E & S BR, E BO, PY, NE AR, UY

NE
Monotypic
HB 6.5–7.3 cm
T tailless
FA 4.3–4.7 cm
W 14–25 g

Gianna's Yellow-shouldered Bat
Sturnira giannae

FR Sturnire de Gianna
DE Gianna-Gelbschulterfledermaus
ES Esturniro de Gianna

below 2000 m

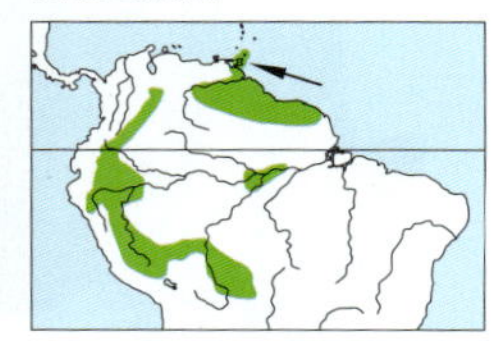

E Andes in CO, VE (E of R Orinoco), TT, Guianas, EC, PE (W of Andes in N), N BO & Amazonian BR

NT
Monotypic
HB 6.3–7.5 cm
T tailless
FA 4.2–4.7 cm

Dominica Yellow-shouldered Bat
Sturnira angeli

FR Sturnire d'Angel
DE Dominica-Gelbschulterfledermaus
ES Esturniro de Ángel

0–610 m

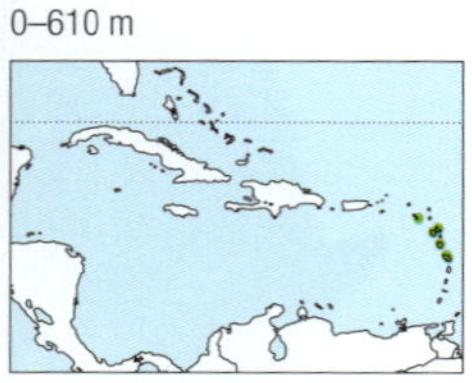

Lesser Antilles (MS, GP, DM, MQ)

NT
Monotypic
HB 6.4–6.7 cm
T tailless
FA 4–4.2 cm

Paulson's Yellow-shouldered Bat
Sturnira paulsoni

FR Sturnire de Paulson
DE Paulson-Gelbschulterfledermaus
ES Esturniro de Paulson

0–300 m

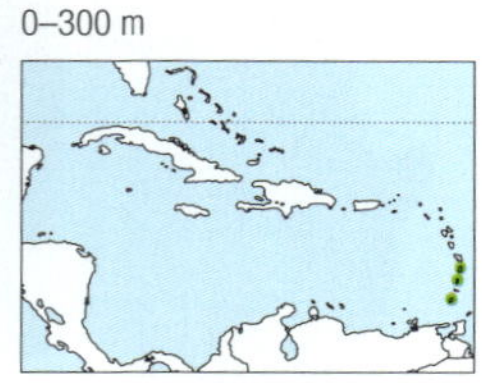

Lesser Antilles (LC, VC, GD)

LC Monotypic
HB 6.7–7.4 cm T tailless
FA 4.1–4.5 cm
W 17–25 g

Luis's Yellow-shouldered Bat
Sturnira luisi

FR Sturnire de Luis
DE Luis-Gelbschulterfledermaus
ES Esturniro de Luis

up to 700 m

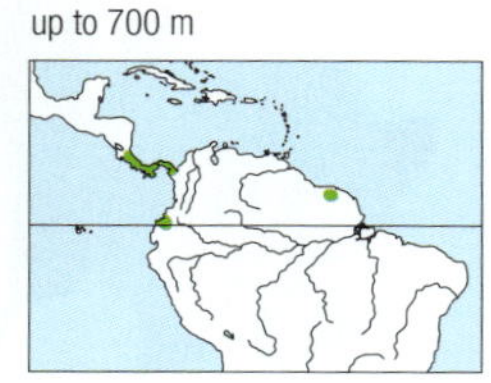

Known with certainty from CR, PA, NW EC & SR

Genus *CHIRODERMA*

(7 species)

LC
Monotypic
HB 5–7 cm
T tailless
FA 3.4–4.3 cm
W 11–22 g

Little Big-eyed Bat
Chiroderma trinitatum

FR Chiroderme orné
DE Kleine Großaugenfledermaus
ES Quiroderma pequeño

0–1000 m

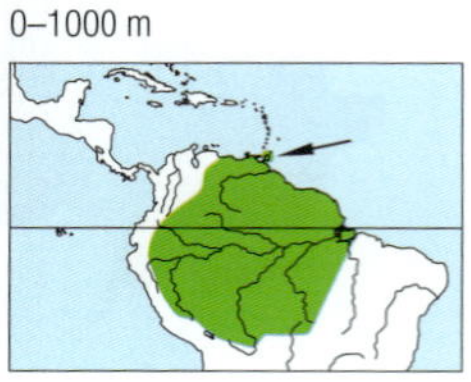

E CO, VE, TT (Trinidad), Guianas, Amazonian BR, E EC, E PE, BO

LC
Monotypic
HB 6.9–8 cm
T tailless
FA 4.5–5.5 cm
W 20–45 g

Brazilian Big-eyed Bat
Chiroderma doriae

FR Chiroderme de Doria
DE Brasilien-Großaugenfledermaus
ES Quiroderma de Brasil

0–1200 m

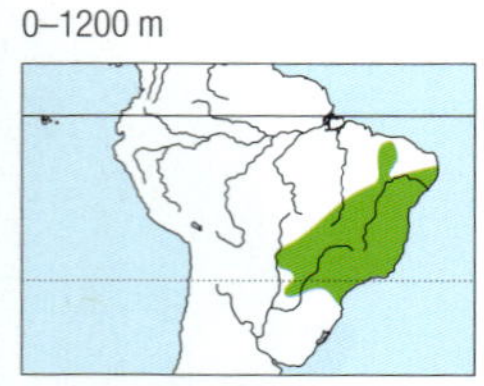

C, E & S BR (from Piauí, Ceará & Rio Grande do Norte S to Santa Catarina), E PY (San Pedro, Cordillera)

NE Monotypic
HB 5.3–6.5 cm T tailless
FA 3.7–3.9 cm W 12–19 g

Handley's Big-eyed Bat
Chiroderma gorgasi

FR Chiroderme de Gorgas
DE Gorgas-Großaugenfledermaus
ES Quiroderma de Gorgas

0–1000 m

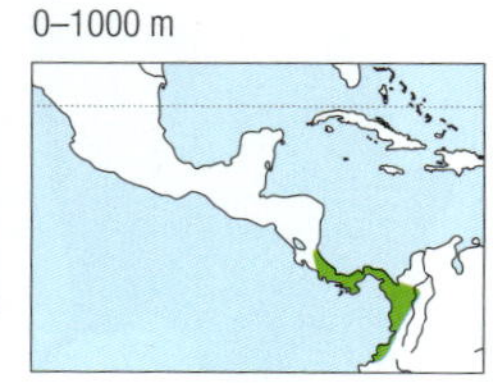

CR, PA, W CO, NW EC

Salvin's Big-eyed Bat
Chiroderma salvini

FR Chiroderme de Salvin
DE Salvin-Großaugenfledermaus
ES Quiroderma de Salvin

LC
Monotypic
HB 6.6–8.9 cm
T tailless
FA 4.5–5.3 cm
W 24–34 g

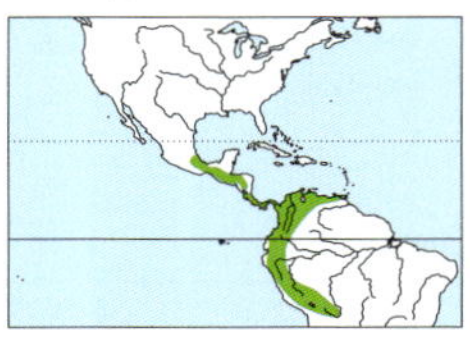

75–2240 m

From E & S MX (S from Veracruz) to N & W VE, W & N CO, W EC, E PE (Andean foothills) & W BO

Mexican Big-eyed Bat
Chiroderma scopaeum

FR Chiroderme du Mexique
DE Mexiko-Großaugenfledermaus
ES Quiroderma de México

NE
Monotypic
HB 6.1–8 cm
T tailless
FA 4.3–4.7 cm
W 19–28 g

0–1720 m

W of isthmus of Tehuantepec (from Chihuahua & Sinaloa E to Veracruz, S to Oaxaca)

Guadeloupean Big-eyed Bat
Chiroderma improvisum

FR Chiroderme de Guadeloupe
DE Guadeloupe-Großaugenfledermaus
ES Quiroderma de Guadalupe

EN
Monotypic
HB 8.5–8.7 cm
T tailless
FA 5.7–5.8 cm
W 34–35 g

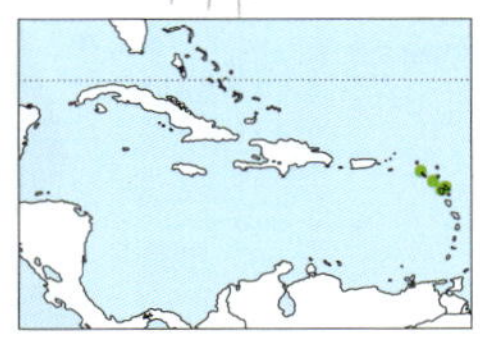

0–355 m

Lesser Antilles (KN, MS, GP)

Hairy Big-eyed Bat
Chiroderma villosum

FR Chiroderme velu
DE Haarige Großaugenfledermaus
ES Quiroderma peludo

LC
2 sspp
HB 5.8–7.9 cm
T tailless
FA 4.1–5.1 cm
W 17–30 g

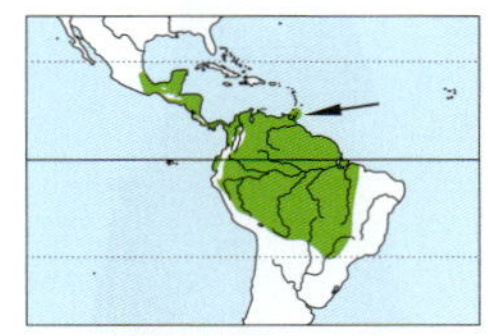

up to 950 m

S MX (S from Hidalgo & Veracruz) to W of Andes (S to PE) & E of Andes S to N BO & BR (SE to Paraná)

Genus *VAMPYRISCUS*

(3 species)

Bidentate Yellow-eared Bat
Vampyriscus bidens

FR Vampyresse bidentée
DE Zweizahn-Gelbohrenfledermaus
ES Vampiresa bidente

LC
Monotypic
HB 5–6 cm
T tailless
FA 3.5–3.7 cm
W 11–13 g

200–1000 m

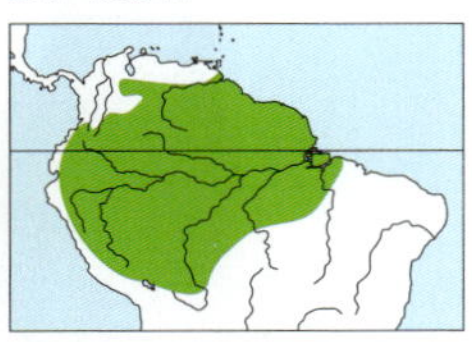

Amazonian Basin & Guianan Shield in CO, VE, Guianas, EC, PE, BR and N & C BO

Brock's Yellow-eared Bat
Vampyriscus brocki

FR Vampyresse de Brock
DE Brock-Gelbohrenfledermaus
ES Vampiresa de Brock

LC Monotypic
HB 4.5–5.5 cm T tailless
FA 3–3.3 cm W 4.8–6 g

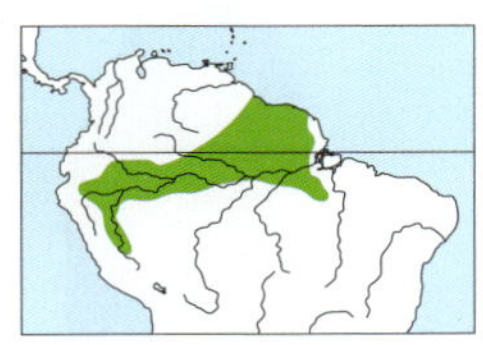

SE CO, Guianas, N BR, PE

Striped Yellow-eared Bat
Vampyriscus nymphaea

FR Grande Vampyresse
DE Streifen-Gelbohrenfledermaus
ES Vampiresa grande

LC
Monotypic
HB 4.7–7 cm
T tailless
FA 3.4–3.8 cm
W 10–13 g

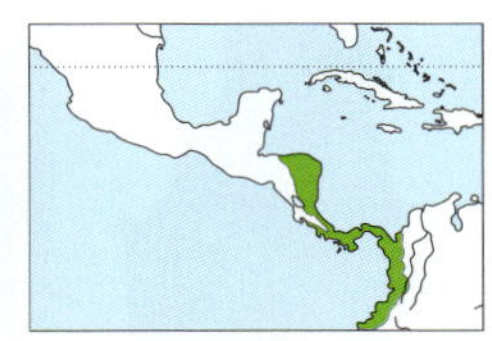

From HN to PA, along coasts of W CO & NW EC

Genus *URODERMA*

(5 species)

Baker's Tent-making Bat
Uroderma bakeri

FR Uroderme de Baker
DE Baker-Zeltfledermaus
ES Urodermo de Baker

NE
Monotypic
HB 5.3–5.8 cm
T tailless
FA 3.6–3.8 cm

500–2500 m

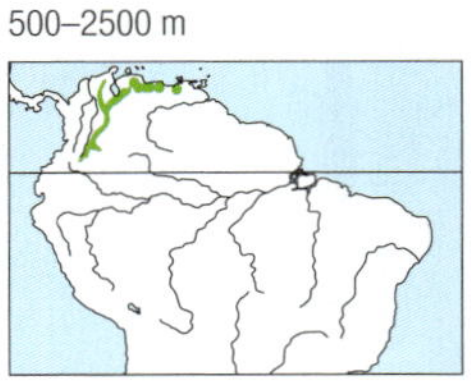

Along E Andes of CO & VE and Coastal Range in N VE

Common Tent-making Bat
Uroderma bilobatum

FR Uroderme bilobé
DE Gemeine Zeltfledermaus
ES Urodermo común

LC
3 sspp
HB 5.5–7.1 cm
T tailless
FA 3.9–4.5 cm
W 13.7–22 g

0–1850 m

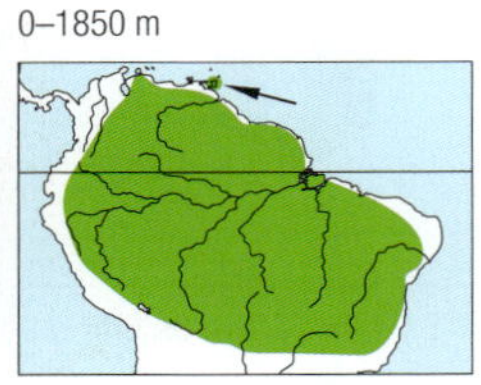

E of Andes in CO, VE, TT (Trinidad), Guianas, EC, PE, BO & BR (S to N Minas Gerais, Mato Grosso do Sul)

Pacific Tent-making Bat
Uroderma convexum

FR Uroderme de Lyon
DE Pazifik-Zeltfledermaus
ES Urodermo de Lyon

NE Monotypic
HB 5.9–6.9 cm T tailless
FA 4–4.4 cm W 13–20 g

below 1200 m

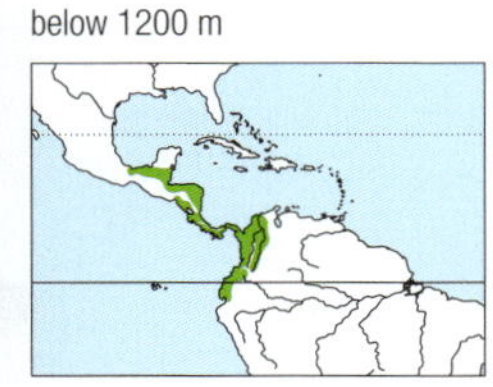

Atlantic side of S MX (except Yucatán Pen), GT & HN S along both sides to W CO & W EC

NE
Monotypic
HB 5.7–6.6 cm
T tailless
FA 4–4.3 cm
W 13–16 g

Davis's Tent-making Bat
Uroderma davisi

FR Uroderme de Davis
DE Davis-Zeltfledermaus
ES Urodermo de Davis

0–1000 m

Pacific side of MX (from E Oaxaca) S to SV; perhaps adjacent HN

LC
Monotypic
HB 5.8–6.5 cm
T tailless
FA 3.6–4.7 cm
W 12–21 g

Brown Tent-making Bat
Uroderma magnirostrum

FR Uroderme brun
DE Braune Zeltfledermaus
ES Urodermo castaño

up to 1000 m

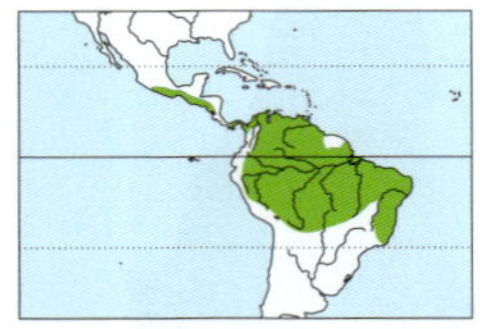

W MX (from Michoacán) S to E PE, N & C BO, BR (S to Rio de Janeiro)

Genus *VAMPYRESSA*

(6 species)

NE Monotypic
HB 5.3–5.8 cm T tailless
FA 3.6–3.8 cm

Kalko's Yellow-eared Bat
Vampyressa elisabethae

FR Vampyresse d'Elisabeth
DE Kalko-Gelbohrfledermaus
ES Vampiresa de Kalko

above 1400 m

Upper R Changena region (Bocas del Toro)

VU
Monotypic
HB 5.8–6.6 cm
T tailless
FA 3.4–3.7 cm
W 14–16 g

Melissa's Yellow-eared Bat
Vampyressa melissa

FR Vampyresse mélissa
DE Melissa-Gelbohrfledermaus
ES Vampiresa de Melissa

1180–2765 m

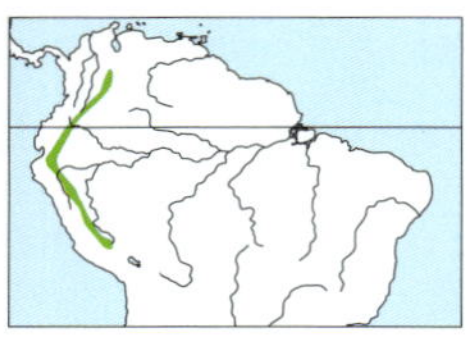

E Andes in CO, E slopes of Andes in EC & PE

NE
Monotypic
HB 5–5.6 cm
T tailless
FA 3.7–3.8 cm
W 11–14 g

Voragine's Yellow-eared Bat
Vampyressa voragine

FR Vampyresse de Chameza
DE Voragine-Gelbohrfledermaus
ES Vampiresa de Chámeza

1300–1600 m

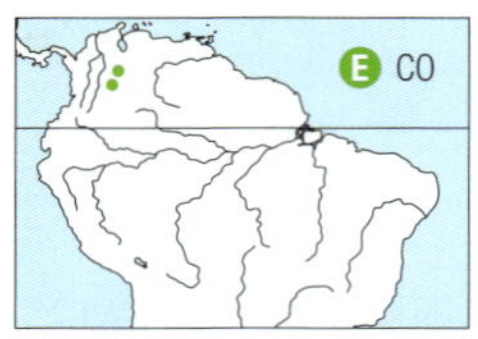

Known from 3 sites in E slope of E Andes (Casanare, Norte de Santander)

NE
Monotypic
HB c. 7 cm
T tailless
FA 3.9–4.1 cm

Quechuan Yellow-eared Bat
Vampyressa sinchi

FR Vampyresse quechua
DE Quechua-Gelbohrfledermaus
ES Vampiresa quechua

above 1600 m

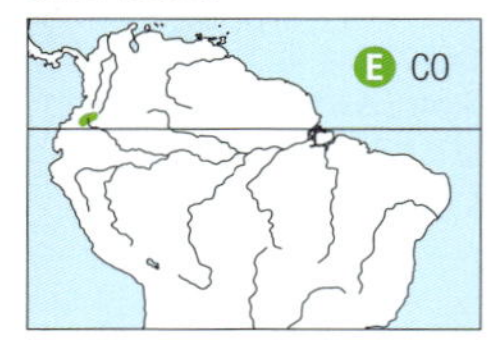

Few sites in S of E slope of E Andes (Nariño, Huila)

LC Monotypic
HB 4.3–5.2 cm T tailless
FA 2.9–3.4 cm W 6–10 g

Northern Little Yellow-eared Bat
Vampyressa thyone

FR Vampyresse thyoné
DE Nördliche Kleine Gelbohrfledermaus
ES Vampiresa de Thyone

up to 1500 m

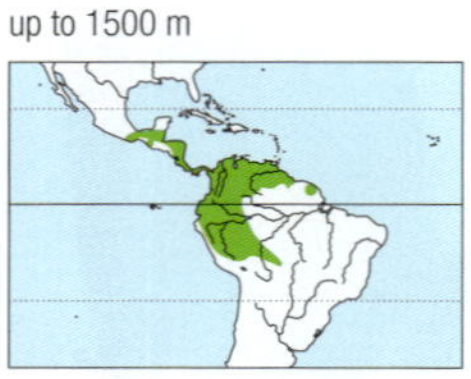

S MX (S from Veracruz & Oaxaca) to GF, N & W BR, EC, PE & BO

DD
Monotypic
HB 5.3–5.4 cm
T tailless
FA 2.9–3.7 cm
W 11–15 g

Southern Little Yellow-eared Bat
Vampyressa pusilla

FR Petite Vampyresse
DE Südliche Kleine Gelbohrfledermaus
ES Vampiresa pequeña

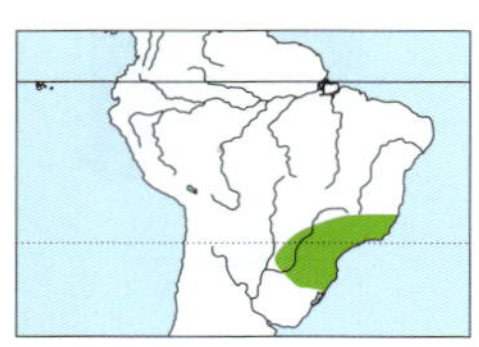

SE & S BR, adjacent E PY, NE AR

Genus *MESOPHYLLA*

(1 species)

LC
Monotypic
HB 4.5–4.9 cm
T tailless
FA 2.9–3.4 cm
W 6–8 g

MacConnell's Bat
Mesophylla macconnelli

FR Vampyresse soufrée
DE MacConnell-Fledermaus
ES Murciélago de MacConnell

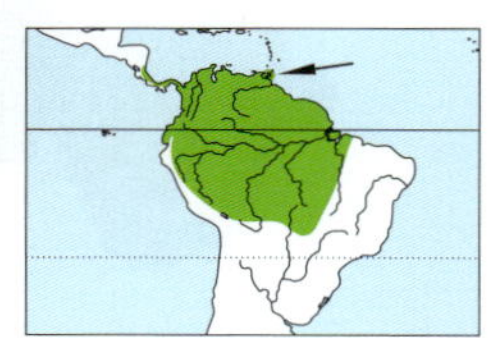

NI, CR, PA, CO, VE, TT (Trinidad), Guianas, BR, EC, E PE, N BO

Genus *VAMPYRODES*

(2 species)

Caracciolo's Stripe-faced Bat
Vampyrodes caraccioli

FR Vampyrode de Caracciolo
DE Caracciolo-Streifengesichtfledermaus
ES Vampirode de Caracciolo

LC
Monotypic
HB 7.3–7.9 cm
T tailless
FA 4.7–5.6 cm
W 29–56 g

300–1000 m

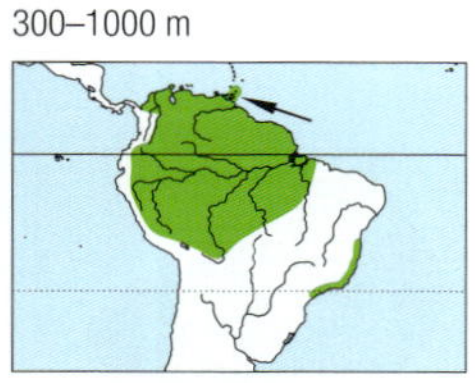

N & E CO, VE, TT, Guianas, BR, E EC, E PE, N BO

LC
Monotypic
HB 7.7–8.9 cm
T tailless
FA 4.9–5.8 cm
W 30–47 g

Great Stripe-faced Bat
Vampyrodes major

FR Grand Vampyrode
DE Große Streifengesichtfledermaus
ES Vampirode grande

0–1400 m

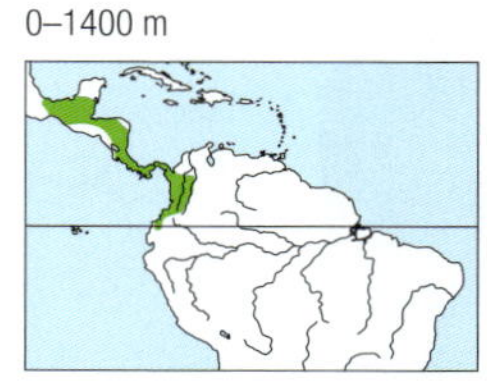

S MX (from Veracruz & Oaxaca) S to W CO & NW EC; probably SV

Genus *PLATYRRHINUS*

(19 species)

White-lined Broad-nosed Bat
Platyrrhinus lineatus

LC
Monotypic
HB 6–7.8 cm
T tailless
FA 4.3–5 cm
W 18–28 g

FR Sténoderme rayé
DE Weißlinien-Breitnasenfledermaus
ES Platirrino rayado

100–1000 m

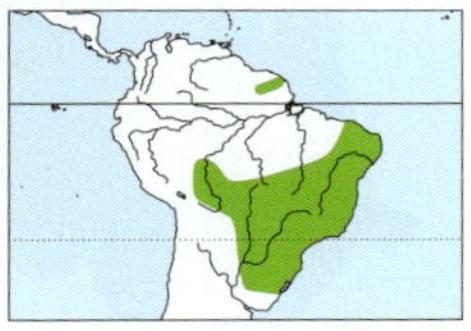

E, C & S BR, N & E BO, PY, N UY, NE AR, disjunct in S SR & GF

Short-headed Broad-nosed Bat
Platyrrhinus brachycephalus

LC
2 sspp
HB 5.4–6.2 cm
T tailless
FA 3.6–4.1 cm
W 12.1–13.9 g

FR Sténoderme à tête large
DE Kurzkopf-Breitnasenfledermaus
ES Platirrino de cabeza ancha

100–700 m

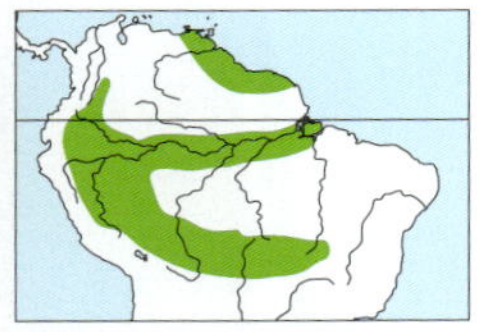

Amazonian Basin in CO, EC, PE, BR & BO, disjunct in NE VE & Guianas

Heller's Broad-nosed Bat
Platyrrhinus helleri

LC Monotypic
HB 5.5–6.5 cm T tailless
FA 3.7–4.1 cm
W 11–21 g

FR Sténoderme de Heller
DE Heller-Breitnasenfledermaus
ES Platirrino de Heller

up to 1500 m

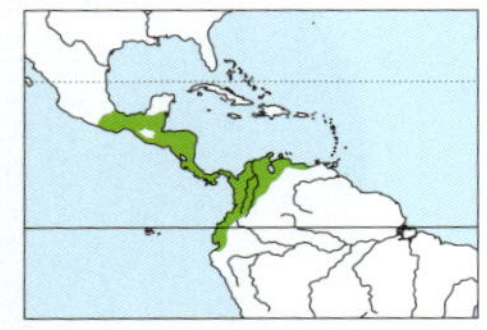

S MX (from Oaxaca & Veracruz) S, except N Yucatán Pen, to W & N CO, NW & N VE and W EC

Matapalo Broad-nosed Bat
Platyrrhinus matapalensis

NT Monotypic
HB 5.6–6.5 cm
T tailless
FA 3.7–3.9 cm
W 16–20 g

FR Sténoderme de Matapalo
DE Matapalo-Breitnasenfledermaus
ES Platirrino de Matapalo

50–680 m

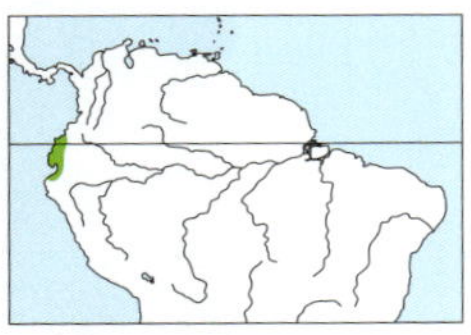

W slope of Andes in W EC (from Esmeraldas to Loja) & NW PE (Tumbes)

Recife Broad-nosed Bat
Platyrrhinus recifinus

LC
Monotypic
HB 5.7–6.3 cm
T tailless
FA 3.6–4.1 cm
W 14–19 g

FR Sténoderme de Recife
DE Recife-Breitnasenfledermaus
ES Platirrino de Recife

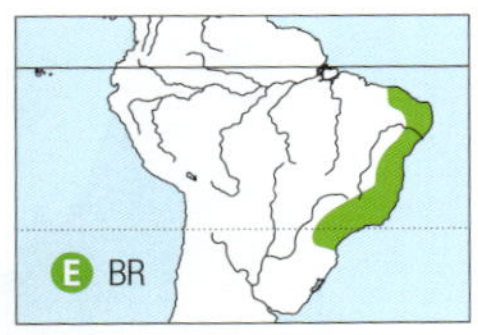

From Ceará S to São Paulo & Paraná

Guianan Broad-nosed Bat
Platyrrhinus guianensis

NE
Monotypic
HB 5.4–5.9 cm
T tailless
FA 3.7–4.1 cm
W 12–15 g

FR Sténoderme de Guyana
DE Guiana-Breitnasenfledermaus
ES Platirrino de Guyana

below 500 m

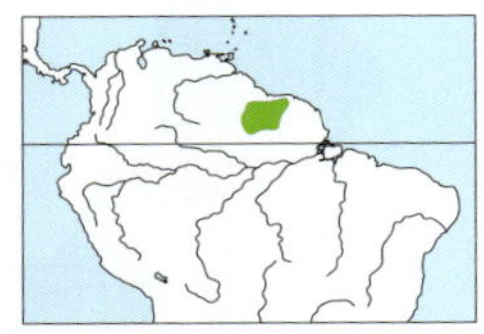

Known from Guiana Shield in GY & SR

Incan Broad-nosed Bat
Platyrrhinus incarum

LC
Monotypic
HB 5.1–6.3 cm
T tailless
FA 3.5–4.1 cm
W 10–18 g

FR Sténoderme inca
DE Inka-Breitnasenfledermaus
ES Platirrino inca

below 1200 m

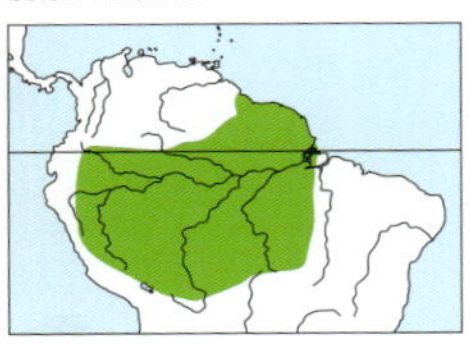

SE CO, Amazonian BR, Guianas, E EC, E PE, N & C BO

Slender Broad-nosed Bat
Platyrrhinus angustirostris

LC
Monotypic
HB 5.2–5.8 cm
T tailless
FA 3.5–3.9 cm
W c. 13 g

FR Sténoderme à nez mince
DE Zierliche Breitnasenfledermaus
ES Platirrino de nariz estrecha

150–1300 m

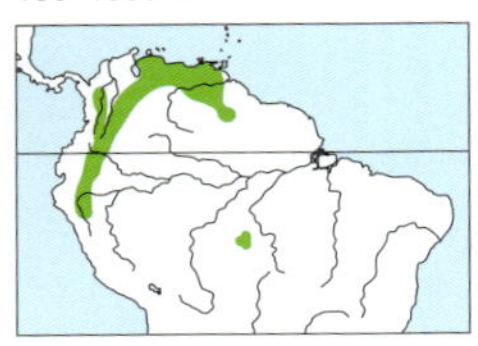

CO, VE, E EC, N PE, C BR

Brown-bellied Broad-nosed Bat
Platyrrhinus fusciventris

LC
Monotypic
HB 5.3–6.4 cm
T tailless
FA 3.5–4.2 cm
W 10–17 g

FR Sténoderme à ventre brun
DE Braunbauchbreitnasenfledermaus
ES Platirrino de vientre castaño

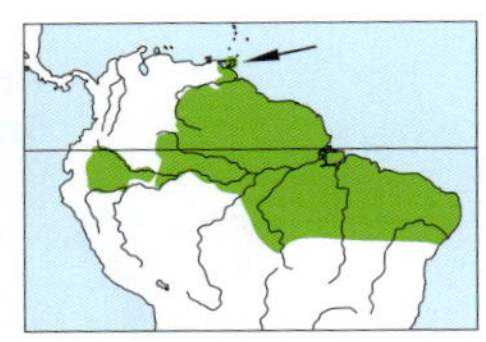

S & E VE, TT (Trinidad), Guianas, N BR, E EC, NE PE

450–3220 m

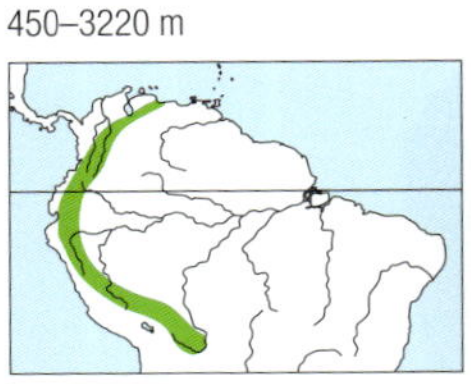

N & W VE, CO, EC, E PE, N BO

Alberico's Broad-nosed Bat
Platyrrhinus albericoi

FR Sténoderme d'Alberico
DE Alberico-Breitnasenfledermaus
ES Platirrino de Alberico

LC
Monotypic
HB 9–10 cm
T tailless
FA 5.7–6.3 cm
W 55–68 g

900–2000 m

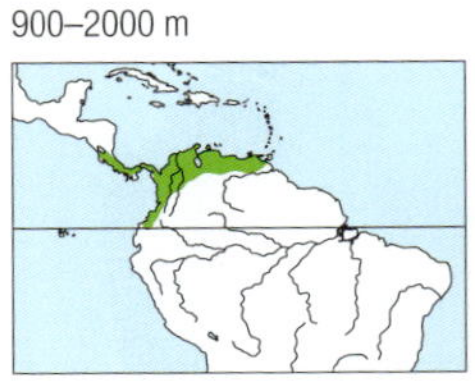

CR, PA, N & W CO, N VE, NW EC

Greater Broad-nosed Bat
Platyrrhinus vittatus

FR Grand Sténoderme
DE Große Breitnasenfledermaus
ES Platirrino grande

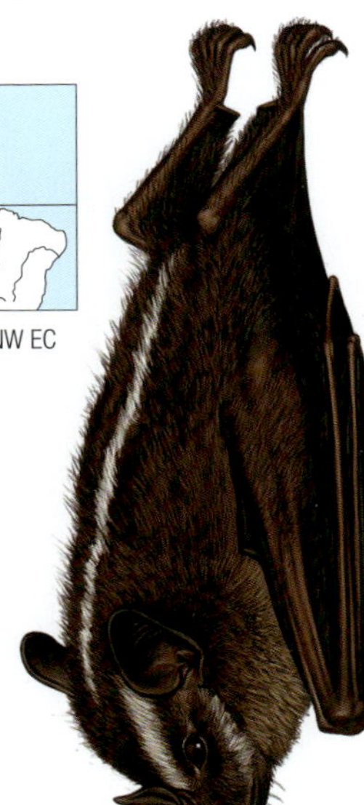

LC
Monotypic
HB c. 10 cm
T tailless
FA 5.7–6.2 cm
W 60–65 g

150–2200 m

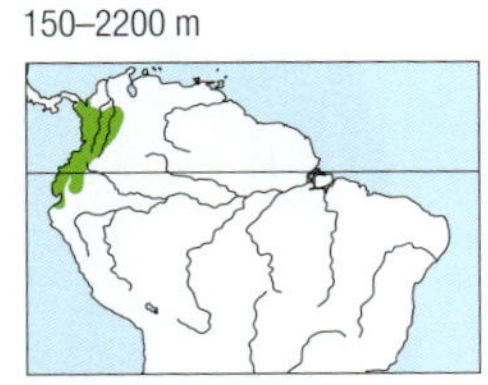

Andean slopes of CO & EC

LC
Monotypic
HB 7.1–8 cm
T tailless
FA 4.6–5 cm
W 30–40 g

Thomas's Broad-nosed Bat
Platyrrhinus dorsalis

FR Sténoderme de Thomas
DE Thomas-Breitnasenfledermaus
ES Platirrino de Thomas

Buffy Broad-nosed Bat
Platyrrhinus infuscus

FR Sténoderme de Peters
DE Gelbbraune Breitnasenfledermaus
ES Platirrino de Peters

LC
Monotypic
HB 7.7–10.5 cm
T tailless
FA 5.5–6.1 cm
W 36–59 g

180–2200 m

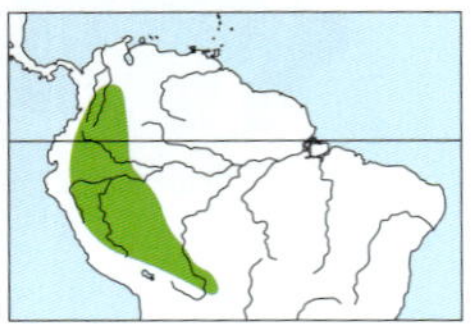

CO, E EC, E PE, NW BR, BO

Darien Broad-nosed Bat
Platyrrhinus aquilus

FR Sténoderme du Darien
DE Darién-Breitnasenfledermaus
ES Platirrino de Darién

DD
Monotypic
HB 7–7.8 cm
T tailless
FA 4.5–4.7 cm
W 16–20 g

1250–1435 m

Cerro Malí & Cerro Pirre (Darién); probably nearby NW Chocó (CO)

Eldorado Broad-nosed Bat
Platyrrhinus aurarius

FR Sténoderme d'El Dorado
DE Eldorado-Breitnasenfledermaus
ES Platirrino de El Dorado

LC
Monotypic
HB 7.1–8.4 cm
T tailless
FA 4.9–5.6 cm
W 24–48 g

400–2100 m

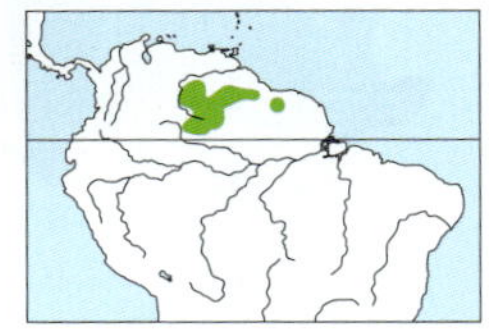

S VE, WC GY, outlying mts in C SR & N BR

Ismael's Broad-nosed Bat
Platyrrhinus ismaeli

FR Sténoderme d'Ismael
DE Ismael-Breitnasenfledermaus
ES Platirrino de Ismael

NT
Monotypic
HB 7.9–8.4 cm
T tailless
FA 5.2–5.6 cm
W 30–40 g

1230–2950 m

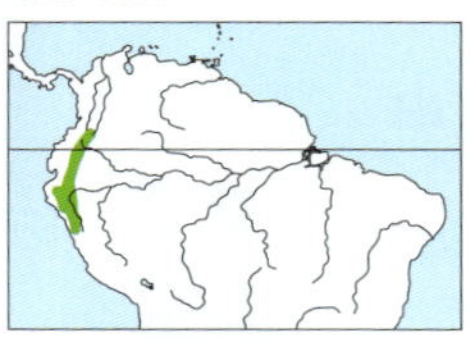

Andean slopes in SW CO, EC & NC PE

Quechuan Broad-nosed Bat
Platyrrhinus masu

FR Sténoderme quechua
DE Quechua-Breitnasenfledermaus
ES Platirrino quechua

LC
Monotypic
HB 7–8.2 cm
T tailless
FA 4.5–5.1 cm
W 23–33 g

650–3350 m

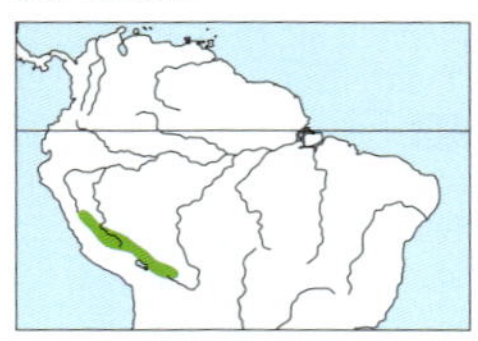

E slope of Andes from C PE to W BO

Shadowy Broad-nosed Bat
Platyrrhinus umbratus

FR Sténoderme ombré
DE Schatten-Breitnasenfledermaus
ES Platirrino sombreado

DD
Monotypic
HB 6–7.4 cm
T tailless
FA 4–4.8 cm
W 19–30 g

400–3150 m

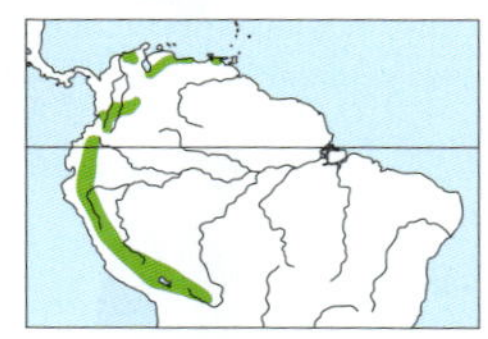

N CO and N & W VE S through EC & PE to C BO

Western Broad-nosed Bat
Platyrrhinus nitelinea

FR Sténoderme à bande dorsale
DE Westliche Breitnasenfledermaus
ES Platirrino ecuatoriano

DD
Monotypic
HB 8.4–9 cm
T tailless
FA 5.3–5.8 cm

10–1100 m

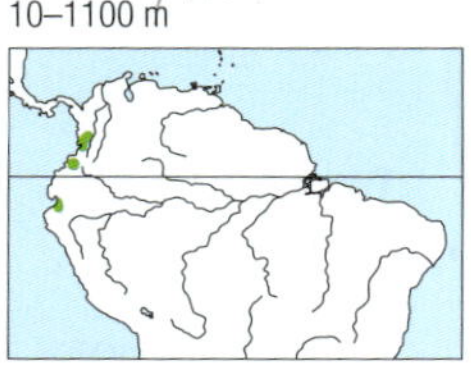

W CO (Chocó, Valle del Cauca, Nariño), SW EC (Guayas, El Oro)

Genus *ENCHISTHENES*

(1 species)

Velvety Fruit-eating Bat
Enchisthenes hartii

FR Sténoderme de Hart
DE Hart-Speernase
ES Platirrino de Hart

LC
Monotypic
HB 5.5–6.8 cm
T tailless
FA 3.8–4.2 cm
W 14–18 g

up to 3540 m

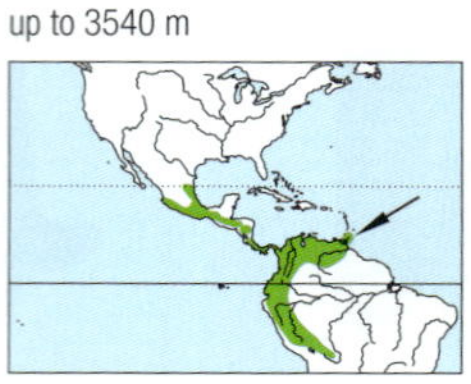

MX (from C Jalisco & S Tamaulipas) S to CO, VE, TT, EC, PE & BO

Genus *ECTOPHYLLA*

(1 species)

Honduran White Bat
Ectophylla alba

FR Sténoderme blanc
DE Honduras-Weißfledermaus
ES Murciélago hondureño blanco

NT
Monotypic
HB 3.7–4.7 cm
T tailless
FA 2.8–2.9 cm
W 5–6 g

up to 750 m

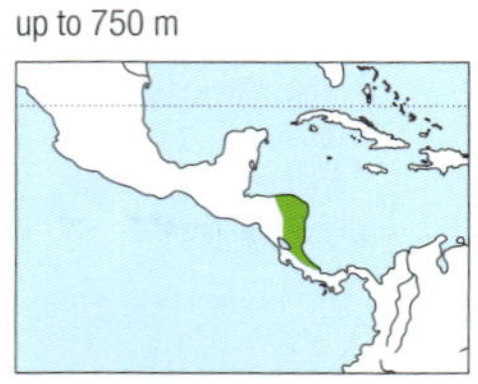

Caribbean slope in E HN, E NI, E CR & extreme NW PA

Genus *KOOPMANIA*

(1 species)

Brown Fruit-eating Bat
Koopmania concolor

FR Artibée unicolore
DE Brauner Fruchtvampir
ES Artibeo unicolor

LC
Monotypic
HB 5.9–6.4 cm
T tailless
FA 4.4–4.7 cm
W 18–20 g

100–1000 m

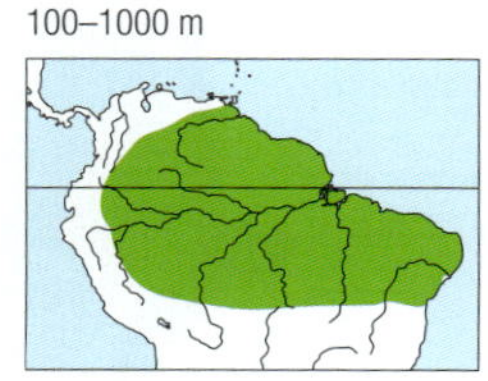

E CO, S & E VE, Guianas, BR, E EC, E PE, N BO

Genus *ARTIBEUS*

(11 species)

Honduran Fruit-eating Bat
Artibeus inopinatus

FR Artibée du Honduras
DE Honduras-Fruchtvampir
ES Artibeo hondureño

DD

Monotypic
T tailless
FA 4.8–5.3 cm
W 25–36 g

up to 1100 m

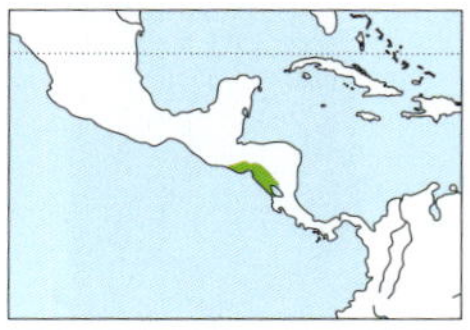

Pacific slopes from SV to NI

Fraternal Fruit-eating Bat
Artibeus fraterculus

FR Artibée d'Anthony
DE Peru-Fruchtvampir
ES Artibeo de Anthony

LC

Monotypic
HB 6.4–7.6 cm
T tailless
FA 5.2–5.9 cm
W 30–55 g

0–2150 m

Pacific coast from W EC (Manabí) to W PE (Lima)

Hairy Fruit-eating Bat
Artibeus hirsutus

FR Artibée hirsute
DE Haariger Fruchtvampir
ES Artibeo piloso

LC

Monotypic
HB 6.9–9 cm
T tailless
FA 5.2–5.8 cm
W 32–47 g

up to 2575 m

Pacific slope from C Sonora to Morelos & S Guerrero

Fringed Fruit-eating Bat
Artibeus fimbriatus

FR Artibée de Gray
DE Fransenfruchtvampir
ES Artibeo de Gray

LC

Monotypic
HB 8.7–10 cm
T tailless
FA 6–7.1 cm
W 48–60 g

0–800 m

E & S BR (from Ceará S to Rio Grande do Sul), E PY, NE AR (Formosa, Chaco, Misiones)

Ecuadorian Fruit-eating Bat
Artibeus aequatorialis

FR Artibée d'Équateur
DE Ecuador-Fruchtvampir
ES Artibeo de Ecuador

LC

Monotypic
HB 7.3–9.4 cm
T tailless
FA 6.5–7.5 cm
W 42–65 g

up to 1100 m

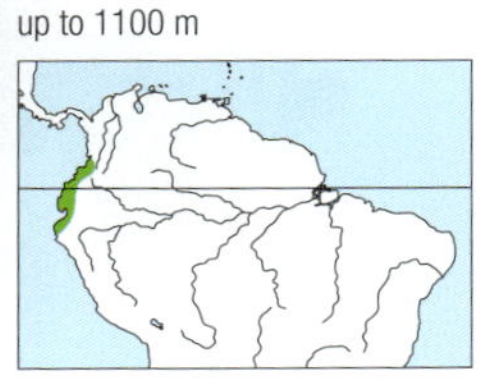

W of Andes in W CO, W EC & extreme NW PE

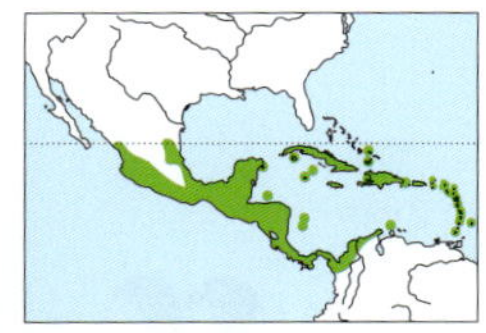

From MX (Sinaloa & Tamaulipas) S to NW CO, many Caribbean Is

LC

6 sspp
HB 7.8–9.3 cm
T tailless
FA 5.2–6.2 cm
W 43–55 g

Jamaican Fruit-eating Bat
Artibeus jamaicensis

FR Artibée fer-de-lance
DE Jamaika-Fruchtvampir
ES Artibeo de Jamaica

Dark Fruit-eating Bat
Artibeus obscurus

FR Artibée noirâtre
DE Dunkler Fruchtvampir
ES Artibeo oscuro

LC

Monotypic
HB 7.2–9 cm
T tailless
FA 5.2–6.5 cm
W 28–52 g

0–1660 m

E of Andes in E CO & VE (S of Llanos), Guianas, EC, PE, N BO & most of BR

Schwartz's Fruit-eating Bat
Artibeus schwartzi

FR Artibée de Schwartz
DE Schwartz-Fruchtvampir
ES Artibeo de Schwartz

DD

Monotypic
HB 8.5–9.6 cm
T tailless
FA 6–6.7 cm
W 42–46 g

0–120 m

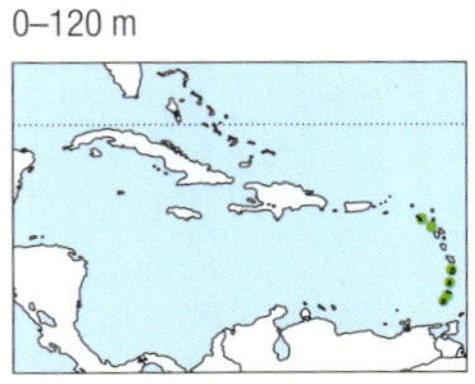

Lesser Antilles (KN, MS, LC, VC, GD)

Great Fruit-eating Bat
Artibeus lituratus

FR Artibée rayée
DE Großer Fruchtvampir
ES Artibeo grande

0–2620 m

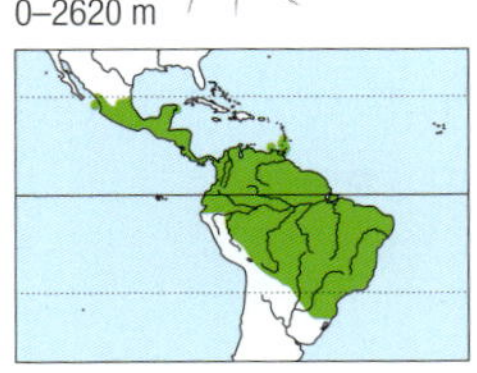

C & S MX (with Cozumel & Marías Is) S to EC, PE, BO, PY, NE AR & S BR, S Lesser Antilles

LC

4 sspp
HB 7.2–10.1 cm
T tailless
FA 5.9–6.8 cm
W 43–59 g

Large Fruit-eating Bat
Artibeus amplus

FR Artibée de Handley
DE Venezuela-Fruchtvampir
ES Artibeo de Handley

LC

Monotypic
HB 8–10.4 cm
T tailless
FA 6.5–7.5 cm
W 46–65 g

25–1700 m

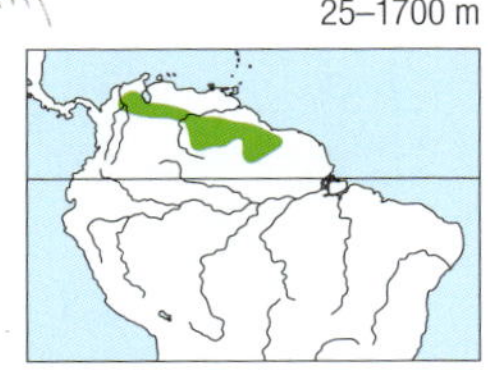

N CO, W & S VE, C & S GY, C SR; doubtful record from N BR (Amazonas)

Flat-faced Fruit-eating Bat
Artibeus planirostris

LC
5 sspp
HB 6.7–9.7 cm
T tailless
FA 6.4–7 cm
W 40–69 g

FR Artibée commune
DE Flachgesicht-Fruchtvampir
ES Artibeo común

0–2000 m

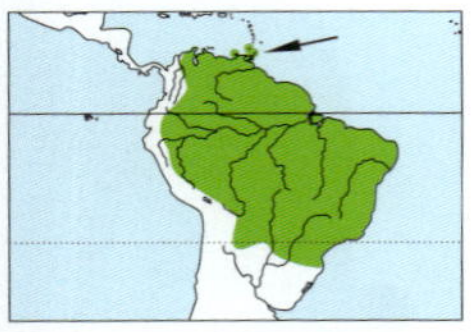

CO, VE (with Margarita I), TT, GD, Guianas, Amazon Basin of BR S to EC, PE, S BO, PY & N AR

Genus *DERMANURA*

(11 species)

Rosenberg's Fruit-eating Bat
Dermanura rosenbergi

DD
Monotypic
HB 5–5.6 cm
T tailless
FA 3.6–4.1 cm
W 9–15 g

FR Dermanure de Rosenberg
DE Rosenberg-Fruchtvampir
ES Dermanura de Rosenberg

up to 1000 m

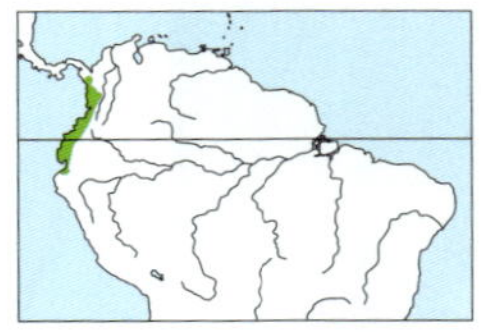

NW CO (Chocó) through Pacific slope of Andes to S EC

Thomas's Fruit-eating Bat
Dermanura watsoni

LC
Monotypic
HB 5–5.8 cm
T tailless
FA 3.5–4.1 cm
W 9–15.6 g

FR Dermanure de Watson
DE Thomas-Fruchtvampir
ES Dermanura de Watson

up to 1500 m

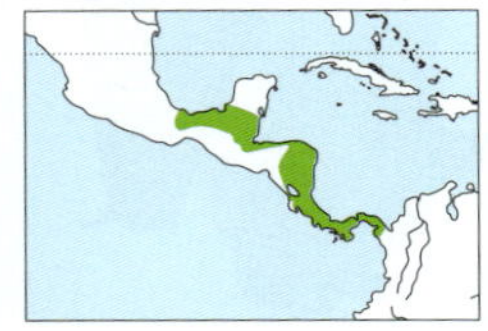

S MX (Veracruz) through mainly Caribbean lowlands to CR & PA

Toltec Fruit-eating Bat
Dermanura tolteca

LC
2 sspp
HB 5.1–6.3 cm
T tailless
FA 3.7–4.2 cm
W 15–20 g

FR Dermanure toltèque
DE Tolteken-Fruchtvampir
ES Dermanura tolteca

600–1500 m

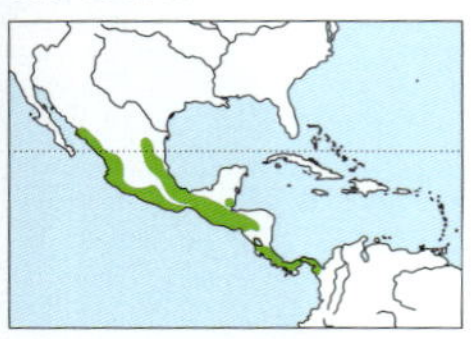

MX (from Sinaloa & Nuevo León) S to NC & SW NI (Ometepe I), CR & PA

Pygmy Fruit-eating Bat
Dermanura phaeotis

LC
3 sspp
HB 5.1–6 cm
T tailless
FA 3.5–4.2 cm
W 8–15.6 g

FR Dermanure pygmée
DE Zwergfruchtvampir
ES Dermanura pigmea

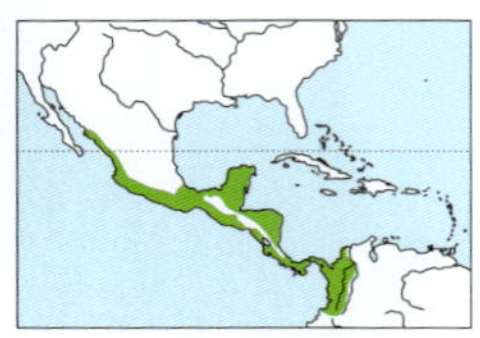

W & S MX (S from Sinaloa) S to W CO; unverified in W EC, N PE, VE, N BR & GY

Gervais's Fruit-eating Bat
Dermanura cinerea

LC
2 sspp
HB 4.7–5.9 cm
T tailless
FA 3.8–4.2 cm
W 10–18 g

FR Dermanure cendrée
DE Gervais-Fruchtvampir
ES Dermanura de Gervais

below 350 m

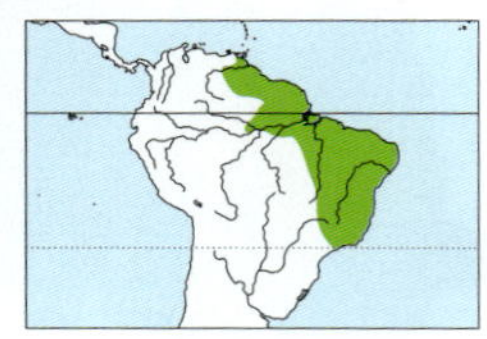

E VE (S of R Orinoco), Guianas, BR (lower Amazon Basin & E)

Andersen's Fruit-eating Bat
Dermanura anderseni

FR Dermanure d'Andersen
DE Andersen-Fruchtvampir
ES Dermanura de Andersen

LC
Monotypic
HB 4.6–5.5 cm
T tailless
FA 3.4–3.8 cm
W 8–9 g

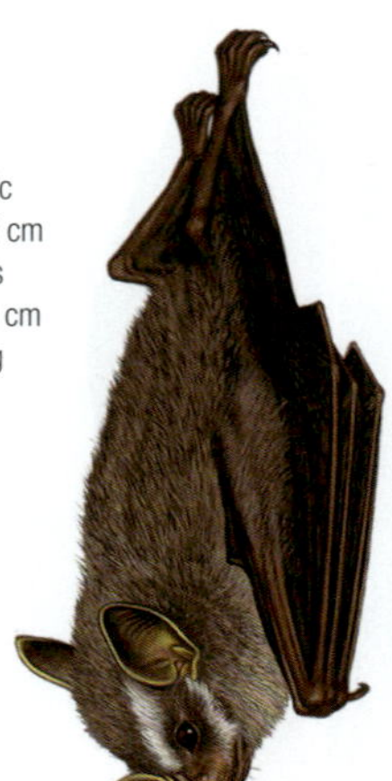

below 1000 m

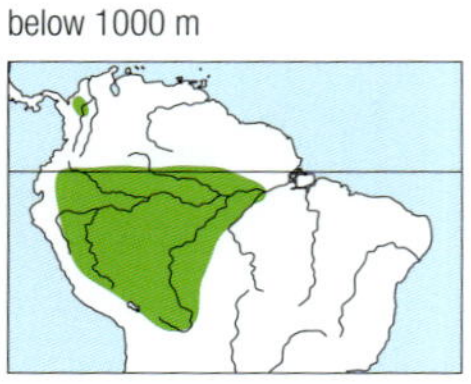

W Amazon Basin in SE CO, NW & W BR, E EC, E PE & BO, isolated in NW CO (Antioquia)

Little Fruit-eating Bat
Dermanura rava

FR Dermanure jaunâtre
DE Kleiner Fruchtvampir
ES Dermanura amarilla

LC
Monotypic
HB 4.4–6.2 cm
T tailless
FA 3.7–4.2 cm
W 10–17 g

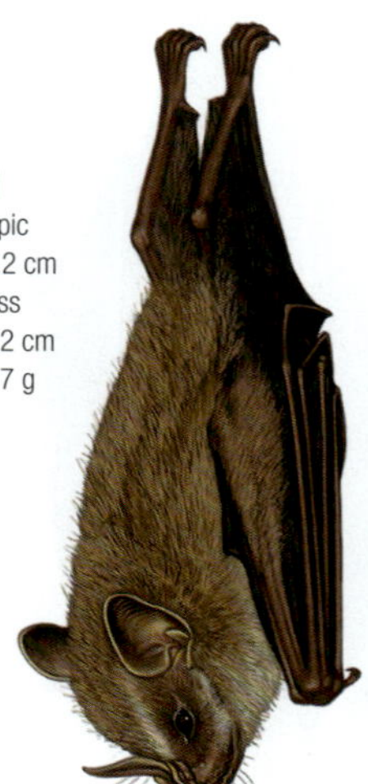

up to 2000 m

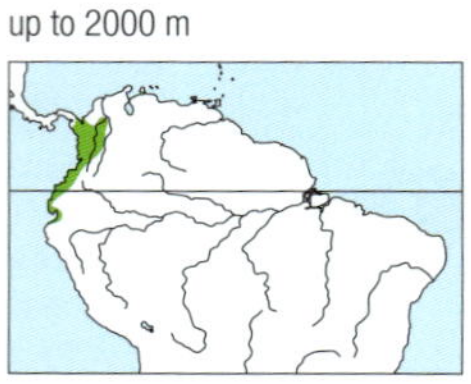

W CO (Chocó, Cauca & Magdalena valleys) S through W slope of Andes to SW EC & NW PE (Tumbes)

Aztec Fruit-eating Bat
Dermanura azteca

FR Dermanure aztèque
DE Azteken-Fruchtvampir
ES Dermanura azteca

LC
3 sspp
HB 5.9–7.5 cm
T tailless
FA 4.1–4.9 cm
W 15–33 g

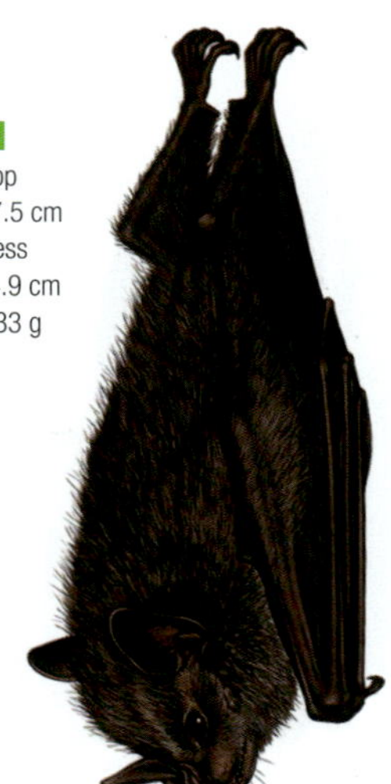

600–3300 m

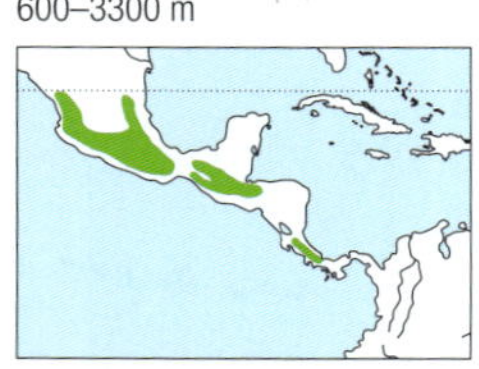

Disjunct in highlands of MX (from Sinaloa & Nuevo León) S to CR & PA

Bogota Fruit-eating Bat
Dermanura bogotensis

FR Dermanure de Bogota
DE Bogotá-Fruchtvampir
ES Dermanura de Bogotá

LC
Monotypic
HB 4.2–6.3 cm
T tailless
FA 3.7–4.3 cm
W 11–16 g

360–2200 m

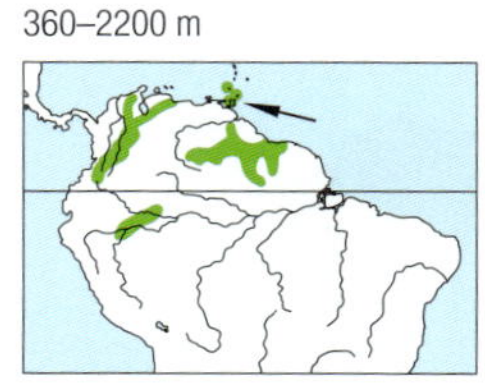

CO, VE, GY, SR, N BR, NE PE, TT, GD

LC
Monotypic
HB 4.3–6.2 cm
T tailless
FA 3.7–4.2 cm
W 10–14 g

Silvery Fruit-eating Bat
Dermanura glauca
FR Dermanure de Thomas
DE Silberfruchtvampir
ES Dermanura de Thomas

360–2600 m

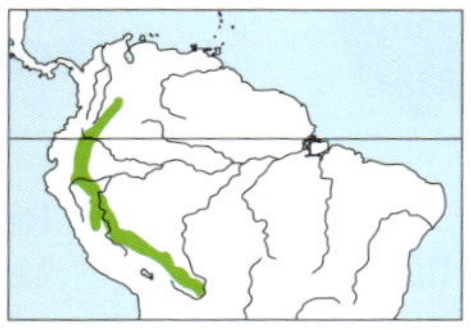

E CO, E EC, E PE, N & C BO; records from SE VE & N BR probably *D. bogotensis*

LC
Monotypic
HB 4.4–5.4 cm
T tailless
FA 3.4–3.8 cm
W 8–13 g

Dwarf Fruit-eating Bat
Dermanura gnoma
FR Dermanure naine
DE Zwergsilberfruchtvampir
ES Dermanura enana

100–680 m

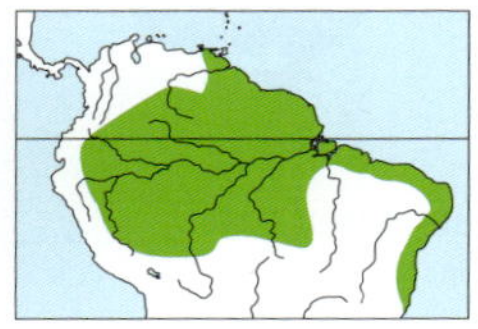

E of Andes in E VE, Guianas, E CO, E EC, E PE, BO & BR

Genus *ARITEUS*

(1 species)

LC
Monotypic
HB 5–6.7 cm
T tailless
FA 3.6–4.4 cm
W 9.2–13.1 g

Jamaican Fig-eating Bat
Ariteus flavescens
FR Arite des figuiers
DE Jamaika-Feigenfledermaus
ES Ariteo de Jamaica

below 1500 m

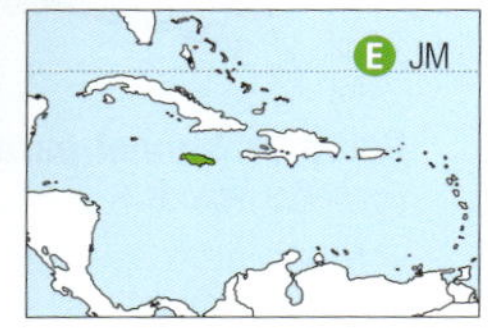

Frequent in natural forest and more disturbed habitats

Genus *ARDOPS*

(1 species)

Tree Bat
Ardops nichollsi
FR Ardops des Petites Antilles
DE Baumfledermaus
ES Ardops de Dominica

LC 5 sspp
HB 6–6.7 cm
T tailless
FA 4.2–4.8 cm
W 15.1–23 g

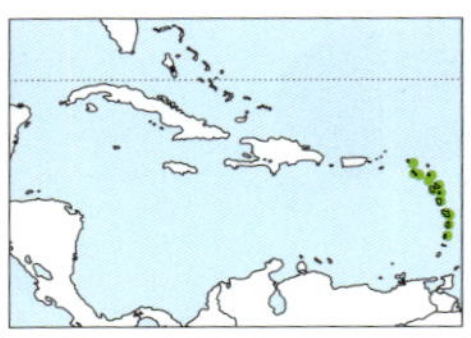

Lesser Antilles, from MS & BQ (Saba, St Eustatius) S to VC (St Vincent)

Genus *STENODERMA*

(1 species)

NT
2 sspp
HB 6–6.9 cm
T tailless
FA 4.8–5.2 cm
W 20–31 g

Red Fruit Bat
Stenoderma rufum
FR Sténoderme roux
DE Roter Fruchtvampir
ES Estenodermo rojizo

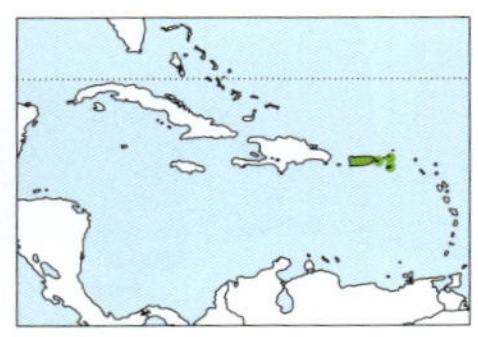

PR (probably Vieques I), VI (St Thomas, St John, St Croix)

Genus *CENTURIO*

(1 species)

LC
2 sspp
HB 5.8–6.5 cm
T tailless
FA 4.1–4.5 cm
W 17–23 g

Wrinkle-faced Bat
Centurio senex
FR Sténoderme ridé
DE Greisengesicht
ES Centurio de cara arrugada

0–1500 m

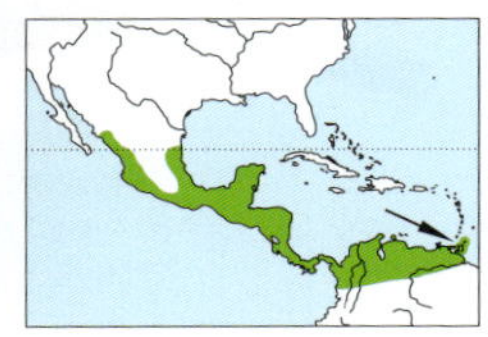

MX (from N Sinaloa & S Tamaulipas) S to N CO & VE, TT

Genus *PYGODERMA*

(1 species)

Ipanema Broad-nosed Bat
Pygoderma bilabiatum
FR Sténoderme d'Ipanema
DE Ipanema-Breitnasenfledermaus
ES Pigoderma de Ipanema

LC
2 sspp
FA 3.6–4.1 cm

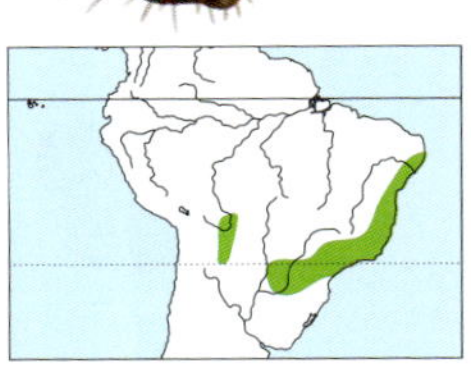

E & S BR, SE BO (Santa Cruz), PY, NW (Salta) & NE AR (Misiones)

Genus *SPHAERONYCTERIS*

(1 species)

LC
Monotypic
HB 5.8–8.5 cm
T tailless
FA 3.6–4.1 cm
W 11–21 g

Visored Bat
Sphaeronycteris toxophyllum
FR Sténoderme à visière
DE Schirmfledermaus
ES Estenodermo de visera

0–3000 m

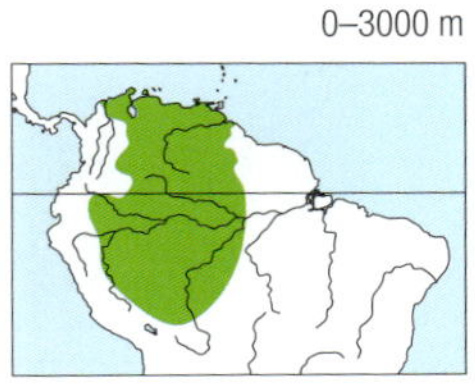

N & E CO, VE, W BR, E EC, E PE, N BO

Genus *AMETRIDA*

(1 species)

Little White-shouldered Bat
Ametrida centurio
FR Centurion amétride
DE Kleine Weißschulterfledermaus
ES Ametrida de cara plana

LC
Monotypic
HB 3.5–5.3 cm
T tailless
FA 2.4–3.3 cm
W 7.8–10 g

up to 2000 m

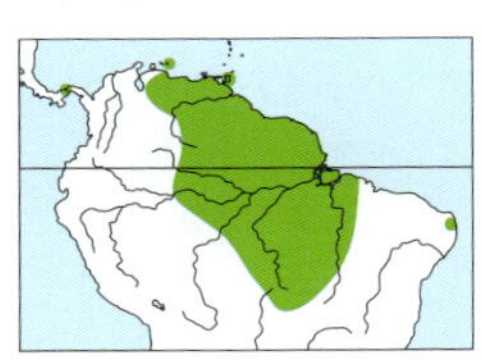

PA (with Barro Colorado I), VE, BQ (Bonaire), TT (Trinidad), Guianas, BR (S to Mato Grosso, 1 record in Paraíba)

Genus *PHYLLOPS*

(1 species)

Cuban Fig-eating Bat
Phyllops falcatus
FR Phyllops de Cuba
DE Kuba-Feigenfledermaus
ES Filops de Cuba

LC
2 sspp
HB 5.5–6.5 cm
T tailless
FA 4–4.8 cm
W 16–23 g

below 700 m

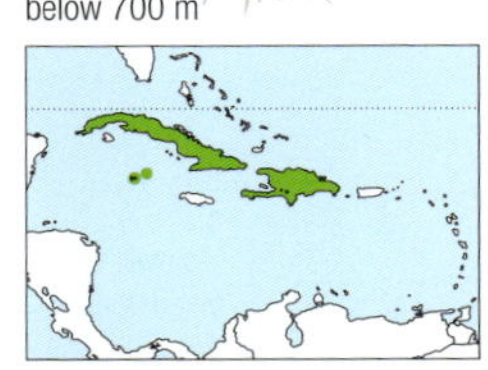

CU, KY (Grand Cayman, Cayman Brac), HT, DO; a record from S Florida Keys may be accidental

Family NATALIDAE – Funnel-eared Bats

3 genera, 12 species. Tropical zone of the New World from deserts through wet forests to mountain areas. Nearctic and Neotropical Regions.

Genus *NYCTIELLUS*

(1 species)

LC
Monotypic
FA 2.7–3.1 cm
W 1.9–2.7 g

Gervais's Funnel-eared Bat
Nyctiellus lepidus

FR Natalide de Gervais
DE Gervais-Trichterohr
ES Natálido de Gervais

0–420 m

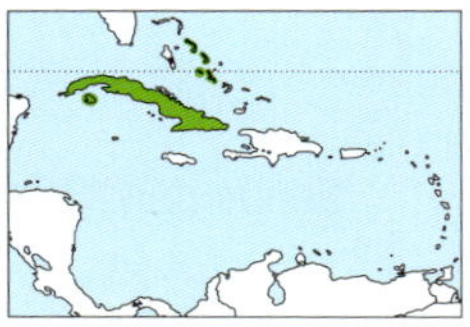

CU (with Isla de la Juventud), BS (Eleuthera, Cat, Great Exuma, Little Exuma, Long)

Genus *CHILONATALUS*

(3 species)

VU
Monotypic
FA 3.1–3.5 cm W 2.6–5 g

Caribbean Lesser Funnel-eared Bat
Chilonatalus micropus

FR Natalide à pattes courtes
DE Kleines Karibisches Trichterohr
ES Natálido del Caribe

0–400 m

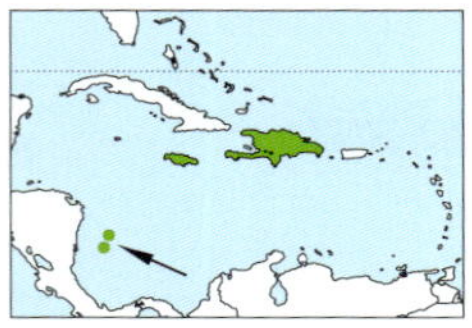

JM, HT, DO, CO (Providencia & San Andrés Is)

DD
Monotypic
FA 3.2–3.4 cm W 2.6–3.3 g

Cuban Lesser Funnel-eared Bat
Chilonatalus macer

FR Natalide de Cuba
DE Kleines Kuba-Trichterohr
ES Natálido menor de Cuba

0–230 m

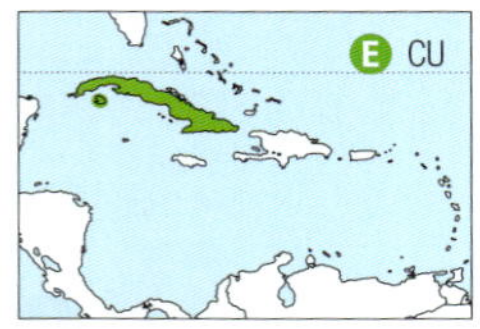

With Isla de la Juventud

NT
Monotypic
FA 3.2–3.6 cm W 3–3.5 g

Bahamian Lesser Funnel-eared Bat
Chilonatalus tumidifrons

FR Natalide des Bahamas
DE Kleines Bahamas-Trichterohr
ES Natálido de las Bahamas

c. 0 m

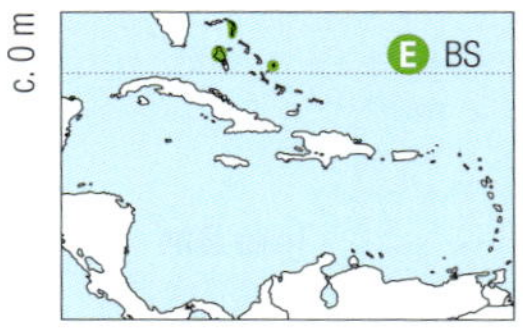

Abaco, Andros, San Salvador

Genus *NATALUS*

(8 species)

VU
Monotypic
FA 4.6–5.1 cm
W 6–12.6 g

Cuban Greater Funnel-eared Bat
Natalus primus

FR Grande Natalide
DE Großes Kuba-Trichterohr
ES Natálido mayor de Cuba

c. 15 m

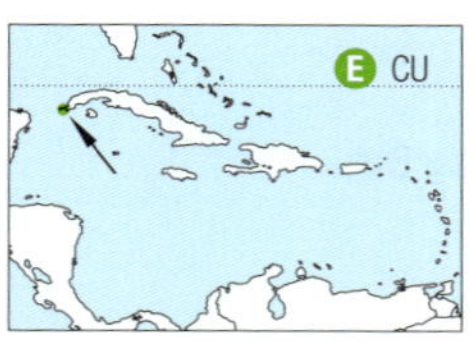

Known only from 1 locality in W

CR
Monotypic
HB 5.1–5.2 cm T 5.7–5.8 cm
FA 4.4–4.7 cm W 5.9–7.3 g

Jamaican Greater Funnel-eared Bat
Natalus jamaicensis

FR Natalide de Jamaïque
DE Großes Jamaika-Trichterohr
ES Natálido de Jamaica

c. 100 m

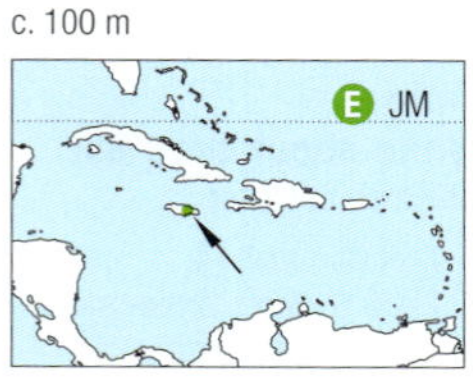

Known only from type locality (St Clair, St Catherine Parish)

NT
Monotypic
FA 4.1–4.5 cm
W 5.5–10 g

Hispaniolan Greater Funnel-eared Bat
Natalus major

FR Natalide d'Hispaniola
DE Großes Hispaniola-Trichterohr
ES Natálido de La Española

0–1000 m

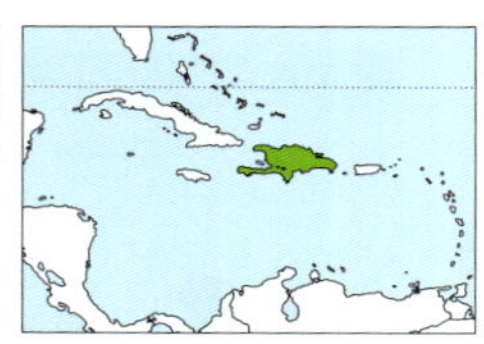

HT, DO

NE
Monotypic
HB c. 4 cm T 4.7–5.5 cm
FA 3.5–3.9 cm W 5–6.5 g

Woolly Funnel-eared Bat
Natalus lanatus

FR Natalide laineuse
DE Wollhaariges Trichterohr
ES Natálido lanudo

500–2000 m

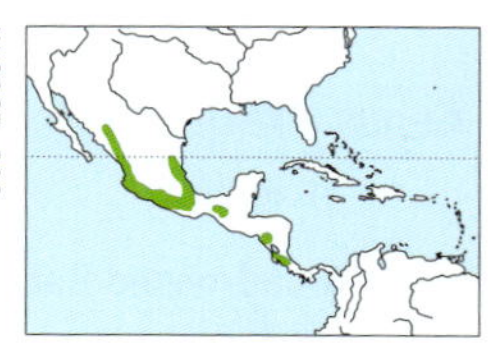

MX (from Chihuahua & N Veracruz S to Chiapas), NW & SW NI, CR

LC
Monotypic
HB 3.8–4.3 cm T 4.7–5.5 cm
FA 3.4–4.1 cm W 3.5–8 g

Mexican Funnel-eared Bat
Natalus mexicanus

FR Natalide du Mexique
DE Mexiko-Trichterohr
ES Natálido mejicano

0–2300 m

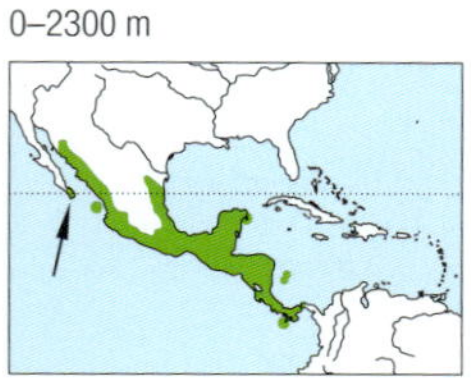

MX (with María Magdalena & Cozumel Is) S to PA (with Coiba I), CO (Providencia & San Andrés Is)

Trinidadian Funnel-eared Bat
Natalus tumidirostris

FR Natalide à nez renflé
DE Trinidad-Trichterohr
ES Natálido de Trinidad

LC
Monotypic
FA 3.5–4.2 cm
W 4.3–8.9 g

0–1400 m

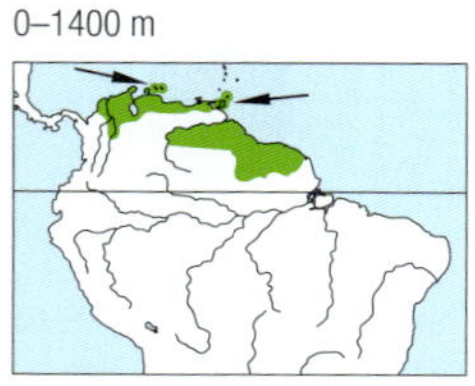

N CO, VE, Guianas, adjacent Is (CW, Bonaire, TT); probably N BR

LC
Monotypic
FA 3.8–4.2 cm
W 4.6–5.3 g

Lesser Antillean Funnel-eared Bat
Natalus stramineus

FR Natalide paillée
DE Kleine-Antillen-Trichterohr
ES Natálido de las Pequeñas Antillas

0–470 m

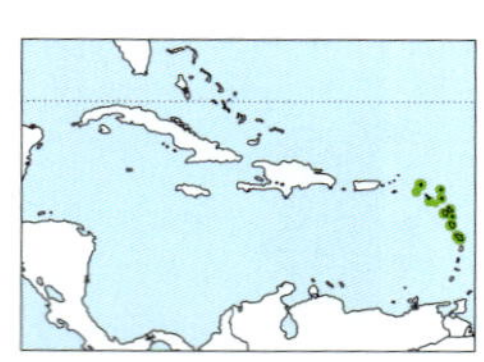

N Lesser Antilles: AI, BQ (Saba), KN (Nevis), AG, MS, GP (with Marie-Galante), DM, MQ

Brazilian Funnel-eared Bat
Natalus macrourus

FR Natalide du Brésil
DE Brasilien-Trichterohr
ES Natálido de Brasil

NE
Monotypic
FA 3.7–4.2 cm
W c. 6 g

0–1000 m

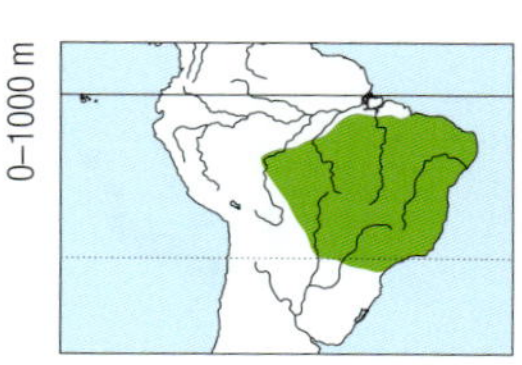

C & E BR S of R Amazon (from Pará S to São Paulo), E BO, PY

Family MOLOSSIDAE – Free-tailed Bats

21 genera, 131 species. From deserts through savanna grasslands and forests. Nearctic, Neotropical, Palearctic, Afrotropical, Indo-Malayan and Australasian Regions.

Genus *TOMOPEAS*

(1 species)

EN

Monotypic
HB 3.6–4.2 cm T 2.9–4.5 cm
FA 3.1–3.6 cm W 2–4 g

Blunt-eared Bat
Tomopeas ravus

FR Murin-molosse
DE Stumpfohrfledermaus
ES Tomopeo de Perú

0–2300 m

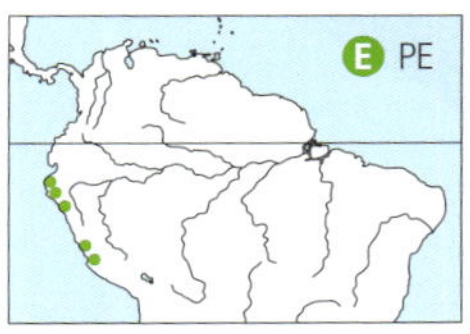

From Piura to Ica

Lesser Naked Bat
Cheiromeles parvidens

FR Chiromèle à dents courtes
DE Sulawesi-Nacktfledermaus
ES Queirómelo pequeño de espalda desnuda

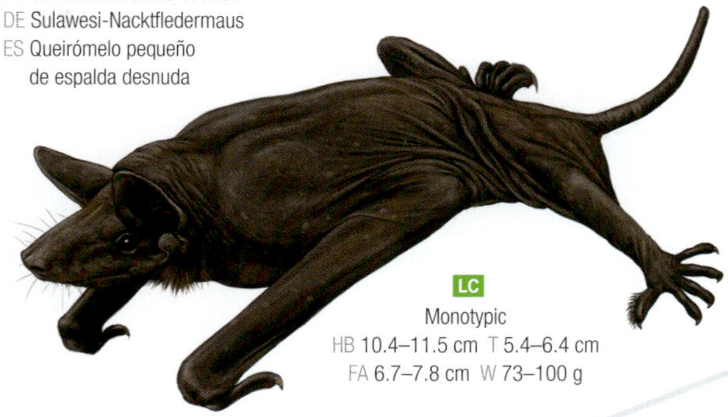

LC

Monotypic
HB 10.4–11.5 cm T 5.4–6.4 cm
FA 6.7–7.8 cm W 73–100 g

0–200 m

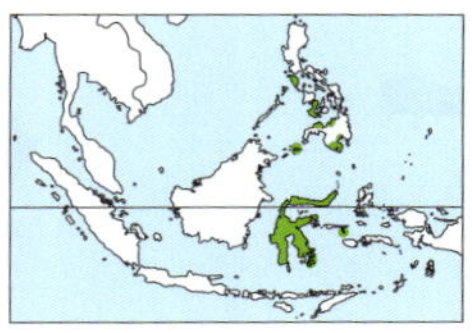

PH (Mindoro, Negros, Mindanao, Jolo, Tawi-Tawi), ID (Sulawesi, with Buton & Sanana Is)

Genus *CHEIROMELES*

(2 species)

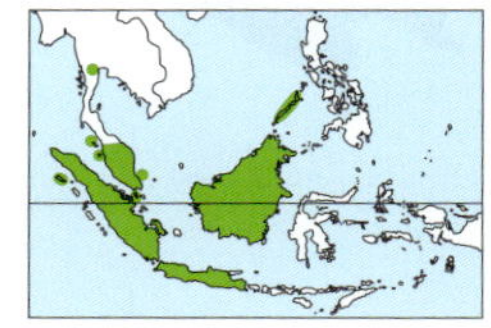

SW & extreme S TH, Pen MY, SG, PH (Palawan), Greater Sundas, several offshore Is

LC

3 sspp
HB 12.5–18 cm
T 6–7.5 cm
FA 6.5–9 cm
W 150–196 g

Greater Naked Bat
Cheiromeles torquatus

FR Chiromèle nu
DE Kragen-Nacktfledermaus
ES Queirómelo grande de espalda desnuda

Genus *MORMOPTERUS*

(7 species)

LC

Monotypic
HB 6–7 cm T 2–3.8 cm
FA 3–3.9 cm W 8.5–17 g

0–1750 m

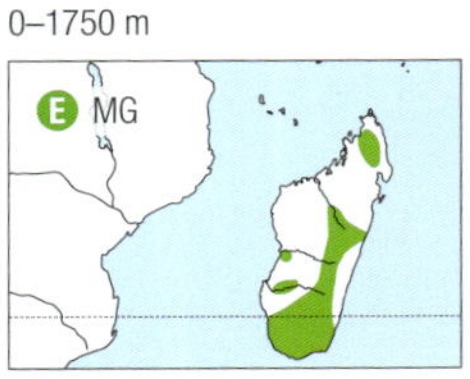

Except most W region

Peters's Little Mastiff Bat
Mormopterus jugularis

FR Molosse de Madagascar
DE Peters-Mastino-Fledermaus
ES Mormóptero de Madagascar

LC

Monotypic
HB 5.1–5.2 cm T 3.8–4.5 cm
FA 3.8–4.2 cm W 5–7.2 g

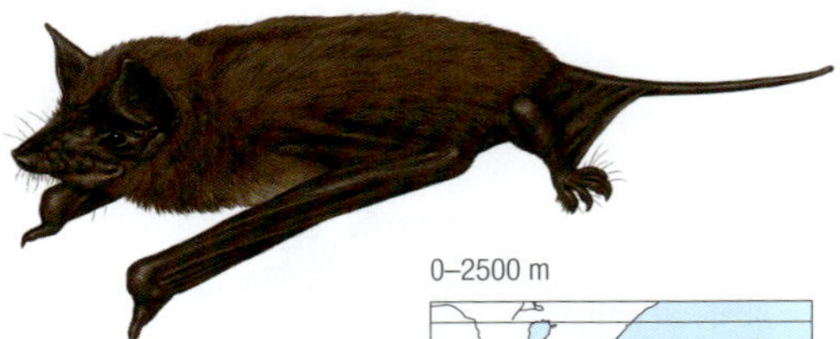

Reunion Little Mastiff Bat
Mormopterus francoismoutoui

FR Molosse de La Réunion
DE Réunion-Mastino-Fledermaus
ES Mormóptero de la Reunión

0–2500 m

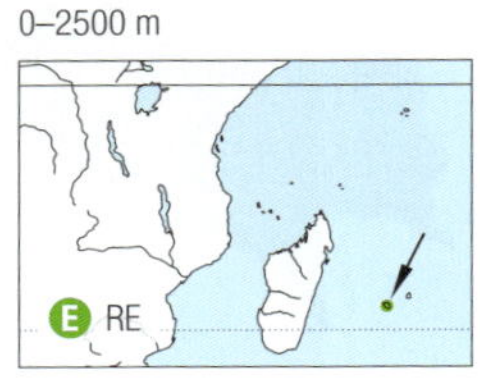

Non-forested habitats

Mauritian Little Mastiff Bat
Mormopterus acetabulosus

FR Molosse de Port-Louis
DE Mauritius-Mastino-Fledermaus
ES Mormóptero de Port-Louis

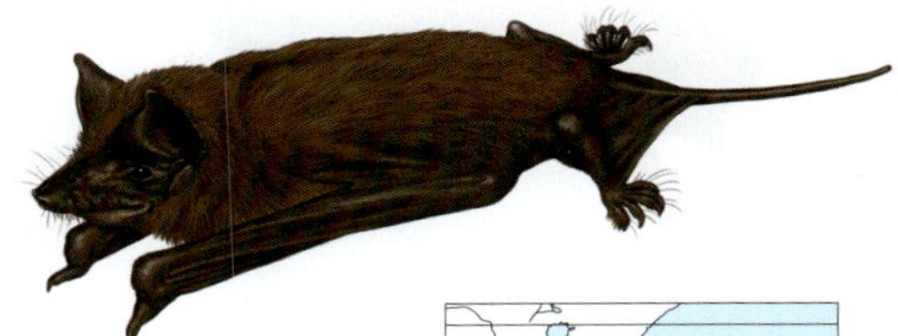

EN

Monotypic
HB 4.9–5.2 cm
T 4.3–5 cm
FA 3.8–4.3 cm
W 6–8.2 g

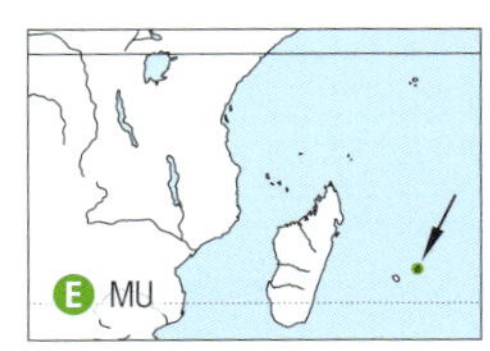

Scrubland, woodland and agricultural areas

DD

Monotypic
T c. 3 cm
FA c. 3.8 cm

below 250 m

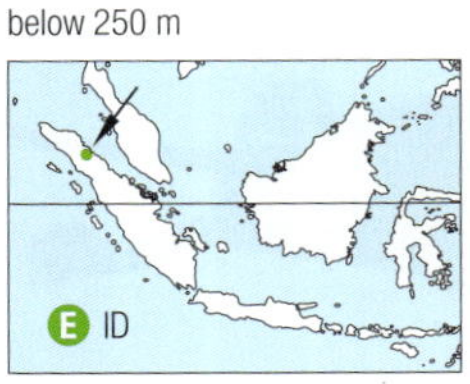

N Sumatra (known only from type locality around Medan)

Sumatran Little Mastiff Bat
Mormopterus doriae

FR Molosse de Sumatra
DE Sumatra-Mastino-Fledermaus
ES Mormóptero de Sumatra

VU

Monotypic
HB 4.6–4.8 cm
T 2.2–3.4 cm
FA 2.8–3.3 cm
W 4–8 g

Mostly in C & E region

Least Little Mastiff Bat
Mormopterus minutus

FR Molosse de Cuba
DE Kuba-Mastino-Fledermaus
ES Mormóptero de Cuba

VU

Monotypic
HB 5–5.1 cm
T 2.9–3.2 cm
FA 3.4–3.5 cm

0–2600 m

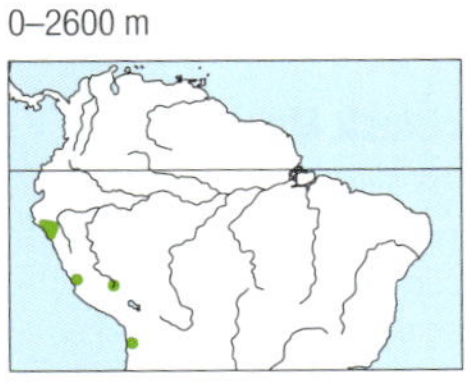

Scattered on dry Pacific slope of Andes from NW PE to NW CL, arid Andean valleys of SE PE

Kalinowski's Little Mastiff Bat
Mormopterus kalinowskii

FR Molosse de Kalinowski
DE Kalinowski-Mastino-Fledermaus
ES Mormóptero de Kalinowski

LC

Monotypic
HB 4–5.3 cm T 2.6–3.8 cm
FA 3.2–3.8 cm W 3–8 g

1800–3000 m

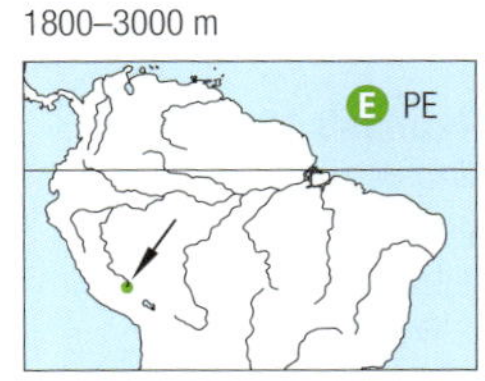

S Andes in Machu Picchu area (R Urubamba, Santuario histórico)

Incan Little Mastiff Bat
Mormopterus phrudus

FR Molosse du Pérou
DE Inka-Mastino-Fledermaus
ES Mormóptero de Perú

Genus *MOLOSSUS*

(15 species)

NE

Monotypic

HB 5.5–6.2 cm T 3–3.7 cm
FA 3.4–3.6 cm W 7–14 g

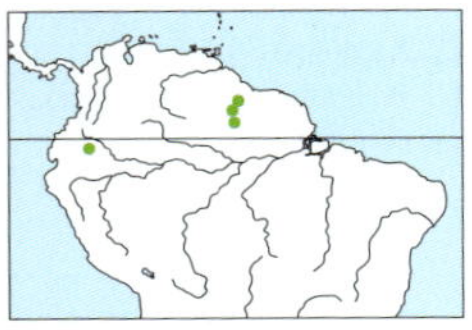

Fenton's Mastiff Bat
Molossus fentoni

FR Molosse de Fenton
DE Fenton-Samtfledermaus
ES Moloso de Fenton

Known from few localities in NE EC & GY; range uncertain

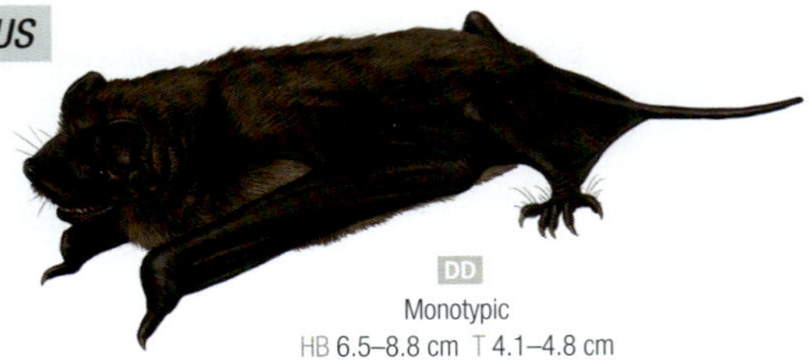

DD

Monotypic

HB 6.5–8.8 cm T 4.1–4.8 cm
FA 4.3–4.7 cm W 20–30 g

up to 2400 m

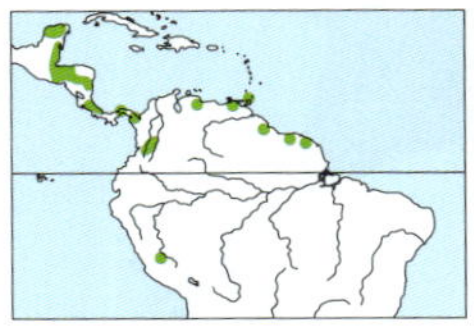

Alvarez's Mastiff Bat
Molossus alvarezi

FR Molosse d'Alvarez
DE Alvarez-Samtfledermaus
ES Moloso de Alvarez

MX (Yucatán Pen), BZ & GT to CO, coastal VE, TT (Trinidad), Guianas, SE PE

NE

Monotypic

HB 5.5–8 cm T 3–4.6 cm
FA 3.7–4.1 cm W 13–20 g

up to 460 m

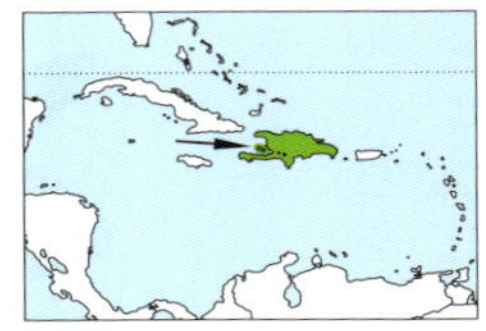

Hispaniolan Mastiff Bat
Molossus verrilli

FR Molosse d'Hispaniola
DE Hispaniola-Samtfledermaus
ES Moloso de La Española

HT (with Gonâve I), DO

NE

Monotypic

HB 5.5–7.1 cm T 3.2–4.2 cm
FA 3.7–4 cm W 13–19 g

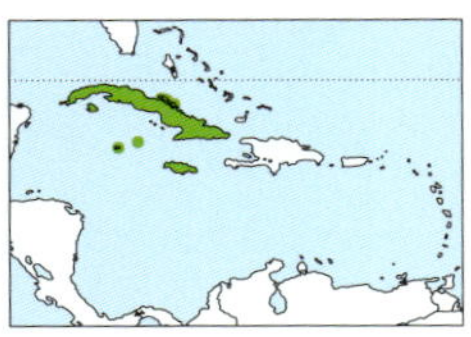

Pug-nosed Mastiff Bat
Molossus milleri

FR Molosse de Jamaïque
DE Stumpfnasen-Samtfledermaus
ES Moloso de Jamaica

CU (with Isla de la Juventud),
KY (Grand Cayman, Cayman Brac), JM

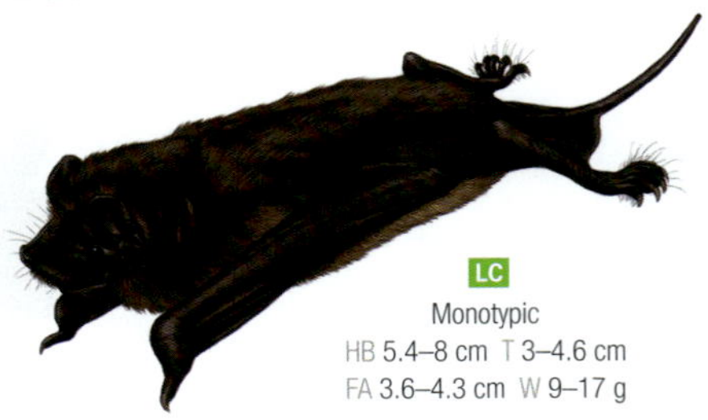

LC

Monotypic

HB 5.4–8 cm T 3–4.6 cm
FA 3.6–4.3 cm W 9–17 g

up to 1250 m

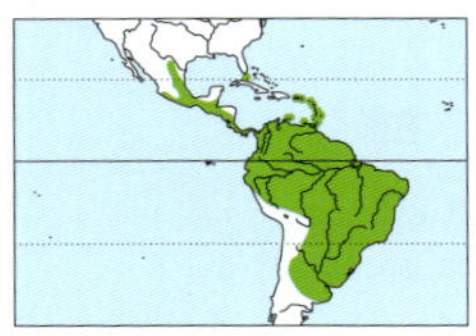

Pallas's Mastiff Bat
Molossus molossus

FR Molosse commun
DE Große Samtfledermaus
ES Moloso común

From NE MX (Coahuila) S to AR & UY,
also SE US (Florida), PR & Lesser Antilles

LC

Monotypic

HB 5.2–7.8 cm T 2.8–4 cm
FA 3.3–3.7 cm W 10–18 g

up to 1300 m

Coiban Mastiff Bat
Molossus coibensis

FR Molosse de Coiba
DE Coiba-Samtfledermaus
ES Moloso de Coiba

Scattered from S MX (Chiapas)
S to NE EC, C PE and E & S BR

NE

Monotypic

HB 7.3–7.6 cm T 4.2–4.8 cm
FA 4.1–4.4 cm W 20–23 g

c. 90 m

Melin's Mastiff Bat
Molossus melini

FR Molosse de Melin
DE Melin-Samtfledermaus
ES Moloso de Melin

Known only from type locality (near Melincué, Santa Fe)

LC

Monotypic

HB 5.4–8.9 cm T 3.1–4.1 cm
FA 3.5–4.2 cm W 12–23 g

up to 1800 m

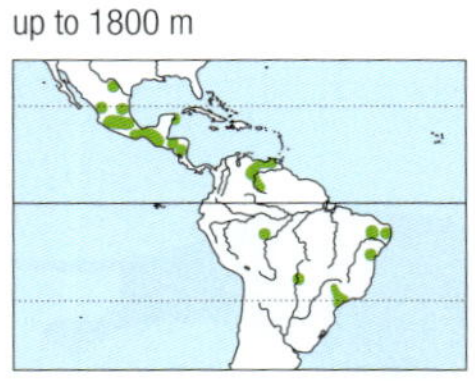

Aztec Mastiff Bat
Molossus aztecus

FR Molosse aztèque
DE Azteken-Samtfledermaus
ES Moloso azteca

Scattered in MX (with Cozumel I), GT, HN, NI, VE & BR

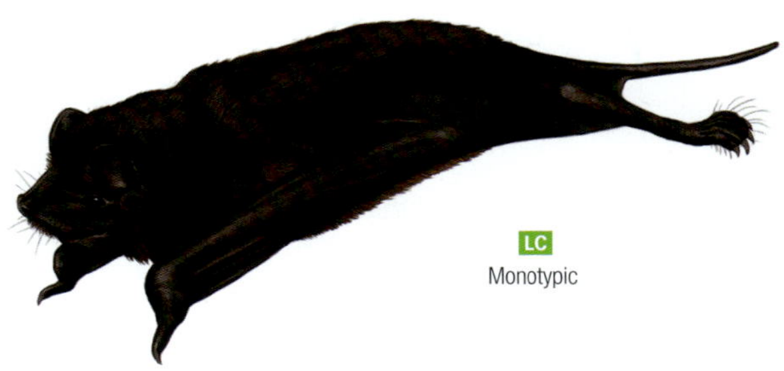

LC

Monotypic

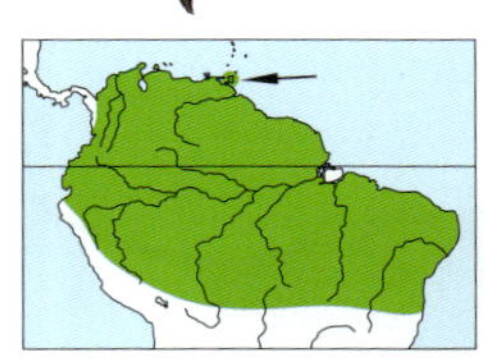

Common Black Mastiff Bat
Molossus rufus

FR Grand Molosse
DE Rote Samtfledermaus
ES Moloso rufo

South America S to C BR & BO, TT (Trinidad)

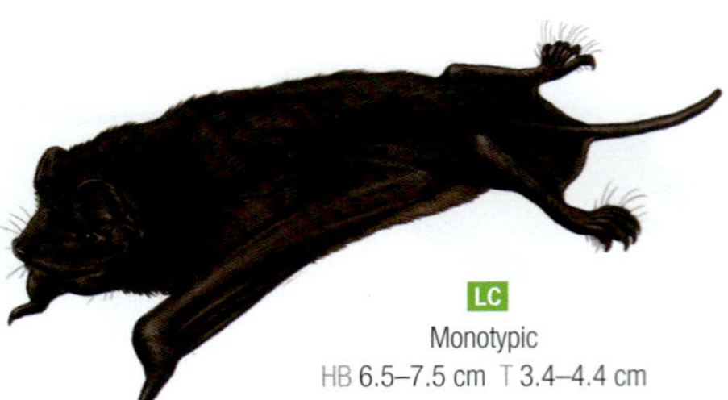

LC

Monotypic

HB 6.5–7.5 cm T 3.4–4.4 cm
FA 3.8–4.4 cm W 13–22 g

below 600 m

Thomas's Mastiff Bat
Molossus currentium

FR Molosse de Thomas
DE Thomas-Samtfledermaus
ES Moloso de Thomas

CW BR, N PY, NE AR, UY; perhaps E BO

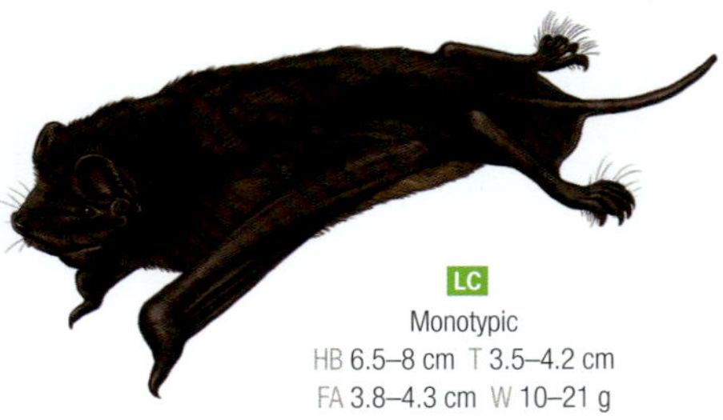

LC

Monotypic

HB 6.5–8 cm T 3.5–4.2 cm
FA 3.8–4.3 cm W 10–21 g

up to 1060 m

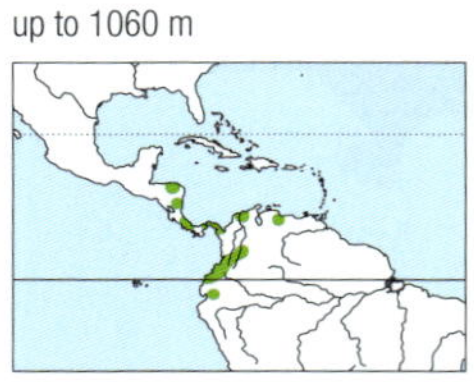

Bonda Mastiff Bat
Molossus bondae

FR Molosse de Bonda
DE Bonda-Samtfledermaus
ES Moloso de Bonda

Central & N South America from HN to CO, NE VE & N EC

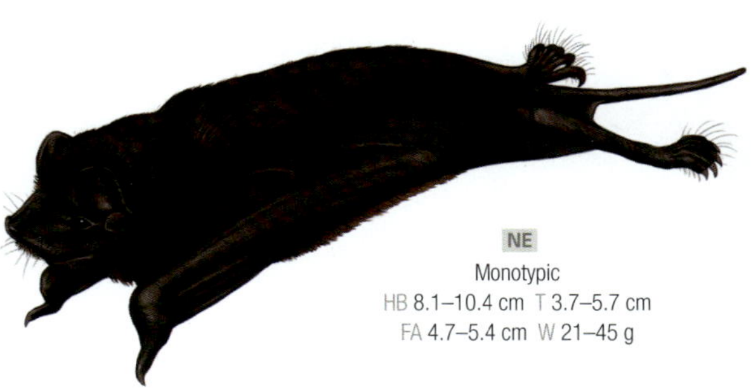

NE

Monotypic

HB 8.1–10.4 cm T 3.7–5.7 cm
FA 4.7–5.4 cm W 21–45 g

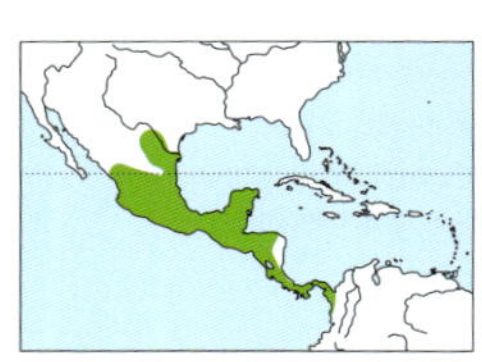

Northern Black Mastiff Bat
Molossus nigricans

FR Molosse noir
DE Nördliche Schwarze Samtfledermaus
ES Moloso negro septentrional

Central America from E & S MX to PA

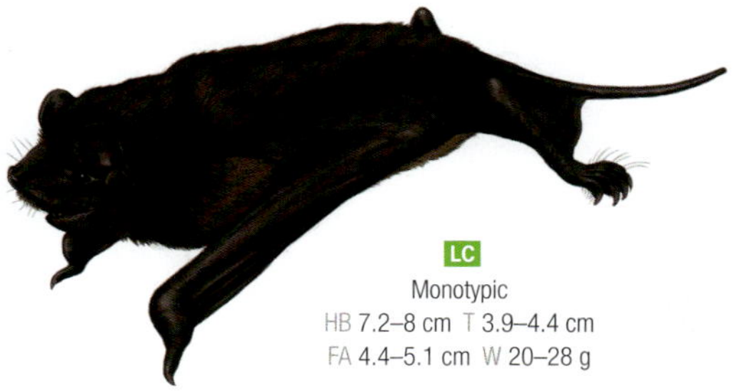

LC
Monotypic
HB 7.2–8 cm T 3.9–4.4 cm
FA 4.4–5.1 cm W 20–28 g

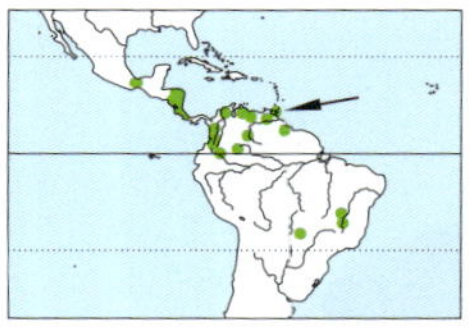

Miller's Mastiff Bat
Molossus pretiosus
FR Molosse de Miller
DE Miller-Samtfledermaus
ES Moloso de Miller

S MX (Guerrero, Oaxaca), E HN, NI, CR, CO, VE, TT (Trinidad), GY, CW & E BR

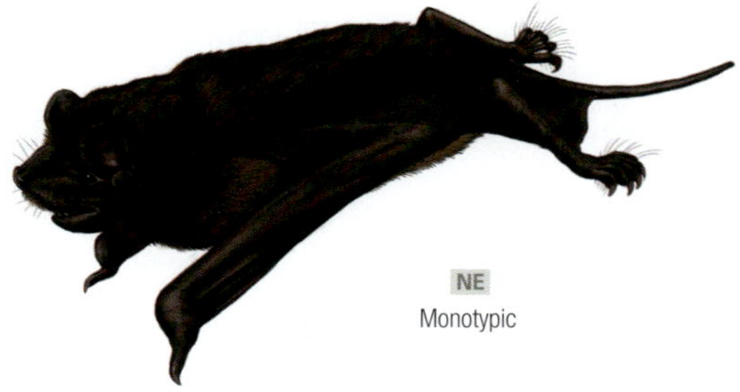

NE
Monotypic

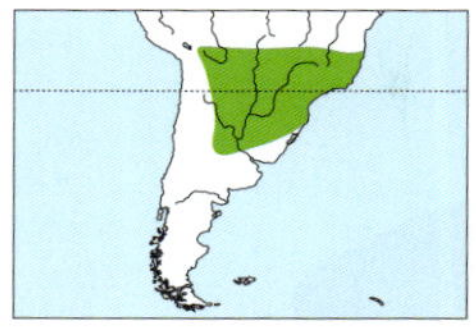

Southern Black Mastiff Bat
Molossus fluminensis
FR Molosse de Rio de Janeiro
DE Südliche Schwarze Samtfledermaus
ES Moloso negro meridional

SE BR, S BO, PY, N AR

LC
Monotypic
HB 7–8.5 cm T 4–5.2 cm
FA 4.5–4.9 cm W 14–28 g

up to 2400 m

Sinaloan Mastiff Bat
Molossus sinaloae
FR Molosse du Sinaloa
DE Sinaloa-Samtfledermaus
ES Moloso de Sinaloa

S Sinaloa, Nayarit, Jalisco, Colima, Michoacán, Guerrero, Morelos, Puebla, Chiapas

Genus *PROMOPS*

(3 species)

0–1800 m

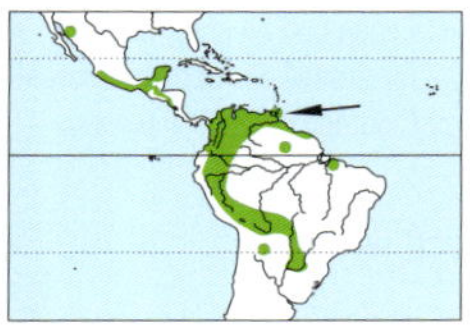

Patchily from MX (Sonora along W coast to Yucatán Pen) to W & N BR, N AR & PY, TT (Trinidad)

Big Crested Mastiff Bat
Promops centralis
FR Grand Promope
DE Schädelkamm-Bulldogfledermaus
ES Promop grande

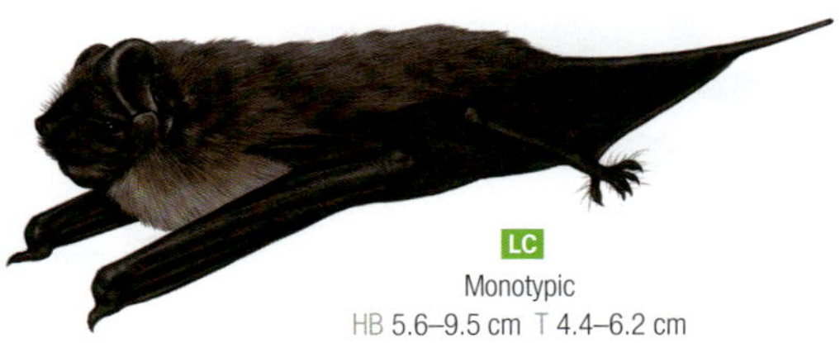

LC
Monotypic
HB 5.6–9.5 cm T 4.4–6.2 cm
FA 4.9–5.7 cm W 18–32 g

Davison's Mastiff Bat
Promops davisoni
FR Promope de Davison
DE Davison-Bulldogfledermaus
ES Promop de Davison

DD
Monotypic
HB 5.4–7.4 cm T 4.9–5.8 cm
FA 4.6–5.2 cm W 10–23 g

0–1320 m

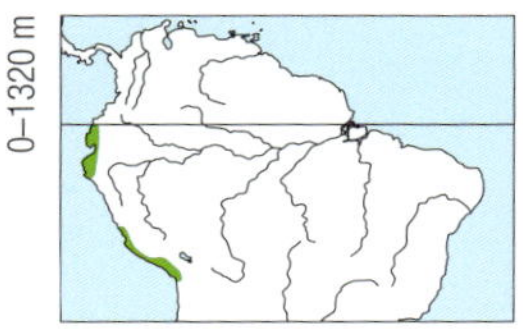

W EC, W PE, NW CL

Brown Mastiff Bat
Promops nasutus
FR Petit Promope
DE Braune Bulldogfledermaus
ES Promop pequeño

LC
Monotypic
HB 6.5–8 cm T 4.7–5.8 cm
FA 4.5–5.2 cm W 13–22 g

up to 2560 m

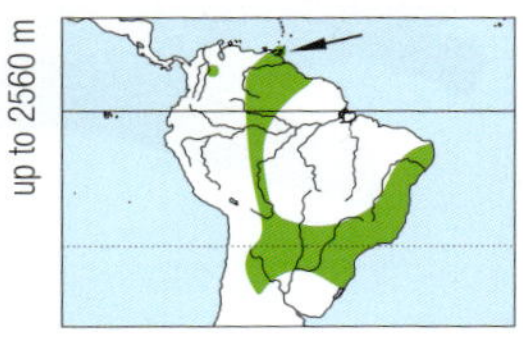

NE CO, VE, TT (Trinidad), GY, SR, W, E & S BR, BO, PY, N AR

Genus *NEOPLATYMOPS*

(1 species)

LC
Monotypic
HB 4.4–5.8 cm
T 2–2.3 cm
FA 2.7–3 cm

75–195 m

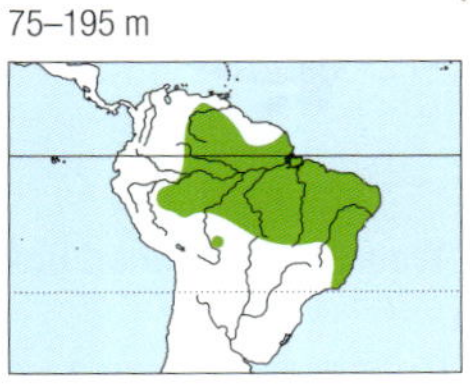

Mato Grosso Dog-faced Bat
Neoplatymops mattogrossensis
FR Molosse du Mato Grosso
DE Mato-Grosso-Bulldogfledermaus
ES Neoplatimop de Mato Grosso

E CO, VE, SW GY, NE BO, BR

Genus *NYCTINOMOPS*

(4 species)

LC
Monotypic
HB 6.3–7.3 cm T 4.1–5.5 cm
FA 4.7–5.3 cm W 13–23 g

0–3150 m

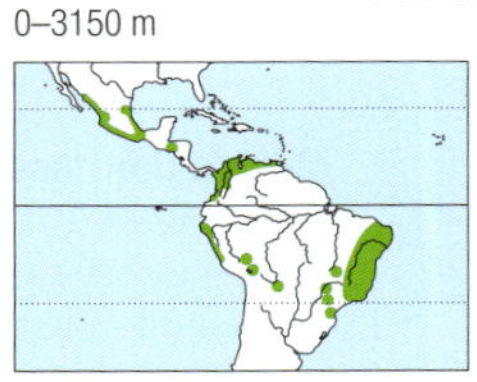

Peale's Free-tailed Bat
Nyctinomops aurispinosus
FR Nyctinomope de Peale
DE Peale-Bulldogfledermaus
ES Nictónimo de Peale

Patchily from N MX (S Sonora & S Tamaulipas) & SW HN to WC & SE BO, E & SE BR; perhaps EC

LC
Monotypic
HB 6–7.5 cm T 3.1–4.7 cm
FA 4.4–4.9 cm W 11–18 g

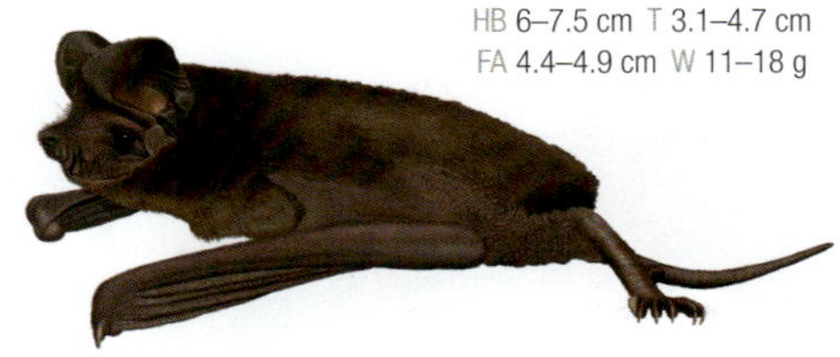

0–2250 m

Pocketed Free-tailed Bat
Nyctinomops femorosaccus
FR Nyctinomope à poche
DE Taschen-Bulldogfledermaus
ES Nictónimo de bolsillos

SW US (S California, C & S Arizona, S New Mexico, SW Texas) S to W MX (Guerrero, Morelos, Puebla)

LC
5 sspp
HB 5–6.8 cm T 3.2–4.8 cm
FA 4–4.7 cm W 9–16 g

0–1700 m

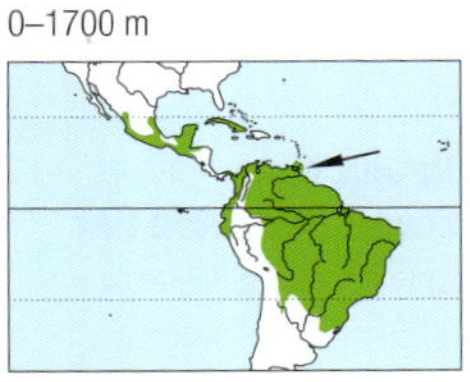

Broad-eared Free-tailed Bat
Nyctinomops laticaudatus
FR Nyctinomope des rochers
DE Großohr-Bulldogfledermaus
ES Nictónimo de orejas anchas

CU, MX (from Durango & S Tamaulipas) S to NW & E PE, S BO, N AR, PY & N UY, TT (Trinidad)

LC
Monotypic
HB 7.3–8.9 cm T 4–6.3 cm
FA 5.4–6.5 cm
W 17–34 g

0–2600 m

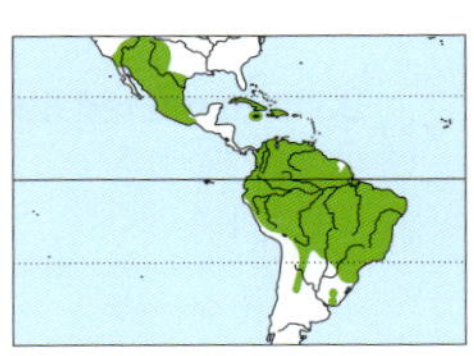

Big Free-tailed Bat
Nyctinomops macrotis
FR Grand Nyctinomope
DE Große Bulldogfledermaus
ES Nictónimo grande

From SW US to S MX, Greater Antilles (CU, JM, HT, DO), South America S to N AR & UY; probably GF

Genus *CABRERAMOPS*

(1 species)

EN
Monotypic
HB 4.4–5.2 cm
T 2.6–3.1 cm
FA 3.1–3.8 cm
W c. 5 g

20–50 m

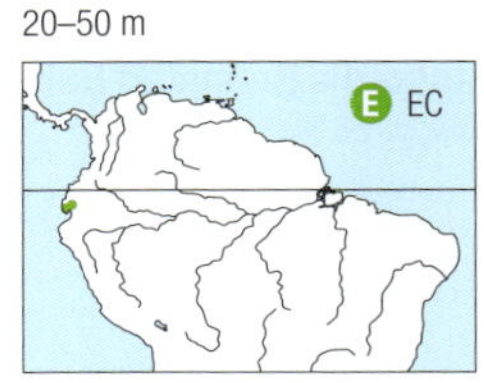

Equatorial Dog-faced Bat
Cabreramops aequatorianus
FR Molosse d'Équateur
DE Ecuador-Bulldogfledermaus
ES Cabreramop de Ecuador

Known from 2 localities in coastal lowlands (Los Ríos, Guayas)

Genus *MOLOSSOPS*

(2 species)

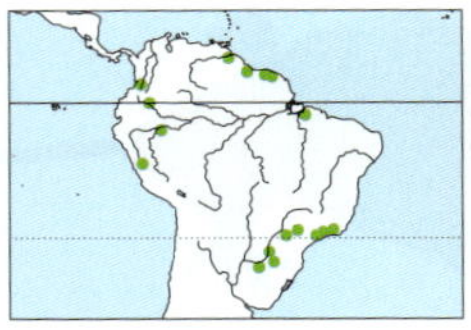

below 300 m

Rufous Dog-faced Bat
Molossops neglectus

FR Molossope négligé
DE Verkannte Bulldogfledermaus
ES Molosop rufo

DD
Monotypic
HB 5.5–6.8 cm T 2.7–4 cm
FA 3.3–4 cm
W 9–16 g

Disjunct in W & S CO, E VE, Guianas, NE & C PE, N, SE & S BR and NE AR (Misiones, Corrientes)

Dwarf Dog-faced Bat
Molossops temminckii

FR Molossope nain
DE Temminck-Bulldogfledermaus
ES Molosop de Temminck

LC
Monotypic
HB 4.8–6 cm
T 1.9–2.7 cm
FA 2.9–3.3 cm
W 4–9 g

0–1000 m

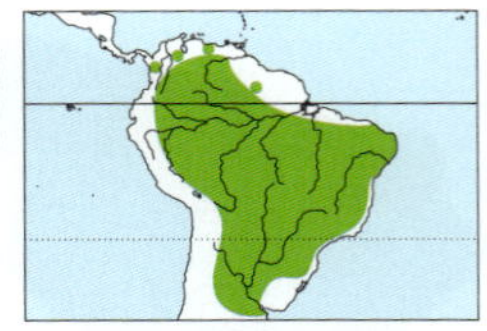

CO, VE, SW GY, NE EC, E PE, BR, BO, PY, N AR, W UY

Genus *CYNOMOPS*

(9 species)

NE
Monotypic
HB 5.6–6.2 cm
T 2.6–3.3 cm
FA 3.3–3.6 cm
W 13–19 g

25–55 m

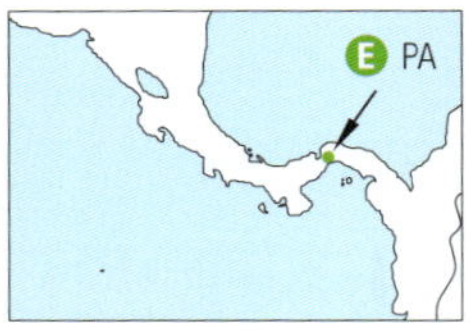

Freeman's Dog-faced Bat
Cynomops freemani

FR Cynomope de Freeman
DE Freeman-Bulldogfledermaus
ES Cinomop de Freeman

Pacific coast of Canal Zone region

LC Monotypic
HB 6.1–7.4 cm T 2.8–3.4 cm
FA 3.3–3.7 cm
W 11–18.3 g

up to 1500 m

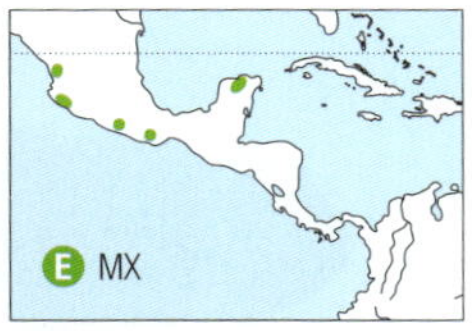

Mexican Dog-faced Bat
Cynomops mexicanus

FR Cynomope du Mexique
DE Mexiko-Bulldogfledermaus
ES Cinomop de México

Scattered sites on Pacific coast & Yucatán Pen

LC Monotypic
HB 5.1–7.5 cm T 2–3.1 cm
FA 2.9–3.7 cm
W 9–15 g

0–2500 m

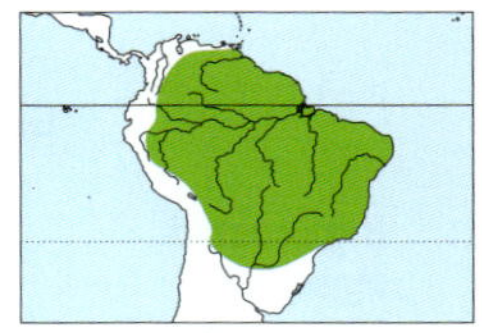

Southern Dog-faced Bat
Cynomops planirostris

FR Cynomope à ventre blanc
DE Amazonas-Bulldogfledermaus
ES Cinomop de vientre blanco

E of Andes in CO, VE, Guianas, PE, BO, BR, PY & N AR

DD Monotypic
HB 7.1–8.6 cm T 3.3–4.2 cm
FA 4.1–4.9 cm
W 24–44 g

0–1100 m

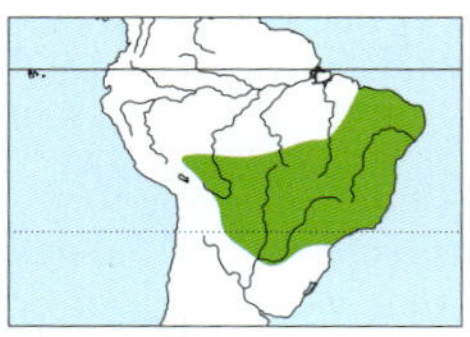

Cinnamon Dog-faced Bat
Cynomops abrasus

FR Cynomope à poil ras
DE Zimtfarbene Bulldogfledermaus
ES Cinomop de pelo corto

E of Andes from S PE through BO, PY & N AR to E & SE BR

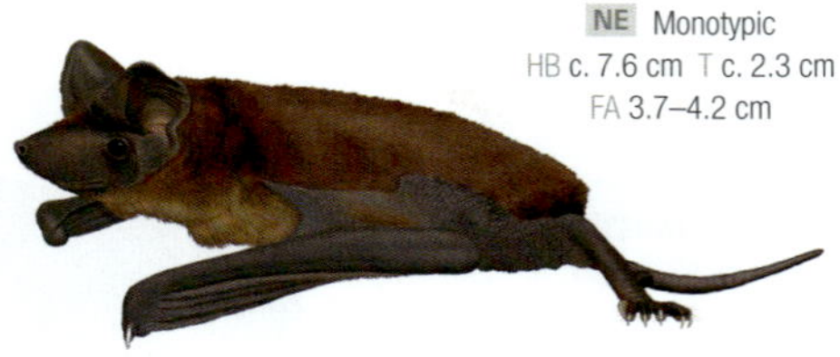

NE Monotypic
HB c. 7.6 cm T c. 2.3 cm
FA 3.7–4.2 cm

0–1000 m

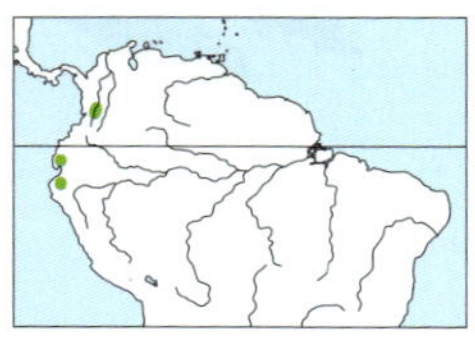

Awa Dog-faced Bat
Cynomops kuizha

FR Cynomope des Awas
DE Awa-Bulldogfledermaus
ES Cinomop de Awa

Lowlands of Cauca Valley in CO, W slope of Andes in EC & NW PE

LC Monotypic
HB 5.5–7.8 cm T 2.4–3.4 cm
FA 3.3–4 cm
W 10.8–24 g

0–975 m

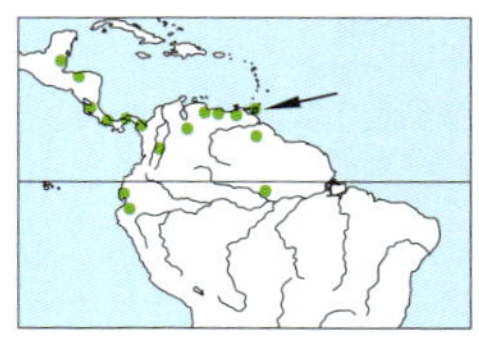

Greenhall's Dog-faced Bat
Cynomops greenhalli

FR Cynomope de Greenhall
DE Greenhall-Bulldogfledermaus
ES Cinomop de Greenhall

Scattered in BZ, N HN, CR, PA, C CO, VE, TT (Trinidad), EC & N BR

NE Monotypic
HB 7.4–8.8 cm T 3.2–4.4 cm
FA 4.2–5.1 cm
W 27–55 g

0–535 m

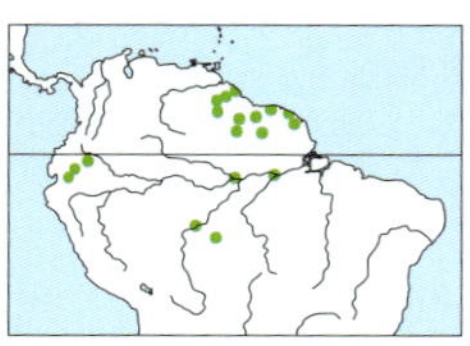

Thomas's Dog-faced Bat
Cynomops mastivus

FR Grand Cynomope
DE Thomas-Bulldogfledermaus
ES Cinomop de Thomas

Scattered on E slopes of Andes in EC, E VE, Guianas & Amazonian BR

LC Monotypic
HB 5.5–8 cm T 2.4–3.1 cm
FA 3–3.7 cm
W 12–20 g

25–240 m

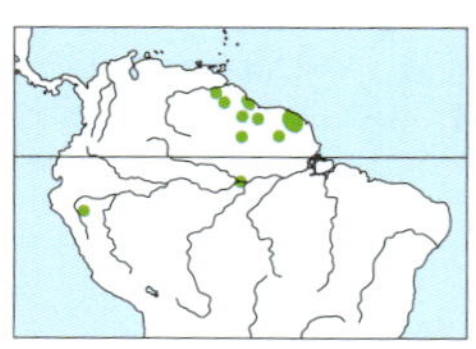

Miller's Dog-faced Bat
Cynomops milleri

FR Cynomope de Miller
DE Miller-Bulldogfledermaus
ES Cinomop de Miller

Scattered in VE, Guianas, N PE & Amazonian BR

NE Monotypic
HB 5.5–6.4 cm T 2.5–3 cm
FA 3.1–3.5 cm
W 12.8–15.3 g

195–530 m

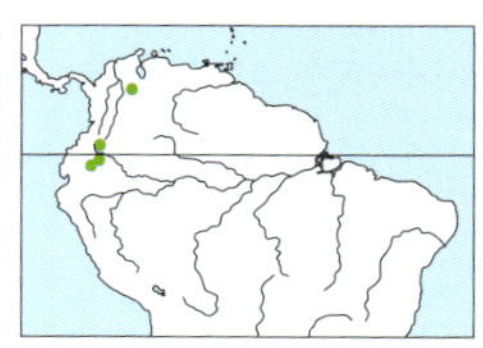

Waorani Dog-faced Bat
Cynomops tonkigui

FR Cynomope des Huaoranis
DE Waorani-Bulldogfledermaus
ES Cinomop de los Huaorani

E slopes of Andes in CO & EC

Genus *EUMOPS*

(17 species)

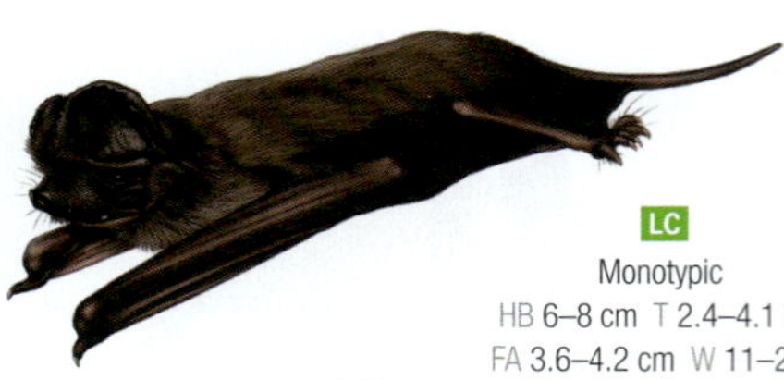

LC
Monotypic
HB 6–8 cm T 2.4–4.1 cm
FA 3.6–4.2 cm W 11–22 g

up to 600 m

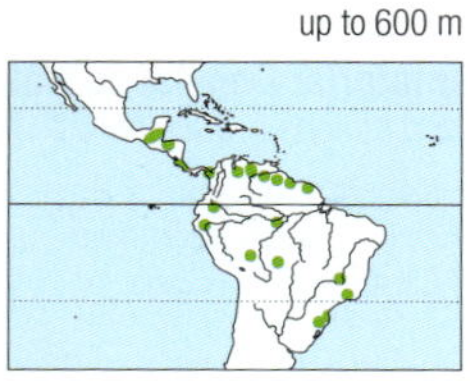

Sanborn's Bonneted Bat
Eumops hansae

FR Eumope de Hansa
DE Sanborn-Bulldogfledermaus
ES Eumop de Hansa

Scattered in S MX, BZ, N HN, CR, E PA, VE, GY, GF, NE EC, N PE, N BO & BR

DD Monotypic
HB 6.5–7.2 cm T 3.4–4.5 cm
FA 4.5–4.8 cm W c. 15 g

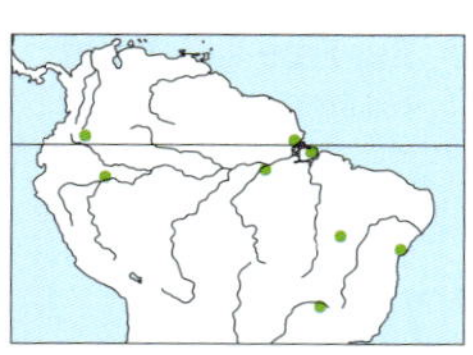

Delta Bonneted Bat
Eumops delticus

FR Eumope du delta
DE Delta-Bulldogfledermaus
ES Eumop del delta

Scattered in S CO (Caquetá), NE PE (Loreto), N & E BR (Amapá, Pará, Bahia, Minas Gerais)

LC Monotypic
HB 5.4–7.4 cm T 2.7–4.3 cm
FA 4–4.7 cm W 7–16 g

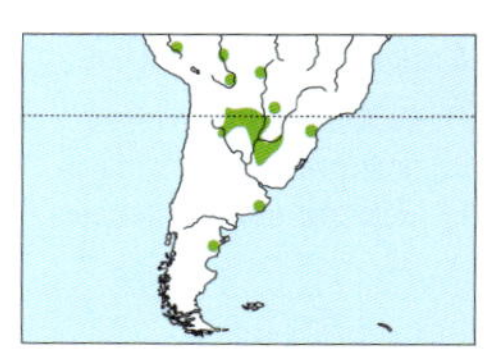

Patagonian Dwarf Bonneted Bat
Eumops patagonicus

FR Eumope de Patagonie
DE Patagonien-Bulldogfledermaus
ES Eumop de la Patagonia

S PE, BO, S BR, PY, AR, extreme NW UY

Northern Dwarf Bonneted Bat
Eumops nanus

LC Monotypic
HB 4.9–6.8 cm
T 2.8–4.7 cm
FA 3.9–4.8 cm
W 7–13 g

FR Eumope nain
DE Zwergbulldogfledermaus
ES Eumop nano

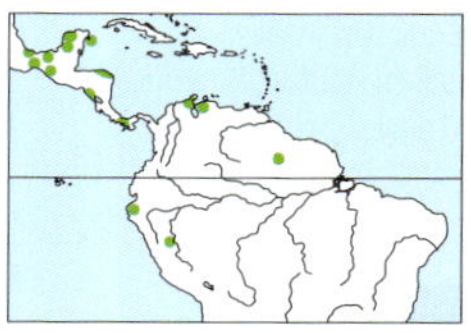

Patchily in S MX (from Veracruz & Oaxaca, Cozumel I) S to N CO, NW VE, S GY & PE

LC
Monotypic
HB 7.2–8.7 cm T 3.3–4.3 cm
FA 4.5–4.9 cm W 15–20 g

PY, S BR, N AR, UY

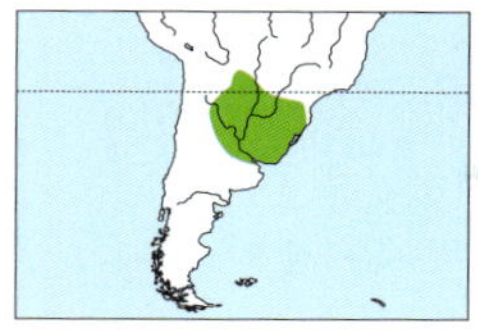

Common Dwarf Bonneted Bat
Eumops bonariensis

FR Eumope de Peters
DE Peters-Bulldogfledermaus
ES Eumop de Peters

NE Monotypic
HB c. 8.2 cm T 5.2–5.3 cm
FA 6.1–6.2 cm W 20–30 g

110–740 m

Chiribaya's Bonneted Bat
Eumops chiribaya

FR Eumope des Chiribayas
DE Chirabaya-Bulldogfledermaus
ES Eumop de Chiribaya

Known from 2 localit es in lowlands of Arequipa & Moquegua

LC 2 sspp
HB 7.4–10 cm T 4.2–6.4 cm
FA 5.5–6.9 cm W 24–38 g

0–2000 m

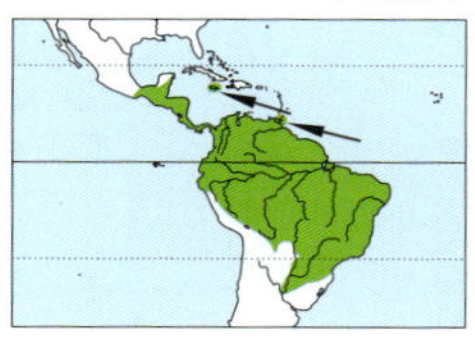

Black Bonneted Bat
Eumops auripendulus

FR Eumope des palmiers
DE Dunkle Bulldogfledermaus
ES Eumop de Shaw

S MX (from Oaxaca & Tabasco) S to PE, BO, S PY, NE AR & S BR, also JM & TT (Trinidad)

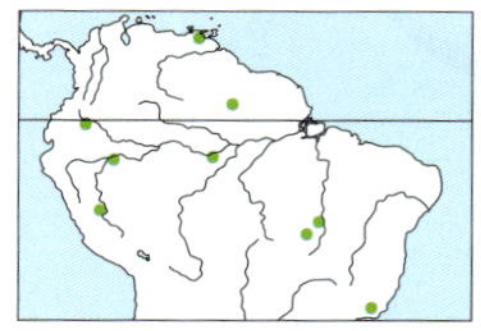

Guianan Bonneted Bat
Eumops maurus

FR Eumope de Thomas
DE Guayana-Bulldogfledermaus
ES Eumop de Thomas

Scattered in NE VE, S GY, SR (unknown site, not mapped), NE EC, E PE & BR

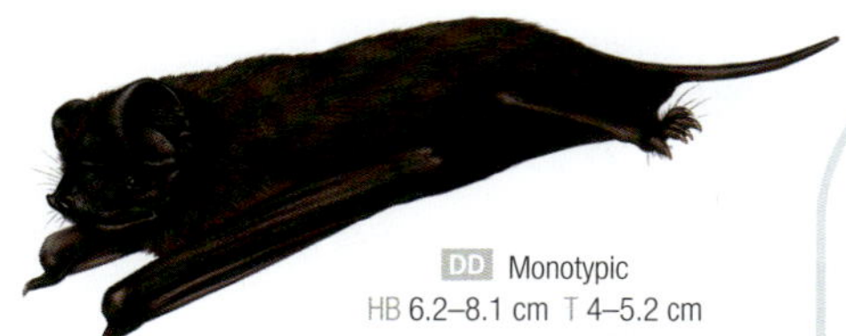

DD Monotypic
HB 6.2–8.1 cm T 4–5.2 cm
FA 5.1–6.1 cm W 22–24 g

Amazon Basin in E CO, S VE, GY, SR, E PE, BR & N BO

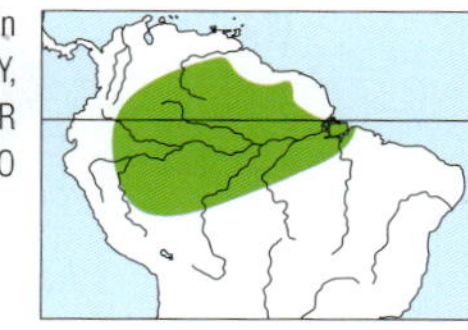

Trumbull's Bonneted Bat
Eumcps trumbulli

below 500 m

FR Eumope de Trumbull DE Trumbull-Bulldogfledermaus
ES Eumop de Trumbull

LC Monotypic
HB 9–13 cm T 4.5–5.7 cm
FA 6.8–7.5 cm

LC 2 sspp
HB 10–12 cm T 5.3–7.1 cm
FA 7.2–8.3 cm W 52–76 g

-60–3200 m

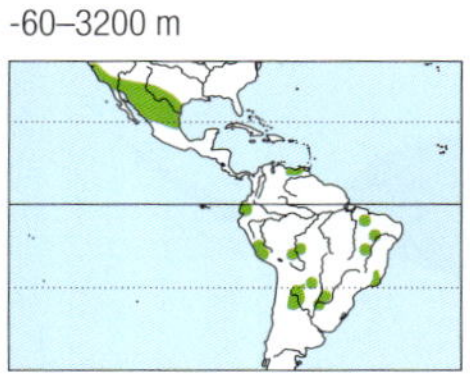

Western Bonneted Bat
Eumops perotis

FR Eumope de Schinz
DE Schinz-Bulldogfledermaus
ES Eumop de Schinz

SW US & N MX, scattered in N VE, SW EC, SW PE, N BO, E BR, PY & N AR

NE Monotypic
HB 7.8–9.1 cm T 5.8–6.1 cm
FA 6.7–6.8 cm

380–650 m

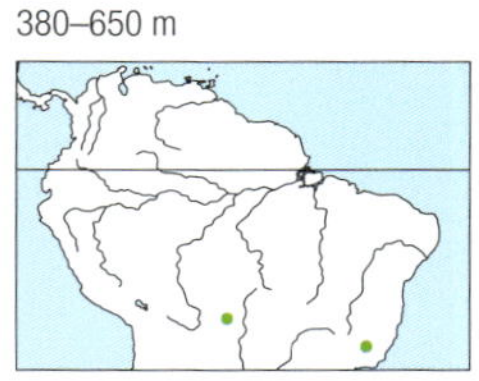

Chimera's Bonneted Bat
Eumops chimaera

FR Eumope chimère
DE Chimära-Bulldogfledermaus
ES Eumop de Chimera

Known from 2 localities in lowlands of E BO & SE BR

DD
Monotypic
HB 6.7–8.4 cm
T 4.5–4.6 cm
FA 5.6–6 cm
W 24–32 g

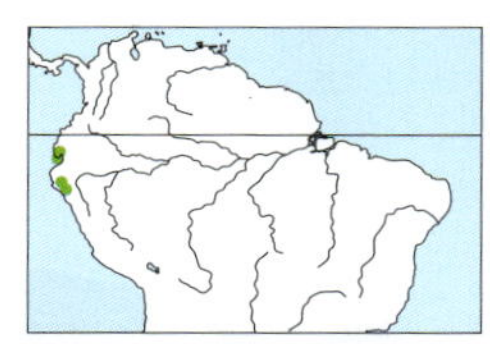

Wilson's Bonneted Bat
Eumops wilsoni

FR Eumope de Wilson
DE Wilson-Bulldogfledermaus
ES Eumop de Wilson

Known from 4 sites in SW EC (Guayas, with Puná I) & NW PE (Lambayeque)

0–1100 m

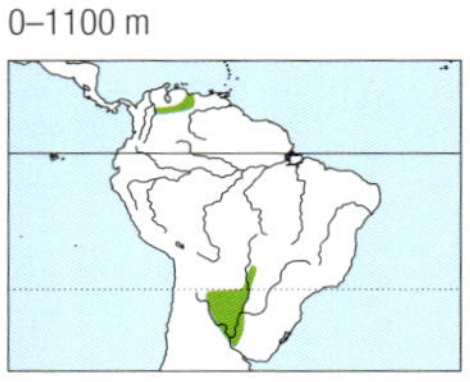

Big Bonneted Bat
Eumops dabbenei

FR Eumope de Dabbene
DE Dabbene-Bulldogfledermaus
ES Eumop de Dabbene

Disjunctly in CO & VE and in BR (Mato Grosso do Sul), PY & N AR

LC
Monotypic
HB 7.1–11.5 cm T 5.5–6.6 cm
FA 7.2–8.6 cm W 74–77 g

LC
2 sspp
HB 8.5–11.2 cm
T 4.8–6.4 cm
FA 6.5–7.5 cm
W 40–68 g

0–2000 m

SW US (S Arizona) & W MX S along Pacific slope to GT, BZ, SV, HN, W NI & NW CR

Underwood's Bonneted Bat
Eumops underwoodi

FR Eumope d'Underwood
DE Underwood-Bulldogfledermaus
ES Eumop de Underwood

LC Monotypic
HB 8.3–9.1 cm T 4.8–6.1 cm
FA 5.4–6.3 cm W 29–42 g

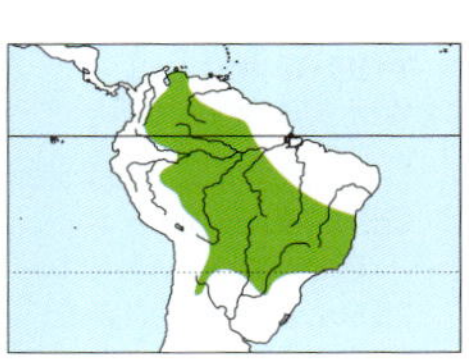

Wagner's Bonneted Bat
Eumops glaucinus

FR Eumope de Wagner
DE Wagner-Bulldogfledermaus
ES Eumop de Wagner

E Andes in CO, VE, S GY, W, C & S BR, N & E BO, PY & N AR

LC Monotypic
HB 7.5–9.5 cm T 4–5.4 cm
FA 5.5–6.3 cm W 34–42 g

up to 900 m

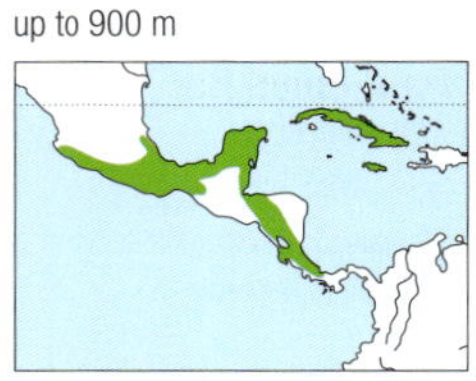

Fierce Bonneted Bat
Eumops ferox

FR Eumope féroce
DE Wilde Bulldogfledermaus
ES Eumop salvaje

Patchily from C MX (S Jalisco, Morelos, N Veracruz) S to CR, also CU & JM

VU
Monotypic
HB 7.7–10 cm T 4–6.4 cm
FA 6.1–6.6 cm W 25–60 g

S Florida

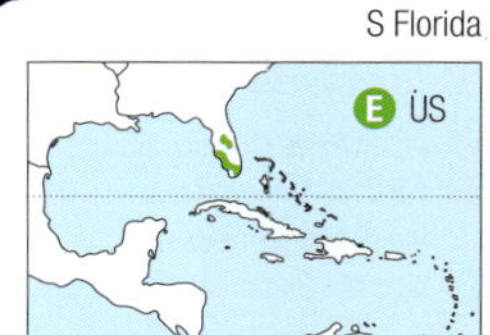

Florida Bonneted Bat
Eumops floridanus

FR Eumope de Floride
DE Florida-Bulldogfledermaus
ES Eumop de Florida

Genus *OTOMOPS*

(8 species)

VU
Monotypic
HB 9–10.9 cm
T 4–5.8 cm
FA 6.4–7.4 cm
W 27–45 g

above 1000 m

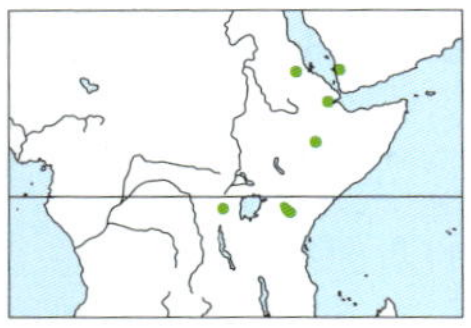

Patchily in NE & E Africa (ER, DJ, ET, UG, RW, KE) & SW Arabia (YE)

Harrison's Giant Mastiff Bat
Otomops harrisoni

FR Tadaride de Harrison
DE Harrison-Riesenbulldogfledermaus
ES Otomop de Harrison

NT
Monotypic
HB 8.7–10.3 cm
T 3.9–5.2 cm
FA 6–6.8 cm
W 25–38 g

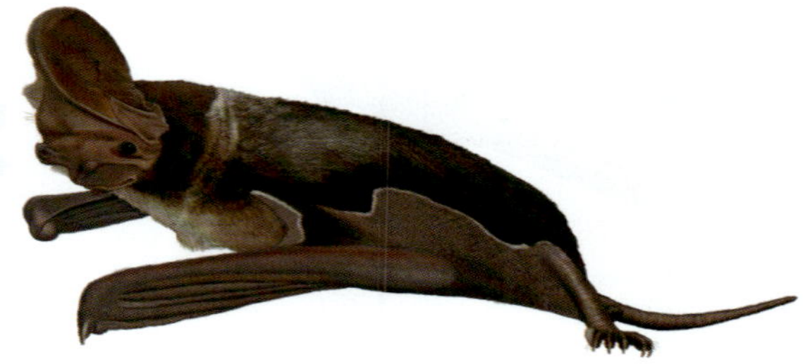

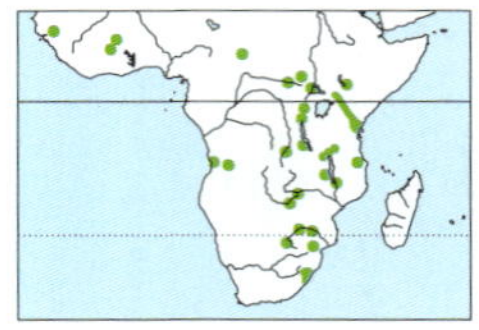

Patchily in GN, CI, GH, CF, CD, W UG, RW, BI, KE, TZ, AO, ZM, MW, ZW & E ZA

Large-eared Giant Mastiff Bat
Otomops martiensseni

FR Tadaride à grandes oreilles
DE Großohrige Riesenbulldogfledermaus
ES Otomop de orejas grandes

LC
Monotypic
HB 8.9–9.6 cm
T 3.5–5 cm
FA 5.7–6.6 cm
W 17.5–29 g

Madagascar Giant Mastiff Bat
Otomops madagascariensis

FR Tadaride de Madagascar
DE Madagaskar-Riesenbulldogfledermaus
ES Otomop de Madagascar

0–1350 m

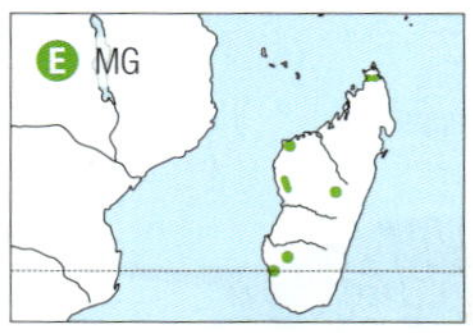

Isolated sites mostly in drier W from Ankarana SR S to Isalo NP & Toliara

DD
Monotypic
HB 8.7–9.9 cm
T 4.1–4.9 cm
FA 6.3–6.7 cm

Wroughton's Giant Mastiff Bat
Otomops wroughtoni

FR Tadaride de Wroughton
DE Wroughton-Riesenbulldogfledermaus
ES Otomop de Wroughton

up to 800 m

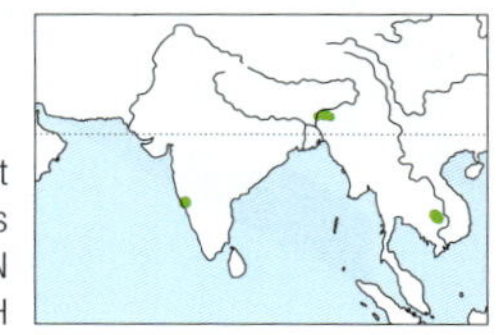

Few disjunct locations in SW & NE IN and NE KH

DD
Monotypic
HB 7.9–8.6 cm T 3–4.3 cm
FA 5.9–6 cm
W 26–27 g

up to 1450 m

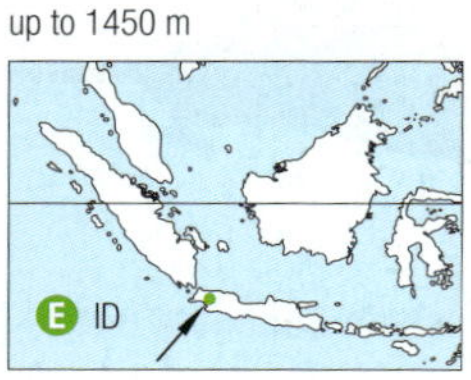

W Java (known from 2 localities, Cibadak & Cibodas, just S of Jakarta)

Java Giant Mastiff Bat
Otomops formosus

FR Tadaride de Java
DE Java-Riesenbulldogfledermaus
ES Otomop de Java

DD
Monotypic
HB c. 7.9 cm T c. 4.4 cm
FA c. 6 cm
W c. 19.5 g

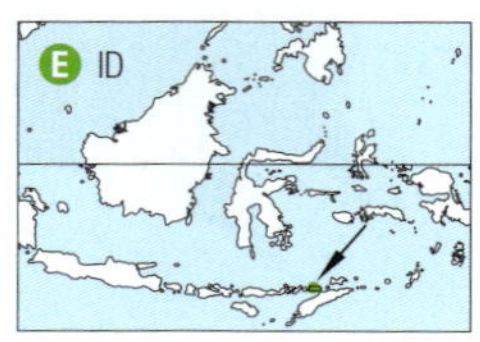

Lesser Sundas (known only from type locality on Alor)

Johnstone's Giant Mastiff Bat
Otomops johnstonei

FR Tadaride de Johnstone
DE Johnstone-Riesenbulldogfledermaus
ES Otomop de Johnstone

DD
Monotypic
HB 7–7.1 cm
T 3.6–3.7 cm
FA c. 5.8 cm

Mantled Giant Mastiff Bat
Otomops secundus

FR Tadaride à cape
DE Mantel-Riesenbulldogfledermaus
ES Otomop con capa

1800–2000 m

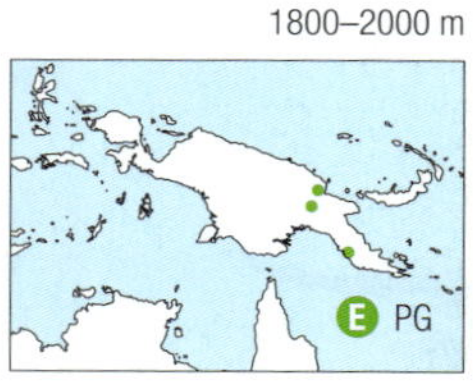

Known from 3 sites (Tapu in Madang, Kimi Creek Camp in Eastern Highlands, Hohola in Port Moresby)

DD
Monotypic
HB 6.7–7.1 cm
T 3.3–3.6 cm
FA 4.8–5.1 cm
W 15.5–17.5 g

Papuan Giant Mastiff Bat
Otomops papuensis

FR Tadaride papoue
DE Papua-Riesenbulldogfledermaus
ES Otomop de Papúa

0–300 m

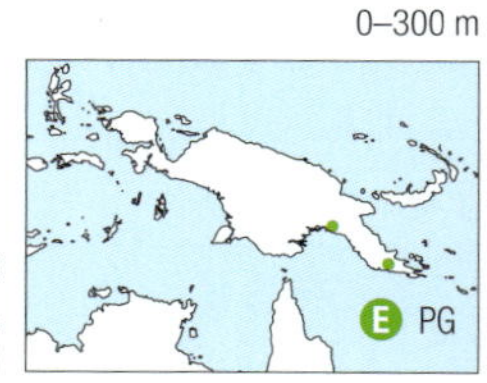

Known from 2 sites (R Vailala near Kerema, R Mai-u near Mt Suckling)

LC Monotypic
HB 7.4–8.6 cm T 3.4–4.6 cm
FA 4.8–5.3 cm W 18–38 g

up to 1200 m

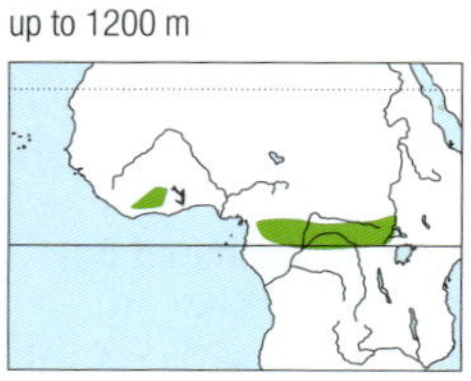

Disjunct in CI & GH and in CM, GA, CF, CD, SS & UG; perhaps CG

Duke of Abruzzi's Free-tailed Bat
Mops aloysiisabaudiae

FR Tadaride du Duc des Abruzzes
DE Fürst-von-Abruzzen-Bulldogfledermaus
ES Mop del Duque de los Abruzos

Genus *MOPS*

(36 species)

LC Monotypic
HB 6.6–8.1 cm T 3.6–4.2 cm
FA 4.3–5.1 cm W 15–23 g

400–2000 m

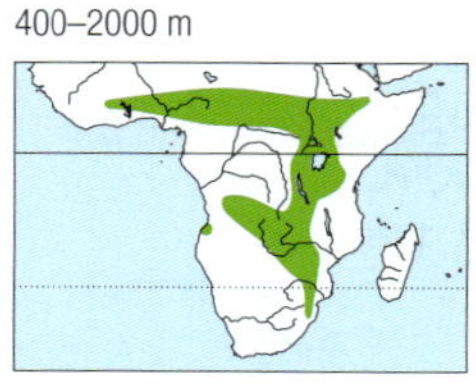

NE CI E to ET, S through E Africa to ZW & E ZA (KwaZulu-Natal), 1 record from W AO

Ansorge's Free-tailed Bat
Mops ansorgei

FR Tadaride d'Ansorge
DE Ansorge-Bulldogfledermaus
ES Mop de Ansorge

LC 2 sspp
HB 6–7 cm T 3.4–4 cm
FA 4.1–4.8 cm W 11–14 g

c. 1600 m

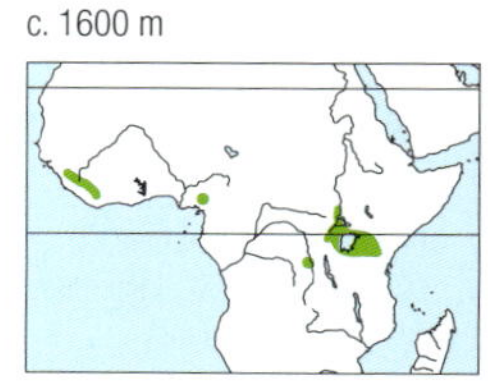

Disjunctly in GN, SL, LR, CI, CM, SC SS, UG, E CD, SW KE, N TZ

Gland-tailed Free-tailed Bat
Mops bemmeleni

FR Tadaride de van Bemmelen
DE Van-Bemmelen-Bulldogfledermaus
ES Mop de van Bemmelen

LC
Monotypic
HB 7.4–8.1 cm
T 3.2–4.8 cm
FA 4.6–5.1 cm
W 15–32 g

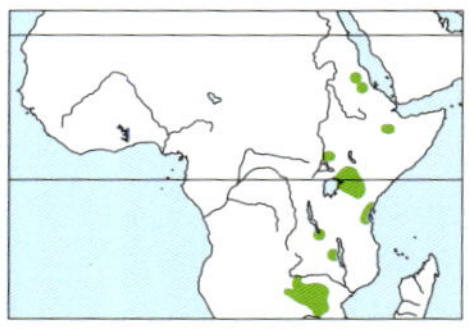

Spotted Free-tailed Bat

Mops bivittatus

FR Tadaride tachetée
DE Flecken-Bulldogfledermaus
ES Mop manchado

Widespread in ZW, marginally to S ZM, E MW & W MZ, isolated from ER & ET S to N ZM

LC
Monotypic
HB 5–6 cm
T 2.7–4.4 cm
FA 3.4–4 cm
W 5–11 g

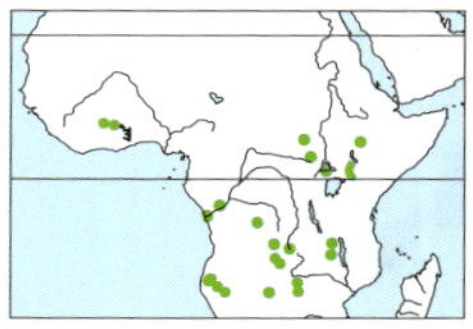

Long-crested Free-tailed Bat

Mops chapini

FR Tadaride de Chapin
DE Chapin-Bulldogfledermaus
ES Mop de Chapin

Patchily over sub-Saharan Africa in CI, GH, SS & ET S to AO, NA & ZW

DD
Monotypic
HB c. 4.9 cm
T c. 2.8 cm
FA c. 3.7 cm

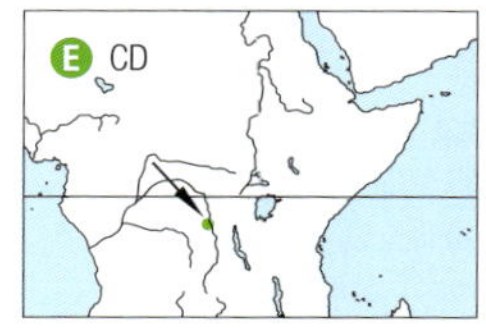

Gallagher's Free-tailed Bat

Mops gallagheri

FR Tadaride de Gallagher
DE Gallagher-Bulldogfledermaus
ES Mop de Gallagher

Known from type locality in Scierie Forest (SW of Kindu)

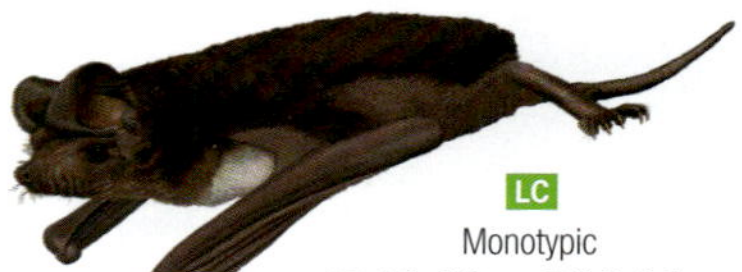

LC
Monotypic
HB 5.3–6.2 cm T 2.6–3.7 cm
FA 3.3–3.7 cm W 6–10 g

0–900 m

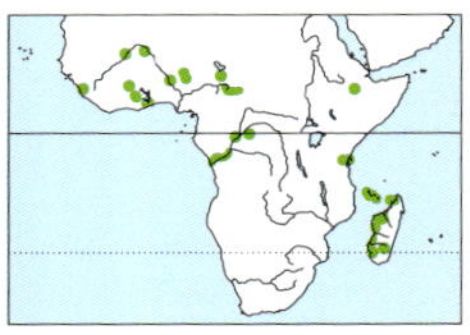

Grandidier's Free-tailed Bat

Mops leucogaster

FR Tadaride de Grandidier
DE Grandidier-Bulldogfledermaus
ES Mop de Grandidier

Scattered from SL & ML E to ET, S to CD, E TZ (with Pemba I), Comoro Is & MG (mainly W)

LC
Monotypic
HB 5.8–7.1 cm T 2.7–4.2 cm
FA 3.9–4.6 cm W 10–28 g

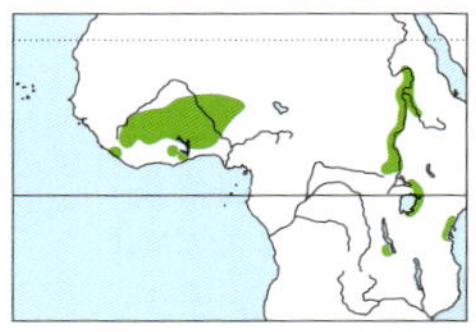

Lappet-eared Free-tailed Bat

Mops major

FR Tadaride à oreillettes
DE Ohrlappen-Bulldogfledermaus
ES Mop de orejas lobuladas

GN & LR E to NE & NG, White Nile valley, L Victoria, coast from SE KE to NE TZ, SW L Tanganyika (NE ZM)

LC
2 sspp
HB 6.8–7.8 cm T 3.1–5.2 cm
FA 4.1–5.1 cm W 10–26 g

up to 1400 m

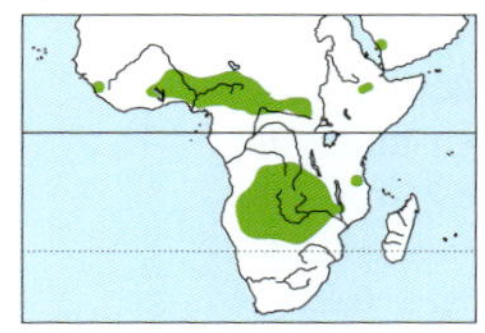

Nigerian Free-tailed Bat

Mops nigeriae

FR Tadaride du Nigeria
DE Nigeria-Bulldogfledermaus
ES Mop de Nigeria

Disjunct in SL E to ET & SW Arabia and from AO E to TZ, S to N NA & N BW

EN
Monotypic
HB 5–6 cm T 2.9–3.4 cm
FA 3.5–3.8 cm W 7–7.2 g

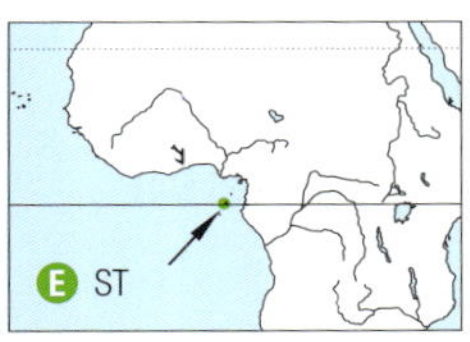

Sao Tome Free-tailed Bat

Mops tomensis

FR Tadaride de Sao Tomé
DE São-Tomé-Bulldogfledermaus
ES Mop de Sao Tomé

São Tomé I (known only from 2 localities)

LC
Monotypic
HB 4.2–6.7 cm
T 2.4–4 cm
FA 3.2–4.2 cm
W 6–16 g

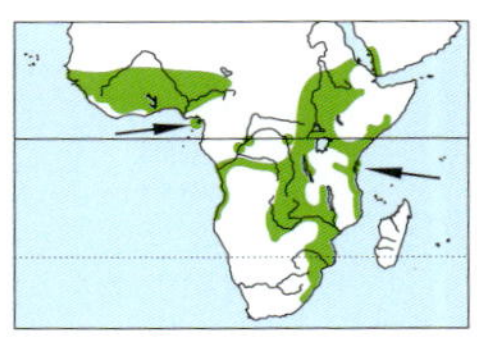

Little Free-tailed Bat

Mops pumilus

FR Petite Tadaride
DE Kleine Bulldogfledermaus
ES Mop pequeño

Disjunct from SN E to ER, ET & SO, S to AO, BW & E ZA, with Bioko, Pemba & Unguja Is, also SW Arabia

DD
Monotypic
HB 6.4–7.3 cm
T 2.9–3.7 cm
FA 4.2–4.6 cm
W 16–19 g

Isolated sites in S GN (Simandou Range), SW CI, S GH, C CM, NW & NE CD and SW KE

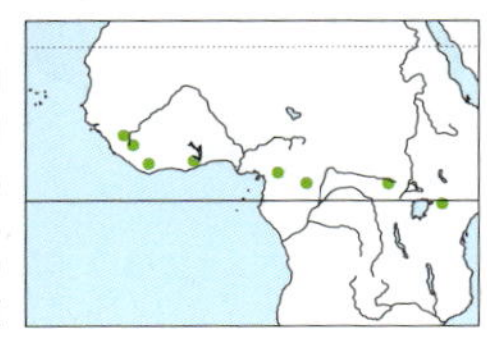

Russet Free-tailed Bat

Mops russatus

FR Tadaride rousse DE Rötliche Bulldogfledermaus ES Mop rojizo

0–400 m

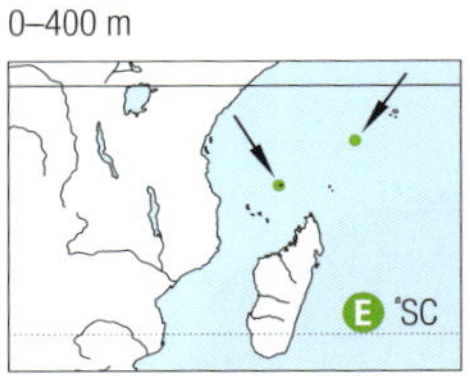

Amirante Is, Aldabra Atoll

Seychelles Free-tailed Bat

Mops pusillus

FR Tadaride des Seychelles
DE Seychellen-Bulldogfledermaus
ES Mop de las Seychelles

VU
Monotypic
HB 5.2–5.7 cm T 2.7–3.6 cm
FA 3.6–3.9 cm W 5–12 g

0–1100 m

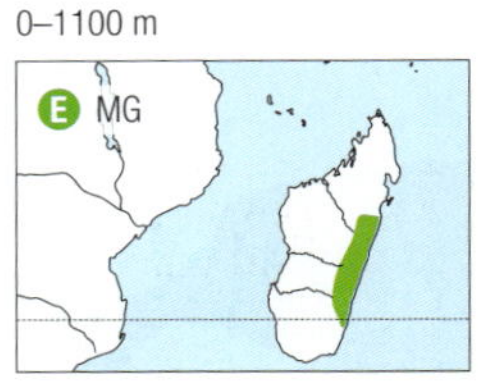

Along E seaboard, from c. 16° S to Vangaindrano (23° S)

Madagascar Free-tailed Bat

Mops atsinanana

FR Tadaride d'Atsinanana
DE Atsinanana-Bulldogfledermaus
ES Mop de Atsinanana

LC
Monotypic
HB 6.1–6.7 cm T 2.7–3.9 cm
FA 3.7–4.2 cm W 9–17 g

LC
Monotypic
HB 7.5–7.9 cm T 3.2–5.1 cm
FA 4.5–4.8 cm W 12–16 g

50–870 m

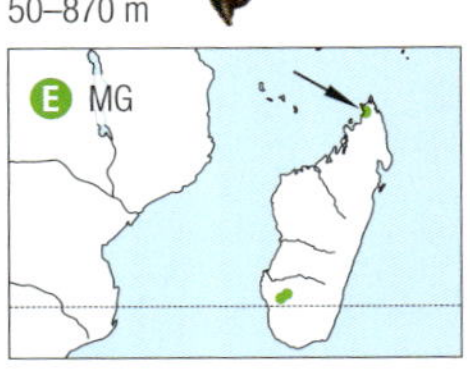

Black-and-red Free-tailed Bat

Mops jobimena

FR Tadaride dimorphe
DE Schwarzrote Bulldogfledermaus
ES Mop bicolor

Ankarana SR (in N), Isalo NP & Zombitse-Vohibasia NP (in S)

VU
Monotypic
HB 6–7.2 cm T 3.6–4.3 cm
FA 4.4–4.9 cm W 15–25 g

below 300 m

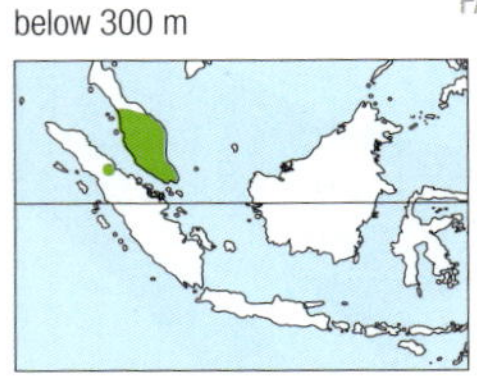

Lesser Northern Free-tailed Bat

Mops johorensis

FR Tadaride de Dobson
DE Johor-Bulldogfledermaus
ES Mop de Dobson

Extreme S TH (Hala-Bala Forest), Pen MY, NE Sumatra (ID)

LC
Monotypic
HB 6.5–7.5 cm T 2.8–4 cm
FA 4–5 cm W 13–31 g

0–950 m

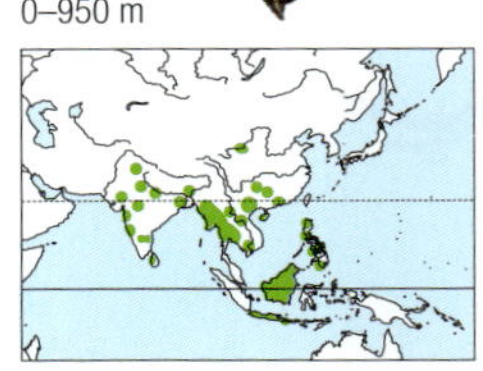

Wrinkle-lipped Free-tailed Bat

Mops plicatus

FR Tadaride à nez plissé
DE Faltenlippen-Bulldogfledermaus
ES Mop de nariz arrugada

Scattered in IN, LK, CN, mainland SE Asia, PH, Borneo, Java, Bali & Lombok

EN Monotypic
HB 6.5–6.7 cm T 4.2–4.5 cm
FA 5.1–5.4 cm W 16–22 g

0–850 m

Fijian Free-tailed Bat
Mops bregullae

FR Tadaride du Pacifique
DE Fidschi-Bulldogfledermaus
ES Mop del Pacífico

Few islands in VU (Espiritu Santo, Malo) & FJ (Vanua Levu, Taveuni)

c. 180 m

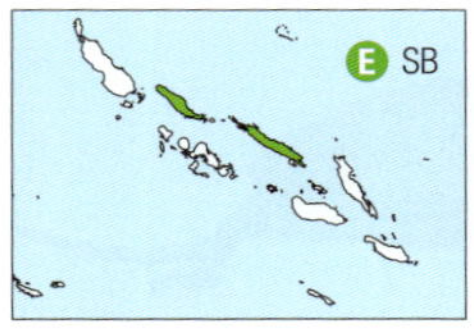

Choiseul, Santa Isabel

Solomons Free-tailed Bat
Mops solomonis

FR Tadaride des Salomon
DE Salomonen-Bulldogfledermaus
ES Mop de las Salomón

EN
Monotypic
HB 5.6–6.3 cm T 3–3.5 cm
FA 4.1–4.3 cm W 11.5–13 g

LC
2 sspp
HB 5.2–9 cm
T 3.1–4.5 cm
FA 4.3–5.2 cm
W 10.5–30 g

Greater Northern Free-tailed Bat
Mops jobensis

FR Tadaride de Miller
DE Yapen-Bulldogfledermaus
ES Mop de Miller

0–1400 m

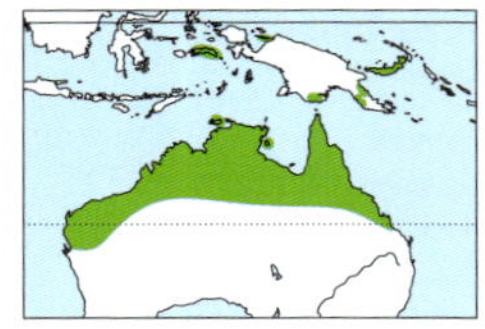

ID (Seram & Yapen Is), PG (E New Guinea, New Britain I), N AU (from N of Geraldton E to Rockhampton)

DD
Monotypic
HB 6.3–7.2 cm T 2.4–3.3 cm
FA 3.4–3.8 cm W 13–18.5 g

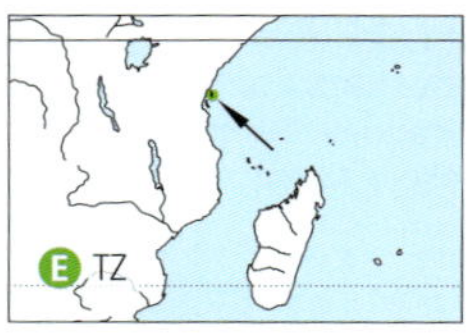

Bakari's Free-tailed Bat
Mops bakarii

FR Tadaride de Bakari
DE Bakari-Bulldogfledermaus
ES Mop de Bakari

Pemba I (Zanzibar Arch)

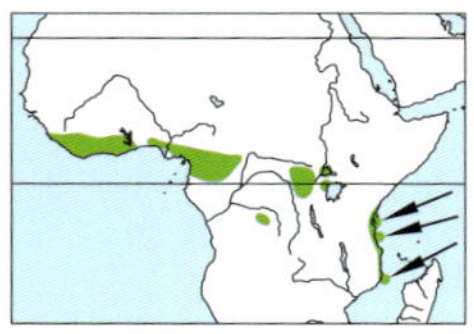

Short-winged Free-tailed Bat
Mops brachypterus

FR Tadaride à ailes courtes
DE Kurzflügel-Bulldogfledermaus
ES Mop de alas cortas

SL E to UG, along coast from SE KE S to NE MZ (with Unguja, Mafia & Mozambique Is)

LC
2 sspp
HB 5.2–6.5 cm
T 2.5–3.5 cm
FA 3.4–4.1 cm
W 12–18 g

LC Monotypic
HB 6.4–7.5 cm T 3.2–5.2 cm
FA 4.5–5.1 cm
W 16–39 g

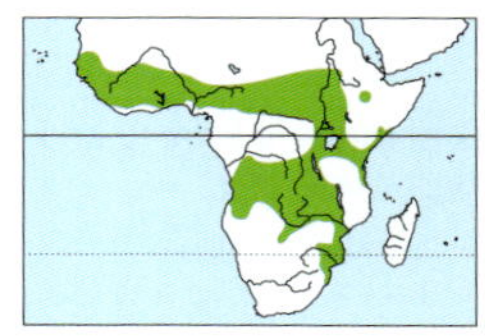

Angolan Free-tailed Bat
Mops condylurus

FR Tadaride d'Angola
DE Angola-Bulldogfledermaus
ES Mop de Angola

S MR & SN E to SW SO, S (except Congo Basin) to SW AO, N BW, SZ & E ZA (KwaZulu-Natal)

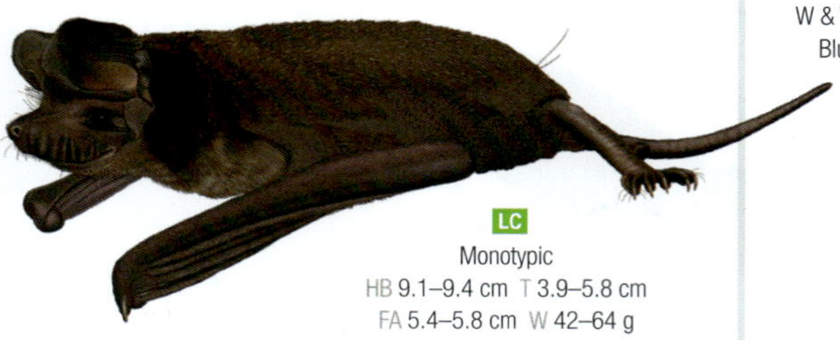

LC
Monotypic
HB 9.1–9.4 cm T 3.9–5.8 cm
FA 5.4–5.8 cm W 42–64 g

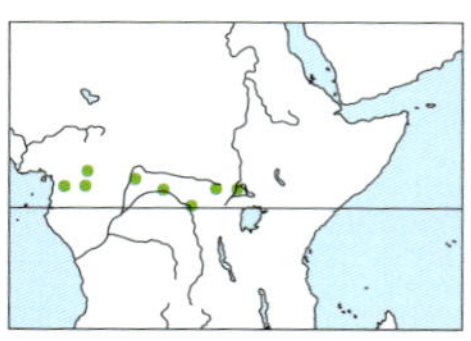

Congo Free-tailed Bat
Mops congicus

FR Tadaride de Medje
DE Kongo-Bulldogfledermaus
ES Mop del Congo

Patchily in C Africa from S CM E to N CD & W UG

W & WC Africa (CI, GH, BF, CM), Blue Nile & White Nile valleys (SD, SS, NE CD, W UG); probably CF

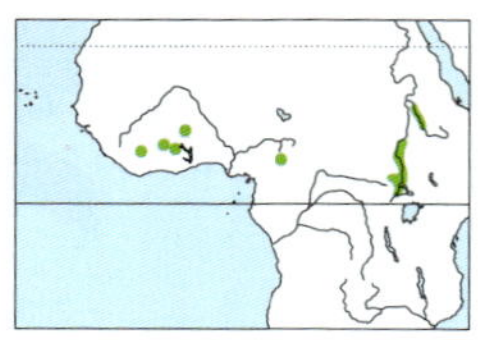

Mongalla Free-tailed Bat
Mops demonstrator

FR Tadaride de Mongalla
DE Mongalla-Bulldogfledermaus
ES Mop de Mongalla

LC
Monotypic
HB 7.6–8.2 cm
T 2.7–3.7 cm
FA 4.1–4.6 cm
W 20–24 g

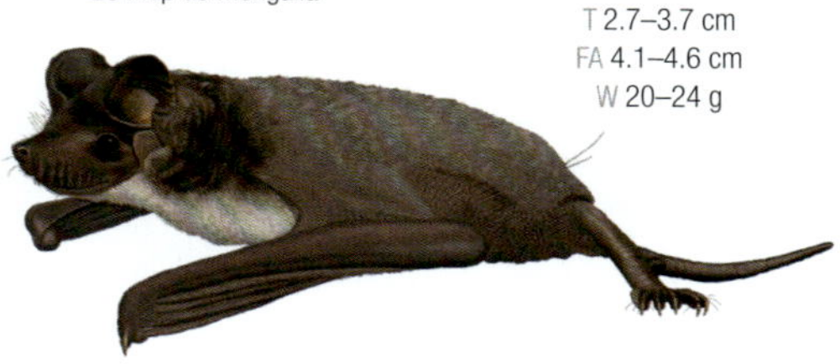

LC Monotypic
HB 8.9–9.4 cm T 3.7–5.6 cm
FA 5.9–6.7 cm W 38–69 g

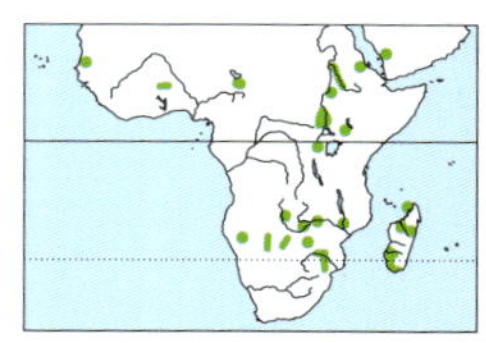

Midas Free-tailed Bat
Mops midas

FR Tadaride midas
DE Midas-Bulldogfledermaus
ES Mop de Midas

Patchily from SN E to ER & ET, S to N NA, N BW, ZW, NE ZA & MG, also SW SA

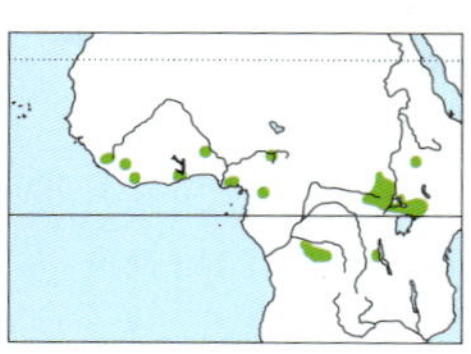

From SL & GN patchily across Congo Basin to SS, WC ET & W KE

Dwarf Free-tailed Bat
Mops nanulus

FR Tadaride naine
DE Zwerg-Bulldogfledermaus
ES Mop nano

LC
Monotypic
HB 5.6–5.9 cm T 1.9–2.9 cm
FA 2.7–3.1 cm W 6–11 g

DD
Monotypic
HB c. 9.1 cm T c. 3.4 cm
FA c. 5.2 cm

Niangara Free-tailed Bat
Mops niangarae

FR Tadaride de Niangara
DE Niangara-Bulldogfledermaus
ES Mop de Niangara

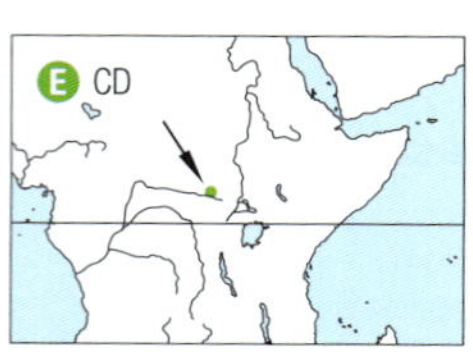

Known from type locality (Niangara) in extreme NE

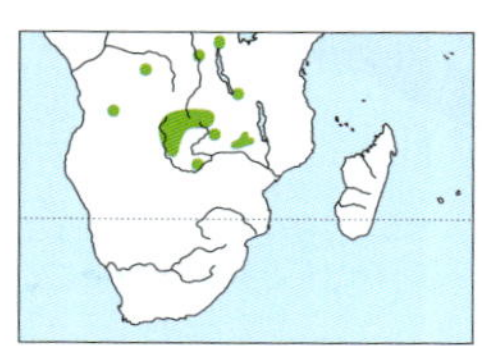

Quite sparsely in SC Africa in E & S CD, RW, BI, SW TZ, C & E AO, ZM & NW MZ; perhaps MW

White-bellied Free-tailed Bat
Mops niveiventer

FR Tadaride à ventre blanc
DE Weißbauch-Bulldogfledermaus
ES Mop de vientre blanco

LC
Monotypic
HB 7.2–7.9 cm T 3.2–3.7 cm
FA 4.4–4.8 cm
W 20–30 g

NT
Monotypic
HB c. 6.6 cm T c. 2.7 cm
FA 3.2–3.5 cm

Peterson's Free-tailed Bat
Mops petersoni

FR Tadaride de Peterson
DE Peterson-Bulldogfledermaus
ES Mop de Peterson

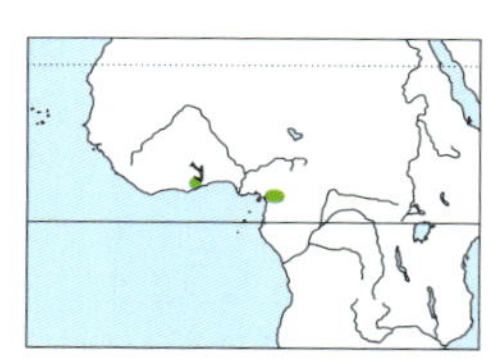

Known from 1 locality in GH and 5 in CM

LC
Monotypic
HB 5.8–6.6 cm
T 2–2.5 cm
FA 2.7–3 cm
W c. 11 g

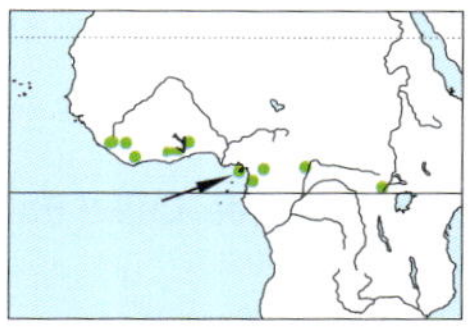

Spurrell's Free-tailed Bat
Mops spurrelli
FR Tadaride de Spurrell
DE Spurrell-Bulldogfledermaus
ES Mop de Spurrell

Patchily in SL, LR, GN, CI, GH, TG, CM, GQ (with Bioko I), CF & NE CD

SL & GN E to CM (with Bioko I) and across Congo Basin to UG, RW & W KE

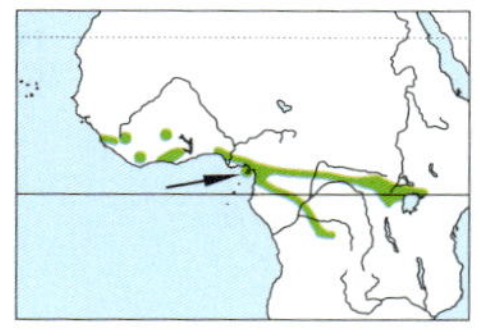

Railer Free-tailed Bat
Mops thersites
FR Tadaride railleuse
DE Thersites-Bulldogfledermaus
ES Mop de Thersites

LC
Monotypic
HB 6.6–7.7 cm T 2.5–3.6 cm
FA 3.5–4.2 cm W 10–34 g

DD Monotypic
HB 8.1–9 cm T 3.4–4.4 cm
FA 5.1–5.5 cm

Trevor's Free-tailed Bat
Mops trevori
FR Tadaride de Trevor
DE Trevor-Bulldogfledermaus
ES Mop de Trevor

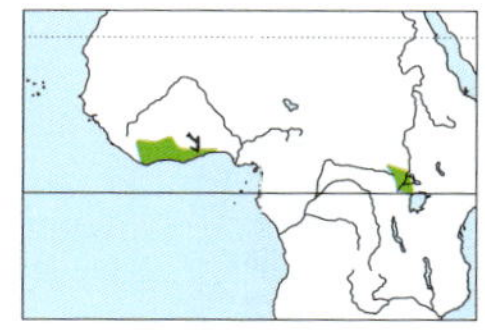

W Africa (extreme SE GN E to extreme SW NG), C Africa (NE CD, W UG)

0–1300 m

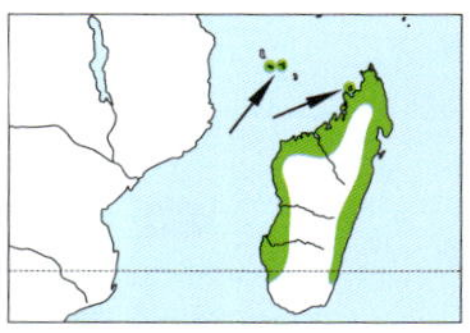

Malagasy White-bellied Free-tailed Bat
Mops leucostigma
FR Tadaride à giron blanc
DE Weißbauch-Madagaskar-Bulldogfledermaus
ES Mop de Madagascar

KM (Mohéli, Anjouan), MG (with Nosy Be) except for C higher-lying areas

LC Monotypic
HB 7–9 cm T 3.2–4.9 cm
FA 4–4.7 cm W 12–28 g

NT
Monotypic
HB 6.5–7.2 cm
T 3.5–3.9 cm
FA 4–4.6 cm
W 17–24 g

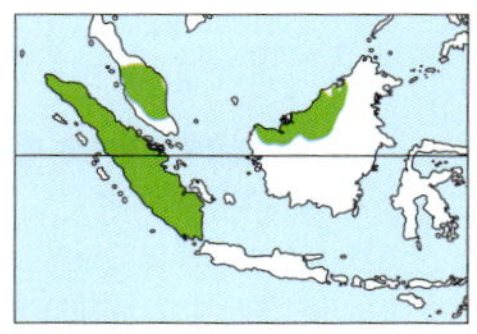

Malayan Free-tailed Bat
Mops mops
FR Tadaride de Malaisie
DE Malaya-Bulldogfledermaus
ES Mop de Malasia

Extreme S TH (Hala-Bala Forest), MY (Peninsula, Sarawak), ID (Sumatra, possibly Java)

DD
2 sspp
HB 6.7–7.2 cm T 3–3.8 cm
FA 3.7–4.5 cm W 16–35 g

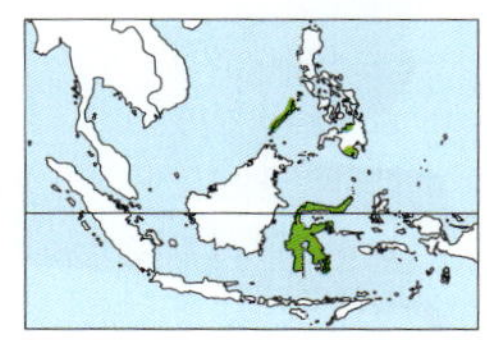

Sulawesian Free-tailed Bat
Mops sarasinorum
FR Tadaride des Célèbes
DE Sulawesi-Bulldogfledermaus
ES Mop de Célebes

ID (Sulawesi, with Peleng & Buton Is), PH (Palawan, Mindanao)

Genus *MYOPTERUS*

(2 species)

Daubenton's Winged-mouse Bat
Myopterus daubentonii
FR Tadaride de Daubenton DE Daubenton-Bulldogfledermaus ES Mióptero de Daubenton

up to 1250 m

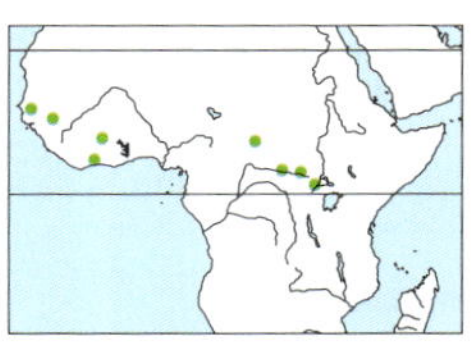

DD
2 sspp
HB 6.7–7.6 cm
T 3.7–4.4 cm
FA 4.8–5.4 cm
W 20–22 g

Patchily in SN, CI, CF & N CD

Bini Winged-mouse Bat
Myopterus whitleyi
FR Tadaride de Whitley DE Whitley-Bulldogfledermaus ES Mióptero de Whitley

LC
Monotypic
HB 5.3–5.6 cm
T 2.5–3.1 cm
FA 3.5–3.7 cm
W 10–12 g

From GH E through NG, CM, CF, GA & CD to S UG, absent from C Congo Basin

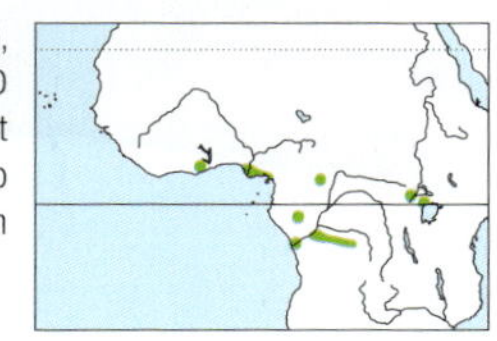

Genus *PLATYMOPS*

(1 species)

up to 2000 m

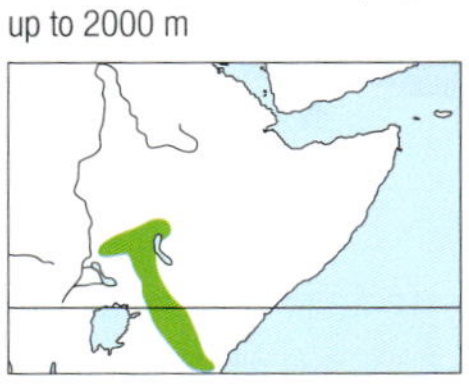

LC
Monotypic
HB 5–7.8 cm
T 2.2–3.6 cm
FA 2.6–3.6 cm
W 7–18 g

SE SS, SW ET, KE; possibly adjacent UG

Peters's Flat-headed Bat
Platymops setiger
FR Tadaride à tête plate
DE Peters-Flachkopf-Bulldogfledermaus
ES Platimop de Peters

Genus *SAUROMYS*

(1 species)

up to 2000 m

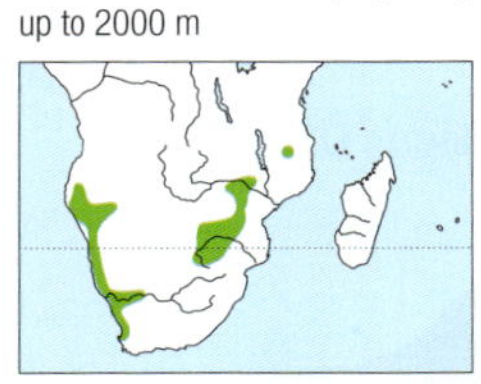

LC
Monotypic
HB 6–8.2 cm
T 2.9–4.9 cm
FA 3.7–5 cm
W 6–22 g

Disjunct in arid SW AO S to NW ZA and from N MZ S to N ZA

Roberts's Flat-headed Bat
Sauromys petrophilus
FR Tadaride de Roberts
DE Roberts-Flachkopf-Bulldogfledermaus
ES Sauromis de Roberts

Genus *TADARIDA*

(8 species)

LC 9 sspp
HB 4.6–6.2 cm T 2.8–4.2 cm
FA 3.6–4.7 cm W 8–15 g

0–3000 m

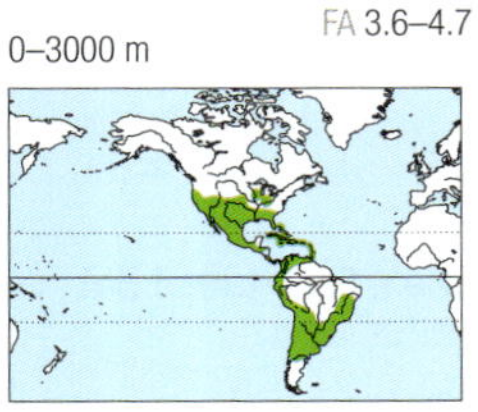

Brazilian Free-tailed Bat
Tadarida brasiliensis
FR Tadaride du Brésil
DE Brasilianische Bulldogfledermaus
ES Murciélago rabudo de Brasil

From US to South America (W of Orinoco Basin, W & S of Amazon Basin), Greater & Lesser Antilles

LC Monotypic
HB 8.2–8.7 cm
T 3.7–5.7 cm
FA 5.6–6.4 cm
W 20–40 g

0–3100 m

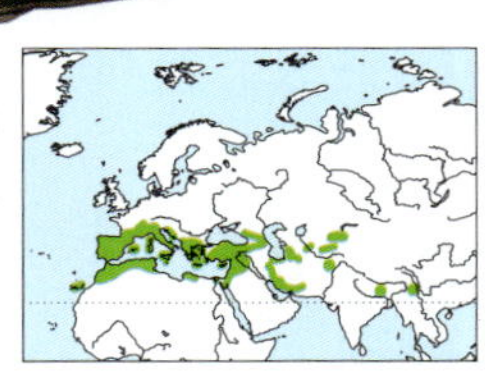

European Free-tailed Bat
Tadarida teniotis
FR Tadaride de Cestoni
DE Europäische Bulldogfledermaus
ES Murciélago rabudo

Mediterranean Basin (with Is) E to Middle East & patchily to IR, Central Asia, AF, NP & MM, also Canary Is

0–2100 m

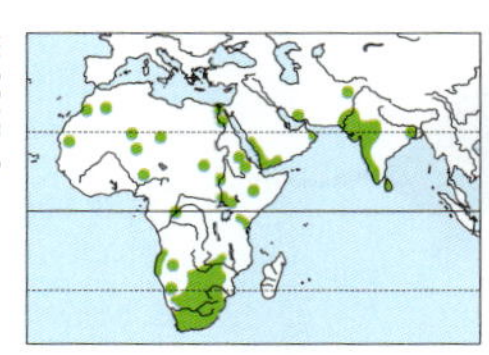

Egyptian Free-tailed Bat
Tadarida aegyptiaca
FR Tadaride d'Égypte
DE Ägyptische Bulldogfledermaus
ES Murciélago rabudo de Egipto

Widespread throughout Africa (very patchy N of R Zambezi), S Arabian Pen, S IR, E AF, PK, IN, LK

LC Monotypic
HB 6–8.7 cm
T 3–5 cm
FA 4.2–5.2 cm
W 9–22 g

LC Monotypic
HB 7.8–8.6 cm T 4.6–6 cm
FA 5.5–6.2 cm W 20–33 g

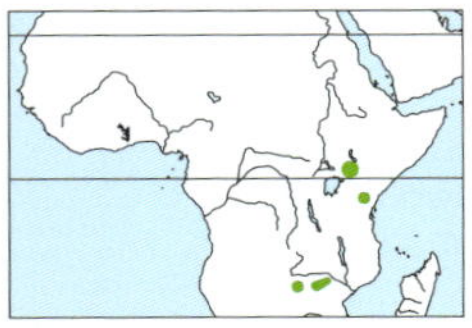

Kenyan Big-eared Free-tailed Bat
Tadarida lobata
FR Tadaride du Kenya
DE Kenia-Bulldogfledermaus
ES Murciélago rabudo de Kenia

Disjunct in W & S KE and N ZW

above 1400 m

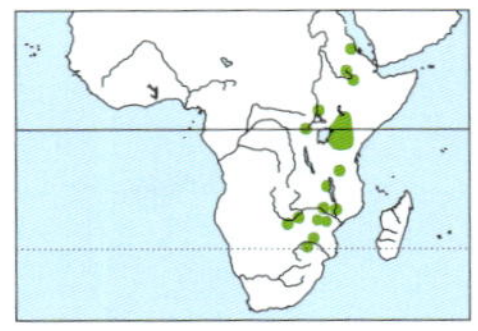

African Giant Free-tailed Bat
Tadarida ventralis
FR Tadaride africaine
DE Große Afrika-Bulldogfledermaus
ES Murciélago rabudo africano

Sparsely but widely in E Africa (ER & ET S to ZW & NE ZA)

DD Monotypic
HB 9.1–10.2 cm T 5.1–6.6 cm
FA 6–6.7 cm W 31–55 g

LC Monotypic
HB 7.9–9.5 cm
T 5.3–6.6 cm
FA 5.6–6.1 cm
W 23–49 g

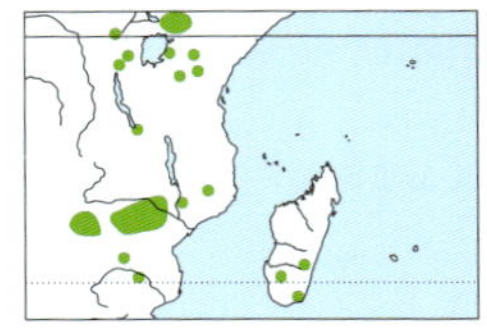

Malagasy Free-tailed Bat
Tadarida fulminans
FR Tadaride de Thomas
DE Madagaskar-Bulldogfledermaus
ES Murciélago rabudo de Thomas

E & S Africa (NE CD, E UG, W & S KE, RW, N TZ, NE & SW ZM, ZW, MZ, S MW, NE ZA), S MG (3 records)

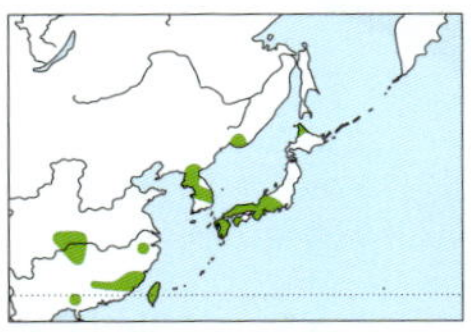

Widespread across C & S CN, TW, Korean Pen, RU (SE Far East), JP

East Asian Free-tailed Bat
Tadarida insignis
FR Tadaride d'Asie
DE Ostasiatische Bulldogfledermaus
ES Murciélago rabudo de Asia

DD
Monotypic
HB 8.2–10.2 cm T 4.8–6 cm
FA 5.7–6.5 cm
W 15–20 g

c. 1500 m

Disjunct in NE CN (Liaoning, Inner Mongolia, Hebei, Beijing, Shandong), N & C LA, NE TH (Loei Prov)

La Touche's Free-tailed Bat
Tadarida latouchei
FR Tadaride de La Touche
DE La-Touche-Bulldogfledermaus
ES Murciélago rabudo de La Touche

EN
Monotypic
HB 6.7–7.2 cm
T 4.1–4.9 cm
FA 5.3–5.7 cm

Genus *AUSTRONOMUS*

(2 species)

LC Monotypic
HB 7.1–8.4 cm T 4.1–4.8 cm
FA 5.4–6 cm W 23–26 g

1900–3100 m

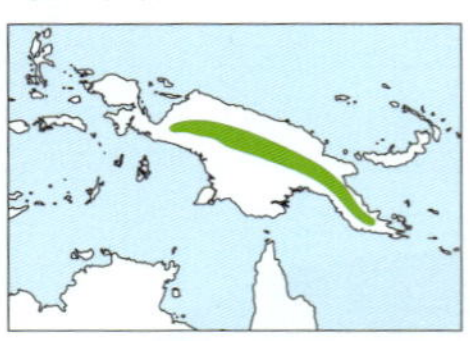

New Guinea Free-tailed Bat
Austronomus kuboriensis
FR Tadaride du Kubor
DE Neuguinea-Bulldogfledermaus
ES Murciélago rabudo de Kubor

ID/PG: New Guinea Central Highlands

LC Monotypic
HB 8.5–10 cm
T 4–5.5 cm
FA 5.7–6.5 cm
W 26–48 g

0–1400 m

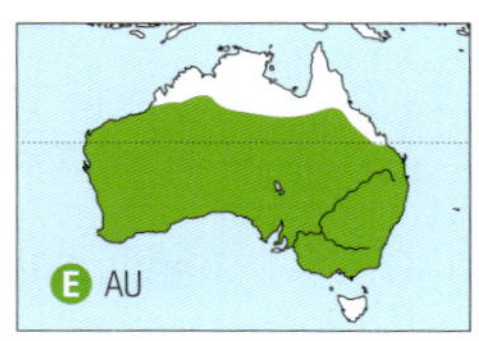

White-striped Free-tailed Bat
Austronomus australis
FR Tadaride australe
DE Weißstreifen-Bulldogfledermaus
ES Murciélago rabudo austral

Except far tropical N and Tas

Genus *SETIROSTRIS*

(1 species)

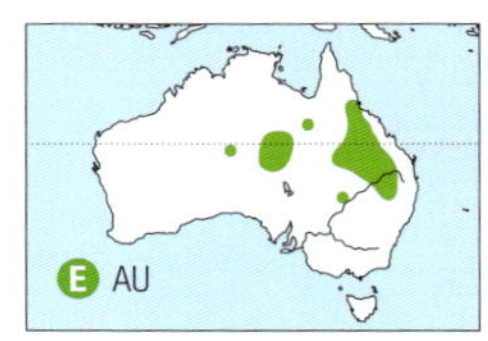

Arid and semiarid regions of C & E to Great Dividing Range in Qld & N NSW

Hairy-nosed Free-tailed Bat
Setirostris eleryi
FR Tadaride d'Elery
DE Nasenhaar-Bulldogfledermaus
ES Murciélago rabudo de Elery

NT
Monotypic
HB 4.1–4.9 cm
T 2.7–3.8 cm
FA 3.1–3.6 cm
W 4.5–7.5 g

Genus *MICRONOMUS*

(1 species)

East Coast Free-tailed Bat
Micronomus norfolkensis
FR Tadaride de Norfolk
DE Ostküsten-Bulldogfledermaus
ES Murciélago rabudo de Norfolk

NT Monotypic
HB 5–5.5 cm T 3.2–4.5 cm
FA 3.5–4.1 cm W 6–12 g

E of Great Dividing Range from Conondale Range (SE Qld) S to Bega (SE NSW)

Genus *OZIMOPS*

(9 species)

DD Monotypic
HB 4.7–5.8 cm T 2.5–3.6 cm
FA 3.1–3.5 cm W 6–9 g

Cape York Free-tailed Bat
Ozimops halli
FR Tadaride de Cape York
DE Kap-York-Bulldogfledermaus
ES Murciélago rabudo de Cape York

Known from 3 locations on Cape York Pen (Qld)

N & C Moluccas, New Guinea (with Kairiru, Kadovar, Mussau, New Britain & Fergusson Is)

0–300 m

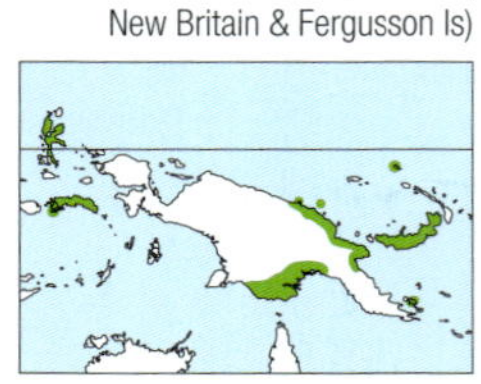

LC 2 sspp
HB 5.4–6.6 cm T 2.7–3.8 cm
FA 3.3–3.6 cm W 13–17 g

Beccari's Free-tailed Bat
Ozimops beccarii
FR Tadaride de Beccari
DE Beccari-Bulldogfledermaus
ES Murciélago rabudo de Beccari

LC
Monotypic
HB 4.5–5.5 cm
T 2.8–3.4 cm
FA 3.2–3.5 cm
W 6–10.5 g

Northern Coastal Free-tailed Bat
Ozimops cobourgianus
FR Tadaride de Cobourg
DE Nordküsten-Bulldogfledermaus
ES Murciélago rabudo de Cobourg

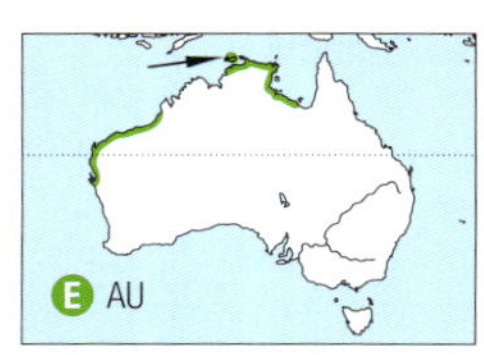

Coastal WA (Exmouth E to Broome) & NT to Qld border (with Melville I)

0–100 m

Few sites in SE coast (in and around National Capital District & Central Prov)

DD Monotypic
HB 4.8–5 cm T c. 2.7 cm
FA 3.2–3.4 cm W 6–13 g

Loria's Free-tailed Bat
Ozimops loriae
FR Tadaride de Loria
DE Loria-Bulldogfledermaus
ES Murciélago rabudo de Loria

LC
Monotypic
HB 5.5–6 cm T 2.8–3.6 cm
FA 3.3–3.5 cm W 8–10.5 g

South-western Free-tailed Bat
Ozimops kitcheneri

FR Tadaride de Kitchener
DE Südwestliche Bulldogfledermaus
ES Murciélago rabudo de Kitchener

S WA

Southern Free-tailed Bat
Ozimops planiceps

FR Tadaride de Sydney
DE Südliche Bulldogfledermaus
ES Murciélago rabudo de Sydney

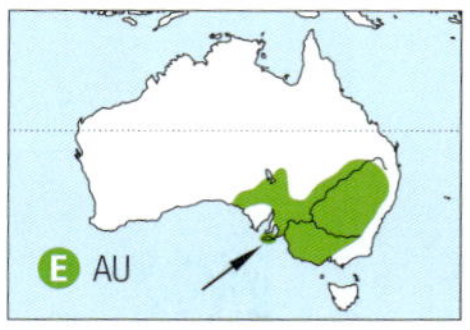

W slopes of Great Dividing Range (NSW) through Vic to S SA (to Flinders & Gawler ranges), Kangaroo I

LC
Monotypic
HB 5–5.6 cm T 2.7–3.5 cm
FA 3.2–3.6 cm W 6.5–12 g

LC
Monotypic
HB 5.9–6.8 cm T 3.3–3.8 cm
FA 3.5–4.1 cm W 11–19 g

Lumsden's Free-tailed Bat
Ozimops lumsdenae

FR Tadaride de Lumsden
DE Lumsden-Bulldogfledermaus
ES Murciélago rabudo de Lumsden

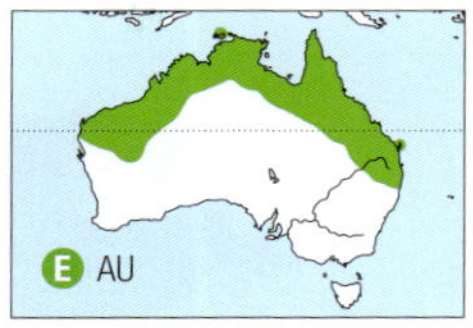

N coastline (inland no more than c. 600 km) from WA (Pilbara region) E to Qld & NE NSW

Inland Free-tailed Bat
Ozimops petersi

FR Tadaride de Peters
DE Inland-Bulldogfledermaus
ES Murciélago rabudo de Peters

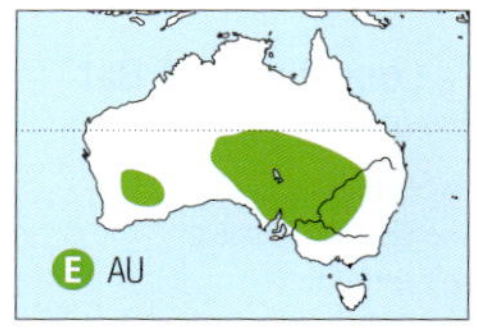

Arid areas of inland S of Tropic of Capricorn

LC
Monotypic
HB 4.5–5.7 cm T 3–3.5 cm
FA 3.2–3.8 cm W 7–12 g

LC
Monotypic
HB 5–6 cm T 2.8–3.4 cm
FA 3–3.5 cm W 5–11 g

Ride's Free-tailed Bat
Ozimops ridei

FR Tadaride de Ride
DE Ride-Bulldogfledermaus
ES Murciélago rabudo de Ride

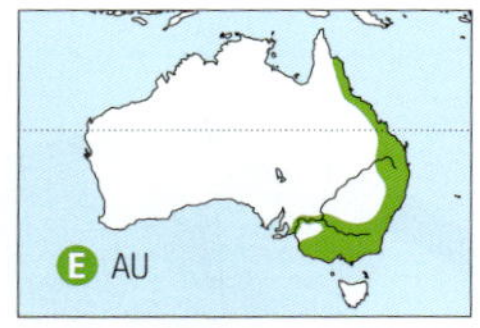

From Qld (Cape Melville, N of Cooktown) S to S Vic, far inland along R Murray basin to SA

Family MINIOPTERIDAE – Long-fingered Bats

1 genus, 40 species. Cave roosting in all biomes, except deserts. Palearctic, Afrotropical, Indo-Malayan and Australasian Regions.

Genus *MINIOPTERUS*

(40 species)

NE
Monotypic
HB 4.7–6.5 cm
T 4.4–6.1 cm
FA 4.5–5 cm
W 12.5–14.7 g

0–2000 m

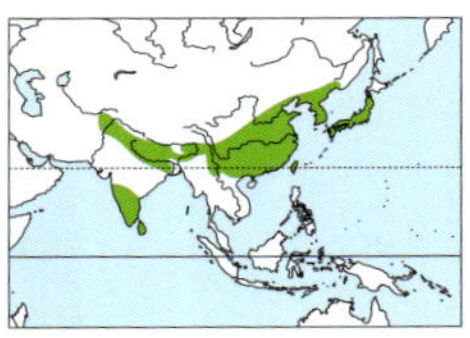

Asian Long-fingered Bat
Miniopterus fuliginosus

FR Minioptère fuligineux
DE Asiatische Langflügelfledermaus
ES Minióptero de Asia

NE AF to N MM, N VN, CN & TW, N to Korean Pen, RU (S Far East) & JP, also S IN & LK

LC
Monotypic
HB 5.8–7.5 cm
T 5.2–6.4 cm
FA 4.8–5.4 cm
W 11.3–21 g

0–2100 m

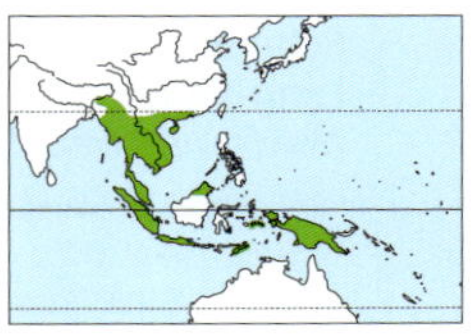

Large Long-fingered Bat
Miniopterus magnater

FR Grand Minioptère
DE Malaiische Langflügelfledermaus
ES Minióptero grande

NE IN, S CN (with Hainan), SE Asia, Sunda Is (Sumatra, Borneo, Timor), C Moluccas, New Guinea

LC
Monotypic
HB 3.8–5 cm
T 4–5.1 cm
FA 3.6–4.4 cm
W 9–11 g

up to 1200 m

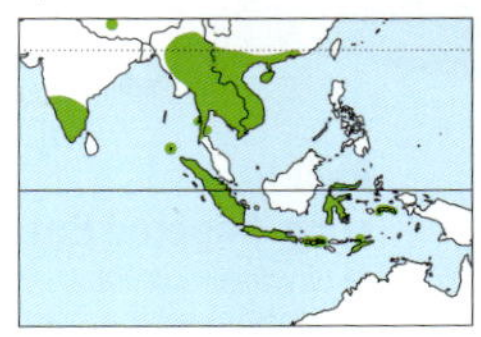

Small Long-fingered Bat
Miniopterus pusillus

FR Minioptère nain
DE Winzige Langflügelfledermaus
ES Minióptero enano

Patchily in S Asia (NP, S IN), mainland & insular SE Asia (from S CN & Nicobar Is E to Moluccas)

0–2600 m

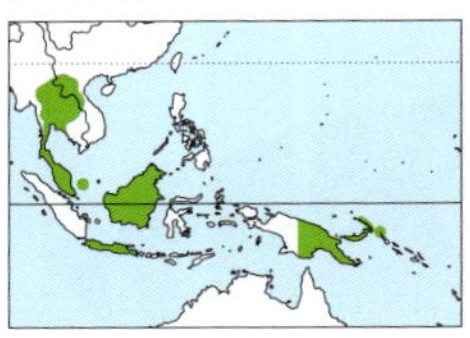

Intermediate Long-fingered Bat
Miniopterus medius

FR Minioptère intermédiaire
DE Mittelgroße Langflügelfledermaus
ES Minióptero mediano

TH, LA, Pen MY, Anambas Is, Borneo, Java, E New Guinea, Bismarck Arch, Buka I

LC
Monotypic
HB 4.2–6 cm T 4.3–5.9 cm
FA 4–4.8 cm W 6–12.1 g

0–1300 m

Ryukyu Long-fingered Bat
Miniopterus fuscus

FR Minioptère des Ryukyu
DE Ryukyu-Langflügelfledermaus
ES Minióptero de Ryukyu

Ryukyu Is (from Amami Oshima S to Ishigakijima & Iriomotejima); apparently extinct from Kii Pen (Honshu)

EN
Monotypic
HB 4.7–5.9 cm
T 5.2–5.5 cm
FA 4.2–4.6 cm

NE
Monotypic
HB c. 5.5 cm
T 4.6–5.7 cm
FA 4.2–4.6 cm
W 8–14 g

0–1500 m

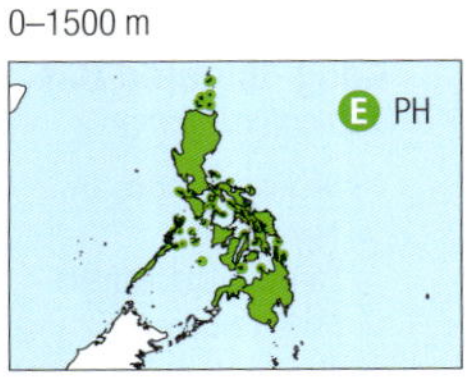

Eschscholtz's Long-fingered Bat
Miniopterus eschscholtzii

FR Minioptère d'Eschscholtz
DE Eschscholtz-Langflügelfledermaus
ES Minióptero de Eschscholtz

Widespread in a wide variety of habitats

LC
3 sspp
HB 4.7–5.1 cm
T 3.4–4.9 cm
FA 3.4–3.9 cm
W 5–6 g

up to 1450 m

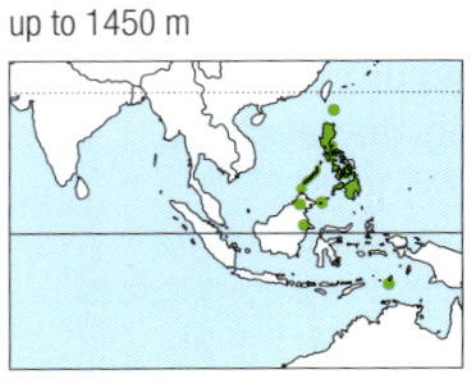

Philippine Long-fingered Bat
Miniopterus paululus

FR Minioptère des Philippines
DE Philippinen-Langflügelfledermaus
ES Minióptero de las Filipinas

PH, Borneo (scattered in N Sabah & East Kalimantan), Tanimbar Is (Selaru)

LC
5 sspp
HB 6.5–7.3 cm
T 5–6.1 cm
FA 4.5–5.7 cm
W 16–22 g

0–1600 m

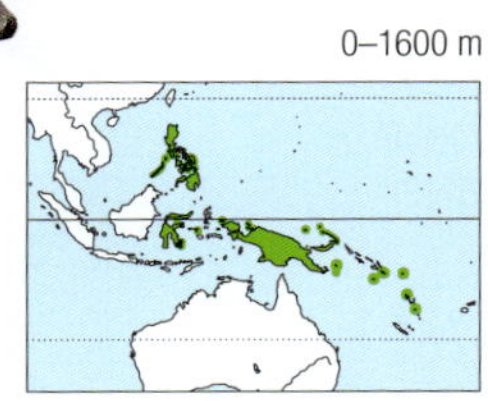

Great Long-fingered Bat
Miniopterus tristis

FR Minioptère triste
DE Große Langflügelfledermaus
ES Minióptero triste

PH, Sulawesi, New Guinea (with nearby Is & arch), SB, VU (Espirito Santo, Efate)

DD
Monotypic
HB 3.3–4.6 cm
T 3.4–4.1 cm
FA 3.3–3.9 cm

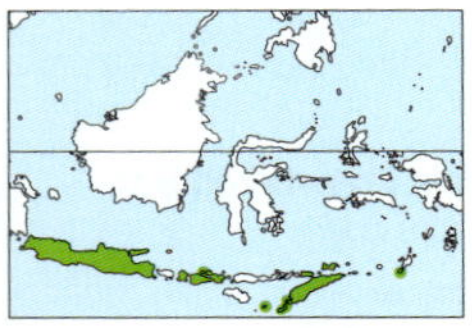

Shortridge's Long-fingered Bat
Miniopterus shortridgei

FR Minioptère de Shortridge
DE Shortridge-Langflügelfledermaus
ES Minióptero de Shortridge

Java, most of Lesser Sundas (from Lombok E to Wetar & Timor), Tanimbar Is (Selaru); perhaps S Sumatra

NE
Monotypic
HB 4.6–6.3 cm
T 3.9–6.2 cm
FA 4.5–5.2 cm
W 8.8–17.5 g

0–2900 m

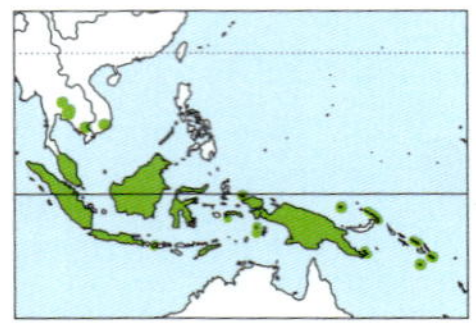

Javanese Long-fingered Bat
Miniopterus blepotis

FR Minioptère blépote
DE Java-Langflügelfledermaus
ES Minióptero de Java

Thai-Malay Pen, Greater & some Lesser Sundas, Moluccas, New Guinea, SB; possibly C TH, S KH, S VN

LC
Monotypic
HB 3.6–4.8 cm
T 3.8–5 cm
FA 3.5–4.1 cm
W 5–6.5 g

0–1500 m

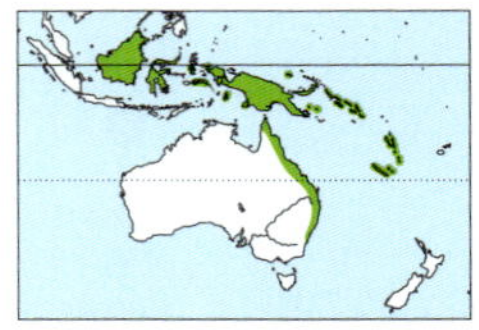

Little Long-fingered Bat
Miniopterus australis

FR Minioptère australasien
DE Kleine Australische Langflügelfledermaus
ES Minióptero australiano

Borneo, Sulawesi, Moluccas, New Guinea (with nearby Is & arch), Solomon Is, VU, NC, E coast of AU

LC
Monotypic
HB 4.2–4.8 cm T 4–5.4 cm
FA 3.9–4.5 cm W 7.5–11 g

0–3200 m

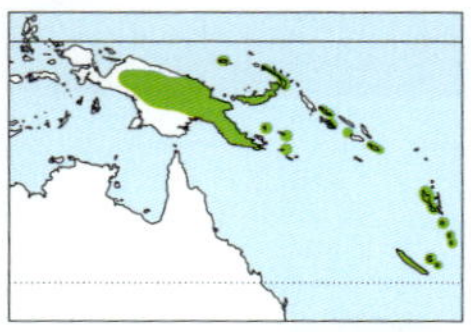

Small Melanesian Long-fingered Bat
Miniopterus macrocneme

FR Minioptère de Mélanésie
DE Kleine Melanesische Langflügelfledermaus
ES Minióptero de Melanesia

New Guinea (with Manus I, Bismarck & Louisiade arch), SB, VU, NC

EN
Monotypic
HB 4.3–4.9 cm
T 3.7–4.6 cm
FA 3.9–4.2 cm

c. 100 m

Loyalty Long-fingered Bat
Miniopterus robustior

FR Minioptère des Loyauté
DE Loyalty-Inseln-Langflügelfledermaus
ES Minióptero de Loyalty

Loyalty Is (Lifou, Maré)

NE
3 sspp
HB 4.7–6.3 cm
T 4.3–5.8 cm
FA 4.3–5 cm
W 8.6–16 g

Australian Long-fingered Bat
Miniopterus orianae

FR Minioptère d'Oriana
DE Australische Langflügelfledermaus
ES Minióptero de Oriana

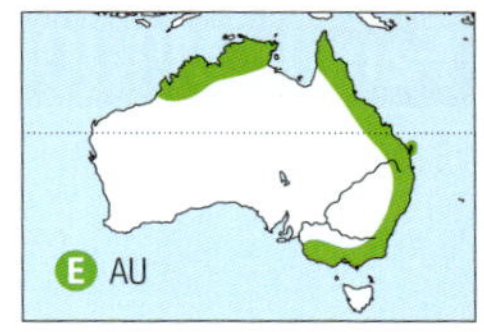

NE WA, N NT, Qld (with Fraser I) to Vict & SE SA (Robe & Naracoorte S to Port MacDonnell)

NT
Monotypic
HB 5.6–6.5 cm
T 5.8–6.6 cm
FA 4.3–4.8 cm
W 14.7–15.2 g

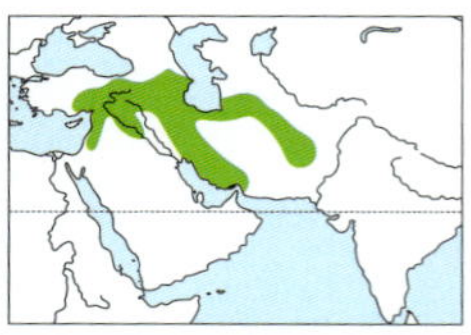

Pallid Long-fingered Bat
Miniopterus pallidus

FR Minioptère pâle
DE Fahle Langflügelfledermaus
ES Minióptero pálido

TR (E Anatolia), S AM, AZ, SY, LB, NW JO, IQ, IR, S TM, S AF; range uncertain

VU
Monotypic
HB 5.1–5.9 cm
T 5.2–6.3 cm
FA 4.3–4.7 cm
W 10–14 g

1000–1400 m

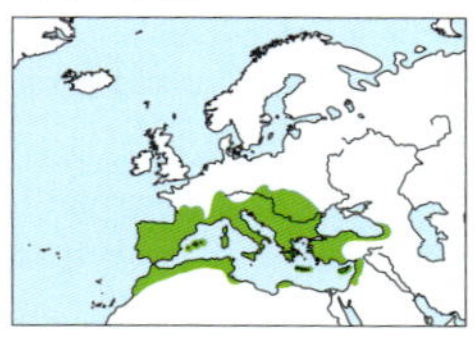

Schreibers's Long-fingered Bat
Miniopterus schreibersii

FR Minioptère de Schreibers
DE Eigentliche Langflügelfledermaus
ES Murciélago de cueva

S Europe E to W GE, TR & Levant, N to C FR & SK, S to N Africa, most Mediterranean Is; range uncertain

NT
Monotypic
HB c. 6 cm
T c. 6.3 cm
FA 4.5–4.8 cm

Maghrebian Long-fingered Bat
Miniopterus maghrebensis

FR Minioptère du Maghreb
DE Maghreb-Langflügelfledermaus
ES Minióptero de Maghreb

300–1200 m

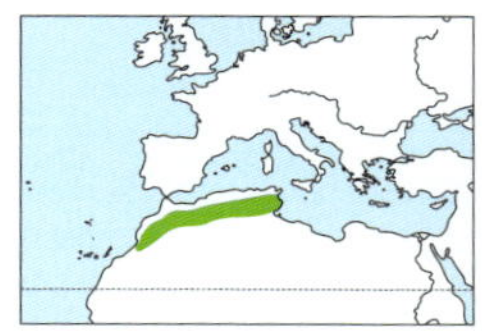

C & SW MA (Middle & High Atlas), N TN; perhaps continuously across most of N Africa, including DZ

200–1350 m

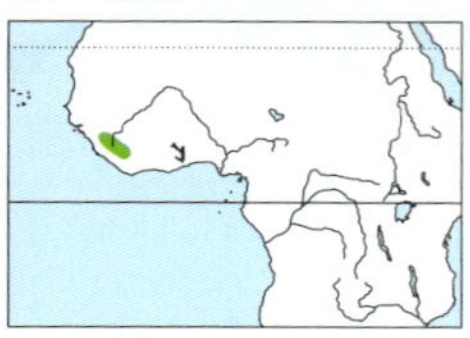

Upper Guinea forest zone in GN, SL & LR

Villiers's Long-fingered Bat
Miniopterus villiersi

FR Minioptère de Villiers
DE Villiers-Langflügelfledermaus
ES Minióptero de Villiers

NE
Monotypic
T 4.8–5.6 cm FA 4.2–4.4 cm

500–2200 m

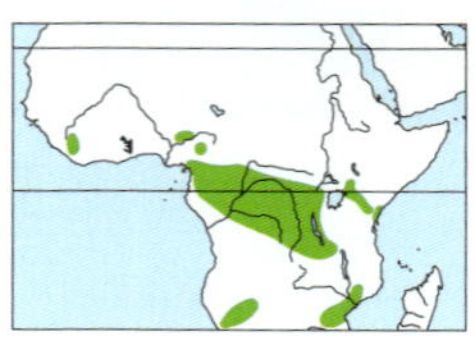

Sub-Saharan Africa in C (CM E to CD), E (UG, RW, BI, KE, TZ), W (GN, LR, NG) & S (NA, ZW, MZ)

Greater Long-fingered Bat
Miniopterus inflatus

FR Minioptère à couronne
DE Größere Langflügelfledermaus
ES Minióptero gigante

LC Monotypic
HB 5.7–6.3 cm T 4.8–5.7 cm
FA 4.5–4.9 cm W 9.9–16 g

NE
Monotypic
HB 6.3–6.9 cm
T 4.7–5.7 cm
FA 4.6–4.9 cm
W 9–10.5 g

720–970 m

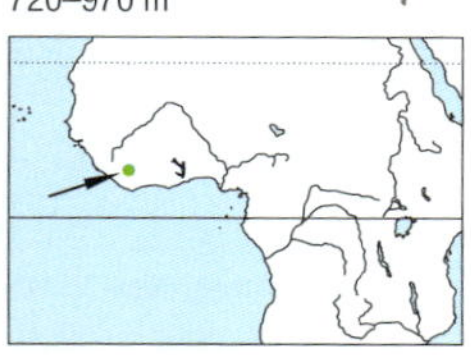

Nimba Long-fingered Bat
Miniopterus nimbae

FR Minioptère du Nimba
DE Nimba-Langflügelfledermaus
ES Minióptero de Nimba

N LR (Mt Nimba, Wonegizi Mts) & SE GN (Mt Béro); perhaps CI

DD
Monotypic
HB c. 4.5 cm
T c. 4.3 cm
FA 3.8–4 cm
W c. 7 g

0–1300 m

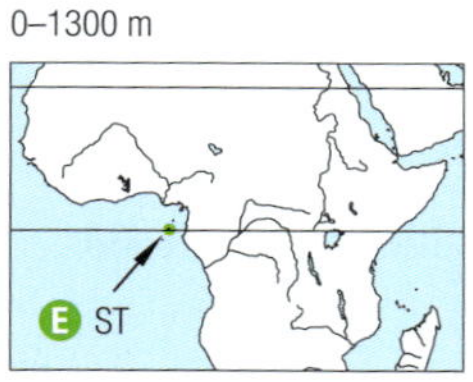

Newton's Long-fingered Bat
Miniopterus newtoni

FR Minioptère de Sao Tomé
DE Newton-Langflügelfledermaus
ES Minióptero de Newton

São Tomé I

DD
2 sspp
HB 5.2–5.3 cm
T 3.7–4.6 cm
FA 3.7–4.1 cm
W 4.6–6.5 g

below 300 m

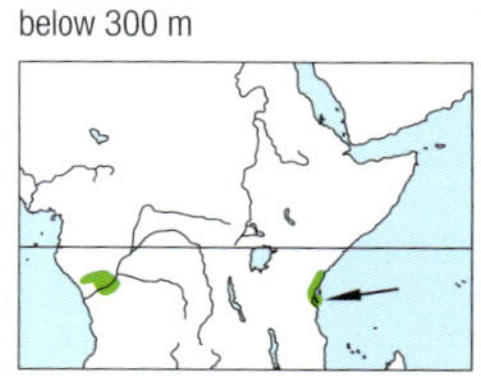

Least Long-fingered Bat
Miniopterus minor

FR Minioptère minuscule
DE Kleinste Langflügelfledermaus
ES Minióptero menor

Disjunctly in lower R Congo (CG, CD) & coastal S KE & NE TZ (with Unguja I)

NE
Monotypic
HB 4.8–5.4 cm
T 4.4–5.1 cm
FA 4.3–4.6 cm
W 6.5–8.5 g

above 1000 m

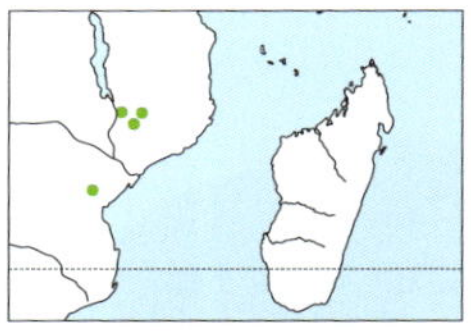

Wilson's Long-fingered Bat
Miniopterus wilsoni

FR Minioptère de Wilson
DE Wilson-Langflügelfledermaus
ES Minióptero de Wilson

N MZ (Gorongosa, Namuli & Mabu mts), S MW (Zomba Massif); perhaps other mts or E ZM

NE
Monotypic
HB c. 5.5 cm
T c. 4.7 cm
FA 4.2–4.6 cm

up to 1400 m

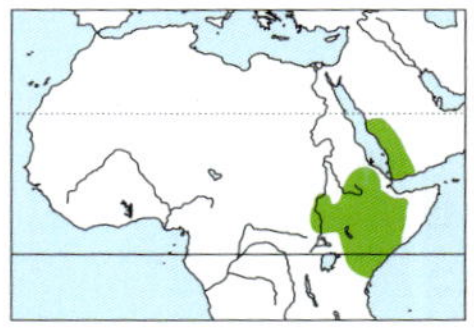

Sandy Long-fingered Bat
Miniopterus arenarius

FR Minioptère des sables
DE Sandfarbene Langflügelfledermaus
ES Minióptero de las arenas

SW Arabian Pen (SW SA, W YE), E Africa (ET, E SS, SO, KE)

DD
Monotypic
FA 4.8–5 cm

1000–3300 m

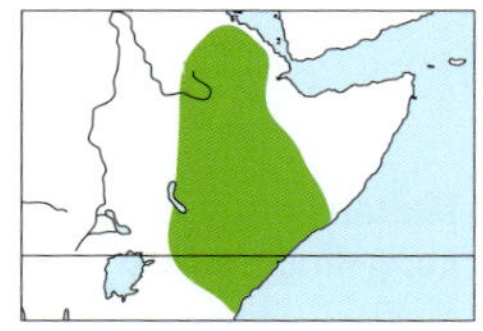

African Long-fingered Bat
Miniopterus africanus

FR Minioptère d'Afrique
DE Ostafrika-Langflügelfledermaus
ES Minióptero de África

E Africa in ER, ET, SO & KE

LC
Monotypic
HB 5.7–6.8 cm
T 4.6–5.5 cm
FA 4.2–4.8 cm
W 9.4–13 g

0–2200 m

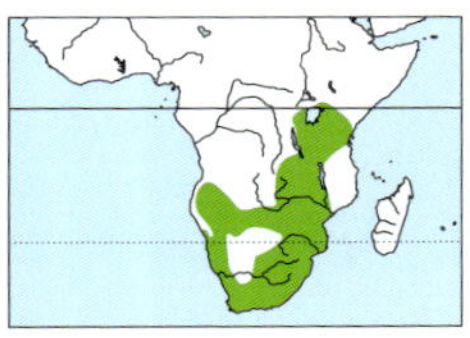

Natal Long-fingered Bat
Miniopterus natalensis

FR Minioptère du Natal
DE Natal-Langflügelfledermaus
ES Minióptero de Natal

E & S Africa (from UG, KE & AO S to ZA)

NE
Monotypic
HB 5.2–5.4 cm
T 4.7–5 cm
FA 4.1–4.5 cm
W 6.7–9 g

420–1800 m

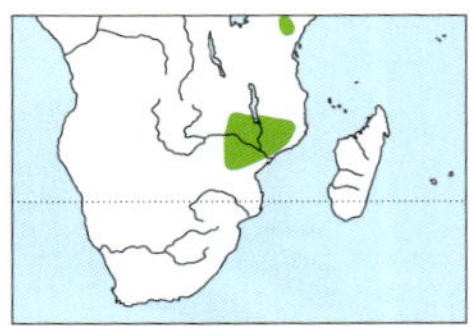

Mozambican Long-fingered Bat
Miniopterus mossambicus

FR Minioptère du Mozambique
DE Mosambik-Langflügelfledermaus
ES Minióptero de Mozambique

S KE (Taita Hills), N & C MZ, neighboring MW, ZM & ZW; possibly TZ

LC
Monotypic
HB 4.6–5.7 cm
T 3.9–5.2 cm
FA 4.2–4.4 cm
W 7–8.5 g

Lesser Long-fingered Bat
Miniopterus fraterculus

FR Petit Minioptère
DE Kleine Langflügelfledermaus
ES Minióptero pequeño

E & S ZA, SZ; perhaps S MZ

LC
Monotypic
HB 4.8–4.9 cm
T 3.9–4.4 cm
FA 3.5–3.8 cm
W 3.7–5.4 g

5–690 m

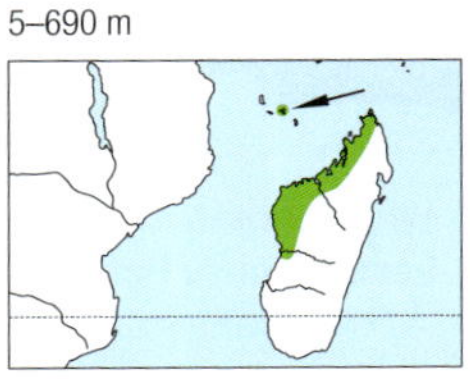

Aellen's Long-fingered Bat
Miniopterus aelleni

FR Minioptère d'Aellen
DE Aellen-Langflügelfledermaus
ES Minióptero de Aellen

N & W MG, KM (Anjouan)

DD
Monotypic
HB 5–5.1 cm
T 3.5–4.3 cm
FA 3.5–3.8 cm
W 4.1–7.1 g

0–890 m

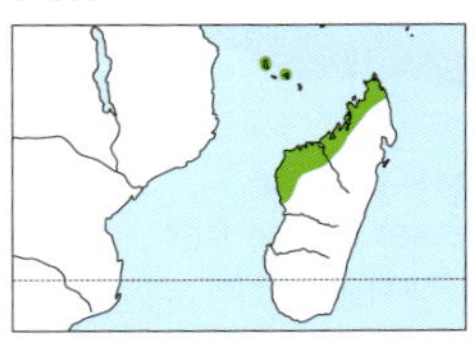

Griveaud's Long-fingered Bat
Miniopterus griveaudi

FR Minioptère des Comores
DE Griveaud-Langflügelfledermaus
ES Minióptero de Griveaud

N & W MG, KM (Grande Comore, Anjouan)

LC
Monotypic
HB c. 6.8 cm
T 5.2–6.3 cm
FA 4.7–5 cm
W 10.5–17.5 g

0–1250 m

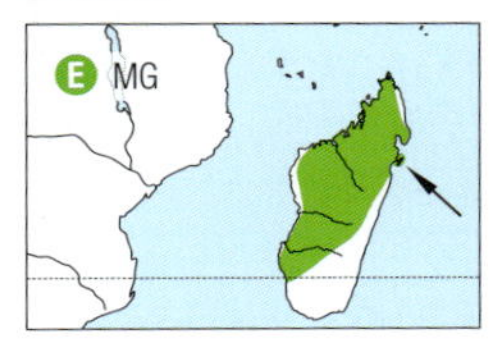

Glen's Long-fingered Bat
Miniopterus gleni

FR Minioptère de Glen
DE Glen-Langflügelfledermaus
ES Minióptero de Glen

Widespread (with Nosy Boraha) except S of R Onilahy

0–600 m

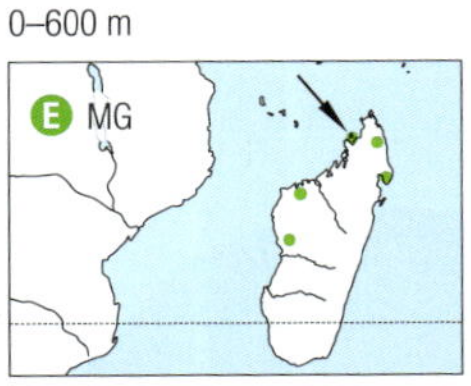

Madagascar Long-fingered Bat
Miniopterus brachytragos

FR Minioptère à oreillons courts
DE Nordwest-Madagaskar-Langflügelfledermaus
ES Minióptero de Madagascar

Scattered in N & W (with Nosy Komba)

LC Monotypic
HB 4.5–4.9 cm T 3.8–4.3 cm
FA 3.5–3.8 cm W 2.9–6.3 g

800–1600 m

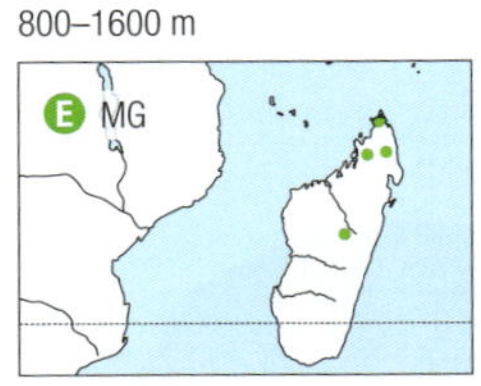

Montagne d'Ambre Long-fingered Bat
Miniopterus ambohitrensis

FR Minioptère d'Ambohitra
DE Montagne-d'Ambre-Langflügelfledermaus
ES Minióptero de Ambohitra

Marojejy NP, Montagne d'Ambre NP, Ambohitantely R & Bemanevika

LC Monotypic
HB 5.3–5.5 cm T 4–4.7 cm
FA 3.7–4.2 cm W 5.3–7.7 g

LC
Monotypic
HB 6–6.1 cm
T 5.1–6 cm
FA 4.3–4.7 cm
W 8.4–12.5 g

0–1000 m

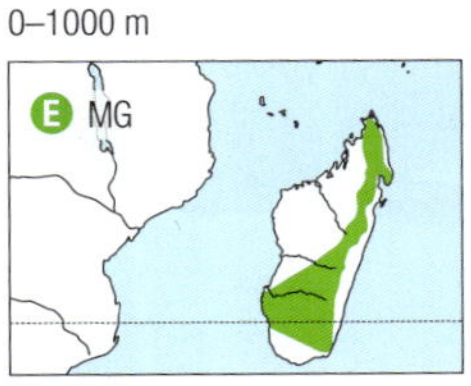

Major's Long-fingered Bat
Miniopterus majori

FR Minioptère de Major
DE Major-Langflügelfledermaus
ES Minióptero de Major

Widely but patchily across E highlands, some records from W & S

LC
Monotypic
HB 4.9–5.1 cm
T 4–4.4 cm
FA 3.7–4 cm
W 4.2–7.6 g

5–1300 m

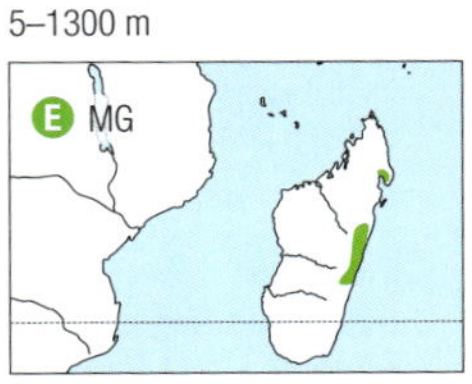

Eger's Long-fingered Bat
Miniopterus egeri

FR Minioptère d'Eger
DE Eger-Langflügelfledermaus
ES Minióptero de Eger

Known from several localities in E

LC
Monotypic
HB c. 5.1 cm
T c. 3.9 cm
FA 3.7–3.9 cm
W c. 6.4 g

900–1500 m

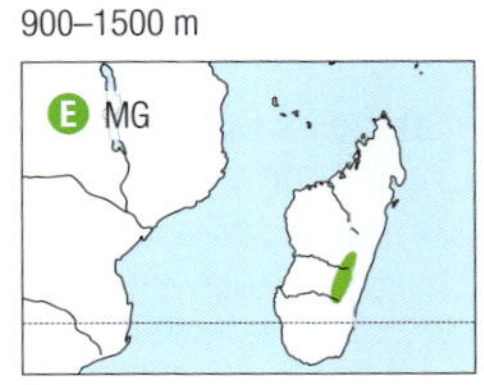

Manavil Long-fingered Bat
Miniopterus manavi

FR Minioptère des Betsileo
DE Zentral-Madagaskar-Langflügelfledermaus
ES Minióptero de Betsileo

C highlands from Fandriana S to Ranomafana (Ifanadiana) & Vinanitelo

40–2200 m

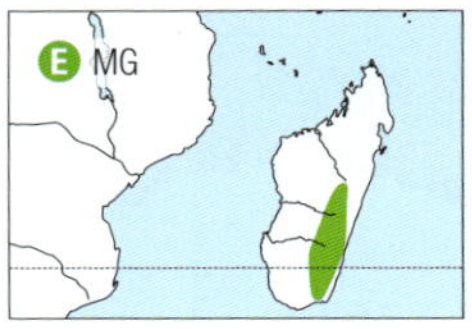

C highlands

Sororcula Long-fingered Bat
Miniopterus sororculus
FR Minioptère soeur
DE Schwesterchen-Langflügelfledermaus
ES Minióptero de Goodman

LC Monotypic
HB 5.4–5.7 cm T 5.1–5.8 cm
FA 4.2–4.5 cm W 7–9.1 g

0–550 m

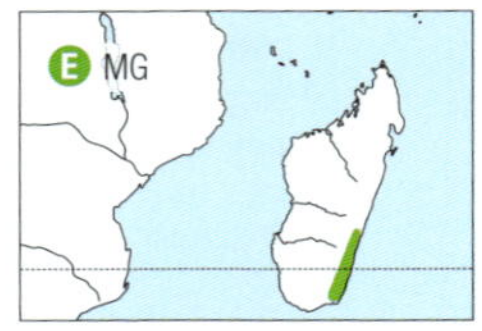

SE from Kianjavato to Tôlanaro

Peterson's Long-fingered Bat
Miniopterus petersoni
FR Minioptère de Peterson
DE Peterson-Langflügelfledermaus
ES Minióptero de Peterson

DD Monotypic
HB 4.6–4.9 cm T 3.9–5 cm
FA 3.8–4.3 cm W 4.2–8.2 g

0–950 m

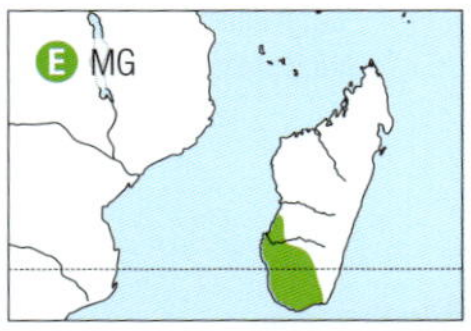

Dry S (S of Kirindy Mitea, Isalo, Ihosy)

Mahafaly Long-fingered Bat
Miniopterus mahafaliensis
FR Minioptère du Mahafaly
DE Mahafaly-Langflügelfledermaus
ES Minióptero de Mahafaly

LC Monotypic
HB 4.8–4.9 cm T 3.8–4.8 cm
FA 3.5–4 cm W 3.8–7.3 g

below 110 m

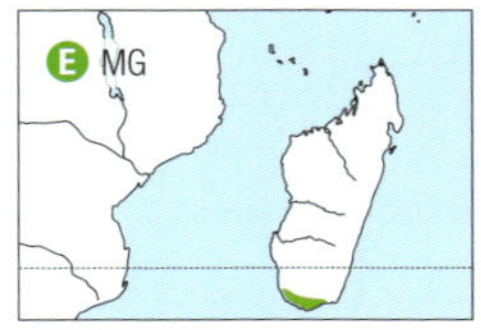

Known from 3 localities S of R Onilahy

Griffiths's Long-fingered Bat
Miniopterus griffithsi
FR Minioptère de Griffiths
DE Griffiths-Langflügelfledermaus
ES Minióptero de Griffiths

DD Monotypic
HB 6.5–6.8 cm T 5.4–6.3 cm
FA 4.8–5 cm W 12–15.5 g

Family CISTUGIDAE – Wing-gland Bats

1 genus, 2 species. Rocky outcrops in arid habitats, grasslands, and shrublands, often close to water. Afrotropical Region.

Genus *CISTUGO*

(2 species)

below 1000 m

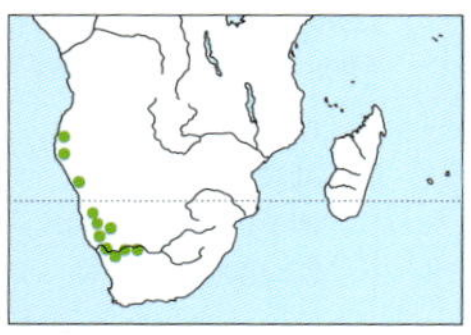

Scattered in SW coast of AO, W NA & ZA (N Northern Cape)

Angolan Wing-gland Bat
Cistugo seabrae
FR Cistugo d'Angola
DE Angola-Mausohr
ES Ratonero cistugo de Angola

LC Monotypic
HB 4–5.4 cm T 3.2–4 cm
FA 3–3.5 cm W 2.5–4.2 g

LC Monotypic
HB 4.4–5.5 cm T 4.3–4.5 cm
FA 3.3–3.8 cm W 5.5–7.9 g

Le Sueur's Wing-gland Bat
Cistugo lesueuri
FR Cistugo de Le Sueur
DE Le-Sueur-Mausohr
ES Ratonero cistugo de Le Sueur

1500–2420 m

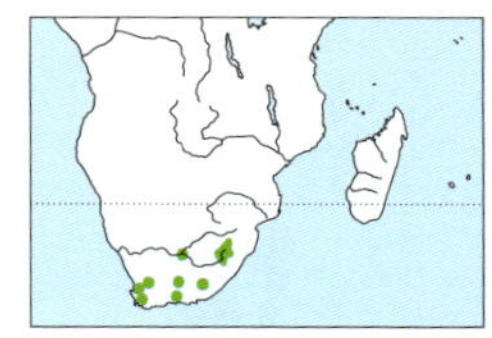

Scattered in ZA & LS

Family VESPERTILIONIDAE – Vesper Bats

57 genera, 525 species. From arid deserts to temperate and tropical forests. Worldwide, except Antarctica.

Genus *SCOTOECUS*

(5 species)

LC Monotypic
HB 4.7–5.9 cm
T 2.7–4.1 cm
FA 3.4–3.9 cm
W 9–14.3 g

0–2500 m

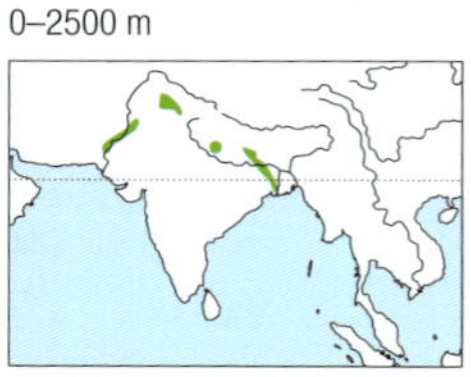

PK (Punjab, Sind), N IN (from Himachal Pradesh E to West Bengal); perhaps Maharashtra (C IN)

Desert Yellow Lesser House Bat
Scotoecus pallidus
FR Scotoèque d'Asie
DE Fahle Hausfledermaus
ES Escotoeco de Asia

LC
Monotypic
HB c. 4.3 cm
FA c. 3.2 cm

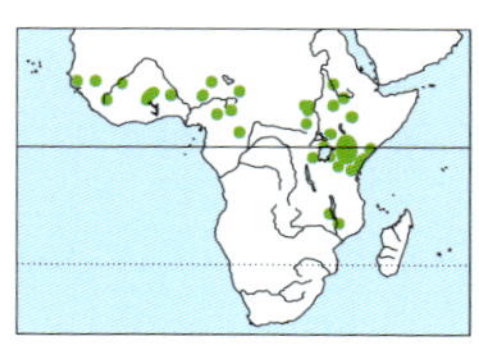

From SN E to E SS, SD & ET, S through UG, KE & TZ to MW

Dark-winged Lesser House Bat
Scotoecus hirundo
FR Scotoèque à ailes noires
DE Schwarzflügel-Hausfledermaus
ES Escotoeco de alas negras

NE
Monotypic
HB c. 5.8 cm
T c. 3.5 cm
FA c. 3.5 cm

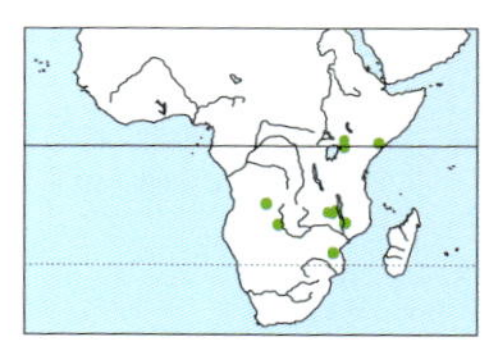

Scattered in SO, UG, KE, AO, ZM, MW, MZ; possibly TZ

White-throated Lesser House Bat
Scotoecus albigula
FR Scotoèque à gorge blanche
DE Weißkehl-Hausfledermaus
ES Escotoeco de garganta blanca

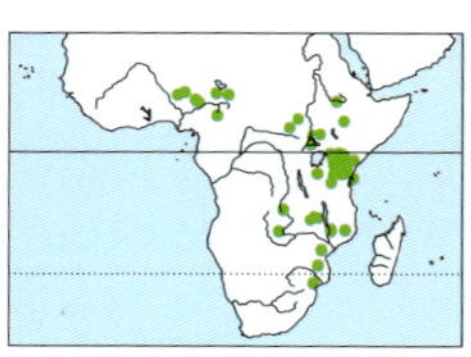

NG & CM, E Africa in SS, ET, CD, UG, KE, TZ, ZM, MW & MZ

Hinde's Lesser House Bat
Scotoecus hindei
FR Scotoèque de Hinde
DE Hinde-Hausfledermaus
ES Escotoeco de Hinde

NE
Monotypic
FA 3.5–3.7 cm

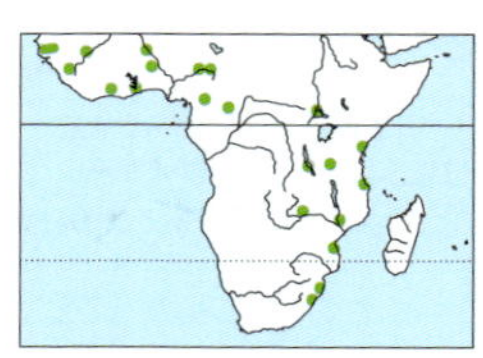

Patchily from SN & GM E to CM and from N UG & S KE S to NE ZA

Light-winged Lesser House Bat
Scotoecus albofuscus
FR Scotoèque à ailes pâles
DE Weißflügel-Hausfledermaus
ES Escotoeco de alas pálidas

DD
2 sspp
HB 4.9–5.4 cm T 2.7–4.1 cm
FA 2.9–3.4 cm W 5–9.5 g

Genus *VANSONIA*

(1 species)

0–1700 m

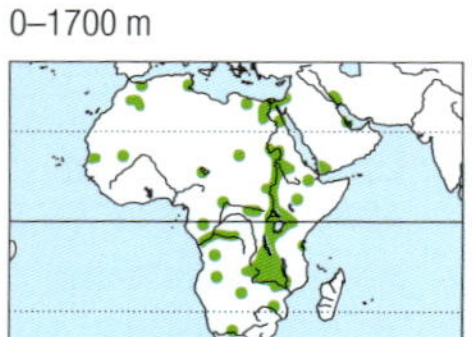

Scattered throughout Africa & Middle East (E to SW IR); probably more continuous

Rüppell's Bat
Vansonia rueppellii
FR Pipistrelle de Rüppell
DE Rüppell-Zwergfledermaus
ES Pipistrela de vientre blanco

LC
Monotypic
HB 4.3–5.6 cm T 2.5–4.1 cm
FA 2.8–3.7 cm
W 4–9 g

Genus *SCOTOZOUS*

(1 species)

0–2000 m

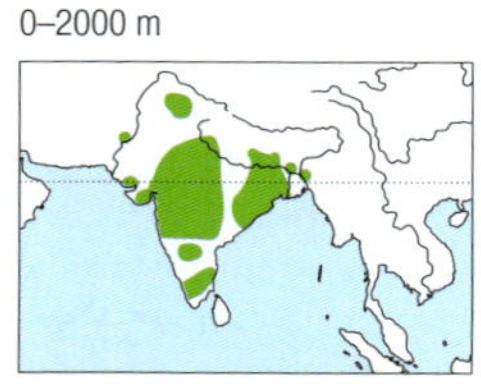

E PK (Punjab, Sind), much of IN, SW NP, W & C BD (Rajshahi, Dhaka); perhaps BT

Dormer's Bat
Scotozous dormeri
FR Pipistrelle de Dormer
DE Dormer-Zwergfledermaus
ES Pipistrela de Dormer

LC
Monotypic
HB 3.9–5.5 cm T 2.7–4.1 cm
FA 3.3–3.6 cm

100–360 m

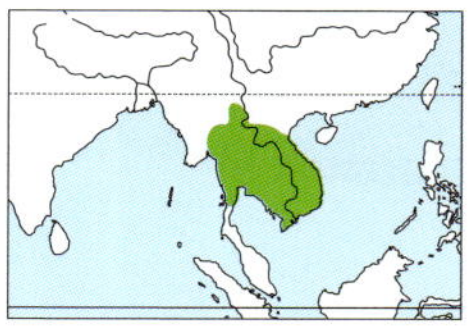

Mainland SE Asia N of Isthmus of Kra (E & S MM, TH, LA, VN & KH)

Indochinese Thick-thumbed Bat
Glischropus bucephalus

FR Pipistrelle bucéphale
DE Indochina-Dickdaumenfledermaus
ES Pipistrela de Indochina

LC Monotypic
HB 4–4.8 cm T 3.8–4.3 cm
FA 3.2–3.6 cm W 4–4.9 g

Genus *GLISCHROPUS*

(4 species)

Dark Thick-thumbed Bat
Glischropus aquilus

FR Pipistrelle sombre
DE Dunkle Dickdaumenfledermaus
ES Pipistrela de Sumatra

NE
Monotypic
HB c. 3.5 cm T c. 3.9 cm
FA c. 3.2 cm W c. 4.8 g

c. 770 m

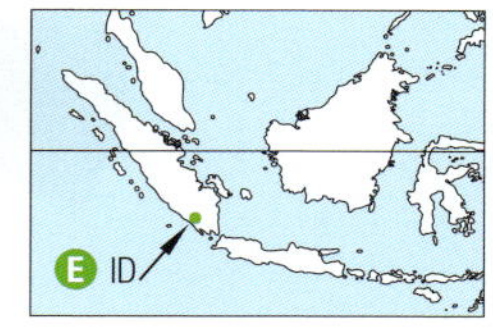

SW Sumatra (known only from Sukabanjar in Lampung)

below 1000 m

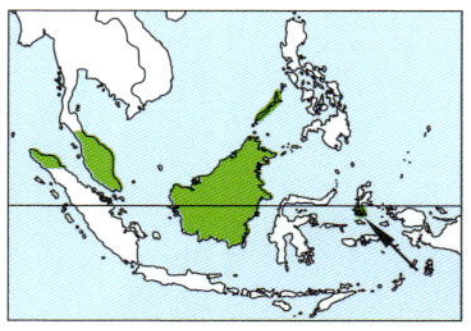

Thai-Malay Pen, N Sumatra, Borneo, PH (Palawan), Moluccas (Bacan)

LC Monotypic
HB c. 3.9 cm T 3.2–3.9 cm
FA 2.7–3.1 W 3.5–5 g

Common Thick-thumbed Bat
Glischropus tylopus

FR Pipistrelle de Dobson
DE Gewöhnliche Dickdaumenfledermaus
ES Pipistrela de Dobson

DD
Monotypic
HB c. 4 cm T c. 4 cm
FA c. 3.2 cm

Javan Thick-thumbed Bat
Glischropus javanus

FR Pipistrelle du Pangrango
DE Java-Dickdaumenfledermaus
ES Pipistrela de Pangrango

900–1000 m

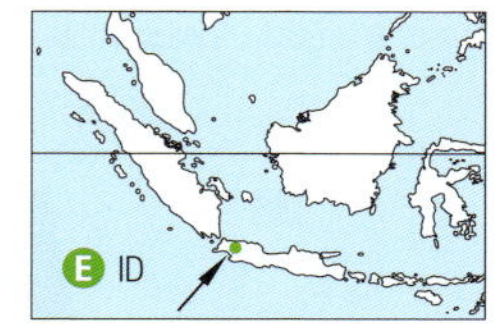

W Java (Mt Pangrango)

Genus *NYCTALUS*

(8 species)

LC
3 sspp
HB 6.5–8.7 cm
T 3.6–5.9 cm
FA 4.7–5 cm
W 19–29 g

0–2310 m

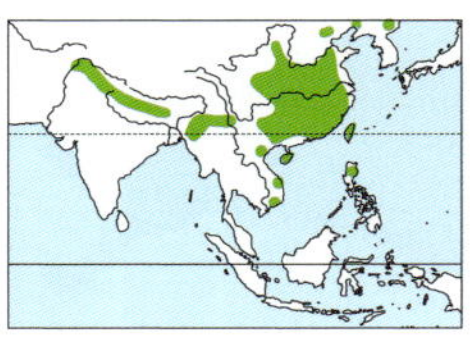

From N PK, N IN & NP E to N MM & CN, TW, VN, PH (Luzon); perhaps TH, N LA & Pen MY

Chinese Noctule
Nyctalus plancyi

FR Noctule de Chine
DE China-Abendsegler
ES Nóctulo de China

NT
Monotypic
HB 8–10.6 cm T 4.5–6.2 cm
FA 5.8–6.4 cm
W 26–61 g

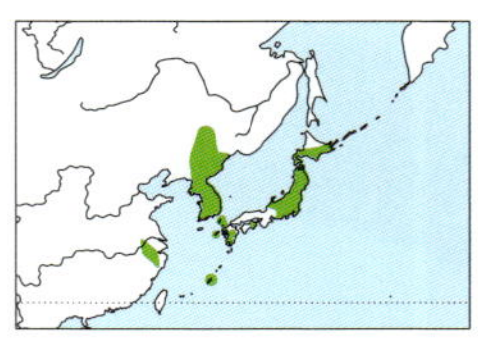

NE & EC CN, Korean Pen, JP; perhaps RU (Far East)

Bird-like Noctule
Nyctalus aviator

FR Noctule aviatrice
DE Großer Ostasien-Abendsegler
ES Nóctulo oriental

EN
Monotypic
HB 7–8.4 cm
T 4.2–5.4 cm
FA 4.8–5.3 cm
W 19–46 g

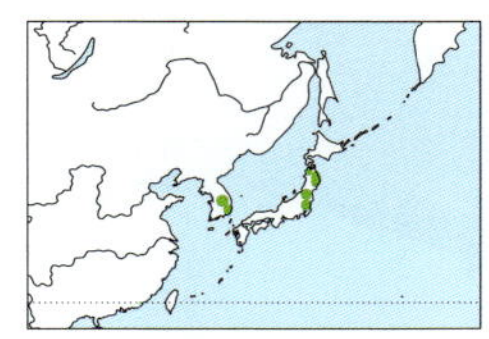

JP (Honshu), C & SW KR

Japanese Noctule
Nyctalus furvus

FR Noctule sombre
DE Japan-Abendsegler
ES Nóctulo de Japón

LC
Monotypic
HB 6–7 cm
T c. 4.3 cm
FA 4.2–4.4 cm

680–2100 m

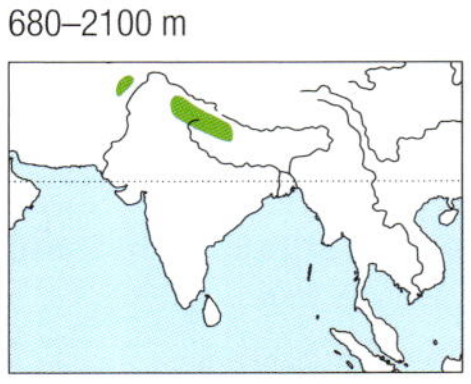

WC AF (Nangarhar, Paktika), S Himalayas of N IN, W NP; perhaps N PK

Mountain Noctule
Nyctalus montanus

FR Noctule des montagnes
DE Bergabendsegler
ES Nóctulo montano

VU
Monotypic
HB 8.4–10.4 cm
T 4.8–7.4 cm
FA 5.9–7 cm
W 33–74 g

0–1900 m

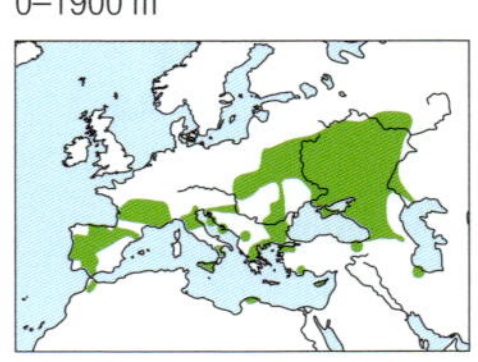

Iberian Pen E to W RU & extreme W KZ, S to MA, NE LY, CY, Caucasus & NW IR

Greater Noctule
Nyctalus lasiopterus

FR Grande Noctule
DE Riesenabendsegler
ES Nóctulo grande

LC
3 sspp
HB 6–8.9 cm
T 4–6.6 cm
FA 4.7–6 cm
W 17–44 g

0–1900 m

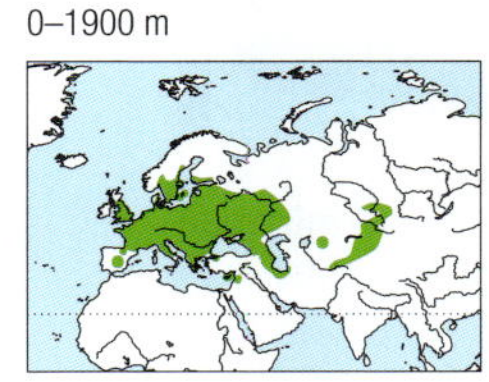

Europe E to NW CN (Xinjiang), S to Levant; perhaps N Africa

Common Noctule
Nyctalus noctula

FR Noctule commune
DE Großer Abendsegler
ES Nóctulo mediano

LC
2 sspp
HB 4.8–7.2 cm
T 3.5–4.8 cm
FA 3.5–4.6 cm
W 8–20 g

0–2400 m

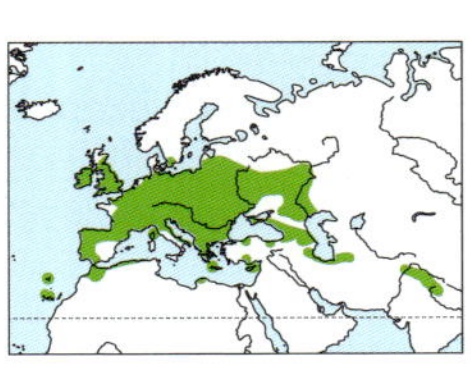

Europe E to SW RU & N IR, N Africa (MA, DZ, NE LY), SC Asia (NE AF to NW IN), Madeira & Canary Is

Leisler's Noctule
Nyctalus leisleri

FR Noctule de Leisler
DE Kleiner Abendsegler
ES Nóctulo pequeño

VU
Monotypic
HB 5.1–5.7 cm
T 4.2–4.3 cm
FA 3.5–4.2 cm
W 6–15 g

0–600 m

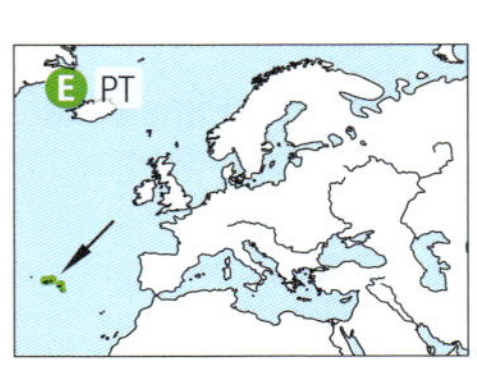

Azores Arch (Faial, Pico, São Jorge, Graciosa, Terceira, São Miguel, Santa Maria)

Azores Noctule
Nyctalus azoreum

FR Noctule des Açores
DE Azoren-Abendsegler
ES Nóctulo de las Azores

Genus *PIPISTRELLUS*

(31 species)

LC
2 sspp
HB 3.5–5.2 cm
T 2.3–3.6 cm
FA 2.7–3.4 cm
W 3–8.5 g

0–2000 m

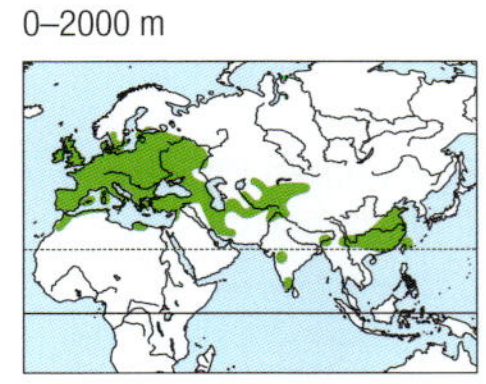

Most of Europe (E to RU), W, C, S & E Asia (with TW), N Africa (mts of MA, DZ, TN & LY)

Common Pipistrelle
Pipistrellus pipistrellus

FR Pipistrelle commune
DE Zwergfledermaus
ES Murciélago enano

LC
2 sspp
HB 3.6–5.1 cm
T 2.3–3.6 cm
FA 2.7–3.3 cm
W 4–7.5 g

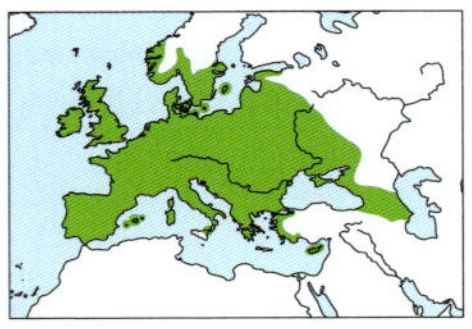

Soprano Pipistrelle
Pipistrellus pygmaeus
FR Pipistrelle pygmée
DE Mückenfledermaus
ES Murciélago de Cabrera

Europe E to RU & Caucasus, major Mediterranean Is

NE
Monotypic
HB 3.7–4.3 cm
T 3.4–3.7 cm
FA 2.9–3 cm
W 3.3–3.9 g

0–940 m

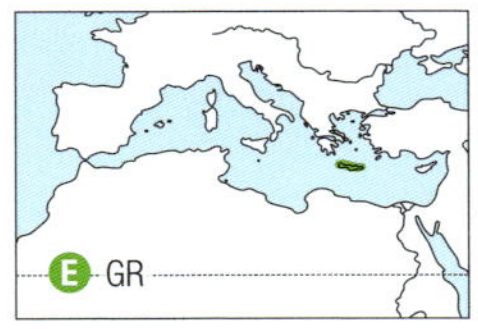

Crete Pipistrelle
Pipistrellus creticus
FR Pipistrelle de Crète
DE Kreta-Zwergfledermaus
ES Pipistrela de Creta

Crete

VU
Monotypic
HB 4.1–4.9 cm
T 3.2–3.9 cm
FA 3.1–3.3 cm
W 3.8–6.5 g

0–500 m

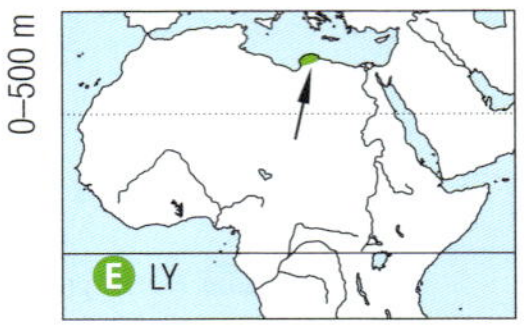

Hanak's Pipistrelle
Pipistrellus hanaki
FR Pipistrelle de Libye
DE Hanák-Zwergfledermaus
ES Pipistrela de Libia

N Cyrenaica region

LC
Monotypic
HB 4.4–5.4 cm
T 3–4 cm
FA 3.2–3.7 cm
W 6–15.5 g

0–2200 m

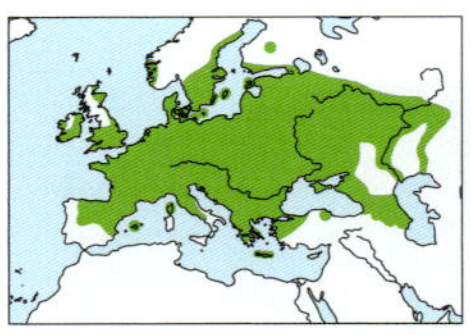

Nathusius's Pipistrelle
Pipistrellus nathusii
FR Pipistrelle de Nathusius
DE Rauhautfledermaus
ES Pipistrela de Nathusius

Most of Europe E to RU & W KZ, TR, Caucasus, Crete & Mallorca

VU
Monotypic
HB 3.9–4.7 cm
T 3–3.6 cm
FA 2.9–3.5 cm
W 5–9 g

0–2150 m

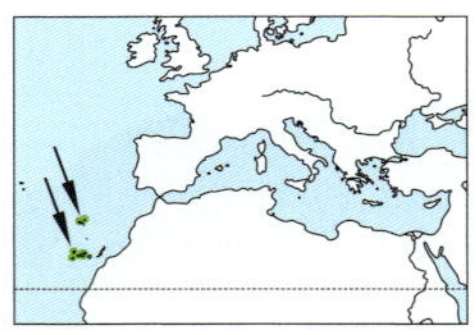

Madeira Pipistrelle
Pipistrellus maderensis
FR Pipistrelle de Madère
DE Madeira-Zwergfledermaus
ES Pipistrela de Madeira

Madeira Arch (Madeira, Porto Santo), W Canary Is (La Palma, La Gomera, El Hierro, Tenerife)

LC
4 sspp
HB 4–5.5 cm
T 3–4.5 cm
FA 3–3.7 cm
W 5–10 g

0–2000 m

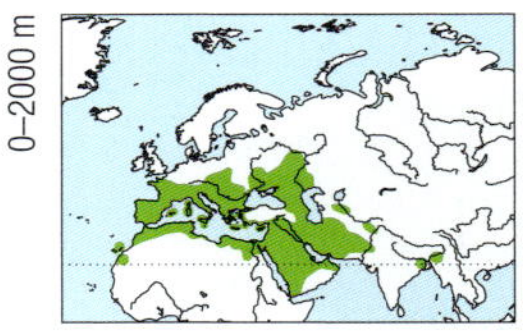

Kuhl's Pipistrelle
Pipistrellus kuhlii
FR Pipistrelle de Kuhl
DE Weißrandfledermaus
ES Pipistrela de borde claro

C & S Europe E to W RU, W KZ, Middle East, UZ, TM, AF, PK & IN, N Africa (from MA E to EG)

LC
3 sspp
HB 4.3–5.5 cm
T 2–4.1 cm
FA 2.7–3.8 cm
W 3.5–9 g

0–3000 m

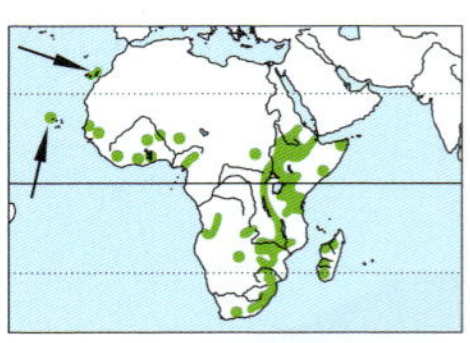

Dusky Pipistrelle
Pipistrellus hesperidus
FR Pipistrelle hespéride
DE Abendstern-Zwergfledermaus
ES Pipistrela africana

Much of sub-Saharan Africa, Canary & Cape Verde Is, MG

LC
2 sspp
HB 4.4–4.6 cm
T 1.9–3.2 cm
FA 2.4–3.1 cm
W 3.4–5.1 g

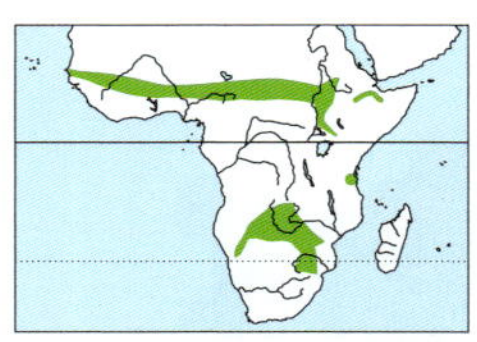

Rusty Pipistrelle
Pipistrellus rusticus
FR Pipistrelle rouille
DE Rostfarbene Zwergfledermaus
ES Pipistrela campestre

From SN to E ET, N UG & WC KE, E TZ, S Africa (SE AO, ZM, NE NA, N BW, ZW & NE ZA); perhaps MW

NE
Monotypic
HB 4.7–6 cm
T 2.6–3.3 cm
FA 3.2–3.4 cm
W 5–8 g

450–1200 m

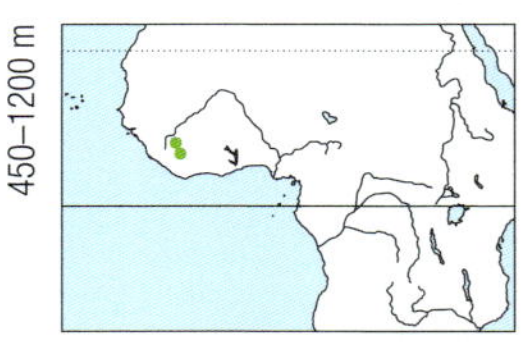

Simandou Pipistrelle
Pipistrellus simandouensis
FR Pipistrelle du Simandou
DE Simandou-Zwergfledermaus
ES Pipistrela de Simandou

Known from N LR (Mt Nimba) & SE GN (Simandou Range)

LC
Monotypic
HB 3.8–4.7 cm
T 2.1–3.2 cm
FA 2.4–3.1 cm
W 2–7 g

0–1835 m

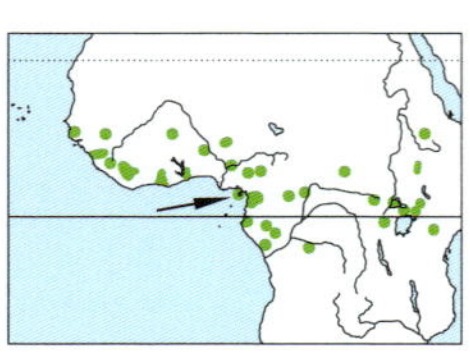

Tiny Pipistrelle
Pipistrellus nanulus
FR Pipistrelle minuscule
DE Kleine Zwergfledermaus
ES Pipistrela minúscula

W, C & E Africa from SN & GM E to W ET and W & S KE, Bioko I

DD
Monotypic
HB c. 4.7 cm
T 3–3.7 cm
FA 3.1–3.3 cm
W 3–5.5 g

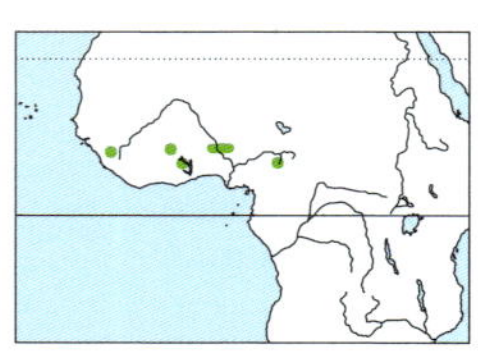

Aellen's Pipistrelle
Pipistrellus inexspectatus
FR Pipistrelle d'Aellen
DE Aellen-Zwergfledermaus
ES Pipistrela de Aellen

W Africa (N SL, SW BF, C GH, N BJ, NW NG, NC CM); perhaps E SD & UG

DD
Monotypic
HB 4–4.9 cm
T 3.2–3.4 cm
FA 3.1–3.4 cm
W 4–5 g

1200–2700 m

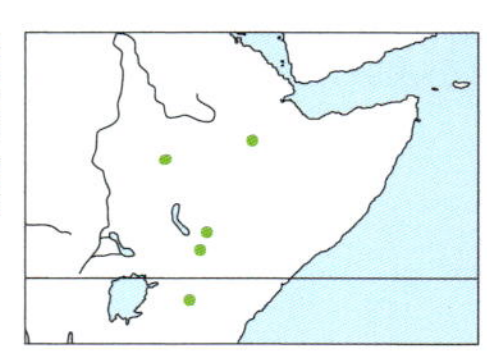

Mount Gargues Pipistrelle
Pipistrellus aero
FR Pipistrelle du Kenya
DE Mount-Gargues-Zwergfledermaus
ES Pipistrela de Kenya

ET (R Beko, R Yamboshi, Chercher Mts), W KE (L Marsabit, Mt Gargues, Ngong)

DD
Monotypic
HB c. 4.2 cm
T c. 3.3 cm
FA c. 3.3 cm

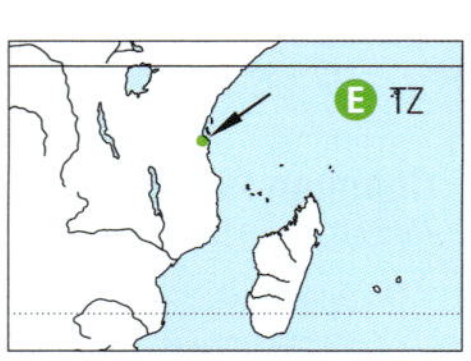

Dar-es-Salaam Pipistrelle
Pipistrellus permixtus
FR Pipistrelle de Tanzanie
DE Dar-es-Salaam-Zwergfledermaus
ES Pipistrela de Tanzania

Known from type specimen at Dar es Salaam

DD
Monotypic
HB 4.6–5.2 cm
T 2.3–3.4 cm
FA 2.7–3.3 cm
W 3.8–5.8 g

10–300 m

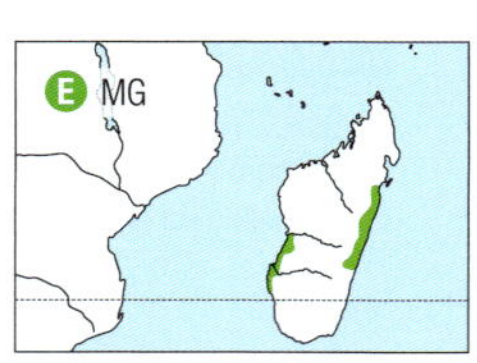

Racey's Pipistrelle
Pipistrellus raceyi
FR Pipistrelle de Racey
DE Racey-Zwergfledermaus
ES Pipistrela de Racey

Lowlands of W & E coasts

NE
Monotypic
HB 4.6–5.7 cm
T 3.1–4.2 cm
FA 3.3–3.8 cm

60–735 m

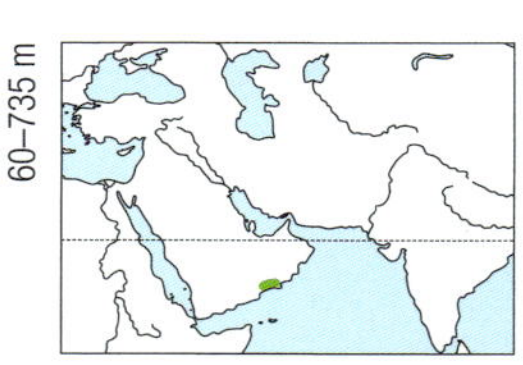

Dhofar Pipistrelle
Pipistrellus dhofarensis
FR Pipistrelle du Dhofar
DE Dhofar-Zwergfledermaus
ES Pipistrela de Dhofar

From Hawf (SE YE) to Mirbat (SW OM)

LC
Monotypic
HB 3.8–6 cm
T 2.7–4.5 cm
FA 2.9–3.6 cm
W 3.8–5.8 g

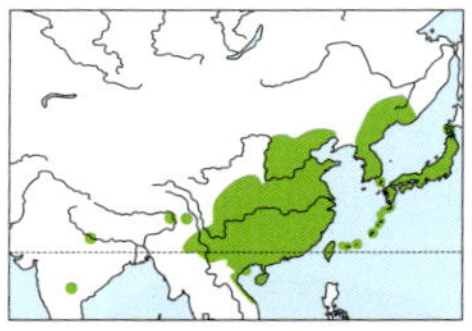

Japanese Pipistrelle
Pipistrellus abramus
FR Pipistrelle abramusi
DE Japan-Zwergfledermaus
ES Pipistrela japonesa

RU (Ussuri), Korean Pen, JP, TW, CN S to N MM, N LA & C VN, scattered in IN

NT
Monotypic
HB 4–4.9 cm
T 2.8–3.8 cm
FA 3–3.4 cm
W 5.6–8.7 g

100–1500 m

Endo's Pipistrelle
Pipistrellus endoi
FR Pipistrelle d'Endo
DE Endo-Zwergfledermaus
ES Pipistrela de Endo

Honshu, Shikoku

LC
5 sspp
HB 4–5.5 cm
T 2.6–4 cm
FA 2.9–3.7 cm
W 4.1–8 g

0–2380 m

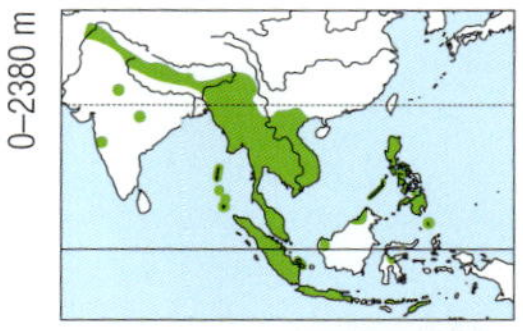

Javan Pipistrelle
Pipistrellus javanicus
FR Pipistrelle de Java
DE Java-Zwergfledermaus
ES Pipistrela de Java

N PK, IN (with Andaman & Nicobar Is), S CN, SE Asia E to PH & Timor

LC
Monotypic
HB 3.4–4.9 cm
T 2.2–3.9 cm
FA 2.5–3.5 cm
W 3.4–8.3 g

100–2000 m

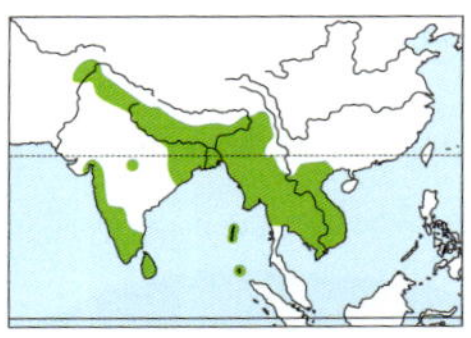

Indian Pipistrelle
Pipistrellus coromandra
FR Pipistrelle de Coromandel
DE Indien-Zwergfledermaus
ES Pipistrela de Coromandel

NE AF E to BD & CN (S Tibet), S to LK, Andaman & Nicobar Is, mainland SE Asia

LC
6 sspp
HB 4.5–6.4 cm
T 3–4.5 cm
FA 3.3–4.2 cm
W 7–10 g

0–2150 m

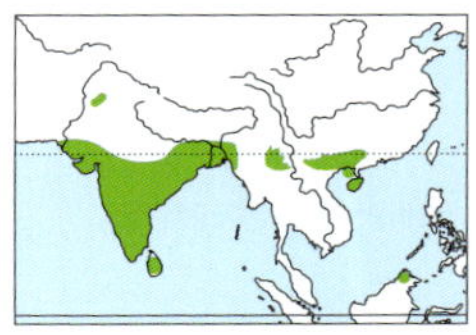

Kelaart's Pipistrelle
Pipistrellus ceylonicus
FR Pipistrelle de Kelaart
DE Sri-Lanka-Zwergfledermaus
ES Pipistrela de Kelaart

E PK (Punjab, Sind), IN, BD, LK, EC MM, N VN, S CN (S Guangxi, Weizhou & Hainan Is), N Borneo

LC
6 sspp
HB 3.3–4.5 cm
T 2–3.5 cm
FA 2.5–3.1 cm
W 2.9–4.2 g

0–2650 m

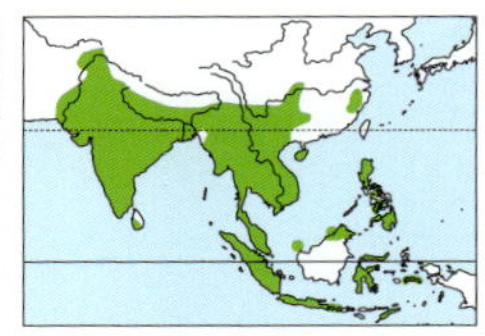

Least Pipistrelle
Pipistrellus tenuis
FR Pipistrelle grêle
DE Winzige Zwergfledermaus
ES Pipistrela diminuta

NE AF, Indian subcontinent, SE CN (with Hainan), SE Asia E to PH, Moluccas (Seram, Ambon) & Timor

LC
2 sspp
HB 4.2–4.8 cm
T 3.1–3.8 cm
FA 2.6–3.4 cm
W 4.3–5.7 g

0–2400 m

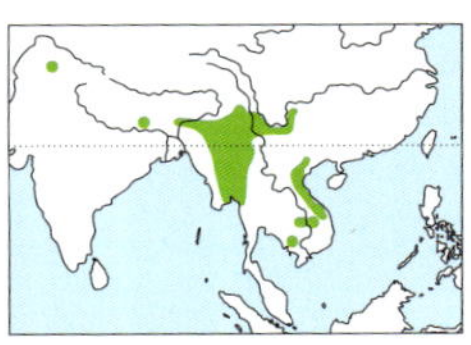

Mount Popa Pipistrelle
Pipistrellus paterculus
FR Pipistrelle du Popa
DE Mount-Popa-Zwergfledermaus
ES Pipistrela de Popa

N & NE IN, MM, SC CN (Yunnan), E LA, VN, N & SW KH; unconfirmed in NW TH

LC
Monotypic
HB 5.1–6.2 cm
T 3.7–5.7 cm
FA 3.7–4.2 cm
W 13–22 g

100–1200 m

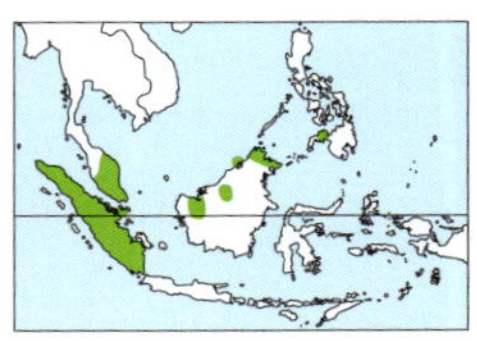

Narrow-winged Pipistrelle
Pipistrellus stenopterus
FR Pipistrelle à ailes étroites
DE Schmalflügel-Zwergfledermaus
ES Pipistrela de alas estrechas

S Pen TH, Pen MY, SG, Sumatra, Riau Arch, N & W Borneo, PH (SW Mindanao)

DD
Monotypic
HB c. 5.9 cm
T c. 3.7 cm
FA c. 3.6 cm

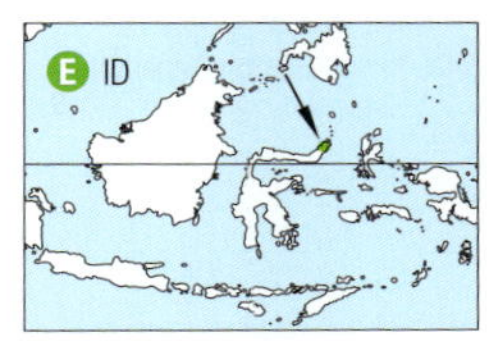

Minahassa Pipistrelle
Pipistrellus minahassae
FR Pipistrelle de Minahassa
DE Minahassa-Zwergfledermaus
ES Pipistrela de Minahasa

N Sulawesi (known with certainty from NE Minahasa Pen)

LC
Monotypic
HB 3.5–5.3 cm
T 3–4.5 cm
FA 3.4–3.8 cm
W 4.3–8.5 g

700–2900 m

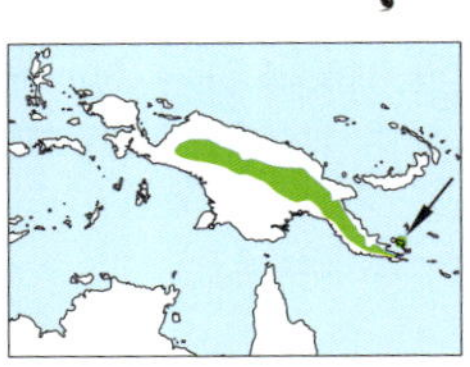

Greater Papuan Pipistrelle
Pipistrellus collinus
FR Pipistrelle des montagnes
DE Große Papua-Zwergfledermaus
ES Pipistrela grande de Papúa

ID/PG: New Guinea (Central Range), D'Entrecasteaux Is (Goodenough)

LC
Monotypic
HB 3.4–5 cm
T 2.2–3.5 cm
FA 2.7–3.2 cm
W 3.9–5.8 g

0–1300 m

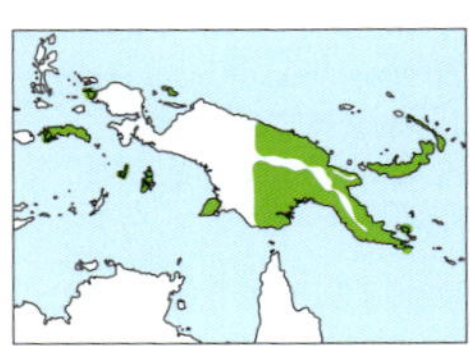

Lesser Papuan Pipistrelle
Pipistrellus papuanus
FR Pipistrelle papoue
DE Kleine Papua-Zwergfledermaus
ES Pipistrela pequeña de Papúa

Moluccas (Ambon, Seram, Kai, Aru), New Guinea (with several nearby Is), Bismarck Arch

LC
2 sspp
HB 3.3–4.8 cm
T 2.7–4.1 cm
FA 3.1–3.7 cm
W 3.2–4.3 g

0–2400 m

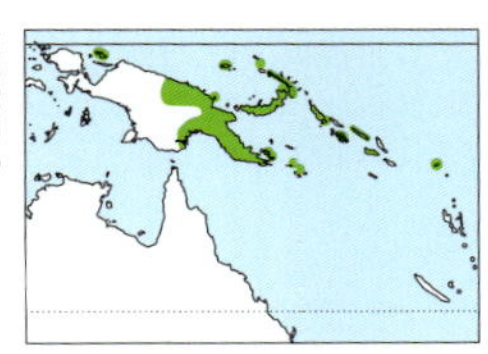

New Guinea Pipistrelle
Pipistrellus angulatus
FR Pipistrelle anguleuse
DE Neu-Guinea-Zwergfledermaus
ES Pipistrela de Nueva Guinea

E New Guinea and nearby Is, Bismarck Arch, Solomon Is (E to Nendö)

LC
Monotypic
HB 3.4–4.1 cm
T 1.9–2.8 cm
FA 2.8–3.3 cm
W c. 3.4 g

0–200 m

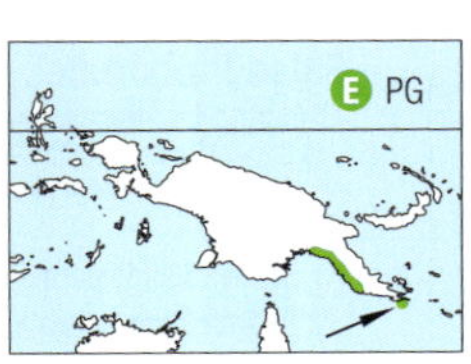

Watts's Pipistrelle
Pipistrellus wattsi
FR Pipistrelle de Watts
DE Watts-Zwergfledermaus
ES Pipistrela de Watts

SE New Guinea, Samarai I

LC
Monotypic
HB 3.4–4.2 cm
T 2.9–3.7 cm
FA 2.7–3.1 cm
W 2.7–4.8 g

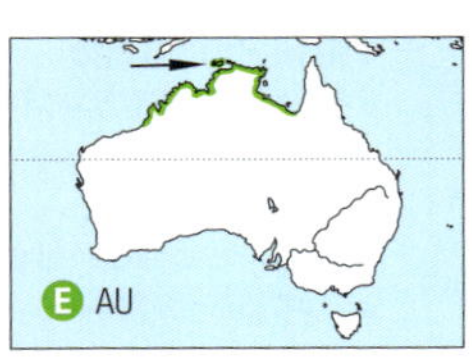

Northern Pipistrelle
Pipistrellus westralis
FR Pipistrelle de Koopman
DE Mangroven-Zwergfledermaus
ES Pipistrela Koopman

N coast from Cape Bossut (WA) to Karumba (W Qld), with Tiwi Is

LC
Monotypic
HB 3.5–4.4 cm
T 2.6–3.5 cm
FA 2.9–3.3 cm
W 3–6.2 g

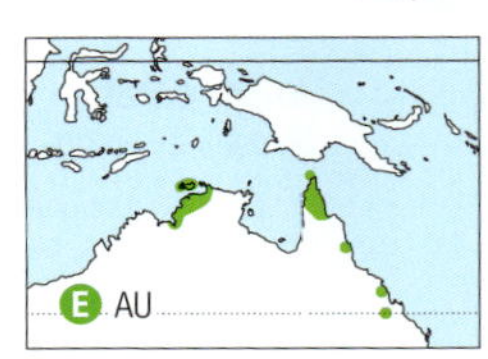

Forest Pipistrelle
Pipistrellus adamsi
FR Pipistrelle d'Adams
DE Kap-York-Zwergfledermaus
ES Pipistrela de Adams

NT (Top End, Tiwi Is), Qld (Cape York, Prince of Wales I, sparsely S to near Blackwater)

Genus *VESPERTILIO*

(2 species)

0–3400 m

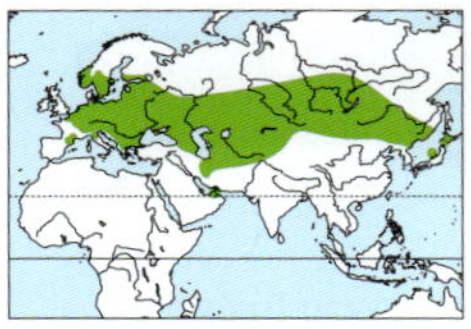

Much of Europe & Asia to NE CN, NE KP & JP, recorded in SE Arabia

LC 2 sspp
HB 4.8–6.4 cm T 3.7–4.4 cm
FA 4.1–5 cm W 10–24 g

Eurasian Particolored Bat
Vespertilio murinus

FR Vespertilion bicolore
DE Zweifarbfledermaus
ES Murciélago bicolor

Asian Particolored Bat
Vespertilio sinensis

FR Vespertilion oriental
DE Asiatische Zweifarbfledermaus
ES Murciélago bicolor asiático

LC Monotypic
HB 5.8–8 cm T 3.4–5.4 cm
FA 4.3–5.5 cm W 14–30 g

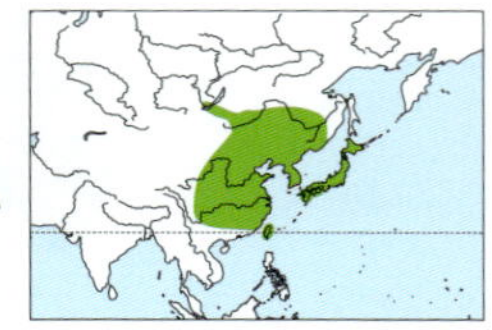

RU (S Siberia, S Far East), E MN, C & E CN, Korean Pen, JP, TW

Genus *TYLONYCTERIS*

(6 species)

DD
Monotypic
HB 2.8–3.4 cm
T 2.3–2.7 cm
FA 2.4–2.6 cm
W 2.5–3.5 g

c. 670 m

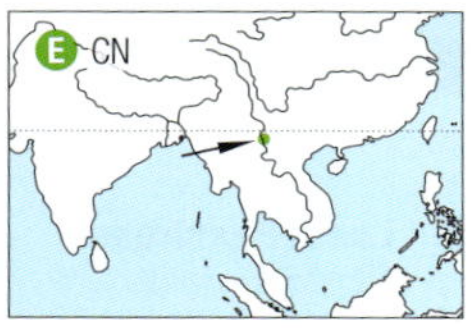

Known only from SC Yunnan

Pygmy Bamboo Bat
Tylonycteris pygmaea

FR Vespertilion pygmée
DE Zwergbambusfledermaus
ES Tilonícteris pigmeo

NE
2 sspp
HB 3.4–4.3 cm T 2–3.2 cm
FA 2.3–2.8 cm W 3.1–3.5 g

0–1260 m

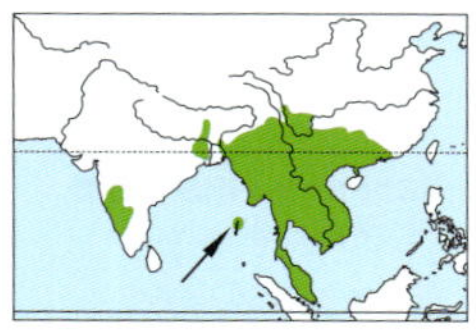

S, E & NE IN, BD, S CN, mainland SE Asia, Andaman Is; possibly BT

Indomalayan Lesser Bamboo Bat
Tylonycteris fulvida

FR Vespertilion fauve
DE Indomalayische Bambusfledermaus
ES Tilonícteris indomalayo

LC
3 sspp
HB 3.1–4.1 cm
T 2.1–3.1 cm
FA 2.4–2.9 cm
W 2.7–4.8 g

0–500 m

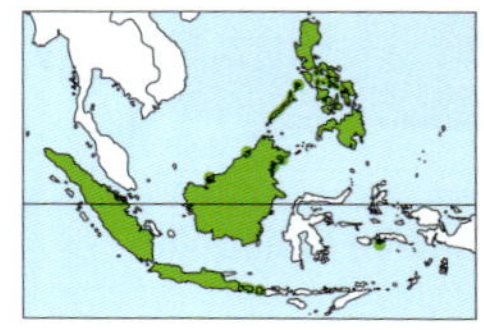

Sunda Lesser Bamboo Bat
Tylonycteris pachypus

FR Vespertilion épaté
DE Sunda-Bambusfledermaus
ES Tilonícteris de la Sonda

Sumatra, Borneo, Java, Bali, Lombok, PH, tentatively Moluccas (Ambon)

NE
Monotypic
HB 3.6–4.5 cm
T 1.9–3.4 cm
FA 2.4–2.9 cm
W 4.1–6.6 g

500–1285 m

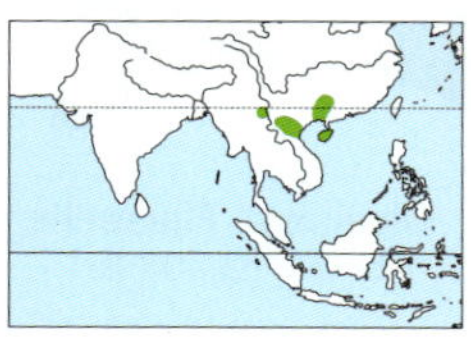

Tonkin Greater Bamboo Bat
Tylonycteris tonkinensis

FR Vespertilion du Tonkin
DE Tonkin-Bambusfledermaus
ES Tilonícteris de Tonkin

S CN (Yunnan, Guizhou, Guangxi, Hainan), N VN, NE LA; perhaps MM

NE
2 sspp
HB 3.6–3.7 cm
T 2.6–2.8 cm
FA 2.3–2.7 cm
W 6.5–8.5 g

up to 1000 m

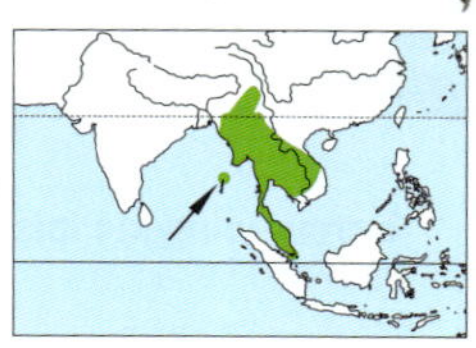

Malayan Greater Bamboo Bat
Tylonycteris malayana

FR Vespertilion de Malaisie
DE Malaya-Bambusfledermaus
ES Tilonícteris de Malasia

NE IN (Mizoram), mainland SE Asia, Andaman Is (North Andaman); perhaps CN (Yunnan)

LC
Monotypic
HB 4–4.5 cm
T 2.8–3.1 cm
FA 2.5–3 cm
W 5–7.8 g

0–1000 m

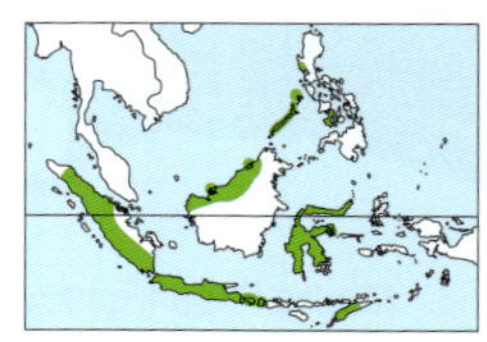

Sumatran Greater Bamboo Bat
Tylonycteris robustula

FR Vespertilion robuste
DE Sumatra-Bambusfledermaus
ES Tilonícteris robusto

PH (Luzon, Negros, Calauit, Palawan), Sumatra, W & NW Borneo, Java, Bali, Lombok, Timor, Sulawesi

Genus *CASSISTRELLUS*

(2 species)

NE
Monotypic
HB 6.1–6.6 cm
T 4.6–5.2 cm
FA 4.3–4.7 cm
W 12–15 g

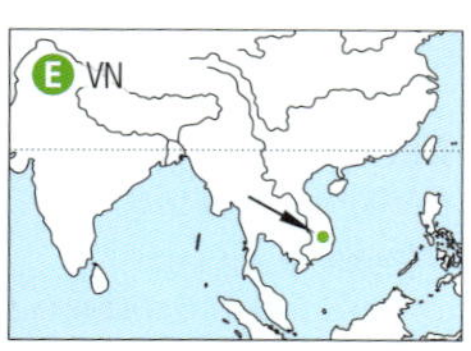

Known only from Yok Don NP

Yok Don Helmeted Bat
Cassistrellus yokdonensis

FR Vespertilion du Yok Don
DE Yok-Don-Helmfledermaus
ES Casistrela de Yok Don

DD
Monotypic
HB 5.9–6.3 cm T 4–4.1 cm
FA 3.9–4.2 cm W c. 15 g

190–675 m

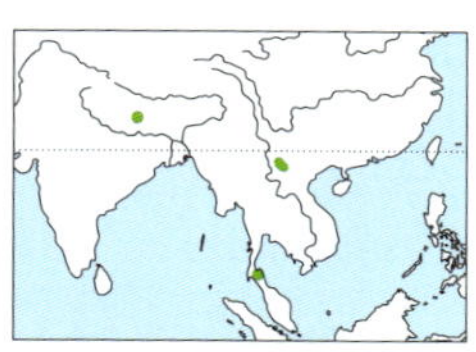

Surat Helmeted Bat
Cassistrellus dimissus

FR Vespertilion de Surat Thani
DE Surat-Helmfledermaus
ES Casistrela de Surat Thani

S NP (Royal Chitwan NP), N LA (Ban Naten) & S TH (Tai Rom Yen NP)

Genus *PHILETOR*

(1 species)

LC
Monotypic
HB 5.2–6.4 cm T 3–3.8 cm
FA 3–3.8 cm W 8–13 g

0–2100 m

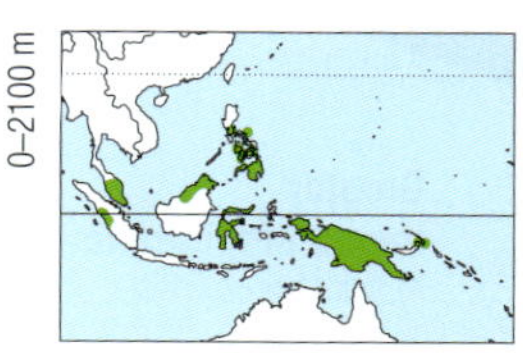

Rohu's Bat
Philetor brachypterus

FR Vespertilion brachyptère
DE Rohu-Fledermaus
ES Vespertiliónido de alas cortas

Pen MY, CW Sumatra, N & NW Borneo, Sulawesi, PH, New Guinea, Bismarck Arch

Genus *MIROSTRELLUS*

(1 species)

DD
Monotypic
HB c. 6.1 cm
T 3.7–4.1 cm
FA 3.8–4 cm

575–2135 m

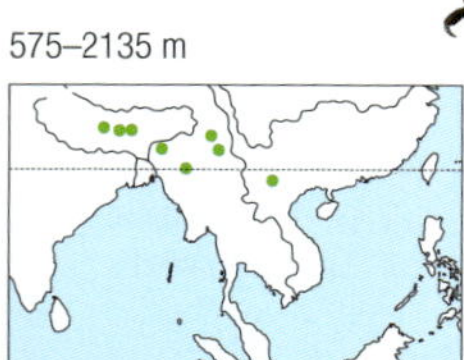

Joffre's Bat
Mirostrellus joffrei

FR Vespertilion de Joffre
DE Joffre-Zwergfledermaus
ES Hipsugo de Joffre

E NP, NE IN (Sikkim, Meghalaya), N MM, NW VN

Genus *FALSISTRELLUS*

(2 species)

NT
Monotypic
HB 5.5–6.7 cm
T 4–5.3 cm
FA 4.8–5.4 cm
W 17–26 g

Western False Pipistrelle
Falsistrellus mackenziei

FR Falsistrelle de McKenzie
DE Westliche Scheinzwergfledermaus
ES Falsistrela de McKenzie

Extreme SW WA (from near Perth to W margin of Wheatbelt)

VU
Monotypic
HB 5.5–7 cm T 4–5.2 cm
FA 4.8–5.4 cm
W 17–28 g

0–1500 m

Eastern False Pipistrelle
Falsistrellus tasmaniensis

FR Falsistrelle de Tasmanie
DE Östliche Scheinzwergfledermaus
ES Falsistrela de Tasmania

SE Qld, NSW, Vic, Tas

Genus *VESPADELUS*

(9 species)

LC

Monotypic

HB 3.5–4.3 cm

T 3.5–4.2 cm

FA 3.4–3.8 cm

W 4.5–6.1 g

Yellow-lipped Cave Bat

Vespadelus douglasorum

FR Vespertilion des Douglas

DE Gelblippen-Waldfledermaus

ES Vespadela de los Douglas

N WA (W Kimberley region & few offshore Is)

LC

Monotypic

HB 3.2–4 cm

T 2.4–3.5 cm

FA 2.7–3.2 cm

W 2.3–4.2 g

Northern Cave Bat

Vespadelus caurinus

FR Vespertilion du Kimberley

DE Nördliche Höhlen-Waldfledermaus

ES Vespadela de Kimberley

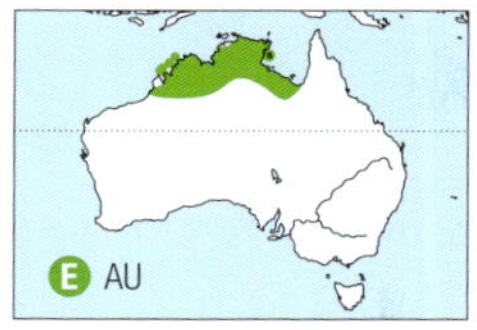

N WA (Kimberley region, Bonaparte Arch), N NT (with Groote Eylandt), NW Qld

LC

Monotypic

HB 3.4–4.6 cm

T 3.1–4.2 cm

FA 3–3.8 cm

W 3–7 g

Finlayson's Cave Bat

Vespadelus finlaysoni

FR Vespertilion de Finlayson

DE Finlayson-Waldfledermaus

ES Vespadela de Finlayson

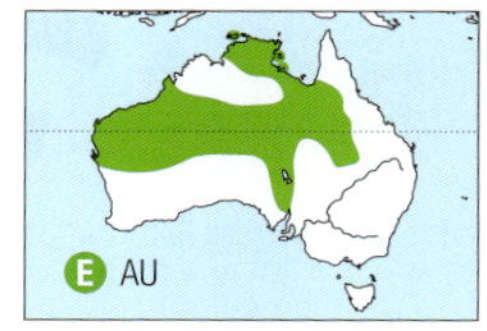

Arid and semiarid areas of N, C & W (with Tiwi, Groote Eylandt & Edward Pellew Is)

LC

Monotypic

HB 3.7–4.4 cm

T 3.1–3.8 cm

FA 3–3.6 cm

W 4.5–6.7 g

Eastern Cave Bat

Vespadelus troughtoni

FR Vespertilion de Troughton

DE Troughton-Waldfledermaus

ES Vespadela de Troughton

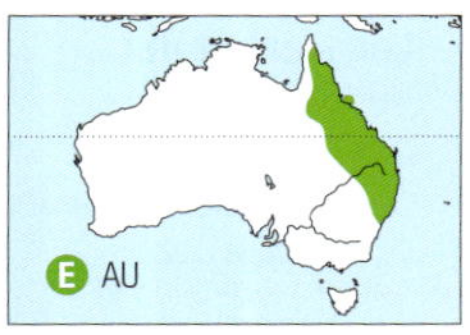

Qld (with Magnetic I), NE NSW

LC

Monotypic

HB 3.5–4.3 cm

T 2.6–3.4 cm

FA 2.6–3.1 cm

W 3.6–5.6 g

Inland Forest Bat

Vespadelus baverstocki

FR Vespertilion de Baverstock

DE Baverstock-Waldfledermaus

ES Vespadela de Baverstock

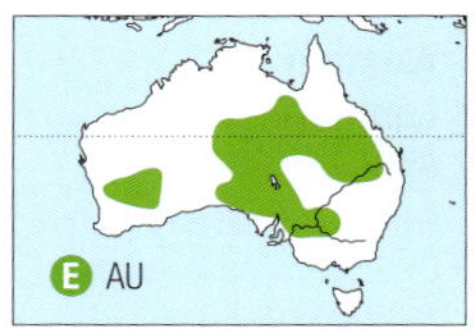

Inland, primarily in C & E, with a more isolated presence in SW WA

LC

Monotypic

HB 3.5–4.4 cm

T 2.8–3.4 cm

FA 2.8–3.3 cm

W 3.5–5.5 g

0–2000 m

Eastern Forest Bat

Vespadelus pumilus

FR Petit Vespertilion

DE Östliche Höhlen-Waldfledermaus

ES Vespadela pequeña

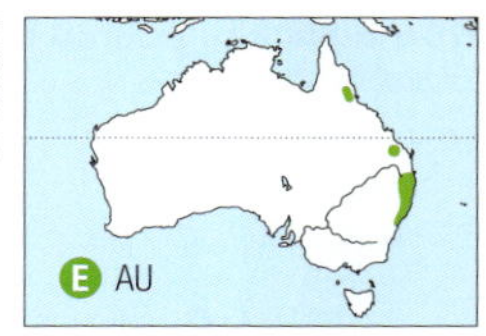

E coast of Qld & NE NSW (from Atherton Tablelands, Kirrama & Kroombit Top to just N of Sydney)

LC

Monotypic

HB 3.5–4.8 cm

T 2.7–3.4 cm

FA 2.3–3.3 cm

W 2.6–6.5 g

0–1100 m

Little Forest Bat

Vespadelus vulturnus

FR Vespertilion sylvestre

DE Kleine Waldfledermaus

ES Vespadela silvestre

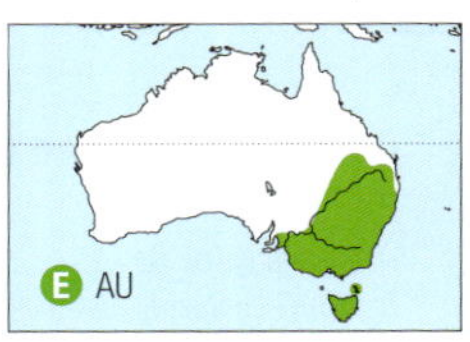

SE Qld, NSW, Vic, SE SA, Flinders I, Tas

LC

Monotypic

HB 3.8–5.2 cm

T 2.9–3.9 cm

FA 3.2–3.9 cm

W 4–10 g

0–1300 m

Large Forest Bat

Vespadelus darlingtoni

FR Vespertilion de Darlington

DE Große Waldfledermaus

ES Vespadela de Darlington

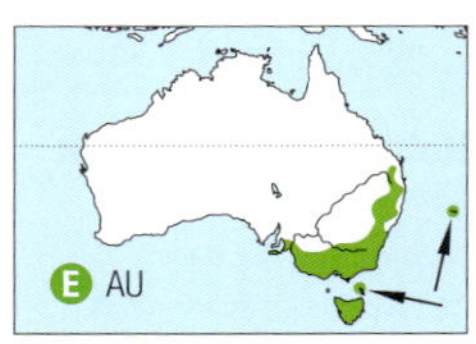

SE Qld, E & SE NSW, Vic, SE SA (with Kangaroo I), Flinders I, Tas, Lord Howe I

LC

Monotypic

HB 3.6–4.6 cm

T 2.8–3.9 cm

FA 2.8–3.4 cm

W 3.6–7 g

0–1700 m

Southern Forest Bat

Vespadelus regulus

FR Vespertilion de King River

DE Südliche Waldfledermaus

ES Vespadela de King River

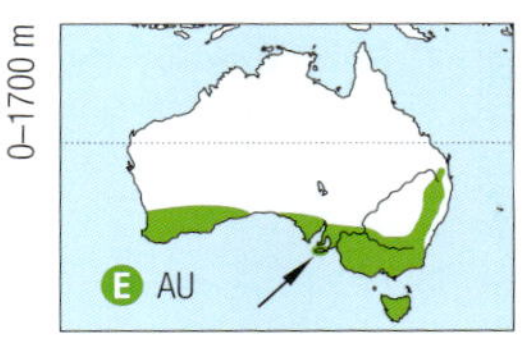

From S WA to Vic, S & E NSW, extreme SE Qld, Kangaroo I, Tas

Genus *CHALINOLOBUS*

(7 species)

VU

Monotypic

HB 4.3–5.6 cm

T 3.4–5 cm

FA 3.7–4.5 cm

W 5.5–12.2 g

above 1500 m

Large-eared Pied Bat

Chalinolobus dwyeri

FR Chalinolobe de Dwyer

DE Langohr-Lappenfledermaus

ES Calinolobo de Dwyer

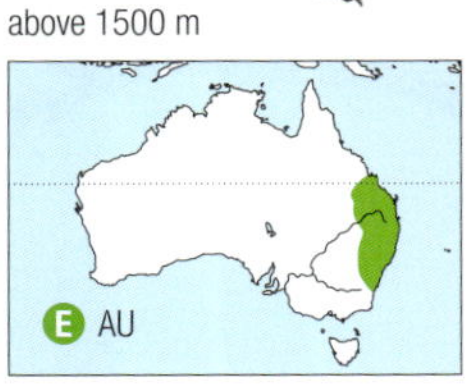

SE Qld, E NSW

NT

Monotypic

HB 4–5 cm T 2.9–3.9 cm

FA 3.1–3.6 cm W 3–8 g

40–400 m

Little Pied Bat

Chalinolobus picatus

FR Chalinolobe pie

DE Kleine Lappenfledermaus

ES Calinolobo pequeño

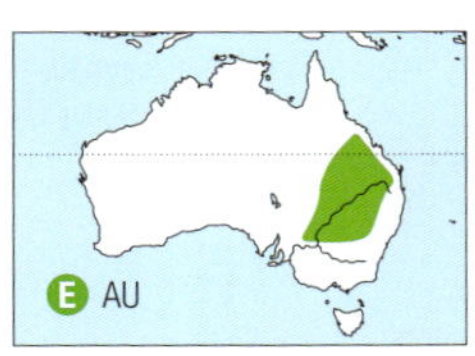

SE Qld, NW NSW, E SA

LC

2 sspp

HB 4–5.8 cm

T 2.6–4.3 cm

FA 3.2–3.9 cm

W 4.2–10 g

0–300 m

Hoary Wattled Bat

Chalinolobus nigrogriseus

FR Chalinolobe cendré

DE Schwarzgraue Lappenfledermaus

ES Calinolobo cenizo

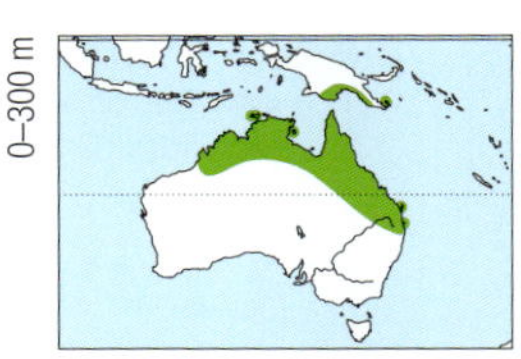

SE New Guinea (with Fergusson I), AU (NE WA, NT, Qld, NE NSW, some offshore Is)

Gould's Wattled Bat

Chalinolobus gouldii

FR Chalinolobe de Gould

DE Gould-Lappenfledermaus

ES Calinolobo de Gould

LC

Monotypic

HB 4.6–7.5 cm T 4–5 cm

FA 3.7–4.8 cm W 8–18 g

0–1500 m

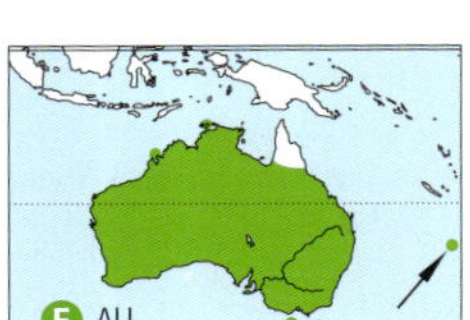

Mainland except Cape York & Nullarbor Plain, offshore Is, Tas, Norfolk I

EN

Monotypic

HB c. 5 cm T 3.3–4.4 cm

FA 3.5–4 cm W 8.3–9.7 g

New Caledonian Wattled Bat

Chalinolobus neocaledonicus

FR Chalinolobe de Nouvelle-Calédonie

DE Neukaledonien-Lappenfledermaus

ES Calinolobo de Nueva Caledonia

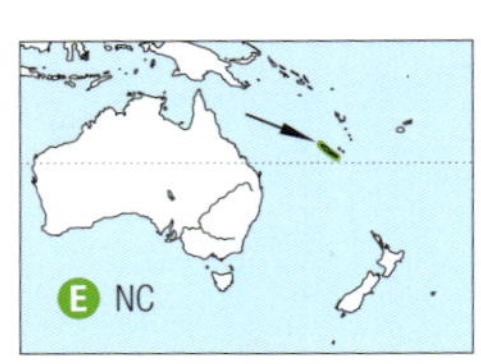

Grande Terre; unconfirmed in Loyalty Is (Lifou)

Chocolate Wattled Bat

Chalinolobus morio

FR Chalinolobe chocolat DE Narren-Lappenfledermaus ES Calinolobo chocolate

0–1570 m

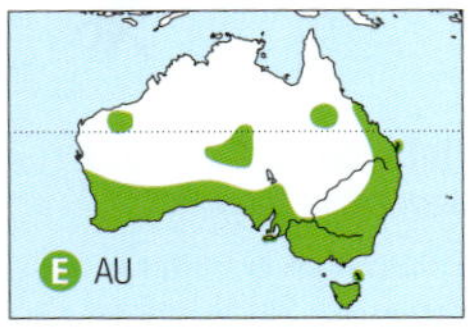

LC

Monotypic
HB 4–6.1 cm
T 3.9–4.9 cm
FA 3.5–4.2 cm
W 5.5–13 g

Mostly S & E (with Kangaroo, Fraser, Flinders & Tas Is), scattered in WC WA, S NT, N SA & Qld

New Zealand Long-tailed Bat

Chalinolobus tuberculatus

FR Chalinolobe de Nouvelle-Zélande DE Neuseeland-Lappenfledermaus ES Calinolobo de Nueva Zelanda

0–1000 m

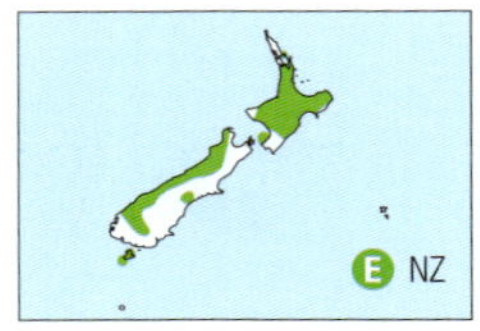

CR

Monotypic
HB 4.2–6.3 cm
T 3–4.6 cm
FA 3.7–4.6 cm
W 7.1–12.5 g

North & South Is, with Great Barrier, Little Barrier, Kapiti & Stewart Is

Genus *PHAROTIS*

(1 species)

up to 100 m

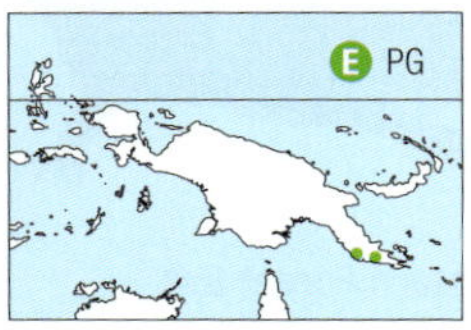

SE New Guinea (known from 2 sites in Central Prov)

CR Monotypic
HB 4.3–5 cm T 4.2–4.4 cm
FA 3.7–4 cm W c. 7.7 g

New Guinea Big-eared Bat

Pharotis imogene

FR Nyctophile imogène
DE Neuguinea-Langohr
ES Pharotis de Nueva Guinea

Genus *NYCTOPHILUS*

(15 species)

LC 2 sspp
HB 5.1–6.5 cm T 4–5.2 cm
FA 3.7–4.8 cm W 11–17.5 g

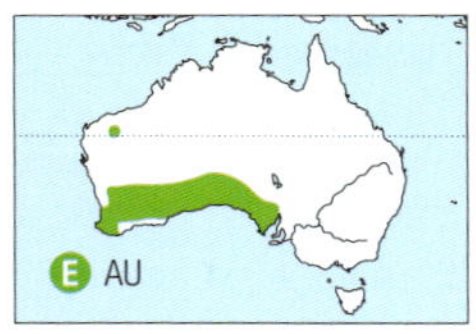

Western Long-eared Bat

Nyctophilus major

FR Nyctophile de Perth
DE Große Langohrfledermaus
ES Nictófila de Perth

SW WA & S SA E to Eyre Pen, 1 record in WC WA

VU

Monotypic
HB 5–7.5 cm
T 3.5–5.4 cm
FA 3.9–4.9 cm
W 11.2–21 g

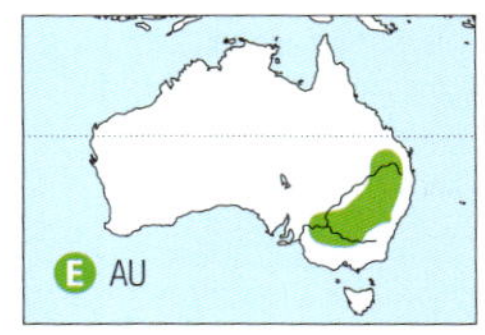

Corben's Long-eared Bat

Nyctophilus corbeni

FR Nyctophile de Corben
DE Corben-Langohrfledermaus
ES Nictófila de Corben

SE Qld, NSW, NW Vic, SE SA

LC

Monotypic
HB 4.5–5.7 cm
T 4–4.5 cm
FA 3.8–4.6 cm
W 7.6–14 g

0–400 m

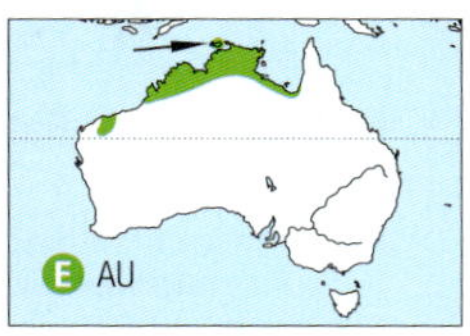

Pallid Long-eared Bat

Nyctophilus daedalus

FR Nyctophile dédale
DE Fahle Langohrfledermaus
ES Nictófila pálido

NW & NE WA, N NT (with Melville I), NW Qld

LC

Monotypic
T c. 4.6 cm
FA c. 4.1 cm
W c. 9 g

Holts' Long-eared Bat

Nyctophilus holtorum

FR Nyctophile des Holt
DE Holts-Langohrfledermaus
ES Nictófila de Holt

Restricted to far SW WA

LC

Monotypic
HB 4.4–5.2 cm
T 3.9–4.1 cm
FA 3.6–4.7 cm
W 5.2–16.5 g

0–1240 m

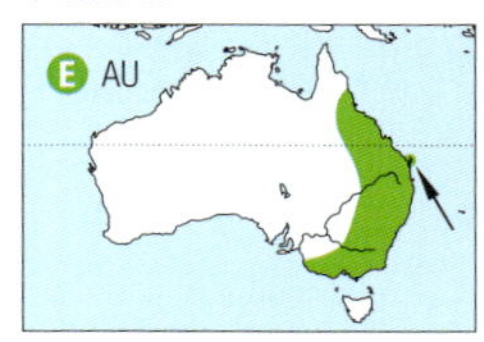

Gould's Long-eared Bat

Nyctophilus gouldi

FR Nyctophile de Gould
DE Gould-Langohrfledermaus
ES Nictófila de Gould

E & SE Qld (with Fraser I), C & E NSW, Vic, SE SA

VU

Monotypic
HB 5.1–5.4 cm
T 3.5–3.7 cm
FA 4.4–4.8 cm
W 9.8–18.9 g

Tasmania

0–740 m

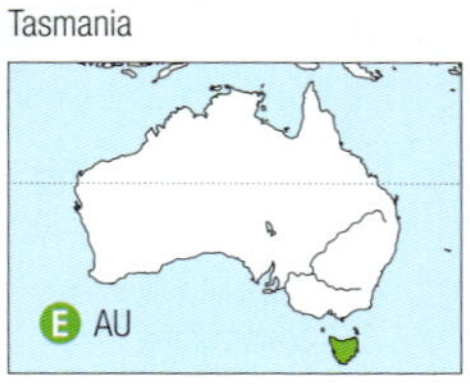

Tasmanian Long-eared Bat

Nyctophilus sherrini

FR Nyctophile de Tasmanie
DE Tasmanien-Langohrfledermaus
ES Nictófila de Tasmania

c. 430 m

S Grande Terre (Mt Koghis)

New Caledonian Long-eared Bat

Nyctophilus nebulosus

FR Nyctophile de Nouvelle-Calédonie
DE Neukaledonien-Langohrfledermaus
ES Nictófila de Nueva Caledonia

CR

Monotypic
HB 5.3–5.5 cm
T 4.8–5 cm
FA 4.2–4.3 cm
W 9.4–12 g

LC

Monotypic
HB 4.2–6.3 cm
T 3.6–4.9 cm
FA 3.7–4.2 cm
W 6–10.5 g

0–2600 m

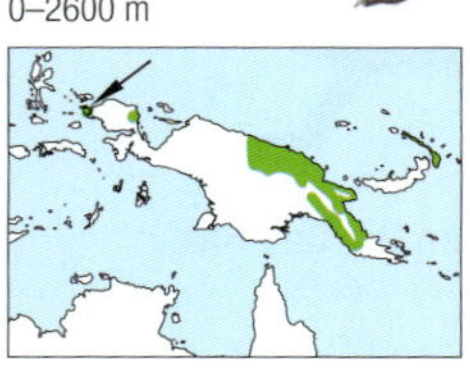

Papuan Long-eared Bat

Nyctophilus microtis

FR Nyctophile à petites oreilles
DE Papua-Langohrfledermaus
ES Nictófila de orejas pequeñas

W & mostly E New Guinea (with Salawati & New Ireland Is); perhaps Manus, New Britain & Sudest Is

LC

Monotypic
HB 3.8–4.4 cm
T 2.6–3.6 cm
FA 3–3.6 cm
W 3–7 g

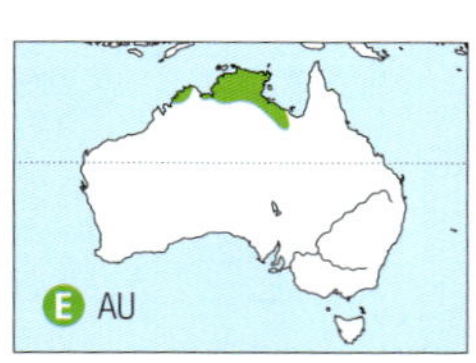

Pygmy Long-eared Bat

Nyctophilus walkeri

FR Nyctophile de Walker
DE Walker-Langohrfledermaus
ES Nictófila de Walker

NE WA, N NT, NW Qld

LC

Monotypic
HB 3.5–5.5 cm
T 3.9–4.6 cm
FA 3.7–4.7 cm
W 5–12 g

0–500 m

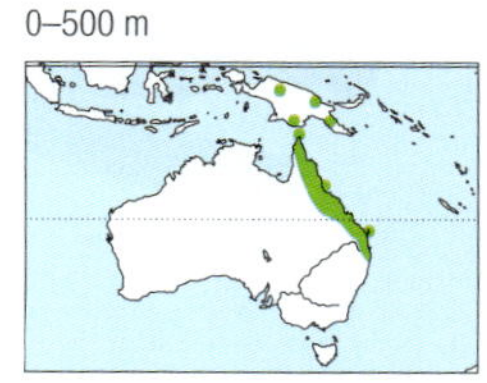

Eastern Long-eared Bat

Nyctophilus bifax

FR Nyctophile du Queensland
DE Queensland-Langohrfledermaus
ES Nictófila de Queensland

E Qld (with Moa, Prince of Wales, Hinchinbrook, Palm & Fraser Is), NE NSW; scattered across New Guinea

LC
Monotypic
HB 4–5.9 cm
T 3.6–4.1 cm
FA 3.3–3.9 cm
W 5.2–9.6 g

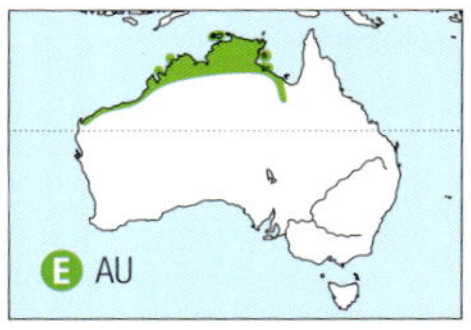

Northern Long-eared Bat
Nyctophilus arnhemensis
FR Nyctophile d'Arnhem
DE Arnhem-Langohrfledermaus
ES Nictófila de Arnhem

N coast of WA (with Bonaparte Arch), N NT (with Tiwi Is, Groote Eylandt & Sir Edward Pellew Is), NW Qld

DD
Monotypic
HB 5.2–6.4 cm
T 4.8–5.2 cm
FA 4.5–4.8 cm
W 11.5–13 g

500–1700 m

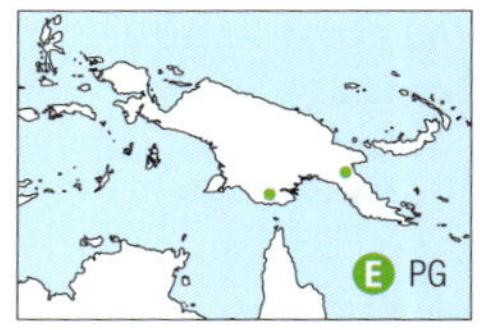

Mount Missim Long-eared Bat
Nyctophilus shirleyae
FR Nyctophile de Shirley
DE Shirley-Langohrfledermaus
ES Nictófila de Shirley

Known from Mt Missim (Morobe Prov) & Trans-Fly region (Western Prov); likely wider range

LC
Monotypic
HB 4.2–6.3 cm
T 3.6–4.9 cm
FA 3.7–4.1 cm
W 5.8–9.5 g

1600–3315 m

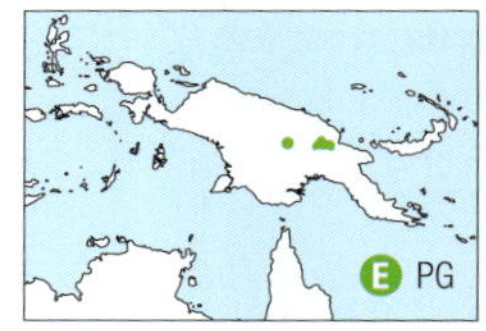

Small-toothed Long-eared Bat
Nyctophilus microdon
FR Nyctophile à petites dents
DE Kleinzähnige Langohrfledermaus
ES Nictófila de dientes pequeños

C highlands (Eastern, Western & Southern Highlands, Chimbu, Enga)

0–1580 m

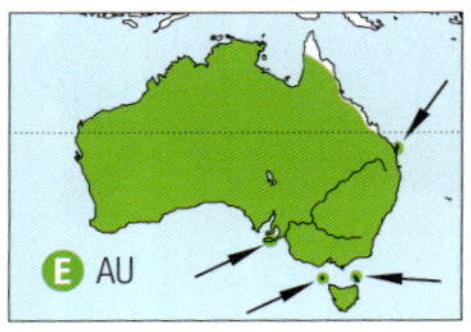

Mainland (except NE coast of Qld), Tas, Fraser, Kangaroo, King & Flinders Is

LC
3 sspp
HB 3.8–5 cm T 3.1–4 cm
FA 3.1–4.2 cm W 3.9–14.5 g

Lesser Long-eared Bat
Nyctophilus geoffroyi
FR Nyctophile de Geoffroy DE Geoffroy-Langohrfledermaus ES Nictófila de Geoffroy

c. 0 m

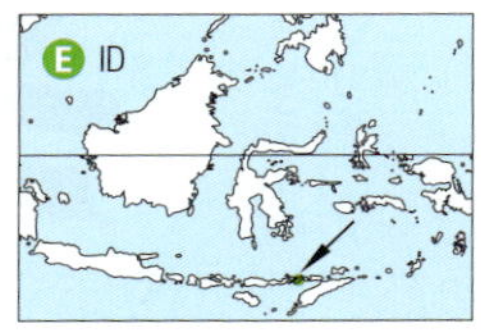

Lesser Sundas (known only in Lembata)

DD Monotypic
HB c. 5.1 cm T c. 4.1 cm
FA c. 3.9 cm W c. 7.6 g

Sunda Long-eared Bat
Nyctophilus heran
FR Nyctophile de la Sonde DE Sunda-Langohrfledermaus ES Nictófila de Sonda

Genus *MIMETILLUS*

(2 species)

up to 2300 m

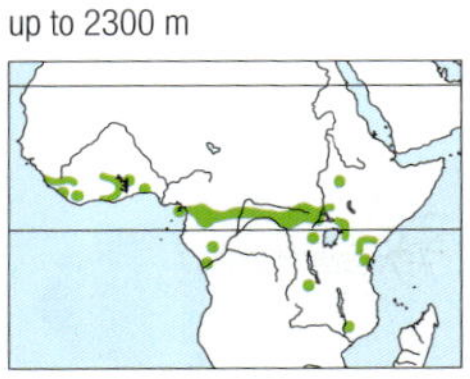

From SL E to S SD & SW ET, KE & NE TZ (with Bioko I), S to NW AO, extreme W CD & WC MZ

LC
Monotypic
HB 5.1–6.3 cm
T 2.3–3.2 cm
FA 2.7–3.1 cm
W 5.5–10 g

Moloney's Mimic Bat
Mimetillus moloneyi
FR Mimétille de Moloney
DE Moloney-Schmalflügelfledermaus
ES Mimétilo de Moloney

Thomas's Mimic Bat
Mimetillus thomasi
FR Mimétille de Thomas
DE Thomas-Schmalflügelfledermaus
ES Mimétilo de Thomas

NE
2 sspp
HB c. 5.6 cm
T c. 3.8 cm
FA c. 3.1 cm

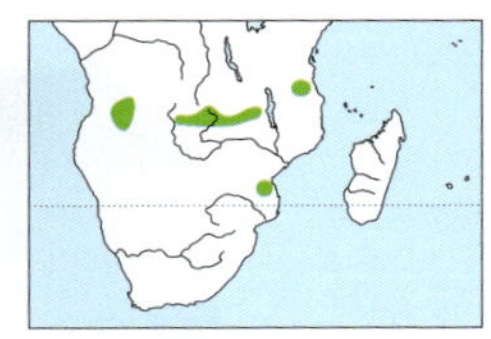

Angolan Plateau, SE CD, S TZ, ZM, S MZ

Genus *HYPSUGO*

(18 species)

DD
Monotypic
HB 3.5–4.5 cm
T 3.2–4.4 cm
FA 2.8–3.4 cm
W 2.3–4.8 g

345–1225 m

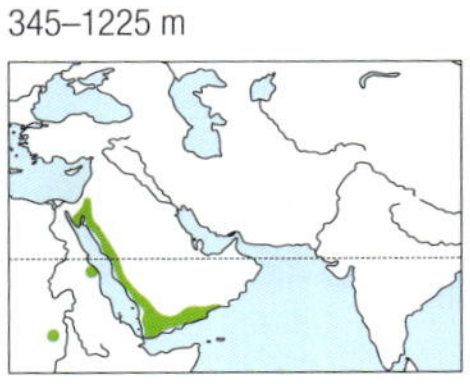

Fairy Pipistrelle
Hypsugo ariel
FR Vespère ariel
DE Elfen-Zwergfledermaus
ES Hipsugo ariel

PS, S IL, SW JO, NE EG (Sinai), W & S Arabian Pen, NE & SC SD; perhaps SE EG

EN
Monotypic
HB c. 4.6 cm
T c. 4.2 cm
FA 3.1–3.3 cm

10–365 m

Lanza's Pipistrelle
Hypsugo lanzai
FR Vespère de Lanza
DE Lanza-Zwergfledermaus
ES Hipsugo de Lanza

Socotra I

DD
Monotypic
HB 3.7–4.8 cm
T 3.7–4.4 cm
FA 2.9–3.4 cm

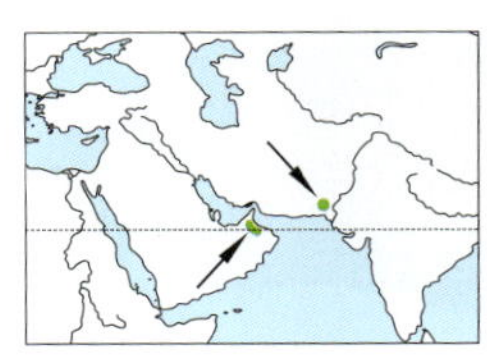

Arabian Pipistrelle
Hypsugo arabicus
FR Vespère d'Oman
DE Arabische Zwergfledermaus
ES Hipsugo de Omán

N OM, SE IR

LC
4 sspp
HB 4–5.5 cm
T 3.1–4.3 cm
FA 3.1–3.8 cm
W 5–9 g

0–3000 m

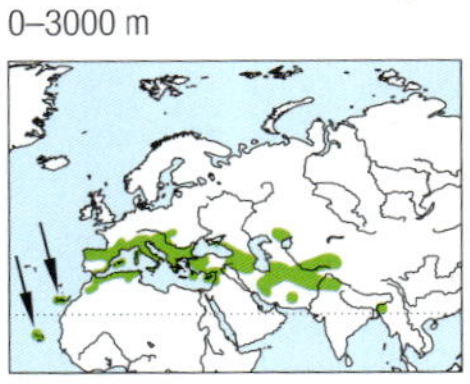

Savi's Pipistrelle
Hypsugo savii
FR Vespère de Savi
DE Alpenfledermaus
ES Murciélago montañero

S Europe E to KZ, W CN & NE IN, Mediterranean Is, N Africa (from N MA to N TN), Canary & Cape Verde Is

LC
Monotypic
HB 3.8–5 cm
T 3–4 cm
FA 3.5–3.8 cm
W 6.3–9.4 g

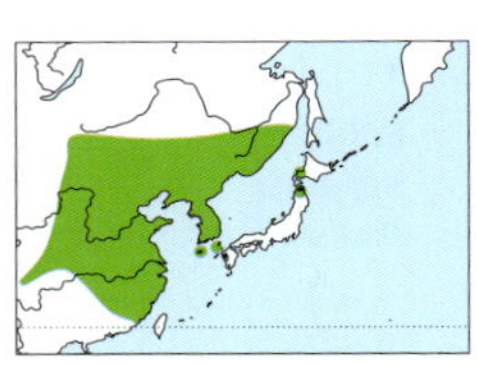

Alashanian Pipistrelle
Hypsugo alaschanicus
FR Vespère de l'Alashan
DE Alaschan-Zwergfledermaus
ES Hipsugo de Ala Shan

SE MN, RU (S Far East), Korean Pen (with Jeju I), C & E CN, JP (S Hokkaido, N Honshu, Tsushima Is)

NE
Monotypic
HB c. 5.5 cm
T c. 4 cm
FA 3.3–3.6 cm
W c. 7 g

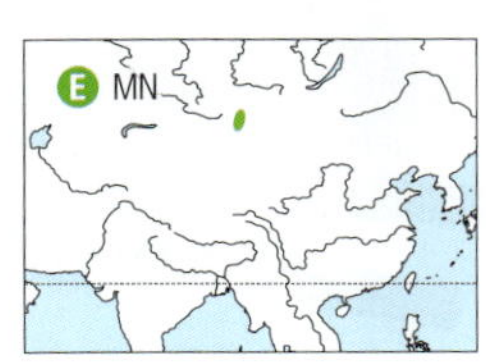

Stubbe's Pipistrelle
Hypsugo stubbei
FR Vespère de Stubbe
DE Stubbe-Zwergfledermaus
ES Hipsugo de Stubbe

Known from 4 sites in far W near Bulgan-gol & Altai Mts

LC
Monotypic
HB 4.6–5.5 cm
T 3.2–4.2 cm
FA 3.4–3.6 cm
W 5.5–7 g

710–1950 m

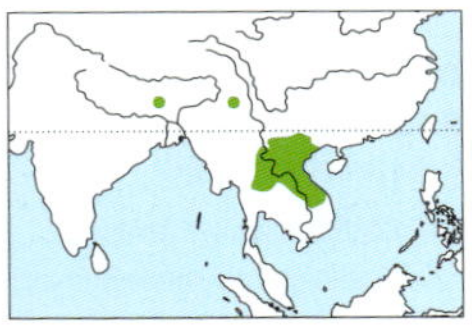

Cadorna's Pipistrelle
Hypsugo cadornae
FR Vespère de Cadorna
DE Cadorna-Zwergfledermaus
ES Hipsugo de Cadorna

NE IN (West Bengal), N MM, N TH, LA, N & C VN, NE KH

LC
Monotypic
HB 3.8–4.9 cm
T 3.2–3.8 cm
FA 3.2–3.7 cm
W 4.3–5.8 g

0–1080 m

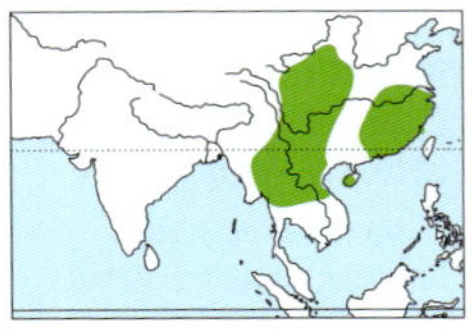

Chinese Pipistrelle
Hypsugo pulveratus
FR Vespère de Chine
DE China-Zwergfledermaus
ES Hipsugo de China

C, SC & SE CN (with Hainan & Matsu Is), C & E MM, N TH, N & C LA, N & C VN

LC
Monotypic
HB 4.3–5.1 cm
T 3–4.1 cm
FA 3.8–4.1 cm

up to 2000 m

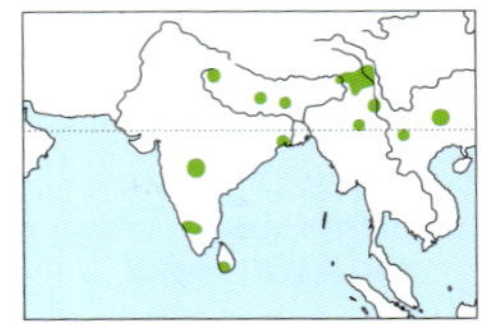

Chocolate Pipistrelle
Hypsugo affinis
FR Vespère chocolat
DE Hochland-Zwergfledermaus
ES Hipsugo chocolatero

IN, C NP, LK, SC CN (SE Tibet, Yunnan, Guangxi), NE MM (Kachin)

DD
Monotypic
HB 4.7–5.6 cm
T 3.7–4.2 cm
FA 3.9–4.2 cm

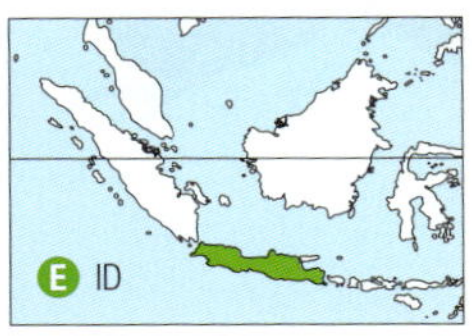

Pungent Pipistrelle
Hypsugo mordax
FR Vespère mordante
DE Bissige Zwergfledermaus
ES Hipsugo mordaz

Known only by few specimens from Java

DD
Monotypic
HB 4.9–5.3 cm
T 3.5–3.9 cm
FA 3.8–4.2 cm
W 6–9 g

1000–2130 m

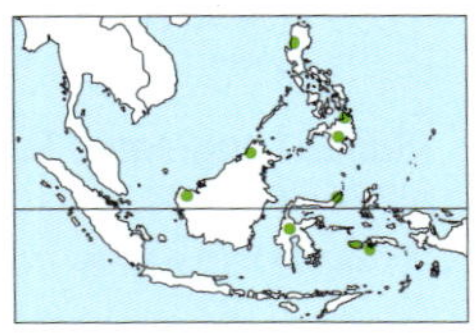

Peters's Pipistrelle
Hypsugo petersi
FR Vespère de Peters
DE Peters-Zwergfledermaus
ES Hipsugo de Peters

PH (Luzon, Mindanao), Borneo (Sabah, Sarawak), NE & C Sulawesi, Moluccas (Buru, Ambon)

DD
Monotypic
FA 3.5–3.8 cm

50–980 m

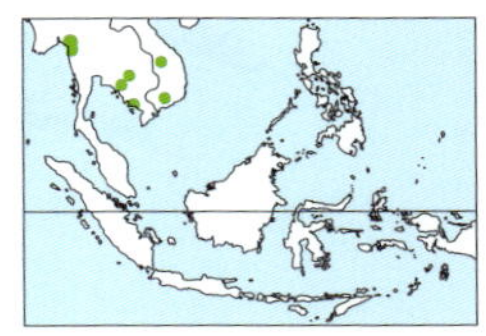

Long-toothed Pipistrelle
Hypsugo dolichodon
FR Vespère à longues dents
DE Langzahn-Zwergfledermaus
ES Hipsugo de dientes largos

Scattered in SE MM, W & S KH, SE LA, S VN

DD
Monotypic
HB c. 5.6 cm
T c. 3.9 cm
FA c. 3.5 cm

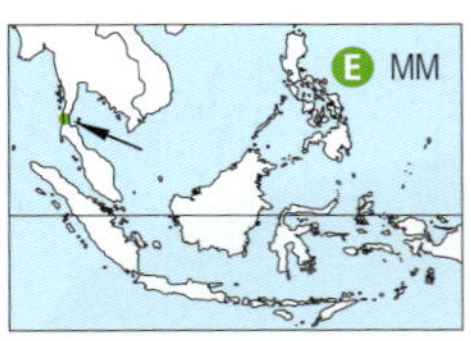

Burmese Pipistrelle
Hypsugo lophurus
FR Vespère de Birmanie
DE Burma-Zwergfledermaus
ES Hipsugo de Birmania

Known only from type locality (Maliwun, S Tenasserim)

DD
Monotypic
HB c. 5.6 cm
T c. 4.1 cm
FA 3.5–3.8 cm

0–1500 m

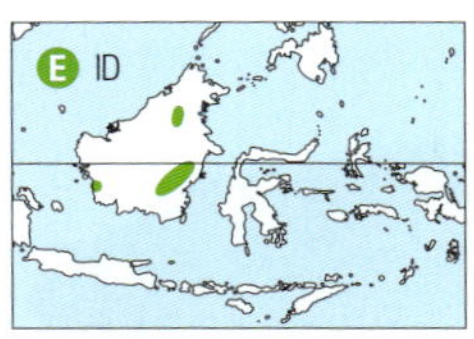

Red-brown Pipistrelle
Hypsugo kitcheneri
FR Vespère de Kitchener
DE Kitchener-Zwergfledermaus
ES Hipsugo de Kitchener

NC, SW & SE Borneo (Kalimantan); perhaps Sabah (MY)

LC
Monotypic
HB 4.1–4.5 cm
T 3.6–4.1 cm
FA 3.2–3.7 cm
W 4.4–5.3 g

0–400 m

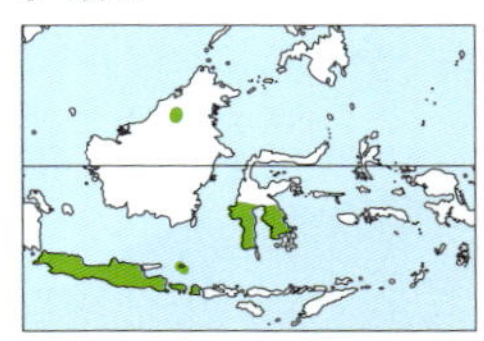

Brown Pipistrelle
Hypsugo imbricatus
FR Vespère imbriquée
DE Braune Zwergfledermaus
ES Hipsugo marrón

MY (Sarawak in NC Borneo), ID (Java, Kangean, Bali, Lombok, S Sulawesi)

DD
Monotypic
HB c. 5 cm
T c. 2 cm
FA 3.2–3.4 cm
W 4–6 g

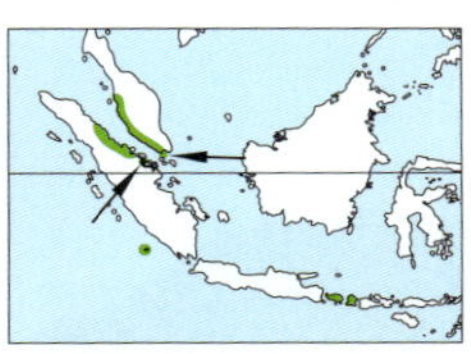

Big-eared Pipistrelle
Hypsugo macrotis
FR Vespère à grandes oreilles
DE Großohr-Zwergfledermaus
ES Hipsugo orejudo

W Pen MY, SG, ID (NE Sumatra, Padang, Enggano, Bali, Lombok)

DD
Monotypic
HB c. 5 cm
T 1.4–1.5 cm
FA 3–3.3 cm
W 4–5 g

0–100 m

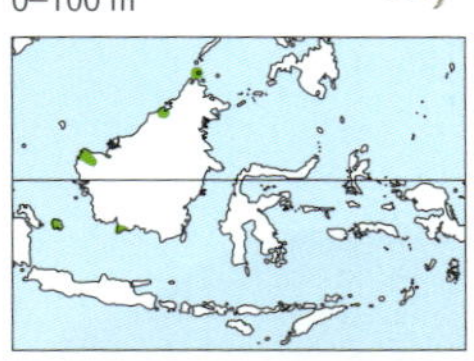

Vordermann's Pipistrelle
Hypsugo vordermanni
FR Vespère de Vordermann
DE Vordermann-Zwergfledermaus
ES Hipsugo de Vordermann

NW, WC & S Borneo (with Banggi & Belitung Is)

DD
Monotypic
HB c. 4 cm
T 2.1–2.6 cm
FA 2.3–2.6 cm
W c. 3.8 g

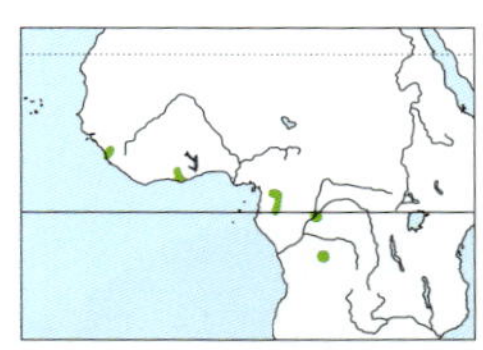

Mouse-like Pipistrelle
Hypsugo musciculus
FR Vespère murine
DE Mäuschen-Zwergfledermaus
ES Hipsugo murino

W SL, SW GH, SC CM, NE GA, 2 records from W CD

Genus *NYCTICEINOPS*

(7 species)

850–2235 m

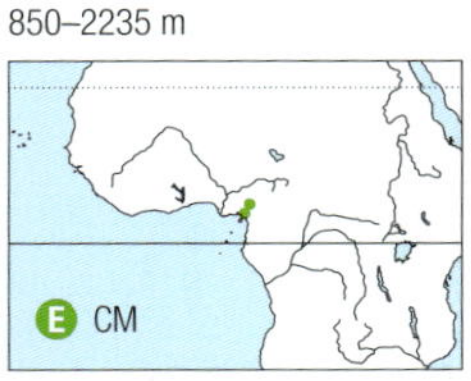

Rumpi Hills, Mt Kupe, Mt Nlonako

Eisentraut's Serotine
Nycticeinops eisentrauti
FR Vespère d'Eisentraut
DE Eisentraut-Zwergfledermaus
ES Nicticenop de Eisentraut

DD Monotypic
T 2.9–3.7 cm FA 3.3–3.6 cm
W 7–8.3 g

Schlieffen's Serotine
Nycticeinops schlieffenii
FR Vespère de Schlieffen
DE Schlieffen-Abendfledermaus
ES Nicticenop de Schlieffen

LC
Monotypic
HB 3.7–4.8 cm T 2.4–3.4 cm
FA 2.8–3.5 cm W 4–9 g

up to 1000 m

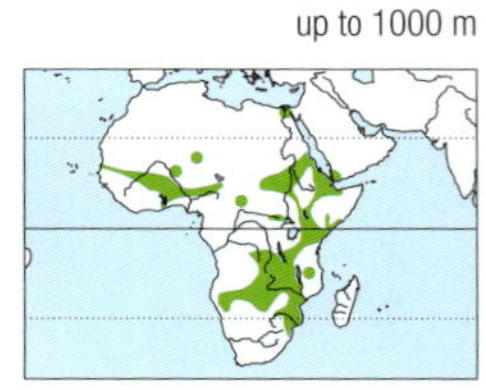

S MR E to N SO (isolated in N EG) S to N NA, N BW & NE ZA, SW Arabia

LC
Monotypic
HB 4.4–5 cm
T 2.7–3.1 cm
FA 2.8–3.2 cm
W 4–7 g

300–1600 m

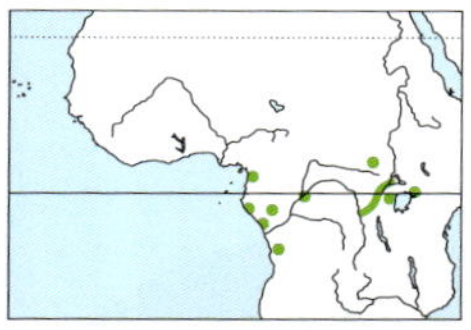

Broad-headed Serotine
Nycticeinops crassulus
FR Vespère à grosse tête
DE Breitköpfige Zwergfledermaus
ES Nicticenop de cabeza ancha

Scattered in S CM, S GA, S CG, NW AO, CD, SW SS, W & S UG, W KE

NE
Monotypic
HB 4.7–5.2 cm
T 2.5–3.2 cm
FA 2.8–3.3 cm
W 4–8 g

70–600 m

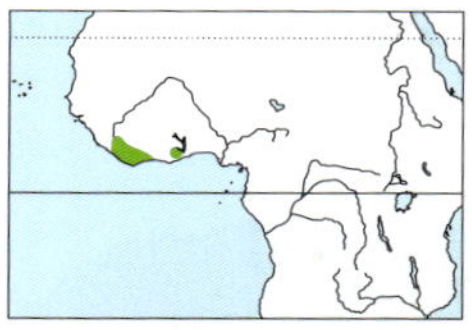

Bellier's Serotine
Nycticeinops bellieri
FR Vespère de Bellier
DE Bellier-Zwergfledermaus
ES Nicticenop de Bellier

W Africa in SE GN, LR, S CI & S GH

NE
Monotypic
HB c. 6.2 cm
T 2.8–3.7 cm
FA 3.5–3.7 cm
W 7.3–9.5 g

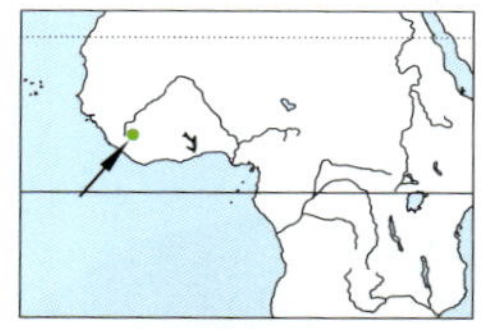

Happolds' Serotine
Nycticeinops happoldorum
FR Vespère des Happold
DE Happold-Zwergfledermaus
ES Nicticenop de Happold

W Africa in SE GN & N LR (Mt Nimba)

Large-headed Serotine
Nycticeinops macrocephalus
FR Vespère à face large DE Großkopf-Zwergfledermaus
ES Nicticenop de cabeza grande

NE
Monotypic
HB c. 5.3 cm
T c. 3.5 cm
FA 3.7–3.8 cm
W 9.8–12 g

1950–2500 m

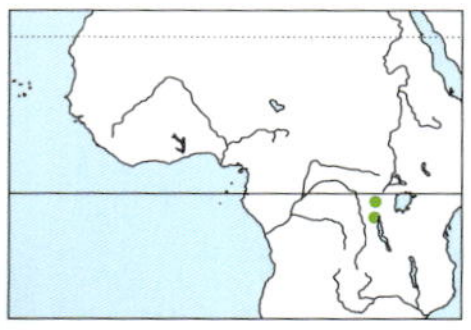

W Africa in RW & EC CD

Grandidier's Serotine
Nycticeinops grandidieri
FR Vespère de Grandidier DE Grandidier-Zwergfledermaus
ES Nicticenop de Grandidier

DD
2 sspp
HB 5.1–5.6 cm
T 3–3.4 cm
FA 3.3–3.8 cm
W 7–8 g

120–1635 m

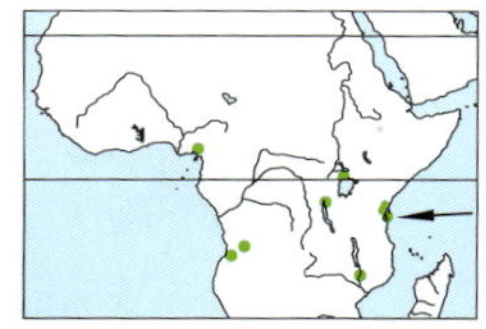

CM, AO, UG, BI, SE KE, NE TZ (with Unguja I), MW; perhaps SO

Genus *PSEUDOROMICIA*

(7 species)

LC
2 sspp
HB 3.7–6.5 cm
T 3–4.1 cm
FA 2.9–3.7 cm
W 4–9 g

0–1480 m

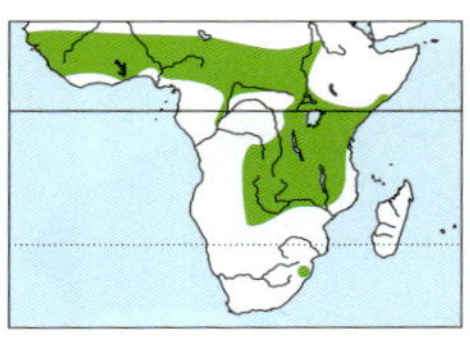

Rendall's Serotine
Pseudoromicia rendalli
FR Vespère de Rendall
DE Rendall-Zwergfledermaus
ES Pipistrela de Rendall

SN E to TD, SD & SO S to BW, MZ & E ZA

400–1470 m

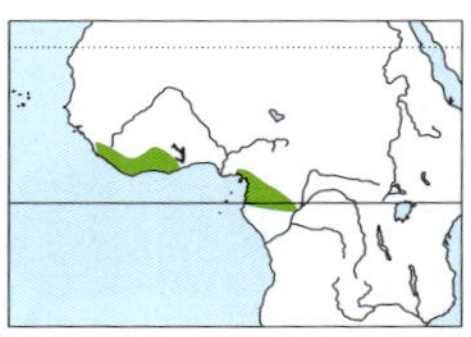

NT Monotypic
HB 4.3–4.8 cm T 3.3–4.6 cm
FA 3.3–3.8 cm W 4–9 g

Dark-brown Serotine
Pseudoromicia brunnea
FR Vespère brun-noir
DE Dunkelbraune Zwergfledermaus
ES Pipistrela oscura

W & WC Africa in SL E to CG

DD
Monotypic
HB 4.4–5 cm
T 2.7–3.6 cm
FA 2.8–3.2 cm
W 4–5.5 g

c. 765 m

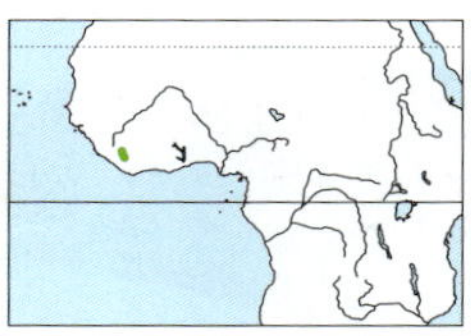

Isabelline Serotine
Pseudoromicia isabella
FR Vespère isabelle
DE Isabelle-Zwergfledermaus
ES Pipistrela isabela

W Africa in GN (Simandou Range) & LR (Mt Nimba)

NE
Monotypic
HB 5.3–5.4 cm
T 3.4–3.6 cm
FA 3.7–3.8 cm
W 7.9–8 g

c. 1130 m

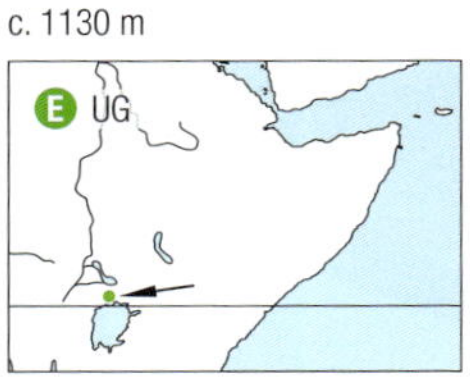

Kityo's Serotine
Pseudoromicia kityoi
FR Vespère de Kityo
DE Kityo-Zwergfledermaus
ES Pipistrela de Kityo

Known only from Mabira Forest R

EN
Monotypic
HB 4.7–5 cm
T 3.9–4.4 cm
FA c. 3.7 cm
W 6–6.3 g

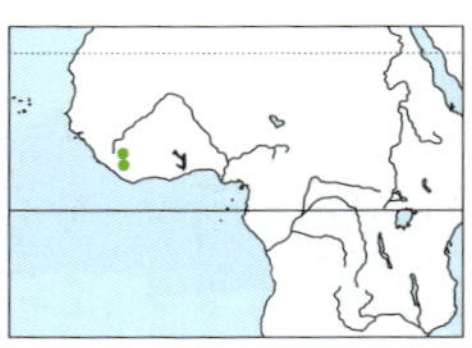

Rosevear's Serotine
Pseudoromicia roseveari
FR Vespère de Rosevear
DE Rosevear-Zwergfledermaus
ES Pipistrela de Rosevear

W Africa in GN (Simandou Range) & LR (Mt Nimba)

LC
Monotypic
HB 3.4–4.8 cm
T 2.8–3.4 cm
FA 2.8–3.3 cm
W 3–7 g

up to 3200 m

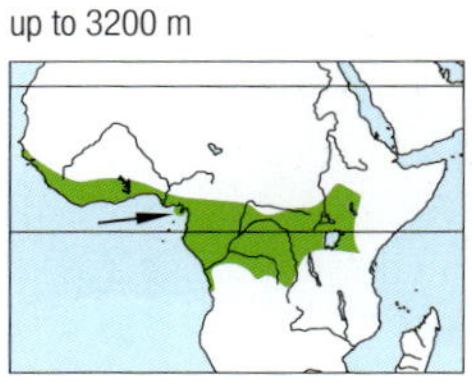

White-winged Serotine
Pseudoromicia tenuipinnis
FR Vespère à ailes blanches
DE Weißflügel-Zwergfledermaus
ES Pipistrela de alas blancas

Widely from SN to ET, W KE & TZ, S to NW AO & S CD, Bioko I (GQ)

NE
Monotypic
HB 4.1–5.1 cm
T 3–3.8 cm
FA 2.9–3.3 cm
W 4.8–8.3 g

c. 1130 m

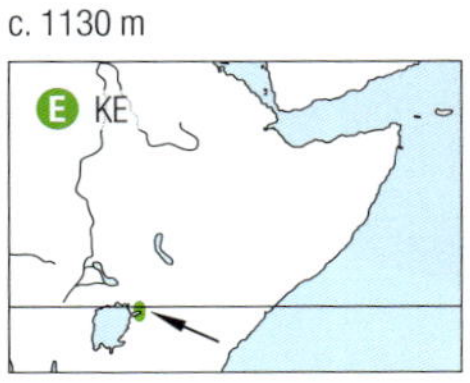

Nyanza Serotine
Pseudoromicia nyanza
FR Vespère du Nyanza
DE Nyanza-Zwergfledermaus
ES Pipistrela de Nyanza

W of Rift Valley; presumably UG

Genus *AFRONYCTERIS*

(2 species)

LC
7 sspp
HB 4–4.4 cm
T 2.4–4 cm
FA 2.5–3.5 cm
W 2.5–6.5 g

0–2500 m

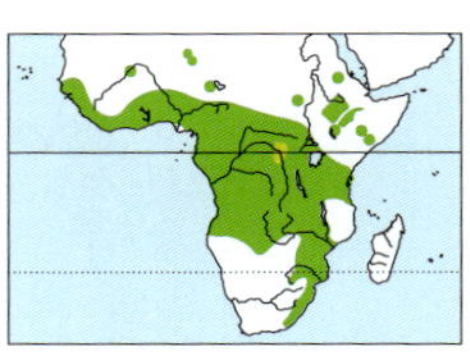

Banana Serotine
Afronycteris nanus
FR Vespère naine
DE Bananen-Zwergfledermaus
ES Pipistrela bananera

Sub-Saharan Africa from SN E to ET, S to S AO, N NA & E ZA, also ML & NE; uncertain in SD, SS & SO

DD
Monotypic
HB 3.4–4.1 cm
T 2.6–3.4 cm
FA 2.6–3 cm
W 2–6 g

400–1335 m

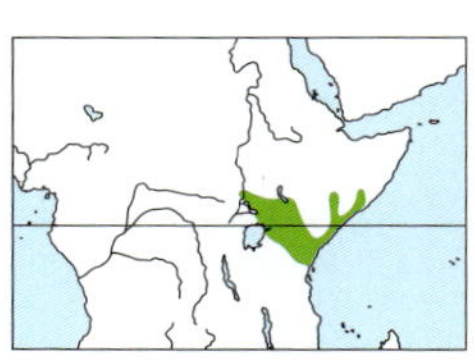

Heller's Serotine
Afronycteris helios
FR Vespère soleil
DE Heller-Zwergfledermaus
ES Pipistrela de Heller

Extreme S SS, S SO, NE & E UG, KE & N TZ; uncertain in DJ

Genus *NEOROMICIA*

(5 species)

0–870 m

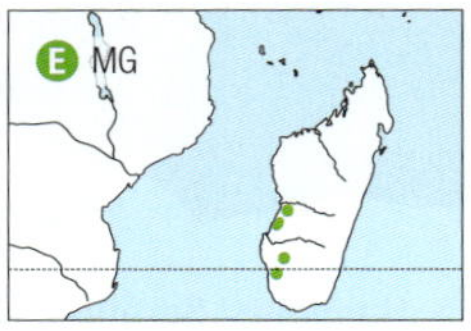

Kirindy Serotine
Neoromicia bemainty

FR Vespère de Kirindy
DE Kirindy-Zwergfledermaus
ES Neoromicia de Kirindy

Few sites in CW & SW mainland

LC Monotypic
HB 4.2–4.7 cm T 3.3–3.6 cm
FA 2.9–3.2 cm W 2.8–4.4 g

Anchieta's Serotine
Neoromicia anchieta

FR Vespère d'Anchieta
DE Anchieta-Zwergfledermaus
ES Neoromicia de Anchieta

LC Monotypic
HB 4.5–4.9 cm T 3.2–3.9 cm
FA 2.9–3.4 cm W 4.2–6 g

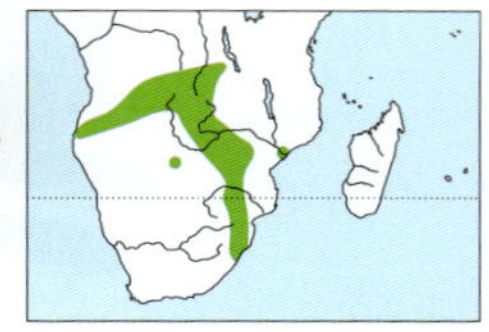

AO, SE CD, W ZM, ZW, C MZ, N BW, E ZA, SZ

LC
Monotypic
HB 3.9–5.3 cm
T 2.4–3.3 cm
FA 2.2–3.2 cm
W 2–5 g

380–2150 m

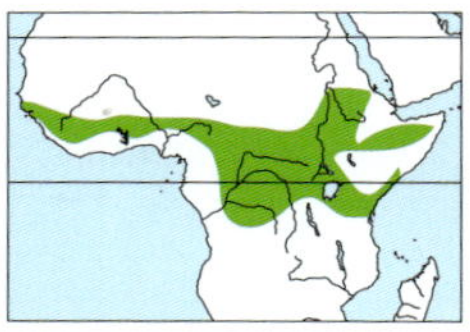

Somali Serotine
Neoromicia somalica

FR Vespère de Somalie
DE Somalia-Zwergfledermaus
ES Neoromicia de Somalia

From SN & GW E to SD, ER & SO, S to S CD & N TZ; perhaps MW

LC
Monotypic
HB 4–4.5 cm
T 2.7–4 cm
FA 2.7–3.3 cm
W 3–6 g

500–2650 m

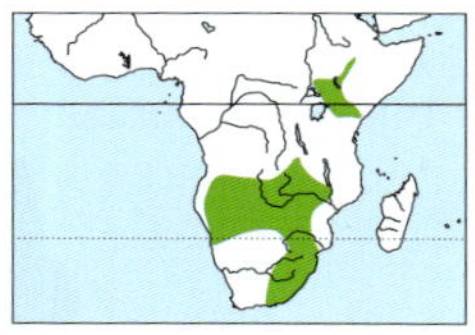

Zulu Serotine
Neoromicia zuluensis

FR Vespère zouloue
DE Zulu-Zwergfledermaus
ES Neoromicia zuluense

Disjunct in ET, SS, UG & KE and C AO, extreme S CD & ZM S to S ZA

LC
2 sspp
HB 3.8–3.9 cm
T 2.1–3.4 cm
FA 2.3–3.5 cm
W 2–5 g

420–2165 m

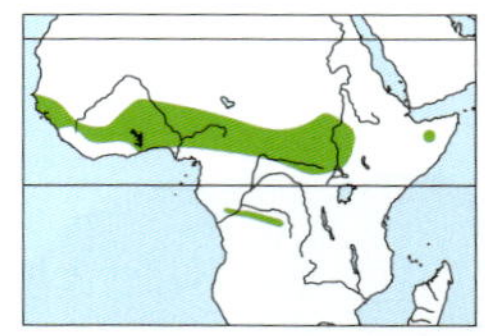

Tiny Serotine
Neoromicia guineensis

FR Vespère de Guinée
DE Guinea-Zwergfledermaus
ES Neoromicia de Guinea

SN E to S SD, SS, ET & NE CD, disjunct in S CG & S CD

Genus *LAEPHOTIS*

(10 species)

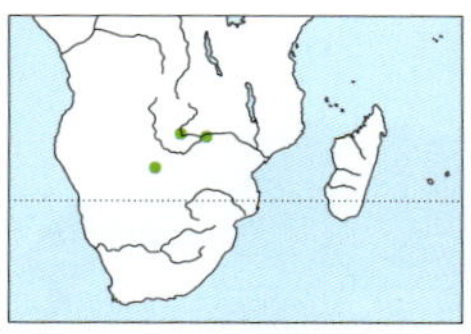

Stanley's Serotine
Laephotis stanleyi

FR Vespère de Stanley
DE Stanley-Langohrfledermaus
ES Laefotis de Stanley

BW (Okavango Delta along R Gomoti), ZM (Kafue NP), ZW (Mana Pools NP); perhaps MZ & ZA

NE Monotypic
HB 4.9–5.3 cm T 3.4–4.8 cm
FA 3.4–3.9 cm W 6.1–6.8 g

Isalo Serotine
Laephotis malagasyensis

FR Vespère de l'Isalo
DE Isalo-Langohrfledermaus
ES Laefotis de Isalo

VU Monotypic
T c. 3 cm FA 3–3.2 cm

450–700 m

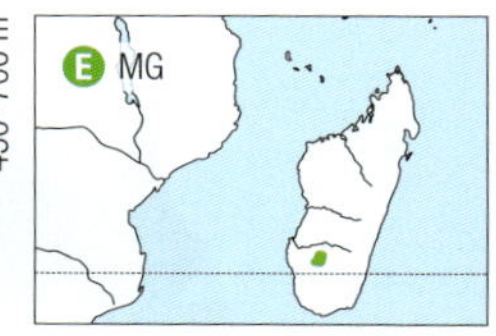

Known from 4 sites around Isalo Massif

DD
Monotypic
HB 5.3–5.8 cm
T 3.1–3.5 cm
FA 3.4–3.8 cm

900–1300 m

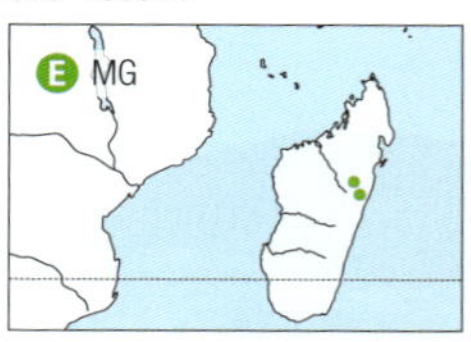

Roberts's Serotine
Laephotis robertsi

FR Vespère de Roberts
DE Roberts-Langohrfledermaus
ES Laefotis de Roberts

Anjozorobe-Angavo & Mantadia-Analamazaotra (Andasibe-Mantadia NP)

LC
Monotypic
HB 5.3–6.4 cm
T 3.8–5 cm
FA 3.7–4.1 cm
W 7.7–11 g

1000–1700 m

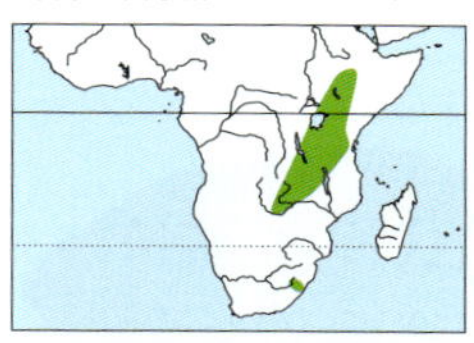

De Winton's Long-eared Serotine
Laephotis wintoni

FR Vespère de De Winton
DE De-Winton-Langohrfledermaus
ES Laefotis de De Winton

C ET S through KE & TZ to S ZM, disjunct in LS & E ZA

LC
Monotypic
HB 4.7–5.5 cm
T 3.9–4.7 cm
FA 3–3.9 cm
W 5–8 g

800–1700 m

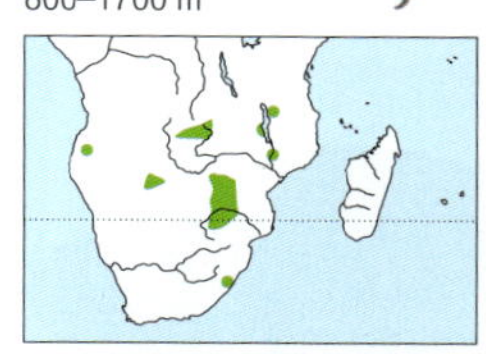

Botswanan Long-eared Serotine
Laephotis botswanae

FR Vespère du Botswana
DE Botswana-Langohrfledermaus
ES Laefotis de Botsuana

SW TZ, extreme S CD, ZM, MW, SW AO, NE NA, NW BW, ZW & NE ZA

LC
Monotypic
HB 4.9–6.5 cm
T 3.8–4.7 cm
FA 3.6–4 cm
W 7.8–11 g

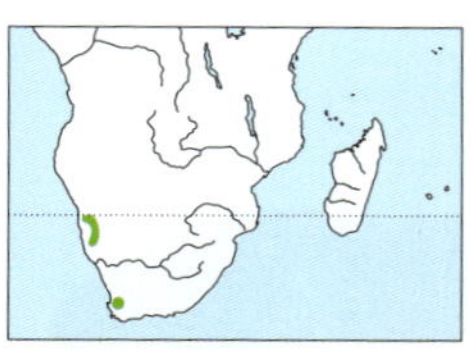

Namib Long-eared Serotine
Laephotis namibensis

FR Vespère de Namibie
DE Namib-Langohrfledermaus
ES Laefotis de Namibia

S Africa in SW NA & W ZA

DD
Monotypic
HB 4.6–5 cm
T 3.6–3.8 cm
FA 3.2–3.6 cm

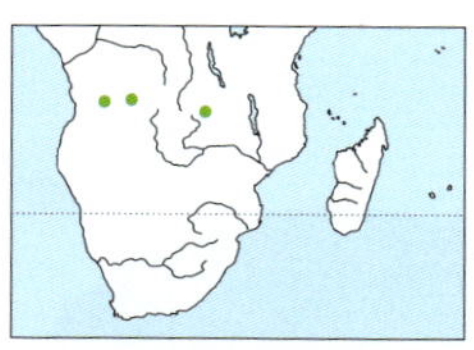

Angolan Long-eared Serotine
Laephotis angolensis

FR Vespère d'Angola
DE Angola-Langohrfledermaus
ES Laefotis de Angola

Known from 3 localities (2 in AO, 1 in CD)

NE Monotypic
HB 4.6–5.8 cm
T 2.8–3.7 cm
FA 3–3.4 cm
W 4–7 g

1155–1700 m

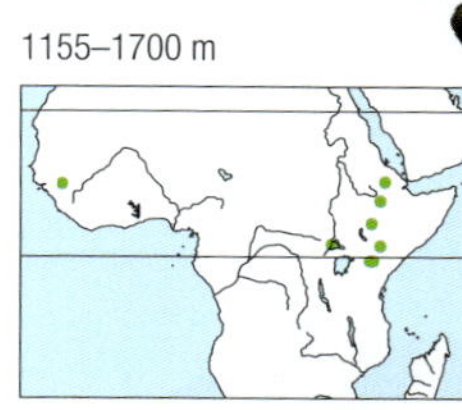

East African Serotine
Laephotis kirinyaga

FR Vespère du Kenya
DE Ostafrika-Langohrfledermaus
ES Laefotis de Kenya

Highlands of KE (both sides of Rift Valley), UG (Murchison Falls NP), ET, out of range in SN

970–1300 m

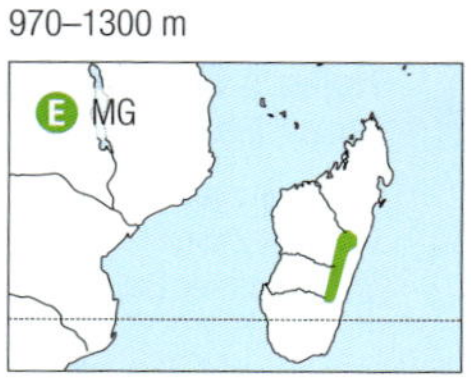

Malagasy Serotine
Laephotis matroka

FR Vespère de Madagascar
DE Madagaskar-Langohrfledermaus
ES Laefotis de Madagascar

Mainly restricted to E highlands

LC
Monotypic
HB 4.6–5.4 cm T 2.8–3.6 cm
FA 2.9–3.5 cm W c. 9 g

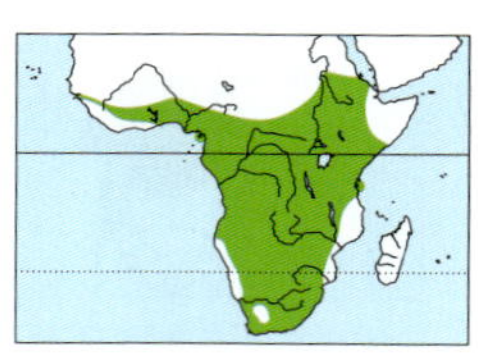

Cape Serotine
Laephotis capensis

FR Vespère du Cap
DE Kap-Langohrfledermaus
ES Laefotis de El Cabo

Sub-Saharan Africa from GW E to SD & ER, S to ZA, also Bioko & Unguja Is

LC
Monotypic
HB 4.4–7.7 cm T 2.5–3.8 cm
FA 2.8–3.8 cm W 6–9 g

Genus *RHYNEPTESICUS*

(1 species)

LC
4 sspp
HB 4–5.6 cm
T 3.8–4.9 cm
FA 3.1–4 cm

0–1110 m

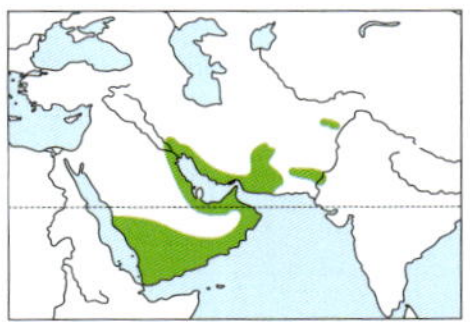

Sind Bat
Rhyneptesicus nasutus

FR Sérotine perse
DE Arabische Breitarmfledermaus
ES Rineptesicus persa

S Arabian Pen, S IQ, S & SE IR, E AF, S PK; perhaps QA & KW

Genus *THAINYCTERIS*

(2 species)

LC Monotypic
HB 6–6.8 cm T 4.3–5.8 cm
FA 4.7–5.2 cm W 13–18 g

200–2000 m

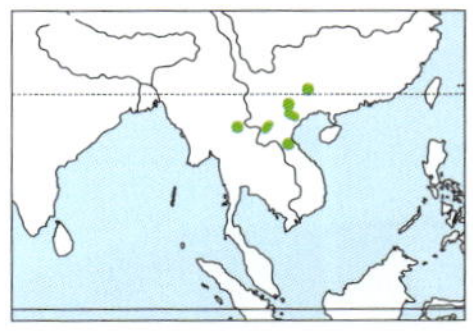

Collared Sprite
Thainycteris aureocollaris

FR Farfadet du Laos
DE Goldkragen-Elfenfledermaus
ES Tainícteris de Laos

NW TH (Chiang Mai Prov), N LA, N & C VN, S CN (Guizhou); perhaps KH

LC Monotypic
HB 4.8–6.5 cm T 3.5–4.8 cm
FA 4.1–5 cm W 12.5–19.5 g

c. 1800 m

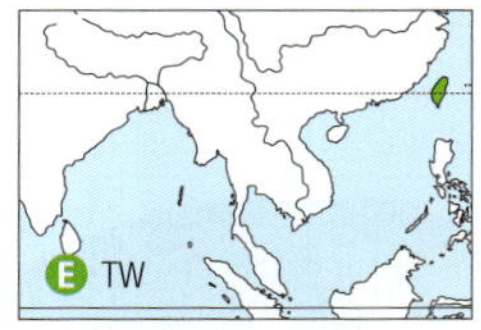

Necklace Sprite
Thainycteris torquatus

FR Farfadet de Taïwan
DE Halsband-Elfenfledermaus
ES Tainícteris de Taiwán

Forests in highlands

LC
Monotypic
HB 4.6–5.8 cm
T 3.6–4.6 cm
FA 3.9–4.4 cm
W 6.2–17.3 g

1300–2100 m

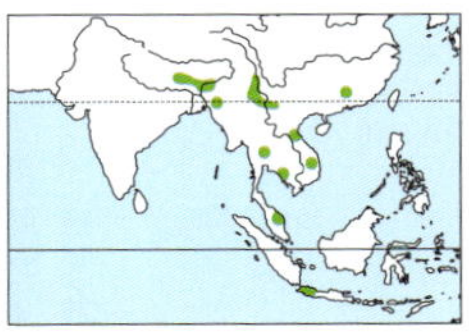

Bronze Sprite
Arielulus circumdatus

FR Farfadet à oreilles bordées
DE Bronzefarbene Elfenfledermaus
ES Arielulus bronceado

Patchily from NP & NE IN E to SC CN (at least Yunnan & Guangdong), S to Pen MY (Fraser's Hill) & Java

Genus *ARIELULUS*

(3 species)

LC Monotypic
HB 4.1–4.4 cm T 3.3–3.7 cm
FA 3.7–4.1 cm W 4.3–8.5 g

c. 265 m

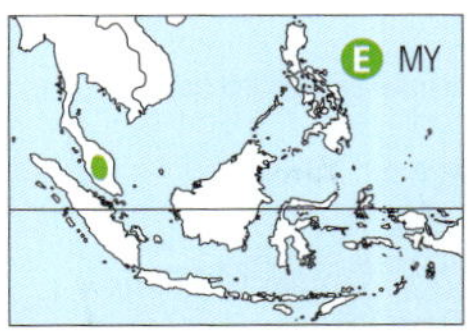

Peninsula (Selangor, Pahang)

Social Sprite
Arielulus societatis

FR Farfadet des sociétés
DE Soziale Elfenfledermaus
ES Arielulus social

VU
Monotypic
HB c. 5.7 cm
T 3.5–4.2 cm
FA 3.4–3.7 cm
W 5.3–6.5 g

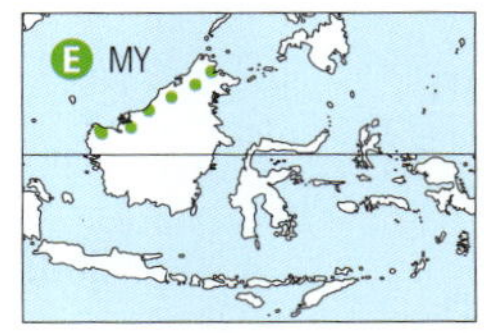

Coppery Sprite
Arielulus cuprosus

FR Farfadet de Bornéo
DE Kupferfarbene Elfenfledermaus
ES Arielulus de Borneo

Few localities in Borneo (Sabah, Sarawak), perhaps Kalimantan

up to 1000 m

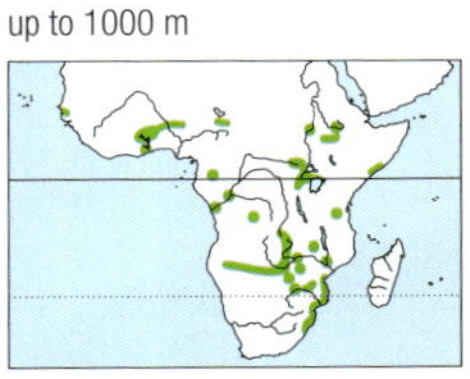

Variegated Butterfly Bat
Glauconycteris variegata

FR Glauconyctère réticulé
DE Genetzte Schmetterlingsfledermaus
ES Glauconícteris reticulado

Patchily from SN E to SC SD, ET & SO, S to NE NA, N BW & NE ZA

LC 2 sspp
HB 4.8–6.2 cm
T 4–5.3 cm
FA 3.8–4.5 cm
W 5–14 g

Genus *GLAUCONYCTERIS*

(13 species)

DD
Monotypic
HB c. 5.7 cm
T c. 4.8 cm
FA c. 4.6 cm

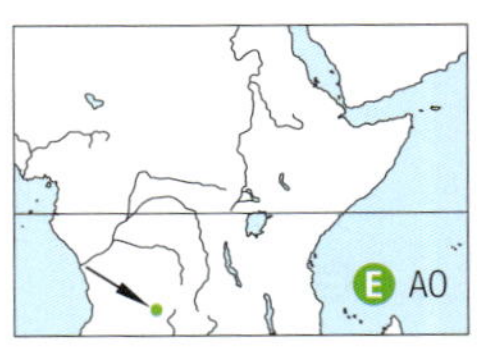

Known only from L Calungo

Machado's Butterfly Bat
Glauconycteris machadoi

FR Glauconyctère de Machado
DE Machado-Schmetterlingsfledermaus
ES Glauconícteris de Machado

DD
Monotypic
HB 5.4–5.6 cm
T 4.4–5 cm
FA 3.8–4.2 cm
W 8.5–15 g

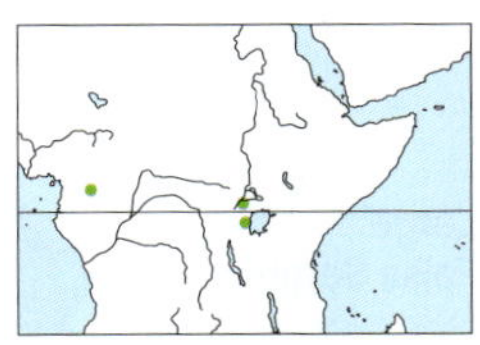

Glen's Butterfly Bat
Glauconycteris gleni

FR Glauconyctère de Glen
DE Glen-Schmetterlingsfledermaus
ES Glauconícteris de Glen

Known only from 3 sites in SE CM, W UG & NW TZ

Pied Butterfly Bat
Glauconycteris superba

FR Glauconyctère pie
DE Schwarzweiße Schmetterlingsfledermaus
ES Glauconícteris pastel

LC
Monotypic
HB 4.9–6.3 cm
T 3.7–4.9 cm
FA 4.3–4.7 cm
W 13.5–19 g

360–720 m

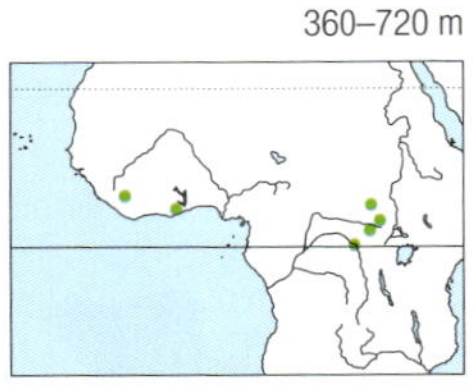

Know from few sites in W CI, S GH, SW SS & NE CD

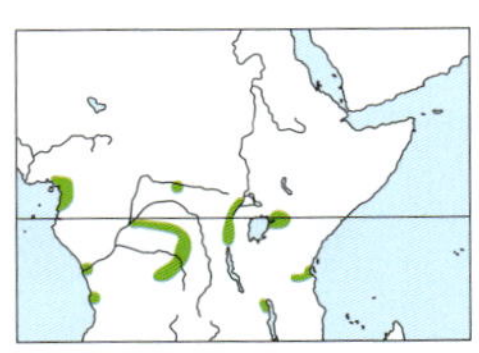

C & E Africa, from CM & GQ (Mbini) to CD, RW, BI, UG, KE, TZ, NW AO & N MW

LC
Monotypic
HB 5.3–6.1 cm
T 4.1–5.3 cm
FA 3.9–4.4 cm
W 6–12 g

Common Butterfly Bat
Glauconycteris argentata

FR Glauconyctère argenté
DE Gewöhnliche Schmetterlingsfledermaus
ES Glauconícteris argentado

c. 1200 m

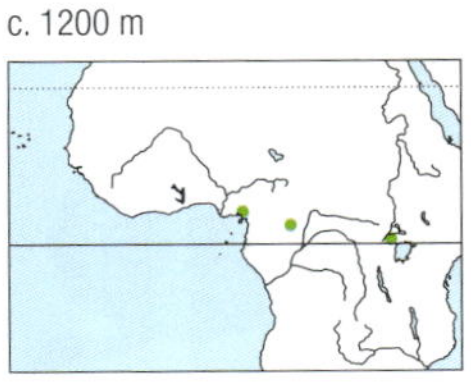

Bibundi Butterfly Bat
Glauconycteris egeria

FR Glauconyctère de Bibundi
DE Bibundi-Schmetterlingsfledermaus
ES Glauconícteris de Bibundi

SW CM (4 sites), SW CF (Dzanga-Sangha SR), W UG (Budongo Forest)

DD
Monotypic
HB 4.6–5 cm T 3.5–4.8 cm
FA 3.7–3.9 cm W 5.5–7.5 g

450–1300 m

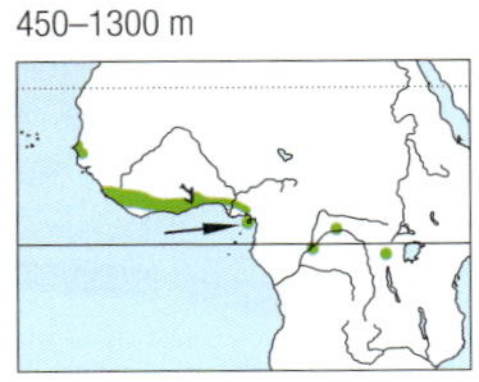

W & C Africa from SN E to CM, Bioko I, isolated in CD; possiby NE TZ

Abo Butterfly Bat
Glauconycteris poensis

FR Glauconyctère d'Abo
DE Abo-Schmetterlingsfledermaus
ES Glauconícteris de Abo

LC
Monotypic
HB 4.3–5.3 cm T 3.4–4.6 cm
FA 3.2–4.1 cm W 15.7–18.9 g

LC
Monotypic
HB 5.1–5.5 cm
T 4.2–4.6 cm
FA 3.8–4.2 cm
W 6–9.5 g

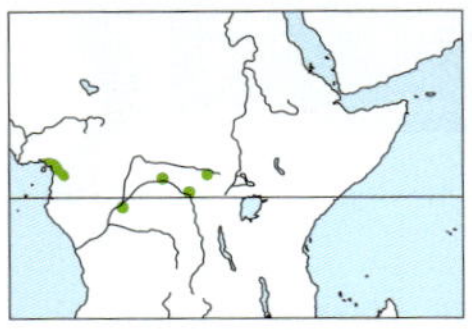

Striped Butterfly Bat
Glauconycteris alboguttata

FR Glauconyctère d'Allen
DE Gestreifte Schmetterlingsfledermaus
ES Glauconícteris de Allen

C Africa in SW CM and W & N CD

LC
Monotypic
HB 3.6–5 cm
T 3.5–5.1 cm
FA 3.5–4.2 cm
W 4–9 g

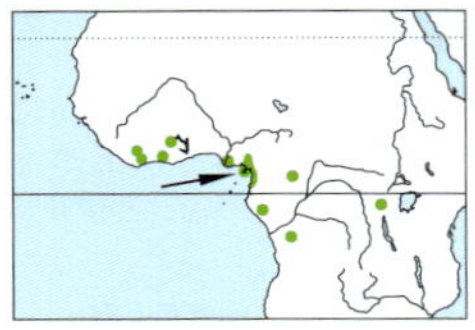

Beatrix's Butterfly Bat
Glauconycteris beatrix

FR Glauconyctère béatrice
DE Kleine Schmetterlingsfledermaus
ES Glauconícteris pequeño

Scattered from CI E to CM, GQ (with Bioko I), S GA, SW CF, EC CD & NW AO

DD
Monotypic
HB 4–4.4 cm
T 3.5–4.1 cm
FA 3.4–3.8 cm
W 4–5 g

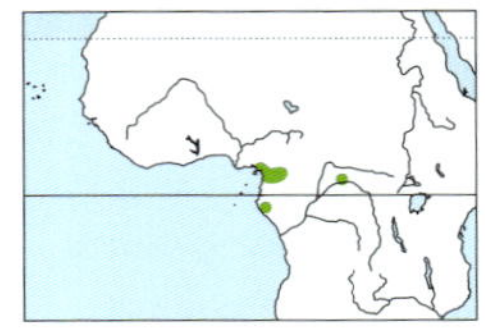

Curry's Butterfly Bat
Glauconycteris curryae

FR Glauconyctère de Curry
DE Curry-Schmetterlingsfledermaus
ES Glauconícteris de Curry

C Africa in S CM, GA & N CD (1 locality)

DD
Monotypic
HB 3.9–4.5 cm
T 3.7–5.9 cm
FA 3.5–4.1 cm
W 4–7 g

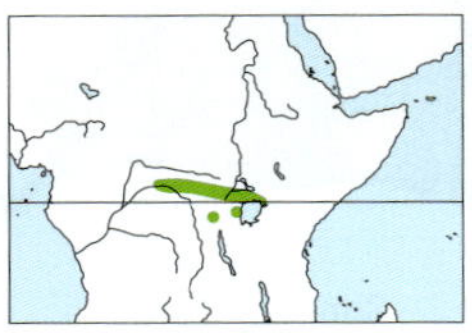

Spotted Butterfly Bat
Glauconycteris humeralis

FR Glauconyctère tacheté
DE Gefleckte Schmetterlingsfledermaus
ES Glauconícteris manchado

C Africa in CD, UG & TZ; perhaps KE

NE
Monotypic
HB 4.6–5.6 cm
T 3.9–4.3 cm
FA 3.4–3.8 cm
W 6.9–8.7 g

400–460 m

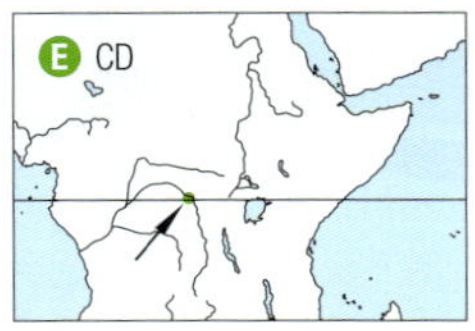

Blackish Butterfly Bat
Glauconycteris atra

FR Glauconyctère sombre
DE Schwärzliche Schmetterlingsfledermaus
ES Glauconícteris negruzco

Known from 3 sites (Yaengo, Yatolema, Yoko) on left bank of R Congo

DD
Monotypic
HB c. 4.9 cm
T c. 4.9 cm
FA c. 4 cm
W c. 7 g

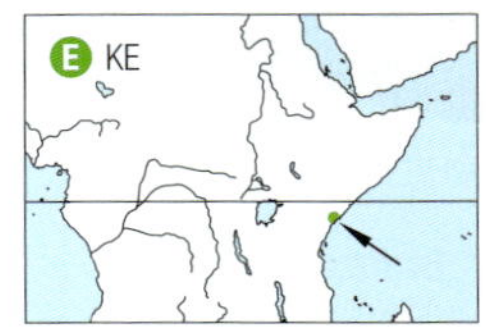

Kenyan Butterfly Bat
Glauconycteris kenyacola

FR Glauconyctère du Kenya
DE Kenia-Schmetterlingsfledermaus
ES Glauconícteris de Kenya

Known only from type locality (near Garsen)

Genus *HESPEROPTENUS*

(5 species)

Doria's False Serotine
Hesperoptenus doriae

FR Sérotine de Doria
DE Doria-Scheinbreitflügel
ES Hesperopteno de Doria

Peninsula and W Borneo (Sarawak)

DD
Monotypic
HB c. 5.1 cm
T c. 4 cm
FA 3.8–4.1 cm
W 8.2–8.3 g

LC
Monotypic
HB 6.1–7.9 cm
T 4.4–6.3 cm
FA 4.9–6 cm
W 15–21 g

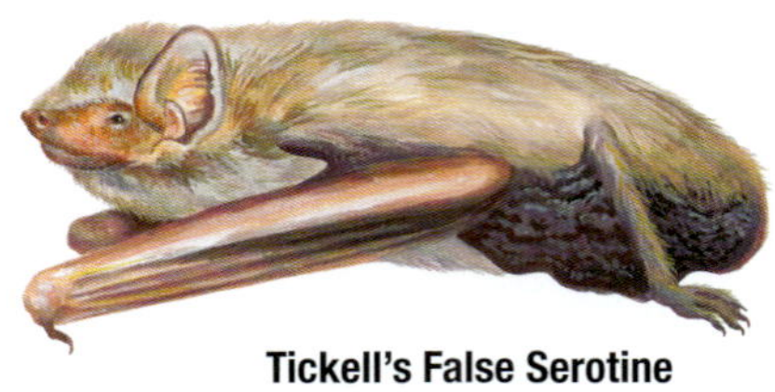

Tickell's False Serotine
Hesperoptenus tickelli

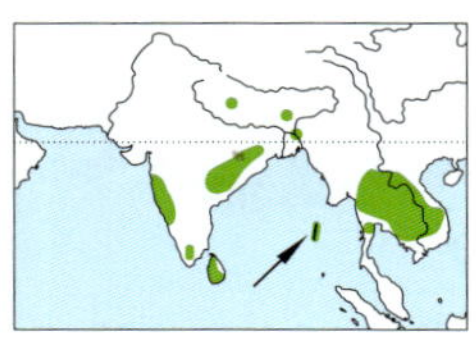

FR Sérotine de Tickell
DE Tickell-Scheinbreitflügel
ES Hesperopteno de Tickell

IN (with Andaman Is), NP, BT, BD, LK, mainland SE Asia; perhaps S CN

Blanford's False Serotine
Hesperoptenus blanfordi

0–1300 m

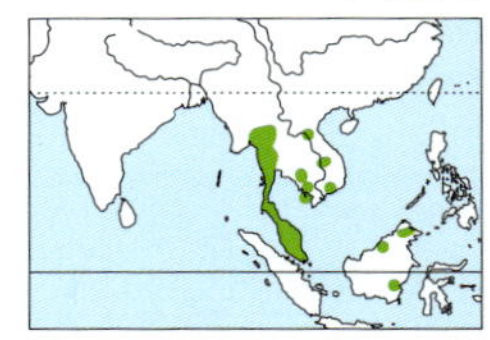

FR Sérotine de Blanford
DE Blanford-Scheinbreitflügel
ES Hesperopteno de Blanford

SE MM (Tenasserim) S to Thai-Malay Pen, scattered in LA, KH, VN & Borneo (MY, ID)

LC
Monotypic
HB 3.7–4.5 cm
T 3.3–3.4 cm
FA 2.4–2.9 cm
W 5.5–8 g

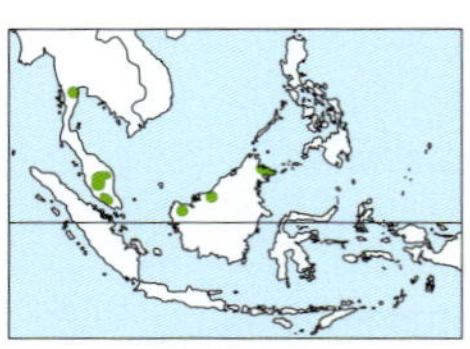

Large False Serotine
Hesperoptenus tomesi

FR Sérotine de Tomes
DE Großer Scheinbreitflügel
ES Hesperopteno de Tomes

VU Monotypic
HB c. 7.4 cm T 4.9–5.3 cm
FA 5–5.3 cm W 30–32 g

S TH (Kaeng Krachan NP), Pen MY, N & W Borneo (Sabah, Sarawak)

c. 10 m

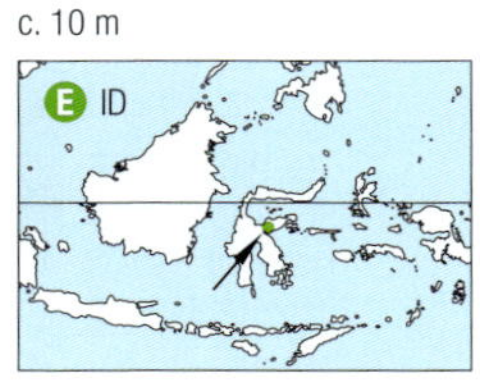

Gaskell's False Serotine
Hesperoptenus gaskelli

FR Sérotine de Gaskell
DE Gaskell-Scheinbreitflügel
ES Hesperopteno de Gaskell

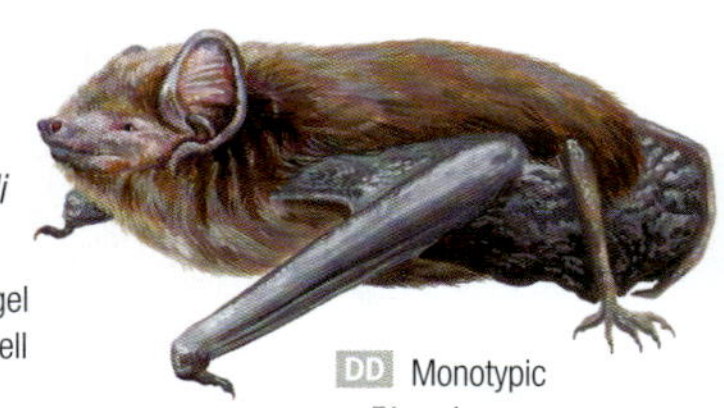

DD Monotypic
FA c. 4 cm

EC Sulawesi (known from type locality in R Ranu)

Genus *IA*

(1 species)

Great Evening Bat
Ia io

200–1700 m

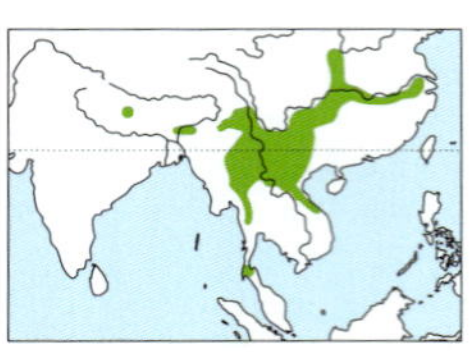

FR Sérotine vespérale
DE Große Abendfledermaus
ES Vespertino mayor

C NP, NE IN (Assam, Meghalaya), N & E MM, S & SE CN, NW & Pen TH, N & C Laos, N & C VN

NT 2 sspp
HB 8.3–10.7 cm T 4.5–8.3 cm
FA 6.6–8.3 cm W 48–54 g

Genus *SCOTOMANES*

(1 species)

Harlequin Bat
Scotomanes ornatus

0–2200 m

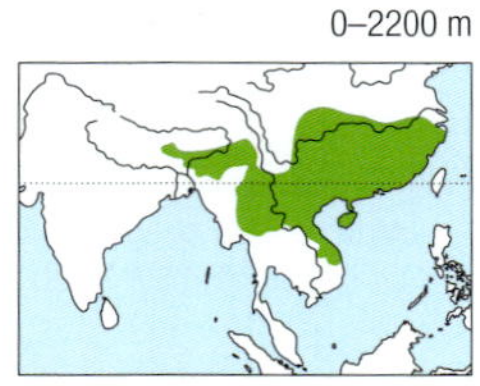

FR Sérotine ornée
DE Harlekin-Fledermaus
ES Escotomano arlequín

E NP, NE IN, N BD, N & E MM, N TH, LA, N & C VN, CN (with Hainan)

LC 2 sspp
HB 6.4–8.5 cm T 5–6.6 cm
FA 5–6.4 cm W 20–30 g

Genus *SCOTEANAX*

(1 species)

Rüppell's Broad-nosed Bat
Scoteanax rueppellii

0–1200 m

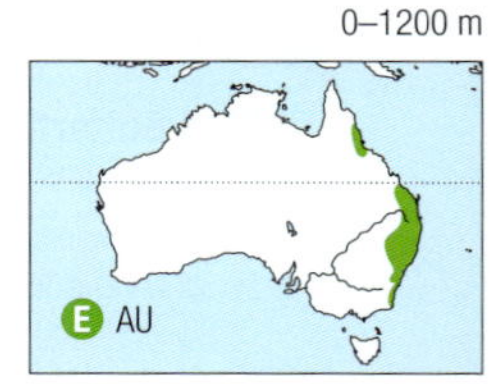

FR Sérotine de Rüppell
DE Rüppell-Breitnasenfledermaus
ES Escoteno de Rüppell

E & SE Qld, E NSW

LC Monotypic
HB 6.3–7.3 cm T 4.5–5.8 cm
FA 5–5.6 cm W 20–40 g

Genus *SCOTOREPENS*

(4 species)

LC Monotypic
HB 3.7–5.7 cm T 2.7–3.9 cm
FA 2.8–3.7 cm W 5.7–9.1 g

Northern Broad-nosed Bat
Scotorepens sanborni

FR Sérotine de Sanborn
DE Sanborn-Breitnasenfledermaus
ES Escotorepo de Sanborn

0–2200 m

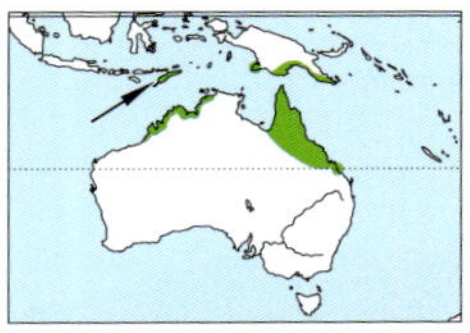

Timor, SE New Guinea (with Dolak I), AU (NE WA, NW NT, NE Qld)

LC Monotypic
HB 3.7–5.3 cm T 2.5–4.8 cm
FA 2.7–3.5 cm W 4–8.5 g

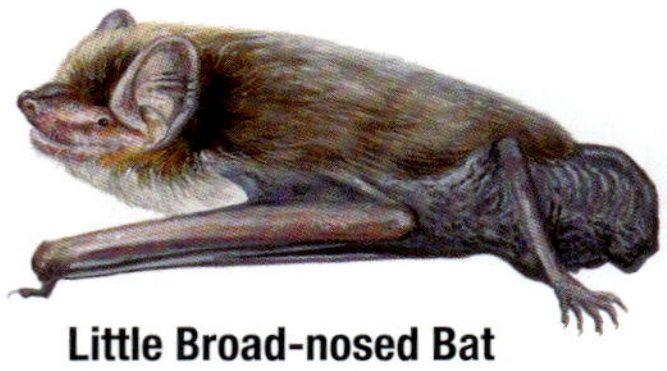

Little Broad-nosed Bat
Scotorepens greyii

FR Sérotine de Grey
DE Grey-Breitnasenfledermaus
ES Escotorepo de Grey

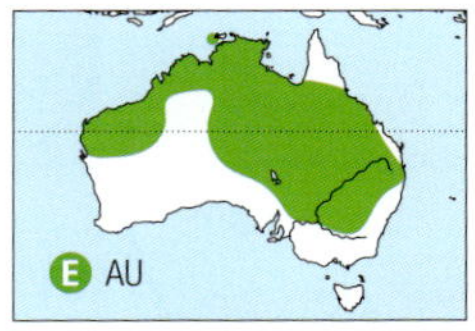

N WA, NT (with Bathurst I), Qld (except Cape York Pen), NSW, N & E SA

LC Monotypic
HB 4.2–6 cm T 2.9–4.2 cm
FA 3.2–4 cm W 6.3–14 g

Inland Broad-nosed Bat
Scotorepens balstoni

FR Sérotine de Balston
DE Balston-Breitnasenfledermaus
ES Escotorepo de Balston

40–480 m

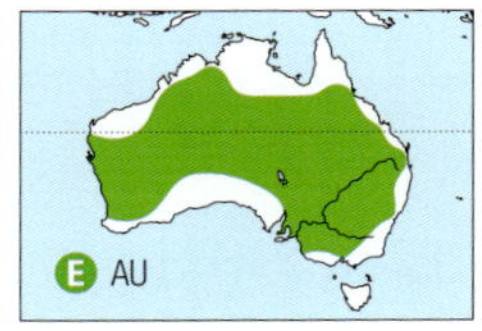

Most of mainland (except N, Cape York Pen and E & S coasts)

LC Monotypic
HB 4.4–5.3 cm T 2.9–3.8 cm
FA 3.2–3.8 cm W 7–14.1 g

Eastern Broad-nosed Bat
Scotorepens orion

FR Sérotine orion
DE Östliche Breitnasenfledermaus
ES Escotorepo de orión

0–720 m

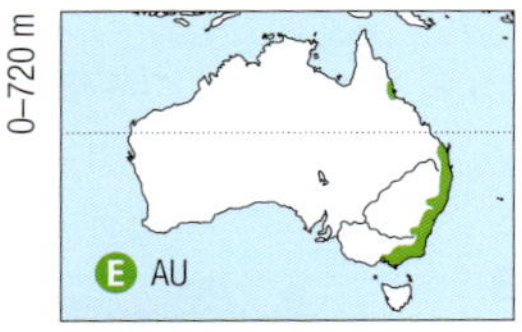

NE & SE Qld, E NSW, E Vic

Genus *LASIONYCTERIS*

(1 species)

LC
Monotypic
HB 5.7–7 cm T 3.5–4.5 cm
FA 3.7–4.4 cm W 8.1–11 g

Silver-haired Bat
Lasionycteris noctivagans

FR Sérotine argentée
DE Silberhaarfledermaus
ES Lasionícteris argentado

150–1965 m

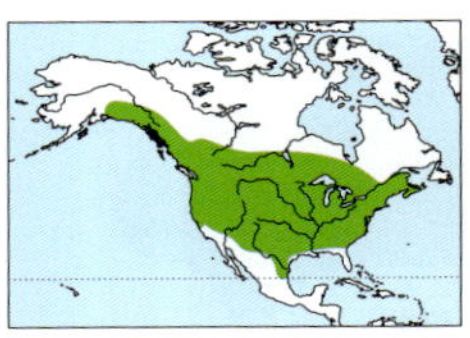

SE Alaska, S CA, widely in US (except on SW & SE coasts) S to NE MX (Sierra de San Carlos)

Genus *HISTIOTUS*

(11 species)

LC
2 sspp
HB 5.1–6.7 cm
T 4–5.5 cm
FA 4.2–5.1 cm
W 9–13 g

Small Big-eared Brown Bat
Histiotus montanus

FR Sérotine des montagnes
DE Kleine Großohrfledermaus
ES Histiotus montano

0–4115 m

PE, CL, AR, UY, S & SE BR; probably W BO

NE
Monotypic
FA c. 4.9 cm

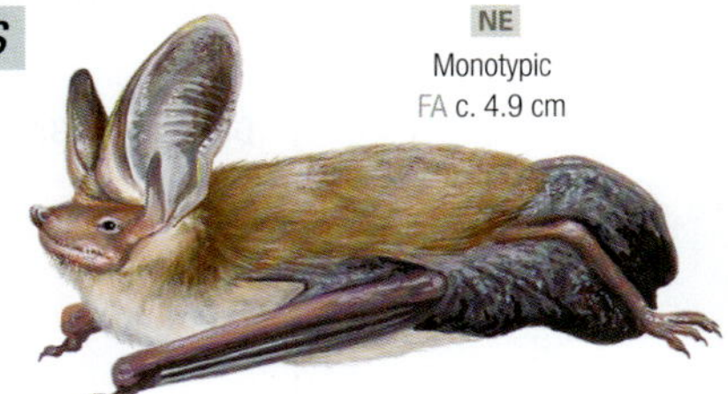

Colombian Big-eared Brown Bat
Histiotus colombiae

FR Sérotine de Colombie
DE Kolumbien-Großohrfledermaus
ES Histiotus de Colombia

2600–3100 m

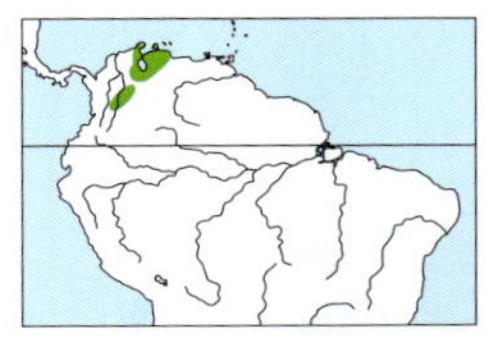

Andes of VE, high plains of E Andes of CO (Boyacá, Cundinamarca)

LC
Monotypic
HB 5.9–6.5 cm
T 4.8–6.1 cm
FA 4.5–4.6 cm
W 14.5–15 g

Southern Big-eared Brown Bat
Histiotus magellanicus

FR Sérotine de Magellan
DE Südliche Großohrfledermaus
ES Histiotus de Magallanes

500–1200 m

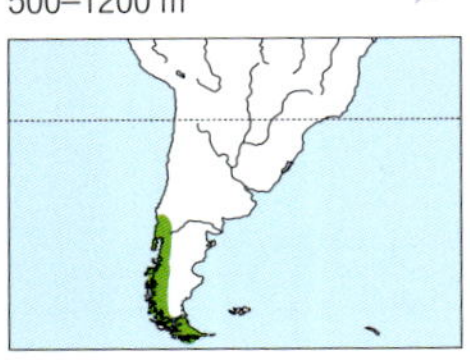

S CL, WC & S AR

NE
Monotypic
HB 5.2–6 cm
T 4.3–5.7 cm
FA 4.2–4.9 cm
W 8–11 g

Transparent-winged Big-eared Brown Bat
Histiotus diaphanopterus

FR Sérotine diaphane
DE Glasflügel-Großohrfledermaus
ES Histiotus de alas transparentes

640–1800 m

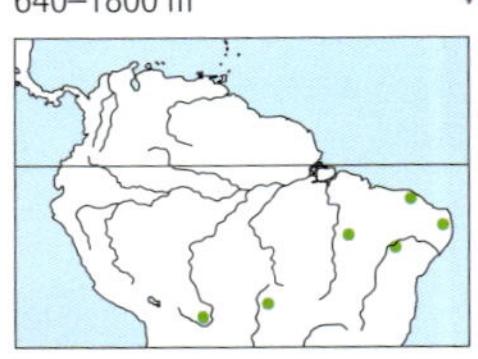

Scattered in NE & CW BR and C BO

DD
Monotypic
HB c. 5.4 cm
T c. 4.5 cm
FA c. 4.5 cm

Strange Big-eared Brown Bat
Histiotus alienus

FR Sérotine de Santa Catarina
DE Merkwürdige Großohrfledermaus
ES Histiotus de Santa Catarina

below 100 m

Santa Catarina (known from type locality in Joinville)

DD
Monotypic
HB 5.4–5.9 cm
T 4.7–5.6 cm
FA 4.4–4.9 cm
W 9–12 g

Humboldt's Big-eared Brown Bat
Histiotus humboldti

FR Sérotine de Humboldt
DE Humboldt-Großohrfledermaus
ES Histiotus de Humboldt

1500–3000 m

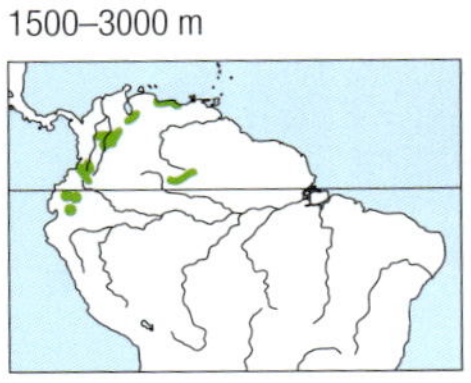

N, W & S VE, NC & SW CO, C EC

NE Monotypic
HB 5.6–5.9 cm T 5.9–6 cm
FA 4.6–4.7 cm W 8–9.5 g

Moche Big-eared Brown Bat
Histiotus mochica

FR Sérotine mochica
DE Moche-Großohrfledermaus
ES Histiotus moche

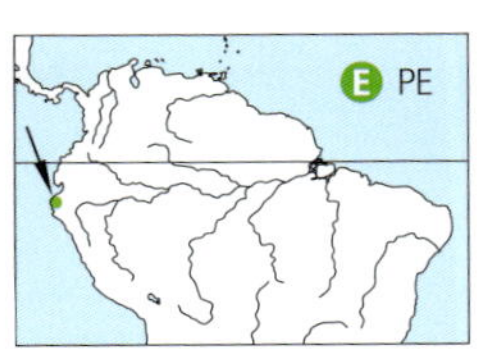

Piura (Quebrada Pariñas Valley)

NE Monotypic
HB 6–6.6 cm T 4.7–5 cm
FA 4.5–5 cm W 10–12 g

Cadena-Garcia's Big-eared Brown Bat
Histiotus cadenai

FR Sérotine de Cadena
DE Cadena-Großohrfledermaus
ES Histiotus de Cadena

2550–4000 m

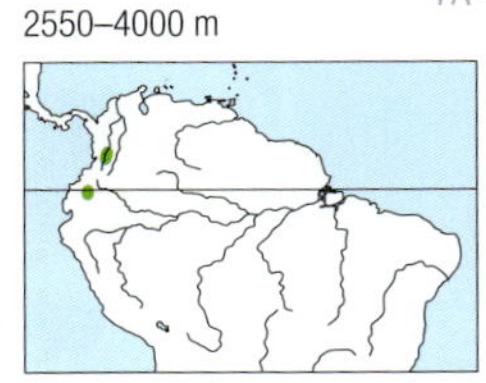

Few localities of C Andes in CO & N Andes in EC

LC
Monotypic
HB 5.7–6.8 cm
T 4.3–5.9 cm
FA 4.4–5 cm
W 9–14 g

350–3685 m

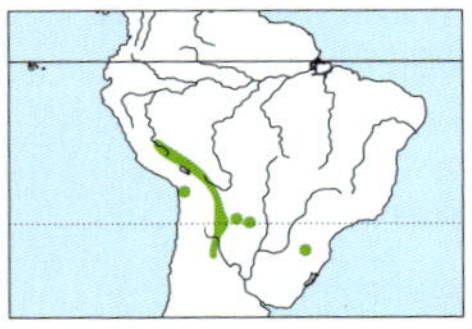

Thomas's Big-eared Brown Bat
Histiotus laephotis

FR Sérotine de Bolivie
DE Thomas-Großohrfledermaus
ES Histiotus de Thomas

S PE, BO, N CL, E slope of Andes in NW AR, W & C PY, S BR (Santa Catarina)

DD
Monotypic
HB 5.8–7.1 cm
T 4.6–6 cm
FA 4.2–5 cm
W 9–17 g

0–2400 m

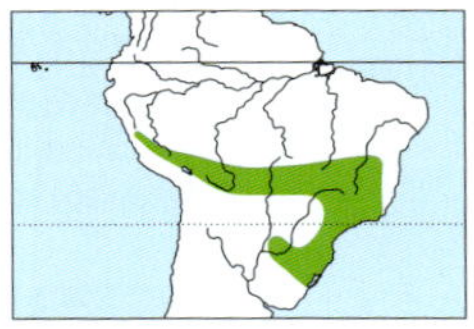

Tropical Big-eared Brown Bat
Histiotus velatus

FR Sérotine voilée
DE Tropische Großohrfledermaus
ES Histiotus tropical

PE, BO, C & S BR, S PY, NE AR

LC Monotypic
HB 5.9–6.5 cm T 4.8–6.2 cm
FA 4.6–5.1 cm W 10.3–20 g

240–3600 m

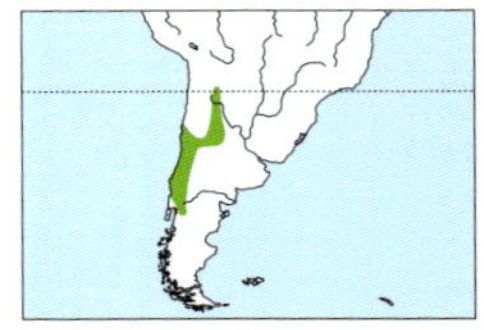

Common Big-eared Brown Bat
Histiotus macrotus

FR Sérotine à grandes oreilles
DE Gewöhnliche Großohrfledermaus
ES Histiotus orejudo

C CL, NW, C & W AR; uncertain in PE

Genus *EPTESICUS*
(26 species)

LC
11 sspp
HB 5.3–8.2 cm
T 3.4–5.6 cm
FA 3.9–5.4 cm
W 11–23 g

300–3465 m

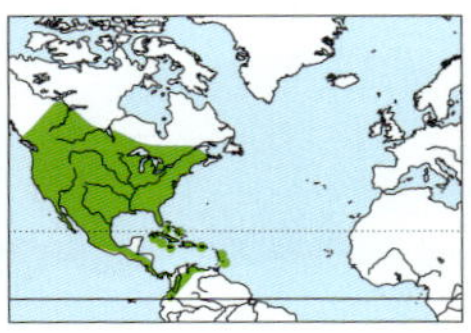

Big Brown Bat
Eptesicus fuscus

FR Sérotine brune
DE Große Braune Breitflügelfledermaus
ES Eptesicus bruno

CA, US, MX, Central America, Andes of W VE, N CO & N EC, BS, KY, Greater & some Lesser Antilles

EN Monotypic
HB 7.3–7.5 cm T 5.4–6 cm
FA 4.9–5.1 cm

0–300 m

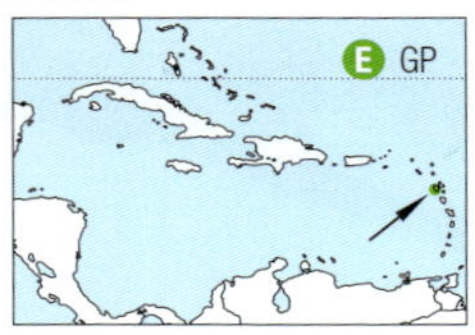

Guadeloupe Serotine
Eptesicus guadeloupensis

FR Sérotine de Guadeloupe
DE Guadeloupe-Breitflügelfledermaus
ES Eptesicus de Guadalupe

Basse-Terre I in Guadeloupe Arch (Lesser Antilles)

NT
Monotypic
HB 4.4–5.5 cm T 3.2–4.1 cm
FA 3.4–4 cm

0–900 m

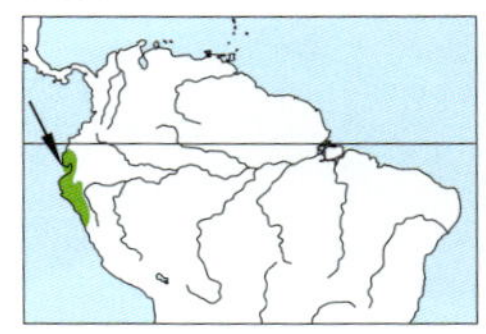

Harmless Serotine
Eptesicus innoxius

FR Sérotine de Gervais
DE Friedliche Breitflügelfledermaus
ES Eptesicus de Gervais

Pacific coast from WC EC (with Puná I) S to NW PE

LC
Monotypic
HB 5.2–6.2 cm T 3.1–3.9 cm
FA 3.7–4.4 cm W 7–10.4 g

100–3600 m

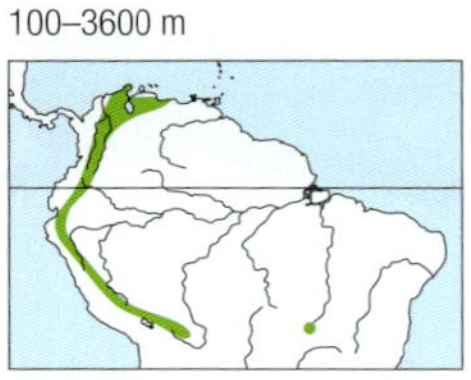

Little Black Serotine
Eptesicus andinus

FR Sérotine des Andes
DE Kleine Schwarze Breitflügelfledermaus
ES Eptesicus de los Andes

N VE S along high Andes of CO, EC, PE & BO, WC BR; recorded in GY

LC
Monotypic
HB 6.2–6.9 cm T 3.5–5.1 cm
FA 4.2–4.9 cm W 9–14 g

25–2840 m

Chiriquinan Serotine
Eptesicus chiriquinus

FR Sérotine du Chiriqui
DE Chiriquí-Breitflügelfledermaus
ES Eptesicus de Chiriquí

From S CR to NW & N Brazilian Amazon, WC BO & NW AR, also E coast of BR (Bahia, Minas Gerais)

LC
4 sspp
HB 5.5–6.2 cm T 3.3–4.9 cm
FA 3.9–4.8 cm W 8–18 g

0–3000 m

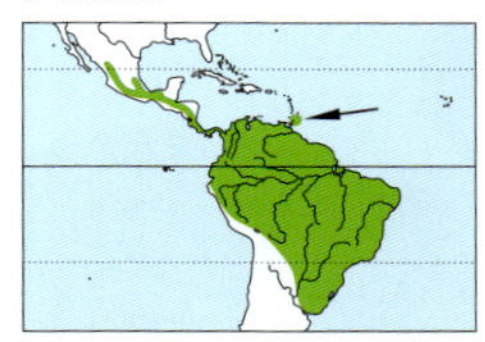

Brazilian Serotine
Eptesicus brasiliensis

FR Sérotine du Brésil
DE Brasilien-Breitflügelfledermaus
ES Eptesicus de Brasil

MX S through Central & South America to E PE, PY, UY & N AR, TT (Tobago); perhaps NE BO

LC 2 sspp
HB c. 5.1 cm T c. 3.7 cm
FA c. 3.6 cm W 4.5–7 g

Diminutive Serotine
Eptesicus diminutus

FR Petite Sérotine
DE Kleine Breitflügelfledermaus
ES Eptesicus pequeño

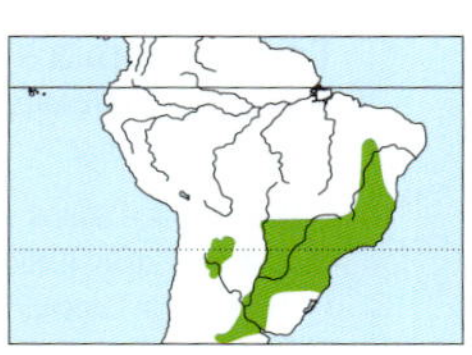

E & S BR, PY, UY, N & C AR

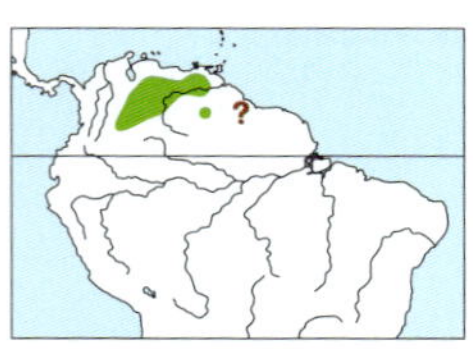

Llanos of Orinoco region in CO & VE; likely in GY

Orinoco Serotine
Eptesicus orinocensis

FR Sérotine de l'Orénoque
DE Orinoco-Breitflügelfledermaus
ES Eptesicus de Orinoco

NE
Monotypic

LC 2 sspp
HB 4.8–6.4 cm T 3–4.9 cm
FA 3.6–4.2 cm W 7–14 g

Argentine Serotine
Eptesicus furinalis

FR Sérotine tropicale
DE Argentinien-Breitflügelfledermaus
ES Eptesicus de Argentina

0–1800 m

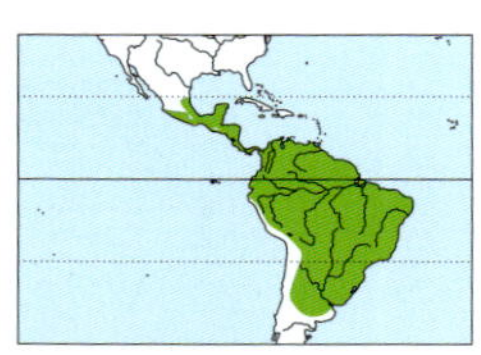

MX S through Central & South America to PE, C & S BO, PY, UY and N & C AR

1800–2020 m

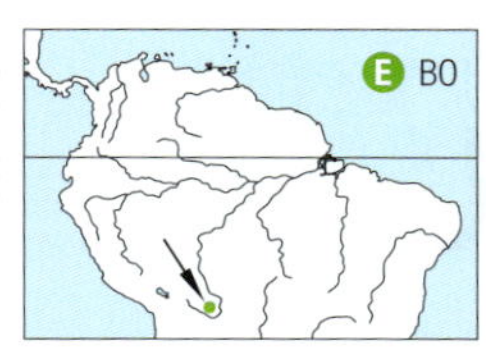

Santa Cruz (known from 3 sites in Florida Prov)

Langer's Serotine
Eptesicus langeri

FR Sérotine de Langer
DE Langer-Breitflügelfledermaus
ES Eptesicus de Langer

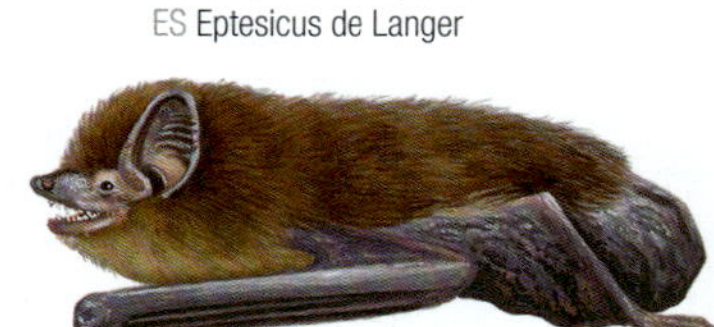

NE Monotypic
HB 5.1–6 cm T 3.9–4.6 cm
FA 4–4.4 cm W 8–12 g

DD
Monotypic
HB 5.6–6.3 cm
T 4.3–5.3 cm
FA 4.4–4.9 cm
W 12–17.5 g

30–1255 m

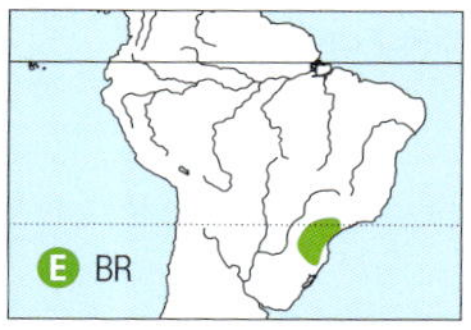

Taddei's Serotine
Eptesicus taddeii

FR Sérotine de Taddei
DE Taddei-Breitflügelfledermaus
ES Eptesicus de Taddei

São Paulo, Paraná, Santa Catarina, Rio Grande do Sul

NE
Monotypic
HB 5.7–6.6 cm
T 3.6–4.9 cm
FA 4.1–4.6 cm
W 9–14 g

495–635 m

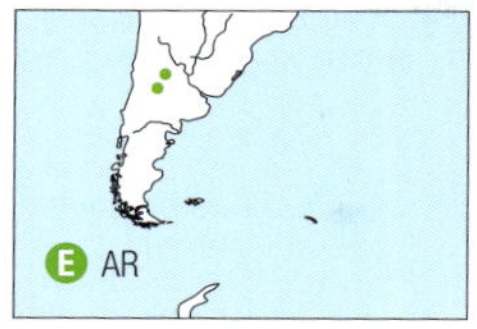

Ulapes Serotine
Eptesicus ulapesensis

FR Sérotine d'Ulapes
DE Ulapes-Breitflügelfledermaus
ES Eptesicus de Ulapes

La Rioja, Mendoza

LC
Monotypic
HB 4.5–4.7 cm
T 3.4–3.6 cm
FA 3.3–3.8 cm

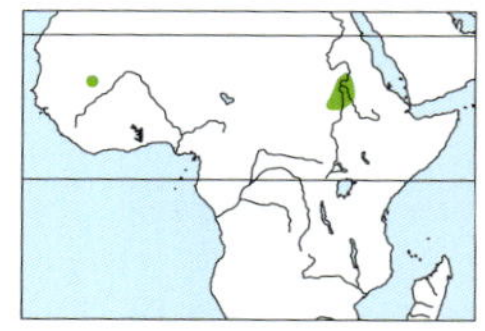

Horn-skinned Serotine
Eptesicus floweri

FR Sérotine de Flower
DE Warzen-Breitflügelfledermaus
ES Eptesicus de Flower

E SD, SE MR (1 record); possibly ML

DD Monotypic
HB 6–7.5 cm T 4–4.9 cm
FA 4.5–5.2 cm

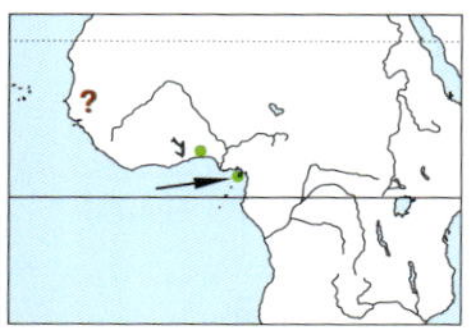

Lagos Serotine
Eptesicus platyops

FR Sérotine de Lagos
DE Lagos-Breitflügelfledermaus
ES Eptesicus de Lagos

SW NG, Bioko I (GQ), SN (unspecified locality)

LC 3 sspp
HB 6.2–7.7 cm T 3.8–5.8 cm
FA 4.5–5.4 cm W 10.8–24 g

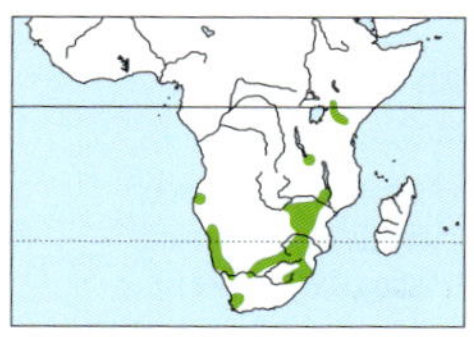

Long-tailed Serotine
Eptesicus hottentotus

FR Sérotine hottentote
DE Hottentotten-Breitflügelfledermaus
ES Eptesicus de cola larga

SW KE, NE ZM, SW AO, W NA, S MW, NW MZ, ZW, W & E ZA, LS; probably SZ

LC Monotypic
HB 6.2–6.5 cm T 4.5–5.2 cm
FA 4.2–4.7 cm

895–3200 m

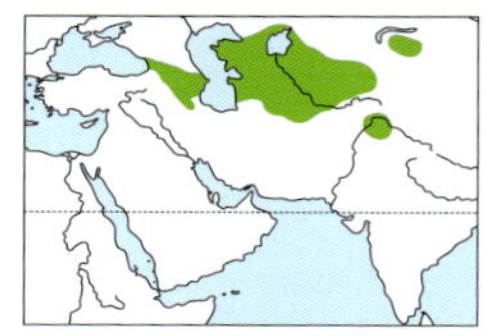

Ognev's Serotine
Eptesicus ognevi

FR Sérotine d'Ognev
DE Ognev-Breitflügelfledermaus
ES Eptesicus de Ognev

Transcaucasia, NW & NE IR, S KZ, UZ, TM, W KG, TJ, NW AF, N PK & NW IN (Jammu and Kashmir)

LC Monotypic
HB 6.6–7.2 cm T 4.9–5.7 cm
FA 4.6–5 cm W c. 14.3 g

0–1850 m

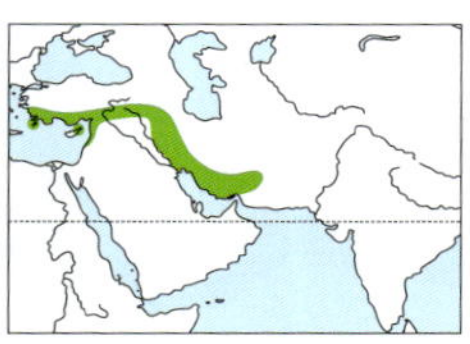

Anatolian Serotine
Eptesicus anatolicus

FR Sérotine d'Anatolie
DE Anatolien-Breitflügelfledermaus
ES Eptesicus de Anatolia

Rhodes I (GR), S TR, NE CY, N SY, LB, N & E IQ, W & S IR

LC
5 sspp
HB 5.6–6.7 cm
T 4.4–5.4 cm
FA 3.8–4.7 cm
W 8–9 g

0–2100 m

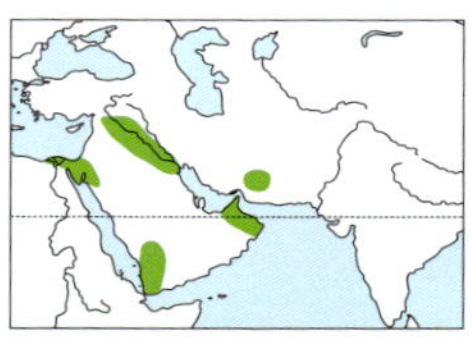

Botta's Serotine
Eptesicus bottae

FR Sérotine de Botta
DE Botta-Breitflügelfledermaus
ES Eptesicus de Botta

NE EG (Nile Delta, Sinai), Levant (S IL, SW JO), Arabian Pen, Mesopotamian region (SY, IQ, KW), SE IR

LC
2 sspp
HB 6.2–7.6 cm
T 4.3–5.6 cm
FA 4.4–5.4 cm
W 15–27 g

0–1800 m

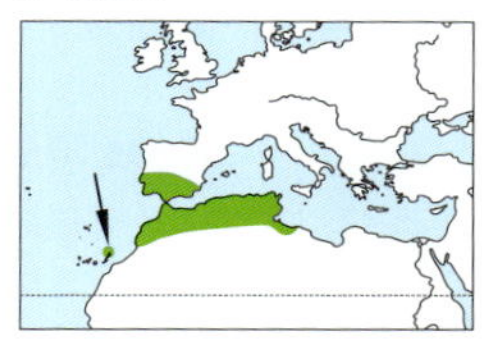

Meridional Serotine
Eptesicus isabellinus

FR Sérotine isabelle
DE Mittelmeer-Breitflügelfledermaus
ES Murciélago hortelano mediterráneo

S PT, S ES, N & C MA, TN, DZ, NW LY, Canary Is (Lanzarote)

460–2340 m

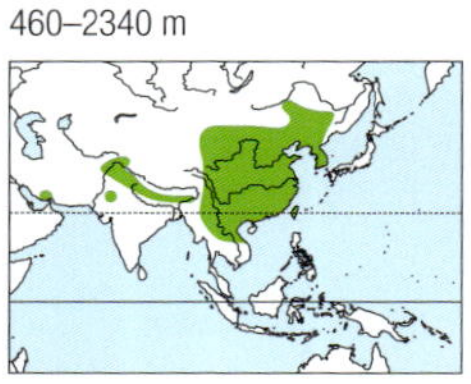

SE IR, NE AF E to Nagaland (NE IN), CN, S MN, Korean Pen, TW, N SE Asia

LC
4 sspp
HB 7.3–8 cm
T 5–6 cm
FA 4.9–5.7 cm
W 15–35 g

Oriental Serotine
Eptesicus pachyomus

FR Sérotine à gros museau DE Orientalische Breitflügelfledermaus
ES Eptesicus de hocico largo

0–1440 m

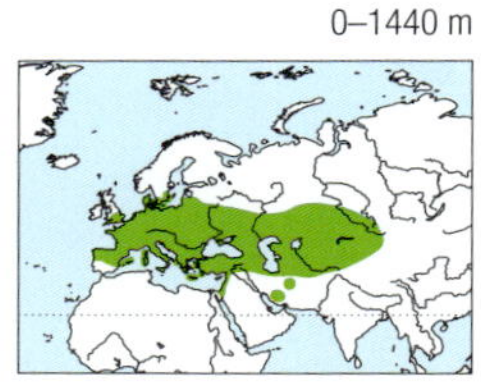

Europe & Mediterranean Is N to S SE, E to C Asia & NW CN (Xinjiang), S to TR, Levant, Caucasus & IR

Eurasian Serotine
Eptesicus serotinus

FR Sérotine commune
DE Eigentliche Breitflügelfledermaus
ES Murciélago hortelano

LC
3 sspp
HB 6.2–8.2 cm
T 3.9–6.5 cm
FA 4.8–5.8 cm
W 18–25 g

LC
Monotypic
HB 5.4–6.8 cm
T 3.5–5 cm
FA 3.7–4.4 cm
W 9–13 g

0–2920 m

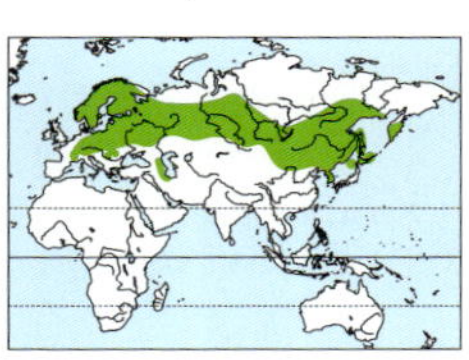

Northern Serotine
Eptesicus nilssonii

FR Sérotine de Nilsson
DE Nordfledermaus
ES Murciélago hortelano norteño

E FR & NO E to N MN, RU (to Kamchatka Pen) & N JP, S to Caucasus, NW IR, N KZ, Korean Pen & N CN

DD
Monotypic
HB c. 4.8 cm
T c. 4.6 cm
FA c. 4.3 cm

Sombre Serotine
Eptesicus tatei

FR Sérotine de Tate
DE Tate-Breitflügelfledermaus
ES Eptesicus de Tate

Known only from type locality in Darjeeling (West Bengal)

LC
2 sspp
HB 4.9–6.7 cm
T 3.8–5.1 cm
FA 3.7–4.4 cm
W 9–15 g

200–3350 m

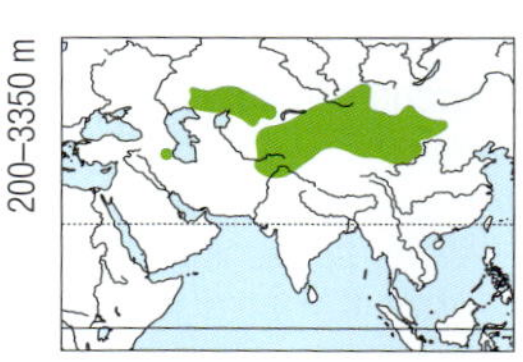

Gobi Serotine
Eptesicus gobiensis

FR Sérotine de Gobi
DE Gobi-Breitflügelfledermaus
ES Eptesicus de Gobi

NW IR, NE AF, N PK (Gilgit), IN (record uncertain), KZ, KG, TJ, S RU (Tuva), W & S MN, NW & N CN

above 700 m

C Honshu; record from Hegurajima I likely a vagrant

Japanese Serotine
Eptesicus japonensis
FR Sérotine du Japon
DE Japan-Breitflügelfledermaus
ES Eptesicus de Japón

VU
Monotypic
HB 5.4–6.8 cm T 3.5–4.3 cm
FA 3.8–4.1 cm

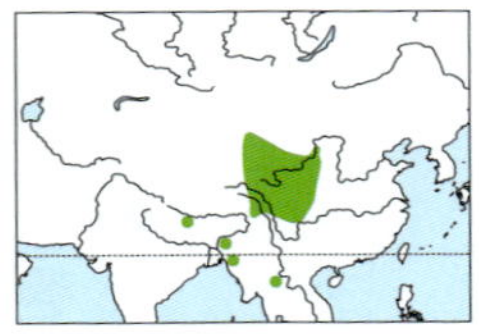

CN (Tibet, Qinghai, Gansu, Ningxia, Sichuan), BD, IN (Meghalaya, Mizoram), NW TH; possibly N MM (mapped)

Thick-eared Serotine
Eptesicus pachyotis
FR Sérotine à oreilles épaisses
DE Dickohr-Breitflügelfledermaus
ES Eptesicus de orejas gruesas

LC
Monotypic
HB 5.5–5.7 cm T 4–4.1 cm
FA 3.8–4.5 cm

Genus *PARASTRELLUS*

(1 species)

LC
Monotypic
HB 3.4–5.3 cm T 2.5–3.6 cm
FA 2.6–3.3 cm W 2–6 g

Canyon Bat
Parastrellus hesperus
FR Parastrelle de l'Ouest
DE Westliche Amerikanische Zwergfledermaus
ES Parastrelo del oeste

up to 2800 m

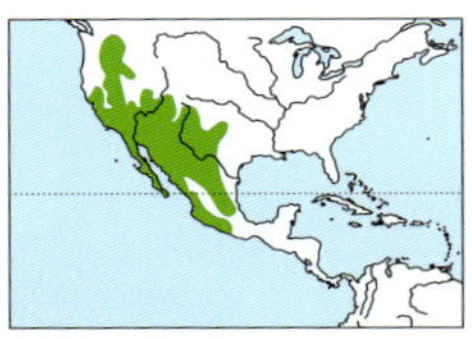

W & SW US (SE Washington S to SW Oklahoma & W Texas), MX (S to Guerrero)

Genus *PERIMYOTIS*

(1 species)

Tricolored Bat
Perimyotis subflavus
FR Blondin de Cuvier
DE Dreifarb-Zwergfledermaus
ES Pipistrela tricolor

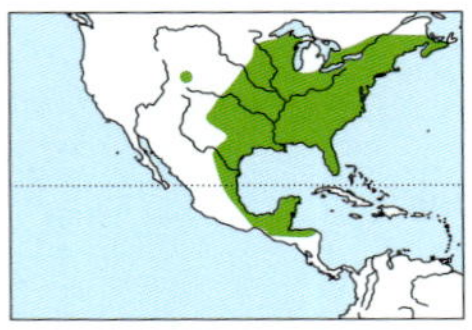

SE CA (S Quebec, New Brunswick, Nova Scotia), E US (isolated record in Colorado), E MX, GT, NE HN

VU
4 sspp
HB 4.3–4.8 cm T 3.4–4.1 cm
FA 3.1–3.4 cm W 5–10 g

Genus *BAUERUS*

(1 species)

NT
Monotypic
HB 5.5–7.8 cm T 4.3–5.7 cm
FA 5–5.7 cm W 18–25 g

Van Gelder's Bat
Bauerus dubiaquercus
FR Antrozon de Van Gelder
DE Van-Gelder-Fledermaus
ES Antrozo de Van Gelder

100–2300 m

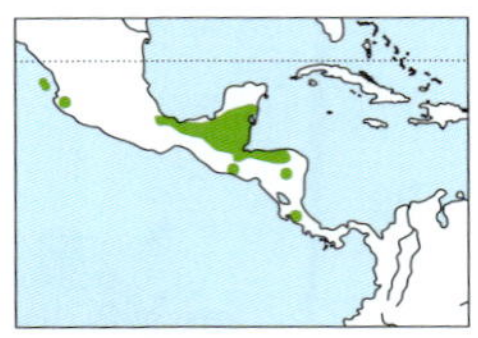

Disjunct in WC (Jalisco, Marías Is) & S MX (S from Veracruz) to GT, BZ & N HN, also SV, N NI & CR

Genus *ANTROZOUS*

(1 species)

Pallid Bat
Antrozous pallidus
FR Antrozon blond
DE Fahlfledermaus
ES Antrozo pálido

up to 2440 m

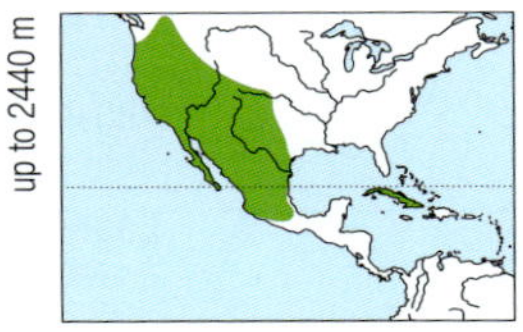

SW CA (British Columbia), W & C US, N & C MX (with Baja California Pen), CU

LC
6 sspp
HB 5.7–8.2 cm T 3.5–5.3 cm
FA 4.5–6 cm W 14–29 g

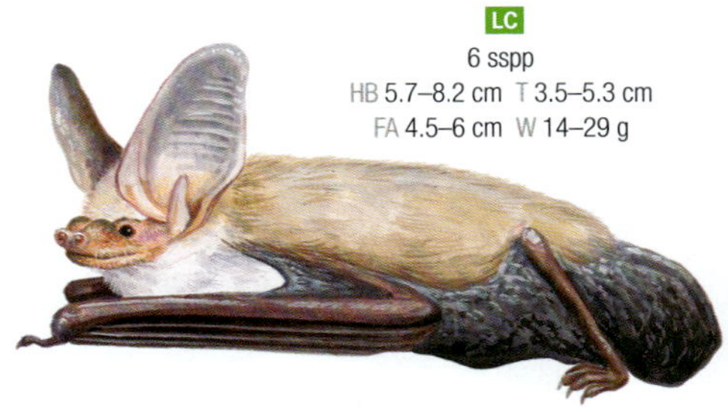

Genus *RHOGEESSA*

(14 species)

LC
Monotypic
HB 3.6–4.4 cm
T 2.7–3.5 cm
FA 2.7–3.2 cm
W 3–5 g

0–1800 m

Black-winged Little Yellow Bat
Rhogeessa tumida
FR Rhogeessa à ailes noires
DE Schwarzflügel-Gelbfledermaus
ES Rogesa de alas negras

Atlantic coast from Tamaulipas S to C Tabasco & N Chiapas

LC
Monotypic
HB 3.7–4.3 cm
T 2.6–3.6 cm
FA 2.6–2.9 cm
W 3–5 g

below 50 m

Yucatan Yellow Bat
Rhogeessa aenea
FR Rhogeessa du Yucatan
DE Yucatán-Gelbfledermaus
ES Rogesa de Yucatán

Yucatán Pen (SE MX, N GT, N BZ)

LC
Monotypic
HB 4.2–4.4 cm
T 2.2–3.6 cm
FA 2.8–3 cm

0–1500 m

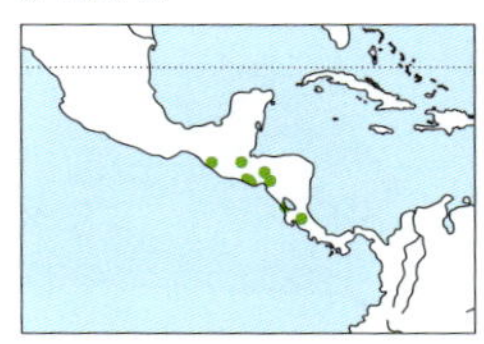

Bickham's Yellow Bat
Rhogeessa bickhami
FR Rhogeessa de Bickham
DE Bickham-Gelbfledermaus
ES Rogesa de Bickham

Pacific coast from S MX (Chiapas) to NW CR (Guanacaste), on Atlantic slope in GT & HN; perhaps PA

DD
Monotypic
HB 4–4.4 cm
T 2.9–3.2 cm
FA 2.9–3 cm

up to 300 m

Menchu's Yellow Bat
Rhogeessa menchuae
FR Rhogeessa de Menchu
DE Menchú-Gelbfledermaus
ES Rogesa de Menchú

Known from 2 localities on Atlantic coast of GT & HN; perhaps N NI

DD
Monotypic
HB 3.9–4.3 cm
T 2.2–3.1 cm
FA 2.7–3.1 cm
W 2.8–4.6 g

0–550 m

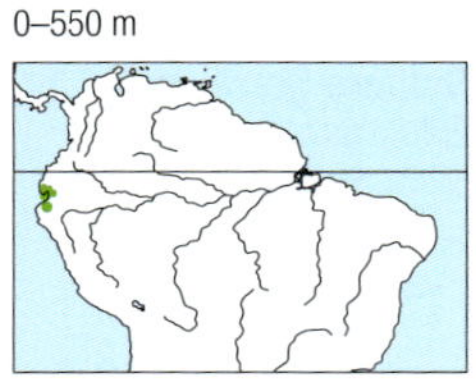

Ecuadorian Little Yellow Bat
Rhogeessa velilla
FR Rhogeessa d'Équateur
DE Ecuador-Gelbfledermaus
ES Rogesa de Ecuador

Scattered localities in W EC (with Puná I) & NW PE

VU
2 sspp
HB 4.2–4.3 cm
T 2.8–3.4 cm
FA 2.5–2.9 cm
W 2.7–4.3 g

0–900 m

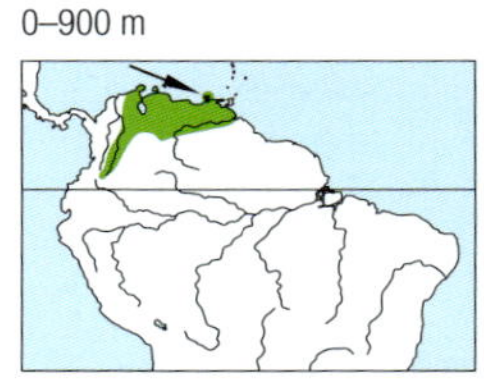

Tiny Yellow Bat
Rhogeessa minutilla
FR Rhogeessa de Miller
DE Winzige Gelbfledermaus
ES Rogesa de Miller

CO (Guajira Pen S to Caquetá & Cauca), NW & N VE (with Margarita I)

EN
Monotypic
HB 3.7–5 cm
T 2.7–3.3 cm
FA 2.8–3 cm
W 3–5 g

20–50 m

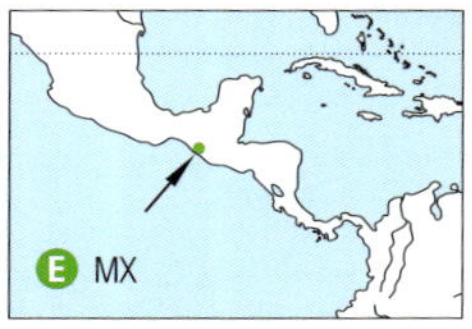

Genoways's Yellow Bat
Rhogeessa genowaysi
FR Rhogeessa de Genoways
DE Genoways-Gelbfledermaus
ES Rogesa de Genoways

S Chiapas (2 nearby localities in Pacific lowlands)

LC
Monotypic
HB 3.7–5 cm
T 2.9–3.3 cm
FA 2.8–3.2 cm
W 3–6 g

0–1500 m

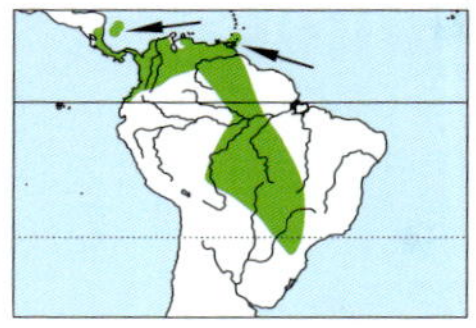

Thomas's Yellow Bat
Rhogeessa io
FR Rhogeessa de Thomas
DE Thomas-Gelbfledermaus
ES Rogesa de Thomas

C & SE NI S to CO (with San Andrés & Providencia Is), NW EC, VE, TT, GY, BR & BO

NE
Monotypic
HB c. 4.7 cm
T c. 3.2 cm

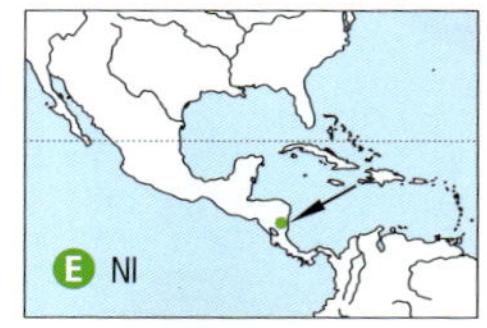

Nicaraguan Little Yellow Bat
Rhogeessa permutandis
FR Rhogeessa du Nicaragua
DE Nicaragua-Gelbfledermaus
ES Rogesa de Nicaragua

Known from type locality near town of Rama

LC
2 sspp
HB 4–4.7 cm
T 2.1–3.2 cm
FA 2.6–3 cm
W 3–8 g

0–1480 m

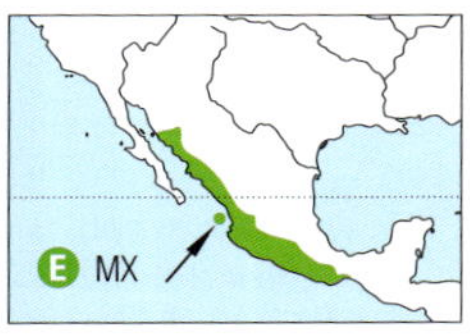

Northern Little Yellow Bat
Rhogeessa parvula
FR Petite Rhogeessa
DE Nördliche Gelbfledermaus
ES Rogesa norteña pequeña

Pacific coast from Sonora S to Oaxaca, with Marías Is (off Nayarit)

VU
Monotypic
HB 3.8–3.9 cm
T 2.6–3.3 cm
FA 2.5–2.8 cm
W c. 3 g

125–340 m

Least Yellow Bat
Rhogeessa mira
FR Rhogeessa menue
DE Kleinste Gelbfledermaus
ES Rogesa menuda

Michoacán (few records between villages of El Infiernillo and Zicuirán)

LC
Monotypic
HB 4.7–4.8 cm
T 3.4–4.2 cm
FA 3.2–3.4 cm
W 5.8–8 g

125–1990 m

Allen's Yellow Bat
Rhogeessa alleni
FR Rhogeessa d'Allen
DE Allen-Gelbfledermaus
ES Rogesa de Allen

N Jalisco to Oaxaca

600–2000 m

Slender Yellow Bat
Rhogeessa gracilis
FR Rhogeessa gracile
DE Schlanke Gelbfledermaus
ES Rogesa grácil

LC Monotypic
HB 4.6–4.8 cm T 3.6–4.3 cm
FA c. 3.3 cm W 3–4 g

From Zacatecas & Jalisco to Oaxaca

285–370 m

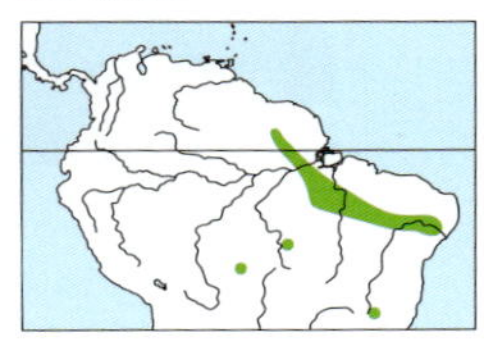

Husson's Yellow Bat
Rhogeessa hussoni
FR Rhogeessa de Husson
DE Husson-Gelbfledermaus
ES Rogesa de Husson

DD
Monotypic
FA c. 3 cm

SR E through N Amazonia to NE BR, scattered records from C & SE BR

Genus *BARBASTELLA*

(6 species)

NT 2 sspp
HB 4.5–6 cm T 3.6–5.2 cm
FA 3.5–4.3 cm W 5–14 g

0–2260 m

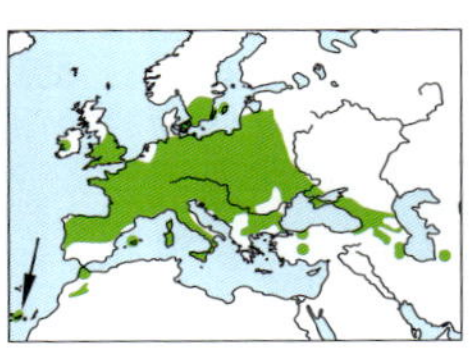

Western Barbastelle
Barbastella barbastellus
FR Barbastelle d'Europe
DE Mopsfledermaus
ES Barbastela

C & S Europe E to LV, BY, UA & Caucasus, major Mediterranean Is, MA, TR (Anatolia), N IR, Canary Is

NE Monotypic
HB 4.5–6.3 cm T 4–5.5 cm
FA 4.1–4.4 cm W 6–18 g

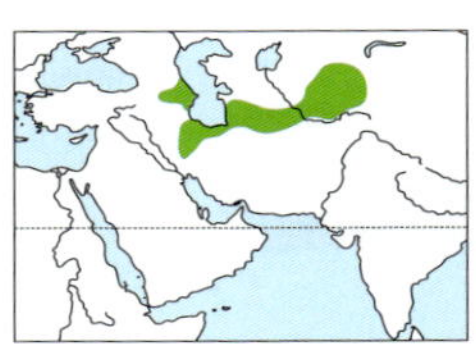

Caspian Barbastelle
Barbastella caspica
FR Barbastelle caspienne
DE Kaspische Mopsfledermaus
ES Barbastela caspiana

E Caucasus & N IR E to W KG; range very poorly delimited

LC
Monotypic
HB c. 5 cm
T 1.9–2 cm
FA 3.9–4.5 cm
W c. 9.5 g

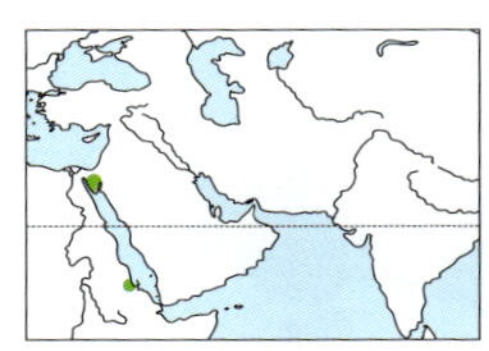

Arabian Barbastelle
Barbastella leucomelas
FR Barbastelle du Sinaï
DE Arabische Mopsfledermaus
ES Barbastela arábica

EG (Sinai), ER

LC Monotypic
HB 4.7–5.1 cm T 4–4.7 cm
FA 3.8–4.2 cm W c. 9 g

0–2500 m

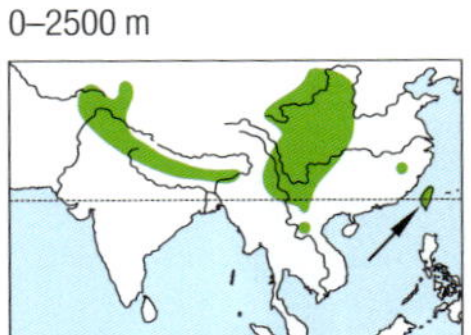

Eastern Barbastelle
Barbastella darjelingensis
FR Barbastelle d'Asie
DE Östliche Mopsfledermaus
ES Barbastela asiática

N AF E through Himalayas to C & E CN, N LA & N VN, TW

NE Monotypic
HB 4.7–6.2 cm T 4.1–5.2 cm
FA 3.9–4.3 cm

Japanese Barbastelle
Barbastella pacifica
FR Barbastelle du Japon
DE Japan-Mopsfledermaus
ES Barbastela de Japón

JP (Hokkaido, scattered on Honshu & Shikoku), Kuril Is (Kunashir)

DD Monotypic
HB 5–5.4 cm T 3.3–4.7 cm
FA 4.1–4.6 cm W 10.5–13.9 g

Beijing Barbastelle
Barbastella beijingensis
FR Barbastelle de Beijing
DE Peking-Mopsfledermaus
ES Barbastela de Pekín

Known only from Beijing Municipality

Genus *PLECOTUS*

(18 species)

Gray Long-eared Bat
Plecotus austriacus

FR Oreillard gris
DE Graues Langohr
ES Orejudo gris

NT
Monotypic
HB 4.1–5.8 cm
T 3.7–5.5 cm
FA 3.6–4.3 cm
W 6–10 g

up to 1390 m

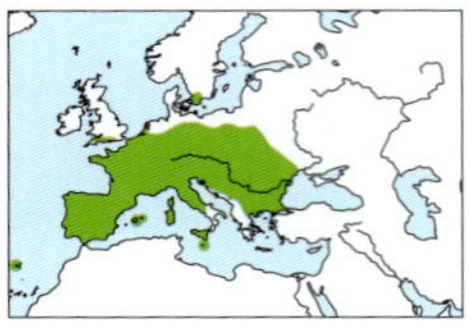

Iberian Pen, S Britain & S SE to W UA & Balkans, major W Mediterranean Is, Madeira I

Canary Long-eared Bat
Plecotus teneriffae

FR Oreillard de Tenerife
DE Teneriffa-Langohr
ES Orejudo canario

VU
Monotypic
HB 4.5–5.5 cm
T 4.4–5.2 cm
FA 4–4.6 cm

up to 2300 m

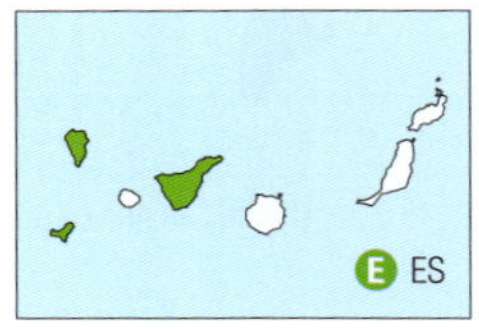

Canary Is (Tenerife, La Palma, El Hierro, probably La Gomera)

Gaisler's Long-eared Bat
Plecotus gaisleri

FR Oreillard de Gaisler
DE Gaisler-Langohr
ES Orejudo de Gaisler

NE
Monotypic
HB 4.5–5.5 cm
T 4.4–5.2 cm
FA 3.7–4.1 cm
W 5–8 g

0–2600 m

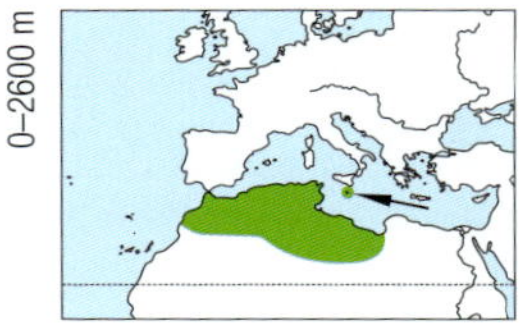

Mediterranean zone from MA E to N LY, tentatively MT

Christie's Long-eared Bat
Plecotus christii

FR Oreillard d'Égypte
DE Christie-Langohr
ES Orejudo de Egipto

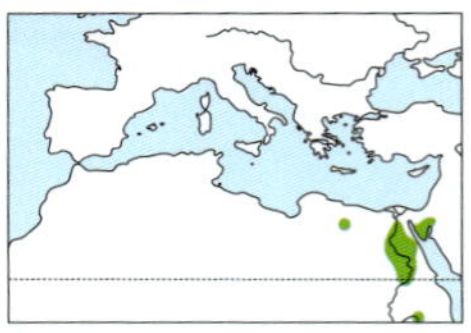

EG, N SD, S IL, SW JO

DD
2 sspp
HB 4.2–5.4 cm
T 4.2–4.8 cm
FA 3.6–4.1 cm
W 6–8 g

Mediterranean Long-eared Bat
Plecotus kolombatovici

FR Oreillard des Balkans
DE Mittelmeer-Langohr
ES Orejudo de los Balcanes

0–800 m

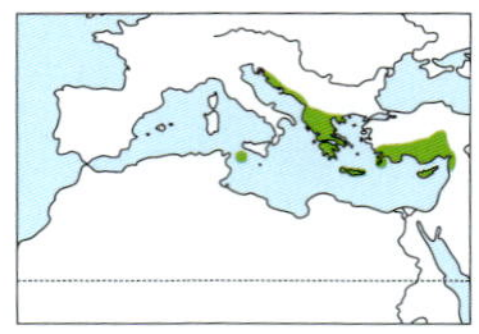

Along Adriatic coast to GR, S TR, NW SY & LB, some E Mediterranean Is; probably IL, PS & JO

LC
Monotypic
HB 4.2–5.4 cm
T 4.2–4.8 cm
FA 3.6–3.9 cm
W c. 7 g

Ethiopian Long-eared Bat
Plecotus balensis

FR Oreillard d'Éthiopie
DE Bale-Langohr
ES Orejudo de Etiopía

2000–3000 m

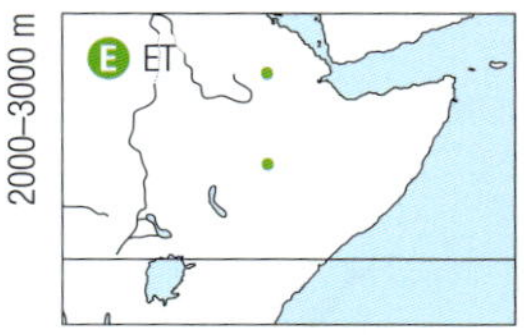

Known with certainty from upper belts of Harenna Forest (Bale Mts NP) & Mt Aune Yosef

DD
Monotypic
HB 4.5–5 cm
T 4.7–5.5 cm
FA 3.6–4.2 cm
W 6.5–8.5 g

Turkmen Long-eared Bat
Plecotus turkmenicus

FR Oreillard turkmène
DE Turkmenistan-Langohr
ES Orejudo de Turkmenistán

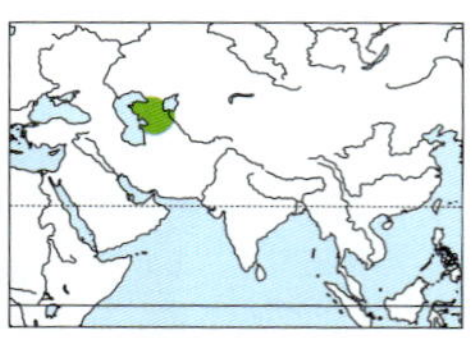

SW KZ, W UZ, W TM; possibly MN

LC
Monotypic
HB c. 5.5 cm
T c. 5 cm
FA c. 4.5 cm
W c. 9 g

Strelkov's Long-eared Bat
Plecotus strelkovi

FR Oreillard de Strelkov
DE Strelkov-Langohr
ES Orejudo de Strelkov

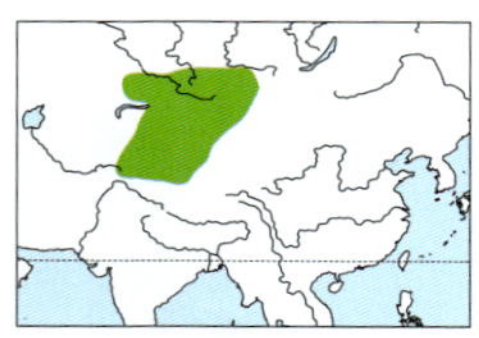

Mts (with Pamir & Tian Shan) in E KZ, NW CN, W MN, E KZ, E TJ & N AF; possibly NE IR

LC
Monotypic
HB 4.9–5.8 cm
T 4.2–5.1 cm
FA 4–4.4 cm
W 6.8–9.3 g

Ognev's Long-eared Bat
Plecotus ognevi

FR Oreillard d'Ognev
DE Ognev-Langohr
ES Orejudo de Ognev

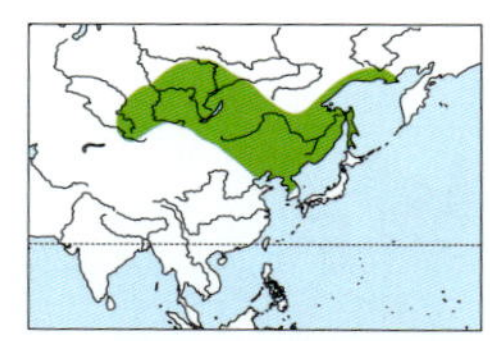

Extreme E KZ & C Asian Mts E through MN & NE CN to Korean Pen & Russian Far East (with Sakhalin I)

LC
Monotypic
HB 4.8–5 cm
T 4.4–4.6 cm
FA 4–4.2 cm
W 7–9 g

Kozlov's Long-eared Bat
Plecotus kozlovi

FR Oreillard de Kozlov
DE Kozlov-Langohr
ES Orejudo de Kozlov

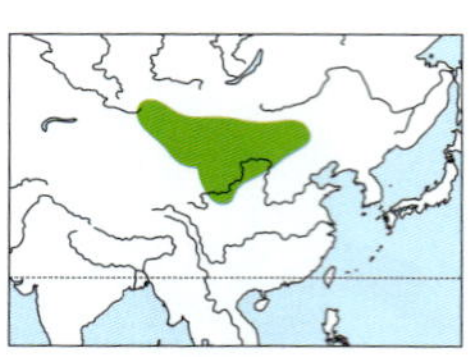

C & S MN, NC CN (Qaidam Basin)

LC
Monotypic
T 4.6–5.2 cm
FA 4.3–4.6 cm

Gobi Long-eared Bat
Plecotus gobiensis

FR Oreillard de Gobi
DE Gobi-Langohr
ES Orejudo de Gobi

Limited to oases in Trans-Altai Gobi

NE
Monotypic
HB 4.4–6.5 cm
T 3.6–6.3 cm
FA 4–6.3 cm
W 6.5–11 g

Himalayan Long-eared Bat
Plecotus homochrous

FR Oreillard de Hodgson
DE Himalaya-Langohr
ES Orejudo de Hodgson

1500–3120 m

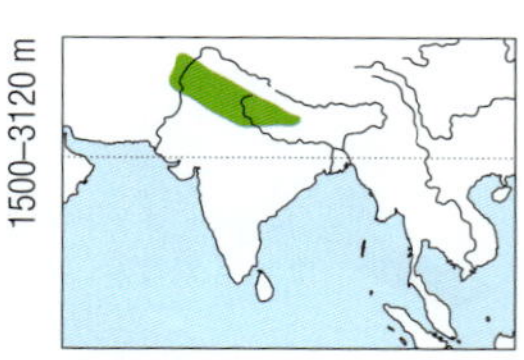

S Himalayas in N PK, NW IN & NP

DD
Monotypic
HB 4–4.5 cm
T 4.8–5 cm
FA 3.6–4.4 cm

Taiwan Long-eared Bat
Plecotus taivanus

FR Oreillard de Taïwan
DE Taiwan-Langohr
ES Orejudo de Taiwán

c. 2250 m

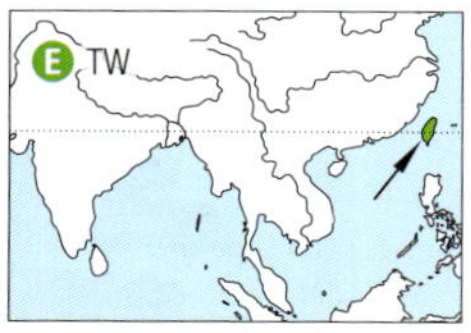

Mountain areas

NT
Monotypic
HB 4.8–5 cm
T 4.7–4.9 cm
FA 3.5–3.8 cm

Japanese Long-eared Bat
Plecotus sacrimontis

FR Oreillard du Japon
DE Japan-Langohr
ES Orejudo de Japón

700–1700 m

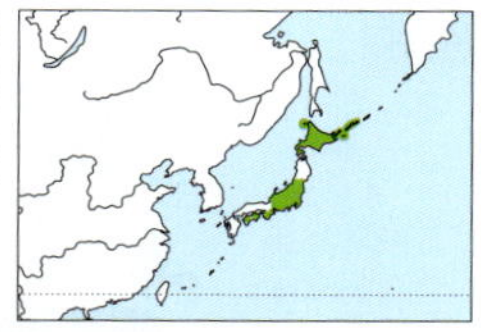

JP (Hokkaido, with Rebun & Rishirito Is, Honshu, Shikoku), S Kuril Is (Iturup, Kunashir, Shikotan)

LC
Monotypic
HB 4.4–5.7 cm
T 4.3–5.4 cm
FA 3.8–4.4 cm
W 7.4–9.2 g

Ward's Long-eared Bat
Plecotus wardi

FR Oreillard de Ward
DE Ward-Langohr
ES Orejudo de Ward

1700–3600 m

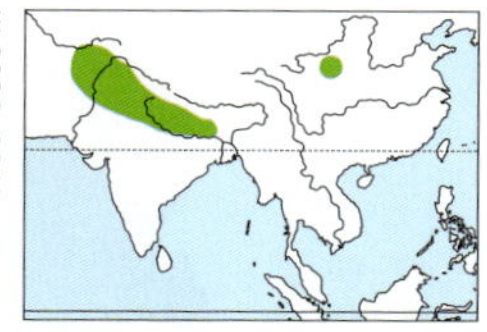

N AF, N PK, N IN, NP, disjunct in CN (W Sichuan)

LC
Monotypic
HB 4.7–5.3 cm
T 4.9–5.4 cm
FA 4.2–4.6 cm

Alpine Long-eared Bat
Plecotus macrobullaris

FR Oreillard alpin
DE Alpen-Langohr
ES Orejudo alpino

0–2800 m

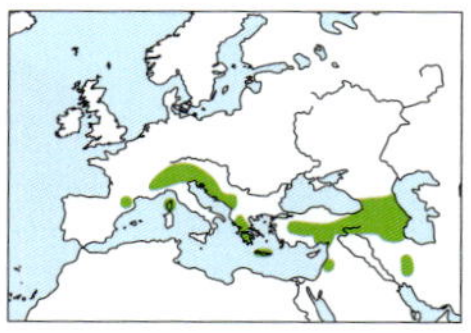

Pyrenees, Alps, Balkans, GR, TR (Anatolia), Caucasus, SE SY, W IR, Corsica & Crete Is

LC
Monotypic
HB 4.6–5.5 cm
T 4.1–5.3 cm
FA 3.7–4.6 cm
W 6–10 g

Sardinian Long-eared Bat
Plecotus sardus

FR Oreillard de Sardaigne
DE Sardinien-Langohr
ES Orejudo de Cerdeña

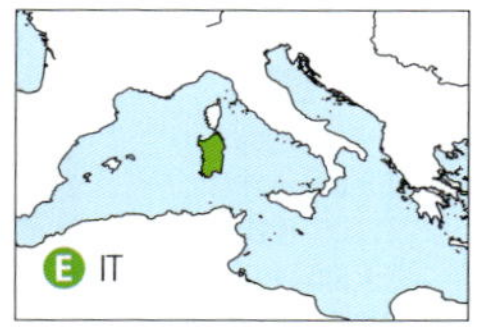

Sardinia

VU
Monotypic
HB c. 4.5 cm
T c. 5.1 cm
FA 4.1–4.2 cm

Brown Long-eared Bat
Plecotus auritus

FR Oreillard roux
DE Braunes Langohr
ES Orejudo dorado

0–2000 m

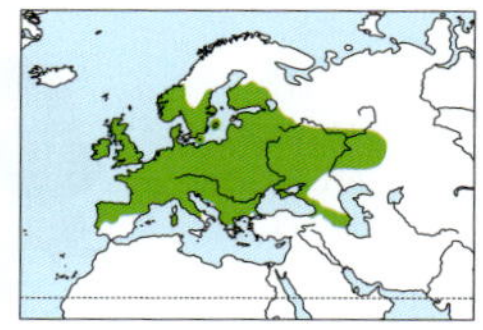

Most of Europe (with British Is, Iberian Pen, Sardinia, S Scandinavia & S FI) E to Urals & Caucasus

LC
2 sspp
HB 3.9–4.6 cm
T 3.9–4.9 cm
FA 3.5–4.3 cm
W 6–9 g

Genus *OTONYCTERIS*

(2 species)

Desert Long-eared Bat
Otonycteris hemprichii

FR Oreillard de Hemprich DE Hemprich-Wüstengroßohr ES Orejudo de Hemprich

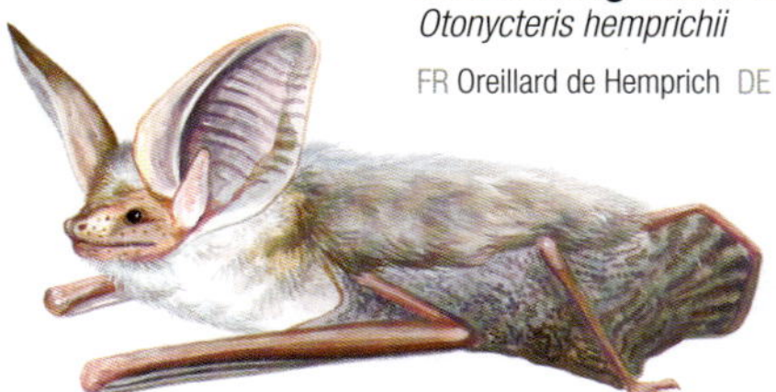

LC
3 sspp
HB 5.7–8.6 cm
T 4–6.4 cm
FA 5–7 cm
W 18–19 g

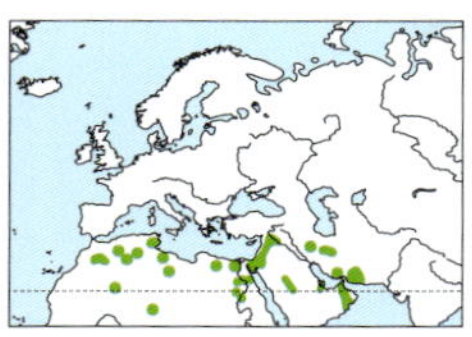

N Africa (MA, DZ, TN, N NE, LY, EG & N SD), SE TR, Levant, C & E Arabia, SE IQ (tentatively), IR

Turkestani Long-eared Bat
Otonycteris leucophaea

FR Oreillard de Severtsov DE Turkestan-Wüstengroßohr ES Orejudo de Severtsov

DD
Monotypic
HB 7.3–7.8 cm
T 4.6–6.5 cm
FA 5.7–6.4 cm

up to 2400 m

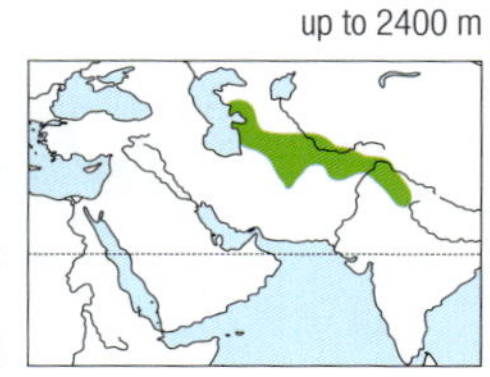

SW KZ, TM, NE IR, SE UZ, KG, TJ, N AF, N PK, NW IN

Genus *CORYNORHINUS*

(3 species)

Rafinesque's Big-eared Bat
Corynorhinus rafinesquii

FR Oreillard de Rafinesque
DE Rafinesque-Langohr
ES Orejudo de Rafinesque

LC
2 sspp
HB 3.8–5.6 cm
T 4.2–5.4 cm
FA 3.8–4.3 cm
W 7.9–13.6 g

0–800 m

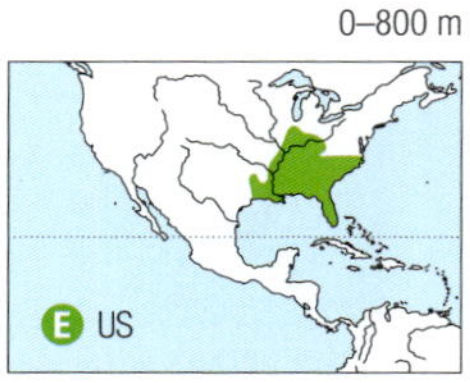

From E & S Illinois, W & S Indiana & S Ohio S to Florida, W to SW & E Arkansas & E Texas

Townsend's Big-eared Bat
Corynorhinus townsendii

FR Oreillard de Townsend
DE Townsend-Langohr
ES Orejudo de Townsend

LC
5 sspp
HB 5.5–5.8 cm
T 3.5–5.4 cm
FA 3.9–4.7 cm
W 5–13 g

450–3300 m

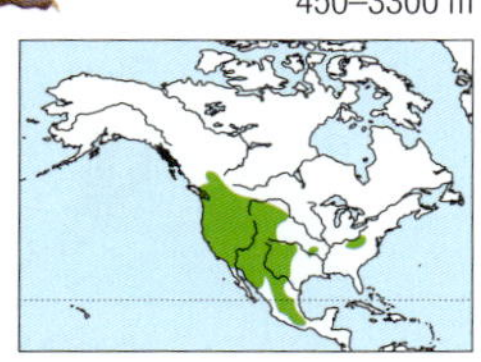

SW CA (British Columbia, with Vancouver I) S across W & C US (and C Appalachian in E) to S MX (Oaxaca)

Mexican Big-eared Bat
Corynorhinus mexicanus

FR Oreillard du Mexique
DE Mexikanisches Langohr
ES Orejudo de México

NT
Monotypic
HB 4.9–5.2 cm
T 4.1–5.1 cm
FA 3.9–4.5 cm
W 5–12 g

1460–3200 m

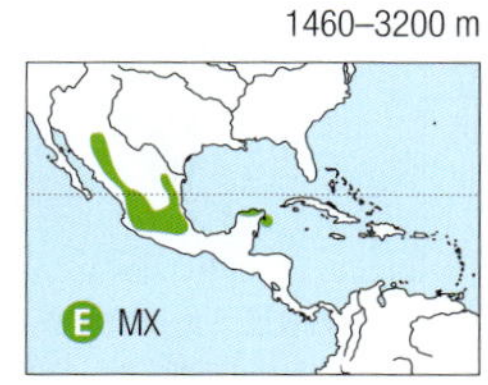

Along Sierra Madre Occidental & Oriental and Sierra Madre del Sur, uncertain in Yucatán Pen & Cozumel I

Genus *IDIONYCTERIS*

(1 species)

Allen's Big-eared Bat
Idionycteris phyllotis

FR Oreillard d'Allen DE Allen-Langohr
ES Orejudo de Allen

405–3225 m

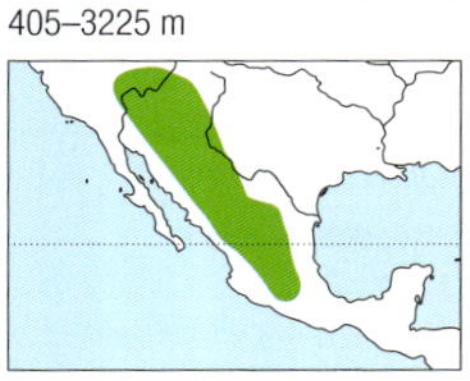

US (extreme S Nevada, S Utah, Arizona, New Mexico) S to SC MX (Michoacán)

LC
2 sspp
HB c. 6.5 cm
T 4–5.3 cm
FA 4.2–4.9 cm
W 8–16 g

Genus *EUDERMA*

(1 species)

Spotted Bat
Euderma maculatum

FR Oreillard maculé DE Gefleckte Fledermaus
ES Orejudo maculado

0–3000 m

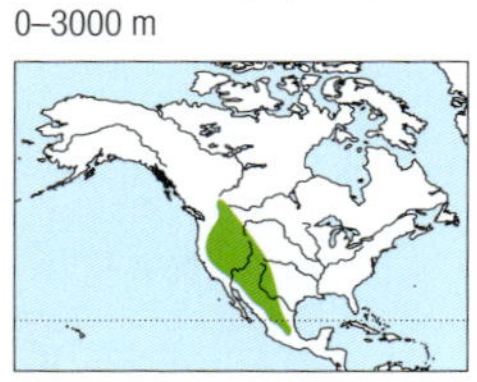

SW CA (British Columbia), NW US (S from Montana & Wyoming) to C MX (Querétaro)

LC
Monotypic
HB 6–6.5 cm
T 4.7–5 cm
FA 4.8–5.1 cm
W 16–20 g

Genus *LASIURUS*

(20 species)

DD
Monotypic
HB 5.7–7 cm
T 4.7–5.7 cm
FA 4.7–4.8 cm
W 11–16 g

25–2900 m

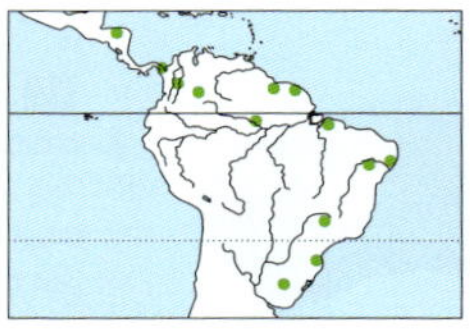

Big Red Bat
Lasiurus egregius
FR Grand Lasiure
DE Brasilien-Haarschwanzfledermaus
ES Lasiuro grande rojo

Scattered in widespread area including HN, PA, CO, SR, GF & BR

NE
4 sspp
HB c. 8 cm
T 3.9–5.9 cm
FA 5–5.7 cm
W 18–22 g

600–3300 m

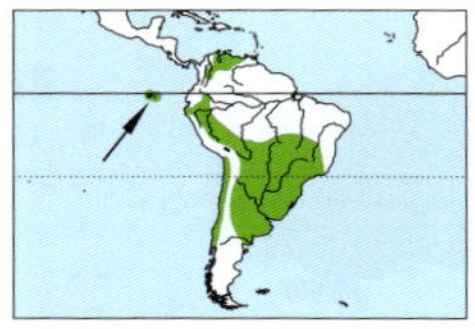

Southern Hoary Bat
Lasiurus villosissimus
FR Lasiure du Paraguay
DE Südamerika-Haarschwanzfledermaus
ES Lasiuro de Paraguay

CO, VE, PE S through BO, PY, S & E BR, UY to C AR, CL, Galápagos Is

LC
Monotypic
HB 8–8.4 cm
T 4.7–6.1 cm
FA 5–5.7 cm
W 20–35 g

500–1900 m

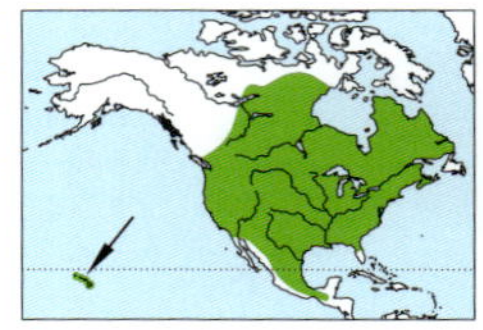

Northern Hoary Bat
Lasiurus cinereus
FR Lasiure cendré
DE Eisgraue Haarschwanzfledermaus
ES Lasiuro cenizo

C & E CA S throughout US & MX to W GT, Hawaii Is

NE
Monotypic
HB c. 5.1 cm
T c. 4.6 cm
FA c. 4 cm
W 16.5–18 g

0–4400 m

Hawaiian Hoary Bat
Lasiurus semotus
FR Lasiure d'Hawaï
DE Hawaii-Haarschwanzfledermaus
ES Lasiuro de Hawai

Hawaii Is

LC
Monotypic
HB 5.5–5.9 cm
T 4.5–6.2 cm
FA 3.6–4.6 cm
W 7–16 g

400–1600 m

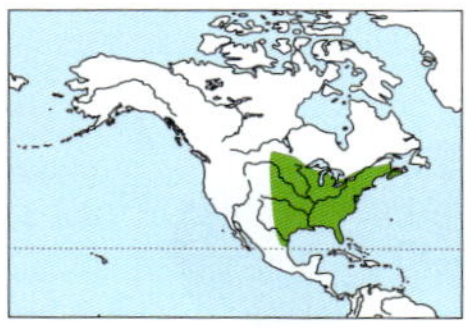

Eastern Red Bat
Lasiurus borealis
FR Lasiure boréal
DE Rote Haarschwanzfledermaus
ES Lasiuro boreal

SC & SE CA S through C & E US to NE MX

NT
Monotypic
HB 5.4–5.5 cm
T 4.8–4.9 cm
FA 4–5 cm
W 12–18 g

50–1000 m

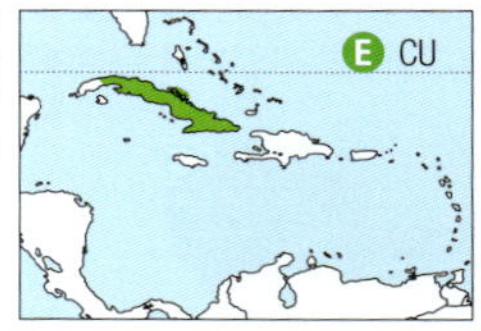

Pfeiffer's Red Bat
Lasiurus pfeifferi
FR Lasiure de Pfeiffer
DE Pfeiffer-Haarschwanzfledermaus
ES Lasiuro de Pfeiffer

Mountainous woodlands and woodland patches

LC
Monotypic
HB 4.7–7.3 cm
T 3.5–5 cm
FA 3.9–4.7 cm
W 7–13.8 g

below 500 m

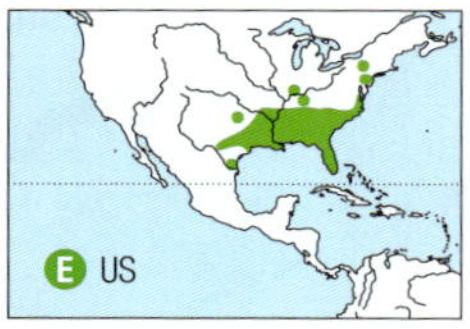

Seminole Bat
Lasiurus seminolus
FR Lasiure séminole
DE Südstaaten-Haarschwanzfledermaus
ES Lasiuro seminola

From SE Oklahoma & SE Virginia S to C & E Texas & Florida, isolated records from New York S to S Texas

LC
Monotypic
HB c. 5.9 cm
T 5.3–5.7 cm
FA 4.5–4.8 cm

100–280 m

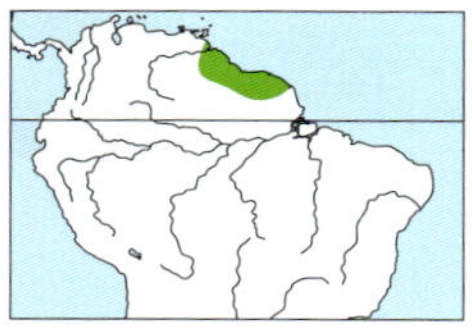

Greater Red Bat
Lasiurus atratus
FR Lasiure deuil
DE Große Haarschwanzfledermaus
ES Lasiuro gigante

E VE, Guianas

NE
Monotypic
HB 5–6 cm
T 5.6–6.4 cm
FA 4.6–4.8 cm
W 12–13 g

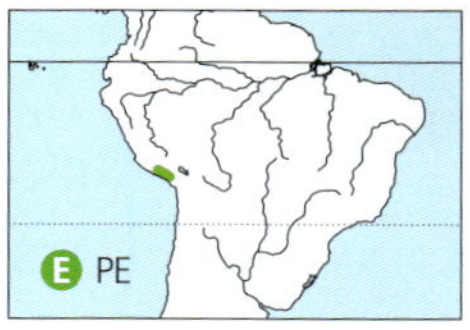

Arequipa Red Bat
Lasiurus arequipae
FR Lasiure d'Arequipa
DE Arequipa-Haarschwanzfledermaus
ES Lasiuro de Arequipa

Arequipa (known from 3 localities)

LC
3 sspp
HB 4.4–6.2 cm
T 3.8–5.8 cm
FA 3.6–4.8 cm
W 6–13.5 g

200–2400 m

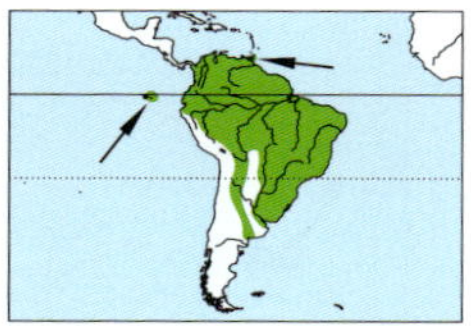

Southern Red Bat
Lasiurus blossevillii
FR Lasiure de Blosseville
DE Blosseville-Haarschwanzfledermaus
ES Lasiuro de Blosseville

From CO & VE S to SE BR, UY & CE AR, TT (Trinidad), Galápagos Is (Santa Cruz, San Cristóbal, Floreana)

NE
2 sspp
HB 4.9–6.5 cm
T 4.5–5.7 cm
FA 3.7–4.3 cm
W 7–12 g

30–3200 m

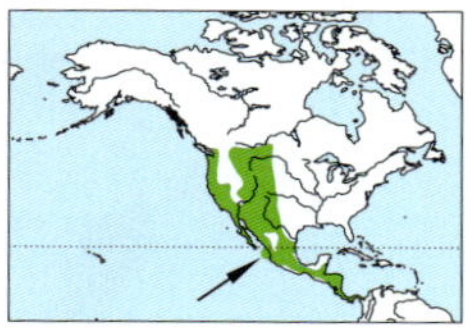

Desert Red Bat
Lasiurus frantzii
FR Lasiure de Frantzius
DE Wüsten-Haarschwanzfledermaus
ES Lasiuro de Frantzius

SW CA, W US, MX (with Marías Is, off Nayarit) S to NI, CR & PA

LC
Monotypic
HB 5.5–6.1 cm
T 4.4–5.8 cm
FA 3.6–4.2 cm
W 7–13 g

up to 1000 m

Cinnamon Red Bat
Lasiurus varius
FR Lasiure du Chili
DE Zimtrote Haarschwanzfledermaus
ES Lasiuro de Chile

C CL & WC AR S to Tierra del Fuego

VU
Monotypic
HB 4.9–7 cm
T 3–4.6 cm
FA 3.7–4.3 cm
W 6–11 g

100–450 m

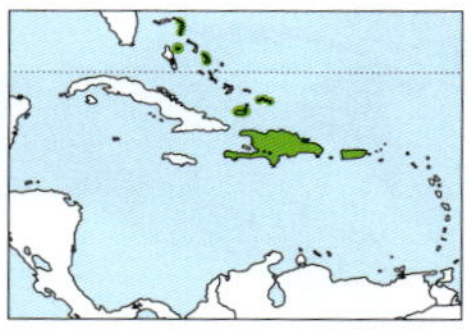

Minor Red Bat
Lasiurus minor
FR Lasiure d'Haïti
DE Kleine Haarschwanzfledermaus
ES Lasiuro de Haití

BS, TC, Hispaniola (HT, DO), PR

DD
Monotypic
HB 5.3–5.6 cm
T 4.7–5 cm
FA 4.1–4.5 cm
W 11.5–17 g

0–1500 m

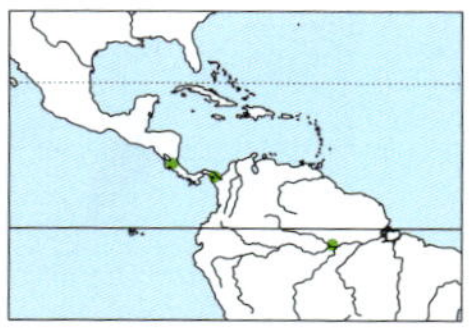

Tacarcunan Bat
Lasiurus castaneus
FR Lasiure de Tacarcuna
DE Panama-Haarschwanzfledermaus
ES Lasiuro de Tacarcuna

Scattered in NW CR, E PA & N BR (Amazonas)

DD
Monotypic
HB 5.7–6.2 cm
T 5–5.8 cm
FA c. 4.6 cm
W 12.5–14 g

40–630 m

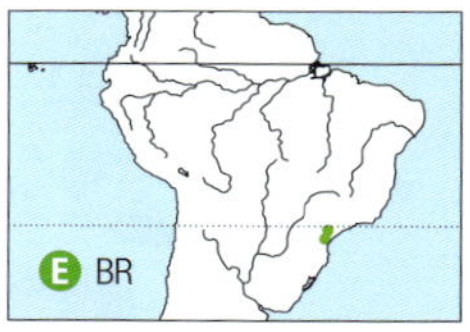

Hairy-tailed Bat
Lasiurus ebenus
FR Lasiure du Brésil
DE São-Paulo-Haarschwanzfledermaus
ES Lasiuro de Brasil

São Paulo (known from 2 localities in SE Atlantic Forest)

VU
Monotypic
HB 5.3–5.7 cm
T 5.2–5.3 cm
FA 4.1–4.7 cm

below 400 m

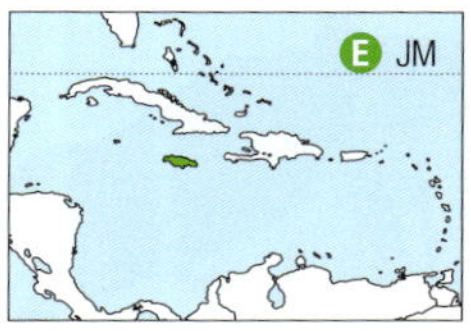

Jamaican Red Bat
Lasiurus degelidus

FR Lasiure de Jamaïque
DE Jamaika-Haarschwanzfledermaus
ES Laisuro de Jamaica

Open areas with low trees, coconuts or disturbed vegetation

LC
Monotypic
HB 6.1–7.6 cm
T 4–5 cm
FA 4.2–4.8 cm
W 12–19 g

25–2400 m

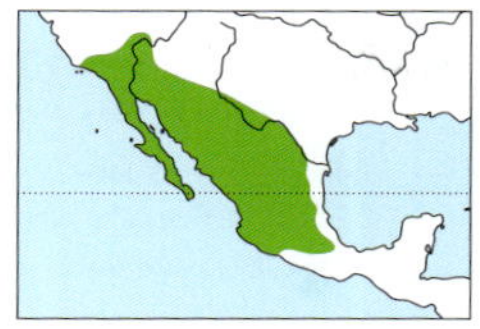

Western Yellow Bat
Lasiurus xanthinus

FR Lasiure de Californie
DE Westliche Haarschwanzfledermaus
ES Lasiuro de California

SW US (S California, S Nevada, Arizona, SW New Mexico, SW Texas) S to W & C MX

LC
2 sspp
HB c. 7.6 cm
T 5.1–7 cm
FA 4.7–5.4 cm
W 14–20 g

0–3000 m

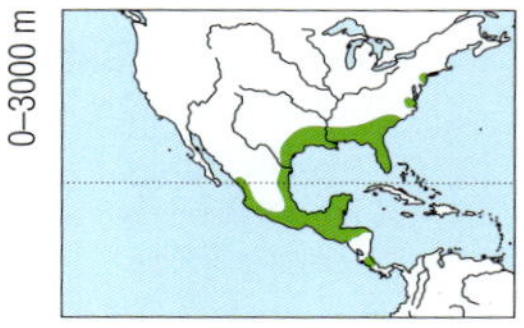

Northern Yellow Bat
Lasiurus intermedius

FR Lasiure du Mexique
DE Nördliche Haarschwanzfledermaus
ES Lasiuro de México

US (New Jersey, SE Virginia S to SE Texas) & MX (S from S Sinaloa & Tamaulipas) to HN, recorded in CR

Cuban Yellow Bat
Lasiurus insularis

FR Lasiure de Cuba DE Kuba-Haarschwanzfledermaus ES Lasiuro de Cuba

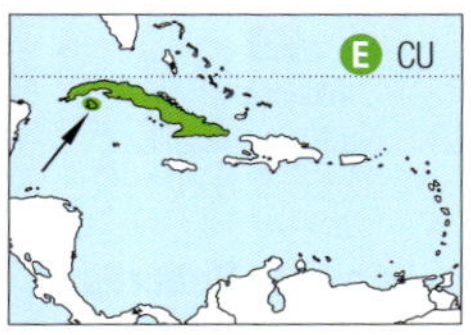

VU
Monotypic
HB 8.3–8.5 cm
T 6.5–8.1 cm
FA 5.7–6.4 cm
W 20–30 g

Palm groves, also in Isla de la Juventud

Southern Yellow Bat
Lasiurus ega

FR Lasiure jaune DE Südliche Haarschwanzfledermaus ES Lasiuro amarillo

0–1860 m

LC
4 sspp
HB 5.7–7.7 cm
T 4.3–6.8 cm
FA 4–5.2 cm
W 13–20 g

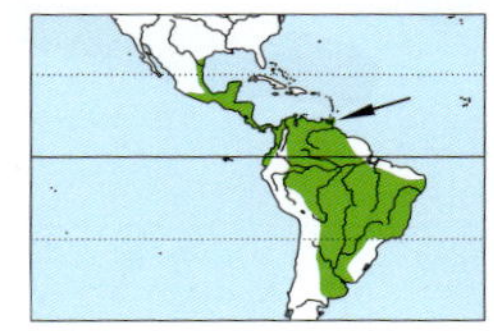

From US (SE Texas) S along E & S MX to W CO & W EC and E of Andes S to PY, UY & EC AR

Genus *SCOTOPHILUS*

(19 species)

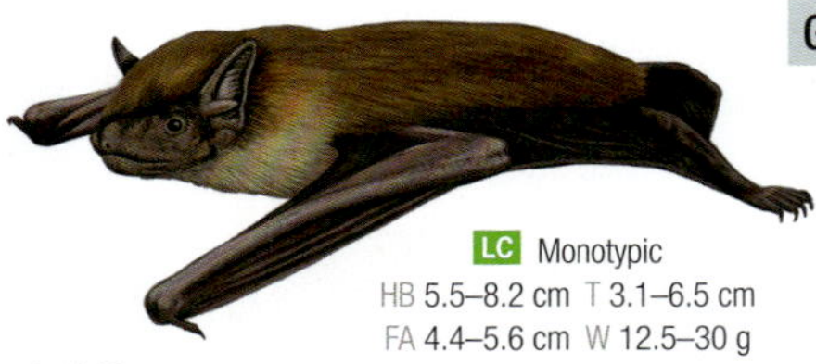

LC Monotypic
HB 5.5–8.2 cm T 3.1–6.5 cm
FA 4.4–5.6 cm W 12.5–30 g

0–1100 m

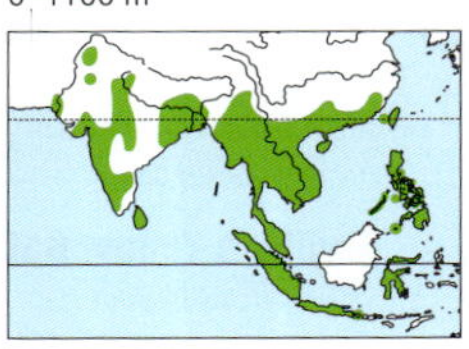

Lesser Asian Yellow Bat
Scotophilus kuhlii

FR Scotophile de Kuhl
DE Kleine Asiatische Hausfledermaus
ES Escotofilo de Kuhl

E PK, IN, LK, SE NP, BD, S & SE CN (with Hainan), TW, SE Asia with Sumatra, Java, Bali, Sulawesi & PH

LC Monotypic
HB 5.6–7.1 cm T 2.8–5.2 cm
FA 4.4–5.2 cm W 18–23 g

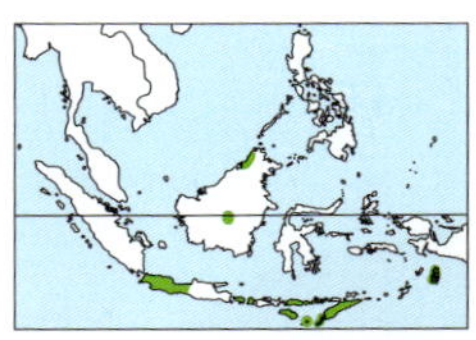

Sody's Yellow Bat
Scotophilus collinus

FR Scotophile de Sody
DE Sody-Hausfledermaus
ES Escotofilo de Sody

NW & C Borneo, W Java, Bali, Lombok, Flores, Lembata, Sumba, Sawu, Timor, Semau, Roti, Banda, Aru Is

LC Monotypic
HB 6.7–9.3 cm T 4.3–7.2 cm
FA 5.5–6.9 cm W 29–53 g

0–1500 m

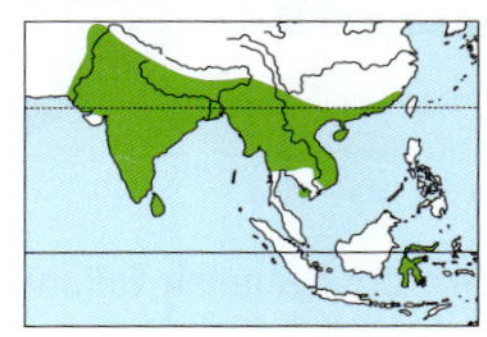

Greater Asian Yellow Bat
Scotophilus heathii

FR Scotophile de Heath
DE Große Asiatische Hausfledermaus
ES Escotofilo de Heath

NE AF, PK, IN & LK E to S CN (with Hainan), mainland SE Asia (with Cat Ba & Phu Quoc Is), Sulawesi

LC Monotypic
HB 6.7–8.9 cm T 4.4–5.4 cm
FA 5.3–6.1 cm W 25–37 g

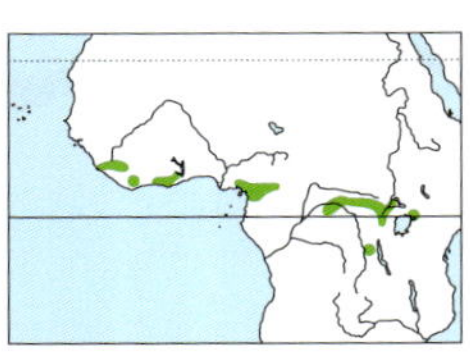

Nut-colored Yellow Bat
Scotophilus nux

FR Scotophile noisette
DE Nussfarbene Hausfledermaus
ES Escotofilo castaño

Patchily in SL, GN, LR, CI, GH, NG, CM, CD, UG & KE; possibly GQ & GA

DD
Monotypic
HB 6.5–6.8 cm
T 4.3–4.6 cm
FA 4.4–4.7 cm
W c. 14 g

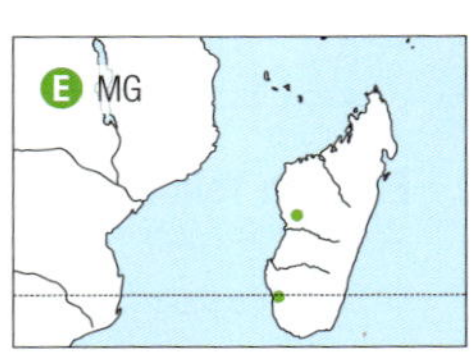

Malagasy Yellow Bat
Scotophilus tandrefana

FR Scotophile d'Andadoany
DE Westmadagaskar-Hausfledermaus
ES Escotofilo de Andadoany

Confirmed from 2 sites near Tsingy de Bemaraha NP & Sarodrano; possibly Kirindy Forest

LC
Monotypic
HB 6.2–6.8 cm
T 3.8–4.5 cm
FA 4.1–4.5 cm
W 12.5–16.8 g

0–200 m

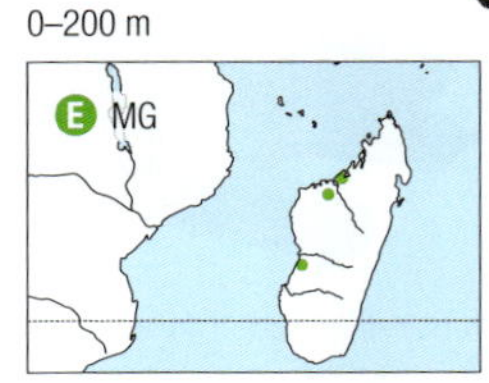

Marovaza Yellow Bat
Scotophilus marovaza

FR Scotophile de Marovaza
DE Marovaza-Hausfledermaus
ES Escotofilo de Marovaza

Known from 3 sites (Marovaza, Ankarafantsika, Mahabo); perhaps across much of W lowland

White-bellied Yellow Bat
Scotophilus leucogaster

FR Scotophile à ventre blanc
DE Weißbauch-Hausfledermaus
ES Escotofilo de vientre blanco

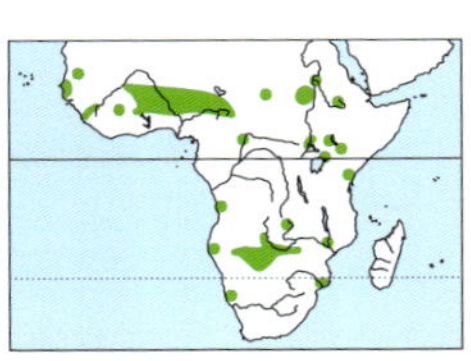

LC
2 sspp
HB 7–7.5 cm T 3.7–5.4 cm
FA 4.3–5.8 cm W 12–27 g

N of equator from MR & SN E to ET & NW KE, S of equator in SE KE & W AO S to NA & S MZ; possibly NE ZA

East African Yellow Bat
Scotophilus altilis

FR Scotophile du Soudan
DE Ostafrika-Hausfledermaus
ES Escotofilo de Sudán

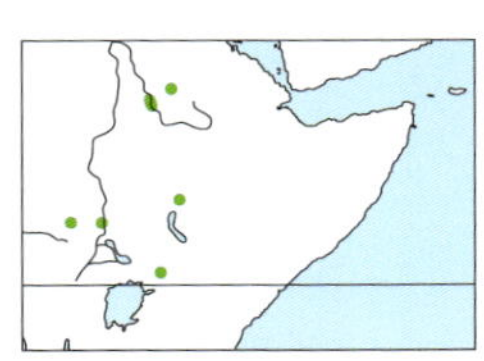

NE
Monotypic
FA c. 4.6 cm

Scattered in SE SD, W ET, SC SS, NE CD & SW KE

LC Monotypic
HB 9.3–9.8 cm T 5.5–7 cm
FA 6.2–6.5 cm W 40–49 g

0–1400 m

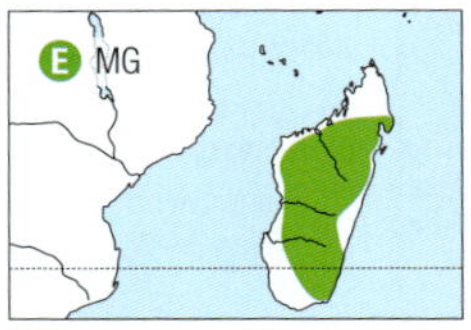

Robust Yellow Bat
Scotophilus robustus

FR Scotophile robuste
DE Robuste Hausfledermaus
ES Escotofilo robusto

Broadly distributed in all distinct biomes (except spiny bush)

LC Monotypic
HB 7.1–8.2 cm T 3.4–4.9 cm
FA 4.1–5.1 cm W 12–20 g

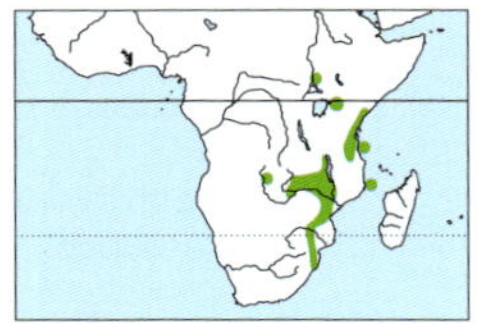

Eastern Greenish Yellow Bat
Scotophilus viridis

FR Scotophile verdâtre
DE Östliche Grünliche Hausfledermaus
ES Escotofilo verdoso

E & SE Africa S to E ZA & SZ, Mafia & Mozambique Is

1390–2100 m

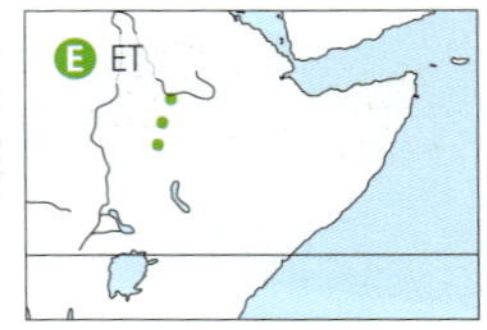

Known only from 3 localities

Ejeta's Yellow Bat
Scotophilus ejetai

FR Scotophile d'Ejeta
DE Ejeta-Hausfledermaus
ES Escotofilo de Etiopía

LC Monotypic
HB 7.3–7.7 cm
T 4–4.5 cm
FA 5–5.1 cm

LC 2 sspp
HB 10.3–12.3 cm T 6.4–9.2 cm
FA 7.8–8.8 cm W 88–91 g

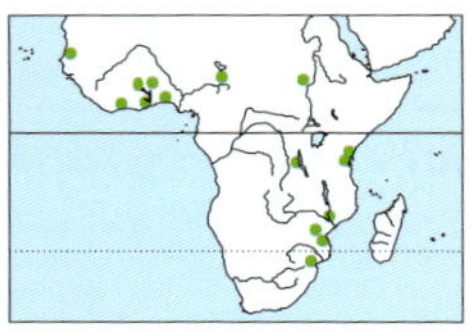

Schreber's Yellow Bat
Scotophilus nigrita

FR Scotophile de Schreber
DE Schreber-Hausfledermaus
ES Escotofilo de Schreber

Scattered in W Africa (SN E to N CM) and E & SE Africa (SC SD, E CD & SE KE S to NE ZA)

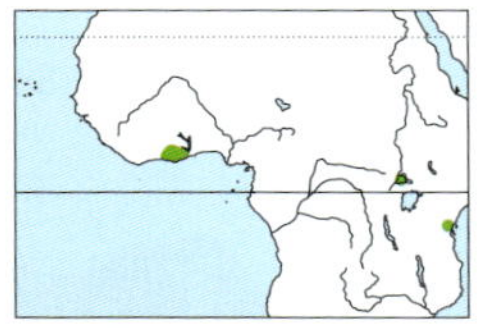

Robbins's Yellow Bat
Scotophilus nucella

FR Scotophile de Robbins
DE Robbins-Hausfledermaus
ES Escotofilo de Robbins

Known from SE CI, S GH, extreme WC UG & TZ (Kambai Forest R in East Usambara Mts)

DD
Monotypic
HB 7.4–7.8 cm
T 4.1–4.7 cm
FA 4.9–5.3 cm

LC
Monotypic
HB 7.5–8.6 cm T 4.3–5 cm
FA 4.6–5.4 cm

330–1905 m

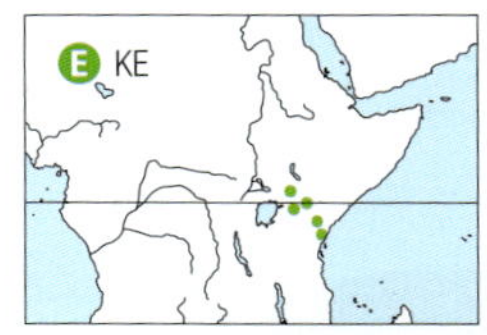

Few localities from W to coast in E

Andrew Rebori's Yellow Bat
Scotophilus andrewreborii

FR Scotophile de Rebori
DE Andrew-Rebori-Hausfledermaus
ES Escotofilo de Rebori

NE Monotypic
HB 7.1–8.2 cm T 3.4–4.9 cm
FA 4.1–5.1 cm W 12–20 g

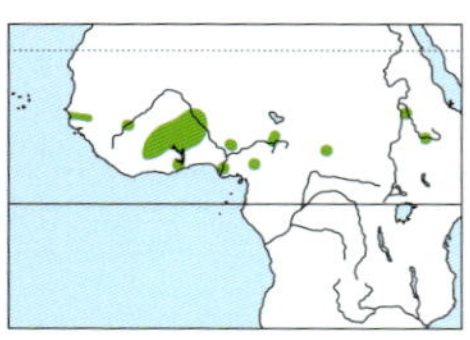

Western Greenish Yellow Bat
Scotophilus nigritellus

FR Scotophile de De Winton
DE Westliche Grünliche Hausfledermaus
ES Escotofilo de Winton

Scattered in SN & GM E to N CM, CF, SD & ET

Livingstone's Yellow Bat
Scotophilus livingstonii

FR Scotophile de Livingstone
DE Livingstone-Hausfledermaus
ES Escotofilo de Livingston

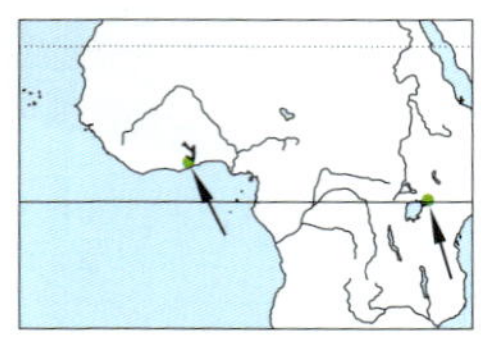

Known from 3 sites in SE GH & SW KE (with a gap of c. 4000 km between them)

LC
Monotypic
HB 7.4–8.9 cm
T 3.2–5.1 cm
FA 5.2–5.7 cm

LC Monotypic
HB 6.5–7.5 cm T 3.7–4.4 cm
FA 4.4–4.6 cm

0–760 m

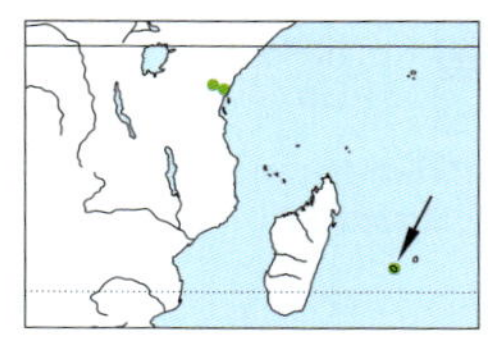

Trujillo's Yellow Bat
Scotophilus trujilloi

FR Scotophile de Trujillo
DE Trujillo-Hausfledermaus
ES Escotofilo de Trujillo

SE KE (3 localities), Réunion I; perhaps S SO & NE TZ, unconfirmed in SW MG

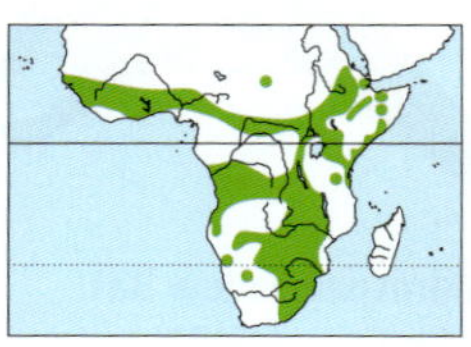

Sub-Saharan Africa from SN & GM E to ER, DJ & ET, S through E & S Africa to E ZA, LS & SZ

African Yellow Bat
Scotophilus dinganii

FR Scotophile de Dingane DE Gelbbauch-Hausfledermaus ES Escotofilo de Dingane

LC Monotypic
HB 7.4–9 cm T 4.6–6.5 cm
FA 4.7–6 cm W 15–36 g

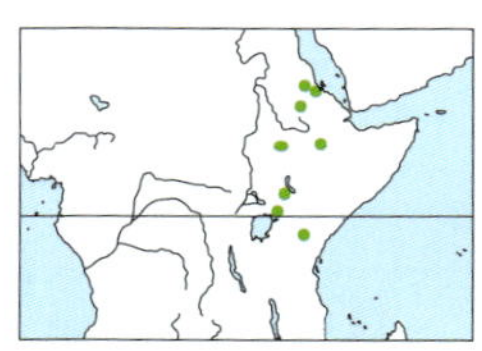

ER, ET, KE

Eritrean Yellow Bat
Scotophilus colias

FR Scotophile soufré DE Eritrea-Hausfledermaus ES Escotofilo de Eritrea

NE
Monotypic
FA 5.5–5.7 cm

Genus *NYCTICEIUS*

(3 species)

LC
3 sspp
HB c. 5.6 cm
T 3.5–4.1 cm
FA 3.3–3.9 cm
W 6.4–14 g

0–900 m

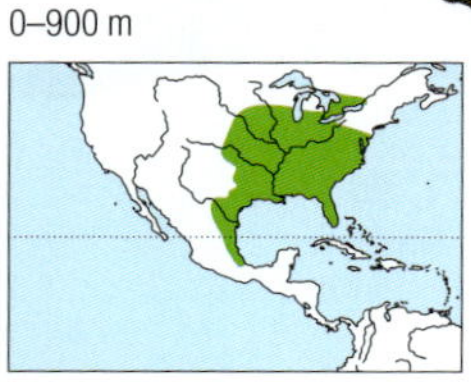

North American Evening Bat
Nycticeius humeralis

FR Nycticée vespérale
DE Neuweltabendsegler
ES Nicticeo vesperino

CA (S Ontario) S across most of E US to E MX (S along Gulf coast to Veracruz)

NT Monotypic
HB c. 5.2 cm T 2.4–3.6 cm
FA 2.9–3.2 cm W 4–7 g

0–250 m

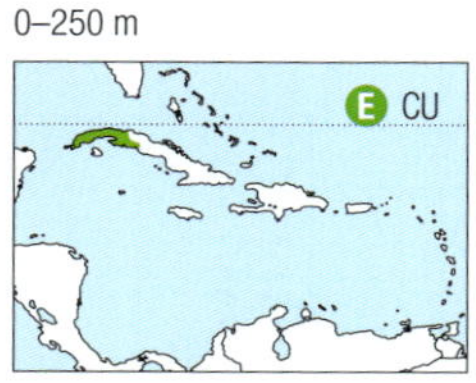

Cuban Evening Bat
Nycticeius cubanus

FR Nycticée de Cuba
DE Kuba-Neuweltabendsegler
ES Nicticeo de Cuba

Known only from W half of island

DD
Monotypic
HB c. 3.7 cm
T c. 2 cm
FA c. 2.9 cm

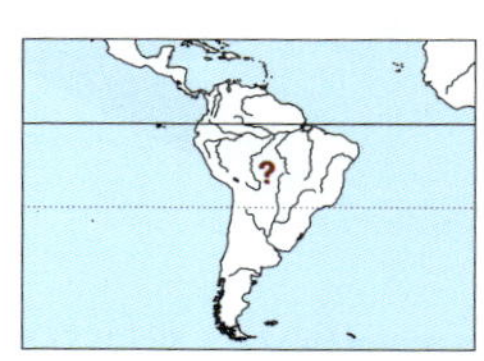

Temminck's Mysterious Bat
Nycticeius aenobarbus

FR Nycticée à barbe rousse
DE Temminck-Neuweltabendsegler
ES Nicticeo de barba roja

According to type locality, it occurs in South America, but this is dubious

Genus *PHONISCUS*

(4 species)

up to 615 m

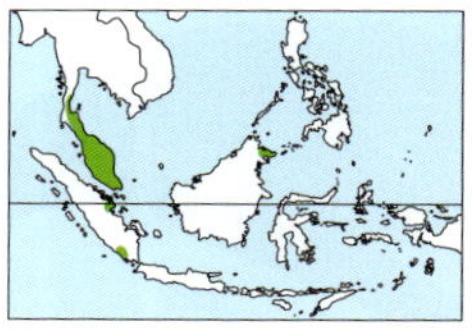

S TH, Pen MY, EC & SW Sumatra, NE Borneo; perhaps widespread

NT Monotypic
HB c. 4.1 cm T 2.9–3.9 cm
FA 3.1–3.5 cm W 3–6.5 g

Groove-toothed Trumpet-eared Bat
Phoniscus atrox

FR Kérivoule de Miller
DE Furchenzahn-Trompetenohr
ES Fonisco de Miller

LC Monotypic
HB 4–4.5 cm T 3.6–4.5 cm
FA 3.5–4.2 cm W 5.8–11 g

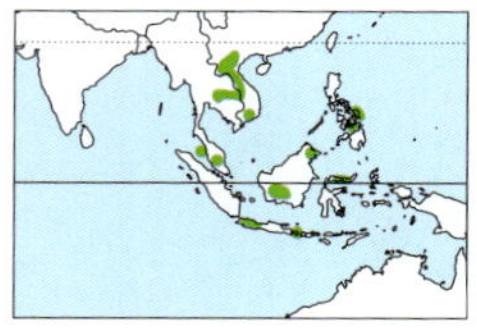

E TH, LA, VN, Pen MY (with Penang I), Borneo, W Java, Bali, N Sulawesi, PH (Samar, Bohol); possibly N KH

Peters's Trumpet-eared Bat
Phoniscus jagorii

FR Kérivoule de Peters
DE Peters-Trompetenohr
ES Fonisco de Peters

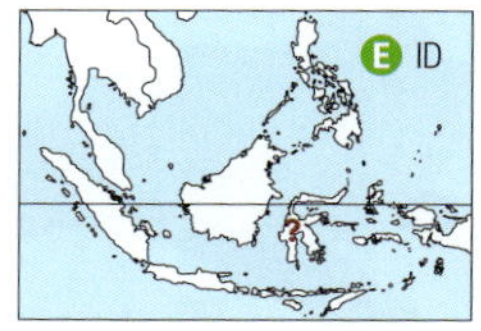

Possibly Sulawesi

DD Monotypic
HB 4.4–5.1 cm T 3.8–4.2 cm
FA c. 3.7 cm

Dubious Trumpet-eared Bat
Phoniscus aerosus

FR Kérivoule rousse
DE Zweifelhaftes Trompetenohr
ES Fonisco rojizo

VU Monotypic
HB 4.1–5.3 cm T 3.7–4.2 cm
FA 3.5–4 cm W 5.3–11.2 g

0–1300 m

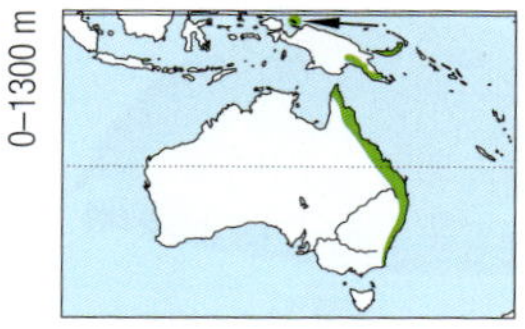

Biak I, E New Guinea, Bismarck Arch (New Britain), E AU (E Qld, E NSW)

Golden-tipped Bat
Phoniscus papuensis

FR Kérivoule papoue
DE Papua-Trompetenohr
ES Fonisco de Papúa

Genus *KERIVOULA*

(28 species)

60–1600 m

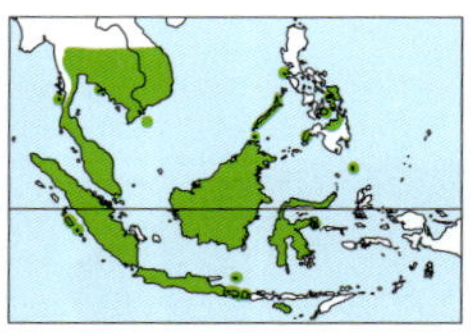

S TH, S LA, S VN, KH, Pen MY, SG, Greater Sundas, Lombok, Sumba & PH, also many offshore Is

Hardwicke's Woolly Bat
Kerivoula hardwickii

FR Kérivoule de Hardwicke
DE Hardwicke-Wollfledermaus
ES Querivoula de Hardwicke

LC Monotypic
HB 3.9–4.1 cm T 3.5–5 cm
FA 2.6–3.6 cm W 2.9–6 g

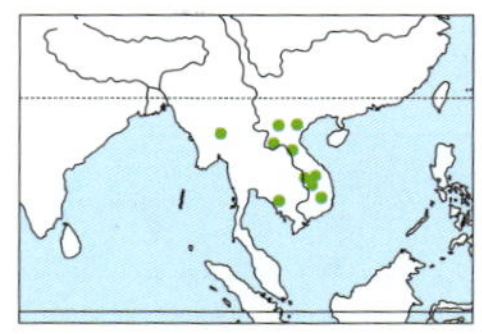

E MM, LA, N & S VN, W & NE KH; likely occurs in much of Indochina

Flat-skulled Woolly Bat
Kerivoula depressa

FR Kérivoule à tête plate
DE Flachkopf-Wollfledermaus
ES Querivoula de cabeza plana

LC Monotypic
HB c. 3.3 cm T c. 4.3 cm
FA 3.2–3.3 cm

Sri Lankan Woolly Bat
Kerivoula malpasi

FR Kérivoule du Sri Lanka
DE Sri-Lanka-Wollfledermaus
ES Querivoula de Sri Lanka

NE Monotypic
HB 3.9–5.5 cm T 3.5–4.3 cm
FA 3.3–3.6 cm

500–1100 m

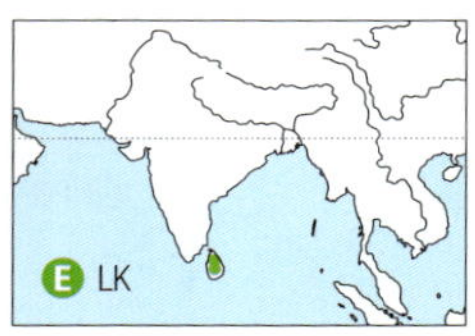

Central highlands, 1 record from NE

LC Monotypic
HB c. 4.4 cm T c. 4.2 cm
FA c. 3.1 cm

Cryptic Woolly Bat
Kerivoula crypta

FR Kérivoule cryptique
DE Kryptische Wollfledermaus
ES Querivoula críptico

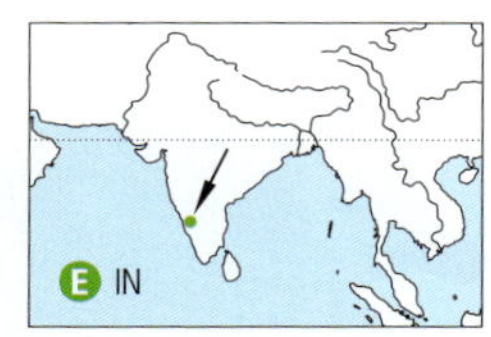

Known from type locality in Shimoga (Karnataka)

LC Monotypic
T 3.5–4.5 cm FA 3–3.7 cm
W 3.2–7 g

up to 2100 m

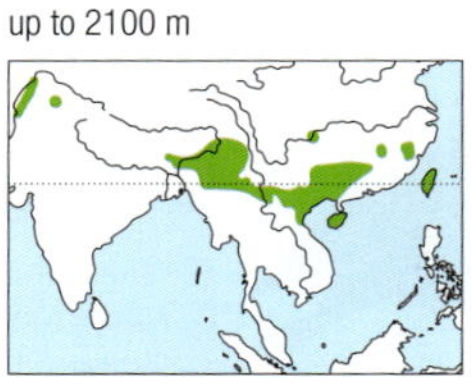

N PK, NW & NE IN, N MM, SC & SE CN (with Hainan), TW, N VN

Dark Woolly Bat
Kerivoula furva

FR Kérivoule sombre
DE Dunkle Wollfledermaus
ES Querivoula oscuro

Kachin Woolly Bat
Kerivoula kachinensis

FR Kérivoule du Kachin
DE Kachin-Wollfledermaus
ES Querivoula de Kachin

LC Monotypic
HB 4.5–5.4 cm T 5.2–6.1 cm
FA 4–4.3 cm W 6.3–9.5 g

100–1300 m

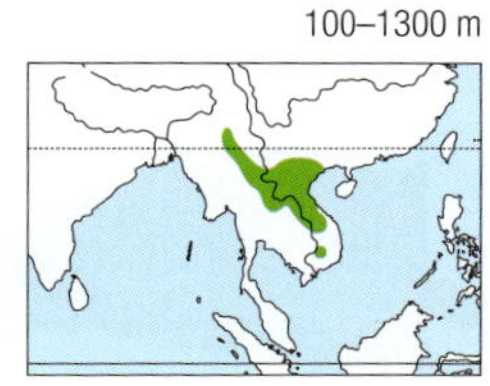

N MM, N TH, LA, N & C VN, SE KH

1115–1535 m

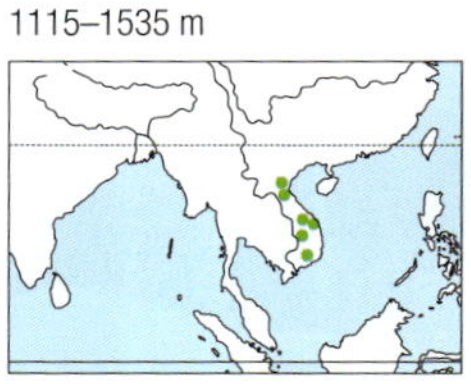

Annamite Mts of E LA, VN & E KH

LC Monotypic
T c. 3.8 cm FA 3–3.3 cm
W c. 4.5 g

Indochinese Woolly Bat
Kerivoula dongduongana

FR Kérivoule d'Indochine
DE Indochina-Wollfledermaus
ES Querivoula de Indochina

0–1600 m

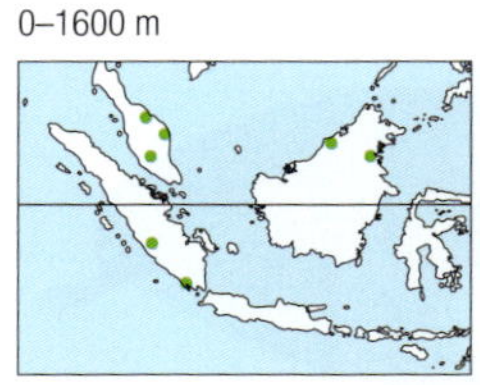

Scattered in extreme S TH, Pen MY, C & S Sumatra and WC & EC Borneo

Krau Woolly Bat
Kerivoula krauensis

FR Kérivoule de Krau
DE Krau-Wollfledermaus
ES Querivoula de Krau

NT Monotypic
HB 3.8–4 cm T 3.4–4.3 cm
FA 2.8–3.4 cm W 2–4.5 g

260–1600 m

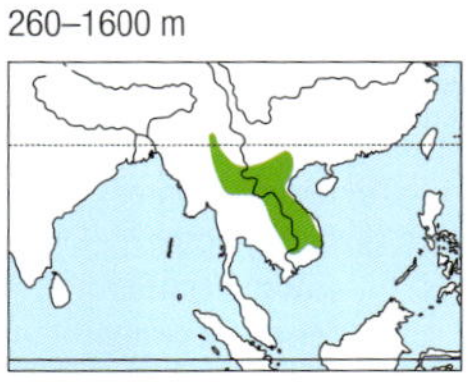

E MM, N TH, LA, VN, E KH

Titania's Woolly Bat
Kerivoula titania

FR Kérivoule titania
DE Titania-Wollfledermaus
ES Querivoula titania

LC Monotypic
HB 3.8–4.1 cm T 4.6–5.3 cm
FA 3.2–3.6 cm W 4–7.9 g

NT Monotypic
HB 2.8–3.9 cm T 2.7–4.3 cm
FA 2.5–3 cm W 2.5–3.5 g

0–500 m

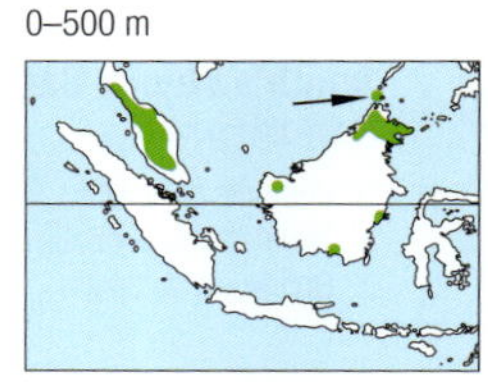

Extreme S TH, Pen MY, N, SW & SE Borneo (with Balambangan I)

Least Woolly Bat
Kerivoula minuta

FR Petite Kérivoule
DE Zwergwollfledermaus
ES Querivoula pequeño

0–500 m

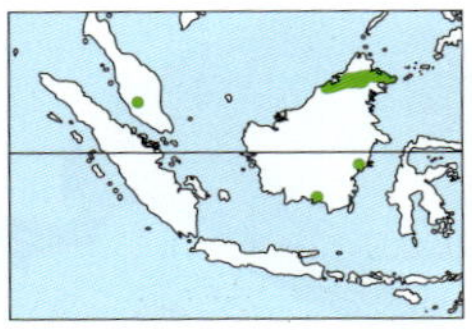

Small Woolly Bat
Kerivoula intermedia

FR Kérivoule de Bornéo
DE Kleine Wollfledermaus
ES Querivoula de Borneo

NT Monotypic
HB 3.4–3.6 cm T 3.3–4.2 cm
FA 2.6–3.1 cm W 2.9–4.2 g

Scattered in Pen MY and N & SE Borneo; perhaps throughout Thai-Malay Pen & Borneo

LC Monotypic
HB 4.8–4.9 cm T 4.7–5.3 cm
FA 3.7–4.1 cm W 6.7–6.9 g

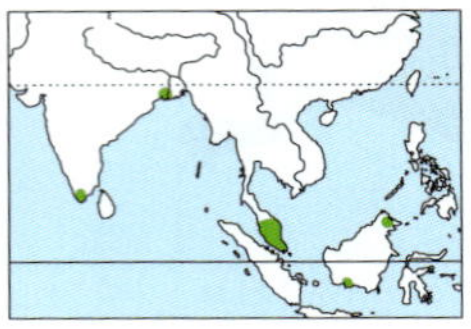

Lenis Woolly Bat
Kerivoula lenis

FR Kérivoule légère
DE Sanfte Wollfledermaus
ES Querivoula de Malasia

NE & S IN (West Bengal, Tamil Nadu), Pen MY, N & SC Borneo

0–1500 m

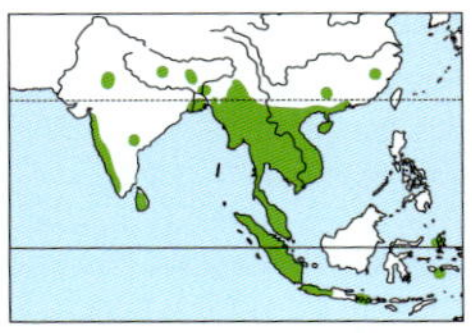

Painted Woolly Bat
Kerivoula picta

FR Kérivoule peinte
DE Bunte Wollfledermaus
ES Querivoula pintado

NT 2 sspp
HB 4–4.8 cm T 3.7–4.8 cm
FA 3.1–3.9 cm W 4.5–5.5 g

IN, LK, C NP, BD, SE CN (with Hainan), mainland SE Asia, Sumatra, Java, Bali, Lombok, N & C Moluccas

Bornean Woolly Bat
Kerivoula pusilla

FR Kérivoule minuscule
DE Borneo-Wollfledermaus
ES Quirovoula enana

NE 2 sspp
HB c. 3.3 cm T c. 3.9 cm
FA c. 2.8 cm

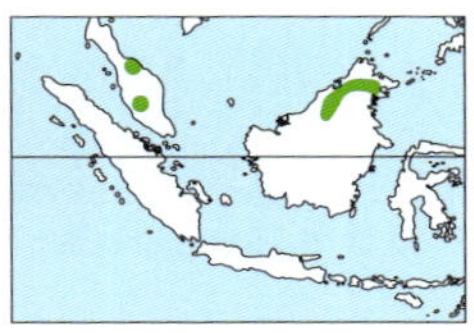

Extreme S TH, C Pen MY, N Borneo

LC Monotypic
HB 3.5–4.3 cm T 3.3–4.4 cm
FA 3.2–3.6 cm W 4–5.4 g

0–1600 m

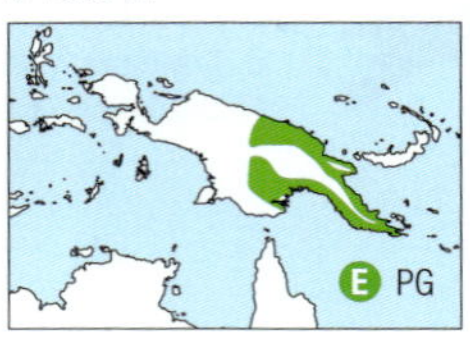

Fly River Woolly Bat
Kerivoula muscina

FR Kérivoule de Fly River
DE Fly-River-Wollfledermaus
ES Querivoula de Fly River

Lowlands; possibly Papua Prov (ID)

VU Monotypic
HB 3.3–4 cm T 3.4–4.6 cm
FA 3.1–3.8 cm W 3.5–6.5 g

Bismarck Woolly Bat
Kerivoula myrella

FR Kérivoule des Bismarck
DE Bismarck-Archipel-Wollfledermaus
ES Querivoula de Bismarck

0–500 m

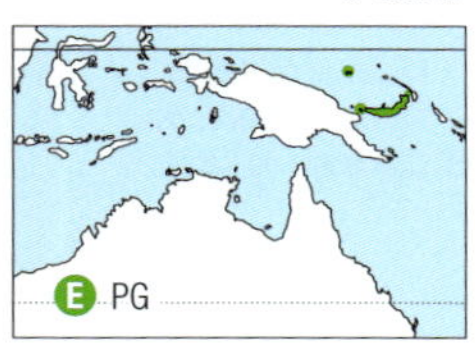

Manus, New Britain, Duke of York & Umboi Is; unconfirmed from Wetar (ID)

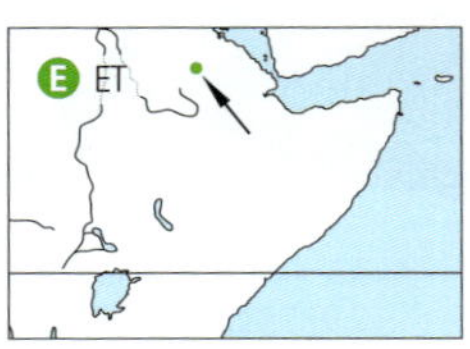

Ethiopian Woolly Bat
Kerivoula eriophora

FR Kérivoule d'Éthiopie
DE Äthiopien-Wollfledermaus
ES Querivoula de Etiopía

DD Monotypic
HB c. 5.3 cm T c. 3.2 cm
FA 2.8–3.2 cm

Known from type locality (Belegaz Valley, Begemder)

900–2800 m

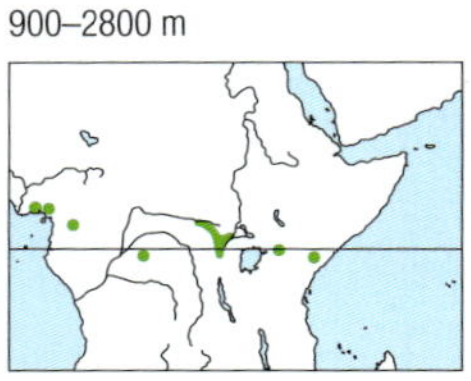

Smith's Woolly Bat
Kerivoula smithii

FR Kérivoule de Smith
DE Smith-Wollfledermaus
ES Querivoula de Smith

LC Monotypic
HB 3.9–4.4 cm T 4–4.9 cm
FA 3.2–3.6 cm W 5.9–7 g

Scattered records from NG & CM E through CD to UG & KE

0–1050 m

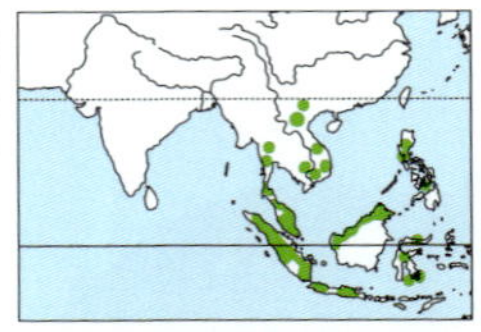

Papillose Woolly Bat
Kerivoula papillosa

FR Kérivoule mamelonnée
DE Warzen-Wollfledermaus
ES Querivoula piloso de Java

LC 2 sspp
HB 4–5.2 cm T 4.7–5.6 cm
FA 3.9–4.9 cm W 6–13 g

WC & S TH, NE LA, VN, C KH, Pen MY, Sumatra, N & W Borneo, Java, N & C Sulawesi, PH (Luzon, Bohol)

0–1200 m

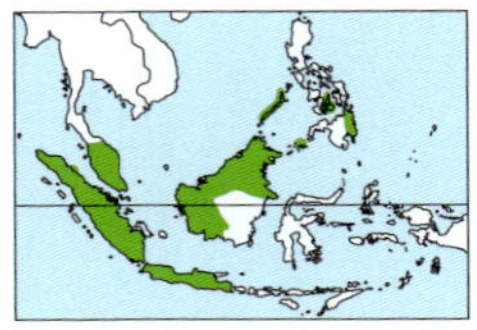

Clear-winged Woolly Bat
Kerivoula pellucida

FR Kérivoule diaphane
DE Transparentflügel-Wollfledermaus
ES Querivoula de alas claras

NT Monotypic
HB 3.5–4.1 cm T 4–5.2 cm
FA 2.9–3.2 cm W 3.5–5.5 g

Thai-Malay Pen, Sumatra, Borneo (except EC & SE), Java, PH (Palawan, Cebu, Bohol, Mindanao, Jolo)

LC Monotypic
HB 3.6–4.9 cm T 2.9–4.3 cm
FA 2.8–3.5 cm W 2.5–5.7 g

0–1490 m

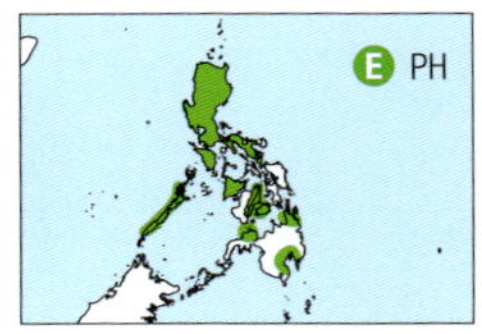

Whitehead's Woolly Bat
Kerivoula whiteheadi

FR Kérivoule de Whitehead
DE Whitehead-Wollfledermaus
ES Querivoula de Whitehead

Luzon, Mindoro, Palawan, Panay, Cebu, Bohol, Mindanao

0–1500 m

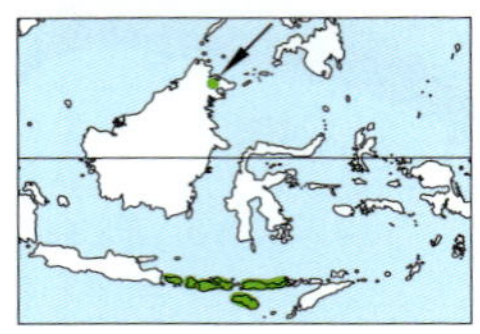

Flores Woolly Bat
Kerivoula flora

FR Kérivoule de Florès
DE Flores-Wollfledermaus
ES Querivoula de Flores

VU Monotypic
HB c. 4.3 cm T c. 4.9 cm
FA 3.8–3.9 cm W c. 6.4 g

N Borneo (Sabah), Lesser Sundas (Bali, Lombok, Sumbawa, Flores, Sumba)

0–700 m

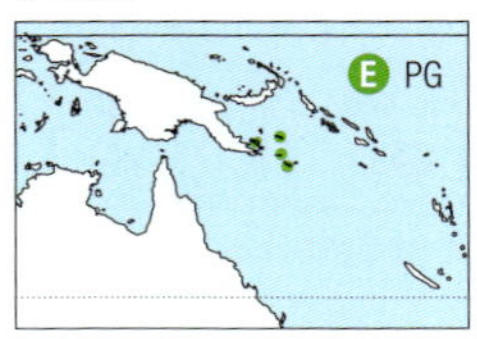

Saint Aignan's Woolly Bat
Kerivoula agnella

FR Kérivoule de Louisiade
DE Misima-Wollfledermaus
ES Querivoula de Louisiade

EN Monotypic
HB 4.2–4.4 cm T c. 4.8 cm
FA 3.4–3.8 cm

Fergusson, Woodlark, Misima & Sudest Is

LC 4 sspp
HB 4.4–4.8 cm T 3–4.2 cm
FA 3–3.4 cm W 4.5–8 g

Lesser Woolly Bat
Kerivoula lanosa

FR Kérivoule laineuse
DE Kleinere Wollfledermaus
ES Querivoula lanudo pequeño

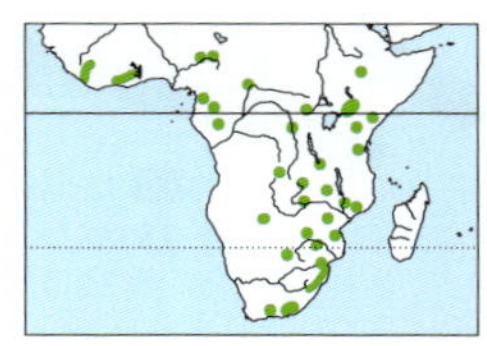

Scattered in GN & LR E to CM, GA & CF and from ET, CD & KE S to SZ & ZA

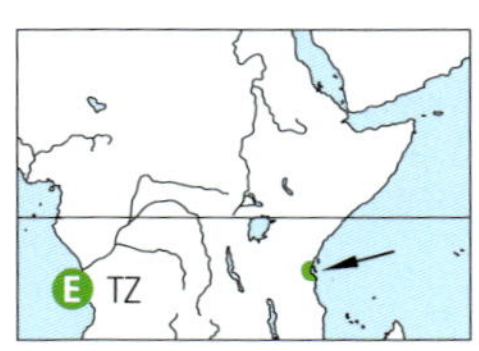

Tanzanian Woolly Bat
Kerivoula africana

FR Kérivoule de Tanzanie
DE Tansania-Wollfledermaus
ES Querivoula de Tanzania

EN Monotypic
HB c. 3.5 cm T c. 3.4 cm
FA c. 2.8 cm

Morogoro, Genda Genda & Tong'omba coastal forests

DD Monotypic
HB 3.4–3.9 cm T 4–4.4 cm
FA 3–3.3 cm W 4–4.5 g

up to 600 m

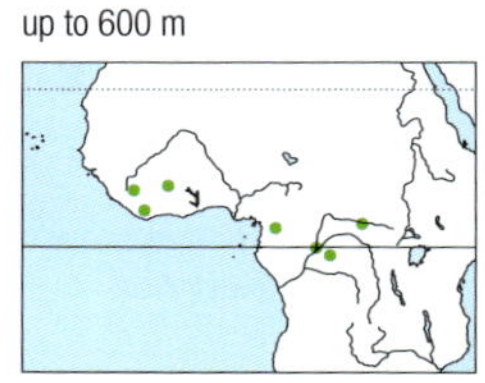

Coppery Woolly Bat
Kerivoula cuprosa

FR Kérivoule cuivrée
DE Kupferfarbige Wollfledermaus
ES Querivoula cobrizo

Scattered in GN, LR, CI, CM & CD; possibly GH

450–2400 m

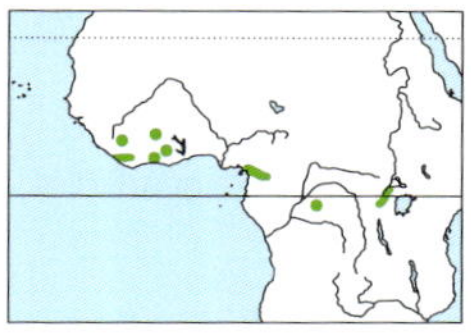

Spurrell's Woolly Bat
Kerivoula phalaena

FR Kérivoule phalène
DE Spurrell-Wollfledermaus
ES Querivoula de Spurrell

LC Monotypic
HB 3.1–3.7 cm T 3.2–4.2 cm
FA 2.5–3 cm W 2.2–5 g

GN, LR, CI, GH, CM, CD, W UG

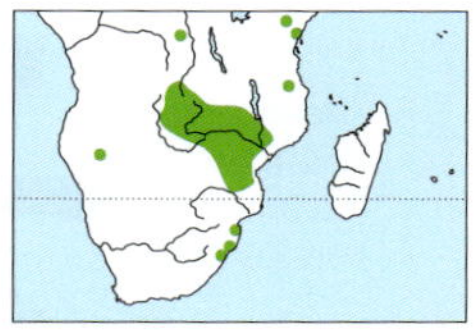

Damara Woolly Bat
Kerivoula argentata

FR Kérivoule argentée
DE Damara-Wollfledermaus
ES Querivoula argentado

LC 3 sspp
HB 3.4–5 cm T 4–5.2 cm
FA 2.8–3.9 cm W 5.5–10.5 g

C & S CD, SE KE, S TZ, ZM, MW, MZ, ZW, E ZA (KwaZulu-Natal), N NA; possibly AO

Genus *HARPIOCEPHALUS*

(1 species)

LC
4 sspp
HB 5.7–7.5 cm
T 4–5 cm
FA 4.4–5.5 cm
W 12–30 g

Hairy-winged Bat
Harpiocephalus harpia

185–2480 m

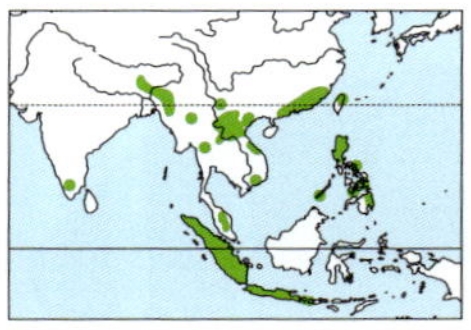

FR Murine à ailes velues
DE Haarflügelfledermaus
ES Ratonero de alas pilosas

S & NE IN, BT, NW & C MM, SC & SE CN, TW, VN, N LA, TH, Pen MY, Sumatra, Java, Bali, Lombok, PH

Genus *HARPIOLA*

(2 species)

DD
Monotypic
HB c. 4.5 cm
T c. 2.5 cm
FA 3.2–3.3 cm

1365–1690 m

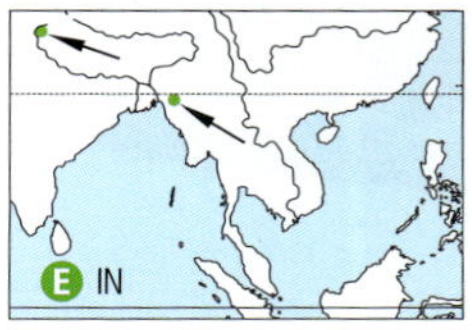

Gray Tube-nosed Bat
Harpiola grisea

FR Murine de l'Himalaya
DE Graue Röhrennase
ES Ratonero del Himalaya

Uttarakhand, Mizoram

LC
Monotypic
HB 4.3–5.3 cm
T 2.4–3.6 cm
FA 3–3.7 cm
W 3.4–8 g

1000–2400 m

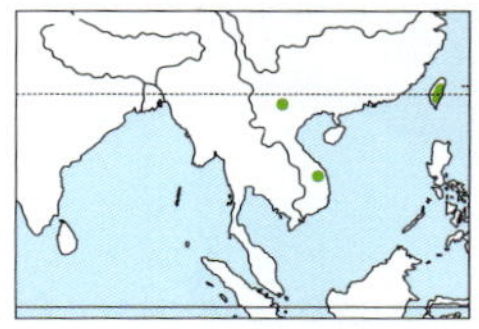

Formosan Tube-nosed Bat
Harpiola isodon

FR Murine isodonte
DE Goldgelbe Formosa-Röhrennase
ES Ratonero isodonte

NW & C VN (Hoang Lien NP, Ngoc Linh R), TW

Genus *MURINA*

(40 species)

NE
Monotypic
HB 4–4.6 cm
T 3.3–3.8 cm
FA 3.2–3.6 cm
W 4.5–7.5 g

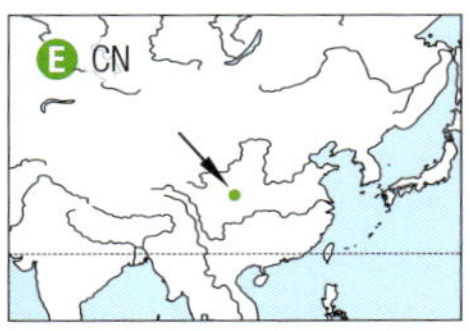

Jinchu's Tube-nosed Bat
Murina jinchui

FR Murine de Jinchu
DE Jinchu-Röhrennase
ES Ratonero narizudo de Jinchu

Known only from type locality (Wolong National NR, Sichuan)

LC 2 sspp
HB 4.7–6.4 cm T 3.3–4.8 cm
FA 3.7–4.7 cm W 7–9 g

700–2135 m

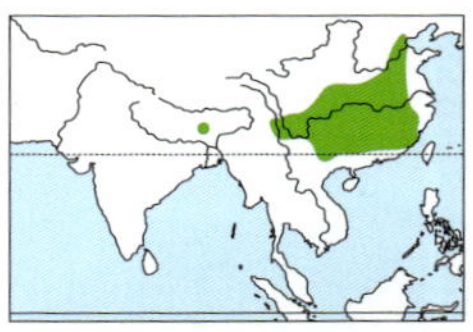

Rufous Tube-nosed Bat
Murina leucogaster

FR Murine à ventre blanc
DE Rötliche Röhrennase
ES Ratonero narizudo de vientre blanco

NE IN (West Bengal), C & E CN, N VN; dubious record from NP

400–3350 m

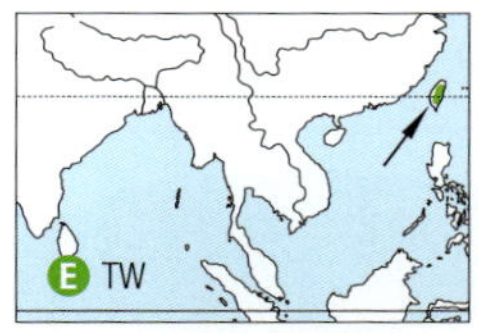

Bicolored Tube-nosed Bat
Murina bicolor

FR Murine bicolore
DE Zweifarbige Röhrennase
ES Ratonero narizudo bicolor

Lowland and montane forests

LC
Monotypic
HB 4.7–4.9 cm
T 4.1–4.5 cm
FA 3.7–4.2 cm

NE
Monotypic
HB 4.3–5.1 cm
T 3.2–4.2 cm
FA 3–3.4 cm
W 3.5–5 g

c. 570 m

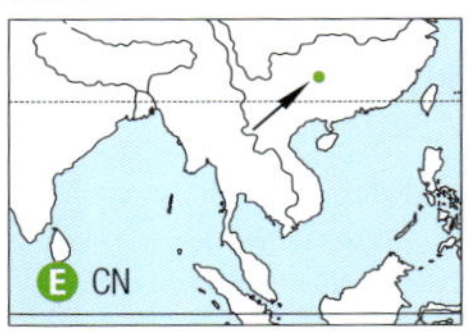

Rongjiang Tube-nosed Bat
Murina rongjiangensis

FR Murine du Rongjiang
DE Rongjang-Röhrennase
ES Ratonero narizudo de Rongjiang

Guizhou (known from type locality in Xinghua town)

1070–1215 m

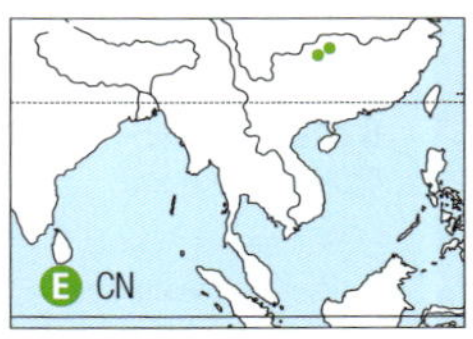

Fang He Tube-nosed Bat
Murina fanjingshanensis

FR Murine du Fanjing
DE Fangjingshan-Röhrennase
ES Ratonero narizudo de Fanjing

Guizhou, Hunan

NE
Monotypic
HB 4.3–5.4 cm
T 3.9–4.7 cm
FA 4–4.1 cm
W c. 9 g

DD
Monotypic
HB 3.4–4.6 cm
T 3.2–4.2 cm
FA 3–3.5 cm
W 4–6.8 g

c. 650 m

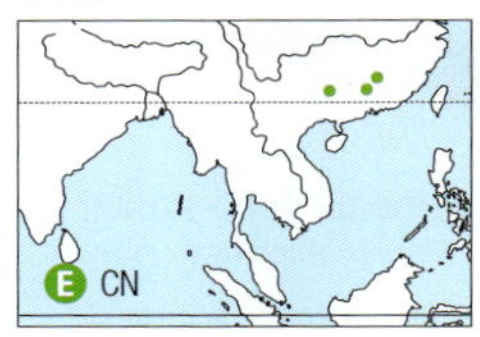

Shuipu Tube-nosed Bat
Murina shuipuensis

FR Murine de Shuipu
DE Shuipu-Röhrennase
ES Ratonero narizudo de Shuipu

Guizhou, Guangdong, Jiangxi

Dusky Tube-nosed Bat
Murina fusca

FR Murine de Mandchourie
DE Dunkle Röhrennase
ES Ratonero narizudo oscuro

Heilongjiang (known from type locality in Imienpo area)

DD
Monotypic
HB c. 5.8 cm
T c. 3.4 cm
FA c. 4 cm

LC
3 sspp
HB 4.6–6.9 cm
T 3.3–4.7 cm
FA 3.9–4.6 cm
W 9–11 g

up to 4000 m

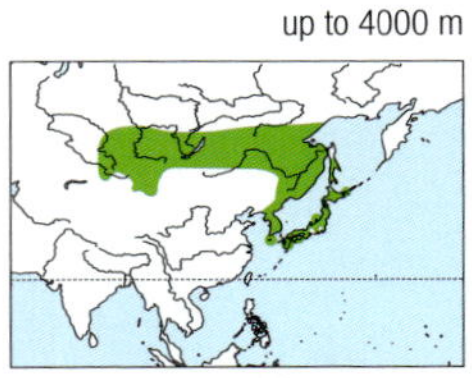

Hilgendorf's Tube-nosed Bat
Murina hilgendorfi

FR Murine de Hilgendorf
DE Hilgendorf-Röhrennase
ES Ratonero narizudo de Hilgendorf

S RU (S Siberia, Far East, Sakhalin), NE KZ, NW MN, NE CN, Korea Pen (with Jeju I), JP, Kuril Is (Kunashir)

1155–4155 m

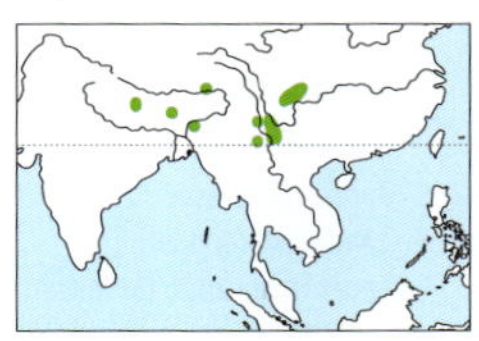

Little Tube-nosed Bat
Murina aurata

FR Murine blonde
DE Kleine Röhrennase
ES Ratonero narizudo dorado

C & E NP, NE IN (Sikkim, Meghalaya), WC & C CN (Gansu, Tibet, Sichuan, Yunnan); possibly BT

DD
Monotypic
HB 3.3–4.5 cm
T 2.7–3.1 cm
FA 2.8–3.2 cm

1000–1500 m

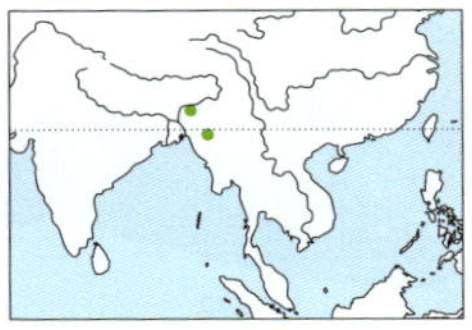

Jaintia Tube-nosed Bat
Murina jaintiana

FR Murine des Jaintias
DE Jaintia-Röhrennase
ES Ratonero narizudo de Jaintias

NE IN (Jaintia Hills in Meghalaya), NW MM (Chin Hills); limits uncertain

DD
Monotypic
HB c. 4 cm
T c. 3.3 cm
FA 2.9–3.1 cm

DD
Monotypic
HB c. 4.4 cm
T c. 3.4 cm
FA c. 3.6 cm

780–1200 m

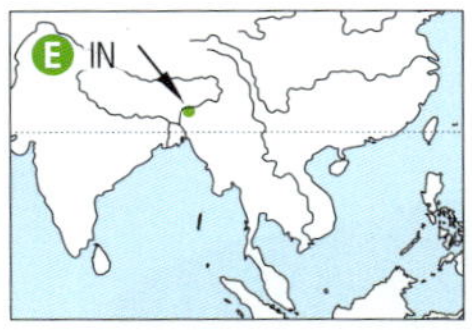

Rainforest Tube-nosed Bat
Murina pluvialis

FR Murine des pluies
DE Regenwald-Röhrennase
ES Ratonero narizudo lluvioso

Meghalaya (Mawphlang & E Khasi Hills)

360–1230 m

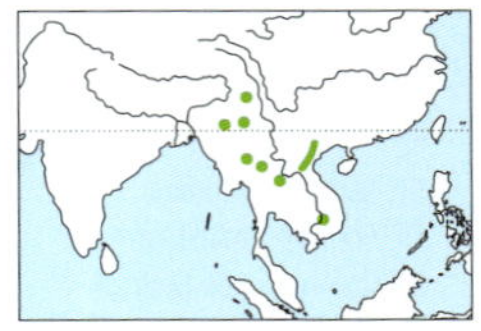

Fea's Tube-nosed Bat
Murina feae

FR Murine de Fea
DE Fea-Röhrennase
ES Ratonero narizudo de Fea

N & E MM, N TH, N LA, N VN, E KH; perhaps NE IN

LC
Monotypic
HB 3.3–4.8 cm
T 2.8–4.2 cm
FA 2.7–3.6 cm
W 3.2–6.4 g

DD
Monotypic
HB 4–4.9 cm
T 3.3–4.5 cm
FA 3.4–3.7 cm
W c. 5.5 g

400–1600 m

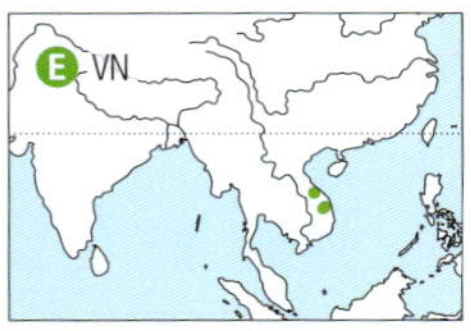

Beelzebub Tube-nosed Bat
Murina beelzebub

FR Murine belzébuth
DE Beelzebub-Röhrennase
ES Ratonero narizudo de belzebú

Bac Huong Hoa NR, Ngoc Linh NR, Ba To, Kon Ka Kinh NP

DD
2 sspp
HB 3.7–4.5 cm
T 3.1–4.1 cm
FA 3.1–3.6 cm
W 4–6.6 g

115–1680 m

Lorelie's Tube-nosed Bat
Murina lorelieae

FR Murine de Lorelie
DE Lorelie-Röhrennase
ES Ratonero narizudo de Lorelie

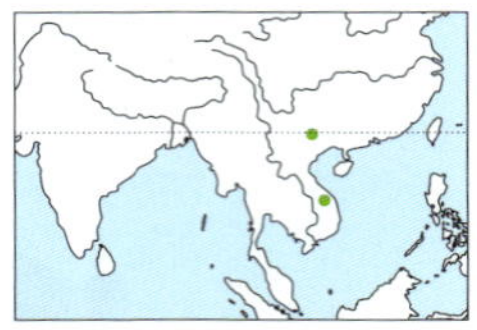

SC CN (Guangxi), SC VN (Ngoc Linh NR)

Vietnamese Tube-nosed Bat
Murina harpioloides

FR Murine de Da Lat
DE Vietnam-Röhrennase
ES Ratonero narizudo de Vietnam

1400–1800 m

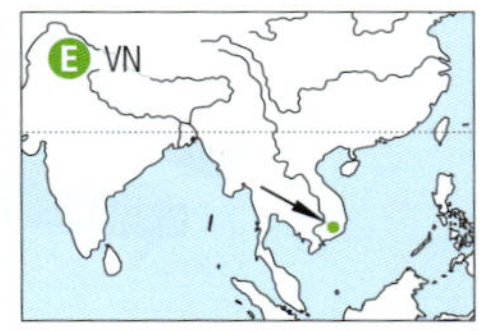

Bi Doup Nui Ba NP, Da Lat Plateau

EN
Monotypic
HB 3.4–3.5 cm
T 2.7–3 cm
FA c. 3 cm
W c. 4.2 g

DD
Monotypic
HB 4–4.1 cm
T 2.4–2.6 cm
FA 2.6–3 cm
W 3–4.4 g

980–1950 m

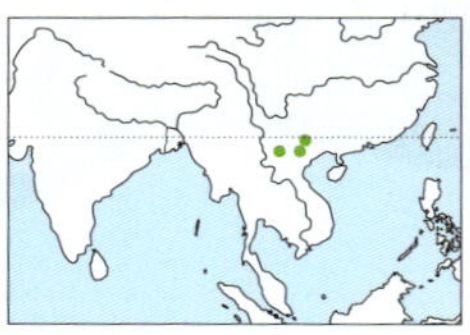

Golden-haired Tube-nosed Bat
Murina chrysochaetes

FR Murine à toison d'or
DE Goldbraune Röhrennase
ES Ratonero narizudo de puntas doradas

SC CN (Guangxi), N VN (Hoang Lien NP, Phia Oac-Phia Den NR)

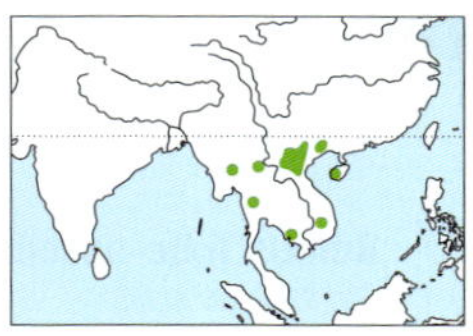

Harrison's Tube-nosed Bat
Murina harrisoni

FR Murine de Harrison
DE Harrison-Röhrennase
ES Ratonero narizudo de Harrison

C MM, N & W TH, N LA, VN, S KH, SE CN (Guangxi, Hainan)

LC
Monotypic
HB 3.9–5.5 cm
T 3.4–4.5 cm
FA 3.4–4 cm
W 8.4–10.5 g

LC
3 sspp
HB 4–4.9 cm
T 2.9–3.8 cm
FA 2.7–3.6 cm
W c. 5.6 g

Ussuri Tube-nosed Bat
Murina ussuriensis

FR Murine de l'Oussouri
DE Ussuri-Röhrennase
ES Ratonero narizudo de Ussuri

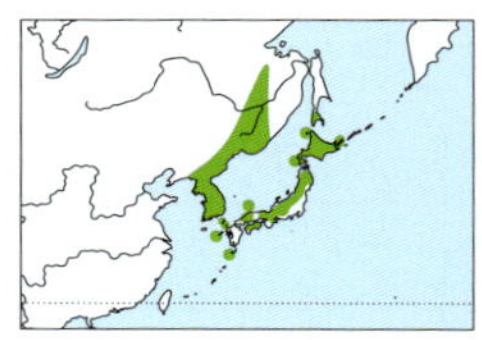

RU (Far East, S Sakhalin), NE CN (Heilongjiang, Jilin), Korea Pen, JP (with offshore Is), Kuril Is (Kunashir)

CR(PE)
Monotypic
HB c. 5 cm
T c. 3.4 cm
FA c. 3.4 cm

Tsushima Is

Gloomy Tube-nosed Bat
Murina tenebrosa

FR Murine sombre
DE Düstere Röhrennase
ES Ratonero narizudo cenizo

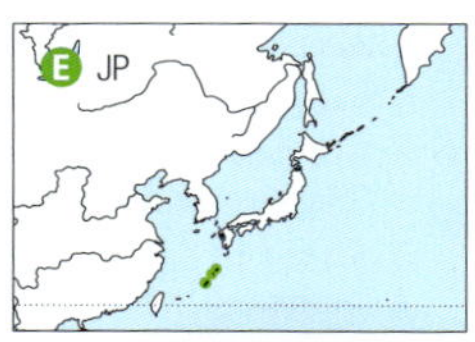

Ryukyu Tube-nosed Bat
Murina ryukyuana

FR Murine des Ryukyu
DE Ryukyu-Röhrennase
ES Ratonero narizudo de Ryukyu

Ryukyu Is (Amami Oshima, Okinawajima)

EN
Monotypic
HB 4.4–4.7 cm
T 3.4–4.5 cm
FA 3.2–3.8 cm
W 7.5–10.6 g

LC
2 sspp
HB 4–5.3 cm
T 3.1–4.2 cm
FA 3.2–3.6 cm
W 5.3–8 g

1140–2460 m

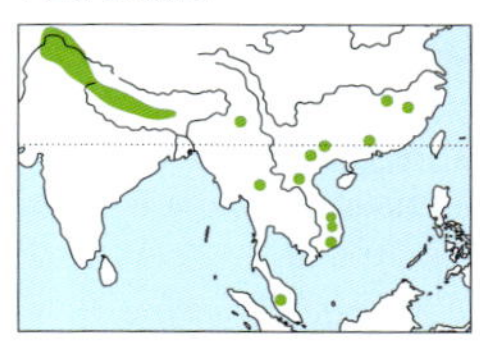

Hutton's Tube-nosed Bat
Murina huttonii

FR Murine de Hutton
DE Hutton-Röhrennase
ES Ratonero narizudo de Hutton

NE PK, N IN, NP, SE CN, N MM, N TH, N LA, VN, Pen MY; possibly BT

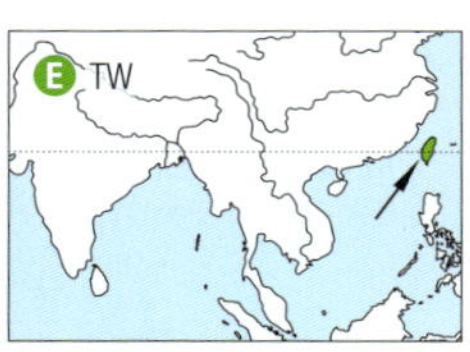

Taiwan Tube-nosed Bat
Murina puta

FR Murine de Taïwan
DE Taiwan-Röhrennase
ES Ratonero narizudo de Taiwán

Restricted to temperate montane forests

LC
Monotypic
HB 5.9–6.1 cm
T 3.2–3.6 cm
FA 3.3–3.9 cm

LC
Monotypic
HB 3.6–5.5 cm
T 3.2–4 cm
FA 2.7–3.5 cm
W 4.3–8 g

Annam Tube-nosed Bat
Murina annamitica

FR Murine annamite
DE Annam-Röhrennase
ES Ratonero narizudo annamita

500–1300 m

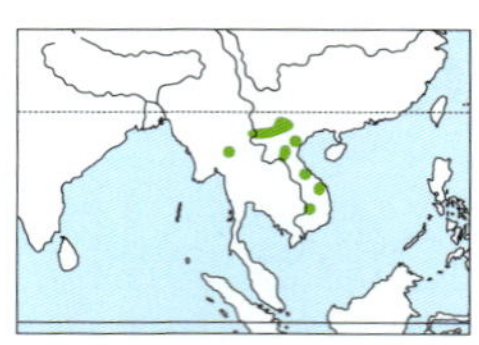

NW TH, VN, N & C LA, EC KH

DD
Monotypic
HB 3.5–4.5 cm
T 2.6–3.2 cm
FA 2.8–3.5 cm
W 4.2–4.6 g

100–400 m

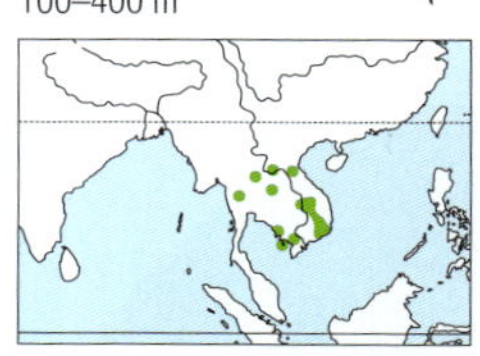

Walston's Tube-nosed Bat
Murina walstoni

FR Murine de Walston
DE Walston-Röhrennase
ES Ratonero narizudo de Walston

EC & WC TH, C & S LA, S VN (with Phu Quoc I), SE & SW KH

LC
2 sspp
HB 3.3–6.2 cm
T 2.2–3.5 cm
FA 2.9–3.3 cm
W 2.5–5.5 g

0–1540 m

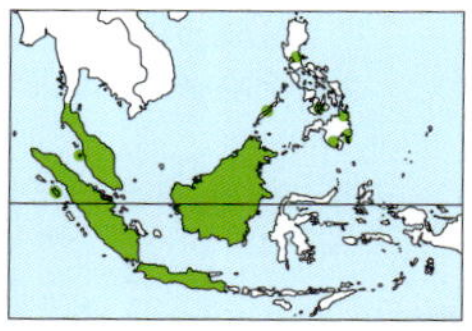

Brown Tube-nosed Bat
Murina suilla
FR Murine pourceau
DE Braune Röhrennase
ES Ratonero narizudo marrón

Thai-Malay Pen (with Langkawi I), Borneo, Sumatra (with Nias I), Java, PH; perhaps Sulawesi & New Guinea

0–2800 m

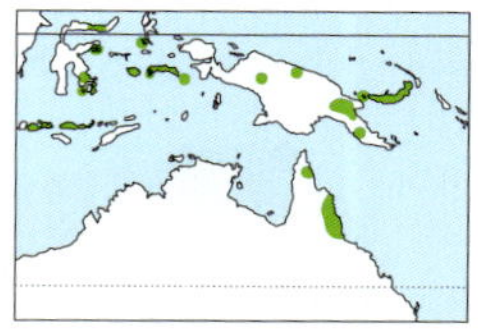

Flores Tube-nosed Bat
Murina florium
FR Murine de Florès
DE Flores-Röhrennase
ES Ratonero narizudo de Flores

N & SE Sulawesi (with offshore Is), Sumbawa, Flores, C Moluccas, New Guinea, Bismarck Arch, AU (NE Qld)

LC
Monotypic
HB 3.8–5.6 cm
T 3.1–3.9 cm
FA 3.2–3.7 cm
W 4.3–10.9 g

VU
Monotypic
HB 4.2–5.2 cm
T 3.2–4.1 cm
FA 3.5–3.8 cm
W 6–8.5 g

0–1200 m

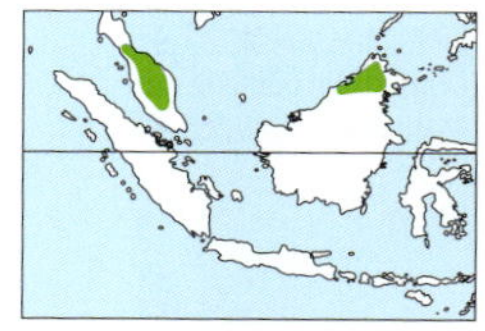

Bronze Tube-nosed Bat
Murina aenea
FR Murine bronzée
DE Bronzefarbige Röhrennase
ES Ratonero narizudo latón

Thai-Malay Pen, N Borneo (Sabah, Sarawak, BN, Kalimantan)

DD
Monotypic
HB c. 4.6 cm
T c. 3.6 cm
FA 3.1–3.6 cm

2150–3000 m

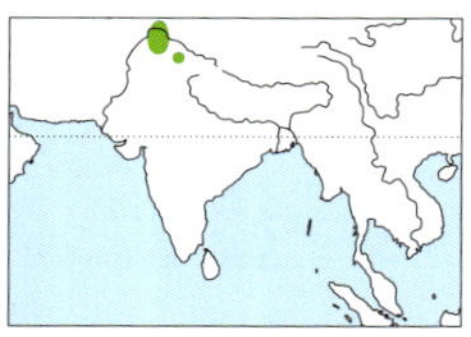

Scully's Tube-nosed Bat
Murina tubinaris
FR Murine de Scully
DE Scully-Röhrennase
ES Ratonero narizudo de Scully

N PK, NW IN (Jammu and Kashmir, Himachal Pradesh)

290–1140 m

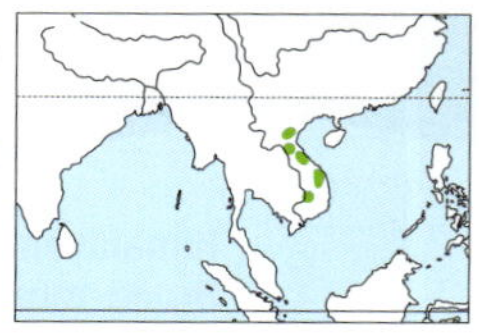

Fiona's Tube-nosed Bat
Murina fionae
FR Murine de Fiona
DE Fiona-Röhrennase
ES Ratonero narizudo de Fiona

N & C VN, C LA, EC KH

LC
Monotypic
HB 4.2–5.1 cm
T 3.7–4.4 cm
FA 3.4–4 cm
W 7.5–12 g

NE
2 sspp
HB 4.2–5.2 cm
T 2.8–4.2 cm
FA 3.2–3.6 cm
W 3–8 g

0–80 m

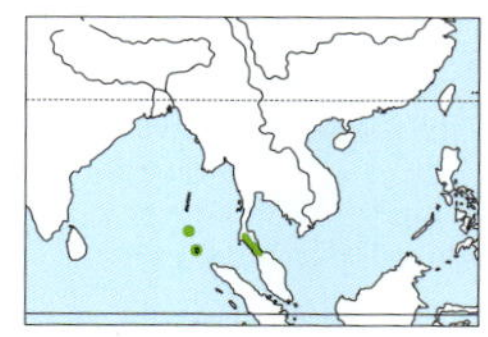

Guillen's Tube-nosed Bat
Murina guilleni
FR Murine de Guillen
DE Guillén-Röhrennase
ES Ratonero narizudo de Guillén

Pen TH, Nicobar Is (IN)

LC
2 sspp
HB 3.8–5 cm
T 2.6–4.3 cm
FA 2.9–3.7 cm
W 4.1–8.2 g

0–1650 m

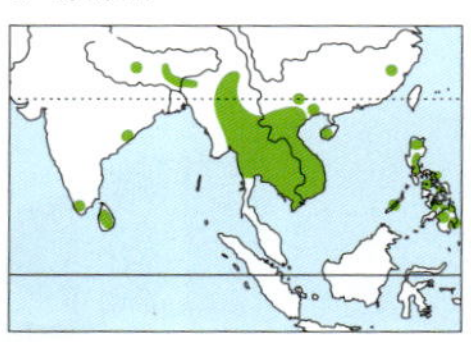

Round-eared Tube-nosed Bat
Murina cyclotis
FR Murine à oreilles rondes
DE Rundohrröhrennase
ES Ratonero narizudo de orejas redondas

C NP, NE & EC IN (tentatively S), LK, SE CN (Jiangxi, Guangxi, S Hainan), mainland SE Asia, PH

NE
Monotypic
HB 4–4.2 cm
T 3.2–4.6 cm
FA 3.4–3.9 cm
W 5.5–11.9 g

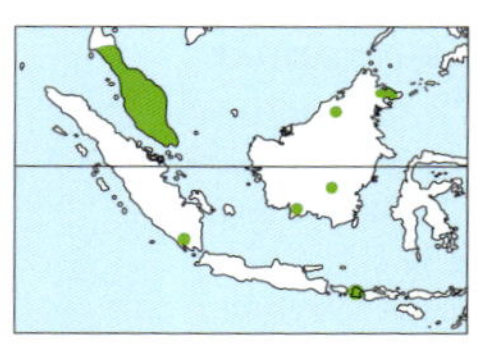

Peninsular Tube-nosed Bat
Murina peninsularis
FR Murine péninsulaire
DE Malaysia-Röhrennase
ES Ratonero narizudo peninsular

Thai-Malay Pen, S Sumatra (Lampung), Borneo, Lombok

c. 685 m

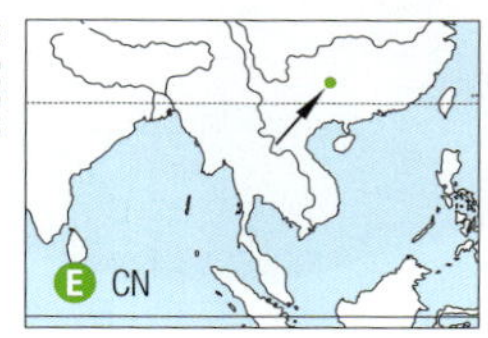

Libo Tube-nosed Bat
Murina liboensis
FR Murine du Libo
DE Libo-Röhrennase
ES Ratonero narizudo de Libo

Guizhou (known from type locality in Libo County)

NE
Monotypic
HB 3.7–4.4 cm
T 3.1–3.8 cm
FA 2.8–3.4 cm
W 4.5–5.5 g

NE
Monotypic
HB c. 4 cm
T c. 3.8 cm
FA c. 3.2 cm
W c. 5 g

c. 1780 m

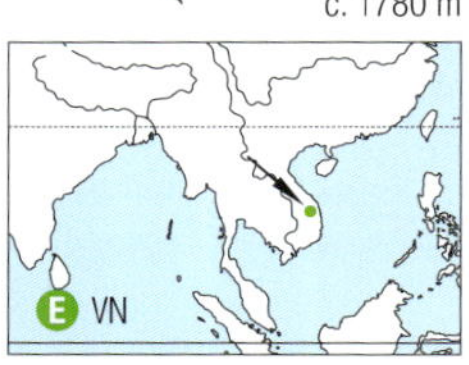

Kon Tum Tube-nosed Bat
Murina kontumensis
FR Murine du Kon Tum
DE Kon-Tum-Röhrennase
ES Ratonero narizudo de Kon Tum

Kon Tum (known from type locality in Dak Glei District)

c. 510 m

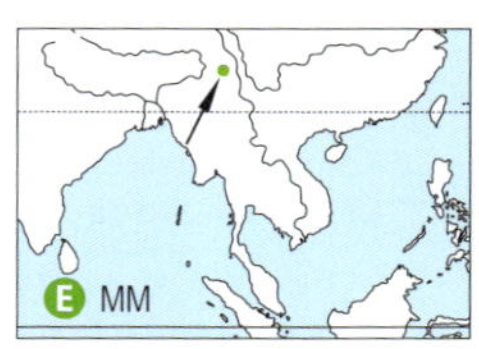

Hkakabo Razi Tube-nosed Bat
Murina hkakaboraziensis
FR Murine du Hkakabo Razi
DE Hkakabo-Razi-Röhrennase
ES Ratonero narizudo de Hkakabo Razi

Kachin (known from type locality in Hkakabo Razi NP)

NE
Monotypic
HB c. 3.5 cm
T c. 3 cm
FA c. 3 cm
W c. 3.4 g

VU
Monotypic
HB 4–4.4 cm
T 2.9–4.1 cm
FA 2.8–3.3 cm
W 3.8–5 g

0–1500 m

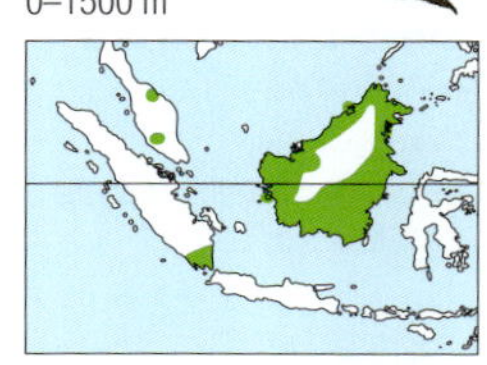

Gilded Tube-nosed Bat
Murina rozendaali
FR Murine de Rozendaal
DE Rozendaal-Röhrennase
ES Ratonero narizudo de Rozendaal

Thai-Malay Pen, lowland Borneo (most records in N), S Sumatra (Lampung)

Elery's Tube-nosed Bat
Murina eleryi
FR Murine d'Elery
DE Elery-Röhrennase
ES Ratonero narizudo de Elery

525–2550 m

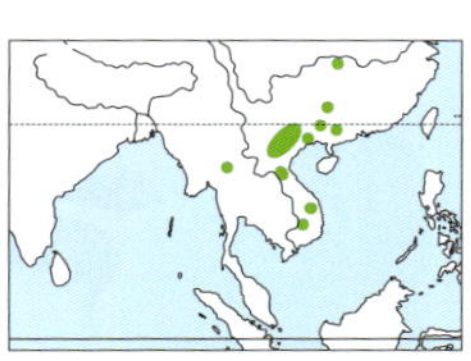

SE CN (Hunan, Guizhou, Guangxi, Guangdong), N TH, C LA, VN, E KH

LC Monotypic
HB 3.1–3.9 cm
T 2.6–3.2 cm
FA 2.7–3.1 cm
W 2.5–5.5 g

CR
Monotypic
HB c. 3.4 cm
T 3–3.1 cm
FA 2.8–3 cm
W 3.5–4 g

340–370 m

Bala Tube-nosed Bat
Murina balaensis
FR Murine de Bala
DE Bala-Röhrennase
ES Ratonero narizudo de Bala

Known from type locality in Hala-Bala WS (Bala Forest)

LC
Monotypic
FA 2.8–3.1 cm

40–2200 m

Hidden Tube-nosed Bat
Murina recondita
FR Murine terne
DE Versteckte Röhrennase
ES Ratonero narizudo oculto

Lowland and montane forests

LC
Monotypic
FA 2.8–3.2 cm

1400–2700 m

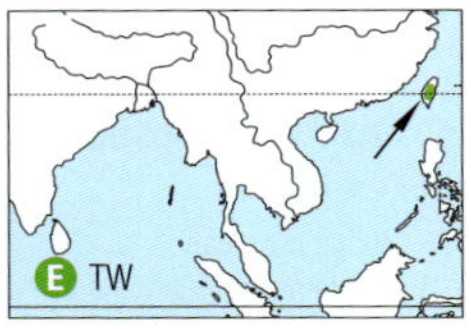

Gracile Tube-nosed Bat
Murina gracilis
FR Murine gracile
DE Zierliche Röhrennase
ES Ratonero narizudo grácil

Broadleaf, mixed and conifer forests

Genus *EUDISCOPUS*

(1 species)

Disk-footed Bat
Eudiscopus denticulus
FR Murin à disques
DE Diskusfüßige Fledermaus
ES Ratonero de ventosas

LC
Monotypic
HB 4.1–4.9 cm
T 3.6–4.5 cm
FA 3.5–3.8 cm
W 4.6–7.8 g

30–1300 m

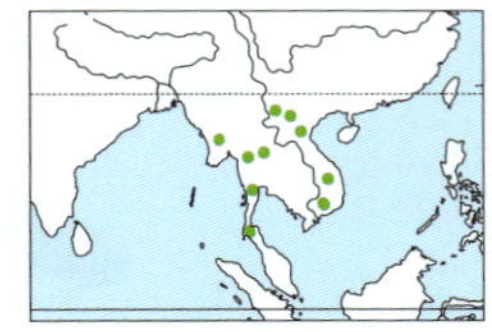

MM (Bago Region), TH, N LA (Phongsaly Prov), VN

Genus *SUBMYOTODON*

(3 species)

LC Monotypic
HB 3.5–4 cm T 3.1–4 cm
FA 3.2–3.5 cm

200–3000 m

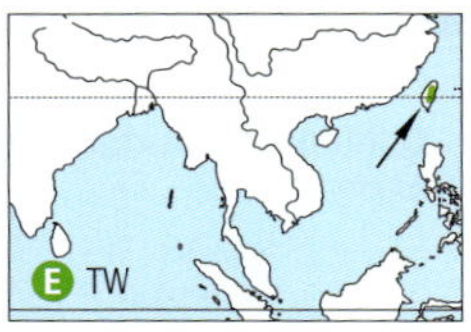

Taiwan Broad-muzzled Bat
Submyotodon latirostris
FR Murin de Kishida
DE Taiwan-Breitmaulfledermaus
ES Ratonero de Taiwán

Broadleaf and mixed coniferous forests and urban areas

NE Monotypic
HB c. 4.1 cm T c. 2.5 cm
FA 3.2–3.4 cm

2000–3250 m

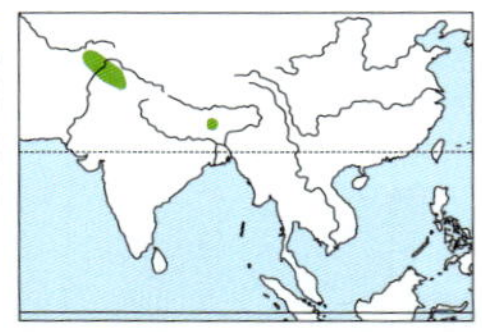

Himalayan Broad-muzzled Bat
Submyotodon caliginosus
FR Murin obscur
DE Himalaya-Breitmaulfledermaus
ES Ratonero oscuro

NE AF, N PK (Khyber Pakhtunkhwa, Punjab), NW & N IN (Jammu and Kashmir, Punjab, Sikkim); perhaps NP

NE
Monotypic
HB c. 4.1 cm
T c. 3.4 cm
FA c. 3.4 cm

Moupin Broad-muzzled Bat
Submyotodon moupinensis
FR Murin du Moupin
DE Moupin-Breitmaulfledermaus
ES Ratonero de Moupin

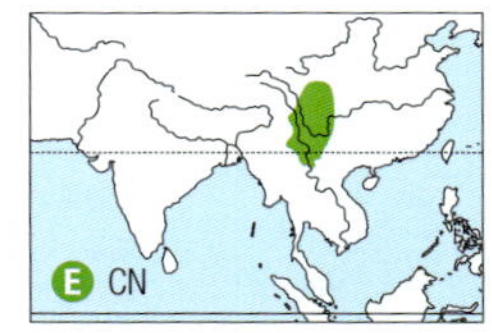

Sichuan, Yunnan; possibly MM

Genus *MYOTIS*

(139 species)

NT
Monotypic
HB 4–4.6 cm
T 3.6–4.3 cm
FA 3.5–3.9 cm
W 4–11 g

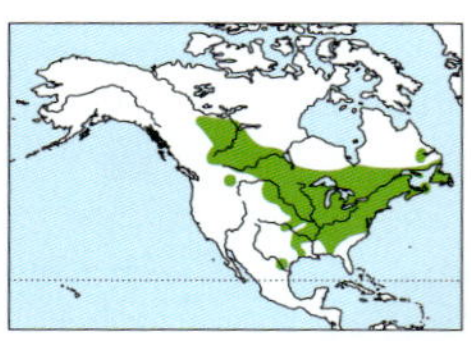

Northern Myotis
Myotis septentrionalis
FR Murin nordique
DE Nordamerika-Mausohr
ES Ratonero nórdico

CA (E from SW Yukon), C & E US (E from E Montana, S to NW Florida, recorded in W Montana & S Texas)

LC 2 sspp
HB 5.1–5.2 cm T 3.4–4.9 cm
FA 3.7–4 cm W 6–8 g

up to 2200 m

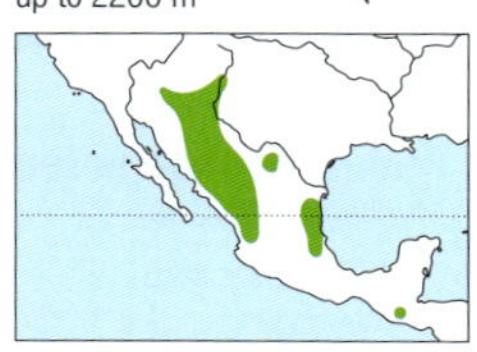

South-western Myotis
Myotis auriculus
FR Murin du Mexique
DE Mexiko-Mausohr
ES Ratonero de México

US (mts of SE Arizona & SW New Mexico) S to MX (to Jalisco, N Coahuila, Tamaulipas & Veracruz)

300–3300 m

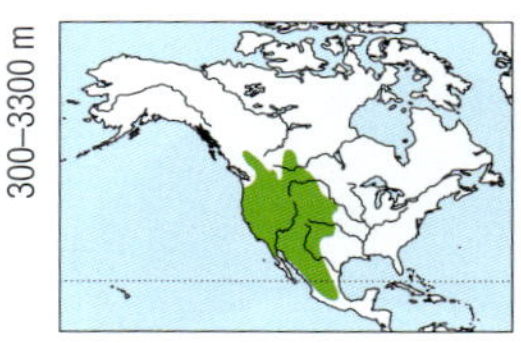

Western Small-footed Myotis
Myotis ciliolabrum
FR Murin pygmée
DE Westliches Kleinfuß-Mausohr
ES Ratonero pigmeo de cara negra

CA (SE British Columbia E to SW Saskatchewan) S through W & C US to C MX

LC
Monotypic
HB 3.9–4.4 cm
T 2.9–4.9 cm
FA 3.1–3.5 cm
W 3–6 g

EN
Monotypic
HB 4–4.4 cm
T 3.3–3.8 cm
FA 3–3.5 cm
W 3–6 g

700–1125 m

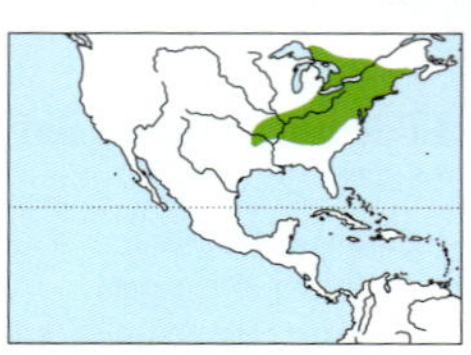

Eastern Small-footed Myotis
Myotis leibii
FR Murin de Leib
DE Östliches Kleinfuß-Mausohr
ES Ratonero de Leib

SE CA (SE Ontario, SW Quebec), E US (from SE Oklahoma E to New England & NW South Carolina)

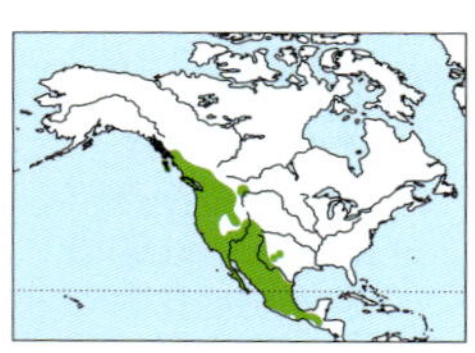

California Myotis
Myotis californicus
FR Murin de Californie
DE Kalifornien-Mausohr
ES Ratonero de California

Extreme SE Alaska, SW CA (British Columbia) S through W US & MX (with Baja California) to C GT

LC
4 sspp
HB 3.8–4.8 cm
T 3.2–4.5 cm
FA 3.1–3.5 cm
W 3.3–5.4 g

EN
Monotypic
HB 3.9–5.1 cm
T 2.9–4.1 cm
FA 3.4–4.1 cm

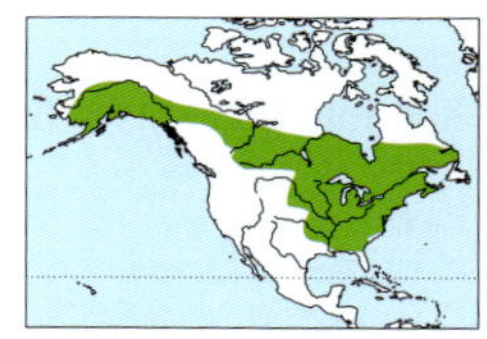

Eastern Little Brown Myotis
Myotis lucifugus
FR Murin brun
DE Kleines Nordamerika-Mausohr
ES Ratonero marrón pequeño

S Alaska, CA (Northwest Territories) throughout E seaboard to N Florida (US)

NE
Monotypic
HB 4.5–5.2 cm
T 3.5–3.9 cm
FA 3.5–3.8 cm

W CA (British Columbia, Alberta) S to Pacific NW US

North-western Little Brown Myotis
Myotis alascensis
FR Murin d'Alaska
DE Nordwestliches Kleines Nordamerika-Mausohr
ES Ratonero marrón pequeño de Alaska

NE
Monotypic
HB 4.1–5 cm
T 3.4–4.2 cm
FA 3.6–4 cm

SW CA, W & C US, except for coastal regions

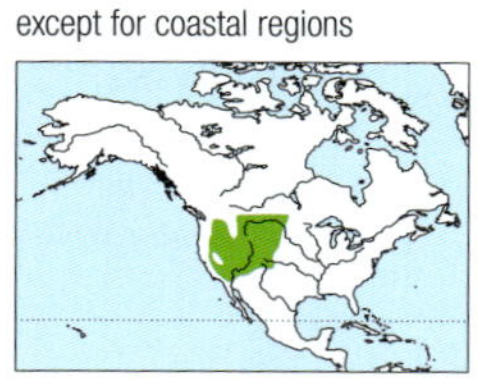

Western Little Brown Myotis
Myotis carissima
FR Murin du Wyoming
DE Westliches Kleines Nordamerika-Mausohr
ES Ratonero marrón pequeño de Wyoming

NE
Monotypic
HB 5.3–5.4 cm
T 3.9–4.4 cm

C border between British Columbia and Alberta

Alberta Little Brown Myotis
Myotis pernox
FR Murin d'Alberta
DE Alberta Kleines Nordamerika-Mausohr
ES Ratonero marrón pequeño de Alberta

NE
Monotypic

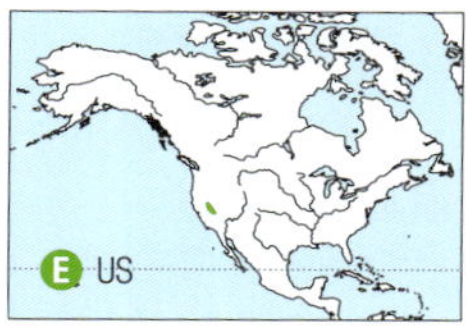

California Mts

Relict Little Brown Myotis
Myotis relictus
FR Murin d'Inyo
DE Reliktäres Kleines Nordamerika-Mausohr
ES Ratonero marrón pequeño de Inyo

LC
Monotypic
HB 4–5.5 cm
T 3.4–4.4 cm
FA 3.5–3.9 cm
W 4–6 g

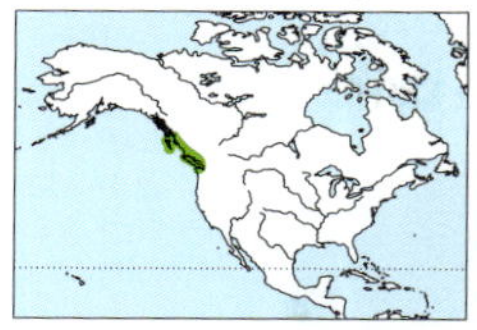

Keen's Myotis
Myotis keenii
FR Murin de Keen
DE Keen-Mausohr
ES Ratonero de Keen

From SE Alaska S along CA (British Columbia, with Haida Gwaii & Vancouver I) to NW Washington (US)

0–2830 m

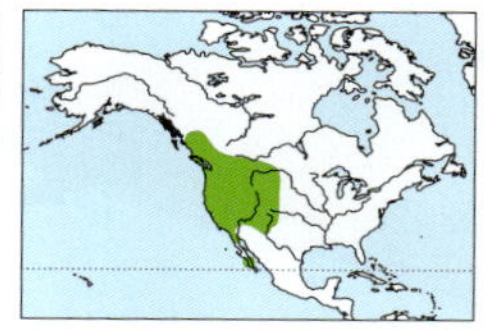

Long-eared Myotis
Myotis evotis
FR Murin à oreilles longues
DE Langohriges Mausohr
ES Ratonero de orejas largas

SW CA (British Columbia E to S Saskatchewan) S through W US to NW Mexico (Baja California Pen)

LC
6 sspp
HB 4.3–5.2 cm
T 3.6–4.5 cm
FA 3.7–4 cm
W 4–9 g

LC
4 sspp
HB 4.5–5.7 cm
T 3.5–4.5 cm
FA 4–4.5 cm
W 6–12 g

0–2850 m

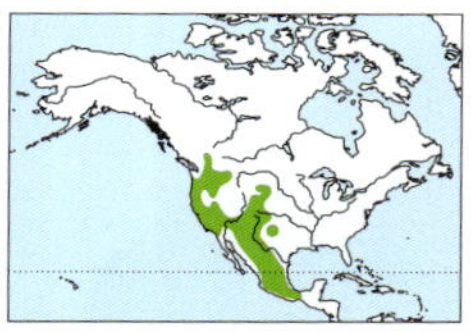

Fringed Myotis
Myotis thysanodes
FR Murin à queue frangée
DE Nordamerika-Fransenfledermaus
ES Ratonero de flecos

SW CA (S British Columbia), W & C US (E to SW South Dakota, NW Nebraska & N Texas) S to S MX

2000–3000 m

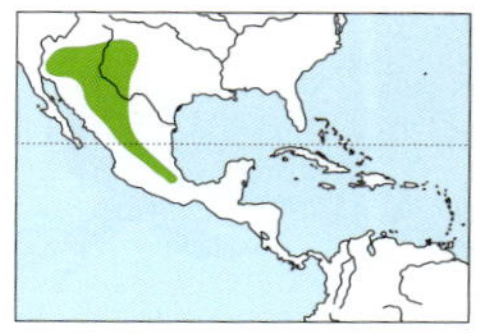

Arizona Myotis
Myotis occultus
FR Murin de Hollister
DE Arizona-Mausohr
ES Ratonero de Arizona

US (SE California, Arizona, S Colorado, New Mexico, perhaps S Utah & W Texas) S to C MX; limits uncertain

LC
Monotypic
HB 4.1–5.4 cm
T 2.9–4.1 cm
FA 3.6–4.2 cm
W 7–9 g

LC
4 sspp
HB 4.3–6.3 cm
T 4–5 cm
FA 3.8–4.2 cm
W 5–10 g

60–3770 m

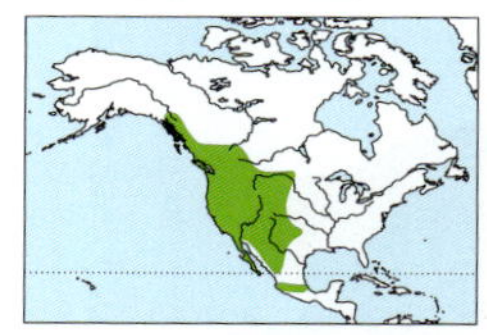

Long-legged Myotis
Myotis volans
FR Murin à longues pattes
DE Langbeiniges Mausohr
ES Ratonero de patas largas

Extreme SE Alaska, W CA (SW Yukon E to S Saskatchewan) & W US S to C MX

1000–3200 m

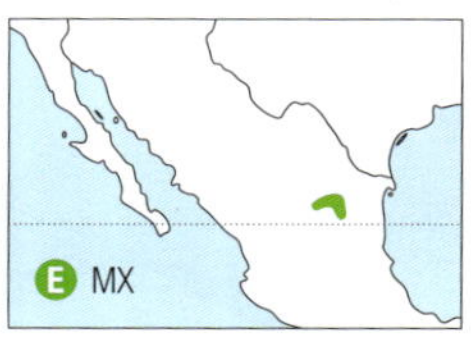

Flat-headed Myotis
Myotis planiceps
FR Murin à tête plate
DE Flachkopf-Mausohr
ES Ratonero de cabeza plana

Very small area in Coahuila, Nuevo León & Zacatecas

EN
Monotypic
HB 3–5.1 cm
T 2.1–2.5 cm
FA 2.6–2.7 cm
W 5–7 g

NT
Monotypic
HB 4.2–4.8 cm
T 2.8–4.5 cm
FA 3.5–4 cm
W 5–10 g

0–1745 m

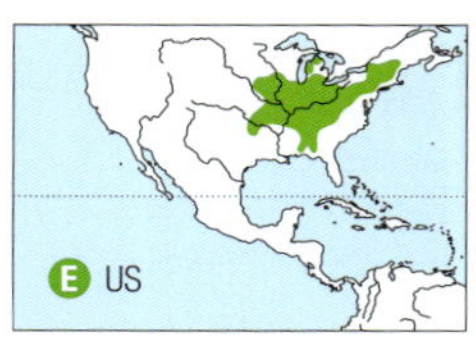

Indiana Myotis
Myotis sodalis
FR Murin d'Indiana
DE Indiana-Mausohr
ES Ratonero de Indiana

From Oklahoma & Iowa N & E to Michigan, New England & N New Jersey, S to Alabama & Arkansas

Cinnamon Myotis
Myotis fortidens
FR Murin à dents fortes
DE Zimtbraunes Mausohr
ES Ratonero dentudo

0–870 m

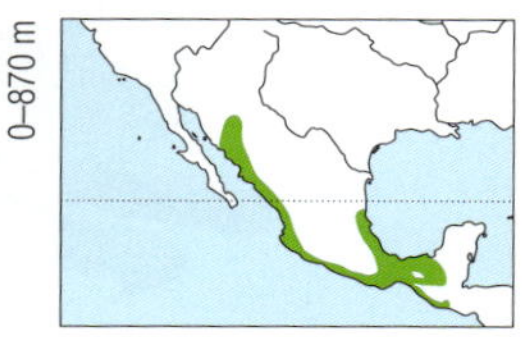

MX (S from Sonora along narrow strip on Pacific coast & from Veracruz S to Chiapas), S GT

LC 2 sspp
HB 4.6–5.4 cm
T 3.1–4 cm
FA 3.5–3.8 cm
W 5–8 g

EN
Monotypic
HB 4.4–4.7 cm
T 2.6–3.3 cm
FA 2.9–3.3 cm
W 2–3 g

0–600 m

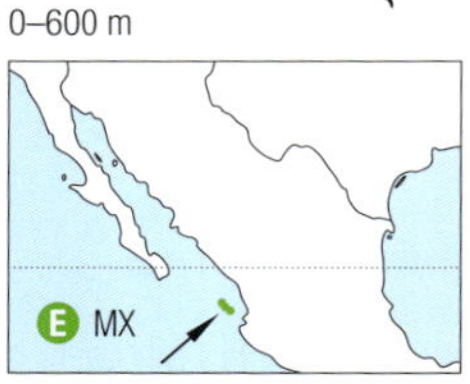

Findley's Myotis
Myotis findleyi
FR Murin de Findley
DE Findley-Mausohr
ES Ratonero de Findley

Marías Is (Madre, Magdalena, Cleofas), off Nayarit

975–2500 m

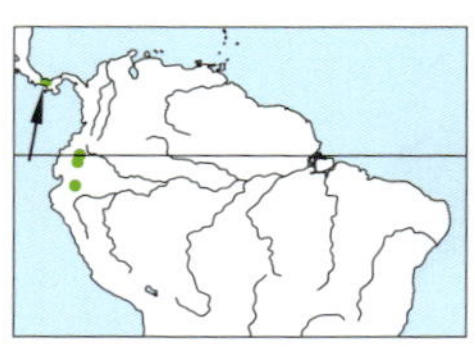

Armien's Myotis
Myotis armiensis
FR Murin d'Armien
DE Armién-Mausohr
ES Ratonero de Armien

PA (Chiriquí Prov, La Amistad International Park), CR (Valle del Silencio), EC (E Andes)

NE
Monotypic
HB 7.7–9.2 cm
T 3.2–4.8 cm
FA 3.6–3.9 cm
W 4.5–5.6 g

0–2500 m

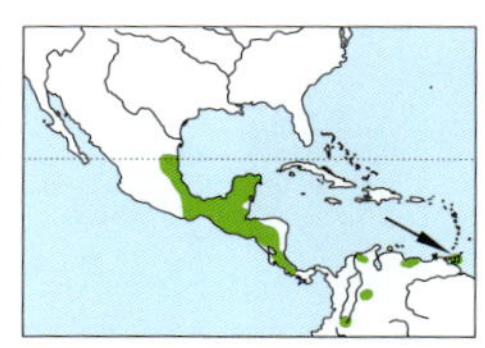

Northern Hairy-legged Myotis
Myotis pilosatibialis
FR Murin à pattes velues
DE Nördliches Haarfüßiges Mausohr
ES Ratonero de patas peludas

EC MX through Central America to CO, N VE & TT (Trinidad)

NE
Monotypic
HB 4.5–5.1 cm
T 3.4–3.9 cm
FA 3.2–3.9 cm
W 4–7 g

550–2500 m

Southern Hairy-legged Myotis
Myotis keaysi
FR Murin de Keays
DE Südliches Haarfüßiges Mausohr
ES Ratonero de Keays

Highlands in VE to Andes of CO, EC, PE, BO & NW AR

LC
Monotypic
HB 5.6–6.1 cm
T 2.9–4.1 cm

NT
Monotypic
HB 4.9–5 cm
T 3.3–4.4 cm
FA 3.4–4.2 cm
W 5–9 g

Red Myotis
Myotis ruber
FR Murin rouge
DE Rotes Mausohr
ES Ratonero rojizo

500–1500 m

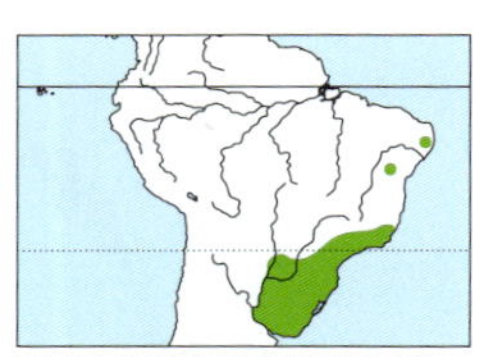

NE, SE & S BR, E PY, NE AR, UY

NE
Monotypic
FA 3.5–3.8 cm

Moratelli's Myotis
Myotis moratellii
FR Murin de Moratelli
DE Moratelli-Mausohr
ES Ratonero de Moratelli

0–150 m

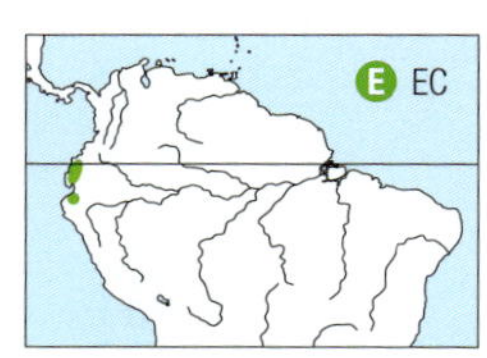

Restricted to W coastal area

LC
Monotypic
HB 4.3–4.5 cm
T 2.8–4.8 cm
FA 3.2–3.9 cm
W 4–7 g

0–2000 m

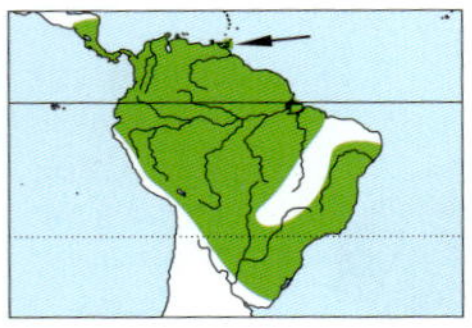

Riparian Myotis
Myotis riparius

FR Murin des ruisseaux
DE Ufermausohr
ES Ratonero ripario

From S HN, NI, CR & PA to South America (except CL), TT (Trinidad)

0–750 m

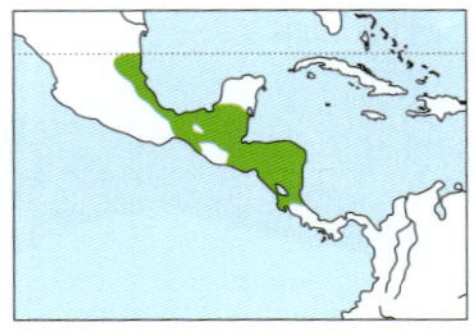

Elegant Myotis
Myotis elegans

FR Murin élégant
DE Elegantes Mausohr
ES Ratonero elegante

Lowlands from E MX (San Luis Potosí) to N CR

LC
Monotypic
HB 3.9–4.5 cm
T 3.1–3.5 cm
FA 3.2–3.4 cm
W 3–5 g

NE
Monotypic
FA 3.4–3.6 cm

Pampas Myotis
Myotis pampa

30–240 m

FR Murin des pampas
DE Pampa-Mausohr
ES Ratonero de la pampa

Known from 2 localities
(Artigas & Tacuarembó dep)

30–600 m

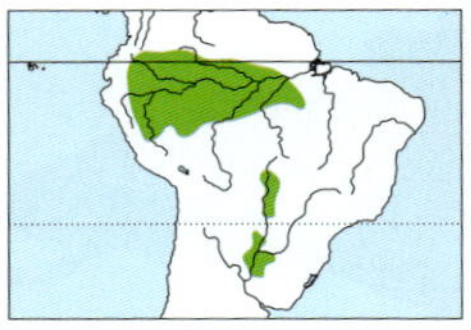

Velvety Myotis
Myotis simus

FR Murin camus
DE Samtiges Mausohr
ES Ratonero aterciopelado

Amazon Basin (CO, EC, PE, BR), Paraná Basin (C BR to NE AR); records in BO & PY assigned to *M. midastactus*

DD
Monotypic
HB 5–5.2 cm
T 3.1–3.7 cm
FA 3.5–4 cm
W 5–11 g

NE
Monotypic
HB 5–5.9 cm
T 3.6–4.1 cm
FA 3.8–4.1 cm
W 6–11 g

150–250 m

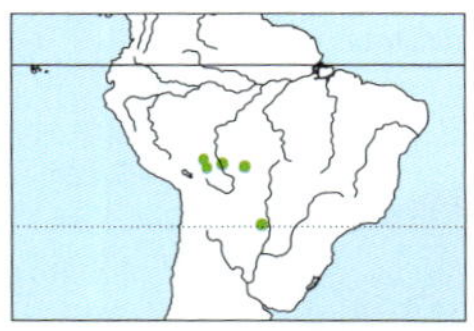

Golden Myotis
Myotis midastactus

FR Murin midas
DE Goldgelbes Mausohr
ES Ratonero dorado

BO (Beni, Santa Cruz), PY (Presidente Hayes)

c. 0 m

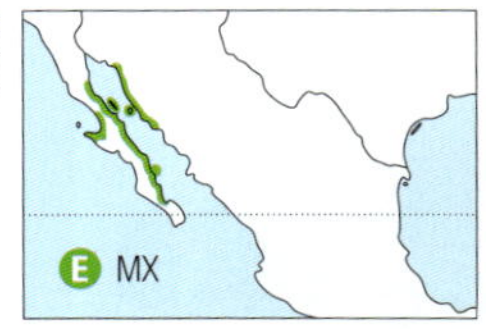

Fish-eating Myotis
Myotis vivesi

FR Murin pêcheur
DE Fischfressendes Mausohr
ES Ratonero pescador

Baja California Pen (Pacific coast & Gulf of California with many Is), Sonora (Bay of San Jorge S to Guaymas)

VU
Monotypic
HB 8.5–8.8 cm
T 5–7.8 cm
FA 5.4–6.3 cm
W 22–28 g

VU
Monotypic
HB 4.7–5.5 cm
T 3.3–4.1 cm
FA 4.1–4.6 cm
W 7–16 g

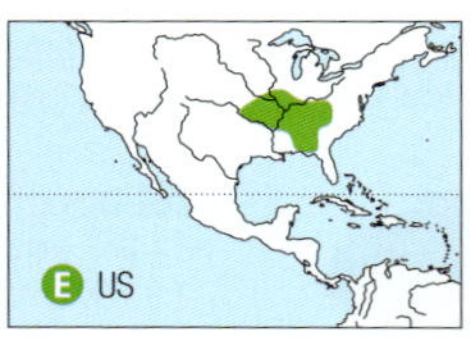

Gray Myotis
Myotis grisescens

FR Murin gris
DE Graues Mausohr
ES Ratonero gris

Kansas, Oklahoma, Missouri & Illinois
E to W Virginia & W North Carolina, S to Alabama & NW Florida

LC
Monotypic
HB 4.5–5.1 cm
T 2.6–4.4 cm
FA 3.3–4 cm
W 5–8 g

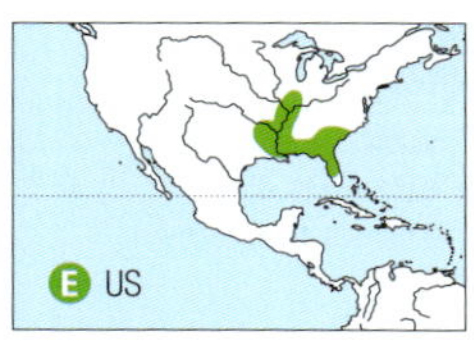

South-eastern Myotis
Myotis austroriparius

FR Murin du Mississippi
DE Südöstliches Mausohr
ES Ratonero del Misisipi

S Illinois & S Indiana S to E Texas & Florida,
E through W Kentucky, W Tennessee & Alabama to Carolinas

Yuma Myotis
Myotis yumanensis

FR Murin de Yuma
DE Yuma-Mausohr
ES Ratonero de Yuma

LC 6 sspp
HB 4.8–4.9 cm T 2.7–4 cm
FA 3.2–3.8 cm W 4–7 g

up to 3350 m

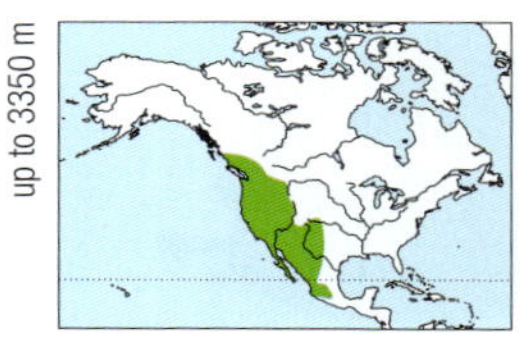

SW CA (British Columbia), W US
(E to W Montana, Oklahoma & Texas),
W & C MX (Baja California, S to Morelos)

Cave Myotis
Myotis velifer

FR Murin des grottes
DE Höhlenmausohr
ES Ratonero de las cavernas

0–3000 m

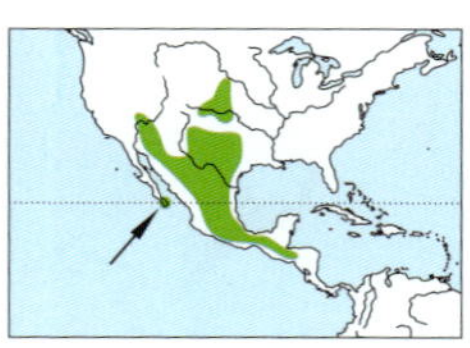

C & S US (S from S Arizona, S Nebraska & Oklahoma) to MX (with S Baja California Pen), GT & HN

LC 4 sspp
HB 4.1–5.4 cm
T 3.5–5.5 cm
FA 3.6–4.7 cm
W 4–11 g

DD
Monotypic
HB c. 3.5 cm
T c. 3.5 cm
FA c. 4.2 cm

c. 1305 m

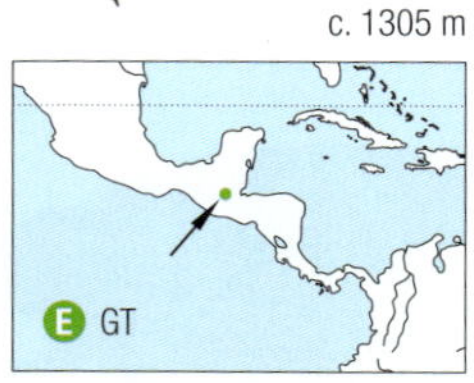

Guatemalan Myotis
Myotis cobanensis

FR Murin du Guatemala
DE Guatemala-Mausohr
ES Ratonero de Guatemala

Known from type locality (Cobán Cathedral, Alta Verapaz)

Chilean Myotis
Myotis chiloensis

FR Murin du Chili
DE Chile-Mausohr
ES Ratonero de Chile

LC Monotypic
HB 4.3–4.5 cm T 2.8–3.9 cm
FA 3.4–4.2 cm W 6–9 g

CL (from Valdivia S to N shore of Navarino I), W & S AR (from Neuquén to Tierra del Fuego); dubious in N CL

Silver-tipped Myotis
Myotis albescens

FR Murin poudré
DE Silberspitzen-Mausohr
ES Ratonero cenizo

0–1500 m

LC Monotypic
HB 4.6–4.9 cm T 3.5–4.6 cm
FA 3.2–3.9 cm W 4–8 g

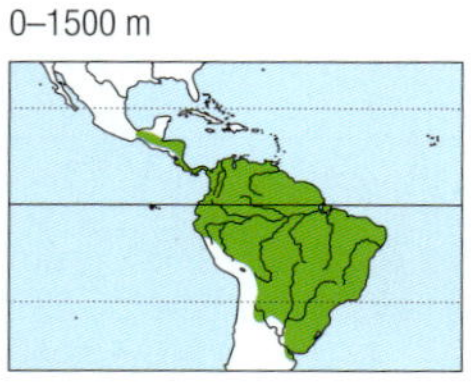

From MX (S Veracruz) S through Central & South America to PE, BO, S BR, PY, UY & N AR

LC
Monotypic
HB 3.8–5.2 cm
T 4.2–4.8 cm
FA 3.6–4.4 cm
W 5–10 g

Yellowish Myotis
Myotis levis

300–1900 m

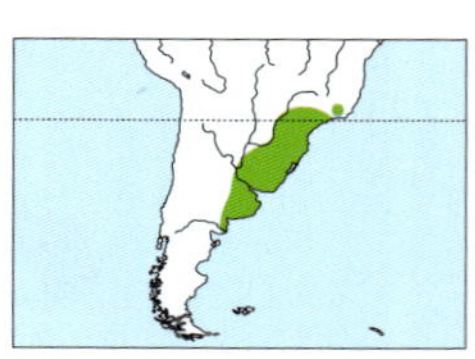

FR Murin léger
DE Gelbliches Mausohr
ES Ratonero amarillento

Atlantic Forest
from SE & S BR to EC AR & UY

350–3890 m

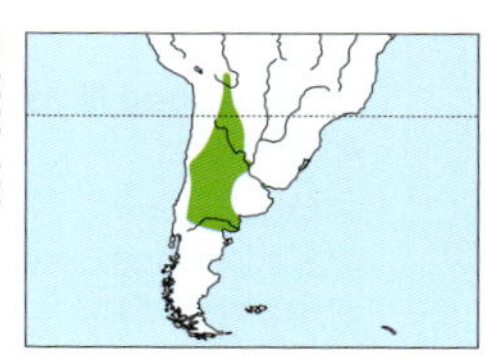

Dinelli's Myotis
Myotis dinellii

FR Murin de Dinelli
DE Dinelli-Mausohr
ES Ratonero de Dinelli

C BO S to C AR & extreme E CL

LC
Monotypic
HB 4.5–5.5 cm
T 2.2–4.4 cm
FA 3.4–3.8 cm
W 4–8 g

Izecksohn's Myotis
Myotis izecksohni
FR Murin d'Izecksohn
DE Izecksohn-Mausohr
ES Ratonero de Izecksohn

250–1200 m

DD Monotypic
HB 4.6–6.4 cm T 3.5–4.2 cm
FA 3.3–3.9 cm W 5–9 g

Atlantic Forest and cerrado in SE & S BR to NE AR

15–900 m

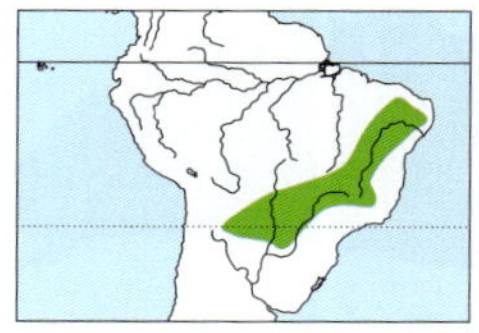

LaVal's Myotis
Myotis lavali
FR Murin de LaVal
DE LaVal-Mausohr
ES Ratonero de LaVal

Diagonal corridor from NE BR S to PY & NW AR

LC
Monotypic
HB 4.1–5 cm
T 3.1–4.2 cm
FA 3.1–3.7 cm
W 3–7 g

Montane Myotis
Myotis oxyotus
FR Murin à oreilles effilées
DE Gebirgsmausohr
ES Ratonero montano

1000–3450 m

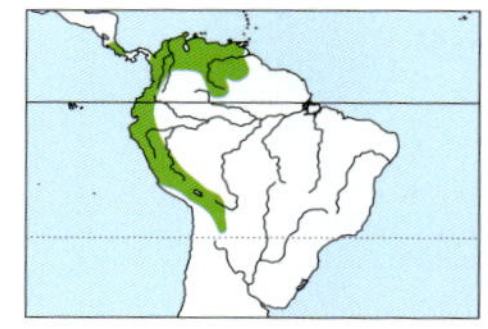

LC 2 sspp
HB 4–5.3 cm T 4.1–4.6 cm
FA 3.7–4.3 cm W 4–7 g

CR & PA (Cordillera Central, Cordillera de Talamanca), CO, VE, EC, PE, BO & NW AR

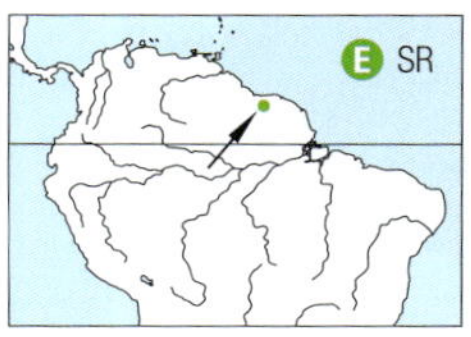

Clyde Jones's Myotis
Myotis clydejonesi
FR Murin de Clyde Jones
DE Clyde-Jones-Mausohr
ES Ratonero de Clyde Jones

Known only from type locality (an island in R Coppername)

NE
Monotypic
HB c. 5 cm
T c. 3.8 cm
FA c. 3.5 cm
W c. 4.3 g

NT
Monotypic
FA 3.6–4 cm
W 5–8 g

Schwartz's Myotis
Myotis martiniquensis
FR Murin de Martinique
DE Schwartz-Mausohr
ES Ratonero de Schwartz

Forested habitats

Dominican Myotis
Myotis dominicensis
FR Murin de la Dominique
DE Dominika-Mausohr
ES Ratonero dominicano

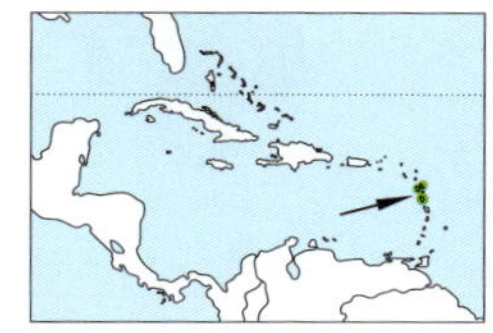

Lesser Antilles (GP, DM)

VU
Monotypic
HB 4.6–4.8 cm
T 2.8–3.6 cm
FA 3.3–3.5 cm
W 3–5 g

VU
Monotypic
HB 4.5–4.7 cm
T 3.5–3.7 cm
FA 3.3–3.7 cm
W c. 5 g

Barbados Myotis
Myotis nyctor
FR Murin de Barbade
DE Barbados-Mausohr
ES Ratonero de Barbados

50–300 m

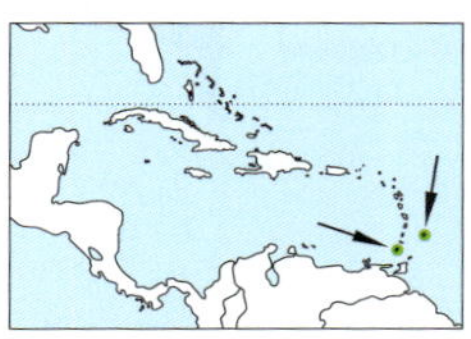

Lesser Antilles (BB, GD)

0–3475 m

Atacama Myotis
Myotis atacamensis
FR Murin d'Atacama
DE Atacama-Mausohr
ES Ratonero de Atacama

CW PE to CW CL

EN
Monotypic
HB 4.1–4.5 cm
T 3.1–4 cm
FA 3.1–3.4 cm
W 3–4 g

NE
Monotypic
HB 4.1–4.9 cm
T 2.6–3.1 cm
FA 2.9–3.2 cm
W 3.3–5 g

200–335 m

Baker's Myotis
Myotis bakeri
FR Murin de Baker
DE Baker-Mausohr
ES Ratonero de Baker

Known from 2 disjunct localities (Lambayeque, Lima)

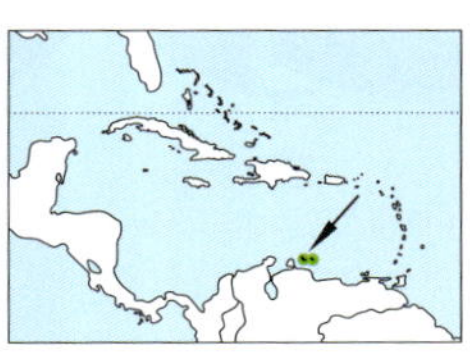

Curaçao Myotis
Myotis nesopolus
FR Murin de Curaçao
DE Curaçao-Mausohr
ES Ratonero de Curaçao

Netherlands Antilles (CW, Bonaire)

LC
Monotypic
HB c. 3.4 cm
T c. 3.6 cm
FA 2.9–3.1 cm

Lara Myotis
Myotis larensis
FR Murin du Lara
DE Lara-Mausohr
ES Ratonero de Lara

50–500 m

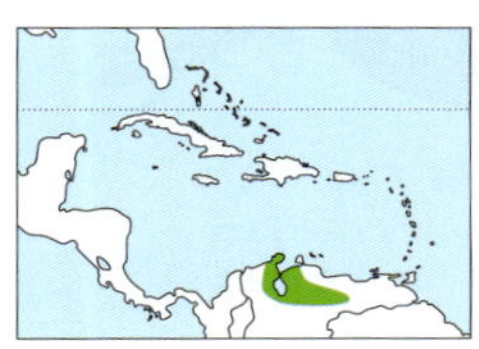

NE CO, NW VE

NE
Monotypic
FA 3.1–3.3 cm
W 3–5 g

Colombian Black Myotis
Myotis caucensis
FR Murin du Cauca
DE Kolumbien-Mausohr
ES Ratonero del Cauca

NE Monotypic
HB 4.7–4.9 cm T 3.4–4.1 cm
FA 3.4–3.8 cm W 4–6.5 g

200–2600 m

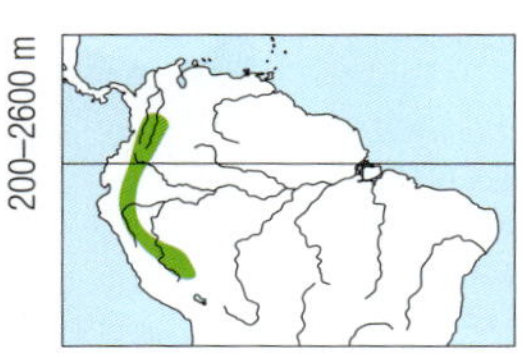

Andes of CO, EC & PE (with itermontane valleys and adjacent Amazon lowlands); perhaps N BO & W BR

NE
Monotypic
HB 3.5–4.3 cm
T 3.2–3.5 cm
FA 3.1–3.3 cm
W c. 4 g

0–260 m

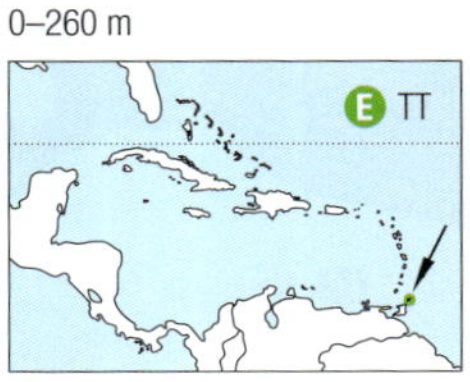

Sir David Attenborough's Myotis
Myotis attenboroughi
FR Murin d'Attenborough
DE David-Attenborough-Mausohr
ES Ratonero de David Attenborough

Tobago

Handley's Myotis
Myotis handleyi
FR Murin de Handley
DE Handley-Mausohr
ES Ratonero de Handley

NE
Monotypic
HB 4–4.6 cm T 3.3–4.2 cm
FA 3.4–3.7 cm W 3.5–6 g

1100–2090 m

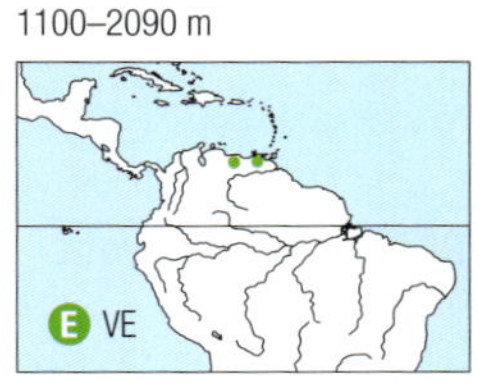

Cordillera de la Costa (Distrito Capital, Aragua, Miranda), Macizo Oriental (Monagas); perhaps other ranges

150–225 m

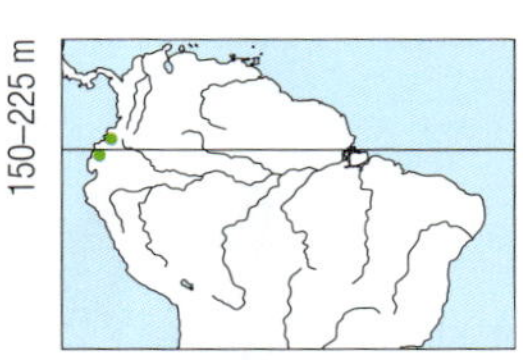

Diminutive Myotis
Myotis diminutus
FR Murin du Choco
DE Winziges Mausohr
ES Ratonero diminuto

W of Andes in SW CO (La Guayacana, Nariño) & WC EC (Los Ríos)

DD
Monotypic
HB c. 4.4 cm
T c. 3.3 cm
FA c. 3.3 cm
W 3.5–4 g

NE

Monotypic
HB 3.9–5.2 cm
T 2.8–3.9 cm
FA 3.3–3.8 cm
W 4–5 g

Carter's Myotis
Myotis carteri
FR Murin de Carter
DE Carter-Mausohr
ES Ratonero de Carter

From Nayarit & Michoacán to Balsas Basin (S México State)

Common Black Myotis
Myotis nigricans
FR Murin noirâtre
DE Schwarzes Mausohr
ES Ratonero hollinoso

LC
3 sspp
HB 3.9–5.2 cm T 2.8–3.9 cm
FA 3.1–3.6 cm W 3–7 g

0–2500 m

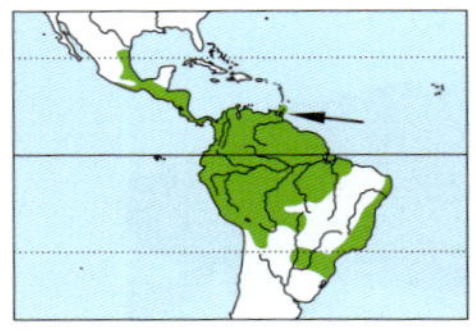

MX (from S Tamaulipas)
S through South & Central America
to PE, BO, N AR, PY & BR, TT

0–1800 m

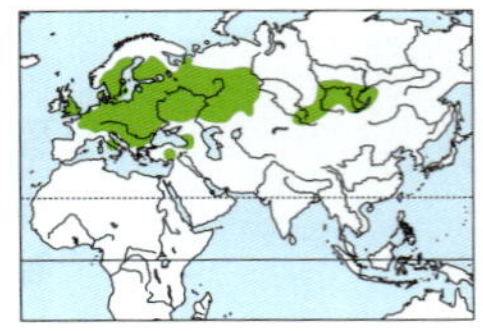

N & C Europe E to W Siberia,
isolated in C IT, TR, Caucasus
& C Asia; range uncertain

LC Monotypic
HB 3.9–5.1 cm T 3.2–4.4 cm
FA 3.3–3.8 cm W 5–7 g

Brandt's Myotis
Myotis brandtii
FR Murin de Brandt
DE Große Bartfledermaus
ES Ratonero de Brandt

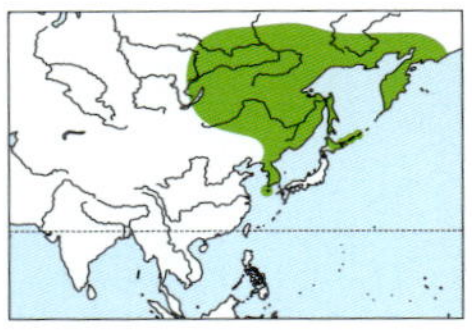

Siberian Whiskered Myotis
Myotis sibiricus
FR Murin de Sibérie
DE Sibirien-Mausohr
ES Ratonero de Siberia

From NE MN to RU (Far East), NE CN,
Korean Pen (with Jeju I), N JP (Hokkaido), Kuril Is

LC
Monotypic
HB 3.8–5.1 cm
T 3.2–4.5 cm
FA 3.2–3.8 cm
W 4–9.5 g

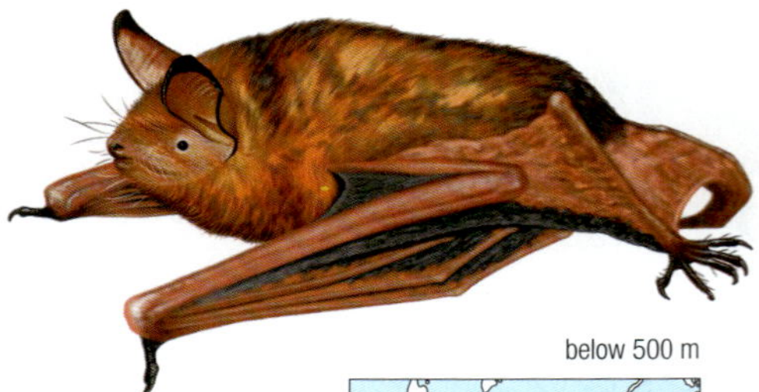

DD
Monotypic
HB 6.4–6.8 cm
T 5.4–5.8 cm
FA 5.6–6 cm
W 16–24 g

below 500 m

Herman's Myotis
Myotis hermani
FR Murin de Herman
DE Herman-Mausohr
ES Ratonero de Herman

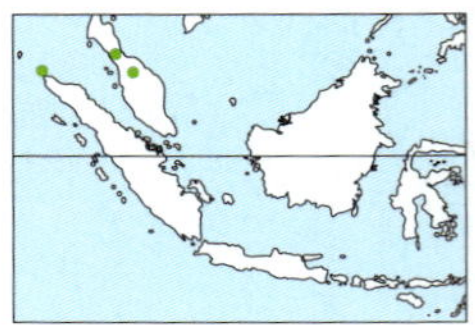

Known from
Songkhla (Pen TH),
Belum-Temenggor R
(Pen MY) & Sabang I
(off NW Sumatra)

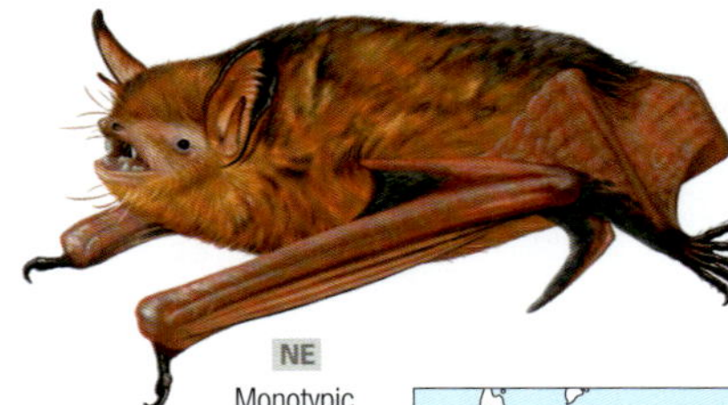

NE
Monotypic
FA c. 5.3 cm

Bartels's Myotis
Myotis bartelsii
FR Murin de Bartels
DE Bartels-Mausohr
ES Ratonero de Bartels

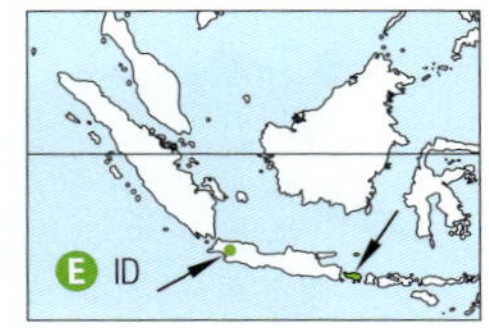

Java (Mt Pangrango), Bali

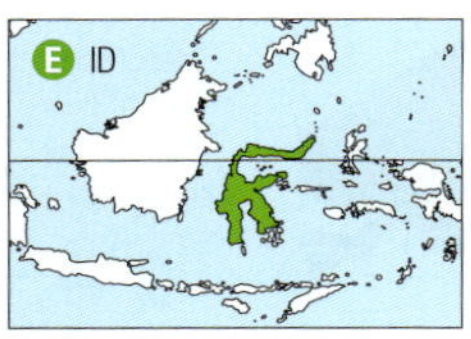

Sulawesi

Weber's Myotis
Myotis weberi
FR Murin de Weber DE Weber-Mausohr ES Ratonero de Weber

DD Monotypic
HB c. 5.7 cm T c. 4.2 cm
FA 4.9–5.3 cm

50–1450 m

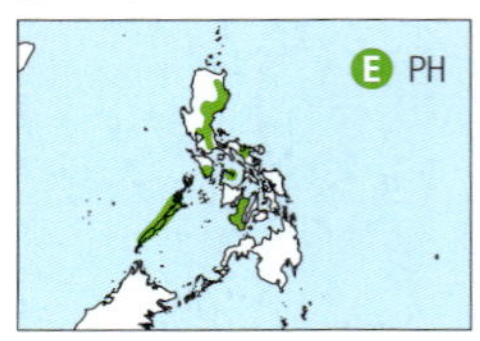

Orange-fingered Myotis
Myotis rufopictus
FR Murin à doigts roux
DE Orangefinger-Mausohr
ES Ratonero de dedos naranjas

Luzon, Mindoro, Sibuyan, Negros, Palawan (with Cuyo Is)

DD Monotypic
HB c. 6.3 cm T 4.3–5.9 cm
FA 4.7–5.8 cm W 9.5–17 g

0–3000 m

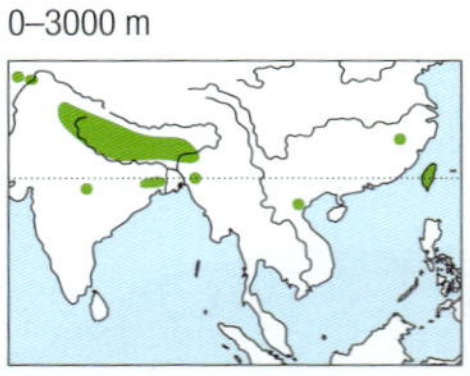

Black-and-orange Myotis
Myotis formosus
FR Murin de Hodgson
DE Hodgson-Mausohr
ES Ratonero de Hodgson

NT 2 sspp
HB 5–5.6 cm T 3.6–5.8 cm
FA 4.4–5.3 cm W c. 13.4 g

NE AF, N PK, N & C IN, NP, BT, N BD,
SW & SE CN (Tibet, Jiangxi), N VN, TW

0–1800 m

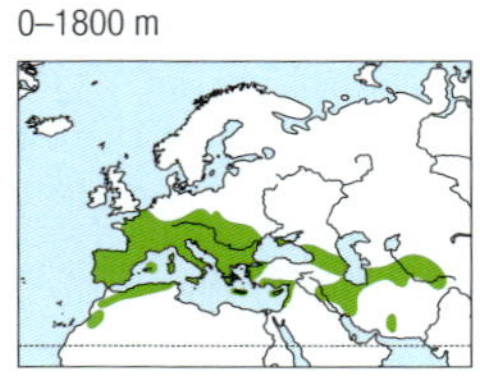

Geoffroy's Myotis
Myotis emarginatus
FR Murin à oreilles échancrées
DE Wimperfledermaus
ES Ratonero pardo

LC 2 sspp
HB 4.1–5.4 cm T 3.8–4.6 cm
FA 3.6–4.5 cm W 5.5–15.5 g

S & C Europe, N Africa (MA, DZ, TN),
larger Mediterranean Is E to Caucasus, Middle East, UZ & TJ

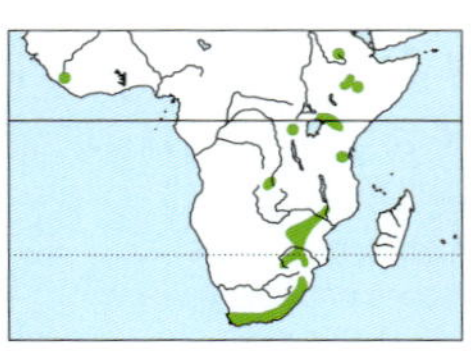

up to 2600 m

Patchily in NW uplands of LR, E & S CD & RW, more widely from ET to S ZA

Temminck's Myotis
Myotis tricolor
FR Murin tricolore DE Temminck-Mausohr ES Ratonero tricolor

LC Monotypic
HB 5.1–6.5 cm T 3.5–5.6 cm
FA 4.7–5.3 cm W 8–16 g

CR Monotypic
HB 6.9–7.5 cm T 4.1–4.7 cm
FA 5.2–5.5 cm W 15.5–17 g

c. 1400 m

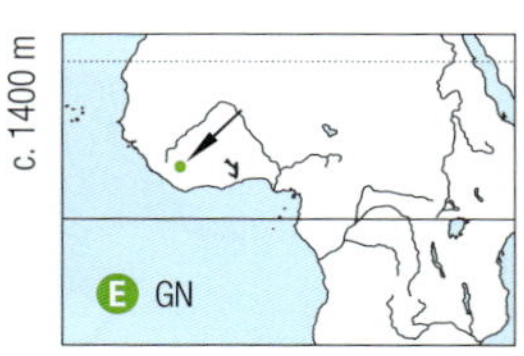

Nimba Mts

Nimba Myotis
Myotis nimbaensis
FR Murin du Nimba DE Nimba-Mausohr ES Ratonero de Nimba

Welwitsch's Myotis
Myotis welwitschii
FR Murin de Welwitsch DE Welwitsch-Mausohr
ES Ratonero de Welwitsch

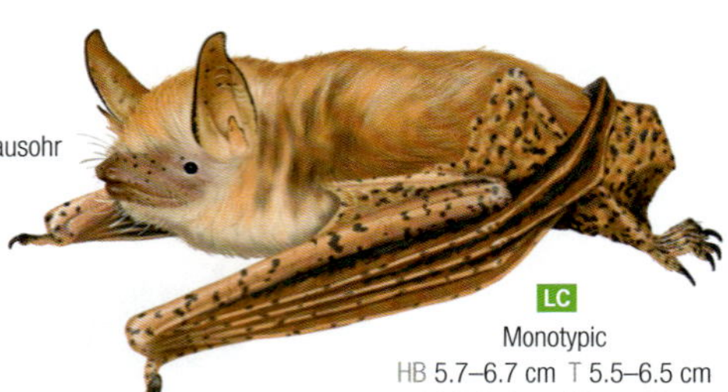

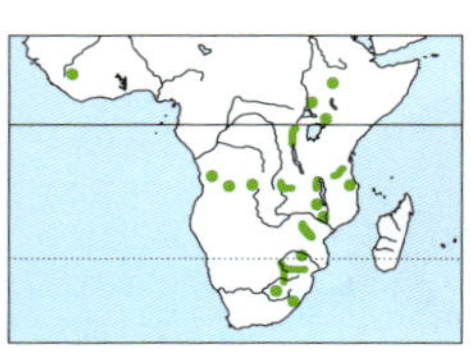

up to 2200 m

LC
Monotypic
HB 5.7–6.7 cm T 5.5–6.5 cm
FA 5.2–6 cm W 14–19 g

Patchily in SE GN
and from ET & SS S through W UG, RW, BI, S CD & AO to N & NE ZA

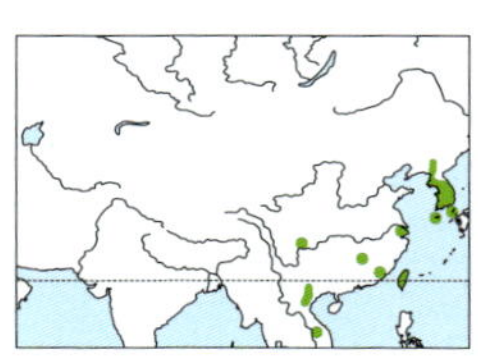

Korean Pen (with Jeju I), JP (Tsushima I), CN (Shanghai, Jiangxi, Fujian, Sichuan), N VN, SE LA, TW

Reddish-black Myotis
Myotis rufoniger
FR Murin rouge-et-noir DE Shanghai-Mausohr ES Ratonero rojinegro

LC Monotypic
HB 5.7–7 cm T 4.3–5.6 cm
FA 4.5–5.6 cm W 12–13 g

Scott's Myotis
Myotis scotti

FR Murin de Scott
DE Scott-Mausohr
ES Ratonero de Scott

1300–2500 m

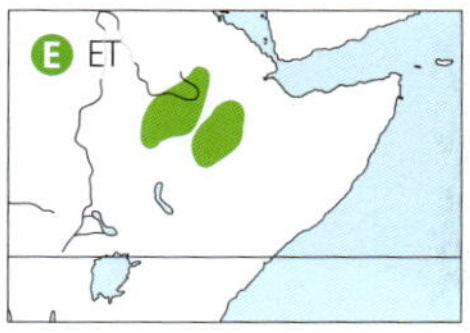

Highlands

VU
Monotypic
HB 4.2–5.1 cm
T 4–4.7 cm
FA 3.7–4.1 cm
W 4–5.6 g

Morris's Myotis
Myotis morrisi

FR Murin de Morris
DE Morris-Mausohr
ES Ratonero de Morris

DD
Monotypic
HB c. 4.8 cm T c. 4.5 cm
FA 4.5–4.6 cm W c. 8 g

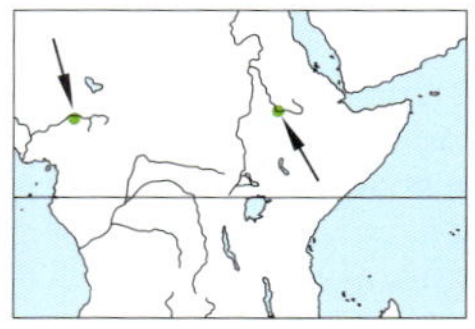

NE NG (Adamawa) & W ET (mouth of R Didessa); perhaps widespread across Sudanian Savanna Biome

Bocage's Myotis
Myotis bocagii

FR Murin de Du Bocage
DE Bocage-Mausohr
ES Ratonero de Bocage

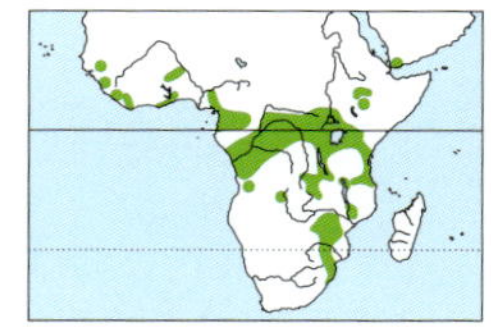

W & C Africa from SN E to CD, most of sub-Saharan Africa from ET S to AO & NE ZA, S YE

LC 2 sspp
HB 5–5.3 cm
T 3–5 cm
FA 3.3–4.2 cm
W 5–10 g

LC
Monotypic
FA 3.7–4 cm
W 5–6 g

Malagasy Myotis
Myotis goudotii

FR Murin de Madagascar
DE Madagaskar-Mausohr
ES Ratonero de Madagascar

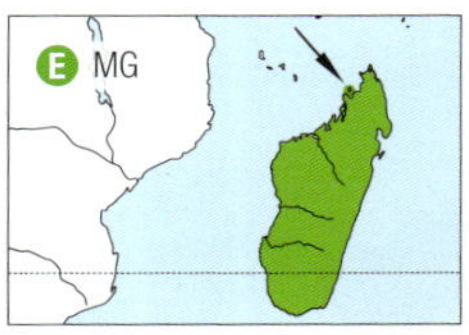

Including Nosy Be & Nosy Komba

Anjouan Myotis
Myotis anjouanensis

FR Murin d'Anjouan
DE Anjouan-Mausohr
ES Ratonero de Anjouan

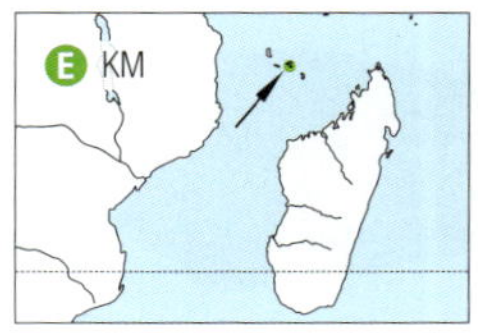

Anjouan

DD
Monotypic
HB c. 4.8 cm
T c. 4.5 cm
FA c. 4.3 cm
W c. 6.6 g

DD
Monotypic
HB c. 4.8 cm
T c. 3.8 cm
FA c. 3.7 cm

Kock's Myotis
Myotis dieteri

FR Murin de Kock
DE Kock-Mausohr
ES Ratonero de Kock

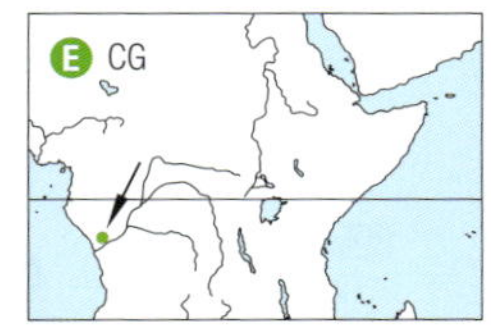

Known from type locality in Loudima

Alcathoe Whiskered Myotis
Myotis alcathoe

FR Murin alcathoé
DE Nymphenfledermaus
ES Ratonero bigotudo pequeño

DD
2 sspp
HB 3.9–4.4 cm T 3.6–3.7 cm
FA 3–3.5 cm W 3.5–5.5 g

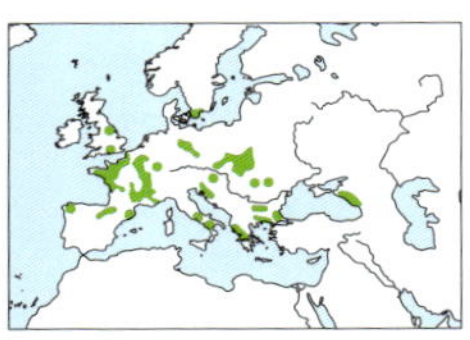

Patchily in Europe N to SE, E to SW UA & NW TR (E Thrace), Greater Caucasus (RU, GE); possibly wider range

Hyrcanian Myotis
Myotis hyrcanicus

FR Murin d'Hyrcanie
DE Hyrkanien-Mausohr
ES Ratonero de Hircania

c. 215 m

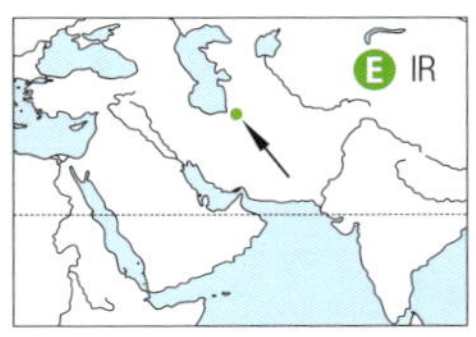

Known from type locality in Golestan

NE
Monotypic
FA c. 3.2 cm

NT
Monotypic
HB 5.7–6.8 cm
T 4.6–5.1 cm
FA 4.3–4.9 cm
W 13–18 g

0–1500 m

Pond Myotis
Myotis dasycneme

FR Murin des marais
DE Teichfledermaus
ES Ratonero lagunero

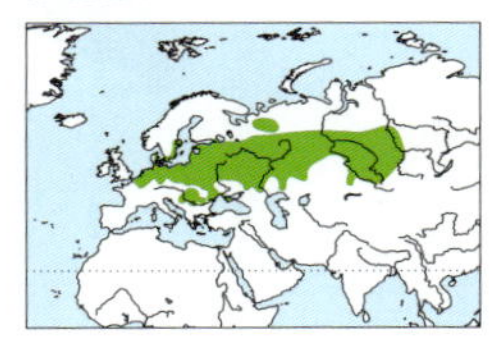

C & E Europe from S SE to N FR, E to N KZ & R Yenisei (C RU)

Ikonnikov's Myotis
Myotis ikonnikovi

FR Murin d'Ikonnikov
DE Kurzohr-Bartfledermaus
ES Ratonero de Ikonnikov

LC
Monotypic
HB 3.6–5.5 cm T 2.9–4.1 cm
FA 3–3.6 cm

200–1000 m

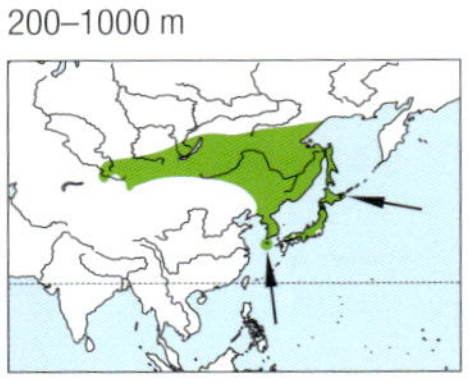

NE KZ, SC & SE RU (with Sakhalin I), N MN, NE CN, Korean Pen (with Jeju I), much of JP, S Kuril Is (Kunashir)

LC
Monotypic
HB 4.7–6 cm
T 3.6–4.9 cm
FA 4.1–4.7 cm
W 8–11 g

c. 1000 m

Sichuan Myotis
Myotis altarium

FR Murin de l'Emei
DE Sichuan-Bartfledermaus
ES Ratonero de Sichuán

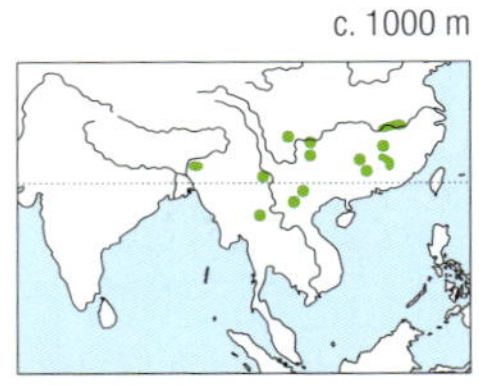

NE IN (Meghalaya), C & E CN, E TH, N VN

0–2000 m

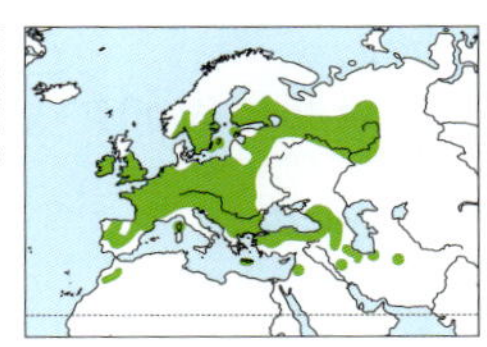

Widespread in Europe N to 64° N (with Corsica & Crete), E to Urals, Caucasus, IR & LB, N MA

LC 3 sspp
HB 3.5–4.8 cm T 3–4.3 cm
FA 3.2–3.6 cm W 4–7 g

Common Whiskered Myotis
Myotis mystacinus

FR Murin à moustaches
DE Kleine Bartfledermaus
ES Ratonero bigotudo

LC
Monotypic
HB 3.8–5 cm
T 3–4.3 cm
FA 3.1–3.7 cm
W 4–7.9 g

David's Myotis
Myotis davidii

FR Murin de David
DE David-Mausohr
ES Ratonero de David

0–3015 m

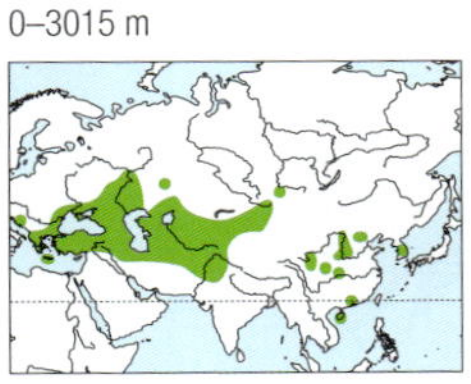

SE Europe from S HR, AL & GR (with Crete) E to N IR, S RU, W MN, Korean Pen & CN (with Hainan)

S slope of Himalayas in IN, NP, BT & MM; range uncertain

LC
Monotypic
HB c. 4.2 cm
T c. 3.4 cm
FA c. 3.4 cm

Nepal Myotis
Myotis nipalensis

FR Murin du Népal
DE Nepal-Mausohr
ES Ratonero de Nepal

NE
Monotypic
HB c. 3.8 cm
T c. 3.7 cm
FA 3.2–3.4 cm
W 3.7–4.7 g

Valley Myotis
Myotis ancricola

FR Murin des vallées
DE Bergtalmausohr
ES Ratonero de los valles

1030–1040 m

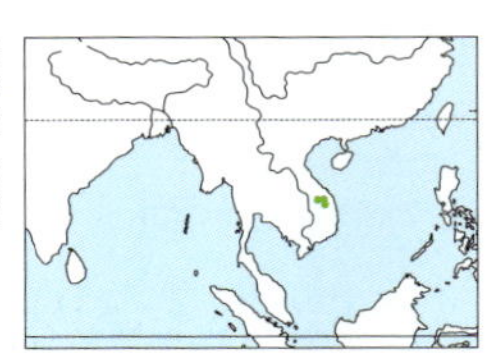

C VN, SE LA

200–1300 m

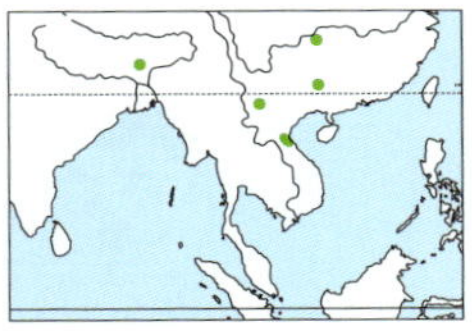

Anna Tess's Myotis
Myotis annatessae
FR Murin d'Anna Tess
DE Anna-Tess-Mausohr
ES Ratonero de Anna Tess

NE IN (West Bengal), C CN (Shaanxi, Guizhou), N VN, C LA

DD
Monotypic
HB 3.7–4.4 cm
T 3.4–4 cm
FA 3.3–3.5 cm
W 2.9–4.8 g

LC
Monotypic
T 3.6–4.5 cm
FA 4–4.3 cm
W 6.4–8.8 g

Gomantong Myotis
Myotis gomantongensis
FR Murin de Gomantong
DE Gomantong-Mausohr
ES Ratonero de Gomantong

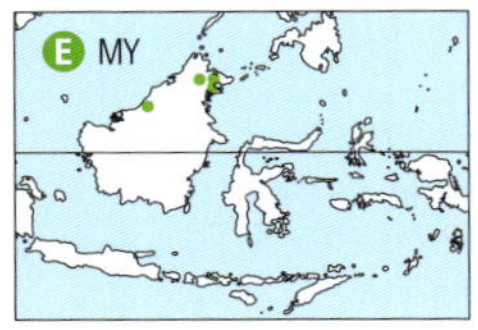

Borneo in Sabah (Gomantong Caves, Baturong Caves, Sinoa) & Sarawak (Bukit Sarang); perhaps widespread

LC
2 sspp
HB 4.6–5.3 cm
T 3.5–4.8 cm
FA 3.4–3.9 cm
W 4.3–8 g

Peters's Myotis
Myotis ater
FR Murin de Peters
DE Peters-Mausohr
ES Ratonero de Peters

0–1100 m

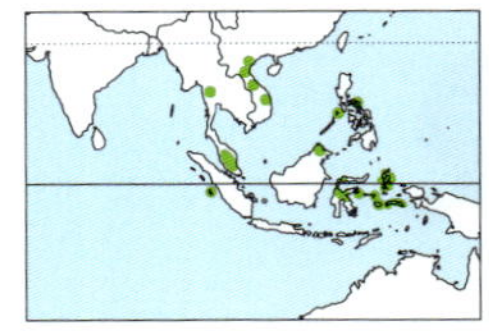

WC TH, VN, Pen MY, Nias I, N Borneo (Sabah), Sulawesi, Sula Is, N & C Moluccas, PH (Luzon, Calamian Is)

0–3200 m

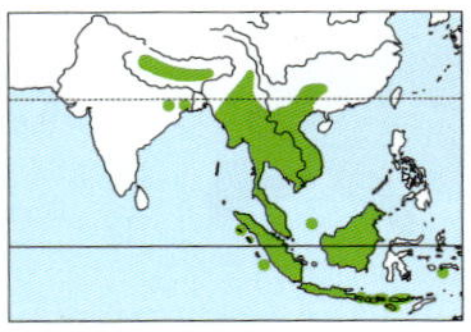

Nepalese Whiskered Myotis
Myotis muricola
FR Murin de Gray
DE Südasien-Bartfledermaus
ES Ratonero de Gray

NP, NE & E IN (with Sikkim), BT, SE CN (Guizhou, Hunan, Guangxi), SE Asia E to Flores & C Moluccas (Ambon)

LC Monotypic
HB 4.1–5.1 cm
T 3.5–4.9 cm
FA 3.1–3.9 cm
W 3.2–6.5 g

NE
Monotypic
HB 3.3–4 cm
T 3–3.6 cm
FA 3–3.4 cm
W 3–4.5 g

0–1600 m

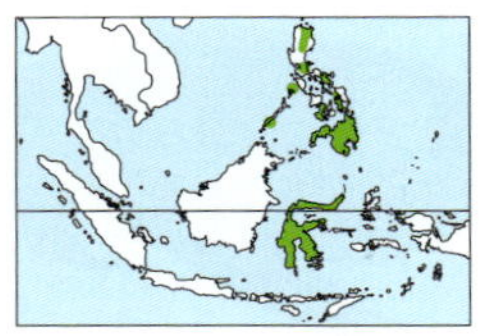

Brown's Whiskered Myotis
Myotis browni
FR Murin de Brown
DE Brown-Bartfledermaus
ES Ratonero de Brown

PH (Luzon, Busuanga, Maripipi, Biliran, Leyte, Negros, dubious in Palawan & Mindanao), ID (Sulawesi)

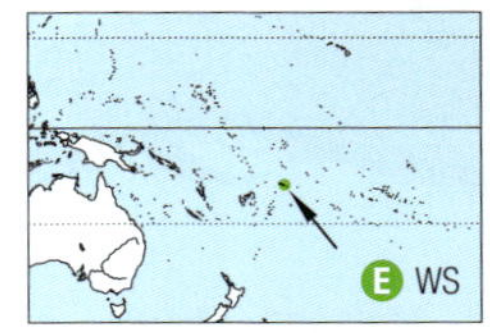

Insular Myotis
Myotis insularum
FR Murin des Samoa
DE Samoa-Wasserfledermaus
ES Ratonero de Samoa

Known from type specimen, distribution highly uncertain

DD
Monotypic
HB c. 4.1 cm
T c. 3.8 cm
FA c. 3.6 cm

LC
Monotypic
HB 5.8–6.3 cm
T 4.1–5 cm
FA 4.3–4.9 cm
W 4–6.5 g

600–1250 m

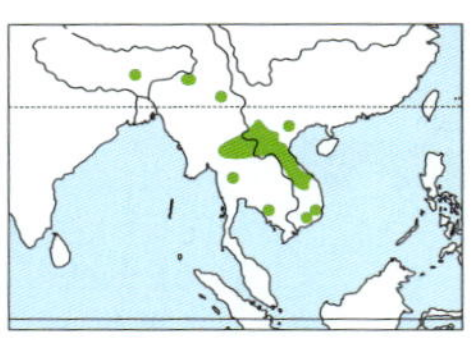

Hairy-faced Myotis
Myotis annectans
FR Murin velu
DE Mittlere Bartfledermaus
ES Ratonero peludo

NE IN (West Bengal, Nagaland), SC CN (Yunnan), N TH, LA, VN, SW KH (Cardamom Mts)

up to 300 m

Frosted Myotis
Myotis pruinosus
FR Murin givré
DE Honshu-Bartfledermaus
ES Ratonero de Japón

Honshu, Shikoku, Kyushu

EN
Monotypic
HB 3.7–4.4 cm
T 3.3–4 cm
FA 3–3.4 cm
W 4–7 g

CR
Monotypic
HB 3.6–4.4 cm
T 3.9–4.6 cm
FA 3.5–3.8 cm
W 4–8 g

Yanbaru Whiskered Myotis
Myotis yanbarensis
FR Murin du Yanbaru
DE Yanbaru-Mausohr
ES Ratonero bigotudo de Yanbaru

Ryukyu Is (Amami Oshima, Tokunoshima, Okinawajima)

40–2600 m

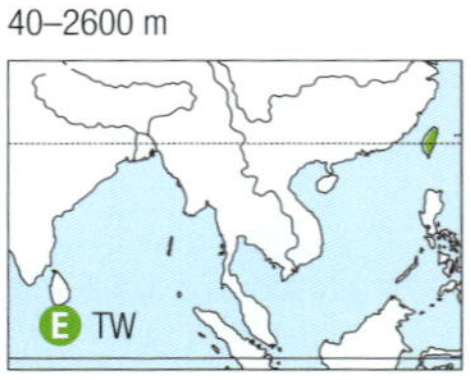

Long-toed Myotis
Myotis secundus
FR Murin second
DE Taiwan-Mausohr
ES Ratonero de pies pequeños

Lowland and montane forest habitats

LC
Monotypic
HB 3.4–4.2 cm
T 3.6–4.6 cm
FA 3.3–3.8 cm
W 3–5 g

DD
Monotypic
T 4.1–4.7 cm
FA 4.2–4.5 cm
W 9–13.5 g

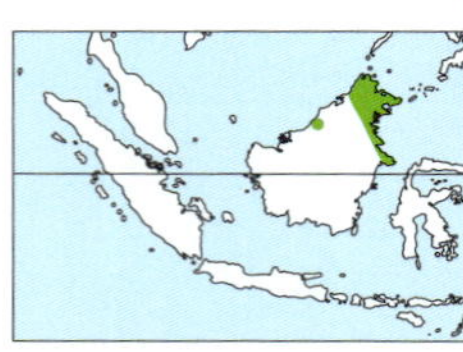

Bornean Whiskered Myotis
Myotis borneoensis
FR Murin de Bornéo
DE Borneo-Bartfledermaus
ES Ratonero bigotudo de Borneo

N Borneo (scattered in Sabah, Sarawak & Kalimantan); unconfirmed in S Sumatra (Bukit Barisan Selatan NP)

Malaysian Whiskered Myotis
Myotis federatus
FR Murin de Malaisie
DE Malaysia-Bartfledermaus
ES Ratonero bigotudo de Malasia

DD
Monotypic
HB c. 5.3 cm T c. 3.5 cm
FA 3.9–4.6 cm
W 8.3–12.8 g

260–900 m

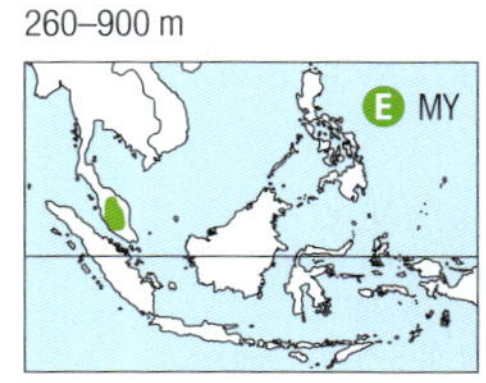

Peninsula (Belum-Temenggor R, Semangko Pass, Ulu Gombak, Genting Highlands, Batu Caves)

DD
Monotypic
FA 4.4–4.6 cm

370–970 m

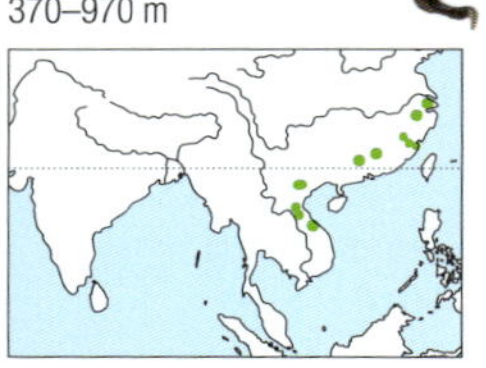

Indochinese Myotis
Myotis indochinensis
FR Murin d'Indochine
DE Laos-Mausohr
ES Ratonero de Indochina

SE CN (Shanghai, Zhejiang, Jiangxi, Fujian, Guangdong), N & C VN, C LA

DD
Monotypic
HB 5.6–6.2 cm
T 4.2–4.8 cm
FA 4.5–4.7 cm

390–1340 m

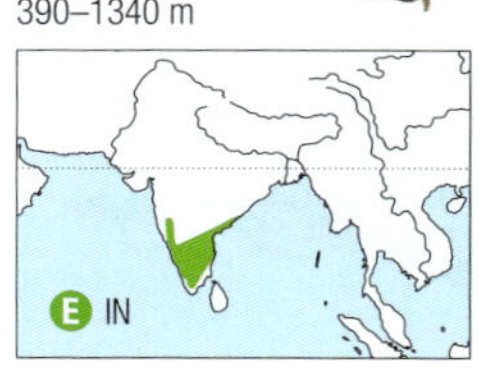

Peyton's Myotis
Myotis peytoni
FR Murin de Peyton
DE Peyton-Mausohr
ES Ratonero de Peyton

Maharashtra, Andhra Pradesh, Karnataka, Kerala, Tamil Nadu

Burmese Whiskered Myotis
Myotis montivagus
FR Murin de montagne
DE Burma-Bartfledermaus
ES Ratonero bigotudo de Birmania

500–1500 m

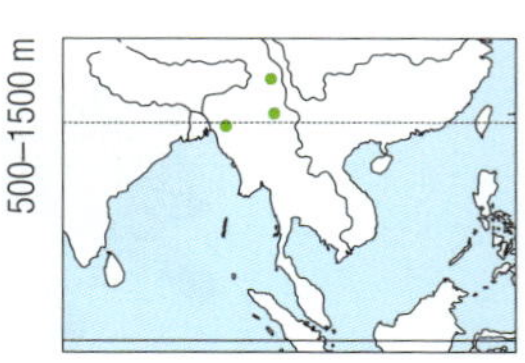

SC CN (Yunnan), N MM, NE IN (Mizoram)

DD
Monotypic
FA 3.9–4.1 cm

VU
2 sspp
HB 4.3–5.4 cm
T 3.4–4.2 cm
FA 3.8–4.4 cm
W 7–10 g

Long-fingered Myotis
Myotis capaccinii
FR Murin de Capaccini
DE Langfußfledermaus
ES Ratonero patudo

up to 1200 m

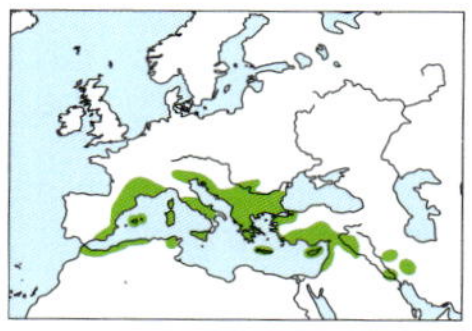

Mediterranean below 46° N, from E Iberian Pen & N Africa E to W & S TR, Levant, IQ & W IR

DD
Monotypic
HB 3.7–4.6 cm
T 3.7–4.2 cm
FA 3.4–3.9 cm

Kashmir Cave Myotis
Myotis longipes
FR Murin du Cachemire
DE Kaschmir-Langfußfledermaus
ES Ratonero de Kashmir

up to 2000 m

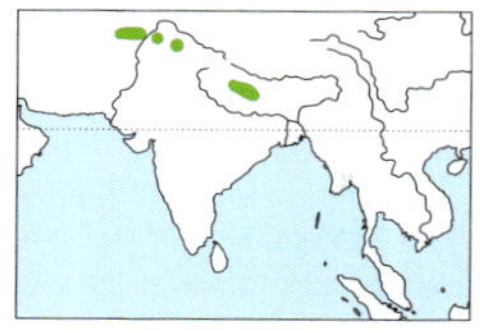

NE AF, N PK, NW IN (Jammu and Kashmir), NP

490–2770 m

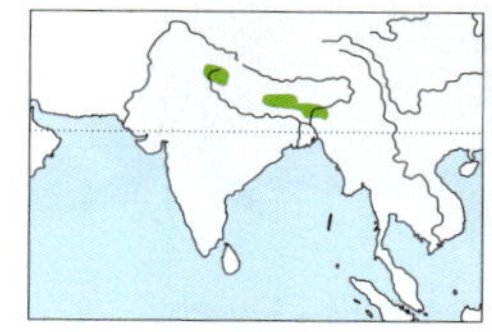

Himalayan Whiskered Myotis
Myotis siligorensis
FR Murin de Siliguri
DE Himalaya-Bartfledermaus
ES Ratonero de Siliguri

N IN (Uttarakhand, Sikkim, West Bengal, Meghalaya), E NP; possibly BT

LC
Monotypic
FA c. 3.2 cm

Sowerby's Whiskered Myotis
Myotis sowerbyi
FR Murin de Sowerby
DE Sowerby-Bartfledermaus
ES Ratonero de Sowerby

c. 915 m

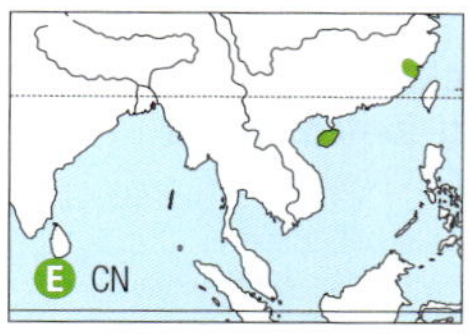

Fujian, Hainan I

NE
Monotypic
HB c. 4 cm
T c. 3.7 cm
FA c. 3.4 cm

LC
Monotypic
HB 3.6–4.4 cm
T 3–4 cm
FA 3.1–3.7 cm
W 3.3–4.5 g

Chinese Water Myotis
Myotis laniger
FR Murin laineux
DE China-Wasserfledermaus
ES Ratonero lanudo

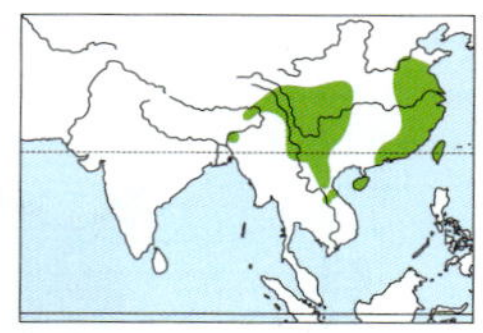

NE IN (Meghalaya), CN (with Hainan), NE & C LA, N & C VN, TW

NE
2 sspp
HB 2.6–4.2 cm
T 2.5–3.8 cm
FA 3–3.8 cm
W 2.3–4 g

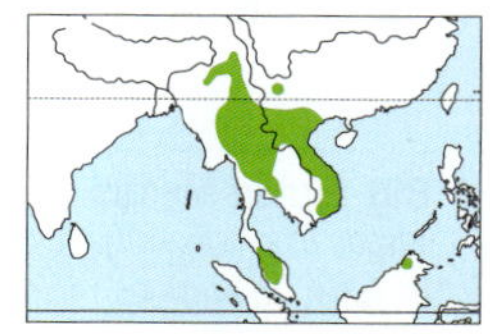

Indochinese Whiskered Myotis
Myotis alticraniatus
FR Murin du Tonkin
DE Indochina-Bartfledermaus
ES Ratonero de Tonkin

SC CN (Yunnan), N & E MM, N & C TH, LA, VN, WC KH, Thai-Malay Pen, N Borneo (Sabah)

NE
Monotypic
HB c. 3.9 cm
T c. 3.5 cm
FA 3.3–3.7 cm
W c. 4 g

900–1500 m

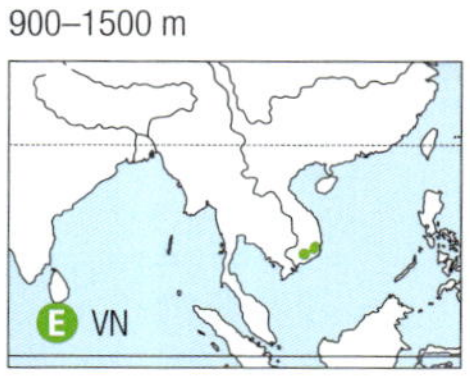

Phan Luong's Myotis
Myotis phanluongi
FR Murin de Phan Luong
DE Vietnam-Langfußfledermaus
ES Ratonero de Phan Luong

Known from 3 localities

Ridley's Myotis
Myotis ridleyi
FR Murin de Ridley
DE Ridley-Langfußfledermaus
ES Ratonero de Ridley

c. 1000 m

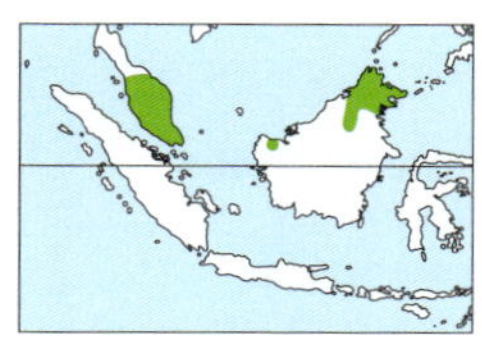

Thai-Malay Pen, W & N Borneo (Sabah, N Kalimantan, Gunung Gading NP in SW Sarawak)

NT
Monotypic
HB 3.6–4.1 cm
T 2.3–3.3 cm
FA 2.7–3.7 cm
W 4–6.5 g

LC
Monotypic
HB 3.6–4.7 cm
T 3.2–4.2 cm
FA 2.7–3.1 cm
W 3.8–6 g

Thick-thumbed Myotis
Myotis rosseti
FR Murin de Rosset
DE Kambodscha-Langfußfledermaus
ES Ratonero de Rosset

up to 500 m

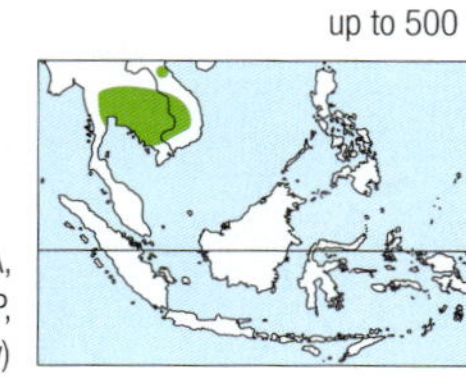

S & SE TH, KH, S LA, C & S VN (Ha Tinh Prov, Cat Tien NP, perhaps Binh Phuoc Prov)

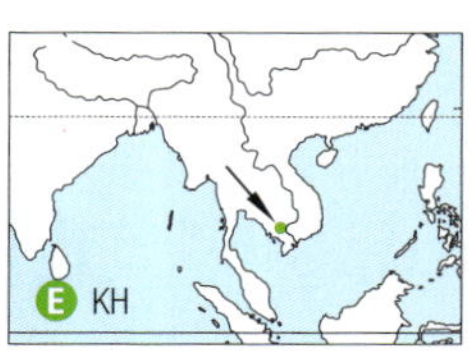

Know from holotype collected in Phnom Penh

Hayes's Thick-thumbed Myotis
Myotis hayesi
FR Murin de Hayes
DE Hayes-Langfußfledermaus
ES Ratonero de Hayes

NE
Monotypic
FA c. 3.4 cm

LC
5 sspp
HB 4.4–5.1 cm
T 3.3–4.3 cm
FA 3.4–4.1 cm
W 5–7.5 g

Horsfield's Myotis
Myotis horsfieldii
FR Murin de Horsfield
DE Horsfield-Langfußfledermaus
ES Ratonero de Horsfield

0–1450 m

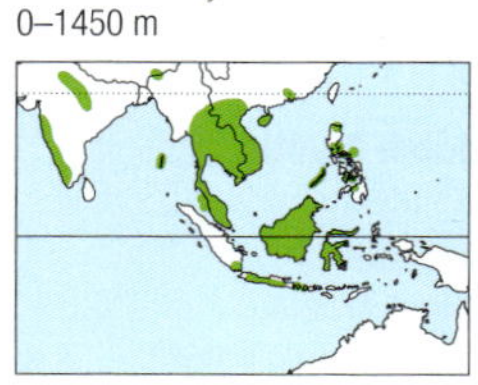

C, W & NE IN, Andaman Is, SE CN (Guangdong, Hainan), HK, SE Asia E to PH, Sulawesi & Lombok

Pallid Large-footed Myotis
Myotis macrotarsus
FR Murin des Philippines
DE Blasse Langfußfledermaus
ES Ratonero de las Filipinas

0–750 m

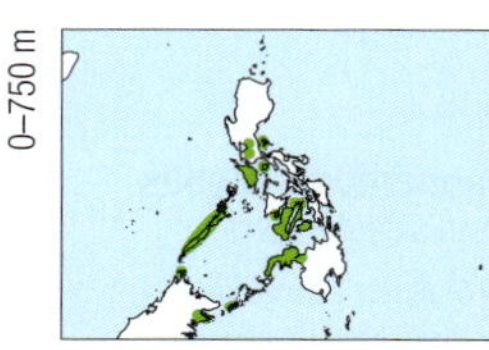

N Borneo in Sabah (Banggi & Balambangan Is, Madai Caves), PH; probably widespread

LC 2 sspp
HB c. 5.5 cm
T 4.5–5.9 cm
FA 4.6–5 cm
W 11–16 g

LC
Monotypic
HB 4.9–7.3 cm
T 5–5.5 cm
FA 4.9–5.1 cm
W 16–18 g

Stalker's Myotis
Myotis stalkeri
FR Murin de Stalker
DE Stalker-Langfußfledermaus
ES Ratonero de Stalker

below 250 m

Moluccas (Gebe & Kai Kecil in Kai Is), Raja Ampat Is (Waigeo, Batanta)

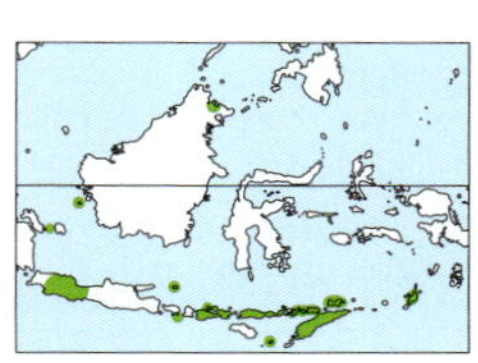

N Borneo (Sabah), Karimata & Mendanau Is, Java, Lesser Sundas, Tanimbar Is (Yamdena)

Gray Large-footed Myotis
Myotis adversus
FR Murin adverse
DE Südasiatische Wasserfledermaus
ES Ratonero adverso

LC
4 sspp
HB 4.3–5.5 cm
T 3.3–4.8 cm
FA 3.8–4.5 cm

LC
Monotypic
HB 4–5.9 cm
T 2.7–4.7 cm
FA 3.6–4.3 cm
W 5.7–11.1 g

Maluku Myotis
Myotis moluccarum
FR Murin des Moluques
DE Maluku-Wasserfledermaus
ES Ratonero de las Molucas

0–1200 m

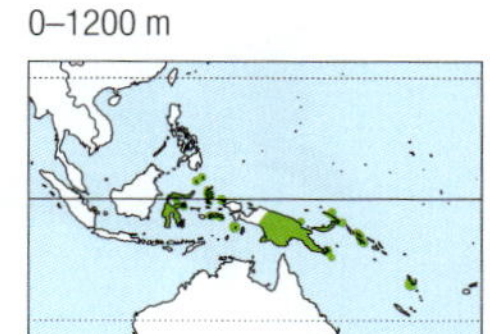

Sulawesi, N & C Moluccas, Kai Is, New Guinea (with Is & arch), Solomon Is, VU (Malakula)

LC
Monotypic
HB 3.5–5 cm
T 3.3–4.2 cm
FA 3.6–4.3 cm
W 5–14.9 g

Southern Myotis
Myotis macropus

FR Murin de Gould
DE Australien-Langfußfledermaus
ES Ratonero de Gould

0–840 m

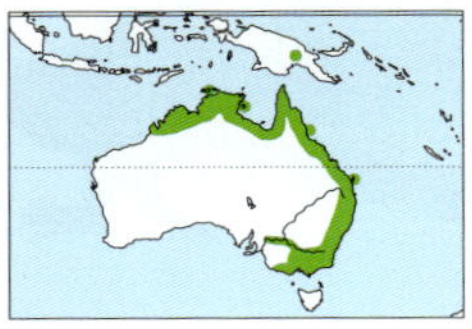

C PG (Yuro), N, E & SE AU (from NE WA to Qld, S to Vic & extreme SE SA, offshore Is)

LC
4 sspp
HB 4.4–6 cm
T 3.5–4.3 cm
FA 3.7–4.3 cm
W 7–12.5 g

Lesser Large-footed Myotis
Myotis hasseltii

FR Murin de Hasselt
DE Van-Hasselt-Mausohr
ES Ratonero de Hasselt

0–1000 m

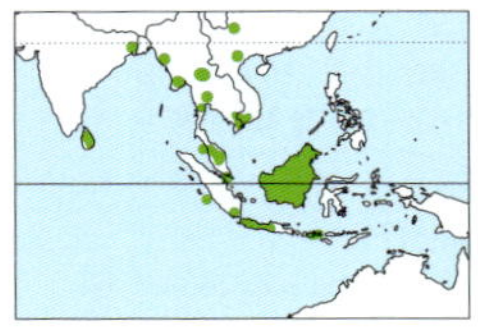

E IN (West Bengal), LK, SC CN (Yunnan), mainland SE Asia, Riau Arch, Sumatra, Mentawai Is, Borneo, Java, Sumbawa

LC 3 sspp
HB 4–5.4 cm
T 2.7–4.1 cm
FA 3.4–3.8 cm

Russian Myotis
Myotis petax

FR Murin de l'Altai
DE Sibirien-Wasserfledermaus
ES Ratonero de Altai

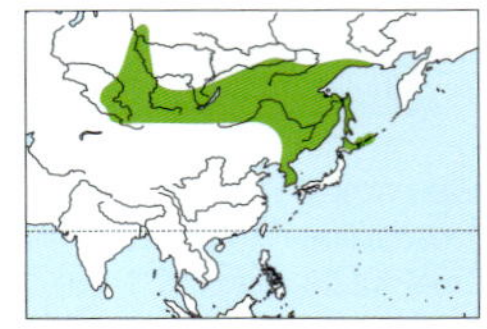

NE KZ, RU (S Siberia, Sakhalin), N MN, NE CN, Korean Pen, JP (Hokkaido), Kuril Is

LC
3 sspp
HB 4–7.1 cm
T 2.9–4.9 cm
FA 3.4–4.2 cm
W 6–8 g

Big-footed Myotis
Myotis macrodactylus

FR Murin macrodactyle
DE Ostasiatische Wasserfledermaus
ES Ratonero macrodáctilo

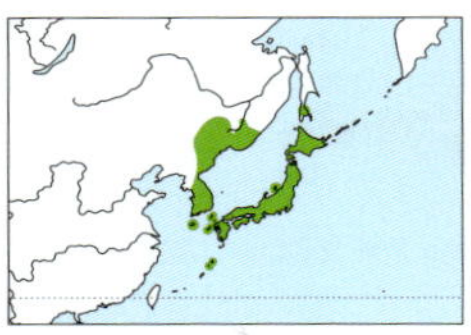

RU (Far East, S Sakhalin), NE CN, N & S Korean Pen (with Jeju I), most of JP

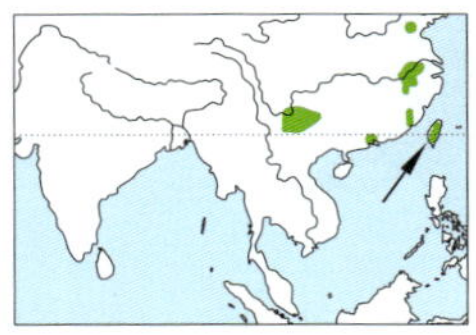

Fringed Long-footed Myotis
Myotis fimbriatus

FR Murin de Swinhoe
DE Haarige Wasserfledermaus
ES Ratonero de Swinhoe

E & SE CN (Shandong, Jiangsu, Anhui, Zhejiang, Jiangxi, Fujian, Guangdong), TW

LC
2 sspp
HB 3.6–5.2 cm
T 3.3–4.8 cm
FA 3.7–4.3 cm

Rickett's Big-footed Myotis
Myotis pilosus

FR Murin de Rickett
DE Rickett-Wasserfledermaus
ES Ratonero de Rickett

1400–3000 m

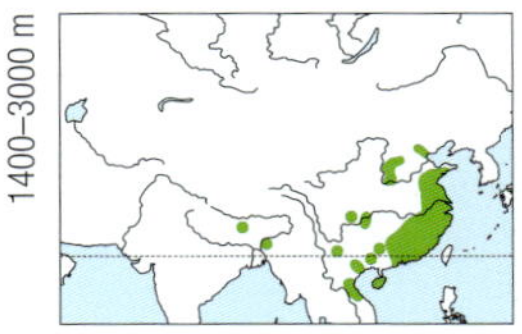

NE IN, E & S CN (with Hainan), C LA (Khammouane Prov), N VN; perhaps MM & TH

VU
Monotypic
HB 5.1–6.5 cm
T 3.5–5.5 cm
FA 5.2–6.3 cm
W 11.7–32 g

VU
Monotypic
HB 5–5.6 cm
T 4.4–5.4 cm
FA 4.8–5.5 cm
W 9.6–11 g

Mandelli's Myotis
Myotis sicarius

FR Murin du Sikkim
DE Mandelli-Wasserfledermaus
ES Ratonero de Sikkim

1350–1500 m

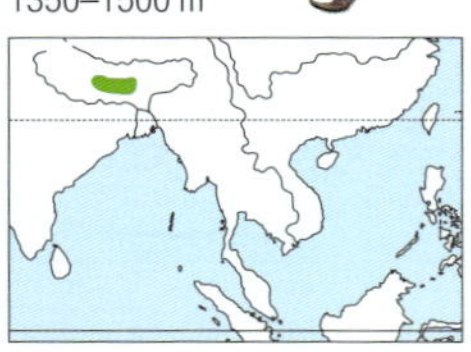

C NP (Annapurna CA, Kathmandu Valley), N IN (Sikkim, West Bengal)

Bechstein's Myotis
Myotis bechsteinii

FR Murin de Bechstein
DE Bechstein-Fledermaus
ES Ratonero forestal

0–1500 m

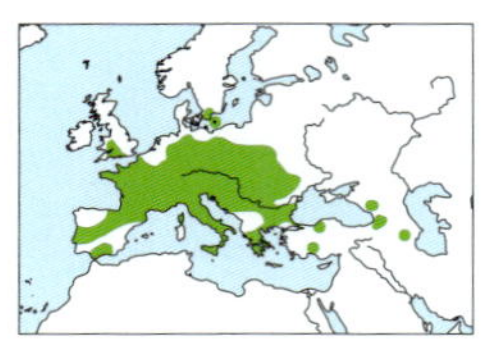

Europe, N to S SE, E to UA & Thrace, Corsica & Sicily Is, isolated in TR (Anatolia) & Caucasus

NT Monotypic
HB 3.8–5.5 cm T 3–4.7 cm
FA 3.9–4.7 cm W 7–14 g

Long-tailed Myotis
Myotis longicaudatus

FR Murin à longue queue
DE Langschwanz-Mausohr
ES Ratonero de cola larga

LC 3 sspp
HB 4.3–5.6 cm
T 3.8–4.7 cm
FA 3.7–4.1 cm
W 6–11 g

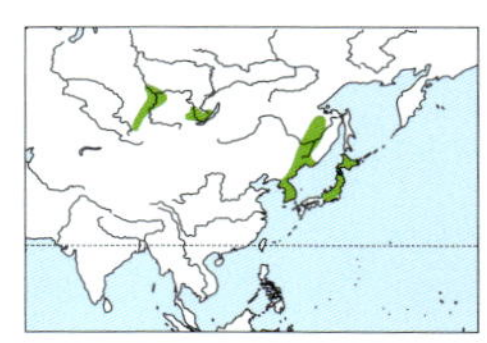

RU (SC Siberia, Far East), NE CN (Heilongjiang), Korean Pen, JP (Hokkaido, Honshu)

Daubenton's Myotis
Myotis daubentonii

FR Murin de Daubenton
DE Wasserfledermaus
ES Ratonero ribereño

0–1400 m

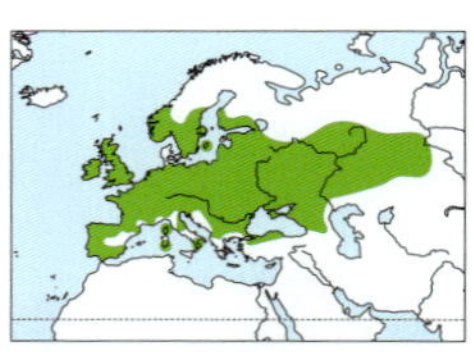

Europe from Iberian Pen, British Is & S Fennoscandia E to W RU (Omsk), Caucasus & N TR

LC
Monotypic
HB 4.5–5.5 cm
T 2.7–5 cm
FA 3.3–4.2 cm
W 6–10 g

DD
Monotypic
HB c. 4.8 cm
T c. 4.1 cm
FA c. 4.2 cm

Reddish Myotis
Myotis soror

FR Murin soeur
DE Rotes Taiwan-Mausohr
ES Ratonero rojo de Taiwán

up to 2600 m

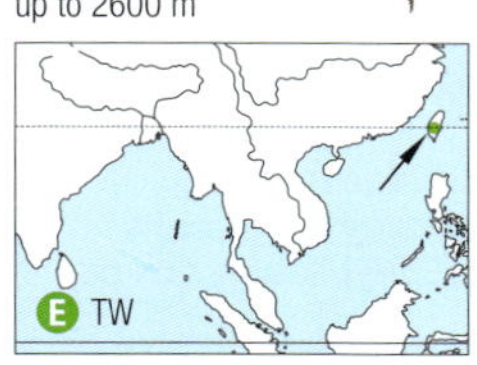

Known from type locality in Nantou County

Fraternal Myotis
Myotis frater

FR Murin frère
DE Korea-Wasserfledermaus
ES Ratonero fraile

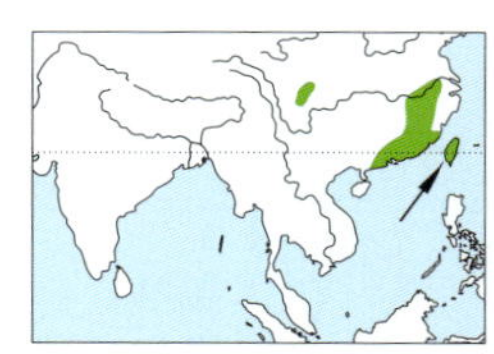

C & SE CN (Sichuan, Anhui, Jiangxi, Fujian, Guangdong), TW

LC
Monotypic
HB 3.8–4.3 cm
T 4.4–5.4 cm
FA 3.8–4.2 cm

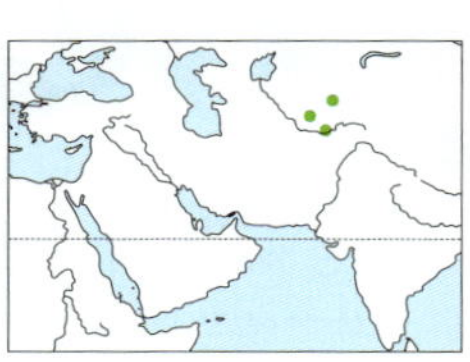

Known from 3 sites in E UZ & SW TJ; erroneously reported from AF

DD Monotypic
HB 4.6–5.1 cm T 4.6–5.5 cm
FA 3.8–4.3 cm

Bokhara Myotis
Myotis bucharensis

FR Murin du Boukhara DE Buchara-Mausohr ES Ratonero de Bujará

Lesser Myotis
Myotis blythii

FR Petit Murin
DE Kleines Mausohr
ES Ratonero mediano

0–2100 m

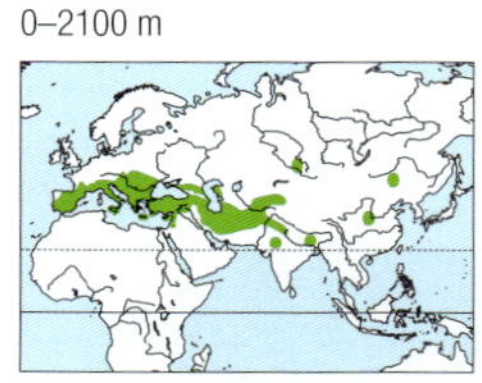

SC & S Europe (with Mediterranean Is), TR, Caucasus & Levant to NP, N IN, Altai Mts and N & C CN

LC
Monotypic
HB 5.4–7.6 cm T c. 5.9 cm
FA 5–6.2 cm W 19–29 g

LC
2 sspp
HB 6.5–8.4 cm T 4–6 cm
FA 5.5–6.7 cm W 20–27 g

0–2000 m

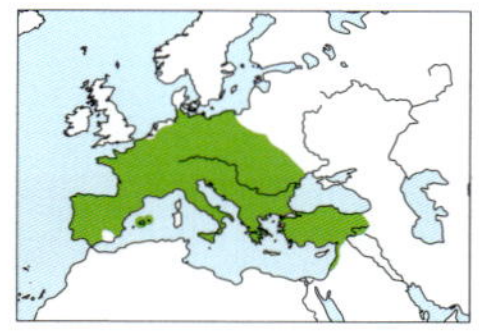

S & C Europe E to TR, Balearic Is & Sicily, N to N DE, S DK, NL, N PL & W UA

Greater Myotis
Myotis myotis

FR Grand Murin
DE Großes Mausohr
ES Ratonero grande

Maghreb Myotis
Myotis punicus

FR Murin du Maghreb
DE Punisches Mausohr
ES Ratonero de Magreb

0–1200 m

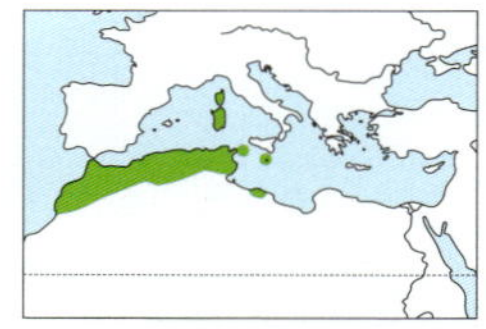

W Mediterranean Is (Corsica, Sardinia, Pantelleria, Gozo, Malta), NW MA, DZ, TN, LY

DD
Monotypic
HB 6.7–7.9 cm
FA 5.4–6.4 cm
W 19–25 g

LC
Monotypic
HB 4.3–4.8 cm
T 3.8–4.7 cm
FA 3.7–4.2 cm
W 7–10 g

Natterer's Myotis
Myotis nattereri

FR Murin de Natterer
DE Natterer-Fransenfledermaus
ES Ratonero de Natterer

0–2000 m

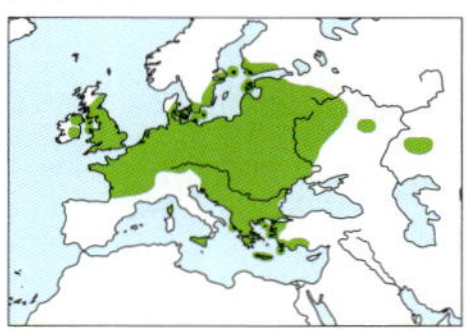

N, C & E Europe N to 60° N, E to W TR & W RU

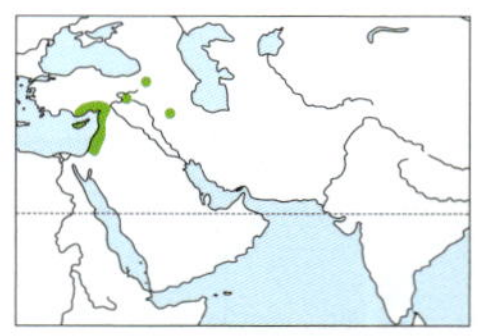

SC TR, W SY, LB, IL, W JO, N IQ, CY

Hovel's Myotis
Myotis hoveli

FR Murin de Hovel
DE Hovel-Fransenfledermaus
ES Ratonero de Hovel

NE
Monotypic
FA 3.8–4.3 cm

Tschuli Myotis
Myotis tschuliensis

FR Murin de Tschuli
DE Tschuli-Fransenfledermaus
ES Ratonero de Tschuli

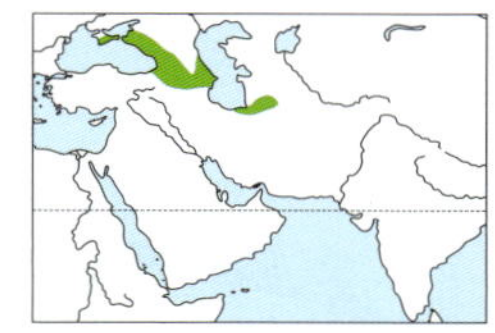

Crimea S to Caucasus, Transcaucasia & N IR to TM

NE
Monotypic
FA 3.9–4.5 cm

NE
Monotypic
HB 4–5 cm
T 3.2–4.6 cm
FA 3.6–4.2 cm
W 5–12 g

Cryptic Myotis
Myotis crypticus

FR Murin cryptique
DE Kryptisches Mausohr
ES Ratonero gris ibérico

up to 2000 m

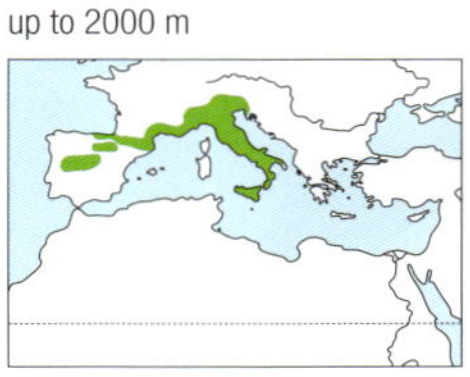

N & C ES, SE FR, IT (with Sicily), SW AT; perhaps Corsica & Sardinia, N limits unknown

Schaub's Myotis
Myotis schaubi

FR Murin de Schaub
DE Schaub-Mausohr
ES Ratonero de Schaub

up to 2590 m

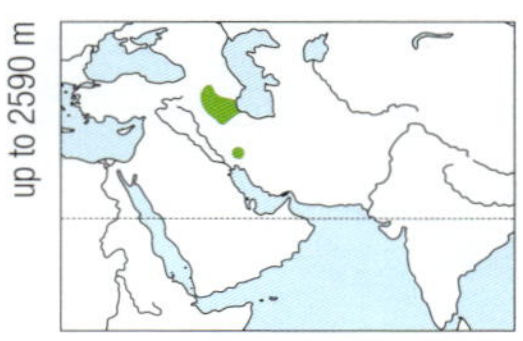

AM, NW & WC IR; perhaps TR & AZ

DD
Monotypic
HB 4.8–5.3 cm
T 4.8–4.9 cm
FA 4.1–4.4 cm

LC
Monotypic
HB 4.2–5 cm
T 4.2–5 cm
FA 3.7–4.2 cm
W 4.5–6.3 g

up to 1400 m

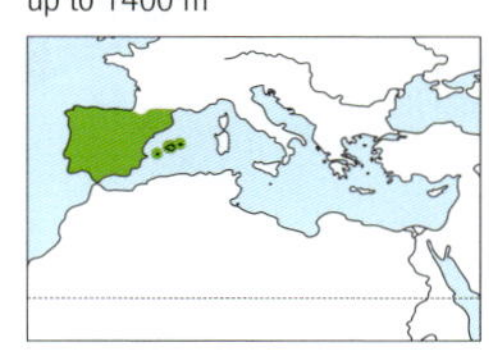

Escalera's Myotis
Myotis escalerai

FR Murin d'Escalera
DE Iberische Fransenfledermaus
ES Ratonero ibérico

Iberian Pen N to French Pyrenees, Balearic Is

NW Africa in MA & DZ; possibly TN

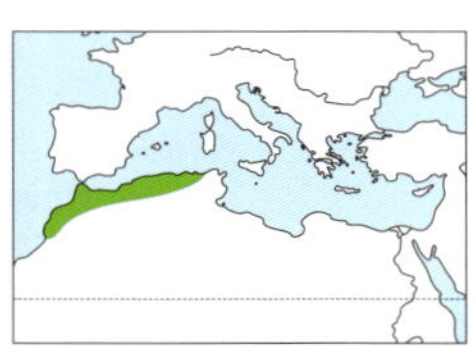

NE
Monotypic
HB 4.4–4.8 cm T 3.8–4.2 cm
FA 3.8–4 cm

Zenati Myotis
Myotis zenatius

FR Murin zénète
DE Zenati-Mausohr
ES Ratonero zenate

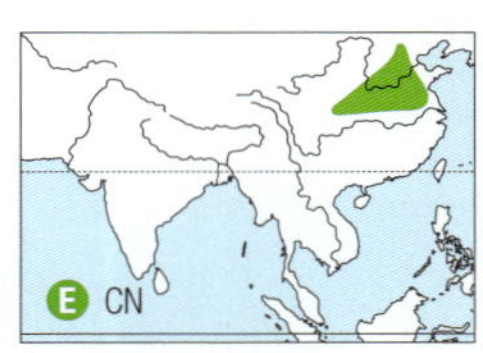

Sichuan E to Beijing & Anhui

Peking Myotis
Myotis pequinius

FR Murin de Pékin
DE Peking-Mausohr
ES Ratonero de Pekín

LC
Monotypic
HB c. 6.2 cm T c. 4.2 cm
FA 4.4–5.3 cm

Far Eastern Myotis
Myotis bombinus

FR Murin bourdon
DE Östliche Fransenfledermaus
ES Ratonero abejorro

NT
2 sspp
HB 4.1–5.2 cm T 3.8–4.5 cm
FA 3.7–4.2 cm W 5–10 g

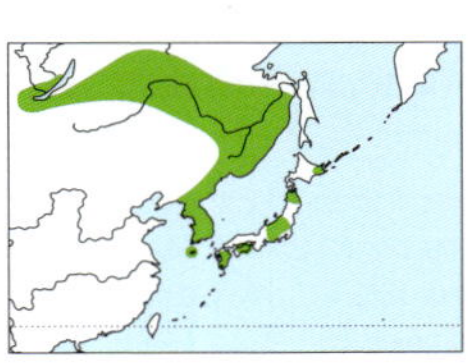

NC MN, SE RU, NE CN (Heilongjiang, Jilin), Korean Pen (with Jeju I), JP (with Kuchinoerabu-jima I)

Large Myotis
Myotis chinensis

FR Murin de Chine
DE China-Mausohr
ES Ratonero de China

50–1000 m

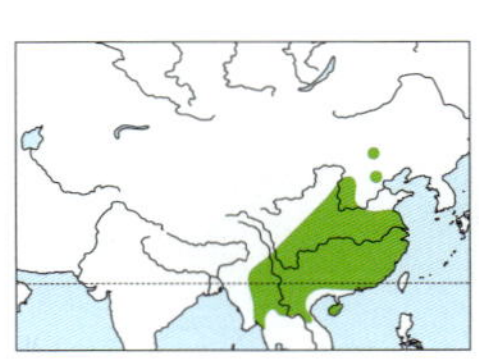

C & SE CN (S from Shanxi & Jiangsu, isolated in Inner Mongolia & Beijing), E MM, N TH, N LA, N & C VN

LC Monotypic
HB 9.1–9.7 cm T 5.3–5.8 cm
FA 6.4–6.9 cm W 25–30 g

Order ARTIODACTYLA (23 families)

Family BALAENIDAE – Right and Bowhead Whales

2 genera, 4 species. Mainly near shore and continental shelf, but also continental slope areas. Arctic, North Pacific, North Atlantic and Southern oceans.

Genus *EUBALAENA*

(3 species)

North Atlantic Right Whale
Eubalaena glacialis

FR Baleine noire
DE Atlantik-Nordkaper
ES Ballena franca del Atlántico

CR
Monotypic
TL 1500–1650 cm
W ≤ 70,000 kg

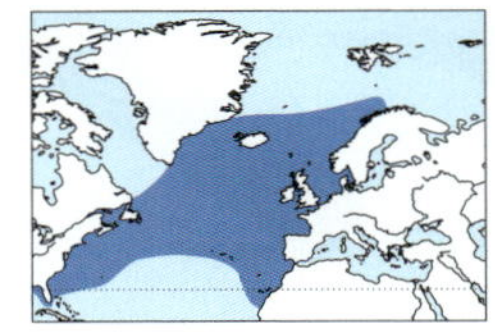

N Atlantic, in W from Nova Scotia to Gulf of Mexico & BM, in E S to 23° N (EH); occasional in GL, IS & NO

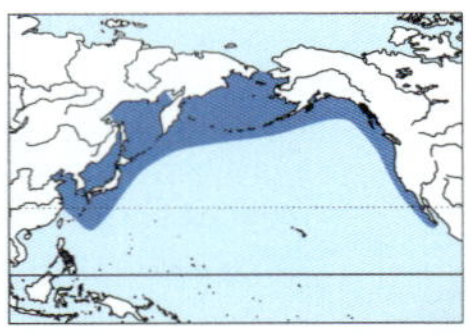

N Pacific, from Bonin & Ryukyu Is (JP) N to Bering Sea & Gulf of Alaska, rarely S to Baja California (MX) in E Pacific

North Pacific Right Whale
Eubalaena japonica

FR Baleine du Japon
DE Pazifik-Nordkaper
ES Ballena franca del Pacífico

EN
Monotypic
TL 1410–1900 cm
W > 80,000 kg

variants

Southern Right Whale
Eubalaena australis

FR Baleine australe
DE Südkaper
ES Ballena franca austral

LC
Monotypic
TL 1500–1650 cm
W c. 60,000 kg

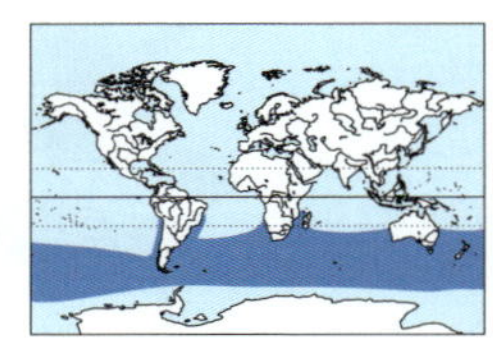

S Hemisphere, mainly at 20–60° S (lower latitudes in South America); in summer in Antarctic as far as 65° S

Genus *BALAENA*

(1 species)

Bowhead Whale
Balaena mysticetus

FR Baleine boréale
DE Grönland-Wal
ES Ballena de Groenlandia

LC
Monotypic
TL 1340–2000 cm
W > 80,000 kg

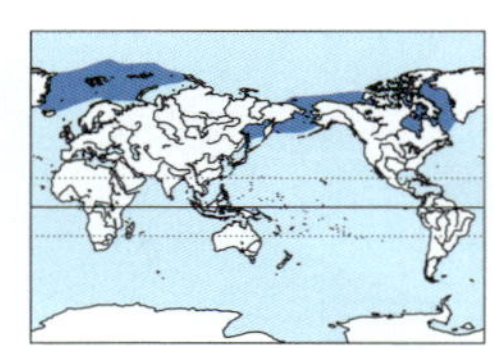

N Atlantic (NE CA, GL, SJ, Barents & Kara seas) & N Pacific (Bering–Chuckchi–Beaufort seas, Sea of Okhotsk)

Family CETOTHERIIDAE – Pygmy Right Whale

1 genus, 1 species. Pelagic and coastal waters, temperate to subantarctic. Southern Hemisphere seas.

Pygmy Right Whale
Caperea marginata

FR Baleine pygmée
DE Zwergglattwal
ES Ballena pigmea

Genus *CAPEREA*

(1 species)

LC
Monotypic
TL 590–650 cm
W 2850–3863 kg

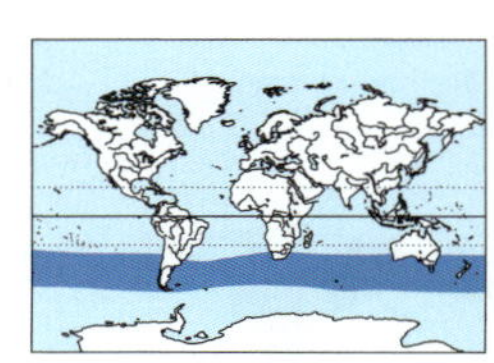

S Hemisphere, circumpolar, generally at 30°–55° S but can occur N to 19° S if cold currents are present

Family BALAENOPTERIDAE – Rorquals

3 genera, 11 species. Coastal continental shelf waters, and mid-latitude temperate and subtropical coastal bays, lagoons, and coastal areas to breed. Cosmopolitan, inhabiting all the world's oceans.

Genus *ESCHRICHTIUS*

(1 species)

Gray Whale
Eschrichtius robustus

FR Baleine grise
DE Grauwal
ES Ballena gris

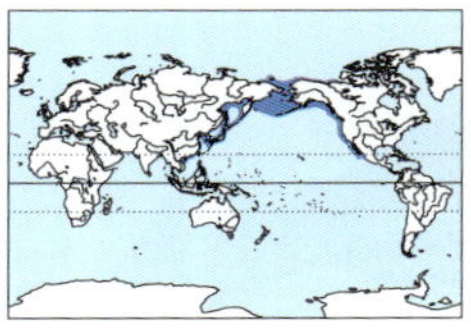

N Pacific, from Arctic Ocean & Bering Sea to SE CN in W & to MX in E; sighted in S Atlantic (NA) & Mediterranean (IL, ES)

LC
Monotypic
TL 1300–1420 cm
W 14,000–35,000 kg

Genus *MEGAPTERA*

(1 species)

Humpback Whale
Megaptera novaeangliae

FR Rorqual à bosse
DE Buckelwal
ES Yubarta

LC
3 sspp
TL 1500–1700 cm
W 30,000–34,000 kg

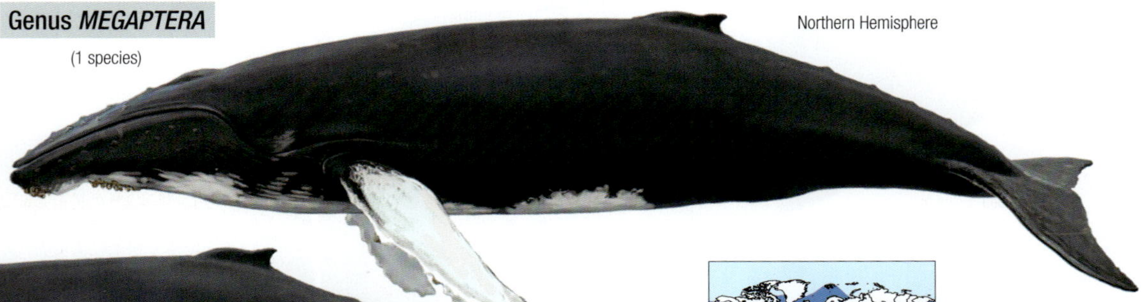

Northern Hemisphere

Southern Hemisphere

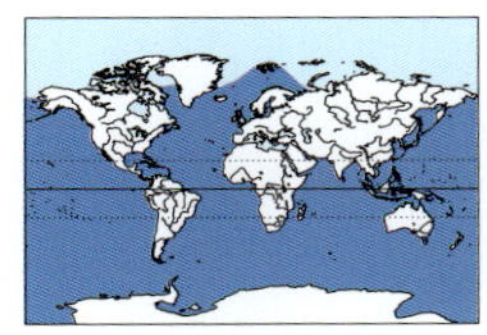

Waters of N Atlantic, N Pacific & S Hemisphere; rarely in Mediterranean Sea

Genus *BALAENOPTERA*

(9 species)

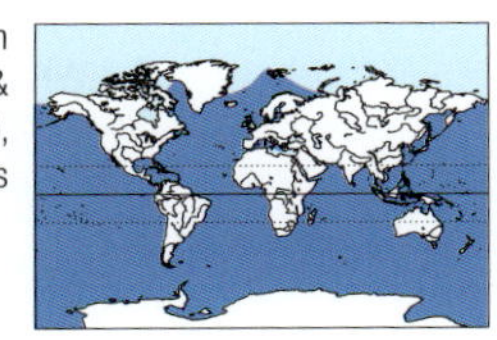

Atlantic (N to c. 80° N in Barents Sea), Arctic, Pacific & Southern oceans (S to c. 80° S), Mediterranean Sea; rare in tropics

Fin Whale
Balaenoptera physalus

FR Rorqual commun
DE Finnwal
ES Rorcual común

VU
4 sspp
TL 2200–2700 cm
W 60,000–90,000 kg

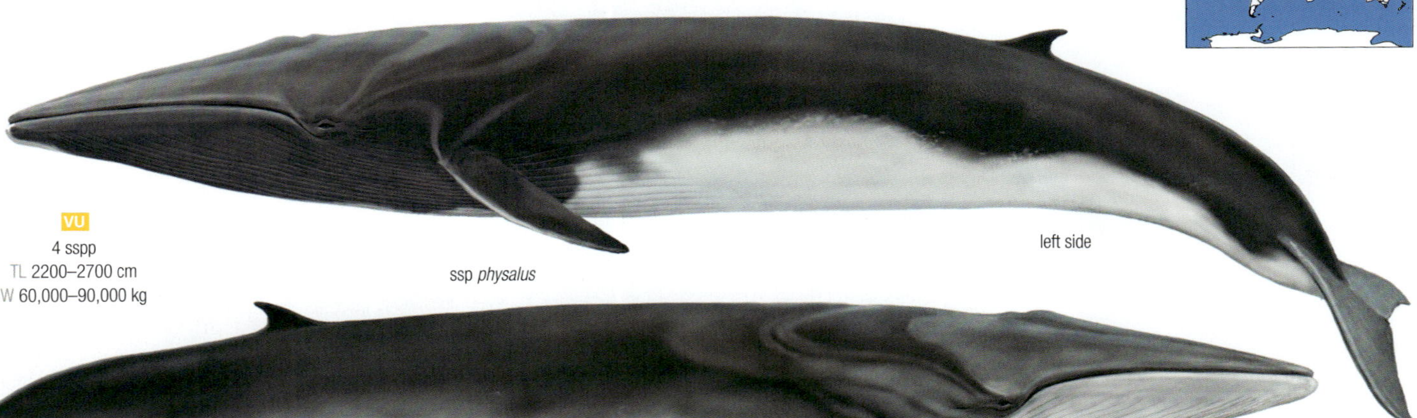

ssp *physalus*

left side

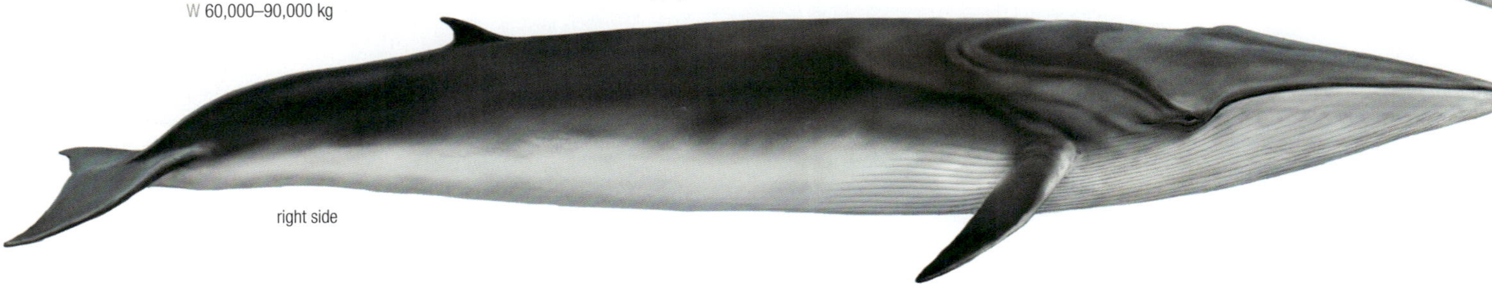

right side

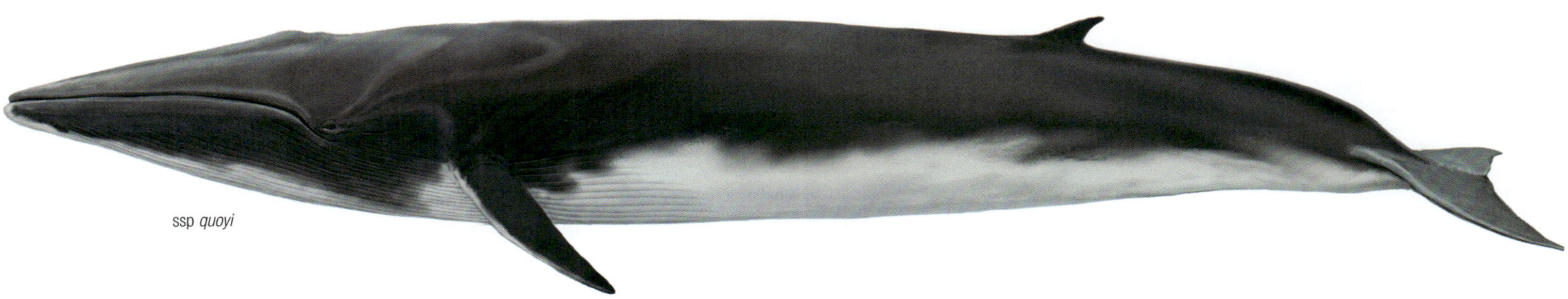

ssp *quoyi*

Blue Whale
Balaenoptera musculus
FR Rorqual bleu
DE Blauwal
ES Rorcual azul

EN
4 sspp
TL 3170–3260 cm
W 113,000–150,000 kg

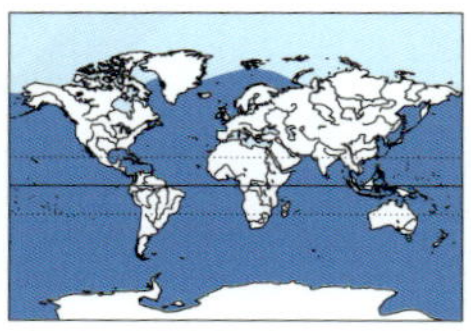

Worldwide in all oceans but absent from some seas such as Mediterranean, Okhotsk & Bering

---

Sei Whale
Balaenoptera borealis
FR Rorqual boréal
DE Seiwal
ES Rorcual boreal

EN
2 sspp
TL 1700–2000 cm
W 22,000–38,000 kg

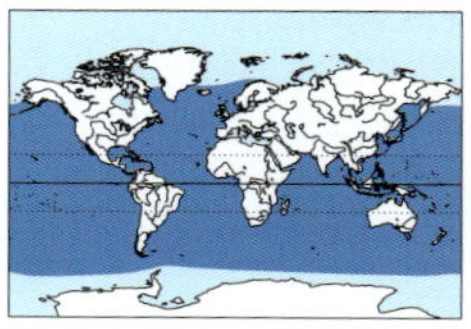

Cosmopolitan in N Atlantic, N Pacific & S Hemisphere; unconfirmed in N Indian Ocean

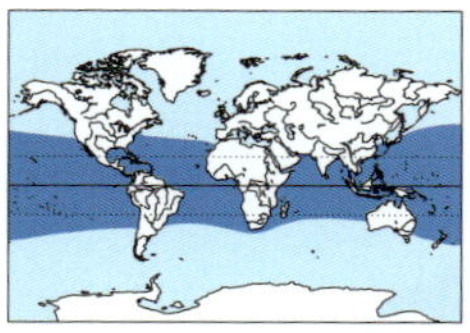

Genetically confirmed records from Atlantic, Pacific & Indian oceans; generally associated with deep & pelagic waters

Bryde's Whale
Balaenoptera brydei
FR Rorqual de Bryde
DE Bryde-Wal
ES Rorcual de Bryde

NE
Monotypic
TL 1040–1550 cm
W 15,000–16,600 kg

Eden's Whale
Balaenoptera edeni

FR Rorqual d'Eden
DE Eden-Wal
ES Rorcual tropical

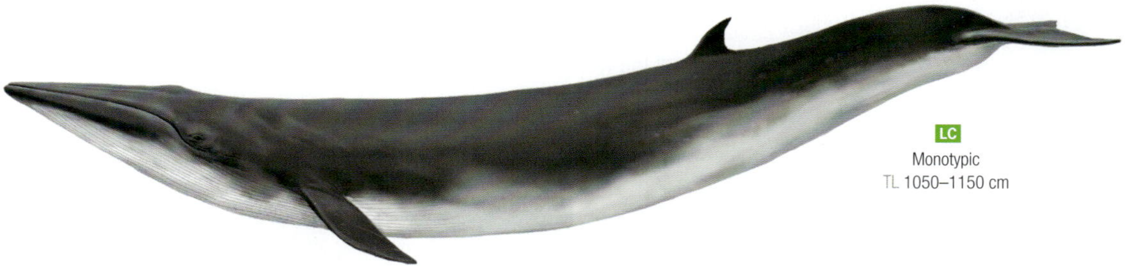

LC
Monotypic
TL 1050–1150 cm

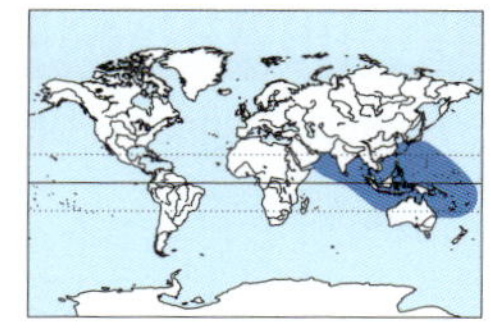

Genetically confirmed records from coastal waters of OM E throughout N Indian Ocean to ID and S & E China seas

Rice's Whale
Balaenoptera ricei

FR Rorqual de Rice
DE Rice-Wal
ES Rorcual de Rice

CR
Monotypic
TL 1126–1250 cm

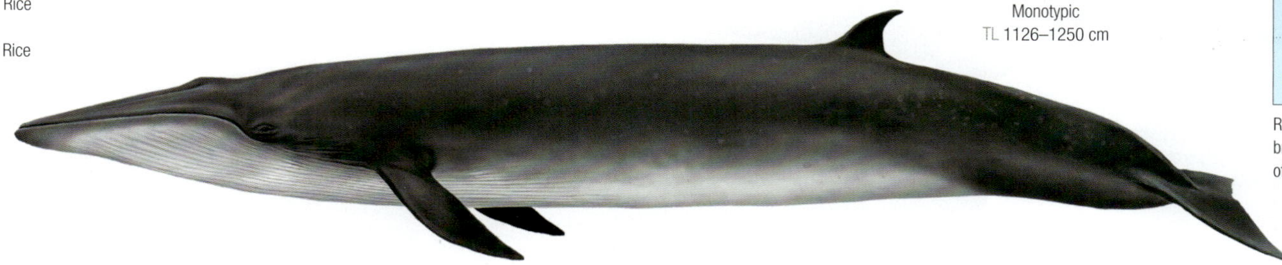

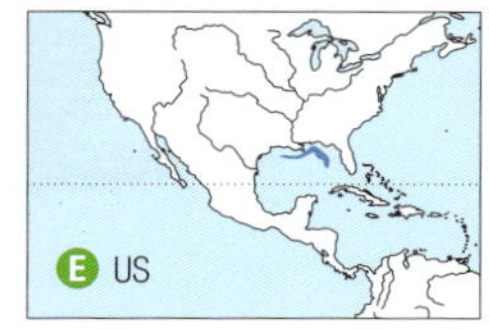

Restricted along continental shelf break near De Soto Canyon area of NE Gulf of Mexico

Omura's Whale
Balaenoptera omurai

FR Rorqual d'Omura
DE Omura-Wal
ES Rorcual de Omura

DD
Monotypic
TL 1000–1200 cm

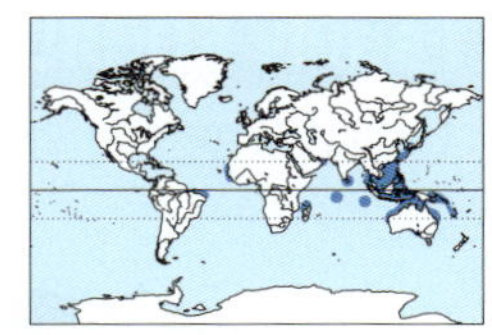

Tropical & subtropical waters in Indo-Pacific region (36° N to 38° S), records W to C Atlantic suggest wider range

Common Minke Whale
Balaenoptera acutorostrata

FR Petit Rorqual
DE Nördlicher Zwergwal
ES Rorcual aliblanco

ssp *acutorostrata*

LC
2 sspp
TL 650–880 cm
W 2000–2700 kg

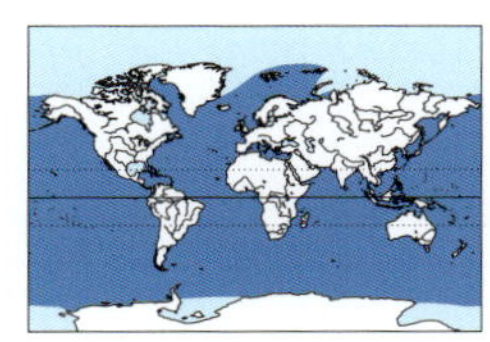

Cosmopolitan in N Atlantic, N Pacific & throughout S Hemisphere, from nearly 70° S to 80° N; possibly N Indian Ocean

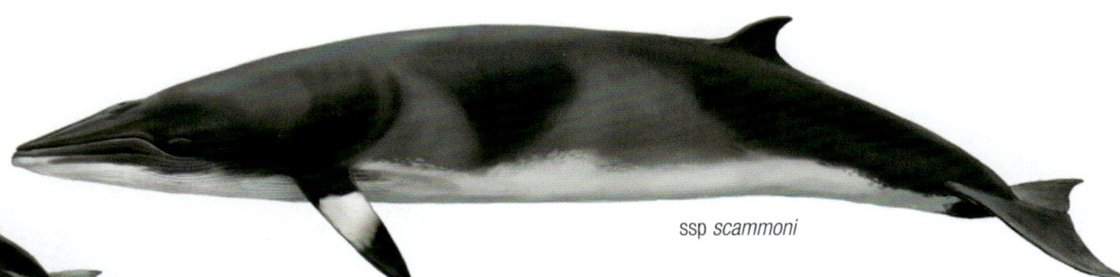

ssp *scammoni*

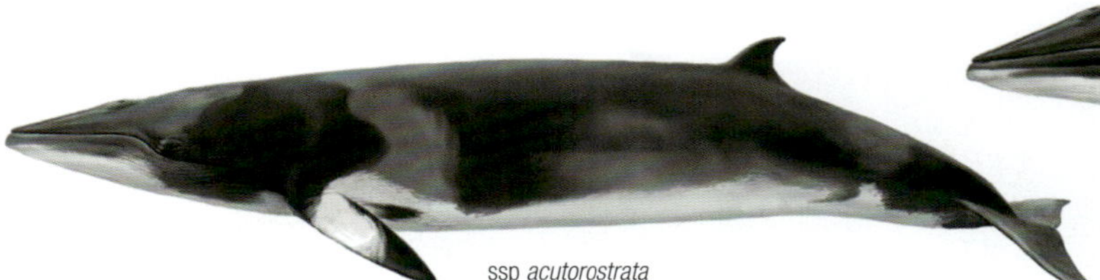

ssp *acutorostrata*
dwarf morph

Antarctic Minke Whale
Balaenoptera bonaerensis

FR Rorqual antarctique
DE Südlicher Zwergwal
ES Rorcual austral

NT
Monotypic
TL 840–1020 cm
W 6800–11,000 kg

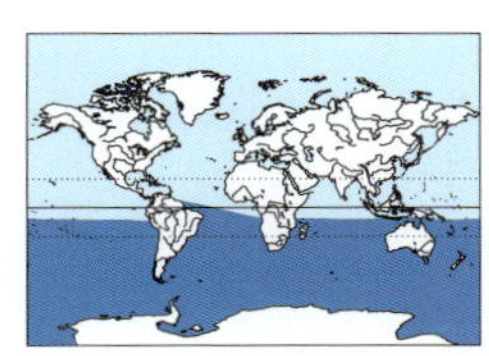

S Hemisphere, from tropics (7° S) to pack ice (65° S) around AQ; recorded off coast of SR (4° N) in N Hemisphere

Family PHYSETERIDAE – Sperm Whale

1 genus, 1 species. Deep, ice-free oceanic waters. Cosmopolitan, inhabiting all the world's oceans.

Genus *PHYSETER*

(1 species)

♂

Sperm Whale
Physeter macrocephalus

FR Grand Cachalot
DE Pottwal
ES Cachalote

VU
Monotypic
TL 1520–1920 cm (M), 1040–1250 cm (F)
W > 70,000 kg (M), > 24,000 kg (F)

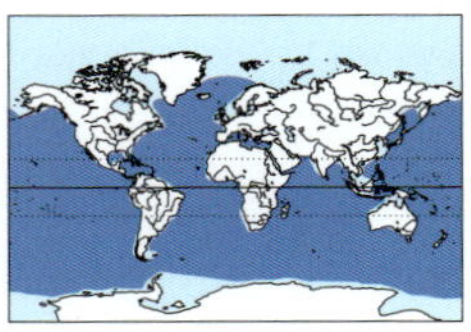

Almost cosmopolitan, found from edges of polar pack ice to equator in S & N hemispheres

♀

Family KOGIIDAE – Pygmy and Dwarf Sperm Whales

1 genus, 2 species. Shoreward and seaward of continental shelves in waters of 400–3500 m or more. Cosmopolitan in temperate and tropical seas.

Genus *KOGIA*

(2 species)

Pygmy Sperm Whale
Kogia breviceps

FR Cachalot pygmée
DE Zwergpottwal
ES Cachalote pigmeo

LC
Monotypic
TL 270–420 cm
W 342–680 kg

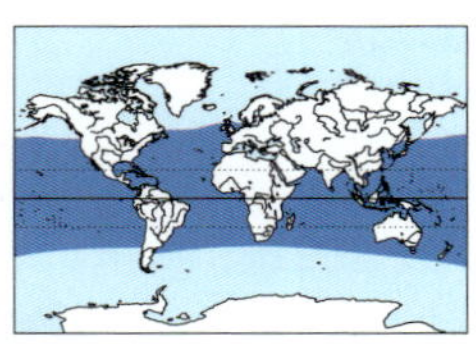

Temperate & tropical waters of Atlantic, Pacific & Indian oceans; strandings mainly along E US, ZA, NZ, FR & ES

♂

Dwarf Sperm Whale
Kogia sima

FR Cachalot nain
DE Kleinstpottwal
ES Cachalote enano

♀

LC
Monotypic
TL 200–270 cm
W 136–280 kg

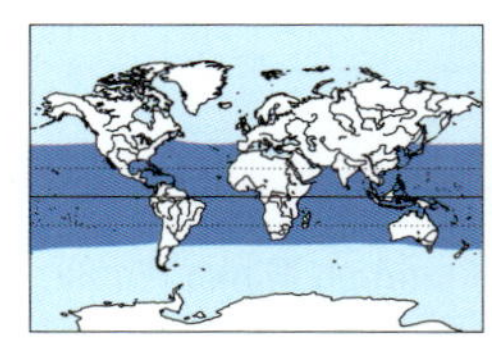

Tropical & temperate waters of Atlantic, Pacific & Indian oceans; strandings mainly on SW & SE US, ZA & S Arabia

Family ZIPHIIDAE – Beaked Whales

6 genera, 24 species. Deep oceanic waters from the ice edges to the tropics. Cosmopolitan, inhabiting all the world's oceans.

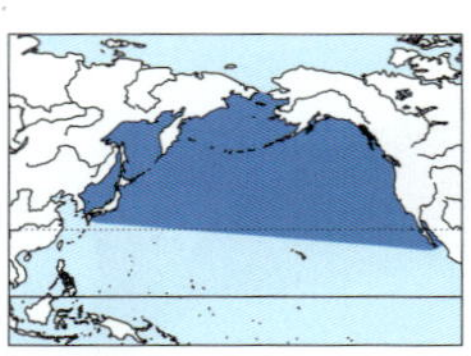

Cooler waters of N Pacific, N to shallow waters of Bering Sea, S to S JP in W & to Baja California (MX) in E; S limit in C Pacific unclear

Baird's Beaked Whale
Berardius bairdii

FR Baleine-à-bec de Baird
DE Baird-Schnabelwal
ES Zifio de Baird

LC
Monotypic
TL 1000–1200 cm
W 10,000–12,000 kg

Genus *BERARDIUS*

(3 species)

Arnoux's Beaked Whale
Berardius arnuxii

FR Baleine-à-bec d'Arnoux
DE Arnoux-Schnabelwal
ES Zifio de Arnoux

LC
Monotypic
TL 800–930 cm
W c. 9000 kg

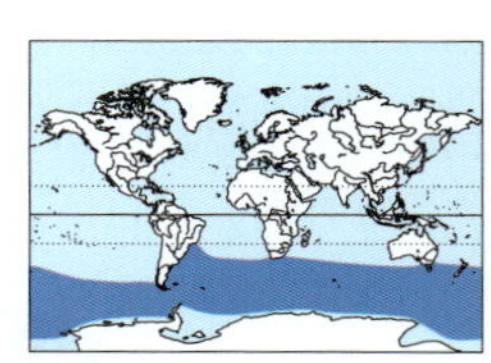

Subantarctic & Antarctic waters from South America, ZA, S AU & N NZ to AQ

Least Beaked Whale

Berardius minimus

FR Petite Baleine-à-bec
DE Kleiner Schnabelwal
ES Zifio pequeño

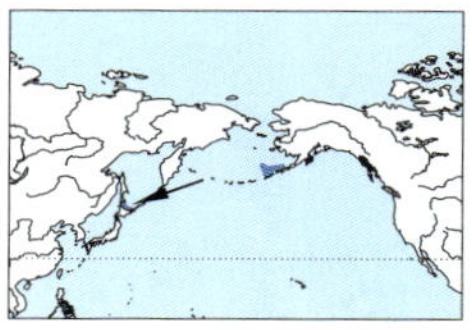

Known from N Pacific off N Hokkaido, JP (40–60° N) & off islands W of Alaska (140° E–160° W); likely further S

NT
Monotypic
TL 630–690 cm

Genus *TASMACETUS*

(1 species)

Shepherd's Beaked Whale

Tasmacetus shepherdi

FR Baleine-à-bec de Shepherd
DE Tasmanien-Schnabelwal
ES Zifio de Shepherd

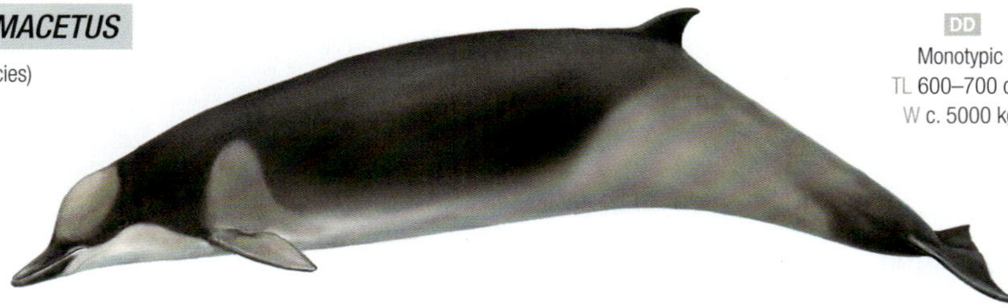

DD
Monotypic
TL 600–700 cm
W c. 5000 kg

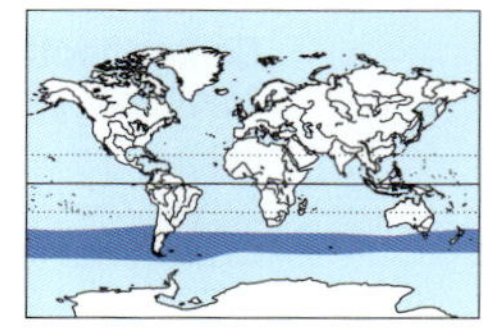

Poorly known but apparently circumpolar in colder waters in S Hemisphere

Genus *ZIPHIUS*

(1 species)

Cuvier's Beaked Whale

Ziphius cavirostris

FR Baleine-à-bec de Cuvier
DE Cuvier-Schnabelwal
ES Zifio de Cuvier

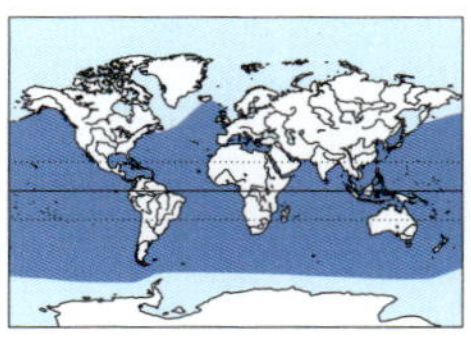

Warm temperate to tropical waters of the world, perhaps cooler waters in N Pacific & Southern oceans; common in Mediterranean

♂

LC
Monotypic
TL 600–700 cm
W 2500–3500 kg

♀

Northern Bottlenose Whale

Hyperoodon ampullatus

FR Baleine-à-bec boréale
DE Nördlicher Entenwal
ES Zifio calderón septentrional

♂

♀

NT
Monotypic
TL 600–1000 cm
W 7500–10,000 kg

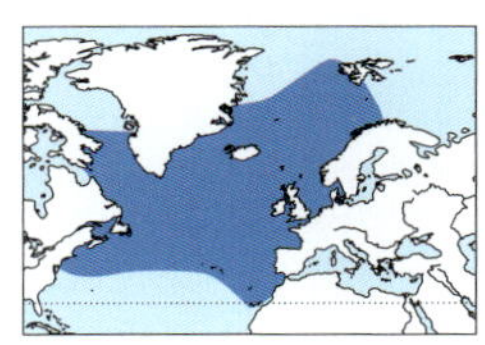

Cooler waters of N Atlantic N to Davis Strait & Spitsbergen, S to NE US, Azores & Canary Is, rare S of Bay of Biscay

Genus *HYPEROODON*

(2 species)

♂

LC
Monotypic
TL 650–750 cm
W 6000–7500 kg

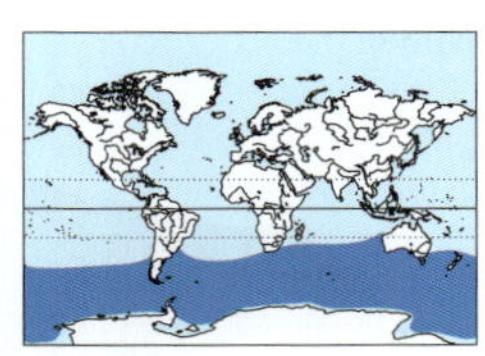

Relatively continuous in Subantarctic & Antarctic waters from South America, ZA, S AU & N NZ to AQ

Southern Bottlenose Whale

Hyperoodon planifrons

FR Baleine-à-bec australe
DE Südlicher Entenwal
ES Zifio calderón meridional

♀

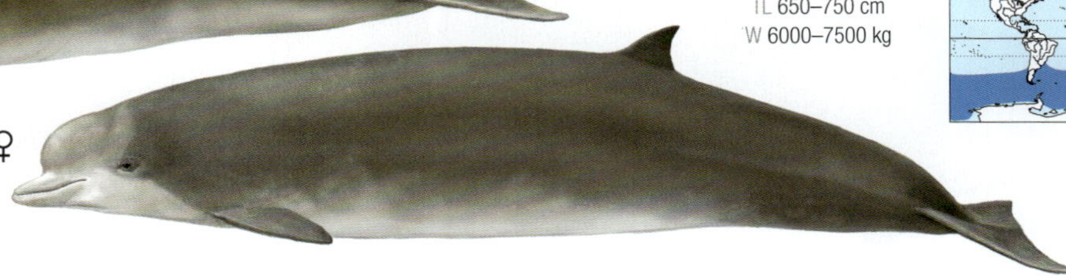

Genus *INDOPACETUS*

(1 species)

Longman's Beaked Whale

Indopacetus pacificus

FR Baleine-à-bec de Longman
DE Indopazifik-Schnabelwal
ES Zifio de Longman

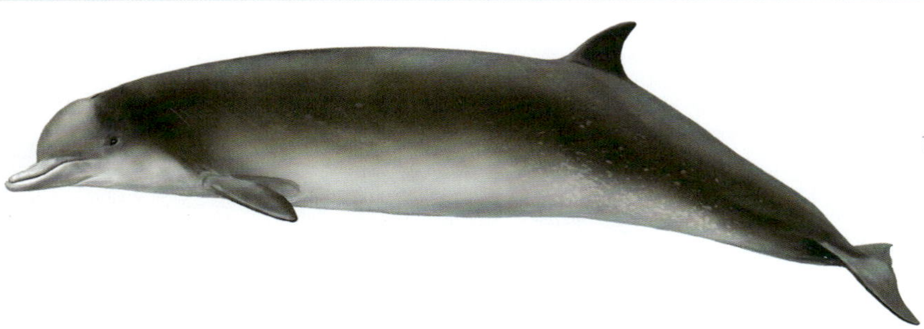

LC
Monotypic
TL 565–650 cm
W c. 7500 kg

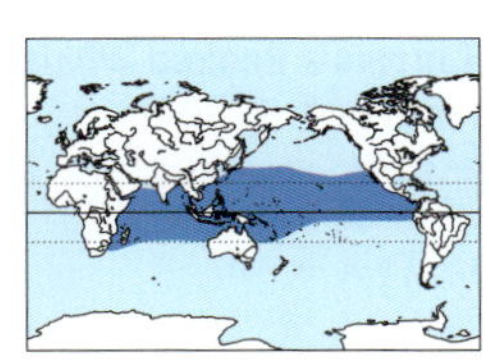

Relatively continuous in warmest waters of Indian & Pacific oceans from Arabian Sea, S IN, JP & MX S to ZA & AU

Sowerby's Beaked Whale
Mesoplodon bidens

FR Baleine-à-bec de Sowerby
DE Sowerby-Zweizahnwal
ES Zifio de Sowerby

LC
Monotypic
TL 450–550 cm
W 1000–1500 kg

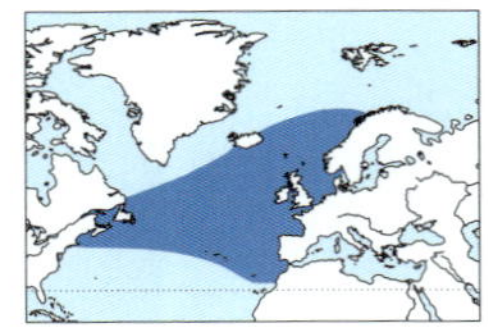

Cooler waters of N Atlantic from Labrador (CA), IS & 71° 30' N in NO S to NE US, Azores & Madeira; vagrant in Gulf of Mexico & E Mediterranean

Genus *MESOPLODON*

(16 species)

Ginkgo-toothed Beaked Whale
Mesoplodon ginkgodens

FR Baleine-à-bec du Japon
DE Japan-Zweizahnwal
ES Zifio de dientes de ginkgo

♂

DD
Monotypic
TL 470–528 cm
W c. 1000 kg

♀

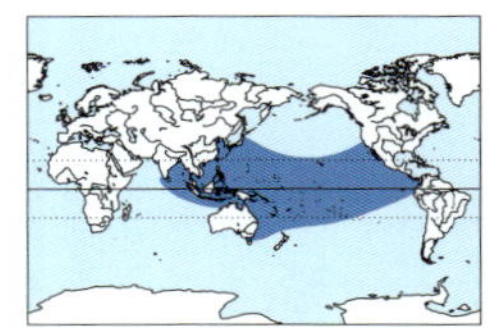

Warmer waters of Indian & Pacific oceans from S IN & LK to TW, JP, SE AU, Galápagos Is & California; limits unclear

Deraniyagala's Beaked Whale
Mesoplodon hotaula

FR Baleine-à-bec de Deraniyagala
DE Deraniyagala-Zweizahnwal
ES Zifio de Deraniyagala

DD
Monotypic
TL 400–450 cm
W c. 1000 kg

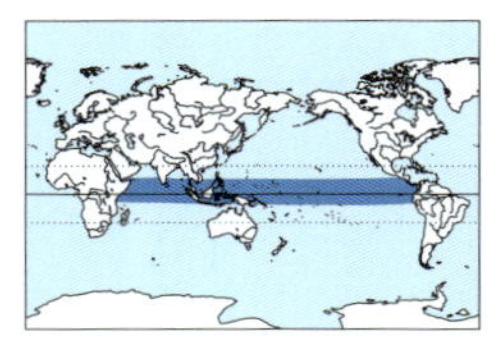

Warmer waters of Indian & Pacific oceans from E Africa & S IN to Galápagos Is, possibly to Central & South America; limits unclear

True's Beaked Whale
Mesoplodon mirus

FR Baleine-à-bec de True
DE True-Zweizahnwal
ES Zifio de True

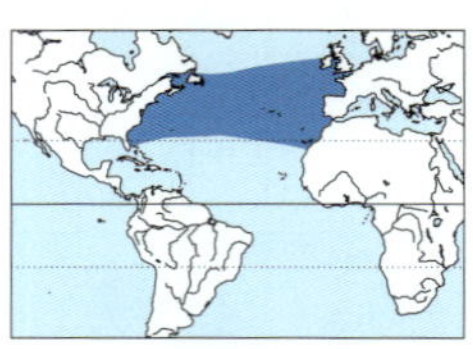

N Atlantic, in W from Nova Scotia (CA) to Florida (US), in E from IE to Canary Is

LC
Monotypic
TL 480–540 cm
W 1000–1400 kg

Ramari's Beaked Whale
Mesoplodon eueu

FR Baleine-à-bec de Ramari
DE Ramari-Zweizahnwal
ES Zifio de Ramari

DD
Monotypic
TL c. 530 cm

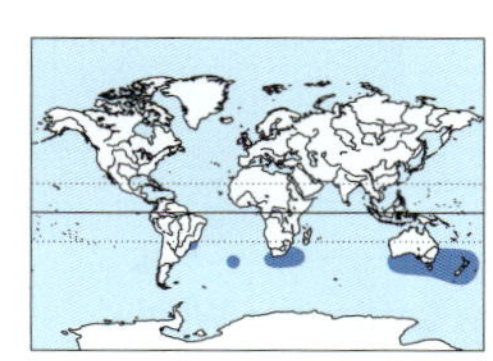

Genetically confirmed in ZA, AU & NZ, also in S BR, Tristan da Cunha, MZ & Walters Shoals (MG); probably throughout temperate S Hemisphere

Gervais's Beaked Whale
Mesoplodon europaeus

FR Baleine-à-bec de Gervais
DE Gervais-Zweizahnwal
ES Zifio de Gervais

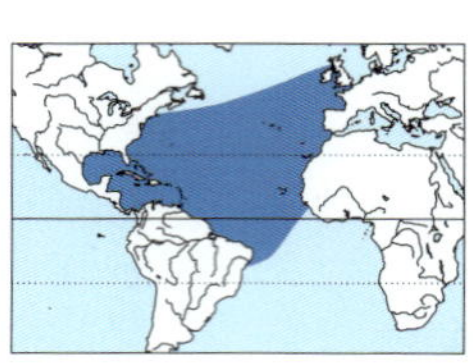

Subtropical & tropical waters of Atlantic, in W from N US to BR (perhaps to UY), in E from IE to GW (perhaps to AO)

LC
Monotypic
TL 420–485 cm
W c. 800 kg

Strap-toothed Whale
Mesoplodon layardii

FR Baleine-à-bec de Layard
DE Layard-Zweizahnwal
ES Zifio de Layard

♂

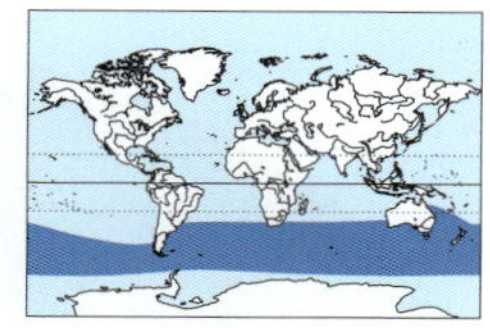

Temperate waters of S Atlantic, Indian & S Pacific oceans to waters of AQ; vagrant (1 record) to MM

LC
Monotypic
TL 550–625 cm
W 1000–1300 kg

♀

Hubbs's Beaked Whale
Mesoplodon carlhubbsi

FR Baleine-à-bec de Hubbs
DE Hubbs-Zweizahnwal
ES Zifio de Hubbs

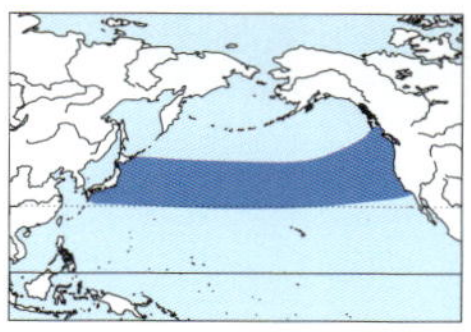

N Pacific mainly in W North America (54° 18' N–32° 42' N), also on coast of JP (41° 42' N–35° N); no records from C Pacific

DD
Monotypic
TL 470–532 cm
W 1000–1500 kg

♂

♀

Andrews's Beaked Whale
Mesoplodon bowdoini

FR Baleine-à-bec d'Andrews
DE Andrews-Zweizahnwal
ES Zifio de Andrews

DD
Monotypic
TL 390–441 cm
W c. 1000 kg

♂

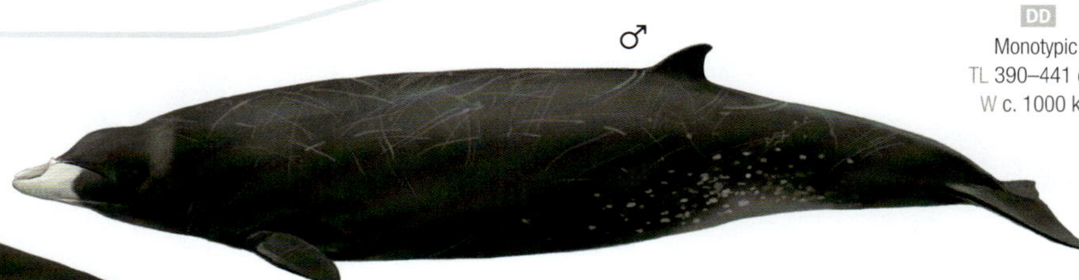

♀

Appears to be restricted to cooler waters of S Hemisphere (32° S to Antarctic Convergence); might be circumpolar

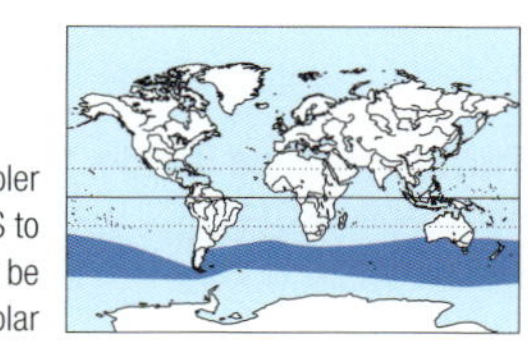

Spade-toothed Whale
Mesoplodon traversii

FR Baleine-à-bec de Travers
DE Travers-Zweizahnwal
ES Zifio de Travers

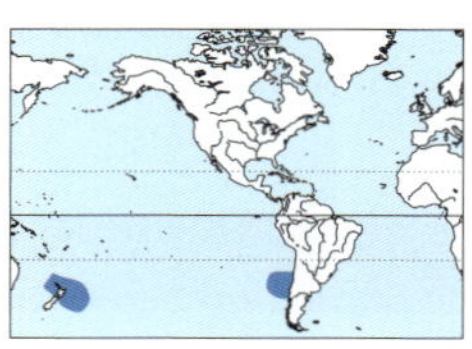

Known from few records in NZ, Chatham Is & CL (Juan Fernández Is); not known how these records reflect range

DD
Monotypic
TL 500–550 cm

Hector's Beaked Whale
Mesoplodon hectori

FR Baleine-à-bec de Hector
DE Hector-Zweizahnwal
ES Zifio de Hector

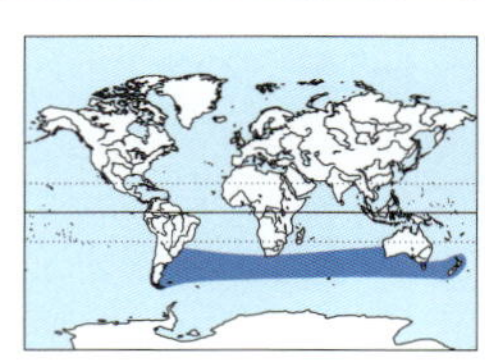

Most records from NZ & Atlantic South America, also recorded in ZA & AU; no records from Pacific E of NZ, but limits unclear

DD
Monotypic
TL 343–443 cm
W c. 500 kg

Gray's Beaked Whale
Mesoplodon grayi

FR Baleine-à-bec de Gray
DE Gray-Zweizahnwal
ES Zifio de Gray

♂

LC
Monotypic
TL 450–500 cm
W c. 900 kg

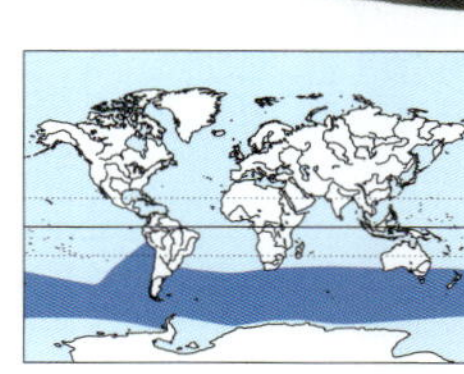

♀

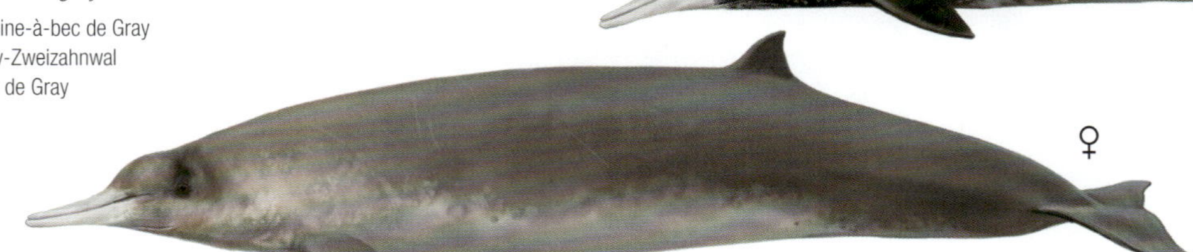

From temperate waters of S Atlantic, Indian & S Pacific oceans to AQ; 1 stranded in NL thought to be vagrant

NT
Monotypic
TL 480–574 cm
W 1000–1300 kg

Stejneger's Beaked Whale
Mesoplodon stejnegeri
FR Baleine-à-bec de Stejneger
DE Stejneger-Zweizahnwal
ES Zifio de Stejneger

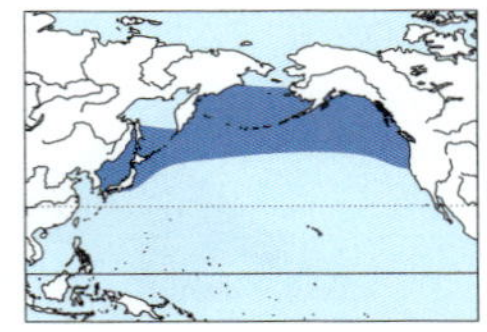

Subpolar & cold temperate waters of N Pacific, from shallow waters of Bering Sea S to S JP in W & to C California in E

Blainville's Beaked Whale
Mesoplodon densirostris
FR Baleine-à-bec de Blainville
DE Blainville-Zweizahnwal
ES Zifio de Blainville

LC
Monotypic
TL 425–475 cm
W c. 800 kg

♂

♀

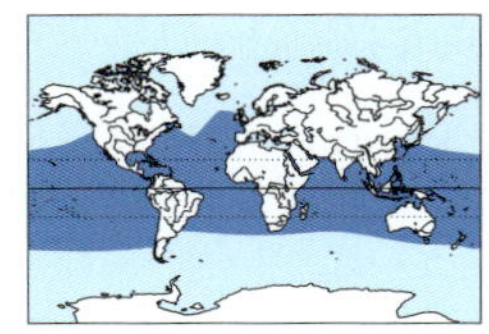

Tropical & subtropical waters throughout the world; occasional strandings in temperate waters might be vagrants

Pygmy Beaked Whale
Mesoplodon peruvianus
FR Baleine-à-bec du Pérou
DE Zwergzweizahnwal
ES Zifio del Perú

E tropical Pacific from 28° N to 29° S; 1 record from NZ currently thought to be a vagrant

♂

LC
Monotypic
TL 370–390 cm
W c. 500 kg

♀

Perrin's Beaked Whale
Mesoplodon perrini
FR Baleine-à-bec de Perrin
DE Perrin-Zweizahnwal
ES Zifio de Perrin

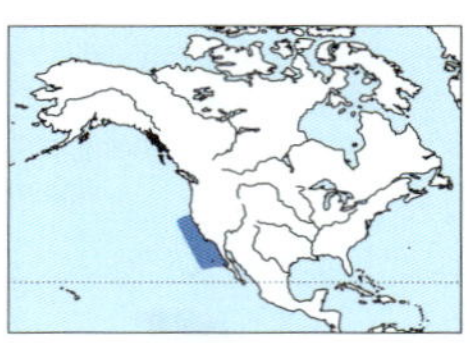

Known from few strandings in California (W US) and a possible sighting at sea off California; range unknown

EN
Monotypic
TL 390–443 cm
W c. 700 kg

Family PLATANISTIDAE – South Asian River Dolphins

1 genus, 2 species. Large and medium-sized fresh or brackish rivers in the Indus, Ganges, and Brahmaputra river basins in South Asia. Indo-Malayan Region.

Ganges River Dolphin
Platanista gangetica
FR Plataniste du Gange
DE Ganges-Delfin
ES Delfín del Ganges

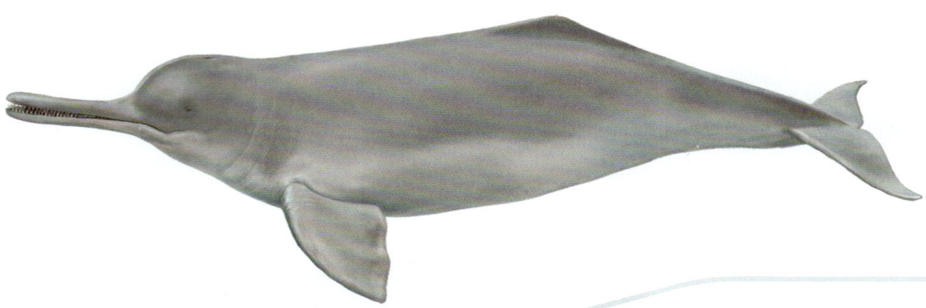

EN
Monotypic
TL 200–400 cm
W 51–89 kg

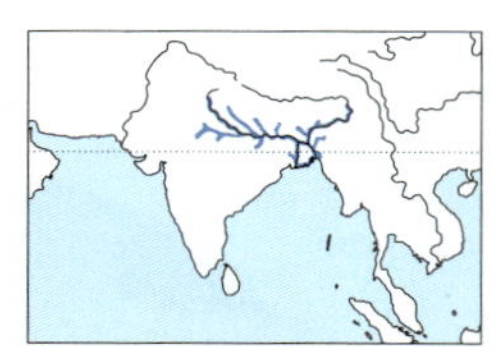

Ganges-Brahmaputra-Meghna & Karnaphuli-Sangu river systems (S NP, N & NE IN, BD)

Genus *PLATANISTA*

(2 species)

Indus River Dolphin
Platanista minor
FR Plataniste de l'Indus
DE Indus-Delfin
ES Delfín del Indo

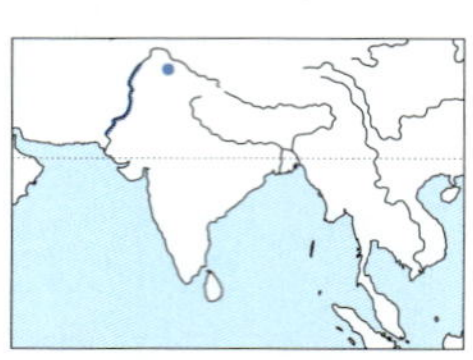

R Indus system of SC PK & NW IN (R Beas)

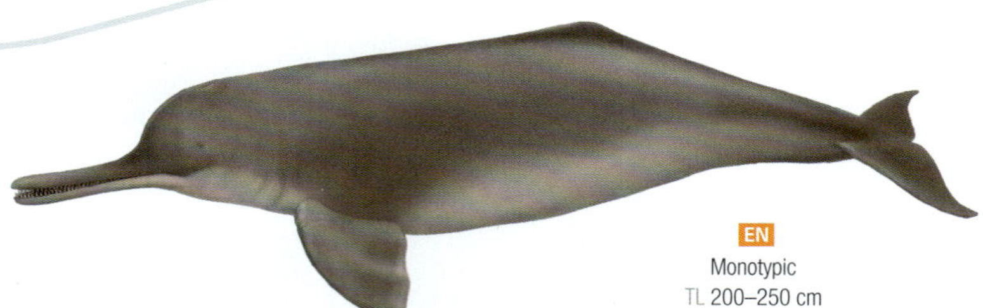

EN
Monotypic
TL 200–250 cm
W c. 84 kg

Family LIPOTIDAE – Baiji

1 genus, 1 species. Middle and lower reaches of the Yangtze River, China. Palearctic Region.

Genus *LIPOTES*

(1 species)

Baiji
Lipotes vexillifer
FR Baiji
DE Jangtse-Delfin
ES Delfín del Yangtsé

CR(PE)
Monotypic
TL 180–253 cm
W ≤ 167 kg

Middle & lower reaches of R Yangtze, including Dongting & Poyang lakes; probably extinct

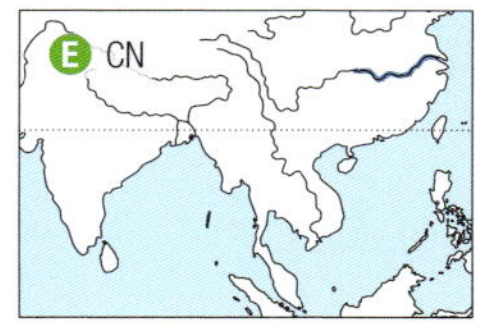

Family INIIDAE – Amazon River Dolphins

1 genus, 4 species. Endemic to the Amazon and Orinoco river basins in northern South America. Neotropical Region.

Amazon River Dolphin
Inia geoffrensis
FR Boto de l'Amazone
DE Amazonas-Delfin
ES Delfín del Amazonas

♂

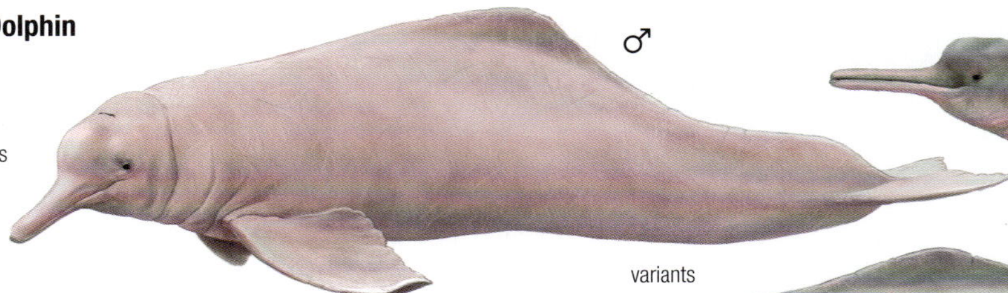

♀

variants

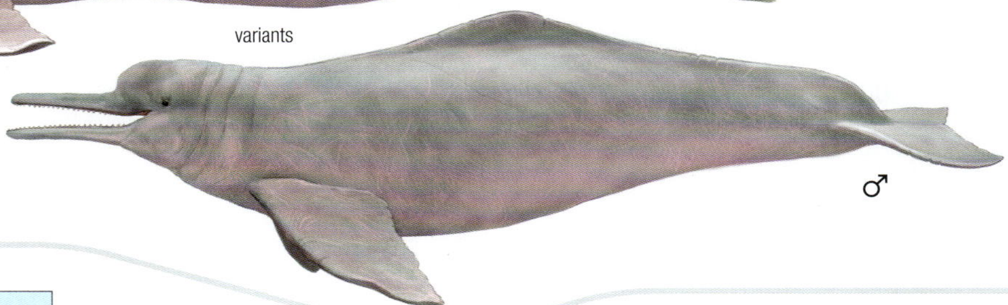

♂

EN
Monotypic
TL 178–228 cm (M), 201–228 cm (F)
W 113–207 kg (M), 72–154 kg (F)

below 100 m

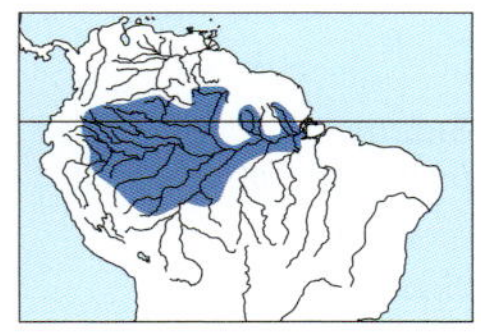

Most of Amazon basins in CO, BR, EC & PE

Genus *INIA*

(4 species)

Orinoco River Dolphin
Inia humboldtiana
FR Boto de l'Orénoque
DE Orinoco-Delfin
ES Delfín del Orinoco

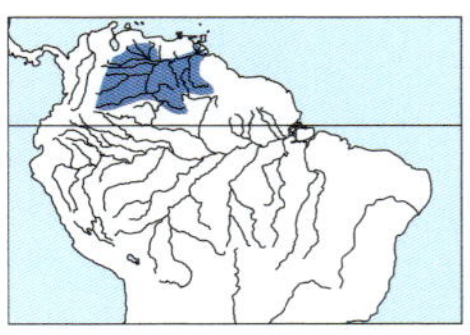

Orinoco basin including Apure & Meta rivers, upstream to rapids at Puerto Ayacucho (VE, CO)

NE
Monotypic
TL 184–214 cm

Bolivian River Dolphin
Inia boliviensis
FR Boto de Bolivie
DE Bolivien-Delfin
ES Delfín del Madeira

NE
Monotypic
TL 208–216 cm (F)

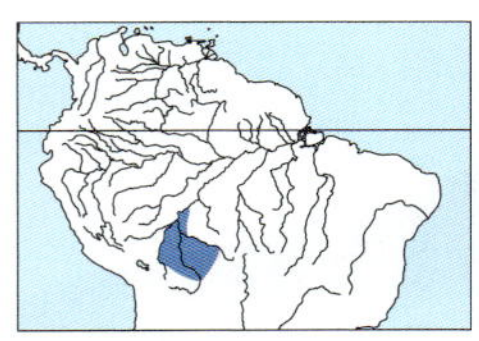

BO lowlands & SW BR in upper R Madeira & tributaries (Beni, Ichilo, Marmoré, Ipurupuru, Iboré, Baré, Abuná, Guaporé)

Araguaian River Dolphin
Inia araguaiaensis
FR Boto de l'Araguaia
DE Araguaia-Delfin
ES Delfín del Araguaia

NE
Monotypic
TL ≤ 252 cm

Araguaia-Tocantins Basin in Goiás, Tocantins, Mato Grosso & Pará

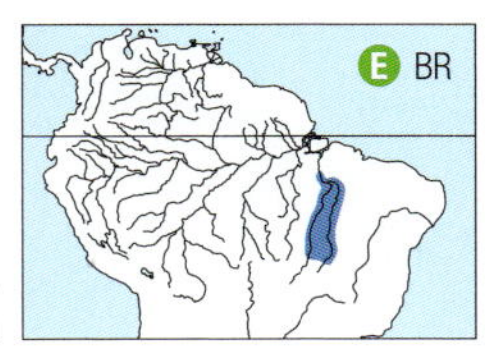

Family PONTOPORIIDAE – Franciscana

1 genus, 1 species. Shoreward of the inner continental shelf, mostly tropical to temperate waters less than 40 m deep. Western South Atlantic.

Genus *PONTOPORIA*

(1 species)

Franciscana
Pontoporia blainvillei
FR Franciscain
DE La-Plata-Delfin
ES Franciscana

SE BR (N to Espírito Santo), UY & N AR (S to Golfo San Matías in N Patagonia); not continuously distributed

VU
Monotypic
TL 117–162 cm
W 20–40 kg

Family MONODONTIDAE – Narwhal and Beluga

2 genera, 2 species. Circumpolar, mainly Arctic and subarctic seas. Arctic and subarctic Regions.

Genus *MONODON*

(1 species)

Narwhal
Monodon monoceros
FR Narval
DE Narwal
ES Narval

LC
Monotypic
TL 370–500 cm
W 700–1800 kg

♂

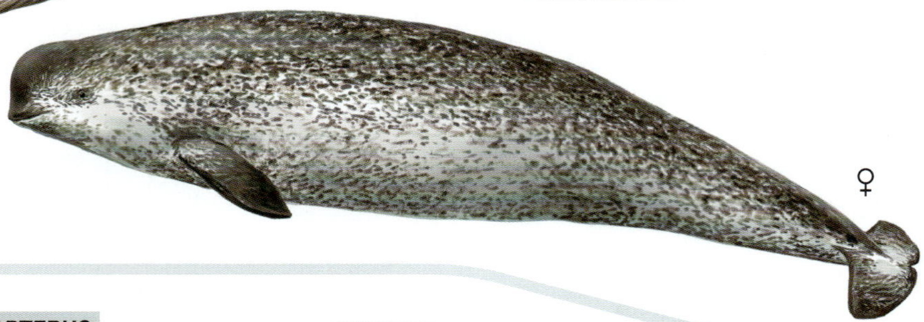

♀

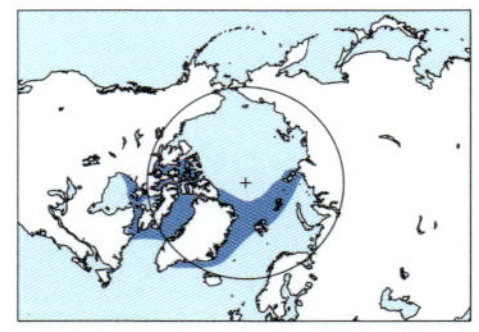

Arctic Ocean N to 85° N, S to c. 60° N (Hudson Bay & Labrador Sea); rare in Siberian, Alaskan and W CA waters

Genus *DELPHINAPTERUS*

(1 species)

Beluga
Delphinapterus leucas
FR Béluga
DE Weißwal
ES Beluga

LC
Monotypic
TL 300–450 cm
W 500–1600 kg

♂
spring

♀

Arctic & subarctic waters S to c. 50° N (in GL & E Europe to c. 64° N); occasionally S to New York (US) & R Seine (FR)

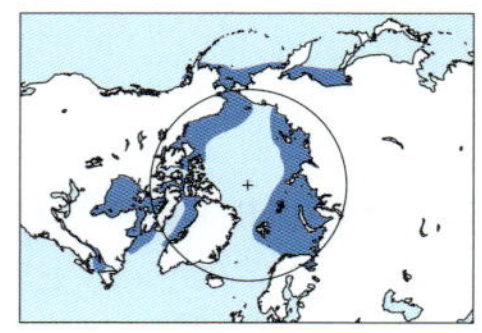

Family DELPHINIDAE – Ocean Dolphins

19 genera, 37 species. Open ocean, coastal, estuarine, and riverine. Cosmopolitan, inhabiting all the world's oceans and some river systems.

Genus *LEUCOPLEURUS*

(1 species)

Atlantic White-sided Dolphin
Leucopleurus acutus
FR Dauphin à flancs blancs
DE Atlantik-Weißseitendelfin
ES Delfín de flancos blancos del Atlántico

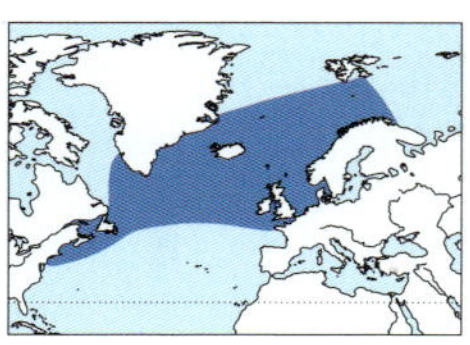

Cold temperate to subarctic waters of N Atlantic (excluding Baltic Sea) S to c. 38° N in W & from S Svalbard to FR in E

LC
Monotypic
TL 250–280 cm
W 182–235 kg

Genus *LAGENORHYNCHUS*

(1 species)

White-beaked Dolphin
Lagenorhynchus albirostris
FR Dauphin à bec blanc
DE Weißschnauzendelfin
ES Delfín de hocico blanco

LC
Monotypic
TL 240–310 cm
W 180–350 kg

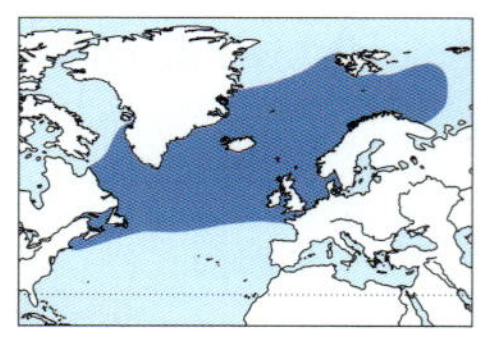

Temperate to subarctic N Atlantic S to Cape Cod (US) in W & to N FR in E; extralimital in Baltic Sea, Bay of Biscay & Iberian Pen

Genus *SAGMATIAS*

(4 species)

Pacific White-sided Dolphin
Sagmatias obliquidens
FR Dauphin de Gill
DE Pazifik-Weißseitendelfin
ES Delfín de flancos blancos del Pacífico

variants

LC
Monotypic
TL 230–250 cm
W c. 198 kg

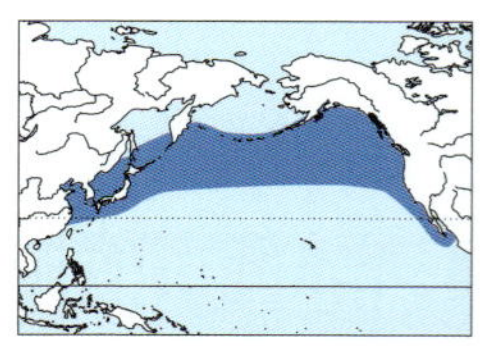

Cool temperate waters of N Pacific (primarily 38–47° N), from S Bering Sea S to Yellow Sea in W & to Gulf of California in E

Genus *ORCINUS*

(1 species)

Killer Whale

Orcinus orca

FR Épaulard
DE Schwertwal
ES Orca

♂ NE Pacific resident

♀ NE Pacific resident

♂ N Atlantic

♂ NE Pacific transient

DD
Monotypic
TL ≤ 980 cm (M), ≤ 770 cm (F)
W ≤ 6600 kg (M), ≤ 4700 kg (F)

♂ Antarctic type C

♂ Antarctic type A

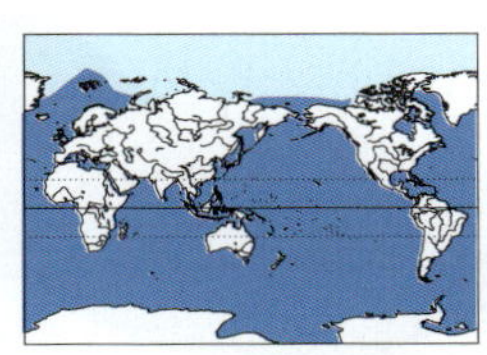

Widely distributed in almost any marine environment from equator to both polar zones, including semi-enclosed seas

♂ Antarctic type B

♀ Antarctic type B

♂ Antarctic type D

New Zealand form

Coastal S Africa (NA, ZA), Prince Edward & Amsterdam Is, S AU, NZ, S South America (N to PE in Pacific, N to AR in Atlantic)

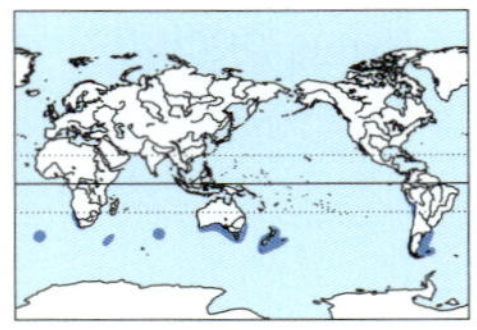

LC

3 sspp

TL ≤ 210 cm

W 70–85 kg

Dusky Dolphin

Sagmatias obscurus

FR Dauphin obscur

DE Dunkler Delfin

ES Delfín de Fitzroy

ssp *fitzroyi*

ssp *obscurus*

LC

Monotypic

TL 140–190 cm

W ≤ 94 kg

Hourglass Dolphin

Sagmatias cruciger

FR Dauphin sablier

DE Sanduhrdelfin

ES Delfín cruzado

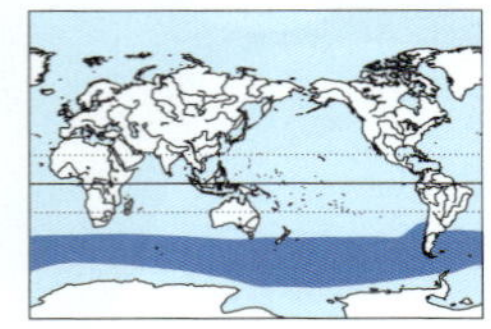

Circumpolar in subantarctic & Antarctic waters, primarily 45–65° S, occasionally N up to 33° S (off CL)

Peale's Dolphin

Sagmatias australis

FR Dauphin de Peale

DE Schwarzkinndelfin

ES Delfín austral

Coastal waters of South America from c. 33° S in Pacific & c. 38° S in Atlantic, S to Drake Passage (c. 59° S) & FK

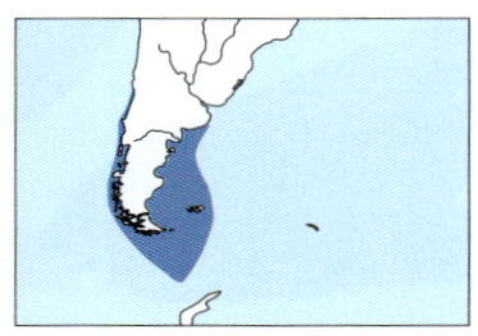

LC

Monotypic

TL ≤ 220 cm

W ≤ 115 kg

Heaviside's Dolphin

Cephalorhynchus heavisidii

FR Dauphin de Heaviside

DE Namibia-Delfin

ES Delfín de Heaviside

NT

Monotypic

TL ≤ 170 cm

W ≤ 75 kg

Coastal waters of SW Africa from c. 17° S (AO) to c. 34° S (Cape Point, ZA)

Genus *CEPHALORHYNCHUS*

(4 species)

Hector's Dolphin

Cephalorhynchus hectori

FR Dauphin de Hector

DE Hector-Delfin

ES Delfín de Héctor

♂ ssp *hectori*

♂

EN

2 sspp

TL c. 144 cm (M), 153–163 cm (F)

W ≤ 65 kg

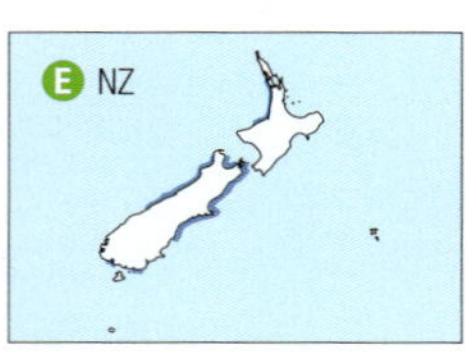

North Island (1 population on NW coast), South Island (3 populations around)

ssp *maui* ♀

ssp *commersonii*

LC

2 sspp

TL ≤ 180 cm

W ≤ 87 kg

ssp *kerguelensis*

Commerson's Dolphin

Cephalorhynchus commersonii

FR Dauphin de Commerson

DE Commerson-Delfin

ES Delfín pío

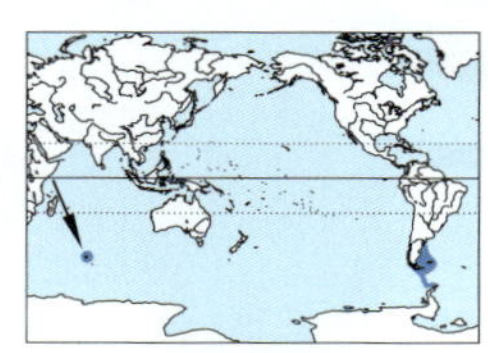

Kerguelen Is (S Indian Ocean), S South America & FK (N to 40–42° S in Atlantic & to c. 53° S in Pacific), S to South Shetland Is

Chilean Dolphin

Cephalorhynchus eutropia

FR Dauphin du Chili

DE Chile-Delfin

ES Delfín chileno

NT

Monotypic

TL ≥ 170 cm

W ≥ 63 kg

Coastal waters of SW South America from c. 33° S to Cape Horn at c. 55° S

Southern Right-whale Dolphin
Lissodelphis peronii

FR Dauphin de Péron
DE Südlicher Glattdelfin
ES Delfín sin aleta meridional

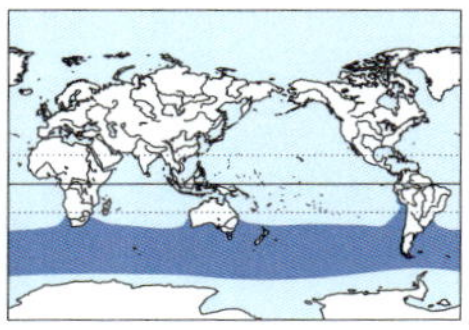

Circumpolar in subantarctic waters (primarily 30–65° S) but N to 23° S in Benguela Current, to 12° S in Humboldt Current & to 25° S in Malvinas Current

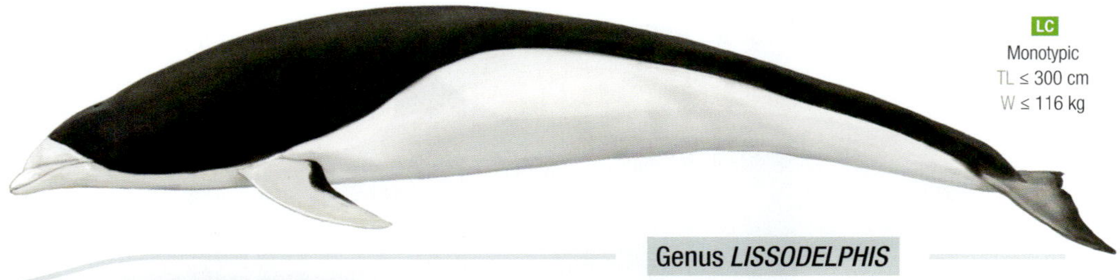

LC
Monotypic
TL ≤ 300 cm
W ≤ 116 kg

Genus *LISSODELPHIS*
(2 species)

Northern Right-whale Dolphin
Lissodelphis borealis

FR Dauphin à dos lisse
DE Nördlicher Glattdelfin
ES Delfín sin aleta septentrional

♂
swirl morph

LC
Monotypic
TL c. 310 cm (M), c. 230 cm (F)
W ≤ 115 kg

♀

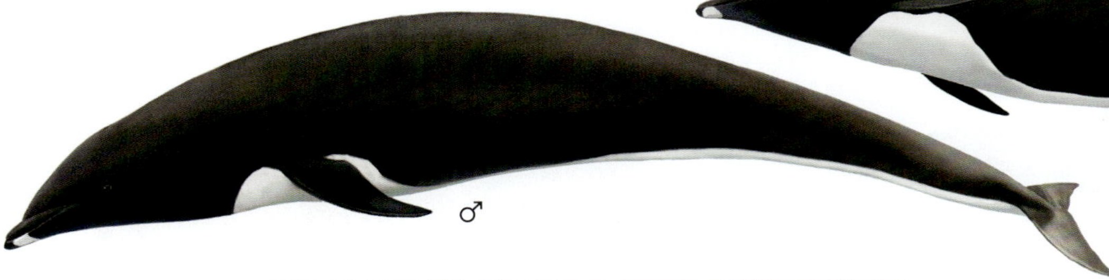

♂

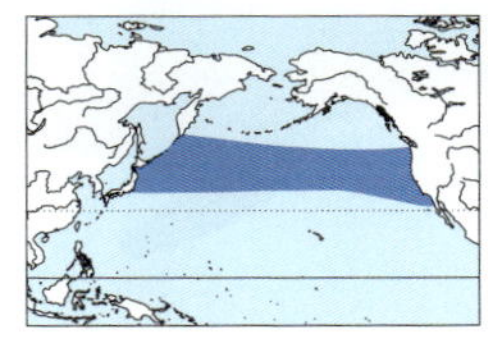

Temperate & subarctic waters of N Pacific (c. 30–50° N)

Irrawaddy Dolphin
Orcaella brevirostris

FR Orcelle de l'Irrawaddy DE Irawadi-Delfin ES Delfín del Irawadi

EN
Monotypic
TL ≤ 280 cm
W 115–130 kg

Genus *ORCAELLA*
(2 species)

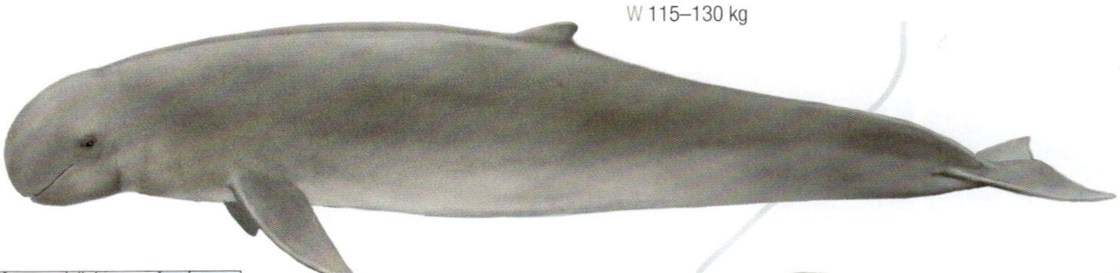

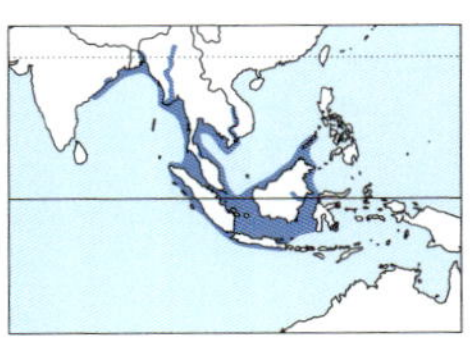

SE Asian coasts (Bay of Bengal, Ganges Delta, Strait of Malacca, Gulf of Thailand, Sunda Shelf, Malampaya Sound), also Irrawady, Mekong & Mahakam rivers and Chilika & Songkhla lakes

Australian Snubfin Dolphin
Orcaella heinsohni

FR Orcelle d'Australie
DE Australischer Stupsflossendelfin
ES Delfín de Heinsohn

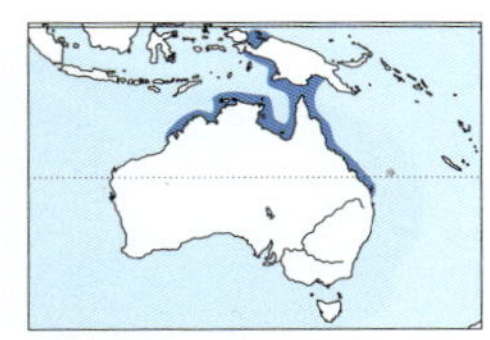

Coastal waters & river mouths on Sahul Shelf of NW & S New Guinea & N AU from Broome (WA) to R Brisbane (Qld)

VU
Monotypic
TL ≤ 270 cm
W 114–135 kg

Genus *STENO*
(1 species)

LC Monotypic
TL ≤ 280 cm
W ≤ 155 kg

Rough-toothed Dolphin
Steno bredanensis

FR Dauphin sténo
DE Rauzahndelfin
ES Delfín de dientes rugosos

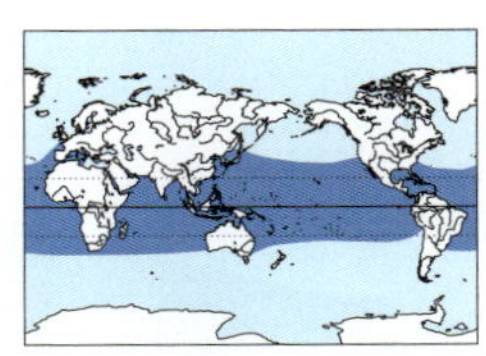

Tropical & subtropical deep, oceanic waters in all major oceans from c. 40° N to c. 35° S; scarce in Mediterranean

Risso's Dolphin
Grampus griseus

FR Grampus
DE Rundkopfdelfin
ES Calderón gris

Genus *GRAMPUS*
(1 species)

variants

LC
Monotypic
TL 380–410 cm
W 400–500 kg

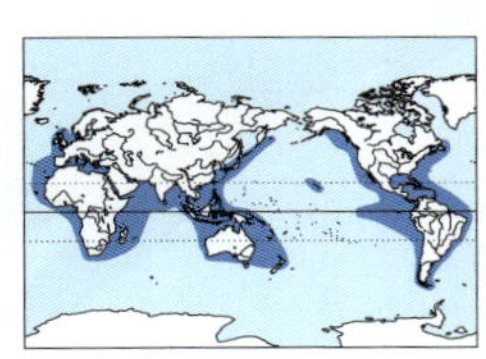

Worldwide in tropical to temperate oceanic waters (c. 60° N–60° S), including North Sea, Mediterranean Sea & N Gulf of Alaska

Genus *PSEUDORCA*

(1 species)

False Killer Whale
Pseudorca crassidens

FR Fausse-orque
DE Kleiner Schwertwal
ES Falsa orca

NT
Monotypic
TL 510–610 cm
W ≤ 2200 kg

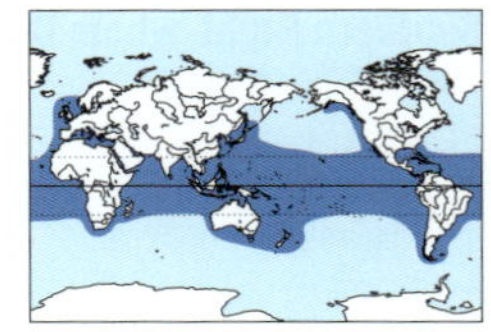

Worldwide in tropical & warm temperate oceanic waters (mainly c. 50° N–50° S), including semi-enclosed seas

Genus *FERESA*

(1 species)

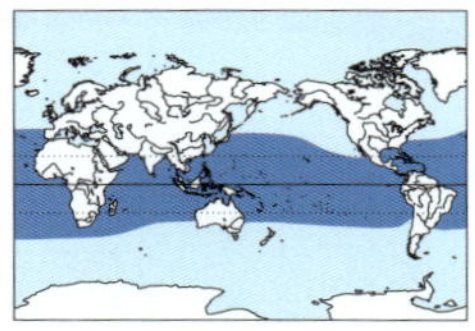

Cosmopolitan tropical & subtropical waters (c. 40° N–35° S), including Gulf of Mexico & Arabian Sea

Pygmy Killer Whale
Feresa attenuata

FR Orque pygmée
DE Zwergschwertwal
ES Orca pigmea

LC
Monotypic
TL 210–260 cm
W ≤ 225 kg

Genus *PEPONOCEPHALA*

(1 species)

Melon-headed Whale
Peponocephala electra

FR Péponocéphale
DE Breitschnabeldelfin
ES Delfín de cabeza de melón

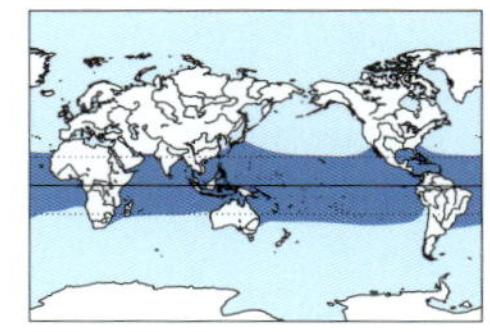

Worldwide in tropical & subtropical waters (c. 40° N–35° S), including Gulf of Mexico, Arabian Sea, Bay of Bengal, S China Sea & Timor Sea

LC
Monotypic
TL ≤ 280 cm
W ≤ 275 kg

Genus *GLOBICEPHALA*

(2 species)

Long-finned Pilot Whale
Globicephala melas

FR Globicéphale noir
DE Langflossen-Grindwal
ES Calderón de aleta larga

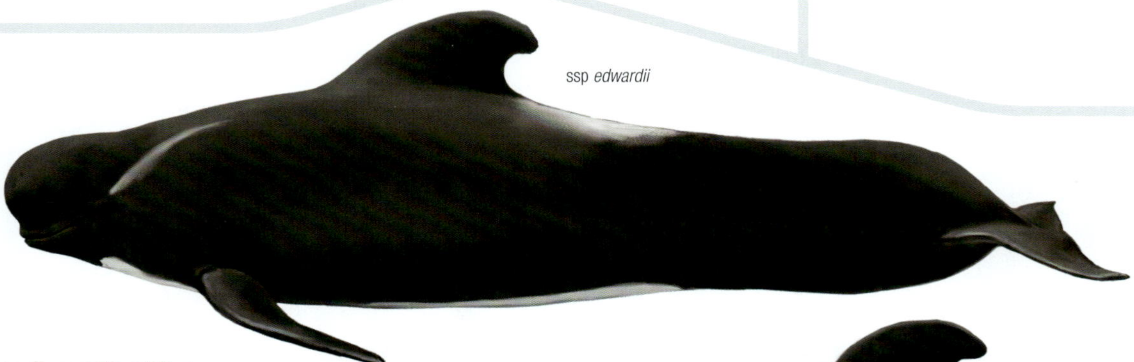

ssp *edwardii*

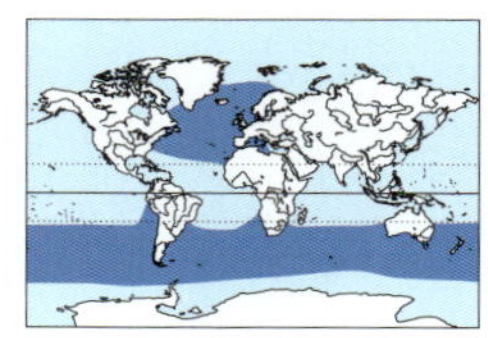

N Atlantic S to about Tropic of Cancer (including Barents, North & W Mediterranean seas, Gulf of Saint Lawrence), circumglobal in Southern Ocean (14–68° S)

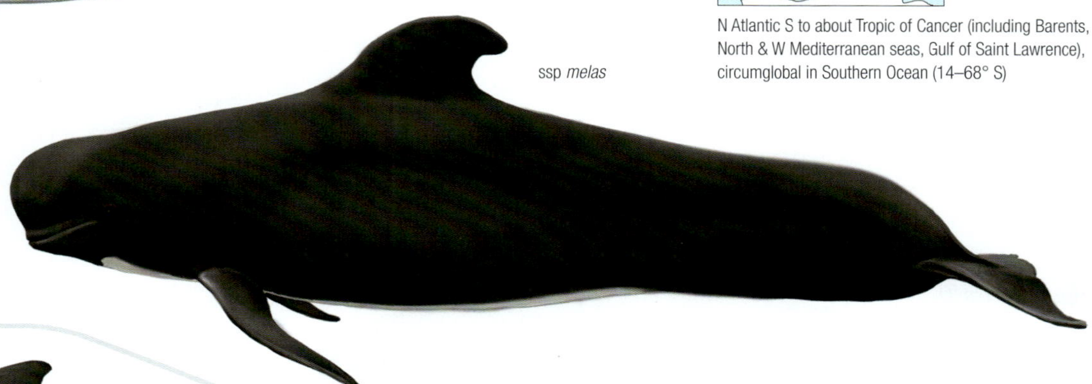

ssp *melas*

LC
2 sspp
TL 570–670 cm
W 1300–2300 kg

Short-finned Pilot Whale
Globicephala macrorhynchus

FR Globicéphale tropical
DE Kurzflossen-Grindwal
ES Calderón de aleta corta

♀

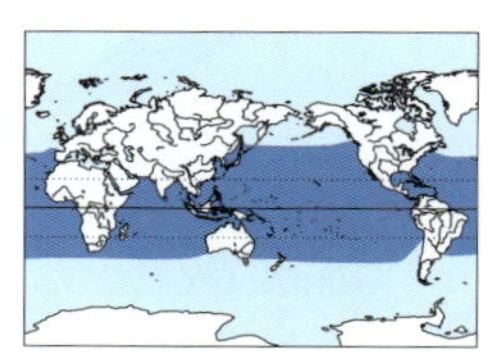

Tropical, subtropical & warm temperate waters worldwide (c. 50° N–40° S), except Mediterranean Sea & Persian Gulf

♂

LC
Monotypic
TL c. 720 cm (M), 550 cm (F)
W c. 3500 kg (M), c. 1500 kg (F)

Genus *SOTALIA*

(2 species)

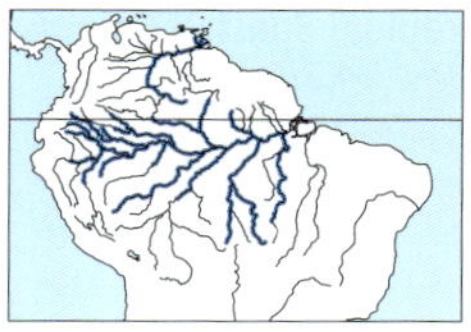

Amazon Basin & tributaries in BR, CO, EC & PE; perhaps Orinoco system in VE (this species or *S. guianensis*)

EN

Monotypic
TL ≤ 152 cm
W ≤ 52 kg

Tucuxi
Sotalia fluviatilis

FR Dauphin tucuxi
DE Amazonas-Sotalia
ES Tucuxi

Guiana Dolphin
Sotalia guianensis

FR Dauphin costero
DE Küsten-Sotalia
ES Delfín costero

NT

Monotypic
TL ≤ 190 cm
W ≤ 40 kg

E coasts of Central & South America, from c. 16° N (HN) to c. 27° S (S BR); perhaps R Orinoco in VE (this species or *S. fluviatilis*)

Genus *SOUSA*

(4 species)

Atlantic Humpback Dolphin
Sousa teuszii

FR Dauphin de Teusz
DE Westafrika-Buckeldelfin
ES Delfín jorobado atlántico

CR

Monotypic
TL 235–260 cm
W ≤ 284 kg (M)

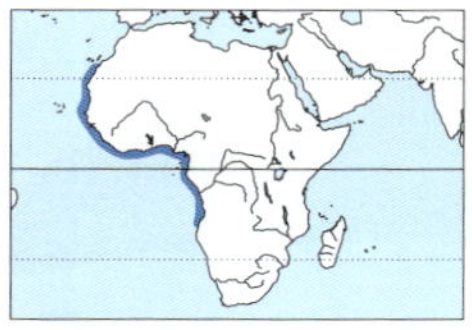

Near-shore waters of tropical & sub-tropical W Africa from EH to S AO

Indian Humpback Dolphin
Sousa plumbea

FR Dauphin de Malabar
DE Bleifarbener Buckeldelfin
ES Delfín jorobado índico

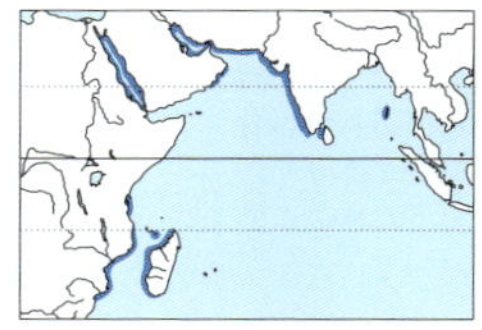

Coastal waters in E Africa (N from False Bay in ZA), MG, Middle East E to W IN, offshore Is (Mayotte, Zanzibar Arch, Andamans)

EN

Monotypic
TL 216–226 cm
W 240–280 kg

Indo-Pacific Humpback Dolphin
Sousa chinensis

FR Dauphin de Chine
DE Indopazifik-Buckeldelfin
ES Delfín jorobado indopacífico

Coastal tropical to subtropical waters of NE Indian Ocean & W Pacific from R Yangtze (C CN) S through SE Asia (E to at least Borneo)

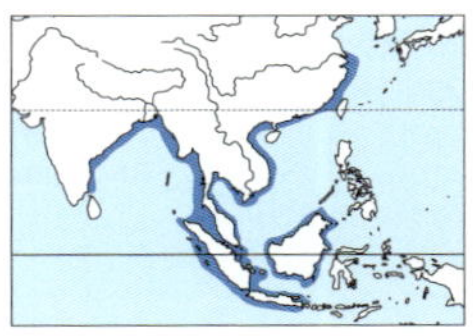

VU

2 sspp
TL ≤ 280 cm
W 240–280 kg

variants

Australian Humpback Dolphin
Sousa sahulensis

FR Dauphin du Sahul
DE Australischer Buckeldelfin
ES Delfín jorobado australiano

VU

Monotypic
TL ≤ 270 cm

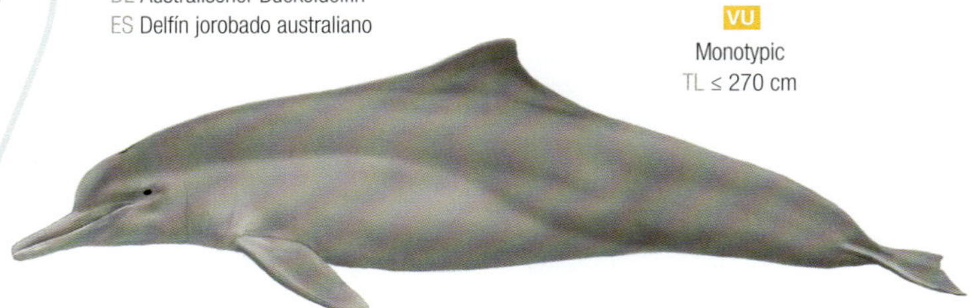

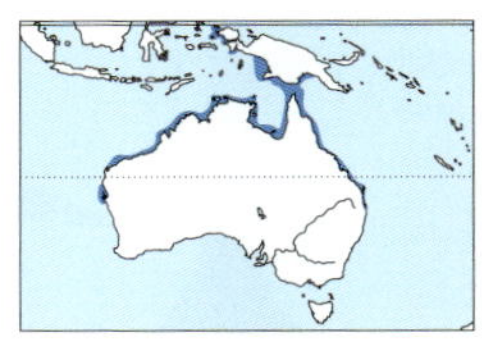

Tropical & subtropical waters of Sahul Shelf from N AU to waters off NW & S New Guinea

Genus *TURSIOPS*

(2 species)

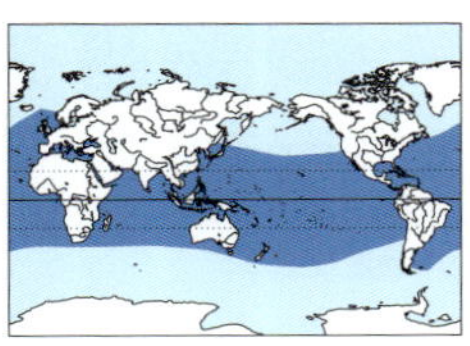

Worldwide temperate & tropical waters N generally to 45° N but reaching FO in N Atlantic, S to NZ & S AR

ssp *truncatus*
coastal form

LC

3 sspp
TL 190–380 cm
W 135–635 kg

ssp *ponticus*

Common Bottlenose Dolphin
Tursiops truncatus

FR Grand Dauphin
DE Großer Tümmler
ES Delfín mular

ssp *truncatus*
offshore form

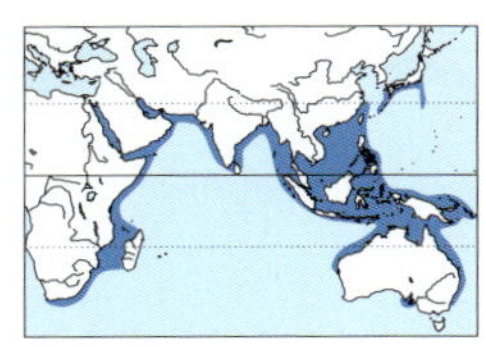

Warm temperate to tropical Indo-Pacific waters, from ZA, Red Sea & Persian Gulf E to Indo-Malay Arch, S JP, Solomon Is & AU

NT

Monotypic
TL c. 270 cm (M)
W c. 230 kg (M)

Indo-Pacific Bottlenose Dolphin
Tursiops aduncus

FR Dauphin indien
DE Indopazifik-Tümmler
ES Delfín mular indopacífico

Genus *STENELLA*

(5 species)

Atlantic Spotted Dolphin

Stenella frontalis

FR Dauphin tacheté
DE Zügeldelfin
ES Delfín manchado atlántico

LC
Monotypic
TL 190–230 cm
W 110–143 kg

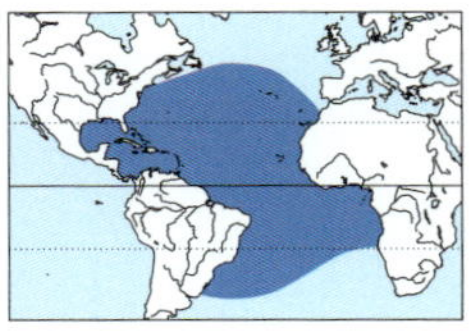

Tropical to temperate coastal & oceanic waters of Atlantic (c. 50° N–25° S)

Pantropical Spotted Dolphin

Stenella attenuata

FR Dauphin bridé
DE Schlankdelfin
ES Delfín manchado tropical

ssp *attenuata* variants

Pantropical in all oceans & semi-enclosed seas (c. 40° N–40° S), excluding Mediterranean Sea

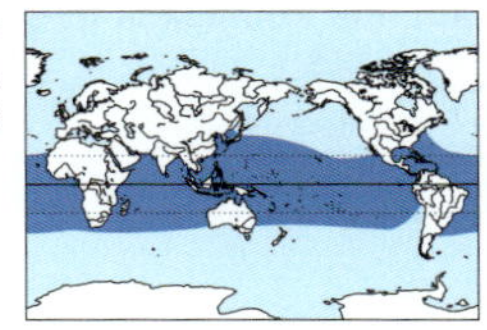

LC
2 sspp
TL 160–260 cm
W ≤ 119 kg

ssp *graffmani*

Striped Dolphin

Stenella coeruleoalba

FR Dauphin bleu et blanc
DE Streifendelfin
ES Delfín listado

LC
Monotypic
TL 216–255 cm
W 130–156 kg

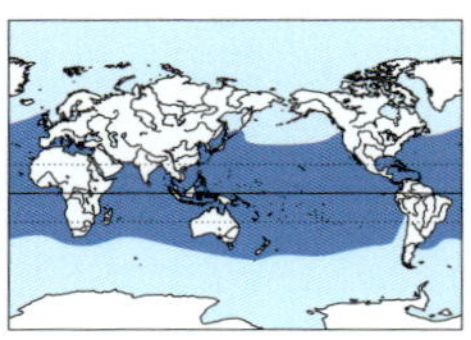

Tropical to warm temperate waters (c. 50° N–40° S) in Indian, Pacific & Atlantic oceans and seas; vagrants to Red Sea & Persian Gulf

Clymene Dolphin

Stenella clymene

FR Dauphin clymène
DE Clymene-Delfin
ES Delfín clímene

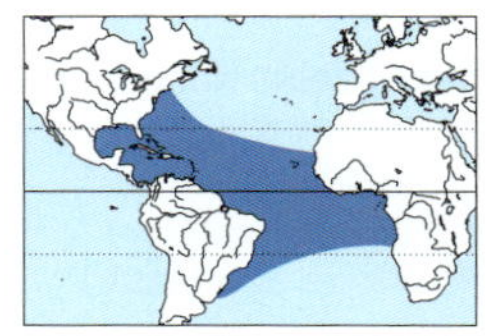

Tropical & subtropical waters of Atlantic including Caribbean & Gulf of Mexico

LC
Monotypic
TL ≤ 197 cm
W ≤ 90 kg

Spinner Dolphin

Stenella longirostris

FR Dauphin à long bec
DE Spinnerdelfin
ES Delfín girador

♂ ssp *centroamericana*

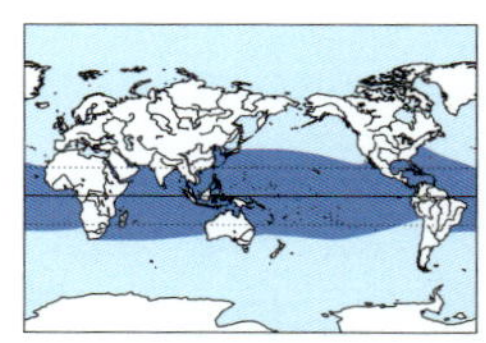

Tropical waters of Atlantic, Indian & Pacific oceans

♀ ssp *centroamericana*

♂ ssp *orientalis*

♂ ssp *longirostris*

LC
4 sspp
TL 129–280 cm (M), 138–211 cm (F)
W ≤ 80 kg (M), ≤ 63 kg (F)

♀ ssp *orientalis*

♂ white belly morph

♂ ssp *roseiventris*

Genus *DELPHINUS*

(1 species)

ssp *tropicalis*

LC

4 sspp

TL 160–260 cm

W ≤ 235 kg

ssp *delphis*
Pacific form

ssp *ponticus*

Common Dolphin

Delphinus delphis

FR Dauphin commun

DE Gewöhnlicher Delfin

ES Delfín de hocico corto

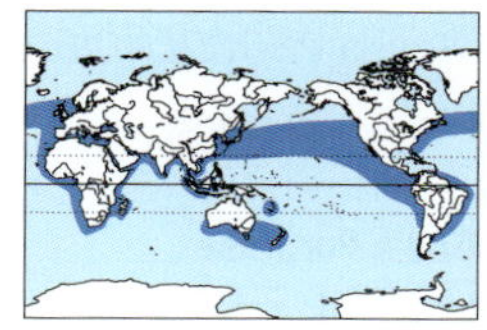

Primarily offshore tropical to temperate waters of Atlantic, Indian & Pacific oceans, excluding Gulf of Mexico & Caribbean Sea

ssp *delphis*
Atlantic form

ssp *delphis*
long-beaked form

Fraser's Dolphin

Lagenodelphis hosei

FR Dauphin de Fraser

DE Borneo-Delfin

ES Delfín de Fraser

Genus *LAGENODELPHIS*

(1 species)

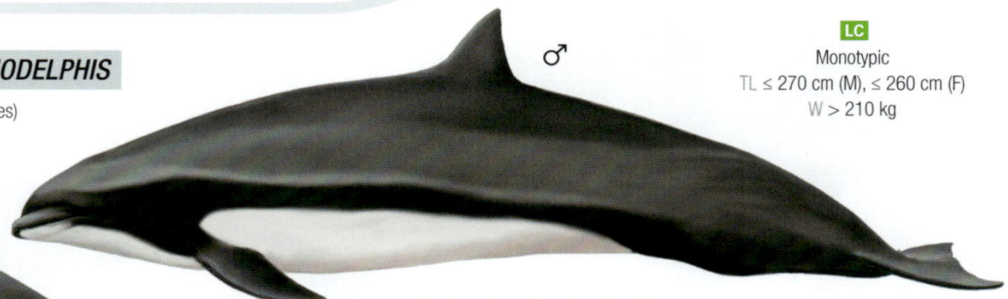

♂

LC

Monotypic

TL ≤ 270 cm (M), ≤ 260 cm (F)

W > 210 kg

♀

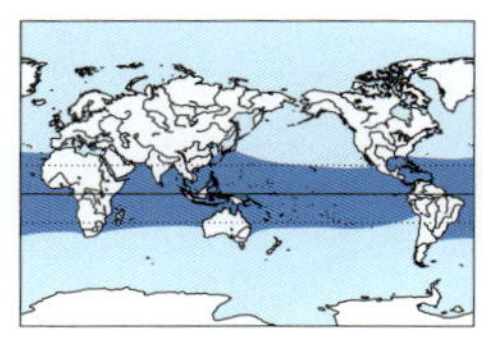

Tropical & subtropical waters in all major oceans from c. 30° N to c. 30° S

Family PHOCOENIDAE – Porpoises

3 genera, 8 species. Primarily continental shelf, less than 180 m deep; some occur in deep, oceanic waters, and two in rivers. All oceans and some rivers in East and South Asia.

Indo-Pacific Finless Porpoise

Neophocaena phocaenoides

FR Marsouin aptère

DE Glattschweinswal

ES Marsopa sin aleta indopacífica

VU

Monotypic

TL 135–171 cm

W 30–55 kg

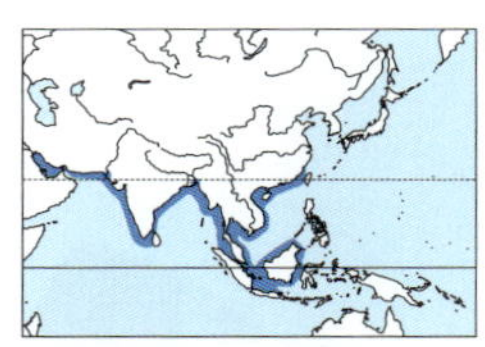

Shallow, coastal waters of tropical & subtropical S & E Asia, from Persian Gulf E to E CN, S to Sunda Is; possibly OM & PH

Genus *NEOPHOCAENA*

(3 species)

East Asian Finless Porpoise

Neophocaena sunameri

FR Marsouin du Japon

DE Ostasiatischer Glattschweinswal

ES Marsopa sin aleta de Japón

NE

Monotypic

TL 130–227 cm

W 40–72 kg

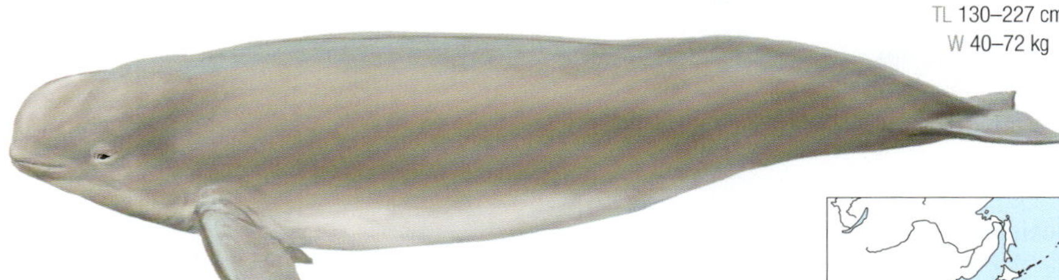

Shallow marine waters of E Asia, records from JP, KR, CN & TW; possibly KP, 1 vagrant stranded on Okinawa I (S JP)

Yangtze Finless Porpoise

Neophocaena asiaeorientalis

FR Marsouin du Yangzi

DE Östlicher Glattschweinswal

ES Marsopa sin aleta asiaticoriental

EN

Monotypic

Middle & lower reaches of R Yangtze; historically occurred in upper reaches but now extinct there

Genus *PHOCOENOIDES*

(1 species)

Dall's Porpoise

Phocoenoides dalli

FR Marsouin de Dall
DE Weißflanken-Schweinswal
ES Marsopa de Dall

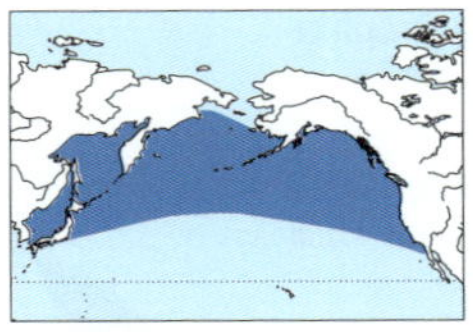

Cool temperate regions of N Pacific S to JP in W & S to N Baja California in E

♀ ssp *dalli*

♂ ssp *dalli*

LC
2 sspp
TL 165–240 cm
W 60–200 kg

♂ ssp *truei*

Harbor Porpoise

Phocoena phocoena

FR Marsouin commun
DE Schweinswal
ES Marsopa común

ssp *phocoena*

LC
3 sspp
TL 130–200 cm
W 45–75 kg

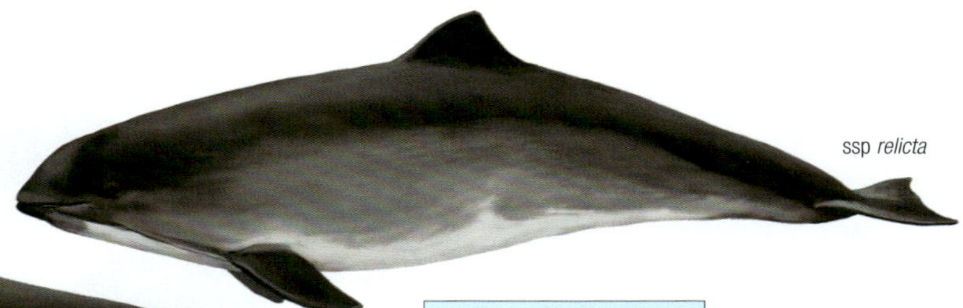

ssp *relicta*

ssp *vomerina*

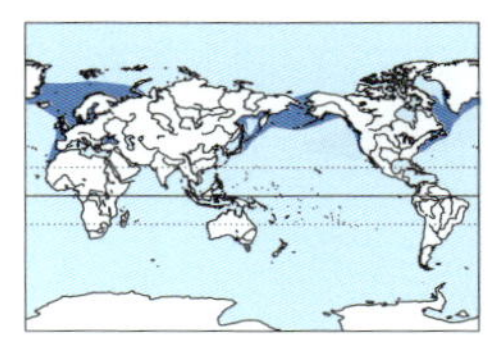

Coastal waters of N Atlantic & N Pacific oceans, isolated in Black, Azov & Marmara seas (some stragglers in Aegean Sea)

Genus *PHOCOENA*

(4 species)

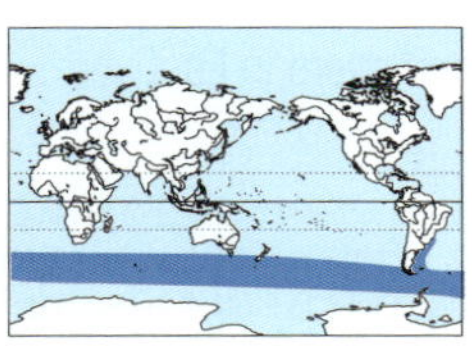

Circumpolar and largely oceanic in S Hemisphere (S to 64° 34' S), in E South America N to S BR

Spectacled Porpoise

Phocoena dioptrica

FR Marsouin à lunettes
DE Brillenschweinswal
ES Marsopa de anteojos

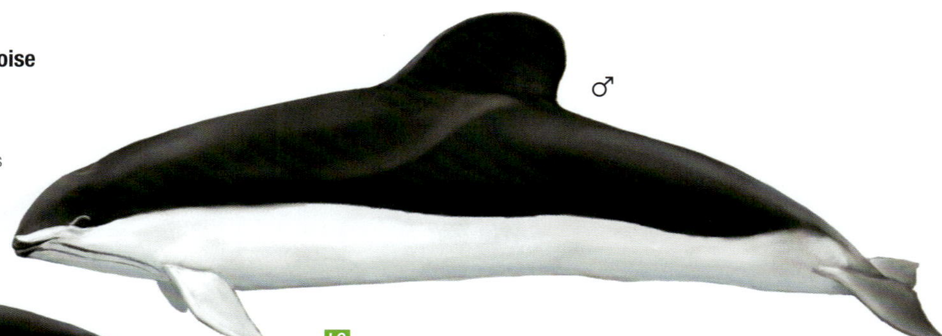

♂

LC
Monotypic
TL 150–225 cm
W 50–115 kg

♀

Burmeister's Porpoise

Phocoena spinipinnis

FR Marsouin de Burmeister
DE Burmeister-Schweinswal
ES Marsopa espinosa

Near-shore waters of South America, from N PE (5° S) S to Cape Horn, up along Atlantic coast to S BR (28° 50' S)

NT
Monotypic
TL 150–200 cm
W 50–105 kg

Vaquita

Phocoena sinus

FR Marsouin de Californie
DE Kalifornien-Schweinswal
ES Vaquita marina

CR
Monotypic
TL 120–150 cm
W 30–48 kg

N Gulf of California; smallest known distribution of any marine cetacean

Family HIPPOPOTAMIDAE – Hippopotamuses

2 genera, 2 species. Aquatic habitats and densely forested areas. Afrotropical Region.

Genus *HIPPOPOTAMUS*

(1 species)

Common Hippopotamus
Hippopotamus amphibius

FR Grand Hippopotame
DE Flusspferd
ES Hipopótamo

VU
Monotypic
HB 290–505 cm
T 40–56 cm
W 1000–4500 kg

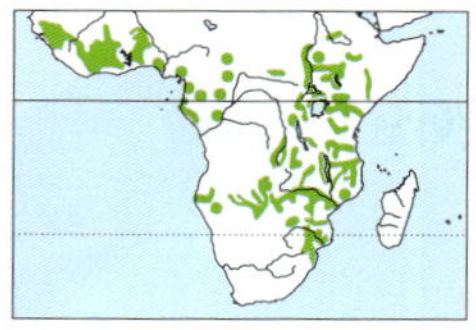

0–2000 m

Scattered localities throughout sub-Saharan Africa; previously continuous across sub-Saharan Africa & along R Nile

Genus *CHOEROPSIS*

(1 species)

Pygmy Hippopotamus
Choeropsis liberiensis

FR Hippopotame nain
DE Zwergflusspferd
ES Hipopótamo pigmeo

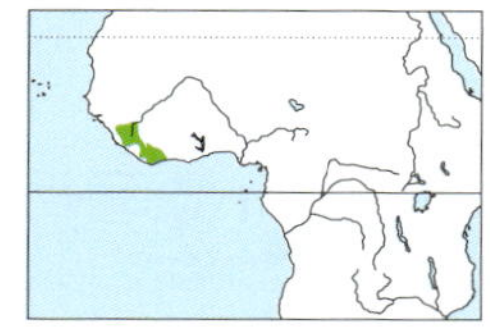

W Africa in GN, SL, LR & CI; population in S NG (swamps of Niger & Cross rivers) presumably extinct

EN
1 extant ssp
HB 150–175 cm
T c. 20 cm
W 160–270 kg

Family TRAGULIDAE – Chevrotains

3 genera, 10 species. Dense wooded habitats, drylands, and riverine subtropical and tropical forests. Afrotropical and Indo-Malayan Regions.

Genus *MOSCHIOLA*

(3 species)

White-spotted Chevrotain
Moschiola meminna

FR Chevrotain meminna
DE Ceylon-Kantschil
ES Ciervo ratón manchado

LC
Monotypic
HB 55–60 cm
T 2–5 cm
W c. 2.4 kg

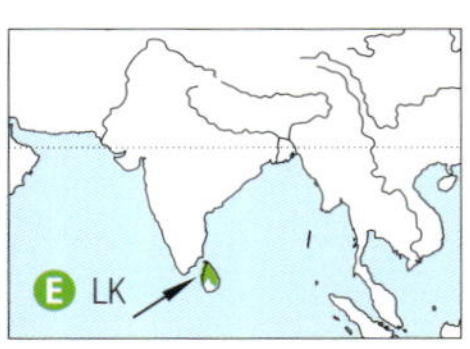

Dry Zone

Yellow-striped Chevrotain
Moschiola kathygre

FR Chevrotain kathygre
DE Ceylon-Gelbstreifenkantschil
ES Ciervo ratón rayado

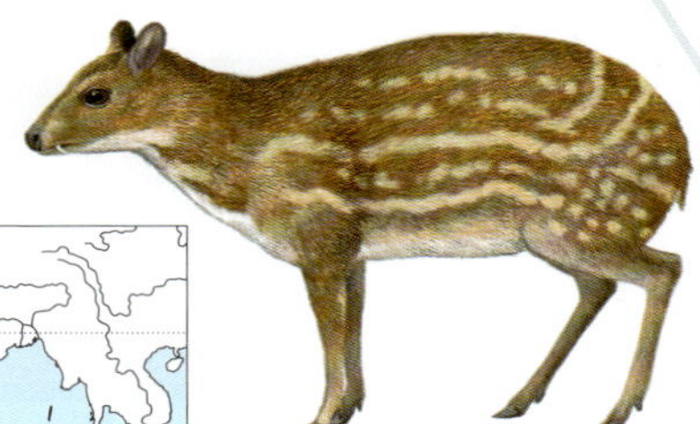

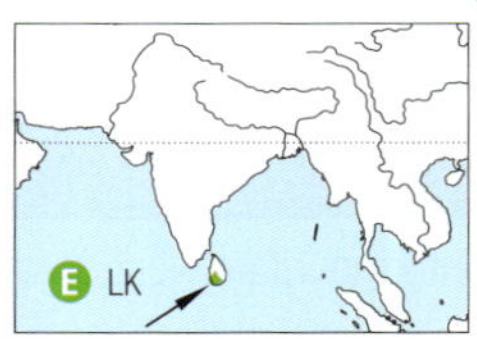

Wet Zone, from Sinharaja R through lowlands N to Katagamuwa, E to highlands at least to Kandy District

LC
Monotypic
HB 43–51 cm
T 1–3 cm

LC
Monotypic
HB 55–59 cm
T c. 2.5 cm
W c. 3 kg

Indian Chevrotain
Moschiola indica

FR Chevrotain indien
DE Indien-Kantschil
ES Ciervo ratón indio

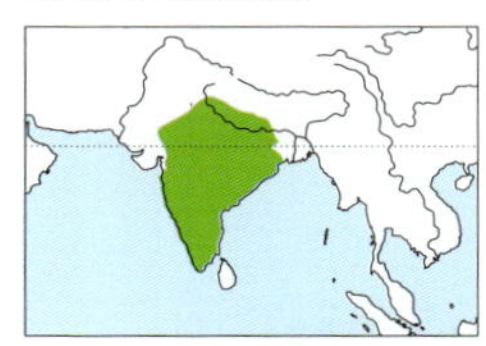

up to 1850 m

Peninsular IN, S NP (where perhaps now extinct)

Genus *TRAGULUS*

(6 species)

Silver-backed Chevrotain
Tragulus versicolor

FR Chevrotain du Vietnam
DE Annam-Kantschil
ES Ciervo ratón plateado

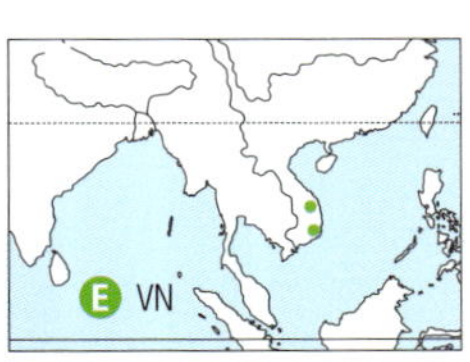

If still extant, likely VN, possibly neighboring KH & LA; limits unknown

DD
Monotypic
HB c. 40 cm
T c. 5 cm
W c. 1.7 kg

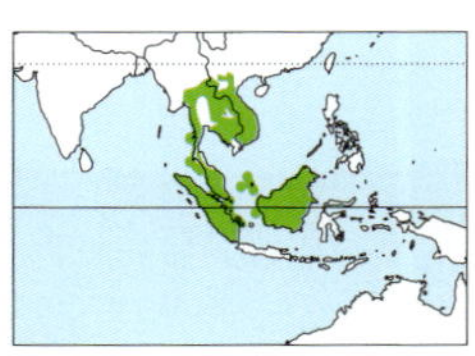

up to 600 m

Mainland SE Asia (S MM, TH, LA, VN, KH, Thai-Malay Pen), Sumatra, Borneo, many Is & arch; perhaps CN (S Yunnan)

Lesser Indo-Malayan Chevrotain
Tragulus kanchil

FR Chevrotain kanchil
DE Kleinkantschil
ES Ciervo ratón pequeño

ssp *klossi*

ssp *ravus*

LC 16 sspp
HB 35–55 cm T 6–9 cm
W 1.5–2.5 kg

Northern Chevrotain
Tragulus williamsoni

FR Chevrotain de Williamson
DE Thailand Kantschil
ES Ciervo ratón de Tailandia

DD
Monotypic

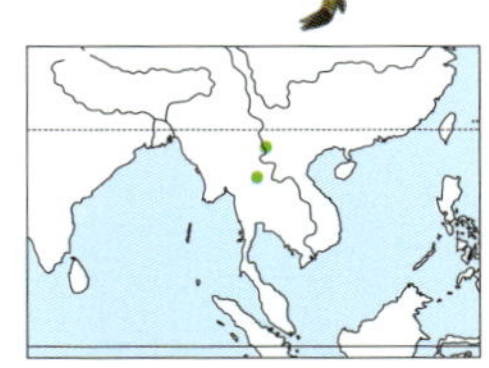

TH, S CN (S Yunnan); perhaps LA, VN & MM

Javan Chevrotain
Tragulus javanicus
FR Chevrotain de Java
DE Java-Kantschil
ES Ciervo ratón de Java

gray-necked form

red-necked form

0–1600 m

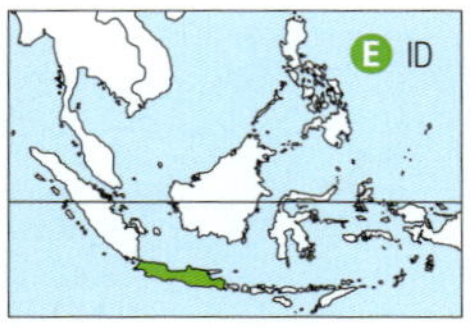

Java; unconfirmed from Bali

DD
Monotypic
HB 50–55 cm
T 4–6 cm
W 1.7–2.1 kg

Greater Indo-Malayan Chevrotain
Tragulus napu
FR Chevrotain napu
DE Großkantschil
ES Ciervo ratón grande

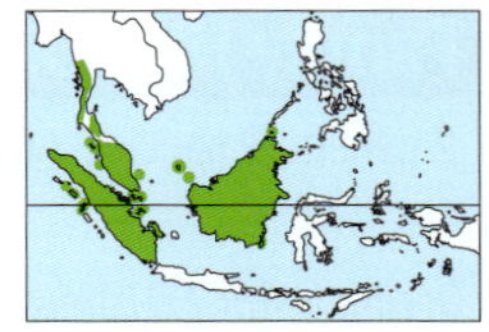
S MM, Thai-Malay Pen, SG, Sumatra, Borneo, many offshore Is & arch

LC
7 sspp
HB 42–68 cm
T c. 8.5 cm
W 3.5–4.5 kg

ssp *rufulus*

ssp *bunguranensis*

ssp *napu*

Malacca Strait Is form

Balabac Chevrotain
Tragulus nigricans
FR Chevrotain de Balabac
DE Balabac-Kantschil
ES Ciervo ratón de Balabac

EN
Monotypic
HB 40–47 cm
T 6.5–8.5 cm

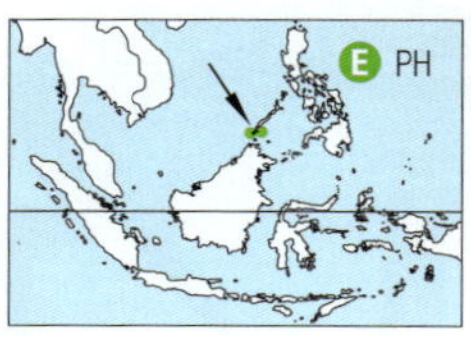

Palawan region (Balabac, Ramos, Bugsuk); introduced to Calauit I & Palawan

Genus *HYEMOSCHUS*
(1 species)

Water Chevrotain
Hyemoschus aquaticus
FR Chevrotain aquatique
DE Hirschferkel
ES Ciervo ratón acuático

LC 3 sspp
HB 60–102 cm
T 7.2–10.9 cm
W 7–16 kg

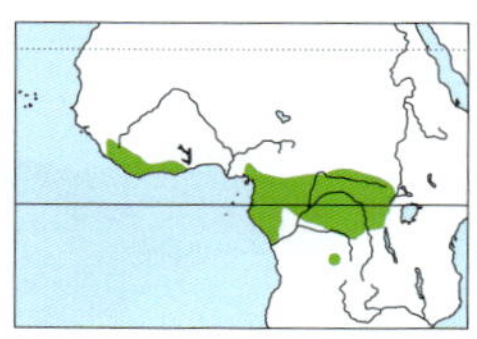
W Africa from GN & SL E to GH, C Africa from NG to CD & marginally UG, S to AO (Cabinda, Lunda Norte)

Family ANTILOCAPRIDAE – Pronghorn

1 genus, 1 species. Grasslands and deserts of the western Great Plains and the intermountain west. Nearctic Region.

Genus *ANTILOCAPRA*
(1 species)

Pronghorn
Antilocapra americana
FR Pronghorn
DE Gabelbock
ES Berrendo

♀

♂

LC
3 sspp
HB 130–140 cm
T 9.7–10.5 cm
W 30–80 kg

0–3350 m

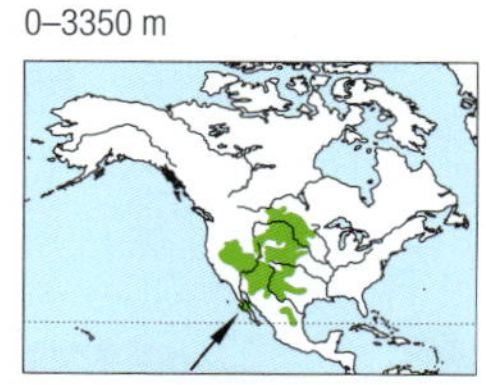
From S CA (S Alberta, S Saskatchewan) through W US to N & C MX (including Baja California Pen)

Family GIRAFFIDAE – Giraffes and Okapi

2 genera, 5 species. Savanna woodlands and dense rainforests. Afrotropical Region.

Genus *GIRAFFA*

(4 species)

VU
4 sspp

ssp *camelopardalis*

ssp *peralta*

ssp *rothschildi*

Northern Giraffe
Giraffa camelopardalis
FR Girafe de Nubie
DE Giraffe
ES Jirafa septentrional

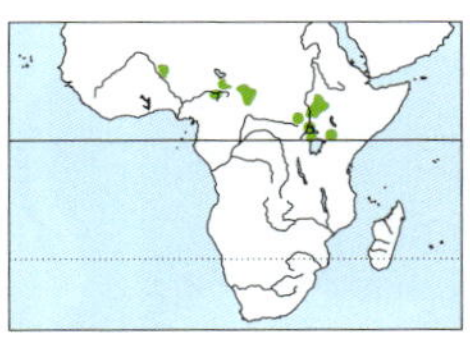

W Africa in NE, S TD, CF, NE CD, SE SD, E & S SS, W ET, N UG, W KE

Reticulated Giraffe
Giraffa reticulata
FR Girafe réticulée
DE Netzgiraffe
ES Jirafa de Somalia

S ET, SW SO, N KE

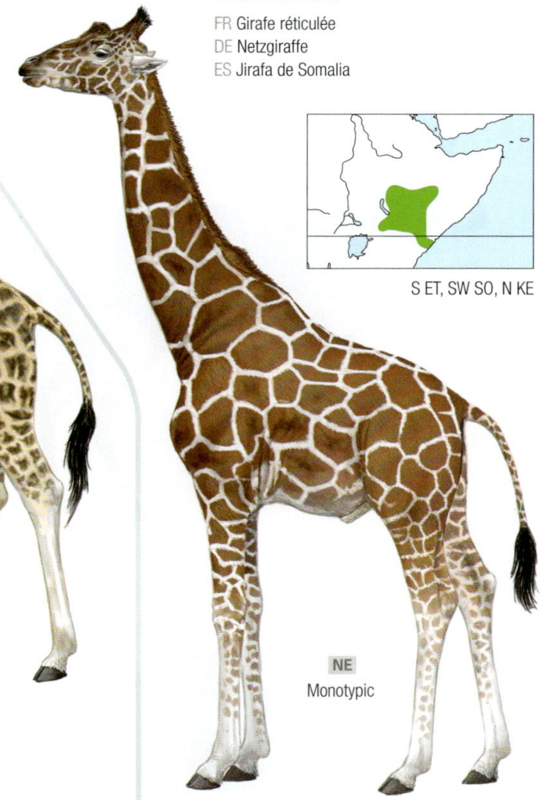

NE
Monotypic

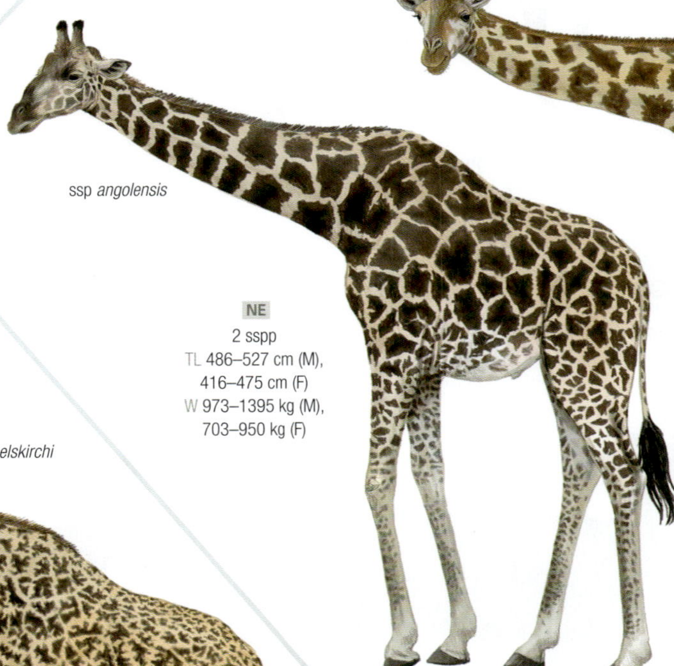

ssp *angolensis*

ssp *giraffa*

Southern Giraffe
Giraffa giraffa
FR Girafe du Cap
DE Südgiraffe
ES Jirafa meridional

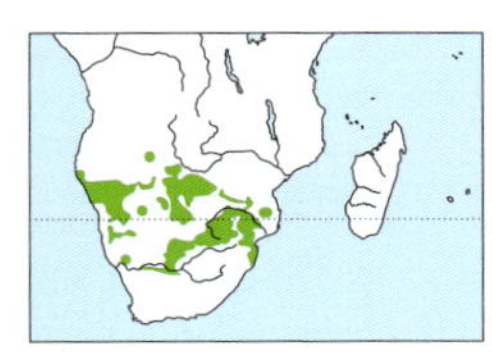

N NA, N BW, S ZM, NW & S ZW, SW MZ, NE ZA

NE
2 sspp
TL 486–527 cm (M), 416–475 cm (F)
W 973–1395 kg (M), 703–950 kg (F)

NE
2 sspp

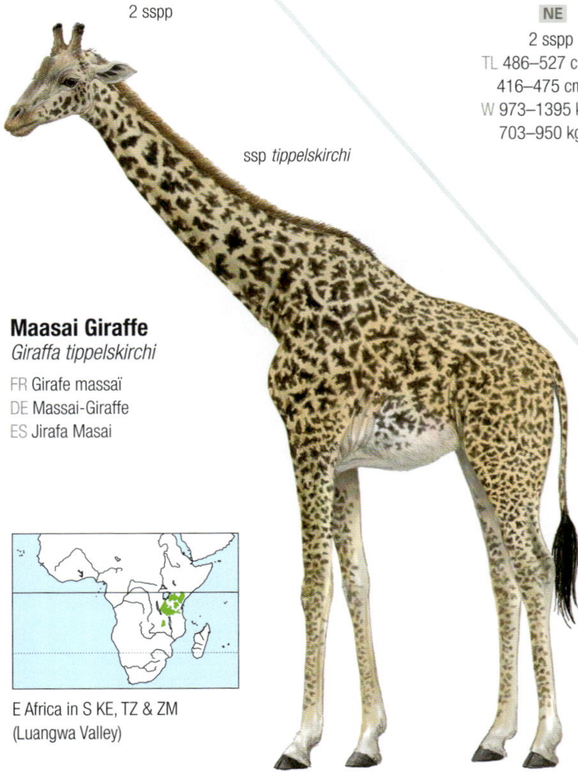

ssp *tippelskirchi*

Maasai Giraffe
Giraffa tippelskirchi
FR Girafe massaï
DE Massai-Giraffe
ES Jirafa Masai

E Africa in S KE, TZ & ZM (Luangwa Valley)

Genus *OKAPIA*

(1 species)

Okapi
Okapia johnstoni
FR Okapi
DE Okapi
ES Okapi

up to 1500 m

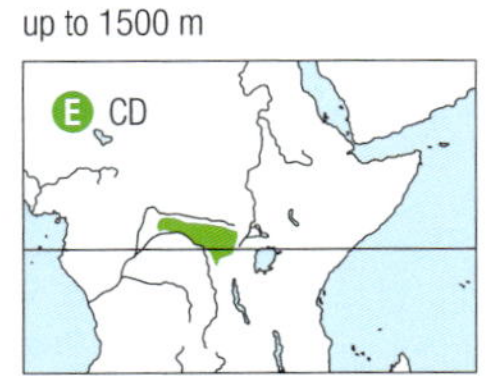

Rainforests of N & NE (Ituri Forest in particular)

EN
Monotypic
HB 200–210 cm
T 30–42 cm
W 180–320 kg

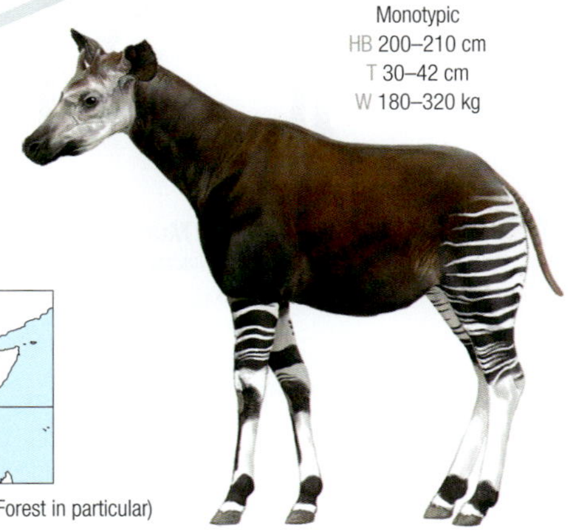

Family CERVIDAE – Deer

18 genera, 54 species. Mainly forested habitats, but also grasslands, swamps, and tundra. Holarctic, Indo-Malayan and Neotropical Regions.

Genus *ELAPHODUS*

(1 species)

up to 4750 m

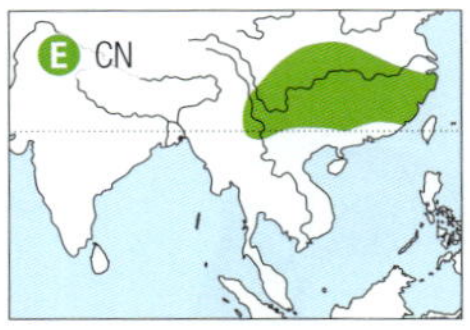

C & S; perhaps N MM, but no recent records

NT
3 sspp
HB 100–120 cm
T 7–13 cm
W 17–30 kg

Tufted Deer
Elaphodus cephalophus
FR Élaphode
DE Schopfhirsch
ES Eláfodo

Genus *MUNTIACUS*

(13 species)

LC
2 sspp
HB 70–80 cm
T 12–13 cm
W 12–15 kg

200–3500 m

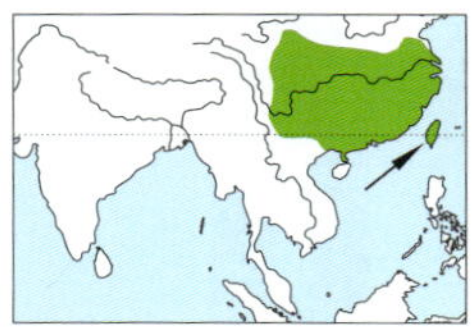

C & SE CN, TW; introduced to England in early 20th century

Reeves's Muntjac
Muntiacus reevesi
FR Muntjac de Chine
DE China-Muntjak
ES Muntíaco de Reeves

up to 1200 m

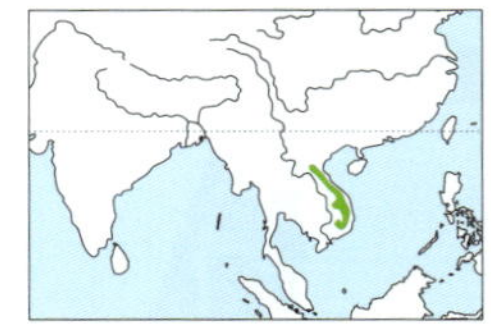

Annamite Mts (LA, VN), E KH

CR
Monotypic
HB 110–115 cm
T c. 17 cm
W c. 34 kg

Giant Muntjac
Muntiacus vuquangensis
FR Muntjac géant
DE Riesenmuntjak
ES Muntíaco gigante

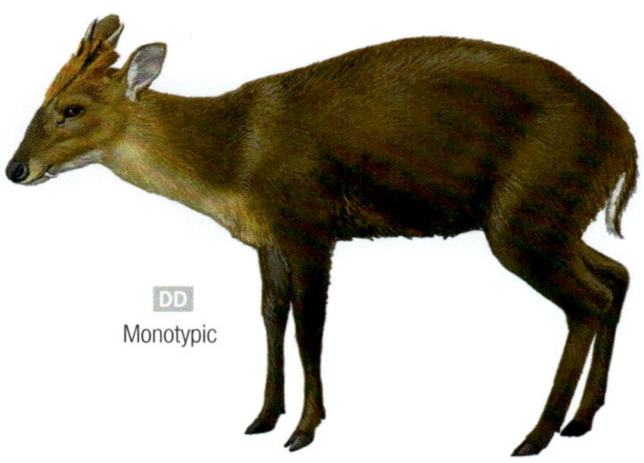

DD
Monotypic

above 1000 m

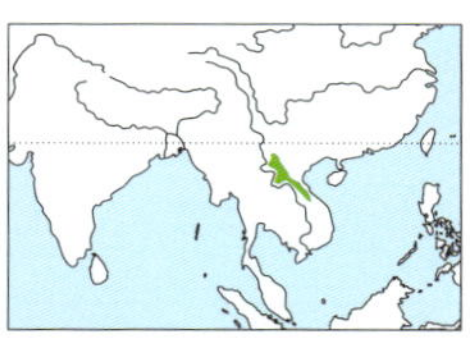

Range very unclear, type locality from LA, but possibly also present in VN & likely S CN

Roosevelts' Muntjac
Muntiacus rooseveltorum
FR Muntjac des Roosevelt
DE Roosevelt-Muntjak
ES Muntíaco de los Roosevelt

DD
Monotypic

above 900 m

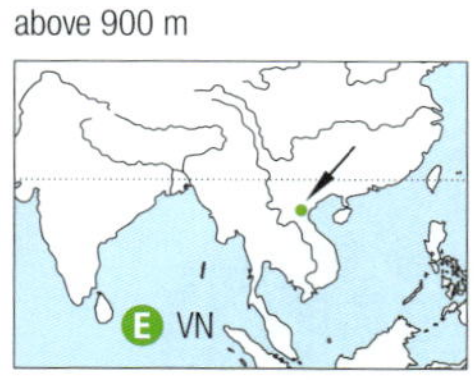

NW VN; perhaps neighboring LA

Puhoat Muntjac
Muntiacus puhoatensis
FR Muntjac de Pu Hoat
DE Vietnam-Muntjak
ES Muntíaco de Pu Hoat

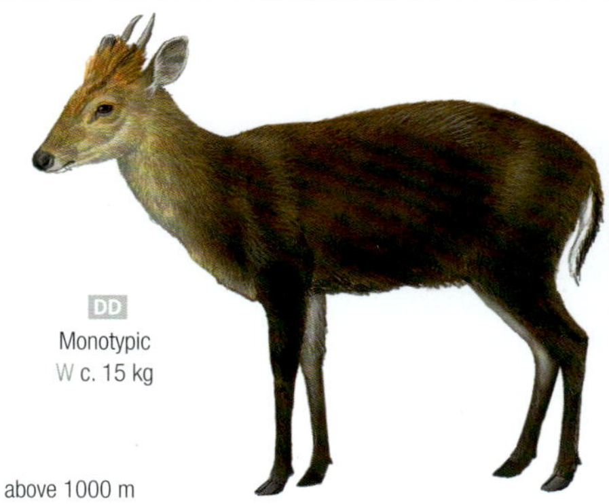

DD
Monotypic
W c. 15 kg

above 1000 m

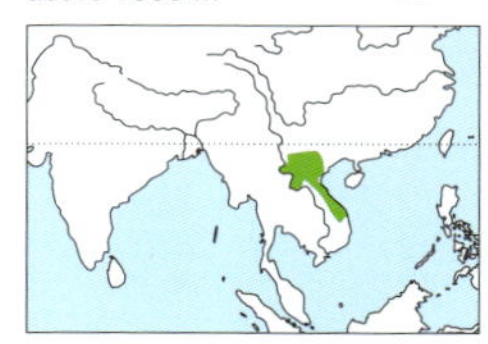

LA, VN; perhaps S CN (Yunnan)

Annamite Muntjac
Muntiacus truongsonensis
FR Muntjac des Truong Son
DE Annam-Muntjak
ES Muntíaco de Truong Son

DD
Monotypic
W c. 12 kg

700–1200 m

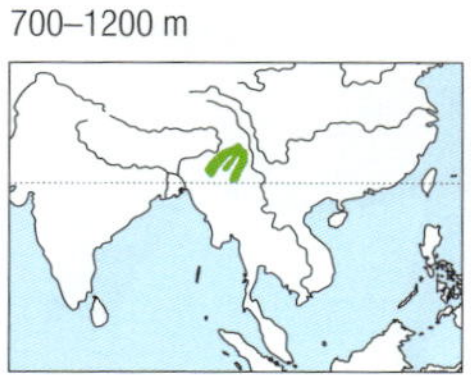

NE IN, N MM

Leaf Muntjac
Muntiacus putaoensis
FR Muntjac de Putao
DE Burma-Muntjak
ES Muntíaco de Putao

DD
Monotypic
HB 90–110 cm
T c. 20 cm
W 20–25 kg

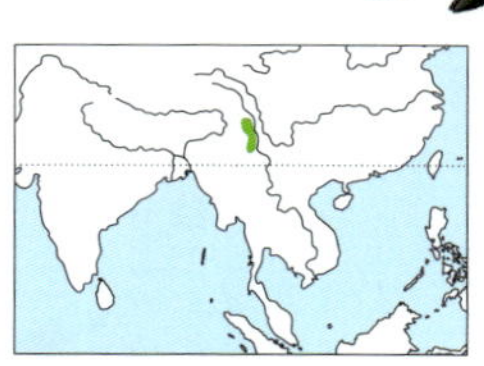

S CN (Yunnan), N MM; perhaps SE Tibet, NE IN & BT

Gongshan Muntjac
Muntiacus gongshanensis
FR Muntjac du Gongshan
DE Gongshan-Muntjak
ES Muntíaco de Gongshan

VU
Monotypic
HB 90–110 cm
T c. 20 cm
W 20–25 kg

up to 1000 m

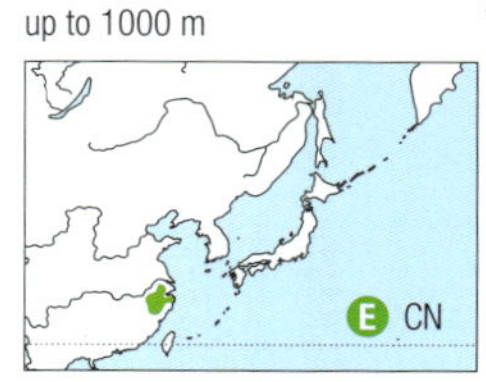

Zhejiang & neighboring areas

Black Muntjac
Muntiacus crinifrons
FR Muntjac noir
DE Schwarzer Muntjak
ES Muntíaco negro

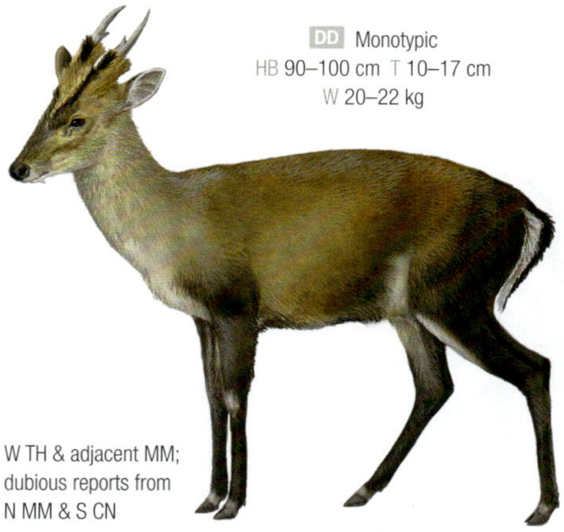

DD Monotypic
HB 90–100 cm T 10–17 cm
W 20–22 kg

W TH & adjacent MM; dubious reports from N MM & S CN

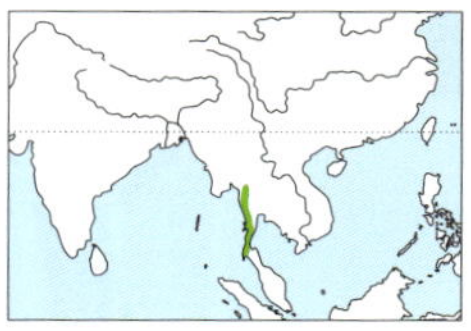

Fea's Muntjac
Muntiacus feae

FR Muntjac de Fea
DE Tenasserim-Muntjak
ES Muntíaco de Fea

NT
Monotypic
HB 85–90 cm
T 15–19 cm
W 14–18 kg

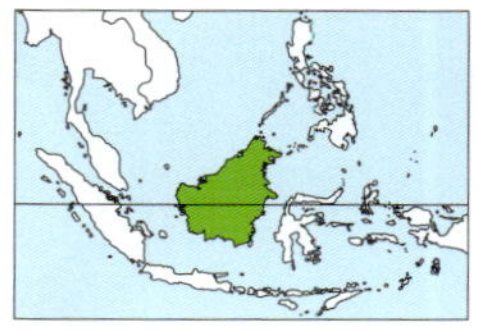

Bornean Yellow Muntjac
Muntiacus atherodes

FR Muntjac de Bornéo
DE Borneo-Muntjak
ES Muntíaco de Borneo

Borneo (Sabah & Sarawak, Kalimantan, presumably BN)

Southern Red Muntjac
Muntiacus muntjak

FR Muntjac de Java
DE Indischer Muntjak
ES Muntíaco de Java

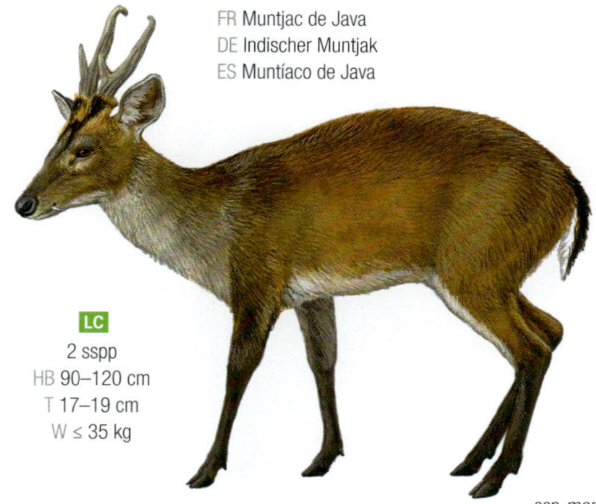

LC
2 sspp
HB 90–120 cm
T 17–19 cm
W ≤ 35 kg

ssp *montanus*

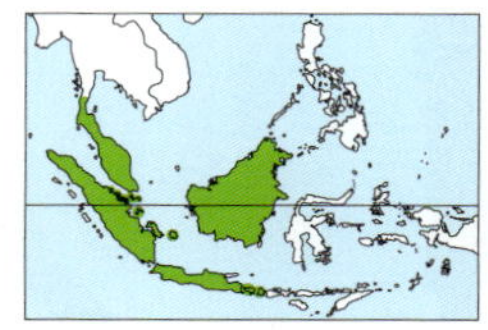

Thai-Malay Pen, Sumatra, Borneo, Java, Bali, several associated smaller Is

Northern Red Muntjac
Muntiacus vaginalis

FR Muntjac du Bengale
DE Nordindischer Muntjak
ES Muntíaco de India

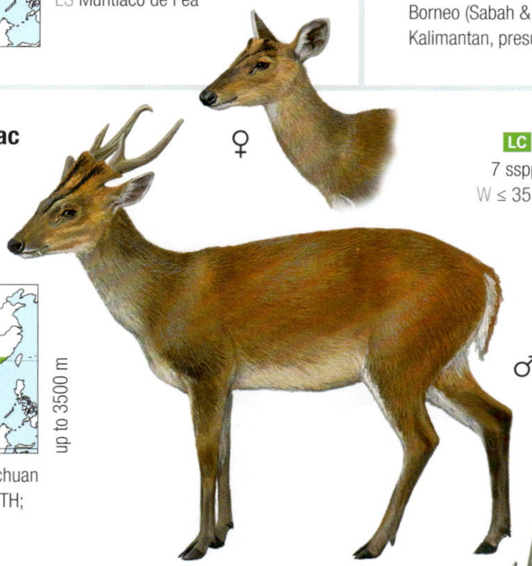

LC
7 sspp
W ≤ 35 kg

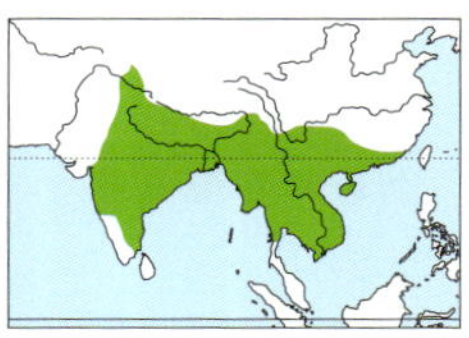

up to 3500 m

PK & IN E to S CN (from S Sichuan E to S Fujian, Hainan), S to S TH; introduced to Andaman Is

NE
Monotypic

Malabar Red Muntjac
Muntiacus malabaricus

FR Muntjac de Malabar
DE Malabar-Muntjak
ES Muntíaco de Malabar

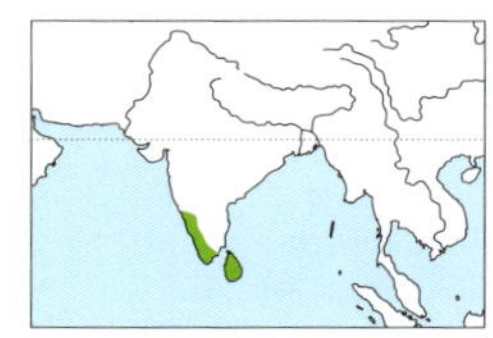

S IN, LK

Chital
Axis axis

FR Cerf axis
DE Axishirsch
ES Axis

Genus *AXIS*
(4 species)

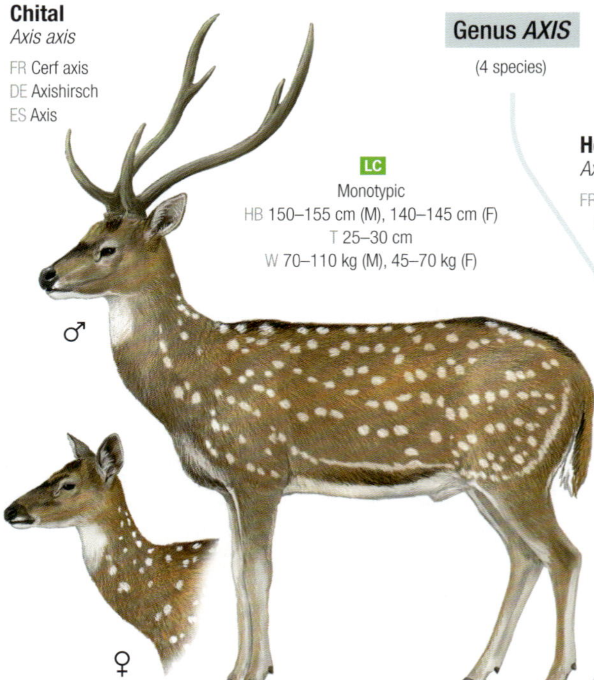

LC
Monotypic
HB 150–155 cm (M), 140–145 cm (F)
T 25–30 cm
W 70–110 kg (M), 45–70 kg (F)

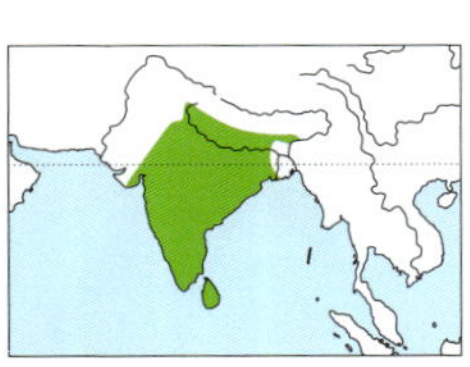

IN, S NP, S BT, BD, LK; introduced to Andaman Is, New Guinea, AU, E Europe, AM, US, BR, UY & AR

Hog Deer
Axis porcinus

FR Cerf cochon
DE Schweinshirsch
ES Axis porcino

EN
2 sspp
HB 130–150 cm
T 17–21 cm
W 30–95 kg

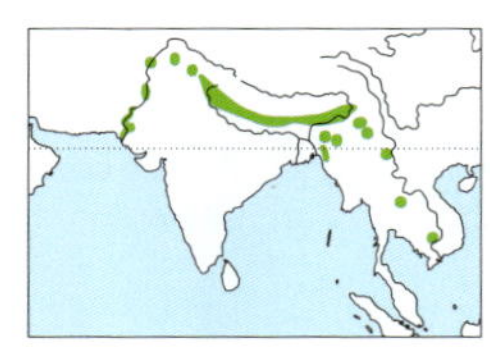

PK, N IN, NP, BT, BD, MM, S CN (Yunnan), TH (reintroduced), KH; extinct in LA & VN; introduced to LK, AU & ZA

Bawean Deer
Axis kuhlii

FR Cerf de Kuhl
DE Bawean-Schweinshirsch
ES Axis de Bawean

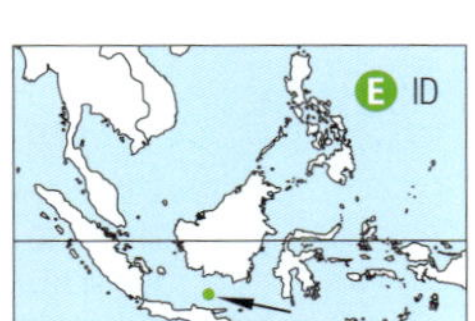

Bawean I, off Java

CR
Monotypic
HB 130–140 cm
T 17–20 cm
W 40–60 kg (M)

Calamian Deer
Axis calamianensis

FR Cerf des Calamian
DE Calamian-Schweinshirsch
ES Axis de Calamianes

EN
Monotypic
HB c. 130 cm
T c. 20 cm
W 35–50 kg

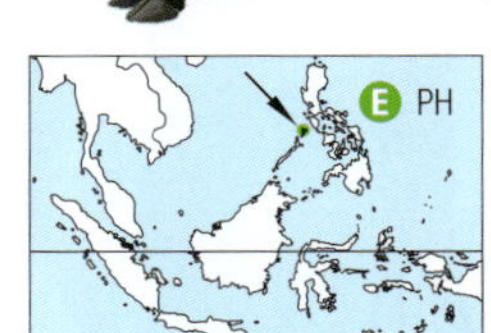

Calamian Is (Busuanga, Calauit, Culion, Marily, Dimaquiat)

ssp *branderi*

♀

♂

♂
ssp *ranjitsinhi*
summer

VU
3 sspp
HB 180–190 cm
T 14–19 cm
W 170–270 kg (M), 140–145 kg (F)

♂
ssp *duvaucelii*
winter

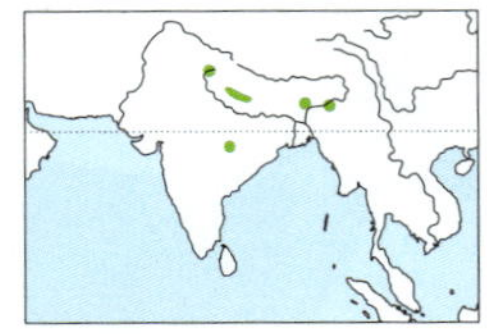

N, C & NE IN, S NP

Barasingha
Rucervus duvaucelii
FR Cerf de Duvaucel
DE Barasingha
ES Ciervo de Duvaucel

Genus *RUCERVUS*

(2 species)

Brow-antlered Deer
Rucervus eldii
FR Cerf d'Eld
DE Leierhirsch
ES Ciervo de Eld

♀

ssp *siamensis*

♂
summer

EN
3 sspp
HB 160–170 cm (M), 140–150 cm (F)
T 22–25 cm
W 90–125 kg (M), 60–80 kg (F)

♂
ssp *eldii*

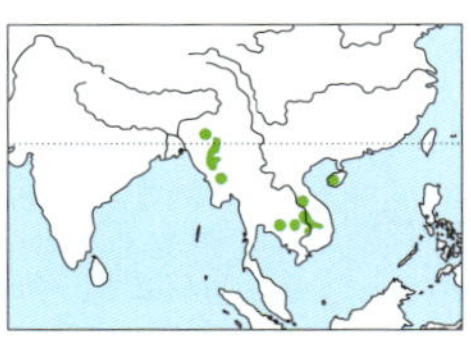

NE IN (Manipur), MM, LA, VN (could be extinct), KH, CN (Hainan)

Common Fallow Deer
Dama dama
FR Daim d'Europe
DE Damhirsch
ES Gamo

♀
summer

♂
winter

LC
Monotypic
HB 145–155 cm (M), 130–145 cm (F)
T 16–19 cm
W 50–80 kg (M), 35–50 kg (F)

♂

up to 1500 m

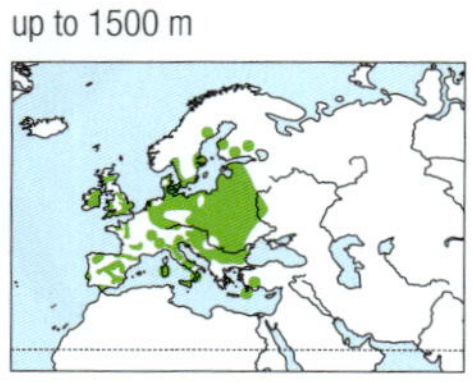

Native to Anatolia (TR), old European introductions mapped; later introductions to North & South America, ZA, AU, NZ & FJ

Genus *DAMA*

(2 species)

Persian Fallow Deer
Dama mesopotamica
FR Daim de Perse
DE Mesopotamien-Damhirsch
ES Gamo persa

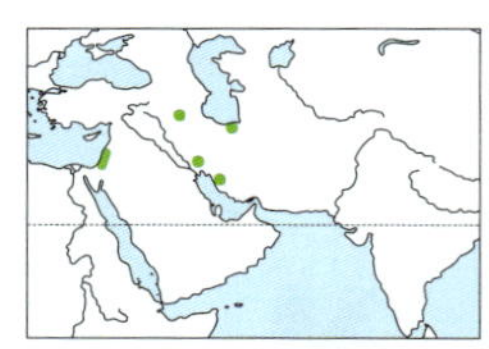

IR & IL (reintroduced); map includes both native relict population and reintroduced

EN
Monotypic
HB 180–190 cm (M), 160–170 cm (F)
T 16–20 cm
W 120–140 kg (M), 70–80 kg (F)

summer

Genus *ELAPHURUS*

(1 species)

♀

♂

EW

Monotypic

HB 195–210 cm (M), 180–190 cm (F)

T c. 35 cm

W 160–220 kg (M), 140–170 kg (F)

Pere David's Deer

Elaphurus davidianus

FR Cerf du Père David
DE Davidshirsch
ES Ciervo del Padre David

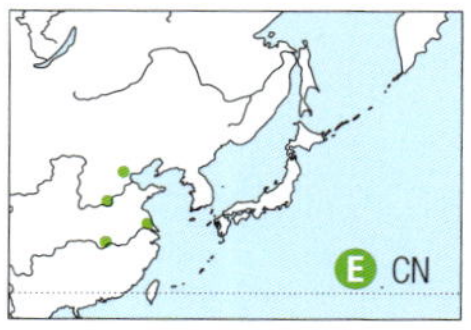

Reintroduced in E since 1985; main introduced populations mapped

Sambar

Rusa unicolor

FR Cerf sambar
DE Sambar
ES Sambar

up to 3800 m

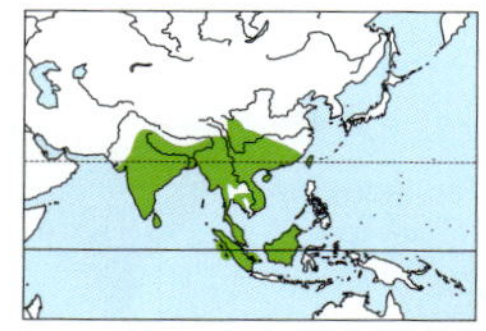

IN, NP & LK E to S CN (including Hainan) & TW, S to Borneo, Sumatra & associated Is; introduced to AU, NZ, ZA & US

♂ ssp *brookei*

♂ ssp *swinhoei*

VU

5 sspp

HB 160–210 cm

T 25–33 cm

W 180–350 kg (M), 90–230 kg (F)

♂ ssp *unicolor*

♀

Genus *RUSA*

(4 species)

Javan Deer

Rusa timorensis

FR Cerf de Timor
DE Mähnenhirsch
ES Sambar de Java

0–900 m

Native to Java & Bali, old introductions mapped; later introduced further E to NZ & NC, also MU & Comoro Is

VU

7 sspp

HB 140–180 cm

T c. 25 cm

W 70–135 kg (M), 50–90 kg (F)

Philippine Spotted Deer

Rusa alfredi

FR Cerf d'Alfred
DE Prinz-Alfred-Hirsch
ES Sambar manchado de Filipinas

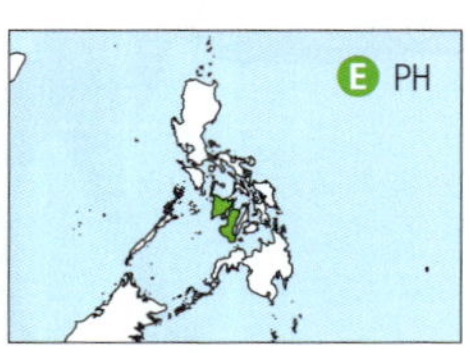

W Visayas (Panay, Negros)

EN

Monotypic

HB c. 130 cm

T c. 12 cm

W c. 40 kg (M)

Philippine Brown Deer

Rusa marianna

FR Cerf des Philippines
DE Philippinen-Hirsch
ES Sambar de Filipinas

VU

3 sspp

HB c. 140 cm

T 8–12 cm

W 40–60 kg

up to 2900 m

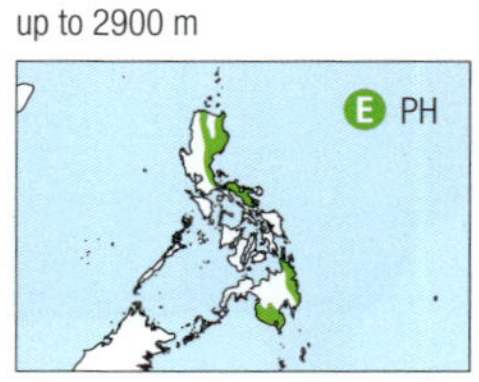

Luzon, Mindoro, Mindanao, Basilan, Samar, Leyte; introduced to Marianas (Guam, Saipan, Rota) & Caroline Is (Pohnpei)

Genus *CERVUS*

(5 species)

White-lipped Deer

Cervus albirostris

FR Cerf à museau blanc
DE Weißlippenhirsch
ES Ciervo hociblanco

3500–5100 m

E Tibetan Plateau (E Tibet, W Gansu, Qinghai, W Sichuan, NW Yunnan)

♂

♀

VU

Monotypic

HB c. 190 cm (M), c. 180 cm (F)

T 12–13 cm

W 190–220 kg (M), 90–150 kg (F)

Sika Deer
Cervus nippon

FR Cerf sika
DE Sikahirsch
ES Sika

up to 3000 m

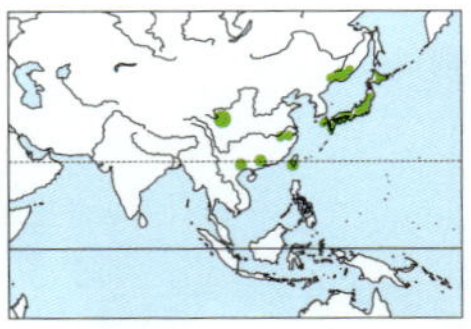

RU (Ussuriland), C, NE & E CN, JP, TW, N VN; possibly KP; introduced to PH, Europe, AM, AZ, MG, AU, NZ & US

♂
ssp *yesoensis*
winter

♀
ssp *yesoensis*
summer

♂
ssp *yesoensis*
summer

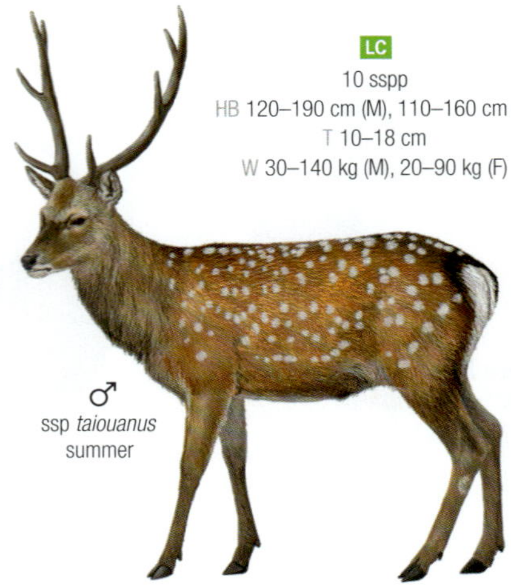

♂
ssp *taiouanus*
summer

LC
10 sspp
HB 120–190 cm (M), 110–160 cm (F)
T 10–18 cm
W 30–140 kg (M), 20–90 kg (F)

Wapiti
Cervus canadensis

FR Cerf wapiti
DE Wapiti
ES Uapiti

♂
ssp *macneilli*
winter

♂
ssp *wallichii*
winter

LC
6 sspp
HB 210–240 cm (M), 190–230 cm (F)
T 10–16 cm
W 220–400 kg (M), 150–250 kg (F)

up to 3300 m

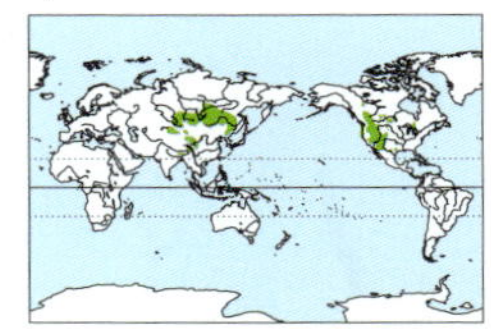

Asia from Tien Shan (E KG) & Altai (NE KZ) through MN & RU (E to Ussuriland) to N, C & S CN, BT, North America (CA, US)

♂
ssp *alashanicus*
winter

♂
ssp *xanthopygus*
winter

♂
ssp *canadensis*
winter

♀
ssp *canadensis*
summer

Central Asian Red Deer
Cervus hanglu

FR Cerf du Cachemire
DE China-Rothirsch
ES Ciervo asiático

4300–4900 m

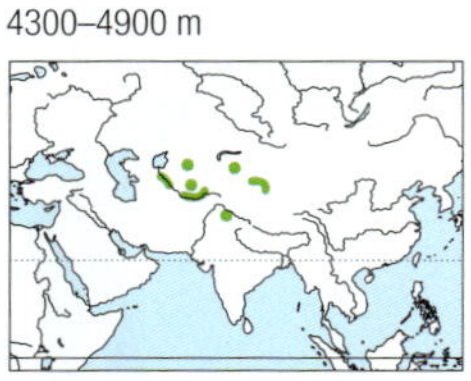

C Asia in KZ, UZ, TJ, N AF, CN (S Xinjiang), N IN (Kashmir)

ssp *hanglu*
winter

LC
3 sspp
HB 190–205 cm (M), 180–195 cm (F)
T 8–12 cm
W 150–240 kg (M), 110–170 kg (F)

ssp *bactrianus*
winter

ssp *yarkandensis*
winter

up to 2800 m

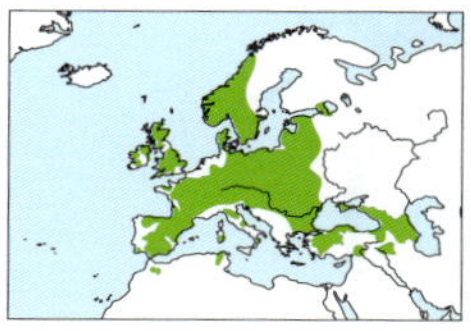

Europe, Corsica (reintroduced), Sardinia, TR, Caucasus, NW IR, Atlas Mts (DZ, TN, reintroduced to N MA); introduced to CL, AR, AU & NZ

Western Red Deer

Cervus elaphus

FR Cerf élaphe
DE Rothirsch
ES Ciervo rojo

LC
5 sspp
HB 180–205 cm (M), 165–180 cm (F)
T 14–16 cm
W 110–220 kg (M), 75–120 kg (F)

♂ ssp *corsicanus* winter

♀ ssp *elaphus* summer

♂ ssp *elaphus* winter

♂ ssp *maral* winter

Genus *HYDROPOTES*

(1 species)

Chinese Water Deer

Hydropotes inermis

FR Hydropote
DE Chinesisches Wasserreh
ES Ciervo acuático

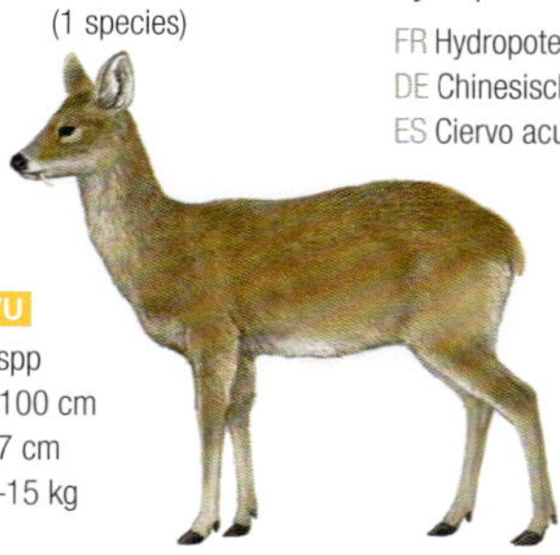

VU
2 sspp
HB 90–100 cm
T 6–7 cm
W 11–15 kg

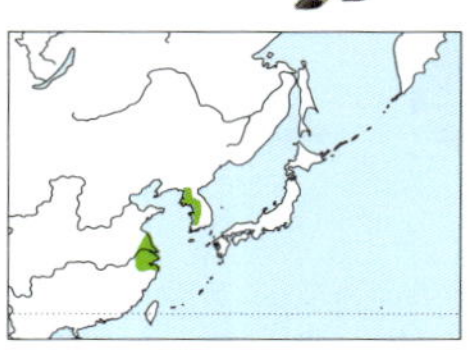

E CN, Korean Pen; introduced to England (GB)

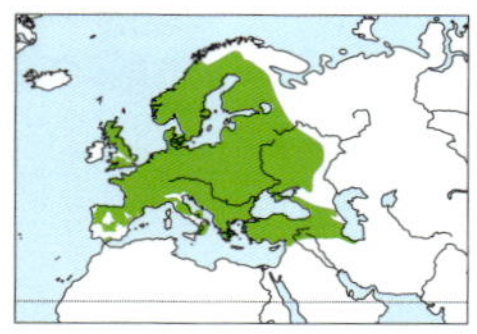

up to 2400 m

Most of Europe (excluding IS, IE & Mediterranean Is), Caucasus, Near East

Western Roe Deer

Capreolus capreolus

FR Chevreuil d'Europe
DE Reh
ES Corzo

♂ summer

♀ winter

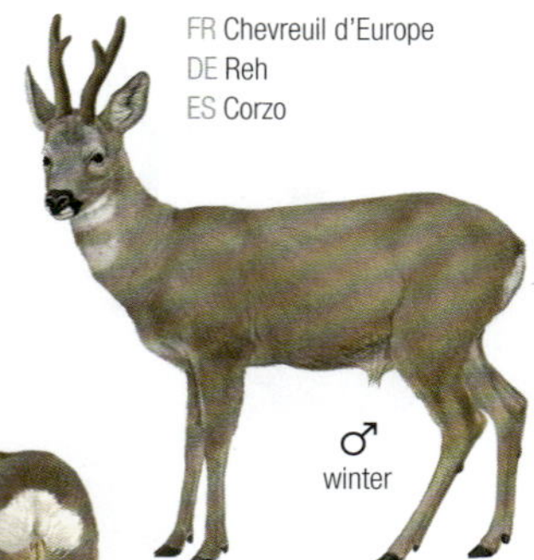

♂ winter

LC
Monotypic
HB 107–127 cm
T 1.5–3 cm
W 20–30 kg (M), 17–29 kg (F)

Genus *CAPREOLUS*

(2 species)

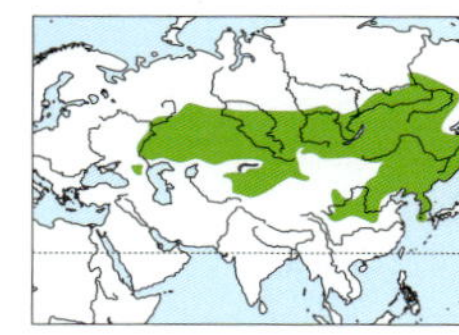

up to 3300 m

From European RU E through C Asia to SE Siberia, C & NE CN and Korean Pen; isolated on N slopes of Caucasus

Eastern Roe Deer

Capreolus pygargus

FR Chevreuil de Sibérie
DE Sibirisches Reh
ES Corzo siberiano

summer

LC 2 sspp
HB 125–145 cm
T 2–4 cm
W 32–50 kg

Genus *ALCES*

(1 species)

♂ ssp *alces*

Moose

Alces alces

FR Élan
DE Elch
ES Alce

♀

ssp *gigas* ♂

♂ ssp *cameloides*

up to 1700 m

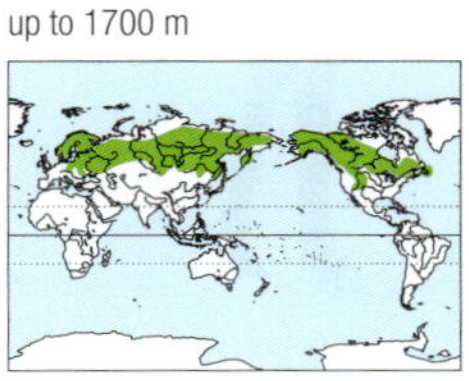

N Eurasia from Scandinavia, PL, N AT & S CZ E to NE Siberia & Kamchatka, S to UA, N KZ, N CN & N MN, North America in Alaska, CA & US S through Rocky Mts to Utah, N Great Lakes & New England

LC
8 extant sspp
HB 250–300 cm (M), 240–290 cm (F)
T 12–16 cm
W 300–600 kg (M), 280–460 kg (F)

♂
ssp *caribou*
winter

♂
ssp *groenlandicus*
summer

♂
ssp *groenlandicus*
winter

Genus *RANGIFER*

(1 species)

Caribou
Rangifer tarandus

FR Renne
DE Ren
ES Reno

VU
10 sspp
HB 185–210 cm (M), 170–195 cm (F)
T 14–16 cm
W 65–170 kg (M), 55–110 kg (F)

♂
ssp *phylarchus*

♀
ssp *platyrhynchus*
winter

♂
ssp *platyrhynchus*
winter

♂
ssp *tarandus*
winter

up to 3000 m

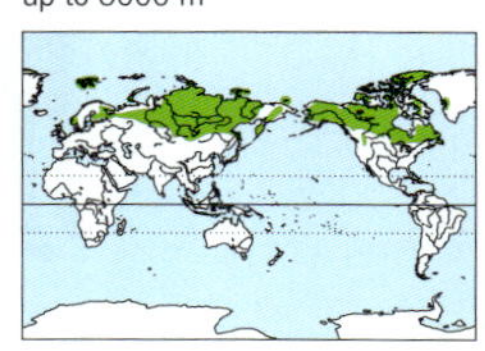

NO, FI, RU through Siberia to Kamchatka & Sakhalin, MN, Svalbard & Novaya Zemlya, Alaska, CA (including Canadian Arctic Is), extreme NW US, W GL; introduced to IS, South Georgia Is & Kerguelen Is

Genus *PUDU*

(2 species)

Northern Pudu
Pudu mephistophiles

FR Poudou d'Équateur
DE Nordpudu
ES Pudu chico

DD
Monotypic
HB c. 75 cm
T c. 3 cm
W 5–6 kg

1700–4500 m

C Andes in C CO, E Andes of EC to PE (S to Junín), natural gap in dry forest S of Huancabamba Depression (N PE)

Southern Pudu
Pudu puda

FR Poudou austral
DE Südpudu
ES Pudu

winter

NT
Monotypic
HB c. 80 cm
T c. 4 cm
W 9–14 kg

0–1700 m

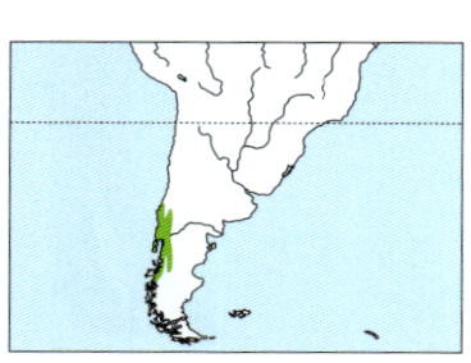

S CL between 35° S and 47° S, W AR from SW Neuquén S along Andes foothills to Los Alerces NP (Chubut)

Genus *OZOTOCEROS*

(1 species)

NT
5 sspp
HB 85–120 cm
T 10–14 cm
W 22–40 kg

Pampas Deer
Ozotoceros bezoarticus

FR Cerf des pampas
DE Pampashirsch
ES Ciervo de la pampa

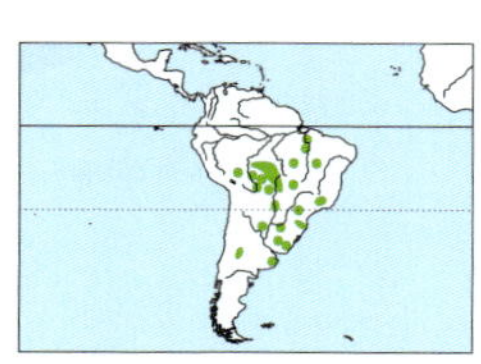

BR (S of Amazonia), BO, PY, UY, N AR (Santa Fe, Corrientes, San Luis, Buenos Aires)

Genus *BLASTOCERUS*

(1 species)

Marsh Deer
Blastocerus dichotomus

FR Cerf des marais
DE Sumpfhirsch
ES Ciervo de los pantanos

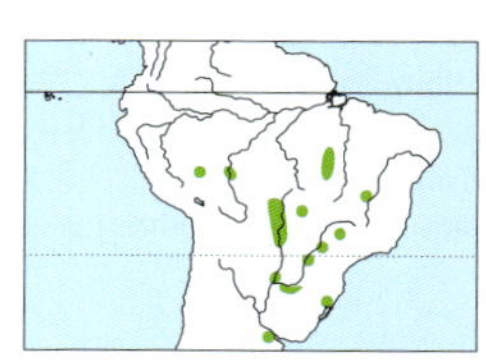

SE PE, BO, BR (S of Amazonia), PY, NE AR

VU
Monotypic
HB 165–180 cm
T 13–15 cm
W 70–130 kg

Genus *HIPPOCAMELUS*

(2 species)

VU
Monotypic
HB 140–145 cm
T 11–12 cm
W 45–60 kg

up to 5000 m

North Andean Huemul
Hippocamelus antisensis

FR Taruca
DE Nördlicher Andenhirsch
ES Huemul septentrional

Andes in PE, BO, N CL & NW AR

EN
Monotypic
HB 155–165 cm (M)
T 13–15 cm
W 60–75 kg

0–3000 m

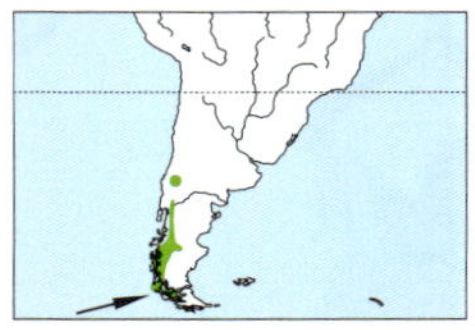

South Andean Huemul
Hippocamelus bisulcus

FR Guémal
DE Südlicher Andenhirsch
ES Huemul meridional

Andes in S CL & S AR

Genus *MAZAMA*

(8 species)

Amazonian Brown Brocket
Mazama nemorivaga

FR Cariacou de Cayenne
DE Amazonien-Mazama
ES Corzuela amazónica

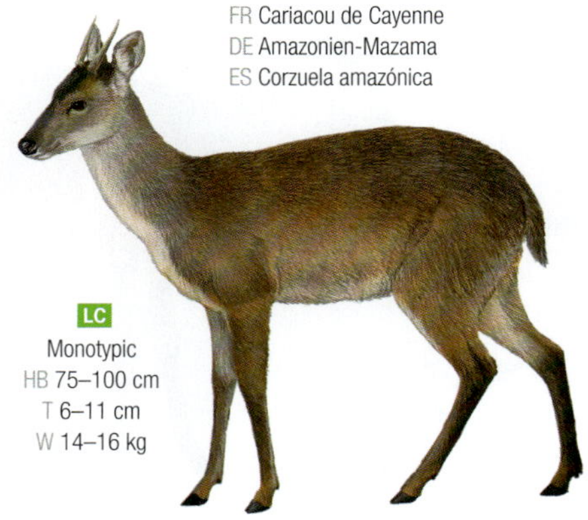

LC
Monotypic
HB 75–100 cm
T 6–11 cm
W 14–16 kg

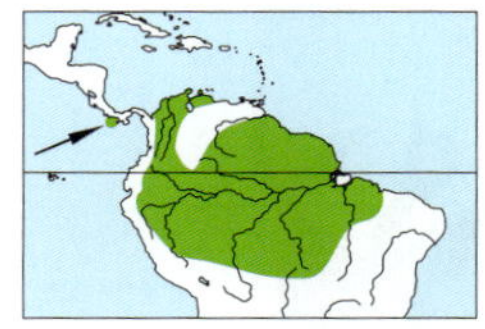

up to 1500 m

N & E CO, VE, Guianas, E EC, E PE, Amazonian BR, N BO; isolated on San José I (PA)

LC
Monotypic
HB 85–105 cm
T 8–9 cm
W 11–25 kg

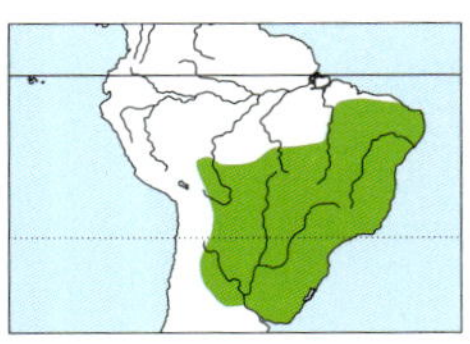

Common Brown Brocket
Mazama gouazoubira

FR Cariacou brun
DE Graumazama
ES Corzuela

E & S BR, C & E BO, PY, UY, N AR

VU
Monotypic
HB 70–75 cm
W c. 11 kg

1000–4000 m

Common Dwarf Brocket
Mazama chunyi

FR Mazame pygmée
DE Südlicher Zwergmazama
ES Corzuela enana

Andes of S PE & N BO

VU
Monotypic
HB 80–95 cm
T 8–9 cm
W 3–15 kg

1400–3600 m

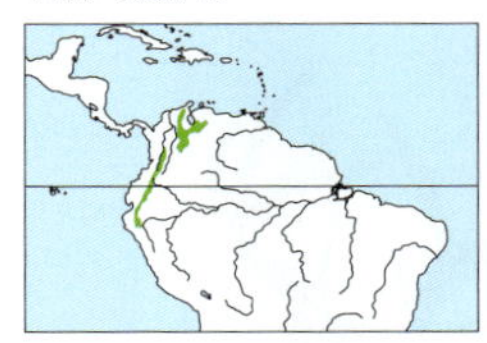

Little Red Brocket
Mazama rufina

FR Mazame d'Équateur
DE Roter Kleinmazama
ES Corzuela roja pequeña

Andes from W VE and NE & C CO to EC & NW PE

DD
Monotypic
HB 90–145 cm
T 12–16 cm
W 30–65 kg

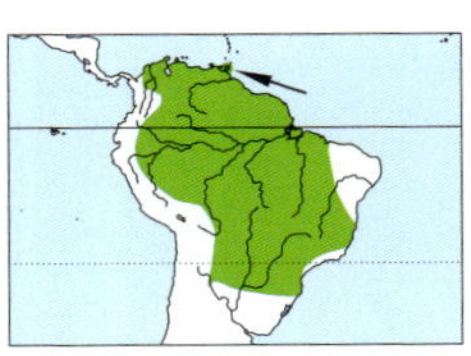

Common Red Brocket
Mazama americana

FR Mazame roux
DE Großmazama
ES Corzuela roja

CO & VE to S BR, PY & N AR, TT (Trinidad, recently extirpated on Tobago)

VU
Monotypic
HB c. 70 cm
T 9–12 cm
W 14–16 kg

Lesser Brocket
Mazama nana

FR Mazame nain
DE Kleinmazama
ES Corzuela menor

SE BR, SE PY, N AR

VU
Monotypic
HB c. 85 cm
T 11–14 cm
W c. 25 kg

up to 1200 m

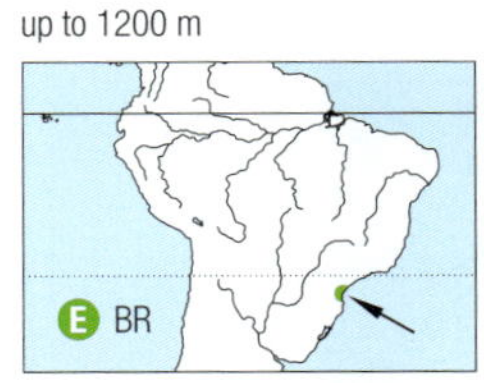

Small Red Brocket
Mazama jucunda

FR Mazame de Sao Paulo
DE Küstenwald-Mazama
ES Corzuela bororo

E BR (SE São Paulo, NE Paraná); smallest range of any Neotropical deer species

Mexican Red Brocket
Mazama temama

FR Mazame de Hernandez
DE Mexico-Großmazama
ES Corzuela temama

DD
3 sspp
HB 80–110 cm
T 10–14 cm
W 12–32 kg

up to 2800 m

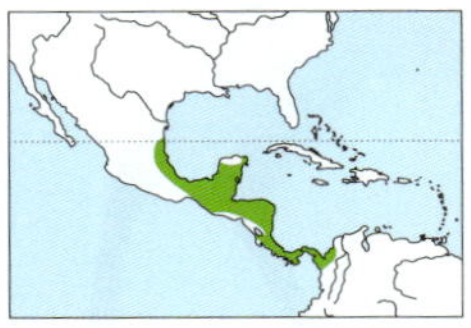

S MX (S from Tamaulipas), GT & BZ to PA & NW CO

Yucatan Brown Brocket
Odocoileus pandora

FR Cariacou du Yucatan
DE Yucatán-Mazama
ES Corzuela de Yucatán

Yucatán Pen; perhaps N GT & N BZ

VU
Monotypic
W 17–21 kg

Mule Deer
Odocoileus hemionus

FR Cerf mulet
DE Maultierhirsch
ES Ciervo mulo

Coastal S Alaska, W CA (from Yukon) S through W & C US (Rocky Mts, Western Plains) to NW & C MX

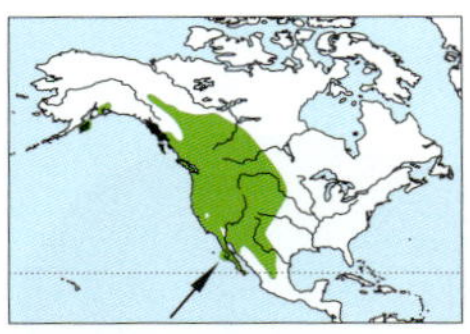

ssp *sitkensis*
summer

LC
8 sspp
HB 135–180 cm
T 15–23 cm
W 35–110 kg

ssp *columbianus*
winter

ssp *hemionus*
winter

Genus *ODOCOILEUS*

(3 species)

♀

♂
ssp *virginianus*
summer

♂
ssp *virginianus*
winter

♂
ssp *borealis*
winter

White-tailed Deer
Odocoileus virginianus

FR Cerf de Virginie
DE Weißwedelhirsch
ES Ciervo de cola blanca

up to 4500 m

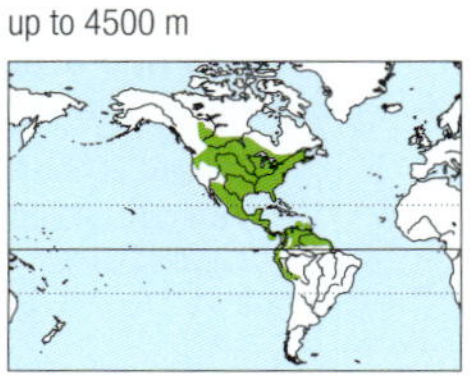

From S CA S through US (except much of SW) & MX to South America (CO, EC, PE, NW BO, VE, Guianas, N BR); introduced to CZ, FI, NZ & CU

♂
ssp *goudotti*

♂
ssp *toltecus*

LC
38 sspp
HB 120–190 cm (M), 115–150 cm (F)
T 14–37 cm
W 30–130 kg (M), 25–85 kg (F)

Family MOSCHIDAE – Musk-deer

1 genus, 7 species. Forest and alpine scrub in mountains and hilly country. Palearctic and Indo-Malayan Regions.

Genus *MOSCHUS*

(7 species)

Siberian Musk-deer
Moschus moschiferus

FR Porte-musc de Sibérie
DE Sibirien-Moschustier
ES Ciervo almizclero siberiano

ssp *parvipes*

ssp *moschiferus*

VU
3 sspp
HB 65–90 cm
T 4–6 cm
W 7–17 kg

300–1600 m

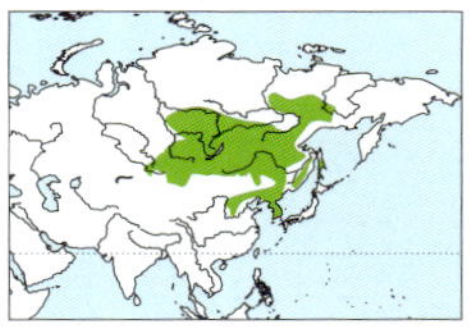

Widely in RU (Siberia, Far East, Sakhalin), extreme E KZ, NW & NE CN, MN, KP, KR

Alpine Musk-deer
Moschus chrysogaster

FR Porte-musc alpin
DE Gelbbauch-Moschustier
ES Ciervo almizclero dorado

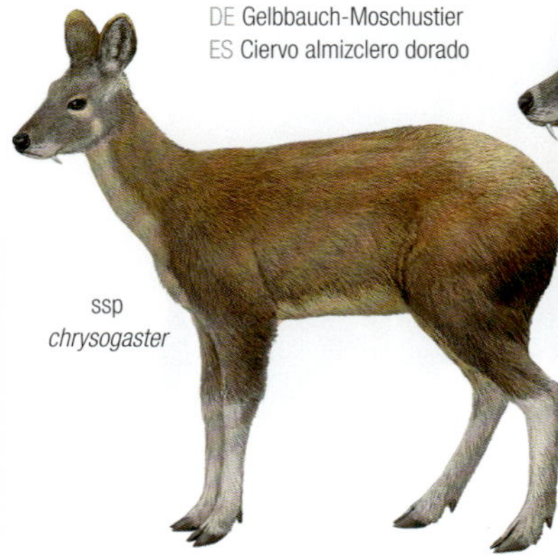

ssp *chrysogaster*

ssp *sifanicus*

2800–4000 m

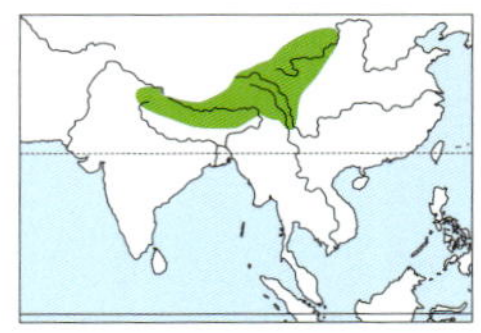

Alpine zones in N IN, NP, BT and C & S CN (Qinghai, S Gansu, S Ningxia, SE Tibet, W Sichuan, NW Yunnan)

EN
2 sspp
HB 85–90 cm
T 4–6 cm
W 11–18 kg

Himalayan Musk-deer
Moschus leucogaster

FR Porte-musc de l'Himalaya
DE Himalaya-Moschustier
ES Ciervo almizclero himalayo

2500–4400 m

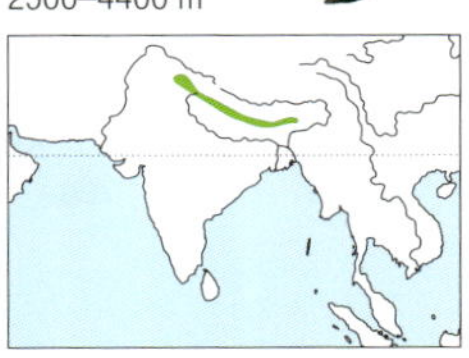

EN
Monotypic
HB 86–100 cm
T 4–6 cm
W 13–18 kg

S slopes of Himalayas in N IN, NP & BT

Kashmir Musk-deer
Moschus cupreus

FR Porte-musc du Cachemire
DE Kaschmir-Moschustier
ES Ciervo almizclero de Cachemira

1500–4000 m

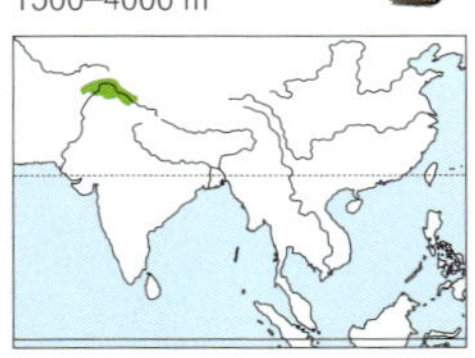

EN
Monotypic
HB 85–100 cm
T 4–6 cm
W 12–17 kg

NE AF (Nuristan), Kashmir (NE PK, NW IN)

Black Musk-deer
Moschus fuscus

FR Porte-musc noir
DE Schwarzes Moschustier
ES Ciervo almizclero negro

EN
Monotypic
HB 73–80 cm
T 4–6 cm
W 10–15 kg

2200–4600 m

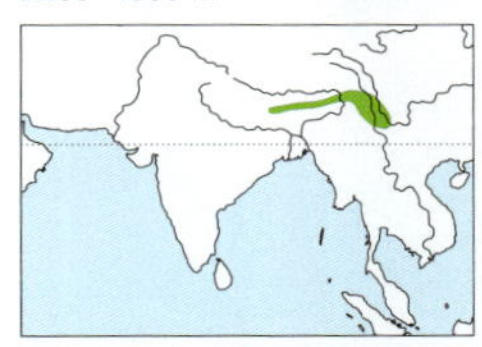

NE NP, NE IN (Sikkim, likely Arunachal Pradesh), BT, S CN (SE Tibet, NW Yunnan), N MM

Anhui Musk-deer
Moschus anhuiensis

FR Porte-musc d'Anhui
DE Anhui-Moschustier
ES Ciervo almizclero de Anhui

below 500 m

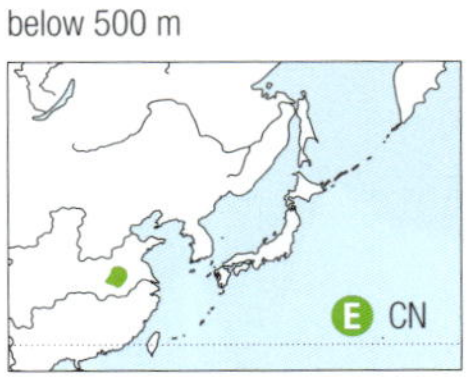

Apparently restricted to SW Anhui

EN
Monotypic
HB 70–76 cm
T 2–4 cm
W 7.1–9.7 kg

Forest Musk-deer
Moschus berezovskii

FR Porte-musc de forêt
DE China-Moschustier
ES Ciervo almizclero de bosque

0–3500 m

C & S CN (from SE Qinghai, S Gansu & S Shaanxi S to S Tibet, Yunnan & Guangxi), N VN (Cao Bang Prov); perhaps LA

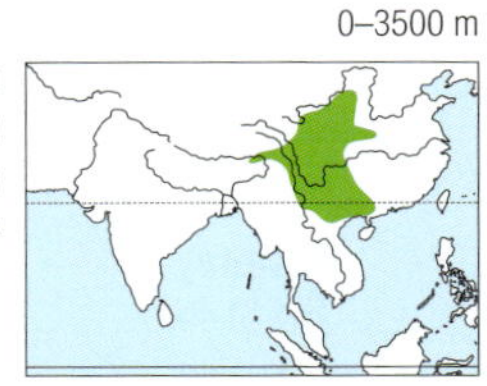

ssp *berezovskii*

ssp *caobangis*

EN
4 sspp
HB 70–80 cm
T 3–4 cm
W 6–9 kg

Family BOVIDAE – Hollow-horned Ruminants

51 genera, 146 species. From deserts and open plains, through forests to mountain areas, from cold temperate zones to tropics. Holarctic, Afrotropical and Indo-Malayan Regions.

Genus *BOS*

(6 species)

Gaur
Bos gaurus

FR Gaur
DE Gaur
ES Gaur

VU
2 sspp
HB 250–330 cm
T 70–105 cm
W 650–1000 kg

♂ ssp *laosiensis*

♂ ssp *gaurus*

♀

0–3000 m

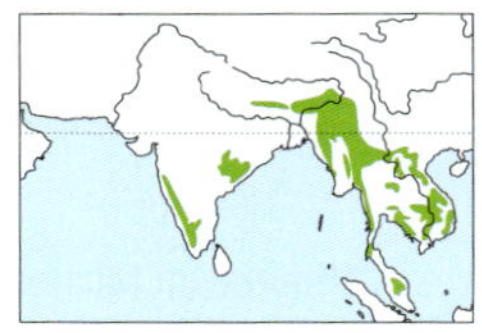

IN, S NP, S BT, MM, TH, LA, VN, KH, Pen MY; possibly BD & SC CN (S Yunnan)

Banteng
Bos javanicus

FR Banteng sauvage
DE Banteng
ES Banteng

EN
3 sspp
HB 190–225 cm
T 65–80 cm
W 600–800 kg

0–2100 m

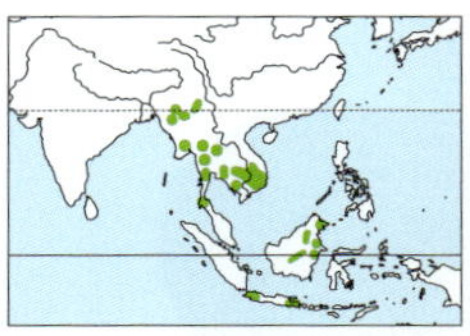

MM, S CN (S Yunnan, current presence uncertain), LA, VN, KH, N Thai-Malay Pen, Borneo, Java; perhaps Bali

♂ ssp *birmanicus*

♂ ssp *javanicus*

♂ ssp *lowi*

♀

Kouprey
Bos sauveli

FR Kouprey
DE Kouprey
ES Kuprey

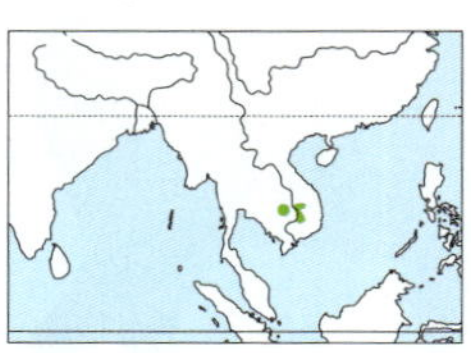

Historically in KH, S LA, SE TH & W VN (perhaps wider), if still extant in the wild, probably restricted to several protected areas in E KH

CR(PE)
Monotypic
HB 210–220 cm
T 100–110 cm
W 700–900 kg

♀

♂

Wisent
Bos bonasus

FR Bison d'Europe
DE Wisent
ES Bisonte europeo

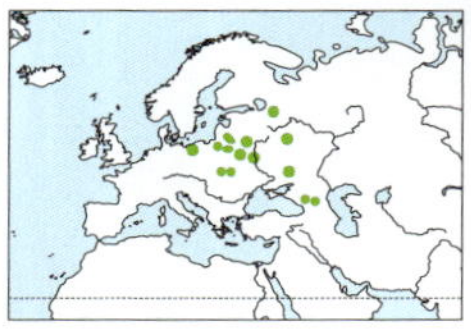

Historically C & E Europe, N Caucasus; now scattered free & semi-free populations in PL, LT, BY, SK, UA & RU

♀

♂

NT
1 extant ssp
HB 210–350 cm
T 30–60 cm
W 350–1000 kg

American Bison
Bos bison

FR Bison d'Amérique
DE Bison
ES Bisonte americano

♂

ssp *bison*

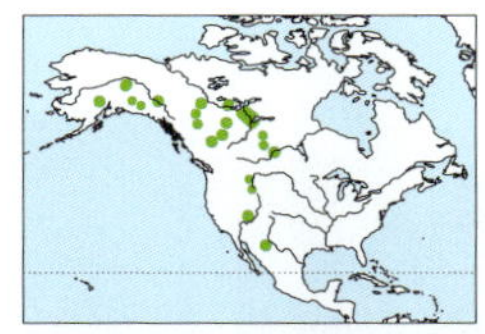

Scattered free-ranging conservation herds in W & C CA, W US (Alaska, Montana, Utah, Wyoming) & extreme N MX (Chihuahua)

♂
ssp *athabascae*

♀

NT
2 sspp
HB 304–380 cm (M), 210–350 cm (F)
T 43–60 cm
W 460–998 kg (M), 360–544 kg (F)

Wild Yak
Bos mutus

FR Yack sauvage
DE Wildyak
ES Yak

4000–6100 m

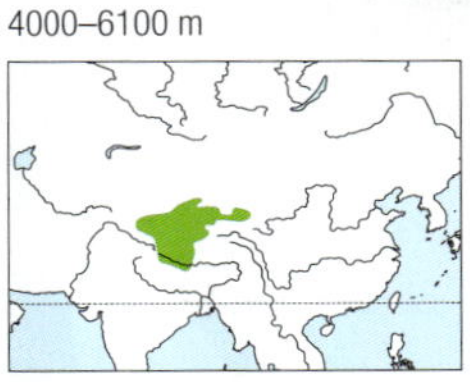

Fragmented on Tibetan Plateau in CN (Xinjiang, Tibet, Qinghai, Gansu), NW IN (Ladakh) & NW NP

♀

♂

VU
Monotypic
HB ≤ 380 cm
T ≤ 100 cm
W 535–1000 kg (M), 300–350 kg (F)

Asian Wild Buffalo
Bubalus arnee

FR Buffle d'Asie
DE Arni
ES Búfalo de agua

S NP, C & NE IN, S LK, N MM, W TH, EC KH

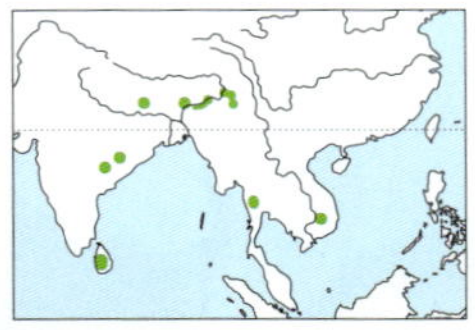

ssp *arnee*

ssp *theerapati*

ssp *fulvus*

EN
4 sspp
HB 240–300 cm
T 60–100 cm
W 250–1200 kg

Genus *BUBALUS*

(4 species)

Tamaraw
Bubalus mindorensis

FR Tamarau
DE Mindoro-Büffel
ES Tamarao

0–2000 m

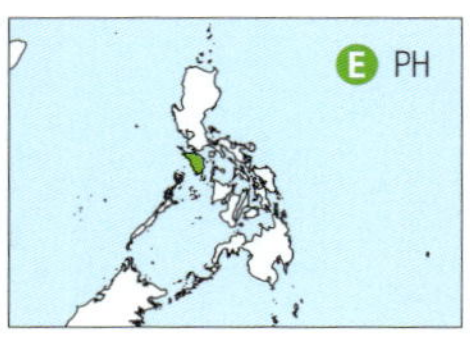

Mindoro

CR
Monotypic
HB c. 220 cm
T c. 60 cm
W 200–300 kg

Lowland Anoa
Bubalus depressicornis

FR Anoa des plaines
DE Anoa
ES Anoa de llanura

EN Monotypic
HB 122–188 cm W < 300 kg

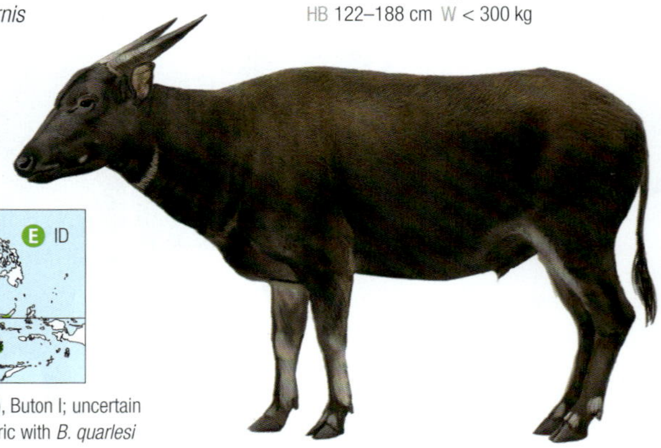

up to 1000 m

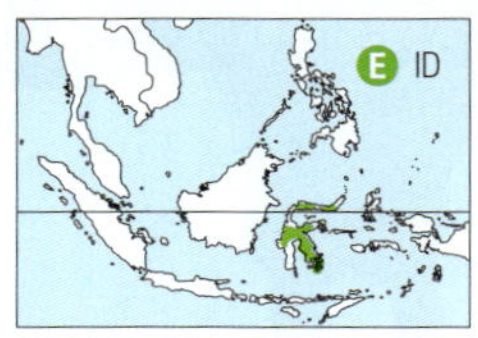

Sulawesi (N, C, E & SE), Buton I; uncertain if sympatric or parapatric with *B. quarlesi*

Mountain Anoa
Bubalus quarlesi

FR Anoa des montagnes
DE Berganoa
ES Anoa de montaña

EN
Monotypic
HB 122–188 cm
W < 300 kg

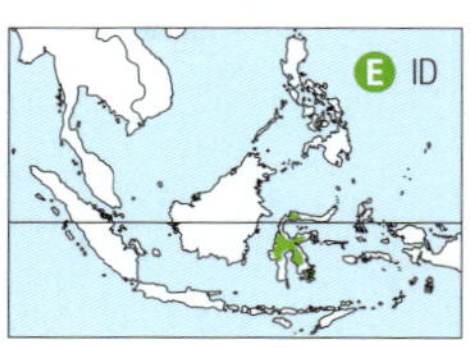

up to 2300 m

Sulawesi (mainly C, also part of N & SE), N Buton I; uncertain if sympatric or parapatric with *B. depressicornis*

ssp *brachyceros*

ssp *nanus*

Genus *SYNCERUS*

(1 species)

Cape Buffalo
Syncerus caffer

FR Buffle d'Afrique
DE Kaffernbüffel
ES Búfalo cafre

ssp *mathewsi*

up to 3600 m

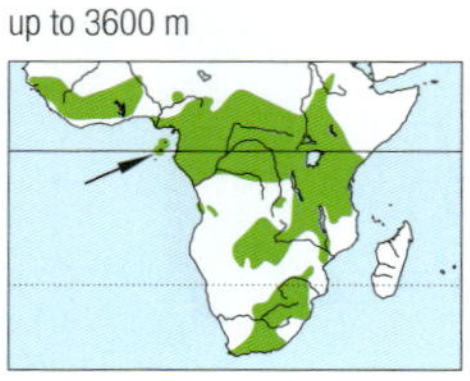

Sub-Saharan Africa, from SE SN & extreme SW ML E to ET & S SO, S through mainly E Africa to ZA

ssp *caffer*

NT
4 sspp
HB 180–340 cm
T 50–110 cm
W 250–900 kg

Genus *PSEUDORYX*

(1 species)

Saola
Pseudoryx nghetinhensis
FR Saola
DE Saola
ES Saola

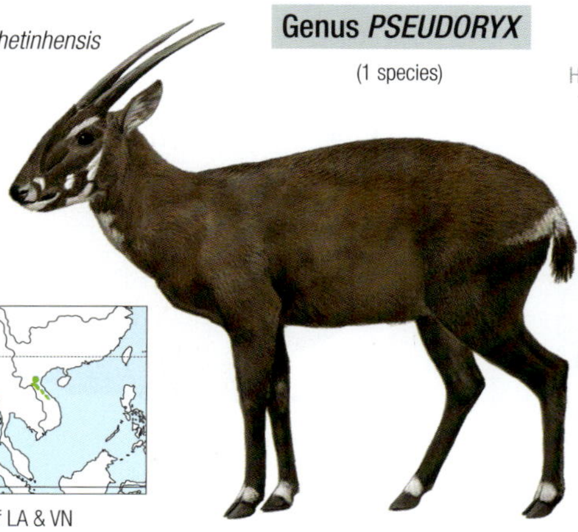

CR
Monotypic
HB 143–150 cm
T 16.5–23 cm
W 70–100 kg

500–1200 m

Annamite Mts of LA & VN

Genus *BOSELAPHUS*

(1 species)

Nilgai
Boselaphus tragocamelus
FR Nilgaut
DE Nilgau
ES Nilgo

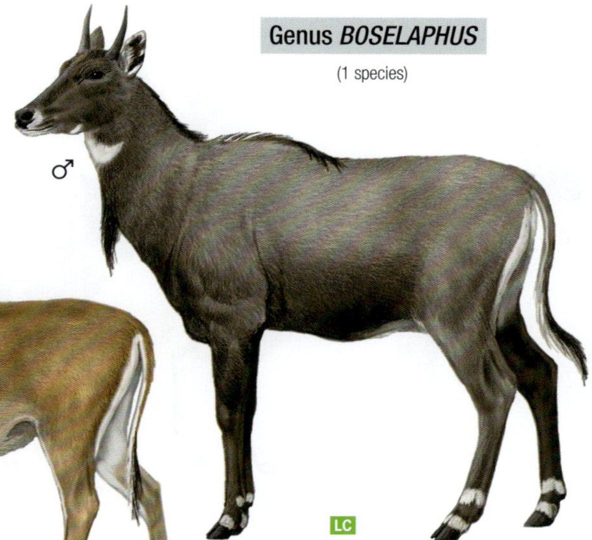

♂

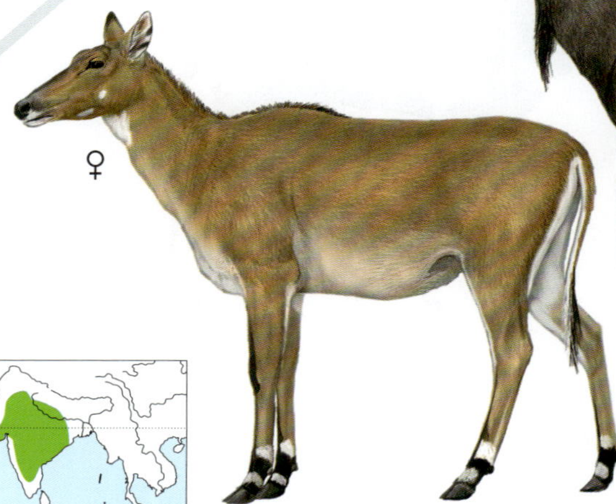

♀

LC
Monotypic
HB 180–210 cm (M), c. 170 cm (F)
T 46–53 cm (M), 45 cm (F)
W 200–288 kg (M), 120–212 kg (F)

NE PK, lowland S NP, Pen IN, extirpated from BD; introduced to S US (Texas), N MX, ZA & IT (now extinct)

Genus *TETRACERUS*

(1 species)

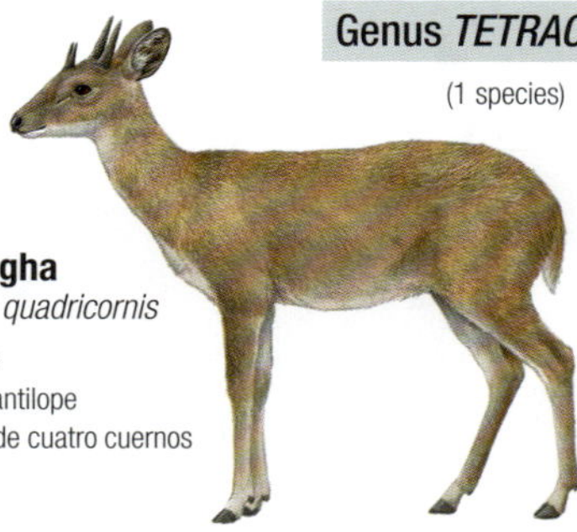

Chowsingha
Tetracerus quadricornis
FR Tétracère
DE Vierhornantilope
ES Antílope de cuatro cuernos

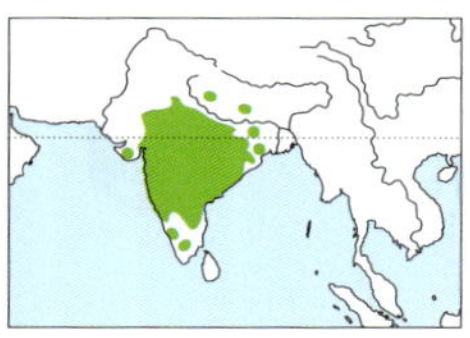

N & Pen IN, S NP

VU
3 sspp
HB 90–110 cm
T 10–15 cm
W 15–25 kg

Genus *TRAGELAPHUS*

(10 species)

Nyala
Tragelaphus angasii
FR Nyala
DE Nyala
ES Niala

LC
Monotypic
HB 159–198 cm (M), 132–146 cm (F)
T 37–47 cm (M), 34–40 cm (F)
W 92–126 kg (M), 55–68 kg (F)

S MW, MZ, N & S ZW, E ZA, SZ (reintroduced); introduced to ranches in NA & ZA from which they have escaped to E BW

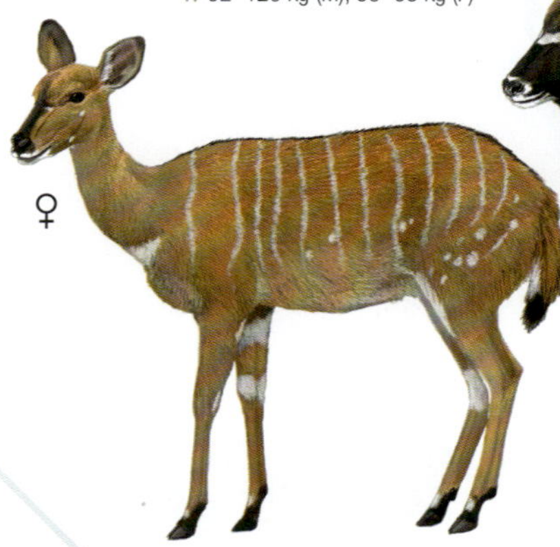

♀

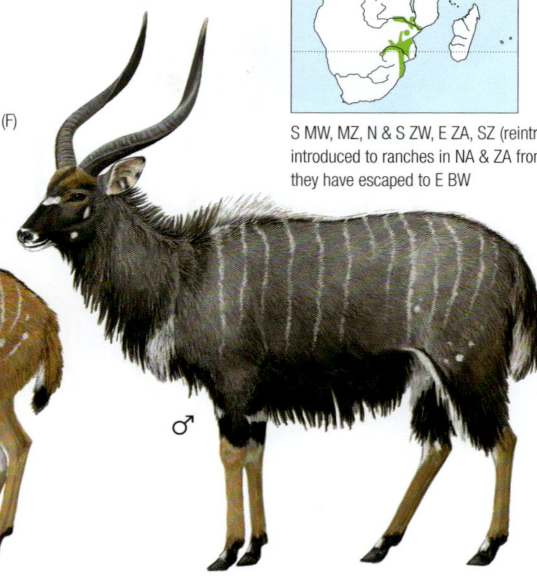

♂

Northern Bushbuck
Tragelaphus scriptus
FR Guib harnaché
DE Senegal-Schirrantilope
ES Bushbuck septentrional

♂

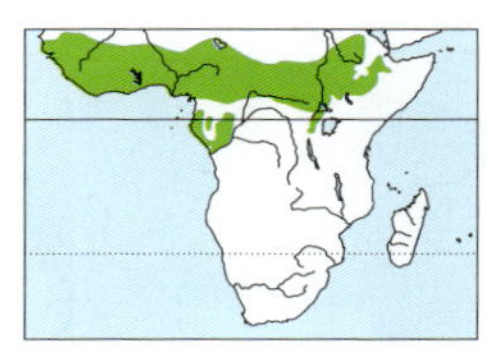

Extreme S MR, S SN & S ML E to ET, S to extreme SW CD (N of R Congo), W UG & RW

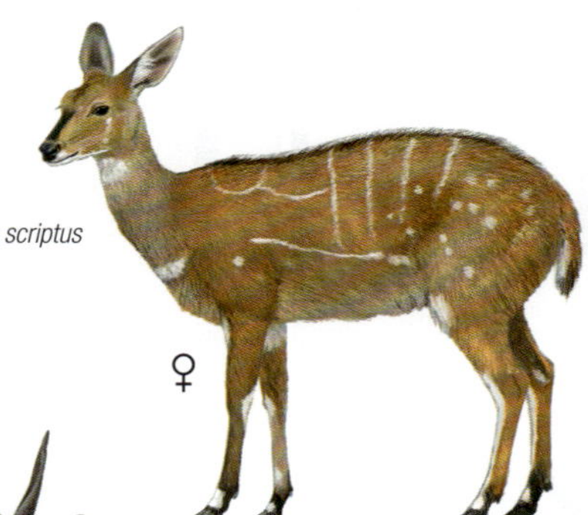

ssp *scriptus*
♀

♂
ssp *phaleratus*

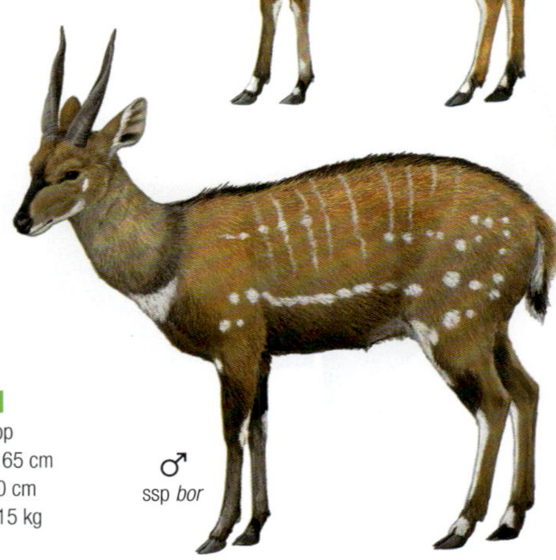

♂
ssp *bor*

♂
ssp *decula*

LC
4 sspp
HB 114–165 cm
T 19–30 cm
W 24–115 kg

Southern Bushbuck

Tragelaphus sylvaticus

FR Guib sylvain
DE Südliche Schirrantilope
ES Bushbuck meridional

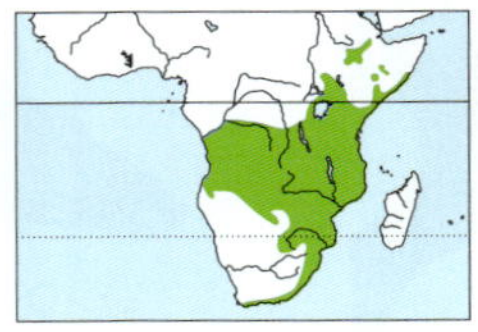

From C ET & SO S through E Africa to S CD, AO (except SW), NE NA (Caprivi Strip), BW and E & S ZA

NE
4 sspp
HB 117–145 cm (M), 114–132 cm (F)
T 19–24 cm
W 40–80 kg (M), 24–60 kg (F)

♀ ♂ ssp *sylvaticus*

♂ ssp *fasciatus*

♀ ♂ ssp *ornatus*

♀ ♂ ssp *meneliki*

Sitatunga

Tragelaphus spekii

FR Sitatunga
DE Ostafrika-Sitatunga
ES Sitatunga del Lago Victoria

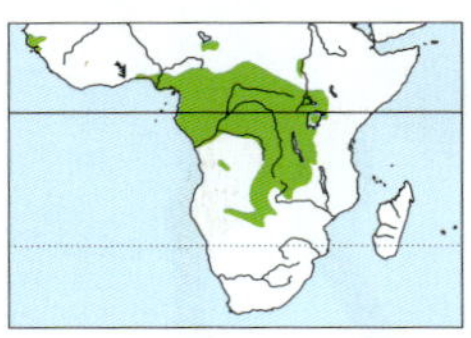

Rare and localized in W Africa (SN, GM, GW, GN), widespread from S GH & BJ through Congo Basin E to KE, S to S AO & BW (Okavango)

LC
5 sspp
HB 115–170 cm
T 20–43 cm
W 50–125 kg

♀ ♂ ssp *spekii*

♀ ♂ ssp *selousi*

♂ ssp *sylvestris*

♂ ssp *larkenii*

♂ ssp *gratus*

Bongo

Tragelaphus eurycerus

FR Bongo
DE Bongo
ES Bongo

NT Monotypic
HB 220–235 cm
T 45–65 cm
W 240–405 kg (M), 210–253 kg (F)

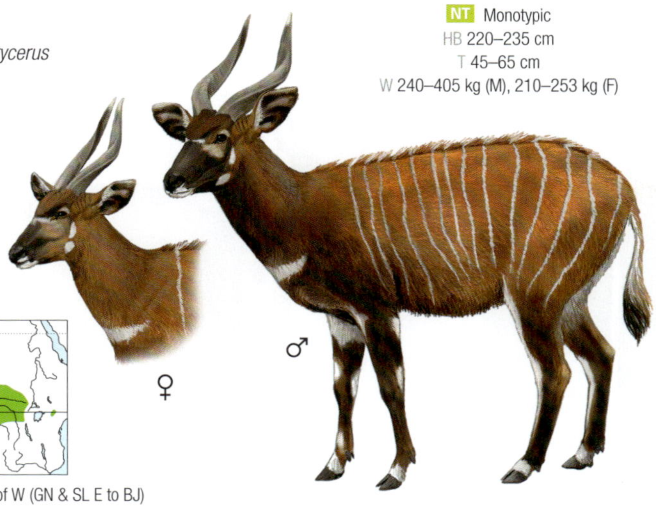

♀ ♂

up to 3000 m

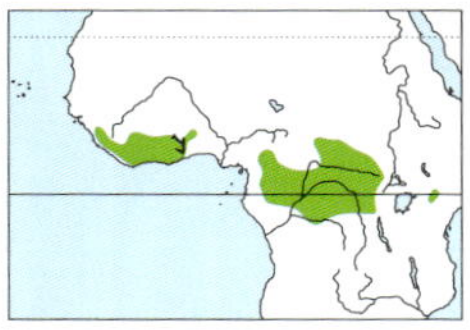

Lowland rainforest belt of W (GN & SL E to BJ) & C Africa (Congo Basin in CM, GA, CG, CD, CF & SW SS), C KE Mts

Gedemsa

Tragelaphus buxtoni

FR Gedemsa
DE Bergnyala
ES Gedemsa

EN Monotypic
HB 190–200 cm
T 20–25 cm
W 180–320 kg (M), 150–200 kg (F)

♀

♂

1600–4300 m

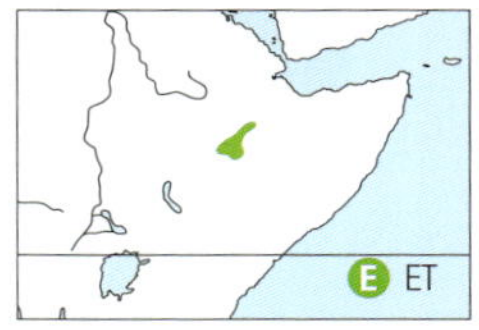

Restricted to 5 isolated localities in Bale & Arsi Mt ranges, S & SE of Rift Valley

Lesser Kudu

Tragelaphus imberbis

FR Petit Koudou
DE Nördlicher Kleinkudu
ES Kudú menor septentrional

NT
2 sspp
HB 140–178 cm
T 28–46 cm

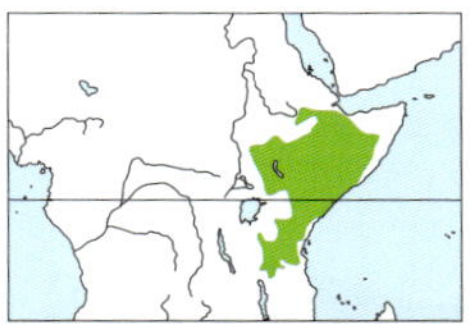

Lowlands of C & S ET, SO, extreme SE SD, extreme NE UG, N, C & S KE, E TZ

♀

♂ ssp *australis*

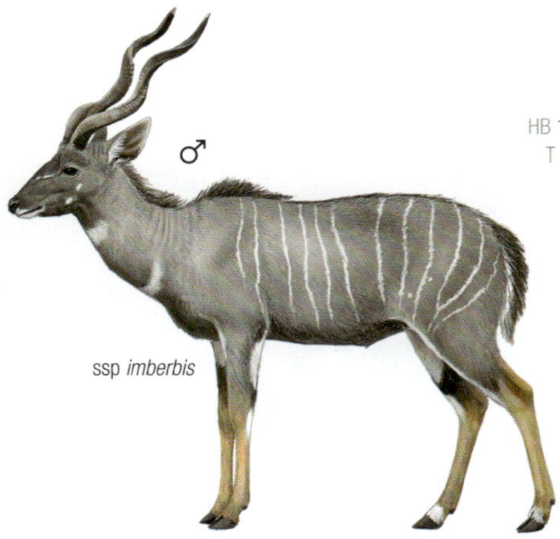

♂ ssp *imberbis*

Greater Kudu

Tragelaphus strepsiceros

FR Grand Koudou
DE Kap-Großkudu
ES Gran kudú meridional

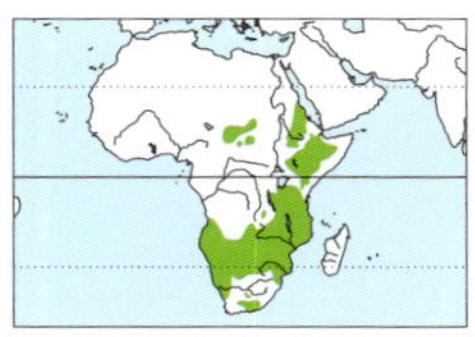

From SE TD, W SD & N CF E to Red Sea, S through E Africa to C AO & ZA (Eastern Cape)

LC
4 sspp
HB 193–248 cm
T 32–75 cm

♀

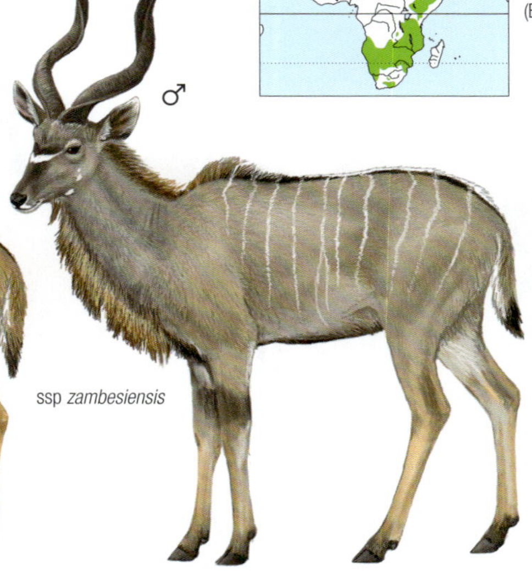

♂ ssp *zambesiensis*

♂ ssp *strepsiceros*

♂ ssp *chora*

♀

♂ ssp *cottoni*

Common Eland
Tragelaphus oryx

FR Éland du Cap
DE Elenantilope
ES Eland de El Cabo

up to 4600 m

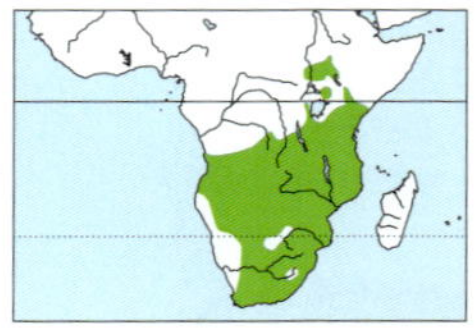

E & S Africa from SE SS, SW ET, UG, RW, KE & S SO S through TZ, S CD & AO to ZA

LC
3 sspp
HB 250–340 cm (M), 200–280 cm (F)
T 54–75 cm
W 400–942 kg (M), 390–600 kg (F)

♀

ssp *livingstonii*

♂

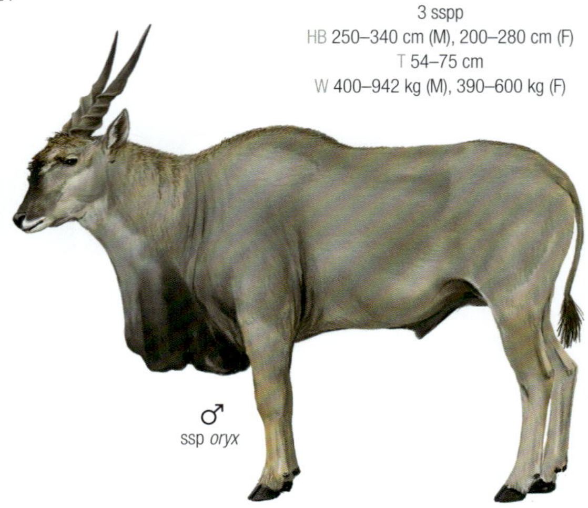

♂
ssp *oryx*

Giant Eland
Tragelaphus derbianus

FR Éland de Derby
DE Riesenelen
ES Eland de Derby

VU Monotypic
HB c. 290 cm (M), 220 cm (F) T c. 90 cm
W 440–950 kg (M), c. 440 kg (F)

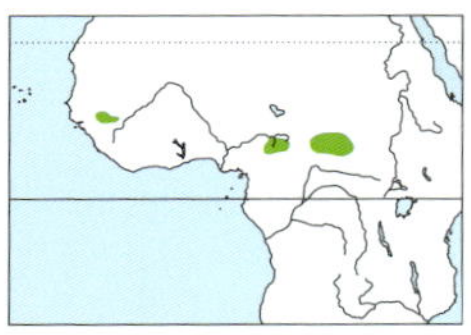

Disjunct in 3 areas: SE SN, N GN, SW ML (maybe GW); N CM, SW TD (maybe NG); CF, SE TD, W SS (maybe CD & UG)

Genus *NEOTRAGUS*

(1 species)

Royal Antelope
Neotragus pygmaeus

FR Néotrague pygmée
DE Kleinstböckchen
ES Antílope enano

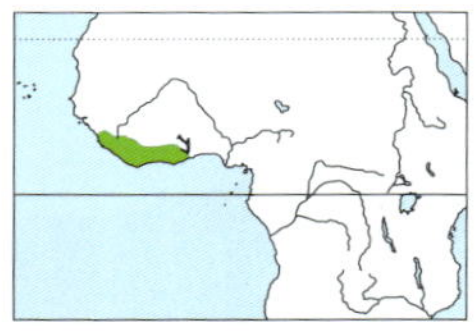

S GN & SL E to GH (W of R Volta)

LC Monotypic
HB 38–51 cm
T 5–8 cm
W 2–3 kg

Genus *NESOTRAGUS*

(2 species)

Dwarf Antelope
Nesotragus batesi

FR Suni de Bates
DE Bates-Böckchen
ES Antílope enano de Bates

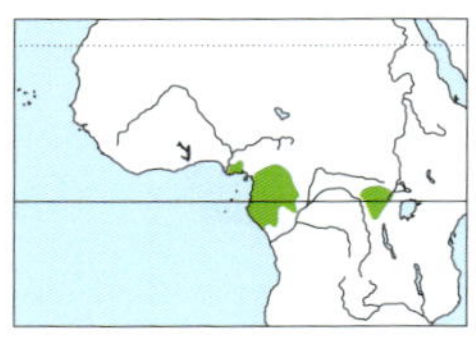

NG (from R Niger to R Cross), CM (S of R Sanaga), SW CF, GQ (Mbini), GA, W CG, NE CD (E of R Congo) to W UG

LC Monotypic
HB 49–62 cm T c. 9 cm
W 2.2–2.8 kg

Suni
Nesotragus moschatus

FR Suni de Zanzibar
DE Moschusböckchen
ES Suni

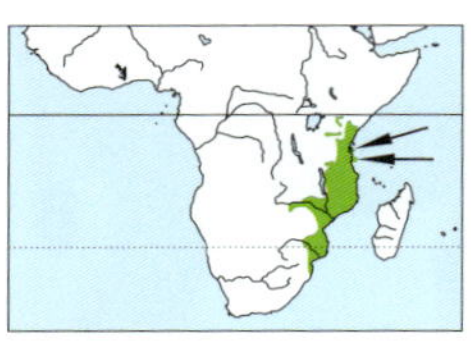

Highlands & coasts of S KE & TZ (including Unguja I & several smaller Is) S to MW, MZ, ZW & NE ZA

ssp *moschatus*

ssp *livingstonianus*

ssp *kirchenpaueri*

LC
3 sspp
HB 58–66 cm
T 9–13 cm
W 4–10 kg

Genus *AEPYCEROS*

(1 species)

♀

ssp *petersi*

♂

♀

ssp *melampus*

♂

Impala
Aepyceros melampus

FR Impala
DE Impala
ES Impala

LC
2 sspp
HB 130–135 cm (M), 125–130 cm (F)
T 25–35 cm
W 57–64 kg (M), 43–52 kg (F)

0–3000 m

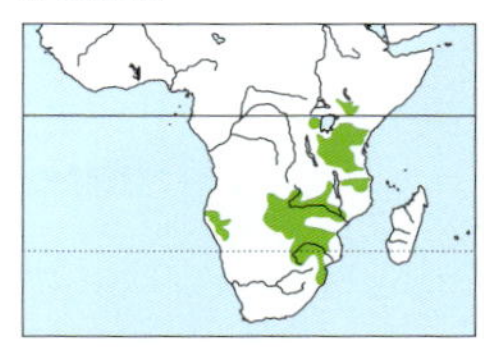

E & S Africa from C & S KE and NE UG S to ZA (N KwaZulu-Natal); small disjunct population in SW AO (R Kunene) & NW NA

Cape Grysbok
Raphicerus melanotis
FR Grysbok du Cap
DE Kap-Greisböckchen
ES Raficero de El Cabo

S of Karoo, from Atlantic coast (S of c. 30° S) E as far as S KwaZulu-Natal

LC
Monotypic
HB 70–75 cm
T 4–7.2 cm
W 8.8–11.4 kg

Genus *RAPHICERUS*

(3 species)

W & S TZ (N almost to L Victoria) S to SE CD, ZA (E of R Zambezi), NA (Caprivi Strip), NE BW, NE ZA & E SZ

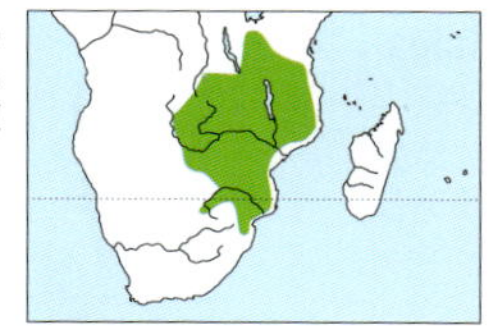

LC
2 sspp
HB 65–80 cm
T 4–8 cm
W 6.4–16 kg

ssp *sharpei*

ssp *colonicus*

Sharpe's Grysbok
Raphicerus sharpei
FR Grysbok de Sharpe
DE Nördliches Greisböckchen
ES Raficero de Sharpe

Steenbok
Raphicerus campestris
FR Steenbok
DE Steinböckchen
ES Raficero

♀

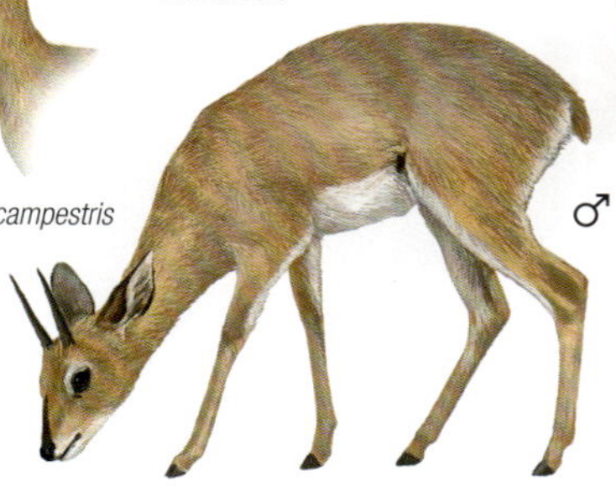

ssp *campestris*

♂ ssp *capricornis*

♂

♂ ssp *neumanni*

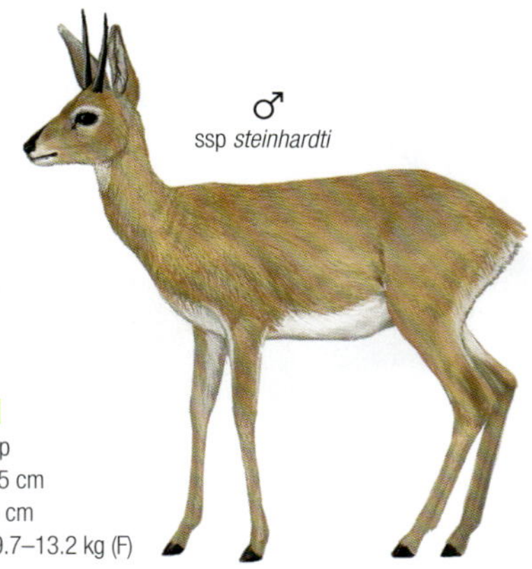

♂ ssp *steinhardti*

LC
6 sspp
HB 74–75 cm
T 5–5.5 cm
W 9–13.2 kg (M), 9.7–13.2 kg (F)

up to 4750 m

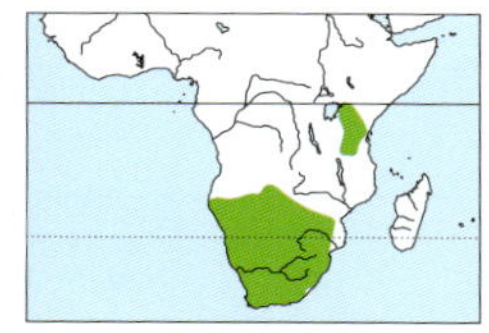

S Africa from S AO, W ZM, ZW & S MZ S to much of ZA, disjunct in E Africa (S KE, N & C TZ)

Springbok
Antidorcas marsupialis
FR Springbok
DE Kap-Springbock
ES Gacela saltarina de El Cabo

Genus *ANTIDORCAS*

(1 species)

From coastal regions of AO (N to Benguela) S to NA, BW, ZA & W LS

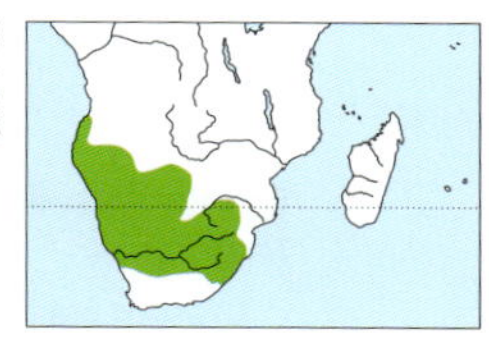

♀

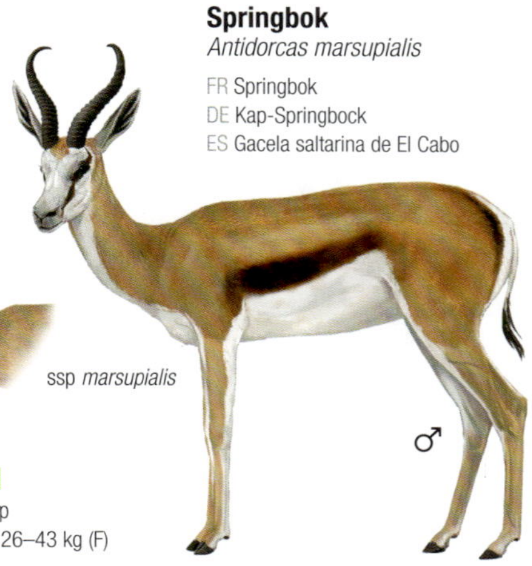

ssp *marsupialis*

♂

♂ ssp *hofmeyri*

♂ ssp *angolensis*

LC
3 sspp
W 31–48 kg (M), 26–43 kg (F)

Genus *AMMODORCAS*

(1 species)

Dibatag
Ammodorcas clarkei
FR Gazelle de Clarke
DE Dibatag
ES Gacela de Clarke

VU
Monotypic
HB c. 117 cm
T c. 36 cm
W c. 27 kg

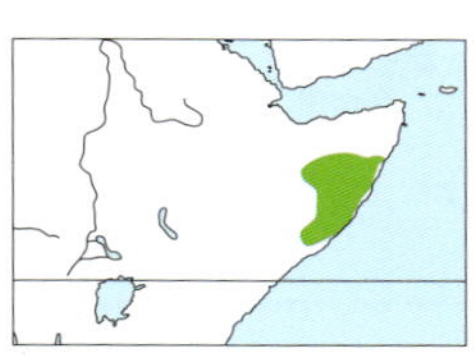

E ET, C SO

Gerenuk
Litocranius walleri
FR Gazelle de Waller
DE Südliche Giraffengazelle
ES Gacela jirafa meridional

Genus *LITOCRANIUS*

(1 species)

NT
2 sspp
HB c. 145 cm (M)
T c. 23 cm (M)
W 31–52 kg (M), 28–45 kg (F)

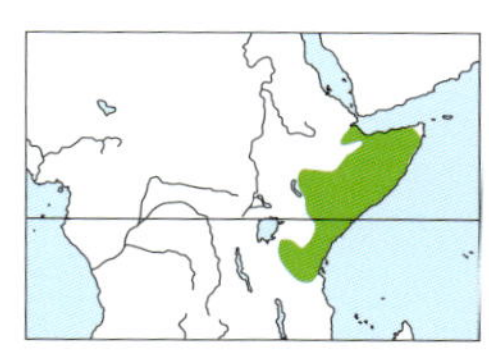

NE Africa from SE DJ, ET & N SO S through KE to NE TZ

♀

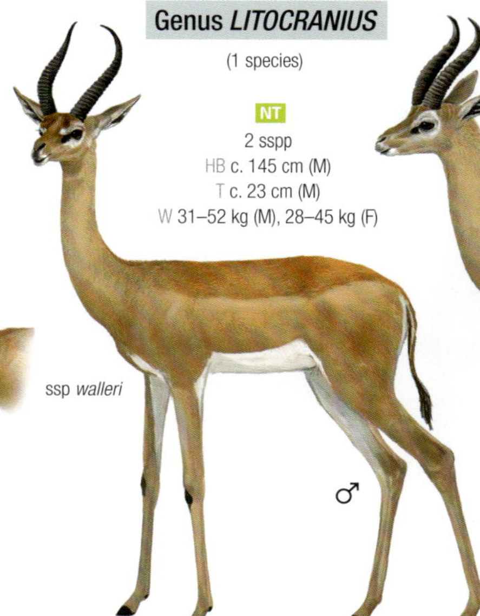

ssp *walleri*

♂

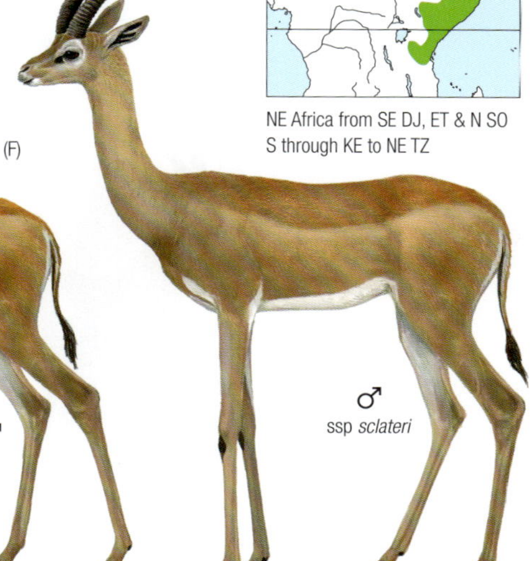

♂ ssp *sclateri*

Saiga
Saiga tatarica
FR Saïga
DE Saiga
ES Saiga

Genus *SAIGA*
(1 species)

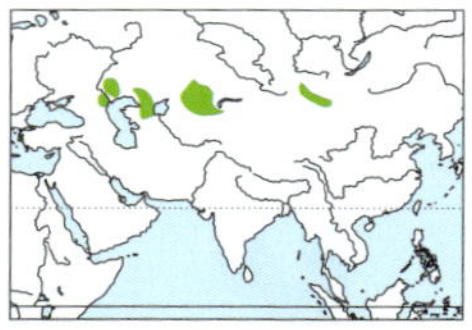

Disjunct in S European RU (Kalmykia), KZ (R Ural, Ustyurt, Betpak-Dala) & MN (E of Altai); perhaps NW UZ & NW TM in winter

♂
ssp *tatarica* summer
♀

♂
ssp *tatarica* winter

♂
ssp *mongolica*

CR 2 extant sspp
HB 123–146 cm (M), 108–125 cm (F)
T 8–12.5 cm (M), 6–10 cm (F)
W 32–51 kg (M), 21–41 kg (F)

Blackbuck
Antilope cervicapra
FR Cervicapre
DE Hirschziegenantilope
ES Antílope indio

Genus *ANTILOPE*
(1 species)

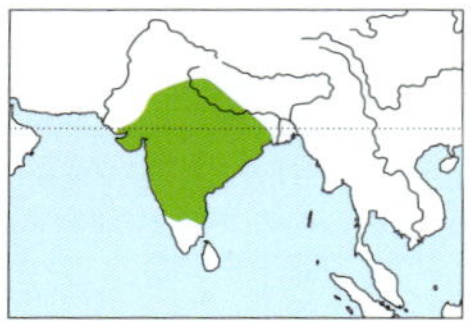

SE PK, SW NP, Pen IN S to Karnataka & Tamil Nadu; introduced to US (Texas) & AR

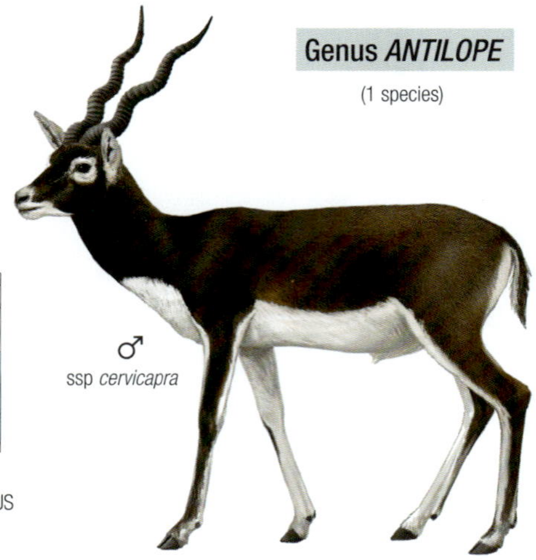

♂
ssp *cervicapra*

LC
2 sspp
HB 120–132 cm
T 8.2–13.5 cm
W 25–35 kg

♀

ssp *rajputanae*
♂

Grant's Gazelle
Nanger granti
FR Gazelle de Grant
DE Südliche Grant-Gazelle
ES Gacela de Grant

♀

♂
♂
variants

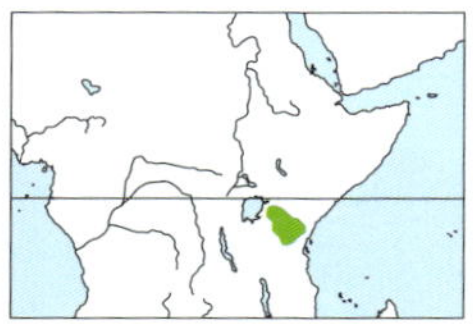

KE (from L Elmenteita in Kenya Rift Valley SE to Tsavo West NP), SW to Serengeti ecosystem (N TZ)

LC
Monotypic
HB 134–153 cm (M), c. 127 cm (F)
T 27–34 cm
W 58–81 kg (M), 38–67 kg (F)

Genus *NANGER*
(5 species)

Bright's Gazelle
Nanger notatus
FR Gazelle de Bright
DE Nördliche Grant-Gazelle
ES Gacela de Bright

NE
Monotypic

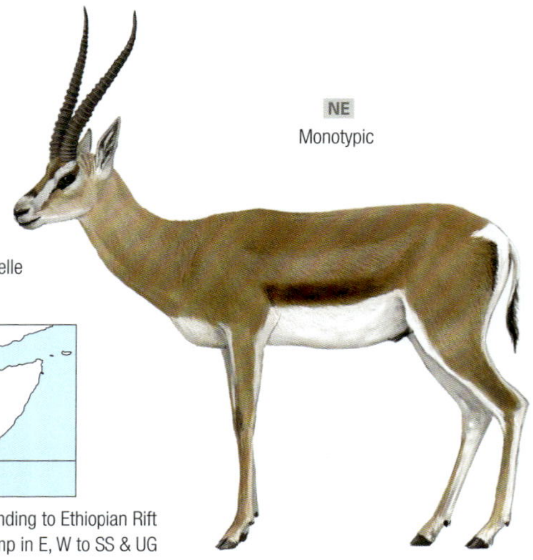

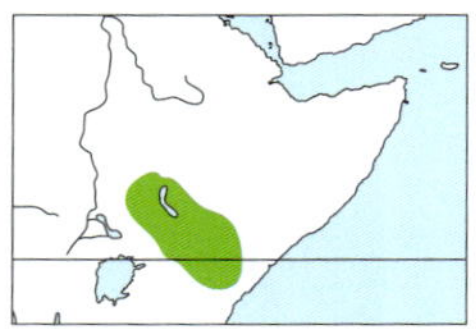

N of KE Highlands, extending to Ethiopian Rift Valley in N, Lorian Swamp in E, W to SS & UG

Peters's Gazelle
Nanger petersii
FR Gazelle de Peters
DE Östliche Grant-Gazelle
ES Gacela de Peters

NE
Monotypic

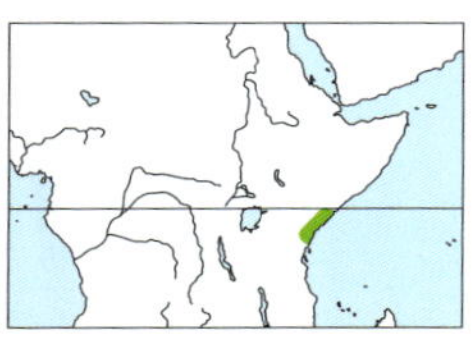

Coast of KE, going N to SE SO

Soemmerring's Gazelle
Nanger soemmerringii
FR Gazelle de Sömmerring
DE Soemmerring-Gazelle
ES Gacela de Soemmerring

VU
2 sspp
HB c. 135 cm
T c. 27 cm

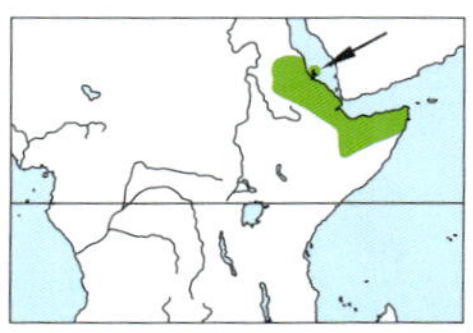

N SD (E of White Nile) E to ER (including Dahlak Arch), N ET, DJ & N SO

Dama Gazelle
Nanger dama

FR Gazelle dama
DE Dama-Gazelle
ES Gacela dama

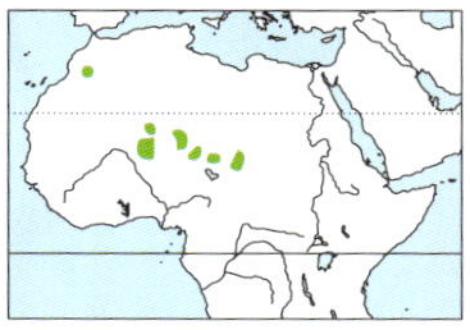

S MA, E ML, S DZ, NE, TD; possibly SD W of Nile, introduced to Safia NR (S MA)

CR
3 sspp
HB 135–165 cm (M), 142–153 cm (F)
T 28–30 cm
W 46–74 kg (M), 42–62 (F)

ssp *dama*

ssp *ruficollis*

ssp *mhorr*

Slender-horned Gazelle
Gazella leptoceros

FR Gazelle leptocère
DE Dünengazelle
ES Gacela de Loder

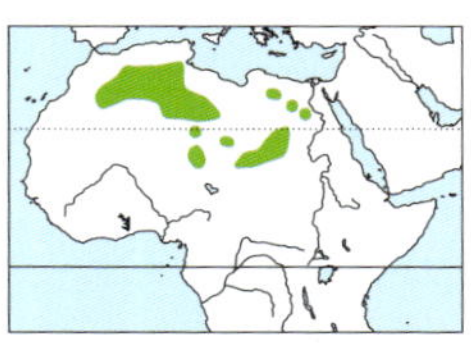

EN
Monotypic
HB 100–110 cm
T 15–20 cm
W 14–18 kg

Patchily in dune regions of Sahara as far E as R Nile, confirmed records from DZ, TN, LY & EG

Cuvier's Gazelle
Gazella cuvieri

FR Gazelle de Cuvier
DE Cuvier-Gazelle
ES Gacela de Cuvier

0–2600 m

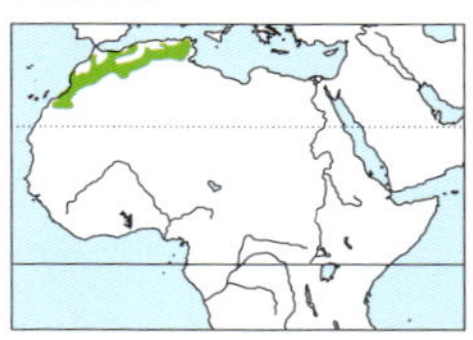

VU
Monotypic
HB 95–105 cm
T 15–20 cm
W 20–35 kg

Atlas Mts from N EH E to N TN

Genus *GAZELLA*
(9 species)

Sand Gazelle
Gazella marica

FR Gazelle marica
DE Sandgazelle
ES Gacela árabe de dunas

VU
2 sspp
HB c. 97 cm
T c. 14.8 cm
W c. 20 kg

ssp *karamii*

ssp *marica*

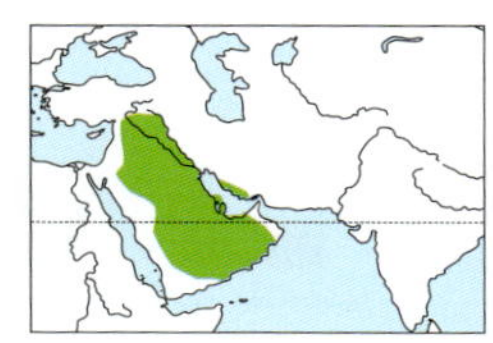

Scattered in SE TR, JO, SA (mainly in Mahazat As-Asyd R & Uruq Bani Ma'arid R), BH, AE, OM, SW IQ, SW IR, uncertain in SY, KW, QA & YE

Goitered Gazelle
Gazella subgutturosa

FR Gazelle à goitre
DE Kropfgazelle
ES Gacela persa

ssp *subgutturosa*

ssp *gracilicornis*

0–3000 m

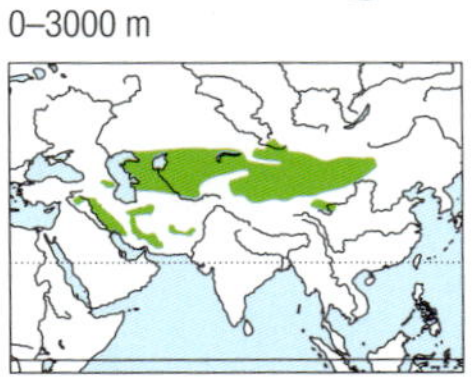

SE TR, AZ, SY, N & E IQ, IR, S AF & W PK through C Asia (KZ, TM, UZ, marginally TJ & KG) to MN & N CN

VU
3 sspp
HB 90–126 cm
T 10–23 cm
W 18–40 kg

ssp *yarkandensis*

Dorcas Gazelle
Gazella dorcas

FR Gazelle dorcas
DE Dorcas-Gazelle
ES Gacela dorcas

VU
6 extant sspp
HB 90–110 cm
T 15–22 cm
W 15–23 kg

ssp *isabella*

ssp *osiris*

ssp *pelzelni*

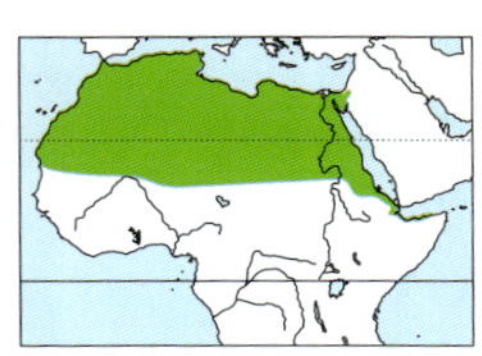

N Africa from Mediterranean to S Sahel & from Atlantic to Red Sea, N to Levant (IL, SY & JO), E along coast to N SO; extinct from Arabian Pen

ssp *farasani*

ssp *erlangeri*

ssp *muscatensis*

ssp *dareshurii*

Arabian Gazelle
Gazella arabica

FR Gazelle d'Arabie
DE Arabische Gazelle
ES Gacela arábiga

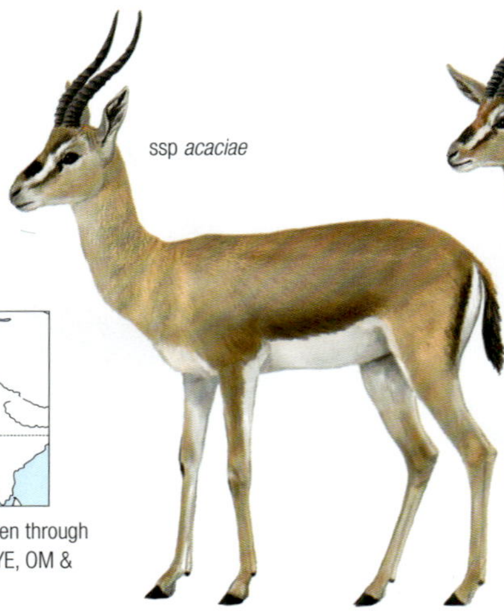

ssp *acaciae*

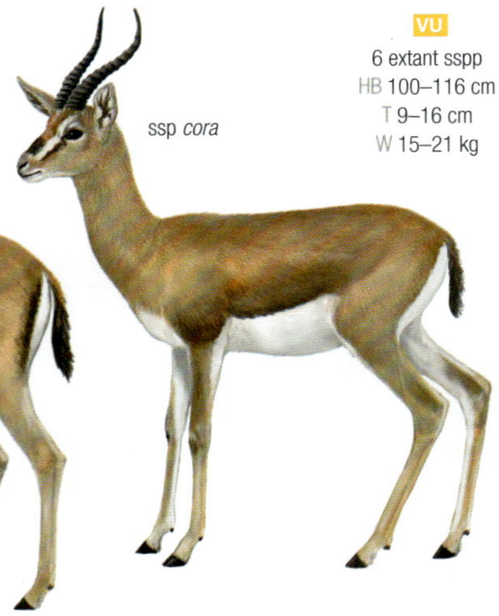

ssp *cora*

VU
6 extant sspp
HB 100–116 cm
T 9–16 cm
W 15–21 kg

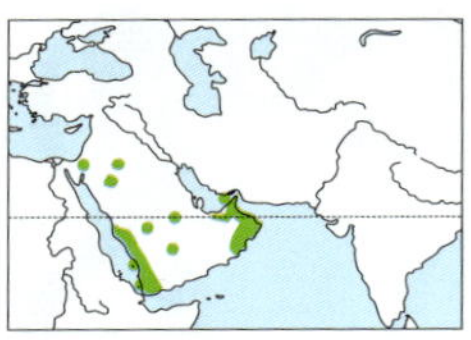

IL (S Arava Valley), Arabian Pen through SA (including Farasan Is) to YE, OM & AE, Farrur I (IR); perhaps JO

Mountain Gazelle
Gazella gazella

FR Gazelle de montagne
DE Echtgazelle
ES Gacela de montaña

EN
Monotypic
HB 101–115 cm
T 8–13 cm
W 17–30 kg

N & C IL, PS (Golan Heights), SC TR (Hatay Prov); unclear if any population survives in LB, JO & SY

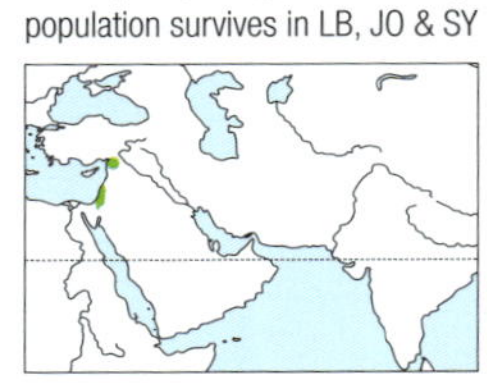

Speke's Gazelle
Gazella spekei

FR Gazelle de Speke
DE Speke-Gazelle
ES Gacela de Speke

EN
Monotypic
HB 95–105 cm
T 15–20 cm
W 15–25 kg

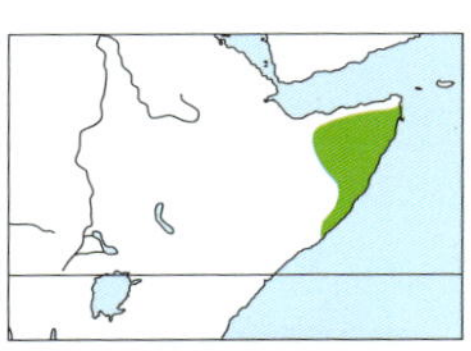

SO N of R Shebelle, except for Red Sea coastal strip, extending to lowlands of E ET

ssp *bennettii*

ssp *fuscifrons*

LC
5 sspp
HB 90–110 cm
T 10–22 cm
W 15–30 kg

Chinkara
Gazella bennettii

FR Gazelle indienne
DE Dekkan-Chinkara
ES Gacela india

ssp *shikarii*

ssp *christii*

ssp *salinarum*

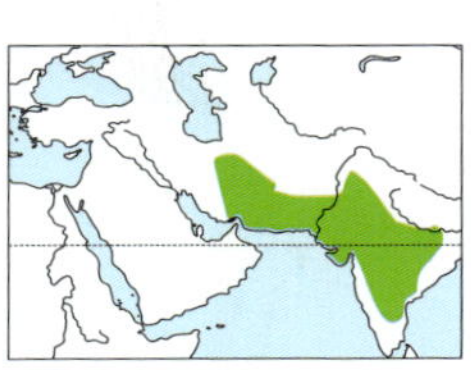

IR (E from Kavir NP), SW AF, PK, much of W & C IN (Thar Desert remains stronghold)

Genus *EUDORCAS*

(4 species)

VU
Monotypic
HB 98–108 cm
T 20–24 cm
W 25–28 kg

Red-fronted Gazelle
Eudorcas rufifrons

FR Gazelle à front roux
DE Rotstirngazelle
ES Gacela de frente roja

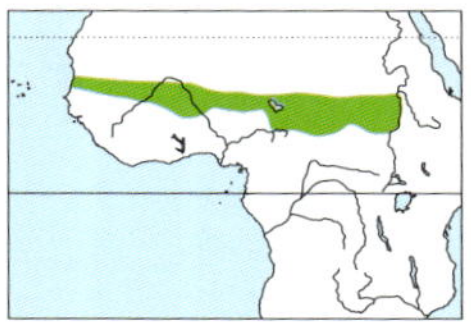

MR & SN E to White Nile and across Nile from Malakal N to Kassala in NE SD

EN
Monotypic
HB 55–68 cm
T 25–29 cm
W 22–25 kg

Eritrean Gazelle
Eudorcas tilonura

FR Gazelle de Heuglin
DE Heuglin-Gazelle
ES Gacela de Eritrea

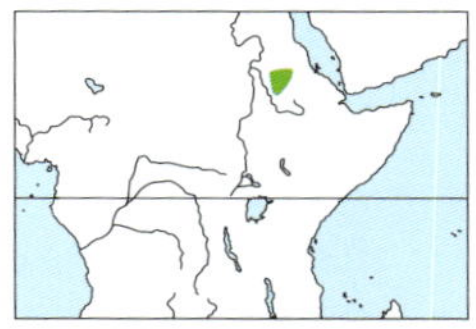

SD (E of R Nile), ER (N of R Setit as far as R Bogos), extreme NW ET

LC
Monotypic
W 23–27 kg (M)

Mongalla Gazelle
Eudorcas albonotata

FR Gazelle du Mongalla
DE Mongalla-Gazelle
ES Gacela de Sudán

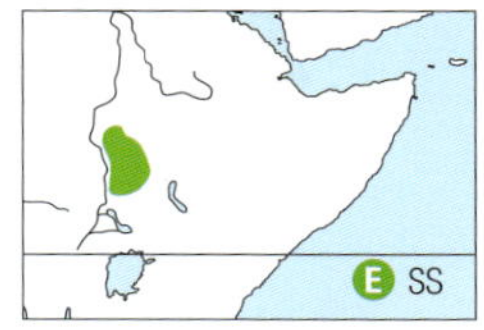

E of White Nile, from about Ugandan border N to R Veveno, E of Bor; small isolated population at Ket (c. 8° N)

Thomson's Gazelle
Eudorcas thomsonii

FR Gazelle de Thomson
DE Östliche Thomson-Gazelle
ES Gacela de Thomson

ssp *thomsonii*

ssp *nasalis*

LC
2 sspp
HB 89–107 cm
T 19–26 cm
W 14–24 kg

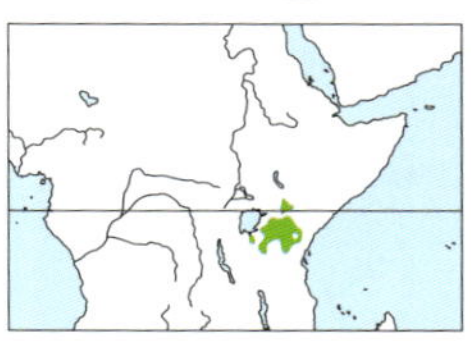

E of Rift Valley in KE & TZ, SW via Serengeti & Arusha district to L Eyasi, Wembere Plains & Shinyanga

Genus *DORCATRAGUS*

(1 species)

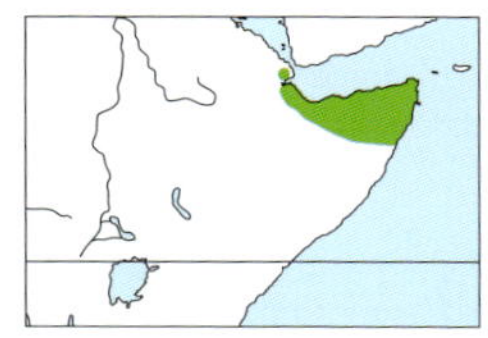

Along Somali coast from Horn of Africa W to just over Ethiopian border, N to SE DJ

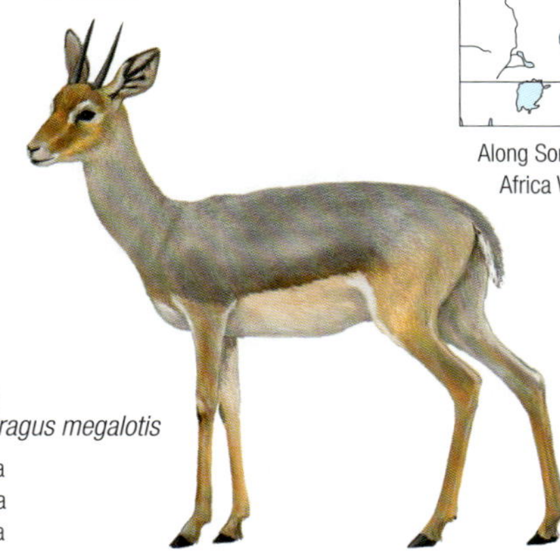

Beira
Dorcatragus megalotis

FR Beira
DE Beira
ES Beira

VU
Monotypic
HB 81–85 cm
T 6–8 cm
W 10–11.2 kg

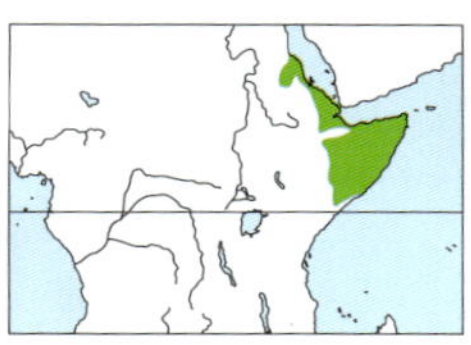

Horn of Africa (ET, ER, DJ, SO), marginaly in SE SD (S from R Atbara); perhaps NE KE

ssp *saltiana*

ssp *swaynei*

ssp *lawrancei*

ssp *phillipsi*

ssp *hararensis*

LC
5 sspp
HB 52–67 cm
T 3.5–8 cm
W 2.3–3 kg

Salt's Dik-dik
Madoqua saltiana

FR Dikdik de Salt
DE Eritrea-Dikdik
ES Dikdik de Salt

Genus *MADOQUA*

(8 species)

Silver Dik-dik
Madoqua piacentinii

FR Dikdik de Piacentini
DE Silber-Dikdik
ES Dikdik de Piacentini

DD
Monotypic
HB c. 47 cm
T c. 3.7 cm
W c. 2.1 kg

Known from few localities on central coast, nearly as far S as R Shebelle; distribution appears to be sporadic

NE
Monotypic

Thomas's Dik-dik
Madoqua thomasi

FR Dikdik de Thomas
DE Thomas-Dikdik
ES Dikdik de Thomas

From Mwanza, Olduvai & L Manyara to Kondoa, Dodoma, Bugogo, Irangi & Tabora districts

Hinde's Dik-dik
Madoqua hindei

FR Dikdik de Hinde
DE Hinde-Dikdik
ES Dikdik de Hinde

NE
Monotypic

Highlands in SE KE and N & NE TZ; N probably to R Tana, S to Mt Kilimanjaro & Pare-Usambara Mts, W to Rift Highlands

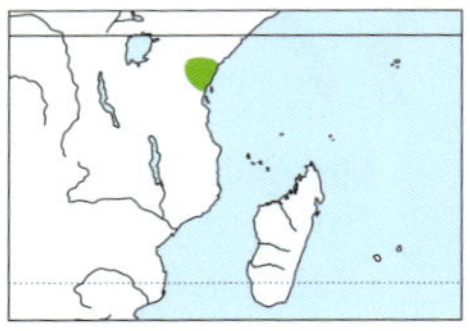

Kirk's Dik-dik
Madoqua kirkii

FR Dikdik de Kirk
DE Kirk-Dikdik
ES Dikdik de Kirk

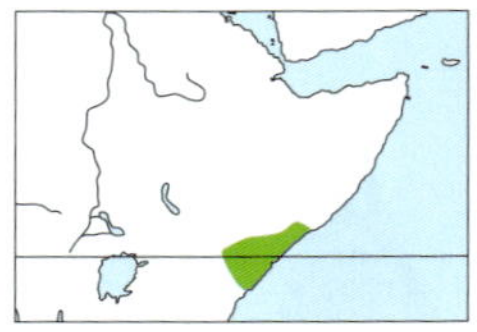

NE KE from N Ewaso Ng'iro to coast at Lamu, N to SO (to Mogadishu), probably E & N of R Tana (or R Athi/ Galana)

LC
Monotypic
HB c. 63 cm
W c. 9.1 kg

Cavendish's Dik-dik
Madoqua cavendishi

FR Dikdik de Cavendish
DE Cavendish-Dikdik
ES Dikdik de Cavendish

NE
Monotypic

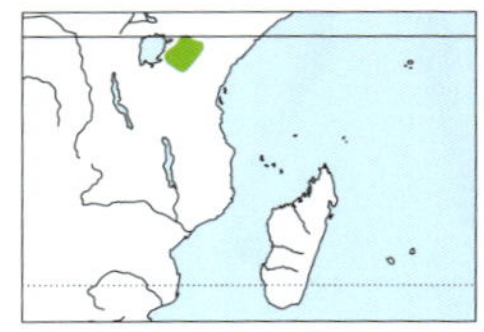

KE Rift Valley, extending to Serengeti-Mara ecosystem in SW KE & NW TZ

Guenther's Dik-dik
Madoqua guentheri

FR Dikdik de Günther
DE Günther-Dikdik
ES Dikdik de Günther

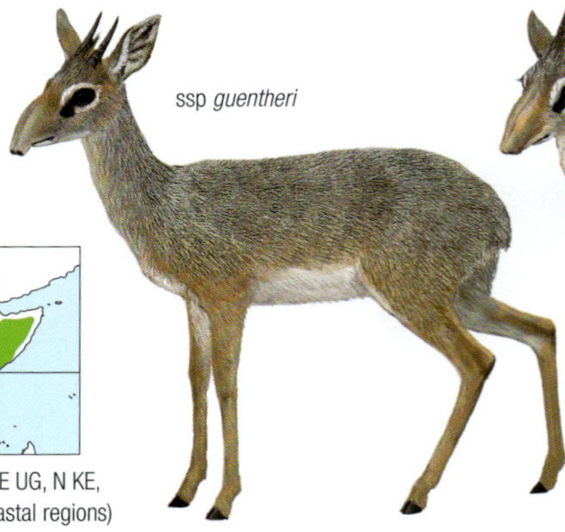

ssp *guentheri*

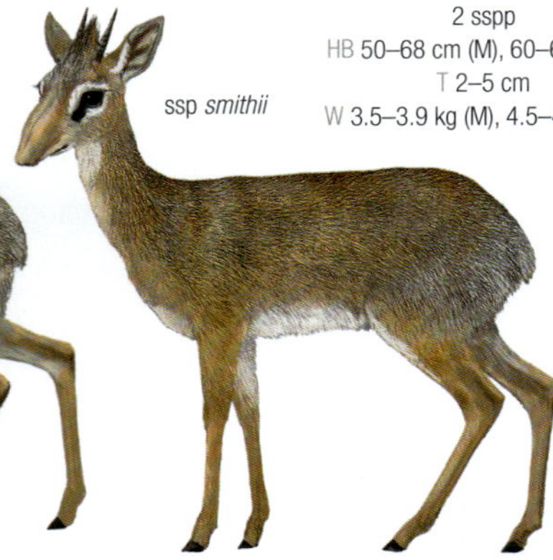

ssp *smithii*

LC
2 sspp
HB 50–68 cm (M), 60–67 cm (F)
T 2–5 cm
W 3.5–3.9 kg (M), 4.5–4.6 kg (F)

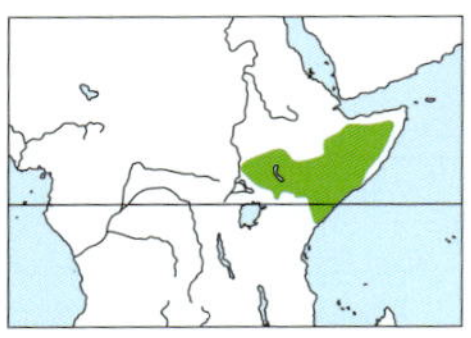

SE SS (E of White Nile), NE UG, N KE, S & SE ET, SO (except coastal regions)

Damara Dik-dik
Madoqua damarensis

FR Dikdik damara
DE Damara-Dikdik
ES Dikdik damara

NE
Monotypic
HB 58–69 cm
T 3.2–5.6 cm
W 4.6–5.5 kg

SW AO, NW & N NA, S as far as Brukkaros Mt at 25° 50' S

Genus *OUREBIA*

(1 species)

Oribi
Ourebia ourebi

FR Ourébi
DE Südliches Oribi
ES Oribí meridional

♀

♂
ssp *ourebi*

♂
ssp *hastata*

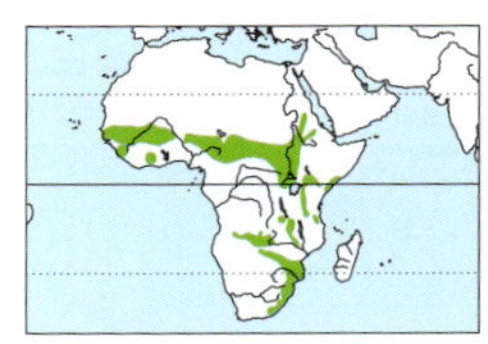

Patchily from SN E to ET, ER & SO, S through E Africa to AO & ZA (Eastern Cape)

LC
4 sspp
HB 89–109 cm
T 6.5–10.5 cm
W 7.5–17 kg

♂
ssp *quadriscopa*

♂
ssp *montana*
dark clay morph

♀
ssp *montana*
orange-buff morph

Genus *PROCAPRA*

(3 species)

Tibetan Gazelle
Procapra picticaudata

FR Gazelle du Tibet
DE Tibet-Gazelle
ES Gacela del Tibet

3000–5750 m

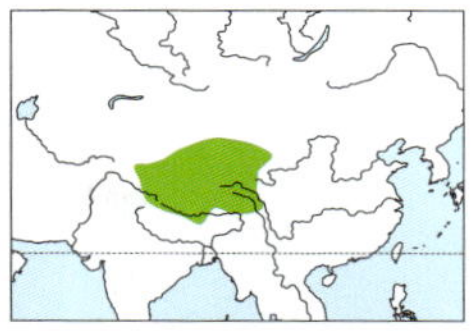

Qinghai–Tibetan Plateau in N IN (Ladakh, Sikkim) & mainly W CN (Xinjiang, Qinghai, Gansu, Tibet, Sichuan)

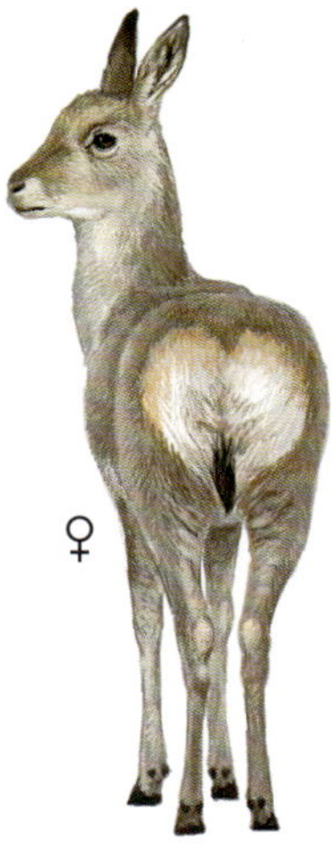

♀

♂

NT
Monotypic
HB 91–105 cm
T 8–10 cm
W 13–16 kg

Przewalski's Gazelle
Procapra przewalskii

FR Gazelle de Przewalski
DE Przewalski-Gazelle
ES Gacela de Przewalski

EN
Monotypic
HB 109–160 cm
T 7–12 cm
W 17–32 kg

below 3000 m

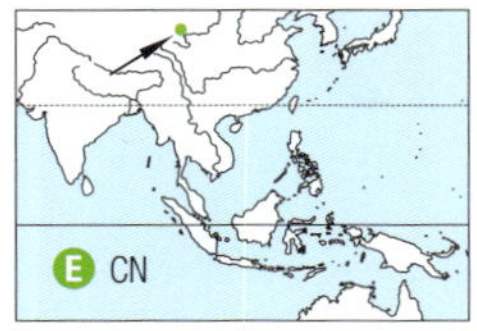

Isolated populations in Qinghai, near L Qinghai & along R Buha; apparently not in Gansu

♀

♂

Mongolian Gazelle
Procapra gutturosa

FR Gazelle de Mongolie
DE Mongolei-Gazelle
ES Gacela de Mongolia

LC
Monotypic
HB 105–148 cm
T 5–12 cm
W 20–39 kg (M)

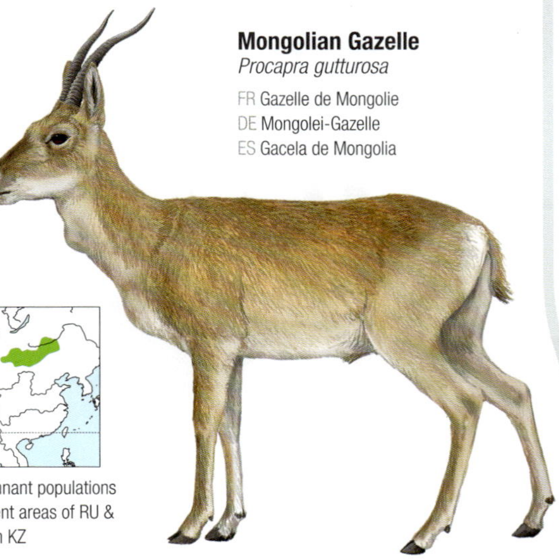

Primarily E MN (remnant populations in S, C & W), adjacent areas of RU & NE CN; extirpated in KZ

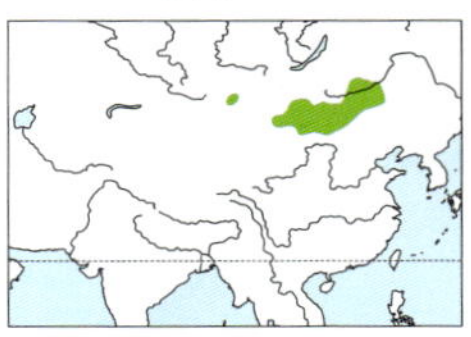

Southern Reedbuck
Redunca arundinum

FR Rédunca des roseaux
DE Großriedbock
ES Redunca meridional

LC
2 sspp
HB 130–160 cm (M), 120–140 cm (F)
T 18–30 cm
W 51–95 kg (M), 39–85 kg (F)

♀

ssp *arundinum*

♂

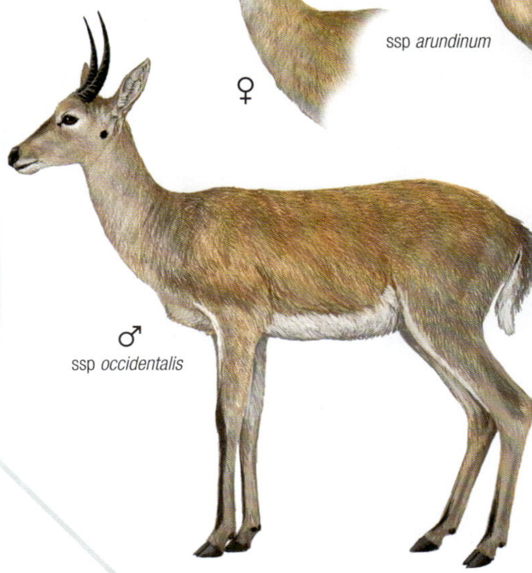

♂
ssp *occidentalis*

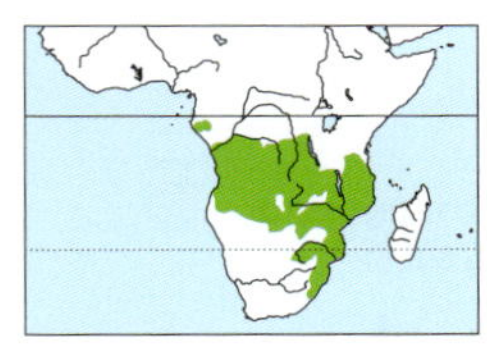

S CD S to N NA & N BW, E to TZ and through MZ & ZW to NE ZA (KwaZulu-Natal) & SZ; isolated in S GA

Genus *REDUNCA*

(3 species)

Bohor Reedbuck
Redunca redunca

FR Rédunca commun
DE Senegal-Riedbock
ES Redunca occidental

ssp *redunca*

LC
4 sspp
HB 125–145 cm (M), 115–130 cm (F)
T 20–25 cm (M), 15–23 cm (F)
W 45–65 kg (M), 35–55 kg (F)

ssp *nigeriensis*

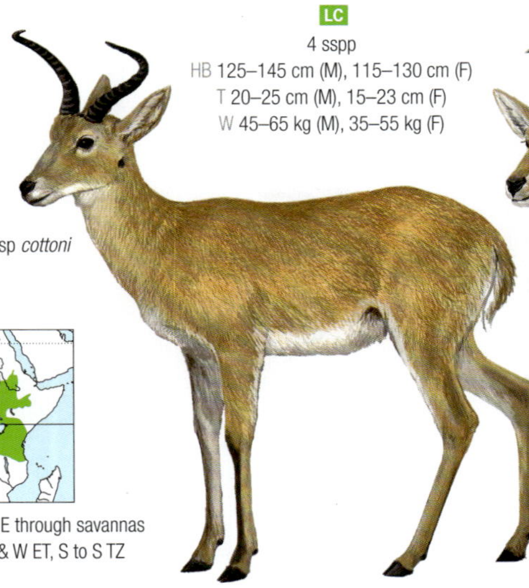

ssp *cottoni*

ssp *bohor*

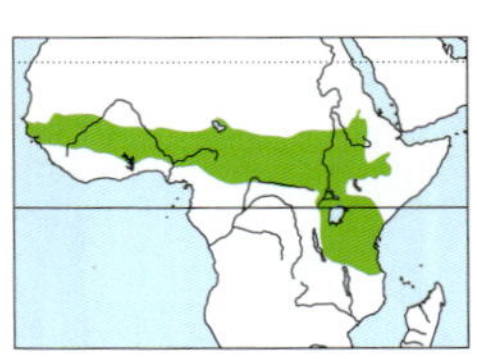

From SN, GM & SW MR E through savannas of W & C Africa to E SD & W ET, S to S TZ (R Rovuma)

Mountain Reedbuck
Redunca fulvorufula

FR Rédunca de montagne
DE Südlicher Bergriedbock
ES Redunca de montaña

EN
3 sspp
HB 97–136 cm
T 12–20 cm
W 19–38 kg

up to 5000 m

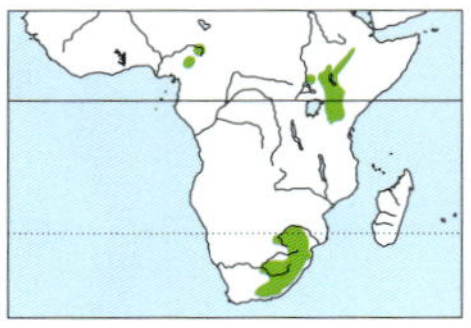

Disjunct in W (E NG & W CM in Adamawa Massif), E (S ET, SE SD, E UG, W KE, N TZ) & S (SE BW, E ZA, SZ, LS) Africa

Southern Lechwe
Kobus leche

FR Cobe lechwe
DE Litschi-Moorantilope
ES Kob rojo

Scattered in S CD (Upemba Swamps), C & SE AO, NE NA (Caprivi Strip), N BW (Okabango Basin), W, C & S ZM

NT
5 sspp
HB 150–180 cm (M), 134–162 cm (F)
T 30–45 cm
W 45–135 kg (M), 40–97 kg (F)

Genus *KOBUS*

(5 species)

Nile Lechwe
Kobus megaceros

FR Cobe du Nil
DE Weißnacken-Moorantilope
ES Kob del Nilo

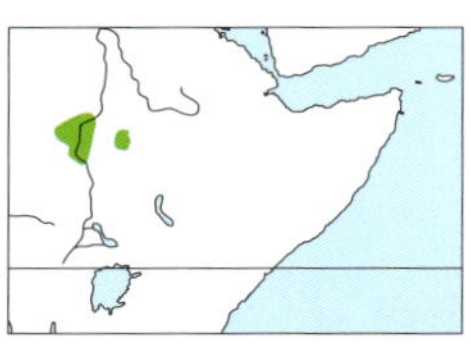

Swamps and seasonally flooded grasslands in SS & W ET

EN
Monotypic
HB 130–180 cm
T 45–50 cm
W 90–120 kg (M), 60–90 kg (F)

Kob
Kobus kob

FR Cobe de Buffon
DE Senegal-Grasantilope
ES Kob de Buffon

SN E to SS & extreme SW ET, extending S to NE CD & UG

LC
4 sspp
HB 160–180 cm
T 10–15 cm
W 40–121 kg

Puku
Kobus vardonii

FR Puku
DE Puku
ES Puku

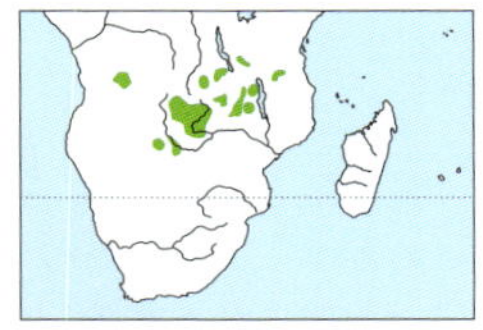

Patchily in AO, E NA, NE BW, ZM, SE CD, N & W MW and SW TZ

NT
Monotypic
HB 126–156 cm
T 28–32 cm
W 67–91 kg (M), 48–78 kg (F)

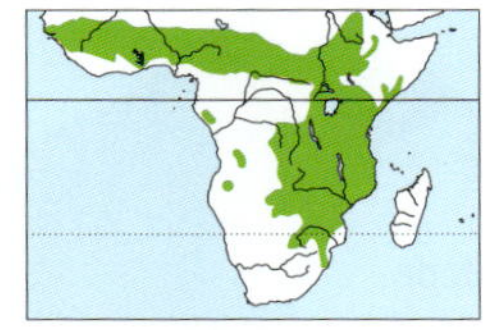

From SN & GM E to N ET, ER & S SO, S through E Africa to E BW & NE ZA; isolated in S GA, W CG & AO

LC
2 sspp
HB 175–235 cm
T 33–40 cm
W 198–275 kg (M), 160–214 kg (F)

Waterbuck
Kobus ellipsiprymnus

FR Cobe à croissant
DE Ellipsen-Wasserbock
ES Kob acuático oriental

ssp *defassa*

ssp *ellipsiprymnus*

Genus *HIPPOTRAGUS*

(2 species)

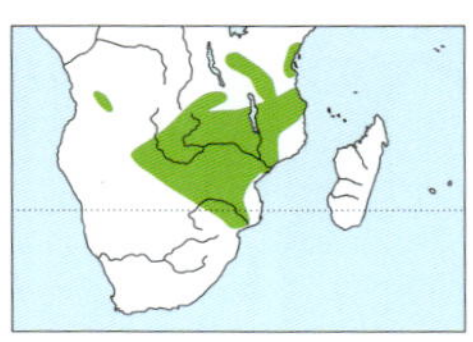

SE KE (Shimba Hills), TZ (Sadani, Songea & Selous Game R), SE CD, ZM, MW, MZ, isolated in C AO

Sable Antelope
Hippotragus niger

FR Hippotrague noir
DE Rappenantilope
ES Hipótrago sable meridional

ssp *anselli*

ssp *roosevelti*

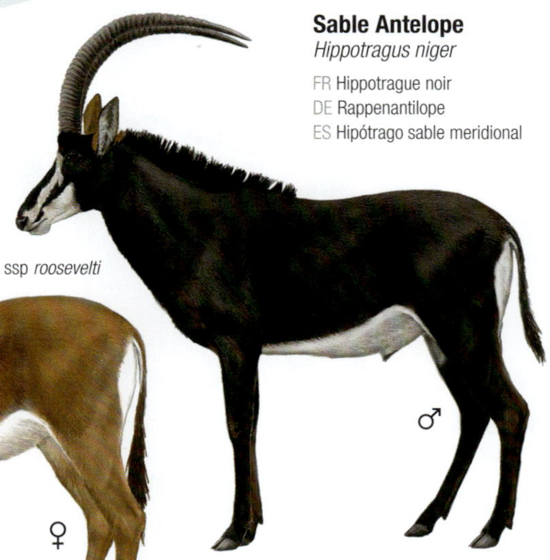

ssp *variani*

ssp *niger*

LC
4 sspp
HB 175–194 cm
T 44–53 cm
W 160–230 kg

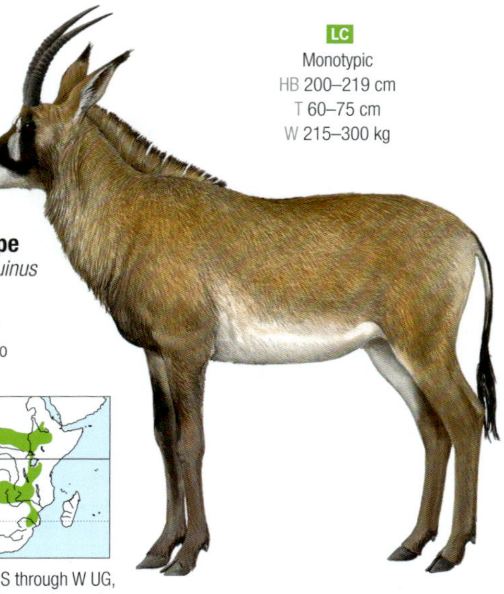

LC
Monotypic
HB 200–219 cm
T 60–75 cm
W 215–300 kg

Roan Antelope
Hippotragus equinus
FR Rouanne
DE Pferdeantilope
ES Hipótrago ruano

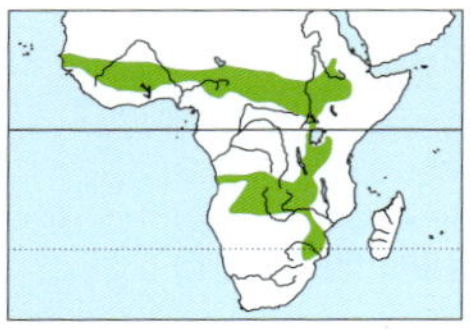

SN & GW to W ET, S through W UG, SW KE, WC TZ (not coastal areas) to AO, NE NA, N BW & NE ZA (Kruger NP)

Genus *PELEA*

(1 species)

Rhebok
Pelea capreolus
FR Péléa
DE Rehantilope
ES Pelea

1400–3300 m

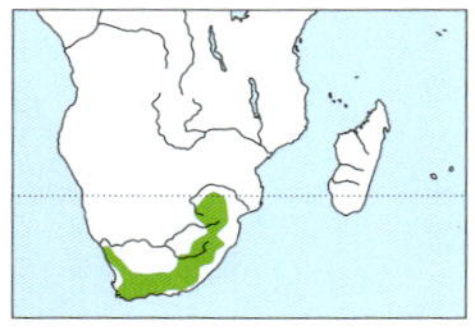

Montane grasslands interspersed with rocky areas in ZA, SZ, LS

NT
Monotypic
HB 105–125 cm
T 10–20 cm
W 18–30 kg

Genus *ADDAX*

(1 species)

Addax
Addax nasomaculatus
FR Addax
DE Addax
ES Adax

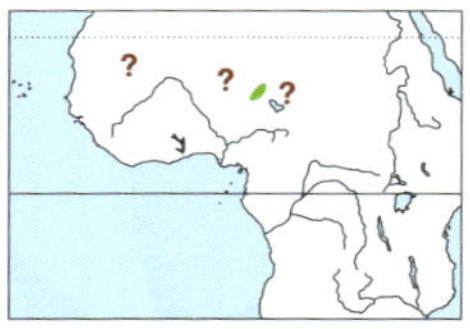

CR
Monotypic
HB 120–130 cm
T 25–35 cm
W 60–125 kg

NE (viable population in Termit-Tin Toumma, recorded in E Aïr Mts), W TD (Djourab), C MR, S DZ, LY, ML–MR border

Genus *ORYX*

(4 species)

Beisa Oryx
Oryx beisa
FR Oryx beisa
DE Beisa-Oryx
ES Oryx beisa

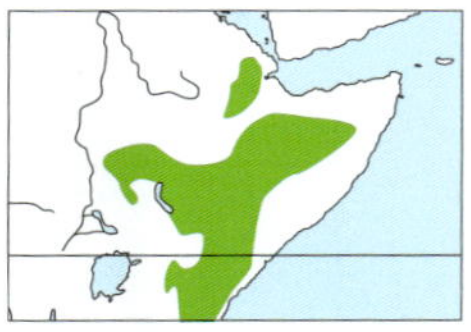

Currently mainly limited to ET, N & E KE, NE TZ and small population in SS (Boma NP)

EN
3 sspp
HB 153–170 cm
T 45–50 cm
W 116–209 kg

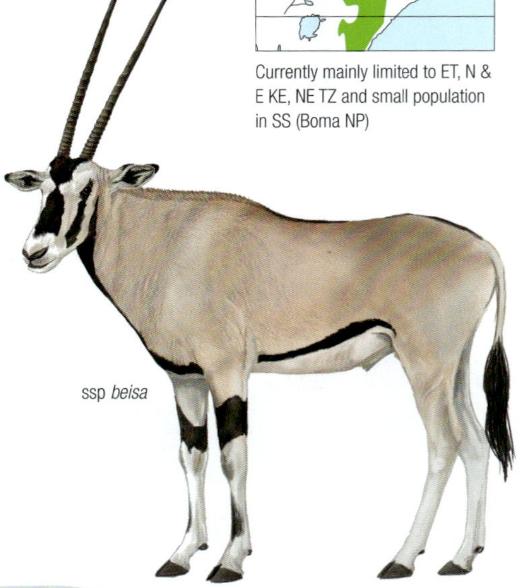

ssp *beisa*

ssp *gallarum*

ssp *callotis*

Gemsbok
Oryx gazella
FR Gemsbok
DE Südliche Oryx
ES Oryx de El Cabo

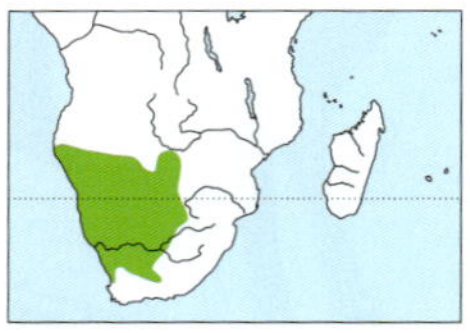

Arid zone in NA, most of BW, ZA (Northern Cape, marginally further E & S), ZW, SW AO (where probably extinct)

LC
Monotypic
HB 180–195 cm
T 48–52 cm
W 190–240 kg

Scimitar-horned Oryx
Oryx dammah
FR Oryx de Libye
DE Säbelantilope
ES Oryx blanco

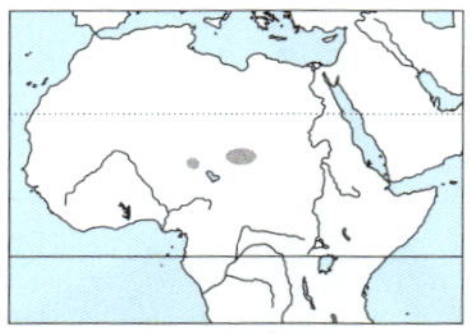

Extinct in the wild; originally occurred on N & S edges of Sahara Desert

EW
Monotypic
HB 159–175 cm
T 37–44 cm
W 180–200 kg (M)

Arabian Oryx
Oryx leucoryx
FR Oryx d'Arabie
DE Arabische Oryx
ES Oryx de Arabia

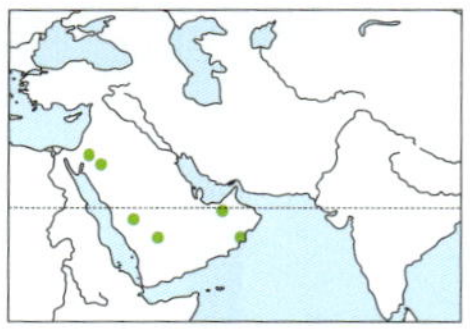

Extinct in the wild by 1972; free-ranging Arabian Oryx have been reestablished in IL, JO, SW SA, AE & WC OM

VU
Monotypic
HB 153–235 cm
T 45–60 cm
W 48–105 kg

Genus *ALCELAPHUS*

(1 species)

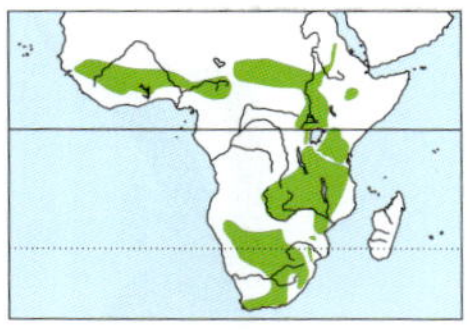

From SN E to ET, N to E SD & ER (where perhaps extinct), S through E Africa to E AO, N & E NA and ZA

Hartebeest

Alcelaphus buselaphus

FR Bubale
DE Nordafrikanische Kuhantilope
ES Alcélafo septentrional

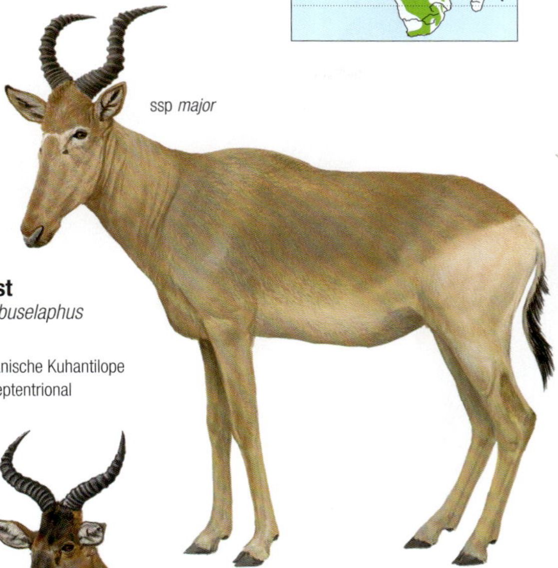

ssp *major*

ssp *lelwel*

ssp *tora*

ssp *swaynei*

ssp *cokii*

ssp *lichtensteinii*

ssp *caama*

LC
7 extant sspp
HB 160–250 cm
T 40–57 cm
W 105–218 kg

Genus *BEATRAGUS*

(1 species)

Herola

Beatragus hunteri

FR Hirola
DE Hirola
ES Hirola

CR
Monotypic
HB 164–200 cm
T 36–45 cm
W 73–118 kg

E KE (between R Tana and R Jubba, on N bank of R Tana, N to Garissa), SW SO (current status unknown)

Genus *DAMALISCUS*

(2 species)

Bontebok

Damaliscus pygargus

FR Bontebok
DE Buntbock
ES Bontebok

LC
2 sspp
HB 140–160 cm
T 30–45 cm
W 56–86 kg

ssp *phillipsi*

ssp *pygargus*

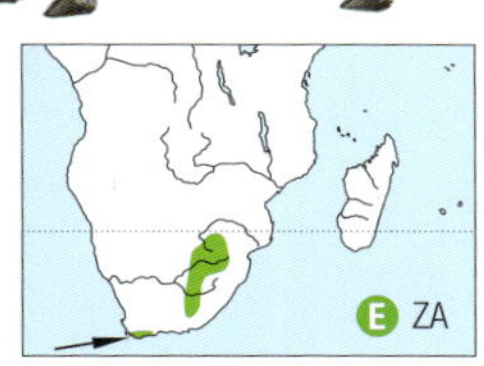

Eastern Cape N of Karoo & KwaZulu-Natal N to BW border, disjunct in S Western Cape; introduced to ZW, NA & E BW

Tsessebe
Damaliscus lunatus

FR Sassaby
DE Sassaby
ES Topi

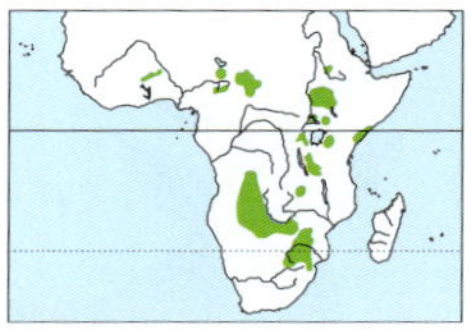

Very fragmented from BF & GH to E TD, N CF, E SD, SW ET & coastal SO S to S CD, E AO, NE NA, N BW & NE ZA

ssp *korrigum*

ssp *superstes*

ssp *lunatus*

ssp *eurus*

ssp *tiang*

ssp *ugandae*

LC
8 extant sspp
HB 185–223 cm
T c. 45 cm
W 90–168 kg

ssp *jimela*

ssp *topi*

Genus *CONNOCHAETES*

(2 species)

Black Wildebeest
Connochaetes gnou

FR Gnou à queue blanche
DE Weißschwanzgnu
ES Ñú negro

Temperate grasslands in ZA, LS & SZ

LC
Monotypic
HB 170–220 cm
T 80–100 cm
W 110–161 kg

Blue Wildebeest
Connochaetes taurinus

FR Gnou bleu
DE Streifengnu
ES Ñú azul

LC
4 sspp
HB 176–203 cm (M), 172–178 cm (F)
T 56–73 cm (M), 55–63 cm (F)
W 171–271 kg (M), 141–242 kg (F)

♂

ssp *taurinus*

♀

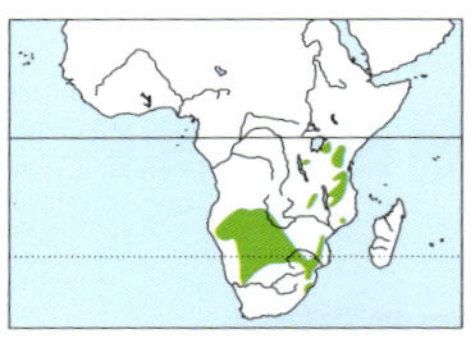

S KE, N & E TZ, ZM, S AO, NA, BW, ZW, MZ, N & E ZA, SZ; introduced to regions outside former range

♂
ssp *johnstoni* variants

♂
ssp *albojubatus*

♂
ssp *mearnsi*

Genus *PANTHOLOPS*

(1 species)

Chiru
Pantholops hodgsonii

FR Chiru
DE Tschiru
ES Antílope tibetano

NT
Monotypic
HB 100–140 cm
T 13–14 cm
W 24–42 kg

3250–5500 m

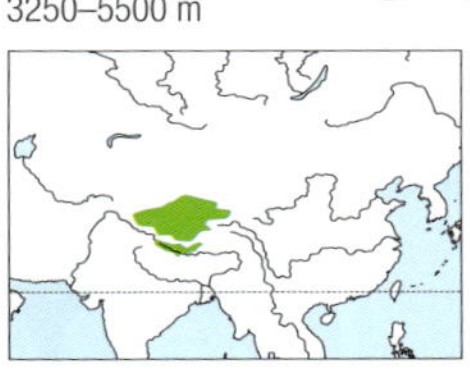

Tibetan Plateau in CN (Xinjiang, Qinghai, Tibet) & IN (small numbers mainly during summer in Ladakh)

Genus *OREAMNOS*

(1 species)

Rocky Mountain Goat
Oreamnos americanus

FR Chèvre des Rocheuses
DE Schneeziege
ES Cabra blanca de las Rocosas

LC
Monotypic
HB 140–180 cm
T 10–20 cm
W 60–115 kg

0–2700 m

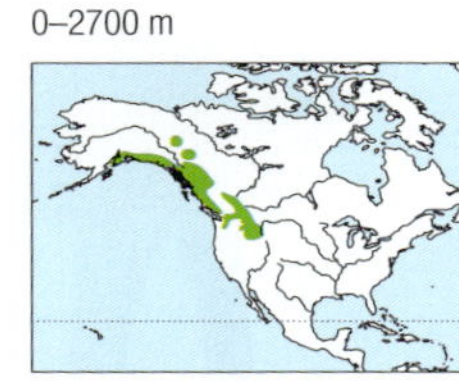

W CA (Yukon S to Alberta), W US (SE Alaska, Washington, W Montana, Idaho); introduced to other US states

Genus *BUDORCAS*

(1 species)

ssp *taxicolor*

ssp *bedfordi*

ssp *whitei*

ssp *tibetana*

Takin
Budorcas taxicolor
FR Takin
DE Mishmi-Takin
ES Takín de Mishmi

VU
4 sspp
HB 170–220 cm
T 7–22 cm
W 150–350 kg

1200–3500 m

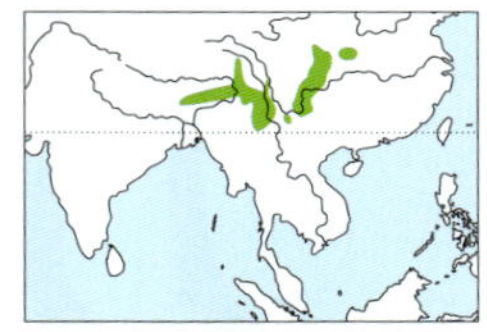

NE IN (Sikkim, Arunachal Pradesh), N BT, N MM, SC CN (SE Tibet, NW Yunnan, W Sichuan, S Gansu, S Shaanxi)

Genus *AMMOTRAGUS*

(1 species)

Aoudad
Ammotragus lervia
FR Aoudad de Barbarie
DE Mähnenschaf
ES Arrui

VU
6 sspp
HB 112–128 cm
T 17.5–20 cm
W 41–86 kg

0–4000 m

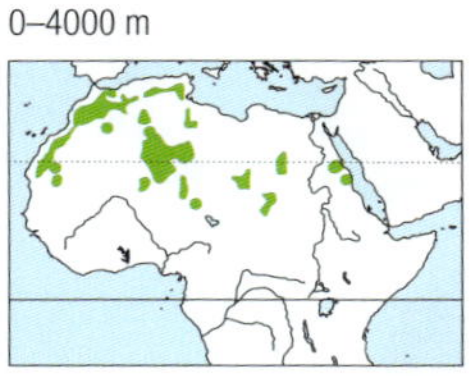

Scattered in MA, EH, DZ, TN, LY, EG, NW MR, NE ML, NE, TD, SD; introduced to ES, Canary Is, US & NE MX

Genus *ARABITRAGUS*

(1 species)

EN
Monotypic
HB 93–95 cm
T 8.4–10 cm
W 17–45 kg

Arabian Tahr
Arabitragus jayakari
FR Aoudad d'Oman
DE Arabien-Tahr
ES Tahr de Omán

900–1800 m

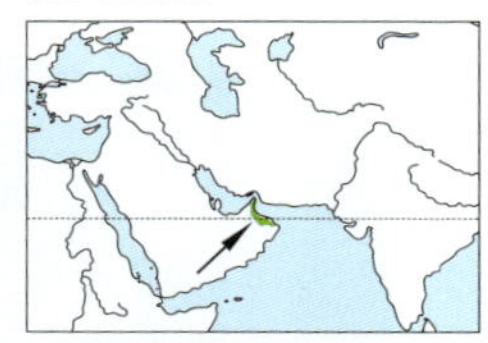

Al Hajar Mts in N OM & extreme NE AE

Genus *HEMITRAGUS*

(1 species)

Himalayan Tahr
Hemitragus jemlahicus
FR Jharal
DE Himalaja-Tahr
ES Tahr del Himalaya

♂

NT
Monotypic
HB 130–170 cm (M), c. 90 cm (F)
T 9–12 cm
W 90–124 kg (M), 55–72 kg (F)

♀

1550–5300 m

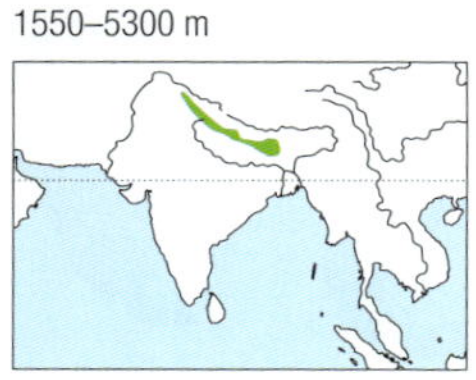

Himalayas in N IN (Jammu and Kashmir, Sikkim), NP, SW CN (S Tibet); introduced to NZ (South Island), ZA, & W US

Genus *PSEUDOIS*

(1 species)

LC 2 sspp
HB 120–140 cm T 13–20 cm
W 25–75 kg

♀

ssp *nayaur*

♂

Blue Sheep
Pseudois nayaur
FR Bharal
DE Blauschaf
ES Cabra azul del Himalaya

♂
ssp *schaeferi*

1200–5500 m

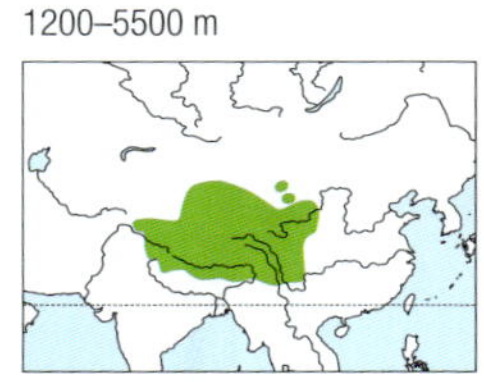

N PK, N IN, NP, BT, N MM, CN (S Xinjiang, Qinghai, Gansu, S Inner Mongolia, Ningxia, Shaanxi, Sichuan, Tibet, N Yunnan)

Genus *CAPRA*

(9 species)

Wild Goat
Capra aegagrus

FR Bouc sauvage
DE Bezoarziege
ES Cabra bezoar

♀

♂

NT
Monotypic
HB 129–152 cm (M)
T 11–13 cm (M)
W 45–90 kg (M), 25–55 kg (F)

0–4000 m

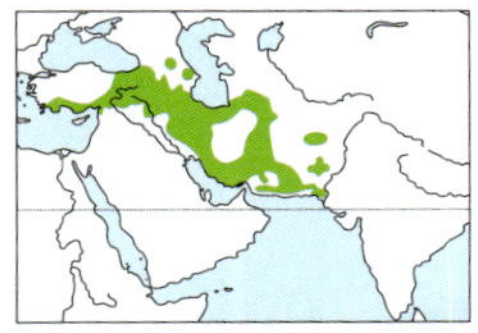

TR, S RU (Daghestan, Chechnya), NE GE, AM, AZ, N IQ, IR, TM, AF & SW PK; introduced to US (SC New Mexico)

Markhor
Capra falconeri

FR Markhor
DE Markhor
ES Marjor

♂
ssp *falconeri* variants

NT
3 sspp
HB 155–170 cm (M)
T 10–20 cm
W 80–108 kg (M), 32–50 kg (F)

600–3600 m

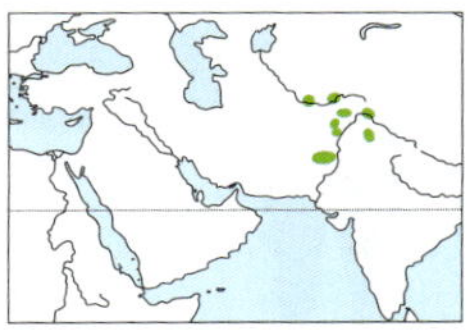

S UZ, SE TM, S TJ, N & NE AF, N & C PK, N IN (SW Jammu and Kashmir)

♀
ssp *megaceros*

♂
ssp *megaceros*

♂
ssp *heptneri*

Nubian Ibex
Capra nubiana

FR Bouquetin de Nubie
DE Nubien-Steinbock
ES Íbice de Nubia

VU
Monotypic
HB 90–160 cm
T 6–17 cm
W 25–85 kg

up to 2600 m

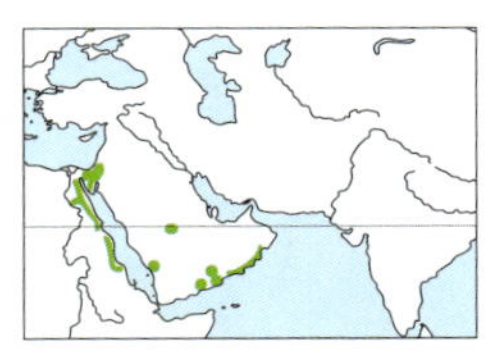

EG E of R Nile, NE SD, IL, W JO, SA, SE YE, SW OM; unconfirmed in ER

Iberian Ibex
Capra pyrenaica

FR Bouquetin d'Espagne
DE Iberischer Steinbock
ES Cabra montés

♂
ssp *hispanica*
summer

♀
ssp *victoriae*

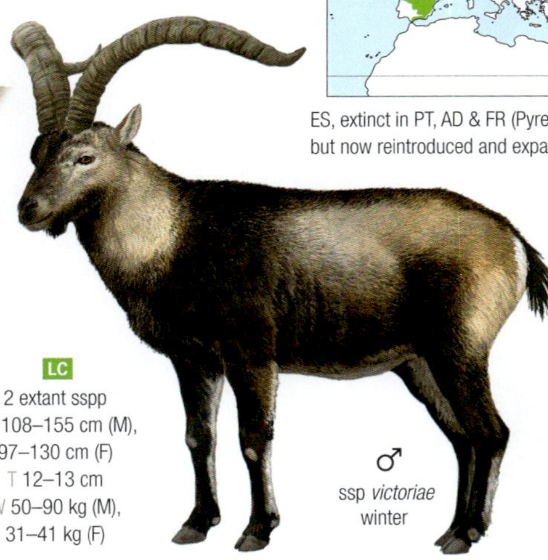

♂
ssp *victoriae*
winter

LC
2 extant sspp
HB 108–155 cm (M), 97–130 cm (F)
T 12–13 cm
W 50–90 kg (M), 31–41 kg (F)

0–3500 m

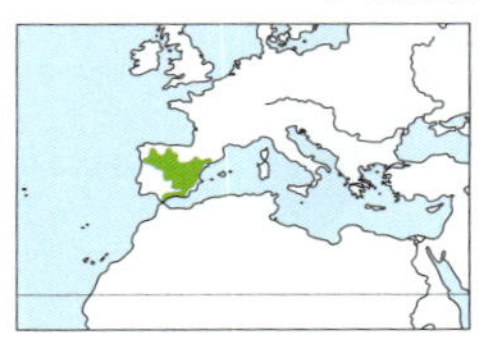

ES, extinct in PT, AD & FR (Pyrenees) but now reintroduced and expanding

Walia Ibex
Capra walie

FR Bouquetin d'Abyssinie
DE Äthiopien-Steinbock
ES Íbice de Etiopía

VU
Monotypic
HB 175–196 cm (M)
T 22–25 cm (M)
W 100–125 kg (M)

2500–4200 m

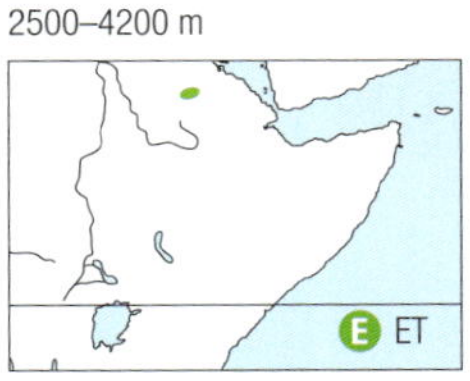

Principally Simien Mts NP along 25 km of N escarpment

Alpine Ibex
Capra ibex

FR Bouquetin des Alpes
DE Alpen-Steinbock
ES Íbice de los Alpes

1600–3200 m

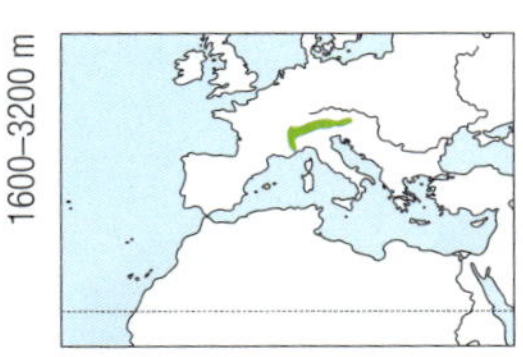

Alps of CH, S DE, LI, AT, N IT & SE FR; introduced to NW SI & BG

♀

♂

LC
Monotypic
HB 115–135 cm (M), 55–100 cm (F)
T 21–29 cm (M), 15–23 cm (F)
W 70–120 kg (M), 40–50 kg (F)

Siberian Ibex
Capra sibirica

FR Bouquetin de Sibérie
DE Sibirien-Steinbock
ES Íbice de Siberia

ssp *sakeen*

ssp *sibirica*

NT
4 sspp
HB 85–132 cm
T 13–22 cm
W 30–130 kg

700–6700 m

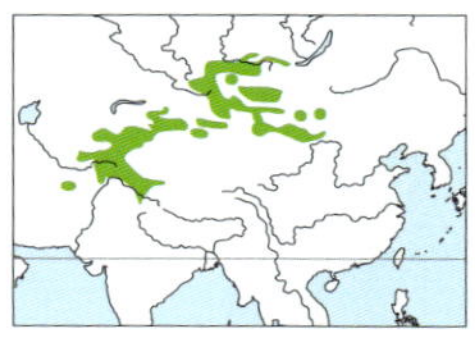

C & N AF, N PK, NW IN, TJ, KG, NE UZ (W Tian Shan), S & E KZ, W & N CN, W & C MN, RU (S Siberia)

Kuban Tur
Capra caucasica

FR Bouquetin du Caucase
DE Kuban-Tur
ES Tur del Cáucaso occidental

EN Monotypic
HB 159–196 cm (M), 136–164 cm (F)
T 12–17 cm (M), 8–14 cm (F)
W 123–155 kg (M), 58–71 kg (F)

♀

♂

1000–3300 m

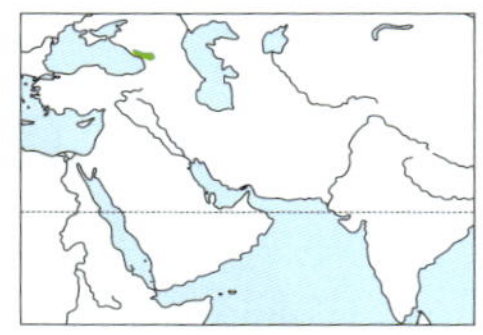

W Great Caucasus Mts (GE, RU)

Daghestan Tur
Capra cylindricornis

FR Bouquetin du Daghestan
DE Dagestan-Tur
ES Tur del Cáucaso oriental

NT
Monotypic
HB 135–192 cm
T 11–15 cm
W 48–143 kg

1000–4000 m

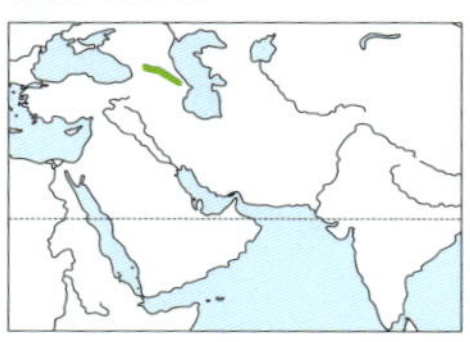

E Caucasus stretching c. 500 km in GE, AZ & RU, W to R Cherek-Bezengi, E to Babadagh Mts

Genus *NILGIRITRAGUS*

(1 species)

Nilgiri Tahr
Nilgiritragus hylocrius

FR Mouflon des Nilgiri
DE Nilgiri-Tahr
ES Tahr del Nilgiri

1200–2600 m

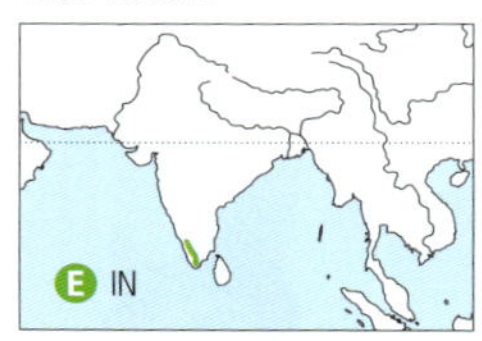

Western Ghats along border of Kerala & Tamil Nadu

EN
Monotypic
W 50–100 kg

Genus *OVIS*

(6 species)

Mouflon
Ovis gmelini

FR Mouflon d'Arménie
DE Armenien-Wildschaf
ES Muflón de Anatolia

♀

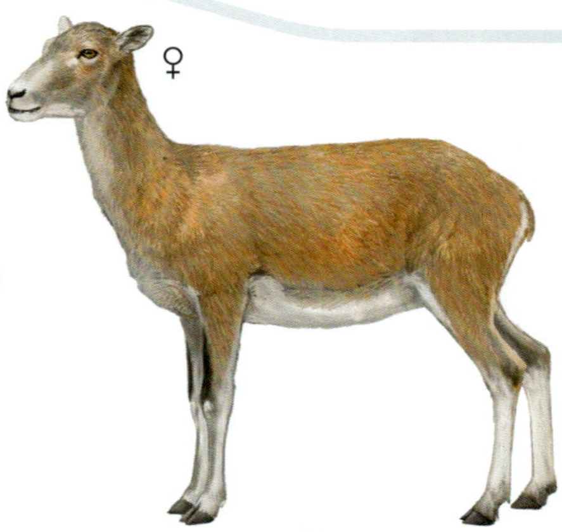

NT
4 sspp
HB 124–140 cm (M), 119–128 cm (F)
T 11–13 cm (M), 11–12 cm (F)
W 49–67 kg (M), 29–45 kg (F)

♂
ssp *laristanica*

ssp *gmelini*

♂

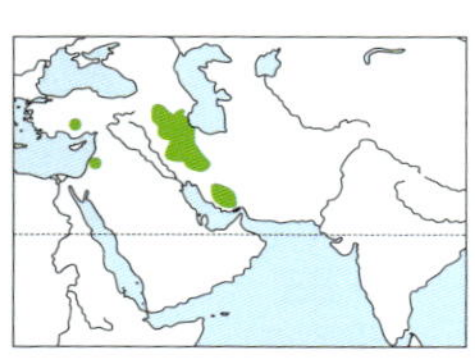

SC & E TR, N IQ, S AM, AZ (Nakhchivan), N IL (Mt Carmel), W & S IR; introduced to Kabudan I in L Urumiyeh (NW IR)

♂
ssp *isphahanica*

Urial
Ovis vignei

FR Mouflon urial
DE Ladakh-Wildschaf
ES Urial de Ladakh

♂ ssp *punjabiensis*

♀

ssp *vignei* ♂

♂ ssp *arabica*

♂ ssp *bochariensis*

♂ ssp *cycloceros*

♀

ssp *arkal* ♂

0–4500 m

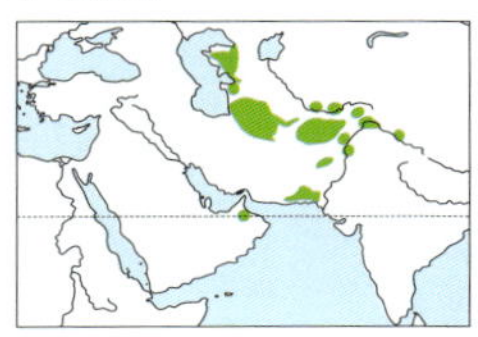

N & NE IR, TM, W KZ, S UZ, S TJ, N & NE AF, PK, NW IN (Ladakh), also OM (where possibly introduced)

VU
6 sspp
HB 109–160 cm (M), 120–134 cm (F)
T 10–13 cm
W 36–66 kg (M), 25–42 kg (F)

Snow Sheep
Ovis nivicola

FR Mouflon des neiges
DE Schneeschaf
ES Carnero nival

LC
4 sspp
HB 126–188 cm
T 7.5–13 cm
W 33–105 kg

ssp *nivicola*

ssp *lydekkeri*
white forehead morph

0–1800 m

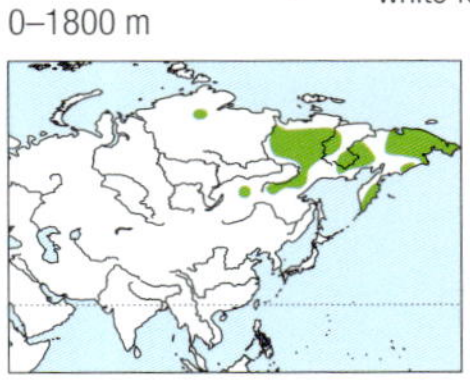

Mts of E Siberia E of R Lena to Chukotski & Kamchatka pen, isolated in Putoran Mts S of Taymyr Pen

Bighorn Sheep
Ovis canadensis

FR Mouflon d'Amérique
DE Dickhornschaf
ES Carnero de las Rocosas

up to 3500 m

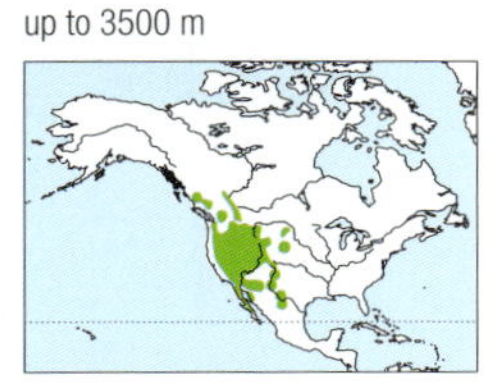

SW CA (S British Columbia, SW Alberta), W US, NW MX (Baja California Pen, NW Sonora, reestablished in Coahuila & Chihuahua)

ssp *canadensis* ♀

♂ ssp *mexicana*

♂

LC
5 sspp
HB 109–169 cm (M), 96–126 cm (F)
T 10.2–15.2 cm (M), 10.2–12.7 cm (F)
W ≤ 145 kg (M), ≤ 104 kg (F)

Argali

Ovis ammon

FR Mouflon argali
DE Altai-Wildschaf
ES Argali del Altai

NT
9 sspp
HB 133–199 cm
T 5.5–18 cm
W 47–175 kg

up to 5700 m

E KZ, RU (Altai Mts) & MN S through W & C CN to E KG, E UZ, TJ, NE AF, N PK, N IN (Ladakh, Sikkim) & N NP; extinct in E CN

Thinhorn Sheep

Ovis dalli

FR Mouflon de Dall
DE Dall-Schaf
ES Carnero de Dall

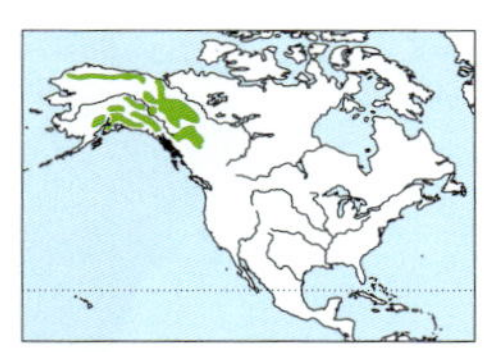

US (Alaska), CA (Yukon, extreme W Northwest Territories, N British Columbia)

LC
2 sspp
HB 130–178 cm (M), 132–162 cm (F)
T 8–13 cm (M), 8–10 cm (F)
W c. 82 kg (M), c. 57 kg (F)

Southern Chamois
Rupicapra pyrenaica

FR Isard
DE Pyrenäen-Gämse
ES Rebeco pirenaico

400–2800 m

Disjunt in 3 areas: Cantabrian Mts (NW ES), Pyrenees (S FR, NE ES, AD), Abruzzi Apennines (C IT)

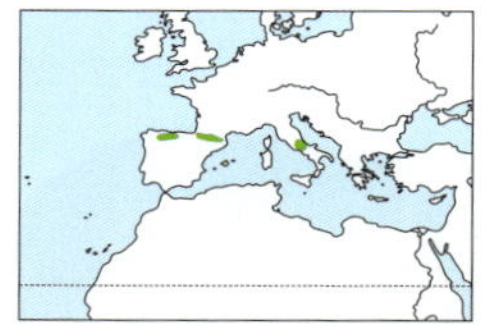

LC
3 sspp
HB 90–130 cm
T 3–4 cm
W 20–35 kg

summer

winter

ssp *pyrenaica*

ssp *parva*

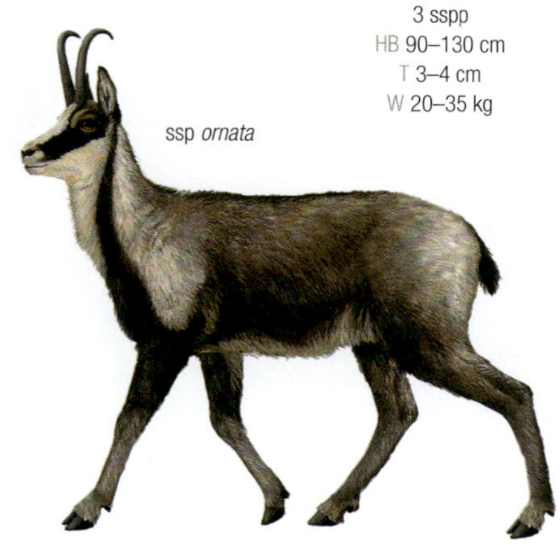

ssp *ornata*

Genus *RUPICAPRA*

(2 species)

Northern Chamois
Rupicapra rupicapra

FR Chamois
DE Alpen-Gämse
ES Rebeco alpino

120–3000 m

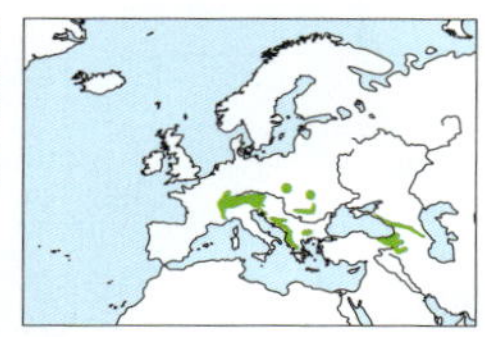

From E FR & S DE through Alps, Tatra, Balkans & Carpathian Mts to NE TR & Caucasus; introduced to CZ, SK, BG & NZ

LC
6 sspp
HB 96–135 cm
T 3–10 cm
W 25–60 kg

summer

ssp *rupicapra*

winter

ssp *carpatica*

ssp *asiatica*

Genus *NAEMORHEDUS*

(6 species)

Himalayan Goral
Naemorhedus goral

FR Goral à queue courte
DE Osthimalaja-Goral
ES Goral marrón del Himalaya

350–4000 m

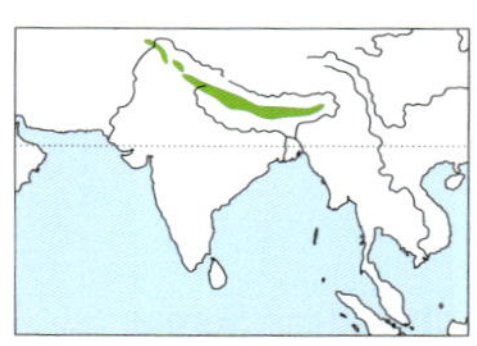

Across Himalayas in PK (Kashmir), N IN, NP, BT & marginally extreme S Tibet (CN)

ssp *goral*

NT
2 sspp
HB 82–120 cm
T 7.5–20 cm
W 35–42 kg

ssp *bedfordi*

Chinese Goral
Naemorhedus griseus

FR Goral gris
DE Westchina-Goral
ES Goral de China

1000–4000 m

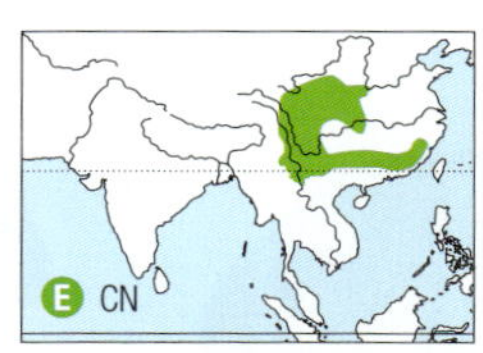

Sichuan NE to Shanxi, Hubei & Henan, E to Fujian, S to E Tibet & Yunnan

NE
Monotypic
HB 80–120 cm
T 8–20 cm
W 20–30 kg

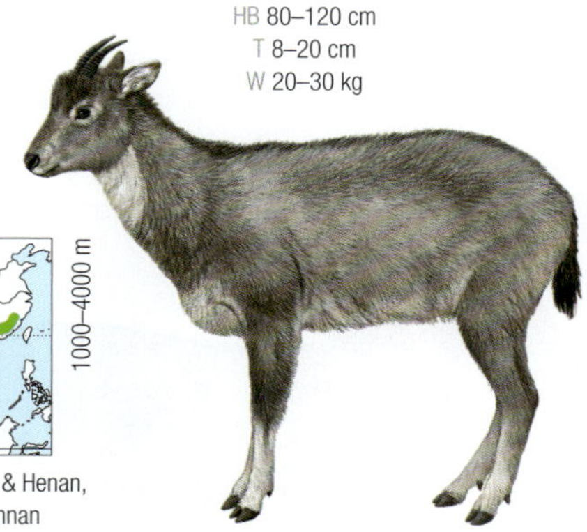

Burmese Goral
Naemorhedus evansi

FR Goral d'Evans
DE Burma-Goral
ES Goral de Birmania

1600–1970 m

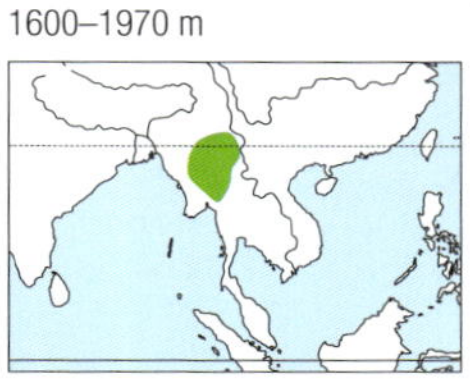

MM, TH (E to Chiang Mai Prov, S to Raheng) N to CN (Yunnan)

NE
Monotypic
HB 50–70 cm
T 12–18 cm
W 20–30 kg

Long-tailed Goral
Naemorhedus caudatus

FR Goral à longue queue
DE Nordchina-Goral
ES Goral colilargo

80–2000 m

NE CN (Inner Mongolia & Hebei E to Heilongjiang), Korean Pen, RU (Primorsky Krai); perhaps W to extreme SE MN

VU
Monotypic
HB 106–135 cm
T 13–18 cm
W 24–47 kg

Red Goral
Naemorhedus baileyi
FR Goral de Bailey
DE Tibet-Goral
ES Goral rojo de Tibet

2000–4500 m

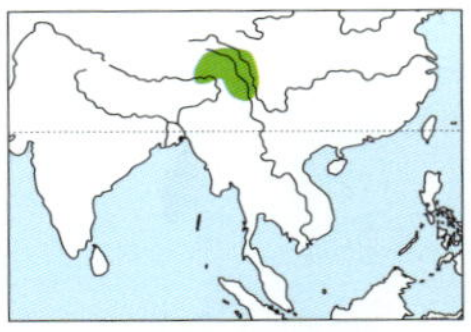

SC CN (SE Tibet, Yunnan), NE IN (Arunachal Pradesh close to CN & MM borders)

VU
Monotypic
HB 90–110 cm
T 7–10 cm
W 20–30 kg

Cranbrook's Goral
Naemorhedus cranbrooki
FR Goral de Cranbrook
DE Cranbrook-Goral
ES Goral rojo de Birmania

2000–4500 m

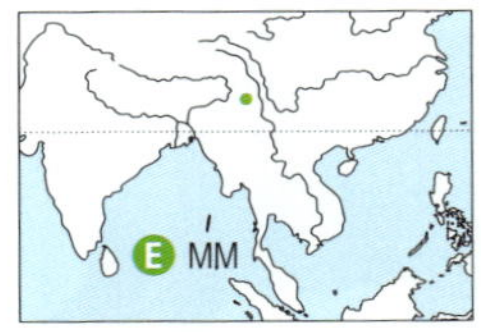

High mountains of N

NE
Monotypic
HB 90–110 cm
T 7–10 cm
W 20–30 kg

Maned Serow
Capricornis sumatraensis
FR Serow de Sumatra
DE Sumatra-Serau
ES Sirao de Sumatra

ssp *sumatraensis*

ssp *thar*

ssp *milneedwardsi*

ssp *maritimus*

0–3500 m

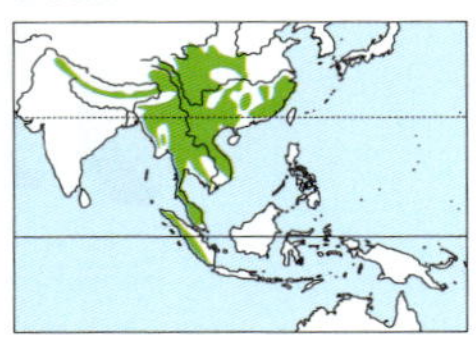

S slopes of Himalayas (N IN, NP, BT, BD) E to C & SE CN, mainland SE Asia, N & W Sumatra (Bukit Barisan Mts)

VU
4 sspp
HB 140–170 cm
T 11–16 cm
W 85–160 kg

Genus *CAPRICORNIS*

(4 species)

Red Serow
Capricornis rubidus
FR Serow rouge
DE Roter Serau
ES Sirao de Birmania

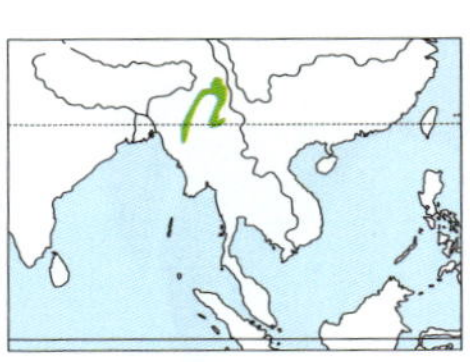

N (perhaps W) MM, adjacent E IN & CN (Yunnan); range uncertain

VU
Monotypic
HB 140–155 cm

Japanese Serow
Capricornis crispus
FR Serow du Japon
DE Japan-Serau
ES Sirao de Japón

0–2700 m

Honshu, Shikoku, Kyushu

LC
Monotypic
T 6–8 cm
W 31–48 kg

Formosan Serow
Capricornis swinhoei
FR Serow de Taïwan
DE Formosa-Serau
ES Sirao de Formosa

200–3870 m

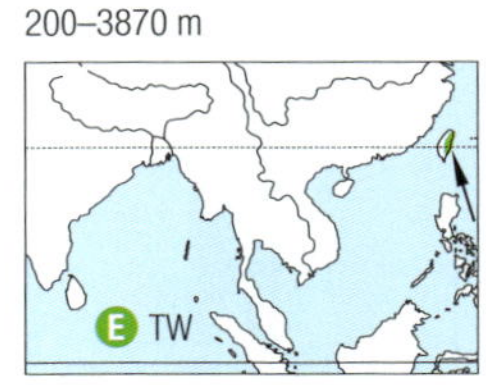

Steep, mountainous terrain in forests

LC
Monotypic
HB 80–114 cm
T 7–12 cm
W 18–30 kg

Genus *OVIBOS*

(1 species)

Muskox
Ovibos moschatus

FR Boeuf-musqué
DE Moschusochse
ES Buey almizclero

Arctic areas of Alaska (US), CA and NE & N GL; introduced to W & SW GL, NO, RU & Alaska outside its range

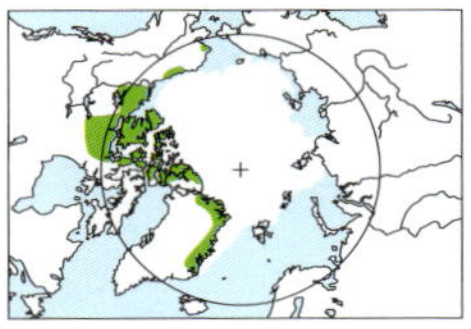

LC
Monotypic
HB 190–230 cm
T 9–10 cm
W 200–410 kg

Genus *SYLVICAPRA*

(1 species)

♂ ssp *splendidula*

LC
13 sspp
HB 70–105 cm (M), 90–115 cm (F)
T 10–20 cm
W 9.7–22 kg (M), 10.3–26 kg (F)

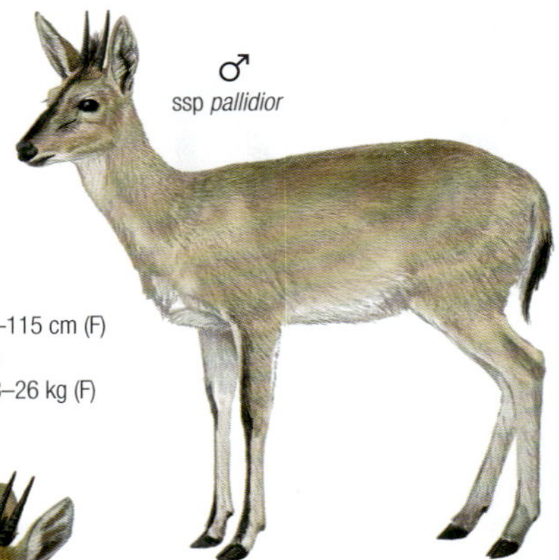

♂ ssp *pallidior*

♂ ssp *altivallis*

Bush Duiker
Sylvicapra grimmia

FR Céphalophe de Grimm
DE Kronenducker
ES Duiker de sabana

♀

ssp *grimmia* ♂

♂ ssp *orbicularis*

0–4300 m

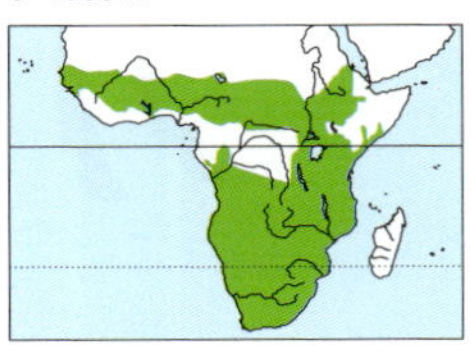

Throughout savanna woodland zones of sub-Saharan Africa, from SN E to W ER & ET, S to ZA

♂ ssp *coronata*

Genus *PHILANTOMBA*

(3 species)

Maxwell's Duiker
Philantomba maxwellii

FR Céphalophe de Maxwell
DE Maxwell-Ducker
ES Duiker de Maxwell

SN & GM to E GH (likely limited in E by R Volta), Yatward & Sherbro Is (SL)

LC
2 sspp
HB 55–75 cm
T 8–16 cm
W 6.5–12.6 kg

Verheyen's Duiker
Philantomba walteri

FR Céphalophe de Verheyen
DE Verheyen-Ducker
ES Duiker de Verheyen

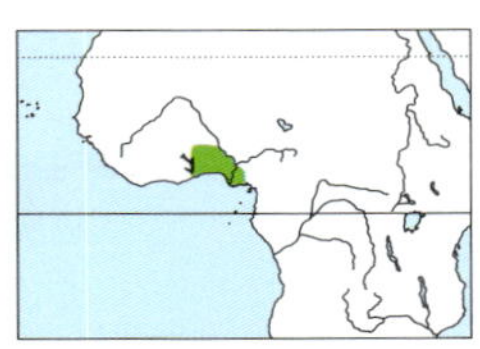

TG, BJ, NG (W of R Cross); perhaps W to R Volta (E GH)

DD
Monotypic
HB 55–75 cm
T ≥ 15 cm
W 6–12 kg

Blue Duiker
Philantomba monticola

FR Céphalophe bleu
DE Kap-Blauducker
ES Duiker azul de El Cabo

ssp *monticola*

ssp *congica*

ssp *melanorhea*

ssp *anchietae*

ssp *bicolor*

ssp *hecki*

LC
10 sspp
HB 55–67 cm
T 8–12 cm
W 3.8–7.3 kg

ssp *aequatorialis*

ssp *lugens*

ssp *defriesi*

ssp *simpsoni*

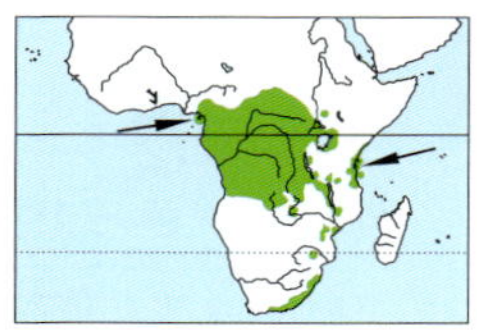

From R Cross (NG) E to SW SS, S to C AO, ZM, E ZW & C MZ, disjunct in S & SE ZA, also Bioko, & Zanzibar Arch (Pemba, Unguja, Mafia)

Banded Duiker
Cephalophus zebra

FR Céphalophe zébré
DE Zebraducker
ES Duiker cebra

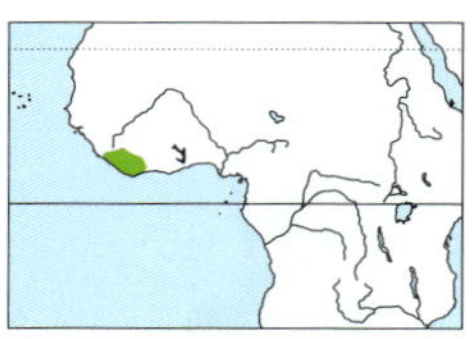

SL (E of R Moa), LR, SE GN, CI (W of R Niouniourou)

VU Monotypic
HB 85–90 cm T c. 15 cm
W 15–23 kg

Bay Duiker
Cephalophus dorsalis

FR Céphalophe à dorsale
DE Schwarzrückenducker
ES Duiker de dorso oscuro occidental

ssp *dorsalis*

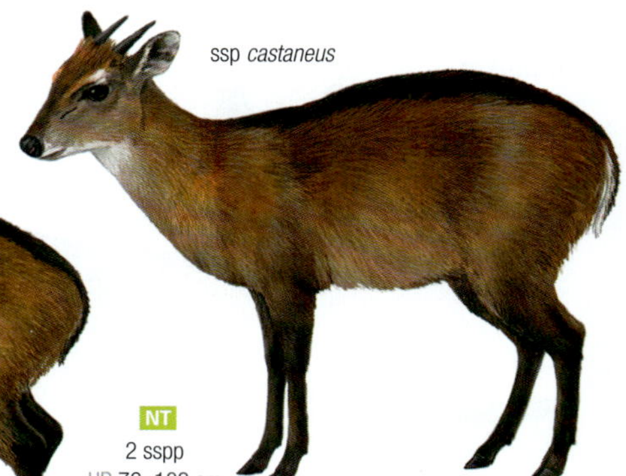
ssp *castaneus*

NT
2 sspp
HB 76–103 cm
T 8.1–12.6 cm
W 17.9–24 kg

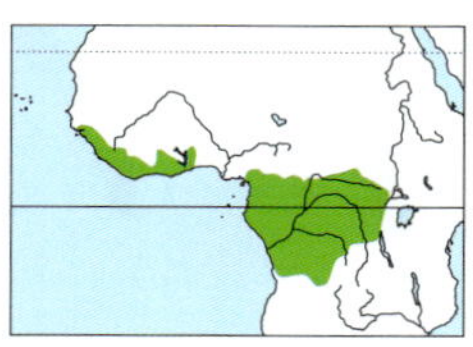

Disjunct in W Africa (from GW E to TG) & C Africa (SE NG, E of R Cross, CM, S CF & E CD S to N AO)

Genus *CEPHALOPHUS*

(16 species)

EN
Monotypic
HB c. 135 cm
T c. 15 cm
W 56–79 kg

Jentink's Duiker
Cephalophus jentinki

FR Céphalophe de Jentink
DE Jentink-Ducker
ES Duiker de Jentink

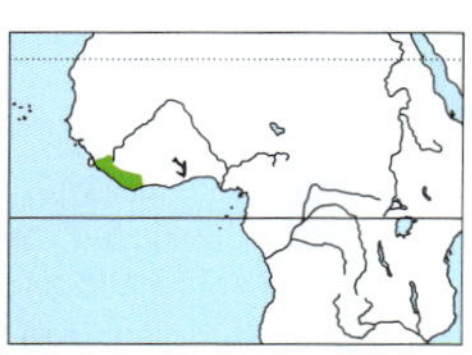

SL, LR, SW CI, bordered in E by R Niouniourou

Yellow-backed Duiker
Cephalophus silvicultor

FR Céphalophe à dos jaune
DE Gelbrückenducker
ES Duiker de dorso amarillo occidental

ssp *curticeps*

ssp *silvicultor*

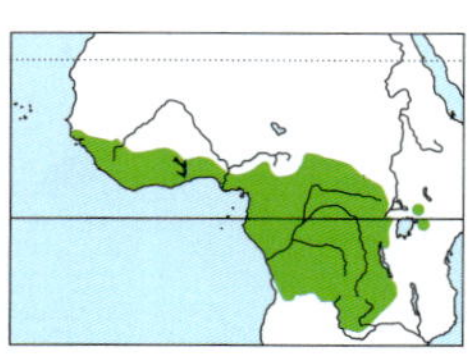

S SN E to S TD, SW SS & W KE, S through Congo Basin to AO & ZM

NT 4 sspp
HB 115–145 cm T 11–18 cm
W 45–80 kg

Abbott's Duiker
Cephalophus spadix

FR Céphalophe d'Abbott
DE Abbot-Ducker
ES Duiker de Abbott

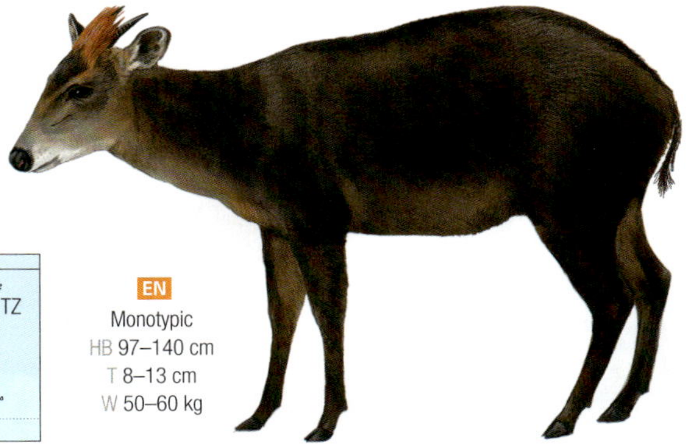

300–4000 m

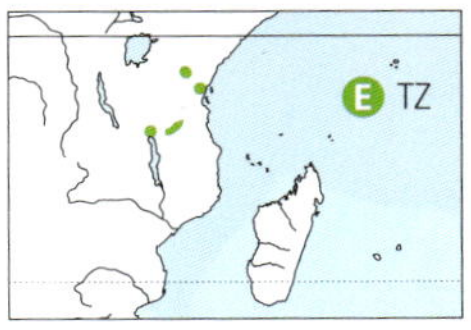

EN
Monotypic
HB 97–140 cm
T 8–13 cm
W 50–60 kg

Mt Kilimanjaro, Mt Meru and Usambara, Uluguru, Udzungwa & Rungwe Mts

Black Duiker
Cephalophus niger

FR Céphalophe noir
DE Schwarzducker
ES Duiker negro

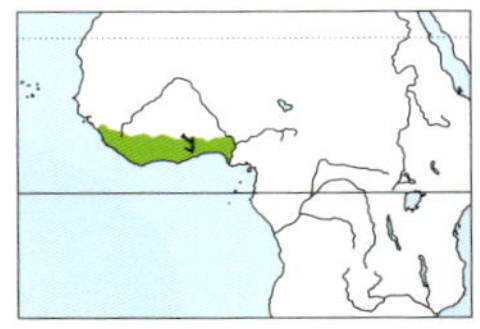

LC
Monotypic
HB 80–100 cm
T 9–14 cm
W 17–26 kg

W Africa from S GN E to W NG (W of R Niger)

Ogilby's Duiker
Cephalophus ogilbyi

FR Céphalophe d'Ogilby
DE Ogilby-Ducker
ES Duiker de Ogilby

LC 2 sspp
HB c. 90 cm T c. 15 cm
W 18–20 kg

ssp *crusalbum*

ssp *ogilbyi*

10–2260 m

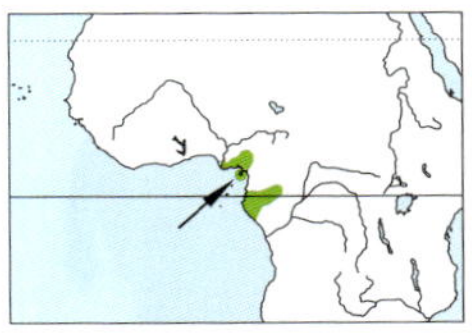

Bioko I (GQ) & mainland Africa from E NG (E of R Niger) to W CM, disjunctly from GA to W CG

Brooke's Duiker
Cephalophus brookei

FR Céphalophe de Brooke
DE Brooke-Ducker
ES Duiker de Brooke

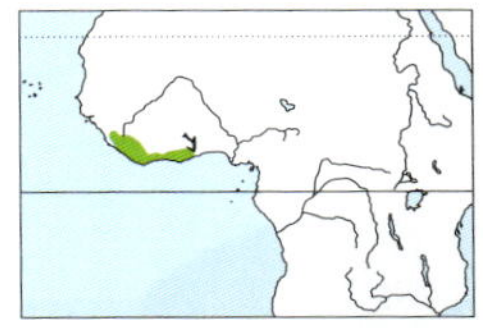

W Africa from SL E to GH (W of R Volta)

NE Monotypic
HB c. 100 cm T c. 12 cm
W 14–20 kg

Peters's Duiker
Cephalophus callipygus

FR Céphalophe de Peters
DE Peters-Ducker
ES Duiker de Peters

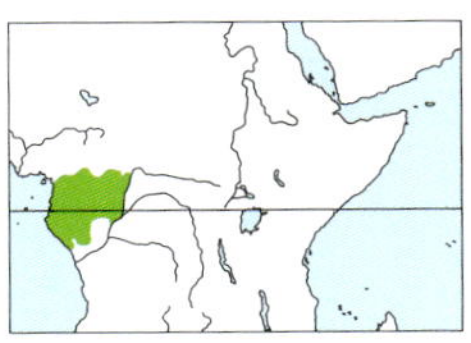

S CM, SW CF, GQ, GA, CG (E to Congo & Ubangi rivers)

LC Monotypic
HB 101–135 cm T 13–16.5 cm
W 17.2–27 kg

ssp *weynsi*

ssp *lestradei*

Weyns's Duiker
Cephalophus weynsi

FR Céphalophe de Weyns
DE Weyns-Ducker
ES Duiker de Weyns

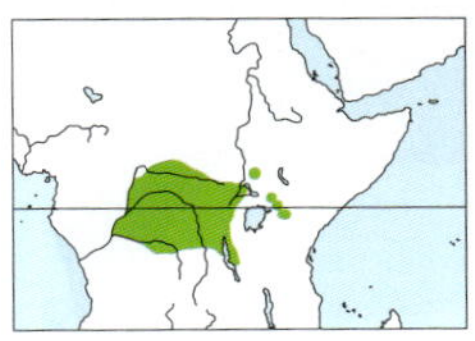

C Africa in S CF, CD, S SS, W KE, UG, W RW & W BI to W TZ

LC
3 sspp
HB 90–100 cm
T 10.5–20 cm
W 14–17.5 kg

ssp *johnstoni*

Aders's Duiker
Cephalophus adersi

FR Céphalophe d'Aders
DE Sansibar-Ducker
ES Duiker de Aders

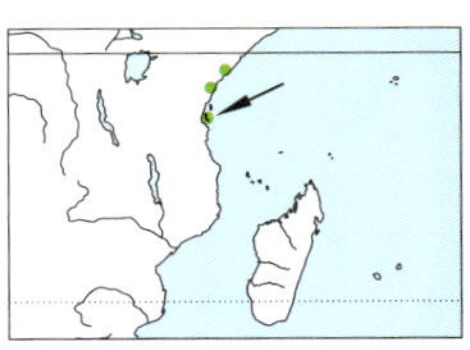

SE KE (Arabuko Sokoke Forest, Boni & Dodori R), TZ (Unguja I in Zanzibar Arch)

VU Monotypic
HB 63–78 cm T 6–13.8 cm
W 6.8–12.4 kg

ssp *leucogaster*

White-bellied Duiker
Cephalophus leucogaster

FR Céphalophe à ventre blanc
DE Weißbauchducker
ES Duiker de vientre blanco

NT
2 sspp
HB 92–105 cm
T 10.5–15 cm
W 14.7–21 kg

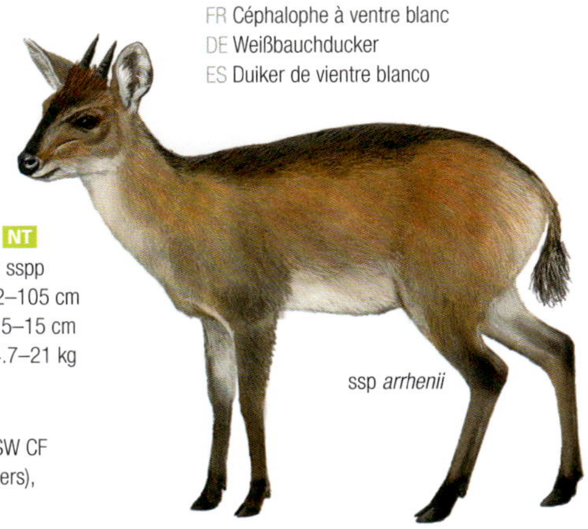

ssp *arrhenii*

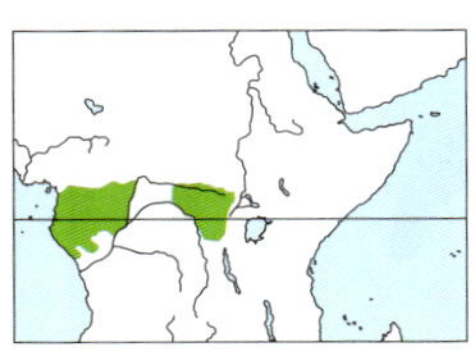

CM (S of R Sanaga), GQ, GA, SW CF & CG (E to Congo & Ubangi rivers), disjunct E CD (NE of R Congo)

Family SUIDAE – Pigs

6 genera, 17 species. From near-desert conditions to closed forests in tropical, subtropical, and boreal regions. Palearctic, Afrotropical, Indo-Malayan and Australasian Regions.

Genus *BABYROUSA*

(3 species)

Sulawesi Babirusa

Babyrousa celebensis

FR Babiroussa des Célèbes
DE Sulawesi-Hirscheber
ES Babirusa de Célebes

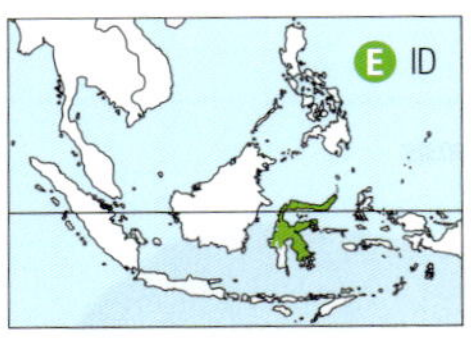

Sulawesi

VU Monotypic
HB 85–110 cm T 20–32 cm
W ≤ 100 kg

Moluccan Babirusa

Babyrousa babyrussa

FR Babiroussa des Moluques
DE Molukken-Hirscheber
ES Babirusa de Molucas

above 800 m

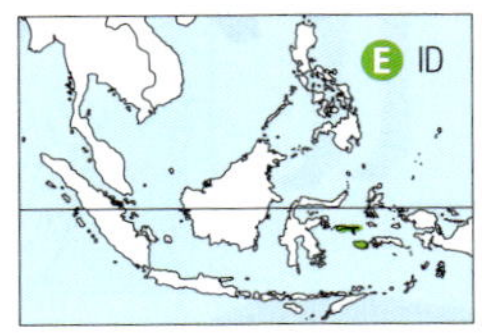

C Moluccas in Sula Is (Mangole, Taliabu) & Buru

VU Monotypic
HB 85–110 cm T 20–32 cm
W 43–100 kg

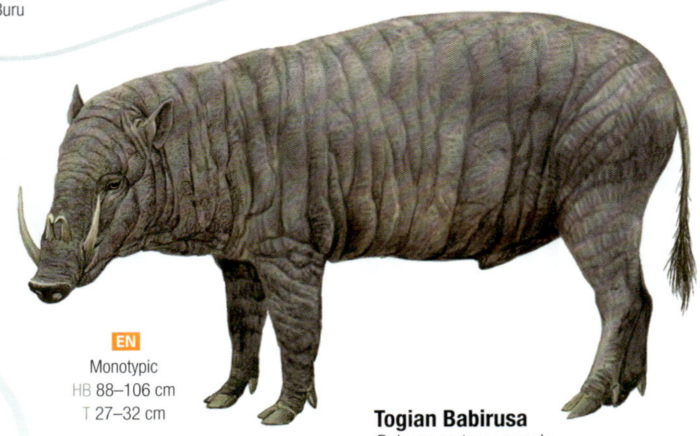

EN
Monotypic
HB 88–106 cm
T 27–32 cm

Togian Babirusa

Babyrousa togeanensis

FR Babiroussa des Togian
DE Togian-Hirscheber
ES Babirusa de Togian

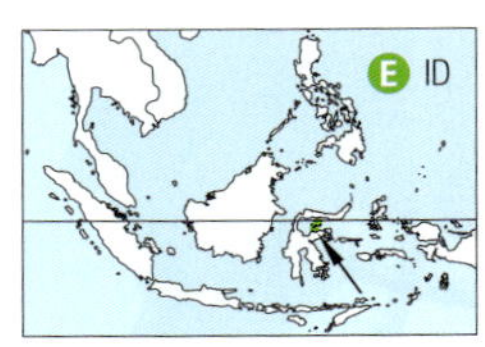

Togian Arch (Batudaka, Togian, Talatakoh, Malenge, tracks reported from Kadidiri, N of Togian I)

♂

Common Warthog

Phacochoerus africanus

FR Phacochère d'Afrique
DE Warzenschwein
ES Facocero de sabana

♀

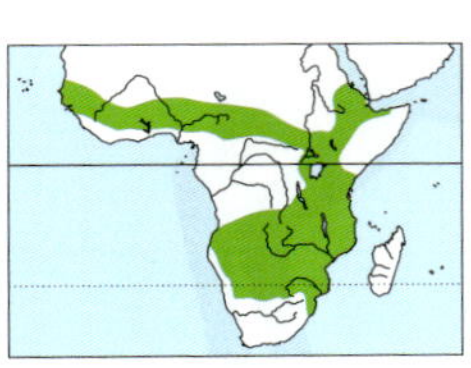

N savanna & Sahel region from MR E to ER, DJ & N SO, S through E Africa to SW AO, NA, N BW & ZA

LC
4 sspp
HB 125–150 cm (M), 105–140 cm (F)
T 35–50 cm
W 60–150 kg (M), 50–75 kg (F)

Genus *PHACOCHOERUS*

(2 species)

Desert Warthog

Phacochoerus aethiopicus

FR Phacochère du désert
DE Wüstenwarzenschwein
ES Facocero de desierto

below 1000 m

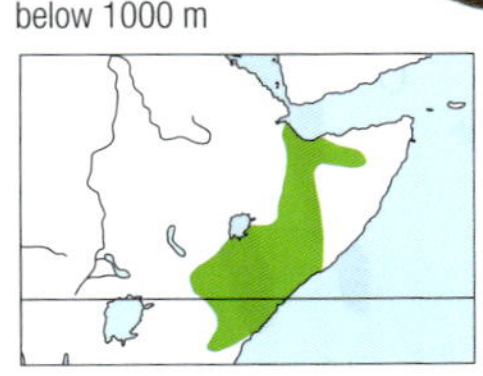

Horn of Africa in NW & S SO, E ET, E KE

LC
1 extant ssp
HB c. 131 cm
T c. 44 cm

Genus *HYLOCHOERUS*

(1 species)

♀

ssp *meinertzhageni*

Giant Forest Hog
Hylochoerus meinertzhageni

FR Hylochère
DE Riesenwaldschwein
ES Hilocero

LC
3 sspp
HB 130–210 cm
T 25–45 cm
W 140–275 kg (M), 100–200 kg (F)

♂
ssp *rimator*

♂

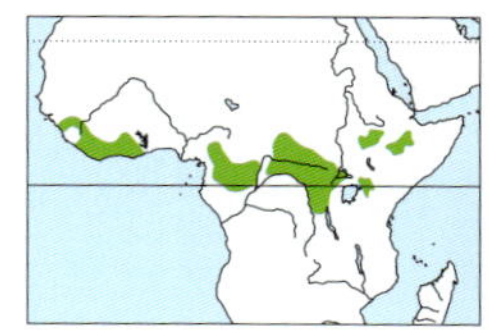

Fragmented from GN to GH; from SE NG & CM to CD, SW SS, UG, RW & BI; from ET to KE, perhaps TZ

Genus *POTAMOCHOERUS*

(2 species)

Bushpig
Potamochoerus larvatus

FR Potamochère du Cap
DE Buschschwein
ES Potamocero de río

LC 3 sspp
HB 100–150 cm T 30–40 cm
W 50–115 kg

0–4000 m

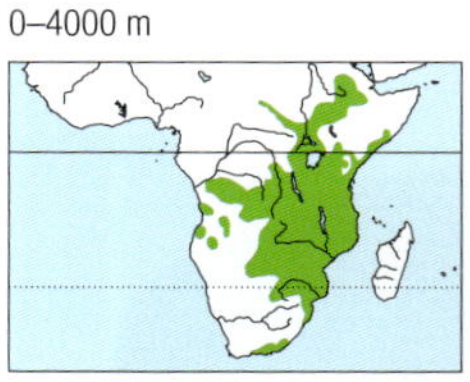

From S SS & ET S through E Africa and E & S CD to BW, MZ & NE ZA; isolated populations in AO & S ZA

Red River Hog
Potamochoerus porcus

FR Potamochère roux
DE Pinselohrschwein
ES Potamocero rojo

LC
Monotypic
HB 100–145 cm
T 30–45 cm
W 45–115 kg

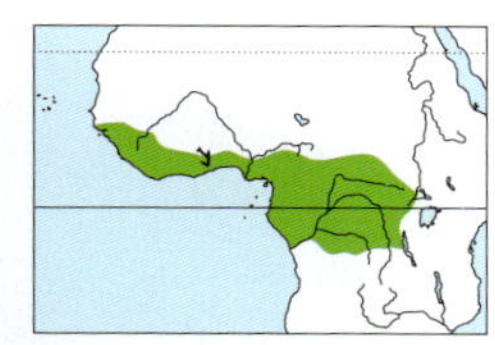

Forest belt from SN to NW UG & E CD, S to N AO (Cabinda); no recent records from GM or TD, perhaps SS & SW ET

Genus *PORCULA*

(1 species)

NW Assam

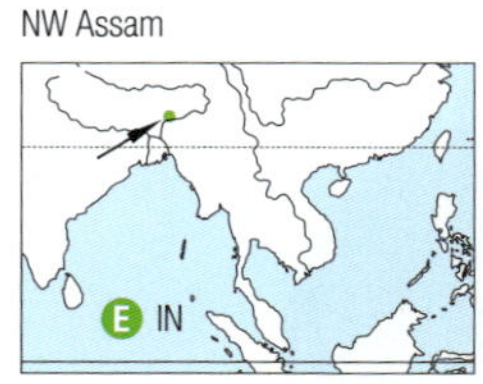

Pygmy Hog
Porcula salvania

FR Sanglier nain
DE Zwergwildschwein
ES Jabalí pigmeo

EN Monotypic
HB 55–71 cm T c. 2.5 cm
W 6.6–9.7 kg

Genus *SUS*

(8 species)

Visayan Warty Pig
Sus cebifrons

FR Sanglier des Visayas
DE Visayas-Pustelschwein
ES Jabalí de Bisayas

0–1600 m

W Visayas (Negros, Panay, possibly Masbate)

♀

CR
1 extant ssp
HB c. 100 cm
T c. 23 cm
W 35–40 kg (M), 20–35 kg (F)

♂

Philippine Warty Pig
Sus philippensis

FR Sanglier des Philippines
DE Philippinen-Pustelschwein
ES Jabalí de Filipinas

ssp *mindanensis*

VU
2 sspp

ssp *philippensis*

925–2150 m

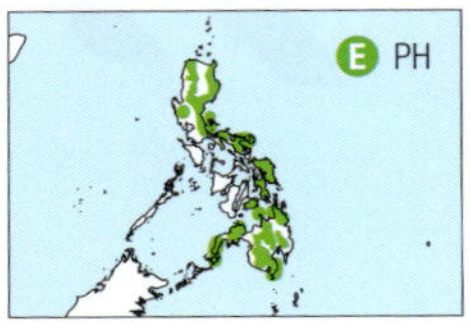

Luzon, Polillo, Catanduanes, Samar, Biliran, Leyte, Bohol, Camiguin, Mindanao, Basilan; formerly Marinduque

Mindoro Warty Pig
Sus oliveri

FR Sanglier de Mindoro
DE Mindoro-Pustelschwein
ES Jabalí de Mindoro

VU
Monotypic

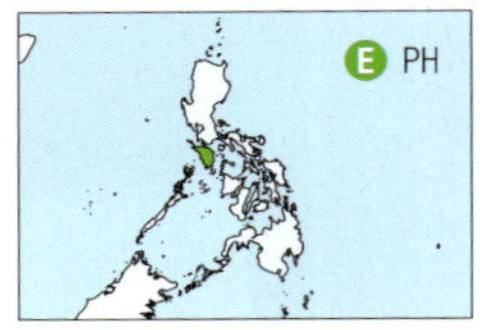

Mindoro

Palawan Bearded Pig
Sus ahoenobarbus

FR Sanglier de Palawan
DE Palawan-Bartschwein
ES Jabalí barbudo de Palawan

0–1500 m

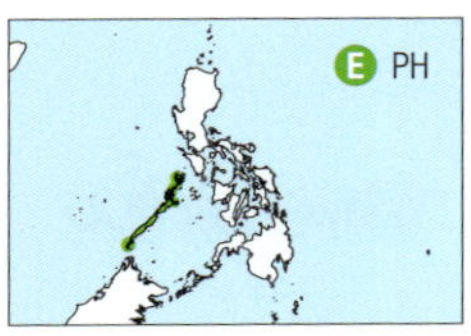

Palawan and associated Is

NT Monotypic
HB 100–160 cm T 15–25 cm
W ≤ 150 kg

Bearded Pig
Sus barbatus

FR Sanglier barbu
DE Bartschwein
ES Jabalí barbudo

VU 2 sspp
HB 122–152 cm T 17–26 cm
W 58–83 kg

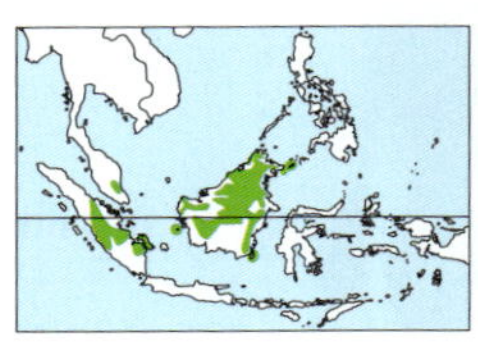

Pen MY, Sumatra, Bangka, Borneo and associated Is, possibly Tawi-Tawi (PH)

Javan Warty Pig
Sus verrucosus

FR Sanglier de Java
DE Java-Pustelschwein
ES Jabalí de Java

below 800 m

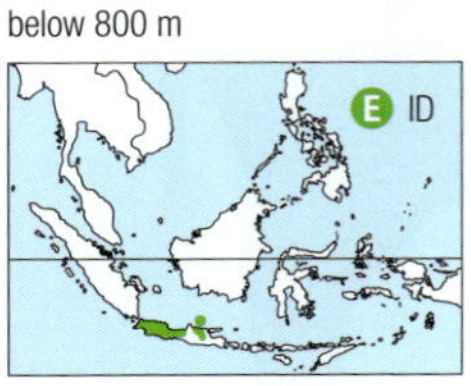

W & C Java, Bawean I

EN
2 sspp
HB 90–190 cm
T 15–25 cm
W 35–150 kg

Sulawesi Warty Pig
Sus celebensis

FR Sanglier des Célèbes
DE Sulawesi-Pustelschwein
ES Jabalí de Célebes

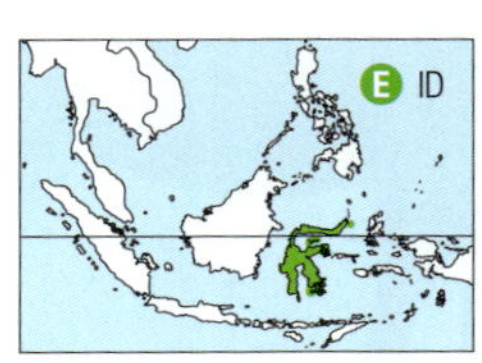

Sulawesi and adjacent Is (Buton, Kabaena, Muna, Peleng, Lembeh, Togian Is); perhaps extinct on Selayar I

NT
Monotypic
HB 80–130 cm
T 25–35 cm
W 40–70 kg

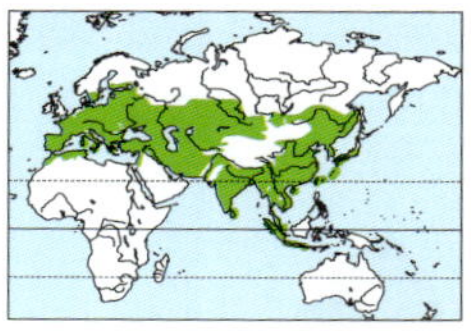

Europe, N Africa, Middle East, C & S Asia to RU Far East, CN, JP, TW & SE Asia; introduced to AU, NZ & America

ssp *cristatus*

ssp *scrofa*

Eurasian Wild Pig
Sus scrofa
FR Sanglier d'Eurasie
DE Wildschwein
ES Jabalí

LC
19 sspp
HB 90–200 cm
T 15–40 cm
W 44–320 kg

ssp *ussuricus*

ssp *vittatus*

Family TAYASSUIDAE – Peccaries

3 genera, 3 species. Variety of habitats, including deserts, thorn scrub, savannas, and dry and moist forests, from temperate to tropical zones. Nearctic and Neotropical Regions.

Genus *CATAGONUS*

(1 species)

Chacoan Peccary
Catagonus wagneri
FR Pécari du Chaco
DE Chaco-Pekari
ES Pecarí del Chaco

EN
Monotypic
HB 90–117 cm
T 2.4–10.2 cm
W 30–40 kg

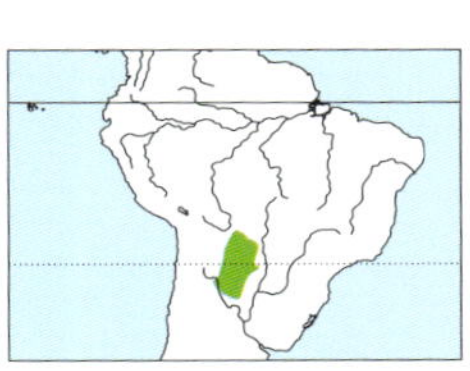

Dry Chaco of SE BO, W PY & N AR (S to Santiago del Estero, possibly NE Tucumán)

Genus *DICOTYLES*

(1 species)

Collared Peccary
Dicotyles tajacu
FR Pécari à collier
DE Halsbandpekari
ES Pecarí de collar

up to 3000 m

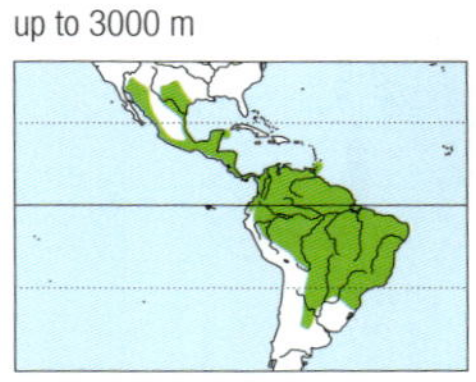

S US (Arizona, New Mexico, S Oklahoma, Texas) & MX S to C AR, also Cozumel I (SE MX) & TT

LC 14 sspp
HB 84–106 cm T 1–10 cm
W 15–42 kg

Genus *TAYASSU*

(1 species)

White-lipped Peccary
Tayassu pecari
FR Pécari à lèvres blanches
DE Weißbartpekari
ES Pecarí barbiblanco

VU
4 sspp
HB 90–139 cm
T 1–6.5 cm
W 25–40 kg

0–1900 m

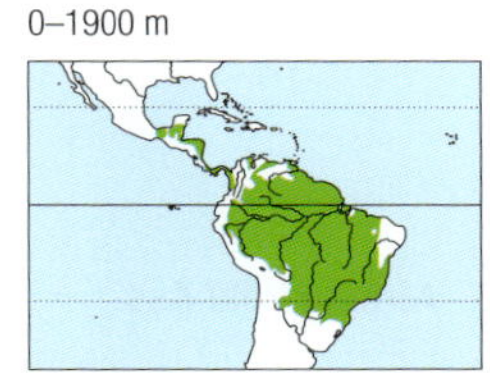

From S MX S through Central & South America to N AR (Santiago del Estero, Misiones) & S BR (Rio Grande do Sul)

Family CAMELIDAE – Camels

2 genera, 3 species. Deserts, semi-arid to arid plains, grasslands, steppes, and shrublands, from sea level to 4800 m. Palearctic, Afrotropical and Neotropical Regions.

Genus *LAMA*

(2 species)

Guanaco
Lama guanicoe

FR Guanaco
DE Guanako
ES Guanaco

0–4500 m

From N PE & BO S to CL & AR (to S Patagonia, Tierra del Fuego & Navarino I); introduced to FK

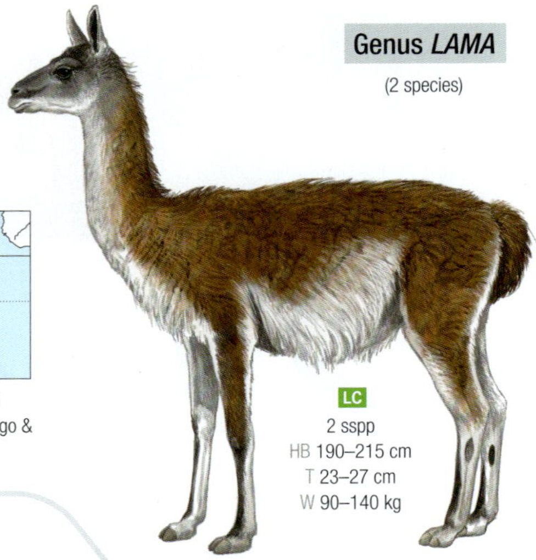

LC
2 sspp
HB 190–215 cm
T 23–27 cm
W 90–140 kg

Vicuña
Lama vicugna

FR Vigogne
DE Vikunja
ES Vicuña

LC
2 sspp
HB 125–190 cm
T 15–25 cm
W 38–45 kg

3200–4800 m

Puna & High Andes in SE PE (S from Ancash), W BO, NE CL & NW AR (S to -29° S); introduced to EC

Genus *CAMELUS*

(1 species)

CR
Monotypic
HB 320–350 cm
T 51–64 cm
W 450–500 kg

Wild Bactrian Camel
Camelus ferus

FR Chameau de Tartarie
DE Wildkamel
ES Camello bactriano

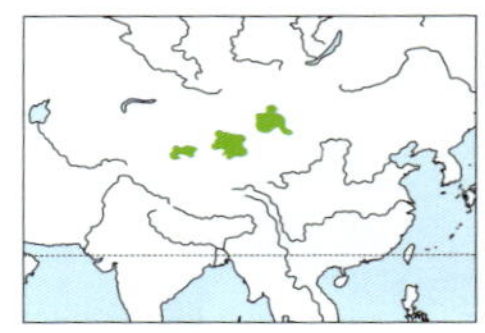

NW CN (C & SE Xinjiang, Inner Mongolia, NW Qinghai, NW Gansu), MN

Order PERISSODACTYLA (3 families)

Family EQUIDAE – Horses and relatives

1 genus, 7 species. Mesic grasslands to semi-arid savannas, arid lowlands, and mountain deserts. Palearctic and Afrotropical Regions.

Genus *EQUUS*

(7 species)

Onager
Equus hemionus

FR Hémione
DE Asiatischer Halbesel
ES Onagro

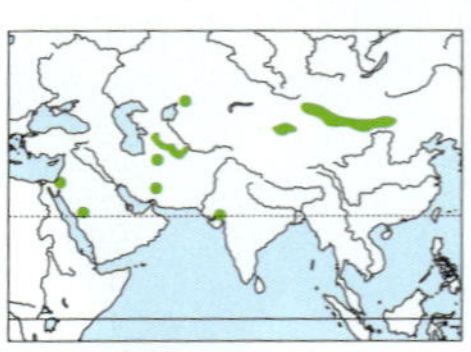

Stronghold in S MN & N CN, also IL (Negev Desert), IR (Qatrouiye, Touran, Bahram-e-Goor), TM (Badkhyz), KZ & IN (Little Rann of Kutch)

NT
4 extant sspp
HB 200–250 cm
T 30–49 cm
W 200–260 kg

Kiang
Equus kiang

FR Kiang
DE Kiang
ES Kiang

2700–5300 m

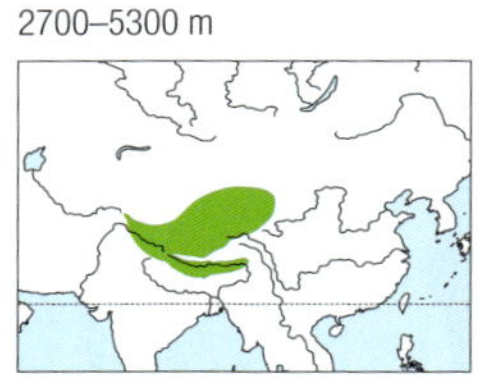

CN (Xinjiang, Qinghai, Gansu, Tibet), NE PK (Khunjerab NP), N IN (Ladakh, Sikkim), N NP (Mustang); perhaps BT

LC
3 sspp
HB 182–214 cm
T 32–45 cm
W 250–400 kg

African Wild Ass
Equus africanus

FR Âne sauvage
DE Afrikanischer Wildesel
ES Asno salvaje

CR
2 extant sspp
HB 195–205 cm
T 40–45 cm
W 270–280 kg

up to 1500 m

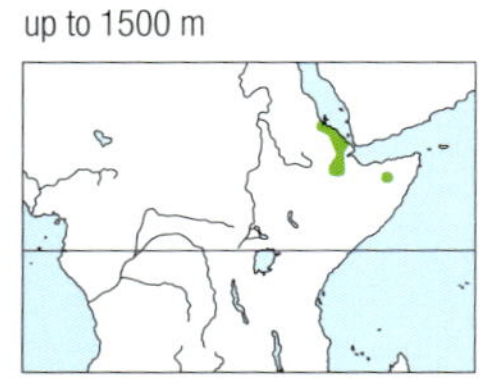

ER, ET, some animals may persist in DJ, SO & E SD; feral on several Caribbean Is

Plains Zebra

Equus quagga

FR Zèbre de Burchell
DE Steppenzebra
ES Cebra de Burchell

0–4300 m

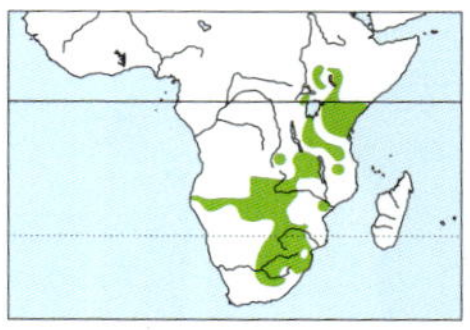

E SS & ET S through E Africa to N & NE ZA and S MZ, W to S AO & N NA; extinct in BI, W ZA & LS

NT
5 extant sspp
HB 217–246 cm
T 47–57 cm
W 175–320 kg

Angola, Namibia and W Botswana form

ssp *burchelli*

ssp *chapmanni*

ssp *boehmi*

ssp *selousi*

ssp *crawshaii*

Grevy's Zebra

Equus grevyi

FR Zèbre de Grévy
DE Grévy-Zebra
ES Cebra de Grévy

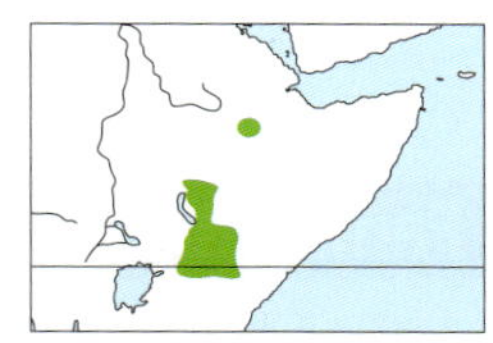

ET, N & C KE; perhaps SS

EN
Monotypic
HB 250–275 cm
T 38–75 cm
W 350–450 kg

Mountain Zebra

Equus zebra

FR Zèbre de montagne
DE Bergzebra
ES Cebra de montaña

ssp *zebra*

ssp *hartmannae*

VU
2 sspp
HB 210–260 cm
T 40–55 cm
W 240–380 kg

up to 2000 m

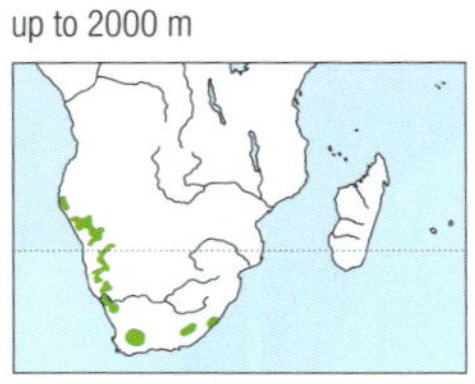

W NA, S ZA (fragmented in W & S edge of Central plateau); perhaps SW AO

Przewalski's Horse

Equus przewalskii

FR Cheval de Przewalski
DE Przewalski-Pferd
ES Caballo de Przewalski

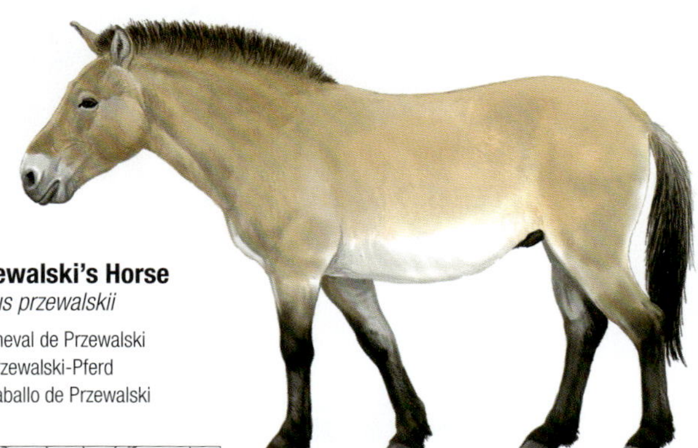

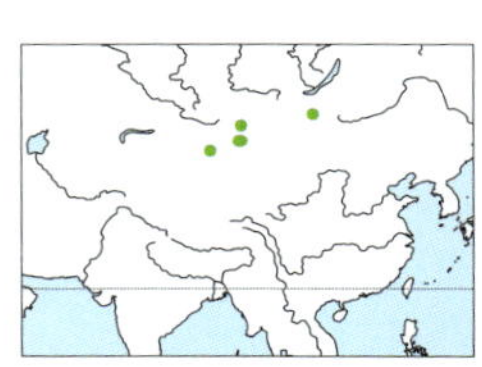

MN (reintroduced to Khustain Nuruu NP and Takhin Tal & Khomin Tal NR), CN (Ka La Mai Li Shan NR)

EN
Monotypic
HB 220–280 cm
T 99–111 cm
W 200–300 kg

Family TAPIRIDAE – Tapirs

1 genus, 4 species. Tropical and subtropical montane, lowland moist, swamp, and cloud forests, páramos, seasonally moist grasslands, and savannas. Indo-Malayan and Neotropical Regions.

Genus *TAPIRUS*

(4 species)

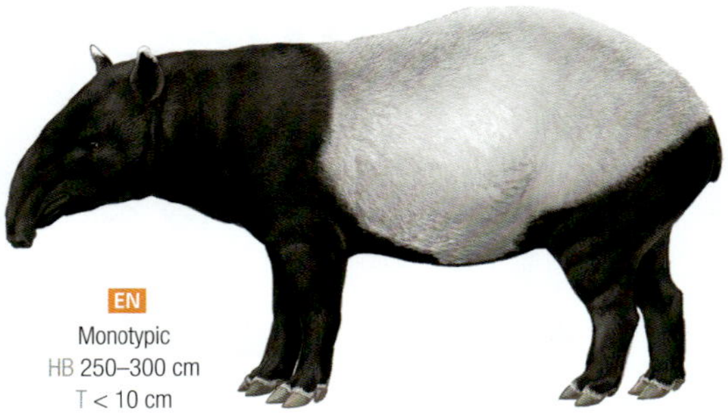

EN
Monotypic
HB 250–300 cm
T < 10 cm
W 280–400 kg

Malayan Tapir
Tapirus indicus
FR Tapir de Malaisie
DE Schabrackentapir
ES Tapir malayo

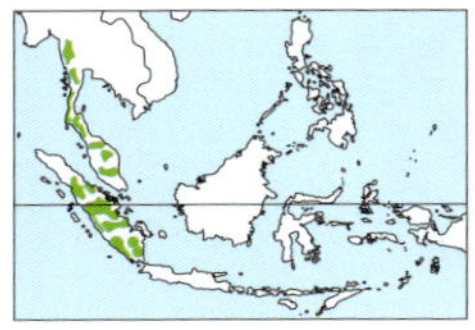

100–2000 m

Tropical lowland moist forests in mainland SE Asia (S MM, S TH, Pen MY) and C & S Sumatra (ID)

EN
Monotypic
HB 200–230 cm
T < 10 cm
W 250–350 kg

Central American Tapir
Tapirus bairdii
FR Tapir de Baird
DE Mittelamerika-Tapir
ES Tapir centroamericano

0–3600 m

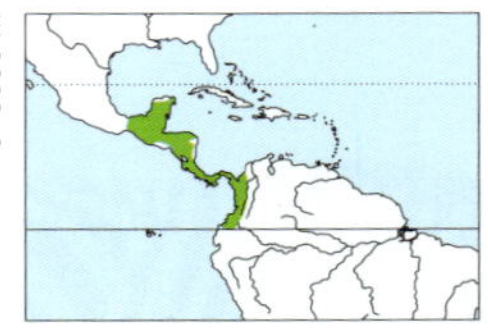

SE MX to NW CO; formerly W EC (S to Gulf of Guayaquil), but status there uncertain

Mountain Tapir
Tapirus pinchaque
FR Tapir des Andes
DE Andentapir
ES Tapir de montaña

EN
Monotypic
HB 180–200 cm
T < 10 cm
W 150–200 kg

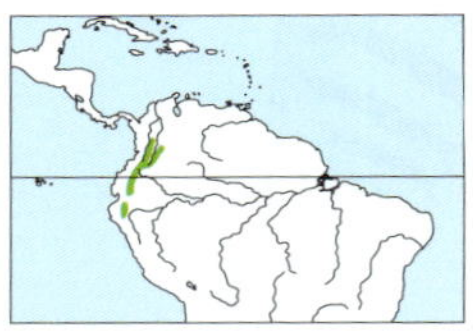

2200–4800 m

Andean areas of CO, EC & PE; historically W VE but now extirpated

Lowland Tapir
Tapirus terrestris
FR Tapir du Brésil
DE Flachlandtapir
ES Tapir amazónico

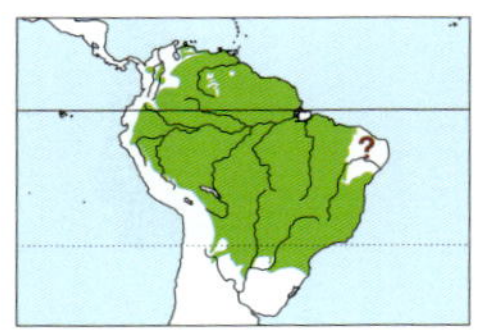

up to 2000 m

Lowlands of South America from CO, VE & Guianas through Amazonian Basin to BO, PY, N AR & SE BR

VU
4 sspp
HB 191–242 cm
T < 10 cm
W 180–300 kg

Family RHINOCEROTIDAE – Rhinoceroses

4 genera, 5 species. Tropical montane, lowland, and mangrove forests, short grasslands, tall alluvial grasslands, savannas, woodlands, semi-deserts, and scrub. Afrotropical and Indo-Malayan Regions.

ssp *cottoni*

♂

White Rhinoceros
Ceratotherium simum
FR Rhinocéros blanc
DE Breitmaulnashorn
ES Rinoceronte blanco

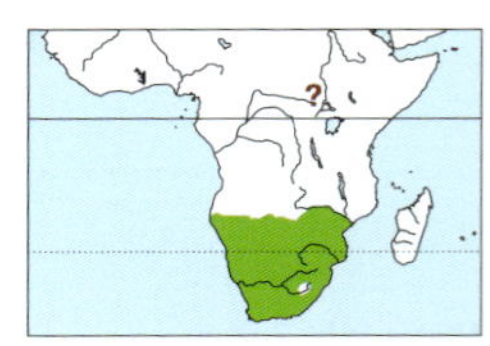

NT
2 sspp
HB 340–420 cm
T 50–70 cm
W 1350–3500 kg

SE AO, NE NA, BW, ZW, S MZ, SZ, ZA, perhaps formerly extreme SW ZM; historically in S TD, E CF, SW SS, NE CD & NW UG, but probably extinct there; introduced to KE, UG & ZM; map vague for conservation reasons

Genus *CERATOTHERIUM*

(1 species)

♀

ssp *simum*

♂

Genus *DICEROS*

(1 species)

Black Rhinoceros
Diceros bicornis

FR Rhinocéros noir
DE Spitzmaulnashorn
ES Rinoceronte negro

CR
3 extant sspp
HB 300–380 cm
T 25–35 cm
W 800–1300 kg

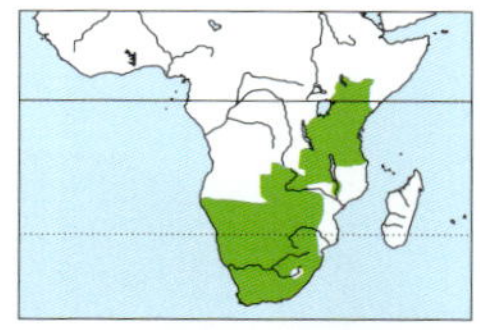

From KE & TZ S to ZA, W to AO (?) & NA; considered extinct in W & C Africa (from NG & CM E to SD, ET, UG & RW); map vague for conservation reasons

Greater One-horned Rhinoceros
Rhinoceros unicornis

FR Rhinocéros indien
DE Panzernashorn
ES Rinoceronte indio

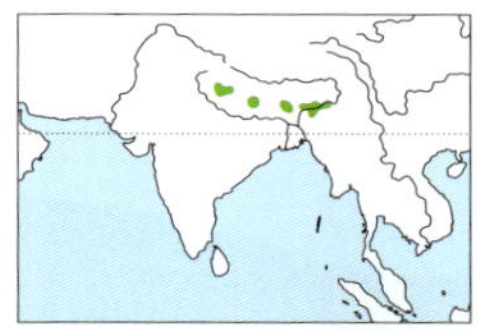

Lowlands of Himalayas in NP & IN; extinct in PK, BT & BD

VU
Monotypic
HB 335–346 cm
T c. 66 cm
W c. 2000 kg

Genus *RHINOCEROS*

(2 species)

♂

♀

Lesser One-horned Rhinoceros
Rhinoceros sondaicus

FR Rhinocéros de Java
DE Java-Nashorn
ES Rinoceronte de Java

CR
1 extant ssp
HB 305–344 cm
W 1200–1500 kg

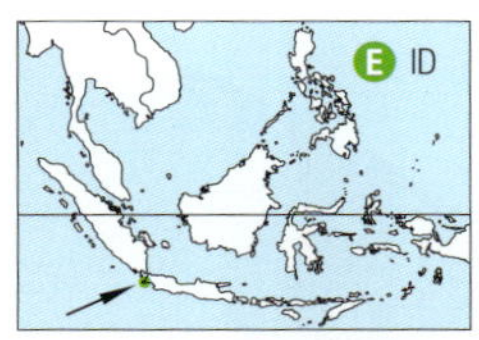

W Java; extinct in mainland S & SE Asia (last individual shot in VN in 2010) & Sumatra

Genus *DICERORHINUS*

(1 species)

Sumatran Rhinoceros
Dicerorhinus sumatrensis

FR Rhinocéros de Sumatra
DE Sumatra-Nashorn
ES Rinoceronte de Sumatra

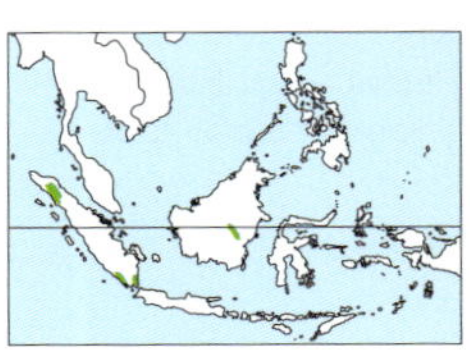

Sumatra & Borneo; extinct in S & SE Asia (may still survive in MM)

ssp *sumatrensis*

ssp *harrissoni*

CR
3 sspp
HB 236–318 cm
W 600–950 kg

Order PHOLIDOTA (1 family)

Family MANIDAE – Pangolins

3 genera, 8 species. Forested and open habitats in tropical and intertropical zones, from sea level to 2300 m. Afrotropical and Indo-Malayan Regions.

Genus *MANIS*

(4 species)

Thick-tailed Pangolin
Manis crassicaudata
FR Pangolin indien
DE Indien-Schuppentier
ES Pangolín coligrueso

EN
Monotypic
HB 51–75 cm
T 33–47 cm
W 10–16 kg

1100–2300 m

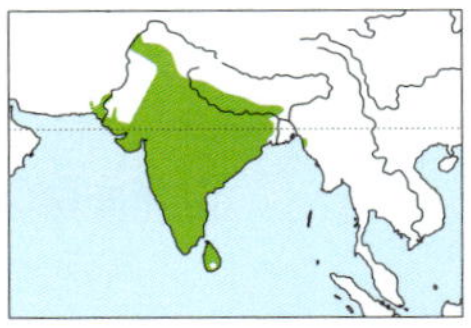

NE & SE PK, IN, S NP, BD (almost extirpated), LK; doubtful records from MM & CN (Yunnan)

Short-tailed Pangolin
Manis pentadactyla
FR Pangolin à queue courte
DE China-Schuppentier
ES Pangolín colicorto

CR
3 sspp
HB 40–58 cm
T 25–38 cm
W 2.5–7 kg

up to 2000 m

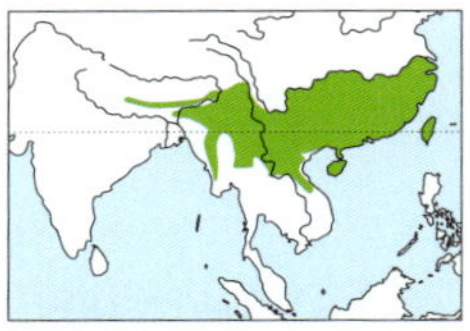

Himalayan foothills in E NP, BT & NE IN to NE BD, MM, N TH, N LA, N VN, CN (S of R Yangtze) & TW

Sunda Pangolin
Manis javanica
FR Pangolin de Malaisie
DE Malaien-Schuppentier
ES Pangolín de la Sonda

CR
Monotypic
HB 40–65 cm
T 35–58 cm
W 3–10 kg

up to 1700 m

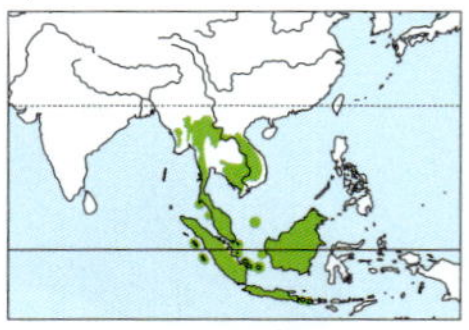

C & S MM, TH, LA, KH, VN, Pen MY, Greater Sundas, Lombok, several small islands; likely extinct in E BD

Palawan Pangolin
Manis culionensis
FR Pangolin de Palawan
DE Palawan-Schuppentier
ES Pangolín de Palawan

CR
Monotypic
HB 45–54 cm
T 39–50 cm
W 2.5–8 kg

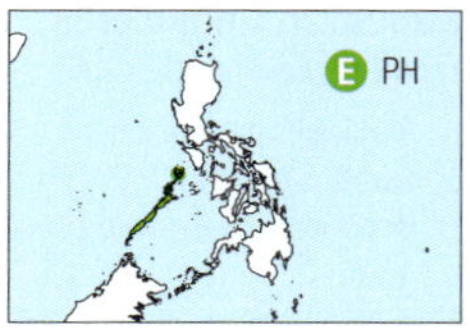

Palawan, Calauit, Busuanga, Culion; introduced to Apulit

Genus *SMUTSIA*

(2 species)

Giant Pangolin
Smutsia gigantea

EN

FR Pangolin géant
DE Riesenschuppentier
ES Pangolín gigante

Monotypic
HB 67–81 cm
T 58–68 cm
W c. 30 kg

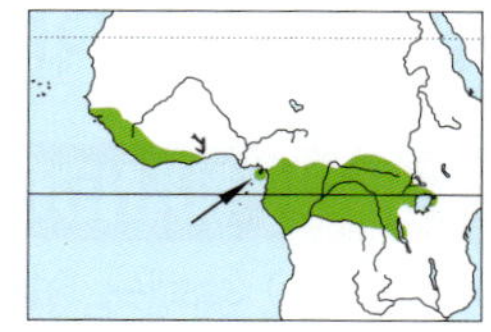

W Africa (from SN E to GH), C Africa (from Bioko & CM E to UG, W KE & W TZ, S to NW AO); perhaps Dahomey Gap & S NG

Ground Pangolin
Smutsia temminckii

VU

FR Pangolin de Temminck
DE Steppenschuppentier
ES Pangolín terrestre

Monotypic
HB 45–55 cm
T 40–52 cm
W 5–20 kg

up to 1700 m

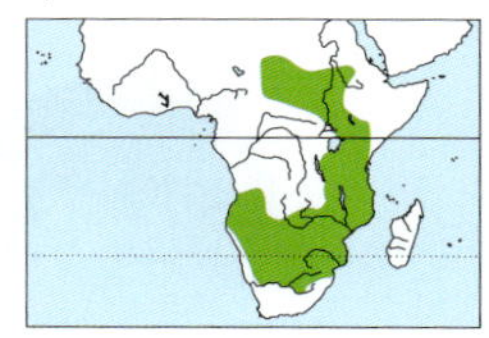

Patchily in open areas from E TD & N CF E to extreme W ET, S through E Africa to C AO, NA & N ZA

Genus *PHATAGINUS*

(2 species)

Long-tailed Pangolin
Phataginus tetradactyla

VU

FR Pangolin à longue queue
DE Langschwanzschuppentier
ES Pangolín colilargo

Monotypic
HB 30–40 cm
T 55–70 cm
W 2–3.5 kg

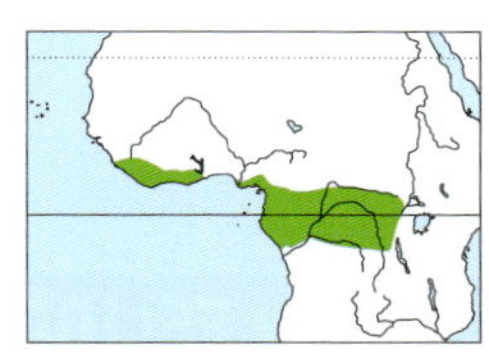

W Africa (from SL E to GH), C Africa (from Niger Delta & Congo Basin to E CD; perhaps NW AO (Cabinda)

Common African Pangolin
Phataginus tricuspis

EN

FR Petit Pangolin
DE Weißbauchschuppentier
ES Pangolín arborícola

Monotypic
HB 25–43 cm
T 35–62 cm
W 1.6–3 kg

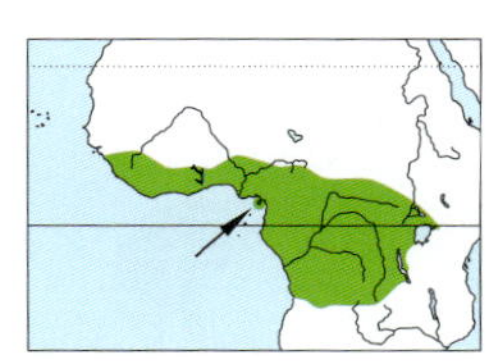

Continuous in W & C African rainforests E to SW KE, S to extreme W TZ, NW ZM & N AO, Bioko I (GQ)

Order CARNIVORA (16 families)

Family NANDINIIDAE – African Palm Civet

1 genus, 1 species. Rain and deciduous forests, from coastal to montane, also savanna woodlands. Tropical Africa.

Genus *NANDINIA*

(1 species)

African Palm Civet
Nandinia binotata
FR Nandinie
DE Pardelroller
ES Nandinia

LC
4 sspp
HB 37–62 cm
T 34–76 cm
W 1.2–3 kg

up to 2500 m

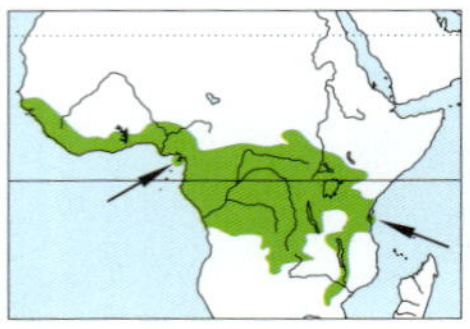

GM E to SW SS, S UG & W KE, S to N AO, NW ZM, TZ, MW, E ZW & MZ, Bioko I (GQ) & Zanzibar Arch

Family PRIONODONTIDAE – Linsangs

1 genus, 2 species. Moist and evergreen forests, including montane forests up to 2700 m. South-east Asia.

Genus *PRIONODON*

(2 species)

Banded Linsang
Prionodon linsang
FR Linsang rayé
DE Bänderlinsang
ES Linsang rayado

LC
2 sspp
HB 38–45 cm
T 33–37 cm
W 590–800 g

up to 2700 m

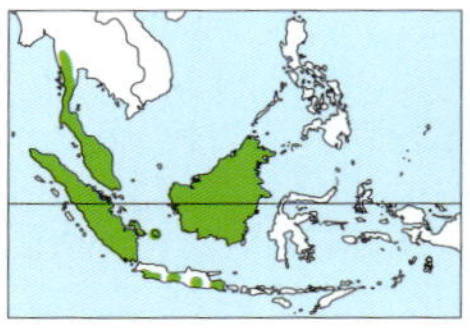

S MM, S TH, Pen MY, Sumatra, Bangka I, Belitung I, Borneo, Java

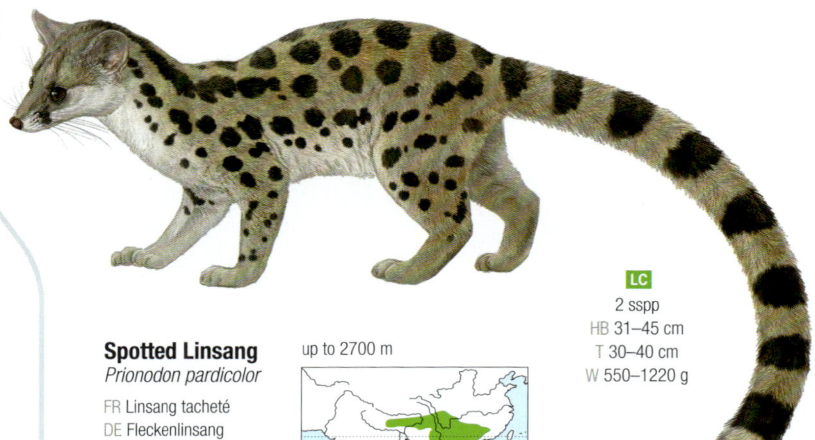

Spotted Linsang
Prionodon pardicolor
FR Linsang tacheté
DE Fleckenlinsang
ES Linsang manchado

LC
2 sspp
HB 31–45 cm
T 30–40 cm
W 550–1220 g

up to 2700 m

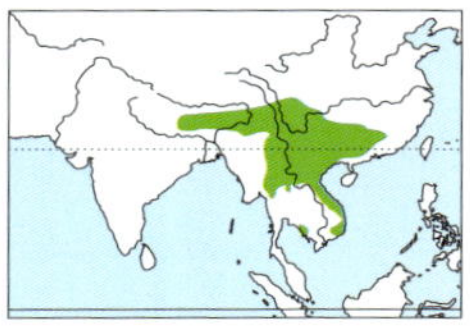

NP, NE IN, BT, N MM, C & S CN, N TH, LA, VN, KH

Family FELIDAE – Cats

14 genera, 45 species. From deserts through forests to mountain areas, from cold temperate zone to tropics. Holarctic, Neotropical, Afrotropical and Indo-Malayan Regions.

Genus *PANTHERA*

(5 species)

Jaguar
Panthera onca
FR Jaguar
DE Jaguar
ES Yaguar

NT
Monotypic
HB 116–170 cm
T 44–80 cm
W 31–121 kg

up to 2000 m

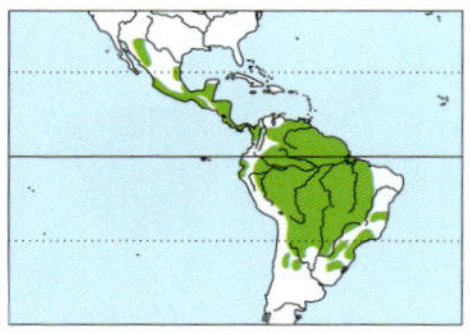

SW US (vagrant in Arizona), scattered from MX through Central & South America S to N AR & SE BR

melanistic morph

Leopard
Panthera pardus

FR Léopard
DE Leopard
ES Leopardo

VU
8 sspp
HB 92–190 cm
T 64–99 cm
W 21–71 kg

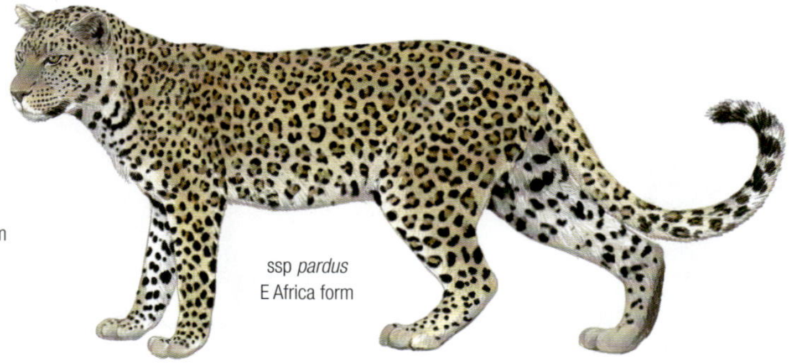
ssp *pardus*
E Africa form

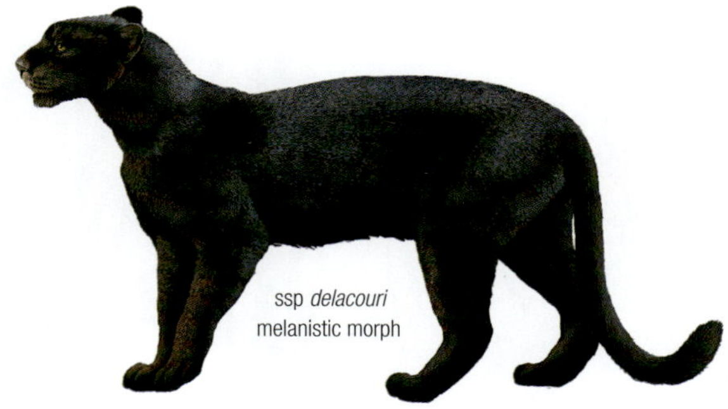
ssp *delacouri*
melanistic morph

ssp *fusca*

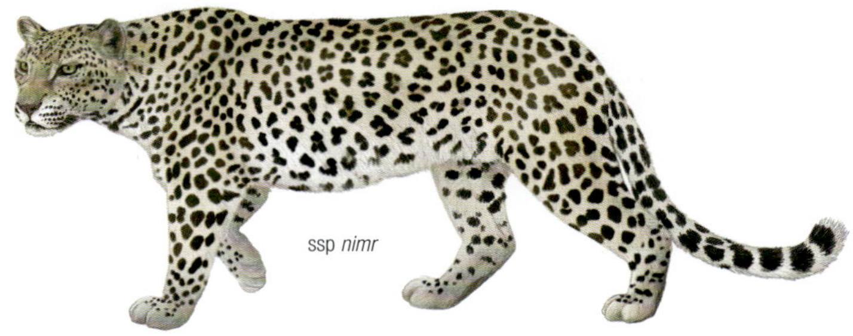
ssp *nimr*

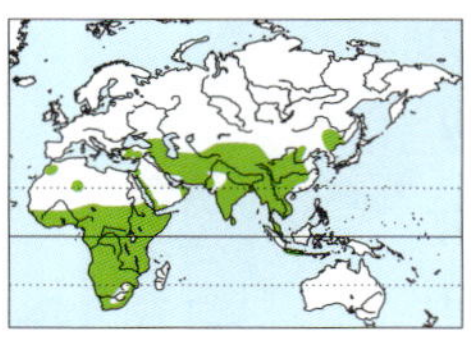
N Africa (MA?, DZ, EG), scattered in sub-Saharan Africa, Middle East & Caucasus E to RU (Far East), KP & CN, S to IN, LK, Pen MY & Java

ssp *orientalis*

ssp *pardus*

VU
2 sspp
HB 172–250 cm (M), 158–192 cm (F)
T 61–100 cm
W 150–272 kg (M), 122–192 kg (F)

♂
India form

NW Africa form
♂

up to 4000 m

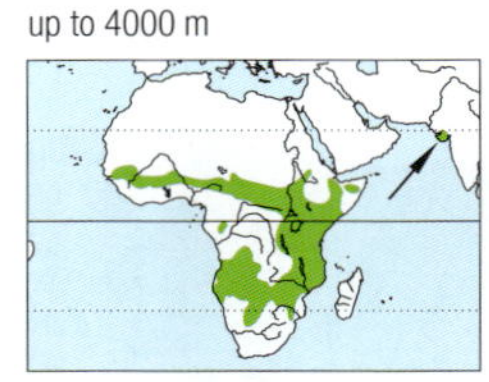
Fragmented throughout W, C, E & S sub-Saharan Africa, small remnant population in IN (Gir Forest); historically N Africa, Arabian Pen, C Asia, Indian subcontinent & S Europe

Lion
Panthera leo

FR Lion
DE Löwe
ES León

♀

♂
NE & E Africa form

ssp *tigris*
Russian Far East
and NE China form

ssp *sondaica*

Tiger
Panthera tigris

FR Tigre
DE Tiger
ES Tigre

EN
2 sspp
HB 146–290 cm
T 72–109 cm
W 75–325 kg

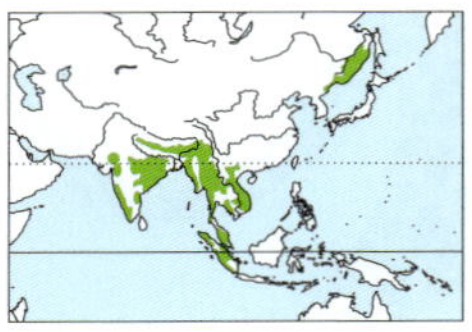

Highly fragmented throughout IN, NP, BT, BD, SC & NE CN, RU (Amur), mainland SE Asia & Sumatra; extinct in a large part of its range, once extended W through C Asia to E TR & Caucasus, N to KZ, also Java & Bali

ssp *tigris*
Indian subcontinent
form

Snow Leopard
Panthera uncia

FR Panthère des neiges
DE Schneeleopard
ES Leopardo de las nieves

VU
2 sspp
HB 86–125 cm
T 80–105 cm
W 22–52 kg

up to 6000 m

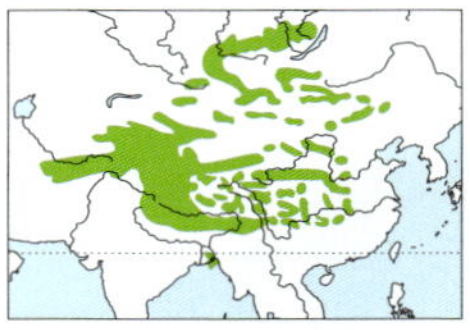

C Asia Mts NE to Altai & Sayan, S to Himalayas (AF, UZ, TJ, KG, KZ, RU, MN, CN, IN, PK, NP, BT, unconfirmed in MM)

Genus *NEOFELIS*

(2 species)

Mainland Clouded Leopard
Neofelis nebulosa

FR Panthère longibande
DE Nebelparder
ES Pantera nebulosa

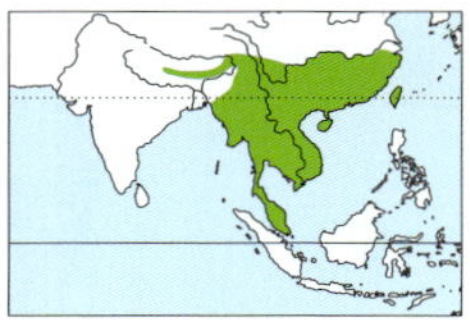

Highly fragmented from sub-Himalayan foothills in NP & NE IN through mainland SE Asia to S CN

VU
Monotypic
HB 69–107 cm
T 61–84 cm
W 11–23 kg

Sunda Clouded Leopard
Neofelis diardi

FR Panthère de Diard
DE Sunda-Nebelparder
ES Pantera de la Sonda

up to 2000 m

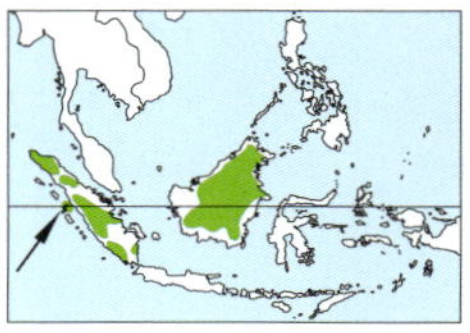

Sumatra (including Batu Is), Borneo

VU
2 sspp
HB 70–105 cm
T 60–85 cm
W 10–25 kg

Genus *PARDOFELIS*

(1 species)

Marbled Cat

Pardofelis marmorata

FR Chat marbré
DE Marmorkatze
ES Gato jaspeado

NT
2 sspp
HB 45–62 cm
T 36–55 cm
W 2–5 kg

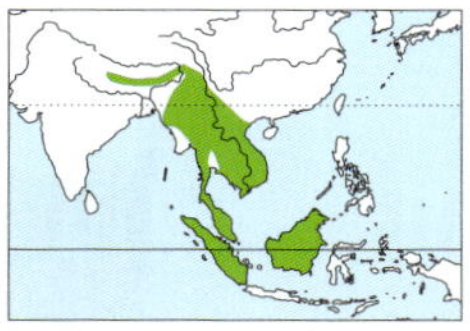

Sub-Himalayan region from NP to MM, SW CN & mainland SE Asia, Sumatra, Borneo

Genus *CATOPUMA*

(2 species)

Bay Cat

Catopuma badia

FR Chat de Bornéo
DE Borneo-Goldkatze
ES Gato badia

EN
Monotypic
HB c. 53 cm (F)
T c. 39 cm
W c. 2 kg

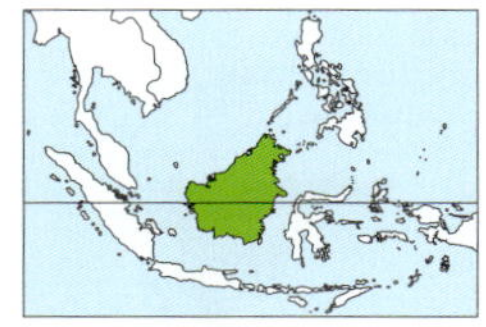

Borneo (Sabah, Sarawak, Kalimantan, uncertain in BN)

Asian Golden Cat

Catopuma temminckii

FR Chat de Temminck
DE Asiatische Goldkatze
ES Gato dorado asiático

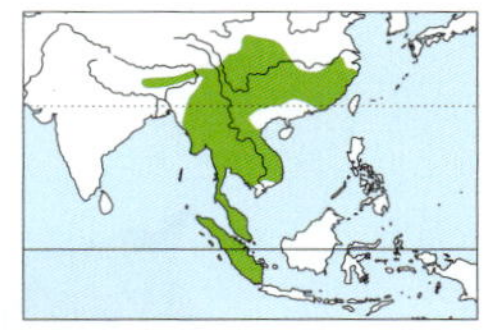

Sub-Himalayan region in NP, IN & BT E to much of CN, mainland SE Asia, Sumatra

ssp *moormensis*

ssp *temminckii*

NT
2 sspp
HB 66–105 cm
T 42–57 cm
W 8.5–15.7 kg

African Golden Cat

Caracal aurata

FR Chat doré
DE Afrikanische Goldkatze
ES Gato dorado africano

Genus *CARACAL*

(2 species)

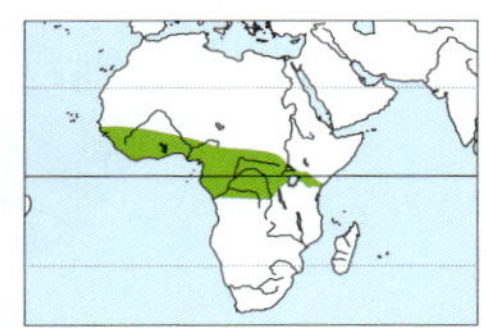

Forests of Equatorial Africa from GN (perhaps GM & GW) to KE & NW TZ, S to N AO

ssp *aurata*

ssp *celidogaster*

VU
2 sspp
HB 62–101 cm
T 16–35 cm
W 6–16 kg

Caracal

Caracal caracal

FR Caracal
DE Karacal
ES Caracal

up to 2500 m

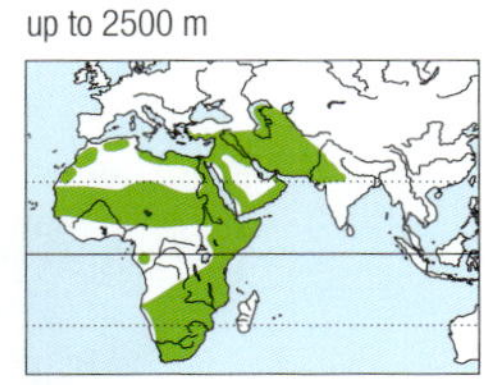

Widely in Africa (from N MA E to NE EG, from SN E to SD, from SE EG S to AO & ZA), Middle East N to TM, E to W IN

ssp *caracal*

ssp *schmitzi*

LC
3 sspp
HB 61–106 cm
T 19–34 cm
W 6.2–20 kg

Genus *LEPTAILURUS*

(1 species)

Serval
Leptailurus serval

FR Serval
DE Serval
ES Serval

LC
3 sspp
HB 59–92 cm
T 20–38 cm
W 7–13.5 kg

up to 3000 m

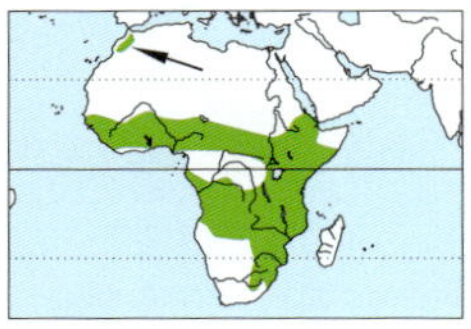

N Africa (N MA, perhaps DZ), sub-Saharan Africa (except tropical rainforest) from SN to ET, S to NA & N ZA; extinct in S ZA

Genus *LEOPARDUS*

(13 species)

NT
3 sspp
HB 42–79 cm
T 30–52 cm
W 2.3–4.9 kg

Margay
Leopardus wiedii

FR Margay
DE Baumozelot
ES Margay

up to 1200 m

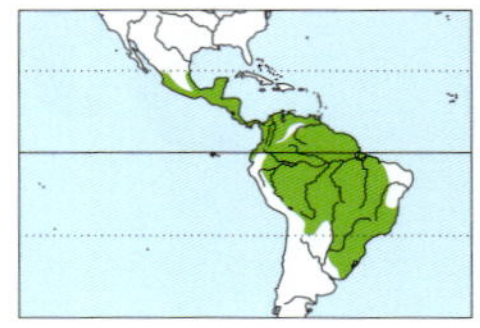

From MX S through Central & South America to PE, BO, PY, N AR & UY; extinct in S Texas (US)

ssp *mitis*

Ocelot
Leopardus pardalis

FR Ocelot
DE Ozelot
ES Ocelote

LC
2 sspp
HB 73–100 cm
T 25–41 cm
W 6.6–15.5 kg

0–1200 m

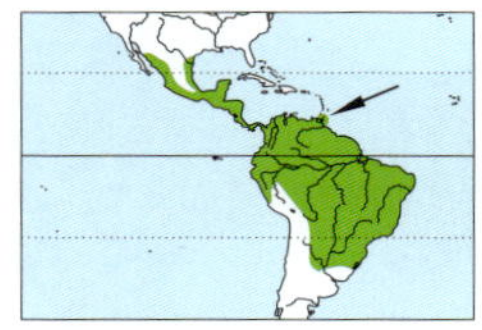

From US (S Arizona, SE Texas) & MX S through Central & South America to N AR & UY, also TT

ssp *pardalis*

Andean Mountain Cat
Leopardus jacobita

FR Chat des Andes
DE Andenkatze
ES Gato andino

EN
Monotypic
HB 58–64 cm
T 41–48 cm
W c. 4 kg (M)

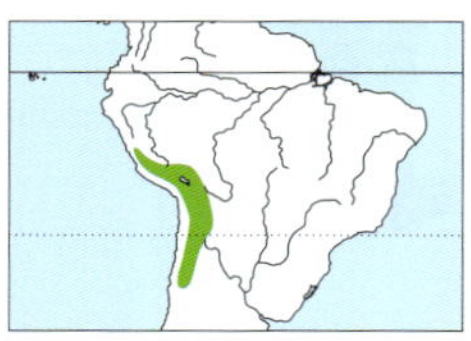

High Andes & Patagonia steppe in C & S PE, SW BO, NE CL, NW AR (S to Neuquén, perhaps Río Negro)

Central Chilean Colocolo
Leopardus colocola

FR Colocolo
DE Pampaskatze
ES Gato de las pampas

NT
Monotypic
HB 56–67 cm
T 28–32 cm

0–1800 m

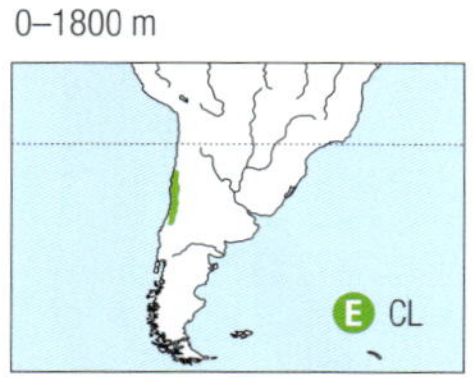

W slope of Andes

Southern Colocolo
Leopardus pajeros

FR Chat des pampas
DE Südliche Pampaskatze
ES Gato de las pampas meridional

NE
Monotypic
HB 54–65 cm
T 23–28 cm

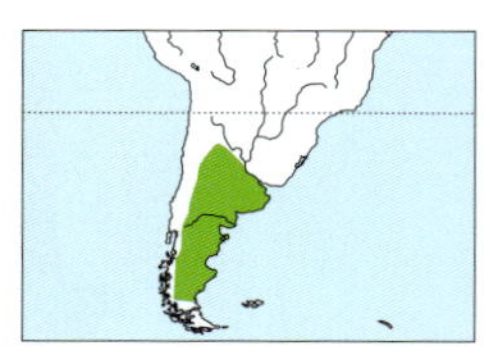

Pampas grasslands in C & S AR (S from Buenos Aires Prov) & CL; uncertain in N AR

Brazilian Colocolo
Leopardus braccatus
FR Chat du Pantanal
DE Pantanal-Pampaskatze
ES Gato del Pantanal

BO (lowlands in Beni), C & E BR (from Pantanal NE to SW Piauí) S to PY & N AR (Formosa)

NE
Monotypic
HB 47–56 cm
T 23–33 cm
W c. 2.9 kg

Northern Colocolo
Leopardus garleppi
FR Chat de Garlepp
DE Garlepp-Pampaskatze
ES Gato de Garlepp

up to 4800 m

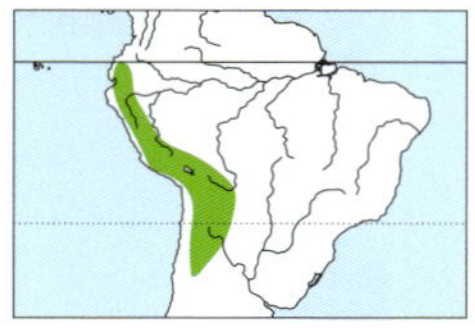

Andes from EC to N CL (Tarapacá) & NW AR (Catamarca, Córdoba); perhaps N to Andes of CO

NE
Monotypic
HB 46–72 cm
T 24–38 cm

Muñoa's Colocolo
Leopardus munoai
FR Chat de Munoa
DE Muñoa-Pampaskatze
ES Gato de Muñoa

S BR (S part of Rio Grande do Sul), UY, NE AR (Corrientes)

NE
Monotypic
HB 38–62 cm
T 24–34 cm
W 3.7–4 kg

ssp *tigrinus*

ssp *oncilla*

Northern Oncilla
Leopardus tigrinus
FR Oncille de Cayenne
DE Ozelotkatze
ES Tigrillo

0–4800 m

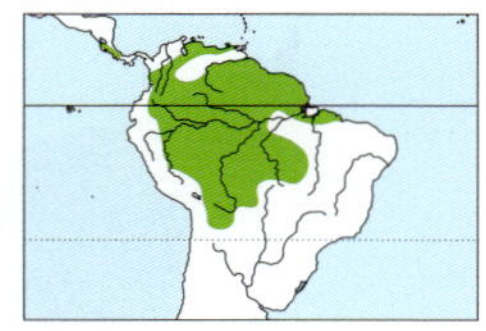

CR, W PA, VE, E Andes & Amazonian Basin S to BO & C BR; S limits not well known

VU
2 sspp
HB 45–56 cm
T 24–34 cm
W c. 2.4 kg

Eastern Oncilla
Leopardus emiliae
FR Oncille d'Emilie
DE Östliche Ozelotkatze
ES Gato de Emilie

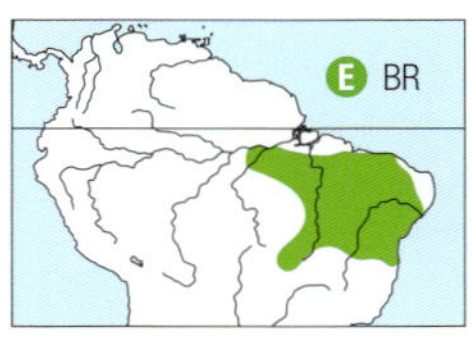

From Pará (S of R Amazon) E to Rio Grande do Norte, S to Bahia & Goiás

NE
Monotypic
HB 41–51 cm
T 26–32 cm
W 1.3–3.5 kg

Southern Oncilla
Leopardus guttulus
FR Oncille du Sud
DE Südliche Ozelotkatze
ES Tirica

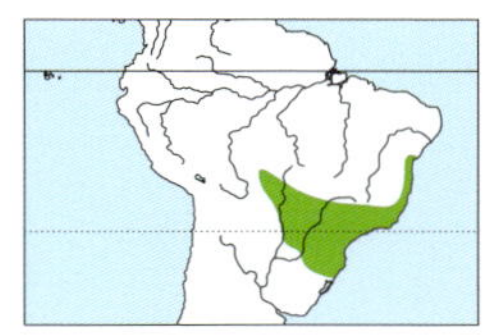

S & E BR (S from Mato Grosso & Espírito Santo), E PY, NE AR (Misiones)

VU
Monotypic
HB 36–54 cm
T 22–35 cm
W 1–4.6 kg

ssp *guigna*

ssp *tigrillo*
melanistic morph

VU
2 sspp
HB 39–51 cm
T 19.5–25 cm
W 1.3–3 kg

ssp *tigrillo*

Kodkod
Leopardus guigna
FR Kodkod
DE Chilenische Waldkatze
ES Huiña

below 2000 m

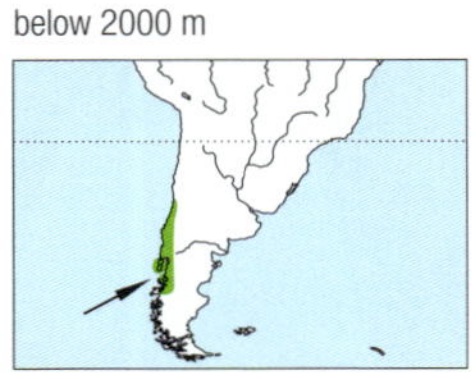

C & S CL, adjacent SW AR

Geoffroy's Cat
Leopardus geoffroyi
FR Chat de Geoffroy
DE Kleinfleckkatze
ES Gato de Geoffroy

0–3300 m

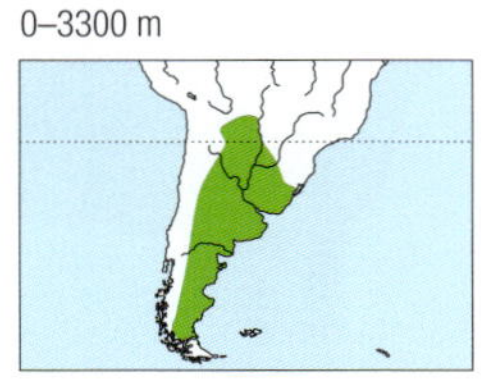

From C BO, PY, S BR & UY S to S CL & S AR

LC
Monotypic
HB 43–88 cm
T 23–40 cm
W 1.8–7.8 kg

Genus *LYNX*

(4 species)

Bobcat

Lynx rufus

FR Lynx roux
DE Rotluchs
ES Lince rojo

LC
4 sspp
HB 65–105 cm
T 9–11 cm
W 5.8–13.3 kg

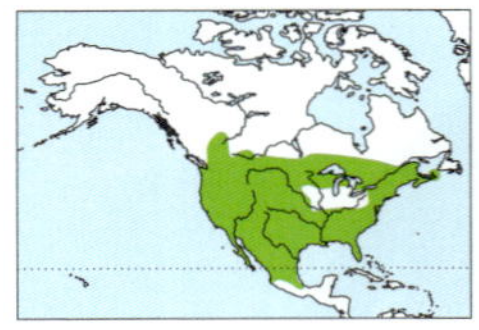

From S CA (from British Columbia E to Quebec) S throughout US to C MX

Canada & N USA form

ssp *rufus*

S USA form

ssp *fasciatus*

summer

winter

Canadian Lynx

Lynx canadensis

FR Lynx du Canada
DE Kanadischer Luchs
ES Lince canadiense

LC
Monotypic
HB 76–107 cm
T 5–12.7 cm
W 5–17.3 kg

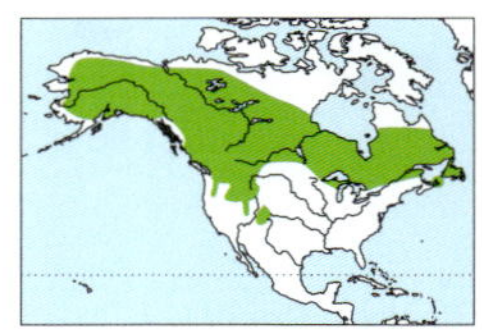

Boreal forests of Alaska, C & S CA (including Newfoundland I) S along Rocky Mts to Colorado (US)

ssp *wrangeli*

ssp *carpathicus*

ssp *dinniki*

summer

ssp *lynx*

winter

Eurasian Lynx

Lynx lynx

FR Lynx d'Europe
DE Eurasischer Luchs
ES Lince boreal

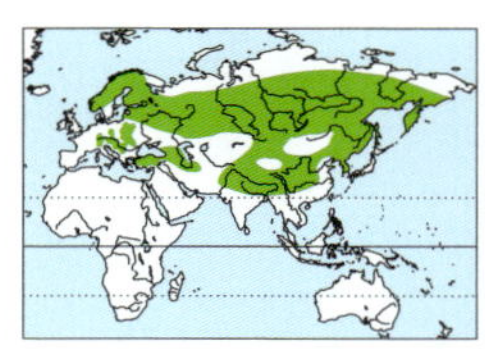

Eurasia from Scandinavia & FR E to E Siberia & CN, S to TR, Caucasus, N IR, Tibetan Plateau & Himalayas; extirpated from parts of W & C Europe

LC
6 sspp
HB 80–110 cm
T 16–23 cm
W 15–29 kg

Iberian Lynx
Lynx pardinus
FR Lynx ibérique
DE Pardelluchs
ES Lince ibérico

EN
Monotypic
HB 68–82 cm
T 12.5–16 cm
W 7–14 kg

SW ES, reintroduced to S PT

color variants

Genus *ACINONYX*
(1 species)

Cheetah
Acinonyx jubatus
FR Guépard
DE Gepard
ES Guepardo

VU
4 sspp
HB 121–145 cm
T 63–76 cm
W 36–59 kg

typical morph
ssp *jubatus*

"*rex*" morph

up to 2000 m
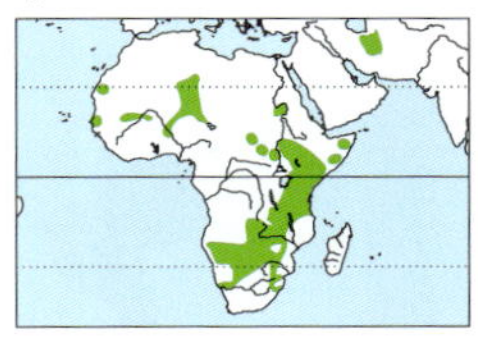
Notably fragmented across Africa, strongholds in E (ET, SS, KE, TZ, N UG) & S (AO, NA, BW, ZM, N ZA, SW MZ), rare in N (mainly SC DZ, NE ML, NE, BJ, BF, SW TD & NE CF), in Asia, surviving only in IR

ssp *venaticus*

ssp *soemmeringii*

Genus *PUMA*

(1 species)

Puma
Puma concolor

FR Couguar
DE Puma
ES Puma

LC
2 sspp
HB 86–155 cm
T 60–97 cm
W 34–72 kg

ssp *cougar*
NW Canada & NW USA form

ssp *concolor*

0–4000 m

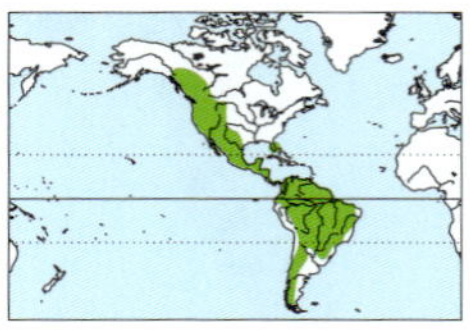

Widely in SW CA, W US (isolated population in Florida), MX, Central & South America S to S CL

ssp *cougar*
SE USA (Florida)
spotted morph

Genus *HERPAILURUS*

(1 species)

Jaguarundi
Herpailurus yagouaroundi

FR Jaguarondi
DE Jaguarundi
ES Yaguarundi

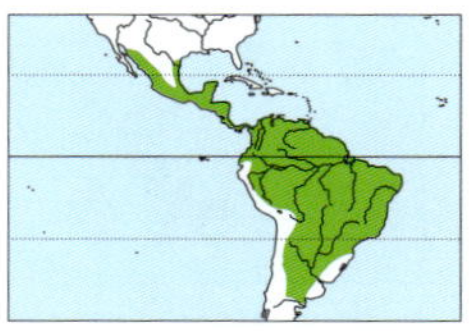

From MX S to C AR (Neuquén, Río Negro); probably extinct in US (S Texas)

red-brown morph

LC
Monotypic
HB 49–83 cm
T 27–59 cm
W 3–7.6 kg

iron-gray morph

Genus *OTOCOLOBUS*

(1 species)

Pallas's Cat
Otocolobus manul

FR Manul
DE Manul
ES Manul

LC
2 sspp
HB 46–65 cm
T 21–31 cm
W 2.5–4.5 kg

up to 4000 m

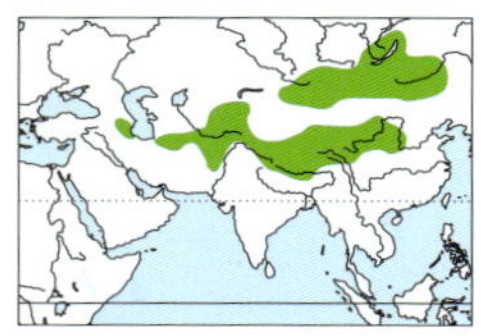

S Caucasus, IR, C Asia (N to KZ), AF, PK, NP, IN, Tibetan Plateau through NW & N CN to MN & RU (L Baikal area)

N form

S form

Genus *PRIONAILURUS*

(5 species)

Rusty-spotted Cat
Prionailurus rubiginosus

FR Chat rougeâtre
DE Rostkatze
ES Gato indio

NT
3 sspp
HB 35–48 cm
T 15–30 cm
W 1.1–1.6 kg

up to 2100 m

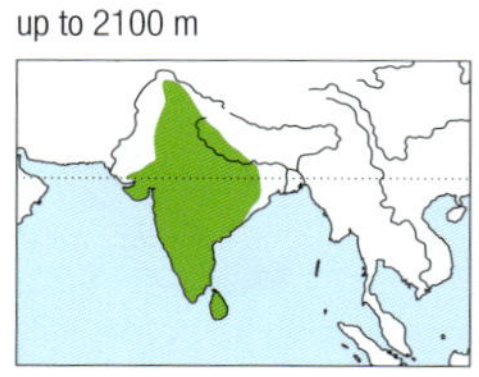

Much of IN, SW NP, LK

Flat-headed Cat
Prionailurus planiceps
FR Chat à tête plate
DE Flachkopfkatze
ES Gato cangrejero

EN
Monotypic
HB 45–52 cm
T 12.8–16.9 cm
W 1.5–2.5 kg

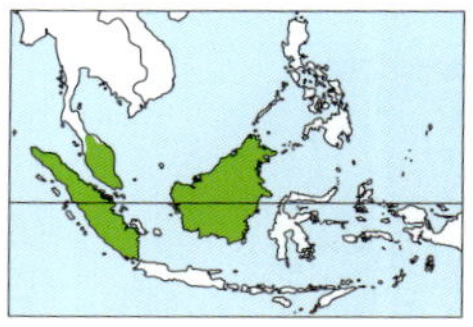
Thai-Malay Pen, Sumatra, Borneo

Fishing Cat
Prionailurus viverrinus
FR Chat viverrin
DE Fischkatze
ES Gato pescador

VU
2 sspp
HB 57–115 cm
T 24–40 cm
W 5–16 kg

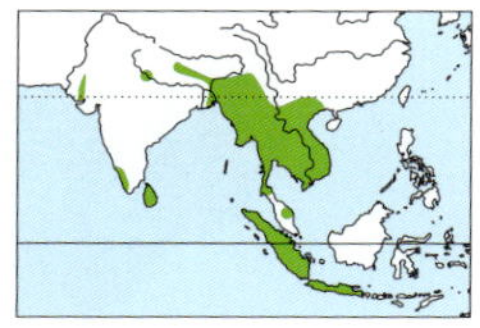
IN, LK, mainland SE Asia, Sumatra, Java

Mainland Leopard Cat
Prionailurus bengalensis
FR Chat du Bengale
DE Bengalkatze
ES Gato de Bengala

0–3000 m

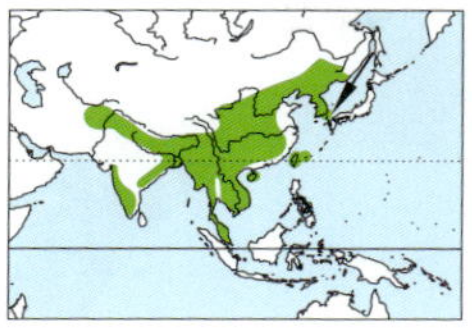
From PK & IN E to mainland SE Asia, N through CN (including Hainan) to RU (Far East), Korea Pen, TW & JP (Iriomote & Tsushima Is)

LC
2 sspp
HB 45–75 cm
T 19.5–31 cm
W 1.7–7.1 kg

ssp *euptilurus*

Iriomote I form

ssp *bengalensis*

Sunda Leopard Cat
Prionailurus javanensis
FR Chat de la Sonde
DE Sunda-Bengalkatze
ES Gato de Sonda

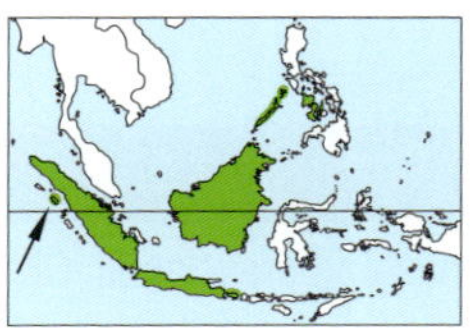
Sumatra (including Nias I), Java, Bali, Borneo, likely introduced to PH (Palawan, Panay, Negros, Cebu)

NE
2 sspp
HB 40–43 cm
T 21–22 cm
W c. 2.5 kg

Genus *FELIS*

(6 species)

Jungle Cat
Felis chaus
FR Chat des marais
DE Rohrkatze
ES Gato marismeño

Egypt form

LC
3 sspp
HB 61–85 cm
T 20–31 cm
W 2.6–12 kg

up to 1000 m

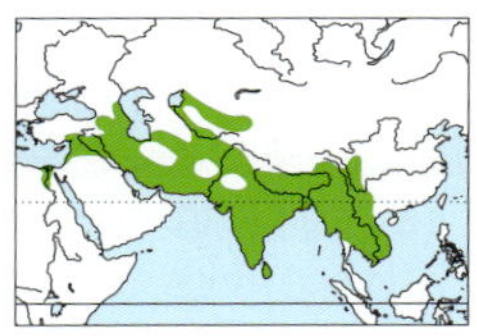
EG, Caucasus, Middle East, C Asia (N to KZ), Indian subcontinent, LK, mainland SE Asia, W & S CN

Black-footed Cat
Felis nigripes
FR Chat à pieds noirs
DE Schwarzfußkatze
ES Gato de pies negros

VU
Monotypic
HB 36–52 cm
T 12.6–20 cm
W 1–2.4 kg

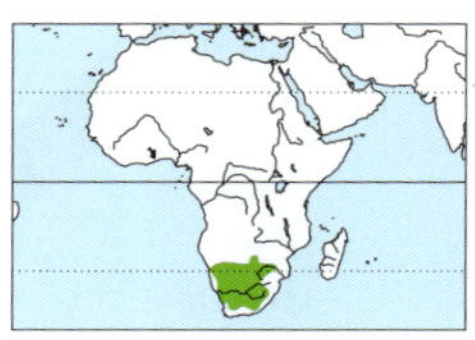
Karoo & Kalahari regions in NA, ZA & BW, marginally ZW; perhaps S AO

ssp *margarita*

ssp *thinobia*

Sand Cat
Felis margarita

FR Chat des sables
DE Sandkatze
ES Gato de las arenas

LC
2 sspp
HB 39–52 cm
T 23–31 cm
W 1.3–3.4 kg

up to 1200 m

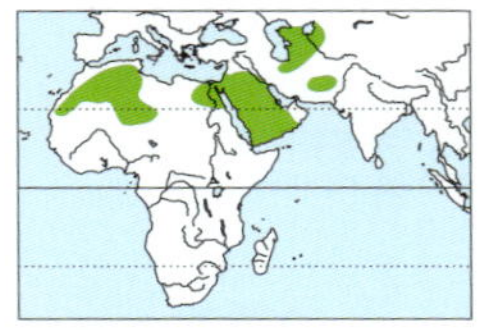

Disjunct and fragmented in deserts of N Africa (Sahara), Arabian Pen & SW Asia; range uncertain

Chinese Mountain Cat
Felis bieti

FR Chat de Biet
DE Graukatze
ES Gato chino de montaña

VU
Monotypic
HB 68–84 cm
T 32–35 cm
W 6.5–9 kg

2500–4100 m

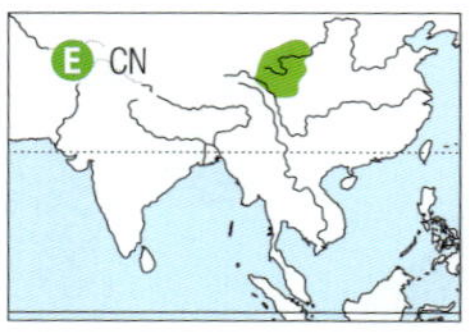

Reportedly Qinghai, Gansu & Sichuan; limits undefined

ssp *lybica*

African Wildcat
Felis lybica

FR Chat ganté
DE Afrikanische Wildkatze
ES Gato montés africano

LC
3 sspp
HB 40–66 cm
T 24–37 cm
W 2.4–6.4 kg

ssp *ornata*

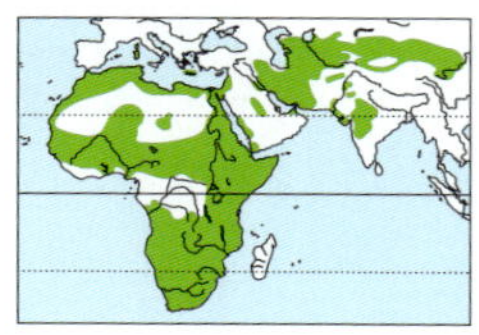

Most of Africa (except rainforests and parts of Sahara), Corsica, Sardinia, Crete, SW & C Asia to IN, CN & MN

European Wildcat
Felis silvestris

FR Chat sauvage
DE Wildkatze
ES Gato montés europeo

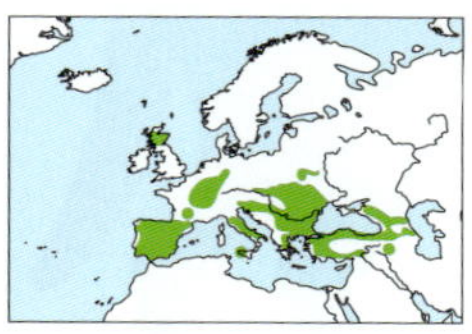

Europe E to Carpathians & R Dnieper, Scotland, Sicily, TR, Caucasus; introduced to Sardinia & Corsica, likely to Crete

LC
2 sspp
HB 47–74 cm
T 22–36 cm
W 2–7 kg

Family VIVERRIDAE – Civets, Genets and Oyans

14 genera, 36 species. Found mainly in forests, although some species also live in more open habitats such as savannas and grasslands. Old World tropics throughout Asia and Africa, also southern Europe.

Genus *POIANA*

(2 species)

Leighton's Oyan
Poiana leightoni

FR Poiane de Leighton
DE Westafrikanischer Linsang
ES Oyán occidental

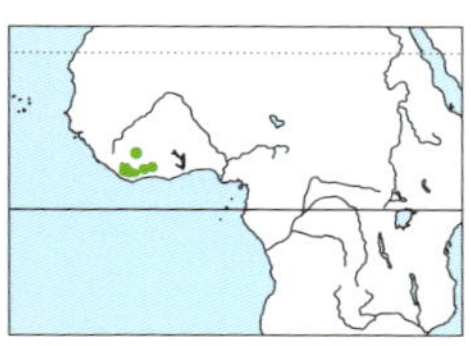

W Africa in LR & CI; unconfirmed in SE GN

VU
Monotypic
HB 30–38 cm
T 35–40 cm
W 500–700 g

Central African Oyan
Poiana richardsonii

FR Poiane de Richardson
DE Zentralafrikanischer Linsang
ES Oyán

LC
2 sspp
HB 32–40 cm
T 34–40 cm
W 455–750 g

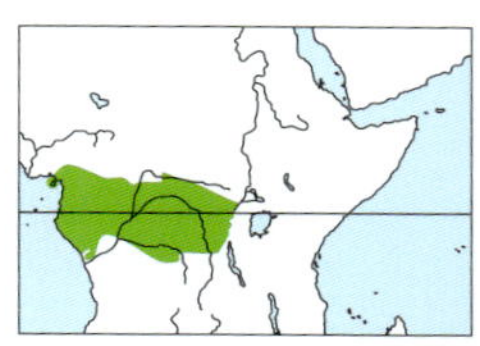

Lowland & montane forests in CM, CF, GQ (including Bioko I), GA, CG & CD

Genus *GENETTA*

(15 species)

Abyssinian Genet
Genetta abyssinica
FR Genette d'Éthiopie
DE Äthiopien-Genette
ES Gineta abisinia

up to 3750 m

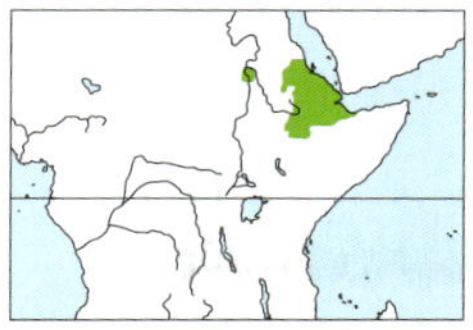

SE SD, ER, ET, DJ, NW SO

DD
Monotypic
HB 41–43 cm
T 38–40 cm
W 1.3–2 kg

Hausa Genet
Genetta thierryi
FR Genette de Thierry
DE Haussa-Genette
ES Gineta Hausa

LC
Monotypic
HB 44–45 cm
T 40–43 cm
W 1.3–1.5 kg

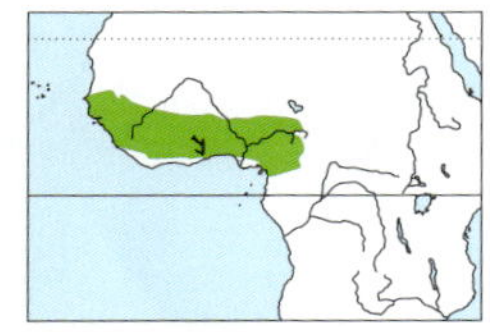

W Africa from SN E to NG & CM

Giant Genet
Genetta victoriae
FR Genette géante
DE Riesengenette
ES Gineta gigante

up to 2000 m

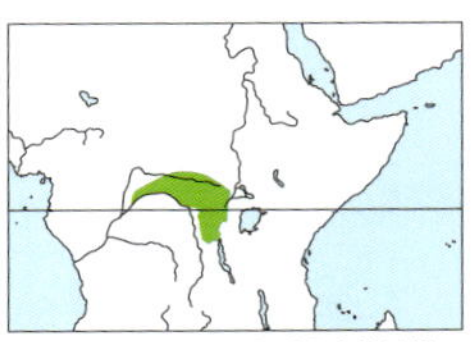

C Africa rainforest in CD & W UG

LC
Monotypic
HB 55–60 cm
T 41–49 cm
W 2.5–3.5 kg

NT
Monotypic
HB 47–51 cm
T 46–49 cm
W 2.2–2.6 kg

Johnston's Genet
Genetta johnstoni
FR Genette de Johnston
DE Liberia-Genette
ES Gineta de Nimba

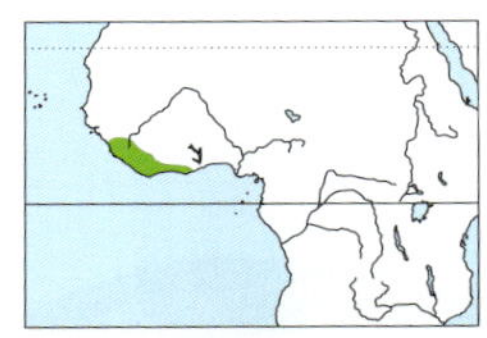

W Africa in GN, SL, LR, CI, GH

Aquatic Genet
Genetta piscivora
FR Genette aquatique
DE Wassergenette
ES Gineta acuática

460–1500 m

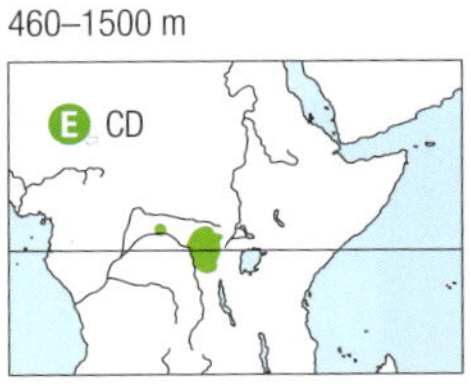

Rainforest, near small rivers

NT
Monotypic
HB 44–49 cm (M)
T 34–41 cm
W c. 1.5 kg

Servaline Genet
Genetta servalina
FR Genette servaline
DE Serval-Genette
ES Gineta servalina

LC
5 sspp
HB 44–51 cm
T 37–48 cm
W c. 2.3 kg (F)

up to 3500 m

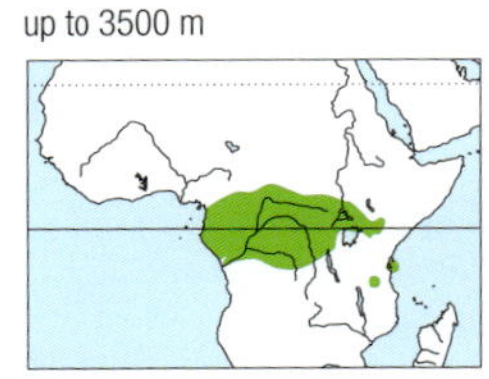

CM, CF, GQ, GA, CG, CD, SS, UG, RW, BI, KE, disjunct in C TZ & Unguja I (Zanzibar Arch)

Crested Genet
Genetta cristata
FR Genette à crête
DE Niger-Genette
ES Gineta crestada

up to 1000 m

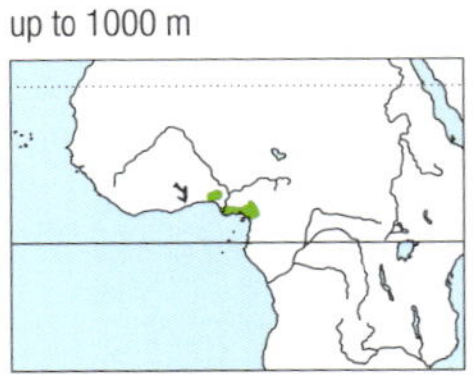

Forests in NG & CM; perhaps GA & CG

VU
Monotypic
HB 49–62 cm
T c. 43 cm
W c. 2.5 kg

ssp *pulchra*

Common Genet
Genetta genetta

FR Genette commune
DE Kleinfleckgenette
ES Gineta común

LC
5 sspp
HB 46–52 cm
T 40–52 cm
W 1.4–2.6 kg

0–2000 m

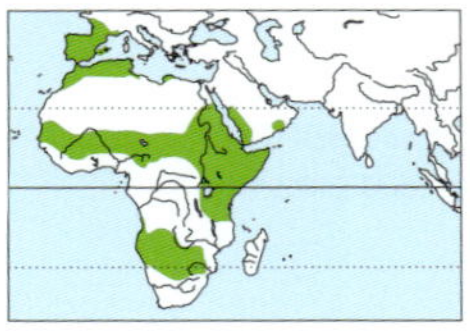

SW Europe (FR, ES, AD, PT), N Africa (MA E to LY), disjunct in W, C & S Africa, Arabian Pen (SW SA, YE, OM)

ssp *dongolana*

ssp *genetta*

Feline Genet
Genetta felina

FR Genette féline
DE Südliche Kleinfleckgenette
ES Gineta felina

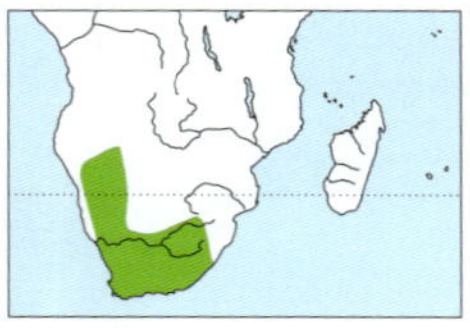

NA, ZA; unconfirmed in AO & ZM

NE
Monotypic
HB 43–56 cm
T 41–49 cm
W 1.4–2.4 kg

LC
2 sspp
HB 43–58 cm
T 38–46 cm
W 1.4–2.1 kg

Cape Genet
Genetta tigrina

FR Genette tigrine
DE Südliche Großfleckgenette
ES Gineta manchada

Higher rainfall areas in ZA (S Western Cape to S KwaZulu-Natal) & adjacent LS

Rusty-spotted Genet
Genetta maculata

FR Genette à grandes taches
DE Großfleckgenette
ES Gineta de manchas grandes

LC
4 sspp
HB 41–52 cm
T 39–54 cm
W 1.3–3.2 kg

280–1470 m

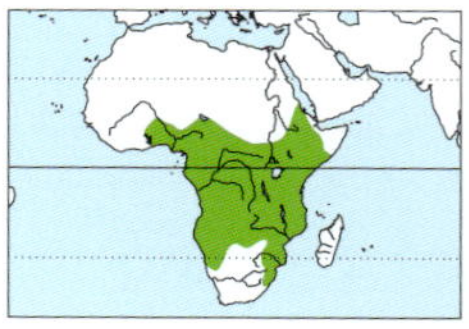

Widely in sub-Saharan Africa, from NG E to ER, ET & SO, S to C NA & NE ZA (KwaZulu-Natal)

Bourlon's Genet
Genetta bourloni

FR Genette de Bourlon
DE Bourlon-Genette
ES Gineta guineana

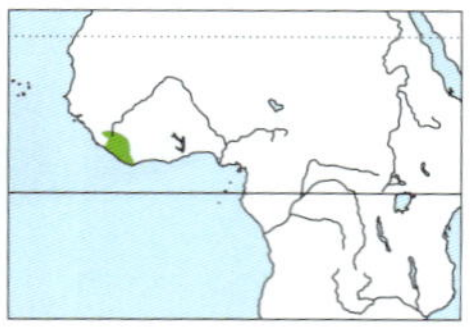

W Africa in SL, SE GN, LR & CI

VU
Monotypic
HB c. 49 cm
T c. 41 cm
W 1.5–2 kg

King Genet
Genetta poensis

FR Genette royale
DE Königsgenette
ES Gineta real

DD
Monotypic
HB c. 60 cm
T c. 41 cm
W 2–2.5 kg

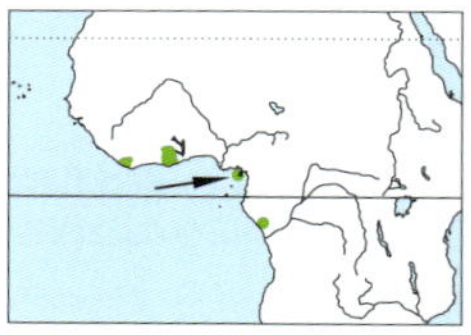

Disjunct in LR, CI, GH, GQ (Bioko I) & CG

Pardine Genet
Genetta pardina
FR Genette pardine
DE Pardelgenette
ES Gineta pardina

LC
Monotypic
HB 41–55 cm
T 39–49 cm
W ≤ 3.1 kg

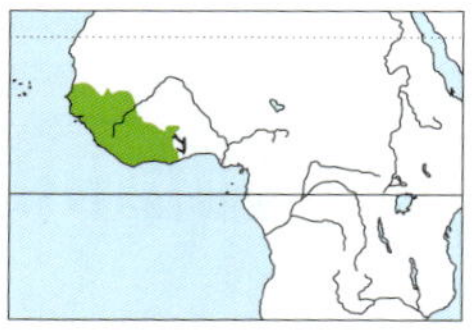

W Africa from SN & ML E to BF & GH

Angolan Genet
Genetta angolensis
FR Genette d'Angola
DE Angola-Genette
ES Gineta angoleña

LC
Monotypic
HB 44–48 cm
T 38–43 cm
W 1.3–2 kg

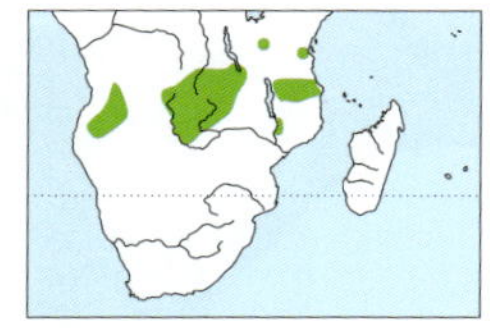

Disjunct in AO, S CD, ZM, TZ, N MZ & S MW

Genus *VIVERRICULA*

(1 species)

Small Indian Civet
Viverricula indica
FR Civette indienne
DE Kleine Indische Zibetkatze
ES Civeta india pequeña

LC
4 sspp
HB 48–68 cm
T 30–43 cm
W 2–4 kg

up to 1200 m

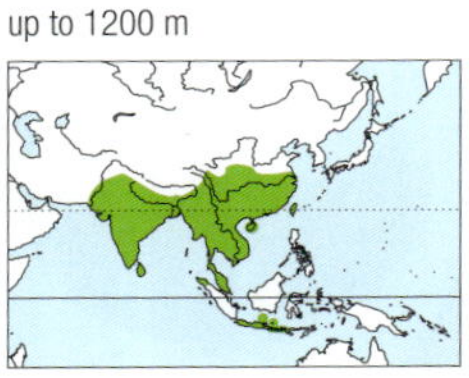

E PK, IN & LK E to S CN (including Hainan), TW, S to N Sumatra, Java & Bali; introduced to MG & Is in E Africa & SE Asia

Genus *CIVETTICTIS*

(1 species)

African Civet
Civettictis civetta
FR Civette d'Afrique
DE Afrikanische Zibetkatze
ES Civeta africana

ssp *civetta*

LC
6 sspp
HB 67–84 cm
T 34–47 cm
W 7–20 kg

ssp *schwarzi*

Widely from SN & MR E to SS, ET, DJ & S SO, S in all countries to NE NA, N & E BW, NE ZA & SZ, Zanzibar Arch

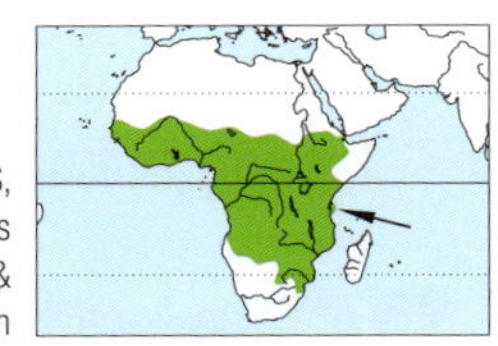

Malay Civet
Viverra tangalunga
FR Civette malaise
DE Malaiische Zibetkatze
ES Civeta malaya

Genus *VIVERRA*

(4 species)

up to 1100 m

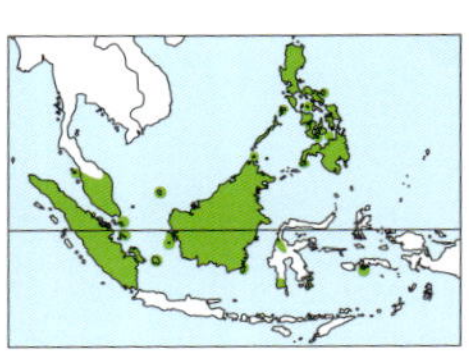

Pen MY, Sumatra, Borneo, Sulawesi, Java? & some smaller Is (E to Ambon), PH; probably introduced in part of range

LC
2 sspp
HB 54–77 cm
T 26–39 cm
W 3–7 kg

Large Indian Civet
Viverra zibetha
FR Grande Civette
DE Indische Zibetkatze
ES Civeta india grande

LC
6 sspp
HB 75–85 cm
T 38–49 cm
W 8–9 kg

up to 1600 m

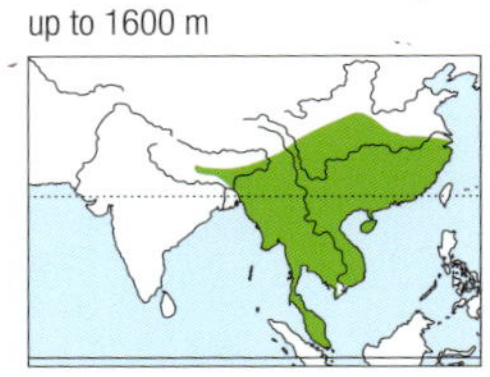

NP, NE IN, C & SE CN (including Hainan I), mainland SE Asia; introduced to Andamans (IN)

Large-spotted Civet
Viverra megaspila
FR Civette à grandes taches
DE Großfleck-Zibetkatze
ES Civeta moteada

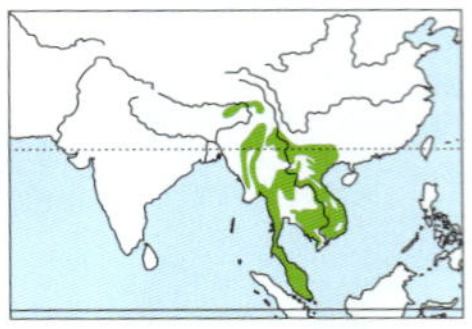

up to 520 m

Patchily in S CN & mainland SE Asia to Pen MY

EN
Monotypic
HB 72–85 cm
T 30–37 cm
W 8–9 kg

Malabar Civet
Viverra civettina
FR Civette de Malabar
DE Malabar-Zibetkatze
ES Civeta malabar

CR(PE)
Monotypic
HB 76–85 cm
T 30–40 cm
W 6.6–8 kg

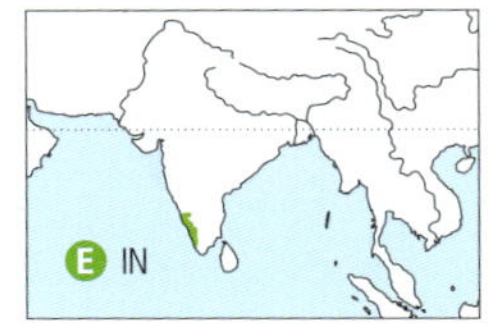

Western Ghats

Genus *ARCTOGALIDIA*

(1 species)

LC
4 sspp
HB 43–53 cm
T 46–66 cm
W 2–2.5 kg

up to 1500 m

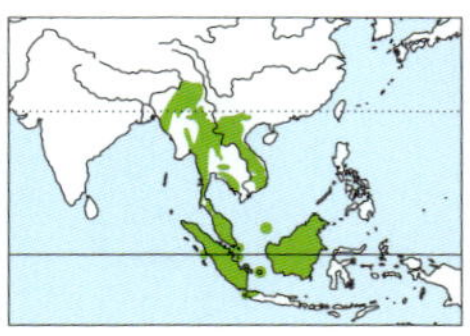

Small-toothed Palm Civet
Arctogalidia trivirgata
FR Civette à trois bandes
DE Streifenroller
ES Galidia

NE IN (Assam), BD, CN (Yunnan), mainland SE Asia, Sumatra, Borneo, W Java, several small Indonesian Is

Genus *ARCTICTIS*

(1 species)

mainland Asia & Sundas form

Binturong
Arctictis binturong
FR Binturong
DE Binturong
ES Binturong

Philippines form

VU
Monotypic
HB 61–96 cm
T 50–84 cm
W 9–20 kg

up to 2500 m

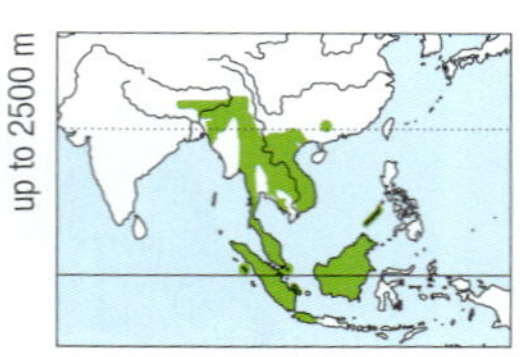

E NP, NE IN, BT, BD, SC CN (Yunnan, Guangxi), mainland SE Asia, Sumatra, Java, several small Is, PH (Calauit, Palawan)

Genus *PAGUMA*

(1 species)

LC
6 sspp
HB 51–87 cm
T 51–64 cm
W 3–5 kg

up to 2500 m

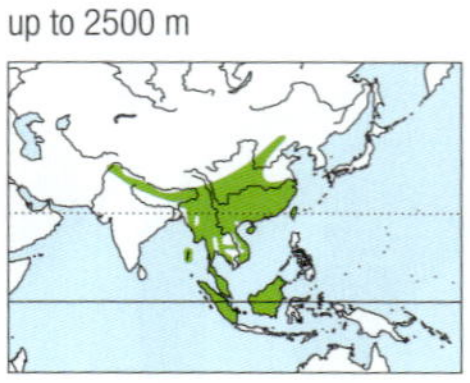

Masked Palm Civet
Paguma larvata
FR Civette masquée
DE Larvenroller
ES Paguma

Himalayas from PK to CN (including Hainan), TW, mainland SE Asia, Borneo, Sumatra, Andaman Is; introduced to JP

Genus *MACROGALIDIA*

(1 species)

up to 2600 m

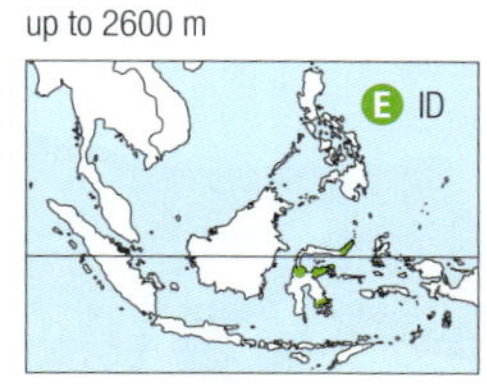

N, C, E & SE Sulawesi

Sulawesi Palm Civet
Macrogalidia musschenbroekii
FR Civette des Célèbes
DE Sulawesi-Roller
ES Civeta de las Célebes

VU
Monotypic
HB 65–71 cm
T 44–54 cm
W 3.9–6.1 kg

Northern Palm Civet
Paradoxurus hermaphroditus

FR Civette hermaphrodite
DE Nördlicher Fleckenmusang
ES Musang septentrional

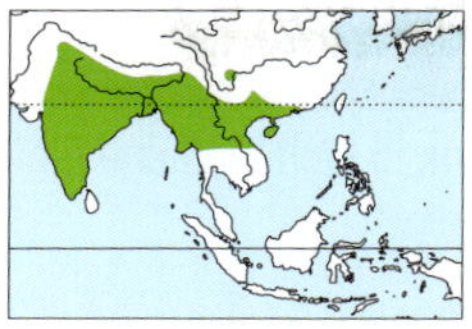

PK, IN, NP, BT, BD, S CN (including Hainan), MM, N VN, N LA, N TH; introduced to JP & LK

LC
2 sspp
HB 42–71 cm
T 33–66 cm
W 2–5 kg

Genus *PARADOXURUS*

(5 species)

Southern Palm Civet
Paradoxurus musangus

FR Civette de Sumatra
DE Südlicher Fleckenmusang
ES Musang de Sumatra

NE
2 sspp

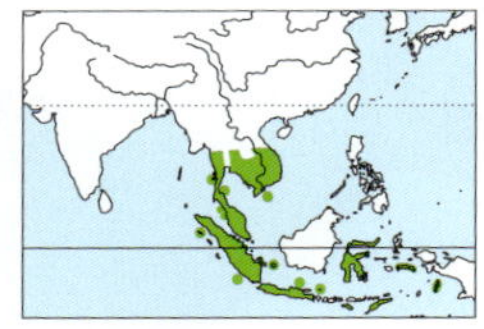

S VN, S LA, KH, S TH, Pen MY, Sumatra & Java, Is further E (to Aru) probably introductions

Philippine Palm Civet
Paradoxurus philippinensis

FR Civette des Philippines
DE Philippinen-Fleckenmusang
ES Musang de Filipinas

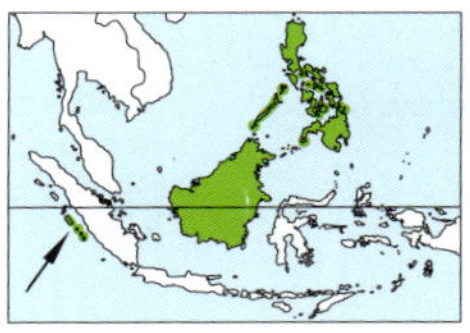

Mentawai Is (Siberut, North Pagai, South Pagai), Borneo, PH

NE
2 sspp

500–1300 m

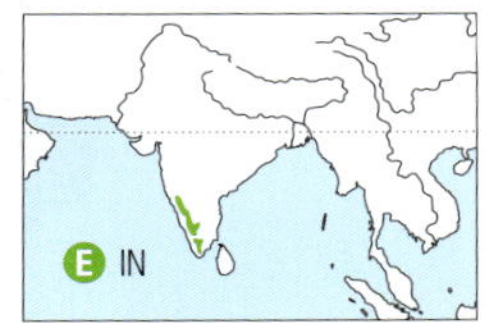

Brown Palm Civet
Paradoxurus jerdoni

FR Civette de Jerdon
DE Jerdon-Musang
ES Musang indio

LC
2 sspp
HB 51–61 cm
T 44–50 cm
W 2–4.3 kg

Western Ghats (Maharashtra S to Kerala)

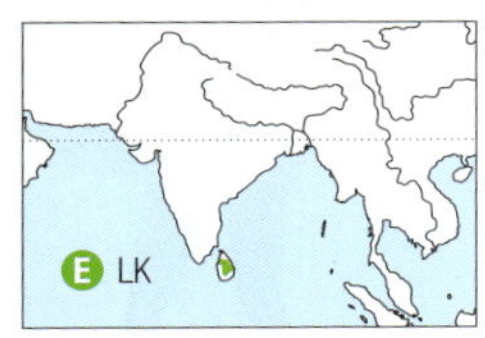

Golden Palm Civet
Paradoxurus zeylonensis

FR Civette de Ceylan
DE Goldmusang
ES Musang dorado

LC
Monotypic
HB 50–58 cm
T 44–52 cm
W c. 3.6 kg

Range includes almost entire country

Genus *CYNOGALE*

(1 species)

Otter Civet
Cynogale bennettii

FR Civette loutre
DE Otterzivette
ES Civeta pescadora

up to 1200 m

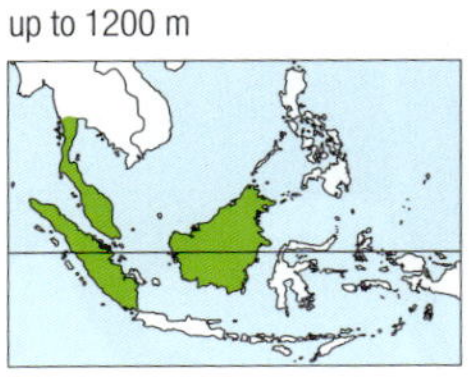

S MM, S TH, Pen MY, SG, Sumatra, Borneo; possible records in N VN (Tonkin) & CN (S Yunnan)

EN
Monotypic
HB 57–68 cm
T 12–20 cm
W 3–5 kg

Genus *CHROTOGALE*

(1 species)

Owston's Palm Civet
Chrotogale owstoni

FR Civette d'Owston
DE Fleckenroller
ES Hemigalo chino

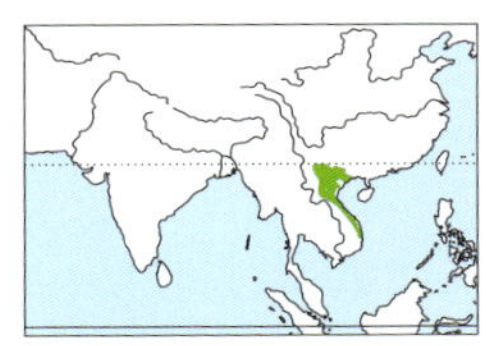

SC CN (Yunnan, Guangxi), LA, VN

EN
Monotypic
HB 56–72 cm T 35–47 cm
W 2.5–4.2 kg

Genus *HEMIGALUS*

(1 species)

Banded Palm Civet
Hemigalus derbyanus

FR Civette de Derby
DE Bänderroller
ES Hemigalo cebrado

NT
3 sspp
HB 41–56 cm
T 23–37 cm
W 1–3 kg

up to 1200 m

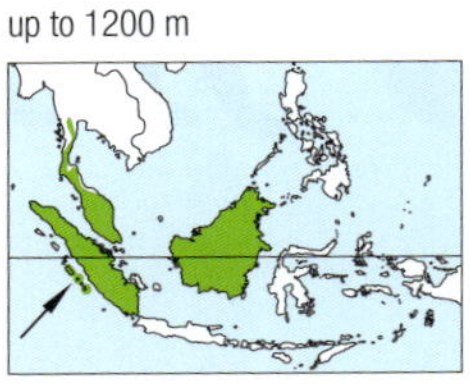

S MM, S TH, Pen MY, Sumatra (including Siberut, South Pagai & Sipora Is), Borneo

Genus *DIPLOGALE*

(1 species)

Hose's Palm Civet
Diplogale hosei

FR Civette de Hose
DE Schlichtroller
ES Hemigalo de Borneo

450–1700 m

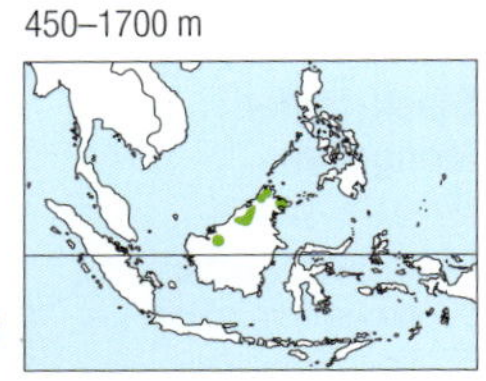

Borneo (Sabah, Sarawak, Kalimantan, BN)

VU
Monotypic
HB 47–54 cm T 30–33 cm
W c. 1.3 kg

Family HERPESTIDAE – Mongooses

15 genera, 35 species. From open areas, such as deserts, savannas, and grasslands, to closed forests, from lowlands to high montane regions. Old World tropics throughout Asia and Africa, also Middle East and southern Europe.

Genus *ICHNEUMIA*

(1 species)

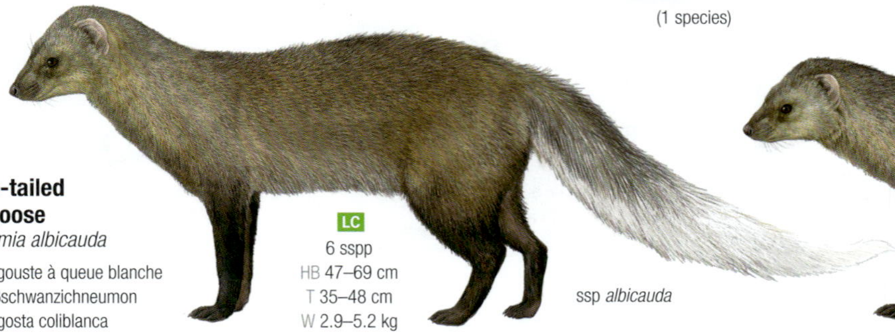

ssp *albicauda*

ssp *loandae*

up to 4000 m

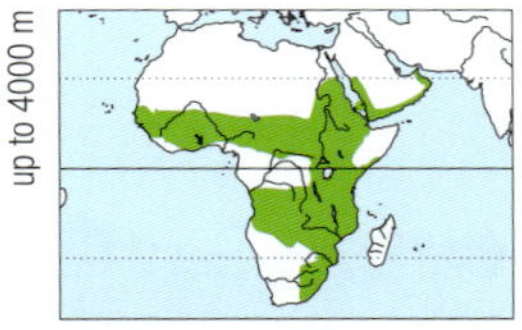

Widespread in sub-Saharan Africa (except rainforest & desert areas), SW SA, YE, OM

White-tailed Mongoose
Ichneumia albicauda

FR Mangouste à queue blanche
DE Weißschwanzichneumon
ES Mangosta coliblanca

LC
6 sspp
HB 47–69 cm
T 35–48 cm
W 2.9–5.2 kg

Yellow Mongoose
Cynictis penicillata

FR Mangouste fauve
DE Fuchsmanguste
ES Mangosta dorada

Genus *CYNICTIS*

(1 species)

LC
3 sspp
HB 26–46 cm T 15–29 cm
W 715–900 g

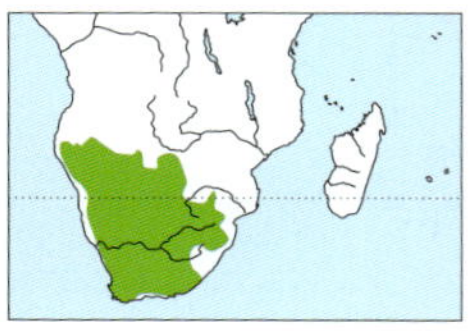

S Africa in S AO, NA, BW, W ZW, ZA & LS

Selous's Mongoose
Paracynictis selousi

FR Mangouste de Selous
DE Trugmanguste
ES Mangosta de Selous

Genus *PARACYNICTIS*

(1 species)

LC
4 sspp
HB 39–47 cm
T 28–43 cm
W 1.4–2.2 kg

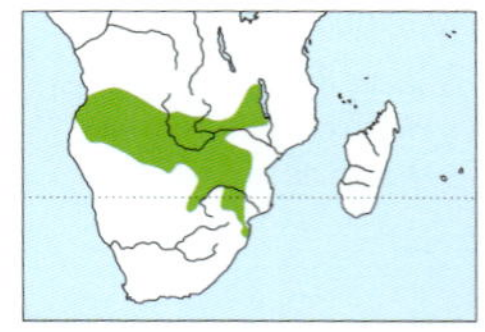

S Africa in S AO, NE NA, N & E BW, ZM, MW, ZW, S MZ, NE ZA

Genus *RHYNCHOGALE*

(1 species)

LC
2 sspp
HB 44–50 cm
T 28–41 cm
W 1.8–2.8 kg

Meller's Mongoose
Rhynchogale melleri

FR Mangouste de Meller
DE Meller-Manguste
ES Mangosta crestada

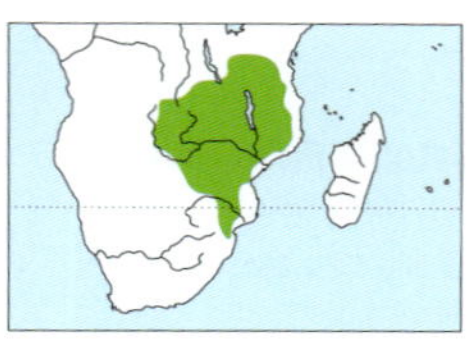

E & SE Africa from C TZ & SE CD through ZM, MW, ZW, MZ to NE ZA & SZ; perhaps NE BW

Genus *BDEOGALE*

(4 species)

LC
4 sspp
HB 38–45 cm
T 22–28 cm
W 1.1–2.1 kg

Bushy-tailed Mongoose
Bdeogale crassicauda

FR Mangouste à queue touffue
DE Buschschwanzmanguste
ES Mangosta coligruesa

up to 1850 m

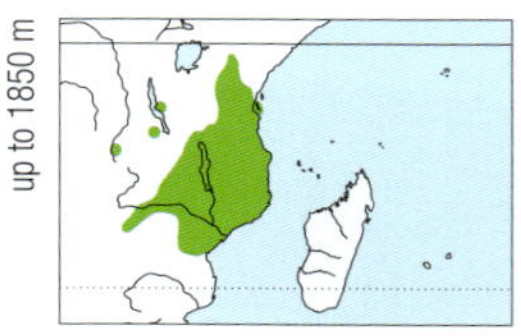

KE (Lukenya Hill), TZ (including Unguja I in Zanzibar Arch), SE CD, N & E ZM, MW, N & C MZ, N ZW; uncorfirmed in YE

Sokoke Mongoose
Bdeogale omnivora

FR Mangouste côtière
DE Sokoke-Buschschwanzmanguste
ES Mangosta de Sokoke

up to 1700 m

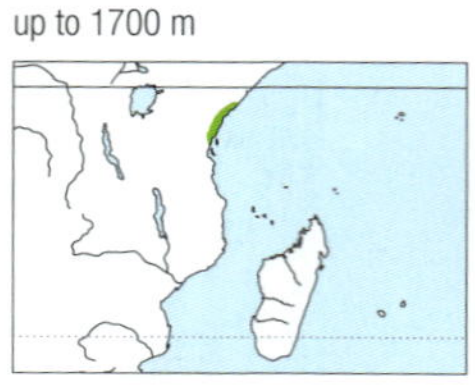

Coastal forests of KE & extreme NE TZ

VU
Monotypic
HB c. 42 cm (F)
T c. 24 cm (F)

LC
Monotypic
HB 45–65 cm
T 29–40 cm
W 2–4.8 kg

Black-footed Mongoose
Bdeogale nigripes

FR Mangouste à pattes noires
DE Schwarzfußmanguste
ES Mangosta calzada

up to 1000 m

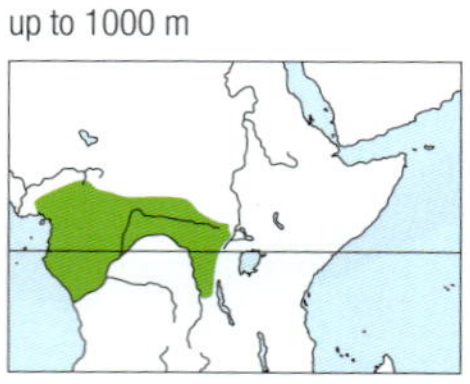

C Africa in SE NG, CM, CF, CD, GQ, GA, CG; uncertain in AO

NT
Monotypic
HB 51–57 cm T 28–32 cm
W 2–3 kg

up to 3300 m

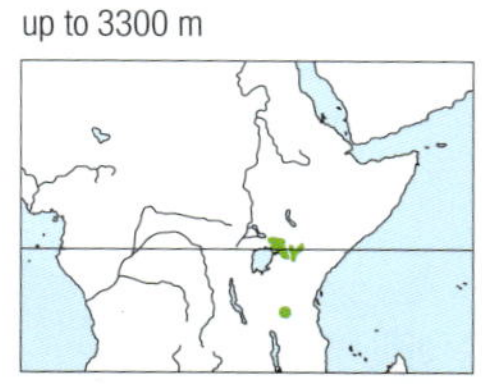

E Africa in SE UG, KE (Aberdare Range, Mt Kenya) & TZ (Udzungwa Mts)

Jackson's Mongoose
Bdeogale jacksoni

FR Mangouste de Jackson
DE Jackson-Manguste
ES Mangosta masai

Egyptian Mongoose
Herpestes ichneumon

FR Mangouste d'Égypte
DE Ichneumon
ES Meloncillo

LC
Monotypic
HB 50–61 cm
T 43–61 cm
W 2.2–4.1 kg

Genus *HERPESTES*
(5 species)

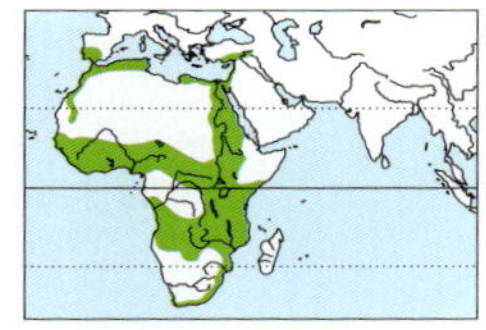

S Iberian Pen, N Africa, S TR, Levant, sub-Saharan Africa from SN E to Red Sea, S to N NA & N BW, along coast to S ZA

Common Slender Mongoose
Herpestes sanguineus

FR Mangouste svelte
DE Rotes Schlankichneumon
ES Mangosta esbelta

0–2700 m

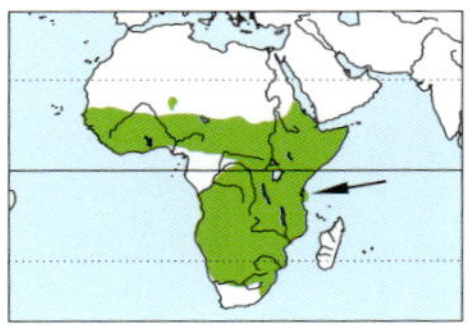

Widely in sub-Saharan belt from SN E to Red Sea (SD), S to ZA, also Unguja I (Zanzibar Arch)

LC
Monotypic
HB 27–34 cm
T 19–31 cm
W 277–789 g

gray morph

reddish morph

dark morph

Somalian Slender Mongoose
Herpestes ochraceus

FR Mangouste d'Abyssinie
DE Somalia-Schlankichneumon
ES Mangosta somalí

LC
Monotypic
HB 25–29 cm
T 22–27 cm (M)

up to 600 m

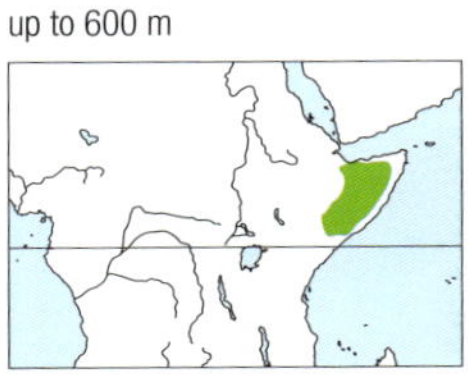

Horn of Africa in ET, SO & marginally II NE KE

dark brown morph

gray morph

Angolan Slender Mongoose
Herpestes flavescens

FR Mangouste flavescente
DE Kaokoveld-Schlankichneumon
ES Mangosta angoleña

LC
Monotypic
HB 31–35 cm
T 32–36 cm

dark morph

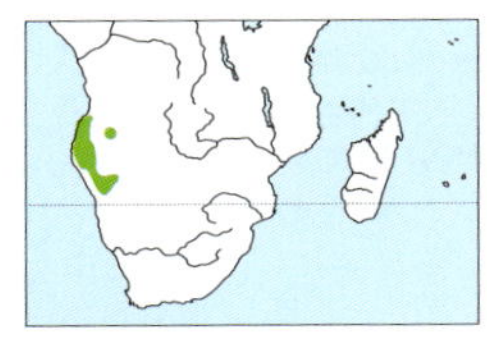

Restricted to SW AO & NW NA

reddish morph

Cape Gray Mongoose
Herpestes pulverulentus

FR Mangouste du Cap
DE Kap-Schlankichneumon
ES Mangosta de El Cabo

LC
3 sspp
HB 33–42 cm
T 20–34 cm
W 0.5–1.2 kg

0–1900 m

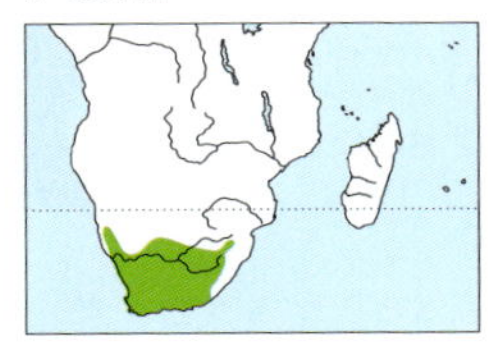

S Africa in S NA, ZA & LS

Genus *ATILAX*
(1 species)

Marsh Mongoose
Atilax paludinosus

FR Mangouste des marais
DE Sumpfmanguste
ES Mangosta negra

LC
10 sspp
HB 49–51 cm
T 31–41 cm
W 2.4–4 kg

0–2500 m

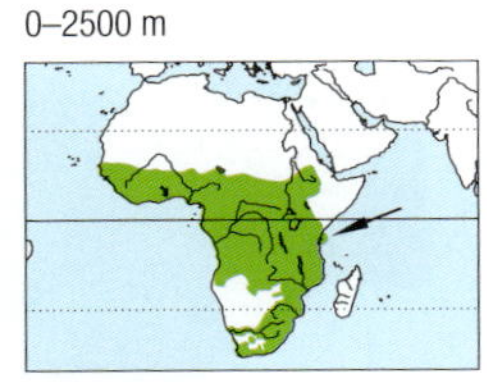

Widely from SN & GW E to S SD & ET, S to ZA (except most of NA, BW & parts of ZA), also Pemba I

Genus *XENOGALE*

(1 species)

Long-nosed Mongoose

Xenogale naso

FR Mangouste à long nez
DE Langnasenichneumon
ES Mangosta hocicuda

LC

Monotypic
HB 40–61 cm
T 32–42 cm
W 1.9–4.5 kg

up to 640 m

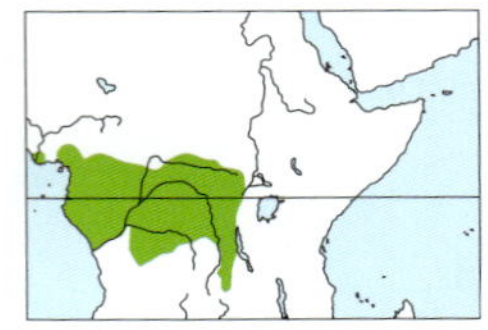

C Africa in SE NG, CM, CF, GQ, GA, CG, CD

Short-tailed Mongoose

Urva brachyura

FR Mangouste à queue courte
DE Kurzschwanzmungo
ES Meloncillo colicorto

up to 1280 m

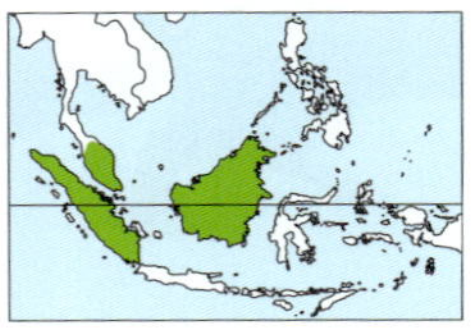

S TH, Pen MY, SG, Sumatra, Borneo

NT

2 sspp
HB 35–49 cm
T 17–25 cm
W 1–3 kg

Genus *URVA*

(9 species)

Collared Mongoose

Urva semitorquata

FR Mangouste à collier
DE Halsbandmungo
ES Meloncillo de collar

NT

2 sspp
HB 40–45 cm
T 26–30 cm
W 3–4 kg

up to 1200 m

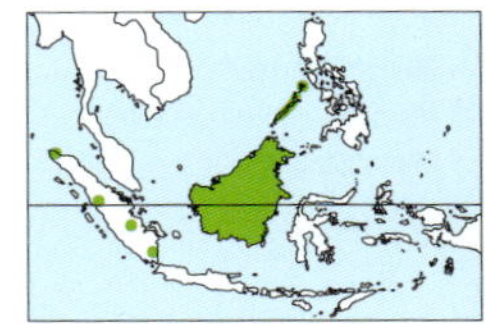

Borneo (Sabah, Sarawak, Kalimantan, perhaps BN), Sumatra (few records), PH (Palawan, Calamian Is)

Crab-eating Mongoose

Urva urva

FR Mangouste crabière
DE Krabbenmanguste
ES Meloncillo cangrejero

up to 2000 m

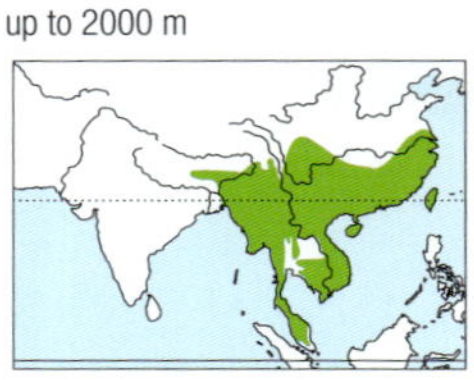

NP, NE IN, BT, BD, mainland SE Asia, TW, SC & SE CN (including Hainan I), TW

LC

4 sspp
HB 44–56 cm
T 26–34 cm
W 3–4 kg

Ruddy Mongoose

Urva smithii

FR Mangouste roussâtre
DE Indischer Rotmungo
ES Meloncillo rojo

LC

2 sspp
HB 39–47 cm
T 35–47 cm
W c. 2.7 kg

up to 2200 m

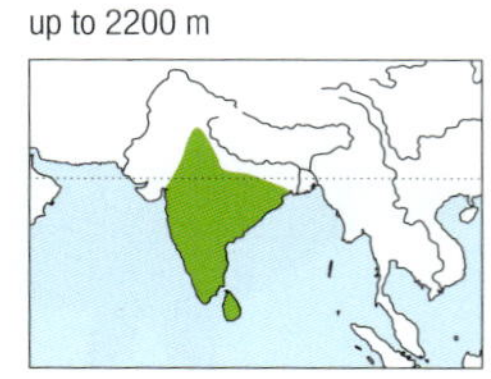

Much of peninsular IN, LK

Stripe-necked Mongoose

Urva vitticolla

FR Mangouste à cou rayé
DE Halsstreifenmungo
ES Meloncillo indio

up to 2135 m

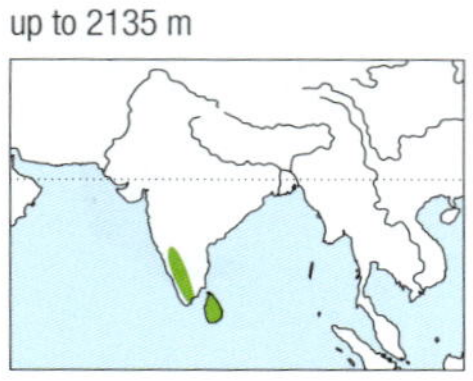

SW IN (Western Ghats), LK

LC

2 sspp
HB 46–53 cm
T 30–32 cm
W ≤ 3.4 kg

Indian Brown Mongoose

Urva fusca

FR Mangouste de Malabar
DE Indischer Kurzschwanzmungo
ES Meloncillo pardo

LC

4 sspp
HB 33–48 cm
T 19.8–34 cm
W c. 2.7 kg

900–1400 m

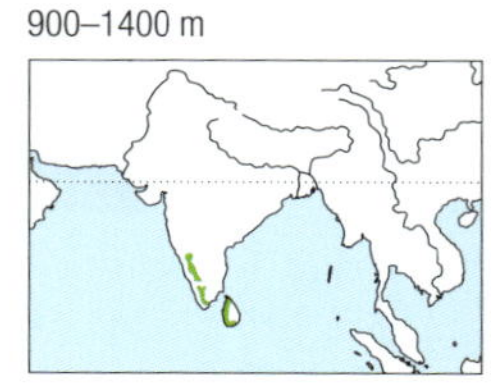

SW IN (Western Ghats), LK

Indian Gray Mongoose
Urva edwardsii

FR Mangouste d'Edwards
DE Indischer Mungo
ES Meloncillo gris

LC
5 sspp
HB 35–45 cm
T 32–45 cm
W c. 1.4 kg

up to 2100 m

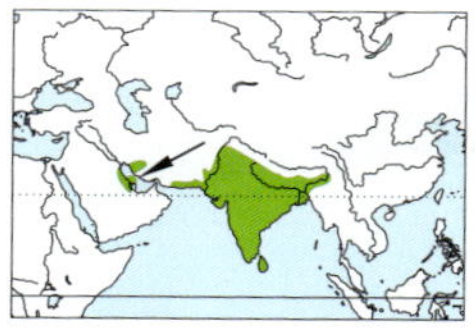

From Arabian Pen E through IR to IN, NP, BT, BD & LK

Small Indian Mongoose
Urva auropunctata

FR Mangouste tachetée
DE Goldstaubmungo
ES Meloncillo chico

LC
Monotypic
HB 25–37 cm
T 19.2–29 cm
W 305–662 g

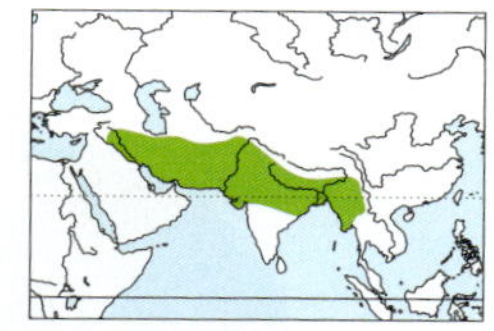

From IQ & IR E to BD & MM; introduced to West Indies, Guianas, HR, Mafia I (TZ), MU, JP, Hawaii & FJ

Javan Mongoose
Urva javanica

FR Mangouste de Java
DE Kleiner Mungo
ES Meloncillo de Java

LC
Monotypic
HB 30–41 cm
T 21–31 cm
W 0.5–1 kg

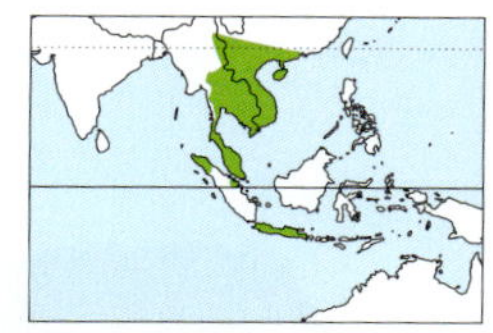

S CN (including Hainan), mainland SE Asia, Sumatra, Java

Genus *SURICATA*

(1 species)

Meerkat
Suricata suricatta

FR Mangouste suricate
DE Erdmännchen
ES Suricata

LC
3 sspp
HB 24–29 cm
T 19–24 cm
W 620–969 g

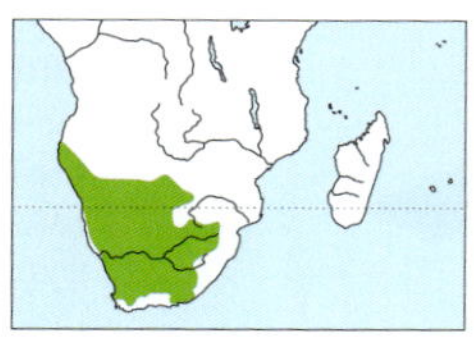

Namib & Kalahari deserts in SW AO, NA, BW & ZA; perhaps LS

Genus *MUNGOS*

(2 species)

Banded Mongoose
Mungos mungo

FR Mangouste rayée
DE Zebramanguste
ES Mangosta rayada

LC
Monotypic
HB 30–40 cm
T 18–31 cm
W 0.9–1.9 kg

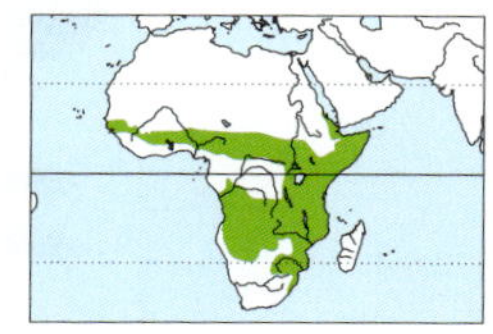

Sub-Saharan Africa (except rainforest & desert areas) in SN & GM E to ER & SO, S to AO, NE NA & E ZA

Genus *LIBERIICTIS*

(1 species)

Liberian Mongoose
Liberiictis kuhni

FR Mangouste du Libéria
DE Liberia-Kusimanse
ES Mangosta de Liberia

VU
Monotypic
HB 43–47 cm
T 19.7–20 cm
W c. 2.3 kg

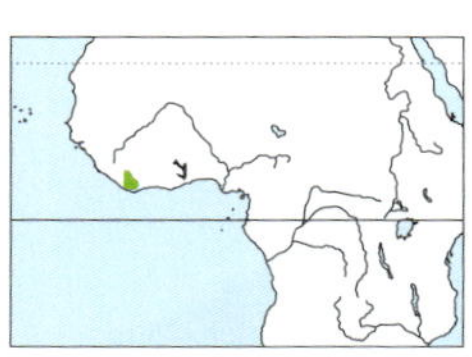

W Africa in LR & CI

Gambian Mongoose
Mungos gambianus

FR Mangouste de Gambie
DE Gambia-Manguste
ES Mangosta de Gambia

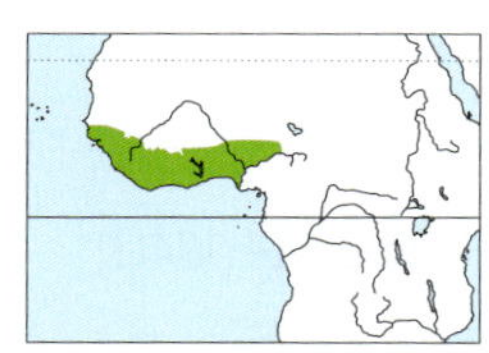

W Africa from SN & GM E to NG

LC
Monotypic
HB 30–45 cm
T 23–29 cm
W 1–2.2 kg

Genus *CROSSARCHUS*

(4 species)

Alexander's Cusimanse
Crossarchus alexandri

FR Mangouste d'Alexander
DE Kongo-Kusimanse
ES Cusimansé del Congo

LC
Monotypic
HB 35–44 cm
T 22–32 cm
W 1–2 kg

1500–2900 m

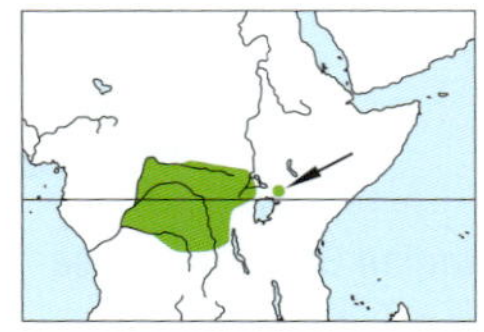

C Africa in CD & UG

Angolan Cusimanse
Crossarchus ansorgei

FR Mangouste d'Ansorge
DE Angola-Kusimanse
ES Cusimansé de Angola

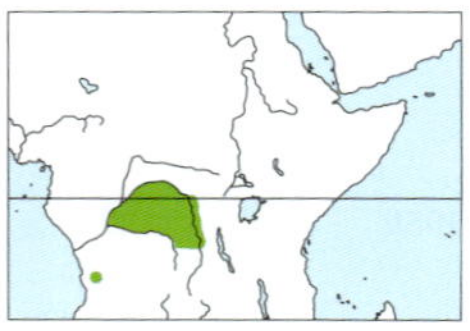

C Africa in CD (R Congo Basin) & N AO

LC
2 sspp
HB 32–36 cm
T 21–22 cm
W 0.6–1.5 kg (M)

LC
Monotypic
HB 30–37 cm
T 14.6–21 cm
W 0.4–1 kg

Common Cusimanse
Crossarchus obscurus

FR Mangouste brune
DE Dunkelkusimanse
ES Cusimansé del Níger

0–1500 m

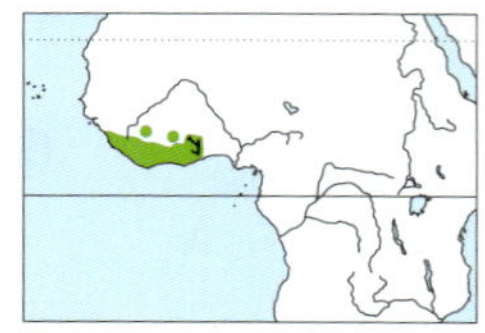

W Africa in GN, SL, LR, CI & GH

Flat-headed Cusimanse
Crossarchus platycephalus

FR Mangouste à tête plate
DE Kamerun-Kusimanse
ES Cusimansé de Camerún

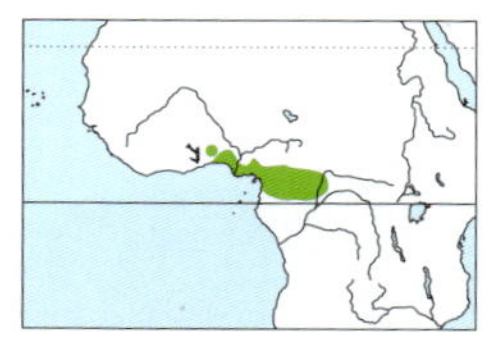

W & C Africa in BJ, NG, CM, CF, GQ & CG; probably GA

LC
Monotypic
HB 30–36 cm
T 15.6–21 cm
W 0.5–1.5 kg

Genus *HELOGALE*

(2 species)

Ethiopian Dwarf Mongoose
Helogale hirtula

FR Mangouste d'Éthiopie
DE Somalia-Zwergmanguste
ES Mangosta etíope

LC
Monotypic
HB 20–27 cm T 15–18 cm
W 220–354 g

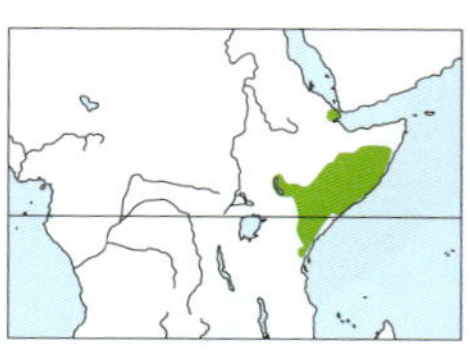

up to 600 m

E Africa in DJ, S ET, C & S SO, N & C KE & NE TZ

Common Dwarf Mongoose
Helogale parvula

FR Mangouste naine
DE Zwergmanguste
ES Mangosta enana

LC
7 sspp
HB 16–23 cm
T 14–19 cm
W 213–415 g

up to 1100 m

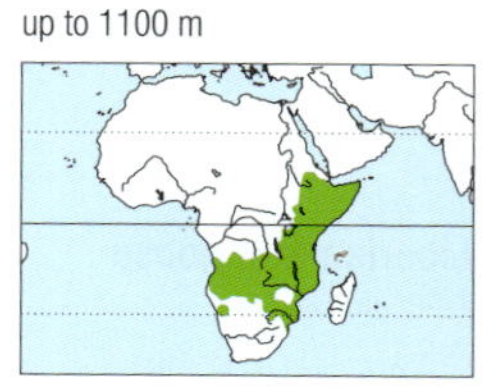

Widely distributed from SD, ET & SO S to NE ZA (KwaZulu-Natal), W to AO & W NA

Genus *DOLOGALE*

(1 species)

Pousargues's Mongoose
Dologale dybowskii

FR Mangouste de Dybowski
DE Listige Manguste
ES Mangosta centroafricana

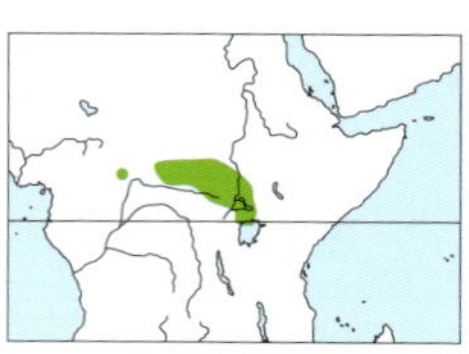

C Africa in CF, NE CD, S SS & W UG; perhaps CG

DD
Monotypic
HB 25–33 cm
T 16–23 cm
W 300–400 g

Family EUPLERIDAE – Madagascar Carnivores

7 genera, 7 species. Forested habitats (dry deciduous, rainforest, and montane); also sub-deserts and above forest line. Madagascar.

Genus *CRYPTOPROCTA*

(1 species)

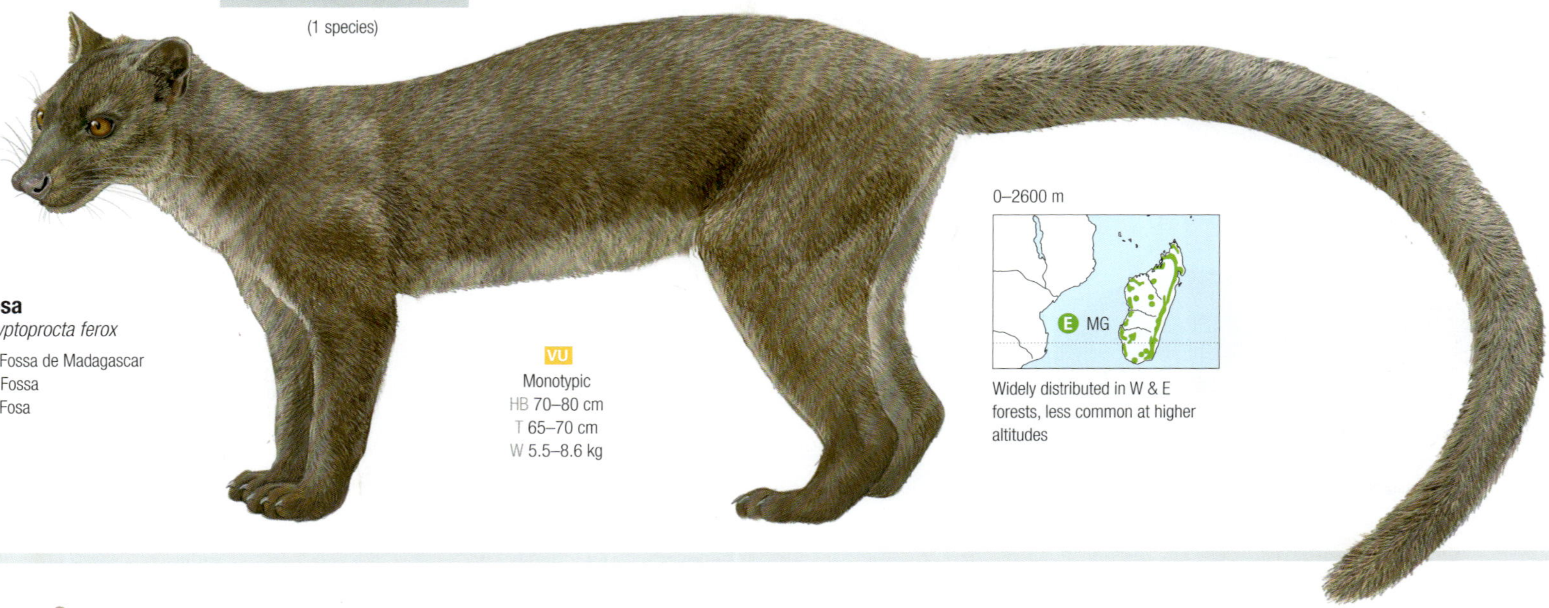

Fosa
Cryptoprocta ferox
FR Fossa de Madagascar
DE Fossa
ES Fosa

VU
Monotypic
HB 70–80 cm
T 65–70 cm
W 5.5–8.6 kg

0–2600 m

Widely distributed in W & E forests, less common at higher altitudes

VU
Monotypic
HB 40–45 cm
T 22–26 cm
W 1.3–1.9 kg

Genus *FOSSA*

(1 species)

Spotted Fanaloka
Fossa fossana
FR Fossane
DE Fanaloka
ES Fosana

0–1300 m

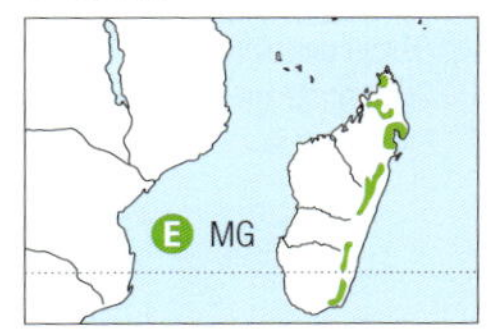

Humid E forests (from Montagne d'Ambre NP S to Andohahela NP)

VU
Monotypic
HB 45–65 cm
T 22–25 cm
W 1.6–4.6 kg

Genus *EUPLERES*

(1 species)

Falanouc
Eupleres goudotii
FR Euplère de Goudot
DE Ameisenschleichkatze
ES Falanuc

50–1600 m

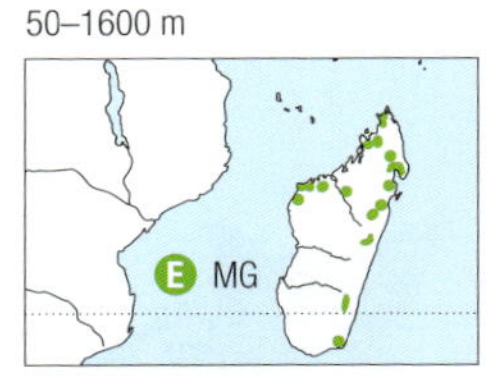

Mainly upland humid forest & marsh areas in NW & E (from Montagne d'Ambre NP S to Andohahela NP)

ssp *dambrensis*

LC
3 sspp
HB 30–38 cm
T 26–29 cm
W 655–965 g

Genus *GALIDIA*

(1 species)

Ring-tailed Vontsira
Galidia elegans
FR Galidie à queue annelée
DE Ringelschwanzmungo
ES Galidia de cola anillada

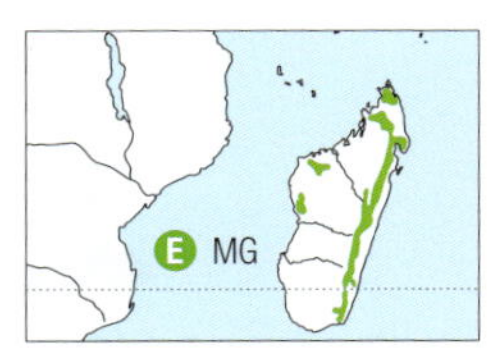

Forests in E (from Montagne d'Ambre NP S to Andohahela NP), limestone areas in W (around Namoroka & Bemaraha)

ssp *elegans*

Genus *GALIDICTIS*

(1 species)

Broad-striped Vontsira
Galidictis fasciata

FR Galidie à bandes larges
DE Breitstreifenmungo
ES Galidia bandeada

up to 1500 m

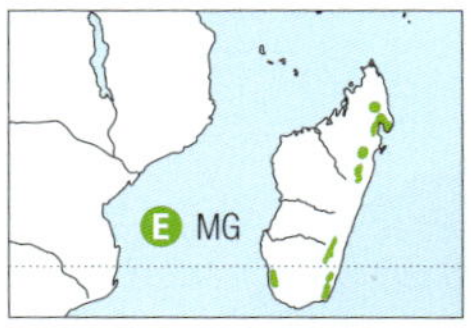

Humid lowlands to montane forests in E & SW (S from Marojejy NP)

ssp *grandidieri*

VU
2 sspp
HB 30–48 cm
T 25–33 cm
W 0.5–1.5 kg

ssp *fasciata*

Genus *SALANOIA*

(1 species)

Brown-tailed Vontsira
Salanoia concolor

FR Galidie unicolore
DE Schlichtmungo
ES Galidia parda

0–1000 m

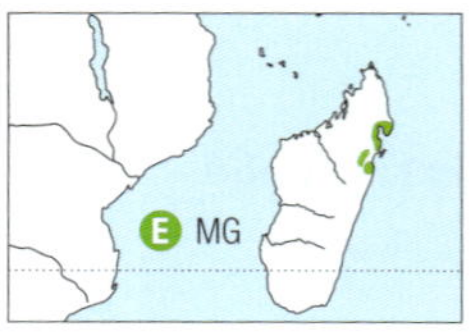

Lowland tropical humid forests in NE

VU
Monotypic
HB 35–38 cm
T 16–20 cm
W c. 780 g

Genus *MUNGOTICTIS*

(1 species)

Narrow-striped Boky
Mungotictis decemlineata

FR Galidie à dix raies
DE Schmalstreifenmungo
ES Galidia rayada

EN
2 sspp
HB 26–33 cm
T 19–21 cm
W 450–740 g

0–400 m

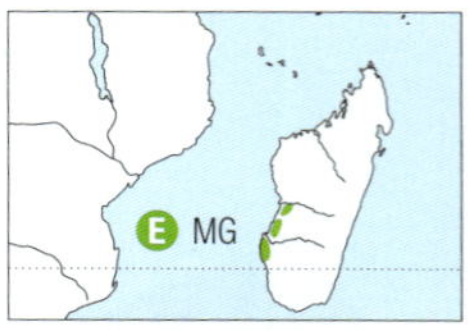

Dry deciduous forests in CW & SW (from R Tsiribihina S to R Manombo)

Family HYAENIDAE – Hyenas

4 genera, 4 species. Mainly savannas, but other semi-arid and desert regions to edges of forests, both tropical and temperate zones. Africa and Middle East to India.

Genus *PROTELES*

(1 species)

Aardwolf
Proteles cristatus

FR Protèle
DE Erdwolf
ES Proteles

LC
2 sspp
HB 55–80 cm
T 20–30 cm
W 8–14 kg

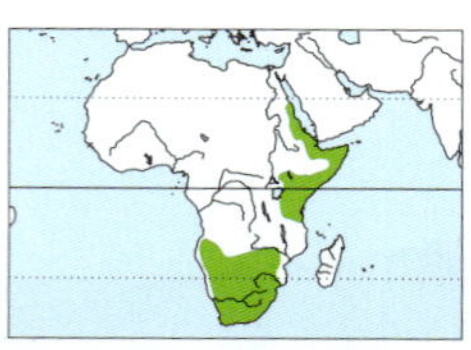

E Africa (EG, SD, ER, ET, DJ, SO, KE, NE UG to C TZ), most of S Africa (S from S AO, S ZM, ZW & SW MZ)

Genus *CROCUTA*

(1 species)

Spotted Hyena
Crocuta crocuta

FR Hyène tachetée
DE Tüpfelhyäne
ES Hiena manchada

LC
Monotypic
HB 125–160 cm
T 22–27 cm
W 45–86 kg

up to 4000 m

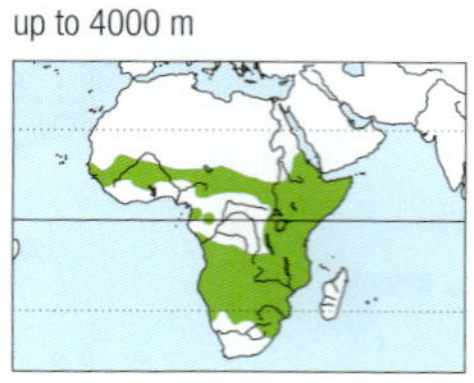

Most of sub-Saharan Africa, except lowland tropical rainforest; extirpated from many areas of S Africa

Genus *HYAENA*

(1 species)

Striped Hyena
Hyaena hyaena

FR Hyène rayée
DE Streifenhyäne
ES Hiena rayada

up to 3300 m

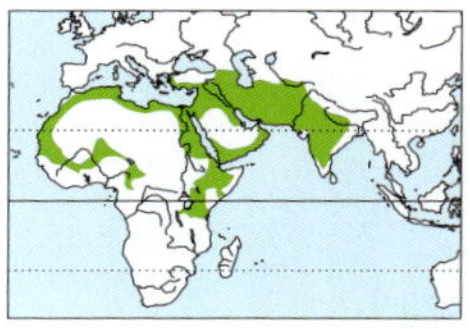

Widely but in isolated populations across N, W & E Africa (S to CM & TZ), Middle East, Caucasus & C Asia (N to UZ & TJ, E to IN & NP)

NT
Monotypic
HB 100–115 cm
T 30–40 cm
W 26–41 kg

Genus *PARAHYAENA*

(1 species)

NT
Monotypic
HB 110–136 cm
T 18.7–26 cm
W 28–47 kg

Brown Hyena
Parahyaena brunnea

FR Hyène brune
DE Braune Hyäne
ES Hiena parda

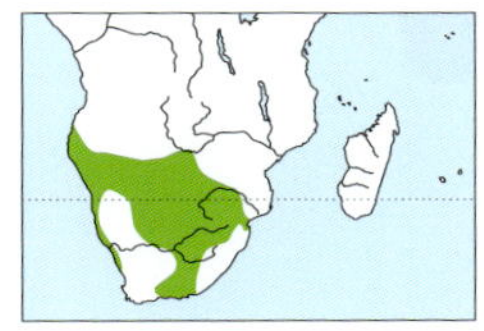

S Africa in NA, BW, W & S ZW, S MZ, SZ, W LS & ZA; old records from SW AO

Family CANIDAE – Dogs

13 genera, 38 species. All major habitats, spanning tropical forests, woodlands, savannas, deserts, alpine heathlands, and the Arctic, from sea level to over 5000 m. Cosmopolitan, all regions except Antarctica and many oceanic islands.

Genus *UROCYON*

(2 species)

LC
16 sspp
HB 54–66 cm
T 28–44 cm
W 2–5.5 kg

Northern Gray Fox
Urocyon cinereoargenteus

FR Renard gris
DE Graufuchs
ES Chacalillo gris

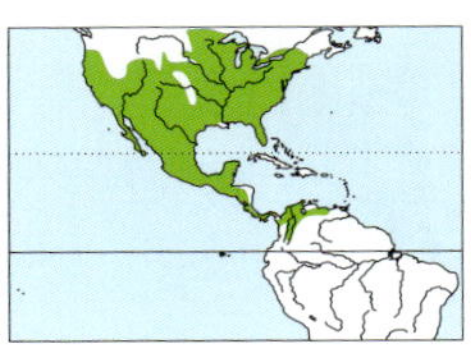

SC & SE CA, US (except N Rocky Mts) & MX S through Central America to N VE & CO

Island Fox
Urocyon littoralis

FR Renard insulaire
DE Insel-Graufuchs
ES Chacalillo isleño

NT
6 sspp
HB 46–63 cm
T 11–31 cm
W 1.3–2.5 kg

California Channel Is (San Miguel, Santa Rosa, Santa Cruz, Santa Catalina, San Clemente, San Nicolas)

Genus *OTOCYON*

(1 species)

Bat-eared Fox
Otocyon megalotis

FR Otocyon
DE Löffelhund
ES Zorro orejudo

LC
2 sspp
HB 46–61 cm
T 23–34 cm
W 3.2–5.4 kg

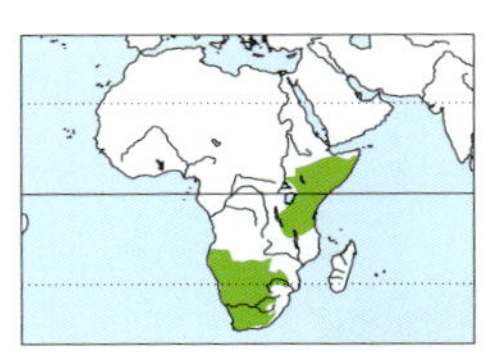

Disjunct in E (from S SD, ET & SO S to UG, KE & SW TZ) & S (from AO, NA & BW to MZ & ZA) Africa

Common Raccoon Dog
Nyctereutes procyonoides

FR Tanuki de Chine
DE Marderhund
ES Perro mapache chino

LC
4 sspp
HB 49–70 cm
T 15–23 cm
W 2.9–12.5 kg

ssp *procyonoides*

ssp *ussuriensis*

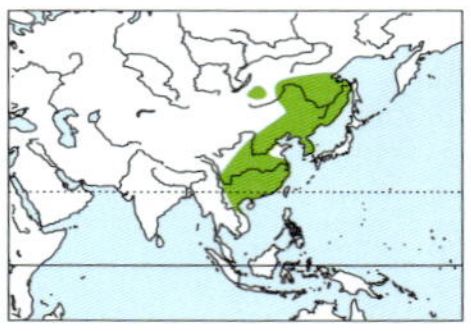
E MN, SE RU, Korean Pen, E CN, NE VN; introduced to many European countries

Genus *NYCTEREUTES*
(2 species)

NE
2 sspp

Japanese Raccoon Dog
Nyctereutes viverrinus

FR Tanuki du Japon
DE Japan-Marderhund
ES Perro mapache japonés

Hokkaido, Honshu, Shikoku, Kyushu, Awaji, Sado & other Is (except S of Kyushu)

Genus *VULPES*
(12 species)

Fennec Fox
Vulpes zerda

FR Renard fennec
DE Fennek
ES Fenec

LC
Monotypic
HB 33–39 cm
T 12.5–25 cm
W 0.8–1.9 kg

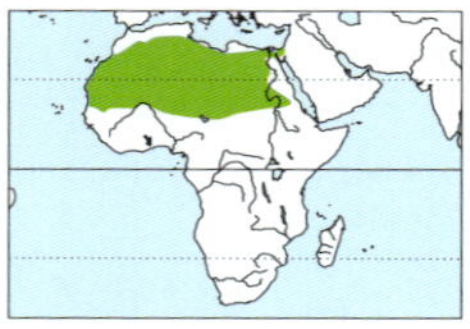
Widespread in sandy deserts & semi-deserts of N Africa (from EH & MR E to EG in N Sinai)

Blanford's Fox
Vulpes cana

FR Renard de Blanford
DE Afghanfuchs
ES Zorro persa

LC
Monotypic
HB 38–80 cm
T 26–35 cm
W 0.8–1.6 kg

below 2000 m
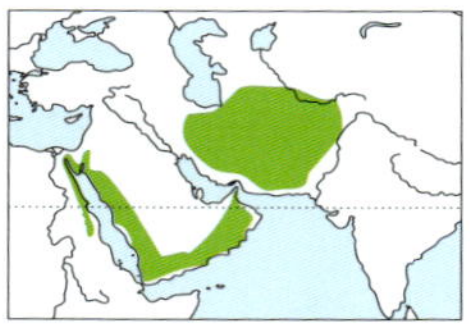
Arid mts of Middle East (E EG, IL, JO, SA, OM, AE), disjunct in IR, AF, PK, TM, UZ & TJ

Cape Fox
Vulpes chama

FR Renard du Cap
DE Kap-Fuchs
ES Zorro de El Cabo

LC
Monotypic
HB 45–62 cm
T 25–41 cm
W 2–4.2 kg

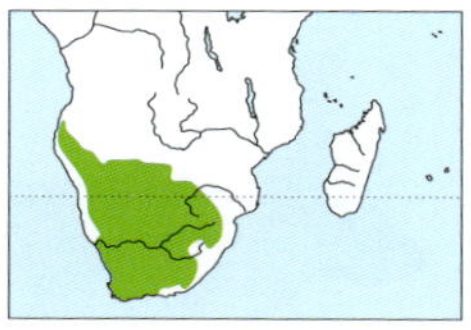
S Africa in SW AO (S from c. 15° N), NA, BW & ZA; perhaps SW SZ & LS

Indian Fox
Vulpes bengalensis

FR Renard du Bengale
DE Bengalfuchs
ES Zorro bengalí

LC
Monotypic
HB 39–57 cm
T 24–32 cm
W 1.8–3.2 kg

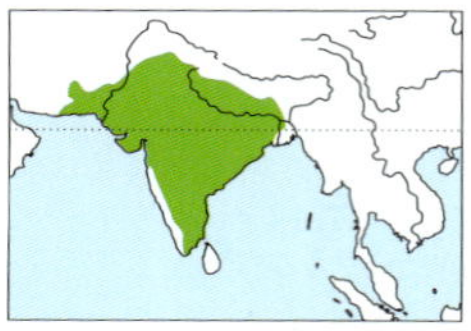
Indian subcontinent, from foothills of Himalayas in NP to S IN (except Western & Eastern Ghats), BD, PK

Pale Fox
Vulpes pallida

FR Renard pâle
DE Blassfuchs
ES Zorro pálido

LC
4 sspp
HB 38–55 cm
T 23–29 cm
W 2–3.6 kg

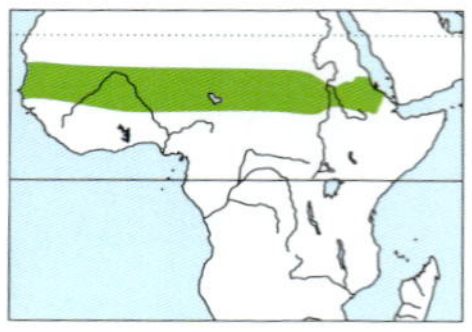
Mainly sub-Saharan desert & semi-desert areas from MR & SN E to SD, ER & ET

Rüppell's Fox
Vulpes rueppellii

FR Renard de Rüppell
DE Sandfuchs
ES Zorro de Rüppell

LC
6 sspp
HB 34–56 cm
T 22–38 cm
W 1.1–2.3 kg

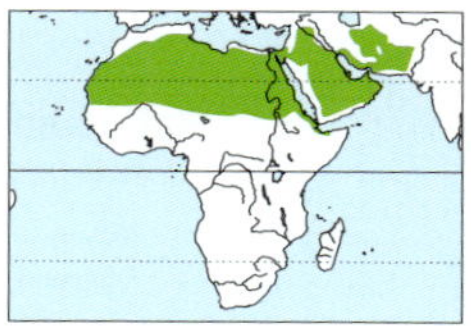
Widespread in desert & semi-desert areas of N Africa (SE to SO), Middle East (N to SY, E to IR), S AF, W PK

ssp *alascensis*

ssp *silacea*

ssp *vulpes*

Red Fox
Vulpes vulpes

FR Renard roux
DE Rotfuchs
ES Zorro rojo

0–4500 m

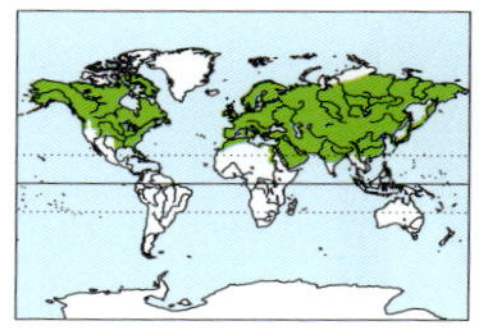

N Hemisphere from Arctic Circle to S North America, N Africa & Eurasia (except IS, Arctic Is, parts of Siberia); introduced to AU & FK

LC
45 sspp
HB 45–90 cm
T 28–49 cm
W 3–14 kg

LC
4 sspp
HB 45–59 cm
T 19–30 cm
W 1.6–3.2 kg

Corsac Fox
Vulpes corsac

FR Renard corsac
DE Steppenfuchs
ES Zorro estepario

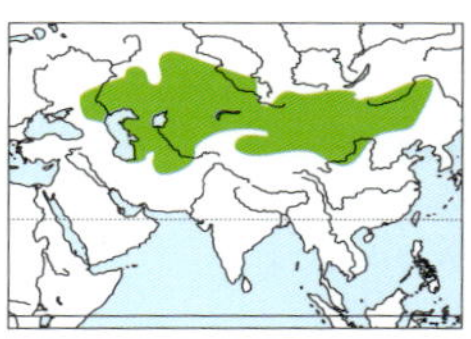

Eurasia from N Caucasus & R Volga E through S Siberia to Transbaikalia, S to NE IR, N AF & C CN

Tibetan Fox
Vulpes ferrilata

FR Renard du Tibet
DE Tibet-Fuchs
ES Zorro tibetano

LC
7 extant sspp
HB 45–53 cm
T 25–34 cm
W 1.6–2.7 kg

2500–5200 m

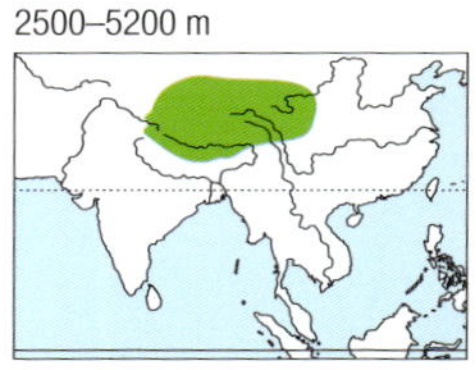

Tibetan Plateau from Ladakh (IN) E across CN (to Gansu, Sichuan & Yunnan), also NP (Mustang area)

Kit Fox
Vulpes macrotis

FR Renard nain
DE Wüsten-Flinkfuchs
ES Zorro chico

LC
Monotypic
HB 51–65 cm
T 22–29 cm
W 3–4.6 kg

400–1900 m

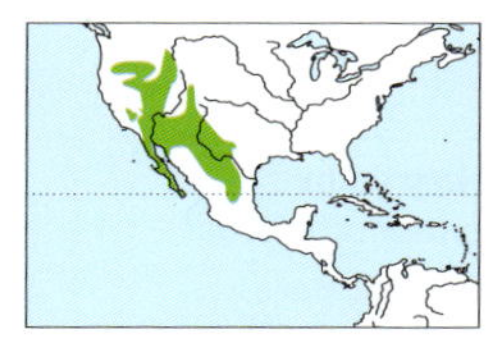

W US (S from S Oregon & SW Idaho, E to W Texas), MX (Baja California Pen, S to N Zacatecas)

Swift Fox
Vulpes velox

FR Renard véloce
DE Prärie-Flinkfuchs
ES Zorro veloz

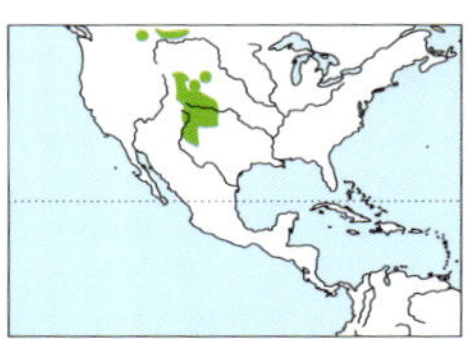

Great Plains in S CA (Alberta, Saskatchewan) & US (S to New Mexico & NW Texas)

LC
Monotypic
HB 47–54 cm
T 25–34 cm
W 1.6–2.5 kg

winter

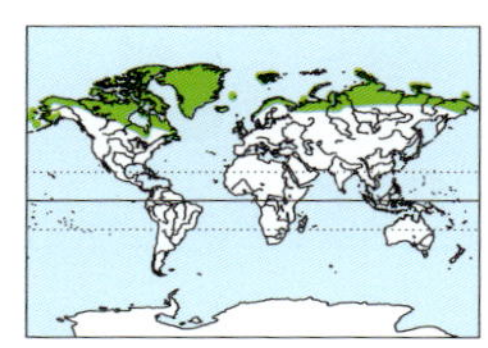

Circumpolar in all Arctic tundra habitats including most Arctic Is & many Is of Bering Sea

Arctic Fox
Vulpes lagopus

FR Renard arctique
DE Polarfuchs
ES Zorro ártico

LC
4 sspp
HB 55–75 cm
T 25–42 cm
W 3.1–4.2 kg

winter

blue morph

summer

typical morph

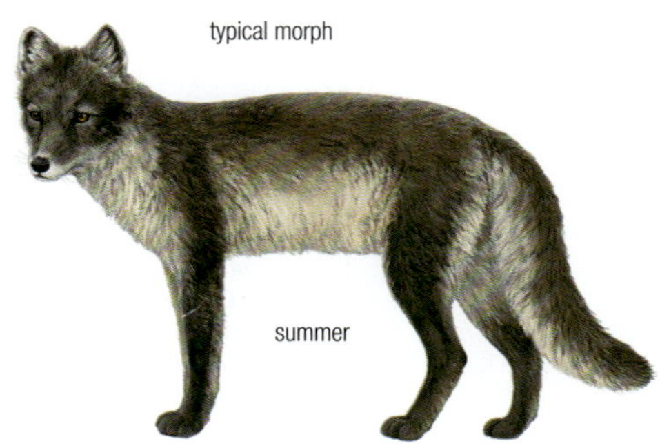

summer

Genus *CHRYSOCYON*

(1 species)

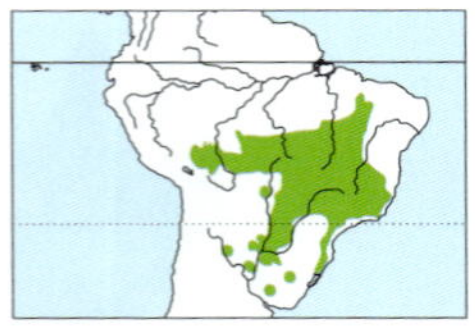

Grasslands & scrub forest from NE BR W to BO & PE border, S to PY (Chaco), UY & N AR (to 30° S)

Maned Wolf

Chrysocyon brachyurus

FR Renard à crinière
DE Mähnenwolf
ES Aguaraguazú

NT
Monotypic
HB 95–115 cm
T 38–50 cm
W 20–30 kg

Bush Dog

Speothos venaticus

FR Chyen-bwa
DE Waldhund
ES Zorro vinagre

NT
3 sspp
HB 57–75 cm
T 12.5–15 cm
W 5–8 kg

Genus *SPEOTHOS*

(1 species)

variants

below 1500 m

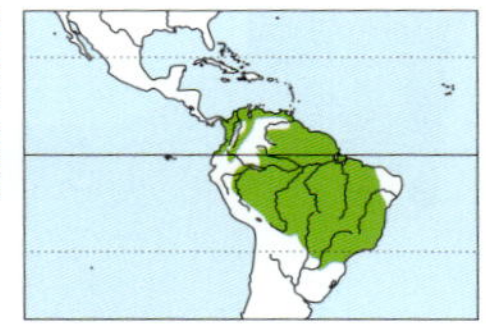

Forests near water sources in PA, CO, VE, Guianas, EC, PE, BR, BO, PY & NE AR (Misiones)

Short-eared Dog

Atelocynus microtis

FR Renard à petites oreilles
DE Kurzohrfuchs
ES Zorro orejicorto

Genus *ATELOCYNUS*

(1 species)

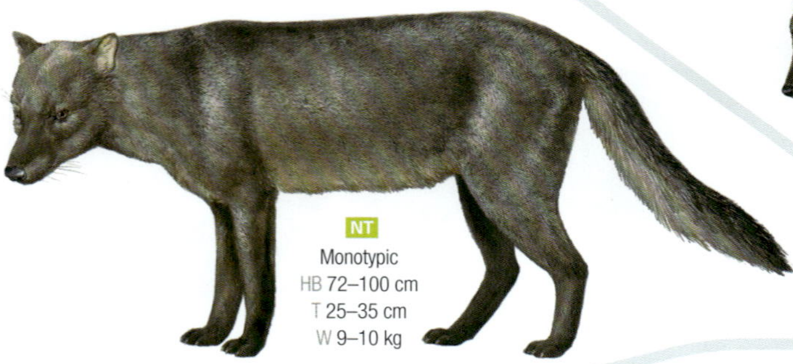

NT
Monotypic
HB 72–100 cm
T 25–35 cm
W 9–10 kg

up to 1200 m

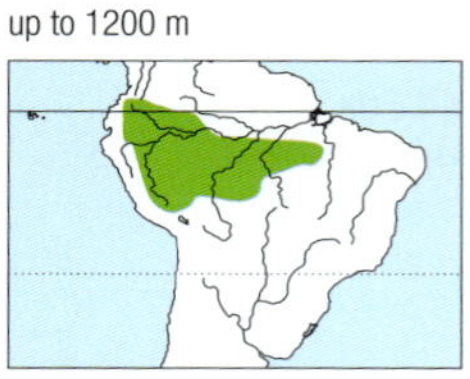

W lowland Amazonia (CO, EC, BR, PE, BO); limits uncertain

Crab-eating Fox

Cerdocyon thous

FR Renard crabier
DE Savannenfuchs
ES Zorro cangrejero

Genus *CERDOCYON*

(1 species)

up to 3000 m

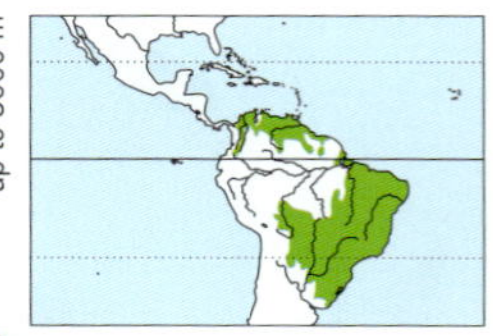

N CO, N & SE VE, Guianas, BR (avoiding Amazon Basin), BO, PY, AR & UY

LC
5 sspp
HB 57–77 cm
T 22–41 cm
W 4.5–8.5 kg

Genus *LYCALOPEX*

(6 species)

Hoary Fox

Lycalopex vetulus

FR Renard chenu
DE Brasilianischer Kampfuchs
ES Zorro brasileño

NT
Monotypic
HB 49–71 cm T 25–38 cm
W 2.5–4 kg

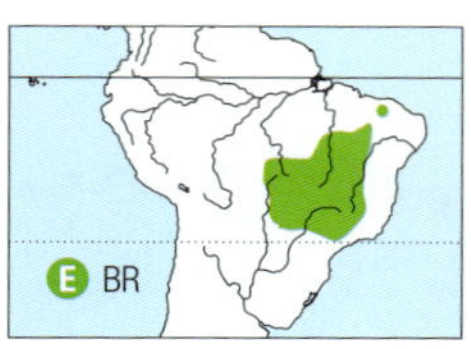

Associated with cerrado habitats of C Plateau

Darwin's Fox

Lycalopex fulvipes

FR Renard de Darwin
DE Darwin-Fuchs
ES Zorro chilote

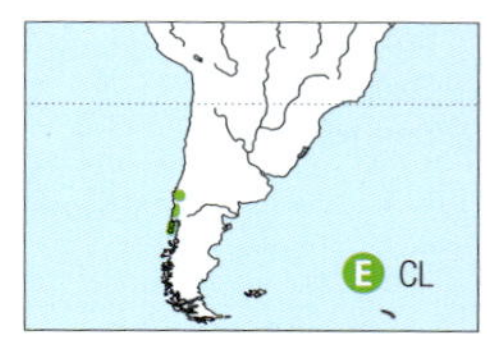

Disjunct range in forests of Chiloé I & mainland coastal mts (Nahuelbuta NP, Valdivian Coastal Range)

EN
Monotypic
HB 48–59 cm
T 19.5–25 cm
W 1.8–3.9 kg

Culpeo

Lycalopex culpaeus

FR Renard des Andes
DE Andenfuchs
ES Zorro andino

LC
6 sspp
HB 44–92 cm
T 30–49 cm
W 3.4–13.8 kg

up to 4800 m

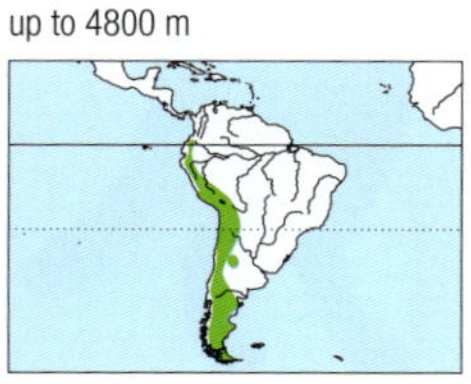

Along Andes & hilly areas from S CO (Nariño), EC, PE, BO, CL & AR (including Córdoba Mts) S to Tierra del Fuego

ssp *culpaeus*

ssp *andinus*

South American Gray Fox
Lycalopex griseus

FR Renard d'Argentine
DE Argentinischer Kampfuchs
ES Zorro de Magallanes

up to 4000 m

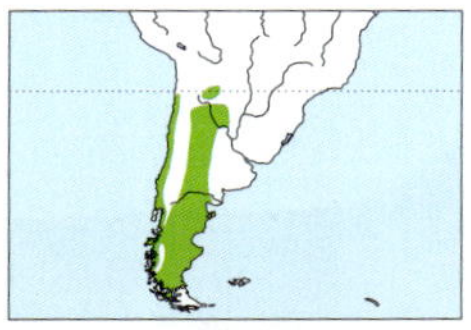

Both sides of Andes in S PE, CL & AR; introduced to Tierra del Fuego

LC
Monotypic
HB 50–66 cm
T 11.5–35 cm
W 2.5–5 kg

Pampas Fox
Lycalopex gymnocercus

FR Renard d'Azara
DE Pampasfuchs
ES Zorro de La Pampa

LC
Monotypic
HB 50–74 cm
T 25–41 cm
W 3–8 kg

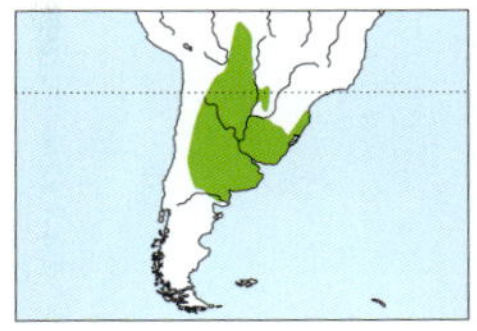

Pampas grasslands in E BO, PY, SE BR, UY and N & C AR

Sechuran Fox
Lycalopex sechurae

FR Renard de Sechura
DE Sechura-Fuchs
ES Zorro del Sechura

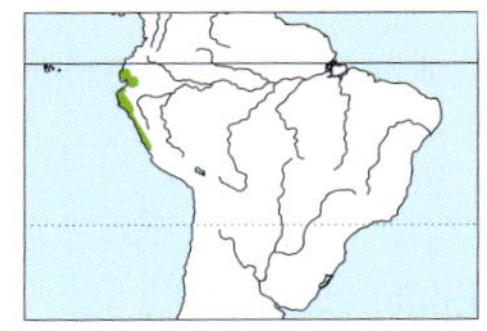

Coastal zones of SW EC & NW PE (3–12° S)

NT
Monotypic
HB 50–78 cm (M)
T 27–34 cm (M)
W 2.6–4.2 kg (M)

Genus *LUPULELLA*

(2 species)

Side-striped Jackal
Lupulella adustus

FR Chacal rayé
DE Streifenschakal
ES Chacal rayado

LC
Monotypic
HB 65–77 cm
T 30–41 cm
W 7.3–12 kg

up to 2700 m

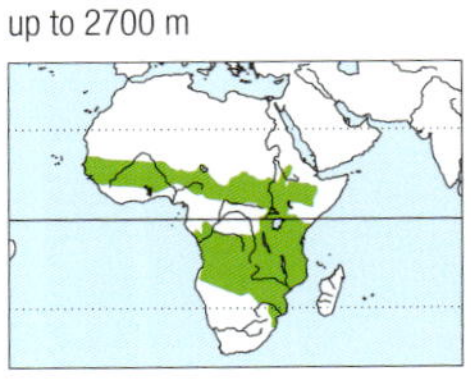

Sub-Saharan Africa from SN & GM through Sahel to ET, S to NE ZA, absent in rainforests & arid SW areas

Black-backed Jackal
Lupulella mesomelas

FR Chacal à dos noir
DE Schabrackenschakal
ES Chacal dorsinegro

LC
2 sspp
HB 65–90 cm
T 26–39 cm
W 5.9–12 kg

0–3000 m

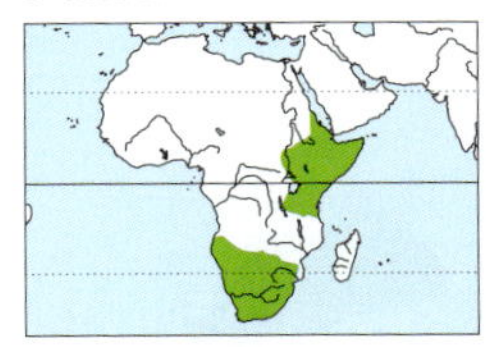

Disjunct in E (Horn of Africa, SS, UG, KE & TZ) & S (S AO, NA, BW, ZW, S MZ, ZA) Africa

Genus *LYCAON*

(1 species)

African Wild Dog
Lycaon pictus

FR Lycaon
DE Afrikanischer Wildhund
ES Licaón

EN
Monotypic
HB 84–141 cm
T 31–42 cm
W 18–34 kg

variants

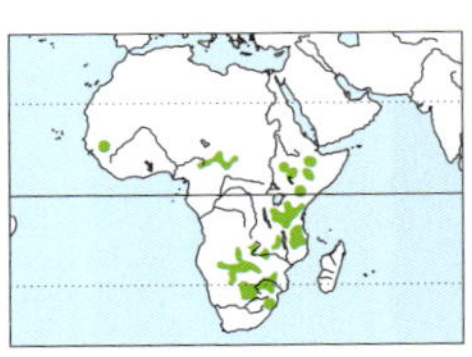

Sub-Saharan Africa, virtually eradicated from W, greatly reduced in C & NE, largest populations in TZ, BW & W ZW

Genus *CUON*

(1 species)

Dhole
Cuon alpinus

FR Dhole
DE Asiatischer Wildhund
ES Cuón

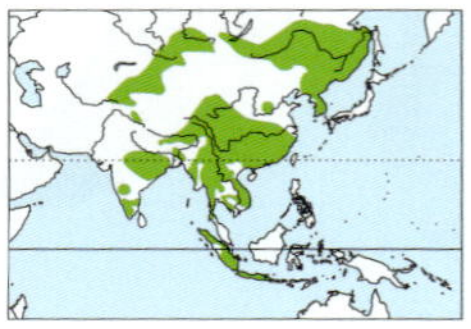

C & E Asian Mts (no recent records), Korean Pen (?), CN, IN, NP, BT, BD, mainland SE Asia, Sumatra, Java; largely extirpated and fragmented

EN
3 sspp
HB 88–135 cm
T 32–50 cm
W 10–20 kg

ssp *alpinus*

ssp *hesperius*

Genus *CANIS*

(7 species)

Ethiopian Wolf
Canis simensis

FR Loup d'Abyssinie
DE Äthiopien-Wolf
ES Lobo etíope

EN 2 sspp
HB 84–101 cm
T 27–40 cm
W 11.2–19.3 kg

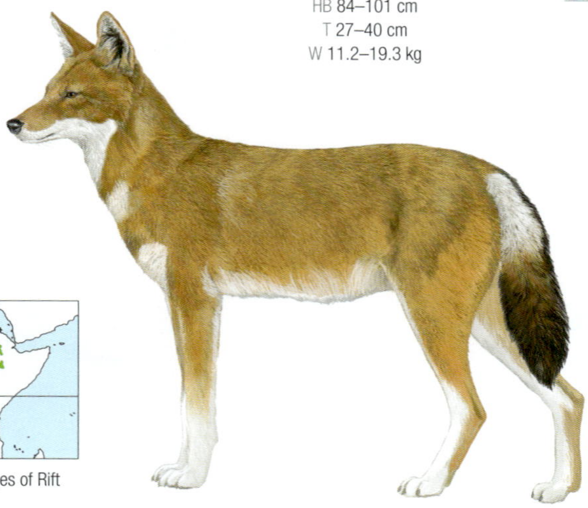

3000–4500 m

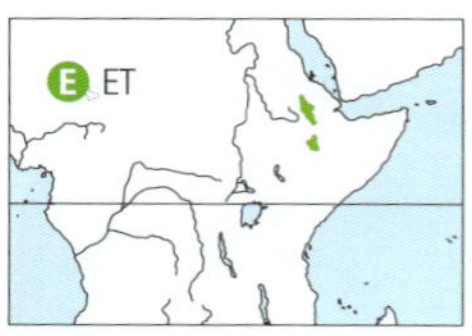

Isolated mountain ranges of Rift Valley

Golden Jackal
Canis aureus

FR Chacal doré
DE Goldschakal
ES Chacal dorado

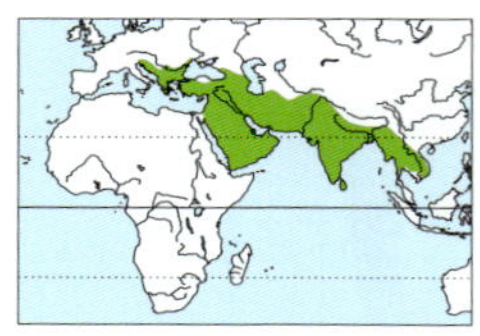

Widespread in SW & S Asia, mainland SE Asia, expanded to SE Europe (N to AT, S SK & S UA, W to NE IT)

LC
Monotypic
HB 74–84 cm
T 20–24 cm
W 6.5–9.8 kg

African Golden Wolf
Canis lupaster

FR Loup doré
DE Afrikanischer Goldwolf
ES Lobo dorado africano

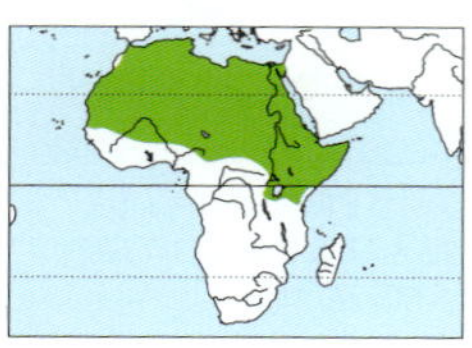

N Africa from MA to EG (including Sinai Pen), S to SN in W & N TZ in E

LC
Monotypic
HB 74–89 cm
T 27–35 cm
W 6.3–15 kg

variants

Gray Wolf
Canis lupus

FR Loup gris
DE Wolf
ES Lobo

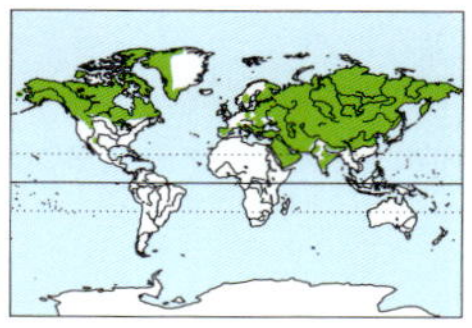

Wilderness and remote areas in Alaska, CA, N US, GL, Europe & Asia from about 75° N to 12° N

ssp *arctos*

ssp *occidentalis*

LC
12 extant sspp
HB 87–130 cm
T 35–52 cm
W ≤ 62 kg

ssp *arabs*

ssp *lupus*

ssp *cubanensis*

Eastern Wolf
Canis lycaon

FR Loup de l'Est
DE Timberwolf
ES Lobo rojo oriental

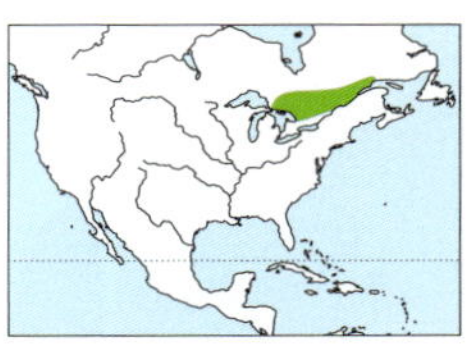

From Great Lakes to SE Quebec (SE CA, NE US)

NE
Monotypic
HB 108–115 cm
W 24–29 kg

Red Wolf
Canis rufus

FR Loup roux
DE Rotwolf
ES Lobo rojo

CR
Monotypic
HB 99–125 cm
T 30–46 cm
W 20–34 kg

Today restricted to E North Carolina, historically from E Texas to S edge of Great Lakes & Maine

Coyote
Canis latrans

FR Coyote
DE Kojote
ES Coyote

ssp *latrans*

ssp *texensis*

LC
19 sspp
HB 74–94 cm
T 26–36 cm
W 7.7–15.8 kg

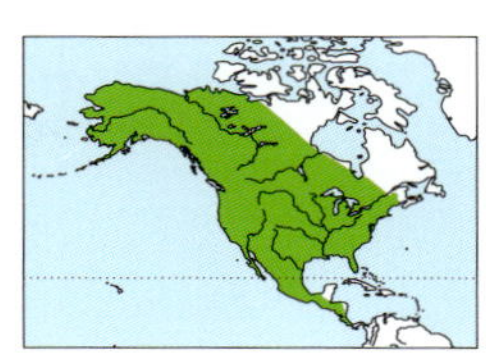

North America (except for polar areas) S through Central America to PA (Darién)

Family URSIDAE – Bears

5 genera, 8 species. Forested environments (boreal, temperate, and tropical) to tundras and semi-deserts; one species in Arctic. Holarctic, Neotropical and Indo-Malayan Regions.

Genus *AILUROPODA*

(1 species)

Giant Panda
Ailuropoda melanoleuca

FR Ours panda
DE Bambusbär
ES Panda

1200–4100 m

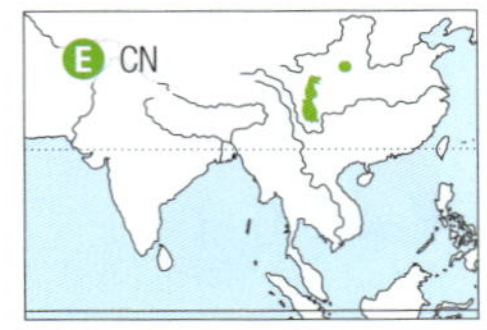

Gansu, Shaanxi, Sichuan

VU
2 sspp
HB 120–180 cm
T 10–16 cm
W 70–125 kg

Genus *TREMARCTOS*

(1 species)

Andean Bear
Tremarctos ornatus

FR Ours à lunettes
DE Brillenbär
ES Oso de anteojos

VU
Monotypic
HB 130–190 cm
T < 10 cm
W 60–175 kg

200–4700 m

Andes of VE, CO, EC, PE & BO; controversial evidence in N AR

variants

Genus *HELARCTOS*

(1 species)

Sun Bear
Helarctos malayanus

FR Ours malais
DE Malaienbär
ES Oso malayo

VU
2 sspp
HB 100–150 cm
T 3–7 cm
W 30–80 kg

0–2100 m

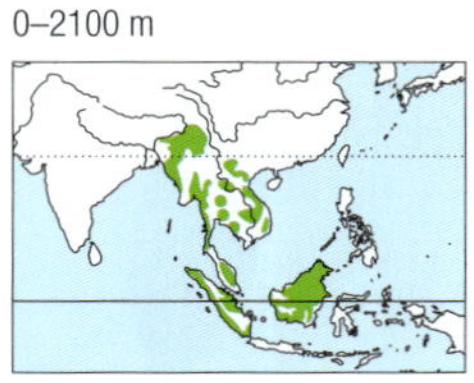

BD, NE IN, SC CN (Yunnan), mainland SE Asia, Sumatra, Borneo

Genus *MELURSUS*

(1 species)

Sloth Bear
Melursus ursinus

FR Ours paresseux
DE Lippenbär
ES Oso bezudo

VU
2 sspp
HB 140–190 cm
T 8–17 cm
W 50–145 kg

up to 2000 m

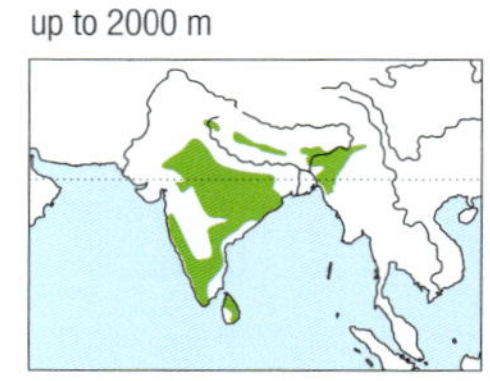

IN, NP, BT, BD, LK

Genus *URSUS*

(4 species)

ssp *thibetanus*

ssp *gedrosianus*

Asian Black Bear
Ursus thibetanus

FR Ours à collier
DE Kragenbär
ES Oso negro asiático

VU
7 sspp
HB 110–190 cm
T < 12 cm
W 35–200 kg

0–4300 m

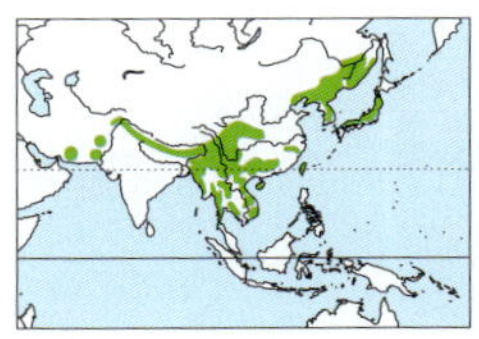

SE IR, N AF, S & N PK, S Himalayas (IN, NP, BT) to mainland SE Asia (S to S TH), CN, RU (Far East), Korean Pen, S JP (Honshu, Shikoku), TW

American Black Bear
Ursus americanus

FR Ours noir
DE Baribal
ES Oso negro americano

white morph

black morph

brown morph

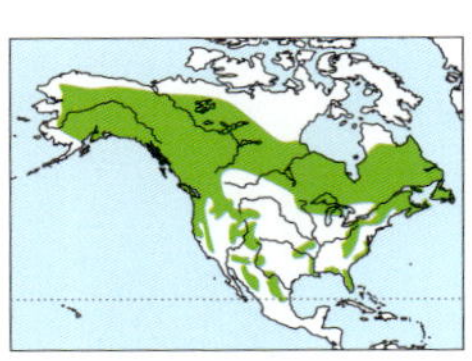

Throughout CA, higly fragmented in continental US to much of N MX; reintroductions & recolonizations across US

LC
Monotypic
HB 120–190 cm
T < 12 cm
W 40–225 kg

Brown Bear
Ursus arctos

FR Ours brun
DE Braunbär
ES Oso pardo

LC
14 extant sspp
HB 150–280 cm
T 6–21 cm
W 80–550 kg

ssp *lasiotus*

ssp *arctos*

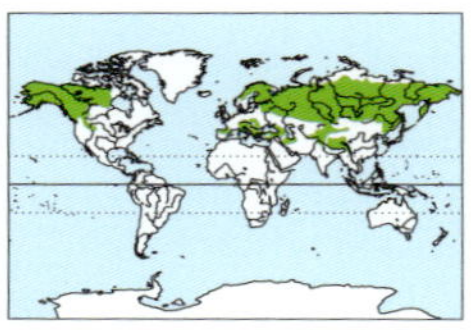
NW North America, Europe, N, W & C Asia, strongholds in RU, Alaska (US) & CA, in the rest small and isolated populations; reintroduced to Pyrenees & Alps

ssp *pruinosus*

ssp *syriacus*

ssp *middendorfi*

ssp *horribilis*

Polar Bear
Ursus maritimus

FR Ours blanc
DE Eisbär
ES Oso polar

VU
Monotypic
HB 180–280 cm
T 6–13 cm
W 150–650 kg

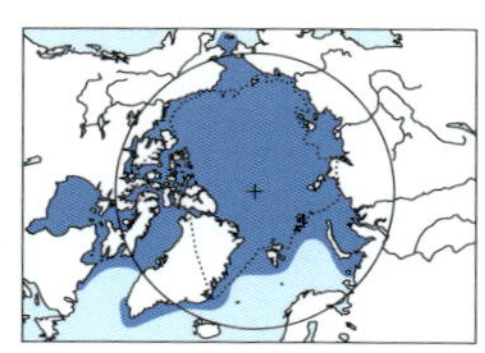
Arctic Ocean in Alaska (US), CA, GL, Svalbard Is & RU; sighted within 0.5° of North Pole

Family OTARIIDAE – Eared Seals

7 genera, 15 species. Mainly temperate and subpolar seas, extending to tropics and polar waters, in estuarine, coastal, and pelagic zones. All oceans except North Atlantic and Arctic oceans.

Genus *CALLORHINUS*

(1 species)

Northern Fur Seal
Callorhinus ursinus
FR Otarie à fourrure
DE Nördlicher Seebär
ES Lobo marino septentrional

VU
Monotypic
TL ≤ 210 cm (M), ≤ 150 cm (F)
W 180–270 kg (M), 40–60 kg (F)

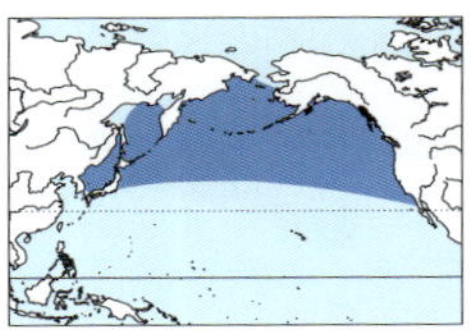

N Pacific, including Bering Sea, Sea of Okhotsk & Sea of Japan, S to JP (Honshu) in W & to NW MX (Baja California) in E

Genus *ARCTOCEPHALUS*

(8 species)

Antarctic Fur Seal
Arctocephalus gazella
FR Otarie antarctique
DE Kerguelen-Seebär
ES Lobo marino antártico

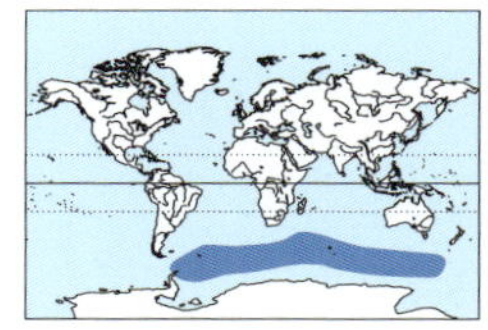

Subantarctic & Antarctic waters (S or just N of Antarctic Convergence) & scattered Is (mainly South Georgia), haul-out areas include Antarctic Pen

LC
Monotypic
TL ≤ 200 cm (M), 120–140 cm (F)
W 130–204 kg (M), 22–51 kg (F)

Juan Fernandez Fur Seal
Arctocephalus philippii
FR Otarie de Juan-Fernandez
DE Juan-Fernández-Seebär
ES Lobo marino de Juan Fernández

LC
Monotypic
TL c. 200 cm (M), 120–140 cm (F)
W c. 140 kg (M), c. 48 kg (F)

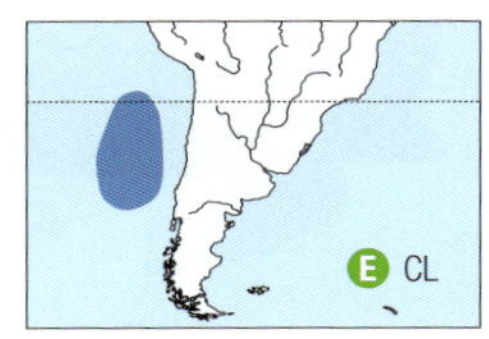

S Pacific in Juan Fernández Is & Desventuradas Is (San Félix), which lie 600 km N

Guadalupe Fur Seal
Arctocephalus townsendi
FR Otarie de Guadalupe
DE Guadalupe-Seebär
ES Lobo marino de Guadalupe

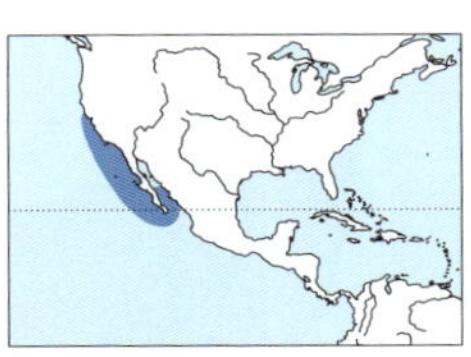

Mainly Guadalupe I, also San Benito Is (W Baja California, MX), Channel Is (San Miguel) & Farallon Is (US), wanders N to Washington & S Alaska

LC
Monotypic
TL c. 200 cm (M), c. 120 cm (F)
W 160–170 kg (M), 40–50 kg (F)

Galapagos Fur Seal
Arctocephalus galapagoensis
FR Otarie des Galapagos
DE Galapagos-Seebär
ES Lobo marino de las Galápagos

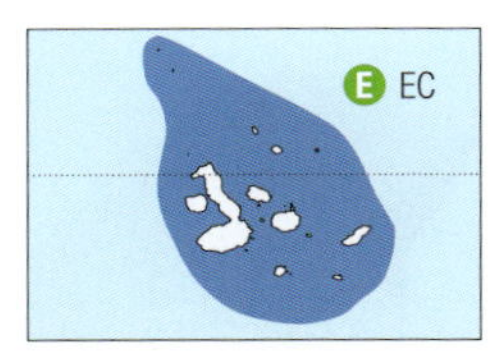

Galápagos Is (on almost all islands)

EN
Monotypic
TL 150–160 cm (M), 110–130 cm (F)
W 60–68 kg (M), c. 27 kg (F)

South American Fur Seal
Arctocephalus australis

FR Otarie des Falkland
DE Südamerika-Seebär
ES Lobo marino de dos pelos

LC
Monotypic
TL ≤ 190 cm (M), c. 150 cm (F)
W 120–160 kg (M), 40–50 kg (F)

South America from UY S to Cape Horn, FK & Magellanic region, regular to S BR, also from C PE to N CL

♀

Atlantic form

♂

C Peru form
♂

S Chile form
♂

New Zealand Fur Seal
Arctocephalus forsteri

FR Otarie de Forster
DE Neuseeland-Seebär
ES Lobo marino de Nueva Zelanda

LC
Monotypic
TL ≤ 190 cm (M), c. 150 cm (F)
W 120–160 kg (M), 30–50 kg (F)

♀

♂

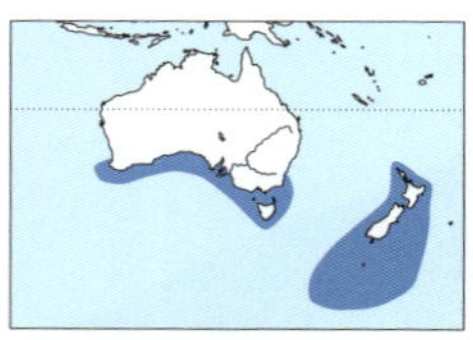

NZ (North & South Is, all subantarctic Is), SW & S AU and Tas

Subantarctic Fur Seal
Arctocephalus tropicalis

FR Otarie d'Amsterdam
DE Subantarktik-Seebär
ES Lobo marino subantártico

♀

♂

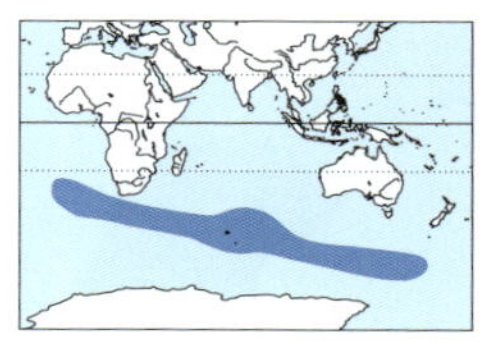

Widespread in subantarctic waters, primarily N of Antarctic Convergence from SE Atlantic & Indian oceans to SW Pacific

LC
Monotypic
TL ≤ 180 cm (M), 119–152 cm (F)
W 70–165 kg (M), 25–67 kg (F)

Afro-Australian Fur Seal
Arctocephalus pusillus

FR Otarie du Cap
DE Südafrika-Seebär
ES Lobo marino de El Cabo

LC
2 sspp
TL 200–230 cm (M), 120–171 cm (F)
W ≤ 360 kg (M), ≤ 110 kg (F)

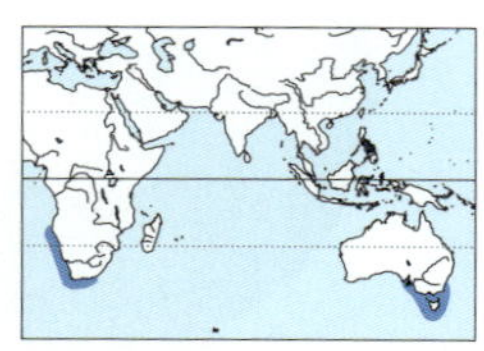

Coastal S Africa, from AO S to Algoa Bay (ZA), coastal SE AU in Vic, NSW, Tas & Bass Strait (including Kangaroo I)

ssp *doriferus*
♂

♂

ssp *pusillus*

♀

Genus *OTARIA*

(1 species)

South American Sea Lion
Otaria flavescens

FR Otarie à crinière
DE Mähnenrobbe
ES León marino de Sudamérica

South America, from S BR to Cape Horn & FK in Atlantic, N to N PE in Pacific

LC
Monotypic
TL c. 260 cm (M), c. 200 cm (F)
W ≤ 350 kg (M), ≤ 144 kg (F)

Genus *NEOPHOCA*

(1 species)

Australian Sea Lion
Neophoca cinerea

FR Otarie d'Australie
DE Australien-Seelöwe
ES León marino de Australia

S AU, from Houtman Albrolhos Is in C WA E along coast to SE SA (including Kangaroo I)

EN
Monotypic
TL c. 250 cm (M), 130–180 cm (F)
W 200–300 kg (M), 61–105 kg (F)

Genus *PHOCARCTOS*

(1 species)

New Zealand Sea Lion
Phocarctos hookeri

FR Otarie de Nouvelle-Zélande
DE Neuseeland-Seelöwe
ES León marino de Nueva Zelanda

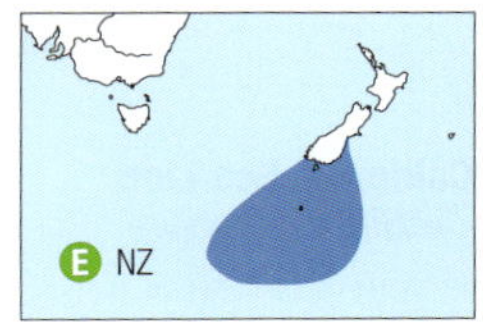

South Island, S Pacific subantarctic Is (mainly Auckland, Campbell & Stewart)

EN
Monotypic
TL 230–270 cm (M), 180–200 cm (F)
W 320–450 kg (M), 90–165 kg (F)

Genus *EUMETOPIAS*

(1 species)

Steller Sea Lion
Eumetopias jubatus

FR Otarie de Steller
DE Steller-Seelöwe
ES León marino de Steller

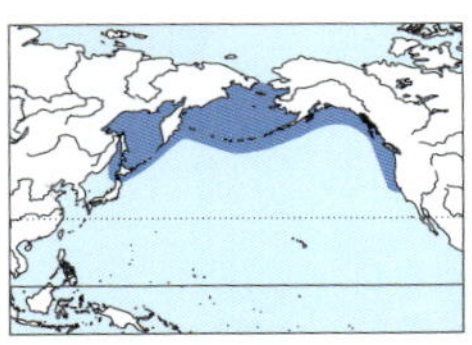

N Pacific, including Bering Sea, Sea of Okhotsk & Sea of Japan, S to N JP in W & to C California (US) in E; occasional to Korean Pen & CN

NT
2 sspp
TL ≤ 330 cm (M), ≤ 250 cm (F)
W ≤ 1120 kg (M), ≤ 273 kg (F)

♂
W form

♀

♂
E form

Genus *ZALOPHUS*

(2 species)

California Sea Lion
Zalophus californianus

FR Otarie de Californie
DE Kalifornischer Seelöwe
ES León marino de California

LC
Monotypic
TL c. 240 cm (M), c. 200 cm (F)
W c. 390 kg (M) c. 110 kg (F)

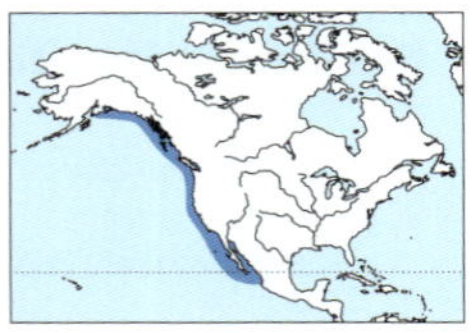

NE Pacific from SC Alaska, throughout coastal CA & US to Marías Is (off Nayarit) & Gulf of California (MX)

Galapagos Sea Lion
Zalophus wollebaeki

FR Otarie de Wollebaek
DE Galapagos-Seelöwe
ES León marino de las Galápagos

EN
Monotypic
TL c. 150 cm (M), c. 120 cm (F)
W ≤ 250 kg (M), 50–100 kg (F)

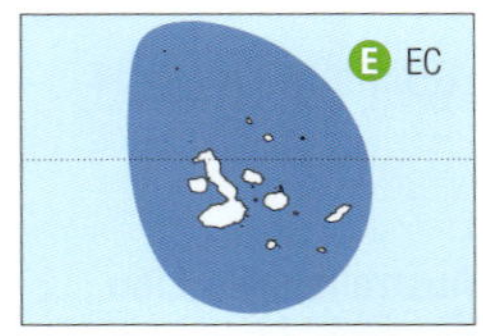

Galápagos Is (all main islands & numerous smaller rocks), irregular on Isla de la Plata (off mainland EC)

Family ODOBENIDAE – Walrus

1 genus, 1 species. Polar and subpolar Arctic seas, mainly in shallow areas. Arctic Region.

Walrus
Odobenus rosmarus

FR Morse
DE Walross
ES Morsa

Genus *ODOBENUS*

(1 species)

VU
2 sspp
TL c. 315 cm (M), c. 260 cm (F)
W 800–1800 kg (M), c. 1000 kg (F)

ssp *rosmarus*

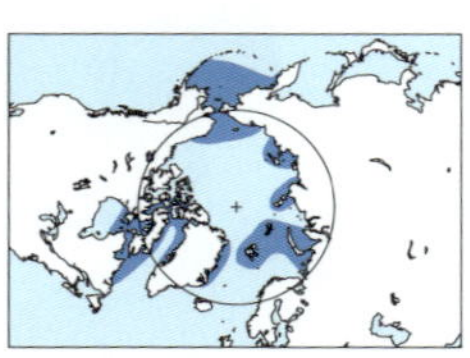

NE CA, W & E GL, Svalbard, N RU to Bering & Chukchi seas, Alaska (US); occasionally IS, stragglers in W Europe (S to Bay of Biscay) & E US (S to New England)

ssp *divergens*
♂

Family PHOCIDAE – Earless Seals

14 genera, 18 species. Mainly polar and subpolar waters, but some temperate to tropical waters, estuaries, and freshwater lakes. All major oceans and a few enclosed seas and lakes.

Genus *MONACHUS*

(1 species)

♀

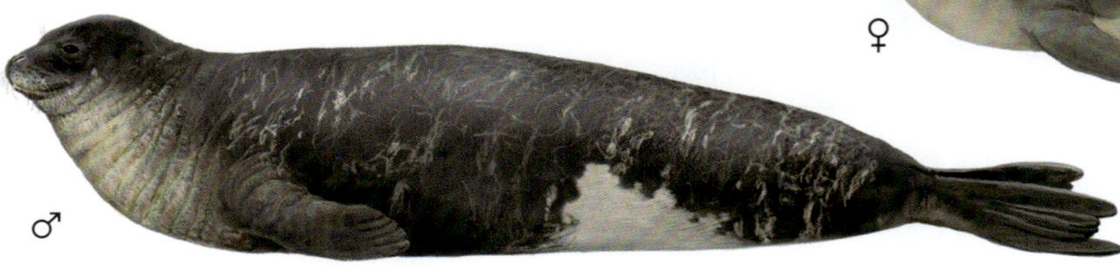

♂

Mediterranean Monk Seal
Monachus monachus

FR Phoque de la Méditerranée
DE Mittelmeer-Mönchsrobbe
ES Foca monje del Mediterráneo

EN
Monotypic
TL 230–280 cm
W 240–300 kg

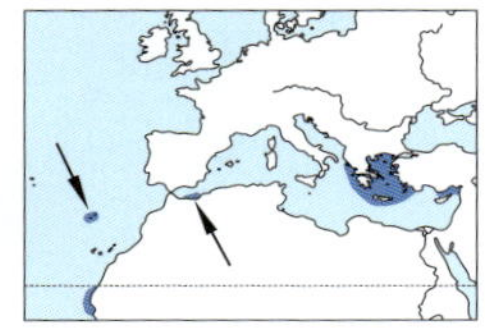

Mediterranean (Is & coasts of GR, W TR, NE MA & NW DZ), E Atlantic at Desertas Is (PT) & Cap Blanc Pen (EH/MR border); occasional in Canary Is, Banc d'Arguin (MR), TN, LY & HR

Genus *NEOMONACHUS*

(1 species)

EN
Monotypic
TL 183–240 cm
W 136–272 kg

Hawaiian Monk Seal
Neomonachus schauinslandi

FR Phoque d'Hawaï
DE Hawaii-Mönchsrobe
ES Foca monje de Hawái

♀

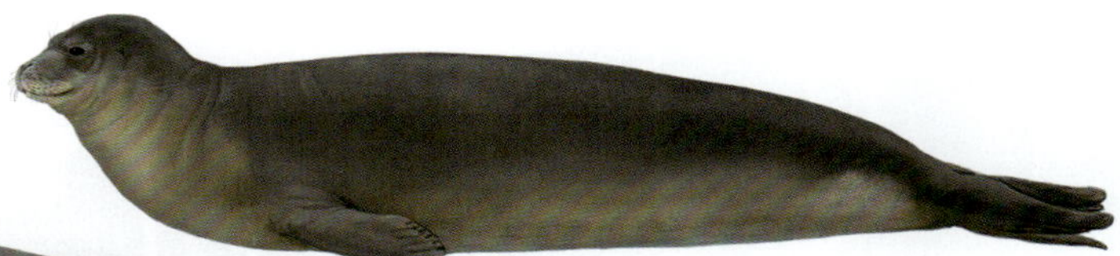

♂

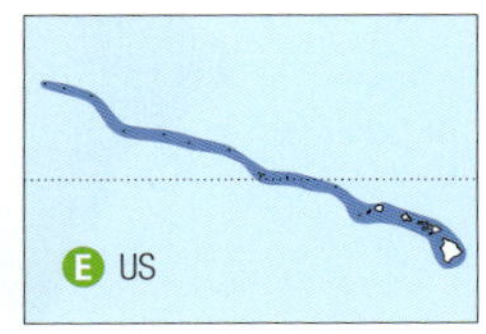

Hawaiian Is, mainly in NW (Kure Atoll, Midway Atoll, Pearl and Hermes Reef, Lisianski I, Laysan I, French Frigate Shoals)

Genus *MIROUNGA*

(2 species)

♀

♂

Northern Elephant Seal
Mirounga angustirostris

FR Éléphant-de-mer boréal
DE Nördlicher See-Elefant
ES Elefante marino septentrional

LC
Monotypic
TL 350–400 cm (M), 215–300 cm (F)
W 1800–2500 kg (M), 300–600 kg (F)

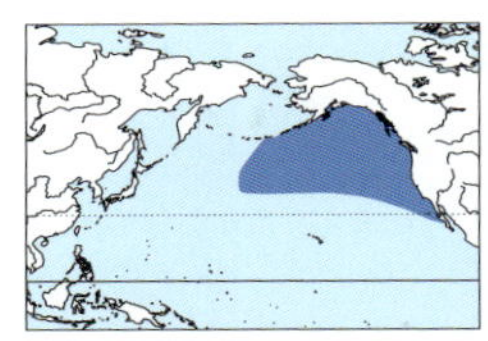

NE Pacific in Aleutian Is, Gulf of Alaska, W coast of North America S to C Baja California (MX); occasional to Hawaii, JP & Bering Sea

♀

♂

Southern Elephant Seal
Mirounga leonina

FR Éléphant-de-mer austral
DE Südlicher See-Elefant
ES Elefante marino meridional

LC
Monotypic
TL 450–500 cm (M), 300–350 cm (F)
W 1500–3000 kg (M), 600–800 kg (F)

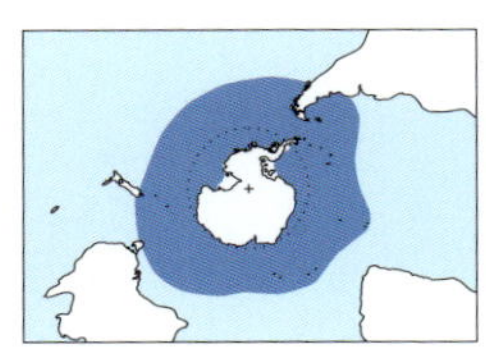

Circumpolar in Southern Ocean; breeding sites scattered on subantarctic Is, Antarctic Pen & coast of S AR

Genus *HYDRURGA*

(1 species)

Leopard Seal
Hydrurga leptonyx

FR Phoque léopard
DE Seeleopard
ES Foca leopardo

LC
Monotypic
TL 241–338 cm
W 200–590 kg

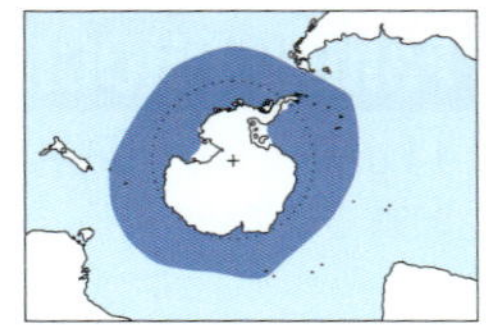

Antarctic & subantarctic cold waters at 50–80° S

Genus *LEPTONYCHOTES*

(1 species)

Weddell Seal
Leptonychotes weddellii

FR Phoque de Weddell
DE Weddell-Robbe
ES Foca de Weddell

LC
Monotypic
TL 280–330 cm
W 400–600 kg

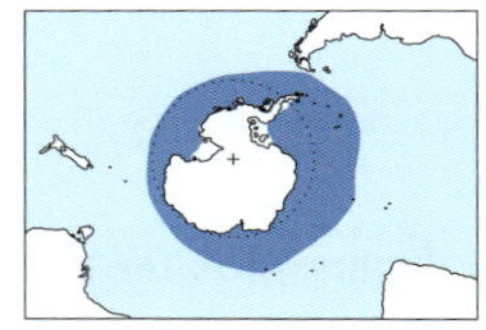

Circumpolar in Southern Ocean; primary breeding colonies in fast-ice habitats near Antarctica, also in South Shetland, South Orkney & South Georgia Is

Genus *LOBODON*

(1 species)

Crabeater Seal
Lobodon carcinophaga

FR Phoque crabier
DE Krabbenfresser
ES Foca cangrejera

LC
Monotypic
TL 219–247 cm
W 182–229 kg

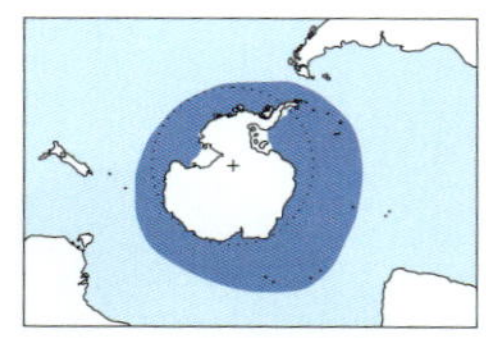

Circumpolar in Southern Ocean, mainly along edges of pack ice and just over edges of Antarctic continental shelf

Genus *OMMATOPHOCA*

(1 species)

Ross Seal
Ommatophoca rossii

FR Phoque de Ross
DE Ross-Robbe
ES Foca de Ross

LC
Monotypic
TL 170–250 cm
W 129–216 kg

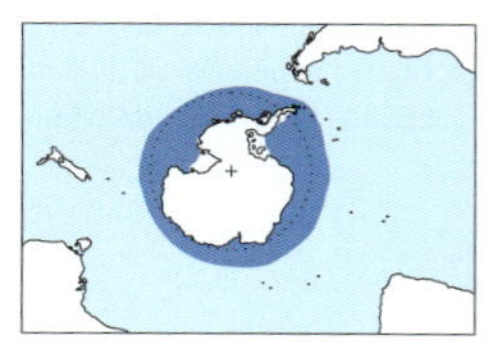

Circumpolar in Southern Ocean, breeding mainly in heavily consolidated pack ice in Ross, Amundsen & Bellingshausen seas (W Antarctica)

Genus *ERIGNATHUS*

(1 species)

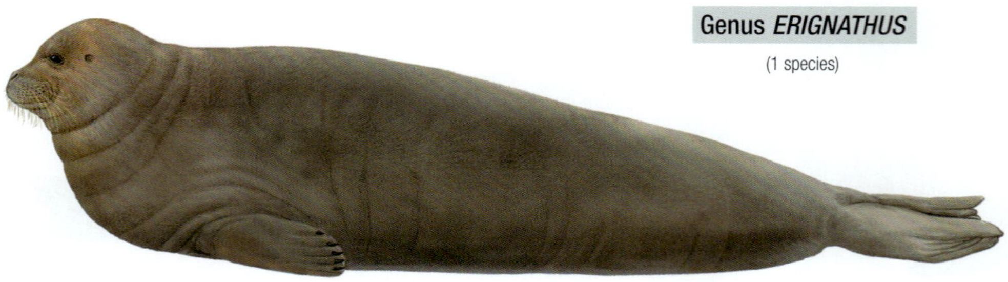

Bearded Seal
Erignathus barbatus

FR Phoque barbu
DE Bartrobbe
ES Foca barbuda

LC
2 sspp
TL 210–250 cm
W 250–450 kg

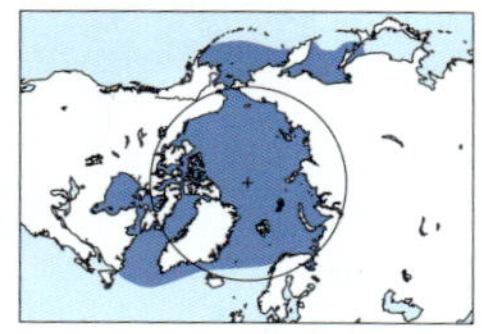

Arctic & subarctic from CA, GL, Svalbard Is, RU & Alaska (US), also Bering Sea & Sea of Okhotsk S to Hokkaido (JP)

Family AILURIDAE – Red Pandas

1 genus, 2 species. Forested mountainous areas; typically deciduous and coniferous forests with bamboo-thicket understory. Mountains of Central Asia.

Genus *AILURUS*

(2 species)

Western Red Panda
Ailurus fulgens
FR Panda roux
DE Kleiner Panda
ES Panda rojo

EN
Monotypic
HB 57–61 cm
T 36–41 cm
W 3.8–4.3 kg

1500–4800 m

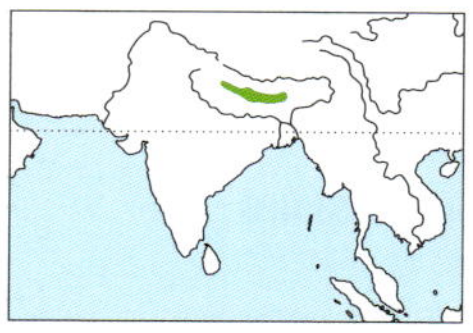

NP, NE IN, BT, perhaps extending to CN; limits uncertain

Eastern Red Panda
Ailurus styani
FR Panda de Styan
DE Styans Kleiner Panda
ES Panda rojo de Styan

NE
Monotypic
HB 59–79 cm
T 43–50 cm
W 3–6 kg

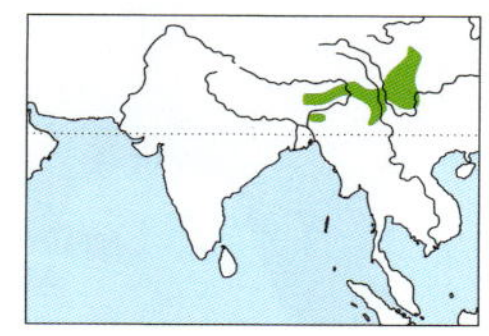

NE IN, N MM, SC CN (W Tibet, W Sichuan, N Yunnan)

Family MEPHITIDAE – Skunks

4 genera, 14 species. Almost all habitats within their range, except for the most arid deserts and the colder alpine and arctic climates. South-east Asia, North, Central and South America.

Genus *MYDAUS*

(2 species)

Sunda Stink Badger
Mydaus javanensis
FR Télagon de Java
DE Sunda-Stinkdachs
ES Melandro indonesio

LC
3 sspp
HB 37–52 cm
T 3.4–7.5 cm
W 1.4–3.6 kg

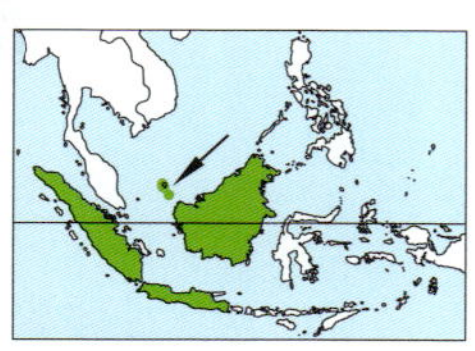

Sumatra, Java, Borneo, Natuna Is

Palawan Stink Badger
Mydaus marchei
FR Télagon de Palawan
DE Palawan-Stinkdachs
ES Melandro de Palawan

LC
Monotypic
HB 32–49 cm
T 1.5–4.5 cm
W 844–2490 g

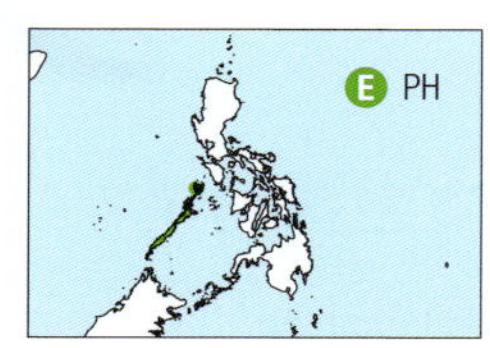

Palawan, Calamian Is

Genus *CONEPATUS*

(3 species)

color variants

American Hog-nosed Skunk
Conepatus leuconotus
FR Moufette à dos blanc
DE Ferkelskunk
ES Zorrino dorsiblanco

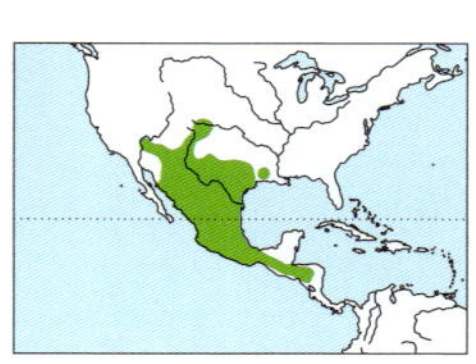

S US (SE Colorado, Arizona, New Mexico, Texas), most of MX (except Baja California & Yucatán Pen) S to NI

Striped Hog-nosed Skunk
Conepatus semistriatus
FR Moufette d'Amazonie
DE Amazonas-Skunk
ES Zorrino amazónico

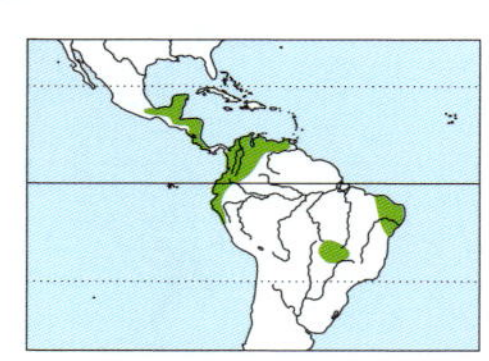

SE MX (Veracruz, Yucatán Pen), BZ & GT S to CO, VE, W EC & N PE, isolated in NE & C BR

LC
8 sspp
HB 33–50 cm
T 13–31 cm
W 1.4–3.5 kg

Molina's Hog-nosed Skunk
Conepatus chinga

FR Moufette de Molina
DE Anden-Skunk
ES Zorrino chileno

color variants

LC
Monotypic
HB 20–49 cm
T 13.3–29 cm
W 0.5–3 kg

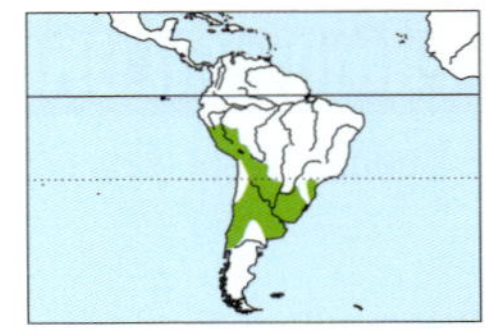

S PE, C & S BO, PY, SE BR, UY, CL, AR

Genus *MEPHITIS*

(2 species)

Striped Skunk
Mephitis mephitis

FR Moufette rayée
DE Streifenskunk
ES Mofeta rayada

typical morph

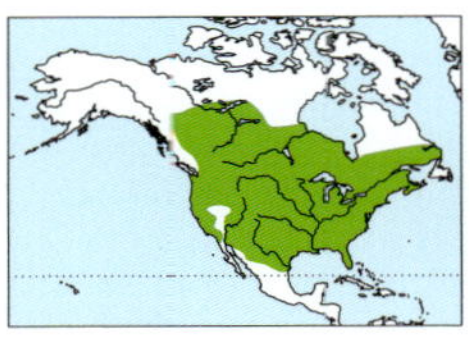

C & S CA (from British Columbia to Hudson Bay & Nova Scotia) throughout US to N MX

color variants

LC
13 sspp
HB 17–40 cm
T 15–47 cm
W 0.6–4.1 kg

Hooded Skunk
Mephitis macroura

FR Moufette à capuchon
DE Haubenskunk
ES Mofeta encapuchada

LC
4 sspp
HB 28–31 cm
T 27–43 cm
W 882–957 g

0–3110 m

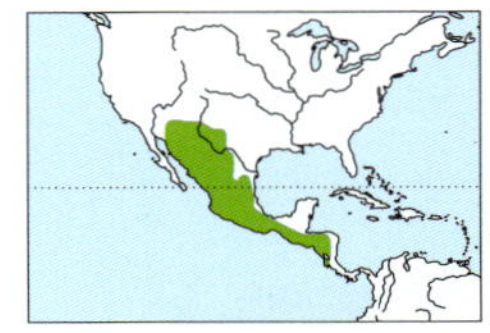

S US (S Arizona, New Mexico, Texas), S throughout MX (absent in Yucatán Pen) to HN, NI & CR

color variants

Genus *SPILOGALE*

(7 species)

Alleghanian Spotted Skunk
Spilogale putorius

FR Moufette tachetée
DE Östlicher Fleckenskunk
ES Mofeta oriental

E of R Mississippi to Atlantic Coastal Plain, N through Appalachian Mts, S to Florida Pen

VU
Monotypic
HB 19–33 cm
T 8–28 cm
W 207–885 g

Plains Spotted Skunk
Spilogale interrupta

FR Moufette des Prairies
DE Plains-Fleckenskunk
ES Mofeta de llanura

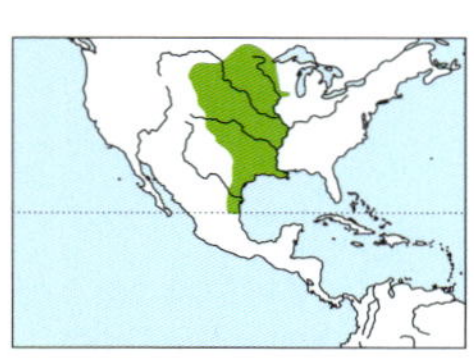

US in Great Plains to E to R Mississippi, N marginally to CA (SE Manitoba, SW Ontario), S to NE MX (Tamaulipas)

NE
Monotypic
HB 19–33 cm
T 8–28 cm
W 207–885 g

Yucatan Spotted Skunk
Spilogale yucatanensis

FR Moufette du Yucatan
DE Yucatán-Fleckenskunk
ES Mofeta de Yucatán

NE
Monotypic
HB 16.6–18.5 cm
T 10.4–11.5 cm

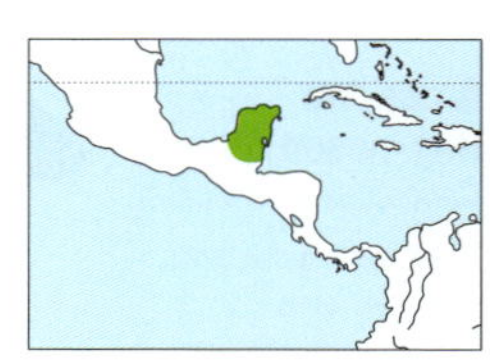

MX (throughout Yucatán Pen), BZ, N GT; S limits uncertain

Western Spotted Skunk
Spilogale gracilis

FR Moufette gracile
DE Westlicher Fleckenskunk
ES Mofeta occidental

LC
Monotypic
HB 24–37 cm
T 8.5–21 cm
W 255–997 g

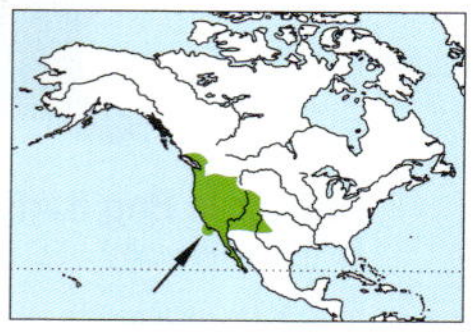

SW CA, along W US to MX (Baja California Pen), inland E to Great Plains, S to Colorado Plateau, Channel Is

Desert Spotted Skunk
Spilogale leucoparia

FR Moufette du désert
DE Wüsten-Fleckenskunk
ES Mofeta de desierto

NE
Monotypic
HB 24–37 cm
T 8.5–21 cm
W 255–997 g

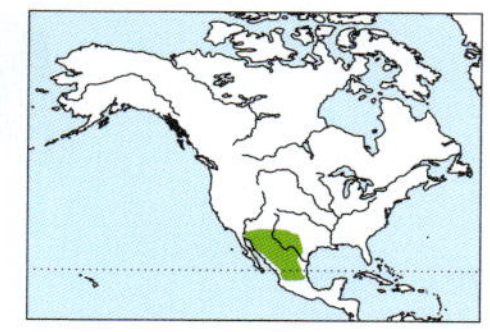

SW US (from S Rocky Mts, E from R Colorado & Mojave Desert to Great Plains) S to C MX

Southern Spotted Skunk
Spilogale angustifrons

FR Moufette de Howell
DE Südlicher Fleckenskunk
ES Mofeta meridional

LC
Monotypic
HB 20–25 cm
T 8–28 cm
W 207–885 g

0–2800 m

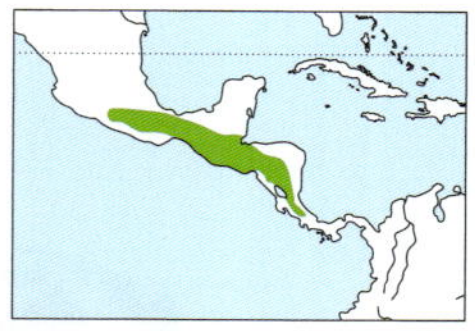

S MX (S from Trans-Mexican Volcanic Belt), GT, SV, HN, NI S to C CR

Pygmy Spotted Skunk
Spilogale pygmaea

FR Moufette naine
DE Zwerg-Fleckenskunk
ES Mofeta chica

VU
3 sspp
HB 19–21 cm
T 5.9–7.3 cm
W 150–320 g

0–1630 m

Pacific coast in Sinaloa, Nayarit, Jalisco, Colima, Michoacán, Guerrero & Oaxaca

Family MUSTELIDAE – Weasels and relatives

23 genera, 65 species. Forests, grasslands, tundras, and oceans, from the Arctic to the tropics. Holarctic, Neotropical, African and Indo-Malayan Regions.

Genus *TAXIDEA*

(1 species)

LC
4 sspp
HB 42–72 cm
T 10–15.5 cm
W 6.3–8.7 kg

American Badger
Taxidea taxus

FR Blaireau d'Amérique
DE Amerikanischer Dachs
ES Tejón americano

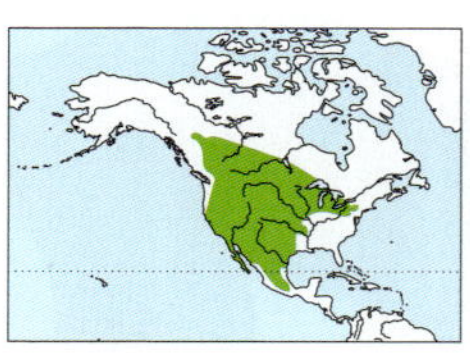

S CA (from British Columbia E to Ontario), most of US, N & C MX (S to Puebla)

Genus *MELLIVORA*

(1 species)

LC
10 sspp
HB 73–96 cm
T 14.3–23 cm
W 6.2–13.6 kg

Honey Badger
Mellivora capensis

FR Ratel
DE Honigdachs
ES Ratel

0–4050 m

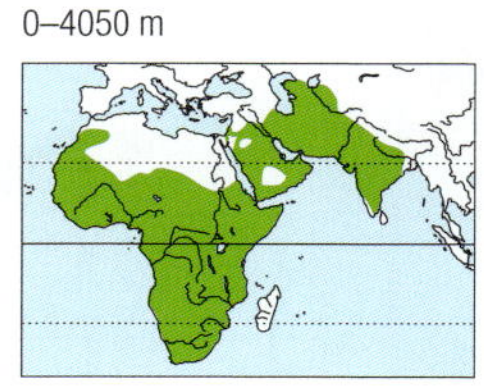

Most of sub-Saharan Africa (from S MA & SW DZ S to ZA), Arabian Pen, IR N to TM & UZ, E to IN & NP

Genus *ARCTONYX*

(3 species)

Northern Hog Badger
Arctonyx albogularis

FR Balisaur à gorge blanche
DE Nördlicher Schweinsdachs
ES Tejón porcino septentrional

LC
Monotypic
HB 55–70 cm
T 11–22 cm

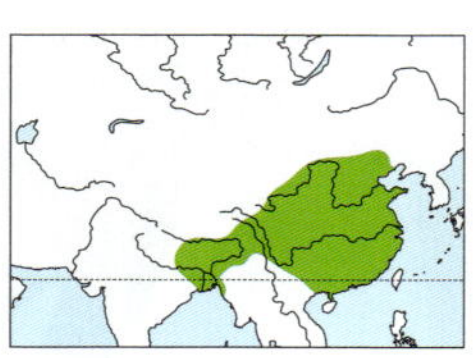

C, E & S CN, E MN to sub-Himalayan zone in BT & NE IN, extreme N MM

Greater Hog Badger
Arctonyx collaris

FR Grand Balisaur
DE Schweinsdachs
ES Tejón porcino

VU
Monotypic
HB 51–71 cm
T 8–18 cm

100–1500 m

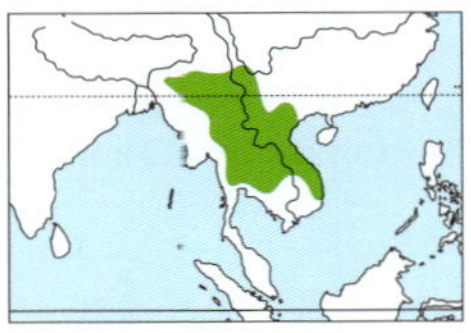

From NE IN & C CN (Yunnan) through MM, TH, LA & VN (perhaps extirpated) to KH

Sumatran Hog Badger
Arctonyx hoevenii

FR Balisaur de Sumatra
DE Sumatra-Schweinsdachs
ES Tejón porcino de Sumatra

LC
Monotypic
HB 51–71 cm
T 8–18 cm

700–3780 m

Sumatra (Barisan Mts)

Genus *MELES*

(4 species)

European Badger
Meles meles

FR Blaireau d'Europe
DE Europäischer Dachs
ES Tejón europeo

LC
Monotypic
HB 56–90 cm
T 11.5–20 cm
W 10–16 kg

up to 1700 m

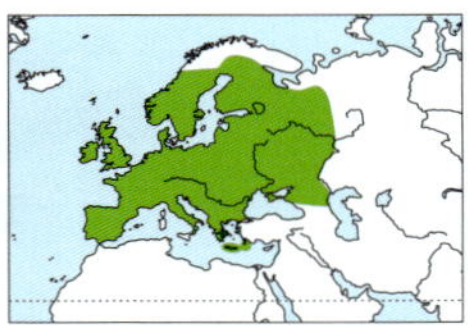

Europe E to R Volga (RU), N of Caucasus Mts, Crete & Rhodes Is

Caucasian Badger
Meles canescens

FR Blaireau de Perse
DE Transkaukasischer Dachs
ES Tejón caucásico

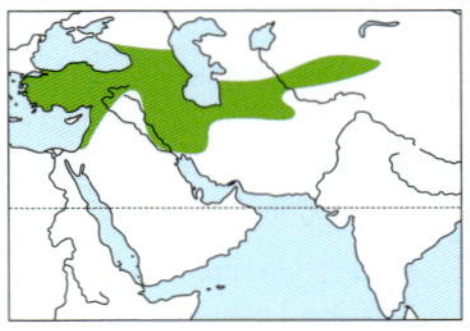

Caucasus & Middle East from TR, IL, LB & SY to IQ, IR & N AF

NE
Monotypic
HB c. 84 cm

Asian Badger
Meles leucurus

FR Blaireau d'Asie
DE Asiatischer Dachs
ES Tejón asiatico

LC
2 sspp
HB 49–70 cm
T 13–20 cm
W 3.5–9 kg

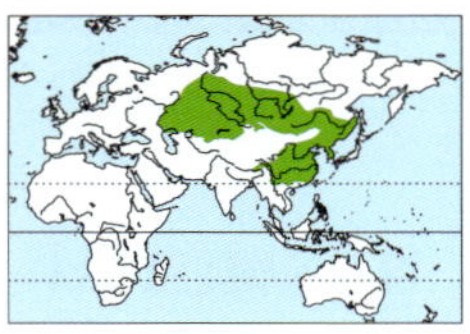

RU (E of R Volga through Siberia to Far East), KZ, MN, N, C, E & SE CN, Korean Pen

Japanese Badger
Meles anakuma

FR Blaireau du Japon
DE Japanischer Dachs
ES Tejón japonés

LC
Monotypic
HB 50–80 cm
T 14–20 cm
W 6–17 kg

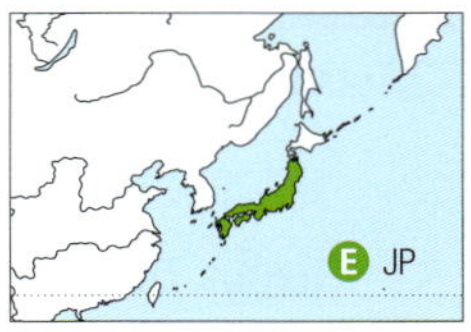

Honshu, Kyushu, Shikoku

Genus *PEKANIA*

(1 species)

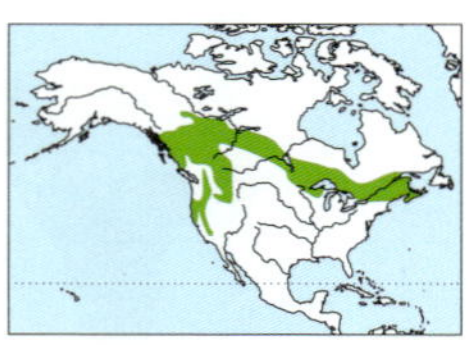

SE Alaska, C & S CA (from Yukon to Quebec), W US (S to California)

Fisher
Pekania pennanti

FR Pékan
DE Fischermarder
ES Marta pekan

LC
3 sspp
HB 45–65 cm
T 30–50 cm
W 2–5.5 kg

Genus *EIRA*

(1 species)

Tayra
Eira barbara

FR Tayra
DE Tayra
ES Taira

color variants

LC
7 sspp
HB 56–71 cm
T 36–46 cm
W 2.7–7 kg

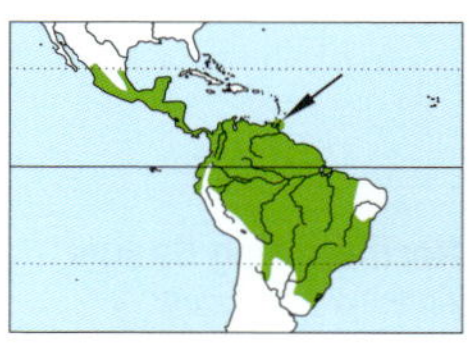

From C MX through Central & South America (except high Andes & caatinga of E BR) to N AR & S BR, TT (Trinidad)

Genus *GULO*

(1 species)

Wolverine
Gulo gulo

FR Carcajou
DE Vielfraß
ES Glotón

LC
2 sspp
HB 65–105 cm
T 21–26 cm
W 6–18 kg

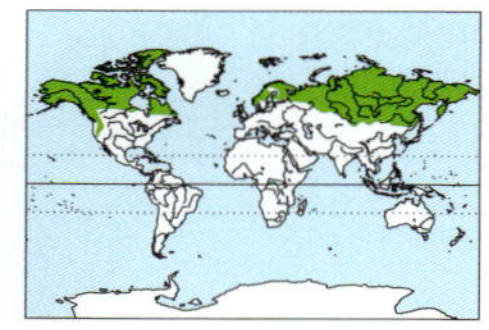

Boreal zone from Scandinavia, RU, MN & N CN to Alaska (US), CA & W US (S to California)

Genus *MARTES*

(8 species)

ssp *flavigula*

ssp *peninsularis*

200–3000 m

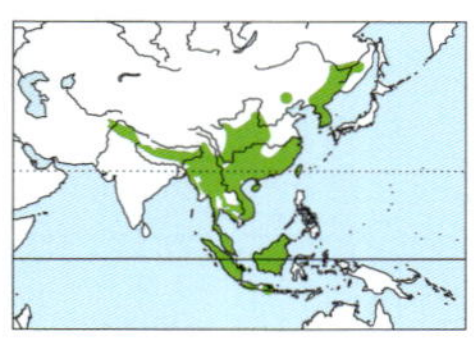

Himalayas (E from PK) E to CN, N to RU Far East & Korean Pen, TW, mainland SE Asia, Sumatra, Java, Borneo

Yellow-throated Marten
Martes flavigula

FR Martre à gorge jaune
DE Buntmarder
ES Marta papigualda

LC
6 sspp
HB 45–65 cm
T 37–45 cm
W 1.3–3 kg

120–2385 m

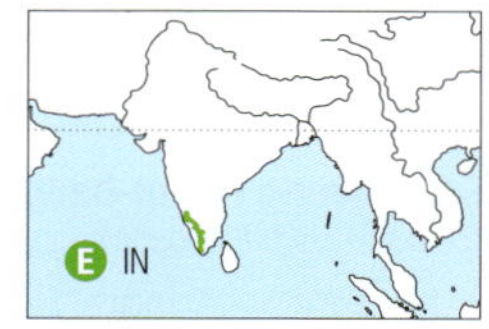

Western Ghats

Nilgiri Marten
Martes gwatkinsii

FR Martre des Nilgiri
DE Südindischer Buntmarder
ES Marta india

VU
Monotypic
HB 50–70 cm (M)
T 35–50 cm (M)
W 1–3 kg (M)

Sable
Martes zibellina

FR Martre zibeline
DE Zobel
ES Marta cibelina

LC
Monotypic
HB 35–56 cm
T 11.5–19 cm
W 700–1800 g

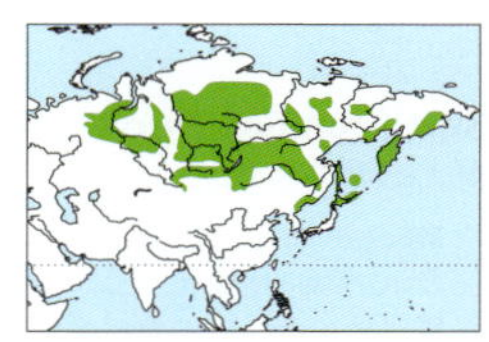

RU (E from W foothills of Urals), MN, CN, KP, JP (Hokkaido)

Stone Marten
Martes foina

FR Fouine
DE Steinmarder
ES Garduña

LC
Monotypic
HB 40–54 cm
T 22–30 cm
W 1.1–2.3 kg

up to 4000 m

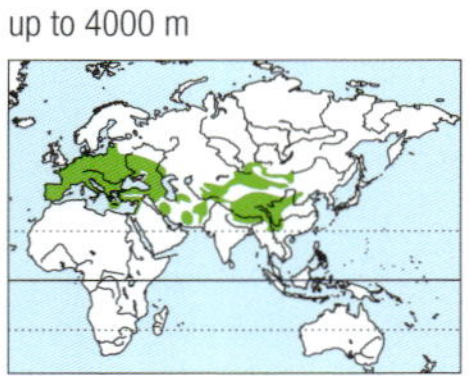

Mainland C & S Europe, Caucasus, Middle East, C Asia to MN, CN & N MM; introduced to US (Wisconsin)

European Pine Marten
Martes martes

FR Martre des pins
DE Baummarder
ES Marta europea

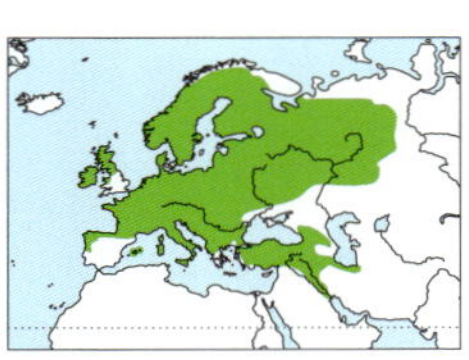

Most of Europe to RU (W Siberia), major Mediterranean Is, Middle East (TR, Caucasus, IQ, IR)

LC
Monotypic
HB 45–58 cm
T 16–28 cm
W 0.8–1.8 kg

Japanese Marten
Martes melampus

FR Martre du Japon
DE Japanischer Marder
ES Marta japonesa

Korea Pen, JP (Honshu, Kyushu, Shikoku, Tsushima); introduced to Sado & Hokkaido Is (JP)

LC
3 sspp
HB 47–54 cm
T 17–22 cm
W 1–1.5 kg

American Marten
Martes americana
FR Martre d'Amérique
DE Fichtenmarder
ES Marta norteamericana

LC
Monotypic
HB 32–45 cm
T 18–23 cm
W 280–1250 g

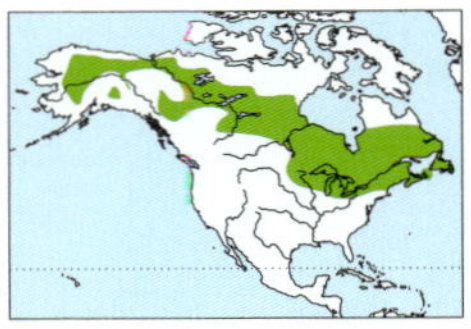

Most of North America from Alaska through much of forested CA to NE US

Pacific Marten
Martes caurina
FR Martre de Merriam
DE Merriam-Marder
ES Marta del Pacífico

NE
Monotypic
HB 32–45 cm
T 18–23 cm
W 280–1250 g

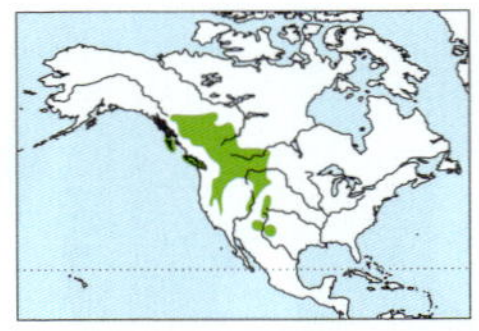

S Alaska, W CA, W US (S to California & Colorado), some Is off Alaska & British Columbia (including Vancouver)

Genus *MELOGALE*

(6 species)

Cuc Phuong Ferret-badger
Melogale cucphuongensis
FR Mélogale de Cuc Phuong
DE Cuc-Phuong-Sonnendachs
ES Melandro de Cuc Phuong

DD
2 sspp
HB c. 35 cm
T c. 18 cm
W c. 800 g

c. 1075 m

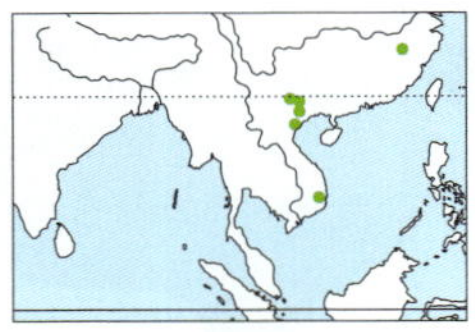

VN (Ninh Binh, Lang Son, Ha Giang, Cao Bang, Dak Lak), SE CN (known from type locality in Fujian)

Large-toothed Ferret-badger
Melogale personata
FR Mélogale indien
DE Burma-Sonnendachs
ES Melandro birmano

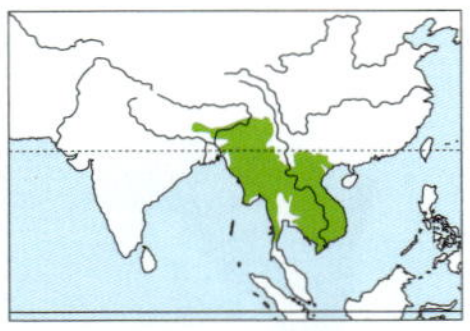

NP, NE IN, MM, CN, TH, LA, VN, KH

LC
3 sspp
HB 35–40 cm
T 15–21 cm
W 1.5–3 kg

Small-toothed Ferret-badger
Melogale moschata
FR Mélogale de Chine
DE China-Sonnendachs
ES Melandro chino

LC
5 sspp
HB 30–40 cm
T 10–15 cm
W 0.8–1.6 kg

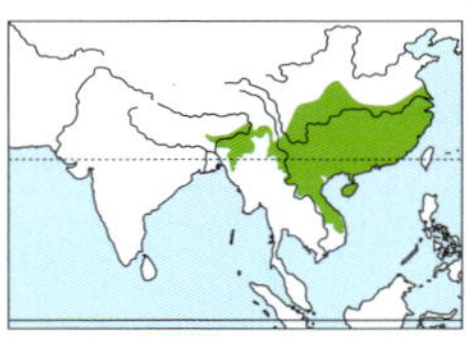

NE IN through N MM to C & SE CN (including Hainan), N LA, VN

Formosan Ferret-badger
Melogale subaurantiaca
FR Mélogale de Taïwan
DE Formosa-Sonnendachs
ES Melandro de Taiwán

NE
Monotypic

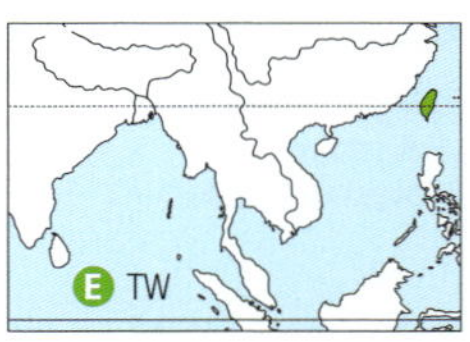

Tropical & subtropical forests, grasslands, cultivated areas

Bornean Ferret-badger
Melogale everetti
FR Mélogale d'Everett
DE Borneo-Sonnendachs
ES Melandro de Borneo

EN
Monotypic
HB 35–40 cm
T 16–17 cm
W 1–2 kg

900–3000 m

Uplands & highlands of N Borneo (currently known only from Sabah)

Javan Ferret-badger
Melogale orientalis
FR Mélogale de Java
DE Java-Sonnendachs
ES Melandro javanés

LC
2 sspp
HB 35–40 cm
T 16–17 cm
W c. 2 kg

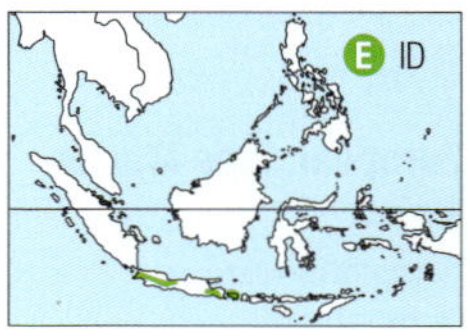

Java, Bali

Lesser Grison
Galictis cuja
FR Petit Grison
DE Kleingrison
ES Grisón chico

LC
4 sspp
HB 28–50 cm
T 12–19.3 cm
W 1–2.5 kg

0–4200 m

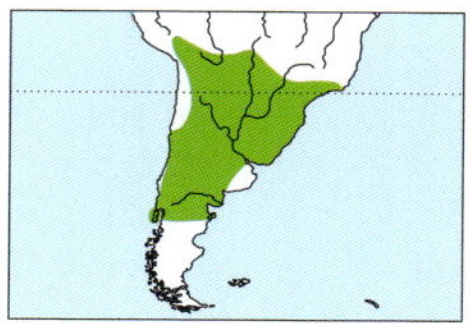

From SE PE, W BO and C & E Brazil through PY & UY to CL & AR

Genus *GALICTIS*

(2 species)

Greater Grison
Galictis vittata
FR Grand Grison
DE Großer Grison
ES Grisón

LC
4 sspp
HB 47–55 cm
T c. 16 cm
W 1.4–3.3 kg

0–2000 m

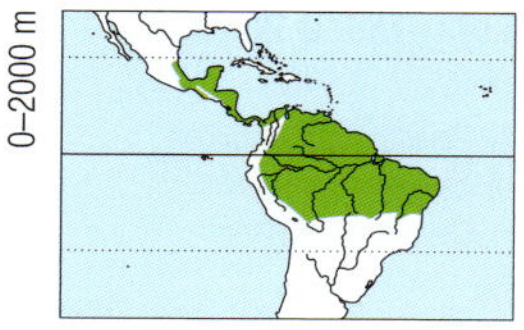

From E MX S through Central & South America to BO & C BR

Patagonian Weasel
Lyncodon patagonicus
FR Belette de Patagonie
DE Patagonisches Wiesel
ES Hurón patagónico

Genus *LYNCODON*

(1 species)

LC
Monotypic
HB 30–35 cm
T 6–9 cm
W 200–250 g

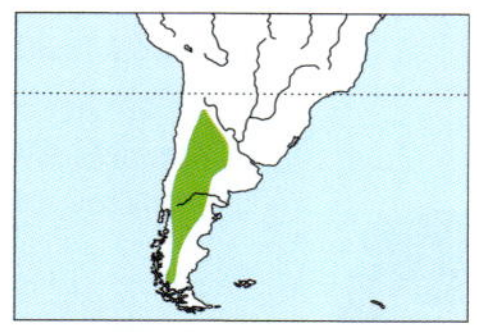

AR (S from Salta), adjacent CL (S to Puerto Prat)

Marbled Polecat
Vormela peregusna
FR Zorille marbrée
DE Tigeriltis
ES Turón jaspeado

Genus *VORMELA*

(1 species)

VU
6 sspp
HB 29–48 cm
T 15.5–17.8 cm
W 295–715 g

up to 2100 m

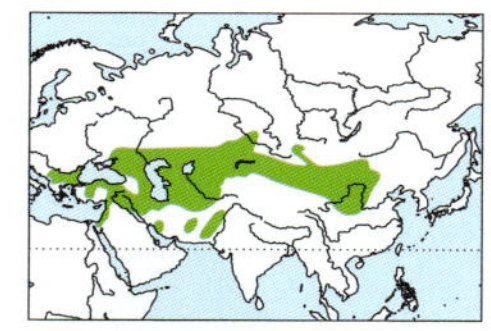

From SE Europe to Middle East, Caucasus, AF, PK & through C Asia to W & N CN and S MN

Genus *ICTONYX*

(2 species)

Saharan Striped Polecat
Ictonyx libycus
FR Zorille de Libye
DE Streifenwiesel
ES Hurón del Sahara

LC
4 sspp
HB 21–26 cm
T 11.4–18 cm
W 200–600 g

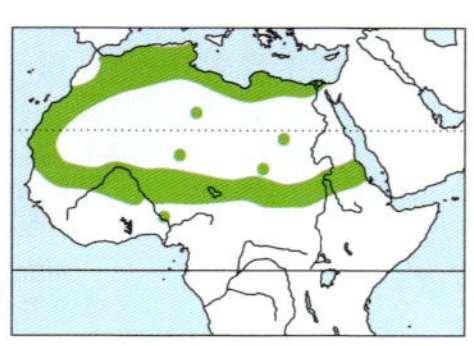

From MA, MR & SN E along N & S edges of Sahara to EG, N SD & ER

Zorilla
Ictonyx striatus
FR Zorille commune
DE Zorilla
ES Hurón estriado

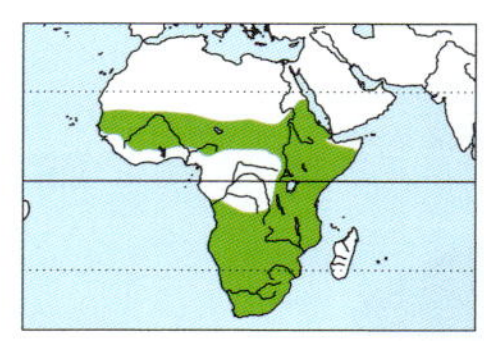

Sub-Saharan Africa from MR & SN E to SD, ET & DJ, S to ZA, absent from W & C rainforests

LC
Monotypic
HB 28–38 cm
T 16.5–28 cm
W 0.4–1.2 kg

Genus *POECILOGALE*

(1 species)

African Striped Weasel
Poecilogale albinucha
FR Zorille à nuque blanche
DE Weißnackenwiesel
ES Hurón de nuca blanca

up to 2200 m

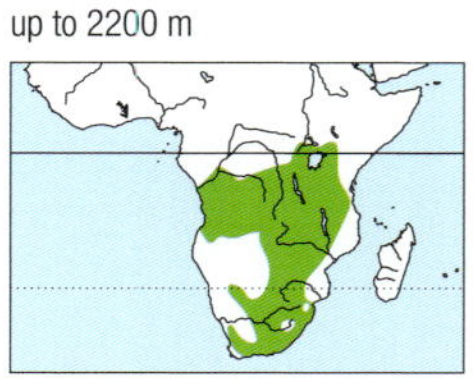

C & S Africa from AO, S CG, S & E CD, UG & KE S to ZA

LC
Monotypic
HB 24–33 cm
T 13.8–20 cm
W 210–380 g

Genus *NEOGALE*

(4 species)

American Mink
Neogale vison
FR Vison d'Amérique
DE Amerikanischer Nerz
ES Visón americano

LC
15 sspp
HB 30–43 cm
T 15.2–20 cm
W 450–1805 g

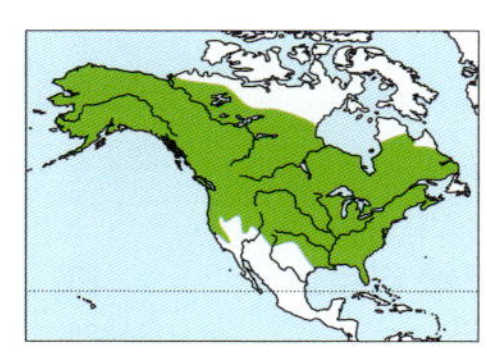

From Alaska & CA through US (except dryer areas in S); introduced to Europe, RU & JP

Long-tailed Weasel
Neogale frenata

FR Belette à longue queue
DE Langschwanzwiesel
ES Comadreja colilarga

LC
42 sspp
HB 20–26 cm
T 7.6–15.2 cm
W 80–450 g

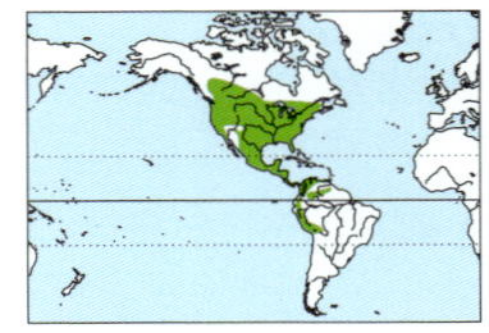

From S CA, US & MX through Central & South America to CO, VE, EC, PE & BO

Colombian Weasel
Neogale felipei

FR Belette de Colombie
DE Kolumbien-Wiesel
ES Comadreja de Don Felipe

VU
Monotypic
HB 21–22 cm
T 11.1–12.2 cm
W c. 138 g

1700–2700 m

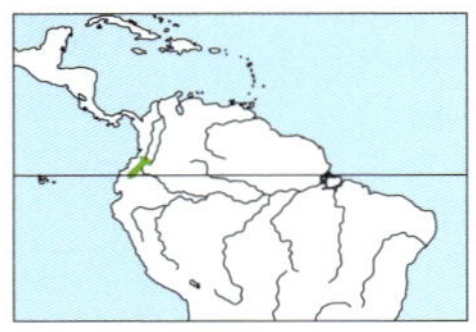

Andes in S CO & N EC

Amazon Weasel
Neogale africana

FR Belette des tropiques
DE Amazonas-Wiesel
ES Comadreja amazónica

LC
Monotypic
HB 24–38 cm
T 16–21 cm

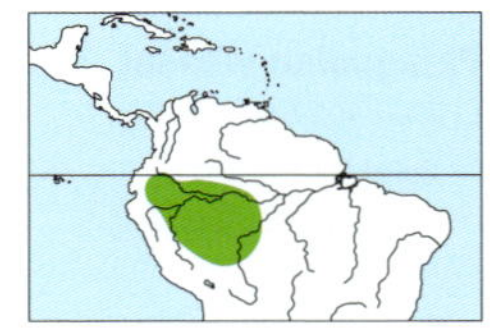

Amazon Basin in CO, EC, PE, BR & BO

Genus *MUSTELA*
(15 species)

Back-striped Weasel
Mustela strigidorsa

FR Putois à dos rayé
DE Rückenstreifenwiesel
ES Comadreja estriada

LC
Monotypic
HB 25–32 cm
T 10.3–20 cm
W 700–2000 g

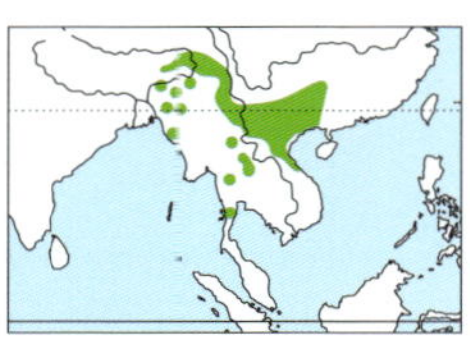

NE IN, N & C MM, S CN (Yunnan, Guizhou, Guangxi), N & S TH, N & C LA, N & C VN

Malay Weasel
Mustela nudipes

FR Putois à pieds nus
DE Nacktfußwiesel
ES Comadreja descalza

LC
Monotypic
HB 30–36 cm
T 24–26 cm
W c. 1000 g

400–1700 m

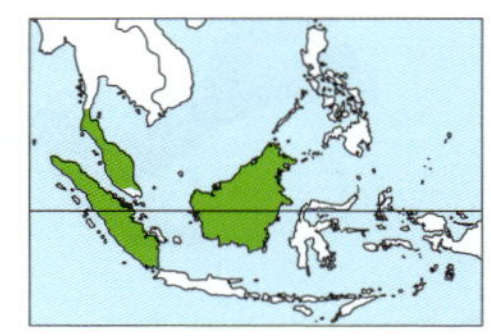

S TH, Pen MY, Sumatra, Borneo

Indonesian Mountain Weasel
Mustela lutreolina

FR Putois d'Indonésie
DE Indonesisches Bergwiesel
ES Comadreja javanesa

1000–2200 m

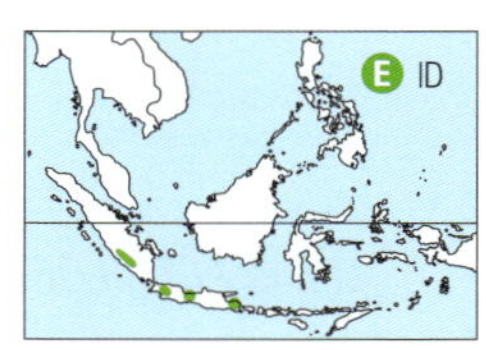

S Sumatra, Java

LC
Monotypic
HB 30–32 cm
T 13.6–17 cm
W 295–340 g

Yellow-bellied Weasel
Mustela kathiah

FR Belette à ventre jaune
DE Gelbbauchwiesel
ES Comadreja china

LC
Monotypic
HB 20–29 cm
T 13–18 cm
W 150–260 g

0–5200 m

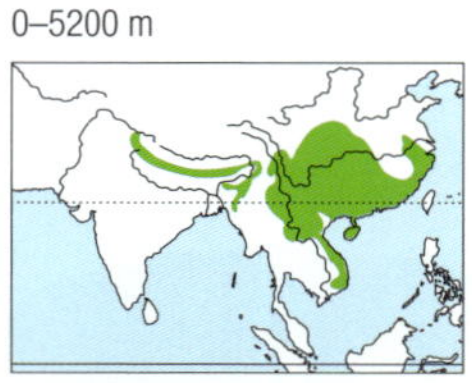

Himalayas in N & NE IN, NP & BT to MM, S & E CN (including Hainan), N TH, LA & VN

Beringian Ermine
Mustela erminea

FR Hermine d'Europe
DE Hermelin
ES Armiño

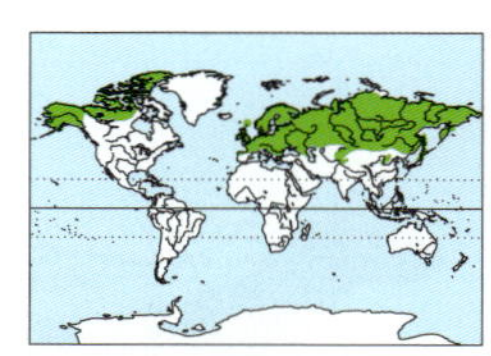

Much of Europe (S to N PT) & Asia (S to W Himalayas), Alaska (US), NW CA

LC
18 sspp
HB 19–34 cm
T 4.2–12 cm
W 134–365 g

NE
3 sspp
HB c. 21 cm
T c. 10 cm

SE Alaska in Alexander Arch (Suemez, Prince of Wales), W CA in Haida Gwaii Arch (Graham, Moresby)

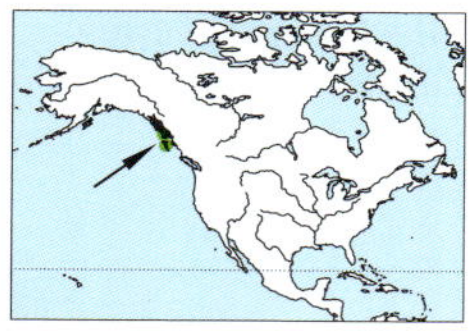

Haida Ermine
Mustela haidarum

FR Hermine des Haïdas
DE Haida-Hermelin
ES Armiño de Haida

NE
13 sspp
HB 17–24 cm
T 5.9–9.8 cm
W 39–154 g

0–3000 m

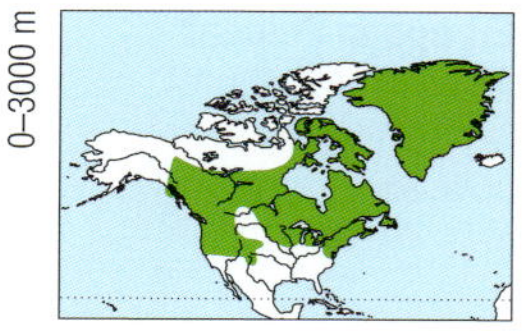

GL, CA, US (S to C California, N New Mexico, Great Lakes & N Virginia, absent in Alaska)

American Ermine
Mustela richardsonii

FR Hermine d'Amérique
DE Amerika-Hermelin
ES Armiño americano

Least Weasel
Mustela nivalis

FR Belette pygmée
DE Mauswiesel
ES Comadreja común

LC
13 sspp
HB 11.4–30 cm
T 7–12.9 cm
W 25–250 g

ssp *subpalmata*

winter

ssp *nivalis*

summer

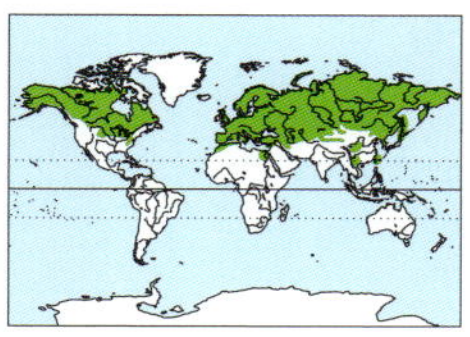

Much of Europe, mainland Asia (S to Levant, IR, MN, C CN, TW, JP), N Africa (MA, DZ, TN, EG); introduced to NZ, MT, Crete, Azores Is & ST

Altai Mountain Weasel
Mustela altaica

FR Belette des montagnes
DE Altai-Wiesel
ES Comadreja de montaña

1500–4000 m

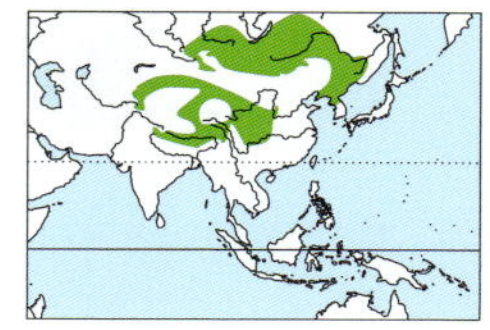

Altai, Tian Shan, Pamir & Himalayan ranges, RU (S & SE Siberia), MN, W, C & NE CN, KP

NT
Monotypic
HB 22–29 cm
T 9–14.5 cm
W 122–350 g

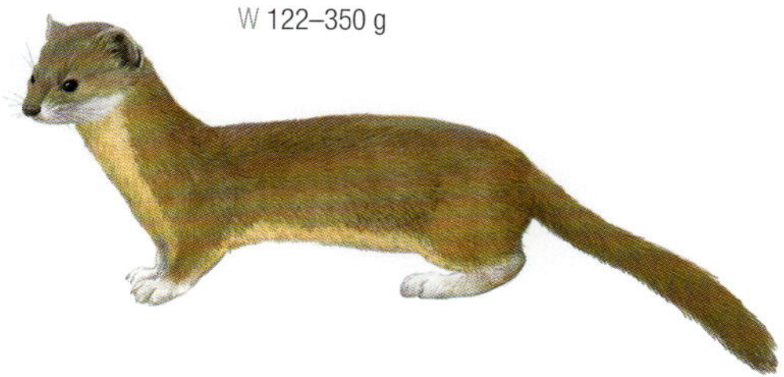

Japanese Weasel
Mustela itatsi

FR Putois du Japon
DE Japan-Wiesel
ES Comadreja japonesa

NT
Monotypic
HB 25–39 cm
T 13.3–21 cm
W 360–820 g

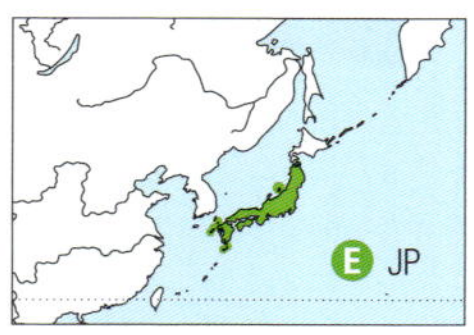

Honshu, Kyushu, Shikoku, Sado & several small Is; introduced to Hokkaido & S Sakhalin I (RU)

Siberian Weasel
Mustela sibirica

FR Putois de Sibérie
DE Sibirisches Feuerwiesel
ES Comadreja siberiana

LC
Monotypic
HB 25–39 cm
T 13.3–21 cm
W 360–820 g

1500–5000 m

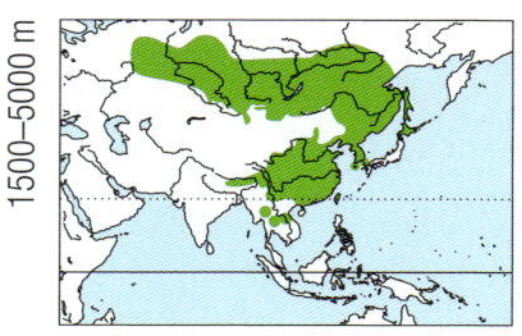

RU (E from Urals), MN, CN, Korea Pen, N JP (Hokkaido), TW, IN, NP, BT, N MM, TH, LA; introduced to JP (Honshu, Kyushu & Shikoku)

European Mink
Mustela lutreola

FR Vison d'Europe
DE Europäischer Nerz
ES Visón europeo

CR
Monotypic
HB 20–36 cm
T 12–17.5 cm
W 500–1000 g

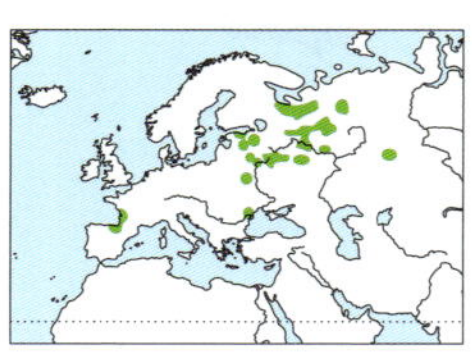

Isolated populations in N ES, SW FR, RO, EE, LV, BY & RU (W of Urals)

Steppe Polecat
Mustela eversmanii

FR Putois d'Eversmann
DE Steppeniltis
ES Turón estepario

LC
Monotypic
HB 29–56 cm
T 7–18.3 cm
W 1.3–2 kg

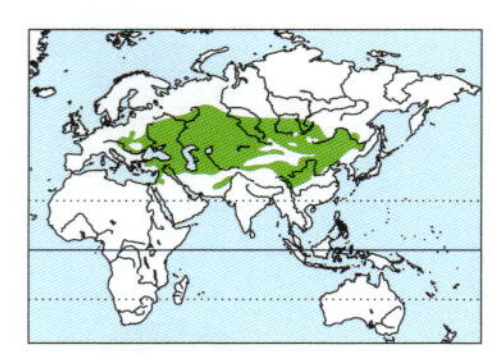

SE Europe, Caucasus, Middle East through C Asia to MN, NE CN & RU Far East

European Polecat
Mustela putorius

FR Putois d'Europe
DE Waldiltis
ES Turón europeo

LC
Monotypic
HB 20–46 cm
T 7–19 cm
W 400–1710 g

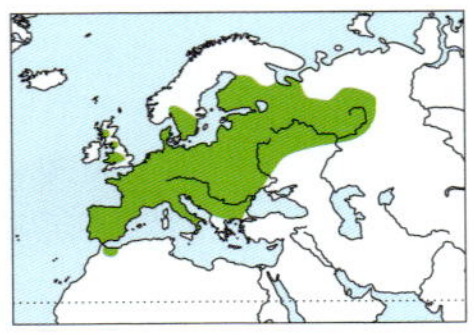

Most of Europe W of Urals (except IE, N Scandinavia, part of Balkans), N MA

Black-footed Ferret
Mustela nigripes

FR Putois d'Amérique
DE Schwarzfußiltis
ES Turón patinegro

EN
Monotypic
HB 40–50 cm
T 11.4–15 cm
W 764–1078 g

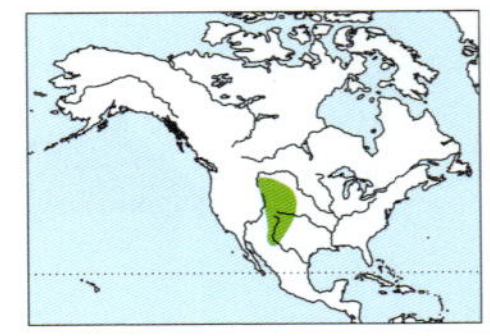

Great Plains of US (from Montana S to Arizona & New Mexico), N MX; extirpated from most of its range

Genus *PTERONURA*

(1 species)

Giant Otter
Pteronura brasiliensis

FR Loutre géante
DE Riesenotter
ES Nutria gigante

EN
Monotypic
HB 100–130 cm
T 45–65 cm
W 22–32 kg

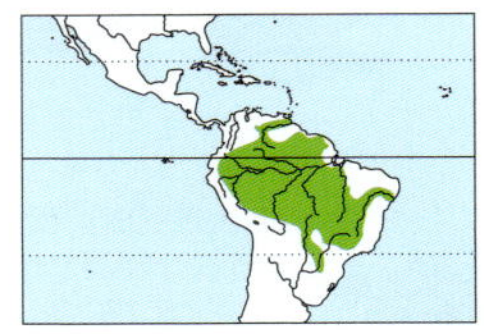

Amazon & Orinoco basins from VE to PY & S BR; formerly AR & UY, but perhaps extinct there

Genus *LONTRA*

(4 species)

North American River Otter
Lontra canadensis

FR Loutre du Canada
DE Nordamerikanischer Fischotter
ES Nutria neártica

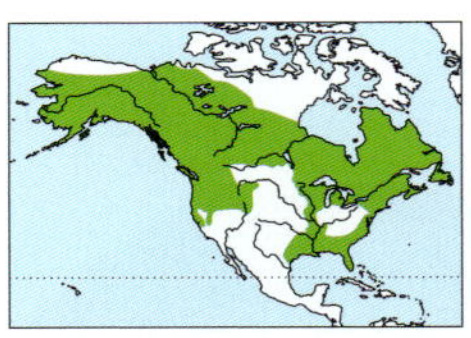

From arctic Alaska & N CA to S US (mainly along coasts of Atlantic, Great Lakes area, NW Pacific & Gulf of Mexico)

LC
7 sspp
HB 58–73 cm
T 32–47 cm
W 7.3–9.4 kg

Neotropical Otter
Lontra longicaudis

FR Loutre néotropicale
DE Südamerikanischer Fischotter
ES Nutria neotropical

NT
3 sspp
HB 36–66 cm
T 37–84 cm
W 5–15 kg

up to 3000 m

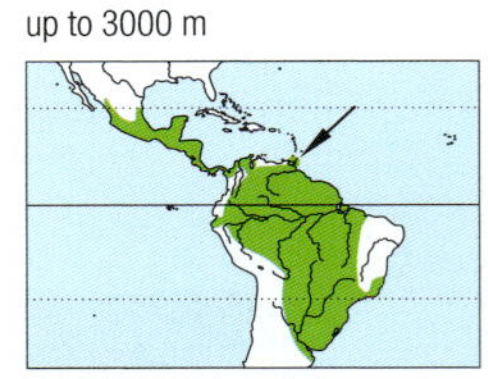

From MX through Central & South America to BR (absent in NE), PY, UY & NE AR, TT (Trinidad)

Southern River Otter
Lontra provocax

FR Loutre du Chili
DE Patagonischer Fischotter
ES Huillín

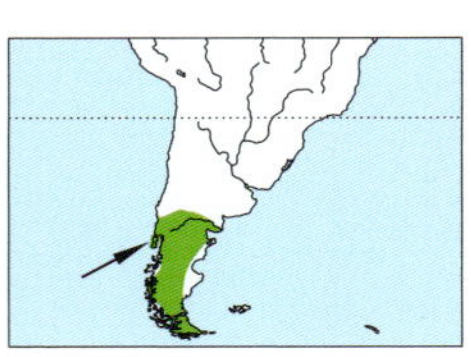

CL (S from Araucanía), AR (from Neuquén to Tierra del Fuego), today strongly extirpated of its range

EN
Monotypic
HB 57–61 cm
T 35–40 cm
W 5–10 kg

Marine Otter
Lontra felina

FR Loutre chungungo
DE Südamerikanischer Meerotter
ES Chungungo

EN
Monotypic
HB 53–79 cm
T 30–36 cm
W 3.2–5.8 kg

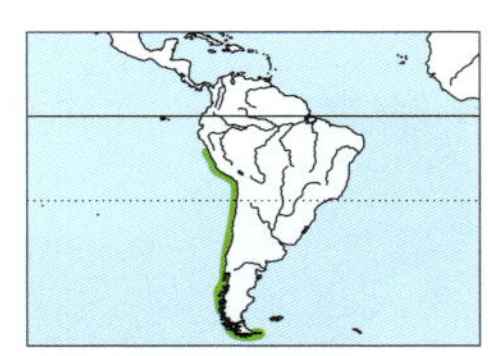

Pacific coast from N PE to S CL, extreme S AR (Patagonia)

Genus *ENHYDRA*

(1 species)

Sea Otter
Enhydra lutris

FR Loutre de mer
DE Meerotter
ES Nutria marina

EN
3 sspp
HB 100–120 cm
T 25–37 cm
W 14–45 kg

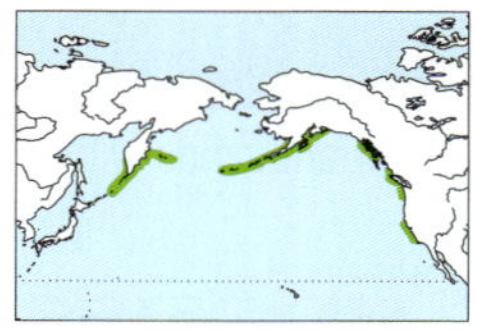

NE RU (Kamchatka, Commander & Kuril Is), Aleutian Is, S Alaska, W CA, W USA (Washington, California)

Genus *HYDRICTIS*

(1 species)

Spotted-necked Otter
Hydrictis maculicollis

FR Loutre à cou tacheté
DE Fleckenhals-Otter
ES Nutria moteada

NT
Monotypic
HB 57–76 cm
T 38–44 cm
W 3.8–6 kg

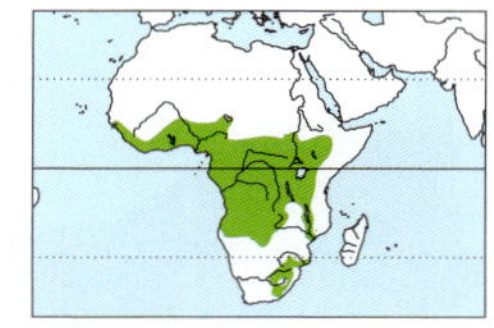

Sub-Saharan Africa from GW E to SW ET & E KE, S to N NA, BW & NW ZW, also MW, MZ & E ZA

Genus *LUTRA*

(2 species)

Eurasian Otter
Lutra lutra

FR Loutre d'Europe
DE Eurasischer Fischotter
ES Nutria paleártica

NT
Monotypic
HB 50–82 cm
T 33–50 cm
W 5–14 kg

0–4120 m

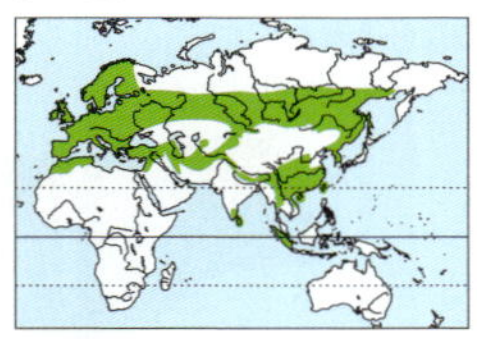

From Europe to RU Far East, Korea Pen & JP, N Africa, Middle East, C, S & E Asia to E CN & TW, Sumatra

Hairy-nosed Otter
Lutra sumatrana

FR Loutre de Sumatra
DE Haarnasen-Otter
ES Nutria indonesia

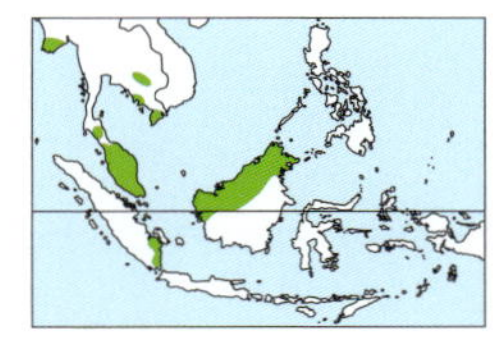

Mainland SE Asia (MM, KH, S VN, Pen MY), S Sumatra, Borneo

EN
Monotypic
HB 50–82 cm
T 37–50 cm
W 3.5–6 kg

Genus *AONYX*

(3 species)

African Clawless Otter
Aonyx capensis

FR Loutre à joues blanches
DE Fingerotter
ES Nutria africana

NT
Monotypic
HB 113–138 cm
T 44–57 cm
W 10–21 kg

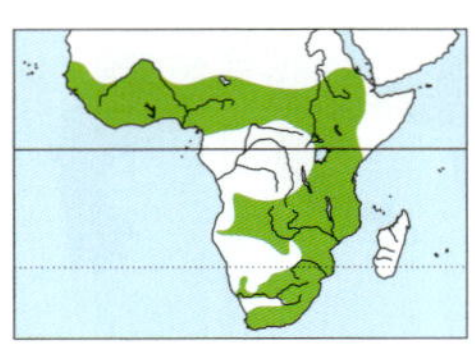

Sub-Saharan Africa from SN E to E ET, S to ZA, except Congo Basin and arid zones of NA, BW & ZA

Congo Clawless Otter
Aonyx congicus

FR Loutre du Congo
DE Kongo-Fingerotter
ES Nutria del Congo

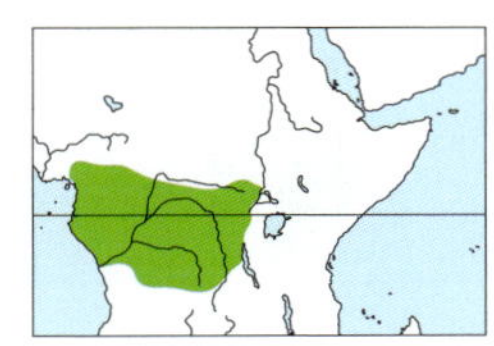

Congo Basin in WC Africa

NT
Monotypic
HB 69–74 cm
T 45–47 cm
W 13–19.4 kg

Asian Small-clawed Otter
Aonyx cinereus

FR Loutre cendrée
DE Zwergotter
ES Nutria chica

VU
3 sspp
HB 36–47 cm
T 22–27 cm
W 2.4–3.8 kg

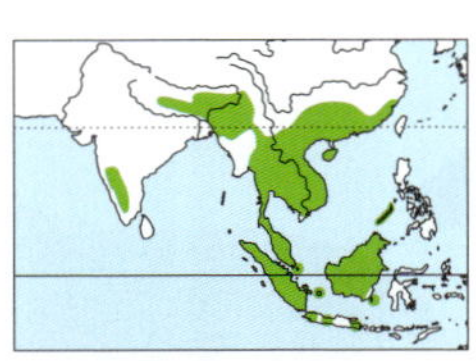

SW IN, sub-Himalayas (NP, BT, NE IN, MM, SW CN), SE CN, mainland SE Asia, Sumatra, Java, Borneo, PH (Palawan)

Genus *LUTROGALE*

(1 species)

Smooth-coated Otter
Lutrogale perspicillata

FR Loutre indienne
DE Indischer Fischotter
ES Nutria lisa

VU
3 sspp
HB 59–75 cm T 37–45 cm
W 7–11 kg

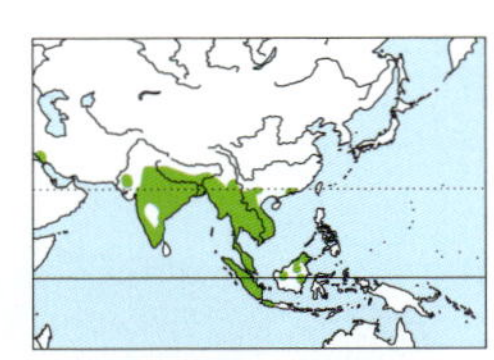

IQ, AF, from PK & IN E to S CN, mainland SE Asia, Sumatra, E Java, Borneo; perhaps IR

Family PROCYONIDAE – Raccoons

5 genera, 13 species. Every forest type within their range; some also use urban areas or arid lands with rocky shelters. North, Central and South America.

Genus *POTOS*

(1 species)

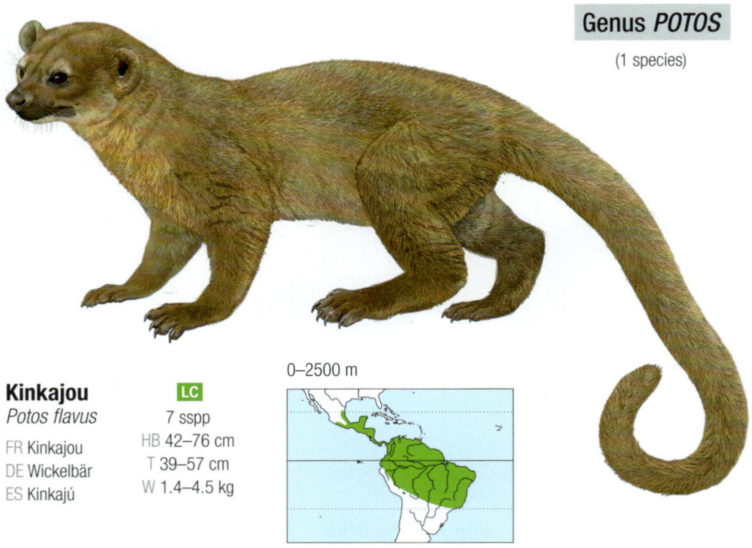

Kinkajou
Potos flavus

FR Kinkajou
DE Wickelbär
ES Kinkajú

LC
7 sspp
HB 42–76 cm
T 39–57 cm
W 1.4–4.5 kg

0–2500 m

MX (E & S of Sierras Madre) S through Central & South America to E BO & SE BR

Genus *BASSARICYON*

(4 species)

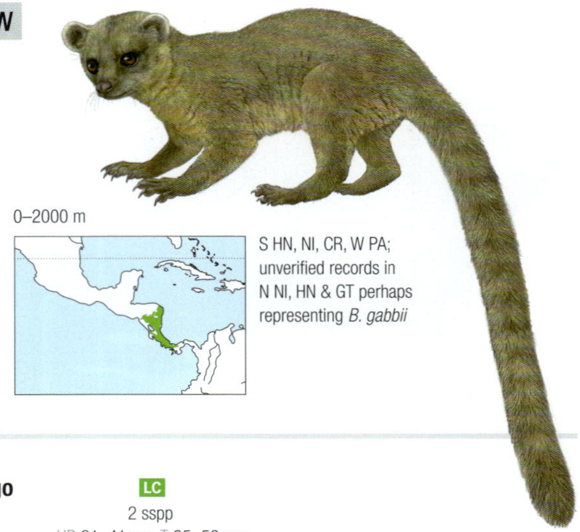

Northern Olingo
Bassaricyon gabbii

FR Olingo de Gapp
DE Mittelamerika-Makibär
ES Olingo de Gabb

LC
Monotypic
HB 35–49 cm
T 40–53 cm
W 1–1.6 kg

0–2000 m

S HN, NI, CR, W PA; unverified records in N NI, HN & GT perhaps representing *B. gabbii*

Western Lowland Olingo
Bassaricyon medius

FR Olingo du Choco
DE Chocó-Makibär
ES Olingo del Chocó

LC
2 sspp
HB 31–41 cm T 35–52 cm
W 0.9–1.6 kg

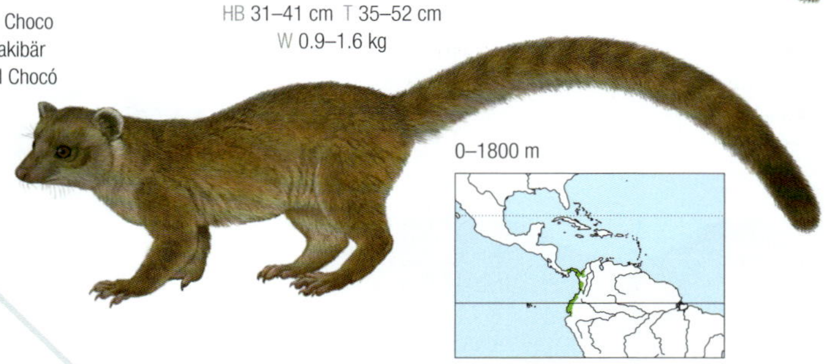

0–1800 m

E PA & NW CO, W of Andes in Chocó region of W CO & EC

Eastern Lowland Olingo
Bassaricyon alleni

FR Olingo d'Allen
DE Makibär
ES Olingo de Allen

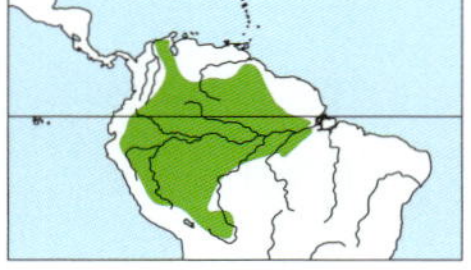

Amazonia E of Andes from VE & GY through N BR to EC, PE and N & C BO

LC
Monotypic
HB 30–45 cm
T 40–53 cm
W 0.9–1.6 kg

Olinguito
Bassaricyon neblina

FR Olinguito
DE Anden-Makibär
ES Olinguito

NT
4 sspp
HB 32–40 cm
T 33–43 cm
W 0.7–1.1 kg

1500–2750 m

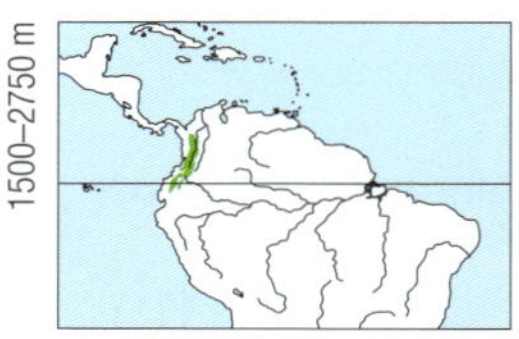

CO (both slopes of W & C Andes), EC (Andes in Pichincha & Cotopaxi)

Genus *NASUA*

(3 species)

South American Coati
Nasua nasua

FR Coati brun
DE Nasenbär
ES Coatí rojo

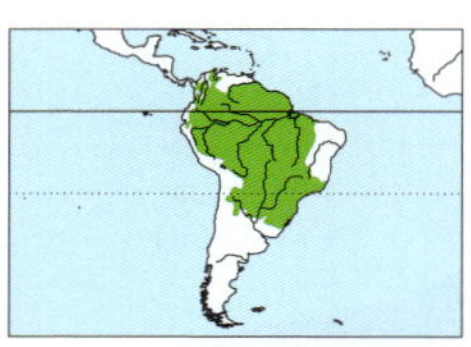

up to 2500 m

From CO & VE (except Llano grasslands) S to BO, PY, N AR & UY

LC
12 sspp
HB 43–58 cm T 42–55 cm
W 2–7.2 kg

White-nosed Coati
Nasua narica

FR Coati à nez blanc
DE Weißrüssel-Nasenbär
ES Coatí pizote

ssp *nelsoni*

SW US (Arizona, S New Mexico), MX (including Cozumel I) S through Central America to W & N CO

ssp *narica*

LC
4 sspp
HB 43–68 cm
T 42–68 cm
W 3.5–5.6 kg

ssp *meridensis*

Mountain Coati
Nasua olivacea

FR Coati des montagnes
DE Bergnasenbär
ES Coatí oliva

NT
3 sspp
HB 36–39 cm
T 20–24 cm
W 1–1.5 kg

ssp *olivacea*

above 2000 m

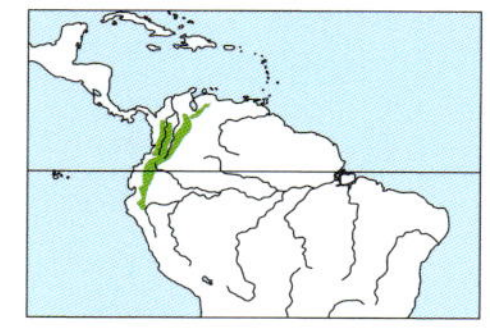

Andes of NW VE, CO & EC

Genus *BASSARISCUS*

(2 species)

Ringtail
Bassariscus astutus

FR Bassaris rusé
DE Nordamerikanisches Katzenfrett
ES Cacomixtle

LC
14 sspp
HB 30–37 cm
T 31–44 cm
W 0.8–1.1 kg

0–2900 m

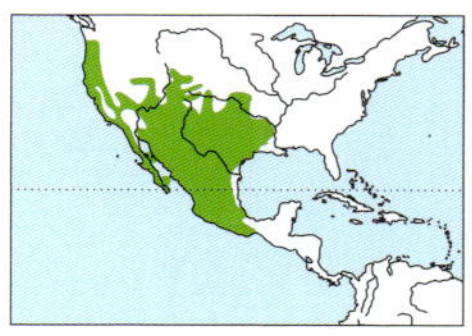

SW US (S from Oregon, S Wyoming & Kansas) to S MX (Oaxaca), Gulf of California Is

Cacomistle
Bassariscus sumichrasti

FR Bassaris de Sumichrast
DE Mittelamerikanisches Katzenfrett
ES Cacomixtle meridional

LC
5 sspp
HB 38–50 cm
T 39–55 cm
W 0.7–1.2 kg

up to 2700 m

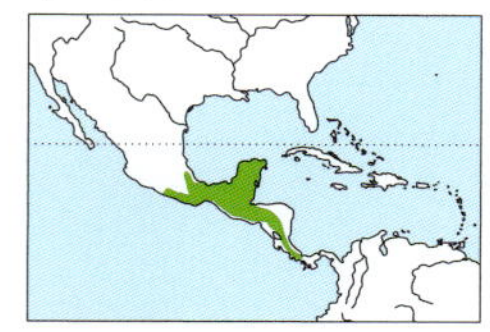

S MX (S from Veracruz & Guerrero), BZ & GT to PA

Genus *PROCYON*

(3 species)

Crab-eating Raccoon
Procyon cancrivorus

FR Raton crabier
DE Krabbenwaschbär
ES Mayuato

LC
4 sspp
HB 54–76 cm
T 25–38 cm
W 3.1–7.7 kg

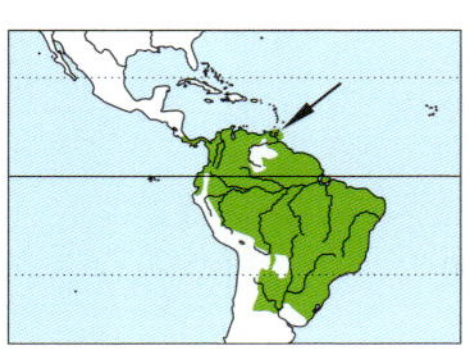

S CR, PA, CO, EC, VE, TT (Trinidad) & Guianas through Amazonia to N AR & N UY

Cozumel Raccoon
Procyon pygmaeus

FR Raton de Cozumel
DE Cozumel-Waschbär
ES Mapache de Cozumel

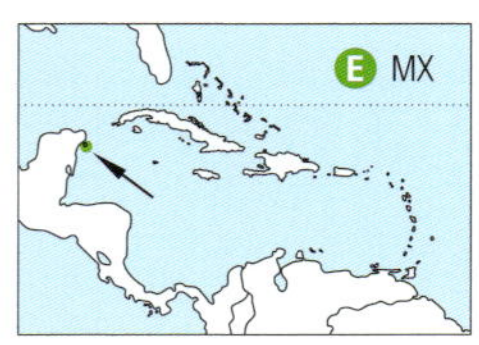

Cozumel I

CR
Monotypic
HB 35–43 cm
T 22–25 cm
W c. 3.5 kg

Northern Raccoon
Procyon lotor

FR Raton laveur
DE Waschbär
ES Mapache común

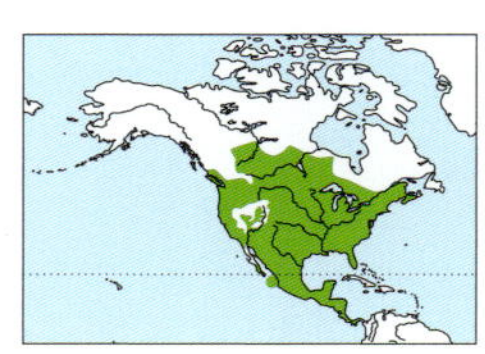

From Central Plains of S CA across US (absent from parts of Rocky Mts) & Central America to PA

LC
20 sspp
HB 44–62 cm
T 19–36 cm
W 2.7–10.4 kg

Appendices

Appendix 1

EXTINCT SPECIES

This Appendix covers the 103 mammal species believed to have become extinct roughly since 1500 AD, which is the date used by The IUCN Red List of Threatened Species to define "recent" extinctions. They are all currently assigned the threat category of Extinct (EX) on the Red List.

This section is divided with headers to indicate the family to which each species belongs, with its scientific name in red and English common name in gray. This way interested readers can find and explore the extant members of the same family, as all scientific family names are listed in the Index, Contents, and endpapers of the book. Unfortunately, there are seven mammal families for which all species are extinct, so they are only found in this Extinct species Appendix: specifically, Thylacinidae (Thylacine), Chaeropodidae (Pig-footed Bandicoots), Palaeopropithecidae (Sloth Lemurs), Megaladapidae (Koala Lemurs), Archaeolemuridae (Monkey Lemurs), Prolagidae (Sardinian Pika), and Nesophontidae (Nesophontes).

As in the section of the book covering extant species, here each species account includes the Latin and vernacular names of the taxon in English, French, German, and Spanish. In addition, each has a short summary describing when each species went extinct and the factors that contributed to its demise. The taxonomy here follows the MDD and all species are monotypic unless otherwise indicated in the text.

The geographic range for each Extinct species is shaded in a gray crosshatch pattern on the map to indicate its approximate former distribution, with solid shading for small ranges and/or islands where the crosshatching would be difficult to see. The map is accompanied by a short phrase to help situate the former range of the species.

Illustrations are not included due to the lack of information and resources for many of these extinct species, including those known exclusively from fossil material.

Data for this section are based on the species accounts appearing in the *Illustrated Checklist of the Mammals of the World,* which were written by Connor J. Burgin, with more recent information compiled by Anna Motis.

Family THYLACINIDAE Thylacine

Thylacine

Thylacinus cynocephalus

FR Thylacine DE Beutelwolf ES Tilacino

Probably became extinct in New Guinea and mainland Australia c. 3200 years ago (but possibly in the 1830s in S Australia); in modern times known only from Tasmania, where it might have survived into the 1960s or 1970s before being declared officially extinct in 1982. Extinction on mainland Australia was most likely caused by factors including increasing human population, potential competition with dingoes (*Canis familiaris*), and bushfires due to climate change. In Tasmania, hunting by farmers and bounty hunters were the primary causes of extinction.

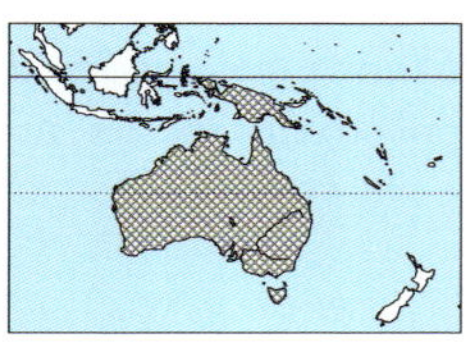

Throughout New Guinea & AU (including Tas)

Family THYLACOMYIDAE Bilbies

Lesser Bilby

Macrotis leucura

FR Petit Bilby DE Kleiner Kaninchen-Nasenbeutler ES Bilbi pequeño

The last living specimens were collected in 1931; a skull (c. 15 years old) found in 1967 in a raptor's nest was the last known specimen. According to reports from Aboriginal people, it probably became extinct in the 1950s or 1960s due to factors such as changes in habitat and fire regimes, although the introduction of foxes (*Vulpes vulpes*) and domestic cats seems to have been the most significant cause.

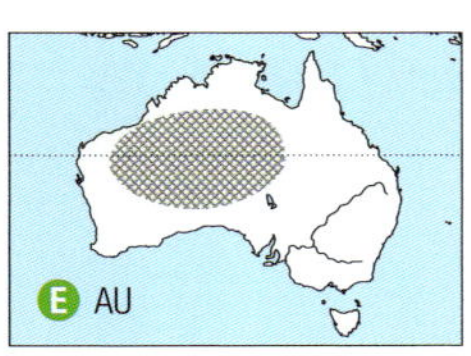

Found in C deserts in N WA, S NT & N SA

Family CHAEROPODIDAE Pig-footed Bandicoots

Northern Pig-footed Bandicoot

Chaeropus yirratji

FR Bandicoot yirratji DE Nördlicher Stützbeutler

ES Bandicut de pie de cerdo septentrional

The last modern specimen was collected in 1907 near Lake Eyre in S Australia but reports from local Aboriginal people suggest it might have survived into the 1950s. Primary cause of extinction was most likely the introduction of foxes (*Vulpes vulpes*) and domestic cats.

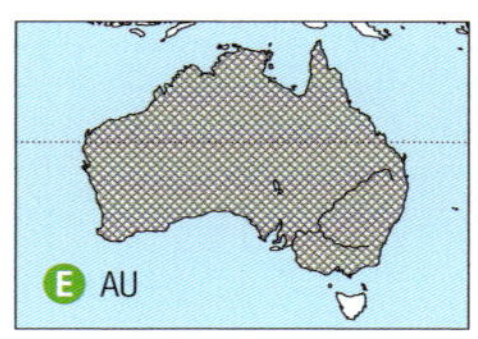

Modern and subfossil remains from the arid zone of the Flinders Ranges (SA), NT & deserts of WA; Pleistocene fossils from Cape York (Qld)

Southern Pig-footed Bandicoot

Chaeropus ecaudatus

FR Bandicoot à pieds de cochon DE Südlicher Stützbeutler

ES Bandicut de pie de cerdo meridional

Two subspecies. Known from modern, subfossil, and Pleistocene fossil remains. The last modern specimen was collected in 1900 in Ozenkadnook (Victoria); it probably became extinct in the 1920s or 1930s due to the introduction of foxes (*Vulpes vulpes*) and domestic cats, although habitat degradation caused by cattle farming also played a part.

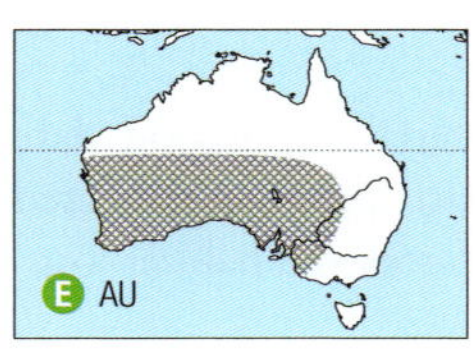

Known from S WA, SW SA, NSW & Vic

Family PERAMELIDAE Bandicoots and Echymiperas

New South Wales Barred Bandicoot

Perameles fasciata

FR Bandicoot de Liverpool DE Westlicher Langnasenbeutler
ES Bandicut listado de Nueva Gales del Sur

Known from several fossil sites in Queensland and New South Wales. Described from Darling Downs (Queensland), it was probably more widespread in the Pleistocene and in the Holocene survived on Liverpool Plains (NSW) until extinction. Last specimen collected in 1846 and last New South Wales record of any subspecies of the *P. bougainville* group (*P. fasciata* was treated as a subspecies) was in 1866, which gives a probable extinction date in the mid-19th century. Probably extirpated by land clearing and introduction of foxes (*Vulpes vulpes*).

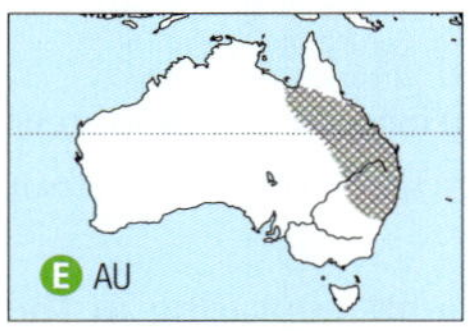

Qld & NSW

South-western Barred Bandicoot

Perameles myosuros

FR Bandicoot de King George DE Südwestlicher Langnasenbeutler
ES Bandicut listado suroccidental

Was noted as disappearing rapidly around 1890 and presumed extinct by c. 1910, probably due to invasive predators and habitat loss.

Formerly locally abundant in Wheatbelt of SW WA; some records from S edge of Nullarbor caves (WA/SA)

Southern Barred Bandicoot

Perameles notina

FR Bandicoot de Saint-Vincent DE Südlicher Langnasenbeutler
ES Bandicut listado meridional

Last observed and collected in 1857 but exact date of extinction unknown. Disappearance probably due to invasive predators and habitat loss.

Specimens collected mainly from SE SA; also noted as common on R Murray (Vic & NSW)

Ooldea Barred Bandicoot

Perameles papillon

FR Bandicoot papillon DE Nullarbor-Langnasenbeutler ES Bandicut listado de Nullarbor

Known from collected specimens and subfossil material; extinction date unknown but believed to be around the late 1930s (last specimen collected in 1928). Extinction was probably due to invasive predators and habitat loss.

Known from caves on Nullarbor Plain and adjacent Roe Plain (either side of WA/SA border)

Desert Bandicoot

Perameles eremiana

FR Bandicoot du désert DE Wüsten-Langnasenbeutler ES Bandicut listado del desierto

Last specimen collected in 1943 along the Canning Stock Route (W Australia). Its ultimate decline probably occurred in 1940s–1960s, with reports from near Lake Mackay up to the 1960s. Predation by introduced foxes (*Vulpes vulpes*) and domestic cats were probably the main causes of its decline and extinction.

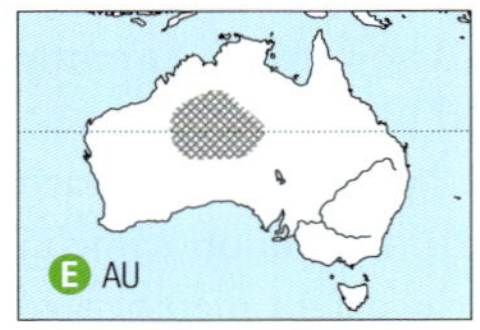

Known from S NT as far N as Tanami Desert, extending into NW SA and WA

Family POTOROIDAE Bettongs and Potoroos

Desert Rat Kangaroo

Caloprymnus campestris

FR Bettong du désert DE Nacktbrustkänguru ES Rata canguro del desierto

Originally known from only three specimens, the last collected in the 1840s. Rediscovered in the 1930s only to disappear again; final specimens collected in 1935 although there were possible sightings in the 1950s, 1970s, and 2010s. Predation by introduced foxes (*Vulpes vulpes*) and domestic cats probably led to its extinction.

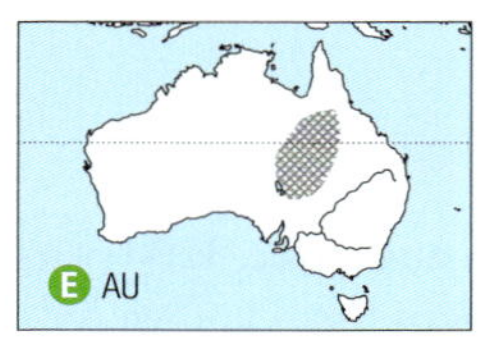

Known from modern specimens in NE SA & SW Qld (N to Coorabulka) and subfossils from Chambers Gorge (Flinders Range)

Nullarbor Dwarf Bettong

Bettongia pusilla

FR Bettong de Nullarbor DE Nullarbor-Bürstenrattenkänguru ES Betong de Nullarbor

Known from Holocene (or late Pleistocene) fossils but may have survived into modern times given Aboriginal familiarity with an unknown macropodoid from the same range whose indigenous names might apply to this species; no modern specimens are known.

Known from Nullarbor region of W SA & WA

Desert Bettong

Bettongia anhydra

FR Bettong du MacKay DE Wüsten-Bürstenrattenkänguru ES Betong del desierto

Known from only two specimens, the last and only living individual being collected (but only skull preserved) in 1933 from type locality. Cause of extinction is not known.

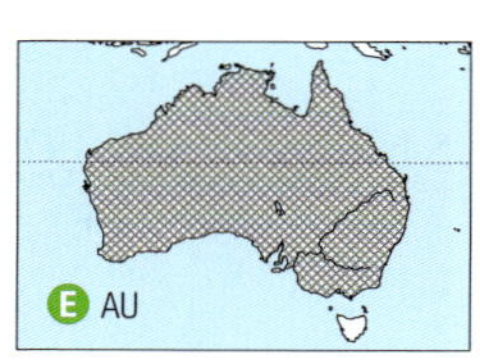

Known from a modern skull collected in WC NT and a second skull from Stegmite Cave (SE WA)

Broad-faced Potoroo
Potorous platyops
FR Potorou à face large DE Breitkopf-Kaninchenkänguru ES Potoro de cara ancha

The last living specimens were collected in 1875 and it was presumed to be extinct by the end of the 18th century, most likely due to introduced predators and changes in bushfire regimes.

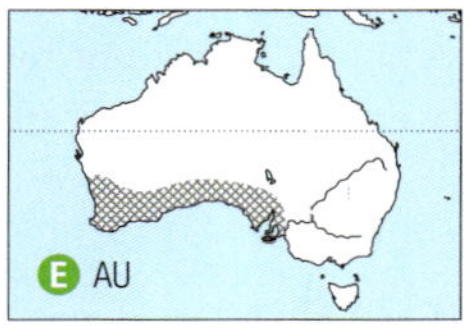

Modern specimens known from SW WA; fossil remains found in coastal WA (from near Albany N through Wheatbelt to at least near Geraldton) & SA (Eyre Pen, Kangaroo I)

Family MACROPODIDAE Kangaroos and Wallabies

Crescent Nail-tailed Wallaby
Onychogalea lunata
FR Wallaby à croissant DE Mondnagelschwanzkänguru
ES Ualabí de cola de uña de medialuna

Last collected in the late 1920s but known to occur in remote arid areas until the 1950s after which it became extinct, probably due to the introduction of foxes (*Vulpes vulpes*).

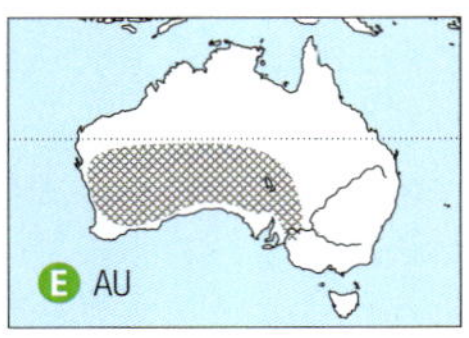

Known in modern times from much of inland S stretching to W coast, including S half of WA, S NT, most of SA, and far W Vic & NSW

Central Hare Wallaby
Lagorchestes asomatus
FR Wallaby du Mackay DE Zentralaustralisches Hasenkänguru
ES Ualabí liebre de Mackay

Only known specimen (a skull extracted from a freshly killed carcass) was collected in 1932. Given widespread Aboriginal familiarity with a small wallaby that was probably this species, it seems to have persisted until the 1940s–1960s across central deserts. Cause of extinction uncertain.

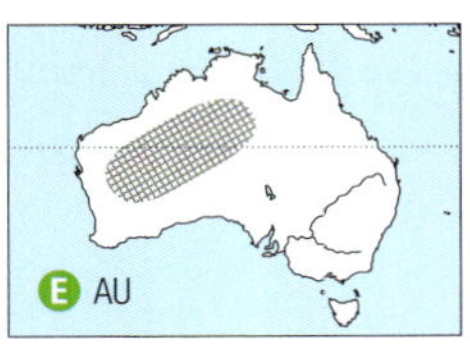

C mainland between Mt Farewell & L Mackay; probably extended across much of C deserts of WA & NT

Eastern Hare Wallaby
Lagorchestes leporides
FR Wallaby lièvre DE Östliches Hasenkänguru ES Ualabí liebre oriental

Ultimately extinct by the late 1800s, the last specimen being collected in 1889 with no reports since. Introduction of foxes (*Vulpes vulpes*), domestic cats, and European rabbits (*Oryctolagus cuniculus*) probably caused a rapid decline leading to extinction, although the exact cause is unknown.

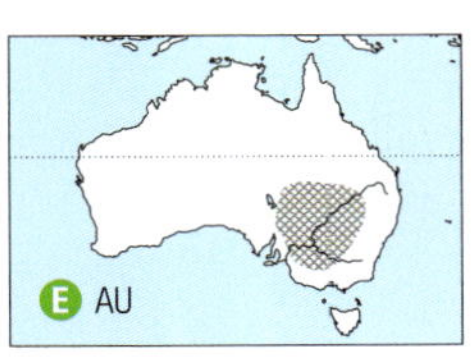

Known in modern times from inland areas, mainly along Murray & Darling rivers, from Qld border through C NSW to W Vic & E SA

Toolache Wallaby
Notamacropus greyi
FR Wallaby de Grey DE Östliches Irma-Wallaby ES Ualabí de Grey

Possibly persisted into the 1940s or 1950s, although the last known individual died in captivity in 1939. Disappearance probably due to changes in land use and unsustainable hunting.

Modern specimens found in a small area in SE SA near Vic border; Holocene remains from Kangaroo I and further E in S Vic & NW Tas

Family DUGONGIDAE Dugong and Steller's Sea Cow

Steller's Sea Cow
Hydrodamalis gigas
FR Rhytine de Steller DE Stellers Seekuh ES Vaca marina de Steller

The last living population, discovered in 1741 and suspected to consist of less than 1500 individuals, became extinct just 27 years later. Based on records from native peoples, who probably hunted them, it may have survived in W Aleutian Is (Attu I) until the 1800s. It became extinct due to overhunting for meat, fat, and hides. Harvesting of sea otters (*Enhydra lutris*) might have also contributed to its demise because these otters control urchin populations that, if left unchecked, deplete the kelp forests this species relied upon.

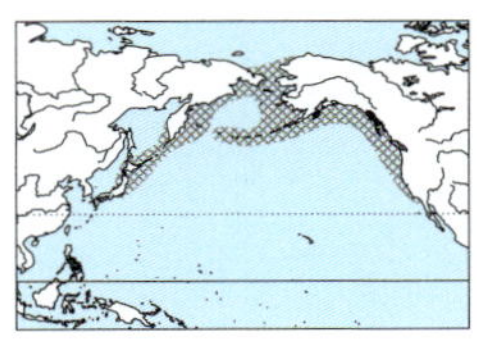

Last population found around Commander Is and probably W Aleutian Is; remains confirm the species was more widespread, ranging N to St Lawrence I, SW to Honshu (JP) & SE to California (US)

Family PALAEOPROPITHECIDAE Sloth Lemurs

Large Sloth Lemur
Palaeopropithecus ingens
FR Lémur-paresseux de Grandidier DE Faultiermaki ES Lémur perezoso gigante

Known from fossil and subfossil remains, based on radio-carbon dating it appears to have survived until as late as 1620, making it the last of the extinct "giant" lemurs to die out. Probably driven extinct by human activities (e.g. overhunting and slash-and-burn farming).

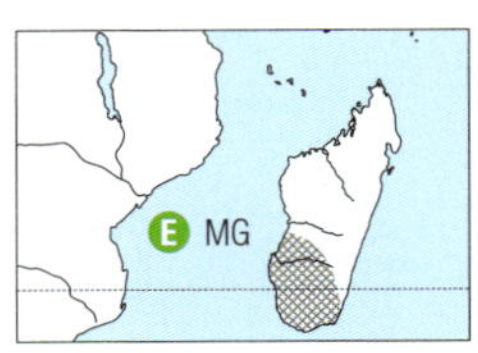

Known from SW & S

Family MEGALADAPIDAE Koala Lemurs

Koala Lemur

Megaladapis madagascariensis

FR Lémur-koala de Madagascar DE Koalamaki ES Lémur gigante

Known from fossil and subfossil material, it appears to have become extinct 500–600 years ago due to human activities such as slash-and-burn farming and overhunting.

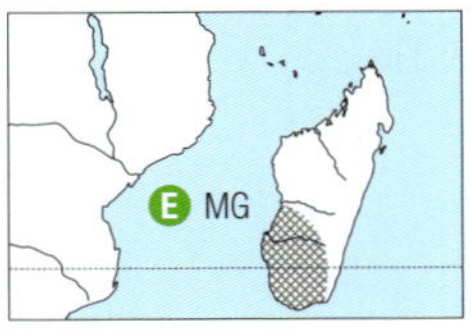

Known from S & SW

Family ARCHAEOLEMURIDAE Monkey Lemurs

Edwards's Monkey Lemur

Archaeolemur edwardsi

FR Archéolémur d'Edwards DE Pavianmaki ES Lémur gigante de Edwards

Known from fossil and subfossil remains, the most recent of which dated 1150 AD but might have survived to more recent times. It probably became extinct due to overhunting, slash-and-burn farming, and other human activities.

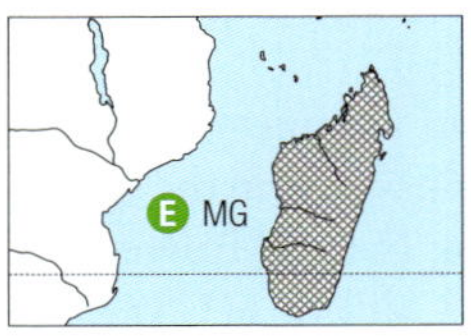

Known from C and possibly from W, N & SE

Family PITHECIIDAE Titis, Sakis and Uacaris

Jamaican Monkey

Xenothrix mcgregori

FR Titi de Jamaïque DE Jamaika-Affe ES Tití de Jamaica

Known only from fossil and subfossil remains, the species is thought to have survived until sometime after 1700 AD but probably became extinct because of human activities on the island.

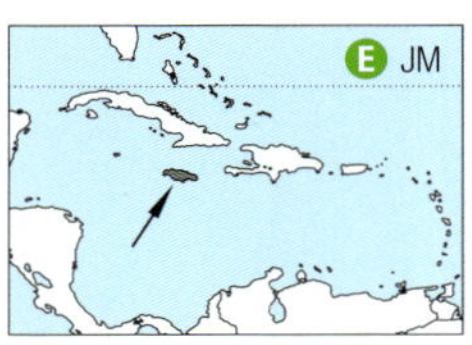

Greater Antilles

Family PROLAGIDAE Sardinian Pika

Sardinian Pika

Prolagus sardus

FR Pika sarde DE Sardischer Pfeifhase ES Pica sarda

Skeletal remains indicate that it was still present 2000 years ago. Large rodent-like mammals reported in historical records suggest that the species survived until the 1700s, a fact not confirmed by specimens. Overhunting and competition with introduced mammals might have contributed to its extinction.

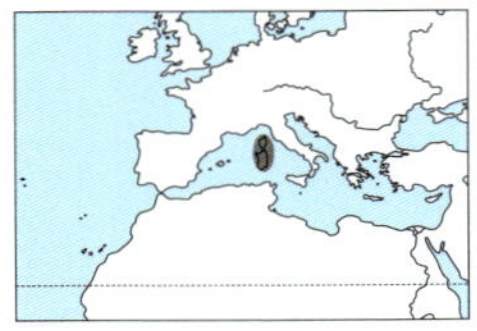

Corsica, Sardinia, and surrounding islands

Family CRICETIDAE New World Rats, New World Mice and relatives

Pemberton's Deermouse

Peromyscus pembertoni

FR Péromyscus de Pemberton DE Pemberton-Hirschmaus ES Ratón ciervo de Pemberton

Last collected in 1931. Most likely driven to extinction by the introduction of domestic cats.

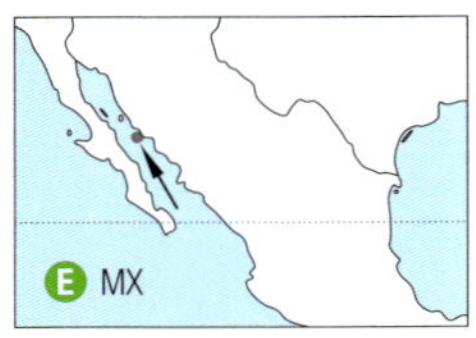

Known from San Pedro Nolasco I (Gulf of California)

Saint Vincent Pygmy Rice Rat

Oligoryzomys victus

FR Colilargo de Saint-Vincent DE Sankt-Vincent-Zwergreisratte
ES Rata arrocera pigmea de San Vicente

Not collected since 1892 and presumed extinct due to introduced rats (*Rattus rattus*) and mongooses (*Urva auropunctata*).

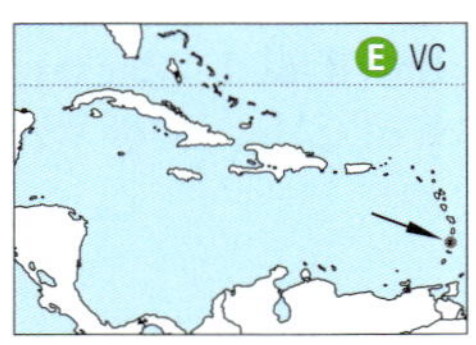

Only known from holotype collected on Saint Vincent I

Vespucci's Rat

Noronhomys vespuccii

FR Oryzomys de Vespucci DE Vespucci-Ratte ES Rata de Noroña

Known from Holocene fossil remains; historical records suggest this species was encountered in 1503 by Amerigo Vespucci but probably became extinct after the introduction of rats (*Rattus rattus*) and mice (*Mus musculus*).

Fernando de Noronha I

Galapagos Giant Rat

Megaoryzomys curioi

FR Oryzomys de Curio DE Galapagos-Riesenratte ES Rata gigante de Galápagos

Known from fossil and subfossil remains but thought to have survived into recent times (due to discovery of its bones alongside *Nesoryzomys* and *Mus*). May have been driven to extinction by the introduction of rats (*Rattus rattus*) and mice (*Mus musculus*).

Santa Cruz I (Galápagos Is)

Tres Marias Rice Rat

Oryzomys nelsoni

FR Oryzomys des Marias DE Tres-Marías-Sumpfreisratte ES Rata arrocera de Tres Marías

Only collected once in 1897. Most likely became extinct due to introduced rodents.

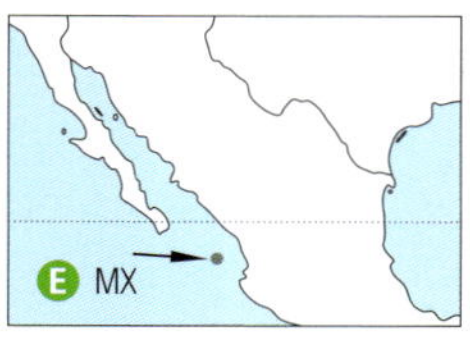

Known only from María Madre I (off Nayarit)

Jamaican Rice Rat

Oryzomys antillarum

FR Oryzomys de Jamaïque DE Jamaika-Sumpfreisratte ES Rata arrocera de Jamaica

Common in subfossil cave faunas and known from three specimens collected alive in the 19th century. Probably became extinct in the 1870s due to introduced mongooses (*Urva javanica*).

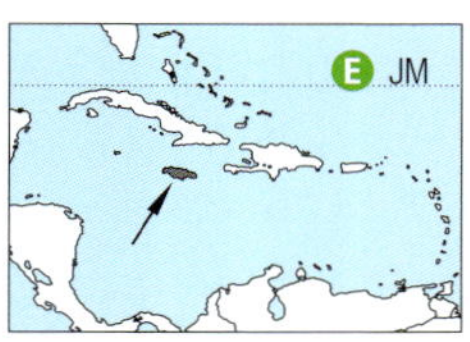

Greater Antilles

Nevis Rice Rat

Pennatomys nivalis

FR Pilori de Niévès DE Sankt-Kitts-Reisratte ES Rata arrocera de Nieves

Known from late Holocene fossil material. Although not known from recent specimens, historical reports may refer to this species; possibly became extinct as late as the early 20th century due to human habitation on the islands and the introduction of non-native rats (*Rattus rattus*) and mice (*Mus musculus*).

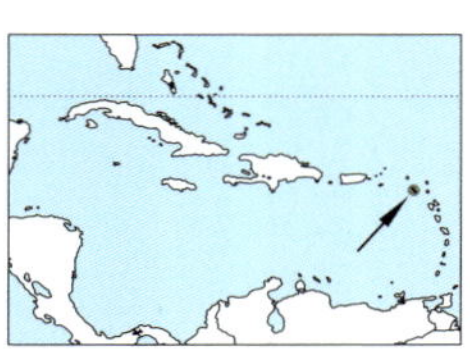

Sint Eustatius, Saint Kitts & Nevis (Lesser Antilles)

Barbados Giant Rice Rat

Megalomys georginae

FR Pilori de Barbade DE Barbados-Riesenreisratte ES Rata arrocera gigante de Barbados

Known from late Quaternary (and possibly late Pleistocene) remains. It might have survived into modern times in light of accounts by 17th century European explorers. Cause of extinction is uncertain but probably linked to the introduction of non-native rats (*Rattus rattus*).

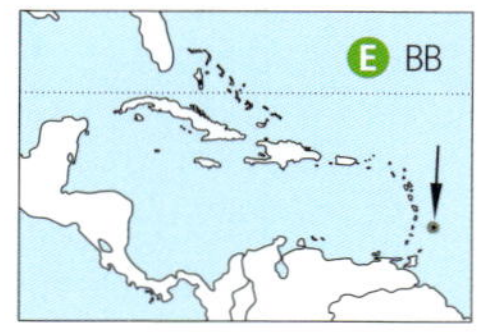

Lesser Antilles

Martinique Giant Rice Rat

Megalomys desmarestii

FR Pilori de Martinique DE Martinique-Riesenreisratte
ES Rata arrocera gigante de Martinica

Attempts were made to exterminate the species in the late 19th century and at the time it was also eaten as food. Although the eruption of Mt Pelée in 1902 is credited with its final demise, predation by introduced mongooses (*Urva auropunctata*) probably also played a key part in its extinction.

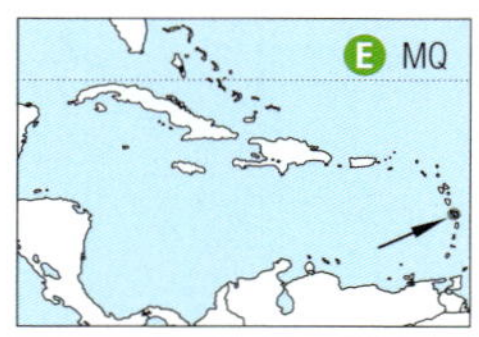

Lesser Antilles

Saint Lucia Giant Rice Rat

Megalomys luciae

FR Pilori de Sainte-Lucie DE Sankt-Lucia-Riesenreisratte
ES Rata arrocera gigante de Santa Lucía

Probably extinct by the late 19th century. The last known specimen died in 1852 in captivity and the last record was in 1881. Introduction of mongooses (*Urva semitorquata*) probably contributed to its extinction.

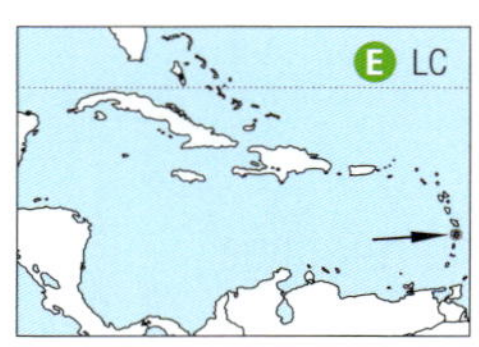

Lesser Antilles

Darwin's Galapagos Mouse

Nesoryzomys darwini

FR Oryzomys de Darwin DE Darwin-Galapagosratte ES Ratón de Galápagos de Darwin

Last collected in 1929 and was probably pushed to extinction by introduced rats.

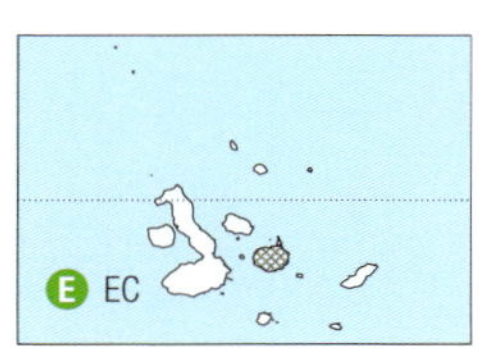

Santa Cruz I (Galápagos Is)

Indefatigable Galapagos Mouse

Nesoryzomys indefessus

FR Oryzomys de Santa Cruz DE Indefatigable-Galapagosratte
ES Ratón de Galápagos de Santa Cruz

Last collected in 1934. Extinction probably due to introduced rats.

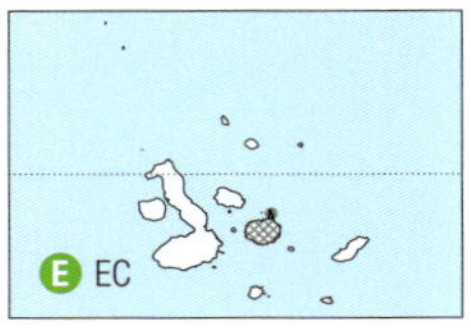

Santa Cruz and Baltra Is (Galápagos Is)

Lesser Antillean Rice Rat

Antillomys rayi

FR Oryzomys de Ray DE Kleine Antillen-Reisratte ES Rata arrocera de Antillas

Became rare and possibly even extinct in the 16th century, shortly after contact with Europeans. Although affected by overexploitation by Amerindians and deforestation by European colonists, the consequences (biological competition, disease, etc.) of the introduction of rats (*Rattus rattus*) represent the most likely explanation for its decline and eventual disappearance.

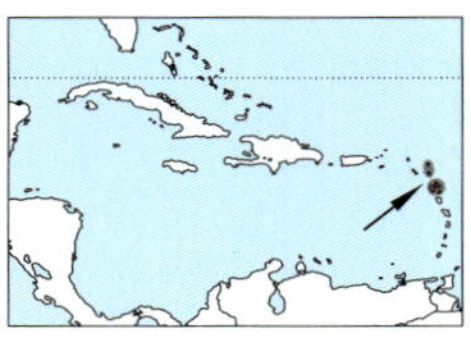

Lesser Antilles in GP (Basse-Terre, Grande-Terre, Marie-Galante) and AG (Antigua, Barbuda)

Candango Burrowing Mouse

Juscelinomys candango

FR Akodon de Brasilia DE Candango-Maulwurfsmaus ES Ratón cavador de Candango

Not recorded since 1960 and probably became extinct shortly after being discovered.

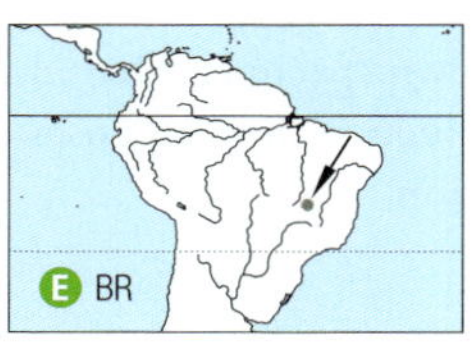

Known only from Federal District (Brasília)

Family MURIDAE True Mice and Rats, Gerbils and relatives

Musser's Giant Rat

Coryphomys musseri

FR Coryphomys de Musser DE Musser-Timor-Riesenratte ES Rata gigante de Musser

Known from subfossil Quaternary remains but might have survived into modern times.

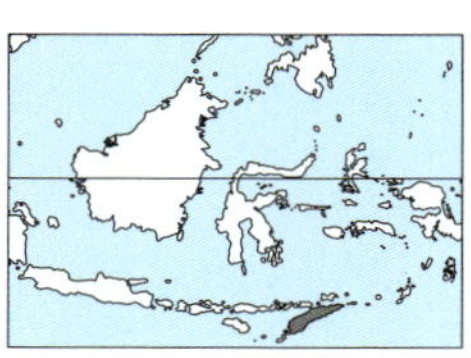

Timor I (Lesser Sundas)

Buehler's Giant Rat

Coryphomys buehleri

FR Coryphomys de Bühler DE Bühler-Timor-Riesenratte ES Rata gigante de Bühler

Known from subfossil Quaternary remains and possibly survived into modern times.

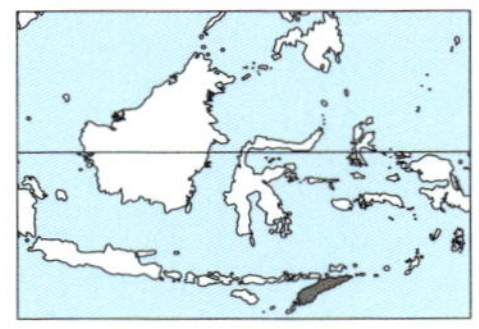

Timor I (Lesser Sundas)

Groves's Giant Rat

Lenomys grovesi

FR Rat de Groves DE Groves-Riesenratte ES Rata gigante de Groves

Known only from late Holocene subfossil remains but might be extant given that collecting efforts in the region have been limited.

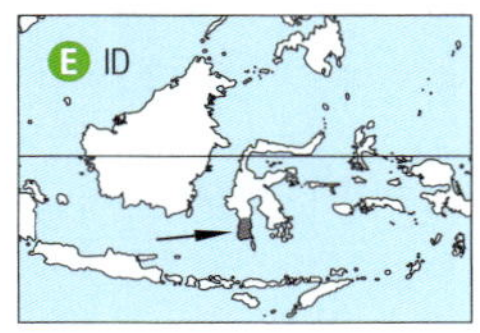

SE portion of SW pen of Sulawesi (Greater Sundas)

Bulldog Rat

Rattus nativitatis

FR Rat de Christmas DE Weihnachtsinsel-Ratte ES Rata de Navidad

Not recorded with certainty since 1897–1898 and probably became extinct sometime in 1900–1904 following the introduction of rats (*R. rattus*) that outcompeted and spread disease to this species.

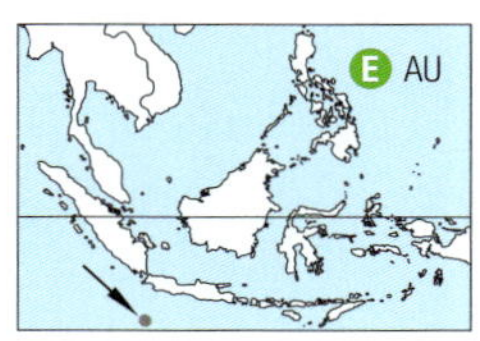

Christmas I (E Indian Ocean)

Maclear's Rat

Rattus macleari

FR Rat de Maclear DE Maclear-Ratte ES Rata de Maclear

Extinct by 1902–1904 after the introduction of roof rats (*R. rattus*), which probably outcompeted and outbred it (there is evidence of hybridization), and also spread diseases to which it had no resistance.

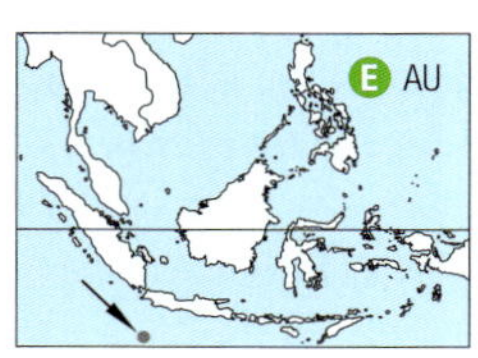

Christmas I (E Indian Ocean)

New Ireland Forest Rat

Rattus sanila

FR Rat de Nouvelle-Irlande DE Neuirland-Ratte ES Rata de Nueva Irlanda

Known from late Pleistocene fossils and late Holocene subfossils (as late as 3000 years ago). Although no recent specimens have been collected, there is some speculation that it could survive in undisturbed and underexplored areas of primary forests on New Ireland I.

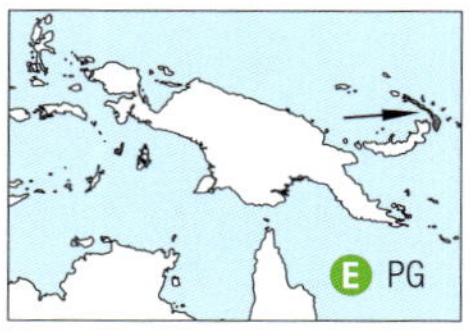

New Ireland I (Bismarck Arch)

Buka Island Giant Rat

Solomys spriggsarum

FR Solomys des Spriggs DE Buka-Insel-Nacktschwanzratte ES Rata gigante de Buka

Known only from late Quaternary subfossil remains. Although no recent specimens have been collected, it could have survived into more recent times.

Buka I (Solomon Is)

Bramble Cay Mosaic-tailed Rat

Melomys rubicola

FR Mélomys de Bramble Cay DE Bramble-Cay-Mosaikschwanzratte
ES Rata de cola moteada de Cayo Bramble

Last seen in 2009 and declared extinct in 2015 after extensive surveys failed to find living individuals. Rising sea levels and increased storm intensity due to climate change have almost certainly led to its extinction.

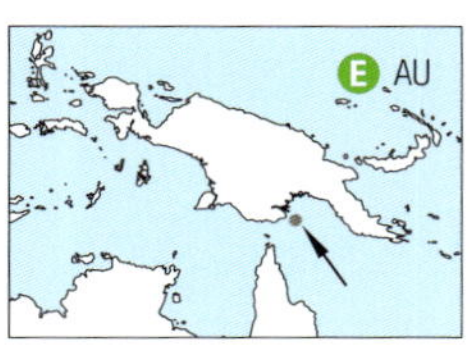

Known only from Bramble Cay in Great Barrier Reef of NE Qld

Buka Island Mosaic-tailed Rat

Melomys spechti

FR Mélomys de Buka DE Buka-Mosaikschwanzratte ES Rata de cola moteada de Buka

Known only from late Quaternary subfossil archeological remains. Although no recent specimens have been collected, it may have survived into more recent times.

Buka I (Solomon Is)

Lesser Stick-nest Rat

Leporillus apicalis

FR Petit Léporille DE Kleine Häschenratte ES Rata constructora menor

Last two specimens were collected on Mt Crombie (NW South Australia) in 1933, although reports continued up to the 1970s with occasional reports of old stick nests with fresh vegetation. Domestic cats were the primary cause of its decline but introduced foxes (*Vulpes vulpes*) probably finished off all remaining subpopulations.

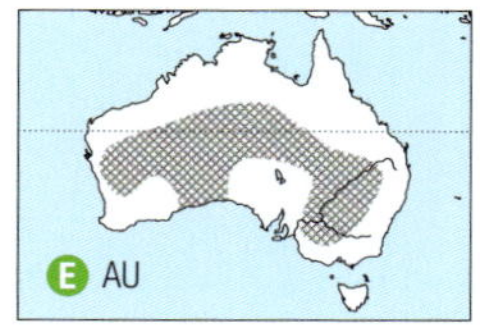

Known from WA, NT, SA, NSW & NW Vic

Capricorn Rabbit Rat

Conilurus capricornensis

FR Conilure du Capricorne DE Steinbock-Kaninchenratte ES Rata conejo de Capricornio

Known from fossil and subfossil remains from late Pleistocene and Holocene. Recent, remarkably well preserved specimens from Broken River area indicate that it might be extant, but more searching is needed.

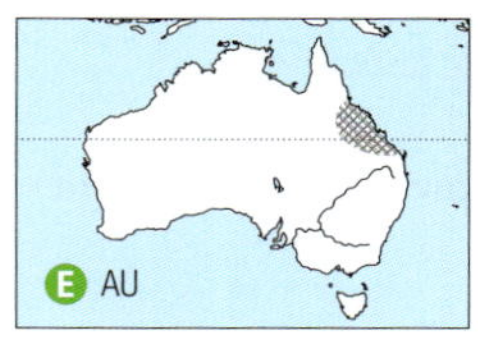

NE & E Qld in Capricorn Caves and Mt Etna (both near Rockhampton) and R Broken and Christmas Creek (both inland from Townsville)

White-footed Rabbit Rat

Conilurus albipes

FR Conilure à pattes blanches DE Weißfuß-Kaninchenratte
ES Rata conejo de pies blancos

Last collected in 1845, although there were reports from 1856–1857 and it may have survived until the 1930s. A combination of competition with introduced rats and predation by domestic cats probably led to its extinction.

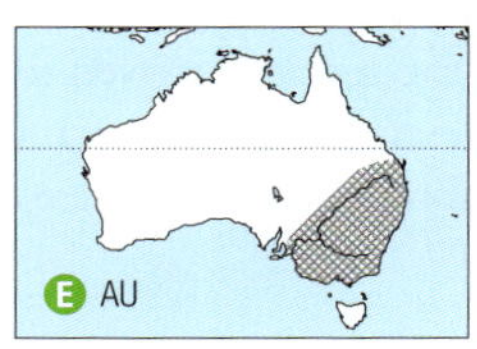

Originally known from SE SA, Vic, NSW & SE Qld

Big-eared Hopping Mouse

Notomys macrotis

FR Notomys oreillard DE Großohr-Australienhüpfmaus
ES Ratón saltador de orejas grandes

Last collected in 1843, although the lack of collecting efforts in the region suggest it could have survived until the 20th century. Cause of extinction unknown but probably due to the introduction of domestic cats and habitat alteration.

Known from 2 living specimens collected near R Moore (WA) and extensive subfossil records from WC Wheatbelt of WA

Long-tailed Hopping Mouse

Notomys longicaudatus

FR Notomys à longue queue DE Langschwanz-Australienhüpfmaus
ES Ratón saltador de cola larga

Last collected in 1901–1902 but skull fragments present in owl pellets until 1977. Extinction was probably caused by a combination of factors, including introduced predators and habitat destruction caused by human settlement.

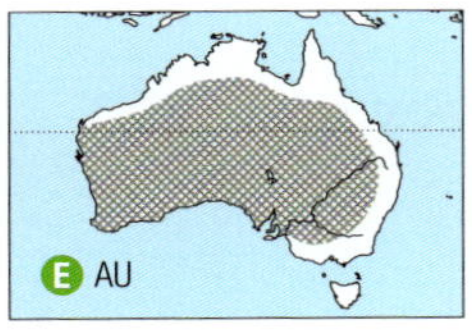

Known from recent specimens in WA & NT and subfossil remains in SA

Darling Downs Hopping Mouse

Notomys mordax

FR Notomys de Darling Downs DE Darling-Downs-Australienhüpfmaus
ES Ratón saltador de Darling Downs

Known from one skull bought from J. Gould in 1846 (probably collected in the 1840s) in Darling Downs. No species of living or fossil *Notomys* has ever been recorded in this region and so uncertainty persists regarding the origin of the specimen, although there is little doubt that Gould did collect it from this region. It may have become extinct due to domestic cats.

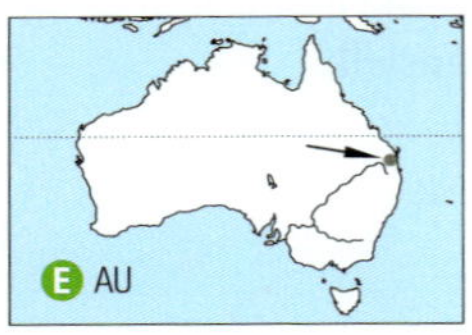

Known only from Darling Downs region (SE Qld)

Broad-cheeked Hopping Mouse

Notomys robustus

FR Notomys costaud DE Breitwangen-Australienhüpfmaus
ES Ratón saltador de mejillas anchas

Known only from a few skulls found in old owl pellets from roosts in 1976 and 1998 (but not thought to be over 200 years old); never collected alive. Pellets also contained remains of house mice, suggesting that the species survived to modern times, possibly up to the mid-1900s. Cause of extinction is uncertain but most likely linked to introduced predators and habitat alteration due to human settlement.

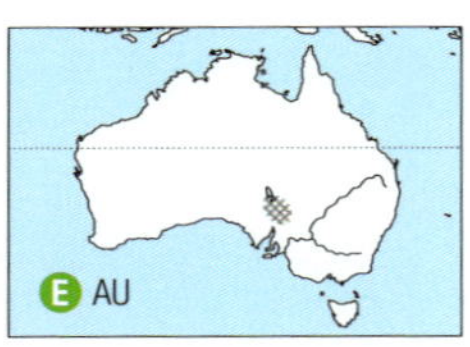

Known from Flinders & Davenport ranges (SA)

Short-tailed Hopping Mouse

Notomys amplus

FR Notomys à queue courte DE Kurzschwanz-Australienhüpfmaus
ES Ratón saltador de cola corta

Not recorded alive since 1896 and probably became extinct due to introduced predators and habitat alteration.

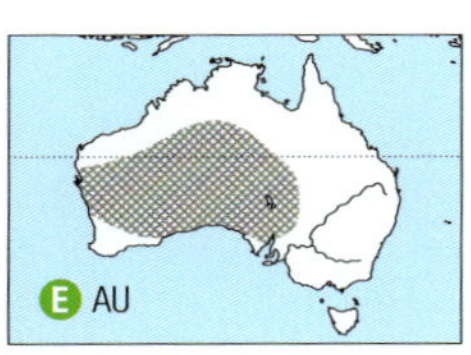

Known from 2 live specimens from type locality (Charlotte Waters) and a single specimen from Burt Plain near Alice Springs (NT); also found in owl pellets from Flinders Ranges (SA)

Long-eared Mouse

Pseudomys auritus

FR Pseudomys oreillard DE Langohr-Australienmaus ES Ratón de orejas largas

Not collected since the 1800s and probably extirpated after the introduction of domestic cats and habitat alteration.

Known from Coorong throughout SE SA to basalt plains of SW Vic (including Kangaroo I)

Blue-grey Mouse

Pseudomys glaucus

FR Pseudomys bleuté DE Blaugraue Australienmaus ES Ratón gris azulado

Known from three specimens, two collected in 1892 and one in 1956 (although most likely collected long before). Predation from domestic cats and habitat destruction probably drove it to extinction.

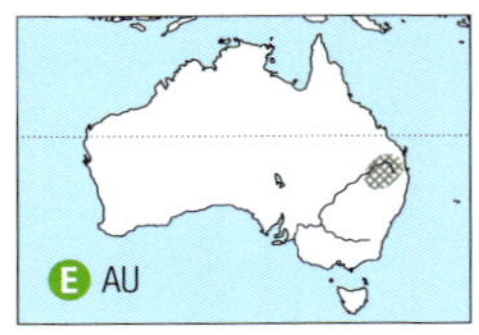

Known from SE Qld & N NSW (Cryon)

Family CHINCHILLIDAE Viscachas and Chinchillas

Peruvian Plains Viscacha

Lagostomus crassus

FR Viscache du Pérou DE Peru-Viscacha ES Vizcacha de Perú

Known from a single skull. Although it might have survived until recent times, no other specimens have ever been collected.

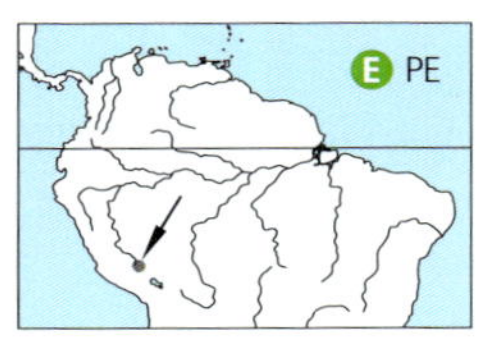

Known only from Santa Ana (Cusco)

Family ECHIMYIDAE Hutias and South American Spiny rats

Small Haitian Hutia

Plagiodontia spelaeum

FR Petit Hutia DE Kleiner Zaguti ES Jutía pequeña de la Española

Known from late Quaternary fossil material but may have survived until more recent times.

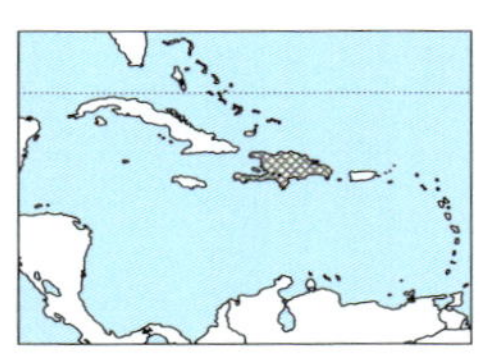

Hispaniola (Greater Antilles)

Samana Hutia

Plagiodontia ipnaeum

FR Hutia de Samana DE Samana-Zaguti ES Jutía de Samana

Known only from recent fossil remains found alongside introduced rat fossils, suggesting that the species lived in recent times. There are also indications based on native people's records that it might have survived to 1530–1540s and possibly even to the 1900s, although it probably became extinct after the European colonization of the island.

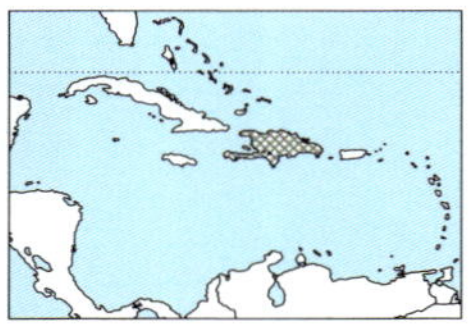

Hispaniola (Greater Antilles)

Wide-toothed Hutia

Hyperplagiodontia araeum

FR Hutia de San Rafael DE San-Rafael-Ferkelratte ES Jutía de San Rafael

Although known only from skeletal material, it is possible that it survived into post-Columbian times.

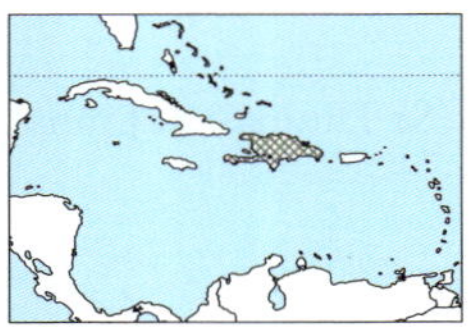

Hispaniola (Greater Antilles)

Lemke's Hutia

Rhizoplagiodontia lemkei

FR Hutia de Lemke DE Lemke-Ferkelratte ES Jutía de Lemke

Known only from skeletal material but may have survived into post-Columbian times.

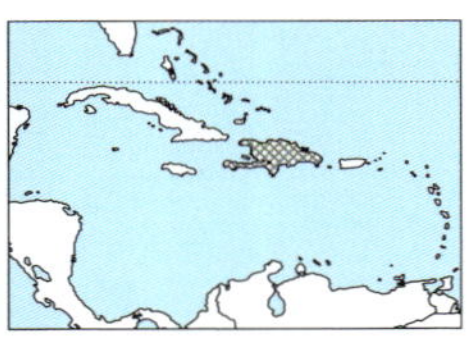

Hispaniola (Greater Antilles)

Imposter Hutia

Hexolobodon phenax

FR Hutia trompeur DE Gaukler-Hutia ES Jutía engañosa

Known from recent subfossil remains that also contain introduced rat skeletons suggesting that it became extinct after European colonization.

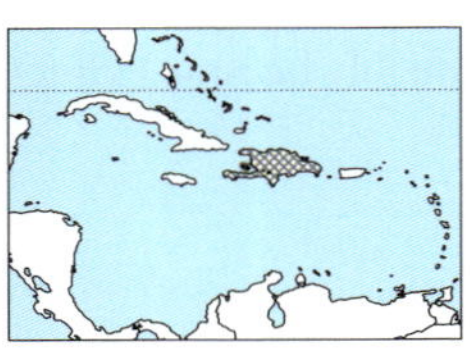

Hispaniola & Gonâve (Greater Antilles)

Puerto Rican Hutia

Isolobodon portoricensis

FR Hutia de Porto Rico DE Puerto-Rico-Hutia ES Jutía de Puerto Rico

Thought to have survived until European colonization (c. 1525), although reports suggest it survived into the 1800s. Cause of extinction is uncertain, especially because it appears to have been eaten by Amerindians who translocated animals to islands other than Hispaniola.

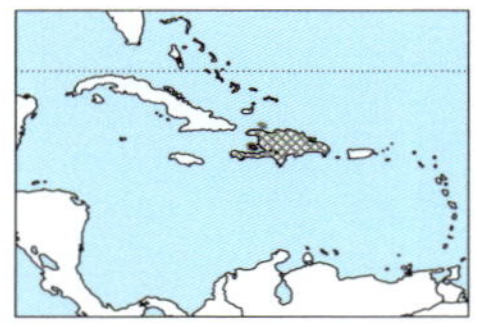

Across several Caribbean Is, probably native to Gonâve, Tortue & Hispaniola; introduced to Mona, PR, Vieques & VI (St Thomas, St Croix)

Montane Hutia

Isolobodon montanus

FR Hutia montagnard DE Berg-Hutia ES Jutía de montaña

Probably survived until European colonization. Cause of extinction unknown.

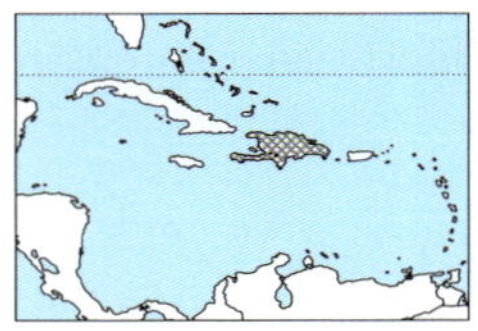

Hispaniola (Greater Antilles)

Little Swan Island Hutia

Geocapromys thoracatus

FR Hutia de Santanilla DE Schwaneninseln-Ferkelratte ES Jutía de Santanilla

Became extinct in 1955 after a severe hurricane and the introduction of domestic cats.

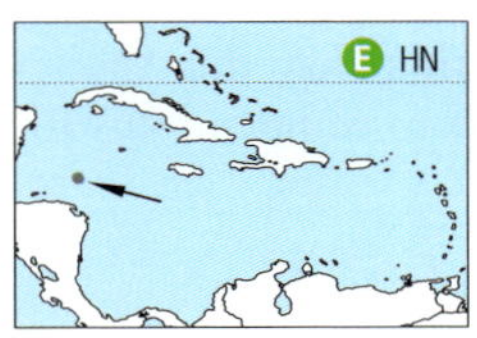

Known only from Little Swan I in Gulf of Honduras (Caribbean)

Cayman Hutia

Geocapromys caymanensis

FR Hutia des Caïmans DE Cayman-Ferkelratte ES Jutía de Caimán

Known only from late Pleistocene and Holocene skeletal remains. Fossils found in association with introduced rats suggest that it survived initial European colonization of the islands but probably became extinct soon after.

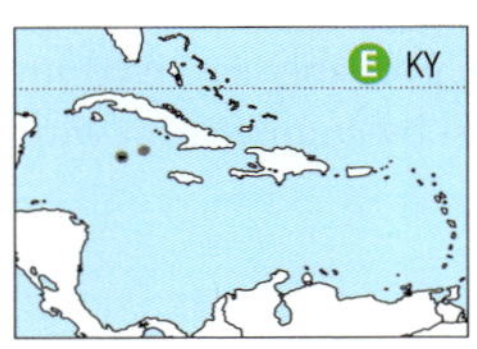

Cayman Brac & Grand Cayman (Caribbean)

Cuban Hutia

Geocapromys columbianus

FR Hutia de Colomb DE Kuba-Ferkelratte ES Jutía de Cuba

It might have survived to recent times as it is known from deposits containing introduced rats, which probably led to the extinction of this species after European colonization.

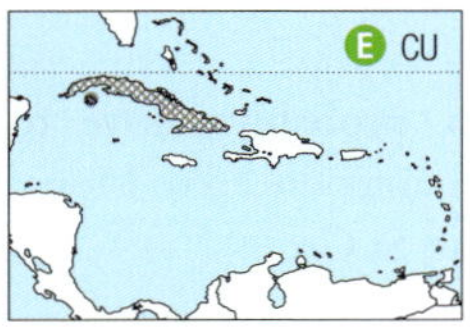

Main island & Isla de la Juventud (Greater Antilles)

Hispaniolan Edible Rat

Brotomys voratus

FR Mohuy DE Mohuy ES Mohuy

Known from recent fossil remains and one written record from Gonzalo Fernández de Oviedo y Valdés who lived on Hispaniola in 1536–1546 and documented a spiny rat-like rodent known as the *mohuy* that probably represented this species. It most likely became extinct shortly after European colonization due to competition with introduced rats.

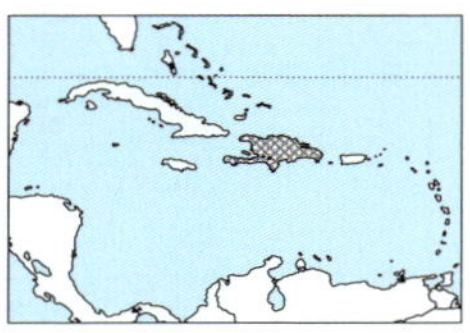

Hispaniola (Greater Antilles)

Oriente Cave Rat

Boromys offella

FR Boromys d'Oriente DE Oriente-Höhlenratte ES Rata de cueva de Oriente

Fossil remains also contain introduced rats suggesting that this species survived up to recent times; possibly disappeared due to competition with introduced rats.

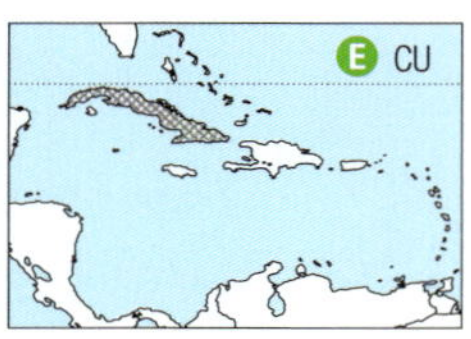

Greater Antilles

Torre's Cave Rat

Boromys torrei

FR Boromys de De la Torre DE Torre-Höhlenratte ES Rata de cueva de De la Torre

Fossil remains contain introduced rats, so this species probably survived up to recent times and was most likely driven to extinction by competition with these rats.

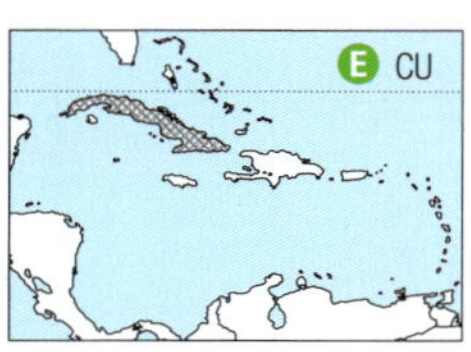

Greater Antilles

Insular Cave Rat

Heteropsomys insulans

FR Rat-épineux de Porto Rico DE Puerto-Rico-Höhlenstachelratte ES Rata de cueva de Puerto Rico

Known from fossil material and probably survived until European colonization.

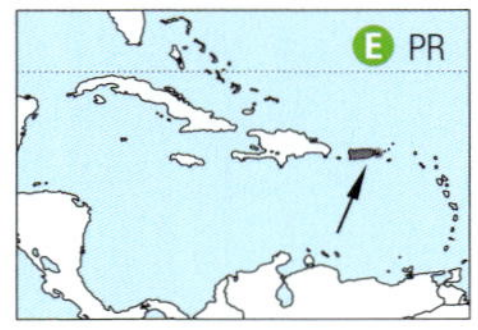

Main island & Vieques I (Greater Antilles)

Family NESOPHONTIDAE Nesophontes

Puerto Rican Nesophontes

Nesophontes edithae

FR Nésophonte de Porto Rico DE Puerto-Rico-Karibikspitzmaus ES Nesofonte de Puerto Rico

Known from relatively recent skeletal remains. It probably survived up to the modern era and became extinct soon after European colonization.

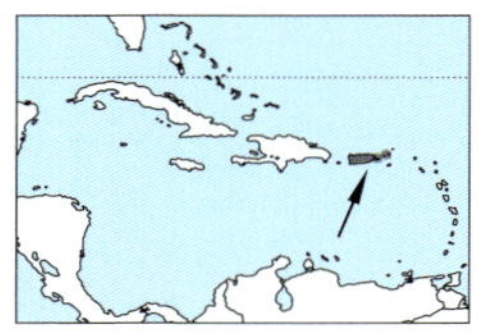

PR (including Vieques I), VI (St Thomas, St Johns)

Saint Michel Nesophontes

Nesophontes paramicrus

FR Nésophonte d'Haïti DE Saint-Michel-Karibikspitzmaus ES Nesofonte de San Miguel

Known from recent skeletal remains found with introduced rats and mice suggesting that it survived into the post-Columbian era and became extinct soon after.

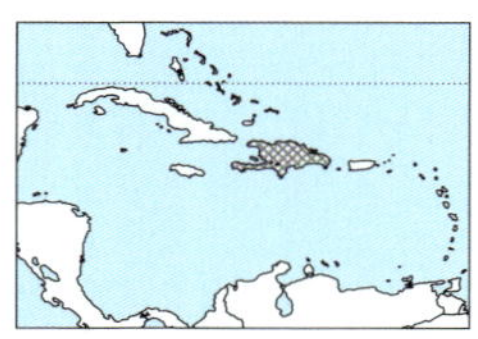

Hispaniola (Greater Antilles)

Haitian Nesophontes

Nesophontes zamicrus

FR Nésophonte nain DE Haiti-Karibikspitzmaus ES Nesofonte de Haití

Known from recent skeletal remains found with introduced rats and mice, so may have survived into the post-Columbian era; it probably became extinct soon after.

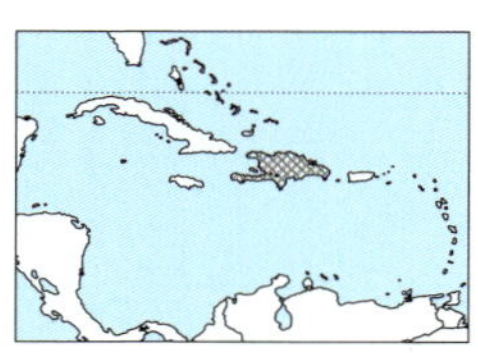

Hispaniola (Greater Antilles)

Atalaye Nesophontes

Nesophontes hypomicrus

FR Petit Nésophonte DE Atalaye-Karibikspitzmaus ES Nesofonte de Atalaya

Known from recent skeletal remains found with introduced rats and mice, evidence that it survived into the post-Columbian era but most likely becoming extinct soon after.

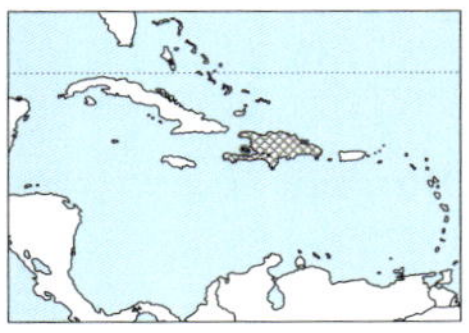

Hispaniola & Gonâve I (Greater Antilles)

Greater Cuban Nesophontes

Nesophontes major

FR Grand Nésophonte DE Große Kuba-Karibikspitzmaus ES Nesofonte grande de Cuba

Known from late Pleistocene and Holocene fossil and subfossil remains and probably became extinct around the time of European colonization.

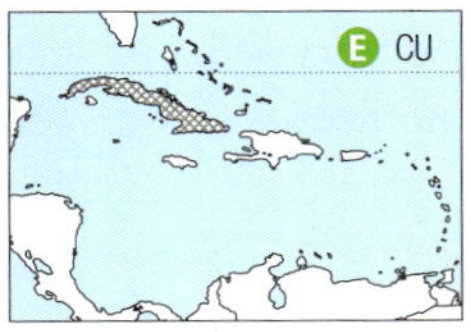

Greater Antilles

Lesser Cuban Nesophontes

Nesophontes micrus

FR Nésophonte de Cuba DE Kleine Kuba-Karibikspitzmaus ES Nesofonte pequeño de Cuba

Known from late Pleistocene and Holocene remains but probably survived into the modern era and went extinct after European colonization.

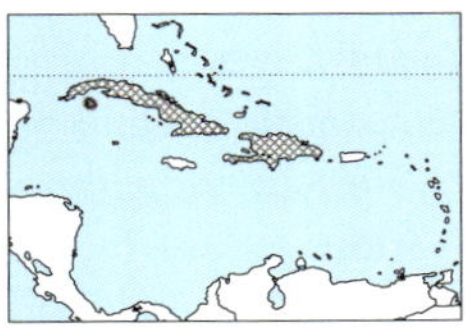

CU (including Isla de la Juventud) & Hispaniola (Greater Antilles)

Cayman Nesophontes

Nesophontes hemicingulus

FR Nésophonte des Caïmans DE Cayman-Karibikspitzmaus ES Nesofonte de Caimán

Known from late Pleistocene and Holocene skeletal remains, this is the only *Nesophontes* found outside Greater Antilles and surrounding islands. Fossils of this species have been found in association with introduced rats, which suggests that it survived the initial European colonization of the islands but became extinct soon after.

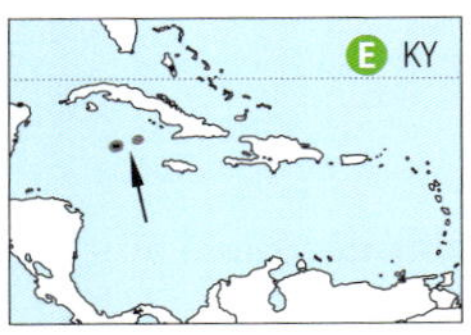

Cayman Brac & Grand Cayman (Caribbean)

Family SOLENODONTIDAE Solenodons

Giant Solenodon

Solenodon arredondoi

FR Solénodonte géant DE Riesenschlitzrüssler ES Solenodonte gigante

Known only from fossil remains from three late Pleistocene sites whose exact age are uncertain. Appears to have been contemporary with *Atopogale cubana* and *Nesophontes micrus*, suggesting that it survived up to recent times. Habitat destruction and predation from introduced dogs brought by pre-Columbian inhabitants is likely to have contributed to its extinction.

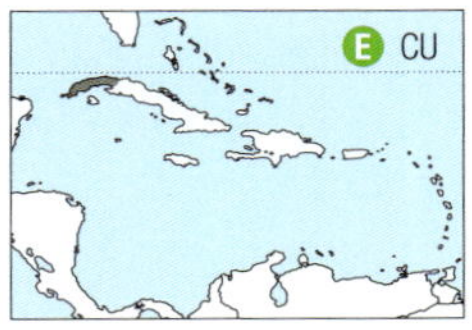

Collected in W part of the island

Marcano's Solenodon

Solenodon marcanoi

FR Solénodonte de Marcano DE Marcano-Schlitzrüssler ES Solenodonte de Marcano

Known from recent skeletal material, it probably survived up to the modern era but became extinct soon after European colonization.

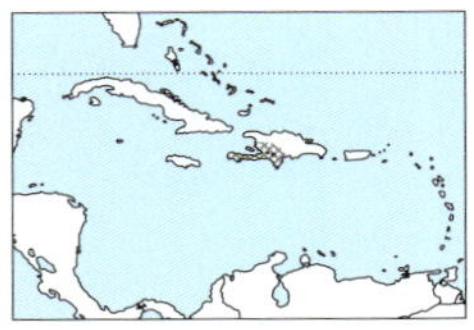

S Hispaniola in HT (La Hotte & La Selle massifs) & DO (Sierra de Neiba)

Family PTEROPODIDAE Old World Fruit Bats

Guam Flying Fox

Pteropus tokudae

FR Roussette de Guam DE Guam-Flughund ES Zorro volador de Guam

The last individual was shot by a hunter in 1968 but only known previously from three specimens. Despite extensive fieldwork, apart from an unconfirmed sighting in the 1970s, no new records have been reported. Local hunting is thought to have been the main factor leading to its extinction, although the introduction of the brown tree snake (*Boiga irregularis*) could have played a part.

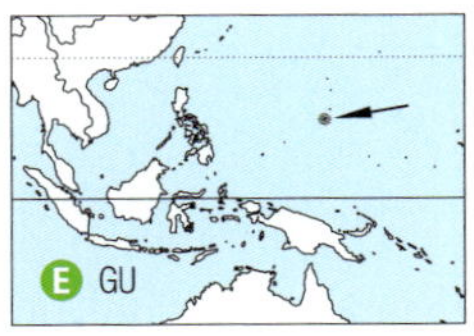

S Marianas Is, NW Pacific

Percy Island Flying Fox

Pteropus brunneus

FR Roussette de Percy DE Percy-Island-Flughund ES Zorro volador de Percy

Known from type specimen collected in 1874 and subsequent records of colonies. It was apparently plentiful near the end of the 1800s but became extinct soon after, possibly due to habitat loss.

Percy Is (off coast of Qld), exact island uncertain

Large Samoan Flying Fox

Pteropus coxi

FR Roussette de Cox DE Großer Samoa-Flughund ES Zorro volador grande de Samoa

Two skulls are known with associated skins that appear to have been lost. Eyewitness reports from the early 1980s indicate that it may have survived until recent decades or even still be extant.

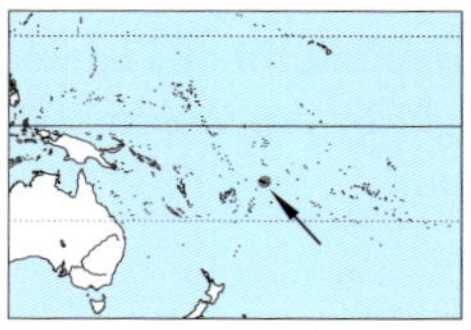

Samoan Is (exact locality uncertain)

Small Samoan Flying Fox

Pteropus allenorum

FR Roussette des Allen DE Kleiner Samoa-Flughund ES Zorro volador pequeño de Samoa

Known only from a single specimen collected in 1856. It is now almost certainly extinct because it has never been collected since. Cause and timing of extinction are unknown.

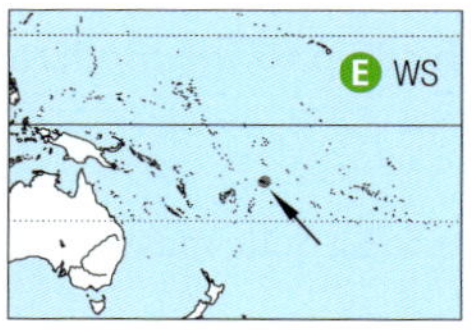

Upolu I

Large Palau Flying Fox

Pteropus pilosus

FR Roussette à poil doux DE Großer Palau-Flughund ES Zorro volador grande de Palau

Known only from two specimens collected prior to 1874 after which it probably became extinct. Cause of its disappearance is uncertain but most likely linked to overhunting by native people and deforestation.

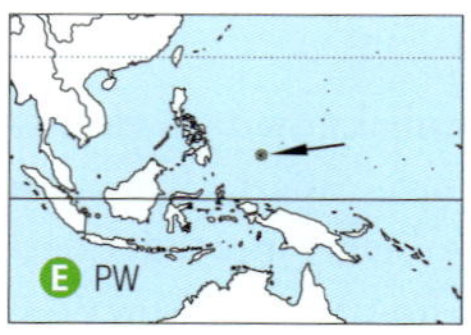

Caroline Is, W Pacific

Lesser Mascarene Flying Fox

Pteropus subniger

FR Roussette à collet rouge DE Kleiner Maskarenen-Flughund

ES Zorro volador menor de las Islas Mascareñas

Remarkably common during the 1700s but last authenticated record on Mauritius in 1859 and believed to have become extinct between 1864 and 1873. No records on Réunion since 1862 and it seems probable that it became extinct in the 1860s. Deforestation and local hunting are thought to have been major factors in its demise.

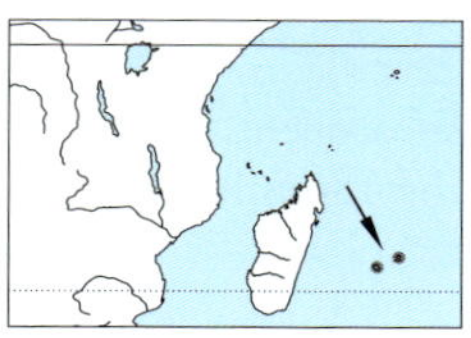

RE & MU (Mascarene Is, W Indian Ocean)

Family PHYLLOSTOMIDAE New World Leaf-nosed Bats

Jamaican Round-eared Bat

Tonatia saurophila

FR Tonatia des lézards DE Kopfstreifen-Rundohrblattnase ES Tonatia de cabeza rayada

Known only from subfossil specimens, all collected in 1919–1920. No recent records but, judging from subfossil status of remains, it probably survived into relatively recent times or might even be extant.

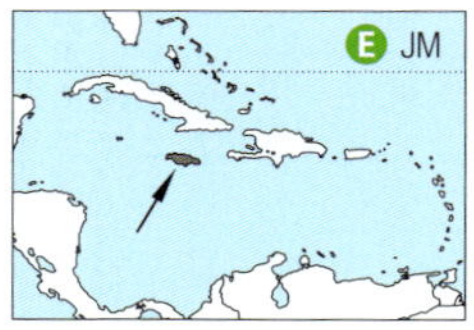

Known from Wallingford Roadside Cave (type locality) & Dairy Cave

Family VESPERTILIONIDAE Vesper Bats

Sturdee's Pipistrelle

Pipistrellus sturdeei

FR Pipistrelle de Sturdee DE Sturdee-Zwergfledermaus ES Pipistrela de Sturdee

Only the type specimen collected in 1889 is known; no further records for over a century despite extensive fieldwork.

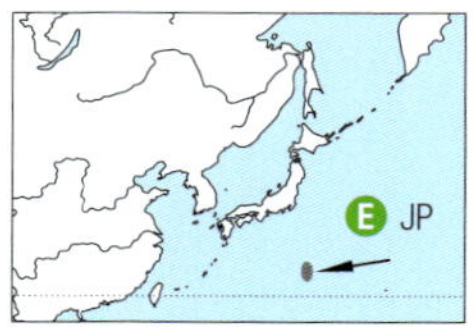

Bonin Is

Christmas Island Pipistrelle

Pipistrellus murrayi

FR Pipistrelle de Christmas DE Weihnachtsinsel-Zwergfledermaus

ES Pipistrela de Christmas

Was previously common across the island but underwent drastic declines in the 1990s and 2000s, ultimately disappearing in 2009; no new individuals have ever been detected despite extensive surveys. Various causes for its extinction have been mooted, including the introduction of oriental wolf snakes (*Lycodon capucinus*), yellow crazy ants (*Anoplolepis gracilipes*), and rats, as well as poisoning and roost disturbance.

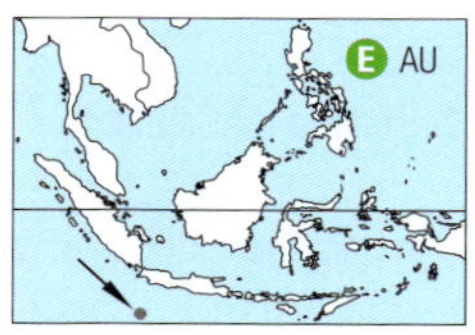

Christmas I (E Indian Ocean)

Lord Howe Long-eared Bat

Nyctophilus howensis

FR Nyctophile de Lord Howe DE Lord-Howe-Langohrfledermaus

ES Nictofila de Lord Howe

Only known specimen (an incomplete skull) was probably deposited 50–100 years ago. Since its description, mammologists have searched unsuccessfully for this species.

Lord Howe I, off NSW

Family HIPPOPOTAMIDAE Hippopotamuses

Lemerle's Dwarf Hippopotamus

Hippopotamus lemerlei

FR Hippopotame de Lemerle DE Lemerle-Flusspferd ES Hipopótamo pigmeo de Lemerle

Last appearance in the fossil record dates to c. 1000 AD. Local oral traditions suggest that it might have survived to more recent times. How humans contributed to its extinction is unclear but hunting probably played a role.

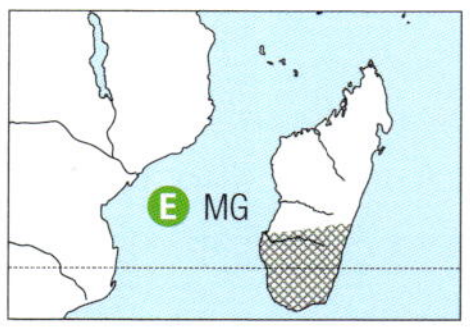

Coastal to lowland S MG

Madagascan Dwarf Hippopotamus

Hippopotamus madagascariensis

FR Hippopotame de Madagascar DE Madagaskar-Flusspferd
ES Hipopótamo pigmeo de Madagascar

Last appearance in the fossil record dates to c. 1000 AD. Local oral traditions indicate that it may have survived to more recent times. The human role in its extinction is uncertain but hunting was probably a factor.

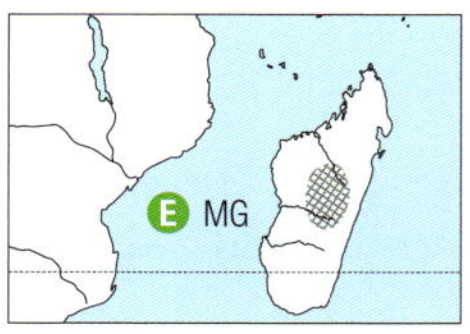

Fossil remains mostly in C highlands

Family CERVIDAE Deer

Schomburgk's Deer

Rucervus schomburgki

FR Cerf de Schomburgk DE Schomburgk-Hirsch ES Ciervo de Schomburgk

The last known individual in captivity died in 1938 and the last known wild individual was killed in 1932. Antlers have been discovered in wildlife trade as recently as 1991 in Laos but, because the dates these antlers were collected cannot be determined, it is still regarded as having become extinct in the 1930s. Overhunting and habitat destruction from increasing rice production are thought to have been implicated in its demise.

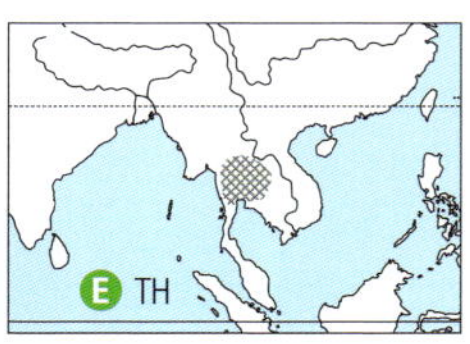

C plains of TH (last individual was killed near Sai Yoke and Kwae Yai); reports from S CN, LA & MM unconfirmed

Family BOVIDAE Hollow-horned Ruminants

Aurochs

Bos primigenius

FR Auroch DE Auerochse ES Uro euroasiático

Three extinct subspecies often recognized, but one (*primigenius*) one survived into recent times. This last subspecies gradually died out due to uncontrolled hunting, habitat loss caused by spread of farming, and disease transmission from domestic cattle. The last individual was a female that died in 1627 from natural causes in Jaktorów Forest (Poland).

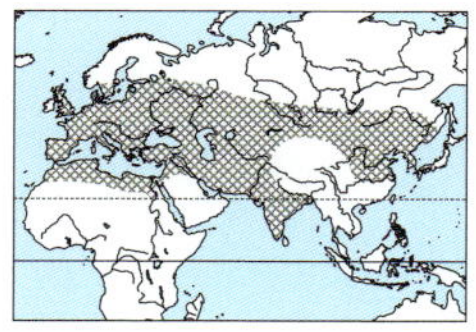

Known from much of Eurasia, including Europe, W & E Asia, also IN (ancient ssp *namadicus*), and N Africa (ancient ssp *africanus*)

Bluebuck

Hippotragus leucophaeus

FR Hippotrague bleu DE Blaubock ES Hipótrago azul

Hunted to extinction during European colonization, the last individual was shot in 1799–1800; a later sighting from 1853 of a "bastard gemsbok" is, however, claimed to be the true final record.

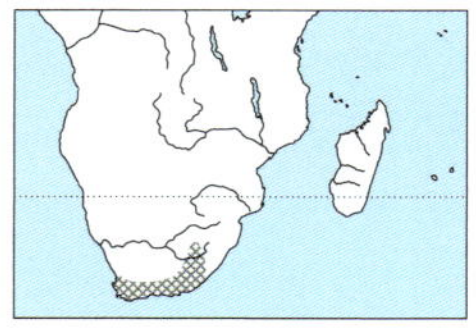

Confined to SW Western Cape of ZA in recent times, although fossils extend it throughout much of S ZA and LS

Family EQUIDAE Horses and relatives

Tarpan

Equus ferus

FR Tarpan DE Tarpan ES Tarpán

During the 1500s it started to disappear from much of mainland Europe but remained common in what is now Russia and less common in E Europe (e.g. Poland). It was present in S Russian Steppe between 1875 and 1890. Last truly wild individual was accidentally killed during an attempted capture in 1879, while the last captive individual died in a Russian zoo in 1909. Commonly hunted during Paleolithic for food and was later targeted by farmers and ranchers to prevent it from interbreeding with their domestic stock.

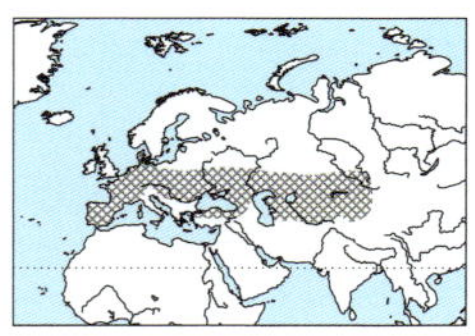

Much of Europe & W Asia from ES to C RU

Family EUPLERIDAE Madagascar Carnivores

Giant Fosa

Cryptoprocta spelea

FR Fossa géant DE Riesenfossa ES Fosa gigante

Known from subfossil remains. Cause of extinction unknown.

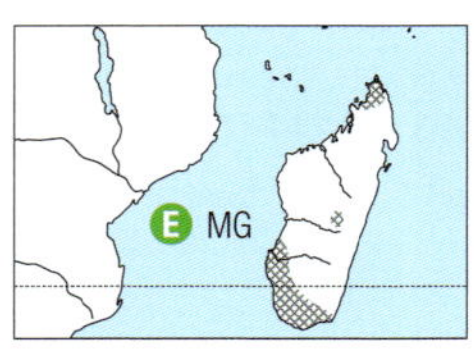

Known in far N (near Antsiranana), along SW coast to S & C highlands (Antsirabe)

Family CANIDAE Dogs

Argentinean Warrah

Dusicyon avus

FR Renard de Burmeister DE Burmeister-Fuchs ES Zorro de Argentina

Known from fossil material from as recently as 700 years ago in Pampean region and 400 years ago in Paraguay. There is speculation about a "big dog" in Tierra del Fuego based on accounts from native people that suggest it could have survived up to the early 1900s. However, it probably became extinct 324–496 years ago for reasons that remain unclear, although it seemed to have ritualistic value for native people. Most likely the anthropogenic impact of European colonization including hunting and introduction of domestic dogs, coupled with habitat change, led to its extinction.

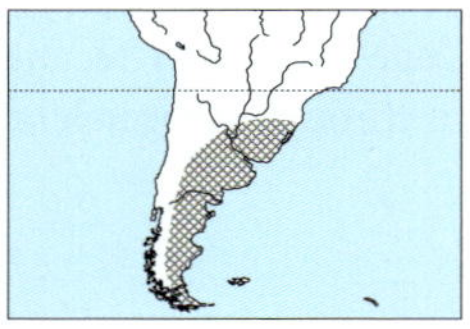

Throughout Pampas and Patagonia of SC & S South America from S BR to Tierra del Fuego

Falkland Warrah

Dusicyon australis

FR Renard des Falkland DE Falkland-Fuchs ES Zorro de Malvinas

The only native terrestrial mammal on the Falkland Is (Malvinas). Driven to extinction by persecution, having declined rapidly in the mid-1800s due to overhunting by Argentinian and Scottish settlers and US fur traders. It was tame and unafraid of humans, making it an easy target. By 1865, it was no longer found on E part of East Falkland I and became extinct in 1876 when the last individual is thought to have been killed at Shallow Bay, West Falkland I.

S Atlantic Ocean

Family OTARIIDAE Eared Seals

Japanese Sea Lion

Zalophus japonicus

FR Otarie du Japon DE Japanischer Seelöwe ES León marino de Japón

Last reliable records are of an individual shot in 1949 on Moneron I and sightings of 50–60 individuals on Takeshima I in 1951. Records persisted until the 1970s but are doubtful because they probably refer to escaped *Z. californianus*.

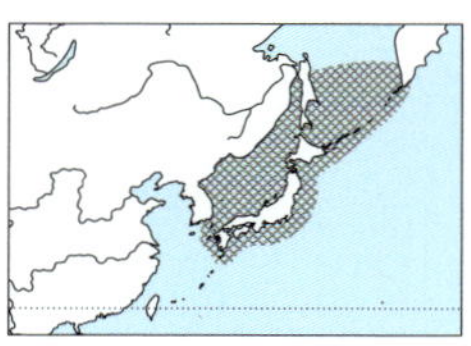

NW Pacific coasts of JP, Korea & RU (Sakhalin I, uncertain in S Kamchatka Pen)

Family PHOCIDAE Earless Seals

Caribbean Monk Seal

Neomonachus tropicalis

FR Phoque des Caraïbes DE Karibik-Mönchsrobbe ES Foca monje del Caribe

The last reliable record was in 1952 of a small colony at Seranilla Bank (between Jamaica and Honduras). Despite numerous surveys, it has never been relocated and is thought to have become extinct. Recent records have been attributed to wandering *Cystophora cristata*.

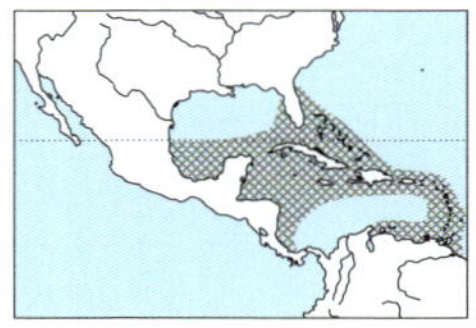

Islands and cays of Caribbean Sea & E Gulf of Mexico (such as BS, Greater & Lesser Antilles), also Yucatán Pen and S along Central & South American coasts to GY and along coasts of Florida & Georgia (US)

Family MUSTELIDAE Weasels and relatives

Sea Mink

Neogale macrodon

FR Vison de mer DE Seenerz ES Visón marino

It became extinct when the last two individuals were killed in 1880 (with a possible killing in 1894, although probably a large *N. vison*) after having become increasingly scarce after 1860.

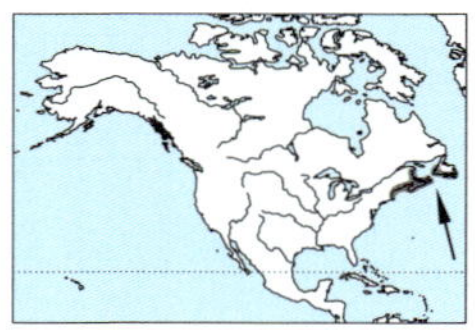

Marine habitats in SE CA (New Brunswick, Newfoundland) & NE US (Massachusetts, Maine)

Japanese Otter

Lutra nippon

FR Loutre du Japon DE Japan-Fischotter ES Nutria japonesa

Common until the 1920s but by the early 1970s its population, estimated at about 20 individuals, had become restricted to Shikoku I. It has not been seen since the 1980s and probably became extinct due to hunting, river pollution, and human development.

Widely distributed in Honshu, Kyushu & Shikoku

Appendix 2

DOMESTIC SPECIES

This Appendix covers the 19 domestic mammal species, as taxonomically defined by the MDD, organized by their pertinent families, which are labelled with their scientific names in red and English common names in gray. This way interested readers can find and explore the wild members of the same family, as all scientific family names are listed in the Index, Contents, and endpapers of the book.

As in the section of the book covering wild species, here each species account includes the Latin name of the taxon and its common names in English, French, German, and Spanish. In addition, each has a short summary describing the domestication, use, and distribution of the species. All the listed domestic species are considered monotypic here.

Maps are not included because domestic species generally share worldwide distributions with humans. Illustrations are not included due to the variation within domestic forms (e.g. domestic dogs and cats).

Data for this section are based on the species accounts appearing in the *Illustrated Checklist of the Mammals of the World*, which were written by Connor J. Burgin, with more recent information compiled by Anna Motis.

Family CAVIIDAE Cavies, Capybaras and Maras

Domestic Guinea Pig

Cavia porcellus

FR Cobaye domestique DE Hausmeerschweinchen ES Cobaya doméstica

Domesticated c. 5000 BC as livestock in South American Andes. Probably originated from *Cavia tschudii* but multiple species might have contributed to its modern gene pool (e.g. *C. aperea*). Now kept as a pet worldwide, although still eaten in parts of South America.

Domesticated worldwide; does not occur naturally in the wild, but feral cavy populations in CO and BR possibly related to this species.

Family BOVIDAE Hollow-horned Ruminants

Gayal

Bos frontalis

FR Gayal DE Gayal ES Gayal

Originally domesticated from *Bos gaurus* but genetic data indicate that it might hybridize or have hybridized with other domestic cattle (*B. taurus*, *B. indicus*). When interbred with *B. taurus* and *B. indicus*, female offspring are likely to be fertile but males typically are not. Also bred with *B. indicus* to produce offspring giving higher milk yields. Revered as a sacrificial animal and a status symbol in some areas.

Found in certain regions of NE IN, BD, MM, and CN (Yunnan).

Bali Cattle

Bos domesticus

FR Banteng domestique DE Bali-Hausrind ES Banteng doméstico

Domesticated from *Bos javanicus* c. 5500 years ago in either Java or on SE Asian mainland and introduced to many regions of Indonesian Archipelago. Will hybridize with *B. indicus* and *B. taurus* to produce fertile offspring (but male offspring of *B. taurus*-crossed animals are sterile). Used primarily for meat but also to a lesser extent as a draught animal.

Domesticated in Bali and on many E Indonesian Is (including Sulawesi, Sumbawa, and Sumba), MY, AU, and New Guinea; feral populations in AU (N NT) and Borneo (East Kalimantan).

Taurine Cattle

Bos taurus

FR Boeuf domestique DE Hausrind ES Toro

Genetic data suggest it was domesticated from Eurasian *Bos primigenius* 10,000–10,500 years ago in W Asia (Mesopotamian region), perhaps originating from a small ancestral stock. Various cattle breeds throughout Asia, Africa, and North and South America are crossbreeds between *B. taurus* and *B. indicus*, although African cattle originated from *B. taurus* that bred with wild *B. primigenius* before being hybridized with *B. indicus*. European breeds appear to have originated from two waves of introduction, the older from W Asia and the more recent an introduction of W African cattle into the Mediterranean. Few breeds are present in S Asia and are often of hybrid origin. *Bos taurus* has been known to have both fertile and infertile hybrids with other *Bos* species including *B. indicus*, *B. bonasus*, *B. bison*, and *B. grunniens*. Used for various purposes, but most commonly for meat and dairy products, and as a draught animal.

Domesticated worldwide; feral and semi-feral populations in much of Europe, US, AU, CO, AR, New Guinea, Hawaii, Galápagos Is, TW, Hispaniola, Aleutian Is, and on many other islands.

Indicine Cattle

Bos indicus

FR Boeuf zébu DE Zebu ES Cebú

Domesticated from *Bos primigenius namadicus* c. 10,000 years ago in what is now Pakistan; its wild form is thought to have disappeared c. 3000 years ago. Many breeds are of hybrid origin, primarily with *B. taurus* but also some with *B. mutus*/*B. grunniens*, *B. gaurus*/*B. frontalis*, or *B. javanicus*/*B. domesticus*. Generally used for labor, riding, and dairy products; hybridized with *B. taurus* to produce individuals giving higher milk yields.

Primarily known from pure breeds in S Asia where it originated, although pure and hybrid breeds are found throughout Asia, Africa, North America, and South America; now found worldwide.

Domestic Yak

Bos grunniens

FR Yack domestique DE Hausyak ES Yak doméstico

Domesticated in S Asia from *Bos mutus* no earlier than c. 10,000 years ago but exact timing uncertain. Hybridization with cattle gives infertile males ("dzo") but fertile females ("dzomo") that are often used to produce larger offspring. Used for meat and dairy products and as a draught animal.

Primarily found throughout Himalayan region, Tibetan Plateau, N MM, CN, MN, and Siberia but also as a domestic animal elsewhere in the world.

Water Buffalo

Bubalus bubalis

FR Buffle domestique DE Wasserbüffel ES Búfalo de agua doméstico

Appears to have been domesticated twice, in India c. 5000 years ago and China c. 4000 years ago, before spreading through W Asia to Egypt and Mediterranean Europe c. 2000 years ago. There are two main forms, the Swamp Buffalo and the River Buffalo, seemingly genetically well separated; this suggests they diverged before being domesticated, although they are not different enough to be considered distinct species. Used for a variety of tasks, particularly meat and dairy production, and as a draught animal, especially for ploughing.

Swamp Buffalo are found from NE IN (Assam) through SE Asia and up to R Yangtze Valley of CN but River Buffalo found throughout Indian subcontinent. Currently feral throughout Asia, with introduced populations in North America, South America, and N AU.

Domestic Goat

Capra hircus

FR Bouc domestique DE Hausziege ES Cabra doméstica

Originally domesticated from *Capra aegagrus* but has probably been hybridizing with other wild goat species in Asia for centuries. Genetic and archeological data suggest that goat domestication originated in Iran c. 10,000 years ago, while other archeological sites indicate that domesticated goats spread across the rest of W Asia 8000–9000 years ago. Used for various purposes, including for hides, meat, and dairy products.

Domesticated worldwide with feral populations in many regions, including the Mediterranean Is (Mallorca, Crete) where "wild" goats are known; becomes feral very quickly after release.

Domestic Sheep

Ovis aries

FR Mouton DE Hausschaf ES Oveja doméstica

Probably domesticated from *Ovis gmelini* (although exact species uncertain) in the Mesopotamian region 9000–11,000 years ago. Originally used for milk, meat, and hides. According to archaeological evidence, selection for woolly sheep did not occur until c. 6000 years ago, when the earliest Europeans apparently started using domestic sheep. Wild sheep on Mediterranean Is (Corsica, Sardinia, Cyprus) are considered to be *O. aries* because genetic data suggest they are derived from domestic sheep released there in the Neolithic.

Domesticated worldwide. European wild sheep (Mouflon) found throughout mainland Europe (N to Scandinavia) deriving from feral sheep from Mediterranean Is; Mouflon also introduced to Canary Is, Kerguelen Is, CL, AR, and US (on some islands and in game ranches).

Family SUIDAE Pigs

Domestic Pig

Sus domesticus

FR Porc DE Hausschwein ES Cerdo doméstico

The earliest evidence of pig domestication based on different stock populations of *Sus scrofa* comes from archeological sites in various regions in the Tigris Basin and Turkey. Genetic data suggest it was domesticated separately in W Asia and China. Domestication is thought to have begun in W Asia 9000–10,000 years ago but perhaps earlier because evidence of pigs on Cyprus from 11,400 years ago suggests introduction there from domesticated animals on mainland. Brought to Europe from W Asia c. 8500 years ago and subsequently interbred with wild *S. scrofa*, making it difficult to track the genetic heritage of European breeds of *S. domesticus*. Domestication in China probably began c. 8000 years ago. Much feral and farm stock is of domestic/wild hybrid origin. Used for many products but most notably for food.

Domesticated worldwide. Introduced feral populations in various parts of the world, including AU, NZ, E Malay Arch, and all the Americas. In all these areas, it is now generally recognized as a major pest. Inadvertently reintroduced into Great Britain and other regions, where it is or was once found in the wild, thanks to hybrid escapees from commercial farms.

Family CAMELIDAE Camels

Llama

Lama glama

FR Lama DE Lama ES Llama

Selectively bred from *Lama guanicoe* 4000–4500 years ago in SC Andes (N Chile, NW Argentina) and 4500–5500 years ago in S Andes (Junín de los Andes, Argentina). Osteological remains and DNA analysis document the origin of domestication within the range of N subspecies of Guanaco, *L. g. cacsilensis*. There are two distinct phenotypic breeds: Short-Woolled and Long-Woolled Llamas. Although indigenous to South America, where kept as pack animals and for food, they now serve worldwide as pets and guard animals for protecting sheep and goats from predators (North America), for trekking and producing home-spun wool in cottage industries, and also appear in livestock shows.

Present in C Andes from C PE to W BO and N AR at 3800–5000 m; now throughout the world for domestic uses.

Alpaca

Lama pacos

FR Alpaca DE Alpaka ES Alpaca

Previously included under *Lama glama* but now considered separate species. Recent DNA studies have established that it was domesticated from *L. vicugna mensalis*, although some introgression between *L. glama* and *L. pacos* occurred in the past, leading to confusion as to whether it came from *L. guanicoe* or *L. vicugna*. Archeological digs at Telarmachay in C Andes of Peru indicate that it was domesticated 5500–6500 years ago by a hunter-gatherer society. Two distinct breeds are recognized: Huacaya and Suri. Not bred as working animals but for their wool, used like sheep's wool for making fabrics and knitwear.

Present in C Andes from C PE to mid-BO and N CL; imported to US, AU, NZ, CA, and Europe in 1980s–1990s.

Domestic Bactrian Camel

Camelus bactrianus

FR Chameau de Bactriane DE Kamel ES Camello

Recent mtDNA studies suggest that wild (*Camelus ferus*) and domestic (*C. bactrianus*) Bactrian Camels derived from separate ancestors, of which the domestic form is extinct. It is estimated to have diverged c. 700,000 years ago in the Pleistocene, long before domestication 4000–6000 years ago. Genetic justification for recognizing the domestic form as a separate species is growing. Used as pack and riding animals and to provide nomads with wool, meat, and milk.

Found in arid lands and deserts of C & S Asia as domesticated livestock.

Dromedary Camel

Camelus dromedarius

FR Dromadaire DE Dromedar ES Dromedario

Based on evidence from archeological sites in S Arabian Peninsula, it was domesticated 4000–5000 years ago, with the wild form becoming extinct 2000–5000 years ago. Used as pack and riding animals, and for ploughing, but also provides meat, milk, wool, and leather.

Present in arid and semi-arid regions of N Africa to Middle East and parts of C Asia. Introduced to E & S Africa, Canary Is (1405), and W & C AU (1840–1907), where a sizeable free-ranging/feral population exists. From 17th to early 20th century, unsuccessful introductions to Caribbean, BO, PE, CO, BR, NA, and SW US.

Family EQUIDAE Horses and relatives

Domestic Ass

Equus asinus

FR Âne domestique DE Hausesel ES Asno

Domesticated from two wild forms of *Equus africanus* (Nubian subspecies *E. a. africanus* and an unknown subspecies) 6000–7000 years ago, probably in Egypt or by Nubian pastoral people. Earliest evidence of domestication from Egypt, dating to 6500–6800 years ago based on remains of small-sized donkeys (although disputed). Remains of donkeys in Yemen dating from 6500 years ago are domesticated individuals or Asiatic populations of *E. africanus*. Present in Levant c. 6000 years ago, although domestication not likely to have occurred there or in Mesopotamia since it coincides with the earliest certain record of domestic individuals in Egypt 6000 years ago. Generally used as a draught or pack animal but also for meat and milk in some regions.

Domesticated worldwide, with feral populations in many regions, including SD, SA, Socotra I, LK, AU, US (including Hawaii), Galápagos Is, Chagos Is, and probably other oceanic islands; most frequent in impoverished regions of Africa, Asia, and Latin America, where demand for asses is greater.

Domestic Horse

Equus caballus

FR Cheval domestique DE Hauspferd ES Caballo

Domesticated from *E. ferus* more than 5500 years ago but probably not fully domesticated until c. 3000 years ago. Domestication probably occurred in E Europe and C Asia, as indicated by archeological evidence from Ukraine and Kazakhstan 5500–6000 years ago; apparently had spread throughout Europe by c. 4000 years ago. Used for various purposes including food, leisure activities, and sports, as well as a working animal; commonly used in warfare and as a primary means of travel throughout much of human history.

Domesticated worldwide with many feral populations, especially in North America, AU, S South America, C Asia, and E & S Africa where large populations exist.

Family FELIDAE Cats

Domestic Cat

Felis catus

FR Chat domestique DE Hauskatze ES Gato doméstico

Domesticated from *Felis lybica* in W Asia (probably Mesopotamian region) c. 9500 years ago. Earliest indication of cat domestication comes from archeological sites on Cyprus where cats are not native, indicating they were brought from mainland Asia. Cats were later domesticated independently in Egypt but did not appear in Europe until c. 3200 years ago, first in Greece. Thought to have lived commensally with humans until they were finally domesticated to eradicate pest mice and other small mammals in and around human habitation, but most often just as companions.

Domesticated worldwide, with accompanying feral populations in just about every region, causing major environmental concerns since cats can be highly invasive and prey on numerous small animals.

Family CANIDAE Dogs

Domestic Dog

Canis familiaris

FR Chien domestique DE Haushund ES Perro doméstico

The earliest domesticated mammal, having been first tamed from a now extinct population of *Canis lupus* 25,000–40,000 years ago. Molecular evidence suggests two origins: one in W Eurasia and another in E Eurasia. Earliest records of probable *C. familiaris* come from Europe (Germany) 35,000–40,000 years ago, although the earliest undisputable occurrence (confirmed by genetic data) comes from 14,200 years ago. Dingos are considered a feral form of *C. familiaris*; the Dingo, New Guinea Singing Dog, and other Malay wild dogs appear to have diverged from other dog breeds c. 8300 years ago, probably originating on mainland Asia and spreading with humans throughout Indonesia and into Australia c. 4000 years ago. Bred for various purposes, including guarding, hunting, and sled transportation but today mostly kept as a pet.

Domesticated worldwide with feral populations in nearly every region.

Family MUSTELIDAE Weasels and relatives

Domestic Ferret

Mustela furo

FR Furet DE Frettchen ES Hurón

Molecular data suggest it was domesticated from N African lineages of *Mustela putorius* c. 2500 years ago and possibly first domesticated in Egypt; however, no archeological evidence is available. Originally domesticated to hunt rabbits but now most often a household pet.

Found worldwide; occasionally inadvertently introduced to regions including NZ and Europe where some cross with *M. putorius* and create hybrid populations.

Appendix 3

TAXONOMIC CHANGES AND CONTROVERSIES IN MAMMALOGY

Given that *All the Mammals of the World* (AMW) is focused primarily on the presentation to a more general audience of the diversity of existing mammalian species, there is less specialized taxonomic information in this work than in its predecessors, the *Handbook of the Mammals of the World* (HMW) and the *Illustrated Checklist of the Mammals of the World* (CMW). However, the taxonomy presented in AMW still represents an updated listing of both these series based on version 1.9 of the *Mammal Diversity Database* (MDD), as discussed in the Introduction to this book. These changes were implemented both to keep up with the constantly changing landscape of mammalian taxonomy and to provide added value to this volume compared to the CMW.

For those who are interested in the taxonomic decisions made in this volume, we have summarized the differences between this book and the CMW in this Appendix. We have also included an overview of certain major taxonomic controversies and highlighted discussions in relation to the taxonomy used in the AMW volume (see *Major Controversies in Mammalian Systematics* below). For more taxonomic and nomenclatural details and the relevant citations for these changes, as well as taxonomic information on the species presented in AMW, we refer readers to the current version of the MDD available at mammaldiversity.org. We also recommend readers consult the HMW and CMW volumes for detailed taxonomic information and a wealth of further information on the species included in those series. In particular, the introduction to the CMW series includes further detailed discussions on mammalian taxonomy and nomenclature. With the exception of some minor spelling changes of specific epithets, the tables presented in this Appendix summarize all the differences between the AMW and CMW treatments. Since AMW is based completely on version 1.9 of the MDD—with the exception of a select few taxonomic changes—we have highlighted these differences in the final table of this Appendix.

TAXONOMIC DIFFERENCES BETWEEN THE CMW, MDD, AND AMW

EPITHET CHANGES. Some of the changes summarized here are purely nomenclatural. These include many epithet changes, which are governed by the rules dictated by the International Commission of Zoological Nomenclature (ICZN), often referred to as "the Code". The ICZN determines the use of scientific names applied to animals, including how taxa are described, how type specimens and type localities are designated, what names are available for nomenclatural use, how these names should be spelt, and which names have priority over others. Although the rules of the ICZN are essential for maintaining stability in the use of scientific names over time, they can be somewhat difficult to interpret due to the vague language used in some articles, which can lead to subjective decisions on how these names are employed. There are many specific epithet spelling changes between the CMW and AMW not listed in the following table, which are often based on particular interpretations of the ICZN Code and on the original descriptions of these names, some of which may have been coined over 200 years ago before the Code even existed. Most of the epithet changes listed in the following table result from the movement of synonyms—names considered to represent the same species as another—between species: a good example is the specific epithet change from *Peromyscus oaxacensis* to *P. cordillerae* due to the name *oaxacensis* being moved to synonymy under *P. aztecus*, thereby making *cordillerae* the oldest name available for that species according to the rules of nomenclatural priority (i.e. the oldest nomenclaturally available name according the ICZN Code applicable to a defined species; Kilpatrick *et al.* 2021). This is especially complex in the case of *Otaria byronia* versus *flavescens*, where both names having been changed back and forth in the pertinent literature depending on whether *flavescens* is considered a *nomen nudum* or not (see discussion in Lucero *et al.* 2019). There is also an ongoing discussion regarding *Chionomys nivalis* versus *syriacus*; this taxon was recently changed to *syriacus* based on the discovery that *syriacus* (which had been described before *nivalis*) applies to this species, although there have been appeals made to the ICZN to preserve *nivalis* over *syriacus* to maintain taxonomic stability given its long history of use (Kryštufek *et al.* 2021, ICZN 2021). These examples highlight the importance of the ICZN Code as well as the complexities of decision-making when determining the applicable name for a given species. There are eight epithet changes between the CMW and AMW reported here.

CMW Species Name	AMW Species Name
Sminthopsis griseoventer	*Sminthopsis fuliginosa*
Lemmus ognevi	*Lemmus paulus*
Chionomys nivalis	*Chionomys syriacus*
Alexandromys gromovi	*Alexandromys shantaricus*
Peromyscus oaxacensis	*Peromyscus cordillerae*
Hipposideros pratti	*Hipposideros swinhoei*
Mazama bororo	*Mazama jucunda*
Otaria byronia	*Otaria flavescens*

GENERIC CHANGES. Taxonomic change can occur at all taxonomic levels. Although most often occurring at species level, genus-level changes are also common and are often proposed as more data accumulate to support claims that a predefined genus is either paraphyletic or polyphyletic (see Figure 2 on p. 20 for examples of these terms). A core philosophy for the definition of taxa above species level is that defined groups must be monophyletic, so when a genus is shown not to be monophyletic, generic revisions need to take place. This is the case of most generic name changes for species between the CMW and AMW, including the inclusion of *Chaerephon* under *Mops* (Gregorin & Cirranello 2016), the movement of *Elephantulus rufescens* and *E. revoilii* to the recently described *Galegeeska* (Heritage *et al.* 2020, Krásová *et al.* 2021), and the erection of *Neogale* from synonymy for various species of American mustelid previously included in the genera *Mustela* and *Neovison* (Patterson *et al.* 2021). However, there are a number of cases in which an already monotypic genus has been split based on subjective judgements regarding the distinctiveness of the specific clades within that genus. These cases can cause great controversy within the taxonomic community because generic changes may prompt disruptive changes to long-standing taxonomic arrangements, which will affect all downstream research on these species. For further discussion of examples of taxonomic issues and debates stemming from this concept, see the primates and bats in the *Major Controversies in Mammalian Systematics* section below. Lastly, there are several generic changes resulting from nomenclatural decisions that resemble the epithet changes discussed above. A good example of nomenclatural generic changes is the switch from *Myodes* to *Clethrionomys*, which is based on interpretations of the type species of *Myodes* (see discussions in Kryštufek *et al.* 2020, who consider the name *Myodes* a synonym of *Lemmus* and use *Clethrionomys* for red-backed voles); this change in particular is highly controversial within the mammalogical community as *Clethrionomys* was used historically but *Myodes* has been used fairly consistently over the last 30 years or so, during which time many papers have argued for the use of one or either. There are 107 generic changes between the CMW and AMW reported here.

CMW Species Name	AMW Species Name
Elephantulus rufescens	*Galegeeska rufescens*
Elephantulus revoilii	*Galegeeska revoilii*
Nycticebus pygmaeus	*Xanthonycticebus pygmaeus*
Callibella humilis	*Mico humilis*
Leontocebus fuscus	*Saguinus fuscus*
Leontocebus nigricollis	*Saguinus nigricollis*
Leontocebus leucogenys	*Saguinus leucogenys*
Leontocebus illigeri	*Saguinus illigeri*
Leontocebus lagonotus	*Saguinus lagonotus*
Leontocebus tripartitus	*Saguinus tripartitus*
Leontocebus fuscicollis	*Saguinus fuscicollis*
Leontocebus nigrifrons	*Saguinus nigrifrons*
Leontocebus cruzlimai	*Saguinus cruzlimai*
Leontocebus weddelli	*Saguinus weddelli*
Brachylagus idahoensis	*Sylvilagus idahoensis*
Dendromus nyikae	*Poemys nyikae*
Dendromus nikolausi	*Megadendromus nikolausi*
Myodes rutilus	*Clethrionomys rutilus*
Myodes californicus	*Clethrionomys californicus*
Myodes gapperi	*Clethrionomys gapperi*
Myodes glareolus	*Clethrionomys glareolus*
Myodes centralis	*Clethrionomys centralis*
Myodes macrotis	*Alticola macrotis*
Aschizomys lemminus	*Alticola lemminus*
Paruromys dominator	*Taeromys dominator*
Bunomys fratrorum	*Frateromys fratrorum*
Mastomys pernanus	*Serengetimys pernanus*
Praomys delectorum	*Montemys delectorum*
Praomys verschureni	*Congomys verschureni*
Praomys lukolelae	*Congomys lukolelae*
Myomyscus yemeni	*Ochromyscus yemeni*
Myomyscus brockmani	*Ochromyscus brockmani*
Myomyscus angolensis	*Mastomys angolensis*
Dolichotis salinicola	*Pediolagus salinicola*
Rousettus celebensis	*Pilonycteris celebensis*
Artibeus concolor	*Koopmania concolor*
Artibeus rosenbergi	*Dermanura rosenbergi*
Artibeus watsoni	*Dermanura watsoni*
Artibeus toltecus	*Dermanura tolteca*
Artibeus phaeotis	*Dermanura phaeotis*
Artibeus cinereus	*Dermanura cinerea*
Artibeus anderseni	*Dermanura anderseni*
Artibeus ravus	*Dermanura rava*
Artibeus aztecus	*Dermanura azteca*
Artibeus bogotensis	*Dermanura bogotensis*

CMW Species Name	AMW Species Name
Artibeus glaucus	*Dermanura glauca*
Artibeus gnomus	*Dermanura gnoma*
Chaerephon aloysiisabaudiae	*Mops aloysiisabaudiae*
Chaerephon ansorgei	*Mops ansorgei*
Chaerephon bemmeleni	*Mops bemmeleni*
Chaerephon bivittatus	*Mops bivittatus*
Chaerephon chapini	*Mops chapini*
Chaerephon gallagheri	*Mops gallagheri*
Chaerephon leucogaster	*Mops leucogaster*
Chaerephon major	*Mops major*
Chaerephon nigeriae	*Mops nigeriae*
Chaerephon tomensis	*Mops tomensis*
Chaerephon pumilus	*Mops pumilus*
Chaerephon russatus	*Mops russatus*
Chaerephon pusillus	*Mops pusillus*
Chaerephon atsinanana	*Mops atsinanana*
Chaerephon jobimena	*Mops jobimena*
Chaerephon johorensis	*Mops johorensis*
Chaerephon plicatus	*Mops plicatus*
Chaerephon bregullae	*Mops bregullae*
Chaerephon solomonis	*Mops solomonis*
Chaerephon jobensis	*Mops jobensis*
Parahypsugo eisentrauti	*Nycticeinops eisentrauti*
Parahypsugo crassulus	*Nycticeinops crassulus*
Parahypsugo bellieri	*Nycticeinops bellieri*
Parahypsugo happoldorum	*Nycticeinops happoldorum*
Parahypsugo macrocephalus	*Nycticeinops macrocephalus*
Neoromicia grandidieri	*Nycticeinops grandidieri*
Neoromicia rendalli	*Pseudoromicia rendalli*
Neoromicia brunnea	*Pseudoromicia brunnea*
Neoromicia isabella	*Pseudoromicia isabella*
Neoromicia roseveari	*Pseudoromicia roseveari*
Neoromicia tenuipinnis	*Pseudoromicia tenuipinnis*
Neoromicia nanus	*Afronycteris nanus*
Neoromicia helios	*Afronycteris helios*
Hypsugo bemainty	*Neoromicia bemainty*
Hypsugo anchieta	*Neoromicia anchieta*
Neoromicia stanleyi	*Laephotis stanleyi*
Neoromicia malagasyensis	*Laephotis malagasyensis*
Neoromicia robertsi	*Laephotis robertsi*
Neoromicia matroka	*Laephotis matroka*
Neoromicia capensis	*Laephotis capensis*
Panolia eldii	*Rucervus eldii*
Nyala angasii	*Tragelaphus angasii*
Ammelaphus imberbis	*Tragelaphus imberbis*

CMW Species Name	AMW Species Name
Strepsiceros strepsiceros	*Tragelaphus strepsiceros*
Taurotragus oryx	*Tragelaphus oryx*
Taurotragus derbianus	*Tragelaphus derbianus*
Neotragus batesi	*Nesotragus batesi*
Neotragus moschatus	*Nesotragus moschatus*
Pecari tajacu	*Dicotyles tajacu*
Vicugna vicugna	*Lama vicugna*
Vicugna pacos	*Lama pacos*
Acrocodia indicus	*Tapirus indicus*

CMW Species Name	AMW Species Name
Tapirella bairdii	*Tapirus bairdii*
Hyaena brunnea	*Parahyaena brunnea*
Mustela vison	*Neogale vison*
Mustela macrodon	*Neogale macrodon*
Mustela frenata	*Neogale frenata*
Mustela felipei	*Neogale felipei*
Mustela africana	*Neogale africana*
Nasuella olivacea	*Nasua olivacea*

NEWLY DESCRIBED SPECIES. Probably the most straightforward taxonomic changes are those involving the description of a new scientific name. All the names listed in this section were described for the first time since the CMW series was released and are included in AMW with illustrations and range maps. Most of these new species are based on previously known populations of existing species that have been found to be sufficiently different to be classified as distinct species: for example, *Sicista talgarica* and *S. terskeica* have been separated based on populations previously attributed to *S. tianschanica* (Lebedev *et al.* 2021). However, there are many species such as *Chingawaemys rarus* (Nicolas *et al.* 2021) and *Alpiscaptulus medogensis* (Chen *et al.* 2021)—described for the first time thanks to newly discovered specimens or populations—that represent previously unknown species or genera. Most recently described mammal species are representatives of species-rich and cryptically diverse mammalian orders such as Rodentia, Chiroptera, and Eulipotyphla, although there are also a few recently described species from more charismatic orders such as the primates, *Mico schneideri* (Costa-Araújo *et al.* 2021) and *Trachypithecus popa* (Roos *et al.* 2020), and cetaceans, *Balaenoptera ricei* (Rosel *et al.* 2021) and *Mesoplodon eueu* (Carroll *et al.* 2021). Although most of these newly described species represent extant populations, there are also a few species included in Appendix 1: Extinct species (see p. 697) such as *Antillomys rayi* (Brace *et al.* 2015, 2023) and *Perameles papillon* (Travouillon & Phillips 2018) that have recently been described on the basis of populations that have unfortunately become extinct since 1500 AD. In all, 107 species newly described since the publication of the CMW series (or omitted from it) are included in AMW; the CMW was only published two years prior to the taxonomic cut-off for the inclusion of taxonomic changes in AMW, which shows how new species are rapidly and consistently being described all the time.

AMW Species Name
Marmosa adleri
Marmosa jansae
Marmosops marina
Perameles papillon
Dendrohyrax interfluvialis
Mico schneideri
Trachypithecus popa
Sicista talgarica
Sicista terskeica
Typhlomys huangshanensis
Eospalax muliensis
Peromyscus purepechus
Peromyscus ensinki
Peromyscus greenbaumi
Nephelomys ricardopalmai
Mindomys kutuku
Pattonimus ecominga
Pattonimus musseri
Oecomys matogrossensis
Oligoryzomys guille
Neacomys serranensis
Neacomys auriventer
Neacomys vossi
Neacomys elieceri
Neacomys xingu
Neacomys marajoara
Neacomys jau
Neacomys aletheia
Neacomys leilae
Neacomys oliveirai
Holochilus oxe

AMW Species Name
Antillomys rayi
Oxymycterus willkaurco
Akodon kadiweu
Rhagomys septentrionalis
Rhipidomys ochoagrateroli
Thomasomys pardignasi
Thomasomys antoniobracki
Phyllotis camiari
Phyllotis pehuenche
Niviventer fengi
Bullimus carletoni
Otomys willani
Mus harennensis
Chingawaemys rarus
Colomys lumumbai
Colomys wologizi
Stenocephalemys zimai
Stenocephalemys sokolovi
Microcavia sorojchi
Ctenomys plebiscitum
Ctenomys bidaui
Ctenomys thalesi
Ctenomys contrerasi
Octodon ricardojeda
Tamiops minshanicus
Eupetaurus nivamons
Eupetaurus tibetensis
Sorex cruzi
Notiosorex tataticuli
Chodsigoa dabieshanensis
Crocidura dongyangjiangensis

AMW Species Name
Crocidura narcondamica
Crocidura pseudorhoditis
Crocidura microelongata
Crocidura pallida
Crocidura caudicrassa
Crocidura brevicauda
Crocidura quasielongata
Crocidura australis
Crocidura ordinaria
Crocidura solita
Crocidura normalis
Crocidura parva
Crocidura mediocris
Crocidura tenebrosa
Crocidura baletei
Crocidura makeda
Crocidura similiturba
Uropsilus dabieshanensis
Alpiscaptulus medogensis
Rhinolophus yonghoiseni
Micronycteris simmonsae
Micronycteris tresamici
Vampyressa voragine
Molossus melini
Cynomops kuizha
Miniopterus nimbae
Miniopterus wilsoni
Pipistrellus simandouensis
Nyctophilus holtorum
Hypsugo stubbei
Pseudoromicia kityoi

AMW Species Name
Pseudoromicia nyanza
Laephotis kirinyaga
Histiotus mochica
Histiotus cadenai
Eptesicus orinocensis
Eptesicus langeri
Plecotus gobiensis
Myotis armiensis
Myotis moratellii
Myotis pampa
Myotis nimbaensis
Myotis hayesi
Balaenoptera ricei
Mesoplodon eueu

SPLIT SPECIES. Along with the recently recognized species that represent entirely new name descriptions, there are also several newly recognized species that have already applicable names. These generally represent names that were previously considered synonyms or subspecies of a parent species but for which fresh data has determined that the population upon which the original name was based actually represents a distinct species. This is often termed a "split" by taxonomists because a single "parent" species is split into multiple "daughter" species based on preexisting names, although this term can also apply to newly described names for a species established from previously known populations of a parent species. Like newly described species, most of these split species are found within the large and cryptically diverse mammalian orders Rodentia, Chiroptera, and Eulipotyphla. However, there are some notable splits within more charismatic taxa including the recognition of *Giraffa reticulata* and the two other previously recognized *Giraffa* species as separate from *G. camelopardalis*, the exact species boundaries of which are under debate in recent literature (see Coimbra *et al.* 2021 and references therein). Another recent taxonomic revision of interest is that of the spotted skunks of the genus *Spilogale* in North America, which have been increased from four to seven species based on molecular genetic data (McDonough *et al.* 2022). Most splits leave the new species in the same genus as their parent species, although in some cases such as *Priapomys leonardi*, split from *Hylopetes alboniger*, the new species are moved to a completely new genus based on phylogenetic inferences (Li *et al.* 2021). There are 106 species split from pre-existing species included in AMW that were not included in the CMW, which underlines how mammalian taxonomy is consistently changing.

AMW Species Name	AMW Parent Species
Marmosa nicaraguae	*Marmosa alstoni*
Marmosa perplexa	*Marmosa phaea*
Marmosa parda	*Marmosa germana*
Philander nigratus	*Philander andersoni*
Marmosops woodalli	*Marmosops pinheiroi*
Dromiciops bozinovici	*Dromiciops gliroides*
Dromiciops mondaca	*Dromiciops gliroides*
Sminthopsis froggatti	*Sminthopsis macroura*
Sminthopsis stalkeri	*Sminthopsis macroura*
Isoodon fusciventer	*Isoodon obesulus*
Isoodon peninsulae	*Isoodon obesulus*
Perameles fasciata	*Perameles bougainville*
Perameles myosuros	*Perameles bougainville*
Perameles notina	*Perameles bougainville*
Phalanger matabiru	*Phalanger ornatus*
Phalanger breviceps	*Phalanger orientalis*
Petaurus notatus	*Petaurus breviceps*
Petaurus ariel	*Petaurus breviceps*
Cabassous squamicaudis	*Cabassous unicinctus*
Cheirogaleus minusculus	*Cheirogaleus sibreei*
Macaca brunnescens	*Macaca ochreata*
Trachypithecus melamera	*Trachypithecus phayrei*
Heteromys bulleri	*Heteromys irroratus*
Sicista cimlanica	*Sicista subtilis*
Sicista severtzovi	*Sicista subtilis*
Poemys pecilei	*Dendromus melanotis*
Dendromus kumasi	*Dendromus messorius*
Lemmus nigripes	*Lemmus trimucronatus*
Alexandromys alpinus	*Alexandromys mongolicus*
Baiomys brunneus	*Baiomys musculus*
Reithrodontomys albilabris	*Reithrodontomys microdon*
Reithrodontomys wagneri	*Reithrodontomys microdon*
Peromyscus chrysopus	*Peromyscus perfulvus*
Peromyscus collinus	*Peromyscus pectoralis*
Peromyscus amplus	*Peromyscus difficilis*
Peromyscus felipensis	*Peromyscus difficilis*
Neacomys carceleni	*Neacomys amoenus*
Holochilus nanus	*Holochilus sciureus*
Necromys lenguarum	*Necromys lasiurus*
Akodon baliolus	*Akodon aerosus*
Akodon oenos	*Akodon spegazzinii*
Calomys fecundus	*Calomys boliviae*
Phyllotis rupestris	*Phyllotis xanthopygus*
Phyllotis vaccarum	*Phyllotis xanthopygus*
Meriones psammophilus	*Meriones meridianus*
Meriones penicilliger	*Meriones meridianus*
Niviventer mekongis	*Niviventer huang*

AMW Species Name	AMW Parent Species
Niviventer sacer	*Niviventer bukit*
Colomys eisentrauti	*Colomys goslingi*
Cryptomys mahali	*Cryptomys hottentotus*
Cryptomys nimrodi	*Cryptomys hottentotus*
Cryptomys natalensis	*Cryptomys hottentotus*
Cryptomys pretoriae	*Cryptomys hottentotus*
Coendou baturitensis	*Coendou prehensilis*
Coendou longicaudatus	*Coendou prehensilis*
Capromys garridoi	*Capromys pilorides*
Dremomys ornatus	*Dremomys rufigenis*
Sundasciurus natunensis	*Sundasciurus lowii*
Sundasciurus robinsoni	*Sundasciurus lowii*
Priapomys leonardi	*Hylopetes alboniger*
Otospermophilus douglasii	*Otospermophilus beecheyi*
Glis persicus	*Glis glis*
Sorex kozlovi	*Sorex thibetanus*
Chimarrogale varennei	*Chimarrogale himalayica*
Crocidura armenica	*Crocidura arispa*
Scapanus occultus	*Scapanus latimanus*
Mogera hainana	*Mogera insularis*
Euroscaptor ngoclinhensis	*Euroscaptor parvidens*
Coelops hirsutus	*Coelops robinsoni*
Hipposideros antricola	*Hipposideros ater*
Rhinolophus hirsutus	*Rhinolophus macrotis*
Rhinolophus imaizumii	*Rhinolophus perditus*
Rhinolophus foetidus	*Rhinolophus luctus*
Glossophaga bakeri	*Glossophaga commissarisi*
Glossophaga mutica	*Glossophaga soricina*
Glossophaga valens	*Glossophaga soricina*
Glossophaga antillarum	*Glossophaga soricina*
Chiroderma scopaeum	*Chiroderma salvini*
Scotoecus albigula	*Scotoecus hirundo*
Scotoecus hindei	*Scotoecus hirundo*
Mimetillus thomasi	*Mimetillus moloneyi*
Histiotus colombiae	*Histiotus montanus*
Kerivoula pusilla	*Kerivoula whiteheadi*
Myotis alascensis	*Myotis lucifugus*
Myotis carissima	*Myotis lucifugus*
Myotis pernox	*Myotis lucifugus*
Myotis relictus	*Myotis lucifugus*
Myotis larensis	*Myotis nesopolus*
Myotis nipalensis	*Myotis davidii*
Myotis sowerbyi	*Myotis siligorensis*
Myotis alticraniatus	*Myotis siligorensis*
Myotis hoveli	*Myotis nattereri*
Myotis tschuliensis	*Myotis nattereri*
Balaenoptera brydei	*Balaenoptera edeni*

AMW Species Name	AMW Parent Species
Platanista minor	*Platanista gangetica*
Inia humboldtiana	*Inia geoffrensis*
Giraffa reticulata	*Giraffa camelopardalis*
Bubalus quarlesi	*Bubalus depressicornis*
Naemorhedus cranbrooki	*Naemorhedus baileyi*
Bdeogale omnivora	*Bdeogale crassicauda*
Spilogale interrupta	*Spilogale putorius*
Spilogale yucatanensis	*Spilogale angustifrons*
Spilogale leucoparia	*Spilogale gracilis*
Mustela haidarum	*Mustela erminea*
Mustela richardsonii	*Mustela erminea*
Lutra nippon	*Lutra lutra*

LUMPED SPECIES. Although new species are constantly being recognized (107 newly described and 106 splits between the CMW and AMW in this work), taxonomic revisions also result in the grouping of previously recognized species under another species, which is often termed as a "lump" by taxonomists. Recent trends in mammalogy have generally favored splitting species over lumping species due to the taxonomic philosophies and species concepts used when interpreting the raw data (see the CMW Introduction for further discussion on this topic), which has generally led to a net increase in mammal species over time (Burgin *et al.* 2018). The guiding philosophy behind the reluctance to lump species derives from the assumption that conservation efforts are often limited to populations recognized as species; thus, if a population is no longer recognized as a separate species, it could end up being ignored by conservation organizations. This is especially true in orders such as primates and ungulates with many endangered species. Thus, more data is often needed to support a lump than a split in taxonomic literature. Nevertheless, between the CMW and AMW, there are still 188 lumped species. The primary cause of this is the use of a specific taxonomic arrangement employed in the HMW and CMW for Bovidae, which was not widely accepted and has been removed in subsequent versions of the MDD and now in AMW (see *Major Controversies in Mammalian Systematics* section below for details on ungulate taxonomic controversies; Groves & Grubb 2011). Of the 188 lumped species between the CMW and AMW, only 46 are from species outside of Bovidae, which illustrates the fact that in the literature lumps are considerably rarer than splits or newly described species. Lastly, there are a few occasions in which lumping two species brings an extinct species back from the grave; this is the case of *Pseudomys gouldii*, which was considered Extinct on The IUCN Red List until recent molecular data indicated that *P. fieldi*, which has living populations, is in fact the same species, thereby re-establishing *P. gouldii* as an extant species (Roycroft *et al.* 2021).

CMW Species Name	Lumped Into AMW Species
Marmosa regina	*Marmosa isthmica*
Sminthopsis aitkeni	*Sminthopsis fuliginosa*
Lepilemur mittermeieri	*Lepilemur dorsalis*
Plecturocebus dubius	*Plecturocebus caligatus*
Colobus caudatus	*Colobus guereza*
Lepus insularis	*Lepus californicus*
Arvicola monticola	*Arvicola amphibius*
Microtus miurus	*Microtus abbreviatus*
Reithrodontomys bakeri	*Reithrodontomys wagneri*
Peromyscus interparietalis	*Peromyscus eremicus*
Peromyscus pseudocrinitus	*Peromyscus fraterculus*
Holochilus vulpinus	*Holochilus brasiliensis*
Euneomys mordax	*Euneomys fossor*
Euneomys petersoni	*Euneomys chinchilloides*
Abrothrix xanthorhina	*Abrothrix olivacea*
Meriones dahli	*Meriones meridianus*
Meriones chengi	*Meriones psammophilus*
Maxomys tompotika	*Maxomys hellwaldii*
Pseudomys fieldi	*Pseudomys gouldii*
Ctenomys goodfellowi	*Ctenomys boliviensis*
Ctenomys mariafarelli	*Ctenomys erikacuellarae*
Ctenomys azarae	*Ctenomys mendocinus*
Ctenomys porteousi	*Ctenomys mendocinus*
Ctenomys coyhaiquensis	*Ctenomys sericeus*
Otospermophilus atricapillus	*Otospermophilus beecheyi*
Dryomys yarkandensis	*Dryomys nitedula*
Sorex cristobalensis	*Sorex salvini*
Crocidura huangshanensis	*Crocidura dongyangjiangensis*
Talpa transcaucasica	*Talpa levantis*
Rhinolophus gorongosae	*Rhinolophus simulator*
Rhinolophus rhodesiae	*Rhinolophus simulator*
Rhinolophus cohenae	*Rhinolophus hildebrandtii*
Rhinolophus mabuensis	*Rhinolophus hildebrandtii*
Rhinolophus smithersi	*Rhinolophus hildebrandtii*
Rhinolophus kahuzi	*Rhinolophus ruwenzorii*
Nycteris madagascariensis	*Nycteris macrotis*
Chiroderma vizottoi	*Chiroderma doriae*
Plecotus ariel	*Plecotus wardi*
Scotophilus celebensis	*Scotophilus heathii*
Scotophilus borbonicus	*Scotophilus trujilloi*
Myotis melanorhinus	*Myotis ciliolabrum*
Myotis peninsularis	*Myotis velifer*
Myotis csorbai	*Myotis longipes*
Myotis annamiticus	*Myotis laniger*
Myotis badius	*Myotis alticraniatus*
Bos caucasicus	*Bos bonasus*
Syncerus brachyceros	*Syncerus caffer*
Syncerus mathewsi	*Syncerus caffer*
Syncerus nanus	*Syncerus caffer*
Tragelaphus bor	*Tragelaphus scriptus*
Tragelaphus decula	*Tragelaphus scriptus*
Tragelaphus phaleratus	*Tragelaphus scriptus*
Tragelaphus fasciatus	*Tragelaphus sylvaticus*
Tragelaphus meneliki	*Tragelaphus sylvaticus*
Tragelaphus ornatus	*Tragelaphus sylvaticus*
Tragelaphus gratus	*Tragelaphus spekii*
Tragelaphus larkenii	*Tragelaphus spekii*
Tragelaphus selousi	*Tragelaphus spekii*
Tragelaphus sylvestris	*Tragelaphus spekii*
Ammelaphus australis	*Tragelaphus imberbis*
Strepsiceros chora	*Tragelaphus strepsiceros*
Strepsiceros cottoni	*Tragelaphus strepsiceros*

Lumped species (Continued)

CMW Species Name	Lumped Into AMW Species
Strepsiceros zambesiensis	*Tragelaphus strepsiceros*
Neotragus kirchenpaueri	*Nesotragus moschatus*
Neotragus livingstonianus	*Nesotragus moschatus*
Aepyceros petersi	*Aepyceros melampus*
Raphicerus colonicus	*Raphicerus sharpei*
Antidorcas angolensis	*Antidorcas marsupialis*
Antidorcas hofmeyri	*Antidorcas marsupialis*
Litocranius sclateri	*Litocranius walleri*
Saiga mongolica	*Saiga tatarica*
Gazella karamii	*Gazella marica*
Gazella gracilicornis	*Gazella subgutturosa*
Gazella yarkandensis	*Gazella subgutturosa*
Gazella pelzelni	*Gazella dorcas*
Gazella saudiya	*Gazella dorcas*
Gazella acaciae	*Gazella arabica*
Gazella bilkis	*Gazella arabica*
Gazella cora	*Gazella arabica*
Gazella dareshurii	*Gazella arabica*
Gazella erlangeri	*Gazella arabica*
Gazella muscatensis	*Gazella arabica*
Gazella christii	*Gazella bennettii*
Gazella fuscifrons	*Gazella bennettii*
Gazella salinarum	*Gazella bennettii*
Gazella shikarii	*Gazella bennettii*
Eudorcas rufina	*Eudorcas rufifrons*
Eudorcas nasalis	*Eudorcas thomsonii*
Madoqua hararensis	*Madoqua saltiana*
Madoqua lawrancei	*Madoqua saltiana*
Madoqua phillipsi	*Madoqua saltiana*
Madoqua swaynei	*Madoqua saltiana*
Ourebia hastata	*Ourebia ourebi*
Ourebia montana	*Ourebia ourebi*
Ourebia quadriscopa	*Ourebia ourebi*
Redunca occidentalis	*Redunca arundinum*
Redunca bohor	*Redunca redunca*
Redunca cottoni	*Redunca redunca*
Redunca nigeriensis	*Redunca redunca*
Redunca adamauae	*Redunca fulvorufula*
Redunca chanleri	*Redunca fulvorufula*
Kobus anselli	*Kobus leche*
Kobus kafuensis	*Kobus leche*
Kobus smithemani	*Kobus leche*
Kobus leucotis	*Kobus kob*
Kobus loderi	*Kobus kob*
Kobus thomasi	*Kobus kob*
Kobus defassa	*Kobus ellipsiprymnus*
Hippotragus roosevelti	*Hippotragus niger*
Oryx callotis	*Oryx beisa*
Oryx gallarum	*Oryx beisa*
Alcelaphus caama	*Alcelaphus buselaphus*
Alcelaphus cokii	*Alcelaphus buselaphus*
Alcelaphus lelwel	*Alcelaphus buselaphus*
Alcelaphus lichtensteinii	*Alcelaphus buselaphus*
Alcelaphus major	*Alcelaphus buselaphus*
Alcelaphus swaynei	*Alcelaphus buselaphus*
Alcelaphus tora	*Alcelaphus buselaphus*
Damaliscus phillipsi	*Damaliscus pygargus*
Damaliscus eurus	*Damaliscus lunatus*
Damaliscus jimela	*Damaliscus lunatus*
Damaliscus korrigum	*Damaliscus lunatus*
Damaliscus selousi	*Damaliscus lunatus*
Damaliscus superstes	*Damaliscus lunatus*
Damaliscus tiang	*Damaliscus lunatus*

CMW Species Name	Lumped Into AMW Species
Damaliscus topi	*Damaliscus lunatus*
Damaliscus ugandae	*Damaliscus lunatus*
Connochaetes albojubatus	*Connochaetes taurinus*
Connochaetes johnstoni	*Connochaetes taurinus*
Connochaetes mearnsi	*Connochaetes taurinus*
Budorcas bedfordi	*Budorcas taxicolor*
Budorcas tibetana	*Budorcas taxicolor*
Budorcas whitei	*Budorcas taxicolor*
Pseudois schaeferi	*Pseudois nayaur*
Ovis isphahanica	*Ovis gmelini*
Ovis laristanica	*Ovis gmelini*
Ovis arabica	*Ovis vignei*
Ovis bochariensis	*Ovis vignei*
Ovis cycloceros	*Ovis vignei*
Ovis punjabiensis	*Ovis vignei*
Ovis collium	*Ovis ammon*
Ovis darwini	*Ovis ammon*
Ovis hodgsoni	*Ovis ammon*
Ovis jubata	*Ovis ammon*
Ovis karelini	*Ovis ammon*
Ovis nigrimontana	*Ovis ammon*
Ovis polii	*Ovis ammon*
Ovis severtzovi	*Ovis ammon*
Rupicapra ornata	*Rupicapra pyrenaica*
Rupicapra parva	*Rupicapra pyrenaica*
Rupicapra asiatica	*Rupicapra rupicapra*
Rupicapra carpatica	*Rupicapra rupicapra*
Nemorhaedus bedfordi	*Naemorhedus goral*
Capricornis maritimus	*Capricornis sumatraensis*
Capricornis milneedwardsi	*Capricornis sumatraensis*
Capricornis thar	*Capricornis sumatraensis*
Sylvicapra coronata	*Sylvicapra grimmia*
Sylvicapra pallidior	*Sylvicapra grimmia*
Philantomba aequatorialis	*Philantomba monticola*
Philantomba anchietae	*Philantomba monticola*
Philantomba bicolor	*Philantomba monticola*
Philantomba congica	*Philantomba monticola*
Philantomba defriesi	*Philantomba monticola*
Philantomba hecki	*Philantomba monticola*
Philantomba lugens	*Philantomba monticola*
Philantomba melanorhea	*Philantomba monticola*
Philantomba simpsoni	*Philantomba monticola*
Cephalophus castaneus	*Cephalophus dorsalis*
Cephalophus curticeps	*Cephalophus silvicultor*
Cephalophus crusalbum	*Cephalophus ogilbyi*
Cephalophus johnstoni	*Cephalophus weynsi*
Cephalophus lestradei	*Cephalophus weynsi*
Cephalophus arrhenii	*Cephalophus leucogaster*
Cephalophus harveyi	*Cephalophus natalensis*
Cephalophus fosteri	*Cephalophus nigrifrons*
Cephalophus hooki	*Cephalophus nigrifrons*
Cephalophus hypoxanthus	*Cephalophus nigrifrons*
Oreotragus aceratos	*Oreotragus oreotragus*
Oreotragus aureus	*Oreotragus oreotragus*
Oreotragus centralis	*Oreotragus oreotragus*
Oreotragus porteousi	*Oreotragus oreotragus*
Oreotragus saltatrixoides	*Oreotragus oreotragus*
Oreotragus schillingsi	*Oreotragus oreotragus*
Oreotragus somalicus	*Oreotragus oreotragus*
Oreotragus stevensoni	*Oreotragus oreotragus*
Oreotragus transvaalensis	*Oreotragus oreotragus*
Oreotragus tyleri	*Oreotragus oreotragus*
Nasuella meridensis	*Nasua olivacea*

ADDED SPECIES. There are two isolated examples of species added to AMW that were not included in the CMW and HMW series and have neither been split from any other species nor recently described: *Homo sapiens* and *Gerbillus burtoni*. The justification for omitting *Homo sapiens* from previous volumes was that humans could never be summarized or illustrated adequately within the confines of these books. However, this species is included here with a distribution map and basic measurements (but no illustrations) for the sake of completeness. *Gerbillus burtoni* was not included in either the HMW or CMW series—despite generally being recognized as a species in other works—and no reference was made therein to any synonymous species. However, this species is included here as a valid species pending further taxonomic revisions.

AMW Species Name
Homo sapiens
Gerbillus burtoni

MDD TAXONOMY NOT APPLIED. Although AMW is based on version 1.9 of the MDD, we decided not to include a handful of taxonomic arrangements from the MDD for various reasons. In the table below, 'NR' is added when a species is Not Recognized in one or either of MDD or AMW but is recognized in the other. MDD data are generally based on the most recent taxonomic arrangement for each taxon to minimize the degree of subjectivity in the curators' decision-making (with some exceptions when the most recent revision is not well supported). Thus, there are several arrangements recognized in the MDD that are considered premature and are not recognized here: *Marmosa limae* and *M. meridae* are not recognized and are lumped here under *M. demerarae* (Voss 2022, Voss *et al.* 2020), the lumping of *Sylvilagus cognatus* and *S. robustus* under *S. holzneri* (itself split from *S. floridanus*) is not accepted and these two species are recognized as distinct here while *S. holzneri* is not (Diersing & Wilson 2021), and the lumping of *Pithecia mittermeieri*, *P. pissinattii*, and *P. rylandsi* under *P. irroratus* is not accepted here and all these species are recognized here (Marsh 2014, Serrano-Villavicencio *et al.* 2019). The data used to support these taxonomic arrangements are presently insufficient to justify these changes given our interpretation of the data. There are also some taxa that are currently too ill-defined geographically or morphologically to be included, even though their specific status is likely to be recognized in the future. Examples of this include taxonomic changes within Neotropical *Sylvilagus* (all taxa illustrated but currently included under either *S. gabbi* or *S. brasiliensis*; Diersing & Wilson 2017, Ruedas 2017, Ruedas *et al.* 2017, 2019), the splits within *Napaeozapus* and *Zapus* (not recognized here; Malaney *et al.* 2017), and the recognition of the recently described *Calomys mattevii* (included under *C. expulsus* here pending further revision given its uncertain distribution and species boundaries; Gurgel-Filho *et al.* 2015, Gutiérrez & Marinho-Filho 2017, Quintela *et al.* 2020). To highlight its distinctiveness from the extinct form *Equus ferus* and despite the fact that most modern publications treat them conspecifically, *E. przewalskii* is included here in AMW despite its exclusion from the MDD. Lastly, recent generic revisions of the genus *Lasiurus* are not recognized here following the substantial pushback from the bat taxonomic community to this change, which is regarded as unnecessary given that the genus was already monophyletic (see *Major Controversies in Mammalian Systematics* below for further details); these changes have subsequently been removed from later versions of the MDD.

AMW Species Name	MDD Species Name
NR	*Marmosa limae*
NR	*Marmosa meridae*
NR	*Sylvilagus andinus*
NR	*Sylvilagus apollinaris*
NR	*Sylvilagus daulensis*
NR	*Sylvilagus fulvescens*
NR	*Sylvilagus holzneri*
NR	*Sylvilagus incitatus*
NR	*Sylvilagus nicefori*
NR	*Sylvilagus parentum*
NR	*Sylvilagus salentus*
NR	*Sylvilagus sanctaemartae*
NR	*Sylvilagus surdaster*
NR	*Sylvilagus tapetillus*
NR	*Napaeozapus abietorum*
NR	*Zapus luteus*
NR	*Zapus montanus*
NR	*Zapus oregonus*

AMW Species Name	MDD Species Name
NR	*Zapus pacificus*
NR	*Zapus saltator*
NR	*Calomys mattevii*
Pithecia mittermeieri	NR
Pithecia pissinattii	NR
Pithecia rylandsi	NR
Sylvilagus cognatus	NR
Sylvilagus robustus	NR
Equus przewalskii	NR
Lasiurus cinereus	*Aeorestes cinereus*
Lasiurus egregius	*Aeorestes egregius*
Lasiurus semotus	*Aeorestes semotus*
Lasiurus villosissimus	*Aeorestes villosissimus*
Lasiurus ega	*Dasypterus ega*
Lasiurus insularis	*Dasypterus insularis*
Lasiurus intermedius	*Dasypterus intermedius*
Lasiurus xanthinus	*Dasypterus xanthinus*

MAJOR CONTROVERSIES IN MAMMALIAN SYSTEMATICS

UNGULATE TAXONOMY. Ungulates represent a charismatic group of herbivorous hooved mammals belonging to the orders Perissodactyla and Artiodactyla. For a long time, their taxonomy was considered to be fairly stable, chiefly because so few taxonomists were actually working in this field given the difficulty in obtaining specimens for many of the species. However, in 2011, a compendium was published that reviewed the taxonomy of these two orders based on currently published literature and on qualitative morphological distinctions (Groves & Grubb 2011). Most suggested taxonomic changes were based entirely on morphological inferences and small sample sizes and more than doubled the number of recognized ungulate species at the time. Although most of the taxonomic changes in this work have been rejected by taxonomists studying these taxa (e.g. Holbrook 2013, Gutiérrez & Garbino 2018), a considerable number of arguments have been put forward in support of this taxonomy, including, in particular, its use for Bovidae in the HMW series (which carried on into the CMW series). The decision to use this taxonomy, based on the most recent revision of this family, in the HMW and CMW series was taken, above all, because Colin Groves was the lead author of the texts on the Bovidae. Groves and Grubb's publication contains what is essentially a fresh point of departure for ungulate taxonomy. Although these taxonomic changes are not followed here due to the lack of data supporting them and the controversies they have stirred up, we would like to acknowledge that Groves and Grubb have made a great leap forwards in the field of mammalian and, in particular, ungulate taxonomy. Their 2011 publication on ungulate taxonomy will ultimately help spur further integrative systematic research on a group that has been neglected for so long. Their book sparked a major debate regarding the use of the Phylogenetic Species Concept (despite its multiple definitions, a simplified summary would be "a species defined as diagnosable or diagnosable and monophyletic"), which was used to justify the taxonomic changes suggested in Groves & Grubb (2011). A number of publications either criticize their philosophy (e.g. Heller *et al.* 2013, Zachos 2013, 2015, 2016, 2018, Zachos & Lovari, 2013, Zachos *et al.* 2013, 2020) or support it (e.g. Gippoliti *et al.* 2013, 2018, Gippoliti 2019, 2020, Groves 2013, Groves *et al.* 2017, Gutiérrez & Garbino 2018, Taylor *et al.* 2019), many of which have serious implications on whether or not these taxonomic changes—i.e. the lumping or splitting of a species—have a negative effect on the conservation of these organisms. Despite the major ongoing debate stirred up within both the mammalogical community and biological systematics as a whole by this taxonomic arrangement, it is now generally considered to have been based on poorly constructed datasets and small sample sizes and so is not recognized here in line with most recent taxonomic authorities and the MDD. Given that the HMW and CMW series followed this taxonomic arrangement specifically for Bovidae, although not for any other families, 142 species of Bovidae were lumped between the CMW and AMW. Even though these proposed changes are not recognized, it should be noted that many ungulates are still in need of more integrative taxonomic revisions.

SPLITTING OF MONOTYPIC GENERA. Most generic revisions become necessary because an already defined genus is shown to be paraphyletic or polyphyletic compared to other genera. However, there have been a number of cases recently in which well-established monophyletic genera have been split into multiple genera based on various different arguments, namely: mammal genera should be considered distinct at specific divergence times (e.g. 4–7 million years, suggested by Groves 2001); there are consistent morphotypes between clades that warrant generic status; and sympatric or biogeographically isolated clades should constitute distinct genera (e.g. Alfaro *et al.* 2012, Byrne *et al.* 2016, Rylands *et al.* 2016). However, there is an alternative argument that this cladistic and biogeographic diversity should be separated into subgenera rather than genera, since splitting a single genus into multiple genera causes taxonomic instability by changing each species' scientific name (Teta 2019). This philosophy has been echoed in other discussions regarding the splitting of both Neotropical primates (Garbino 2015, Gutiérrez & Marinho-Filho 2017) and the bat genus *Lasiurus* (Ziegler *et al.* 2016, Novaes *et al.* 2018, Teta 2019, Francis *et al.* 2023). This debate has primarily been applied to Neotropical primates and involves the proposed splitting of *Sapajus* from *Cebus* (recognized here; Alfaro *et al.* 2012), *Leontocebus*, *Oedipomidas*, and *Tamarinus* from *Saguinus* (not recognized here; Rylands *et al.* 2016, Brcko *et al.* 2022), and *Cheracebus* and *Plecturocebus* from *Callicebus* (recognized here; Byrne *et al.* 2016); and in bats, the splitting of *Dermanura* and *Koopmania* from *Artibeus* (*Dermanura* but not *Koopmania*, recognized here; Hoofer *et al.* 2008) and *Aeorestes* and *Dasypterus* from *Lasiurus* (not recognized here; Baird *et al.* 2015, 2017, 2021). Although this debate has been settled for these genera, there is still inconsistency regarding when the splitting of a monophyletic genus is allowed or followed; this is the case of the splitting of *Neotamias* and *Eutamias* from *Tamias* (Patterson & Norris 2016), which is recognized here, unlike a number of the splits listed above, but still inconsistently accepted in the literature. This is an ongoing discussion within mammalian systematics and is one that is currently applied unevenly across the MDD and AMW, as some splits are recognized based on their acceptance in the literature, while others are rejected based on their rejection in the literature.

Connor J. Burgin
PhD Candidate, Department of Biology,
University of New Mexico, Albuquerque, New Mexico, USA

References

Alfaro, J.W.L., Silva-Júnior, J.S. & Rylands, A.B. (2012). How different are robust and gracile capuchin monkeys? An argument for the use of *Sapajus* and *Cebus*. *Am. J. Primatol.* **74(4)**: 273–286.

Baird, A.B., Braun, J.K., Engstrom, M.D., Holbert, A.C., Huerta, M.G., Lim, B.K., Mares, M.A., Patton, J.C. & Bickham, J.W. (2017). Nuclear and mtDNA phylogenetic analyses clarify the evolutionary history of two species of native Hawaiian bats and the taxonomy of Lasiurini (Mammalia: Chiroptera). *PLoS ONE* **12(10)**: e0186085.

Baird, A.B., Braun, J.K., Engstrom, M.D., Lim, B.K., Mares, M.A., Patton, J.C. & Bickham, J.W. (2021). On the utility of taxonomy to reflect biodiversity: the example of Lasiurini (Chiroptera: Vespertilionidae). *Therya* **12(2)**: 283–289.

Baird, A.B., Braun, J.K., Mares, M.A., Morales, J.C., Patton, J.C., Tran, C.Q. & Bickham, J.W. (2015). Molecular systematic revision of tree bats (Lasiurini): doubling the native mammals of the Hawaiian Islands. *J. Mammal.* **96(6)**: 1255–1274.

Brace, S., Turvey, S.T., Weksler, M., Hoogland, M.L.P. & Barnes, I. (2015). Unexpected evolutionary diversity in a recently extinct Caribbean mammal radiation. *Proc. Roy. Soc. London B, Biol. Sci.* **282(1807)**: 20142371.

Brace, S., Turvey, S.T., Weksler, M., Hoogland, M.L.P. & Barnes, I. (2023). Establishing the availability of the mammalian genus name *Antillomys* and species name *Antillomys rayi* (Rodentia, Cricetidae). *Therya* **14(2)**. In Press.

Brcko, I.C., Carneiro, J., Ruiz-García, M., Boubli, J.P., Silva-Júnior, J.S., Farias, I., Hrbek, T., Schneider, H. & Sampaio, I. (2022). Phylogenetics and an updated taxonomic status of the Tamarins (Callitrichinae, Cebidae). *Mol. Phylogenet. Evol.* **17**: 107504.

Burgin, C.J., Colella, J.P., Kahn, P.L. & Upham, N.S. (2018). How many species of mammals are there? *J. Mammal.* **99(1)**: 1–14.

Byrne, H., Rylands, A.B., Carneiro, J.C., Alfaro, J.W.L., Bertuol, F., da Silva, M.N.F., Messias, M., Groves, C.P., Mittermeier, R.A., Farias, I., Hrbek, T., Schneider, H., Sampaio, I. & Boubli, J.P. (2016). Phylogenetic relationships of the New World titi monkeys (*Callicebus*): first appraisal of taxonomy based on molecular evidence. *Front. Zool.* **13**: 10.

Carroll, E.L., McGowen, M.R., McCarthy, M.L., Marx, F.G., Aguilar, N., Dalebout, M.L., Dreyer, S., Gaggiotti, O.E., Hansen, S.S., van Helden, A., Onoufriou, A.B., Baird, R.W., Baker, C.S., Berrow, S., Cholewiak, D., Claridge, D., Constantine, R., Davison, N.J., Eira, C., Fordyce, R.E., Gatesy, J., Hofmeyr, G.J.G., Martín, V., Mead, J.G., Mignucci-Giannoni, A.A., Morin, P.A., Reyes, C., Rogan, E., Rosso, M., Silva, M.A., Springer, M.S., Steel, D. & Olsen, M.T. (2021). Speciation in the deep: genomics and morphology reveal a new species of beaked whale *Mesoplodon eueu*. *Proc. Roy. Soc. London B, Biol. Sci.* **288(1961)**: 20211213.

Chen, Z.Z., He, S.W., Hu, W.H., Song, W.Y., Onditi, K.O., Li, X.Y. & Jiang, X.L. (2021). Morphology and phylogeny of scalopine moles (Eulipotyphla: Talpidae: Scalopini) from the eastern Himalayas, with descriptions of a new genus and species. *Zool. J. Linn. Soc.* **193(2)**: 432–444.

Coimbra, R.T.F., Winter, S., Kumar, V., Koepfli, K.P., Gooley, R.M., Dobrynin, P., Fennessy, J. & Janke, A. (2021). Whole-genome analysis of giraffe supports four distinct species. *Curr. Biol.* **31(13)**: 2929–2938.

Costa-Araújo, R., Silva-Júnior, J.S., Boubli, J.P., Rossi, R.V., Canale, G.R., Melo, F.R., Bertuol, F., Silva, F.E., Silva, D.A., Nash, S.D., Sampaio, I., Farias, I. & Hrbek, T. (2021). An integrative analysis uncovers a new, pseudo-cryptic species of Amazonian marmoset (Primates: Callitrichidae: *Mico*) from the arc of deforestation. *Sci. Rep.* **11**: 15665.

Diersing, V.E. & Wilson, D.E. (2017). Systematic status of the rabbits *Sylvilagus brasiliensis* and *S. sanctaemartae* from northwestern South America with comparisons to Central American populations. *J. Mammal.* **98(6)**: 1641–1656.

Diersing, V.E. & Wilson, D.E. (2021). Systematics of the mountain-inhabiting cottontails (*Sylvilagus*) from southwestern United States and northern Mexico (Mammalia: Lagomorpha: Leporidae). *Proc. Biol. Soc. Washington* **134(1)**: 42–79.

Francis, C.M., Simmons, N.B., Cakenberghe, V.V., Upham, N.S. & Burgin, C.J., on behalf of the Global Bat Taxonomy Working Group of the IUCN SSC Bat Specialist Group. (2023). On the taxonomy of *Lasiurus*. *Zenodo*: 1–11. DOI: https://doi.org/10.5281/zenodo.7696845

Garbino, G.S.T. (2015). Defining genera of New World monkeys: the need for a critical view in a necessarily arbitrary task. *Int. J. Primatol.* **36(6)**: 1049–1064.

Gippoliti, S. (2019). Species delimitation in mammals: A comment on Zachos (2018). *Mammal. Biol.* **94**: 127–131.

Gippoliti, S. (2020). Everything mammal conservation biologists always wanted to know about taxonomy (but were afraid to ask). *J. Nat. Conserv.* **54**: 125793.

Gippoliti, S., Cotterill, F.P.D. & Groves, C.P. (2013). Mammal taxonomy without taxonomists: a reply to Zachos and Lovari. *Hystrix, Ital. J. Mammal.* **24(2)**: 145–147.

Gippoliti, S., Cotterill, F.P.D., Zinner, D. & Groves, C.P. (2018). Impacts of taxonomic inertia for the conservation of African ungulate diversity: an overview. *Biol. Rev.* **93(1)**: 115–130.

Gregorin, R. & Cirranello, A. (2016). Phylogeny of Molossidae Gervais (Mammalia: Chiroptera) inferred by morphological data. *Cladistics* **32(1)**: 2–35.

Groves, C.P. (2001). *Primate Taxonomy*. Smithsonian Institution Press, Washington, D.C.

Groves, C.P. (2013). The nature of species: A rejoinder to Zachos et al. *Mammal. Biol.* **78(1)**: 7–9.

Groves, C.P. & Grubb, P. (2011). *Ungulate Taxonomy*. The Johns Hopkins University Press, Baltimore, Maryland.

Groves, C.P., Cotterill, F.P.D., Gippoliti, S., Robovský, J., Roos, C., Taylor, P.J. & Zinner, D. (2017). Species definitions and conservation: a review and case studies from African mammals. *Conserv. Genet.* **18(6)**: 1247–1256.

Gurgel-Filho, N.M., Feijó, A. & Langguth, A. (2015). Pequenos mamíferos do Ceará (marsupiais, morcegos e roedores sigmodontíneos) com discussão taxonômica de algumas espécies. *Rev. Nordest. Biol.* **23(2)**: 3–150.

Gutiérrez, E.E. & Garbino, G.S.T. (2018). Species delimitation based on diagnosis and monophyly, and its importance for advancing mammalian taxonomy. *Zool. Res.* **39(5)**: 301–308.

Gutiérrez, E.E. & Marinho-Filho, J. (2017). The mammalian faunas endemic to the Cerrado and the Caatinga. *ZooKeys* **644**: 105–157.

Heller, R., Frandsen, P., Lorenzen, E.D. & Siegismund, H.R. (2013). Are there really twice as many bovid species as we thought? *Syst. Biol.* **62(3)**: 490–493.

Heritage, S., Rayaleh, H., Awaleh, D.G. & Rathbun, G.B. (2020). New records of a lost species and a geographic range expansion for sengis in the Horn of Africa. *PeerJ* **8**: e9652.

Holbrook, L.T. (2013). Taxonomy Interrupted: UNGULATE TAXONOMY. By Colin Groves and Peter Grubb. Baltimore: The Johns Hopkins University Press. 2011. 317 pp., $100.00 (cloth). ISBN 978-1-4214-0093-8. *J. Mammal. Evol.* **20(2)**: 153–154.

Hoofer, S.R., Solari, S., Larsen, P.A., Bradley, R.D. & Baker, R.J. (2008). Phylogenetics of the fruit-eating bats (Phyllostomidae: Artibeina) inferred from mitochondrial DNA sequences. *Occ. Pap. Mus. Texas Tech. Univ.* **277**: 1–15.

ICZN (International Commission on Zoological Nomenclature). (2021). Notice of new applications to the Commission (Case 3858–3862). *Bull. Zool. Nomen.* **78**: 166.

Kilpatrick, C.W., Pradhan, N. & Norris, R.W. (2021). A re-examination of the molecular systematics and phylogeography of taxa of the *Peromyscus aztecus* species group, with comments on the distribution of *P. winkelmanni*. *Therya* **12(2)**: 331–346.

Krásová, J., Mikula, O., Šumbera, R., Horáková, S., Robovský, J., Kostin, D.S., Martynov, A.A., Lavrenchenko, L.A. & Bryja, J. (2021).

The Rufous Sengi is not *Elephantulus*—Multilocus reconstruction of evolutionary history of sengis from the subfamily Macroscelidinae. *J. Zool. Syst. Evol. Res.* **59(4)**: 918–932.

Kryštufek, B., Shenbrot, G. & Janžekovič, F. (2021). Long-standing taxonomic confusion over the identity of *Hypudaeus syriacus* Brants, 1827, at last resolved. *Mammalia* **85(6)**: 603–611.

Kryštufek, B., Tesakov, A.S., Lebedev, V.S., Bannikova, A.A., Abramson, N.I. & Shenbrot, G. (2020). Back to the future: the proper name for red-backed voles is *Clethrionomys* Tilesius and not *Myodes* Pallas. *Mammalia* **84(2)**: 214–217.

Lebedev, V.S., Kovalskaya, Y., Solovyeva, E.N., Zemlemerova, E.D., Bannikova, A.A., Rusin, M.Yu. & Matrosova, V.A. (2021). Molecular systematics of the *Sicista tianschanica* species complex: a contribution from historical DNA analysis. *PeerJ* **9**: e10759.

Li, Q., Cheng, F., Jackson, S.M., Helgen, K.M., Song, W.Y., Liu, S.Y., Sanamxay, D., Li, S., Li, F., Xiong, Y., Sun, J., Wang, H.J. & Jiang, X.L. (2021). Phylogenetic and morphological significance of an overlooked flying squirrel (Pteromyini, Rodentia) from the eastern Himalayas with the description of a new genus. *Zool. Res.* **42(4)**: 389–400.

Lucero, S., Rodríguez, S.M., Teta, P., Cassini, G. & D'Elía, G. (2019). Solving a long-standing nomenclatorial controversy: designation of a neotype for the southern sea lion *Otaria flavescens* (Shaw, 1800). *Zootaxa* **4555(2)**: 296–300.

Malaney, J.L., Demboski, J.R. & Cook, J.A. (2017). Integrative species delimitation of the widespread North American jumping mice (Zapodinae). *Mol. Phylogenet. Evol.* **114**: 137–152.

Marsh, L.K. (2014). A taxonomic revision of the saki monkeys, *Pithecia* Desmarest, 1804. *Neotrop. Primates* **21(1)**: 1–165.

McDonough, M.M., Ferguson, A.W., Dowler, R.C., Gompper, M.E. & Maldonado, J.E. (2022). Phylogenomic systematics of the spotted skunks (Carnivora, Mephitidae, *Spilogale*): Additional species diversity and Pleistocene climate change as a major driver of diversification. *Mol. Phylogenet. Evol.* **167**: 107266.

Nicolas, V., Mikula, O., Lavrenchenko, L.A., Šumbera, R., Bartáková, V., Bryjová, A., Meheretu, Y., Verheyen, E., Missoup, A.D., Lemmon, A.R., Lemmon, E.M. & Bryja, J. (2021). Phylogenomics of African radiation of Praomyini (Muridae: Murinae) rodents: First fully resolved phylogeny, evolutionary history and delimitation of extant genera. *Mol. Phylogenet. Evol.* **163**: 107263.

Novaes, R.L.M., Garbino, G.S.T., Cláudio, V.C. & Moratelli, R. (2018). Separation of monophyletic groups into distinct genera should consider phenotypic discontinuities: the case of Lasiurini (Chiroptera: Vespertilionidae). *Zootaxa* **4379(3)**: 439–440.

Patterson, B.D. & Norris, R.W. (2016). Towards a uniform nomenclature for ground squirrels: the status of the Holarctic chipmunks. *Mammalia* **80(3)**: 241–251.

Patterson, B.D., Ramírez, H.E., Vilela, J.F., Soares, A.E.R. & Grewe, F. (2021). On the nomenclature of the American clade of weasels (Carnivora: Mustelidae). *J. Anim. Divers.* **3(2)**: 1–8.

Quintela, F.M., da Rosa, C.A. & Feijó, A. (2020). Updated and annotated checklist of recent mammals from Brazil. *An. Acad. Bras. Ciênc.* **92(Suppl. 2)**: e20191004.

Roos, C., Helgen, K.M., Portela, R., Thant, N.M.L., Lwin, N., Lin, A.K., Lin, A., Yi, K.M., Soe, P., Hein, Z.M., Myint, M.N.N., Ahmed, T., Chetry, D., Urh, M., Veatch, E.G., Duncan, N., Kamminga, P., Chua, M.A.H., Yao, L., Matauschek, C., Meyer, D., Liu, Z.J., Li, M., Nadler, T., Fan, P.F., Quyet, L.K., Hofreiter, M., Zinner, D. & Momberg, F. (2020). Mitogenomic phylogeny of the Asian colobine genus *Trachypithecus* with special focus on *Trachypithecus phayrei* (Blyth, 1847) and description of a new species. *Zool. Res.* **41(6)**: 656–669.

Rosel, P.E., Wilcox, L.A., Yamada, T.K. & Mullin, K.D. (2021). A new species of baleen whale (*Balaenoptera*) from the Gulf of Mexico, with a review of its geographic distribution. *Mar. Mamm. Sci.* **37(2)**: 577–610.

Roycroft, E., MacDonald, A.J., Moritz, C., Moussalli, A., Portela, R. & Rowe, K.C. (2021). Museum genomics reveals the rapid decline and extinction of Australian rodents since European settlement. *Proc. Natl. Acad. Sci. USA* **118(27)**: e2021390118.

Ruedas, L.A. (2017). A new species of cottontail rabbit (Lagomorpha: Leporidae: *Sylvilagus*) from Suriname, with comments on the taxonomy of allied taxa from northern South America. *J. Mammal.* **98(4)**: 1042–1059.

Ruedas, L.A., Silva, S.M., French, J.H., Platt II, R.N., Salazar-Bravo, J., Mora, J.M. & Thompson, C.W. (2017). *A Prolegomenon to the Systematics of South American Cottontail Rabbits (Mammalia, Lagomorpha, Leporidae:* Sylvilagus*): Designation of a Neotype for* S. brasiliensis *(Linnaeus, 1758), and Restoration of* S. andinus *(Thomas, 1897) and* S. tapetillus *Thomas, 1913.* Miscellaneous Publications **205**. Museum of Zoology, University of Michigan, Ann Arbor, Michigan.

Ruedas, L.A., Silva, S.M., French, J.H., Platt II, R.N., Salazar-Bravo, J., Mora, J.M. & Thompson, C.W. (2019). Taxonomy of the *Sylvilagus brasiliensis* complex in Central and South America (Lagomorpha: Leporidae). *J. Mammal.* **100(5)**: 1599–1630.

Rylands, A.B., Heymann, E.W., Alfaro, J.W.L., Buckner, J.C., Roos, C., Matauschek, C., Boubli, J.P., Sampaio, R. & Mittermeier, R.A. (2016). Taxonomic review of the New World tamarins (Primates: Callitrichidae). *Zool. J. Linn. Soc.* **177(4)**: 1003–1028.

Serrano-Villavicencio, J.E., Hurtado, C.M., Vendramel, R.L. & Nascimento, F.O. (2019). Reconsidering the taxonomy of the *Pithecia irrorata* species group (Primates: Pitheciidae). *J. Mammal.* **100(1)**: 130–141.

Taylor, P.J., Denys, C. & Cotterill, F.P.D. (2019). Taxonomic anarchy or an inconvenient truth for conservation? Accelerated species discovery reveals evolutionary patterns and heightened extinction threat in Afro-Malagasy small mammals. *Mammalia* **83(4)**: 313–329.

Teta, P. (2019). The usage of subgenera in mammalian taxonomy. *Mammalia* **83(3)**: 209–211.

Travouillon, K.J. & Phillips, M.J. (2018). Total evidence analysis of the phylogenetic relationships of bandicoots and bilbies (Marsupialia: Peramelemorphia): reassessment of two species and description of a new species. *Zootaxa* **4378(2)**: 224–256.

Voss, R.S. (2022). An annotated checklist of Recent opossums (Mammalia: Didelphidae). *Bull. Am. Mus. Nat. Hist.* **455(1)**: 1–76.

Voss, R.S., Giarla, T.C., Díaz-Nieto, J.F. & Jansa, S.A. (2020). A revision of the didelphid marsupial genus *Marmosa*. Part 2. Species of the *Rapposa* Group (Subgenus *Micoureus*). *Bull. Am. Mus. Nat. Hist.* **439(1)**: 1–62.

Zachos, F.E. (2013). Species splitting puts conservation at risk. *Nature* **494(7435)**: 35.

Zachos, F.E. (2015). Taxonomic inflation, the Phylogenetic Species Concept and lineages in the Tree of Life – a cautionary comment on species splitting. *J. Zool. Syst. Evol. Res.* **53(2)**: 180–184.

Zachos, F.E. (2016). *Species Concepts in Biology: Historical Development, Theoretical Foundations and Practical Relevance*. Springer International Publishing, Cham, Switzerland.

Zachos, F.E. (2018). Mammals and meaningful taxonomic units: the debate about species concepts and conservation. *Mammal Rev.* **48(3)**: 153–159.

Zachos, F.E. & Lovari, S. (2013). Taxonomic inflation and the poverty of the Phylogenetic Species Concept – a reply to Gippoliti and Groves. *Hystrix, Ital. J. Mammal.* **24(2)**: 142–144.

Zachos, F.E., Apollonio, M., Bärmann, E.V., Festa-Bianchet, M., Göhlich, U., Habel, J.C., Haring, E., Kruckenhauser, L., Lovari, S., McDevitt, A.D., Pertoldi, C., Rössner, G.E., Sánchez-Villagra, M.R., Scandura, M. & Suchentrunk, F. (2013). Species inflation and taxonomic artefacts—A critical comment on recent trends in mammalian classification. *Mammal. Biol.* **78(1)**: 1–6.

Zachos, F.E., Christidis, L. & Garnett, S.T. (2020). Mammalian species and the twofold nature of taxonomy: a comment on Taylor et al. 2019. *Mammalia*, **84(1)**: 1–5.

Ziegler, A.C., Howarth, F.G. & Simmons, N.B. (2016). A second endemic land mammal for the Hawaiian Islands: a new genus and species of fossil bat (Chiroptera: Vespertilionidae). *Am. Mus. Novit.* **3854**: 1–52.

Appendix 4

COUNTRY CODES

This resource is provided as a convenient guide to the internationally recognized two-letter codes ("ISO 3166-1 alpha-2") published by the International Organization for Standardization (ISO) that represent countries, dependent territories, and special areas of geographical interest. The names of countries and their subdivisions employed by ISO are those used by the United Nations for daily operations and are not necessarily their official names. The assigned designations do not imply the expression of any opinion on the part of the United Nations, ISO, or the present book concerning the legal status of any country, territory, or area, of its authorities, or regarding the delimitation of its boundaries or frontiers. Note that in the ISO system some islands and archipelagos have their own, individual codes (e.g. Norfolk I = NF), while others are assigned the code of the country to which they are considered to belong administratively (e.g. Lord Howe I = AU).

The list is first presented in alphabetical order of the ISO codes, each with the corresponding country/territory/area name following on the same line. Footnotes with further information are included for clarification in relevant cases. The same list is then presented in alphabetical order of country/territory/area names, each followed by the corresponding ISO code on the same line.

A large selection of these countries and codes can also be found in the collection of 34 full-page maps provided in Appendix 5: Reference maps (see p. 729).

Additional abbreviations used in this book for states and territories of Australia are provided in the Introduction (see p. 26).

Country codes: list in alphabetical order by code

Code	Country name
AD	Andorra
AE	United Arab Emirates
AF	Afghanistan
AG	Antigua and Barbuda
AI	Anguilla
AL	Albania
AM	Armenia
AO	Angola
AQ	Antarctica[1]
AR	Argentina
AS	American Samoa
AT	Austria
AU	Australia
AW	Aruba
AZ	Azerbaijan
BA	Bosnia and Herzegovina
BB	Barbados
BD	Bangladesh
BE	Belgium
BF	Burkina Faso
BG	Bulgaria
BH	Bahrain
BI	Burundi
BJ	Benin
BL	Saint Barthélemy
BM	Bermuda
BN	Brunei[2]
BO	Bolivia[3]
BQ	Bonaire, Sint Eustatius and Saba[4]
BR	Brazil
BS	Bahamas
BT	Bhutan
BW	Botswana
BY	Belarus
BZ	Belize
CA	Canada
CC	Cocos (Keeling) Islands
CD	Congo, Democratic Republic of the
CF	Central African Republic
CG	Congo
CH	Switzerland[5]
CI	Côte d'Ivoire[6]
CK	Cook Islands
CL	Chile
CM	Cameroon
CN	China
CO	Colombia
CR	Costa Rica
CU	Cuba
CV	Cabo Verde[7]
CW	Curaçao
CX	Christmas Island
CY	Cyprus
CZ	Czechia
DE	Germany
DJ	Djibouti
DK	Denmark
DM	Dominica
DO	Dominican Republic
DZ	Algeria[8]
EC	Ecuador
EE	Estonia
EG	Egypt
EH	Western Sahara
ER	Eritrea
ES	Spain

Code	Country name
ET	Ethiopia
FI	Finland
FJ	Fiji
FK	Falkland Islands (Malvinas)[9]
FM	Micronesia (Federated States of)
FO	Faroe Islands
FR	France
GA	Gabon
GB	United Kingdom[10]
GD	Grenada
GE	Georgia
GF	French Guiana
GH	Ghana
GI	Gibraltar
GL	Greenland
GM	Gambia
GN	Guinea
GP	Guadeloupe
GQ	Equatorial Guinea
GR	Greece
GS	South Georgia and the South Sandwich Islands
GT	Guatemala
GU	Guam
GW	Guinea-Bissau
GY	Guyana
HK	Hong Kong
HM	Heard Island and McDonald Islands
HN	Honduras
HR	Croatia
HT	Haiti
HU	Hungary
ID	Indonesia
IE	Ireland
IL	Israel
IN	India
IO	British Indian Ocean Territory
IQ	Iraq
IR	Iran[11]
IS	Iceland
IT	Italy
JM	Jamaica
JO	Jordan
JP	Japan
KE	Kenya
KG	Kyrgyzstan
KH	Cambodia
KI	Kiribati
KM	Comoros
KN	Saint Kitts and Nevis
KP	North Korea[12]
KR	South Korea[13]
KW	Kuwait
KY	Cayman Islands
KZ	Kazakhstan
LA	Laos[14]
LB	Lebanon
LC	Saint Lucia
LI	Liechtenstein
LK	Sri Lanka
LR	Liberia
LS	Lesotho
LT	Lithuania
LU	Luxembourg
LV	Latvia
LY	Libya
MA	Morocco

Code	Country name
MC	Monaco
MD	Moldova[15]
ME	Montenegro
MF	Saint Martin (French part)
MG	Madagascar
MH	Marshall Islands
MK	North Macedonia
ML	Mali
MM	Myanmar
MN	Mongolia
MO	Macao
MP	Northern Mariana Islands
MQ	Martinique
MR	Mauritania
MS	Montserrat
MT	Malta
MU	Mauritius
MV	Maldives
MW	Malawi
MX	Mexico
MY	Malaysia
MZ	Mozambique
NA	Namibia
NC	New Caledonia
NE	Niger
NF	Norfolk Island
NG	Nigeria
NI	Nicaragua
NL	Netherlands
NO	Norway
NP	Nepal
NR	Nauru
NU	Niue
NZ	New Zealand
OM	Oman
PA	Panama
PE	Peru
PF	French Polynesia
PG	Papua New Guinea
PH	Philippines
PK	Pakistan
PL	Poland
PM	Saint Pierre and Miquelon
PN	Pitcairn
PR	Puerto Rico
PS	Palestine[16]
PT	Portugal
PW	Palau
PY	Paraguay
QA	Qatar
RE	Réunion
RO	Romania
RS	Serbia
RU	Russia[17]
RW	Rwanda
SA	Saudi Arabia
SB	Solomon Islands
SC	Seychelles
SD	Sudan
SE	Sweden
SG	Singapore
SH	Saint Helena, Ascension and Tristan da Cunha
SI	Slovenia
SJ	Svalbard and Jan Mayen
SK	Slovakia
SL	Sierra Leone

Code	Country name
SM	San Marino
SN	Senegal
SO	Somalia
SR	Suriname
SS	South Sudan
ST	São Tomé and Príncipe
SV	El Salvador
SX	Sint Maarten (Dutch part)
SY	Syria[18]
SZ	Eswatini[19]
TC	Turks and Caicos Islands
TD	Chad
TF	French Southern Territories[20]
TG	Togo
TH	Thailand
TJ	Tajikistan

Code	Country name
TK	Tokelau
TL	Timor-Leste
TM	Turkmenistan
TN	Tunisia
TO	Tonga
TR	Turkey
TT	Trinidad and Tobago
TV	Tuvalu
TW	Taiwan[21]
TZ	Tanzania[22]
UA	Ukraine
UG	Uganda
UM	United States Minor Outlying Islands[23]
US	United States of America
UY	Uruguay
UZ	Uzbekistan

Code	Country name
VA	Vatican City[24]
VC	Saint Vincent and the Grenadines
VE	Venezuela[25]
VG	Virgin Islands (British)
VI	Virgin Islands (U.S.)
VN	Vietnam
VU	Vanuatu
WF	Wallis and Futuna
WS	Samoa
XK	Kosovo
YE	Yemen
YT	Mayotte
ZA	South Africa[26]
ZM	Zambia
ZW	Zimbabwe

1 Covers the territories south of 60° south latitude.
2 UN designated country name: Brunei Darussalam.
3 UN designated country name: Bolivia (Plurinational State of).
4 Consists of three Caribbean "special municipalities", which are part of the Netherlands proper.
5 Code taken from name in Latin: Confoederatio Helvetica.
6 Country name follows UN designation (common name: Ivory Coast).
7 Country name follows UN designation (common name and previous ISO country name: Cape Verde).
8 Code taken from name in Arabic الجزائر al-Djazā'ir.
9 Country name follows UN designation due to the Falkland Islands sovereignty dispute.
10 UN designated country name: United Kingdom of Great Britain and Northern Ireland.
11 UN designated country name: Iran (Islamic Republic of).
12 UN designated country name: Korea (Democratic People's Republic of).
13 UN designated country name: Korea, Republic of.
14 UN designated country name: Lao People's Democratic Republic.
15 UN designated country name: Moldova, Republic of.
16 UN designated country name: Palestine, State of.
17 UN designated country name: Russian Federation.
18 UN designated country name: Syrian Arab Republic.
19 Previous ISO country name: Swaziland.
20 Covers the French Southern and Antarctic Lands except Adélie Land.
21 UN designated country name (due to political status of Taiwan within the UN): Taiwan, Province of China.
22 UN designated country name: Tanzania, United Republic of.
23 Consists of nine minor insular areas of the United States: Baker I, Howland I, Jarvis I, Johnston Atoll, Kingman Reef, Midway Is, Navassa I, Palmyra Atoll and Wake I.
24 UN designated country name: Holy See.
25 UN designated country name: Venezuela (Bolivarian Republic of).
26 Code taken from name in Dutch: Zuid-Afrika.

Country codes: list in alphabetical order by country name

Country name	Code
Afghanistan	**AF**
Albania	**AL**
Algeria	**DZ**
American Samoa	**AS**
Andorra	**AD**
Angola	**AO**
Anguilla	**AI**
Antarctica	**AQ**
Antigua and Barbuda	**AG**
Argentina	**AR**
Armenia	**AM**
Aruba	**AW**
Australia	**AU**
Austria	**AT**
Azerbaijan	**AZ**
Bahamas	**BS**
Bahrain	**BH**
Bangladesh	**BD**
Barbados	**BB**
Belarus	**BY**
Belgium	**BE**
Belize	**BZ**
Benin	**BJ**
Bermuda	**BM**
Bhutan	**BT**
Bolivia	**BO**
Bonaire, Sint Eustatius and Saba	**BQ**
Bosnia and Herzegovina	**BA**
Botswana	**BW**
Brazil	**BR**
British Indian Ocean Territory	**IO**

Country name	Code
Brunei	**BN**
Bulgaria	**BG**
Burkina Faso	**BF**
Burundi	**BI**
Cabo Verde	**CV**
Cambodia	**KH**
Cameroon	**CM**
Canada	**CA**
Cayman Islands	**KY**
Central African Republic	**CF**
Chad	**TD**
Chile	**CL**
China	**CN**
Christmas Island	**CX**
Cocos (Keeling) Islands	**CC**
Colombia	**CO**
Comoros	**KM**
Congo	**CG**
Congo, Democratic Republic of the	**CD**
Cook Islands	**CK**
Costa Rica	**CR**
Côte d'Ivoire	**CI**
Croatia	**HR**
Cuba	**CU**
Curaçao	**CW**
Cyprus	**CY**
Czechia	**CZ**
Denmark	**DK**
Djibouti	**DJ**
Dominica	**DM**
Dominican Republic	**DO**

Country name	Code
Ecuador	**EC**
Egypt	**EG**
El Salvador	**SV**
Equatorial Guinea	**GQ**
Eritrea	**ER**
Estonia	**EE**
Eswatini	**SZ**
Ethiopia	**ET**
Falkland Islands (Malvinas)	**FK**
Faroe Islands	**FO**
Fiji	**FJ**
Finland	**FI**
France	**FR**
French Guiana	**GF**
French Polynesia	**PF**
French Southern Territories	**TF**
Gabon	**GA**
Gambia	**GM**
Georgia	**GE**
Germany	**DE**
Ghana	**GH**
Gibraltar	**GI**
Greece	**GR**
Greenland	**GL**
Grenada	**GD**
Guadeloupe	**GP**
Guam	**GU**
Guatemala	**GT**
Guinea	**GN**
Guinea-Bissau	**GW**
Guyana	**GY**

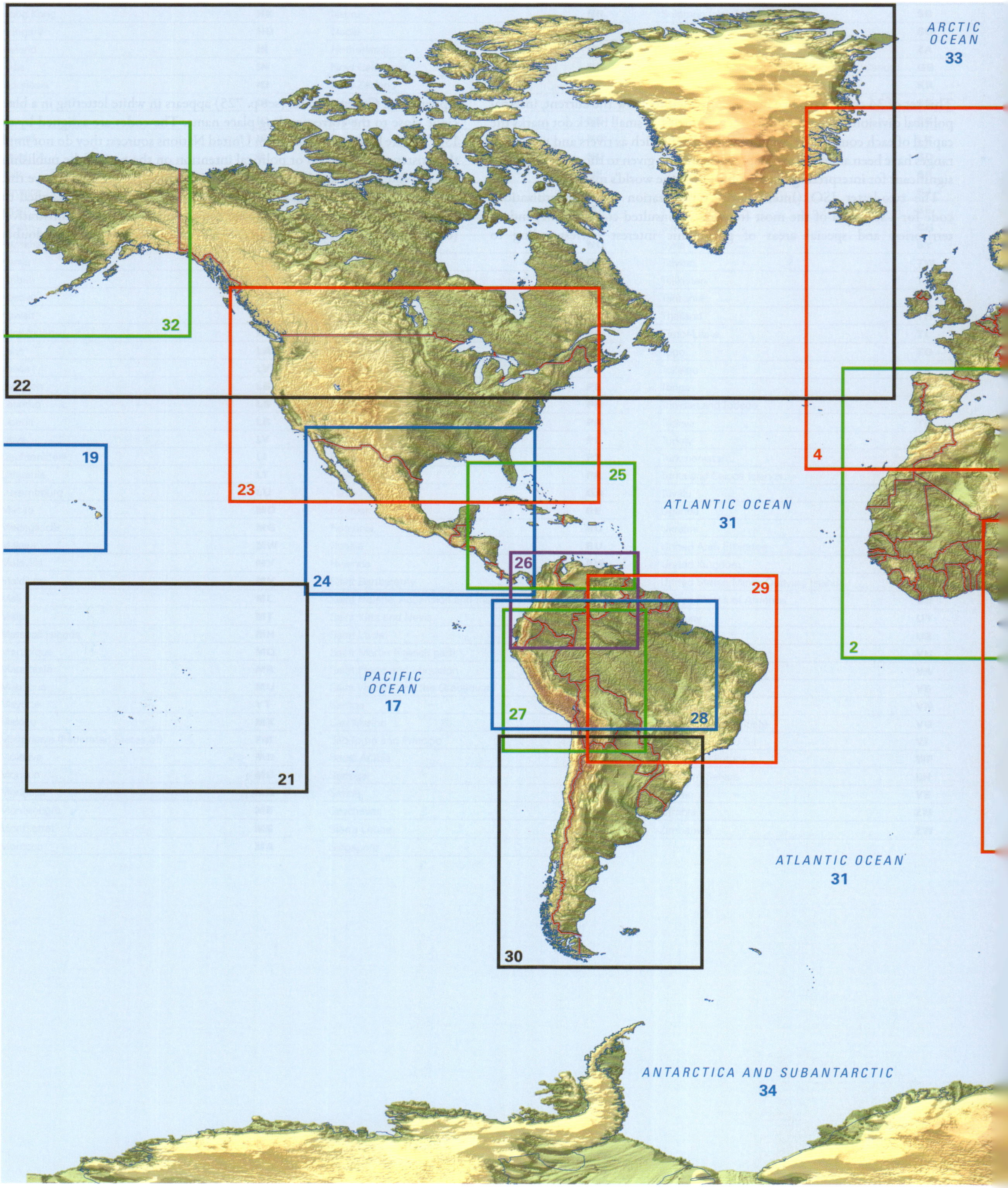
ARCTIC OCEAN
33
22
32
23
19
24
25
4
ATLANTIC OCEAN
31
26
29
21
PACIFIC OCEAN
17
27
28
2
30
ATLANTIC OCEAN
31
ANTARCTICA AND SUBANTARCTIC
34

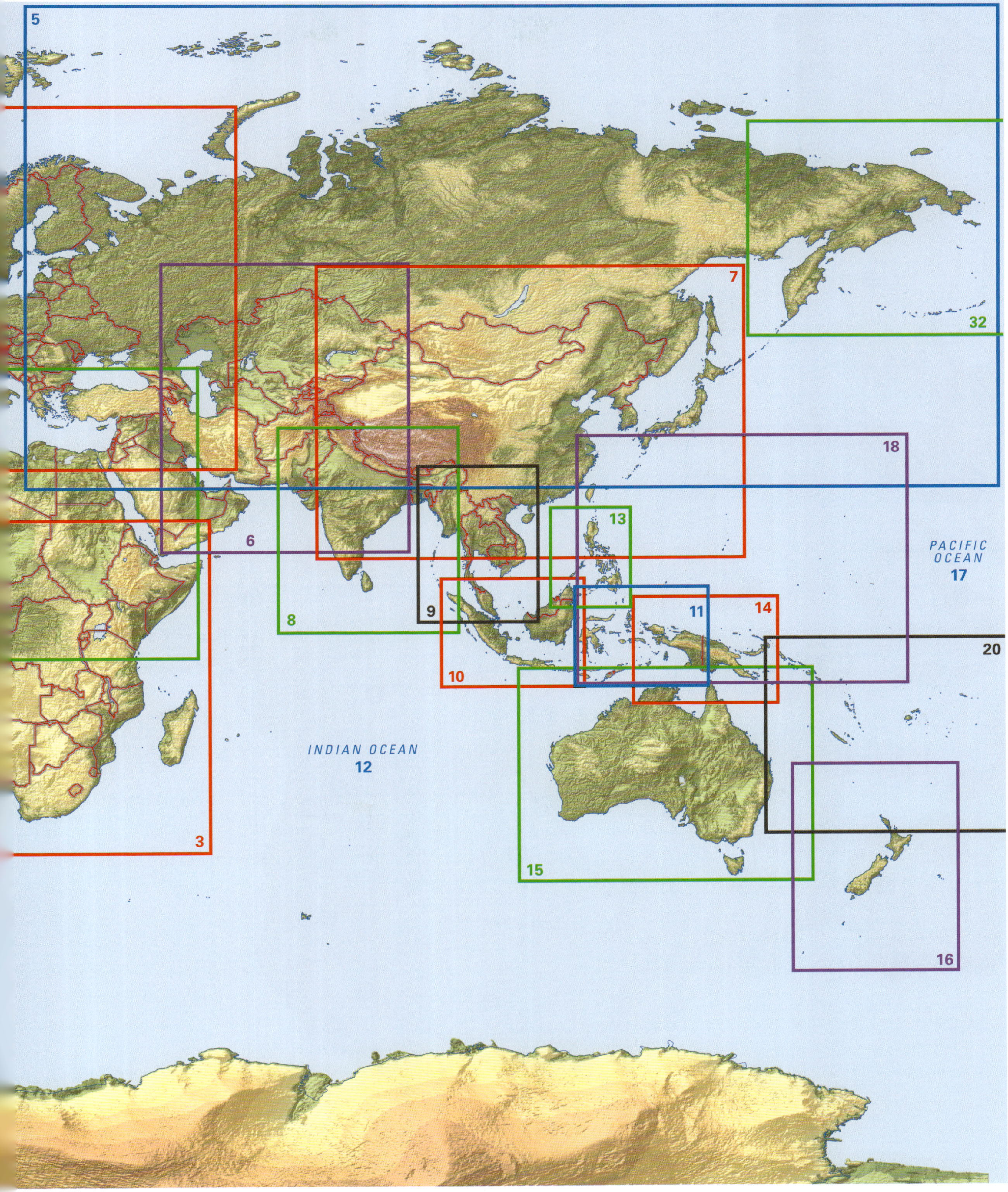
5
7
32
18
6
13
9
8
10
11
14
20
PACIFIC OCEAN
17
INDIAN OCEAN
12
3
15
16

BLACK SEA
MEDITERRANEAN SEA
RED SEA
Gulf of Aden
Gulf of Guinea
Strait of Gibraltar
SAHARA DESERT
NUBIAN DESERT
WESTERN SAHARA
ATLAS MTS
Russia
Georgia
Armenia
Azerbaijan
Iran
Iraq
Kuwait
Saudi Arabia
Yemen
Syria
Lebanon
Jordan
Palestine
Israel
Sinai
Turkey
Cyprus
Crete
Bulgaria
Serbia
Macedonia
Montenegro
Albania
Greece
Italy
Sicily
Sardinia
Corsica
Malta
France
Balearic Is
Mallorca
Spain
Portugal
Madeira
Desertas Is
Canary Is
Morocco
Algeria
Tunisia
Libya
Egypt
Sudan
South Sudan
Eritrea
Djibouti
Ethiopia
Somalia
Kenya
Uganda
Tanzania
Rwanda
Burundi
Pemba
Unguja
Chad
Niger
Mali
Mauritania
Senegal
Gambia
Guinea-Bissau
Guinea
Sierra Leone
Liberia
Côte d'Ivoire
Burkina Faso
Ghana
Togo
Benin
Nigeria
Cameroon
Central African Republic
Equatorial Guinea
Gabon
Congo
Democratic Republic of the Congo
Cabinda
Bioko
Príncipe
São Tomé
Annobón
Tigris
Euphrates
Nile
Blue Nile
White Nile
Lake Tana
Lake Chad
Lake Turkana
Lake Victoria
Niger
Benue
Congo
Ubangi
Ahaggar
Aïr
Tibesti
Ennedi
Darfur
Kordofan
Bale Mts
Mt Kenya
Aberdares
Kilimanjaro
Rwenzori Mts
Mt Oku
Mt Cameroon
IR
IQ
SA
YE
SY
JO
LB
IL
TR
GR
XK
IT
ES
PT
MA
DZ
TN
LY
EG
SD
SS
ER
DJ
ET
SO
KE
UG
TZ
RW
BI
TD
NE
ML
MR
EH
SN
GM
GW
GN
SL
LR
CI
BF
GH
TG
BJ
NG
CM
CF
GQ
GA
CG
CD
ST

Niger
NE
Chad
TD
Sudan
SD
Eritrea
ER
RED SEA
Saudi Arabia
YE
Yemen
Nigeria
NG
Djibouti
DJ
Gulf of Aden
Somaliland
Blue Nile
White Nile
Nile
Benue
Niger
CF
Central African Republic
SS
South Sudan
Ethiopia
ET
Bale Mts
Cameroon
CM
SO
Somalia
Gulf of Guinea
GQ
Equatorial Guinea
UG
Uganda
Lake Turkana
Kenya
KE
Ubangi
Congo
São Tomé and Príncipe
ST
Gabon
GA
Congo
CG
CD
Democratic Republic of the Congo
Rwenzori Mts
Lake Victoria
Aberdares
Mt Kenya
Rwanda
RW
Burundi
BI
Kilimanjaro
Lake Tanganyika
TZ
Tanzania
Eastern Arc Mts
Pemba
Unguja
Mafia
Zanzibar Arch
Cabinda
Zaire
Uige
Lukaga
1. Kinshasa
2. Cuanza Norte
3. Luanda
Bengo
Malanje
Lunda Norte
Lunda Sul
Lualaba
SC
Aldabra
KM
Comoros
Lake Malawi
Cuanza Sul
Benguela
Huambo
Bie
AO
Angola
Moxico
Zambezi
Zambia
ZM
Malawi
MW
Mozambique
MZ
Namibe
Huíla
Cunene
Cuando Cubango
Shire
Caprivi Strip
Okavango Delta
Zimbabwe
ZW
Mt Gorongosa
Namibia
NA
Namib Desert
Botswana
BW
Kalahari Desert
Limpopo
Madagascar
MG
North West
Gauteng
Mpumalanga
Eswatini
SZ
Free State
KwaZulu-Natal
Lesotho
LZ
Orange
Northern Cape
South Africa
ZA
Eastern Cape
Western Cape
Cape of Good Hope
Cape Agulhas

Greenland
GL
Iceland
IS
Jan Mayen
SJ
Faroe Is
FO
Novaya Zemlya
BARENTS SEA
Bear I
NO
NORWEGIAN SEA
ATLANTIC OCEAN
Azores
PT
Madeira
PT
Canary Is
ES
Outer Hebrides
Orkney Is
Shetland Is
Ireland
IE
United Kingdom
GB
Great Britain
English Channel
NORTH SEA
Norway
NO
Sweden
SE
Finland
FI
SCANDINAVIAN MTS
Gulf of Bothnia
Kola Peninsula
WHITE SEA
Russia
RU
Pechora
Kama
Ural
Volga
Don
Dnieper
Lake Onega
Lake Ladoga
Neva
Svir
Estonia
EE
Latvia
LV
Lithuania
LT
Kaliningrad
BALTIC SEA
Belarus
BY
Ukraine
UA
Moldova
MD
SEA OF AZOV
BLACK SEA
CASPIAN SEA
Kazakhstan
KZ
Turkmenistan
TM
Azerbaijan
AZ
Armenia
AM
Georgia
GE
CAUCASUS
Iran
IR
Iraq
IQ
Tigris
Euphrates
Firat
Murat
Saudi Arabia
SA
Syria
SY
Cyprus
CY
Turkey
TR
Rhodes
Crete
Egypt
EG
Libya
LY
Denmark
DK
Netherlands
NL
Belgium
BE
Luxembourg
LU
Germany
DE
Rhine
Danube
Poland
PL
Czechia
CZ
Slovakia
SK
Hungary
HU
Austria
AT
Switzerland
CH
Slovenia
Croatia
HR
Bosnia Herzegovina
BA
Serbia
RS
Kosovo
XK
Montenegro
Albania
AL
North Macedonia
Bulgaria
BG
Romania
RO
CARPATHIANS
Greece
GR
AEGEAN SEA
Ionian Is
ADRIATIC SEA
Italy
IT
ALPS
APENNINES
Corsica
Sardinia
Sicily
Malta
MT
MEDITERRANEAN SEA
Tunisia
TN
France
FR
PYRENEES
Andorra
Balearic Is
Mallorca
Spain
ES
Portugal
PT
Strait of Gibraltar
Morocco
MA
Algeria
DZ

ARCTIC OCEAN
BARENTS SEA
KARA SEA
LAPTEV SEA
EAST SIBERIAN SEA
CHUKCHI SEA
BERING SEA
SEA OF OKHOTSK
SEA OF JAPAN
EAST CHINA SEA
PACIFIC OCEAN
WHITE SEA
CASPIAN SEA
BLACK SEA
MEDITERRANEAN SEA
Svalbard (Spitsbergen)
SJ
Franz Josef Land
Novaya Zemlya
Severnaya Zemlya
New Siberian Is
Medvezhyi Is
Wrangel
Herald
Alaska
Chukchi Pen.
St Lawrence
Pribilof Is
Aleutian Is
US
Commander Is
Karaginsky
Kuril Is
Sakhalin
Japan
JP
Kolguyev
Kanin Pen.
Kola Pen.
Yamal Pen.
Gydan Pen.
Taymyr Pen.
Norway
Finland
FI
Estonia
Latvia
Lithuania
Kaliningrad
Belarus
Ukraine
UA
Murmansk
Karelia
Arkhangel'sk
Vologda
Moskva
Kirov
Komi
Nenets
Pechora
URAL MTS
Sverdlovsk
Ob
Yamal-Nenets
Khanty-Mansiy
Tomsk
Novosibirsk
Altay
Gorno-Altay
Irtysh
Yenisey
Taymyr
Evenk
Russia
RU
Krasnoyarsk
Tuva
Angara
Irkutsk
Lake Baikal
Buryat
Chita
SIBERIA
Vilyuy
Lena
Aldan
Amur
Sakha (Yakutia)
Khabarovsk
Primorye
Maga Buryatdan
Kolyma
Chukotka
Koryak
Kamchatka
Kazakhstan
KZ
Lake Balkhash
Syr Darya
Uzbekistan
Turkmenistan
Kyrgyzstan
Tajikistan
Afghanistan
AF
Pakistan
PK
Indus
India
IN
HIMALAYAS
Plateau of Tibet
Brahmaputra
Salween
Yangtze
Huang He
China
CN
Mongolia
MN
North Korea
South Korea
Iran
IR
Iraq
IQ
Saudi Arabia
Turkey
TR
Georgia
Armenia
Azerbaijan
Egypt
Nile

Volga
Kama
Russia
RU
Tobol
Irtysh
Ob
Ural
Don
Kazakhstan
KZ
ALTAI MTS
LAKE BALKHASH
TARBAGATAI MTS
ARAL SEA
Syr Darya
Caucasus
Georgia
CASPIAN SEA
Transcaspia
Uzbekistan
UZ
Issyk Kul
Kyrgyzstan
KG
TIAN SHAN
Armenia
Azerbaijan
AZ
ALAI MTS
Tarim Basin
Turkmenistan
TM
Amu Darya
Tajikistan
TJ
China
CN
TAKLAMAKAN DESERT
ALTUN SHAN
LAKE URUMIA
PAMIRS
ALBORZ MTS
HINDU KUSH
KARAKORAM
KUNLUN SHAN
Tigris
Afghanistan
AF
Iran
IR
ZAGROS MTS
SULAIMAN MTS
Indus
HIMALAYAS
Plateau of Tibet
Euphrates
Indo-Gangetic Plain
Iraq
IQ
KW
Kuwait
Brahmaputra
Persian Gulf
Nepal
NP
Pakistan
PK
Bahrain
BH
Qeshm
Baluchistan
THAR DESERT
Ganges
Qatar
QA
Gulf of Oman
India
IN
United Arab Emirates
AE
Saudi Arabia
SA
Oman
OM
Kathiawar Peninsula
Masirah
Dhofar
ARABIAN SEA
Deccan Plateau
Yemen
YE
Socotra

Russia
RU
S I B E R I A
Transbaikalia
LAKE BAIKAL
Lena
Angara
Yenisey
Aldan
Amur
Ussuri
SIKHOTE ALIN MTS
XIAO HINGGAN LING
DA HINGGAN LING
Sakhalin
Kuril Is
Hokkaido
Honshu
Japan
JP
Izu Is
Shikoku
Kyushu
Tanegashima
Amami-Oshima
Tokunoshima
Okinawa
Ryukyu Is
Sakishima
Ogasawara (Bonin) Is
Volcano Is
Northern Mariana Is
MP
Saipan
Guam
PACIFIC OCEAN
SEA OF JAPAN
EAST CHINA SEA
YELLOW SEA
SOUTH CHINA SEA
North Korea
KP
South Korea
KR
Heilongjiang
Jilin
Manchuria
Liaoning
Beijing
Tianjin
Hebei
Shandong
Jiangsu
Shanghai
Anhui
Zhejiang
Fujian
Taiwan
TW
Lanyu
Philippines
PH
Hong Kong
HK
Macao
MO
Guangdong
Jiangxi
Henan
Shanxi
Inner Mongolia
Hubei
Hunan
Guangxi
Hainan
Xi Jiang
Yangtze
Chongqing
Guizhou
Shaanxi
Ningxia Hui
Huang He (Yellow)
Sichuan
Yunnan
Vietnam
VN
Laos
LA
Thailand
TH
Cambodia
Myanmar
MM
Irrawaddy
Mekong
Salween
Brahmaputra
Bhutan
Bangladesh
BD
Bay of Bengal
India
IN
Ganges
Indus
Nepal
NP
HIMALAYAS
Tibet (Xizang)
Plateau of Tibet
China
CN
Qinghai
Qaidam Basin
Gansu
Qilian Shan
Kuya Nur
ALTUN SHAN
KUNLUN SHAN
TAKLAMAKAN DESERT
Tarim Basin
Xinjiang Uygur
TIAN SHAN
BOSTEN LAKE
G O B I   D E S E R T
Mongolia
MN
Dornod
Sühbaatar
Hentiy
Dornogovi
Dundgovi
Govisümber
Töv
Ulaanbaatar
Darhan-Uul
Selenge
Orhon
Bulgan
Övörhangay
Ömnögovi
Arhangay
Hövsgöl
Bayanhongor
Dzavhan
Govi-Altay
Uvs
Hovd
Bayan-Ölgiy
ALTAI MTS
Irtysh
TARBAGATAI MTS
Kazakhstan
KZ
LAKE BALKHASH
Issyk Kul
Kyrgyzstan
Tajikistan
Pakistan
PK

HINDU KUSH
KARAKORAM
China
CN
Afghanistan
AF
Jammu and Kashmir
Indus
Plateau of Tibet
Himachal Pradesh
Punjab
Chandigarh
Punjab
Pakistan
PK
Uttarakhand
HIMALAYAS
Brahmaputra
Arunachal Pradesh
Haryana
Delhi
Nepal
NP
Sikkim
Bhutan
BT
Baluchistan
Uttar Pradesh
Assam
Nagaland
Rajasthan
Meghalaya
Manipur
Bihar
Ganges
Sind
Bangladesh
BD
Tripura
Mizoram
Jharkhand
Myanmar
MM
West Bengal
Rann of Kutch
Gulf of Kutch
Gujarat
Madhya Pradesh
Sundarbans
Mouths of the Ganges
Kathiawar Pen.
Chhattisgarh
ARABIAN SEA
Gulf of Khambhat
India
IN
Odisha
Maharashtra
EASTERN GHATS
Munaung
Bay of Bengal
Andhra Pradesh
Mouths of the Irrawaddy
Preparis
INDIAN OCEAN
Great Coco
Goa
WESTERN GHATS
Narcondam
Coromandel Coast
Andaman Is
Barren
Malabar Coast
Karnataka
Pondicherry
Little Andaman
ANDAMAN SEA
Tamil Nadu
IN
Car Nicobar
Laccadive Is
Kerala
Palk Strait
Nicobar Is
Gulf of Mannar
Great Nicobar
Sri Lanka
LK
Male Atoll
Maldives
MV
Hoang He
Yangtze
Mekong
Salween
Irrawaddy

Sichuan
Hunan
Guizhou
China
CN
Arunachal Pradesh
Brahmaputra
India
IN
Nagaland
Assam
Meghalaya
Manipur
Kachin
Irrawaddy
Salween
Yangtze
Yunnan
Mekong
Xi Jiang
Guangxi
Sagaing
Myanmar
MM
Tripura
Mizoram
Tonkin
Bangladesh
BD
Chin
Shan State
Mandalay
Magwe
Arakan
Laos
LA
Gulf of Tonkin
Hainan
Kayah
Doi Inthanon
Pegu
Annam
Vietnam
Karen
Dawna Range
Khorat Plateau
ANNAMITE MTS
Irrawaddy
Yangon
Mon
Thailand
TH
Bolovens Plateau
Mouths of the Irrawaddy
Vietnam
VN
Ratanakiri
Preparis
Great Coco
Cambodia
KH
Mondulkiri
Narcondam
Tenasserim
Cardamom Mts
Da Lat Plateau
Andaman Is
Barren
Mergui Arch.
Gulf of Thailand
Cochinchina
SOUTH CHINA SEA
ANDAMAN SEA
Little Andaman
IN
Isthmus of Kra
Car Nicobar
Nicobar Is
Tarutao
Thai-Malay Peninsula
Great Nicobar
Langkawi
Kedah
Kelantan
Trengganu
1. Perlis
2. Selangor
3. Negeri Sembilan
4. Melaka (Malacca)
Pulau Pinang
Cameron Highlands
Natuna Is
Malaysia
Perak
MY
INDIAN OCEAN
Strait of Malacca
Pahang
Tioman
Anambas Is
Johor
Sumatra
Simeulue
Singapore
Indonesia
ID
Borneo

SOUTH CHINA SEA
SULU SEA
CELEBES SEA
JAVA SEA
FLORES SEA
BALI SEA
INDIAN OCEAN
Makassar Strait
Strait of Malacca
Thailand
TH
Thai-Malay Peninsula
Malaysia
MY
Brunei Darussalam
BN
Singapore
SG
Indonesia
ID
Christmas
CX
Borneo
Sumatra
Java
Kalimantan
Sulawesi
Sabah
Sarawak
Mt Kinabalu
Mt Mulu
Mt Murud
Mt Duit
Labuan
Balabac
Balambangan
Banggi
Mapun
Sibutu
Sebatik
Bunyo
Tarakan
Maratua
Kalimantan Timur
Kalimantan Selatan
Kalimantan Tengah
Kalimantan Barat
Sebuku
Laurot Is
Karamian
Masalembu
Bawean
Kangean Is
Madura
Lombok
Bali
Sumbawa
Komodo
Sumba
Nusa Tenggara (Lesser Sundas)
Jawa Timur
Jawa Tengah
Yogyakarta
Jawa Barat
Banten
Jakarta Raya
Seribu Is
Karimunjawa Is
Belitung
Mendanau
Bangka
Karimata Is
Natuna Besar
Natuna Is
Laut
Tambelan Is
Anambas Is
Tioman
Bintan
Riau Arch.
Lingga Arch.
Krakatau
Lampung
Sumatera Selatan
Jambi
Bengkulu
Enggano
Riau
Sumatera Barat
Sumatera Utara
Aceh
Mentawai Is
Sipura
North Pagai
South Pagai
Siberut
Batu Is
Pini
Musala
Nias
Banyak Is
Babi
Simeulue
We
Trengganu
Kelantan
Pahang
Johor
Negeri Sembilan
Melaka
Selangor
Perak
Cameron Highlands
Kedah
Perlis
Pulau Pinang
Langkawi
Tarutao

PACIFIC OCEAN
New Guinea
Papua
New
Guinea
PG
Papua
Snow Mts
Yapen
Biak
Numfor
Num
Geelvink Bay
Aru Is
ARAFURA SEA
Arfak Mts
Papua Barat
Vogelkop Pen.
Onin Pen.
Bombera
Pen.
Adi
Kai
Besar
Kai Is
Kai
Kecil
Watubela
Kur
Tayandu Is
Gorong
Larat
Yamdena
Tanimbar Is
Selaru
Australia
West Papuan Is
Waigeo
Batanta
Salawati
Misool
Gebe
Gag
Kofiau
Northern
Moluccas
Morotai
Halmahera
Widi Is
SERAM
SEA
Seram
Saparua
Haruku
Ambon
Boano
Banda Is
Southern Moluccas
Nila
Teun
Babar
Sermata
Damar
TIMOR SEA
BANDA SEA
Moa
Lakor
Leti
Romang
Kisar
Nanusa Is
Talaud Is
Kabaruang
Karakelong
Salibabu
Sangihe Is
Sangihe
Siau
Loloda Is
Ternate
Tidore
Kasiruta
Bacan
Obi
Mayu
Lembeh
Bunaken Is
Mangole
Sanana
Taliabu
Sula Is
Buru
Ambelan
Indonesia
ID
Wetar
Atauro
Timor-Leste
TL
Timor
Alor
Pantar
Lomblen
Ili Boleng
Solor
Semau
Rote
Sawu
Philippines
PH
Mindanao
CELEBES SEA
Sulu Arch.
Minahassa Pen.
Sulawesi Utara
Gorontalo
Banggai
Peleng
Banggai Is.
Manui
Wowoni
Butung
Tukangbesi Is
Batuata
Muna
Kabaena
Sulawesi Tenggara
Togian Is
Sulawesi Tengah
Sulawesi
Sulawesi Selatan
Sulawesi Barat
Makassar Strait
Borneo
Mt Lompobattang
Salayar
Tanahjampea
Kalao
Bonerate
Kalaotoa
Madu
FLORES SEA
Paloe
Flores
Komodo
Sangeang
Sumbawa
Sumba
Nusa Tenggara (Lesser Sundas)

INDIAN OCEAN
India
IN
Pakistan
Oman
Saudi Arabia
Yemen
Somalia
Ethiopia
Kenya
Tanzania
South Sudan
Central African Republic
Democratic Republic of the Congo
Angola
Zambia
Zimbabwe
Mozambique
Botswana
Namibia
South Africa
Madagascar
MG
Comoros
KM
Myanmar
Thailand
Malaysia
Sri Lanka
LK
Maldives
MV
Laccadive Is
Andaman Is
Nicobar Is
Socotra
YE
Caseyr
Seychelles
SC
Bird I
Amirante Is
Farquhar Group
Agalega Is
Cosmoledo
Glorieuses Is
Aldabra
Assumption
Europa
Tromelin
Mauritius
MU
Réunion
RE
Cargados Carajos Shoals
Rodrigues
Chagos Arch.
Diego Garcia
IO
Cocos (Keeling) Is
CC
Christmas
CX
Sumatra
Greater Sundas
Java
Borneo
Indonesia
ID
Sulawesi
Philippines
Moluccas
Wallacea
Nusa Tenggara (Lesser Sundas)
New Guinea
Australia
Amsterdam
St Paul
TF
Kerguelen
Heard
HM
Crozet Is
Prince Edward
Marion
ZA
A
Comoros
KM
Ngazidja (Grand Comoro)
Mwali (Mohéli)
Nzwani (Anjouan)
Mayotte
YT
B
Seychelles
SC
Aride
Curieuse
Les Soeurs
Félicité
Marianne
Cousin
Cousine
Praslin
La Digue
Frégate
North
Silhouette
Conception
Thérèse
Ste-Anne
Île au Cerf
Mahé
C
Andaman and Nicobar Is
IN
Narcondam
North Andaman
Middle Andaman
Havelock
South Andaman
North Sentinel
Little Andaman
Car Nicobar
Tillangchong
Teressa
Camorta
Katchall
Little Nicobar
Great Nicobar

Batan Is
Babuyan
Calayan
Camiguin (north)
CORDILLERA CENTRAL
Cagayan Valley
SIERRA MADRE
PHILIPPINE SEA
Luzon
Polillo Is
Calagua Is
Manila Bay
Alabat
Catanduanes
Lubang
Marinduque
Burias
Mindoro
Romblon
Ticao
SOUTH CHINA SEA
Calamian Group
Tablas
Sibuyan
Masbate
Samar
Busuanga
Culion
Biliran
Visayan Is
Panay
Leyte
Calicoan
Cuyo Is
Cebu
Poro
Dumaran
Guimaras
Dinagat
Negros
Bohol
Panaon
Siargao
Palawan
Cagayan Is
Camiguin (south)
Siquijor
Philippines
PH
Mt Malindang
Mt Katanglad
Balabac
SULU SEA
Zamboanga Peninsula
Mt Ragang
Mindanao
Banggi
Cagayan de Tawitawi
Moro Gulf
Mt Apo
Davao Gulf
Pangutaran Group
Basilan
Jolo
Siasi
Sulu Arch.
Tawitawi
Sarangani
Balut
Brunei Darussalam
BN
Malaysia
MY
Sibutu
Talaud Is
Indonesia
ID
CELEBES SEA
Sangihe Is
Indonesia
ID

PACIFIC OCEAN
BISMARCK SEA
SOLOMON SEA
CORAL SEA
ARAFURA SEA
TIMOR SEA
BANDA SEA
Papua New Guinea
PG
Indonesia
ID
Australia
AU
Bismarck Arch.
St Matthias Is
New Hanover
Djaul
Tabar Is
Tabga Is
Lihir Is
Feni Is
Green Is
Nissan
New Ireland
Duke of York
New Britain
Witu Is
Rooke
Long
Manus
Admiralty Is
Kaniet Is
Hermit Is
Ninigo Is
Tarawai Is
Manam
Karkar
Astrolabe Bay
Huon Pen.
Huon Gulf
Trobriand Is
Woodlark
Goodenough
Fergusson
Normanby
D'Entrecasteaux Is
Milne Bay
Misima
Louisiade Arch.
Rossel
Tagula
OWEN STANLEY MTS
HERZOG MTS
WHARTON RANGE
ADELBERT MTS
BISMARCK MTS
VICTOR EMANUEL MTS
SNOW MTS
WEYLAND MTS
ARFAK MTS
Ramu
Purari
Jimi
Kikori
Sepik
Fly
Bensbach
Gulf of Papua
Daru
Saibai
Boigu
Moa
Badu
Torres Strait
Prince of Wales
Cape York Pen.
Humboldt Bay
Cyclops Mts
Idenburg
Mamberamo
Baliem
Digul
Lorentz
Kolepom I.
Gulf of Carpentaria
Groote Eylandt
Arnhem Land
Melville
Biak
Yapen
Numfor
Num
Geelvink Bay
Etna Bay
Triton Bay
Bintuni Bay
Onin Pen.
Bomberai Pen.
Adi
Aru Is
Kai Is
Tanimbar Is
Babar
Vogelkop Pen.
West Papuan Is
Waigeo
Batanta
Salawati
Gag
Kofiau
Misool
Gebe
Seram
Gorong
Watubela
Banda Is

Indonesia
ID
Timor-Leste
TL
Papua
New Guinea
PG
Solomon
Islands
SB
Australia
AU
INDIAN OCEAN
TIMOR SEA
ARAFURA SEA
CORAL SEA
TASMAN SEA
Ashmore Reef
Melville
Arnhem Land
Groote
Eylandt
Gulf of Carpentaria
Wellesley Is
Torres Strait
Cape York
Pen.
Great Barrier Reef
Fraser I
GREAT DIVIDING RANGE
GREAT DIVIDING RANGE
Queensland
Darling
New South Wales
Australian
Capital
Territory
Murray
Victoria
Bass Strait
King
Flinders
Cape Barren
Tasmania
Kangaroo
Spencer Gulf
LAKE
EYRE
South Australia
Great Australian Bight
Nullarbor Plain
Northern Territory
MACDONNELL
RANGES
MUSGRAVE
RANGES
GREAT VICTORIA
DESERT
GIBSON
DESERT
GREAT SANDY
DESERT
Kimberley
Western Australia
Barrow
Shark Bay
Houtman
Abrolhos

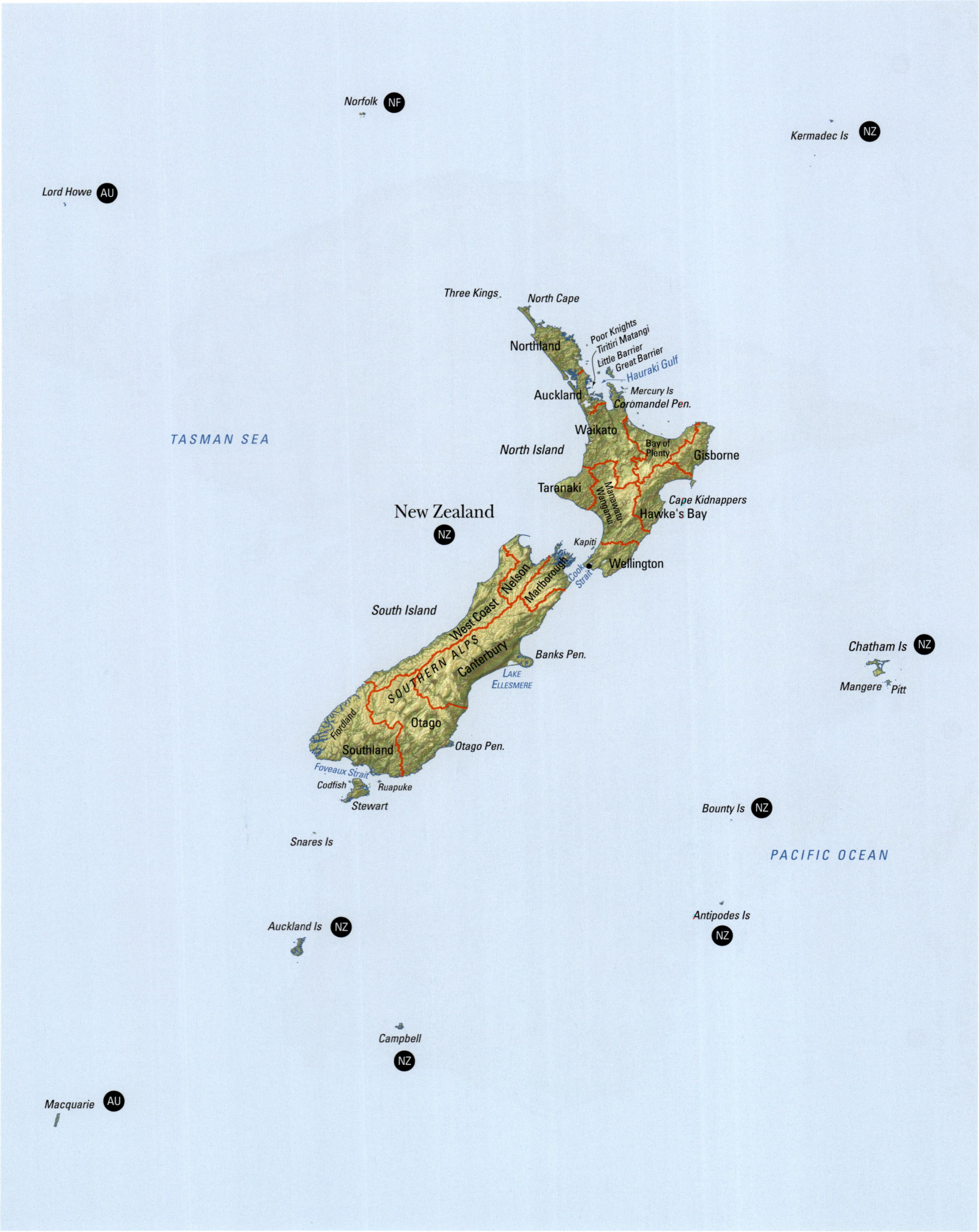
Norfolk
NF
Kermadec Is
NZ
Lord Howe
AU
Three Kings
North Cape
Poor Knights
Tiritiri Matangi
Little Barrier
Great Barrier
Hauraki Gulf
Northland
Mercury Is
Auckland
Coromandel Pen.
TASMAN SEA
Waikato
Bay of Plenty
North Island
Gisborne
Taranaki
Manawatu-Wanganui
Cape Kidnappers
New Zealand
NZ
Hawke's Bay
Kapiti
Wellington
Cook Strait
Nelson
Marlborough
South Island
West Coast
SOUTHERN ALPS
Canterbury
Chatham Is
NZ
Banks Pen.
Lake Ellesmere
Mangere
Pitt
Fiordland
Otago
Southland
Otago Pen.
Foveaux Strait
Codfish
Ruapuke
Stewart
Bounty Is
NZ
Snares Is
PACIFIC OCEAN
Antipodes Is
NZ
Auckland Is
NZ
Campbell
NZ
Macquarie
AU

Galápagos Is EC
B
Pinta
Marchena
Genovesa
Isabela
Fernandina
San Salvador (Santiago)
Bartolomé
Daphne
Seymour
Baltra
Rábida
Santa Cruz
Pinzón
Cuatro Hermanos
Santa Fe
Tortuga
San Cristóbal
Floreana
Española
Juan Fernández Arch. CL
A
Alejandro Selkirk (Más Afuera)
Robinson Crusoe (Más a Tierra)
Santa Clara
Mongolia
China
Russia
Alaska
Canada
United States
Mexico
Cuba
Gulf of Mexico
Costa Rica
Colombia
Ecuador
Peru
Chile
Malpelo
Cocos
Galápagos Is
EC
Juan Fernández Arch.
CL
Desventuradas Is
Sala y Gómez
Easter
Pitcairn Is
PN
Clipperton
PF
Revillagigedo Is
MX
Guadalupe
Marquesas Is
French Polynesia
Tuamotu Arch.
Gambier Is
Austral Is
Society Is
Cook Is
CK
Line Is
Hawaii
US
POLYNESIA
American Samoa
AS
Tokelau
TK
Phoenix Is
Niue
NU
Samoa
WS
Tonga
TO
Kermadec
NZ
Chatham
New Zealand
Kiribati
KI
Tuvalu
TV
Wallis and Futuna
WF
Fiji
FJ
Johnston Atoll
UM
Laysan
Midway
Wake
Marshall Is
MH
Gilbert Is
Nauru
NR
Santa Cruz Is
Loyalty Is
Vanuatu
VU
Norfolk
NF
New Caledonia
NC
Lord Howe
AU
TASMAN SEA
Tasmania
Australia
CORAL SEA
MELANESIA
Solomon Is
SB
Papua New Guinea
MICRONESIA
Caroline Is
Federated States of Micronesia
FM
Northern Mariana Islands
MP
Guam
GU
Bonin Is
Volcano Is
JP
Yap
Palau
PW
Japan
PACIFIC OCEAN
CHUKCHI SEA
BERING SEA
St Lawrence
Pribilof Is
Aleutian Is
US
Commander Is
RU
Kurils
SEA OF OKHOTSK
Kodiak
Gulf of Alaska
Queen Charlotte Is
SEA OF JAPAN
EAST CHINA SEA
Taiwan
Philippines
SOUTH CHINA SEA
Vietnam
Malaysia
Indonesia

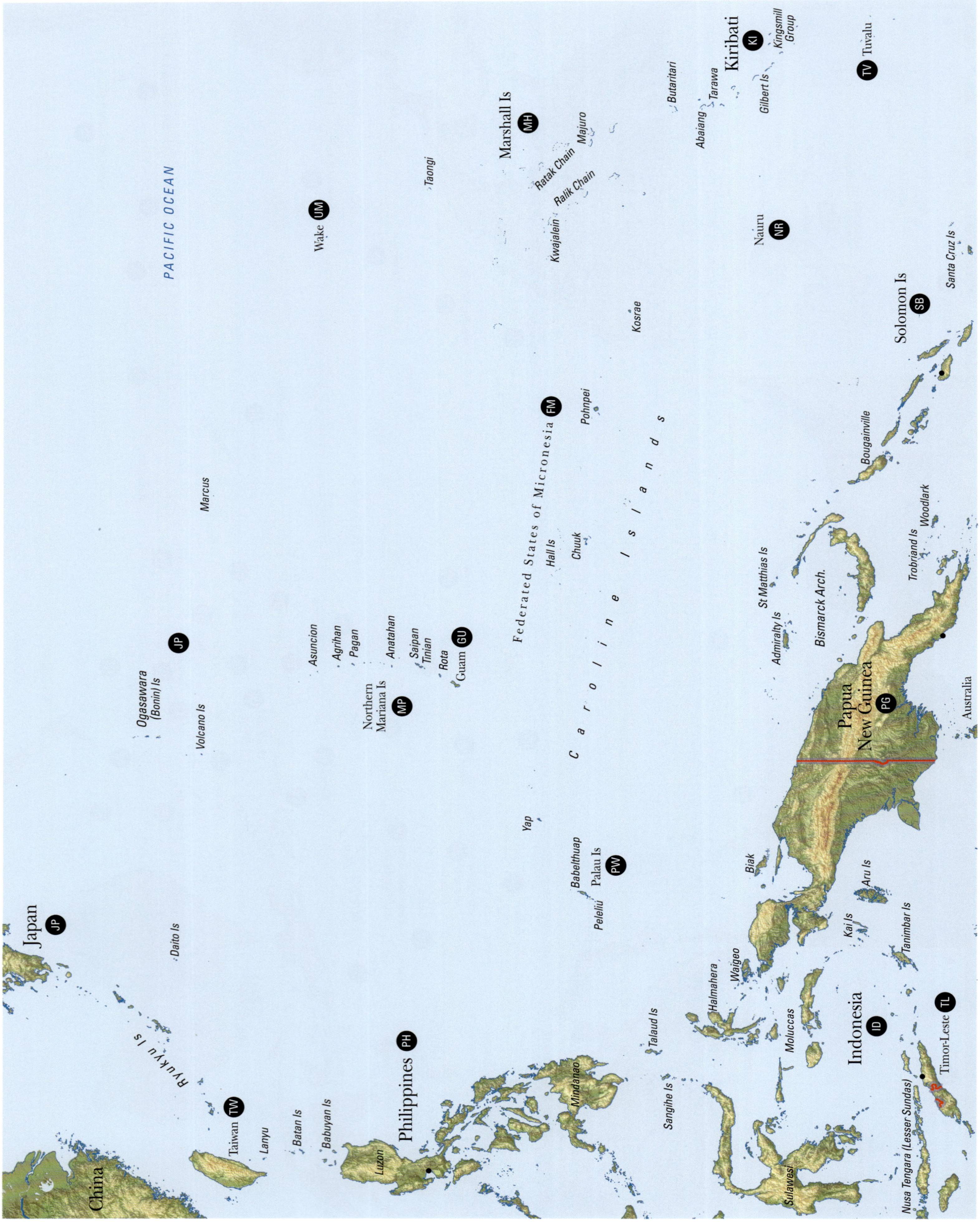
PACIFIC OCEAN
China
Japan
JP
Taiwan
TW
Lanyu
Ryukyu Is
Daito Is
Ogasawara (Bonin) Is
Volcano Is
Marcus
Wake
UM
Taongi
Marshall Is
MH
Ratak Chain
Ralik Chain
Majuro
Kwajalein
Butaritari
Abaiang
Tarawa
Gilbert Is
Kiribati
KI
Kingsmill Group
Tuvalu
TV
Nauru
NR
Santa Cruz Is
Solomon Is
SB
Kosrae
Pohnpei
Federated States of Micronesia
FM
Caroline Islands
Chuuk
Hall Is
Asuncion
Agrihan
Pagan
Anatahan
Saipan
Tinian
Rota
Guam
GU
Northern Mariana Is
MP
Yap
Babelthuap
Palau Is
PW
Peleliu
Batan Is
Babuyan Is
Luzon
Philippines
PH
Mindanao
Sangihe Is
Talaud Is
Halmahera
Waigeo
Moluccas
Sulawesi
Indonesia
ID
Timor-Leste
TL
Nusa Tengara (Lesser Sundas)
Kai Is
Tanimbar Is
Aru Is
Biak
Papua New Guinea
PG
Australia
Admiralty Is
St Matthias Is
Bismarck Arch.
Trobriand Is
Woodlark
Bougainville

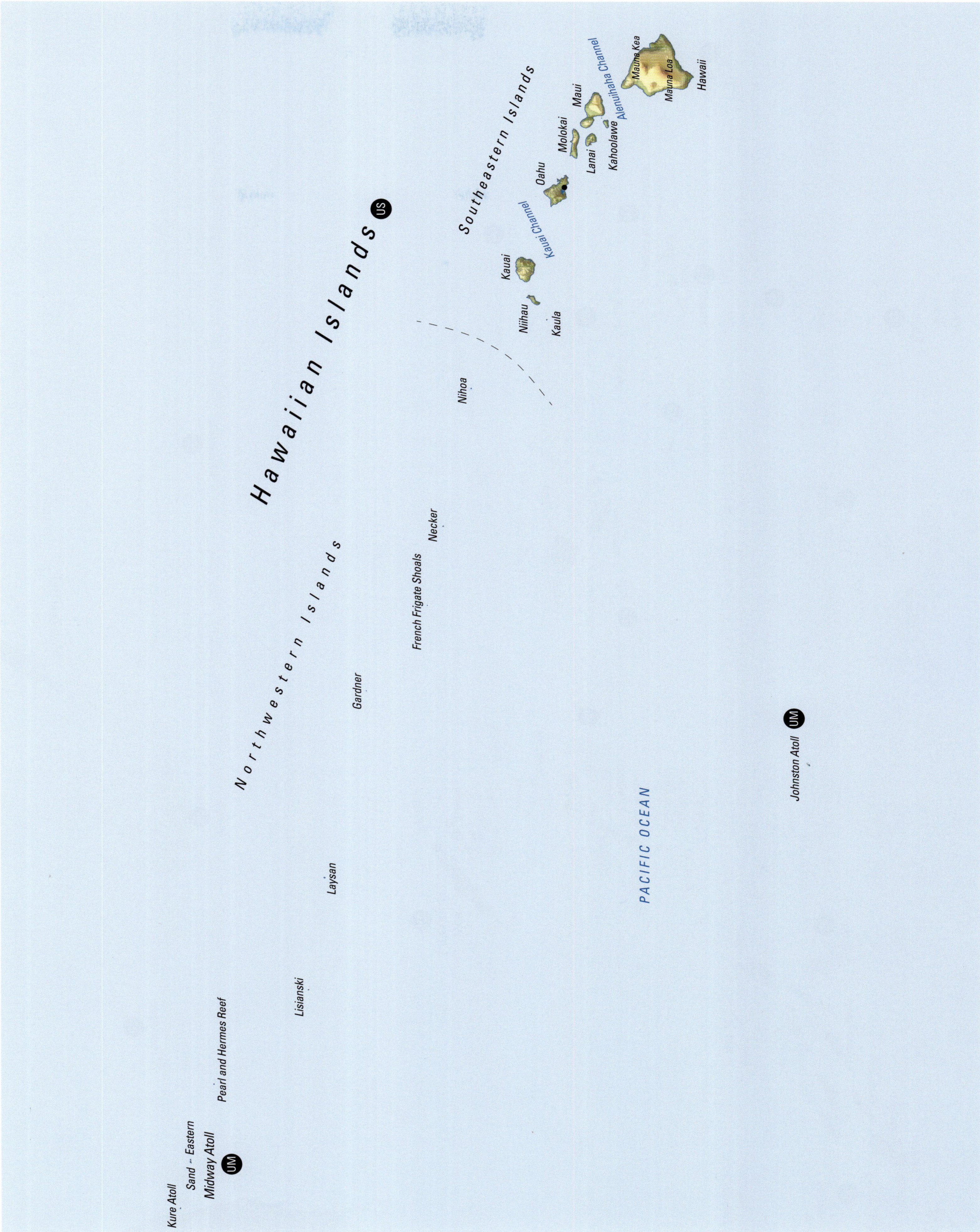
Hawaiian Islands
US
Southeastern Islands
Northwestern Islands
Hawaii
Mauna Kea
Mauna Loa
Alenuihaha Channel
Maui
Kahoolawe
Lanai
Molokai
Oahu
Kauai Channel
Kauai
Niihau
Kaula
Nihoa
Necker
French Frigate Shoals
Gardner
Laysan
Lisianski
Pearl and Hermes Reef
Midway Atoll
UM
Sand - Eastern
Kure Atoll
Johnston Atoll
UM
PACIFIC OCEAN

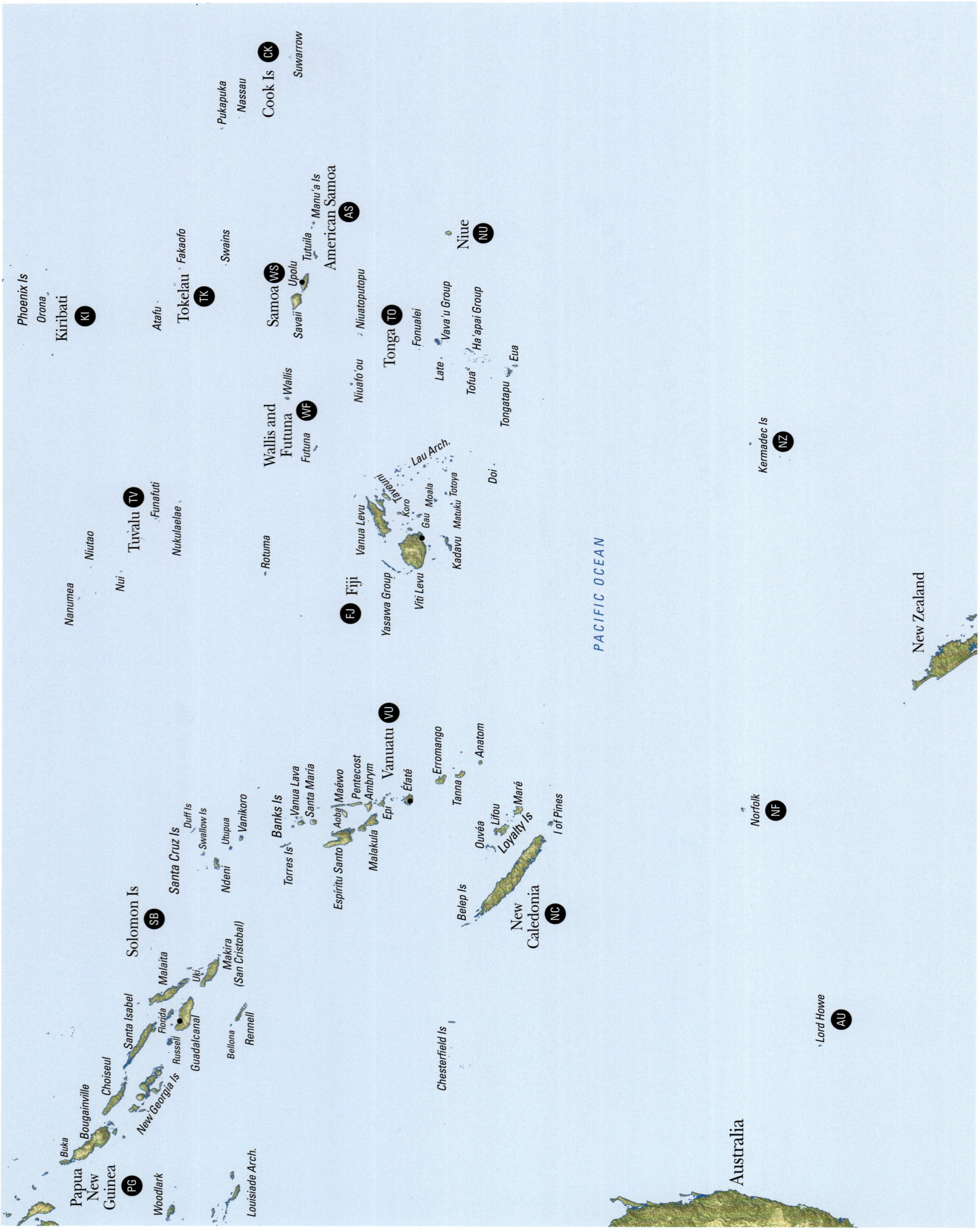
Papua New Guinea
PG
Woodlark
Louisiade Arch.
Buka
Bougainville
Choiseul
New Georgia Is
Santa Isabel
Russell
Florida
Guadalcanal
Malaita
Bellona
Rennell
Uki
Makira (San Cristobal)
Solomon Is
SB
Santa Cruz Is
Duff Is
Swallow Is
Utupua
Vanikoro
Ndeni
Banks Is
Torres Is
Vanua Lava
Santa María
Espíritu Santo
Aoba
Maéwo
Pentecost
Ambrym
Malakula
Epi
Éfaté
Vanuatu
VU
Erromango
Tanna
Anatom
Chesterfield Is
Belep Is
New Caledonia
NC
Ouvéa
Lifou
Maré
Loyalty Is
I of Pines
Norfolk
NF
Lord Howe
AU
Australia
Phoenix Is
Orona
Kiribati
KI
Nanumea
Niutao
Nui
Tuvalu
TV
Funafuti
Nukulaelae
Rotuma
Fiji
FJ
Yasawa Group
Vanua Levu
Viti Levu
Taveuni
Koro
Gau
Moala
Kadavu
Matuku
Totoya
Lau Arch.
Wallis and Futuna
WF
Wallis
Futuna
Atafu
Tokelau
TK
Fakaofo
Swains
Samoa
WS
Savaii
Upolu
Tutuila
Manu'a Is
American Samoa
AS
Pukapuka
Nassau
Cook Is
CK
Suwarrow
Niuatoputapu
Niuafo'ou
Tonga
TO
Fonualei
Late
Vava'u Group
Ha'apai Group
Tofua
Tongatapu
Eua
Niue
NU
Doi
Kermadec Is
NZ
New Zealand
PACIFIC OCEAN

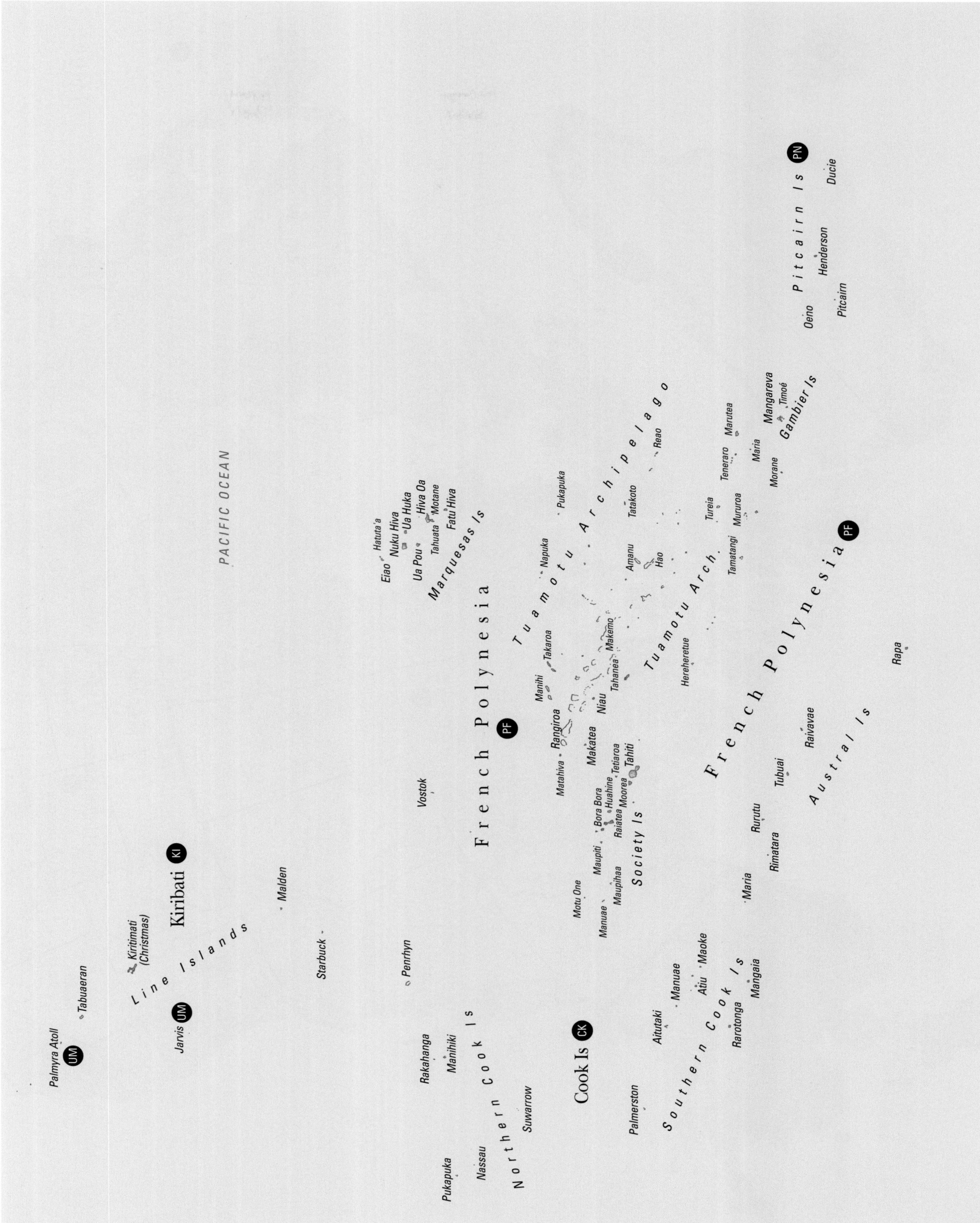
PACIFIC OCEAN
Palmyra Atoll
UM
Tabuaeran
Jarvis
UM
Kiritimati (Christmas)
Line Islands
Kiribati
KI
Malden
Starbuck
Penrhyn
Vostok
Pukapuka
Nassau
Rakahanga
Manihiki
Suwarrow
Northern Cook Is
Cook Is
CK
Palmerston
Aitutaki
Manuae
Atiu
Maoke
Rarotonga
Mangaia
Southern Cook Is
Motu One
Manuae
Maupihaa
Maupiti
Bora Bora
Raiatea
Huahine
Moorea
Tetiaroa
Tahiti
Society Is
Maria
Rimatara
Rurutu
Tubuai
Raivavae
Austral Is
Rapa
French Polynesia
PF
Matahiva
Rangiroa
Makatea
Manihi
Takaroa
Niau
Tahanea
Makemo
Tuamotu Archipelago
Tuamotu Arch.
Hereheretue
Napuka
Pukapuka
Amanu
Hao
Tatakoto
Reao
Tureia
Mururoa
Tamatangi
Tenararo
Maria
Morane
Marutea
Mangareva
Timoé
Gambier Is
French Polynesia
PF
Eiao
Hatuta'a
Nuku Hiva
Ua Huka
Ua Pou
Hiva Oa
Tahuata
Motane
Fatu Hiva
Marquesas Is
Oeno
Pitcairn
Henderson
Ducie
Pitcairn Is
PN

Iceland
IS
Greenland
GL
Tunu
Kitaa
Denmark Strait
Davis Strait
Baffin Bay
Nares Strait
LABRADOR SEA
ARCTIC OCEAN
BEAUFORT SEA
CHUKCHI SEA
Bering Strait
PACIFIC OCEAN
ATLANTIC OCEAN
Russia
St Lawrence
Nunivak
Alaska
ALASKA RANGE
BROOKS RANGE
Kenai Pen.
Alaska Pen.
Kodiak
Alexander Arch.
Queen Charlotte
Vancouver
Yukon
MACKENZIE MTS
Mackenzie
Northwest Territories
GREAT BEAR LAKE
GREAT SLAVE LAKE
Slave
Peace
LAKE ATHABASCA
Canada
CA
Amundsen Gulf
Banks
Victoria
Melville
Parry Channel
Bathurst
Queen Elizabeth Is
Ellesmere
Devon
Somerset
Prince of Wales
King William
Nunavut
Melville Peninsula
Baffin
Prince Charles
Southampton
Coats
Mansel
Hudson Strait
Hudson Bay
Belcher Is
James Bay
Akimiski
British Columbia
Alberta
Saskatchewan
Manitoba
Ontario
Quebec
Newfoundland and Labrador
Newfoundland
Saint Pierre and Miquelon
PM
Anticosti
Gulf of St Lawrence
Prince Edward
New Brunswick
Nova Scotia
REINDEER LAKE
Saskatchewan
Nelson
LAKE WINNIPEG
LAKE NIPIGON
LAKE SUPERIOR
LAKE HURON
LAKE MICHIGAN
LAKE ERIE
LAKE ONTARIO
Columbia
ROCKY MOUNTAINS
CASCADE RANGE
Snake
Colorado
Missouri
Mississippi
United States
US
Washington
Oregon
California
Nevada
Idaho
Montana
Utah
Wyoming
Colorado
North Dakota
South Dakota
Nebraska
Minnesota
Iowa
Wisconsin
Illinois
Michigan
Indiana
Ohio
Pennsylvania
New York
New Jersey
Vermont
New Hampshire
Massachusetts
Rhode Island
Connecticut
Maine

United States
US
Canada
CA
Mexico
MX
Bahamas
BS
Cuba
CU
GREAT PLAINS
ROCKY MOUNTAINS
APPALACHIAN MTS
CASCADE RANGE
SIERRA NEVADA
COASTAL RANGE
Washington
Oregon
California
Idaho
Nevada
Montana
Wyoming
Utah
Arizona
Colorado
New Mexico
North Dakota
South Dakota
Nebraska
Kansas
Oklahoma
Texas
Minnesota
Iowa
Missouri
Arkansas
Louisiana
Wisconsin
Illinois
Michigan
Indiana
Ohio
Kentucky
Tennessee
Mississippi
Alabama
Georgia
Florida
South Carolina
North Carolina
Virginia
West Virginia
Pennsylvania
New York
Maryland
Delaware
New Jersey
Connecticut
Rhode Island
Massachusetts
Vermont
New Hampshire
Maine
New Brunswick
Quebec
Ontario
Manitoba
Saskatchewan
Alberta
British Columbia
Baja California
Baja California Sur
Sonora
Chihuahua
Sinaloa
Durango
Coahuila
Nuevo León
Tamaulipas
Zacatecas
San Luis Potosí
Lake Superior
Lake Michigan
Lake Huron
Lake Erie
Lake Ontario
Lake Winnipeg
Lake Nipigon
St. Clair
Great Salt Lake
Salton Sea
James Bay
Chesapeake Bay
San Francisco Bay
Gulf of Mexico
Gulf of California
Straits of Florida
Mississippi Delta
Cape Hatteras
Long Island
Florida Keys
Everglades
Channel Is
Guadalupe
Cedros
Vancouver
St. Lawrence
Ohio
Mississippi
Missouri
Arkansas
Rio Grande
Colorado
Snake
Columbia
Saskatchewan
Nelson

North Carolina
South Carolina
Georgia
Florida
Alabama
Mississippi
Louisiana
Arkansas
Oklahoma
Texas
United States
US
New Mexico
Arizona
California
Mississippi
Rio Grande
Rio Grande
Colorado
Bahamas
Florida Keys
Straits of Florida
Cuba
CU
I of Pines
Cayman Is
KY
Jamaica
CARIBBEAN SEA
Yucatan Channel
Gulf of Mexico
Mujeres
Cozumel
Yucatán
Quintana Roo
Campeche
Tabasco
Chiapas
Veracruz
Oaxaca
Puebla
Guerrero
Hidalgo
México
Michoacán
Guanajuato
San Luis Potosí
Tamaulipas
Nuevo León
Coahuila
Mexico
MX
Durango
Zacatecas
Jalisco
Nayarit
Colima
Chihuahua
Sinaloa
Sonora
SIERRA MADRE OCCIDENTAL
Baja California
Baja California Sur
Gulf of California
Tiburón
Ángel de la Guarda
Cedros
Magdalena
Santa Margarita
Cerralvo
Cape San Lucas
Tres Marías Is
Revillagigedo Arch.
Socorro
Clarión
Gulf of Tehuantepec
1. Aguascalientes
2. Querétaro
3. Tlaxcala
4. Distrito Federal
5. Morelos
PACIFIC OCEAN
Belize
BZ
Gulf of Honduras
Bay Is
Roatán
Guanaja
Utila
Guatemala
GT
El Salvador
SV
Honduras
HN
Nicaragua
NI
Lake Nicaragua
Cayos Miskitos
Providencia
San Andrés
Costa Rica
CR
Gulf of Nicoya
Gulf of Chiriquí
Bocas del Toro Arch.
Escudo de Veraguas
Panama
PA
Pearl Is
Gulf of Panama
Coiba
Azuero Pen.
Colombia

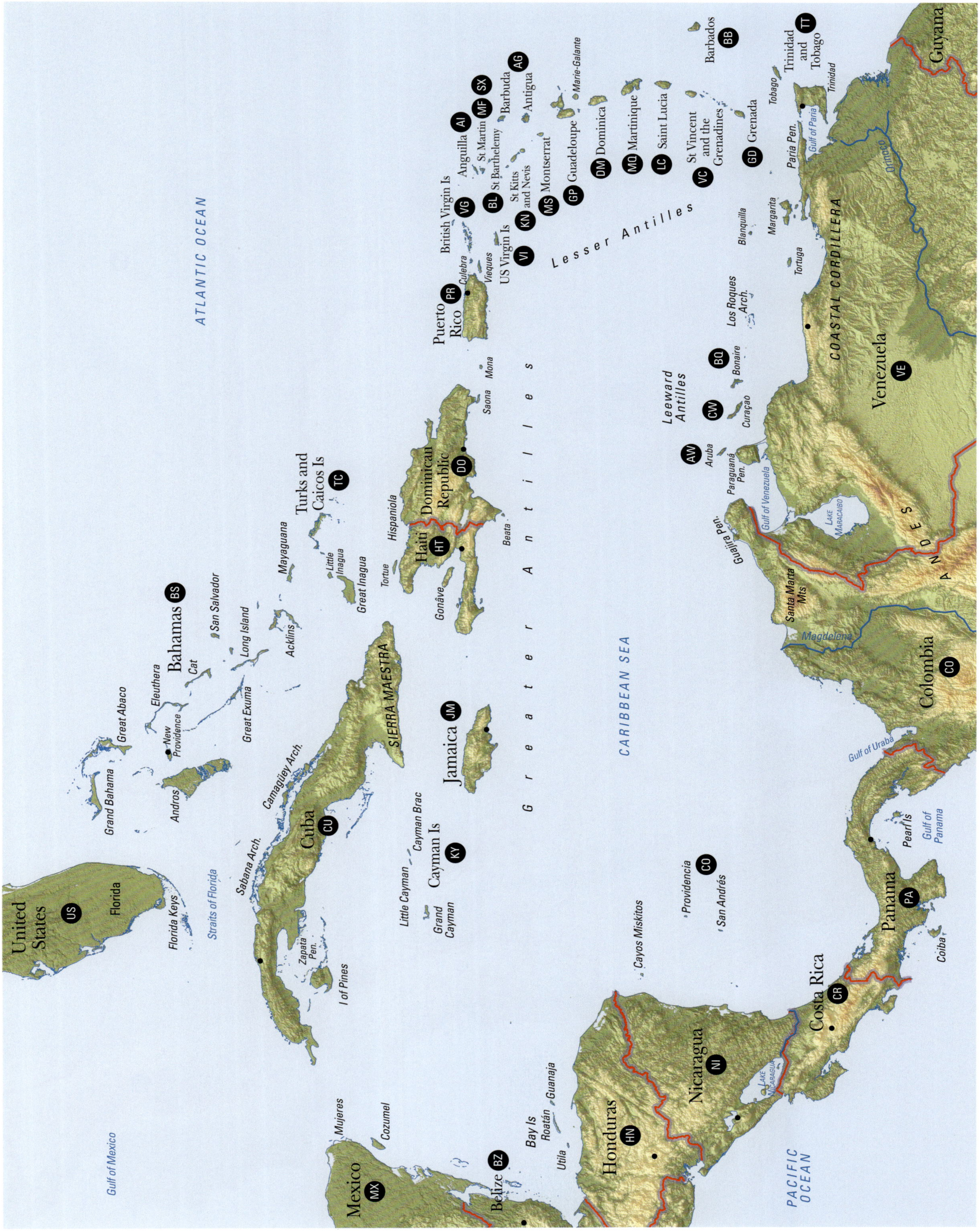
ATLANTIC OCEAN
CARIBBEAN SEA
PACIFIC OCEAN
Gulf of Mexico
Straits of Florida
United States
US
Florida
Florida Keys
Bahamas
BS
Grand Bahama
Great Abaco
Andros
New Providence
Eleuthera
Cat
San Salvador
Long Island
Great Exuma
Acklins
Mayaguana
Little Inagua
Great Inagua
Turks and Caicos Is
TC
Cuba
CU
Sabana Arch.
Camagüey Arch.
Zapata Pen.
I of Pines
SIERRA MAESTRA
Cayman Is
KY
Little Cayman
Cayman Brac
Grand Cayman
Jamaica
JM
Hispaniola
Tortue
Gonâve
Haiti
HT
Dominican Republic
DO
Beata
Saona
Mona
Puerto Rico
PR
Culebra
Vieques
British Virgin Is
VG
US Virgin Is
VI
Anguilla
AI
St Martin
MF
SX
St Barthelemy
BL
Barbuda
Antigua
AG
St Kitts and Nevis
KN
Montserrat
MS
Guadeloupe
GP
Marie-Galante
Dominica
DM
Martinique
MQ
Saint Lucia
LC
St Vincent and the Grenadines
VC
Grenada
GD
Barbados
BB
Tobago
Trinidad
Trinidad and Tobago
TT
Lesser Antilles
Greater Antilles
Leeward Antilles
Aruba
AW
Curaçao
CW
Bonaire
BQ
Los Roques Arch.
Blanquilla
Margarita
Tortuga
Paraguaná Pen.
Guajira Pen.
Gulf of Venezuela
LAKE MARACAIBO
Gulf of Paria
Paria Pen.
Orinoco
COASTAL CORDILLERA
Venezuela
VE
Guyana
ANDES
Santa Marta Mts
Magdalena
Colombia
CO
Gulf of Uraba
Panama
PA
Pearl Is
Gulf of Panama
Coiba
Costa Rica
CR
Nicaragua
NI
LAKE NICARAGUA
Providencia
San Andrés
Cayos Miskitos
Honduras
HN
Bay Is
Utila
Roatán
Guanaja
Belize
BZ
Mexico
MX
Mujeres
Cozumel

CARIBBEAN SEA
PACIFIC OCEAN
Panama
PA
Gulf of Panama
Pearl Is
Darién
Gulf of Urabá
Gorgona
Colombia
CO
Venezuela
VE
Guyana
GY
Brazil
BR
Peru
PE
Ecuador
EC
Aruba
AW
Curaçao
CW
Bonaire
BQ
Grenada
GD
Trinidad and Tobago
TT
Tobago
Trinidad
Paria Gulf
Paria Pen.
Los Roques Arch.
Tortuga
Blanquilla
Margarita
Paraguaná Pen.
Guajira Pen.
Gulf of Venezuela
Lake Maracaibo
Falcón
Lara
Yaracuy
Carabobo
Aragua
Vargas
Miranda
Cojedes
Portuguesa
Trujillo
Barinas
Mérida
Táchira
Zulia
Guárico
Anzoátegui
Monagas
Sucre
Delta Amacuro
Bolívar
Apure
Amazonas
Llanos
COASTAL CORDILLERA
ANDES
SIERRA DE PERIJA
Serranía de Imataca
Merume Mts
Mt. Roraima
Pacaraima Mts
La Gran Sabana
Auyan Tepui
Cerro Guaiquinima
PANTEPUI
Cerro Yavi
Cerro Duida
Serranía Guayapo
Serra Parima
Cerro de la Neblina
Roraima
Branco
Negro
Orinoco
La Guajira
Santa Marta Mts
Cesar
Magdalena
Atlántico
Sucre
Córdoba
Antioquia
Norte de Santander
Santander
Boyacá
Cundinamarca
Caldas
Risaralda
Quindío
Tolima
Huila
Chocó
Valle del Cauca
Cauca
Nariño
Putumayo
Caquetá
Meta
Sierra de Macarena
Arauca
Casanare
Vichada
Guainía
Guaviare
Vaupés
Amazonas
Caquetá
Japurá
Putumayo
Amazon
Amazonia
Gulf of Guayaquil

Ecuador
EC
Colombia
CO
Venezuela
Guyana
Peru
PE
Brazil
BR
Bolivia
BO
Chile
CL
Argentina
AR
Paraguay
PY
Amazonia
ANDES
PACIFIC OCEAN
Gulf of Guayaquil
I. de la Plata
I. San Gallán
Paracas Pen.
Lake Titicaca
Cerro de la Neblina
Esmeraldas
Nariño
Cauca
Meta
Guaviare
Caquetá
Vaupés
Putumayo
Sucumbíos
Pichincha
Manabí
Cotopaxi
Los Ríos
Orellana
Pastaza
Guayas
Morona Santiago
Azuay
El Oro
Loja
Zamora Chinchipe
Tumbes
Piura
Lambayeque
Cajamarca
Amazonas
Loreto
San Martín
La Libertad
Ancash
Huánuco
Pasco
Ucayali
Junín
Lima
Huancavelica
Ica
Ayacucho
Apurímac
Cusco
Madre de Dios
Puno
Arequipa
Moquegua
Tacna
Roraima
Pará
Acre
Rondônia
Mato Grosso
Mato Grosso do Sul
Pando
Beni
La Paz
Cochabamba
Santa Cruz
Oruro
Potosí
Chuquisaca
Tarija
Tarapacá
Antofagasta
Jujuy
Salta
Formosa
Alto Paraguay
Boquerón
Presidente Hayes
Concepción
Japurá
Amazon
Negro
Branco
Purus
Madeira
Juruena
Teles Pires
Tapajós
Guaporé
Beni
Ucayali
Marañón
Paraguay
1. Carchi
2. Imbabura
3. Tungurahua
4. Bolívar
5. Chimborazo
6. Cañar

Colombia
CO
Venezuela
VE
Orinoco
Magdalena
Vaupés
Apaporis
Uaupes
Negro
Caquetá
Japurá
Solimoes (Amazon)
Putumayo
Napo
Curaray
Tigre
Pastaza
Amazon
Ecuador
EC
ANDES
Marañón
Javary
Ucayali
Marañón
Huallaga
Juruá
Purus
Acre
Peru
PE
Purus
Acre
Madeira
Mantaro
Ene
Madre de Dios
Beni
Mamoré
Urubamba
Apurimac
A N D E S
PACIFIC OCEAN
Lake Titicaca
Chapare
Chile
CL

Guyana
GY
Suriname
SR
French Guiana
GF
ATLANTIC OCEAN
Roraima
Maracá
Amapá
Ilha de Marajó
Amazonas
Pará
Maranhão
Brazil
BR
Rondônia
Tocantins
Mato Grosso
Distrito Federal
Goiás
Bolivia
BO
Minas Gerais
Branco
Negro
Mapuera
Trombetas
Paru
Jari
Amazon
Madeira
Tapajós
Iriri
Xingu
Tocantins
Araguaia
Juruena
Teles Pires
Machado
Arinos

Orinoco
Bolívar
Venezuela
VE
Guyana
GY
Suriname
SR
French Guiana
GF
ATLANTIC OCEAN
Amazonas
Roraima
Negro
Maracá
Amapá
Branco
Ilha de Marajó
Japurá
Amazonas
Negro
Amazon
Solimões (Amazon)
Madeira
Tocantins
Maranhão
Ceará
Rio Grande do Norte
Purus
Tapajós
Pará
Paraíba
Piauí
Brazil
BR
Rio Teles Pires
Araguaia
Pernambuco
Alagoas
Tocantins
São Francisco
Pando
Rondônia
Guaporé
Rio Juruena
Sergipe
Bahia
Mato Grosso
Xingu
Beni
Planalto do Brasil
Distrito Federal
Goiás
Cochabamba
Bolivia
BO
Paraguay
Pantanal
Santa Cruz
Minas Gerais
Abrolhos Arch.
Rio Paranaíba
Potosí
Chuquisaca
Mato Grosso do Sul
Espírito Santo
Tarija
Paraguay
PY
São Paulo
Rio de Janeiro
Paraná
Jujuy
Salta
Formosa
Paraná
Argentina
AR
Chaco

Bolivia
Paraguay
PY
Mato Grosso do Sul
São Paulo
Brazil
BR
Paraná
Jujuy
Antofagasta
ANDES
Salta
Formosa
Río Paraguay
Río Paraná
Desventuradas Is
Chaco
Tucumán
Misiones
Santa Catarina
Atacama
Catamarca
Santiago del Estero
Corrientes
Santa Fe
Rio Grande do Sul
La Rioja
Coquimbo
San Juan
Río Paraná
Lagoa dos Patos
Córdoba
Entre Ríos
Uruguay
UY
Valparaíso
San Luis
Juan Fernández Arch.
Santiago
O'Higgins
Mendoza
Río de la Plata
Maule
Argentina
AR
La Pampa
Buenos Aires
Bío-Bío
I. Mocha
Neuquén
Araucanía
Bahía Blanca
Los Lagos
Río Negro
Chile
CL
Golfo San Matías
Península Valdés
Chiloé I.
Golfo de Corcovado
Chubut
PATAGONIA
Golfo de San Jorge
Aisén
Golfo de Penas
Santa Cruz
Falkland Is (Malvinas)
FK
Bahía Grande
Magallanes y Antártica Chilena
West Falkland
East Falkland
Straits of Magellan
Tierra del Fuego
Staten I
Cape Horn

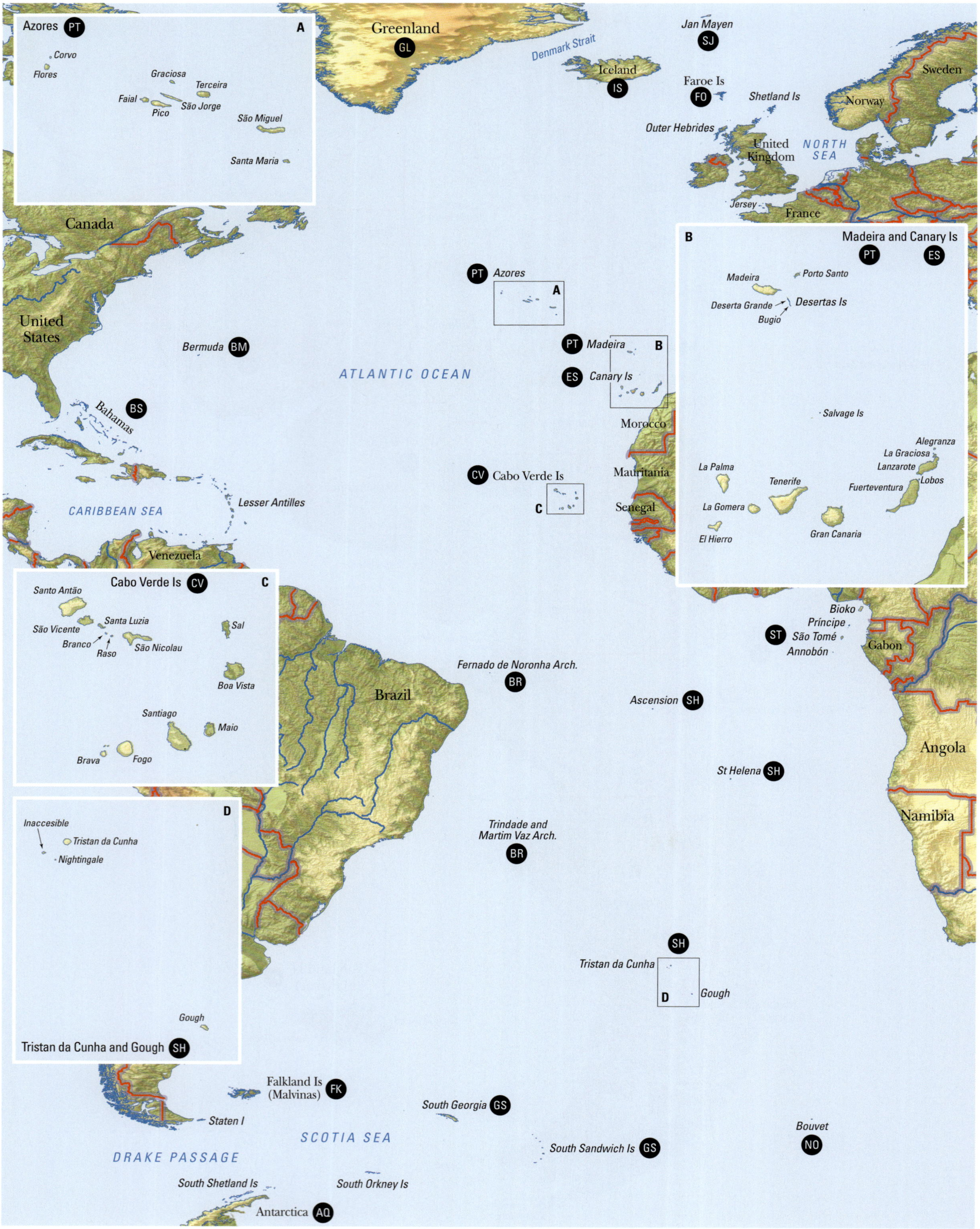
Azores PT
A
Corvo
Flores
Graciosa
Terceira
Faial
São Jorge
Pico
São Miguel
Santa Maria
Greenland GL
Denmark Strait
Jan Mayen SJ
Iceland IS
Faroe Is FO
Shetland Is
Outer Hebrides
Sweden
Norway
United Kingdom
NORTH SEA
Jersey
France
Canada
United States
Bermuda BM
Azores PT
Madeira PT
Canary Is ES
ATLANTIC OCEAN
Bahamas BS
Morocco
Mauritania
Senegal
Cabo Verde Is CV
CARIBBEAN SEA
Lesser Antilles
Venezuela
B
Madeira and Canary Is PT ES
Madeira
Porto Santo
Deserta Grande
Desertas Is
Bugio
Salvage Is
Alegranza
La Graciosa
Lanzarote
Lobos
La Palma
Tenerife
Fuerteventura
La Gomera
El Hierro
Gran Canaria
Cabo Verde Is CV
C
Santo Antão
Santa Luzia
Sal
São Vicente
Branco
Raso
São Nicolau
Boa Vista
Santiago
Maio
Brava
Fogo
Bioko
Príncipe
ST São Tomé
Annobón
Gabon
Fernando de Noronha Arch. BR
Brazil
Ascension SH
Angola
St Helena SH
Namibia
Trindade and Martim Vaz Arch. BR
D
Inaccessible
Tristan da Cunha
Nightingale
Gough
Tristan da Cunha and Gough SH
SH
Tristan da Cunha
Gough
Falkland Is (Malvinas) FK
South Georgia GS
Staten I
SCOTIA SEA
South Sandwich Is GS
Bouvet NO
DRAKE PASSAGE
South Shetland Is
South Orkney Is
Antarctica AQ

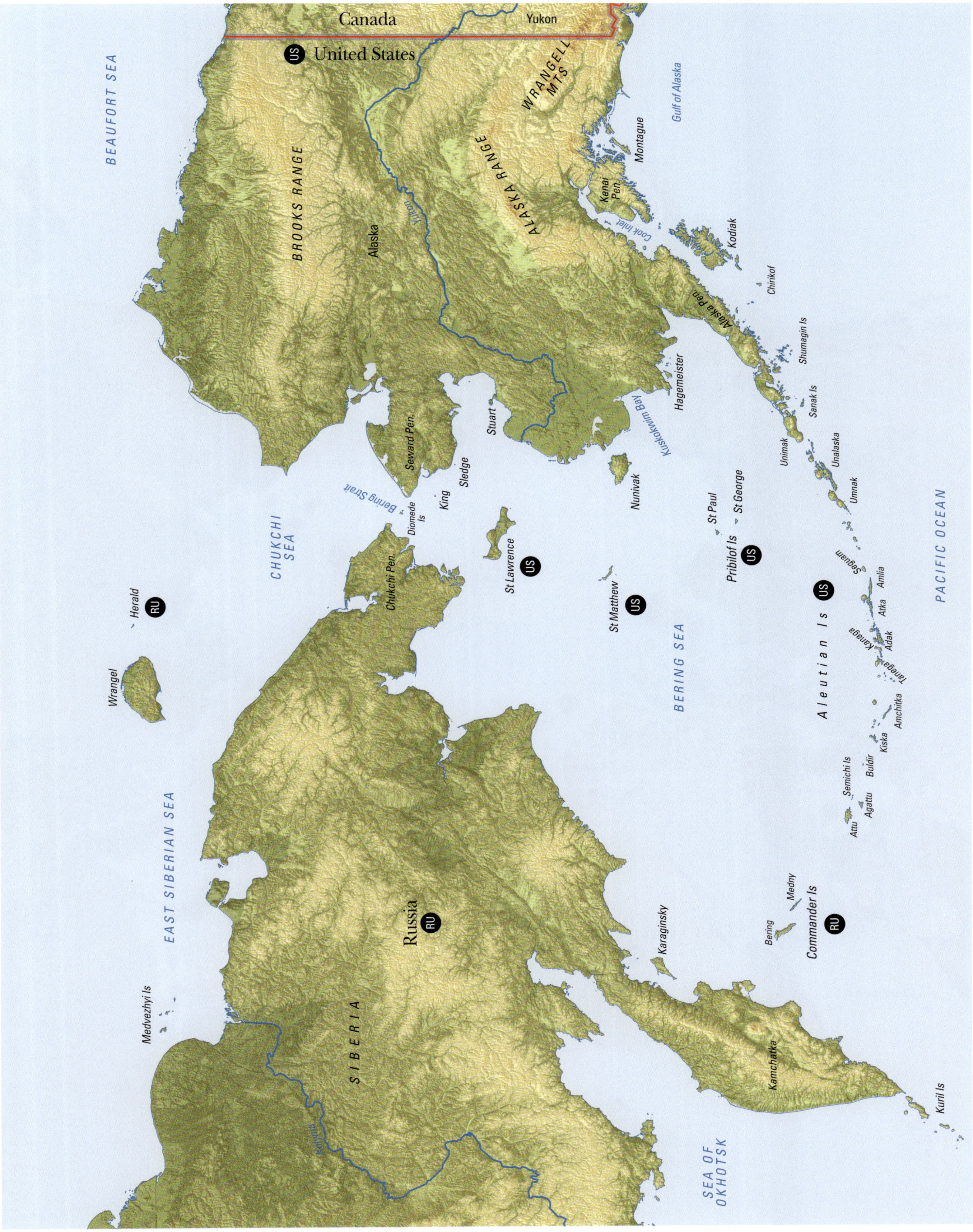
Canada
Yukon
United States
US
BEAUFORT SEA
BROOKS RANGE
Alaska
Yukon
ALASKA RANGE
WRANGELL MTS
Montague
Gulf of Alaska
Kenai Pen.
Cook Inlet
Kodiak
Chirikof
Alaska Pen.
Shumagin Is
Hagemeister
Kuskokwim Bay
Sanak Is
Unimak
Unalaska
Umnak
Stuart
Seward Pen.
Sledge
King
Bering Strait
Diomede Is
Nunivak
St Paul
St George
Pribilof Is
PACIFIC OCEAN
CHUKCHI SEA
Chukchi Pen.
St Lawrence
St Matthew
Seguam
Amlia
Atka
Kanaga
Adak
Tanaga
Aleutian Is
BERING SEA
Herald
RU
Wrangel
Amchitka
Kiska
Buldir
Semichi Is
Agattu
Attu
EAST SIBERIAN SEA
Russia
Mednyy
Commander Is
Bering
Karaginsky
Medvezhyi Is
SIBERIA
Kolyma
Kamchatka
Kuril Is
SEA OF OKHOTSK

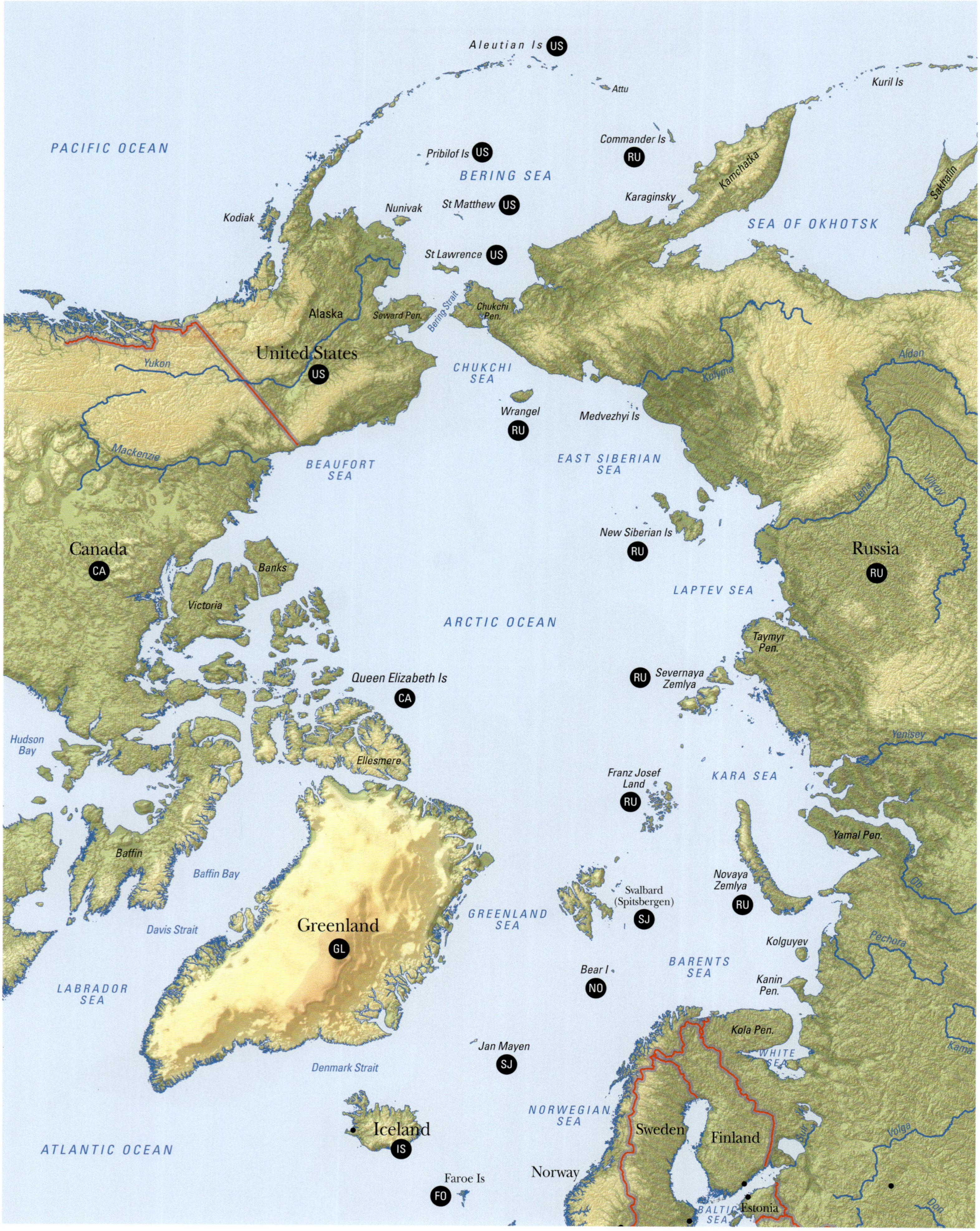
Aleutian Is US
Attu
Kuril Is
PACIFIC OCEAN
Pribilof Is US
Commander Is RU
BERING SEA
Kamchatka
Sakhalin
Karaginsky
Nunivak
St Matthew US
Kodiak
SEA OF OKHOTSK
St Lawrence US
Alaska
Seward Pen.
Bering Strait
Chukchi Pen.
United States US
Yukon
CHUKCHI SEA
Kolyma
Aldan
Wrangel RU
Medvezhyi Is
Mackenzie
BEAUFORT SEA
EAST SIBERIAN SEA
Lena
Vilyuy
New Siberian Is RU
Canada CA
Banks
Russia RU
Victoria
LAPTEV SEA
ARCTIC OCEAN
Taymyr Pen.
Queen Elizabeth Is CA
Severnaya Zemlya RU
Hudson Bay
Yenisey
Ellesmere
Franz Josef Land RU
KARA SEA
Yamal Pen.
Baffin
Baffin Bay
Svalbard (Spitsbergen) SJ
Novaya Zemlya RU
Ob
Davis Strait
Greenland GL
GREENLAND SEA
Pechora
Kolguyev
BARENTS SEA
LABRADOR SEA
Bear I NO
Kanin Pen.
Kola Pen.
Jan Mayen SJ
WHITE SEA
Kama
Denmark Strait
NORWEGIAN SEA
Iceland IS
Sweden
Finland
Volga
ATLANTIC OCEAN
Norway
Faroe Is FO
Estonia
BALTIC SEA
Don

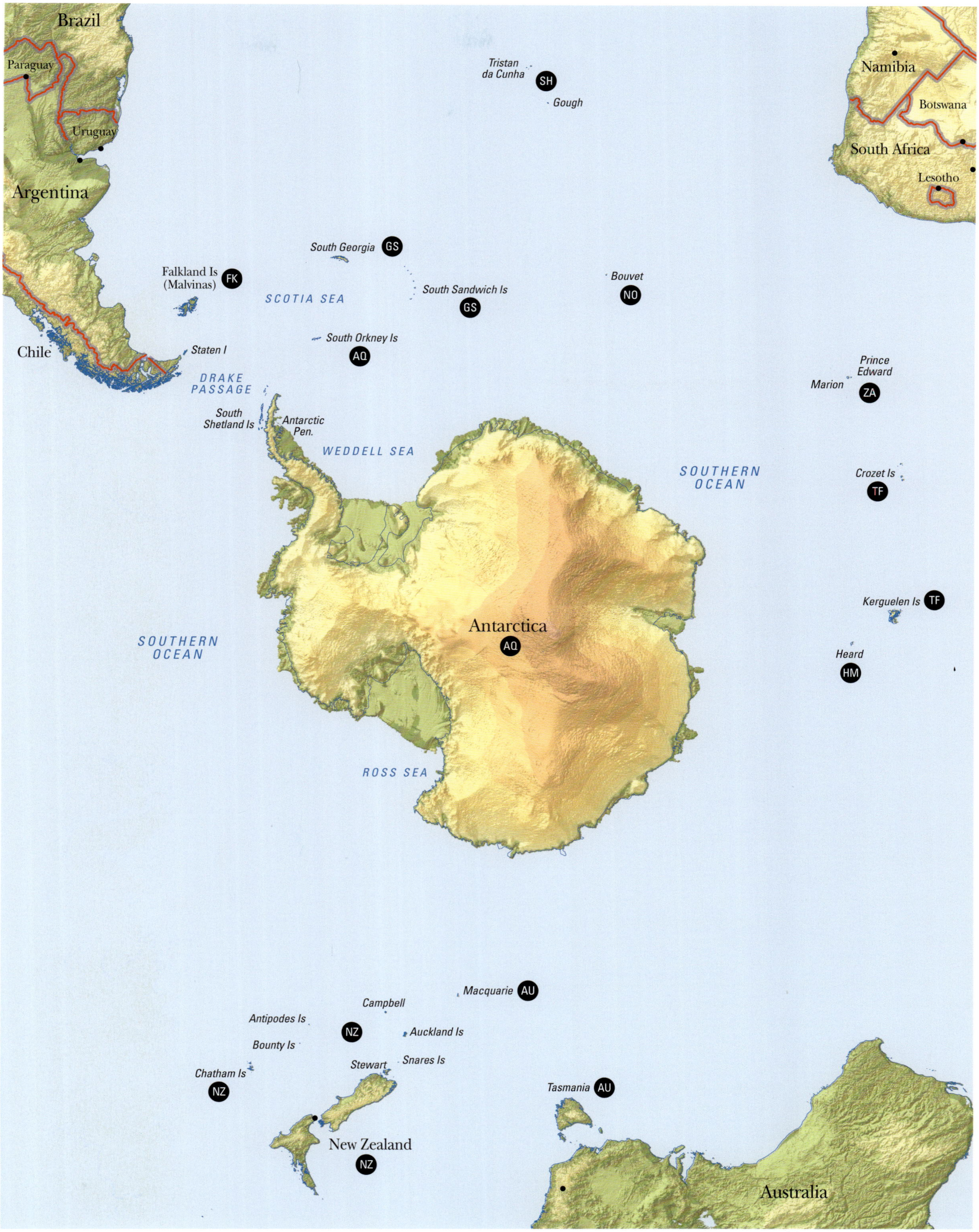
Brazil
Paraguay
Uruguay
Argentina
Chile
Tristan da Cunha
SH
Gough
Namibia
Botswana
South Africa
Lesotho
South Georgia
GS
Falkland Is (Malvinas)
FK
SCOTIA SEA
South Sandwich Is
GS
Bouvet
NO
South Orkney Is
AQ
Staten I
DRAKE PASSAGE
Prince Edward
Marion
ZA
South Shetland Is
Antarctic Pen.
WEDDELL SEA
SOUTHERN OCEAN
Crozet Is
TF
Kerguelen Is
TF
Antarctica
AQ
SOUTHERN OCEAN
Heard
HM
ROSS SEA
Macquarie
AU
Campbell
Antipodes Is
NZ
Auckland Is
Bounty Is
Stewart
Snares Is
Chatham Is
NZ
Tasmania
AU
New Zealand
NZ
Australia

Appendix 6

ONE-COUNTRY ENDEMICS

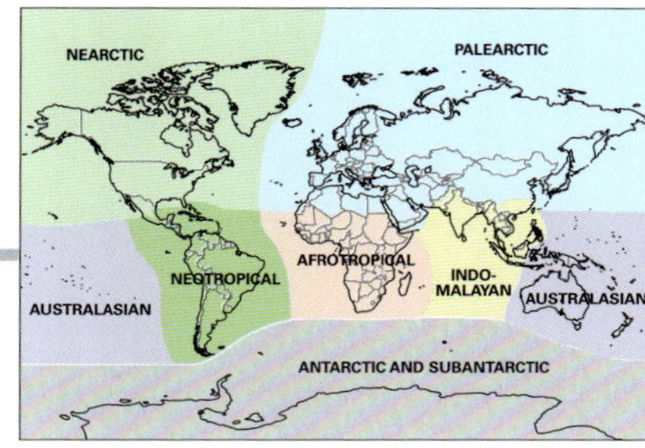

List of countries (based on names used by the United Nations) with their corresponding two-letter ISO codes (see Appendix 4: Country codes, p. 725), the number of extant endemic species in each, and the zoogeographical region (see map here) to which each country belongs, ordered from highest to lowest number of endemic species (and by zoogeographical region when the numbers are the same).

Country (code)	Nº extant Endemics	Zoogeopraphical region
Indonesia (ID)	323	Indo-Malayan Australasian
Australia (AU)	248	Australasian
Brazil (BR)	235	Neotropical
Madagascar (MG)	210	Afrotropical
Mexico (MX)	194	Nearctic Neotropical
China (CN)	145	Palearctic Indo-Malayan
Philippines (PH)	144	Indo-Malayan
United States of America (US)	115	Nearctic
Argentina (AR)	89	Neotropical
Papua New Guinea (PG)	87	Australasian
Peru (PE)	84	Neotropical
Ethiopia (ET)	57	Afrotropical
Colombia (CO)	56	Neotropical
India (IN)	56	Indo-Malayan
Japan (JP)	47	Palearctic
Ecuador (EC)	45	Neotropical
Congo, Democratic Rep. of the (CD)	45	Afrotropical
Tanzania (TZ)	43	Afrotropical
South Africa (ZA)	42	Afrotropical
Malaysia (MY)	33	Indo-Malayan
Venezuela (VE)	31	Neotropical
Bolivia (BO)	29	Neotropical
Vietnam (VN)	28	Indo-Malayan
Taiwan (TW)	26	Indo-Malayan
Russia (RU)	22	Palearctic
Cameroon (CM)	21	Afrotropical
Kenya (KE)	21	Afrotropical
Chile (CL)	20	Neotropical
Sri Lanka (LK)	20	Indo-Malayan
Solomon Islands (SB)	19	Australasian
Panama (PA)	17	Neotropical
Guatemala (GT)	16	Neotropical
Cuba (CU)	15	Neotropical
Angola (AO)	13	Afrotropical
Somalia (SO)	9	Afrotropical
Myanmar (MM)	9	Indo-Malayan
Canada (CA)	7	Nearctic
Costa Rica (CR)	7	Neotropical
Honduras (HN)	7	Neotropical
Italy (IT)	7	Palearctic
Iran (IR)	7	Palearctic
Libya (LY)	7	Palearctic
Kazakhstan (KZ)	7	Palearctic
Sudan (SD)	7	Afrotropical
Zambia (ZM)	7	Afrotropical
Jamaica (JM)	6	Neotropical
Turkey (TR)	6	Palearctic
Nigeria (NG)	6	Afrotropical
São Tomé and Príncipe (ST)	6	Afrotropical
New Caledonia (NC)	6	Australasian
Morocco (MA)	5	Palearctic
Congo (CG)	5	Afrotropical
Thailand (TH)	5	Indo-Malayan
Laos (LA)	5	Indo-Malayan
New Zealand (NZ)	5	Australasian
Seychelles (SC)	4	Afrotropical
Nicaragua (NI)	3	Neotropical
Paraguay (PY)	3	Neotropical
Uruguay (UY)	3	Neotropical
Greece (GR)	3	Palearctic
Mongolia (MN)	3	Palearctic
Spain (ES)	3	Palearctic
Yemen (YE)	3	Palearctic
Pakistan (PK)	3	Palearctic Indo-Malayan
Comoros (KM)	3	Afrotropical
Equatorial Guinea (GQ)	3	Afrotropical
Gabon (GA)	3	Afrotropical
Guinea (GN)	3	Afrotropical
Namibia (NA)	3	Afrotropical
South Sudan (SS)	3	Afrotropical
Uganda (UG)	3	Afrotropical
Micronesia (Federated States of) (FM)	3	Australasian
Bahamas (BS)	2	Neotropical
Algeria (DZ)	2	Palearctic
Armenia (AM)	2	Palearctic
Cyprus (CY)	2	Palearctic
Egypt (EG)	2	Palearctic
Romania (RO)	2	Palearctic
Ukraine (UA)	2	Palearctic
Central African Republic (CF)	2	Afrotropical
Malawi (MW)	2	Afrotropical
Mauritania (MR)	2	Afrotropical
Mauritius (MU)	2	Afrotropical
Rwanda (RW)	2	Afrotropical
Nepal (NP)	2	Indo-Malayan
Vanuatu (VU)	2	Australasian
Guadeloupe (GP)	1	Neotropical
Martinique (MQ)	1	Neotropical
Puerto Rico (PR)	1	Neotropical
Saint Vincent and the Grenadines (VC)	1	Neotropical
Suriname (SR)	1	Neotropical
Trinidad and Tobago (TT)	1	Neotropical
Afghanistan (AF)	1	Palearctic
Iraq (IQ)	1	Palearctic
Kyrgyzstan (KG)	1	Palearctic
Portugal (PT)	1	Palearctic
Syria (SY)	1	Palearctic
Tunisia (TN)	1	Palearctic
Burundi (BI)	1	Afrotropical
Ivory Coast (CI)	1	Afrotropical
Mali (ML)	1	Afrotropical
Mozambique (MZ)	1	Afrotropical
Réunion (RE)	1	Afrotropical
Togo (TG)	1	Afrotropical
Cambodia (KH)	1	Indo-Malayan
Timor-Leste (TL)	1	Australasian
Fiji (FJ)	1	Australasian
Samoa (WS)	1	Australasian

INDEX

A

C

G

M

V

W

Class Mammalia (continued)

EULIPOTYPHLA p. 404

Erinaceidae
Hedgehogs and Gymnures
10 genera • 26 spp
p. 404

Soricidae
Shrews
29 genera • 476 spp
p. 406

Talpidae
Moles, Desmans, Star-nosed Mole and Shrew Moles
19 genera • 60 spp
p. 442

Nesophontidae
Nesophontes
1 genus (†1) • 7 spp (†7)
p. 706

Solenodontidae
Solenodons
2 genera • 4 spp (†2)
p. 447

CHIROPTERA p. 448

Pteropodidae
Old World Fruit Bats
47 genera • 197 spp (†6)
p. 448

Rhinopomatidae
Mouse-tailed Bats
1 genus • 6 spp
p. 465

Craseonycteridae
Hog-nosed Bat
1 genus • 1 sp
p. 465

Megadermatidae
False-vampire Bats
6 genera • 6 spp
p. 466

Rhinonycteridae
Trident Bats
4 genera • 9 spp
p. 466

Hipposideridae
Old World Leaf-nosed Bats
7 genera • 90 spp
p. 467

Rhinolophidae
Horseshoe Bats
1 genus • 107 spp
p. 475

Emballonuridae
Sheath-tailed Bats
14 genera • 55 spp
p. 484

Nycteridae
Slit-faced Bats
1 genus • 14 spp
p. 489

Myzopodidae
Madagascar Sucker-footed Bats
1 genus • 2 spp
p. 490

Mystacinidae
New Zealand Short-tailed Bats
1 genus • 2 spp
p. 490

Noctilionidae
Bulldog Bats
1 genus • 2 spp
p. 491

Furipteridae
Smoky Bat and Thumbless Bat
2 genera • 2 spp
p. 491

Thyropteridae
Disk-winged Bats
1 genus • 5 spp
p. 491

Mormoopidae
Ghost-faced Bats, Naked-backed Bats and Mustached Bats
2 genera • 18 spp
p. 492

Phyllostomidae
New World Leaf-nosed Bats
62 genera • 227 spp (†1)
p. 494

Natalidae
Funnel-eared Bats
3 genera • 12 spp
p. 514

Molossidae
Free-tailed Bats
21 genera • 131 spp
p. 515

Miniopteridae
Long-fingered Bats
1 genus • 40 spp
p. 525

Cistugidae
Wing-gland Bats
1 genus • 2 spp
p. 528

Vespertilionidae
Vesper Bats
57 genera • 528 spp (†3)
p. 528

ARTIODACTYLA p. 566

Balaenidae
Right and Bowhead Whales
2 genera • 4 spp
p. 566

Cetotheriidae
Pygmy Right Whale
1 genus • 1 sp
p. 566

Balaenopteridae
Rorquals
3 genera • 11 spp
p. 567

Physeteridae
Sperm Whale
1 genus • 1 sp
p. 570

Kogiidae
Pygmy and Dwarf Sperm Whales
1 genus • 2 spp
p. 570

Ziphiidae
Beaked Whales
6 genera • 24 spp
p. 570

Platanistidae
South Asian River Dolphins
1 genus • 2 ssp
p. 574

Lipotidae
Baiji
1 genus • 1 sp
p. 575

Iniidae
Amazon River Dolphins
1 genus • 4 spp
p. 575

Pontoporiidae
Franciscana
1 genus • 1 sp
p. 575